PHYSIOLOGY

PHYSIOLOGY

Edited by

ROBERT M. BERNE, M.D., D.Sc. (Hon.)

Alumni Professor of Physiology,
Department of Physiology,
University of Virginia School of Medicine,
Charlottesville, Virginia

MATTHEW N. LEVY, M.D.

Chief of Investigative Medicine, Mount Sinai Medical Center,
Professor of Physiology and Biophysics and of Biomedical Engineering,
Case Western Reserve University,
Cleveland, Ohio

THIRD EDITION

with 963 *illustrations*

St. Louis Baltimore Boston Chicago London Philadelphia Sydney Toronto

Editor: Robert Farrell
Developmental Editor: Emma D. Underdown
Project Manager: John A. Rogers
Senior Production Editor: Shauna Burnett Sticht
Designer: Gail Morey Hudson

THIRD EDITION

Printed in the United States of America

Mosby–Year Book, Inc.
11830 Westline Industrial Drive, St. Louis, Missouri 63146

Library of Congress Cataloging in Publication Data

Physiology / edited by Robert M. Berne, Matthew N. Levy.—3rd ed.
p. cm.
ISBN 0-8016-6465-9
1. Human physiology. I. Berne, Robert M., 1918- . II. Levy, Matthew N., 1922- .
QP34.5.P496 1992
612—dc20 92-19105
CIP

94 95 96 97 C/VH 9 8 7 6 5 4 3 2

Contributors

ROBERT M. BERNE, MD
Alumni Professor of Physiology
Department of Physiology
University of Virginia Health Sciences Center
Charlottesville, Virginia
Section V The Cardiovascular System

SAUL M. GENUTH, MD
Professor, Department of Medicine
School of Medicine
Case Western Reserve University
Cleveland, Ohio
Section IX The Endocrine System

BRUCE M. KOEPPEN, MD, PhD
Associate Professor
Department of Medicine and Physiology
University of Connecticut Health Center
Farmington, Connecticut
Section VIII The Kidney

HOWARD C. KUTCHAI, PhD
Professor, Department of Physiology
University of Virginia Health Sciences Center
Charlottesville, Virginia
Section I Cellular Physiology
Section VII The Gastrointestinal System

MATTHEW N. LEVY, MD
Chief, Investigative Medicine
Mount Sinai Hospital Medical Center
Cleveland, Ohio
Section V The Cardiovascular System

RICHARD A. MURPHY, PhD
Professor, Department of Physiology
University of Virginia School of Medicine
Charlottesville, Virginia
Section III Muscle

OSCAR D. RATNOFF, MD
Professor, School of Medicine
Department of Medicine
Case Western Reserve University
Cleveland, Ohio
Section IV Blood

BRUCE A. STANTON, PhD
Associate Professor, Department of Physiology
Dartmouth College Medical School
Hanover, New Hampshire
Section VIII The Kidney

NORMAN C. STAUB, SR., MD
Professor
Department of Physiology
Cardiovascular Research Institute
University of California School of Medicine
San Francisco, California
Section VI The Respiratory System

WILLIAM D. WILLIS, JR., MD, PHD
Ashbel Smith Professor and Chairman
Department of Anatomy & Neurosciences
Director, Marine Biomedical Institute
The University of Texas Medical Branch at Galveston
Galveston, Texas
Section II The Nervous System

Dedicated to
Alex, Ari, Chris, Daniel, Maggie,
Molly, Sarah, Todd, and Tracy

Preface

The third edition of this text, like the previous editions, is designed to emphasize broad concepts and to minimize the compilation of isolated facts. Each chapter has been altered significantly to make the text as accurate, clear, and current as possible. To help attain this goal, we have revised many illustrations and added a large number of new ones to assist the reader in grasping difficult physiological information. In keeping with our emphasis on broad principles, we have grouped homologies of important mechanisms wherever possible. We have used italics to emphasize important concepts, and we have used boldface to define new terms. Finally, we have added summaries to highlight the key points in each chapter. We hope these changes will enhance the teaching value of the text.

At the beginning of the book and of several of its sections, many of the important physicochemical principles of physiology are analyzed in detail. Among these principles we have included considerable information about relevant advances in molecular biology. When important principles could be represented profitably by equations, the bases of the equations and the major underlying assumptions have been discussed. This approach provides students with a more quantitative understanding of these principles, so that a mastery of certain topics will involve a minimum of pure memorization.

Throughout the section on muscle physiology, we have emphasized that the basic mechanisms of contraction are very similar in skeletal, cardiac, and smooth muscles, and that the differences lie mainly in the relative importance of certain components of the underlying process. We have also added a separate chapter on smooth muscle.

The section on hematology covers blood composition and affords a complete and modern discussion of blood coagulation and its aberrations.

To provide a clearer understanding of cardiovascular physiology, we have dissected the entire system initially into its major components, and we have examined the functions of these individual components in detail. We then reconstructed the system and considered it as a whole to indicate how the various parts of this closed loop system interact in certain physiological and pathophysiological states.

In the endocrine section, homologies again influence the presentation of material. Discussions of the male and female gonads have been included in the same chapter to highlight the similarities between the Sertoli cell functions in spermatogenesis and the granulosa cell functions in oogenesis.

The sections on the central nervous, renal, and respiratory systems have been entirely rewritten by new authors, all of whom are authorities in their respective fields. The section on the nervous system provides a functional neuroanatomical framework for its presentation of contemporary cellular neurophysiology. Substantial attention has been directed toward the sensory and motor systems because of their relevance to clinical problems. The theoretical foundation common to all sensory systems has been constructed so as to facilitate the learning of the various components.

In the renal physiology section, homologies influence the presentation of material. The mechanisms whereby the kidneys handle a few important solutes have been described in detail. The specific details of the transport of the myriad substances that pass through the kidneys have been presented briefly.

The respiratory section emphasizes the physical principles that underly the mechanics of breathing and the processes of gas exchange between the blood and alveoli and between the blood and peripheral tissues. Also, the various neural and chemical processes that regulate respiration have been described in detail.

Again, the framework of this textbook comprises firmly established facts and principles. Isolated phenomena generally are ignored unless they are considered to be highly significant, and few experimental methods are described unless they are essential for the comprehension of a specific topic. We have not documented in the text the assertions we have made throughout the book, but we have provided references at the end of each chapter. These references have been selected because they provide a current and comprehensive review of the topic, a clear and detailed description of important mechanisms, or a complete and current bibliography of the subject.

Although controversies exist in virtually all areas of physiology, such controversies are not described unless they provide a deeper understanding of the subject. The authors of each section have described what they believe to be the most likely mechanism responsible for the phenomenon under consideration. Although we recognize that future advances may prove many of our conjectures wrong, we have made this compromise to achieve brevity, clarity, and simplicity.

We wish to express our appreciation to the colleagues and students who have provided constructive criticism during the revision of this book. We also give special thanks to Frances S. Langley, who skillfully drew the illustrations for the entire volume.

Robert M. Berne
Matthew N. Levy

Contents

SECTION VI

THE RESPIRATORY SYSTEM

Norman C. Staub, Sr.

SECTION VII

THE GASTROINTESTINAL SYSTEM

Howard C. Kutchai

SECTION VIII

THE KIDNEY

Bruce A. Stanton
Bruce M. Koeppen

SECTION IX

THE ENDOCRINE SYSTEM

Saul M. Genuth

Introduction

As the reader embarks on the study of human physiology, it is instructive to consider briefly what physiology is and how it relates to other disciplines.

For centuries physiology and anatomy were the only recognized basic biomedical sciences, having originated in the western world with the ancient Greeks. Comparatively recently, physiology and anatomy, together wtih biology, chemistry, physics, psychology, and other sciences, have given rise to other biomedical disciplines, whose areas of inquiry overlap to considerable degrees. Among the newer biomedical sciences are biochemistry, genetics, pharmacology, biophysics, biomedical engineering, molecular biology, cell biology, and neuroscience. Because the overlap among the interests of the different biomedical sciences is so extensive, it is often difficult, and sometimes not even useful, to decide where one discipline begins and the other ends.

Physiology may be distinguished from the other basic biomedical sciences by its concern with the function of the intact organism and its emphasis on the processes that control and regulate important properties of living systems. In the healthy human many variables are actively maintained within relatively narrow physiological limits. The list of controlled variables is long; it includes body temperature, blood pressure, the ionic composition of blood plasma, blood glucose levels, the oxygen and carbon dioxide content of blood, and a host of other properties. The tendency to maintain the relative constancy of certain variables, even in the face of significant environmental changes, is known as **homeostasis.** A central goal of physiological research is the elucidation of the mechanisms responsible for homeostasis.

In studying a homeostatic mechanism, physiologists attempt to characterize the components of the control system. What is the **sensor** that detects the difference between a physiological variable and its **set point?** How does the sensor operate? What is the **integrating center** that receives information from the sensor via an **afferent pathway** and communicates with **effectors** by **efferent pathways?** The afferent and efferent pathways frequently involve nerves or hormones. What are the effectors that function to bring the level of the controlled variable closer to its set point? The **steady-state** value of a controlled variable typically results from a dynamic balance between certain effectors that function to increase the value of the variable and other effectors that act to decrease it.

We are in the midst of an explosion of knowledge of biological systems. This information expansion is characterized by a progressively greater understanding of the behavior of individual cells and of the molecules that make up those cells. The precise nucleotide sequences of many genes have been determined. The physiological regulation of the oxygen affinity of the blood is understood in terms of precise knowledge of conformational changes in the hemoglobin molecule. The structure of certain ion channels and the nature of the processes by which the channels open and close are being unraveled. The mechanisms that control important variables in individual cells, such as the level of free Ca^{++} in the cytosol, are being elucidated. Many physiologists now study biological processes at the level of single cells or even single molecules in order to explain their in vivo functions. Scientists in related disciplines also concentrate their inquiries on the behavior of cells and biological molecules. What distinguishes physiology as a scientific discipline is its emphasis on homeostatic mechanisms, at all levels of organization, and its concern with synthesizing knowledge of the functions of tissues, cells, and molecules to achieve a more complete understanding of the behavior of the intact organism.

This textbook describes what is now known about the function of the major organ systems of mammals and the mechanisms that control and regulate their behavior. It may be emphasized that physiological knowledge is being continually broadened, and especially deepened. "The fabric of knowledge is being continuously woven, and altered by the introduction of new threads, the texture and device even are mobile, yet it is the object of a comprehensive work to attempt to capture the fleeting pattern. That the design is of growing complexity, and less and less easy of discernment, may be matter for regret though not for surprise."* In our attempt to "capture the fleeting pattern" we have tried to integrate the descriptions of individual organ systems and homeostatic mechanisms in order to provide a broad appreciation for the physiological function of the whole organism. The authors hope that this text will stimulate its readers to actively pursue the study of physiology.

*Evans CL: Preface. In Evan CL, Hartridge H, editors: *Starling's principles of human physiology*, ed 7, Philadelphia, 1936, Lea & Febiger.

SECTION

I

CELLULAR PHYSIOLOGY

Howard C. Kutchai

CHAPTER

1

Cellular Membranes and Transmembrane Transport of Solutes and Water

■ *Cellular Membranes*

Each cell is surrounded by a plasma membrane that separates it from the extracellular milieu. The plasma membrane serves as a permeability barrier that allows the cell to maintain a cytoplasmic composition far different from the composition of the extracellular fluid. The plasma membrane contains enzymes, receptors, and antigens that play central roles in the interaction of the cell with other cells and with hormones and other regulatory agents in the extracellular fluid.

The membranes that enclose the various organelles divide the cell into discrete compartments and allow the localization of particular biochemical processes in specific organelles. Many vital cellular processes take place in or on the membranes of the organelles. Striking examples are the processes of electron transport and oxidative phosphorylation, which occur on, within, and across the mitochondrial inner membrane.

Most biological membranes have certain features in common. However, in keeping with the diversity of membrane functions, membrane composition and structure differ from one cell to another and among the membranes of a single cell.

■ *Membrane Structure*

Proteins and phospholipids are the most abundant constituents of cellular membranes. A phospholipid molecule has a polar head group and two very nonpolar, hydrophobic fatty acyl chains. In an aqueous environment it is most energetically stable for phospholipids to form structures that allow the fatty acyl chains to be kept from contact with water. One such structure is the **lipid bilayer** (Fig. 1-1). Many phospholipids, when dispersed in water, spontaneously form lipid bilayer structures. Most of the phospholipid molecules in biological membranes have a lipid bilayer structure.

The proteins of biological membranes are associated with the membrane phospholipids in two major ways: (1) by charge interactions between the polar head groups of the phospholipids and acidic or basic amino acid residues of the protein, and (2) by hydrophobic interactions of the phospholipid acyl chains with hydrophobic amino acid residues of the proteins.

Fig. 1-2 depicts the "fluid mosaic" model of membrane structure. This model is consistent with many of the properties of biological membranes. Note the bilayer structure of most of the membrane phospholipids. The membrane proteins can be divided into two major classes: (1) **integral** or **intrinsic** membrane proteins that are embedded in the phospholipid bilayer, and (2) **peripheral** or **extrinsic** membrane proteins that are associated with the surface of the phospholipid bilayer. The peripheral membrane proteins interact with membrane lipids predominantly by charge interactions with integral membrane proteins. Thus sometimes they may be removed from the membrane by altering the ionic composition of the medium. Integral membrane proteins have important hydrophobic interactions with the interior of the membrane. These hydrophobic interactions can be disrupted by detergents that solubilize the integral proteins by forming their own hydrophobic interactions with nonpolar amino acid side chains.

Cellular membranes are fluid structures in which many of the constituent molecules are free to diffuse in the plane of the membrane. Most lipids and proteins are free to move in the bilayer plane, but they "flip-flop" from one phospholipid monolayer to the other at much slower rates. This is less likely to occur when a large

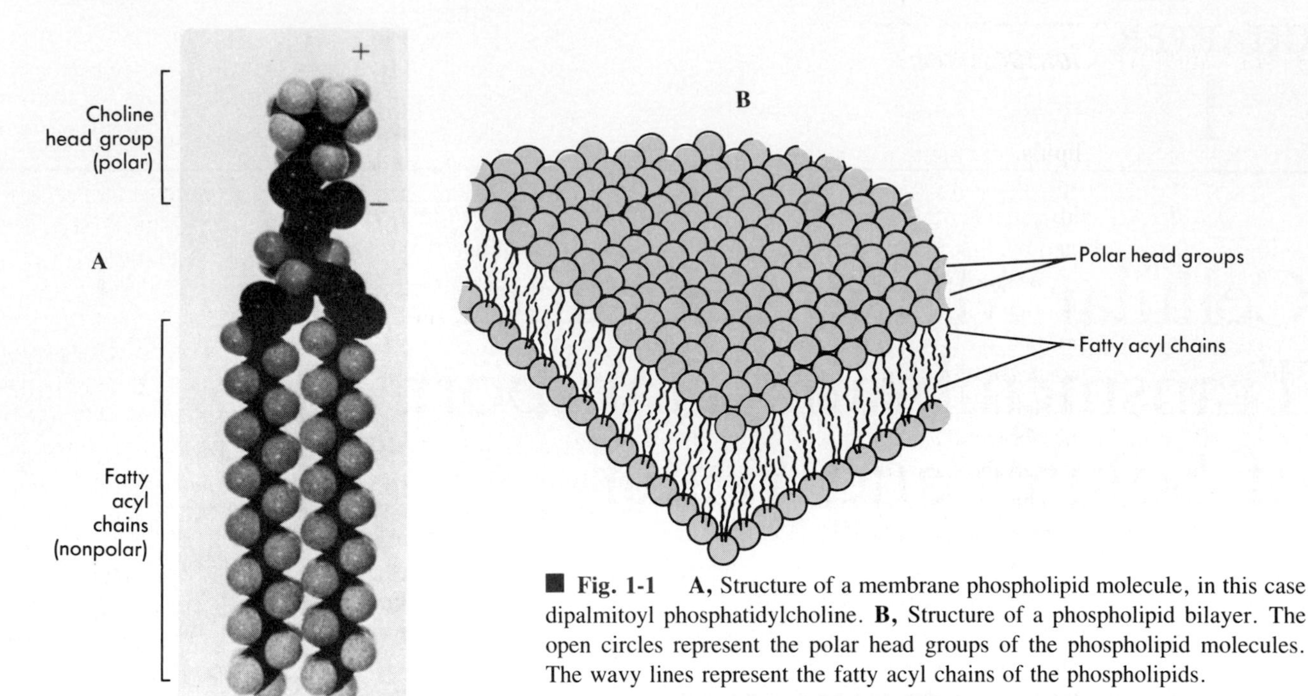

■ **Fig. 1-1** **A,** Structure of a membrane phospholipid molecule, in this case dipalmitoyl phosphatidylcholine. **B,** Structure of a phospholipid bilayer. The open circles represent the polar head groups of the phospholipid molecules. The wavy lines represent the fatty acyl chains of the phospholipids.

hydrophilic moiety must be dragged through the nonpolar interior of the lipid bilayer.

In some cases membrane components clearly are not free to diffuse in the plane of the membrane. Examples of this motional constraint are the sequestration of acetylcholine receptors (integral membrane proteins) at the motor endplate of skeletal muscle and the presence of different membrane proteins in the apical and basolateral plasma membranes of epithelial cells. The cytoskeleton helps to anchor certain membrane proteins and restrain them from lateral diffusion.

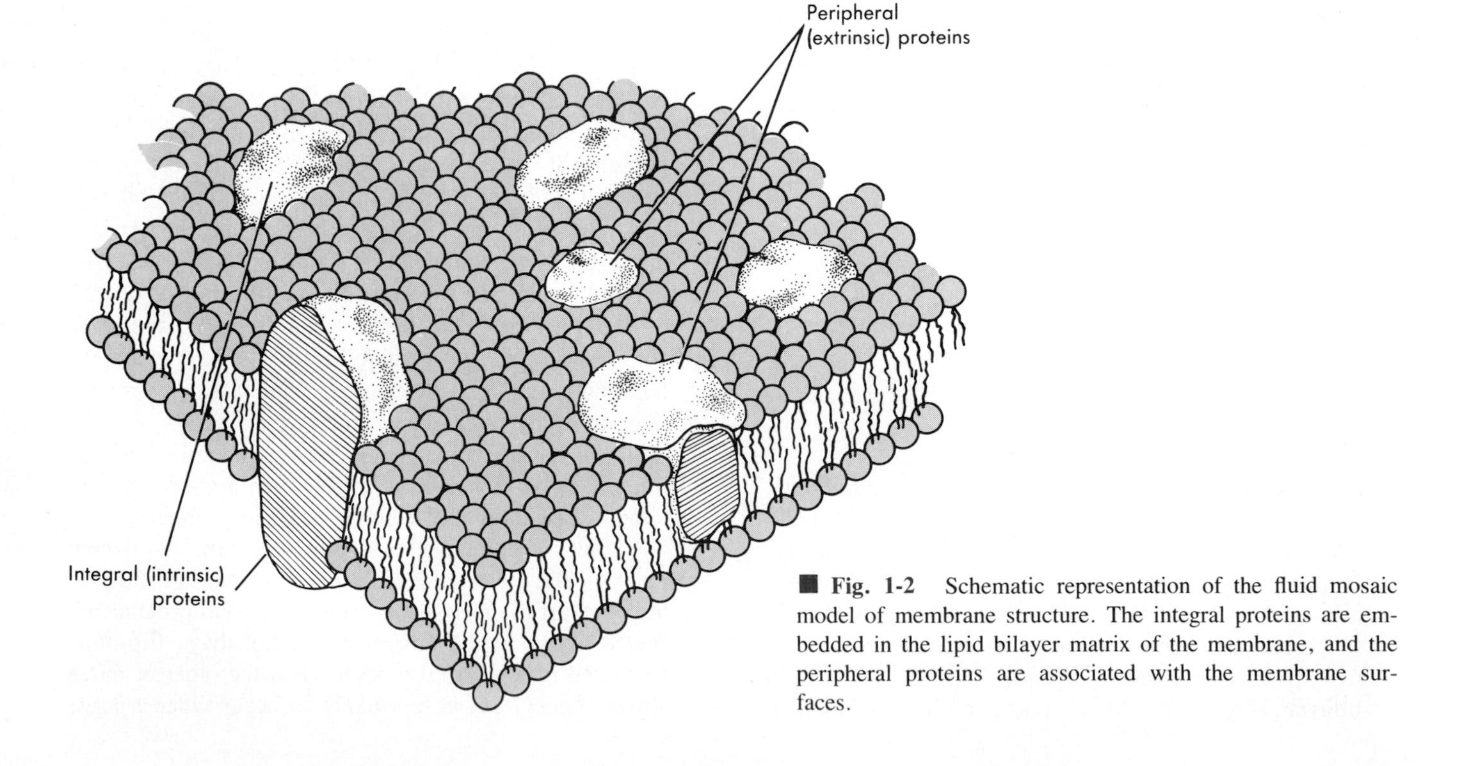

■ **Fig. 1-2** Schematic representation of the fluid mosaic model of membrane structure. The integral proteins are embedded in the lipid bilayer matrix of the membrane, and the peripheral proteins are associated with the membrane surfaces.

Membrane Composition

Lipid Composition

Major phospholipids. In animal cell membranes the most abundant phospholipids are often the choline-containing phospholipids: the lecithins (phosphatidylcholines) and the sphingomyelins. Next in abundance are usually the amino phospholipids: phosphatidylserine and phosphatidylethanolamine. Other important phospholipids that are present in smaller amounts are phosphatidylglycerol, phosphatidylinositol, and cardiolipin.

Cholesterol. Cholesterol is a major constituent of animal cell plasma membranes. The steroid nucleus of cholesterol lies parallel to the fatty acyl chains of membrane phospholipids. Thus cholesterol alters the molecular packing of the membrane phospholipids. In natural membranes cholesterol diminishes the lateral mobility of the lipids and proteins of the membrane.

Glycolipids. Glycolipids are present in small quantities, but they have important functions. Glycolipids are found mostly in plasma membranes, where their carbohydrate moieties protrude from the external surface of the membrane. The blood group antigens and certain other antigens are the carbohydrate side chains of specific glycolipids or glycoproteins.

Asymmetry of lipid distribution in the bilayer. In many membranes the lipid components are not distributed uniformly across the bilayer. As just mentioned, the glycolipids of the plasma membrane are located exclusively in the outer monolayer and thus show absolute asymmetry. Asymmetry of phospholipids occurs but is not absolute. In the red blood cell membrane, for example, the outer monolayer contains most of the choline-containing phospholipids, whereas the inner monolayer is enriched in the amino phospholipids.

Membrane Proteins

The protein composition of membranes may be simple or complex. The highly specialized membranes of the sarcoplasmic reticulum of skeletal muscle and the disks of the rod outer segment of the retina contain only a few different proteins. Plasma membranes perform many functions and may have more than 100 different protein constituents. Membrane proteins include enzymes (such as adenylyl cyclase), transport proteins (such as the Na, K-ATPase), hormone receptors, receptors for neurotransmitters, and antigens.

Glycoproteins. Some membrane proteins are glycoproteins with covalently bound carbohydrate side chains. As with glycolipids, the carbohydrate chains of glycoproteins are located almost exclusively on the external surfaces of plasma membranes. Cell surface carbohydrate has important functions. The negative surface charge of cells is almost entirely the result of the negatively charged sialic acid of glycolipids and glycoproteins. Receptors for viruses may involve surface carbohydrate. Certain surface antigenic determinants reside in carbohydrate moieties on the cell surface. Surface carbohydrate participates in cellular aggregation phenomena and other forms of cell-cell interactions.

Asymmetry of membrane proteins. The absolute asymmetry of glycolipids and glycoproteins was mentioned earlier. The Na, K-ATPase of the plasma membrane and the Ca^{++} pump protein (Ca^{++}-ATPase) of the sarcoplasmic reticulum membrane are other examples of the asymmetrical functions of membrane proteins. In both cases ATP is split on the cytoplasmic face of the membrane, and some of the energy liberated is used to pump ions in specific directions across the membrane. In the case of the Na, K-ATPase, K^+ is pumped into the cell, and Na^+ is pumped out, whereas the Ca^{++}-ATPase actively pumps Ca^{++} into the sarcoplasmic reticulum. Integral membrane proteins are inserted into the membrane lipid bilayer of the endoplasmic reticulum during protein synthesis. The configuration of the protein in the membrane is the result of this specific insertion and the secondary and tertiary structures the protein assumes in the membrane.

Membranes as Permeability Barriers

Biological membranes serve as permeability barriers. Most of the molecules present in living systems have high solubility in water and low solubility in nonpolar solvents. Hence, such molecules have low solubility in the nonpolar environment in the interior of the lipid bilayer of biological membranes. As a consequence, *biological membranes pose a formidable permeability barrier to most water-soluble molecules*. The plasma membrane is a permeability barrier between the cytoplasm and the extracellular fluid. This permeability barrier allows for the maintenance of cytoplasmic concentrations of many substances that differ greatly from their concentrations in the extracellular fluid. The localization of various cellular processes in certain organelles depends on the barrier properties of cellular membranes. For example, the inner mitochondrial membrane is impermeable to the enzymes and substrates of the tricarboxylic acid cycle, allowing the localization of the tricarboxylic cycle in the mitochondrial matrix. The spatial organization of chemical and physical processes in the cell depends on the barrier functions of cellular membranes.

The passage of important molecules across membranes at controlled rates plays a central role in the life of the cell. Examples include the uptake of nutrient molecules, the discharge of waste products, and the release of secreted molecules.

In some cases molecules move from one side of a membrane to another without actually moving through the membrane itself. **Endocytosis** and **exocytosis** are examples of processes that transfer molecules across, but not through, biological membranes. In other cases molecules cross a particular membrane by actually *moving through* the membrane by either passing through or between the molecules that make up the membrane.

Transport Across, But Not Through, Membranes

Endocytosis. Endocytosis allows material to enter the cell without passing through the membrane (Fig. 1-3). The uptake of particulate material is termed **phagocytosis** (Fig. 1-3, *A*). The uptake of soluble molecules is called **pinocytosis** (Fig. 1-3, *B*). Sometimes special regions of the plasma membrane, whose cytoplasmic surface is covered with bristles made primarily of a protein called **clathrin,** are involved in endocytosis. These bristle-covered regions are called **coated pits,** and their endocytosis gives rise to **coated vesicles** (Fig. 1-3, *C*). The coated pits appear to be involved primarily in **receptor-mediated endocytosis.** Specific proteins to be taken up are recognized and bound by specific membrane receptor proteins in the coated pits. The binding often leads to aggregation of receptor-ligand complexes, and the aggregation triggers endocytosis in ways that are not yet understood. Endocytosis is an active process that requires metabolic energy. It also can occur in regions of the plasma membrane that do not contain coated pits.

Exocytosis. Molecules can be ejected from cells by exocytosis, a process that resembles endocytosis in reverse. The release of neurotransmitters, which is considered in more detail in Chapter 4, takes place by exocytosis. Exocytosis is responsible for the release of secretory proteins by many cells; the release of pancreatic zymogens from the acinar cells of the pancreas is a well-studied example. In such cases the proteins to be secreted are stored in secretory vesicles in the cytoplasm. A stimulus to secrete causes the secretory vesicles to fuse with the plasma membrane and to release the vesicle contents by exocytosis.

Fusion of membrane vesicles. The contents of one type of organelle can be transferred to another organelle by fusion of the membranes of the organelles. In some cells secretory products are transferred from the endoplasmic reticulum to the Golgi apparatus by fusion of endoplasmic reticulum vesicles with membranous sacs of the Golgi apparatus. Fusion of phagocytic vesicles with lysosomes allows intracellular digestion of phagocytosed material to proceed.

Transport of Molecules Through Biological Membranes

The traffic of molecules through biological membranes is vital for most cellular processes. Some molecules move through biological membranes simply by diffusing among the molecules that make up the membrane, whereas the passage of other molecules involves the mediation of *specific transport proteins* in the membrane.

Oxygen, for example, is a small molecule with fair solubility in nonpolar solvents. It crosses biological membranes by diffusing among membrane lipid molecules. Glucose, on the other hand, is a much larger molecule with low solubility in the membrane lipids. Glucose enters cells via a specific glucose transport protein in the plasma membrane.

Diffusion

Diffusion is the process whereby atoms or molecules intermingle because of their random thermal (Brownian) motion. Imagine a container divided into two compartments by a removable partition. A much larger number of molecules of a compound is placed on side A than on side B, and then the partition is removed. Every molecule is in random thermal motion. It is equally probable that a molecule that begins on side A will move to side B in a given time as it is that a molecule beginning on side B will end up on side A. Because many more molecules are present on side A, the total *number* of molecules moving from side A to side B will be greater than the number moving from side B to side A. In this way the number of molecules on side A will decrease, whereas the number of molecules on side B will increase. This process of *net diffusion* of molecules will

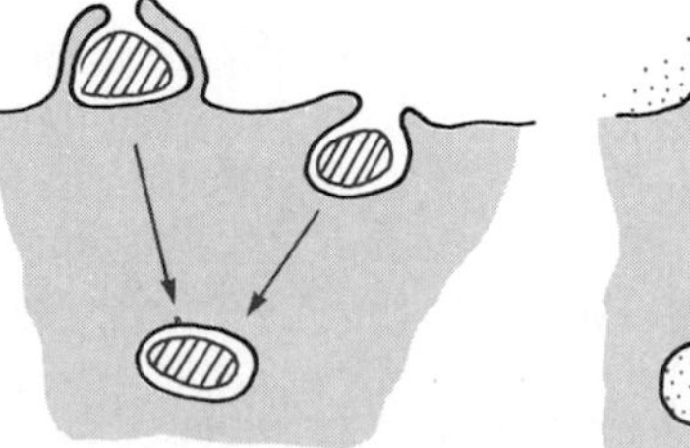

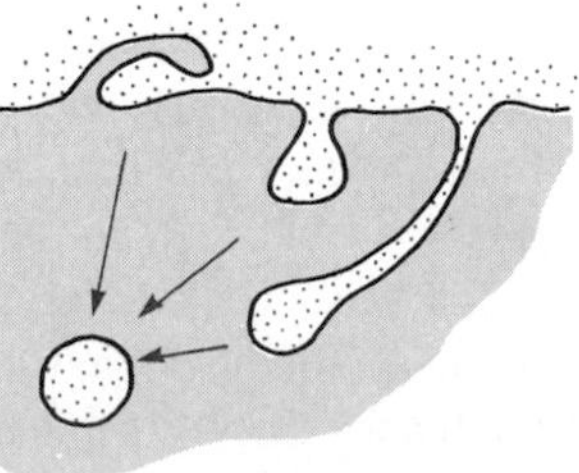

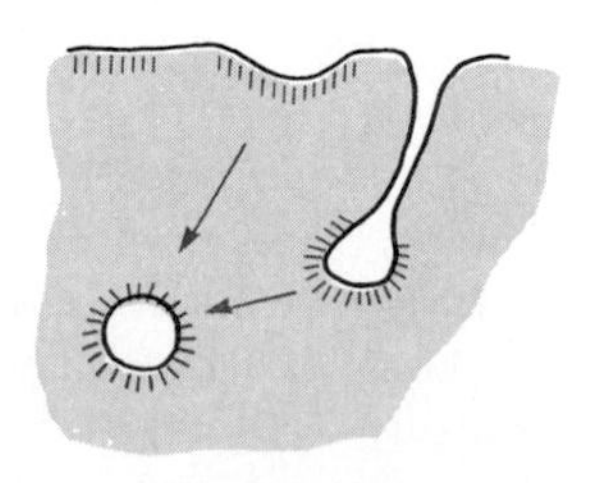

Fig. 1-3 Schematic depiction of endocytotic processes. **A,** Phagocytosis of a solid particle. **B,** Pinocytosis of extracellular fluid. **C,** Receptor-mediated endocytosis by coated pits. (Redrawn from Silverstein SC et al: *Annu Rev Biochem* 46:669, 1977. With permission by Annual Reviews.)

continue until the number of molecules on side A equals that on side B. Thereafter the rate of diffusion of molecules from A to B will equal that from B to A, so no further net movement will occur. A state of *dynamic equilibrium* exists when concentrations on side A and side B are equal.

Diffusion leads to a state in which the concentration of the diffusing species is constant in space and time. Thus diffusion across cellular membranes tends to equalize the concentrations on the two sides of the membrane. The diffusion rate across a particular planar surface is proportional to the area of the plane and to the difference in concentration of the diffusing substance on the two sides of the plane. **Fick's first law of diffusion** states that

$$J = -DA\frac{dc}{dx} \quad (1)$$

where

J = net rate of diffusion in moles or grams per unit time
A = area of the plane
dc/dx = concentration gradient across the plane
D = constant of proportionality called **diffusion coefficient**

The nature of the concentration gradient dc/dx in Fick's first law requires further explanation. The concentration profile of the diffusing substance may take many forms. In Fig. 1-4 two possible forms of the concentration profile are shown, and the locations of planes across which the rates of diffusion are to be determined are represented by x_1 and x_2. In Fig. 1-4, *A*, the concentration profile is a straight line. In this case the value of the concentration gradient is simply the slope $\Delta c/\Delta x$ of the line, and the rate of flux is the same at both planes, that is:

$$J = -DA\frac{dc}{dx} = -DA\frac{\Delta c}{\Delta x} \quad (2)$$

In Fig. 1-4, *B*, however, the concentration profile is nonlinear. At each plane, x_1 and x_2, the value of dc/dx is equal to the slope of the tangent to the curve at that point. Because the slopes of those tangents are different, the rates of diffusion across planes placed at x_1 and x_2 will be different at this time.

Note also that the equation for Fick's first law contains a minus sign. The minus sign indicates the *direction* of diffusion. In Fig. 1-4 the slope of the concentration profile is *negative,* but the direction of diffusion is in the *positive* x direction. The need for the minus sign in the equation arises because molecules flow down a concentration gradient (i.e., from higher to lower concentration).

The diffusion coefficient. The diffusion coefficient, D, has units of square centimeters per second. Solving equation 1 for D and expressing all the variables on the right-hand side of the resulting equation in cgs units, we obtain

$$D = \frac{J}{A\frac{dc}{dx}} = \frac{\text{moles/second}}{\text{cm}^2\,\frac{\text{moles/cm}^3}{\text{cm}}} = \text{cm}^2\text{/second} \quad (3)$$

D can be thought of as proportional to the speed with which the diffusing molecule can move in the surrounding medium. D is smaller the larger the molecule and the more viscous the medium.

For spherical solute molecules that are much larger than the surrounding solvent molecules Albert Einstein obtained the following equation:

$$D = kT/(6\pi r\eta) \quad (4)$$

where

k = Boltzmann's constant
T = absolute temperature (kT is proportional to the average kinetic energy of a solute molecule)
r = molecular radius
η = viscosity of the medium

The equation is called the **Stokes-Einstein relation,** and the molecular radius defined by this equation is known as the Stokes-Einstein radius.

For large molecules equation 4 predicts that D will be inversely proportional to the radius of the diffusing molecule. Because the molecular weight (MW) is approximately proportional to r^3, D should be inversely proportional to $MW^{1/3}$, so that a molecule that is ⅛ the mass of another molecule will have a diffusion coeffi-

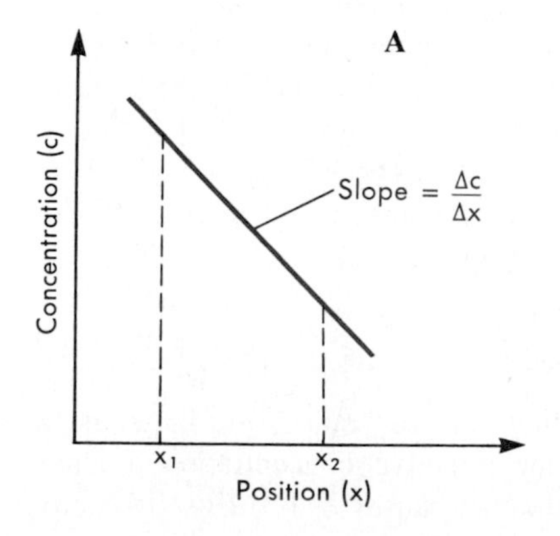

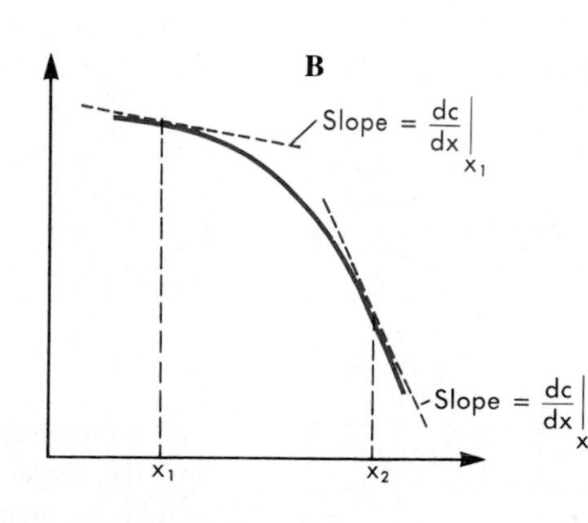

Fig. 1-4 **A,** Linear concentration profile. **B,** Nonlinear concentration profile. The concentration gradient at any point is the slope of the concentration profile at that point.

cient only twice as large as the other molecule. For smaller solutes, with a molecular weight less than about 300, D is inversely proportional to $MW^{1/2}$ rather than $MW^{1/3}$.

Diffusion is a rapid process when the distance over which it must take place is small. This can be appreciated from another relation derived by Einstein. He considered the random movements of molecules that are originally located at $x = 0$. Because a given molecule is equally likely to diffuse in the $+x$ or $-x$ direction, the *average* displacement of all the molecules that begin at $x = 0$ will be zero. The average displacement squared, $\overline{(\Delta x)^2}$, which is a positive quantity, is represented by

$$\overline{(\Delta x)^2} = 2\ Dt \tag{5}$$

where t is the time elapsed since the molecules started diffusing. The **Einstein relation** (equation 5) tells us how far the average molecule will diffuse in time (t), and provides us with a rough estimate of the time scale of a particular diffusion process.

Einstein's relation shows us that the time required for a diffusion process increases with the square of the distance over which diffusion occurs. Thus a tenfold increase in the diffusion distance means that the diffusion process will require about 100 times longer to reach a given degree of completion. Table 1-1 shows the results of calculations using Einstein's relation for a typical, small, water-soluble solute. It can be seen that diffusion is extremely rapid on a microscopic scale of distance. For macroscopic distances diffusion is rather slow. A cell that is 100 μm away from the nearest capillary can receive nutrients from the blood by diffusion with a time lag of only 5 seconds or so. This is sufficiently fast to satisfy the metabolic demands of many cells. However, a nerve axon that is 1 cm long cannot rely on diffusion for the intracellular transport of vital metabolites, because the 14 hours it would take for diffusion over the 1 cm distance is too long on the time scale of cellular metabolism. Some nerve fibers are as long as 1 m; it is no wonder then that intracellular axonal transport systems are involved in transporting important molecules along nerve fibers. Because of the slowness of diffusion over macroscopic distances, it is not surprising that even rather small multicellular organisms have evolved circulatory systems to bring the individual cells of the organism within reasonable diffusion range of nutrients.

■ **Table 1-1** The time required for diffusion to occur over various diffusion distances*

Diffusion distance (μm)	*Time required for diffusion*
1	0.5 msec
10	50 msec
100	5 seconds
1000 (1 mm)	8.3 minutes
10,000 (1 cm)	14 hours

*The time required for the "average" molecule (with diffusion coefficient taken to be 1×10^{-5} cm^2/second) to diffuse the required distance was computed from the Einstein relation.

■ *Diffusive Permeability of Cellular Membranes*

Permeability of lipid-soluble molecules. The plasma membrane serves as a diffusion barrier enabling the cell to maintain cytoplasmic concentrations of many substances that differ markedly from their extracellular concentrations. As early as the turn of the century, the relative impermeability of the plasma membrane to most water-soluble substances was attributed to its "lipoid nature."

The hypothesis that the plasma membrane has a lipoid character is supported by experiments showing that compounds that are soluble in nonpolar solvents (such as ether or olive oil) enter cells more readily than do water-soluble substances of similar molecular weight. Fig. 1-5 shows the relationship between membrane permeability and lipid solvent solubility for a number of different solutes. The ratio of the solubility of the solute in olive oil to its solubility in water is used as a measure of solubility in nonpolar solvents. This ratio is called the olive oil/water partition coefficient. The permeability of the plasma membrane to a particular substance increases with the "lipid solubility" of the substance. For compounds with the same olive oil/water partition coefficient, permeability decreases with increasing molecular weight. As described previously, the fluid mosaic model of membrane structure envisions the plasma membrane as a lipid bilayer with proteins embedded in it. The data of Fig. 1-5 support the idea that the lipid bilayer is the principal barrier to substances that permeate the membrane by simple diffusion.

The positive correlation between lipid solvent solubility and membrane permeability suggests that lipid-soluble molecules can dissolve in the plasma membrane and diffuse across it. Consider a substance that dissolves in the lipid bilayer and then diffuses across the plama membrane to the other side (Fig. 1-6). If the substance equilibrates at the edges of the lipid bilayer, its concentration at the outer face of the bilayer, $C_m(o)$, will be βC_o: the partition coefficient (β) times the concentration in the extracellular medium. Its concentration at the inner face of the membrane, $C_m(i)$, will be βC_i: β times the concentration in the cytoplasm. The concentration difference within the membrane itself is not $C_o - C_i$, but $C_m(o) - C_m(i)$. Because $C_m(o) - C_m(i) = \beta[C_o - C_i]$, the concentration gradient relevant to diffusion across the membrane is

$$\frac{\Delta C_m}{\Delta x} = \beta \frac{C_o - C_i}{\Delta x} \tag{6}$$

The more lipid soluble the substance, the larger is β, and the larger is the effective concentration gradient within the membrane that causes it to diffuse. Accord-

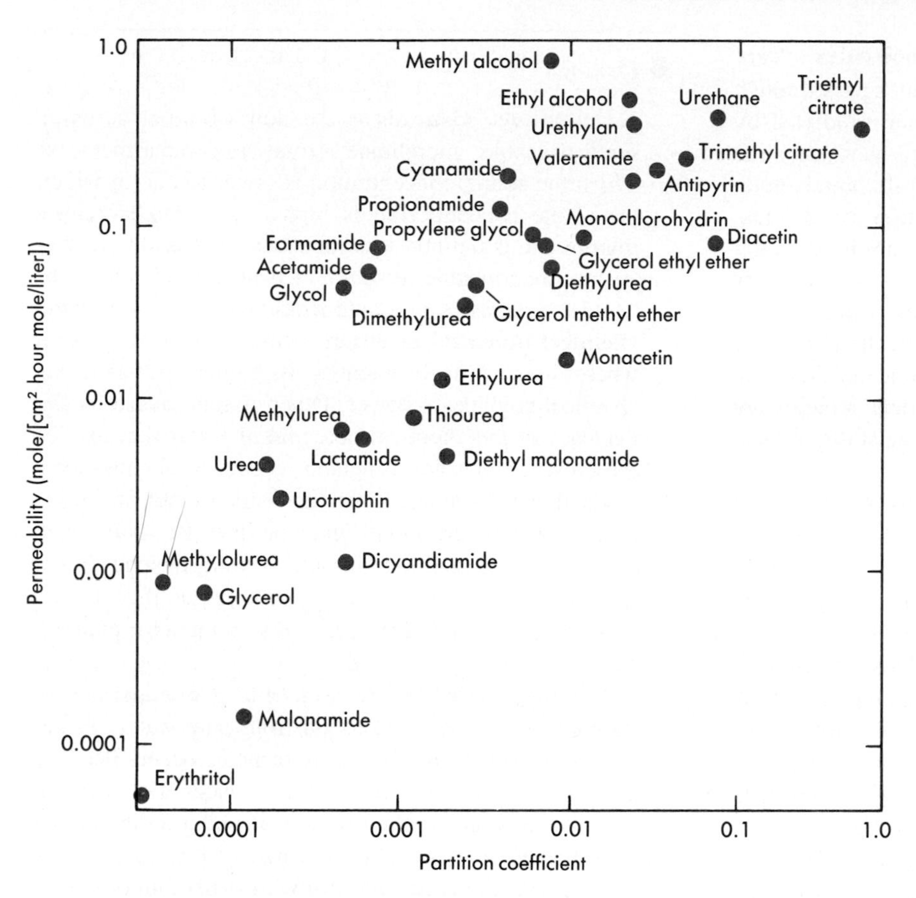

Fig. 1-5 The permeability of the plasma membrane of the alga *Chara ceratophylla* to various nonelectrolytes as a function of the lipid solubility of the solutes. Lipid solubility is represented on the abscissa by the olive oil/water partition coefficient. (Redrawn from Christensen HN: Biological transport, ed 2, Menlo Park, Calif, 1975, WA Benjamin. Data from Collander R: *Trans Faraday Soc* 33:985, 1937.)

ing to Fick's first law the flux of the substance across the membrane is

$$J = -DA\frac{\Delta C_m}{\Delta x} = -DA\beta\frac{C_o - C_i}{\Delta x} = -\frac{D\beta}{\Delta x}A(C_o - C_i) \quad (7)$$

where D is the diffusion coefficient *within the membrane*. $D\beta/\Delta x$ is a constant for a particular substance and a particular membrane, and it is called the **permeability coefficient** (k_p). In terms of the permeability coefficient:

$$J = -k_pA(C_o - C_i) \quad (8)$$

Because $k_p = \frac{D\beta}{\Delta x}$, β is dimensionless, the units of D are expressed in square centimeters per second, the units of k_p are centimeters per second.

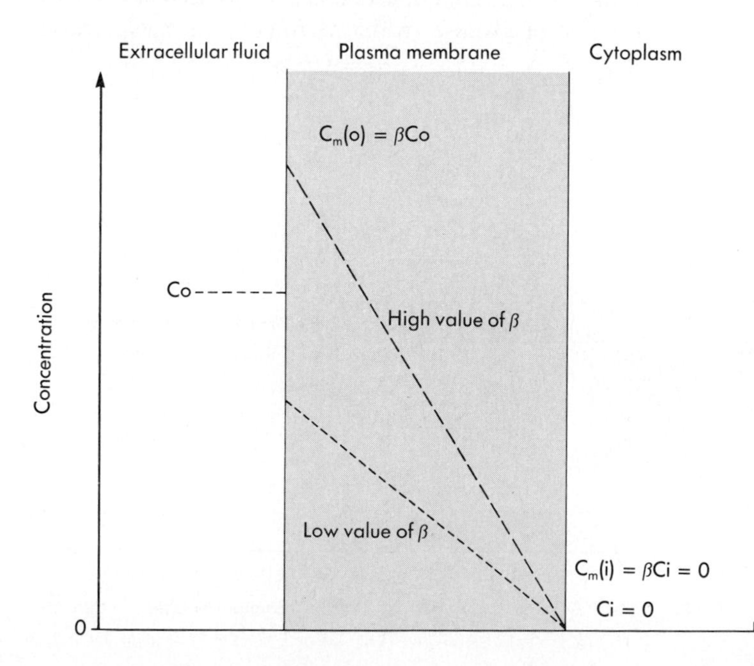

Fig. 1-6 Influence of the membrane lipid/water partition coefficient β of a solute on the concentration profile of the solute within the membrane. The cytoplasmic concentrations of the solutes are assumed to be zero for the sake of illustration. The higher the value of β, the steeper the intramembrane diffusion gradient.

Permeability of water-soluble molecules. Very small, uncharged, water-soluble molecules pass through cell membranes much more rapidly than predicted by their lipid solubility. For example, water permeates the cell membrane much more readily than do larger molecules with similar olive oil/water partition coefficients. The reason for the unusually high permeability of water is controversial. Some investigators think that very small water-soluble molecules can pass between adjacent phospholipid molecules without actually *dissolving* in the region occupied by the fatty acid side chains. Some experiments suggest that membrane proteins are responsible for the high membrane permeability of water.

As the size of uncharged, water-soluble molecules increases, their membrane permeability decreases. Most plasma membranes are essentially impermeable to water-soluble molecules whose molecular weights are greater than about 200.

Because of their charge, ions are relatively insoluble in lipid solvents and therefore lipid membranes have low permeability to most ions. Ionic diffusion across membranes occurs through protein "channels" that span the membrane. Some channels are highly specific with respect to the ions allowed to pass, whereas others allow the passage of all ions below a certain size. Some ion channels are controlled by the voltage difference across the membrane, whereas others are regulated by neurotransmitters or certain other molecules (see Chapters 3 and 4).

Certain water-soluble molecules such as sugars and amino acids, which are essential for cellular survival, do not cross plasma membranes at appreciable rates by simple diffusion. Plasma membranes have specific proteins that allow the transfer of vital metabolites into or out of the cell. The characteristics of membrane protein-mediated transport are discussed later in this chapter.

■ *Osmosis*

Definitions. Osmosis is the flow of water across a semipermeable membrane from a compartment in which the solute concentration is lower to one in which the solute concentration is higher. A **semipermeable membrane** is defined as a membrane permeable to water but impermeable to solutes. Osmosis takes place because the presence of solute results in a decrease of the **chemical potential** of water. Water tends to flow from where its chemical potential is higher to where its chemical potential is lower. Other effects caused by the decrease of the chemical potential of water (because of the presence of solute) include reduced vapor pressure, lower freezing point, and higher boiling point of the solution as compared with pure water. Because these properties, and osmotic pressure as well, depend on the concentration of the solute present rather than on its chemical properties, they are called **colligative properties.**

Osmotic pressure. In Fig. 1-7 a semipermeable membrane separates a solution from pure water. Water flow from side B to side A by osmosis occurs because the presence of solute on side A reduces the chemical potential of water in the solution. Pushing on the piston will increase the chemical potential of the water in the solution on side A and slow down the net rate of osmosis. If the force on the piston is increased gradually, a pressure at which net water flow stops is eventually reached. Application of still more pressure will cause water to flow in the opposite direction, from side A to side B. The pressure on side A that is just sufficient to keep pure water from entering is called the **osmotic pressure** of the solution on side A.

The osmotic pressure of a solution depends on the number of particles in solution. Thus the degree of ionization of the solute must be taken into account. A 1 M solution of glucose, a 0.5 M solution of NaCl, and a

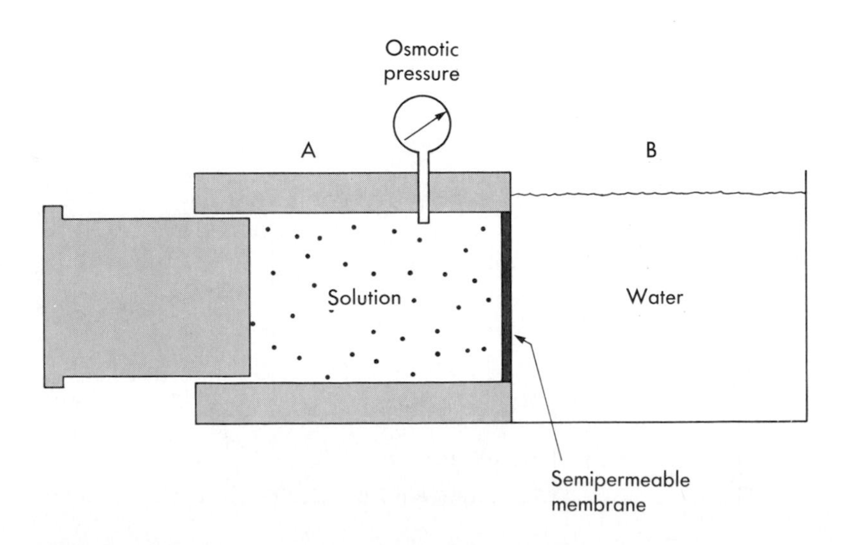

■ **Fig. 1-7** Schematic representation of the definition of osmotic pressure. When the hydrostatic pressure applied to the solution in chamber *A* is equal to the osmotic pressure of that solution, there will be no net water flow across the membrane.

0.333 M solution of $CaCl_2$ theoretically should have the same osmotic pressure. (Actually their osmotic pressures will differ because of the deviations of real solutions from ideal solution theory.) Important equations that pertain to osmotic pressure and the other colligative properties were derived by van't Hoff. One form of **van't Hoff's Law** for calculation of osmotic pressure is

$$\pi = iRTm \tag{9}$$

where

π = osmotic pressure
i = number of ions formed by dissociation of a solute molecule
R = ideal gas constant
T = absolute temperature
m = molal concentration of solute (moles of solute per kilogram of water)

This equation applies more exactly as the solution becomes more dilute.

In the biological sciences molar concentrations are used more frequently than molal concentrations, and van't Hoff's law is approximated by

$$\pi = iRTc \tag{10}$$

where c is the molar concentration of solute (moles of solute per liter of solution). The discrepancy between theory and reality is a bit larger for this form of van't Hoff's law than for equation 9.

Equations 9 and 10 do not predict precisely the osmotic pressures of real solutions. At the concentrations of many substances in cytoplasm and extracellular fluids, the deviations from ideality may be substantial. For example, sodium is the principal cation of the extracellular fluids, and chloride is the main anion. Na^+ is present at about 150 mEq/L and Cl^- at about 120 mEq/L. NaCl solutions in this concentration range differ considerably in their osmotic pressures from the predictions of van't Hoff's law.

One way of correcting for the deviations of real solutions from the predictions of van't Hoff's law is to use a correction factor called the **osmotic coefficient** (ϕ). Including the osmotic coefficient, equation 10 becomes

$$\pi = RT\phi ic \tag{11}$$

The osmotic coefficient may be greater or less than one. It is less than one for electrolytes of physiological importance, and for all solutes it approaches one as the solution becomes more and more dilute. The term **ϕic** can be regarded as the osmotically effective concentration, and ϕic often is referred to as the **osmolar concentration,** with units in osmoles per liter. Values of the osmotic coefficient depend on the concentration of the solute and on its chemical properties. Table 1-2 lists osmotic coefficients for several solutes. These values apply fairly well at the concentrations of these solutes in the extracellular fluids of mammals. The value of ϕ may be highly concentration dependent. More precise values of ϕ can be obtained from tables in handbooks that list values of ϕ for different substances as functions of concentration. Solutions of proteins deviate greatly from van't Hoff's law, and different proteins may deviate to different extents. The deviations from ideality are frequently more concentration dependent for proteins than for smaller solutes.

Table 1-2 Osmotic coefficients (ϕ) of certain solutes of physiological interest

Substance	*i*	*Molecular weight*	ϕ
NaCl	2	58.5	0.93
KCl	2	74.6	0.92
HCl	2	36.6	0.95
NH_4Cl	2	53.5	0.92
$NaHCO_3$	2	84.0	0.96
$NaNO_3$	2	85.0	0.90
KSCN	2	97.2	0.91
KH_2PO_4	2	136.0	0.87
$CaCl_2$	3	111.0	0.86
$MgCl_2$	3	95.2	0.89
Na_2SO_4	3	142.0	0.74
K_2SO_4	3	174.0	0.74
$MgSO_4$	2	120.0	0.58
Glucose	1	180.0	1.01
Sucrose	1	342.0	1.02
Maltose	1	342.0	1.01
Lactose	1	342.0	1.01

Reproduced with permission from Lifson N, Visscher MB: *Osmosis in living systems*. In Glasser O, editor: *Medical physics,* vol 1. 1944, Chicago, Year Book Medical Publishers, Inc.

Sample calculations

1. What is the osmotic pressure (at 0° C) of a 154 mM NaCl solution?

$$\pi = RT\phi ic \tag{12}$$

Taking ϕ for NaCl from Table 1-2, we obtain

$$\pi = 22.4 \text{ L-atm/mol} \times 0.93 \times 2 \times 0.154 \text{ mol/L} = 6.42 \text{ atm} \tag{13}$$

2. What is the osmolarity of this solution?

$$\text{Osmolarity} = \phi ic = 0.93 \times 2 \times 0.154 = 0.286 \text{ osmole/L} = 286 \text{ mosmolar}$$

Measurement of osmotic pressure. The osmotic pressure of a solution can be obtained by determining the pressure required to prevent water from entering the solution across a semipermeable membrane (Fig. 1-7). However, this method is time consuming and technically difficult. Consequently the osmotic pressure more often is estimated from another colligative property, such as depression of the freezing point. The relation that describes the depression of the freezing point of water by a solute is

$$\Delta T_f = 1.86\ \phi ic \tag{14}$$

where ΔT_f is the freezing point depression in degrees centigrade. Thus the effective osmotic concentration (in osmoles per liter) is

$$\phi ic = \frac{\Delta T_f}{1.86} \tag{15}$$

When the freezing point depression of a multicomponent solution is determined, the effective osmolar concentration (in osmoles per liter) of the solution as a whole can be obtained.

If the total osmotic pressures of two solutions (as measured by freezing point depression or by the osmotic pressure developed across a true semipermeable membrane) are equal, the solutions are said to be **isosmotic** (or isoosmotic). If solution A has greater osmotic pressure than solution B, A is said to be **hyperosmotic** with respect to B. If solution A has less total osmotic pressure than solution B, A is said to be **hypoosmotic** to B.

Osmotic Swelling and Shrinking of Cells

The plasma membranes of most of the cells of the body are relatively impermeable to many of the solutes of the interstitial fluid but are highly permeable to water. Therefore when the osmotic pressure of the interstitial fluid is increased, water leaves the cells by osmosis, the cells shrink, and cellular solutes become more concentrated until the effective osmotic pressure of the cytoplasm is again equal to that of the interstitial fluid. Conversely, if the osmotic pressure of the extracellular fluid is decreased, water enters the cells, and the cells swell until the intracellular and extracellular osmotic pressures are equal.

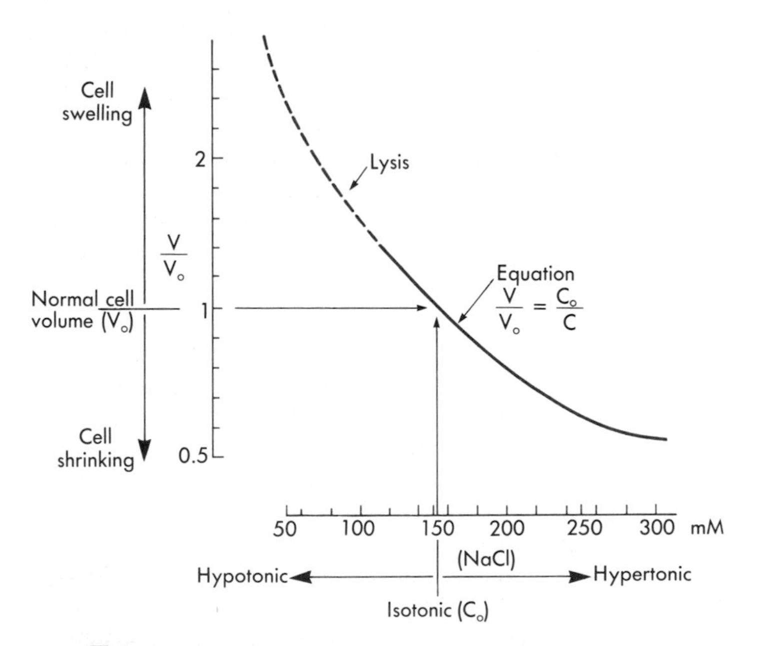

Fig. 1-8 The osmotic behavior of human red blood cells in NaCl solutions. At 154 mM NaCl (isotonic) the red cell has its normal volume. It shrinks in more concentrated (hypertonic) solutions and swells in more dilute (hypotonic) solutions.

Red blood cells often are used to illustrate the osmotic properties of cells because they are readily obtained and are easily studied. Within a certain range of external solute concentrations the red cell behaves as an osmometer, became its volume is inversely related to the solute concentration in the extracellular medium. In Fig. 1-8 the red cell volume, as a fraction of its normal volume in plasma, is shown as a function of the concentration of NaCl solution in which the red cells are suspended. At a NaCl concentration of 154 mM (308 mM particles) the volume of the cells is the same as their volume in plasma; this concentration of NaCl is said to be **isotonic** to the red cell. A concentration of NaCl greater than 154 mM is called **hypertonic** (cells shrink), and a solution less concentrated than 154 mM is termed **hypotonic** (cells swell). When red cells have swollen to about 1.4 times their original volume, some cells **lyse** (burst). At this volume the properties of the red cell membrane abruptly change, and hemoglobin leaks out of the cell. The membrane becomes transiently permeable to other large molecules at this point as well.

The intracellular substances that produce an osmotic pressure that just balances that of the extracellular fluid include hemoglobin, K^+, organic phosphates (such as ATP and 2,3-diphosphoglycerate), and glycolytic intermediates. Regardless of the chemical nature of its contents, the red cell behaves as though it were filled with a solution of **impermeant** molecules with an osmotically effective concentration of 286 milliosmolar (154 mM NaCl = 286 milliosmolar).

Osmotic effects of permeant solutes. Permeating solutes eventually equilibrate across the plasma membrane. For this reason permeating solutes exert only a transient effect on cell volume.

Consider a red blood cell placed in a large volume of 0.154 M NaCl, containing 0.050 M glycerol. Because of the extracellular NaCl and glycerol, the osmotic pressure of the extracellular fluid will initially exceed that of the cell interior, and the cell will shrink. With time, however, glycerol will equilibrate across the plasma membrane of the red cell, and the cell will swell back toward its original volume. The steady-state volume of the cell will be determined only by the impermeants in the extracellular fluid. In this case the impermeant (NaCl) has a concentration that is isotonic, so that the final volume of the cell will be equal to the normal red cell volume. Because the red cell ultimately returns to its normal volume, the solution (0.050 M glycerol in 0.154 M NaCl) is isotonic. Because the red cell initially shrinks when put in this solution, the solution is hyperosmotic with respect to the normal red cell. The transient changes in cell volume depend on equilibration of glycerol across the membrane. Had we used urea (a more rapidly permeating substance), the cell would have reached steady-state volume sooner.

The following rules help predict the volume changes a cell will undergo when suspended in solutions of permeant and impermeant solutes:

1. *The steady-state volume of the cell is determined only by the concentration of impermeant solutes in the extracellular fluid.*
2. *Permeant solutes cause only transient changes in cell volume.*
3. *The time course of the transient changes is more rapid the greater the permeance of the permeant solute.*

The magnitudes of osmotic flows caused by permeating solutes. In the preceding example it was explained that permeants, such as glycerol, exert only a transient osmotic effect. It is sometimes important to determine the magnitude of the osmotic effect exerted by a particular permeant.

When a difference of hydrostatic pressure (ΔP) causes water flow across a membrane, the rate of water flow ($\dot{V}_w$) is

$$\dot{V}_w = L\Delta P \qquad (16)$$

where L is a constant of proportionality called the **hydraulic conductivity.**

Osmotic flow of water across a membrane is directly proportional to the osmotic pressure difference ($\Delta\pi$) of the solutions on the two sides of the membrane; thus

$$\dot{V}_w = L\Delta\pi \qquad (17)$$

Equation 17 holds only for osmosis caused by impermeants; permeants cause less osmotic flow. The greater the permeance of a solute, the less the osmotic flow it causes. Table 1-3 shows the osmotic water flows induced across a porous membrane by solutes of different molecular size. The solutions have identical freezing points, so the total osmotic pressures are the same. Note that the larger the solute molecule, and thus the more impermeant it is to the membrane, the greater the osmotic water flow it causes.

Table 1-3 Osmotic water flow across a porous dialysis membrane caused by various solutes*

Gradient producing the water flow	*Net volume flow (μl/minute)**	*Solute radius (Å)*	*Reflection coefficient (σ)*
D_2O	0.06	1.9	0.0024
Urea	0.6	2.7	0.024
Glucose	5.1	4.4	0.205
Sucrose	9.2	5.3	0.368
Raffinose	11	6.1	0.440
Inulin	19	12	0.760
Bovine serum albumin	25.5	37	1.02
Hydrostatic pressure	25		

Data from Durbin RP: J Gen Physiol 44:315, 1960. Reproduced from *The Journal of General Physiology* by copyright permission of The Rockefeller University Press.

*Flow is expressed as microliters per minute caused by a 1 M concentration difference of solute across the membrane. The flows are compared with the flow caused by a theoretically equivalent hydrostatic pressure.

Equation 17 can be rewritten to take solute permeance into account by including σ, the **reflection coefficient.**

$$\dot{V}_w = \sigma L\Delta\pi \qquad (18)$$

σ is a dimensionless number that ranges from 1 for completely impermeant solutes down to 0 for extremely permeant solutes. σ is a property of a particular solute and a particular membrane and represents the osmotic flow induced by the solute as a fraction of the theoretical maximum osmotic flow (Table 1-3).

Protein-Mediated Membrane Transport

Certain substances enter or leave cells by way of specific carriers or channels that are intrinsic proteins of the plasma membrane. Transport via such protein carriers or channels is called **protein-mediated transport,** or simply **mediated transport.** Specific ions or molecules may cross the membranes of mitochondria, endoplasmic reticulum, and other organelles by mediated transport. Mediated transport systems include **active transport** and **facilitated transport** processes. As discussed later, active transport and facilitated transport have a number of properties in common. The principal distinction between these two processes is that active transport is capable of "pumping" a substance against a gradient of concentration (or electrochemical potential), whereas facilitated transport tends to equilibrate the substance across the membrane.

Properties of mediated transport. The basic properties of mediated transport are the following:

1. Transport is more rapid than that of other molecules of similar molecular weight and lipid solubility that cross the membrane by simple diffusion.
2. The transport rate shows **saturation kinetics:** as the concentration of the transported compound is increased, the rate of transport at first increases; but eventually a concentration is reached above which the transport rate increases no further. At this point the transport system is said to be saturated with the transported compound.
3. The mediating protein has **chemical specificity:** only molecules with the requisite chemical structure are transported. The specificity of most transport systems is not absolute, and, in general, it is broader than the specificity of most enzymes. Glucose is transported into red blood cells by facilitated transport. Glucose is the preferred sub-

strate, but mannose, galactose, xylose, L-arabinose, and certain other sugars also are transported by the same membrane protein. D-Arabinose, however, is a poor substrate for the glucose system, and sorbitol and mannitol do not enter red cells at all. Mediated processes have **stereospecificity** as well. In the red cell sugar transport system, D-glucose is transported well, but L-glucose is barely transported at all.

4. Structurally related molecules may compete for transport. Typically the presence of one transport substrate will decrease the transport rate of a second substrate by competing for the transport protein. The competition is analogous to **competitive inhibition** of an enzyme.
5. Transport may be inhibited by compounds that are not structurally related to transport substrates. An inhibitor may bind to the transport protein in a way that decreases its affinity for the normal transport substrate. The compound phloretin does not resemble a sugar molecule, yet it strongly inhibits red cell sugar transport. Active transport systems, which require some link to metabolism, may be inhibited by metabolic inhibitors. The rate of Na^+ transport out of cells by the Na^+, K^+-ATPase is decreased by substances that interfere with ATP generation.

Facilitated Transport

Facilitated transport, sometimes called facilitated diffusion, occurs via a transport protein that is not linked to metabolic energy. Facilitated transport has the properties discussed previously, except that facilitated transport is not generally depressed by metabolic inhibitors. Because they have no link to energy metabolism, facilitated transport processes cannot move uncharged substances against concentration gradients or ions against electrochemical potential gradients. Facilitated transport systems act to equalize concentrations (or electrochemical potentials of ions) in the cytoplasm and extracellular fluid of the substances they transport.

Monosaccharides enter muscle cells by a facilitated transport process. Glucose, galactose, arabinose, and 3-O-methylglucose compete for the same carrier. The rate of transport shows saturation kinetics. The nonphysiological stereoisomer L-glucose enters the cells very slowly, and nontransported sugars such as mannitol or sorbose enter muscle cells very slowly if at all. Phloretin inhibits sugar uptake. The sugar transport system in muscle cells is stimulated by insulin. In the absence of insulin, glucose transport is rate limiting for glucose use, so that insulin is a major physiological regulator of muscle glucose metabolism.

The molecular details of protein-mediated transport processes are not well understood. Current evidence suggests that most transport proteins span the membrane and are multimeric. Fig. 1-9 depicts a model that has been proposed for the monosaccharide transport protein of the membrane of the human red blood cell.

Active Transport

Active transport processes have most of the properties of facilitated transport and, in addition, can concentrate their substrates against electrochemical potential gradients. This requires energy; hence active transport processes must be linked to energy metabolism in some way. Active transport systems may use ATP directly, or they may be linked more indirectly to metabolism. Because of their dependence on metabolism, active transport processes may be inhibited by any substance that interferes with energy metabolism.

One possible general mechanism for active transport. Mediated transport is analogous in many respects to an enzyme-catalyzed chemical reaction. The rate (V)

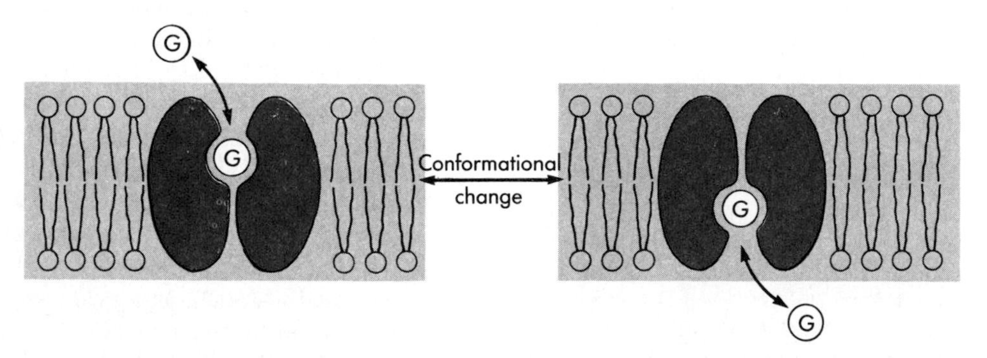

Fig. 1-9 An illustration of an alternating conformation model of glucose transport across the membrane of the human red blood cell. At a given instant a glucose binding site is accessible from only one side of the membrane. The binding of glucose from one side of the membrane is postulated to promote a conformational change of the glucose transporter that exposes the glucose binding site to the other side of the membrane. The glucose transporter is illustrated as a dimer, but the state of aggregation of the functional carrier is controversial.

of transport of substrate (S) is described by a Michaelis-Menten equation:

$$V = \frac{V_{max}[S]}{K_m + [S]} \quad (17)$$

where $\mathbf{V_{max}}$ *is the maximum rate of transport and* $\mathbf{K_m}$ *is equal to the concentration of substrate for which the rate of transport is equal to* $V_{max}/2$. When the transport process is in the steady state across the plasma membrane, the unidirectional flux from cytoplasm to extracellular fluid is exactly equal to the unidirectional flux from extracellular fluid to cytoplasm. Because influx equals efflux, there is no net flux across the plasma membrane and the cellular concentration is constant in the steady state.

For a facilitated transport process across the plasma membrane, influx and efflux are often symmetrical processes. In that case, in the terms of enzyme kinetics, influx and efflux have the same K_m and V_{max}. Imagine that the K_m of a glucose transport system is 1 mM. When the glucose concentration in cytoplasm and extracellular fluid are 1 mM, influx and efflux will be equal (each being equal to $V_{max}/2$). This example illustrates that the steady state of a facilitated transport system is characterized by equilibration of substrate across the membrane.

With an active transport system the influx and efflux processes are not symmetrical. Consider an imaginary transport system for glycine that has a K_m for influx of 0.5 mM, a K_m for efflux of 5 mM, and the same V_{max} for both influx and efflux. When the intracellular glycine concentration is 5 mM and the extracellular glycine concentration is 0.5 mM, influx of glycine will equal glycine efflux (both being equal to $V_{max}/2$), and the steady state will be achieved. Note that the system is in steady state with a glycine concentration 10 times higher inside the cell than outside; therefore this is an active transport system.

In creating the tenfold concentration difference of glycine the transport system does work. The ultimate source of the energy to do transport work is metabolic energy. How is metabolic energy harnessed to do transport work? One way might involve the cyclical phosphorylation and dephosphorylation of the transport protein. In the glycine transport system just described imagine that phosphorylation of the protein at the inner aspect of the membrane (by a protein kinase using ATP) converts the protein to the form in which K_m at the inner surface of the plasma membrane is 5 mM. If dephosphorylation occurs, the protein "reverts" to the form in which K_m at the outer face of the plasma membrane is 0.5 mM. In this way the K_m for efflux is made greater than that for the influx, allowing a tenfold concentration difference to be created. Because the imaginary glycine active transport system described uses ATP directly, this is termed a **primary active transport system,** as would be any transport system that is rather directly linked to any high-energy metabolic intermediate.

Another way in which active transport is powered involves a less direct link to metabolic energy. The transmembrane concentration difference of a second molecule that itself is transported actively may be used to allosterically modify the affinity of the transport protein for its substrate. Sodium is actively extruded from most cells by the Na, K pump (Na^+, K^+-ATPase) by primary active transport, so that the extracellular concentration of Na^+ is often about tenfold higher than its intracellular concentration. In the hypothetical glycine transport system described previously, imagine that at the high Na^+ concentrations present in the extracellular fluid, Na^+ binds to the transport carrier at the external face of the membrane and that Na^+ binding converts the protein to the form in which K_m at the outer face of the membrane is 0.5 mM. At the inner surface of the membrane the Na^+ concentration is much lower. Therefore much less Na^+ will bind to the transport protein, and it will be predominantly in the form in which K_m at the inner membrane surface is 5 mM. In this way the transmembrane gradient of Na^+ (which is created by primary active transport) is harnessed to bring about a transmembrane concentration difference of glycine (in this hypothetical example). In this case glycine transport is said to occur by **secondary active transport.**

Mechanisms of active transport processes

Transport powered by the phosphorylation of a transport protein. In the cytoplasm of most animal cells the concentration of Na^+ is much less and the concentration of K^+ much more than the extracellular concentrations of these ions. This situation is brought about by the action of a Na^+, K^+ pump in the plasma membrane that pumps Na^+ out of the cell and K^+ into the cell. The Na^+, K^+ pump activity is the result of an integral membrane protein called the Na^+, K^+-ATPase. The Na^+, K^+-ATPase consists of a "catalytic" α-subunit of about 100,000 daltons and a glycoprotein β-subunit of about 50,000 daltons. When operating near its capacity for ion transport, the Na^+, K^+-ATPase transports three sodium ions out of the cell and transports two potassium ions into the cell for each ATP hydrolyzed. The cyclical phosphorylation and dephosphorylation of the protein causes it to alternate between two conformations, E1 and E2. In E1 the ion-binding sites of the protein have high affinity for Na^+ and face the cytoplasm. In E2 the ion-binding sites favor the binding of K^+ and face the extracellular fluid.

A simplified view of the ion transporting cycle of the Na^+, K^+-ATPase is as follows (Fig. 1-10). The E1 conformer has high affinities for intracellular Na^+, Mg^{++}, and ATP. When these three ligands are bound, the protein is phosphorylated at an aspartate residue near the ATP binding site, and the Na^+ binding site becomes "occluded," (i.e., inaccessible from either side

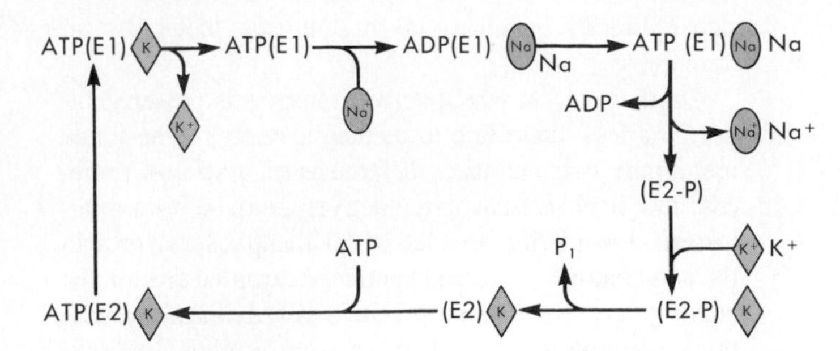

■ **Fig. 1-10** A simplified view of the reaction cycle of the Na^+, K^+, ATPase. Ligands shown on the inside of the cycle bind or are released on the inside of the plasma membrane; ligands shown on the outside of the cycle bind or are released on the outside of the plasma membrane. (Adapted from Post RL: A perspective on sodium and potassium ion transport adenosine triphosphatase. In Mukohata Y, Packer L, editors: *Cation fluxes across biomembranes,* New York, 1979, Academic Press.)

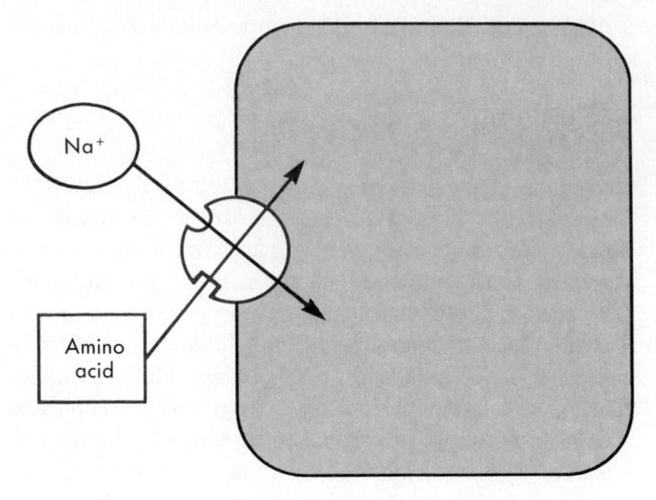

■ **Fig. 1-11** Many cells take up certain neutral amino acids by secondary active transport. The transport protein binds both Na^+ and the amino acid. Na^+ is transported down its electrochemical potential gradient, and the protein uses the energy released by Na^+ flux to transport the amino acid against a concentration gradient.

of the membrane). The ATPase then is thought to undergo a conformational change that makes the ion binding sites accessible to the extracellular fluid and reduces their affinity for Na^+. Na^+ then dissociates on the extracellular side of the membrane, and the protein undergoes a conformational change to E2-P. E2-P has a high affinity for K^+ at the extracellular side of the membrane and proceeds to bind two K^+ ions. The binding of K^+ promotes the hydrolysis of the acylphosphate bond, and the K^+ binding sites then move to an occluded position. Then the binding of ATP to the $E2{\cdot}K_2$ promotes the conformational change back to the E1 state, and dissociation of K^+ occurs at the cytoplasmic face of the membrane. In this way, the Na^+, K^+-ATPase is thought to alternate between the E1 and the E2 conformations, a process termed molecular peristalsis.

Because the Na^+, K^+-ATPase uses the energy in the terminal phosphate bond of ATP to power its transport cycle, it is said to be a **primary active transport** system. A transport process powered by some other high-energy metabolic intermediate or linked directly to a primary metabolic reaction would also be classified as primary active transport.

Transport powered by the gradient of another species: secondary active transport. The last section emphasized that energy is required to create a concentration gradient of a transported substance. Once created, a concentration gradient (or the electrochemical potential gradient of an ion) represents a store of chemical potential energy that can be used to perform work. When the substance flows down its gradient, it releases energy that can be harnessed to do work. For example, in the mitochondrion, the electrochemical potential gradient of H^+ ions created across the inner mitochondrial membrane by electron transport is used to perform the

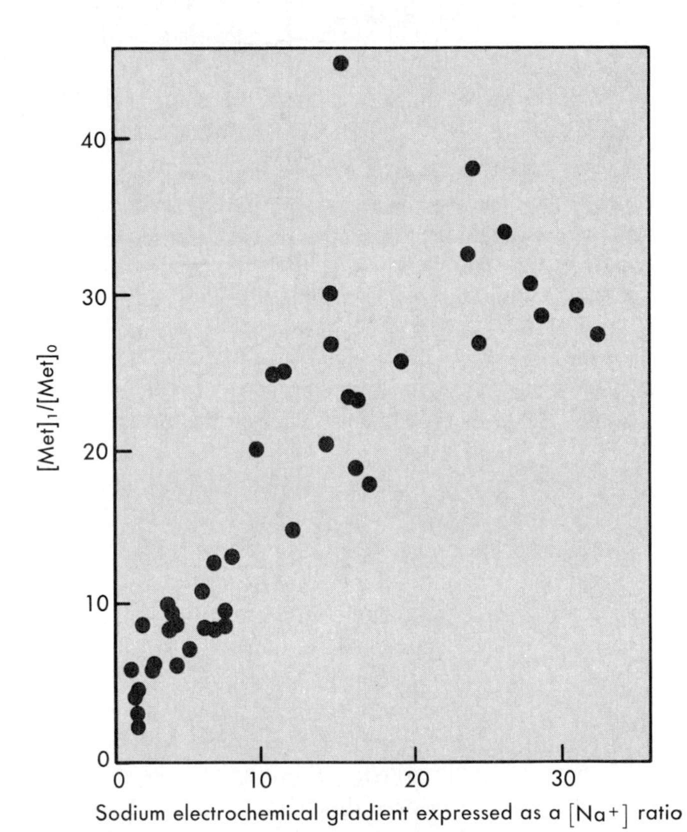

■ **Fig. 1-12** The extent to which methionine can be accumulated by mouse ascites-tumor cells depends on the gradient of the electrochemical potential of Na^+ across the plasma membrane. The ratio of steady-state intracellular to extracellular concentrations of methionine is plotted against the steady-state ratio of the electrochemical gradient of the Na^+ ion across the membrane. (Modified from Philo R, Eddy AA: *Biochem J* 174:811, Biochemical Society, London.)

work of ATP synthesis. In many cell types the electrochemical potential gradient of Na^+, created by the Na^+, K^+-ATPase, is used to actively transport certain other solutes into the cell. Many cells take up neutral, hydrophilic amino acids by a membrane transport protein that links the inward transport of Na^+ down its electrochemical potential gradient to the inward transport of amino acids against their gradients of electrochemical potential (Fig. 1-11). The energy for the transport of the amino acid is not provided directly by ATP or some other high-energy metabolite but indirectly from the gradient of another species that is itself actively transported. Hence the amino acid is said to be transported by **secondary active transport.** In the secondary active transport of amino acids, both the rate of amino acid transport and the extent to which the amino acid is accumulated (Fig. 1-12) are dependent on the electrochemical potential gradient of Na^+.

■ *Other Important Membrane Transport Processes*

Cells contain numerous transport systems, both active and facilitated, that function to transport solutes across the plasma membrane and across the membranes that bound intracellular organelles. The application of molecular cloning techniques, which started in the middle 1980s, has enhanced our knowledge of membrane transport protein molecules. However, the detailed molecular mechanisms of transport remain to be elucidated. The scope of this book permits the discussion of only a few of the many important transport processes and transport proteins.

Ion-transporting ATPases. Ion-transporting ATPases are central to the lives of all cells from *Archebacteria* to *Homo sapiens*. There are three classes of ion transporting ATPases (Fig. 1-13).

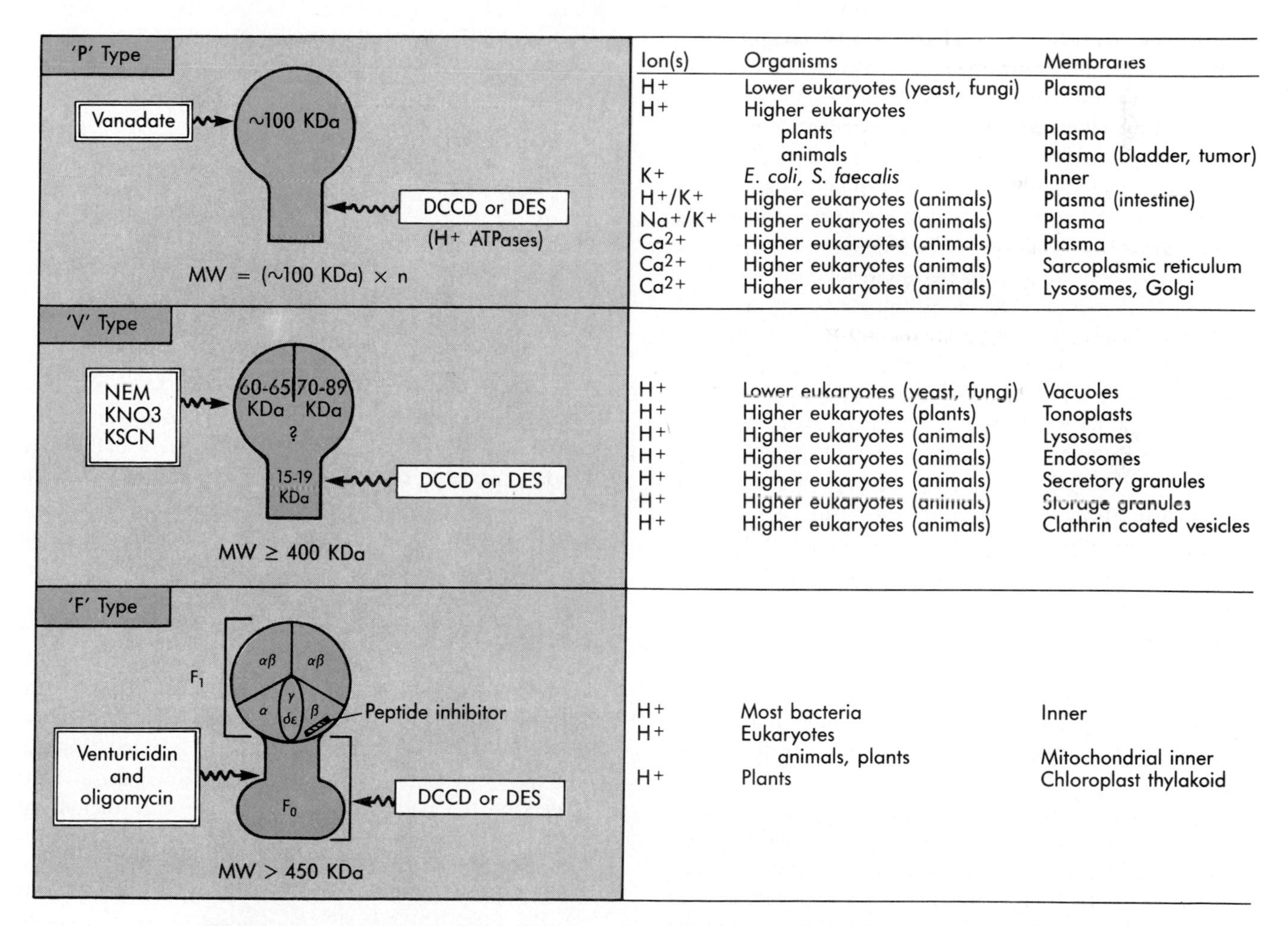

Ion(s)	Organisms	Membranes
H+	Lower eukaryotes (yeast, fungi)	Plasma
H+	Higher eukaryotes	
	plants	Plasma
	animals	Plasma (bladder, tumor)
K+	*E. coli, S. faecalis*	Inner
H+/K+	Higher eukaryotes (animals)	Plasma (intestine)
Na+/K+	Higher eukaryotes (animals)	Plasma
Ca2+	Higher eukaryotes (animals)	Plasma
Ca2+	Higher eukaryotes (animals)	Sarcoplasmic reticulum
Ca2+	Higher eukaryotes (animals)	Lysosomes, Golgi
H+	Lower eukaryotes (yeast, fungi)	Vacuoles
H+	Higher eukaryotes (plants)	Tonoplasts
H+	Higher eukaryotes (animals)	Lysosomes
H+	Higher eukaryotes (animals)	Endosomes
H+	Higher eukaryotes (animals)	Secretory granules
H+	Higher eukaryotes (animals)	Storage granules
H+	Higher eukaryotes (animals)	Clathrin coated vesicles
H+	Most bacteria	Inner
H+	Eukaryotes	
	animals, plants	Mitochondrial inner
H+	Plants	Chloroplast thylakoid

■ **Fig. 1-13** The three classes of ion-transporting ATPases. The left panel represents the subunit structures of P, V, and F types ATPases and shows inhibitors of each class (*NEM*, N-ethylmaleimide; *DES*, diethylstilbesterol; *DCCD*, dicyclohexylcarbodiimide). The number of different subunits of F type ATPases is also indicated. The right panel lists examples of ATPases of each class. (Modified from Pedersen PL, Carafoli E: *Trends Biochem Sci* 12:146, 1987.)

P type ATPases. These ATPases are so named because their transport cycle involves a phosphorylated intermediate (see Fig. 1-10). The Na^+, K^+-ATPase of animal cell plasma membranes, discussed above, is a P type ATPase. So are the Ca^{++}-ATPases of plasma membrane, sarcoplasmic reticulum, and endoplasmic reticulum, and the H^+, K^+-ATPase of the gastric parietal cell that is responsible for acid secretion in the stomach. P type ATPases are frequently called E1-E2 ATPases because their transport cycle involves two distinct conformation states (see Fig. 1-10).

V type ATPases. The membranes of various intracellular organelles such as lysosomes, endosomes, secretory vesicles, and storage granules contain V type ATPases. These ATPases are H^+-ATPases that acidify the interior of these organelles; their activity is essential for the function of lysosomes and for the storage of neurotransmitters, such as epinephrine, norepinephrine, serotonin, and acetylcholine.

F type ATPases. The mitochondrial inner membrane contains F type ATPases, also known as F_1F_0 ATPases. Whereas the normal functions of P and V type ATPases are to use the energy of ATP to create an ion concentration gradient, the mitochondrial F type ATPase usually uses the energy of the H^+ gradient across the inner mitochondrial membrane that is established by electron transport to synthesize ATP. (Because all chemical reactions are reversible, under appropriate conditions P and V type ATPases can use an ion energy gradient to synthesize ATP, and an F type ATPase can use ATP to pump protons.) In the economy of the cell the mitochondrial F type ATPase is the major source of ATP, and the P and V type ion-transporting ATPases are major consumers (sinks) of ATP (Fig. 1-14).

Structures of P type ATPases. We have previously summarized how the Na^+, K^+-ATPase uses the energy of ATP to actively transport Na^+ out of the cell and to actively accumulate K^+ in the cytosol. We have also previously mentioned the role of the Na^+, K^+-ATPase in regulation of cellular volume and the use of the Na^+ gradient produced by the Na^+, K^+-ATPase as a source of energy for transporting other substances, such as sugars and amino acids. The next few chapters explain how the Na^+ and K^+ gradients that result from the activity of the Na^+, K^+-ATPase help create the resting membrane potential and the ionic currents that are responsible for action potentials and for synaptic transmission.

The Na^+, K^+-ATPase is a prominent member of a family of ion-transporting ATPases. In animal cells this family includes (a) the plasma membrane and sarcoplasmic and endoplasmic reticulum Ca^{++}-ATPases that help regulate the level of cytosolic Ca^{++}, and (b) the K^+, H^+-ATPase of gastric parietal cells that secrete HCl into gastric juice. The members of this family of ion transporting ATPases have much in common.

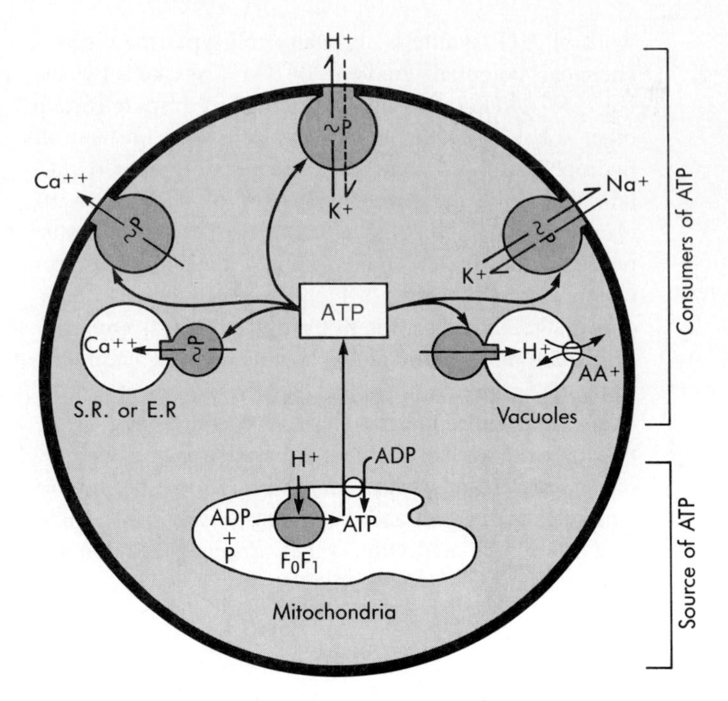

■ **Fig. 1-14** Ion-transporting ATPases are the major source of ATP and important consumers of ATP in cells. (Modified from Pedersen PL, Carafoli E: *Trends Biochem Sci* 12:146, 1987.)

Fig. 1-15 compares hydropathy plots of the Na^+, K^+-ATPase with plots of the sarcoplasmic reticulum Ca^{++}-ATPase. In these plots an upward deflection indicates a stretch of hydrophobic amino acids and a downward deflection represents a stretch of hydrophilic amino acids. A stretch of 20 to 25 predominantly hydrophobic amino acids can form an α-helix that is embedded in and spans the membrane. Many integral membrane proteins have several **membrane-spanning helices.** The Na^+, K^+-ATPase and Ca^{++}-ATPase are each believed to have 8 to 10 membrane-spanning helices. The hydropathy profiles of the Na^+, K^+-ATPase and Ca^{++}-ATPase are strikingly similar, suggesting that the number of transmembrane helices and the overall shape of the proteins are similar.

The amino acid sequence of gastric H^+, K^+-ATPase is 64% homologous to Na^+, K^+-ATPase. By contrast, the degree of amino acid sequence homology of sarcoplasmic reticulum Ca^{++}-ATPase to Na^+, K^+-ATPase is less than 30%, which is much less striking than the similarity of their hydropathy plots. However, as indicated by the parallelograms in Fig. 1-15, regions of high homology between the two proteins exist; such regions include the ATP binding domain and the site where the proteins are phosphorylated. A stretch of 10 amino acids that includes the phosphorylated aspartate residue is identical among the Na^+, K^+, Ca^{++}, and H^+, K^+-ATPases.

Fig. 1-16 is a current model of the secondary structure of the Ca^{++}-ATPase of sarcoplasmic reticulum.

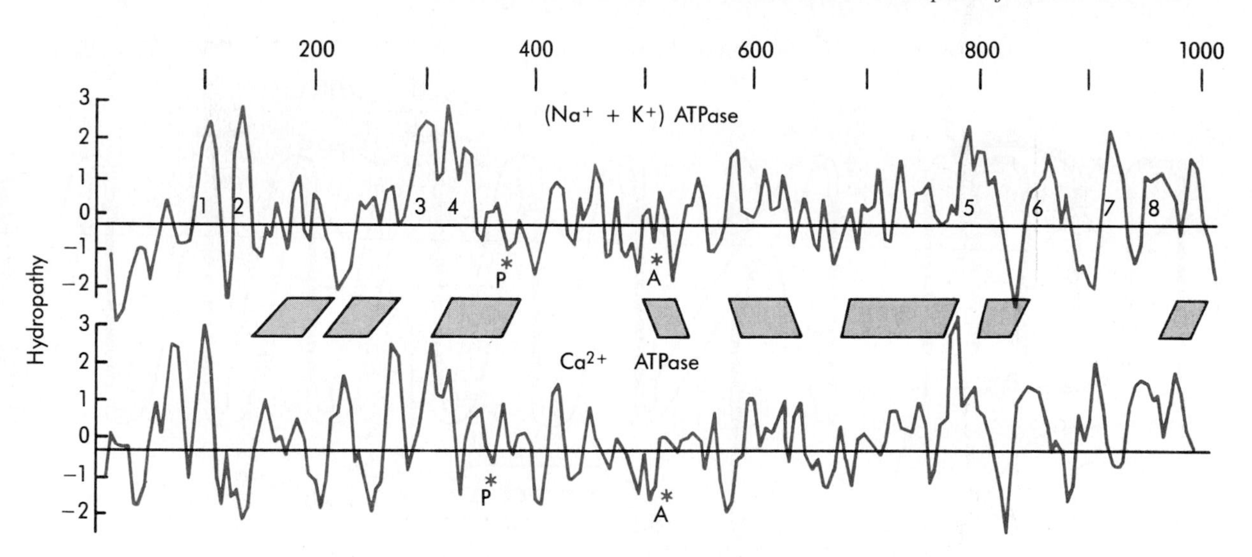

Fig. 1-15 Hydropathy plots of Na^+, K^+-ATPase and Ca^{++}-ATPase are strikingly similar. An upward deflection indicates a stretch of predominantly hydrophobic amino acids and downward deflection represents a sequence that is predominantly hydrophilic. Eight putative transmembrane helices are indicated by 1 to 8. The ATP-binding sites *(A)* and the aspartate that is phosphorylated *(P)* are marked by asterisks. The parallelograms indicate regions of high sequence homology between the two ATPases. (Redrawn from Shull GE, Schwartz A, Lingrel JB: *Nature* 316:691, 1985.)

Mutation of the anionic and hydrophilic residues (indicated by colored arrows) interferes with the binding of Ca^{++} or weakens the effect of Ca^{++} on the reaction cycle of the Ca^{++}-ATPase. The details of how the energy of ATP is harnessed to actively transport Ca^{++} remain obscure.

Calcium transport. Under most circumstances the concentration of Ca^{++} in the cytosol of cells is maintained at rather low levels, below 10^{-7} M, whereas the concentration of Ca^{++} in extracellular fluids is of the order of a few mM. The plasma membranes of most cells contain a Ca^{++}-ATPase that helps to establish the large gradient of Ca^{++} across the plasma membrane. The plasma membrane Ca^{++}-ATPase is a P type ATPase that is distinct from the Ca^{++}-ATPase of the sarcoplasmic reticulum of muscle.

A plasma membrane Ca^{++}-ATPase is present in red blood cells, cardiac muscle, smooth muscle, neurons, liver cells, kidney tubule cells, intestinal epithelial cells, and several other cell types. The Ca^{++}-ATPase of the plasma membrane of the human red blood cell is the best characterized plasma membrane Ca^{++}-ATPase, and it serves as a model of the Ca^{++}-ATPases of the plasma membranes of most other cells. The plasma membrane Ca^{++}-ATPase is an integral membrane protein of 130,000 to 140,000 molecular weight. The number of Ca^{++} ions pumped for each ATP hydrolyzed is still controversial; estimates range from 1 to 2 Ca^{++}/ATP. In keeping with the vital role of Ca^{++} as a controller of cellular processes, the activity of the Ca^{++}-ATPase is highly regulated. Intracellular **calmodulin,** the ubiquitous calcium-dependent regulatory protein, in the presence of micromolar levels of Ca^{++}, increases the affinity of the ATPase for Ca^{++} at the inner surface of the plasma membrane and increases its rate of turnover. The activity of the Ca^{++}-ATPase is also regulated by the phosphatidylinositol cycle: phosphatidylinositol-4,5-bisphosphate enhances the activity of the Ca^{++}-ATPase.

The Na^+-Ca^{++} exchanger. In most cells the rate at which Ca^{++} leaks into the cell down its electrochemical potential gradient is rather slow, so that the energy cost of maintaining a low intracellular level of Ca^{++} is not high. This is in contrast to the cost of pumping Na^+ and K^+; running the Na^+, K^+-ATPase is a major item in the energy budget of many cells.

Many excitable cells, those of the heart for example, have an additional mechanism for controlling the level of intracellular Ca^{++}. A sodium/calcium exchange protein in the plasma membrane uses the energy in the Na^+ gradient to extrude Ca^{++} from the cell. In heart cells it appears that rapid, transient changes in intracellular Ca^{++} are mediated by the sodium/calcium exchange protein, whereas the resting level of intracellular Ca^{++} is set primarily by the Ca^{++}-ATPase.

The secondary structure of the Na^+-Ca^{++} exchanger from canine cardiac sarcolemma has been deduced from its amino acid sequence (Fig. 1-17, *A*). It has numerous (12) transmembrane α-helices, as is also the case for many other transport proteins. The Na^+-Ca^{++} exchanger has a large cytoplasmic domain that contains a calmodulin-binding site and that accounts for about half

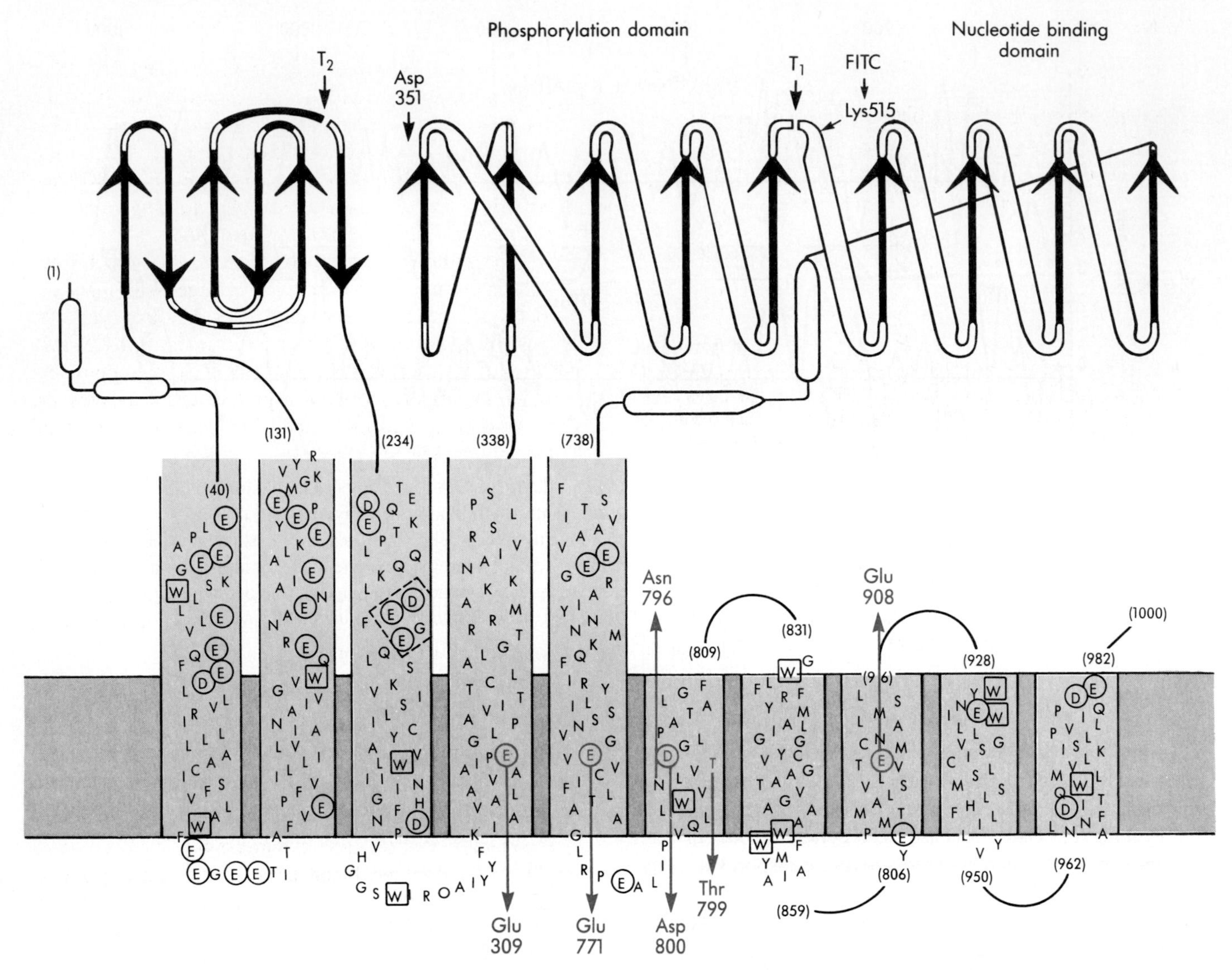

■ **Fig. 1-16** A representation of the secondary structure of the Ca^{++}-ATPase of sarcoplasmic reticulum. T_1 and T_2 are sensitive trypsin cleavage sites. Asp351 is the residue that is phosphorylated as part of the transport cycle. Lys515 is labeled by FITC (fluorescein isothiocyanate). Tryptophans are indicated by squares and acidic residues by circles. Mutations of the anionic or hydrophilic residues indicated by colored arrows diminish Ca^{++} binding or interfere with the effect of Ca^{++} on the reaction cycle. (Redrawn from Inesi G, Sumbilla C, Kirtley ME: *Physiol Rev* 70:749, 1990.)

of its molecular weight. Binding of the Ca^{++}-calmodulin complex to its binding domain stimulates the activity of the sodium-calcium exchanger.

The Na^+-H^+ exchanger. Most cells contain a protein that catalyzes the one-for-one exchange of a sodium ion for a proton. This protein functions to prevent acidification of the cytosol. When the pH of the cytosol is nearly neutral, the Na^+-H^+ exchanger has a very low affinity for H^+ ions and is almost inactive. Acidification of the cytoplasm increases the affinity of the protein for H^+. Na^+ flows into the cell in exchange for H^+ and the cell pH rises toward neutrality. Treatment of cells with certain growth factors, tumor promoters, and mitogens results in phosphorylation of the Na^+-H^+ exchanger and increases its affinity for H^+. This causes the exchanger to be active at neutral pH and leads to persistent alkalinization of the cytosol. The mechanisms whereby alkalinization of the cytosol leads to stimulation of cell division are not understood.

The sodium-hydrogen exchanger (Fig. 1-17, *B*) has 10 transmembrane α-helices that account for about half its molecular weight. A large C-terminal, cytoplasmic domain accounts for the other half.

Anion exchange proteins. Essentially all cells contain an anion exchange protein in their plasma membranes. Three members of this protein family have been

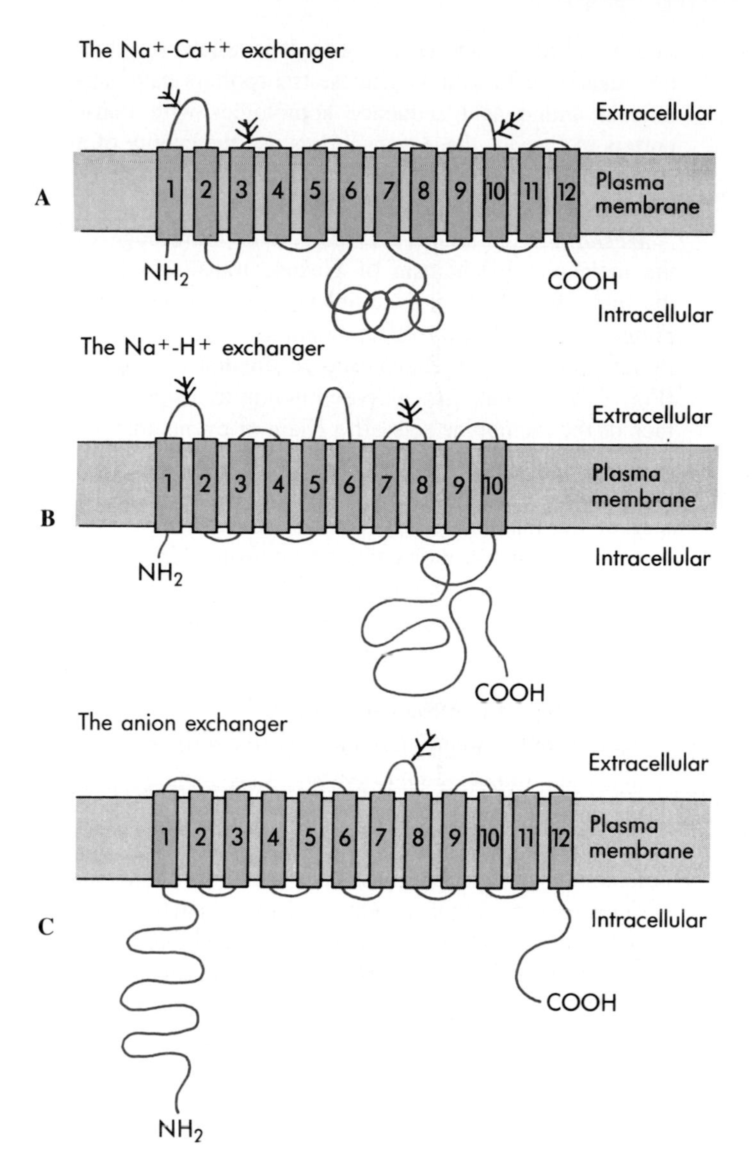

■ **Fig. 1-17** Secondary structure of three ion exchange proteins. **A,** The sodium-calcium exchanger. **B,** The sodium-hydrogen exchanger. **C,** The anion exchanger.

discovered in animal cells. The best-characterized anion exchanger is the Band 3 protein of the erythrocyte membrane. This protein exchanges an extracellular univalent anion for an intracellular one; it will transport various univalent anions. Physiologically the anions present in greatest concentrations are Cl^- and HCO_3^-. Hence the protein is also known as the **chloride-bicarbonate exchanger.** *In the mammalian red blood cell the chloride-bicarbonate exchanger plays an important role in the transport of CO_2 from the tissues to the lungs and the unloading of CO_2 from the blood in the lungs* (Chapter 36).

The anion exchanger helps to prevent the pH of the cytosol from becoming basic. An increase in cytosolic pH shifts the equilibrium of carbonic acid toward elevated bicarbonate:

$$H_2CO_3 \rightleftharpoons H^+ + HCO_3^-$$

The elevated cytosolic bicarbonate activates the efflux of bicarbonate in exchange for chloride and shifts the cytosolic pH toward neutral. *The cytosolic pH is thus kept near neutral by the combined actions of the anion exchanger, (which responds to changes in an alkaline direction) and the sodium-hydrogen exchanger, (which counteracts changes toward an acid pH).*

The erythrocyte anion exchanger (Fig. 1-17, *C*) has 12 transmembrane α-helices and a large N-terminal cytoplasmic domain. The latter has binding sites for ankyrin, a protein which connects it to the cytoskeleton.

Na-K-Cl cotransporters. Many cells, both epithelial and nonepithelial, contain a protein that catalyzes the cotransport of Na^+, K^+, and Cl from extracellular fluid to cytosol. The stoichiometry is 1 Na: 1 K: 2 Cl, and thus the transport is electroneutral. This cotransport can be specifically inhibited by the diuretic drug, bumetanide. Na-K-Cl cotransport is stimulated in many cell types by cell shrinkage, which results in an influx of all three ions that generates an osmotic force to restore cell volume. Na-K-Cl cotransporters play a major role in the regulation of cell volume.

Facilitated transport of glucose. Glucose is poorly soluble in hydrophobic solvents and is sufficiently large so that it very slowly enters cells by simple diffusion. In keeping with the importance of glucose as a metabolic fuel, almost all types of cells take up glucose, galactose, and certain other sugars by facilitated transport. More than 50 years ago measurements of glucose transport in human red blood cells and in sheep placenta demonstrated that simple diffusion could not account for the rate of glucose uptake or for the occurence of substrate saturation, competitive inhibition by certain other sugars, stereospecificity, and other properties of glucose transport. W.F. Widdas invoked a model of a membrane transport protein in 1952 to explain these research findings, and that model remains plausible today.

Species of glucose transporters. Biochemical and molecular biological studies have defined five closely related species of glucose facilitated transporters (GLUT) in eukaryotic cells:

GLUT1: Red cells, muscle, brain, thymocytes, and adipocytes
GLUT2: Liver, pancreatic islet cells, and the basolateral membranes of kidney and intestinal epithelial cells
GLUT3: Fetal skeletal muscle
GLUT4: Insulin-responsive glucose transporter in muscle and fat cells
GLUT5: Intestine

Table 1-4 shows the tissue distribution of GLUT1, GLUT2, and GLUT4 in the adult rat.

Each member of this protein family, GLUT1 to GLUT5, has a high degree of amino acid sequence homology with the others and a similar postulated struc-

■ **Table 1-4** Distribution of different species of glucose transporters in various tissues. Northern blot analysis of mRNAs isolated from the listed rat tissues using cDNA probes specific for GLUT1, GLUT2, and GLUT4

	Glucose carrier type		
Tissue	***GLUT1***	***GLUT2***	***GLUT4***
Brain	+	−	−
Testis	+	−	−
Spleen	+	−	−
Thymus	+	−	−
Muscle			
Soleus	+	−	+
Diaphragm	+	−	+
Heart	+	−	+
Gastrocnemius	+	−	+
Aorta and posterior vena cava	+	±	+
Fat pad	+	−	+
Kidney	+	+	±
Urinary bladder	+	−	±
Liver	−	+	−
Pancreas			
Total	−	±	−
Islets	−	+	−
Stomach	+	−	−
Small intestine	±	+	−
Adrenal gland	±	−	−

± means that a low level of that mRNA was detected. (Modified from Charron MJ, Brosius FC III, Lodish HF: *Proc Nat Acad Sci USA* 86:2535, 1989.)

ture in the membrane (Fig. 1-18). It is noteworthy that the eukaryotic facititated glucose transporters have significant amino acid sequence homologies with transporters in *E. coli;* these transporters use the energy of a H^+ gradient to power the secondary active uptake of sugar from the medium.

Mechanism of sugar transport. Our knowledge of the molecular mechanism of glucose translocation by the transporter remains incomplete. Whether the transporter exists primarily as a monomer, dimer, or tetramer is not settled. **Alternating conformation models** (Fig. 1-9) postulate that glucose binding to a site at one face of the membrane sets off a chain of events that are mediated by conformational changes of the transporter that expose that same glucose binding site to the other face of the membrane. By contrast, **fixed-site models** permit two glucose molecules on opposite sides of the membrane to bind simultaneously to the transporter. Presently it cannot be determined which class of models explains the function of glucose transporters.

Regulation of glucose transporters

Insulin-mediated stimulation of glucose transport

In certain tissues, notably skeletal muscle and adipose tissue, insulin dramatically accelerates glucose uptake. In these cell types a large pool of glucose transporters, predominantly the GLUT4 class, is present in membrane vesicles in the cytosol. On insulin stimulation the vesicles fuse with the plasma membrane and thereby insert their glucose transporters into the plasma membrane (Fig. 1-19), where they function to enhance the rate of glucose uptake. When insulin is removed, glu-

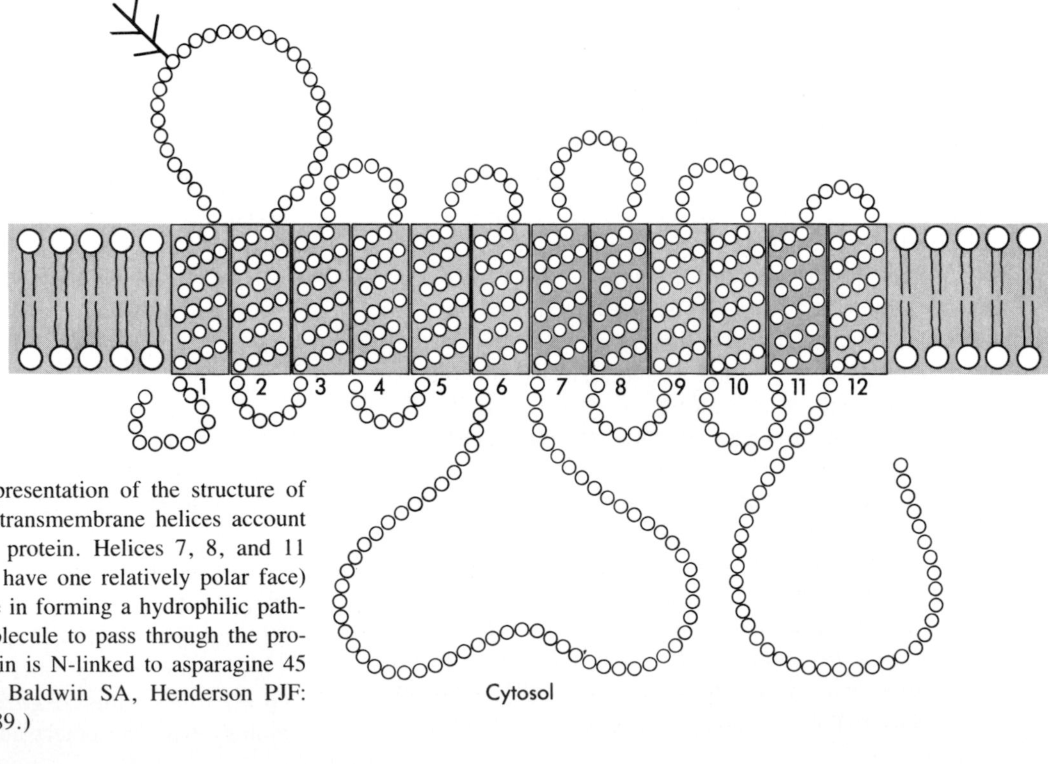

■ **Fig. 1-18** Schematic representation of the structure of glucose transporters. The 12 transmembrane helices account for much of the mass of the protein. Helices 7, 8, and 11 (color) are amphipathic (they have one relatively polar face) and are believed to participate in forming a hydrophilic pathway that allows a glucose molecule to pass through the protein. A carbohydrate side chain is N-linked to asparagine 45 as indicated. (Modified from Baldwin SA, Henderson PJF: *Annu Rev Physiol* 51:459, 1989.)

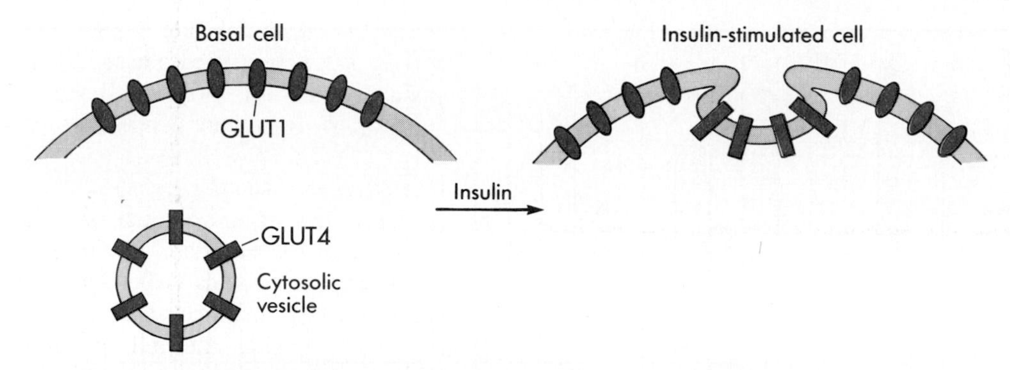

■ **Fig. 1-19** Schematic depiction of the stimulation of glucose transport in a muscle or fat cell in response to insulin. In the basal state GLUT1 transporters are present in the plasma membrane and GLUT4 transporters predominate in a pool of cytosolic vesicles. Insulin causes the cytosolic vesicles to fuse with the plasma membrane thereby increasing the total number of glucose transporters in the plasma membrane. Insulin may also act to increase the transport rate of individual transporters.

cose transporters are again internalized to diminish the number in the plasma membrane. Insulin stimulates sugar uptake more than is predicted from the number of additional glucose transporters that are added to the membrane. The basis for this additional effect of insulin is not settled. The GLUT4 transporters that are inserted into the plasma membrane may be more active than the GLUT1 transporters that are responsible for basal glucose uptake. Alternatively, other effects of insulin, probably mediated by second messenger mechanisms (see Chapter 5), may enhance the functions of glucose transporters in the plasma membrane. The ability of insulin to stimulate glucose transport requires the tyrosine kinase activity of the activated insulin receptor (see Chapter 5).

Metabolic regulation of glucose transport

The glucose uptake capacity of a number of different cell types is modulated in accordance with metabolic requirements. In human red blood cells and molluscan nerve cells, adenine nucleotides mediate the regulation of glucose transport; apparently the nucleotides act directly on glucose transporters that are already present in the plasma membrane. In these cells ATP inhibits glucose transport, but ADP and AMP stimulate the activity of glucose transporters.

Anoxia in the heart and exercise in skeletal muscle dramatically stimulate the rate of glucose uptake. In both tissues these effects are additive to those of insulin and for that reason are judged to occur by separate mechanisms. These stimulatory effects in cardiac and skeletal muscle may involve some recruitment of new transporters to the plasma membrane, but a larger component of the response is caused by enhanced activity of transporters that are already present in the plasma membrane. The stimulation of glucose transport in skeletal muscle on exercise occurs before significant changes have occurred in cytosolic nucleotide concentrations; the mediator of the changes in transport remains unknown.

In adipocytes glucose transport and lipolysis are regulated reciprocally. Lipolytic stimuli, such as the elevation of cellular cyclic AMP by β-adrenergic agonists, diminish glucose uptake by inhibiting the activity of transporters already present in the plasma membrane. This effect is mimicked by the activation of protein kinase C by phorbol esters (see Chapter 5). Antilipolytic agonists, such as adenosine, which lowers cyclic AMP, enhance glucose transport.

Amino acid transport. Most of the cells in the body synthesize proteins and therefore require amino acids. A number of different amino acid transport systems are present in plasma membranes. Among the amino acid transport systems present in the plasma membrane of most cell types are three distinct systems for neutral amino acids, a system for basic amino acids, and a system for acidic amino acids. Amino acid transport systems overlap significantly in specificities, and the distribution of the different systems varies from one cell type to another.

The three transport systems for neutral amino acids (Fig. 1-20) have been named the A system (prefers α-aminoisobutyric acid and other small, hydrophilic amino acids), the L system (prefers leucine and the more hydrophobic amino acids), and the ASC system (prefers alanine, serine, and cysteine). Systems A and ASC cotransport Na^+ along with a neutral amino acid and use the energy in the electrochemical potential gradient of Na^+ to actively transport the neutral amino acids into the cell against significant concentration gradients. The A and ASC systems are thus secondary active transport systems that use the gradient of Na^+ created by the Na^+, K^+-ATPase (a primary active transport system). The L amino acid transport system, on the other hand, does not depend on the Na^+ gradient or on

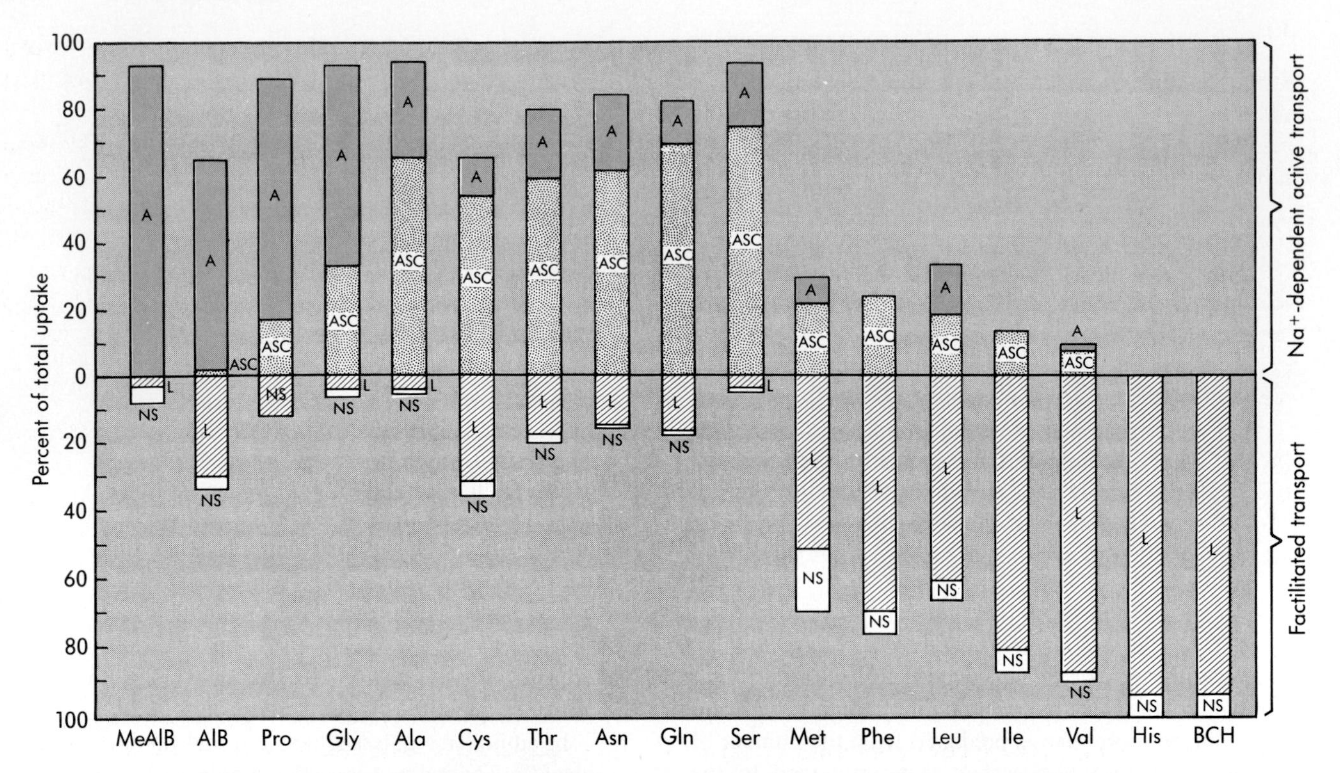

■ **Fig. 1-20** The overlapping specificities of the 3 major transporters for neutral amino acids in Chinese hamster ovary (CHO) cells. Properties of the A, ASC, and L transport systems are described in the text. NS (for nonsaturable) indicates the proportion of transport that is attributed to simple diffusion of the amino acid. MeAIB, AIB, and BCH are model amino acids. AIB, α-amino isobutyric acid; MeAIB, α-(methylamino)isobutyric acid; and BCH, 2-endoamino-bicycloheptane-2-carboxylic acid. (From Shotwell MA, et al: *J Biol Chem* 256:5422, 1981.)

metabolic energy. The L system is a facilitated transport system and functions to equilibrate the concentrations of the transported amino acids across the plasma membrane. The L system prefers the bulkier and more hydrophobic neutral amino acids, namely leucine, isoleucine, valine, phenylalanine, and methionine. At least 10 other transport systems for neutral amino acids exist in particular cell types, but these other systems are less well characterized and less widely distributed than the three just described.

Basic and acidic amino acids are not transported by the neutral amino acid transporters. A basic amino acid facilitated transporter that is present in almost all cells is the **y^+ transporter.** The y^+ transporter is a facilitated transporter with no link to metabolic energy. Multiple transporters for acidic amino acids have been discovered; the most widespread of these is the **X^-_{AG} transporter,** which is an Na^+-powered secondary active transporter. X^-_{AG} is present in nerve terminals of the central nervous system; these terminals employ glutamate or aspartate as neurotransmitter. X^-_{AG} functions to actively reaccumulate glutamate or aspartate that has been released into the synaptic clefts (See Chapter 4).

Transport of nucleosides and of nucleic acid bases. Nucleotides enter most cells very slowly. Nucleosides and nucleic acid bases enter mammalian cells by facilitated transport. After entering the cell, nucleosides are phosphorylated, and bases are phosphoribosylated. The rates of transport are greatly in excess of the rates of the intracellular reactions. A single facilitated transport system, with broad specificity, may be responsible for the transport of nucleosides. The transport system has higher affinities for purine nucleosides than for pyrimidine nucleosides; cytosine nucleosides have especially low affinity for the transport system.

The transport of the bases of nucleic acids is less well characterized. The number of distinct transport systems for nucleobases is a matter of controversy. Purines are transported much more rapidly than pyrimidines. Cytosine is the most poorly transported base; its transport may not be mediated by a membrane protein at all.

The uptake of purines and purine nucleosides by cells is vital for those cells that are incapable of de novo synthesis of purines. Membrane transport may be viewed as the first step in the salvage pathways for purine nucleotides. As mentioned, the rates at which extracellular purines and purine nucleosides are incorporated by cultured cells are *not* limited by the rate of membrane transport but rather by the rates of the intracellular phosphorylation or phosphoribosylation reactions.

Transport across epithelia. Epithelial cells are polarized with respect to their transport properties; (i.e., the transport properties of the plasma membrane facing one side of the epithelial cell layer are different from those of the membrane facing the other side).

The epithelial cells of the small intestine and the proximal tubule of the kidney are good examples of this polarity. The complement of membrane transport proteins in the brush border that faces the lumen of the small bowel or the renal tubule differs from the transport protein composition of the basolateral plasma membrane of the cell. The polarity of the transport proteins of epithelial cells is complete in that no transport protein is known to be present in the apical and basolateral plasma membrane of the same epithelial cell type. Moreover, epithelial cells contain transport proteins that are not found in nonepithelial cells. The brush border membranes of epithelial cells of the renal proximal tubule and small intestine contain Na^+-coupled transporters for neutral amino acids; these transporters are distinct from those of nonepithelial cells. Most epithelial cells contain an electrogenic Cl^- transporter (the one that is deficient in **cystic fibrosis).** In some epithelia this transporter is present in the apical membrane, but in other epithelia, it is in the basolateral membrane.

The tight junctions that join the epithelial cells side to side prevent mixing of the transport proteins of the luminal and basolateral plasma membranes. The brush border plasma membranes of these epithelia do not contain Na^+, K^+-ATPase molecules, which reside in the basolateral plasma membrane. Glucose (and galactose) and neutral amino acids enter these epithelial cells at the brush border by Na^+ gradient-driven secondary active transport systems but leave the cells at the basolateral membrane via facilitated transport systems (Fig. 1-21).

The tight junctions are leaky to water and small water-soluble molecules and ions. There are thus two pathways for transport across the epithelia—**transcellular,** through the cells, and **paracellular,** in between the cells (see Fig. 1-21). The nature and significance of transport across the epithelia of the intestine and the renal tubule are discussed in Chapters 40 and 41, respectively.

■ *Summary*

1. Membranes, consisting primarily of phospholipids and proteins, separate the cytosol from the extracellular fluid and enclose the various organelles of the cell. The selective permeabilities of membranes subserve the compartmentation of the various functions of the cell and the organelles.

2. Lipid-soluble solutes and very small water-soluble substances can cross membranes by simple diffusion. Larger water-soluble molecules require the mediation of particular membrane transport proteins.

3. Water can move across biological membranes by osmosis. The transport of Na^+ out of the cell by the Na^+, K^+-ATPase plays an important role in maintaining the osmotic balance of the cell. Other ion transport processes are also involved.

4. Protein-mediated transport may occur by facilitated transport or by active transport. Facilitated transport can

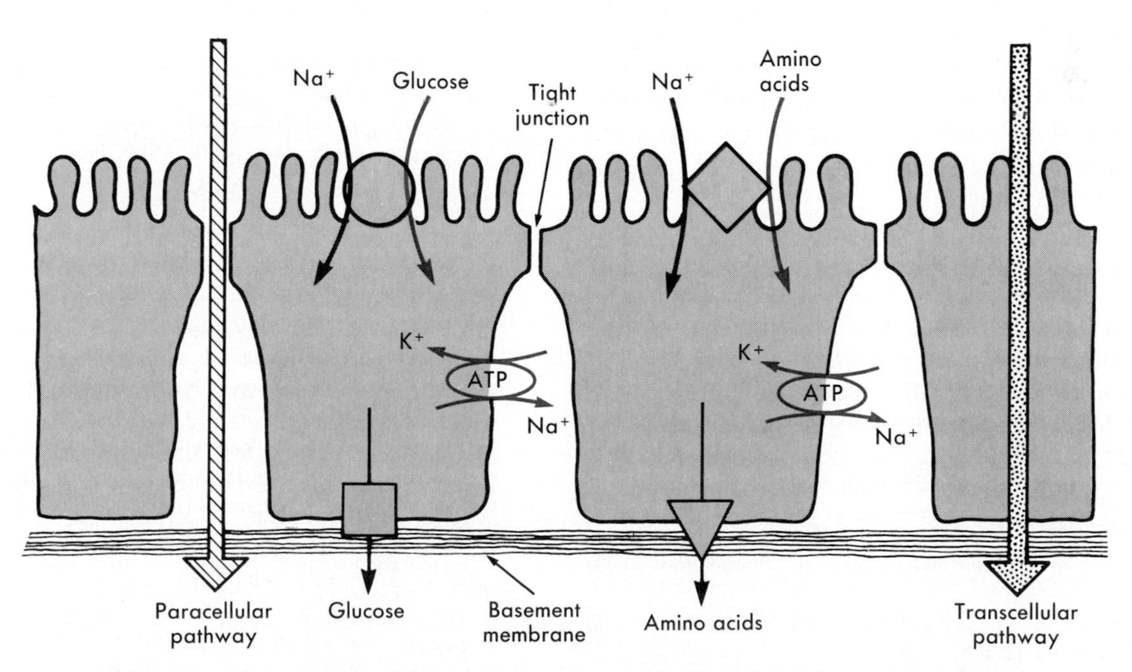

■ **Fig. 1-21** Epithelial transport processes that occur in the small intestine and renal proximal tubules. Epithelia are polarized such that the transport processes that take place on one side of the cell differ from those on the other side. Active fluxes are shown in color.

only transport a substance down its concentration gradient. Active transport is linked to metabolic energy and can transport a solute against its concentration gradient. Primary active transport proteins are directly linked to metabolic energy, most commonly by phosphorylation by ATP. Secondary active transport is driven by the concentration gradient of another actively transported solute, most frequently Na^+ ions.

5. Ion-transporting ATPases play a central role in the economy of the cell. The F_1F_0 ATPase of mitochondria is the major source of ATP. The P type ATPases of the plasma membrane and endoplasmic reticulum and the V type ATPases of lysosomes and other organelles are among the major consumers of ATP.

6. The exchange of ions across the plasma membrane is mediated by the members of a family of ion exchange proteins. Among them are the Na^+-Ca^{++} exchanger, the Na^+-H^+ exchanger, and the anion exchanger.

7. A family of glucose transporters promotes rapid cellular uptake of glucose from the extracellular fluid. In muscle and fat cells insulin markedly stimulates the uptake of glucose by fusing cytosolic vesicles, which contain preformed glucose transporters, with the plasma membrane.

8. Plasma membranes contain three or more different transporters that mediate the uptake of neutral amino acids and separate transport proteins for basic and acidic amino acids. Some are facilitated transporters, whereas others can actively accumulate amino acids powered by the energy of the Na^+ gradient.

9. Epithelia are tissues specialized for the vectorial transport of water and solutes. The apical and basolateral membranes of epithelia have completely different populations of transporters. The tight junctions that join epithelial cells side to side are not completely impermeable to water and small solutes.

Bibliography

Journal articles

Alper S: The band 3-related anion exchanger (AE) gene family, *Annu Rev Physiol* 53:549, 1991.

Baldwin SA, Henderson PJF: Homologies between sugar transporters from eukaryotes and prokaryotes, *Annu Rev Physiol* 51:459, 1989.

Carruthers A: Facilitated diffusion of glucose, *Physiol Rev* 70:1135, 1990.

Christensen HN: Role of amino acid transport and countertransport in nutrition and metabolism, *Physiol Rev* 70:43, 1990.

Fambrough DM: The sodium pump becomes a family, *Trend in Neural Sci* 11:325, 1988.

Haas M: Properties and diversity of (Na-K-Cl) cotransporters, *Annu Rev Physiol* 51:443, 1989.

Handler JS: Overview of epithelial polarity, *Annu Rev Physiol* 51:729, 1989.

Hediger MA, Coady MJ, Ikeda TS, Wright, EM: Expression cloning and cDNA sequencing of the Na^+/glucose co-transporter, *Nature* 330:379, 1987.

Horisberger JD, Lemas V, Kraehenbuhl JP, Rossier BC: Structure-function relationship of Na, K-ATPase, *Annu Rev Physiol* 53:565, 1991.

Inesi G, Sumbilla C, Kirtley ME: Relationships of molecular structure and function in Ca^{++}-transport ATPase, *Physiol Rev* 70:749, 1990.

Jay D, Cantley L: Structural aspects of the red cell anion exchange protein, *Annu Rev Biochem* 55:511, 1986.

Kopito H, Lodish HF: Primary structure and transmembrane orientation of the murine anion exchange protein, *Nature* 316:234, 1985.

Krapf R: Physiology and molecular biology of the renal Na/H antiporter, *Klin Wochenschr* 67:847, 1989.

MacLennan DH et al: Amino acid sequence of a Ca^{++} + Mg^{2+}-dependent ATPase from rabbit muscle sarcoplasmic reticulum, deduced from its complementary DNA sequence, *Nature* 316:696, 1985.

Mueckler M et al: Sequence and structure of a human glucose transporter, *Science* 229:941, 1985.

Nicoll DA, Longoni S, Philipson KD: Molecular cloning and functional expression of the cardiac sarcolemmal Na^+-Ca^{++} exchanger, *Science* 250:562, 1990.

Pedersen PL, Carafoli E: Ion motive ATPases. I. Ubiquity, properties, and significance to cell function, *Trends in Biochem Sci* 12:146, 1987.

Sardet C, Franchi A, Pouyssegur J: Molecular cloning, primary structure, and expression of the human growth factor-activatable Na^+/H^+ antiporter, *Cell* 56:271, 1990.

Shull GE, Schwartz A, Lingrel JB: Amino acid sequence of the catalytic subunit of the (Na^+ + K^+)ATPase deduced from a complementary DNA, *Nature* 316:691, 1985.

Walmsley AR: The dynamics of the glucose transporter, *Trends in Biochem Sci* 13:226, 1988.

Books and monographs

Aaronson PS, Boron W, editors: Na^+-H^+ exchange, intracellular pH, and cell function. In *Current topics in membrane transport, vol* 26, New York, 1986, Plenum Medical.

Andreoli TF, editor: *Physiology of membrane disorders,* New York, 1986, Plenum Medical.

Finkelstein A: *Water movement through lipid bilayers, pores, and plasma membranes: theory and reality,* New York, 1987, John Wiley.

Kaplan JH, De Weer P, editors: *The sodium pump: structure, mechanism, and regulation, 44th Symposium of the Society of General Physiologists,* New York, 1990, Rockefeller Press.

Klausner RD, Kemp C, Renswoude J van, editors: Membrane structure and function. In *Current topics in membranes and transport, vol 29,* Orlando, 1987, Academic Press.

Kotyk A, Janacek K, Koryta J: *Biophysical chemistry of membrane functions,* New York, 1988, Wiley-Interscience.

Poste G, Crooke ST, editors: *New insights into cell and membrane transport processes,* New York, 1986, Plenum Press.

Stein WH: *Channels, carriers, and pumps: an introduction to membrane transport,* San Diego, 1990, Academic Press.

CHAPTER

2

Ionic Equilibria and Resting Membrane Potentials

Most animal cells have an electrical potential difference (voltage) across their plasma membranes. The cytoplasm is usually electrically negative relative to the extracellular fluid. Because the electrical potential difference across the plasma membrane is present even in resting cells, it sometimes is referred to as the **resting membrane potential.** The resting membrane potential plays a central role in the excitability of nerve and muscle cells and in certain other cellular responses.

The major purpose of this chapter is to discuss the ways that electrochemical potential gradients of certain ions across the plasma membrane generate the resting membrane potential. The first part of the chapter deals with some fundamental definitions and concepts that describe the flow of ions across membranes.

Ionic Equilibria

Electrochemical Potentials of Ions

A membrane separates aqueous solutions in two chambers (A and B). Na^+ is at a higher concentration on side A than on side B. If there is no electrical potential difference between side A and side B, Na^+ will tend to diffuse from side A to side B, just as if it were an uncharged molecule. If, however, side A is electrically negative with respect to side B, the situation is more complex. The tendency for Na^+ to diffuse from side A to side B because of the concentration difference remains, but now Na^+ also tends to move in the opposite direction (from B to A) because of the electrical potential difference across the membrane. The direction of net Na^+ movement depends on whether the effect of the concentration difference or the effect of the electrical potential difference is larger. By comparing the two tendencies—concentration and electrical—one can predict the direction of net Na^+ movement.

The quantity that allows us to compare the relative contributions of ionic concentration and electrical potential is called the **electrochemical potential** (μ) of an ion. The electrochemical potential is defined as

$$\mu = \mu^\circ + RT\ln C + zFE \quad (1)$$

where

μ° = electrochemical potential of the ion at some reference state (say 1M concentration at 0° C with zero electrical potential)
R = gas constant
T = absolute temperature
lnC = natural log of concentration
z = charge number of the ion (+ 2 for Ca^{++}, −1 for Cl^-, etc.)
F = Faraday's number
E = electrical potential

The units of μ and of each term in equation 1 are energy per mole. The electrochemical potential represents the **chemical potential energy** possessed by a mole of ions resulting from their concentration and the electrical potential.

The net flow of an ion will be from where its electrochemical potential is higher to where its electrochemical potential is lower. Consider a membrane separating two chambers (A and B), each of which contains ion x in solution. The tendency of the ion to move from A to B is proportional to $\mu(x)$ on side A, and the tendency of the ion to move from B to A is proportional to $\mu(x)$ on side B. The *net* tendency for x to flow from A to B is $\mu_A(x) - \mu_B(x)$. This difference is the **electrochemical potential difference** of x across the membrane ($\Delta\mu$).

$$\Delta\mu(x) = \mu_A(x) - \mu_B(x) \quad (2)$$

Substituting into equation 2 the values of $\mu_A(x)$ and $\mu_B(x)$ derived from equation 1, we obtain

$$\mu_A(x) = \mu^\circ(x) + RT\ln[x]_A + zFE_A$$

$$\mu_B(x) = \mu^\circ(x) + RT\ln[x]_B + zFE_B$$

so that

$$\Delta\mu(x) = \mu_A(x) - \mu_B(x) = RT\ln\frac{[x]_A}{[x]_B} + zF(E_A - E_B) \quad (3)$$

The first term ($RT\ln [x]_A/[x]_B$) on the right-hand side of equation 3 is the tendency for the ion x to move from A to B because of the concentration difference, and the second term [$zF(E_A - E_B)$] is the tendency for the ion to move from A to B because of the electrical potential difference. The first term represents the chemical potential difference between a mole of x ions on side A and a mole of x ions on side B as a result of the concentration difference. The second term represents the chemical potential difference between a mole of x ions on side A and a mole of x ions on side B caused by the electrical potential difference between A and B. Thus $\Delta\mu$ describes the difference in chemical potential between a mole of x ions on side A and a mole of x ions on side B resulting from *both* concentration and electrical potential difference.

The x ions will tend to move from higher to lower electrochemical potential. $\Delta\mu$ is defined as the electrochemical potential of the ion on side A minus that on side B. If $\Delta\mu$ is positive, the ions will tend to move from A to B; if $\Delta\mu$ is zero, there is no net tendency for the ions to move at all; and if $\Delta\mu$ is negative, the ions will tend to move from side B to side A.

Electrochemical Equilibrium and the Nernst Equation

$\Delta\mu$ may be thought of as the *net* force on the ion, whereas $RT\ln [x]_A/[x]_B$ is the force caused by the concentration difference, and $zF(E_A - E_B)$ is the force caused by the electrical potential difference. When the two forces are equal and opposite, $\Delta\mu = 0$, there is no net force on the ion. When there is no net force on the ion, there will be no net movement of the ion, and the ion is said to be in **electrochemical equilibrium** across the membrane. At equilibrium $\Delta\mu = 0$. From equation (3), therefore

$$RT\ln\frac{[x]_A}{[x]_B} + zF(E_A - E_B) = 0 \quad (4)$$

Solving for $E_A - E_B$, we obtain

$$E_A - E_B = \frac{-RT}{zF}\ln\frac{[x]_A}{[x]_B} = \frac{RT}{zF}\ln\frac{[x]_B}{[x]_A} \quad (5)$$

Equation 5 is called the **Nernst equation.** The condition of equilibrium was assumed in its derivation, and *the Nernst equation is valid only for ions at equilibrium.* It allows one to compute the electrical potential difference, $E_A - E_B$, required to produce an electrical force, $zF(E_A - E_B)$, that is equal and opposite to the concentration force, $RT\ln ([x]_A/[x]_B)$, that tends to move the ion from A to B.

Use of the Nernst equation. It is often convenient to convert the Nernst equation to a form involving $\log_{10}$ rather than natural logarithms ($\ln x = 2.303 \log_{10} x$). Because biological potentials are usually expressed in millivolts (mV), the units of R are selected so RT comes out in millivolts. At 29.2° C the quantity 2.303 RT/F is equal to 60 mV. Because this quantity is proportional to the absolute temperature, it changes roughly by only 1/273 for each centigrade degree. Thus the value of 60 mV for 2.303 RT/F holds approximately for most experimental conditions, and a useful form of the Nernst equation is

$$E_A - E_B = \frac{-60\text{ mV}}{z}\log_{10}\frac{[x]_A}{[x]_B} = \frac{60\text{ mV}}{z}\log_{10}\frac{[x]_B}{[x]_A} \quad (6)$$

Examples of uses of the Nernst equation

Example 1. In Fig. 2-1 K^+ is 10 times more concentrated in chamber A than in chamber B. Following is a calculation of the electrical potential difference that must exist between the chambers for K^+ to be in equilibrium across the membrane.

Because we have specified that K^+ should be *in equilibrium,* the Nernst equation will hold.

$$E_A - E_B = \frac{-60\text{ mV}}{+1}\log_{10}\frac{[K^+]_A}{[K^+]_B} = -60\text{ mV}\log_{10}\frac{0.1}{0.01} = -60\text{ mV}\log(10) = -60\text{ mV} \quad (7)$$

The Nernst equation tells us that at equilibrium, side A must be 60 mV negative relative to side B. We can see that this polarity is correct because K^+ will tend to move from B to A due to the electrical force, which will counteract the tendency for it to move from A to B because of the concentration difference.

This example shows that *an electrical potential difference of about 60 mV is required to balance a tenfold concentration difference of a univalent ion.* This is a useful rule of thumb.

Example 2. In Fig. 2-2 the Nernst equation can help decide whether HCO_3^- is in equilibrium. If HCO_3^- is

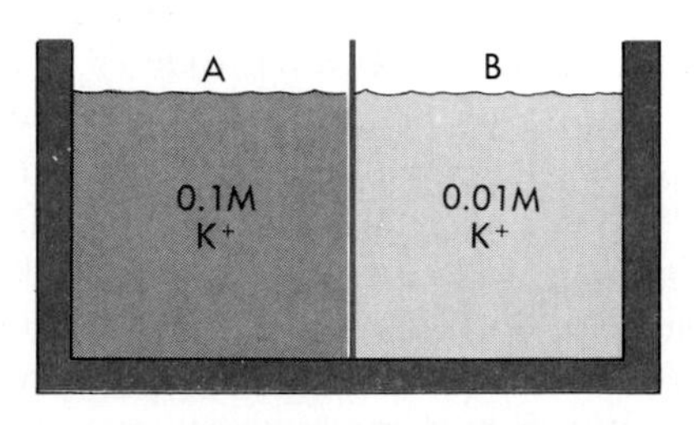

Fig. 2-1 A membrane separates chambers containing different K^+ concentrations. At an electrical potential difference ($E_A - E_B$) of −60 mV, K^+ is in electrochemical equilibrium across the membrane.

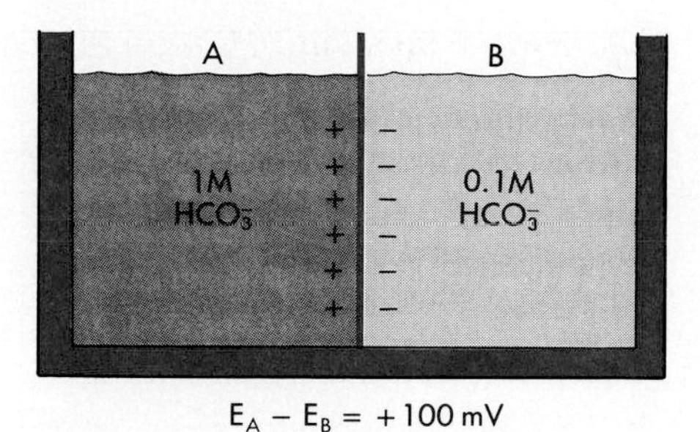

■ **Fig. 2-2** A membrane separates chambers that contain different HCO_3^- concentrations. $E_A - E_B = +100$ mV. HCO_3^- is not in electrochemical equilibrium. If $E_A - E_B$ were +60 mV, HCO_3^- would be in equilibrium. $E_A - E_B$ is stronger than it needs to be to just balance the tendency for HCO_3^- to move from *A* to *B* because of its concentration difference. Thus net movement of HCO_3^- from *B* to *A* will occur.

not in equilibrium, the Nernst equation determines the direction of net flow of HCO_3^-.

The Nernst equation tells us the electrical potential difference, $E_A - E_B$, that will *just balance* the concentration difference of HCO_3^- across the membrane.

$$E_A - E_B = \frac{-60 \text{ mV}}{z} \log_{10} \frac{[HCO_3^-]_A}{[HCO_3^-]_B}$$

$$= \frac{-60 \text{ mV}}{-1} \log_{10} \frac{1}{0.1} \quad (8)$$

$$= 60 \text{ mV} \log_{10} (10) = 60 \text{ mV}$$

Thus a potential difference of +60 mV between A and B would just balance the tendency of HCO_3^- to move from A to B because of its concentration difference. Because $E_A - E_B$ is actually +100 mV, the electrical potential is of the right sign to oppose the concentration force, but it is 40 mV larger than it needs to be to just balance the concentration force. Because the electrical force on HCO_3^- is larger than the concentration force, it will determine the net direction of HCO_3^- movement. Net HCO_3^- flow will occur from B to A.

In brief, the Nernst equation can be used to predict the direction that ions will tend to flow:

1. If the potential difference *measured* across a membrane is equal to the potential difference *calculated* from the Nernst equation for a particular ion, then *that particular ion* is in electrochemical equilibrium across the membrane, and there will be no net flow of that ion across the membrane.
2. If the measured electrical potential is of the same sign as that calculated from the Nernst equation for a particular ion but is larger in magnitude, then the electrical force is larger than the concentration force; net movement of that particular ion will tend to occur in the direction determined by the electrical force.
3. When the electrical potential difference is of the same sign but is numerically less than that calculated from the Nernst equation for a particular ion, then the concentration force is larger than the electrical force; net movement of that ion tends to occur in the direction determined by the concentration difference.
4. If the electrical potential difference measured across the membrane is of the sign opposite to that predicted by the Nernst equation for a particular ion, then the electrical and concentration forces are in the same direction. Thus that ion cannot be in equilibrium, and it will tend to flow in the direction determined by both electrical and concentration forces.

■ *Ionic Mechanisms*

In the discussion on electrical activity of nerve and muscle cells that follows, the principles of ionic equilibrium are used to explain the ionic mechanisms of the resting membrane potential and the action potential.

■ *The Gibbs-Donnan Equilibrium*

Cytoplasm typically contains proteins, organic polyphosphates, and other ionized substances that cannot permeate the plasma membrane. Cytoplasm also contains Na^+, K^+, Cl^-, and other ions to which the plasma membrane is somewhat permeable. The steady-state properties of this mixture of permeant and impermeant ions are described by the **Gibbs-Donnan equilibrium.**

Consider a membrane separating a solution of KCl from a solution of KY, where Y^- is an anion to which the plasma membrane is completely impermeable (Fig. 2-3). The membrane is permeable to water, K^+, and Cl^-. Suppose that *initially* on side A there is a 0.1M solution of KY, and an equal volume of 0.1M KCl is on side B. Because $[Cl^-]_B$ exceeds $[Cl^-]_A$, there will be a net flow of Cl^- from chamber B to chamber A. Negatively charged Cl^- ions flowing from side B to side A will create an electrical potential difference (side A negative) that will cause K^+ also to flow from side B to side A. Essentially the same number of K^+ ions as Cl^- ions will flow from side B to side A to preserve *electroneutrality*. The **principle of electroneutrality** states that any macroscopic region of a solution must have an equal number of positive and negative charges. In reality slight separation of charges does occur in certain situations, but in chemical terms the imbalance between positive and negative charges is very small and essentially impossible to measure by chemical techniques.

If enough time passes, those components of the system that can permeate the membrane—K^+ and Cl^- in this example—will come to equilibrium. At equilib-

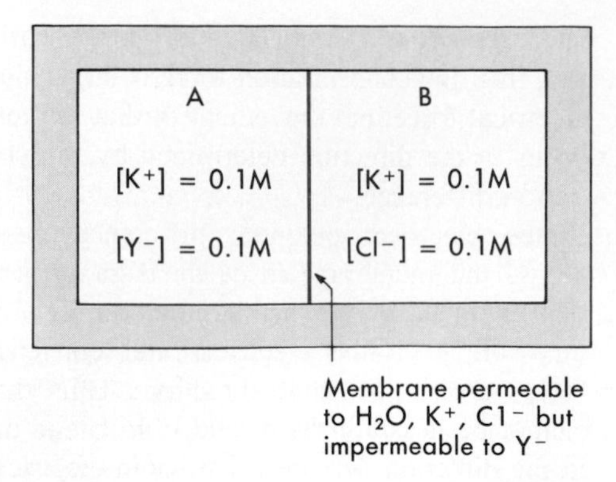

■ **Fig. 2-3** Before a Gibbs-Donnan equilibrium is established, a membrane separates two aqueous compartments. The membrane is permeable to H_2O, K^+, and Cl^- but impermeable to Y^-.

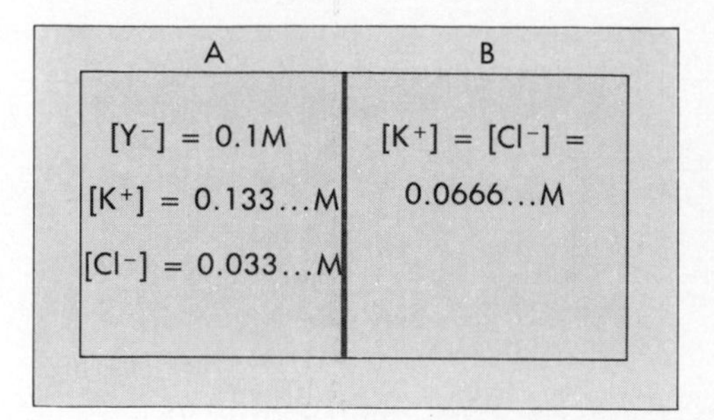

■ **Fig. 2-4** Ion concentrations after Gibbs-Donnan equilibrium has been attained. The initial ion concentrations were as shown in Fig. 2-3.

rium both $\Delta\mu_{K+}$ and $\Delta\mu_{Cl^-}$ must equal zero. From equation 3

$$\Delta\mu_{K+} = RT\ln\frac{[K^+]_A}{[K^+]_B} + F(E_A - E_B) = 0$$

$$\Delta\mu_{Cl^-} = RT\ln\frac{[Cl^-]_A}{[Cl^-]_B} - F(E_A - E_B) = 0 \qquad (9)$$

Adding these two equations and dividing the result by RT gives

$$\ln\frac{[K^+]_A}{[K^+]_B} + \ln\frac{[Cl^-]_A}{[Cl^-]_B} = 0 \qquad (10)$$

This gives

$$\ln\frac{[K^+]_A}{[K^+]_B} = -\ln\frac{[Cl^-]_A}{[Cl^-]_B} = \ln\frac{[Cl^-]_B}{[Cl^-]_A} \qquad (11)$$

Thus

$$\frac{[K^+]_A}{[K^+]_B} = \frac{[Cl^-]_B}{[Cl^-]_A} \qquad (12)$$

Cross multiplying gives

$$[K^+]_A[Cl^-]_A = [K^+]_B[Cl^-]_B \qquad (13)$$

Equation 13 is called the **Donnan relationship** (or the Gibbs-Donnan equation) and holds for any pair of univalent cation and anion in equilibrium between the two chambers. If other ions that could attain an equilibrium distribution were present, the same reasoning and an equation similar to equation 13 would apply to them as well.

Example of a Gibbs-Donnan equilibrium. The Donnan relationship and the principle of electroneutrality make it possible to determine the equilibrium concentrations of the components in the problem posed at the beginning of this section. The initial situation is shown in Fig. 2-3. If b represents the change in $[Cl^-]$ when Cl^- moves from B to A, the equilibrium value of $[Cl^-]_B$ can be denoted as 0.1 − b. By the electroneutrality principle $[K^+]_B = [Cl^-]_B = 0.1 - b$. If the volumes of A and B are kept the same, at equilibrium, $[Cl^-]_A = b$, and $[K^+]_A = 0.1 + b$. Substituting these concentrations into the Donnan relationship:

$$[K^+]_A\,[Cl^-]_A = [K^+]_B[Cl^-]_B$$
$$(0.1 + b)\,(b) = (0.1 - b)\,(0.1 - b) \qquad (14)$$

Solving this equation for b gives b = 0.0333 so that at equilibrium we obtain the concentrations shown in Fig. 2-4.

In this Gibbs-Donnan equilibrium both K^+ and Cl^- (but *not* Y^-) are in electrochemical equilibrium. This means that both K^+ and Cl^- must satisfy the Nernst equation, so that the equilibrium transmembrane potential difference can be computed from the Nernst equation for *either* K^+ or Cl^-.

$$\begin{aligned} E_A - E_B &= \frac{-60\text{ mV}}{+1}\log_{10}\frac{[K^+]_A}{[K^+]_B} \\ &= -60\text{ mV}\log_{10}\frac{0.1333}{0.0666} \\ &= \frac{-60\text{ mV}}{-1}\log_{10}\frac{[Cl^-]_A}{[Cl^-]_B} \qquad (15) \\ &= 60\text{ mV}\log_{10}\frac{0.0333}{0.0666} \\ &= -60\text{ mV}\log_{10}2 = -60\text{ mV}\,(0.3) \\ &= -18\text{ mV} \end{aligned}$$

Note that only the permeant ions attain equilibrium. The impermeant ion, Y^-, *cannot* reach an equilibrium distribution. What may not yet be evident is that water also will not achieve equilibrium, unless provision is made for that to occur. The total number of K^+ and Cl^- ions on side A in the preceding example exceeds that on side B. This is a general property of Gibbs-Donnan equilibria.

Taking the impermeant Y into account as well, the total concentration of osmotically active ions is considerably greater on side A than on side B. Because K^+ and Cl^- are present at equilibrium, they will have their full osmotic force even though the membrane is quite

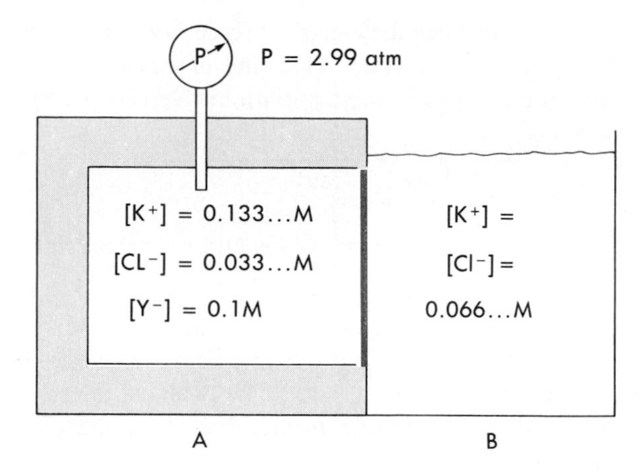

■ **Fig. 2-5** A hydrostatic pressure of 2.99 atm is required to prevent water from flowing from *B* to *A* in the Gibbs-Donnan equilibrium in Fig. 2-4.

permeable to them. Water will tend to flow by osmosis from side B to side A until the total osmotic pressure of the two solutions is equal. But then ions will flow to set up a new Gibbs-Donnan equilibrium, and that requires there be more osmotically active ions on the side with Y^-. All the water from side B will end up on side A unless water is restrained from moving.

Water can be restrained by enclosing the solution on side A in a rigid container (Fig. 2-5). Then, as fluid flows from side B to side A, pressure will build up in A and will oppose further osmotic water flow. The pressure in chamber A at equilibrium is equal to the difference between the total osmotic pressures of the solutions in chambers A and B. In this example the approximate hydrostatic pressure (P) in chamber A at equilibrium (at 0° C) is

$$\begin{aligned} P &= \Delta\pi_{K^+} + \Delta\pi_{Cl^-} + \Delta\pi_{Y^-} \\ &= RT\,(\Delta[K^+] + \Delta[Cl^-] + \Delta[Y^-]) \\ &= RT\,(0.06667 - 0.03333 + 0.1) \qquad (16) \\ &= (22.4 \text{ atm})\,(0.13333) = 2.99 \text{ atm} \end{aligned}$$

■ *Regulation of Cell Volume*

K^+ and Cl^- are nearly in equilibrium across many plasma membranes, and their distribution is influenced by the predominantly negatively charged impermeant ions, such as proteins and nucleotides, in the cytoplasm. K^+ and Cl^- approximately satisfy the Donnan relationship. This being the case, why does the osmotic imbalance discussed previously not cause the cell to swell and finally burst? One reason is that cells actively pump Na^+ out of the cytoplasm to the extracellular fluid, decreasing the osmotic pressure of the cytoplasm and increasing that of the extracellular fluid. Much of the pumping of Na^+ is done by the Na^+ pump, the Na^+, K^+-ATPase, in the plasma membrane. The Na^+, K^+-ATPase (p. 15) splits an ATP and uses some of the energy released to extrude 3 Na^+ from the cytoplasm and to pump 2 K^+ into the cell. Whereas K^+ is only slightly removed from an equilibrium distribution, Na^+ is pumped out against a large electrochemical potential difference.

When the ATP production of the cell is compromised (in the presence of metabolic inhibitors or low O_2 levels), or when the Na^+, K^+-ATPase is specifically inhibited (by cardiac glycosides), cells swell. Some investigators believe that the irreversible brain damage that occurs after acute oxygen deprivation is caused by the injury that results from inhibition of Na^+ pumping and the resulting osmotic swelling of brain cells. Consistent with this hypothesis, when brain cells are "preshrunk" with hypertonic solutions, they may have a greater ability to survive periods of low oxygen.

While the Na^+, K^+-ATPase is involved in volume regulation, other mechanisms play important roles in controlling cell volume. When cells are osmotically shrunk or osmotically swollen, ion transport processes are activated that return cell volume toward normal. In response to an initial osmotic shrinkage, cells gain Na^+ and Cl^-, water enters the cells osmotically, and the cells swell back toward normal volume. In response to an initial osmotic swelling, K^+ and Cl^- exit from the cells, and consequently the cells lose water.

Our knowledge of these volume-regulating ion fluxes is still incomplete. Some may be modulated by volume-dependent alterations in intracellular Ca^{++} concentration by calmodulin-dependent processes.

■ *Resting Membrane Potentials*

Communication between nerve cells depends on an electrical disturbance that is propagated in the plasma membrane and is called an **action potential.** In striated muscle an action potential propagates rapidly over the entire cell surface, allowing the cell to contract synchronously. The action potential and the ionic mechanisms that account for its properties are discussed in Chapter 3. All cells that are able to produce action potentials have sizable **resting membrane potentials** across their plasma membranes. Most inexcitable cells also have a resting membrane potential.

The resting membrane potential of many cells can be measured using glass microelectrodes that have tip diameters of about 0.1 μm and can puncture the plasma membrane of some cells without greatly injuring the cell. The electrical potential difference between the tip of a microelectrode inside a skeletal muscle cell and a reference electrode in the extracellular fluid is about −90 mV. The resting membrane potential is necessary for the cell to fire an action potential. If the resting membrane potential is decreased to −50 mV or less, the cell is no longer able to produce an action potential.

Ions that are actively transported are not in electrochemical equilibrium across the plasma membrane. It is shown later that the flow of ions across the plasma membrane, down their electrochemical potential gradients, is directly responsible for generating most of the resting membrane potential. To understand how the electrochemical potential gradient of an ion can give rise to a transmembrane difference in electrical potential, let us first consider a model system known as a **concentration cell.**

The Concentration Cell

In Fig. 2-6, *A*, the membrane that separates chambers A and B is permeable to cations but not to anions. Initially no electrical potential difference exists across the membrane. Na^+ will flow from A to B because of the concentration force acting on it. Cl^- has the same force on it but cannot flow from A to B because the membrane is impermeable to anions. The flow of Na^+ from A to B will transfer net positive charge to side B and leave a very slight excess of negative charges behind on side A, causing A to become electrically negative to side B (Fig. 2-6, *B*). This electrical force is oppositely directed to the concentration force on Na^+. The more Na^+ that flows, the larger the opposing electrical force. Net Na^+ flow will stop when the electrical force just balances the concentration force (i.e., when the electrical potential difference is equal to the equilibrium (Nernst) potential for Na^+):

$$E_A - E_B = \frac{-60 \text{ mV}}{+1} \log_{10} \frac{0.1}{0.01} = -60 \text{ mV} \log (10) = -60 \text{ mV} \quad (17)$$

Note that *only a very small amount of* Na^+ *flows* from A to B before equilibrium is reached. This is because the separation of positive and negative charges requires a large amount of work. The potential difference that builds up to oppose further Na^+ movement is a manifestation of that work.

The Na^+ *concentration difference in this example acts much like a battery.* The natural tendency for any ion that can flow is to seek equilibrium; thus Na^+ tends to flow until its equilibrium potential difference is established. As is explained later, in a system such as a cell membrane with more than one permeant ion, *each* ion "strives" to make the transmembrane potential difference equal to its equilibrium potential. The more permeant the ion, the greater its ability to force the electrical potential difference toward its equilibrium potential.

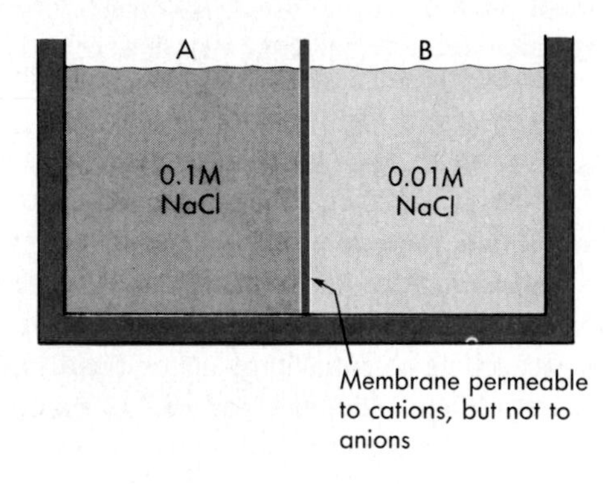

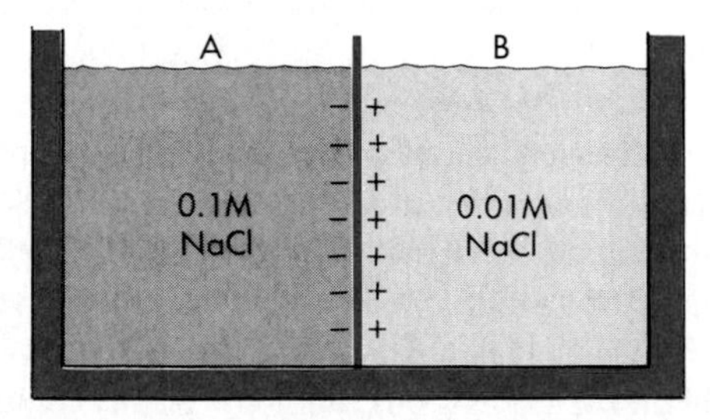

Fig. 2-6 *Top,* A concentration cell. The membrane, which is permeable to cations but not to anions, separates NaCl solutions of different concentrations. *Bottom,* The concentration cell after electrochemical equilibrium has been established. The flow of an infinitesimal amount of Na^+ generated an electrical potential difference across the membrane.

The Distribution of Ions across Plasma Membranes

In most tissues a number of ions are not in equilibrium between the extracellular fluid and the cytoplasm. Table 2-1 gives the concentrations of Na^+, K^+, and Cl^- in the extracellular fluid and in the cytoplasmic water of frog skeletal muscle and squid giant axon. Intracellular ion concentrations for mammalian muscle are similar to those for frog muscle.

Chloride is nearly in equilibrium across the plasma membrane of both frog muscle and squid axon. This is known because chloride's potential difference for equilibrium, as calculated from the Nernst equation, is about equal to the measured transmembrane potential difference. In both tissues K^+ has a concentration force tending to make it flow out of the cell. The electrical force on K^+ is oppositely directed to the concentration force. If the $E_{in} - E_{out}$ in frog muscle were -105 mV, electrical and concentration forces on K^+ would exactly balance. Because $E_{in} - E_{out}$ is only -90 mV, the concentration force is greater than the electrical force, and K^+ therefore has a net tendency to flow out of the cell. In frog muscle and in squid axon *both* the concentration and electrical forces on Na^+ tend to cause it to flow into the cell. Na^+ is the ion furthest from an equilibrium distribution. The larger the difference between the measured membrane potential and the equilibrium potential for an ion, the larger the net force tending to make that ion flow.

Table 2-1 Distribution of Na^+, K^+, and Cl^- across the plasma membranes of frog muscle and squid axon

	Extracellular fluid (mM)	*Cytoplasm (mM)*	*Approximate equilibrium potential (mV)*	*Actual resting potential (mV)*
Frog muscle				
$[Na^+]$	120	9.2	+67	
$[K^+]$	2.5	140	−105	
$[Cl^-]$	120	3 to 4	−89 to −96	−90
Squid axon				
$[Na^+]$	460	50	+58	
$[K^+]$	10	400	−96	
$[Cl^-]$	540	About 40	About −68	−70

Data from Katz B: *Nerve, muscle, and synapse,* New York, 1966, McGraw-Hill Book Co. With permission of McGraw-Hill Book Co.

Active Ion Pumping and the Resting Potential

The Na^+, K^+-ATPase, located in the plasma membrane, uses the energy of the terminal phosphate ester bond of ATP to actively extrude Na^+ from the cell and to actively take K^+ into the cell. The Na^+, K^+ pump is responsible for the high intracellular K^+ concentration and the low intracellular Na^+ concentration. Because the pump moves a larger number of Na^+ ions out than K^+ ions in (3 Na^+ to 2 K^+), it causes a net transfer of positive charge out of the cell and thus contributes to the resting membrane potential. Because it brings about net movement of charge across the membrane, the pump is termed **electrogenic.**

The size of the pump's contribution to the resting potential can be estimated by completely inhibiting the pump with a cardiac glycoside, such as ouabain. Such studies show that in some cells the electrogenic Na^+, K^+ pump is responsible for a large fraction of the resting potential. In most vertebrate nerve and skeletal muscle cells, however, the direct contribution of the pump to the resting potential is small under most circumstances—less than 5 mV. The resting membrane potential in nerve and skeletal muscle results primarily from the diffusion of ions down their electrochemical potential gradients. The ionic gradients are maintained by active ion pumping. In other types of excitable cells electrogenic pumping of ions may contribute more to the resting membrane potential. In certain vertebrate smooth muscle cells, for example, the electrogenic effect of the Na^+, K^+ pump may be responsible for 20 mV or more of the resting membrane potential.

Generation of the Resting Membrane Potential by the Ion Gradients

The earlier section on concentration cells shows how an ion gradient can act as a battery. When a number of ions are distributed across a membrane, all being removed from electrochemical equilibrium, *each* ion will tend to force the transmembrane potential toward *its own* equilibrium potential, as calculated from the Nernst equation. The more permeable the membrane to a particular ion, the greater strength that ion will have in forcing the membrane potential toward its equilibrium potential. In frog muscle (Table 2-1) the Na^+ concentration difference can be regarded as a battery that tries to make $E_{in} - E_{out}$ equal to +67 mV. The K^+ concentration difference is like a battery that attempts to make $E_{in} - E_{out}$ equal to −105 mV. The Cl^- concentration difference resembles a battery trying to make $E_{in} - E_{out}$ equal to −90 mV.

We can draw an equivalent electrical circuit for the plasma membrane. In Fig. 2-7 each ion gradient is represented by a battery of the appropriate polarity. The resistor in series with each battery represents the resistance *(R)* to the passage of that ion through the membrane. The reciprocal of each resistance is the conductance *(g)* of the membrane for that ion. C_m represents the membrane capacitance that stores the transmembrane potential difference. In this circuit if the transmembrane resistance to a particular ion is decreased, the total transmembrane electrical potential difference will move toward the battery potential of that ion.

The Chord Conductance Equation

The way in which the interplay of ion gradients creates the resting membrane potential is also illustrated by a simple mathematical model. If the transmembrane electrical potential difference is equal to the equilibrium potential for a particular ion, there is no net force on that ion, and there will be no net flow of that ion. However, if the membrane potential is *not* equal to the equilibrium potential for a given ion, then the difference between the membrane potential and the ion's equilibrium potential can be regarded as the driving force for that ion. Because ions bear charge, ionic flow is equivalent to electrical current. The net current (I) of an ion across the membrane is equal to the driving force on the ion

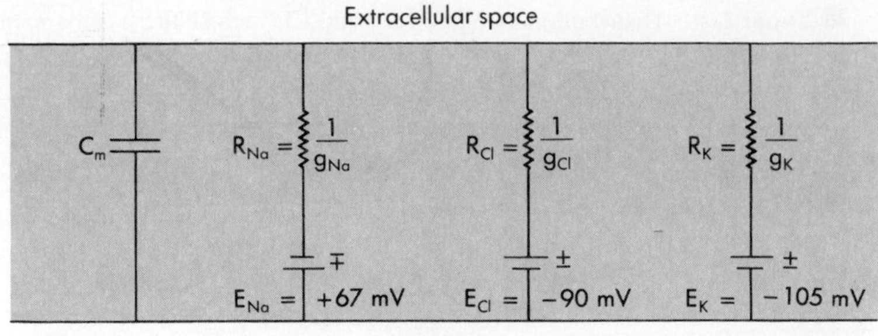

■ **Fig. 2-7** An electrical equivalent circuit model of the plasma membrane of a skeletal muscle cell. The equilibrium potentials of Na^+, K^+, and Cl^- are represented as batteries. The resistances *(R)* and conductances (*g*, reciprocal of resistance) to each ion are shown. C_m represents the membrane capacitance.

times the conductance of the membrane for that ion. For Na^+, K^+, and Cl^-

$$I_K = g_K(E_m - E_K)$$
$$I_{Na} = g_{Na}(E_m - E_{Na}) \quad (18)$$
$$I_{Cl} = g_{Cl}(E_m - E_{Cl})$$

where E_K, and E_{Na}, and E_{Cl} are the equilibrium (Nernst) potentials for the indicated ion.

In the steady state, when the transmembrane electrical potential difference is constant, the sum of all the ionic currents across the membrane must be zero. This is so because, if net current *were* flowing across the membrane, the membrane capacitor would be charging or discharging, and the transmembrane potential difference would be changing. Any net charge transfer across the membrane leads to a change in the degree of charge separation across the membrane and hence in the membrane potential. If K^+, Na^+, and Cl^- are the only important ions, then the algebraic sum of their currents must be zero in the steady state.

$$I_K + I_{Na} + I_{Cl} = 0 \quad (19)$$

Substituting from equation 18 for the currents, we obtain

$$g_K (E_m - E_K) + g_{Na} (E_m - E_{Na}) + g_{Cl} (E_m - E_{Cl}) = 0 \quad (20)$$

For a cell membrane across which chloride is in equilibrium, $E_m = E_{Cl}$, and $E_m - E_{Cl} = 0$, so the third term can be dropped. However, it should be kept in mind that Cl^- is not in equilibrium for all excitable cells. Also, whenever E_m moves away from E_{Cl}, Cl^- will exert a restoring force to bring E_m back toward E_{Cl} just as any ion tries to force E_m toward its equilibrium potential. Without the chloride term equation 10 becomes

$$g_K (E_m - E_K) + g_{Na} (E_m - E_{Na}) = 0 \quad (21)$$

Solving for E_m yields

$$E_m = \frac{g_K}{g_K + g_{Na}} E_K + \frac{g_{Na}}{g_K + g_{Na}} E_{Na} \quad (22)$$

Equation 22 is one form of the **chord conductance equation.** It expresses the transmembrane electrical potential difference as a weighted average of the equilibrium potentials of K^+ and Na^+. The weighting factor for each ion is that fraction of the total ionic conductance caused by that particular ion. To consider more ions, we need only add appropriate terms to the chord conductance equation. In a cell in which Cl^- and Ca^{++} play important roles, the chord conductance equation becomes

$$E_m = \frac{g_K}{g_T} E_K + \frac{g_{Na}}{g_T} E_{Na} + \frac{g_{Cl}}{g_T} E_{Cl} + \frac{g_{Ca}}{g_T} E_{Ca} \quad (23)$$

where

$$g_T = g_K + g_{Na} + g_{Cl} + g_{Ca}$$

For the frog muscle fiber discussed earlier, $E_{in} - E_{out} = -90$ mV. The membrane potential is much closer to E_K (−105 mV) than to E_{Na} (+67 mV). The chord conductance equation predicts that in resting muscle g_K is 10.5 times larger than g_{Na}. This has been confirmed by ion flux measurements with radioactive tracers. In resting frog sartorius muscle, g_K is about 10 times g_{Na}. In resting squid axon ($E_m = -70$ mV), the chord conductance equation predicts that g_K is about five times larger than g_{Na}. In other types of excitable cells the relationship between g_K and g_{Na} is somewhat different. Other ions also may play a role in generating the resting membrane potential. Resting membrane potentials vary from about −7 mV or so in human erythrocytes to −30 mV in some types of smooth muscle and up to −90 mV in vertebrate skeletal muscle and cardiac ventricular cells.

The chord conductance equation shows that, if g_{Na} were suddenly increased, the membrane potential would move toward E_{Na} (toward +67 mV in frog muscle). This is what occurs during an action potential when g_{Na} transiently increases. The ionic mechanism of the action potential is discussed in Chapter 3.

■ *The Constant Field Equation: an Alternative to the Chord Conductance Equation*

The chord conductance equation is derived by considering the current of each individual ion as the product of its conductance and its driving force, where the driving force is equal to the difference between the membrane potential and the equilibrium potential for the ion. By setting the sum of the ionic currents equal to zero (steady-state assumption), the chord conductance equation can be obtained.

Another way of dealing with the fluxes of individual ions across the membrane is treating the flux of each ion as the product of its permeability coefficient (p. 9) and a driving force that is a function of the membrane potential and the intracellular and extracellular concentrations of the ion. By assuming that, in the steady state, the sum of the fluxes of K^+, Na^+, and Cl^- is zero, the following expression for the membrane potential is derived:

$$E_m = \frac{RT}{F} \ln \frac{k_p^{K^+}[K^+]_o + k_p^{Na^+}[Na^+]_o + k_p^{Cl^-}[Cl^-]_i}{k_p^{K^+}[K^+]_i + k_p^{Na^+}[Na^+]_i + k_p^{Cl^-}[Cl^-]_o} \quad (24)$$

where the k_p's are permeability coefficients and the subscripts *o* and *i* represent extracellular and intracellular concentrations, respectively. Equation 24 is known as the **constant field equation** because the expressions for the driving forces of the individual ions are derived by assuming that the electric field within the membrane is of constant strength.

If one of the permeability coefficients in equation 24 becomes very large relative to the other two, then the predicted membrane potential approaches the equilibrium potential for the highly permeable ion. The more permeable a particular ion, the larger its contribution to E_m. Thus the constant field equation and the chord conductance equation lead to qualitatively similar explanations of the membrane potential.

Summary of the Ionic Mechanism of the Resting Membrane Potential

The Na^+, K^+ pump establishes gradients of Na^+ and K^+ across the plasma membranes of the cells. Because the amount of Na^+ pumped out is larger than the amount of K^+ pumped in, the pump transfers net charge across the membrane and contributes to the resting membrane potential. The pump is therefore said to be electrogenic. In vertebrate skeletal and cardiac muscle and in nerve the electrogenic activity of the pump is directly responsible for only a small fraction of the resting membrane potential. The major portion of the resting membrane potential in these tissues is a result of the diffusion of Na^+ and K^+ down their electrochemical potential gradients, with Na^+ flowing into the cell and K^+ flowing out. The principal role of the pump in these tissues is to maintain the ion gradients it has established. K^+ and Na^+ each tend to force the transmembrane potential toward their own equilibrium potential. The resulting E_m is a weighted average of E_K and E_{Na}, with the weighting factor for each ion being the fraction of the total membrane conductance caused by that ion. In these cells the resting membrane potential is *directly* a result of the diffusion of K^+ and Na^+ down their respective electrochemical potential gradients. The resting membrane potential is *indirectly* caused by the Na^+, K^+ pump, which maintains the gradients. In mammalian smooth muscle cells and in certain other cell types the electrogenic effect of the active pumping of Na^+ and K^+ may contribute a substantial fraction of the resting membrane potential.

Summary

1. When an ion is distributed in equilibrium across a membrane, any tendency of the ion to diffuse across the membrane caused by a concentration difference is exactly balanced by the effect on the ion of the electrical potential difference across the membrane.

2. The Nernst Equation describes the relationship between the concentration ratio of an ion across a membrane and the electrical potential difference across the membrane when the ion is distributed in equilibrium.

3. Cells contain impermeant molecules with net negative charge. The Gibbs-Donnan Equilibrium defines the consequences of the fixed negative charge on the distribution of cations and anions that can permeate the plasma membrane.

4. The activity of the Na^+, K^+-ATPase causes animal cells to have a higher intracellular concentration of K^+ and a lower intracellular concentration of Na^+ than the extracellular concentrations of these ions. Because the Na^+, K^+-ATPase is electrogenic, it makes a direct contribution to the resting membrane potential of the cell. In addition Na^+ flowing down its electrochemical potential gradient to enter the cell and K^+ flowing out of the cell down its electrochemical potential gradient also contribute to the resting membrane potential.

5. Each ion that can permeate a membrane tends to bring the resting membrane potential toward its equilibrium potential. The ability of an ion to do this increases as the permeability of the membrane to the ion increases. The chord conductance equation states that the resting membrane potential is a weighted average of the equilibrium potentials of all permeant ions; the weighting factor for each ion is the fraction of the total ionic conductance of the membrane contributed by that ion.

Bibliography

Books and monographs

Aidley DJ: *The physiology of excitable cells,* ed 3, Cambridge, 1990, Cambridge University Press.

Junge D: *Nerve and muscle excitation,* ed 2, Sunderland, Mass, 1981, Sinauer Associates.

Kandel ER, Schwartz JH: *Principles of neural science,* ed 3, New York, 1991, Elsevier Science Publishing.

Katz B: *Nerve, muscle, and synapse,* New York, 1966, McGraw-Hill.

Keynes RD, Aidley DJ: *Nerve and muscle,* ed 2, New York, 1991, Cambridge University Press

Kuffler SW, Nicholls JG, Martin AR: *From neuron to brain,* ed 2, Sunderland, Mass, 1984, Sinauer Associates.

Shepherd GM: *Neurobiology,* ed 2, New York, 1988, Oxford University Press.

CHAPTER

3

Generation and Conduction of Action Potentials

An **action potential** is a rapid change in the membrane potential followed by a return to the resting membrane potential (Fig. 3-1). The size and shape of action potentials differ considerably from one excitable tissue to another. An action potential is propagated with the same shape and size along the whole length of a nerve axon or muscle cell. The action potential is the basis of the signal-carrying ability of nerve cells. It allows all parts of a long muscle cell to contract almost simultaneously. This chapter discusses the ionic currents that generate action potentials, the ion channels that permit the currents, and the ways in which action potentials are propagated and conducted.

Experimental Observations of Membrane Potentials

Our knowledge of the ionic mechanism of an action potential was first obtained from experiments on the squid giant axon. The large diameter (up to 0.5 mm) of the squid giant axon makes it a convenient object for electrophysiological research with intracellular electrodes. The frog sartorius muscle is another useful preparation. The sartorius muscle can be removed from the frog without damage to the muscle. The surface muscle cells can be visualized with a microscope and penetrated with one or more microelectrodes.

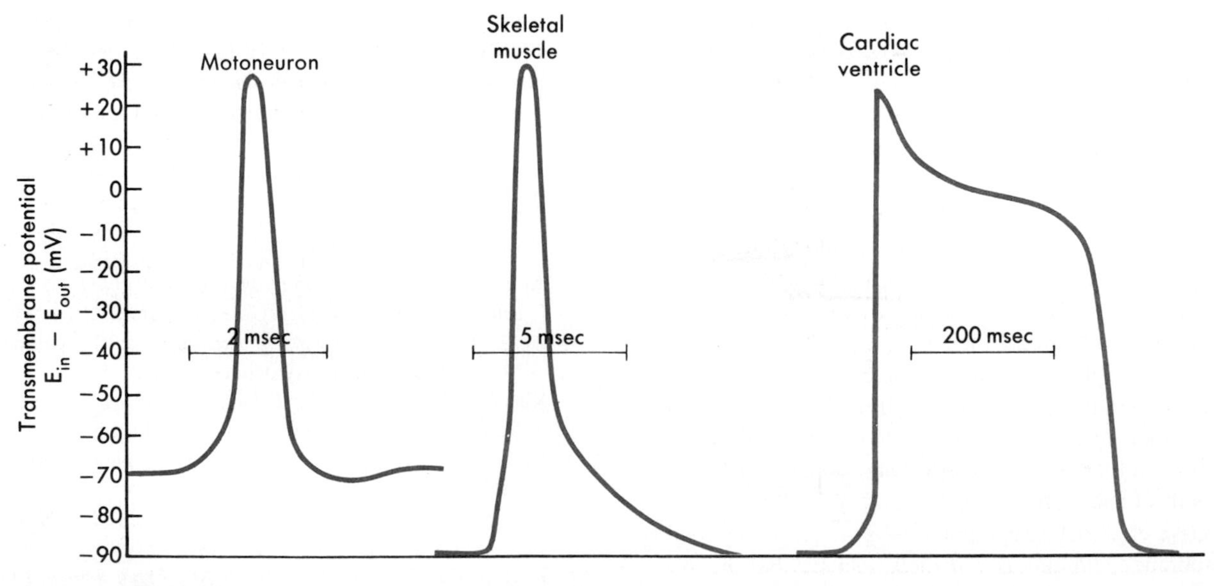

■ **Fig. 3-1** Action potentials (note the different times scales) from three vertebrate cell types. (Redrawn from Flickinger CJ et al: *Medical cell biology,* Philadelphia, 1979, WB Saunders.)

The following experiment illustrates some basic aspects of the resting membrane potential. A sartorius muscle is removed from a frog and put in a dish of fluid with composition similar to the frog's extracellular fluid. Two microelectrodes (tip diameter less than 0.5 μm) are placed in the extracellular fluid. When both microelectrodes are placed in the extracellular fluid, no electrical potential difference between them is observed. One electrode then is moved slowly toward a muscle cell until it penetrates the plasma membrane. At the instant the electrode pops through the membrane, an abrupt change of the potential difference between the two electrodes is observed. The intracellular electrode suddenly becomes about 90 mV negative with respect to the external electrode. This −90 mV potential difference is the **resting membrane potential** of the muscle fiber. Another microelectrode placed in the same cell also will register −90 mV relative to the external solution. At rest there is no net internal current in the muscle cell and thus no potential difference between these two intracellular electrodes. In the absence of perturbing influences the resting membrane potential remains at −90 mV.

■ *Subthreshold Responses: The Local Response*

Fig. 3-2 illustrates the results of an experiment in which the membrane potential of an axon of a shore crab is perturbed by passing rectangular pulses of current across the plasma membrane. Current pulses are depo-

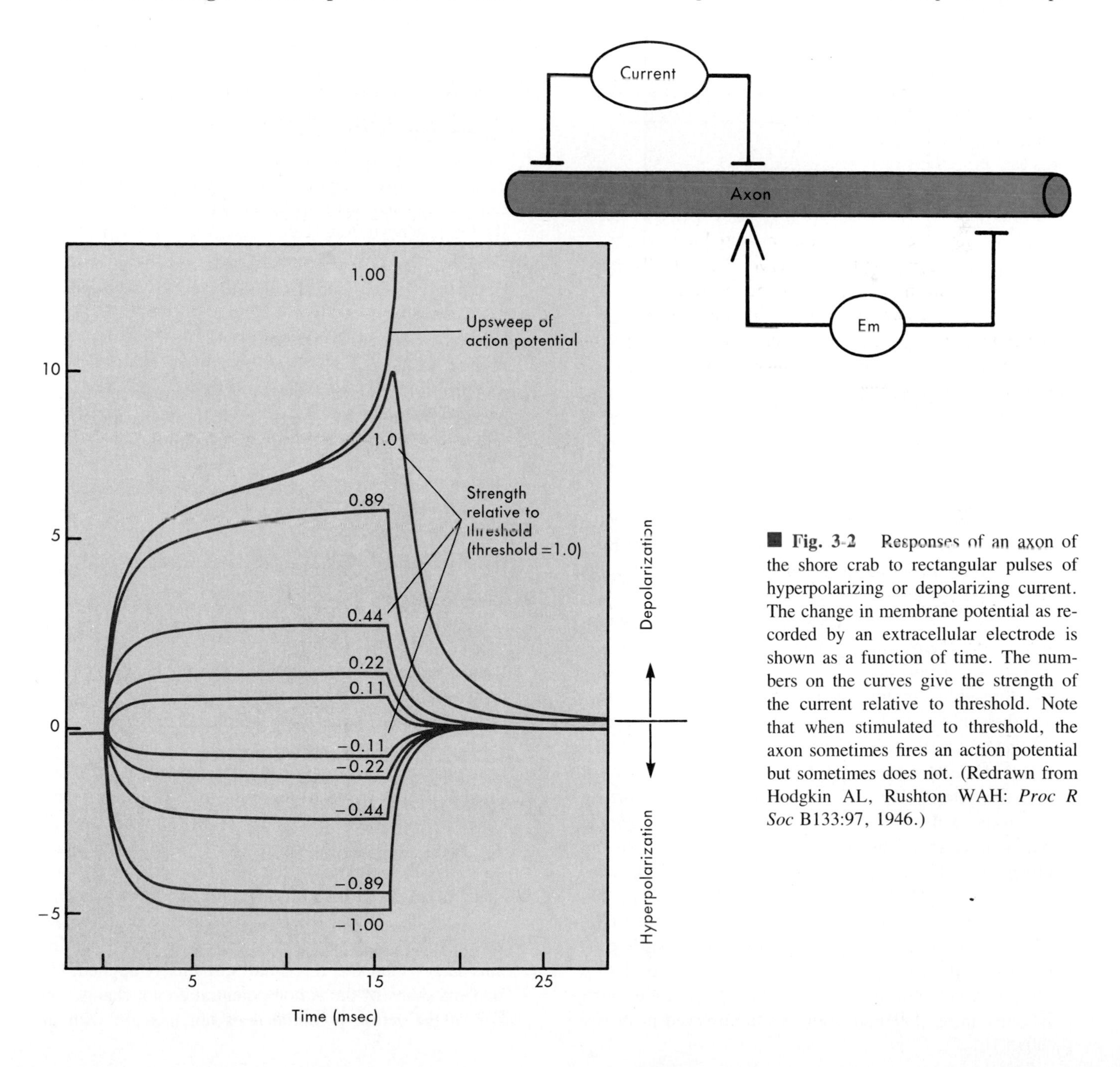

■ **Fig. 3-2** Responses of an axon of the shore crab to rectangular pulses of hyperpolarizing or depolarizing current. The change in membrane potential as recorded by an extracellular electrode is shown as a function of time. The numbers on the curves give the strength of the current relative to threshold. Note that when stimulated to threshold, the axon sometimes fires an action potential but sometimes does not. (Redrawn from Hodgkin AL, Rushton WAH: *Proc R Soc* B133:97, 1946.)

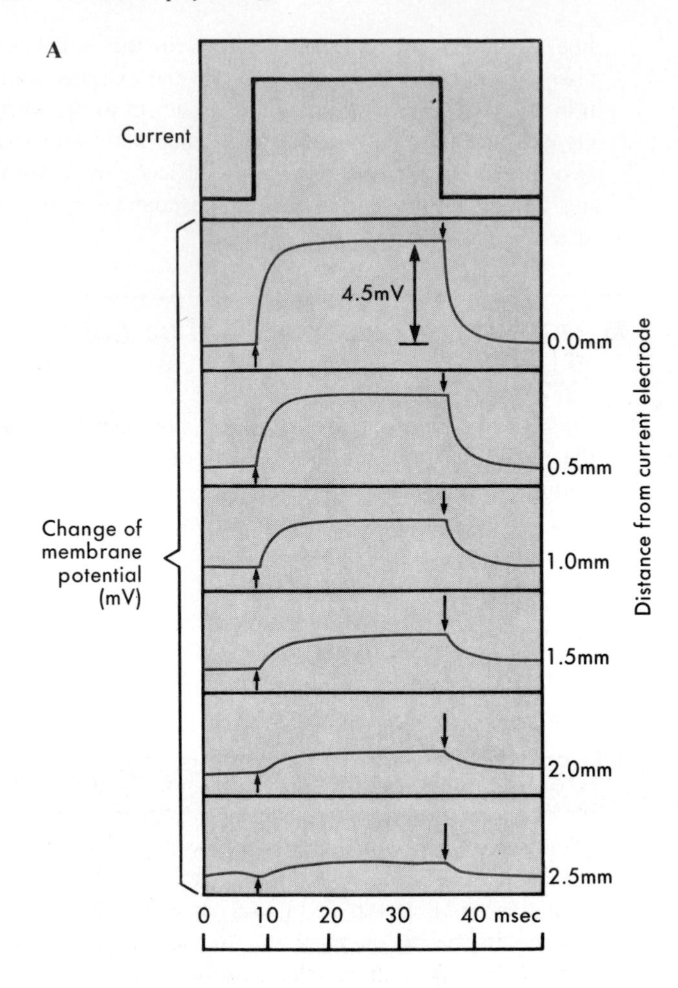

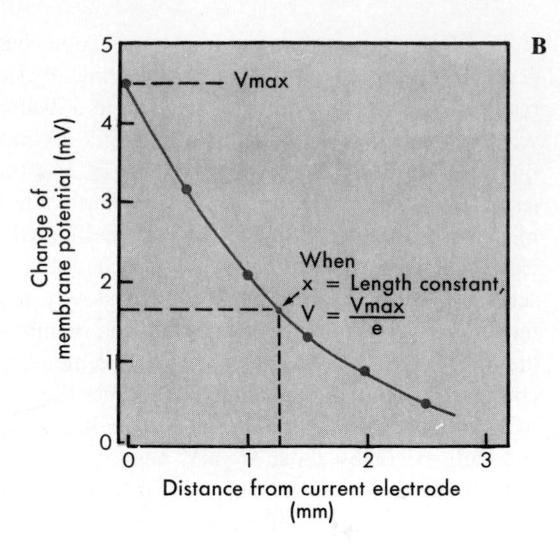

■ Fig. 3-3 **A,** Responses of an axon of a shore crab to a rectangular pulse of current recorded extracellularly by an electrode located different distances from the current-passing electrode. As the recording electrode is moved farther from the point of stimulation, the response of the membrane potential is slower and smaller. **B,** The change in membrane potential from A is plotted versus distance from the point of current passage. The distance over which the response falls to 1/e (37%) of the maximal response is called the length constant. (Part *A* is redrawn from Hodgkin AL, Rushton WAH: *Proc R Soc* B133:97, 1946.)

larizing or hyperpolarizing depending on the direction of current flow. The terms **depolarizing** and **hyperpolarizing** may be confusing. A change of the membrane potential from −90 mV to −70 mV is a depolarization because it is a decrease of the potential difference, or polarization, across the cell membrane. If the membrane potential changes from −90 mV to −100 mV, the polarization across the membrane has increased; this is hyperpolarization.

The larger the current passed, the larger the perturbation of the membrane potential. As shown in Fig. 3-2, in response to depolarizing current pulses above a certain threshold strength, the cell fires an action potential.

When subthreshold current pulses are passed, the size of the potential change observed depends on the *distance* of the recording microelectrode from the point of current passage (Fig. 3-3, *A*). The closer the recording electrode to the site of current passage, the larger the potential change observed. The size of the potential change is found to decrease exponentially with distance from the site of current passage (Fig. 3-3, *B*). The distance over which the potential change decreases to 1/e (37%) of its maximal value is called the **length constant** (or **space constant**). A length constant of 1 to 3 mm is typical for mammalian nerve or muscle cells. Because these potential changes are observed primarily near the site of current passage and the changes are not propagated along the length of the cell (as are action potentials), they are called **local responses.**

■ *Action Potentials*

If progressively larger depolarizing current pulses are applied, a point is reached at which a different sort of response, the **action potential,** occurs (Figs. 3-2 and 3-4). An action potential is triggered when the depolarization is sufficient for the membrane potential to reach a **threshold value,** which is near −60 mV for frog sartorius muscle. The action potential differs from the local depolarizing response in two important ways: (1) it is a much larger response, with the polarity of the membrane potential actually reversing (the cell interior becoming positive with respect to the exterior), and (2) the action potential is *propagated without decrement* down the entire length of the muscle fiber. The size and shape of an action potential remain the same as it travels along the muscle fiber; it does not decrease in size with distance as does the local response. When a stimulus larger than the threshold stimulus is applied, the size and shape of the action potential do not change; the size of the action potential does not increase with in-

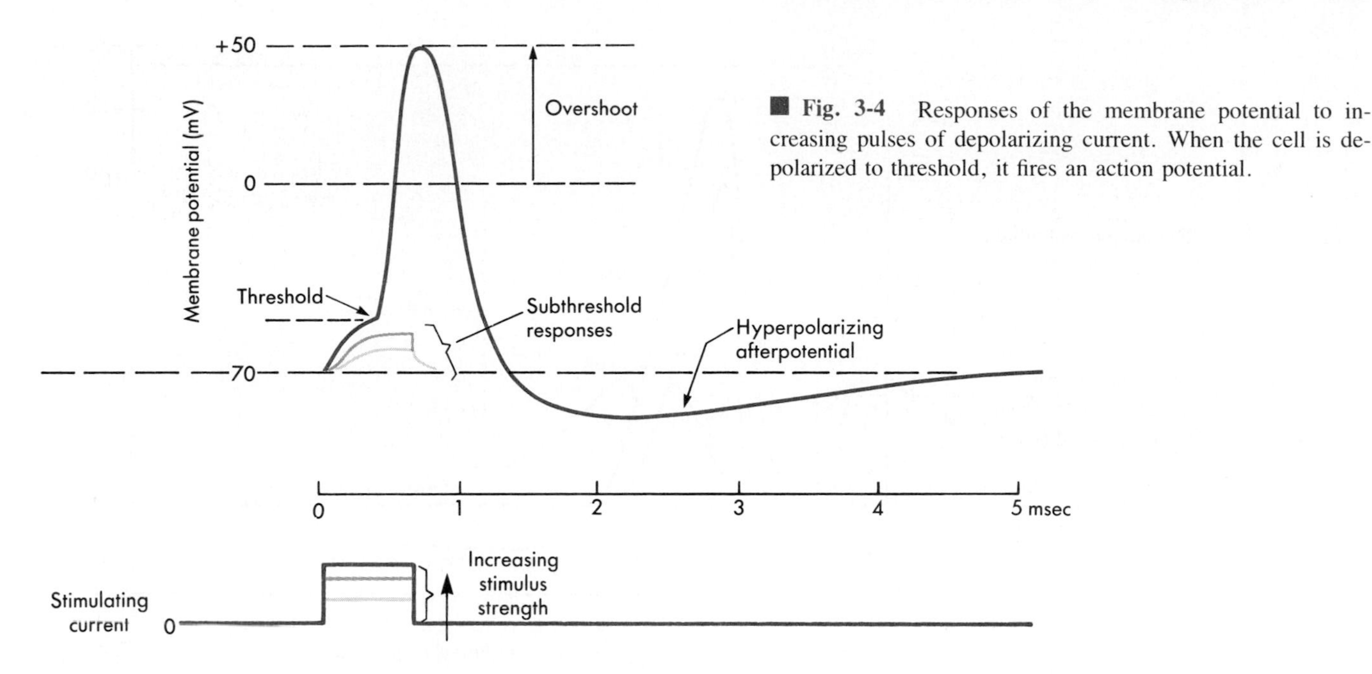

■ **Fig. 3-4** Responses of the membrane potential to increasing pulses of depolarizing current. When the cell is depolarized to threshold, it fires an action potential.

creased stimulus strength. A stimulus either fails to elicit an action potential (a subthreshold stimulus), or it produces a full-size action potential. For this reason the action potential is an **all-or-none response.**

■ *The Shape of the Action Potential*

The form of an action potential of a squid giant axon is shown in Fig. 3-4. Once the membrane is depolarized to the threshold, an explosive depolarization occurs, which completely depolarizes the membrane and even *overshoots* so that the membrane becomes polarized in the reverse direction. The peak of the action potential reaches about +50 mV. The membrane potential then returns toward the resting membrane potential almost as rapidly as it was depolarized. After repolarization a transient hyperpolarization that is known as the **hyperpolarizing afterpotential** occurs. It persists for about 4 msec. The following section discusses the ionic currents that cause the various phases of the action potential.

■ *Ionic Mechanisms of the Action Potential*

In Chapter 2 the resting membrane potential was seen to be a weighted sum of the equilibrium potentials for Na^+, K^+, Cl^-, etc., with the weighting factor for each ion being the fraction it contributes to the total ionic conductance of the membrane (the chord conductance equation). In squid giant axon the resting membrane potential (E_m) is about −70 mV. E_K is about −100 mV in squid axon, so an increase in g_K would hyperpolarize the membrane. E_{Cl} is about −70 mV, so an increase in g_{Cl} would stabilize E_m at −70 mV. An increase in g_{Na} of sufficient magnitude would cause depolarization and reversal of the membrane polarity because E_{Na} is about +60 mV in squid axon. A decrease in g_K also would have the effect of depolarizing the membrane.

It has been known for a long time that Na^+ is involved in the action potential. It was shown in 1902 that Na^+ in the extracellular fluid is required for excitability. Some 50 years later Hodgkin and Huxley showed that the action potential of squid giant axon is caused by successive conductance increases to sodium and potassium ions. They found that the conductance to Na^+, g_{Na}, increases very rapidly during the early part of the action potential (Fig. 3-5). The sodium conductance reaches a peak about the same time as the peak of the action potential, then it decreases rather rapidly. The K^+ conductance, g_K, increases more slowly, reaches a peak at about the middle of the repolarization phase, and then returns more slowly to resting levels.

As described in Chapter 2, the chord conductance equation shows that the membrane potential is a result of the opposing tendencies of the K^+ gradient to bring E_m toward the equilibrium potential for K^+ and the Na^+ gradient to bring E_m toward the equilibrium potential for Na^+. Increasing the conductance of either ion will increase its ability to pull E_m toward its equilibrium potential. The rapid increase in g_{Na} during the early part of the action potential causes the membrane potential to move toward the equilibrium potential for Na^+ (+60 mV). The peak of the action potential reaches only about +50 mV. This is because the conductance of K^+ also increases, albeit more slowly, providing an opposing tendency, and also because g_{Na} quickly decreases toward resting levels. The rapid return of the membrane potential toward the resting potential is caused by the

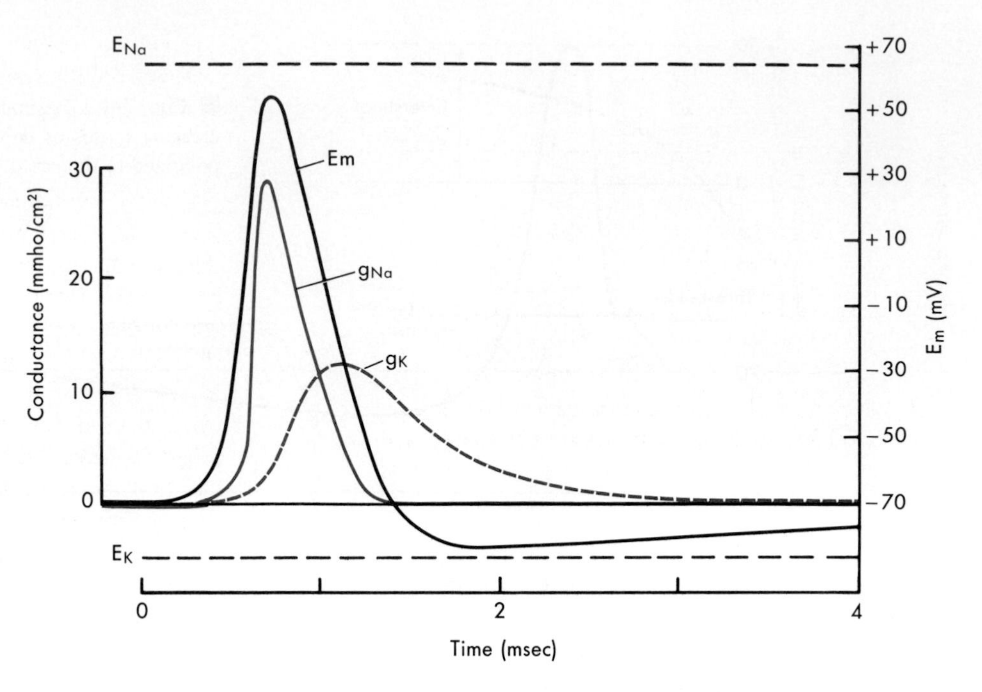

■ **Fig. 3-5** The action potential (E_m) of a squid giant axon is shown on the same time scale with the changes in the conductance of the axon membrane to sodium and potassium ions that occur during the action potential. (Redrawn from Hodgkin AL, Huxley AF: *J Physiol* 117:500, 1952.)

rapid decrease of g_{Na} and the continued increase in g_K, both of which decrease the size of the Na^+ term in the chord conductance equation and increase the size of the K^+ term. During the hyperpolarizing afterpotential, when the membrane potential is actually more negative (more polarized) than the resting potential, g_{Na} has returned to baseline levels, but g_K remains elevated above resting levels. Thus E_m is pulled closer to the K^+ equilibrium potential (−100 mV) for a short time.

■ *The Basis for Knowledge of the Action Potential*

Early in this century it was hypothesized that the action potential was caused by a transient increase of the permeability of the plasma membrane to all ions. Later evidence indicated that the entry of Na^+ into the cell plays a central role in generating the action potential. Removal of Na^+ from the extracellular fluid was found to abolish the action potential. Both the height of the action potential and the maximum rate of depolarization during the action potential depend on the extracellular concentration of Na^+, increasing with increasing external Na^+.

By repetitively stimulating the squid giant axon in a bath containing radioactive $^{22}Na^+$, it was found that 3 to 4×10^{-12} mole of Na^+ enter the axon through each square centimeter of surface area during one action potential. This amount of Na^+ is in good agreement with calculations of the amount of Na^+ that should enter the axon based on electrical considerations. The capacitance (C) of the membrane is defined as dq/dv, the amount of charge that must flow to cause a 1 V change in potential difference. It is known that the capacitance of 1 cm^2 of squid axon membrane is about 1 microfarad. The amount of charge that must flow to discharge each cm^2 of the membrane capacitor by 100 mV (the approximate height of the action potential) is C times the change in V, or 10^{-6} farad $\times$ 10^{-1} V = 10^{-7} coulombs. Dividing this by the amount of charge on 1 mole of Na^+ ions (96,500 coulombs/mole) gives about 10^{-12} mole of Na^+ ions that must enter each square centimeter of axonal membrane to depolarize the membrane by 100 mV.

If only Na^+ and K^+ are involved, the amount of K^+ that leaves the cell during the repolarization phase must be equal to the amount of Na^+ that enters during depolarization. The Na^+, K^+-ATPase pumps out the Na^+ that enters and reaccumulates the lost K^+. Because of the small number of ions that cross the membrane during each impulse, the Na^+, K^+ pump is not required in the short run. A squid axon that has been poisoned with cyanide or ouabain so that ion pumping is abolished can fire nearly 100,000 action potentials before failing. Smaller axons, with a much smaller ratio of volume to surface, will fail after fewer impulses if the Na^+ pump is poisoned.

■ *The Voltage Clamp Technique*

Much of the current knowledge of the ionic mechanism of the action potential first came from experiments using the **voltage clamp** technique. This technique uses electronic feedback to rapidly set the transmembrane potential difference to some chosen value. The voltage clamp circuit holds the membrane potential at this level and measures the net ionic current that flows across the membrane during the clamp (Fig. 3-6).

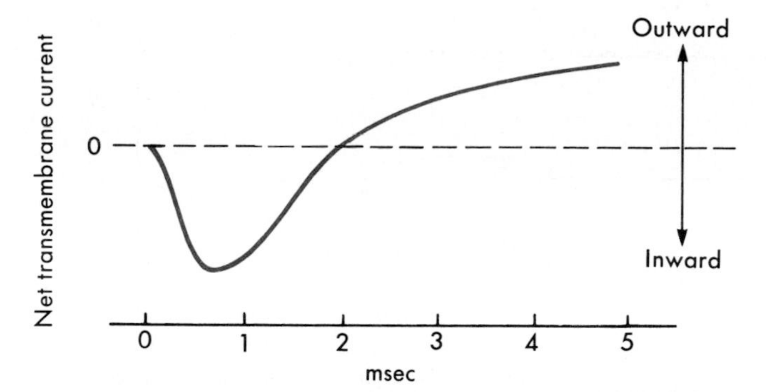

■ **Fig. 3-6** Net transmembrane current flow in a squid giant axon after clamping the membrane potential to zero.

When the membrane of the squid giant axon is depolarized rapidly to some point beyond the threshold, there is an almost instantaneous current that is caused by discharge of the membrane capacitance. This is followed by currents that flow over the same time course as a normal action potential (see Fig. 3-6). The inward phase of the current results from Na^+ entering the axon, and the outward current results from K^+ leaving the axon. The Na^+ current can be separated from the K^+ current as follows (Fig. 3-7).

1. Enough of the external Na^+ is replaced by choline$^+$ (an impermeant cation), so that the Na^+ concentration inside the axon equals that in the extracellular fluid. If the membrane potential is then clamped to zero, there will be no net force causing Na^+ to flow because there is no concentration difference across the membrane and no electrical potential difference. When this is done, the inward current component disappears (see Fig. 3-7, curve B), indicating that the inward current is normally caused by Na^+. Because curve A is the sum of I_K and I_{Na} and curve B is I_K alone, subtracting B from A yields curve C in Fig. 3-7, which represents I_{Na}.

2. The involvement of Na^+ in the inward current during the action potential also can be shown without changing the external $[Na^+]$. If E_m is clamped to +58 mV, which is the Na^+ equilibrium potential, no net flow of Na^+ occurs, and the inward current disappears. Clamping the axon to positive E_m greater than +58 mV causes a net outward Na^+ current. The potential at which the Na^+ current reverses direction (+58 mV) is called the **reversal potential** for the Na^+ current. The fact that the reversal potential for the inward current phase of the action potential is equal to the equilibrium potential for Na^+ supports the contention that Na^+ carries most of the current in the inward phase.

In their pioneering studies of action potentials in squid giant axons Hodgkin and Huxley found that rapidly depolarizing the membrane caused a rapid increase in the flow of Na^+ into the axon and a slower increase in the flow of K^+ out of the axon. By separating the Na^+ current from the K^+ current, they were able to determine the conductances, g, for Na^+ and K^+ as a function of time during voltage clamp. [If I and E_m are known, g can be calculated because $I_{Na} = g_{Na}(E_m - E_{Na})$, and $I_K = g_K(E_m - E_K)$.] When the membrane potential was clamped to zero, the conductances shown in Fig. 3-8 were obtained. The increase in g_{Na} shuts off after a short delay even though the clamp is maintained. The increase in g_K, however, remains and does not decrease to resting levels until the clamp is released. Smaller depolarizations cause smaller changes in g_K and g_{Na}.

■ **Fig. 3-7** **A,** The net ionic current that flows when a squid giant axon in sea water is clamped from its resting potential to a transmembrane potential of 0 mV. This current consists primarily of sodium and potassium ions flowing across the membrane. **B,** The net ionic current when the axon is placed in artificial sea water with most of the sodium replaced by choline (an impermeant cation) so that the intracellular and extracellular sodium concentrations are equal. The current is assumed to be caused by potassium only. **C,** Curve *A* minus curve *B*. This curve is taken to represent the sodium current. (Redrawn from Hodgkin AL, Huxley, AF: The conduction of the nervous impulse, *J Physiol* 116:449, 1952. Courtesy Charles C Thomas, Publisher, Springfield, Illinois.)

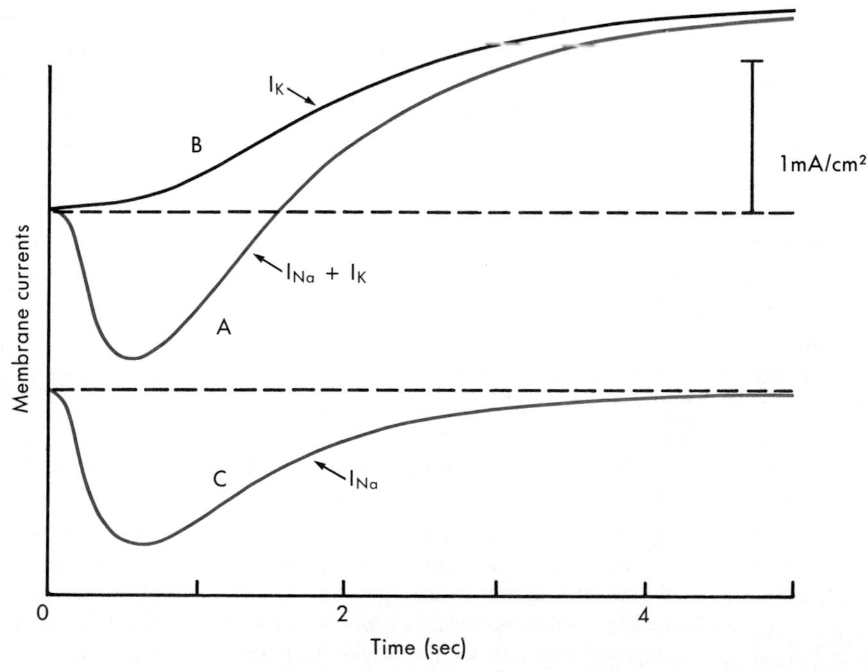

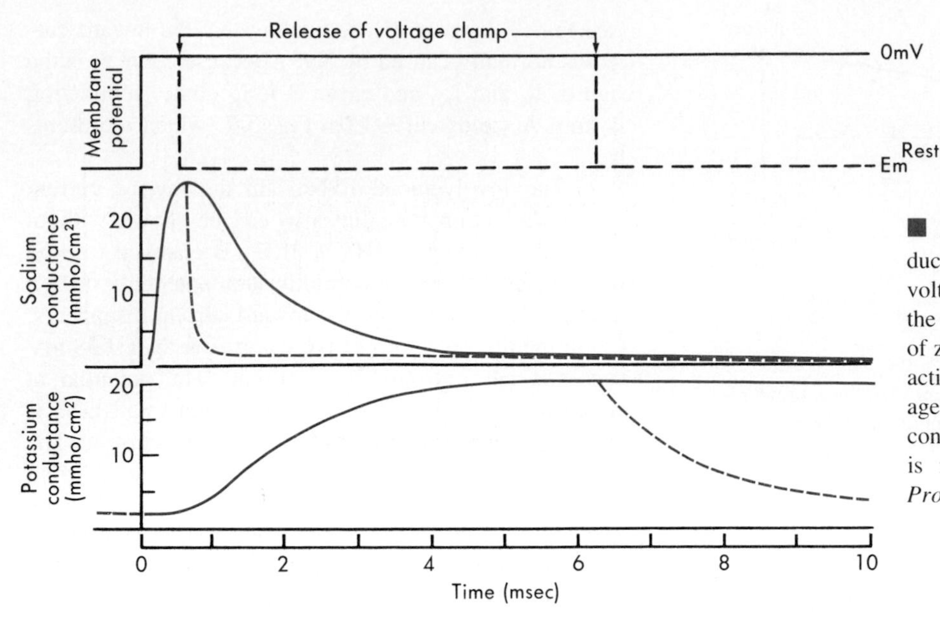

■ **Fig. 3-8** Sodium and potassium conductance changes are shown in response to voltage clamp of the squid giant axon from the resting potential to a membrane potential of zero. Note that the sodium conductance inactivates after a slight delay even if the voltage clamp is maintained, but the potassium conductance remains elevated until the clamp is released. (Redrawn from Hodgkin AL: *Proc Roy Soc* B148:1, 1958.)

In brief, there are three basic events in the action potential: the increase of g_{Na}, the return of g_{Na} to resting levels (called inactivation of Na^+ channels), and the increase of g_K. The rate and extent of each of these depends on the level of the membrane potential. By doing voltage clamp experiments at different E_m levels, the changes in g_{Na} and g_K that must occur during the normal action potential can be reconstructed.

■ *Ion Channels and Gates*

■ *Behavior of Individual Ion Channels*

It has become possible to study the behavior of individual ion channels. One way is to incorporate either purified ion channel proteins or bits of membrane into planar lipid bilayers that separate two aqueous compartments. Then electrodes placed in the compartments can be used to monitor or impose currents and voltages across the membrane. Under some conditions only one, or a few, ion channels of a particular type may be present in the planar membrane. Each ion channel spontaneously oscillates between two conductance states, an open state and a closed state (Fig. 3-9).

Another way to study individual ion channels involves the use of so-called patch electrodes. A fire-polished microelectrode is placed against the surface of a cell and suction is applied to the electrode (Fig. 3-10). A high-resistance seal is formed around the tip of the electrode. The sealed patch electrode can then be used to monitor the activity of whatever channels happen to be trapped inside the seal. The patch of membrane can either be studied in situ or removed from the cell so that the composition of the solution in contact with the intracellular face of the membrane may be manipulated, or even turned inside out if desired! Sometimes the patch trapped inside the electrode contains more than one functional ion channel of a particular type (Fig. 3-11).

Patch electrodes have been used to study Na^+ channels in rat muscle cells. In response to a step depolarization of the muscle cell membrane, some of the Na^+ channels show an opening event, some do not open at all, and some open more than once (Fig. 3-12). When the currents of a large number of channels are summed (Fig. 3-12, *B*), the resulting behavior resembles the macroscopic behavior of Na^+ channels discussed earlier. That is to say the "average channel" opens promptly in response to depolarization; then after a short time delay, the channel closes (inactivates), even though the depolarization is maintained.

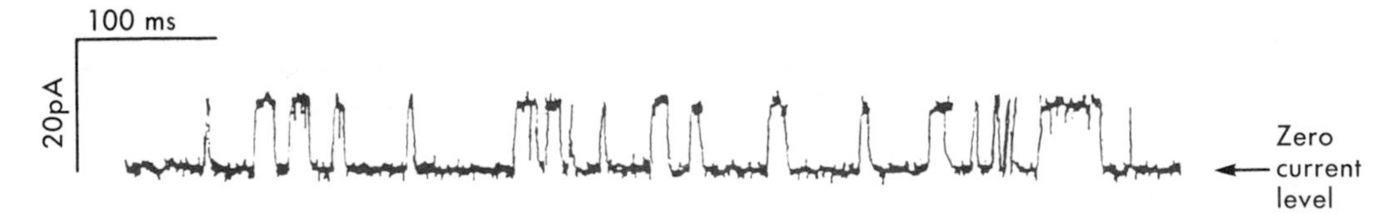

■ **Fig. 3-9** Ionic current through a single ion channel from rat muscle incorporated into a planar lipid bilayer membrane. The channel opens and closes spontaneously. The fraction of time this channel spends in the open state is a function of calcium ion concentration and membrane potential. (Reproduced from Moczydlowski E, Latorre R: *J Gen Physiol,* 82:511, 1983. Copyright permission of the Rockefeller University Press.)

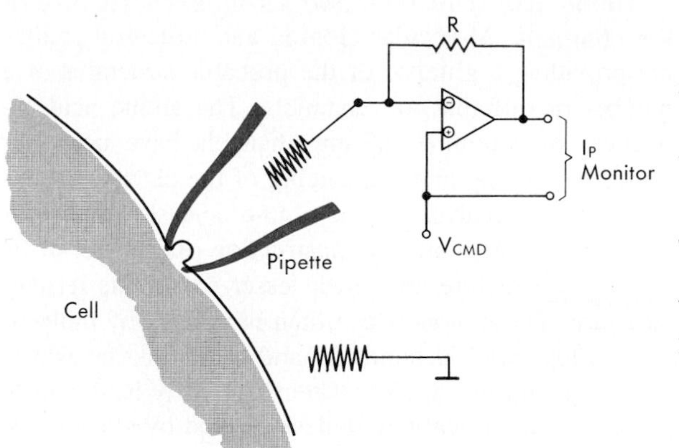

■ **Fig. 3-10** A patch electrode and circuitry required to record the currents that flow through the small number of ion channels isolated in the electrode. (Redrawn by permission from Sigworth FJ, Neher E: *Nature,* 287:447, 1980. Copyright Macmillan Journals.)

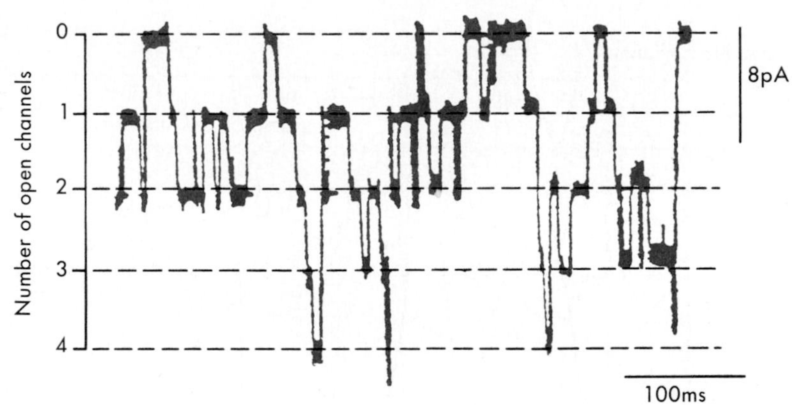

■ **Fig. 3-11** A current recording from a patch electrode on a muscle cell plasma membrane. The five different current levels show that this particular patch of membrane contains four different ion channels. (Redrawn from Hammill OP et al: *Pflugers Arch* 391:85, 1981.)

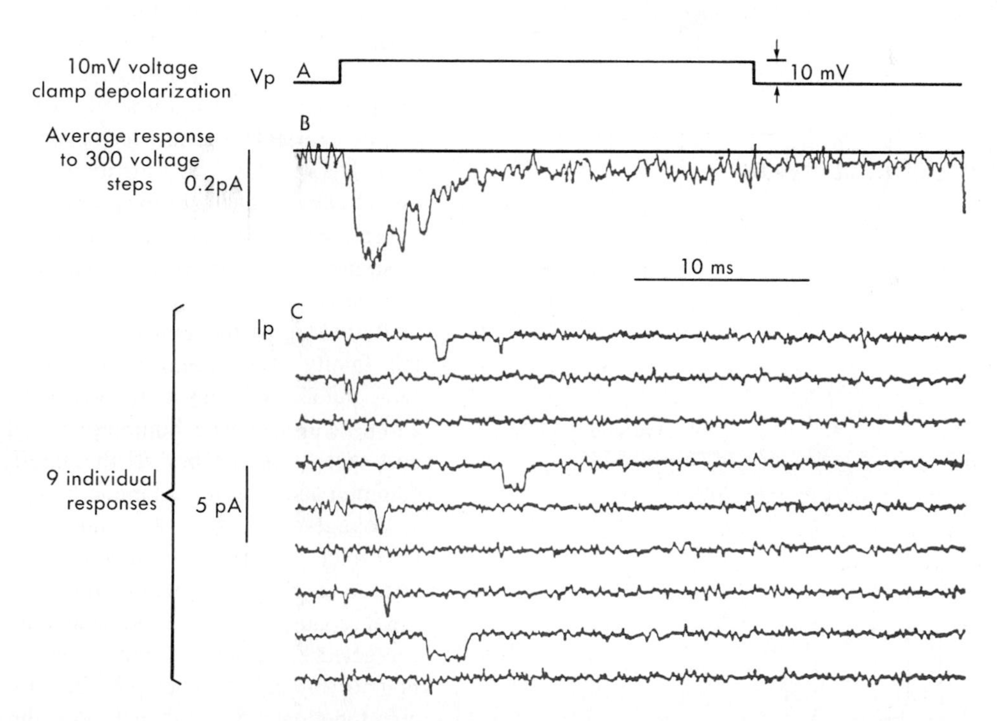

■ **Fig. 3-12** A patch electrode records the currents that flow in a small patch of rat muscle membrane in resonse to a 10 mV depolarization (trace *A*). Tetraethylammonium is used to block potassium channels that may be present in the patch. The traces in panel *C* show responses to nine individual 10 mV depolarizations. The tracing in *B* is the average of 300 individual responses. Note that this average response resembles the response of large numbers of sodium channels as seen in conventional voltage clamp experiments (see Fig. 3-7). (Redrawn by permission from Sigworth FJ, Neher E: *Nature* 287:447, 1980. Copyright Macmillan Journals.)

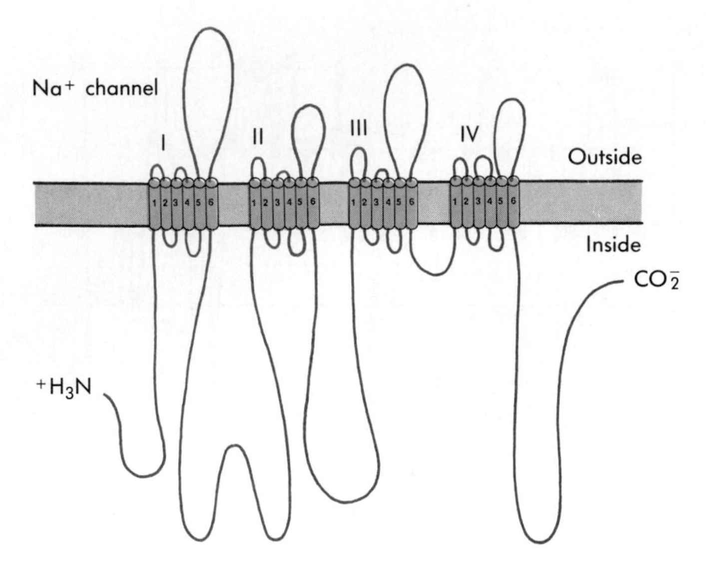

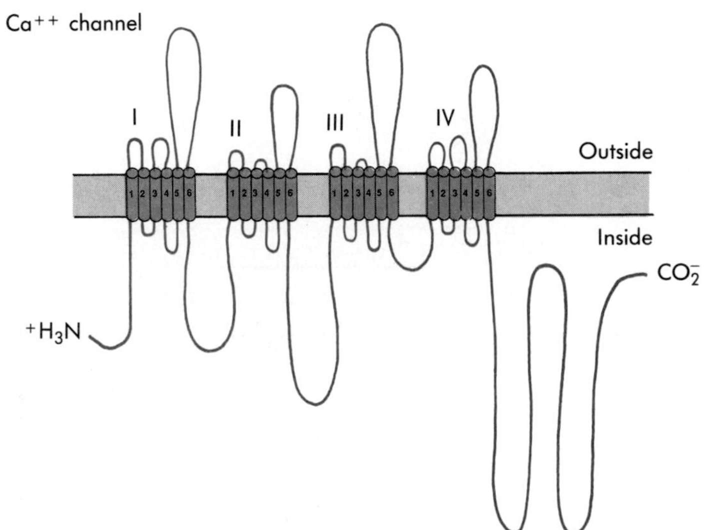

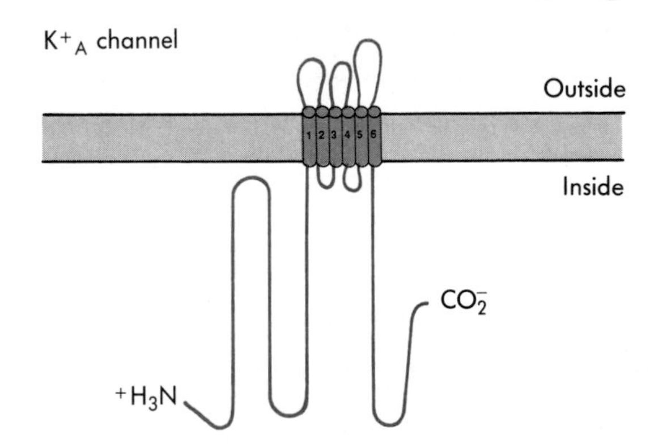

■ **Fig. 3-13** A family of voltage-gated cation channels. Models of the structures of voltage-gated Na^+, Ca^{++}, and K^+ channels. The channels have motifs of six transmembrane α helices (indicated by the numbered cylinders). The motif is repeated 4 times in the Na^+ and Ca^{++} channels. Four of the K^+ channel proteins are required to make a single K^+ channel. (Modified from Caterall W: *Science* 242:50, 1988.)

■ *Molecular Structures of Ion Channels*

Amino acid sequences and secondary structure of ion channels. Molecular cloning and structural studies are providing a glimpse of the probable structures of a number of different ion channels. The amino acid sequences of a number of ion channels have been obtained by cloning and sequencing of the cDNA that encodes them. Analysis of the amino acid sequence of a membrane protein allows construction of a model of its secondary structure and, to a lesser extent, its tertiary structure. These models can then be refined by molecular, biological, biochemical, and structural investigations. Integral membrane proteins typically have one or more **transmembrane α helices** formed by stretches of about 20 predominantly hydrophobic amino acids. The number and disposition of these transmembrane helices are important features of the structure of the protein. Although the structure of no vertebrate ion channel is known in complete detail, the amino acid sequences and some of the important structural features of a number of ion channels are emerging.

Ion channels are often classified based on the gating stimulus to which they are most responsive. Molecular cloning studies have revealed that ion channels with similar gating properties are frequently members of families of proteins with significant homologies in their primary and secondary structures. In this chapter we deal primarily with the **voltage-gated cation channels** that are responsible for action potentials. The structure and function of certain **ligand-gated ion channels** that are responsive to particular neurotransmitter substances will be described in Chapter 4.

Structure of voltage-gated cation channels. Certain voltage-gated cation channels, from insects to mammals, belong to the same protein family. Members of this family of voltage-gated cation channels have the same number of transmembrane helices and significant amino acid sequence homologies. These features suggest that each member of the family derives from a common ancestral ion channel.

Voltage-gated Na^+, K^+, and Ca^{++} channels have important features in common (Fig. 3-13). The Na^+ and Ca^{++} channels contain four repeated motifs; each motif contains six membrane-spanning helices. Voltage-gated K^+ channels contain only one of the motifs. Four of the K^+ channel proteins are required to form one functional K^+ channel. All three channel types have considerable amino acid sequence homology, particularly in the transmembrane helices.

The fourth transmembrane helix (S4) in each motif has seven repeats of arginine or lysine followed by two hydrophobic amino acids (Fig. 3-14). The movement of the positively charged arginine and lysine residues in response to changes in the transmembrane voltage may underlie the gating currents that correspond to activation of the channel.

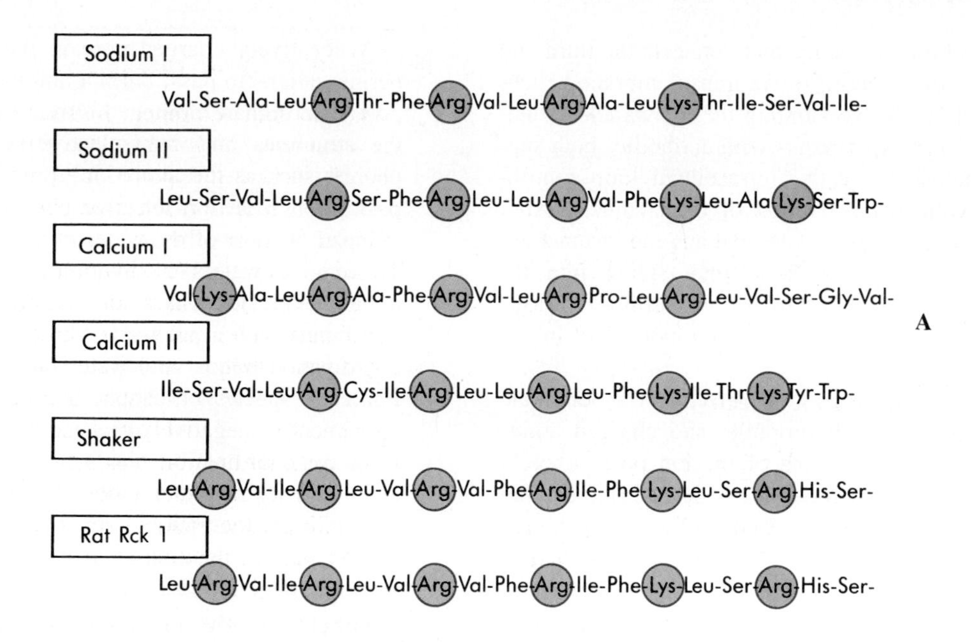

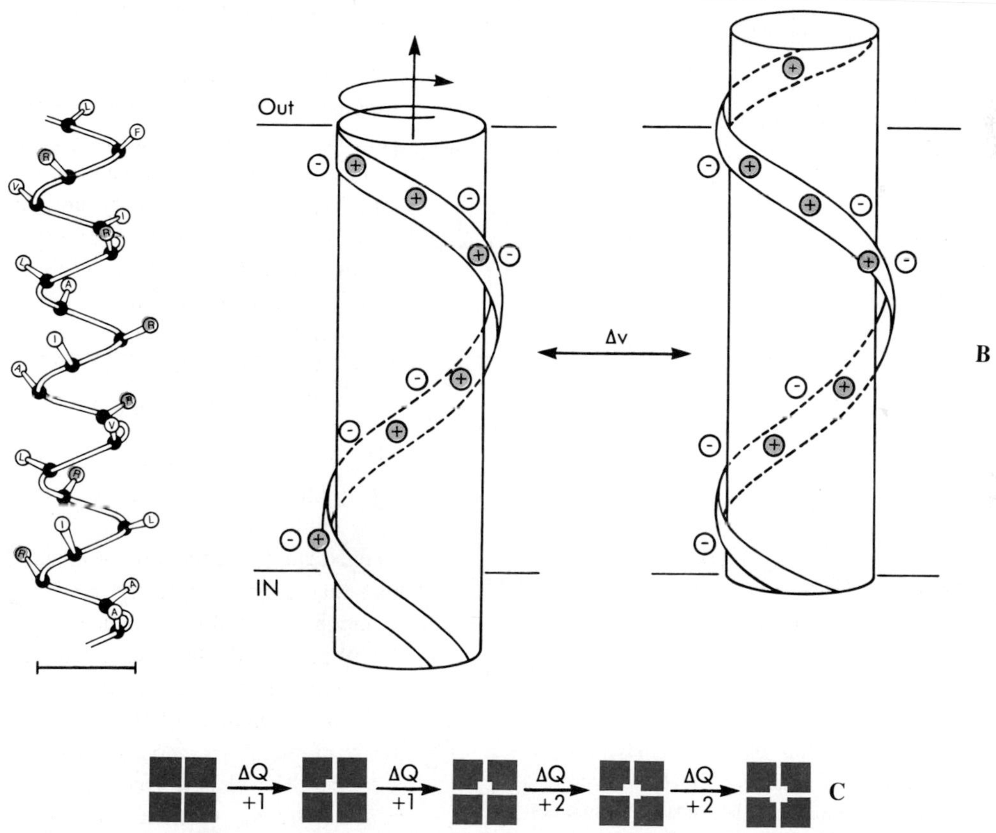

■ **Fig. 3-14** **A,** The S4 helices are the voltage-sensors of voltage-gated cation channels. Amino acid homology in the S4 helices of voltage-gated Na^+, Ca^{++}, and K^+ channels is extensive. Aligned sequences are shown from the S4 helices of domains I and II of the Na^+ and Ca^{++} channels and from the shaker K^+ channel from *Drosophila* and the Rckl K^+ channel from rat. (Modified from Levitan IB, Kaczmarek LK: The neuron—cell and molecular biology, New York, 1991, Oxford University Press.) **B,** Model of the function of the S4 helix as the voltage-sensor. When the membrane is depolarized the S4 helix is postulated to rotate and to move in the direction shown. **C,** Model illustrating that a conformational change in each of the four domains is required to produce an open channel. A total of 6 charge equivalents (shown as ΔQ) move across the membrane to open one ion channel. (Modified from Catterall W: *Trends Neurosci* 9:7, 1986.)

The intracellular sequence that connects the third and fourth repeats of the motif of six transmembrane helices is significantly conserved among these channels. Mutational studies and experiments with antibodies both support the conclusion that this intracellular loop contributes to the voltage inactivation of the channels. After the membrane has been depolarized and the channel has opened, this intracellular loop may swing into the mouth of the channel and thus prevent further ion flow. This is known as the *"ball and chain model"* of inactivation.

The mechanism of ionic selectivity. The ion channel pore is lined with hydrophilic and charged amino acid residues. The structures of the ion pores of voltage-gated ion channels have not been determined, but the pore is believed to be surrounded by one of the helices from each of the repeated motifs, so that charged residues (predominantly negative for a cation channel) line the channel interior (Fig. 3-15). Which of the helices lines the channel is not yet clear, but transmembrane helices S2 and S3 are candidates.

A negatively charged pore in the membrane would permit cations to pass, but not anions, but how does the pore discriminate among different cations? Studies of the structures and ion selectivities of various ionophores, such as the antibiotic gramicidin, show that to pass through an ion-selective channel, an ion must be stripped of most of the water molecules that hydrate it. To remove waters of hydration, negatively charged amino acid residues that line the channel form coordination bonds with a positively charged ion, and the ion's coordination bonds with water molecules are replaced. For each species of cation, a precise geometrical arrangement of negatively charged residues is required to form the coordination bonds that will remove a sufficient number of water molecules to allow that ion to pass through the channel. In this way the geometrical arrangement of the charges that line the pore determine the ionic selectivity of a channel.

Other subunits of voltage-gated cation channels. The α-subunits of Na^+, K^+, and Ca^{++} channels can form functional ion channels by themselves. Neverthe-

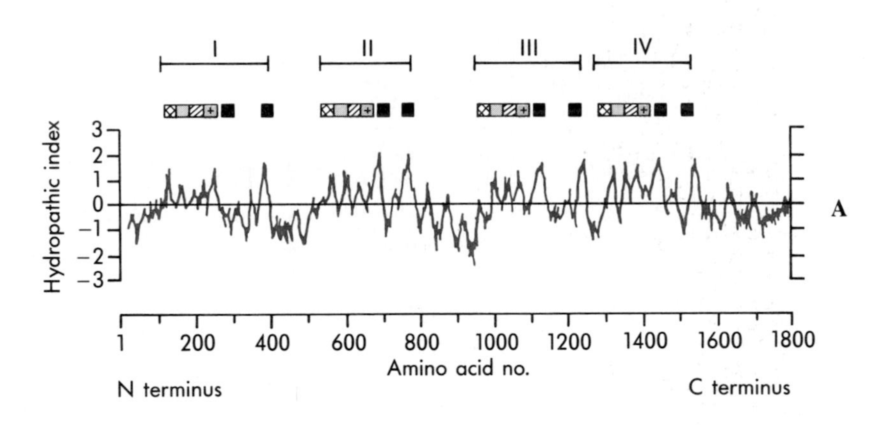

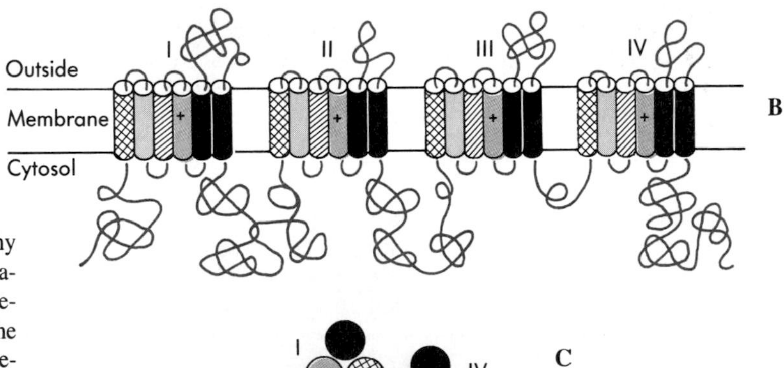

■ **Fig. 3-15** A model of the Na^+ channel. **A,** Hydropathy plot of the α subunit of the Na^+ channel. Above the hydropathy plot the boxes mark the location of the transmembrane helices of domains I to IV. **B,** Two-dimensional model of the Na^+ channel. The S4 helices, marked with + signs, are believed to function as the voltage-sensor for activation of the channel. The intracellar loop connecting domains III and IV plays a key role in voltage inactivation. (Modified from Noda M et al: *Nature* 312:121, 1984.) **C,** Top view of the Na^+ channel showing how the central ion channel is proposed to be lined by one of the helices from each domain. Which helix lines the channel is still not known. (Modified from Levitan IB, Kaczmarek LK: The neuron—cell and molecular biology, New York, 1991, Oxford University Press.)

less, voltage-gated cation channels often possess additional protein subunits. For example, the Na^+ channel from rat brain has two additional subunits, called β_1 and β_2 (Fig. 3-16). The functions of the β subunits remain to be elucidated, but they may influence the gating properties of the channel and the β_1 subunit may contribute to the structural stability of the α subunit. The Na^+ channel from mammalian skeletal muscle has α and β_1 subunits (it lacks the β_2 subunit), and the Na^+ channel from eel electric organ contains only an α subunit.

Ion channel isoforms. Different isoforms of an ion channel may predominate in different cell types. The mRNAs that encode three different isoforms of the α subunit of the Na^+ channels of rat brain have been cloned and sequenced. The proteins they encode, called *isoforms I, II, and III,* are highly homologous to one another. The functional consequences of the presence of one isoform versus the other may be subtle. As mentioned previously, ion channels may have different numbers of subunits in different tissues.

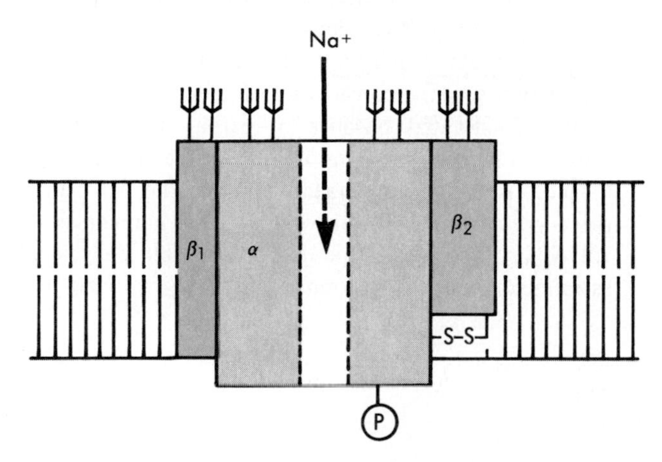

■ **Fig. 3-16** Subunits of the Na^+ channel of rat brain. The functions of the β_1 and β_2 subunits remain to be elucidated. All voltage-gated Na^+ channels have an α subunit, but the presence of the β subunits is variable from one cell type to another. (Modified from Catterall W: *Science* 242:50, 1988).

Different types of voltage-gated Ca^{++} channels. The flow of Ca^{++} ions through Ca^{++} channels is not only important in electrophysiological phenomena such as action potentials, but it also modulates the cytosolic level of Ca^{++}, a second messenger that controls a host of cellular processes. Three different types of voltage-gated Ca^{++} channels have been identified: L (long lasting), T (transient), and N (neuronal). These channels have different single-channel conductances and different activation voltages, and they differ in the extent and time-course of voltage-dependent inactivation (Table 3-1).

The L-type Ca^{++} channel has specific receptors for three different classes of drugs that block Ca^{++} channels: the phenylalkylamines (e.g., verapamil), the benzothiazepines (e.g., diltiazem), and the dihydropyridines (e.g., nifedipine). These Ca^{++} channel antagonists are widely used in treating certain cardiovascular disorders.

■ *Modulation of Ion Channels by Second Messengers*

A plethora of cellular processes are regulated via mechanisms involving second messengers (see Chapter 5). Ion channels (both voltage-gated and the ligand-gated ion channels discussed in Chapters 3 and 4) may be modulated by second messenger mechanisms. Second messenger-mediated mechanisms of modulating ion channels are described more fully after the discussion of ligand-gated ion channels in Chapter 4.

■ *Action Potentials in Cardiac and Smooth Muscle*

■ *Cardiac Muscle*

An action potential in a cardiac ventricular cell is schematically shown in Fig. 3-1. The initial rapid depolarization and overshoot is caused by the rapid entry of

■ **Table 3-1** Properties of different classes of voltage-gated Ca^{++} channels

	Channel type		
Property	***T***	***N***	***L***
Activation voltage range	More positive than −70 mV	More positive than −10 mV	More positive than −10 mV
Inactivation voltage range	−100 to −60 mV	−100 to −40 mV	Little inactivation
Single channel conductance	8-10 pS	11-15 pS	23-27 pS
Ca^{++} current kinetics	Long latency, brief current, moderate rate of current decrease	Long lasting current, moderate rate of current decrease	Longest lasting current, very slow rate of current decrease
Blocked by dihydropyridines	No	No	Yes
Blocked by cadmium	No	Yes	Yes
Blocked by cobalt	Strongly	Weakly	Weakly

(Modified from Tsien RW. In Kaczmarek LK, Levitan IB, editors: *Neuromodulation—the biochemical control of neuronal excitability,* New York, 1987, Oxford University Press.

Na^+ through channels that are very similar to the Na^+ channels of nerve and skeletal muscle. Because of the rapid kinetics of the opening and closing of these channels, they are called **fast Na^+ channels.**

After the initial depolarization and overshoot the cardiac ventricular action potential has a *plateau phase*. The plateau is caused by another set of channels distinct from the fast Na^+ channels. These channels open and close much more slowly than the fast Na^+ channels and are called **slow channels.** The slow channels are L-type Ca^{++} channels and they conduct both Ca^{++} and Na^{++}. The Ca^{++} that enters the cell during the plateau phase helps to initiate contraction of the ventricular cell. The repolarization of the ventricular cell is brought about by the closing of the slow channels and by a much delayed opening of K^+ channels. The ionic mechanism of the cardiac action potential is discussed in more detail in Chapter 23.

Smooth Muscle

Action potentials vary considerably among different types of smooth muscle. Characteristically action potentials in smooth muscle have slower rates of depolarization and repolarization and less overshoot than skeletal muscle action potentials. Smooth muscle cells lack fast Na^+ channels. The depolarizing phase of smooth muscle action potentials is caused primarily by channels that resemble the cardiac slow channels in their kinetics and in conducting both Na^+ and Ca^{++}. The Ca^{++} that enters via the slow channels is often vital for excitation-contraction coupling in smooth muscle. Repolarization is caused by the closing of the slow Na^+/Ca^{++} channels and a simultaneous opening of K^+ channels.

Properties of Action Potentials

Voltage Inactivation of the Action Potential

If a neuron or skeletal muscle cell is depolarized, for example, by increasing the concentration of K^+ in the extracellular fluid, its action potential has a slower rate of rise and a smaller overshoot. This is a result of two factors: (1) a smaller electrical force driving Na^+ into the depolarized cell, and (2) **voltage inactivation** of some of the Na^+ channels. The increase in g_{Na} during the action potential is self-inactivating. Once the Na^+ channels are inactivated, the membrane must be repolarized toward the normal resting membrane potential before the channels can be reopened. As the membrane potential is restored toward normal resting levels, more and more of the Na^+ channels again become capable of being activated. Because the action potential mechanism requires a critical density of open Na^+ channels, an action potential may not be generated in response to stimulation when a considerable fraction of Na^+ channels is inactivated because of partial depolarization. This is called **voltage inactivation** of the action potential because of voltage inactivation of the Na^+ channels. Voltage inactivation of the Na^+ channels is involved in important properties of excitable cells, such as refractoriness and accommodation.

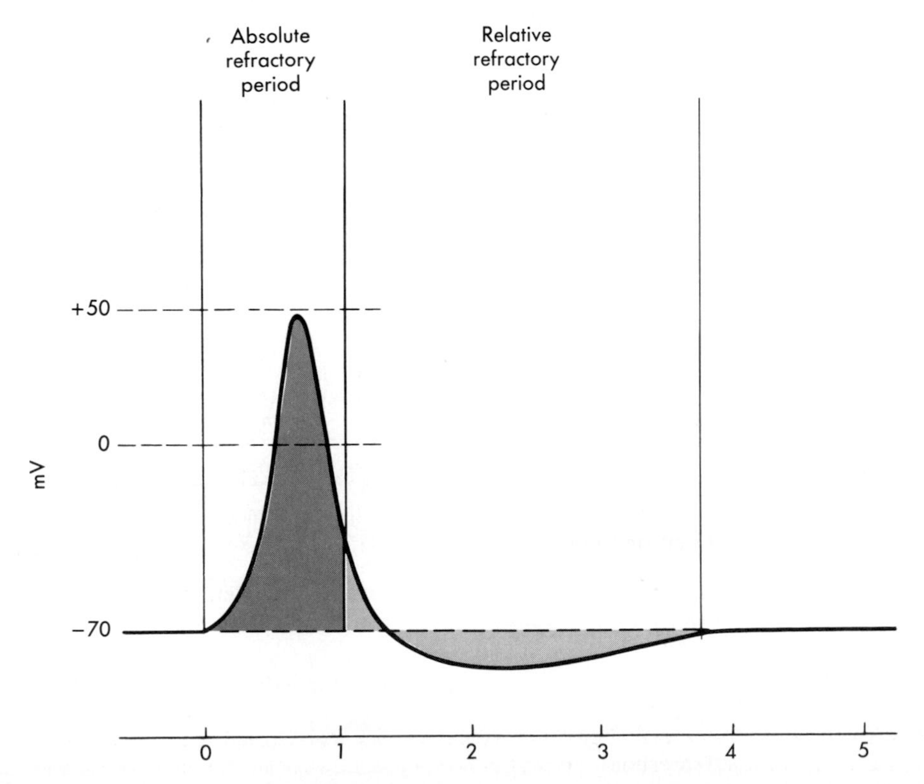

Fig. 3-17 The action potential of nerve showing the absolute and relative refractory periods.

The Refractory Periods

During much of the action potential the membrane is completely refractory to further stimulation. This means that, no matter how strongly the cell is stimulated, it is unable to fire a second action potential. This is called the **absolute refractory period** (Fig. 3-17). The cell is refractory because a large fraction of its Na^+ channels is voltage inactivated and cannot be reopened until the membrane is repolarized.

During the last part of the action potential the cell is able to fire a second action potential, but a stronger than normal stimulus is required. This is the **relative refractory period.** Early in the relative refractory period, before the membrane potential has returned to the resting potential level, some Na^+ channels are voltage inactivated, so a stronger than normal stimulus is needed to open the critical number of Na^+ channels necessary to trigger an action potential. Throughout the relative refractory period the conductance to K^+ is elevated, which results in increased opposition to depolarization of the membrane. This also contributes to the refractoriness.

Accommodation to Slow Depolarization

When a nerve or muscle cell is depolarized slowly, the threshold may be passed without an action potential being fired. This is called **accommodation.** Na^+ and K^+ channels are both involved in accommodation. During slow depolarization some of the Na^+ channels that are opened by depolarization have enough time to become voltage inactivated before the threshold potential is attained. If depolarization is slow enough, the critical number of open Na^+ channels required to trigger the action potential may never be achieved. In addition, K^+ channels open in response to the depolarization. The increased g_K tends to repolarize the membrane, making it still more refractory to depolarization.

Conduction of the Action Potential

A principal function of neurons is to transfer information by the conduction of action potentials. The axons of the motor neurons of the ventral horn of the spinal cord conduct action potentials from the cell body of the neuron in the spinal cord to a number of skeletal muscle fibers. The distance from the motor neuron to one of the muscle fibers it innervates may be longer than 1 m. The mechanism of action potential conduction and the factors that determine the speed of conduction are considered next.

Action potentials are conducted along a nerve or muscle fiber by local current flow, just as occurs in electrotonic conduction of subthreshold potential changes. Thus the same factors that govern the velocity of electrotonic conduction also determine the speed of action potential propagation.

The Local Response

Fig. 3-18, *A*, shows the membrane of an axon or muscle fiber that has been depolarized in a small area. In the depolarized region the external aspect of the membrane is negative relative to the adjacent membrane, and the internal face of the depolarized membrane is positively charged relative to neighboring internal areas. The potential differences cause local currents to flow (Fig. 3-18, *B*), which depolarize the membrane adjacent to the initial site of depolarization. These newly depolarized areas then cause current flows that depolarize other segments of the membrane that are still further removed from the initial site of depolarization. This spread of depolarization is called the **local response.** This mechanism of conduction is known as **electrotonic conduction.**

RC Circuits

The speed of electrotonic conduction is determined by the passive electrical properties of the cell involved. In particular, the capacitance of the plasma membrane and the resistance through which charge must flow to charge or discharge the membrane capacitor help to determine the conduction velocity.

The hydrophobic core of the plasma membrane is a good electrical insulator. The surfaces of the plasma membrane and the electrolyte solutions of the cytosol

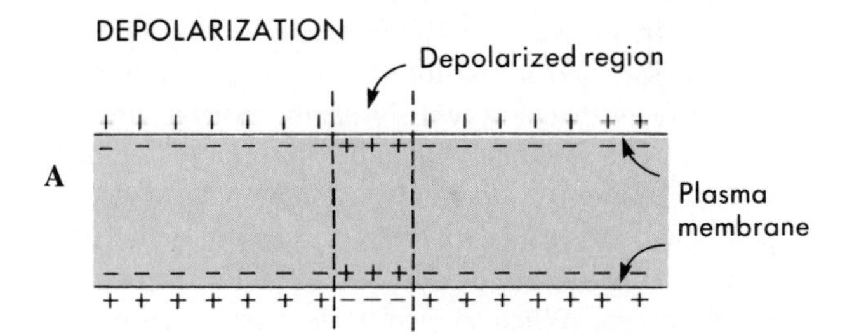

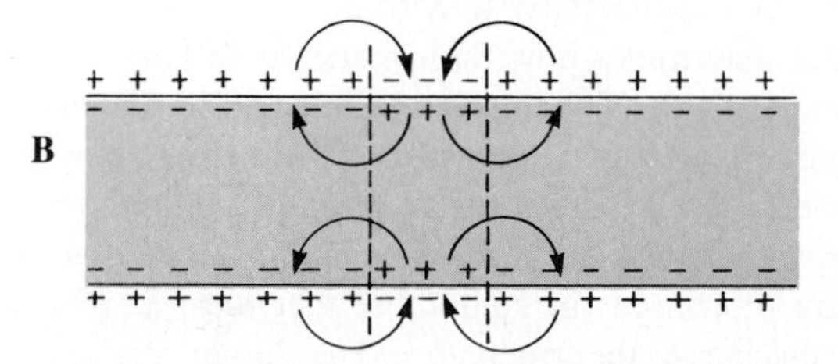

Fig. 3-18 Mechanism of electrotonic spread of depolarization. **A,** The reversal of membrane polarity that occurs with local depolarization. **B,** The local currents that flow to depolarize adjacent areas of the membrane and allow spreading of the depolarization.

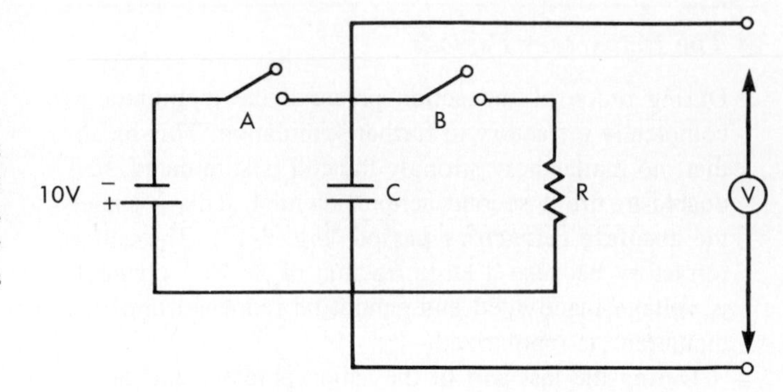

■ **Fig. 3-19** An electrical circuit that illustrates the charging and discharging of a capacitor (C). When switch A is closed, currents flow to charge the capacitor to the battery voltage (10 v). When switch A is opened, the charge remains on the capacitor because there is not a path for current to flow to discharge it. When switch B is closed, current flows through resistor (R) to discharge the capacitor.

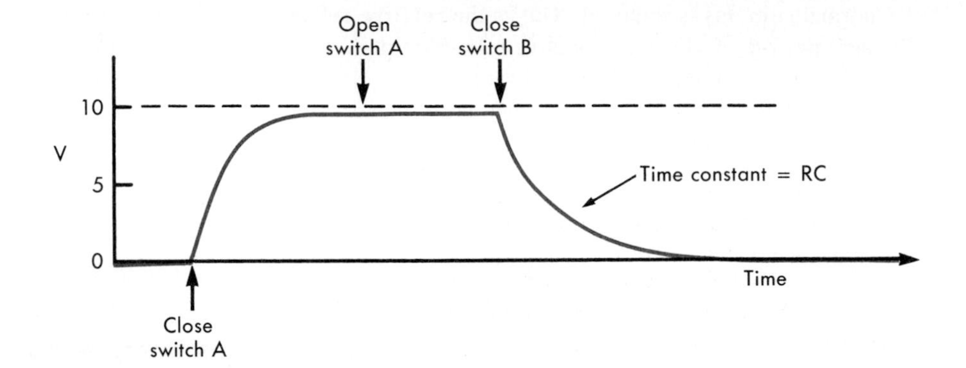

and extracellular fluid are better conductors of electrical current. Capacitors, devices that can store electrical charge, are constructed from a sheet of insulating material sandwiched between two conductors. The plasma membrane has a structure analagous to a capacitor and plasma membranes have significant capacitance (about 1 microfarad/cm^2). Before considering how the capacitance of the plasma membrane helps to determine the velocity of electrotonic conduction, let us consider the electrical circuit shown in Fig. 3-19.

The circuit in Fig. 3-19 consists of a 10-volt battery, a capacitor, and a resistor. When switch A is closed, the currents that flow will bring the voltage difference between the two plates of the capacitor to the battery voltage (10 volts). If switch A is now opened, the voltage across the capacitor will remain constant because there is no conductive path that will allow the capacitor to discharge. When switch B is closed, current will flow through the resistor and discharge the capacitor. As shown in the figure, the voltage across the capacitor will decay exponentially to zero.

What determines how rapidly the current will decay? The capacitance of the capacitor is equal to the amount of charge stored per unit voltage difference across the capacitor. The capacitance (C) times the initial voltage difference across the capacitor is then equal to the amount of charge (in coulombs) that must flow to totally discharge the capacitor. The larger the capacitance, the more charge must flow, and the more time will be required to discharge the capacitor. The resistance (R) in the circuit (and the voltage difference V_C) will determine the current flow that discharges the capacitor. The higher the resistance, the smaller the current, and the slower the discharge of the capacitor. Large values of R and large values of C both increase the time required to discharge the capacitor. The time required for the capacitor to discharge from its original voltage (V_o) to V_o/e (37% of V_o) is equal to R times C; RC is called the **time constant** for the circuit. The smaller the time constant, the more rapidly can the capacitor be discharged.

■ *Conduction Velocity*

The speed of electrotonic conduction along a nerve or muscle fiber is determined by the membrane capacitance and the electrical resistance to the flow of current. The membrane potential is a measurable manifestation of the charge stored by the membrane capacitor. The amount of charge that must flow to depolarize 1 cm^2 of membrane is proportional to the membrane capacitance (the coulombs of charge stored per volt of potential difference across the membrane) per unit area. A typical value of membrane capacitance (c_m) is about 10^{-6} farad/cm^2 membrane. To depolarize the membrane from -100 mV (-0.1 V) to 0 mV, 10^{-7} coulombs of charge must flow across each square centimeter of membrane:

$$\begin{aligned}\text{Charge flow/cm}^2 &= \text{Capacitance/cm}^2 \times \text{Voltage change} \\ &= 10^{-6}\ \text{farad/cm}^2 \times 0.1\ \text{V} \qquad (1)\\ &= 10^{-7}\ \text{coulombs/cm}^2\end{aligned}$$

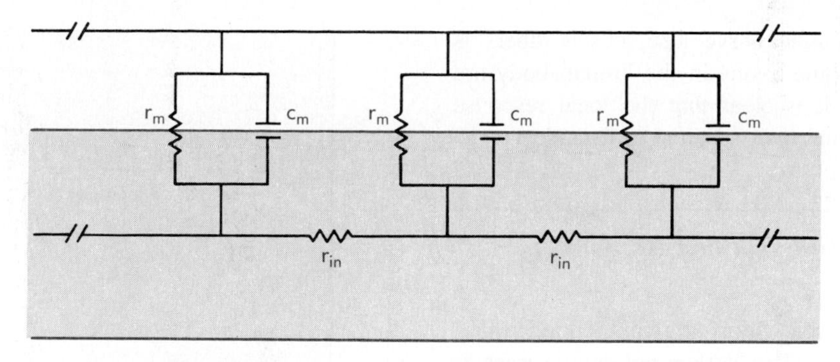

■ **Fig. 3-20** The equivalent electrical circuit for electrotonic conduction. The membrane capacitor (c_m) can be discharged by currents flowing through the membrane resistance (r_m) and the internal longitudinal resistance (r_{in}). The effective time constant for this circuit is $\sqrt{r_m r_{in}}\ c_m$.

The membrane capacitance thus determines *how much charge* must flow to depolarize the membrane. The larger the membrane capacitance, the greater the amount of charge that must flow and the slower the rate of electrotonic spread.

The resistance to electrotonic current flow determines *how rapidly* charge can flow. The resistance to electrotonic current flow depends on the resistance to current flow across the membrane (r_m) and the resistance to longitudinal current flow in the cytoplasm (r_{in}) (Fig. 3-20). The effective resistance is proportional to the geometric mean of r_m and r_{in} ($\sqrt{r_m r_{in}}$). The larger $\sqrt{r_m r_{in}}$, the slower will electrotonic current flow and the slower will be the rate of electrotonic conduction.

The product $\sqrt{r_m r_{in}}\ c_m$ is the time constant for electrotonic conduction. The smaller $\sqrt{r_m r_{in}}\ c_m$, the more rapidly electrotonic conduction can occur and vice versa.

Effect of fiber size on conduction velocity. Consider a cylindrical nerve or muscle cell. The surface area of the cell increases with increasing radius ($A_s = 2\pi rl$). The cross-sectional area of the cell increases with the square of the radius ($A_x = \pi r^2$). The membrane capacitance increases in direct proportion to membrane area. The membrane resistance decreases in proportion to an increase in membrane area. The internal resistance is inversely proportional to the cross-sectional area. The effect of doubling the radius of the cell thus will increase c_m by a factor of 2, decrease r_m by a factor of 2, and decrease r_{in} by a factor of 4. Thus c_m will increase twofold, but $\sqrt{r_m r_{in}}$ will decrease by a factor of $\sqrt{2 \cdot 4}$, or $2\sqrt{2}$. The product $\sqrt{r_m r_{in}}\ c_m$ thus will decrease to $1/\sqrt{2}$ of its former value with a doubling of cell radius, and the velocity of electrotonic conduction will increase to $\sqrt{2}$ times the conduction velocity of the smaller fiber. Thus larger fibers have larger conduction velocities.

■ *Electrotonic Conduction Involves Decrement*

Earlier in this chapter it was noted that the local response dies away to almost nothing over the course of several millimeters (see Fig. 3-3). A nerve or muscle fiber has some of the properties of an electrical cable. In a perfect cable the insulation surrounding the core conductor prevents all current loss so that a signal is transmitted along the cable with undiminished strength (Fig. 3-21). The plasma membrane of an unmyelinated nerve or muscle fiber serves as the insulation. The membrane has a resistance much higher than the resistance of the cytoplasm, but (partly because of its thinness) the plasma membrane is not a perfect insulator. The higher the ratio of r_m to r_{in}, the better the cell can function as a cable and the longer the distance that a signal can be transmitted electrotonically without significant decrement. $\sqrt{r_m/r_{in}}$ determines the **length constant** (see Fig. 3-3) of a cell. The length constant is the distance over which an electrotonically conducted signal falls to 37% (1/e) of its initial strength. A typical length constant for

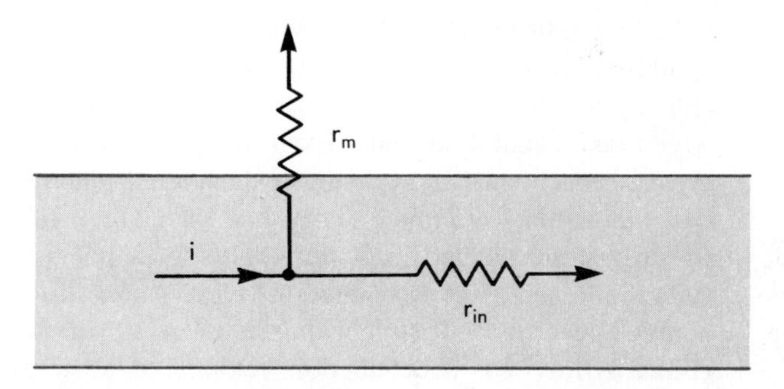

■ **Fig. 3-21** An axon or a muscle fiber resembles an electrical cable. Currents that flow across the membrane resistance (r_m) are lost from the cable. Currents that flow through the longitudinal resistance (r_{in}) carry the electrical signal along the cable.

unmyelinated mammalian nerve and muscle fibers is about 1 to 3 mm. Some axons in the human body are about 1 m long, so it is clear that the local response cannot conduct a signal over so great a distance.

The Action Potential as a Self-Reinforcing Signal

Many nerve and muscle fibers are much longer than their length constants. The action potential serves to conduct an electrical impulse with undiminished strength along the full length of those fibers. To do this, the action potential *reinforces itself* as it is propagated along the fiber. The propagation of the action potential occurs by the mechanism depicted in Fig. 3-18. When the areas on either side of the depolarized region reach threshold, these areas also fire action potentials, which locally reverses the polarity of the membrane potential. By local current flow the areas of the fiber adjacent to these areas are brought to threshold, and then these areas in turn fire action potentials. There is a cycle of depolarization by local current flow followed by generation of an action potential in a restricted region that then travels along the length of the fiber, with "new" action potentials being generated as they spread. In this way the action potential propagates over long distances, keeping the same size and shape.

Because the shape and size of the action potential are ordinarily invariant, only variations in the frequency of the action potentials can be used in the code for information transmission along axons. The maximum frequency is limited by the duration of the absolute refractory period (about 1 msec) to about 1000 impulses/second in large mammalian nerves.

Effect of Myelination on Conduction Velocity

A squid giant axon with 500 μm diameter has a conduction velocity of 25 m/second and is unmyelinated. If conduction velocity were directly proportional to fiber radius, a human nerve fiber with a 10μm diameter would conduct at 0.5 m/second. With this conduction velocity a reflex withdrawal of the foot from a hot coal would take about 4 seconds. Even though our nerve fibers are much smaller in diameter than squid giant axons, our reflexes are much faster than this. The myelin sheath that surrounds certain vertebrate nerve fibers results in a much greater conduction velocity than that of unmyelinated fibers of similar diameters. A 10 μm **myelinated fiber** has a conduction velocity of about 50 m/second, which is twice that of the 500 μm squid giant axon. The high conduction velocity permits reflexes that are fast enough to allow us to avoid dangerous stimuli. Fig. 3-22 shows the large increase in conduction velocity caused by myelination. A myelinated axon has a greater conduction velocity than an unmyelinated fiber that is 100 times larger in diameter. As discussed below, the myelin sheath increases the velocity of action potential conduction by decreasing the capacitance of the axon and by allowing action potentials to be generated only at the **nodes of Ranvier.**

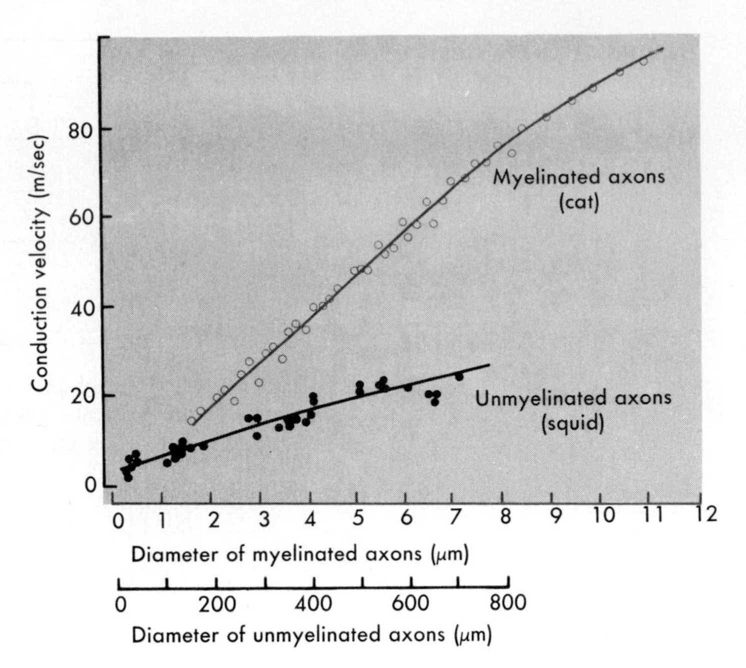

Fig. 3-22 Conduction velocities of myelinated and unmyelinated axons as functions of axon diameter. Myelinated axons *(colored curve* and *abscissa)* are from cat saphenous nerve at 38 C. Unmeylinated axons *(black curve* and *abscissa)* are from squid and cuttlefish at 20 to 22 C. Note that the myelinated axons have faster conduction velocities than unmyelinated axons 100 times greater in diameter. (Data for colored curve and abscissa from Gasser H, Grundfest H: *Am J Physiol* 127:393, 1939; data from black curve and abscissa from Pumphrey RJ, Young JZ: *J Exp Biol* 15:453, 1938.)

The evolution of myelinated fibers in vertebrates is salutary. If each of our peripheral nerve fibers were as large as a squid giant axon, then the nerve trunks (each containing thousands of nerve fibers) would be so large that the peripheral nerves would alter the human form.

Myelination is caused by the wrapping of Schwann cell plasma membrane around a nerve axon (Fig. 3-23). Thus the myelin sheath consists of several to over 100 layers of plasma membrane. Gaps that occur in the sheath every 1 to 2 mm are known as nodes of Ranvier. Nodes of Ranvier are about 1 μm wide and are the lateral spaces between different Schwann cells along the axon.

Myelination greatly alters the electrical properties of the axon. The many wrappings of membrane around the axon increase the effective *membrane resistance,* so that r_m/r_{in} and thus the length constant is much greater. Less of a conducted signal is lost through the electrical insulation of the myelin sheath, so that the amplitude of a conducted signal declines less with distance along the axon.

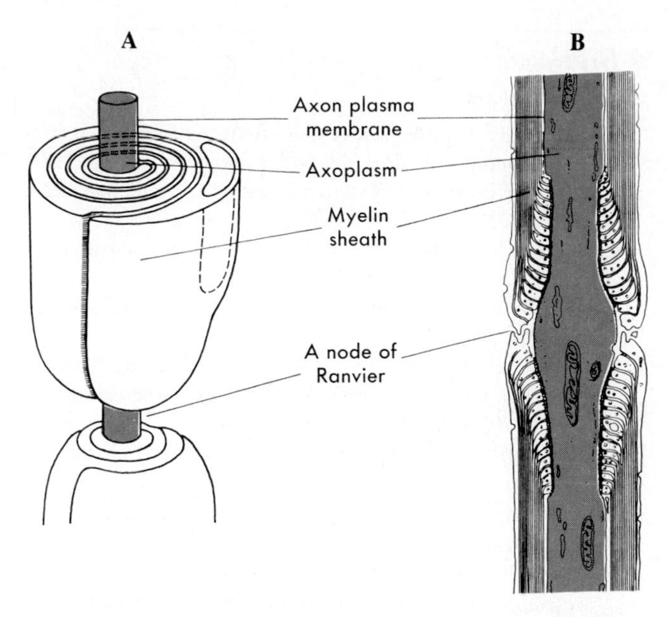

■ **Fig. 3-23** The myelin sheath. **A,** Schematic drawing of Schwann cells wrapping around an axon to form a myelin sheath. **B,** Drawing of a cross-section through a myelinated axon near a node of Ranvier. (Redrawn from Elias H et al: Histology and human microanatomy, ed 4, New York, 1978, John Wiley & Sons.)

Each Schwann cell membrane has a capacitance similar to that of the plasma membrane of the axon. The capacitances of the membranes in the myelin sheath act as though they were connected in series. Capacitors in series are added according to the equation $1/C_t = 1/C_1 + 1/C_2 + 1/C_3 + \ldots$ etc., where C_t is the effective overall capacitance, and C_1, C_2, C_3, etc., are the individual capacitances. If there are 50 identical capacitances in series (25 Schwann cell wraps), then $1/C_t = 50/C$, and $C_t = C/50$. A myelin sheath with 50 membranes thus *lowers* the membrane capacitance by a factor of 50. Resistances in series, however, add directly, so the myelination will increase the membrane resistance by fiftyfold. Earlier in this chapter we saw that the time constant $\sqrt{r_m r_{in}}\, c_m$ determines the electrotonic conduction velocity. The smaller this product, the greater the conduction velocity. The 25 Schwann cell wraps should have no effect on r_{in} but will lower c_m by a factor of 50 and increase r_m by a factor of 50. Thus $\sqrt{r_m r_{in}}\, C_m$ will decrease by $50/\sqrt{50}$, resulting in a sevenfold increase in electrotonic conduction velocity because of myelination.

Another property of conduction in myelinated fibers that enhances conduction velocity is called **saltatory conduction,** because the impulse "jumps" from one node of Ranvier to the next. Saltatory conduction occurs because the action potential is regenerated only at the nodes (1 to 2 mm apart). The action potential in myelinated fibers is not regenerated at each place along the axon as the impulse is propagated. The internodal plasma membrane cannot produce action potentials, because the depolarization of the internodal membrane is divided among 50 or so membranes of the myelin sheath. The resulting depolarization of the internodal plasma membrane is only 1 or 2 mV—not nearly sufficient to reach threshold. Moreover, Na^+ channels are concentrated at the nodes of Ranvier and are relatively scarce in the internodal plasma membrane.

In brief, conduction in myelinated axons is characterized by rapid electrotonic conduction (because of the decreased time constant for conduction) with little decrement (because of the increased length constant) between the nodes of Ranvier. Only at the nodes is the action potential regenerated.

Myelinated axons are also more efficient metabolically than nonmyelinated axons. The sodium-potassium pump extrudes the sodium that enters and reaccumulates the potassium that leaves the cell during action potentials. In a myelinated axon, ionic currents are restricted to the small fraction of the membrane surface at the nodes of Ranvier. For this reason fewer Na^+ and K^+ ions traverse a unit area of membrane, and less ion pumping is required to maintain Na^+ and K^+ gradients.

■

1. A hyperpolarization or a subthreshold depolarization of an axon or skeletal muscle fiber is conducted with decrement along the length of the cell. The distance, typically 1 to 2 mm, over which the conducted signal decreases to 1/e (about 38%) of its maximum strength, is called the length constant of the cell.

2. Depolarization of an electrically excitable cell to threshold elicits an action potential. The action potential is propagated over the entire plasma membrane of the cell without a significant decrease in its strength.

3. Action potentials are caused by the opening and closing of particular populations of voltage-gated ion channels in the plasma membrane of an excitable cell. Cell types that have different types of ion channels have action potentials that differ in shape and time course. In nerve and skeletal muscle cells action potentials are initiated by a rapid opening of fast Na^+ channels, which causes the depolarizing phase of the action potential. The depolarization brings about voltage-inactivation of the Na^+ channels followed by a delayed opening of K^+ channels, which cause the repolarization of the cell.

4. An ion channel typically oscillates irregularly between a low-conductance state and a high conductance state. For a voltage-gated ion channel the probability that the channel is in the high-conductance state is a function of the membrane potential.

5. The voltage-gated ion channels for Na^+, K^+, and Ca^{++} are members of a family of structurally related proteins. The Na^+ and Ca^{++} channel proteins have four repeated motifs, each with six transmembrane heli-

ces. The ion channel is apparently lined by the relatively hydrophilic faces of one helix from each motif. The K^+ channel protein has only one of the motifs that is repeated in the Na^+ and Ca^{++} channels; four of the K^+ channel proteins are required to form one K^+ channel.

6. Three types of voltage-gated Ca^{++} channels have been identified: L-type (long lasting), T-type (transient), and N-type (neuronal). These channels differ in their voltage dependence, single-channel conductance, and time course. The L-type Ca^{++} channels are blocked by three different classes of Ca^{++} channel-blocking drugs.

7. Action potentials in cardiac muscle cells have a plateau phase caused by the movement of Ca^{++} through slow channels that are L-type Ca^{++} channels. In cardiac and smooth muscle cells the Ca^{++} that enters the cytosol via slow channels during the action potential plays an important role in excitation-contraction coupling.

8. During most of the action potential spike, the cell is absolutely refractory to further stimulation. This is caused by voltage-inactivation of such a large fraction of Na^+ channels that a sufficient number cannot be recruited to fire another action potential.

9. An action potential or a subthreshold perturbation of membrane potential is conducted by local circuit currents, a mechanism called electrotonic conduction. The rate of electrotonic conduction is determined by the electrical properties of the plasma membrane and the cytosol, namely by the capacitance and resistance of the membrane and by the longitudinal resistance of the cytosol.

10. Large diameter axons conduct more rapidly than do thin axons. A myelinated axon conducts more rapidly than an unmyelinated axon that is 100 times larger in diameter.

11. The propagation of an action potential along a myelinated axon occurs by saltatory conduction. The action potential jumps (is conducted rapidly and with little decrement) from one node of Ranvier to the next and the action potential is regenerated at each node of Ranvier.

Bibliography

Journal articles

Barchi RL: Probing the molecular structure of the voltage-dependent sodium channel, *Annu Rev Neurosci* 11:455, 1988.

Bean BP: Classes of calcium channels in vertebrate cells, *Annu Rev Physiol* 51:367, 1989.

Catterall WA: Structure and function of voltage-sensitive ion channels, *Science* 242:50, 1988.

Catterall WA: Genetic analysis of ion channels in vertebrates, *Annu Rev Physiol* 50:395, 1988.

Hamill OP et al: Improved patch-clamp techniques for high-resolution recording from cells and cell-free membrane patches, *Pflugers Arch* 391:85, 1981.

Jan LY, Jan YN: Voltage-sensitive ion channels, *Cell* 56:13, 1989.

Maelicke A: Structural similarities between ion channel proteins, *Trends in Biochem Sci* 13:199, 1988.

Steinbach JH: Structural and functional diversity in vertebrate skeletal muscle nicotinic acetylcholine receptors, *Annu Rev Physiol* 51:353, 1989.

Stuhmer W: Structure-function studies of voltage-gated ion channels, *Annu Rev Biophys Biophys Chem* 20:65, 1991.

Trimmer JS, Agnew WS: Molecular diversity of voltage-sensitive Na channels, *Annu Rev Physiol* 51:401, 1989.

Books and monographs

Aidley DJ: *The physiology of excitable cells,* ed 3, Cambridge, 1990, Cambridge University Press.

Hille B: *Ionic channels of excitable membranes,* Sunderland, Mass, 1984, Sinauer Associates.

Hodgkin AL: *The conduction of the nervous impulse,* Springfield, Ill, 1964, Charles C Thomas Publishing.

Kandel ER, Schwartz JH: *Principles of neural science,* ed 3, New York, 1991, Elsevier.

Katz B: *Nerve, muscle, and synapse,* New York, 1966, McGraw-Hill.

Kuffler SW, Nicholls, JG, Martin, AR: *From neuron to brain,* ed 2, Sunderland, Mass, 1984, Sinauer Associates.

Levitan IB, Kaczmarek LK: *The neuron: cell and molecular biology,* New York, 1991, Oxford University Press.

Miller C: *How ion channel proteins work.* In Kazamarek LK, Levitan, IB, editors: *Neuromodulation: the biochemical control of neural excitability,* New York, 1987, Oxford University Press.

Stevens CF: *Neurophysiology: a primer,* New York, 1966, John Wiley & Sons.

CHAPTER

4

Synaptic Transmission

A synapse is a site at which an impulse is transmitted from one cell to another. There are two types of synapses: electrical synapses and chemical synapses. At an **electrical synapse** two excitable cells communicate by the direct passage of electrical current between them. This is called **ephaptic** or **electrotonic** transmission. **Gap junctions** link electrotonically coupled cells and are low-resistance pathways for current flow directly between the cells.

Information is also transmitted between excitable cells by means of **chemical synapses.** Chemical synapses may be somewhat better suited for the complex modulation of synaptic activity and the integration that occurs at synapses in vertebrate central nervous systems. At a chemical synapse an action potential causes a transmitter substance to be released from the presynaptic neuron. The transmitter diffuses across the extracellular synaptic cleft and binds to receptors on the membrane of the postsynaptic cell to cause a change in its electrical properties. Chemical synapses have **synaptic delay**—the time required for these events to occur. The neuromuscular junction is a particularly well-studied vertebrate chemical synapse.

Although the nature of the presynaptic and postsynaptic cells, the structure of the synapse, and the transmitter substance vary, there are certain characteristics that chemical synapses have in common.

Neuromuscular Junctions

The synapses between the axons of motoneurons and skeletal muscle fibers are called **neuromuscular junctions** or **motor endplates.** The neuromuscular junction was the first vertebrate synapse to be well characterized. The neuromuscular junction serves as a model chemical synapse that provides a basis for understanding more complex synaptic interactions among neurons in the central nervous system.

Structure of the Neuromuscular Junction

Near the neuromuscular junction the motor nerve loses its myelin sheath and divides into fine terminal branches (Fig. 4-1). The terminal branches of the axon lie in synaptic troughs on the surfaces of the muscle cells (Fig. 4-1). The plasma membrane of the muscle cell lining the trough is thrown into numerous **junctional folds.** The axon terminals contain many 400 Å smooth-surfaced synaptic vesicles that contain acetylcholine. The

General characteristics of transmission at chemical synapses

Action potential in presynaptic cell

↓

Depolarization of the plasma membrane of the presynaptic axon terminal

↓

Entry of Ca^{++} into presynaptic terminal

↓

Release of the transmitter by the presynaptic terminal

↓

Chemical combination of the transmitter with specific receptors on the plasma membrane of the postsynaptic cell

↓

Transient change in the conductance of the postsynaptic plasma membrane to specific ions

↓

Transient change in the membrane potential of the postsynaptic cell

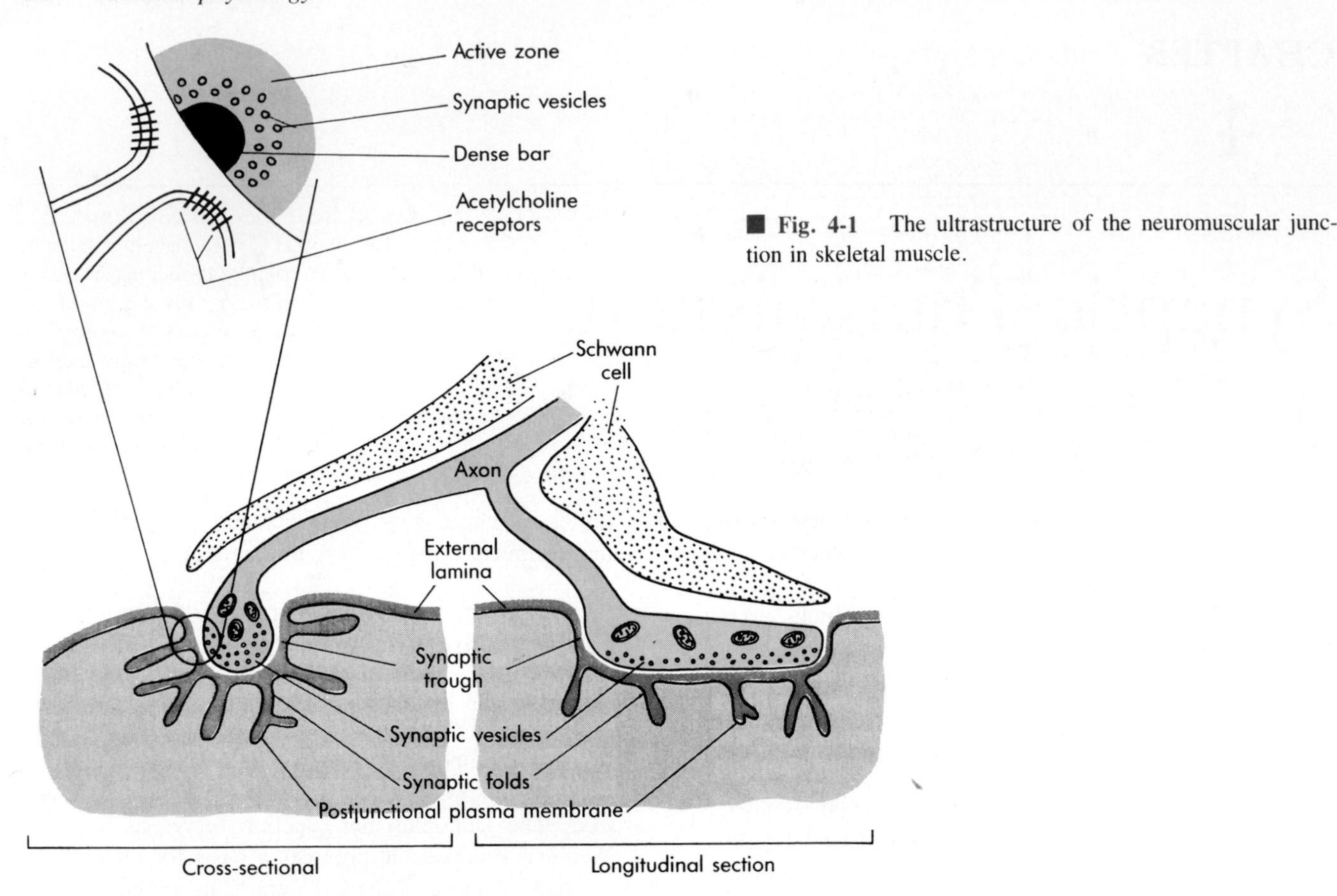

■ **Fig. 4-1** The ultrastructure of the neuromuscular junction in skeletal muscle.

axon terminal and the muscle cell are separated by the **synaptic cleft,** which contains a carbohydrate-rich amorphous material.

Acetylcholine receptor molecules are concentrated near the mouths of the junctional folds. **Acetylcholinesterase** appears to be evenly distributed on the external surface of the postsynaptic membrane. The synaptic vesicles in the nerve terminals and specialized release sites on the presynaptic membrane, called **active zones,** are concentrated opposite the mouths of the junctional folds.

■ *Overview of Neuromuscular Transmission*

The action potential is conducted down the motor axon to the presynaptic axon terminals. Depolarization of the plasma membrane of the axon terminal brings about a transient increase in its calcium conductance. Ca^{++} flows down its electrochemical potential gradient into the axon terminal. The influx of Ca^{++} causes synaptic vesicles to fuse with the plasma membrane and to empty their acetylcholine into the synaptic cleft by exocytosis. Acetylcholine diffuses across the synaptic cleft and combines with a specific acetylcholine receptor protein on the external surface of the muscle plasma membrane of the motor endplate. The combination of acetylcholine with the receptor protein causes a transient increase in the conductance of the postjunctional membrane to Na^+ and K^+. Ionic currents (Na^+ and K^+) result in a transient depolarization of the endplate region. The transient depolarization is called the **endplate potential** or **EPP** (Fig. 4-2). The EPP is transient because the action of acetylcholine is ended by the hydrolysis of acetylcholine to form choline and acetate. The hydrolysis is catalyzed by the enzyme acetylcholinesterase, which is present in high concentration on the postjunctional membrane.

The postjunctional plasma membrane of the neuromuscular junction is not electrically excitable and does not fire action potentials. After it is depolarized, adjacent regions of the muscle cell membrane are depolarized by electrotonic conduction (see Fig. 4-2). When those regions reach threshold, action potentials are generated. Action potentials are propagated along the muscle fiber at high velocity and induce the muscle cells to contract. The steps involved in neuromuscular transmission are listed below, and some are considered next in more detail.

■ *Synthesis of Acetylcholine*

Motoneurons and their axons synthesize acetylcholine. Most other cells are not able to make acetylcholine. The enzyme choline-O-acetyltransferase in the motoneuron

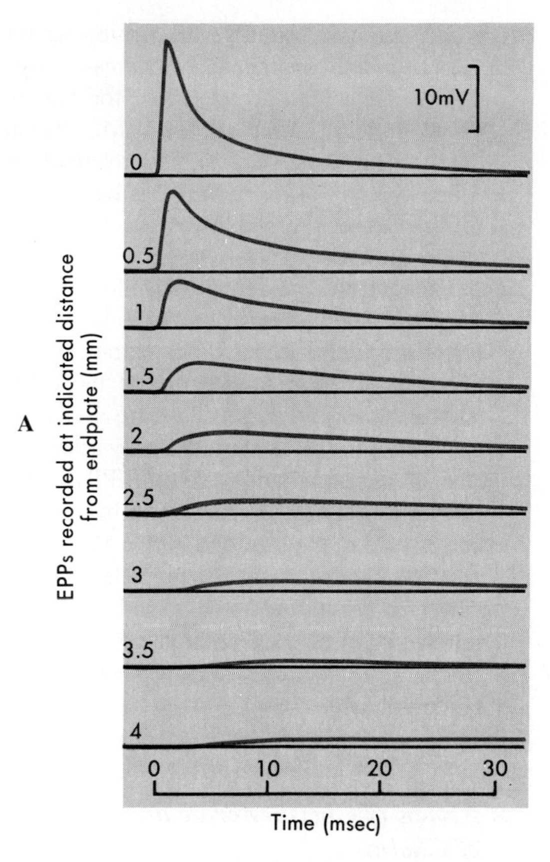

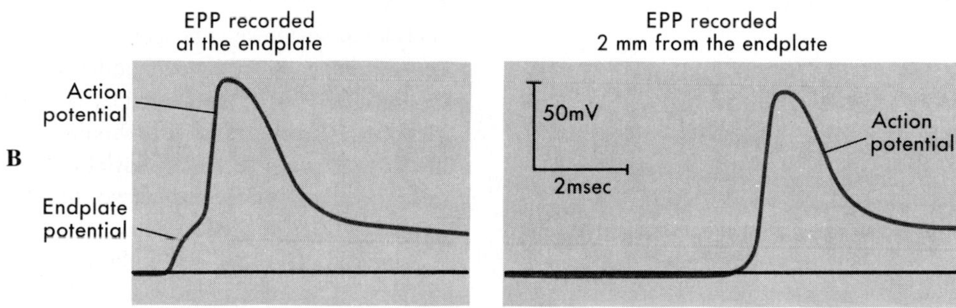

■ **Fig. 4-2** **A,** EPPs recorded with a microelectrode in a frog sartorius muscle. The preparation was treated with curare to bring the EPP just below threshold for eliciting an action potential. The EPP, recorded at increasing distances from the neuromuscular junction, decreases in amplitude and rate of rise. **B,** Intracellular recordings made at the motor endplate *(left panel)* and 2 mm away *(right panel)* in a muscle fiber of frog extensor digitorum longus. When the motor nerve was stimulated, an EPP occurred, which triggered an action potential. Both the EPP and the resulting action potential can be recorded at the endplate, but 2 mm away from the endplate only the action potential can be seen because the EPP is conducted with decrement and has substantially decayed before reaching this point on the muscle fiber. (**A** redrawn from Fatt P, Katz B: *J Physiol* 115:320, 1951; **B** redrawn from Fatt P, Katz B: *J Physiol* 117:109, 1952.)

Summary of events that occur during neuromuscular transmission

Action potential in presynaptic motor axon terminals

↓

Increase in Ca^{++} permeability and influx of Ca^{++} into axon terminal

↓

Release of acetylcholine from synaptic vesicles into the synaptic cleft

↓

Diffusion of acetylcholine to postjunctional membrane

↓

Combination of acetylcholine with specific receptors on postjunctional membrane

↓

Increase in permeability of postjunctional membrane to Na^+ and K^+ causes EPP

↓

Depolarization of areas of muscle membrane adjacent to endplate and initiation of an action potential catalyzes the condensation of acetyl coenzyme A (acetyl CoA) and choline. Acetyl CoA is produced by the neuron (as well as by most cells). Choline cannot be synthesized by the motoneuron and is accumulated by active uptake from the extracellular fluid. The plasma membrane of the motoneuron has a transport system, powered by the Na^+ gradient, that can accumulate choline against a large electrochemical potential gradient. About half the choline that is freed in the synaptic cleft when acetylcholine is hydrolyzed is actively taken back up into the motoneuron to be used in the resynthesis of acetylcholine.

Quantal Release of Transmitter

Even if the motoneuron is not stimulated, small depolarizations of the postjunctional muscle cell occur spontaneously. These small spontaneous depolarizations are known as **miniature endplate potentials,** or **MEPPs** (Fig. 4-3). They occur at random times with a frequency that averages about 1 per second. Each MEPP depolarizes the postjunctional membrane by only about 0.4 mV on average. The MEPP has the same time course as an EPP that is evoked by an action potential in the nerve terminal. The MEPP is similar to the EPP in its response to most drugs. The EPP and MEPP are both prolonged by drugs that inhibit acetylcholinesterase, and both are similarly depressed by compounds that compete with acetylcholine for binding to the receptor protein. The frequency of MEPPs varies in time, but their amplitudes are within a relatively narrow range (Figs. 4-3 and 4-4). A MEPP is caused by the spontaneous release of a single vesicle of transmitter into the synaptic cleft.

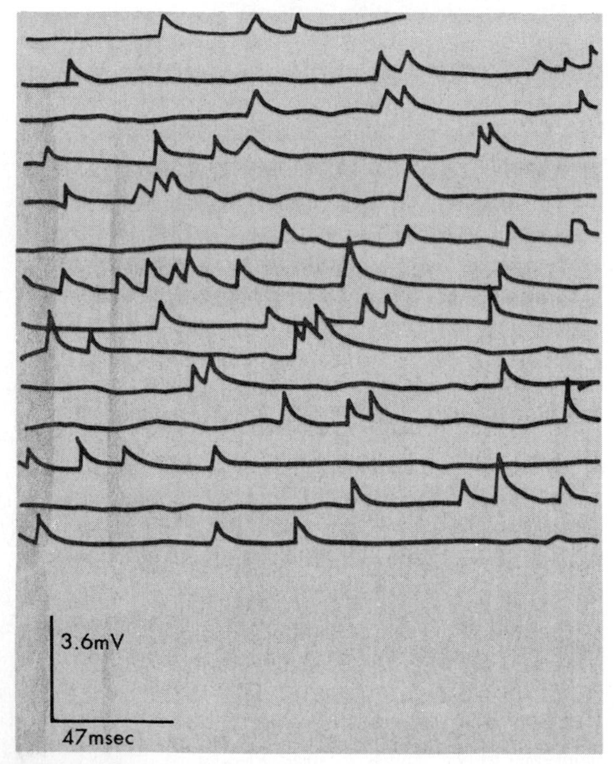

■ **Fig. 4-3** Spontaneous MEPPs recorded at a neuromuscular junction in a fiber of frog extensor digitorum longus. (Redrawn from Fatt P, Katz B: *Nature* 166:597, 1950.)

The quantal nature of transmitter release has been shown in another way. Extracellular Ca^{++} is required for transmitter release. If the extracellular Ca^{++} is reduced to low levels, the sizes of the EPPs evoked by stimulation of the motoneuron are greatly reduced. Under these conditions spontaneous variations occur in the size of the stimulation-evoked EPPs. The size of the EPP does not vary continuously, but in small steps that correspond to the size of a single MEPP (see Fig. 4-4).

Acetylcholine may also be released via a pathway that does not involve membrane-bounded vesicles. The relative roles of transmitter release from synaptic vesicles and via the vesicle-independent pathway remain to be determined.

Action of Cholinesterase and Reuptake of Choline

Acetylcholinesterase is concentrated on the external surface of the postjunctional membrane and in the basal lamina. The EPP is terminated by the hydrolysis of acetylcholine. Eserine and edrophonium are inhibitors of the enzyme and are called **anticholinesterases.** In the presence of an anticholinesterase, the EPP is larger and dramatically prolonged.

Approximately half of the choline released by the hydrolysis of acetylcholine is taken back up into the prejunctional nerve terminal by a Na^+-mediated secondary active transport system. The motoneuron cannot synthesize choline, so the reuptake provides choline needed for the resynthesis of acetylcholine. Hemicholiniums are drugs that block the choline transport system and inhibit choline uptake. Prolonged treatment with hemicholiniums results in depletion of the store of transmitter and ultimately causes a decrease in the acetylcholine content of the quanta.

The Ionic Mechanism of the EPP

The cation channels that acetylcholine causes to open in the postjunctional membrane differ from the cation channels of nerve and muscle by being independent of the membrane potential. The postjunctional channels are gated by the action of acetylcholine, rather than by the transmembrane potential. These channels are thus called **ligand-gated.** The acetylcholine-dependent cation channels of the postjunctional membrane are perme-

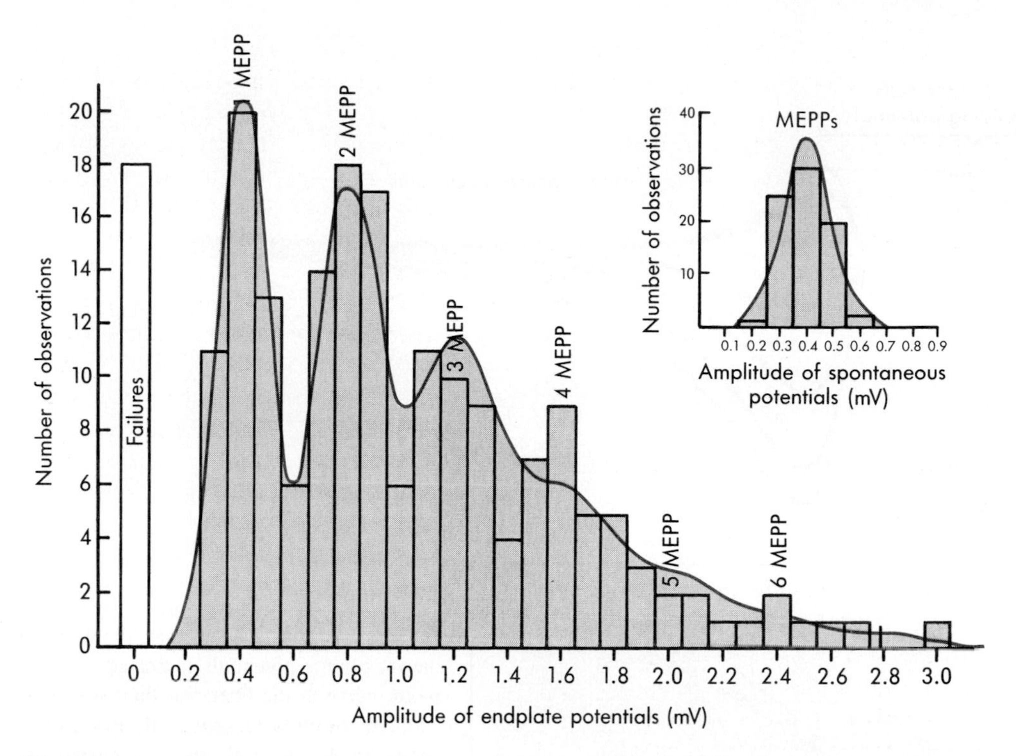

■ **Fig. 4-4** Histogram of the amplitudes of EPPs elicited by stimulating the motor nerve to a cat tenuissimus muscle fiber. Neuromuscular transmission was severely inhibited by bathing the muscle in a solution containing 12.5 mM Mg^{++}. The insert shows an amplitude histogram for MEPPs in the same fiber. Note that the distribution of EPPs has peaks that occur at integral multiples of the mean amplitude of the MEPPs. (Redrawn from Boyd IA, Martin AR: *J Physiol* 132:74, 1956.)

able only to cations, but the channels are not very selective among small cations. Na^+, K^+, Rb^+, and NH_4^+ pass through these channels with roughly equal ease.

The membrane potential may be determined primarily by the membrane conductances to K^+ and Na^+, as shown by the chord conductance equation (short form):

$$E_m = \frac{g_K}{g_K + g_{Na}} E_K + \frac{g_{Na}}{g_K + g_{Na}} E_{Na} \qquad (1)$$

Because acetylcholine causes the postsynaptic membrane to become permeable to both Na^+ and K^+, the membrane potential of the endplate in the presence of acetylcholine should tend toward a value in between E_K and E_{Na}. This has been shown by experiments in which the membrane potential of the postjunctional muscle cell is voltage clamped to various values. The prejunctional motor axon is then stimulated, and the resulting endplate currents are measured (Fig. 4-5, *A*). As the postjunctional cell is depolarized by the voltage clamp, the endplate currents diminish. For positive membrane potentials the net endplate current becomes outward. The potential of the postjunctional membrane at which no net current flows (about 0 mV) is called the **reversal potential** of the neuromuscular junction (Fig. 4-5, *B*).

■ *The Acetylcholine Receptor Protein*

The acetylcholine receptor protein has been studied intensively. Development of methods for isolating and purifying hydrophobic membrane proteins and the availability of snake venom neurotoxins that bind very tightly to the acetylcholine receptor have been essential in these studies. α-Bungarotoxin, from the venom of the Formosan krait (a relative of the cobra), is a useful toxin that binds to the acetylcholine receptor almost irreversibly.

Binding of radioactively labeled α-bungarotoxin by neuromuscular junctions suggests that there are 10^7 to 10^8 binding sites per motor endplate. At mouse diaphragm neuromuscular junctions the acetylcholine binding sites concentrated near the mouths of the postjunctional folds have a density of about 20,000 per μm^2. This suggests that the receptor molecules are quite tightly packed, since the maximum density possible has been estimated to be about 50,000 per μm^2.

The acetylcholine receptor protein is an integral membrane protein and is deeply embedded in the hydrophobic lipid matrix of the postjunctional membrane. Cholinesterase, on the other hand, is loosely associated

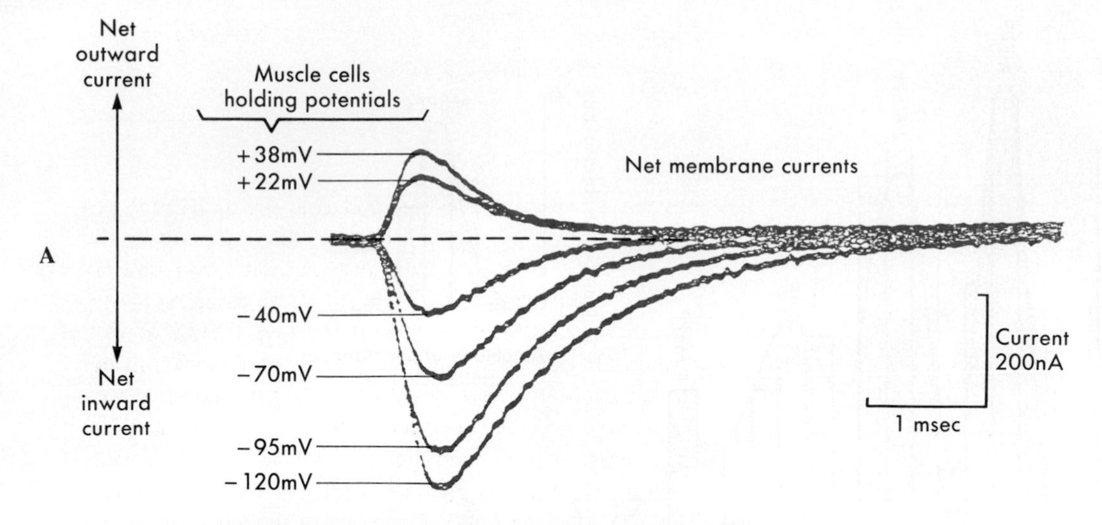

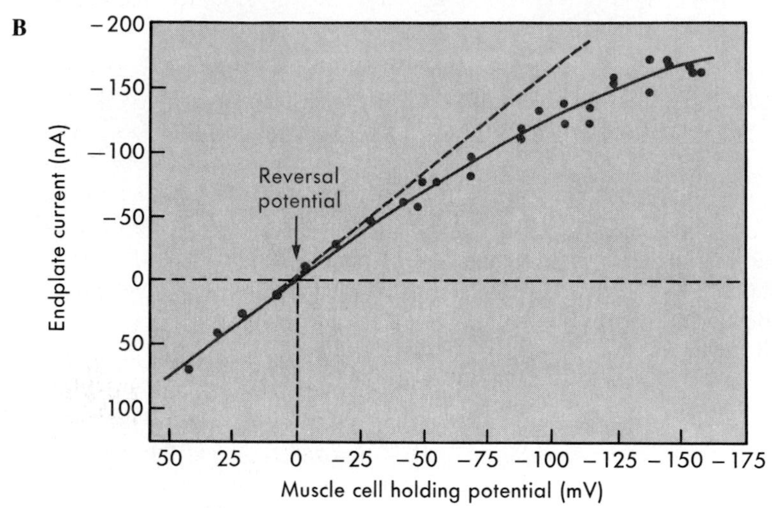

Fig. 4-5 **A,** A frog muscle cell was voltage clamped to the membrane potentials indicated as holding potentials. The motor nerve to the fiber was then stimulated and the resulting endplate currents recorded. **B,** For experiments of the type shown in **A,** the peak endplate current is plotted versus the holding potential. By interpolation the current is seen to approach zero at a holding potential about 0 mV. (Redrawn from Magelby KL, Stevens CF: *J Physiol* 223:173, 1972.)

with the surface of the postjunctional membrane by hydrophilic interactions.

Acetylcholine receptor protein has been isolated from certain neuromuscular junctions and from the electroplax of electrical fish, where the receptor concentration is especially high. The receptor protein has been extensively purified by affinity chromatography. Purified receptor protein has been reconstituted into a lipid bilayer model membrane. In such systems binding of acetylcholine by the receptor protein leads to an increase in the Na^+ and K^+ conductance of the lipid bilayer.

The acetylcholine receptor protein consists of five subunits, two of which are identical, so that there are four different polypeptide chains. The duplicated subunit is the α subunit (molecular weight 40,000). The acetylcholine binding sites are located on the α subunits. The other subunits are β (48,000), γ (58,000), and δ (64,000). In some species the pentamer $\alpha_2\beta\gamma\delta$ is the predominant form of the acetylcholine receptor. In other species two pentamers are linked by a disulfide bridge between their γ subunits. The function of these two forms is apparently identical. The genes for the α,β,γ, and δ subunits have been cloned and sequenced. The amino acid sequences of the α,β,γ, and δ subunits have extensive sequence homology, and thus the four subunits have evolved from a common precursor. Each of the subunits has four transmembrane helical domains. One of these, called M2, has charged amino acid residues (negative, on balance) on one face of the helix. The ion channel pore is believed to be lined by these charges with an M2 helix from each of the five subunits surrounding the ion channel.

X-ray diffraction and image analysis of electron micrographs of negatively stained preparations have revealed a good deal about the three-dimensional structure of the acetylcholine receptor in the membrane. The five subunits surround a central ion channel (Fig. 4-6). All of the subunits span the membrane, with more of their mass protruding from the extracellular face of the membrane than from the intracellular face. The subunits form a funnel around the mouth of the ion channel on the external side of the membrane. The binding of acetylcholine is postulated to cause a conformational change in the α subunit that results in alterations in the conformations of the other subunits and opens the central ion channel.

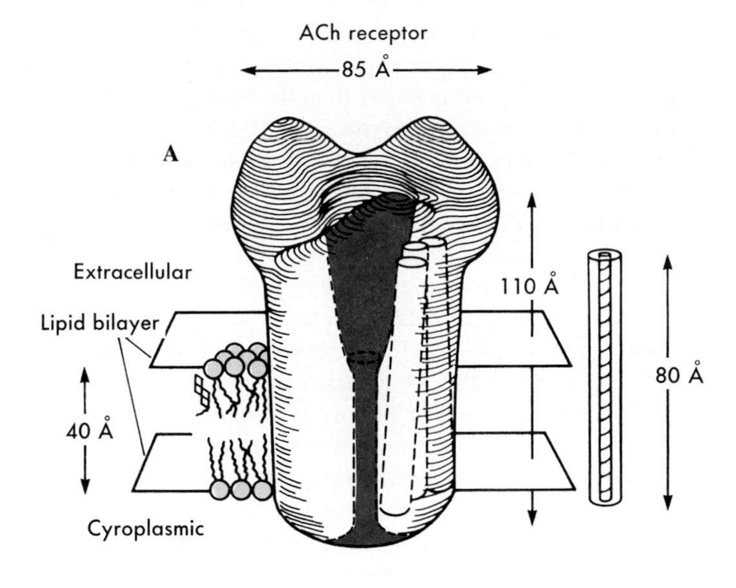

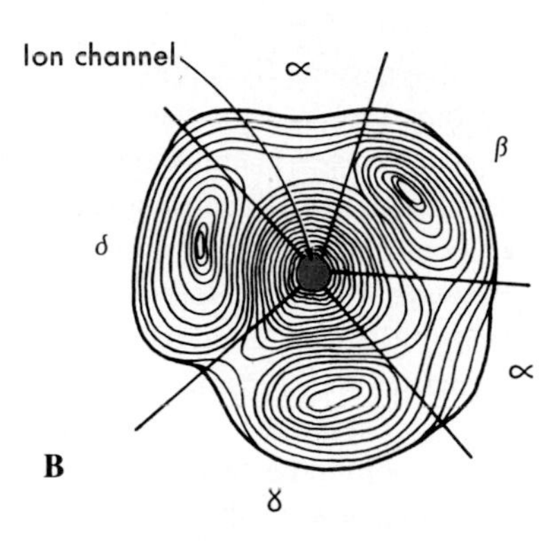

■ **Fig. 4-6** A model of the structure of the acetylcholine receptor protein. **A,** Viewed from the side and, **B,** viewed looking down on the acetylcholine receptor from the cytoplasmic surface. The closed curves are electron density profiles. The five subunits surround a central ion channel. (Redrawn from Kistler J et al: *Biophys J* 37:371, 1982.)

Defects in neuromuscular transmission are involved in myasthenia gravis. Experiments involving tritiated bungarotoxin binding suggest that in this disease there is decreased density of acetylcholine receptors on the postjunctional membrane. Most myasthenic patients have circulating antibodies to acetylcholine receptor protein. Whether these antibodies are the cause of myasthenia gravis or are produced secondarily as the result of endplate degeneration remains to be determined.

■ *Synapses between Neurons*

Chemical transmission between neurons has many of the same properties that characterize the neuromuscular junction. Electrical synapses have been described in the central nervous systems of animals from invertebrates to mammals. The prevalence of electrical synapses in the nervous systems of higher animals and their physiological roles are still poorly defined.

■ *Electrical Synapses*

At an electrical synapse a change in the membrane potential of one cell is transmitted to the other cell by the direct flow of current. Because current flows directly between the two cells that make an electrical synapse, there is essentially no synaptic delay. Chemical synapses typically have a synaptic delay of about 0.5 msec. In general, electrical synapses allow conduction in both directions. In this respect they differ from chemical synapses, which are obligatorily unidirectional. Certain electrical synapses conduct more readily in one direction than in the other; this property is called **rectification.**

Cells that form electrical synapses are joined by **gap junctions.** Gap junctions are plaquelike structures in which the plasma membranes of the coupled cells are very close (less than 30 Å). In freeze-fracture electron micrographs of gap junctions there are regular arrays of intramembrane protein particles. The intramembrane particles consist of six subunits surrounding a central channel that is accessible to water. The hexagonal array is called a **connexon.** Each of the six subunits is a single protein (one polypeptide chain) called **connexin** (molecular weight about 25,000). At the gap junction the connexons of the coupled cells are aligned to form channels (Fig. 4-7, *A*). The channels allow the passage of water-soluble molecules up to molecular weights of 1200 to 1500 from one cell to the other. These channels are the pathways for electrical current flow between the cells.

Cells that are electrically coupled may become uncoupled by closing of the connexon channels. The channels may close in response to increased intracellular Ca^{++} or H^+ in one of the cells or in response to depolarization of one or both of the cells. A model for the mechanism of closing the channels is shown in Fig. 4-7, *B*.

Numerous electrical synapses have been described in the peripheral and central nervous systems of invertebrates and vertebrates. Electrical synapses appear to be particularly useful in reflex pathways in which rapid transmission between cells (little synaptic delay) is necessary or when the synchronous response of a number of neurons is required. Among the many nonneuronal cells that are coupled by gap junctions are hepatocytes, myocardial cells, intestinal smooth muscle cells, and the epithelial cells of the lens.

■ *Chemical Synapses*

When one neuron makes a chemical synapse with another, the presynaptic nerve terminal characteristically

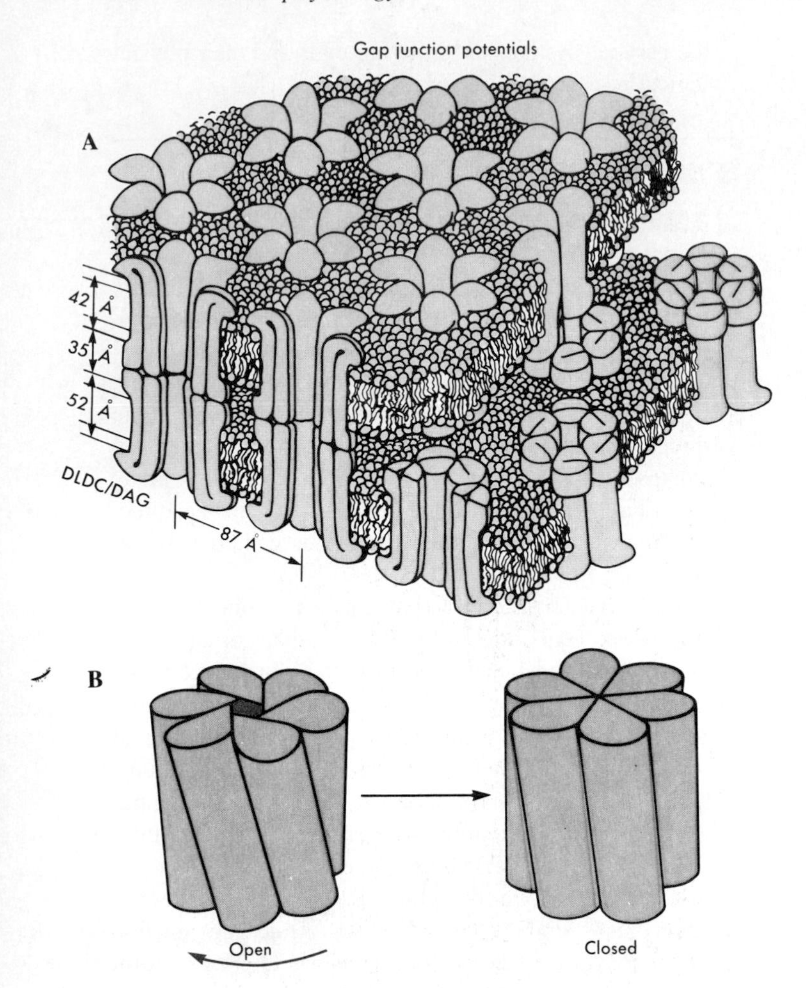

■ **Fig. 4-7** **A,** A model for the structure of gap junction channels. Each plasma membrane contains connexons, each of which consists of an hexagonal array of six connexin polypeptides. The connexons of the two membranes are aligned at the gap junction to form channels between the cytosolic compartments of the two cells. **B,** A model of the opening and closing of the gap junction channel. The individual subunits of the connexon are proposed to twist relative to one another to open and close the central channel. (**A** redrawn from Makowski L et al: *J Cell Biol* 74:629, 1977. From The Rockefeller University Press; **B** redrawn from Unwin PNT, Zampighi G: *Nature* 283:545, 1980. From *Nature*. Copyright 1980 Macmillan Journals.)

broadens to form a **terminal bouton.** At the synapse itself the presynaptic and postsynaptic membranes come more or less into close apposition and lie parallel to one another. Substantial structures stabilize the synaptic association, so that when nervous tissue is disrupted, the relationship of the presynaptic and postsynaptic membranes at the synapse often is preserved.

Electron micrographs of synapses in the central nervous system often show areas of high electron density subjacent to the plasma membranes in the region of synaptic contact. Synapses where this electron-dense region is of roughly similar extent and intensity on both presynaptic and postsynaptic sides of the synapse are called **symmetrical synapses** (Fig. 4-8, *A*). Synapses characterized by a postsynaptic electron density that is greater in extent and intensity than the presynaptic electron density are called **asymmetrical synapses** (Fig. 4-8, *B*). These classifications may indicate the two ends of a spectrum of synaptic structure rather than two distinct classes. The presynaptic nerve terminals at asymmetrical synapses often contain spherical synaptic vesicles, whereas symmetrical synapses are characterized by flattened or ellipsoidal vesicles. The asymmetrical synapses (round vesicles) may involve excitatory transmitters, whereas the symmetrical synapses (flattened vesicles) may involve inhibitory transmitters. This interpretation is, however, controversial.

Because of the structure and organization of chemical synapses, conduction is necessarily one way. **One-way conduction** of chemical synapses contributes to the organization of central nervous systems of vertebrates. A **synaptic delay** of about 0.5 msec is a property of transmission at chemical synapses. Synaptic delay is primarily caused by the time required for the release of transmitter. The time required for calcium channels to open in response to depolarization of the presynaptic terminal appears to be a major component of synaptic delay. In polysynaptic pathways synaptic delay accounts for a significant fraction of the total conduction time.

At chemical synapses the transmitter released by the presynaptic neurons alters the conductance of the postsynaptic plasma membrane to one or more ions. A change of the conductance of the postsynaptic membrane to an ion that is not in equilibrium across the membrane brings about a change in the rate of flow of that ion, which causes a change in the membrane potential of the postsynaptic cell. In most cases transmitters produce their effect by increasing the conductance of the postsynaptic membrane to one or more ions, but a few act by decreasing the postsynaptic conductances to specific ions.

Typically, the presynaptic nerve terminal helps to terminate the action of the transmitter substance by rapidly reaccumulating much of the transmitter that has been released into the synaptic cleft. This is accomplished by a transport protein in the plasma membrane of the nerve terminal. This protein uses the energy in the electrochemical potential gradient of Na^+ to actively take up the transmitter.

The part of the membrane of the postsynaptic neuron that forms the synapse is specialized for **chemical sensitivity** rather than electrical sensitivity. Action potentials are not produced at the synapse. The change in membrane potential, be it depolarization or hyperpolarization, that occurs at the synapse is conducted electrotonically over the membrane of the postsynaptic neuron to the **axon hillock–initial segment** region (Fig. 4-9). The axon hillock–initial segment of certain neurons has a lower threshold than the rest of the plasma membrane of the postsynaptic cell, and an action potential will be

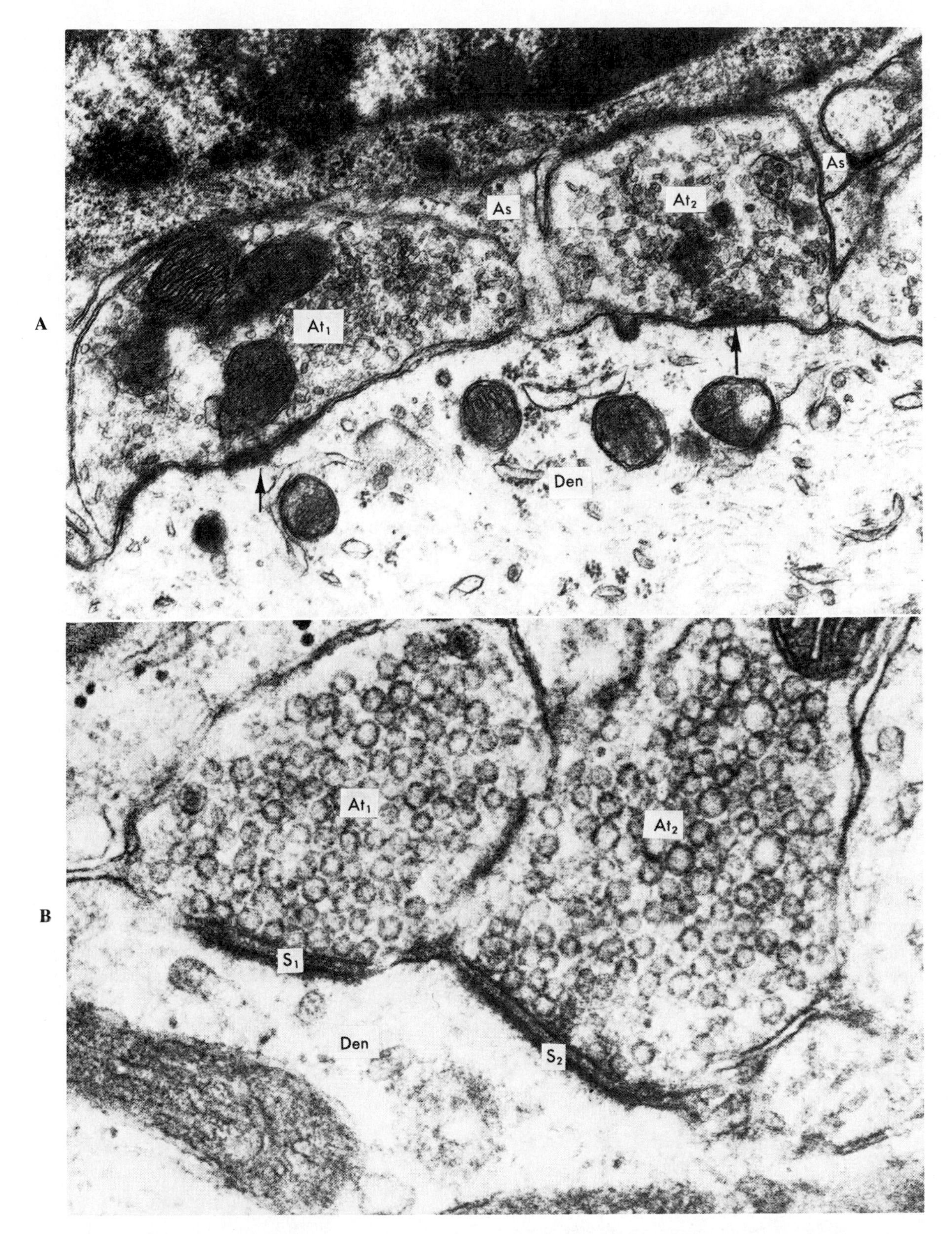

■ **Fig. 4-8** **A,** Symmetrical synapses. Two axon terminals *(At_1 and At_2)* synapse with a large dendrite *(Den)* in the anterior horn of the spinal column. Note that the presynaptic and postsynaptic electron densities *(arrow)* are similar in extent. Astroglial cells *(As)* invest the axon terminals. **B,** Asymmetrical synapses *(S_1 and S_2)* in the cerebral cortex. Two axon terminals synapse with dendrite of a stellate cell. Note the greater extents of the electron densities on the postsynaptic (dendrite) side of the two synapses. (From Peters A, Palay SL, Webster H de F: The fine structure of the nervous system, Philadelphia, 1976, WB Saunders.)

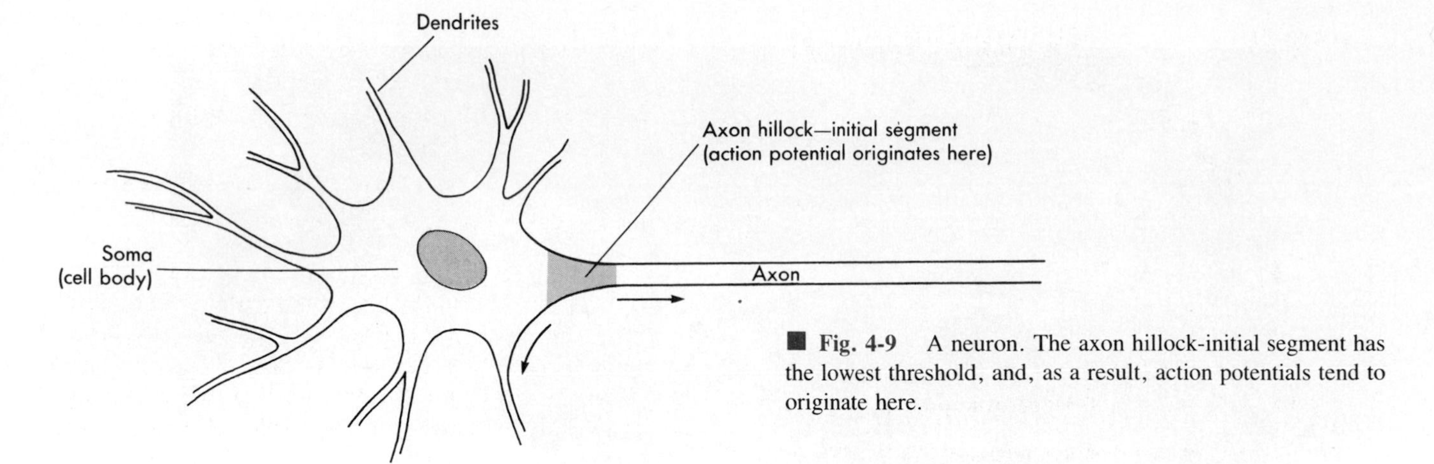

■ **Fig. 4-9** A neuron. The axon hillock-initial segment has the lowest threshold, and, as a result, action potentials tend to originate here.

generated here if the sum of all the inputs to the cell exceeds threshold. Once the action potential has been generated, it is conducted back over the surface of the soma of the postsynaptic cell and is propagated along its axon.

■ *Input-Output Relations*

The neuromuscular junction is representative of a particularly simple type of synapse in which one action potential in the presynaptic cell (the input) results in a single action potential in the postsynaptic cell (the output). In other types of synapses the output may differ from the input. Synapses can be classified as one-to-one, one-to-many, or many-to-one, based on the relationship between input and output.

In a **one-to-one synapse,** like the neuromuscular junction, the input and output are the same. A single action potential in the presynaptic cell evokes a single action potential in the postsynaptic cell. Because the output is the same as the input, no integration can occur at this type of snyapse.

In a **one-to-many synapse** a single action potential in the presynaptic cell elicits many action potentials in the postsynaptic cell. One-to-many synapses are not common, one example being the synapse of motoneurons on Renshaw cells in the spinal cord. One action potential in the motoneuron induces the Renshaw cell to fire a burst of action potentials.

In a **many-to-one synaptic arrangement** one action potential in the presynaptic cell is not enough to make the postsynaptic cell fire an action potential. The nearly simultaneous arrival of presynaptic action potentials in several input neurons that synapse on the postsynaptic cell is necessary to depolarize the postsynaptic cell to threshold. The spinal motoneuron has this type of synaptic organization. One hundred or more presynaptic axons synapse on each spinal motoneuron (Fig. 4-10). Some of these are excitatory inputs that *depolarize* the postsynaptic cell and bring it closer to its threshold. Other inputs are inhibitory and *hyperpolarize* the motoneuron, taking it farther away from threshold. The changes in postsynaptic potential caused by an action potential in a single input are about 100 μV. Thus no one excitatory input is capable of bringing the motoneu-

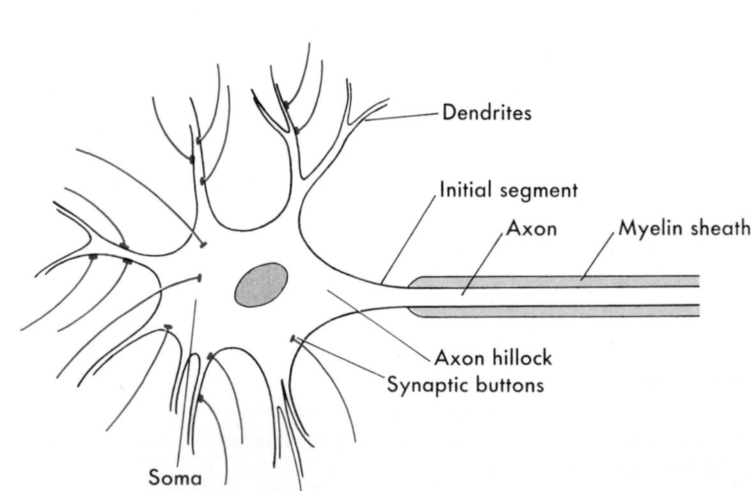

■ **Fig. 4-10** A spinal motor neuron with multiple synapses on both soma and dendrites.

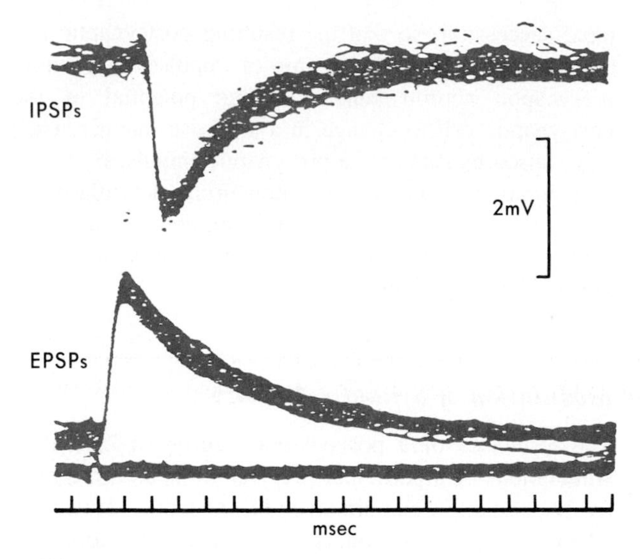

■ **Fig. 4-11** Inhibitory postsynaptic potentials *(IPSPs)* and excitatory postsynaptic potentials *(EPSPs)* recorded with a microelectrode in a cat spinal motor neuron in response to stimulation of appropriate peripheral afferent fibers. Forty traces are superimposed. (Redrawn from Curtis DR, Eccles JC: *J Physiol* 145:529, 1959.)

ron to threshold. A transient depolarization of the postsynaptic neuron as the result of an action potential in a presynaptic cell is called an **excitatory postsynaptic potential (EPSP)** (Fig. 4-11). The transient hyperpolarization caused by an action potential in an inhibitory input is called an **inhibitory postsynaptic potential (IPSP)** (Fig. 4-11). At any instant the postsynaptic cell *integrates* the various inputs. If the momentary sum of the inputs depolarizes the postsynaptic cell to its threshold, it will fire an action potential. This is integration at the level of a single postsynaptic neuron.

■ *Summation of Synaptic Inputs*

The summation (or integration) of inputs can occur by either **spatial summation** or **temporal summation** (Fig. 4-12). *Spatial* summation occurs when two separate inputs arrive simultaneously. The two postsynaptic potentials are added so that two simultaneous excitatory inputs will depolarize the postsynaptic cell about twice as much as either input alone. One EPSP and one IPSP that occur simultaneously will tend to cancel one another. Even inputs that synapse at opposite ends of the postsynaptic body cell will act in this way. The postsynaptic potentials (EPSPs and IPSPs) are conducted rapidly over the entire cell membrane of the postsynaptic cell body with almost no decrement. This is because cellular dimensions (about 100 μm) are much smaller than the length constant (about 1 to 2 mm) for electrotonic conduction. Synaptic potentials that originate in fine dendritic branches decrease in magnitude as they are conducted to the cell body. The finer the branches and the greater their distance from the cell body, the greater the decrement.

Temporal summation occurs when two or more action potentials in a single presynaptic neuron are fired in

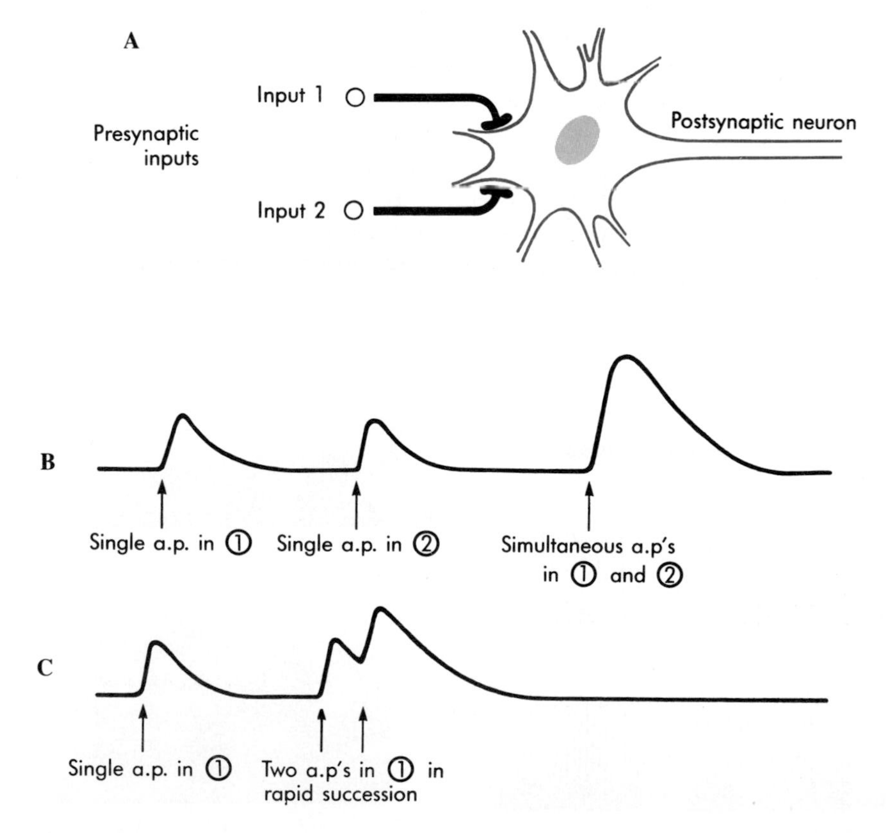

■ **Fig. 4-12** **A,** Spatial and temporal summation at a postsynaptic neuron with two synaptic inputs *(1* and *2)*. **B,** Spatial summation. The postsynaptic potential in response to single action potentials in inputs *1* and *2* occurring separately and simultaneously. **C,** Temporal summation. The postsynaptic response to two impulses in rapid succession in the same input.

■ **Fig. 4-13** **A,** Facilitation at a neuromuscular junction. EPPs at a neuromuscular junction in toad sartorius muscle were elicited by successive action potentials in the motor axon. Neuromuscular transmission is depressed by 5 mM Mg^{++} and 2.1 μm curare, so that action potentials do not occur. **B,** EPPs at a frog neuromuscular junction elicited by repetitively stimulating the motor axon at different frequencies. Note that facilitation fails to occur at the lowest frequency of stimulation (1 per second) and that the degree of facilitation increases with increasing frequency of stimulation in the range of frequency employed. Neuromuscular transmission was inhibited by bathing the preparation in 12 to 20 mM Mg^{++}. **C,** Posttetanic potentiation at a frog neuromuscular junction. The top two traces indicate control EPPs in response to single action potentials in the motor nerve. The subsequent traces indicate the responses to single action potentials following tetanic stimulation (50 impulses/sec for 20 sec) of the motor nerve. The time interval between the end of tetanic stimulation and the single action potential is shown on each trace. The muscle was treated with tetrodotoxin to prevent generation of action potentials. (**A** redrawn from Belnave RJ, Gage PW: *J Physiol* 266:435, 1977; **B** redrawn from Magelby KL: *J Physiol* 234:327, 1973; **C** redrawn from Weinrich D: *J Physiol* 212:431, 1971.)

rapid succession, so that the resulting postsynaptic potentials overlap in time. A train of impulses in a single presynaptic neuron can cause the potential of the postsynaptic cell to change in a stepwise manner, each step caused by one of the presynaptic impulses.

Integration at the spinal motoneuron takes place because many positive and negative inputs impinge on a single motoneuron. This permits fine control of the firing pattern of the spinal motoneuron.

■ *Modulation of Synaptic Activity*

The responses of a postsynaptic neuron to individual stimulations of a particular presynaptic neuron are relatively constant in magnitude and time course. However, when a presynaptic cell is stimulated repeatedly, the postsynaptic response may depend on the frequency and duration of the presynaptic stimulation. When a presynaptic axon is stimulated repeatedly, the postsynaptic response may grow with each stimulus. This phenomenon is called **facilitation** (Fig. 4-13, *A*). As shown in Fig. 4-13, *B*, the extent of facilitation depends on the frequency of presynaptic impulses. Facilitation dies away rapidly, within tens to hundreds of milliseconds.

When a presynaptic neuron is stimulated tetanically (many stimuli at high frequency) for several seconds, a longer-lived enhancement of postsynaptic response occurs called **posttetanic potentiation** (Fig. 4-13, *C*). Posttetanic potentiation persists much longer than facilitation; it lasts tens of seconds to several minutes after cessation of tetanic stimultion.

An enhancement of synaptic efficacy may occur that is intermediate in time course between facilitation and posttetanic potentiation. This enhancement is called

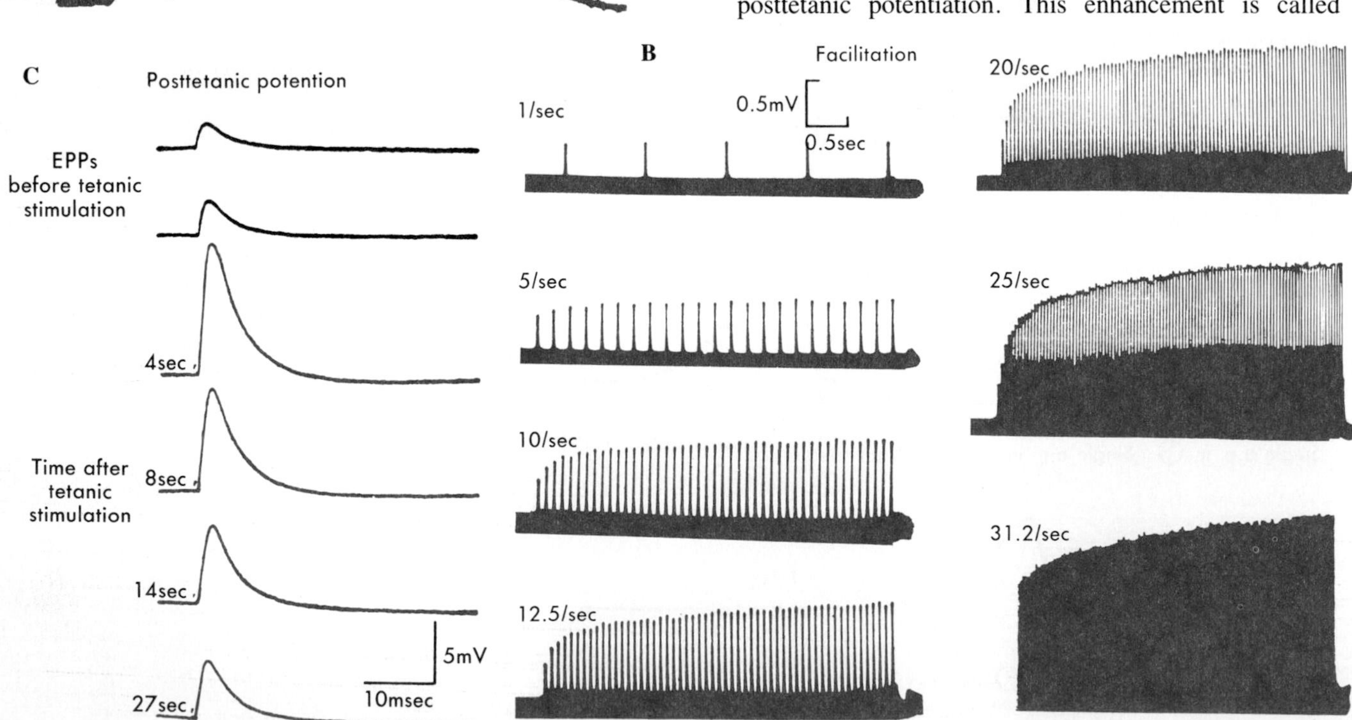

augmentation, and it persists for about 10 seconds after repetitive stimulation is ended.

Facilitation, augmentation, and posttetanic potentiation are the result of the effects of repeated stimulation on the presynaptic neuron. These phenomena do not involve a change in the sensitivity of the postsynaptic cell to transmitter. With repeated stimulation, an increased number of quanta of transmitter is released. Increased levels of intracellular calcium may play a role in the enhancement of transmitter release with repetitive stimulation, but other intracellular events are involved as well. The details of the mechanisms that bring about facilitation, augmentation, and posttetanic potentiation are unknown.

At some synapses in the central nervous system, potentiation of the postsynaptic response after strong tetanic stimulation of a presynaptic input persists for hours or days, a phenomenon known as **long-term potentiation.** Long-term potentiation is one of the few models of the phenomena of memory that can be investigated at the cellular level, but at present the cellular and molecular mechanisms of long-term potentiation are not well understood. Protein synthesis is required for long-term potentiation, but not for posttetanic potentiation.

When a synapse is repetitively stimulated for a long time, a point is reached at which each successive presynaptic stimulation elicits a smaller postsynaptic response. This phenomenon is called **synaptic fatigue** or **depression** (neuromuscular depression at the motor endplate). The postsynaptic cell at a fatigued synapse responds normally to transmitter applied from a micropipette, so the defect appears to be presynaptic. In some cases a decrease in quantal content (the amount of transmitter per synaptic vesicle) has been implicated in synaptic fatigue. A fatigued synapse typically recovers in a few seconds.

Ionic Mechanisms of Postsynaptic Potentials in Spinal Motoneurons

Much of our current knowledge of synaptic mechanisms in the mammalian central nervous system is derived from studies of cat spinal motoneurons.

The EPSP. The EPSP (see Fig. 4-11) of the cat spinal motoneuron is caused by a transient increase of the conductance of the postsynaptic membrane to both Na^+ and K^+. One way in which this was demonstrated was by injecting Na^+ or K^+ into the cell to raise the intracellular concentration of that particular ion. Injection of either Na^+ or K^+ results in a smaller EPSP because, when the conductance increase occurs, there is a smaller tendency for Na^+ to flow in and depolarize the cell after Na^+ injection and a greater tendency for K^+ to flow out and oppose depolarization after K^+ injection. If the postsynaptic membrane is progressively depolarized, the EPSP progressively decreases in size because of a decreased tendency for Na^+ to enter and an increased tendency for K^+ to leave the cell. When E_m reaches about zero, the EPSP disappears, and if E_m is made positive, the EPSP changes direction. Thus zero is the **reversal potential** for EPSP. If Na^+ were the only ion flowing during the EPSP, the reversal potential for the EPSP should equal the equilibrium potential for Na^+ (about +65 mV). If K^+ were the only ion in-

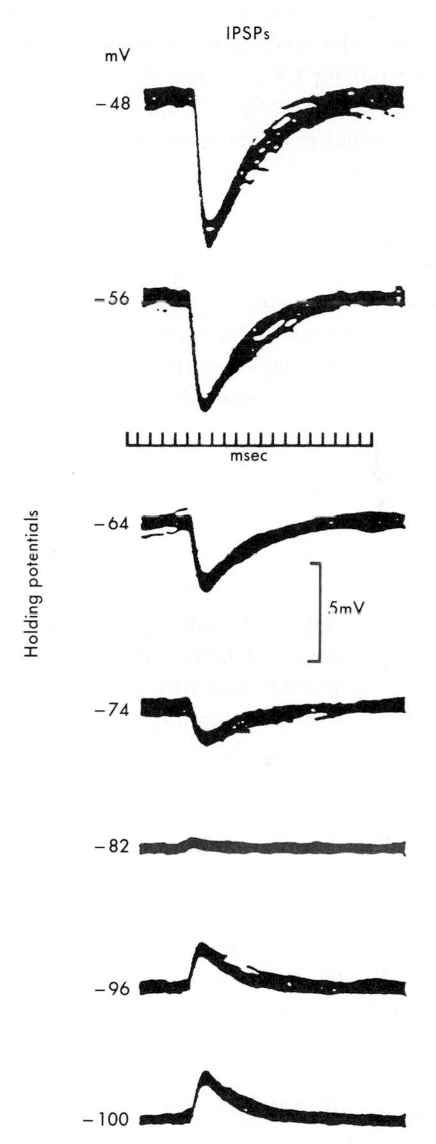

■ **Fig. 4-14** A cat spinal motor neuron was voltage-clamped to the membrane potentials shown, and then IPSPs were elicited by stimulating a peripheral afferent input to the spinal motor neuron. Each record was formed by superimposing 40 individual records. As the "resting" potential was made more negative, the IPSP first diminished in amplitude and, finally, at about −80 mV, reversed direction. The reversal potential (about −80 mV) of the IPSP is close to the value of the equilibrium potential for chloride. This supports the hypothesis that chloride currents are responsible for the IPSP in cat spinal motor neurons. (Redrawn from Coombs JS et al: *J Physiol* 130:326, 1955.)

volved in the EPSP, the reversal potential should equal the K^+ equilibrium potential (about -100 mV). The reversal potential for the EPSP is a weighted average of E_{Na} and E_K. Injection of Cl^- into the cell does not alter the EPSP, so Cl^- apparently is not involved in the EPSP.

The IPSP. The IPSP (see Fig. 4-11) of cat spinal motoneurons is caused by an increased chloride conductance of the postjunctional membrane. The reversal potential for the IPSP is more negative than the normal resting membrane potential of the postsynaptic neuron (Fig. 4-14). The reversal potential is near the equilibrium potential for Cl^-. At rest there is a net tendency for Cl^- to enter the cell. The increase in chloride conductance, as the result of transmitter release at the inhibitory synapse, allows Cl^- to enter the postsynaptic cell and hyperpolarize it. Injecting Cl^- into the cell or hyperpolarizing it decreases the net tendency for Cl^- to enter the cell and decreases the size of the IPSP. Injection of Na^+ or K^+ produces no change in the IPSP, suggesting that neither Na^+ nor K^+ is involved in the IPSP of spinal motoneurons. In certain other cell types the IPSP appears to be caused by an increased K^+ conductance.

Presynaptic inhibition. Inhibitory interactions are vital in stabilizing the central nervous system. Another type of inhibition is called **presynaptic inhibition.** If an inhibitory input to a spinal motoneuron is stimulated tetanically and then an excitatory input is stimulated, the EPSP elicited by stimulating the excitatory input may be reduced in magnitude after the inhibitory volley. This is believed to occur by a mechanism in which axon collaterals of the inhibitory axons synapse on the excitatory nerve terminals (Fig. 4-15). An action potential in the inhibitory nerve produces a depolarization (rather long lived) of the excitatory nerve terminal. This brings the excitatory nerve terminal closer to threshold. What is important here is that the partly depolarized excitatory terminal *will release less transmitter in response to an action potential*. The smaller release of transmitter results in a smaller EPSP. The phenomenon of decreased transmitter release from a partially depolarized nerve terminal is well known at the neuromuscular junction.

The presumed anatomical arrangement that underlies presynaptic inhibition is diagramed in Fig. 4-15. The mechanisms described for presynaptic inhibition have been worked out in invertebrates. They are believed to apply in mammalian nervous systems, but they have not been as fully characterized in mammals.

Transmitters in the Nervous System

Identification of Transmitter Substances

A number of compounds have been proposed to function as neurotransmitters. Such compounds are called **candidate neurotransmitters** or **putative neurotransmitters.** Candidate neurotransmitters usually are concentrated in specific neurons or in specific neuronal pathways. Microapplication of the putative transmitter to particular areas of the central nervous system may evoke specific responses. Correlation of information about the localization of the putative transmitter with knowledge of the location of neurons that respond to the candidate transmitter and the ways in which they respond allows intelligent speculation about the functions of a putative neurotransmitter substance.

To *prove* that a substance is the transmitter at a particular synapse is often difficult. A putative transmitter (X) must satisfy the following criteria before it is accepted as a proven transmitter at a particular synapse:

1. The presynaptic neurons must contain X and must be able to synthesize it.
2. X must be released by the presynaptic neurons on appropriate stimulation.
3. Microapplication of X to the postsynaptic membrane must mimic the effects of stimulation of the presynaptic neuron.
4. The effects of presynaptic stimulation and of microapplication of X should be altered in the same way by pharmacological agents.

Our knowledge of neurotransmitters has increased greatly in recent years. Some transmitters have rapid and transient effects on the postsynaptic cell. Other transmitters have effects that are much slower in onset and may last for minutes or even hours. Most of the candidate neurotransmitters that have been discovered

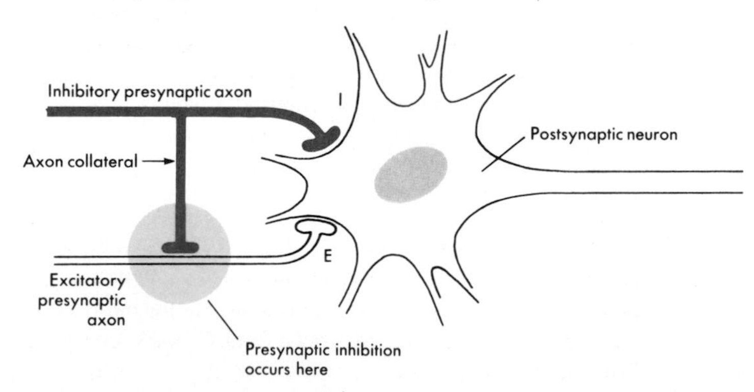

Fig. 4-15 Presynaptic inhibition. Axon collateral of the inhibitory axon *(I)* synapses on the excitatory axon terminal *(E)*. An action potential in the inhibitory axon depolarizes the excitatory axon terminal. The depolarized excitatory axon terminal will release less transmitter in response to an action potential in the excitatory neuron.

so far fall into three major chemical classes: amines, amino acids, and oligopeptides.

Transmitters and Putative Transmitters in the Central Nervous System

Acetylcholine. As discussed previously, acetylcholine is the transmitter used by all motor axons that arise from the spinal cord. Acetylcholine plays a central role in the autonomic nervous system, being the transmitter for all preganglionic neurons and also for postganglionic parasympathetic fibers. Acetylcholine may be the transmitter in a large number of pathways in the central nervous system.

Biogenic Amine Transmitters

Among the amines that may serve as neurotransmitters are norepinephrine, epinephrine, dopamine, serotonin and histamine.

Dopamine, norepinephrine, and **epinephrine** are catecholamines and share a common biosynthetic pathway that starts with the amino acid tyrosine. Tyrosine is converted to L-dopa by tyrosine hydroxylase. L-dopa is converted to dopamine by a specific decarboxylase. In dopaminergic neurons the pathway stops here. Noradrenergic neurons have another enzyme, dopamine β-hydroxylase, that converts dopamine to norepinephrine. Other cells add a methyl group to norepinephrine to produce epinephrine. S-adenosylmethionine is the methyl donor, and the reaction is catalyzed by phenylethanolamine-*N*-methyltransferase.

Norepinephrine is the primary transmitter for postganglionic sympathetic neurons. In the brain norepinephrine-containing cell bodies are found in several locations.

Neurons that contain high levels of dopamine are prominent in certain midbrain regions. The degeneration of dopaminergic synapses occurs in Parkinson's disease and is believed to be a major cause of the muscular tremors and rigidity that characterize this disease. Antipsychotic drugs such as chlorpromazine and related compounds are dopamine-receptor antagonists and thus tend to diminish the effects of endogenous dopamine.

Serotonin (5-hydroxytryptamine)-containing neurons are present in high concentration in many nuclei of the brainstem.

Histamine is present in certain neurons in the hypothalamus.

Amino Acid Transmitters

Glycine, the simplest amino acid, is an inhibitory neurotransmitter released by certain spinal interneurons.

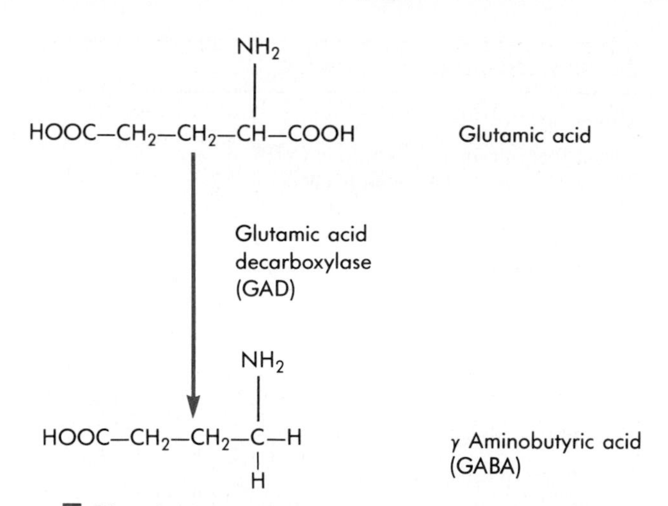

Fig. 4-16 A single reaction converts glutamate, the most important excitatory neurotransmitter, into GABA, the most important inhibitory neurotransmitter. The enzyme that catalyzes this reaction, glutamic acid decarboxylase, is present only in GABAergic neurons.

Glutamate and aspartate, dicarboxylic amino acids, have strong excitatory effects on many neurons in the brain. Glutamate and aspartate are the most prevalent excitatory transmitters in the brain.

γ-Aminobutyric acid (GABA) is not incorporated into proteins, nor is it present in all cells (as are the other naturally occurring amino acids). GABA is produced from glutamate by a specific decarboxylase present only in the central nervous system (Fig. 4-16). GABA functions as an inhibitory transmitter. It is the most common transmitter in the brain.

GABA is believed to be important in many different central control pathways. A deficit in the level of GABA and other neurotransmitters in the corpus striatum occurs in patients with Huntington's chorea. The uncontrolled movements that characterize this disease may be partly due to diminished effectiveness of GABA-mediated inhibition in central motor pathways.

Neuroactive Peptides

Some cells release peptides that act at very low concentrations to excite or inhibit neurons. To date, more than 25 of these so-called neuropeptides, ranging from 2 amino acids to about 40 amino acids long, have been identified. Some of the neuropeptides that have been discovered so far are listed in the following box. Many other neuropeptides could be added to this list.

Neuropeptides typically affect their target neurons at lower concentrations than the classical neurotransmitters discussed previously, and the action of neuropeptides usually lasts longer. A number of the neuropeptides listed in the box are more familiar as hormones. Neuropeptides may act as hormones, as neurotransmit-

Some neuroactive peptides

Gut-brain peptides

Vasoactive intestinal polypeptide (VIP)
Cholecystokinin octapeptide (CCK-8)
Substance P
Neurotensin
Methionine enkephalin
Leucine enkephalin
Motilin
Insulin
Glucagon

Hypothalamic-releasing hormones

Thyrotropin-releasing hormone (TRH)
Luteinizing hormone-releasing hormone (LHRH)
Somatostatin (growth hormone releasing-inhibiting factor, or SRIF)

Pituitary peptides

Adrenocorticotropin (ACTH)
β-Endorphin
α-Melanocyte-stimulating hormone (α-MSH)

Others

Dynorphin
Angiotensin II
Bradykinin
Vasopressin
Oxytocin
Carnosine
Bombesin

Modified from Snyder SH: *Science* 209:976, 1980. Copyright 1980 by American Association for the Advancement of Science.

ters, or as neuromodulators. A hormone is a substance that is released into the blood and that reaches its target cell via the circulation. A neurotransmitter or neuromodulator is released near the surface of its target cell and simply diffuses to the target cell. Neurotransmitters, as discussed earlier, act to change the conductance of the target cell to one or more ions, and in that way they change the membrane potential of the target cell. A neuromodulator modulates synaptic transmission. The neuromodulator may act presynaptically to change the amount of transmitter released in response to an action potential or it may act on the postsynaptic cell to modify its response to transmitter. Whether a neuropeptide acts as a true synaptic transmitter or as a synaptic modulator is often difficult to determine.

In a number of instances, some of which are listed in Table 4-1, neuropeptides coexist in the same nerve terminals with classical transmitters. In some of these cases, the neuropeptide is released along with the transmitter in response to nerve stimulation. The transmitter and the neuropeptide are packaged in separate vesicles. Different proportions of neurotransmitter and neuropeptide may be released depending on the nature of the stimulus to the cell.

Table 4-1 Examples of the coexistence within the same nerve terminal of a classical transmitter and a neuropeptide*

Transmitter	*Peptide*
Acetylcholine	Vasoactive intestinal peptide (VIP)
Norepinephrine	Somatostatin
	Enkephalin
	Neurotensin
Dopamine	Cholecystokinin (CCK)
	Enkephalin
Adrenalin	Enkephalin
Serotonin	Substance P
	Thyrotropic-releasing hormone (TRH)

Reprinted by permission of the publisher from Chemical Messengers: small molecules and peptides by Schwartz JH in Kandel ER, Schwartz JH, editors: *Principles of neural science*. Copyright 1981 by Elsevier Science Publishing.

*Evidence for the coexistence of a classical transmitter substance with a neuroactive peptide has been reported for these combinations. With the information thus far available, it is not yet possible to determine the specificity of the pairs and their physiological significance.

Neuropeptides are very important physiologically. They are being intensively investigated, but our knowledge of the function of neuropeptides is still fragmentary. This section briefly summarizes our current knowledge of neuropeptides; additional information is presented in Chapters 8 and 38.

Synthesis of neuropeptides. Classical neurotransmitters are synthesized in nerve terminals by pathways that involve soluble enzymes and simple precursors. Neuropeptides are synthesized in the cell body. They are encoded by the cell's DNA and transcribed into messenger RNA, which is translated on polyribosomes bound to the endoplasmic reticulum. As is the case for other secretory proteins, the nascent peptide of neuropeptides begins with a hydrophobic signal sequence of 18 to 25 amino acids. The signal sequence leads the nascent chain into the lumen of the endoplasmic reticulum, so that the completely translated polypeptide chain ends up in a cistern of the endoplasmic reticulum. The signal sequence is cleaved off, and smooth vesicles containing the polypeptide bud off the endoplasmic reticulum. These vesicles fuse with the forming face of the Golgi complex and are moved toward the mature face of the Golgi complex. Secretory vesicles, containing the neuropeptides, emerge from the mature face of the Golgi complex. Secretory vesicles are moved by **fast axonal transport** (Fig. 4-17) to the axon terminals, where they are known as synaptic vesicles.

A number of neuropeptides are synthesized as preprohormones. Cleavage of the signal sequence converts the preprohormone to a prohormone. Proteolytic cleavage of the prohormone may then release one or more active peptides. In some cases one prohormone may contain several active peptide sequences. As shown in

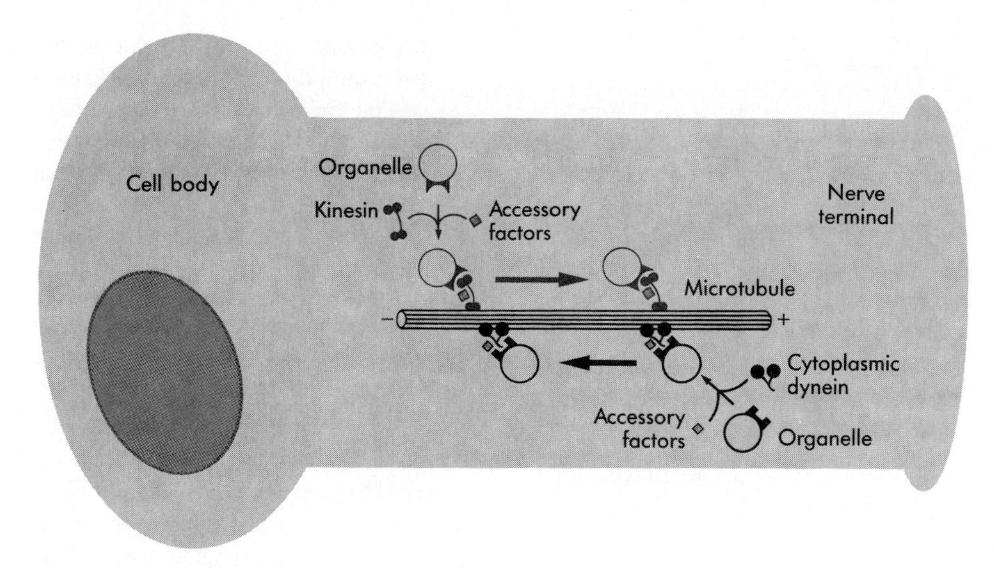

■ **Fig. 4-17** Fast axonal transport of membrane vesicles. The network of microtubules that runs the length of the axon serves as the substrate for fast axonal transport of vesicular organelles from the cell body to the nerve terminal and in the reverse direction. Most of the microtubules are oriented with their + (rapidly growing) ends toward the nerve terminal. Kinesin and dynein are microtubule-associated motors that function in the transport of vesicles toward the nerve terminal and the cell body, respectively. Other microtubule-associated proteins, designated as accessory factors, are required for transport of vesicles. (Modified from Sheetz MP, Steuer ER, and Schroer TA: The mechanism of fast axonal transport, *Trends in Neurosci* 12:474, 1990.)

Fig. 4-18, the prohormone of the opioid peptide, β-endorphin, is a 31,000-dalton polypeptide that contains a number of active sequences. One cleavage of the prohormone releases adrenocorticotrophic hormone (ACTH) and β-lipotropin. Cleavage of ACTH releases a hormone, melanocyte-stimulating hormone (α-MSH). Cleavages of β-lipotropin release β-MSH and a number of active β-endorphins. The following box contrasts certain properties of nonpeptide and peptide neurotransmitters.

Opioid peptides. Opiates are drugs that are derived from the juice of the opium poppy. Opiates are useful therapeutically as powerful analgesics. They exert their analgesic effect by binding to specific opiate receptors. The binding of opiates to their receptors is stereospecifically inhibited by a morphine derivative called *naloxone*. Compounds that do not derive from the opium poppy but that exert direct effects by binding to opiate receptors are called **opioids.** Operationally, opioids are defined as directly acting compounds whose effects are stereospecifically antagonized by naloxone.

The three major classes of endogenous opioid peptides in mammals are enkephalins, endorphins, and dynorphin. Enkephalins are the simplest opioids; they

Distinctions between classical nonpeptide neurotransmitters and peptide neurotransmitters.

Nonpeptide transmitters	Peptide transmitters
Synthesized and packaged in nerve terminal	Synthesized and packaged in cell body; transported to nerve terminal by fast axonal transport
Synthesized in active form	Active peptide formed when it is cleaved from a much larger polypeptide that contains several neuropeptides
Present in small, clear vesicles	Present in large, electron-dense vesicles
Released into a synaptic cleft	May be released some distance from the postsynaptic cell There may be no well-defined synaptic structure
Action terminated by uptake by presynaptic terminal by Na^+-powered active transport	Action terminated by proteolysis or by the peptide diffusing away
Typically, action has short latency and short duration (msec)	Action may have long latency and may persist for many seconds

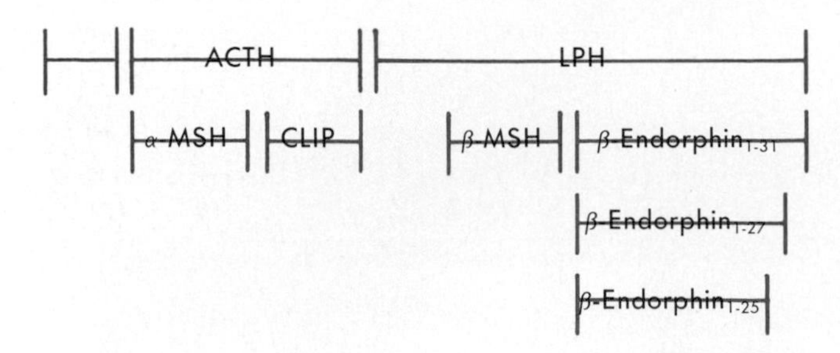

■ **Fig. 4-18** Endorphins are secreted in an inactive prohormone form that contains other important peptides. Specific proteases are presumed to release the active peptides. *ACTH,* adrenocorticotrophic hormone; *LPH,* β-lipotropin; *MSH,* melanocyte-stimulating hormone; *CLIP,* corticotrophin-like intermediate lobe peptide. (Redrawn from Smyth DG: *Br Med Bull* 39:25, 1983. With permission of Churchill Livingstone, Edinburgh.)

are pentapeptides. Met-enkephalin is Tyr-Gly-Gly-Phe-Met. Leu-enkephalin (Tyr-Gly-Gly-Phe-Leu) has leucine in place of methionine. Dynorphin and the endorphins are somewhat longer peptides that share one or the other of the enkephalin sequences at their *N*-terminal ends.

Opioid peptides are widely distributed in neurons of the central nervous system and intrinsic neurons of the gastrointestinal tract. Opioid peptides are found in vesicles that resemble synaptic vesicles. The endorphins are discretely localized in particular structures of the central nervous system, whereas the enkephalins and dynorphins are more widely distributed.

Calcium-dependent, depolarization-induced release of enkephalins from brain slices has been demonstrated, which suggests that enkephalins serve as neurotransmitters in the brain. When opioids are directly applied to particular groups of neurons, the effects may resemble those of transmitters (altered membrane conductance to particular ions) or those of peptide hormones (changes in intracellular levels of cyclic nucleotides or calcium). Opioids hyperpolarize certain neurons of sympathetic ganglia and locus ceruleus by increasing the K^+ conductance of the plasma membrane. Opioids may be inhibitory transmitters in these cells. In some cases opioids appear to act presynaptically to diminish the release of synaptic transmitter (perhaps by decreasing calcium influx in response to stimulation). Opioids may act presynaptically to decrease the release of acetylcholine by preganglionic nerve endings in sympathetic ganglia. **Substance P** is the suspected transmitter at synapses made by primary sensory neurons (their cell bodies are in the dorsal root ganglia) with spinal interneurons in the dorsal horn of the spinal column. Enkephalins act to decrease the release of substance P at these synapses, thereby inhibiting the pathway for pain sensation at the first synapse in the pathway. Opioids appear to have inhibitory effects on cells in the brain involved in the perception of pain.

In certain cell types intracellular levels of cyclic AMP promptly decrease in response to opioids. In such cases the opioid receptor may be linked in an inhibitory fashion to the adenylate cyclase of the plasma membrane. In other cases responses that are triggered by increases in intracellular calcium are inhibited by opioids. In these instances opioids are believed to block calcium influx or calcium release from internal stores.

Nonopioid neuropeptides. As shown in the list of neuroactive peptides on p. 70, most of the known neuropeptides are not opioids. This section briefly summarizes current knowledge of nonopioid neuropeptides.

Substance P. **Substance P,** a peptide of 11 amino acids, is present in specific neurons in the brain, in primary sensory neurons, and in plexus neurons in the wall of the gastrointestinal tract. Substance P is a member of a group of neuropeptides called tachykinins (most of the other known tachykinins are present in amphibians). Substance P was the first so-called gut-brain peptide to be discovered. The wall of the gastrointestinal tract is richly innervated with neurons that form networks or plexuses. The intrinsic plexuses of the gastrointestinal tract exert primary control over its motor and secretory activities (Chapters 38 to 39). These enteric neurons contain many of the neuropeptides, including substance P, that are found in the brain and spinal column.

The postulated role of substance P as the neurotransmitter at the synapses between primary sensory neurons and dorsal horn neurons was discussed earlier. Substance P may also be a transmitter in the brain. Calcium-dependent release of substance P in response to depolarization has been found in brain slices or in synaptosomes from hypothalamus, substantia nigra, and other regions of the brain.

Vasoactive intestinal peptide (VIP). VIP is a member of a family of neuropeptides related to **secretin,** which was first discovered as a gastrointestinal hormone but is now known to be a neuropeptide as well. Secretin (27 amino acids) and **glucagon** (29 amino acids) have 14 amino acids in common at similar positions. VIP (28 amino acids) and **gastric inhibitory peptide** (GIP, 43 amino acids) have extensive sequence homology with secretin and glucagon. These four neuropeptides probably have a common ancestor peptide and arose in evolution by gene duplication.

VIP is widely distributed in the central nervous system and in the intrinsic neurons of the gastrointestinal tract. In neurons in the brain it has been localized in synaptic vesicles. VIP appears to function as an inhibitory transmitter to vascular and nonvascular smooth muscle and as an excitatory transmitter to glandular epithelial cells. VIP and acetylcholine coexist in parasympathetic nerve terminals that innervate pancreatic acinar cells. VIP and acetylcholine potentiate one another in their effects on pancreatic acinar cells. When applied locally, VIP excites a wide variety of cells in the central nervous system.

Secretin, glucagon, and GIP are molecules whose function as hormones has been well characterized. These peptides have also been found in particular neurons in the central nervous system, but their functions in the CNS remain to be determined.

Cholecystokinin (CCK). Cholecystokinin is a member of a group of neuropeptides that includes **gastrin** and **cerulein,** which have similar C-terminal sequences. CCK is a well-known gastrointestinal hormone that elicits contraction of the gallbladder (Chapter 39). CCK(39) is cleaved near its N-terminus to produce CCK(33), the physiological form in the gastrointestinal tract. The N-terminal octapeptide, CCK(8), is present in particular neurons of the central nervous system.

Neurotensin. Neurotensin is one of the most recently discovered gut-brain neuropeptides and is present in enteric neurons and in the brain. When injected into cerebrospinal fluid at low concentrations, neurotensin elicits hypothermia. Thus neurotensin may be involved in the regulation of body temperature.

Neuropeptides are being investigated intensively. Information about their localization, effects, and mechanisms of action is accumulating rapidly.

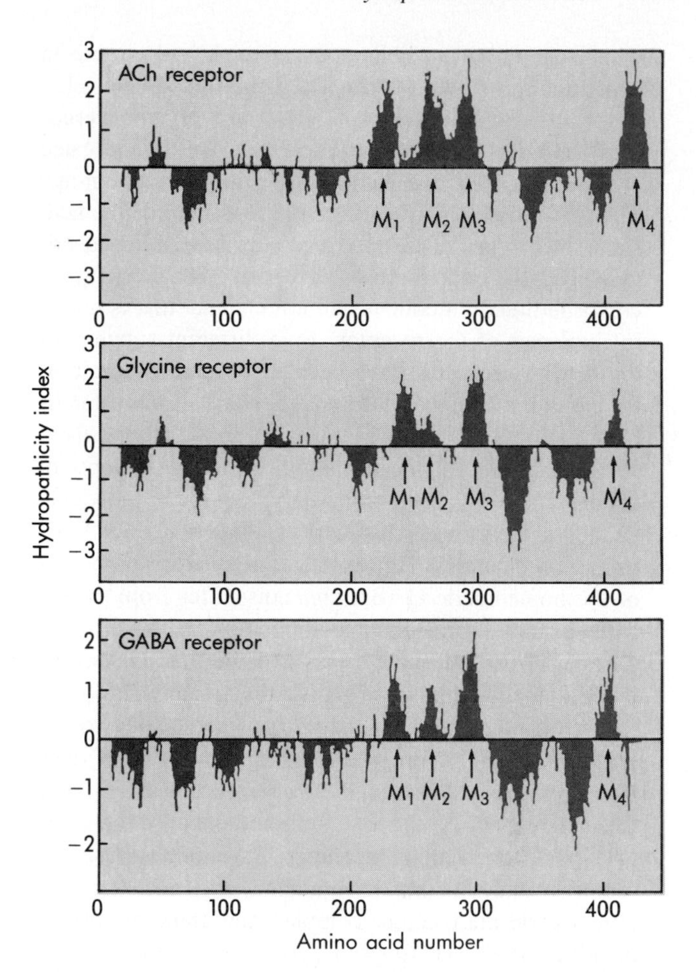

Fig. 4-19 Aligned hydropathy profiles of three ligand-gated ion channels with putative transmembrane helices indicated by M1 through M4. Shown are hydropathy profiles of α subunits of nicotinic acetylcholine receptor (AChR), glycine receptor, and GABA receptor. Positive deflections indicate stretches of hydrophobic amino acids; negative deflections indicate hydrophilic amino acids. (Modified from Maelicke A: *Trends Biochem Sci* 13:199, 1988.)

Other Neuromodulators

There are important neuromodulators that are not peptides. Purines and purine nucleotides and nucleosides function as neuromodulators in the central, autonomic, and peripheral nervous systems. Substances that serve as classical neurotransmitters may also act as neuromodulators. In some cases the transmitter binds to receptors on the presynaptic neuron that released it, with the result that the transmitter acts as a modulator to regulate its own release. Other neuromodulators remain to be discovered and their mechanisms of action elucidated.

Neurotransmitter Receptors

GABA and Glycine Receptors

The amino acid sequences and basic structures of the GABA, glycine, and nicotinic acetylcholine receptors are significantly homologous. Hydropathy plots for α subunits of a nicotinic ACh receptor, a glycine receptor, and a GABA (γ-aminobutyric acid) receptor are shown in Fig. 4-19. Note that the positions of the four putative transmembrane helices (M1 through M4) are similar in all three molecules and that there are also similarities in the sizes and positions of the hydrophilic parts of the three receptors. These three ligand-gated ion channels have evolved on a similar pattern, even though the ACh receptor is a cation channel and the other two transmit anions.

Another common feature of these ligand-gated channels is that they are constructed of different subunits that surround a central ion channel; typically five subunits surround the ion channel (see Fig. 4-6, *B*). The different subunits have a high degree of amino acid sequence homology with one another and have similar hydropathy plots. Moreover, functional, ligand-responsive ion channels can be formed from different combinations of the subunits. Only one or two different subunits are required to form an ion channel with properties similar to those of channels formed from four or five dissimilar subunits. Different cell types in the brain frequently have different receptor subtypes that consist of different combinations of subunit isoforms. Evolution appears to have created a smorgasbord of subunit isoforms from which an individual cell type can pick and choose in constructing its own multimeric ligand-gated channel.

Inhibitory function of GABA and glycine receptors. GABA and glycine receptors are ligand-gated chloride channels that mediate Cl^- influx into neurons. The Cl^- influx hyperpolarizes and thus inhibits the neurons. Two types of GABA receptors have been identified—$GABA_A$ and $GABA_B$ receptors. The $GABA_B$ receptor indirectly modulates an ion channel that is not an integral part of the receptor. Five different subunits of the $GABA_A$ receptor have been discovered by molecular cloning techniques. Moreover, some of the subunits have multiple (as many as four) isoforms. Mixing these subunits allows for the formation of a large number of different $GABA_A$ receptor subtypes. Each individual subunit α through ε, when expressed alone, forms functional ion channels. However, the regulatory properties of the homeomeric GABA channels differ from the native channel. The subunit composition of the $GABA_A$ channel in particular cell types and the functions of the individual subunits is being investigated actively.

$GABA_A$ receptors are the sites of action of two important classes of drugs. Benzodiazepines (like diazepam), widely used antianxiety and relaxant drugs, bind to a specific site on the γ subunit and enhance the effect of GABA on channel opening. Barbiturates, used as sedatives and anticonvulsive agents, also bind to a specific site on the receptor complex and enhance channel opening.

Glycine receptors in spinal cord and brainstem. Although GABA receptors predominate in most parts of the brain, glycine receptors are more numerous in the spinal cord and brainstem. Like $GABA_A$ receptors, glycine receptors are ligand-gated Cl^- channels that hyperpolarize neurons when activated. The glycine receptor is a pentameric complex of three different subunits: α (53 kDa), β (57 kDa), and γ subunits. The probable stoichiometry of the pentameric complex is $\alpha_2\beta_2\gamma$. The subunits have considerable amino acid sequence homology with one another and with nicotinic ACh and $GABA_A$ receptor subunits. The subunits have the same pattern of four transmembrane helices with particularly high sequence homology with other members of this family of ligand-gated receptors in the transmembrane segments (see Fig. 4-19).

Excitatory Amino Acid Receptors

Glutamate is the major neurotransmitter mediating synaptic excitation in the central nervous system. Receptors for glutamate are called **excitatory amino acid (EAA) receptors.** At the present time five subtypes of EAA receptors are recognized (Table 4-2). These receptors are classified principally on the synthetic amino acid analogues they bind tightly and specifically. Four of these are classes of ligand-gated ion channels, and the fifth is a receptor (called the **metabotropic EAA receptor**) that is indirectly linked to an ion channel. The four EAA-gated ion channels are the **NMDA** (N-methyl-D-aspartate) receptor, the **AMPA** (α-amino-3-hydroxy-5-methyl-4-isoxazolepropionic acid) receptor (formerly known as the quisqualate receptor), the **kainate** (salt of kainic acid) receptor, and the **L-AP4** (L-2-amino-4-phosphonobutanoic acid) receptor.

Table 4-2 Different classes of excitatory amino acid receptors*

Receptor class	*Distribution and function*
AMPA	Channel-selective for Na^+ and K^+; widely distributed in CNS
NMDA	Channel-selective for Na^+, K^+, and Ca^{++}; distributed similarly to AMPA receptors
Kainate	Present in a few specific areas of CNS; function not clear
L-AP4	Not widely distributed; may function as a presynaptic glutamate receptor to inhibit glutamate release
Metabotropic	Mobilizes IP_3 and thus Ca^{++}

Modified from Watkins JC, Krogsgaard-Larsen P, Honore T: *Trends Pharmacol Sci* 11:25, 1990.

*Glutamate is the major physiological agonist for all the classes of excitary amino acid receptors.

NMDA and AMPA receptors are widely distributed in the CNS and are especially prevalent in the cerebral cortex and hippocampus. Kainate receptors are concentrated in a few specific areas and the distribution of L-AP4 receptors remains to be elucidated. AMPA receptors contain a ligand-gated channel that permits Na^+ and K^+ to flow, leading to depolarization of the postsynaptic neuron. The NMDA receptor channel allows Na^+, K^+, and Ca^{++} flow. The NMDA receptor is blocked by extracellular Mg^{++} in a voltage-dependent fashion so that near the resting potential the NMDA receptor is unresponsive to glutamate. Only after considerable depolarization of the cell do the NMDA receptors become unblocked. This appears to permit a graded response of a postsynaptic cell to glutamate release from a presynaptic nerve ending. A small amount of glutamate would lead to opening of AMPA channels and a small depolarization. Greater amounts of glutamate would first open AMPA channels, leading to depolarization of the postsynaptic cell and then to opening of unblocked NMDA channels and consequent Ca^{++} entry and greater depolarization. Some neurons in the CNS have relatively low resting membrane potentials at which the block of NMDA receptors by Mg^{++} is only partial. Hence the NMDA receptors of these cells can respond to glutamate in the absence of prior depolarization.

Partial cloning of the kainate receptor shows that it has four membrane-spanning helices, spaced similarly to those in the subunits of the nicotinic ACh receptor subunits. The kainate receptor has only limited amino acid sequence homology with the nicotinic ACh receptor, primarily in the transmembrane segments.

The NMDA receptor has a binding site for glycine that is distinct from the site for NMDA and glutamate. Binding of glycine does not, by itself, alter the function of the NMDA channel, but it allosterically enhances current flow in response to NMDA or glutamate by diminishing the average closed time of the channel. This effect is maximal at low μM levels of glycine. The physiological significance of glycine-modulation of NMDA channels remains to be elucidated.

Control of Ion Channels by Second Messenger Mechanisms

The interaction of acetylcholine with muscarinic ACh receptors opens K^+ channels and hyperpolarizes the cell. In contrast to the neurotransmitter receptors discussed so far, the muscarinic ACh receptor is not an ion channel itself. Rather, the binding of ACh to the muscarinic receptor sets in motion a chain of biochemical events that affects another protein that is an ion channel. The muscarinic ACh receptor is the member of the family of receptors that is linked to GTP-binding proteins (G proteins). A G protein is activated when it interacts with a muscarinic ACh receptor that has ACh bound to it. Activated G proteins then interact with K^+ channels to increase their open times. More often G proteins produce their effects by changing cellular levels of second messenger compounds, such as cyclic AMP, cyclic GMP, Ca^{++}, and inositol trisphosphate. For a number of ion channels, the primary physiological gating stimulus is a second messenger compound. In addition many voltage- and ligand-gated ion channels can be modulated by second messenger mechanisms (see Chapter 5).

Summary

1. The neuromuscular junction is the best-characterized vertebrate chemical synapse. Acetylcholine released by the prejunctional nerve terminal interacts with acetylcholine receptor proteins in the postjunctional membrane to open a cation-selective ion channel. This results in a depolarization (the endplate potential) of the postjunctional membrane.

2. The endplate potential is terminated by hydrolysis of acetylcholine by the enzyme acetylcholinesterase. Choline liberated in the junctional cleft is actively transported back into the nerve terminal.

3. The release of acetylcholine is quantal. A quantum corresponds to the amount of acetylcholine in a single prejunctional vesicle.

4. Direct electrical transmission between adjacent cells is mediated by gap junctions.

5. Spinal motoneurons have hundreds of synaptic inputs. An action potential in an excitatory input causes an excitatory postsynaptic potential that depolarizes the motoneuron and brings it closer to threshold. An action potential in an inhibitory input causes an inhibitory postsynaptic potential that hyperpolarizes the motoneuron.

6. The efficacy of synaptic transmission depends on the timing and frequency of action potentials in the presynaptic neuron. Facilitation, augmentation, posttetanic potentiation, and long-term potentiation are examples of increased efficacy of synaptic transmission in response to multiple stimulations of the presynaptic neuron.

7. Acetylcholine, biogenic amines, and the amino acids glutamate, glycine, and γ-aminobutyric acid are important neurotransmitters in the central nervous system. A large number of neuropeptides function as neurotransmitters or as neuromodulators in the central nervous system.

8. The receptors for acetylcholine, glycine, and γ-aminobutyric acid are ligand-gated ion channels that belong to the same family of related membrane proteins.

9. Glycine and γ-aminobutyric acid are the major transmitters at inhibitory synapses in the central nervous system. Glutamate, acting on five different classes of excitatory amino acid receptors, is a major mediator of synaptic excitation.

10. Some neurotransmitter receptors are not ion channels themselves but control other proteins, which are ion channels, by second messenger mechanisms.

Bibliography

Journal articles

Augustine GJ, Charlton MP, Smith SJ: Calcium action in synaptic transmitter release, *Ann Rev Neurosci* 10:633, 1987.

Bartfai T, Iverfeldt K, Fisone G: Regulation of the release of coexisting neurotransmitters, *Ann Rev Pharmacol Toxicol* 28:285, 1988.

Changeux J-P: The nicotinic acetylcholine receptor: an allosteric protein prototype of ligand-gated ion channels, *Trends Pharmacol Sci* 11:485, 1990.

Cotman CW, Ganong AH: Excitatory amino acid neurotransmission: NMDA receptors and Hebb-type synaptic plasticity, *Annu Rev Neurosci* 11:61, 1988.

Hughes J, editor: Opiod Peptides, *Br Med Bull* 39:1, 1983.

Kow L-M, Pfaff DW: Neuromodulatory actions of peptides, *Annu Rev Pharmacol Toxicol* 28:163, 1988.

Kupfermann I: Functional studies of cotransmission, *Physiol Rev* 71:683, 1991.

LLinas R: Calcium in synaptic transmission, *Sci Am* 247:56, 1982.

Maelicke A: Structural similarities between ion channel proteins, *Trends Biochem Sci* 13:199, 1988.

Monaghan DT, Bridges RJ, Cotman CW: The excitatory amino acid receptors: their classes, pharmacology, and distinct properties in the function of the central nervous system, *Annu Rev Pharmacol Toxicol* 29:365, 1989.

Nicoll RA, Malenka RC, Kauer JA: Functional comparison of neurotransmitter receptor subtypes in mammalian central nervous system, *Physiol Rev* 70:513, 1990.

Sheetz MP, Steuer ER, Schroer TA: The mechanism of fast axonal transport, *Trends Neural Sci* 12:474, 1990.

Watkins JC, Krogsgaard-Larsen P, Honore T: Structure-activity relationships in the development of excitatory amino acid receptor agonists and competitive antagonists, *Trends Pharmacol Sci* 11:25, 1990.

Wroblewski JT, Danysz W: Modulation of glutamate receptors: molecular mechanisms and functional implications, *Annu Rev Pharmacol Toxicol* 29:441, 1989.

Young AB, Fagg GE: Excitatory amino acid receptors in the brain: membrane binding and receptor autoradiographic approaches, *Trends Pharmacol Sci* 11:126, 1990.

Zucker RS: Short-term synaptic plasticity, *Annu Rev Neurosci* 12:13, 1989.

Books and monographs

Aidley DJ: *The physiology of excitable cells,* ed 3, Cambridge, 1990, Cambridge University Press.

Cooper JR, Bloom FE, Roth RH: *The biochemical basis of neuropharmacology,* ed 4, New York, 1982, Oxford University Press.

Gregory RA, editor: Regulatory peptides of gut and brain, *Br Med Bull* 38:219, 1982.

Hille B: *Ionic channels of excitable membranes,* Sunderland, MA, 1984, Sinauer Associates.

Kandel ER, Schwartz JH: *Principles of neural science,* ed 3, New York,

Katz B: *Nerve, muscle, and synapse,* New York, 1966, McGraw-Hill.

Kuffler SW, Nicholls JG, Martin AR: *From neuron to brain,* ed 2, Sunderland, Mass, 1984, Sinauer Associates.

Levitan IB, Kaczmarek LK: *The neuron: cell and molecular biology,* New York, 1991, Oxford University Press.

Stevens CF: *Neurophysiology: a primer,* New York, 1966, John Wiley & Sons.

CHAPTER

5

Membrane Receptors, Second Messengers, and Signal Transduction Pathways

Basic cellular processes are regulated by a host of hormones, neurotransmitters, and paracrine substances. Some regulatory substances, such as steroid hormones, enter the cell and influence the transcription of certain genes. Many regulatory substances, however, exert their effects from outside the cell. The first step in the action of such substances is to bind to specific receptors in the plasma membrane of the target cells. The neurotransmitters discussed in Chapter 4 and their receptors are examples. For most of the neurotransmitters discussed so far, the receptor is a ligand-gated ion channel and the response of the cell is a ligand-induced ionic current. In such cases the ligand-gated ion channel is both the receptor and the effector for the action of the neurotransmitter. For most regulatory molecules, however, a more complex series of events links the binding of the substance by its specific membrane receptor to its final effects on cellular function. It is the purpose of this chapter to succinctly describe the major **signal transduction pathways** by which regulatory molecules, such as hormones and neurotransmitters, exert their effects on cells. Even though many regulatory substances exist, there are only a few signal transduction pathways by which binding of a regulatory substance to its plasma membrane receptor elicits a cellular response.

Signal Transduction Pathways

G Protein-mediated Signal Transduction Pathways

The majority of known hormones, neurotransmitters, and other regulatory molecules that alter cellular processes do so by signal transduction pathways that involve **GTP-binding proteins (G proteins).** A G protein can exist in two states. In its activated state a G protein has a higher affinity for GTP, whereas the inactivated G protein preferentially binds GDP over GTP. Some membrane receptors, when they have regulatory molecules bound to them, can interact with a G protein to promote conversion of the G protein to its activated state and its binding of GTP. The activated G protein can then interact with many **effector proteins,** most notably enzymes or ion channels, and alter their activities. The activated G protein has GTPase activity, so that eventually GTP is hydrolyzed to GDP and the G protein reverts to its inactive state (Fig. 5-1).

Among the most important targets of activated G proteins are molecules that change the cellular concentrations of the **second messengers cyclic AMP (cAMP), cyclic GMP (cGMP),** and **Ca^{++}** (Fig. 5-2). Adenylyl cyclase and cGMP phosphodiesterase are the enzymes responsible for the synthesis of cAMP and for the breakdown of cGMP. These are powerfully modulated by G protein-mediated mechanisms. Ca^{++} channels may be modulated directly by G proteins or indirectly by second messenger-dependent protein kinases. Other effectors that are modulated by G proteins include certain K^+ channels, phospholipase C, and phospholipase A_2.

Protein Kinases and Protein Phosphatases in Signal Transduction Pathways

Frequently the final step in a signal transduction pathway involves the phosphorylation (or dephosphorylation) of a particular protein that plays a central role in a biological process. When this protein is phosphorylated, its activity may be enhanced or suppressed. **Protein kinases** in the cell are responsible for phosphory-

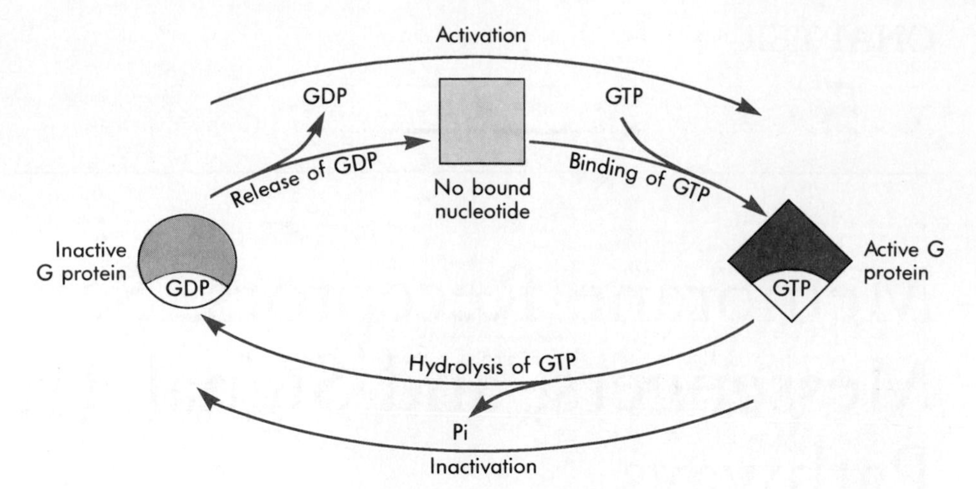

■ **Fig. 5-1** The activity cycle of a G protein. The inactive form of the G protein *(colored circle)* binds GDP. Interaction of the G protein with a ligand-bound membrane receptor promotes a conformational change leading to the release of GDP and the binding of GTP. The GTP bound form of the G protein *(colored diamond)* is the active form that interacts with proteins such as adenylyl cyclase and ion channels to alter their activity. The G protein has an intrinsic GTPase activity; hydrolysis of GTP converts the G protein back to its inactive state.

lating particular proteins, and **protein phosphatases** are responsible for cleaving phosphates from proteins.

A signal transduction pathway frequently involves altered activity of a protein kinase in response to the binding of the regulatory molecule, often called an **agonist,** to its membrane receptor. Among the signals that regulate the activities of protein kinases are the second messengers cAMP, cGMP, Ca^{++}, and diglycerides. Cells contain protein kinases that are modulated by each of these second messengers. Binding of the regulatory molecule to its membrane receptor often changes the intracellular level of one of the second messengers, which then modulates the activity of a protein kinase (Table 5-1). Cells contain protein kinases whose activities are enhanced by cAMP and cGMP, called, respectively, **cAMP-dependent protein kinase** and **cGMP-dependent protein kinase.** Cells also contain protein kinases whose activities are enhanced when they bind the com-

■ **Fig. 5-2** A schematic depiction of the signal transduction cascade by which an extracellular ligand such as a peptide hormone can bind to its receptor so as to activate a G protein and, via the cascade, lead to the activation or inactivation of an ion channel, a protein kinase, a phospholipase, or an ion channel. Amplification can occur at each level of the cascade. (For the sake of simplicity, the figure indicates that occupation of receptor by ligand leads to an increase in a second messenger, which causes an activation of an enzyme or ion channel. In reality, in numerous cases receptor occupation leads to a decreased concentration of second messenger and in other cases increased second messenger results in inactivation of an enzyme or ion channel.)

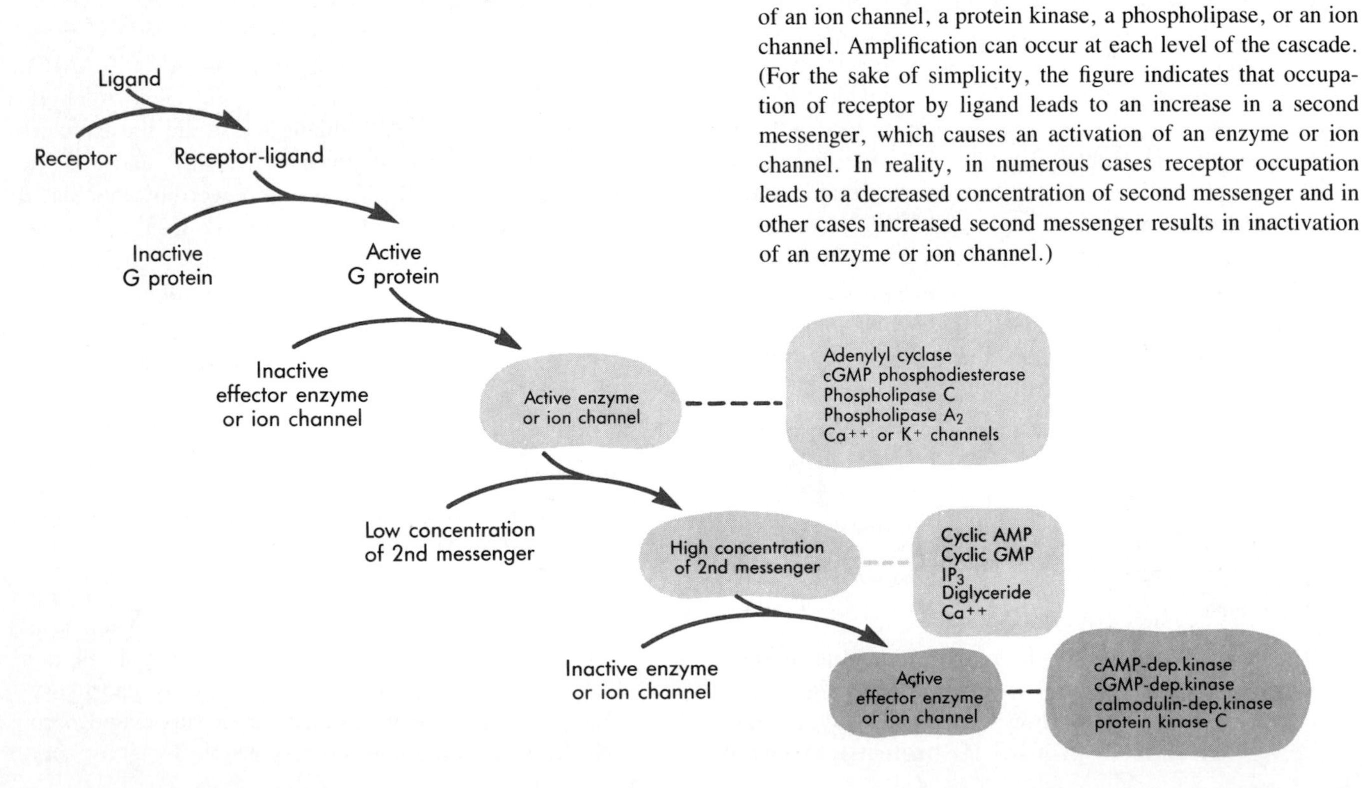

Table 5-1 Second messenger-dependent protein kinases.*

Protein kinase	*Activator*
Cyclic AMP-dependent protein kinase	Cyclic AMP
Cyclic GMP-dependent protein kinase	Cyclic GMP
Calmodulin-dependent protein kinase	Ca^{++}-calmodulin complex
Protein kinase C	Ca^{++} and diglyceride

*Frequently the final step in a signal transduction pathway is the phosphorylation of an important protein, which modifies its activity. The phosphorylation of that protein is catalyzed by a second messenger-dependent protein kinase.

plex of Ca^{++} with a protein called **calmodulin** (MW 16,700). Calmodulin is present in all cells, and it binds four Ca^{++} ions. The complex of Ca^{++} and calmodulin then regulates a host of other intracellular proteins, many of which are not kinases. Protein kinases that are regulated by Ca^{++}-calmodulin are called **calmodulin-dependent protein kinases.**

Inositol Lipids and Signal Transduction Pathways

Another class of extracellular agonists binds to receptors that activate, via a G protein, a specific **phospholipase C** that cleaves phosphatidylinositol bisphosphate (a phospholipid present in minute quantities in the plasma membrane) into **inositol-1,4,5-trisphosphate (IP_3)** and **diglyceride** (Fig. 5-3). Both IP_3 and diglyceride are second messengers. IP_3 binds to specific ligand-gated Ca^{++} channels in the endoplasmic reticulum. It thereby releases Ca^{++} to increase its cytosolic level. The Ca^{++} channel of the endoplasmic reticulum has a similar structure to the Ca^{++} channel of the sarcoplasmic reticulum of skeletal and cardiac muscle. These latter channels are involved in excitation-contraction coupling (Chapters 18 and 24). Diglyceride, together with Ca^{++}, activates another important protein kinase called **protein kinase C.** Among the substrates of protein kinase C are proteins involved in the control of cellular proliferation.

Phospholipase A_2 (PLA_2) is also activated by some agonists via a G protein-dependent pathway. PLA_2 cleaves the number 2 fatty acid from membrane phospholipids. Because some of the phospholipids, particularly inositol phospholipids, are species with arachidonic acid esterified to the number 2 carbon of the glycerol backbone, PLA_2 releases significant amounts of **arachidonic acid,** which is the precursor for the cellular synthesis of **prostaglandins and leukotrienes.** These substances are important classes of potent regulatory molecules.

To summarize the characteristics of a general G protein-protein kinase-mediated signal transduction pathway (see Fig. 5-2):

1. A hormone or other regulatory molecule binds to its plasma membrane receptor.
2. The ligand-bearing receptor interacts with a G protein and activates it. The activated G protein binds GTP.
3. The activated G protein interacts with one or more of the following: adenylyl cyclase, cGMP phosphodiesterase, Ca^{++} channels, or phospholipase C to activate or inhibit them.
4. An increase or decrease occurs in the cellular level of one or more of the following second messengers: cAMP, cGMP, Ca^{++}, or diglyceride.
5. The increase in the second messenger changes the activity of one or more of the second messenger-dependent protein kinases: cAMP-dependent protein kinase, cGMP-dependent protein kinase,

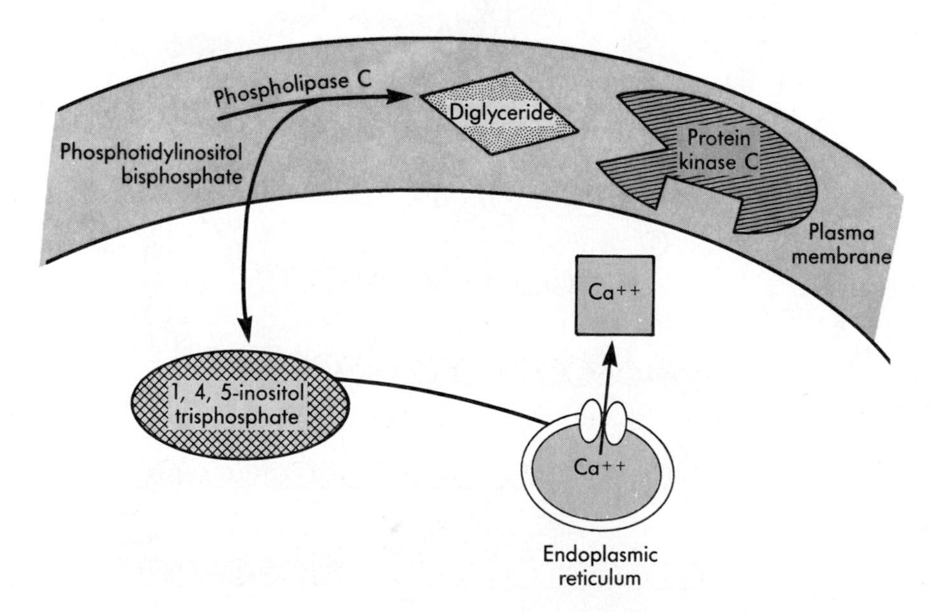

Fig. 5-3 Second messengers that result from the turnover of inositol phospholipids of the plasma membrane. A specific phospholipase C that hydrolyzes the minor membrane phospholipid phosphatidylinositol bisphosphate may be stimulated by activated G proteins. This cleavage releases IP_3 and diglyceride, both of which are second messengers. IP_3 binds to a specific Ca^{++} channel in the endoplasmic reticulum membrane, causing Ca^{++} to be released from the endoplasmic reticulum. Diglyceride, together with Ca^{++}, activates protein kinase C to phosphorylate important cellular effector proteins.

calmodulin-dependent protein kinase, or protein kinase C.

6. The level of phosphorylation of an enzyme or an ion channel is altered, or an ion channel activity changes and brings about the final cellular response.

Protein Tyrosine Kinases

A class of membrane receptors, not linked to G proteins, comprises proteins with intrinsic protein-tyrosine kinase activity. When these receptors bind agonist, their tyrosine kinase activity is turned on and they phosphorylate specific effector proteins on particular tyrosine residues. The other protein kinases that have been discussed phosphorylate proteins on serine and threonine residues exclusively. The receptor for the hormone insulin and receptors for some growth factors are tyrosine kinases.

Membrane Receptors for Regulatory Molecules

The membrane receptors that mediate agonist-dependent activation of G proteins are members of a protein

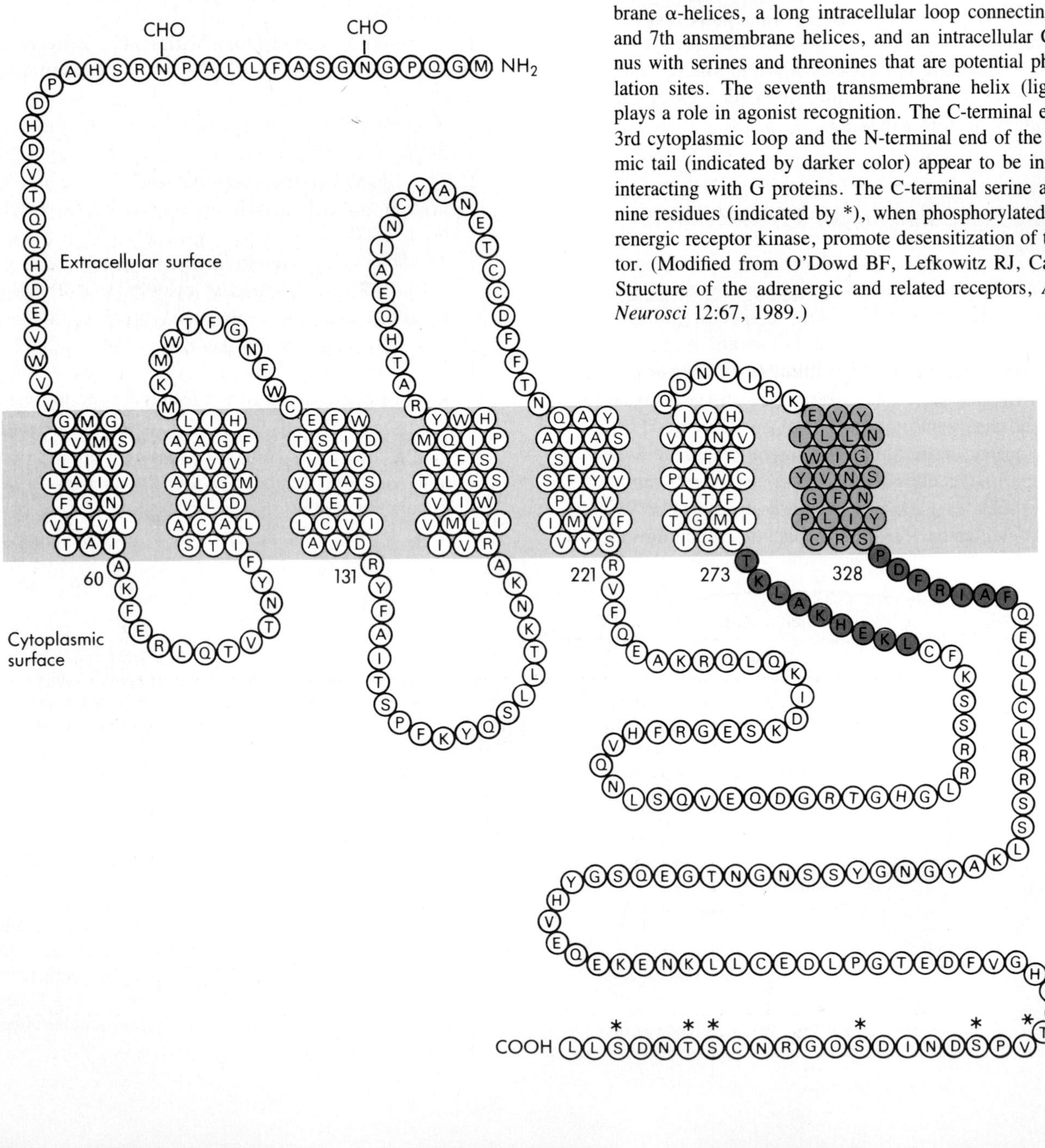

Fig. 5-4 Proposed structure of the human β_2 adrenergic receptor. This receptor is a member of the family of G protein-linked receptors that includes the other adrenergic receptors, the muscarinic acetylcholine receptor, the serotonin receptor, rhodopsin, and others. Among the common structural features of this protein family are an extracellular N-terminus with multiple N-linked glycosylation sites, seven transmembrane α-helices, a long intracellular loop connecting the 6th and 7th ansmembrane helices, and an intracellular C-terminus with serines and threonines that are potential phosphorylation sites. The seventh transmembrane helix (light color) plays a role in agonist recognition. The C-terminal end of the 3rd cytoplasmic loop and the N-terminal end of the cytoplasmic tail (indicated by darker color) appear to be involved in interacting with G proteins. The C-terminal serine and threonine residues (indicated by *), when phosphorylated by β-adrenergic receptor kinase, promote desensitization of the receptor. (Modified from O'Dowd BF, Lefkowitz RJ, Caron MG: Structure of the adrenergic and related receptors, *Annu Rev Neurosci* 12:67, 1989.)

family. This family includes α-adrenergic and β-adrenergic receptors, muscarinic acetylcholine receptors, serotonin receptors, adenosine receptors, receptors for peptide hormones, and rhodopsin. The common structural features of the G protein-coupled receptor family (Fig. 5-4) include 7 transmembrane helices of 22 to 28 amino acids that are predominantly hydrophobic, an extracellular N terminus with sites for N-linked glycosylation, and an intracellular C terminus with several serines and threonines that may be phosphorylated.

Molecular biological studies, particularly those involving construction of chimeric receptors (combinations of parts of different receptor types), are beginning to define the functions of certain regions of the receptor proteins. The C-terminal end of the third cytoplasmic loop, which connects the sixth and seventh transmembrane helices and the N-terminal end of the C-terminal peptide, is involved in interacting with G proteins (see Fig. 5-4). The seventh transmembrane helix plays a role in determining agonist specificity. Phosphorylation of certain serine and threonine residues in the C-terminal tail promotes **desensitization** of the receptor-mediated response. The densensitized receptor has a diminished response to an agonist. In the cases of the β-adrenergic receptor and rhodopsin, specific protein kinases catalyze phosphorylation of these serine and threonine residues.

G Proteins

G proteins, also known as GTP-binding proteins, bind and hydrolyze GTP. They serve as molecular switches that regulate a host of intracellular processes. The active form of a G protein has a high affinity for GTP. Hydrolysis of GTP (promoted by the intrinsic GTPase activity of the G protein) converts the G protein to its inactive form. Active G proteins bind to and modify the activity of various enzymes and ion channels, and in this way modulate many vital cellular activities. Two classes of G proteins are known: **heterotrimeric G proteins** and **low molecular weight (small) G proteins.**

Heterotrimeric G Proteins

A heterotrimeric G protein has three subunits: an α subunit (40,000 to 45,000 daltons), a β subunit (about 37,000 daltons), and a γ subunit (8,000 to 10,000 daltons). Currently we know of at least 16 different genes that encode α subunits, four genes that encode β subunits, and several genes that encode γ subunits. The function and specificity of a G protein is usually determined by its α subunit; in most cases the β and γ subunits are interchangeable. Some of the best-characterized G proteins and some of the properties of the signal transduction pathways they mediate are shown in Table 5-2.

Heterotrimeric G proteins function as intermediaries between the plasma membrane receptors for over 100 different extracellular regulatory substances (hormones, neurotransmitters, and others) and the intracellular processes they control. Binding of the regulatory substance to its receptor activates the G protein, and the activated G protein then either stimulates or inhibits an enzyme or an ion channel.

The α subunit is usually the "business end" of the heterotrimeric G protein (Fig. 5-5). The inactive G protein exists primarily as the αβγ heterotrimer, with GDP in its nucleotide binding site. The interaction of the heterotrimeric G protein with a ligand-bearing receptor

Table 5-2 Major classes of heterotrimeric G proteins*

G protein	*Receptors for*	*Effectors*	*Signalling pathways*
G_s	Adrenaline, noradrenaline, histamine, glucagon, ACTH, luteinizing hormone, follicle-stimulating hormone, thyroid-stimulating hormone, and others	Adenylyl cyclase Ca^{++} channels	↑ cAMP ↑ Ca^{++} influx
G_{olf}	Odorants	Adenylyl cyclase	↑ cAMP (olfaction)
G_{t1} (rods)	Photons	cGMP phosphodiesterase	↓ cGMP (vision)
G_{t2} (cones)	Photons	cGMP phosphodiesterase	↓ cGMP (colour vision)
G_{i1}, G_{i2}, G_{i3}	Noradrenaline, prostaglandins, opiates, angiotensin, many peptides	Adenylyl cyclase Phospholipase C Phospholipase A_2 K^+ channels	↓ cAMP ↑ IP_3, diacylglycerol, Ca^{++} Arachidonate release Membrane polarization
G_0	Probably many, but not yet defined	Phospholipase C Ca^{++} channels	↑ IP_3, diacylglycerol, Ca^{++} ↓ Ca^{++} influx

Modified from Bourne HR, Sanders DA, McCormick F: *Nature* (London) 348:125, 1990.

*The classification is primarily based on the structure of their α subunits. In general there is more than one isoform of each type of α subunit.

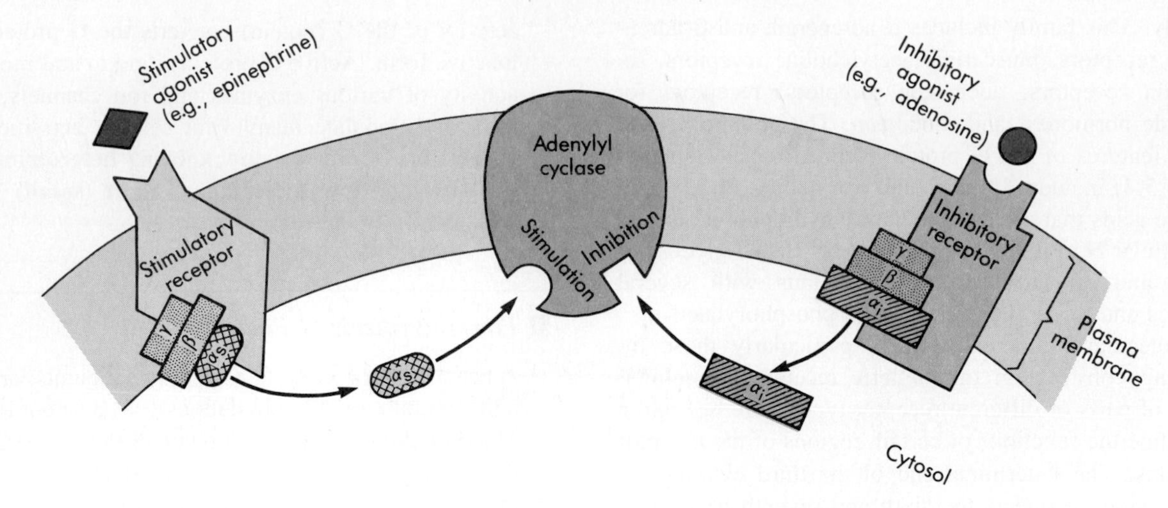

■ **Fig. 5-5** Adenylyl cyclase may be stimulated or inhibited by signal transduction pathways. Receptors for agonists that result in stimulation of adenylyl cyclase activate $G_s\alpha$, which interacts with adenylyl cyclase to stimulate it. Receptors for agonists that produce inhibition of adenylyl cyclase activate $G_i\alpha$, which causes inhibition of adenylyl cyclase.

causes a conformational change in the α subunit to the active form, which has a high affinity for GTP and a low affinity for the βγ pair. Therefore the activated α subunit exchanges GTP for GDP and dissociates from βγ. In some cases, however, the βγ dimer may be responsible for all or some of the receptor-mediated response.

Regulation of adenylyl cyclase. cAMP was the first of the second messengers to be discovered. The regulation of adenylyl cyclase, the enzyme that produces cAMP, is the prototype for G protein-mediated signal transduction pathways. Adenylyl cyclase is subject to both positive and negative control by G protein-mediated pathways (see Fig. 5-5). The binding of a stimulatory ligand, such as epinephrine acting through β-adenergic receptors, results in activation of heterotrimeric G proteins with α subunits of the type called $G_s\alpha$. Activation of the G proteins by the ligand-bearing receptor causes $G_s\alpha$ to dissociate from βγ and to bind GTP. $G_s\alpha$ then interacts with adenylyl cyclase to activate it.

When a regulatory substance, such as adenosine (which inhibits adenylyl cyclase), binds to its receptor, it activates G proteins with α-subunits of a different type, called $G_i\alpha$. Binding of the inhibitory ligand to its receptor activates the G protein and causes $G_i\alpha$ to dissociate from the βγ dimers. $G_i\alpha$ inhibits adenylyl cyclase. However, the mechanism of inhibition remains controversial. $G_i\alpha$ may bind to adenylyl cyclase to inhibit it. Alternatively the βγ dimers may bind to $G_s\alpha$ and thus inhibit the stimulation of adenylyl cyclase by stimulatory ligands. Both of these mechanisms may be operative.

Low molecular weight G proteins. Cells contain another type of GTP-binding protein called low molecular weight G proteins or **small G proteins** (MW 20,000 to 35,000 daltons). Small G proteins, like their heterotrimeric cousins, operate via the cycle of activation and inactivation shown in Fig. 5-1. The small G proteins do not mediate the signals of extracellular regulatory molecules but rather are involved in intracellular control. Among the processes that small G proteins help regulate are polypeptide chain elongation in protein synthesis, proliferation and differentiation of cells, neoplastic transformation of cells, transport of vesicles among different organelles, and exocytotic secretion. Table 5-3 lists some of the major classes of small G proteins.

■ **Table 5-3** Some of the major subfamilies of small GTP-binding proteins*

	Approximate number		
Subclass	***Mammals***	***Yeast***	***Representative proteins***
Ras and ras-like	6	3	Ha-, Ki-, N-ras (mammals) R-ras, Rap proteins (mammals) Ras1, Ras2 *(S. cerevisiae)*
Ypt1/Sec4	8	2	Rab3 (mammals) Ypt1 *(S. cerevisae)* Sec4 *(S. cerevisae)*
Rho	3	3	Rho C (mammals) CDC42 *(S. cerevisae)*
ARF-like	1	3	ADP-ribosylation factor (ARF)

Modified from Bourne HR, Sanders DA, McCormick F: *Nature* (London) 348:125, 1990.

*Other subfamilies (not listed) are known. The small GTP-binding proteins are classified largely on the basis of similarities of primary structure in domains believed to participate in binding GTP.

Ion Channels Modulated Directly by G Proteins

In Chapter 4 several ligand-gated ion channels that are modulated directly by an extracellular agonist were discussed. Other ion channels are regulated by second messenger-mediated mechanisms that involve G proteins in the second step of the signal transduction cascade. Some ion channels, however, are directly modulated by G proteins without the involvement of a second messenger. The binding of acetylcholine to M_2 muscarinic receptors in the heart and in certain neurons leads to the activation of K^+ channels. Acetylcholine binding to the muscarinic receptor leads to activation of a G protein of the G_i subclass. The activated α subunit, denoted $G_i\alpha$, then directly interacts with a particular class of K^+ channels to increase their probability of opening. In some cases the βγ dimer, rather than the α subunit, appears to alter the ion channel gating.

The L-type (dihydropyridine-sensitive) Ca^{++} channel in the heart and in skeletal muscle is activated in response to agonists that bind to the β-adrenergic receptor. The L-type channel is regulated both directly by G proteins and indirectly by a second messenger signal transduction cascade involving cAMP-dependent protein kinase. For both the direct and indirect effects of β-adrengergic agonists on the L-type Ca^{++} channels, the G protein involved is $G_s\alpha$.

Second Messenger-Dependent Protein Kinases

Rall and Sutherland first identified cAMP as a second messenger in investigations of the mechanisms involved in the hormonal control of glycogen synthesis and breakdown. They showed that the phosphorylation of rate-determining enzymes in these metabolic pathways by cAMP-dependent protein kinases was responsible for the hormonal regulation of glycogen metabolism.

It is now recognized that many other enzymes and effector molecules are regulated by cAMP-dependent protein kinase and that some protein kinases depend on the concentrations of other second messengers (see Table 5-1).

Cyclic AMP–Dependent Protein Kinase

In the absence of cAMP, cAMP-dependent protein kinase is composed of four subunits: two regulatory subunits and two catalytic subunits. Most cell types contain the same catalytic subunit, but the regulatory subunits differ significantly. The presence of the regulatory subunits greatly inhibits the enzymatic activity of the complex. In the presence of micromolar levels of cAMP, each regulatory subunit binds two molecules of cAMP. The binding of cAMP causes a conformational change in the regulatory subunits, and this change diminishes their affinity for binding the catalytic subunits. The regulatory subunits dissociate from the catalytic subunits, and in this way the catalytic subunits become activated (Fig. 5-6). The active catalytic subunit phosphorylates target proteins on particular serine and threonine residues.

Comparison of the amino acid sequence of cAMP-dependent protein kinase with representatives of the other classes of protein kinases shows that, in spite of vast differences in their regulatory properties, the different classes of protein kinases share a common core with high amino acid homology (Fig. 5-7). The core structure includes the ATP binding domain and the active center of the enzyme where the transfer of phosphate from ATP to the acceptor protein occurs. The parts of the kinases outside the catalytic core are involved in regulation of the kinase activities. Recently the crystal

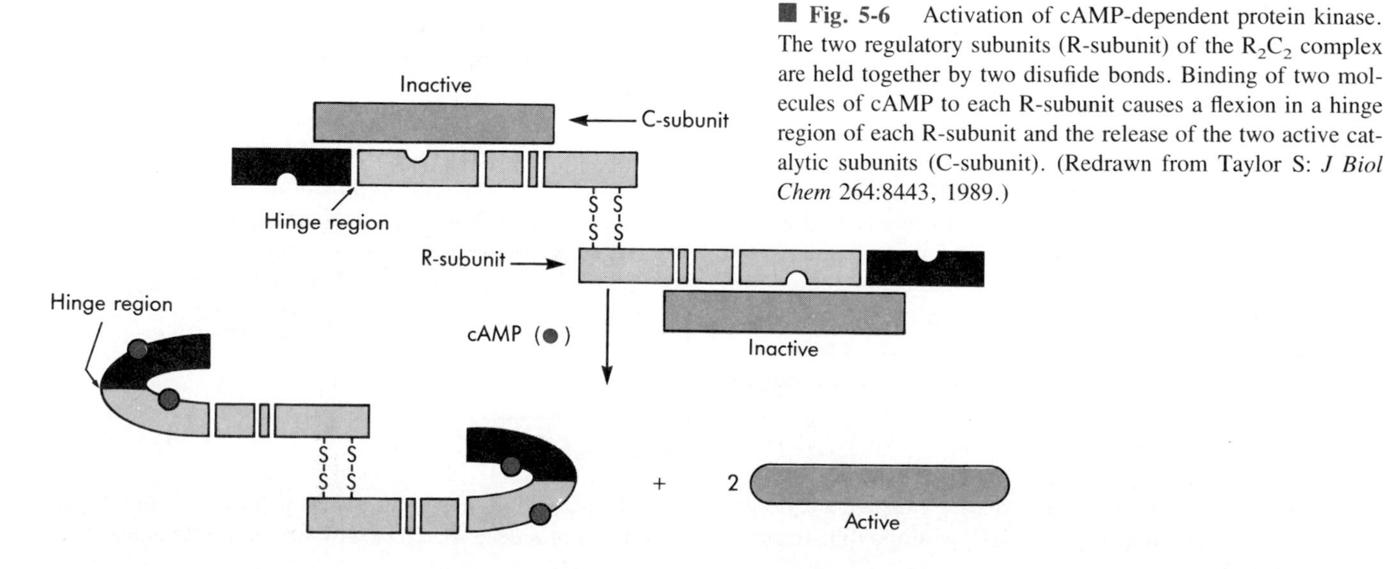

Fig. 5-6 Activation of cAMP-dependent protein kinase. The two regulatory subunits (R-subunit) of the R_2C_2 complex are held together by two disufide bonds. Binding of two molecules of cAMP to each R-subunit causes a flexion in a hinge region of each R-subunit and the release of the two active catalytic subunits (C-subunit). (Redrawn from Taylor S: *J Biol Chem* 264:8443, 1989.)

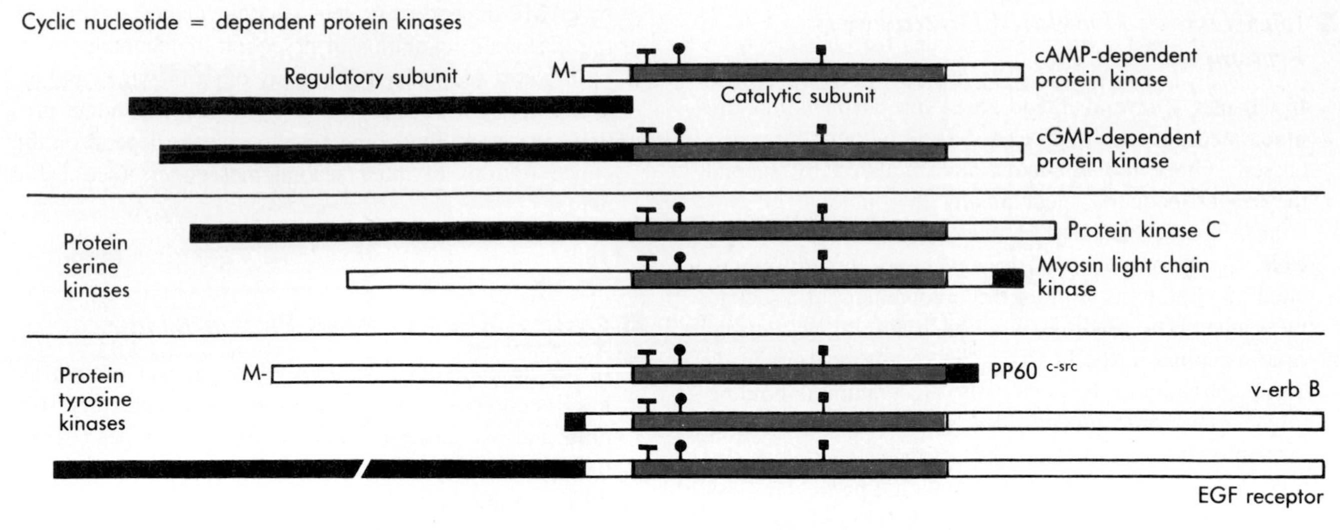

■ **Fig. 5-7** The protein kinase family. All known protein kinases share a common catalytic core (solid color) that contains ATP and peptide binding domains and the active site where phosphoryl transfer occurs. Conserved residues are aligned with lysine 72 (●), aspartate 184 (■), and glycines 50, 52, and 55 ▬ of the catalytic subunit of cAMP-dependent protein kinase. Regions important for regulation are shown as black bars. (Modified from Taylor S: *J Biol Chem* 264:8443, 1989.)

structure of the catalytic subunit of cAMP-dependent protein kinase has been determined. The catalytic core that is conserved among all known protein kinases consists of two lobes. The smaller of the two lobes contains an unusual ATP binding site, whereas the larger lobe contains the peptide binding site.

Many protein kinases also contain a regulatory region, known as a **pseudosubstrate domain,** whose amino acid sequence resembles the phosphorylation sites of substrate proteins. The pseudosubstrate region binds to the active site of the protein kinase and inhibits the phosphorylation of true substrates of the protein kinase. Activation of the kinase may involve phosphorylation or a noncovalent allosteric modification of the protein kinase to remove the inhibition.

■ *Calcium-Calmodulin–Dependent Protein Kinases*

A host of vital cellular processes, including release of neurotransmitters, secretion of hormones, and muscle contraction, are regulated by the cytosolic level Ca^{++}. One way that Ca^{++} exerts control is by binding to calmodulin. The complex of Ca^{++} and calmodulin then can influence the activity of many different proteins, among them a group of protein kinases known as Ca^{++}-calmodulin–dependent protein kinases (Table 5-4).

Ca^{++}-calmodulin–dependent protein kinase II may play a central role in the mechanism by which an increase in Ca^{++} in a nerve terminal causes exocytotic release of neurotransmitter. Ca^{++}-calmodulin–dependent protein kinase II is one of the most abundant proteins in the nervous system; it may account for as much as 1% of total protein in certain regions of the brain. Its favorite substrate is a protein called synapsin I that is present in nerve terminals and binds to the external surface of synaptic vesicles. A current hypothesis is that bound synapsin I prevents vesicles from participating in exocytosis, and that phosphorylation of synapsin I causes it to dissociate from the vesicles and to allow exocytosis to proceed.

Ca^{++}-calmodulin–dependent protein kinase II is a large complex of two different subunits—α (50 kDa) and β (60 kDa). The size of the complex and the proportion of α and β subunits varies from one cell type to another. As indicated in Table 5-4, the types and sizes of the subunits that compose the different types of Ca^{++}-calmodulin–dependent protein kinases and in the molecular weights of the active complexes are very diverse.

Another Ca^{++}-calmodulin–dependent protein kinase is myosin light chain kinase, which plays a central role in the regulation of contraction of smooth muscle (Chapter 19. Elevation of the cytosolic Ca^{++} concentration in a smooth muscle cell stimulates the activity of myosin light chain kinase. The resulting phosphorylation of the regulatory light chains of myosin allows contraction to proceed.

■ *Protein Kinase C*

The activation of protein kinase C by Ca^{++} and diglycerides produced as a result of receptor activation of the

Table 5-4 The family of Ca^{++}-calmodulin–dependent protein kinases*

Enzyme	*Substrate*	*Structure*
Myosin light-chain kinase	Myosin P light chain	M_r 80-150,000
Phosphorylase kinase	Phosphorylase	Native M_r 1.3 × 10^6 (α, β, γ, δ) Subunit M_r α, 145,000 Subunit β, 128,000 Subunit γ, 45,000 Subunit δ, 17,000
Ca^{++}/calmodulin kinase I	Synapsin I (site I), protein III	Native M_r 48,000 Subunit M_r 37,000-42,000
Ca^{++}/calmodulin kinase II	Synapsin I (site II), glycogen synthase, MAP-2 tyrosine hydroxylase	Native M_r 250,000-600,000 Subunit M_r α 50,000 Subunit β 60,000
Ca^{++}/calmodulin kinase III	Elongation factor 2	Native M_r 140,000

From Blackshear PJ, Nairn AC, Kuo FJ: *FASEB J* 2:2957, 1988.
*Note the considerable diversity in the numbers and types of subunits that compose these protein kinases and the total size of the active multisubunit complexes.

hydrolysis of phosphatidylinositol bisphosphate was briefly described earlier. The best-known pathway of activation of protein kinase C is as follows. In an unstimulated cell much of the protein kinase C is present in the cytosol and is inactive. Ca^{++} binds to protein kinase C when the cytosolic levels of Ca^{++} rise. This causes protein kinase C to bind to the inner surface of the plasma membrane where it can be activated by the diglyceride that is produced by the hydrolysis of phosphatidylinositol bisphosphate. Membrane phosphatidylserine is also a potent activator of protein kinase C once the enzyme has bound to the membrane.

Even though we are ignorant of many of the most important proteins that are phosphorylated by protein kinase C, this enzyme certainly plays a vital role in the control of certain cellular processes. The primary action of certain lipophilic tumor promoting substances, most notably the **phorbol esters,** is to directly activate protein kinase C. Its activation is a powerful stimulus to cell division in many cell types and to the conversion of normal cells with controlled growth properties to transformed cells that resemble tumor cells in their uncontrolled growth.

At present seven different subtypes of protein kinase C have been discovered (Fig. 5-8, Table 5-5). Although some of the subtypes are present in many or most mammalian cells, the γ and ϵ subtypes appear to be confined to certain cells of the central nervous system. Besides being differentially distributed among the cells and tissues of the body, the subtypes of protein kinase C appear to be differentially regulated (see Table 5-5). Some of the subtypes may be bound to the plasma membrane in unstimulated cells, and thus they do not require elevated Ca^{++} for activation. At least two of the subtypes are activated by arachidonic acid, by other unsaturated fatty acids, and by lysophospholipids. Thus receptor activated phospholipase A_2 may play a role in activation of protein kinase C.

Tyrosine Kinases as Modulators of Cellular Processes

The receptors for certain peptide hormones and growth factors are proteins with a glycosylated extracellar domain, a single transmembrane sequence, and an intracellular domain with protein tyrosine kinase activity. Members of this superfamily (Fig. 5-9) of peptide receptors include the receptors for insulin and related growth factors, epidermal growth factor (EGF), platelet-derived growth factor (PDGF), colony-stimulating factor (CSF), and fibroblast growth factor (FGF). The binding of hormone or growth factor to its receptor triggers multiple cellular responses, including Ca^{++} influx, increased Na^+/H^+ exchange, stimulation of the uptake of sugars and amino acids, and stimulation of phospholipase C and hydrolysis of phosphatidylinositol bisphosphate with all the consequences thereof.

The known protein tyrosine kinase receptors fall into four subclasses (see Fig. 5-9), based on their structures. Within a subclass there is high amino acid sequence homology. Binding of ligand to the receptor results in dimerization of the receptor-ligand complexes. The dimerization enhances binding affinity and activates the protein-tyrosine kinase activity. In subclass II receptors, the insulin receptor family, the unliganded receptor exists as a disulfide-linked dimer, and ligand binding causes a conformational change of both "monomers." This change enhances ligand binding and tyrosine kinase activity.

The tyrosine kinase activity of the receptors is required for the signal transduction function of the receptors. Protein-tyrosine kinase receptors phosphorylate certain of their own tyrosine residues. The tyrosines that are phosphorylated may reside in the kinase insert regions of subclasses III and IV or in the carboxyl terminal tail. The autophosphorylation enhances tyrosine kinase activity partly because the unphosphorylated ty-

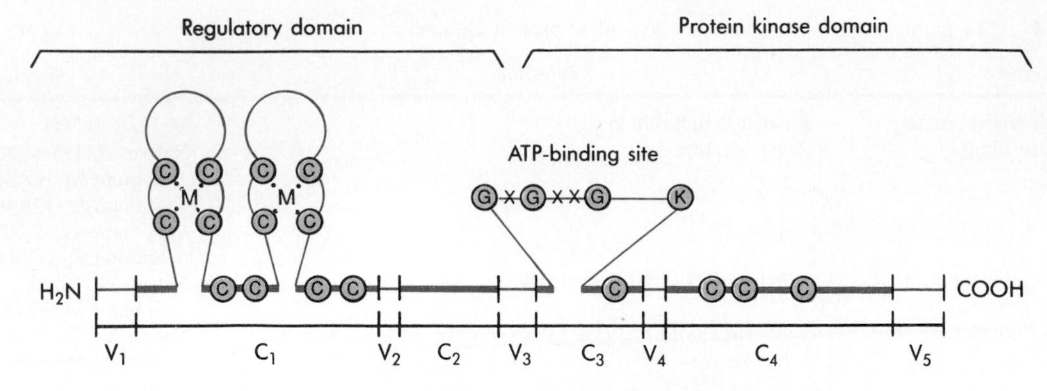

■ **Fig. 5-8** The structure of the members of the protein kinase C family. There are 4 conserved region (denoted by C_1 to C_4) and 5 variable segments (V_1 to V_5). Certain subtypes lack the C_2 domain. The letters C, G, K, X, and M represent cysteine, glycine, lysine, any amino acid, and metal, respectively. The regulatory domain contains two metal-binding zinc-fingerlike domains, but the functional significance of these is not clear. The sites for binding Ca^{++}, diglyceride, and phosphatidylserine remain to be determined. (Modified from Nishizuka Y: *Nature* 334:661, 1988.)

■ **Table 5-5** Properties of the known isozymes of protein kinase C in mammals

Isozyme	α	β_I	β_{II}	γ	δ	ε	ζ
Amino-acid residues	672	671	673	697	673	737	592
Calculated molecular weight	76,799	76,790	76,933	78,366	77,517	83,474	67,740
Chromosome location (human)	17	16		19	?	?	?
Activators*	PS + DG + Ca^{++} AA + Ca^{++}	PS + DG + Ca^{++}		PS + DG + Ca^{++} AA	PS + DG + (Ca^{++})	PS + DG + (Ca^{++})	PS + (DG + Ca^{++})
Tissue expression	Universal	Some tissues and cells	Many tissues and cells	Brain and spinal cord only	Many tissues	Brain only?	Many tissues

Modified from Nishizuka Y: *Nature* 334:661, 1988.
*PS, DG, and AA represent phosphatidylserine, diglyceride, and arachidonic acid, respectively. Activators in parentheses are of questionable physiological significance.

rosine residues of the kinase compete with phosphorylation sites on substrate proteins. By contrast, phosphorylation of the receptor protein on serines or threonines by other kinases, such as protein kinase C, may diminish the tyrosine kinase activity of the receptor.

Protein tyrosine kinases that are out of control play a central role in cell transformation and cancer. In some cell types, mutation of the receptor renders it active in phosphorylating tyrosines, regardless of the presence or absence of a growth factor. Other tumor cells secrete a growth factor *and* overexpress its receptor; these processes lead to abnormally high protein-tyrosine kinase activity.

■ *Protein Phosphatases and Their Modulation*

Phosphorylation of proteins is one the the most significant means of regulating their activities. The extent of phosphorylation of a regulated protein at a given instant is the result of the opposing activities of the protein kinase that phosphorylates that protein and of the protein phosphatase that dephosphorylates it. In addition to the different types of protein kinases discussed earlier, all cells also contain **protein phosphatases** whose task is to reverse the effects of protein phosphorylation. Protein phosphatases are classified, in keeping with the classification of protein kinases, as **serine-threonine protein phosphatases** and **tyrosine protein phosphatases.**

■ *Serine-threonine protein phosphatases*

The serine-threonine protein phosphatases are a large family of structurally related molecules. They are classified as type 1 **(PP-1)** or type 2 **(PP-2),** based on which subunit of phosphorylase kinase they dephosphorylate. PP-1s act on the β-subunit, but PP-2s act on the α-subunit of phosphorylase kinase (Table 5-6). The

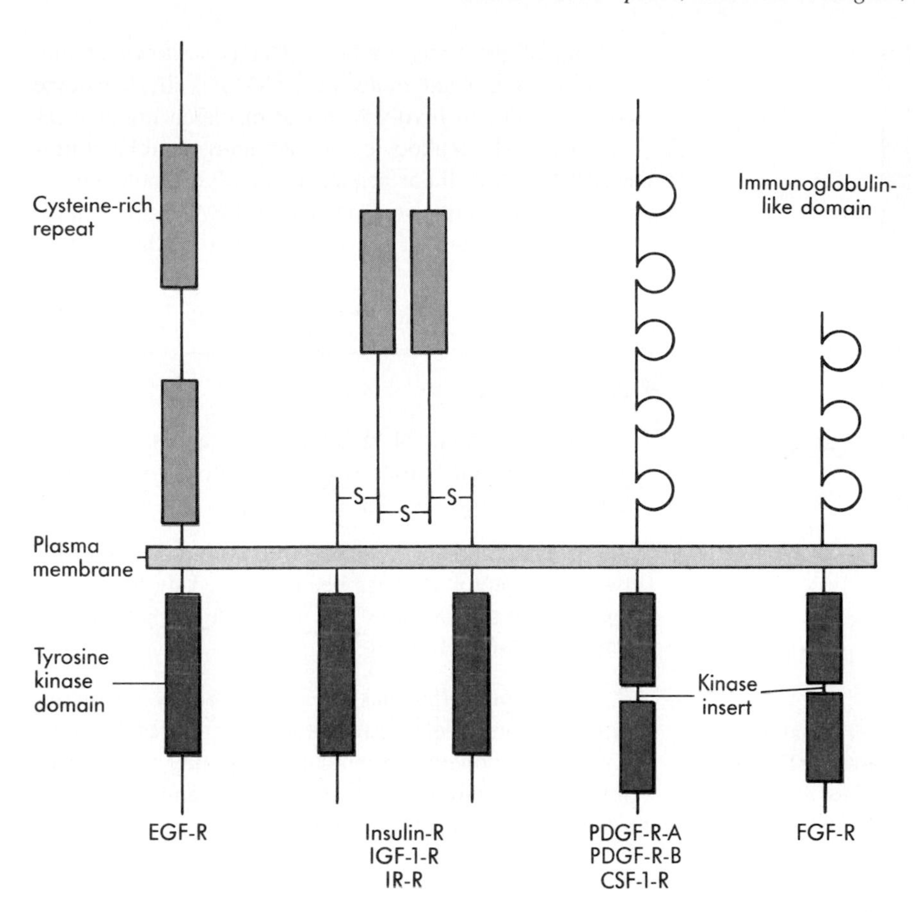

■ **Fig. 5-9** The structures of the different subclasses of receptor protein-tyrosine kinases. Subclasses I and II have extracellular domains with cysteine-rich repeat sequence domains *(light color)*, whereas the extracellular domains of subclasses III and IV have immunoglobulin-like regions (loops). The protein-tyrosine kinase domains *(solid color)* are the most conserved sequences. The short intracellular region just inside the membrane, the kinase insert region (quite variable in length), and the carboxyl terminal tail are sites of regulation of protein kinase activity. *EGF*, epidermal growth factor; *IGF-1*, insulinlike growth factor-1; *IR*, insulin-related protein; *PDGF*, platelet-derived growth factor; *CSF-1*, colony stimulating factor-1; *FGF*, fibroblast growth factor. (Modified from Ullrich A, Schlessinger J: *Cell* 61:203, 1990.)

PP-2s are subclassified into PP-1A, PP-1B, and PP-1C, based on their regulation by divalent cations. PP-2A does not require divalent cations for activity. PP-2B has an absolute requirement for the Ca^{++}-calmodulin complex. PP-2C absolutely requires Mg^{++}. PP-1 and the PP-2s can also be distinguished by their inhibition by **okadaic acid,** a complex fatty acid produced by marine dinoflagellates. Okadaic acid is as potent a tumor promoter as the phorbol esters, presumably because both okadaic acid and phorbol esters enhance the phosphorylation of certain substrates of protein kinase C. PP-1s and PP-2s contain a catalytic core with a significantly homologous amino acid sequence.

The PP-2s are mainly cytosolic enzymes, but PP-1 in muscle and liver cells is inactive in the cytosol. PP-1 in liver is bound to glycogen particles, and in muscle PP-1 is bound to glycogen, to sarcoplasmic reticulum, and to the myofibrils (the contractile proteins). Cytosolic PP-1 is relatively inactive. Protein subunits of PP-1 are responsible for the binding of PP-1 to these specific cellular structures, and this binding appears to direct the activity of PP-1 toward particular physiological substrates.

The activity of PP-1 is also regulated by two classes of **endogenous inhibitor proteins: I-1 and I-2.** I-1 is an effective inhibitor only when it is phosphorylated by protein kinase A. Apparently the phosphorylated I-1 has a high affinity for PP-1 and is able to bind PP-1, thereby removing it from the subunit that binds it to glycogen or other cellular structure and, in this way, inactivating PP-1. The complex regulation of PP-1 sup-

■ **Table 5-6** Properties of subtypes of serine-threonine protein phosphatases*

Subtype	*PP-1*	*PP-2A*	*PP-2B*	*PP-2C*
Preference for the α or β subunit of phosphorylase kinase	β subunit	α subunit	α subunit	α subunit
Inhibition by I-1 and I-2	Yes	No	No	No
Absolute requirement for divalent cations	No	No	Yes (Ca^{++})	Yes (Mg^{++})
Stimulation by calmodulin	No	No	Yes	No
Inhibition by okadaic acid (K_i)	Yes (20 nM)	Yes (0.2 nM)	Yes (5 μM)	No
Phosphorylase phosphatase activity	High	High	Very low	Very low

Modified from Cohen P: *Annu Rev Biochem* 58:453, 1989.

*The phosphatases are classified as type 1 or type 2 based on which subunit of phosphorylase kinase they prefer to dephosphorylate. Subtypes 2A, 2B, and 2C can be distinguished based on the nature of their dependence on divalent cations.

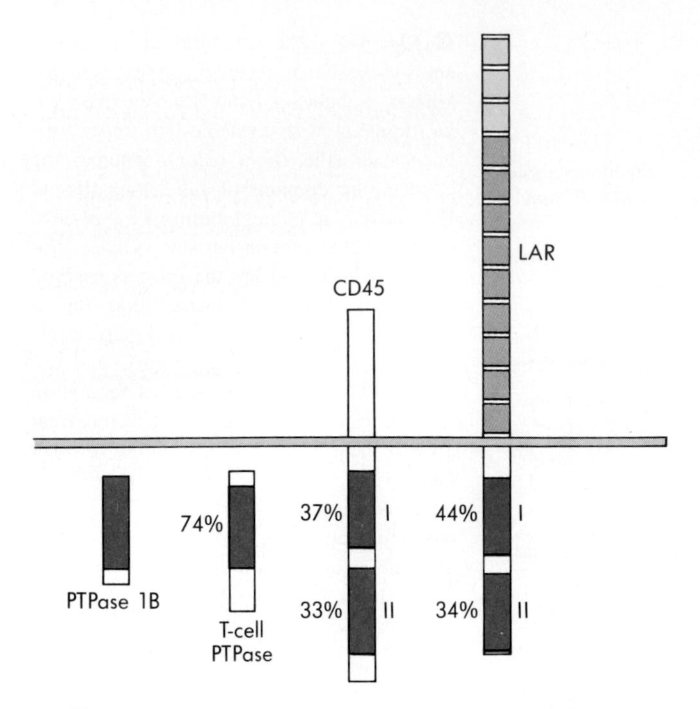

■ **Fig. 5-10** Protein tyrosine phosphatases (PTPases) schematically depicted. Shown are two small cytosolic PTPases: PTPase 1B from human placenta, and T-cell PTPase from human T-lymphocytes. Also shown are two transmembrane PTPases: CD45, the leukocyte common antigen, and LAR (leukocyte common antigen related protein). The solid-colored cytosolic segments of each protein are the PTPase catalytic domains. The extracellular domains of LAR are homologous to N-CAM (neural cell adhesion molecule). The lighter colored domains are homologous to the IgG-like domains, and the intermediate colored domains are homologous to the non-IgG-like domains of N-CAM. The intramembranous segments of CD45 and LAR have the seven-transmembrane helix motif that characterizes receptor proteins. (Modified from Tonks NK, Charbonneau H: *Trends in Biochem Sci* 14:497, 1989.)

ports the view that it plays a key role in cellular regulatory processes.

■ *Protein tyrosine phosphatases*

Protein tyrosine phosphatases (PTPases) have been less extensively studied than the serine-threonine protein phosphatases. PTPases are not structurally homologous to serine-threonine protein phosphatases. Recall that, by contrast, all the protein kinases appear to derive from a common ancestor protein kinase. Fig. 5-10 depicts four of the PTPases that are currently known. Note that whereas two of the PTPases are small cytosolic proteins, two other PTPases are larger transmembrane proteins whose intramembrane segments appear to have the seven transmembrane helices that are characteristic of receptor proteins (see Fig. 5-4). This structure suggests that the transmembrane PTPases are receptors whose PTPase activity will be modulated by extracellular ligands that have yet to be identified.

Both of the transmembrane PTPases shown in Fig. 5-10 are significant molecules. CD45 is the leukocyte common antigen involved in cell-mediated immune responses. LAR (leukocyte common antigen–related protein) has extracellular sequences highly homologous to the neural cell adhesion molecule (N-CAM). This molecule is important in the development of the nervous system.

■ *Summary*

1. A large number of regulatory substances exert their effects on cellular processes via a relatively small number of signal transduction processes.

2. Heterotrimeric G proteins serve as intermediaries between a receptor that has been activated by binding an agonist and the enzymes and ion channels that are activated by agonist binding.

3. A G protein that has been activated by interacting with an agonist-bearing receptor then changes the activity of an enzyme or an ion channel to alter the intracellular concentration of a second messenger, such as cAMP, cGMP, Ca^{++}, or diacylglycerol.

4. An increased level of one or more second messengers increases the activity of a second messenger-dependent protein kinase, such as cAMP-dependent protein kinase, cGMP-dependent protein kinase, calmodulin-dependent protein kinase, or protein kinase C.

5. Certain membrane receptors for hormones and growth factors are protein tyrosine kinases that are activated directly by binding of the agonist.

6. Myriad cellular processes are regulated by the phosphorylation of enzymes and ion channels.

7. Protein phosphatases that are subject to complex regulation by agonists and second messengers reverse the effects of protein phosphorylation.

■ *Bibliography*

Journal articles

Birnbaumer L, Brown AM: G proteins and the mechanism of action of hormones, neurotransmitters, and autocrine and paracrine regulatory factors, *Am Rev Respir Dis* 141:S106, 1990.

Blackshear PJ, Nairn AC, Kuo JF: Protein kinases 1988: a current perspective, *FASEB J:* 2:2957, 1988.

Bourne HR, Sanders DA, McCormick F: The GTPase superfamily: a conserved switch for diverse cell functions, *Nature* 348:125, 1990.

Brown AM, Birnbaumer L: Ionic channels and their regulation by G protein subunits, *Annu Rev Physiol* 52:197, 1990.

Cohen P: The structure and regulation of protein phosphatases, *Annu Rev Biochem* 58:453, 1989.

Freissmuth M, Casey PJ, Gilman AG: G proteins control diverse pathways of transmembrane signaling, *FASEB J* 3:2125, 1989.

Hall A: The cellular functions of small GTP-binding proteins, *Science* 249:635, 1990.

Hunter T: A thousand and one protein kinases, *Cell* 50:823, 1987.

Nathanson NM: Molecular properties of the muscarinic acetylcholine receptor, *Annu Rev Neurosci* 10:195, 1987.

Nishizuka Y: The molecular heterogeneity of protein kinase C and its implications for cellular regulation, *Nature* 334:661, 1988.

O'Dowd BF, Lefkowitz RJ, Caron MG: Structure of the adrenergic and related receptors, *Annu Rev Neurosci* 12:67, 1989.

Ross EM: Signal sorting and amplification through G protein-coupled receptors, *Neuron* 3:141, 1989.

Simon MI, Strathmann MP, Gautam N: Diversity of G proteins in signal transduction, *Science* 252:802, 1991.

Szabo G, Otero AS: G protein-mediated regulation of K^+ channels in heart, *Annu Rev Physiol* 52:293, 1990.

Taylor SS: cAMP-dependent protein kinase: model for an enzyme family, *J Biol Chem* 264:8443, 1989.

Tonks NK, Charbonneau H: Protein tyrosine dephosphorylation and signal transduction, *Trends Biochem Sci* 14:497, 1989.

Ullrich A, Schlessinger J: Signal transduction by receptors with tyrosine kinase activity, *Cell* 61:203, 1990.

Books and monographs

Boyer PD, Krebs EG, editors: Control by phosphorylation, parts A and B. In *The enzymes, vols 17* and *18,* New York, 1986 and 1987, Academic Press.

Cohen P, Klee C, editors: Calmodulin. In *Molecular aspects of cellular regulation, vol 5,* New York, 1988, Elsevier.

Conn P, editor: *The receptors,* 4 vols, New York, 1984, Academic Press.

Cheung WY, editor: *Calcium and cell function, vol* 4, New York, 1984, Academic Press.

Evered D, Nugent J, Whelan J, editors: Growth factors in biology and medicine, Ciba Foundation Symposium No 116, London, 1985, Pitman.

Houslay MD, Milligan G, editors: *G proteins as mediators of cellular signalling processes,* New York, 1990, John Wiley & Sons.

Merlevede W, DiSalvo J, editors: *Advances in protein phosphatases, vol 5,* Leuven, 1989, University of Leuven Press.

Michell RH, Drummond AH, Downes CP, editors: *Inositol lipids in cell signaling,* New York, 1989, Academic Press.

SECTION
II

THE NERVOUS SYSTEM

William D. Willis, Jr.

CHAPTER

6

The Nervous System and Its Components

The nervous system is a communications network that allows an organism to interact in appropriate ways with its environment. In a broad sense the environment includes both the external environment (the world outside the body) and the internal environment (the contents of the body). The nervous system includes sensory components that detect environmental events, integrative components that process and store sensory and other data, and motor components that generate movements and glandular secretions. The nervous system can be divided into peripheral and central parts, each with further subdivisions.

■ *Organization of the Nervous System*

The nervous system consists of a highly complex aggregation of cells, part of which forms a communications network and part of which forms a supportive matrix. The communications network is composed of neural circuits. **Neurons** are cells specialized for receiving and making decisions on information and for transmitting signals to other neurons or to effector cells. The supportive cells are the **neuroglia** (meaning **"nerve glue"**). These cells help maintain an appropriate local environment for neurons, or they ensheath axons to increase the speed of propagation of action potentials. We will discuss these microscopic components of the nervous system, but first let us present an overview of the structures in which these cells are found.

■ *The Peripheral Nervous System*

The peripheral nervous system (PNS) provides an interface between the central nervous system and the environment. It includes sensory components formed by sensory receptors and primary afferent neurons and motor components formed by somatic and autonomic motor neurons.

Sensory receptors serve as transducers, sensing interactions of various forms of environmental energy with the body. Information obtained by the sensory receptors is then transmitted to the central nervous system by **primary afferent neurons** by way of **dorsal roots** or **cranial nerves.** These cells have their somas in **dorsal root ganglia** or **cranial nerve ganglia.** *A ganglion in the peripheral nervous system is an aggregation of neuronal cell bodies with a similar function.*

Somatic motor neurons have their cell bodies in the spinal cord or brainstem, and they innervate skeletal muscle fibers. They have long dendrites (see p. 99) and receive many synaptic connections. Motor neurons that supply a given muscle are located in a motor nucleus. *A* ***nucleus*** *is a collection of neurons in the central nervous system (CNS) with a similar function* (as distinguished from the nucleus found in an individual cell). The axons of somatic motor neurons leave the central nervous system either through a ventral root in the spinal cord or a cranial nerve.

Autonomic motor neurons include both **preganglionic** and **postganglionic neurons** of the **sympathetic** and **parasympathetic** nervous systems (see Chapter 15). This terminology reflects the fact that the autonomic outflow passes through the **autonomic ganglia.** Preganglionic neurons are located in the central nervous system either in the spinal cord or in the brainstem. The cell bodies of postganglionic neurons are in autonomic ganglia. Preganglionic neurons synapse on postganglionic neurons, and postganglionic neurons synapse on autonomic **effectors** (cardiac muscle, smooth muscle, or glands).

The neural components of the PNS are described in more detail in Chapter 7.

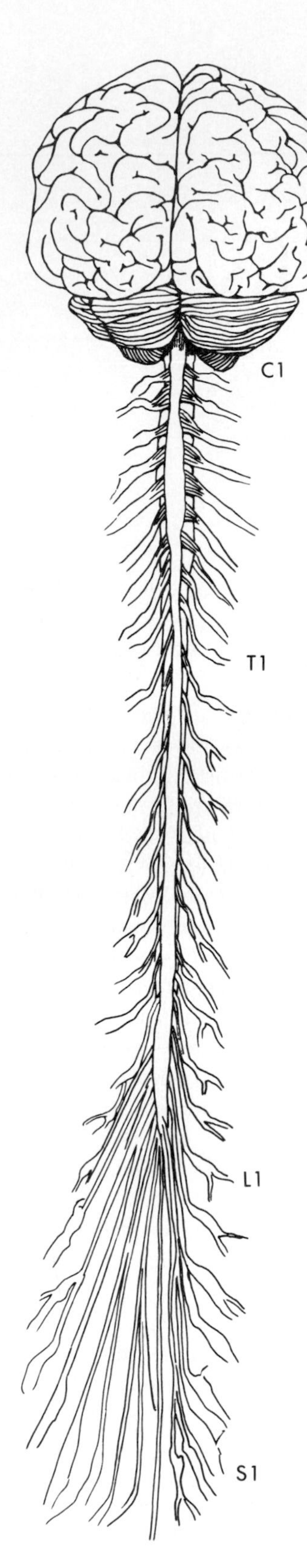

■ **Fig. 6-1** Brain and spinal cord with attached spinal nerves. Note the relative size of various components. *C1, T1, L1,* and *S1,* First cervical, thoracic, lumbar, and sacral segments, respectively. (Redrawn from Williams PL, Warwick R: *Functional neuroanatomy of man,* Edinburgh, 1975, Churchill Livingstone.)

■ *The Central Nervous System*

The central nervous system (CNS), among other functions, gathers information about the environment from the PNS, processes this information and perceives part of it, organizes reflex and other behavioral responses, is responsible for cognition, learning, and memory, and plans and executes voluntary movements. The CNS includes the **spinal cord** and the **brain** (Fig. 6-1). The spinal cord can be subdivided into a series of regions, each composed of a number of segments (in humans, 8 cervical, 12 thoracic, 5 lumbar, 5 sacral, and 1 coccygeal). The brain can be subdivided into five regions, based on embryologic development: the **myelencephalon,** the **metencephalon,** the **mesencephalon,** the **diencephalon,** and the **telencephalon** (Table 6-1). In the adult brain the myelencephalon includes the **medulla oblongata** (or medulla); the metencephalon includes the **pons** and **cerebellum;** the mesencephalon is the **midbrain;** the diencephalon includes the **thalamus** and **hypothalamus;** and the telencephalon includes the **basal ganglia** and the **cerebral cortex** (Figs. 6-2 and 6-3). The cerebral cortex is further divided into lobes (named after the overlying bones of the skull): **frontal, parietal, temporal,** and **occipital.** The **cerebral hemispheres** on the two sides are connected across the midline by a massive bundle of axons, the **corpus callosum** ("hard body").

Some of the functions of different parts of the CNS are given in Table 6-1.

■ *Environment of the Neuron*

The local environment of most neurons is controlled so that neurons are normally protected from extreme variations in the composition of the extracellular fluid that bathes them. This control is provided by regulation of the CNS circulation (see Chapter 31), the presence of a blood-brain barrier, the buffering function of neuroglia, and the exchange of substances between the CSF and the extracellular fluid of the CNS.

The cranial cavity contains the brain, blood, and CSF (Fig. 6-4). The human brain weighs about 1350 g; approximately 15%, or 200 ml, is extracellular fluid. The intracranial blood volume is about 100 ml as is the cranial volume of CSF. Thus the extracellular fluid space in the cranial cavity totals approximately 400 ml.

■ *The Blood-Brain Barrier*

The movement of large molecules and highly charged ions from the blood into the brain and spinal cord is severely restricted (see Fig. 6-4). The restriction is at least partly caused by the barrier action of tight junctions between the capillary endothelial cells of the

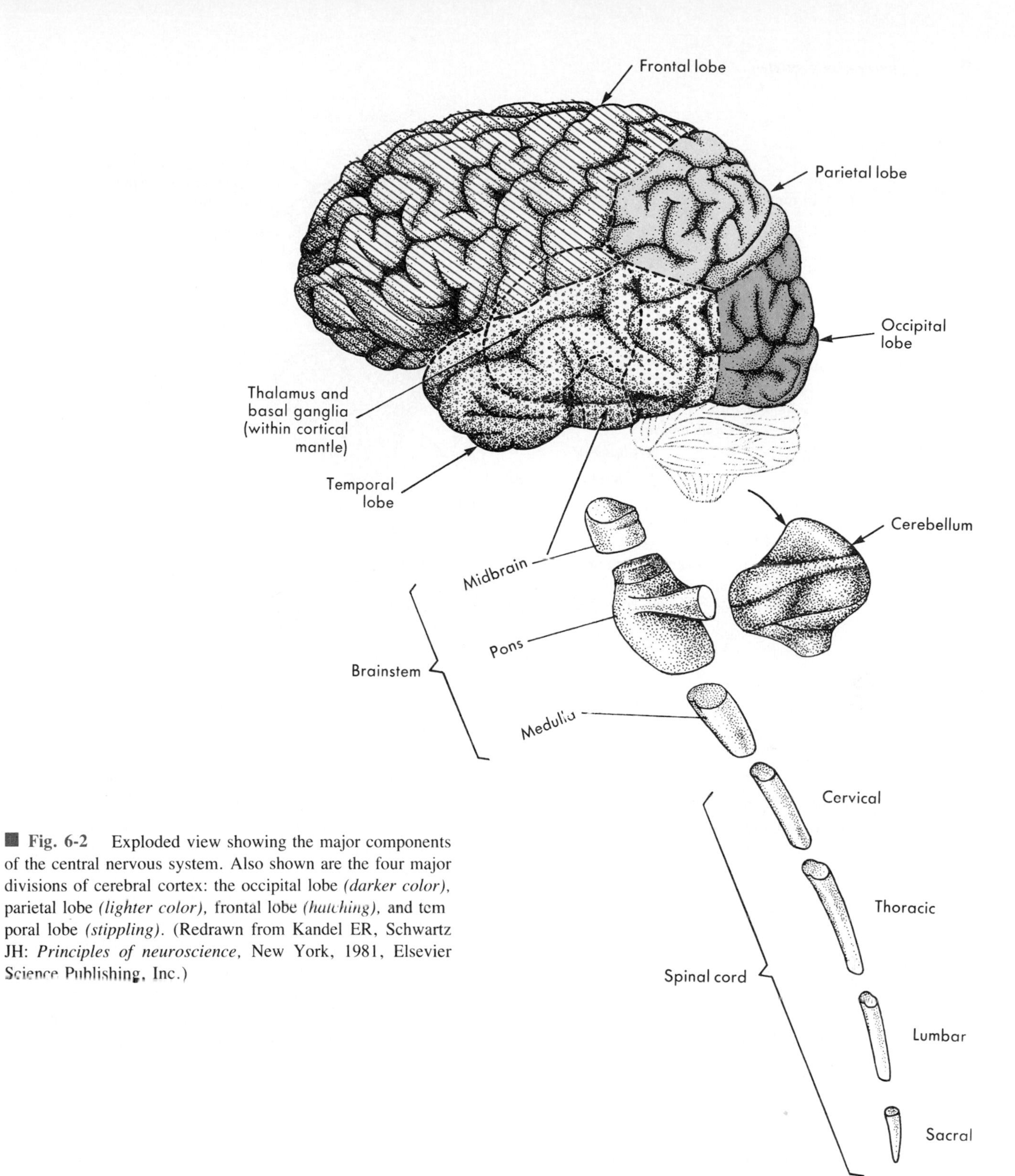

■ **Fig. 6-2** Exploded view showing the major components of the central nervous system. Also shown are the four major divisions of cerebral cortex: the occipital lobe *(darker color)*, parietal lobe *(lighter color)*, frontal lobe *(hatching)*, and temporal lobe *(stippling)*. (Redrawn from Kandel ER, Schwartz JH: *Principles of neuroscience*, New York, 1981, Elsevier Science Publishing, Inc.)

■ **Table 6-1** Parts of the CNS and functions

Region	*Subdivision*	*Function*
Spinal cord		Sensory input; reflex organization; somatic and autonomic motor output
Myelencephalon	Medulla	Cardiovascular control; respiratory control; brainstem reflexes
Metencephalon	Pons	Respiratory and urinary bladder control; vestibular control of eye movements
	Cerebellum	Motor control; motor learning
Mesencephalon	Midbrain	Acoustic relay; control of eye movements; motor control
Diencephalon	Thalamus	Sensory and motor relay to cerebral cortex
	Hypothalamus	Autonomic and endocrine control
Telencephalon	Basal ganglia	Motor control
	Cerebral cortex	Sensory perception; cognition; learning and memory; motor planning and voluntary movement

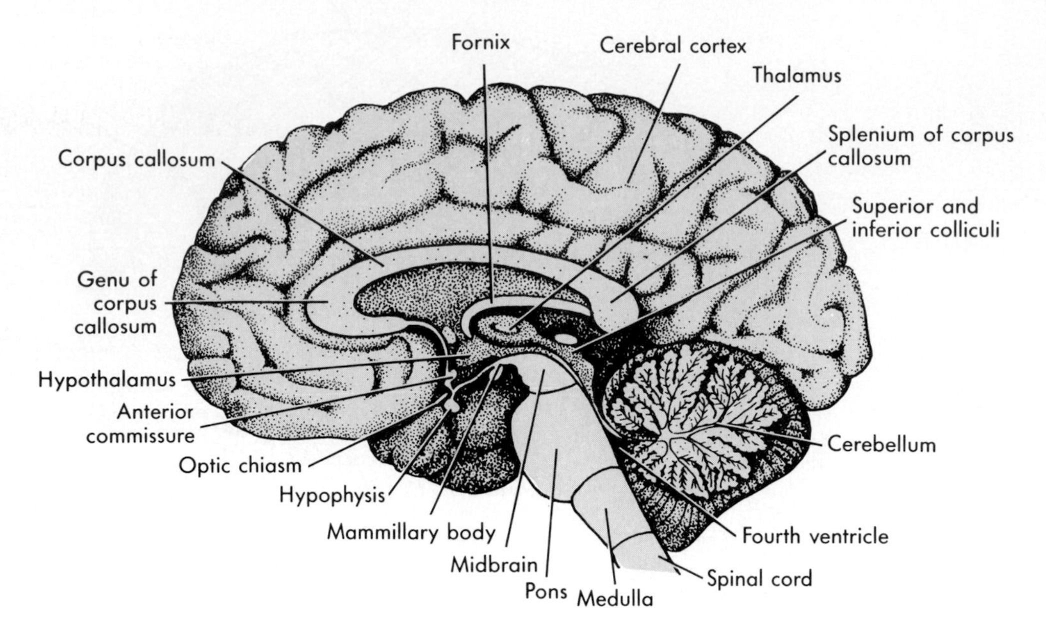

■ **Fig. 6-3** Midsagittal section of the brain. Note the relationships among the cerebral cortex, cerebellum, thalamus, and brainstem plus the location of various commissures. (Redrawn from Kandel ER, Schwartz JH: *Principles of neuroscience,* New York, 1981, Elsevier North-Holland.)

CNS. Neuroglia called astrocytes (see p. 100) may also help limit the movement of certain substances. For example, astrocytes can take up potassium ions and thus regulate the K^+ concentration in the extracellular space. Some substances such as penicillin are removed from the CNS by transport mechanisms.

■ *Cerebrospinal Fluid*

Within the substance of the brain (and spinal cord) is the **ventricular system,** a series of spaces filled with **cerebrospinal fluid** (Fig. 6-5). CSF cushions the brain and regulates the extracellular environment of neurons. The CSF is formed largely by the **choroid plexuses,** which are covered by specialized ependymal cells. The choroid plexuses are located in the lateral, third, and fourth ventricles. The **lateral ventricles** are found within the two cerebral hemispheres. These connect with the **third ventricle** through the **interventricular foramena** (of Monro). The third ventricle lies in the midline between the diencephalon on the two sides. The **cerebral aqueduct** (of Sylvius) traverses the midbrain and connects the third ventricle with the **fourth ventricle.** The fourth ventricle is interposed between the pons and medulla below and the cerebellum above. The **central canal** of the spinal cord continues caudally from the fourth ventricle (in adult humans it is generally not patent).

The CSF escapes from the ventricular system into the **subarachnoid space** through three apertures in the roof of the fourth ventricle: the **medial aperture** (of Magendie) and the two **lateral apertures** (of Luschka). After leaving the ventricular system, the CSF circulates through the **subarachnoid space** that surrounds the brain and spinal cord. Regions where these spaces are distended are called **subarachnoid cisterns.** An example is the **lumbar cistern,** which surrounds the lumbar and sacral spinal roots below the level of termination of the spinal cord. *The lumbar cistern is the target for lumbar puncture,* a procedure used clinically to sample the CSF. A large part of the CSF is removed by bulk flow through the valvular **arachnoid villi** into the dural venous sinuses in the cranium.

The volume of the CSF within the cerebral ventricles is approximately 35 ml and that in the subarachnoid spaces is about 100 ml. About 0.35 ml of CSF is produced each minute. This rate allows the CSF to be turned over approximately 4 times daily.

The pressure in the CSF column is about 120 to 180 mm H_2O when a person is recumbent. The rate at which CSF is formed is relatively independent of the pressure in the ventricles and subarachnoid space, as well as of the systemic blood pressure. However, the absorption rate of CSF is a direct function of CSF pressure.

The extracellular fluid within the CNS communicates directly with the CSF. Thus the composition of the CSF is an indication of the composition of the extracellular environment of neurons in the brain and spinal cord. The main constituents of CSF in the lumbar cistern are listed in Table 6-2. For comparison the concentrations of the same constituents in the blood are also given. The CSF has a lower concentration of K^+, glucose, and

■ **Fig. 6-4** The structural and functional relationships involved in the blood-brain and blood-CSF barriers. Substances entering the neurons and glial cells (i.e., intracellular compartment) must pass through the cell membrane. Arrows indicate direction of fluid flow under normal conditions.

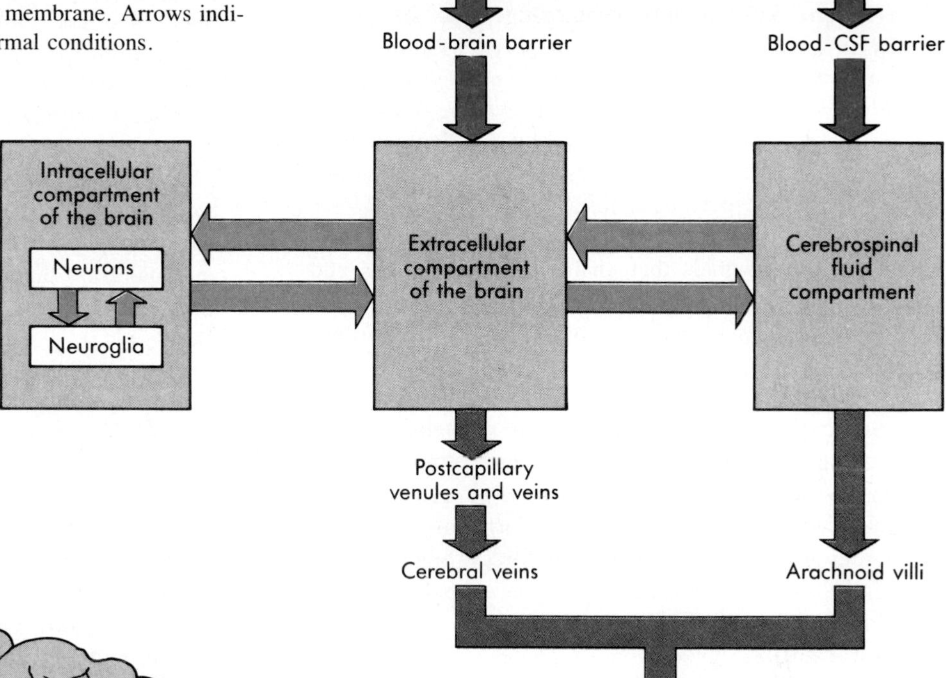

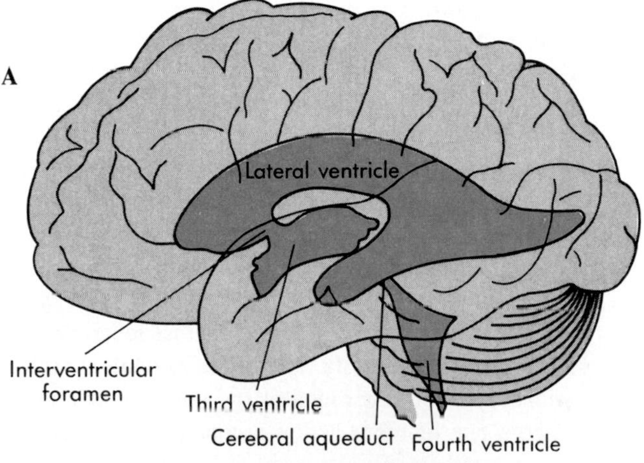

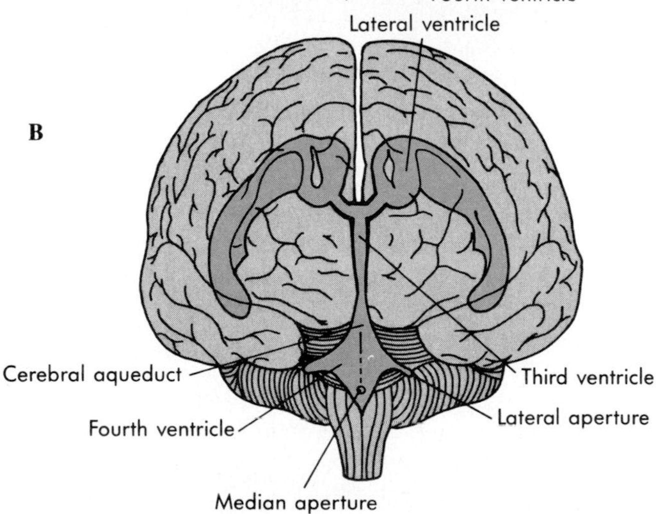

■ **Fig. 6-5** The ventricular system in situ as seen from the side **A** and from the front **B.**

protein, but a greater concentration of Na^+ and Cl^- than does blood. Furthermore CSF contains practically no blood cells. The increased concentration of Na^+ and Cl^- enables the CSF to be isotonic to blood, despite the much lower concentration of protein in the CSF.

Obstruction of the circulation of CSF leads to increased CSF pressure and **hydrocephalus,** an abnormal accumulation of fluid in the cranium. In hydrocephalus the ventricles become distended, and, if the pressure increase is sustained, brain substance is lost. When the obstruction is within the ventricular system or in the roof of the fourth ventricle, the condition is called **noncommunicating hydrocephalus.** If the obstruction is in the subarachnoid space or arachnoid villi, it is known as **communicating hydrocephalus.**

■ **Table 6-2** Constituents of CSF and blood

Constituent	*Lumbar CSF*	*Blood*
Na^+ (mEq/l)	148	136-145
K^+ (mEq/l)	2.9	3.5-5
Cl^- (mEq/l)	120-130	100-106
Glucose (mg/dl)	50-75	70-100
Protein (mg/dl)	15-45	$6\text{-}8 \times 10^3$
pH	7.3	7.4

From Willis WD, Grossman RG: *Medical neurobiology,* ed 3, St Louis, 1981, CV Mosby.

General Functions of the Nervous System

The functions of the nervous system include **sensory detection, information processing,** and **behavior.** *Learning and memory are special forms of information processing that permit behavior to change appropriately in response to environmental challenges based on past experience.* Other systems, such as the endocrine and immune systems, share these functions, but the nervous system is specialized for them.

Excitability is a cellular property of neurons involving electrical signals that enable them to receive and transmit information. Excitability is manifested by such electrical events as **action potentials, receptor potentials,** and **synaptic potentials** (see Section I). Chemical events often accompany these electrical ones.

Sensory detection is the process whereby neurons **transduce** environmental energy into neural signals. Sensory detection is accomplished by special neurons called **sensory receptors.** Various forms of energy can be sensed, including mechanical forces, light, sound, chemicals, temperature, and in some animals electrical fields.

Information processing includes, among other events, the following:

1. The transmission of information in neural networks
2. The transformation of signals by combining them with other signals **(neural integration)**
3. Storage of information in and retrieval of information from memory
4. Use of sensory information for **perception**
5. Thought processes
6. Learning
7. Planning and implementation of motor commands
8. Emotions

Information processing, including learning and memory, depends on **intercellular communication** in neural circuits. The mechanisms involve both electrical and chemical events.

Behavior consists of the totality of the organism's responses to its environment. Behavior may be covert, as in **cognition,** but it is often readily observable as a motor act, such as a **movement** or an **autonomic response.** In humans a particularly important set of behaviors are those involved in **language.**

How are these highly complex functions carried out? Whether simple or complex, each response is communicated by neurons organized into neural pathways. The remainder of this chapter is devoted to the cellular mechanisms that allow neurons to interact (see Section I).

Cellular Components of the Nervous System

For its communicative activities the functional unit of the nervous system is the **neuron** (Fig. 6-6). The typical neuron has a receptive surface consisting of a **cell body,** or **soma,** and several branchlike **dendrites** that receive **synapses** or neuron-to-neuron connections. Its **axon** makes synaptic connections with other neurons or with effector cells. The nervous system forms a communications network through **neural circuits** made up of synaptically interconnected neurons. Neural activity is generally coded by sequences of **action potentials** propagated along the axons in the neural circuits (see Chapter 3). The coded information is passed from one neuron to the next by **synaptic transmission.** In synaptic transmission the action potentials that reach the **presynaptic ending** usually release a **neurotransmitter** substance, which either **excites** the **postsynaptic cell** to discharge one or more action potentials or **inhibits** the

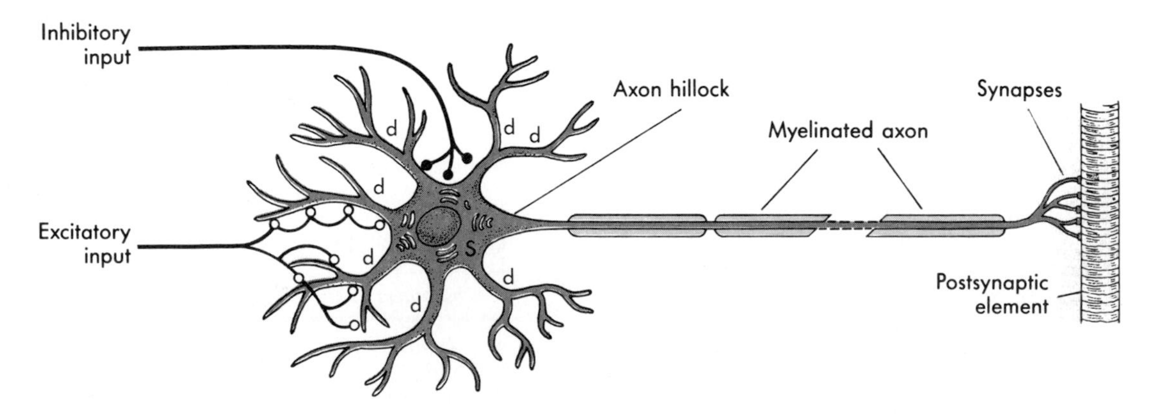

Fig. 6-6 Schematic diagram of an idealized neuron and its major components. Most afferent input from axons of other cells terminates in synapses on the dendrites *(d),* although some may terminate on the soma *(S).* Excitatory terminals tend to terminate more distally on dendrites than do inhibitory ones, which often terminate on the soma. (Redrawn from Williams PL, Warwick R: *Functional neuroanatomy of man,* Edinburgh, 1975, Churchill Livingstone.)

activity of the postsynaptic cell. Axons not only transmit information in neural circuits but also convey chemical substances toward the synaptic terminals by **axonal transport** (see p. 103).

The other cellular elements of the nervous system are the **neuroglia** (Fig. 6-7), or supportive cells. Neuroglial cells in the human CNS outnumber neurons by an order of magnitude: about 10^{13} neuroglia and 10^{12} neurons. Neuroglia do not participate directly in the short-term communication of information through the nervous system, but they do provide important assistance in that function. For example, some types of neuroglial cells provide many axons with **myelin sheaths** that speed up the conduction of action potentials along axons (see p.100). This increase in conduction velocity allows some axons to communicate rapidly with other cells over relatively long distances.

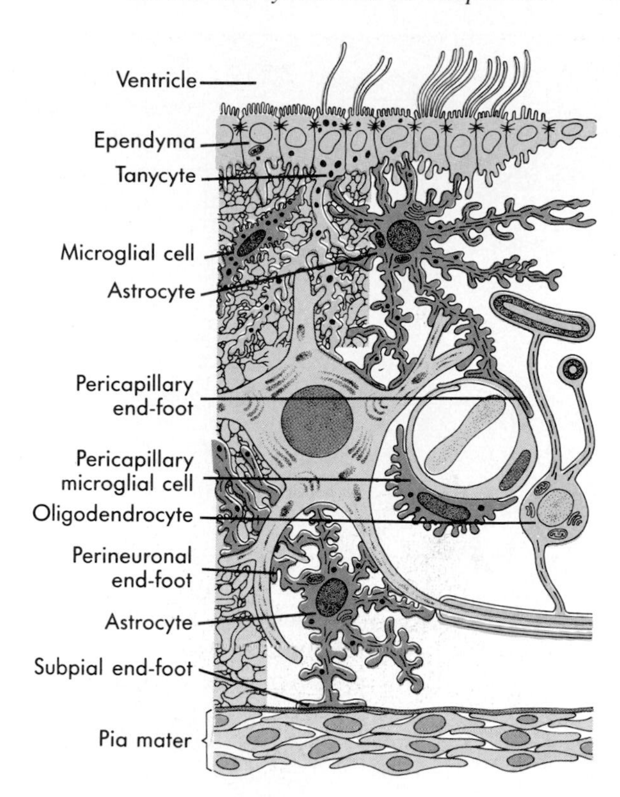

Fig. 6-7 Schematic representation of nonneural elements in the central nervous system. Two astrocytes *(darker color)* are shown ending on a neuron's soma and dendrites. They also contact the pial surface and/or capillaries. An oligodendrocyte *(lighter color)* provides the myelin sheaths for axons. Also shown are microglia *(darker color)* and ependymal cells *(lighter color)*. (Redrawn from Williams PL, Warwick R: *Functional neuroanatomy of man,* Edinburgh, 1975, Churchill Livingstone.)

The Soma

The soma contains the **nucleus** and **nucleolus** of the neuron (Fig. 6-8). It also possesses a well-developed biosynthetic apparatus for manufacturing membrane constituents, synthetic enzymes, and other chemical substances needed for the specialized functions of nerve cells. The neuronal biosynthetic apparatus includes **Nissl bodies,** which are stacks of rough endoplasmic reticulum, and a prominent **Golgi apparatus.** The soma also contains numerous **mitochondria** and cytoskeletal elements, including **neurofilaments** and **microtubules. Lipofuscin** is a pigment formed from incompletely degraded membrane components; it accumulates with aging in some neurons. A few groups of neurons in the brainstem (e.g., in the **substantia nigra** and the **locus coeruleus**) contain **melanin** pigment.

Dendrites

Dendrites are extensions of the cell body. In some neurons the dendrites are more than 1 mm long, and they account for more than 90% of the surface area. The proximal dendrites (near the cell body) contain Nissl bodies and parts of the Golgi apparatus. However the main cytoplasmic organelles in dendrites are microtubules and neurofilaments. Traditionally dendrites have not been regarded as electrically excitable. However, we now know that the dendrites of many neurons have voltage-dependent conductances. These often depend on calcium channels that, when activated, produce calcium spikes.

The Axon

The axon arises from the soma (or sometimes from a dendrite) in a specialized region called the **axon hillock.** The axon hillock and axon differ from the soma and proximal dendrites in that they lack rough endoplasmic reticulum, free ribosomes, and Golgi apparatus. The axon contains smooth endoplasmic reticulum and a prominent cytoskeleton. Axons may be short and, as with dendrites, terminate near the soma **(Golgi type 1 neurons);** or they may be long **(Golgi type 2 neurons)** and extend for centimeters or even more than a meter.

Types of Neurons and Neuroglia

Nervous tissue consists largely of neurons and neuroglia (as well as vascular and some connective tissue elements). Neurons are responsible for communications. Many different kinds of neurons exist, as reflected both in their specific functions and in their morphology (Fig. 6-9). For example, the **dorsal root ganglion cells** receive information from sensory endings in receptor organs rather than by synaptic transmission. Hence they have a cell body that lacks dendrites (Fig. 6-9, *E*) and

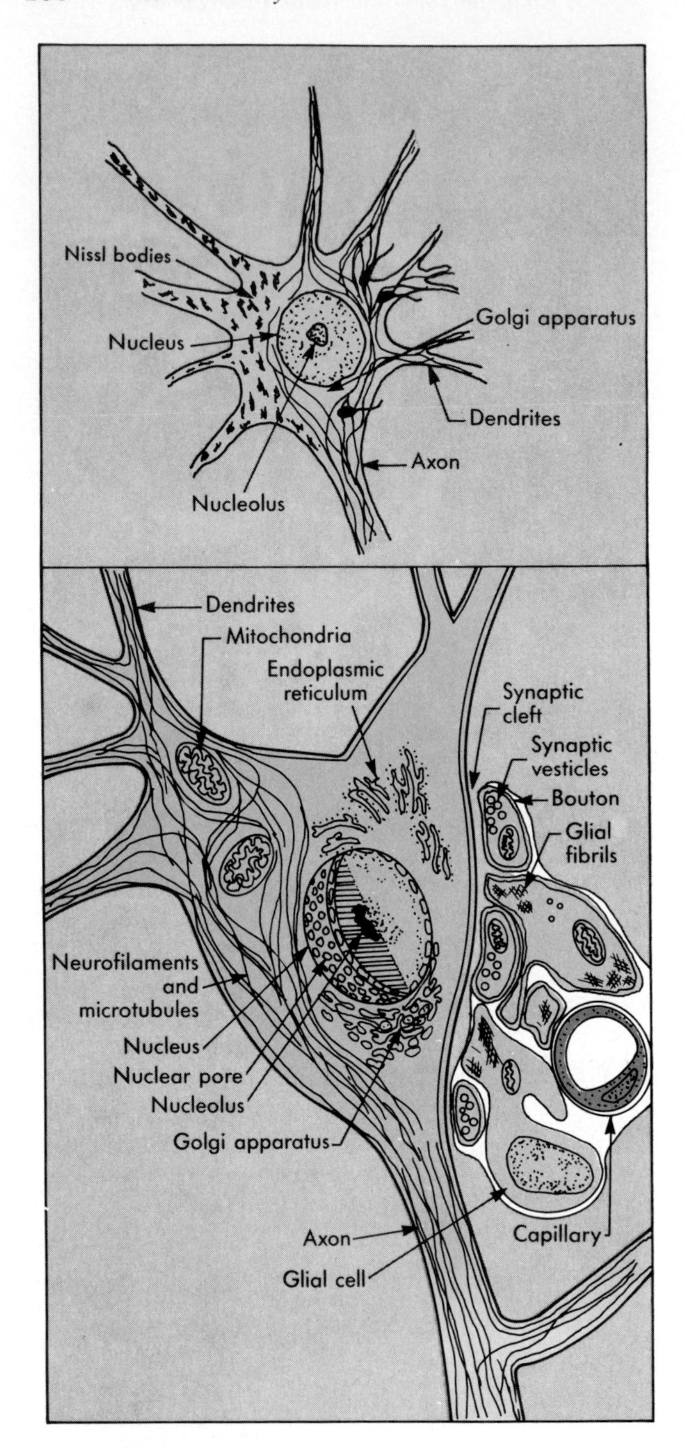

■ **Fig. 6-8** Organelles of the neuron. The smaller drawing on top shows the organelles typical of a neuron, as seen with the light microscope. The portion of the illustration to the left of the soma represents structures seen with a Nissl stain. These include the nucleus and nucleolus, Nissl bodies in the cytoplasm of the cell body and proximal dendrites, and as a negative image, the Golgi apparatus. The absence of Nissl bodies in the axon hillock and axon is also shown. To the right of the soma are structures seen with a heavy-metal stain: these include neurofibrils. The appropriate heavy-metal stain may demonstrate the Golgi apparatus (not shown). On the surface of the neuron several synaptic endings are indicated, as stained by the heavy metal. The large drawing shows structures visible at the electron microscopic level. The nucleus, nucleolus, chromatin, and nuclear pores are represented. Mitochondria, rough endoplasmic reticulum, Golgi apparatus, neurofilaments, and microtubules are in the cytoplasm. Along the surface membrane are such associated structures as synaptic endings and astrocytic processes.

receives no synaptic endings. The axon branches near the cell body and one branch (the **peripheral process**) traverses a peripheral nerve to supply a sensory receptor; the other branch (the **central process**) reaches the spinal cord through a **dorsal root** or the brainstem through a **cranial nerve.**

Other neurons are engaged in information processing, such as the **pyramidal cells** of the cerebral cortex and the **Purkinje cells** of the cerebellar cortex (Fig. 6-9, *A* and *B*). These neurons have greatly expanded dendritic surfaces covered with dendritic spines, and they receive enormous numbers of synaptic endings.

Neuroglial cells provide support for the activity of neurons (Fig. 6-10). These include **astrocytes** and **oligodendroglia** in the CNS and **Schwann cells** and **satellite cells** in the PNS. **Microglia** and **ependymal cells** can also be considered central neuroglial cells.

Astrocytes (so named because they are star-shaped) regulate the microenvironment of neurons in the CNS although they contact only a part of the surfaces of central neurons (see Fig. 6-7). However, their processes surround groups of synaptic endings and isolate these from adjacent synapses. Astrocytes have **foot processes** that contact capillaries and the connective tissue at the surface of the CNS, the **pia mater** (see Fig. 6-7). These foot processes may help limit the free diffusion of substances into the CNS. Astrocytes can actively take up K^+ ions and neurotransmitter substances, which they can metabolize. Thus they serve to buffer the extracellular environment of neurons with respect both to ions and neurotransmitters. The cytoplasm of astrocytes contains glial filaments, which help provide mechanical support for CNS tissue. After injury, astrocytic processes containing these glial filaments hypertrophy, forming a glial "scar."

Axons may be ensheathed or bare. Many axons are surrounded by a spiral multilayered wrapping of glial cell membrane called a **myelin sheath;** other **unmyelinated axons** lack a myelin sheath but may nevertheless be surrounded by a neuroglial cell. In the CNS myelinated axons are ensheathed by the membranes of **oligodendroglia** (Fig. 6-11, *A*), and unmyelinated axons are bare. In the PNS **Schwann cells** spirally wrap the axons (Fig. 6-11, *B*). In the PNS unmyelinated axons are embedded in Schwann cells although they are not ensheathed by myelin (Fig. 6-12). Myelin increases the speed of conduction of the action potential, in part by limiting the flow of ionic current during action poten-

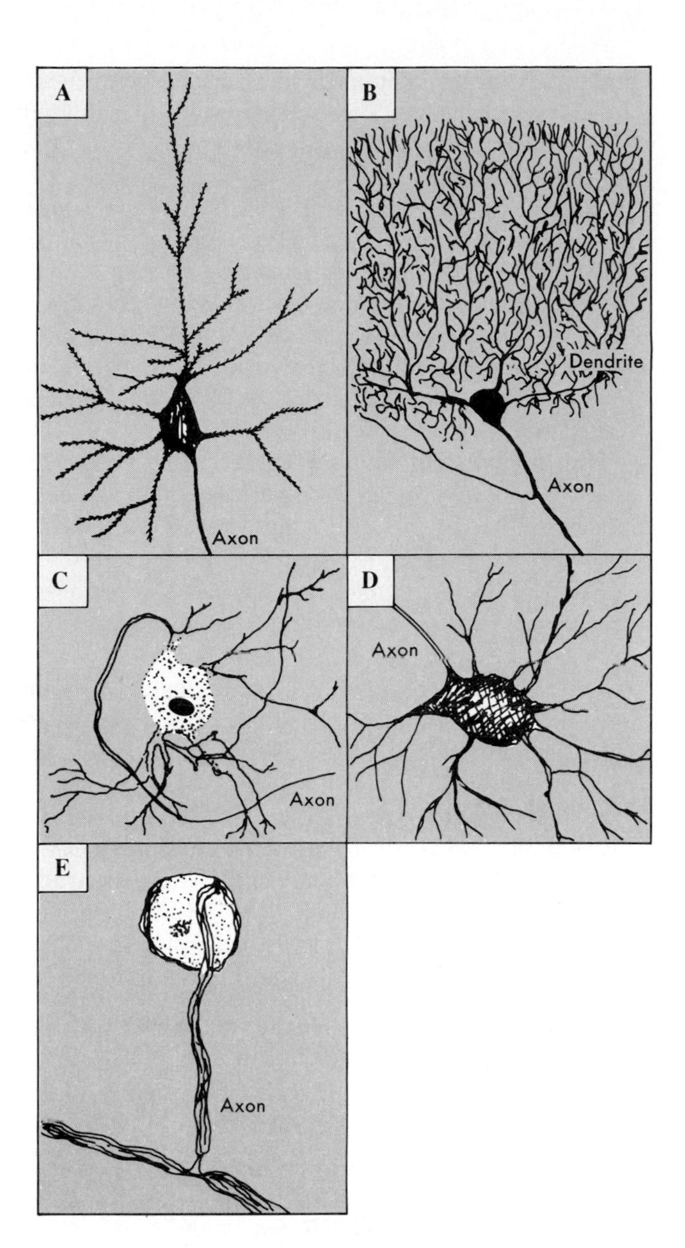

■ **Fig. 6-9** Various forms of neurons. **A,** Neuron characterized by a cell body that has a roughly pyramidal shape. This type of neuron, called a pyramidal cell, is typical of the cerebral cortex. Note the many spinous processes lining the surface of the dendrites. **B,** Cell type first described by the Czechoslovakian neuroanatomist Purkinje and since known as the Purkinje cell. Purkinje cells are characteristic of the cerebellar cortex. The cell body is pear shaped, with a rich dendritic plexus originating from one end and the axon from the other. The fine branches of the dendrites are covered with spines (not shown). **C,** A sympathetic postganglionic motoneuron. **D,** An α-motoneuron of the spinal cord. Both **C** and **D** are multipolar neurons with radially arranged dendrites. **E,** A sensory dorsal root ganglion cell; no dendrites are present. The axon branches into a central and a peripheral process. Because the axon results from fusion of two processes during embryonic development, these cells are described as pseudounipolar neurons rather than unipolar.

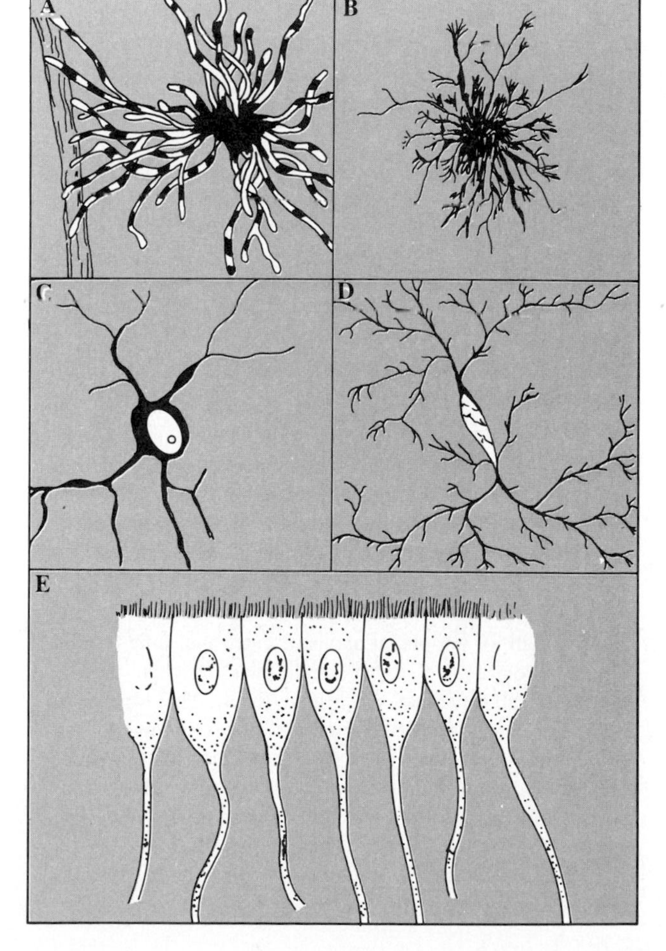

■ **Fig. 6-10** Different types of neuroglial cells of the central nervous system. **A,** Fibrous atrocyte; **B,** protoplasmic astrocyte. Note the glial foot processes in association with a capillary in **A. C,** An oligodendrocyte. Each of the processes is responsible for the production of one or more myelin sheath internodes about central axons. **D,** Microglial cell. **E,** Ependymal cells.

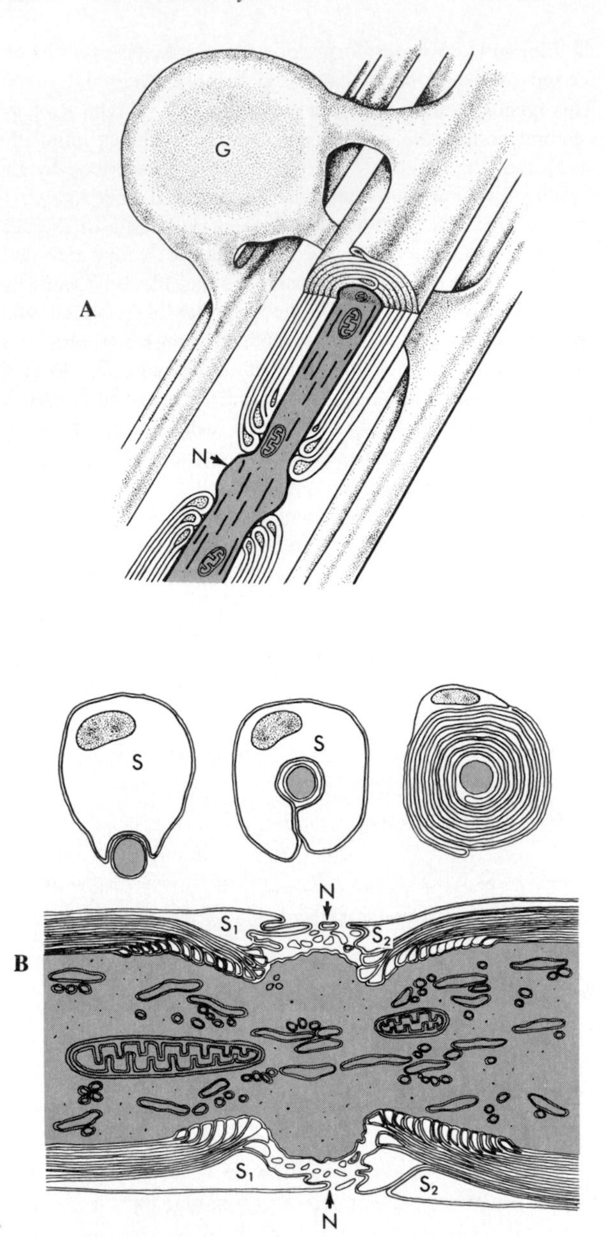

■ **Fig. 6-11** Myelin sheaths of axons. **A,** Myelinated axons in the central nervous system. A single oligodendrocyte *(G)* emits several processes, each of which winds in a spiral fashion around an axon to form the myelin sheath. The axon *(color)* is shown in cutaway. The myelin from a single oligodendrocyte ends before the next wrapping from another oligodendrocyte. The bare axon between sheaths is the node of Ranvier *(N)*. Conduction of action potentials is saltatory down the axon, skipping from node to node. **B,** Myelinated axon in the peripheral nervous system. A Schwann cell forms a myelinated sheath for peripheral axons in much the same fashion as oligodendrocytes do for central ones, except that each Schwann cell myelinates a single axon. The top shows a cross-sectional view of progressive stages in myelin sheath formation by a Schwann cell *(S)* around an axon *(color)*. The bottom shows a longitudinal view of a myelinated axon *(color)*. The node of Ranvier *(N)* is shown between adjacent sheaths formed by two Schwann cells (S_1 and S_2). (Redrawn from Patton HD et al: *Introduction to basic neurology*, Philadelphia, 1976, WB Saunders.)

tials to the **nodes of Ranvier** (the junction between adjacent sheath cells). This action results in **saltatory conduction,** the skipping of nerve impulses from node to node.

Satellite cells encapsulate dorsal root and cranial nerve ganglion cells and regulate their microenvironment in a fashion similar to astrocytes.

Microglia are latent phagocytes. When the CNS is damaged, the microglia help remove the cellular products of the damage. They are assisted by neuroglia and by other phagocytes that invade the CNS from the circulation.

Ependymal cells form the epithelium that separates the CNS from CSF in the ventricles (see Fig. 6-7). Many substances diffuse readily across the ependyma between the extracellular space of the brain and the CSF. The CSF is secreted in large part by specialized ependymal cells of the choroid plexuses, located in the ventricular system (see p. 96).

Nutrients are delivered and wastes removed by the vascular system. Capillaries and other blood vessels are abundant in nervous tissue. Diffusion of many substances between the blood and the CNS is limited by the **blood-brain barrier** (see p. 94). The external surface of the CNS is covered by several layers of connective tissue. These layers form the **pia mater,** the **arachnoid,** and the **dura mater.** These layers protect the CNS. The space between the pia mater and arachnoid, the **subarachnoid space,** contains CSF (see p. 96).

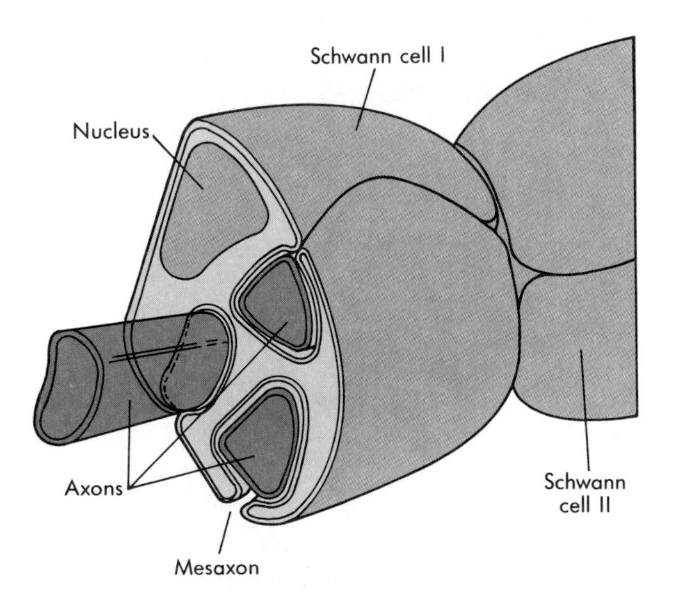

■ **Fig. 6-12** Three-dimensional impression of the appearance of a Remak's bundle. The cut face of the bundle is seen to the left. One of the three unmyelinated axons is represented as protruding from the bundle. A mesaxon is indicated, as is the nucleus of the Schwann cell. To the right, the junction of adjacent Schwann cells is depicted.

Transmission of Information

A major role of axons is to transmit information from the region of the cell body and dendrites of a neuron to synapses on other neurons or effector cells. The information is generally transmitted as a series of nerve impulses.

The conduction velocity of an axon is the speed at which an action potential is propagated (see Chapter 3). The conduction velocity depends on the diameter of the axon and whether the axon is myelinated or unmyelinated. Unmyelinated axons are generally less than 1 μm in diameter and conduct at less than 2.5 m/sec^{-1}. It would take about 1 sec for a signal from a sensory receptor in a person's foot to reach the spinal cord via an unmyelinated axon having a conduction velocity of 1 m/sec^{-1}. Myelinated axons have diameters of 1 to 20 μm and conduct at 3 to 120 m/sec^{-1}. A spinal motoneuron with an axon that conducts at 100 m/sec^{-1} would be able to trigger the contraction of a toe muscle in about 10 msec.

In the CNS certain neurons that lack axons **(amacrine cells)** send information to the synaptic terminals by intracellular electrical current flow rather than by generating action potentials. This current flow produces a **local potential,** which decays over a short distance (millimeters to hundreds of micrometers, depending on the **length constant;** see Chapter 3) of the neuron involved. Local potentials differ from action potentials in that they are nonpropagating and therefore cannot spread over long distances. By contrast, action potentials can propagate over long distances along axons.

Signaling of local potentials is also characteristic of sensory receptors, which produce **receptor potentials,** and of communications between nerve cells by **synaptic potentials.**

Coding

Information conveyed by axons may be encoded in several ways. A **labeled line** is a set of neurons dedicated to a general function, such as a particular sensory modality. For example, the visual pathway includes neurons in the retina, the lateral geniculate nucleus of the thalamus, and the visual areas of the cerebral cortex. Fiber tracts carrying visual signals include the optic nerve and optic tract and the optic radiation. The normal means for activating the visual system is light striking the retina. Neurons of the retina process the information and then transmit signals along the visual pathway. However, mechanical or electrical stimulation of neurons in the visual pathway will also produce a visual sensation, although a distorted one. Thus neurons of the visual system can be regarded as a labeled line, which when activated by whatever means, causes a visual sensation. Motor pathways also provide examples of labeled lines. For example, certain neurons of the cerebral cortex when activated cause motor neurons to the muscles of the hand to discharge and the hand muscles to contract, whereas other cortical neurons cause movements of the foot.

A second way in which information is encoded by the nervous system is through neural maps. A **somatotopic map** is formed by arrays of neurons in the sensory or motor systems that (1) receive information from corresponding locations on the body surface or (2) issue motor commands to move particular parts of the body. In the visual system, points on the retina are represented by neuronal arrays that form **retinotopic maps.** In the auditory system the frequency of sounds is represented in **tonotopic maps.**

A third method for encoding information is by **patterns of nerve impulses,** which are sequences of nerve impulses that result in synaptic transmission of information to a new set of neurons. The communicated information is coded in terms of the structure of the nerve impulse trains. Several different types of nerve impulse codes have been proposed. A commonly used code depends on the **mean discharge frequency.** For example, in many sensory systems increases in the intensity of a stimulus cause a greater frequency of discharge of the sensory neurons. Other candidate codes depend on the **time of firing,** the **temporal pattern,** and the **duration of bursts.** Other codes have also been proposed.

Synaptic Transmission

Neurons communicate with each other at specialized junctions called **synapses** (see Chapter 4). Typically, synapses are formed between the terminals of the axon of one neuron and the dendrites of another; these are called **axodendritic.** However, other types of synapses occur including **axosomatic, axoaxonal,** and **dendrodendritic.** The synapse between a motoneuron and a skeletal muscle fiber is called an **endplate** or **neuromuscular junction.**

Axonal Transport

Many axons are too long to allow efficient movement of substances from the soma to the synaptic endings simply by diffusion. Membrane and cytoplasmic components that originate in the biosynthetic apparatus of the soma and proximal dendrites must be distributed along the axon (especially to the presynaptic elements of synapses) to replenish secreted or inactivated materials. A special transport mechanism, called **axonal transport,** accomplishes this distribution.

Several types of axonal transport exist. Membrane-bound organelles and mitochondria are transported relatively rapidly by **fast axonal transport.** Substances (e.g., proteins) that are dissolved in cytoplasm are moved by **slow axonal transport.** In mammals fast axonal transport proceeds as rapidly as 400 mm/day, whereas slow axonal transport occurs at about 1 mm/day. Synaptic vesicles, which travel by fast axonal transport, can travel from the soma of a motoneuron in the spinal cord to a neuromuscular junction in a person's foot in about 2½ days. In comparison the movement of many soluble proteins over the same distance takes nearly 3 years.

Axonal transport requires metabolic energy and involves calcium ions. The cytoskeleton, particularly the microtubules, provides a system of guide-wires along which membrane-bound organelles move (Fig. 6-13). These organelles may attach to microtubules through a linkage similar to that between the thick and thin filaments of skeletal muscle fibers; calcium triggers the movement of the organelles along the microtubules.

Axonal transport occurs in both directions. Transport from the soma toward the axonal terminals is called **anterograde axonal transport** (Fig. 6-14, *A*). This process allows the replenishment of synaptic vesicles and enzymes responsible for neurotransmitter synthesis in synaptic terminals. Transport in the opposite direction is **retrograde axonal transport** (Fig. 6-14, *B*). This process returns synaptic vesicle membrane to the soma for lysosomal degradation. Marker substances that can be transported anterogradely or retrogradely are used experimentally to trace neural pathways. For example, when the plant lectin **Phaseolus vulgaris leucoagglutinin** is injected near the somas of neurons, it is taken up by the neurons and is transported anterogradely. Therefore this lectin is employed to demonstrate the axons and synaptic endings of CNS neurons. The enzyme **horseradish peroxidase (HRP)** is taken up by synaptic terminals and is transported retrogradely. For this reason HRP is often used to demonstrate the somas of neurons that project to a particular region of synaptic termination.

Nervous Tissue Reactions to Injury

Injury to nervous tissue elicits responses by neurons and neuroglia. Severe injury causes cell death. Once a neuron is lost, it cannot be replaced because neurons are postmitotic cells.

Degeneration

When an axon is transected, the soma of the neuron may show the **axonal reaction.** Normally, Nissl bodies stain well with basic aniline dyes that attach to the ribonucleic acid of the ribosomes (Fig. 6-15, *A*). After injury to the axon (Fig. 6-15, *B*), the neuron attempts to repair the axon by making new structural proteins. During the axonal reaction the cisterns of the rough endoplasmic reticulum become distended with the products of protein synthesis. The ribosomes appear disorganized, and the Nissl bodies are stained weakly by basic aniline dyes. This process, called **chromatolysis,** produces an alteration in staining (Fig. 6-15, *C*). The soma may swell and become rounded, and the nucleus may assume an eccentric position. These morphological

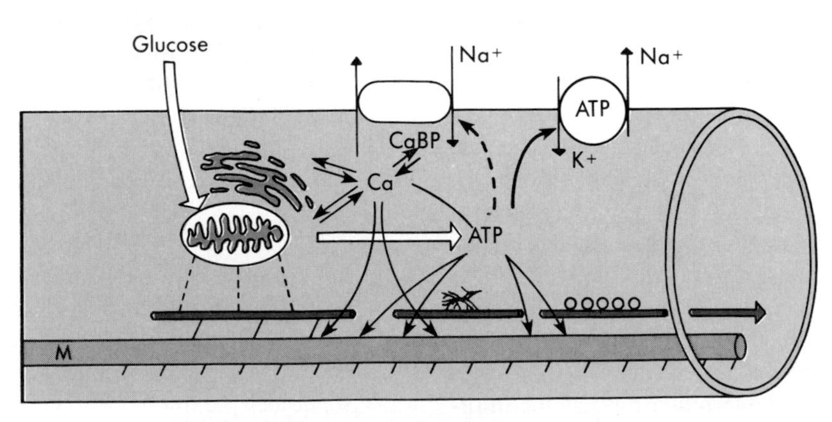

Fig. 6-13 Axonal transport has been proposed to depend on the movement of transport filaments. Energy is required and is supplied by glucose. Mitochondria control the level of cations in the axoplasm by supplying adenosine triphosphate *(ATP)* to the ion pumps. An important cation for axonal transport is calcium. Transport filaments (red bars at bottom of drawing) move along the cytoskeleton (microtubules, *M,* or neurofilament, *NF*) by means of crossbridges. Transported components attach to the transport filaments.

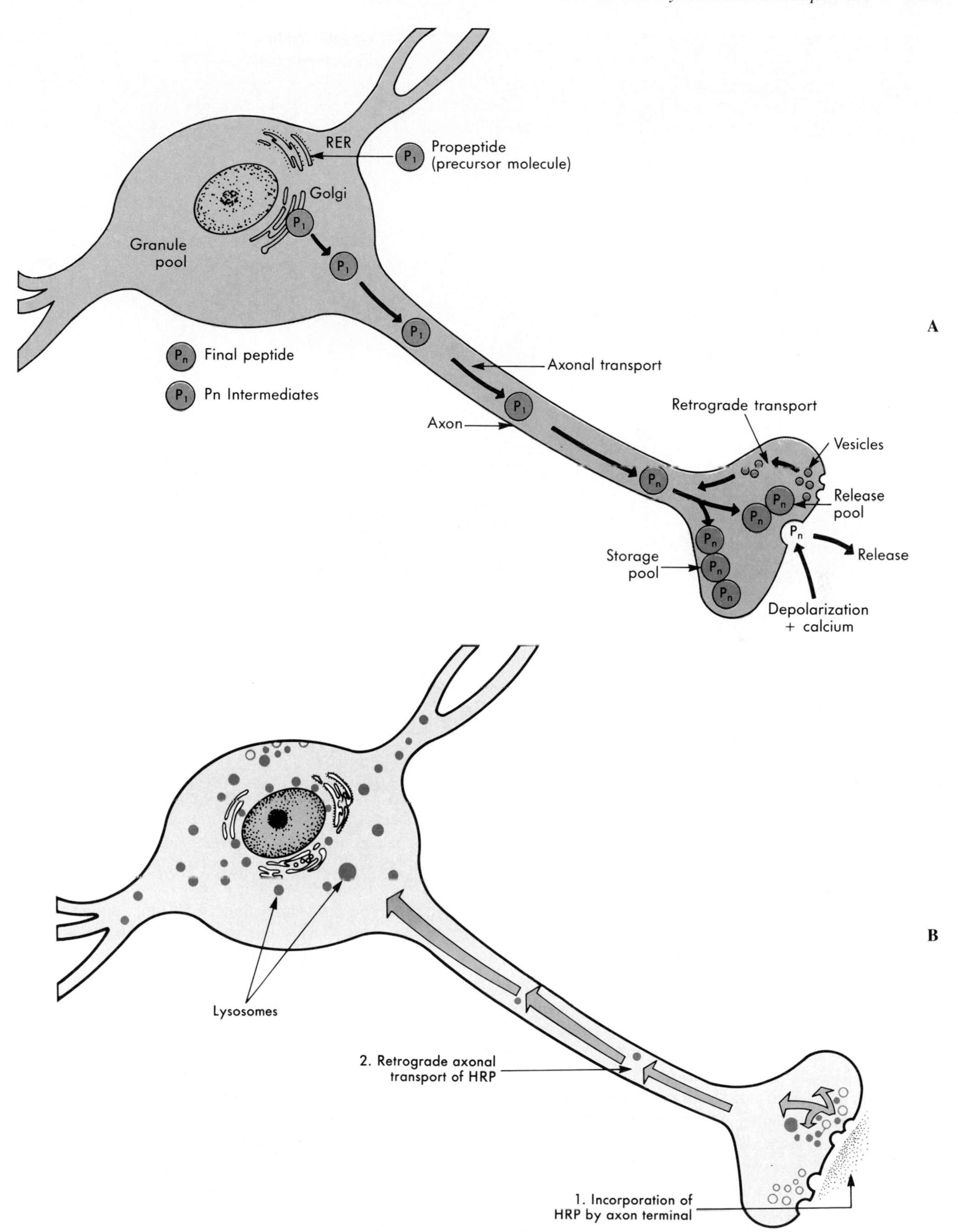

■ **Fig. 6-14** **A,** Axonal transport and its relation to the synthesis of peptides in the cell body and their release from terminals. *RER,* Rough endoplasmic reticulum. **B,** Schematic summary of incorporation, retrograde axonal transport, and lysosomal accumulation of horseradish peroxidase *(HRP)* in neurons. Anterograde axonal transport of HRP from the soma is not illustrated.

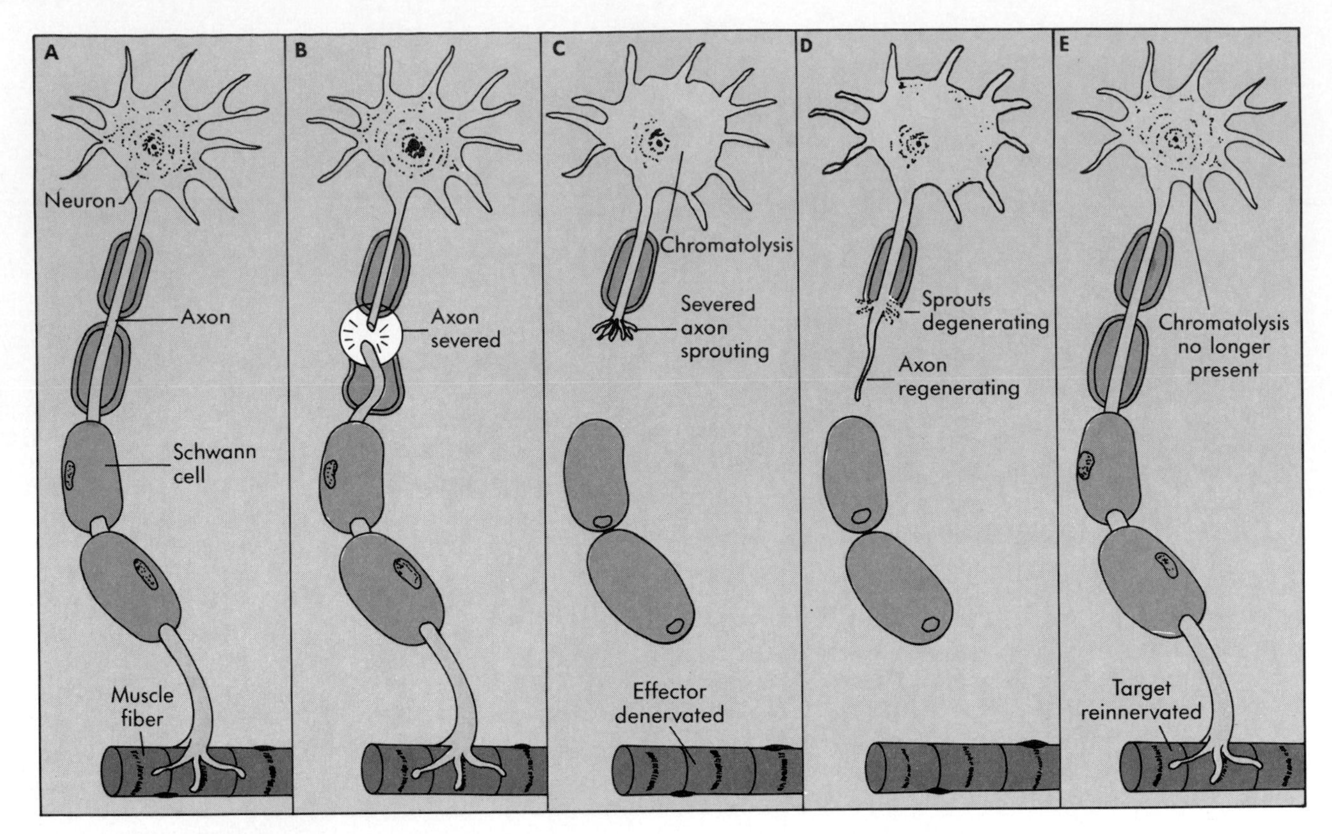

■ **Fig. 6-15** **A,** Normal motoneuron innervating a skeletal muscle fiber. **B,** Motor axon has been severed, and the motoneuron is undergoing chromatolysis. **C,** This is associated in time with sprouting and, in **D,** with regeneration of the axon. The excess sprouts degenerate. **E,** When the target cell is reinnervated, chromatolysis is no longer present.

changes reflect the cytological processes that accompany increased protein synthesis.

The axon distal to the transection dies (Fig. 6-15, *C*). Within a few days the axon and all of the synaptic endings formed by the axon disintegrate. If the axon had been myelinated, the myelin sheath fragments and is eventually phagocytized and removed. However, the neuroglial cells that had formed the myelin sheath remain viable. This sequence of events was originally described by Waller and is called **wallerian degeneration.**

If the axons that provide the sole or predominant synaptic input to a neuron or an effector cell are interrupted, the postsynaptic cell may undergo degeneration and even death. The best-known example of this is the atrophy of skeletal muscle fibers following interruption of their innervation by motoneurons.

These pathological changes have been useful in neuroanatomical investigations to trace neural pathways. For example, retrograde chromatolysis has been used to reveal groups of neurons whose axons have been deliberately interrupted. The projection target of axons can be determined by following the course of interrupted axons undergoing wallerian degeneration. Synaptic targets can also be mapped if neurons undergo transneuronal degeneration after an axonal bundle is transected.

■ *Regeneration*

After an axon is lost through injury, many neurons can regenerate a new axon. The proximal stump of the damaged axon develops **sprouts** (see Fig. 6-15, *C*). In the PNS these sprouts elongate and grow along the path of the original nerve if this route is available (Fig. 6-15, *D*). The Schwann cells in the distal stump of the nerve not only survive the wallerian degeneration, but they also proliferate and form rows along the course previously taken by the axons. **Growth cones** of the sprouting axons find their way along the rows of Schwann cells and may eventually reinnervate the original peripheral target structures (Fig. 6-15, *E*). The Schwann cells then remyelinate the axons. The rate of regeneration is limited by the rate of slow axonal transport to about 1 mm/day^{-1}.

In the CNS transected axons also sprout. Proper guidance for the sprouts is lacking, however, because the oligodendroglia do not form a path along which the sprouts can grow. This limitation may occur because a single oligodendroglial cell myelinates many central axons, whereas a single Schwann cell provides myelin for only a single axon in the periphery. Alternatively, different chemical signals may affect peripheral and central attempts at regeneration differently. Another obstacle is the formation of glial scars by astrocytes.

Trophic factors. A number of proteins are now known to affect the growth of axons and the maintenance of synaptic connections. The best studied of these substances is **nerve growth factor** (NGF). NGF was initially thought to enhance the growth and maintain many sensory and autonomic postganglionic neurons. However, we now believe that NGF affects the CNS and the PNS.

NGF is secreted by target cells and binds to NGR receptors; these receptors are located on neurons that synapse with the target cells. The bound NGF and receptor are internalized, and the NGF is transported retrogradely to the soma. NGF may directly act on the nucleus and affect the production of enzymes responsible for neurotransmitter synthesis and axonal growth.

Summary

1. Sensory, integrative, and motor components of the nervous system permit the body to communicate with the environment.

2. The neuron is the functional unit of the nervous system. Information is conveyed through neural circuits by action potentials in neurons and by synaptic transmission between neurons.

3. Neuroglial cells regulate the microenvironment of neurons and provide myelin sheaths that speed conduction velocities.

4. The PNS includes sensory receptors, primary afferent neurons, somatic motor neurons, and autonomic pre- and postganglionic neurons.

5. The CNS includes the spinal cord and brain. The brain includes the medulla, pons, cerebellum, midbrain, thalamus, hypothalamus, basal ganglia, and cerebral cortex.

6. CNS extracellular fluid composition is regulated by the cerebrospinal fluid, the blood-brain barrier, and the astrocytes.

7. Choroid plexuses form CSF. CSF leaves the ventricles through the roof of the fourth ventricle, traverses the subarachnoid space, and returns to the circulation through arachnoid villi.

8. CSF differs from blood in having a lower concentration of K^+, glucose, and protein and a higher concentration of Na^+ and Cl^-; CSF normally lacks blood cells. Its production is relatively independent of ventricular and blood pressure, but its absorption is a function of CSF pressure.

9. General functions of the nervous system include excitability, sensory detection, information processing, and behavior. Different types of neurons are specialized for different functions.

10. Neuroglial cells include astrocytes (regulate CNS microenvironment), oligodendroglia (form CNS myelin), Schwann cells (form PNS myelin), ependymal cells (line the ventricles), and microglia (CNS macrophages).

11. Neurons contain a nucleus and nucleolus, Nissl bodies (rough endoplasmic reticulum), Golgi apparatus, mitochondria, neurofilaments, and microtubules. Most neurons have dendrites and an axon; dendrites receive synaptic contacts from other neurons, and the axon makes synaptic contacts with other neurons.

12. Neurons encode information by labeled lines, neural maps, and patterns of nerve impulses.

13. Chemical substances are distributed along axons by fast or by slow axonal transport; the direction of axonal transport may be anterograde or retrograde.

14. Damage to the axon of a neuron causes an axonal reaction in the cell body (chromatolysis) and wallerian degeneration of the axon distal to the injury. Regeneration of PNS axons is more likely than CNS axons.

15. The growth and maintenance of axons is affected by trophic factors, such as the nerve growth factor.

Bibliography

Journal articles

Bray GM, Rasminsky M, Aguayo AJ: Interactions between axons and their sheath cells, *Annu Rev Neurosci* 4:127, 1981.

Fawcett JW, Keynes RJ: Peripheral nerve regeneration, *Annu Rev Neurosci* 13:43, 1990.

Gerfen CR, Sawchenko PE: A method for anterograde axonal tracing of chemically specified circuits in the central nervous system: combined Phaseolus vulgaris-leucoagglutinin (PHA-L) tract tracing and immunohistochemistry, *Brain Res* 343:144, 1985.

Partridge WM: Brain metabolism: a perspective from the blood-brain barrier, *Physiol Rev* 63:1481, 1983.

Udin SB, Fawcett JW: Formation of topographic maps, *Annu Rev Neurosci* 11:289, 1988.

Vallee RB, Bloom GS: Mechanisms of fast and slow axonal transport, *Annu Rev Neurosci* 14:59, 1991.

Books and monographs

Bullock TH, Orkand R, Grinnell A: *Introduction to nervous systems,* San Francisco, 1977, WH Freeman.

Cajal SR: *Degeneration and regeneration of the nervous system,* New York, 1959, Hafner Publishing.

Cajal SR: *The neuron and the glial cell,* Springfield, Ill, 1984, Charles C Thomas.

Carpenter MB, Sutin J: *Human neuroanatomy,* Baltimore, 1983, Williams & Wilkins.

Eccles JC: *The physiology of nerve cells,* Baltimore, 1957, Johns Hopkins University Press.

Heimer L, Robards MJ: editors: *Neuroanatomical tract-tracing methods,* New York, 1981, Plenum Press.

Kandel ER, Schwartz JH: *Principles of neural science,* ed 3, New York, 1991, Elsevier.

Kuffler SW, Nicholls JG, Martin AR: *From neuron to brain,* ed 2, Sunderland, Mass, 1976, Sinaer Associates.

Millen JW, Woollam DHM: *The anatomy of the cerebrospinal fluid,* New York, 1962, Oxford University Press.

Paxinos G, editor: *The human nervous system,* San Diego, 1990, Academic Press.

Peters A, Palay SL, Webster HdeF: *The fine structure of the nervous system: the neurons and supporting cells,* Philadelphia, 1976, WB Saunders.

Shephard GM, editor: *The synaptic organization of the brain,* ed 3, New York, 1990, Oxford University Press.

Whitfield IC: *Neurocommunications: an introduction,* New York 1984, John Wiley & Sons.

Willis WD, Grossman RG: *Medical neurobiology,* ed 3, St Louis, 1981, CV Mosby.

CHAPTER

7

The Peripheral Nervous System

The central nervous system (CNS) analyzes sensory information transmitted from sensory receptors by way of primary afferent neurons. It then uses this information to produce motor commands that are transmitted (1) by motor axons from somatic motoneurons to skeletal muscle fibers or (2) by autonomic preganglionic and postganglionic neurons to cardiac muscle, smooth muscle or glands. The nervous system thus senses and analyzes the environment in order to generate behavior appropriate to that environment.

The axons of primary afferent neurons, somatic motoneurons, and autonomic motoneurons all pass through the peripheral nervous system (Fig. 7-1). The peripheral nervous system (PNS) serves as a bridge between the environment and the CNS. Before discussing the various sensory and motor systems of the CNS, we will consider the sensory and somatic motor components of the PNS. However, for detail about the composition and destination of particular peripheral nerves, a standard textbook of gross anatomy should be consulted. The peripheral and central components of the autonomic nervous system will be discussed in Chapter 15.

■ *Sensory Components of the PNS*

Sensory receptors are specialized neurons that serve as transducers of environmental energy. They provide information to the organism about its external and internal environment. In general, that information is then trans-

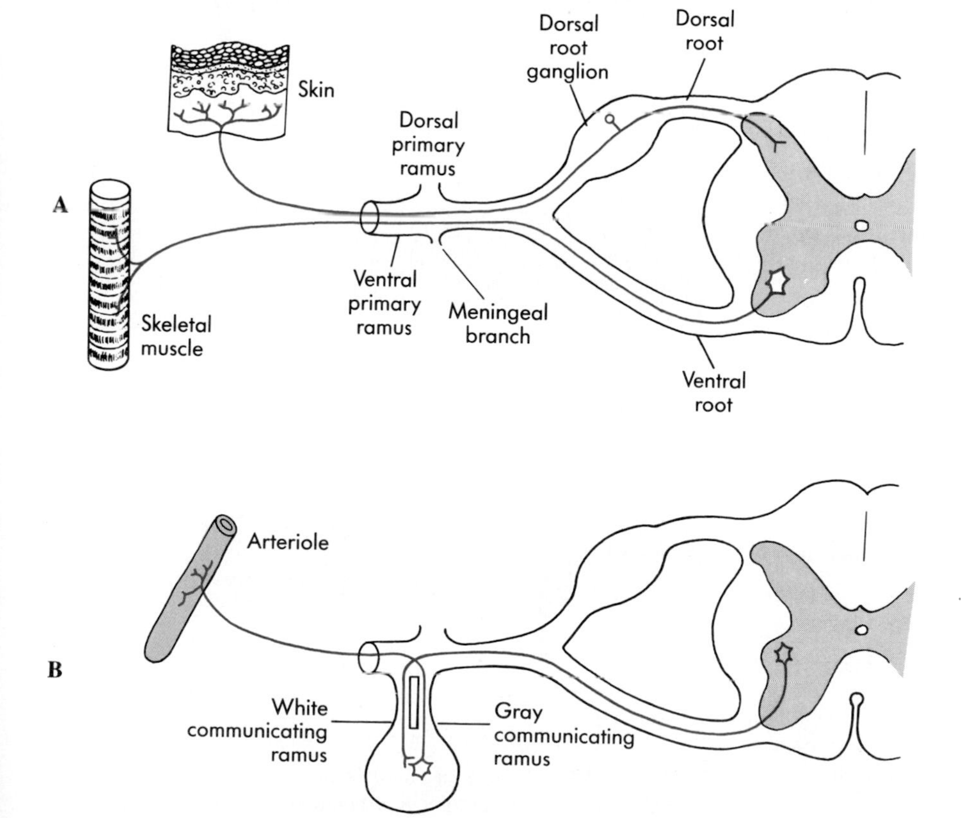

■ **Fig. 7-1** In **A** is a diagram of the spinal cord, spinal roots, and spinal nerve. A primary afferent neuron is shown with its cell body in the dorsal root ganglion and its central and peripheral processes distributing, respectively, to the spinal cord gray matter and to a sensory receptor in the skin. A motoneuron is shown to have its cell body in the spinal cord gray matter and to project its axon out of the ventral root to innervate a skeletal muscle fiber. In **B,** a sympathetic preganglionic neuron is shown to have its cell body in the spinal cord gray matter. The axon courses out of the ventral root and enters a sympathetic ganglion through a white communicating ramus. The preganglionic axon synapses on a ganglion cell, and the ganglion cell sends its process, the postganglionic axon, through a gray communicating ramus into the spinal nerve. The postganglionic axon is shown to terminate on an arteriole in the body wall.

mitted to the CNS by trains of nerve impulses in primary afferent neurons. As mentioned in Chapter 6, the cell bodies of primary afferent neurons are located in dorsal root and cranial nerve ganglia. Each primary afferent neuron has a peripheral process that extends distally through a peripheral nerve to reach the appropriate sensory receptor(s) and a central process that enters the CNS through a dorsal root or a cranial nerve (Fig. 7-1, *A*).

Types of Sensory Receptors

Sensory receptors can be classified according to whether they provide information about the external environment **(exteroceptors),** the internal environment **(interoceptors),** or the position of the body in space **(proprioceptors).** Another more detailed classification is given in Table 7-1.

Functions of Sensory Receptors

Transduction. Sensory systems are designed to respond to the environment. *Stimulation is the action of environmental energy on the body through activation of one or more sensory receptors.* A **stimulus** is the environmental event that excites sensory receptors, which then provide information about the stimulus to the CNS. The **response** to the stimulus is the effect the stimulus has on the organism. Responses can be recognized at several levels; these include receptor potentials in sensory receptors, the transmission of action potentials along axons in sensory pathways, synaptic events in sensory neural networks, motor activity triggered by sensory stimulation, and ultimately behavioral events. The process that enables a sensory receptor to respond usefully to a stimulus is called **sensory transduction.**

Environmental events that involve sensory transduction can be mechanical, thermal, chemical, or other forms of energy, depending on the sensory apparatus that serves as a transducer. Although humans cannot sense electrical or magnetic fields, other animals can respond to such stimuli. For example, many fish have electroreceptors, and both fish and birds may use the earth's magnetic field for orientation during migration.

Fig. 7-2 shows how different types of stimuli can alter the membrane properties of sensory receptor neurons specialized to transduce such stimuli. In Fig. 7-2, *A,* a **chemoreceptor** responds when a molecule of a *chemical stimulant* reacts with **receptor molecules** on the plasma membrane of the sensory receptor. (Note the distinction between a sensory receptor, which includes one or more cells, and a receptor molecule, which is a protein inserted into the membrane of a cell.) The reaction between the chemical stimulant and the receptor molecules opens ion channels, which enables the influx of ionic current that depolarizes the sensory receptor. In Fig. 7-2, *B,* the ion channel of a **mechanoreceptor** is opened in response to the application of a *mechanical force* along the membrane; this opens the ion channel and allows an influx of current that depolarizes the sensory receptor. In Fig. 7-2, *C,* the ion channel of a **photoreceptor** cell (responds to *light*) is open in the dark and closed when a photon is absorbed by pigment on the disc membrane. In this case, an influx of current occurs in the dark (called the **dark current**); the current

Table 7-1 Classification of sensory receptors

Special	Vision, audition, taste, olfaction, balance
Superficial	Touch, pressure, flutter, vibration, tickle, warmth, cold, pain, itch
Deep	Position, kinesthesia, deep pressure, deep pain
Visceral	Hunger, nausea, distention, visceral pain

From Willis WD, Grossman RG: *Medical neurobiology,* ed 3, St Louis, 1981, Mosby–Year Book.

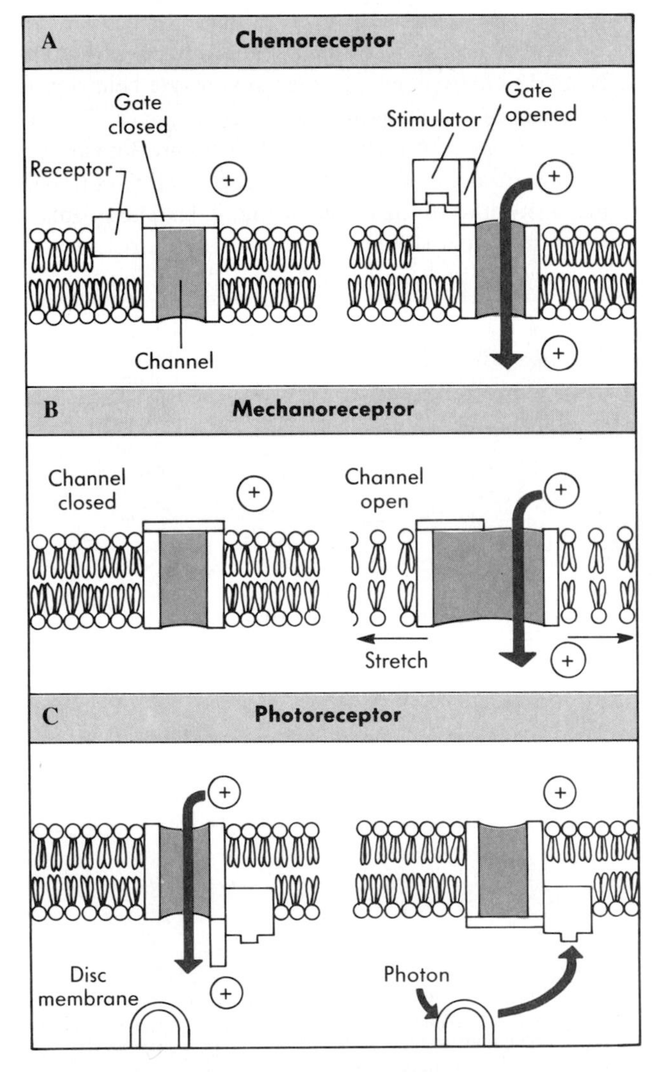

Fig. 7-2 Conceptual models of transducer mechanisms in three types of receptors. **A,** Chemoreceptor; **B,** mechanoreceptor; **C,** vertebrate photoreceptor. (See text.)

ceases when light is applied. When the current stops, the photoreceptor hyperpolarizes.

As illustrated in Fig. 7-3, sensory transduction generally produces a **receptor potential** in the peripheral end of the sensory afferent neuron. The **receptor potential** is usually a depolarizing event that results from inward current flow, and it brings the membrane potential of the sensory receptor toward or past the threshold needed to trigger an action potential. For example, in Fig. 7-3 a mechanical stimulus distorts the ending of a mechanoreceptor and causes inward current flow at the end of the axon and longitudinal and outward current flow along the axon. The outward current produces a depolarization (the receptor potential), which may or may not exceed threshold for an action potential. In this case the action potential is generated in a trigger zone at the first node of Ranvier of the afferent fiber. However, in photoreceptors, as mentioned above, the receptor potential is hyperpolarizing. Information transmission in the retina will be discussed in Chapter 9.

In some sensory receptors, the primary afferent fiber terminates on a separate, peripherally located sensory cell. For example, in the **cochlea,** primary afferent fibers end on **hair cells.** Sensory transduction in such sense organs is made more complex by this arrangement. In the cochlea, a receptor potential arises in the hair cells in response to sound (see Chapter 10). The receptor potential is oscillatory; but in each oscillatory cycle when the membrane of the hair cell is depolarized, the hair cell releases an excitatory neurotransmitter onto the primary afferent terminal. The result is a **generator potential,** which in turn depolarizes the primary afferent fiber. The generator potential brings the membrane potential of the primary afferent fiber toward or beyond the threshold to fire nerve impulses.

Adaptation is a characteristic property of sensory receptors; it suits them better to signal particular kinds of sensory information. In **slowly adapting receptors,** a long-lasting stimulus produces a prolonged repetitive discharge in the primary afferent neurons that supply the sensory receptors. However, the same stimulus produces only a short-lived response (one or a few discharges) in **rapidly adapting receptors.** These different rates of adaptation occur because a prolonged stimulus may produce either a maintained or a transient receptor potential in the sensory receptor. The functional implication of the adaptation rate is that different temporal features of a stimulus can be analyzed by receptors with different adaptation rates. For example, during an indentation of the skin, a slowly adapting receptor may respond repetitively at a rate proportional to the amount of indentation (Fig. 7-4). On the other hand, rapidly adapting receptors in the skin respond best to transient mechanical stimuli. The information signaled may reflect stimulus velocity or acceleration rather than the amount of skin indentation.

Receptive fields. The relationship between the location of a stimulus and the activation of particular sensory neurons is a major theme in sensory physiology.

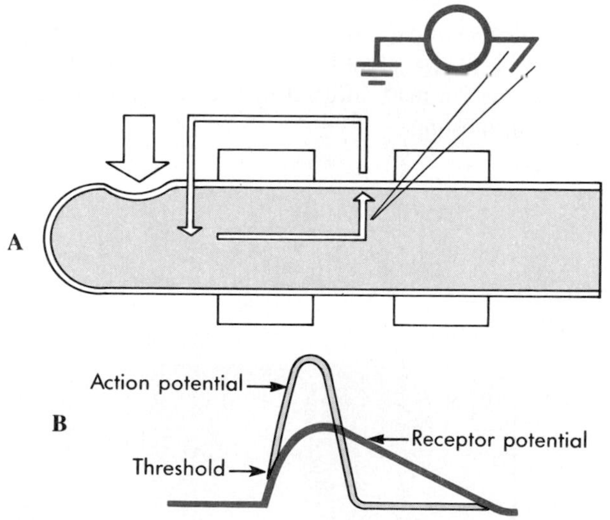

■ **Fig. 7-3** **A,** The current flow produced by stimulation of a mechanoreceptor at the site indicated by the arrow and an intracellular recording from a node of Ranvier. **B,** The receptor potential produced by the current and an action potential that may be superimposed on the receptor potential if the latter exceeds threshold.

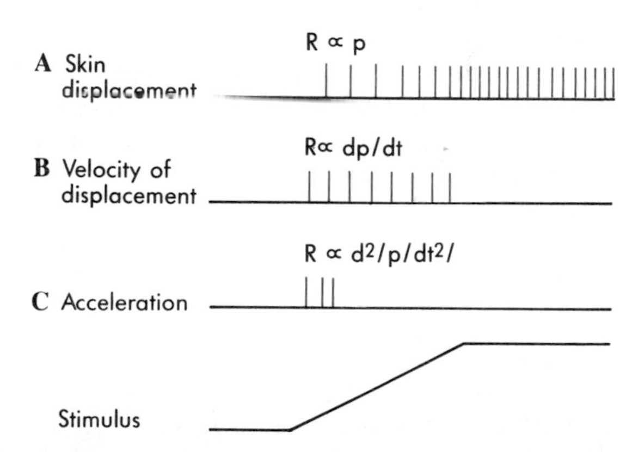

■ **Fig. 7-4** Responses of slowly and rapidly adapting mechanoreceptors to displacement of the skin. The discharges of the primary afferent fibers supplying the receptors in response to the ramp and hold stimulus (shown at the *bottom*) are termed the response *(R)*. **A,** *R* is proportional to skin position *(p)*. The receptor is slowly adapting and signals skin displacement. **B,** *R* is a function of the velocity of displacement *(dp/dt)*. **C,** *R* is a function of the acceleration *(d^2p/dt^2)*. These receptors are rapidly adapting, but signal different, dynamic features of the stimulus.

The **receptive field** of a sensory neuron is the region that, when stimulated, affects the discharge of the neuron. For example, a sensory receptor might be activated by indentation of only a small area of skin. That area is the **excitatory receptive field** of the sensory receptor. A neuron in the CNS might be excited by stimulation of a receptive field several times as large as that of a sensory receptor. The reason that CNS sensory neurons often have larger receptive fields than do sensory receptors is that CNS sensory neurons may receive information from many sensory receptors, each with a slightly different receptive field. The receptive field of the CNS neuron is the sum of the receptive fields of the sensory receptors that influence the CNS neuron. The location of the receptive field is determined by the location of the sensory transduction apparatus responsible for signaling information about the stimulus to the sensory neuron.

Generally, the receptive fields of sensory receptors are excitatory. However, a central sensory neuron can have either an excitatory or an **inhibitory receptive field.** For example, the somatosensory neuron illustrated in Fig. 7-5 had both excitatory and inhibitory receptive fields. This cell was located in the primary somatosensory cerebral cortex (SI cortex). Inhibition results from data processing in sensory neural circuits and is mediated by inhibitory interneurons.

Sensory coding. Sensory neurons encode stimuli. In the process of sensory transduction, one or more aspects of the stimulus must be encoded in a way that can be interpreted by the CNS. The encoded information is an abstraction based on (1) which sensory receptors are activated, (2) the responses of sensory receptors to the stimulus, and (3) information processing in the sensory pathway. Some of the aspects of stimuli that may be encoded include the **sensory modality, spatial location, threshold, intensity, frequency,** and **duration.** Other aspects of stimuli that are encoded are presented in relation to particular sensory systems in later chapters.

A **sensory modality** is a readily identified class of sensation. For example, maintained mechanical stimuli applied to the skin result in sensations of **touch** or **pressure,** and transient mechanical stimuli may evoke sensations of **flutter** or **vibration.** Other cutaneous modalities include cold, warmth, and pain. Vision, audition, position, sense, taste, and smell are examples of noncutaneous sensory modalities. Coding for modality is signaled by labeled-line sensory channels in most sensory systems (see Chapter 6). A **labeled-line** sensory channel consists of a set of neurons devoted to a particular sensory modality.

The **spatial location** of a stimulus is often signaled by the activation of the particular population of sensory neurons whose receptive fields are affected by the stimulus (Fig. 7-6, *A*). In some cases an inhibitory receptive field or a contrasting border between an excitatory and an inhibitory receptive field can have localizing value. Resolution of two different adjacent stimuli may depend on excitation of partially separate populations of neurons and on inhibitory interactions (Fig. 7-6, *B*).

A **threshold stimulus** is the weakest that can be reliably detected. For detection, a stimulus must produce receptor potentials that are large enough to activate one or more primary afferent fibers. Weaker intensities of stimulation can produce subthreshold receptor potentials; however, such stimuli would not excite central sensory neurons and so could not be perceived. Furthermore, the number of primary afferent neurons that need to be excited for sensory detection depends on the requirements for **spatial** and **temporal summation** in the sensory pathway (see Chapter 4). In some sensory systems a stimulus at threshold for detection must be much greater than the threshold for activation of the most responsive primary afferent neurons. Thus a stimulus that excites some primary afferent neurons may not be perceived. On the other hand, if a stimulus is perceived, at least one primary afferent neuron has to be excited beyond threshold.

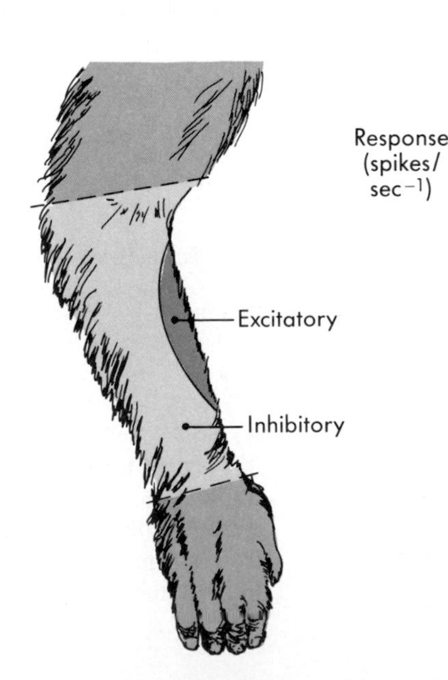

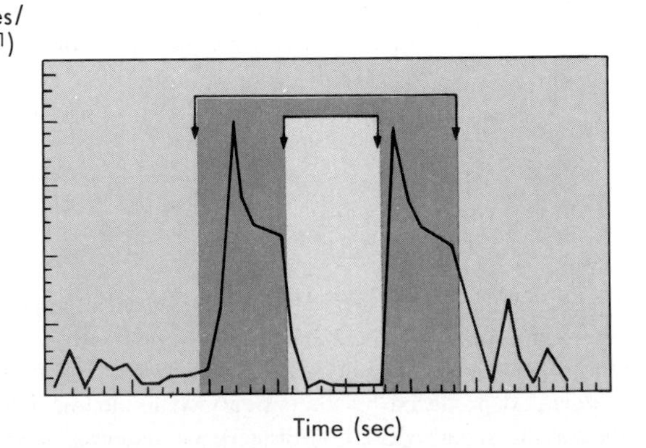

■ **Fig. 7-5** Excitatory and inhibitory receptive fields of a central somatosensory neuron located in the SI (primary) somatosensory cerebral cortex. The excitatory receptive field is on the forearm and is surrounded by an inhibitory receptive field. The graph shows the response to an excitatory stimulus and the inhibition of that response by a stimulus applied in the inhibitory field.

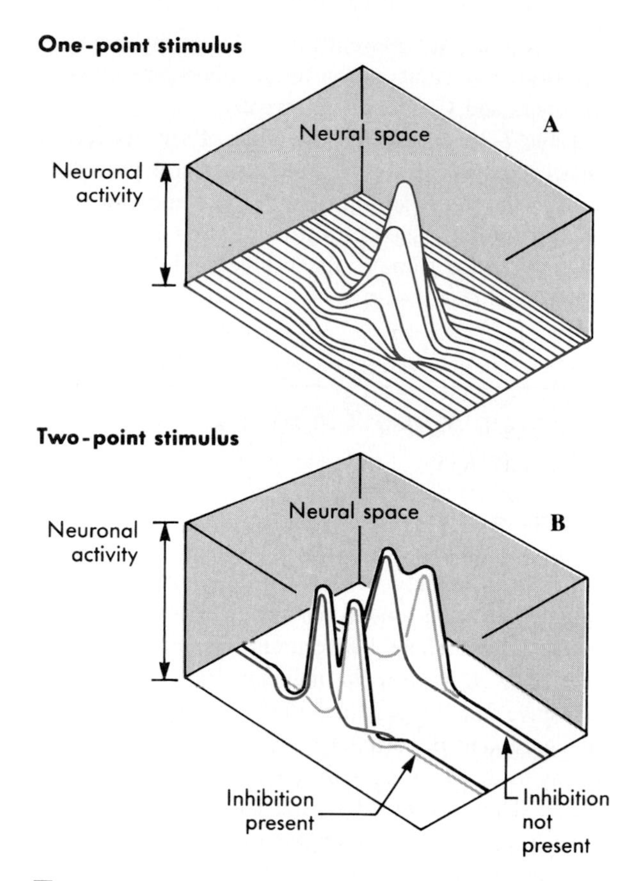

■ **Fig. 7-6** **A,** Representation of the activity of a large population of neurons distributed three-dimensionally in neural space. The activity is in response to stimulation of a point on the skin. Note that an excitatory peak is surrounded by an inhibitory trough; these are determined by the excitatory and inhibitory fields of sensory neurons in the central pathways. **B,** The activity in response to stimulation of two adjacent points on the skin. Note that the sum of the activity *(black line)* is separated better into two peaks when inhibition is present than when it is not.

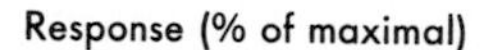

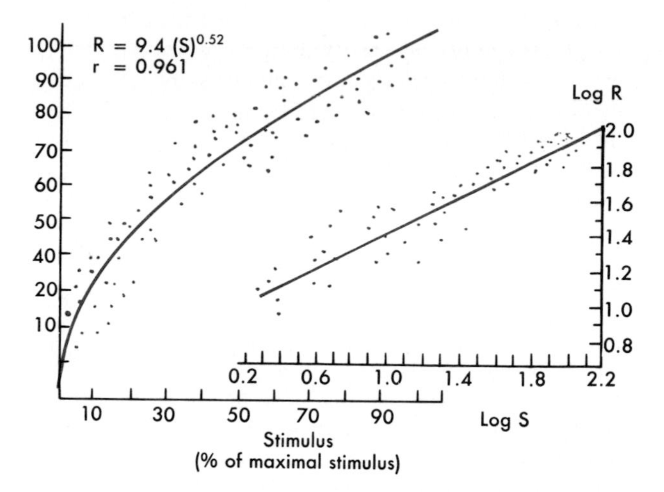

■ **Fig. 7-7** Stimulus-response function for slowly adapting cutaneous mechanoreceptors. The rate of discharge is plotted against stimulus strength (normalized to maximal). The plots are on linear and on log-log scales. The stimulus-response function is $R = 9.4(S).^{0.52}$

Stimulus intensity may be encoded by the mean frequency of discharge of sensory neurons. The relationship between stimulus intensity and response can be plotted as a stimulus-response function. For many sensory neurons, the stimulus-response function approximates an exponential curve (Fig. 7-7). The general equation for such a curve is

$$\text{Response} = \text{constant} \times (\text{stimulus} - \text{threshold stimulus})^{n}$$

The exponent, n, can be less than, equal to, or greater than 1. Stimulus-response functions with fractional exponents are found for many mechanoreceptors (see Fig. 7-7). **Thermoreceptors,** which detect changes in *temperature,* have linear stimulus-response curves (exponent of 1). **Nociceptors,** which detect *painful* stimuli, may have linear or positively accelerating stimulus-response functions; i.e., the exponent for these curves is 1 or more.

Another way in which stimulus **intensity** is encoded is by the number of sensory receptors activated. A stimulus at the threshold for perception may activate just one or a few primary afferent neurons of an appropriate class, whereas a strong stimulus of the same type may excite many similar receptors. Central sensory neurons that receive input from this particular class of sensory receptor would be more powerfully activated as more primary afferent neurons discharge. Greater activity in central sensory neurons is perceived as a stronger stimulus.

Stimuli of different intensities may also activate different sets of sensory receptors. For example, a weak mechanical stimulus applied to the skin might activate only mechanoreceptors, whereas a strong mechanical stimulus might activate both mechanoreceptors and nociceptors. In this case the sensation evoked by the stronger stimulus would be more intense, and the quality would be different.

Stimulus **frequency** can be encoded by the intervals between the discharges of sensory neurons. Sometimes the interspike intervals correspond exactly to the intervals between stimuli (Fig. 7-8). In other cases a given neuron may discharge at intervals that are multiples of the interstimulus interval.

Stimulus **duration** may be encoded in slowly adapting sensory neurons by the duration of enhanced firing. The beginning and end of a stimulus may be signaled by transient discharges of rapidly adapting sensory receptors.

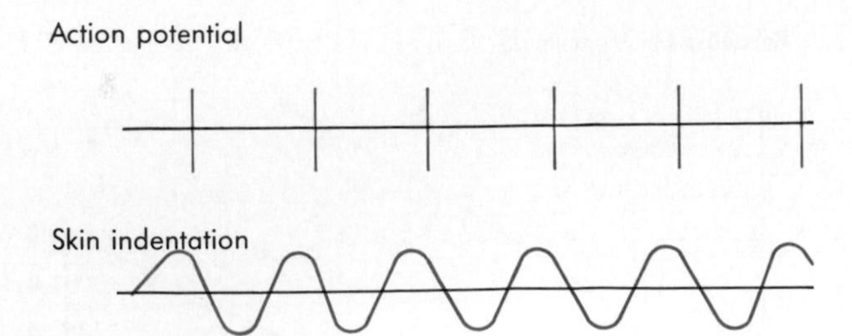

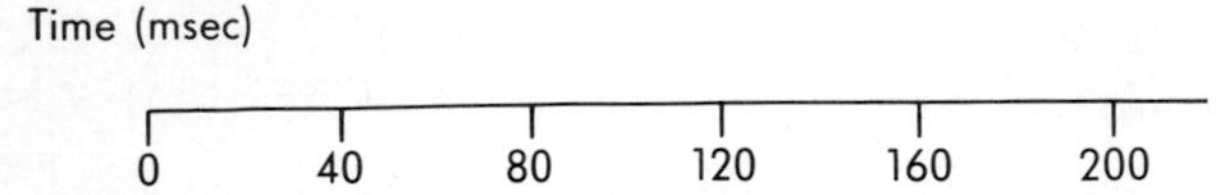

■ **Fig. 7-8** Coding for the frequency of stimulation. Discharge of a rapidly adapting cutaneous mechanoreceptor in phase with a sinusoidal stimulus. The action potentials are shown at the top and the stimulus in the middle trace.

■ *Primary Afferent Neurons*

The peripheral processes of primary afferent neurons that supply different types of sensory receptors have characteristic ranges of conduction velocity. For example, some sensory receptors in muscle are supplied by the largest myelinated axons in the PNS, the **group I** afferent fibers. Other sensory receptors in muscle are innervated by medium-sized **(group II)** or small **(group III)** myelinated axons or by unmyelinated **(group IV)** axons. The largest myelinated axons that supply cutaneous sensory receptors are comparable in size to the medium-sized muscle afferent fibers. Sometimes they are also called group II fibers, but more commonly they are referred to as **Aβ** afferent fibers. Small myelinated and unmyelinated cutaneous afferent fibers are classed as **Aδ** fibers and **C** fibers, respectively.

Table 7-2 lists some of the types of sensory receptors associated with different sized peripheral processes of primary afferent neurons that supply muscle or skin. The terminology used for different sized axons that supply joints is the same as for muscle, and that for viscera is the same as for skin.

■ *Somatic Motor Components of the PNS*

Contractions of skeletal muscle fibers are responsible for movements of the body. Skeletal muscle fibers are innervated by large neurons in the ventral horn of the spinal cord or in cranial nerve nuclei called **alpha motor neurons.** These large, multipolar neurons range in size up to 70 microns in diameter (Fig. 7-9). They usually have 7 to 11 dendrites that may each exceed 1 mm in length. The dendrites are oriented radially and extend as far in the rostrocaudal direction as in the transverse plane. Their axons leave the spinal cord through the ventral roots. As the axon passes ventrally, it may give off one or more recurrent collaterals that synapse on **Renshaw cells** (see Chapter 12), which are inhibitory interneurons in the ventral horn and named after their discoverer. The motor axons distribute to the appropriate skeletal muscles through peripheral nerves and terminate synaptically on skeletal muscle fibers at **neuromuscular junctions** or **endplates.**

A given skeletal muscle is supplied by a group of alpha motoneurons located in a **motor nucleus** (Fig. 7-10, *A*). In the ventral horn, a motor nucleus is typi-

■ **Table 7-2** Types of sensory receptors

Type	*Group*	*Subgroup*	*Diameter (μm)*	*Conduction velocity (m/sec)*	*Tissue supplied*	*Function*
Afferent						
A	I	Ia	12-20	72-120	Muscle	Afferents from muscle spindle primary endings
		Ib			Muscle	Afferents from Golgi tendon organs
	II		6-12	36-72	Muscle	Afferents from muscle spindle secondary endings
	Beta				Skin	Afferents from pacinian corpuscles, touch receptors
	III		1-6	6-36	Muscle	Afferents from pressure-pain endings
	Delta				Skin	Afferents from touch, temperature, and pain receptors
C	IV		<1	0.5-2	Muscle	Afferents from pain receptors
	Dorsal root				Skin	Afferents from touch, pain and temperature receptors

From Willis WD, Grossman RG: *Medical neurobiology*, ed 3, St Louis, 1981, Mosby–Year Book.

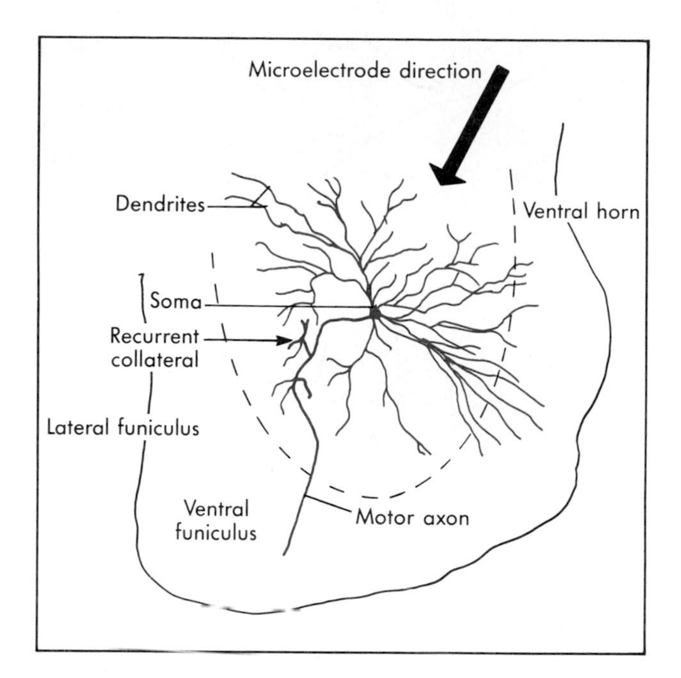

■ **Fig. 7-9** α-Motoneuron injected intracellularly with horseradish peroxidase. A small arrow indicates a recurrent collateral of the motor axon. The large arrow shows the direction followed by the microelectrode.

cally a sausage-shaped array of motoneurons extending over several spinal cord segments (Fig. 7-10, *B*). Motor nuclei that supply different muscles are arranged somatotopically in the ventral horn (Fig. 7-11). Motor nuclei that supply the axial muscles of the body are in the medial part of the ventral horn in the cervical and lumbosacral enlargements and in the most ventral part of the ventral horn in the upper cervical, thoracic, and upper lumbar segments of the spinal cord. Motor nuclei that innervate the limb muscles are in the lateral part of the ventral horn in the cervical and lumbosacral enlargements. The most distal muscles are supplied by motor nuclei located in the dorsolateral part of the ventral horn, whereas more proximal muscles are innervated by motor nuclei in the ventrolateral ventral horn. The set of α-motoneurons that innervates a muscle is called the **motor neuron pool** of the muscle.

Each skeletal muscle fiber in mammals is supplied by just one alpha motoneuron. However, a given alpha motoneuron may innervate a variable number of skeletal muscle fibers, depending on how fine a control of the muscle is required. For highly regulated muscles, such as the eye muscles, an alpha motoneuron may supply only a few skeletal muscle fibers. However, in a proximal limb muscle, such as the quadriceps femoris, a single alpha motoneuron may innervate thousands of skeletal muscle fibers.

A ***motor unit*** *is an alpha motoneuron, its motor axon, and all of the skeletal muscle fibers that it supplies*. The motor unit can be regarded as the basic unit of movement. When an alpha motoneuron discharges, under normal circumstances all of the muscle fibers of the motor unit contract. A given alpha motoneuron may participate in a variety of reflexes and in voluntary movements. Because decisions about whether or not the synaptic input from a variety of sources will cause particular muscle fibers to contract is made at the level of

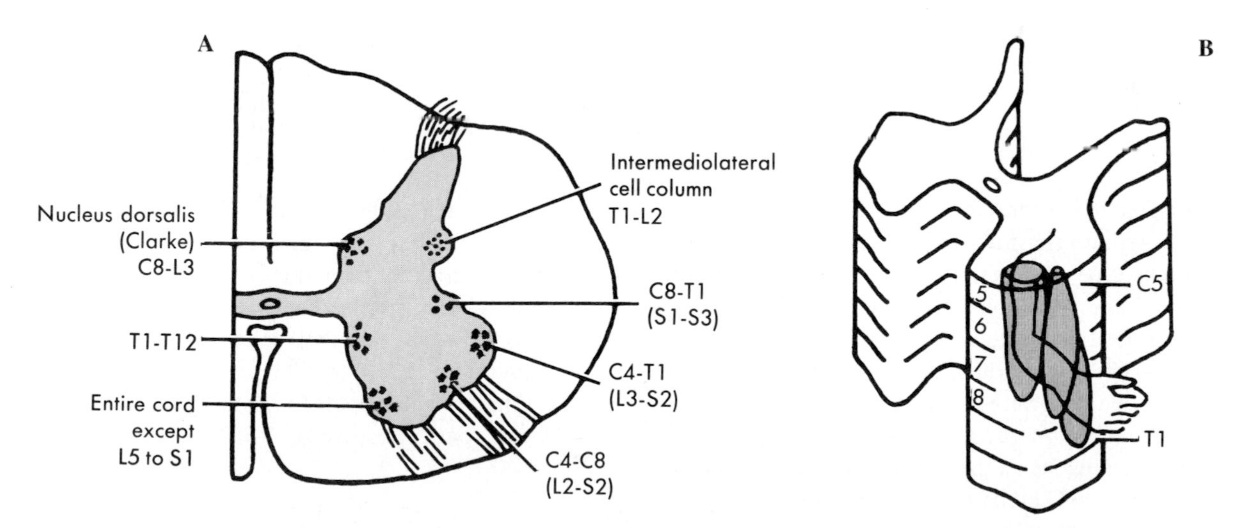

■ **Fig. 7-10** Diagrammatic illustration of the topographical organization of the motoneurons and the muscles they innervate. **A** shows a transverse section of the spinal cord with the positions of various cellular columns and their longitudinal extents. **B** shows three longitudinal columns of motoneurons *(color)* at cervical levels in relation to the parts of the arm they innervate. It should be appreciated that the more proximal muscles of the arm are innervated by more ventromedially situated motoneurons and the more distal muscles by more dorsolaterally situated motoneurons. (Redrawn from Brodal A: *Neurological anatomy,* ed 3, 1981, New York, Oxford University Press.)

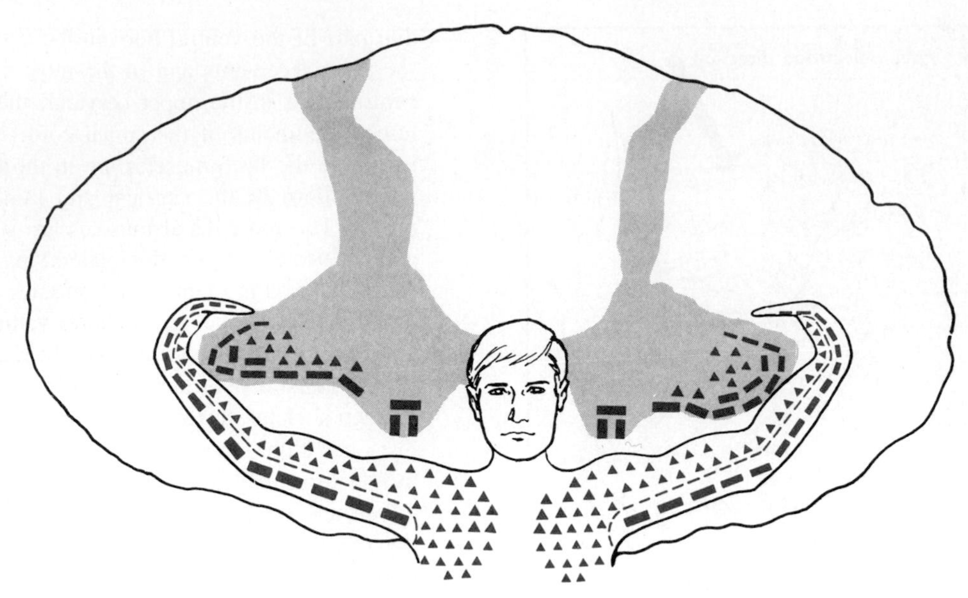

■ **Fig. 7-11** Somatotopic organization of spinal cord motoneurons. Motoneurons to axial muscles are in the medial ventral horn. In the lateral part of the motor nucleus, motoneurons to more proximal muscles are indicated by the larger symbols. Extensor muscles are supplied by motoneurons indicated by solid rectangles and flexor muscles by motoneurons indicated by triangles.

■ **Table 7-3** Characteristics of axons of somatic motoneurons

Type	*Group*	*Subgroup*	*Diameter (μm)*	*Conduction velocity (m/sec)*	*Tissue supplied*	*Function*
Motor						
A	Alpha		12-20	72-120	Muscle	Motor supply of extrafusal skeletal muscle fibers
	Gamma		2-8	12-48	Muscle	Motor supply of intrafusal muscle fibers

From Willis WD, Grossman RG: *Medical Neurobiology,* ed 3, St Louis, 1981, Mosby–Year Book.

the alpha motoneuron (in mammals), the motoneuron has been termed **the final common pathway.**

Another type of motoneuron is called the **γ-motorneuron.** Gamma motoneurons are smaller than alpha motoneurons, and have a soma diameter of about 35 microns. Their dendrites are simpler and oriented mostly in the transverse plane. The gamma motoneurons that project to a particular muscle are located in the same motor nucleus as the alpha motoneurons that supply that muscle. Gamma motoneurons do not supply ordinary skeletal muscle fibers. Instead, they synapse on the specialized striated muscle fibers, the **intrafusal muscle fibers,** that are found within muscle spindles (see Chapter 12).

Table 7-3 shows the sizes of the axons of somatic motoneurons and their conduction velocities.

The skeletal muscle fibers that belong to a given motor unit are called a ***muscle unit.*** All of the muscle fibers in a muscle unit are of the same histochemical type; i.e., they are either all type I, type IIB, or IIA. The contractile properties of these muscle fiber types are summarized in Table 7-4. The motor units that twitch slowly and resist fatigue are classified as **S** (slow) and have type I fibers. S motor units depend on oxidative metabolism for their energy supply and have weak contractions (Fig. 7-12). The muscle units with

■ **Table 7-4** Muscle fiber contractile properties

Type	*Speed*	*Strength*	*Fatigability*	*Motor unit*
I	Slow	Weak	Fatigue resistant	S
IIB	Fast	Strong	Fatigable	FF
IIA	Fast	Intermediate	Fatigue resistant	FR

From Berne RM, Levy MN, editors: *Principles of physiology,* St Louis, 1990, Mosby–Year Book.

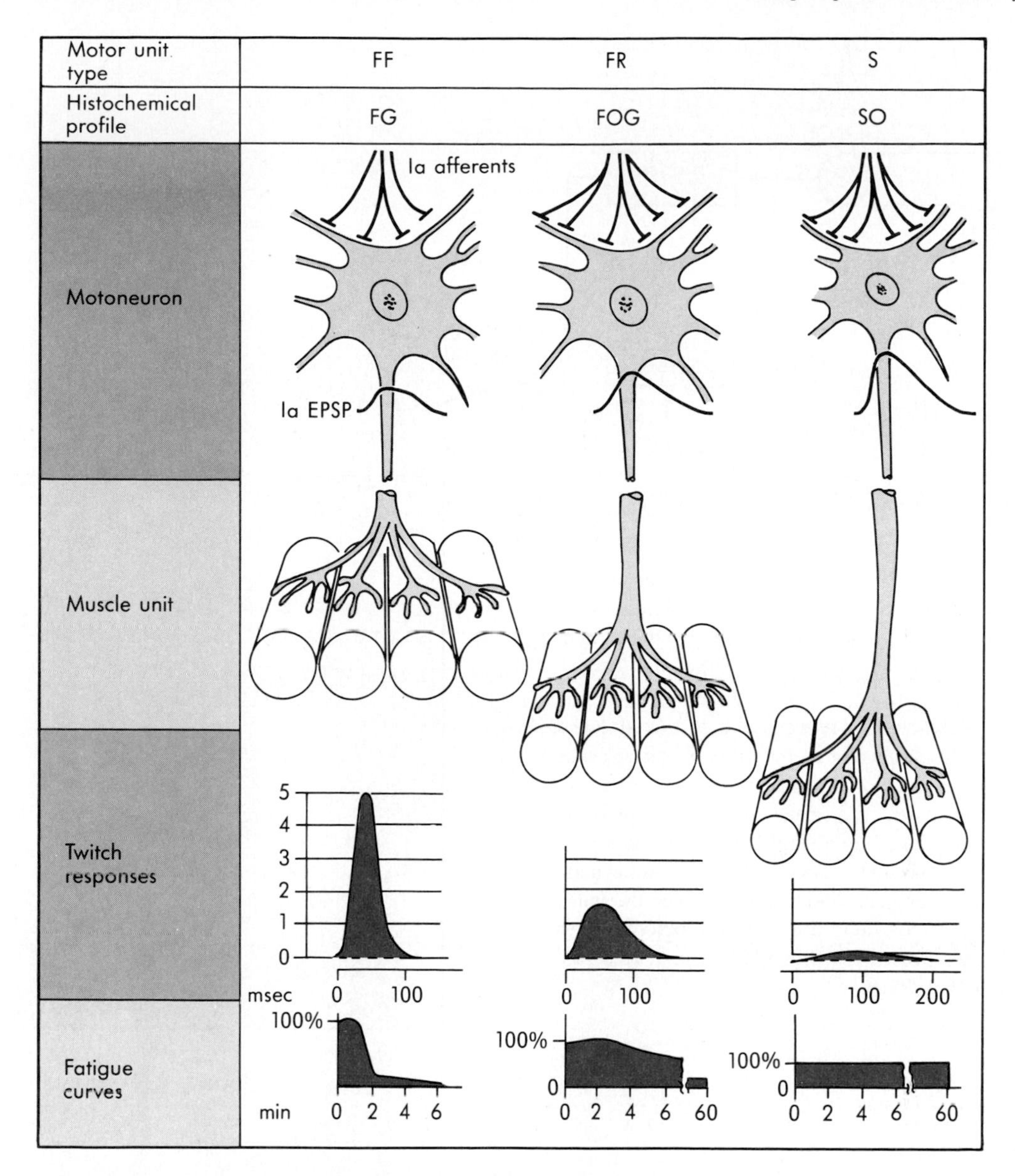

■ **Fig. 7-12** Summary of features of motor units in a mixed muscle (medial gastrocnemius of cat). Relative sizes are shown for motoneurons, muscle fibers, monosynaptic excitatory postsynaptic potentials evoked by volleys in group Ia afferent fibers, and twitch responses. *EPSP*, Excitatory postsynaptic potential; *FG*, fast glycolytic; *FOG*, fast oxidative-glycolytic; *SO*, slow oxidative.

fast twitches are **FF** (fast, fatigable) and **FR** (fast, fatigue resistant). FF muscle units have type IIB fibers, use glycolytic metabolism, and have strong contractions, but they fatigue easily. FR muscle units have type IIA fibers and rely on oxidative metabolism; their contractions are of intermediate force, and these muscle units resist fatigue.

A clinically useful way to monitor the activity of motor units is **electromyography.** An electrode is placed within a skeletal muscle to record the summed action potentials of the skeletal muscle fibers of a muscle unit (Fig. 7-13). If no spontaneous activity is noted, the patient can be asked to contract the muscle voluntarily to increase the activity of motor units in the muscle. As the force of voluntary contraction increases, more motor units are **recruited.** The first motor units to be activated, either by voluntary effort or during reflex action (see Chapter 12), are those with the smallest motor axons (Fig. 7-14); these motor units generate the smallest contractile forces and allow the initial contraction to be finely graded. As more motor units are recruited, the motoneurons with progressively larger axons become

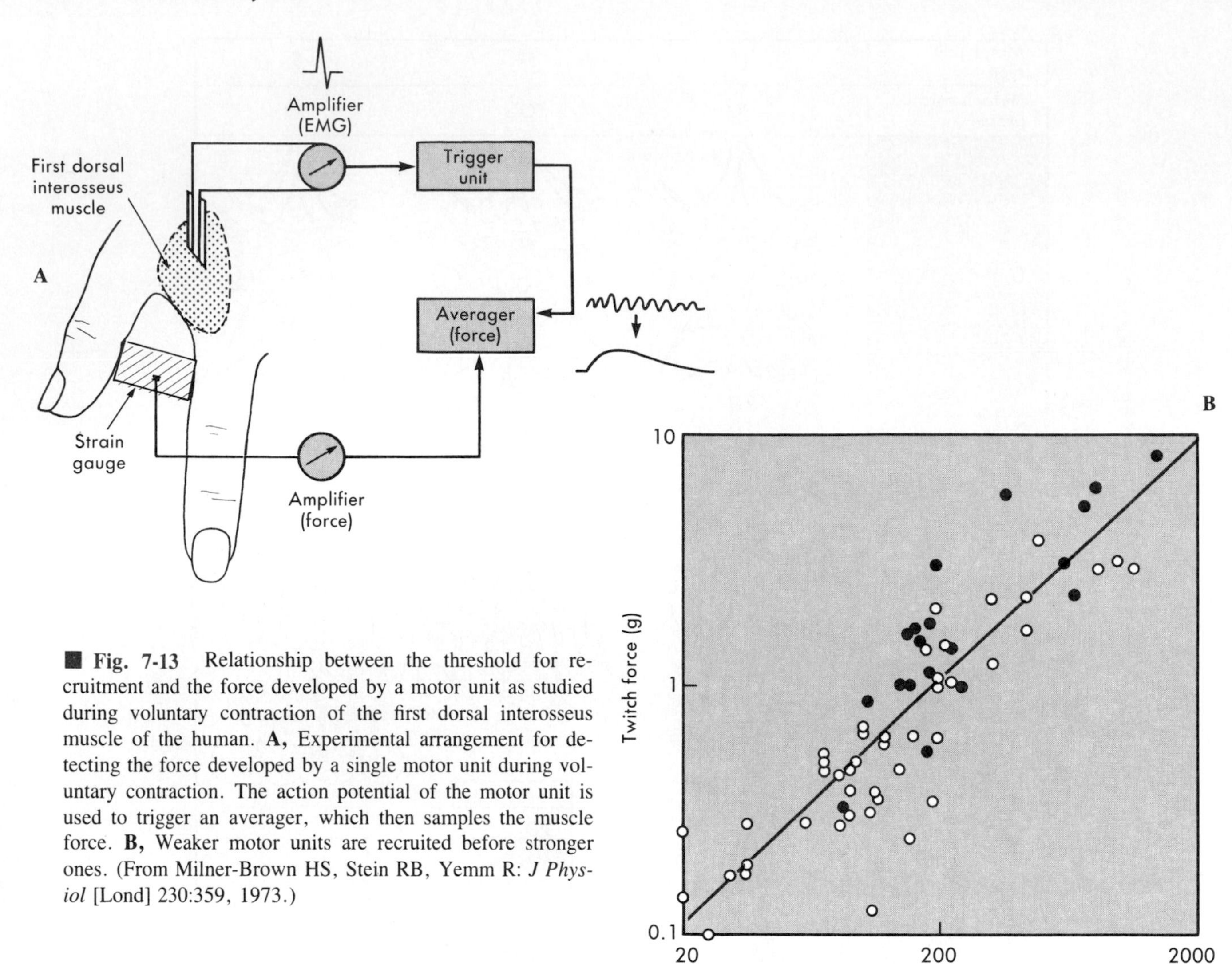

■ **Fig. 7-13** Relationship between the threshold for recruitment and the force developed by a motor unit as studied during voluntary contraction of the first dorsal interosseus muscle of the human. **A,** Experimental arrangement for detecting the force developed by a single motor unit during voluntary contraction. The action potential of the motor unit is used to trigger an averager, which then samples the muscle force. **B,** Weaker motor units are recruited before stronger ones. (From Milner-Brown HS, Stein RB, Yemm R: *J Physiol* [Lond] 230:359, 1973.)

involved and generate progressively larger amounts of tension. This orderly recruitment of motor units is called the **size principle,** because the motor units are recruited in order of motor axon size. The size principle depends on the fact that small alpha motoneurons are activated by excitatory postsynaptic potentials that are greater than those in large alpha motoneurons (Fig. 7-12). In addition to the recruitment of more motoneurons, contractile strength is affected by the rate of discharge of active alpha motoneurons.

■ *Summary*

1. The peripheral nervous system (PNS) contains the axons of primary afferent neurons, somatic motoneurons, and autonomic preganglionic and postganglionic neurons.

2. The cell bodies of primary afferent neurons are in dorsal root or cranial nerve ganglia.

3. Sensory receptors include exteroceptors, interoceptors, and proprioceptors. Stimuli are environmental events that excite sensory receptors; responses are the effects of stimuli; sensory transduction is the process by which stimuli are detected.

4. Sensory transduction is accomplished in different ways by different sensory receptors, but in general it involves the production of a receptor potential.

5. Sensory receptors may be slowly or rapidly adapting.

6. A receptive field is the region that, when stimulated, causes a response in sensory neurons.

7. Sensory receptors encode modality, spatial location, threshold, intensity, frequency, and duration of stimuli.

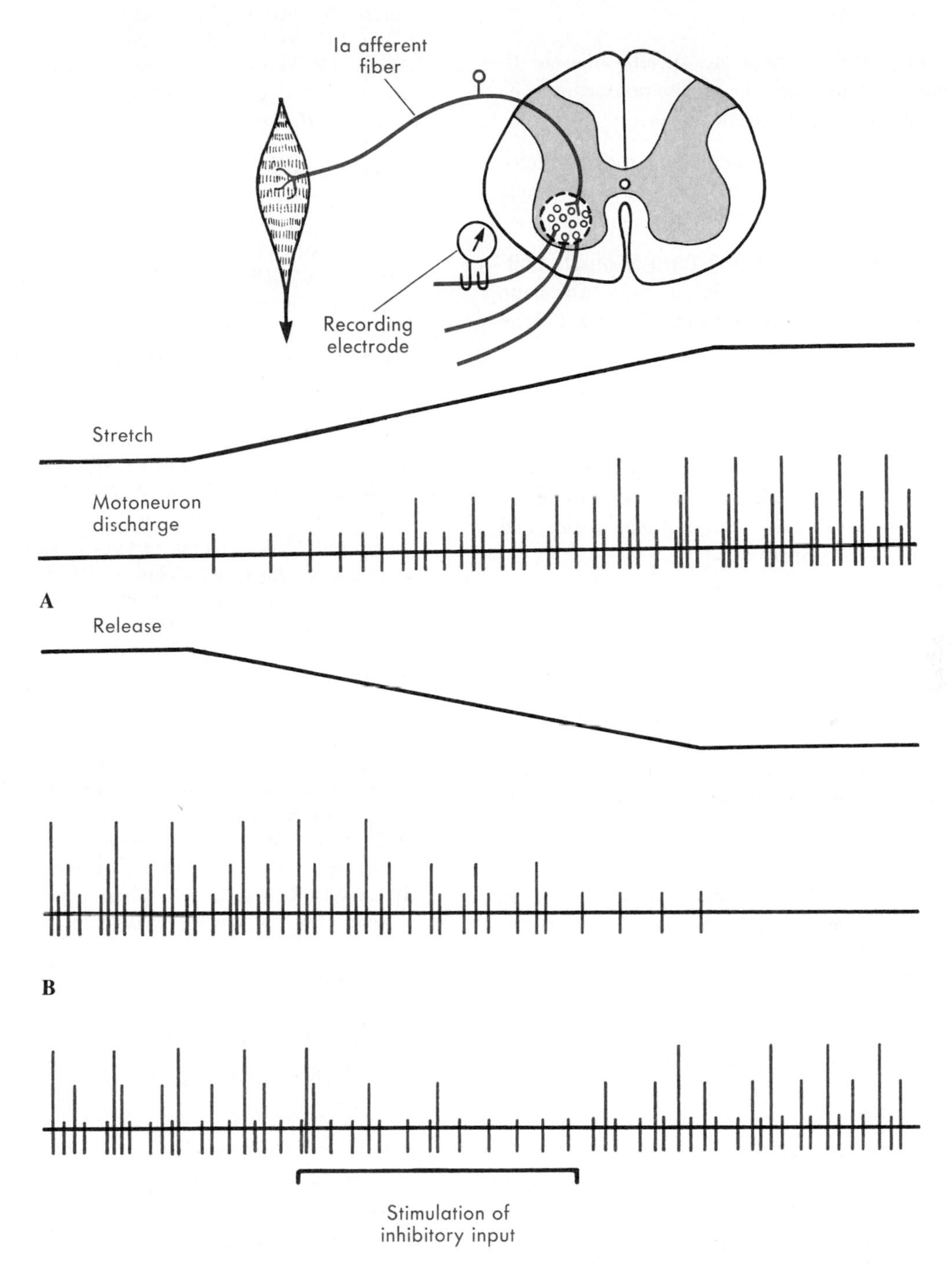

■ **Fig. 7-14** The size principle in the recruitment of motoneurons. The schematic at the top shows Ia afferent fiber from a muscle spindle and recording electrodes on a dissected ventral root filament arising from an homonymous motoneuron. **A,** Excitation. As a muscle is stretched, the increased activity of the Ia afferent fibers first recruits the smaller motoneurons. As the stretch increases, successively larger motoneurons are recruited. On release from stretch, the larger motoneurons stop discharging first and the smaller motoneurons last. **B,** Inhibition. Activating an inhibitory input to the motoneuron pool first silences the larger motoneurons and then successively smaller motoneurons. (From Eyzaguirre C, Fidone SJ: *Physiology of the nervous system: an introductory text,* ed 2, Chicago 1975, Mosby–Year Book.

8. Primary afferent fibers can be subdivided according to the different receptor types that they supply and by their size.

9. Alpha motoneurons innervate skeletal muscle fibers; a motor unit includes an alpha motoneuron, its axon, and a set of skeletal muscle fibers.

10. Gamma motoneurons are smaller than alpha motoneurons and supply intrafusal muscle fibers in muscle spindles.

11. Muscle fibers in a motor unit form a muscle unit; all of the muscle fibers of a muscle unit are of the same histochemical type (slow; fast, fatigable; or fast, fatigue resistant).

12. The activity of motor units can be monitored by electromyography.

13. Motor units are recruited in an orderly fashion during voluntary or reflex activity; the first units to be recruited are those with the smallest motor axons and contractile force (size principle).

■ *Bibliography*

Journal articles

Burke RE: Motor unit properties and selective involvement in movement, *Exercise sports science rev,* 3:31, 1975.

Westbury DR: A comparison of the structures of alpha- and gamma-spinal motoneurones of the cat, *J Physiol* 325:79, 1982.

Books and monographs

Akoev GN, Andrianov GN: Synaptic transmission in the mechano-and electroreceptors of the acousticolateral system. In Ottoson D, editor: *Progress in sensory physiology 9,* Berlin, 1989, Springer-Verlag.

Basmajian JV: *Muscles alive: their functions revealed by electromyography,* Baltimore, 1967, Williams & Wilkins.

Binder MD, Mendell LM, editors: *The segmental motor system,* New York, 1990, Oxford University Press.

Burgess PR, Perl ER: Cutaneous mechanoreceptors and nociceptors. In Iggo A, editor: *Handbook of sensory physiology, vol II, Somatosensory system,* Berlin, 1973, Springer-Verlag.

Burke RE: Motor units: anatomy, physiology and functional organization. In Brookhart JM, Mountcastle VB, editors: *Handbook of physiology: the nervous system,* vol II, Bethesda, Md, 1981, American Physiological Society.

Henneman E, Mendell LM: Functional organization of motoneuron pool and its inputs. In Brookhart JM, Mountcastle VB, editors: *Handbook of physiology: the nervous system,* vol 2, Bethesda, Md, 1981, American Physiological Society.

Lindauer M, Martin H: The biological significance of the earth's magnetic field. In Ottoson D, editor: *Progress in sensory physiology 5,* Berlin, 1985, Springer-Verlag.

Loewenstein O, editor: *Principles of receptor physiology,* Berlin, 1971, Springer-Verlag.

Mountcastle VB: Central nervous system mechanisms in mechanoreceptive sensibility. In Brookhart JM, Mountcastle VB, editors: *Handbook of physiology, the nervous system,* vol III, Bethesda, Md, 1984, American Physiological Society.

Shepherd GM: *Neurobiology,* New York, 1983, Oxford University Press.

Sherrington CS: *The integrative action of the nervous system,* ed 2, New Haven, Con, 1947, Yale University Press.

Uttal WR: *The psychobiology of sensory coding,* New York, 1973, Harper & Row.

Willis WD, Coggeshall RE: *Sensory mechanisms of the spinal cord,* ed 2, New York, 1991, Plenum Press.

Willis WD, Grossman RG: *Medical neurobiology,* ed 3, St Louis, 1981, Mosby–Year Book.

CHAPTER

8

The Somatosensory System

As the mammalian brain evolved, the role of the cerebral cortex and thalamus in sensory processing expanded enormously. In primitive brains, subcortical and extrathalamic sensory structures were crucial. Comparable structures continue to be very important in the advanced brains of modern mammals. For example, the reticular formation in the brainstem was one of the major sensory-motor integration systems in nonmammalian vertebrates. In mammals it continues to be involved in sensory processing, and it contributes to the arousal mechanism, to selective attention, and motor control. Another example is the optic tectum, which is the most important structure in the visual system of nonmammalian vertebrates. The tectum still has important visual functions in mammals, even though the highly evolved visual cortex has been added.

In this and the following chapters, the sensory systems of mammals will be considered. Emphasis will be placed on the more recently evolved components of most of these sensory systems that depend on thalamocortical interactions. However, reference will also be made to the more primitive components.

■ *Sensory Pathways*

A sensory pathway can be viewed as a set of sensory neurons arranged in series (Fig. 8-1). First-, second-, third-, and higher-order neurons serve as sequential elements in a given sensory pathway. Furthermore, several parallel sensory pathways are often involved in transmitting similar sensory information.

The **first-order neuron** in a sensory pathway is the primary afferent neuron. The peripheral endings of this neuron form a sensory receptor (or receive input from an accessory sensory cell, such as a hair cell). Thus the first-order neuron responds to a stimulus, transduces it, and then transmits encoded information to the central nervous system (CNS). The primary afferent neuron often has its soma in a dorsal root or cranial nerve ganglion.

The **second-order neuron** is generally located in the spinal cord or brainstem. It receives information from first-order neurons (usually from several or even many) and transmits information to the thalamus. The information may be transformed at the level of the second-order neurons by local neural processing circuits and by the biophysical properties of the second-order neurons. The axon of a second-order neuron typically crosses the midline to ascend to the thalamus. Thus sensory information that originates on one side of the body reaches the contralateral thalamus.

The **third-order neuron** resides in one of the sensory nuclei of the thalamus. Again, local circuits in the thalamus and intrinsic membrane properties of the third-order neurons may transform information received from the second-order neurons before the signals are transmitted to the cerebral cortex.

Fourth-order neurons in the appropriate sensory receiving areas of the cerebral cortex and **higher-order neurons** in the same and other cerebral cortical areas process the information further. At some undetermined site the sensory information results in **perception,** which is a conscious awareness of the stimulus.

■ *Somatovisceral Sensory System*

The somatosensory system, or appropriately the somatovisceral sensory system, transmits information from sensory receptor organs in the skin, muscles, joints, and viscera. Information arising from these sensory receptors reaches the CNS by way of first-order neurons, which are the primary afferent neurons. The cell bodies of the primary afferent neurons are located in dorsal root or cranial nerve ganglia. Each ganglion cell gives off a neurite that branches into a peripheral process and a central process. The **peripheral process** has the structure of an axon and terminates peripherally as a sensory receptor (or on an accessory cell). The **central process** is also an axon, and it enters the spinal cord through a dorsal root or it enters the brainstem through a cranial nerve. The central process typically gives rise to numerous collateral branches that end synaptically on several second-order neurons.

The processing of somatovisceral sensory informa-

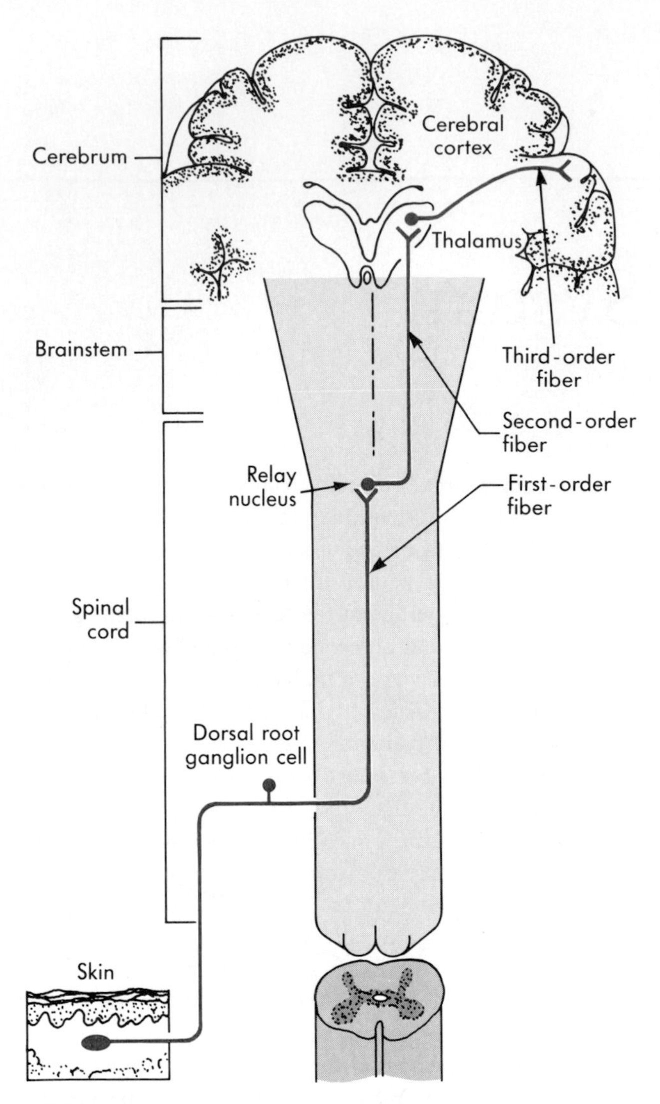

■ **Fig. 8-1** General arrangement of sensory pathways. First-, second-, and third-order neurons are shown. Note that the axon of the second-order neuron crosses the midline, so that sensory information from one side of the body is transmitted to the opposite side of the brain.

tion involves a number of CNS structures, including the spinal cord, brainstem, thalamus, and cerebral cortex. The ascending pathways arise from second-order neurons that are located in the spinal cord and brainstem and that project to the contralateral thalamus. The most important ascending somatosensory pathways that carry somatovisceral information from the body are the dorsal column–medial lemnisicus path and the spinothalamic tract. The main somatosensory projection that represents the face is the trigeminothalamic tract. Ancillary somatosensory pathways include the spinocervicothalamic path, the postsynaptic dorsal column path, the dorsal spinocerebellar tract, the spinoreticular tract, and the spinomesencephalic tract.

The sensory modalities that are mediated by the somatovisceral sensory system include touch-pressure, flutter-vibration, proprioception (position sense and joint movement), thermal sense (warmth and cold), pain, and visceral distention.

■ *Sensory Receptors*

■ *Cutaneous Receptors*

Cutaneous receptors can be subdivided according to the type of stimulus to which they respond. The major types of receptors include mechanoreceptors, thermoreceptors, and nociceptors (there are also recently discovered chemoreceptors).

Mechanoreceptors. Mechanoreceptors respond to such mechanical stimuli as stroking or indenting the skin, and they can be rapidly adapting or slowly adapting. Rapidly adapting cutaneous mechanoreceptors include **hair follicle receptors** in the hairy skin, **Meissner's corpuscles** in the glabrous (nonhairy) skin, and **Pacinian corpuscles** in subcutaneous tissue (Fig. 8-2, *A*). Hair follicle receptors and Meissner's corpuscles respond best to stimuli repeated at rates of about 30 to 40 Hz, whereas Pacinian corpuscles prefer stimuli repeated at about 250 Hz. Slowly adapting cutaneous mechanoreceptors include **Merkel cell endings** and **Ruffini endings** (Fig. 8-2, *B*). Merkel cell receptors have punctate receptive fields, whereas Ruffini endings can be activated by stretching the skin, even some distance away from the receptor terminals.

The axons of all of these receptor types are myelinated. Most of the axons are Aβ fibers, although one class of hair follicle receptor, the down hair receptor, is supplied by Aδ fibers. Some mechanoreceptors are innervated by unmyelinated axons. These are the C mechanoreceptors, which respond best to very slowly moving stimuli, such as stroking. The C mechanoreceptors have only recently been found in humans, although they are common in other mammals, such as cats.

Thermoreceptors. Thermoreceptors signal the temperature of the skin. The two types of thermoreceptors in the skin are **cold** and **warm receptors.** Both classes are slowly adapting, although they also discharge phasically when skin temperature is changing rapidly (Fig. 8-3, *A*). These are among the few receptor types that discharge spontaneously under normal circumstances. The receptors are active over a broad range of temperatures (Fig. 8-3, *B*). At a moderate skin temperature, such as 35° C, both types of receptor may be active. However, as the skin is warmed, the cold receptors become silent; conversely, as the skin is cooled, the warm receptors become inactive. Note that the warm receptors also stop discharging as the temperature extends into the **noxious** (damaging) range (above 45° C). Therefore these receptors cannot signal heat pain.

The stimulus-response curve for cold receptors in Fig. 8-3, *B,* shows that the mean discharge frequencies of these fibers would not provide a way for the CNS to

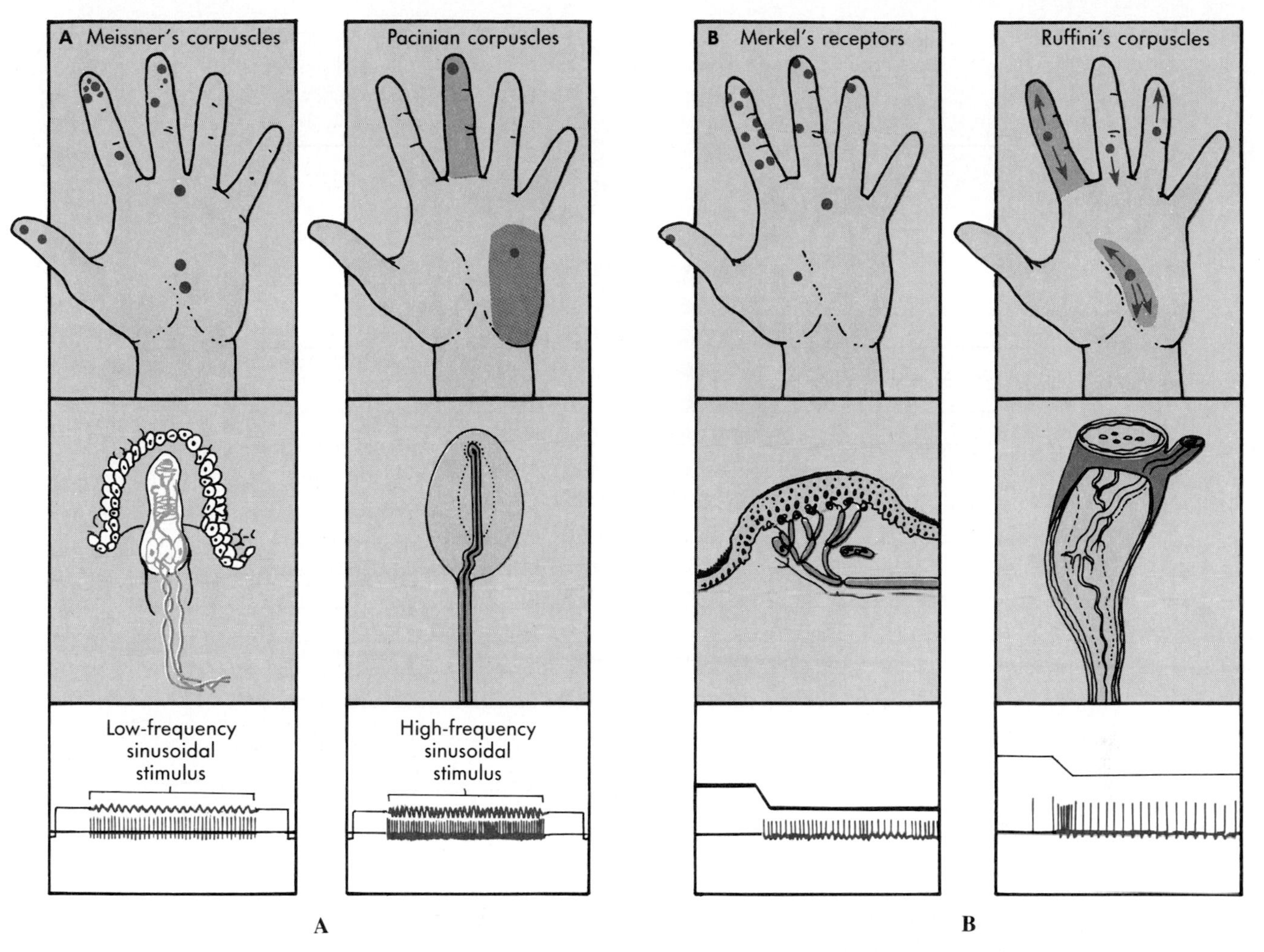

Fig. 8-2 The receptive fields of several types of cutaneous mechanoreceptors are shown in the top row of drawings. **A,** Rapidly adapting mechanoreceptors: Meissner's corpuscles and Pacinian corpuscles. **B,** Slowly adapting mechanoreceptors: Merkel's receptors and Ruffini's corpuscles. The second row of drawings shows the morphology of the receptors; the third row, the responses to sinusoidal stimuli **(A)** or to step indentations of the skin **(B).**

discriminate between temperatures at points above and below the peak of the curve. However, cold receptors behave unusually in that, over a certain range of temperatures, their discharges occur in bursts (Fig. 8-3, *A*). These bursts may provide information that allows the CNS to distinguish between the activity of cold receptors exposed to higher and lower temperatures. Alternatively, an additional class of high threshold cold receptors is activated when the temperature of the skin is lowered sufficiently. Most cold fibers are supplied by Aδ fibers and warm receptors by C fibers.

Nociceptors. Nociceptors respond to stimuli that may produce damage. The two major classes of cutaneous nociceptors are the **Aδ mechanical nociceptors** and the **C-polymodal nociceptors,** although several other types also exist. As their names suggest, the Aδ mechanical nociceptors are supplied by finely myelinated afferent fibers and the C-polymodal nociceptors by unmyelinated fibers. The Aδ mechanical nociceptors respond to strong mechanical stimuli, such as pricking the skin with a needle or crushing the skin with forceps. They typically do not respond to noxious thermal or chemical stimuli unless they have previously been sensitized (see below). C-polymodal nociceptors, on the other hand, respond to several types of noxious stimuli, including mechanical, thermal (Fig. 8-4) and chemical stimuli.

Sensitization of nociceptors is a process that causes these afferent fibers to become more responsive. Sensitized nociceptors discharge more vigorously after a given noxious stimulus and their threshold for activation is lower (Fig. 8-4). This can lead to **hyperalgesia,** an increase in the pain produced by stimulation at a given intensity and a decrease in the pain threshold. The nociceptors may also develop a background discharge and therefore produce spontaneous pain.

Sensitization occurs when chemical products, such as

A

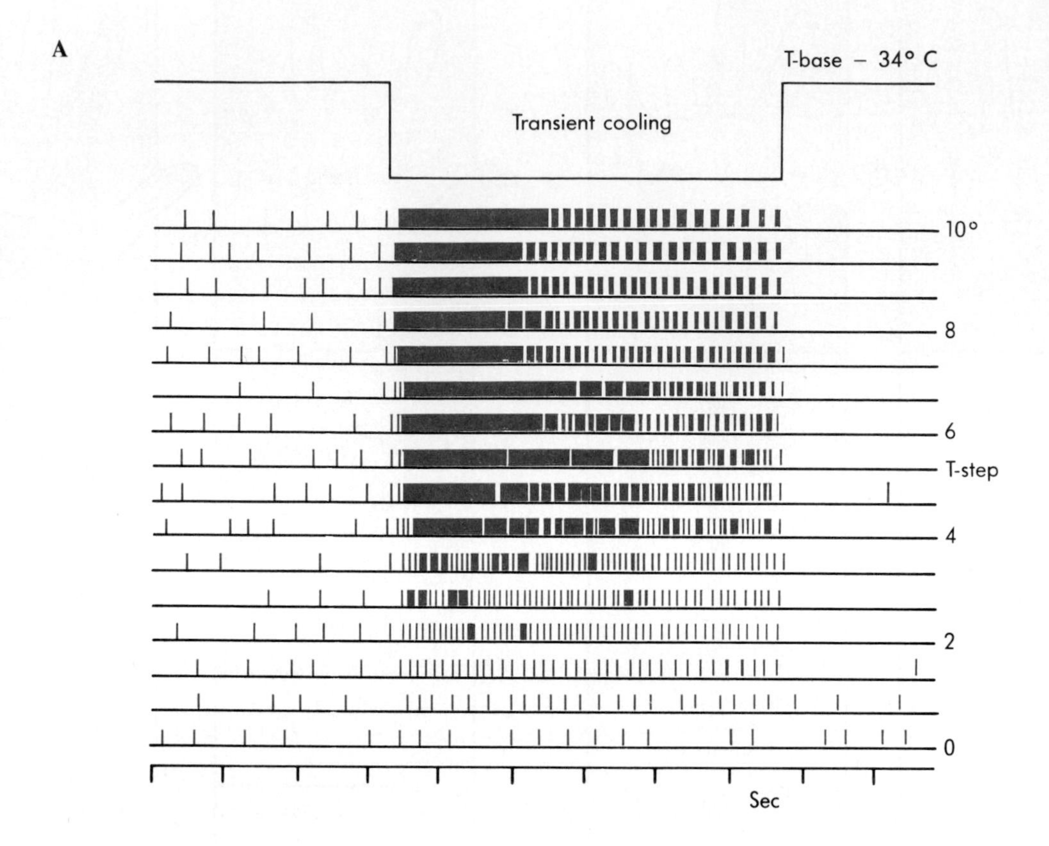

B

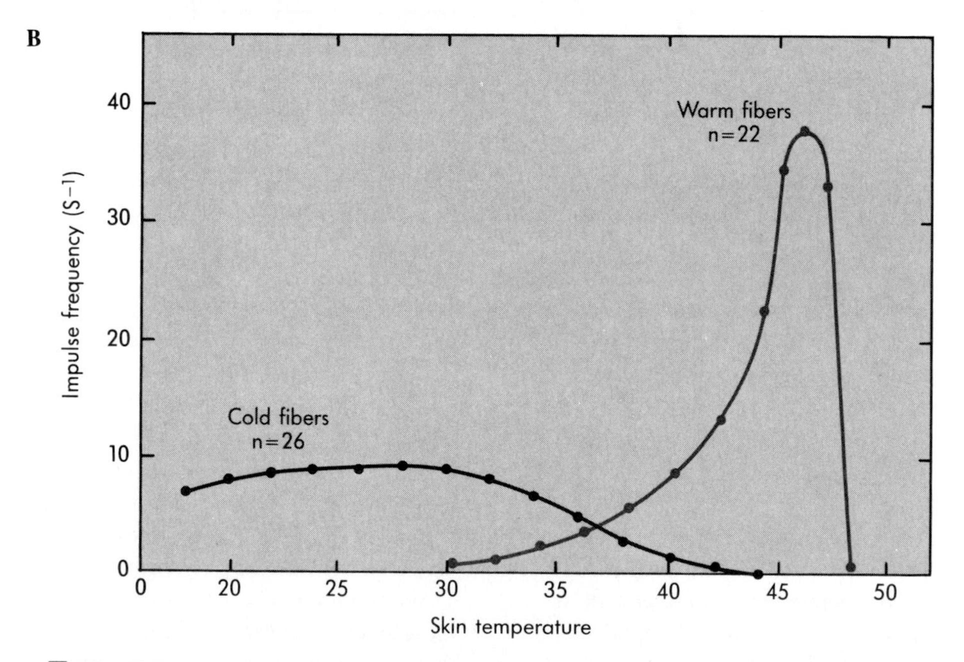

■ **Fig. 8-3** **A,** Responses recorded from the afferent supplying a cold receptor after graded cooling pulses. **B,** Average static discharge rates for populations of cold and warm receptors. (**A,** From Darian-Smith I, Johnson KD, Dykes R: "Cold" fiber population innervating palmar and digital skin of the monkey: response to cooling pulses *J Neurophysiol* 36:325, 1973. (**B,** From Hensel H, Kenshalo DR: *J Physiol* 204:99, 1969.)

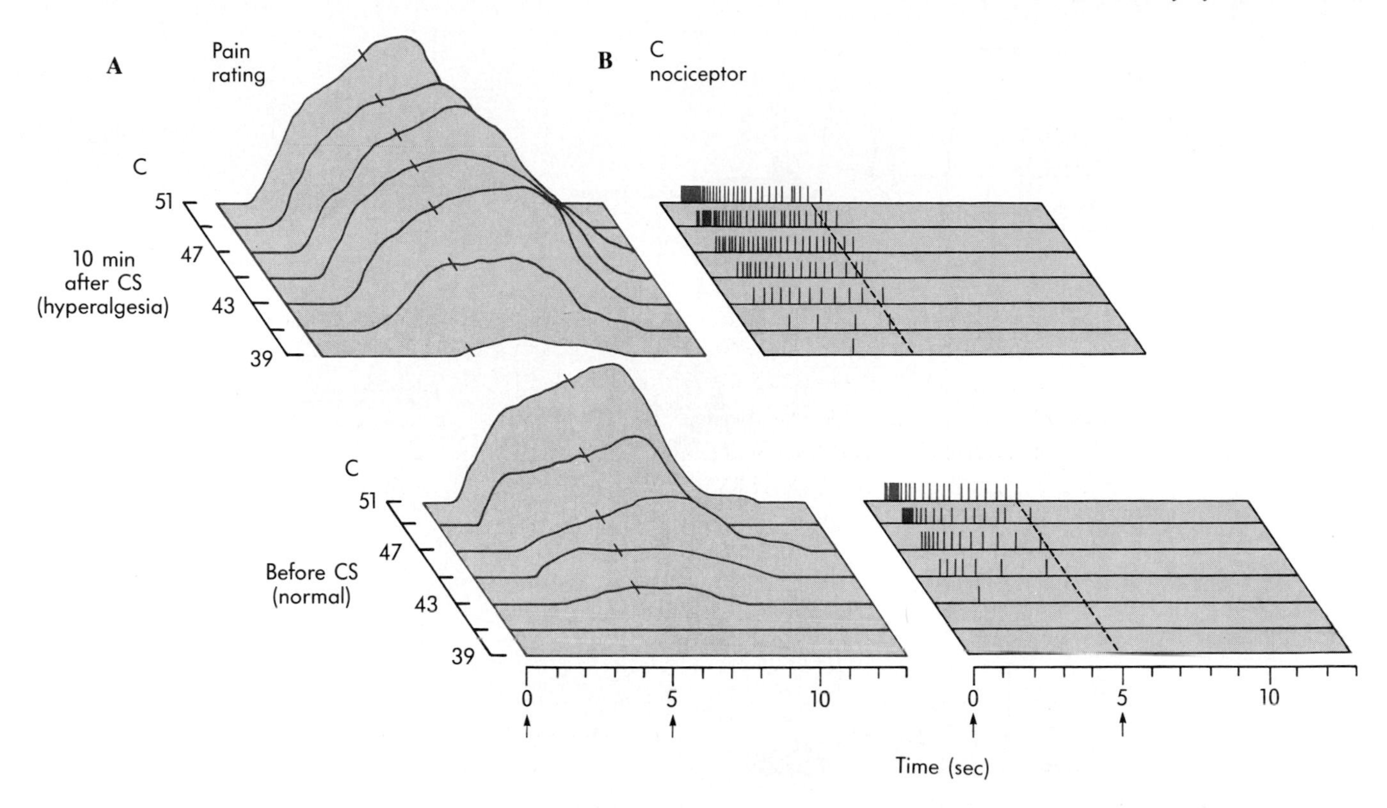

■ **Fig. 8-4** **A,** Curves show magnitude ratings of pain in a human in response to heat pulses applied to the hairy skin before and after a mild burn. Arrows indicate the beginning and end of the stimuli. **B,** The responses of a C polymodal nociceptor innervating the hairy skin in a monkey in response to the same heat pulses before and after the mild burn. (From LaMotte RH, Thalhammer JG, Robinson CJ: *J Neurophysiol* 50:1, 1983.

K^+, bradykinin, serotonin, histamine, and eicosanoids (prostaglandins and leukotrienes), are released near nociceptor terminals after damage or during inflammation. For example, a noxious stimulus applied to the skin may destroy cells near a nociceptor (Fig. 8-5, *A*). The dying cells release K^+, which will depolarize the nociceptor. They also release proteolytic enzymes that react with circulating globulins to form bradykinin. The bradykinin then binds to a receptor on the membrane of the nociceptor and activates a second messenger system, which in turn sensitizes the ending. Other chemical agents, such as serotonin released from platelets, histamine from mast cells, and eicosanoids from various cellular elements also contribute to sensitization, either by opening ion channels or by activating second messenger systems. Many of these substances also act on blood vessels, immune cells, platelets and other effectors that participate in inflammation.

Activation of a nociceptor terminal can also release chemicals, such as the peptide substance P (SP), from other terminals of the same nociceptor through an **axon reflex** (Fig. 8-5, *A*). A nerve impulse established in one branch of a nociceptor conducts centrally through the parent axon, but in addition it spreads antidromically into other branches of the axon in the skin. As the nerve impulse invades these peripheral branches of the nociceptor, SP is released into the skin (Fig. 8-5, *B*). The SP has several effects, including vasodilation and increased capillary permeability. These add to the effects of other agents released from damaged cells and from platelets, mast cells, and invading leucocytes. The resulting inflammation causes a typical series of changes in the skin, such as reddening and warming due to the increased blood flow, swelling due to the neurogenic edema, and pain and tenderness due to the sensitization of nociceptors. These responses constitute the classical combination in inflammation of **rubor** (redness), **calor** (heat), **tumor** (swelling), and **dolor** (pain).

■ *Muscle, Joint, and Visceral Receptors*

Skeletal muscle also contains several types of sensory receptors. These are chiefly mechanoreceptors and nociceptors, although some muscle receptors may possess thermosensitivity or chemosensitivity. The best studied muscle receptors are the **stretch receptors,** which include muscle spindles and Golgi tendon organs. Although these are important in proprioception, they may be even more important in motor control. Their structure and function are discussed in Chapter 12. Nociceptors in muscle respond to pressure applied to the muscle

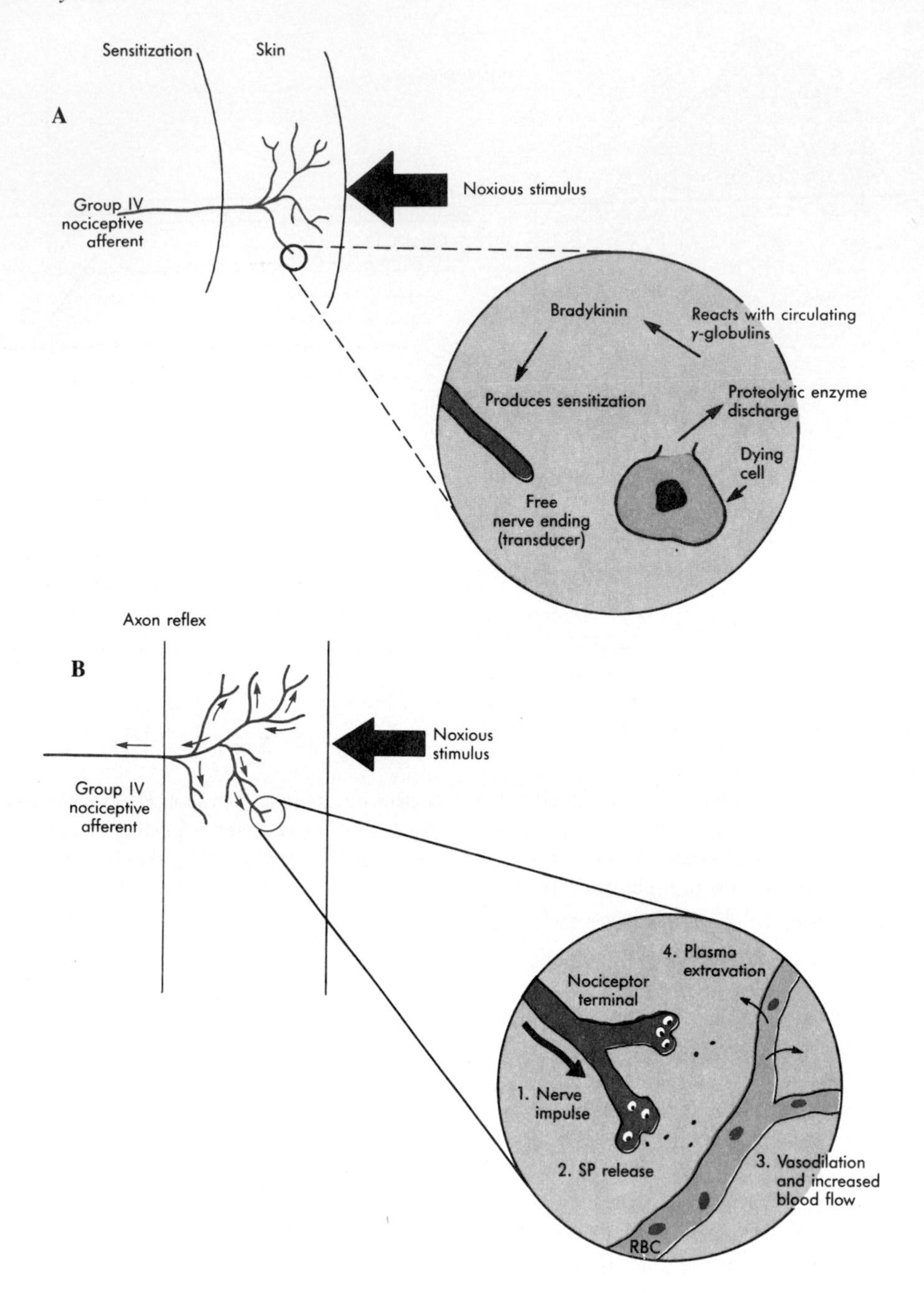

■ **Fig. 8-5** **A,** sensitization of the terminals of a nociceptor. A noxious stimulus causing damage to cells results in the local release of proteolytic enzymes that react with circulating proteins to produce bradykinin. Bradykinin then binds to a receptor on the membrane of the nociceptive afferent fiber and sensitizes it (after activation of a second messenger system). The nociceptor is now more responsive to further stimulation. Other agents that would have similar effects include prostaglandins, serotonin, histamine, leukotrienes, and K^+ ions, although these have different actions at the membrane level (e.g., serotonin would act on a receptor that opens an ion channel). **B,** release of substance P *(SP)* from a nociceptor following an axon reflex causes changes in the local environment. One action is vasodilation, resulting in reddening of the skin and warming. Another is increased capillary permeability, resulting in plasma extravasation.

and to release of metabolites, especially during ischemia. Muscle nociceptors are supplied by medium-sized and small myelinated (group II and III) axons or by unmyelinated (group IV) afferent fibers. Other muscle receptors with fine afferent fibers are regarded as **ergoreceptors,** which signal muscle work.

Joints are associated with several types of sensory receptors, such as rapidly and slowly adapting mechanoreceptors and nociceptors. The rapidly adapting mechanoreceptors are Pacinian corpuscles, which respond to mechanical transients, including vibration. The slowly adapting receptors are Ruffini endings, which respond best to extreme movements of a joint. These endings signal pressure or torque applied to the joint. Joint mechanoreceptors are innervated by medium-sized (group II) afferent fibers. Joint nociceptors are activated by hyperextension or hyperflexion, although many articular nociceptors fail to respond to joint movements under normal conditions. If sensitized by inflammation, however, they respond to movements or to weak pressure stimuli, which are normally innocuous. Joint nociceptors are innervated by finely myelinated (group III) or unmyelinated (group IV) primary afferent fibers.

Viscera are sparsely supplied with sensory receptors. These are usually involved in reflexes and have little to do with sensory experience. However, some visceral mechanoreceptors are responsible for the sensation of distension, and visceral nociceptors signal visceral pain. Pacinian corpuscles are present in the mesentery and in the capsules of visceral organs such as the pancreas. These presumably signal mechanical transients. Whether some forms of visceral pain result from overactivity of mechanoreceptor afferent fibers is still controversial. Some viscera, however, clearly have specific nociceptors. It is likely that some visceral nociceptors are silent under normal circumstances but become active after sensitization by damage or inflammation.

Microneurography

The sensory functions of various cutaneous sensory receptors have been examined in human subjects by a technique known as microneurography. A fine metal microelectrode is inserted into a nerve trunk in the arm or leg. When a recording can be made from a single sensory axon, the receptive field of the fiber is mapped. Many of the same types of sensory receptors that have been studied in experimental animals have also been found in humans.

In some humans it has been possible to activate what seems to be the same sensory axon by stimulating through the microelectrode. The subject is asked to locate the perceived receptive field of the sensory axon, and this turns out to be identical to the mapped receptive field. Repetitive stimulation is usually required to evoke a sensation, but the sensations produced by individual nerve fibers are perceived as pure sensations that match the properties of the receptors. The quality of the sensations remains the same no matter what the frequency of stimulation. For example, stimulation of the afferent fiber of a Meissner corpuscle causes a sensation of flutter, of a Pacinian corpuscle a sensation of vibration, and of a Merkel receptor a sensation of maintained touch. If the frequency of stimulation is increased, the perceived frequency of flutter or of vibration increases in the case of Meissner or Pacinian corpuscles. However, for Merkel endings, an increase in the frequency of stimulation augments the intensity of the touch or pressure sensation. In no case does the quality of the sensation change, for example, to pain.

The axons of most Ruffini endings so far tested do not produce a sensation when stimulated, but a few have recently been shown to cause sensations of either touch or position of a finger joint. Central summation is probably required for most Ruffini endings to produce a sensation, and the sensation can be either touch-pressure or proprioception.

Activation of individual Aδ nociceptors causes pricking pain. Stimulation of C-polymodal nociceptors produces either burning pain or in some cases itch. The sensory effects of C fibers probably result from activation of more than one afferent fiber.

Dermatomes, Myotomes, and Sclerotomes

Primary afferent fibers in the adult are distributed systematically both peripherally and centrally. The pattern of innervation is determined during embryological development. The mammalian embryo becomes segmented, and each body segment is called a **somite.** A somite is innervated by an adjacent segment of the spinal cord or, in the case of a somite of the head, by a cranial nerve. The portion of a somite destined to form skin is called a **dermatome.** Similarly, the part of a somite that will form muscle is a **myotome,** and the part that will form bone, a **sclerotome.** Viscera are also supplied by particular segments of the spinal cord or particular cranial nerves.

Many dermatomes become distorted during development, chiefly because of the rotation of the upper and lower extremities as they are formed and also because humans maintain an upright posture. However, the sequence of dermatomes can readily be understood if depicted on the body in a quadrupedal position (Fig. 8-6).

Although a dermatome receives its densest innervation from the corresponding spinal cord segment, the dermatome is also supplied by several adjacent spinal segments. Thus transection of a single dorsal root causes little sensory loss in the corresponding dermatome. Anesthesia of any given dermatome requires interruption of several successive dorsal roots.

Spinal Roots

As shown in Fig. 8-7, axons of the PNS enter or leave the CNS through the spinal roots (or through cranial nerves). The **dorsal root** on one side of a given spinal segment is composed entirely of the central processes of dorsal root ganglion cells. The **ventral root** consists chiefly of motor axons, including α-motor axons, γ-motor axons, and at certain segmental levels, autonomic preganglionic axons. Ventral roots also contain many primary afferent fibers, whose role is still unclear.

Just before they penetrate the spinal cord, the large myelinated primary afferent fibers assume a medial position in the dorsal root, whereas the fine myelinated and unmyelinated fibers shift to a lateral position (see

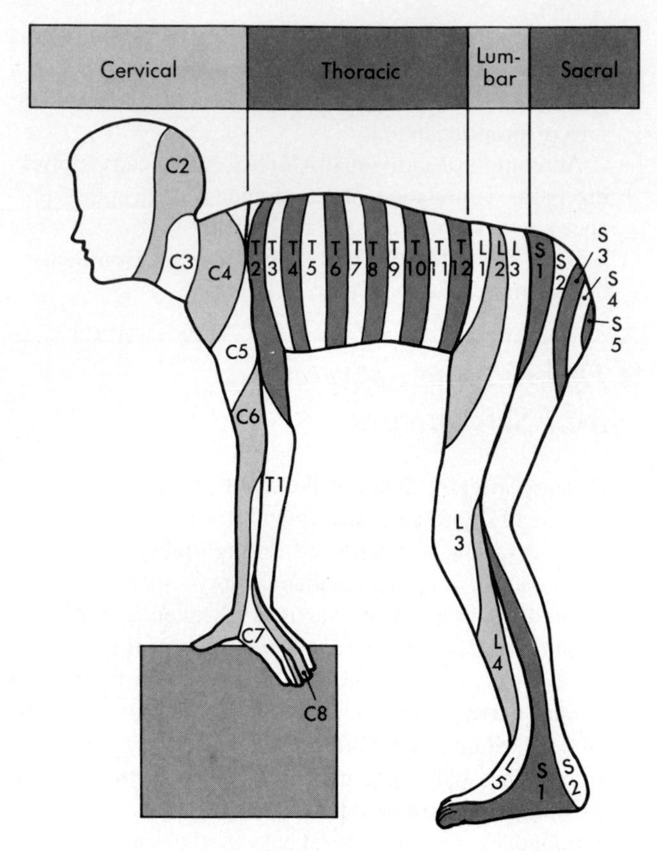

■ **Fig. 8-6** Dermatomes represented on a drawing of a person assuming a quadrupedal position.

Fig. 8-7). The large, medially placed afferent fibers enter the dorsal column, where they bifurcate to send one branch rostrally and another branch caudally. These branches travel through several segments, and some ascend to the medulla as part of the dorsal column-medial lemniscus pathway (see p. 129). The axons in the dorsal funiculus give off collaterals that pass ventrally into the gray matter of the spinal cord. These transmit sensory information to neurons in the dorsal horn and also provide the afferent limb of reflex pathways.

The fine myelinated and unmyelinated primary afferent fibers enter the dorsolateral (Lissauer's) fasciculus, where they bifurcate and send branches rostrally and caudally for a short distance. Collaterals from these branches enter and terminate in the spinal cord gray matter. The endings of the large and small primary afferent fiber projections to the spinal cord are located in distinctly different regions of the gray matter.

The spinal cord gray matter can be subdivided into 10 **laminae** or layers. The dorsal horn is composed of laminae I to VI. The intermediate region is equivalent to parts of laminae VI and VII. The ventral horn in the cervical and lumbar enlargements is subdivided into a medial zone, lamina VIII, and a lateral zone, lamina IX, which is the motor nucleus that supplies the limb muscles. A component of lamina IX is also found medially (the motor nucleus of the axial musculature). Lamina VII extends ventrally between lamina VIII and

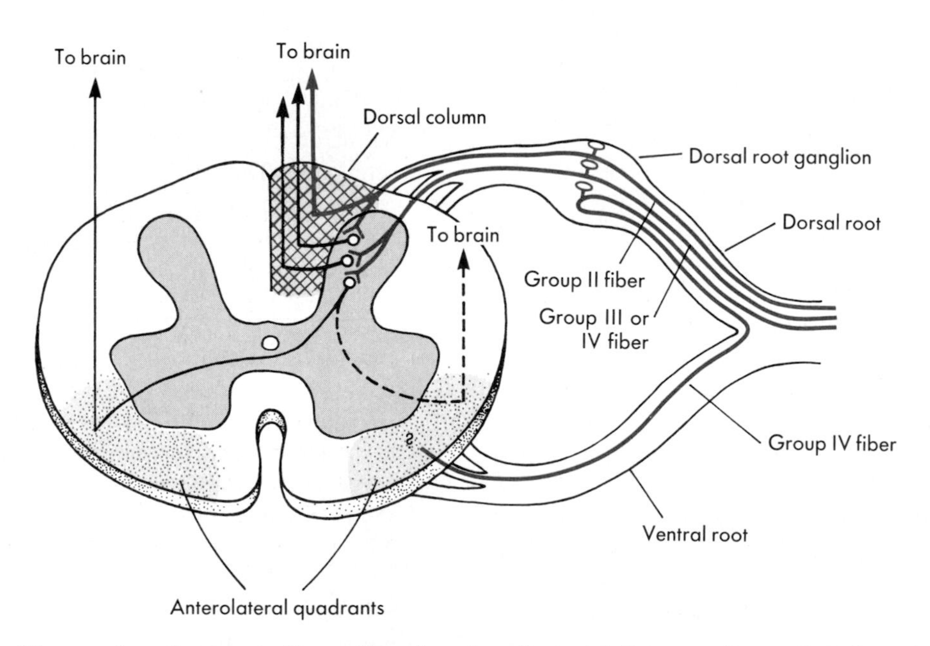

■ **Fig. 8-7** Entry of large and small primary afferent fibers into the spinal cord. Primary afferent fibers have their cell bodies in the dorsal root ganglion. The central processes of the large fibers enter the spinal cord through the medial part of the dorsal root and join the dorsal funiculus. Collaterals synapse in the spinal cord gray matter. The central processes of small afferent fibers enter the cord through the lateral part of the dorsal root (or in some cases through the ventral root). Collaterals from primary afferent axons in the dorsolateral fasciculus synapse in the dorsal horn. Some of the second order neurons project to the brain, either through the contralateral ventrolateral funiculus or through ipsilateral pathways in the dorsal or lateral funiculus.

the lateral component of lamina IX in the cervical and lumbosacral enlargements. Lamina X surrounds the central canal.

The large myelinated primary afferent fibers that enter through the medial part of the dorsal root terminate in the deeper laminae of the dorsal horn (III to VI), the intermediate region, and the ventral horn. The small myelinated and unmyelinated fibers that enter through the lateral part of the dorsal root terminate in laminae I, II, and part of V.

The Trigeminal Nerve

The arrangement of primary afferent fibers that supply the face is comparable to that of fibers that supply the body. Peripheral processes of neurons in the trigeminal ganglion pass through the ophthalamic, maxillary, and mandibular divisions of the trigeminal nerve to innervate dermatome-like regions of the face. The trigeminal nerve also innervates the oral and nasal cavities and the cranial dura mater.

The large myelinated fibers that supply mechanoreceptors of the face and structures of the oral and nasal cavities synapse in the main sensory nucleus of the trigeminal nerve. Small myelinated and unmyelinated primary afferent fibers of the trigeminal nerve descend through the brainstem in the spinal tract of the trigeminal nerve and terminate in the spinal nucleus. Primary afferent fibers from stretch receptors in muscles of the head have their cell bodies in the mesencephalic nucleus of the trigeminal nerve. This is an exceptional arrangement, because all other primary afferent cell bodies in the somatovisceral sensory system are in peripheral ganglia. The central processes synapse in the motor nucleus of the trigeminal nerve.

The Dorsal Column–Medial Lemniscus Pathway

The ascending branches of many large myelinated primary afferent nerve fibers travel rostrally in the dorsal funiculus all the way to the medulla. Axons that innervate sensory receptors of the lower extremity and the lower trunk ascend in the gracile fasciculus, whereas fibers from receptors of the upper extremity and upper trunk ascend in the cuneate fasciculus. These axons are the first-order neurons of the dorsal column-medial lemniscus pathway (Fig. 8-8).

Second-order neurons that receive synaptic input from the ascending branches of primary afferent fibers in the dorsal funiculus are located in the dorsal column nuclei, which include the gracile and cuneate nuclei. Neurons in the dorsal column nuclei respond similarly

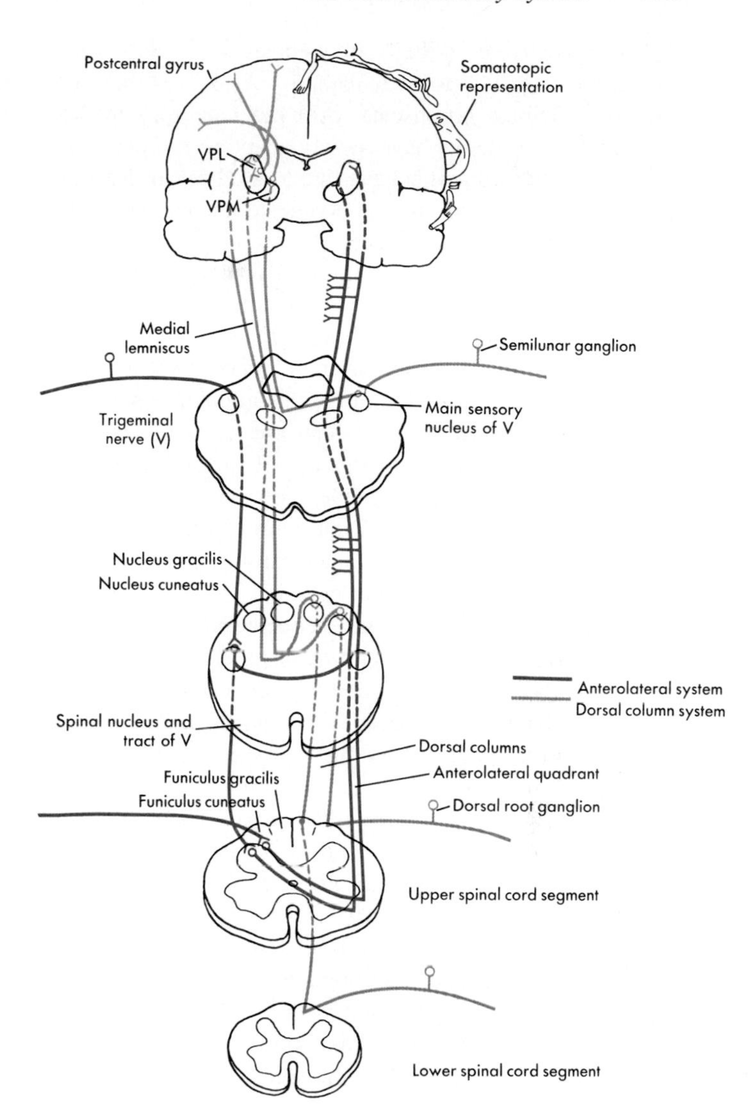

Fig. 8-8 Schematic representation of some central connections in the dorsal column and anterolateral systems. For simplicity, only the primary fibers are shown in the dorsal columns. No obvious difference has yet emerged between the central connections of the primary and secondary fibers that ascend. Both trigeminal and spinal components are shown. The dorsal column system is shown in lighter color, and the anterolateral system is shown in darker color. Note the somatotopic arrangement throughout the neuraxis for the dorsal column system. Fibers that enter dorsal roots at higher spinal segments ascend in the dorsal columns more laterally, and this relationship is preserved throughout; furthermore, the trigeminal component merges with the spinal component in the medial lemniscus next to fibers that represent the highest cervical spinal segments. Consequently an accurate map of the body surface exists throughout, including the cortical areas of the postcentral gyrus. *VPL,* Ventral posterolateral nucleus; *VPM,* ventral posteromedial nucleus.

to the primary afferent fibers that synapse on them. Some behave like rapidly adapting receptors that respond to hair movement or to mechanical transients applied to the glabrous skin, which are stimuli that activate hair follicle afferents and Meissner corpuscles.

Others discharge at high frequencies when vibratory stimuli are applied to their receptive fields and thus resemble Pacinian corpuscles. Still other neurons in the dorsal column nuclei have slowly adapting responses to cutaneous stimuli and behave like Merkel cell or Ruffini endings. Many neurons in the cuneate nucleus are activated by stretching muscles.

The main differences between the responses of dorsal column neurons and primary afferent neurons are these: (1) dorsal column neurons have larger receptive fields because multiple primary afferent fibers synapse on a given dorsal column neuron; (2) dorsal column neurons sometimes respond to more than one class of sensory receptor because of convergence of several different types of primary afferent fibers on the second-order neurons; and (3) dorsal column neurons often have inhibitory receptive fields mediated through interneuronal circuits in the dorsal column nuclei.

Neurons in the dorsal column nuclei project their axons through the medial lemniscus to the contralateral thalamus (see Fig. 8-8). The third-order neurons are in the ventral posterior lateral (VPL) nucleus. The third-order neurons in the thalamus then project to the somatosensory areas of the cerebral cortex.

The pathway in the trigeminal system that is equivalent to the dorsal column–medial lemnisicus path involves a relay from primary afferent fibers that supply the face in the main sensory nucleus of the trigeminal nerve (see Fig. 8-8). Second-order neurons in this nucleus send their axons through the trigeminothalamic tract to the ventral posterior medial (VPM) nucleus of the contralateral thalamus, and third-order neurons of the VPM nucleus project to the somatosensory cerebral cortex.

Other Somatosensory Pathways of the Dorsal Spinal Cord

Three other pathways that carry somatosensory information ascend in the dorsal part of the spinal cord on the same side as the afferent input: (1) the spinocervical tract; (2) the postsynaptic dorsal column pathway; and (3) the dorsal spinocerebellar tract.

The cells of origin of the **spinocervical tract** are in the dorsal horn of the spinal cord. They receive input largely from hair follicle afferent fibers, although many of these cells are activated by nociceptors as well. These cells project to the lateral cervical nucleus, a relay nucleus in the upper cervical spinal cord. Cells in the lateral cervical nucleus project to the contralateral VPL nucleus in the thalamus, and information is relayed from there to the somatosensory cerebral cortex.

Neurons that give rise to the **postsynaptic dorsal column pathway** are also located in the dorsal horn. These cells may receive input from mechanoreceptors of various kinds, including Pacinian corpuscles and slowly adapting cutaneous mechanoreceptors, and from nociceptors. The axons ascend in the dorsal funiculus to the dorsal column nuclei. From here, the sensory pathway is through the medial lemniscus to the contralateral VPL nucleus and then to the somatosensory cerebral cortex.

The **dorsal spinocerebellar tract** (DSCT) responds to input from muscle and joint receptors of the lower extremity. The cells of the DSCT are in Clarke's column, a nucleus found in lamina VII at segmental levels T1 to L3. Other DSCT cells are found in the dorsal horn in the lumbosacral spinal cord. The main destination of the DSCT is the cerebellum. However, it also provides proprioceptive information from the leg to the VPL nucleus of the thalamus by way of a relay in a medullary nucleus known as nucleus z, which is located just rostral to the gracile nucleus. Proprioceptive information from the upper extremity is signalled by the dorsal column pathway.

Sensory Functions of the Dorsal Spinal Cord Pathways

The sensory qualities mediated by dorsal spinal cord pathways include flutter-vibration, touch-pressure, and proprioception. Each of these qualities of sensation depends on the activity in a set of sensory neurons that collectively form a labeled-line sensory channel. A sensory channel may involve several parallel ascending pathways, and it includes certain primary afferent neurons and sensory processing mechanisms at spinal cord, brainstem, thalamic, and cerebral cortical levels.

Flutter-vibration is a complex sensation. Flutter refers to recognition of mechanical stimuli that have low-frequency components. In clinical testing, this often involves sensory responses to transient applications of a wisp of cotton or a brief tap on the skin. The sensory receptors that detect flutter include hair follicles and Meissner's corpuscles. Several parallel ascending sensory tracts convey information used for flutter sensation. These tracts include the dorsal column–medial lemniscus pathway, the spinocervical tract, and the postsynaptic dorsal column path. In addition, the spinothalamic tract in the ventral part of the cord is partly responsible for flutter sensation, as discussed later in the chapter.

Vibratory sense involves recognition of stimuli with high-frequency components. A common clinical test of vibratory sense is discrimination of whether a tuning fork placed against a bony prominence is vibrating. High-frequency vibration is detected primarily by Pacinian corpuscles. The information is transmitted through the dorsal column–medial lemniscus pathway.

Touch-pressure involves the recognition of maintained skin indentation. The receptors include Merkel cell and Ruffini endings. The ascending pathways that convey information from these receptors include the dorsal column–medial lemniscus and the postsynaptic dorsal column paths.

Proprioception can be separated into the senses of **joint movement** and **joint position.** The sensory information arises from receptors in muscle, joints, and skin. For proximal joints, such as the knee, the most important information is derived from the activity of muscle spindles in the muscles that move the joint. In distal joints, such as those of the fingers, Ruffini endings in skin and joint receptors also contribute. All of the information required for proprioception in the upper extremity ascends in the dorsal column–medial lemniscus pathway. However, a major part of the information needed for proprioception in the lower extremity depends on the DSCT and its relay in the nucleus z.

Visceral distention is the sense of filling of such viscera as the urinary bladder. Information about visceral distention arises from stretch receptors in the wall of the viscus and is transmitted by way of the dorsal column–medial lemniscus pathway.

Effects of Interruption of Dorsal Cord Pathways

A lesion that interrupts pathways that ascend in the dorsal part of the spinal cord will result in deficits in tactile discrimination, vibratory sense, and position sense. Particularly affected is the ability to recognize figures drawn on the skin **(graphesthesia)** and tactile recognition of objects placed in the hand **(stereognosis).** Some tactile function remains that allows localization of a tactile stimulus and flutter sensation. Pain and temperature sensations are unaffected.

Spinothalamic Tract

The spinothalamic tract is the most important sensory pathway for pain and thermal sensations. It also contributes to tactile sensation. The first-order neurons are primary afferent fibers from nociceptors, thermoreceptors, and mechanoreceptors. The second-order neurons are in the spinal cord (rather than in the medulla, as in the dorsal column–medial lemniscus pathway). The axons of the spinothalamic tract cells cross to the opposite side within the spinal cord and ascend to the brain in the ventral part of the lateral funiculus (see Fig. 8-8), and they terminate on third-order neurons in the thalamus.

The cells of origin of the spinothalamic tract are found chiefly in spinal cord laminae I and V. Most spinothalamic tract cells receive an excitatory input from nociceptors in the skin, but many can also be excited by noxious stimulation of muscle or viscera. Effective stimuli include noxious, mechanical, thermal, and chemical stimuli. Some spinothalamic tract cells are excited by activity in cold or warm thermoreceptors or in sensitive mechanoreceptors. Thus different spinothalamic tract cells respond in a manner appropriate for signaling noxious, thermal, or mechanical events.

Some nociceptive spinothalamic tract cells receive a convergent excitatory input from several different classes of sensory receptors. For example, a given spinothalamic neuron may be activated weakly by tactile stimuli but more powerfully by noxious stimuli (Fig. 8-9). Such neurons are called **wide-dynamic range cells** because they are activated by stimuli with a wide range of intensities. Wide-dynamic range neurons mainly signal noxious events, the weak responses to tactile stimuli perhaps being ignored by higher centers. However, in pathological conditions these neurons may be activated sufficiently to evoke a sensation of pain. This would explain some pain states in which the activation of mechanoreceptors causes pain **(mechanical**

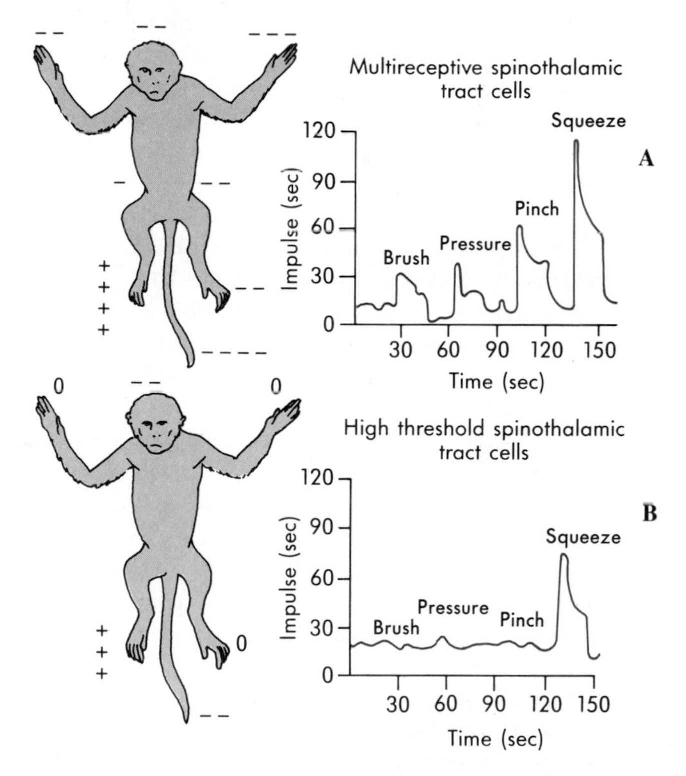

■ **Fig. 8-9** **A,** Responses of a wide-dynamic-range or multireceptive spinothalamic tract cell. **B,** Responses of a high-threshold spinothalamic tract cell. The figures indicate the excitatory *(plus signs)* and inhibitory *(minus signs)* receptive fields. The graphs show the responses to graded intensities of mechanical stimulation. *Brush* is with a camel's hair brush repeatedly stroked across the receptive field. *Pressure* is applied by attachment of an arterial clip to the skin. This is a marginally painful stimulus to a human. *Pinch* is by attachment of a stiff arterial clip to the skin and is distinctly painful. *Squeeze* is by compressing a fold of skin with forceps and is damaging to the skin.

allodynia). Other spinothalamic tract cells are activated only by noxious stimuli. Such neurons are often called **nociceptive-specific** or **high-threshold cells** (see Fig. 8-9).

The neurotransmitters released by nociceptors that activate spinothalamic tract cells include the excitatory amino acid, glutamate, and any of several peptides, such as substance P (SP), calcitonin gene-related peptide (CGRP), vasoactive intestinal polypeptide (VIP), and others. Glutamate appears to act as a fast transmitter by its action on non-N-methyl-d-aspartic acid excitatory amino acid (non-NMDA) receptors. However, with repetitive stimulation, glutamate can also cause a build-up of activity or wind-up through an action mediated by NMDA receptors. SP appears to act as a neuromodulator. Through a combined action with an excitatory amino acid, such as glutamate, SP can produce a long-lasting increase in the responses of spinothalamic tract cells. CGRP seems to increase SP release and to prolong the action of SP by inhibiting its enzymatic degradation.

The spinothalamic tract projects mainly to the contralateral thalamus, although some ipsilateral projections are present as well. The decussation of the axon of a spinothalamic tract cell is within the same segment as the cell body, and the axon ascends in the ventrolateral funiculus. The spinothalamic tract terminates in several nuclei of the thalamus, including the VPL nucleus and the central lateral nucleus (one of the intralaminar nuclei). A given spinothalamic tract cell may terminate only in the VPL nucleus, in both the VPL and central lateral nuclei, or just in the central lateral nucleus. Spinothalamic tract cells that project just to the VPL or to both the VPL and central lateral nuclei have small receptive fields located on the contralateral side of the body. The activity of these cells could help signal stimulus location. Those that project just to the central lateral nucleus have very large receptive fields that often include much of the surface of the body and face bilaterally. The very large receptive fields suggest that these cells trigger motivational-affective responses to painful stimuli, rather than participate in sensory discrimination.

Spinothalamic tract cells often have inhibitory receptive fields. Inhibition may result from weak mechanical stimuli, but usually the most effective inhibitory stimuli are noxious ones. The nociceptive inhibitory receptive fields may be very large and may include most of the body and face (see Fig. 8-9). Such receptive fields may account for the ability of various physical manipulations, including transcutaneous electrical nerve stimulation and acupuncture, to suppress pain. Neurotransmitters that can inhibit spinothalamic tract cells include the inhibitory amino acids, GABA and glycine, as well as monoamines and the endogenous opioid peptides (see p. 139).

The **gate control theory of pain** explains how innocuous stimuli may inhibit the responses of dorsal horn neurons that transmit information about painful stimuli to the brain. In this theory pain transmission is prevented by innocuous inputs mediated by large myelinated afferent fibers, whereas pain transmission is enhanced by inputs carried over fine afferent fibers. The inhibitory interneurons of lamina II serve as the gating mechanism. The circuit diagram originally proposed has been criticized, but the basic notion of a gating mechanism is still viable.

Noxious stimuli applied to large areas of the body can also inhibit the discharges of nociceptive dorsal horn neurons by activating inhibitory pathways that descend from the brainstem (see later description of the endogenous analgesia system).

Other Somatosensory Pathways of the Ventral Spinal Cord

Two other pathways that transmit somatosensory information ascend in the ventrolateral part of the spinal cord: the spinoreticular tract and the spinomesencephalic tract.

The cells of origin of the **spinoreticular tract** are often difficult to activate. However, when receptive fields are found, they are generally large, sometimes bilateral, and the effective stimuli include noxious ones. The reticular formation is involved in attentional mechanisms and arousal (see Chapter 16). The reticular formation projects to the intralaminar complex of the thalamus and thence to wide areas of the cerebral cortex. Reticulospinal projections also contribute to the descending systems that control pain transmission.

Many cells of the **spinomesencephalic** tract respond to noxious stimuli, and the receptive fields may be small or large. The terminations of this tract are in several midbrain nuclei, including the **periaqueductal gray,** which is an important component of the endogenous analgesia system (see p. 138). Motivational-affective responses may also result from activation of the periaqueductal gray and reticular formation. For example, stimulation in the periaqueductal gray can cause vocalization and aversive behavior. Information from the midbrain is relayed not only to the thalamus, but also to the amygdala, a part of the limbic system. This provides one of several pathways by which noxious stimuli can trigger emotional responses.

Sensory Functions of the Ventral Spinal Cord Pathways

The sensory modalities mediated by ventral spinal cord pathways include a contribution to flutter, but the most important functions are pain and thermal sensations. Although the most essential pathway for flutter-vibration is the dorsal column–medial lemniscus pathway, the spinothalamic tract can provide sufficient information for flutter, but not for vibratory, sensation. However, the spinothalamic tract alone is insufficient for the recognition of the direction of a tactile stimulus, and discrimination is much less accurate than when the dorsal column–medial lemniscus path is intact. This can be tested clinically by drawing numbers on the fingertip (graphesthesia). Graphesthesia is lost when the dorsal column–medial lemniscus pathway is interrupted but not when the spinothalamic tract is interrupted.

Thermal sense includes the two submodalities of cold and warm. Thermal sense appears to depend on input from cold and warm receptors to spinothalamic tract cells in lamina I of the dorsal horn. Section of the spinothalamic tract causes a loss of thermal sense on the contralateral side of the body.

Pain that results from stimulation of nociceptors is mediated partly by spinothalamic tract cells and partly by the spinoreticular and spinomesencephalic tracts. Pain is a complex phenomenon and includes both sensory-discriminative and motivational-affective components. That is, pain is a sensory experience that is accompanied by emotional responses and by somatic and autonomic motor adjustments. Presumably the sensory-discriminative component of pain depends on the spinothalamic tract projection to the VPL nucleus and further transmission of nociceptive information to the SI and SII regions of the cerebral cortex (see p. 135). Sensory processing at these and higher levels of the cortex results in the perception of the quality of pain (e.g., pricking, burning, or aching), the location of the painful stimulus, the intensity of the pain, and its duration.

The motivational-affective responses to painful stimuli include attention and arousal, somatic and autonomic reflexes, endocrine responses, and emotional changes. These collectively account for the unpleasant nature of painful stimuli. The motivational-affective responses depend on activity transmitted in several ascending pathways, including the component of the spinothalamic tract that projects to the medial thalamus, the spinoreticular tract, and the spinomesencephalic tract. In addition, several recently discovered pathways connect the spinal cord directly with the hypothalamus (spinohypothalamic tract) and the basal forebrain (spinotelencephalic tract).

Effects of Interruption of the Spinothalamic Tract

Both the sensory-discriminative and the motivational-affective components of pain are lost on the contralateral side of the body when the spinothalamic tract is interrupted. This motivated the development of the surgical procedure known as **anterolateral cordotomy,** which formerly was used to treat pain in many individuals, especially those suffering from cancer pain. This operation is now used infrequently because of improvements in drug therapy and because pain often returns months to years after an initially successful cordotomy. The return of pain may reflect either an extension of the disease or the development of a central pain state (see later discussion). In addition to the loss of pain sensation, anterolateral cordotomy produces a loss of cold and warmth sensations on the contralateral side of the body. Careful testing may reveal a minimal tactile deficit as well, but the intact sensory pathways of the dorsal part of the spinal cord provide sufficient tactile information that any loss caused by interruption of the spinothalamic tract is insignificant.

Pain

Referred Pain

Pain that originates from the skin is generally well localized, presumably because spinothalamic tract cells have relatively discrete cutaneous receptive fields. Also, the ascending system through which they signal is somatotopically organized. However, pain that originates from deep structures, including muscle and viscera, is poorly localized and is often mistakenly attributed (referred pain) to superficial structures. An example of this is **angina pectoris,** the pain that originates from ischemia of the heart. Frequently, ischemic heart pain is referred to the inner aspect of the left arm. This area of pain referral is in the T1 dermatome, which corresponds to the spinal cord level that provides the main sensory innervation of the heart.

One explanation of referred pain is that many spinothalamic neurons receive excitatory input not only from the skin but also from muscle and viscera. The spinal cord segments that innervate the dermatomes containing the cutaneous receptive field of the cell correspond to the segments that innervate the muscle or viscus. The activity in a population of spinothalamic tract cells may be interpreted as originating in somatic structures, based on learning this association during childhood. Subsequently, activation of these neurons by pathologic input from visceral nociceptors is misinterpreted as resulting from stimulation of superficial parts of the body.

Central Pain

Pain sometimes occurs in the absence of nociceptor stimulation. This is most likely to happen after damage to peripheral nerves or to parts of the CNS involved in transmitting nociceptive information. Examples of pain secondary to neural damage are phantom limb pain and "thalamic pain." **Phantom limb pain** follows amputation in some individuals and is clearly not caused by activation of nociceptors, because these are no longer present in the area in which pain is felt. Similarly, lesions of the thalamus or at other levels of the spinothalamocortical pathway may cause severe spontaneous pain; yet interruption of the nociceptive pathway by the same lesion may prevent or reduce pain evoked by peripheral stimulation. The mechanism of such pain caused by neural damage is poorly understood, but the pain appears to depend on changes in the activity and response properties of neurons upstream in the nociceptive system.

Trigeminal Nociceptive System

Sensory processing of nociceptive and thermoreceptive information that originates from the face, oral cavity, and dura mater is organized in a fashion similar to that for the trunk and limbs. Pain in the trigeminal distribution is of particular importance, because this includes both tooth and headache pain.

The primary afferent fibers that supply nociceptors and thermoreceptors in the head enter the brainstem through the trigeminal nerve (some also enter through the facial, glossopharyngeal, and vagus nerves) and descend through the brainstem to the upper cervical spinal cord through the spinal tract of the trigeminal nerve (see Fig. 8-8). Some mechanoreceptive afferent fibers also join the spinal tract. Axons in the spinal tract synapse on second-order neurons in the spinal nucleus. These neurons transmit information concerning pain and temperature sensations that originate from the face, oral cavity, and dura mater to the contralateral VPM nucleus through the trigeminothalamic tract. The VPM nucleus in turn projects to the somatosensory cerebral cortex. The spinal nucleus also projects to the central lateral nucleus of the intralaminar complex.

Higher Processing of Somatosensory Information

Thalamus

The medial lemniscus and the spinothalamic tract synapse in the ventral posterior lateral (VPL) nucleus of the thalamus. Several other parallel sensory tracts, such as the spinocervical tract and the pathway through nucleus z, also terminate in the VPL nucleus. The trigeminothalamic tracts from the main sensory and spinal nuclei of the trigeminal nerve synapse in the ventral posterior medial (VPM) nucleus of the thalamus.

The responses of many neurons in the VPL and VPM nuclei resemble those of first- and second-order neurons in the ascending tracts. The responses may be dominated by a particular type of sensory receptor, and the receptive field may be small, although generally larger than that of a primary afferent fiber. The receptive fields are contralateral to the thalamic neuron, and the location of the thalamic neuron is systematically related to that of the receptive field. That is, there is a **somatotopic organization** of the VPL and VPM nuclei. The lower extremity is represented by neurons in the lateral part of the VPL nucleus, the upper extremity in the medial part of the VPL nucleus, and the face in the VPM nucleus (Fig. 8-10).

Thalamic neurons often have inhibitory, as well as excitatory, receptive fields. The inhibition may actually take place in the dorsal column nuclei or in the dorsal horn of the spinal cord, but there are also inhibitory circuits within the thalamus. The VPL and VPM nuclei contain inhibitory interneurons (in primates, but not in rodents), and some of the inhibitory interneurons in the reticular nucleus of the thalamus project into the VPL and VPM nuclei. The inhibitory neurons intrinsic to the VPL and VPM nuclei and in the reticular nucleus use GABA as their inhibitory neurotransmitter.

A notable difference between neurons in the VPL and VPM nuclei and sensory neurons at lower levels of the somatosensory system is that the excitability of the thalamic neurons depends on the stage of the sleep-wake cycle and on the presence or absence of anesthesia. This may reflect the level of excitation of thalamic neurons by excitatory amino acids acting at non-NMDA and NMDA receptors. It may also reflect the inhibition of the thalamic neurons by recurrent pathways through the reticular nucleus. During a state of drowsiness or during barbiturate anesthesia, thalamic neurons tend to undergo an alternating sequence of excitatory and inhibitory postsynaptic potentials. The alternating bursts of discharges in turn intermittently excite neurons in the cerebral cortex. Such excitation results in an alpha rhythm or in spindling in the electroencephalogram (see Chapter 16).

The spinothalamic tract and the part of the trigeminothalamic tract that originates in the spinal nucleus of the trigeminal nerve project to the central lateral nucleus of the intralaminar complex of the thalamus. The intralaminar nuclei are not somatotopically organized, and they project diffusely to the cerebral cortex, as well as to the basal ganglia. The projection of the central lateral nucleus to the SI cortex may be involved in arousal of this part of the cortex and in selective attention.

Effects of lesions of the somatosensory thalamus. *Destruction of the VPL or VPM nuclei diminishes sensation on the contralateral side of the body or face.* The loss reflects mainly those sensory qualities transmitted by the dorsal column–medial lemniscus pathway and its trigeminal equivalent. The sensory-discriminative component of pain sensation is also lost, but the motivational-affective component of pain is still present if the medial thalamus is intact, presumably because of the medial spinothalamic and spinoreticulothalamic projection. In some individuals, a lesion of the somatosen-

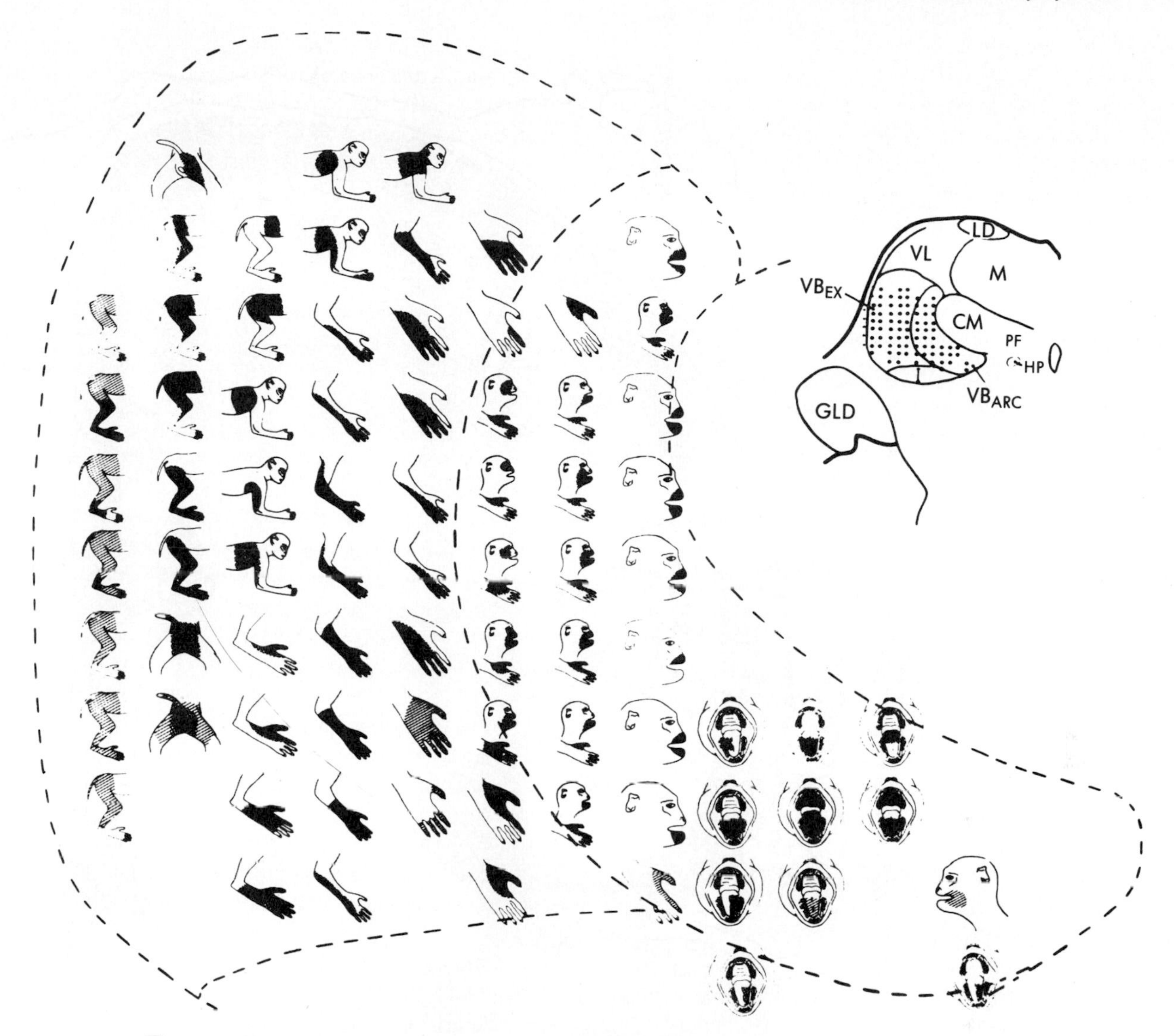

■ **Fig. 8-10** Somatotopic map of the VPL and VPM nuclei in the monkey thalamus. Recordings were made from small groups of neurons and the receptive fields were mapped using weak mechanical stimuli. The figurines show the receptive fields for multiunit activity in the VPL and VPM nuclei (demarcated by the dashed lines). The locations of the recording sites are shown by the dots on the drawing of a section through the thalamus on one side. VB_{EX}, VPL: VB_{ARC}, VPM; *VL*, ventral lateral; *LD*, lateral dorsal; *M*, medial dorsal; *CM*, centre median; *PF*, parafascicular; *GLD*, dorsal lateral geniculate; *HP*, habenulopeduncular tract. (From Mountcastle VN, editor: Medical physiology, ed 14, St Louis, 1979, CV Mosby.)

sory thalamus results in a central pain state known as "thalamic pain." However, pain indistinguishable from thalamic pain can also be produced by brainstem or cortical lesions.

■ *Somatosensory Cortex*

Third-order sensory neurons in the VPL and VPM nuclei of the thalamus project to the somatosensory cortex. The main somatosensory receiving areas of the cortex are called the SI and SII areas. The SI cortex occupies much of the postcentral gyrus, and the SII cortex is in the superior bank of the lateral fissure.

The **SI cortex,** like the somatosensory thalamus, has a somatotopic organization (Fig. 8-11). The face is represented in the lateral part of the postcentral gyrus, above the lateral fissure. The hand and the rest of the upper extremity are represented in the dorsolateral part of the postcentral gyrus and the lower extremity on the medial surface of the hemisphere. The map of the surface of the body and face of a human on the postcentral gyrus is called a **sensory homunculus.** The map is distorted, because the greatest volume of neural tissue is

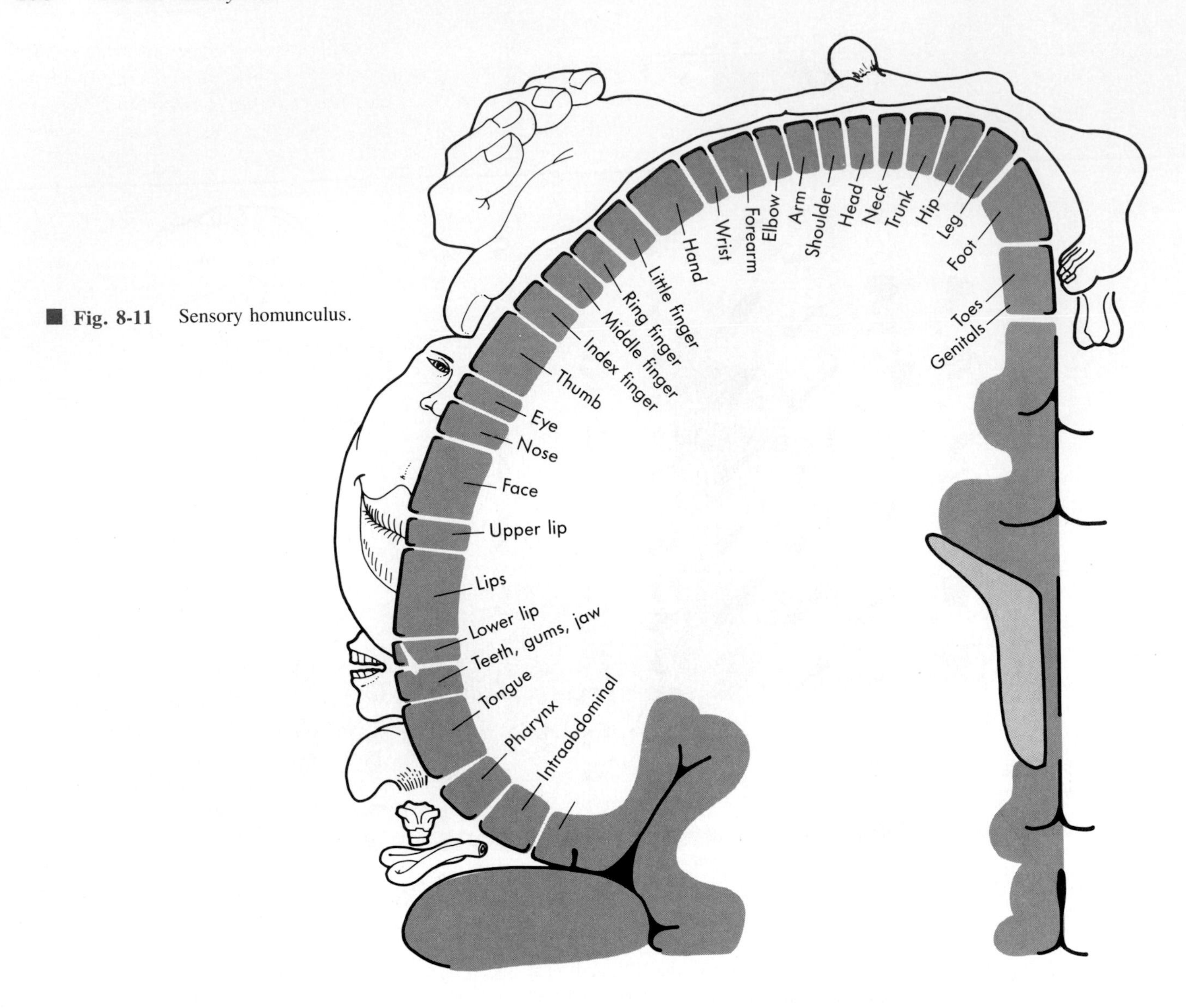

■ **Fig. 8-11** Sensory homunculus.

devoted to the mostly densely innervated regions, such as the perioral area and the thumb and other digits. The sensory homunculus is an expression of place coding of somatosensory information. A locus in the SI cortex encodes the location of a somatosensory stimulus on the surface of the body or face. For example, the way in which the brain knows that a certain part of the body has been stimulated is that certain neurons in the dorsolateral part of the postcentral gyrus are activated.

The **SII cortex** also contains a somatosensory map, as do several other less understood areas of cortex.

The SI cortex has several morphological and functional subdivisions, each of which has a somatotopic map like that shown in Fig. 8-11. These subdivisions were originally described by Brodmann, based on the arrangements of neurons in the various layers of the cortex as seen in Nissl-stained preparations. The subdivisions are known as Brodmann's areas 3a, 3b, 1, and 2. Cutaneous input dominates in areas 3b and 1, whereas muscle and joint input dominates in areas 3a and 2. Thus separate cortical zones are specialized for the processing of tactile and proprioceptive information. The inputs to these cortical areas are from distinct parts of the VPL and VPM nuclei (Fig. 8-12). For example, the cutaneous input from the extremities is from the core of the VPL nucleus, whereas the muscle and joint input is from a "shell" region.

Within any particular area of the SI cortex, all of the neurons along a line perpendicular to the cortical surface have similar response properties and receptive fields. The SI cortex is thus said to have a **columnar organization.** A comparable columnar organization has also been demonstrated for other primary sensory receiving areas, including the primary visual and auditory cortices (see Chapters 9 and 10). Nearby cortical columns in the SI cortex may process information for different sensory modalities. For example, the cutaneous information that reaches one cortical column in area 3b may come from rapidly adapting mechanoreceptors, whereas the information that reaches a neighboring col-

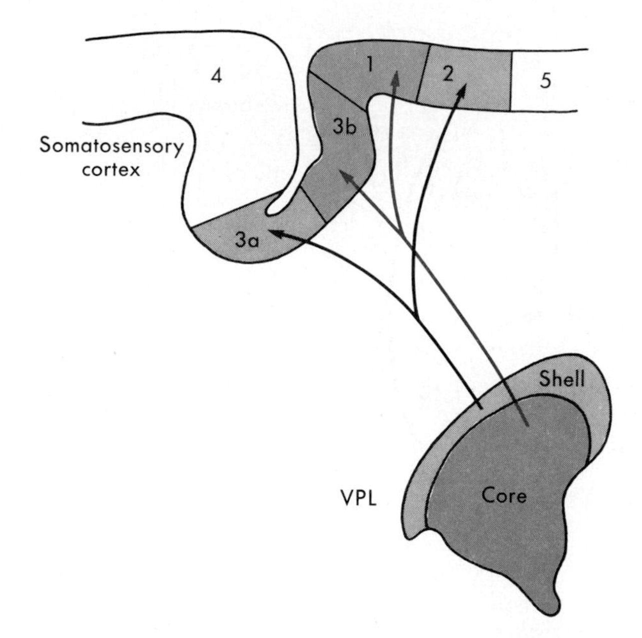

■ **Fig. 8-12** Schematic relationships in thalamocortical projection from VPL to areas of somatosensory cortex. A core region *(color)* of VPL projects primarily to cortical areas 1 and 3b, and a shell region projects primarily to areas 2 and 3a. (Redrawn from Jones EG, Friedman DP: *J Neurophysiol* 48:521, 1982.

umn might be from slowly adapting mechanoreceptors.

Besides being responsible for the initial processing of somatosensory information, the SI cortex also begins higher-order processing, such as feature extraction. For example, certain neurons in area 1 respond preferentially to a stimulus that moves in one direction across the receptive field but not in the opposite direction (Fig. 8-13). Such neurons presumably contribute to the perceptual ability to recognize the direction of an applied stimulus.

■ *Association Cortex*

The SI cortex is connected with many other cortical areas, such as the SII cortex, the motor cortex, supplementary sensory and motor cortices, and also the parietal association cortex. The parietal association cortex also receives input from other sensory systems, notably from the visual system. A major function of the parietal association cortex is the relationship of the body to extrapersonal space. For example, this part of the cortex on one side helps to coordinate hand and eye movements on the contralateral side. In humans the posterior parietal cortex of the nondominant hemisphere is particularly involved in spatial relations, whereas that in the dominant hemisphere is concerned with language (see Chapter 16).

Effects of cortical lesions. A lesion of the SI cortex in humans causes sensory changes like those produced by a lesion of the somatosensory thalamus. However, usually only a part of the cortex is involved, and so *the sensory loss may be confined, for example, to the face or to the leg, depending on the location of the lesion with respect to the sensory homunculus.* The sensory modalities most affected are discriminative touch and position sense. Graphesthesia and stereognosis are par-

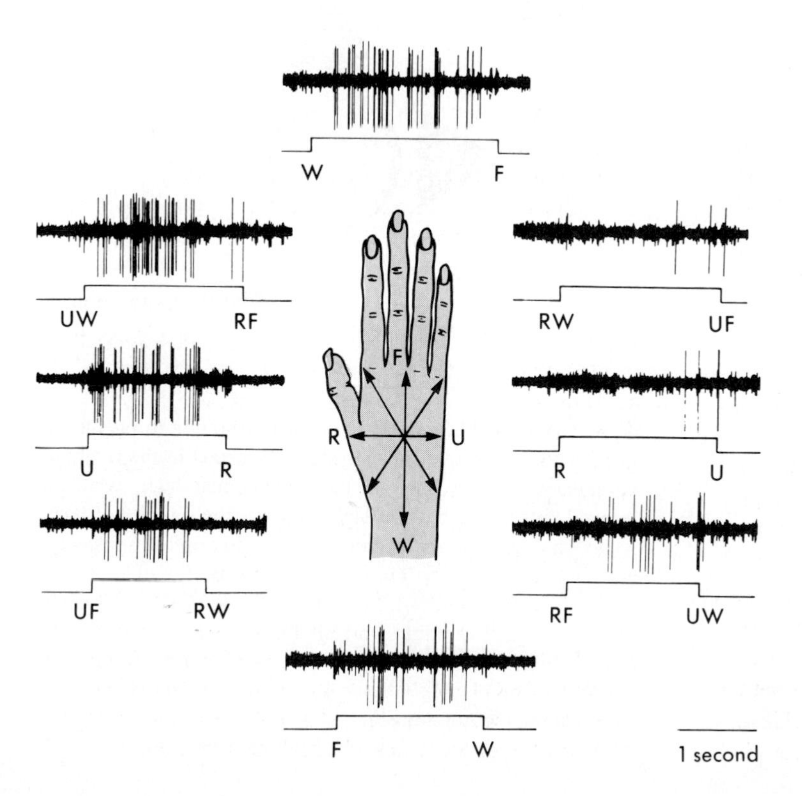

■ **Fig. 8-13** Feature extraction by cortical neurons. The responses were recorded from a neuron in the somatosensory cortex of a monkey. The direction of a stimulus was varied, as shown by the arrows in the drawing. Note that the responses were greatest when the stimulus moved in the direction from *UW* to *RF* and least from *RW* to *UF*. (From Costanzo RM, Gardner EP: *J Neurophysiol* 43:1319, 1980.)

ticularly disturbed. Pain and thermal sensation may be relatively unaffected, although a loss of pain may follow cortical lesions. Conversely, cortical lesions can result in a central pain state that resembles thalamic pain.

A lesion of the parietal association cortex on the nondominant side causes deficits in the ability to relate to extrapersonal space. When an affected person is asked to copy geometric figures, the figures are distorted. For example, the numbers of a clock face may all be drawn on the side of the clock that corresponds to the normal side of the person with the lesion **(constructional apraxia).** The person may deny the contralateral side of the body **(neglect syndrome)** and have difficulty in dressing on that side. The individual with a lesion may also have difficulty reading maps, driving, or other activities that require spatial orientation.

Centrifugal Control of Somatovisceral Sensation

Sensory experience is not just the passive detection of environmental events. Instead, it more often depends on exploration of the environment. Tactile cues are sought by moving the hand over a surface. Visual cues result from scanning visual targets with the eyes. Thus sensory information is often received as the result of activity in the motor system. Furthermore, transmission in pathways to the sensory centers of the brain is regulated by descending control systems. This allows the brain to control its input by filtering the incoming sensory messages. Important information can be attended to and unimportant information ignored.

The tactile and proprioceptive somatosensory pathways are regulated by descending pathways that originate in the SI and motor regions of the cerebral cortex. For example, corticobulbar projections to the dorsal column nuclei help control sensory input that is transmitted by the dorsal column–medial lemniscus path.

Of particular interest is the descending control system that regulates the transmission of nociceptive information. This system presumably suppresses excessive pain under certain circumstances. For example, it is well known that soldiers on the battlefield, accident victims, and athletes in competition often feel little or no pain at the time a wound occurs or a bone is broken. At a later time, pain may develop and become severe. Although the descending regulatory system that controls pain is part of a more general centrifugal control system that modulates all forms of sensation, the pain control system is so important medically that it is distinguished as a special system called the **endogenous analgesia system.**

Several centers in the brainstem and descending pathways from these contribute to the endogenous analgesia

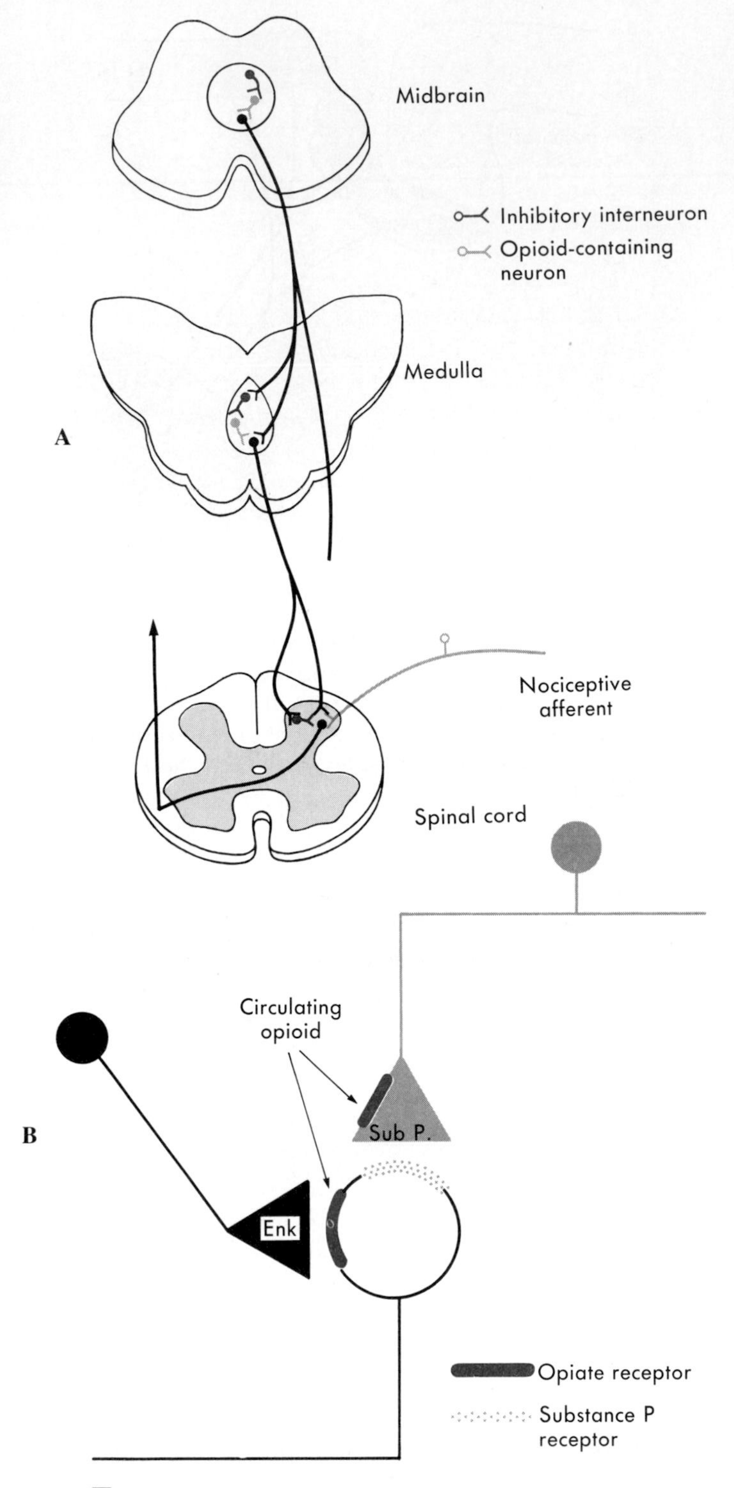

■ **Fig. 8-14** **A,** Some of the neurons thought to play a role in the endogenous analgesia system. Neurons in the midbrain periaqueductal gray activate the raphe-spinal tract, which in turn inhibits nociceptive spinal neurons, such as those of the spinothalamic tract. Interneurons containing opioid substances are involved in the system at each level. **B,** Possible presynaptic and postsynaptic sites of action of enkephalin *(Enk).* The presynaptic action might prevent the release of substance P *(Sub P)* from nociceptors. (Redrawn from Henry, JL. In Porter R, O'Connor M, editors: *Ciba Foundation Symposium 91,*London, 1982, Pitman.)

system. For example, stimulation in the periaqueductal gray, the locus coeruleus, or the medullary raphe nuclei inhibits nociceptive neurons at spinal cord and brainstem levels, including spinothalamic tract and trigeminothalamic tract cells (Fig. 8-14, *A*). Other inhibitory pathways originate in the sensorimotor cortex, the hypothalamus, and the reticular formation.

The endogenous analgesia system can be subdivided into components that use one of the endogenous **opioid peptides** as neurotransmitters or modulators and those that do not. The endogenous opioids are neuropeptides that activate one of several types of opiate receptors. Some of the endogenous opioids include enkephalin, dynorphin and β-endorphin. Opiate analgesia can generally be prevented or reversed by the narcotic antagonist, naloxone. Therefore naloxone is often used as a test of whether or not analgesia is mediated by an opioid mechanism.

The opioid-mediated endogenous analgesia system can be activated by the exogenous administration of morphine or other opiate drugs. Thus one of the oldest medical treatments for pain depends on the triggering of a sensory control system. Opiates typically inhibit neural activity in nociceptive pathways. Two sites of action have been proposed for opiate inhibition, presynaptic and postsynaptic (Fig. 8-14, *B*). The presynaptic action of opiates on nociceptive afferent terminals is thought to prevent the release of excitatory transmitters, such as substance P. The postsynaptic action produces an inhibitory postsynaptic potential. How can an inhibitory neurotransmitter activate descending pathways? One hypothesis is that the descending analgesia system is under tonic inhibitory control by inhibitory interneurons in both the midbrain and the medulla. The action of opiates would inhibit the inhibitory interneurons and thereby disinhibit the descending analgesia pathways.

Some endogenous analgesia pathways operate by neurotransmitters other than opioids and thus are unaffected by naloxone. One way of engaging a nonopioid analgesia pathway is through certain forms of stress. The analgesia so produced is a form of **stress-induced analgesia.**

Many neurons in the raphe nuclei use serotonin as a neurotransmitter. Serotonin can inhibit nociceptive neurons and presumably plays an important role in the endogenous analgesia system. Other brainstem neurons release catecholamines, such as norepinephrine and epinephrine, in the spinal cord. These catecholamines also inhibit nociceptive neurons; therefore catecholaminergic neurons may contribute to the endogenous analgesia system. Furthermore, these monoamine neurotransmitters interact with endogenous opioids. Undoubtedly, many other substances are involved in the analgesia system. In addition, there is evidence for the existence of endogenous opiate antagonists that can prevent opiate analgesia.

Summary

1. Sensory pathways include serially connected neurons. First-order neurons have cell bodies in sensory nerve ganglia; they connect peripherally to a sensory receptor and centrally to second-order neurons. Higher-order neurons are in the thalamus and cortex.

2. Skin contains mechanoreceptors, thermoreceptors, and nociceptors. Mechanoreceptors may be rapidly or slowly adapting. Thermoreceptors include cold and warm receptors. $A\delta$ and C nociceptors may be sensitized by release of chemical substances from damaged cells. Muscle, joints, and viscera have mechanoreceptors and nociceptors.

3. Large primary afferent fibers enter the dorsal funiculus through the medial dorsal root; collaterals synapse in the deep dorsal horn, intermediate zone, and ventral horn. Small primary afferent fibers enter the dorsolateral fasciculus through the lateral dorsal root; collaterals synapse in the dorsal horn.

4. Trigeminal primary afferent fibers supply the face, oral and nasal cavities, and dura. The cell bodies are in the trigeminal ganglion or, if proprioceptive, the mesencephalic nucleus.

5. Ascending branches of large primary afferent fibers synapse on second-order neurons in the dorsal column nuclei, which project in the medial lemniscus to the contralateral thalamus and synapse on third-order neurons of the ventral posterior lateral (VPL) nucleus. The equivalent trigeminal pathway relays in the main sensory nucleus and contralateral ventral posterior medial (VPM) nucleus.

6. The spinocervical tract, the postsynaptic dorsal column path, and the part of the dorsal spinocerebellar tract that relays in nucleus z act in parallel with the dorsal column–medial lemniscus path.

7. The dorsal spinal cord pathways signal the sensations of flutter-vibration, touch-pressure, and proprioception.

8. The spinothalamic tract includes nociceptive, thermoreceptive, and tactile neurons; its cells of origin are mostly in the dorsal horn, and the axons cross, ascend in the ventrolateral funiculus, and synapse in the VPL and intralaminar nuclei of the thalamus. The equivalent trigeminal pathway relays in the spinal nucleus and projects to the contralateral VPM and intralaminar nuclei.

9. The spinothalamic relay in the VPL nucleus helps account for the sensory-discriminative aspects of pain. Parallel nociceptive pathways in the ventrolateral funiculus are the spinoreticular and spinomesencephalic tracts; these and the spinothalamic projection to the me-

dial thalamus contribute to the motivational-affective aspects of pain.

10. Referred pain is explained by convergent input to spinothalamic tract cells from the body wall and from viscera. Damage to a peripheral nerve or to the central nociceptive pathway may result in neurogenic pain, such as phantom limb pain or thalamic pain.

11. The VPL and VPM nuclei are somatotopically organized and contain inhibitory circuits. The somatosensory cortex includes the SI and SII regions; these are also somatotopically organized and the SI cortex is further subdivided into regions that process different kinds of information. The SI cortex contains columns of neurons with similar receptive fields and response properties. Some SI neurons are involved in feature extraction. The association cortex of the parietal lobe is concerned with extrapersonal space.

12. Transmission in somatosensory pathways is regulated by descending control systems. The endogenous analgesia system regulates nociceptive transmission, using such transmitters as the endogenous opioid peptides, norepinephrine and serotonin.

■ Bibliography

Journal articles

Akil H et al: Endogenous opioids: biology and function, *Annu Rev Neurosci* 7:223, 1984.

Basbaum AI, Fields HL: Endogenous pain control systems: brainstem spinal pathways and endorphin circuitry, *Annu Rev Neurosci* 7:309, 1984.

Besson JM, Chaouch A: Peripheral and spinal mechanisms of nociception, *Physiol Rev* 67:67, 1987.

Boivie J, Leijon G, Johansson I: Central post-stroke pain—a study of the mechanisms through analysis of the sensory abnormalities, *Pain* 37:173, 1989.

Constanzo RM, Gardner, EP: A quantitative analysis of responses of direction-sensitive neurons in somatosensory cortex of awake monkeys, *J Neurophysiol* 43:1319, 1980.

Darian-Smith I, Johnson KO, Dykes R: "Cold" fiber population innervating palmar and digital skin of the monkey: response to cooling pulses, *J Neurophysiol* 36:325, 1973.

Hensel H, Kenshalo DR: Warm receptors in the nasal region of cats, *J Physiol* 204:99, 1969.

Jones EG, Friedman DP: Projection pattern of functional components of thalamic ventrobasal complex on monkey somatosensory cortex, *J Neurophysiol* 48:521, 1982.

LaMotte RH, Thalhammer JG, Robinson CJ: Peripheral neural correlates of magnitude of cutaneous pain and hyperalgesia: a comparison of neural events in monkey with sensory judgements in human, *J Neurophysiol* 50:1, 1983.

Mountcastle VB, Henneman E: The representation of tactile sensibility in the thalamus of the monkey, *J Comp Neurol* 97:409, 1952.

Mountcastle VB, Powell, TPS: Neural mechanisms subserving cutaneous sensibility, with special reference to the role of afferent inhibition in sensory perception and discrimination, *Bull Johns Hopkins Hosp* 105:201, 1959.

Mountcastle VB et al: Posterior parietal association cortex of the monkey: command functions for operations within extrapersonal space, *J Neurophysiol* 38:871, 1975.

Northcutt RG: Evolution of the telencephalon in nonmammals, *Annu Rev Neurosci* 4:301, 1981.

Schaible HG, Schmidt RF: Effects of an experimental arthritis on the sensory properties of fine articular afferent units, *J Neurophysiol* 54:1109, 1985.

Tanji J, Wise SP: Submodality distribution in sensorimotor cortex of the unanesthetized monkey, *J Neurophysiol* 45:467, 1981.

Torebjork HE, Vallbo AB, Ochoa J: Intraneural microstimulation in man: its relation to specificity of tactile sensations, *Brain* 110:1509-1529, 1987.

Books and monographs

Darian-Smith I: The trigeminal system. In Iggo A, editor: *Handbook of sensory physiology, Vol II, somatosensory system,* New York, 1973, Springer-Verlag.

Fields HL, Besson JM, editors: Pain modulation. *Progress in brain research,* Vol 77, Elsevier, Amsterdam, 1988.

Fitzgerald M: The course and termination of primary afferent fibers. In Wall PD, Melzack R, editors: *Textbook of pain,* New York, 1989, Churchill Livingstone.

Foreman RD: Organization of the spinothalamic tract as a relay for cardiopulmonary sympathetic afferent fiber activity. In Ottoson D, editor: *Progress in sensory physiology 9,* Berlin, 1989, Springer-Verlag.

Gilman AG et al: The pharmacological basis of therapeutics, ed 7, New York, 1985, Macmillan Publishing.

Mountcastle VB: Central nervous system mechanisms in mechanoreceptive sensibility. In Brookhart JM, Mountcastle VB, editors: *Handbook of physiology,* Section 1: The nervous system; vol 3, Sensory processes, part 2, Bethesda, MD, 1984, American Physiological Society.

Penfield W, Jasper H: *Epilepsy and the functional anatomy of the human brain,* Boston, 1954, Little, Brown & Co.

Uttal WR: *The psychobiology of sensory coding,* New York, 1973, Harper & Row.

Willis WD: Control of nociceptive transmission in the spinal cord, Berlin, 1982, Springer-Verlag.

Willis WD: *The pain system,* Basel, 1985, Karger.

Willis WD, Coggeshall RE: *Sensory mechanisms of the spinal cord,* ed 2, New York, 1991, Plenum Press.

CHAPTER

9

The Visual System

Encephalization is an evolutionary trend in which special sensory organs develop in the heads of animals, along with corresponding neural systems in the brain. These special sensory systems, which include the visual, auditory, vestibular, olfactory, and gustatory systems, allow the animal to detect and analyze light, sound, and chemical signals in the environment, as well as to signal the position of the head. This chapter is concerned with the visual system. The following chapters will consider the functions mediated by the eighth cranial nerve (audition, vestibular functions) and the chemical senses.

■ *Functions of the Visual System*

Vision is one of the most important special senses in humans. We depend on vision and audition for much of human communication. Vision evolved as a dominant sense in arboreal primates, and concomitantly olfaction became less significant; the converse trend occurred in rodents.

The visual system detects and interprets photic stimuli. In vertebrates, effective photic stimuli are electromagnetic waves between 400 and 750 nm long; such wavelengths constitute **visible light.** The eye can distinguish two aspects of light, its **brightness** (or luminance) and its **wavelength** (or color). Light enters the eye and impinges on **photoreceptors** of a specialized sensory epithelium, the **retina.** The photoreceptors include rods and cones. **Rods** have low thresholds for detecting light and operate best under conditions of reduced lighting **(scotopic vision).** However, rods neither provide well-defined visual images nor contribute to color vision. **Cones,** by contrast, are not as sensitive as rods to light and so operate best under daylight conditions **(photopic vision).** Cones are responsible for high visual acuity and color vision.

Information processing within the retina is performed by retinal interneurons, and the output signals are carried to the brain by the axons of retinal ganglion cells. The axons travel in the optic nerves; some cross to the opposite side of the brain in the optic chiasm, and others continue on the same side. Posterior to the optic chiasm, the axons of retinal ganglion cells pass through the optics tracts and synapse in nuclei of the brain (Fig. 9-1). The main visual pathway in humans is through the dorsal lateral geniculate nucleus of the thalamus. This nucleus projects through the optic radiation to the visual receiving areas of the cerebral cortex (see p. 158). Other extrageniculate visual pathways relay in the superior colliculus, pretectum, and hypothalamus (see p. 163). These participate in orientation of the eyes, control of pupil size, and circadian rhythms, respectively.

■ *Structure of the Eye*

The wall of the eye is formed of three concentric layers (Fig. 9-2). The outer layer is the fibrous coat, which includes the transparent **cornea,** with its epithelium, the **conjunctiva,** and the opaque **sclera.** The middle layer is the vascular coat, which includes the iris and the choroid. The **iris** contains both radially and circularly oriented smooth muscle fibers, which make up the pupillary dilator and sphincter muscles (Fig. 9-3). The **choroid** is rich in blood vessels that supply the outer layers of the retina, and it also contains pigment. The inner coat of the eye is the neural layer or retina. The functional part of the retina covers the entire posterior eye except for the blind spot, which is the **optic nerve head** or **optic disc** (see Fig. 9-2). Visual acuity is highest in the central part of the retina, the **macula lutea.** The **fovea** is a pitlike depression in the middle of the macula where visual targets are focused. The inner retinal layers are nourished by tributaries of the central artery and vein of the retina; these vessels course with the optic nerve and can be seen (with an ophthalmoscope) to diverge from the optic nerve head and to spread over the inner surface of the retina (Fig. 9-10). However, the vessels avoid the macula.

Besides the retina, the eye contains a **lens** to focus light on the retina, **pigment** to reduce light scatter, and fluids called **aqueous** and **vitreous humor.** These fluids

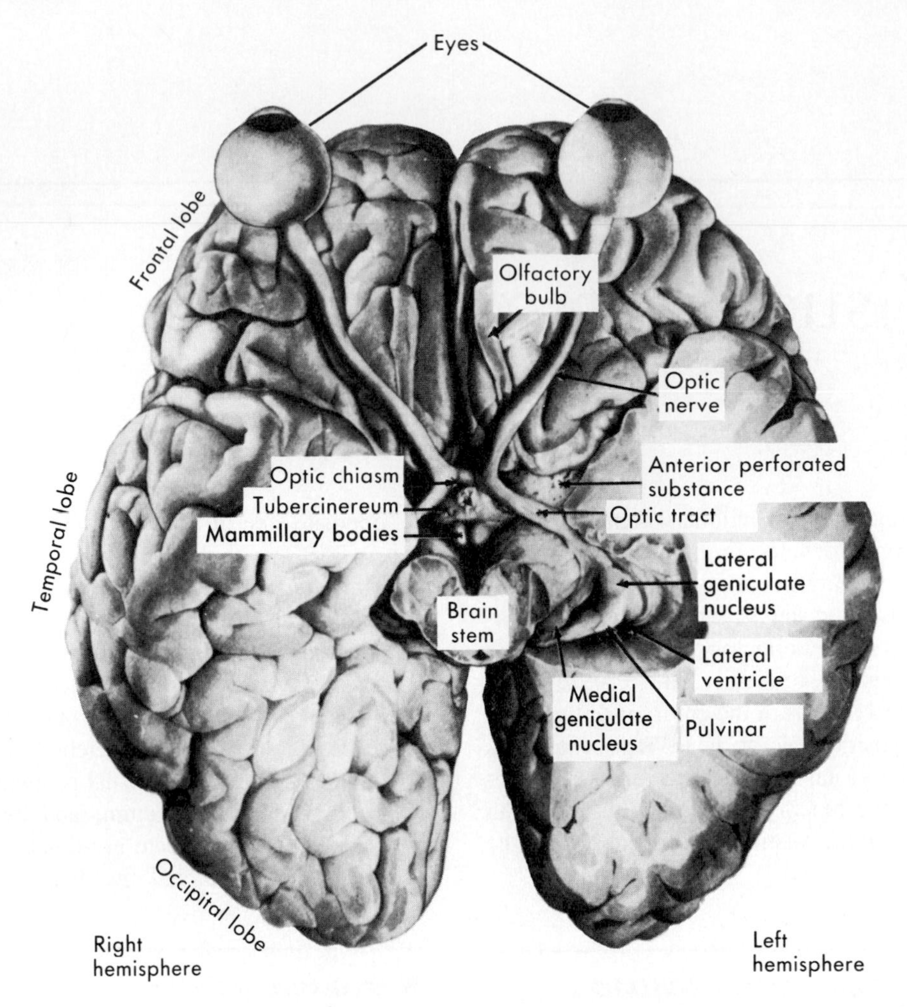

■ **Fig. 9-1** Structures associated with the visual system as seen in a view of the ventral surface of the brain. Visual structures include the eyes, optic nerve, optic chiasm, optic tracts, and lateral geniculate nucleus. (From Polyak S: *The vertebrate visual system,* Chicago, 1957, University of Chicago Press.)

are located in the anterior and posterior chambers and the space behind the lens, and they help maintain the shape of the eye. The aqueous humor is secreted by the ciliary epithelium into the posterior chamber of the eye. It then circulates through the pupil and into the anterior chamber, where it is drained by the canal of Schlemm into the venous system (Fig. 9-3). The pressure in the aqueous humor determines the pressure within the eye. This is normally less than 22 mm Hg. A failure of absorption of aqueous humor leads to an increase in intraocular pressure, a condition known as **glaucoma,** which can cause blindness by impeding blood flow to the retina. The vitreous humor is a gel composed of extracellular fluid containing collagen and hyaluronic acid; it turns over very slowly.

A number of functions of the eyes are under muscular control. Externally attached extraocular muscles aim the eyes toward an appropriate visual target (see Chapter 13). These are innervated by the oculomotor, trochlear, and abducens nerves. There are also several intraocular muscles. The pupillary dilator and sphincter muscles allow the iris to serve as a diaphragm to control pupil size; the dilator is activated by the sympathetic nervous system and the sphincter by the parasympathetic nervous system (through the oculomotor nerve). The shape of the lens is also affected by muscular action. The lens is held in place behind the iris by the suspensory ligaments or zonule fibers, which attach to the wall of the eye at the ciliary body (see Fig. 9-3). When the **ciliary muscles** are relaxed, the tension exerted by the suspensory ligaments flattens the lens. The ciliary muscles act as a sphincter, because they surround the eye. When the ciliary muscles contract, they reduce the tension on the suspensory ligaments, which allows the lens to assume a more spheric shape because of its elastic properties. The ciliary muscles are activated by the parasympathetic nervous system (by way of the oculomotor nerve).

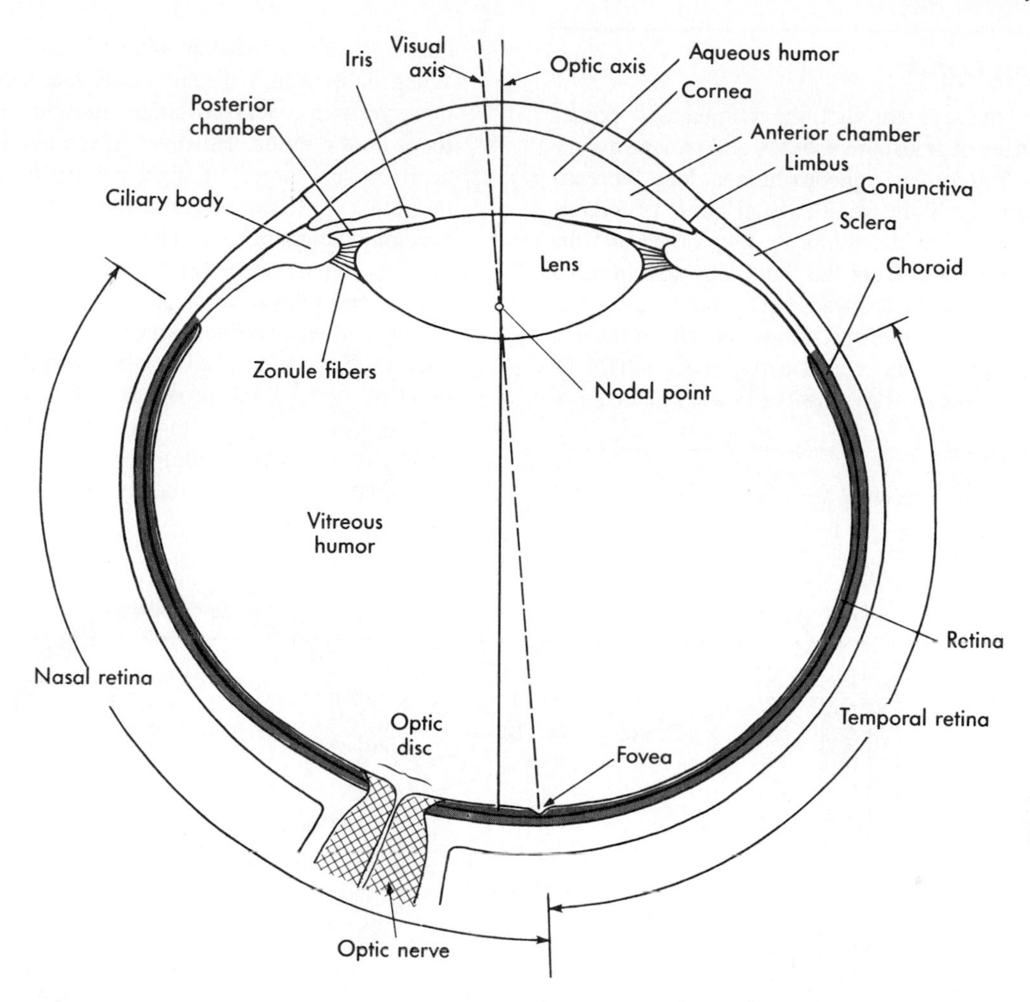

■ **Fig. 9-2** View of a horizontal section of the right eye. (Redrawn from Wall GL: *The vertebrate eye and its adaptive radiation,* Bloomfield Hills, Mich, 1942, Cranbrook Institute of Science.)

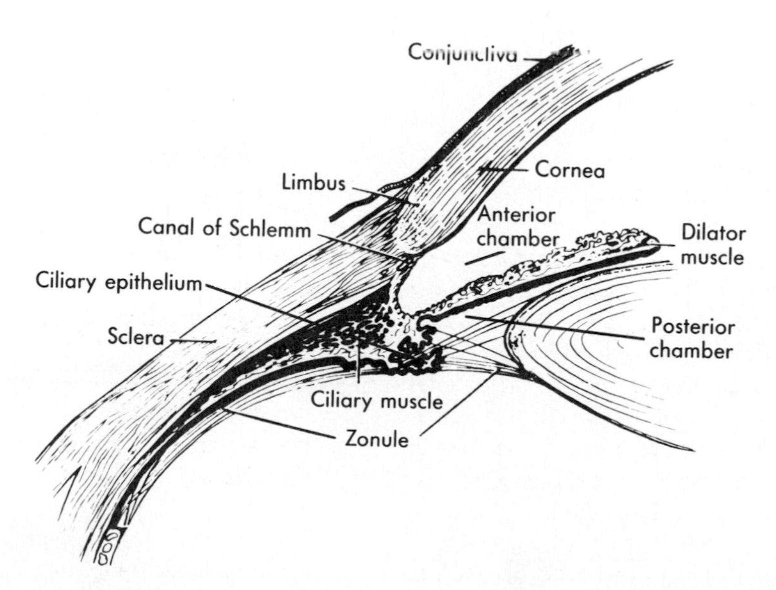

■ **Fig. 9-3** Structures in the anterior part of the eye. Details are shown of the area of the limbus (junction of the cornea and sclera), ciliary body, and lens. (From Davson H: *The eye,* ed 2, vol 1, New York, 1969, Academic Press.)

■ *Physiological Optics*

Light enters the eye through the cornea and passes through a series of transparent fluids and structures, the **dioptric media** (cornea, aqueous humor, lens, vitreous humor). Normally, light from a visual target of interest is focused sharply on the retina by the cornea and the lens, which bend or **refract** the light. The cornea has a refractive power of 43 **diopters*** (D) and is the major refractive element of the eye. However, the refractive power of the lens can be varied between 13 and 26 D; thus the lens is able to allow the eye to accommodate so that both near or distant objects can be seen clearly. When light from a distant visual target enters the normal eye with a relaxed ciliary muscle, the target is in focus on the retina. However, if the eye is directed at a nearby visual target, the light is initially focused behind the retina (i.e., the image at the retina is blurred) until accommodation occurs. The image is sharpened when the lens rounds up because of contraction of the ciliary muscle and relaxation of the zonule fibers.

The cornea and lens serve as a compound convex lens. Light from a visual target passes through the nodal point of the lens and produces a reversed image on the retina, just as in a camera (Fig. 9-4). The retina can be regarded as similar to film in the sense of capturing visual images. However, the retina is far more complex

*The diopter describes the refractile power of a lens and equals the reciprocal of the focal length of the lens in meters.

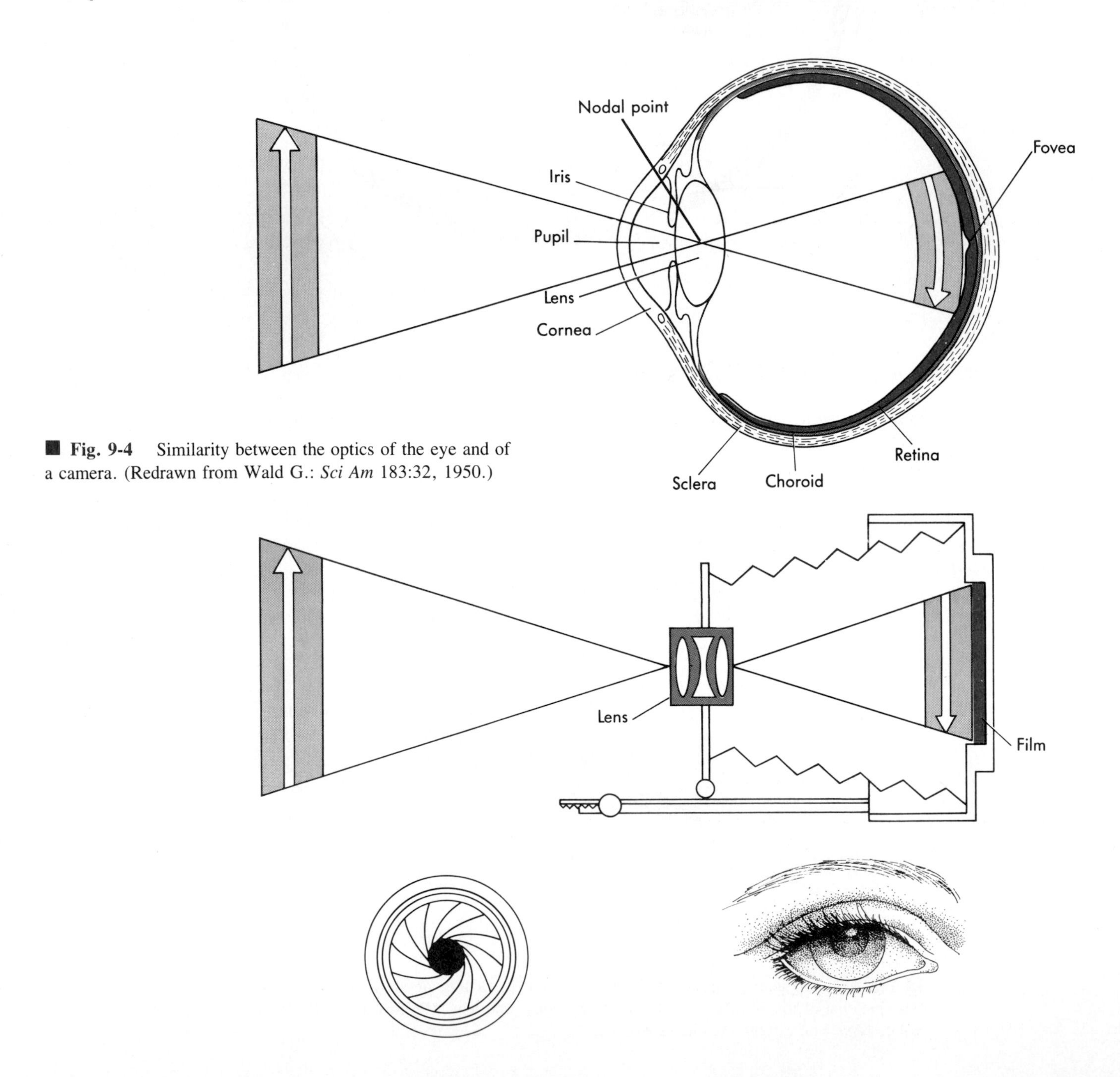

■ **Fig. 9-4** Similarity between the optics of the eye and of a camera. (Redrawn from Wald G.: *Sci Am* 183:32, 1950.)

than film; it processes a continuous sequence of images, and it provides the brain with clues about the movement of visual targets, threats, the light-dark cycle, and other features of the visual environment.

Although the optic axis of the human eye passes through the nodal point of the lens and reaches the retina at a point between the fovea and the optic disc (see Fig. 9-2), the eye is directed by the oculomotor system at a point on the visual target called the **fixation point.** Light from the fixation point passes through the nodal point and is focused on the fovea; this course of the light is along the visual axis. Light from the remainder of the visual target is focused on the retina surrounding the fovea (see Fig. 9-4).

Proper focus of light on the retina depends not only on the lens but also on the iris. As mentioned earlier, the iris acts like the diaphragm in a camera not only in regulating the amount of light entering the eye, but more importantly in controlling the depth of field of the image and the amount of spherical aberration produced by the lens. When the pupil is constricted, the depth of field is increased, and the light is directed through the central part of the lens where spherical aberration is minimal. Pupillary constriction occurs reflexly when the eye accommodates for near vision. Thus the quality of the image is improved by the optical system of the eye when a person reads or does other fine visual work. Another factor that affects the quality of the image is light scatter. This is minimized in the eye by restricting the light path and also by absorption of stray light by pigment in the choroid and the retinal pigment layer. In a camera, light scatter is also prevented by limiting the light path and by covering the interior of the camera with black paint.

As an individual ages, the elasticity of the lens is gradually reduced. The result is that accommodation of

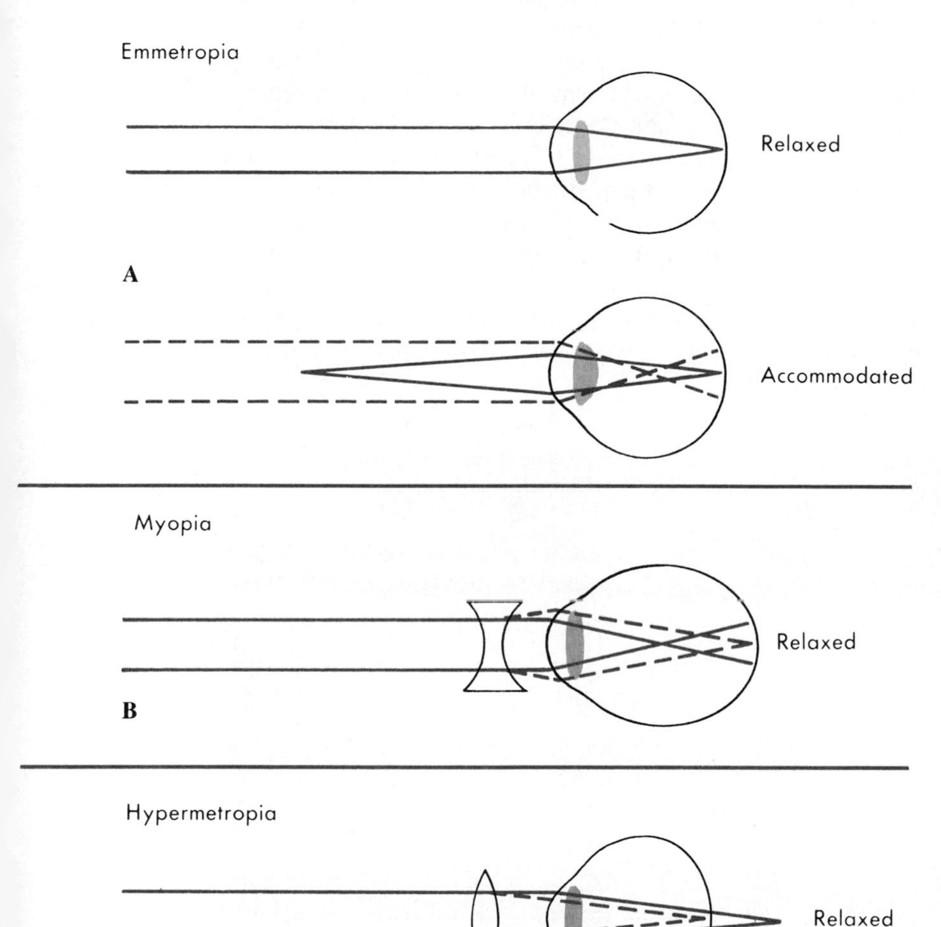

■ **Fig. 9-5** Accommodation and disorders of accommodation. **A,** The eye is emmetropic (has normal accommodation). Light rays from a distant visual target are focused on the retina *(above),* and rays from a nearby target can be focused when the eye is accommodated *(below).* In **B,** the image of a distal visual target is focused in front of the retina unless a concave lens is used. In **C,** the image is focused behind the retina unless a convex lens is employed *(above)* or the eye is accommodated *(below).*

the lens for near vision becomes progressively less effective, a condition called **presbyopia.** A young person can change the power of the lens by as much as 14 D. However, by the time a person reaches 40 years of age, the amount of accommodation halves, and after 50 years it decreases to 2 D or less. Presbyopia can be corrected by convex lenses.

Difficulty in focusing can also occur in young individuals because of a discrepancy between the size of the eye and the refractive power of the dioptric media. For example, in **myopia** (near-sightedness), the images of distant objects are focused in front of the retina (Fig. 9-5). Correction is by concave lenses. Conversely, in **hypermetropia** (far-sightedness), the images of distant objects are focused behind the retina, a problem that can be corrected with convex lenses (Fig. 9-5). Temporary focusing is also possible by accommodation, but this may result in eyestrain caused by fatigue of the ciliary muscles. **Astigmatism** is a condition in which there is an asymmetry in the radii of curvature of different meridians of the cornea or lens (or sometimes of the retina). Astigmatism can often be corrected with lenses with appropriately matched radii of curvature.

■ The Retina

■ Layers of the Retina

The retina is a layered structure. The ten layers of the retina are shown in Fig. 9-6. The retina begins just inside the choroid with the **pigment epithelium** (layer 1). As mentioned earlier, the pigment epithelium absorbs stray light. The pigment cells have tentacle-like processes that extend into the photoreceptor layer and that surround the outer segments of the rods and cones. These processes prevent transverse scatter of light between photoreceptors. They probably also serve a mechanical function to maintain contact between layers 1 and 2. Pigment cells also have other important functions that include phagocytosis of the ends of the outer segments of the rods, which are continuously shed. Another major function of the pigment cells is to reduce **all-trans-retinal** that is taken up from the photoreceptors, and then to transform it back to 11-**cis-retinal,** the form of retinal that combines with opsin in the photoreceptors (see p. 149). The 11-cis-retinal is transported back to the photoreceptors.

The **outer** and **inner segments** of the photoreceptors form layer 2 of the retina (see Fig. 9-6). The structure of photoreceptors is described on p. 147.

The junction between layers 1 and 2 of the retina in adults represents the surface of fusion between the anterior and posterior walls of the embryonic optic cup. This junction is structurally weak and is the locus of **retinal detachment.** A retinal detachment not only results in loss of vision because of displacement of the retina from the focal plane of the eye but also can lead to death of the photoreceptor cells, which are maintained by the blood supply of the choroid (the photoreceptor layer itself is avascular).

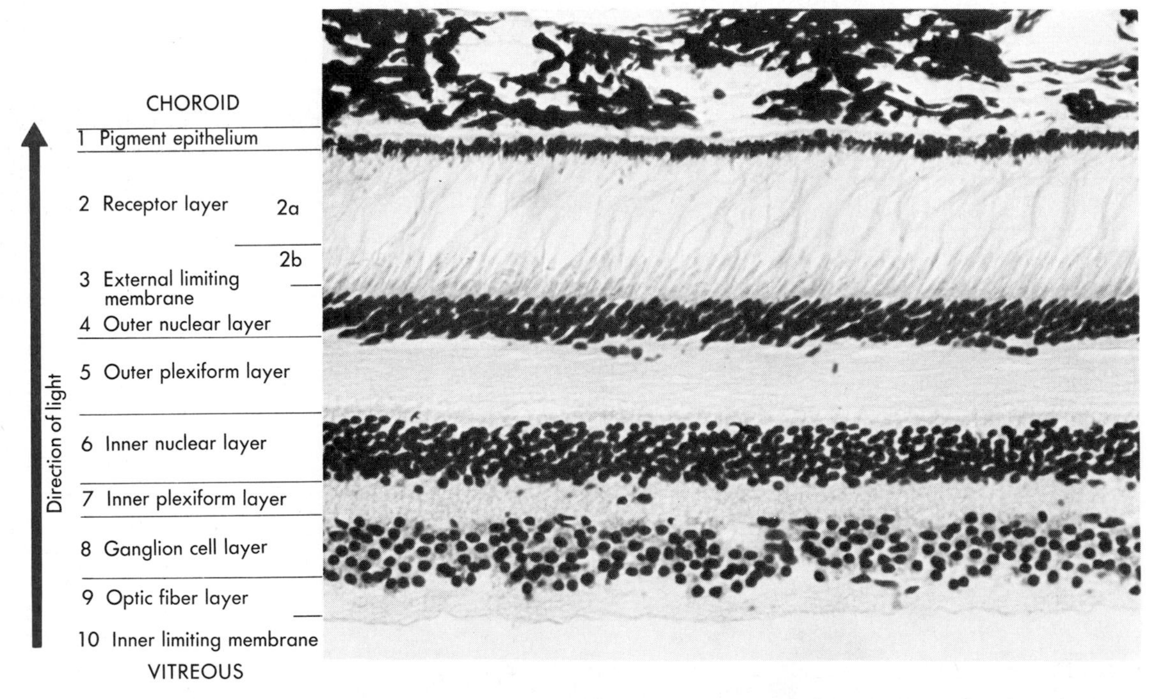

■ **Fig. 9-6** Layers of the retina (from a macaque). The arrow at the left shows the direction of light impinging on the retina. (Nissl stain; courtesy of R.E. Weller.)

The photoreceptors form an organized surface upon which light impinges. Light rays that originate from different parts of the visual target correspond point-to-point to particular photoreceptors. Therefore the geometry of the retina and in particular that of the photoreceptors must be maintained. Retinal glial cells known as **Mueller cells** play an important role in maintaining the geometry of the retina. The Mueller cells are oriented radially, parallel to the light path through the retina. The outer ends of the Mueller cells form tight junctions with the inner segments of the photoreceptors. The numerous connections made between Mueller cells and inner segments give the appearance of a continuous layer in the light microscope (see Fig. 9-6). This layer is called the **outer limiting membrane** (layer 3 of the retina).

Inside the outer limiting membrane is a layer of nuclei (see Fig. 9-6), the **outer nuclear layer** (layer 4 of the retina). These are the nuclei of the rods and cones.

The next layer of the retina (layer 5) is called the **outer plexiform layer** (see Fig. 9-6). This is a synaptic zone that contains presynaptic and postsynaptic elements of synapses between the photoreceptors and retinal interneurons, including bipolar cells and horizontal cells.

Deep to the outer plexiform layer is the **inner nuclear layer** (layer 6 of the retina). This layer contains the cell bodies and nuclei of a number of cell types, including the retinal interneurons (bipolar cells, horizontal cells, amacrine cells, and interplexiform cells) and the Mueller cells.

The next layer is the **inner plexiform layer** (layer 7 of the retina). This is another synaptic zone that contains the presynaptic and postsynaptic elements of synapses between some of the retinal interneurons, including the bipolar and amacrine cells, and the ganglion cells.

Layer 8 of the retina is the **ganglion cell layer.** As mentioned earlier, the ganglion cells are the output cells of the retina. They transmit visual information to the brain.

The axons of the retinal ganglion cells form the **optic fiber layer** (layer 9 of the retina). The axons pass across the vitreous surface of the retina and avoid the macula, and they enter the optic disc. They leave the eye in the optic nerve. The portions of the ganglion cell axons that are in the optic fiber layer remain unmyelinated, but the axons become myelinated after they reach the optic disc and nerve. The lack of myelin where the axons cross the retina is a specialization that helps permit light to pass through the inner retina with minimal distortion.

The innermost layer of the retina is the **inner limiting membrane** (layer 10 of the retina). This is formed by projections of the Mueller cells.

Structure of Photoreceptors

The photoreceptors include the rods and the cones. Each photoreceptor cell includes a cell body, an inner and an outer segment that extends into layer 2, and a synaptic terminal (Fig. 9-7). The outer segments of rods are longer than those of cones, and they contain stacks of freely floating membrane disks rich in rhodopsin molecules (10^8 per rod outer segment). The outer segments of cones also contain membranous disks associated with photopigment, but the outer segments of cones are not as long as those of rods and the disk membranes consist of infoldings of the surface membrane. Rods contain much more photopigment than do cones. This accounts in part for their greater sensitivity to light. A single photon can elicit a rod response, whereas several hundred photons are required for a cone response.

The inner segments of the photoreceptors are connected to the outer segments by a modified cilium that contains nine pairs of microtubules but lacks the two central pairs of microtubules that are found in most cilia. The inner segments contain a number of organelles, including numerous mitochondria.

The photopigment is synthesized in the inner segment and incorporated into the membranes of the outer segment. In rods the pigment is inserted into new membranous disks (three new disks are formed per hour) that

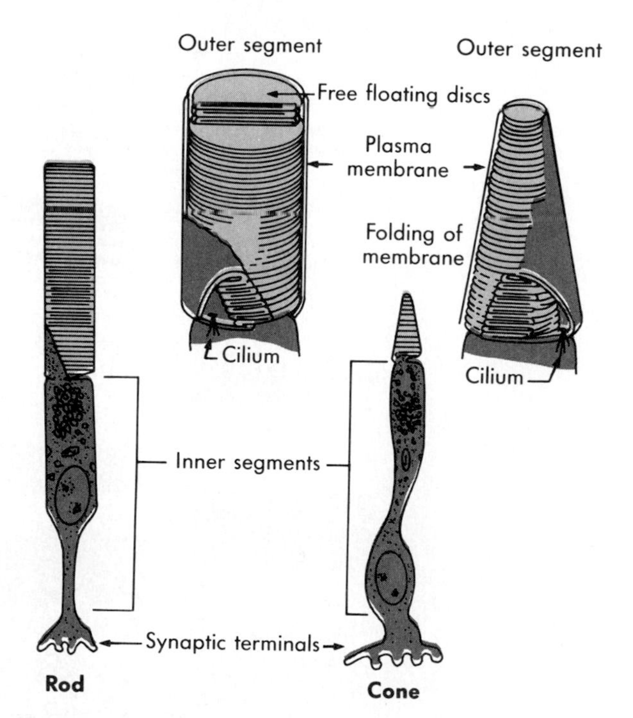

■ **Fig. 9-7** Rods and cones. The drawings at the bottom show the general features of a rod and a cone. The insets show the outer segments in more detail.

are then displaced distally until they are eventually shed at the apex of the outer segment to be phagocytized by the pigment cell epithelium. This process determines the rodlike shape of the outer segments of rods. In cones, the photopigment is inserted randomly into the membranous folds of the outer segment and there is no shedding comparable with that seen in rods.

Regional Variations in the Retina

Fovea. The fovea is a depression in the macula lutea; it is the region of the retina that has the highest visual resolution. Correspondingly, the image from the fixation point is focused on the fovea. The fovea has an unusual layering pattern in that several of the retinal layers appear to be pushed aside (Fig. 9-8). This allows light to reach the photoreceptors without having to pass through the inner layers of the retina; this arrangement reduces distortion of the image. An additional specialization here is the unusually long and thin cone outer segments. This shape permits a high packing density. In fact, cone density is maximum in the fovea (see Fig. 9-9). The cones provide high visual resolution, which matches the high quality of the image provided to the fovea.

Optic disk. As mentioned earlier, the axons of the retinal ganglion cells cross the retina in the optic fiber layer (layer 9) and enter the optic nerve at the optic disk. The axons in the optic fiber layer pass around the macula and avoid the fovea, as do the blood vessels that supply the inner layers of the retina (Fig. 9-10). The optic disk can be visualized on physical examination by use of an ophthalmoscope. The normal optic disk has a slight depression in its center. Changes in the appearance of the optic disk are of considerable clinical importance. For example, the depression may be exaggerated by loss of ganglion cell axons (optic atrophy) or the optic disk may protrude into the vitreous space because of edema (papilledema) due to increased intracranial pressure.

The optic disk lacks photoreceptors and therefore photosensitivity. Thus the optic disk is a blind spot in the visual surface of the retina. The blind spot is normally ignored by the subject, both because the corresponding part of the visual field can be seen by the contralateral eye and because of a psychological process by which incomplete visual images tend to be completed perceptually. However, the blind spot can be mapped, as demonstrated in Fig. 9-11.

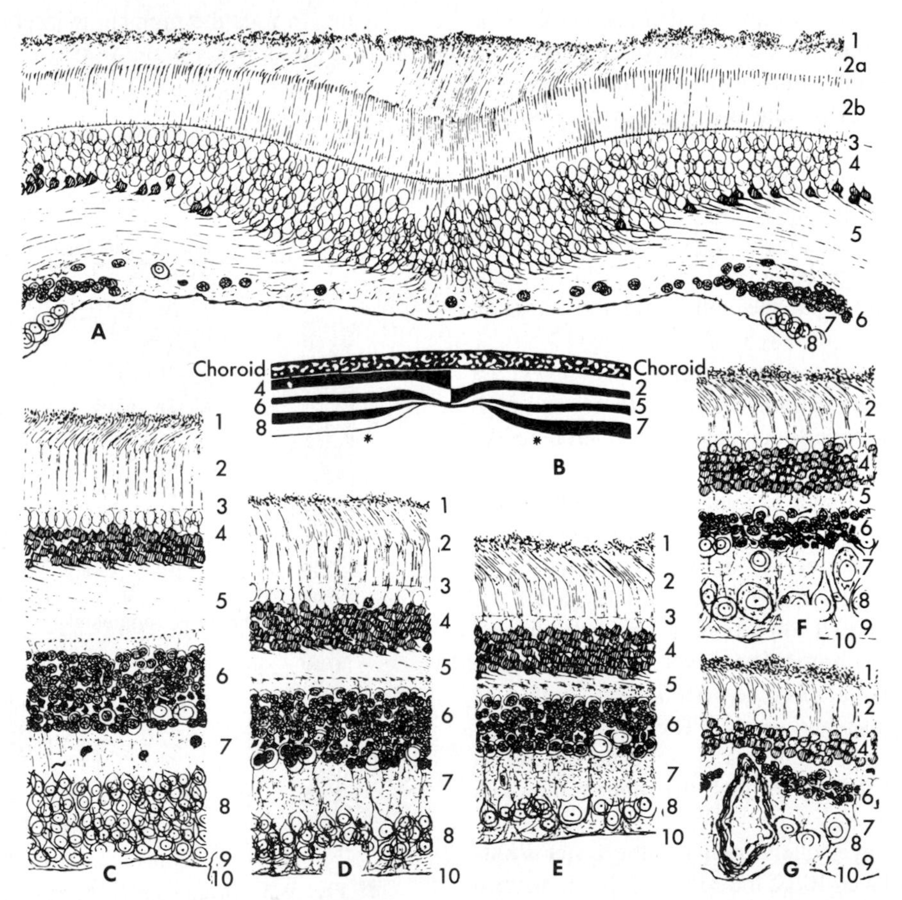

Fig. 9-8 Section through the fovea of a retina from a macaque monkey. The layers of the retina are numbered. (From Polyak S: *The vertebrate visual system,* Chicago, 1957, University of Chicago Press.)

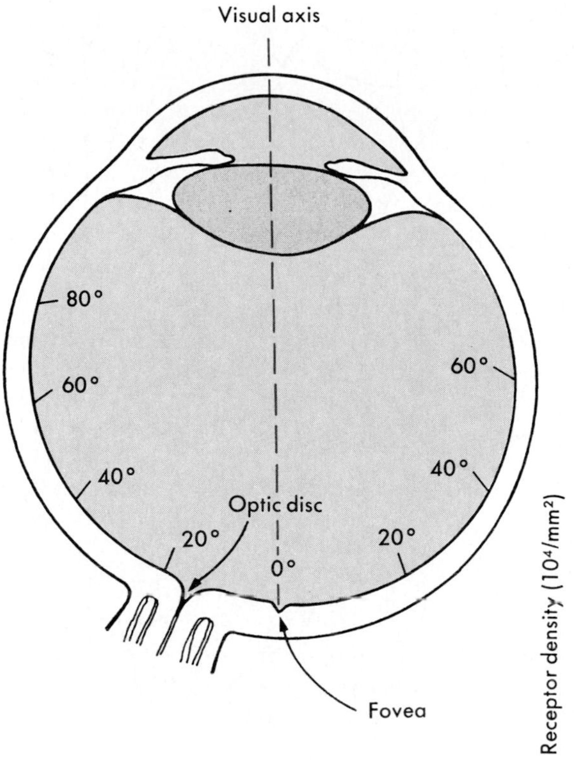

■ **Fig. 9-9** The drawing on the left shows the location of the fovea (0°) and different parts of the retina at varying degrees of eccentricity from the fovea. The graph on the right plots the density of cones and rods as a function of retinal eccentricity from the fovea. Note that cone density peaks at the fovea, rod density peaks at an eccentricity of about 20°, and no photoreceptors occur in the optic disc. (Redrawn from Cornsweet TN: *Visual perception,* New York, 1970, Academic Press.)

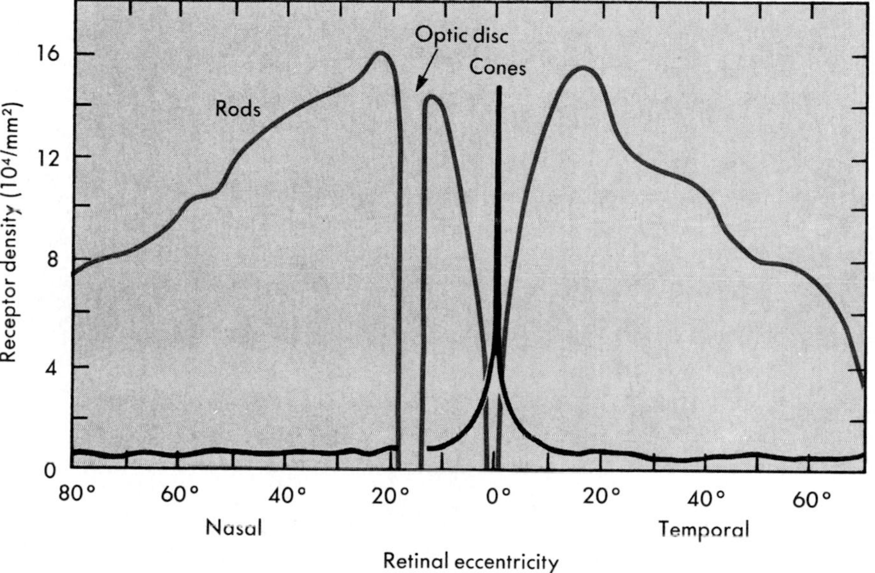

■ *Visual Pigments*

Light must be absorbed for it to be detected by the retina. Light absorption is accomplished by the visual pigments, which are located in the outer segments of rods and cones. The pigment found in the outer segments of rods is **rhodopsin,** or visual purple (because it has a purple appearance after green and blue light have been absorbed). Three variants of visual pigment are found in different cone types. Rhodopsin absorbs light best at a wavelength of 500 nm, whereas the cone pigments absorb best at 420 nm (blue), 531 nm (green), or 558 nm (red). However, the absorption spectra of these visual pigments overlap considerably.

Rhodopsin contains a chromophore called retinal, which is the aldehyde of **retinol,** or vitamin A. Retinol is derived from carotenoids, such as β-carotene, the orange pigment found in carrots. Like other vitamins, retinol cannot be synthesized by humans; instead it is derived from food sources. Individuals with a severe vitamin A deficiency suffer from "night blindness," a condition in which vision in poor illumination is defective.

An isomer of retinal known as 11-cis-retinal, is combined with a glycoprotein known as opsin, to form **rhodopsin.** When rhodopsin absorbs light, the higher energy level causes a series of chemical changes that lead to isomerization of 11-cis-retinal to all-trans-retinal, release of the bond with opsin, and conversion of the retinal to retinol. The separation of all-trans-retinal from opsin causes the visual pigment to be bleached; i.e., it loses its purple color.

Visual adaptation. Reduction in the amount of rhodopsin is associated with a process known as **light adaptation,** which leads to a reduction in visual sensitivity. Light adaptation occurs rapidly, within seconds, and favors cone vision because the rhodopsin in rods bleaches more readily than do the cone pigments. (Cones have a special mechanism by which a reduction in Ca^{++} concentration occurs during light adaptation and permits cAMP synthesis and the reopening of Na^{+} channels; see p. 151).

As all-trans-retinal is formed, it is transported to the retinal pigment cell layer, where it is reduced, isomerized, and esterified. As all-cis-retinal is converted back to 11-cis retinal, it is transported back to the photoreceptor layer, taken up by outer segments, and recombined to opsin. In this way, bleached rhodopsin is regenerated. The regeneration of photopigment is one

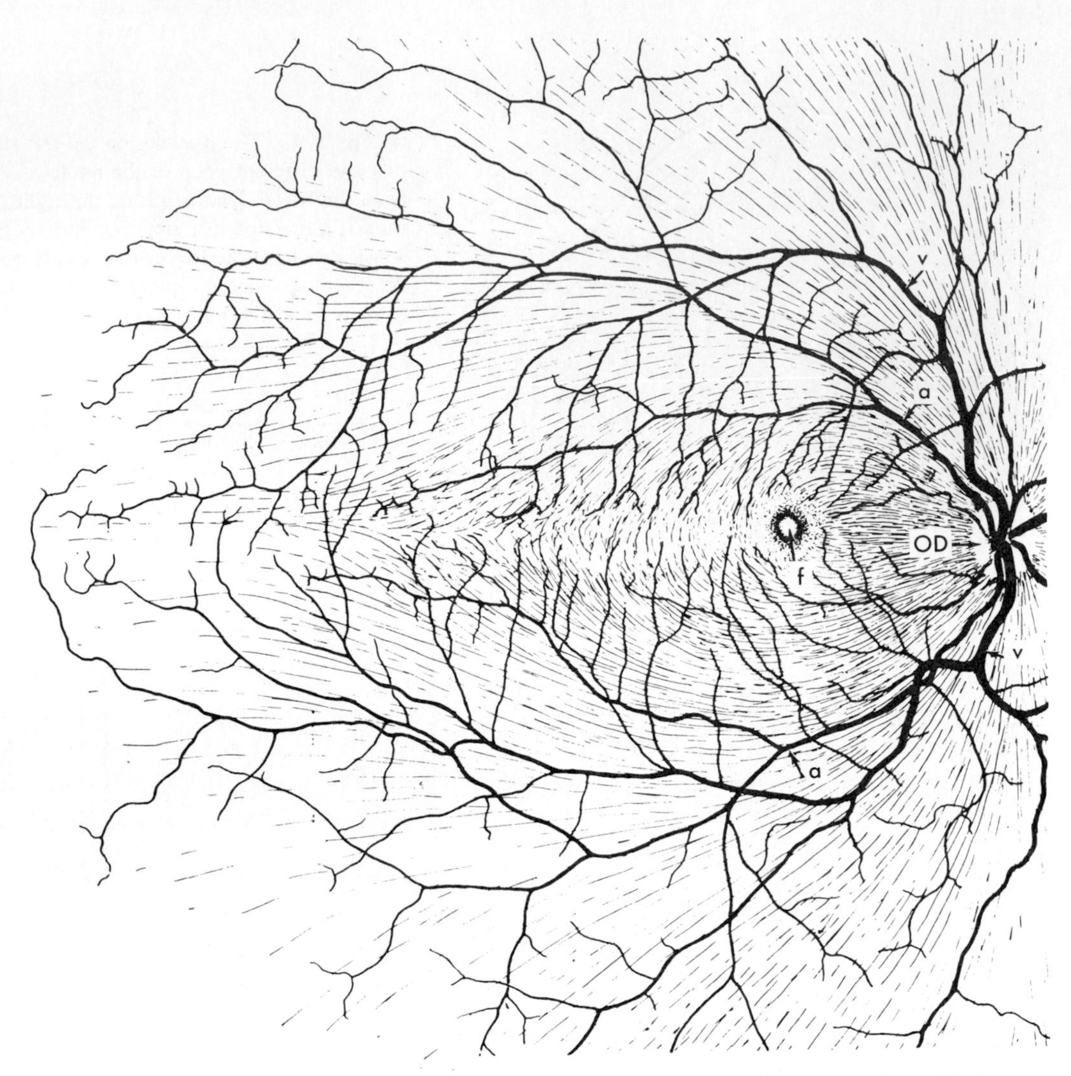

■ **Fig. 9-10** View of the posterior surface of the human eye as seen through an ophthalmoscope. Branches of the central artery *(a)* and vein (v) are seen as they leave the optic disc *(OD)*. The fovea *(f)* is located a short distance to the temporal side of the optic disc. Note the pattern formed by the ganglion cell axons *(fine lines)* as they sweep towards the optic disc. They and the blood vessels avoid the area of the fovea. (From Polyak S: *The vertebrate visual system,* Chicago, 1957, University of Chicago Press.)

mechanism involved in **dark adaptation,** a process that results in an increase in visual sensitivity. Cones adapt more rapidly to darkness than do rods, but their adapted threshold is relatively high. Thus cones do not function when the ambient light level is low. By contrast, rods adapt to darkness slowly, but the sensitivity of rods is high. Within 10 minutes in a dark room, rod vision is more sensitive than cone vision.

Dark adaptation is very familar to moviegoers who must wait several minutes after entering the darkened theater before they can see an empty seat. While the theater is dark and rod vision is in use, visual acuity is low and colors are not distinguished (scotopic vision). When the movie is projected, light adaptation allows cone function to resume (photopic vision) and visual acuity and color vision are restored.

Color vision. The visual pigments in cone outer segments have opsins that differ from that in rhodopsin. The result is that the three types of cone pigments absorb light best in the blue, green, or red parts of the visible light spectrum (Fig. 9-12). The differences in ab-

■ **Fig. 9-11** If one eye is closed and the page is held about 30 cm from the open eye, when one of the symbols is fixated, the other will disappear as the page is moved back and forth. For the right eye, the fixation point should be the circle and for the left eye the cross.

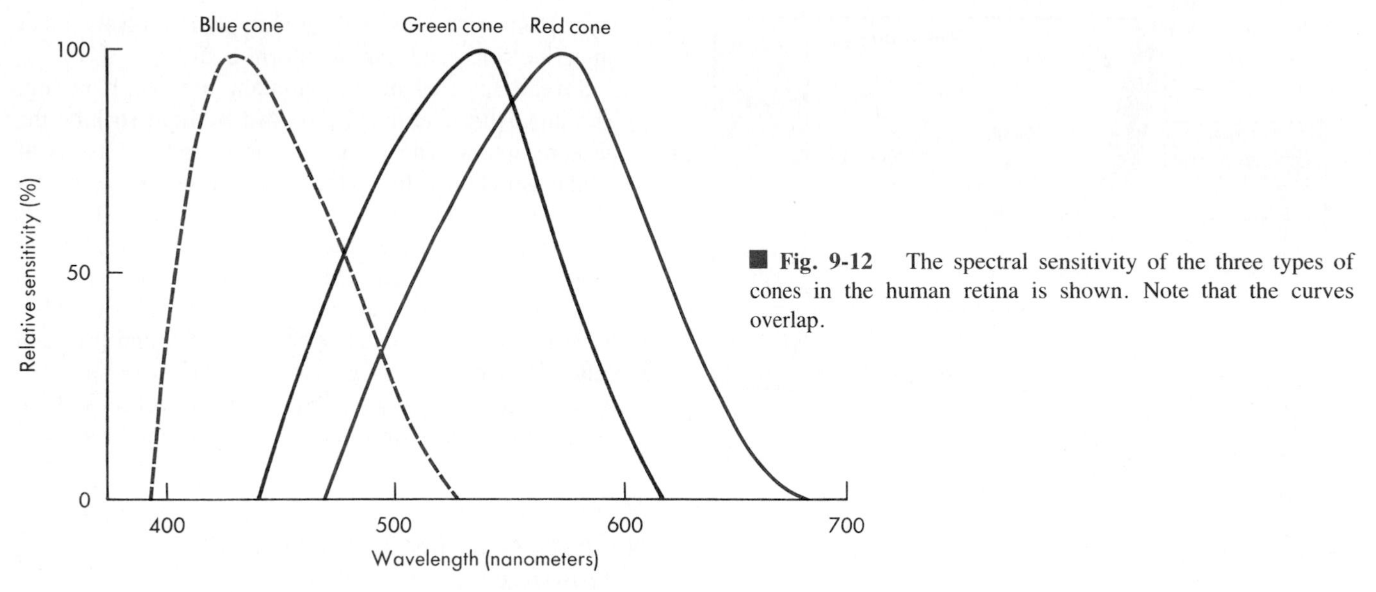

Relative spectral sensitivity of the three cone types in humans.

■ **Fig. 9-12** The spectral sensitivity of the three types of cones in the human retina is shown. Note that the curves overlap.

sorption are presumed to account for color vision, according to the **trichromacy theory.** The basis of the trichromacy theory is that any color can be produced by a suitable mixture of three other colors. Because three types of cone pigments exist, it has been proposed that these pigments in some way provide for a neural analysis of color mixing. However, there must also be a neural system for the analysis of color brightness, because absorption of light by a visual pigment depends in part on the wavelength and in part on the intensity of the light. A given wavelength of light at a particular intensity may be absorbed by two or three of the visual pigments found in cones. However, absorption by one of the pigments will be greater than that by the others. If the intensity of the light is changed, but not the wavelength, the ratio of absorption will remain constant. The relationship in the effectiveness of aborption of light of different wavelengths by the different types of cones provides clues used by the visual system to distinguish different colors. At least two different kinds of cones are required for color vision. The presence of three kinds decreases the ambiguity in distinguishing colors when all three absorb light, and it assures that at least two types of cones will absorb most wavelengths of visible light.

Consistent with the trichromacy theory are observations on color blindness. In **color blindness,** a genetic defect (sex-linked recessive) results in the loss of one or more cone mechanisms. Normal people are *trichromats,* because they have three cone mechanisms. Loss of one of the cone mechanisms results in individuals called *dichromats.* When the long wavelength cone mechanism is lost, the resultant condition is called **protanopia;** loss of the medium wavelength system causes **deuteranopia;** loss of the short wavelength system, **tritanopia.** *Monochromats* have lost all three cone mechanisms (or in some cases, two of these).

Note that other theories of color vision have been proposed. The **opponent-process theory** is based on the observation that certain pairs of colors are perceived as if these colored lights activate opposing neural processes. Green and red are opposed, as are yellow and blue and black and white. For example, if a gray area is surrounded by a green ring, the gray area appears to acquire a reddish color. Furthermore, a greenish-red or a bluish-yellow color do not exist. These observations lead to the proposal that neurons activated by green would be inhibited by red. Similarly, neurons excited by blue would be inhibited by yellow. Neurons with these characteristics are found both in the retina and at higher levels of the visual pathway.

A recent theory, the **retinex** theory, attributes color vision to the combined action of neural activity at several levels of the visual system, including the retina and cerebral cortex. All of these theories of color vision may be applicable.

■ *Visual Transduction*

When light is absorbed by rhodopsin, the signal is amplified by a special transduction mechanism in the rods. This amplification mechanism, along with the large amount of photopigment in rod outer segments, accounts for the extraordinary sensitivity of rods, which can detect a single photon after full dark adaptation. In the dark, rods have open sodium channels (Fig. 9-13). A net influx of Na^+ results in a continuous current, called the **dark current.** This causes a maintained depolarization of the rods (to a resting potential of about −40 mV). As a consequence of the depolarization, neurotransmitter (considered to be glutamate) is tonically released at the rod synapses on bipolar and horizontal cells. The intracellular Na^+ concentration is kept at a

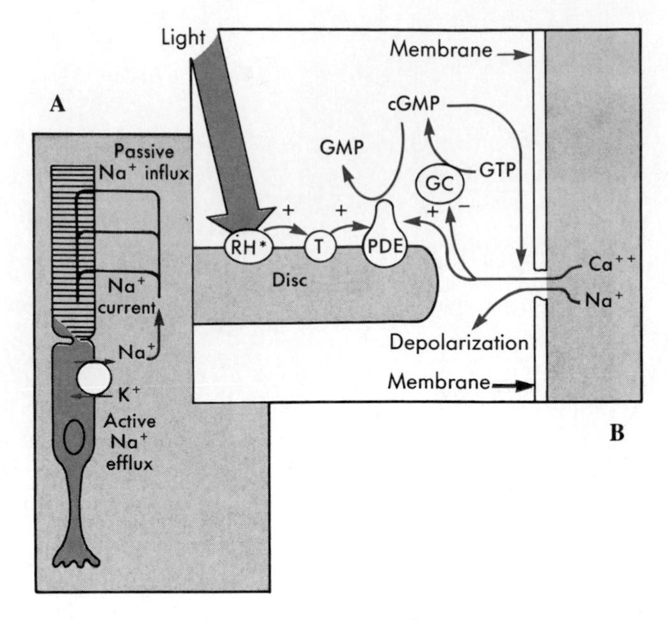

■ **Fig. 9-13** **A,** A drawing of a rod. The flow of dark current is indicated, as well as the Na pump. In **B,** the sequence of second messenger events that follow absorption of light is shown. *Rh,* rhodopsin; *T,* transducin; *PDE,* phosphodiesterase; *GC,* guanylate cyclase; *cGMP,* cyclic guanosine monophosphate; *GTP,* guanosine triphosphate.

steady-state level by the pumping action of Na^+-K^+ ATPase.

Absorption of light activates a G-protein, called transducin. This in turn activates cyclic guanosine monophosphate (cGMP) phosphodiesterase, which is associated with the rhodopsin-containing discs. cGMP phosphodiesterase hydrolyzes cGMP to 5′-GMP and lowers the cGMP concentration in the rod cytoplasm. cGMP normally keeps the sodium channels open, so that a reduction in cGMP concentration causes the channels to close and the membrane to hyperpolarize. Amplification results from the fact that a single rhodopsin molecule activates hundreds of transducin molecules, and each phosphodiesterase molecule hydrolyzes thousands of cGMP molecules per second.

Similar events occur in cones, but the membrane hyperpolarization occurs more quickly than in rods, perhaps because the intracellular distances are shorter in cones.

■ *Retinal Circuitry*

A diagram of the basic circuitry of the retina is shown in Fig. 9-14. Photoreceptors (R) synapse on the dendrites of bipolar cells (B) and horizontal cells (H) in the outer plexiform layer. The horizontal cells make transverse connections with bipolar cells, and they receive input from interplexiform cells (I). Bipolar cells synapse on the dendrites of ganglion cells (G) and the processes of amacrine cells (A) in the inner plexiform layer. Amacrine cells connect with ganglion cells, other amacrine cells, and interplexiform cells.

Several features of this circuitry are worth noting. The input to the retina is provided by light striking the photoreceptors. The output is carried by the axons of retinal ganglion cells to the brain. Information is processed within the retina by the interneurons. The most direct pathway through the retina is from the photoreceptors to the bipolar cells and then to the ganglion cells. Another more indirect pathway involves the photoreceptors, bipolar cells, amacrine cells, and ganglion cells. Horizontal cells provide lateral interactions between adjacent pathways. Interplexiform cells allow interactions to occur from the inner to the outer retina.

■ *Contrasts in Rod and Cone Pathway Functions*

Rod and cone pathways have several important functional differences, based partly on the differences in their phototransduction mechanisms and partly on retinal circuitry.

Rods have more photopigment and a better signal amplification system than do cones, and there are many more rods than cones. Thus rods function better in dim light (scotopic vision). However, they contain a single photopigment and so cannot signal color differences. Furthermore, many rods converge on single bipolar cells. Therefore rods cannot provide high resolution vision, because the effective receptive field for the rod pathway is large. In bright light, rhodopsin is bleached. Hence, rods no longer function under photopic conditions because of light adaptation. Loss of rod function results in night blindness.

Cones have a higher threshold to light and so are not activated in dim light after dark adaptation. However, they operate very well in daylight. They provide high resolution vision, because only a few cones converge on bipolar cells in the cone pathways, and in the fovea there is no convergence (there is a one-to-one connection from cones to bipolar cells). This means that cone pathways have very small receptive fields. Therefore, the cone pathways can resolve stimuli that originate from sources that are very close to each other. Cones also respond to sequential stimuli with good temporal resolution. Finally, cones have three different cone photopigments. Thus, they can discriminate among wavelengths and therefore can participate in color vision. Loss of cone function results in functional blindness (rod vision is not sufficient for normal activities).

■ *Synaptic Interactions*

The distances within the retina are short. Hence most of the activity in retinal circuits involves just receptor and synaptic potentials and not action potentials. Only the

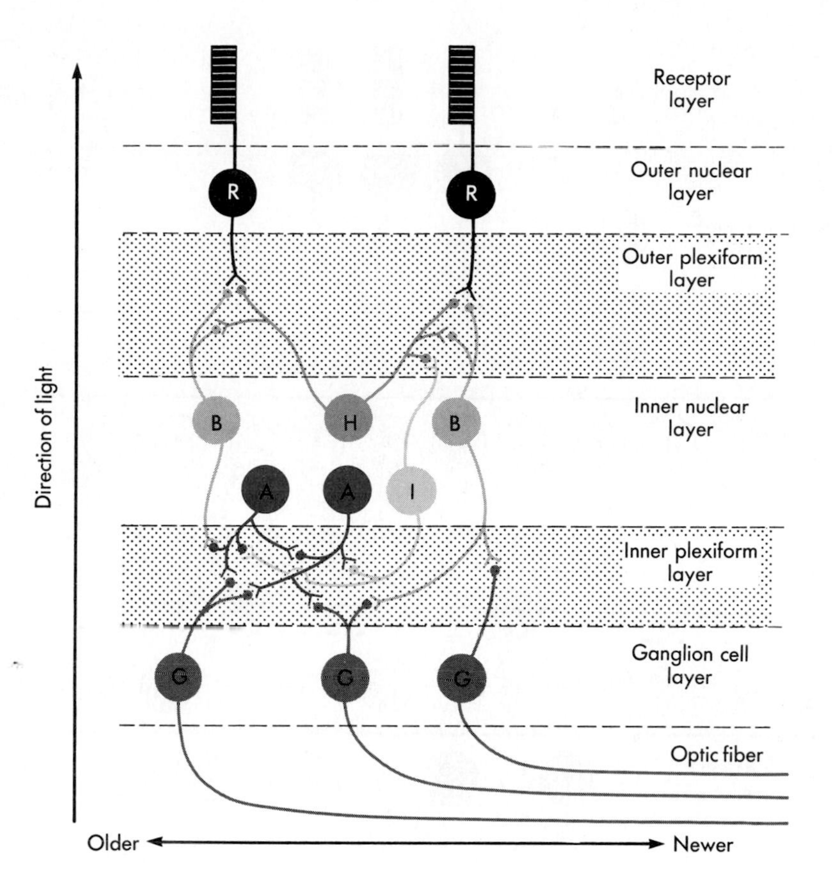

■ **Fig. 9-14** Basic retinal circuitry. The arrow at the left indicates the direction of light through the retina. *R,* photoreceptors; *B,* bipolar cells; *H,* horizontal cells; *A,* amacrine cells; *I,* interplexiform cells; *G,* ganglion cells.

ganglion cells and some amacrine cells can generate action potentials. It is unclear why amacrine cells have action potentials, but ganglion cells must have them to transmit information to the brain.

The receptor potential in photoreceptors is hyperpolarizing, and synaptic potentials in the retina can be either hyperpolarizing or depolarizing. Hyperpolarizing events reduce neurotransmitter release from the synaptic terminals of a retinal interneuron, whereas depolarizing events increase neurotransmitter release.

■ *Receptive Field Organization*

The receptive fields of retinal ganglion cells constitute an important step in visual information processing, because these fields reflect the characteristics of visual signals that are conveyed to the brain. Fig. 9-15 shows how the receptive fields of photoreceptors and retinal interneurons determine the receptive fields of the retinal ganglion cells.

In Fig. 9-15, *A,* the receptive field of a photoreceptor is indicated by a circle with a minus sign. This indicates that the receptive field is small and circular and that light in the receptive field will hyperpolarize the photoreceptor cell.

The receptive field of a horizontal cell is shown in Fig. 9-15, *B* as a larger circle with minus signs. Light reaching any of the photoreceptors converging on this horizontal cell will hyperpolarize it. The sequence of events would be as follows: light hyperpolarizes one or more photoreceptors; these release less excitatory neurotransmitter; the horizontal cell is hyperpolarized because of the reduction in tonic excitatory drive (i.e., it is disfacilitated).

The receptive fields of two different types of bipolar cells are shown in Fig. 9-15, *C*. The bipolar cell on the left (B1) has a centrally located excitatory receptive field (shown by the white circle containing a plus sign), surrounded by an inhibitory receptive field (gray annulus containing minus signs). This type of receptive field is described as having a **center-surround organization,** and this particular arrangement is called an **on-center, off-surround** receptive field (abbreviated "on-center"). The bipolar cell at the right in Fig. 9-15, *C* labeled B2 has an "off-center" receptive field.

The responses of bipolar cells depend on input from one or more photoreceptors and from horizontal cells. The response to stimulation of the center of the receptive field reflects direct connections from one or a few photoreceptors. If the neurotransmitter tonically released by the photoreceptor hyperpolarizes the bipolar cell, then light striking the photoreceptor and hyperpolarizing it will reduce the release of the neurotransmitter and the bipolar cell will be depolarized (disinhibited). On the other hand, if the neurotransmitter tonically released by the photoreceptor is depolarizing, the bipolar cell will be disfacilitated (just like the horizontal cell in Fig. 9-15, *B*). The surround response results when light impinges on adjacent photoreceptors and changes the

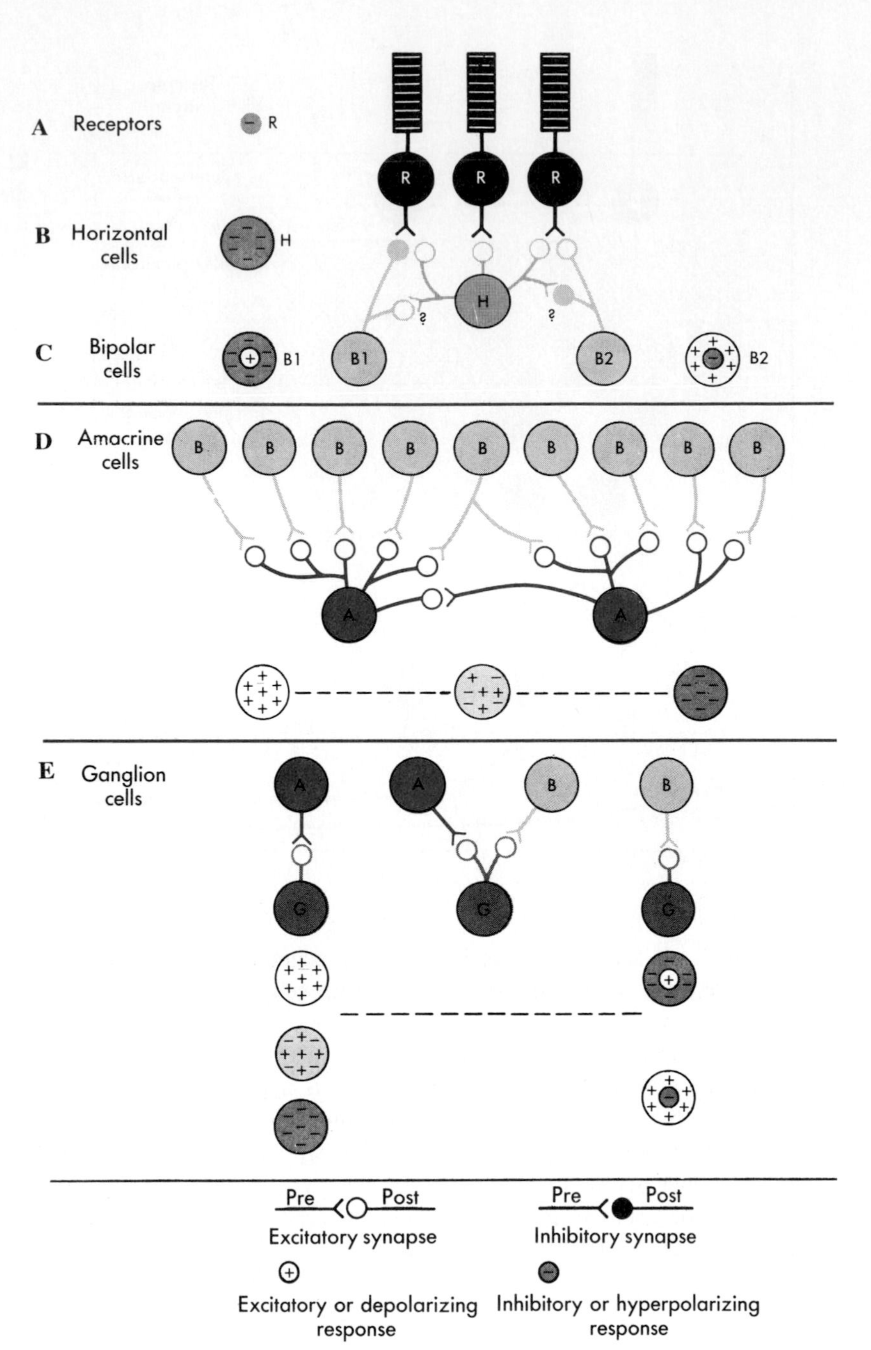

■ **Fig. 9-15** **A-E** show the receptive fields and neural circuitry determining them for photoreceptors *(R)*, horizontal cells *(H)*, bipolar cells *(B)*, amacrine cells *(A)*, and ganglion cells *(G)*. Hyperpolarizing events are shown by minus signs and depolarizing events by plus signs.

activity of horizontal cells. The pathway through the horizontal cells results in a response that is opposite in sign to that produced directly by the photoreceptor(s) that mediate the center response.

If light strikes not only the photoreceptors responsible for the center response but also those that cause the surround response, the bipolar cell may not respond at all because of the opposing actions from the center and surround. On the other hand, light moving across the receptive field will sequentially cause dramatic changes in the activity of the bipolar cell as it crosses the receptive field from surround to center and then again to surround.

The receptive fields of amacrine cells are shown in Fig. 9-15, *D*. The amacrine cells receive input from different combinations of "on-center" and "off-center" bipolar cells. Thus their receptive fields are mixtures of "on-center" and "off-center" regions. There are many different types of amacrine cells, and at least eight neurotransmitters are known to be released by the amacrine cells.

Fig. 9-15, *E* shows the receptive fields of ganglion cells. Ganglion cells may receive a dominant input from amacrine cells (left), a mixed input from amacrine and bipolar cells (middle), or a dominant input from bipolar cells (right). When the amacrine cell input dominates, the receptive fields of ganglion cells tend to be diffuse and either excitatory or inhibitory. On the other hand, when the input is dominated by bipolar cells, the ganglion cells have a center-surround organization similar to that of bipolar cells.

■ *X-, Y-, and W-Cells*

Experimental studies have shown that, based on a number of features, retinal ganglion cells can be subdivided into three general types, called X-cells, Y-cells, and W-cells. X- and Y-cells are fairly homogeneous groups, whereas W-cells are heterogeneous. X- and Y-cells have center-surround receptive fields; hence they are presumably controlled predominantly by bipolar cells. Some W-cells also have center-surround receptive fields, but many have diffuse receptive fields. Thus they are probably influenced chiefly through amacrine cell pathways.

The different types of ganglion cells have strikingly different morphology (Fig. 9-16). X-cells have medium-sized cell bodies and axons and restricted dendritic trees. On the other hand, Y-cells have larger cell bodies and axons and much more extensive dendritic trees than those of X-cells. W-cells have small cell bodies and axons but extensive dendritic trees.

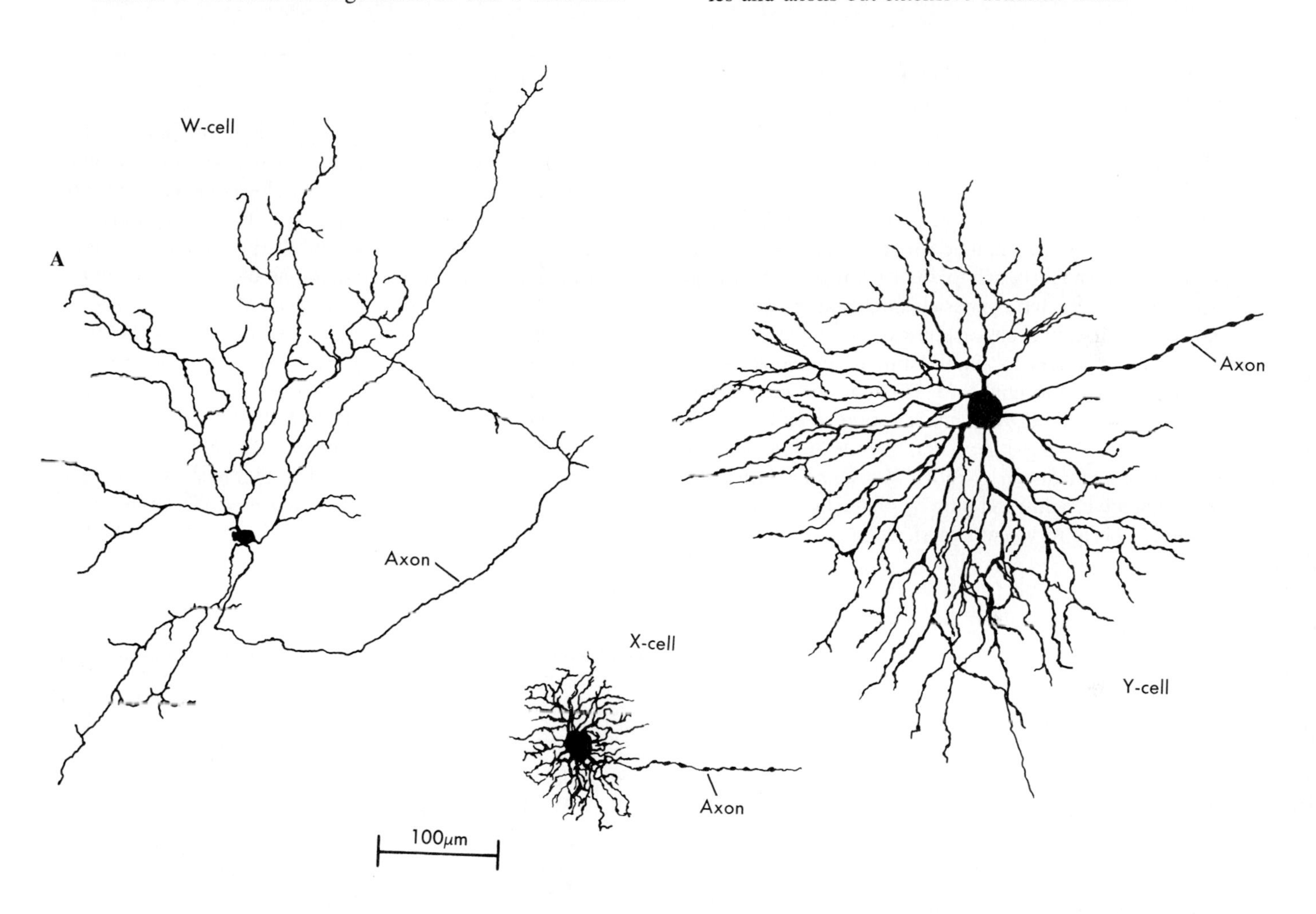

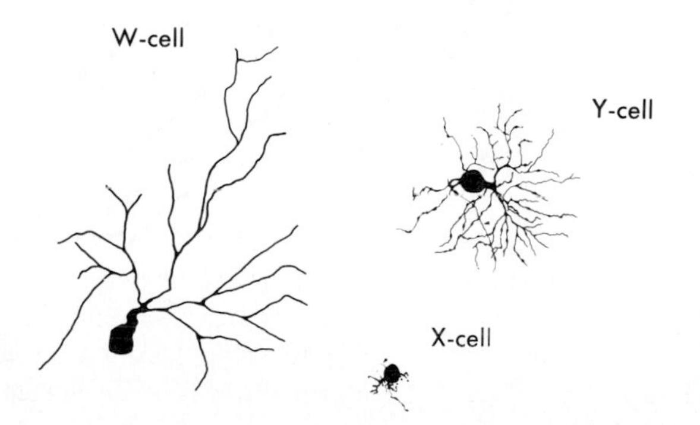

■ **Fig. 9-16** Y- and W-cells. **A,** The cells were in the retina of cat. **B,** Cells were in monkey retina. (**A** is from Stanford LR, Sherman SM: *Brain Res* 297:381, 1984; **B** redrawn from Perry VH et al: *Neuroscience* 12:1101 1984, and Perry VH, Cowey A: *Neuroscience* 12:1125, 1984.)

■ Table 9-1 Properties of x-, y-, and w-cells

Properties	X-cells	Y-cells	W-cells
Cell body and axon	Medium-sized	Large	Small
Dendritic tree	Restricted	Extensive	Extensive
Receptive field			
Size	Small	Medium	Large
Organization	Center-surround	Center-surround	Diffuse
Adaptation	Tonic	Phasic	Poorly responsive
Linearity	Linear	Nonlinear	Insensitive
Wavelength	Sensitive	Insensitive	Insensitive
Luminance	Insensitive	Sensitive	Sensitive

Several of the physiological differences among these cell types correspond to the morphological differences (Table 9-1). For example, X-cells have smaller receptive fields (corresponding to the smaller dendritic trees) and more slowly conducting axons than Y-cells. In addition, X-cells have tonic responses to visual stimuli, show more linear summation of responses than do Y-cells, and respond better to small stimuli than to large stimuli. Y-cells have phasic, nonlinear responses to complex stimuli; such responses cannot be predicted from the responses to simple stimuli. In monkeys, X-cells respond differently to different wavelengths of light, whereas Y-cells are not sensitive to differences in wavelength. On the other hand, Y-cells are more sensitive to luminance than X-cells. W-cells have large, diffuse receptive fields and slowly conducting axons; they respond poorly to visual stimuli.

■ *The Visual Pathway*

The retinal ganglion cells transmit information to the brain by way of the optic nerve, optic chiasm, and optic tract (see Fig. 9-1). Figure 9-17 shows the relationships between a visual target (arrow), the retinal images of the target in the two eyes, and the projections of retinal ganglion cells to the two hemipheres. The eyes and the optic nerves, chiasm, and tract are viewed from above.

The visual target, an arrow, is in the **visual fields** of both eyes (see Fig. 9-17). The visual target in this case is so long that it extends into the monocular segments of each retina (i.e., one end of the target can be seen only by one eye and the other end only by the other eye). The fixation point is shown by the shaded circle at the center of the target. The image of the target is reversed on the retinas by the lens system. The left half of

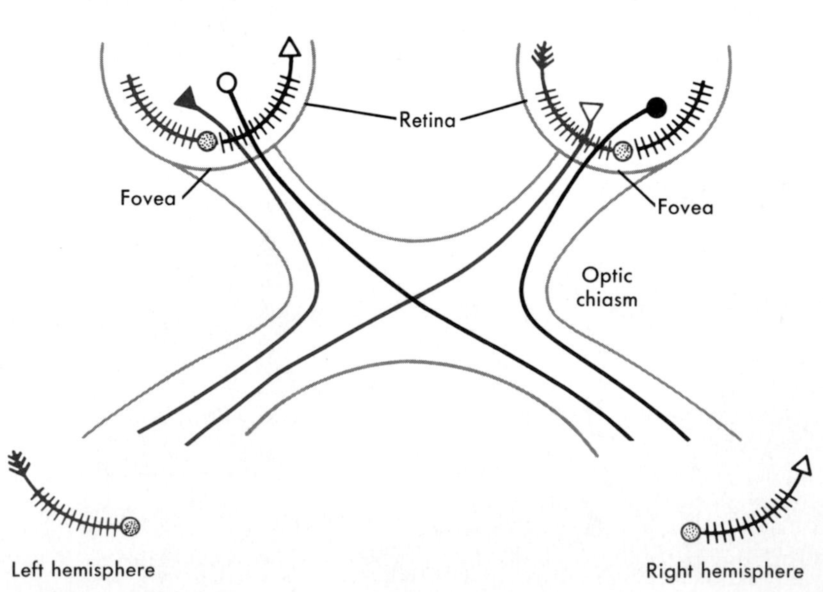

■ Fig. 9-17 Relationships between a visual target, images on the retinas of the two eyes, and the projections of the ganglion cells carrying visual information about these images. The image is so large that it extends into the monocular segments of the eyes where the image is seen in only one eye.

the visual target is imaged on the nasal retina of the left eye and the temporal retina of the right eye. Thus the left visual field is seen by the left nasal retina and the right temporal retina. Similarly, the right half of the visual target is imaged on and seen by the left temporal retina and the right nasal retina.

The projections of retinal ganglion cells may be uncrossed or crossed, depending on the location of the ganglion cell. For example, a given axon from the left retina may pass through the left optic nerve, the left side of the optic chiasm, and the left optic tract to terminate in the brain on the left side. Alternatively, an axon in the left retina may pass through the left optic nerve, cross to the opposite side in the optic chiasm, and then pass through the right optic tract to end in the right side of the brain. *Axons that remain uncrossed arise from ganglion cells in the temporal retina, whereas axons that do cross arise from ganglion cells in the nasal retina.*

This arrangement results in the representation of the left field of vision in the right side of the brain and of the right field of vision in the left side of the brain (see Fig. 9-17).

The axons of retinal ganglion cells can synapse in any of several nuclei of the brain. The main pathway for vision relays in the dorsal lateral geniculate nucleus (LGN), one of the sensory nuclei of the thalamus. The LGN in turn projects to the primary visual cortex or striate cortex by way of the optic radiations or geniculostriate tract (Fig. 9-18). As the optic radiation passes caudally, it fans out, and some of the fibers loop forward in the temporal lobe as Meyer's loop. The axons in Meyer's loop carry information derived from the lower half of the appropriate hemiretinas. Thus the axons in Meyer's loop represent the contralateral upper visual field. Axons that pass directly caudally through the parietal lobe in the optic radiation represent the contralateral lower visual field.

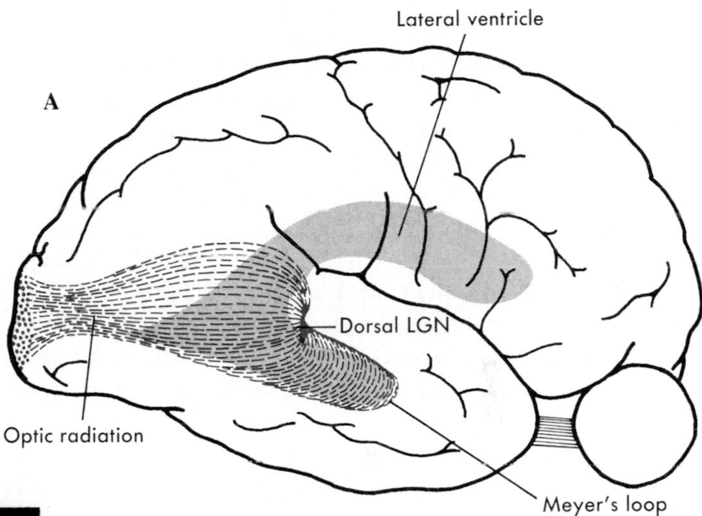

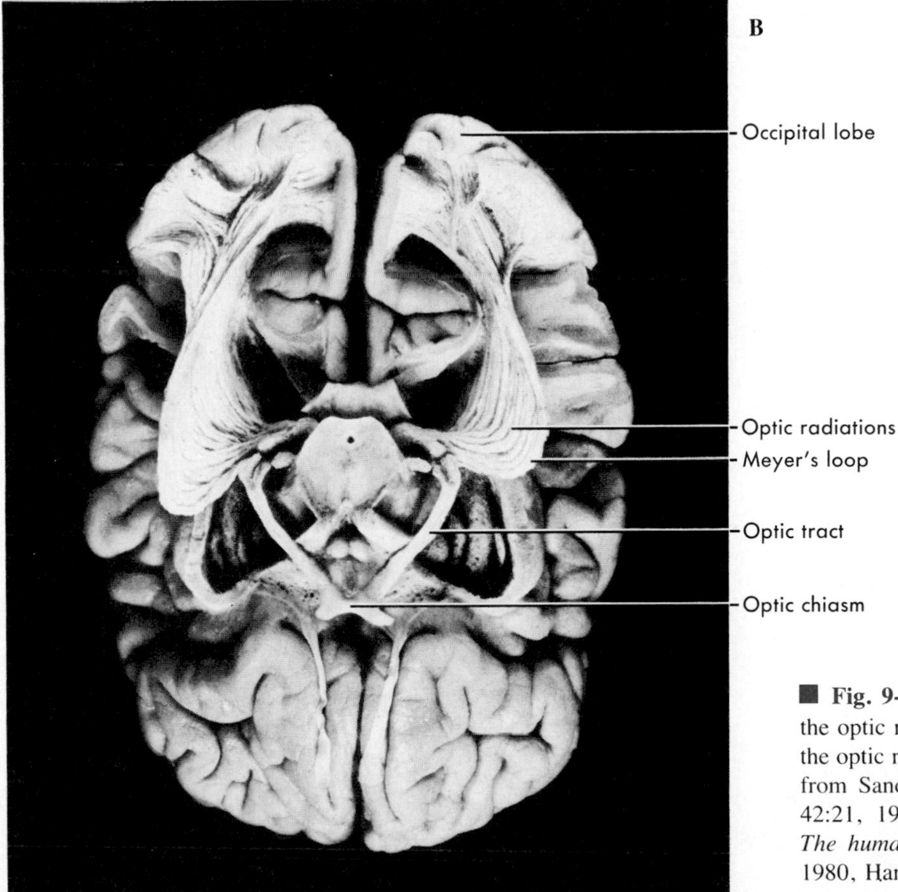

■ **Fig. 9-18** Geniculostriate tract. **A,** The course of the optic radiation in a lateral view. **B,** A dissection of the optic radiation, including Meyer's loop. (**A** redrawn from Sandford HS, Blair JHL: *Arch Neurol Psychiat* 42:21, 1939; **B** from Gluhbegovic N, Williams TH: *The human brain: a photographic guide,* New York, 1980, Harper & Row.)

The optic radiation ends in the striate cortex, which is located dorsal and ventral to the calcarine fissure in the occipital lobe. The gyrus dorsal to the calcarine fissure is the *cuneus,* and that ventral is the **lingual gyrus.** The cuneus receives information from the upper part of the appropriate hemiretinas, and the lingual gyrus receives information from the lower hemiretinas. Thus the cuneus represents the contralateral lower visual field and the lingual gyrus the upper visual field.

Visual Field Defects

Interruption of the visual pathway at any level will cause a defect in the appropriate part of the visual field (Fig. 9-19). For example, a lesion of the retina in one eye produces localized blindness, a **scotoma,** in that eye. Complete interruption of the optic nerve produces complete blindness, and partial interruption of the optic nerve a scotoma. Damage to the optic nerve fibers as they cross in the optic chiasm results in loss of vision in both temporal fields of vision, a condition known as **bitemporal hemianopsia.** This occurs because the crossing fibers originate from ganglion cells in the nasal halves of each retina. A lesion of the entire optic tract, LGN, optic radiation, or visual cortex on one side causes loss of vision in the entire contralateral visual field, or **homonymous hemianopsia.** Partial lesions result in partial visual field defects. For example, a lesion of Meyer's loop causes a loss of vision in the contralateral upper visual field, an upper **homonymous quadrantanopsia** ("pie in the sky" visual defect). A lesion of the striate cortex may not destroy all of the neurons that represent the macula, and the result is sometimes a homonymous hemianopsia with **macular sparing.**

Lateral Geniculate Nucleus

The dorsal lateral geniculate nucleus (LGN) is a layered structure (Fig. 9-20). The first two layers contain large size neurons and are called the magnocellular layers. The other four layers are the parvocellular layers. There is a point-to-point projection from the retina to the LGN. The LGN thus has a retinotopic map. Cells that represent a particular retinal location are aligned along projection lines that can be drawn across the LGN (see Fig. 9-20).

The projection from one eye is distributed to three of the layers of the LGN, to one of the magnocellular lay-

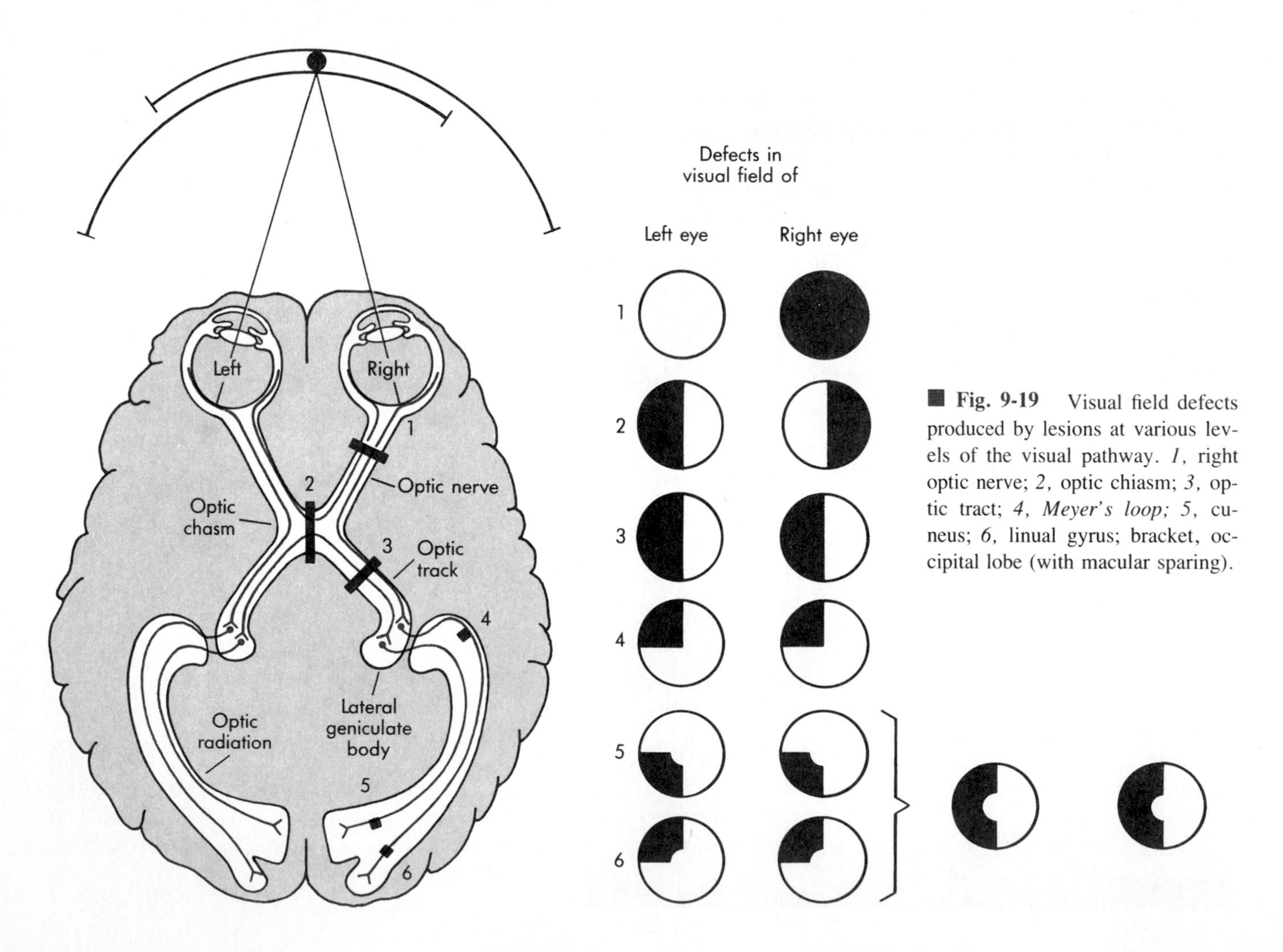

Fig. 9-19 Visual field defects produced by lesions at various levels of the visual pathway. *1,* right optic nerve; *2,* optic chiasm; *3,* optic tract; *4, Meyer's loop; 5,* cuneus; *6,* linual gyrus; bracket, occipital lobe (with macular sparing).

A

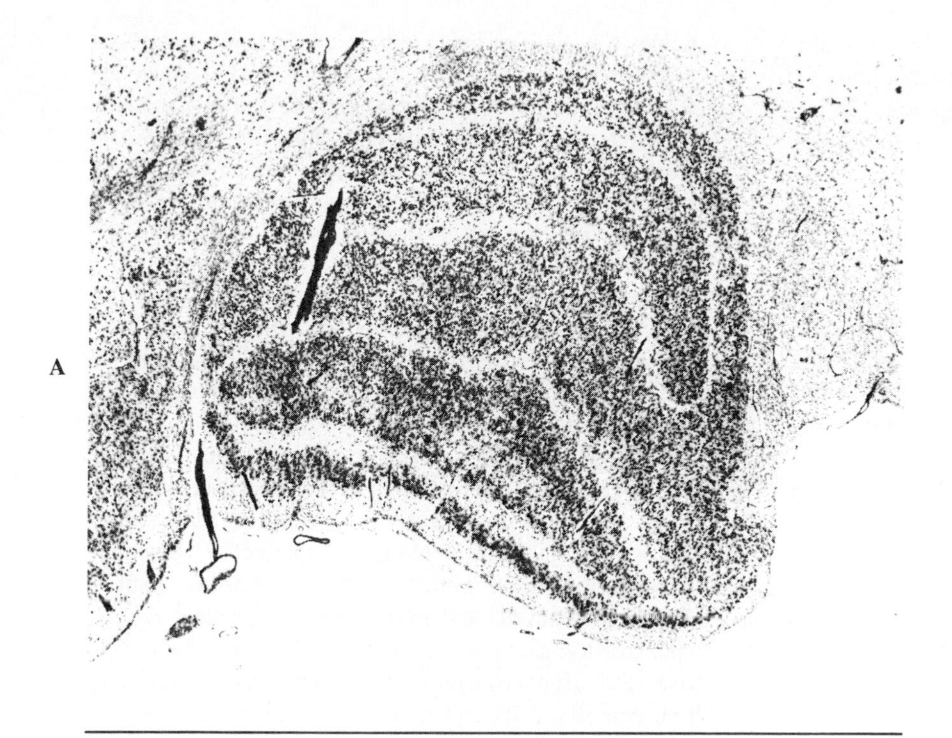

B

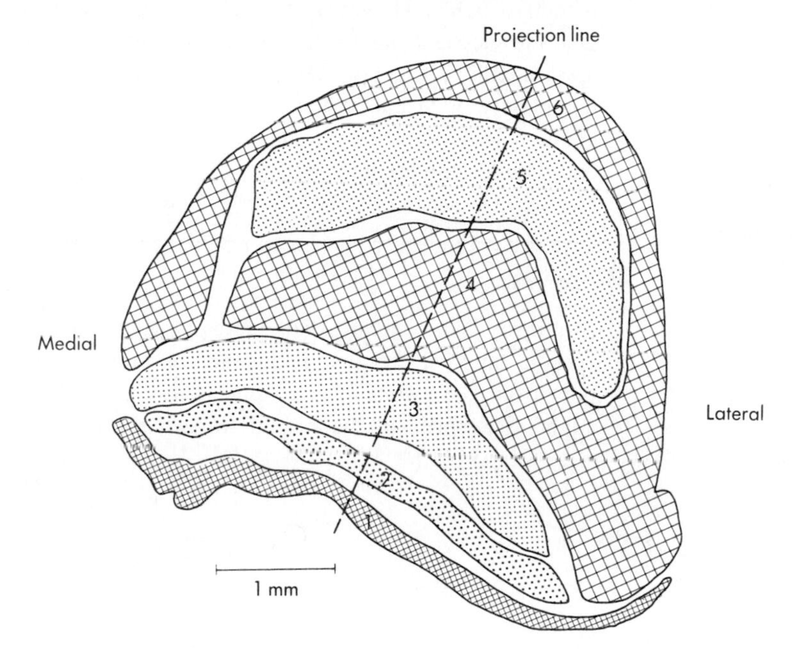

■ **Fig. 9-20** Section through the LGN of a human infant. **A,** a micrograph of the Nissl-stained LGN. **B,** A drawing. The layers are numbered, *1-6*. Cross-hatching indicates the layers innervated by the contralateral eye and the dots the layers supplied by the ipsilateral eye. The projection line shows the location of cells in all the layers that map a point in the visual field. (**A** courtesy TL Hickey and RW Guillery.)

ers, and to two parvocellular layers. The contralateral eye projects to layers 1, 4, and 6, whereas the ipsilateral eye projects to layers 2, 3, and 5. Another basis for the diversion of retinal input to different layers of the LGN depends on the subdivision of retinal ganglion cells into X- and Y-cells. Y-cells innervate layers 1 and 2, whereas X-cells supply layers 3-6. Furthermore, off-center X-cells tend to end in layers 3 and 4 and on-center X-cells in layers 5 and 6.

Most of the neurons in the LGN project to the striate cortex. However, about a quarter of the cells are interneurons, which are inhibitory in function. Each LGN neuron receives an input from only a limited number of retinal ganglion cells. Consequently, LGN neurons have properties that are very similar to those of ganglion cells. For example, LGN neurons can be classified as X- or Y-cells, and they have on-center or off-center receptive fields.

The LGN also receives input from the visual areas of the cerebral cortex, the thalamic reticular nucleus, and several nuclei of the brainstem reticular formation. The activity of LGN projection neurons is inhibited by interneurons both in the LGN and in the thalamic reticular nucleus. These cells use γ aminobutyric acid (GABA) as their inhibitory neurotransmitter. In addition, the activity of LGN neurons is influenced by brainstem neurons that use monoamine transmitters and by corticofugal pathways. These control systems serve to filter visual information and are likely to be important for selective attention.

Striate Cortex

The geniculostriate pathway ends chiefly in layer 4 of the striate cortex. A dense band of axons from the optic radiation in layer 4 forms the **stripe of Gennari,** which accounts for the name of this part of the cerebral cortex. Axons that represent one eye or the other terminate alternately in patches known as **ocular dominance columns.** Cortical neurons in an ocular dominance column respond preferentially to input from the appropriate eye. Near the border between two ocular dominance columns, neurons respond about equally to inputs from the two eyes.

Like the LGN, the striate cortex contains a retinotopic map (actually, two interlaced retinotopic maps, one for each eye). The macula is represented by a region that is relatively large compared with that of the remainder of the retina (Fig. 9-21). The macular representation extends forward from the occipital pole for about a third of the length of the striate cortex.

The receptive fields of neurons in the striate cortex are more complex than are those of LGN neurons. For one thing, they may respond to stimulation of both eyes, although often the input from one eye dominates (see above). Also, some neurons in layer 4 that receive direct input from the LGN may be activated by stimulation of just one eye. Another important difference is that cortical neurons generally have **orientation selectivity;** i.e., they respond best when the stimulus is elongated, such as a bar or an edge, and is oriented in a particular way with respect to the horizontal position (Fig. 9-22). Neurons in a particular zone of the cortex all tend to have the same orientation selectivity and are considered to form an orientation column. (Fig. 9-22). Cortical neurons may also display direction selectivity; i.e., they may respond when the stimulus is moved in one direction, but not when it is moved in the opposite direction.

Several theories have been proposed to account for these properties of neurons in the visual cortex. One idea is that the patterns of convergence of neurons at different serial levels of the visual pathway cause the receptive fields to become progressively more complex. **Simple cells** have on- and off-zones in their receptive fields. However, the receptive fields are rectangular and have an orientation selectivity. **Complex cells** reflect the convergent input of several simple cells. They also have an orientation preference, but there are no distinct on- and off-zones, and many cells are particularly responsive to movement of the stimulus across the receptive field. **Hypercomplex cells** receive inputs from several complex cells and thus have even more elaborate receptive fields. However, this classification does not take into account the X- and Y-cell pathways. Presumably, parallel X- and Y- cell pathways contribute to the complexity of visual cortical organization. Cortical receptive field organization may depend on both serial and parallel processing.

Stereopsis

Stereopsis is binocular depth perception. It must be a cortical function, because it depends on convergent inputs from the two eyes. It appears to be caused by slight differences in the retinal images formed in the

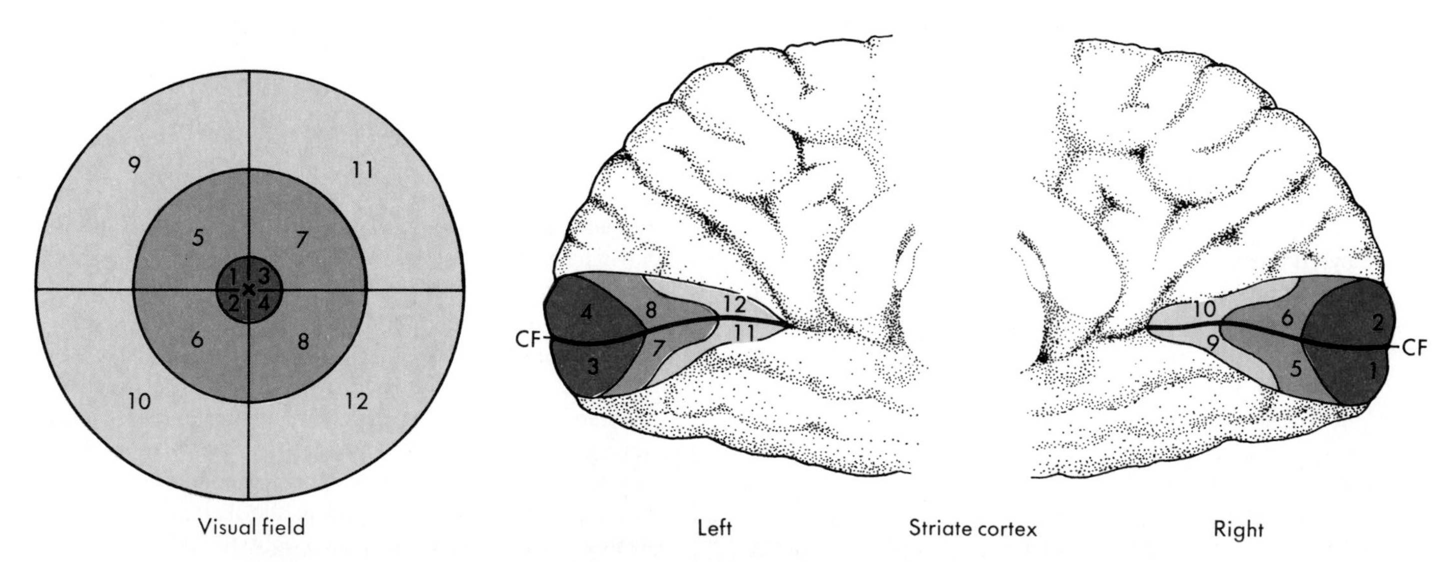

Fig. 9-21 The representation of different parts of the visual field (left) in the retinotopic map of the visual cortex is shown by the corresponding numbers.

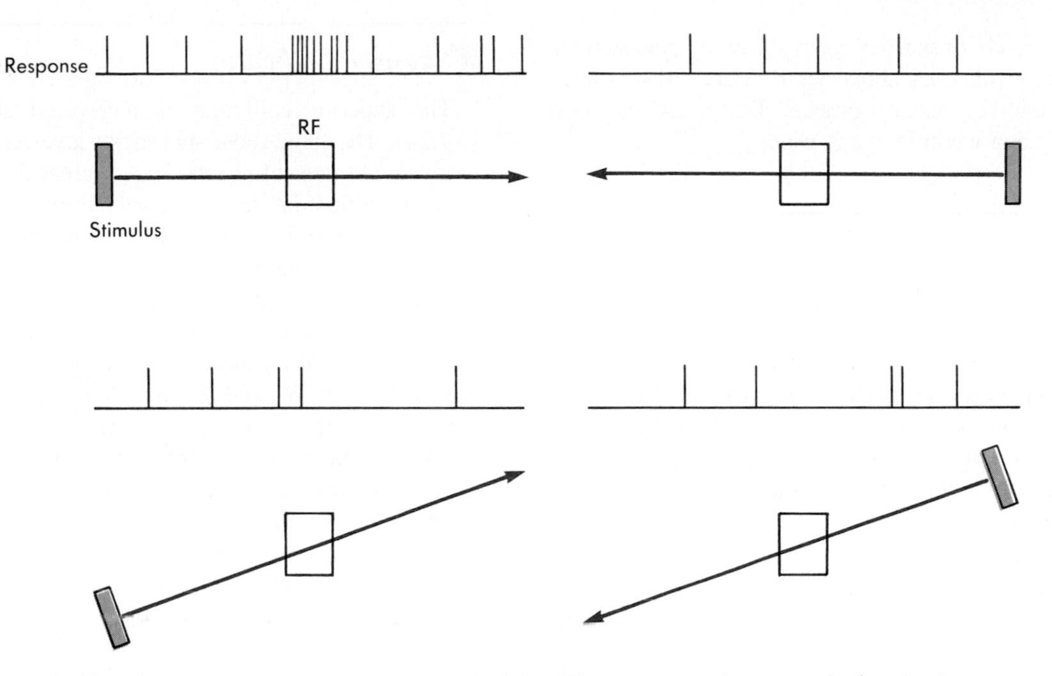

■ **Fig. 9-22** Orientation and direction selectivity. The responses of a neuron in the visual cortex are shown for stimulus oriented vertically (upper) or obliquely (lower) and moved to the right or to the left *(arrows)*.

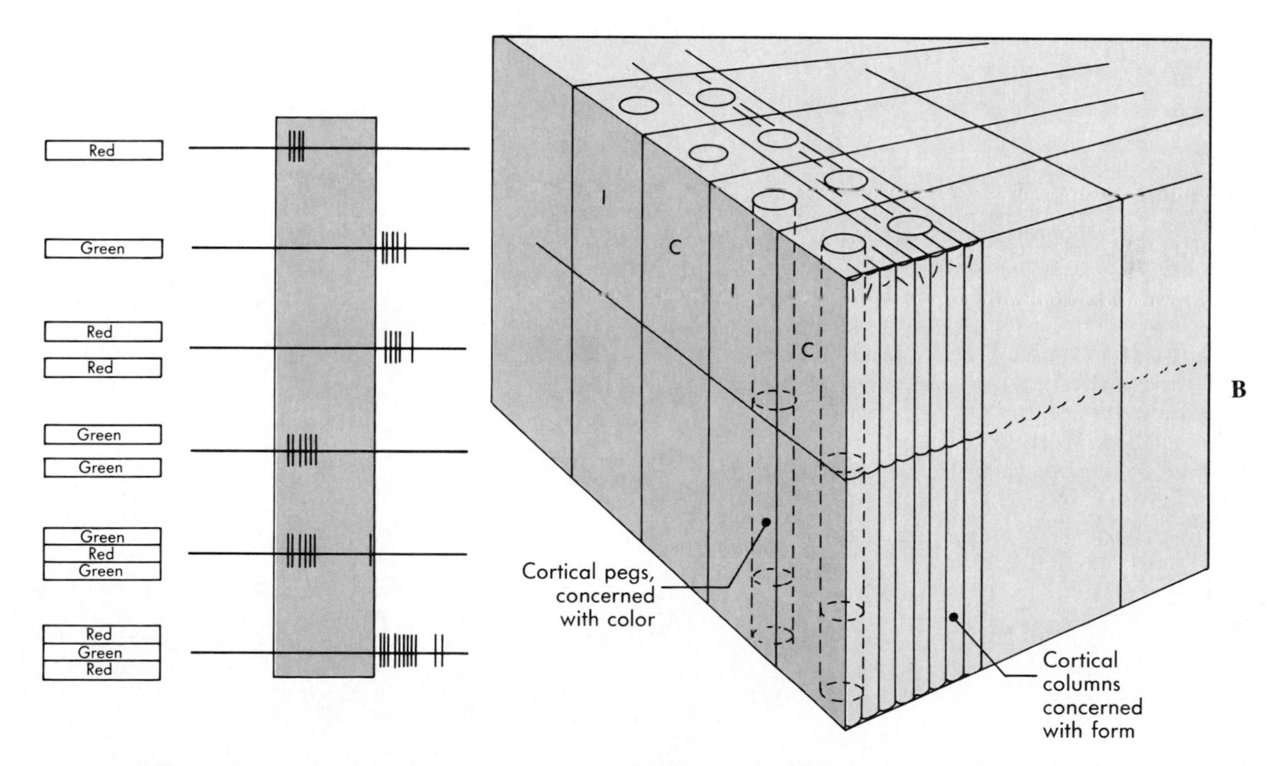

■ **Fig. 9-23** **A,** The responses of a neuron in the striate cortex that responds to various combinations of red and green bars. The best response was to a red bar flanked by 2 green bars. **B,** A diagram of the columnar arrangement of the visual cortex. Ocular dominance columns are indicated by *I* (for ipsilateral) and *C* (for contralateral). Orientation columns are indicated by the short bars at various angles. The cortical pegs contain neurons like that of **A** with spectral opponent receptive fields.

two eyes. Such disparities give different perspectives that lead to visual cues about depth. Stereopsis is useful only for relatively nearby objects. Depth cues are also available when a single eye is used.

Color Vision

As already discussed, color vision may depend on the presence in the retina of cones of three different types, as well as on neurons in the visual pathway that show spectral opposition. Retinal ganglion cells, LGN neurons, and visual cortical neurons have been found that show spectral opponent properties (Fig. 9-23, *A*). These are the X-cells. Other neurons, the Y-cells, respond to luminance but not to spectral opposition. The spectral opponent neurons in the visual cortex are found in cortical "pegs" or "blobs." The relationship between the ocular dominance and orientation columns and the cortical color pegs is shown in Fig. 9-23, *B*.

Superior Colliculus

The superior colliculus is a layered structure (Fig. 9-24). The three most superficial layers are exclusively involved in visual processing, whereas the deeper layers have multimodal inputs not only from the visual system but also from the somatosensory and auditory systems.

Neurons in the superficial layers of the superior colliculus receive a projection from retinal ganglion cells. The axons reach the superior colliculus through the brachium of the superior colliculus. The ganglion cells include both W- and Y-cells (but not X-cells) and are located chiefly in the contralateral nasal retina. There is also a projection from the visual cortex, including the striate cortex. The cortical loop involves neurons activated by Y-cells. The superficial layer of the superior colliculus in turn projects to several thalamic nuclei (pulvinar, LGN) and is therefore indirectly connected with large areas of visual cortex.

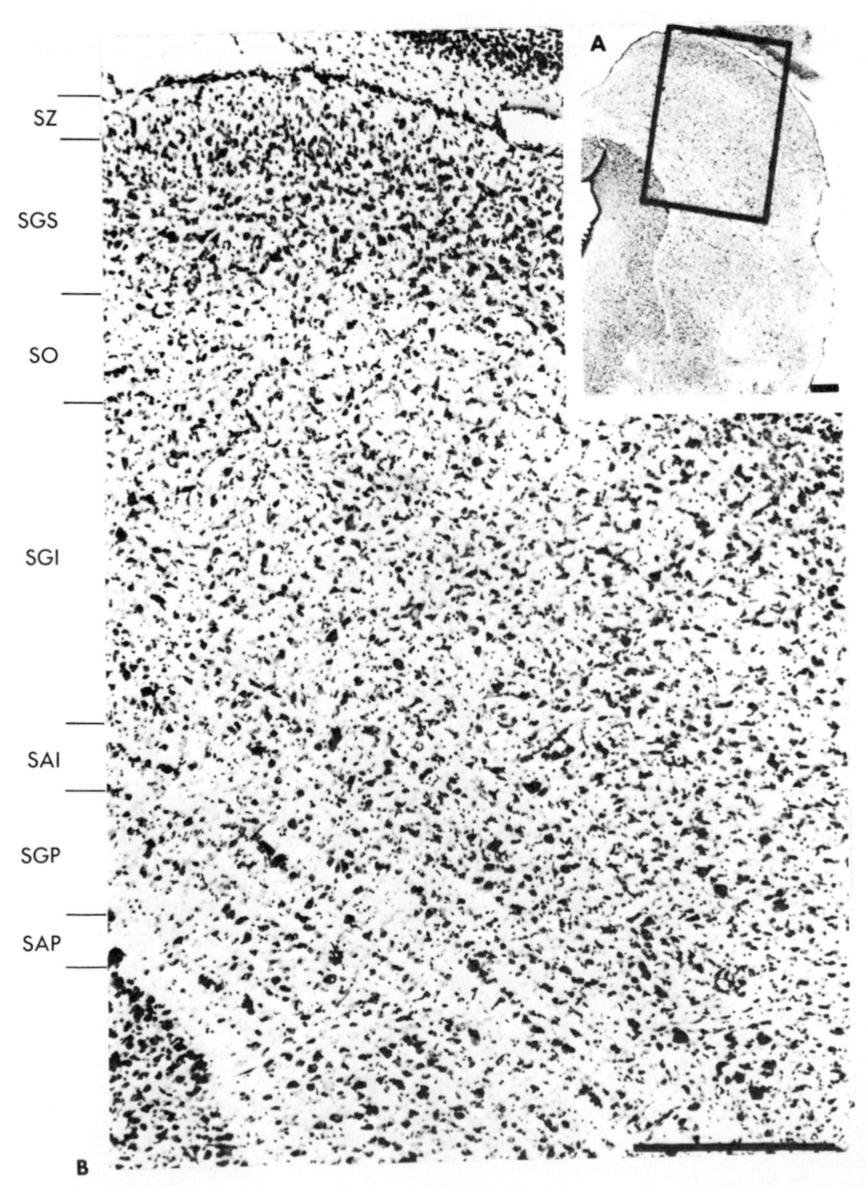

Fig. 9-24 Coronal section through the superior colliculus in a primate. **A,** A low power micrograph; **B,** A higher power view of the region indicated by the box in **A.** *SZ,* stratum zonale; *SGS,* st. griseum superficiale; *SO,* st. opticum; *SGI,* st. griseum intermedium; *SAI,* st. album intermedium; *SGP,* st. griseum profundum; *SAP,* st. album profundum. (Courtesy D Raczkowski.)

The superior colliculus contains a retinotopic map. Collicular neurons are particularly sensitive to rapid stimulus motion in a particular direction. Most of the cells have binocular inputs, but they lack orientation selectivity.

Studies on cats demonstrate that the superior colliculus plays an important role in visual perception in these animals. Bilateral destruction of the striate cortex of cats impairs visual performance only slightly; some visual acuity is lost. The superior colliculus in cats may be particularly important for determining the location of objects in visual space. It is unclear how important "collicular vision" is in humans.

The deep layers of the superior colliculus receive connections from somatosensory and auditory pathways, as well as visual input from the superficial layers. Thus the deep layers of the superior colliculus contain somatotopic and retinotopic maps, as well as a map of sound in space. Corresponding parts of these are overlaid. For example, an area that receives information about the contralateral visual field will also receive information about sounds that originate from the contralateral auditory space and about somatic stimuli applied to the contralateral surface of the body. In addition, there is a motor map that functions to control eye and head position. For instance, activation of neurons in the superior colliculus by a visual target causes movement of the eyes to center the visual target on the fovea. In this way, the superior colliculus is involved in reflex responses to the sudden appearance of a novel or threatening object in the visual field. Similarly, a sound or a sudden contact with the body will elicit appropriate eye and head movements to visualize the source of the stimulus. The descending pathways include connections to the oculomotor control system through tectoreticular connections and to the spinal cord through the tectospinal tract.

Extrastriate Visual Cortex

In animal studies at least 13 different visual areas have been identified in the cerebral cortex. One of these, V4, may be crucial in color vision. Another, MT, is specialized for the analysis of visual motion. X-cells influence the inferotemporal areas, which are involved in recognition and analysis of fine spatial detail (Fig. 9-25). Y-cells in the parietal cortex may affect activity that is involved in motion analysis and the relative location of objects in visual space.

Other Visual Pathways

The visual pathways include connections to nuclei that serve functions other than vision. For example, there is a retinal projection to the suprachiasmatic nucleus of the hypothalamus. This nucleus controls circadian rythmicity.

Another retinal projection is to the pretectum. This pathway activates parasympathetic preganglionic neurons in the Edinger-Westphal nucleus, which in turn causes pupillary constriction in the pupillary light reflex. The pretectal areas are interconnected through the posterior commissure, and thus the reflex causes both ipsilateral (direct) and contralateral (consensual) pupillary constriction.

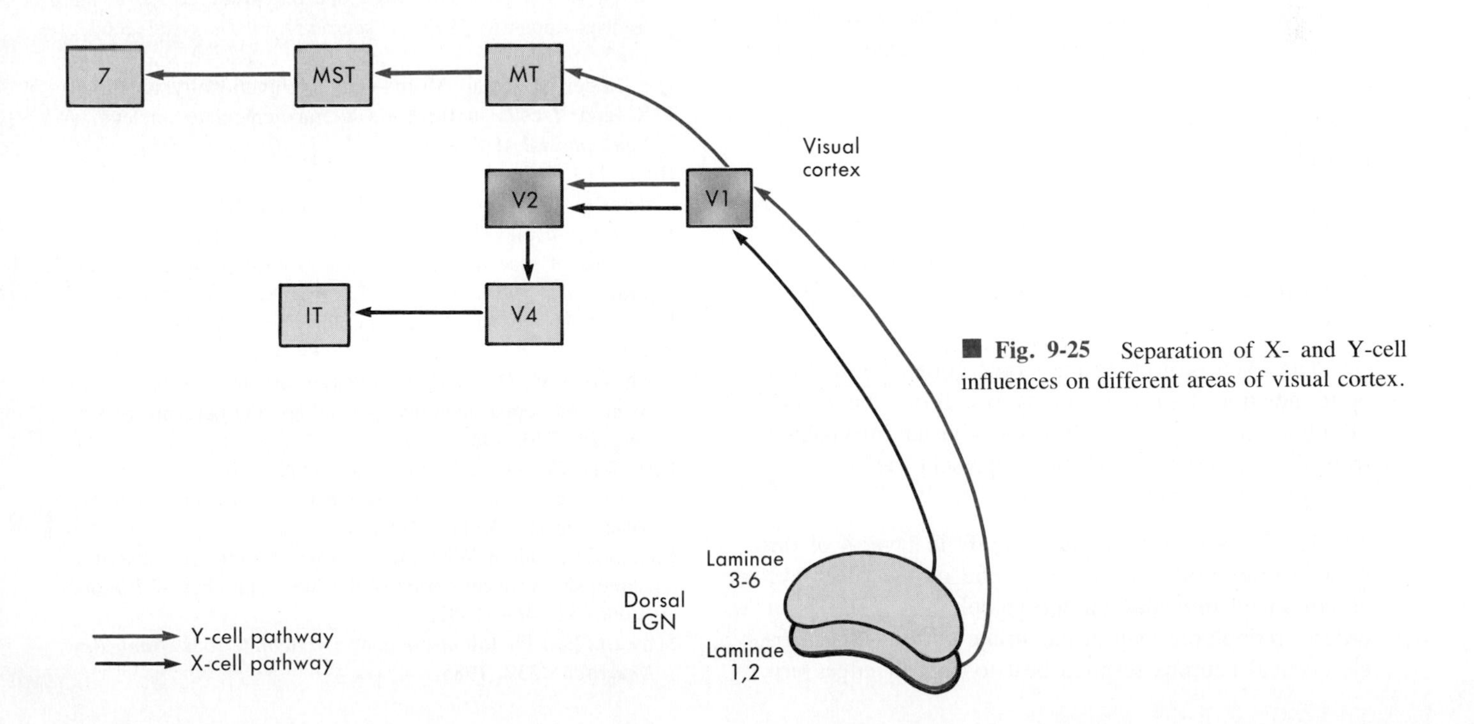

Fig. 9-25 Separation of X- and Y-cell influences on different areas of visual cortex.

Summary

1. The eye has three main layers. The fibrous layer includes the cornea and sclera; the vascular layer includes the choroid, iris, and ciliary body; the nervous layer is the retina.

2. The cornea is the most powerful refractive surface, but the lens has a variable power that allows images from near objects to be focused on the retina. Depth of field is adjusted by the iris. Stray light is absorbed by pigment.

3. The retina has 10 layers. The photoreceptor layer absorbs light. Photoreceptors synapse on retinal interneurons, which in turn synapse on each other and on ganglion cells. The latter project to the brain through the optic nerve.

4. The fovea is specialized for high resolution and color vision and contains only cones. The visual fixation point is imaged on the fovea.

5. The optic disk contains no photoreceptors and therefore is a blind spot.

6. Photoreceptors transduce visual signals. They are hyperpolarized by light. Pathways through the retina involve relays by retinal interneurons. Bipolar pathways are more direct than amacrine ones. Horizontal cells mediate lateral inhibition. Bipolar cells and many ganglion cells have receptive fields with an "on-center, off surround" or "off-center, on-surround" organization. Amacrine cells and some ganglion cells have large, diffuse receptive fields.

7. Many ganglion cells can be classified as X-, Y-, or W-cells. X-cells have small receptive fields, tonic and linear responses, and signal fine detail and wavelength. Y-cells have nonlinear responses and signal motion. Most W-cells are difficult to activate.

8. The axons of ganglion cells in the temporal retina project ipsilaterally; those in the nasal retina cross in the optic chiasm. Crossed axons end in layers 1, 4, and 6 of the lateral geniculate nucleus (LGN); uncrossed axons end in layers 2, 3, and 5. Layers 1 and 2 receive Y-cell projections; the other layers receive X-cell projections.

9. The LGN projects to the striate cortex through the optic radiation. Part of the axons extend into the temporal lobe in Meyer's loop. These carry visual information from the lower retinas and thus represent the contralateral upper visual field quadrants.

10. The LGN projection ends largely in layer 4 of the striate cortex. Information from one or the other eye predominates in ocular dominance columns. There is an orderly retinotopic map in the striate cortex. Most striate cortical neurons respond best to bars or edges oriented in a certain way. Cells that prefer a particular stimulus orientation are grouped in orientation columns.

11. Stereopsis involves differences in the retinal images in the 2 eyes.

12. Color vision depends on wavelength discrimination, based on the three types of cone pigment and also on color opponent neurons.

13. The upper layers of the superior colliculus are involved in visual processing. The deep layers produce eye movements directed at visual targets that move into the field of vision or its somatosensory or auditory stimuli.

14. Many cortical visual areas have different functions. Some are influenced chiefly by X-cells and others by Y-cells.

Bibliography

Journal articles

Baylor DA, Lamb TD, Yau KW: Responses of retinal rods to single photons, *J Physiol* (Lond) 288:265, 1979.

Bowling DB, Michael CR: Terminal patterns of single physiologically characterized optic tract fibers in the cat lateral geniculate nucleus, *J Neurosci* 4:198, 1984.

Daw NW: The psychology and physiology of colour vision, *Trends Neurosci* 7:330, 1984.

Derrington AM, Lennie P: Spatial and temporal contrast sensitivities of neurones in lateral geniculate nucleus of macaque, *J Physiol* (Lond) 311:623, 1984.

Ferster D: A comparison of depth mechanisms in areas 17 and 18 of the cat visual cortex, *J Physiol* (Lond) 311:623, 1981.

Fitzpatrick D, Itoh K, Diamond IT: The laminar organization of the lateral geniculate body and the striate cortex in the squirrel monkey *(Saimiri sciureus), J Neurosci* 3:673, 1983.

Friedlander MJ et al: Morphology of functionally identified X- and Y-cells in the cat's lateral geniculate nucleus, *J Neurophysiol* 46:80, 1981.

Hubel DH, Wiesel TN: Functional architecture of macaque monkey visual cortex, *Proc R Soc Lond* B 198:1, 1977.

Kaplan E, Shapley RM: X and Y cells in the lateral geniculate nucleus of macaque monkeys, *J Physiol* (Lond) 330:125, 1982.

Land EH: The retinex theory of color vision, *Sci Am* 237:108, 1977.

Livingstone M, Hubel D: Segregation of form, color, movement, and depth: anatomy, physiology and perception, *Science* 240:740, 1988.

Perry VH, Cowey A: Retinal ganglion cells that project to the superior colliculus and pretectum in the macaque monkey, *Neuroscience* 12:1125, 1984.

Poggio GF, Talbot WH: Mechanisms of static and dynamic stereopsis in foveal cortex of the rhesus monkey, *J Physiol* (Lond) 315:469, 1981.

Schwartz EA: Phototransduction in vertebrate rods, *Annu Rev Neurosci* 8:339, 1985.

Stryer L: Cyclic GMP cascade of vision, *Annu Rev Neurosci* 9:87, 1986.

Sur M, Sherman SM: Retinogeniculate terminations in cats: morphological differences between X- and Y-cell axons, *Science* 218:389, 1982.

Van Essen DC, Maunsell JHR: Hierarchical organization and functional streams in the visual cortex, *Trends Neurosci* 6:370, 1983.

Books and monographs

Boynton RM: *Human color vision,* New York, 1979, Holt, Rinehart, & Winston.

Dowling JE: *The retina: an approachable part of the brain,* Cambridge, Mass, 1987, Belknap Press.

Kandel ER, Schwartz JH, Jessell TM: *Principles of neural science,* ed 3, New York, 1991, Elsevier.

Rodieck RW: *The vertebrate retina,* San Francisco, 1973, WH Freeman.

Schmidt RF, Thews G, editors: *Human physiology,* Heidelberg, 1983, Springer-Verlag.

Sherman SM: Functional organization of the W-, X-, and Y-cell pathways in the cat: a review and hypothesis. In Sprague JM, Epstein AN, editors: *Progress in psychobiology and physiological psychology,* vol 11, New York, 1985, Academic Press.

Stone J: *Parallel processing in the visual system,* New York, 1983, Plenum Publishing.

Tusa RJ: Visual cortex: multiple areas and multiple functions. In Morrison AR, Strick PL, editors: *Changing concepts of the nervous system,* New York, 1982, Academic Press.

CHAPTER

10

The Auditory and Vestibular Systems

The peripheral parts of the auditory and vestibular systems share components of the bony and membranous labyrinths, use hair cells as mechanical transducers, and transmit information to the CNS through the eighth cranial nerve. However, their CNS processing and sensory functions are quite distinct. The auditory system allows us to transduce sound to recognize environmental cues and to communicate with other organisms. The most complex auditory functions are those involved in language. The vestibular system provides the CNS with information related to the position and movements of the head in space. The control of eye movements by the vestibular system is discussed in Chapter 13.

■ *Audition*

■ *Sound*

Sound is produced by waves of compression and decompression transmitted in air or in other elastic media such as water. Sound propagates at about 335 m/s in air. The waves are associated with changes in pressure called sound pressure. The unit of sound pressure is N/m^2, but more commonly sound pressure is expressed as the **sound pressure level** (SPL) in **decibels (dB):**

$$SPL = 20 \log P/P_R$$

where P is the sound pressure and P_R is a reference pressure (either 0.0002 $dyne/cm^2$, the absolute threshold for human hearing, or 1 $dyne/cm^2$).

Sound frequency is measured in cycles per second or **Hertz (Hz).** A sound can be described in terms of a mixture of pure tones. A pure tone results from sinusoidal waves at a particular frequency and is characterized not only by its frequency but also by its amplitude and phase (Fig. 10-1).

Most sounds are complex mixtures of pure tones. The composition of a particular sound can be broken down into a set of pure tones by **Fourier analysis.** Conversely, Fourier synthesis permits the construction of a sound by mixing pure tones. Fig. 10-2 shows how a complex sound wave can be characterized using Fourier analysis to determine its frequency composition. This particular sound had most of its energy concentrated at a frequency slightly greater than 100 Hz, its fundamental frequency, but a number of other frequencies (up to about 3,000 Hz) were also present. **Noise** is a sound composed of many unrelated frequencies, and **white noise** is a mixture of all audible frequencies.

The normal human ear is sensitive to pure tones with frequencies between about 20 and 20,000 Hz. The threshold for detection of a pure tone varies with its frequency (Fig. 10-3). The lowest thresholds for human hearing are for pure tones of about 1,000 to 3,000 Hz (see Fig. 10-3). By definition, threshold at these frequencies is approximately 0 dB (reference pressure, 0.0002 $dynes/cm^2$). A sound with an intensity 10 times greater would be 20 dB, and one 100 times greater would be 40 dB. Sounds that exceed 100 dB can damage the peripheral auditory apparatus, and those over 120 dB can cause pain.

By this same scale, speech has an intensity of about 65 dB. The main frequencies used in speech are in the range of 300 to 3,500 Hz. As people age, their ability to hear high frequencies declines, a condition called presbycusis.

■ *The Ear*

The peripheral auditory apparatus is the ear, which can be subdivided into the external ear, the middle ear, and the inner ear (Fig. 10-4).

External ear. The external ear includes the pinna, the external auditory meatus, and the auditory canal. The latter contains glands that secrete **cerumen,** a waxy protective substance. The pinna may help direct sounds into the auditory canal, at least in animals. The auditory canal transmits sound waves to the tympanic membrane. In humans the auditory canal has a resonant frequency of about 3,500 Hz and limits the frequencies that reach the tympanic membrane.

Middle ear. The external ear is separated from the middle ear by the **tympanic membrane** (Fig. 10-5, *A*).

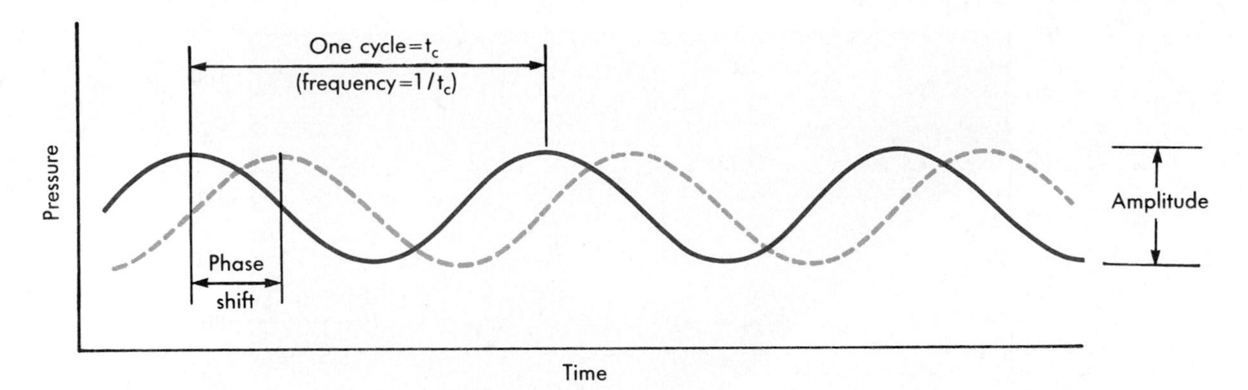

■ Fig. 10-1 Two pure tones are shown by the solid and dashed lines. Frequency is determined from the wavelength as indicated. Amplitude is the peak-to-peak change in sound pressure. Both of the tones have the same frequency and amplitude but differ in phase.

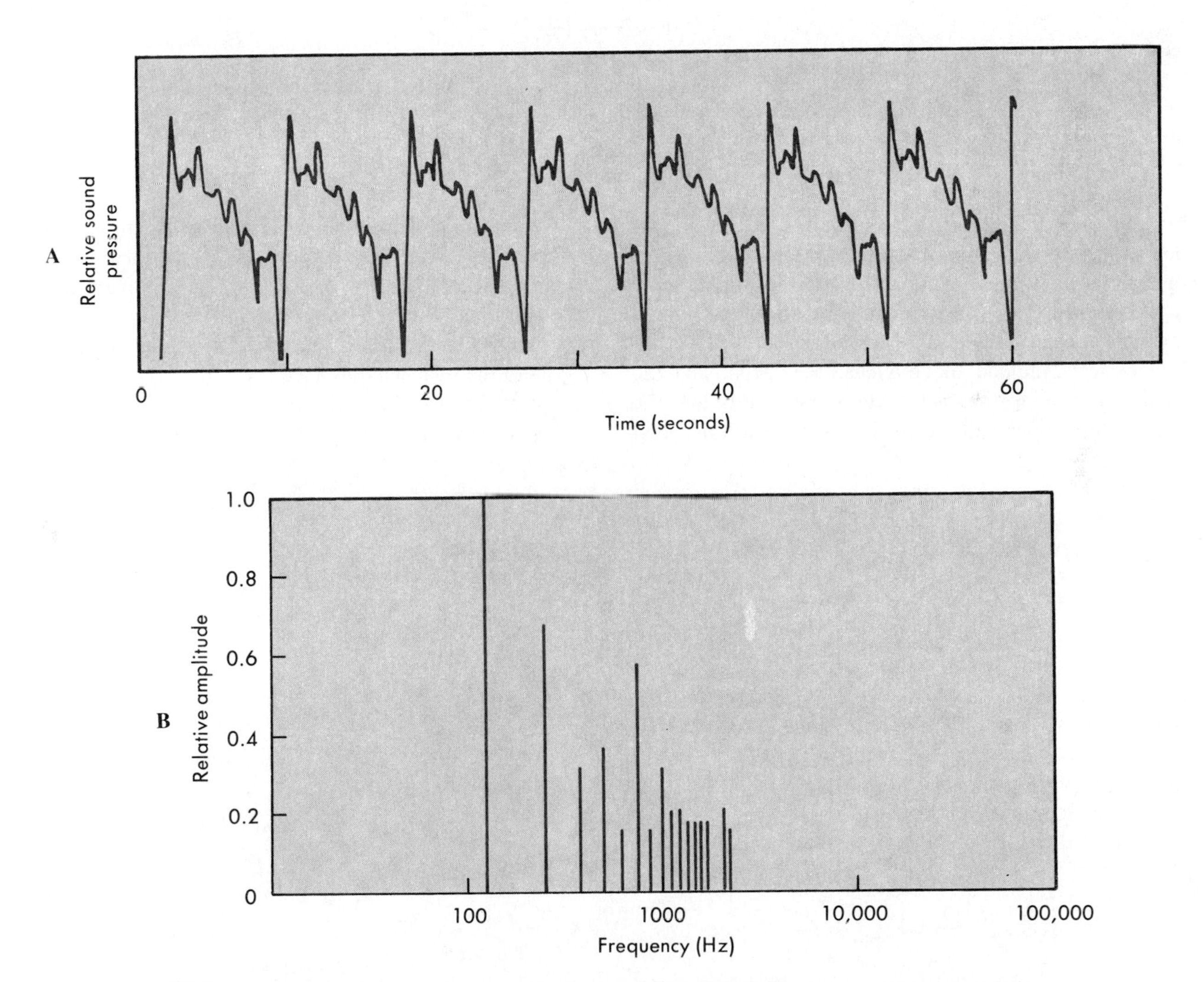

■ Fig. 10-2 Fourier analysis of sound. **A,** A complex sound. The waves can be regarded as the sum of a number of pure tones. **B,** A Fourier spectrum indicating the component pure tones that make up the sound, as well as their relative strengths. (From Cornsweet TN: *Visual perception,* New York, 1970, Academic Press.)

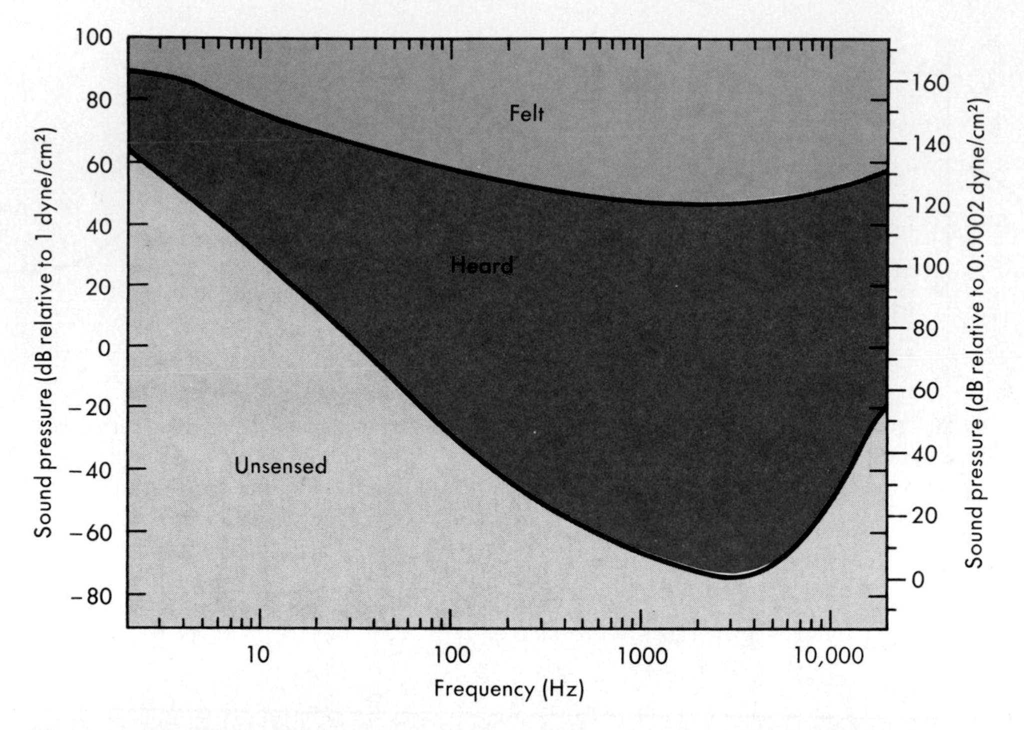

■ **Fig. 10-3** Sound intensities required for hearing at different frequencies. The gray area is subthreshold for hearing. The dark area is the normal range for audition. The light-colored area is the range at which sound is painful.

The middle ear contains air. A chain of ossicles connects the tympanic membrane to the oval window, an opening into the inner ear. Adjacent to the oval window is the round window, another membrane-covered opening between the middle and inner ears (Fig. 10-5, *B*). The ossicles include the malleus, the incus, and the stapes. The stapes has a footplate that inserts into the oval window. Beneath the oval window is a fluid-filled component of the cochlea. This component is called the vestibule, which is continuous with a tubular structure known as the scala vestibuli. Inward movement of the tympanic membrane by a sound pressure wave causes the chain of ossicles to push the footplate of the stapes into the oval window (see Fig. 10-5, *B*). This, in turn, displaces the fluid within the scala vestibuli. The pressure wave is transmitted through the basilar membrane

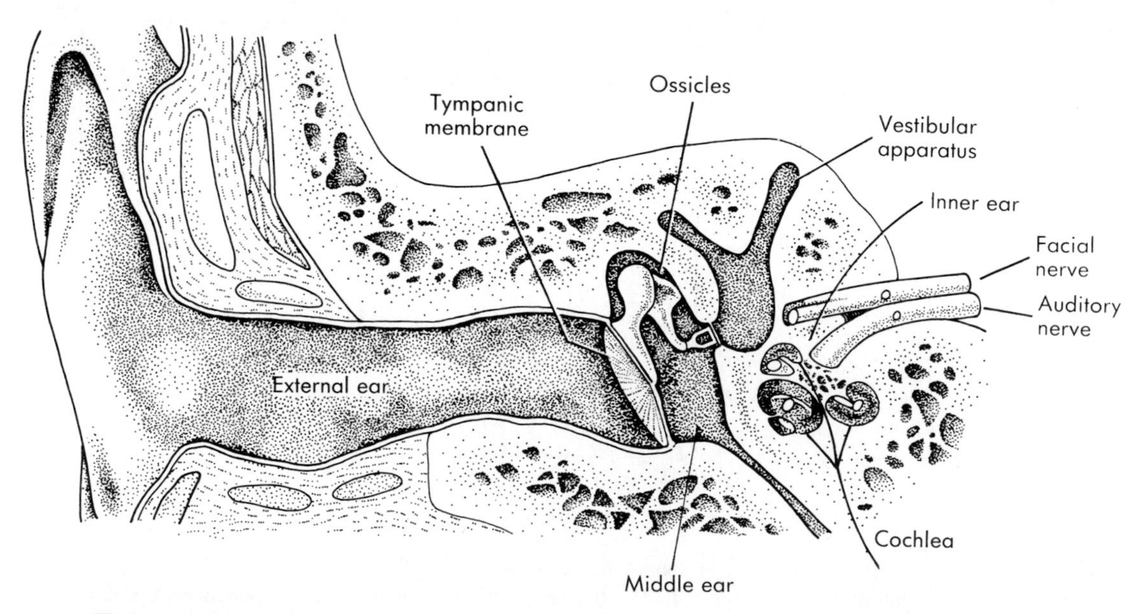

■ **Fig. 10-4** Division of the ear into the external ear, middle ear, and inner ear. (Redrawn fron von Bekesy G: *Sci Am* 197:66, 1957.)

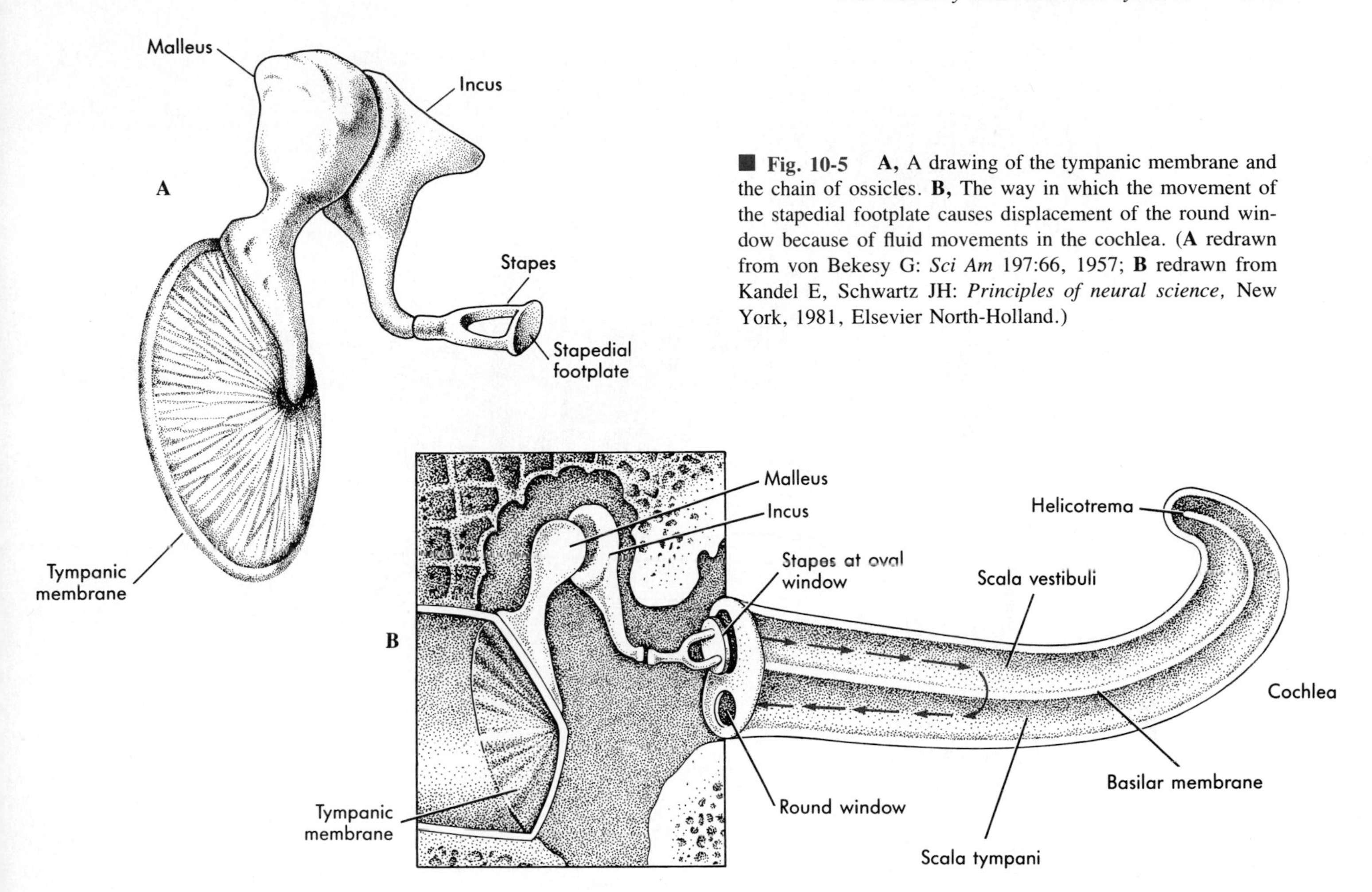

■ **Fig. 10-5** **A,** A drawing of the tympanic membrane and the chain of ossicles. **B,** The way in which the movement of the stapedial footplate causes displacement of the round window because of fluid movements in the cochlea. (**A** redrawn from von Bekesy G: *Sci Am* 197:66, 1957; **B** redrawn from Kandel E, Schwartz JH: *Principles of neural science,* New York, 1981, Elsevier North-Holland.)

of the cochlea to the scala tympani (see below) and causes the round window to bulge into the middle ear.

The tympanic membrane and the chain of ossicles serve as an impedance matching device. The ear must detect sound waves traveling in air, but the neural transduction mechanism depends on movements established in the fluid column within the cochlea. Thus, pressure waves in air must be converted into pressure waves in fluid. The acoustic impedance of water is much higher than that of air; and so without a special device for impedance matching, most sound reaching the ear would simply be reflected. Impedance matching in the ear depends on (1) the ratio of the surface area of the tympanic membrane to that of the oval window and on (2) the mechanical advantage of the lever system formed by the ossicle chain. The efficiency of the impedance match is sufficient to improve hearing by 10 to 20 dB.

The middle ear also serves other functions. Two muscles are found in the middle ear: the tensor tympani (supplied by the trigeminal nerve) and the stapedius (supplied by the facial nerve). These muscles attach, respectively, to the malleus and stapes. When they contract, they dampen movements of the ossicular chain and so decrease the sensitivity of the acoustic apparatus. This action can be protective for sounds that can be anticipated, such as vocalization. However, a sudden explosion can still damage the acoustic apparatus, because reflex contraction of the middle ear muscles cannot occur quickly enough.

The middle ear connects to the pharynx through the Eustachian tube. Pressure differences between the external and middle ears can be equalized through this passage. The Eustachian tube may be blocked, for example, by fluid collection during a middle ear infection. The resulting pressure difference between the external and middle ears can produce pain by displacement of the tympanic membrane and, in extreme cases, by rupture of the tympanic membrane. Other causes for a pressure difference are flying and diving.

Inner ear. The inner ear includes the bony and membranous labyrinths. The cochlea and the vestibular apparatus are formed from these. The cochlea is described first.

The cochlea is a spiral-shaped organ (Fig. 10-6). The spiral in humans consists of 2¾ turns, and it starts from a broad base and extends to a narrow apex. The cochlea forms from the rostral end of the bony and membranous labyrinths. The apex of the cochlea faces laterally (Fig. 10-6, *A*).

The bony labyrinth component of the cochlea in-

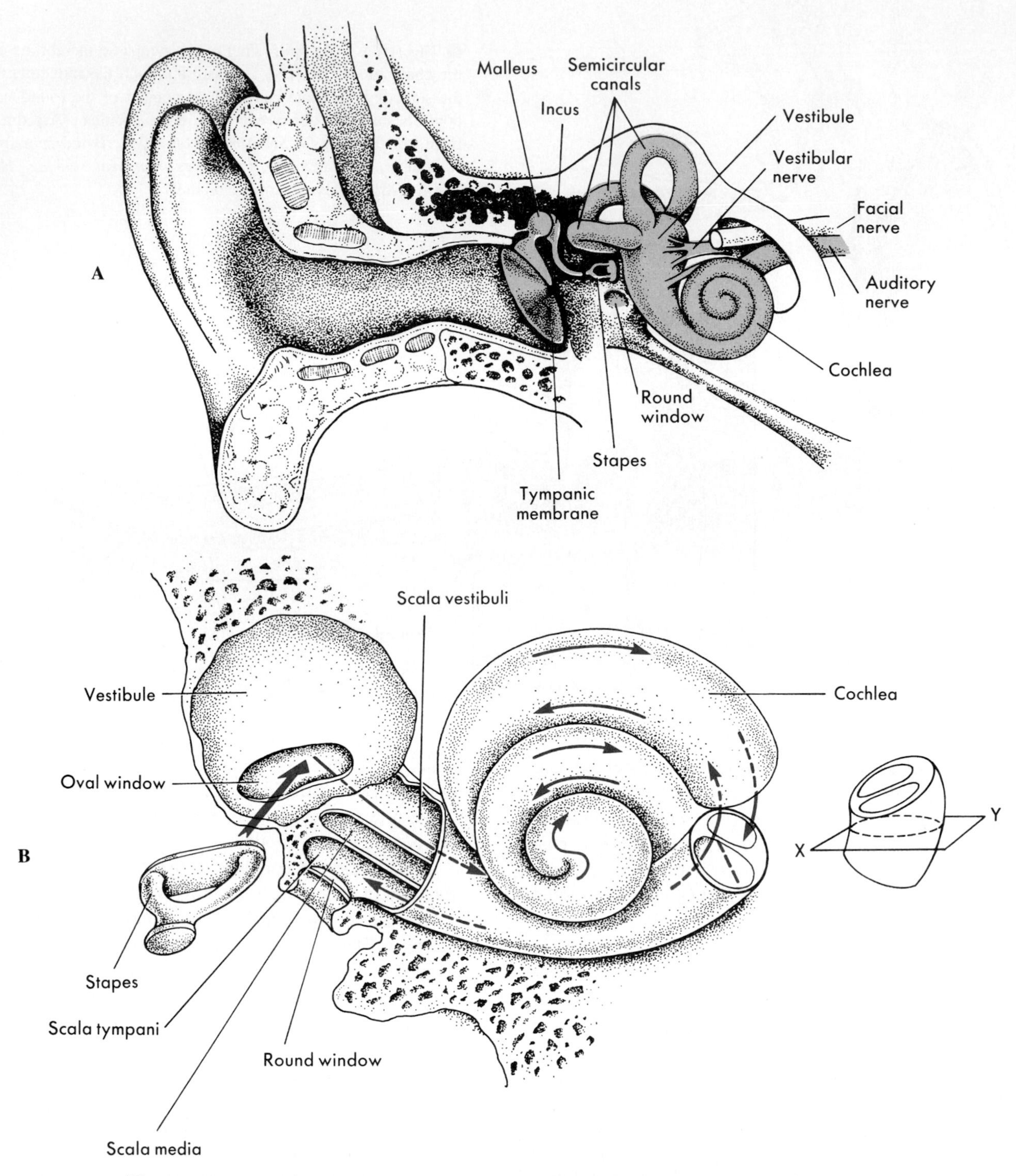

■ **Fig. 10-6** Cochlear structure. **A,** The location of the right human cochlea in relation to the vestibular apparatus and middle and external ears. **B,** The relationships between the spaces within the cochlea. **C,** A drawing of a cross-section through the cochlea in the plane indicated by the inset to the right of **B. D,** An expanded view of the organ of Corti. (**B** redrawn from Gulick WL: *Hearing: physiology and psychophysics,* New York, 1971, Oxford University Press; after Maloney.)

C

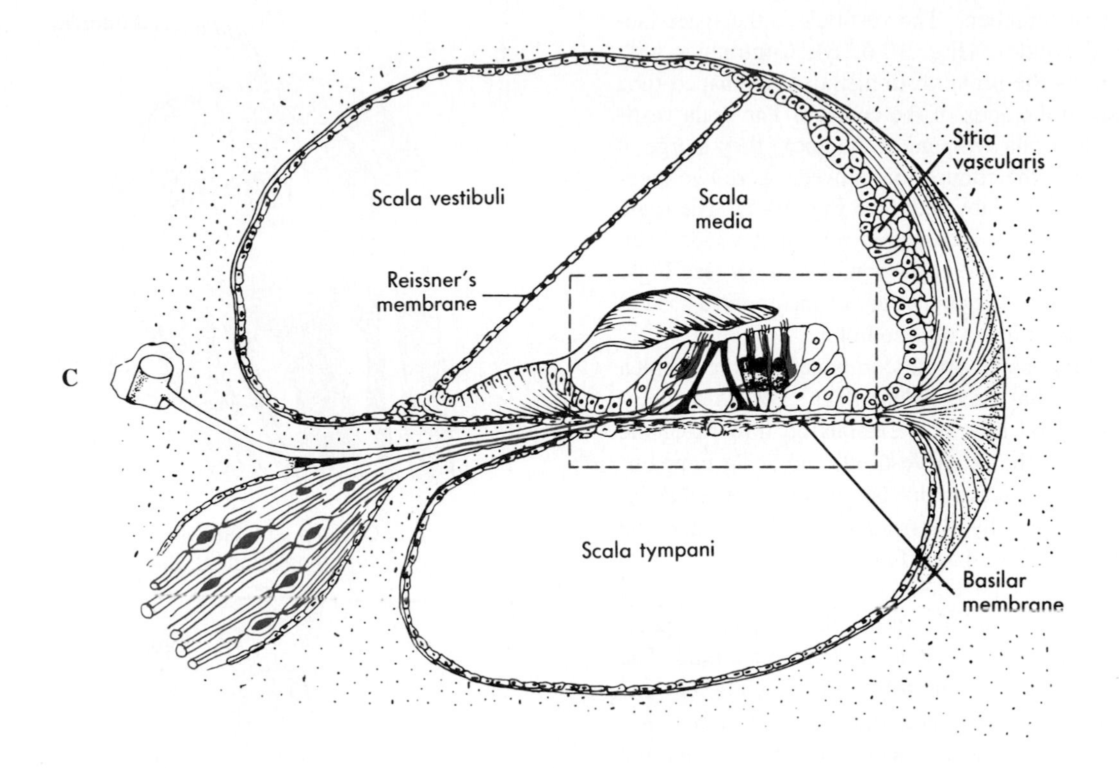

D

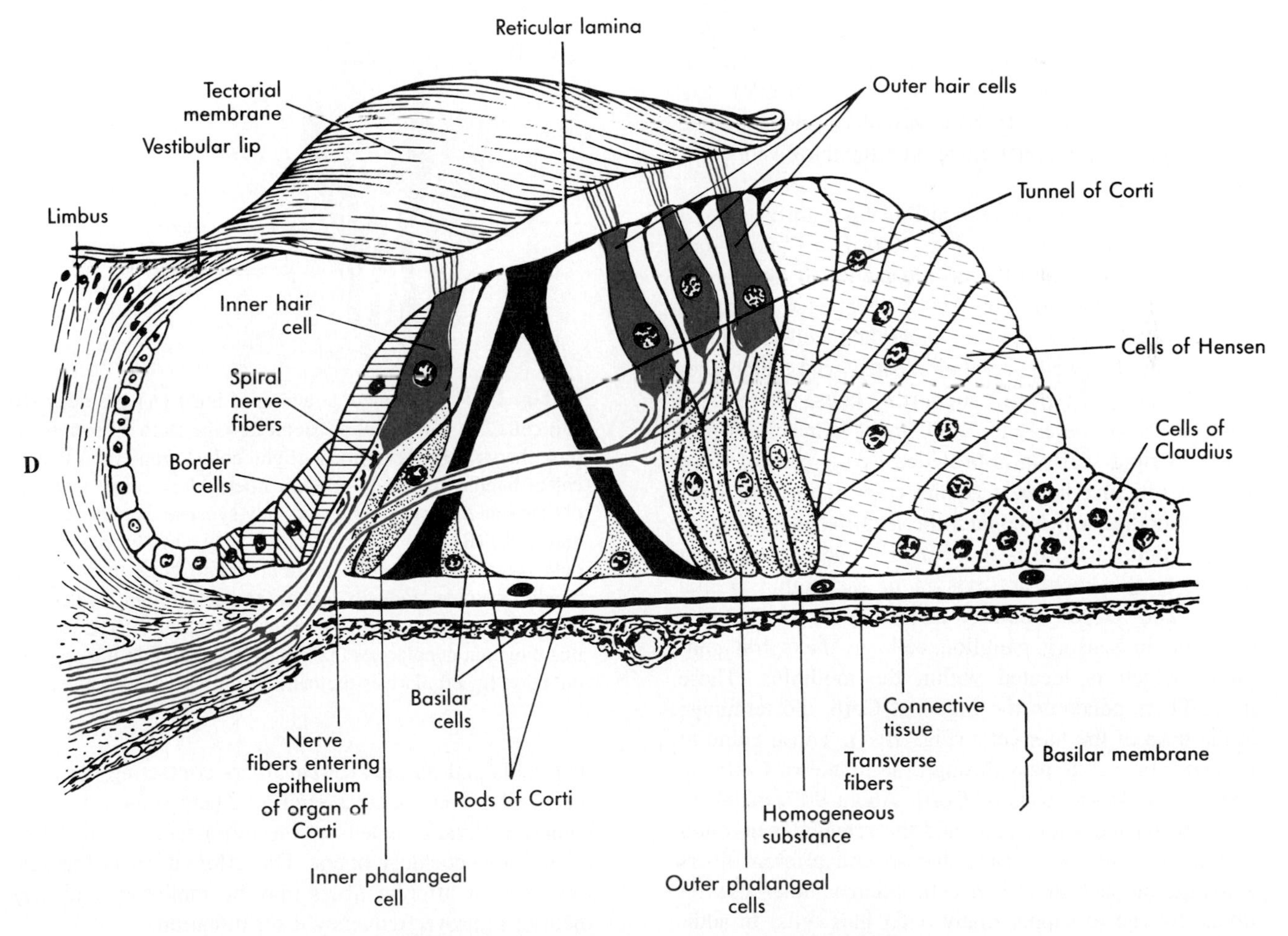

■ **Fig. 10-6, cont'd** For legend see opposite page.

cludes several chambers. The vestibule is the space facing the oval window (Fig. 10-6, *B*). Continuous with the vestibule is the scala vestibuli, a spiral-shaped tube that extends to the apex of the cochlea. The scala vestibuli meets the scala tympani at the apex; they merge at the helicotrema, the connection between these two components of the bony labyrinth (see Fig. 10-5). The scala tympani is another spiral-shaped tube that winds back down the cochlea to end at the round window (see Figs. 10-5 and 10-6, *B*). The bony core of the cochlea around which the scalae turn is the modiolus.

The membranous labyrinth component of the cochlea is the scala media, or cochlear duct. This is a membrane-bound spiral tube that extends 35 mm along the cochlea between the scala vestibuli and scala tympani. One wall of the scala media is formed by the basilar membrane, another by Reissner's membrane, and the third by the stria vascularis (Fig. 10-6, *C*).

The spaces within the cochlea are filled with fluid. The fluid in the scala vestibuli and scala tympani is **perilymph,** which closely resembles cerebrospinal fluid. The fluid in the scala media is **endolymph,** which is very different. Endolymph contains a high concentration of K^+ (about 145 mM) and a low concentration of Na^+ (about 2 mM); it resembles intracellular fluid in this respect. Endolymph has a positive potential (about +80 mV), which results in a large potential gradient across the membranes of the cochlear hair cells (about 140 mV). Endolymph is secreted by the stria vascularis and is drained through the endolymphatic duct into the dural venous sinuses.

The neural apparatus responsible for transduction of sound is the organ of Corti (Fig. 10-6, *D*). The organ of Corti is located within the cochlear duct. It lies on the basilar membrane and consists of several components, which include three rows of outer hair cells, a single row of inner hair cells, a gelatinous tectorial membrane, and a number of types of supporting cells. There are a total of 15,000 outer and 3,500 inner hair cells in humans. A rigid scaffold is provided by the rods of Corti and the reticular lamina. At the apex of the hair cells are stereocilia, which contact the tectorial membrane.

The organ of Corti is innervated by nerve fibers that belong to the cochlear division of the eighth cranial nerve. The 32,000 auditory afferent fibers in humans originate in sensory ganglion cells in the spiral ganglion, which is located within the modiolus. These nerve fibers penetrate the organ of Corti and terminate at the base of the hair cells (Fig. 10-7). Those going to the outer hair cells pass through the tunnel of Corti, an opening below the rods of Corti. About 90% of the fibers end on inner hair cells and the remainder on outer hair cells. This fact means that several afferent fibers converge on each inner hair cell, whereas other afferent fibers diverge to supply many outer hair cells. In addition to afferent fibers, the organ of Corti is supplied by cochlear efferent fibers, which terminate on the outer

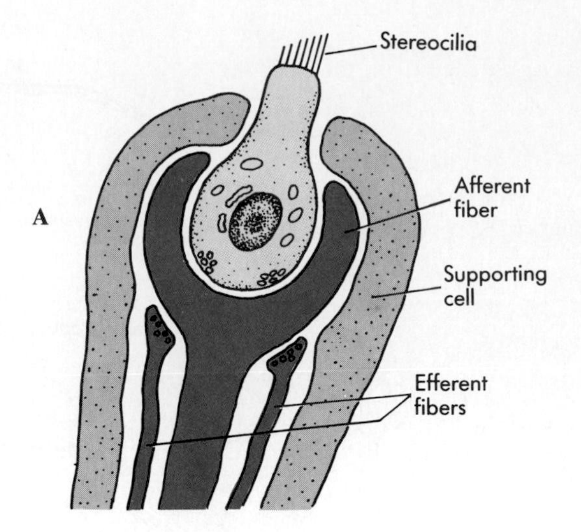

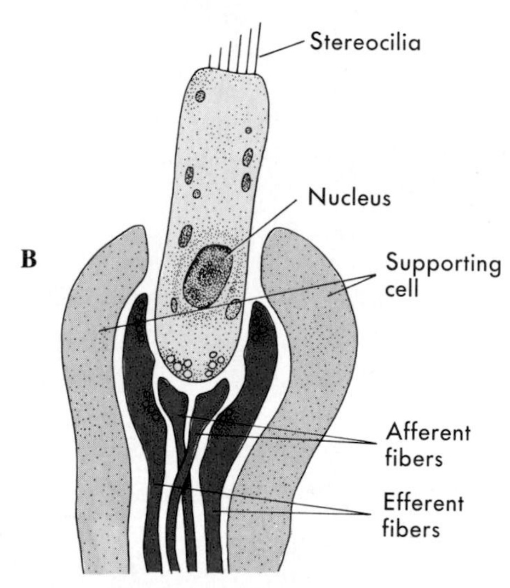

■ **Fig. 10-7** Schematic drawing of inner **(A)** and outer **(B)** hair cell. The hair bundles extend into the tectorial membrane (not shown). Efferent fibers (light color) from the olivocochlear bundle synapsion afferent fibers (dark color) that supply the inner hair cells (**A**), and they synapse directly on outer hair cells (**B**). The efferent fibers may also form synapses onto the terminals of the afferent fibers. Note that the hair bundles become gradually longer from left to right, which sets up an axis of polarization *(arrows)*. Bending of the hairs in this direction depolarizes the hair cell; bending in the opposite direction hyperpolarizes the cell.

hair cells and on the afferent fibers contacting the inner hair cells. The cochlear efferent fibers originate in the superior olivary nucleus of the brainstem and are often called olivocochlear fibers. The efferent fibers that end on cochlear afferent fibers may be inhibitory and may help to improve frequency discrimination.

The inner hair cells clearly provide most of the neural information about acoustic signals that the CNS uses for hearing. The function of the outer hair cells is less clear. The length of the outer hair cells is variable. This suggests that changes in outer hair cell length may affect the sensitivity or tuning of the inner hair cells. Cochlear efferent fibers may control outer hair cell length. This could be a mechanism whereby the brain influences which sounds are recognized.

Sound Transduction

Sound waves are transduced by the organ of Corti in the following manner. Sound waves cause the tympanic membrane to oscillate. These oscillations result in fluid movements within the scala vestibuli and scala tympani (see Fig. 10-5). Part of the hydraulic energy is used to displace the basilar membrane and, with it, the organ of Corti (Fig. 10-8). The stereocilia of the hair cells are bent because of the shear forces set up by the relative displacements of the basilar membrane and the tectorial membrane. When the stereocilia of a hair cell are moved in the direction toward the tallest cilium, the hair cell is depolarized; when the stereocilia are bent in the opposite direction, the hair cell is hyperpolarized (see Fig. 10-7).

The changes in membrane potential result from changes in cation conductance in membranes at the apical ends of the hair cells. The potential gradient that affects ion movement into the hair cells includes both the resting potential of the hair cells and the positive potential of the endolymph. The total gradient is about 140 mV. A change in membrane conductance in the apical membranes of the hair cells therefore results in a large current flow, which produces the receptor potential in these cells. This current flow can be recorded extracellularly as the **cochlear microphonic potential,** an oscillatory event that has the same frequency as the acoustic stimulus; the cochlear microphonic potential represents the sum of the receptor potentials of a number of hair cells.

Hair cells, like retinal photoreceptors, release an excitatory neurotransmitter (probably glutamate or aspartate) when they are depolarized. The transmitter produces a generator potential, which excites the cochlear afferent nerve fibers with which the hair cell synapses. Thus, oscillatory movements of the basilar membrane cause intermittent discharges of cochlear afferent nerve fibers. The activity of a large number of cochlear afferent fibers can be recorded extracellularly as a compound action potential.

However, most cochlear afferent fibers fail to discharge in response to a particular sound frequency. One factor that influences which afferent fiber discharges is its location along the organ of Corti because a given

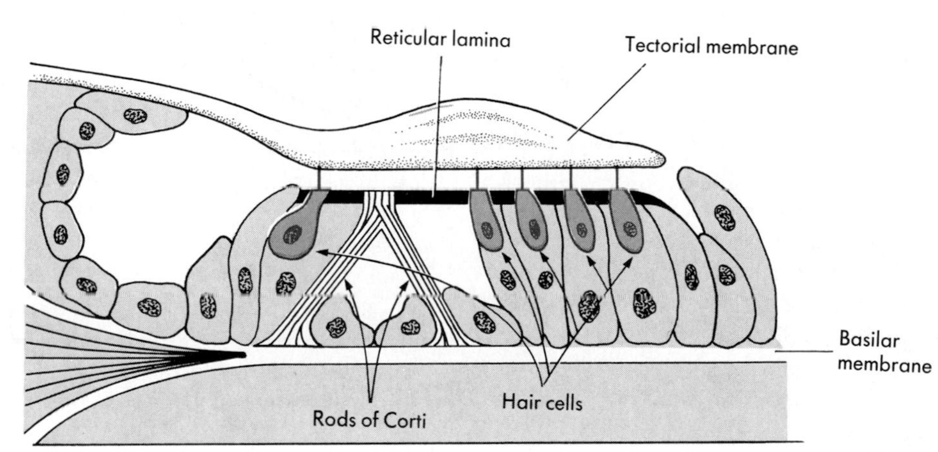

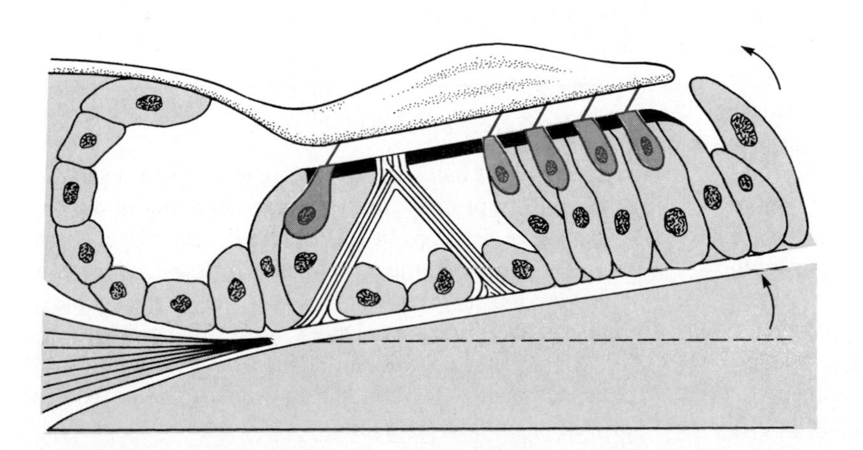

Fig. 10-8 How a movement of the basilar membrane will cause the stereocilia to bend because of shear forces produced by the relative displacement of the hair cells and the tectorial membrane.

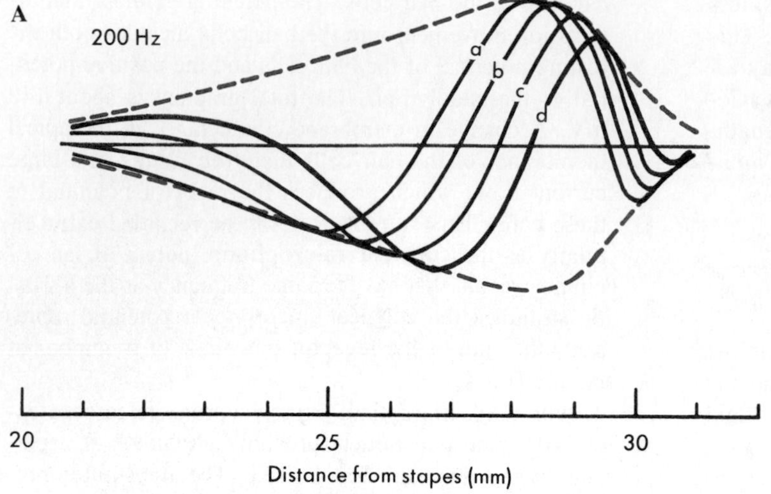

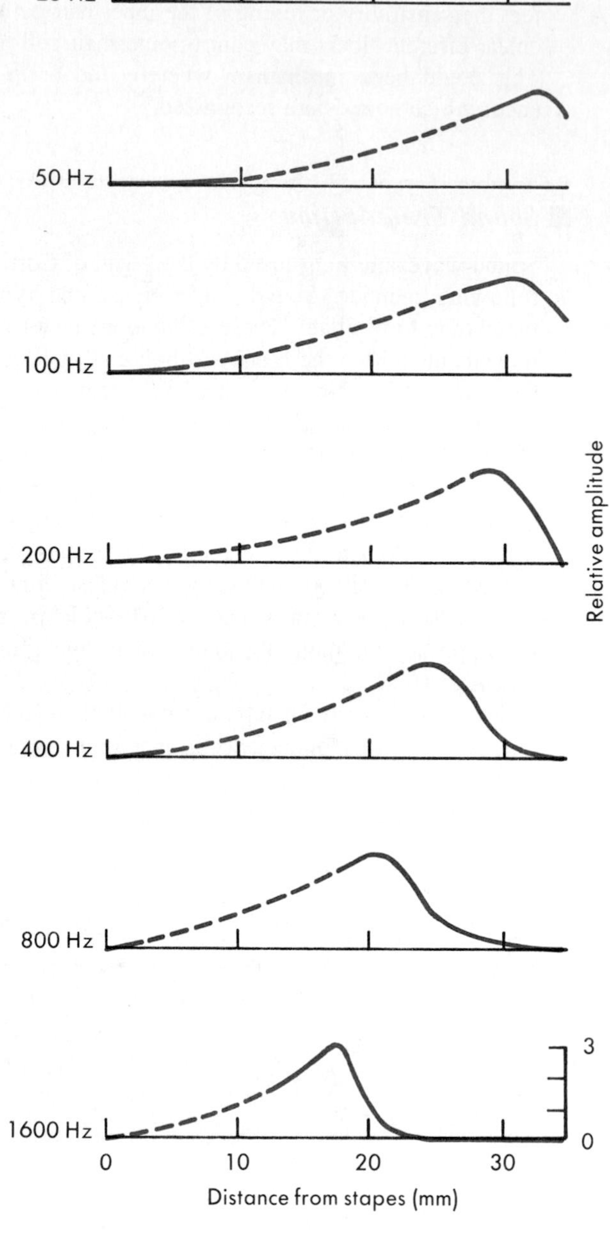

■ **Fig. 10-9** Different frequencies of sound result in different amplitudes of displacement of the basilar membrane at different sites along the organ of Corti. **A,** A traveling wave produced in the basilar membrane by a sound of 200 Hz. The curves at *a, b, c,* and *d* represent the displacements of the basilar membrane at different times, and the dashed line is the envelope formed by the peaks of the wave at different times. The maximum deflection occurs at about 29 mm from the oval window. **B,** The envelopes of traveling waves produced by several frequencies of sound. Note that the maximum displacement varies with frequency and is closest to the oval window when the frequency is highest. (Redrawn from von Bekesy G: *Experiments in hearing,* New York, McGraw-Hill.)

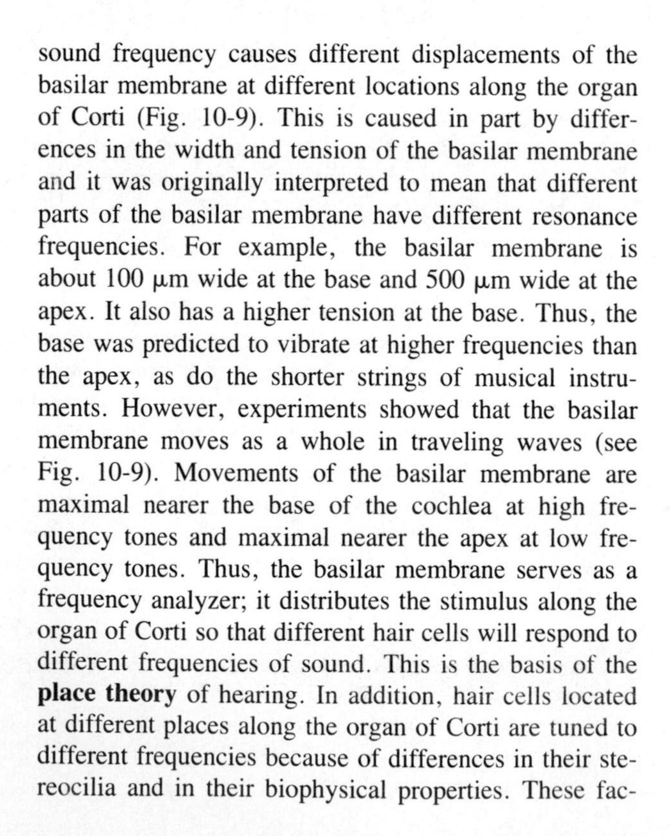

sound frequency causes different displacements of the basilar membrane at different locations along the organ of Corti (Fig. 10-9). This is caused in part by differences in the width and tension of the basilar membrane and it was originally interpreted to mean that different parts of the basilar membrane have different resonance frequencies. For example, the basilar membrane is about 100 μm wide at the base and 500 μm wide at the apex. It also has a higher tension at the base. Thus, the base was predicted to vibrate at higher frequencies than the apex, as do the shorter strings of musical instruments. However, experiments showed that the basilar membrane moves as a whole in traveling waves (see Fig. 10-9). Movements of the basilar membrane are maximal nearer the base of the cochlea at high frequency tones and maximal nearer the apex at low frequency tones. Thus, the basilar membrane serves as a frequency analyzer; it distributes the stimulus along the organ of Corti so that different hair cells will respond to different frequencies of sound. This is the basis of the **place theory** of hearing. In addition, hair cells located at different places along the organ of Corti are tuned to different frequencies because of differences in their stereocilia and in their biophysical properties. These factors cause the basilar membrane and organ of Corti to have a tonotopic map (Fig. 10-10).

■ *Cochlear Nerve Fibers*

The activity of hair cells in the organ of Corti results in discharges in primary afferent fibers traveling in the cochlear nerve. The cell bodies of these nerve fibers are in the spiral ganglion; the peripheral processes end on hair cells, and the central processes terminate in the cochlear nuclei of the brainstem. Unlike most other primary afferent neurons, those of the eighth cranial nerve are bipolar cells with a myelin sheath around the cell bodies as well as around the axons.

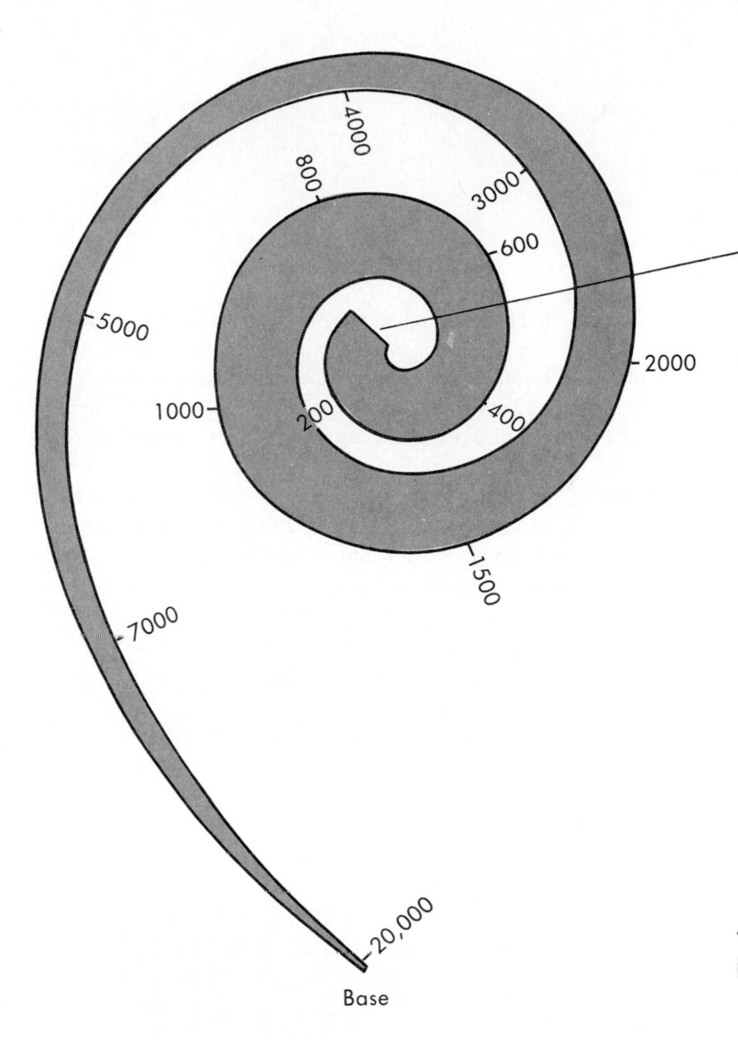

Fig. 10-10 The layout of the tonotopic map of the cochlea. (Redrawn from Stuhlman O: *An introduction to biophysics*, New York, 1943, John Wiley & Sons.)

A cochlear afferent fiber discharges maximally when stimulated by a particular sound frequency, called the **characteristic frequency** of that fiber. The characteristic frequency can be determined from a tuning curve for the fiber (Fig. 10-11). A **tuning curve** plots the threshold for activation of the nerve fiber by different sound frequencies. Typically, tuning curves are sharp near threshold but broad at high sound pressure levels. Not only excitatory but also inhibitory areas can be included in a tuning curve (Fig. 10-11, *A*). The sharpness of some tuning curves may reflect inhibitory processes.

Different aspects of an acoustic stimulus are encoded in the discharges of cochlear nerve fibers. Duration is signaled by the duration of activity, and intensity is signaled both by the amount of neural activity and by the number of fibers that discharge. For low frequency sounds (up to 4,000 Hz), the frequency is signaled by the tendency of an afferent fiber to discharge in phase with the stimulus **(phase locking).** Phase locking is required for sounds with periods shorter than the absolutely refractory period, which limits neural discharges to rates lower than about 500 Hz. Therefore, a given fiber cannot discharge in each cycle for tones of 500 Hz or more. However, frequency information can be detected by the CNS from the activity of a population of afferent fibers, each of which discharges in phase with the stimulus. This is the **frequency theory** of hearing. For high frequency sounds the place theory applies. The CNS interprets sounds that activate afferent fibers that supply hair cells near the base of the cochlea as being of high frequency. Thus, both the place and the frequency theories are required to explain the frequency coding of sound **(duplex theory).**

Central Auditory Pathway

The cochlear afferent fibers synapse in the most rostral part of the medulla on neurons of the dorsal and ventral cochlear nuclei (Fig. 10-12). These neurons give rise to axons that contribute to the central auditory pathways. Some of the axons cross to the contralateral side and ascend in the lateral lemniscus, the main ascending auditory tract. Others connect with the ipsi- or contralateral superior olivary nuclei. The superior olivary nuclei project through the ipsi- and contralateral lateral lemnisci. Many of the auditory fibers cross in the trapezoid body, although some cross in the pontine tegmentum. The lateral lemniscus ends in the inferior colliculus. The two inferior colliculi are interconnected through the commissure of the inferior colliculus. Axons from neurons of the inferior colliculus ascend through the brachium of the inferior colliculus to terminate in the medial geniculate nucleus of the thalamus. The medial geniculate nucleus gives rise to the auditory radiation, which ends in the auditory cortex, located in the transverse temporal gyrus in the temporal lobe. Projections from the auditory cortex also descend to the medial geniculate nucleus and inferior colliculus, and projections from the inferior colliculus descend to the superior olivary complex and cochlear nuclei.

It is noteworthy that a mixture of ascending auditory system fibers represent both ears at the level of the lateral lemniscus. Thus the representation of auditory

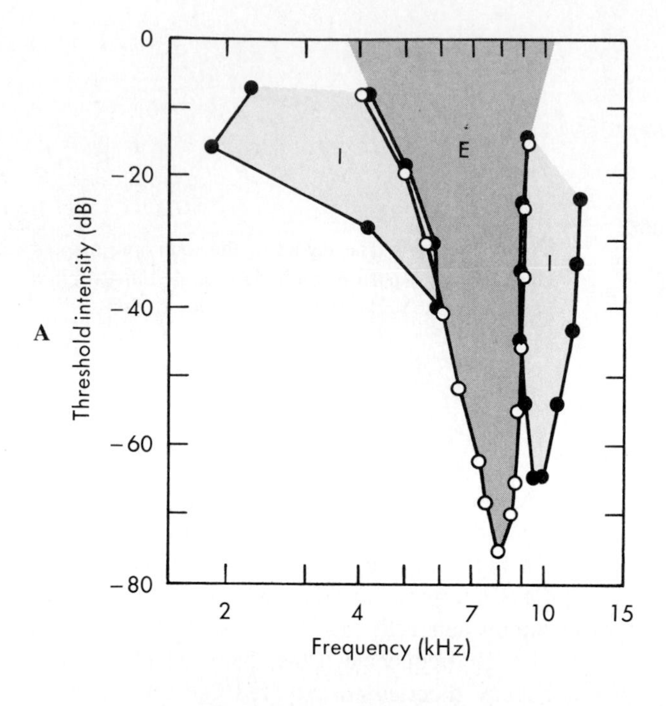

■ **Fig. 10-11** Tuning curves of neurons in the auditory system. Tuning curves can be considered as receptive field plots. **A,** A tuning curve with excitatory *(E)* and inhibitory *(I)* regions. **B,** Tuning curves for cochlear nerve fibers *(upper left),* neurons in the inferior colliculus *(upper right),* trapezoid body *(lower left),* and medial geniculate nucleus *(lower right).* (**A** redrawn from Arthur RM et al: *J Physiol* [Lond]. 212:593, 1971; **B** redrawn from Katsui Y. In Rosenblith WA, editor: *Sensory communication,* Cambridge, Mass, 1961, MIT Press.)

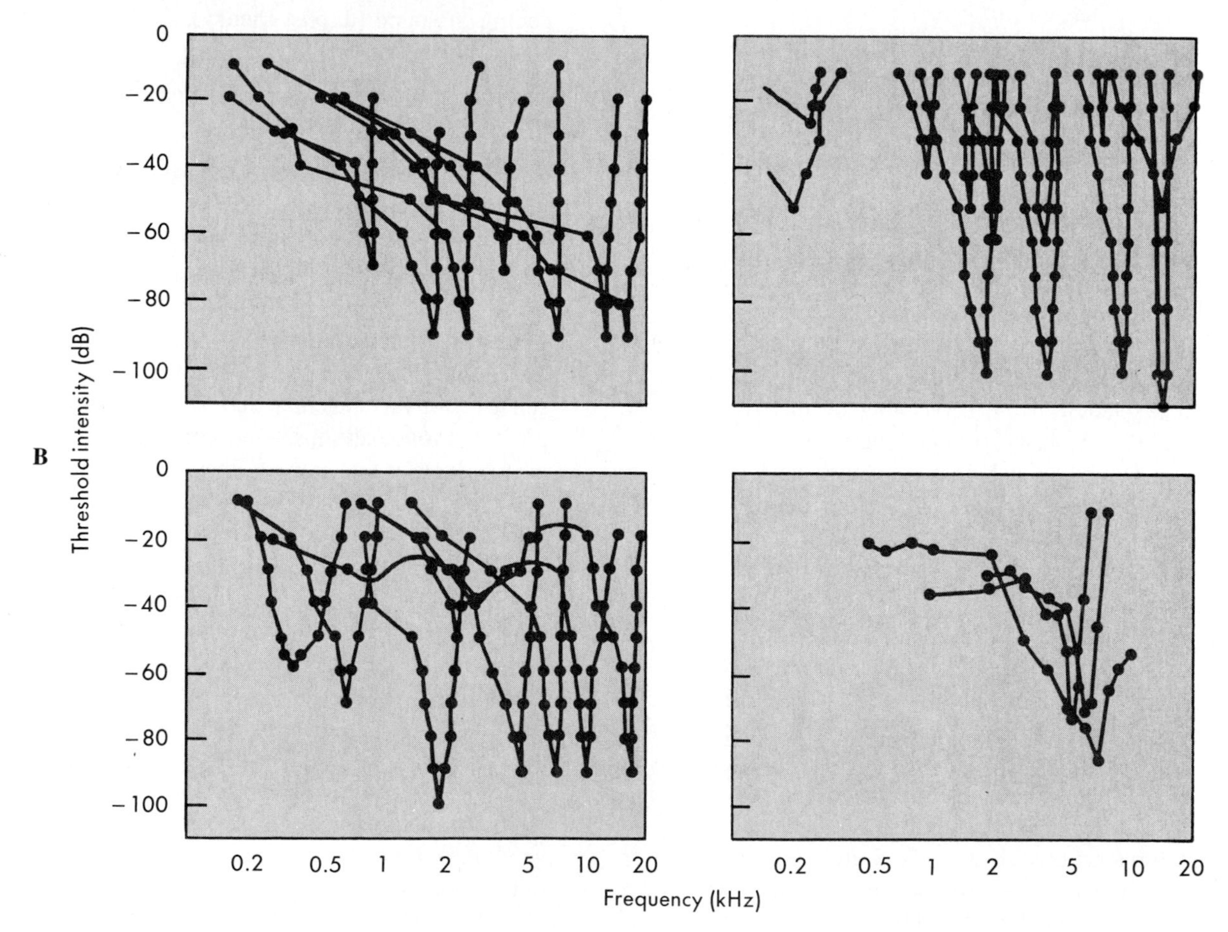

space is complex even at the brainstem level. Consequently, unilateral deafness may occur with lesions only of the cochlear nuclei or more peripheral structures. Central lesions do not cause unilateral deafness, although they may interfere with sound localization or discrimination of tones.

Fig. 10-12 shows a distinction between core and belt regions of the inferior colliculus, medial geniculate nucleus, and auditory cortex. The **core regions** include the central nucleus of the inferior colliculus, the laminated part of the medial geniculate nucleus, and the primary auditory cortex (AI). The **belt regions** include the pericentral nucleus of the inferior colliculus; the nonlaminated ventral, dorsal, and magnocellular divisions of the medial geniculate nucleus; and nonprimary parts of the auditory cortex. The core regions may represent a more recently evolved component of the auditory system and may be responsible for frequency representation. The belt regions, on the other hand, receive not only auditory input but also a convergent input from the

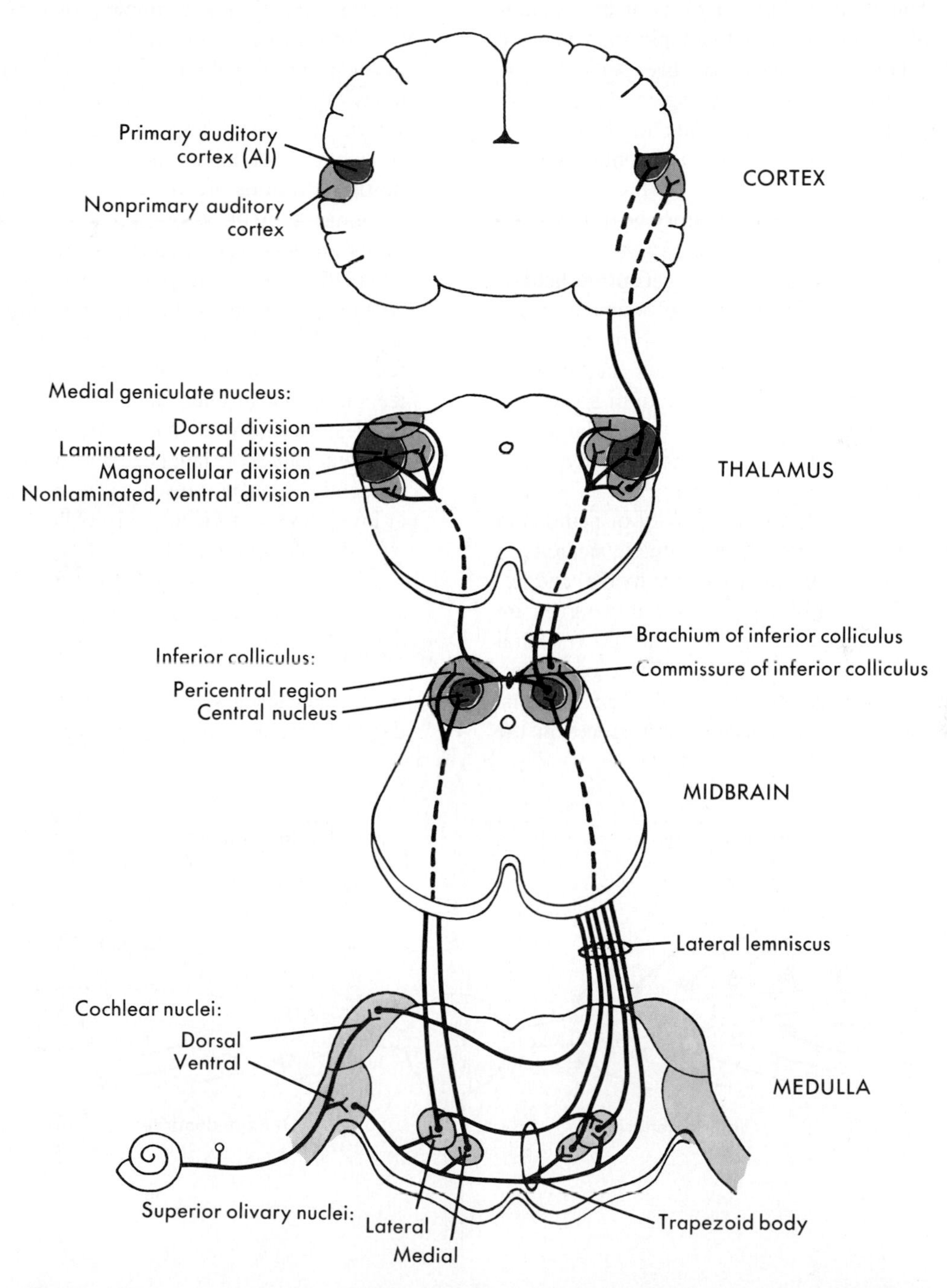

■ **Fig. 10-12** Central auditory pathway. Subdivision into core (darker color) and belt (lighter color) regions is shown for the inferior colliculus, medial geniculate nucleus, and auditory cortex.

visual and somatosensory systems. Hence, the belt system may be involved in activities, such as speech, which require integration across several modalities.

Functional Organization of Central Auditory System

Receptive fields and tonotopic maps. The responses of neurons in several structures belonging to the auditory system can be described by tuning curves (Fig. 10-11, *B*). By plotting the distribution of the characteristic frequencies of neurons within a nucleus or in the auditory cortex, the presence of a tonotopic map may be demonstrated. Tonotopic maps have been found in the cochlear nuclei, superior olivary complex, inferior colliculus, medial geniculate nucleus, and auditory cortex. A given auditory structure may, in fact, contain several tonotopic maps.

Binaural interactions. Most auditory neurons at levels above the cochlear nuclei respond to stimulation of either ear, (i.e., they have **binaural receptive fields**). Binaural receptive fields contribute to sound localization. A human can distinguish sounds originating from sources separated by as little as 1°. Clues that are used by the auditory system to judge the origin of sounds include differences in the time (or phase) of sound arrival at the two ears and differences in sound intensity on the two sides of the head.

These factors can influence the activity of neurons in the superior olivary complex. For example, neurons in the medial superior olivary nucleus have medial and lateral dendrites. The synapses on the medial dendrites are largely excitatory and originate from the contralateral ventral cochlear nucleus (Fig. 10-13). Those on the lateral dendrites are mostly inhibitory and come from the ipsilateral ventral cochlear nucleus. Differences in the phase of the sound reaching the two ears affect the strength and timing of the excitation and inhibition reaching a particular medial olivary neuron. The activity of that neuron can then provide information about sound localization. The lateral superior olivary nucleus uses differences in sound intensity reaching the two ears to provide information about the source of the sound.

Cortical organization. The primary auditory cortex has several features that resemble those of other primary sensory-receiving areas. Not only are sensory maps, in this case tonotopic maps, present in the auditory cortex, but this cortical region also performs feature extractions. For example, some neurons are selective for the direction of frequency modulation. Neurons in the primary auditory cortex form isofrequency columns (all neurons in the column have the same characteristic frequency) and alternating columns known as summation and suppression columns. Neurons in **summation columns** are more responsive to binaural than to monaural input, whereas neurons in **suppression columns** are less responsive to binaural than to monaural stimulation and the response to one ear is dominant.

Bilateral lesions of the auditory cortex have little effect on the ability to distinguish the frequencies of different sounds or sound intensity, but the ability to localize sound and to understand speech is reduced. Unilateral lesions, however, have little effect, especially if the nondominant (for language) hemisphere is involved. Evidently frequency discrimination depends on activity at lower levels of the auditory pathway, presumably the inferior colliculus.

Deafness

As already discussed, unilateral deafness is caused by damage to the peripheral auditory apparatus or to the

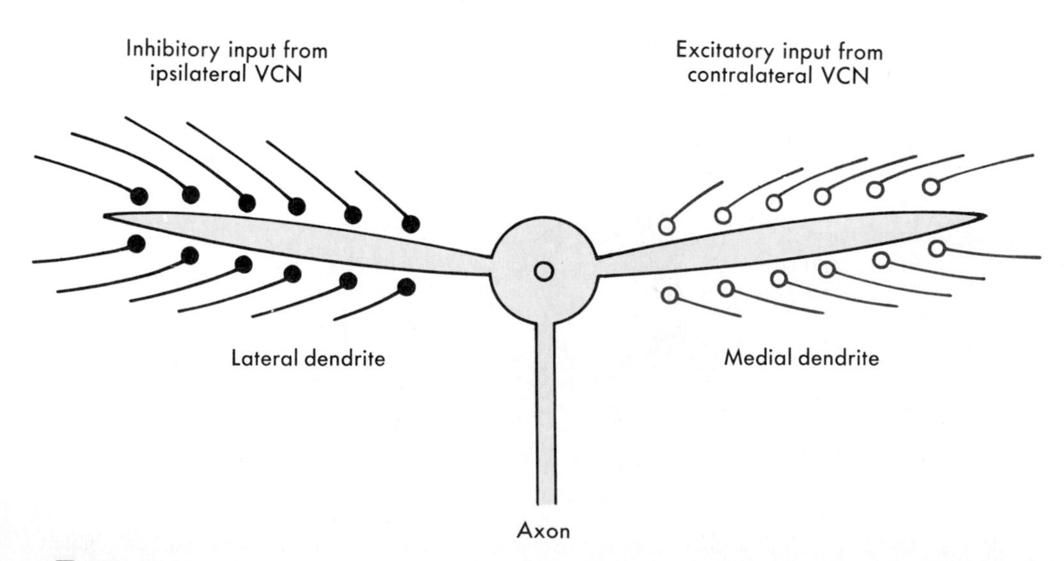

Fig. 10-13 Diagram of the synaptic input to a neuron in the medial superior olivary nucleus.

cochlear nuclei, but not by CNS lesions. A discrete loss of hearing for particular frequencies can result from damage to a part of the organ of Corti (e.g., by exposure to intense sound, such as particularly loud rock music or industrial noise).

The degree of deafness can be quantitated for different frequencies by **audiometry.** In this test each ear is presented with tones of different frequencies and intensities. An **audiogram** is plotted that shows the thresholds of each ear for representative frequencies of sound. Comparison with the audiogram of normal individuals shows the auditory deficit (in dB). The pattern of deficit aids in the diagnosis of the cause of the hearing loss.

Two simple tests are often used clinically to distinguish the most important types of deafness, *conduction loss* and *sensorineural loss*. These tests involve a tuning fork (generally one that vibrates at 256 Hz) and are called the Weber test and the Rinne test.

In the **Weber test** the base of the vibrating tuning fork is placed against the middle of the forehead, and the subject is asked if the sound is localized to one ear. Normally the sound is not localized to a particular ear. However, if the person has a conductive hearing loss (because of a punctured tympanic membrane, fluid in the middle ear, otosclerosis, loss of continuity of the ossicular chain), the sound is localized to the deaf ear. The reason the sound can be heard on the deaf side is that the sound is conducted to the cochlea through bone. Bone-conducted sound can activate the organ of Corti, although not as well as sound that can use the normal conduction mechanism through the tympanic membrane and ossicle chain. One reason why the sound in the Weber test is not localized to the normal ear may be that hearing in the normal ear is inhibited by the ambient sound level **(auditory masking).** The localization of sound in the Weber test to the ear deafened by a loss of the normal conduction mechanism can easily be demonstrated in normal subjects asked to place a finger in one ear. If a subject has a sensorineural hearing loss (because of damage to the organ of Corti, cochlear nerve, or cochlear nuclei), the sound is localized to the normal side.

The **Rinne test** also involves a vibrating tuning fork. In this case the base of the tuning fork is placed against the mastoid process, and the subject is asked to indicate when the sound dies out. The tuning fork is then held near the external auditory meatus. In normal subjects the sound is again heard. If the conduction mechanism is damaged, the sound is not heard. Bone conduction in this case is better than air conduction. If the hearing loss is sensorineural, the sound is heard again.

The Vestibular System

The vestibular system detects angular and linear accelerations of the head. Signals from the vestibular system trigger head and eye movements to provide the retina with a stable visual image and to make adjustments in posture to maintain balance. The following description of the vestibular system emphasizes the sensory aspects of vestibular function and introduces the central vestibular pathways. The role of the vestibular apparatus in motor control is discussed in Chapter 13.

The Vestibular Apparatus

Structure of the vestibular labyrinth. The vestibular apparatus, like the cochlea, consists of a component of the membranous labyrinth located within the bony labyrinth (Fig. 10-14). The vestibular apparatus on each side is composed of three semicircular ducts and two otolith organs. These are surrounded by perilymph and contain endolymph. The semicircular ducts include the horizontal, superior, and posterior ducts. The otolith organs include the utricle and the saccule. A swelling called an ampulla is found on each semicircular duct. The semicircular ducts all connect with the utricle. The utricle joins the saccule through the ductus reuniens. The endolymphatic duct originates from the ductus reuniens and ends in the endolymphatic sac. The saccule has a connection with the cochlea through which endolymph produced by the stria vascularis of the cochlea can reach the vestibular apparatus.

The three semicircular ducts on one side are matched with corresponding coplanar semicircular ducts on the other side. This arrangement allows the sensory epithelia in pairs of ducts on the two sides to sense movements of the head in all planes. Fig. 10-15 shows the orientation of the ducts on the two sides of the head; note that the cochlea is positioned rostrally to the vestibular apparatus and that the coil of the cochlea points laterally. The pairs of semicircular ducts that correspond are the two horizontal ducts, and the superior and posterior ducts on the two sides. The horizontal ducts can be placed in the horizontal plane with respect to the horizon if the head is tilted down 30°. The utricle is oriented nearly horizontally and the saccule, vertically.

Each of the semicircular ducts contains sensory epithelium in its ampulla. The sensory epithelium in a semicircular duct is called a crista ampullaris or ampullary crest (Fig. 10-16). An ampullary crest consists of a ridge in which are located vestibular hair cells. These are innervated by primary afferent fibers of the vestibular nerve, which is a subdivision of the eighth cranial nerve. Vestibular hair cells resemble cochlear hair cells in that each has a set of stereocilia on its apical surface. However, unlike cochlear hair cells, there is also a single kinocilium. The cilia on ampullary hair cells are embedded in a gelatinous structure called the cupula. The cupula crosses the ampulla and occludes its lumen completely. Movements of endolymph produced by an-

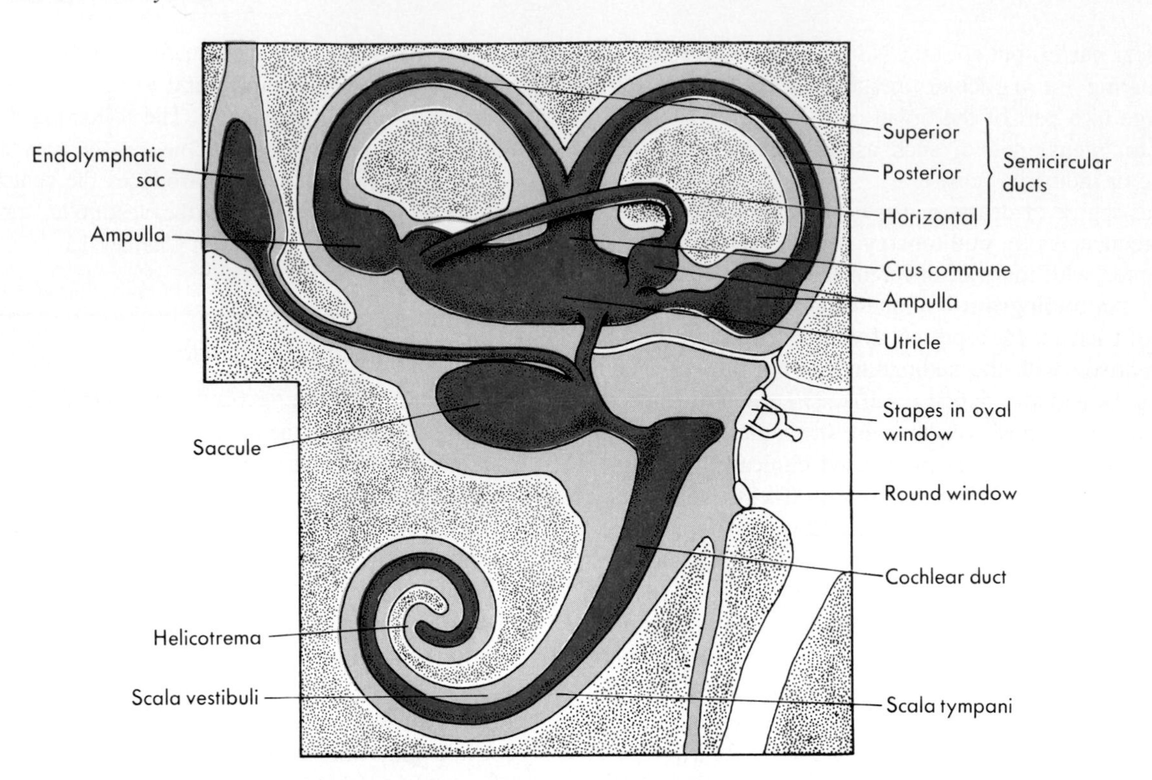

■ **Fig. 10-14** Vestibular apparatus. (Redrawn from Kandel ER, Schwartz JH: *Principles of neural science,* New York, 1981, Elsevier.)

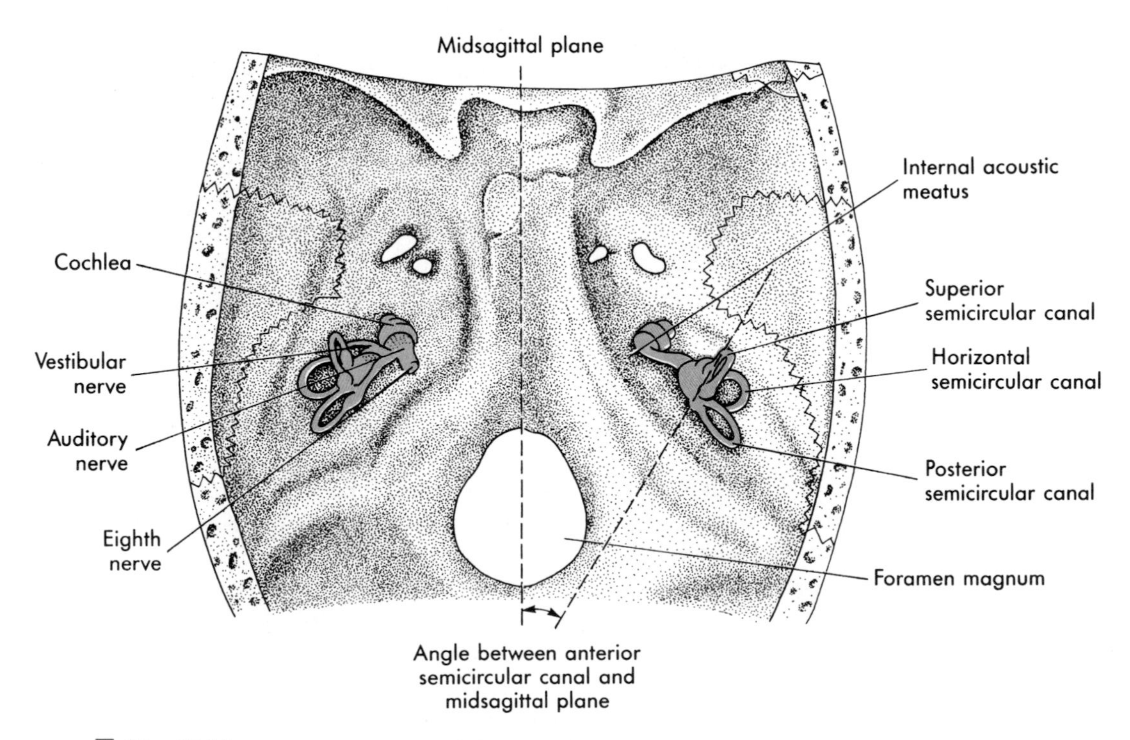

■ **Fig. 10-15** View of the base of the skull showing the orientation of structures of the inner ear. Coplanar pairs of semicircular ducts include the horizontal ducts, as well as the superior and contralateral posterior ducts. (Redrawn from Kandel ER, Schwartz JH: *Principles of neural science,* New York, 1981, Elsevier.)

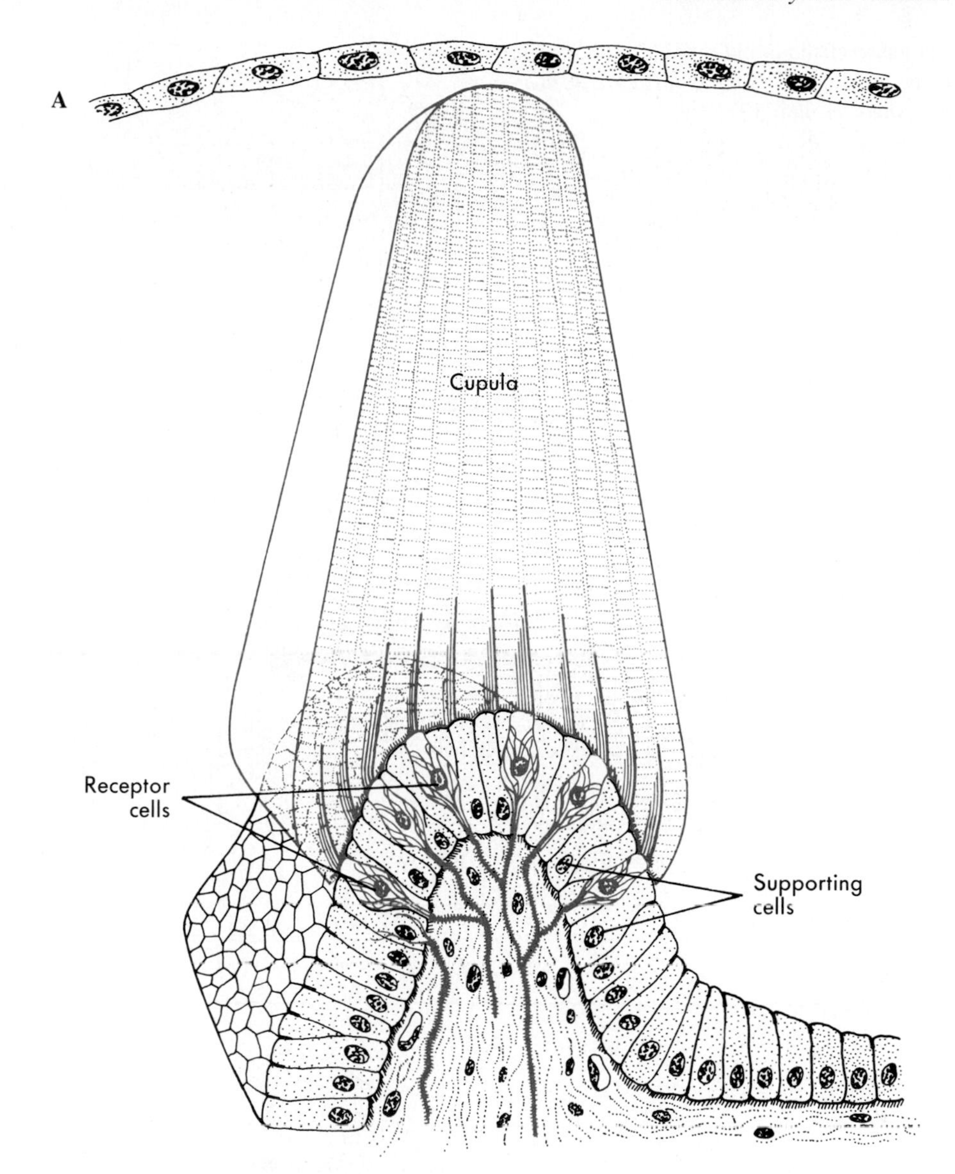

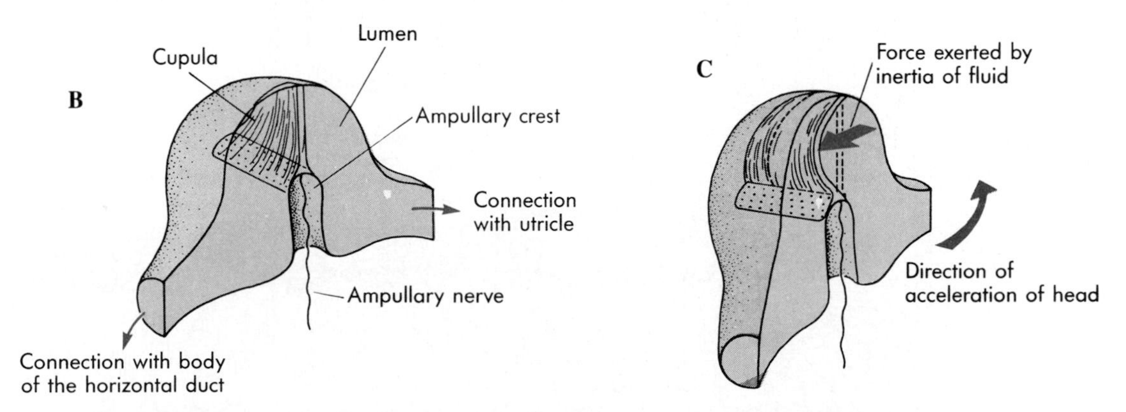

■ **Fig. 10-16** **A,** Drawing of an ampullary crest. The stereocilia and the kinocilium of each hair cell extend into the cupula. **B** and **C** show the distortion of the cupula that is produced when the head is rotated; **B** is before and **C** is during head rotation. (**A** redrawn from Wersäll J: *Acta Otolaryngol* (Stockholm) suppl 162:1, 1956; **B** and **C** redrawn from Kandel ER, Schwartz JH, Jessell TM: *Principles of neural science,* ed 3, New York, 1991, Elsevier.)

■ **Fig. 10-17** Structure of the otolith organs. The saccule is shown in **A** and the utricle in **B.** (Redrawn from Lindeman HH: *Adv Otorhinolaryngol* 20:405, 1973.)

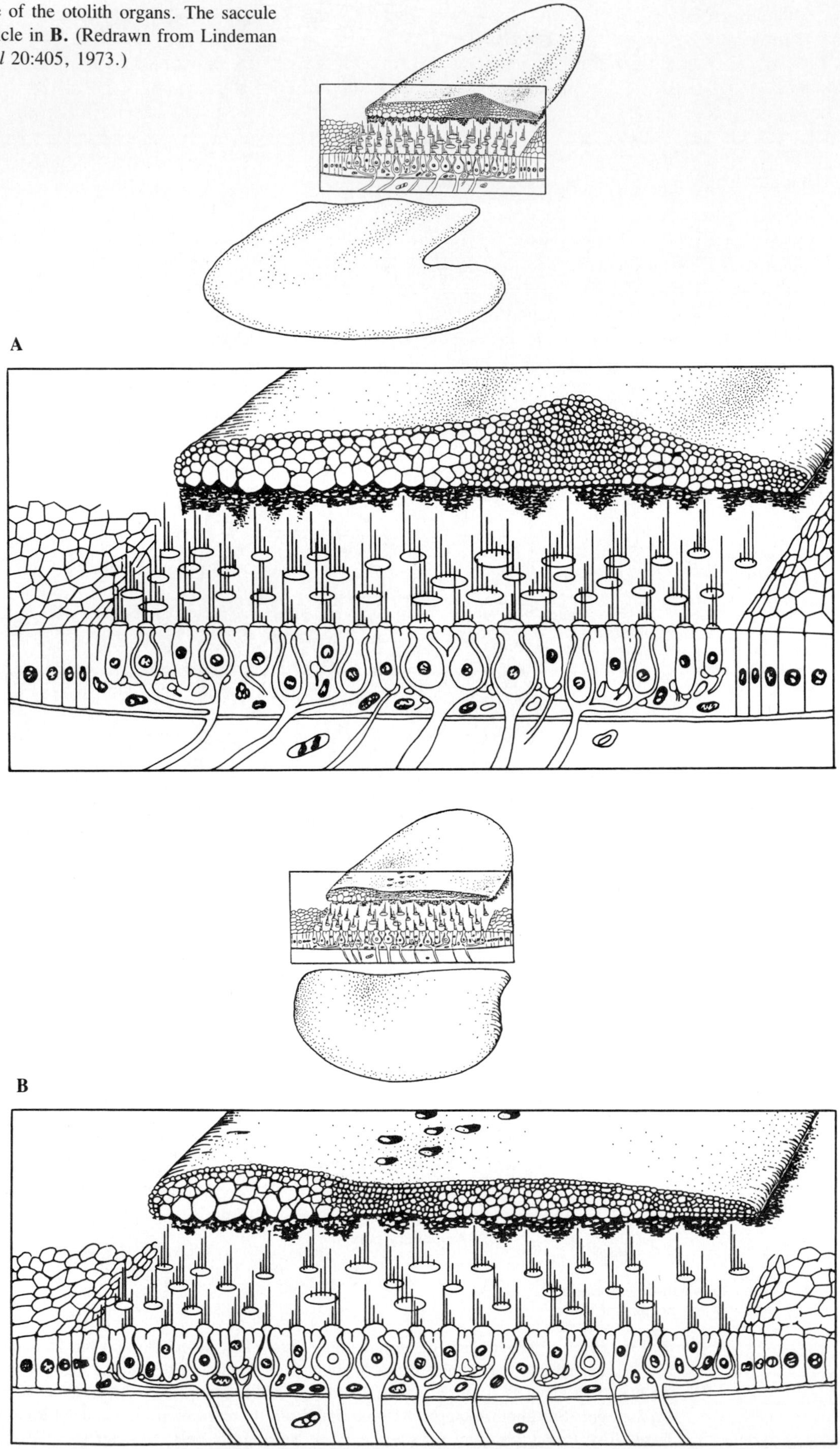

A

B

gular accelerations of the head deflect the cupula and consequently bend the cilia on the hair cells. The cupula has the same specific gravity as the endolymph, and so it is unaffected by linear acceleratory forces, such as that exerted by gravity.

The sensory epithelia of the otolith organs are called the macula utriculi and the macula sacculi (Fig. 10-17). Vestibular hair cells underlie each macula. As in the ampullary crests, the stereo- and kinocilia of these hair cells are embedded in a gelatinous mass. However, the gelatinous mass in this case contains numerous otoliths composed of crystals of calcium carbonate. The gelatinous mass with otoliths is called an **otolithic membrane.** The otoliths increase the specific gravity of the otolithic membrane to about twice that of the endolymph. Hence, the otolithic membrane tends to move when it is affected by a linear acceleration, such as that produced by gravity. Angular accelerations of the head do not affect the otolithic membranes, which do not protrude substantially into the lumen of the membranous labyrinth.

Innervation of sensory epithelia of vestibular apparatus. The primary afferent fibers of the vestibular nerve have their cell bodies in Scarpa's ganglion. As in the spiral ganglion, the neurons are bipolar, and their cell bodies, as well as axons, are myelinated. The vestibular nerve gives off separate branches to each of the sensory epithelia (Fig. 10-18). The vestibular nerve is accompanied by the cochlear and facial nerves in the internal auditory meatus of the skull.

Vestibular hair cells. There are two types of vestibular hair cells, called type I and type II hair cells (Fig. 10-19). The **type I hair cells** are pear-shaped, and they make synaptic contacts with chalice-like primary afferent terminals of vestibular afferent fibers. **Type II hair cells** are more rectangular, and they also synapse on vestibular afferent fibers. Vestibular efferent fibers synapse on the primary afferent fibers in relation to type I hair cells but directly on type II hair cells. Note the similarity between these arrangements and those of the cochlear afferent and efferent fibers on the inner and outer hair cells of the organ of Corti (see Fig. 10-7). The efferent endings on type II hair cells may account for the tendency of the afferent fibers that receive synapses from these cells to discharge irregularly.

Vestibular transduction. Like cochlear hair cells, vestibular hair cells are functionally polarized. When the stereocilia are bent towards the longest cilium (in this case, the kinocilium), the conductance of the apical membrane increases for cations and the vestibular hair cell is depolarized (Fig. 10-20). Conversely, when the cilia are bent away from the kinocilium, the hair cell is hyperpolarized. The hair cell releases an excitatory neurotransmitter (either glutamate or aspartate) tonically, so that the afferent fiber on which it synapses has a resting discharge. When the hair cell is depolarized, more

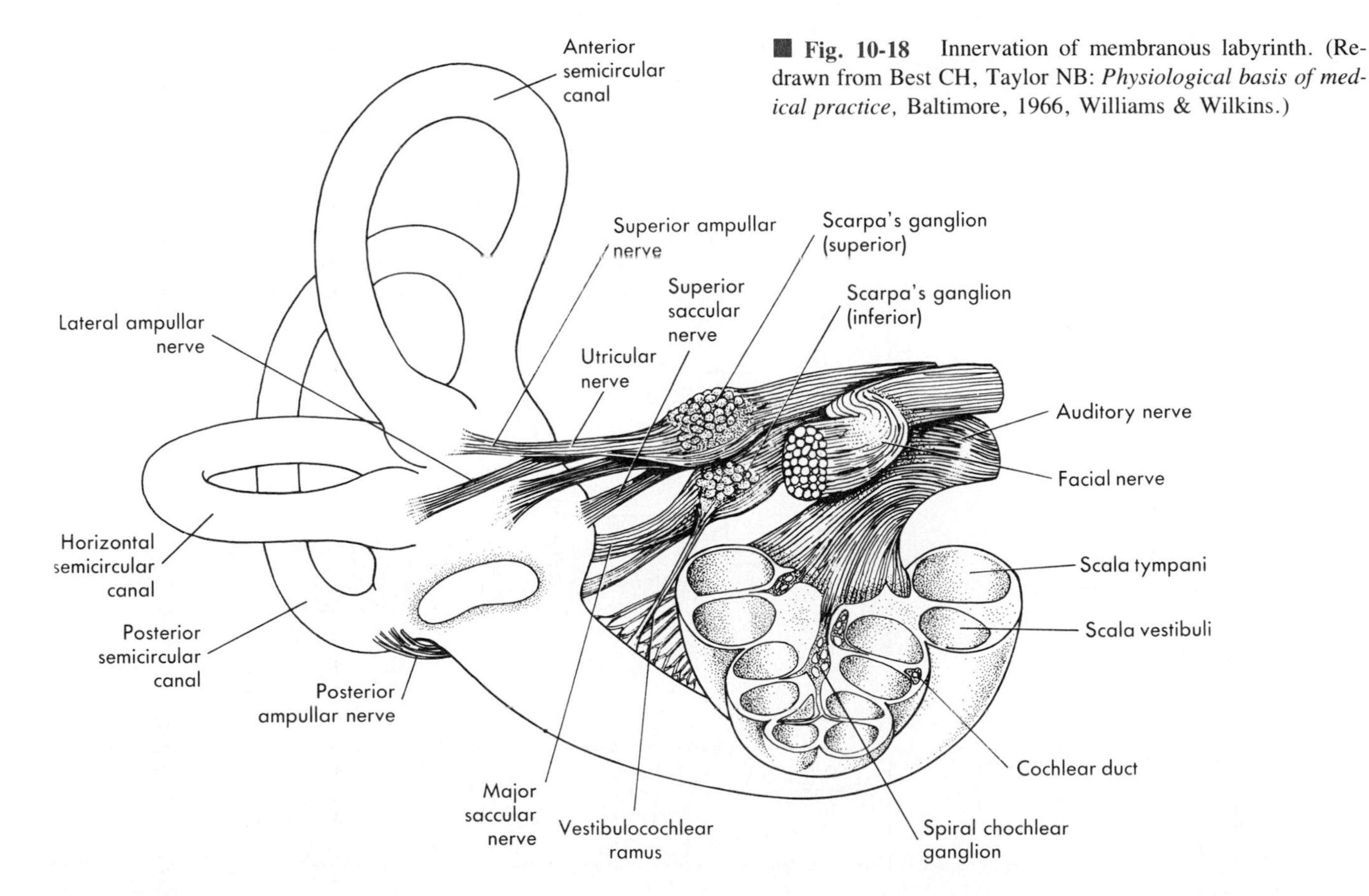

■ **Fig. 10-18** Innervation of membranous labyrinth. (Redrawn from Best CH, Taylor NB: *Physiological basis of medical practice,* Baltimore, 1966, Williams & Wilkins.)

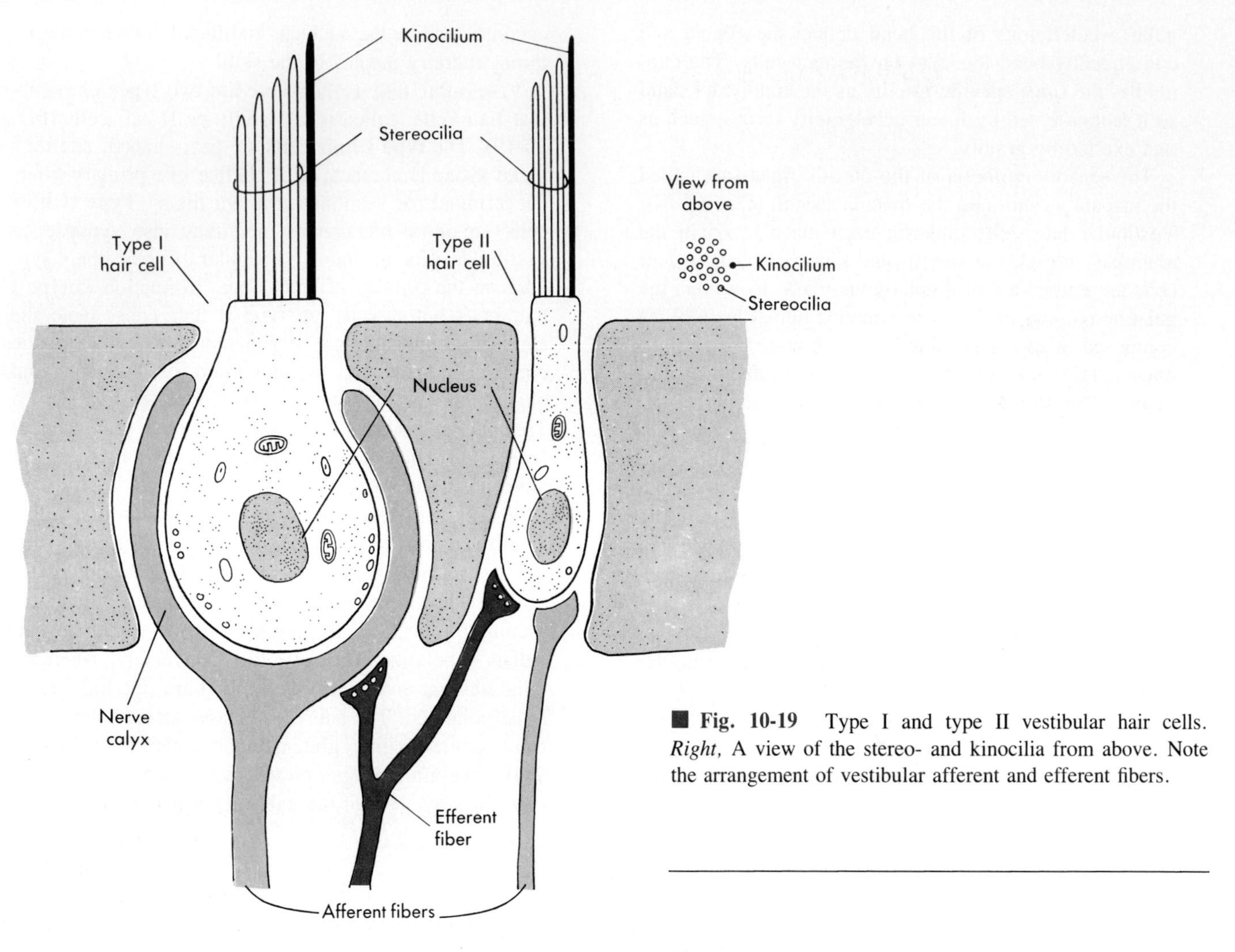

■ **Fig. 10-19** Type I and type II vestibular hair cells. *Right,* A view of the stereo- and kinocilia from above. Note the arrangement of vestibular afferent and efferent fibers.

■ **Fig. 10-20** Functional polarization of vestibular hair cells. When the stereocilia are bent towards the kinocilium, the hair cell is depolarized and the afferent fiber is excited. When the stereocilia are bent away from the kinocilium, the hair cell is hyperpolarized and the afferent discharge slows or stops. (Redrawn from Kandel ER, Schwartz JH: *Principles of neural science,* New York, 1981, Elsevier.)

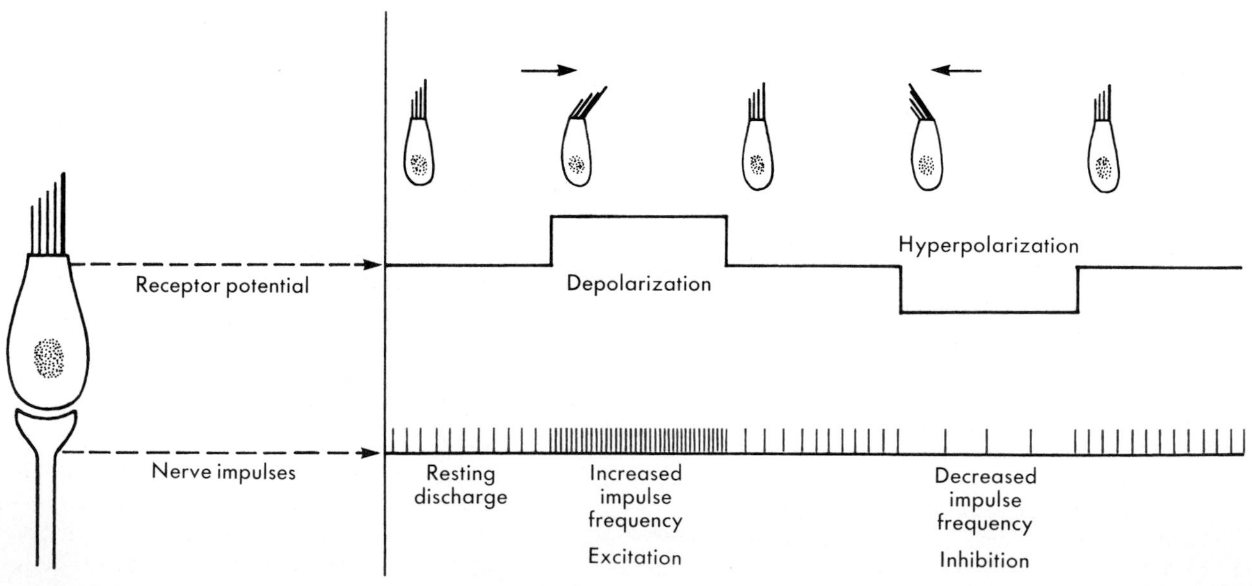

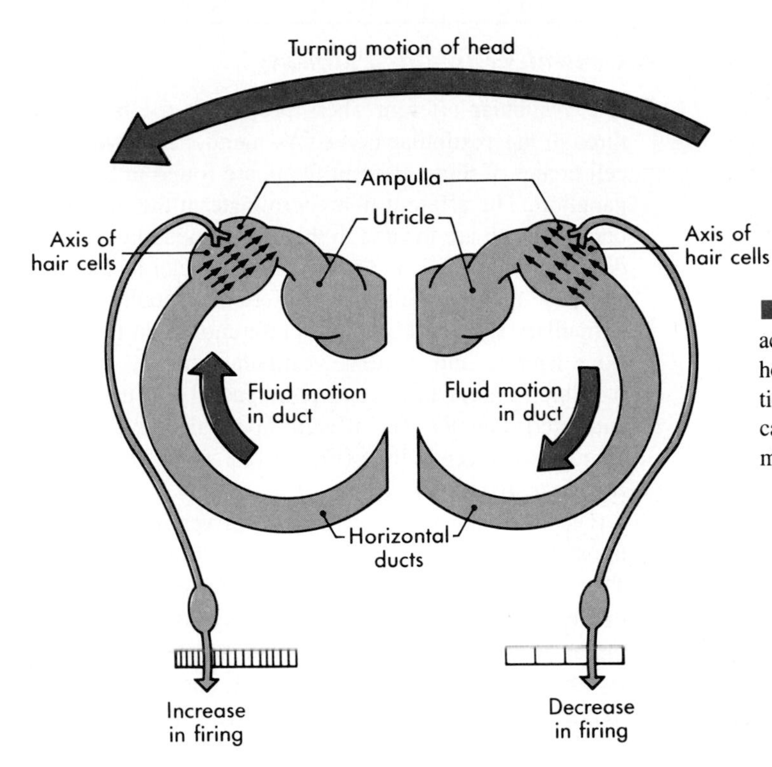

■ **Fig. 10-21** Effect of head movement to the left on the activity of vestibular afferent fibers supplying hair cells in the horizontal semicircular ducts. *Small arrows* indicate the functional polarity of the hair cells. The *large arrow, top,* indicates movement of the head; *Open arrows,* relative movements of the endolymph.

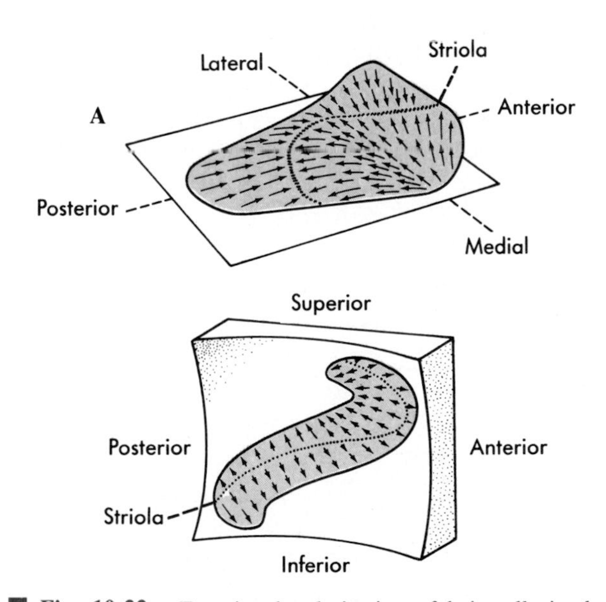

■ **Fig. 10-22** Functional polarization of hair cells in the otolith organs. **A,** The utricle. **B,** The saccule. The striola in each case is indicated by the dotted line. (Redrawn from Spoendlin HH. In Wolfson RJ, editor: *The vestibular system and its diseases,* Philadelphia, 1966, University of Pennsylvania Press.)

transmitter is released; and the discharge rate of the afferent fiber increases. Conversely, when the hair cell is hyperpolarized, less transmitter is released; and the firing rate of the afferent fiber slows or stops.

Semicircular ducts. Movements of the cilia on the hair cells of the ampullary crests of the semicircular ducts are caused by angular accelerations of the head. These accelerations produce relative movements of the endolymph. The inertia of the endolymph causes it to shift in relation to the fixed wall of the membranous labyrinth. This shift distorts the cupula and causes the cilia to bend. All of the cilia in a given ampullary crest are oriented in the same direction. In the horizontal duct, the cilia are oriented toward the utricle, and in the other ampulla they are oriented away from the utricle.

The way in which an angular acceleration of the head affects the discharges of vestibular afferent fibers can be exemplified by the activity that originates from the horizontal ducts. Fig. 10-21 shows the horizontal ducts and utricle as seen from above. The hair cells in these ducts are polarized toward the utricle. Thus, movement of the cilia toward the utricle will increase the discharge rates of the afferent fibers, whereas movements of the cilia away from the utricle will reduce the discharge rate. In the illustration the head is rotated to the left. This causes a relative movement of the endolymph in the horizontal ducts to the right. This movement of endolymph bends the cilia on the hair cells of the ampulla of the left horizontal duct toward the utricle and those of the right duct away from the utricle. These effects on the cilia increase the firing rate in horizontal duct afferent fibers on the left and decrease the firing rate on the right.

Otolith organs. The hair cells in the otolith organs, unlike those in the ampullary crests, are not all oriented in the same direction. Instead, they are oriented with respect to a ridge, called the striola, along the otolith organ (Fig. 10-22). In the utricle the hair cells on either side of the striola are polarized toward the striola, whereas in the saccule they are polarized away from the striola. Because the striola in each otolith organ is curved, the hair cells have diverse orientations. When the head is tilted so that gravity produces a different linear acceleration, the otolithic membranes shift and the cilia of the hair cells bend in a new way. This changes the pattern of input from the otolith organs to the CNS. Similarly, a linear acceleration caused by other forces, such as might occur in a space launch or a free fall, will change the output from the otolith organs.

Central Vestibular Pathways

The vestibular afferent fibers project to the brainstem through the vestibular nerve. As mentioned earlier, the cell bodies of these afferent fibers are found in Scarpa's ganglion. The afferent fibers terminate in the vestibular nuclei, which are located in the rostral medulla and caudal pons (Fig. 10-23). The vestibular nuclei include the superior, lateral, medial, and inferior vestibular nuclei. Ampullary afferent fibers end preferentially in the superior, lateral, and medial vestibular nuclei, whereas otolithic afferent fibers end preferentially in the lateral and inferior nuclei. The afferent fibers also give off collaterals to the cerebellum (Fig. 10-23, *right, upward-directed arrows*).

The vestibular nuclei give rise to a variety of projections. Some of these are illustrated in Fig. 10-23, *left*. The superior and medial vestibular nuclei project through the medial longitudinal fasciculus to the oculomotor nuclei. Therefore, it is not surprising that the vestibular nuclei exert a powerful control over eye movements (the vestibulo-ocular reflex). The lateral and medial vestibular nuclei give rise to the lateral and medial vestibulospinal tracts. These pathways provide, respectively, for the activation of postural and neck muscles and thereby contribute to balance and to head movements (vestibulo-collic reflex). There are vestibular projections to the cerebellum, the reticular formation, and the contralateral vestibular complex (arrows extending leftwards in Fig. 10-23), as well as to the thalamus. The latter mediate conscious sensations related to vestibular activity. The vestibular efferent fibers also originate from the vestibular nuclei.

Vestibular reflexes and clinical tests of vestibular function are described in Chapter 13.

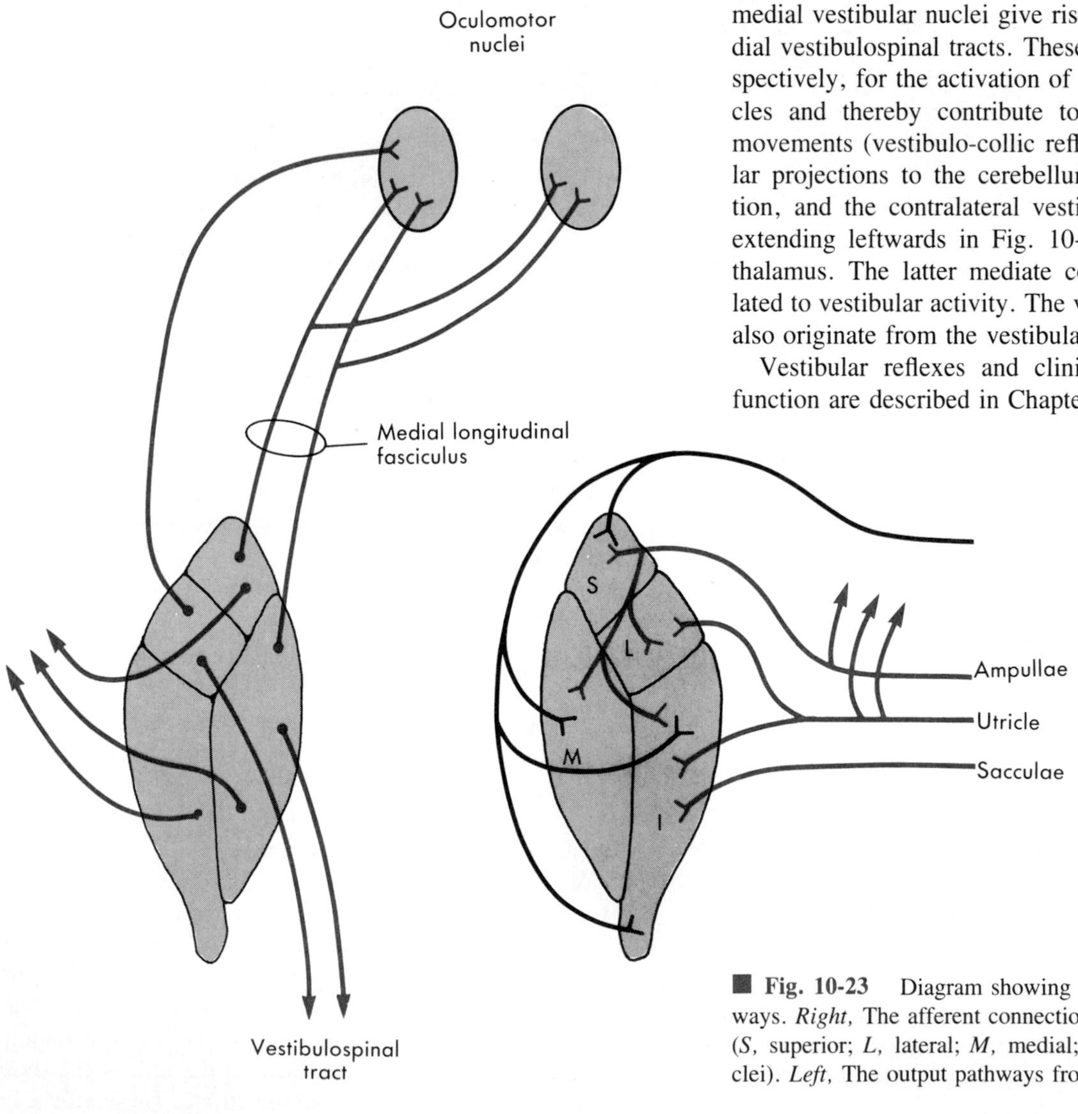

Fig. 10-23 Diagram showing the major vestibular pathways. *Right,* The afferent connections to the vestibular nuclei (*S,* superior; *L,* lateral; *M,* medial; *I,* inferior vestibular nuclei). *Left,* The output pathways from the vestibular nuclei.

Summary

1. Sound waves are combinations of pure tones and the composition of a sound can be determined by Fourier analysis. A pure tone is characterized in terms of its amplitude, frequency, and phase.

2. The unit of sound pressure is the decibel. Hearing is most sensitive at about 3,000 Hz.

3. The main parts of the ear are the external, middle, and inner ears. The external ear includes the pinna and auditory canal. The middle ear includes the tympanic membrane and a chain of ossicles that ends at the oval window. It is separated from the inner ear by the oval and round windows. The middle ear apparatus serves as an impedance matching device for energy transfer between air and the fluid in the inner ear.

4. The inner ear includes the cochlea and vestibular apparatus. The cochlea has three main compartments: the scala vestibuli and scala tympani, which are parts of the bony labyrinth, and the scala media, which is part of the membranous labyrinth. The bony labyrinth contains perilymph and the membranous labyrinth, endolymph.

5. The cochlear duct is bounded on one side by the basilar membrane, on which lies the organ of Corti, the sound transduction mechanism. Hair cells of the organ of Corti synapse with cochlear afferent fibers. They are also controlled by efferent fibers that originate in the superior olivary complex. When the basilar membrane oscillates, the stereocilia of the hair cells are subjected to shear forces at their contacts with the tectorial membrane. This results in a membrane conductance change that causes neurotransmitter release and thus a generator potential in cochlear afferent fibers.

6. Hair cells near the base of the cochlea are best activated by high frequency sounds and those near the apex by low frequency sounds. A tonotopic organization is also found in central auditory structures, including the cochlear nuclei, superior olivary complex, inferior colliculus, medial geniculate nucleus, and primary auditory cortex. Auditory processing in the central auditory pathway contributes to sound localization, frequency and intensity analysis, and speech recognition.

7. The vestibular apparatus is part of the membranous labyrinth and includes three semicircular ducts (horizontal, superior, and posterior) and two otolith organs (utricle and saccule) on each side. These transduce angular and linear accelerations of the head.

8. The sensory epithelium, the crista ampullaris, of a semicircular duct is found in a dilatation called the ampulla. Stereocilia and a single kinocilium extend from each hair cell into the cupula. Angular head movements displace the endolymph and distort the cupula, bending the cilia. If the stereocilia bend toward the kinocilium, the hair cell is depolarized; this causes a greater firing rate in the afferent fiber.

9. In the otolith organs, the cilia project into an otolithic membrane. Linear acceleration of the head displaces the otolithic membrane, which is sensitive to gravity because of the otoliths. The hair cells have various orientations, but displacements of the head are coded by the patterned input from the afferent fibers.

10. Central vestibular pathways include afferent connections to four vestibular nuclei and to the cerebellum. Vestibular nuclei project (a) in the medial longitudinal fasciculus to the oculomotor nuclei (to control eye position); (b) in the lateral vestibulospinal tract, which excites motoneurons to postural muscles (to control balance); and (c) in the medial vestibulospinal tract, which activates motoneurons to neck muscles (to control head position).

Bibliography

Journal articles

Allen JB: Cochlear micromechanics—a physical model of transduction, *J Acoust Soc Am* 68:1660, 1981.

Allon N et al: Responses of single cells in the medial geniculate body of awake squirrel monkeys, *Exp Brain Res* 41:222, 1981.

Brownell WE et al: Evoked mechanical responses of isolated cochlear outer hair cells, *Science* 227:194, 1985.

Brugge JF, Merzenich MM: Responses of neurons in auditory cortex of the macaque monkey to monaural and binaural stimulation, *J Neurophysiol* 36:1138, 1973.

Calford MB, Aitkin LM: Ascending projections to the medial geniculate body of the cat: evidence for multiple, parallel auditory pathways through thalamus, *J Neurosci* 3:2365, 1983.

Carelton SC, Carpenter MB: Afferent and efferent connections of the medial, inferior, and lateral vestibular nuclei, *Brain Res* 278:29, 1983.

Flock A, Strelioff D: Graded and nonlinear mechanical properties of sensory hairs in the mammalian hearing organ, *Nature* 310:597, 1984.

Goldberg JM, Brown PB: Response of binaural neurons of dog superior olivary complex to dichotic tonal stimuli: some physiological mechanisms of sound localization, *J Neurophysiol* 32:613, 1969.

Hudspeth AJ: The cellular basis of hearing: the biophysics of hair cells, *Science* 230:745, 1985.

Imig TJ, Morel A: Organization of the thalamocortical auditory system in the cat, *Ann Rev Neurosci* 6:95, 1983.

Jenkins WM, Merzenich MM: Role of cat primary cortex for sound-localization behavior, *J Neurophysiol* 52:819, 1984.

Shotwell SL et al: Direction sensitivity of individual vertebrate hair cells to controlled deflection of their hair bundles, *Ann NY Acad Sci* 374:1, 1981.

Tomko DL et al: Responses to head tilt in cat eighth nerve afferents, *Exp Brain Res* 41:216, 1981.

Books and monographs

Brodal A: *Neurological anatomy in relation to clinical medicine,* ed 3, New York, 1981, Oxford University Press.

Kiang NYS: *Discharge patterns of single fibers in the cat's auditory nerve,* Cambridge, Mass, 1965, MIT Press.

Kornhuber HH, editor: *Vestibular system.* I. Basic mechanisms: handbook of sensory physiology, vol VI/1, Springer-Verlag, 1974, Heidelberg.

Von Bekesy G: *Experiments in hearing,* New York, 1960, McGraw-Hill.

Wilson VJ, Melville JG: *Mammalian vestibular physiology,* New York, 1979, Plenum Press.

Yin TCT, Kuwada S: Neuronal mechanisms of binaural interaction. In Edelman GM, Cowan WM, editors: *Dynamic aspects of neocortical function,* New York, 1984, John Wiley & Sons.

CHAPTER

11

The Chemical Senses

The senses of **gustation** (taste) and **olfaction** (smell) depend on chemical stimuli that are present either in food and drink or in the air. In humans these senses do not have the survival value of some of the other senses, but the chemical senses contribute considerably to the quality of life and are important stimulants of digestion. In other animals the chemical senses do have survival value and evoke a number of social behaviors, including mating, territoriality, and feeding.

■ *Taste*

Tastes can be viewed as mixtures of four elementary taste qualities: salty, sweet, sour, and bitter. Taste stimuli that are particularly effective in eliciting these sensations include NaCl, sucrose, hydrochloric acid, and quinine.

■ *Taste Receptors*

Tastes depend on the activation of chemoreceptors located in taste buds. A **taste bud** consists of a group of 50 to 150 receptor cells, of two types, as well as supporting and basal cells (Fig. 11-1). The chemoreceptor cells synapse at their bases with primary afferent nerve fibers. The receptor cells can be distinguished by differences in their synaptic vesicle content; one type has dense core vesicles, and the other, clear round vesicles. The apexes of the cells have microvilli that extend toward a taste pore.

Chemoreceptor cells have a short lifespan, about 10 days. They are constantly being replaced by new chemoreceptor cells that differentiate from basal cells located near the base of the taste bud.

Chemoreceptive molecules on the microvilli detect stimulatory molecules that diffuse into the taste pore from the overlying fluid layer. Part of this fluid originates from glands located adjacent to the taste buds. A change in membrane conductance of a chemoreceptor cell leads to a receptor potential and to the release of an excitatory neurotransmitter. The neurotransmitter evokes a generator potential in the primary afferent nerve fiber and causes a discharge that is transmitted to the CNS.

Coding for the four primary taste qualities is not based on complete selectivity of the chemoreceptors for the different qualities. Instead, a given chemoreceptor will respond to stimuli that can evoke several different taste qualities, although perhaps most vigorously to one. Recognition of taste quality appears to depend on the patterned input from a population of chemoreceptors. The intensity of the stimulus is reflected in the total amount of evoked activity.

■ *Distribution and Innervation of Taste Buds*

Taste buds are located on different types of taste papillae located on the tongue, palate, pharynx, and larynx (Fig. 11-2, *C*). The types of taste papillae include fungiform and foliate papillae on the anterior and lateral tongue and circumvalate papillae on the base of the tongue. The latter may contain several hundred taste buds, and there are a total of several thousand taste buds in humans.

The sensitivity of the tongue for different taste qualities varies with the region of the tongue (Fig. 11-2, *A*). Sweet tastes are detected best at the tip of the tongue, salty and sour along the sides, and bitter at the base.

The taste buds are innervated by three cranial nerves; two of these are shown in Fig. 11-2, *B*. The chorda tympani branch of the facial nerve supplies taste buds on the anterior two thirds of the tongue, and the glossopharyngeal nerve supplies taste buds on the posterior one third of the tongue (Fig. 11-3, *B*). The vagus nerve supplies a few taste buds in the larynx and upper esophagus.

■ *Central Pathways*

The taste fibers in cranial nerves VII, IX, and X have their cell bodies in the geniculate, petrosal, and nodose

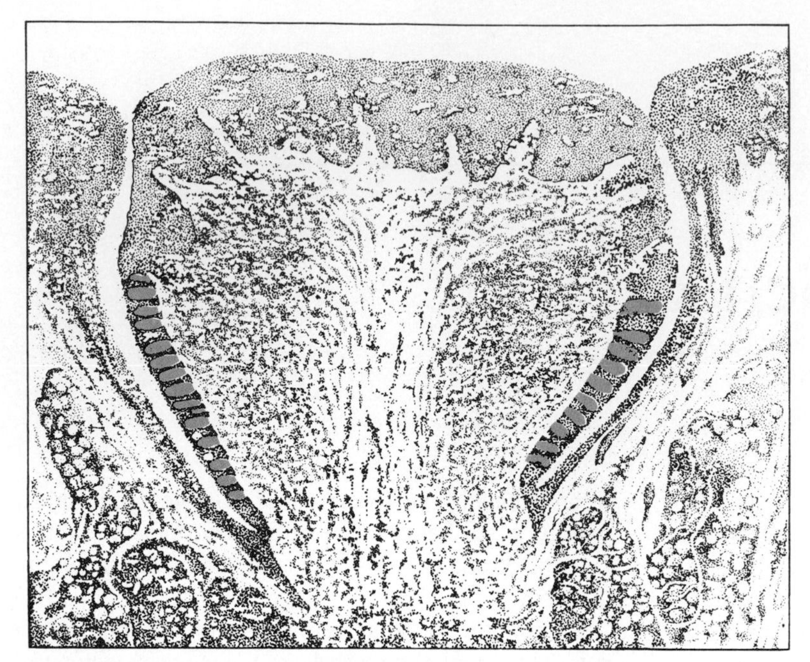

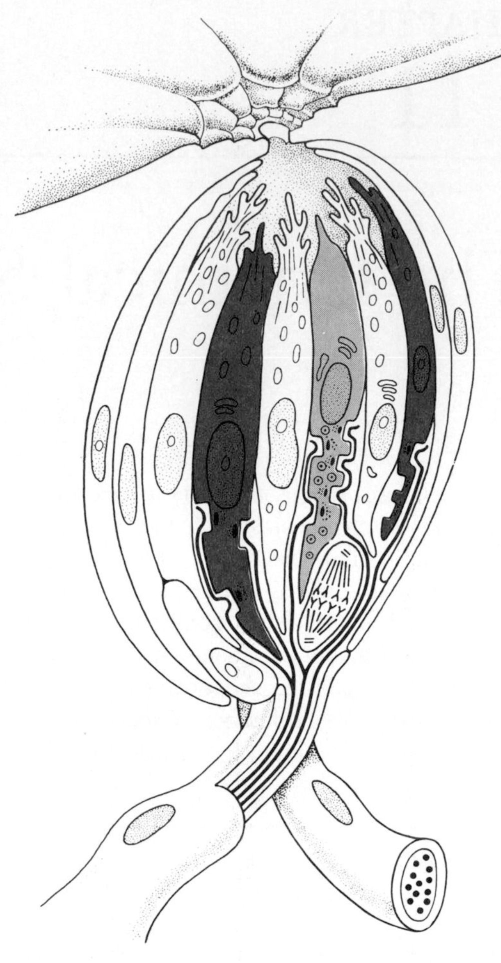

■ **Fig. 11-1** Taste bud. The two types of chemoreceptor cells are shown in color and the supporting cells are uncolored. *Above,* a circumvallate papilla is shown with its taste buds indicated in pink. *Right,* a taste bud is shown with the taste pore at the top and its innervation below. (Redrawn from Williams PL, Warwick R: *Functional neuroanatomy of man,* Philadelphia, 1975, WB Saunders.)

■ **Fig. 11-2** **A,** Distribution of sensitivity for the four taste qualities. **B,** The innervation of the anterior two-thirds and posterior one-third of the tongue by the facial and glossopharyngeal nerves. **C,** The arrangement of taste buds in the three types of papillae.

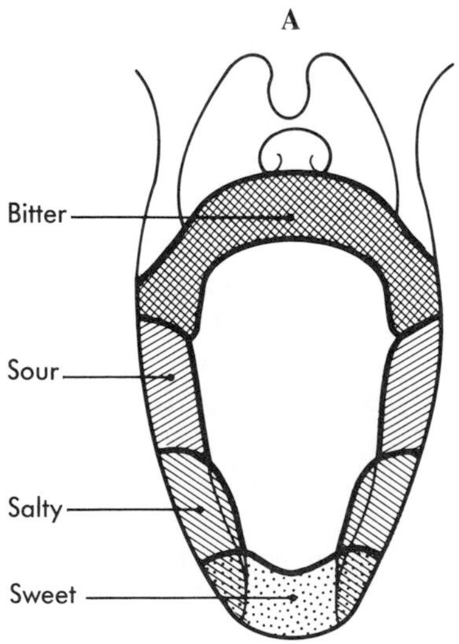

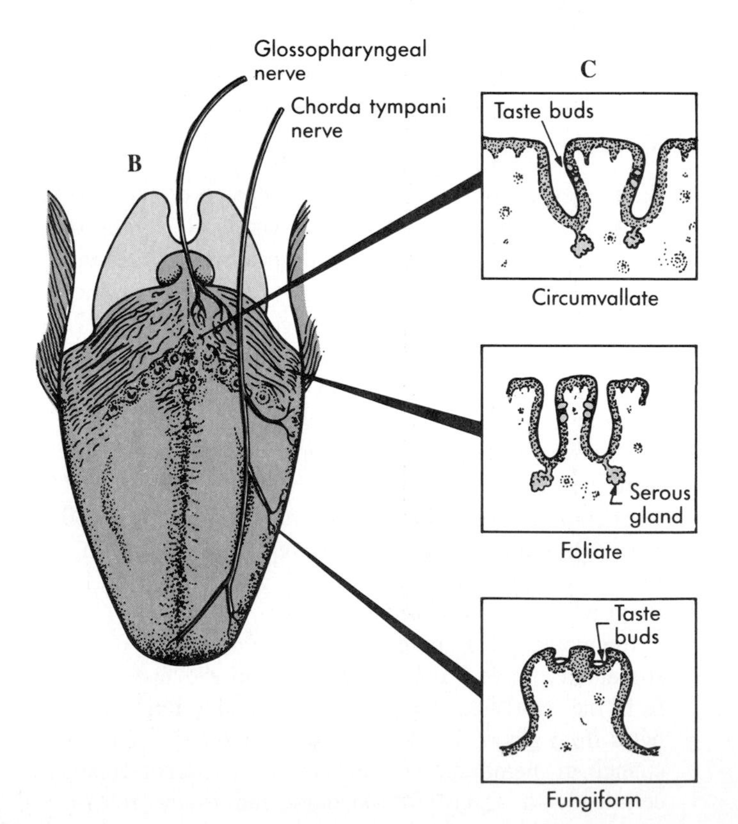

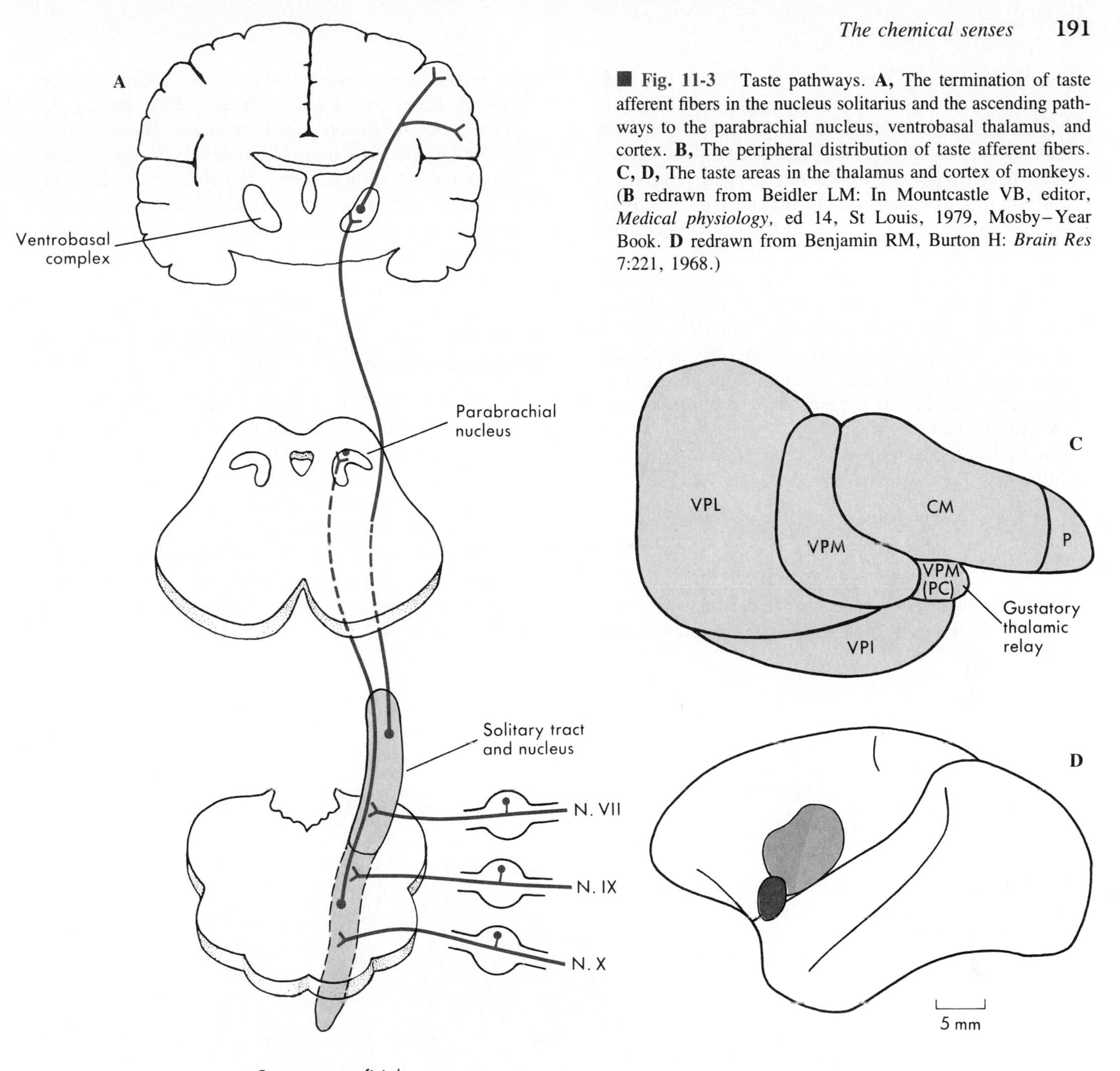

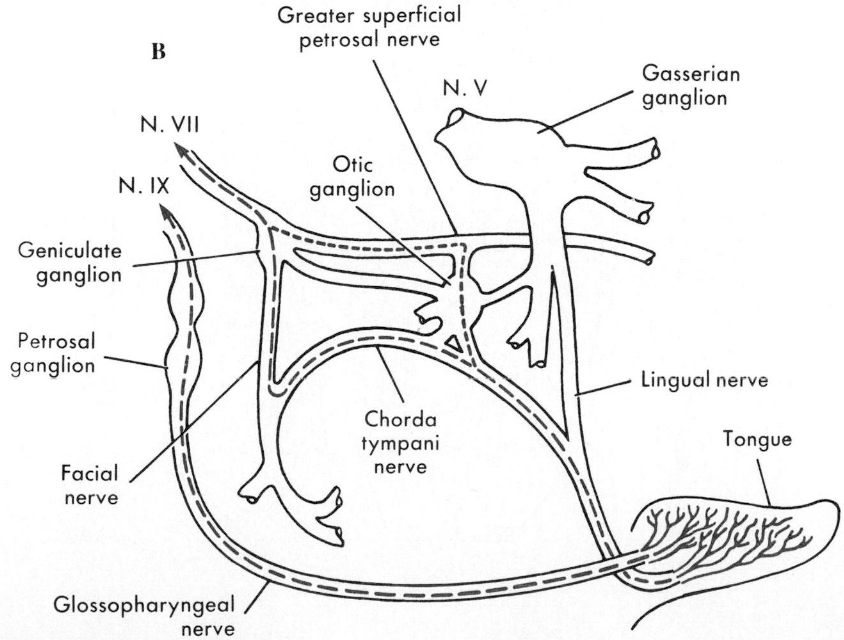

■ **Fig. 11-3** Taste pathways. **A,** The termination of taste afferent fibers in the nucleus solitarius and the ascending pathways to the parabrachial nucleus, ventrobasal thalamus, and cortex. **B,** The peripheral distribution of taste afferent fibers. **C, D,** The taste areas in the thalamus and cortex of monkeys. (**B** redrawn from Beidler LM: In Mountcastle VB, editor, *Medical physiology,* ed 14, St Louis, 1979, Mosby–Year Book. **D** redrawn from Benjamin RM, Burton H: *Brain Res* 7:221, 1968.)

ganglia. The central processes of the afferent fibers enter the medulla, join the solitary tract, and synapse in the nucleus of the solitary tract (Fig. 11-3, *A*). In some animals, including several rodent species, the second order taste neurons of the solitary nucleus project rostrally to the ipsilateral parabrachial nucleus. The parabrachial nucleus then projects to the small-celled (parvocellular) part of the ventral posterior medial (VPM_{pc}) **nucleus** of the thalamus (Fig. 11-3, *C*). In monkeys the solitary nucleus projects directly to the VPM_{pc} nucleus. The VPM_{pc} nucleus is connected with two different gustatory areas of the cerebral cortex, one in the face area of the SI cortex and another in the insula (Fig. 11-3, *D*). An unusual feature of the central gustatory pathway is that it is predominantly an uncrossed pathway (unlike the central somatosensory, visual, and auditory pathways, which are predominantly crossed).

Olfaction

The sense of smell is much better developed in other animals (macrosmatic animals) than in primates, including humans (microsmatic animals). The ability of dogs to track based on odor is legendary, as is the use of pheromones by insects to attract mates. However, olfaction contributes to our emotional life, and odors can effectively call up memories. The absence of olfaction in disease **(anosmia)** can be an important diagnostic symptom, as can olfactory hallucinations in the **uncinate fits** observed in people with temporal lobe epilepsy.

Olfactory Receptors

An olfactory chemoreceptor is a bipolar cell (Fig. 11-4). On its apical surface it has immobile cilia that detect odorants dissolved in the overlying mucus layer, and it gives off an unmyelinated axon from its basal surface. This axon joins others in olfactory nerve filaments that penetrate the base of the skull through openings in the cribriform plate of the ethmoid bone. The olfactory nerves make synaptic connections in the olfactory bulb, a CNS structure at the base of the cranial cavity, just below the frontal lobe. The olfactory chemoreceptor

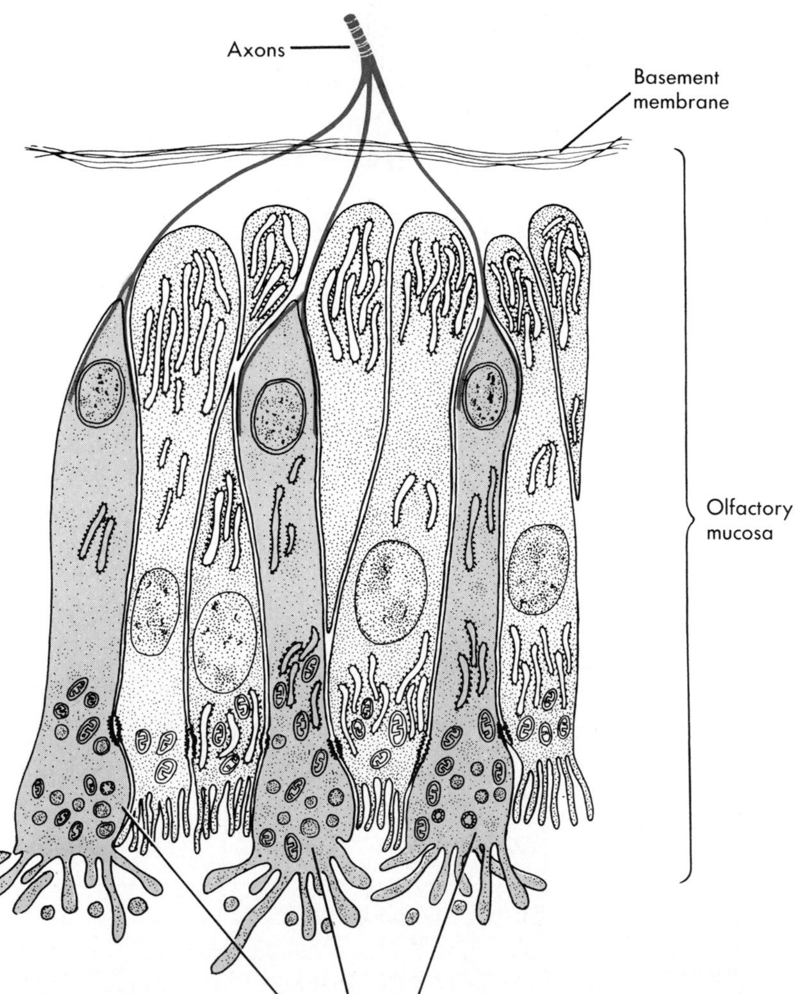

Fig. 11-4 Olfactory chemoreceptors are shown in color and supporting cells are uncolored. (Redrawn from de Lorenzo AJD. In Zotterman Y, editor: *Olfaction and taste,* Elmsford, New York, 1963, Pergamon Press.

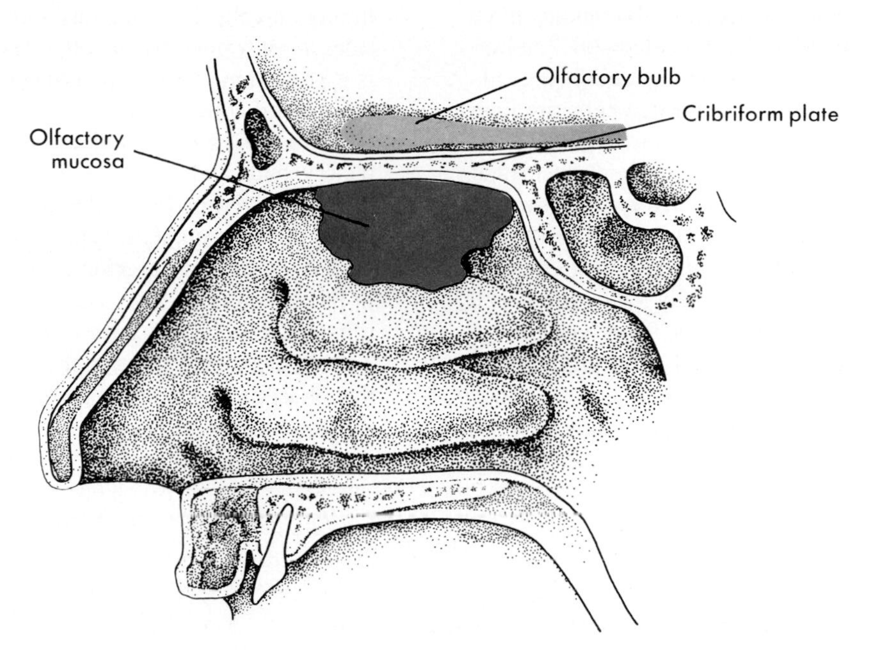

■ **Fig. 11-5** Location of the olfactory mucosa is shown by the dark colored region in the nasopharynx. Overlying the olfactory mucosa is the cribriform plate of the ethmoid bone and above this is the olfactory bulb. The olfactory mucosa also spreads onto the lateral aspect of the nasopharynx.

cells are located in the olfactory mucosa, a specialized part of the nasopharynx with a total surface area on the two sides of about 10 cm^2 (Fig. 11-5). In humans there are about 10^7 receptors. Like taste cells, olfactory chemoreceptors are continuously replaced, with a life span of about 60 days.

Odorant molecules are introduced to the olfactory mucosa by ventilatory air currents or from the oral cavity during feeding. Sniffing increases the influx of odorants. The odorants are temporarily bound in the mucus to an olfactory binding protein, which is secreted by a gland in the nasal cavity.

Odor has more primary qualities than does taste. There are at least 6 odor qualities: **floral, ethereal, musky, camphor, putrid,** and **pungent.** Natural stimuli with these odors are roses, pears, musk, eucalyptus, rotten eggs, and vinegar, respectively. The olfactory mucosa also contains somatosensory receptors of the trigeminal nerve. Clinical testing of olfaction must avoid activating these somatosensory receptors with noxious or thermal stimuli.

A few odorant molecules that reach an olfactory chemoreceptor cell produce a depolarizing receptor potential, which triggers a neural discharge. However, behavioral responses require the activation of a number of olfactory chemoreceptors. The receptor potential probably results from an increased conductance for Na^+. However, a G-protein is activated, and a cascade of second messengers is also involved in olfactory transduction.

Olfactory coding resembles taste coding in that an individual olfactory chemoreceptor responds to more than one odorant class. Coding for a particular olfactory quality depends on a population response, and the strength of the odorant is represented by the overall amount of afferent neural activity.

Central Pathways

The initial relay of the olfactory pathway is in the olfactory bulb, which is a cortical structure. It contains three main cell types, the mitral cells, the tufted cells, and interneurons (granule cells; periglomerular cells) (Fig. 11-6). The dendrites of the mitral and tufted cells are long and branch to form the postsynaptic components of the olfactory glomeruli. The olfactory afferent fibers that reach the olfactory bulb from the olfactory mucosa ramify as they approach the olfactory glomeruli and then synapse on the dendrites of the mitral and tufted cells. Olfactory axons converge extensively onto mitral cell dendrites; as many as 1,000 afferent fibers synapse on the dendrites of a single mitral cell. The granule and periglomerular cells are inhibitory interneurons. They form dendrodendritic reciprocal synapses with the dendrites of the mitral cells. Evidently, activity in a mitral cell depolarizes the inhibitory cells that synapse with it and release inhibitory neurotransmitter that acts back on the mitral cell. The olfactory bulb has other inputs besides those formed by the olfactory nerves; among these is a projection from the contralateral olfactory tract via the anterior commissure.

The axons of the mitral and tufted cells leave the olfactory bulb and enter the olfactory tract (Figs. 11-6 and 11-7). From here, the olfactory connections become highly complex. Within the olfactory tract is a nucleus called the **anterior olfactory nucleus.** Neurons in this structure receive synaptic connections from neurons of the olfactory bulb and project to the contralateral olfactory bulb through the anterior commissure. As the olfactory tract approaches the anterior perforated substance at the base of the brain, it splits into the lat-

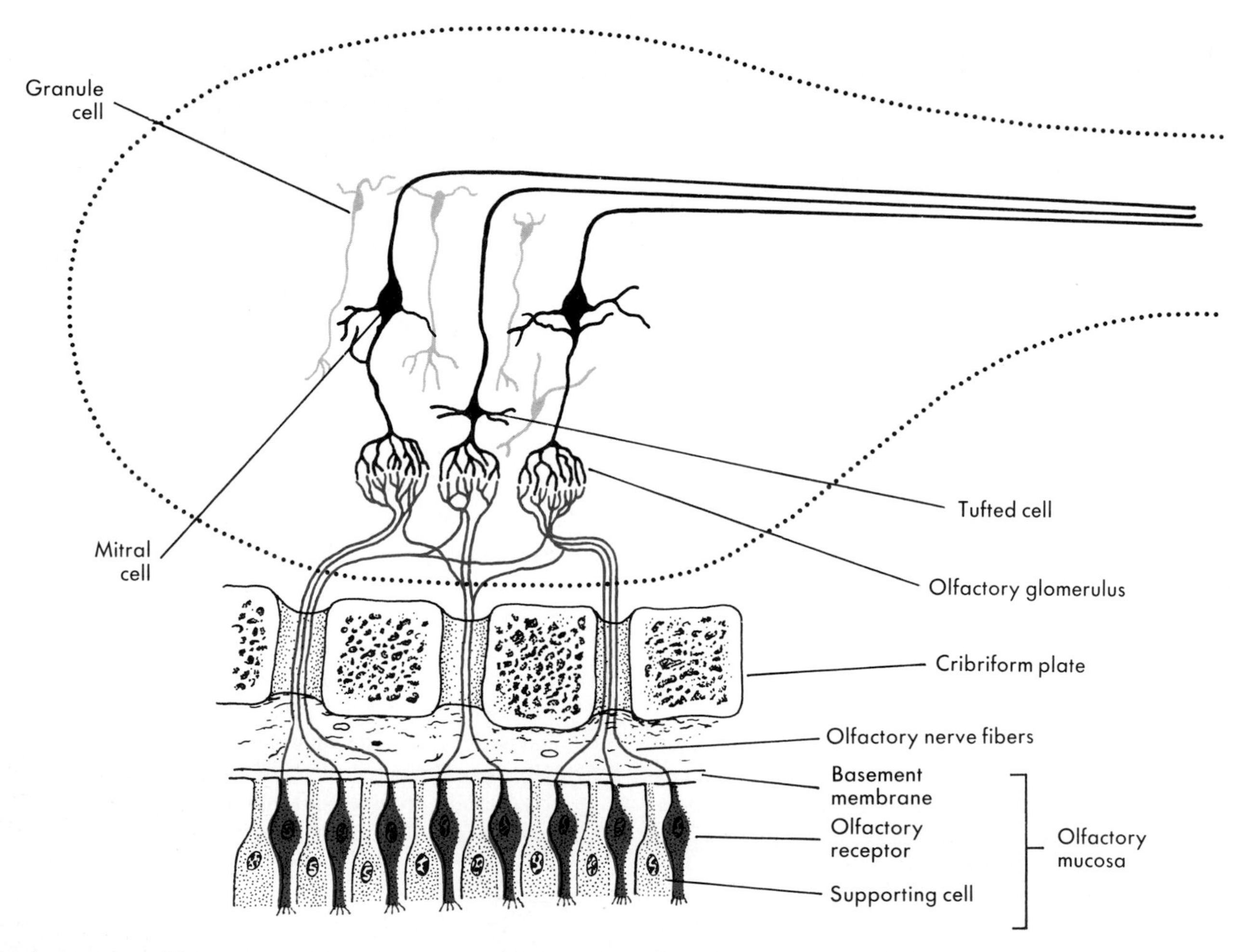

■ **Fig. 11-6** Drawing of a sagittal section through an olfactory bulb, showing the terminations of the olfactory chemoreceptor cells in the olfactory glomerulis and the intrinsic neurons of the olfactory bulb. The axons of the mitral and tufted cells are shown exiting in the olfactory tract to the right. (Redrawn from House EL, Pansky B: *A functional approach to neuroanatomy,* ed 2, New York, 1967, McGraw-Hill. Used with permission.)

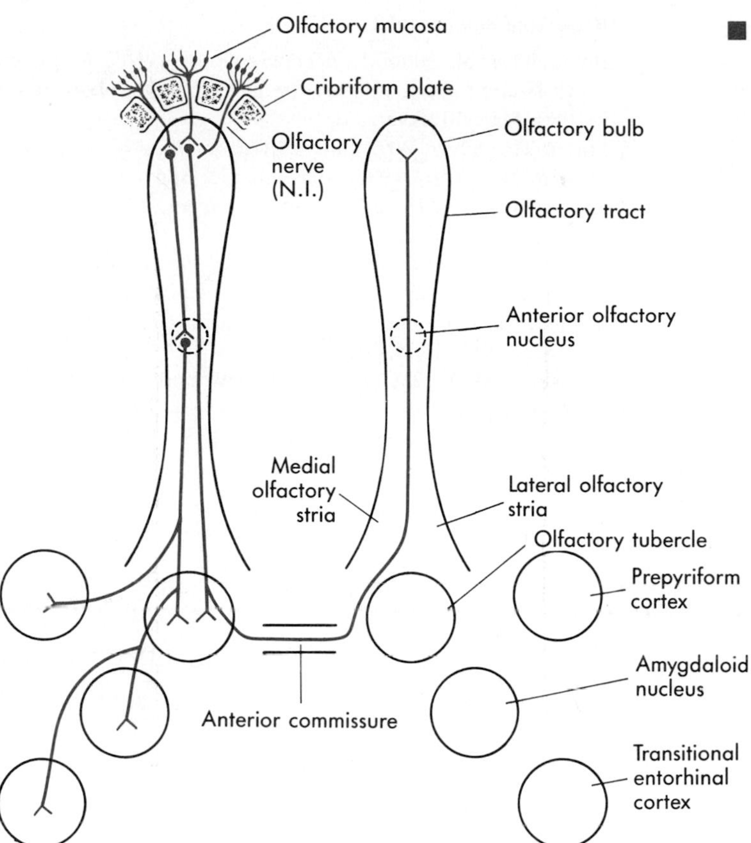

■ **Fig. 11-7** Major olfactory pathways.

eral and medial olfactory striae. Axons of the lateral olfactory stria synapse in the primary olfactory receiving area, which includes the prepyriform cortex (and, in animals, the piriform lobe). The medial olfactory stria includes projections to the amygdaloid nucleus, as well as part of the cortex of the basal forebrain (Fig. 11-7).

Note that the olfactory pathway is the only sensory system that does not have an obligatory synaptic relay in the thalamus. This may reflect the phylogenetic primitiveness of the olfactory system. However, olfactory information does reach the mediodorsal nucleus of the thalamus, and olfactory information is then transmitted to the prefrontal and orbitofrontal cortex.

■ *Summary*

1. Taste buds detect gustatory stimuli. A population code is used to signal the four elementary qualities of taste: salty, sweet, sour, and bitter.

2. Taste buds are located on several kinds of papillae on the tongue and in the pharynx and larynx. Taste buds contain chemoreceptor cells arranged around a taste pore; these cells are innervated by taste afferent fibers of cranial nerves VII, IX, and X.

3. The taste afferent fibers synapse in the nucleus of the solitary tract. Higher pathways differ in different species. The parabrachial nucleus is included in rodents, but not in primates. The thalamus gustatory nucleus is the small-celled part of the VPM nucleus, and there are taste-receiving areas in the SI cortex and the insula.

4. Odors are detected by olfactory chemoreceptor cells in the olfactory mucosa. At least six elementary odor qualities are encoded by the population of olfactory afferent nerves.

5. The olfactory chemoreceptor cells project to the olfactory bulb, where they synapse in olfactory glomeruli on the dendrites of mitral and tufted cells. Local inhibitory dendrodendritic circuits are formed by the granule cells. The mitral and tufted cells project to the olfactory tract.

■ *Bibliography*

Journal articles

Beckstead RM et al: The nucleus of the solitary tract in the monkey: projections to the thalamus and brainstem nuclei, *J Comp Neurol* 190:259, 1980.

Hamilton RB, Norgren R: Central projections of gustatory nerves in the rat, *J Comp Neurol* 222:560, 1984.

Macrides F, Schneider SP: Laminar organization of mitral and tufted cells in the main olfactory bulb of the adult hamster. *J Comp Neurol* 208:419, 1982.

Pfaffman C et al: Neural mechanisms and behavioral aspects of taste, *Ann Rev Psychol* 30:283, 1979.

Price JL, Slotnick BN: Dual olfactory representation in the rat thalamus: an anatomical and electrophysiological study, *J Comp Neurol* 205:63, 1983.

Roper SD: The cell biology of vertebrate taste receptors, *Ann Rev Neurosci* 12:329, 1989.

Smith DV et al: Gustatory neuron types in hamster brain stem, *J Neurophysiol* 50:522, 1983.

Yamamoto T et al: Gustatory responses of cortical neurons in rats. I. Response characteristics, *J Neurophysiol* 51:616, 1984.

Books and monographs

Brand JG et al, editors: *Chemical senses,* vol *1, Receptor events and transduction in taste and olfaction,* New York, 1989, Marcel Dekker.

Finger TE, Silver WL, editors: *Neurobiology of taste and smell,* New York, 1987, John Wiley & Sons.

Kandel ER et al: *Principles of neural science,* ed 3, Elsevier, 1991, New York.

Pfaff DW, editor: *Taste, olfaction, and the central nervous system,* New York, 1985, Rockefeller University Press.

Schmidt RF, Thews G, editors: *Human physiology,* Springer-Verlag, 1983, Heidelberg.

Shepherd GM: *Neurobiology,* ed 2, New York, 1988, Oxford University Press.

CHAPTER

12

Spinal Organization of Motor Function

Movements and posture depend on the correct balance of contraction of muscles operating around joints. These muscular contractions in turn depend on the amount and timing of the discharges of motoneurons to the appropriate muscles and the lack of discharges of motoneurons to inappropriate muscles. The motor system is designed to coordinate muscle actions through control of motoneuronal activity. Motor control is in part voluntary, but it occurs mainly by reflex action and by subconscious mechanisms.

Spinal reflexes are an important substrate for motor activity. Many subconscious actions depend largely on simple reflexes that are triggered by the activation of sensory receptors that excite interneurons and motoneurons in the spinal cord. Such excitation then triggers the contraction or relaxation of particular muscles. Spinal reflexes can be observed in **spinalized** individuals (i.e., below the level of a complete transection of the spinal cord). Therefore these reflexes do not depend on motor commands that originate in the brain and conveyed to the spinal cord by way of descending motor pathways. Furthermore, many of the motor acts that do originate from motor commands issued by the brain depend on spinal reflex circuitry for their implementation. Only a few of the descending pathways synapse directly on spinal cord motoneurons. Instead, most of the descending projections influence the activity of interneurons that are interposed in reflex circuits and thus alter ongoing spinal reflex activity.

This chapter describes a number of spinal cord reflex pathways. Motor pathways that descend from the brain will be discussed in Chapter 13. The motor control systems of the brain will be the topic of Chapter 14.

■ *Spinal Cord Transection*

Spinal cord reflexes can be investigated at levels of the spinal cord below a complete transection; thus descending motor commands are absent. However, there are important differences in the effects of transection of the spinal cord in human and animal subjects.

In humans an abrupt transection of the spinal cord results initially in a condition called **spinal shock,** which is characterized by a flaccid paralysis, areflexia, loss of autonomic function, and loss of all sensation below the level of the transection. In a flaccid paralysis, the joint offers no resistance to the passive movement when an examiner bends the joint. This results from the absence of muscle stretch reflexes, which normally cause muscle contractions that oppose changes in the position of the joint (see p. 203).

Spinal shock generally lasts 3 to 4 weeks, after which reflexes gradually return and then become hyperactive. At this time, an examiner finds that passive movements of a joint result in much greater resistance than in normal subjects. Voluntary movement and sensation never return, and the paralysis changes from a flaccid to a spastic type. Furthermore, pathological reflexes appear (e.g., the sign of Babinski; see Chapter 13), muscle tone increases, and bowel and bladder functions return, but in altered form. The reflexes that are hyperactive include the muscle stretch reflexes and the flexion reflexes. Hyperactive stretch reflexes are associated not only with an increased resistance to passive stretch, but also often with clonus (an alternating contraction of agonist and antagonist muscles about a joint, such as the ankle, after an initial quick passive flexion of the joint). Hyperactive flexion reflexes in response to a noxious stimulus applied to a foot may include not only flexion of one or both lower extremities but also urination, defecation, and sweating. The posture is often one of maintained flexion of the lower extremities.

In animals spinal transection produces similar changes, but the period of spinal shock is usually very brief. This allows experimental work to be done on spinal reflexes in the absence of descending controls.

Decerebration

Another experimental preparation that has been very useful for the study of reflexes is the decerebrate preparation. Surgical decerebration is done by transecting the midbrain, often at an intercollicular level. Decerebrate animals no longer have sensation, and their motor control system is profoundly altered. Some descending pathways, such as those that originate in the cerebral cortex, are interrupted, whereas others, such as those that originate in the brainstem, remain intact. In fact, activity in some descending pathways becomes hyperactive because of a change in the balance of excitatory and inhibitory control systems. The result is that some spinal reflexes, such as the flexion reflex, are suppressed, whereas others, such as the stretch reflex, are exaggerated, a condition called **decerebrate rigidity.** In fact, decerebrate animals maintain a posture that has been called exaggerated standing. Decerebrate preparations are thus ideal for the study of the stretch reflex and also the inverse myotatic reflex (see p. 205). Human patients with brainstem damage may also develop a decerebrate state with many of the same reflex features that are shown by animals. The prognosis is poor when the signs of decerebration appear.

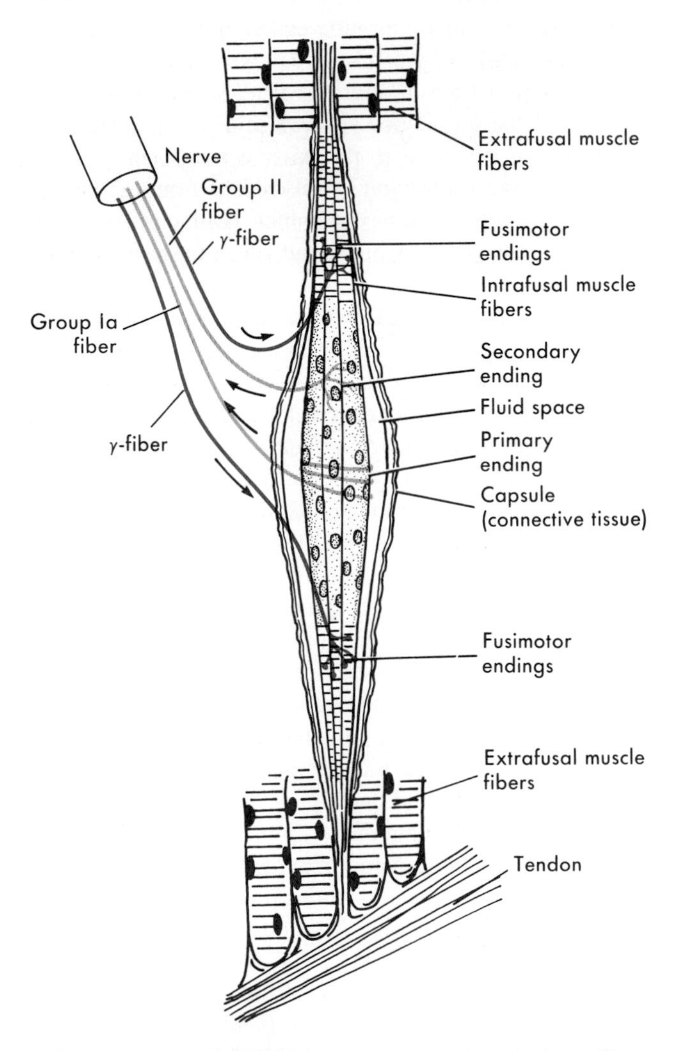

Fig. 12-1 Drawing of a muscle spindle. (Redrawn from Brodal A: *Neurological anatomy,* ed 3, New York, 1981 Oxford University Press.)

Sensory Receptors Responsible for Eliciting Spinal Reflexes

Several important spinal reflexes are elicited by muscle stretch receptors, which include muscle spindles and Golgi tendon organs. These reflexes are the **muscle stretch reflex** (or myotatic reflex) and the **inverse myotatic reflex.** These reflexes are very important in the maintenance of posture, and alteration in the activity in the circuits for these reflexes is an important means by which descending pathways produce movements. Furthermore, abnormalities in the stretch reflexes are prominent in disorders of the motor system, as already indicated for the syndrome associated with complete spinal cord transection.

The flexion reflex can be evoked by various sensory receptors, including nociceptors in the skin, muscles, joints, and viscera. The flexion component of locomotion is also influenced by various receptors, including relatively low threshold receptors in the skin, as well as the higher threshold receptors. The spectrum of different receptors that collectively can evoke a flexion reflex is often referred to as the **flexion reflex afferents.**

Only the reflexes just mentioned will be discussed in this chapter. Many other spinal reflexes, including both somatic (e.g., scratch reflex) and visceral (e.g., micturition) reflexes, will not be discussed here.

Muscle Stretch Receptors

Muscle stretch receptors are important both for spinal reflexes and for proprioception (see Chapter 8). Their structure and functional properties will be described in this chapter, because an understanding of their operation will be useful for the description of the mechanisms that underlie the spinal reflexes.

Muscle spindle. The muscle spindles are very complex. They are found in most skeletal muscles, but they are particularly concentrated in muscles that exert fine motor control (e.g., the small muscles of the hand). In large muscles they are most abundant in muscles that are rich in slow twitch (type I) muscle fibers.

A **muscle spindle** (or neuromuscular spindle) is a spindle-shaped organ composed of a bundle of modified muscle fibers innervated both by sensory and motor axons (Fig. 12-1). The muscle spindle is about 100 μm in diameter and up to 10 mm long. The innervated part of the muscle spindle is encased in a connective tissue capsule that contains fluid in a so-called lymph space. The muscle spindle lies freely in the space between the regular muscle fibers. The distal ends of the spindle are attached to the connective tissue within the muscle (endomysium). The arrangement of the muscle spindle with respect to the muscle is described as being in parallel, because the muscle spindle lies in parallel to the regular muscle fibers. This arrangement has important functional implications, which will be made clear below.

The modified muscle fibers within a muscle spindle are called **intrafusal muscle fibers,** to distinguish them from the regular or **extrafusal** muscle fibers. Individual intrafusal fibers are much narrower than extrafusal fibers and are too weak to contribute to muscle tension. The two main types of intrafusal muscle fibers have been named nuclear bag and nuclear chain fibers (Fig. 12-2). The names derive from the arrangement of nuclei in the two kinds of intrafusal fibers. **Nuclear bag fibers** are larger than nuclear chain fibers, and the nuclei are accumulated like a bag full of oranges in the central or equatorial region of the fiber. More than one nucleus may be seen in a cross-section through a nuclear bag fiber. In **nuclear chain fibers,** the nuclei are arranged in a row, and only one can be seen in a cross-section through the equatorial region. There are generally two nuclear bag fibers and about five or six nuclear chain fibers per muscle spindle. One of the nuclear bag fibers gives rise to dynamic and the other to static responses; the latter will not be distinguished from nuclear chain fibers in the discussion to follow.

Muscle spindles receive a complex innervation. The sensory supply includes a single **group Ia afferent** and a variable number (often one) of **group II afferent** fibers (see Fig. 12-2). Group Ia fibers belong to the largest diameter class of sensory nerve fibers and conduct at 72 to 120 m/sec; group II fibers are intermediate in size and conduct at 36 to 72 m/sec (see Table 7-2). A group Ia afferent fiber forms a **primary ending,** which consists of a spiral-shaped terminal made by branches of the group Ia fiber on each of the intrafusal muscle fibers. It is functionally important that the primary ending is on both nuclear bag and nuclear chain fibers. The group II afferent fiber forms a **secondary ending,** which is chiefly on the nuclear chain fibers, although a small spriglike ending may occasionally contact a nuclear bag fiber.

The motor supply to the muscle spindle consists of two types of γ-motor axons (see Fig. 12-2). **Dynamic γ-motor axons** end with plate endings on nuclear bag fibers. **Static γ-motor axons** end with trail endings on nuclear chain fibers. γ-motor axons are smaller in diameter than the α-motor axons to extrafusal muscle. Hence they conduct more slowly, at 12 to 48 m/s (see Table 7-3). α-motor axons conduct at 72 to 120 m/sec, which is the same velocity range as the group I afferent fibers.

Operation of the muscle spindle. As the name stretch receptor implies, muscle spindles respond to muscle stretch. Fig. 12-3 shows the changes in the activity of an afferent fiber from a muscle spindle when the muscle spindle is shortened (unloaded) by contraction of the extrafusal muscle fibers and when it is lengthened by stretching the muscle. Contraction of the extrafusal muscle fibers results in shortening of the muscle spindle because of the parallel arrangement of the muscle spindle described earlier. The transduction mechanism by which the muscle spindle afferent fibers are activated involves mechanical stretch of the afferent terminals on the intrafusal fibers. When the muscle spindle is unloaded, the spacing between the coils of the afferent terminals is reduced, and the discharge rate of the afferent fiber decreases. Conversely, when the whole muscle is stretched, the muscle spindle is also stretched (because its ends are attached to the connec-

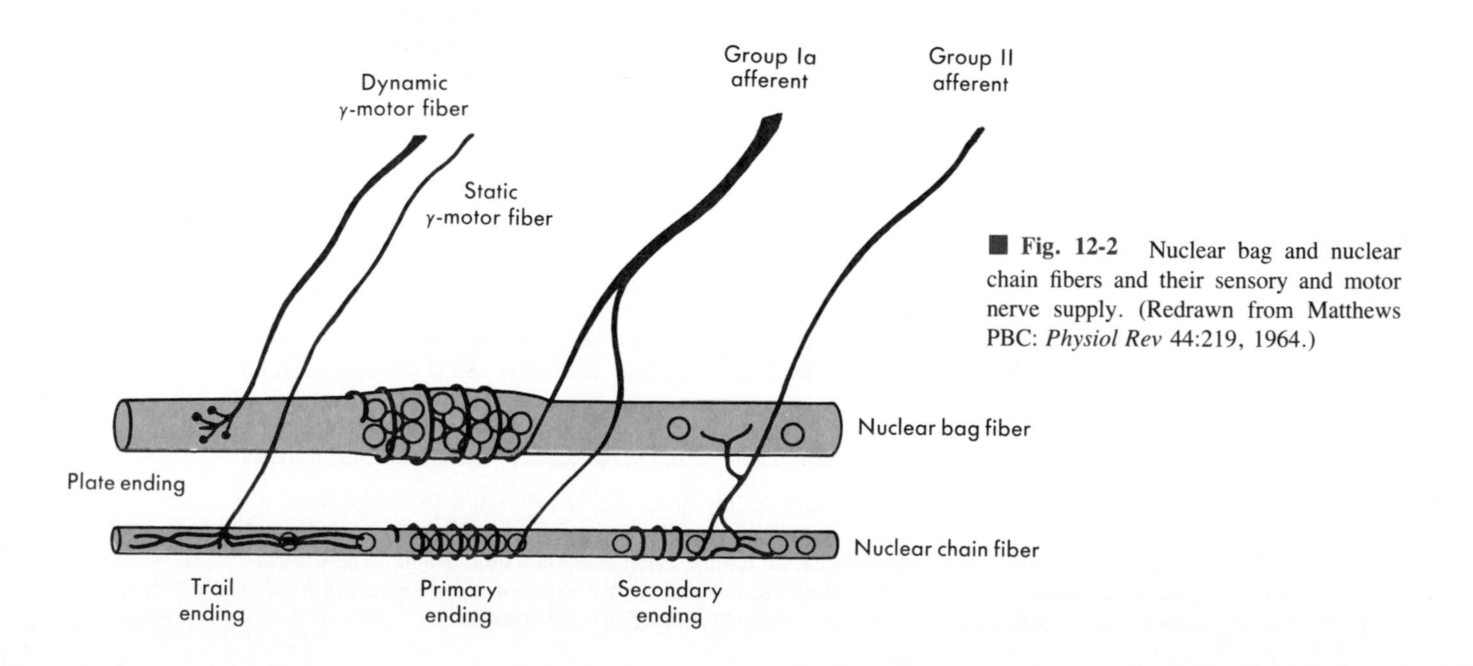

■ **Fig. 12-2** Nuclear bag and nuclear chain fibers and their sensory and motor nerve supply. (Redrawn from Matthews PBC: *Physiol Rev* 44:219, 1964.)

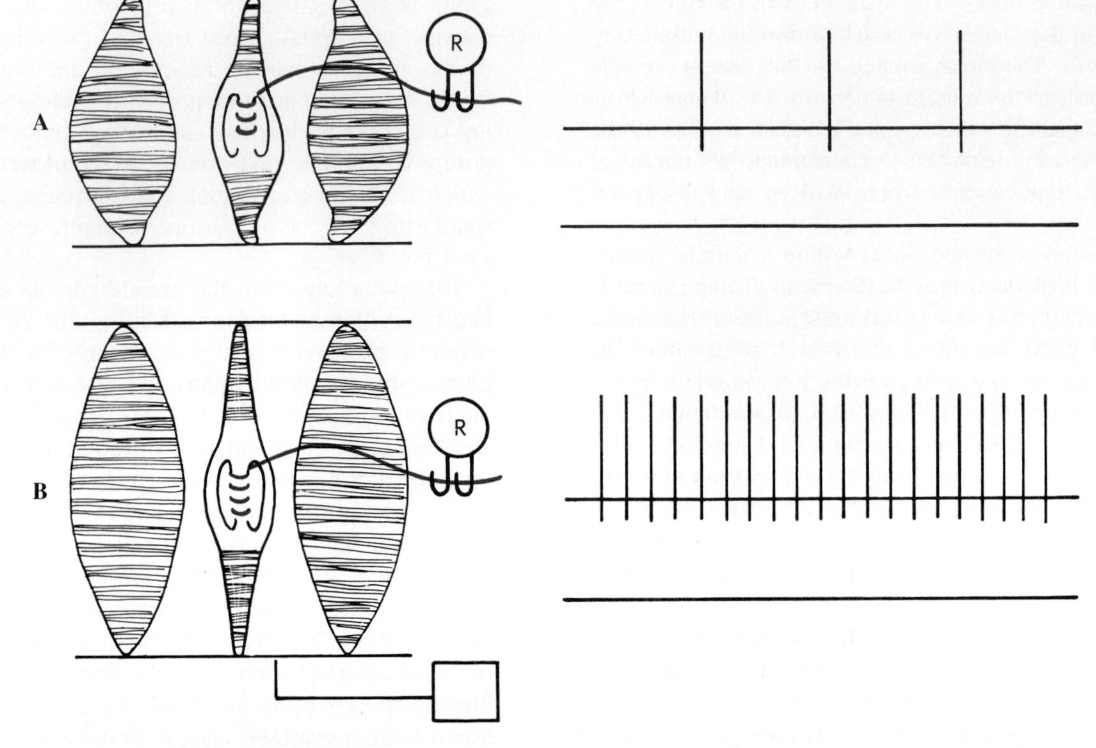

■ **Fig. 12-3** Changes in the discharge rate of a muscle spindle afferent fiber during muscle shortening (due to a contraction of the muscle) and during muscle lengthening (due to muscle stretch). During the contraction, **A,** the muscle spindle is unloaded as a result of its parallel arrangement within the muscle. In **B,** the muscle spindle is stretched along with the muscle. (Redrawn from Eyzaguirre C, Fidone SJ: *Physiology of the nervous system,* ed 2, Chicago, 1975, Year Book Medical Publishers; modified from Ruch TC, Patton HD: *Physiology and biophysics,* ed 19, Philadelphia, 1965, WB Saunders; and Hunt CC, Kuffler SW: *J Physiol* 113:298, 1951.)

tive tissue framework of the muscle), and the elongation of the receptor terminals increases the discharge rate.

The primary and secondary endings respond differently to stretch. The primary ending is sensitive both to the amount of stretch and to its rate, whereas the secondary ending responds chiefly to the amount of stretch (Fig. 12-4). The firing rates of both the group Ia fiber and the group II fiber are in proportion to the length of the muscle spindle, as shown for linear-stretch at the left in Fig. 12-4 and when the stretch is released, as shown at the right of the illustration. This type of response is called the **static response** of muscle spindle afferent fibers. The difference in the behavior of the

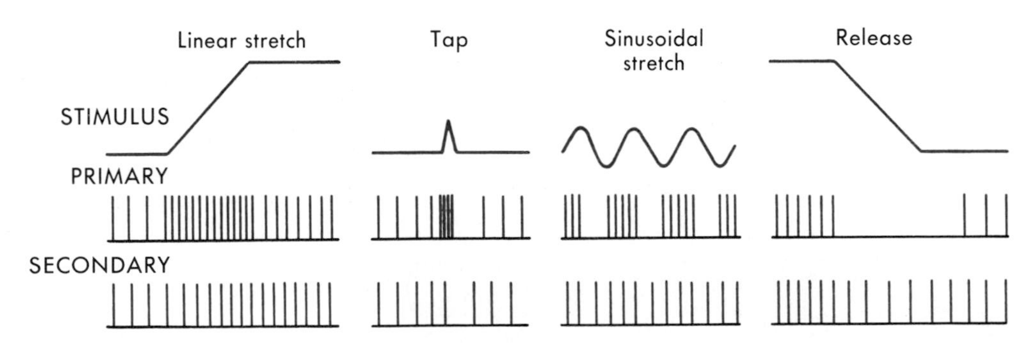

■ **Fig. 12-4** The responses of a primary ending and of a secondary ending to various types of changes in muscle length are shown to illustrate the difference in dynamic and static responsiveness of these endings. The waveforms at the top are the changes in muscle length. The vertical lines in the middle and bottom parts of the figure show the discharges of a primary and of a secondary ending. (Redrawn from Matthews PBC: *Physiol Rev* 44:219, 1964.)

two endings is that the activity of the primary ending overshoots during muscle stretch, and it stops firing when the stretch is first released. These are called **dynamic responses** of the group Ia afferent fiber. The responses in the center of Fig. 12-4 show further examples of dynamic responses of the primary ending. Tapping the muscle or its tendon and sinusoidal stretch are much more effective in causing discharges of the primary than the secondary ending.

These responses show that the primary endings signal both the length and the rate of change in length of the muscle, whereas the secondary endings signal only the length of the muscle. The mechanism for these differences appears to depend largely on the mechanical differences between the nuclear bag and nuclear chain fibers. Nuclear bag fibers lack contractile protein in the equatorial region because of the accumulation of nuclei in this region. Therefore the nuclear bag fibers are readily stretched in their mid-region. However, immediately after they are stretched, the equatorial region of nuclear bag fibers tends to return toward its original length as the polar regions lengthen. This phenomenon is called **creep** and is caused by the viscoelastic properties of these intrafusal fibers. The result is an overshoot in activity followed by a reduction in activity toward a new static level of firing.

The length of the nuclear chain fibers more closely parallels that of the extrafusal muscle fibers because the nuclear chain fibers contain contractile proteins even in their equatorial regions. Hence they have more uniform viscoelastic properties throughout their length. Therefore they do not display creep, and the secondary ending has only a static response.

Up to this point, the behavior of muscle spindles in which the γ-motoneurons were not active has been described. However, the efferent innervation of muscle spindles is extremely important, because it determines the sensitivity of muscle spindles to stretch. For example, in Fig. 12-5, *A*, the activity of a muscle spindle afferent is shown during a steady stretch. As already discussed, when the extrafusal part of the muscle contracts (Fig. 12-5, *B*), the muscle spindle is unloaded, and the muscle spindle afferent may stop discharging. However, this effect of muscle unloading can be counteracted if γ-motoneurons are stimulated. This causes the muscle spindle to shorten along with the extrafusal muscle fibers (Fig. 12-5, *C*). Actually, the γ-motor axons cause both polar regions of the muscle spindle to

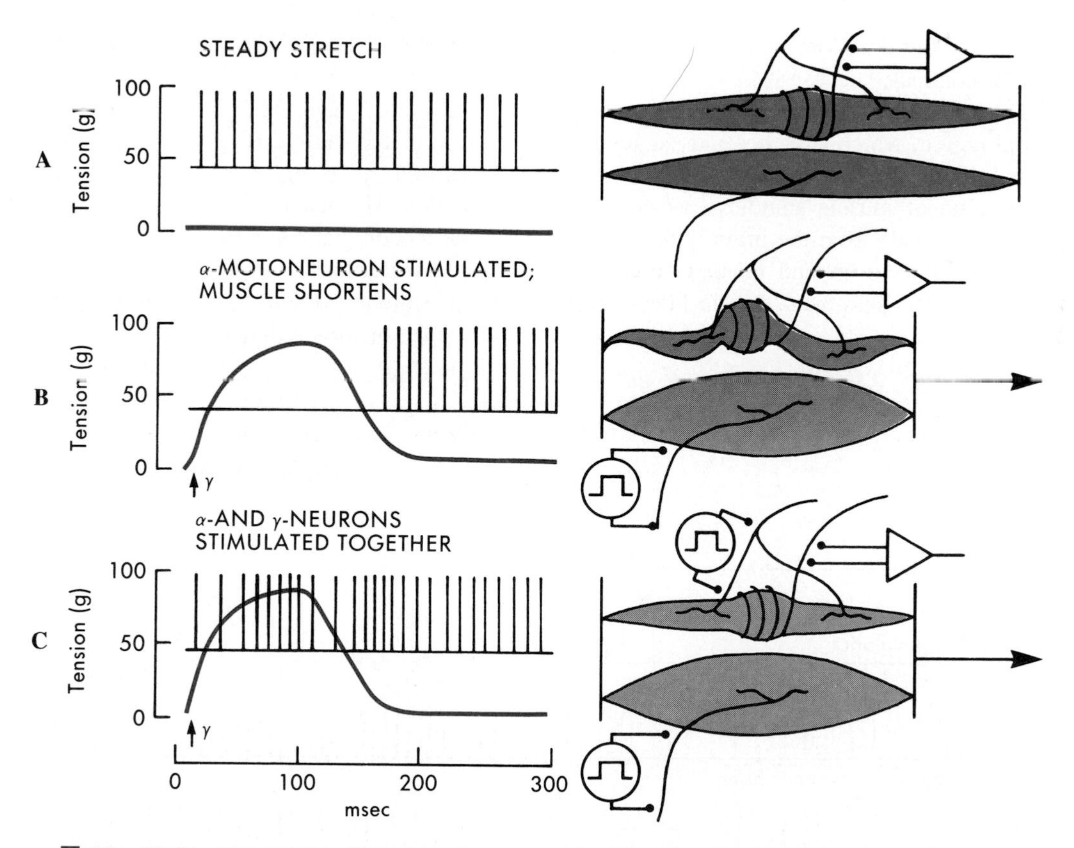

Fig. 12-5 The activity of gamma motor axons can counteract the effects of unloading on the discharges of a muscle spindle afferent. **A,** The activity of a muscle spindle afferent is shown during steady stretch. **B,** The afferent stops firing when the extrafusal muscle fibers contract, due to unloading of the muscle spindle. **C,** Activation of aγ-motoneuron causes shortening of the muscle spindle, counteracting the effects of unloading. (Redrawn from Kuffler SW, Nicholls JG: *From neuron to brain,* Sunderland, Mass, 1976, Sinauer Associates.)

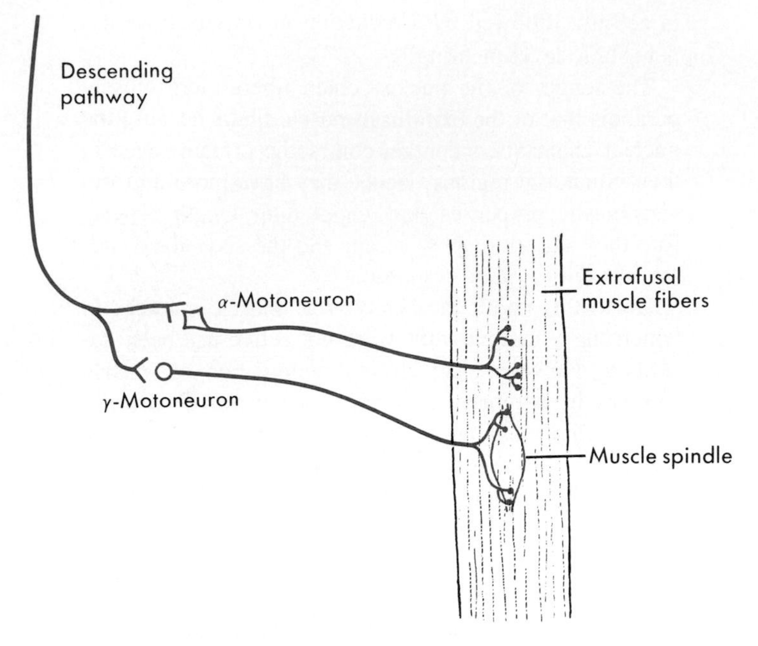

■ **Fig. 12-6** Coactivation of an α- and a γ-motoneuron by a descending motor pathway.

contract, but the equatorial region, where the nuclei are located, does not contract, because it has little contractile protein. This arrangement results in an elongation of the equatorial region, which stretches and excites the afferent terminals. This mechanism is very important in the normal operation of muscle spindles, because descending motor commands from the brain typically coactivate α- and γ-motoneurons and thus cause cocontraction of extrafusal and intrafusal muscle fibers (Fig. 12-6).

As mentioned earlier, there are two types of γ-motoneurons called dynamic and static γ-motoneurons (see Fig. 12-2). Dynamic γ-motor axons end on nuclear bag fibers and static γ-motor axons synapse on nuclear chain fibers. When a dynamic γ-motoneuron is activated, the dynamic response of the group Ia afferent fiber is enhanced (Fig. 12-7, *D*). When a static γ-motoneuron discharges, the static responses of both group Ia and II afferent fibers are increased (Fig. 12-7, *C*); at the same time, the dynamic response may be reduced. Different descending pathways can preferentially influence dynamic or static γ-motoneurons and thereby alter the nature of reflex activity in the spinal cord.

Golgi tendon organ. The other type of stretch receptor found in skeletal muscle is the Golgi tendon organ (Fig. 12-8). A **Golgi tendon organ** is formed by the terminals of a group Ib afferent fiber. The diameter of a Golgi tendon organ is about 100 μm and its length about 1 mm. The group Ib fiber has a large diameter and conducts in the same velocity range as the group Ia fiber (see Table 7-2). The terminals are wrapped about bundles of collagen fibers in the tendon of a muscle (or in tendonous inscriptions within the muscle). The sensory ending is arranged in series with the muscle, in contrast to the parallel arrangement of the muscle spindle.

Operation of Golgi tendon organs. Because of their arrangement in series with muscle, Golgi tendon organs can be activated either by muscle stretch or by contraction of the muscle (Fig. 12-9). However, muscle contraction is a more effective stimulus than muscle stretch. The actual stimulus is the force that develops in the tendon that contains the Golgi tendon organ. Therefore Golgi tendon organs signal force, unlike the muscle spindle, which signals muscle length and rate of change of muscle length.

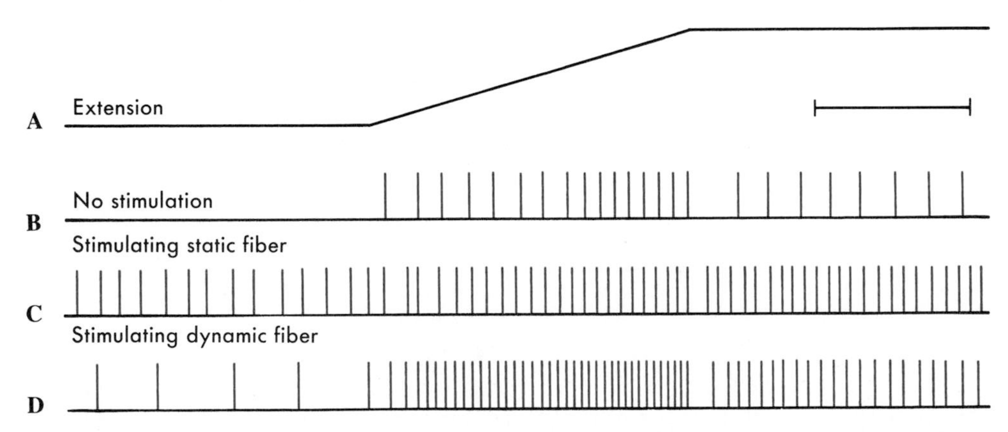

■ **Fig. 12-7** Effects of static and dynamic γ-motoneurons on the responses of a primary ending to muscle stretch. The upper trace is the time course of stretch, **A. B** shows the discharge of the group Ia fiber in the absence of γ-motoneuron activity. In **C,** a static γ-motor axon was stimulated, and in **D** a dynamic γ-motor axon was stimulated. (Redrawn from Crowe A, Matthews PBC: *J Physiol* 174:109, 1964.)

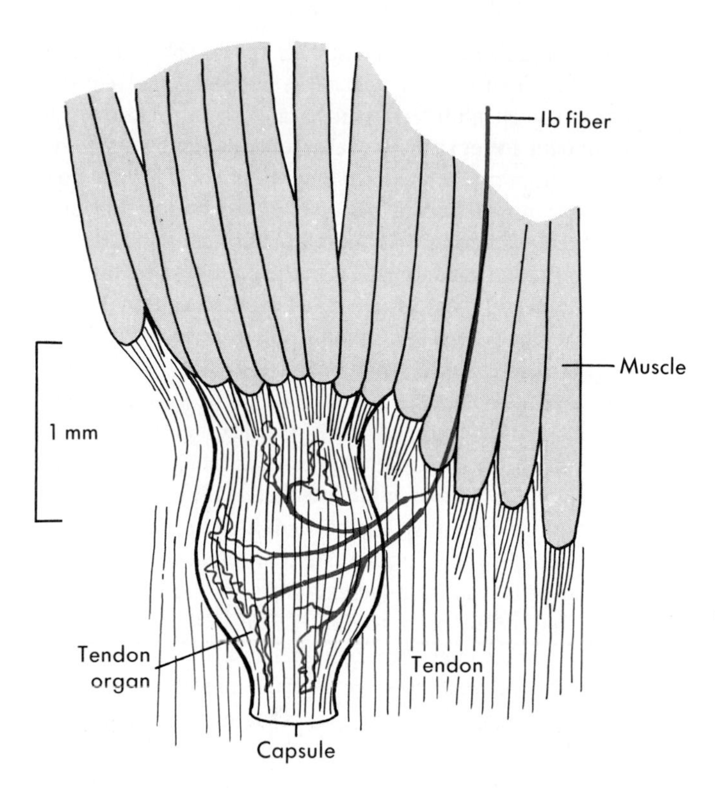

■ **Fig. 12-8** Drawing of a Golgi tendon organ. (Redrawn from Barker D: *Muscle receptors,* Hong Kong, 1962, Hong Kong University Press.)

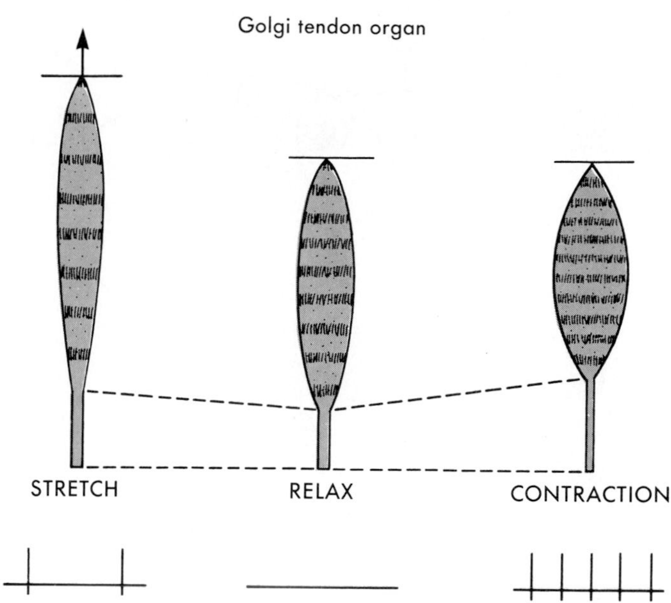

■ **Fig. 12-9** Activation of a Golgi tendon organ by muscle stretch *(left)* or by contraction of the muscle *(right).* (Redrawn with permission from Eyzaguirre C, Fidone SJ: *Physiology of the nervous system,* ed 2, Chicago, 1975, Year Book Medical Publishers.)

■ *Spinal Reflexes*

A reflex is a simple, relatively stereotyped motor response to a specific type of stimulus. The **reflex arc** is the neuronal circuit responsible for the reflex. Typically, a reflex arc includes a set of sensory receptors of a particular kind that, when excited, elicit the reflex by exciting a set of interneurons and motoneurons (Fig. 12-10). Other interneurons and motoneurons may be inhibited, so that an appropriate pattern of muscle contractions occurs about one or more joints. In spinal reflexes, all of the components of the reflex circuit are present at the spinal cord level. There are also brainstem reflexes, in which the reflex arc is located in the brainstem, and supraspinal or long reflexes, in which sensory receptors activate long ascending pathways to the brain to elicit stereotyped motor commands transmitted by descending pathways to the spinal cord.

■ *Myotatic or Stretch Reflex*

The stretch reflex is a key reflex that helps maintain posture. In addition, changes in this reflex are involved in actions commanded by the brain, and pathological alterations are important signs of neurological disease. This reflex actually has two forms: the phasic stretch reflex and the tonic stretch reflex. The **phasic stretch reflex** is elicited by primary endings of muscle spindles, whereas the **tonic stretch reflex** depends on both primary and secondary endings. These reflexes are best studied in decerebrate animals, although they are also quite active in spinalized animals that have recovered from spinal shock.

The reflex arc responsible for the phasic stretch reflex

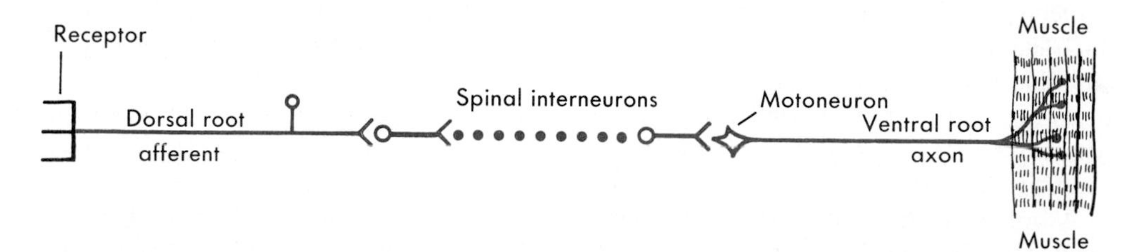

■ **Fig. 12-10** Diagram of a reflex arc, including a sensory receptor, spinal interneurons, a motoneuron, and muscle.

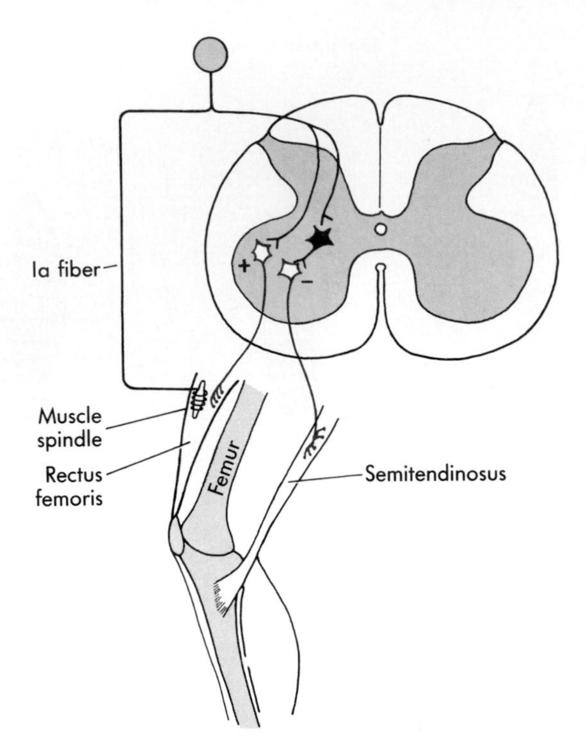

■ **Fig. 12-11** Reflex arc of the stretch reflex. The interneuron shown in black is a group Ia inhibitory interneuron.

is shown in Fig. 12-11. A group Ia afferent fiber from a muscle spindle in the rectus femoris muscle branches as it enters the spinal cord. One branch ascends in the dorsal column; this branch is destined for Clarke's column, the nucleus of origin of the dorsal spinocerebellar tract. The pathway through Clarke's column provides sensory information to the cerebellum for its function in motor regulation and to the cerebral cortex (through relays in nucleus z and the thalamus) for proprioception.

The group Ia fiber also sends branches into the spinal cord gray matter. Some of these synapse directly or monosynaptically on α-motoneurons that supply the rectus femoris muscle (and its synergists, such as the vastus intermedius muscle), which extend the leg at the knee. The group Ia fiber and others like it produce a monosynaptic excitation of the α-motoneuron. If the excitation is powerful enough, the motor neuron discharges and causes a contraction of the muscle.

Other collaterals of the group Ia fiber end on group Ia inhibitory interneurons, such as the one shown in black in Fig. 12-11. These inhibitory interneurons end on α-motoneurons that innervate the hamstring muscles, including the semitendinosus muscle, which are antagonists and flex the knee. The activity of the Ia inhibitory interneurons inhibits the motor neurons to the antagonist muscles. A volley in the group Ia afferent fibers from muscle spindles in the quadriceps muscle evokes a quick contraction of this muscle and a concomitant relaxation of the hamstring muscles.

The organization of the stretch reflex arc guarantees that one set of α-motoneurons is activated and the opposing set is inhibited. This arrangement is known as **reciprocal innervation.** Many reflexes have such a reciprocal innervation. However, this is not the only possible organization of a motor control system. In some instances, a motor command will cause cocontraction of synergists and antagonists. This happens, for example, when a person makes a fist. The muscles that extend and flex the wrist contract and allow the wrist to resist motion.

In a clinical setting, a volley in group Ia afferent fibers is elicited by a tap with a reflex hammer on the tendon of a muscle, such as the quadriceps. The result is normally a brief contraction that is quickly damped. When the excitability of the α-motoneurons is altered pathologically, the phasic stretch reflex may be depressed or it may be hyperexcitable. In the past, such a reflex has often been termed a "deep tendon reflex." This is a misnomer, however, because the receptors responsible for the reflex are in the muscle, not the tendon. The tendon only provides for a quick stretch of the muscle.

The other type of stretch reflex is the tonic stretch reflex. This is elicited by passively bending a joint. The reflex circuit is the same as that illustrated in Fig. 12-11, except that the receptors involved include both group Ia and group II afferent fibers from muscle spindles. Many group II fibers make monosynaptic excitatory connections with α-motoneurons. Therefore the tonic stretch reflex is largely a monosynaptic reflex, like the phasic stretch reflex. The tonic stretch reflex contributes to muscle tone, which is judged by the resistance that a joint offers to bending. However, its importance lies in its contribution to posture. When an individual stands, the joints of the leg must maintain a particular position to avoid falling. The tonic stretch reflex helps maintain posture, because a slight extension or flexion will elicit a stretch reflex in the muscles required to oppose the movement. For example, if the knee of a soldier standing at attention begins to flex because of fatigue, the quadriceps muscle will be stretched, a tonic stretch reflex will be elicited, and the quadriceps will contract more, thereby opposing the flexion and restoring the posture.

γ-motoneurons help set the sensitivity of the stretch reflexes. Muscle spindle afferent fibers have no direct influence on γ-motoneurons. These are affected more by the flexion reflex pathways at a spinal cord level and by descending commands. In many clinical disorders of motor control, the activity of γ-motoneurons is inappropriate because of a change, for instance, in the activity of descending pathways. As already mentioned, spinal cord transection results initially in spinal shock, with a loss of the stretch reflexes. Spinal shock may be caused in part by the loss of descending excitation of γ-motoneurons. Conversely, increased activity in γ-motoneurons may underlie **spasticity,** in which phasic stretch reflexes are hyperactive, and **hypertonia,** in which tonic stretch reflexes are hyperactive.

Clinical testing with a reflex hammer is a common way for a physician to assess the level of excitability of spinal motor neurons. However, another approach is to stimulate the axons of group Ia fibers electrically and to examine changes in the reflex responses to a synchronous volley in muscle spindle afferent fibers. The reflex can be monitored objectively by electromyography. The monosynaptic reflex elicited by stimulating the tibial nerve at the popliteal fossa and recording from the triceps surae muscles is called an **H reflex** (for Hoffmann).

Inverse Myotatic Reflex

Activation of Golgi tendon organs has a reflex effect that seems to oppose the stretch reflex (it actually complements the stretch reflex, as discussed in the next paragraph). This is called the **inverse myotatic reflex,** and its reflex arc is shown in Fig. 12-12. In this case, the receptor organ is a Golgi tendon organ located in the rectus femoris muscle. The afferent fiber branches as it enters the spinal cord. One branch ascends to Clarke's column. Like the muscle spindle afferent fibers that follow this same route, it contributes information used by the cerebellum and cerebral cortex for motor control and for proprioception. Other collaterals end on interneurons in the spinal cord. There are no monosynaptic connections to α-motoneurons. The Golgi tendon organ pathway involves inhibitory interneurons that inhibit α-motoneurons that supply the rectus femoris muscle and excitatory interneurons that activate α-motoneurons to the antagonistic hamstring muscles. Another pathway through commissural interneurons affects the activity of contralateral motor neurons.

The organization of the inverse myotatic reflex is contrary to that of the stretch reflexes, which explains the name given to this reflex. However, the function of this reflex is actually complementary to that of the stretch reflex. The Golgi tendon organs monitor force in the tendon that they supply. During maintained posture, such as standing at attention, if the rectus femoris muscle begins to fatigue, the force in the patellar tendon will decline. This will reduce the activity of Golgi tendon organs in this tendon. Because these receptors normally inhibit the α-motoneurons to the rectus femoris muscle, a reduced discharge in the Golgi tendon organs will enhance the excitability of the α-motoneurons and increase the force. Thus a coordinated reflex change will occur, involving both muscle spindle and Golgi tendon organ afferent fibers, that causes a greater contraction of the rectus femoris muscle and maintenance of the posture.

When reflexes are hyperactive it may be possible to

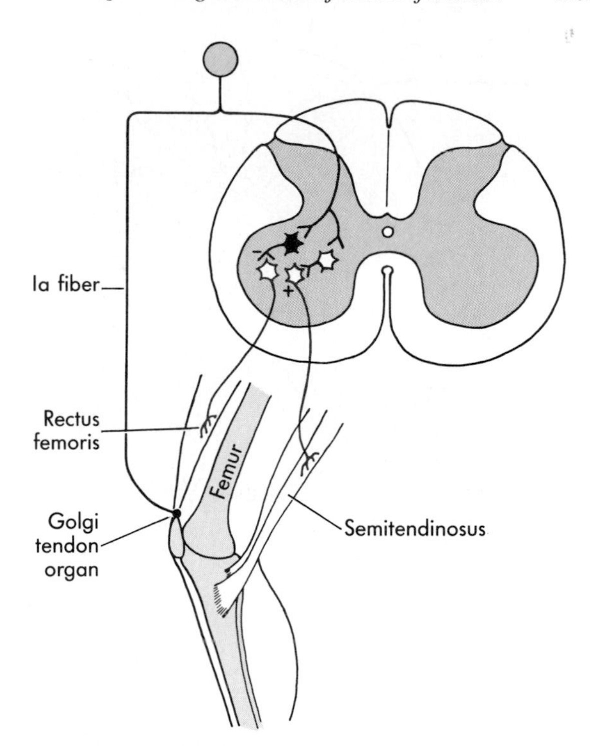

■ **Fig. 12-12** Reflex arc for the inverse myotatic reflex. The interneurons include both excitatory (clear) and inhibitory (black) interneurons.

demonstrate a clasp-knife reflex. When a joint is passively bent, resistance to the passive movement increases initially. However, as the bending is continued, the resistance suddenly decreases and the joint movement is readily completed. This change is attributed to reflex inhibition. The clasp-knife reflex was once attributed to the activation of Golgi tendon organs, because these were initially thought to have a high threshold to muscle stretch. However, it is now thought that the clasp-knife reflex is caused by activation of high threshold muscle receptors that supply the fascia around the muscle.

Flexion Reflexes

The afferent limb of flexion reflexes is furnished by a variety of sensory receptors, which collectively give rise to the **flexion reflex afferents** (FRA). In the flexion reflexes, afferent volleys (a) cause excitatory interneurons to activate α-motoneurons that supply flexor muscles in the ipsilateral limb, and (b) they cause inhibitory interneurons to prevent the activation of α-motoneurons that supply the antagonistic extensor muscles (Fig. 12-13). This pattern of activity causes one or more joints to flex. In addition, commissural interneurons evoke the opposite pattern of activity in the contralateral side of the spinal cord. This opposite pattern results in extension, the **crossed extension reflex.** The contralateral effect helps the subject maintain balance.

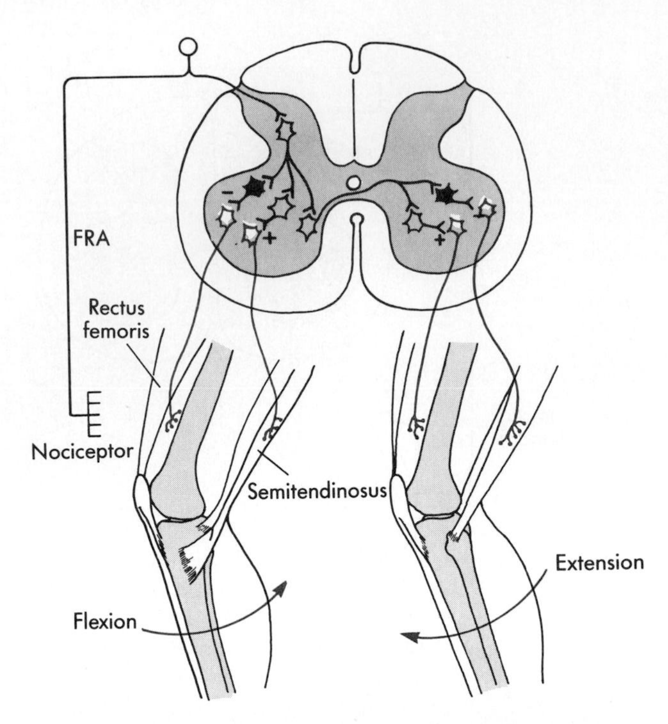

■ **Fig. 12-13** The reflex arc of the flexor reflex. Black interneurons are inhibitory and clear ones are excitatory.

There are several different types of flexion reflexes, although they produce similar patterns of muscle contraction. Locomotion has a flexion phase that involves a pattern of muscle contraction that can be regarded as a flexion reflex. This is controlled predominantly by a neural circuit, called the locomotor pattern generator, in the spinal cord. However, afferent input can alter the locomotor pattern so that it is suited to moment-by-moment changes in the terrain.

The most powerful flexion reflex is the **flexor withdrawal reflex.** This reflex takes precedence over other reflexes, including those associated with locomotion, presumably because flexor withdrawal protects the limb from further damage. This can readily be observed in a dog with a hurt paw that is held away from the ground while the animal walks. Nociceptors form the afferent limb of this reflex. These nociceptors include cutaneous, muscle, joint, and visceral nociceptors. Other, lower threshold FRA may also contribute.

In the flexor withdrawal reflex, a strong noxious stimulus results in a withdrawal of the limb from the stimulus. In Fig. 12-13, the neural circuit of the flexion reflex is shown for neurons that affect only the knee joint. However, there is actually considerable divergence of the primary afferent and interneuronal pathways in the flexion reflex (Fig. 12-14); in fact, all of the major joints of a limb (e.g., hip, knee, ankle) may be involved in a strong flexor withdrawal reflex. However, the details of the flexor withdrawal reflex vary, depending on the nature and location of the stimulus. Fig. 12-15 shows differences in the amount of flexion at the hip, knee, and ankle that occurs when different nerves

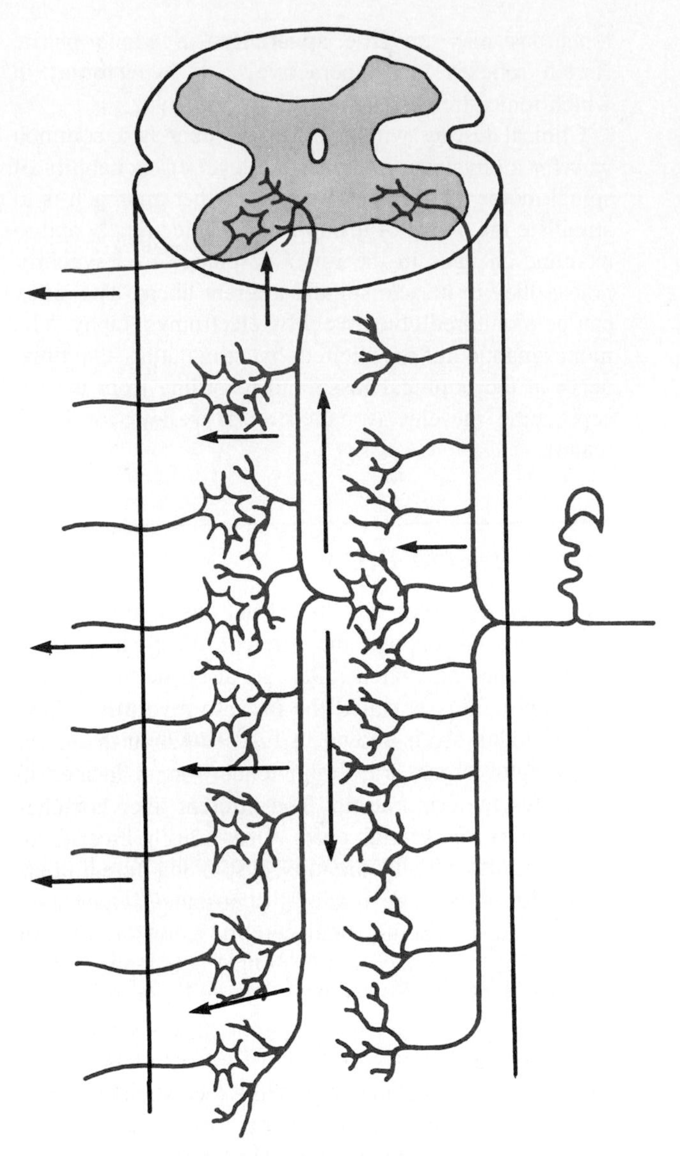

■ **Fig. 12-14** Drawing showing the divergence of the FRA pathways in the spinal cord. (Redrawn with permission from Eyzaguirre C, Fidone SJ: *Physiology of the nervous system,* ed 2, Chicago, 1975, Year Book Medical Publishers. Slightly modified from Cajal SR: *Histologie du systeme nerveux,* Paris, 1909, Maloine.)

of the hindlimb are stimulated electrically. This variability of the flexion reflex is called the **local sign.** Flexor withdrawal reflexes also occur in areas other than the limbs; e.g., visceral disease may cause contractions of the muscles in the chest wall or abdomen, and thereby decrease the mobility of the trunk.

■ *Comparison of the Stretch and Flexion Reflexes*

The flexion reflexes have a number of properties that differ strikingly from those of the stretch reflex. These are listed in Table 12-1.

Table 12-1 Comparison of the stretch and flexion reflexes

	Stretch reflex	*Flexion reflex*
Afferent limb	Group Ia (and II) muscle spindle afferents	FRA
Latency	Short (monosynaptic)	Long (polysynaptic)
Divergence	Some	Widespread
Target muscle	Same and synergists of same side	Flexor muscles on same side; extensor muscles on opposite side
Reciprocal innervation	Yes	Yes (double)
Linearity	Linear	Nonlinear
Duration	Same as stimulus	May persist because of after-discharges
Specificity	Specific to set of muscles	Less specific, involves many muscles

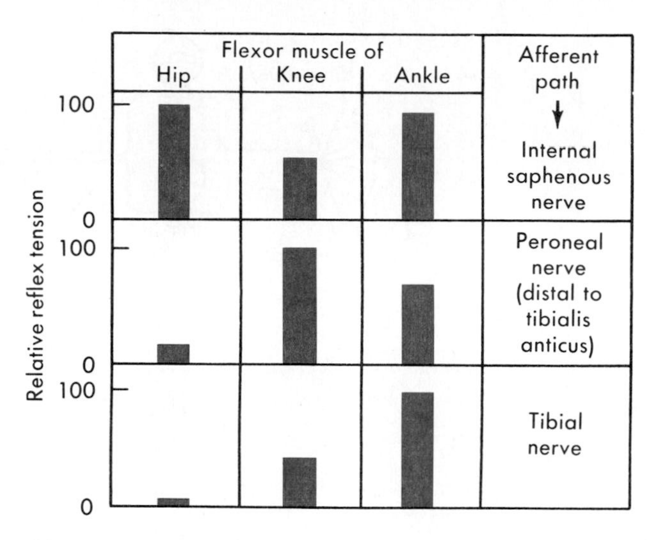

Fig. 12-15 Variations in the amount of flexion at the hip, knee, and ankle produced by electrical stimulation of nerves innervating different parts of the hindlimb. These variations are called local sign. (Redrawn from Patton HD: *Reflex regulation of movement and posture*. In Ruch T, Patton HD, editors: *Physiology and biophysics,* vol IV, Philadelphia, 1982, WB Saunders.)

The stretch reflex is activated by stimulation of group Ia (and II) muscle spindle afferent fibers. It has a short latency because the excitation is monosynaptic. The afferent pathway shows some divergence because it affects all the α-motoneurons that supply the muscle, plus some of those that supply synergistic muscles. In addition, the pathway activates inhibitory interneurons to the antagonistic α-motoneurons, an example of reciprocal innervation. The stretch reflex terminates when the afferent volley ceases, and it exerts a graded, specific, and discrete control over the muscles that operate across a joint, such as the knee or the ankle.

By contrast, the flexion reflex can be evoked by a diversity of receptor types supplied by the FRA. The latency is long, because the reflex arc is polysynaptic. There is substantial divergence in the reflex pathway, which may involve interneurons that influence α-motoneurons that supply muscles at all of the joints of a limb. In addition, the reflex can activate extensor motoneurons of the opposite limb. Thus the flexion reflex involves double reciprocal innervation. The reflex is nonlinear. Weak stimuli have little or no effect, but when stimuli of a certain level of intensity is reached, a powerful flexor withdrawal reflex may be elicited, and it dominates other reflexes. The flexion may persist long after the stimulus ends, presumably because of after-discharges of interneurons in the reflex arc.

Principles of Spinal Organization

As discussed in relation to the stretch and especially the flexion reflex, divergence is an important aspect of reflex pathways (see Fig. 12-14). Convergence is another important organizational feature of reflex arcs. **Convergence** refers to the termination of several neurons on another neuron. For example, all of the group Ia afferent fibers from the muscle spindles of a particular hindlimb muscle have convergent monosynaptic terminals on a given α-motoneuron to that muscle. This convergent input accounts for the phenomenon of **spatial facilitation** in the stretch reflex. The flexion reflex varies in its details, depending on the particular afferent input that triggers it (see Fig. 12-15). This variability is called a local sign. The function of a local sign in the flexion reflex is to allow the reflex to withdraw the limb from a noxious stimulus. Slight variations in the reflex are required to do this, depending on the exact location of the stimulus.

An example of spatial facilitation is shown in Fig. 12-16. A monosynaptic reflex is elicited by electrical stimulation of the group Ia fibers in each of two branches of a muscle nerve (Fig. 12-16, *A*). Instead of observing muscle contractions, the reflex is recorded from the appropriate ventral root as the discharges of α-motor axons. When muscle nerve branch A is stimulated, a small compound action potential is recorded as reflex A. Similarly, when muscle nerve branch B is stimulated, reflex B is recorded. These reflex discharges have a low electrical threshold because the group Ia fibers in the muscle nerve are large axons. Also, the latency of the reflex discharge is short, because the reflex pathway is monosynaptic and the conduction velocities of the afferent and motor axons are high. The drawing of the motor neurons contained within the motor nucleus in Fig. 12-16, *B* shows the set of α-motoneurons that are activated when each muscle nerve branch is stimulated separately (two smaller black areas that represent two motoneurons each). When the two nerves are

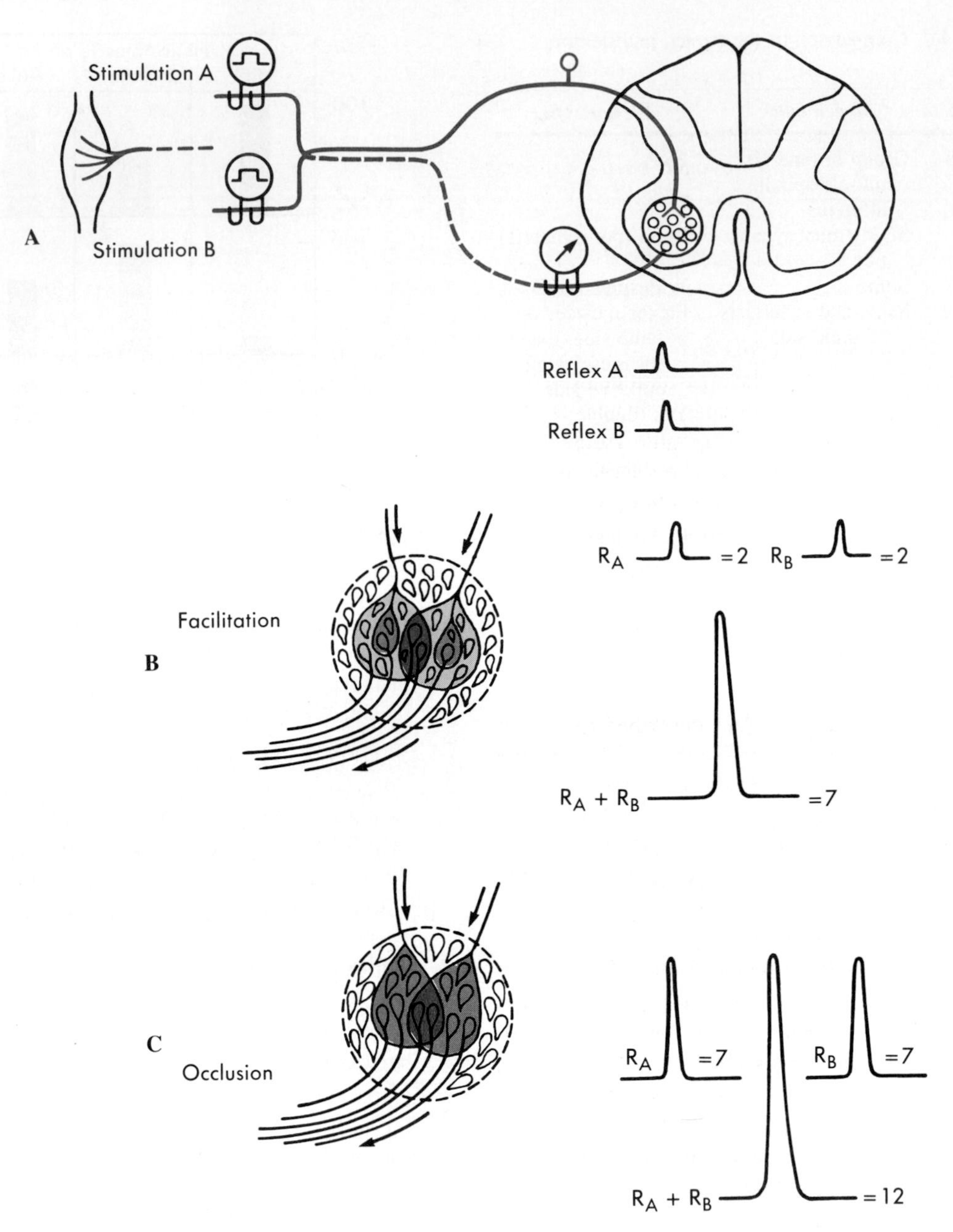

■ **Fig. 12-16** In **A** is shown the arrangement for using electrically evoked afferent volleys and recordings from motor axons in a ventral root to study reflexes. **B** represents an experiment in which combined stimulation of two muscle nerves resulted in spatial summation. In **C** the combined volleys caused occlusion. (Redrawn with permission from Eyzaguirre C, Fidone SJ: *Physiology of the nervous system,* ed 2, Chicago, 1975, Year Book Medical Publishers.)

stimulated at the same time, a much larger reflex discharge is recorded (Fig. 12-16, *B,* right). This reflex represents the discharges of seven α-motoneurons, as represented by the three black areas in the diagram in Fig. 12-16, *B,* left; i.e., three additional motor neurons discharged when the two muscle nerves were stimulated simultaneously.

The explanation for this spatial summation is that all of the α-motoneurons to the muscle were excited by either muscle afferent volley. However, when only one nerve was stimulated, the excitation was powerful enough to activate only two motoneurons. These motoneurons were in the discharge zone, whereas those that were excited, but not enough to reach threshold, were in the subliminal fringe. However, the combined excitation produced by volleys in the two nerves reached threshold in three additional motoneurons, and a reflex discharge resulted in a total of seven motor neurons.

A similar effect could be elicited by repetitive stimulation of one of the muscle nerves, provided that the stimuli occur close enough together in time so that some of the excitatory effect of the first volley still persists after the time of arrival of the second volley. This effect is called temporal summation. Both spatial and temporal summation depend on the properties of the excitatory postsynaptic potentials evoked by the group Ia afferent fibers in α-motoneurons (see Chapter 4).

The number of α-motoneurons that supply each muscle is limited, and not all can be activated even by a large peripheral input. For these reasons, the availability of motoneurons for summation is limited. If a volley in one of the two muscle nerves of Fig. 12-16 reached the motor nucleus at a time when the motoneurons were highly excitable, the reflex discharge would be relatively large (Fig. 12-16, *C*). A similar volley in the other muscle nerve might also produce a large reflex response. However, when the two muscle nerves are excited simultaneously, the reflex might be less than the sum of the two independently evoked reflexes. In this case, each muscle nerve activated seven α-motoneurons, but the volleys in the two nerves together caused only twelve motor neurons to discharge. This phenomenon is called **occlusion.**

Reflex testing by the techniques described to demonstrate spatial and temporal summation and occlusion can also be employed to demonstrate inhibition. A monosynaptic reflex discharge can be evoked (see Fig. 12-16) by stimulating the group Ia afferent fibers in a muscle nerve. This tests the reflex excitability of a population of α-motoneurons. The discharges of either extensor or flexor α-motoneurons can be recorded by proper choice of the muscle nerve to be stimulated. However, instead of using another group Ia volley in a nerve to the same muscle, other kinds of afferent fibers can be stimulated. For example, stimulation of the group Ia afferent fibers in the nerve to the antagonist muscles will produce **reciprocal inhibition.** If the small afferent fibers of a cutaneous nerve are stimulated

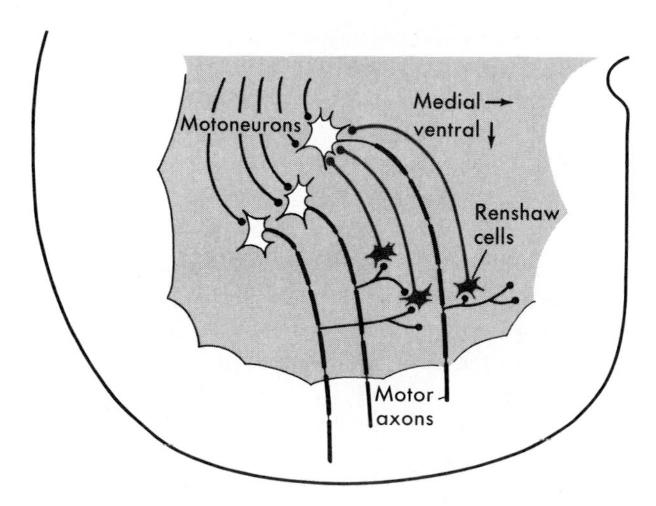

■ **Fig. 12-17** Recurrent inhibitory pathway. (Redrawn from Eccles JC: *The physiology of synapses,* New York, 1964, Academic Press.)

to evoke a flexion reflex, the α-motoneurons to extensor muscles will be inhibited (and those to flexor muscles excited). Stimulation of a ventral root causes the excitation of Renshaw cells by way of the recurrent collaterals of α-motor axons (Fig. 12-17). The **Renshaw cells** inhibit monosynaptic reflexes (and also group Ia inhibitory interneurons) and produce recurrent inhibition (or facilitation).

Reflex testing of this kind has been used to determine the circuitry involved in the various reflexes described in this chapter. As mentioned earlier, similar reflex testing can be done in human subjects, using the H-reflex as a test of the excitability of α-motoneurons.

■ *Summary*

1. Transection of the spinal cord reveals the reflexes that depend only on spinal cord circuits. These include the stretch reflex and the flexion reflex.

2. Muscle spindles are complex sensory receptors found in muscle in parallel to regular muscle fibers. They contain nuclear bag and nuclear chain intrafusal muscle fibers. Group Ia afferent fibers form primary endings on nuclear bag and chain fibers, and group II fibers form secondary endings on nuclear chain fibers. Dynamic γ-motor axons end on nuclear bag fibers and static γ-motor axons on nuclear chain fibers.

3. Primary endings have static and dynamic responses, which signal muscle length and rate of change in muscle length; secondary endings have just static responses and signal only muscle length. γ-motoneurons cause muscle spindles to shorten, preventing the unloading effect of muscle contraction.

4. Golgi tendon organs are located in the tendons of muscles and are arranged in series. They are supplied by group Ib afferent fibers and are excited both by stretch and by contraction of the muscle.

5. Reflexes are simple, stereotyped motor responses to a stimulus. A reflex arc includes the afferent fibers, interneurons, and motoneurons responsible for the reflex. The reflex arcs of many reflexes are located in the spinal cord or brainstem.

6. The stretch reflex includes a monosynaptic excitatory pathway from group Ia (and II) muscle spindle afferent fibers to α-motoneurons that supply the same and synergistic muscles, and a disynaptic inhibitory pathway to antagonistic motoneurons. Phasic stretch reflexes are triggered by the dynamic responses of group Ia fibers and tonic stretch reflexes by the static responses of group Ia and II afferents.

7. The inverse myotatic reflex is evoked by Golgi tendon organs. Afferent volleys cause a disynaptic inhibition of α-motoneurons to the same muscle and excita-

tion of antagonist muscles. However, because Golgi tendon organs monitor force, their reflex action actually complements that of the stretch reflex.

8. The flexion reflex is evoked by volleys in flexion reflex afferent fibers that supply a variety of receptors, including nociceptors. In the flexion reflex, ipsilateral flexor motoneurons are excited and extensor motoneurons are inhibited through polysynaptic pathways. The converse pattern may occur contralaterally. There is a flexion component of locomotion, but the most powerful flexion reflex is the flexor withdrawal reflex. This varies somewhat with the location of the stimulus (local sign).

9. Spinal reflexes can be studied by electrically stimulating the afferent fibers in nerves and recording volleys in α-motor axons from ventral roots. Principles of reflex organization that are derived from such studies include spatial and temporal summation, occlusion, and various patterns of inhibition.

Bibliography

Journal articles

Boyd IA: The isolated mammalian muscle spindle, *Trends Neurosci* 3:258, 1980.

Crowe A, Matthews PBC: The effects of stimulation of static and dynamic fusimotor fibres on the response to stretching of the primary endings of muscle spindles, *J Physiol* 174:109, 1964.

Eccles JC, Fatt P, Koketsu K: Cholinergic and inhibitory synapses in a pathway from motor-axon collaterals to motoneurones, *J Physiol* 126:524, 1954.

Houk J, Henneman E: Responses of Golgi tendon organs to active contractions of the soleus muscle of the cat, *J Neurophysiol* 30:466, 1967.

Hunt CC, Kuffler SW: Stretch receptor discharges during muscle contraction, *J Physiol* 113:298, 1951.

Mendell LM, Henneman E: Terminals of single Ia fibers: location, density, and distribution within a pool of 400 homonymous motoneurons, *J Neurophysiol* 34:171, 1971.

Nichols TR, Houk JC: Improvement in linearity and regulation of stiffness that results from actions of stretch reflex, *J Neurophysiol* 39:119, 1976.

Renshaw B: Central effects of centripetal impulses in axons of spinal ventral roots, *J Neurophysiol* 9:191, 1946.

Sherrington CS: Flexion-reflex of the limb, crossed extension-reflex, and reflex stepping and standing, *J Physiol* 40:28, 1910.

Swett JE, Schoultz TW: Mechanical transduction in the Golgi tendon organ: a hypothesis, *Arch Ital Biol* 113:374, 1975.

Books and monographs

Baldissera F et al: *Integration in spinal neuronal systems*. In Brooks VB, editor: *Handbook of physiology,* sect 1, The nervous system, vol II, Motor control, part 1, Bethesda, Md, 1981, American Physiological Society.

Creed RS et al: *Reflex activity of the spinal cord,* London, 1932, Oxford University Press.

Eccles JC: *The physiology of synapses,* New York, 1964, Academic Press.

Houk JC, Rymer WZ: *Neural control of muscle length and tension*. In Brooks VB, editor: *Handbook of physiology,* sect 1. The nervous system, vol II. Motor control, part 1, Bethesda, Md, 1981, American Physiological Society.

Matthews PBC: *Mammalian muscle receptors and their central actions,* London, 1972, Arnold.

Matthews PBC: *Muscle spindles: their messages and their fusimotor supply*. In Brooks VB, editor: *Handbook of physiology,* Sect 1, The nervous system, vol II, Motor control, part 1, Bethesda, Md, 1981, American Physiological Society.

Sherrington CS: *The integrative action of the nervous system,* ed 2, New Haven, 1947, Yale University Press.

Stein RB, Lee RG: *Tremor and clonus*. In Brooks VB, editor: *Handbook of physiology,* sect 1, The nervous system, vol II, Motor control, part 1, Bethesda, Md, 1981, American Physiological Society.

CHAPTER

13

Descending Pathways Involved in Motor Control

■ *Topographical Organization of Spinal Motor System*

As discussed in Chapter 7 (see Fig. 7-11), spinal cord motoneurons are organized topographically in the ventral horn. Motoneurons that supply the axial musculature form a column of cells extending the length of the spinal cord. In the cervical and lumbosacral enlargements, they are located in the most medial part of the ventral horn. Motoneurons that supply the limb muscles form columns that extend for one or two segments in the lateral part of the ventral horn in the cervical and lumbar enlargements. The motoneurons to the muscles of the distal limb are located most laterally, whereas those that innervate more proximal muscles are located more medially in the lateral ventral horn. It should be noted that the α- and γ-motoneurons to a given muscle are found side by side in the same motoneuron column.

The interneurons that connect with the motoneurons in the enlargements are also topographically organized. In general, interneurons that supply the limb muscles are located mainly in the lateral parts of the deep dorsal horn and intermediate region, whereas those that supply the axial muscles are in the medial part of the ventral horn (Fig. 13-1). Many of these interneurons are found in spinal reflex arcs and in the descending motor control systems, because they receive synaptic connections from primary afferent fibers and from the axons of pathways descending from the brain. An important aspect of the interneuronal systems is that the laterally placed interneurons project ipsilaterally to motoneurons that supply the distal or the proximal limb muscles, whereas the medial ones project bilaterally (see Fig. 13-1). This arrangement of the lateral interneurons is related to the organization of limb movements in higher mammals, because the limbs must be controlled independently to allow manipulations. On the other hand, the axial muscles are controlled bilaterally because they provide postural support of the trunk and neck.

Fig. 13-2 shows some of the sites in the reflex circuitry that can be influenced by descending motor control pathways. Synapses can be made by descending pathways on interneurons that cause presynaptic inhibition of primary afferent fibers or that participate in postsynaptic excitation or inhibition of other interneurons or of motoneurons. Alternatively, some descending pathways make direct excitatory connections to α- or γ-motoneurons.

■ *Cranial Nerve Motor Nuclei*

A similar arrangement can be described for cranial nerve motor nuclei. Most of the cranial nerve motor nuclei can be regarded as similar to the axial motor nuclei of the spinal cord because they supply muscles near the midline. The oculomotor nuclei, the trigeminal motor outflow, part of the facial motor nuclei, and the nucleus ambiguus all supply muscles that are part of a bilateral motor control system. For example, the trigeminal motor nuclei supply the muscles of mastication, such as the temporalis, masseter, and pterygoid muscles. These operate bilaterally during chewing. Similarly, facial motoneurons that supply the corrigator muscle of the forehead and the orbicularis oculi muscles, which close the eyes, must operate bilaterally (e.g., the blink reflex causes both eyelids to close simultaneously). The head muscles that resemble in function the distal limb muscles are those that control facial expression and the tongue. Thus neurons in the facial motor nuclei that supply the lower face elicit changes in facial expression that can be unilateral, and hypoglossal motoneurons can cause the tongue to move toward one side. Presumably, the interneuronal systems in the brainstem are organized

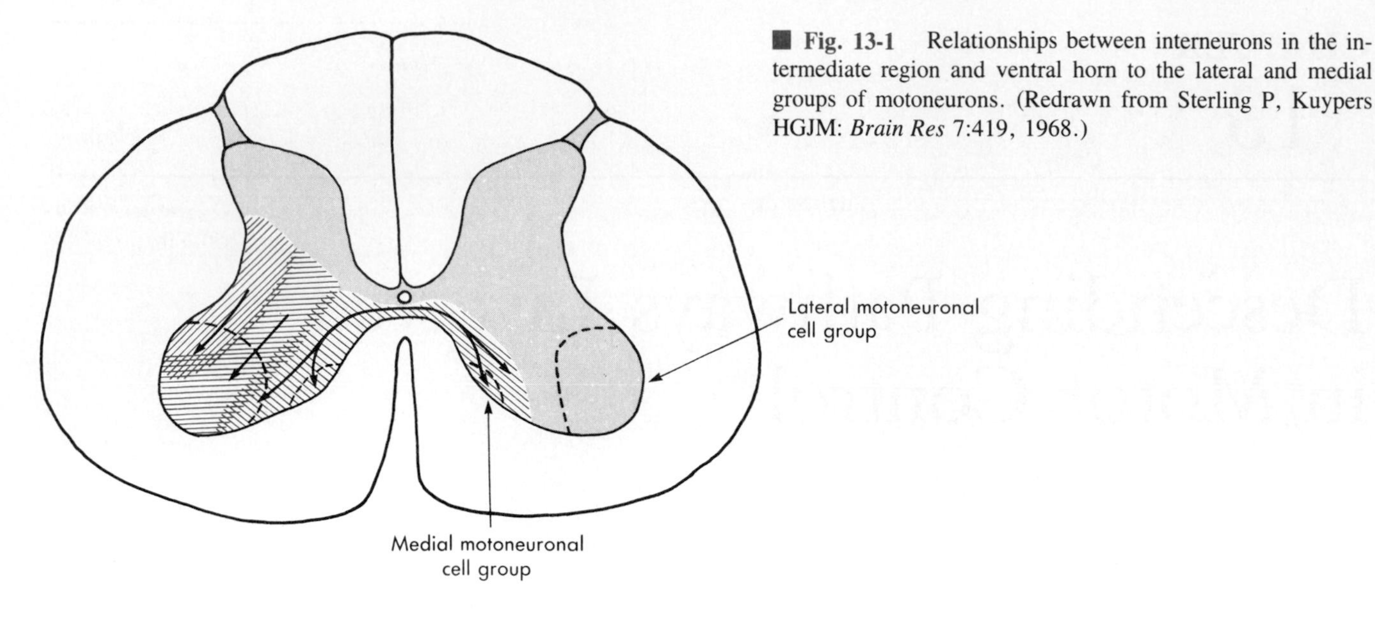

■ **Fig. 13-1** Relationships between interneurons in the intermediate region and ventral horn to the lateral and medial groups of motoneurons. (Redrawn from Sterling P, Kuypers HGJM: *Brain Res* 7:419, 1968.)

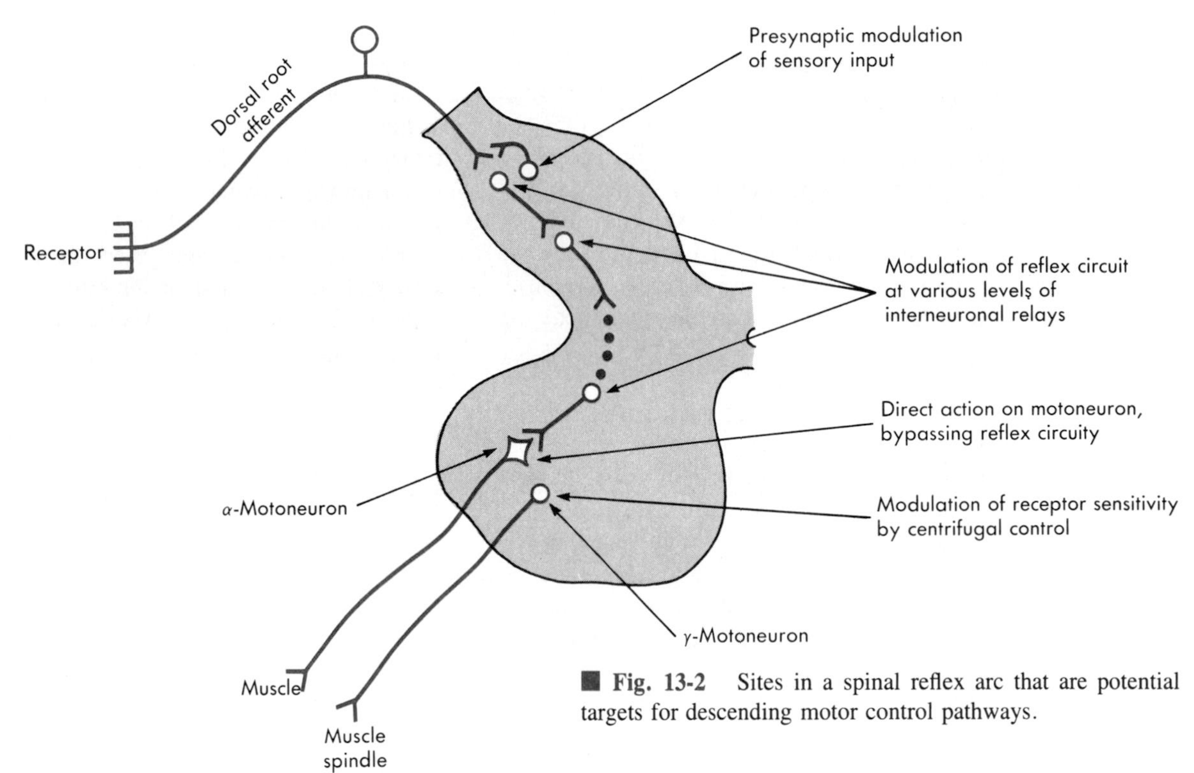

■ **Fig. 13-2** Sites in a spinal reflex arc that are potential targets for descending motor control pathways.

like those in the spinal cord to support these bilateral or unilateral motor activities.

■ *Classification of Descending Motor Pathways*

Descending motor pathways have traditionally been subdivided into the pyramidal tract and the extrapyramidal pathways. This terminology reflects the important clinical dichotomy between pyramidal tract disease and extrapyramidal disease. In **pyramidal tract disease,** the corticospinal tract is interrupted. Hence the signs of the disease were originally attributed to the loss of function of the pyramidal tract (so named because the corticospinal tract passes through the medullary pyramid). However, in many cases, the functions of other pathways are also altered, and pyramidal tract signs are not necessarily only caused by the loss of the corticospinal tract (see p. 219).

The term **extrapyramidal** raises even more difficulties. The extrapyramidal tracts would presumably include all descending motor pathways except the pyramidal tract. The term extrapyramidal disease generally signifies one of several diseases of the basal ganglia. However, the main motor pathway involved in basal ganglion diseases is the corticospinal tract (Chapter 14). Other extrapyramidal motor pathways, such as the reticulospinal tract, play a prominent role in cerebellar and other motor disorders, as well as in basal ganglion disease.

Lateral Versus Medial Motor Systems

Another way of classifying the motor pathways is based on their sites of termination in the spinal cord and the consequent differences in their roles in the control of manipulation and posture. The **lateral pathways** are those that terminate directly on motoneurons or on the

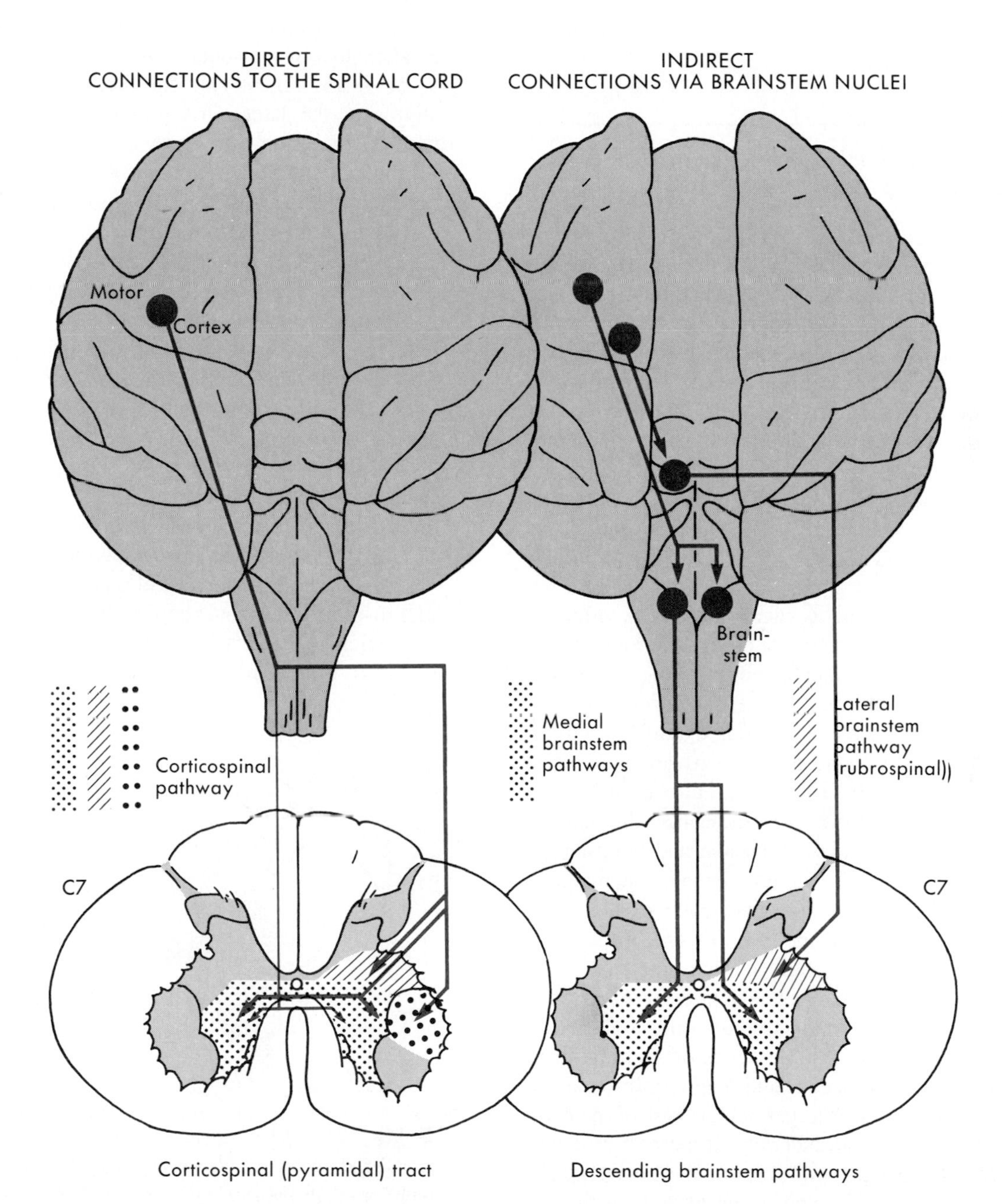

Fig. 13-3 Subdivision of the pathways descending from the cerebral cortex and brainstem to the spinal cord into a lateral and a medial system, based on the terminations of these pathways in the spinal cord gray matter. The lateral pathways end on motoneurons to distal muscles *(dots)* and interneurons projecting to these motor neurons *(hatched area)*. The medial pathways end on interneurons supplying motoneurons to the axial muscles *(stippled area)*. (Redrawn from Brinkman C: Split-brain monkeys: cerebral control of contralateral and ipsilateral arm, hand, and finger movements, doctoral dissertation, Rotterdam, The Netherlands, 1974, Erasmus University.)

interneuronal groups in the lateral parts of the spinal cord gray matter (Fig. 13-3). They are in a position to excite motor neurons directly and to influence reflex arcs that control the fine movements of the distal limbs, as well as to activate supporting musculature in the proximal limbs. The **medial pathways** end in the medial ventral horn on the medial group of interneurons (see Fig. 13-3). These interneurons connect with motoneurons that control the axial musculature bilaterally and thereby contribute to balance and postural support. They also contribute to the control of proximal limb muscles.

Lateral System: Lateral Corticospinal Tract and Rubrospinal Tract

The pathways that are most important for control of manipulative ability of the limbs are the **lateral corticospinal tract** and, in animals, the **rubrospinal tract** (see Fig. 13-3). The ventral corticospinal tract belongs to the medial system (see p. 217). A comparable subdivision can be made of the **corticobulbar tract,** which projects to the cranial nerve motor nuclei. One component of the corticobulbar tract ends in the part of the facial motor nucleus that innervates the muscles of facial expression in the lower face and in the hypoglossal nucleus that supplies the tongue muscles. This component of the corticobulbar tract is equivalent to the lateral corticospinal tract.

The corticospinal and corticobulbar tracts originate from a wide area of the cerebral cortex that includes the motor, premotor, and supplementary motor areas and the somatosensory cortex. The cells of origin of these tracts include both large and small pyramidal cells of layer V of the cortex and the giant pyramidal cells of Betz, which are a characteristic feature of the motor cortex. The corticospinal tract descends through the posterior limb of the internal capsule and brainstem to the level of the medulla, where it occupies the pyramid. At the caudalmost medulla, about 80% of the axons cross to the opposite side and then descend in the dorsal lateral funiculus as the lateral corticospinal tract. The remaining axons continue caudally in the ventral funiculus on the same side as the ventral corticospinal tract. The corticobulbar tract also descends through the internal capsule, but in the genu, and then terminates in the brainstem at a level near target nuclei. Part of the corticobulbar tract ends contralaterally in the part of the facial nucleus that supplies muscles of the lower face, in the hypoglossal nucleus, and in the nucleus of the spinal accessory nerve. This component of the corticospinal tract is similar in organization to the lateral corticospinal tract. The remainder of the corticobulbar tract ends bilaterally and, in that respect, is equivalent to the ventral corticospinal tract.

The corticospinal projections from the motor and sensory areas of the cerebral cortex terminate in different parts of the spinal cord gray matter. The motor areas project to the intermediate region and ventral horn (Fig. 13-4), whereas the sensory areas project to the dorsal horn. The part of the motor area projections that descends in the lateral corticospinal tract terminates in part directly on motoneurons and in part on interneurons of the lateral group (see Fig. 13-4). The ventral corticospinal tract, on the other hand, ends bilaterally on the medial group of interneurons.

The motor cortex has a topographic organization that parallels that of the somatosensory cortex (see also Fig. 13-5, *A,* and Fig. 8-11). The face is represented laterally, near the lateral fissure; the hand is represented more medially on the convexity of the cerebrum; and the lower extremity is represented largely on the medial aspect of the hemisphere. The figurine in the illustration is called a **motor homunculus.** The various body parts are distorted so as to indicate approximately how much of the cortex is devoted to their motor control.

The other cortical motor areas also contain somatotopic representations. Fig. 13-5, *B,* shows the motor areas of the cerebral cortex of a monkey. The somatotopic map of the motor cortex is shown along the precentral gyrus. Note how closely it resembles that of the human shown in Fig. 13-5, *A*. There are other somatotopic maps in the SII cortex, which is in the cortex of the roof of the lateral fissure (part of which is shown in Fig. 13-5, *B*), and in the supplementary motor cortex, which is located on the medial aspect of the hemisphere just rostral to the motor cortex. It should be noted that the digital representation in monkeys (and presumably humans) is toward the central sulcus, whereas the

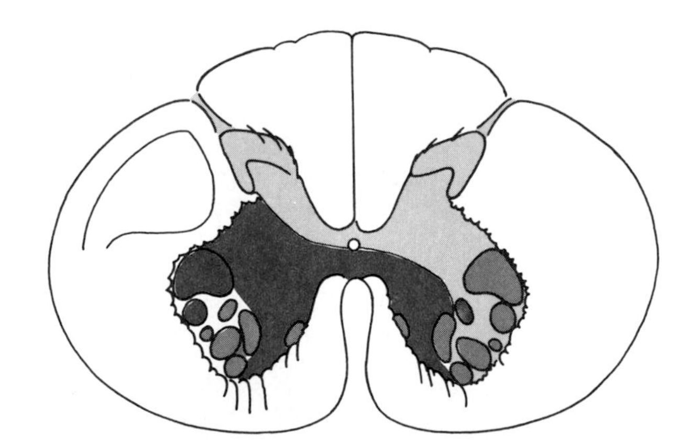

Fig. 13-4 The region of termination of the part of the lateral and ventral corticospinal tracts that originate from the motor areas of the cerebral cortex is shown. Part of the projection, that of the lateral corticospinal tract, is directly on motoneurons and part on the lateral interneurons. By contrast, the ventral corticospinal tract ends bilaterally on the medial interneurons. (Redrawn from Brinkman C: Split-brain monkeys: cerebral control of contralateral and ipsilateral arm, hand and finger movements, doctoral dissertation, Rotterdam, The Netherlands, 1974, Erasmus University.)

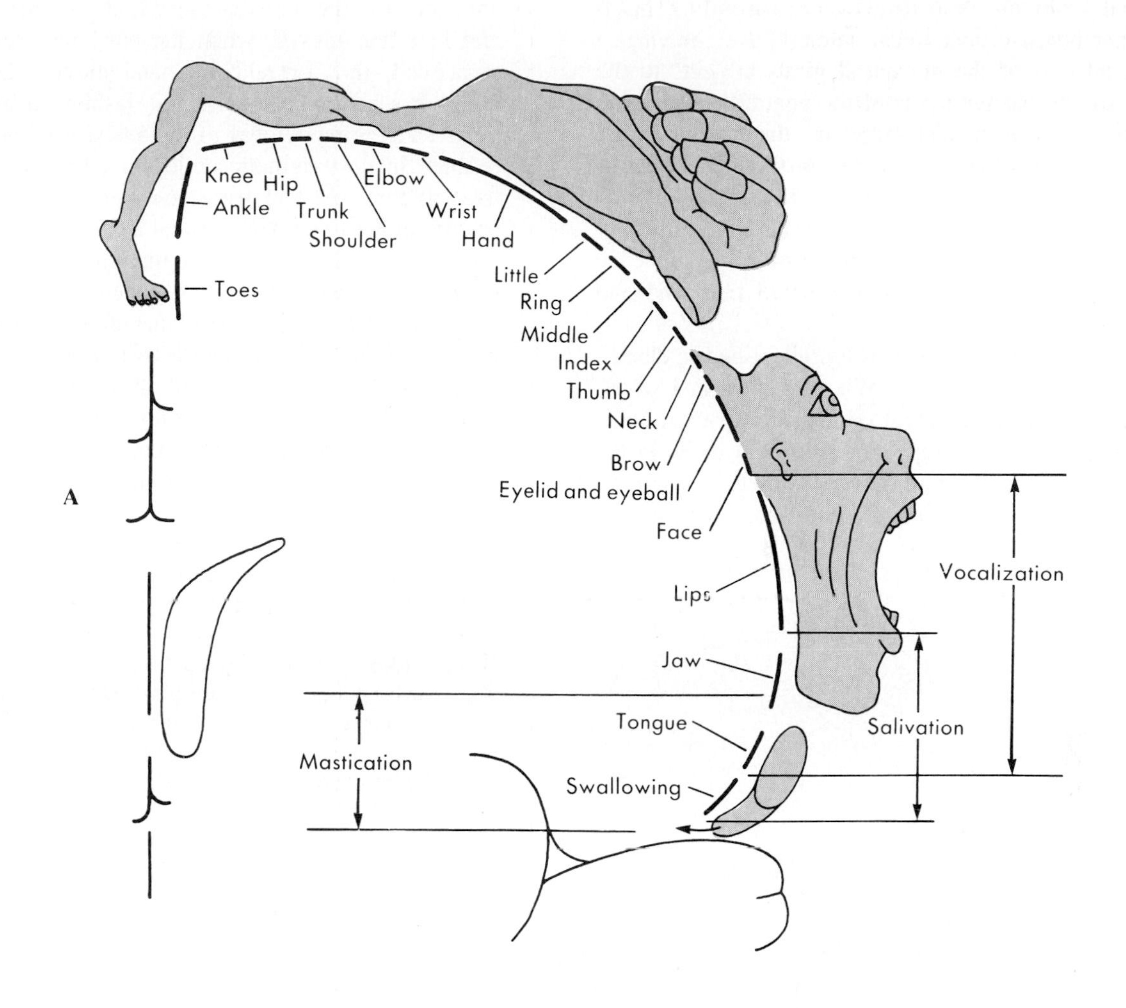

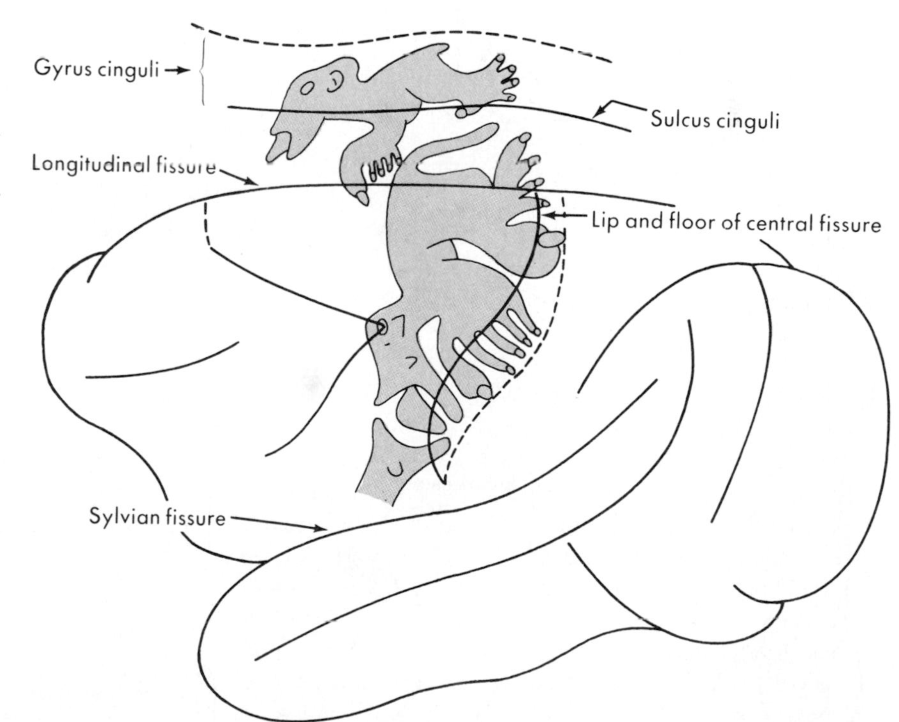

■ **Fig. 13-5** Topographic organization of the motor cortex. **A,** The cortex is cut in a coronal section, and the somatotopic map of the body and face is shown as a "motor homunculus." The sizes of different body parts indicate the amount of cortex devoted to motor control of that part. **B,** The somatotopic organization of the motor cortex, supplementary motor cortex, and SII areas of the cerebral cortex of a monkey. (Redrawn from Eyzaguirre C, Fidone SJ: *Physiology of the nervous system,* ed 2, Chicago, 1975, Year Book Medical Publishers. **A** slightly modified from Penfield W, Rasmussen T: *The cerebral cortex of man,* New York, 1950, The Macmillan Co; **B** slightly modified from Woolsey CN et al: *Res Publ Assoc Res Nerv Ment Dis* 30:238, 1952.)

proximal limbs are represented more rostrally. This is important because anatomical studies have shown that the caudal part of the precentral gyrus projects to the area of the dorsolateral part of the ventral horn, which contains motoneurons that supply the distal musculature of the extremities. The rostral part of the precentral gyrus, however, projects to the intermediate region, which contains interneurons of the lateral group. Thus the caudal part of the motor cortex is in a position to influence directly the activity of motoneurons to the hand and digit muscles.

The motor cortex proper was first identified in electrical stimulation experiments. When an electrical stimulus is applied to the cerebral cortex, the lowest threshold points for a motor response are found to lie in the motor cortex. Such stimuli evoke discrete movements of distal muscles on the contralateral side. When stimuli are applied in the face representation, it is the contralateral face that moves; when the hand representation is stimulated, the contralateral hand moves. The motor cortex has been explored in this fashion in human patients undergoing surgery to remove scar tissue that had resulted in posttraumatic epilepsy. Often, resection of the scar cures the epilepsy. However, the surgeon must be careful not to damage normal areas of the motor cortex, because this will paralyze the affected musculature.

The direct innervation of motoneurons by the lateral corticospinal tract is of particular interest, because the most important function that is lost after interruption of this pathway is the fine control of the digits (and loss of voluntional movements of the lower face and tongue in the case of interruption of the corticobulbar tract). The direct innervation of motoneurons by corticospinal fibers is most prominent in humans and the great apes

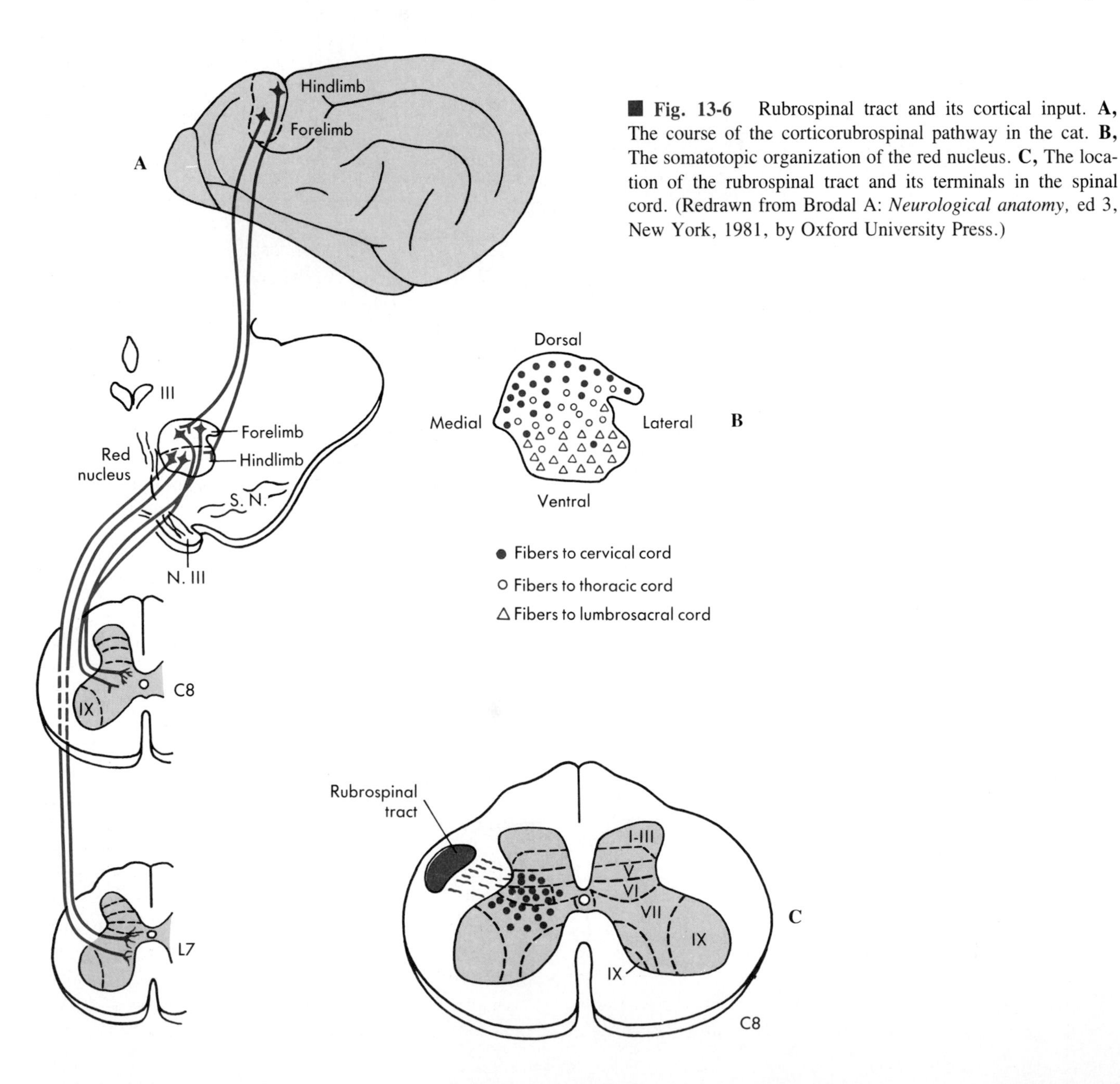

■ **Fig. 13-6** Rubrospinal tract and its cortical input. **A,** The course of the corticorubrospinal pathway in the cat. **B,** The somatotopic organization of the red nucleus. **C,** The location of the rubrospinal tract and its terminals in the spinal cord. (Redrawn from Brodal A: *Neurological anatomy,* ed 3, New York, 1981, by Oxford University Press.)

and is substantial in monkeys. However, there are no direct connections in most mammals, including cats and rats. This distribution of direct corticospinal projections correlates with the capability of various animals for fine digital control.

Another lateral system pathway is the rubrospinal tract (Fig. 13-6). It is unclear how important this pathway is in humans. However, it is a major descending motor control pathway in animals like the cat. The rubrospinal tract originates from neurons in the red nucleus in the midbrain, and it receives a projection from the motor cortex (Fig. 13-6, *A*). In addition, the cells of origin of the rubrospinal tract receive a powerful input from the deep cerebellar nuclei (Chapter 14). The neurons within the nucleus have a somatotopic organization (Fig. 13-6, *B*). Their axons leave the ventral aspect of the nucleus and cross to the opposite side of the midbrain; from this position, they descend through the brainstem and pass down the spinal cord in the dorsal lateral funiculus adjacent to the lateral corticospinal tract and terminate on the lateral group of interneurons (Fig. 13-6, *C*). The rubrospinal tract in cats does not control the most distal muscles, but rather more proximal muscles of the limb.

Medial System Pathways

As already mentioned, the ventral corticospinal tract and much of the corticobulbar tract can be regarded as medial system pathways. These tracts end on the medial group of interneurons in the spinal cord and on equivalent medial system neurons in the brainstem. The muscles controlled by these pathways are the axial muscles, and they often contract bilaterally to provide postural support or to participate in some other bilateral function, like chewing or wrinkling of the brow.

Other medial system pathways originate in the brainstem. Some of these include the lateral and medial vestibulospinal tracts, the pontine and medullary reticulospinal tracts, and the tectospinal tract.

The lateral vestibulospinal tract originates from the lateral vestibular nucleus (Fig. 13-7, *A*). Like the red nucleus, the lateral vestibular nucleus has a somatotopic organization (Fig. 13-7, *B*). The lateral vestibulospinal tract descends ipsilaterally through the brainstem, then down the ventral funiculus of the spinal cord, and ends on interneurons of the medial group (Fig. 13-7, *A* and *C*). The lateral vestibulospinal tract excites motoneurons that supply proximal postural muscles. The input to the lateral vestibular nucleus is from both the semicircular ducts and the otolith organs. An important function of the lateral vestibulospinal tract is to assist with postural adjustments after angular and linear accelerations of the head. In decerebrate experimental preparations, the lateral vestibulospinal tract becomes hyperactive, presumably because of the loss of descending inhibitory controls. This hyperactivity is largely responsible for the extensor hypertonus seen in these animals.

The medial vestibulospinal tract originates from the medial vestibular nucleus (Fig. 13-8). It descends in the ventral funiculus of the spinal cord to cervical and midthoracic levels and ends on the medial group of interneurons (see Fig. 13-7, *C*). The input to the medial vestibular nucleus from the labyrinth is chiefly from the semicircular ducts. Hence this pathway mediates adjustments in head position in response to angular accelerations of the head.

The cells that give rise to the pontine reticulospinal tract are in the medial pontine reticular formation (nuclei reticularis pontis caudalis and pontis oralis). The tract descends in the ipsilateral ventral funiculus and ends on the medial group of interneurons (Fig. 13-9, *A*). Its function is, like that of the lateral vestibulospinal tract, to excite motoneurons to the proximal extensor muscles to support posture.

The medullary reticulospinal tracts arise from neurons of the medial medulla (nuclei reticularis gigantocellularis and ventralis). The tracts descend bilaterally in the ventral lateral funiculus and end mostly on the medial group of interneurons, although some also end on lateral interneurons (Fig. 13-9, *B*). The function of the pathway is largely inhibitory. Some of the inhibitory connections end directly on motor neurons.

The tectospinal tract originates in the deep layers of the superior colliculus (see Fig. 9-24). The axons cross to the contralateral side just below the periaqueductal gray, and they descend in the ventral funiculus of the spinal cord to terminate on the medial group of interneurons in the upper cervical spinal cord. The function of the tectospinal tract is to cause contralateral movement of the head in response to visual, auditory and somatic stimuli.

Other medial system pathways originate in the interstitial nucleus of Cajal and the solitary nucleus.

Monoaminergic Pathways

In addition to the lateral and medial systems, less specifically organized systems descend from the brainstem to the spinal cord. These include several pathways that use monoamines as synaptic transmitters.

The locus coeruleus and the nucleus subcoeruleus are nuclei located in the upper pons and are composed of norepinephrine-containing neurons. These nuclei project widely to the spinal cord through the lateral funiculi. The terminals are on interneurons and motoneurons. The dominant effect of the pathway is inhibitory.

The raphe nuclei of the medulla give rise to several raphe-spinal projections to the spinal cord. Many of the raphe-spinal cells contain serotonin. Terminals on dorsal horn interneurons produce inhibition, whereas terminals on motoneurons have excitatory actions. The dorsal horn projection may help reduce nociceptive trans-

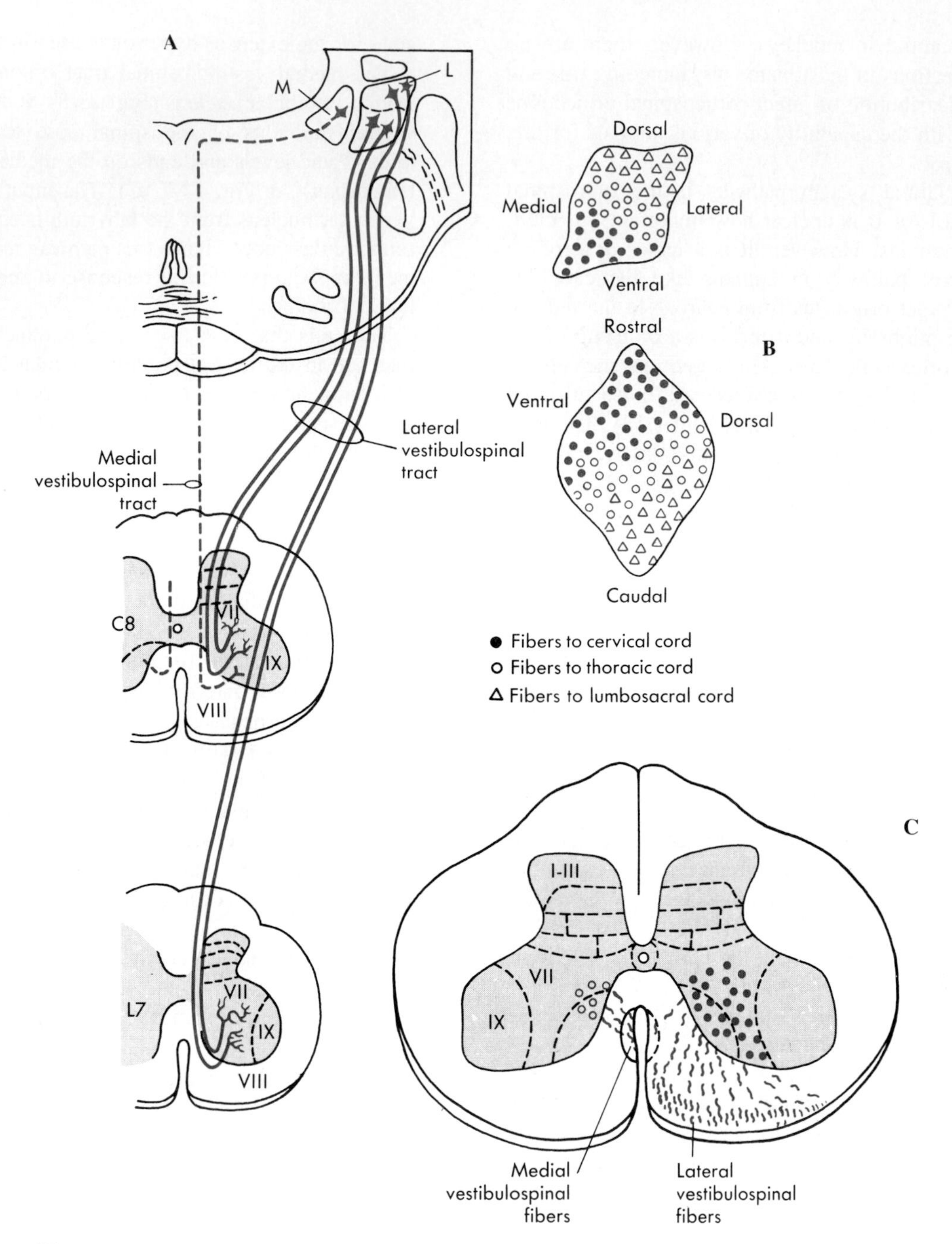

■ **Fig. 13-7** Organization of the vestibulospinal tract of the cat. **A,** The projection from the lateral vestibulospinal tract. **B,** The somatotopic organization of the lateral vestibular nucleus. **C,** The zone of termination on the medial group of interneurons. The termination of the medial vestibulospinal tract in also shown. (Redrawn from Brodal A: *Neurological anatomy,* ed 3, New York, 1981, Oxford University Press.)

mission, whereas the ventral horn projection may enhance motor activity.

Other monoamine pathways project to the spinal cord and contain dopamine or epinephrine. Their functions are less clear than those of the norepinephrine- and serotonin-containing pathways.

Generally, the monoaminergic pathways may alter the responsiveness of spinal cord circuits, including the reflex arcs. They would be important in causing widespread changes in excitability rather than in producing discrete movements or specific changes in behavior.

■ *Effects of Interruption of the Lateral or Medial Systems*

A common cause of motor disorder in human patients is interruption of the corticospinal tract as it traverses the

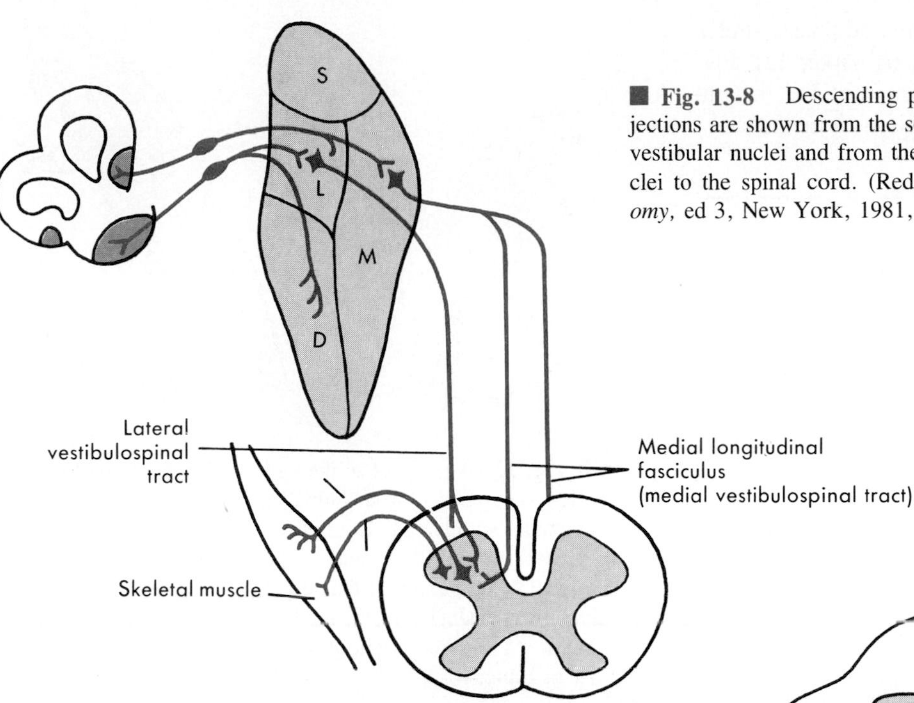

■ **Fig. 13-8** Descending pathways from the vestibular nuclei. Projections are shown from the semicircular ducts and otolith organs to the vestibular nuclei and from the lateral *(L)* and medial *(M)* vestibular nuclei to the spinal cord. (Redrawn from Brodal A: *Neurological anatomy,* ed 3, New York, 1981, Oxford University Press.)

internal capsule, which occurs in capsular strokes. The result is often termed a **pyramidal tract syndrome** or **upper motoneuron disease.** The motor changes include (1) increased phasic and tonic stretch reflexes (spasticity); (2) weakness; (3) pathological reflexes, including the **sign of Babinski** (dorsiflexion of the big toe and fanning of the other toes when the sole of the foot is stroked); and (4) a reduction in superficial reflexes, such as the abdominal and cremasteric reflexes. The weakness is most evident in the distal muscles, especially in the finger muscles.

However, if the corticospinal tract alone is interrupted, as can occur with a lesion of the medullary pyramid, most of these signs are missing. The most prominent deficit in this case is weakness of the distal muscles, especially of the fingers, and a positive sign of Babinski. There is no spasticity, but rather a decrease in muscle tone. Evidently, the spasticity depends on disordered function of the corticospinal tract and other pathways, such as the reticulospinal tracts.

Interruption of the medial system pathways has quite different effects from those produced by corticospinal tract lesions. The main deficits are an initial reduction in the tone of postural muscles and loss of righting reflexes (see p. 220). Longer term effects are impairment of locomotion and frequent falling. However, manipulation of objects by the fingers is perfectly normal.

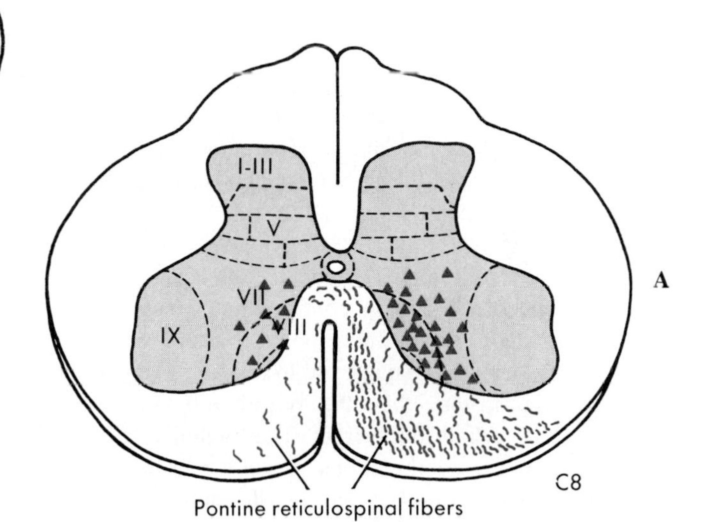

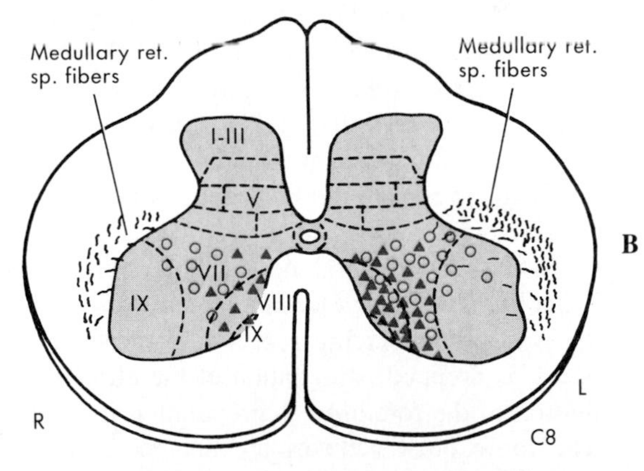

■ **Fig. 13-9** **A,** The course and terminations of the pontine reticulospinal tract. **B,** The course and terminations of the medullary reticulospinal tracts; in addition, the terminations of the pontine reticulospinal tract are indicated. (Redrawn from Brodal A: *Neurological anatomy,* ed 3, New York, 1981, Oxford University Press.)

■ *Brainstem Control of Posture and Movement*

The importance of pathways that originate in the brainstem in motor control is evident from observations of the extensor hypertonus and the increased phasic stretch reflexes seen in decerebrate animals (Chapter 12). Particular brainstem systems have been identified that influence posture, locomotion, and eye movements.

Postural Reflexes

Several reflex mechanisms are brought into action when the head is moved or the neck is bent. The sensory receptors responsible for these reflexes include the vestibular apparatus, which is stimulated by head movements, and stretch receptors in the neck.

The **vestibular reflexes** constitute one class of postural reflexes. Rotation of the head, which produces an angular acceleration, activates sensory receptors of the semicircular ducts (Chapter 10). In addition to eye movements (see p. 221), the input to the vestibular nuclei results in postural adjustments. These are mediated by commands transmitted to the spinal cord through the lateral and medial vestibulospinal tracts (and the reticulospinal tracts). The lateral vestibulospinal tract activates extensor muscles that support posture. For instance, if the head is rotated to the left, the postural support is increased on the left side. This increased support prevents the tendency for the subject to fall to the left as the head rotation continues. A disease process that eliminates labyrinthine function in the left ear will cause the person to tend to fall to the left. Conversely, a disease that irritates the left labyrinth will cause the person to tend to fall to the right. The medial vestibulospinal tract causes contractions of neck muscles that oppose the induced movement **(vestibulocollic reflex).**

Tilting the head will change linear acceleration and will activate the otolith organs of the vestibular apparatus. This can produce eye movements (see p. 222) and postural adjustments. For example, tilting the head and body forward (without bending the neck and consequently without evoking the tonic neck reflexes) in a quadriped, such as a cat, results in extension of the forelimbs and flexion of the hindlimbs. This vestibular action tends to restore the body toward its original posture. Conversely, if the head and body are tilted backward (without bending the neck), the forelimbs flex and the hindlimbs extend. Otolithic organs also contribute to the **vestibular placing reaction.** If an animal, such as a cat, is dropped, stimulation of the utricles leads to extension of the forelimbs in preparation for landing.

The **tonic neck reflexes** are another type of positional reflex. These reflexes are activated by the muscle spindles found in the neck muscles. These muscles contain the largest concentration of muscle spindles of any muscle in the body. If the neck is bent (without tilting the head), the neck muscle spindles evoke the tonic neck reflexes without interference from the vestibular system. When the neck is extended relative to the body, the forelimbs extend and the hindlimbs flex (Fig. 13-10). The converse occurs when the neck is flexed. Note that these effects are the opposite of those evoked by the vestibular system. Furthermore, if the neck is bent to the left, the extensor muscles in the limbs on the left contract more and the flexor muscles in the limbs on the right side will relax.

Fig. 13-10 Effect of the tonic neck reflexes on limb position. The head is in a normal position in the cat, avoiding vestibular stimulation. Dorsiflexion of the neck causes extension of the forelimbs and flexion of the hindlimbs. Conversely, ventriflexion of the neck results in flexion of the forelimbs and extension of the hindlimbs. (Redrawn from Roberts TDM: *Neurophysiology of postural mechanisms,* London, 1979, Butterworth.)

The third class of postural reflexes are the **righting reflexes.** These tend to restore an altered position of the head and body toward normal. The receptors responsible for righting reflexes include not only the vestibular apparatus and neck stretch receptors, but also mechanoreceptors of the body wall.

Locomotion

The spinal cord contains neural circuits that serve as a pattern generator for locomotion. There are actually several pattern generators, one for each limb involved in locomotion. These permit independent movements of the limb. However, the pattern generators are interconnected to ensure coordination of limb movements. The pattern generator for locomotion is an example of a biologic oscillator. Similar oscillators are responsible for activities like scratching, chewing, and respiration.

The pattern generator for locomotion is normally activated by commands that descend from the brain. The **midbrain locomotor center** is thought to organize commands to initiate locomotion. Voluntary activity that originates in the motor cortex can trigger locomotion by an action of corticobulbar fibers on the midbrain locomotor center. The commands are relayed through the pontomedullary reticular formation through the reticulospinal tracts. Locomotion is also influenced by afferent activity. This ensures that the pattern generator adapts to changes in the terrain as locomotion proceeds. Such changes may occur rapidly during running, and locomotion must then be adjusted to ensure proper coordination.

An important requirement for locomotion is adequate postural support. This is normally provided by the postural muscles in response to activity in the reticulospinal tract. After a spinal cord transection, postural activity is lost during spinal shock. In humans no recovery of locomotion occurs even after spinal shock is over. However, in animals some locomotion is possible in spinalized animals, especially if some postural support is produced by stimulation of afferent fibers or by pharmacological activation of spinal cord interneuronal circuits.

Control of Eye Position

The position of the eyes is controlled by several neural systems. **Conjugate eye movements** (movements of both eyes in the same direction and amount) are controlled by the vestibuloocular reflex and by the optokinetic, saccadic, and pursuit movement systems. The eyes can also converge or diverge under the control of the vergence system.

Vestibuloocular reflex. The vestibuloocular reflex is designed to maintain a stable image on the retina during rapid head rotation. This reflex produces conjugate movements of the eyes in the direction opposite to and in an amount equal to the head movement.

Optokinetic reflex. The optokinetic reflex is another means by which the nervous system compensates for head movements to maintain a visual target. In this case, the movements are slow and the sensory input for the reflex is provided by the visual system. The optokinetic system can be activated by having a subject view a rotating drum painted with vertical stripes. A more familiar stimulus is provided by viewing telephone poles from a moving car.

Smooth pursuit. In contrast to the vestibuloocular and optokinetic reflexes, which allow the eyes to remain fixated on the visual target during head movements, the smooth pursuit system elicits eye movements to maintain fixation on a moving visual target even though the head may be still. Smooth pursuit occurs only when there is a moving stimulus; it cannot result from a command.

Saccadic system. A **saccade** is a stereotyped rapid eye movement that allows foveation of a visual target. Saccades can be made voluntarily, or they can be a part of several reflexes. The velocity of a saccade is too rapid for visual processing. Hence there is no visual feedback during the movement. Saccades are corrected by smaller saccades that occur after the initial large change in eye position.

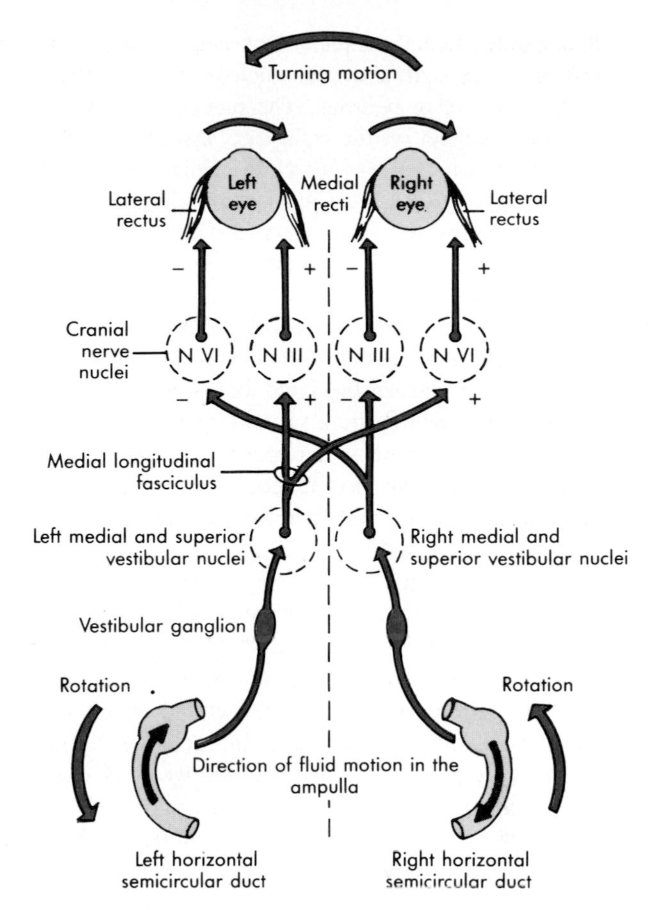

Fig. 13-11 Neural circuit for the vestibulo-ocular reflex. The eyes, brainstem pathways, and horizontal semicircular ducts are viewed as from above. Rotation of the head to the left is indicated by the large arrows and the relative movement of endolymph to the right by the smaller arrows.

Vergence system. The vergence system allows the two eyes to converge or diverge. This allows fixation on nearby or distant objects and on objects that approach or move away. During convergence movements, accommodation of the lens for near vision and pupillary constriction occur (see Chapter 9).

Neural circuit for the vestibuloocular reflex. Fig. 13-11 provides an overview of the neural pathways for the vestibuloocular reflex. The reflex is triggered by angular acceleration of the head. The sensory receptors that are involved are the semicircular ducts (Chapter 10). When the head is rotated, for instance, to the left, the endolymph distorts the cupulas in the horizontal ducts. Activity in vestibular afferent fibers to the left horizontal duct increases and that in afferent fibers to the right duct decreases. The afferent fibers project to the medial and superior vestibular nuclei. The increased activity in the vestibular nuclei on the left side activates ascending fibers (a) that project to the left oculomotor nucleus (through the medial longitudinal fasciculus) and that excite medial rectus motor neurons and (b) that project to the right abducens nucleus and that excite lateral rectus motor neurons. The reduced activity in the vestibular nuclei on the right side has the converse effect on motoneurons to the right medial rectus and left lateral rectus. Reciprocal innervation also occurs in the ascending vestibular pathways; this innervation results in active inhibition of the motor neurons to antagonistic muscles.

As the head continues to rotate, the eyes reach the limit of their excursion. When this happens the eyes perform a saccade in the same direction as the head rotation. They then fixate a visual target and again begin to rotate in the direction opposite to the head movement. Saccadic eye movements are so rapid that visual images are blurred and therefore the visual process is minimally disrupted.

The alternation of slow and fast eye movements as the head turns is called **vestibular nystagmus.** Under these conditions of vestibular stimulation, the nystagmus is normal. However, vestibular nystagmus can also be produced by diseases that either reduce or increase vestibular afferent discharges. In such cases, the eye movements produce a sense that the world is spinning, and the subject may report dizziness.

Another reflex that affects eye position and that originates with vestibular signals is ocular counter-rolling. When the head is tilted, activation of the otolith organs causes rotation of the eyes in the opposite direction. This movement tends to keep the retinal image aligned with the horizon.

Gaze Centers

In addition to the vestibular nuclei, brainstem centers for control of eye movement include the horizontal and vertical gaze centers. The **horizontal gaze center** consists of neurons in the reticular formation (paramedian pontine reticular formation) in the vicinity of the abducens nucleus. The **vertical gaze center** is in the reticular formation of the midbrain. The circuitry and operation of the horizontal gaze center is better understood than that of the vertical gaze center, and so the discussion here will emphasize the horizontal gaze center. The horizontal gaze center controls both saccadic eye movements and smooth pursuit movements.

Fig. 13-12 is an overview of the neural circuitry by which the horizontal gaze center elicits conjugate eye movements, and Fig. 13-13 shows the activity of some of the types of neurons that form this circuitry. The right horizontal gaze center has excitatory connections with motoneurons in the ipsilateral abducens nucleus

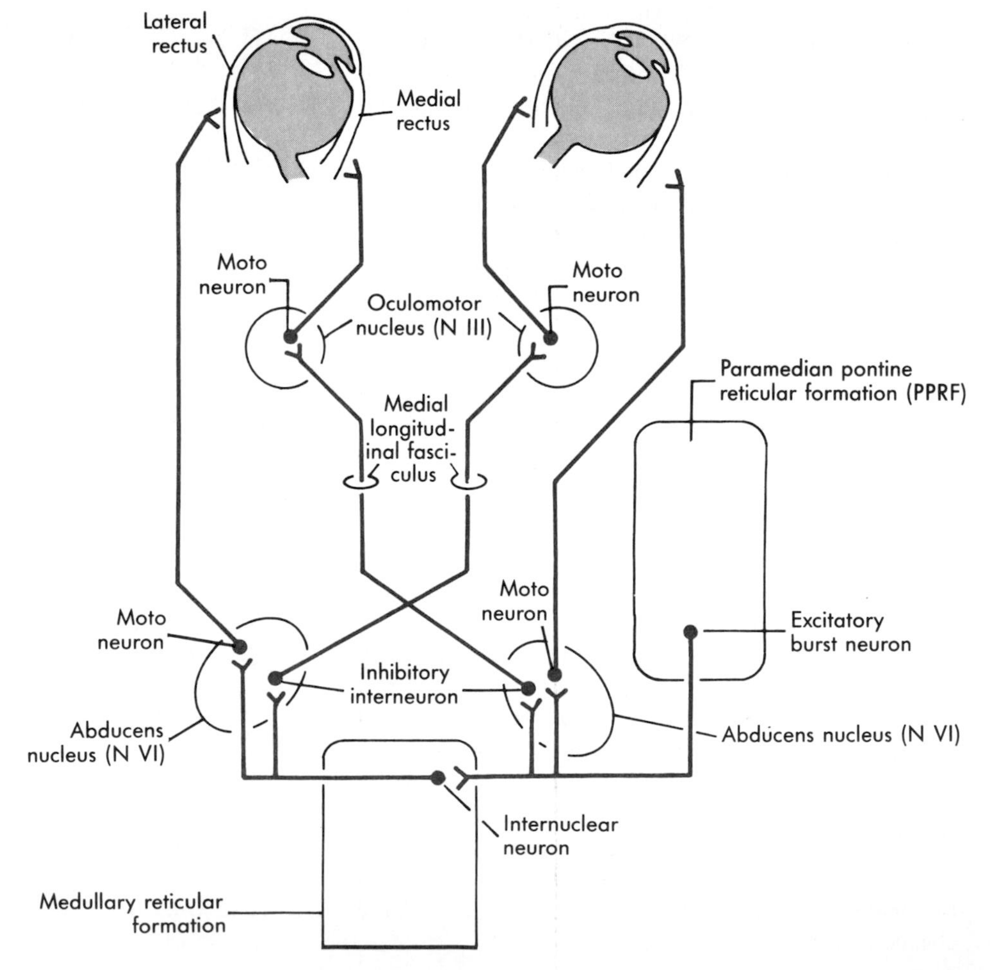

Fig. 13-12 Neural circuit by which the horizontal gaze center elicits conjugate eye movements. Excitation of burst neurons of the right horizontal gaze center causes activation of abducens motoneurons on the right and medial rectus motoneurons on the left. The ascending pathway to the oculomotor nucleus is through the MLF. The left horizontal gaze center is simultaneously inhibited by way of the reticular formation.

and with medial rectus motoneurons in the contralateral oculomotor nucleus. It also has inhibitory connections with the contralateral horizontal gaze center by way of the reticular formation.

Burst neurons in one of the horizontal gaze centers may initiate saccadic eye movements. At the same time, pause cells, which are inhibitory interneurons located in the nucleus of the dorsal raphe, cause contralateral burst cells to stop discharging. The activity of burst neurons and the inhibition of pause neurons would be triggered by commands that originate elsewhere, such as the frontal eye field located in the premotor area of the contralateral frontal lobe or the superior colliculus (Chapter 14). Tonic cells discharge during smooth pursuit movements, and burst-tonic neurons fire both during saccades and during smooth pursuit movements. Presumably, the burst helps initiate the movement and the tonic activity maintains the new eye position. Tonic activity may arise as a consequence of activity in a neural integration circuit located in the nucleus praepositus hypoglossi, a nucleus found near the midline just rostral to the hypoglossal nucleus. Motoneurons to the eye muscles involved in a saccade show a burst-tonic form of discharge to initiate movement and then to maintain eye position.

A lesion of the brainstem that destroys one of the horizontal gaze centers tonically deviates the eyes toward the opposite side. This is partly because of the paralysis of the ipsilateral lateral rectus muscle, but also because of tonic activity of the contralateral horizontal gaze center that is no longer compensated for by the center that has a lesion.

The role of the cerebellum and the cerebral cortex in the control of eye movements will be discussed in Chapter 14.

Superior colliculus. Neurons in the deep layers of the superior colliculus can trigger conjugate eye movements that cause the eyes to target novel or threatening visual, auditory or somatosensory stimuli. The superior colliculus sends impulses to the horizontal or vertical gaze centers to organize the conjugate eye movement.

Summary

1. Spinal cord motoneurons are organized topographically. Those in the lateral ventral horn supply the limb muscles, and those in the medial ventral horn supply the axial muscles. Lateral and medial groups of interneurons synapse on the respective lateral and medial groups of motoneurons. The latter project bilaterally.

2. Descending pathways can be subdivided into (a) a lateral system, which ends on motor neurons to limb muscles and on the lateral group of interneurons, and (b) a medial system, which ends on the medial group of interneurons.

3. The lateral system includes the lateral corticospinal tract and part of the corticobulbar tract, as well as the rubrospinal tract. These pathways influence contralateral motoneurons that supply the musculature of the limbs, especially the digits and muscles of the lower face and the tongue.

4. The medial system includes the ventral corticospinal, lateral and medial vestibulospinal, pontine and medullary reticulospinal, and tectospinal tracts. These pathways mainly affect posture and provide the motor background for movements of the limbs and digits.

5. Descending monoaminergic pathways affect the gen-

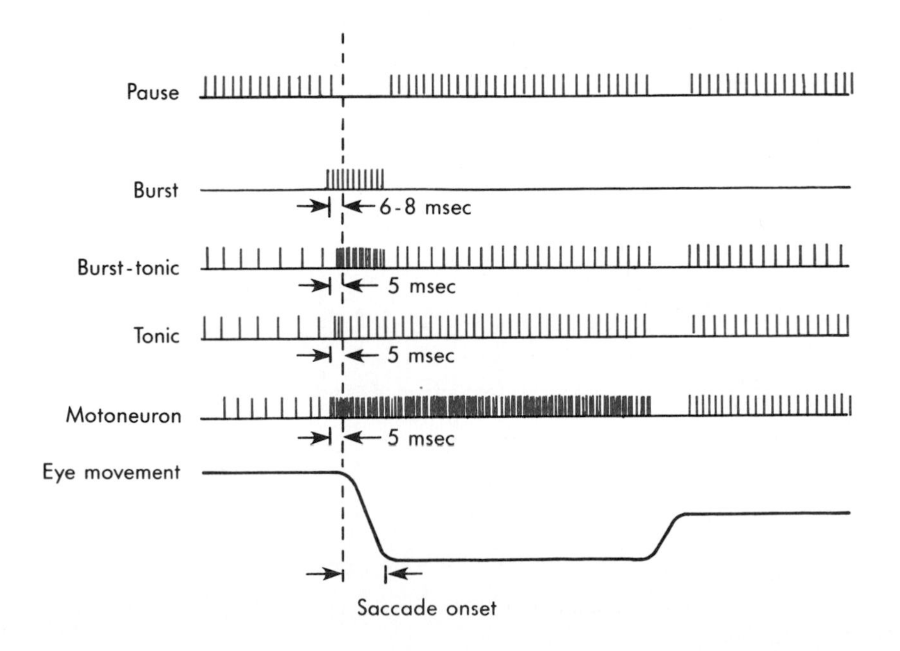

Fig. 13-13 Several types of neurons contribute to the circuitry by which the horizontal gaze center elicits conjugate eye movements. These include pause, burst, burst-tonic, and tonic neurons, in addition to motoneurons.

eral level of excitability of spinal cord reflex circuits and ascending pathways.

6. Pathways that originate in the brainstem influence posture, locomotion, and eye movements. Postural reflexes include several vestibular reflexes (activation of extensor motoneurons as a result of head movements, the vestibulocollic reflex, the vestibular placing reaction, and ocular counter-rolling), the tonic neck reflexes, and the righting reflexes.

7. Locomotion is triggered by commands relayed through the midbrain locomotor center. However, locomotor activity is organized by central pattern generators formed by spinal cord circuits and influenced by afferent input.

8. Conjugate eye movements are produced by several different control systems (vestibuloocular reflex, optokinetic reflex, smooth pursuit system, and saccadic system). Vergent eye movements are controlled by the vergence system.

9. The brainstem centers for control of eye movement include the vestibular nuclei, the horizontal and vertical gaze centers, and the superior colliculi.

Journal articles

Armstrong DM: Review lecture: the supraspinal control of mammalian locomotion, *J Physiol* 405:1, 1988.

Brown TG: The intrinsic factors in the act of progression in the mammal, *Proc R Soc Lond* B84:308, 1911.

Coulter JD, Ewing L, Carter C: Origin of primary sensorimotor cortical projections to lumbar spinal cord of cat and monkey, *Brain Res* 403:366, 1976.

Fuchs AF et al: Brainstem control of saccadic eye movements, *Annu Rev Neurosci* 8:307, 1985.

Grillner S, Wallen P: Central pattern generators for locomotion, with special reference to vertebrates, *Annu Rev Neurosci* 8:233, 1985.

Landau WM, Clare MH: The plantar reflex in man, with special reference to some conditions where the extensor response is unexpectedly absent, *Brain* 82:321, 1959.

Lisberger SG et al: Visual motion processing and sensory-motor integration for smooth pursuit eye movements, *Annu Rev Neurosci* 10:97, 1987.

Sherrington CS: Decerebrate rigidity, and reflex coordination of movements, *J Physiol* 22:319, 1898.

Shik ML, Orlovsky GN: Neurophysiology of locomotor automatism, *Physiol Rev* 56:465, 1976.

Sparks DL: Translation of sensory signals into commands for control of saccadic eye movements: role of primate superior colliculus, *Physiol Rev* 66:118, 1986.

Books and monographs

Asanuma H: *The pyramidal tract.* In Brooks VB, editor: *Handbook of physiology, sect I, The nervous system, vol II, Motor control, part 1,* Bethesda, Md, 1981, American Physiological Society.

Becker W: The neurobiology of saccadic eye movements, *Reviews of oculomotor research, vol 3,* Amsterdam, 1989, Elsevier.

Brodal A: *Neurological anatomy,* ed 3, New York, 1981, Oxford University Press.

DeJong RN: *The neurologic examination,* New York, 1979, Harper & Row.

Gelfand IM et al: Locomotion and scratching in tetrapods. In Cohen AH et al, editors: *Neural control of rhythmic movements in vertebrates,* New York, 1988, John Wiley & Sons.

Kandel ER et al: *Principles of neural science,* New York, 1991, Elsevier.

Kuypers HGJM: *Anatomy of the descending pathways.* In Brooks VB, editor: *Handbook of physiology, sect I, The nervous system, vol II, Motor control, part 1,* Bethesda, Md, 1981, American Physiological Society.

Kuypers HGJM: *A new look at the organization of the motor system.* In Kuypers HGJM, Martin GF, editors: *Descending pathways to the spinal cord,* Prog Brain Res 57:390, 1982.

Lundberg A: Control of spinal mechanisms from the brain. In Tower DB, editor: *The nervous system, Vol 1, The basic neurosciences,* New York, 1975, Raven Press.

Phillips CB, Porter R: *Corticospinal neurones; their role in movement,* London, 1977, Academic Press.

Roberts TDM: *Neurophysiology of postural mechanisms,* London, 1979, Butterworth.

Wilson VJ, Melville Jones G: *Mammalian vestibular physiology,* New York, 1979, Plenum Press.

CHAPTER

14

Motor Control by the Cerebral Cortex, Cerebellum, and Basal Ganglia

■ *Voluntary Movements*

In Chapter 12 emphasis was placed on reflexes, which are simple, stereotyped motor acts that occur in response to specific stimuli. This chapter will be concerned with the neural basis for voluntary movements. These are often complex, they may vary when repeated, and they are often initiated as a result of cognitive processes rather than in response to an external stimulus.

The most important descending pathway for fine movements that use distal muscles, such as those that control the hand and fingers, is the lateral corticospinal tract (a portion of the corticobulbar tract serves an equivalent function for the head). However, many other pathways are also engaged to activate more proximal and axial muscles, as described in Chapter 13.

Commands carried by descending motor pathways must first be organized in the brain. The target of the movement is identified by pooling sensory information in the posterior parietal cerebral cortex. This information is transmitted to the supplementary motor and premotor areas, so that a motor plan can be developed. The plan includes what muscles need to be contracted, by how much, and in what sequence. Then the motor plan is implemented by commands transmitted from the primary motor cortex through the descending pathways. However, successful execution of the motor commands needs to be monitored, with feedback provided to the motor cortex through the ascending pathways to the somatosensory cortex, as well as through the visual pathway. During both the planning stage and the execution of a movement, motor processing is also provided by two major motor control systems, the cerebellum and the basal ganglia.

■ *Motor Control by the Cerebral Cortex*

■ *Cortical Motor Areas*

As discussed in Chapter 13, the primary motor cortex was originally defined on the basis of experiments in which electrical stimuli applied to the cortex evoked discrete, contralateral movements. However, other cortical areas can also evoke movements when stimulated with more intense stimuli. On the basis of such studies, the deficits produced by lesions, anatomical experiments, electrophysiological recordings, and modern imaging studies in humans, it is possible to recognize several "motor" areas of the cerebral cortex (Fig. 14-1). These include the primary motor cortex in the precentral gyrus, the premotor area just rostral to this, the secondary somatosensory cortex in the roof of the lateral fissure (usually called SII), and the supplementary motor cortex on the medial aspect of the hemisphere. In a part of the premotor area just rostral to the face representation in the motor cortex are found the frontal eye fields.

Stimuli applied to the surface of the **primary motor cortex** evoke discrete contralateral movements that involve several muscles. However, microstimulation within the cortex with microelectrodes can elicit contractions of individual muscles. Mapping studies reveal that the motor cortex can be considered to be made up of a mosaic of motor points related to particular muscles or sets of muscles. These points are termed **cortical efferent zones,** which can be regarded as motor columns. They are organized somatotopically and in sum produce the motor homunculus (see Fig. 13-5).

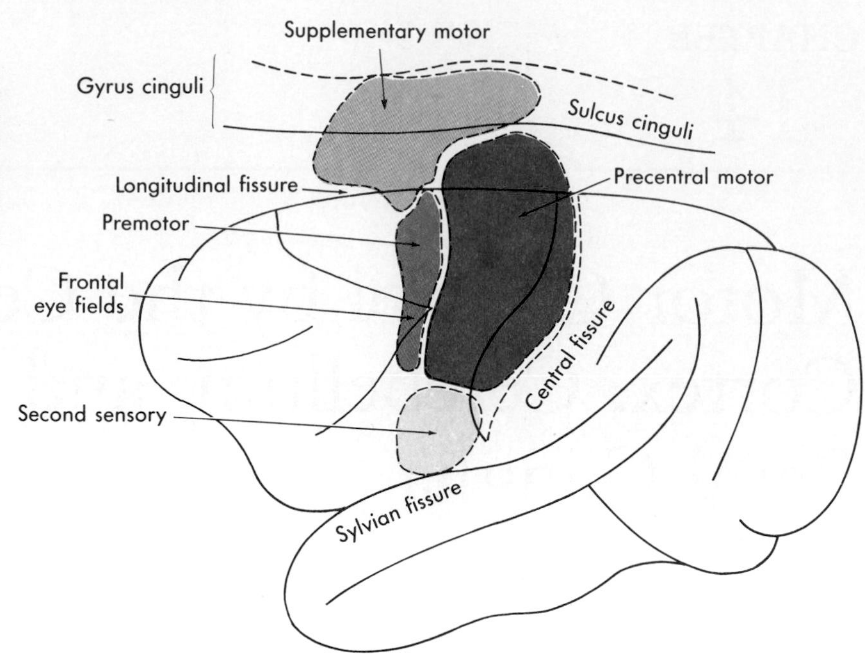

■ **Fig. 14-1** Motor regions of the cerebral cortex of a monkey. The motor cortex, premotor area, secondary motor cortex, and supplementary motor cortex are indicated with different colors. The dashed lines around the secondary and supplementary cortex are meant to indicate that these areas are hidden from surface view. (Redrawn from Eyzaguirre C, Fidone SJ: *Physiology of the nervous system,* ed 2, Chicago, 1975, Year Book Medical Publishers.

Stimulation in the **supplementary motor cortex** can produce vocalization or complex postural movements, such as a slow movement of the contralateral hand in an outward, backward, and upward direction, accompanied by movement of the head and eyes toward the hand. The postural movements can be bilateral. Rhythmic movements can sometimes be elicited from this area of cortex, but stimulation can also have the opposite result, a temporary arrest of movement or speech. Removal of the supplementary motor cortex produces slowed movements of the opposite extremities and a tendency to make forced grasping movements with the opposite hand.

Stimulation of the **secondary somatosensory (SII) cortex** generally evokes sensation. However, a sensation of movement may occur, and voluntary movements are often inhibited; this indicates a role of SII in motor function. Ablation of SII results in neither a sensory nor a motor deficit.

Stimulation of the **frontal eye fields** in one hemisphere causes a contralateral saccadic conjugate deviation of the eyes. Vertical saccades require bilateral stimulation of the frontal eye fields. Ablation of the frontal eye field on one side transiently weakens the contralateral gaze, and in humans the eyes may deviate toward the side of the lesion. Memory-guided saccades are eliminated, but visually evoked saccades persist. However, bilateral lesions of the frontal eye fields and the superior colliculi eliminate saccadic eye movements.

■ *Connections of the Motor Regions of the Cortex*

The motor areas of the cortex receive input from a number of sources (Fig. 14-2). Ascending pathways that relay in the thalamus provide information about somatosensory events. This information can reach the motor cortex directly from the thalamus (from the ventral lateral nucleus) or indirectly by way of the SI somatosensory cortex. Both somatosensory and visual information is conveyed to the motor areas from the posterior parietal cortex. The frontal eye fields receive visual input from the occipital lobe (connection not shown). The motor areas of the cortex also receive information through circuits that interconnect with the cerebellum and basal ganglia. In addition, the motor regions of the cortex are interconnected.

The output of the motor regions of the cortex to the spinal cord and brainstem is through several descending pathways, including not only direct projections through the corticospinal and corticobulbar tracts, but also indirect projections (by way of corticorubral and corticoreticular fibers) through the red nucleus and reticular formation. The motor regions also contribute to the cerebellar and basal ganglion circuits. The frontal eye fields project to the superior colliculus and also to the pontine and mesencephalic reticular formation.

■ *Role of Premotor and Supplementary Motor Areas in Motor Programming*

Preparation for volutional movements requires several hundred milliseconds, the exact time depending on the difficulty of the task. The ability to prepare for voluntary movements is hampered by lesions of the premotor, supplementary motor, and posterior parietal areas. When such a lesion is produced experimentally, the result resembles **apraxia,** the failure of patients with frontal or parietal lobe lesions to perform complex movements despite retention of sensation and the ability to make simple movements.

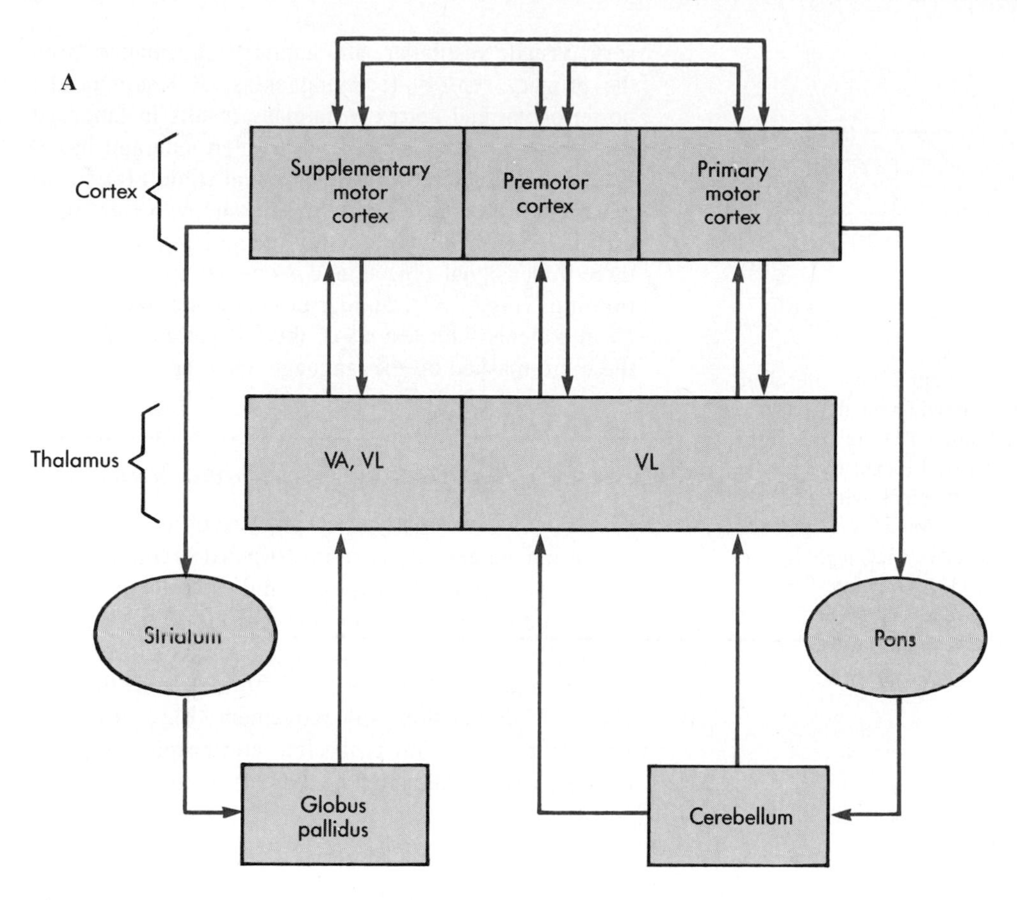

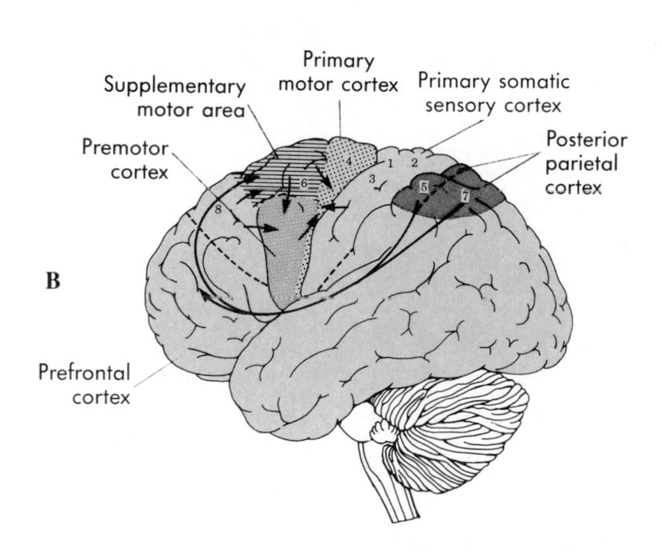

■ **Fig. 14-2** **A,** Some of the connections of the motor areas of the cortex with subcortical structures. *VA,* Ventral anterior thalamic nucleus; *VL,* ventral lateral thalamic nucleus. **B,** Some of the cortical projections involving the motor areas. Arrows indicate transmission in one direction, but actually there are also connections in the opposite direction. (Redrawn from Kandel ER, Schwartz JH, Jessell TM: *Principles of neural science,* ed 3, New York, Elsevier, 1991.)

The supplementary motor cortex is involved in motor programming and is active during both the planning and the execution of complex (but not simple) movements. Its actions are partly mediated by direct corticospinal connections, but they partly depend on a relay to the primary motor cortex. When the regional blood flow of the cerebral cortex is monitored during motor tasks, it increases when the individual is just thinking about a movement, as well as when the movement is executed. By contrast, the blood flow in the primary motor cortex increases only when the movement is executed. Besides its role in motor planning, the supplementary motor cortex may assist in the coordination of posture and voluntary movements.

The premotor cortex receives a major input from the posterior parietal cortex, and its output influences chiefly the medial system of descending pathways. These connections suggest that this region of cortex controls the axial muscles. Neurons in this area seem to discharge during preparation for a movement.

The posterior parietal cortex is often called the parietal association cortex. This region receives somatosen-

■ **Fig. 14-3** Drawing made by a patient 2 days after damage to the right parietal lobe. The drawing on the left was made by the physician to show a house. The drawing in the middle was made by the patient and was meant to copy the physician's drawing. The drawing of the clock with all of the numbers on the right indicates neglect of the left extrapersonal space by the patient. (From Cotman CW, McGaugh JL: *Behavioral neuroscience,* New York, 1980, Academic Press.

sory, visual, vestibular, and auditory information from the primary sensory receiving areas. A lesion of the posterior parietal cortex in humans results in language disorders when the lesion is on the left side and in neglect of contralateral somatic or visual stimuli **(agnosia)** when the lesion is on the right. Patients with right parietal lobe lesions have difficulty recognizing or drawing three-dimensional objects and recognizing spatial relationships (Fig. 14-3). Similar problems are likely to exist in patients with lesions of the left parietal lobe, but these are masked by the language disorder.

■ *Activity of Individual Corticospinal Neurons*

The role of individual corticospinal neurons in the control of movements has been investigated in trained monkeys by recording discharges from these cells in the primary motor cortex during movements. The monkey learns to execute a simple movement, such as wrist flexion, and recordings are made from neurons that discharge in association with movement (Fig. 14-4). An important observation is that the corticospinal neurons discharge before the onset of the movement, which sug-

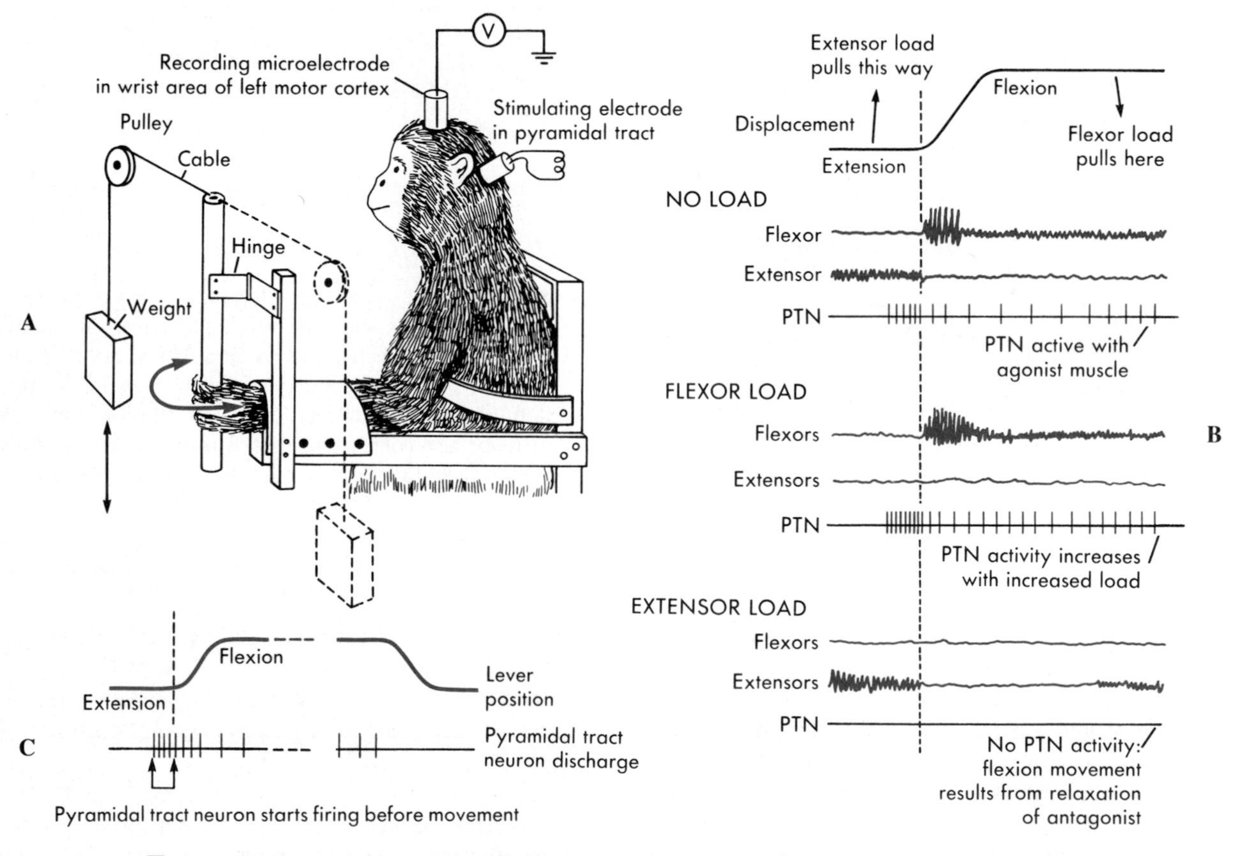

■ **Fig. 14-4** **A,** The experimental arrangement for recording from a corticospinal neuron during trained movements of the wrist. **B** and **C,** The cell discharges before the movement. With an extensor load (**B,** lower trace), the cell fails to discharge, indicating that it encodes force rather than displacement. (Redrawn from Kandel ER, Schwartz JH: *Principles of neural science,* New York, 1981, Elsevier Science Publishing.)

gests that these cells cause the movement. Furthermore, spike-triggered averaging of the electromyogram from the appropriate muscle reveals that a particular corticospinal neuron excites a particular motoneuron monosynaptically. The discharges of corticospinal neurons appear to relate to the contractile force that generates the movement or to the rate of change in force, rather than to the position of the joint.

Corticospinal neurons that influence more proximal muscles inhibit extensor and excite flexor muscles. Apparently the corticospinal output in this case is directed at postural changes that support fine movements of the distal muscles. Other corticospinal neurons trigger the fine movements through excitatory connections to motoneurons that supply the distal muscles.

A given corticospinal neuron may discharge before movements in various directions. However, it tends to discharge most vigorously when the movement occurs in a preferred direction. Neurons in a motor column have the same preferred direction of movement. Commands from a population of corticospinal neurons that have somewhat different preferred directions will determine the actual direction of a movement.

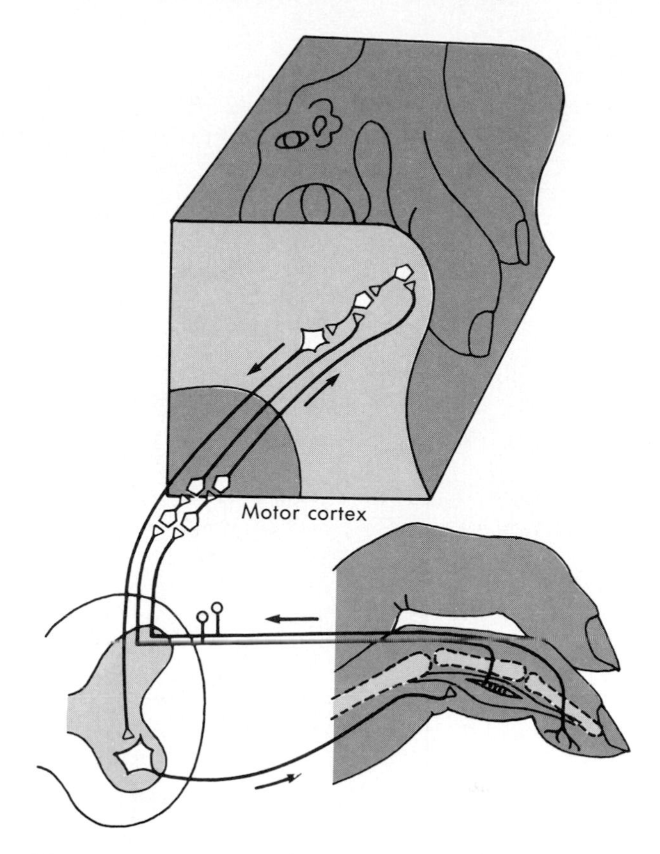

■ **Fig. 14-5** Sensory input to a corticospinal neuron that causes flexion of a digit. (Redrawn from Asanuma H: *Physiologist* 16:143, 1973.)

■ *Sensory Feedback*

As already emphasized, the corticospinal neurons of the primary motor cortex receive sensory information through the thalamus, as well as from sensory areas of the cerebral cortex. This information is used by the motor cortex to ensure that the movements it evokes are appropriate.

Fig. 14-5 shows how somatosensory information influences corticospinal neurons. Cutaneous receptors and also proprioceptive receptors, such as muscle spindles, transmit information about mechanical contact of the fingertip with a surface and about the position of the finger joints by way of the ascending somatosensory pathways, such as the dorsal column–medial lemniscus system (Chapter 8). In this case, the corticospinal neuron activates motoneurons that flex the digit. As the fingertip moves into contact with a surface, cutaneous receptors in the skin of the ventral surface of the fingertip discharge and thereby excite the corticospinal neuron through somatosensory projections. Muscle spindle afferent fibers activated by stretch of the flexor muscle as the finger contacts a surface will also activate the corticospinal neuron. Thus sensory feedback from both skin and muscle facilitates the activity of the corticospinal neuron and enhances the movement.

Another example of sensory feedback to the motor cortex is illustrated in Fig. 14-6. In this preparation the corpus callosum and the optic chiasm of the monkey were previously transected surgically, so that visual information from either eye could reach only the ipsilateral cortex, and no information could be transmitted from one hemisphere to the other. The monkey's task was to remove a pellet of food from one of the wells on a board. This requires fine movements of the digits and thus corticospinal control originating on the side of the brain contralateral to the hand used. When the right eye was open, the monkey could retrieve the food pellets with its left hand (Fig. 14-6, *A*). However, if the right eye was masked, the monkey could no longer retrieve the food with its left hand (Fig. 14-6, *B*), although it could with its right hand. It did so because visual information that reached the left hemisphere through the left eye influenced corticospinal neurons in the left motor cortex, which in turn controlled fine movements of the digits in the right hand (Fig. 14-7). The pathway that conveys the visual information to the motor cortex passes through the posterior parietal lobe.

■ *Motor Control by the Cerebellum*

The cerebellum helps regulate movements and posture, and it plays a key role in certain forms of motor learning. It is noteworthy that removal of the cerebellum affects neither sensation nor muscle strength.

The aspects of movements that are influenced by the cerebellum include their rate, range, force, and direc-

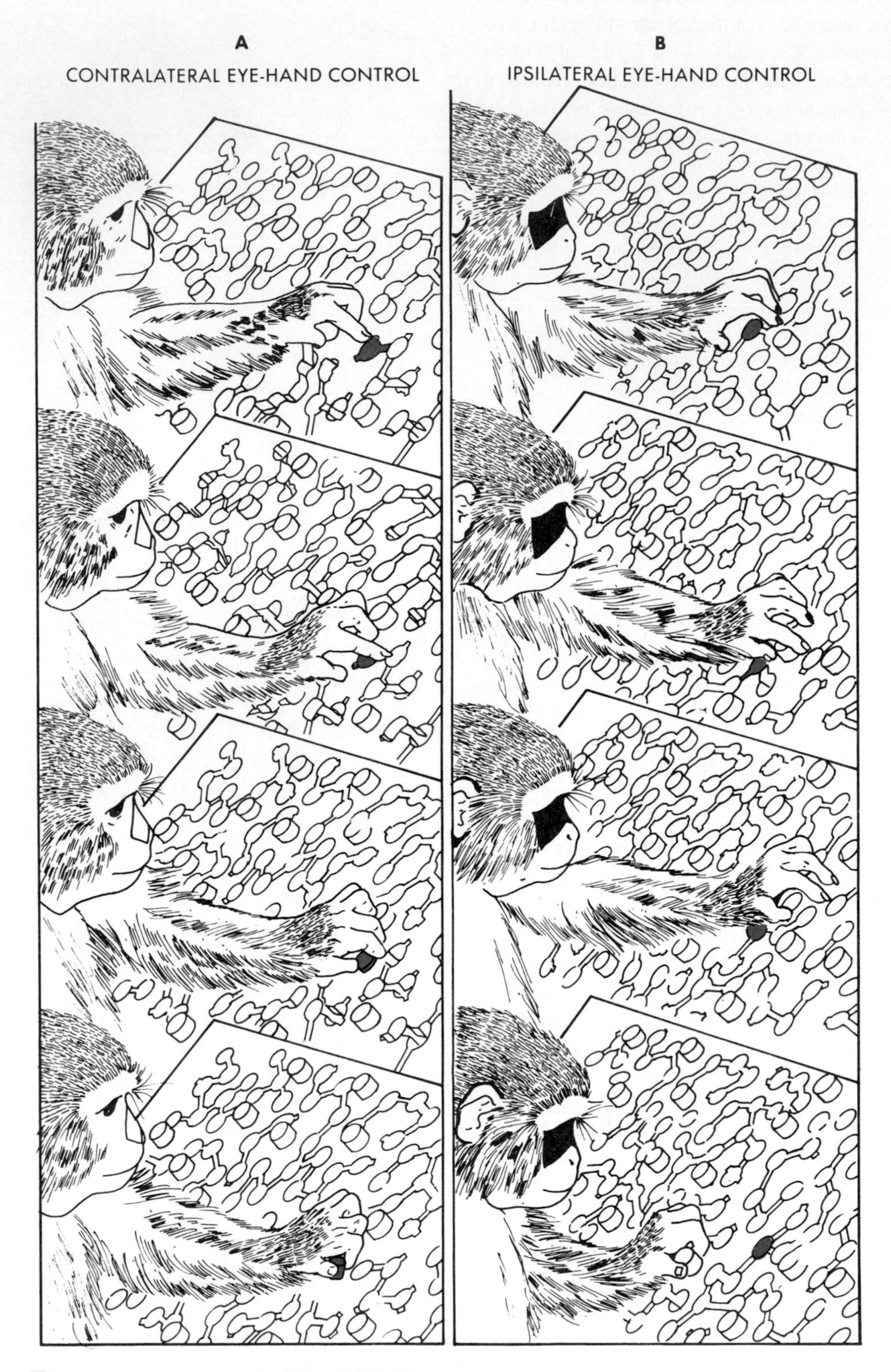

■ **Fig. 14-6** Drawings showing hand and finger movements by a monkey subjected to a complete commissurotomy. **A,** The monkey is able to retrieve a food pellet from a well, using a precision grip involving hand and digit muscles. **B,** The task cannot be performed by the left hand when the right eye is masked. (Redrawn from Brinkman C: *Split-brain monkeys: cerebral control of contralateral and ipsilateral arm, hand, and finger movements,* doctoral dissertation, Rotterdam, The Netherlands, 1974, Erasmus University.)

tion. Damage to the cerebellum results in incoordination of movements because of the loss of this regulation.

The role of the cerebellum in posture can be demonstrated in animals in which (1) stimulation of the cerebellum inhibits decerebrate rigidity, and (2) removal of

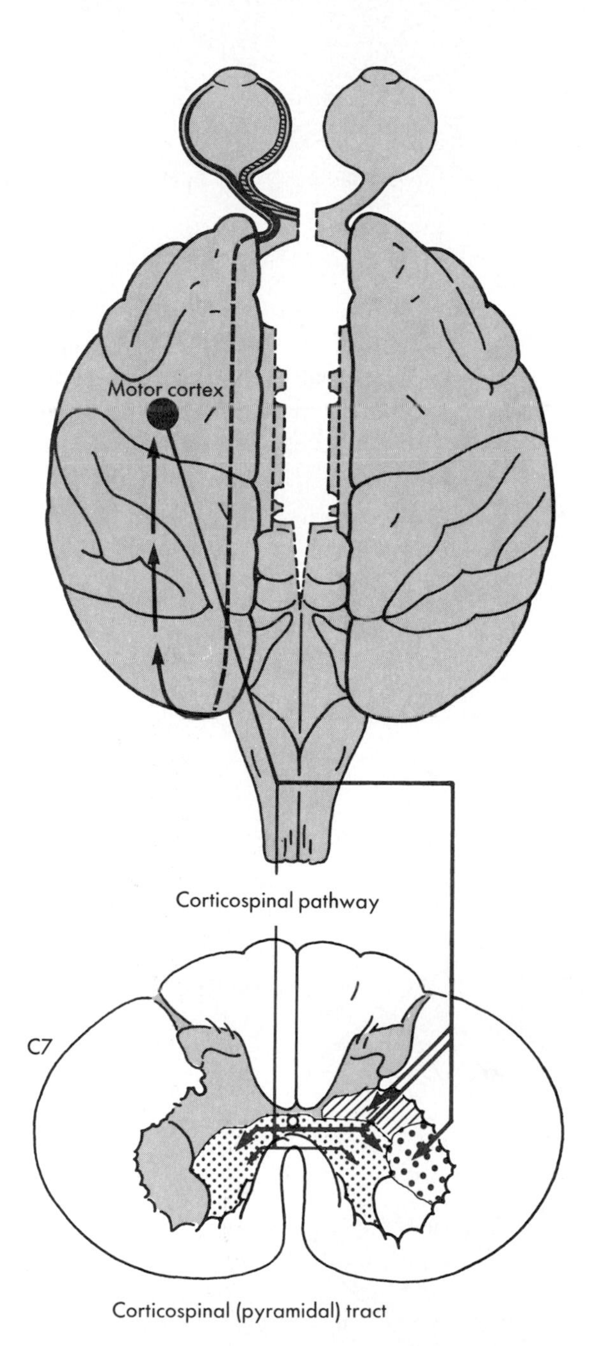

■ **Fig. 14-7** Split-brain preparation, as in Fig. 14-6. Drawing above shows the relationship of pathways carrying visual information to the left hemisphere from the left eye to the corticospinal neurons on the left side that control hand and digit movement on the right. (Redrawn from Brinkman C: *Split-brain monkeys: cerebral control of contralateral and ipsilateral arm, hand, and finger movements,* doctoral dissertation, Rotterdam, The Netherlands, 1974, Erasmus University.)

the cerebellum enhances the rigidity. In humans the dominant postural effect of cerebellar action is opposite that in animals, because cerebellar damage in humans usually results in hypotonia. Disequilibrium can also result from cerebellar damage.

The cerebellum helps mediate changes in the gain of the vestibuloocular reflex, and it participates in the improvements in motor performances that result from practice of motor skills. It is therefore thought to be involved in motor learning.

The cerebellum exerts its motor control on a moment-by-moment basis. It uses sensory information from a variety of sources, but most prominently from the proprioceptive system, to update its computations of body position, muscle length, and muscle tension. The cerebellum may be able to compare this sensory feedback with motor commands that are transmitted to the cerebellum from the motor areas of the cerebral cortex. Errors are corrected by output signals from the cerebellum to other components of the motor system.

■ *Cerebellar Organization*

The cerebellum ("little brain") is located in the posterior fossa of the cranium just below the occipital lobe. It is connected to the brainstem. In a surface view, only the cortex is visible. The cerebellar cortex is subdivided into three rostrocaudally arranged lobes, the anterior lobe, the posterior lobe, and the flocculonodular lobe (Fig. 14-8). The cerebellar lobes are separated by major fissures, the **primary fissure** and the **posterolateral fissure,** and each lobe is made up of one or more **lobules.** Each of the lobules has an esoteric name, but the lobules can also be assigned numbers. In this scheme, the anterior lobe consists of lobules I to V, the posterior lobe of lobules VI to IX, and the flocculonodular lobe of lobule X (Fig. 14-9). Each lobule of the cerebellar cortex is composed of a series of transverse folds, called folia.

The cerebellum has not only this rostrocaudal organization, but also a sagittal organization. At the midline is the **vermis** (named for its segmented appearance, which resembles an earthworm). Extending laterally are the **hemispheres.** In the paravermal region, between the hemispheres and vermis, is the **intermediate region.**

Both the cerebellum and the cerebrum are cortical structures, and in both parts of the brain white matter is found beneath the cortex. Deep nuclei are buried within the white matter. When the cerebellum is sectioned, as in Fig. 14-10, the white matter of the cerebellum can be seen beneath the cortex. Buried in the white matter are the deep cerebellar nuclei. The dentate nucleus, shown in Fig. 14-10, is the most lateral of these. The other deep nuclei include the emboliform, globose, and fastigial nuclei. (In cats the emboliform and globose nuclei combine to form the interpositus nucleus.)

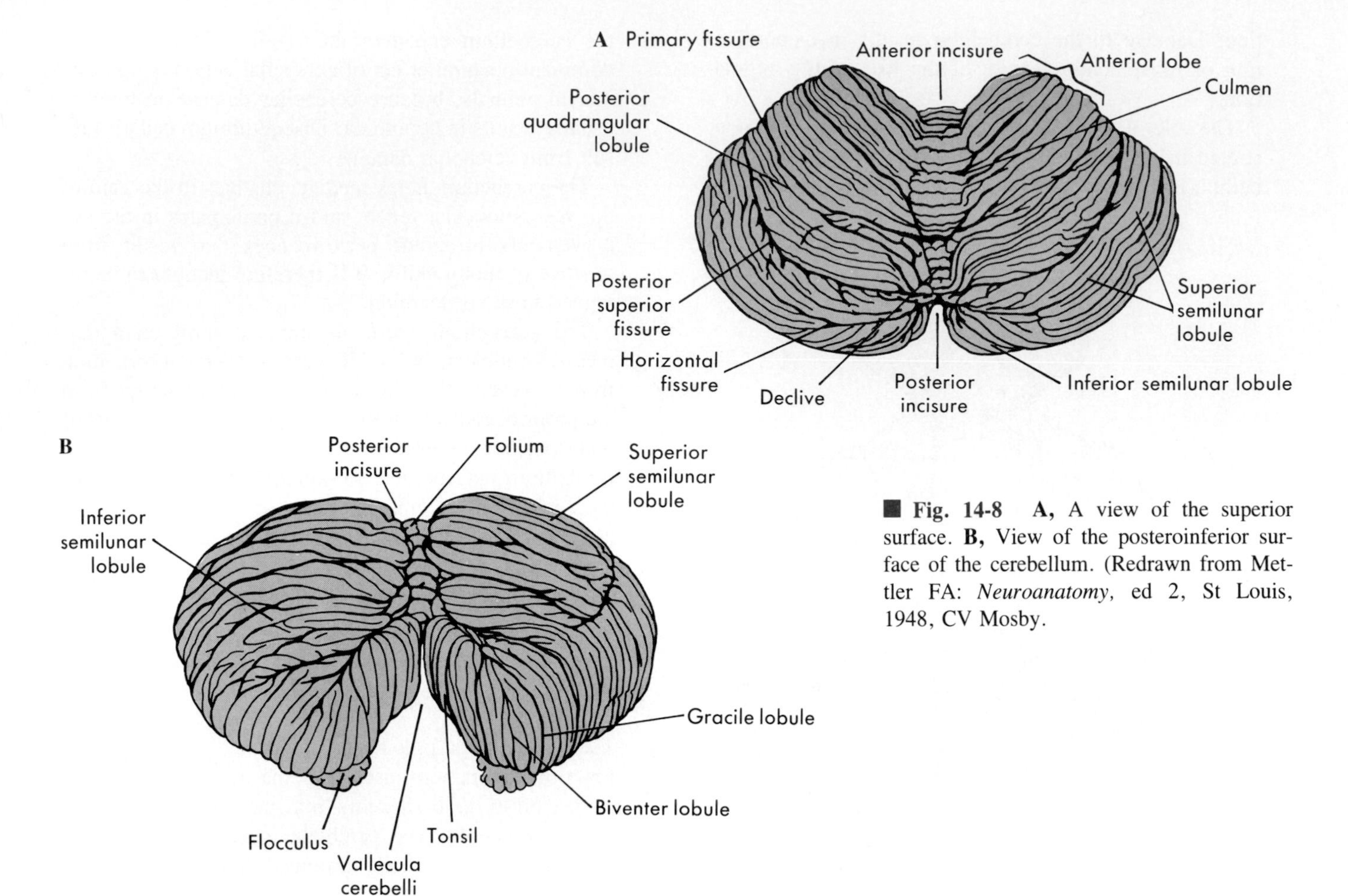

■ **Fig. 14-8** **A,** A view of the superior surface. **B,** View of the posteroinferior surface of the cerebellum. (Redrawn from Mettler FA: *Neuroanatomy,* ed 2, St Louis, 1948, CV Mosby.

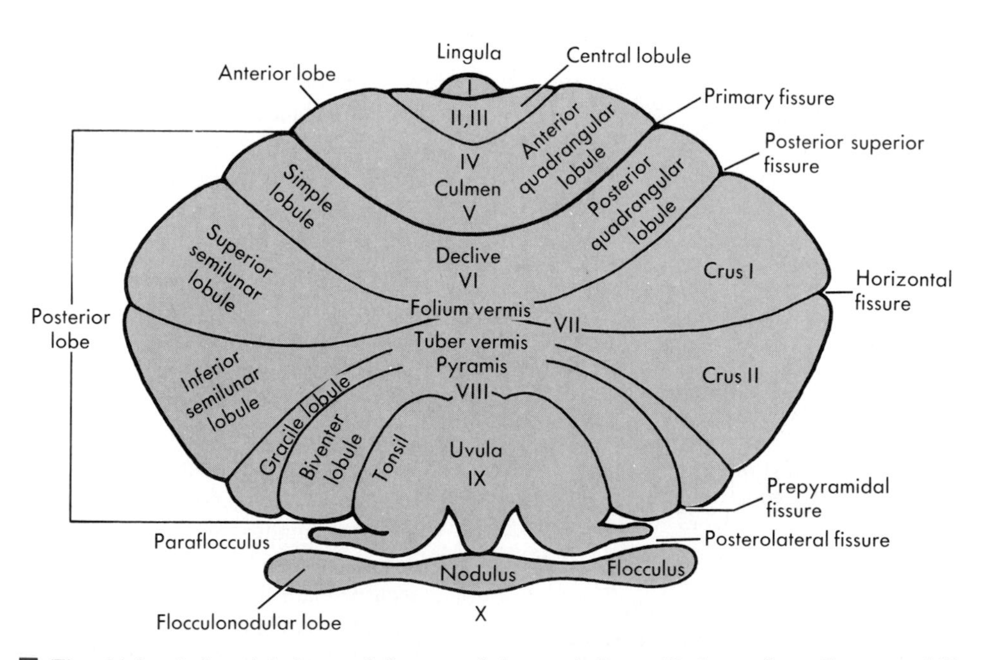

■ **Fig. 14-9** Lobes, lobules, and fissures of the cerebellum. (Redrawn from Carpenter MB: *Human neuroanatomy,* ed 7, Baltimore, 1976, Williams & Wilkins; modified from Larsell O: *Anatomy of the nervous system,* ed 2, New York, 1957, Appleton–Century–Croft, and Angervine JB et al: *The human cerebellum,* Boston, 1961, Little, Brown, & Co.)

The cerebellum is connected to the brainstem by three pairs of cerebellar peduncles (Fig. 14-10). The inferior cerebellar peduncle (or restiform body) connects the cerebellum with the medulla; the middle cerebellar peduncle connects the cerebellum with the pons; and the superior cerebellar peduncle connects the cerebellum with the midbrain. The cerebellar peduncles contain axons that are either cerebellar afferent or cerebellar efferent nerve fibers. Most of the axons in the inferior cerebellar peduncle and all of those in the middle cerebellar peduncle are cerebellar afferent fibers, whereas most of the axons in the superior cerebellar peduncle are cerebellar efferent fibers.

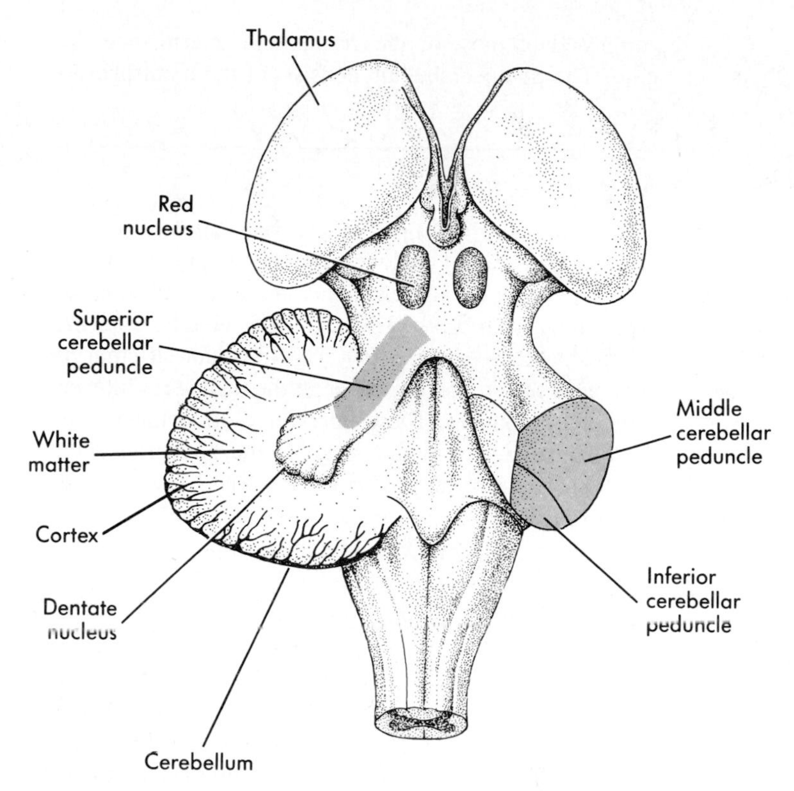

■ **Fig. 14-10** Cerebellar peduncles. (Redrawn from Carpenter MB: *Human neuroanatomy,* ed 7, Baltimore, 1976, Williams & Wilkins.)

■ *Subdivisions of the Cerebellum*

Based on phylogenetic grounds, the cerebellum can be subdivided into the archicerebellum, paleocerebellum, and neocerebellum (Fig. 14-11). These subdivisions correspond to regions of the cerebellum that are dominated by vestibular input (the **vestibulocerebellum**), by spinal cord input (the **spinocerebellum**), and by indirect input from the cerebrum by way of the pontine nuclei (the **pontocerebellum**). The vestibulocerebellum consists chiefly of the flocculonodular lobe, but it also includes a small part of the vermis and intermediate cortex of the posterior lobe. The spinocerebellum is

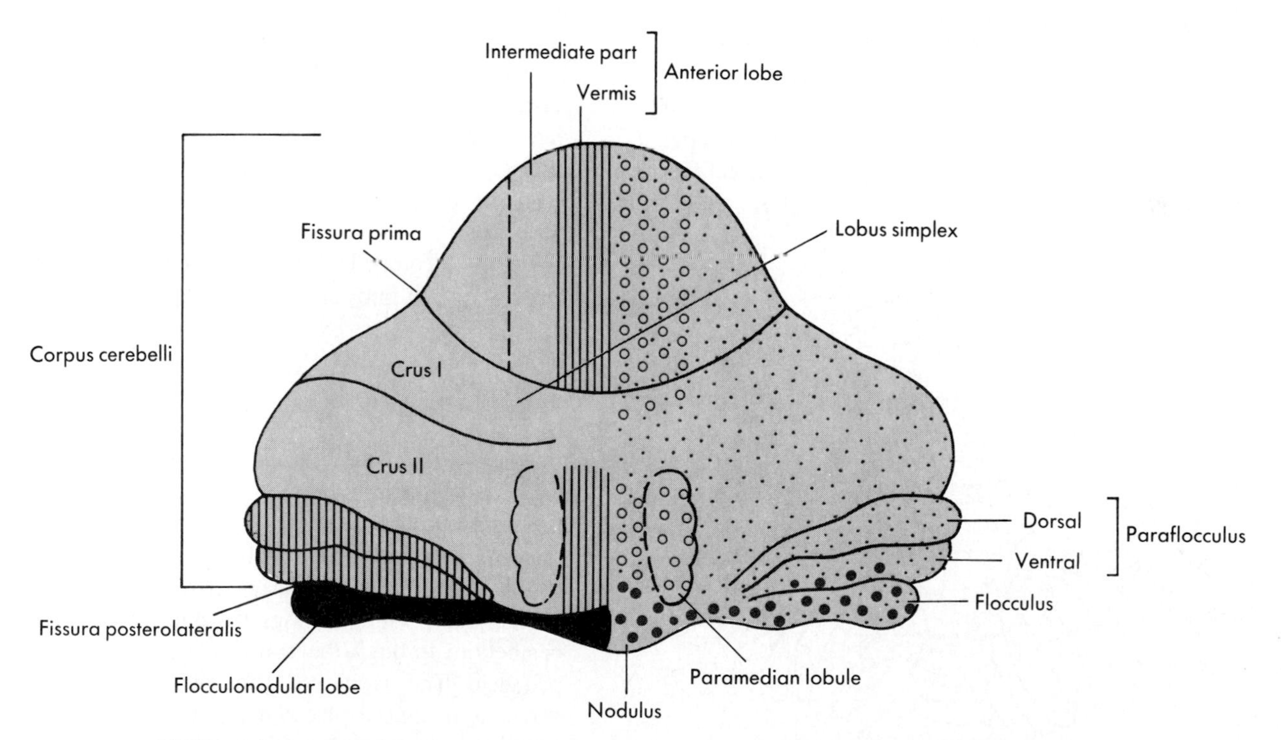

■ **Fig. 14-11** The subdivision of the cerebellum into archicerebellum, paleocerebellum, and neocerebellum is shown by the black, hatched, and gray areas. Terminations of vestibulocerebellar fibers are indicated by the large dots, spinocerebellar fibers by the open circles, and pontocerebellar fibers by the small dots. (Redrawn from Brodal A: *Neurological anatomy,* ed 3, New York, 1981, Oxford University Press.)

composed of most of the vermis and intermediate region. The pontocerebellum consists of the hemispheres.

Afferent Pathways

Vestibulocerebellum. The afferent pathways of the vestibulocerebellum include both direct projections of primary vestibular afferent fibers and also second-order projections from the vestibular nuclei, chiefly from the inferior nucleus (Chapter 10). The vestibular afferent fibers enter the cerebellum through the ipsilateral inferior cerebellar peduncle. These fibers give off collaterals to the fastigial nuclei before reaching the flocculonodular lobe and parts of the posterior lobe. The vestibulocerebellum regulates eye movements, stance, and gait.

Spinocerebellum. The spinocerebellum has several somatotopic maps, one in the anterior lobe and another in the posterior lobe (Fig. 14-12). The trunk is represented in the vermis and the extremities in the intermediate zone. The head is oriented toward the primary fissure in both maps. Thus the two maps are inverted with respect to each other. Visual and auditory information is received in the areas that represent the head. Stimulation of the cerebellar cortex results in movements of parts of the body that correspond to the sensory maps.

The ascending pathways that convey somatosensory information to the spinocerebellum include the dorsal and ventral spinocerebellar tracts from the lower extremities and trunk and the cuneocerebellar and rostral spinocerebellar tracts from the upper extremities and trunk. The dorsal spinocerebellar tract originates from **Clarke's column,** a nucleus found in the thoracic and upper lumbar spinal cord. This nucleus receives input from muscle stretch receptors, as well as from cutaneous receptors, through the dorsal funiculus, and it projects its axons to the cerebellum by way of the ipsilateral dorsal lateral funiculus and the inferior cerebellar peduncle. It gives off collaterals to the emboliform and globose nuclei en route to the cortical representations of the lower extremities and trunk.

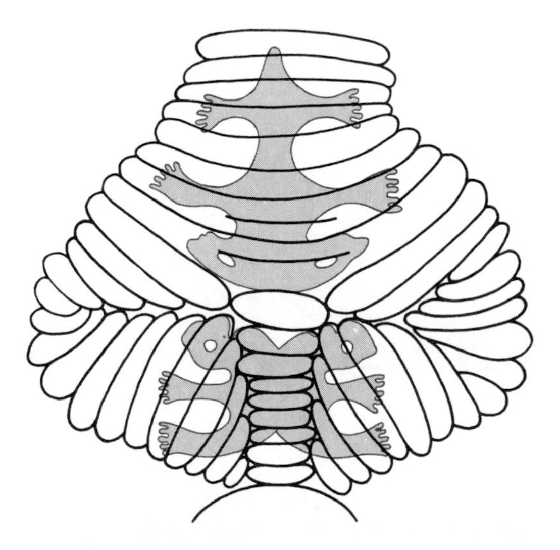

Fig. 14-12 Somatotopic maps in the spinocerebellum. (Redrawn from Snider R: The cerebellum, *Sci Am* 199:4, 1958.)

The cuneocerebellar tract originates from the lateral cuneate nucleus, which is located in the caudal medulla just lateral to the main cuneate nucleus. The neurons of the lateral cuneate nucleus receive proprioceptive information and project to the cerebellum through the adjacent inferior cerebellar peduncle. Again, collaterals are provided to the emboliform and globose nuclei, and the projection ends in the forelimb and upper trunk representations. The dorsal and cuneocerebellar pathways are responsible for providing discrete, moment-by-moment information to the cerebellum about limb position and muscle actions.

The ventral and rostral spinocerebellar tracts have a more complex organization, provide less discrete information, project bilaterally in the cerebellum, and operate under the control of descending systems. There are also indirect projections from the spinal cord to the cerebellum through relays in the dorsal column nucleus and in the lateral reticular nucleus of the medulla.The spinocerebellum regulates truncal and proximal limb movements.

Pontocerebellum. Wide areas of the cerebral cortex, including parts of the frontal, parietal, and temporal lobes, provide input to the pontocerebellum indirectly by connections to the pontine nuclei. The pontine nuclei also receive connections from the spinal cord and from brainstem pathways. The pontocerebellar tract originates from the pontine nuclei, crosses the midline, and enters the cerebellum through the middle cerebellar peduncle. This tract gives off collaterals to the dentate nucleus on its way to the cortex of the cerebellar hemisphere.

The pontocerebellum regulates movements of the distal parts of the limbs and participates in motor planning.

Inferior Olive

The inferior olivary nucleus is a massive structure located in the rostral medulla. This nucleus receives input from the vestibular system, the spinal cord, and the cerebral cortex through a number of pathways. The cells of the inferior olivary nucleus give rise to the olivocerebellar tract, which crosses the midline and enters the cerebellum through the contralateral inferior cerebellar peduncle. The axons are distributed to all parts of the cerebellum, and send collaterals to the deep cerebellar nuclei *and* to the cortex. Their terminals form a special type of cerebellar afferent fibers called climbing fibers; by contrast, all of the other cerebellar afferent pathways terminate in the cerebellar cortex as mossy fibers (see p. 235).

Cerebellar Cortex

The cerebellar cortex is organized in reference to its output cell, the Purkinje cell. The cell bodies of Purkinje cells form the middle of the three layers of the cerebellar cortex. The other layers are the granular layer and the molecular layer (Fig. 14-13).

The granular layer is adjacent to the white matter and contains many interneurons, including numerous granule cells (there are about 10^{11} granule cells, about half the number of neurons in the brain!) and Golgi cells. The mossy fibers form excitatory terminals in the cerebellar cortex on the dendrites of granule cells. A given granule cell receives convergent inputs from many mossy fibers. The terminal zones are called cerebellar glomeruli. The Golgi cells provide inhibitory projections to the glomeruli.

The axons of the granule cells ascend through the Purkinje cell layer to the molecular layer, where they bifurcate and form **parallel fibers.** The parallel fibers pass along the long axis of the folium and form excitatory synapses with the dendrites of the Purkinje cells and the Golgi cells and also with interneurons of the molecular layer, the stellate and basket cells. A given parallel fiber synapses with about 50 Purkinje cells, and a given Purkinje cell receives synapses from about 200,000 parallel fibers.

The stellate and basket cells are inhibitory interneurons that synapse with Purkinje cells. The stellate cells form terminals on the dendrites of the Purkinje cells,

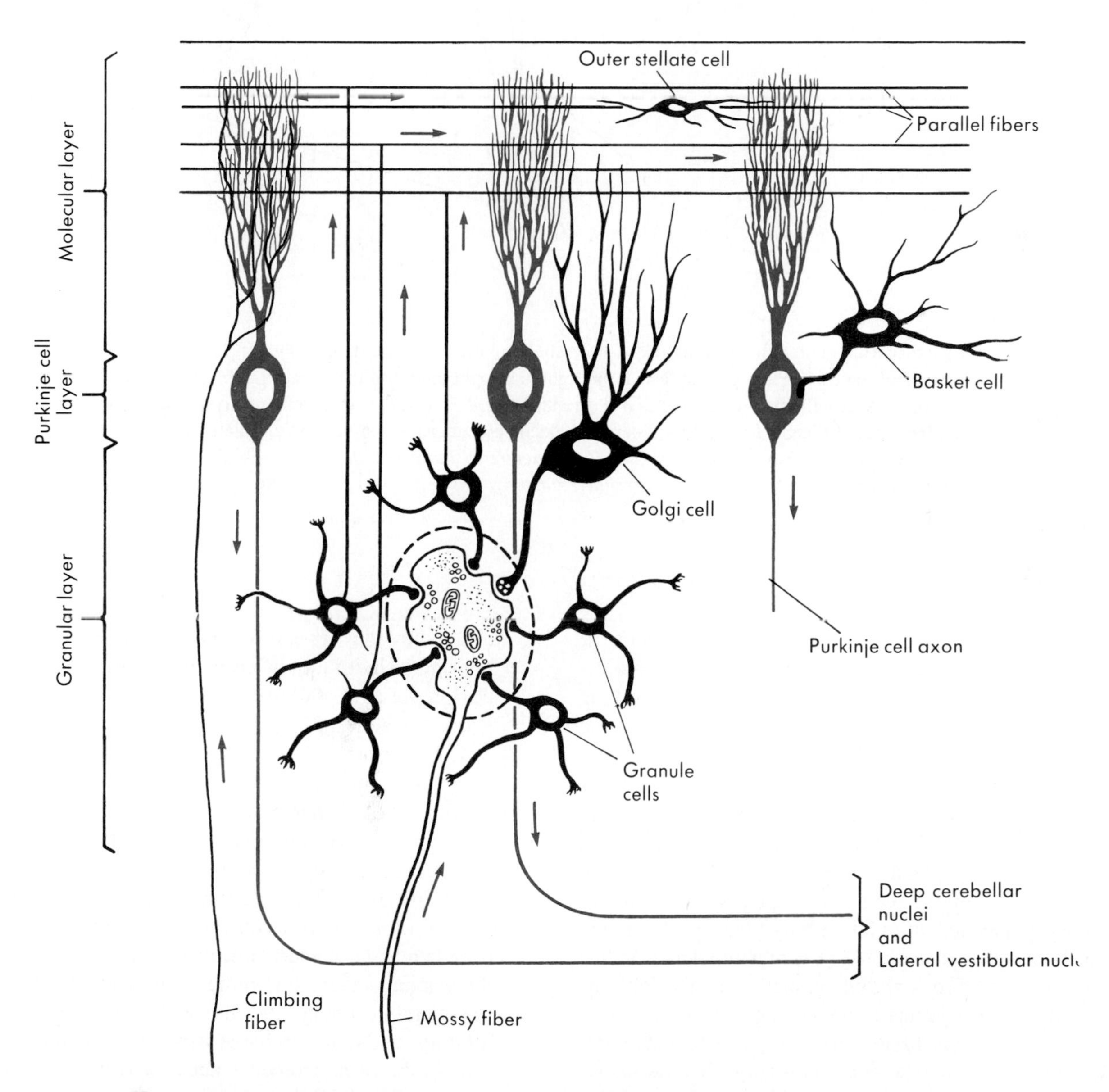

Fig. 14-13 Cerebellar cortex cut along the long axis of a folium. (Redrawn from Carpenter MB: *Human neuroanatomy,* ed 7, Baltimore, 1976, Williams & Wilkins; based on Gray EG: *J Anat,* 95:345, 1961, and Eccles JC et al: *The cerebellum as a neurorial machine,* New York, 1967, Springer-Verlag.)

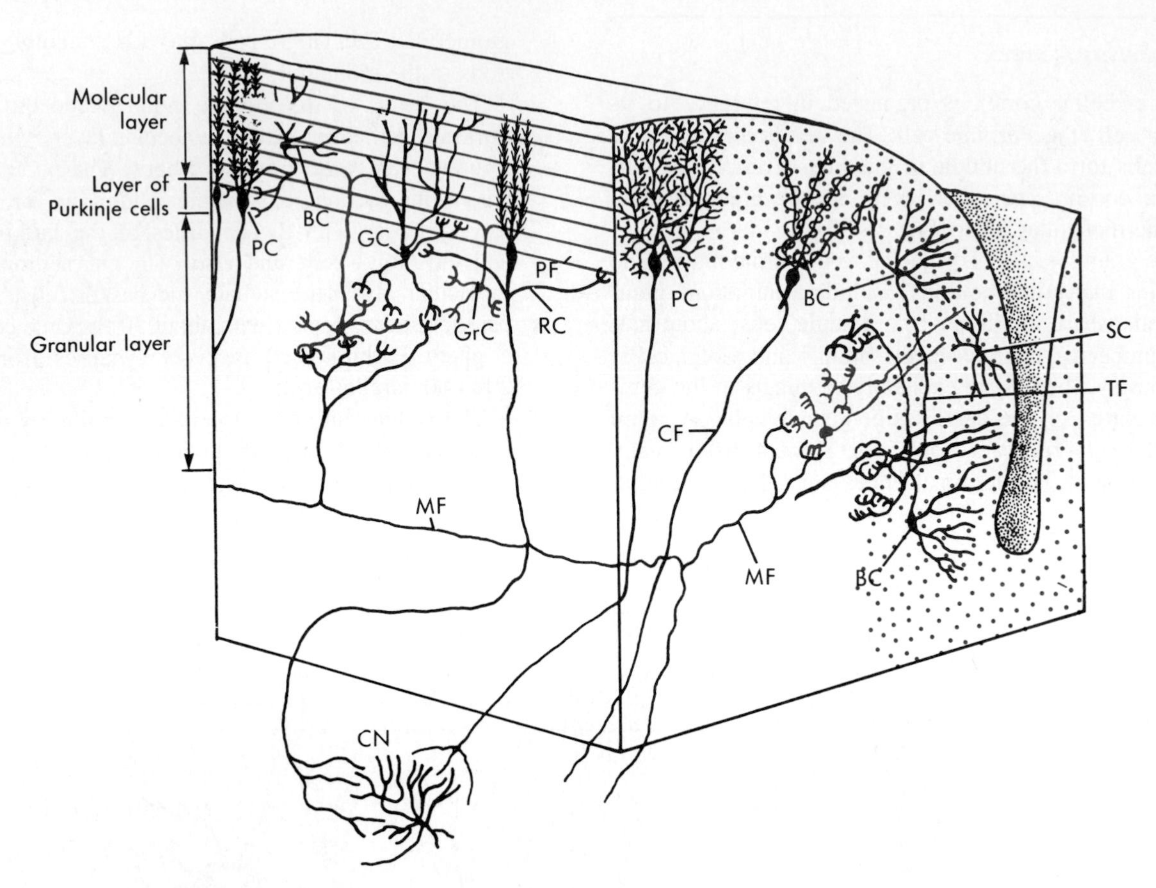

■ **Fig. 14-14** Three-dimensional view of cerebellar cortex. The cut face at the left is along the long axis of the folium and the cut face at the right is at right angles to the long axis. *PC,* Purkinje cell; *BC,* basket cell; *GC,* Golgi cell; *GrC,* granule cell; *PF,* parallel fiber; *RC,* recurrent collateral; *MF,* mossy fiber; *CF,* climbing fiber; *CN,* deep cerebellar nuclear cell; *SC,* stellate cell; *TF,* transverse fiber. (Redrawn from Fox CA: *The structure of the cerebellar cortex.* In Crosby EC, Humphrey TH, Lauer EW, editors: *Correlative anatomy of the nervous system,* New York, 1962, Macmillan.)

and the basket cells form terminals on the somas of the Purkinje cells. The orientation of the basket cell projections to the Purkinje cells is at right angles to the long axis of the folium. These axons are called **transverse fibers** (Fig. 14-14).

As mentioned, the Purkinje cells are the output cells of the cerebellar cortex. They receive numerous excitatory synapses on their dendrites from the parallel fibers formed by the granule cells (see Figs. 14-13 and 14-14). Each Purkinje cell also receives a powerful excitatory connection from a climbing fiber. The climbing fiber branches repeatedly as it ascends the dendritic tree and makes numerous active contacts with the Purkinje cell. Mossy fiber inputs to the cerebellar cortex cause a Purkinje cell to discharge single action potentials **(simple spikes),** whereas a single climbing fiber causes repetitive discharges of the Purkinje cell **(complex spikes)** (Fig. 14-15). The climbing fibers generate complex spikes at a low frequency, and so they do not change the average firing rates of Purkinje cells. However, they do appear to alter the responsiveness of Purkinje cells to mossy fiber inputs. These changes can be long-lasting and hence may play a role in motor learning (see p. 238).

The Purkinje cell also receives inhibitory connections from the basket and stellate cells (see Figs. 14-13 and 14-14). These connections produce inhibitory postsynaptic potentials in Purkinje cells (see Fig. 14-15). The axon of the Purkinje cell descends through the granule layer and enters the cerebellar white matter (see Fig. 14-14). Most Purkinje cell axons terminate in one of the deep cerebellar nuclei. However, some Purkinje cells in the vestibulocerebellum project out of the cerebellum to the lateral vestibular nucleus. The Purkinje cells are inhibitory, and their discharge inhibits the activity of neurons in the deep cerebellar nuclei and lateral vestibular nucleus. The inhibitory neurotransmitter used by Purkinje cells is GABA.

Another source of afferent input to the cerebellum is from monoaminergic neurons in the brainstem. Nor-

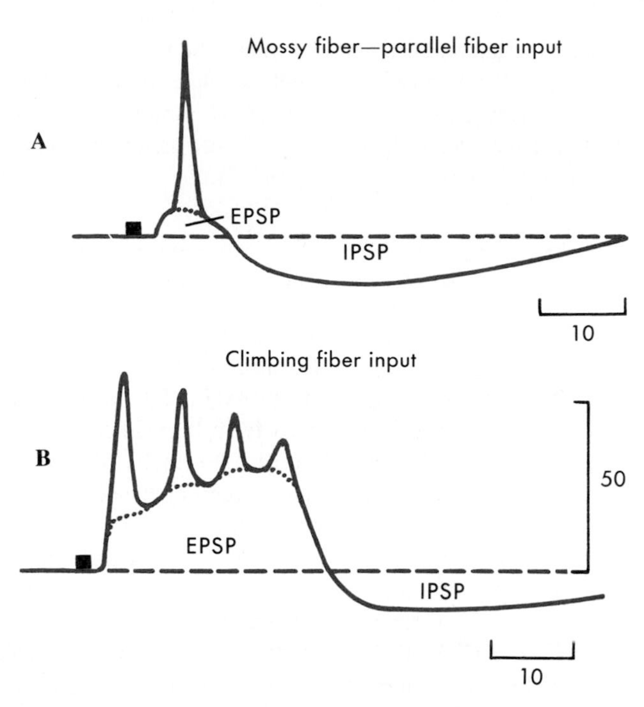

■ **Fig. 14-15** Responses of a Purkinje cell to a mossy fiber input, **A,** and to a climbing fiber input, **B.**

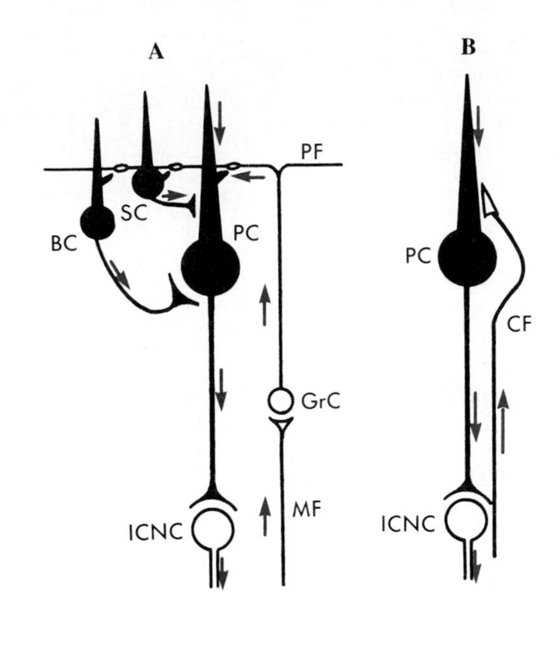

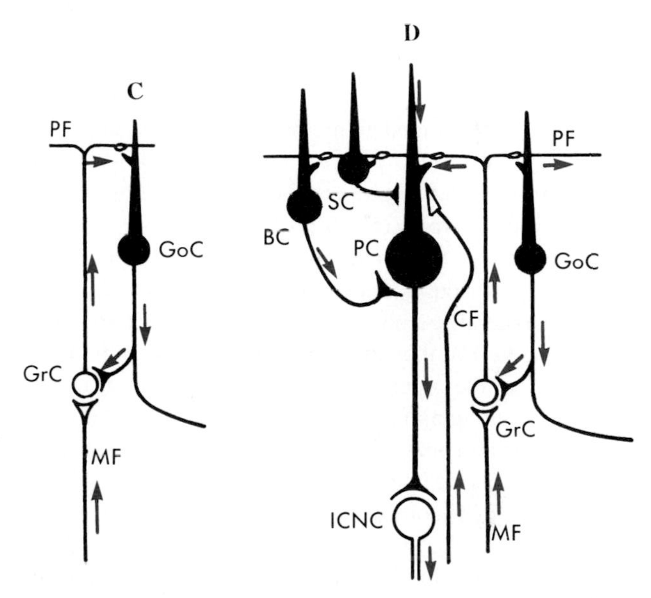

■ **Fig. 14-16** Neural circuits of the cerebellar cortex. Inhibitory neurons are black and excitatory ones clear. *PC,* Purkinje cell; *SC,* stellate cell; *BC,* basket cell; *PF,* parallel fiber; *GrC,* granule cell; *MF,* mossy fiber; *ICNC,* deep cerebellar nuclear cell; *CF,* climbing fiber; *GoC,* Golgi cell. (Redrawn from Eccles JC: *Nobel Symposium I: Muscular afferents and motor control,* New York, 1966, John Wiley & Sons.)

adrenergic neurons of the locus coeruleus and serotonergic neurons of the raphe nuclei project to the cerebellar cortex and modulate the synaptic activity of cerebellar neurons.

Some of the circuits that are formed by the cerebellar afferent fibers, interneurons, and Purkinje cells are shown in Fig. 14-16. The mossy fiber pathway can activate granule cells and Purkinje cells, but it can also produce a feedforward inhibition of Purkinje cells through basket and stellate cells (Fig. 14-16, *A*). The climbing fiber pathway (Fig. 14-16, *B*) provides a powerful excitation of Purkinje cells. Mossy fiber inputs are regulated by inhibitory feedback from Golgi cells (Fig. 14-16, *C*). Complicated interactions can occur between elements of the mossy fiber and climbing fiber pathways (Fig. 14-16, *D*).

That the output of the cerebellar cortex is inhibitory may seem paradoxical. However, recall that all of the input pathways to the cerebellum send collaterals to the deep cerebellar nuclei. Therefore it should not be surprising that the cells in the deep nuclei are very active. The role of the cerebellar cortex is to modulate this activity in the appropriate direction. The real output of the cerebellum is formed by the projections made from the deep cerebellar nuclei.

■ *Projections of the Deep Cerebellar Nuclei*

The deep cerebellar nuclei receive topographically organized projections from the Purkinje cells of the different parasagittal zones of the cerebellum (Fig. 14-17).

Vestibulocerebellum and vermal spinocerebellum. Purkinje cells in the flocculonodular lobe and the remainder of the vermis project to the fastigial nucleus or directly to the lateral vestibular nucleus. The fastigial nucleus, in turn, projects to the lateral vestibular nucleus and to the pontine reticular formation. Thus the output of these regions of the cerebellum influence axial and proximal limb muscles by way of the lateral vestibulospinal and pontine reticulospinal tracts, which

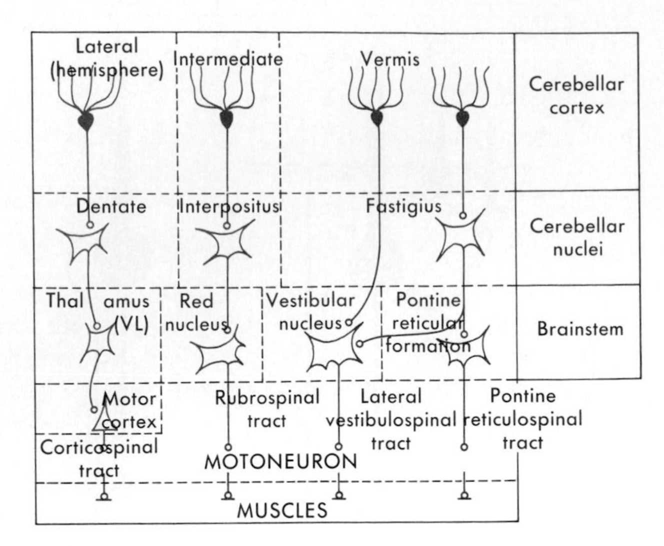

■ **Fig. 14-17** Cerebellar output. Inhibitory neurons are shown in solid red. (Redrawn from Bell CC, Dow RS: *Cerebellar circuitry.* In Schmitt FO et al, editors: *Neuroscience research summaries,* vol 2, Cambridge, Mass, 1967, MIT Press.)

belong to the medial motor projection system (Chapter 13).

Intermediate spinocerebellum. The Purkinje cells of the paravermal region project to the emboliform and globose nuclei (in cats, these are fused into the interpositus nucleus). These nuclei are connected with the contralateral red nucleus by way of fibers that leave the cerebellum in the superior cerebellar peduncle and then cross to the opposite side in the decussation of the brachium conjunctivum. This pathway allows the paravermal spinocerebellum to influence the discharges of neurons of the rubrospinal tract (which converges with the corticorubral tract; see Chapter 13). It should be recalled that the rubrospinal tract crosses at the midbrain level. Thus the paravermal region on one side of the cerebellum influences motor activity on the same side of the body, because the cerebellar output crosses the midline, and the red nucleus output then recrosses. The rubrospinal tract controls movements of proximal muscles of the limbs and belongs to the lateral motor system (Chapter 13).

Pontocerebellum. Purkinje cells of the cerebellar hemisphere project to the dentate nucleus. Neurons in the dentate nucleus project contralaterally through the superior cerebellar peduncle and the decussation of the brachium conjunctivum to the contralateral thalamus, and they end in the ventral lateral (VL) nucleus (see Fig. 14-2). The thalamic neurons distribute their axons to the premotor and primary motor cortex, where they can influence the planning and initiation of voluntary movements. Like the spinocerebellum, the pontocerebellum on one side influences movements on the same side of the body because of its crossed connection with a contralaterally projecting motor output system, in this case the lateral corticospinal tract. The pontocerebellum affects distal muscles through the lateral corticospinal tract, which belongs to the lateral motor system and which terminates in part directly on motor neurons to the distal muscles (see Chapter 13).

■ *Role of Cerebellum in Motor Learning*

As mentioned earlier, the activity of climbing fibers may participate in motor learning by influencing the effectiveness of mossy fibers in exciting Purkinje cells. Evidence for this idea comes from experiments on the gain of the vestibuloocular reflex. Normally, the gain of this reflex is one. For instance, when the head rotates to the left, the eyes will rotate by an equal angle in the opposite direction. However, if the visual fields are reversed by appropriate lenses, after a learning period, the eye movements are reversed. Thus the reflex gain is reduced and finally changed in sign. This alteration in gain does not occur after a lesion of the vestibulocerebellum.

In monkeys trained to perform a new task, the frequency of recorded complex spikes initially increases. Once the task is learned, the rate of complex spikes gradually decreases. The mechanism appears to be a form of long-lasting inhibition of the Purkinje cells.

■ *Cerebellar Disease*

Damage to the cerebellum results in motor dysfunction on the ipsilateral side of the body. The particular deficits that result depend on which functional component of the cerebellum is most affected. If the flocculonodular lobe is damaged, the motor disorders resemble those produced by a lesion of the vestibular apparatus and include difficulty in balance and gait and often nystagmus. When the vermis is affected, the motor disturbance affects the trunk, and when the intermediate region or hemisphere is involved, motor disorders occur in the limbs. The part of the limbs affected depends on the site of damage; the distal muscles are affected more by hemispheric lesions than by paravermal lesions.

Particular types of motor dysfunction in cerebellar disease include disorders of coordination, equilibrium, and muscle tone. Cerebellar incoordination is one form of **ataxia.** It is often expressed as **dysmetria,** a condition in which there are errors in the direction and force of movements that prevent a limb from being moved smoothly to a desired position. Ataxia may also be expressed as **dysdiadochokinesia,** in which there is difficulty in making repeated supinations and pronations of the arm. When more complicated movements are attempted, there is **decomposition of movement,** in which the movement is accomplished in a series of discrete steps, rather than as a smooth sequence. An **inten-**

tion tremor appears when the subject is asked to touch a target; the affected hand (or foot) develops a tremor that increases in magnitude as the target is approached. When equilibrium is disturbed, balance is difficult; the individual tends to fall toward the affected side and may walk with a wide-based stance. Speech may be slow and slurred, a defect called **scanning speech.** When muscle tone is involved, there is **hypotonia.** This may be associated with a **pendular knee jerk** when a leg is affected; when a phasic stretch reflex of the quadriceps muscle is elicited by striking the patellar tendon, the leg continues to swing back and forth because of the hypotonia, in contrast to the highly damped oscillation in a normal person.

■ *Motor Control by Basal Ganglia*

The basal ganglia are the deep nuclei of the cerebrum. In association with other nuclei in the diencephalon and midbrain, the basal ganglia differ from the cerebellum in the way they regulate motor activity. The basal ganglia do not have an input from the spinal cord like the cerebellum does, but they do receive direct input from the cerebral cortex, unlike the cerebellum. The main action of the basal ganglia is on the motor areas of the cortex by way of the thalamus. In addition to their role in motor control, the basal ganglia contribute to affective and cognitive functions. Lesions of the basal ganglia produce abnormal movements and posture.

■ *Organization of the Basal Ganglia and Related Nuclei*

The basal ganglia include the caudate nucleus, the putamen, and the globus pallidus (Fig. 14-18). Some authorities also include the claustrum. The term **corpus striatum** refers to these four structures, whereas the term **striatum** means just the caudate nucleus and putamen. These nuclei are separated by the anterior limb of the internal capsule; cellular bridges exist between the nuclei that give the structure a striated appearance. The globus pallidus has two parts, the external and the internal segments. The combination of putamen and globus pallidus is often referred to as the lentiform nucleus.

Associated with the basal ganglia are several thalamic nuclei. These include the ventral anterior (VA) and ventral lateral (VL) nuclei and several components of the intralaminar complex, including the centre median (CM) nucleus. Other associated nuclei are the subthalamic nucleus of the diencephalon and the substantia nigra of the midbrain. The substantia nigra ("black substance") derives its name from its pigment content. Many of the neurons in this nucleus contain melanin, a biproduct of dopamine synthesis. The substantia nigra can be subdivided into a dorsal pars compacta and a ventral pars reticulata. The melanin-containing cells are in the pars compacta.

■ *Connections and Operation of the Basal Ganglia*

Figs. 14-19 to 14-21 show the main connections of the basal ganglia. The circuitry of the basal ganglia is very complex, and how the circuits actually operate is still unknown. The neurons of the striatum do not discharge before activation of neurons in the motor cortex by a sensory input or to movements. Hence they do not appear to be concerned with the initiation of stimulus-trig-

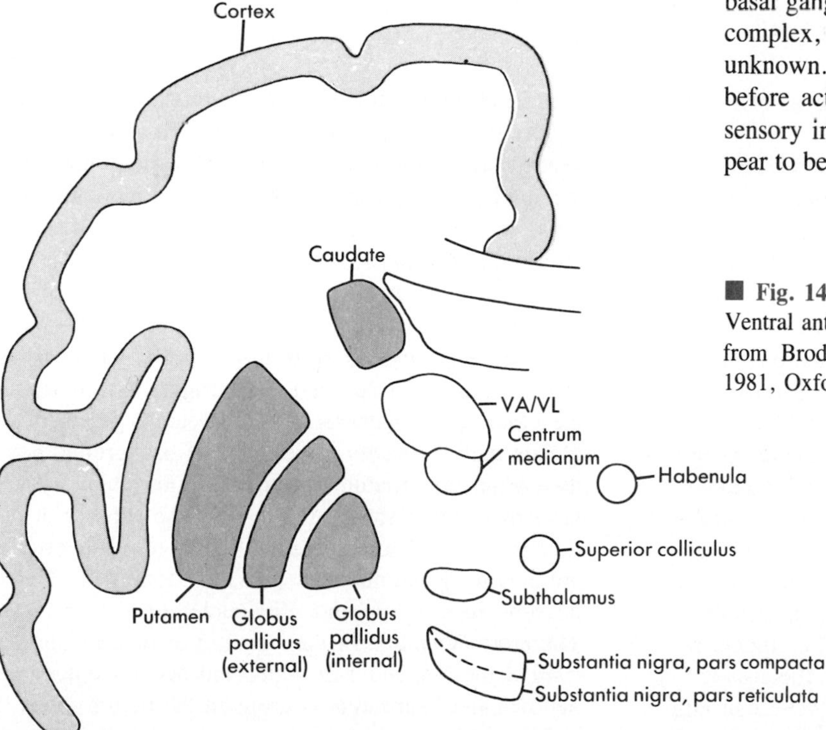

■ **Fig. 14-18** Basal ganglia and associated nuclei. *VA/VL,* Ventral anterior and ventral lateral thalamic nuclei. (Redrawn from Brodal A: *Neurological anatomy,* ed 3, New York, 1981, Oxford University Press.)

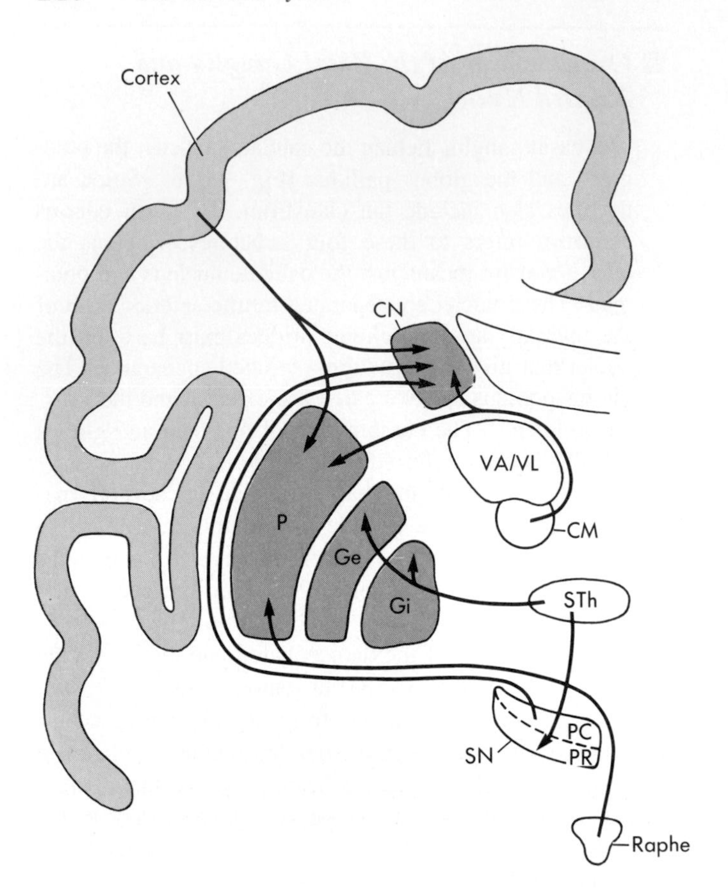

■ **Fig. 14-19** Afferent pathways to the basal ganglia. *CN,* Caudate nucleus; *VA/VL,* ventral anterior and ventral lateral thalamic nuclei; *CM,* centre median; *STh,* subthalamic nucleus; *PC,* pars compacts; *PR,* pars reticulata; *SN,* substantia nigra; *Ge, Gi,* external and internal segments of globus pallidus; *P,* putamen. (Redrawn from Brodal A: *Neurological anatomy,* ed 3, New York, 1981, Oxford University Press.)

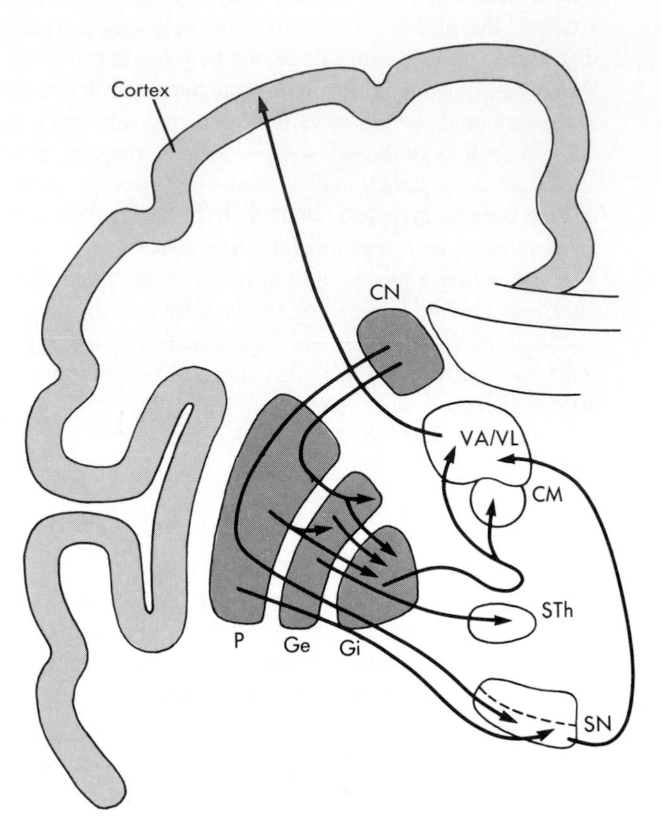

■ **Fig. 14-20** Efferent connections of the basal ganglia, including intrinsic connections. *CN,* Caudate nucleus; *VA/VL,* ventral anterior and ventral lateral thamalic nuclei; *CM,* centre median; *STh,* subthalamic nucleus; *SN,* substantia nigra; *Ge, Gi,* external and internal segments of the globus pallidus; *P,* putamen. (Redrawn from Brodal A: *Neurological anatomy,* ed 3, New York, 1981, Oxford University Press.)

gered movements. However, they may be involved in internally generated motor commands.

All regions of the cerebral cortex project topographically to the striatum (caudate nucleus and putamen). An important component of the cortical input to the striatum originates in the motor cortex. The corticostriatal projection arises from neurons in layer V of the cortex and appears to use glutamate as its excitatory neurotransmitter. The striatum then influences neurons in the VA and VL nuclei of the thalamus by two pathways, a direct pathway and an indirect one (Fig. 14-21).

In the direct pathway, the striatum projects to the internal segment of the globus pallidus and to the pars reticulata of the substantia nigra. This projection is inhibitory, and the transmitters include both GABA and substance P. The internal segment of the globus pallidus and the pars reticulata of the substantia nigra project to the VA and VL nuclei of the thalamus. The efferent fibers from the globus pallidus pass through the ansa lenticularis and the lenticular fasciculus. These connections are also GABAergic and inhibitory. The VA and VL nuclei send excitatory connections to the prefrontal, premotor, and supplementary motor cortex. This input to the cortex influences motor planning and eventually the discharges of corticospinal and corticobulbar neurons. The pars reticulata also influences eye movements by a projection to the superior colliculus.

The direct pathway appears to function as follows. Neurons in the striatum have little background activity, and they are activated only during movements by their inputs from the cortex and the intralaminar nuclei. On the other hand, neurons in the internal segment of the globus pallidus have a high level of background activity. When the striatum is activated, its inhibitory projections to the globus pallidus slow the activity of pallidal neurons. However, the pallidal neurons themselves are inhibitory and normally provide a tonic inhibition of neurons in the VA and VL nuclei of the thalamus. Therefore activation of the striatum disinhibits the neurons of the VA and VL nuclei and hence excites these neurons and their target neurons in the motor cortex.

The indirect pathway involves a connection from the

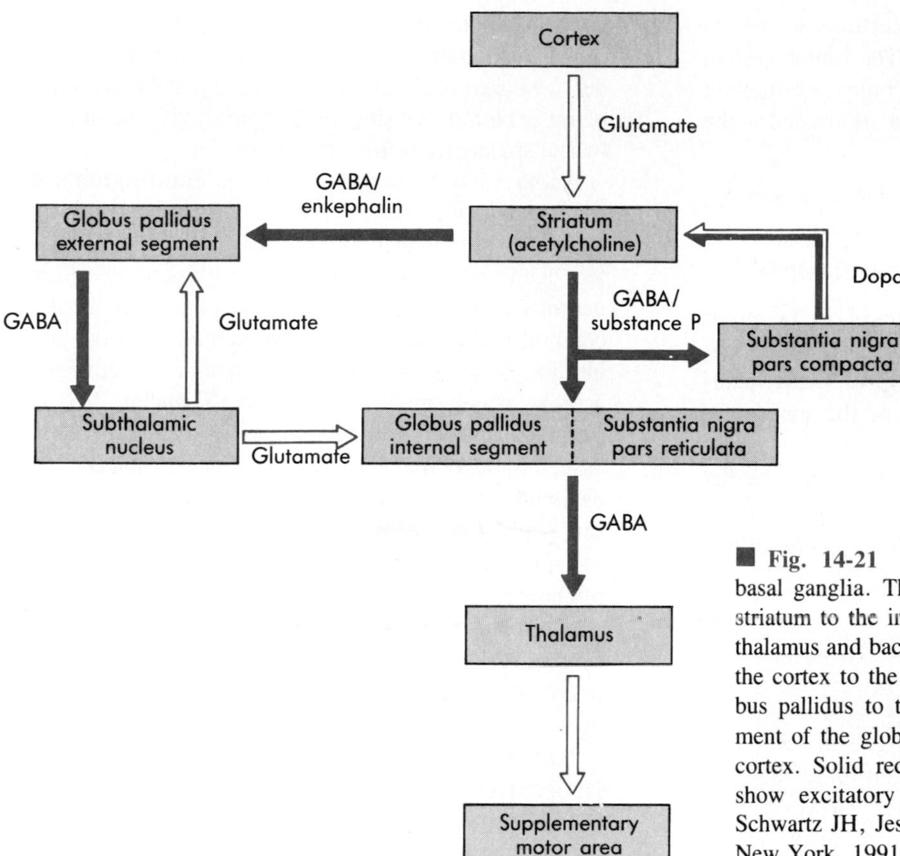

■ **Fig. 14-21** Direct and indirect pathways through the basal ganglia. The direct pathway is from the cortex to the striatum to the internal segment of the globus pallidus to the thalamus and back to the cortex. The indirect pathway is from the cortex to the striatum to the external segment of the globus pallidus to the subthalamic nucleus to the internal segment of the globus pallidus to the thalamus and back to the cortex. Solid red arrows show inhibitory, and open arrows show excitatory connections. (Redrawn from Kandel ER, Schwartz JH, Jessell TM: *Principles of neural science,* ed 3. New York, 1991, Elsevier.)

striatum to the external segment of the globus pallidus, which projects to the subthalamic nucleus, which in turn projects back to the internal segment of the globus pallidus. In this pathway, pallidal neurons in the external segment are inhibited by GABA and enkephalin released from striatal terminals in the globus pallidus. The GABA and enkephalin disinhibit neurons of the subthalamic nucleus. The subthalamic neurons become more active because of the disinhibition, and they release glutamate in the internal segment of the globus pallidus. This transmitter excites neurons that project to the VA and VL thalamic nuclei. The pallidal action is inhibitory, and consequently the activity of the thalamic neurons decreases, as do the cortical neurons that they influence.

The direct and indirect pathways thus have opposing actions; an increase in the activity of either one of these pathways might lead to an imbalance in motor control. Such imbalances could be in the direction of either an increase or a decrease in the motor output of the cortex and are typical of basal ganglion diseases (see p. 242).

An important connection exists from the pars compacta of the substantia nigra to the striatum. Dopamine is the neurotransmitter for this pathway; it appears to have an excitatory action on the direct pathway and an inhibitory action on the indirect pathway. Both actions facilitate activity in the cerebral cortex.

Other projections to the striatum come from the CM nucleus and from the dorsal raphe nucleus. The latter pathway is serotonergic.

■ *Differences Between the Basal Ganglion and Cerebellar Motor Loops*

The organization of the motor loops that connect the basal ganglia and cerebellum with the motor regions of the cerebral cortex differ in several ways. The basal ganglia receive input from all areas of the cerebral cortex, whereas the inputs to the cerebellum from the cortex are more restricted. The output from the basal ganglia is also more widespread, and reaches the prefrontal cortex, as well as all of the premotor areas; the cerebellar circuit influences only the premotor and motor cortex. Finally, the basal ganglia do not receive somatosensory information from ascending pathways in the spinal cord and they have few connections with the brainstem; the cerebellum is the target of several somatosensory pathways and it has rich connections with brainstem nuclei.

■ *Subdivision of the Striatum into Striosomes and Matrix*

By use of a variety of markers for neurotransmitters, the striatum has been subdivided into zones called striosomes and matrix. The cortical projections related to motor control end in the matrix area. The limbic system projects to the striosomes. The striosomes are thought to synapse in the pars compacta and to influence the dopaminergic nigrostriatal pathway.

Role of the Basal Ganglia in Motor Control

The main influence of the basal ganglia is on the motor cortex. This suggests that the basal ganglia have an important influence on the lateral system of motor pathways which is consistent with some of the movement disorders observed in diseases of the basal ganglia. However, the basal ganglia also relate to the medial motor pathways, because diseases of the basal ganglia can also affect posture and tone of the proximal muscles.

Basal Ganglion Disease

Experimental lesions of the basal ganglia in carnivores or rodents have so far not been very helpful in shedding light on human basal ganglion diseases. However, Parkinson's disease can be mimicked in monkeys with a neurotoxin called MTPT, which destroys dopamine-containing neurons. This discovery will help further our understanding of the abnormalities in parkinsonism.

The deficits seen in the various basal ganglion diseases include abnormal movements **(dyskinesias),** increases in muscle tone **(cogwheel rigidity),** and slowness in initiating movements **(bradykinesia).** The abnormal movements include tremor, athetosis, chorea, ballism, and dystonia. The tremor of basal ganglion disease is a "pill-rolling" tremor that occurs when the limb is at rest. **Athetosis** consists of slow, writhing movements of the distal parts of the limbs, whereas **chorea** is rapid, flicking movements of the extremities and facial muscles. **Ballism** is associated with violent, flailing movements of the limbs (ballistic movements). Finally, **dystonic movements** are slow truncal movements that result in distorted body positions.

Parkinson's disease is a common disorder that is characterized by tremor, rigidity, and bradykinesia. The main pathologic condition is loss of neurons in the pars compacta of the substantia nigra and consequently a severe loss of dopamine in the striatum. Loss of neurons of the locus coeruleus and the raphe nuclei, as well as of other monoaminergic nuclei, also occurs. Presumably, the loss of dopamine results in overactivity of the inhibitory pathway from the striatum to the globus pallidus and therefore a disinhibition of neurons in the VA and VL nuclei. The neurons in the VA and VL nuclei would then activate neurons of the motor cortex and would increase the discharges of motor neurons, including γ-motor neurons. Before complete loss of the dopaminergic neurons occurs, administration of L-dopa, a precursor of dopamine that can cross the blood-brain barrier, relieves some of the motor disorders in Parkinson's disease. Currently, the possibility is being explored that dopamine-synthesizing neurons can be transplanted into the striatum.

Another basal ganglion disease is **Huntington's disease.** This results from a genetic defect that involves an autosomal dominant gene on chromosome 4. This defect leads to the loss of GABAergic and cholinergic neurons of the striatum (and also degeneration of the cerebral cortex, with a resultant dementia). Loss of inhibition of the globus pallidus presumably reduces the activity of neurons in the VA and VL nuclei. This may cause the choreiform movements of Huntington's disease, although the exact mechanism is unclear. It appears that the rigidity of Parkinson's disease is in a sense the opposite of chorea, because overtreatment of patients with Parkinson's disease with L-dopa can result in chorea.

Cerebral palsy is another common motor disorder associated with a lesion of the basal ganglia. In cerebral palsy, athetosis often occurs in association with lesions of both the striatum and globus pallidus.

Ballism is produced by a partial lesion of the subthalamic nucleus.

In all of these disorders of the basal ganglia, the motor dysfunction is contralateral to the diseased component. This is understandable because the main final output of the basal ganglia is mediated by the corticospinal tract.

Summary

1. Voluntary movements depend on interactions among motor areas of the cerebral cortex, the cerebellum, and the basal ganglia.

2. Motor areas of the cerebral cortex include (a) the primary motor cortex, which controls distal muscles of the extremities; (b) the premotor area, which helps control proximal and axial muscles; (c) the supplementary motor cortex, which participates in motor planning and in coordination; and (d) the frontal eye fields, which help initiate saccadic eye movements.

3. Individual corticospinal neurons discharge before voluntary contraction of related muscles. The discharges are related to contractile force, rather than to joint position.

4. Cortical motoneurons receive feedback from the sensory systems by way of the somatosensory cortex and the posterior parietal lobe; this feedback helps correct motor commands.

5. The cerebellum influences the rate, range, force, and direction of movements. It also influences muscle tone and posture, as well as eye movements and balance. The cerebellum is involved in motor learning.

6. The vestibulocerebellum connects with the vestibular system and influences eye movements and also balance by connections with the vestibulospinal and reticulospinal tracts.

7. The spinocerebellum receives input from spinal cord pathways. It controls (a) the axial musculature through the medial system of descending pathways and (b) the proximal limb muscles through the rubrospinal tract of the lateral system.

8. The pontocerebellum receives information from the cerebral cortex by way of the pontine nuclei. It controls the distal muscles of the limbs by connections to the motor cortex (via the thalamus) and lateral corticospinal tract. The pontocerebellum is also involved in motor planning.

9. Most of the input to the cerebellum is through pathways that end as mossy fibers. These evoke single-action potentials called simple spikes in Purkinje cells. However, the inferior olive projections to the cerebellum end as climbing fibers. A climbing fiber produces repetitive discharges of a Purkinje cell. Complex spikes are thought to alter the effectiveness of simple spikes and to play a role in motor learning.

10. The basal ganglia include several deep telencephalic nuclei (including the caudate, putamen, globus pallidus). They interact with the cerebral cortex, subthalamic nucleus, substantia nigra, and thalamus.

11. Activity transmitted from the cortex through the basal ganglia can either facilitate or inhibit thalamic neurons that project to motor areas of the cortex. The basal ganglia thus affect the output of the motor cortex.

12. Diseases of the cerebellum and basal ganglia affect motor behavior profoundly, but in ways that can be rationalized based on the interconnections of the cerebellum and basal ganglia with the lateral and medial motor systems.

Bibliography

Journal articles

Alben RL et al: The functional anatomy of basal ganglia disorders, *Trends Neurosci* 12:366, 1989.

Alexander GE, Crutcher MD: Functional architecture of basal ganglia circuits: neural substrates of parallel processing, *Trends Neurosci* 13:266, 1990.

Alexander GE et al: Parallel organization of functionally segregated circuits linking basal ganglia and cortex, *Annu Rev Neurosci* 9:357, 1986.

Allen GL, Tsukahara N: Cerebrocerebellar communication systems, *Physiol Rev* 54:957, 1974.

Asanuma H, Rosen I: Topographical organization of cortical efferent zones projecting to distal muscles in the monkey, *Exp Brain Res* 14:243, 1972.

Asanuma H et al: Relationship between afferent input and motor outflow in cat motorsensory cortex, *J Neurophysiol* 31:670, 1968.

Brinkman C: Supplementary motor area of the monkey's cerebral cortex: short- and long-term deficits after unilateral ablation and the effects of subsequent callosal section, *J Neurosci* 4:918, 1984.

Cheney PD, Fetz EE: Functional classes of primate corticomotoneuronal cells and their relation to active force, *J Neurophysiol* 44:773, 1980.

DeLong MR: Primate models of movement disorders of basal ganglia origin, *Trends Neurosci* 13:281, 1990.

Evarts EV: Relation of pyramidal tract activity to force exerted during voluntary activity, *J Neurophysiol* 31:14, 1968.

Fetz EE et al: Corticomotoneuronal connections of precentral cells detected by post-spike averages of EMG activity in behaving monkeys, *Brain Res* 114:505, 1976.

Glickstein M, Yeo C: The cerebellum and motor learning, *J Cogn Neurosci* 2:69, 1990.

Holmes G: The cerebellum of man, *Brain* 62:1, 1939.

Kopin IJ, Markey SP: MPTP toxicity: implications for research in Parkinson's disease, *Ann Rev Neurosci* 11:81, 1988.

Roland PE et al: Supplementary motor area and other cortical areas in organization of voluntary movements in man, *J Neurophysiol* 43:118, 1980.

Snider RS, Stowell A: Receiving areas of the tactile, auditory and visual system in the cerebellum, *J Neurophysiol* 7:331, 1944.

Tanji J, Evarts EV: Anticipatory activity of motor cortex neurons in relation to direction of an intended movement, *J Neurophysiol* 39:1062, 1976.

Thach WT: Correlation of neural discharge with pattern and force of muscular activity, joint position, and direction of intended next movement in motor cortex and cerebellum, *J Neurophysiol* 41:654, 1978.

Wise SP: The primate premotor cortex: past, present, and preparatory, *Ann Rev Neurosci* 8:1, 1985.

Yurek DM, Sladek JR: Dopamine cell replacement: Parkinson's disease, *Ann Rev Neurosci* 13:415, 1990.

Books and monographs

Brodal A: *Neurological anatomy,* ed 3, New York, 1981, Oxford University Press.

Brooks VB, editor: *Handbook of physiology, Sect 1, The nervous system, Vol II, Motor control, Part 2,* Bethesda, MD, 1981, American Physiological Society.

Brooks VB: *The neural basis of motor control,* New York, 1986, Oxford University Press.

Carpenter MB, Sutin J: *Human neuroanatomy,* ed 8, Baltimore, 1983, Williams & Wilkins.

Cotman CW, McGaugh JL: *Behavioral neuroscience,* New York, 1980, Academic Press.

DeJong RN: *The neurological examination,* ed 4, Hagerstown, Mass, 1979, Harper & Row, Publishers.

Eccles JC et al: *The cerebellum as a neuronal machine,* New York, 1967, Springer-Verlag.

Ito M: *The cerebellum and motor control,* New York, 1984, Raven Press.

Kandel ER et al: *Principles of neural sciences,* ed 3, New York, 1991, Elsevier.

Palay SL, Chan–Palay V: *Cerebellar cortex,* New York, 1974, Springer-Verlago

Phillips CG, Porter R: *Corticospinal neurones,* London, 1977, Academic Press.

Schmidt RF, Thews G: *Human physiology,* New York, 1983, Springer-Verlag.

CHAPTER

15

The Autonomic Nervous System and its Central Control

The autonomic nervous system can be regarded as a part of the motor system, but in this case the effectors are smooth muscle, cardiac muscle, and glands, rather than skeletal muscle. The autonomic nervous system is, therefore, the visceral motor system. It is also sometimes called the vegetative nervous system. However, this terminology does not seem appropriate for a system that is important for all levels of activity, including aggressive behavior.

One important function of the autonomic nervous system is to assist the body in maintaining a constant internal environment **(homeostasis).** The central nervous system and its autonomic outflow take compensatory actions when internal stimuli signal that regulation is required. For example, a sudden increase in systemic blood pressure activates the baroreceptors, which in turn reflexly adjust the autonomic nervous system and restore the blood pressure towards its previous level (see Chapter 29).

The autonomic nervous system also participates in appropriate and coordinated responses to external stimuli. One instance of this is regulation of pupil size in response to different intensities of ambient light. An extreme example arises when a threat causes massive activation of the sympathetic nervous system and results in adrenal medullary secretion, increases in heart rate and blood pressure, dilation of the bronchioles, inhibition of intestinal motility and secretion, increased glucose metabolism, pupillary dilation, piloerection, vasoconstriction of cutaneous and splanchnic vessels, and dilation of vessels in skeletal muscle. However, this "fight or flight response" is an uncommon event and does not represent the ordinary mode of operation of the sympathetic nervous system. On the other hand, the combination of changes seen in such a reaction serves as a useful reminder of some of the functions of the autonomic nervous system.

Accompanying the autonomic motor fibers in peripheral nerves are visceral afferent fibers that originate from sensory receptors in the viscera. Many of these trigger reflexes, but the activity of some receptors cause sensory experiences, such as pain, a sense of visceral distention, hunger, thirst, and nausea. The chemical senses can also be considered visceral senses (see Chapter 11).

The term **autonomic nervous system** generally refers to the sympathetic and parasympathetic nervous systems. We will also include the enteric nervous system, although this is sometimes considered as a separate entity. However, because the autonomic nervous system is under central nervous system (CNS) control, those parts of the CNS that are components of the central autonomic control system also are discussed in this chapter. This includes the hypothalamus and higher levels of the limbic system, which are associated with the emotions and with many visceral behaviors that have survival value (e.g., feeding, drinking, thermoregulation, reproduction, defense, aggression).

■ *The Autonomic Nervous System*

■ *Organization of the Autonomic Nervous System*

The autonomic nervous system has three major subdivisions: the sympathetic nervous system, the parasympathetic nervous system, and the enteric nervous system. The **sympathetic** and **parasympathetic nervous systems** consist of a two neuron motor pathway, including preganglionic neurons, whose cell bodies are located in the CNS, and postganglionic neurons, whose cell bodies are located in one of the autonomic ganglia. The **enteric nervous system** includes the neurons and nerve fibers in the wall of the gastrointestinal tract, including the myenteric and submucosal plexuses.

The sympathetic preganglionic neurons are located in the thoracic and upper lumbar segments of the spinal cord. For this reason, the sympathetic nervous system is sometimes referred to as the thoracolumbar division of the autonomic nervous system. By contrast, the parasympathetic preganglionic neurons are found in the brainstem and in the sacral spinal cord. Hence, this part of the autonomic nervous system can be called the craniosacral division. Sympathetic postganglionic neurons are generally found in the paravertebral or the prevertebral ganglia, which are located at some distance from the target organs. On the other hand, parasympathetic postganglionic neurons are found in parasympathetic ganglia near or actually in the walls of the target organs.

The sympathetic and parasympathetic nervous systems have often been described as antagonistic in their control of many organs. However, it is more appropriate to consider these different parts of the autonomic control system as working in a coordinated way, sometimes acting reciprocally and sometimes synergistically, to regulate visceral function. Furthermore, it is important to recognize that not all visceral structures are innervated by both systems. For example, the smooth muscles and glands in the skin receive only a sympathetic innervation. The parasympathetic nervous system

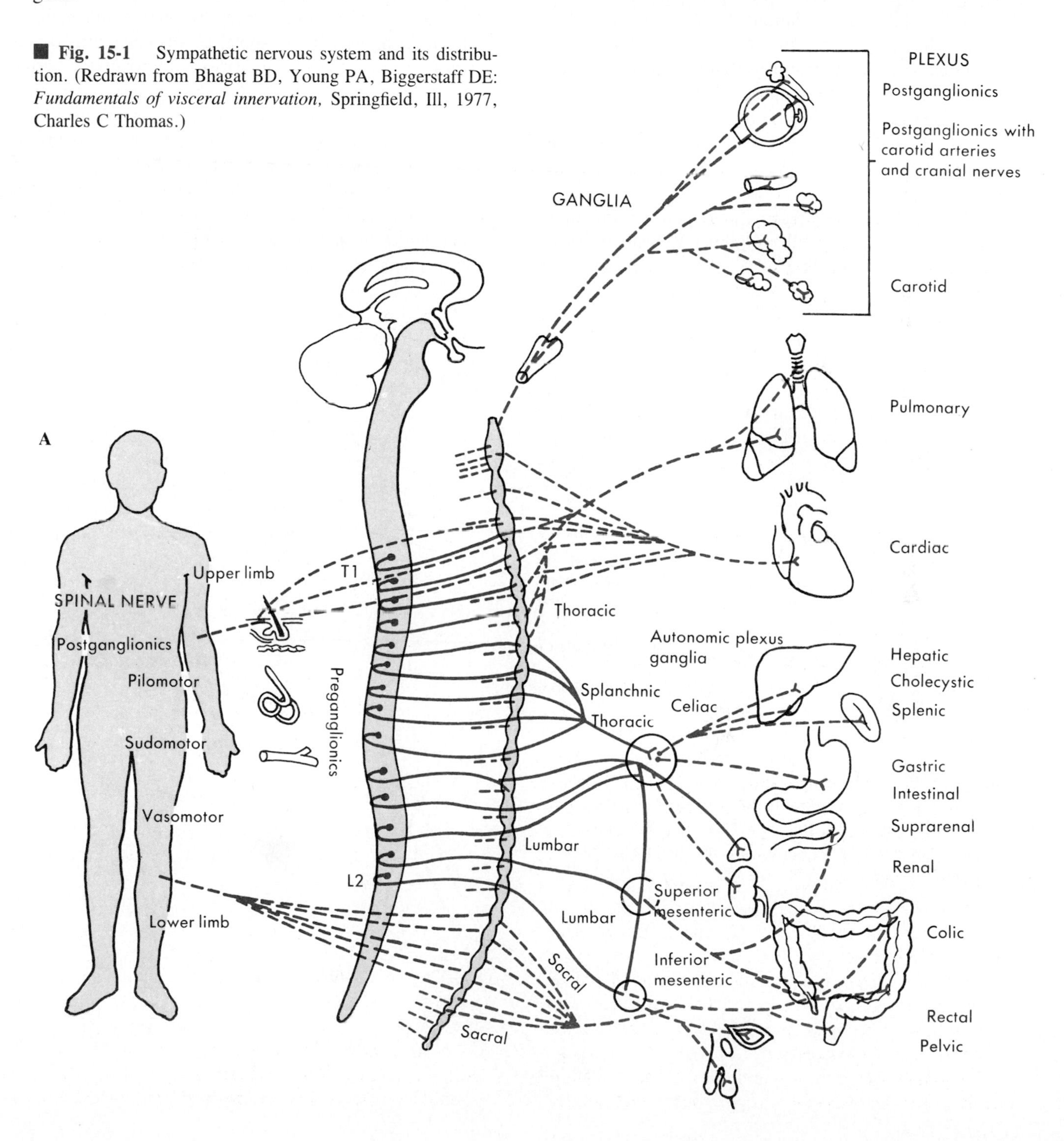

■ **Fig. 15-1** Sympathetic nervous system and its distribution. (Redrawn from Bhagat BD, Young PA, Biggerstaff DE: *Fundamentals of visceral innervation,* Springfield, Ill, 1977, Charles C Thomas.)

does not distribute to the body wall, only to structures in the head and the thoracic, abdominal, and pelvic cavities; and few blood vessels have a parasympathetic innervation.

The sympathetic nervous system. Sympathetic preganglionic neurons are concentrated in the **intermediolateral cell column** in the thoracic and upper lumbar segments of the spinal cord (Figs. 15-1 and 15-2). Some may also be found in the C8 segment. In addition to the intermediolateral cell column, groups of sympathetic preganglionic neurons are found in other locations, including the lateral funiculus, the intermediate region, and the part of lamina X dorsal to the central canal. The cells located medially are periodically concentrated and then sparse at different rostrocaudal levels. The arrangement resembles a ladder, with the intermediolateral cell columns on the two sides forming the sides of the ladder and the strands of neurons in the intermediate region and in the lamina X forming the rungs of the ladder.

The axons of the preganglionic neurons are often small myelinated nerve fibers known as B fibers. However, some are unmyelinated C fibers. They leave the spinal cord in the ventral root and enter the paravertebral ganglion at the same segmental level through a white communicating ramus. White rami are found only from T1 to L2. The preganglionic axon may synapse on postganglionic neurons in this ganglion, or they may pass through the ganglion and enter either the sympathetic chain or a splanchnic nerve (see Fig. 15-2).

Preganglionic axons in the sympathetic chain, which connects the paravertebral ganglia, may travel rostrally or caudally to a nearby or distant paravertebral ganglion and then synapse. If the synapse is in a paravertebral ganglion, the postganglionic axon often passes through a gray communicating ramus to enter a spinal nerve. Each of the 31 pairs of spinal nerves has a gray ramus. Postganglionic axons are distributed through the peripheral nerves to effectors in the skin, muscle, and joints. The target structures include piloerector muscles, blood vessels, and sweat glands. Postganglionic axons are generally unmyelinated (C fibers), although there are some exceptions. The distinction between white and gray rami is based on the relative content of myelinated and unmyelinated axons in these rami.

Preganglionic axons in a splanchnic nerve often

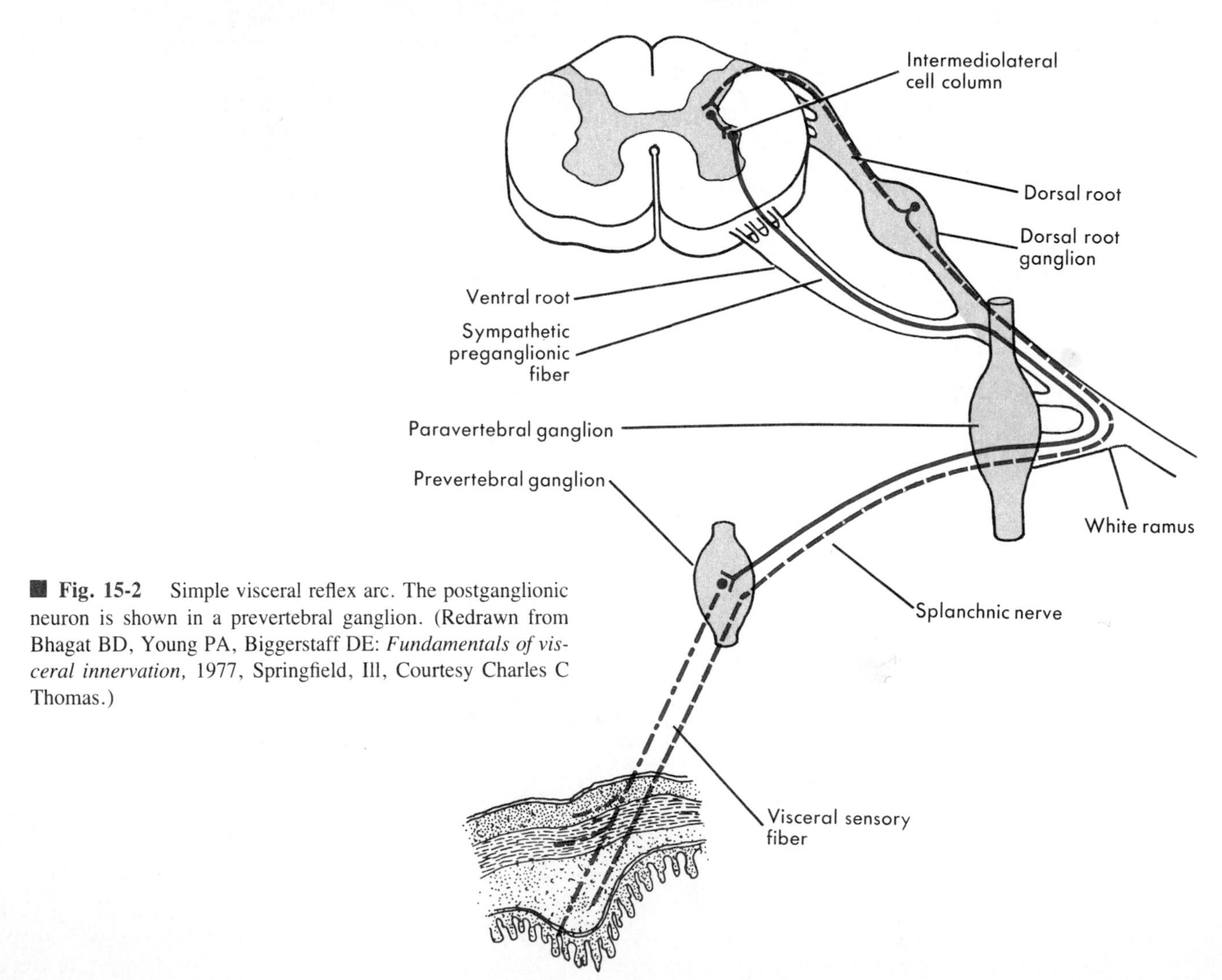

■ **Fig. 15-2** Simple visceral reflex arc. The postganglionic neuron is shown in a prevertebral ganglion. (Redrawn from Bhagat BD, Young PA, Biggerstaff DE: *Fundamentals of visceral innervation,* 1977, Springfield, Ill, Courtesy Charles C Thomas.)

travel to a prevertebral ganglion and synapse, or they may pass through the ganglion and an autonomic plexus and end in a more distant ganglion. Some preganglionic axons pass through a splanchnic nerve and end directly on cells of the adrenal medulla.

The sympathetic chain extends from cervical to coccygeal levels. This arrangement serves as a distribution system so that preganglionic neurons, which are limited to thoracic and upper lumbar segments, can activate postganglionic neurons that innervate all body segments. However, there are fewer paravertebral ganglia than there are spinal segments, because some of the segmental ganglia fuse during development. For example, the superior cervical sympathetic ganglion represents the fused ganglia of C1 through C4; the middle cervical sympathetic ganglion is the fused ganglia of C5 and C6; and the inferior cervical sympathetic ganglion is the combination of the ganglia at C7 and C8. The term **stellate ganglion** refers to a fusion of the inferior cervical sympathetic ganglion with the ganglion of T1. The superior cervical sympathetic ganglion provides postganglionic innervation to the head and neck; and the middle cervical and stellate ganglia innervate the heart, lungs, and bronchi.

Generally, the sympathetic preganglionic neurons are distributed to ipsilateral ganglia and thus control autonomic function on the same side of the body. However, there are important exceptions to this. For example, the sympathetic innervation of the intestine and of the pelvic viscera is bilateral. As with motor neurons to skeletal muscle, sympathetic preganglionic neurons that control a particular organ are spread over several segments. For example, the sympathetic preganglionic neurons that control sympathetic functions in the head and neck region are distributed in C8 to T5, whereas those controlling the adrenal gland are in T4 through T12.

The parasympathetic nervous system. The parasympathetic postganglionic neurons are located in several cranial nerve nuclei in the brainstem, as well as in the intermediate region of the S3 and S4 segments of the sacral spinal cord (Fig. 15-3). The cranial nerve nuclei that contain parasympathetic preganglionic neurons are the **Edinger-Westphal nucleus** (cranial nerve III), the **superior** (cranial nerve VII) and **inferior salivatory nuclei** (cranial nerve IX), and the **dorsal motor nucleus** and **nucleus ambiguus** (cranial nerve X).

Many postganglionic parasympathetic cells are located in cranial ganglia, including the ciliary ganglion, the pterygopalatine and submandibular ganglia, and the otic ganglion. Preganglionic parasympathetic neurons to the ciliary ganglion originate in the Edinger-Westphal nucleus; those to the pterygopalatine and submandibular ganglia arise from the superior salivatory nucleus, and those that reach the otic ganglion originate from the inferior salivatory nucleus. The ciliary ganglion innervates the pupillary sphincter and ciliary muscles in the eye. The pterygopalatine ganglion supplies the lacrimal gland, as well as glands in the nasal and oral pharynx. The submandibular ganglion projects to the submandibular and sublingual salivary glands and glands in the oral cavity. The otic ganglion innervates the parotid salivary gland and glands in the mouth.

Other parasympathetic postganglionic neurons arise near or in the walls of visceral organs in the thoracic, abdominal, and pelvic cavities. Neurons of the enteric plexus include cells that can also be considered parasympathetic postganglionic neurons. These cells receive input from the vagus or pelvic nerves. The vagus nerves innervate the heart, lungs, bronchi, liver, pancreas, and all of the gastrointestinal tract from the esophagus to the splenic flexure of the colon. The remainder of the colon and rectum, as well as the urinary bladder and reproductive organs, are supplied by sacral parasympathetic preganglionic neurons that distribute through the pelvic nerves to postganglionic neurons in the pelvic ganglia.

The parasympathetic preganglionic neurons that project to the viscera of the thorax and part of the abdomen are in the dorsal motor nucleus of the vagus and the nucleus ambiguus. The dorsal motor nucleus is thought to be largely **secretomotor** (activating glands) and the nucleus ambiguus **visceromotor** (modifying the activity of cardiac muscle). The dorsal motor nucleus supplies visceral organs in the neck (pharynx, larynx), thoracic cavity (trachea, bronchi, lungs, heart, esophagus), and abdominal cavity (including much of the gastrointestinal tract, liver, and pancreas). Electrical stimulation of the dorsal motor nucleus results in gastric acid secretion, as well as in secretion of insulin and glucagon by the pancreas. Although projections to the heart have been described, their function is uncertain. The nucleus ambiguus contains two groups of neurons: (1) a dorsal group that activates striated muscle in the soft palate, pharynx, larynx, and esophagus (branchiomotor); and (2) a ventrolateral group that innervates and slows the heart.

Visceral afferent fibers. The visceral motor fibers in the autonomic nerves are accompanied by visceral afferent fibers. Most of these afferent fibers supply information that originates in sensory receptors in the viscera. The activity of these sensory receptors never reaches the level of consciousness. Instead, these afferent fibers form the afferent limb of reflex arcs. Both viscero-visceral and viscero-somatic reflexes are elicited by these afferent fibers. Visceral reflexes operate at a subconscious level but are very important for homeostatic regulation and adjustment to external stimuli.

The fast-acting neurotransmitters used by visceral afferent fibers are not well documented, although many of these neurons may use an excitatory amino acid transmitter, such as glutamate or aspartate. However, visceral afferent fibers contain many neuropeptides, or combinations of these, including angiotensin II, arginine-vasopressin, bombesin, calcitonin gene-related peptide, cholecytstokinin, galanin, substance P, enkepha-

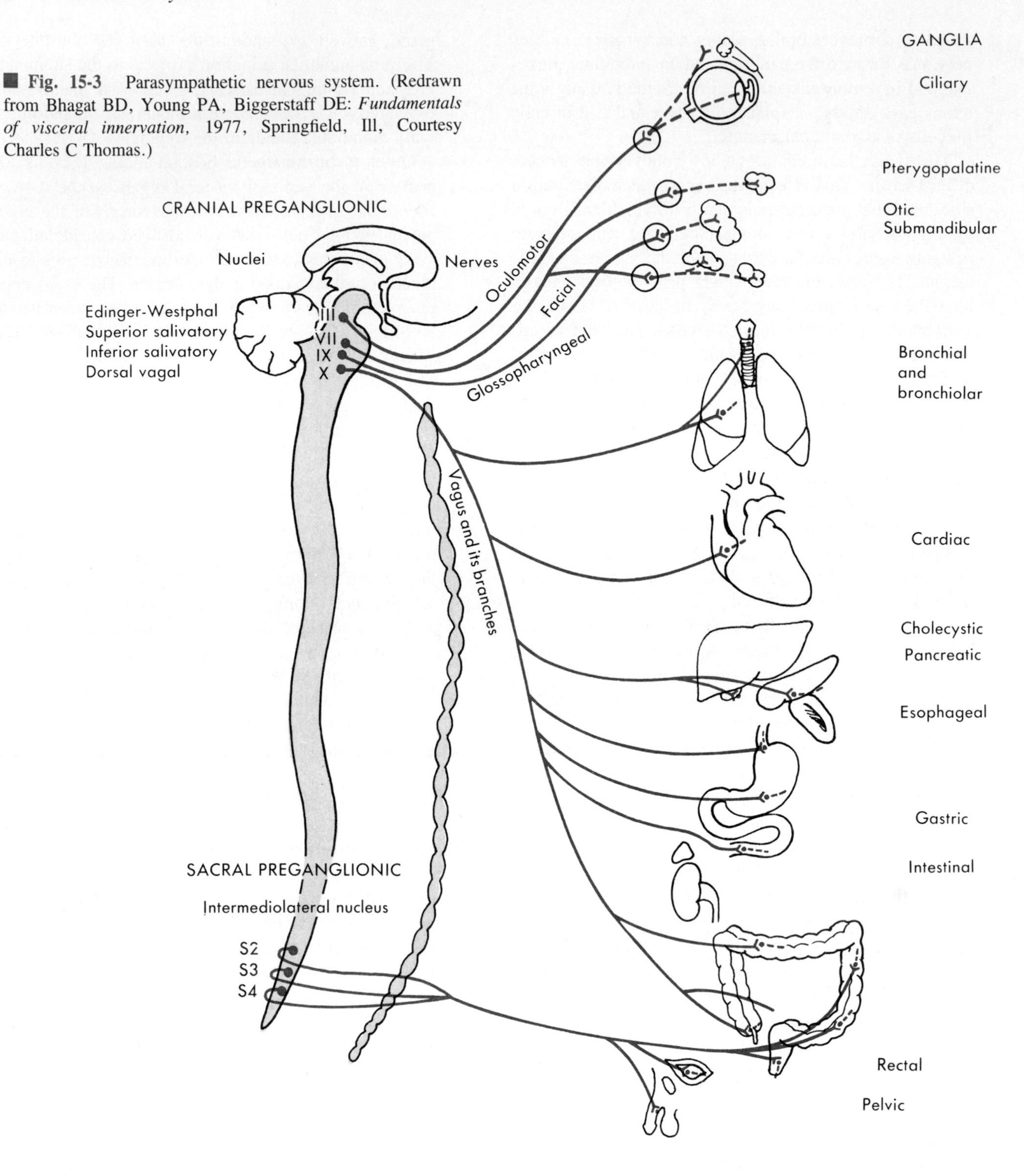

■ **Fig. 15-3** Parasympathetic nervous system. (Redrawn from Bhagat BD, Young PA, Biggerstaff DE: *Fundamentals of visceral innervation,* 1977, Springfield, Ill, Courtesy Charles C Thomas.)

lin, oxytocin, somatostatin, and vasoactive intestinal polypeptide.

Visceral afferent fibers that provide a sensory experience include nociceptors that travel in sympathetic nerves, such as the splanchnic nerves. Visceral pain is caused by excessive distention of hollow viscera, contraction against an obstruction, or ischemia. The origin of visceral pain is often difficult to identify because of its diffuse nature and its tendency to be referred to somatic structures (see Chapter 8). Visceral nociceptors in sympathetic nerves reach the spinal cord by way of the sympathetic chain, white rami, and dorsal roots. The terminals of the nociceptive afferent fibers distribute widely in the superficial dorsal horn and also in laminae V and X. They activate not only local interneurons, which participate in reflex arcs, but also projection cells, which include spinothalamic tract cells that signal pain to the brain.

Other visceral afferent fibers travel in parasympathetic nerves. These are generally involved in reflexes rather than in sensation (except for the taste afferent fibers; see Chapter 11). For example, the baroreceptor afferent fibers that innervate the carotid sinus are in the glossopharyngeal nerve. They enter the brainstem and pass through the solitary tract to terminate in the nucleus of the solitary tract. Neurons in the nucleus of the solitary tract connect with interneurons in the brainstem reticular formation. These interneurons in turn project to the autonomic preganglionic neurons that control heart rate and blood pressure (see Chapter 29). The nucleus of the solitary tract receives information from all visceral organs, except those in the pelvis. The nucleus is subdivided into a number of areas that receive information from specific visceral organs, as well as a zone of convergence of input from several organs (called the commissural nucleus and medial nucleus of the solitary tract). The specific regions are thought to be in reflex arcs that regulate that organ, whereas the region with convergent inputs project to the forebrain, probably in a viscerotopic manner.

The enteric nervous system. There are about 100 million neurons in the enteric nervous system, which is located in the walls of the gastrointestinal tract. The enteric nervous system is subdivided into the myenteric plexus, which lies between the longitudinal and circular muscle layers of the gut, and the submucosal plexus, which lies in the submucosa of the gut. The neurons of the myenteric plexus control gastrointestinal motility, whereas those in the submucosal plexus regulate body fluid homeostasis (see Chapter 38).

The types of neurons found in the myenteric plexus include not only excitatory and inhibitory motor neurons (which can be considered as parasympathetic postganglionic neurons), but also interneurons and primary afferent neurons. The afferent neurons supply mechanoreceptors within the wall of the gastrointestinal tract. These mechanoreceptors form the afferent limb of reflex arcs within the enteric plexus. The reflex processing is done by the local excitatory and inhibitory interneurons, and the output is through the motor neurons to the smooth muscle cells. Excitatory motor neurons contain acetylcholine and substance P, and inhibitory motor neurons contain dynorphin and vasoactive intestinal polypeptide. The circuitry of the enteric plexus is sufficient to allow coordinated movements of an intestine that is completely removed from the body. However, normal function requires innervation by the autonomic preganglionic neurons and regulation by the CNS.

Activity in the enteric nervous system is modulated by the sympathetic nervous system. Sympathetic postganglionic neurons that contain norepinephrine inhibit intestinal motility; those that contain norepinephrine and neuropeptide Y regulate blood flow; and those that contain norepinephrine and somatostatin control intestinal secretion. Feedback is provided by intestinofugal neurons that project back from the myenteric plexus to the sympathetic ganglia.

The submucosal plexus regulates ion and water transport across the intestinal epithelium and glandular secretion. It also communicates with the myenteric plexus to ensure coordination of the functions of the two components of the enteric nervous system. The neurons and neural circuits of the submucosal plexus are not as well understood as are those of the myenteric plexus, but many of the neurons contain neuropeptides and the neural networks are well organized.

Autonomic Ganglia

The main type of neuron in autonomic ganglia is the postganglionic neuron. These cells receive synaptic connections from preganglionic neurons, and they project to autonomic effector cells. However, many autonomic ganglia also contain interneurons. These interneurons allow some information processing to occur within the autonomic ganglia. (The enteric plexus can be regarded as an elaborate example of this.) A type of interneuron found in some autonomic ganglia can be shown by fluorescence histochemistry to contain a high concentration of catecholamines. Hence these interneurons have been called **small, intensely fluorescent cells,** or **SIF cells.** The SIF cells are believed to be inhibitory.

Neurotransmitters in Autonomic Ganglia

The classical neurotransmitter of autonomic ganglia, whether sympathetic or parasympathetic, is acetylcholine. There are two classes of acetylcholine receptors in autonomic ganglia, nicotinic and muscarinic, so named because of their responses to the plant alkaloids, nicotine and muscarine. Nicotinic acetylcholine receptors can be blocked by agents such as curare or hexamethonium, and muscarinic receptors can be blocked by atropine. Nicotinic receptors in autonomic ganglia differ somewhat from those on skeletal muscle cells.

Autonomic ganglia also contain neuropeptides that appear to act as neuromodulators. Besides acetylcholine, sympathetic preganglionic neurons may contain enkephalin, substance P, luteinizing hormone-releasing hormone, neurotensin, or somatostatin.

Catecholamines, such as norepinephrine or dopamine, serve as the neurotransmitters of the SIF cells in autonomic ganglia.

Stimulation of preganglionic neurons elicits a fast excitatory postsynaptic potential (EPSP), followed by a slow EPSP. The fast EPSP results from activation of nicotinic receptors, which cause the opening of ion channels. The slow EPSP is mediated by muscarinic receptors that inhibit the M current, a current produced by a potassium conductance.

Table 15-1 Responses of effector organs to autonomic nerve impulses

Effector organs	*Receptor type*	*Adrenergic impulses[1] Responses[2]*	*Cholinergic impulses[1] Responses[2]*
Eye			
Radial muscle, iris	α	Contraction (mydriasis) ++	—
Sphincter muscle, iris		—	Contraction (miosis) +++
Ciliary muscle	β	Relaxation for far vision +	Contraction for near vision +++
Heart			
SA node	β_1	Increase in heart rate ++	Decrease in heart rate; vagal arrest +++
Atria	β_1	Increase in contractility and conduction velocity ++	Decrease in contractility, and (usually) increase in conduction velocity ++
AV node	β_1	Increase in automaticity and conduction velocity ++	Decrease in conduction velocity; AV block +++
His-Purkinje system	β_1	Increase in automaticity and conduction velocity +++	Little effect
Ventricles	β_1	Increase in contractility, conduction velocity, automaticity, and rate of idioventricular pacemakers +++	Slight decrease in contractility
Arterioles			
Coronary	α,β_2	Constriction +; dilation[3] ++	Dilation ±
Skin and mucosa	α	Constriction +++	Dilation[4]
Skeletal muscle	α,β_2	Constriction ++; dilation[3,5] ++	Dilation[6] +
Cerebral	α	Constriction (slight)	Dilation[4]
Pulmonary	α,β_2	Constriction +; dilation[3]	Dilation[4]
Abdominal viscera; renal	α,β_2	Constriction +++; dilation[5] +	—
Salivary glands	α	Constriction +++	Dilation ++
Veins (systemic)	α,β_2	Constriction ++; dilation ++	—
Lung			
Bronchial muscle	β_2	Relaxation +	Contraction ++
Bronchial glands	?	Inhibition (?)	Stimulation +++
Stomach			
Motility and tone	α_2,β_2	Decrease (usually)[7] +	Increase +++
Sphincters	α	Contraction (usually) +	Relaxation (usually) +
Secretion		Inhibition (?)	Stimulation +++
Intestine			
Motility and tone	α_2,β_2	Decrease[7] +	Increase +++
Sphincters	α	Contraction (usually) +	Relaxation (usually) +
Secretion		Inhibition (?)	Stimulation ++
Gallbladder and ducts		Relaxation +	Contraction +
Kidney	β_2	Renin secretion ++	—
Urinary bladder			
Detrusor	β	Relaxation (usually) +	Contraction +++
Trigone and sphincter	α	Contraction ++	Relaxation ++
Ureter			
Motility and tone	α	Increase (usually)	Increase (?)
Uterus	α,β_2	Pregnant: contraction (α); nonpregnant: relaxation (β)	Variable[8]
Sex organs, male	α	Ejaculation +++	Erection +++
Skin			
Pilomotor muscles	α	Contraction ++	—
Sweat glands	α	Localized secretion[9] +	Generalized secretion +++
Spleen capsule	α,β_2	Contraction +++; relaxation +	—
Adrenal medulla		—	Secretion of epinephrine and norepinephrine
Liver	α,β_2	Glycogenolysis, gluconeogenesis[10] +++	Glycogen synthesis +
Pancreas			
Acini	α	Decreased secretion +	Secretion ++
Islets (β cells)	α	Decreased secretion +++	—
	β_2	Increased secretion +	—
Fat cells	α,β_1	Lipolysis[10] +++	—
Salivary glands	α	Potassium and water secretion +	Potassium and water secretion +++
	β	Amylase secretion +	—
Lacrimal glands		—	Secretion +++
Nasopharyngeal glands		—	Secretion ++
Pineal gland	β	Melatonin synthesis	—

*See footnotes on p. 251.

Neurotransmitters Between Postganglionic Neurons and Autonomic Effectors

Sympathetic postganglionic neurons. Sympathetic postganglionic neurons typically release norepinephrine, which can excite some effector cells and inhibit other effector cells. The receptors on the target cells may be either α or β adrenergic receptors. These can be further subdivided into α_1 and α_2, as well as β_1 and β_2 receptors. The distribution of these types of receptors and the actions that they mediate when activated by sympathetic postganglionic neurons are listed for various target organs in Table 15-1.

α_1 receptors are located postsynaptically, but α_2 receptors may be either pre- or postsynaptic receptors. Receptors located presynaptically are generally called **autoreceptors;** they are thought to inhibit transmitter release. The effects of agents that excite α_1 or α_2 receptors can be distinguished by using antagonists to block these receptors specifically. For example, prazosin is a selective α_1 blocker, and yohimbine is a selective α_2 blocker. The effects of α_1 receptors are mediated by activation of the inositol triphosphate/diacylglycerol second messenger system (see Chapter 5). On the other hand, α_2 receptors decrease the rate of synthesis of cyclic AMP through an action on a G-protein.

β receptors are subdivided into β_1 and β_2 receptors on the basis of the ability of antagonists to block them. The two types of β receptors are similar proteins, with seven membrane spanning regions connected by intracellular and extracellular domains (see Chapter 5). Agonist drugs that work on β receptors activate a G-protein, which stimulates adenylyl cyclase to increase the cyclic AMP concentration. This action is terminated by the buildup of GDP. It can also be antagonized by action of α_2 receptors. The number of β receptors is regulated on a short term basis by the process of desensitization, in which β receptors are phosphorylated and in some cases internalized, thereby decreasing the number of available receptors. β receptors can also be increased in number (up-regulation), e.g., by denervation. The number of α receptors is also regulated.

In addition to norepinephrine, sympathetic postganglionic neurons contain neuropeptides, such as somatostatin or neuropeptide Y. For example, cells that contain both norepinephrine and somatostatin supply the mucosa of the gastrointestinal tract, and others that contain both norepinephrine and neuropeptide Y innervate blood vessels in the gut and the limb. Another chemical mediator in sympathetic postganglionic neurons is ATP.

The endocrine cells of the adrenal medulla are similar in many respects to sympathetic postganglionic neurons. They receive input from sympathetic preganglionic neurons, are excited by acetylcholine, and release catecholamines. However, the cells of the adrenal medulla differ from sympathetic postganglionic neurons in that they release catecholamines into the circulation, rather than synaptically. Also, the main catecholamine released is epinephrine, not norepinephrine (in humans 80% of the catecholamine released by the adrenal medulla is epinephrine and 20%, norepinephrine).

Some sympathetic postganglionic neurons release acetylcholine rather than norepinephrine as their neurotransmitter. For example, sympathetic postganglionic neurons that innervate eccrine sweat glands are cholinergic. The acetylcholine receptors involved are muscarinic and therefore are blocked by atropine. Similarly, some blood vessels are innervated by cholinergic sympathetic postganglionic neurons. In addition to acetylcholine, the postganglionic neurons that supply the sweat glands also contain neuropeptides, including calcitonin gene-related peptide and vasoactive intestinal polypeptide.

Parasympathetic postganglionic neurons. The neurotransmitter for parasympathetic postganglionic neurons is acetylcholine. The effects of these neurons on various target organs are listed in Table 15-1. Parasympathetic postganglionic actions are mediated by muscarinic receptors. Based on binding studies, the action of selective antagonists, and molecular cloning, there now appear to be several types of muscarinic receptors. At least two types of muscarinic receptors, M_1 and M_2, can be distinguished on the basis of the action of the antagonist, pirenzepine. M_1 receptors have a high affinity

Table 15-1 footnotes

From Goodman LS, Gilman A: *The pharmacological basis of therapeutics,* ed 6, New York, 1980, Macmillan Publishing.

[1]A long dash signifies no known functional innervation.

[2]Responses are designated 1+ to 3+ to provide an approximate indication of the importance of adrenergic and cholinergic nerve activity in the control of the various organs and functions listed.

[3]Dilation predominates *in situ* due to metabolic autoregulatory phenomena.

[4]Cholinergic vasodilatation at these sites is of questionable physiological significance.

[5]Over the usual concentration range of physiologically released, circulating epinephrine, β-receptor response (vasodilatation) predominates in blood vessels of skeletal muscle and liver, α-receptor response (vasoconstriction), in blood vessels of other abdominal viscera. The renal and mesenteric vessels also contain specific dopaminergic receptors, activation of which causes dilatation, but their physiological significance has not been established.

[6]Sympathetic cholinergic system causes vasodilatation in skeletal muscle, but this is not involved in most physiological responses.

[7]It has been proposed that adrenergic fibers terminate at inhibitory β-receptors on smooth muscle fibers and at inhibitory α-receptors on parasympathetic cholinergic (excitatory) ganglion cells of Auerbach's plexus.

[8]Depends on stage of menstrual cycle, amount of circulating estrogen and progesterone, and other factors.

[9]Palms of hands and some other sites ("adrenergic sweating").

[10]There is significant variation among species in the type of receptor that mediates certain metabolic responses.

for pirenzepine; and their activation enhances the secretion of gastric acid, whereas M_2 receptors have a low affinity for pirenzepine and their activation slows the heart. A subtype of the M_2 receptor activate glands, such as the lacrimal and submaxillary glands.

Muscarinic receptors, like adrenergic ones, have diverse actions. Some of their effects are mediated by specific second messenger systems. For example, cardiac M_2 muscarinic receptors may act by way of the inositol trisphosphate (IP3) system, and they may also inhibit adenylyl cyclase and thus cAMP synthesis. Muscarinic receptors also open or close ion channels, particularly K^+ or Ca^{++} channels. This action on ion channels is likely to occur through activation of G-proteins, because (1) the action of acetylcholine on ion channels through muscarinic receptors is much slower than that through nicotinic receptors, (2) intracellular GTP is required and (3) the effect is blocked by pertussis toxin (which inactivates certain G-proteins). A third action of muscarinic receptors is to relax vascular smooth muscle by an effect on endothelial cells, which produce endothelium-derived relaxing factor (EDRF). EDRF is now thought to be nitric oxide, a compound that is synthesized from arginine. Nitric oxide induces relaxation of vascular smooth muscle by stimulating guanylate cyclase and thereby increasing levels of cGMP, which in turn activates a cGMP-dependent protein kinase (see Chapter 5).

The number of muscarinic receptors is regulated, and exposure to muscarinic agonists apparently decreases the number of receptors by internalization of the receptors.

■ Central Control of Autonomic Function

The discharges of autonomic preganglionic neurons are controlled by pathways that synapse on autonomic preganglionic neurons. The pathways that influence autonomic activity include spinal cord or brainstem reflex pathways and also descending control systems that originate at higher levels of the nervous system, such as the hypothalamus.

■ Examples of Autonomic Control of Particular Organs

Several examples are given of autonomic control systems that affect different target organs and that depend on local reflex circuitry or on signals from other parts of the CNS (see Table 15-1).

Pupil. The size of the pupils is controlled by reflex pathways and by activity of higher centers. The sphincter and dilator muscles of the iris determine the size of the pupil. Activation of the sympathetic innervation of the eye dilates the pupil **(mydriasis).** This happens during emotional excitement and also in response to painful stimulation (e.g., of the skin). The sympathetic preganglionic neurons that control pupil size are located in the intermediolateral cell column in the T1 and T2 segments. The axons exit through ventral roots, join the paravertebral ganglia after passing through white rami, and then ascend in the sympathetic chain to the superior cervical sympathetic ganglion, where they synapse on postganglionic neurons. These project through the carotid plexus, and then follow the ophthalmic artery and ciliary nerves into the eye. The neurotransmitter at the postganglionic synapses is norepinephrine, which acts at α receptors.

Sympathetic control of the pupil is sometimes affected by disease. For example, interruption of the sympathetic innervation of the head and neck results in *Horner's syndrome* (pupillary constriction, loss of sweating on the face, vasodilation of facial skin, and withdrawal of the eye into the orbit [enophthalmos]). Horner's syndrome can be produced by a lesion that (1) destroys the sympathetic preganglionic neurons in the upper thoracic spinal cord, (2) interrupts the cervical sympathetic chain, or (3) damages the lower brainstem in the region of the reticular formation through which pathways descend to the spinal cord to activate sympathetic preganglionic neurons.

The parasympathetic nervous system has an opposite action on pupillary size to that of the sympathetic nervous system. Whereas the sympathetic system elicits pupillary dilation, the parasympathetic system constricts the pupil **(meiosis).** The preganglionic parasympathetic neurons are located in the Edinger-Westphal nucleus, and they project to the ciliary ganglion through the oculomotor nerve. The postganglionic neurons then project through short ciliary nerves into the eye and synapse on the pupillary sphincter muscle. The main neurotransmitter at this synapse is acetylcholine, which acts on muscarinic receptors. However, peptides may also serve as neuromodulators for some neurons.

Pupil size is reduced by the pupillary light reflex and during accommodation for near vision. In the **pupillary light reflex,** light that strikes the retina is processed by retinal circuits that excite W type retinal ganglion cells (see Chapter 9). These cells are sensitive to luminance. The axons of some of the W cells project through the optic nerve and tract to the pretectal area. There they synapse in several nuclei, including the olivary pretectal nucleus, which contains neurons that are sensitive to luminance. Activity of the luminance-detection neurons of the olivary pretectal nucleus causes pupillary constriction by means of bilateral connections with parasympathetic preganglionic neurons in the Edinger-Westphal nuclei. Fibers cross to the contralateral side through the posterior commissure. The reflex results in contraction of the pupillary sphincter muscle.

In the **accommodation response,** information from

Y cells of the retina is transmitted to the striate cortex through the geniculate-striate visual pathway. The stimulus that triggers accommodation is thought to be a blurred retinal image and binocular disparity. After the information is processed in the visual cortex, signals are transmitted directly or indirectly to the middle temporal cortex, where they activate neurons in a visual area known as MT. By an unknown pathway, MT neurons transmit signals to the midbrain that activate parasympathetic preganglionic neurons in the Edinger-Westphal nuclei bilaterally, which results in pupillary constriction. At the same time signals are transmitted to the ciliary muscle, causing it to contract. This ciliary muscle contraction allows the lens to round up and to increase its refractile power.

The pupillary light reflex is sometimes absent in patients with syphilis that affects the CNS (i.e., in tabes dorsalis). Although the pupil fails to respond to light, it has a normal accommodation response. This condition is known as the **Argyll-Robertson pupil.** The exact mechanism is controversial, but one explanation is that the brachium of the superior colliculus is interrupted, as are the fibers in the brachium that pass from the optic tract to the pretectal area. Thus, the input to the olivary pretectal nucleus is interrupted, but the optic tract connection to the lateral geniculate nucleus is maintained.

Urinary bladder. The urinary bladder is controlled by reflex pathways in the spinal cord and also by a supraspinal center.

The sympathetic innervation originates from preganglionic sympathetic neurons in the upper lumbar segments of the spinal cord. The axons of these neurons leave the spinal cord through ventral roots and gain access to the sympathetic chain through the white rami. The preganglionic axons either descend in the sympathetic chain to lumbosacral paravertebral ganglia, or they project through splanchnic nerves to synapse in a prevertebral ganglion, the inferior mesenteric ganglion. Postganglionic sympathetic axons pass through the pelvic nerves or descend from the inferior mesenteric ganglion in the hypogastric plexus and end in the bladder on the detrusor muscle (which they inhibit) and on muscles of the trigone region and of the internal urethral sphincter (which they excite). The detrusor muscle is tonically inhibited during filling of the bladder. The inhibition of the detrusor muscle is mediated by norepinephrine acting on β receptors, whereas the excitation of the trigone and internal urethral sphincter is elicited by norepinephrine acting on α receptors. Some sympathetic preganglionic axons reach the ganglia located on the bladder wall and synapse on postganglionic sympathetic neurons in this location. The ganglionic connections produce either excitation by α_1 receptors or inhibition by α_2 receptors. The main effect of the sympathetic innervation is to prevent voiding until the bladder is full.

The external sphincter of the urethra also helps prevent voiding. This sphincter is a striated muscle, and it is innervated by motor axons in the pudendal nerves, which are somatic nerves. The motor neurons are located in Onuf's nucleus in the ventral horn of the sacral spinal cord.

The parasympathetic preganglionic neurons that control the bladder are in the sacral spinal cord (S2 and S3 or S3 and S4 segments). They project through the ventral roots into the pelvic nerves and are distributed to ganglia in the pelvic plexus and in the bladder wall. These neurons are cholinergic. Postganglionic parasympathetic neurons in the bladder wall innervate the detrusor muscle, as well as the trigone and sphincter. The parasympathetic actions are contraction of the detrusor and relaxation of the trigone and sphincter. This results in **micturition,** or urination. Some of the postganglionic neurons are cholinergic and others, purinergic (they release ATP).

Micturition is normally controlled by the **micturition reflex** (Fig. 15-4). Mechanoreceptors in the bladder wall are supplied by Aδ and C visceral afferent fibers that reach the sacral spinal cord through the pelvic nerves and that synapse in laminae I, V, VII, and X.

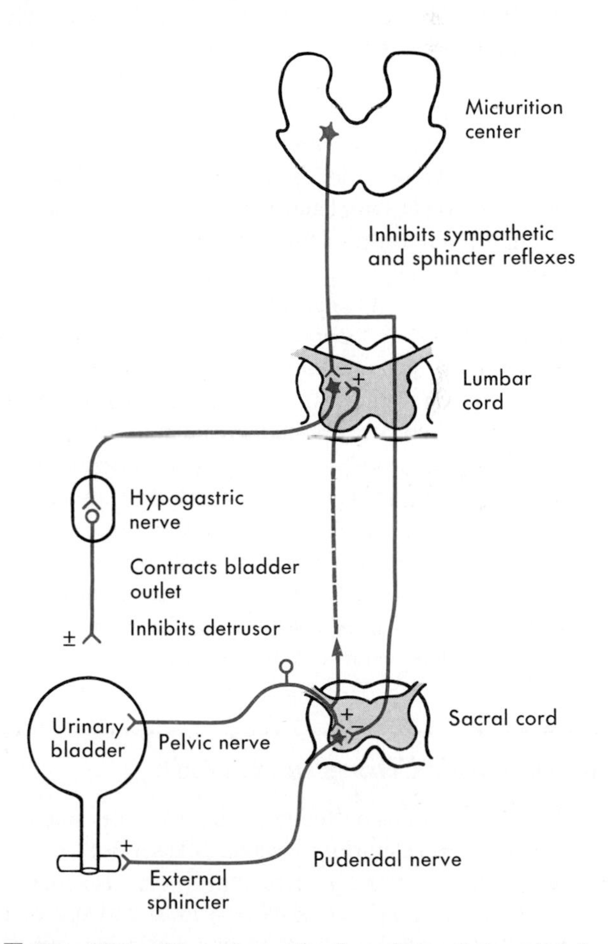

■ **Fig. 15-4** The pathway for the reflexes that control the urinary bladder. (Redrawn from de Groat WC, Booth AM: *Autonomic systems to bladder and sex organs.* In Dyck PJ et al, editors: *Peripheral neuropathy,* ed 2, Philadelphia, 1984, WB Saunders.)

These receptors are excited by both stretch and contraction of the muscles in the bladder wall. Thus, as the bladder distends because of the accumulation of urine, the mechanoreceptors begin to discharge. The threshold for activating the Aδ afferent fibers is 5 to 15 mm Hg, which is also threshold for a sensation of bladder distention in humans. Other bladder afferent fibers are found in sympathetic nerves; these are likely to transmit information from nociceptors and so mediate bladder pain. Still other afferent fibers from the bladder are in the pudendal nerve; these transmit information about temperature and pain sensations, as well as the sense of urine flow.

The pressure in the urinary bladder is low during filling (5 to 10 cm H_2O), but it increases abruptly when micturition begins. Micturition can be triggered either reflexly or voluntarily. In reflex micturition bladder afferent fibers excite neurons that project to the brainstem and that activate the micturition center in the rostral pons (Barrington's center). The ascending projections also inhibit sympathetic preganglionic neurons that prevent voiding. When there is a sufficient level of activity in this ascending pathway, micturition is triggered by the micturition center. Commands reach the sacral spinal cord through a reticulospinal pathway. Activity in the sympathetic projection to the bladder is inhibited and the parasympathetic supply of the bladder is activated. Contraction of the muscle in the wall of the bladder causes a vigorous discharge of the mechanoreceptors that supply the bladder wall and thereby further activate the supraspinal loop. The result is complete emptying of the bladder.

There is also a spinal reflex pathway for micturition. This pathway is operational in the newborn infant. However, with maturation the supraspinal control pathways take on a dominant role in triggering micturition. After spinal cord injury human adults lose bladder control during the period of spinal shock (urinary incontinence). As the spinal cord recovers from spinal shock, some degree of bladder function is recovered because of an enhancement of the spinal cord micturition reflex. However, the bladder has an increased muscle tone, and it fails to empty completely. These circumstances frequently lead to urinary infections.

Autonomic Centers in the Brain

The neural circuit in the pons that regulates micturition is called the micturition center. Many other autonomic centers with a variety of functions, are also located in the brain. A center consists of a local network of neurons that respond to inputs from a particular source and that influence distant neurons by way of long efferent pathways. Activity of a center is recognized by a characteristic set of autonomic actions. The details of the organization of these centers are currently under investigation.

Vasomotor and vasodilator centers are located in the medulla, and respiratory centers are located in the medulla and pons. Thus, autonomic centers are present at several levels of the brainstem. Perhaps the greatest concentration of autonomic centers is in the hypothalamus.

Hypothalamus

The hypothalamus is a part of the diencephalon. Some of the nuclei of the hypothalamus are shown in Figs. 15-5 and 15-6. In the rostrocaudal dimension the hypothalamus can be subdivided into three regions: the suprachiasmatic region, the tuberal region, and the mammillary region. Some important nuclei include the supraoptic, paraventricular, tuberal, and mammillary nuclei. Continuing anteriorly from the hypothalamus are telencephalic structures, the preoptic region and septum. Both the preoptic and septal regions help regulate autonomic function. Important fiber tracts that course through the hypothalamus are the fornix, the medial forebrain bundle, and the mammillothalamic tract. The fornix is used as a landmark to divide the hypothalamus into the medial and the lateral hypothalmus. Some of the connections of the hypothalamus are shown in Fig. 15-7.

The hypothalamus has many functions. Its control of autonomic function is emphasized here. See Chapter 48 for a discussion of hypothalamic control of endocrine function.

Temperature regulation. Homeothermic animals regulate their body temperature. When the environmental temperature decreases, the body adjusts by reducing heat loss and by increasing heat production. Conversely, when the temperature rises, the body increases its heat loss and reduces heat production.

Information about the external temperature is provided by thermoreceptors in the skin (and probably other organs, such as muscle). Internal temperature is monitored by central thermoreceptive neurons in the anterior hypothalamus. The central thermoreceptors monitor the temperature of the blood. The system acts as a servomechanism with a set point at the normal body temperature. Error signals, which represent a deviation from the set point, evoke responses that tend to restore body temperature toward the set point. These responses are mediated by the autonomic, somatic, and endocrine systems.

Cooling causes shivering, which consists of asynchronous muscle contractions that increase heat production. Increases in thyroid gland activity and in sympathetic activity tend to increase heat production metabolically. Heat loss is reduced by piloerection and by cuta-

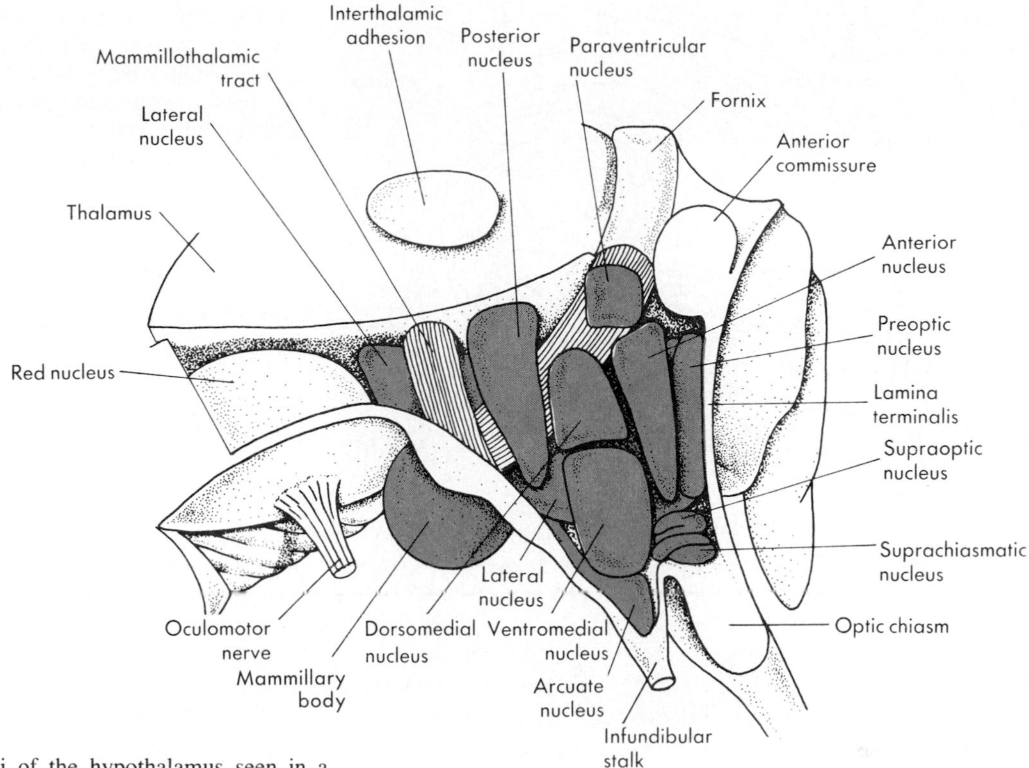

■ **Fig. 15-5** Main nuclei of the hypothalamus seen in a view from the third ventricle. (Redrawn from Nauta WJH, Haymaker W: *The hypothalamus,* Springfield, Ill, 1969, Charles C Thomas.)

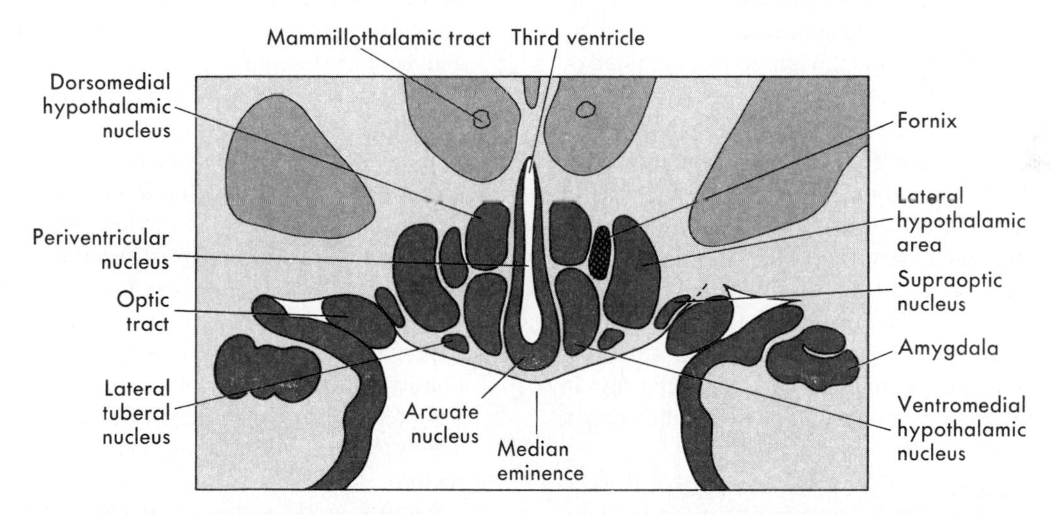

■ **Fig. 15-6** Nuclei of the hypothalamus as seen in a frontal section. The fornix is a fiber bundle that divides the hypothalamus into lateral and medial regions. *A,* amygdala; *H,* hypothalamus; *M,* mammillary body; *MD,* dorsomedial thalamic nucleus; *S,* septus; *VM,* ventromedial hypothalamic nucleus. (Redrawn by permission from the publisher of Kandel ER, Schwartz JH: *Principles of neural science,* ed 2. New York: 1985, Elsevier.)

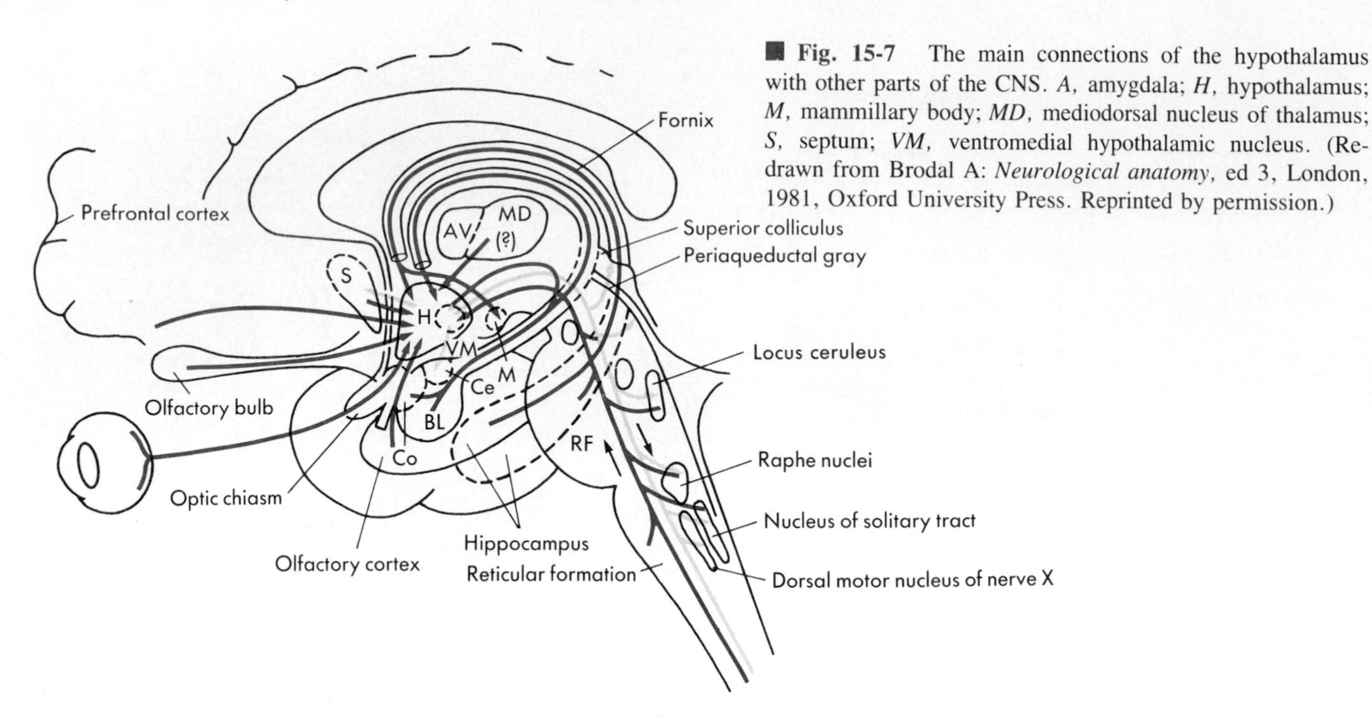

■ **Fig. 15-7** The main connections of the hypothalamus with other parts of the CNS. *A*, amygdala; *H*, hypothalamus; *M*, mammillary body; *MD*, mediodorsal nucleus of thalamus; *S*, septum; *VM*, ventromedial hypothalamic nucleus. (Redrawn from Brodal A: *Neurological anatomy*, ed 3, London, 1981, Oxford University Press. Reprinted by permission.)

neous vasoconstriction. Piloerection is effective in animals with fur, but not in humans; in the latter the result is goose-bumps.

Warming the body causes changes in the opposite direction. The activity of the thyroid gland diminishes, which leads to reduced metabolic activity and less heat production. Heat loss is increased by sweating and cutaneous vasodilation.

The hypothalamus serves as the temperature servomechanism. The heat loss responses are organized by the heat loss center, which is thought to be composed of neurons in the preoptic region and anterior hypothalamus. Lesions here prevent sweating and cutaneous vasodilation and cause **hyperthermia** when the individual is placed in a warm environment. Conversely, electrical stimulation of the heat loss center causes cutaneous vasodilation and inhibits shivering. Neurons in the posterior hypothalamus form a heat production and conservation center. Lesions in the area dorsolateral to the mammillary body eliminate heat production and conservation. This results in hypothermia when the subject is in a cold environment. Electrical stimulation in this region evokes shivering.

Thermoregulatory responses are also produced when the hypothalamus is locally warmed or cooled. These responses are consistent with the presence of central thermoreceptive neurons in the hypothalamus. Many details of the interconnections of the hypothalamus involved in the heat loss and in the heat production and conservation centers are unknown.

In fever the set point for body temperature is elevated. This can be caused by the release of a pyrogen by microorganisms. The pyrogen changes the set point, and this leads to an increased heat production by shivering and to heat conservation by cutaneous vasoconstriction.

Regulation of food intake. Food intake is also regulated by a servomechanism. However, the set point is affected by many factors.

Sensory signals that help regulate food intake operate both on a short-term basis to control ingestion and on a long-term basis to control body weight. Glucoreceptors in the hypothalamus sense blood glucose and use this information to control food intake. Their main action occurs when blood glucose levels decrease. Opioid peptides and pancreatic polypeptide stimulate food intake, and cholecystokinin inhibits food intake. Insulin and adrenal glucocorticoids also affect food intake.

Lesions of the lateral hypothalamus lead to a lack of food intake or **aphagia,** which can cause starvation and death. Electrical stimulation in the lateral hypothalamus causes eating. These observations suggest that the lateral hypothalamus contains a feeding center. Converse effects are produced by manipulations of the ventromedial nucleus of the hypothalamus. A lesion here causes hyperphagia, an increased food intake that can result in obesity, whereas electrical stimulation stops feeding behavior. This area of the hypothalamus is known as the satiety center. The feeding and satiety centers operate reciprocally.

Further work is needed to clarify the role of other parts of the nervous system in feeding behavior. Some structures that are involved are shown in Fig. 15-8.

Regulation of water intake. Water intake also depends on a servomechanism. Fluid intake is influenced by blood osmolality and volume (Fig. 15-9).

With water deprivation the extracellular fluid becomes hyperosmotic; this, in turn, causes intracellular fluid to become hyperosmotic. The brain contains neurons that serve as osmoreceptors, which detect increases in the osmotic pressure of the extracellular fluid. The osmoreceptors appear to be located in one of the cir-

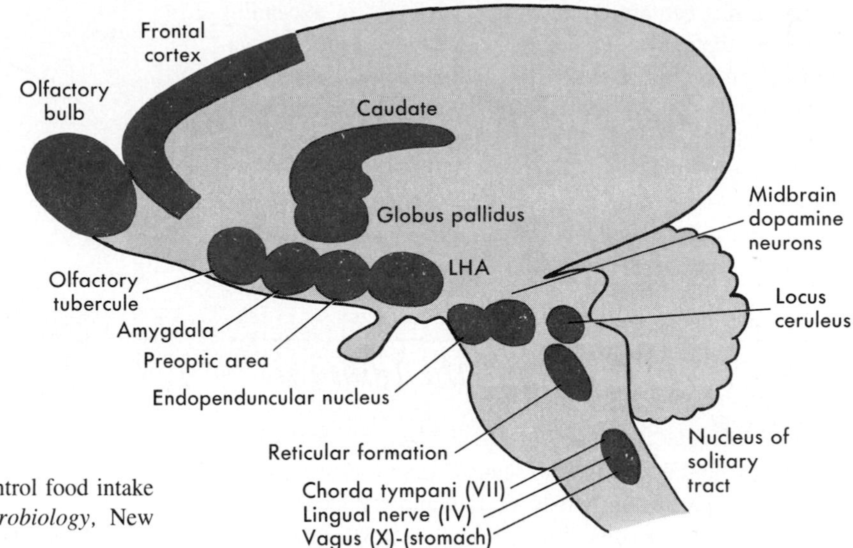

■ **Fig. 15-8** Structures thought to help control food intake in rats. (Redrawn from Shepherd GM: *Neurobiology,* New York, 1983, Oxford University Press.)

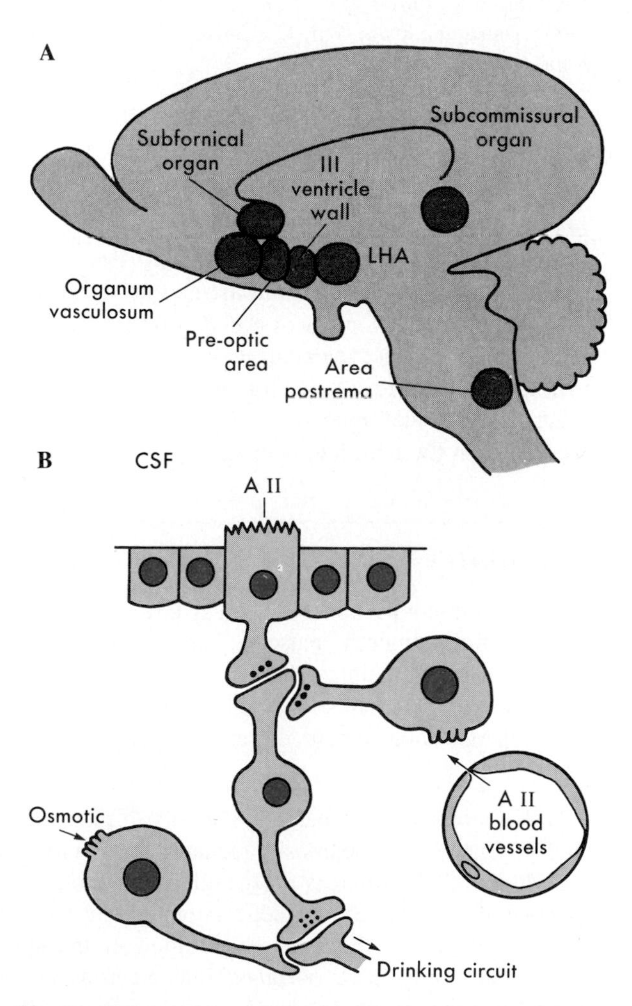

■ **Fig. 15-9** **A,** Structures thought to play a role in regulation of water intake in rats. **B,** The neural circuits that signal changes in blood osmolality and volume. (**A,** Redrawn from Shepherd GM: *Neurobiology,* New York, 1983, Oxford University Press.)

cumventricular organs, the organum vasculosum of the lamina terminalis. Circumventricular organs are found around the cerebral ventricles and lack a blood-brain barrier. The subfornical organ and the organum vasculosum are involved in thirst. The area postrema serves as a chemosensitive zone that triggers vomiting.

Water deprivation also causes a decrease in blood volume. This is sensed by receptors in the low pressure side of the vasculature, including the right atrium. In addition, a decreased blood volume triggers the release of renin by the kidney. Renin breaks down angiotensinogen into angiotensin I, which is then hydrolyzed to angiotensin II (see Chapter 42). This peptide stimulates drinking by an action on angiotensin II receptors in another one of the circumventricular organs, the subfornical organ. Angiotensin II also causes vasoconstriction and release of aldosterone and ADH.

Insufficient water intake, not excess water intake, is usually the problem. When more water is taken in than is required, it is easily eliminated by inhibition of the release of antidiuretic hormone (ADH) from neurons in the supraoptic nucleus at their terminals in the posterior pituitary gland (see Chapter 48). As already mentioned, signals that inhibit ADH release include an increased blood volume and a decreased osmolality of the extracellular fluid (see Chapter 48). Other areas of the hypothalamus, particularly the preoptic region and lateral hypothalamus, help regulate water intake, as do several structures outside the hypothalamus.

■ *Other Autonomic Control Structures*

Several regions of the forebrain other than the hypothalamus also play a role in autonomic control. These include the central nucleus of the amygdala and bed nucleus of the stria terminalis, as well as a number of areas of cerebral cortex. Information reaches these higher autonomic centers from viscera through an ascending system that involves the nucleus of the solitary tract, the parabrachial nucleus, the periaqueductal gray, and the hypothalamus. Descending pathways that help control autonomic activity originate from a number of structures, including the paraventricular nucleus of the hypothalamus, the A5 noradrenergic cell group, the rostral ventrolateral medulla, and the raphe nuclei and adjacent structures of the ventromedial medulla.

Neural Influences on the Immune System

Environmental stress can cause immunosuppression, with a reduction in helper T cells and reduced activity of natural killer cells. Immunosuppression can even be the result of classical conditioning. One mechanism for such an effect involves the release of corticotropin-releasing factor (CRF) from the hypothalamus. CRF causes the release of ACTH from the pituitary gland; and this stimulates the secretion of adrenal corticosteroids, which produce immunosuppression (see Chapter 50). Other mechanisms include direct neural actions on lymphoid tissues. The immune system also may influence neural activity.

Emotional Behavior

The limbic system is thought to control emotional behavior, in part by an influence on the hypothalamus. The limbic lobe is phylogenetically the oldest part of the cerebral cortex. A circuit that connects the limbic lobe with the hypothalamus (the Papez circuit) possibly regulates emotional behavior. The neural components of this circuit were termed the limbic system (Fig. 15-10). The Papez circuit connects many areas of the neocortex to the hypothalamus. Information passes from the cingulate gyrus to the entorhinal cortex and hippocampus, and from there to the mammillary bodies in the hypothalamus. The mammillothalamic tract then connects the hypothalamus with the anterior thalamic nuclei, which projects back to the cingulate gyrus. Other structures that were later added to the limbic system circuitry include the amygdala and the bed nucleus of the stria terminalis.

Evidence that supports a role of the limbic system in emotional behavior came from lesion and stimulation experiments. For example, stimulation in the amygdala can result in aggression, and lesions of the amygdala may cause docility. Hypothalamic stimulation can cause defensive behavior. Surgery that separates the hypothalamus from higher centers results in **sham rage,** a condition in which an animal shows behavioral signs similar to those of an enraged intact animal. Bilateral temporal lobe lesions can produce the **Kluver-Bucy syndrome,** which is characterized by loss of the ability to detect and recognize the meaning of objects from visual cues **(visual agnosia),** a tendency to examine objects orally, attention to irrelevant stimuli, hypersexuality, change in dietary habits, and a decreased emotionality. The components of this syndrome can be attributed to damage to different parts of the neocortex and limbic cortex. For instance, the changes in emotional behavior are largely the result of lesions of the amygdala, whereas the visual agnosia is caused by damage to visual areas in the temporal neocortex.

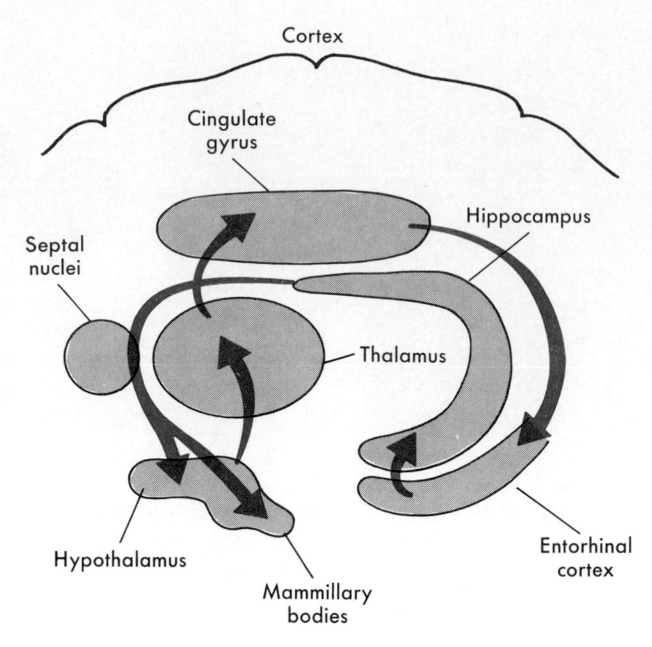

Fig. 15-10 The Papez circuit. (From Groves PM, Schlesinger K: *Introduction to biological psychology,* ed 2, 1982, Dubuque, Iowa, Wm C Brown. All rights reserved. Reprinted by permission.)

Summary

1. The autonomic nervous system is a motor system that controls smooth muscle, cardiac muscle, and glands. It helps maintain homeostasis and coordinates responses to external stimuli. Its components are the sympathetic, parasympathetic, and enteric nervous systems.

2. Autonomic motor pathways involve preganglionic and postganglionic neurons. Preganglionic neurons reside in the CNS, whereas postganglionic neurons lie in peripheral ganglia. Sympathetic preganglionic neurons are located in the thoracolumbar region of the spinal cord, and sympathetic postganglionic neurons are located in paravertebral and prevertebral ganglia. Parasympathetic preganglionic neurons are located in cranial nerve nuclei or in the sacral spinal cord, and parasympathetic postganglionic neurons reside in ganglia located in or near the target organs.

3. Visceral afferent fibers supply sensory receptors in

the viscera. Some have a sensory function, such as visceral pain and taste, but most activate reflexes.

4. The enteric nervous system includes the myenteric and submucosal plexuses in the wall of the gastrointestinal tract. The myenteric plexus regulates motility and the submucosal plexus regulates ion and water transport and secretion. The neural circuitry of the enteric plexuses permits coordinated activity in the isolated intestine, but normal function depends on intact autonomic control.

5. Neurotransmitters at the synapses of preganglionic neurons in autonomic ganglia include acetylcholine (acting at both nicotinic and muscarinic receptors) and a number of neuropeptides. Interneurons use catecholamines. Sympathetic postganglionic neurons generally use norepinephrine (acting at α_1, α_2, β_1, or β_2 adrenergic receptors) as their neurotransmitter, although often neuropeptides are also released. Sympathetic postganglionic neurons that supply sweat glands use acetylcholine. Parasympathetic postganglionic neurons use acetylcholine (acting on M_1 or M_2 muscarinic receptors).

6. The pupil is controlled reciprocally by the sympathetic and parasympathetic nervous systems. Sympathetic activity causes mydriasis and parasympathetic activity causes meiosis. The sympathetic pathway can be activated by excitement or pain. Parasympathetic pathways are involved in the pupillary light reflex and in accommodation.

7. Emptying of the urinary bladder depends on parasympathetic outflow during the micturition reflex. Sympathetic constriction of the external sphincter of the urethra prevents voiding. The micturition reflex is triggered by stretch receptors and is controlled in normal adults by a micturition center in the pons.

8. The hypothalamus contains several centers that control autonomic and other activity. These include the heat loss and the heat production and conservation centers, the feeding and satiety centers, and centers that regulate fluid intake.

9. Information about visceral activities reaches other autonomic control centers outside the hypothalamus. A number of descending pathways control visceral function through activation of the autonomic nervous system.

10. The limbic system consists of several cortical and other structures. The limbic system controls emotional behavior, in part by activation of the autonomic nervous system.

Bibliography

Journal articles

Anand BK, Brobeck JR: Localization of a "feeding center" in the hypothalamus of the rat, *Proc Soc Exp Biol Med* 77:323, 1951.

Andersson B: Regulation of water intake, *Physiol Rev* 58:482, 1978.

Burnstock G: The changing face of autonomic neurotransmission, *Acta Physiol Scand* 126:67, 1986.

Bylund DB, Prichard DC: Characterization of α_1- and α_2-adrenergic receptors, *Int Rev Neurobiol* 243:343, 1983.

Cabanac M: Temperature regulation, *Annu Rev Physiol* 37:415, 1975.

deGroat WC et al: Organization of the sacral parasympathetic reflex pathways to the urinary bladder and large intestine, *J Auton Nerv Syst* 3:135, 1981.

Gershon MD: The enteric nervous system, *Annu Rev Neurosci* 4:227, 1981.

Greenwood B, Davison JS: The relationship between gastrointestinal motility and secretion, *Amer J Physiol* 252:G1, 1987.

Johnson AK, Cunnigham JT: Brain mechanisms and drinking: the role of lamina terminalis-associated systems and extracellular thirst, *Kidney Int* 32:S35, 1987.

Kalia M: Brain stem localization of vagal preganglionic neurons, *J Auton Nerv Syst* 3:451, 1981.

Lundberg JM, Hokfelt T: Multiple coexistence of peptides and classical transmitters in peripheral autonomic and sensory neuron—functional and pharmacological implications, *Prog Brain Res* 68:241, 1986.

Moncada S et al: Endothelium-derived relaxing factor: identification as nitric oxide and role in the control of vascular tone and platelet function, *Biochem Pharmacol* 37:2495, 1988.

Nathanson NM: Molecular properties of the muscarinic acetylcholine receptor, *Annu Rev Neurosci* 10:195, 1987.

Papez JW: A proposed mechanism of emotion, *Arch Neurol Psychiat* 38:725, 1937.

Petras JM, Faden AI: The origin of sympathetic preganglionic neurons in the dog, *Brain Res* 144:3563, 1978.

Smith OA, DeVito JL: Central neural integration for the control of autonomic responses associated with emotion, *Annu Rev Neurosci* 7:43, 1984.

Books and monographs

Bjorklund A et al, editors: *Handbook of chemical neuroanatomy, vol 6, The peripheral nervous system,* Amsterdam, 1988, Elsevier.

Brodal A: *Neurological anatomy,* ed 3, New York, 1981, Oxford University Press.

Cannon WB: *The wisdom of the body,* ed 2, New York, 1939, WW Norton & Co.

Furness JB, Costa M: *The enteric nervous system,* Edinburgh, 1987, Churchill Livingstone.

Gross PM: *Circumventricular organs and body fluids, vols. I-III,* Boca Raton, 1987, CRC Press.

Johnson LR, editor: *Physiology of the gastrointestinal tract,* ed 2, New York, 1987, Raven Press.

Karczmar AG et al: *Autonomic and enteric ganglia: transmission and its pharmacology,* New York, 1986, Plenum Press.

Loewy AD, Spyer KM, editors: *Central regulation of autonomic functions,* New York, 1990, Oxford University Press.

Pick J: *The autonomic nervous system: morphological, comparative, clinical and surgical aspects,* Philadelphia, 1970, JB Lippincott.

Schmidt RF, Thews G: *Human physiology,* Heidelberg, 1983, Springer-Verlag.

Torrens M, Morrison JFB: *The physiology of the lower urinary tract,* Berlin, 1987, Springer-Verlag.

CHAPTER

16

The Cerebral Cortex and Higher Functions of the Nervous System

Interactions between different parts of the cerebral cortex and between the cerebral cortex and other parts of the brain are responsible for the higher functions that characterize humans. The neural basis for some of these higher functions are discussed in this chapter.

The Cerebral Cortex

The cerebral cortex in humans occupies a volume of about 600 cm^3 and a surface area of 2,500 cm^2. The surface of the cortex is highly folded into ridges, known as **gyri,** separated by grooves called **sulci** (if shallow) and **fissures** (if deep). This folding increases the surface area of the cortex, and it hides much of the cortex from a surface view (Fig. 6-2).

The cerebral cortex can be subdivided into a number of lobes, including the frontal, parietal, temporal, and occipital lobes, which are named for the overlying bones of the skull (see Fig. 6-2). The frontal and parietal lobes are separated by the central sulcus; they are separated from the temporal lobe by the lateral fissure. The occipital and parietal lobes are separated (on the medial surface of the hemisphere) by the parietooccipital fissure. Buried within the lateral fissure is another lobe, the insula. The final lobe, the limbic lobe, is formed by the cortex on the medial aspect of the hemisphere that borders with the brainstem. The hippocampal formation, which is a part of the limbic lobe, is folded into the temporal lobe and cannot be seen from the surface of the brain (Fig. 16-1). On the base of the brain can be seen an area of olfactory cortex, which includes the olfactory tubercle in the anterior perforated substance and the prepiriform lobe (see Chapter 11).

Activity in the cerebral cortex in the two hemispheres is coordinated by interconnections through the cerebral commissures. The bulk of the neocortex on the two sides is connected through the massive **corpus callosum** (Fig. 6-3). Part of the temporal lobes connect through the anterior commissure, and the hippocampal formations on the two sides communicate through the hippocampal commissure (which is formed between the fornices on the two sides as they pass under the corpus callosum).

Functions of the Lobes of the Cerebral Cortex

Although disputed in the past, it is now clear that particular functions can be associated with the different lobes of the cerebral hemispheres.

One of the main general functions of the **frontal lobes** is motor behavior. As discussed in Chapter 14, the motor, premotor, and supplementary motor areas are in the frontal lobe, as is the frontal eye field. These areas are responsible for planning and executing voluntary motor acts. In addition, the motor speech area (Broca's area) is located in the inferior frontal gyrus in the dominant hemisphere for human language (almost always the left hemisphere; see the following discussion). In addition, the prefrontal cortex in the rostral part of the frontal lobe appears to play a major role in personality and emotional behavior. Bilateral lesions of this part of the brain may be produced either by disease or therapeutically by frontal lobotomy. Such lesions produce deficits in attention, difficulty in problem-solving, and inappropriate social behavior. Also aggressive behavior is reduced, and the motivational-affective component of pain is lost, even though pain sensation remains. Frontal lobotomies are rarely performed today, because improved drug therapies have become available for mental illness and for chronic pain.

The **parietal lobe** contains the somatosensory cortex and the adjacent parietal association cortex (see Chapter 8). It is involved in the processing and perception of somatosensory information. Connections with the frontal lobe allow somatosensory information to affect voluntary motor activity. Visual information from the occipi-

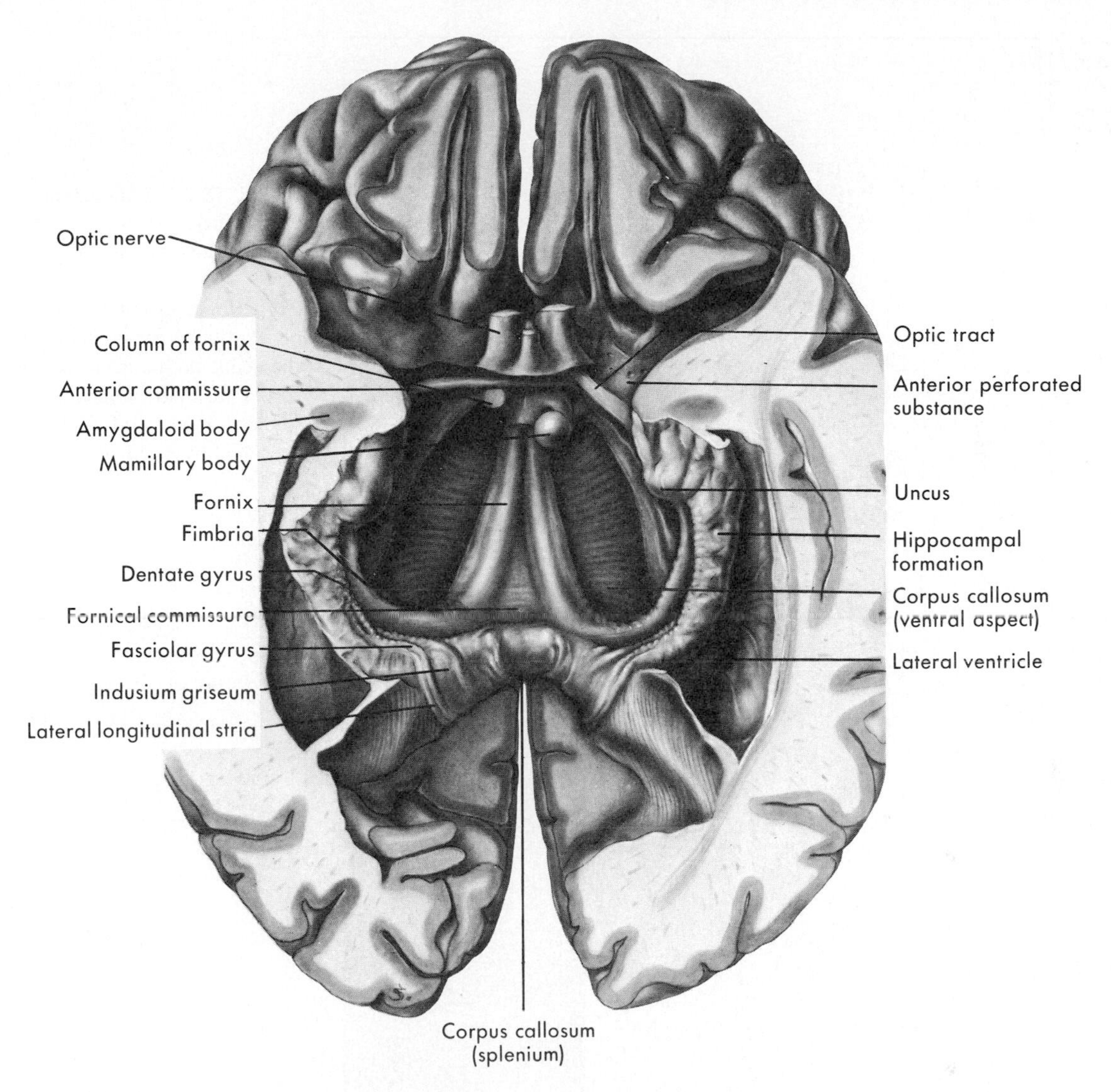

■ **Fig. 16-1** View of the base of the brain after dissection to reveal the hippocampal formation and related structures from below. (From Mettler FA: *Neuroanatomy,* St. Louis, 1948, Mosby–Year Book.)

tal lobe reaches the parietal association cortex and also the frontal lobe to assist in visual guidance of voluntary movements. Somatosensory information can also be transferred to the language centers, such as Wernicke's area, in the dominant hemisphere (see the following discussion). The parietal lobe in the nondominant hemisphere is involved in spatial analysis, as shown by the effects of lesions (see Chapters 8 and 14).

The primary function of the **occipital lobe** is visual processing and perception (see Chapter 9). Occipital eye fields affect eye movements, and a projection to the midbrain assists in the control of convergent eye movements, pupillary constriction, and accommodation, all of which occur when the eyes adjust for near vision.

The **temporal lobe** has many different functions. One of these involves hearing, which depends on the processing and perception of information related to sounds (see Chapter 10). Another is processing of vestibular information. Several visual areas have been discovered in the temporal lobe; hence, this lobe is involved in higher order visual processing (see Chapter 9). For example, the inferior temporal gyrus is involved in the recognition of faces. In addition, Meyer's loop passes through the temporal lobe, which means that part of the optic radiation can be damaged by temporal lobe lesions. Some of Wernicke's area is in the posterior region of the temporal lobe (see the following); damage to the temporal lobe in the dominant hemisphere can cause language disorders.

The medial temporal lobe belongs to the limbic system, which helps control emotional behavior and the autonomic nervous system (see Chapter 15). The hippocampal formation is thought to be involved in learning and memory (see below).

The functions of the different lobes of the cerebral cortex have been defined in part from the effects of lesions produced either by disease or by surgical interventions to treat disease in humans, as well as from experiments on animal subjects. Another approach has been to correlate the findings in **epilepsy** with the brain loca-

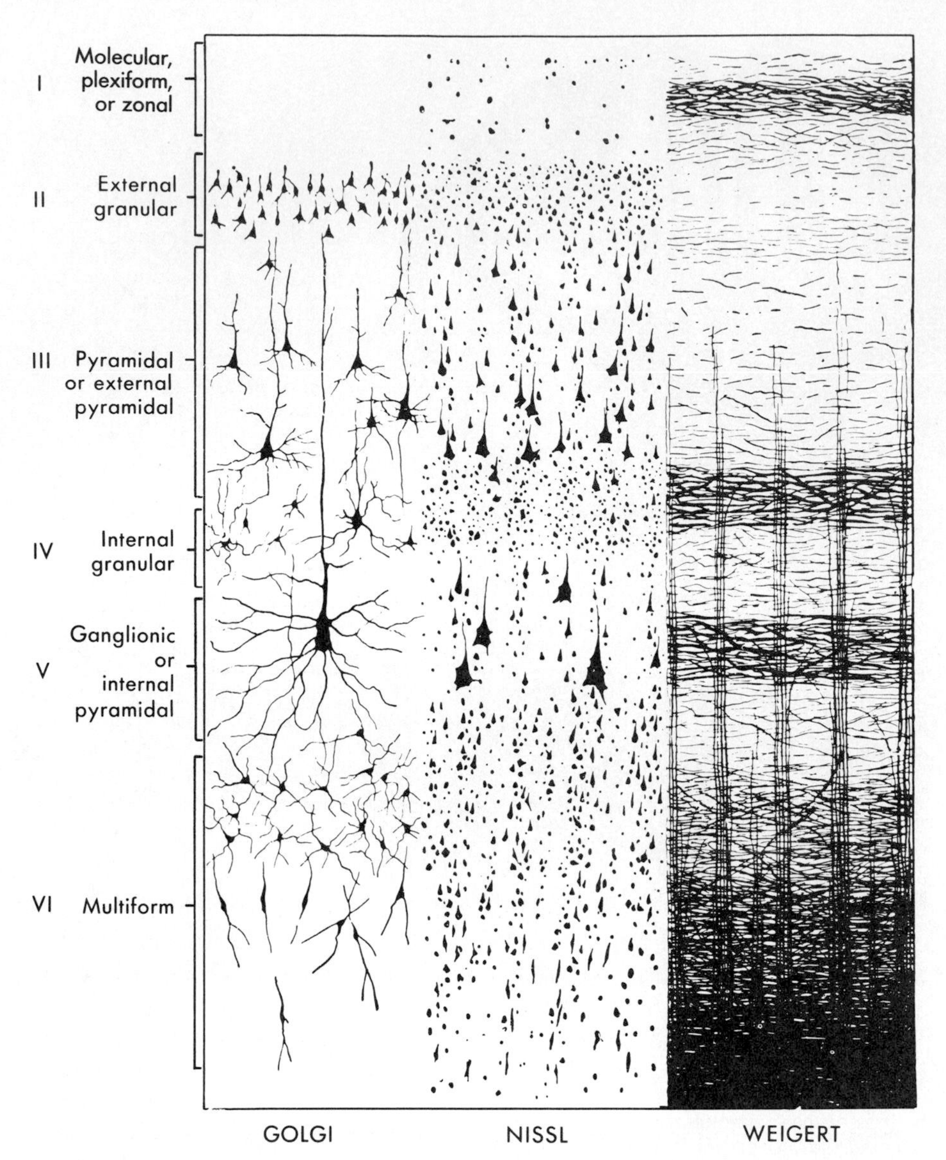

■ **Fig. 16-2** Lamination of the neocortex. Neurons in the different layers are demonstrated by the Golgi stain *(left column),* and the Nissl stain *(middle column),* while myelinated fibers are shown by the myelin sheath stain *(right column).* The layers are numbered at the left. (From Brodal A: *Neurological anatomy,* ed 3, London, 1981, Oxford University Press. Reprinted by permission.)

tions **(epileptic seizure foci)** that give rise to epileptic seizures. For example, epileptic foci in the motor cortex cause movements on the contralateral side; the exact movements relate to the somatotopic location of the seizure focus. Seizures that originate in the somatosensory cortex cause an **epileptic aura** in which a sensation is experienced. Similarly, seizures that start in the visual cortex cause a visual aura (scintillations, colors); those in the auditory cortex, an auditory aura (humming, buzzing, ringing); and those in the vestibular cortex, a feeling of spinning. Complex behaviors result from seizures that originate in the temporal lobe; in addition, there may be a malodorous aura if the olfactory cortex is involved.

■ *Neocortical Layering and Subdivisions*

The cerebral cortex can be subdivided on phylogenetic grounds into the **archicortex** (or allocortex), **paleocortex** (or juxtaallocortex), and **neocortex** (also called isocortex). In humans 90% of the cortex is neocortex.

The different phylogenetic subdivisions of the cerebral cortex can be recognized on the basis of their layering pattern (Fig. 16-2). The cytoarchitecture of the cortical layers can be demonstrated by a cell stain, such as a Golgi or a Nissl stain, and the myeloarchitecture can be demonstrated by a myelin sheath stain. With either technique the cortex is revealed as a layered structure. The neocortex is characterized by the presence of six cortical layers. On the other hand, the allocortex has only three layers, and the juxtaallocortex has four to five layers.

Cell types in neocortex. A number of different cell types have been described in the neocortex. The most abundant cell types are the pyramidal cells, "stellate cells" (various types of nonpyramidal cells), and fusiform cells (see Fig. 16-3).

Pyramidal cells have a large, triangular cell body, a long apical dendrite, and several basal dendrites. The

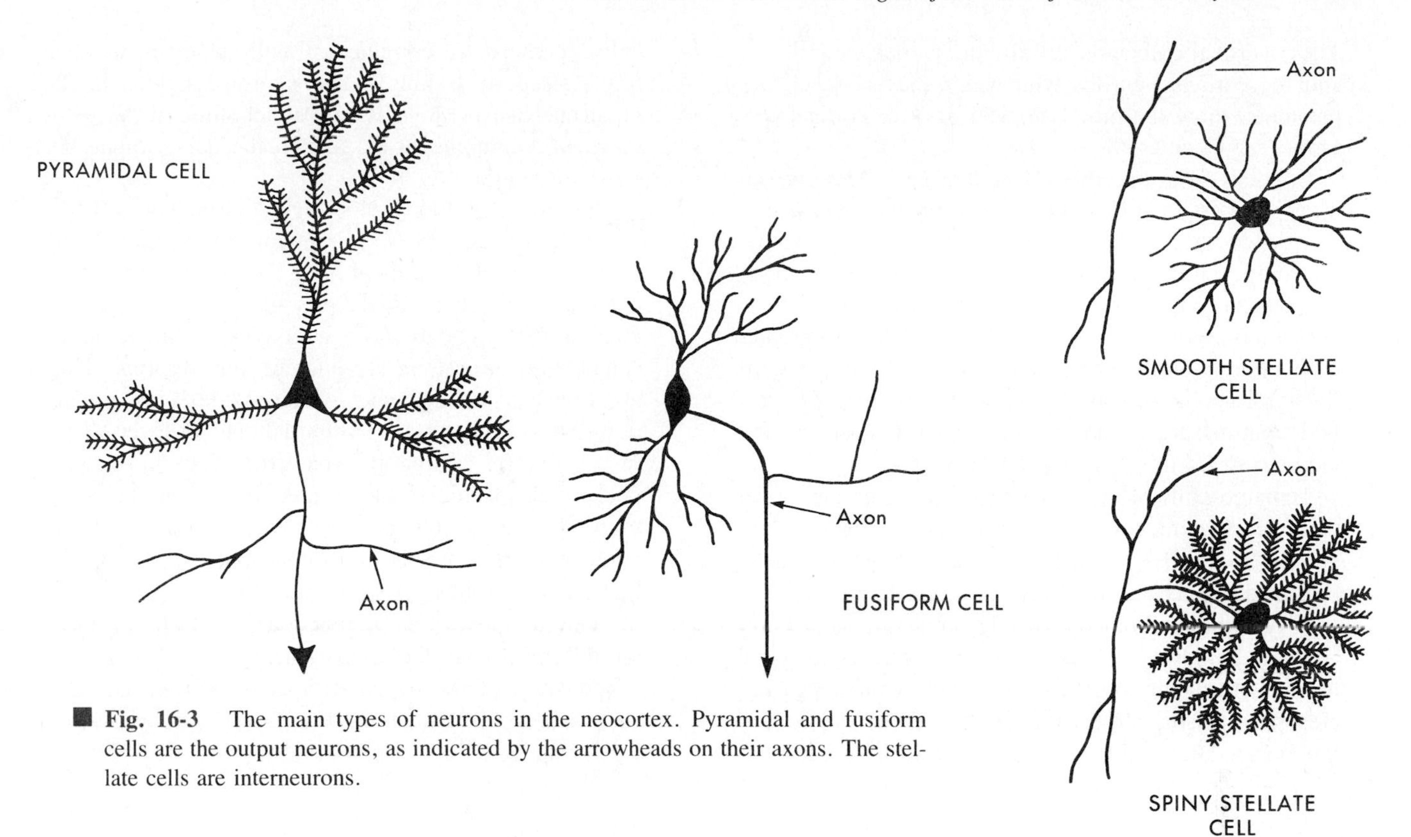

■ **Fig. 16-3** The main types of neurons in the neocortex. Pyramidal and fusiform cells are the output neurons, as indicated by the arrowheads on their axons. The stellate cells are interneurons.

apical dendrite often ascends to near the cortical surface and then branches. The basal dendrites branch near the soma. The dendrites are covered with numerous dendritic spines. Synapses contact the synaptic spines and also the surfaces of the dendritic shafts and the soma. The pyramidal cells are the main cortical efferent cells. The axon emerges from the cell body opposite the apical dendrite, and it projects into the subcortical white matter. The axon may give off collaterals as it descends through the cortex. An excitatory amino acid (glutamate, aspartate) may be used by pyramidal cells as their neurotransmitter.

"Stellate" cells, often called **granule cells,** are interneurons. They have a small soma and numerous branched dendrites. Some of these neurons have dendrites that radiate in all directions and can properly be called stellate cells. These have dendritic spines and are called **spiny stellate cells.** The spiny stellate cells are thought to be excitatory interneurons and are abundant in layer IV of the cortex (see above). Their axons ascend toward the supragranular layers. The smooth "stellate" cells appear to be GABAergic inhibitory interneurons. There are several other types of GABAergic inhibitory interneurons in the cortex, including basket and chandelier cells, and probably also neurogliaform and double bouquet cells. Many of these cells also contain neuropeptides.

Fusiform cells are less common. The cell body is elongated, and it gives off dendrites from either end. The orientation of the cell is vertical to the cortical surface.

Cytoarchitecture of cortical layers. Each of the six layers of the neocortex has a characteristic cellular content (see Fig. 16-2). Layer I (molecular layer) has few neuronal cell bodies; instead it contains mostly axon terminals and synapses on dendrites. Layer II (external granular layer) contains mostly "stellate" cells, although there are some pyramidal cells. Layer III (external pyramidal layer) consists mostly of small pyramidal cells. Layer IV (internal granular layer) includes mostly "stellate" cells, including spiny stellate cells. Layer V (internal pyramidal layer) is dominated by large pyramidal cells. Layer VI (multiform layer) contains pyramidal, fusiform, and other types of cells.

Layers I through III are often called the supragranular layers and layers V and VI, the infragranular layers, in reference to layer IV, the main granular layer. The supragranular layers are concerned chiefly in cortico-cortical pathways; the infragranular layers, in connections with subcortical structures.

Myeloarchitecture of cortical layers. The cortex contains concentrations of myelinated axons that are oriented either horizontally or vertically.

Prominent horizontal sheets of axons can be found in layers I, IV, and V (see Fig. 16-1). Those in layer IV and V are called the outer and inner lines of Baillarger. In the visual cortex a particularly prominent outer line of Baillarger, known as the stripe of Gennari, gives this part of the cortex the name "striate" cortex.

Vertical collections of axons can also be seen crossing the lower layers of the cortex (see Fig. 16-2). These are formed by cortical afferent and efferent fibers.

These vertical collections of afferent and efferent fibers, and the cortical neurons with which they connect, are presumed to be the morphological basis of cortical columns (see Chapters 8-10, 14).

Cortical afferent and efferent fibers. The cortical afferent fibers tend to synapse in particular cortical layers, depending on the source of the fibers (Fig. 16-4). Similarly, cortical efferent fibers that originate from particular layers project to particular destinations.

Cortico-cortical afferent fibers originate from other cortical areas in the same hemisphere or from the contralateral hemisphere (after crossing in one of the cerebral commissures, such as the corpus callosum), and they terminate in the supragranular layers.

Thalamo-cortical afferent fibers from thalamic nuclei that have specific cortical projections end chiefly in layers III, IV, and VI. Neurons in other thalamic nuclei project diffusely and terminate in layers I and VI.

Several nonthalamic, diffusely projecting nuclei, including the basal nucleus of Meynert, the parabrachial nucleus, the locus coeruleus, and the dorsal raphe nucleus, project to all cortical layers. These projections are nonspecific, and they modulate cortical activity globally, perhaps in conjunction with changes in state (e.g., sleep or waking). The neurotransmitter in the basal nucleus of Meynert is acetylcholine; in the locus coeruleus, norepinephrine; and in the dorsal raphe nucleus, serotonin.

The cortical efferent fibers originate from layers II, III, V, and VI (see Fig. 16-4). Pyramidal cells of layers II and III project to other cortical areas, either ipsilaterally or contralaterally. Pyramidal cells of layer V project in many descending pathways, with synaptic targets in the spinal cord, brainstem, red nucleus, and striatum. They also project to thalamic nuclei that provide diffuse projections back to the cortex. Pyramidal cells of layer VI are responsible for cortico-thalamic projections to thalamic nuclei with specific cortical projections and to the claustrum. Reciprocal thalamo-cortical and cortico-thalamic interconnections are likely to contribute importantly to the electroencephalogram (see below).

Regional variations in neocortical structure. Based on differences in the cytoarchitecture, a number of subdivisions of the neocortex can be recognized. One such scheme is shown in Fig. 16-5. The characteristic features of five types of neocortex are shown in

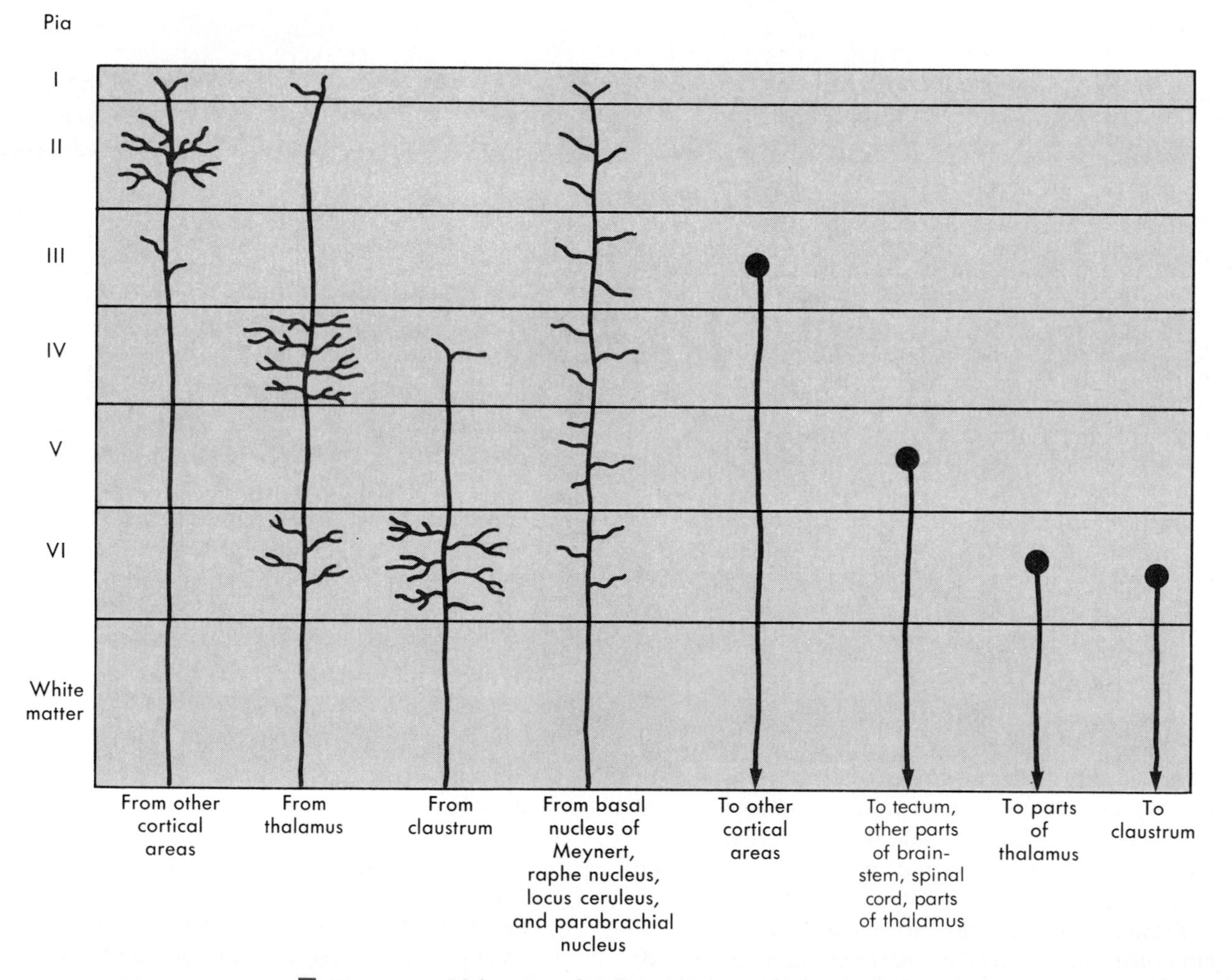

■ **Fig. 16-4** Main types of cortical afferent and efferent fibers.

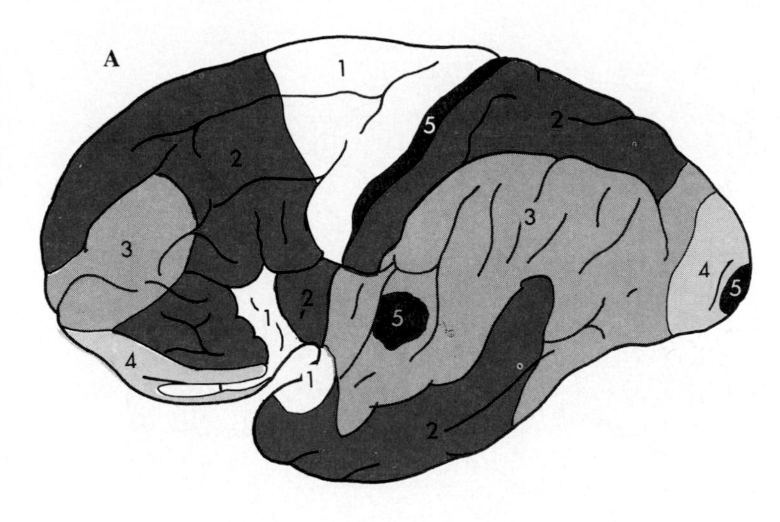

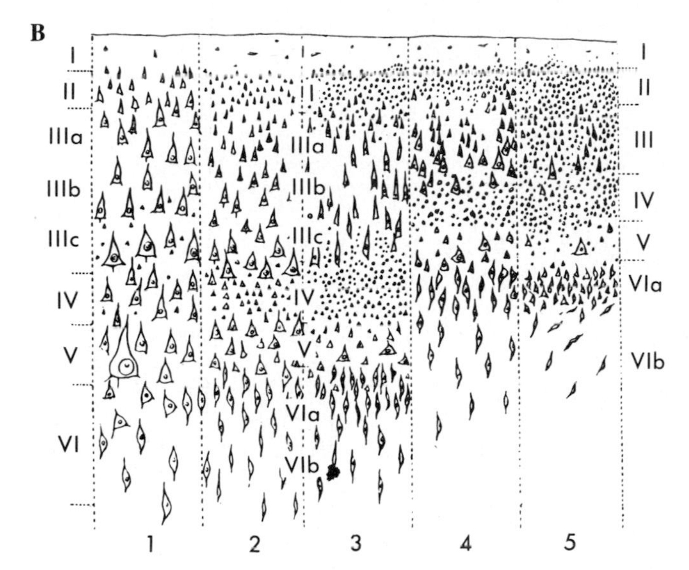

■ **Fig. 16-5** The five main types of neocortex. **A,** The distribution of the five types of cortex on the lateral aspect of the cerebrum. **B,** The cell types and laminar pattern of the five types of cortex. (Redrawn from Kornmüeller AE, Janzen R: *Arch Psychiatr Nervenkr* 110:224, 1939.)

Fig. 16-5, *B,* and the distribution of some of the different subtypes are shown in Fig. 16-5, *A*. Cortex constructed of six readily distinguishable layers is called **homotypical cortex;** variants are considered **heterotypical cortex.**

Type 1 cortex is a heterotypical type called **agranular cortex.** This term is an exaggeration, because many nonpyramidal cells are found in this type of cortex, but the prominence of the pyramidal cell content is the basis for the name. Evidently this kind of cortex specializes in output cells, hence finding type 1 cortex in the motor and premotor areas is not surprising.

Type 5 cortex, on the other hand, has a relatively small number of pyramidal cells and is dominated by nonpyramidal cells. This type of cortex is called **granular cortex** (or **koniocortex,** for "dustlike"). Evidently it is specialized for processing afferent input. Therefore, finding this kind of cortex in the primary sensory receiving areas, the somatosensory cortex (SI), the primary auditory cortex, and the primary visual (striate) cortex is reasonable.

Cortex of types 2 through 4 show less dramatic variations. These types of cortex are homotypical cortex, because they have six well demarcated layers. Homotypical cortex is found in much of the cortex.

The cortex was subdivided to an even greater extent by Brodmann (Fig. 16-6), based on an extensive cytoarchitectural analysis, in humans, monkeys, and many other species. Brodmann divided the human cortex into 44 discrete areas. Important ones include: Brodmann's areas 3, 1, and 2, which form the SI cortex; area 4, the primary motor cortex; area 6, the premotor cortex; areas 41 and 42, the primary auditory cortex; and area 17, the primary visual cortex (striate cortex). Detailed studies have confirmed that the Brodmann areas are in fact distinctly different, both with respect to their interconnections and with respect to their functions.

■ *Allocortex*

About 10% of the human cerebral cortex is allocortex and juxtaallocortex. The allocortex has a three-layered structure, and the juxtaallocortex has four to five layers. The juxtaallocortex is at the border between the allocortex and neocortex.

Hippocampal formation. In humans the hippocampal formation is folded into the temporal lobe and it can only be viewed after dissection (see Fig. 16-1). The hippocampal formation consists of several parts, including the hippocampus (Ammon's horn or Cornu ammonis), the dentate gyrus, and the subiculum. These are well demarcated in a cross-section through the hippocampal formation (Fig. 16-7).

The name for the hippocampus is derived from that of the seahorse (*Hippocampus* sp.). The synonym, Ammon's horn, is from Ammon, an Egyptian god with a ram's head. The hippocampus forms an interlocking C with the dentate gyrus when these are viewed in cross-section (Figs. 16-7 and 16-8). The hippocampus has three layers: the molecular layer, the pyramidal cell layer, and the polymorphic layer. These resemble layers I, V, and VI in the neocortex. The folding of the hippocampus results in an inverted appearance, because the white matter is at the surface of the lateral ventricle (see Figs. 16-7 and 16-8). The white matter over the hippocampus is called the alveus. The alveus contains hippocampal afferent and efferent fibers. The axons in the alveus continue into a nerve fiber bundle called the fimbria, and the fimbria is continuous with the fornix.

The dentate gyrus is also a three-layered cortex. However, instead of a pyramidal cell layer, the middle

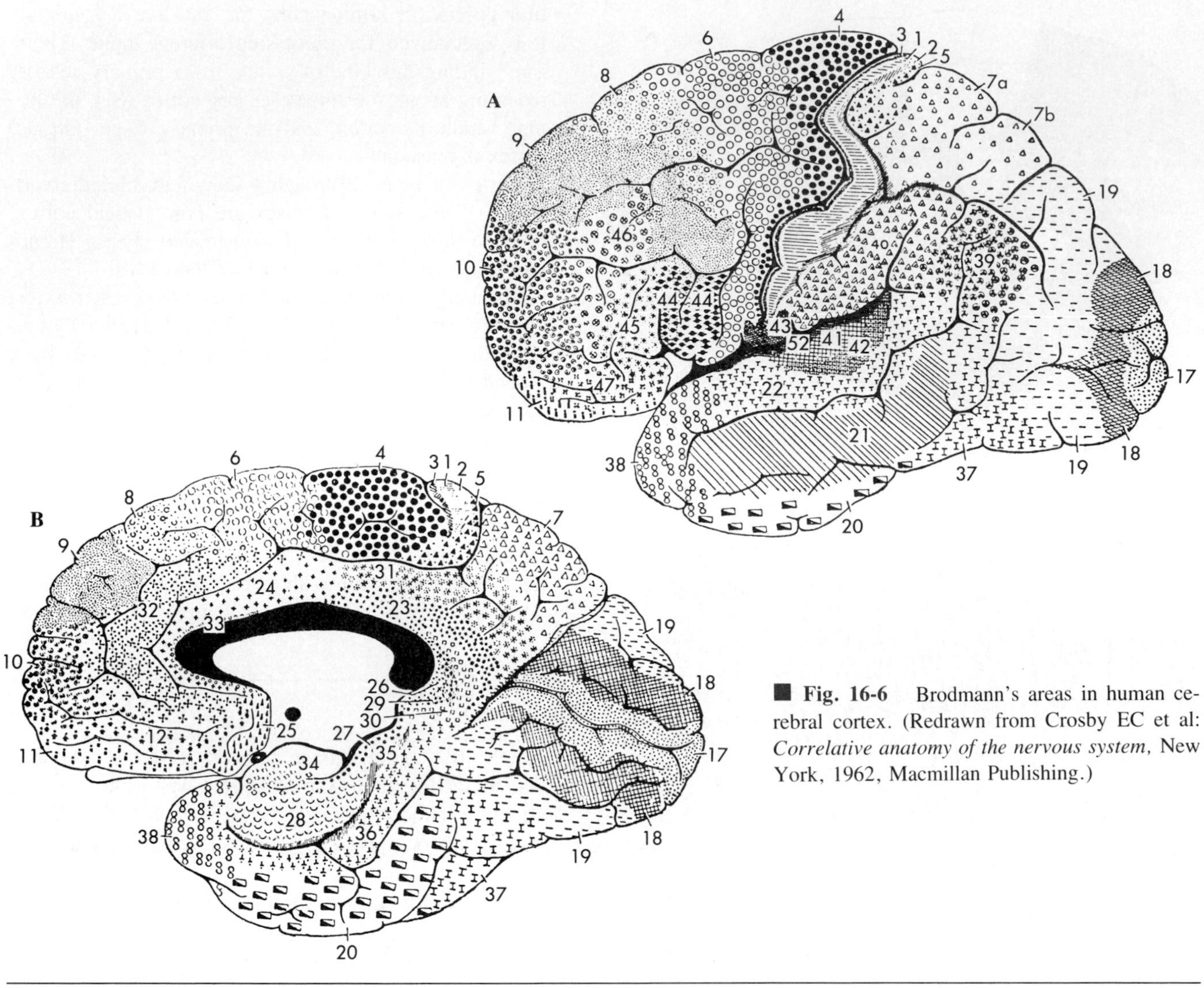

■ **Fig. 16-6** Brodmann's areas in human cerebral cortex. (Redrawn from Crosby EC et al: *Correlative anatomy of the nervous system,* New York, 1962, Macmillan Publishing.)

■ **Fig. 16-7** Section taken through the hippocampal formation in the human brain. (From Carpenter MB: *Human neuroanatomy,* ed 7, Baltimore, 1976, Williams & Wilkins.)

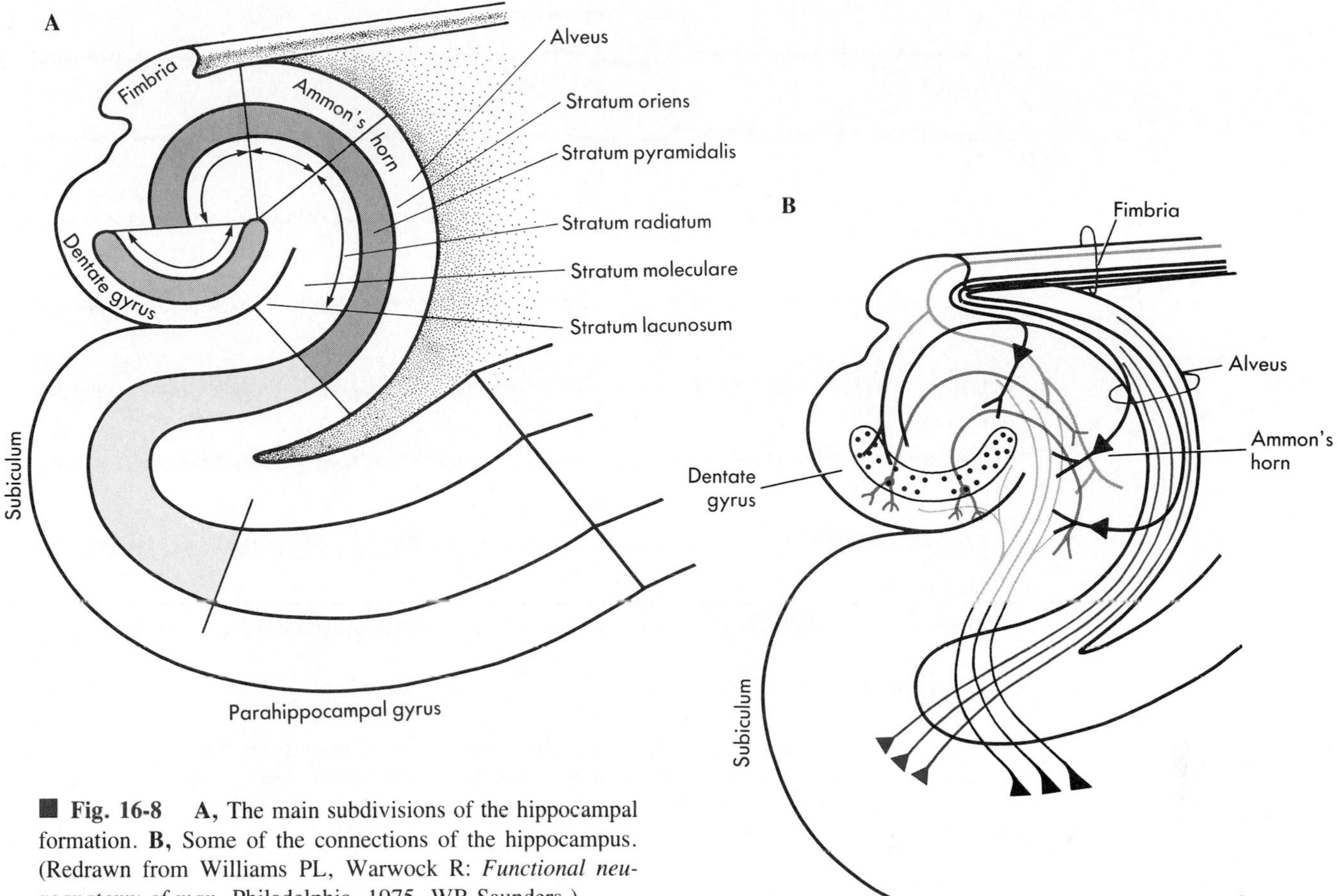

■ **Fig. 16-8** **A,** The main subdivisions of the hippocampal formation. **B,** Some of the connections of the hippocampus. (Redrawn from Williams PL, Warwock R: *Functional neuroanatomy of man,* Philadelphia, 1975, WB Saunders.)

layer in the dentate gyrus is the granule cell layer. The axons of the granule cells do not leave the hippocampal formation. Instead, they project to Ammon's horn.

The subiculum is part of the parahippocampal gyrus. It is composed of juxtaallocortex and merges with the entorhinal area.

The hippocampal formation receives its main input from the entorhinal cortex of the parahippocampal gyrus through two main projections, the perforant path and the alvear path (Fig. 16-8, *B*). Afferent fibers also enter the hippocampus through the fornix. These come from several sources, including the septal nuclei and the contralateral hippocampal formation (by way of the hippocampal commissure). The granule cell layer of the dentate gyrus also projects to the hippocampus. The pyramidal cells are the main output cells of the hippocampal formation. These project through the fornix to several targets, including the septal nuclei, the mammillary body, and the contralateral hippocampus.

■ *Higher Functions of the Nervous System*

■ *The Electroencephalogram*

Rhythmic electrical activity can be recorded from the cerebral cortex. This activity is known as the **electroencephalogram** (EEG) when the activity is recorded from the surface of the skull, or the **electrocorticogram** when it is recorded from the surface of the brain. For human studies the EEG is recorded from a grid of standard recording sites. Thus, EEGs can be recorded consistently from approximately the same sites at different times from one individual or from different subjects (Fig. 16-9). The EEG is an important diagnostic tool in clinical neurology and is particularly useful in patients with epilepsy.

The normal EEG consists of waves of various frequencies. The dominant frequencies depend on several factors, including the state of wakefulness, the age of the subject, the location of the recording electrodes, and the absence or presence of drugs or disease. When a normal awake adult is relaxed and his eyes are closed, the dominant frequencies of the EEG recorded over the parietal and occipital lobes are about 8 to 13 Hz, the alpha rhythm (see Fig. 16-9). If the subject is asked to open his eyes, the EEG becomes less synchronized and the dominant frequency increases to 13 to 30 Hz, the beta rhythm. The delta (0.5 to 4 Hz) and theta (4 to 7 Hz) rhythms are observed during sleep (see the following discussion).

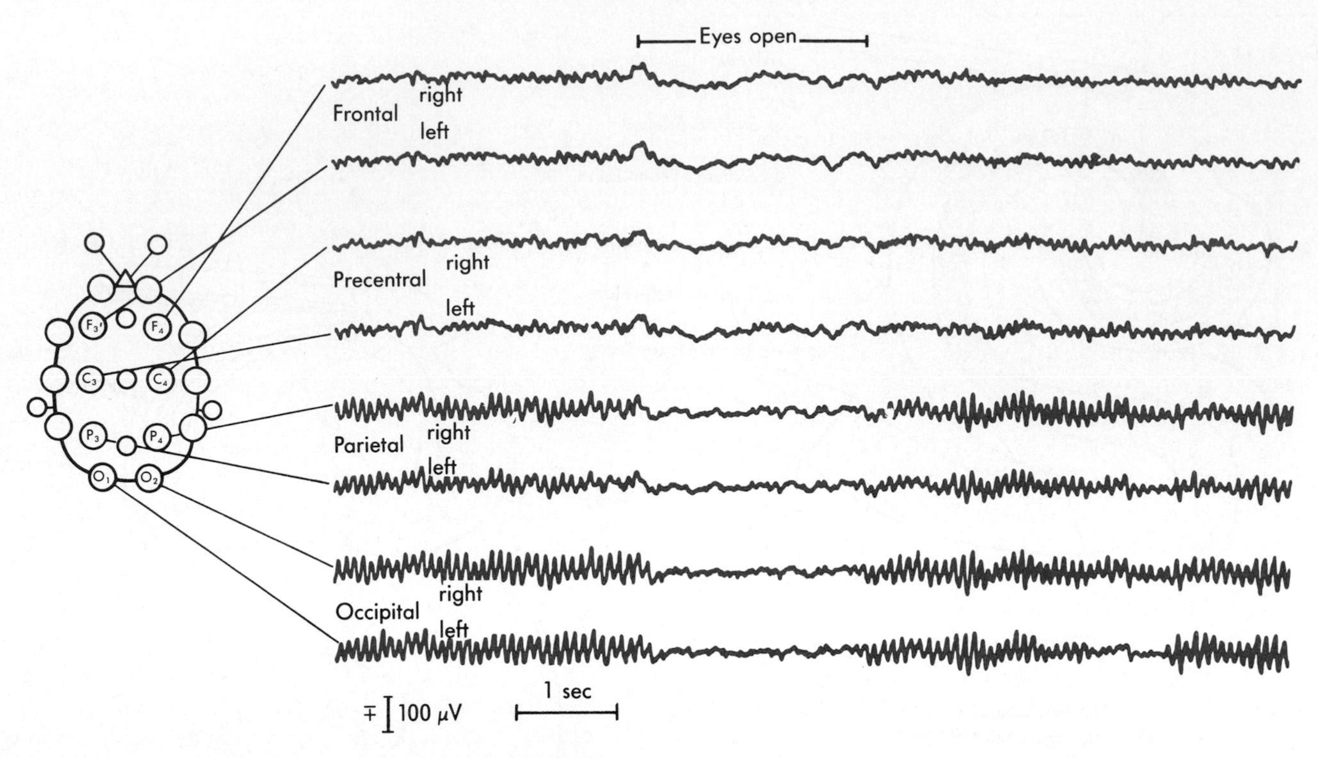

■ **Fig. 16-9** EEG in a normal, resting awake human. The recordings were made from eight channels at the same time. The electrode positions are indicated. When the eyes were opened, the alpha rhythm was blocked. (From Schmidt RF, editor: *Fundamentals of neurophysiology,* ed 2, New York, 1978, Springer-Verlag.)

The source of EEG waves is alternating excitatory and inhibitory synaptic potentials that occur in cortical neurons as a result of thalamocortical and other input. The potentials are produced chiefly by extracellular currents that flow vertically across the cortex during the generation of synaptic potentials in the pyramidal cells. The extracellular currents associated with action potentials are too small, fast, and asynchronous to be recorded with EEG electrodes. The potentials that are recorded are relatively large (around 100 μV), because the many pyramidal cells with their apical dendrites aligned in parallel form a dipole sheet with one pole oriented toward the cortical surface and the other toward the subcortical white matter. Sometimes the term "spike" is used for a sharp EEG wave, but this term does not refer to action potentials. Note that the sign of an EEG wave does not in itself indicate whether pyramidal cells are being excited or inhibited. For instance, a negative EEG potential may be generated at the surface of the skull (or cortex) by excitation of apical dendrites or by inhibition near the somas. Conversely, a positive EEG wave can be produced by inhibition of the apical dendrites or by excitation near the somas.

■ *Evoked Potentials*

An EEG change, called a **cortical evoked potential,** can be evoked by a stimulus. A cortical evoked potential is best recorded from the part of the skull located over the cortical area that is activated. For example, a visual stimulus results in an evoked potential that can be recorded best over the occipital bone, whereas a somatosensory evoked potential is recorded most effectively near the junction of the frontal and parietal bones. Evoked potentials reflect the synaptic potentials evoked in large numbers of cortical neurons. They may also show components generated by subcortical structures.

Evoked potentials are small relative to the size of the EEG waves, but the apparent size can be enhanced by a process called **signal averaging.** Repeated stimulation causes the evoked potential to occur at a fixed interval after the stimulus on each trial. However, the EEG may show a positive or a negative deflection on different trials during the time of occurrence of the evoked potential. In signal averaging evoked potentials are electronically averaged. The EEG waves average out, whereas the evoked potentials sum.

Evoked potentials are useful clinically, because they assess the integrity of a sensory pathway, at least to the level of the primary sensory receiving area. These potentials can be recorded in comatose individuals, as well as in infants too young to permit a sensory examination. The initial parts of the auditory evoked potential actually reflect activity in the brainstem; therefore, this evoked potential can be used to assess the function of brainstem structures.

■ *Sleep-wake Cycle*

Sleep and wakefulness are among the many functions of the body that show **circadian** (about one day) periodicity. The sleep-wake cycle has an endogenous periodicity of about 25 hours, but it normally becomes entrained to the day-night cycle. However, the entrainment can be disrupted by isolating the subject from the environment or when he shifts time zones (jet lag).

Characteristic changes in the EEG can be correlated with the changes in behavioral state during the sleep-wake cycle. **Beta wave** activity dominates in the awake, aroused individual (see Fig. 16-9). The EEG is said to be desynchronized; it displays low voltage, high frequency activity. In relaxed individuals with eyes closed, the EEG is dominated by **alpha waves.** As the person falls asleep, he passes sequentially through four stages of **slow wave sleep** (called stages 1 through 4) over a period of 30 to 45 minutes (Fig. 16-10). In stage 1 alpha waves are interspersed with lower frequency waves (3 to 7 Hz) called **theta waves.** In stage 2 the EEG slows further, but the slow wave activity is interrupted by **sleep spindles,** which are bursts of activity at 12 to 14 Hz, and by large K complexes. Stage 3 sleep is associated with **delta waves,** which occur at frequencies of 0.5 to 2 Hz, and with occasional sleep spindles. Stage 4 is characterized by delta waves.

During slow wave sleep the muscles of the body relax, but the posture is adjusted intermittently. The heart rate and blood pressure decrease and gastrointestinal motility increases. The ease with which the individual can be awakened decreases progressively as he passes through these sleep stages. As the person awakens, he passes through the sleep stages in reverse order.

About every 90 minutes slow wave sleep changes to a different form of sleep, called **rapid eye movement,** or **REM** sleep. In REM sleep the EEG becomes desynchronized, with low voltage, fast activity like that seen in the EEG from an aroused subject (Fig. 16-10, *bottom trace*). The similarity of the EEG to that seen in awake individuals and the difficulty in awaking the person leads to another term for this type of sleep—**paradoxical sleep.** Muscle tone is completely lost, but phasic contractions occur in a number of muscles, most notably of the eye muscles. The resulting rapid eye movements are the basis of the name for this type of sleep. Many autonomic changes also take place. Temperature regulation is lost, and meiosis occurs. Penile erection also occurs during this type of sleep. Heart rate, blood pressure, and respiration change intermittently. REM

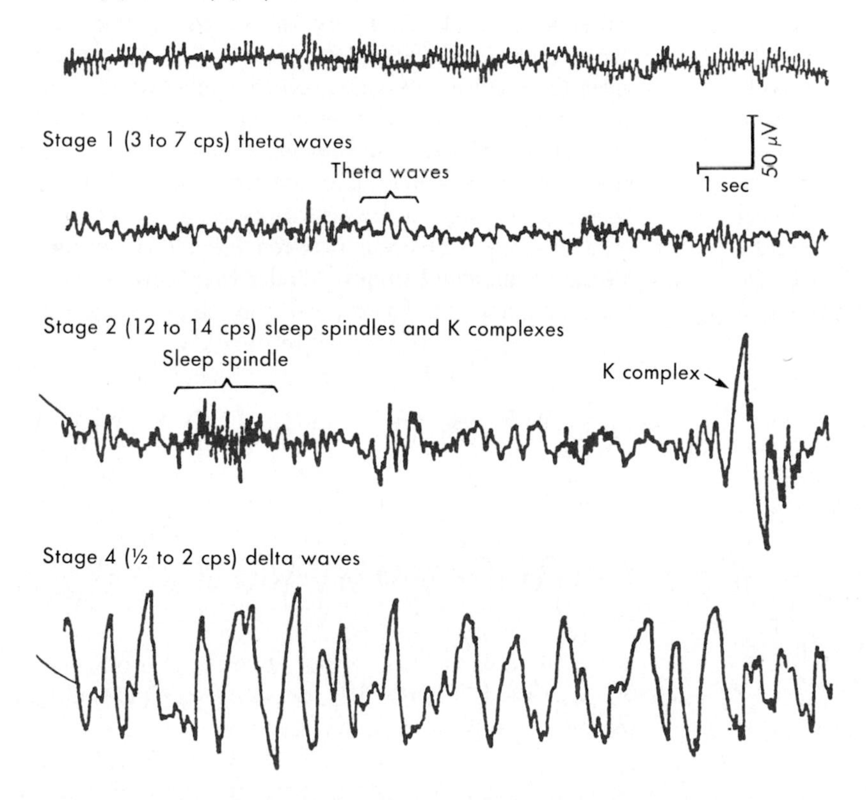

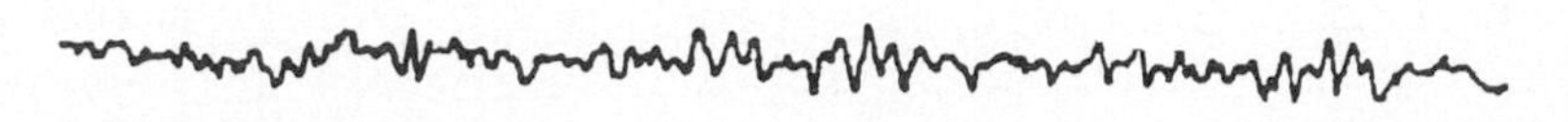

■ **Fig. 16-10** EEG during drowsiness and stages 1, 2, and 4 of slow wave (nonREM) sleep and in REM sleep. (Modified from Shepherd GM: *Neurobiology*, London, 1983, Oxford University Press.)

sleep occurs in several episodes during a given night. Although arousing a person from REM sleep is difficult, internal arousal is common. Most dreams occur during REM sleep.

The proportion of slow wave (nonREM) and REM sleep varies with age. Newborn children spend about half of their sleep time in REM sleep, whereas the elderly have little REM sleep. About 20% to 25% of the sleep of young adults is REM sleep.

The purpose of sleep is still unclear. However, it must have a high value, because so much of life is spent in sleep and because lack of sleep can be debilitating. Medically important disorders of the sleep-wake cycle include insomnia, bedwetting, sleepwalking, sleep apnea, and narcolepsy.

The mechanism of sleep is incompletely understood. Stimulation in the brainstem reticular formation in a wide region known as the **reticular activating system** causes arousal and low voltage, fast EEG activity. Sleep was once thought to be caused by a reduced level of activity in the reticular activating system; however, a number of arguments can be made against this idea. These arguments include the observations that anesthesia of the lower brainstem results in arousal, and stimulation in the medulla near the nucleus of the solitary tract can induce sleep. These and other observations suggest that sleep is an active process. Investigators have tried to relate sleep mechanisms to brainstem networks that use particular neurotransmitters, including serotonin, norepinephrine, and acetylcholine, because manipulations of the levels of these transmitters in the brain can affect the sleep-wake cycle. However, a detailed neurochemical explanation of the neural mechanisms of sleep is not yet available.

The source of circadian periodicity in the brain appears to be the suprachiasmatic nucleus of the hypothalamus. This nucleus receives projections from the retina, and its neurons seem to form a biological clock. Destruction of the suprachiasmatic nucleus disrupts a number of biological rhythms, including the sleep-wake cycle.

■ *Abnormal EEG*

The EEG can become abnormal under a variety of pathological circumstances. For example, during coma the EEG is dominated by delta activity. **Brain death** is defined by a maintained flat EEG.

Epilepsy is a common cause of EEG abnormalities. There are several forms of epilepsy, and examples of EEG patterns from some of these are shown in Fig. 16-11. Epileptic seizures can be either partial or generalized. One form of partial seizures originates in the motor cortex and results in localized contractions of contralateral muscles. The contractions may then spread to other muscles; the spread follows the somatotopic sequence of the motor cortex (see Fig. 13-5). Complex partial seizures may occur in psychomotor epilepsy. These originate in the limbic lobe and result in illusions and semipurposeful motor activity. During and between focal seizures, scalp recordings may reveal EEG spikes (see Figs. 16-11, *C* and *D*).

Generalized seizures involve wide areas of the brain and loss of consciousness. Two major types are petit mal and grand mal seizures. In **petit mal epilepsy** consciousness is lost transiently and the EEG displays spike and wave activity (see Fig. 16-11, *B*). In **grand mal seizures** consciousness is lost for a longer period, and the individual will fall to the ground if he is standing when the seizure starts. The seizure begins with a generalized increase in muscle tone (tonic phase), followed by a series of jerky movements (clonic phase). The bowel and bladder may be evacuated. The EEG shows widely distributed seizure activity (see Fig. 16-11, *A*).

EEG spikes that occur between full blown seizures are called **interictal spikes.** Similar events can be studied experimentally. These arise from abrupt, long-last-

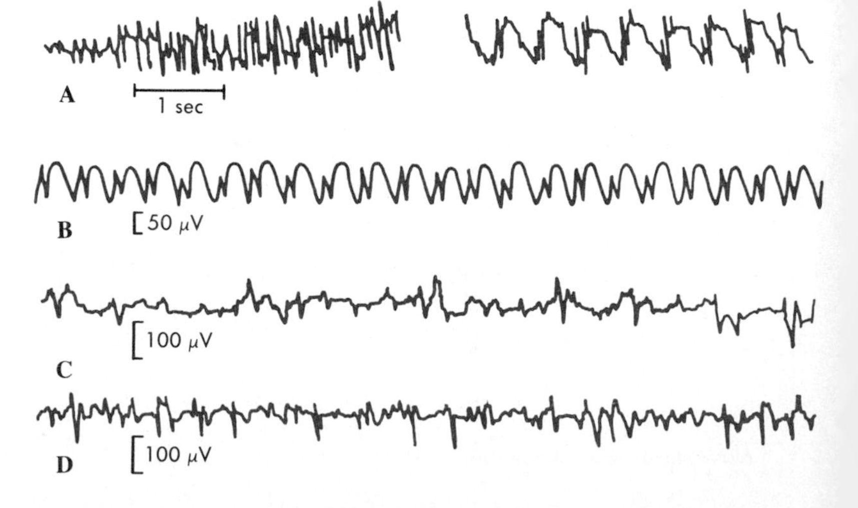

■ **Fig. 16-11** Abnormalities of the EEG in several forms of epilepsy. **A,** The EEG during the tonic *(left)* and clonic *(right)* phases of a grand mal seizure. **B,** The spike and wave components of a petit mal seizure. **C,** The EEG in temporal lobe epilepsy. **D,** A focal seizure. (Redrawn from Eyzaguirre C, Fidone SJ: *Physiology of the nervous system,* ed 2, Chicago, 1975, Yearbook Medical Publishers.)

ing depolarizations called **depolarization shifts,** which trigger repetitive action potentials in cortical neurons. Depolarization shifts may reflect several changes in epileptic foci. Such changes include regenerative Ca^{++}-mediated dendritic action potentials in cortical neurons and a reduction in inhibitory interactions in cortical circuits. Electrical field potentials and the release of K^{+} and excitatory amino acids from hyperactive neurons may also contribute to the increased cortical excitability.

Cerebral Dominance and Language

In most people the left cerebral hemisphere is said to be the **dominant hemisphere** with respect to language. This dominance has been demonstrated (a) by the effects of lesions of the left hemisphere, which may produce deficits in language function **(aphasia),** and (b) by the transient aphasia that results when a short-acting anesthetic is introduced into the left carotid artery. Lesions of the right hemisphere and the injection of anesthetic into the right carotid artery do not usually affect language substantially. The right hemisphere is dominant for functions other than language. For example, left-handedness reflects a dominance of the right hemisphere. However, in most left-handed people the left hemisphere is still dominant for language. Differences in the size of an area, called the planum temporale, in the floor of the lateral fissure, correlate with language dominance. The left planum temporale is usually larger than that of the right hemisphere.

Several areas in the left hemisphere are involved in language. **Wernicke's area** is a large area centered on the posterior part of the superior temporal gyrus near the auditory cortex. Another important language area, called **Broca's area,** is in the posterior part of the inferior temporal gyrus, close to the face representation of the motor cortex. Damage to Wernicke's area results in a **sensory aphasia,** in which the person has difficulty in understanding spoken or written language. However, speech production remains fluent. On the other hand, a lesion of Broca's area causes a **motor aphasia.** Individuals with motor aphasia have difficulty in speech and in writing, although they can understand language relatively well. In sensory aphasia there may be no auditory or visual impairment, and in motor aphasia the motor control of the muscles responsible for speech or writing may be intact. Thus, aphasia does not depend on an alteration in sensation or in the motor system; rather it is a deficit in the reception or planning of language expression. However, lesions in the dominant hemisphere may be large enough to result in mixed forms of aphasia, as well as in sensory changes or paralysis of some of the muscles used to express language.

Interhemispheric Transfer

The two cerebral hemispheres can function relatively independently, as in the case of language function. However, information must be transferred between the hemispheres, so that activity on the two sides of the body can be coordinated. That is, each of the two hemispheres must know what the other is doing. Much of the information transferred between the two hemispheres is transmitted through the corpus callosum, although some is transmitted through other commissures (e.g., anterior commissure, midbrain commissures).

An experiment that shows the importance of the corpus callosum for interhemispheric transfer of information is illustrated in Fig. 16-12. An animal with an intact optic chiasm and corpus callosum and with the left eye closed learns a visual discrimination task (Fig. 16-12, *A*). The information is transmitted to both hemispheres through the bilateral connections made by the optic chiasm or through the corpus callosum or both. When the animal is tested with the left eye open and the right eye closed (Fig. 16-12, *center*), the task can still be performed, because both hemispheres have learned the task. If the optic chiasm is transected before the animal is trained, the same result is found (Fig. 16-12, *B*). Therefore, information is presumably transferred between the two hemispheres through the corpus callosum. This can be confirmed by cutting both the optic chiasm and corpus callosum before training (Fig. 16-12, *C*). Now the information is not transferred, and each hemisphere must learn the task independently.

A similar experiment has been done in human patients who have had a surgical transection of the corpus callosum to prevent the interhemispheric spread of epilepsy (Fig. 16-13). The optic chiasm remained intact, but directing visual information to one or the other hemisphere was possible by having the patient fix his vision on a point on a screen. Then a picture of an object was projected to one side of the fixation point. Visual information about the picture therefore reached only the contralateral hemisphere. An opening beneath the screen allowed the patient to manipulate objects that could not be seen. The objects included those shown in the projected pictures. Normal individuals would be able to locate the correct object with either hand. However, the patients with a transected corpus callosum could locate the correct object only with the hand ipsilateral to the projected image (contralateral to the hemisphere that received the visual information). The visual information had to have access to the motor areas of the cortex for the hand to explore and recognize the correct object. With the corpus callosum cut, the visual and motor areas are interconnected only on the same side of the brain.

Another test was to ask the patient to identify verbally what object was seen in the picture. The patient

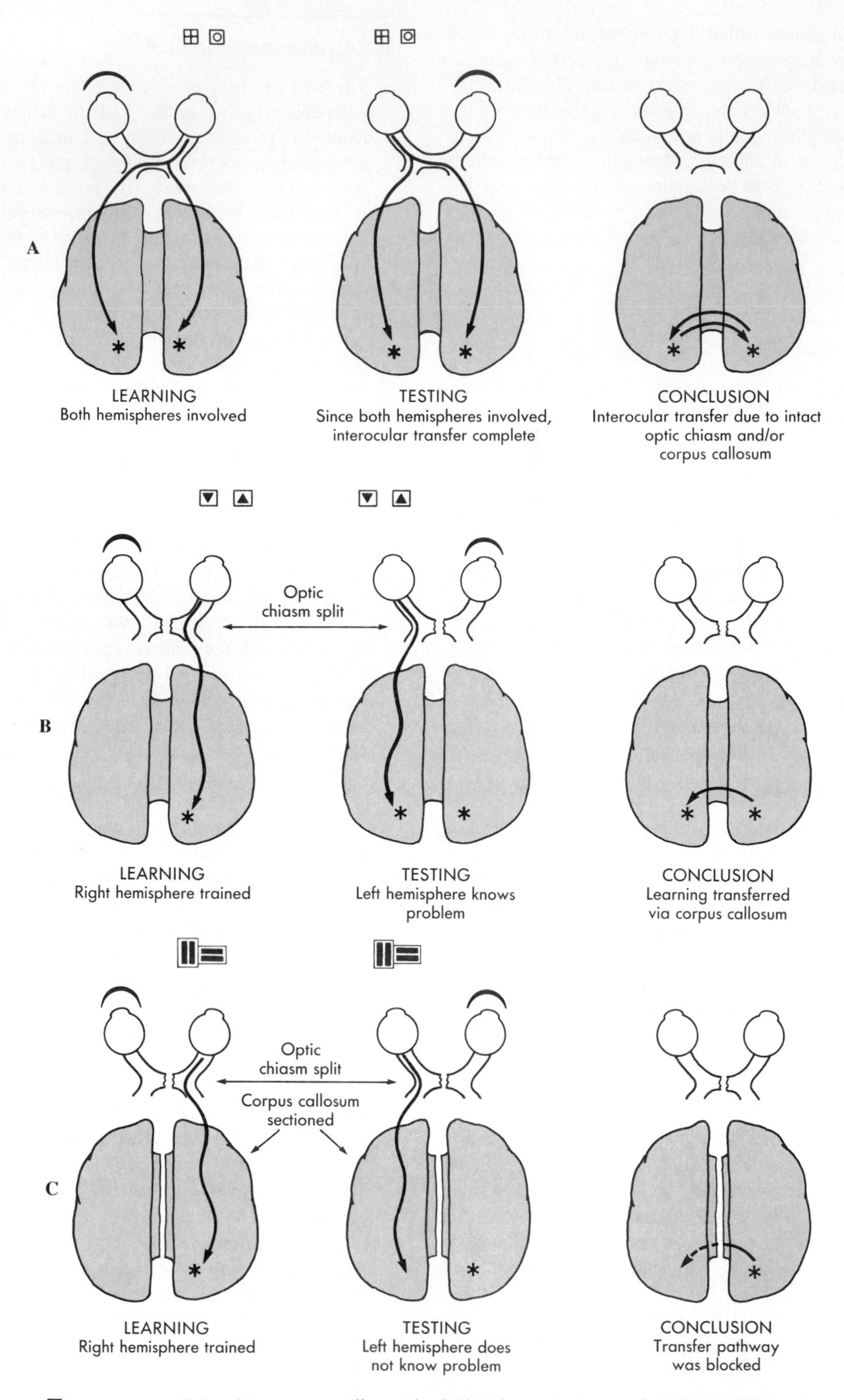

■ **Fig. 16-12** Role of the corpus callosum in the interhemispheric transfer of visual information. **A,** Learning involves one eye. The discrimination depends on distinguishing between a cross and a circle. **B,** Discrimination is between triangles oriented with apex up or down. **C,** Discrimination is between vertical and horizontal bars.

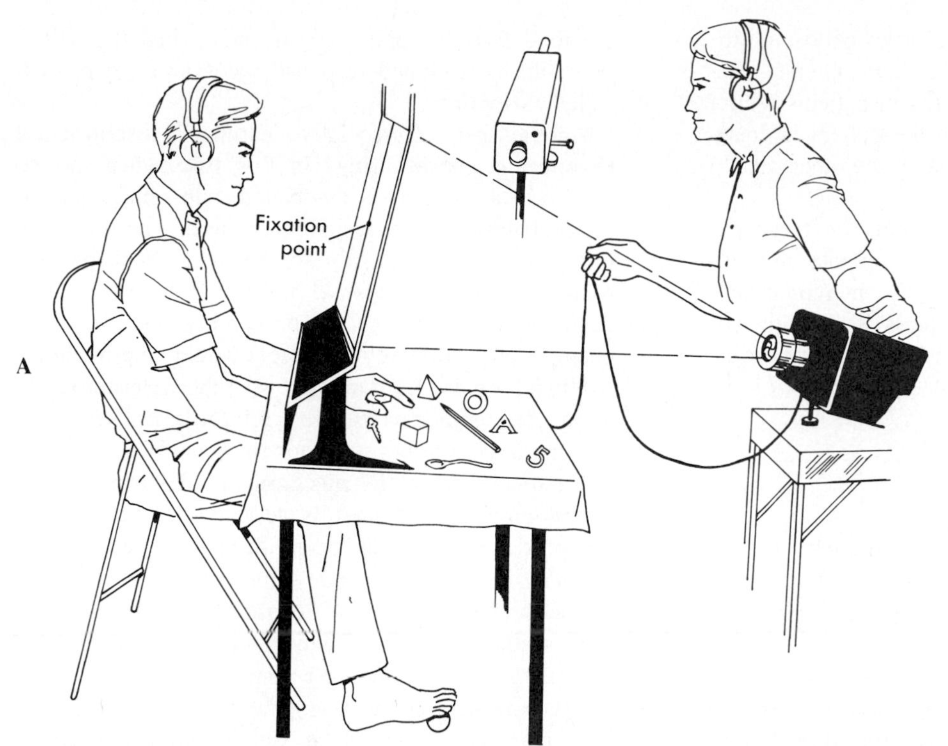

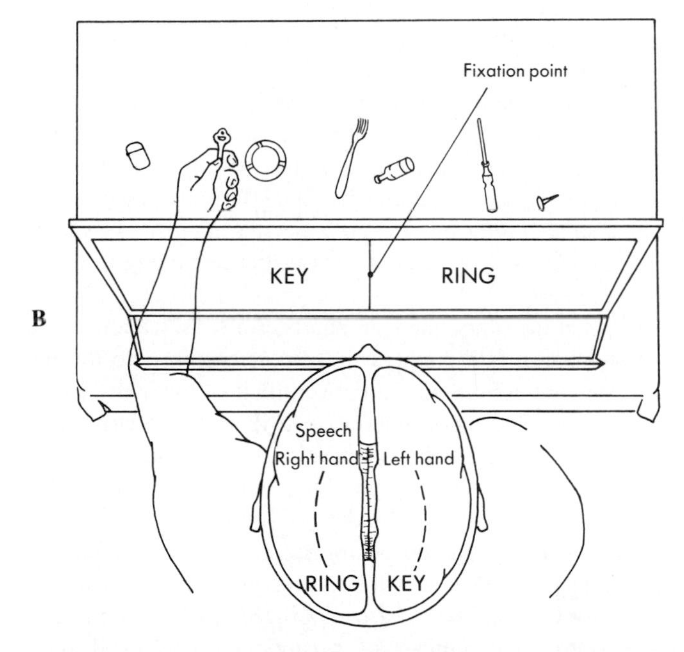

■ **Fig. 16-13** Tests on a patient with a transected corpus callosum. **A,** The patient fixes on a point on a rear projection screen, and pictures are projected to either side of the fixation point. The hand can palpate objects that correspond to the projected pictures, but these cannot be seen. **B,** A response to a picture of a key in the left field of view by the left hand. However, the verbal response is that the patient sees a picture of a ring. (Redrawn from Sperry RW. In Schmitt FO, Worden FG, editor: *The neurosciences third study program,* Cambridge, Mass, 1974, MIT Press.)

would make a correct verbal response to a picture that was projected to the right of the fixation point so that the visual information reached only the left (language dominant) hemisphere. However, the patient could not verbally identify a picture that was presented to the left hemifield so that visual information reached the right hemisphere.

Similar observations can be made in patients with a transected corpus callosum when different forms of stimuli are used. For example, such a patient given a verbal command to raise the right arm will do so without difficulty. The language centers in the left hemisphere send signals to the motor areas on the same side, and these produce the movement of the right arm. However, the patient cannot respond to a command to raise his left arm. The language areas on the left side cannot influence the motor areas on the right unless the corpus callosum is intact. The result is a type of apraxia.

Somatosensory stimuli applied to the right side of the body can be described by patients with a transected corpus callosum, but the same stimuli applied to the left side of the body cannot. Again, information that reaches the right somatosensory areas of the cortex cannot access the language centers if the corpus callosum is not intact.

The functional capabilities of the two hemispheres can be compared by exploring the capabilities of individuals with a transected corpus callosum. For example, such patients solve three-dimensional puzzles better with the right than with the left hemisphere; this finding suggests that the right hemisphere specializes in spatial tasks. Other functions that seem to be more associated with the right than the left hemisphere are facial expression, body language, and speech intonations.

Earlier it was suggested that the corpus callosum promotes coordination between the two hemispheres. This is shown by the lack of coordination in patients with a transected corpus callosum. When the patient is dressing, one hand may button a shirt while the other tries to unbutton it.

A striking conclusion of experiments on these patients is that the two hemispheres can operate quite independently when they are no longer interconnected. However, one hemisphere can express itself with language, whereas the other communicates only nonverbally.

Learning and Memory

Major functions of the higher levels of the nervous system are learning and memory. *Learning is a neural mechanism by which the individual changes behavior as the result of experience. Memory refers to the storage mechanism for what is learned.*

Types of learning. The two broad classes of learning are nonassociative and associative learning. In **nonassociative learning,** the learning does not depend on a particular relationship between what is learned and some other stimulus. For example, in habituation a repeated stimulus causes a response that gradually diminishes. This presumably happens because the individual is learning that the stimulus is not important. A familiar example is the change that typically occurs in attention to a new clock. At first the ticking noise may be annoying and may cause some difficulty in sleeping. However, after several nights the clock is no longer noticed. The same can happen with the clock's morning alarm. Another type of nonassociative learning is **sensitization.** In this case the stimulus is strong and consequently threatening. When the stimulus is first given, the response may be minimal. However, with repetition, the response increases. This indicates that learning is occurring in the opposite direction to that seen in habituation, presumably so that the behavior becomes directed toward escape from the stimulus.

Associative learning occurs when learning involves some relationship between stimuli. **Classical conditioning** involves a temporal association between a neutral conditioned stimulus and an unconditioned stimulus that elicits an unlearned response. When this combination of stimuli is repeated, provided that the timing relationship is maintained, the association between these stimuli is learned, at which point the conditioned stimulus alone will elicit the unlearned response. An example of this is the behavior of dogs in Pavlov's experiments on conditioned reflexes. For example, food presented to a hungry dog elicits an unconditioned response, salivation. If a bell is rung just before the food is presented, the dog learns to associate the bell with the food. Eventually, ringing the bell alone causes salivation. Of course, if the food fails to appear consistently when the bell is rung, the conditioned response fades away, a process called **extinction.**

Another form of associative learning is **instrumental** or **operant conditioning.** In this case when the response to a stimulus is associated with reinforcement, the probability of the response changes. The reinforcement can be either positive or negative. With positive reinforcement the response probability increases; with negative reinforcement the probability decreases. An example of positive reinforcement would be providing a fish to a porpoise for jumping out of the water through a hoop. An instance of negative reinforcement would be sending a child to his room for misbehaving.

Experiments on the mechanisms of learning. The neural circuitry involved in learning in mammals is complex; hence it has been difficult to study these mechanisms. An alternative approach has been to examine the cellular basis of learning in the simpler nervous systems of invertebrates, such as the marine mollusc, *Aplysia*. By isolating a connection between a single sensory neuron and a motoneuron responsible for a particular motor response, modeling habituation, sensitization, and even conditioning has been possible.

These examples of learning at a cellular level revealed that the presynaptic endings of the sensory neuron could change the amount of neurotransmitter released (Fig. 16-14). For example, during short-term habituation, the amount of transmitter released in successive responses gradually diminishes. The change involves an alteration in the Ca^{++} current that triggers the release. The cause of this change is repeated action potentials that lead to a reduction in the number of available Ca^{++} channels. Long-term habituation can also be produced. In this case the number of synaptic endings and of active zones in the remaining terminals decrease.

On the other hand, in short-term sensitization, an interneuron is thought to release serotonin onto the presynaptic terminal, and the serotonin stimulates adenylyl cyclase and the buildup of cAMP intracellularly. This, in turn, causes phosphorylation of K^+ channels, a decrease in K^+ current, and broadening of the presynaptic action potential. Broadened action potentials cause more transmitter to be released and therefore increase synaptic efficacy, which leads to a larger response. More transmitter is also mobilized. In long-term sensitization the number of synaptic terminals and active zones formed by the presynaptic neurons increases, and the dendrites of the postsynaptic neurons expand. Similar kinds of events also may occur in associative learning.

Long-term potentiation. Another model of learning is provided by a synaptic phenomenon called long-term potentiation (LTP). This has been studied most intensively in in vitro slices of the hippocampus. However, LTP has also been described in the neocortex and in

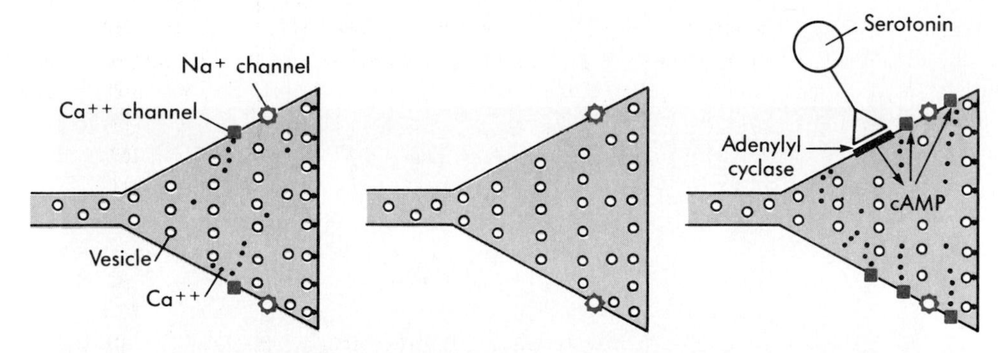

Fig. 16-14 Aplysia model of short-term habituation and sensitization. *Left,* A synapse in the control state. Invasion of the action potential into the ending causes Na^+ influx through Na^+ channels, resulting in the opening of Ca^{++} channels and transmitter release. *Middle,* Repeated activity has reduced the number of available Ca^{++} channels and thus a decrease in transmitter release. *Right,* Repeated strong stimulation activates a serotonergic interneuron, which releases serotonin onto the terminal, causing an increase in cAMP, a reduction in K^+ channel opening, and spike broadening. The broadened spikes cause increased transmitter release. (Redrawn from Kandel ER, Schwartz JH: *Principles of neural science,* New York, 1981, Elsevier/North Holland.)

other parts of the nervous system. Repetitive activation of an afferent pathway to the hippocampus or of one of the intrinsic connections increases the responses of the pyramidal cells. The increased responses last for hours in vitro (and even days to weeks in vivo). The forms of LTP differ, depending on the particular synaptic system. The mechanism of the enhanced synaptic efficacy seems to involve both pre- and postsynaptic events, and the neurotransmitters include excitatory amino acids that act on NMDA receptors. In some cases LTP is associative, in which case a weak synaptic input is strengthened when it is temporally associated with a stronger input. In other cases LTP is nonassociative. The hypothesis that LTP is involved in at least some aspects of memory is reasonable.

Memory. With regard to the stages of memory storage, a distinction between short-term memory and long-term memory is useful. Recent events appear to be stored in short-term memory by ongoing neural activity, because **short-term memory** persists for only minutes. **Long-term memory** can be subdivided into an intermediate form, which can be disrupted, and a long-lasting form, which is difficult to disrupt. Memory loss can be caused by a disruption of memory per se, or it can be a result of interference with the mechanism for recovering information from memory. Long-term memory is likely to involve structural changes in the nervous system, because this form of memory can remain intact even after events that disrupt short-term memory.

The temporal lobes appear to be particularly important for memory, because bilateral removal of the hippocampal formation severely and permanently disrupts recent memory. Short-term and long-term memories are unaffected, but new long-term memories can no longer be stored.

Neural plasticity. Damage to the nervous system can induce remodeling of neural pathways and thereby alter behavior. Such remodeling is said to reflect the **plasticity** of the nervous system. The CNS is much more plastic than was once believed. For example, altering the progress of development of neural connections by certain manipulations, such as lesions of the brain or sensory deprivation, is possible. Plasticity is greatest in the developing brain, but some degree of plasticity remains in the adult brain.

A change in the capability for developmental plasticity may occur for some neural systems at a time referred to as a **critical period.** For example, it may be possible to alter the connections formed in the visual pathways during their development, but only up to a particular developmental period. In visually deprived animals the visual connections may be abnormal (Fig. 16-15). However, visual deprivation that occurs after several postnatal months does not result in abnormal connections. Nor does restoration of vision after this time repair the abnormal connections in previously visually deprived animals. The plastic changes seen in such experiments may reflect a competition between fibers for synaptic connections with postsynaptic neurons in the developing nervous system. If a developing neural pathway loses in such a competition, the result may be a neurological deficit in the adult. For example, in the case of visual deprivation during development of the visual pathways, the consequence may be amblyopia of the deprived eye. **Amblyopia** is reduced visual capacity, and it can occur, for example, in children with strabismus (cross-eyed) because of a relative weakness of one of the extraocular muscles. Similar effects can be produced by a cataract or myopia if it is not corrected.

Plastic changes can also occur after injury to the

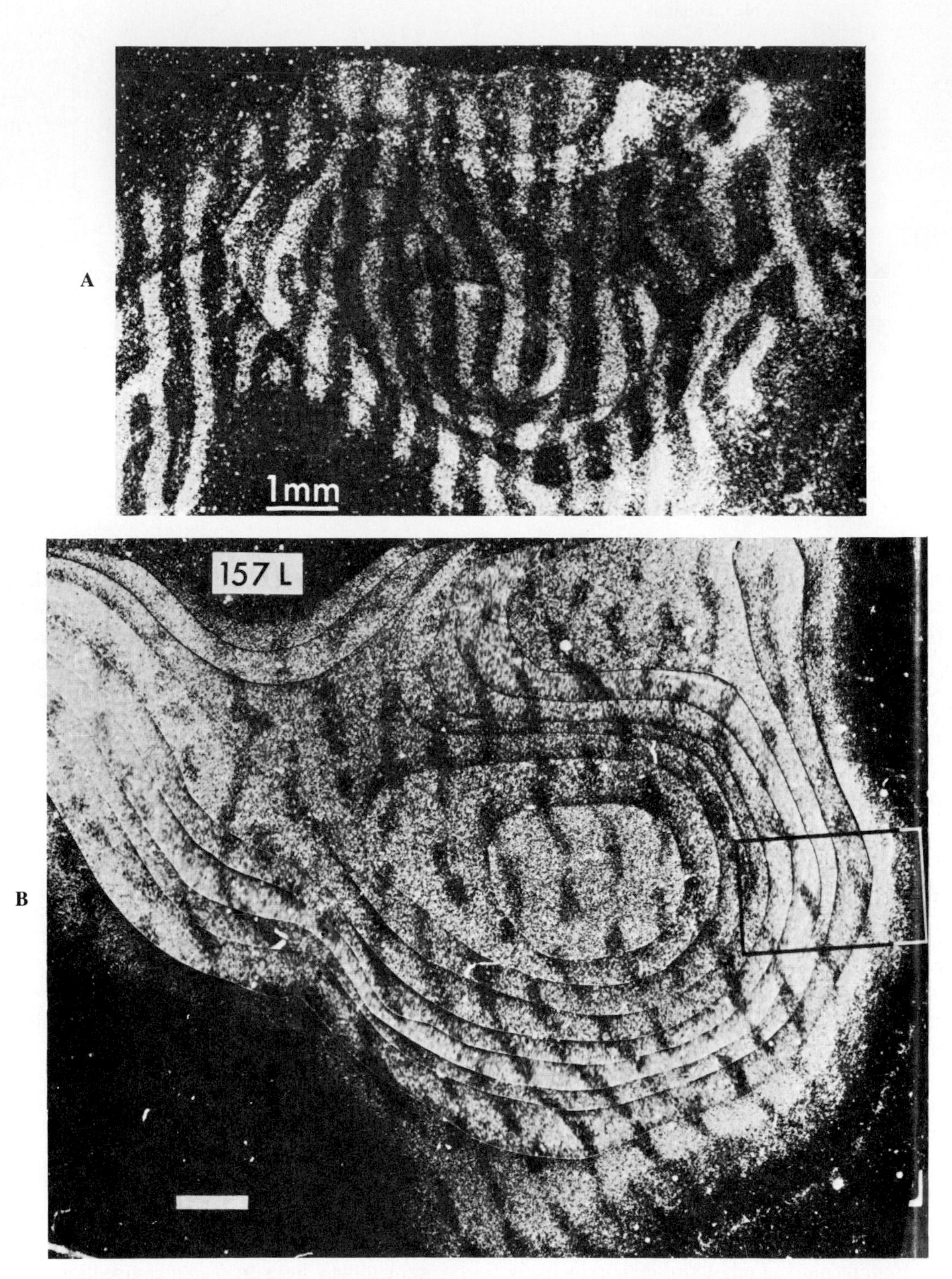

■ **Fig. 16-15** Plasticity in the visual pathway as a result of sensory deprivation during development. The ocular dominance columns are demonstrated by autoradiography following injection of a radioactive tracer into one eye. The tracer is transported to the lateral geniculate nucleus and then transneurally transported to the striate cortex. The cortex is labeled in bands that alternate with unlabeled bands whose input is from the uninjected eye. **A,** The normal pattern; **B,** The changed pattern in an animal raised with monocular visual deprivation. The injection was into the nondeprived eye, and the ocular dominance columns for this eye were clearly expanded. In other experiments it could be shown that the ocular dominance columns for the deprived eye contracted. (**A** from Hubel DH, Wiesel TN: Functional architecture of macaque monkey visual cortex. *Proc Roy Soc B* 198:1, 1977; **B** from LeVay S, Hubel DH, Weisel TN: *J Comp Neurol* 191:1, 1980.)

brain in adults. Axonal sprouting does occur in the damaged CNS. However, the sprouts do not necessarily restore normal function, and many neural pathways do not appear to sprout. Improvements in our knowledge concerning neural plasticity in the adult nervous system are vital if we are to improve medical therapy in many diseases of the nervous system and after neural trauma.

Summary

1. The cerebral cortex can be subdivided into lobes, based on the pattern of gyri and sulci. Each lobe has special functions, as shown by the effects of lesions or seizures.

2. The cerebral cortex can be subdivided into neocortex, allocortex, and juxtaallocortex. The neocortex typically has six layers, whereas the other types of cortex have fewer layers.

3. Neocortex contains a number of cell types, including pyramidal cells, which serve as the output cells, and several kinds of interneurons. The pyramidal cells appear to use an excitatory amino acid neurotransmitter, whereas the inhibitory interneurons are GABAergic.

4. Cortico-cortical afferent fibers end in the upper cortical layers; specific thalamocortical afferent fibers terminate in the middle layers and layer VI; diffuse thalamocortical afferent fibers synapse in layers I and VI. Cortical efferent fibers from layers II and III project to other areas of the cortex; those from layer V project to many subcortical targets, including the spinal cord, brainstem, and striatum, as well as to diffuse thalamic nuclei; layer VI distributes to the appropriate specific thalamic nucleus.

5. Cortical structure varies in different regions. Agranular (type 1) cortex is found in the motor areas, whereas granular cortex (koniocortex; type 5 cortex) occurs in the primary sensory receiving areas. Several types of homotypical cortex are found elsewhere in the neocortex. Brodmann's areas reflect these variations of cortical structure and correlate with functionally discrete areas. Allocortex has three layers, as typified by the hippocampus and dentate gyrus of the hippocampal formation.

6. The electroencephalogram, or EEG, varies with the state of the sleep-wake cycle, disease, and other factors. EEG rhythms include alpha, beta, theta, and delta waves. The EEG reflects synaptic activity of pyramidal cells. Cortical evoked potentials are stimulus-triggered changes in the EEG, and they are useful clinical tests of sensory transmission.

7. Sleep can be divided into slow wave and rapid eye movement (REM) forms. Slow wave sleep progresses through stages 1 through 4, each with a characteristic EEG pattern. Most dreams occur in REM sleep. Sleep is produced actively by a brainstem mechanism, and its circadian rhythmicity is controlled by the suprachiasmatic nucleus.

8. The EEG helps in the recognition of the various forms of epilepsy. Seizures are associated with depolarization shifts in pyramidal cells. Such shifts are caused by dendritic Ca^{++} spikes and a reduction in inhibitory processing.

9. The left cerebral hemisphere is dominant for language in most individuals. Wernicke's area is responsible for the understanding of language and Broca's area for its expression.

10. Information is transferred between the two hemispheres through the corpus callosum. This structure coordinates the two sides of the brain. The right hemisphere is more capable than the left in spatial tasks, facial expression, body language, and speech intonation.

11. Learning includes nonassociative and associative forms. Two kinds of nonassociative learning are habituation and sensitization. Associative learning includes classical and operant conditioning. The mechanisms of learning have been explored in animals that have simple nervous systems. Short-term changes include changes in synaptic efficacy, whereas long-term changes involve alterations in the number of synapses.

12. Long-term potentiation is mediated by an increased synaptic efficacy that lasts hours to weeks and that involves both pre- and postsynaptic changes.

13. Memory involves short-term (minutes), recent, and long-term storage processes and a retrieval mechanism. The hippocampal formation is important for recent memory.

14. Changes in neural pathways may occur during development of the nervous system because of damage. These may become permanent after a critical period and may result in neurological deficits in adults.

Bibliography

Journal articles

Amral DG, Cowan WM: Subcortical afferents to the hippocampal formation in the monkey, *J Comp Neurol* 189:573, 1980.

Bekklers JM, Stevens CF: Presynaptic mechanism for long-term potentiation in the hippocampus, *Nature* 346:724, 1990.

Benowitz LI et al: Hemispheric specialization in nonverbal communication, *Cortex* 19:5, 1983.

Damaasio AR, Geschwind N: The neural basis of language, *Ann Rev Neurosci* 7:127, 1984.

Dichter MA, Ayala GF: Cellular mechanisms of epilepsy: a status report, *Science* 237:157, 1987.

Geschwind N: Specializations of the human brain, *Sci Amer* 241:180, 1979.

Kandel ER, Schwartz JH: Molecular biology of learning: modulation of transmitter release, *Science* 218:433, 1982.

Milner B: Some cognitive effects of frontal-lobe lesions in man, *Phil Trans R Soc London B* 298:211, 1982.

Moruzzi G: The sleep-waking cycle, *Rev Physiol Biochem Exp Pharmacol* 64:1, 1972.

Moruzzi G, Magoun HW: Brain stem reticular formation and activation of the EEG. *Electroencephalogr Clin Neurophysiol* 1:455, 1949.

Prince DA: Neurophysiology of epilepsy, *Ann Rev Neurosci* 1:395, 1978.

Rosenwasser AM: Behavioral neurobiology of circadian pacemakers: a comparative perspective, *Prog Psychobiol Physiol Psychol* 13:155, 1988.

Sherman SM, Spear PD: Organization of visual pathways in normal and visually deprived cats, *Physiol Rev* 62:738, 1982.

Sperry RW: Mental unity following surgical disconnection of the cerebral hemispheres, *Harvey Lecture* 62:293, 1964.

Tsukahara N: Synaptic plasticity in the mammalian central nervous system, *Ann Rev Neurosci* 4:351, 1981.

Books and monographs

Andersen RA: *Inferior parietal lobe function in spatial perception and visuomotor integration.* In Plum F, editor: *Handbook of physiology, sect 1, The nervous system, vol 6, Higher functions of the brain, part 2,* Bethesda, 1987, American Physiological Society.

Benson DF: *Aphasia, alexia, and agraphia,* New York, 1979, Churchill Livingstone.

Bergamini L, Bergamasco B: *Cortical evoked potentials in man,* Springfield, 1967, Charles C Thomas.

Dudai Y: *The neurobiology of memory: concepts, findings, trends,* Oxford, 1989, Oxford University Press.

Fulton JF: *Frontal lobotomy and affective behavior: a neurophysiological analysis,* New York, 1951, Norton.

Fuster JM: *The prefrontal cortex: anatomy, physiology, and neuropsychology of the frontal lobe,* ed 2, New York, 1989, Raven Press.

Gazzaniga MS, LeDoux JE: *The integrated mind,* New York, 1978, Plenum Press.

Kandel ER et al: *Principles of neural science,* ed 3, New York, 1991, Elsevier.

Kryger MH, Roth T, Dement WC, editor: *Principles and practice of sleep medicine,* Philadelphia, 1989, Saunders.

Milner B: *Hemispheric specialization: scope and limits.* In Schmitt FO, Worden FG, editor: *The neurosciences: third study program,* Cambridge, Mass, 1974, MIT Press.

Pavlov IP: *Conditioned reflexes: an investigation of the physiological activity of the cerebral cortex,* GV Anrep, trans, London, 1927, Oxford University Press.

Pedley TA, Traub RD: Physiological basis of the EEG. In Daly DD, Pedley TA, editors: Current practice of clinical electroencephalography, ed 2, New York, Raven Press.

Penfield W, Jasper H: *Epilepsy and the functional anatomy of the human brain,* Boston, 1954, Little, Brown & Co.

Peters A, Jones EG, editors: *Cerebral cortex, vol 1, Cellular components of the cerebral cortex,* New York, 1984, Plenum Press.

Peters A, Jones EG, editors: *Cerebral cortex, vol 2, Functional properties of cortical cells,* New York, 1984, Plenum Press.

Squire LR: *Memory and brain,* New York, 1987, Oxford University Press.

Steriade M, McCarley RW: *Brainstem control of wakefulness and sleep,* New York, 1990, Plenum Press.

SECTION
III

MUSCLE

Richard A. Murphy

CHAPTER 17

Contractile Mechanism of Muscle Cells

■ *Functional Classification of Muscle*

Muscles are traditionally classified in terms of their anatomy (striated versus smooth). Another distinction is made between voluntary muscles under conscious control and involuntary muscles in the internal organ systems with autonomic innervation. However, the properties of different muscle cells are more readily understood in terms of their functional roles. *The basic distinction is between muscle cells attached to the skeleton and those in the walls of hollow organs.*

Cells attached to a skeleton (these are all voluntary and striated) are often very long and bridge the attachment points of the muscle. Thus the individual cells are anatomically and mechanically arranged in parallel. This means that the cells function independently, and the total force produced by muscle equals the sum of the forces generated by its cells. The musculoskeletal system is arranged so that most gravitational loads are borne by the skeleton and ligaments. Skeletal muscle cells are normally relaxed and are usually recruited to generate force and movement.

Muscle cells in the walls of hollow organs cannot function independently. In a continuous sheet of muscle, cells must be connected in series with each other as well as in parallel. Cells mechanically linked in series are like links in a chain: all must bear the same stress and contract uniformly. Furthermore, all cells are interdependent because pressures are transmitted throughout hollow organs, and contraction of some cells necessarily alters the load on the others. Cells in hollow organs (typically involuntary and smooth) have two functional roles. They must be capable not only of generating force and movement, like skeletal muscle cells, but also of maintaining organ dimensions against applied loads. For example, vascular smooth muscle must bear the load imposed by the blood pressure to regulate blood flow. Smooth muscle cells have more complex contractile and regulatory systems to carry out this latter function economically.

This chapter describes the cellular and molecular processes of contraction and the mechanisms involved in control of contraction. The basic apparatus for conversion of chemical energy into mechanical energy (chemomechanical transduction) in the form of force development or movement is considered first in a discussion of skeletal muscle cells. Chapter 18 is devoted to integrated tissue function and the capacity of skeletal muscles to adapt to changing requirements. The heart is a hollow organ, and cardiac muscle has many differences from skeletal muscle, although it is also striated. These characteristics are described in detail in Chapters 22 through 25, but the information on the contractile system presented in this chapter is also applicable to the heart. Smooth muscle cells found in other hollow organs have features not found in skeletal muscles, as described in Chapter 19. Smooth muscle physiology is remarkably diverse and reflects its specialized roles in many different organs. Specific information on the properties of smooth muscle and its innervation is given in the chapters describing the vascular, airway, gastrointestinal, and other organ systems.

■ *Structure of the Contractile Apparatus in Skeletal Muscle*

A skeletal muscle is composed of bundles of enormous, multinucleated cells up to 80 μm in diameter and sometimes many centimeters long (Fig. 17-1). These cells, like those in cardiac muscle, have a striking banding pattern responsible for their classification as *striated* muscles. The striations arise from a highly organized arrangement of subcellular structures. There are few instances in biology in which ultrastructure provides a clearer basis for understanding cell function than in stri-

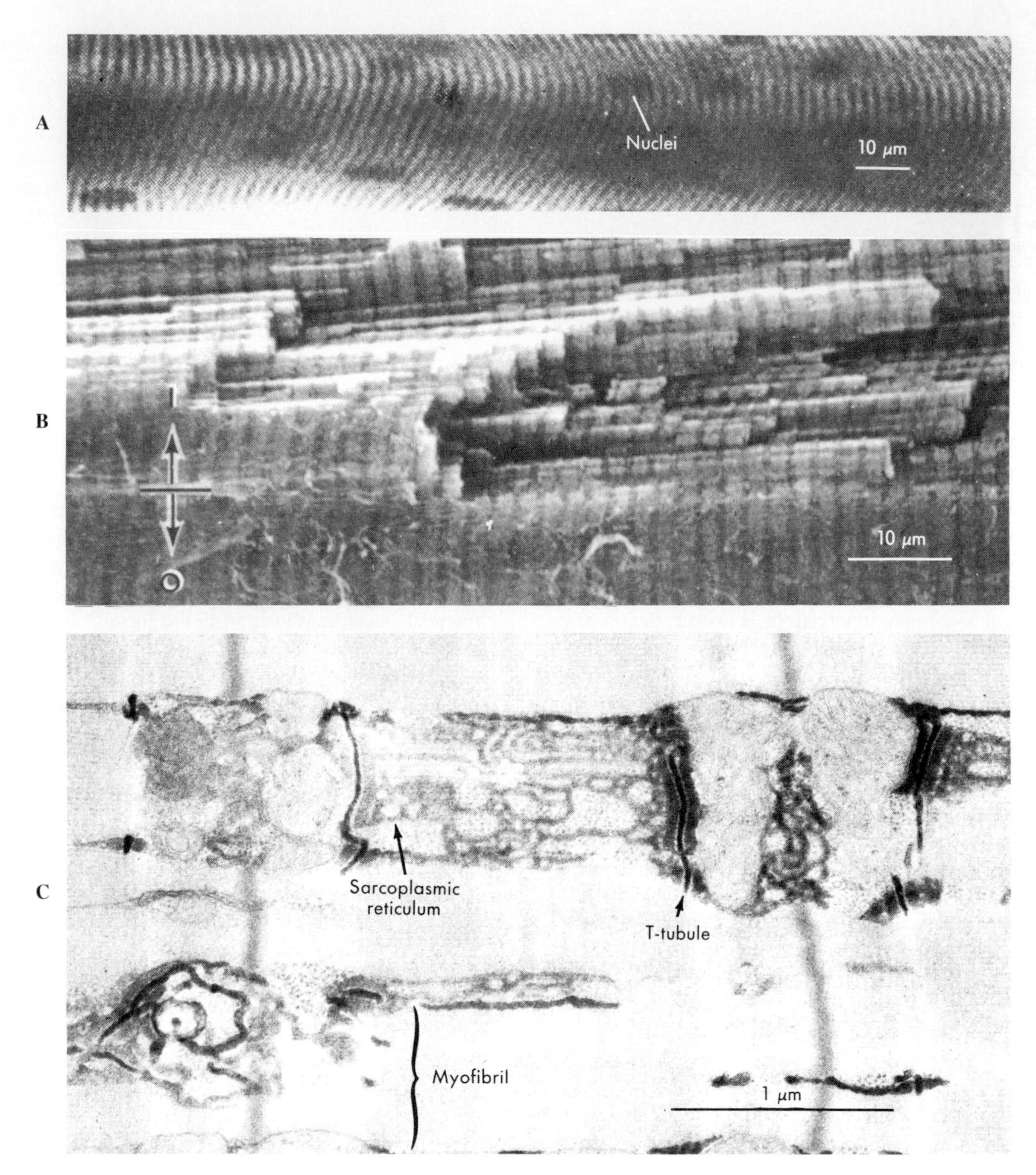

■ **Fig. 17-1** **A,** Segment of a single cell from human gastrocnemius muscle. Cross striations of alternating dark *(A)* and light *(I)* bands are visible along with nuclei located peripherally. **B,** Scanning electron micrograph of a skeletal muscle cell shows the cell surface *(o)* and interior *(i)* with tightly packed myofibrils. **C,** Electron micrograph of a longitudinal section through a mouse skeletal muscle cell. The tissue is stained to show the internal membrane system of the sarcoplasmic reticulum as a very dark network of interconnecting tubules. Invaginations of the cell membranes (T-tubules) are visible as dark elements extending into the cell at the level of the junction between the *A* and *I* bands. (**A** redrawn from Leeson CR, Leeson TS: *Histology,* ed 3, Philadelphia, 1976, WB Saunders. **B** from Sawada H, Ishikawa H, Yamada E: *Tissue Cell* 10:183, 1978; **C** courtesy Dr. Michael S. Forbes.)

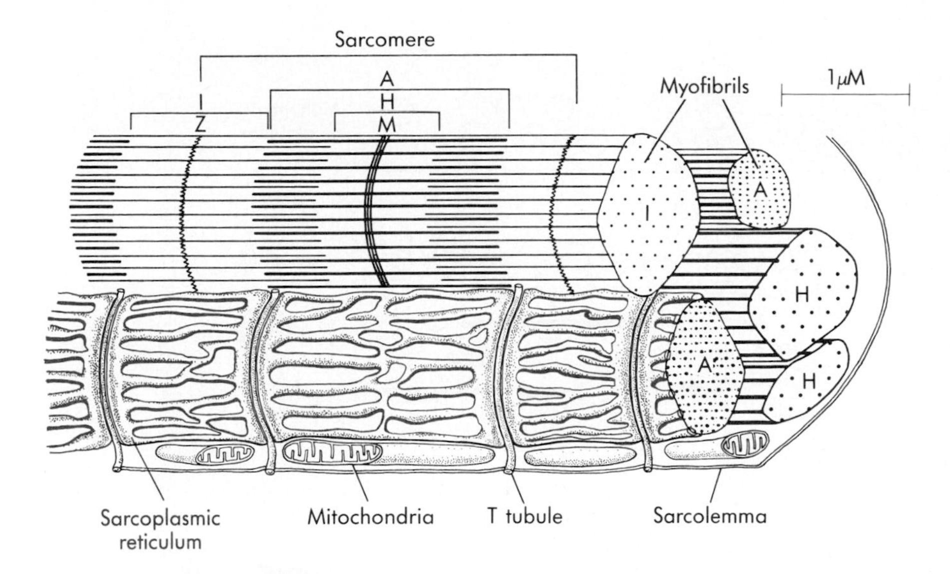

■ **Fig. 17-2** Drawing illustrating the three-dimensional relationships between membrane elements and the filament lattice. (Redrawn from Leeson CR, Leeson TS: *Histology,* ed 3, Philadelphia, 1976, WB Saunders.)

ated muscle. Therefore careful study of Figs. 17-1 and 17-2 is warranted. Electron micrographs reveal bundles of filaments running along the axis of the cell. These bundles are termed **myofibrils** (Fig. 17-1, *B*). The gross striation pattern of the cell arises from a repeating pattern in the myofibrils that is in transverse register across the whole cell (Fig. 17-2).

The banding pattern arises from two sets of filaments in the myofibrils. Their organization is most clearly seen in a drawing that exaggerates the cross-sectional dimensions (Fig. 17-3). The dark striations are a region containing a lattice of **thick filaments.** This region of the myofibril is termed the **A band** because it appears dark (anisotropic) when viewed in a microscope using polarized light. A second lattice consists of **thin filaments** that are attached to a transverse, darkly staining structure termed the **Z line** or **Z disk.** The thin filaments extend from two adjacent Z lines to interdigitate with the thick filaments. Areas of the myofibril or cell containing only thin filaments and Z disks are termed **I bands** because they are light (isotropic) when viewed under polarized light. *The thick and thin filament arrays form the contractile system, and the repeating unit in each myofibril delimited by the Z disks is the basic contractile unit called a* ***sarcomere.*** Each sarcomere contains half of two I bands with a central A band. The latter has a less dense central region (H zone) where there is no overlap of thin filaments. The H zone is bisected by a darkly staining M line containing proteins that link the thick filaments together. Cross sections of the myofibril reveal the relationship between thin and thick filaments (Fig. 17-3). The thin filaments form a hexagonal array around each thick filament, whereas each thin filament is equidistant from three thick filaments. This arrangement reflects the presence of two thin filaments for each thick filament per half sarcomere in vertebrate striated muscle. The complex membrane systems associated with each sarcomere are considered later.

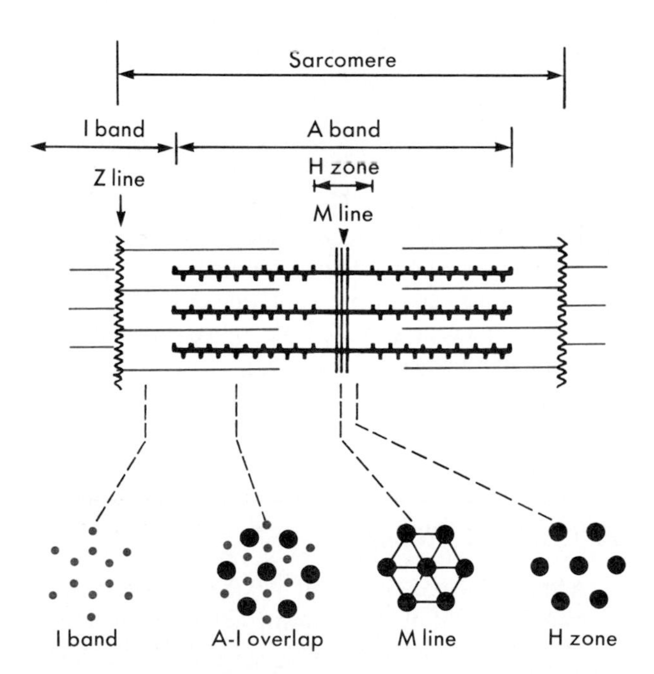

■ **Fig. 17-3** Longitudinal *(top)* and cross-sectional *(bottom)* diagrams showing the relationships between thick *(black)* and thin *(color)* filaments of a sarcomere. (Redrawn from Squire JM: *The structural basis of muscular contraction,* New York, 1981, Plenum Press.)

■ *The Thin Filament*

Composition and structure. Thin filaments are ubiquitous constituents of all nucleated cells and are a dominant feature in muscle. All thin filaments contain two major proteins. A globular protein called **actin,** with a molecular weight of 42,000 daltons, polymerizes under conditions existing in the cytoplasm to form twisted, two-stranded filaments (Fig. 17-4). Rod-shaped molecules of **tropomyosin** stretch along each strand of the thin filament. Each molecule of tropomyosin is as-

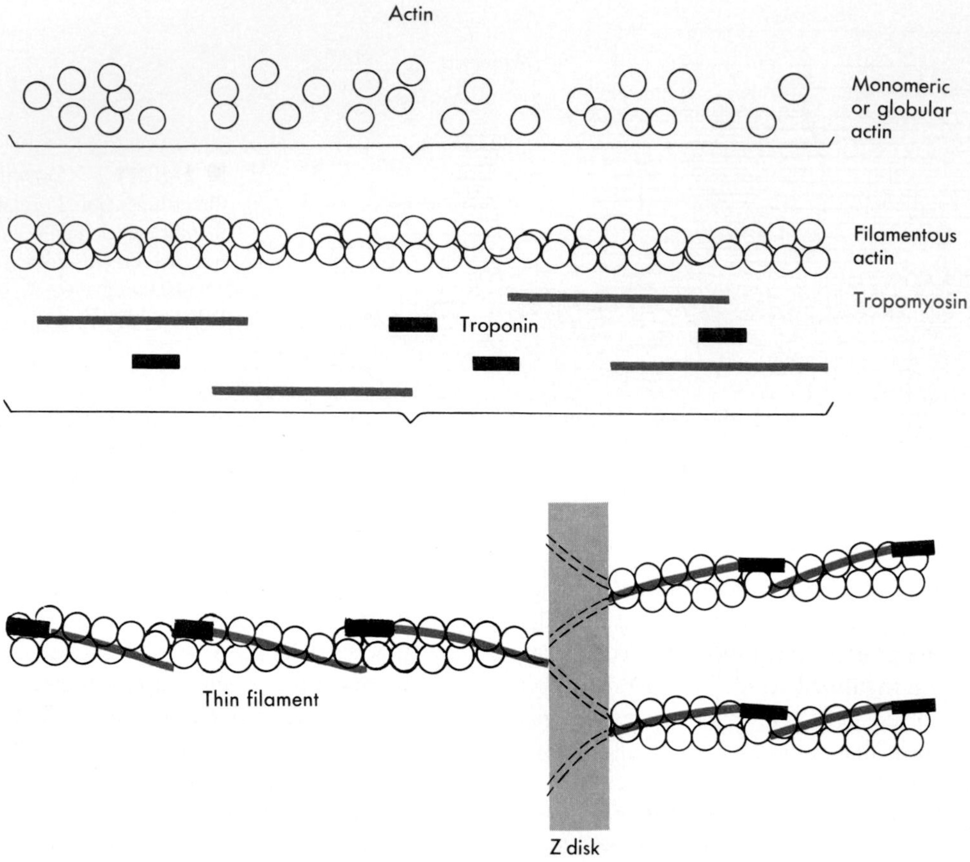

Actin

Monomeric or globular actin

Filamentous actin

Tropomyosin

Troponin

Thin filament

Z disk

■ **Fig. 17-4** Composition and structure of thin filaments in muscle. Globular actin monomers *(top)* polymerize into a two-stranded helical filament. The thin filament structure is completed with the addition of stiff, rod-shaped tropomyosin molecules *(color)*. Troponin *(black rectangles)* is a regulatory protein bound to the tropomyosin component of the thin filament in vertebrate striated muscles *(bottom)*. For clarity the tropomyosin-troponin complexes associated with only one strand of the actin helix are illustrated. Thin filaments are anchored to Z disks in striated muscles. Filaments on each side of a Z disk have opposite polarities. (Redrawn from Murray JM, Weber A: *Sci Am* 230:58, 1974.)

sociated with six or seven actins in one strand. Tropomyosin is composed of two separate polypeptide chains. The individual polypeptides have a basic α-helical structure, and the two helical peptides are wound around each other to form a supercoil. Molecules of this type are long, rigid, and insoluble. Other proteins, present in smaller amounts, are associated with the thin filaments. These minor proteins are involved in the attachment of thin filaments to Z disks and probably contribute to the remarkably uniform thin filament length of 1 μm in vertebrate striated muscles. Finally, additional thin-filament proteins are involved in regulation of the interaction of actin with the thick filament. The regulatory protein in striated muscles is **troponin,** which is bound to tropomyosin (Fig. 17-4). Regulatory proteins are considered later.

Lattice organization and polarity. Thin filaments are anchored at one end. The filaments have a polarity, although this is not apparent in electron micrographs. Thus thin filaments on each side of the Z disk "point" in opposite directions. Fig. 17-4 indicates one view of how this may arise from a splitting of the two strands of the thin filament at the Z disk so that each strand forms half a thin filament of opposite polarity. In cross section the thin filaments can be seen to form a hexagonal lattice around each thick filament (Fig. 17-3).

■ *The Thick Filament*

A very large protein (about 470,000 daltons) called **myosin** forms the thick filament, although small amounts of other proteins are present. The myosin molecule contains six different polypeptides. These peptides, which are not covalently linked, can be dissociated by detergents or denaturing agents (Fig. 17-5) and separated into three pairs: one set of large **heavy chains** and two sets of **light chains.** Most of the heavy chain has an

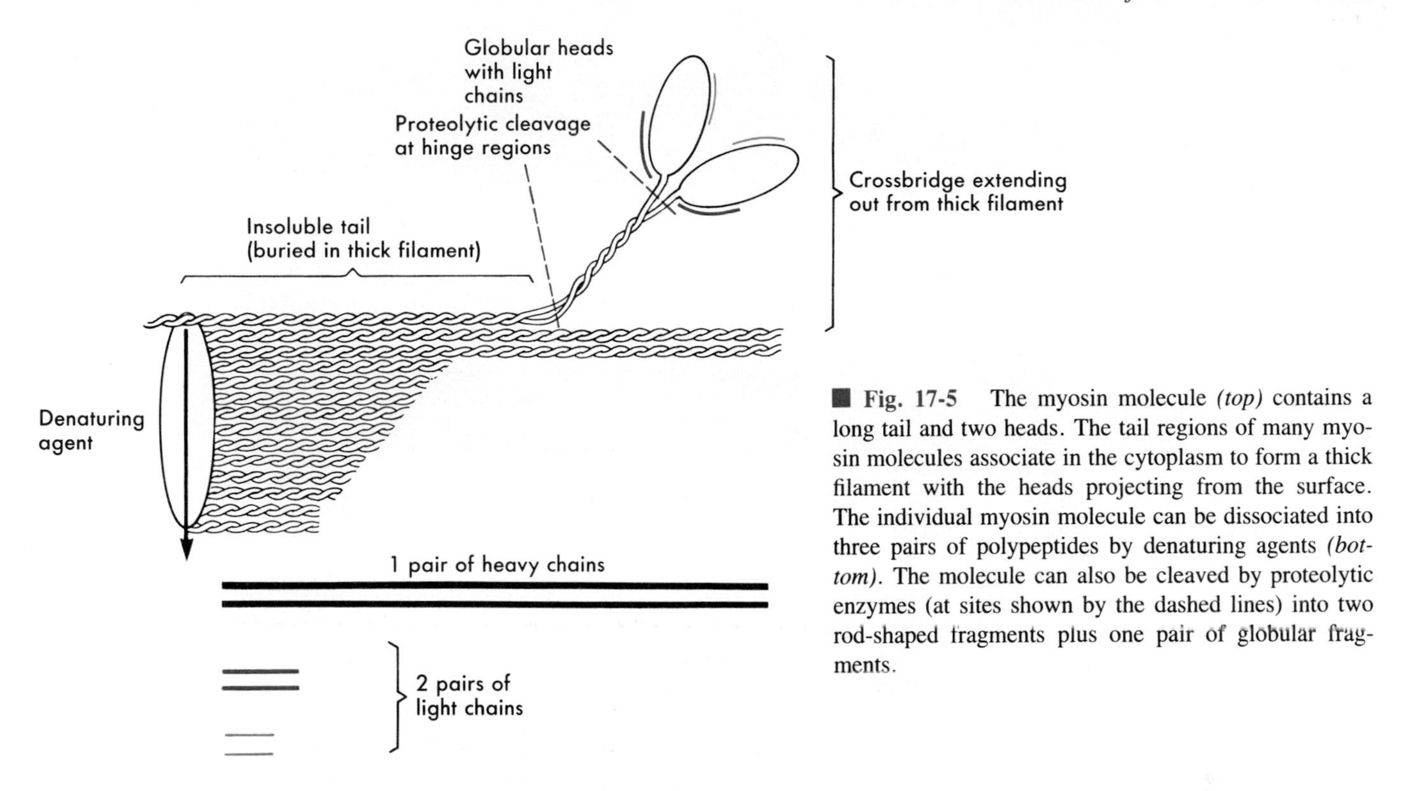

■ **Fig. 17-5** The myosin molecule *(top)* contains a long tail and two heads. The tail regions of many myosin molecules associate in the cytoplasm to form a thick filament with the heads projecting from the surface. The individual myosin molecule can be dissociated into three pairs of polypeptides by denaturing agents *(bottom)*. The molecule can also be cleaved by proteolytic enzymes (at sites shown by the dashed lines) into two rod-shaped fragments plus one pair of globular fragments.

α-helical structure, and the two strands are twisted around each other in a supercoil that forms a long, rigid, insoluble "tail." However, one end of the heavy chains has a globular tertiary structure. Thus each myosin molecule has two "heads" attached to one end of the long tail (Fig. 17-5). One polypeptide of each set of light chains is associated with each head of the molecule.

If myosin is briefly exposed to proteolytic enzymes such as trypsin or papain, it is preferentially cleaved into one long and one short rod segment and two globular fragments (Fig. 17-5). The points of cleavage occur at regions where the basic supercoiled structure is perturbed, exposing the peptide backbone to enzymatic attack. Cleavage points indicate flexible regions that serve as hinges for the molecule. Studies of these proteolytic fragments show the functions of each part of the myosin molecule, as described later.

Filament formation and structure. Myosin aggregates in the cytoplasm to form thick filaments. The basic filament represents aggregates of the tail segment of the molecules, and smooth filaments lacking projecting heads can form from the tail segments obtained by proteolytic cleavage. The remainder of the molecule, including the globular heads and the rod portion between the hinges, projects laterally from the filaments. These projections are visible in electron micrographs and are termed **crossbridges** because they can link adjacent thick and thin filaments. The process of filamentogenesis is extraordinarily specific. Each filament begins with an end-to-end association of the tails of the myosin molecules (Fig. 17-6, *A*). This produces a central thick filament segment lacking crossbridge projections. The crossbridges in each half of the filament are consequently oriented in the opposite direction. Therefore the molecules of the thick and thin filaments in each half of the sarcomere have the same relative orientation as a result of the thin filament polarity. The crossbridges project in groups of three from the filament (Fig. 17-6, *B*). Successive crowns of crossbridges are rotated, producing a helical arrangement of crossbridges along the thick filament. Thick filaments in striated muscle are 1.6 μm long and contain an estimated 300 to 400 crossbridges. Minor protein constituents probably contribute to the highly ordered and regular structure of thick filaments and their basic triangular lattices (Fig. 17-3) in cells. Adjacent thick filaments in the sarcomere are linked by protein connections between the central bare zones, contributing to the stability of the filament lattice.

■ *Crossbridge Interactions with the Thin Filament*

■ *Crossbridge Properties*

Each crossbridge consists of two identical heads that exhibit a remarkable set of properties. Most studies suggest that the two heads act largely independently in the reactions discussed in this section. However, the behavior of one head may be somewhat influenced by reac-

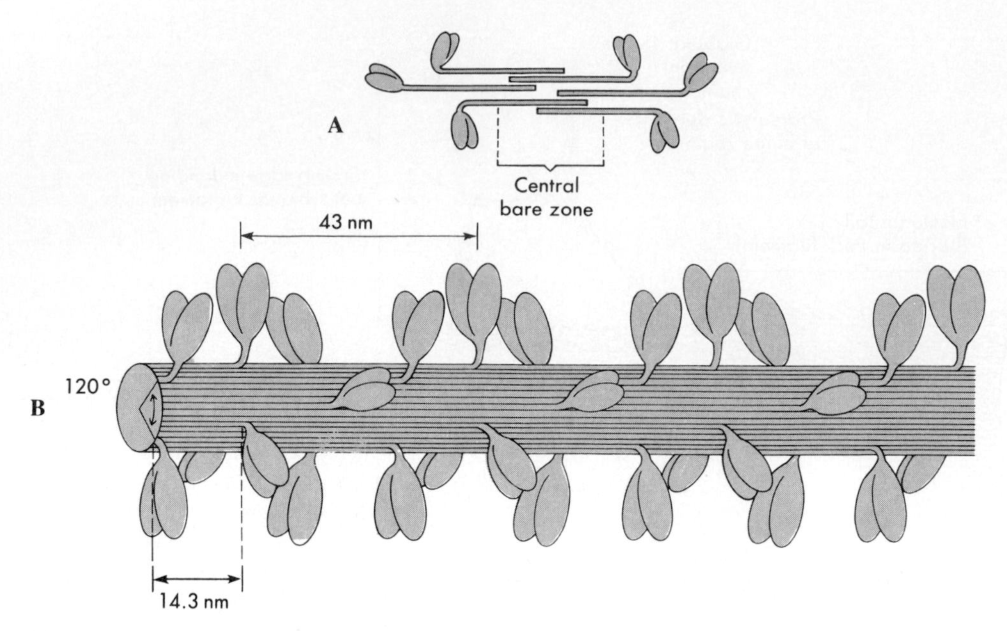

■ **Fig. 17-6** Proposed structure of the thick filament. **A,** Initiation of filament. Filament formation begins with an end-to-end association of the tails of myosin molecules. **B,** Segment of filament. "Crowns" of three crossbridges project at intervals of 14.3 nm along the thick filament, and successive crowns are rotated. The result is a thick filament with nine rows of crossbridges along its length. (Redrawn from Murray JM, Weber A: *Sci Am* 230:58, 1974.)

tions involving the other (cooperativity). The properties of the myosin head depend on the intact globular part of the heavy chain and the two associated light chains.

Myosin can catalyze the hydrolysis of adenosine triphosphate (ATP), producing adenosine diphosphate (ADP) and inorganic phosphate (P_i). However, the ATPase activity of myosin is inhibited by the high magnesium concentrations that exist in cells. Myosin can also bind to actin, forming an **actomyosin** complex. Actomyosin is a very active ATPase in the muscle cells. *The interaction between actin and myosin, associated with ATP hydrolysis, represents the fundamental* ***chemomechanical transduction*** *process, considered later, in which chemical energy is converted into mechanical energy by the muscle.*

The splitting of ATP by actomyosin is a complex cycle involving many steps. The more important steps are illustrated in Fig. 17-7, *A*. In a relaxed muscle, regulatory systems prevent actin-myosin interactions (Fig. 17-7, *A, top*). This inhibition is overcome on stimulation by Ca^{++} ions and is considered later. In the presence of ATP, myosin has ADP and P_i bound to each head and, in this state, exhibits a high affinity for actin. The ADP and P_i are released when the myosin heads bind to actin. Product release allows an ATP molecule to bind to myosin, and the affinity of myosin for actin (as A—M) is greatly reduced. The bound ATP is hydrolyzed with dissociation of myosin and actin, although the products (ADP and P_i) remain bound to myosin in the succeeding step. The energy released by splitting of ATP is stored in the myosin molecule, which is now in a high-energy state and has a renewed high affinity for actin.

■ *The Crossbridge Cycle*

The cycle associated with ATP hydrolysis by isolated actomyosin releases the energy in ATP without any conversion into mechanical work. Chemomechanical transduction involves conformational changes in the crossbridges, which lead to filament movements with shortening and force development (Fig. 17-7). In a resting muscle the crossbridge is not attached to the thin filament and is oriented perpendicularly to the myosin filament. When a muscle is stimulated, a rise in myoplasmic Ca^{++} concentration produces changes in the myofilament structure, allowing crossbridge binding to the thin filament (considered in the section on control mechanisms; step 1). The hinge regions in the crossbridge permit the head to swing toward the thin filament. Crossbridge attachment occurs because of the high affinity of the crossbridge-ADP-P_i complex for actin. *Crossbridges preferentially assume a conformation that minimizes their free energy*. This preferred conformation for attachment corresponds to a 90-degree orientation with respect to the filaments (Fig. 17-7, *B*). The chemical structure of the crossbridge alters during step 2 when ADP and P_i are released. The resulting AM complex can have a lower free energy, but only by

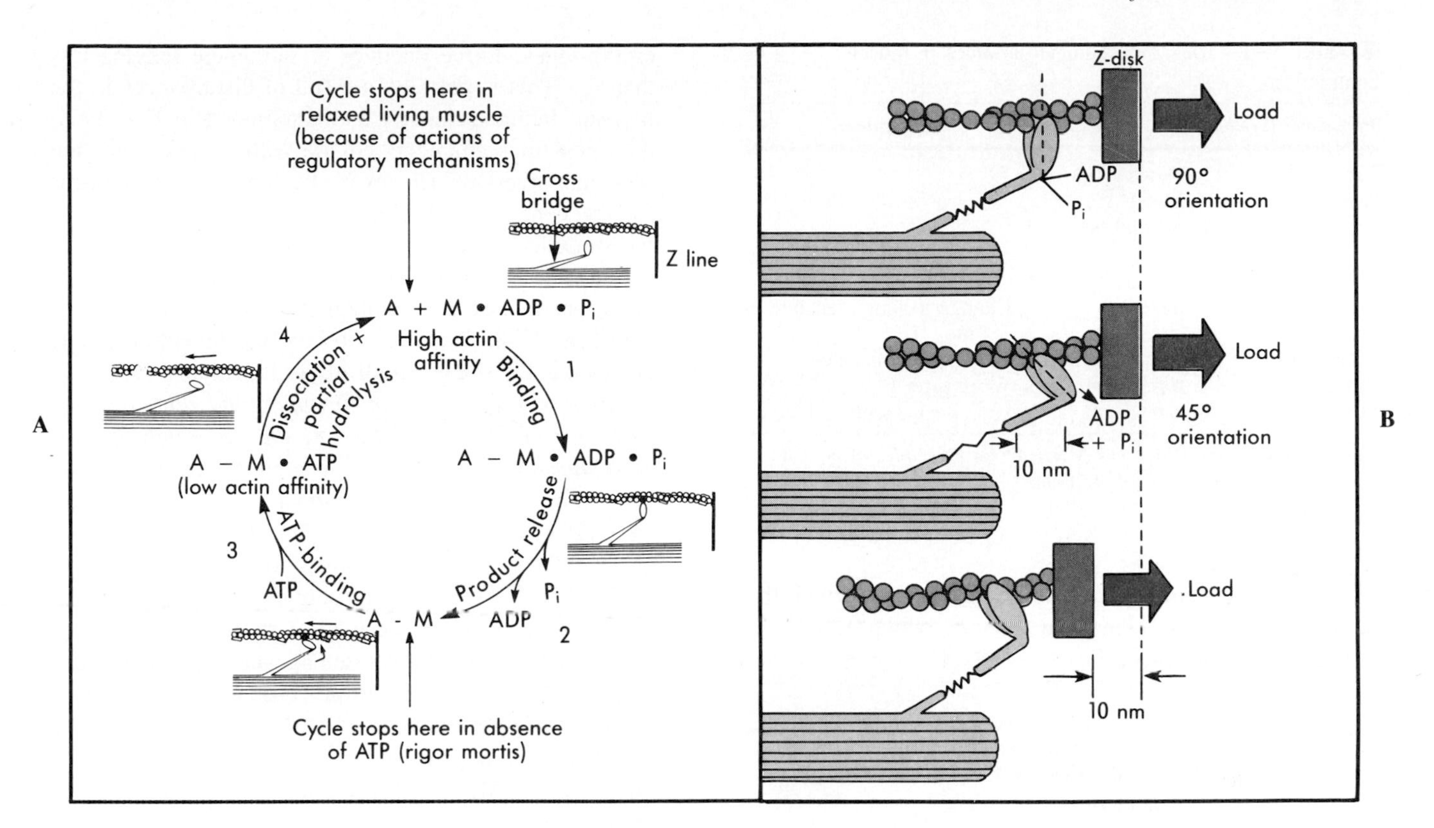

■ **Fig. 17-7** **A,** The mechanism of ATP hydrolysis by actomyosin and the major steps in the crossbridge cycle. Actin in the thin filament *(A)* and the myosin crossbridge projecting from the thick filament *(M)* interact cyclically. This interaction involves a number of steps during which ATP is hydrolyzed, and the energy released is harnessed to induce conformational changes in the crossbridge. Each cycle causes the thick and thin filaments to interdigitate by about 10 nm. **B,** The preferred or minimal free energy conformation of the crossbridge with bound ADP + P_i is 90 degrees *(top)*. This changes to 45 degrees after the release of ADP + P_i *(middle)*. The transition from 90 degrees to 45 degrees generates a force on the thin filament represented by the stretched spring *(middle)* that causes the sarcomere to shorten (*bottom*). (From Berne RM, Levy MN, editors: *Principles of physiology,* St Louis, 1990, Mosby-Year Book, Inc.)

changing its conformation from 90 degrees to 45 degrees. *Part of the free energy made available from ATP hydrolysis is captured in this conformational change, which minimizes the free energy and thereby generates a force that causes the thick and thin filaments to interdigitate*. The resulting protein complex has a high affinity for ATP, whose binding induces crossbridge detachment (Fig. 17-7, step 3). Partial ATP hydrolysis to regenerate the crossbridge-ADP-P_i complex raises the free energy and the 90-degree conformation is favored (step 4).

Each cycle can move the filaments about 10 nm relative to each other. The way in which enormous numbers of such cycles generate muscular contraction is considered in the following section. The illustrated cycle will continue until interrupted in the detached state by control systems (which remove Ca^{++} from the myoplasm and produce relaxation) or until the ATP is exhausted. ATP depletion is an abnormal situation in muscle cells and arrests the cycle with the formation of permanent actomyosin complexes as shown in Fig. 17-7. Death causes muscular rigidity **(rigor mortis)** because ATP depletion leads to permanent crossbridge attachment.

■ *Biophysics of the Contractile System*

Quantitative estimates of the mechanical output of the muscles have provided important inferences about the mechanism of chemomechanical transduction and its control mechanisms. The measurement of muscle contraction also provides a way of assessing the effects of neurotransmitters, drugs, and hormones and of quantifying pathological changes. The important mechanical variables are listed in Table 17-1. The basic approach in muscle mechanics is to control all of these factors except the measured dependent variable. Such measure-

■ **Table 17-1** Basic mechanical variables in muscle contraction

Parameter (symbol)	*Units*	*Definition*
Force (F)	Newton (n)	
Length (L)	Meter (m)	
Time (T)	Second (sec)	
Derived variables		
Velocity (V)	m/sec	Change in length/change in time
Work (W)	n·m = joules	Force times distance
Power (P)	n·m/sec = watts	Work/time
Stress (S)	n/m^2	Force/cross-sectional area

Adapted from Berne RM, Levy MN, editors: *Principles of physiology*, St Louis, 1990, Mosby-Year Book.

ments should employ muscle preparations in which the cells are aligned along the axis in which force and length are determined.

■ *Force-Length Relationships and the Sliding Filament Mechanism*

Force generation depends on the length of the muscle (Fig. 17-8). Points on these curves are determined under conditions in which the length of the muscle is fixed, and the maximum force is measured during a contraction. Such a contraction at constant length is termed **isometric.** Force will vary with the size of the muscle and is best expressed as a stress (force/cross-sectional area of the muscle) to allow comparisons with other tissues. A relaxed muscle is elastic and it requires a force to stretch it to an increased length. The resulting **passive force-length** or **stress-length relationship** primarily reflects the properties of connective tissue in intact muscles. When a muscle is stimulated, the total isometric stress is greater than the passive stress at any length. The difference between the stress-length curves for contracting and relaxed muscles is the **active stress-length relationship** that characterizes the contractile system (Fig. 17-8, *B*).

The correlation of structural and mechanical information reveals that active stress is proportional to the overlap between thick and thin filaments in a sarcomere (Fig. 17-8, *C*). As the muscle is stretched beyond the length (L_o) at which maximum active force (F_o) is developed, *active stress decreases linearly with the overlap between thick and thin filaments—showing that stress is proportional to the number of active crossbridges that can interact with the thin filament.* This number includes all the interacting crossbridges in each half sarcomere. There is a small segment at the peak of the stress-length curve at which stress does not change with length. At these lengths the thin filaments move past the central thick filament bare zones that lack crossbridges. Force declines at sarcomere lengths less than L_o. This is partially a result of disturbances in the filament lattice geometry as diagrammed in Fig. 17-8, *C*, when thick filaments collide with Z disks and thin filaments overlap. However, the control mechanisms for the contractile machinery become less sensitive to the stimulus at short muscle lengths, and partial inactivation contributes to the low stress at short lengths. Slight variations in sarcomere lengths in whole muscles (compare Fig. 17-8, *B* and *C*) obscure the inflection points detected in the stress-length curves for sarcomeres.

Because the lengths of the thick and thin filaments and their packing density are similar in vertebrate striated muscles, they all generate similar maximal stresses at L_o. The maximal stress is about 3×10^5 n/m^2 (or up to 3 kg/cm^2) of cell cross-sectional area.

Velocity-stress relationships and the crossbridge cycle. Another basic relationship characterizing the output is obtained when the load or stress on a muscle is held constant and the shortening velocity is measured (Fig. 17-9). A contraction at constant load is termed **isotonic.** *Shortening velocity is determined by the load on the muscle as illustrated by the hyperbolic velocity-stress curve.* Muscles lift a heavy load slowly but can shorten rapidly when lightly loaded. Fig. 17-9 also shows that a contracting muscle can withstand (briefly) a heavier stress when forcibly stretched than it can develop isometrically. The strength of the crossbridge attachment to the thin filament is greater than the force generated by its movement. Consequently, *muscles can bear a load about 1.6 times F_o before the crossbridge attachment is mechanically broken and rapid lengthening occurs.* This situation is important physiologically when a muscle is contracted to decelerate the body, as when a person is walking downhill.

The contractile system operates most efficiently (i.e., the greatest mechanical work output for the chemical energy used) with optimal loading. This is illustrated by the power curve. Power, or work/time, is simply the product of force and velocity, and power can be calculated by multiplying these values at any point on the velocity-force curve (Fig. 17-9). Maximal power is obtained with a load of about 0.3 of the maximal force that can be developed. A muscle contracting isometrically does no work (force × distance = 0), nor does a muscle shortening with no load. The mechanical efficiency of such contractions is necessarily zero. Inefficient contractions are often physiologically important when either high speed or maximal force is appropriate. *The optimal efficiency of the contractile system in converting chemical energy into mechanical work is about 40% to 45%.* The remaining energy appears as heat that can raise the temperature of working muscles by several degrees.

If a sarcomere is to shorten more than the 10 nm associated with a single crossbridge cycle, each crossbridge must detach and then reattach at a new site on

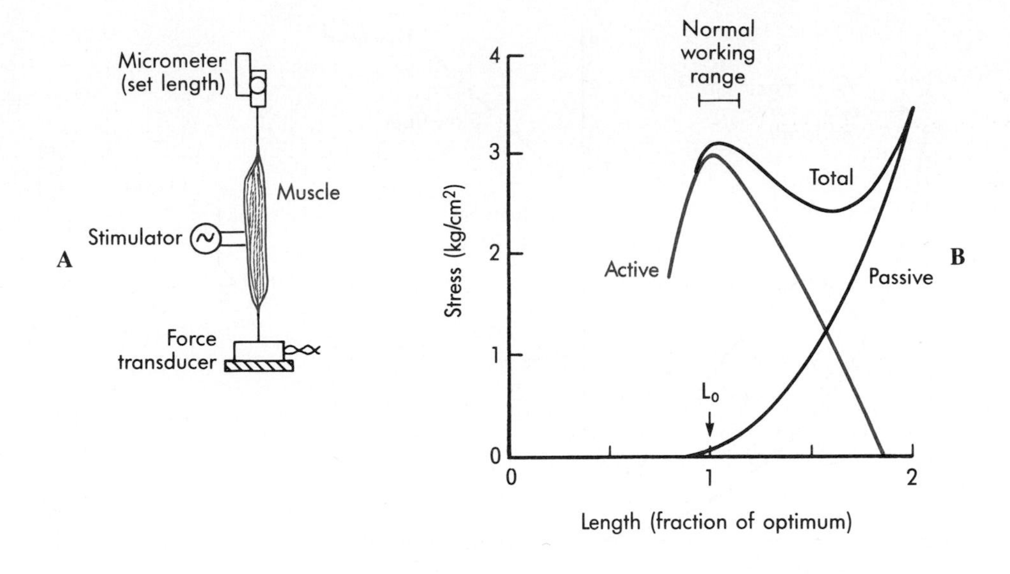

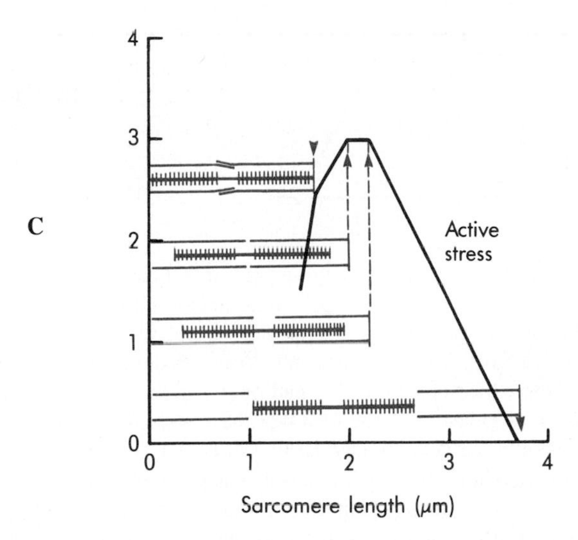

■ **Fig. 17-8** Stress-length relationships in muscle. **A,** Schematic diagram of experimental apparatus in which the tissue is attached to a micrometer to set the length and a transducer to measure force. Force is normalized as a stress, to allow comparisons of muscles of differing size (i.e., force/cross-sectional area of the muscle cells). Values of stress at different lengths are obtained and plotted as shown in **B. B,** Three stress-length curves are shown for skeletal muscle: (1) the passive stress exerted as a function of the length of the relaxed muscle, (2) the total stress exerted by the maximally stimulated muscle (passive + active), and (3) the active stress-length curve for the contractile machinery obtained as the difference between the total and passive stresses at any length *(colored curve)*. All muscles have an inherent maximal force-generating capacity, which is obtained at the optimal length (L_o). **C,** Precise studies of the stress-length behavior of the sarcomeres in single skeletal muscle cells reveal the dependence of stress on the overlap of thick and thin filaments. Diagrams of four sarcomeres at lengths where the slope of the stress-length curve changes show how filament interactions and active stress depend on sarcomere length. (Data from Gordon AM, Huxley AF, Julian FJ: *J Physiol* [London] 184:170, 1966.)

the thin filament closer to the Z line. The crossbridges must cycle asynchronously to maintain a constant force and permit continuous shortening. This is probably facilitated by the fact that the distances between successive crossbridges along a thick filament are different from the repeat distances on the thin filament.

Shortening velocities in a muscle depend on several factors. First, the velocity varies with the number of sarcomeres in a cell. Total shortening and shortening velocity are the sum of the movements of thin filaments past thick filaments times the number of half sarcomeres in the cell. Velocities can be calculated in terms of μm per second per half sarcomere to allow comparisons between different muscles. However, it is more

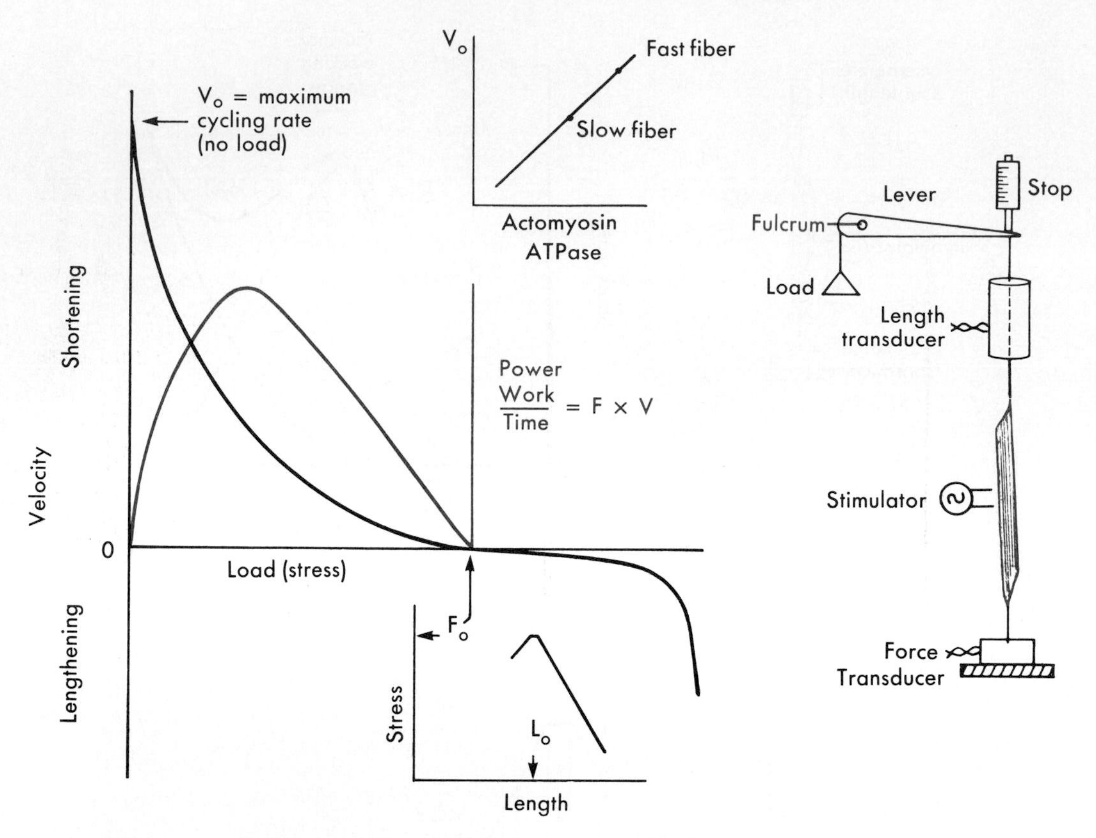

■ **Fig. 17-9** The dependence of shortening velocity on the load (stress) on a muscle. Shortening velocity may be measured using a lever system, which permits a muscle to shorten against a constant load. Velocity is measured using a length transducer (velocity = Δlength/Δtime). The velocity-load curve is constructed from points obtained in a series of contractions against different loads. (The initial length in these experiments is at L_o, and only a slight shortening is permitted.) If the load placed on the muscle is greater than the load that the active crossbridges can bear, the muscle will lengthen. Consequently the velocity-load curve can be extended to describe this situation. Insets emphasize that the maximal stress (F_o) a muscle can develop depends on the number of interacting crossbridges, whereas the maximal shortening velocity (V_o) is limited by the rate at which a particular isoenzymatic form of myosin synthesized in a cell can interact with actin and release the energy stored in ATP. The power output of a muscle *(colored curve)* is the mechanical work (force times distance shortened) per unit of time and can be calculated as the product of load times shortening velocity.

common to normalize shortening velocities by reporting values in terms of optimal muscle lengths (L_o) per second, which accounts for the number of sarcomeres in a cell. The velocity also depends on the load on the muscle; the velocity-stress relationship shows that crossbridge cycling rates fall as the stress on the crossbridges increases. This effect can be understood by referring to Fig. 17-7, *B*. The conformational change in the crossbridge, which causes shortening between attachment and detachment, is opposed by the load. Heavier loads progressively increase the average time for this to occur and allow a new cycle to take place. *An unloaded crossbridge can cycle at a maximal rate, indicated by V_o. This maximal rate depends only on the molecular properties of the myosin isoform synthesized within a cell.* The direct proportionality between the ATPase activity of myosin isolated from a cell and V_o for that cell (inset, Fig. 17-9) illustrates this molecular diversity, which is responsible for physiological differences in the speed of contraction of muscle cells from different sources.

■ *Summary*

1. The basic contractile unit is the sarcomere, which consists of a centrally located array of thick filaments that interdigitate with thin filaments attached to Z disks at each end of the sarcomere. Myofibrils contain many sarcomeres in series, and cells contain large numbers of myofibrils in parallel.

2. In vertebrate striated muscle, thin filaments consist of polymers of actin and tropomyosin, plus the Ca^{++}-binding regulatory protein, troponin. Thick filaments are composed of myosin. The "head" regions of individual myosin molecules project laterally from the filament. These projections contain the actin- and ATP-binding sites and form crossbridges.

3. A high free energy state of the crossbridges occurs after ATP binding and hydrolysis to form the myosin-ADP-P_i complex, which has a high affinity for actin. This crossbridge state rapidly binds to the thin filament in a preferred 90-degree conformation. However, release of bound P_i and ADP leads to a complex whose minimum free energy occurs after a conformational change to 45 degrees. This conformational change produces a force on the thin filament and movement toward the center of the sarcomere. ATP binding reduces the affinity of myosin for actin, the crossbridge detaches, and ATP is split to regenerate the high free energy, 90-degree myosin-ADP-P_i complex and completes the crossbridge cycle.

4. In relaxed muscles the crossbridge cycle is interrupted in the detached, high energy state by regulatory mechanisms that are controlled by Ca^{++}. The fall in cell content of ATP after death leads to the accumulation of low energy actin-myosin complexes and stiffness or muscular rigidity, termed rigor mortis.

5. The active force-length relationship reveals that force generation is a function of the number of crossbridges that can interact with the thin filaments. Force/cross-sectional area (or stress) reaches a very high value of 3 kg/cm^2 at the optimum length in isometric contractions.

6. The velocity-stress or velocity-load relationship shows that, in isotonic contractions, muscles shorten more slowly the heavier the load. Power output is maximized at 30% of the maximum force a muscle can develop. This occurs when the most free energy of ATP hydrolysis is converted into mechanical work with a maximum efficiency of 40% to 45%.

7. Contracting muscles often lengthen when opposing forces are very high. This situation is termed negative work, because work is done on, rather than by, the muscle. Stresses on muscle, tendons, or skeleton can be very high under such conditions, because crossbridges can transiently bear loads that are about 1.6-fold higher than they develop isometrically.

8. Velocity depends on the number of sarcomeres in a cell. Sarcomere-shortening velocities are a function of crossbridge cycling rates and load. Maximum cycling rates at zero load are determined by the isoform of myosin expressed in a cell.

Bibliography

Journal articles

Brenner B: Mechanical and structural approaches to correlation of cross-bridge action in muscle with actomyosin ATPase in solution, *Annu Rev Physiol* 49:655, 1987.

Eisenberg E, Hill TL: Muscle contraction and free energy transduction in biological systems, *Science* 227:999, 1985.

Emerson CP Jr: Molecular genetics of myosin, *Annu Rev Biochem* 56:695, 1987.

Epstein HF, Fischman DA: Molecular analysis of protein assembly in muscle development, *Science* 251:1039, 1991.

Goldman YE: Kinetics of the actomyosin ATPase in muscle fibers, *Annu Rev Physiol* 49:637, 1987.

Homsher E: Muscle enthalpy production and its relationship to actomyosin ATPase, *Annu Rev Physiol* 49:673, 1987.

Homsher E, Millar NC: Caged compounds and striated muscle contraction, *Annu Rev Physiol* 52:875, 1990.

Peters SE: Structure and function in vertebrate skeletal muscle, *Am Zool* 29:221, 1989.

Seow CY, Ford LE: Shortening velocity and power output of skinned muscle fibers from mammals having a 25,000-fold range of body mass, *J Gen Physiol* 97:541, 1991.

Taylor EW: Actomyosin ATPase mechanism and muscle contraction, *Prog Clin Biol Res* 315:9, 1989.

Books and monographs

Huxley A: *Reflections on muscle,* Princeton, NJ, 1980, Princeton University Press.

Needham DM: *Machina carnis: the biochemistry of muscular contraction in its historical development,* Cambridge, 1971, Cambridge University Press.

Peachey LD, Adrian RH, editors: *Handbook of physiology,* section 10, Skeletal muscle, Bethesda, Md, 1983, American Physiological Society.

Pollack GH: *Muscles and molecules: uncovering the principles of biological motion,* Seattle, Wash, 1990, Ebner & Sons.

Squire JM: *Molecular mechanisms in muscular contraction,* Boca Raton, Fl, 1990, CRC Press.

CHAPTER

18

Skeletal Muscle Physiology

Control of Skeletal Muscle

Contraction of skeletal muscles is voluntary and controlled by the motor pathways in the central nervous system (see Chapters 12-14). Each mammalian muscle cell is innervated by one branch of a motor nerve. The process of neuromuscular transmission, whereby an action potential in the motor nerve elicits an action potential in the sarcolemma, was described in Chapter 4. This chapter will focus on control at the cellular and tissue levels: (1) signal transduction at the sarcolemma that generates a transient increase in the myoplasmic Ca^{++} concentration, which is the second messenger that triggers crossbridge cycling; (2) regulation of the output of a muscle by coordinating the mechanical responses of many cells; (3) linkage of energy consumption to supply; and (4) adaptive changes in muscle associated with growth and use.

Crossbridge Regulation by Ca^{++} and Troponin

Vertebrate skeletal and cardiac muscles contain a regulatory protein, termed **troponin,** in the thin filament. One molecule of troponin is bound to the end of each tropomyosin molecule (Figs. 17-4 and 18-1). Troponin binds up to four Ca^{++} ions in a cooperative manner such that all or none of the sites are usually occupied. In the presence of micromolar concentrations of Ca^{++} the binding sites on troponin are occupied, and the conformation of the thin filament changes (Fig. 18-1). This shift allows crossbridges adjacent to that portion of the thin filament to strongly attach and cycle. In effect, *Ca^{++} binding to troponin acts as a switch that regulates the number of crossbridges that are interacting with the thin filament.*

In vertebrate skeletal and cardiac muscles the contraction-relaxation cycle involves five steps: (1) action potentials in the cell membrane cause the myoplasmic Ca^{++} to rise above 0.1 μM; (2) binding of Ca^{++} to troponin changes the thin filament conformation; (3) crossbridges strongly bind and cycle; (4) cessation of stimulation is followed by Ca^{++} removal from the myoplasm and dissociation of Ca^{++} from troponin; (5) the thin filament returns to a configuration in which further crossbridge cycles are inhibited.

The binding of Ca^{++} ions to troponin exhibits a steep response curve so that relatively small changes in Ca^{++} produce large changes in the number of active crossbridges and force development (Fig. 18-1). Each crossbridge cycle is associated with ATP hydrolysis; therefore the myofibrillar ATPase activity is also proportional to the Ca^{++} concentration.

Regulation of Cellular Ca^{++}

The myoplasmic Ca ion concentration in relaxed muscle is very low (less than 0.1 μM). This is maintained by active transport mechanisms against very large concentration gradients. *Contraction is initiated by release of Ca^{++} into the myoplasm, and relaxation follows Ca^{++} removal. The Ca^{++} involved in initiating contraction is localized in an internal compartment called the* **sarcoplasmic reticulum** (see Figs. 17-1, *C* and 17-2).

The **sarcolemma,** or plasma membrane, separates the extracellular space from the **myoplasm,** or the intracellular space that contains the contractile apparatus. The sarcolemma is an excitable membrane and propagates an action potential, which is the first event in the excitation of striated muscle cells (Chapter 3). Small **transverse tubules (T-tubules)** open to the extracellular space at the sarcolemma and form a reticulum in the interior of the cell (Fig. 18-2). This network of T-tubules around the myofibrils forms a grid across the cell at the level of the junction between the A and the I bands in each sarcomere of mammalian striated muscles (see Figs. 17-2 and 18-2). Despite the extent of T-tubule development, the volume of this system that is continuous with the extracellular fluid is only 0.1% to 0.5% of the cell volume. Depolarization of the sarcolemma spreads down the T-tubule into the interior of

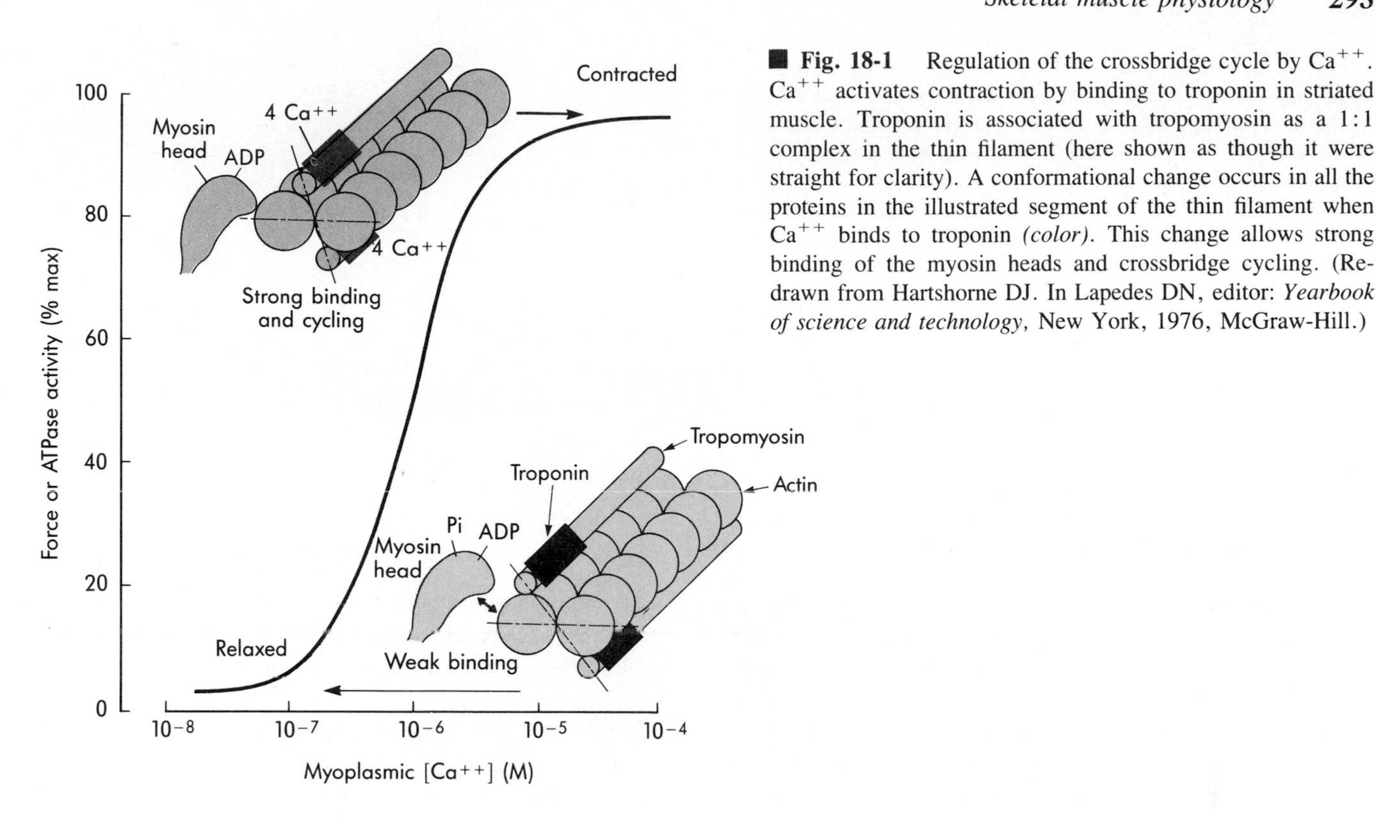

■ **Fig. 18-1** Regulation of the crossbridge cycle by Ca^{++}. Ca^{++} activates contraction by binding to troponin in striated muscle. Troponin is associated with tropomyosin as a 1:1 complex in the thin filament (here shown as though it were straight for clarity). A conformational change occurs in all the proteins in the illustrated segment of the thin filament when Ca^{++} binds to troponin *(color)*. This change allows strong binding of the myosin heads and crossbridge cycling. (Redrawn from Hartshorne DJ. In Lapedes DN, editor: *Yearbook of science and technology*, New York, 1976, McGraw-Hill.)

the cell. The T-tubules are in close association with the fenestrated sheath of sarcoplasmic reticular membranes surrounding the myofibrils (see Figs. 17-1, *C* and 17-2). This sheath consists of repeating units with (1) expanded elements (or terminal cisternae) that surround the T-tubules and with (2) narrower elements that run along the myofibrils.

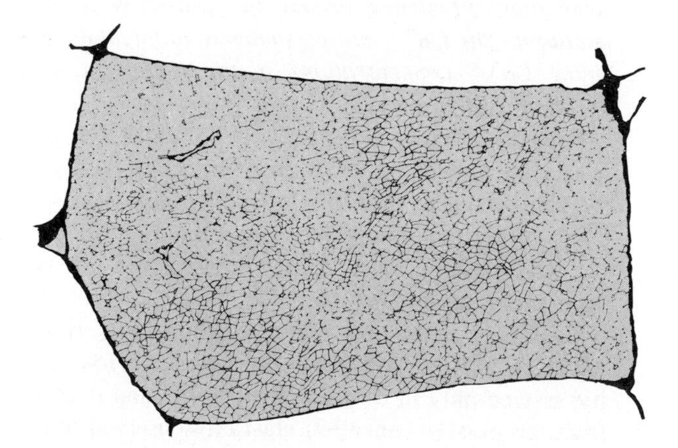

■ **Fig. 18-2** The T-tubular system in a skeletal muscle, reconstructed from high-voltage electron micrographs of serial transverse sections. The extensive network of T-tubules across the fiber has many openings to the extracellular space. (From Peachey LD, Eisenberg BR. Reproduced from the *Biophysical Journal,* 22:145, 1978, by permission of the Biophysical Society.)

■ *Ca^{++} Regulation by the Sarcoplasmic Reticulum*

The membranes of the sarcoplasmic reticulum form an intracellular compartment occupying about 1% to 5% of the volume of mammalian skeletal muscle cells, which contain large amounts of calcium. The sarcoplasmic reticular membrane is highly specialized and consists almost entirely of transport pumps that have a higher affinity for Ca^{++} than does troponin. Through the action of this active transport system, 2 moles of Ca^{++} are sequestered in the sarcoplasmic reticulum for each mole of ATP hydrolyzed. These pumps maintain the low resting myoplasmic Ca^{++} concentration. The transit of an action potential along the sarcolemma causes Ca^{++} release from the sarcoplasmic reticulum into the myoplasm. The nature of the coupling between the sarcolemma T-tubular system to induce Ca^{++} release from the sarcoplasmic reticulum is poorly understood. However, it appears to involve (1) a brief, graded depolarization of T-tubular membranes during the passage of an action potential along the sarcolemma; (2) charge-coupling, such that Ca^{++}-channels in the closely apposed sarcoplasmic reticulum membrane are opened; and (3) release of a pulse of Ca^{++} into the myoplasm. Most of the released Ca^{++} binds to troponin to initiate contraction. However, the increased myoplasmic Ca^{++} concentration also activates sarcoplasmic reticulum pumps that rapidly restore the relaxed state, unless another action potential is elicited.

The interior of the sarcoplasmic reticulum contains a

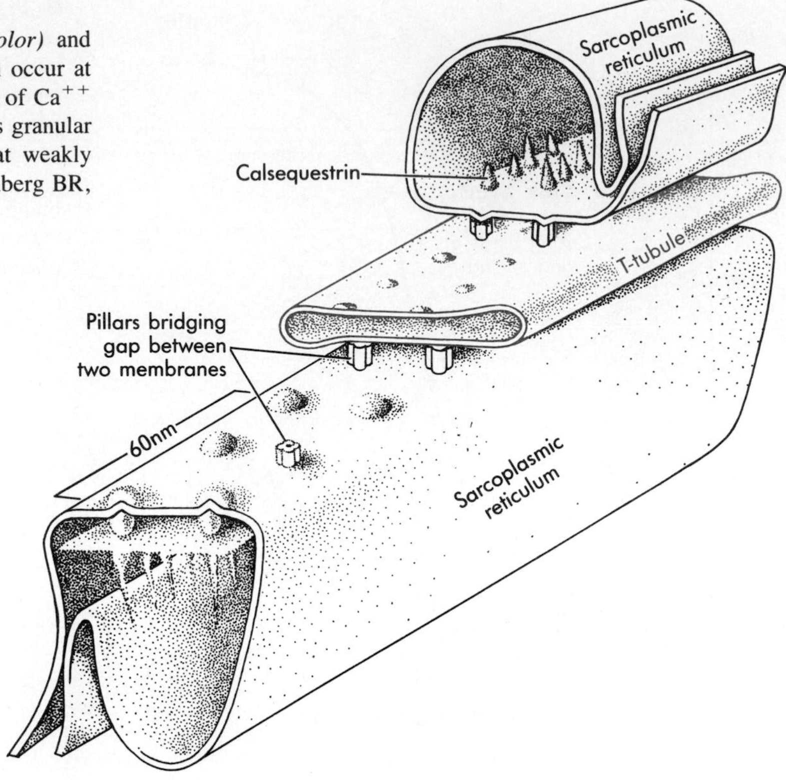

■ **Fig. 18-3** Connections between T-tubules *(color)* and the terminal cisternae of the sarcoplasmic reticulum occur at discrete sites or "pillars" that also appear to be sites of Ca^{++} release. The lumen of the terminal cisternae contains granular material believed to be calsequestrin, a protein that weakly binds large amounts of Ca^{++}. (Redrawn from Eisenberg BR, Eisenberg RS: *J Gen Physiol* 79:1, 1982.)

protein called **calsequestrin** (Fig. 18-3). Each molecule of calsequestrin can weakly bind about 43 Ca^{++} ions. Calsequestrin acts to reduce the sarcoplasmic reticular concentration of Ca^{++} from perhaps 20 mM, if all were free, to an estimated 0.5 mM. This action reduces the concentration gradient against which the pump must act.

■ *Excitation-Contraction Coupling in Skeletal Muscle*

The overall process by which depolarization of the sarcolemma causes Ca^{++} release into the myoplasm and Ca^{++} binding to regulatory sites to initiate crossbridge cycling is termed **excitation-contraction coupling** (E-C coupling). The response of the contractile system depends on the amount of Ca^{++} released into the myoplasm. Ca^{++} release from the sarcoplasmic reticulum depends on the value of the membrane potential, and it occurs when the potential becomes less negative than about −50 mV. Maximal Ca^{++} release with full activation of the cell occurs when the cell is depolarized to a level of about −20 mV (Fig. 18-4). The more negative value is termed the **mechanical threshold** and is similar whether it is achieved during a normal action potential or when the membrane is experimentally depolarized (see Fig. 18-4).

Action potentials in skeletal muscle cells are quite uniform. Hence the electrical signal (Fig. 18-4, *B*) for muscle activation is constant and leads to the release of a reproducible pulse of Ca^{++} (Fig. 18-5). A single action potential may release sufficient Ca^{++} to fully activate the contractile machinery in skeletal muscles. However, the Ca^{++} is very rapidly pumped back into the sarcoplasmic reticulum before the muscle has time to develop its maximal force (Fig. 18-5). *The resulting submaximal response to a single action potential is termed a* ***twitch.*** *Repetitive action potentials can cause summation of twitches, producing a partial or complete* ***tetanus*** *as the Ca^{++} pulses summate to maintain saturating Ca^{++} concentrations in the myoplasm* (Figs. 18-5 and 18-6).

Force development is much slower than Ca^{++} release and binding to troponin, because a number of crossbridge cycles must produce some sarcomere shortening before maximal force is registered on the skeleton. This behavior is a result of the fact that the structural elements of the muscle, including the crossbridges, are somewhat elastic and lengthen when a force is generated. This elasticity (termed **series elasticity**) is seen only in a contracting muscle and is distinct from the **passive (parallel) elasticity** observed in a relaxed muscle (Fig. 17-8), which is largely caused by the presence of connective tissue.

■ *Energy Use and Supply*

Muscular contraction demands a constant supply of ATP at a rate proportional to consumption. In this sec-

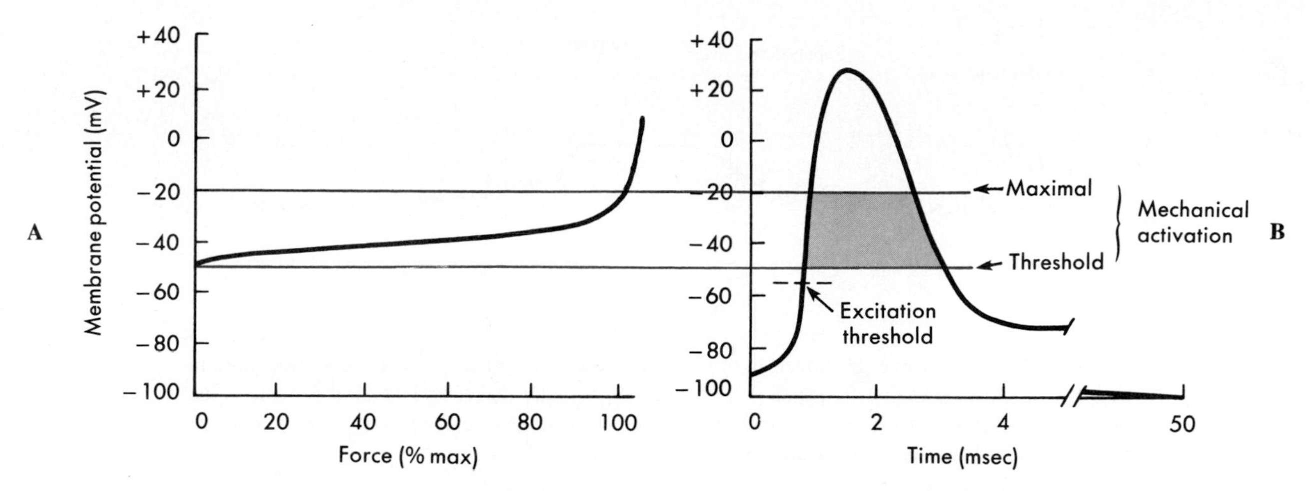

■ **Fig. 18-4** **A,** The relationship between steady state force development and membrane potential (E_m) obtained by increasing the K^+ concentration in the bathing solution. The mechanical threshold for force development occurs at an E_m of about −50 mV when Ca^{++} release from the sarcoplasmic reticulum leads to myoplasmic concentrations exceeding threshold values of about 0.1 μM. At an E_m of −20 mV, myoplasmic Ca^{++} concentrations are maximal for activation. **B,** Under physiological conditions the transit of an action potential along a muscle fiber leads to release of a Ca^{++} pulse. Ca^{++} release is proportional to the time the membrane is depolarized within the values producing mechanical activation, as indicated by the colored area in the illustrated action potential. (C Redrawn from Sandow A: *Arch Phys Med Rehabil* 45:62, 1964.)

tion the ATP requirements for various types of contractions (muscle energetics) are considered. The ways in which cells can adequately provide ATP are then reviewed. Finally, the matching of ATP production and use is described for particular fiber types. Muscle cells are specialized for specific contractile activities and a knowledge of fiber types is needed to understand many physiological aspects of muscle function in the body. Muscle makes up about 45% to 50% of the total body mass. Thus the enormous increase in energy expendi-

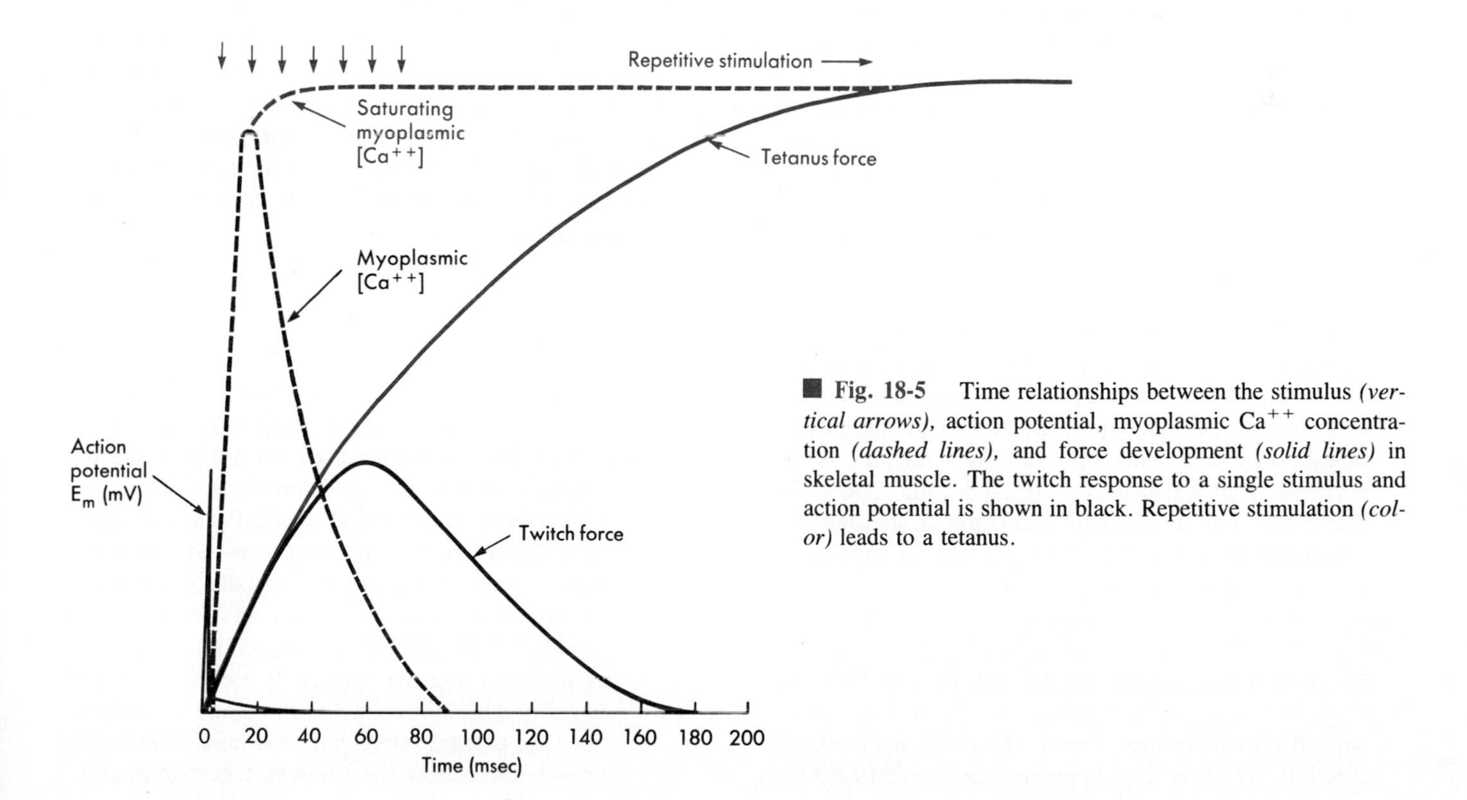

■ **Fig. 18-5** Time relationships between the stimulus *(vertical arrows),* action potential, myoplasmic Ca^{++} concentration *(dashed lines),* and force development *(solid lines)* in skeletal muscle. The twitch response to a single stimulus and action potential is shown in black. Repetitive stimulation *(color)* leads to a tetanus.

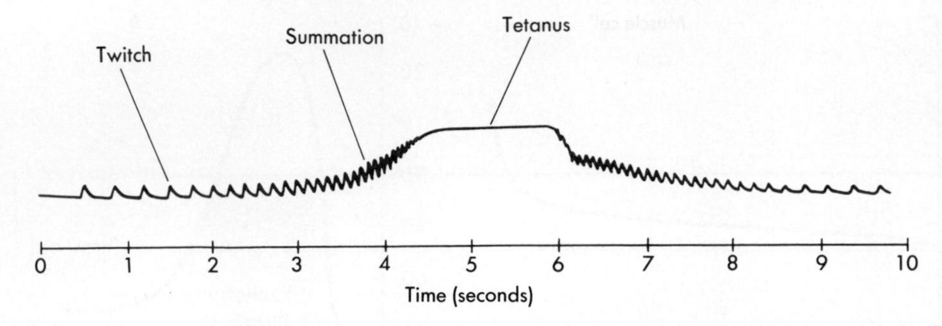

■ **Fig. 18-6** The force developed by a skeletal muscle fiber in response to repetitive stimulation at gradually increasing frequencies, followed by gradually decreasing frequencies. Individual twitch responses start to fuse and sum as the frequency increases, leading to a complete or fused tetanus. (From Buchthal F: *Dan Biol Med* 17:1, 1942.)

ture during sustained exercise dictates many of the characteristics of the cardiovascular and respiratory systems.

■ *Energetics*

The mechanical output of muscle can be readily determined in terms of work performed (force × distance shortened) or the force-time integral (force × time) in isometric contractions. Measurement of energy use by the contractile system is more difficult. Heat production can be determined very accurately and reflects ATP use. On the basis of assumptions about the total energy released during ATP hydrolysis and the efficiency of its conversion to mechanical work, energy consumption can be calculated from heat production. There are potential errors in these assumptions, and the method is not well suited for estimating rapid events. The same is true for metabolic measurements in which oxygen consumption (or lactate production when oxidative metabolism is blocked) is used to estimate ATP use. However, more difficult direct chemical measurements of ATP and creatine phosphate use (in muscles where metabolic resynthesis of these compounds is inhibited) have confirmed the basic relationships in muscular energetics.

Muscle has an ATP consumption or metabolism associated with basal cellular activities, such as maintaining ion gradients and synthesizing and degrading cellular constituents. Muscles are unusual in that this **resting metabolism** represents only a small fraction of the maximal ATP use that is associated with contraction. The energy requirements for activation, associated with action potentials and Ca^{++} release into the myoplasm, are also comparatively small.

Most energy consumption is a consequence of the fact that each complete cycle of the enormous numbers of crossbridges in a muscle cell requires one ATP molecule (see Fig. 17-7). Cycling rates in skeletal muscle depend on two factors. One is the load on the muscle (see Fig. 17-9). A muscle contracting isometrically has a low crossbridge cycling rate and a comparatively moderate rate of ATP use. This rate is reduced even more in a muscle subjected to an imposed stretch. This situation is termed **negative work,** because it is characterized by the product of force times the distance stretched. Such an imposed stretch occurs when muscles are used to decelerate the body. Crossbridge cycling rates increase when a muscle can shorten, and ATP consumption rises with the mechanical work done. Low forces and high velocities are associated with extremely high rates of ATP consumption. Shortening velocity is ultimately limited by the myosin isoenzyme present. Fast fibers, which express a form of myosin that is capable of high rates of ATP hydrolysis, have much greater requirements for rapid ATP synthesis than do slow fibers. Thus the type of myosin is the second factor that dictates metabolic requirements.

Relaxation is also associated with above-resting levels of ATP use. In part this increased ATP use provides energy for the Ca^{++} pumps in the sarcoplasmic reticulum or the sarcolemma. The creatine pool is also rephorphorylated if it were depleted during the contraction, and glycogenesis contributes to ATP use in resynthesis of glycogen stores.

■ *Metabolism*

Muscle shares the same ATP-generating mechanisms that are found in all nucleated cells (Fig. 18-7), although the relative importance of the different mechanisms varies in different muscle cell types.

1. **Direct phosphorylation** of ADP to regenerate ATP from creatine phosphate is an extremely rapid reaction. This pathway usually functions as a buffer to maintain the normal ATP levels of 3 to 5 mM at the beginning of contraction, while other systems for regenerating ATP are being turned on. Myoplasmic creatine phosphate concentrations are about 20 mM, which is only sufficient to provide the energy for a few twitches. In the sec-

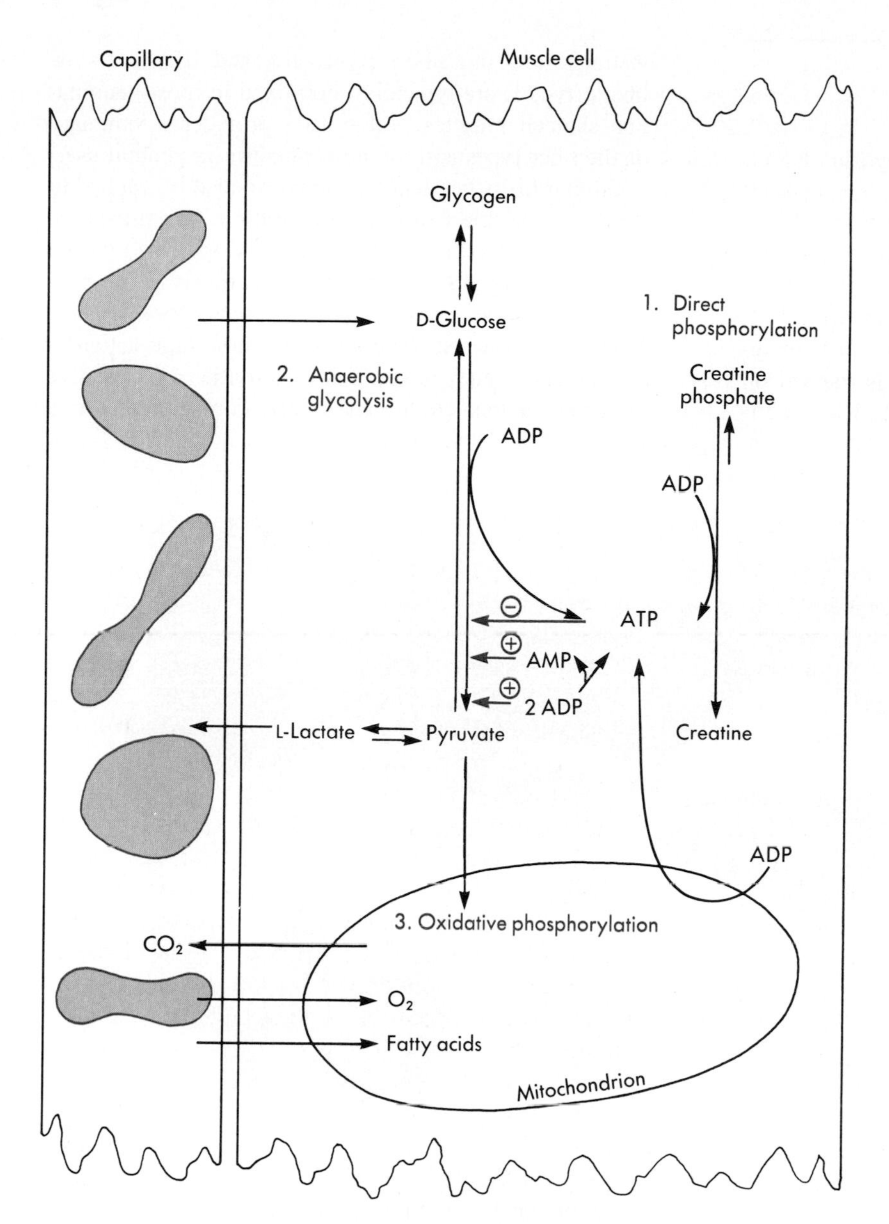

■ **Fig. 18-7** Metabolic pathways in muscle. ATP is supplied via direct phosphorylation of ADP *(1)*, glycolysis *(2)* and oxidative phosphorylation *(3)*.

ond direct phosphorylation reaction, **adenylate kinase** (often termed **myokinase** in muscle) transfers a phosphate group from one ADP to another to form ATP and AMP. This reaction has little metabolic significance, but it plays an important regulatory role in glycolysis (*colored arrows,* Fig. 18-7). Phosphofructokinase, the rate-limiting enzyme in glycolysis, is inhibited by ATP and stimulated by ADP and AMP. Thus ADP and AMP formed on activation of the contractile apparatus will stimulate glycolysis.

2. **Anaerobic glycolysis** is very rapid and readily meets the ATP demands, even of very fast muscle cells. Consequently it is important in cells of this type and in all muscle cells when the oxygen supply is inadequate. However, this pathway has a net yield of only 2 moles of ATP per mole of glucose (or 3 if the glucose is derived from cell glycogen), and it is comparatively inefficient. In the presence of oxygen, pyruvate is converted to CO_2 instead of lactate (aerobic glycolysis), and the net yield of ATP is improved threefold. ATP production by glycolysis may also be limited by the cellular stores of glycogen, which can be rapidly depleted.
3. **Oxidative phosphorylation** of fatty acids is the primary source of energy in muscles that are frequently active. Oxidative phosphorylation is not only efficient (36 moles ATP/mole glucose), but it can operate continuously when the circulation is adequate. However, oxidative phosphorylation is a slow process. It cannot meet the maximal ATP consumption rates of rapidly contracting skeletal muscle fibers unless a major fraction of the fibers consists of mitochondria in close proximity to a capillary.

■ *Matching of ATP Production and Consumption*

One of the two isoenzymatic forms of myosin characteristic of skeletal muscle cells has a higher rate of ATP hydrolysis than the other. Cells in which the "fast" myosin is synthesized have faster shortening velocities. With appropriate histochemical methods, thin sections of frozen muscles can be cut and incubated with ATP under conditions where only the slow myosin is enzymatically active. The phosphate released is trapped on the section, "staining" the slow fibers. Fig. 18-8, *A*, illustrates how fast fibers (types IIA and IIB) and slow fibers (type I) are typically intermixed in most mammalian skeletal muscles. Table 18-1 provides a summary of the fiber types and the nomenclature in general use.

Similar histochemical reactions can also be applied to serial sections cut from the same muscle to estimate the activities of enzymes in the oxidative (Fig. 18-8, *B*) and glycolytic (Fig. 18-8, *C*) metabolic pathways. The metabolic capacities in muscle fibers vary considerably. However, most fast (type II) fibers show high activities of glycolytic enzymes and low activities of oxidative enzymes, a fact confirmed by the comparatively few

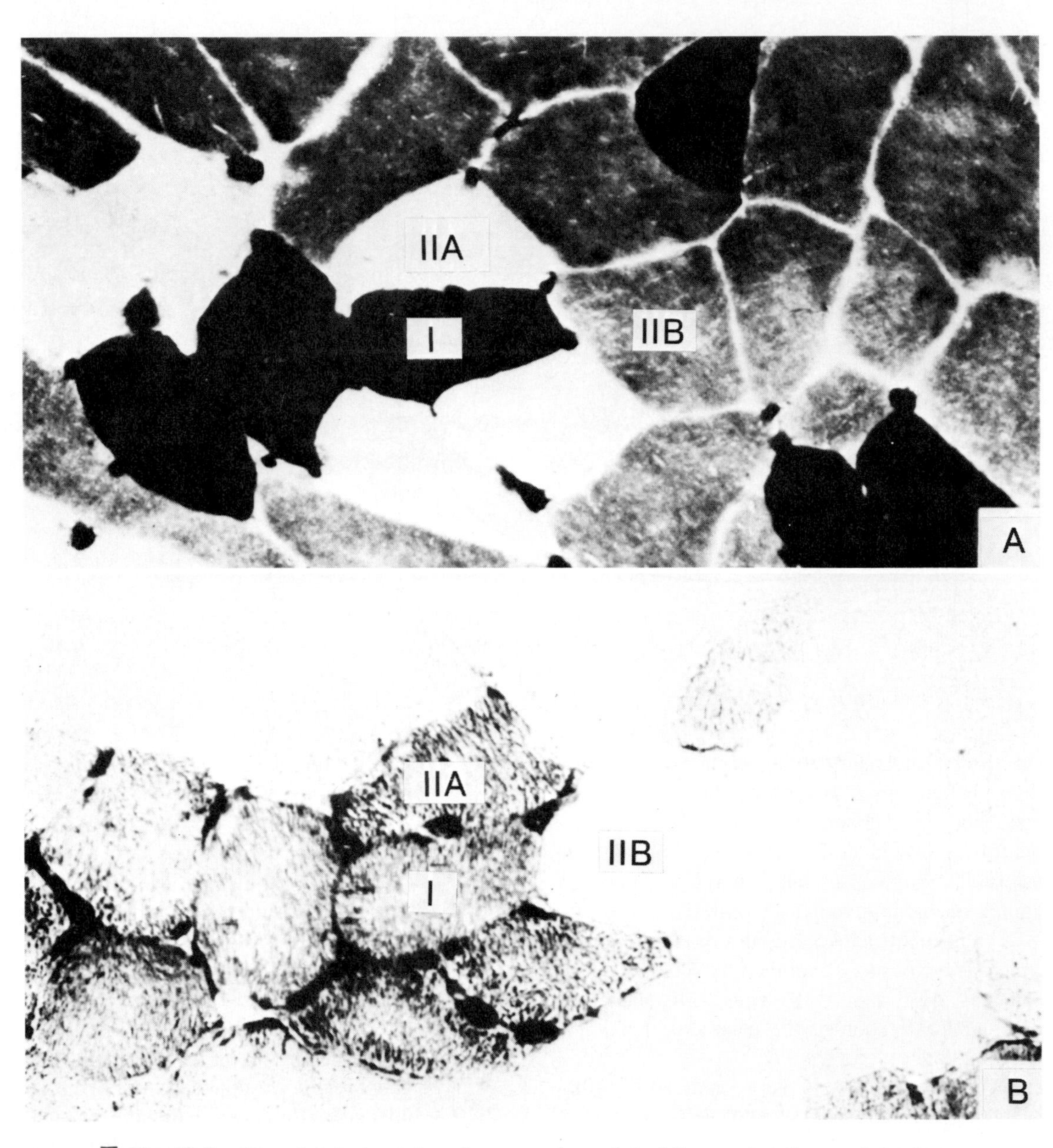

■ **Fig. 18-8** Histochemical staining of cross sections of the feline semitendinosus skeletal muscle. **A,** Staining for the ATPase activity of the slow myosin isoenzyme of type I fibers. Type IIB fibers stain more than type IIA fibers in this species, and the myosins in these two types of fast fibers may differ. **B,** Staining for succinic dehydrogenase activity, an enzyme associated with oxidative phosphorylation.

■ **Table 18-1** Basic classification of skeletal muscle fiber types

	Type I: slow oxidative (red)	*Type IIB: fast glycolytic (white)*	*Type IIA*: fast oxidative (red)*
Myosin isoenzyme (ATPase rate)	Slow	Fast	Fast
Sarcoplasmic reticular Ca^{++} pumping capacity	Moderate	High	High
Diameter (diffusion distance)	Moderate	Large	Small
Oxidative capacity: mitochondrial content, capillary density, myoglobin	High	Low	Very high
Glycolytic capacity	Moderate	High	High

*Comparatively infrequent in humans and other primates. In the text a simple designation of type II fiber refers to a fast-glycolytic (type IIB) fiber.

mitochondria that can be observed in electron micrographs. Fast fibers with high contraction velocities have a much more extensive sarcoplasmic reticulum with high pumping rates to quickly activate and inactivate the contractile machinery than do slow fibers. *Because the ATP consumption rates in the myofibrils and sarcoplasmic reticulum of such fast fibers can most readily be matched by glycolysis, a metabolism based primarily on glycolysis is appropriate, provided that the fibers are recruited only occasionally for brief efforts.* This is true, as is shown below. Theoretically a fiber of small diameter with very high mitochondrial and surrounding capillary densities could oxidatively synthesize ATP at rates used by the fast myosin isoenzyme. Some fast fibers with high glycolytic and oxidative capacities are found in mammals and have led to the subclassification of fast, type II fibers shown in Table 18-1.

Slow, type I fibers can meet relatively modest metabolic demands with oxidative phosphorylation. Molecules associated with oxygen binding (e.g., hemoglobin, myoglobin, cytochromes) contain iron and are red. Therefore the red color of an oxidative muscle cell or a tissue containing mostly type I (or IIA) fibers has led to the designation of such types as **red fibers.**

Cardiac muscle resembles slow skeletal muscle in expressing myosin isoenzymes with low ATPase activities. As might be expected, this continuously active muscle is almost entirely oxidative and is highly sensitive to interruptions of its blood supply.

The oxidation of fatty acids provides most of the ATP used by the muscles in the body. However, all muscles have a significant capacity to use other substrates, including carbohydrates, certain amino acids, and ketone bodies.

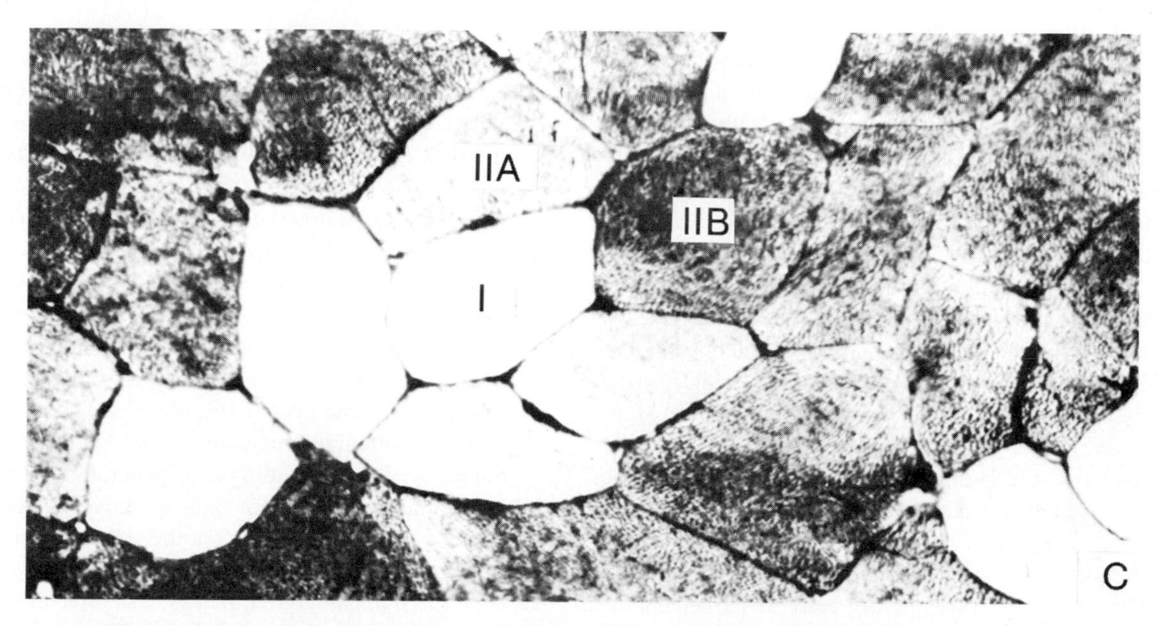

■ **Fig. 18-8, cont'd** **C,** PAS stain indicating glycolytic capacity. Three distinct fiber types can be differentiated in these serial sections: slow oxidative or type I, fast glycolytic or type IIB, and fast oxidative (and glycolytic) or type IIA fibers. (×450.) (From Hoppeler H et al: *Respir Physiol* 44:94, 1981.)

■ *Musculoskeletal Relationships*

Individual muscle cells are encased in a connective tissue layer called the **endomysium.** Groups of skeletal muscle cells form *fascicles* that are bounded by the **perimysium.** These fascicles are grouped into the definitive muscle, which is covered by the **epimysium.** The three connective tissue layers are composed primarily of elastin and collagen fibrils. An estimated 250 million cells are found in the more than 400 skeletal muscles in humans. Each of these muscles exerts specific movements via tendons that are attached to the skeleton in most cases.

The output of a muscle depends on the size of the muscle's cells and their anatomical arrangement. Increasing the diameter of a fiber by synthesis of new myofibrils **(hypertrophy)** will increase the force-generating capacity of a cell (Fig. 18-9). The formation of more cells **(hyperplasia)** also increases tissue force output. However, differentiated skeletal muscle has only a limited capacity to form new cells. The force-generating capacity is unchanged if the length of the cells is increased by adding more sarcomeres in series without increasing the cross-sectional area of the cells. However, the absolute velocity of contraction and the total shortening capacity of the cell increase with the addition of more sarcomeres (see Fig. 18-9).

The output of muscles not only depends on the number of sarcomeres and myofibrils in the cells but also on the orientation of the cells within the tissue. Measurements on intact muscles directly reflect cellular properties only when the cells are arranged parallel with each other and also with the axes of the tendons that transmit the force (Fig. 18-10). This parallel arrangement maximizes shortening capacity and velocity for a muscle. However, force-generating capacity can be enhanced at the expense of shortening capacity and velocity by arrangements placing the muscle fibers at an angle to the tendons. Examples of such **pennate** (featherlike) arrangements are illustrated in Fig. 18-10.

Some skeletal muscles, including those surrounding the mouth and anus, serve as sphincters. Striated muscles are also found in the upper portions of the esophagus, and they are active during swallowing. Others may be attached to the skin. However, most skeletal muscles are attached to the skeleton. The attachments involve the highly inextensible protein collagen, which forms tendons, or flattened sheets termed aponeuroses. Muscles are described in terms of an origin at the proximal or relatively fixed point and an insertion on the bone that is moved. (The distinction between origin and insertion is sometimes arbitrary.) Muscles bridge one or, more frequently, two joints. The contraction of an individual muscle can consequently lead to movement of more than one bone.

Specific coordinated movements require the actions of two or more muscles. These may be **synergists,** when the muscles act together, or **antagonists,** when the actions of the muscles are opposed. **Kinesiology** is the study of the interactions of groups of muscles. The complex nature of coordinated movements has been revealed by electromyographic techniques, in which the summed electrical activity of a muscle is detected from electrodes placed on the overlying skin. Another technique can detect the activity of a single cell with needle electrodes inserted into the muscle. These techniques reveal that a specific movement may involve contractions of antagonistic muscles and may not involve contractions of some synergistic muscles. The activity patterns are often unpredictable and can only be determined by direct recordings. In general, the interactions of the muscles serve not only to move a specific bone but also to fix or stabilize another bone or joint.

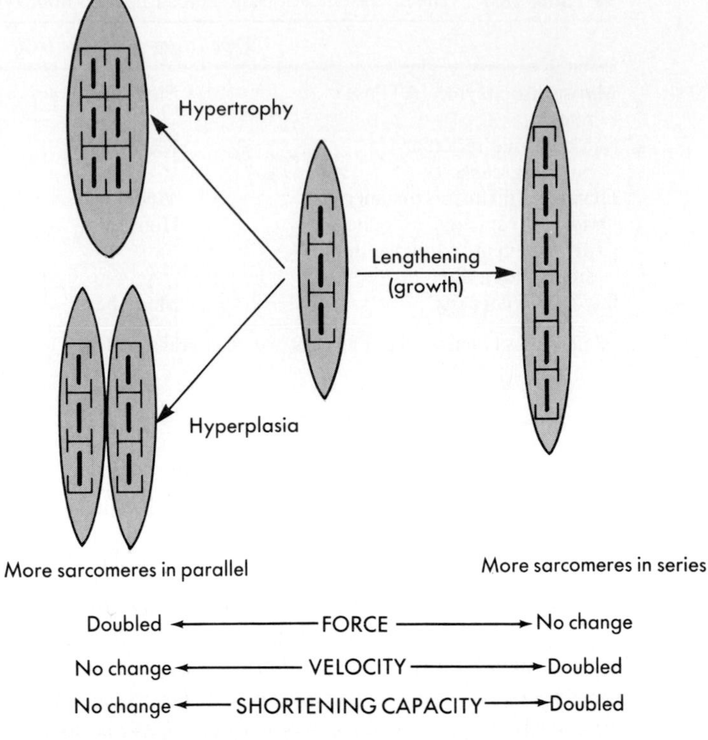

■ **Fig. 18-9** Effects of growth on the mechanical output of a muscle cell. Growth may consist of adding new myofibrils (depicted as a series of model sarcomeres) within a cell (hypertrophy), formation of new cells (hyperplasia), or adding more sarcomeres in series as the muscle cells lengthen along with skeletal growth. The effects of the illustrated cell growth on the absolute force (newtons), shortening velocity (m/sec), and shortening capacity (m) of the muscle are summarized in the table.

The skeleton also acts as a lever system with significant mechanical consequences. As illustrated in Fig. 18-11, the biceps muscle of a person holding a 20-kg weight at approximately right angles must develop an isometric force of 140 kg. Enormous forces can be placed on tendons by large muscles such as the gastroc-

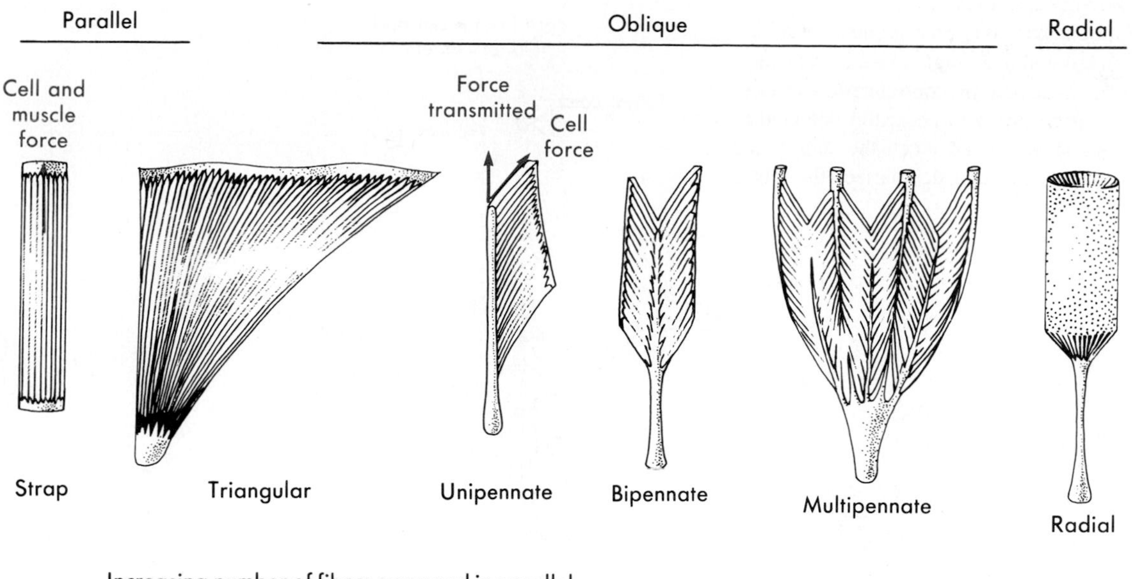

■ **Fig. 18-10** Some arrangements of skeletal muscle fibers. The force generated by all of the individual cells is fully transmitted to the skeleton in only a few straplike muscles. The cells in most muscles are arranged at an angle to the axis of the muscle. This allows more fibers to be attached to a tendon and increases the total force-generating capacity. However, not all of the force generated by each cell is usefully transmitted to the tendon with an oblique geometry, and the overall shortening velocity and shortening capacity of the muscle are less than that of the individual cells. (Redrawn from *Gray's anatomy,* ed 35 [British], Philadelphia, 1973, WB Saunders.)

nemius. This can lead to rupture of the tendon. The greatest stress on a tendon occurs upon gravitational loading of a contracting muscle because the cross-bridges can bear more force than they can develop. Injuries usually occur as the result of a fall, which subjects a contracting muscle to a sudden large increase in stress. Another consequence of the skeletal lever system is that large movements of the limbs can occur with far less shortening of the muscles (Fig. 18-11). With this arrangement sarcomere lengths remain close to their optimum for force development in most movements (see Fig. 17-8).

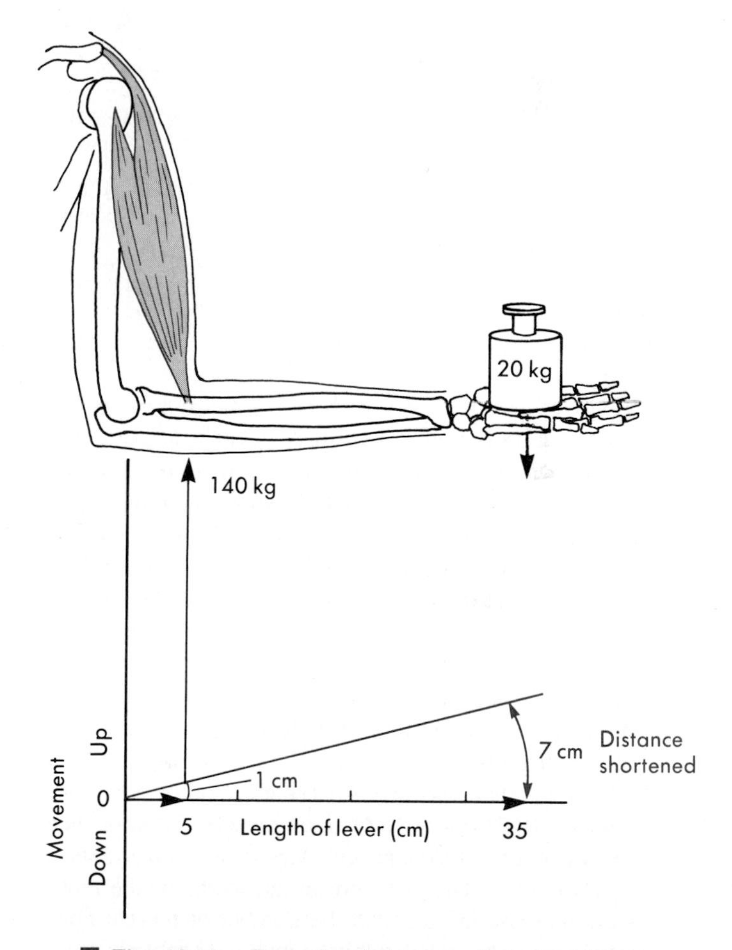

■ **Fig. 18-11** Example of the musculoskeletal lever system. The biceps muscle operates at a 1:7 mechanical disadvantage in this situation and must generate high forces to support a weight on the hand. However, little shortening of the biceps muscle is required to produce large displacements of the hand. (Redrawn from Guyton AC: *Textbook of medical physiology,* ed 6, Philadelphia, 1981, WB Saunders.)

■ *Coordination of Muscular Activity*

■ *Motor Nerves and Motor Units*

The cell bodies of the motor nerves (α-motor axons) are in the ventral horn of the spinal cord (Fig. 18-12). The axon exits via the ventral root and reaches the muscle through a mixed peripheral nerve. The motor nerves branch in the muscle, with each terminal innervating a single muscle cell in mammals. The specialized cholin-

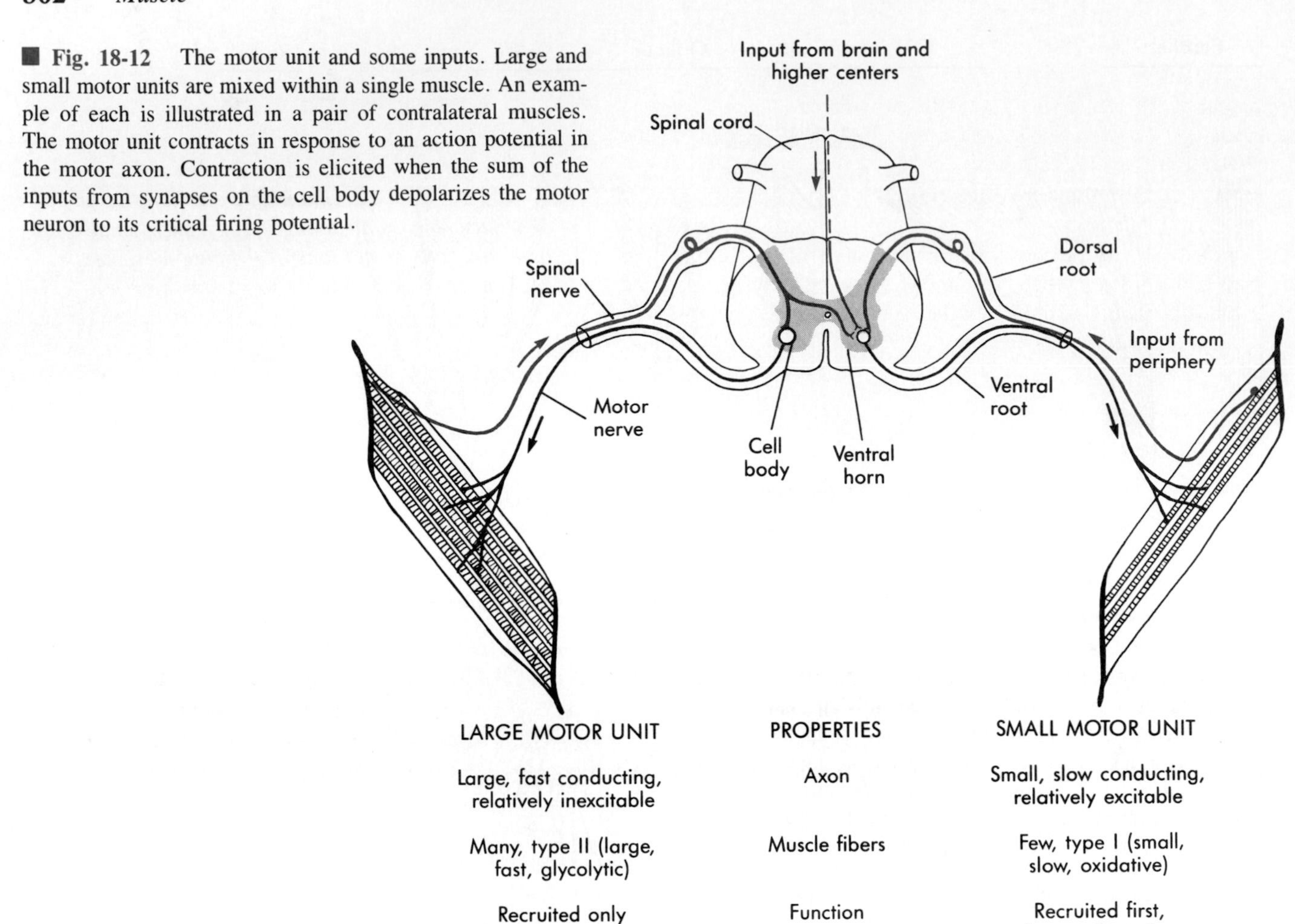

■ **Fig. 18-12** The motor unit and some inputs. Large and small motor units are mixed within a single muscle. An example of each is illustrated in a pair of contralateral muscles. The motor unit contracts in response to an action potential in the motor axon. Contraction is elicited when the sum of the inputs from synapses on the cell body depolarizes the motor neuron to its critical firing potential.

ergic synapse that forms the neuromuscular junction and the neuromuscular transmission process that generates an action potential in the muscle fiber are described in Chapter 4. A **motor unit** consists of the motor nerve and all of the muscle fibers innervated by that nerve. The motor unit (and not the individual cell) is the functional contractile unit because all of the cells within a motor unit contract synchronously when the motor nerve fires. The muscle cells of a motor unit are not segregated anatomically into distinct groups, and considerable intermixture of cells occurs among neighboring motor units.

Motor units exhibit considerable specialization. They may consist of only two or three muscle fibers, or there may be over a thousand cells in some motor units in large muscles. The cell bodies and axons of the motor nerve increase in size with the number of muscle fibers in the unit. This relationship is understandable in terms of the metabolic requirements for synthesis and release of acetylcholine. A distinction between slow oxidative (type I) and fast glycolytic (type II) muscle fibers is shown in Table 18-1. This classification also applies to motor units because all the fibers in one motor unit are of the same type (Table 18-2). The smaller motor units normally consist of type I cells (Table 18-2 and Fig. 18-12). An important point is that only the contraction velocity or myosin ATPase activity clearly distinguishes the fiber types. These two parameters reflect the presence of different isoenzymatic forms of myosin in the cells. Metabolic differences arise from different cellular contents of mitochondria, for example, and may vary considerably between cells of the same type.

■ **Table 18-2** Properties of motor units

	Motor unit classification	
Characteristics	***Type I***	***Type II***
Properties of nerve		
Cell diameter	Small	Large
Conduction velocity	Fast	Very fast
Excitability	High	Low
Properties of muscle cells		
Number of fibers	Few	Many
Fiber diameter	Moderate	Large
Force of unit	Low	High
Metabolic profile	Oxidative	Glycolytic
Contraction velocity	Moderate	Fast
Fatigability	Low	High

The inputs to the motor nerve are both excitatory and inhibitory. The inputs involve (1) neurons from the brain, (2) neurons from elsewhere in the spinal cord, and (3) neurons originating from a variety of receptors within a muscle, from its antagonists and synergists, and from the same group of contralateral muscles (Fig. 18-12). These inputs to the motor neuron and their integrated activity are described in Chapters 12-14. A motor neuron fires when the sum of the excitatory and inhibitory inputs depolarizes the cell to its critical membrane potential.

Recruitment of Motor Units

The functional importance of the variations among motor units can be illustrated by considering how the motor units in a muscle are progressively recruited in graded contractions. Increasing excitatory or decreasing inhibitory input to the motor neuron pool in the ventral horn will depolarize the cell bodies. However, a given level of excitatory input will produce more depolarization of the smallest neurons because of their smaller membrane areas. Thus the first axons to fire are those of the smallest motor units. The conduction velocity of the action potential will be relatively low, reflecting the cable properties of the axons of small diameter. The total force developed by the muscle will be small because there are only a few cells of moderate diameter in these units.

Fig. 18-13 shows the basic relationships between motor unit recruitment and total force generated by a muscle. In the left column 31 indicates the number of motor units among those sampled in the muscle that developed up to 10 g force (average, 5 g) when tetanized. If all 31 motor units fired together, the total force generated would be about 0.15 kg, assuming an average force of 5 g/unit. The important point is that *these units are recruited first, and they remain active as long as any part of the muscle is contracting. Many such small motor units permit fine gradations of delicate movements.*

With increasing excitatory input somewhat larger axons fire. In this sample there were fewer motor units that could generate between 10 and 20 g force, but their summed contributions to the total force in the muscle were comparable, i.e., 10 units × 15 g (average force) = 0.15 kg, shown by the height of the colored column. The total force generated by the muscle with the increased level of excitatory input is now the sum of 31 units averaging 5 g force plus 10 units averaging 15 g force, or 0.31 kg. A smooth gradation in contractile force is still maintained because the percent increase in force produced by a larger motor unit remains small when added to the force already being generated. Fig. 18-13 shows that smaller numbers of larger motor units are successively recruited until all motor units are contracting. There is some evidence that the largest motor units are so inexcitable that most persons cannot recruit them voluntarily. They may account for the exceptional displays of strength exhibited by persons under stress when increased excitatory activity occurs in the central nervous system.

The pattern of increasing force illustrated in Fig. 18-13 is a consequence of recruitment by the **size principle.** Recruitment by size and the simultaneous increase in firing rates, which allows each unit to increase

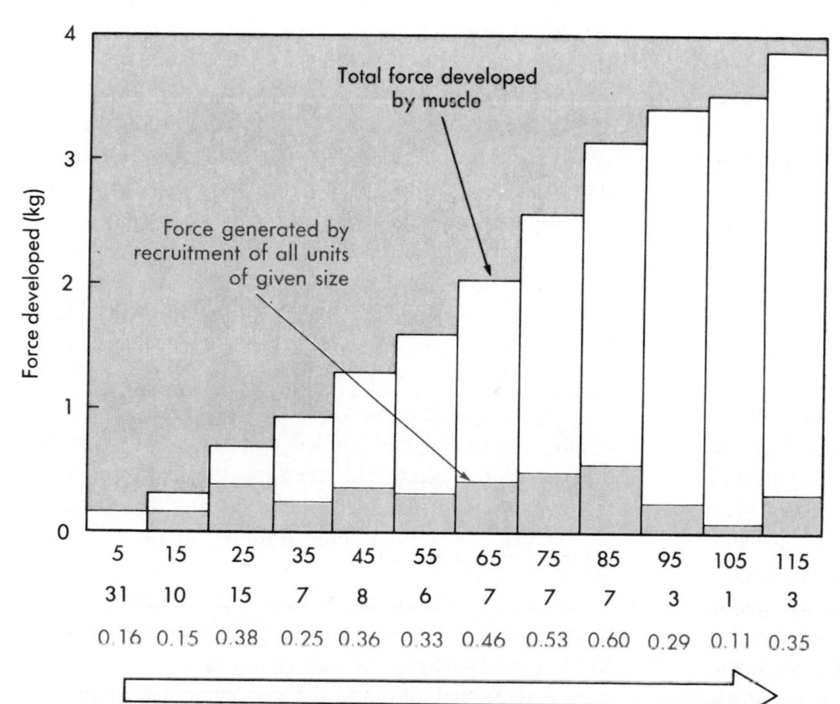

Fig. 18-13 Relationships between the size of a motor unit, the number of motor units of different size classes in a muscle, the contributions of each class of motor units to force development *(colored numbers)*, and the order of recruitment of motor units of different sizes. See text for further explanations. (Data from Henneman E, Olson CB: *J Neurophysiol* 28:581, 1965.)

its force by tetanization, are responsible for gradations in contractile force. Because the larger motor units are also the faster units, whole muscle contraction velocities can be increased by recruiting more motor units. Recruiting more units also contributes to the increase in speed by reducing the effective load on each cell, allowing faster crossbridge cycling rates.

The size principle also explains the metabolic profile pattern of the motor units. *Highly oxidative units are those that are used most. Maximal efforts, in which fast motor units are recruited, cannot be sustained because of the rapid depletion of glycogen.* This intracellular source of energy is needed to supply the glycolytic pathway, which is the principal way in which the high rates of ATP consumption can be met.

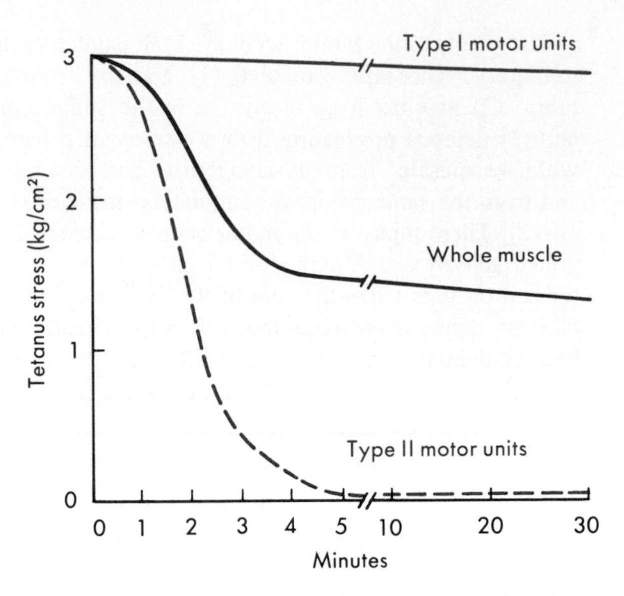

■ **Fig. 18-14** Fatigue of a skeletal muscle in which half of the cross-sectional area is composed of slow, oxidative type I motor units and the other half of fast, glycolytic type II motor units. The muscle was briefly tetanized once every second by stimulation of its motor nerve in situ. With this regimen type II motor units exhibit rapid cellular fatigue and failure to contract, whereas type I motor units maintain almost normal contractile responses.

■ *Tone in Skeletal Muscle*

The skeletal system supports the body mass efficiently when the posture is normal. The amount of energy expended for the muscular contraction that is required to maintain a standing posture is remarkably small. However, muscles normally exhibit some level of contractile activity. Isolated, unstimulated muscles are relaxed and quite flaccid. "Relaxed" muscles in the body are comparatively firm. This **tone** is apparently a result of low levels of contractile activity in some motor units driven by reflex arcs from receptors in the muscles; tone is abolished by dorsal root section. Tone in skeletal muscles should be distinguished from "tone" in vascular smooth muscles. The latter refers to the normal tonic or continuous contraction of all muscle cells that maintains vascular resistance.

■ *Fatigue*

Remarkably little is known about the factors responsible for fatigue. Fatigue may potentially occur in any of the steps involved in muscular contraction, from the brain to the muscle cells, as well as in the systems involved in maintaining energy supplies, including cardiovascular and respiratory functions.

Cellular fatigue. Fatigue of motor units can be assessed experimentally by recording the maximal stress maintained during prolonged contraction or during a series of brief tetani elicited by direct stimulation of the motor nerve to the muscle. The latter regimen allows adequate perfusion when the circulation is intact. Tetanic stress decays rapidly to a level that can be maintained for long periods (Fig. 18-14). This decay represents the rapid and almost total failure of fast motor units. The decline in tetanic stress is paralleled by glycogen and creatine phosphate depletion and by lactic acid production. This implies that ATP depletion leads to the failure of contraction. However, the decline in stress occurs when the ATP pool is not greatly reduced, and the muscle fibers do not go into rigor. Slow motor units, which can meet the energy demands of the stimulus regimen, do not exhibit significant fatigue for many hours. Evidently some factor associated with energy metabolism can inhibit contraction, but this factor has not been clearly identified. There is some evidence that **neuromuscular fatigue** can occur in the largest fast motor units, in which the ability of the motor nerve to synthesize and release acetylcholine may be limiting.

General fatigue. Most persons tire and cease exercise long before there is any motor unit fatigue of the type illustrated in Fig. 18-14. ***General physical fatigue*** *may be defined as a state of disturbed homeostasis produced by work. The basis for the perceived discomfort or even pain probably involves many factors.* These factors may include a lowering of plasma glucose levels and accumulation of metabolites. Motor system function in the central nervous system is not impaired. Highly motivated and trained athletes are able to bear the discomfort and will exercise to the point where some motor unit fatigue occurs. Part of the enhanced performance observed after training involves motivational factors.

Recovery. Muscle blood flow and oxygen uptake remain elevated for some time after exercise. **Oxygen debt** (Fig. 18-15) is the excess amount of oxygen consumed over that required for resting metabolism when energy use by the contractile system has ceased. Some

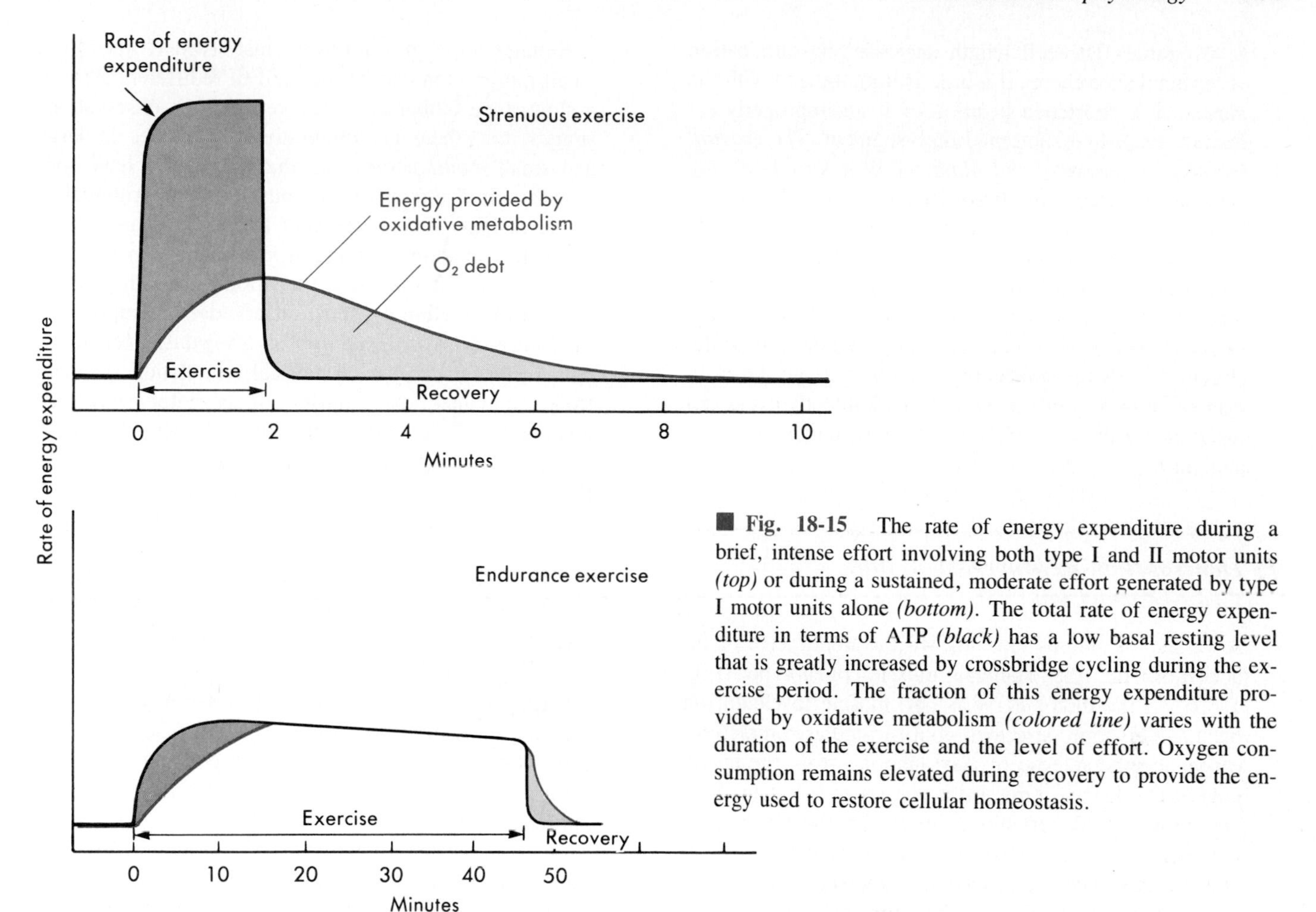

■ **Fig. 18-15** The rate of energy expenditure during a brief, intense effort involving both type I and II motor units *(top)* or during a sustained, moderate effort generated by type I motor units alone *(bottom)*. The total rate of energy expenditure in terms of ATP *(black)* has a low basal resting level that is greatly increased by crossbridge cycling during the exercise period. The fraction of this energy expenditure provided by oxidative metabolism *(colored line)* varies with the duration of the exercise and the level of effort. Oxygen consumption remains elevated during recovery to provide the energy used to restore cellular homeostasis.

oxygen debt occurs even at low levels of exercise, because slow oxidative motor units consume considerable ATP (derived from creatine phosphate or glycolysis) before oxidative metabolism can increase ATP production to the steady-state requirements. The oxygen debt is much greater in strenuous exercise, when fast glycolytic motor units are used. The oxygen debt is approximately equal to the energy consumed during exercise minus that supplied by oxidative metabolism (i.e., the dark- and light-colored areas in Fig. 18-15 are roughly equal). The additional oxygen used during recovery represents the energy requirements for restoring normal cellular metabolite levels.

■ *Trophic Responses of Skeletal Muscle*

Trophic factors are responsible for the long-term development and maintenance of specific characteristics of a tissue. *Skeletal muscle exhibits considerable plasticity (variability in the phenotype or properties of the cells)* as shown by a variety of experimental studies.

■ *Growth and Development*

Skeletal muscle fibers differentiate before innervation (neuromuscular junctions may be formed well after birth). Before innervation the fibers physiologically resemble slow, type I cells. These uninnervated fibers have acetylcholine receptors distributed throughout the sarcolemma and are supersensitive to that neurotransmitter. An endplate is formed when the first growing nerve terminal establishes contact. The fiber forms no further association with nerves, and the receptors to acetylcholine become concentrated in the endplate membranes. Fibers that are innervated by a small motor neuron form slow oxidative motor units. Fibers innervated by large motor nerves develop all the characteristics of fast, type II motor units. Thus innervation produces major changes in the cells, including the synthesis of the fast and slow myosin isoenzymes, which replace the embryonic variant.

An increase in muscle strength and size occurs during maturation. The cells must lengthen with skeletal growth. This is accomplished by formation of additional sarcomeres at the ends of the cells. This process is reversible. The cell length decreases by elimination of terminal sarcomeres if a limb is immobilized with the muscle in a shortened position or if an improperly set fracture leads to a shortened limb segment. *The gradual increase in strength and diameter of a muscle during growth is achieved mainly by hypertrophy (Fig. 18-9). Skeletal muscles have a limited capacity to form new fibers (hyperplasia) by differentiation of satellite cells that are present in the tissues.* Injured cell segments that contain nuclei can grow and fuse with other segments to regenerate a cell. However, major cellular destruction leads to replacement with scar tissue. Overall, skeletal muscle cells exhibit remarkable dynamic adjustments of their morphology to the demands of the organism.

■ *Denervation, Reinnervation, and Cross-innervation*

A variety of studies have shown the importance of innervation for the skeletal muscle phenotype (Fig. 18-16). If the motor nerve is cut, muscle **fasciculation** occurs. This term describes small, irregular contractions caused by the release of acetylcholine from the terminals of the degenerating distal portion of the axon. Several days after denervation, muscle **fibrillation** begins. Fibrillation is characterized by spontaneous, repetitive contractions. These originate after the spread of cholinergic receptors occurs over the entire cell membrane (a return to the preinnervation embryonic characteristic) and reflect a supersensitivity to acetylcholine. **Atrophy** also occurs, with a decrease in the size of the muscle and its cells. Atrophy is progressive in humans, with degeneration of some cells after 3 or 4 months. Most of the muscle fibers are replaced by fat and connective tissue after 1 to 2 years. These changes are all reversed if reinnervation occurs within a few months. Reinnervation is normally achieved by growth of the peripheral stump of the motor nerve axons along the old nerve sheath.

Reinnervation of a formerly fast, type II fiber by a small motor axon causes that cell to redifferentiate into a slow, type I fiber and vice versa. Such observations suggest that there are qualitative differences in large and small motor nerves and that the nerves have specific "trophic" effects on the muscle fibers. Although it is increasingly recognized that nerve terminals release substances in addition to the primary neurotransmitter at synapses, an alternative explanation of the trophic effects of innervation fits most observations. Simply, *it is the frequency of contraction that determines fiber development and phenotype*. Electrical stimulation via electrodes implanted in the muscle or its motor nerves can ameliorate denervation atrophy. More strikingly, chronic low-frequency stimulation of fast motor units (by a pacemaker with electrodes placed on their motor nerves) causes fast motor units to be converted to slow units. Some shifts toward a typical fast-fiber phenotype can be observed when the frequency of contraction in slow units is greatly decreased by reducing the excitatory input in the ventral horn. This may be achieved through appropriate spinal or dorsal root section or by severing the tendon, which functionally inactivates peripheral mechanoreceptors (Fig. 18-16) (Chapter 12). In brief, fibers that undergo frequent contractile activity form many mitochondria and synthesize the slow myosin isoenzyme. Fibers innervated by large, inexcitable axons contract infrequently. Such relatively inactive fibers form few mitochondria and have large concentrations of glycolytic enzymes. The fast myosin isoenzyme is synthesized in such cells.

■ *Response to Exercise*

Exercise physiologists identify three categories of training regimens and responses (Table 18-3). In practice, most athletic endeavors involve elements of all three. The learning aspect can involve motivational factors, as well as neuromuscular coordination. This aspect of

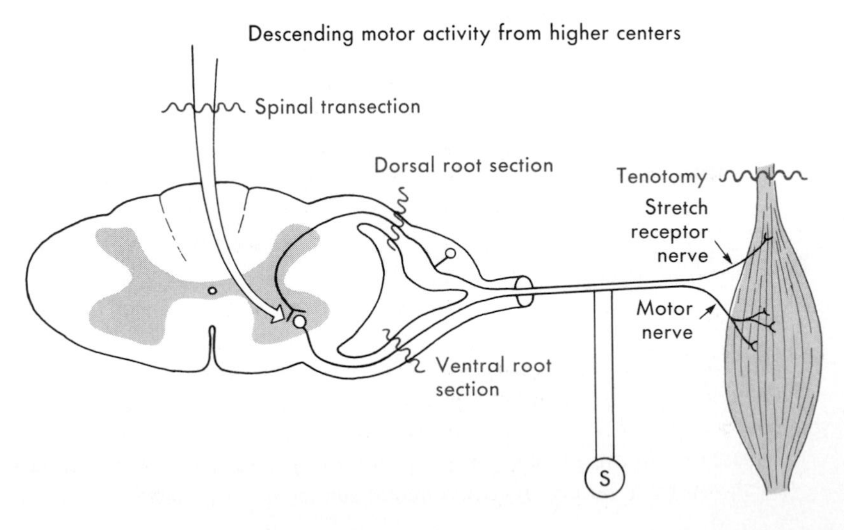

■ **Fig. 18-16** Experimental or pathological situations that modify the phenotype of skeletal muscles. Normal contractile activity in the muscle can be abolished by ventral root section that cuts the motor nerves (peripheral denervation). In a muscle with normal innervation, contractile activity can be increased by pacemakers *(S)* implanted on the motor nerves. Activity can be decreased by blocking excitatory pathways from higher centers (spinal transection) or by interrupting the reflex arcs that modify muscle activity through dorsal root section. If the muscle tendon is severed (tenotomy), stretch receptors within the tendon and muscle become inactive. Tenotomy also decreases excitatory input to the motor nerves.

Table 18-3 Effects of exercise

Type of training	*Example*	*Major adaptive response*
Learning/coordination	Typing	Increases the rate and accuracy of motor skills (central nervous system)
Endurance (submaximal, sustained efforts)	Marathon running	Increased oxidative capacity in all involved motor units with limited cellular hypertrophy
Strength (brief, maximal efforts)	Weight lifting	Hypertrophy and enhanced glycolytic capacity of motor units employed

training does not involve adaptive changes in the muscle fibers per se. However, motor skills can persist for years without regular training, unlike the responses of muscle cells to exercise.

All healthy persons can maintain some level of continuous muscular activity that is supported by oxidative metabolism. This level can be greatly increased by a regular exercise regimen that is sufficient to induce adaptive responses. *The adaptive response of skeletal muscle fibers to endurance exercise is mainly an increase in the metabolic capacity of the motor units involved.* This demand places an increased load on the cardiovascular and respiratory systems and increases the capacity of the heart and respiratory muscles. The latter effects are responsible for the principal health benefits associated with endurance exercise.

Muscle strength can be increased by regular massive efforts that involve all motor units. Such efforts recruit fast glycolytic motor units and are brief. The blood supply may be interrupted as the tissue pressures rise above the intravascular pressures during maximal contractions, further limiting the duration of the contraction. *Regular maximal strength exercise, such as weight lifting, induces synthesis of more myofibrils and hypertrophy of the active muscle cells. The increased stress also induces growth of tendons and bones.*

The precise cellular adaptive responses to exercise are difficult to determine. However, most studies indicate that the effects of exercise on muscle are quantitative. Endurance exercise does not cause fast motor units to become slow, with the synthesis of the slow myosin isoenzyme, nor do maximal muscular efforts produce a shift from slow to fast motor units. Experiments involving cross-innervation and other techniques show that such shifts are possible. It seems likely that any practical exercise regimen, when superimposed on normal daily activities, does not provide sufficient changes in the pattern of activation of motor units to shift the myosin isoenzyme distribution.

Two additional factors that affect the motor unit phenotype and athletic performance are hormones and the genetic endowment of the person. Androgens stimulate muscle hypertrophy and may account for the greater average muscle mass in males. Anabolic steriods are increasingly used to promote meat production in agriculture or athletic performance in humans. It is increasingly clear that the long-term ingestion of these drugs can have serious side effects.

Summary

1. Excitation-contraction coupling involves (1) binding of acetylcholine released from the motor nerve to receptors in the endplate membrane (which increases endplate conductance and generates an action potential that propagates in both directions along the muscle cell); (2) mobilization of the second messenger, Ca^{++}; (3) a thin filament conformational change caused by the allosteric binding of Ca^{++}; and (4) crossbridge attachment and cycling.

2. The sarcoplasmic reticulum surrounding each myofibril contains the pool of Ca^{++} that is mobilized when the propagation of an action potential along the sarcolemma depolarizes transverse tubules. This depolarization briefly opens Ca^{++} channels in the opposing sarcoplasmic reticulum membrane. A transient increase in the myoplasmic Ca^{++} concentration follows the action potential.

3. The binding of four Ca^{++} to troponin induces the conformational change in the thin filament that enables crossbridges to strongly bind and cycle until the myoplasmic Ca^{++} is returned to low values by active transport into the sarcoplasmic reticulum. Ca^{++} dissociates from troponin, the thin filament is "switched off," and relaxation ensues. The mechanical response to a single action potential is a twitch.

4. Sufficient Ca^{++} is released by an action potential to activate all the thin filaments. The force of a twitch is much less than the maximum that can be developed because the brief Ca^{++} transient does not allow sufficient crossbridge cycling to generate the maximum force.

5. The force of contraction is graded in a cell by increasing the frequency of action potentials, thereby maintaining the thin filaments in an on state with prolonged crossbridge cycling in a tetanus. In a muscle, force is also increased by recruiting more cells. Changing filament overlap has minimal effects in most skeletal muscles because the skeleton constrains changes in lengths to values near the optimum for force development.

6. Skeletal muscles have a high power output when shortening and consume ATP at a rapid rate. The cost is minimized by a very high efficiency in converting chemical to mechanical energy. Less ATP is consumed during isometric contractions, but considerable cycling still occurs, which results in poor economy in force maintenance in isometric contractions. The high, free energy of the myosin-ADP-P_i complex is never released during negative work because the transformation from a 90-degree to a 45-degree conformation cannot occur when sarcomeres are lengthening: therefore no ATP is consumed by crossbridge cycling.

7. Contraction requires ATP production and consumption to be matched in a muscle cell. Oxidative phosphorylation meets ATP consumption in slow muscle fibers that are phenotypically red as a result of the iron in molecules that bind oxygen (hemoglobin, myoglobin, cytochromes).

8. Fast fibers are recruited only during maximal efforts, because their high rates of ATP consumption are typically matched by glycolysis. Depletion of the cellular stores of glycogen and other factors lead to rapid fatigue.

9. Motor nerves branch in a muscle and may form neuromuscular junctions with hundreds of muscle cells. Such groupings are termed motor units because firing of the nerve causes all the cells to contract simultaneously. The smaller motor units in a muscle are recruited first because of the high excitability of their motor nerve and therefore are most frequently active. All the muscle fibers in such units are slow, oxidative cells most suited for sustained activity.

10. Skeletal muscle exhibits considerable phenotypic plasticity. Normal growth is associated with cellular hypertrophy caused by the addition of more myofibrils and more sarcomeres at the ends of the cell to match skeletal growth. Strength training induces cellular hypertrophy, whereas endurance training increases the oxidative capacity of all involved motor units. Training regimens are not sufficient to alter fiber type and the expression of myosin isoforms.

11. Skeletal muscle cells atrophy after denervation and depend on the activity of their motor nerves for maintenance of the differentiated phenotype. Reinnervation by axon growth along the original nerve sheath can reverse these changes. This may result in a phenotypic shift from a fast to slow fiber if a small, excitable motor nerve reinnervates the cell. Skeletal muscle has a very limited capacity to replace cells lost as a result of trauma or disease.

Bibliography

Journal articles

Alexander RM: Optimization and gaits in the locomotion of vertebrates, *Physiol Rev* 69:1199, 1989.

Bandman E: Myosin isoenzyme transitions in muscle development, maturation, and disease, *Int Rev Cytol* 97:97, 1985.

Biewener AA: Biomechanics of mammalian terrestrial locomotion, *Science* 250:1097, 1990.

Booth FW, Thomason DB: Molecular and cellular adaptation of muscle in response to exercise: perspectives of various models, *Physiol Rev* 71:541, 1991.

Buckingham ME: The control of muscle gene expression: a review of molecular studies on the production and processing of primary transcripts, *Br Med Bull* 45:608, 1989.

Fleischer S, Inui M: Biochemistry and biophysics of excitation-contraction coupling, *Annu Rev Biophys Biophys Chem* 18:333, 1989.

Florini JR: Hormonal control of muscle growth, *Muscle Nerve* 10:577, 1987.

Freund H-J: Motor unit and muscle activity in voluntary motor control, *Physiol Rev* 63:387, 1983.

Huang CL-H: Intramembrane charge movements in skeletal muscle, *Physiol Rev* 68:1197, 1988.

Krier J, Adams T: Properties of sphincteric striated muscle, *News Physiol Sci* 5:263, 1990.

Martonosi AN et al: The Ca^{2+} pump of skeletal muscle sarcoplasmic reticulum. In Stein WD, editor: *The ion pumps: structure, function, and regulation,* New York, 1988, Alan R Liss.

Payne MR, Rudnick SE: Textbook error: regulation of vertebrate striated muscle contraction, *TIBS* 14:357, 1989.

Rios E, Pizarro G: Voltage sensor of excitation-contraction coupling in skeletal muscle, *Physiol Rev* 71:849, 1991.

Swynghedauw B: Developmental and functional adaptation of contractile proteins in cardiac and skeletal muscles, *Physiol Rev* 66:710, 1986.

Westerblad H et al: Cellular mechanisms of fatigue in skeletal muscle, *Am J Physiol Cell Physiol* 261:C195, 1991.

Books and monographs

Fernandez HL, Donoso JA, editors: *Nerve-muscle cell trophic communication,* Boca Raton, Fl, 1988, CRC Press.

Kedes LH, Stockdale FE, editors: *Cellular and molecular biology of muscle development,* New York, 1988, Alan R Liss.

McMahon TA: *Muscles, reflexes, and locomotion,* Princeton, NJ, 1984, Princeton University Press.

Netter FH: *The CIBA collection of medical illustrations,* vol 8, Musculoskeletal system, West Caldwell, NJ, 1987, CIBA-GEIGY.

Peachey LD, Adrian RH, editors: *Handbook of physiology: skeletal muscle,* section 10, Bethesda, Md, 1983, American Physiological Society.

Rüegg JC: *Calcium in muscle activation,* Berlin, 1986, Springer-Verlag.

Woledge RC et al: *Energetic aspects of muscle contraction,* London, 1985, Academic Press.

CHAPTER
19

Smooth Muscle

■ *Muscle in Hollow Organs*

Muscle plays important roles in the function of most organs, and nonstriated (or smooth) muscle cells are a major component of the airways, vasculature, alimentary canal, and other systems. *Because smooth muscle does not attach directly to the skeleton, it has two major roles. It must develop force or shorten to provide motility or to alter the shape of an organ. However, it must also be capable of sustained or tonic contractions to maintain organ dimensions against imposed loads, such as the ability of blood vessels to withstand the blood pressure. In hollow organs, muscle cells are mechanically coupled like links in a chain, and all the cells must respond in a highly coordinated fashion.* In keeping with these roles, smooth muscle is structurally and functionally diverse.

To understand the many differences between smooth and skeletal muscle, consider the complications inherent in the absence of a skeleton.

1. Coordinated cellular responses require complex neural and hormonal control systems that act on the smooth muscle cells and require extensive communication between the smooth muscle cells.
2. The fibrillar contractile apparatus and its force-transmitting cytoskeleton must be able to operate in nonlinear configurations.
3. ATP consumption should be minimized in muscle cells that are often continuously active.

■ *Types of Smooth Muscle*

Many attempts have been made to classify smooth muscle into general categories. Although this can be done with respect to specific characteristics (structure, electrophysiology, innervation, and so on), smooth muscle is too diverse to allow rigorous generalizations. A useful functional distinction is whether the smooth muscle cells in an organ are normally relaxed or contracted. Examples of the former are the esophagus and the urinary bladder. The cells in the walls of these organs contract intermittently or **phasically** in response to an increase in the volume of the organ. By contrast, smooth muscle in the sphincters at the ends of these organs are normally **tonically** contracted, as is the muscle in the walls of blood vessels and airways. Typically, phasic smooth muscles contract in response to action potentials that propagate from cell to cell, whereas tone is not associated with trains of action potentials. However, all smooth muscle cells appear capable of both phasic and tonic behavior. Both types of contraction are based in large part on common molecular mechanisms for contraction and its control.

■ *Structure*

Smooth muscle cells (along with many other cell types) typically form layers in tubular or saclike structures. The simplest structure is tubular, as found in blood vessels or airways. The smooth muscle cells are arranged circumferentially so that contraction reduces the diameter of the tube; such contraction increases the resistance to the flow of blood or air. A more complex structure occurs in the gastrointestinal tract, whose role is to mix or propel the contents. Layers of smooth muscle in both circumferential and longitudinal orientations provide the mechanical actions. Coordination depends on a very complex system of autonomic nerves linked by plexuses that are characteristically found between the two muscle layers (see Chapter 38).

Smooth muscle in the walls of sacular structures, such as the urinary bladder or rectum, passively allows the organ to increase in size with the accumulation of urine or feces. The varied arrangement of cells in the walls contributes to their ability to reduce the internal volume to zero during urination or defecation. The smooth muscle in all hollow organs is separated from the contents by other cellular elements as simple as the vascular endothelium or as complex as the mucosa of the digestive tract. The walls also contain large amounts of connective tissue that bear an increasing share of the wall stress as the organ volume increases.

Cell-to-Cell Contacts

A variety of specialized contacts between involuntary muscle cells serves two functions: mechanical linkages and communication. Smooth (and cardiac) muscle cells are connected to each other rather than spanning the distance between two tendons. Thus cells anatomically arranged in series not only should be linked mechanically but also should be activated simultaneously and to the same degree. If this were not true, contraction in one region would simply stretch another region without a substantial decrease in radius or increase in pressure. The mechanical connections are provided by sheaths of connective tissue and by specific junctions between muscle cells.

In smooth muscle there are several types of junctions (Fig. 19-1). One is the **gap junction** in which adjacent plasma membranes are separated by only 2 or 3 nm. Gap junctions often exhibit a typical five-layered appearance in electron micrographs. Other specialized contacts are the **intermediate junction** and the **attachment plaque** or **desmosome-like attachment.** The membranes of these attachments are separated by a space of 20 to 60 nm, often marked by a central dense line. A variety of simple appositions between areas of cell membranes that are separated more widely are also common between smooth muscle cells. Some of these are quite elaborate with protrusions of one cell into another.

All these junctions may subserve the roles of both mechanical linkage and cell-to-cell communication. However, gap junctions are of particular importance in forming low-resistance pathways through which currents generated by action potentials in one cell can fire adjacent cells. In certain tissues, such as the outer longitudinal layer of smooth muscle in the intestine, large numbers of such junctions exist. A wave of depolarization is readily propagated from cell to cell through such tissues. Intermediate junctions and desmosomes are presumed to subserve mechanical functions.

Cells and Membranes

Embryonic smooth muscle cells do not fuse, and each individual cell has a single centrally located nucleus (Fig. 19-2, *B*). Although dwarfed by skeletal muscle cells, smooth muscle cells are large: typically 100 to 400 μm long at their optimum lengths for force generation. They are 2 to 5 μm in diameter at the level of the nucleus and taper toward their ends. Cross-sections may

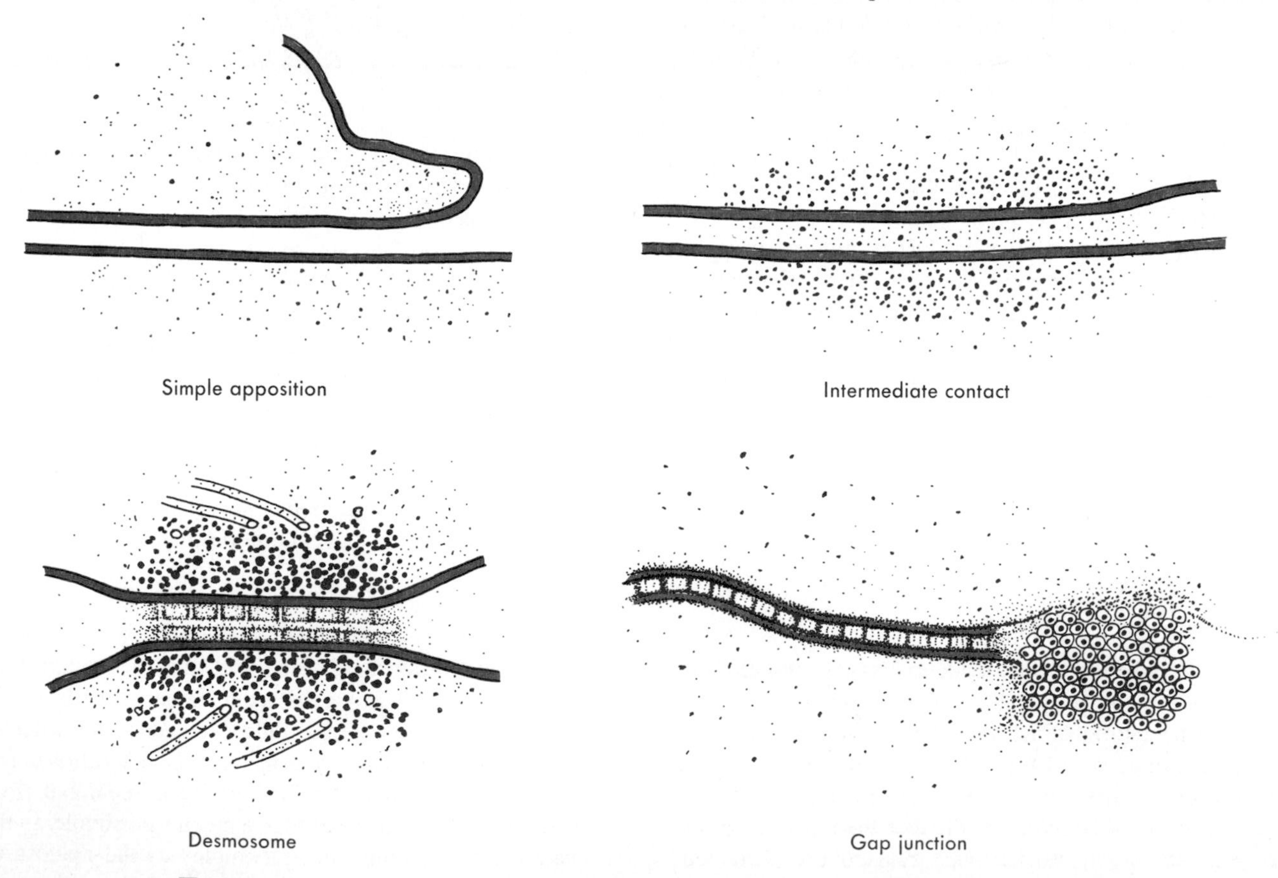

Fig. 19-1 Junctions between smooth muscle cells. The types and frequency of junctions vary widely among different types of smooth muscles. (Courtesy Dr. Michael S. Forbes.)

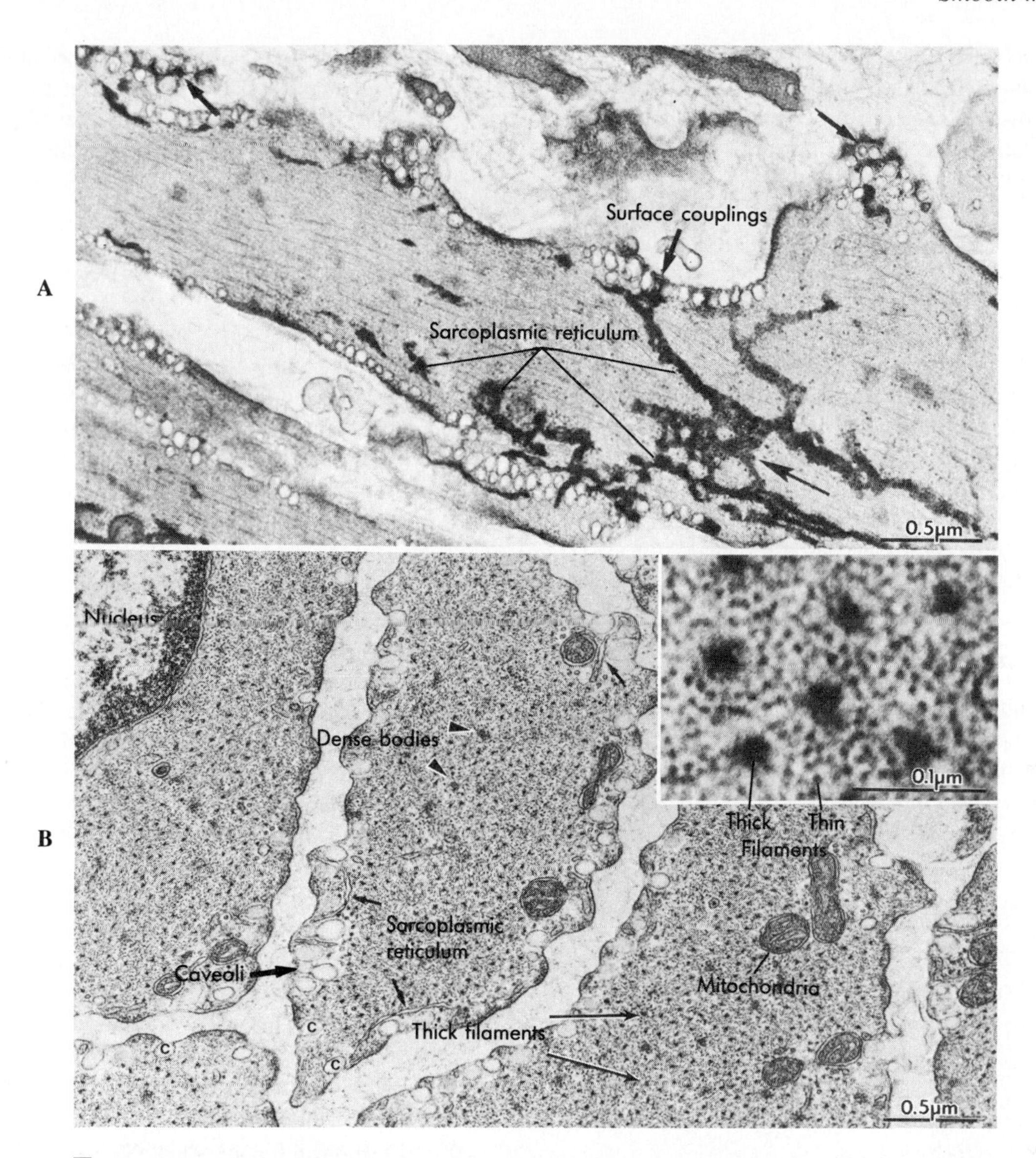

■ **Fig. 19-2** **A,** Longitudinal view of a portion of a rabbit main pulmonary artery smooth muscle cell. The sarcoplasmic reticulum is stained with osmium ferricyanide. The sarcoplasmic reticulum appears to form a continuous network throughout the cell consisting of tubules, fenestrated sheets (long arrows), and surface couplings at the cell membrane (short arrows). **B,** Transverse section of a bundle of smooth muscle fibers of rabbit portal-anterior mesenteric vein, illustrating the regular spacing of thick filaments (large arrows) and the relatively large number of surrounding thin (actin) filaments *(insert)*. Dense bodies (arrow heads) are sites of attachment for the thin actin filaments and equivalent to the Z-discs of striated muscles. Elements of sarcoplasmic reticulum (small arrows) occur at the periphery of these cells. (Redrawn from Somlyo AP, Somlyo AV: *Smooth muscle structure and function*. In Fozzard HA et al, editors: *The heart and cardiovascular system,* ed 2, New York, 1992, Raven Press.)

be very irregular, and during contraction they become quite distorted as a result of forces exerted on the cell by attachments to other cells or to the extracellular matrix (Fig. 19-2).

Smooth muscle cells lack T-tubules, the tiny invaginations of the sarcolemma that provide electrical links to the sarcoplasmic reticulum of skeletal muscle. However, the sarcolemma has longitudinal rows of small saclike inpocketings called **caveoli** (see Fig. 19-2, *B*). Caveoli increase the already large surface-to-volume ratio of the cells, but their functional role is uncertain.

The average diameter of a smooth muscle cell is not much greater than that of a striated muscle myofibril, and it is the sarcolemma, rather than an elaborate sarcoplasmic reticulum, that surrounds the myofilaments. *Ca^{++} diffuses into the cell through channels in the sar-*

colemma during activation, and the extracellular fluid contains an important Ca^{++} pool for regulation of contraction. However, smooth muscle has a sarcoplasmic reticulum (with connections to the surface) that contains an intracellular Ca^{++} pool (Fig. 19-2, *A*). This Ca^{++} pool can be mobilized when stimulatory neurotransmitters, hormones, or drugs bind to receptors on the sarcolemma. The volume of sarcoplasmic reticulum varies (2% to 6% of cell volume), but it can approximate that of skeletal muscle. Chemical signals link the sarcolemma and the sarcoplasmic reticulum.

A smooth muscle cell contains a prominent rough endoplasmic reticulum and Golgi apparatus, which are located centrally at each end of the nucleus. These reflect significant protein synthetic and secretory functions. The scattered mitochondria (see Fig. 19-2, *B*) are sufficient for oxidative phosphorylation to generate the ATP consumed during contraction.

The Fibrillar Contractile Apparatus

Thick and thin filaments are about 10,000 times longer than their diameter and are tightly packed in muscle. This means that the probability of observing an intact filament by electron microscopy is extremely low. Only the precise transverse alignment of thick and thin filaments in striated muscles makes it possible to discern their three-dimensional packing from two-dimensional images or by diffraction methods that require a quasicrystalline regularity in filament spacings.

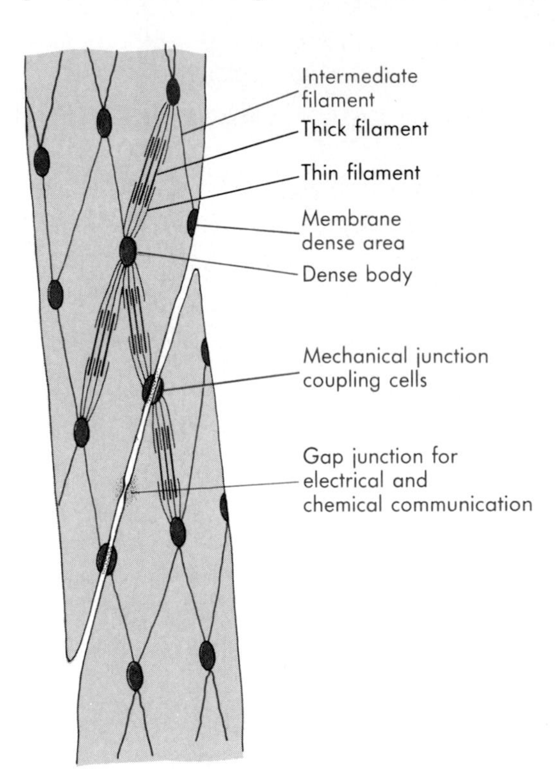

Fig. 19-3 Apparent organization of the cytoskeleton *(color)* and myofilaments in smooth muscles. Small contractile elements functionally equivalent to a sarcomere presumably underlie the similarities in mechanics between smooth and skeletal muscles. Linkages consisting of specialized junctions or interstitial fibrillar material functionally couple the contractile apparatus of adjacent cells.

The cytoskeletal and contractile filaments in smooth muscle are not all aligned transversely. These considerations have two consequences. First, the lack of striations accounts for the anatomical designation of plain or smooth muscle (see Fig. 19-2). Second, the absence of apparent structure does not imply a lack of order, but simply that a highly organized filamentous array may not be established until improved three-dimensional imaging technologies have been developed.

Cell cytoskeleton. The cytoskeleton in muscle cells serves as attachment points for the thin filaments and permits force transmission to the ends of the cell. The contractile apparatus in smooth muscle is not organized into myofibrils, and Z-disks are lacking. The functional equivalents of the Z-disks are centrally located ellipsoidal **dense bodies** and **dense areas** that form bands along the sarcolemma (Figs. 19-2 and 19-3). These structures serve as attachment points for the thin filaments and contain α-actinin, a protein also found in the Z-disks of striated muscle.

Intermediate filaments with diameters between those of thin filaments (7 nm) and thick filaments (15 nm) are prominent in smooth muscle and link the dense bodies and areas into a cytoskeletal network (see Fig. 19-3). All smooth muscles contain intermediate filaments consisting of polymers of a protein termed **desmin.** Smooth muscle cells in some tissues also contain intermediate filaments composed of **vimentin.**

Myofilaments. Little is known about the organization of the contractile apparatus into a contractile unit functionally equivalent to a sarcomere. Fig. 19-3 synthesizes some information about myofilament and cytoskeletal arrangements; important details are lacking about filament lengths and overlap at various points on the force-length relationship.

The thin filaments of smooth and striated muscle have the same basic structure (see Fig. 17-4), except that the regulatory protein, troponin, is not present in smooth muscle. However, the cellular content of actin and tropomyosin in smooth muscle is about twice that of striated muscle, and most of the myoplasm is filled with thin filaments approximately aligned along the long axis of the cell. The thin filaments exhibit the same polarity with respect to their attachments as those in striated muscle. In contrast, the myosin content of smooth muscle is only one-fourth that of striated muscle. Small groups of three to five thick filaments are aligned, surrounded by many thin filaments. Such groups of thick filaments with interdigitating thin filaments connected to dense bodies or areas (see Fig. 19-3) may be the equivalent of the sarcomere.

The contractile apparatus of adjacent cells is mechanically coupled by the links between membrane dense areas. This is an anatomical correlate of the fact that the cells are not functionally independent contractile units.

Control Systems

Neuromuscular relationships. Some smooth muscles show little electrical coupling via junctions, and these cells may not normally fire action potentials. Synchrony of cellular activation occurs by diffusion of hormones or neurotransmitters through the tissue. Neural control of contraction in involuntary muscle is more complex than in skeletal muscle. Three factors must be considered: (1) the types of innervation and neurotransmitters, (2) the proximity of the nerves to the muscle cells, and (3) the type and distribution of the receptors for neurotransmitters in the muscle cell membranes (Fig. 19-4).

The types of innervation can be divided into three categories. **Extrinsic innervation** is derived from axons of the autonomic nervous system. In arteries this is usually limited to sympathetic nerves, but both sympathetic and parasympathetic innervation commonly are present in other tissues. **Intrinsic nerves** contained in plexuses may occur within the smooth muscle tissue, particularly in the gastrointestinal tract. Finally, **afferent sensory neurons** that mediate various reflexes are found in the plexuses. The nervous system for the gastrointestinal tract alone is larger than the total motor system for all the skeletal muscles. A few smooth muscle tissues have no innervation.

Neuromuscular junctions in involuntary muscle are functionally comparable to those in skeletal muscle: presynaptic transmitter release, diffusion across the "junction," and combination with a postsynaptic receptor. However, elaborate neuromuscular contacts at axon terminals are not found. Autonomic nerves have a series of swollen areas or varicosities that are spaced at intervals along the axon. These varicosities contain the vesicles in which the neurotransmitters are found (see Fig. 19-4). Each varicosity functions as a neuromuscular junction, although the adjacent muscle membranes exhibit little specialization. The varicosities are closely apposed to the muscle cell membranes with a gap of 6 to 20 nm in tissues with a rich neural regulation. The average gap may be 80 to 120 nm and, occasionally, considerably more in tissues in which neural control is less extensive. In arteries the nerves tend to be concen-

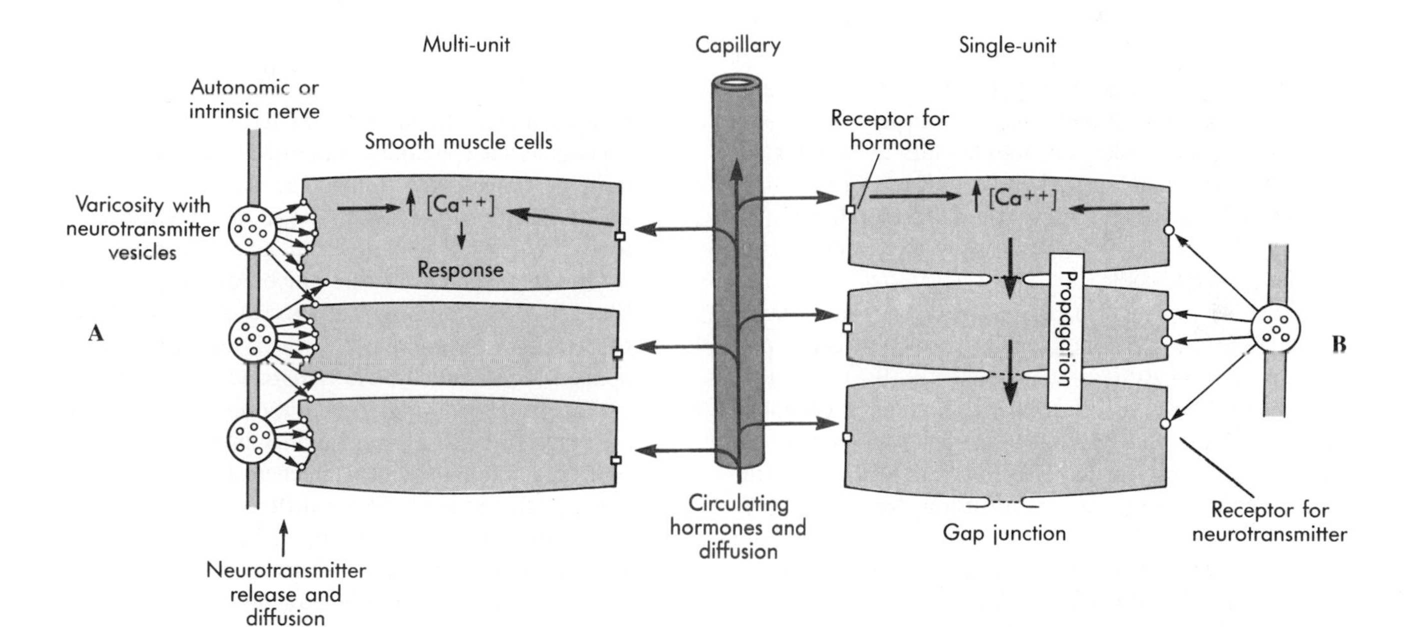

Fig. 19-4 Control systems of smooth muscle. Contraction (or inhibition of contraction) of smooth muscles can be initiated by (1) intrinsic activity of pacemaker cells, (2) neurally released transmitters, or (3) circulating hormones. The combination of a neurotransmitter, hormone, or drug with specific receptors activates contraction by increasing cell Ca^{++}. The response of the cells depends on the concentration of the transmitters or hormones at the cell membrane and the nature of the receptors present. Hormone concentrations depend on diffusion distance, release, reuptake, and catabolism. Consequently cells lacking close neuromuscular contacts will have a limited response to neural activity unless they are electrically coupled so that depolarization is transmitted from cell to cell. **A,** Multiunit smooth muscles resemble striated muscles in that there is no electrical coupling, and neural regulation is important. **B,** Single-unit smooth muscles are like cardiac muscle, and electrical activity is propagated throughout the tissue. Most smooth muscles probably lie between the two ends of the single unit-multiunit spectrum.

trated in the outer layer of smooth muscle cells just under the adventitial sheath of connective tissue. In such cases neurally mediated activation will be greatest at the periphery, and it may be nonexistent for smooth muscle cells in the luminal layers.

The neurotransmitters that are released and the presynaptic and postsynaptic effects of the neurohormones are highly variable. Marked individuality of the responses of different tissues that contain involuntary muscle is achieved by differences in their innervation, the types of transmitter released, and the nature of the receptors for each transmitter.

Relationships with other cell types. Contractile function in smooth muscle can be modulated by cells other than nerves. For example, specialized junctional regions can be observed between the membranes of the smooth muscle cells and the endothelial cells that line blood vessels. The actions of some circulating hormones or drugs can be mediated by the endothelial cells. Circulating acetylcholine, acting on endothelial cell receptors, produces arterial smooth muscle relaxation. Nitric oxide (NO) has been identified as the endothelium-derived relaxing factor that mediates the vasodilation. Acetylcholine causes contraction when it combines with cholinergic receptors in the arterial smooth muscle cell membrane.

Another example of a local control mechanism for vascular smooth muscle contraction is the formation of adenosine as a consequence of the increased metabolism during contraction of skeletal muscle. Adenosine diffuses from the skeletal muscle cells to receptors on the vascular smooth muscle cells of the arteries and induces smooth muscle relaxation and vasodilation. The result is an increase in blood flow to the contracting skeletal muscle cells.

Patterns of function. In a classic grouping of involuntary muscles, a distinction is made between **multiunit** and **single-unit** tissues (see Fig. 19-4). Multiunit tissues are those in which each cell does not communicate with other muscle cells through junctions. Contraction of multiunit smooth muscle is controlled by extrinsic innervation or hormonal diffusion. Skeletal muscles are an example of this pattern, which is also typical of a few smooth muscles (i.e., those in the vas deferens and iris). However, all smooth muscle cells that are anatomically arranged in series must be part of the same functional contractile unit, although many nerves may be involved. Single-unit tissues (Fig. 19-4, *B*) are exemplified by the heart and by a number of smooth muscles that undergo rhythmical contractile activity. All the cells in such tissues are electrically coupled through cell-to-cell junctions. Single-unit tissues can maintain fairly normal contractile activity without extrinsic innervation. The activity pattern in denervated single-unit muscles is caused by pacemaker cells and intrinsic reflex pathways.

The distinction between single-unit and multiunit tissues is overly simplified. Most smooth muscles are controlled and coordinated by a combination of neural elements, by some degree of coupling, and by locally produced activators or inhibitors. One generalization is that tonic muscles, such as arteries and sphincters, which maintain more or less continuous levels of tone, approach the multiunit end of the spectrum. Characteristically, such tissues do not exhibit action potentials when stimulated. Tissues that undergo phasic (rhythmical) activity, such as peristalsis, usually generate action potentials that are propagated from cell to cell. Such tissues more fully meet the criteria characterizing single-unit muscles.

Contractile Function in Smooth Muscle

The evidence that the sliding filament/crossbridge mechanism underlies contraction largely rests on similarities in the mechanics of smooth and striated muscles.

The Force-Length Relationship

Smooth muscle tissues contain large amounts of connective tissue containing extensible **elastin** fibrils and inextensible **collagen** fibrils. This matrix is responsible for the passive force-length curve measured in relaxed tissues. Highly distending forces or loads are withstood by this extracellular matrix, which limits organ volume. The active force developed on stimulation depends on tissue length. *When lengths are normalized to* L_o *(the optimal length for force development), the force-length curves for smooth and skeletal muscle are very similar* (see Fig. 17-8, *B*). This similarity provides strong support for a sliding filament mechanism in smooth muscle. The force-length curves differ quantitatively, however. Smooth muscle cells often shorten more in vivo than do striated muscle cells, although this is not caused by differences in their contractile systems. Smooth muscles are characteristically only partially activated, and the peak isometric forces attained vary with the stimulus. Smooth muscles can generate active stresses comparable to or even somewhat greater than striated muscle. This is accomplished with only one fourth of the myosin (and crossbridge) content of striated muscle. This does not imply that crossbridges in smooth muscle have a greater force-generating capacity. Active crossbridges in smooth muscle are much more likely to be in an attached force-generating configuration because of their slow cycling kinetics.

Velocity-Stress Relationships and the Crossbridge Cycle

Smooth and striated muscles exhibit the same hyperbolic dependence of shortening velocity on load, the same optimal power output at a load of 0.3 F_o, and the same ability to bear (transiently) an applied load greater than the developed active stress (see Fig. 17-9). Contraction velocities are far slower in smooth muscle than in striated muscle. One factor underlying these slow velocities is a myosin isoenzyme with a low ATPase activity.

Smooth muscles exhibit variable velocity-stress curves. Peak stress, S_o, varies with the stimulus and the level of activation of the contractile apparatus (Fig. 19-5, *D*). This is consistent with control of the numbers of crossbridges that are turned on and that interact with the thin filament. The values of V_o also differ (Fig. 19-5, *C*), depending on the level of activation of the muscle. *Such results imply that both crossbridge cycling rates and the number of active crossbridges are regulated, in marked contrast to striated muscle. This behavior reflects properties of a different type of regulatory system for the crossbridge rather than basic differences between crossbridges in smooth and striated muscle.*

Biophysics of Hollow Organs

The mechanical output of a skeletal muscle is readily calculated from the stress-length and velocity-stress re-

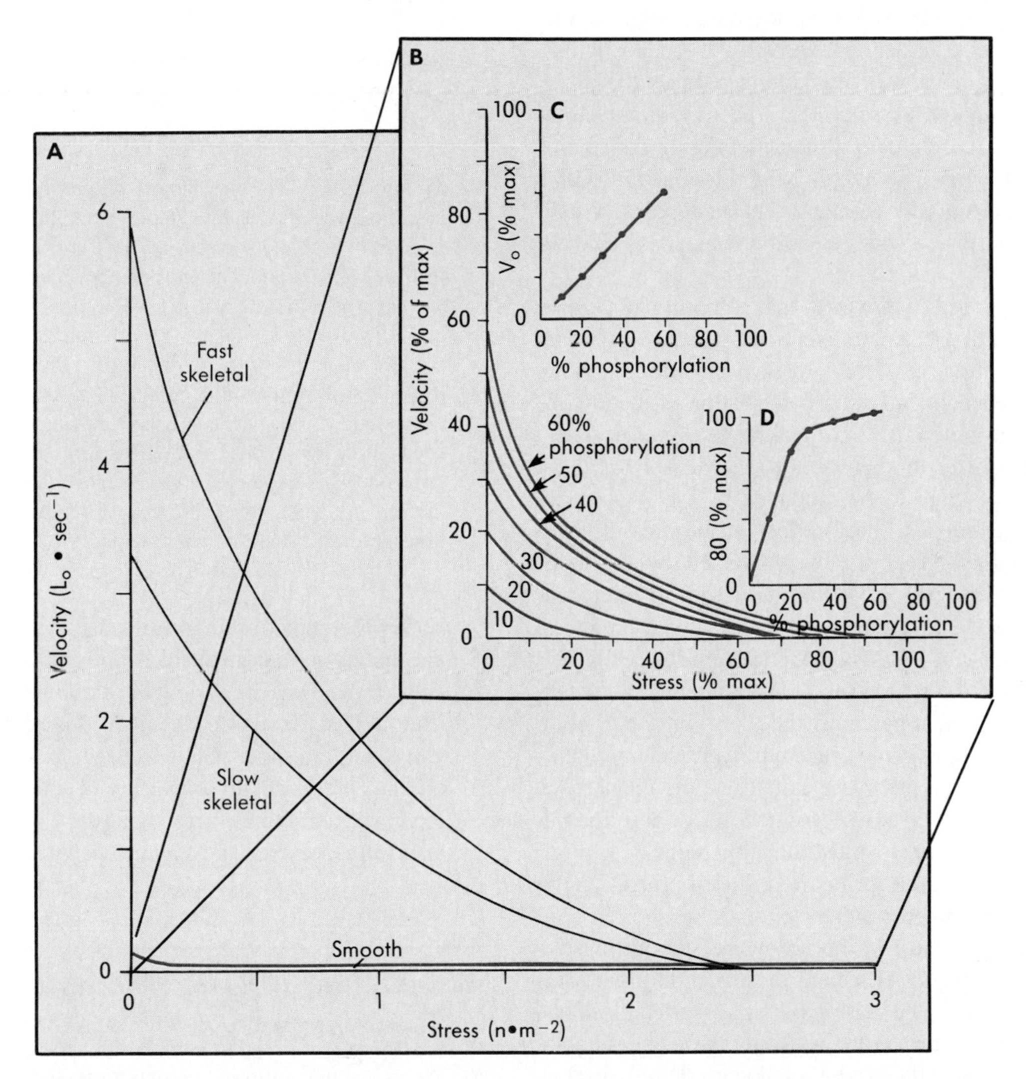

Fig. 19-5 **A,** Velocity-stress curves for fast and slow human skeletal fibers and smooth muscle. **B,** Smooth muscles have variable velocity-stress relationships that are determined by the level of Ca^{++}-stimulated crossbridge phosphorylation. **C,** Maximal shortening velocities with no load (intercepts on the ordinate) are directly dependent on phosphorylation. **D,** Active stress (abscissa intercepts) rises rapidly with phosphorylation, and near maximal stress may be generated with only 25% to 30% of the crossbridges in the phosphorylated state. (From Berne RM, Levy MN, editors: *Principles of physiology,* St Louis, 1990, Mosby–Year Book.)

lationships of the fibers (Figs. 17-8 and 17-9) and the angle formed between the fibers and the axis of the tendon. However, additional factors affect the mechanics of hollow organs.

Structural factors. Arteries have a comparatively simple geometry in which the smooth muscle cells run circumferentially around the vessel. However, the heart and many organs that contain smooth muscle have multiple layers of cells with different orientations. Such structural arrangements can have a profound effect on the pressure-volume relationships of the tissue. A moderate shortening of smooth muscle can produce a disproportionately large volume change. This is illustrated for an arteriole with a lumen diameter of 18 μm and a wall composed of a 1 μm thick endothelium surrounded by a 2 μm thick smooth muscle cell (Fig. 19-6). A 50% shortening will halve the mean radius of the smooth muscle cylinder. However, the cells are not compressible. Hence wall thickening reduces the inner radius of the smooth muscle layer to 7 μm. Contraction of the smooth muscle cells produces bulging of the endothelial cell layer. The overall volume reduction is 36-fold for an arteriolar segment that undergoes a cell shortening of 50%. This reduction can stop red cell flow.

Physical factors. The stress that is borne by the organ wall resulting from transmural pressure (e.g., the internal pressure minus the pressure outside a hollow organ) can be calculated from a derivation of the **law of Laplace** (considered further in Chapter 28). It states that the stress (σ) in the wall of the vessel (*double-headed arrows,* Fig. 19-6) produced by the transmural pressure (P) is equal to the product of the pressure and the mean radius (r) of the wall divided by the wall thickness (w), or $\sigma = Pr/w$, with units of force per unit area. Obviously a given pressure in the lumen produces a smaller stress on shortened smooth muscle cells because of the reduced vessel radius and the increased wall thickness. At larger radii the stress on the arterial wall produced by a given transmural pressure is withstood in part by the passive elasticity of the connective tissue and in part by active contraction of the arterial smooth muscle. This is illustrated by point *X* in Fig. 19-6, where the muscle cells are at their optimal length for force development. After the muscle shortens by 50%, all the load must be borne by the smooth muscle cells. Their ability to withstand the distending pressure is reduced (point *Y*) because of their force-length properties. However, the actual wall stress on the smooth muscle cells (neglecting endothelial cell compression) is now significantly lower because of the Laplace relationship (point *Z*). The smooth muscle cells would have to "relax" to maintain the reduced diameter if the compressive forces were insignificant and would be operating on a lower stress-length curve (colored curve, Fig. 19-6).

The relatively simple example illustrated in Fig. 19-6 shows the general relationship between muscle function and hollow organ mechanics. Pressure-volume relationships rather than stress-length relationships are normally used to describe the mechanical properties of hollow organs. The pressure-volume curve of an organ depends on the stress-length properties of the smooth muscle. However, structural factors and the Laplace relationship also influence pressure-volume behavior.

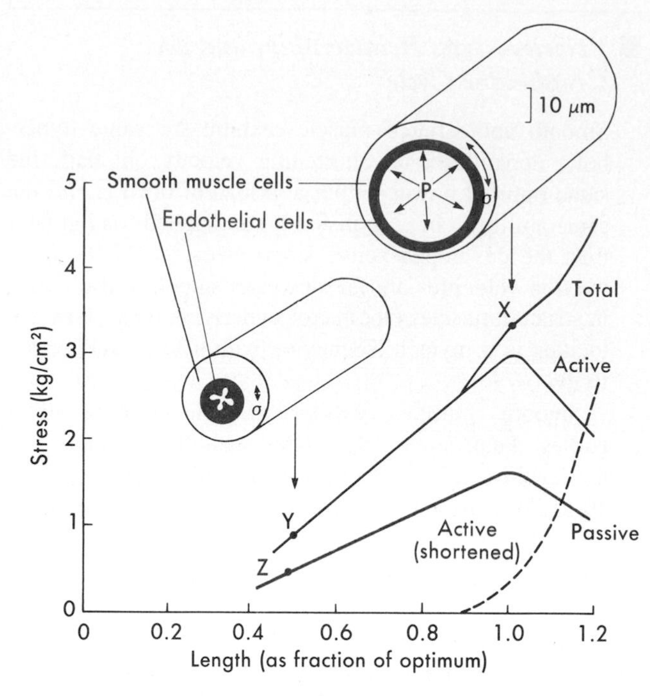

■ **Fig. 19-6** The mechanical properties of smooth muscle in hollow organs. In this simple example the smooth muscle of a small arteriole shortens by 50% from its optimum length for force generation. The stress-length relationships for the arteriolar wall before constriction are illustrated in black. The actual wall stress (σ) in the arteriole is governed by the Laplace relationship (text). If the pressure remained constant, σ would fall from a value proportional to *X* to a value proportional to *Z* during the constriction. Thus the smooth muscle needs to develop considerably less force to maintain the constriction. This geometrical factor more than compensates for the fall in active stress *(Y)* because of the length-stress relationship of the smooth muscle cells.

■ *Calcium and Crossbridge Regulation*

As in striated muscle, *contraction in smooth muscle depends on an increase in the myoplasmic Ca^{++} concentration. However, smooth muscles lack troponin and Ca^{++} regulates crossbridge attachment and cycling indirectly.* There are two aspects of regulation. The first mechanism whereby Ca^{++} regulates crossbridge interactions with the thin filament is discussed in this section. The second aspect, which concerns the mechanisms that regulate the myoplasmic Ca^{++} in response to the inputs to the cell membrane, will be discussed in the following section.

Ca^{++}-stimulated Crossbridge Phosphorylation

The current understanding of the link between Ca^{++} and contraction stems from the discovery of a pair of enzymes, myosin kinase and myosin phosphatase. These enzymes phosphorylate and dephosphorylate, respectively, a specific site on the myosin regulatory light chains that forms part of the crossbridge. Phosphorylation of crossbridges is the covalent linkage of PO_4 derived from the hydrolysis of ATP to a serine residue (Fig. 19-7). Phosphorylation is Ca^{++}-dependent because the active form of myosin kinase is a complex with the small cytoplasmic Ca^{++}-binding protein, **calmodulin.** Calmodulin binds 4 Ca^{++} ions cooperatively (much like troponin in striated muscle).

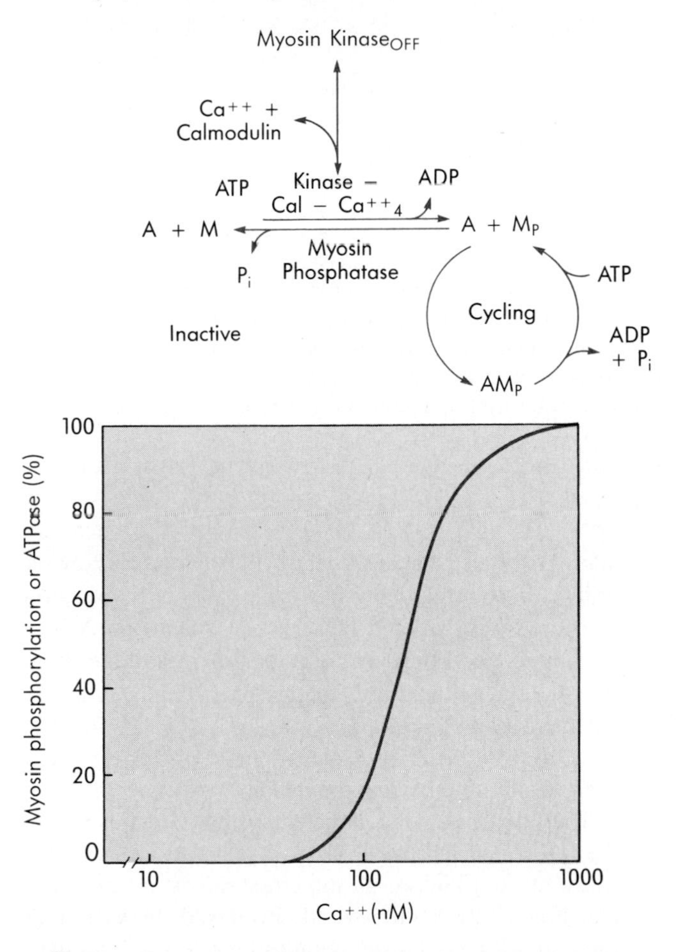

Fig. 19-7 Regulation of smooth muscle myosin ATPase activity by Ca^{++}-stimulated phosphorylation. The active myosin kinase-calmodulin-Ca_4^{++} complex transfers P_i from ATP to the regulatory light chains of myosin. Phosphorylated myosin (M_p) is able to interact with actin (A). Removal of Ca^{++} inactivates myosin kinase, and myosin is dephosphorylated by myosin phosphatase.

Phosphorylation is both necessary and sufficient for activation of ATP hydrolysis by myosin and actin isolated from smooth muscle (see Fig. 19-7) because there is no additional requirement for Ca^{++}. These biochemical data led to the hypothesis that Ca^{++} induces contraction by activating myosin kinase and that relaxation would follow Ca^{++} removal, inactivation of myosin kinase, and dephosphorylation by myosin phosphatase (although many phosphatases are regulated, direct evidence for regulation of crossbridge dephosphorylation has not been obtained). A covalent regulatory mechanism, whereby phosphorylation rather than the reversible allosteric binding of Ca^{++} to a myofibrillar regulatory site, has two inherent disadvantages. First, phosphorylation is slow and maximal crossbridge phosphorylation requires almost a second in smooth muscles. Second, ATP is used for both phosphorylation and to power the crossbridge cycle. This greater use of ATP reduces the efficiency of contraction. However, these disadvantages are more than offset by reductions in ATP consumption that are possible with covalent regulation of crossbridge cycling. These are illustrated by the relationships between Ca^{++}, crossbridge phosphorylation. and crossbridge cycling in smooth muscle cells.

Crossbridge Phosphorylation and Contraction

The scheme depicted in Fig. 19-7 has two crossbridge states (free and attached) as does skeletal muscle, and force should increase in proportion to the Ca^{++} concentration and phosphorylation. A proportionality between Ca^{++} concentration and phosphorylation is observed in smooth muscles exposed to a brief period of stimulation (Fig. 19-8, *A*). Stimulation transiently increases cell Ca^{++}. Changes in the Ca^{++} concentration are followed by changes in crossbridge phosphorylation and by a phasic contraction. However, muscles that are tonically stimulated for long periods behave as shown in Fig. 19-8, *B*. Instead of rising to a sustained value proportional to the stimulatory input, the initial peak Ca^{++} concentration falls to moderate steady-state levels as does crossbridge phosphorylation. Nevertheless, force rises to a high, sustained value. Parameters that reflect crossbridge cycling, such as V_o and ATP consumption, are closely correlated with changes in crossbridge phosphorylation (see Fig. 19-8).

Force maintenance with reduced crossbridge cycling confers a clear physiological advantage for a smooth muscle contracting tonically. Nevertheless, high force with moderate phosphorylation means that attached, dephosphorylated crossbridges must contribute to the force if the sliding filament-crossbridge mechanism is valid for smooth muscle. In fact, the properties illus-

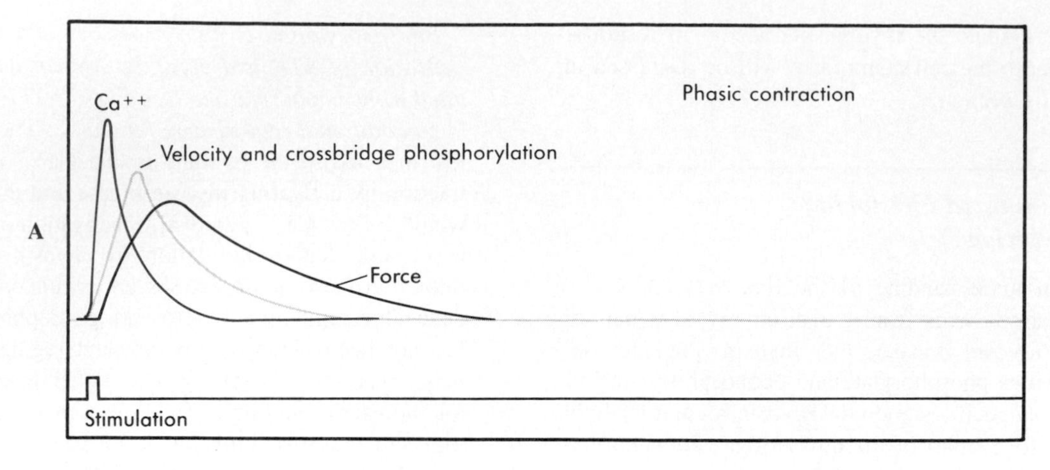

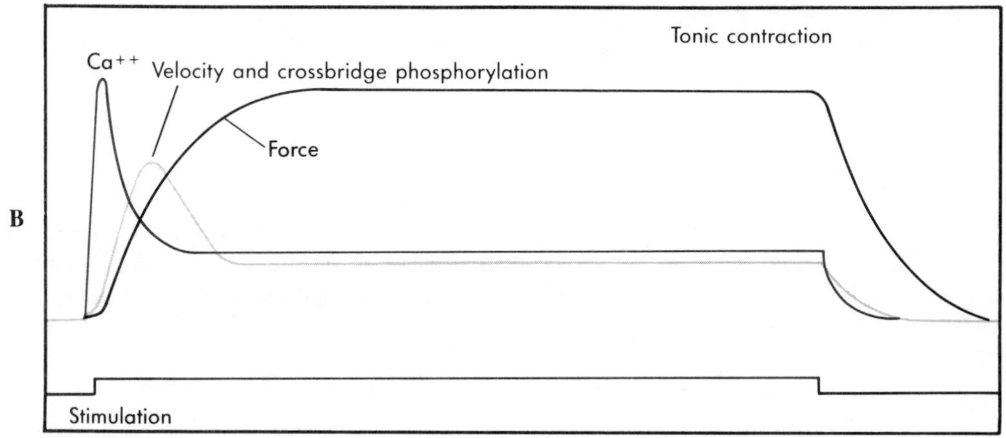

■ **Fig. 19-8** Time course of events in crossbridge activation and contraction in smooth muscle. **A,** A brief period of stimulation is associated with Ca^{++} mobilization followed by crossbridge phosphorylation and cycling to produce a phasic, twitchlike contraction. **B,** In a tonic contraction produced by sustained stimulation, the Ca^{++} and phosphorylation levels typically fall from an initial peak (allowing rapid force development). Force (tone) is maintained at reduced Ca^{++}, and phosphorylation with lower crossbridge cycling rates manifested by lower shortening velocities and ATP consumption.

trated in Figs. 19-5 and 19-8 are conferred by a covalent regulatory mechanism, because this allows four, rather than two, crossbridge states. The existence of four crossbridge states makes it possible to regulate not only the numbers of attached crossbridges (force), but also their cycling rates (velocity and ATP consumption).

■ *Covalent Crossbridge Regulation*

Despite many uncertainties about crossbridge regulation, there is strong evidence that contracting smooth muscle has four crossbridge states: free, attached, phosphorylated, and dephosphorylated (Fig. 19-9). This is possible because myosin kinase and myosin phosphatase can act on free or attached crossbridges. A simple hypothesis predicts many of the properties of smooth muscle (see Fig. 19-9). Four crossbridge states allow two crossbridge cycles. Phosphorylated crossbridges have a relatively fast cycle (although very slow compared with striated muscles) in which one ATP is consumed. A slow cycle occurs via crossbridge phosphorylation, attachment, dephosphorylation, and detachment (via states 1, 2, 3, 4, to 1, Fig. 19-9). This cycle uses two ATP molecules, one to power the crossbridge and the other for regulation.

The output of the contractile system is a function of the Ca^{++} concentration. At Ca^{++} levels too low to activate myosin kinase, all the crossbridges are in state 1 (see Fig. 19-9) and the muscle is relaxed. If stimulation raises the Ca^{++} concentration to high levels, the activity of myosin kinase will be high and most of the crossbridges will be phosphorylated to give rapid force development or shortening. If the cell Ca^{++} concentration then falls to moderate levels (see Fig. 19-8, *B*), phosphorylation falls, but force is sustained by the accumulation of crossbridges in state 4 (see Fig. 19-9), because

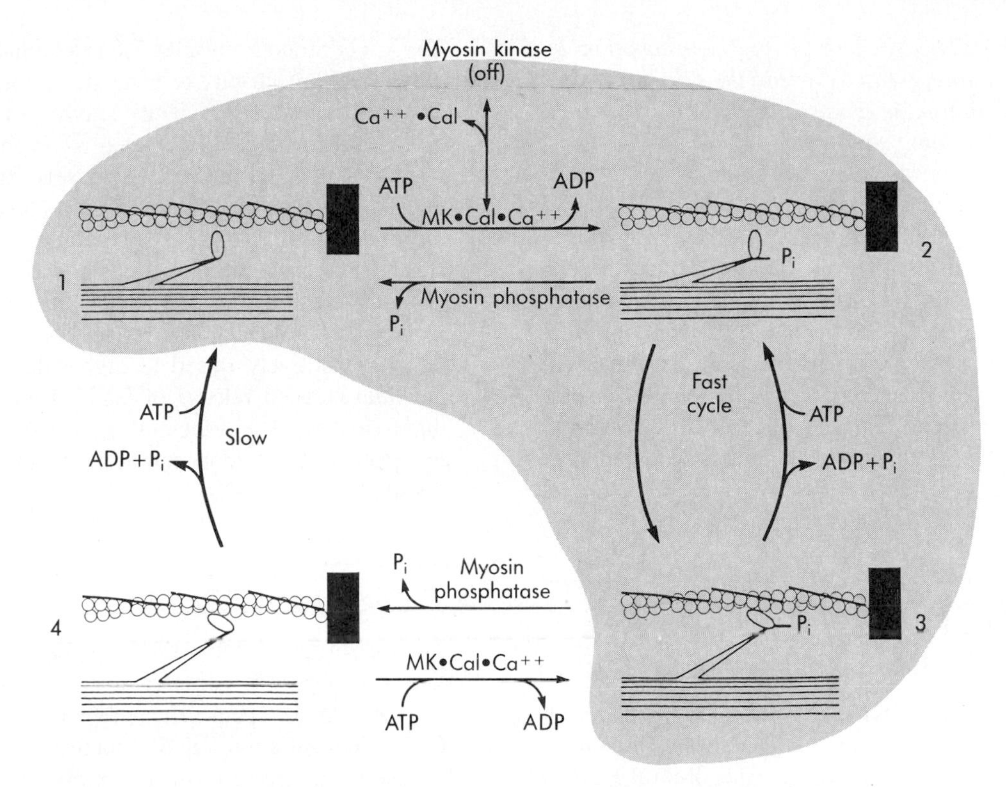

■ **Fig. 19-9** Hypothesis for regulation of crossbridge interactions in smooth muscle. Ca^{++}-stimulated myosin phosphorylation ($—P_i$) is obligatory for crossbridge attachment to the thin filament and cycling. Both free and attached crossbridges are substrates for myosin kinase and phosphatase. Dephosphorylation of an attached crossbridge (state 3) yields a modified crossbridge (state 4) that contributes equally to force but cycles more slowly because its detachment rate is reduced. Ca^{++} regulates myosin kinase activity and phosphorylation as shown in Fig. 19-7.

their detachment rates are slow. Shortening velocity depends on load and on the phosphorylation rates that determine the fraction of the crossbridges that cycle rapidly or slowly. This enables smooth muscle to undergo fairly fast phasic contractions and also to withstand imposed loads during sustained tonic contractions.

Smooth muscles are quite diverse and uncertainties exist about whether the regulatory scheme depicted in Fig. 19-9 is complete or generally applicable to all smooth muscles. However, this scheme can predict quantitatively the mechanical evidence for regulation of crossbridge cycling rates and force (see Fig. 19-5), just as allosteric regulation and two crossbridge states quantitatively predict the unique velocity-stress relationship that characterizes a skeletal muscle cell.

■ *Energetics and Metabolism*

As in skeletal muscle, crossbridge cycling in smooth muscle increases ATP consumption. Thus energy expenditure is minimal during imposed stretches and negative work, modest during isometric contractions, and larger during shortening. However, the energetics of smooth muscle are quantitatively very different from striated muscle for two reasons. First, smooth muscle myosin has a very slow detachment rate, even when phosphorylated. The second factor is the further slowing of cycling rates caused by decreases in Ca^{++}-dependent phosphorylation rates. Because detachment is the rate-limiting step in crossbridge cycling, almost all active crossbridges will be generating force in smooth muscle, whereas many will be in the detached portions of the cycle in skeletal muscle. The result is high forces with low rates of ATP consumption in smooth muscle because of crossbridge cycling.

Both smooth and striated muscle require energy for ion pumping associated with maintenance of membrane potentials and Ca^{++} sequestration. However, smooth muscle has an additional cost associated with regulation. ATP consumption for crossbridge phosphorylation may approach that used for crossbridge cycling under isometric conditions. This explains the low efficiency of smooth muscle (mechanical work/ATP hydrolyzed); efficiency is estimated as 15% to 20%, rather than 40% in skeletal muscle. However, sustained, isometric contractions are important for muscles in hollow organs, and the efficiency is zero in such cases (no work is done). Under such conditions some smooth muscles can sustain the same force as a skeletal muscle but use 300-

fold less ATP. *The savings in ATP consumption by crossbridge cycling greatly exceeds the additional cost of covalent regulation in a muscle with no physiological requirements for rapid shortening.*

The metabolic needs of smooth muscle during contraction are readily met by oxidative phosphorylation because consumption rates are low. Basal or resting rates of ATP consumption are typical of muscle cells generally, but basal ATP consumption is a significant fraction of the total ATP consumption by smooth muscle cells. Even when oxygen is plentiful, aerobic glycolysis with lactate production normally supports membrane ion pumps. This coupling of aerobic glycolysis to ion pumps indicates functional compartmentalization of metabolism.

Regulation of Myoplasmic Calcium Concentration

The mechanisms that couple activation to contraction in smooth muscle involve two Ca^{++} pools. The sarcolemma regulates Ca^{++} influx and efflux from the extracellular Ca^{++} pool (the extracellular fluid). The sarcoplasmic reticulum membranes determine Ca^{++} movements between the myoplasm and the intracellular pool. Several factors explain the presence of the numerous mechanisms that interact to determine the myoplasmic Ca^{++}: (1) Smooth muscles are functionally diverse and must have the capacity to generate various types of mechanical activity. (2) The activity of mechanically linked cells must be coordinated while enabling discrete responses of individual blood vessels, airways, and so on. (3) Covalent regulation requires precise control of the myoplasmic Ca^{++} concentration, which allows steady-state regulation of myosin phosphorylation. By contrast, the myoplasmic Ca^{++} concentration is not regulated in skeletal muscles, but the Ca^{++} concentration is repetitively raised to high values by the action potential-induced release of Ca^{++} from the sarcoplasmic reticulum. The principal mechanisms governing the myoplasmic Ca^{++} concentration are illustrated in Fig. 19-10.

The Sarcoplasmic Reticulum

The role of the sarcoplasmic reticulum with its intracellular Ca^{++} pool is comparable with skeletal muscle in that activation opens channels and the myoplasmic Ca^{++} concentration rapidly increases. However, this release is not linked to membrane potential changes but to binding of a second messenger, **inositol 1,4,5-trisphosphate (IP_3),** to receptors in the sarcoplasmic reticulum. IP_3 is generated by a stimulus that acts on sarcolemmal receptors coupled via a guanine nucleotide binding protein (G-protein) to activate phospholipase C.

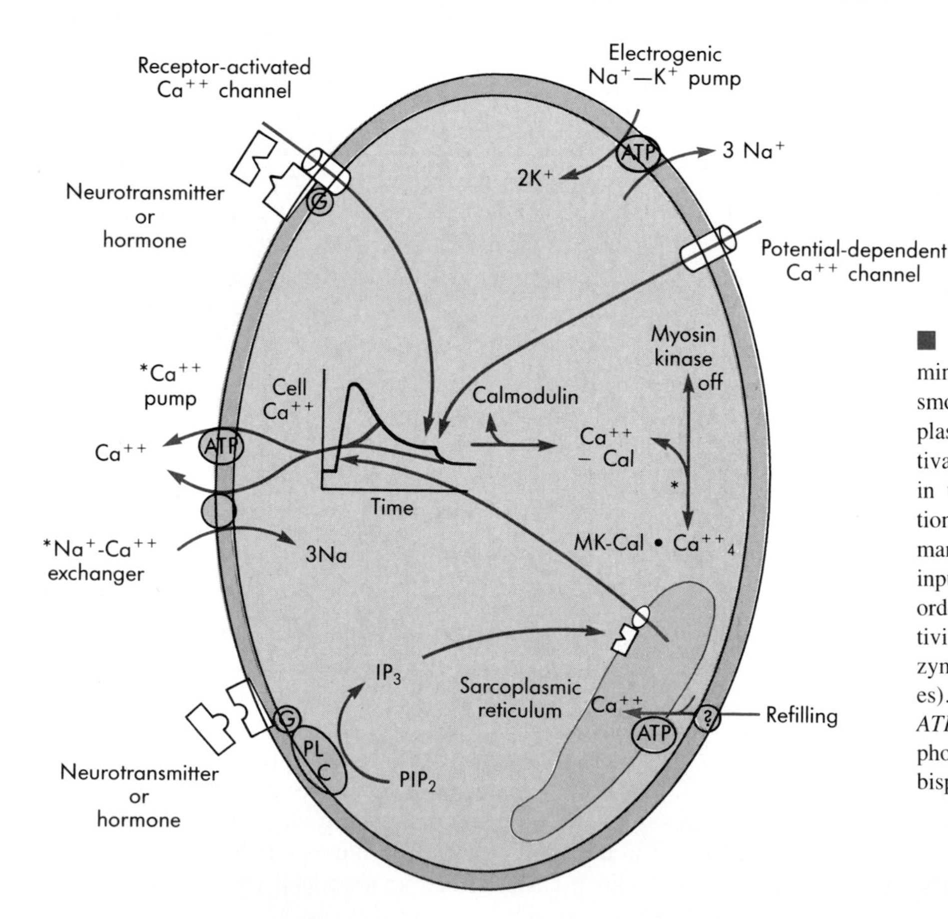

Fig. 19-10 Principle mechanisms determining the myoplasmic Ca^{++} concentration in smooth muscle. Ca^{++} release from the sarcoplasmic reticulum is a rapid initial event in activation, whereas the sarcolemma is involved in the subsequent stimulus-dependent regulation of cell Ca^{++}. The sarcolemma integrates many simultaneous excitatory and inhibitory inputs to govern the cellular response. Higher order regulatory mechanisms can alter the activity of various pumps, exchangers, or enzymes (* designates well-established instances). *G,* guanine nucleotide binding proteins; *ATP,* process requires ATP hydrolysis; *PL C,* phospholipase C; *PIP_2,* phosphatidyl inositol bisphosphate; *IP_3,* inositol 1,4,5-trisphosphate.

Phospholipase C hydrolyzes phosphatidyl inositol bisphosphate (PIP_2). IP_3 is one of the products of the reaction. This apparently complex process may permit a graded Ca^{++} release from the sarcoplasmic reticulum.

The Sarcolemma

Smooth muscles have effective methods to lower Ca^{++} concentrations from the initial peaks reached with contraction. Reduction in the Ca^{++} concentration is achieved by pumping Ca^{++} out of the cell by active transport through the sarcolemma or by a passive exchange coupled to the influx of 3 Na^+ ions (see Fig. 19-10). Refilling of the sarcoplasmic reticulum depends on extracellular Ca^{++}, but the details of this process remain uncertain.

Sustained contraction of smooth muscle is totally dependent on the extracellular Ca^{++} pool. The steady-state myoplasmic Ca^{++}, and thereby crossbridge phosphorylation, is regulated by the sum of the stimulus-dependent processes that govern Ca^{++} exchange with the extracellular pool. Although not shown in Fig. 19-10, the inputs to the cell membrane also can be inhibitory and reduce myoplasmic Ca^{++} concentration and induce relaxation.

Two categories of Ca^{++} channels are operationally distinguished, receptor activated and potential dependent (see Fig. 19-10). The conductance of receptor activated Ca^{++} channels is linked to receptor occupancy. Neurotransmitters or hormones can induce contractions with very little change in membrane potential in a process termed **pharmaco-mechanical coupling** (Fig. 19-11, *D*). Such channels can also be linked by G-proteins to receptors for inhibitory neurotransmitters or hormones.

Membrane potentials in smooth muscle cells characteristically consist of a component arising from a Donnan potential, based on the differential permeabilities of Na^+ and K^+ ions, as in skeletal muscle. However, the pump that transports these ions is electrogenic and expels 3 Na^+ ions in exchange for 2 K^+ ions (see Fig.

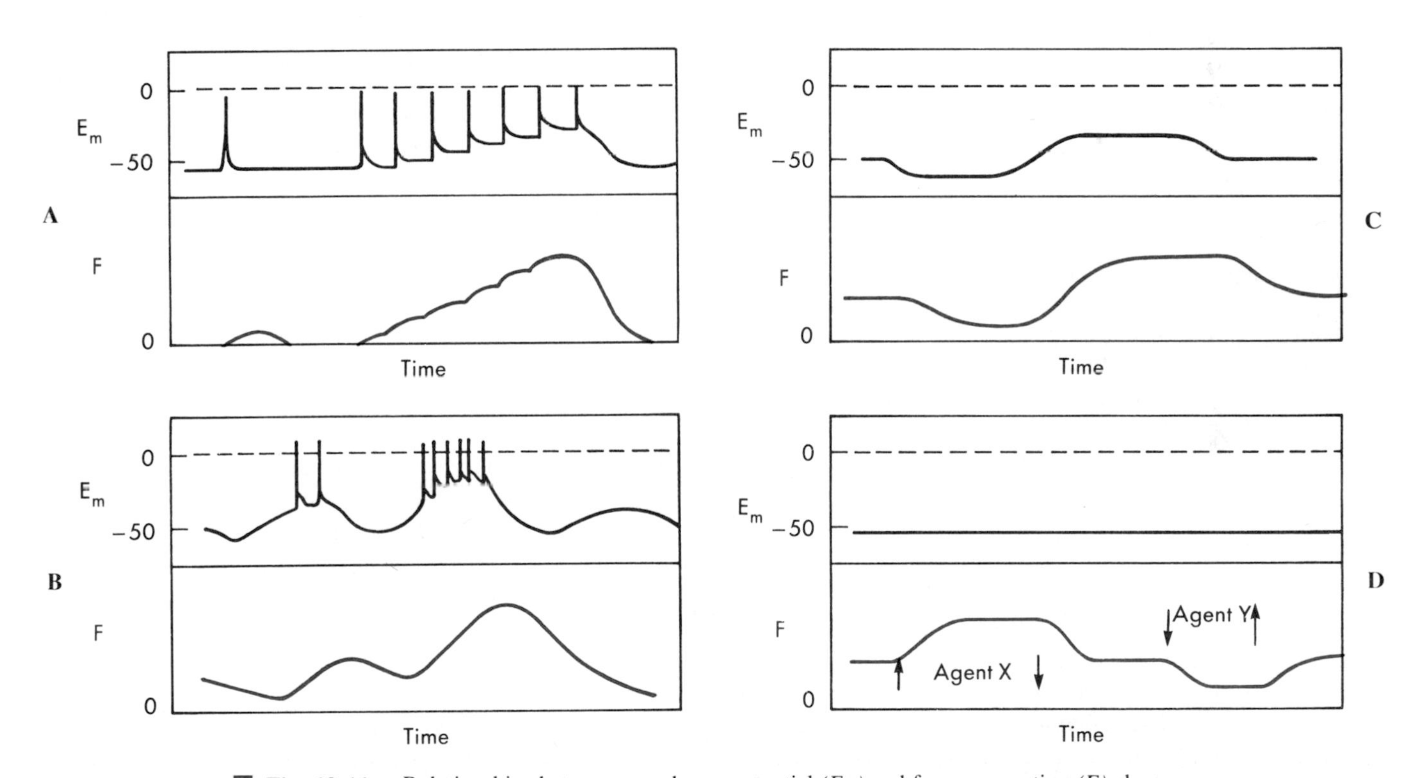

Fig. 19-11 Relationships between membrane potential *(E_m)* and force generation *(F)* characteristic of different types of smooth muscle. **A,** Action potentials may be generated and lead to a twitch or larger summed mechanical responses. Action potentials are characteristic of single-unit smooth muscles (many visceral). Gap junctions permit the spread of action potentials throughout the tissue. **B,** Rhythmical activity produced by slow waves that trigger action potentials. The contractions are usually associated with a burst of action potentials. Slow oscillations in membrane potential usually reflect the activity of electrogenic pumps in the cell membrane. **C,** Tonic contractile activity may be related to the value of the membrane potential in the absence of action potentials. Graded changes in E_m are common in multiunit smooth muscles (e.g.,vascular), where action potentials are not propagated from cell to cell. **D,** Pharmacomechanical coupling; changes in force produced by the addition or removal *(arrows)* of drugs or hormones that have no significant effect on the membrane potential.

19-10). One positive charge is removed from the cell for each cycle of the pump, and the membrane potential becomes more negative.

The sarcolemma also contains potential dependent Ca^{++} channels whose summed conductance increases with depolarization (see Fig. 19-10). Thus (1) action potentials, (2) graded depolarization caused by slowing of Na^+-K^+ exchange, or (3) depolarization propagated via gap junctions from adjacent cells increase Ca^{++} influx and force. Fig. 19-11 illustrates some common relationships between membrane potential and contractile force.

Other Second Messengers and Relaxation

A variety of drugs and hormones relax smooth muscles by increasing the cellular concentrations of cyclic AMP or cyclic guanosine monophosphate (cGMP). The protein kinases activated by these second messengers can phosphorylate a variety of enzymes and modify their activities. Relaxation can result from enhanced Ca^{++} extrusion or sequestration, presumably by phosphorylation of Ca^{++} pumps.

Functional Adaptations

Neurotrophic Relationships

Unlike skeletal muscles, *involuntary muscles show little dependence on their extrinsic innervation for maintaining a normal phenotype*. This may be related to the fact that smooth muscles can maintain contractile activity by other mechanisms when denervated. However, supersensitivity to neurotransmitters occurs in smooth muscle after denervation.

Development and Hypertrophy

During development and growth the number of smooth muscle cells increases (hyperplasia) (Fig. 19-12, *A*). In adults tissue mass increases if an organ is subjected to a sustained increase in mechanical work. This is termed **compensatory hypertrophy.** A striking example of this occurs in the arterial media in hypertension. The increased mechanical load on the muscle cells appears to be the common factor that induces hypertrophy in involuntary muscles. Recent studies indicate that tissue hypertrophy is sometimes caused by cellular hypertrophy. Chromosomal replication (which may or may not be followed by nuclear replication) results in significant numbers of polyploid muscle cells. The polyploid cells, with multiples of the normal sets of chromosomes, synthesize more contractile proteins and increase in size with each chromosomal replication (Fig. 19-12, *B*).

The myometrium, which is the smooth muscle component of the uterus, exhibits another striking example of hypertrophy as parturition approaches. Hormones play an important role in this response. During pregnancy (when the hormone progesterone predominates) the smooth muscle is quiescent, and there are few gap junctions electrically coupling the smooth muscle cells. At term, under the dominant influence of estrogen, there is a marked hypertrophy of the myometrium. A great increase in the number of gap junctions occurs just before birth and converts the myometrium into a single-unit tissue to coordinate contraction during parturition.

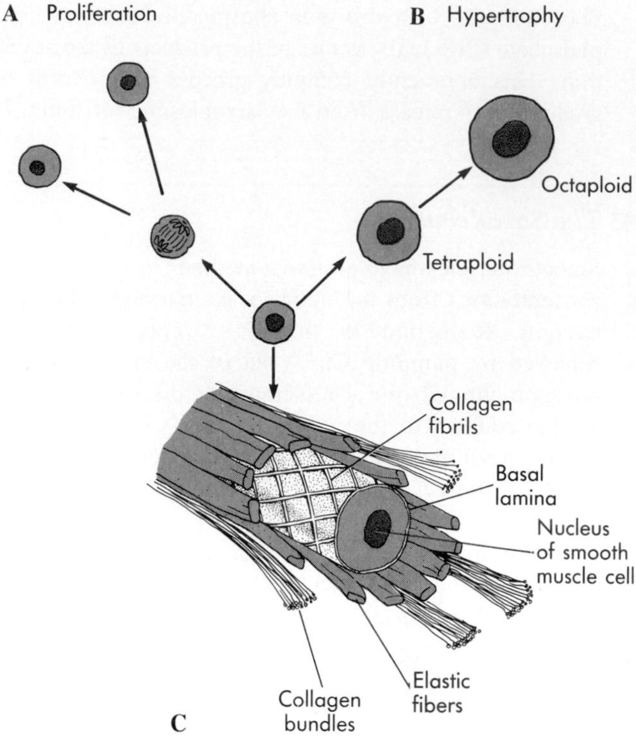

Fig. 19-12 Smooth muscle cells carry out many activities. **A,** They retain the capacity to divide during normal growth or in certain pathological responses such as formation of atherosclerotic plaques. **B,** Cells also may hypertrophy in response to increased loads. Chromosomal replication, not followed by cell division, yields cells with a greater content of contractile proteins. **C,** Smooth muscle cells also synthesize and secrete the constituents of the connective tissue matrix.

Synthetic and Secretory Functions

Growth and development of tissues containing smooth muscles are associated with increases in the connective tissue matrix. Smooth muscle cells can synthesize and secrete the materials that make up this matrix. These constituents include collagen, elastin, and proteoglycans (Fig. 19-12, *C*). The synthetic and secretory capacities are evident when smooth muscle cells with extensive contractile filament arrays are isolated and placed in tissue culture. The cells rapidly modulate and lose thick myosin filaments and much of the thin filament lattice. Their places are filled by a greatly expanded rough endoplasmic reticulum and Golgi apparatus (cellular structures that are associated with protein synthesis and secretion). The modulated cells multiply and lay down connective tissues in the culture plate. This process is reversible, and some degree of redifferentiation with formation of thick filaments is possible after cessation of cell replication. The determinants of the smooth muscle cell phenotype are largely unknown, but hormones and growth factors in the blood, as well as the mechanical loads on the cells, are implicated in the control of phenotypic modulation.

Summary

1. Smooth muscle in hollow organs has two roles: (1) to develop force or shorten like skeletal muscle and (2) to contract tonically to maintain organ dimensions against imposed loads.

2. Smooth muscle cells are comparatively small because embryonic myoblasts do not fuse. They are always part of a tissue in which they are linked by a variety of junctions, which serve both mechanical and communication roles. These linkages are essential in cells that must contract uniformly.

3. Smooth muscle cells have a high ratio of surface area to volume, and the sarcolemma plays an active role in Ca^{++} exchange between the extracellular fluid and the myoplasm. The sarcoplasmic reticulum contains an intracellular Ca^{++} pool that can be mobilized to give transient increases in the myoplasmic Ca^{++}.

4. Smooth muscles contain contractile units that consist of small groups of thick, myosin filaments, which interdigitate with large numbers of thin filaments attached to Z-disk equivalents, termed dense bodies or membrane dense areas. A three-dimensional organization that does not arrange thick filaments in transverse register contributes to an absence of striations.

5. The extracellular control systems include (1) extrinsic and intrinsic nerves that release a variety of transmitters that may be excitatory or inhibitory, (2) circulating hormones, (3) locally generated signaling substances, (4) junctions with other smooth muscle cells that allow electrical or chemical communication, and (5) junctions with other cell types that mediate signals.

6. Contraction appears to result from a common sliding filament-crossbridge mechanism. The stress-length relationship, hyperbolic velocity-load relationships, power output curves, and the ability to resist imposed loads are comparable with those of skeletal muscle. Quantitatively, shortening velocities and ATP consumption rates are very low in smooth muscle, in keeping with expression of a myosin isoform with low activity. Uniquely, smooth muscles have variable velocity-stress relationships that reflect regulation of both the numbers of active crossbridges (determining force) and their average cycling rates for a given constant load (determining velocity).

7. The response to sustained or tonic stimulation is a rapid contraction (for this muscle type) followed by sustained force maintenance with reduced crossbridge cycling rates and ATP consumption. This behavior is advantageous for muscles that may need to withstand continuous external forces, such as the ability of blood vessels to withstand blood pressure.

8. Smooth muscle lacks troponin, and a covalent mechanism regulates crossbridge cycling. Phosphorylation of crossbridges by a Ca^{++}-dependent myosin kinase is obligatory for attachment to the thin filament. Dephosphorylation of an attached crossbridge slows its detachment. Thus covalent regulation allows four crossbridge states (free, attached, phosphorylated, and dephosphorylated) and two crossbridge cycles that differ in their average turnover rate.

9. Phosphorylation consumes a significant fraction of the total ATP consumption of smooth muscle, and it contributes to a low efficiency. This disadvantage is more than offset by the savings in ATP during tonic contractions. Some smooth muscles maintain the same force as striated muscle, even though the rates of ATP consumption are 300-fold lower.

10. Covalent regulation requires precise regulation of the myoplasmic Ca^{++} concentration that determines myosin kinase activity. This requirement and the ability to generate varied patterns of contractile activity underlie complex mechanisms that regulate cell Ca^{++}. In the sarcolemma these include (1) receptor-activated Ca^{++} channels, (2) membrane potential gated Ca^{++} channels, (3) Ca^{++} pumps, (4) electrogenic Na^+-K^+ pumps, and (5) Na^+-Ca^{++} exchangers.

11. The Ca^{++} channels in the sarcoplasmic reticulum open in response to a chemical rather than an electrical signal. Neurotransmitters or hormones that act via receptors in the sarcolemma can activate phospholipase C, followed by generation of the second messenger, inositol 1,4,5-trisphosphate (IP_3), and diffusion of IP_3 to receptors on the sarcoplasmic reticulum.

12. The relationships between membrane potential and contractile activity range from action potentials that elicit small "twitches," which can be summed to tonic responses that are correlated with slow changes in

membrane potential and are unaccompanied by action potentials.

13. Smooth muscle is also a synthetic and secretory cell with a major role in the formation of the extensive extracellular matrix that surrounds and links the cells. Cellular hypertrophy occurs in response to physiological needs, but smooth muscle also retains the potential to divide.

Bibliography

Journal articles

Cooke PH et al: Molecular structure and organization of filaments in single, skinned smooth muscle cells, *Prog Clin Biol Res* 245:1, 1987.

Hai C-M, Murphy RA: Ca^{++}, crossbridge phosphorylation, and contraction, *Annu Rev Physiol* 51:285, 1989.

Hartshorne DJ, Ito M, Ikebe M: Myosin and contractile activity in smooth muscle, *Adv Exp Med Biol* 255:269, 1989.

Lefkowitz RJ, Caron MG: Adrenergic receptors: models for the study of receptors coupled to guanine nucleotide regulatory proteins, *J Biol Chem* 263:4993, 1988.

Paul RJ: Smooth muscle energetics, *Annu Rev Physiol* 51:331, 1989.

Somlyo AP: Excitation-contraction coupling and the ultrastructure of smooth muscle, *Circ Res* 57:497, 1985.

Somlyo AP, Himpens B: Cell calcium and its regulation in smooth muscle, *FASEB J* 3:2266, 1989.

Stull JT et al: Second messenger effects on the myosin phosphorylation system in smooth muscle, *Adv Exp Med Biol* 255:279, 1990.

Thyberg J et al: Regulation of differentiated properties and proliferation of arterial smooth muscle cells, *Arteriosclerosis* 10:966, 1990.

Triggle DJ et al: Calcium channels in smooth muscle: properties and regulation, *Ann NY Acad Sci* 560:215, 1989.

van Breemen C, Saida K: Cellular mechanisms regulating $[Ca^{++}]_i$ smooth muscle, *Annu Rev Physiol* 51:315, 1989.

Books and monographs

Blaustein MP: *Sodium-calcium exchange in cardiac, smooth, and skeletal muscles: key to control of contractility.* In Hoffman JF, Giebisch G, Schultz SC, editors: *Current topics in membranes and transport, vol 34, Cellular and molecular biology of sodium transport,* New York, 1989, Academic Press.

Bohr DF, Somlyo AP, Sparks HV Jr, editors: *Handbook of physiology,* section 2, *The cardiovascular system,* vol 2, Vascular smooth muscle, Bethesda, Md., 1980, American Physiological Society.

Driska SP: Mechanical properties and regulation of vascular smooth muscle contraction. In Sperelakis N, editor: *Physiology and pathophysiology of the heart,* 1989, Kluwer Academic Publishers. Lancaster, England.

Grover AK, Daniel EE: *Calcium and contractility: smooth muscle,* Clifton, NJ, 1985, Humana Press.

Motta PM, editor: *Ultrastructure of smooth muscle,* 1990, Lancaster, England, Kluwer Academic Publishers.

Rüegg JC: *Calcium in muscle activation,* Berlin, 1986, Springer-Verlag.

Sperelakis N, Wood JD: *Frontiers in smooth muscle research* (Prog in Clin Biol Res 327), New York, 1990, Wiley-Liss.

Wood JD, editor: *Handbook of physiology,* section 6, *The gastrointestinal system,* vol I, *Motility and circulation,* Bethesda, Md, 1989, American Physiological Society.

SECTION
IV

BLOOD

Oscar D. Ratnoff

CHAPTER 20

Blood Components

Blood is the vehicle of transportation that makes possible the specialization of structure and function characteristic of all but the lowest organisms. Blood, which makes up about 6% of body weight, is a suspension of various types of cells in a complex aqueous medium, the **plasma.** The elements of blood serve multiple functions essential for metabolism and the defense of the body against injury.

■ *Plasma*

To study plasma and its constituents, venous or arterial blood must be mixed with an anticoagulant to inhibit coagulation, or clot formation. Two types of anticoagulant are in general use: (1) heparin, a complex sulfated mucopolysaccharide that is usually derived from lung or intestinal mucosa and inhibits the enzymes that induce coagulation (see Chapter 21), or (2) agents such as sodium citrate or salts of ethylenediamine tetraacetic acid (EDTA) that reduce the concentration of calcium ions that are needed for coagulation. Plasma is separated from the blood cells by centrifugation. Blood drawn without adding an anticoagulant clots; after coagulation, the cell-depleted, fluid phase of blood, devoid of fibrinogen, is called **serum.**

The normal adult has an average of 25 to 45 ml of plasma per kg of body weight (i.e., a total volume of about 3 L). Many substances are dissolved in plasma, including electrolytes, proteins, lipids, carbohydrates (particularly glucose), amino acids, vitamins, hormones, nitrogenous breakdown products of metabolism (such as urea and uric acid), and gaseous oxygen, carbon dioxide, and nitrogen (Tables 20-1 and 20-2). The concentrations of these constituents are influenced by diet, metabolic demand, and the levels of hormones and vitamins. Normally the composition of blood is maintained at biologically safe and useful levels by a variety of homeostatic mechanisms. The balance may be upset by impaired function in a multitude of disorders, particularly those involving the lungs, the cardiovascular system, kidneys, liver, and endocrine organs.

Atmospheric oxygen, needed for the oxidative metabolic processes of the body, diffuses from the pulmonary alveoli into the plasma circulating in the pulmonary capillaries and thence into red blood cells, where it combines with hemoglobin, the major carrier of oxygen in blood (see below and also Chapter 36). Similarly, carbon dioxide, elaborated by the tissues through oxidation of carbon-containing compounds, diffuses into peripheral capillaries and is carried by the blood to the

■ **Table 20-1** Ionic constituents of plasma in adults

Cations	
Sodium (mEq/L)	135-145
Potassium (mEq/L)	3.5-5.0
Calcium* (mEq/L)	2.2-2.5
Magnesium (mEq/L)	1.5-2.0
Hydrogen (pH)	7.35-7.45
Anions	
Chloride (mEq/L)	95-107
Bicarbonate (mEq/L)	22-26
Lactate (mEq/L)	1.0-1.8
Sulfate (mEq/L)	1.0
Phosphate† (mEq/L)	2.0

*Total plasma calcium is 8.5-10.5 mg/dl.
†Total inorganic phosphorus is 2.5-4.5 mg/dl.

■ **Table 20-2** Some constituents of serum in adults

Proteins (gm/dl)	**6.0-8.0**
Albumin (g/dl)	3.4-5.0
Total globulin (g/dl)	2.2-4.0
Transferrin (mg/dl)	250
Haptoglobulin (mg/dl)	30-205
Hemopexin (mg/dl)	50-100
Ceruloplasmin (mg/dl)	25-45
Ferritin (μg/l)	15-300
Nonproteins	
Cholesterol (mg/dl)	140-250
Glucose (mg/dl)	70-110
Urea nitrogen (mg/dl)	6-23
Uric acid (mg/dl)	4.1-85
Creatinine (mg/dl)	0.7-1.4
Iron (μg/dl)	50-150

lung, where it is excreted. Carbon dioxide is transported in several ways: in solution, as bicarbonate ions, and as carbaminohemoglobin (see Chapter 36).

The ionic constituents of plasma maintain the pH of blood within physiological limits. Along with other, nonionic solutes, the ions also maintain the osmolarity of plasma; the normal osmolarity is 280 to 300 mOsm/kg water (see Chapters 1 and 42). The chief inorganic cation of plasma is sodium, present normally at an average concentration of 142 mEq/L (Table 20-1). Plasma also contains small amounts of ionic potassium, calcium, magnesium, and hydrogen. The principal anion of plasma is chloride, at an average concentration of 103 mEq/L. Ionic equilibrium is maintained by the presence of other anions, including bicarbonate, plasma protein, and (to a lesser degree) phosphate, sulfate, and organic acids. Elaborate mechanisms preserve the normal ionic composition and pH of plasma.

Literally hundreds of different proteins are dissolved in plasma. In all, plasma normally contains about 7 gm of protein per dl. The bulk of protein belongs to two groups, albumin and the various immunoglobulins. Plasma (or serum) proteins have been characterized by their migration in an electric field at pH 8.6. Albumin migrates most rapidly toward the anode; globulin species described as α-1, α-2, and β migrate successively less rapidly; gamma globulin migrates very slowly.

Albumin, a protein synthesized by hepatic parenchymal cells, is normally present at an average concentration of about 4 gm/dl. Because it diffuses poorly through intact vascular endothelium, *albumin provides the critical colloid osmotic or oncotic pressure that regulates the passage of water and diffusible solutes through the capillaries* (see p. 472.) When the concentration of albumin is severely reduced, excess extracellular fluid may accumulate in extravascular tissues, producing the phenomenon of **edema.** In closed body cavities, fluid accumulation is described as **ascites** in the peritoneal cavity or as **effusions** in the pleural or pericardial cavities. A decreased concentration of albumin may be the result of impaired synthesis in hepatic disease, urinary loss in renal disorders described as the nephrotic syndrome, or, more rarely, by gastrointestinal loss in protein-losing enteropathy.

Albumin is also the carrier for substances that are adsorbed to it. Such substances include normal components of blood, such as bilirubin and fatty acids, and certain exogenous agents, such as drugs. Albumin also furnishes some of the anions needed to balance the cations of plasma.

Among the proteins, the antibodies (immunoglobulins) are normally second to albumin in concentration. Antibodies arise on stimulation of lymphocytes in response to their exposure to antigens, which are agents usually foreign to the normal body that evoke formation of specific antibodies (see p. 335). Immunoglobulins are synthesized by plasma cells in the lymphoid organs and are critical as a defense against infection. The immunoglobulins constitute the bulk of gamma globulin in plasma. Other plasma proteins include (1) the clotting factors needed for blood coagulation, of which the most plentiful is fibrinogen; (2) the components of complement, a group of proteins that mediate the biological effects of immune reactions; (3) many enzymes or their precursors; (4) enzyme inhibitors; (5) specific carriers of such constituents as iron (transferrin), copper (ceruloplasmin), hormones, and vitamins; and (6) scavengers of agents inadvertently released into plasma (e.g., haptoglobin, which binds free hemoglobin, and hemopexin, which binds heme). Plasma lipids, the chief of which are triglycerides, cholesterol, and phospholipids, are transported as complexes with plasma proteins, the apolipoproteins.

Blood Cells

The cellular constituents of blood include red blood cells **(erythrocytes),** a variety of white blood cells **(leu-**

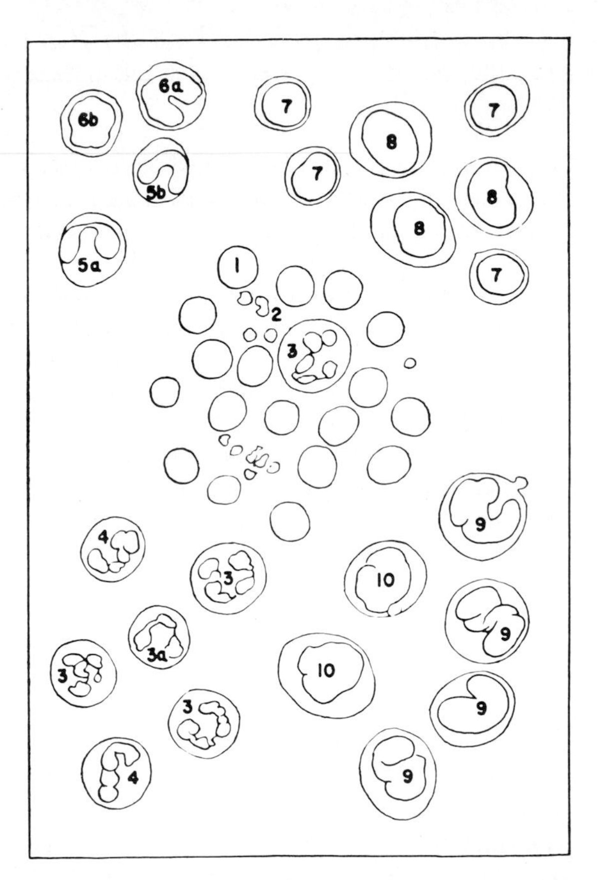

Fig. 20-1 The morphology of blood cells. *1,* Normal red cells. *2,* Platelets. *3,* Neutrophil, adult. *3a,* Neutrophil, adult (two lobes). *4,* Neutrophil, band form. *5a,* Eosinophil, two lobes. *5b,* Eosinophil, band form. *6a,* Basophil, band form. *6b,* Metamyelocyte, basophilic. *7,* Lymphocyte, small. *8,* Lymphocyte, large. *9,* Monocyte, mature. *10,* Monocyte, young. (See color plate between pgs. 338 and 339. From Daland GA: *A color atlas of morphologic hematology,* Cambridge, Mass, 1951, Harvard University Press.)

kocytes), and **platelets,** which play a major role in blood coagulation. Five classes of leukocytes are recognized: neutrophils, eosinophils, basophils, monocytes, and lymphocytes. The various classes are distinguished in blood smears by their morphological and tinctorial characteristics when stained with a mixture of dyes (Wright-Romanowsky stain) (Fig. 20-1, Plate I, and Table 20-3).

Erythrocytes

The mature erythrocyte is an anuclear cell surrounded by a deformable membrane well adapted to the need to traverse narrow capillaries (see Fig. 20-1 and Plate I). The red cells are biconcave disks, each with a diameter of about 8 μm, a thickness of 2 μm at its edge, and a volume of about 94 μm^3. In normal adults, the red cells occupy, on the average, about 48% of the volume of blood in males and about 42% in females. The percentage of the volume of blood made up by erythrocytes is defined as the **hematocrit.** The **red blood cell count** (i.e., the concentration of red cells in blood) normally averages about 5.2 million/μl in adult men and 4.8 million/μl in women.

The principal protein constituent of the cytoplasm of the mature erythrocyte is **hemoglobin.** Normal blood has about 15 g hemoglobin/dl in adult men and about 13.5 g hemoglobin/dl in adult women. The cytoplasm also contains enzymes that are needed to provide enough energy to preserve the integrity of the cell's membrane, to maintain the intracellular concentration of potassium above that in the surrounding plasma, to convert carbon dioxide to bicarbonate ion, and to prevent the oxidative transformation of hemoglobin to a nonfunctional protein, methemoglobin.

Hemoglobin, a protein synthesized in the marrow by the nucleated precursors of erythrocytes, is a complex molecule with a molecular weight of 68,000. It is composed of two dissimilar pairs of polypeptide "globin" subunits, two alpha chains and two nonalpha chains; the latter are either β, γ or δ chains. In the fetus and newborn, most of the hemoglobin is composed of two α and two γ chains, a configuration designated as fetal hemoglobin or hemoglobin F; the remainder consists of two α and two β chains, hemoglobin A. The proportion of hemoglobin F declines in early childhood. By the age of 4 years, hemoglobin F has virtually disappeared and more than 95% of the hemoglobin is hemoglobin A. A small proportion, up to 3% of hemoglobin, is made up of two α and two δ chains (hemoglobin A_2).

Each globin subunit is attached covalently to a prosthetic group consisting of a tetrapyrrole, heme. **Heme** is synthesized from glycine and succinyl-coenzyme A by a sequence of steps that lead to the formation of a pyrrole, porphobilinogen, and then to a tetrapyrrole ring compound, protoporphyrin IX. A mitochondrial enzyme, ferrochelatase, inserts an atom of ferrous iron into protoporphyrin IX to form heme. Oxygen binds reversibly to the iron incorporated into the heme unit. The combination of oxygen with hemoglobin in pulmonary capillaries and release of oxygen from hemoglobin in the capillaries of other tissues are described in Chapter 36.

The average life-span of the normal erythrocyte in the circulation is 120 days. Perhaps 10% to 20% of senescent red cells break up within the blood stream, where the liberated hemoglobin is bound to a specific carrier protein, **haptoglobin.** Some plasma hemoglobin is cleaved intravascularly into globin and heme; the latter binds to another carrier protein, **hemopexin.** Both the hemoglobin-haptoglobin and heme-hemopexin complexes are cleared from the circulation by the liver and catabolized by hepatic parenchymal cells.

The great bulk of senescent red cells are engulfed by the macrophages of the reticuloendothelial system, particularly in the liver and spleen. In the macrophages, hemoglobin is freed from the cells and catabolized into globin and heme. The globin is broken down by cellular proteases into its constituent amino acids, which join the pool of amino acids in plasma and are reused for protein synthesis. The heme tetrapyrrole is split enzymatically, releasing its iron atom and forming a linear tetrapyrrole, biliverdin. The released iron is reused for the most part in erythroblasts to resynthesize heme. The biliverdin is reduced to **bilirubin,** which is liberated into the plasma where, bound to albumin, it is transported to the parenchymal cells of the liver. There bilirubin is coupled to glucuronic acid and forms a water-soluble conjugate that is excreted into the bile. The bilirubin loses its glucuronic acid in the process. Some bilirubin is reabsorbed as such into the blood stream, to be reexcreted by the liver. Most, however, undergoes reduction by bacterial enzymes, successively forming tetrapyrolles, the urobilinogens, which are colorless, and stercobilin and urobilin, which provide the brown color of the stool. In turn, some urobilinogen and urobilin are reabsorbed from the gut and are either reexcreted by the liver or excreted into the urine.

Nearly all the iron needed for hemoglobin synthesis comes from this catabolism of heme-containing compounds. The iron is transported from the macrophages into the plasma, where it is bound in the ferric state to a carrier protein, **transferrin** (iron-binding globulin). Transferrin is normally about one third saturated with iron, and its concentration in plasma is about 250 mg/dl. Transferrin carries the iron to the cells that need it for heme synthesis. These cells are almost entirely bone marrow erythroid precursors, the erythroblasts and reticulocytes. The transferrin binds to specific membrane receptors on these cells and is internalized. The iron, freed from transferrin and reduced to the ferrous state, is either incorporated into heme or stored as ferritin. **Ferritin** is a complex of a water-soluble protein (apoferritin) and ferrous hydroxide. The macrophages of the reticuloendothelial system also store iron as ferritin. Such stores in the macrophages provide a reserve of this essential metal. Some of the ferritin is degraded into an insoluble form, hemosiderin, which also can furnish iron when needed. Ferritin is found in normal plasma, as well as in such organs as the liver, spleen, and heart.

Although the recycling of iron is highly efficient, small amounts are continually lost, largely through the desquamation of intestinal mucosa. A normal adult must ingest about 1 mg of iron daily to compensate for this loss. In women, additional iron must be absorbed to make up for blood loss during menstruation and for use of iron by the developing fetus. Most of the iron in the diet is derived from the heme moiety of meat, but a small amount is also provided by inorganic iron that is reduced by the acid of gastric juice.

Absorption of iron takes place in the mucosal cells of the duodenum and jejunum. Through mechanisms that are disputed, just enough iron is absorbed to make up for that lost to the body. In the mucosal cells the iron attached to heme is liberated and reduced to the ferrous form, and it joins the inorganic ferrous iron that has also been absorbed. The ferrous iron is then transported through the cytoplasm of the mucosal cells to the underlying small blood vessels. Transport through the cells is facilitated by its binding to an intracellular transferrin-like molecule. Other iron molecules are bound to a mucosal cell ferritin, and they may be lost to the body during the desquamation of these cells. The iron that reaches the plasma is oxidized to the ferric state and bound to transferrin. This bound iron joins the plasma pool of iron that is used for the formation of heme.

Table 20-3 Peripheral blood cells in adults

Blood cells	*Content*
Erythrocytes (red blood cells/μl)	
Men	$4.3\text{-}5.9 \times 10^6$
Women	$3.5\text{-}5.5 \times 10^6$
Hematocrit (%)	
Men	39-55
Women	36-48
Hemoglobin (g/dl)	
Men	13.9-16.3
Women	12.0-15.0
Leukocytes (white blood cells/μl)	$4.8\text{-}10.8 \times 10^3$
Neutrophils (% leukocytes)	50-70
Bands (% leukocytes)	0.6
Lymphocytes (% leukocytes)	20-40
T cells (% lymphocytes)	70
B cells (% lymphocytes)	10-20
Monocytes (% leukocytes)	1-10
Eosinophils (% leukocytes)	0-3
Basophils (% leukocytes)	0-1
Platelets (/μl)	$150\text{-}350 \times 10^3$

Leukocytes

Normal blood contains between 4000 and 10,000 leukocytes/μl (Table 20-3). Of these cells, about 40% to 75% are neutrophils, 20% to 45% are lymphocytes, 2% to 10% are monocytes, 1% to 6% are eosinophils, and less than 1% are basophils.

Neutrophils, eosinophils, and basophils, described collectively as **granulocytes,** are distinguished by the nature of the granules in their cytoplasm (see Fig. 20-1 and Plate I). These cells average about 12 to 15 μm in diameter, have small, multilobed nuclei in their mature forms, and have abundant cytoplasm.

Neutrophils. Neutrophils have nuclei with two to five lobes and, in stained blood smears, fine purple granules in a pink cytoplasm. A small fraction of the neutrophils in peripheral blood have a single, sausage-shaped nonlobulated nucleus and are described as band or juvenile cells. Primary granules, which make up a minority of the neutrophils' granules, contain a variety of enzymes, among them lysozyme, an enzyme that digests the walls of certain bacteria, and a peroxidase that reduces hydrogen peroxide. Much more numerous in stained preparations are secondary or specific granules that contain, among other biologically active agents, an iron-binding protein, (lactoferrin), a cationic bactericidal protein, and a vitamin B_{12}-binding protein. Besides the neutrophils detected in samples of peripheral blood, about an equal number appear to be distributed along the margins of the smallest blood vessels, particularly in the spleen and liver. This "marginal pool" of neutrophils can be mobilized into the circulation by various stimuli, e.g., the injection of epinephrine.

Within 12 hours after the neutrophils have been discharged from the marrow into the blood stream, they migrate into extravascular tissues where they survive 4 or 5 days. There they are attracted to sites of injury in extravascular tissues by "chemotactic" agents; among them are agents released by microorganisms, injured tissues, macrophages, granulocytes themselves, components of the clotting system, and components of complement. Neutrophils provide a major defense against infection by bacteria, which they can ingest and destroy.

The mechanisms for the destruction of bacteria begin with their attachment to the neutrophil surface. This process is mediated by an adhesive protein, fibronectin, and by antibody and complement opsonins fixed to the bacterial surface. Pseudopodia extruded by the neutrophils encircle the adherent bacteria and enclose them in newly formed vacuoles, the **phagosomes;** this process is called **phagocytosis.** The bacteria are destroyed in these vacuoles by enzymes released from the neutrophil's granules that fuse to the membranes lining the phagosomes. Among these enzymes are lysozyme, which disrupts the outer membrane of certain microorganisms, and enzymes that induce a sudden increase in the consumption of oxygen by the neutrophil. This **respiratory burst** results in the elaboration of hydrogen peroxide, superoxide ions (O_2^-), hydroxyl radicals (·OH) and singlet oxygen ($^1O_2$). Of these species, the principal bactericidal agent is **hydrogen peroxide,** which oxidizes components of the bacteria by the generation of hypochlorous acid (HOCl) through the interaction of chloride ions and myeloperoxidase, an enzyme discharged from the primary granules. Lactoferrin, a component of the specific granules, contributes to bacterial killing by binding iron that would be used by microorganisms and by catalyzing the generation of hydroxyl radicals. The processes through which bacteria are phagocytized and killed may lead to spillage of the neutrophil's enzymes and oxygen metabolites into the surrounding milieu. There they induce changes associated with the inflammatory reaction and may bring about proteolytic injury of tissue cells and of the neutrophil itself. In a similar manner, the engulfment of particulate matter by the neutrophil leads to its disruption and the release of agents that induce local inflammatory changes.

A sharp increase in the number of circulating neutrophils is a characteristic response to infection with certain microorganisms, such as pneumococcal, streptococcal, or staphylococcal infection. The increase is caused by mobilization of marginated neutrophils and by stimulation of production of these cells. A similar **leukocytosis** (increase in white blood cells) accompanies nonbacterial tissue injury, e.g., myocardial infarction.

Eosinophils. Eosinophils most often have bilobed nuclei; in stained preparations their cytoplasm contains large red granules (Fig. 20-1). Once released into the blood stream, most eosinophils migrate within 30 minutes into extravascular tissues where they survive for weeks. Like neutrophils, the eosinophils are mobile cells whose movement is directed by chemotactic agents derived from a variety of sources, including mast cells and lymphocytes. The eosinophils are phagocytic and destroy organisms through oxidative mechanisms similar but not identical to those of neutrophils. Eosinophils are increased in number in peripheral blood in many situations, including such parasitic infestations as trichinosis and schistosomiasis. Although the eosinophils are incapable of ingesting these organisms, they do exert cytotoxic effects. Eosinophils are also increased in number in patients with allergic or hypersensitivity states, such as asthma, in which exposure to abnormal exogenous or endogenous antigens leads to an immediate immunological reaction mediated by release of leukotrienes and platelet-activating factor.

Basophils. Basophils have a multilobulated nucleus and, in stained blood smears, large, deep blue cytoplasmic granules (Fig. 20-1). Like other granulocytes, basophils are motile cells with phagocytic properties. Basophils may migrate into extravascular tissues, where they may be stimulated, e.g., by complexes of antigens that are bound to IgE. These complexes react with specific IgE receptors on the surface of basophils. The stimulated cells may discharge histamine from their granules into the surrounding tissues. This brings about either an explosive systemic anaphylactic response, or a local dilation and increased permeability of blood vessels. The latter reactions result in local edema.

The relationship between blood basophils and morphologically similar "mast cells" that are found scattered throughout extravascular tissues is not certain, but they may have different origins. Like the basophils, the mast cells may be responsible for some of the phenomena associated with localized immunological reactions. On appropriate stimulation mast cells may release histamine from their granules and bring about acute immunological reactions or localized wheals.

Monocytes. A fourth class of leukocytes, the monocytes, are larger than other leukocytes; they have an average diameter of about 15 to 20 μm. In stained smears they have an indented, often kidney-shaped nucleus and fine pink cytoplasmic granules that are readily differentiated from those of the granulocytes. The monocytes are released into the blood stream from the marrow, where they mature. After a day or two they migrate into the tissues, particularly the liver, spleen, lymph nodes, and lungs, where they may reside for days or even years. There they comprise the macrophages of the reticuloendothelial system. The **macrophages** are cells with multiple functions that can replicate themselves in situ. The monocytes and macrophages are motile cells that are actively phagocytic. They ingest particulate matter, including microorganisms, injured or dead cells, and denatured proteins. The killing of microorganisms is brought about by processes similar to those used by neutrophils.

Monocytes also participate in immune responses. They appear to do this by ingesting antigens, which they then "process" so that they are recognized by T and B lymphocytes (described later). The monocytes are stimulated to secrete an agent, **interleukin-1**, that promotes the proliferation and maturation of T lymphocytes. Interleukin-1 is also a **pyrogen**, i.e., an agent that induces fever. A subset of T-cells, designated T helper or CD4 cells, are stimulated to secrete **interleukin-2**, which activates T-cells. The activated T-cells then release **interleukin-4** and **interleukin-5**, which signal the B lymphocytes to proliferate and differentiate into antibody-producing plasma cells (see following section).

Monocytes have many other biological properties, including the elaboration of tissue thromboplastin, activators of plasminogen, proteolytic enzymes, and other biologically active agents.

Lymphocytes. The lymphocytes are a heterogeneous group of cells with large nuclei (see Fig. 20-1 and Plate

I). The amount of cytoplasm depends on their size, which varies from 6 to 20 μm in diameter. The cytoplasm of most lymphocytes is devoid of granules in stained preparations; an exception is the natural killer cell (described later), which has azurophilic cytoplasmic granules. Some lymphocytes, called **B-cells**, are recognized by the presence of immunoglobulins on their surface. B-cells are transformed on stimulation with antigen into **plasma cells**, which synthesize and secrete the specific immunoglobulin antibodies. Other lymphocytes, called **T-cells**, can be identified because their surfaces have receptors for sheep erythrocytes. They participate in cell-mediated immune responses, such as sensitivity to tuberculin, that do not depend on the presence of circulating antibodies. Certain T-cells, designated as **helper** or **suppressor cells**, respectively, either abet or inhibit the transformation of B lymphocytes to antibody-producing cells. Other T-cells, designated as **cytotoxic lymphocytes**, bring about lysis of tissue cells when sensitized by antigens on the surface of tissue cells. In this way they participate in such phenomena as the rejection of grafted tissues that are incompatible with those of the host. Some lymphocytes with characteristics of neither T or B lymphocytes are described as **null cells.** Some null cells seem capable of destroying certain tumor cells, virus-infected cells, or tissue cells that have been coated with antibody; these are described as **natural killer cells.**

Platelets

Platelets are anuclear cytoplasmic fragments of large polyploid cells called **megakaryocytes** found in the bone marrow (Fig. 20-1 and Plate I). Platelets have an important role in the control of bleeding and the genesis of **thrombosis,** i.e., the formation of clots within blood vessels. The platelets are discussed in detail in Chapter 21.

Hematopoiesis

Our understanding of the origin of peripheral blood cells derives from observations in animals, and it is fortified by knowledge gained from patients with disorders of **hematopoiesis** (the process of generating blood cells). Current evidence suggests that blood cells arise through a succession of steps from primitive **totipotent stem cells.** The totipotent stem cell is derived, ultimately, from blood islands in the embryonal yolk sac whose cells colonize in the liver, spleen, and marrow. In the second trimester of fetal life, the generation of erythrocytes, granulocytes, monocytes, and megakaryocytes (the precursors of platelets) is centered largely in the liver and spleen. Blood cell production gradually shifts to the marrow, where cellular development takes place in endothelium-lined sinuses in the extravascular spaces, the marrow stroma. By the time of birth, the development of these cells is confined to marrow tissues. During childhood the generation of all these cell lines gradually becomes restricted to the marrow of the calvarium, pelvis, ribs, sternum, vertebrae, and the ends of the long bones. Lymphocytes, too, are derived from the totipotent stem cells, first in the liver and spleen and later in the marrow. However, their subsequent generation and differentiation occur not only in the marrow but also in the thymus and peripheral lymphatic organs.

The most primitive totipotent stem cell in the marrow is capable of reproducing itself indefinitely. Under conditions that are not yet clear, it can differentiate into stem cells that are the precursors either of lymphocytes or of pluripotential stem cells (Fig. 20-2). The **pluripotential stem cell,** although able to perpetuate itself, additionally may differentiate further into cells that are committed to the development of erythrocytes, megakaryocytes, granulocytes, or monocytes (the last two probably derived from a common precursor). The differentiation of the pluripotential stem cells into one or another **committed cell** is apparently determined by environmental conditions local to the growing cell colony.

The evolution of the mature blood cells that will enter the blood stream takes place through the successive division and differentiation of the committed or progenitor cells; these steps are described as representing the **mitotic compartment of hematopoiesis.** The earliest forms recognized, the **blast** cells, have large nuclei with prominent nucleoli and relatively scant cytoplasm. Further division and differentiation of these cells is accompanied by shrinkage of the nucleus, loss of visible nucleoli, and expansion of the cytoplasm, which takes on the characteristics of the peripheral blood cells. Ultimately, cell division ceases. Maturation to the functional cells to be released into the blood continues in what is described as the **postmitotic compartment of hematopoiesis.**

Following this general pattern, the earliest committed erythroid precursor cells differentiate into nucleated **erythroblasts (pronormoblasts)** that begin the synthesis of the oxygen-carrying protein, hemoglobin. Further cell division and differentiation bring about the formation of the most mature nucleated cells, the **normoblasts,** in which the nuclear chromatin is condensed. Eventually, the normoblasts extrude their nuclei and are released as erythrocytes through the capillaries of the marrow into the blood stream.

Maturation of red blood cells from the earliest committed precursor cell takes place over 3 to 4 days. For the first 2 days of their life in the blood stream, the anuclear erythrocytes contain remnants of RNA and other intracellular organelles and continue to synthesize hemoglobin; such cells are termed **reticulocytes.** For the rest of their life-span, the erythrocytes lack protein syn-

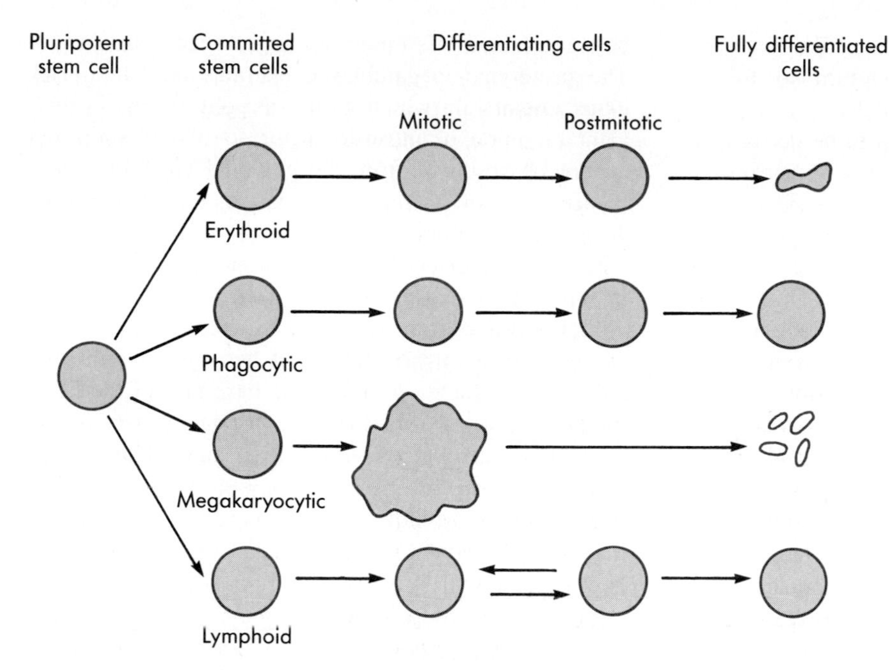

Fig. 20-2 An overall look at hematopoieseis. (Modified from Babior BM, Stossel TP: *Hematology: a pathophysiological approach,* New York, 1984, Churchill Livingstone.)

thetic ability and can no longer synthesize hemoglobin. The proportion of reticulocytes in peripheral blood may rise when the rate of erythropoesis is increased to compensate for severe blood loss or for destruction of red cells by **hemolysis** i.e., disruption of the outer membrane of the erythrocyte with loss of its intracellular contents.

Differentiation of the erythroid precursors is under the influence of a polypeptide hormone, **erythropoietin.** This hormone is elaborated chiefly by the peritubular interstitial cells of the kidney and attaches to receptors on the surface of erythrocyte precursors. The synthesis of erythropoietin is stimulated by tissue hypoxia, which may be brought about by anemia, residence at high altitudes, impaired oxygenation of hemoglobin as the result of cardiac or pulmonary disease, or other stimuli.

Other committed cells differentiate through similar steps from blast cells to mature cells with the characteristics of neutrophils, eosinophils, basophils, or monocytes. The differentiation and maturation of these cells are stimulated by agents called, in the most confusing way, **colony-stimulating factors** (CSF) and **interleukins.** These agents are small polypeptide hormonelike substances that exert their effect by stimulating specific receptors of their target cells. For example, the differentiation and maturation of cell lines other than those of lymphocytes is initiated by **interleukin-3,** which stimulates a multipotential precursor of these various cells. The subsequent development of granulocytes, monocytes (macrophages), and megakaryocytes is stimulated by other peptides such as granulocyte-monocyte (macrophage) colony-stimulating factor (GM-CSF) and interleukin-1. CSFs are produced by many tissues, among them macrophages or monocytes, T lymphocytes, and fibroblasts and endothelial cells within the stroma of the marrow. CSFs are found in the plasma and in lung and placental tissues.

The normal differentiation and maturation of marrow cells depends on adequate supplies of vitamin B_{12} and folic acid. These agents are required for the formation of DNA, a requisite for cell division.

Vitamin B_{12} (cyanocobalamin) is a cobalt-containing compound synthesized by bacteria and molds. This vitamin cannot be synthesized by man but is furnished in the diet by foods of animal origin; approximately 5 μg are needed daily. In the stomach vitamin B_{12} is bound to a specific protein, intrinsic factor, that is secreted by the parietal cells in the body and fundus of the stomach. The complex of vitamin B_{12} and intrinsic factor at taches to and is incorporated into mucosal cells in the lower portion of the ileum. In these cells vitamin B_{12} is split from the complex. Vitamin B_{12} is thence carried to the blood-forming tissues bound to a specific plasma transport protein, transcobalamin II.

Folic acid is also furnished in the diet, particularly by vegetables and dairy products; the daily requirement is about 100 to 200 μg. Dietary folate is conjugated to a polyglutamate chain from which it is cleaved as it is absorbed from the intestine. It is reduced in intestinal epithelial cells and other cells to its biologically active form, tetrahydrofolate.

Lymphocytes are generated from the primitive hematopoietic stem cells in marrow from which they enter the blood stream and colonize in the lymphoid organs, principally the thymus, spleen, lymph nodes, tonsillar tissue, and the submucosal Peyer's patches of the intestines. In these peripheral lymph organs, the lymphocytes can reproduce themselves and undergo further differentiation, thereby sustaining their population. Lymphocytes in these peripheral sites can reenter the circulating blood, where they can again return to the lymphoid tissues via the lymphatic channels.

T lymphocytes (or T-cells) are thought to be derived from lymphocytes that have migrated from the marrow to the cortex of the thymus gland. There they proliferate and migrate to the medulla of the thymus where they mature and then migrate to other lymphoid tissues. The origin of human B-cells is less certain. Their designation derives from the observation that in birds antibodies are synthesized by lymphocytes derived from the bursa of Fabricius. In mammals B-cells probably mature in bone marrow and then migrate to the various lymphoid tissues where they can reproduce themselves. Synthesis of immunoglobulins takes place in those lymphocytes that have migrated to the lymphoid organs, where they are transformed on antigenic stimulation to antibody-producing plasma cells. In the lymphoid organs, the T-cells and B-cells segregate in separate zones. Both types of cells freely recirculate between the lymphoid organs and peripheral blood.

Cells committed to the development of megakaryocytes are stimulated to mature under the influence of a hormone-like substance, thrombopoietin; the platelets are shed from the cytoplasm of mature megakaryocytes.

Anemia

Anemia is a decrease in the circulating mass of erythrocytes. It may result from decreased generation of these cells or their premature destruction or loss through hemorrhage. Failure of synthesis can arise in many ways: (1) as a consequence of hypocellularity of the marrow; (2) from replacement of the marrow by tumor tissue; (3) by suppression of hematopoiesis (as occurs in renal failure or from deficiencies of agents, such as vitamin B_{12} or folic acid that are essential for the maturation of red cells); or (4) from a deficiency of iron needed for the formation of heme. Premature destruction of red cells can come about from such diverse causes as hereditary defects in their outer membranes or direct chemical, physical, or immunological injury. Of the innumerable causes of anemia, only a few are discussed here.

Iron-deficiency anemia, is undoubtedly the most common anemia in Western countries. It is almost invariably caused either by blood loss, such as occurs with menstruation or from lesions in the gastrointestinal tract, or by the use of iron by the fetus during pregnancy. The red cells are characteristically smaller than normal (microcytosis) and are pallid from the resultant deficiency of hemoglobin (hypochromia).

Deficiencies of vitamin B_{12} and folic acid impair DNA synthesis and bring about a characteristic anemia in which the erythrocytes are larger than normal (macrocytosis). In the marrow the erythroblasts are abnormally large, and the nuclei of these "megaloblasts" are both larger and less mature in appearance than normal. The prototypic megaloblastic anemias are (1) **pernicious anemia,** in which gastric atrophy results in deficient secretion of intrinsic factor so that vitamin B_{12} cannot be absorbed from the gut, and (2) **folate deficiency,** most often from inadequate dietary intake, such as occurs in alcoholics, but also from increased demand when the rate of erythropoiesis is greatly increased, as in some hemolytic anemias.

Sickle cell anemia is particularly instructive because it represents a host of defects of hemoglobin synthesis in which an alteration in a single base pair of the DNA directing synthesis results in the replacement of a specific amino acid in the globin molecule. This genetically determined abnormal hemoglobin is designated as hemoglobin S. In individuals who are homozygotes for the sickle hemoglobin abnormality, this single amino acid substitution brings about aggregation of hemoglobin molecules into polymers when the oxygen tension is low. This aggregation distorts the normally pliable discoid red cells into a characteristic sickled shape. In this configuration the cells are no longer flexible and may occlude small blood vessels, resulting in tissue damage. The distorted red cells have a shortened life span, bringing about a **hemolytic anemia,** i.e., an anemia because of premature destruction of red cells. Numerous other abnormal hemoglobins have been described.

Another instructive group of hereditary hemoglobin disorders are the **thalassemias,** which are characterized by deficient synthesis of the α or β chains. The thalassemias are inherited as autosomal recessive traits in which the hemoglobin of homozygotes lacks either α or β chains, whereas heterozygotes synthesize varying proportions of the two chains.

The therapy of anemia depends on its pathogenesis. For example, iron may be prescribed to correct iron-deficiency anemia, or vitamin B_{12} to treat pernicious anemia. In some cases, replacement of red cells by transfusion may be needed. Also, administration of erythropoietin is helpful in certain previously refractory anemias.

Blood Groups

Blood transfusion has been made feasible by the recognition of hereditary differences in the chemical structure of the red cell membranes. For example, individuals can be separated into those with blood groups O, A, B, or AB; the separation depends on the genetically determined polysaccharide groups of the membrane glycoprotein. Furthermore, those individuals with blood group O have plasma antibodies directed against group A and B cells, those of group A have antibodies directed against B cells, and vice versa, whereas those of group AB do not have antibodies directed at any of these blood groups. These "natural antibodies" or **isohemagglutinins** are, for the most part, IgM molecules.

For transfusion to be successful, the recipient must lack antibodies directed against the infused cells. Thus individuals of blood group O are "universal donors," and those of blood groups AB are "universal recipients." In common practice, however, blood of the donor is carefully matched to that of the recipient.

The ABO system is only one of many inherited blood group systems. Of special note are the complex Rh blood groups. Ordinarily, individuals whose red cells lack the Rh factor do not have antibodies against this substance in their plasma. Women who are Rh-negative, however, may develop such antibodies if they carry a fetus who has inherited the Rh factor from its father. The maternal antibodies can cross the placenta to the fetus, and a devastating destruction of its erythrocytes can result. This disorder is described as **hemolytic disease of the newborn.** Sensitization to the Rh antigen can also occur if an Rh-negative individual receives a transfusion containing Rh-positive blood cells. In such individuals a subsequent transfusion of Rh-positive blood may lead to premature destruction of the infused erythrocytes.

Immunoglobulins

An **antigen** is a substance that is foreign to the normal body and is immunogenic; i.e., an antigen can induce the formation of an **antibody** (or immunoglobulin) that can combine in a specific way with that antigen. Antigens are usually proteins, but some antigens are polysaccharides, lipopolysaccharides (such as the endotoxin produced by certain gram-negative bacteria), or nucleic acids. Much smaller molecules, called haptens, may bring about antibody formation, but only when they are combined with larger "carrier" molecules that are immunogenic. Subsequently, haptens alone can bind to the specific immunoglobulins that they have evoked. Antibodies are proteins synthesized by plasma cells (derived from B lymphocytes), in conjunction with monocytes and T lymphocytes.

The combination of antigen and antibody, an **immune complex,** is held together by noncovalent bonds. Both antigens and antibodies may have more than one combining site, i.e., they are polyvalent, and hence they may form a latticework of proteins. In the test tube, the lattice of soluble antigens and antibodies may precipitate out of solution, an event called the **precipitin reaction.** If the antigen is a component of a particle, such as a bacterium, the polyvalent antibodies serve as bridges that make the particles stick together, the phenomenon of **agglutination.** Antibodies may also enhance the phagocytosis of particulate antigens, a process described as **opsonization.** In the presence of a group of serum proteins known collectively as **complement,** bacteria or blood cells that have combined with their specific antibodies may be disrupted, a process called **immune lysis** or **cytolysis.**

Antibodies provide a major defense against infectious agents and they may neutralize toxic agents, such as the toxins elaborated by certain organisms. The reaction between antigen and antibody need not be beneficial, and under some circumstances it may injure the host. For example, tissue damage may be evoked by immune complexes, by the destruction of blood or tissue cells, or by immediate reactions that may threaten life by severe bronchoconstriction or a fall in blood pressure that results in shock **(anaphylaxis).**

The formation of antibodies may be induced experimentally by the parenteral administration of antigen. If the host has not previously been exposed to the antigen, antibodies appear in peripheral blood only after several days have elapsed. The titer of antibodies then rises; it reaches a peak after about 2 or 3 weeks, and then it usually gradually decreases. Subsequent exposure to the same antigen induces a much more rapid increase in titer, the so-called anamnestic response (anamnesis means "memory"). This secondary rise in titer, which usually lasts longer than the primary response, comes about by the stimulation of lymphocytes that have previously been exposed to the antigen, so-called memory cells.

Unstimulated B lymphocytes synthesize immunoglobulins that adhere to their surfaces but are not secreted. The B lymphocytes are highly diverse, synthesizing one or another of innumerable immunoglobulins, each with unique variable regions. The steps leading to the secretion of antibody begin with the transport of antigen to the lymph nodes and spleen via lymphatic and blood vascular channels respectively. Antigen is engulfed there by monocytes or macrophages that "present" the antigen, attached to the surface of these cells, to B lymphocytes in the lymphoid organs. Antigens adhere only to those lymphocytes that have on their surface antibodies specific for the presented antigen. This signal is ordinarily not sufficient in itself to induce the production of antibodies. An additional signal is required and is furnished by several protein hormones, ***interleukin−4, and interleukin−5.*** These agents are probably released by CD4 (T helper) lymphocytes that have been similarly exposed to macrophages coated with the specific antigen. Stimulated by their matching antigen, the B lymphocytes proliferate and transform into plasma cells that secrete immunoglobulins specific for that particular antigen into the surrounding milieu. Inherent in this concept is that immunogenic substances in some way stimulate proliferation of only those clones of B lymphocytes that are committed to the synthesis of the antibody proteins matching the specific antigen.

Five classes of immunoglobulins have been recognized, designated IgG, IgA, IgM, IgD, and IgE (Table 20-4). Common to all is a basic unit with a molecular weight of about 150,000 daltons (Fig. 20-3). Each unit is composed of two large (or heavy) and two small (or light) polypeptide chains, held together by disulfide

Table 20-4 Immunoglobulins

Characteristics	*IgG*	*IgA*	*IgM*	*IgD*	*IgE*
Physical					
Molecular weight ($\times 10^3$)	150	150; 400	900	180	190
Basic units/molecule	1	1, 2*	5*	1	1
Heavy chain†	gamma	alpha	mu	delta	epsilon
Concentration in serum (mg/dl)	600-1500	85-380	50-400	<15	0.01-0.03
Biological half-disappearance time (days)	21-23	6	5	2-8	1-5
Crosses placenta	+	−	−	−	−
Principal Ig in secretions	−	+	−	−	−
Functional					
Toxin neutralization	+	+	+	−	−
Agglutination of particulate antigens	+	+	+	−	−
Opsonization	+	?	−	−	−
Bacterial lysis	+	−	+	−	−
Virus inactivation	+	+	+	−	−
Macrophage (monocyte) and neutrophil binding	+	±	−	−	−
Binding to and degranulation of mast cells and basophils with histamine release	−	−	−	−	+
Damage to host tissues	+	−	+	−	+
Complement fixation	+‡	−	+	−	−

Data are from various sources.
*Polymeric form joined by a J chain.
†All immunoglobulins may have kappa or lambda chains but not both.
‡Complement fixed by IgG of subclasses 1, 2, and 3, but not 4.

bonds and by noncovalent forces. Digestion of the basic unit by papain, a proteolytic enzyme, splits the basic unit into two "Fab" fragments and one "Fc" fragment. Cleavage takes place at a portion of the heavy chain known as the hinge region, an area thought to provide a flexibility that allows room for contact of the molecule with antigen. Each Fab fragment contains one of the light chains and the amino-terminal portion of one of the heavy chains. The Fc fragment is a dimer of the carboxyterminal portion of the two heavy chains.

The striking ability of antibodies to react with specific antigens depends on the genetically determined structural diversity of the aminoterminal ends of the light and heavy chains of the Fab fragment. These portions of the Fab fragment are therefore designated as the variable regions of the immunoglobulin; that of the light chain is V_L and that of the heavy chain is V_H. The variable regions of the Fab fragments contain the sequence of amino acids that bind to antigen. Each of the two Fab fragments thus has one combining site for antigen. The remainder of the Fab fragments and the Fc fragments do not display this variability and are therefore referred to as the constant regions of the immunoglobulin. The Fc fragment of the immunoglobulins does not combine with antigens, but it is the site of receptors for complement and for such cells as neutrophils, monocytes, and mast cells.

The IgG, IgD, and IgE molecules are each composed of a single basic unit. The IgA molecules may be monomeric or dimeric. Those of the IgM class have five units that are joined at the carboxyterminal ends of the Fc fragments. The dimeric IgA and pentameric IgM molecules are each joined by an additional polypeptide, the J chain. Each class of immunoglobulins has a distinctive heavy chain, denoted by the corresponding Greek letter.

At different stages in the immune response, lymphocytes that have been stimulated by antigens to differentiate into plasma cells secrete different classes of immunoglobulin. For example, some clones of plasma cells first synthesize IgM and later IgG antibodies directed against the same antigen; in this switch the proteins retain the same variable regions.

The different composition of the heavy chains accounts for the different biological properties of the five classes of immunoglobulins. **IgG** is the dominant antibody in plasma that reacts with bacteria and probably viruses. It can cross the placenta to enter the fetal circulation. IgG is secreted into the colostrum, the initial milk fed to the newborn infant, and may thereby serve as a source of immunity against infection.

IgA is elaborated by plasma cells localized to secretory tissues, where two IgA molecules joined by a J chain may combine with secretory piece, a protein that

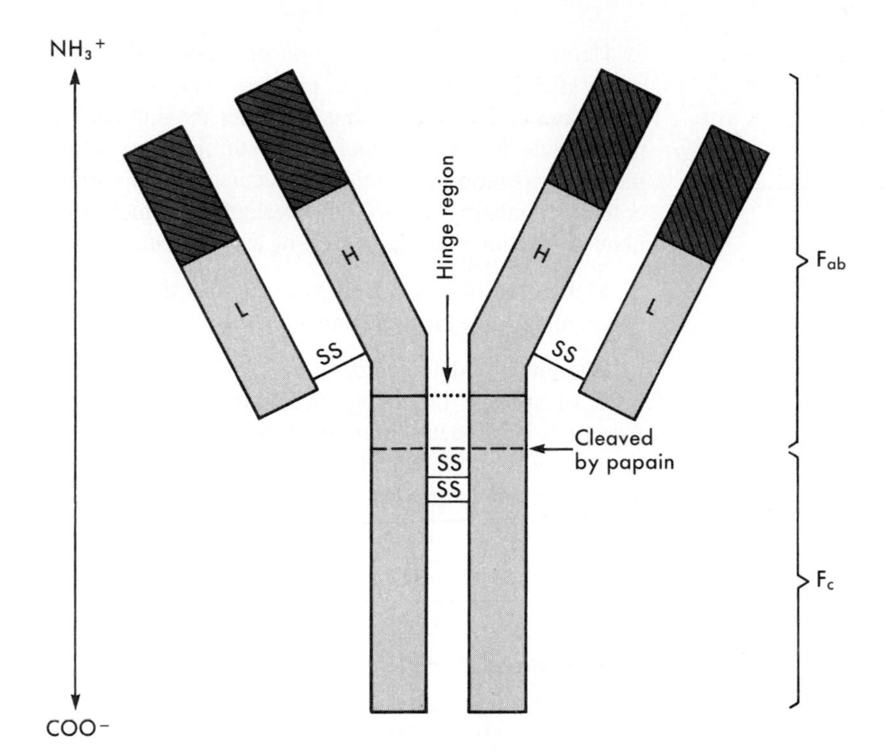

■ **Fig. 20-3** The basic unit of the immunoglobulins is a tetramer of 2 light (L) and two heavy (H) chains, joined by disulfide (SS) bonds. The amino terminals of the light and heavy chains (crosshatching) are the variable regions of the molecule that combine with specific antigen. The remainder of the molecule is designated as the constant region. Papain splits the molecule into three fragments: two Fab (antigen-binding) fragments and one Fc (crystallizable) fragment. Neutrophils, monocytes, and the C1q component of complement bind to the Fc region. The hinge region is indicated by dotted line.

is synthesized by epithelial cells. Combined with secretory piece, IgA is secreted into the tears, saliva, and milk and into respiratory, intestinal, and cervical secretions. The complex forms a barrier to infection by organisms to which the IgA molecules are specifically directed.

IgM is the most prominent immunoglobulin on the surface of lymphocytes. It is the antibody first synthesized by plasma cells after an initial exposure to antigen. The natural antibodies against blood cell antigens are of the IgM class.

IgE binds avidly to mast cells. Exposure of the mast cells to specific antigens may then trigger reactions that lead to discharge of the contents of intracellular granules, including histamine, leukotrienes (oxidation derivatives of arachidonic acid), and other agents. These substances induce vasodilation, increase vascular permeability, and contract certain smooth muscles such as those of the respiratory tree; thus they may bring about attacks of bronchial asthma. In instances that are fortunately rare, the reaction betwen antigen and IgE may be so explosive as to cause a life-threatening fall in blood pressure (from vasodilation) and bronchoconstriction (from contraction of bronchial smooth muscle). These **anaphylactic reactions** may be triggered in previously sensitized individuals by the sting of venomous insects, such as bees or wasps, or by the injection of animal serum or pharmacological agents. The more banal allergic reactions, such as hives (**urticaria**), hay fever, and some types of bronchial asthma, are more localized. The mast cells are degranulated in tissues exposed to ingested or inhaled antigens. IgE may be important as a defense against parasitic infection. The biological function of **IgD** is not yet clarified.

■ *Complement*

The term complement encompasses a group of plasma proteins that, when appropriately activated, bring about responses that defend the body against injury. Among the functions of complement are (1) **chemotaxis,** the attraction of leukocytes to the site of injury; (2) **opsonization,** the presentation of insoluble immune complexes for phagocytosis by granulocytes or monocytes; (3) the **release of anaphylatoxins** from mast cells that increase vascular permeability; and (4) the **disruption of the exterior membranes** of certain microorganisms and blood cells.

The complement system is complex, involving the participation of many different plasma proteins. Two convergent chains of chemical reactions have been delineated: the classic and alternative (or properdin) pathways. The **classic complement pathway** involves the interaction of nine distinct proteins, C1, i.e., the first component of complement, to C9. The components are not numbered in their order of action but rather on historical grounds. The agents participating in the **alternative complement pathway** include proteins designated B, D, and P (for properdin), as well as components C3 to C9 of the classic pathway. A multitude of cell membrane and plasma inhibitors modulate the complement system, among them $C\bar{1}$-inactivator (which has multiple functions), factors H and I (C3b inhibitor), decay accelerating factor (DAF), homologous restriction factor (CD 59), and C4b-binding protein (which circulates in plasma bound to a clotting factor, protein S).

Summary

1. Blood, which makes up about 6% of body weight, is a suspension of various types of cells in a complex aqueous medium, the plasma.

2. Substances dissolved in plasma include electrolytes, proteins, lipids, carbohydrates, amino acids, vitamins, nitrogenous breakdown products of metabolism, and gaseous oxygen, carbon dioxide, and nitrogen. Normally the composition of blood is maintained at safe levels by a variety of homeostatic mechanisms.

3. The cellular elements of blood include erythrocytes (red blood cells), platelets, and five classes of white blood cells (leukocytes): neutrophils, eosinophils, basophils, monocytes, and lymphocytes.

4. Erythrocytes are anuclear cells whose principal protein constituent is hemoglobin. Hemoglobin is composed of four polypeptide chains, to each of which is attached an iron-containing tetrapyrrole, heme. Heme binds oxygen in the pulmonary capillaries and releases it in the capillaries of other tissues. Iron needed for synthesis of hemoglobin is partly recycled from catabolism of heme compounds and partly replenished by iron-containing foodstuffs.

5. Neutrophils provide a major defense against infection by bacteria, which they can ingest and destroy. Eosinophils are cytotoxic for certain parasitic organisms. Basophils, suitably stimulated, release their granular contents to cause either an explosive anaphylactic response or localized vascular dilation and edema. Monocytes are actively phagocytic and participate in other defense reactions including abetting the formation of antibodies by lymphocytes. Lymphocytes, a heterogeneous group of cells, also participate in cell-mediated immune responses.

6. Platelets, which are anuclear fragments of the cytoplasm of megakaryocytes, participate in the formation of clots and in the generation of thrombi.

7. Hematopoiesis, the generation of blood cells, takes place after birth in the bone marrow. The blood cells arise from precursor cells that are under the influence of polypeptide hormones, such as erythropoietin (which induces generation of erythroid precursors) and various colony-stimulating factors and interleukins (which stimulate generation and maturation of leukocytes).

8. Maturation of blood cells also requires the presence of vitamin B_{12} (cyancobalamin) and folic acid.

9. The immune defenses of the body depend on generation of specific immunoglobulins that can neutralize foreign toxic agents. Immunoglobulins may be cytotoxic as well; they can disrupt invading foreign cells and, under some circumstances disrupt cells of the host itself.

Bibliography

Journal articles

Klebanoff SJ: Oxygen metabolism and the toxic properties of phagocytes, *Ann Int Med* 93:480, 1980.

Mayer MM: Complement. Historical perspectives and some current issues, *Complement* 1:2, 1984.

Sieff CA: Hematopoeietic growth factors, *J Clin Invest* 79:1549, 1987.

Weller PF: The immunobiology of eosinophils, *N Engl J Med* 324:1110, 1991.

Books and monographs

Babior BM, Stossel TP: *Hematology: a pathophysiological approach,* New York, 1984, Churchill Livingstone.

Benjamini E, Leskowitz S: *Immunology: a short course,* ed 2, New York, 1991, Wiley-Liss.

Donaldson VH: Complement. In Frohlich ED, editor: *Pathophysiology. Altered regulation, mechanisms in disease,* ed 3, Philadelphia, 1984, JB Lippincott.

Jandl JH: *Blood: pathophysiology,* Boston, 1991, Blackwell Scientific Publications.

Rapaport SI: Introduction to hematology, ed 2, Philadelphia, 1987, JB Lippincott.

Reichlin M, Harley JB: Adaptive immunity. *In* Frohlich ED, editor: *Pathophysiology. Altered Regulation, Mechanisms in Disease,* ed 3, Philadelphia, 1984, JB Lippincott.

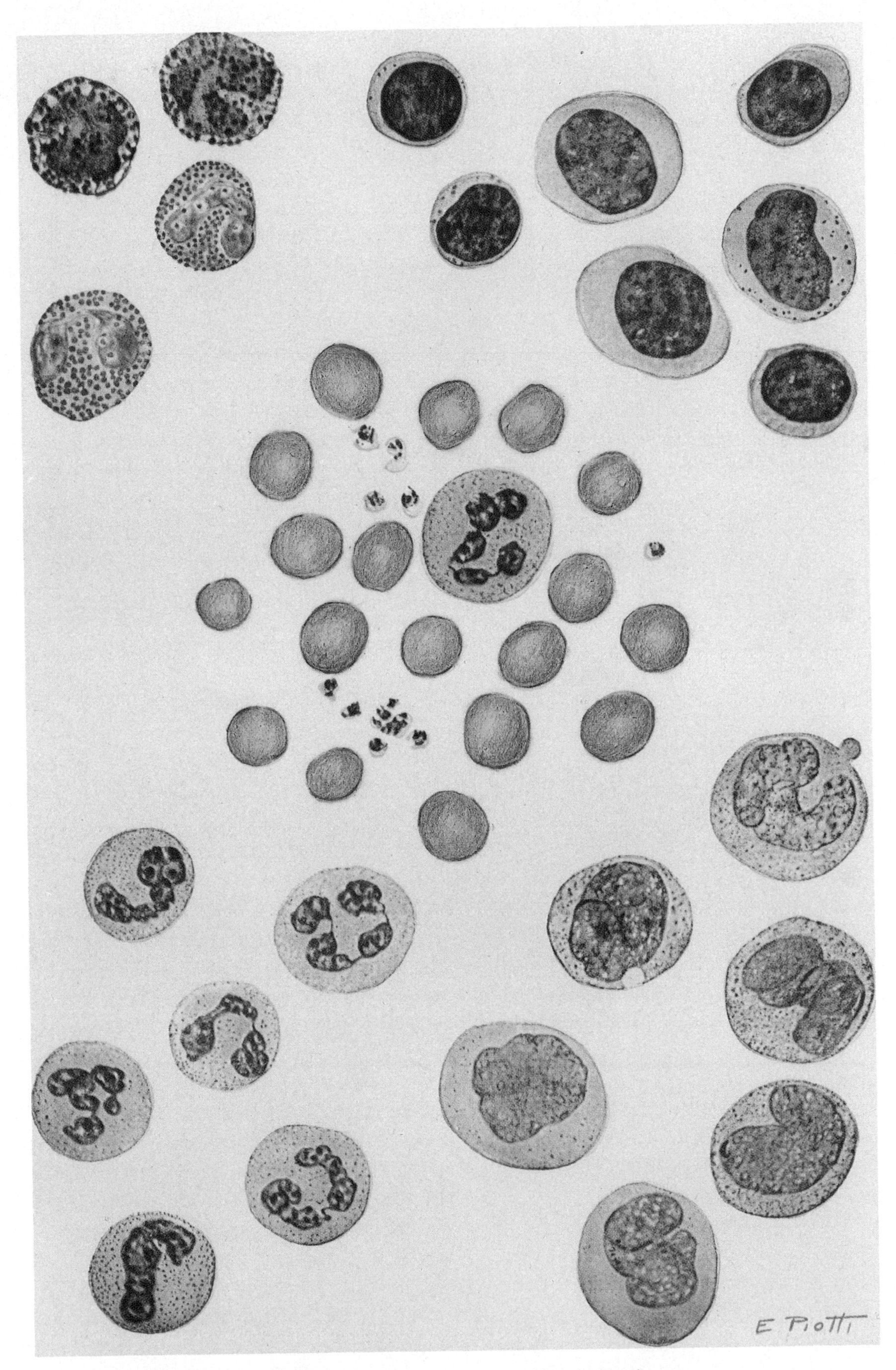

■ **Plate I.** Typical cells of normal human blood.

CHAPTER

21

Hemostasis and Blood Coagulation

The life of every organism depends on preservation of its internal environment. Even one-cell creatures can seal disruptions of their external membranes. Invertebrates protect themselves from loss of their vital hemolymph by processes analogous to vertebrate clotting that use circulating blood cells, hemolymph, or both.

Vertebrates have evolved complex mechanisms to stem hemorrhage after injury. Failure of the hemostatic mechanisms may lead to fatal exsanguination. In mammals an early event after trauma is transitory contraction of damaged blood vessels, but this furnishes little to **hemostasis,** i.e., the arrest of bleeding. Within seconds after vascular injury, however, **platelets,** small anuclear circulating blood cells, adhere to the site of damage and pile up, one on another, to provide a mechanical plug that effectively stops bleeds from minor injury. Hemorrhage from more formidable wounds is stanched by coagulation (or clotting) of blood. When the skin is unbroken, bleeding may be checked by external compression of the injured vessels by extravasated blood. This back pressure is highly effective where the skin is tightly bound to the underlying tissues, as in the finger tip, but it is essentially useless where the skin is distensible. For example, hemorrhage around the eye is unchecked by the counterpressure of extravasated blood, leading to a "black eye." After childbirth hemostasis in the uterus is aided by contraction of the uterine musculature, which compresses the blood vessels that feed the site from which the placenta was wrenched.

Clotted blood is a meshwork of insoluble protein fibers called **fibrin** that traps blood cells and **serum** (i.e., plasma that has undergone coagulation). Fibrin evolves during the clotting process from a dissolved plasma protein, **fibrinogen** (factor I), a transformation catalyzed by a proteolytic enzyme, **thrombin** (see below).

Thrombin is not normally present in circulating blood. When blood is shed, thrombin is cleaved from its precursor in plasma, **prothrombin** (factor II). The release of thrombin is the final step in one or both of two convergent and intertwined chains of chemical reactions, the extrinsic and intrinsic pathways (see below). When trauma disrupts the vascular endothelial lining, blood comes into contact with subendothelial structures and other exposed injured tissues. This sets in motion a succession of catalytic events through either or both pathways. At each step, a proenzyme clotting factor is changed to its enzymatic form, in which it can activate the next proenzyme in the chain. The series of enzymatic steps magnifies the original disturbance of the blood until, in the end, thrombin is released explosively.

The enzymes that are activated via the extrinsic or intrinsic pathways are endopeptidases (i.e., they cleave specific peptide bonds that are not located at the extreme ends of the substrate molecules). Cleavage exposes the site in the substrate clotting factor that is responsible for its biological function. In most cases, the broken bond is located within a folded portion of the polypeptide chain that is held together by internal disulfide cross-links. With the disruption of the peptide bond the polypeptide chain is split into two parts that are held together by a disulfide bridge. The clotting enzymes of the intrinsic and extrinsic pathways are serpins (i.e., proteases whose catalytic sites include a serine residue). Three clotting factors, high-molecular–weight kininogen, antihemophilic factor (AHF, factor VIII), and proaccelerin (factor V), have *not* been shown to have catalytic activity and appear to act as nonenzymatic cofactors.

Reactions of the **extrinsic pathway** of thrombin formation are initiated by contact of blood with injured tissues. The damaged cells furnish a clot-promoting agent, **tissue thromboplastin** (tissue factor; factor III). Tissue thromboplastin interacts with a plasma protein, **factor VII,** to start a chain of events leading to thrombin formation. Thrombin formation can also occur in the absence of tissue thromboplastin via reactions of the **intrinsic pathway** of thrombin formation when plasma comes into contact with a surface different from the normal, intact vascular endothelial lining. Under these conditions a plasma proenzyme, **Hageman factor** (factor XII), is converted to its enzymatically active form by contact with a suitable "foreign" surface. Activated Hageman factor (factor XII_a) then initiates a succession of enzymatic events that involve at least eight other proteins that lead to the elaboration of thrombin. The final steps of the intrinsic and extrinsic pathways are identical.

Clotting is modulated by inhibitory agents within the plasma. The first such substance to be delineated was an inhibitor of thrombin, now called **antithrombin III,** whose action is greatly potentiated by **heparin,** a mucopolysaccharide that is now widely used therapeutically (p. 350).

Blood also has the capacity to redissolve clots that might inadvertently form within blood vessels. This property is attributed for the most part to the elaboration of a plasma proteolytic enzyme, **plasmin.** Plasmin can be generated from its precursor, **plasminogen,** in many ways, including the process of coagulation itself.

Integrity of blood vessel walls is critical in hemostasis. When supporting structures surrounding the blood vessels are defective, as in scurvy (vitamin C deficiency), a hemorrhagic tendency may ensue. Blood flow itself affects hemostasis. The formation of intravascular clots (thrombi) is fostered by stasis of flow and impeded by rapid blood flow.

Blood Coagulation

Nomenclature

As each new clotting factor was discovered, it acquired a trivial name that reflected its supposed function or was derived from the name of a patient whose plasma appeared to be deficient in the particular agent. Often several names were used to designate the same substance. To give order to this chaos, a roman numeral has been assigned to most of the clotting factors. Currently most of the scientific literature uses only the numerical nomenclature. Inevitably this has led to numerous errors, of little consequence in a published article, but of great consequence if the patient or physician is confused. Table 21-1 provides a key to commonly used synonyms. This chapter will avoid the pitfalls of pied type by retaining the admittedly archaic trivial nomenclature, followed, where clarity is needed, by the appropriate roman numeral. Familiarity with both systems of nomenclature will aid in understanding the relationships between basic knowledge and clinical application.

The Synthesis of Clotting Factors

With the exception of the factor VIII/von Willebrand factor complex (VIII/vWf), the clotting factors in plasma are synthesized mainly in the liver. The factor VIII/von Willebrand factor complex can be dissociated into two components of unequal size. The larger subcomponent, designated von Willebrand factor (vWf), is synthesized in vascular endothelial cells and megakaryocytes under direction of an autosomal gene. The site of synthesis of the smaller subcomponent, factor VIII or factor VIII:C, is probably the liver; synthesis is directed by a gene on the X chromosome. The two subcomponents of the factor VIII/vWf complex are held together by noncovalent bonds. Megakaryocytes also synthesize the a (or α) chains of **fibrin-stabilizing factor** (factor

Table 21-1 Blood clotting factors

Roman numeral	*Trivial names*	*Activated or altered state*
Factor I	Fibrinogen	Fibrin
Factor II	Prothrombin	Thrombin
Factor III	Tissue thromboplastin; tissue factor	—
Factor IV	Calcium ions	—
Factor V	Proaccelerin; Ac-globulin	Altered proaccelerin (factor V_a)
Factor VII	—	Factor VII_a
Factor VIII/vWf*	Antihemophilic factor/von Willebrand factor complex	
Factor VIII; factor VIII:C	Coagulant subcomponent of AHF	Altered antihemophilic factor (factor VIII_a)
vWf	Von Willebrand factor subcomponent of factor VIII/vWf complex; von Willebrand factor	
Factor IX	Christmas factor	Factor IX_a
Factor X	Stuart factor	Factor X_a
Factor XI	Plasma thromboplastin antecedent (PTA)	Factor XI_a
Factor XII	Hageman factor (HF)	Factor XII_a
Factor XIII	Fibrin-stabilizing factor (FSF)	Factor XIII_a (fibrinoligase)
—	Plasma prekallikrein (Fletcher factor)	Plasma kallikrein
—	High-molecular–weight (HMW) kininogen (Fitzgerald, Williams, or Flaujeac factor)	—

*Historically the term *factor VIII* was first applied to the coagulant property of antihemophilic factor (factor VIII:C) and later to the noncovalent complex of coagulant AHF and what is now called von Willebrand factor. The von Willebrand factor has been termed vWf, factor VIII:vWf, factor VIIIR:Ag, or ristocetin cofactor (factor VIII:RCo), depending on the context (see text). Some authors now restrict the meaning of the term *factor VIII* to the coagulant part of the factor VIII/vWf complex and designate von Willebrand factor as vWf.

XIII) and a form of fibrinogen whose identity with plasma fibrinogen is disputed. Proaccelerin (factor V) appears to be synthesized by the liver and by vascular endothelial cells. Endothelial cells are also a site of synthesis of tissue thromboplastin (tissue factor), high-molecular-weight kininogen, thrombomodulin, and protein S.

Vitamin K. Hepatocytes synthesize four clotting factors: Christmas factor (factor IX), factor VII, Stuart factor (factor X), and prothrombin (factor II) only if vitamin K is present. Vitamin K is the generic name for a group of fat-soluble quinone derivatives that are plentiful in leafy vegetables. In the newborn infant vitamin K is provided by milk; cow's milk is a much richer source than human milk. Within a few days after birth bacteria that have begun to grow within the lumen of the gut become an important source of the vitamin. Because vitamin K is fat-soluble, its absorption from the gut depends on (1) the presence of bile salts that are excreted by the liver into the duodenum and (2) normal digestive and absorptive mechanisms for fat. Some experiments suggest that synthesis of the vitamin K–dependent clotting factors may be controlled by humoral factors.

Synthesis of the vitamin K–dependent clotting factor proceeds in two stages. First hepatocytes synthesize polypeptide progenitors of each factor, a process not requiring the vitamin. In the second step vitamin K, reduced to its hydroquinone form in the liver, acts as a cofactor for a specific microsomal carboxylase (Fig. 21-1). This carboxylase inserts a second carboxyl group into the γ-carbon of certain glutamic acid residues in the polypeptide chains. These unique tricarboxylic glutamic acid residues serve as points of attachment for the calcium ions that are needed to transform the vitamin K–dependent factors to their enzymatically active states. During carboxylation the reduced vitamin K is oxidized to an epoxide form, from which vitamin K is then regenerated by enzymatic reduction.

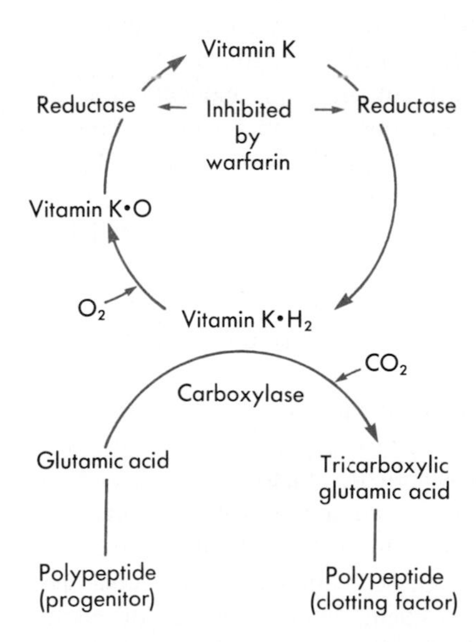

■ **Fig. 21-1** The role of vitamin K is the insertion of a second carboxyl group into the γ-carbon of glutamic acid residues of the vitamin K-dependent clotting factors. *Vitamin K·O,* vitamin K epoxide.

Synthesis of the vitamin K–dependent factors is inhibited by **dicumarol** and its congeners, notably **warfarin.** These agents have had long usage as "anticoagulants" in the prevention and treatment of thrombosis. Dicumarol and similar agents competitively inhibit the enzymes that reduce vitamin K and its epoxide (see Fig. 21-1). In this way they suppress the carboxylation of the progenitors of the vitamin K–dependent factors so that synthesis of these factors is not completed.

Vitamin K is also required for synthesis of other proteins found, for example, in plasma, bone, kidney, lung, spleen, and placenta. Presumably in each case the function of these proteins depends on attachment of calcium ions to tricarboxylic glutamic acid residues formed through the action of vitamin K. Of pertinent interest are two vitamin K–dependent plasma proteins, **protein C** and **protein S.** Protein C, when activated by thrombin, inhibits the coagulant properties of antihemophilic factor (factor VIII) and proaccelerin (factor V), whereas protein S enhances this property of protein C (p. 351).

That four distinct clotting factors require vitamin K to complete their synthesis suggests that they have a common evolutionary origin. In agreement with this, extensive similarities in deoxyribonucleic acid (DNA) and amino acid sequences have been found. A reasonable explanation is that the different factors arose as the result of duplication and subsequent mutation of the genes responsible for the synthesis of an ancestral protein. Similarly sequence similarities between factor VIII and proaccelerin (both nonenzymatic cofactors) and between plasma thromboplastin antecedent (PTA, factor XI) and plasma prekallikrein (both substrates of activated Hageman factor ([factor XII_a]) suggest common evolutionary origins.

■ *Plasma Fibrinogen and the Formation of Fibrin*

Blood clotting is the visible result of the conversion of a soluble plasma protein, **fibrinogen** (factor I), into an insoluble meshwork of fibrin. Fibrinogen is a dimeric glycoprotein with a molecular weight of 340,000 daltons. Each half of the dimer is composed of three polypeptide chains, designated Aα, Bβ, and γ, respectively. The six chains are held together by disulfide bonds.

The conversion of fibrinogen to **fibrin,** (which forms the essential part of blood clots,) takes place in three stages (Fig. 21-2). First fibrinogen undergoes limited proteolysis by thrombin that has evolved during the ear-

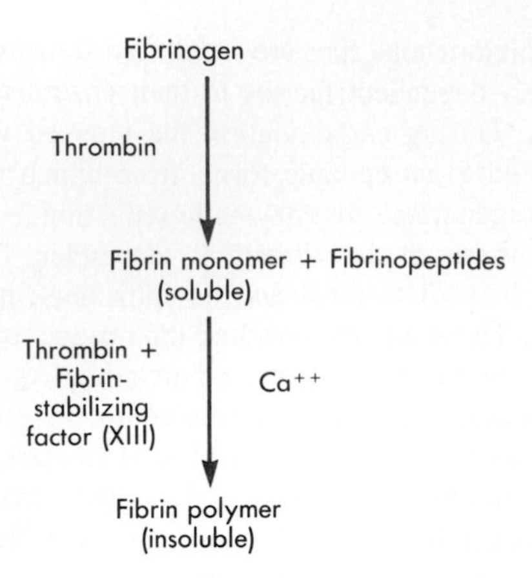

■ **Fig. 21-2** The formation of fibrin. (Redrawn from Ratnoff OD: Hemorrhagic disorders: coagulation defects. In Beeson PB, McDermott W, Wyngaarden JB, editors: *Cecil textbook of medicine,* ed 15, Philadelphia, 1979, WB Saunders.)

lier steps of the clotting process. The partially digested fibrinogen molecules, called fibrin monomers, polymerize by electrostatic forces into insoluble strands of fibrin. Finally the constituent fibrin monomers of the fibrin strands are cross-linked covalently by a plasma enzyme, **activated fibrin-stabilizing factor** (fibrinoligase; factor XIII_a).

In the first step of fibrin formation, thrombin cleaves four small polypeptide fragments, each with a molecular weight of about 1,500 daltons, from each fibrinogen molecule, reducing its weight by about 2%. One fragment, **fibrinopeptide A,** is released from the amino-terminal end of each Aα chain, and another, *fibrinopeptide B,* is released from the amino-terminal end of each Bβ chain. The residue, fibrin monomer, retains its dimeric structure, each half having three chains, designated α, β, and γ. Fibrinopeptide A is released by thrombin more rapidly than fibrinopeptide B, and with its separation polymerization begins. Polymerization depends primarily on the separation of fibrinopeptide A.

As the first fibrin monomers are generated, they are surrounded by unaltered fibrinogen molecules with which they form loose complexes. As more fibrin monomers accumulate, the equilibrium shifts so that fibrin monomers now polymerize with each other. At the same time the complexes of fibrinogen and fibrin monomer dissociate, making more fibrin monomers available for polymerization. The monomers aggregate both side to side and end to end, gradually building insoluble polymers that thicken and lengthen to form the visible fibrous strands. Some of the polymers appear branched so that the fibrin strands build into a complicated meshwork that traps blood cells and serum. In addition platelets bind to the polymerizing strands of fibrin (Fig. 21-3).

Calcium ions are not required for the transformation of fibrinogen to fibrin monomer, but at the concentration present in plasma, they greatly accelerate the polymerization process.

■ **Fig. 21-3** Human blood clot showing red blood cells immobilized within a network of fibrin threads. The small spheres are platelets. Scanning electron micrography (× 9000). (From Shelly WB: Of red cell bondage, *JAMA,* 249:3089, 1983, American Medical Association.)

Clots formed in purified mixtures of thrombin and fibrinogen are held together by noncovalent forces and have low tensile strength. Such clots dissolve readily in solutions of urea or monochloroacetic acid, which dissociate noncovalent bonds. In contrast clots formed in normal human plasma are insoluble in these dispersing agents and have high tensile strength. Plasma contains a proenzyme, **fibrin-stabilizing factor** (factor XIII). When activated, this factor acts as a transaminase to forge covalent links between the γ-carboxyl groups of glutamic acid residues of one fibrin monomer and the ε-amino groups of lysyl residues of another. Fibrin-stabilizing factor is a tetramer composed of two a or α chains and two b or β chains; one half of the fibrin-stabilizing factor in blood is found in platelets, where it is composed only of α chains.

Activation of fibrin-stabilizing factor is brought about by thrombin, which cleaves a small polypeptide from each α chain. In the presence of calcium ions, fibrin-stabilizing factor then induces covalent links among fibrin monomers. It creates dimers between the γ chains of two adjacent fibrin monomers and, at a slower rate, polymers among several α chains. Fibrin resists dissolution by urea or monochloroacetic acid only after these α-chain polymers are formed.

Fibrin-stabilizing factor has other actions besides covalent cross-linking of fibrin. For example, it forges links between molecules of contractile proteins of muscles or platelets; it bonds fibronectin (cold-insoluble globulin) to itself and to fibrin or collagen and it binds α_2-plasmin inhibitor to fibrin (p. 350). In the absence of fibrin-stabilizing factor, experimental wound healing is retarded. This phenomenon may be related to stimulation of the growth of fibroblasts by activated fibrin-stabilizing factor.

The concentration of fibrinogen in normal human plasma averages 270 to 300 mg/dl. Thus about 10 g of fibrinogen is present in the circulating blood; perhaps an additional 5 g is present in extravascular fluid. An increased concentration of fibrinogen is commonplace in normal pregnancy and in innumerable disease states, particularly those associated with inflammation, tissue damage, or neoplasm. In these disease states fibrinogen is thought to be an **acute phase reactant,** (i.e., a plasma component whose concentration is increased under the stress of disease).

An increased concentration of fibrinogen accelerates the settling of red blood cells when blood is allowed to stand. This increased **erythrocyte sedimentation rate** was described by Hippocrates, who interpreted the rapid settling of red cells as a basic cause of disease. Settling is fostered by the higher density of red cells than that of plasma and is resisted by the surface area that the cells present to the surrounding medium. Normally these forces are almost balanced so that red cells sediment only very slowly. Rapid settling comes about because the increased amounts of fibrinogen neutralize the electrostatic forces that exist on the surface of red cells and normally prevent these cells from sticking to each other. Under these conditions the red cells form stacks of cells that resemble piles of coins. The stacked cells, in contrast to individual erythrocytes, present relatively less surface area to the surrounding plasma to oppose the downward pull of gravity, and therefore the cells sediment more rapidly.

The biological half-life of fibrinogen under normal conditions is about 3 to 4 days (i.e., half of the fibrinogen present in the plasma at any given time will have disappeared in this interval) (Table 21-2). Thus about 15% of plasma fibrinogen must be replaced daily by its continual synthesis, which takes place in the parenchymal cells of the liver. Among the agents proposed as stimuli to synthesis are growth hormone, thyroxin, corticotropin, hormones released from inflammatory lesions, and degradation products released by the digestion of fibrinogen and fibrin by plasmin, a plasma proteolytic enzyme (p. 349).

The normal site and mechanism of the catabolism of fibrinogen are unknown. Fibrinogen is used through its conversion to fibrin both in hemostasis and in formation of inflammatory lesions. Perhaps some fibrinogen is digested by plasmin, particularly when this enzyme has been activated by severe stress. But these mechanisms participate only marginally in the normal catabolism of fibrinogen. The hypothesis that fibrinogen catabolism is the result of continual slow intravascular clotting has little to support it.

Table 21-2 Molecular weight, concentration, and biological half-life of plasma hemostatic factors

Factor	*Molecular weight (× 1,000)*	*Approximate plasma concentration (mg/dl)*	*Biological half-life**
Fibrinogen (I)	340	200-400	3-4 days
Prothrombin (II)	72	12	3 days
Proaccelerin (V)	330	0.4-1.4	15-30 days
Factor VII	48	0.05-0.06	4-5 hr
Antihemophilic factor (VIII)	1,000-12,000	0.5-1	10-12 hr
(VIII:C)	265	0.005-0.010	12 hr
(VIII:vWf)	500-20,000	0.5-1.0	24 hr
Christmas factor (IX)	57	0.4-0.5	18-24 hr
Stuart factor (X)	59	0.7-1.2	24-36 hr
Plasma thromboplastin antecedent (XI)	160	0.4-0.6	2-3 days
Hageman factor (XII)	80	1.5-4.5	2 days
Plasma prekallikrein	88	3.5-4.5	35 hr
HMW kininogen	120	8-9	6 days
Fibrin-stabilizing factor (XIII)	320	1-2	11-12 days
Plasminogen	92	20	2-2½ days

*The biologic half-life of a clotting factor is measured by infusing it into a patient with a congenital deficiency of the factor or by infusing radiolabeled factor into a normal individual. Ordinarily the curve describing the rate of disappearance of the factor from plasma has two components. First the concentration of the infused substance decreases relatively rapidly as it diffuses into extravascular spaces. Thereafter the concentration decreases more slowly, reflecting the catabolism of the factor. The biological half-life, (i.e., the length of time until the titer of the infused factor decreases by 50%) is calculated from the second component of the disappearance curve.

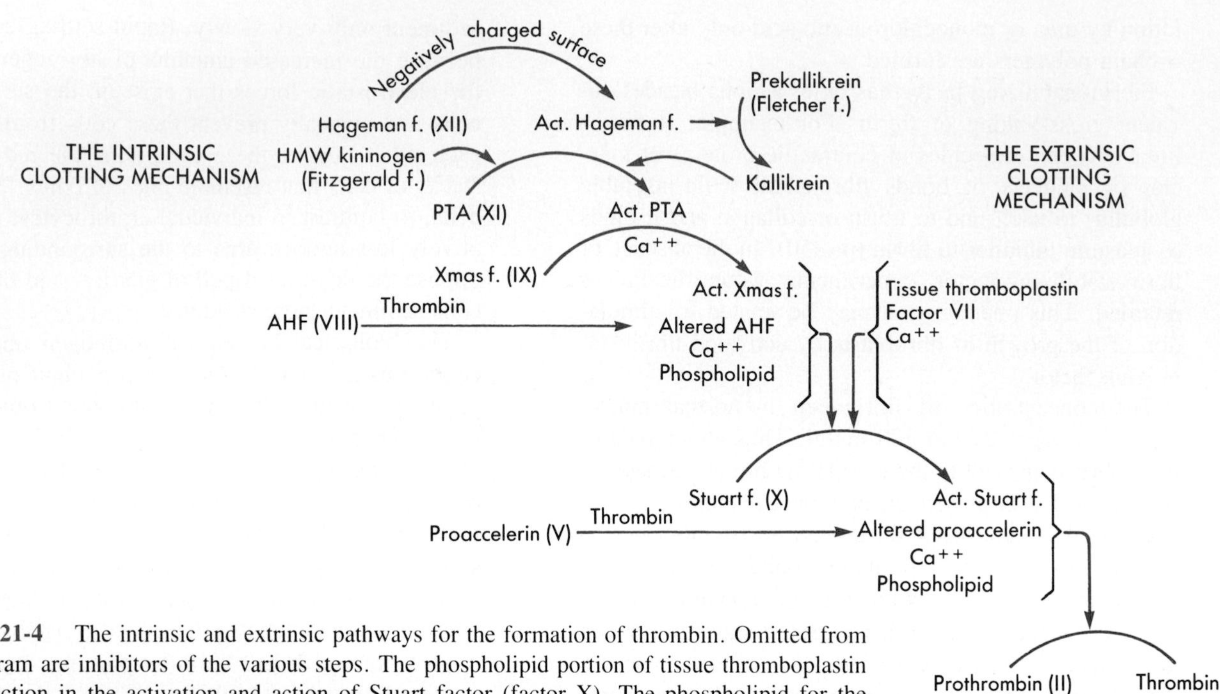

■ **Fig. 21-4** The intrinsic and extrinsic pathways for the formation of thrombin. Omitted from the diagram are inhibitors of the various steps. The phospholipid portion of tissue thromboplastin may function in the activation and action of Stuart factor (factor X). The phospholipid for the intrinsic pathway is furnished by platelets and by the plasma itself. Augmentation of the action of factor VII by thrombin and by the activated forms of Hageman factor (factor XII_a), Christmas factor (factor IX_a), and Stuart factor (factor X_a), and the activation of Christmas factor by the factor VII-tissue thromboplastin complex are not depicted. *Act,* activated; *HMW,* high molecular weight, *PTA,* plasma thromboplastin antecedent; *X-mas,* Christmas, *AHF,* antihemophilic factor. (From Ratnoff OD: Hemorrhagic disorders: coagulation defects. In Beeson PB, McDermott W, Wyngaarden, JB, editors: *Cecil textbook of medicine,* ed 15, Philadelphia, 1979, WB Saunders.)

■ *The Formation of Thrombin*

Thrombin, the enzyme ultimately responsible for the formation of fibrin monomers, is generated by the catalytic scission of prothrombin by activated **Stuart factor** (factor X_a). Under physiological conditions activation of Stuart factor (factor X) takes place through two series of enzymatic steps, the intrinsic and extrinsic pathways (Fig. 21-4). These reactions are so intertwined that the reader may find it helpful to read the next few sections twice to appreciate the nature of the feedback mechanisms that are involved.

■ *Surface-Mediated Reactions Initiating the Intrinsic Pathway*

Activation of Stuart factor (factor X) by the intrinsic pathway is the result of a series of reactions that begins when blood comes into contact with negatively charged surfaces. When venous blood is drawn into a polystyrene or silicone-coated tube and centrifuged to sediment its cells, the separated plasma clots readily when placed in glass tubes, but much more slowly in polystyrene or silicone-coated tubes. Such experiments suggest that glass has qualities that induce a change in plasma leading to elaboration of thrombin. The clot-promoting properties of glass have been related to its negative surface charge. Many similarly charged insoluble substances, such as kaolin (clay), diatomaceous earth (Celite), and talc, have a similar effect. These substances are foreign to the body, but drug addicts may inject themselves with drugs adulterated with kaolin or talc and in this way may self-induce thrombosis.

Uncertainty exists about the natural analogues of the clot-promoting solids. Sebum (the oily secretion of skin), some forms of collagen or basement membrane proteins, sulfatides (such as are found in brain tissue), disrupted vascular endothelial cells, and endotoxins derived from gram-negative bacteria have all been suggested.

The events triggered by negatively charged surfaces include not only the generation of thrombin, but also the mediation of other defense mechanisms of the body against injury, including inflammation, immune responses, and **fibrinolysis** (i.e., dissolution of clots). Four plasma proteins are involved in the initiation of one or another of these "surface-mediated defense reactions": **Hageman factor** (factor XII), **plasma thromboplastin antecedent** (factor XI), **plasma prekallikrein** (Fletcher factor), and **high-molecular-weight kininogen** (Fitzgerald, Williams, or Flaujeac factor). All four

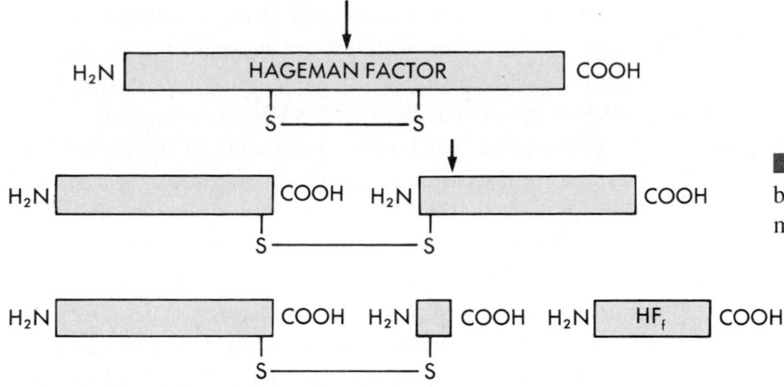

Fig. 21-5 The activation of Hageman factor. What brings about scission (arrows) is uncertain. *HF_f* Carboxy-terminal fragment.

factors are readily adsorbed from normal plasma to negatively charged surfaces such as glass. It is apparently in this adsorbed state that they exert their clot-promoting functions.

The first step in the various surface-mediated defense reactions, including the initiation of events of the intrinsic pathway, is conversion of Hageman factor to an enzymatic form that can activate PTA. When **Hageman factor** is adsorbed to clot-promoting, negatively charged surfaces, it undergoes a change in shape that exposes amino acid residues buried within the native molecule. Whether this shape change is enough to induce clot-promoting activity is disputed. Adsorbed to such surfaces from normal plasma, Hageman factor is split internally within a disulfide loop (Fig. 21-5). The two-chain species that results can transform PTA to its enzymatically active state. Hageman factor is then further cleaved into two polypeptide fragments. The carboxy-terminal fragment, designated HF_f, includes the sequence of amino acids that is responsible for Hageman factor's enzymatic activity, but this fragment is a much weaker activator of PTA than the two-chain species. Recent evidence suggests that thrombin, as well as activated Hageman factor, can activate PTA.

The conversion of PTA to its activated state by activated Hageman factor requires the presence of high-molecular-weight kininogen. The conversion takes place on the surface of clot-promoting, negatively charged surfaces to which the three clotting factors are adsorbed. Activation is accelerated by similarly adsorbed plasma prekallikrein, but the presence of this factor is not an absolute requirement for this step. Activation of PTA may be enhanced because this clotting factor and plasma prekallikrein are both loosely bound to high-molecular-weight kininogen in circulating plasma.

Long before their role as clotting factors was discovered, both **plasma prekallikrein** and **high-molecular–weight kininogen** had been identified in plasma by their participation in experimental inflammatory reactions. In the presence of high-molecular–weight kininogen, activated Hageman factor (factor XII_a) changes plasma prekallikrein to its enzymatic form, plasma kallikrein. In turn plasma kallikrein releases small polypeptide *kinins,* notably a nonapeptide called bradykinin, from plasma proteins described as **kininogens.** Plasma contains two groups of kininogens, distinguished by their molecular weight. The high molecular weight kininogens are much more avid substrates for plasma kallikrein than those of lower molecular weight; only the high-molecular-weight kininogens promote clotting. Bradykinin, released through the action of plasma kallikrein, can dilate small blood vessels, increase their permeability, and induce pain. In this way bradykinin can bring about the warmth, redness, swelling, and pain of inflammatory lesions. Bradykinin also contracts certain smooth muscles, a property useful in its biological assay. Human (but not bovine) high-molecular-weight kininogen retains its clot-promoting properties after the bradykinin sequence has been removed.

Activated Hageman factor participates in at least three other surface-mediated defense reactions. Directly or indirectly it augments the clot-promoting properties of factor VII (p. 347), it converts plasminogen to plasmin (p. 349), and it transforms C1, the first component of complement (a group of proteins participating in immune reactions; see Chapter 20), to its enzymatically activated state ($C\bar{1}$) (p. 337). In the test tube activated Hageman factor also converts prorenin to renin, an enzyme that brings about the formation of angiotensin, a polypeptide that contracts arterioles.

Activation of Christmas factor (factor IX) via the intrinsic pathway. The function of activated PTA (factor XI_a) in the intrinsic pathway is the activation of Christmas factor (factor IX), the first step that requires calcium ions. **Christmas factor** is a vitamin K–dependent protein synthesized by the liver under the direction of a gene on the X chromosome (p. 341). Activated PTA cleaves Christmas factor at two points (Fig. 21-6). First the single chain of Christmas factor is split within an internal disulfide loop. The two-chain molecule that results acquires enzymatic properties only after activated PTA goes on to sever a small "activation" polypeptide from the longer of the two chains.

In the test tube plasma kallikrein, activated Stuart factor (factor Xa), and activated factor VII (factor VII_a) can activate Christmas factor. Two of these agents, Stu-

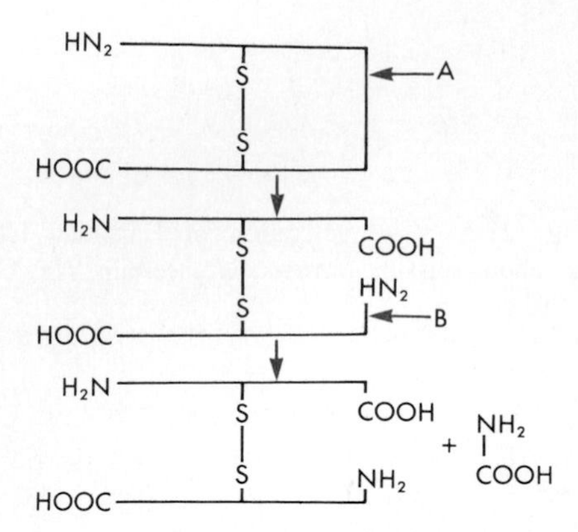

Fig. 21-6 The activation of Christmas factor (factor IX). Activated plasma thromboplastin antecedent (factor XI_a) splits Christmas factor successively at points *A* and *B*. Christmas factor acquires enzymatic properties only with the release of the "activation polypeptide."

art factor and factor VII, participate in the extrinsic pathway of thrombin formation (p. 347). Perhaps this explains the benign nature of deficiencies of Hageman factor, PTA, plasma prekallikrein, or high molecular weight kininogen. Activation of the extrinsic pathway may circumvent the steps of the intrinsic pathway before the participation of Christmas factor.

Antihemophilic factor is part of a complex molecule with a variety of defined properties that is designated as the **factor VIII/vWf complex.** It is functionally deficient in several hereditary bleeding disorders, the most common of which are classic hemophilia and von Willebrand's disease. The factor VIII/vWf complex is readily dissociated into two subcomponents of unequal size. The larger, high molecular weight subcomponent, vWf, is synthesized in vascular endothelial cells and megakaryocytes. It has a molecular weight that varies from about 500,000 to 20,000,000 daltons. The vWf subcomponent forms precipitates when mixed with antiserum raised in goats or rabbits that have been immunized with the factor VIII/vWf complex. For this reason it has been described as factor VIIIR:Ag, (i.e., an antigen [Ag] related [R] to the factor VIII/vWf complex); (recall from Chapter 20 that an antigen is an agent that induces formation of specific antibodies).

The vWf subcomponent of the factor VIII/vWf complex acts as a carrier for antihemophilic factor (factor VIII), and it enhances the adhesion of platelets to subendothelial structures. In this way vWf may be important for hemostasis when subendothelial structures are exposed to circulating blood by vascular injury. In vitro the higher molecular weight forms of vWf bind to platelets in the presence of the antibiotic ristocetin and cause the platelets to clump ("agglutinate"). This property of the von Willebrand factor portion of the factor VIII/vWf complex has therefore been described as factor VIII:RCo (*RCo* represents ristocetin cofactor). The designation vWf is based on its absence in von Willebrand's disease; vWf is not deficient in classic hemophilia. Additionally, when normal blood is filtered from a column of glass beads, most of its platelets are retained within the column, a phenomenon that depends on the presence of the vWf component of the factor VIII/vWf complex.

The second, smaller subcomponent of the factor VIII/vWf complex contains the amino acid sequences needed for blood coagulation and has therefore been named factor VIII:C or, less clearly, factor VIII. This subcomponent has a molecular weight of about 270,000 daltons and is probably synthesized mainly in the liver.

The titer of antihemophilic factor is under hormonal control. Exercise, stress, pregnancy, and administration of estrogens, epinephrine, vasopressin, or its synthetic derivative, 1-desamino-8-D-arginine vasopressin (DDAVP), all increase the titers of both subcomponents of the factor VIII/vWf complex.

Classic hemophilia (hemophilia A) is the most common hereditary disorder of the intrinsic pathway. In this disease the coagulant titer of antihemophilic factor (factor VIII:C) is reduced, but both the concentration of vWf and the capacity to agglutinate platelets in the presence of ristocetin are normal. Classic hemophilia is the prototype of X chromosome–linked disorders. It affects only males, who pass the abnormal gene to their daughters, all of whom are carriers. In turn half of a carrier's sons have hemophilia and half of her daughters are carriers. Women who carry the abnormal gene are usually asymptomatic. However, they can be recognized in about 95% of cases because their plasma contains relatively less coagulant antihemophilic factor (factor VIII:C) than vWf. About the same degree of precision in recognition can be obtained by molecular genetic techniques.

Von Willebrand's disease is a hereditary disorder of both sexes inherited as an autosomal dominant trait in which affected individuals have symptoms suggestive of mild hemophilia. In women menorrhagia and bleeding after childbirth may be troublesome. In the usual forms of Von Willebrand's disease plasma is deficient in both factor VIII: C and vWf, but there are many variant forms of this disorder.

Activation of Stuart factor (factor X) via the intrinsic pathway. Activated Christmas factor (factor IX_a) converts Stuart factor to its activated state. Under physiological conditions the activation of Stuart factor depends on the interaction of activated Christmas factor with antihemophilic factor (factor VIII) and calcium ions on a phospholipid surface; the phospholipids are derived principally from platelets but also from plasma.

Stuart factor (factor X) is a vitamin K–dependent plasma proenzyme (p. 341), which is the precursor of activated Stuart factor, the enzyme immediately responsible for the release of thrombin from prothrombin. Stuart factor is activated by proteolytic separation of a polypeptide fragment from one of its two chains. In the intrinsic pathway this cleavage is brought about by a complex of activated Christmas factor (factor IX_a), antihemophilic factor (factor VIII), and calcium ions, all adsorbed to phospholipid micelles. The protease responsible for cleavage of Stuart factor is activated Christmas factor, whereas antihemophilic factor appears to act as a nonenzymatic cofactor. Native factor VIII has little or no clot-promoting activity. Cleavage by thrombin or activated Stuart factor (factor X_a) releases a molecular weight (MW) 92,000 dalton polypeptide that serves as a cofactor for the activation of Stuart factor by activated Christmas factor, a process that optimally requires the presence of phospholipids (or platelets) and calcium ions. Further proteolysis vitiates this property of factor VIII.

Activation of Stuart factor (factor X) via the extrinsic pathway. The early events of the intrinsic pathway are short-circuited when plasma is exposed to injured tissues. Such tissues contain one or more powerful agents known generically as **tissue thromboplastin** (tissue factor). Tissue thromboplastin is found principally in cell membranes of almost every cell, including peripheral blood monocytes; platelets are an important exception. Tissue thromboplastin is a lipoprotein complex composed of heat-stable phospholipids and a heat-labile glycoprotein with a molecular weight that varies from 50,000 to 330,000 daltons, depending on the source of the thromboplastin. Removal of the phospholipid from tissue thromboplastin inactivates its clot-promoting properties, which can be restored by readdition of the phospholipid portion.

The clot-promoting properties of tissue thromboplastin are mediated through its action on a trace plasma protein, **factor VII,** a single-chain protein that is synthesized when vitamin K is available (p. 341). By itself factor VII has no activity. Complexed stoichiometrically with tissue thromboplastin, factor VII is transformed to an enzymatically active form (factor VII_a) that cleaves Stuart factor, converting it to its activated state (factor X_a). In turn factor X_a activates molecules of factor VII bound to tissue thromboplastin that have not yet been activated, increasing the amount of factor VII_a available. Indeed it has been proposed that factor VII bound to tissue thromboplastin is activated only after some factor X_a has been generated, perhaps via the intrinsic pathway. The subsequent steps of the extrinsic and intrinsic pathways are identical.

As was noted earlier, the factor VII–tissue thromboplastin complex also activates Christmas factor (factor IX). *Thus reactions of the extrinsic pathway affect the intrinsic pathway and vice versa.*

The generation of thrombin. Thrombin, the protease that is ultimately responsible for the formation of a fibrin clot, is separated from its parent molecule, prothrombin, by activated Stuart factor (factor X_a) through reactions that are augmented by proaccelerin (factor V), calcium ions, and phospholipid.

Prothrombin (factor II) is a vitamin K–dependent glycoprotein (p. 341). The molecular structure has been dissected enzymatically into several distinct parts (Fig. 21-7). The functional anatomy of prothrombin has been clarified through the use of various enzymes. Thus thrombin splits prothrombin into two parts, an amino-terminal segment, fragment 1, and a carboxy-terminal portion, prethrombin 1. Fragment 1 contains the tricarboxylic glutamic acid residues that serve as points of attachment for calcium ions. The prethrombin 1 segment can be further cleaved by activated Stuart factor (factor X_a) into an amino-terminal portion, fragment 2, and a carboxy-terminal fragment, prethrombin 2, that contains an internal disulfide loop.

Thrombin formation proceeds through several stages. The first activated Stuart factor that forms via reactions of the intrinsic or extrinsic pathways binds via calcium links to tricarboxylic glutamic acid residues in fragment 1 of prothrombin (see Fig. 21-7). There it slowly cleaves prothrombin in two places, between fragment 2 and prethrombin 2 and within the internal disulfide loop in prethrombin 2. This second split converts prethrombin 2 to thrombin, a molecule with a molecular weight of 39,000 daltons.

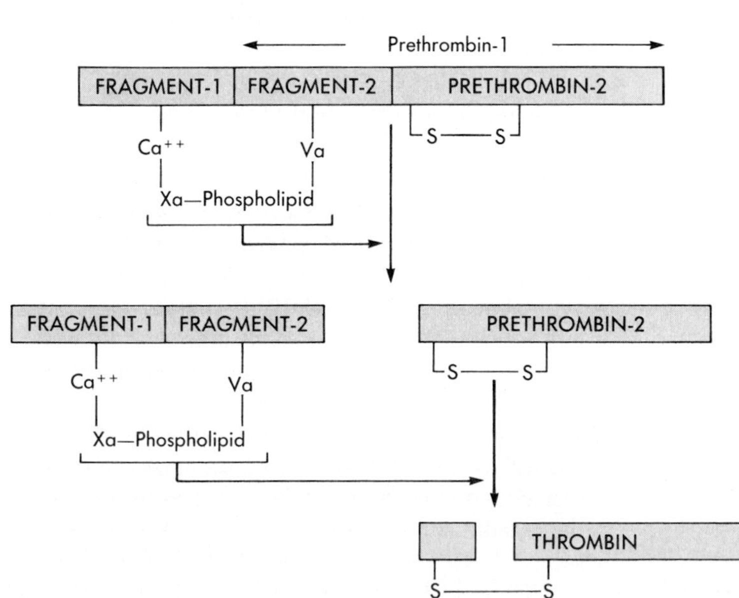

Fig. 21-7 The release of thrombin from prothrombin.

As thrombin is liberated, it alters **proaccelerin** (factor V) by cleavage of its polypeptide chains. Proaccelerin in its native state has no clot-promoting properties. This alteration of proaccelerin is also brought about by the activated form of Stuart factor (factor X_a). In this altered form, described as factor V_a, proaccelerin attaches to the fragment 2 portion of prothrombin molecules. There it is a nonenzymatic accelerator of the action of activated Stuart factor, bringing about a rapid release of thrombin. Both activated Stuart factor and proaccelerin are adsorbed to phospholipid, which appears to augment their activity. Phospholipid for the intrinsic pathway is furnished by platelets and (to a lesser extent) plasma, and for the extrinsic pathway it is furnished by tissue thromboplastin. Thus the events that lead to the activation of Stuart factor and to the formation of thrombin are parallel. Indeed, as noted earlier, homologies exist in the amino acid sequences of the coagulant portion of the antihemophilic factor complex (factor VIII:C) and proaccelerin (factor V).

Thrombin converts fibrinogen to fibrin monomer (p. 341); activates fibrin-stabilizing factor (factor XIII) (p. 342), factor VII (p. 347), and protein C (p. 351); and aggregates platelets. Other properties of thrombin include chemotactic activity for monocytes and neutrophils, release of tissue plasminogen activators from endothelial cells, contraction of vascular smooth muscle, and mitogenic activity. Which of these actions of thrombin are of physiological relevance is unclear. Thrombin binds to fibrin, and in this way fibrin contributes in a minor way to the inactivation of this enzyme.

The role of vascular endothelium in blood coagulation. The endothelium that lines blood vessels has long been thought to provide a surface that lacks the capacity to initiate blood clotting. Normally the luminal surface of vascular endothelial cells does not induce clotting or attract platelets. Indeed these cells appear to retard clot formation. The endothelial surface is rich in heparinlike proteoglycans, notably heparan sulfate. These complex polysaccharides can serve as cofactors for two plasma inhibitors of thrombin, antithrombin III and heparin cofactor II; antithrombin III, bound to heparinlike proteoglycans, also inhibits other serine proteases of the clotting mechanism (p. 350). Endothelial cells under appropriate circumstances may have the reverse effect, promoting the clotting process. Thus activated PTA (factor XI_a), Christmas factor (factor IX), activated Christmas factor (factor IX_a), Stuart factor (factor X), thrombin, and antithrombin III can bind to the surface of endothelial cells grown in tissue culture. Furthermore, endothelial cells synthesize the high-molecular–weight subcomponent (vWf) of the factor VIII/vWf complex and proaccelerin (factor V).

The significance of these observations is only now beginning to be understood. When thrombin is added to endothelial cells grown in culture to confluence (i.e., to the point that each cell is in contact with the next), the endothelial cells appear to shrink, and gaps appear between them. If this phenomenon were to occur in vivo, one might anticipate that the circulating blood could come into contact with subendothelial structures that might foster local thrombosis.

When activated PTA (factor XI_a) binds to endothelial cells, it can activate Christmas factor that has been similarly bound. This activation can then lead to the formation of fibrin via the reactions of the intrinsic pathway. If instead the endothelial surface is perturbed, as can be brought about by thrombin or by endotoxin (a lipopolysaccharide released by gram-negative bacteria), the endothelial cells acquire tissue thromboplastinlike activity, and they can then interact with factor VII. In this way endothelial cells bring about formation of fibrin both by reactions of the extrinsic pathway and by activation of Christmas factor that is adsorbed to the endothelial surface. Factor V for these various reactions is furnished by the plasma, by the endothelial cells themselves, and much more effectively by its release from platelets. Thus under appropriate conditions in vivo endothelial cells may provide a template for the assembly of clotting factors and in this way promote localized coagulation.

Fibrinolysis and Related Phenomena

Under appropriate conditions clotted blood may reliquefy. The agent responsible for fibrinolysis, the dissolution of fibrin, is **plasmin,** a typical serine protease that is the activated form of a plasma glycoprotein proenzyme, **plasminogen.** Plasminogen is a single-chain polypeptide that is synthesized in the liver and perhaps elsewhere as well. It can be changed to plasmin by many activators, not all of which are physiological.

Streptokinase, a protein elaborated by certain β-hemolytic streptococci, is of particular interest as an activator of plasminogen because it is used clinically to dissolve thrombi and emboli. Streptokinase combines stoichiometrically with plasminogen to form a complex that transforms plasminogen to plasmin by cleavage within an internal disulfide loop. Staphylococci synthesize a similar activator of plasminogen called staphylokinase.

Streptokinase and staphylokinase are foreign to the body, but many physiological activators also exist. Normal urine contains a serine protease, **urokinase,** that is elaborated and excreted by the kidney. This enzyme converts plasminogen to plasmin through the same cleavages as the streptokinase/plasminogen complex. Urokinase, too, is used for the clinical dissolution of thrombi and emboli.

Activators of plasminogen have also been identified in human plasma, milk, tears, saliva, seminal fluid, and many tissues, including vascular endothelium, blood monocytes, and tumor cells. The **tissue plasminogen activators** are serine proteases that cleave plasminogen to form plasmin. The tissue activators of plasminogen fall into two groups, depending on whether or not they resemble urokinase as determined by immunological and functional tests. Those tissue plasminogen activators that are not urokinaselike bind more avidly to fibrin, and they lyse clots much more effectively than they hydrolyze fibrinogen. Tissue plasminogen activators are now widely used to dissolve thrombi, particularly in coronary arteries.

A tissue plasminogen activator of considerable physiological interest can be demonstrated in blood drawn from a vein distal to a tourniquet that has been applied to the upper arm for several minutes. Clots formed from the plasma of such blood dissolve at an increased rate. Activation of plasminogen under these conditions appears to be initiated by an agent that is released from venous endothelium and that is similar to the nonurokinase-like tissue plasminogen activator.

Fibrinolysis, as measured in vitro, is greatly enhanced by emotional stress, strenuous physical activity, or injection of epinephrine, pyrogens, or vasopressin or its synthetic analogue, DDAVP. This phenomenon is probably the consequence of the release of the venous endothelial plasminogen activator. Vascular plasminogen activator can also be released by the vitamin K–dependent plasma protease activated protein C (p. 351).

Plasma itself can convert plasminogen to plasmin "spontaneously" through one of several pathways. Three enzymes activated through surface-mediated reactions (p. 345), plasma kallikrein, activated PTA (factor XI_a), and (much more weakly) activated Hageman factor (factor XII_a), can activate plasminogen. Thus activation of the intrinsic pathway of thrombin formation initiates reactions that lead to fibrinolysis.

Plasmin is not very specific. It digests not only fibrin but numerous other substances as well, including fibrinogen, antihemophilic factor (factor VIII), proaccelerin (factor V), and other clotting factors. Plasmin may participate in the immune defenses of the body, for it converts the first component of complement (C1) to its enzymatically activated state ($C\bar{1}$). It also separates a peptide fragment from the fifth component of complement (C5) that has chemotactic properties (i.e., the capacity to attract leukocytes). It fragments Hageman factor, liberating its enzymatically active carboxy-terminal fragment (HF_f), and it releases kinins from kininogens, both directly and possibly by converting prekallikrein to kallikrein (p. 345). In addition plasmin digests casein and certain synthetic amino acid esters and small polypeptide amides, all of which have been used to examine and assay its function.

The most notable property of plasmin is the digestion of fibrinogen and fibrin. In purified systems plasmin attacks these two proteins with equal avidity. In plasma, however, generation of plasmin occurs more readily on the surface of a clot. One explanation for this phenomenon is that plasminogen and its activators adhere to fibrin. In this situation plasmin can digest fibrin relatively unhampered by the inhibitors of this enzyme in plasma (p. 350). The practical result is that the therapeutic injection of streptokinase, urokinase, or tissue plasminogen activator may bring about dissolution of intravascular clots without a major effect on circulating fibrinogen. Any plasmin that leaks into the circulation after its activation on the surface of a clot is readily inactivated by its inhibitors in plasma, notably α_2-plasmin inhibitor (p. 350).

The digestion of fibrinogen takes place in steps (Fig. 21-8). First plasmin separates fragments sequentially from the carboxy-terminal ends of the α-chains and the amino-terminal ends of the β chains of fibrinogen. The principal residue, fragment X, is still coagulable by thrombin. Fragment X is then further digested by plasmin into fragments designated Y, D, and E. The degra-

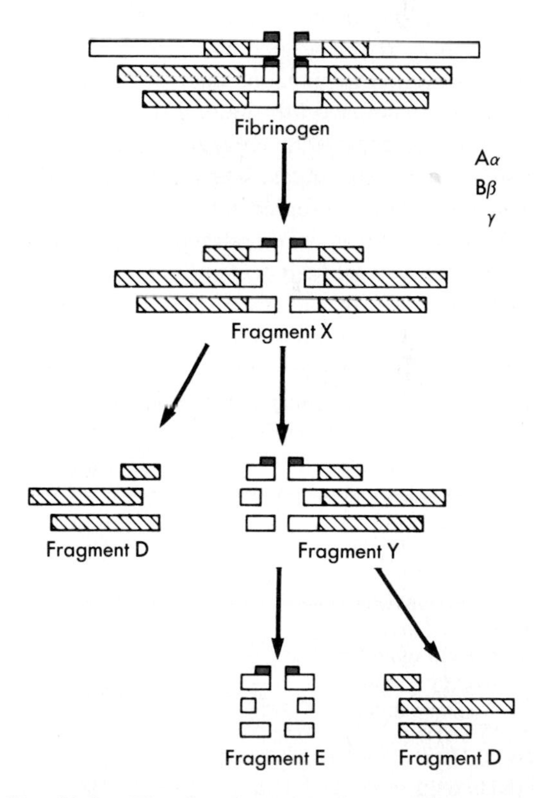

■ **Fig. 21-8** The digestion of fibrinogen by plasmin. The three chains of fibrinogen are Aα, Bβ, and γ; the solid black areas represent fibrinopeptides A and B. The first cleavage of fibrinogen removes a carboxy-terminal segment of the Aα chain; the residue (fragment X) is still coagulable. Fragment X is then cleaved successively to fragment Y, two fragments D, and fragment E. (Redrawn from Sherry S: Mechanisms of fibrinolysis. In Williams WJ et al, editors: Hematology, ed 2, New York, 1977, McGraw-Hill.)

dation products of fibrinogen are inhibitors of clotting. These products interfere with the formation and action of thrombin and with the polymerization of fibrin monomers. In part this inhibition of clotting may be caused by the formation of soluble complexes of fragments X and Y with fibrinogen or fibrin monomer. The fragmentation of fibrin proceeds along similar lines, but fragment X is not coagulable.

Fibrinogen and fibrin can be digested not only by plasmin but also by proteases released by leukocytes. The proteases responsible have been identified as elastases, collagenases, and chymotrypsin-like enzymes.

Inhibitors of Clotting and Fibrinolysis

Plasma is rich in substances that can inhibit activated clotting and fibrinolytic enzymes (Table 21-3). Presumably these serve to restrict the growth of intravascular clots and to minimize the deleterious effects of unbridled fibrinolysis. Most of these inhibitors are members of the **serpin** family, (i.e., proteins that form complexes with and thereby inactivate serine proteases). An exception is α_2-macroglobulin, which first undergoes partial digestion by its target enzyme and then forms an irreversible complex with the enzyme. In this configuration α_2-macroglobulin prevents the enzyme from digesting protein substrates, whereas the catalytic activity of the enzyme is preserved.

Antithrombin III is the principal inhibitor of thrombin and activated Stuart factor (factor X_a). To a lesser extent it blocks the action of the enzymatically active forms of each of the other plasma serine proteases involved in the formation and dissolution of clots; it is without effect on fibrin-stabilizing factor (factor XIII). Inhibition results from the formation of a stoichiometric complex between antithrombin III and the activated enzymes. The importance of antithrombin III is evident from the frequency of thrombosis in individuals with a partial deficiency of this inhibitor.

The inhibitory properties of antithrombin III are greatly potentiated by the addition of heparin, with which it forms a complex. **Heparin** is a negatively charged sulfated polysaccharide that is synthesized principally in mast cells. It is found in many tissues, including lung, liver, and intestinal mucosa, but it is not a normal constituent of blood. Without antithrombin III heparin is not a significant anticoagulant. *Heparin is widely used in the prevention and treatment of thrombotic states,* in which its enhancement of the action of antithrombin III limits the growth of intravascular clots. Overdosage of heparin is readily controlled by administration of protamine sulfate, a highly positively charged polypeptide derived from fish sperm.

A second plasma cofactor for the inhibition of thrombin by heparin has been detected. This agent, variously called **heparin cofactor II** does not react with other proteases of the clotting mechanism.

The major inhibitor of plasmin is **α_2-plasmin inhibitor** (α_2-antiplasmin), with which plasmin combines stoichiometrically and probably covalently in an irreversible link. α_2-Plasmin inhibitor also inhibits the adsorption of plasminogen to fibrin and in this additional way inhibits fibrinolysis. Inhibition of fibrinolysis is enhanced by cross-linkage of α_2-plasmin inhibitor to fibrin by activated fibrin-stabilizing factor (factor XIII_a).

Table 21-3 Molecular weight, concentration, and biological half-life of plasma inhibitors of clotting and fibrinolysis

Factor	*Molecular weight (× 1,000)*	*Approximate plasma concentration (mg/dl)*	*Biological half-life**	*Deficiency state*
Antithrombin III	62	15-30	45-60 hr	Thrombotic tendency
Heparin cofactor II	66	6-12		Thrombotic tendency
α_1-Antitrypsin (α_1-proteinase inhibitor)	55	130	5-6 days	Chronic obstructive pulmonary disease
α_2-Macroglobulin	725	285-350	5 days	—
α_2-Plasmin inhibitor (α_2-antiplasmin)	67	5-8	2½ days	Bleeding tendency
C1-esterase inhibitor ($C\bar{1}$-INH)	105	18-25	3-4 days	Hereditary angioneurotic edema
Protein C	62	0.3-0.6	9-15 hr	Thrombotic tendency
Protein S	69	1-4	11-21 hr	Thrombotic tendency
Histidine-rich glycoprotein	81	9-17	3 days	Thrombotic tendency
Activated protein C inhibitor	57	0.4-0.6	—	—
Tissue factor pathway inhibitor†	38	0.01	—	—
Plasminogen activator inhibitor I	50	0.005	—	—
Plasminogen activator inhibitor II‡	48-70	0.3	—	—

*See note Table 21-2.
†Lipoprotein-associated coagulation inhibitor (LACI) or extrinsic pathway inhibitor (EPI).
‡Detected during pregnancy.

Patients with a hereditary deficiency of this inhibitor have a severe hemorrhagic disorder in which the rapid dissolution of clots by the unchecked action of plasmin impedes hemostasis. α_2-Plasmin inhibitor inactivates activated Hageman factor, plasma kallikrein, activated Stuart factor (factor X_a), and thrombin, but these reactions are probably not significant.

Plasma also contains inhibitors of the activation of plasminogen by tissue plasminogen activators and urokinase. At least three such plasma inhibitors have been described. **Tissue plasminogen activator inhibitor 1** is found in all normal plasmas. **Tissue plasminogen activator inhibitor 2** is synthesized by placental tissues and is detected in the plasma of pregnant women. Tissue plasminogen activator inhibitor 3 is apparently identical with the plasma inhibitor of **activated protein C** (see below).

Histidine-rich glycoprotein is a plasma protein that circulates complexed to plasminogen, inhibits the activation of plasminogen by reducing its binding to fibrin, and neutralizes the anticoagulant properties of heparin.

α_2-Macroglobulin is a plasma protein that slowly inactivates the proteolytic properties of plasmin, thrombin, plasma kallikrein, and tissue plasminogen activator. It is a major inhibitor of plasma kallikrein and the chief back-up system for the inhibition of plasmin when supplies of α_2-plasmin inhibitor are exhausted. As noted earlier, its mode of action is unusual. The enzyme binds to and partially digests the inhibitor. As a consequence the shape of the enzyme is so changed that its ability to attack protein substrates is greatly reduced.

α_1-Antitrypsin (**α_1-proteinase inhibitor**) is a glycoprotein that has broad specificity and is synthesized in the liver. Among enzymes that it inactivates in the test tube are activated PTA (factor XI_a), plasmin, activated protein C, and probably thrombin and plasma kallikrein.

C1 esterase inhibitor (C$\bar{1}$-INH) is a plasma protein that inhibits the activated form of the first component of complement (C1). C$\bar{1}$-INH also inhibits activated HF, plasma kallikrein, activated PTA, plasmin, and tissue plasminogen activator. These properties are probably of no clinical importance because patients with hereditary angioneurotic edema, deficient in C1-INH, do not have defective hemostasis or fibrinolysis.

Protein C is a two-chain vitamin K–dependent plasma proenzyme that can be changed to an enzymatically active state (activated protein C) by thrombin or by activated Stuart factor (factor X_a). Activation by thrombin is greatly accelerated on the surface of endothelial cells that furnish a cofactor, **thrombomodulin,** which binds thrombin to the cell surface (Fig. 21-9). Activated proaccelerin (factor V_a) also enhances activation of protein C by thrombin. Activated protein C inhibits the coagulant properties of antihemophilic factor (factor VIII) and proaccelerin (factor V), particularly after their alteration or activation by thrombin. This may explain in part the low titers of factors VIII and V in some cases of widespread intravascular coagulation. Activated protein C also enhances fibrinolysis, probably by neutralizing the inhibitor of the activation of plasminogen described as plasminogen activator inhibitor 1 (PAI-1), with which it forms complexes.

Protein C inhibitor is a protease inhibitor that forms complexes with and inhibits the formation of activated protein C, a property enhanced by heparin. Its action is nonspecific; it blocks the actions of tissue plasminogen activator, urokinase, plasma kallikrein, activated Stuart factor (factor X_a), activated plasma thromboplastin antecedent (factor XI_a), and thrombin.

Protein S is a nonenzymatic vitamin K–dependent plasma protein that is a cofactor for activated protein C, enhancing the inactivation of the coagulant forms of antihemophilic factor (factor VIII) and proaccelerin (factor V). Perhaps two thirds of protein S circulates in a reversible complex with C4b-binding protein, an inhibitor of the complement system (p. 338).

Plasma tissue factor pathway inhibitor [lipoprotein-associated clotting inhibitor (LACI) or extrinsic path-

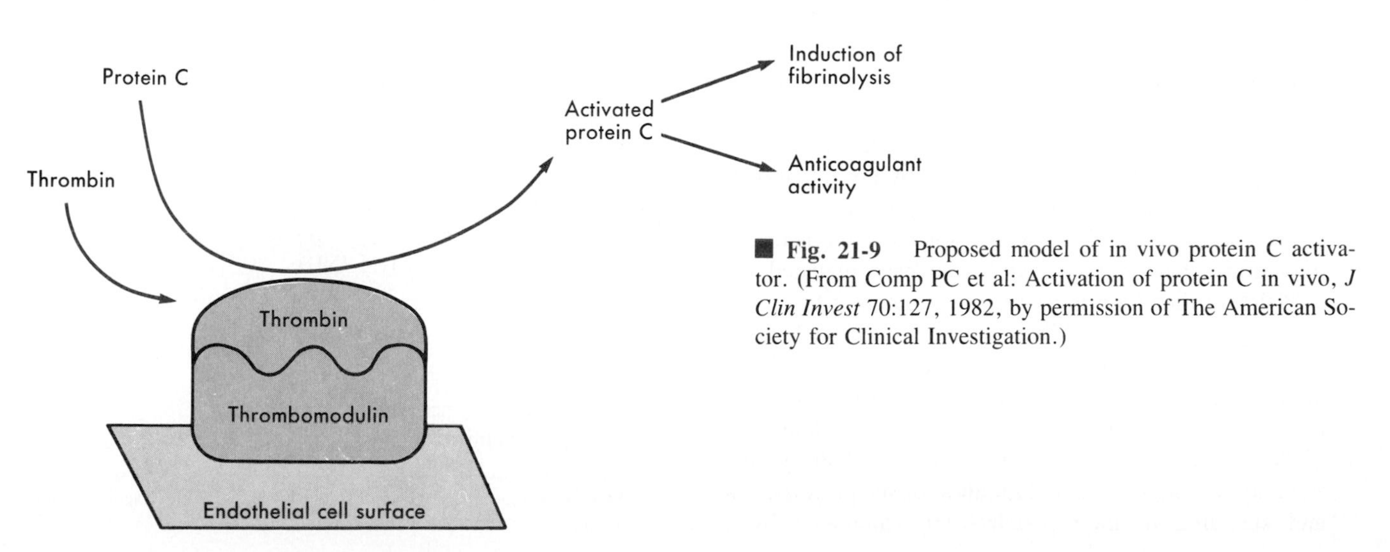

Fig. 21-9 Proposed model of in vivo protein C activator. (From Comp PC et al: Activation of protein C in vivo, *J Clin Invest* 70:127, 1982, by permission of The American Society for Clinical Investigation.)

way inhibitor (EPI)] is a trace plasma protein that modulates the extrinsic pathway of thrombin formation by blocking the clot-promoting effect of factor VII_a that is bound to tissue thromboplastin. In vitro tissue factor pathway inhibitor is synthesized by hepatocytes and vascular endothelial cells. This inhibitor likely plays an active role in maintaining the fluidity of the blood. Its capacity to inhibit the extrinsic pathway may help to explain the bleeding tendency of patients with classic hemophilia (factor VIII deficiency) and Christmas disease (factor IX deficiency) in whom this pathway appears to be intact.

Protease nexins are cell membrane-associated serpin inhibitors that are not present in plasma in significant concentrations. Protease nexin I is said to inhibit thrombin, and protease nexin II inhibits activated plasma thromboplastin antecedent (factor XI_a). Protease nexin II is of particular interest because it is the precursor of a protein found in the central nervous system lesions of Alzheimer's disease.

Platelets

The **platelets** are small, disk-shaped anuclear cells with an average diameter of about 2 to 4 μm. Platelets serve many functions in the hemostatic and defensive mechanisms of the body. They arise by budding from the cytoplasm of their progenitors, the **megakaryocytes,** which are large polyploid cells with 4, 8, or 16 nuclei and are derived from the primitive hematopoietic stem cells. In the normal adult megakaryocytes are found almost exclusively in bone marrow. In the newborn and under certain pathologic conditions in the adult megakaryocytes are also found in the liver, lungs, and other organs. Maturation of megakaryocytes to the point at which they shed their platelets takes 4 or 5 days; it has been estimated that each megakaryocyte gives rise to 1,000 to 1,500 platelets.

Normal blood contains between 150,000 and 350,000 platelets/μl. Not all platelets are in the circulating blood; as many as a third are sequestered elsewhere, principally in the spleen. The life of an individual platelet is about 8 to 12 days; their destruction appears to be a consequence of senescence. Aged or damaged platelets are removed from the circulation by the reticuloendothelial system. The spleen is an important site of removal, and in individuals whose spleens have been removed the platelet count may be modestly elevated.

Platelet production is probably regulated at least partially by the total number of platelets destroyed rather than by their concentration in circulating blood, as though the metabolites of destroyed platelets stimulated thrombopoiesis. One signal for platelet production is probably interleukin 3 (IL-3), a small protein synthesized by T lymphocytes. Maturation of megakaryocytes and shedding of their platelets are stimulated by a poorly defined hormone, thrombopoietin, but this substance is probably not required for platelet production.

Platelet Structure

The anatomy of the platelets is complex (Fig. 21-10). The exterior coat or glycocalyx is rich in glycoproteins that may be responsible for adhesion of platelets to subendothelial structures and for aggregation of platelets one to another when these cells are appropriately stimulated. Separated from plasma, the platelets have a loose covering of plasma proteins. Among these proteins are fibrinogen, antihemophilic factor (factor VIII), and proaccelerin (factor V). The surface glycoproteins serve as receptors for many agents, including fibrinogen, wVf, thrombin, adenosine diphosphate (ADP), catecholamines, serotonin (5-hydroxytryptamine), collagen, and immune complexes (i.e., complexes of antigen and antibody).

Just within the exterior coat is a phospholipid-rich "unit membrane" that is the principal source of clot-promoting phospholipids. This layer is the site of many enzymes, among them phospholipases, glycogen synthetase, and adenylate cyclase. The innermost layer of the platelet membrane contains fibrillar microtubular

Platelet organelles*

Electron-dense bodies (dense granules)
- Serotonin
- Catecholamines
- Nonmetabolic "storage pool" of ATP, ADP
- Calcium ions
- Pyrophosphate
- α Granules
- Albumin
- Platelet factor 4
- β-Thromboglobulin
- Fibronectin
- Plasminogen
- Platelet fibrinogen
- Platelet-derived growth factor
- High-molecular-weight kininogen
- α_2-Plasmin inhibitor
- Proaccelerin (factor V)
- von Willebrand factor (vWf)
- Thrombospondin
- C1-esterase inhibitor ($C\bar{1}$-INH)
- Protease nexin II

Lysosomes
- Acid hydrolases

Mitochondria

Golgi apparatus (inconstant)

Glycogen granules

*The localization of certain substances in the specific granules is only tentative.

structures and microfilaments that contain actin and myosin and maintain the shape of the platelet.

The cytoplasm of the platelets is equally complex. At its periphery is an open canalicular system whose channels appear to be invaginations of the external platelet membranes. Through these tubular structures the contents of platelet granules can be discharged to the exterior of the cell. Scattered through the cytoplasm is a second series of channels, the dense tubular system, which does not connect with the open canalicular system. These structures, which represent smooth endoplasmic reticulum, are probably the sites of prostaglandin synthesis and calcium ion sequestration. The cytoplasm also contains a potent contractile system that includes an actin-binding protein and proteins that resemble muscle actin and myosin. Within the cytoplasm are the a or α chains of fibrin-stabilizing factor (factor XIII) and a metabolic pool of adenosine triphosphate (ATP) and ADP that furnish energy for the cell's metabolism.

Dispersed throughout the cytoplasm are numerous organelles, including small electron-dense bodies, lysosomes, α granules, mitochondria, glycogen granules, and, inconstantly, a Golgi apparatus (see platelet organelles on p. 352).

Platelet Adhesion, Aggregation, and the Release Reaction (Fig. 21-11)

Within seconds after the endothelial lining of blood vessels is disrupted, some circulating platelets adhere to the exposed subendothelial structures, such as collagen. **Von Willebrand factor** (vWf) (p. 346), which is synthesized in endothelial cells and megakaryocytes, appears to enhance platelet adhesion by forming a bridge between collagen fibers and a specific platelet surface glycoprotein, glycoprotein Ib. The adherent platelets become more spherical. By active contraction that involves polymerization of microfibrillar actin, the platelets send out pseudopods or spicules that spread out like

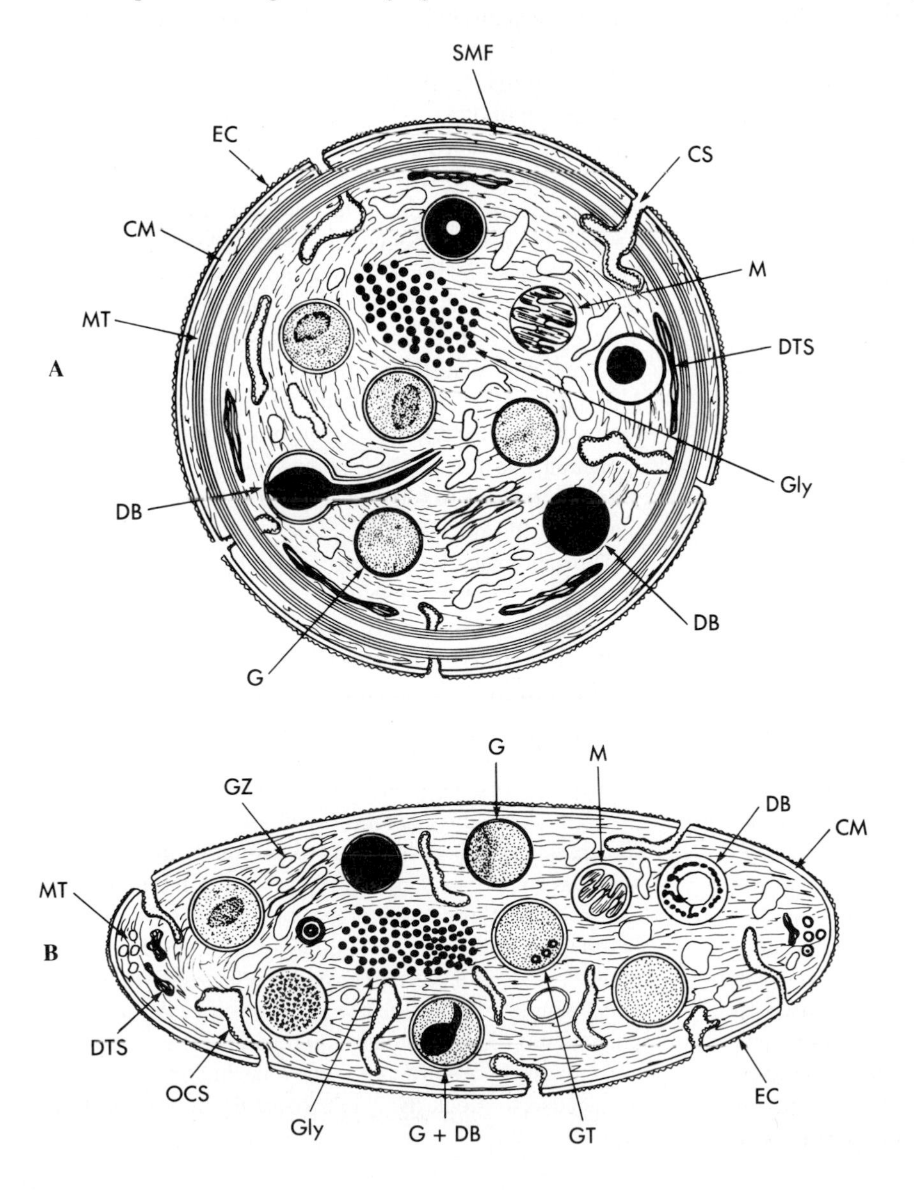

Fig. 21-10 Diagrammatic representation of a platelet in an equatorial plane **A**, and in cross section **B.** Components of the peripheral zone include the exterior coat *(EC)*, trilaminar unit membrane *(CM)*, and submembrane area containing specialized filaments *(SMF)*, which form the platelet wall and line channels of the surface-connected canalicular system *(CS)*. The matrix of platelet cytoplasm contains active microfilaments, structured filaments, the circumferential band of microtubules *(MT)*, and glycogen *(Gly)*. Organelles embedded in the sol-gel zone include mitochondria *(M)*, granules *(G)*, and electron-dense bodies *(DB)*; some α-granules appear to contain tubular structures *(GT)*. The membrane systems include the surface-connected open canalicular system *(CS* or *OCS)* and the dense tubular system *(DTS)*. An occasional Golgi apparatus *(GZ)* is found in some platelets. (Redrawn from White JG: *Am J Clin Pathol* 71:363, 1979.)

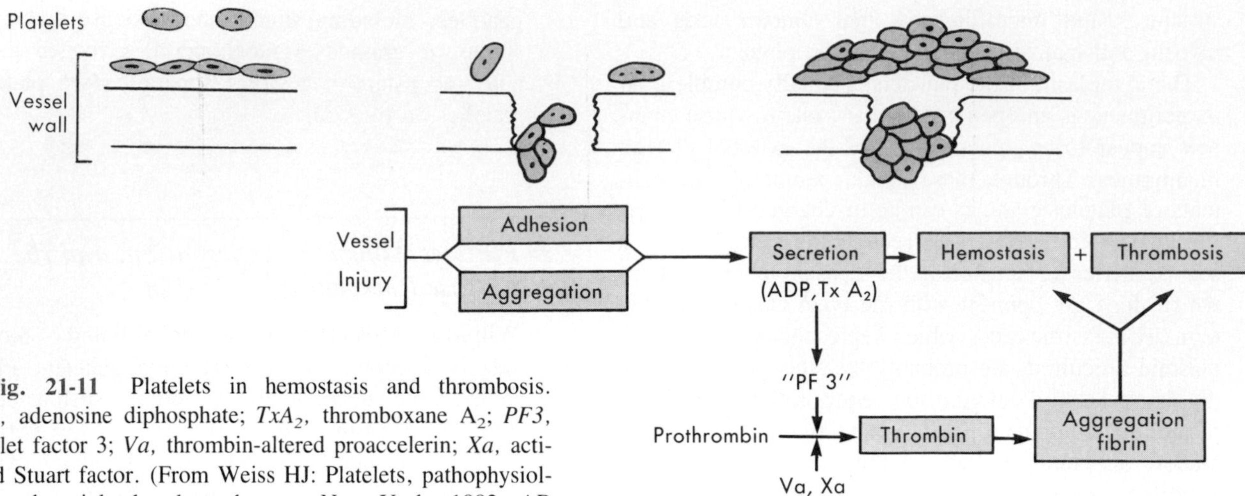

■ **Fig. 21-11** Platelets in hemostasis and thrombosis. *ADP,* adenosine diphosphate; *TxA_2,* thromboxane A_2; *PF3,* platelet factor 3; *Va,* thrombin-altered proaccelerin; *Xa,* activated Stuart factor. (From Weiss HJ: Platelets, pathophysiology and antiplatelet drug therapy, New York, 1982, AR Liss.)

the legs of a spider along the subendothelial fibrils. Simultaneously contraction of the platelet's microfibrils forces the cytoplasmic granules to migrate toward the center of the cell and to discharge the contents of the granules into the surrounding blood via the open canalicular system. This process is called the **release reaction.** Substances extruded from the granules, including the nonmetabolic pool of ADP in the dense bodies, attract other platelets to those adherent to the injured vessel walls. At the same time the clotting mechanisms are activated as blood comes into contact with the damaged vascular surfaces. Thrombin, generated locally in this way, promotes further platelet aggregation and release. It also brings about the formation of strands of fibrin that bind the platelet aggregates into a firm **hemostatic plug** that may control bleeding from small vascular injuries. The plug may also serve as a nidus for a more formidable intravascular clot or thrombus.

How platelets stick to subendothelial structures and to each other is under intensive study. Platelet aggregation and the release reaction can be induced in the test tube by addition of such diverse agents as collagen, thrombin, ADP, catecholamines, serotonin, arachidonic acid, aggregated immunoglobulins, antigen/antibody complexes, and antibodies directed against platelets.

Platelet aggregation takes place only if the surrounding milieu contains fibrinogen, which adheres to specific receptors, the glycoprotein GPIIb/IIIa complex, on the surface of platelets that have been stimulated by the aggregating agent. The platelets of subjects who have ingested aspirin within the preceding few days do not undergo the release reaction. Additionally platelet aggregation is incomplete and reversible. In normal individuals the effect of aspirin is reflected only by a slight prolongation of the bleeding time, (i.e., the duration of bleeding from a deliberately incised wound, but in patients with disorders of hemostasis the tendency to hemorrhage may be enhanced.

The mechanisms underlying platelet aggregation and release have been examined intensively (Fig. 21-12). Collagen, thrombin, ADP, or catecholamines bind to receptors in the platelet membrane where they activate membrane phospholipases, perhaps by releasing membrane-bound calcium ions. The activated phospholipases hydrolyze membrane phosphatidylcholine and phosphatidylinositol. Arachidonic acid released from these phosphatides is converted to cyclic endoperoxides (prostaglandins G_2 and H_2) by a cyclooxygenase that uses molecular oxygen. The endoperoxides are transformed by a microsomal enzyme, thromboxane synthase, to a short-lived compound, **thromboxane A_2,** that can aggregate platelets, bring about the release reaction, and constrict small blood vessels. The release of arachnodonic acid from membrane phospholipids is blocked by elevated concentrations of cyclic adenosine monophosphate (cAMP). Aspirin inhibits platelet aggregation and release by inactivating the cyclooxygenase, which is irreversibly acetylated, preventing the formation of thromboxane A_2.

Thromboxane A_2 is one of a family of arachidonate derivatives known collectively as **eicosanoids.** Other eicosanoids of biological importance are prostacyclin (prostaglandin I_2) and the leukotrienes. **Prostacyclin** is synthesized by endothelial cells that, like platelets, convert arachidonic acid to cyclic endoperoxides through the action of a cyclooxygenase. But in the endothelial cell the cyclic endoperoxides are changed by a microsomal enzyme, prostacyclin synthase, to prostacyclin (prostaglandin I_2) rather than to thromboxane A_2. One stimulus for prostacyclin synthesis is thrombin. Prostacyclin inhibits platelet aggregation, perhaps by increasing the level of cAMP, and dilates blood vessels. Endothelial cells can use endoperoxides released from stimulated platelets to synthesize prostacyclin. Perhaps this phenomenon limits platelet aggregation at sites of vascular damage. Aspirin, at somewhat higher doses than are needed to block the synthesis of thromboxane A_2 in

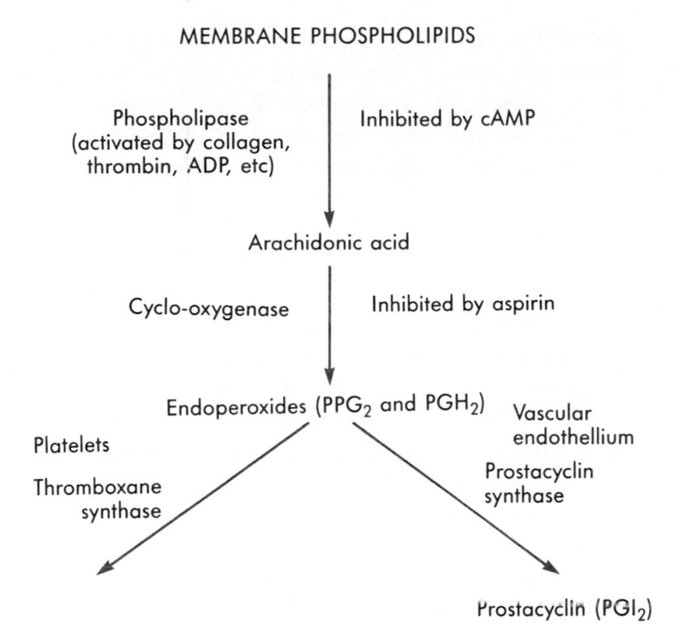

Fig. 21-12 Pathways to the formation of thromboxane A_2 in platelets and of prostacyclin in vascular endothelial cells.

platelets, inhibits prostacyclin formation, suggesting that this drug may have a double-edged effect. The antagonistic effects of thromboxane A_2 and prostacyclin suggest that platelet aggregation and vascular constriction are critical functions. Among the regulatory devices that have evolved current interest focuses on the elaboration of vascular **endothelium-derived relaxing factor**, which dilates small blood vessels and inhibits platelet aggregation (see Chapter 28). This agent appears to be nitric oxide (NO) derived from arginine residues.

The third class of eicosanoids, the **leukotrienes,** are synthesized by granulocytes through the action of cytoplasmic enzymes described as lipoxygenases. The leukotrienes play no direct role in hemostasis, but they are chemotactic (i.e., they bring about migration of leukocytes to the site of their release) and foster adhesion of leukocytes to vascular endothelium. They also increase vascular permeability locally and thereby contribute to the formation of wheals. Furthermore they constrict the smallest bronchioles; such bronchoconstriction is characteristic of bronchial asthma and anaphylaxis (Chapter 20).

Through the thromboxane pathway thrombin brings about aggregation of platelets and discharge of their granular contents. However, thrombin still aggregates platelets and induces the release reaction when the thromboxane pathway is blocked, suggesting that some additional mechanism may be involved, perhaps the elaboration of another mediator. One candidate for this role is a protein component of the α granules, **thrombospondin,** which on stimulation of the platelets by thrombin migrates to the surface of the cells, where it binds to the platelets' membrane glycoproteins. There thrombospondin can form a complex with fibrinogen, and this complex may serve as a bridge between adjacent platelets. Fibrinogen bound to thrombospondin on the platelet membrane may stabilize platelet aggregates that are formed when these cells are stimulated by thrombin or ADP. Thrombospondin binds to other plasma proteins, including plasminogen, fibronectin, histidine-rich glycoprotein, and collagen, and to mucopolysaccharides. Thrombospondin is synthesized not only by platelets but also by fibroblasts, monocytes, smooth muscle cells, and endothelial cells. Hence this protein may play a more general role in the adhesion of cell to cell and of cells to the extracellular matrix.

A minor constituent of the α granules that is discharged during the release reaction is fibronectin, which is also synthesized by fibroblasts, hepatocytes, endothelial cells, and other tissues. Fibronectin is localized to many cellular surfaces, where it may serve as a glue that fosters the adhesion of cells to one another. It also serves as an **opsonin** that enhances the engulfment of microorganisms and particulate matter by phagocytes. It may play a role in promoting platelet aggregation by thrombin. Fibronectin is an important component of the matrix of connective tissue, where it is bound to collagen by an enzyme similar to activated fibrin-stabilizing factor (factor XIII_a).

Fibronectin is also present in plasma, where it has been called cold-insoluble globulin. Plasma fibronectin may participate during disseminated intravascular coagulation in reactions that inhibit the polymerization of fibrin monomers. During in vitro clotting it may be covalently linked to fibrin by activated fibrin-stabilizing factor. Perhaps this property enhances the adhesion of cells to fibrin and in this way promotes wound healing.

Platelets and Blood Coagulation

In the test tube stimulation of platelets by ADP or other agents (or contact of platelets with such negatively charged substances as kaolin) makes available membrane phospholipids (platelet factor 3 [PF3]) for the intrinsic pathway of thrombin formation. Presumably similar stimulation of platelets occurs in vivo. In addition activated Stuart factor (factor X_a) binds to receptor sites on the surface of thrombin-stimulated platelets. In this situation activated Stuart factor converts prothrombin to thrombin much more effectively. The receptor for activated Stuart factor has been identified as proaccelerin (factor V) that has been altered to its clot-promoting state by thrombin. The function of platelet fibrinogen is not clear, although this clotting factor makes up as much as 15% of platelet protein. Both proaccelerin and fibrinogen are found in the platelets' α granules and are secreted during the release reaction.

Platelets may alter coagulation and fibrinolysis

through yet other mechanisms. As noted previously, when platelets are stimulated by thrombin, the α granules release a protein, thrombospondin. This protein migrates to the surface of these platelets, where it can bind heparin, fibrinogen, fibronectin, histidine-rich glycoprotein, plasminogen, and collagen. These properties of thrombospondin may inhibit the anticoagulant properties of heparin and the activation of plasminogen by tissue plasminogen activator. Platelets also contain an inhibitor of activated protein C that is released when these cells are activated.

■ *Clot Retraction*

After normal blood coagulates, the clot gradually shrinks, extruding clear serum. This phenomenon of clot retraction depends on the presence of viable, metabolically active platelets. Retraction is initiated by an action of thrombin on these platelets, but the subsequent steps are only vaguely appreciated. Perhaps 15% or more of platelet protein consists of actin and myosin, which are similar but not identical to that in muscle (see Chapter 17). One construction of the mechanism of clot retraction is that thrombin releases calcium ions from intracellular storage sites. The calcium ions then inactivate regulatory proteins that are similar to muscle troponin. When this takes place, platelet myosin, an adenosine triphosphatase (ATPase), is phosphorylated, and it combines with actin through cross-bridges. The complex then contracts in a manner similar to the contraction of muscle; the energy required comes from the splitting of ATP by myosin ATPase. The contractile process causes the extrusion of pseudopods and migration of granules to the center of the cell. The platelet spicules stick to polymerizing fibrin in the surrounding blood, whereupon the contractile process pulls the fibrin strands toward the platelet. This decreases the volume of the clot so that the entrapped serum is extruded; the fibrin strands themselves are not shortened. Clot retraction draws the platelets together, and these cells gradually assume an amorphous appearance that has been described as platelet metamorphosis, a term now seldom used. Clot retraction is not inhibited by aspirin and is thus not dependent on the release reaction.

The utility of clot retraction has been puzzling. Perhaps the retraction of fibrin strands can bring together the edges of a wound filled with clot, or the shrinkage of the clot allows room for vascular walls to approximate.

■ *Thrombosis*

A **thrombus** (plural **thrombi**) is a clot within a blood vessel. The conditions that foster the formation of thrombi include damage to the vascular wall, impaired blood flow *(stasis),* and alterations in the blood composition that render it more readily coagulable. The initial event is probably a disruption of the normal vascular surface that induces localized platelet adhesion and aggregation and localized blood coagulation; such lesions are more evident in arterial than venous thrombosis. Thereafter a blood clot extends from this nidus into the lumen of the vessel. The anatomy of the thrombus is conditioned by its location. In arteries, in which blood flow is swift, the adherent portion of the thrombus is rich in platelets, whereas in veins, in which flow is sluggish, the base of the thrombus may resemble a typical blood clot. In veins thrombi are particularly likely to begin at the site of valves (a region of stagnant flow).

The degree to which changes in the blood itself foster thrombosis is debated. Experimentally the injection of activated clotting factors induces intravascular clots but only in areas in which blood flow has been halted by the subsequent application of ligatures. Thus stasis and introduction of activated clotting factors foster this type of experimental thrombosis. The role of increased concentrations of clotting factors is uncertain. A familial tendency to recurrent thrombosis has been described in individuals with partial deficiencies of antithrombin III, heparin cofactor II, protein C, or protein S or with qualitatively abnormal species of fibrinogen (dysfibrinogenemia) as if intravascular clotting were fostered by impairment of normal mechanisms that limit the generation or action of thrombin.

Embolism is the process through which thrombi break off from their vascular attachment and are carried by the flowing blood until they lodge in a blood vessel too small to allow their passage. The most frequent example is the often lethal disorder *pulmonary embolism,* in which a venous thrombus that originates in the veins of the legs or the pelvis becomes wedged in the pulmonary arteries.

■ *Summary*

1. Hemostasis, the arrest of bleeding, is mainly produced by circulating platelets that plug small defects in vascular walls and by coagulation of blood.

2. The basic structure of a clot is a meshwork of fibrin, which is a protein derived during coagulation from its precursor, fibrinogen, by the action of a plasma-derived protease, thrombin.

3. Thrombin circulates normally as a precursor, prothrombin. When blood is shed, prothrombin is changed to thrombin via one or both of two intertwining chemical pathways. Blood that comes into contact with a "foreign" surface such as glass clots rapidly, a process mediated by activation of a plasma proenzyme, Hageman factor (factor XII). The steps of this intrinsic pathway of thrombin formation involve the participation of

additional plasma proteins, including plasma thromboplastin antecedent (PTA, factor XI), plasma prekallikrein, high-molecular–weight kininogen, Christmas factor (factor IX), and antihemophilic factor (factor VIII). The product, activated Christmas factor (factor IX_a) converts another plasma protein, Stuart factor (factor X), to an enzymatically active form that releases thrombin from prothrombin; this reaction requires a protein cofactor, proaccelerin (factor V).

4. Stuart factor can be activated through reactions of the extrinsic pathway of thrombin formation. These reactions are initiated when blood comes into contact with injured tissues. Tissue thromboplastin, furnished by injured tissue, combines with and activates a plasma proenzyme, factor VII; factor VII_a then transforms Stuart factor (factor X) to its proteolytic form. The subsequent steps of the intrinsic and extrinsic pathways are identical.

5. Factor VIII (antihemophilic factor) circulates as a loose, noncovalent complex with another plasma protein, vWf, a protein that promotes adhesion of platelets to injured blood vessel walls and thereby enhances hemostasis.

6. Except vWf, which is synthesized by vascular endothelial cells and megakaryocytes, clotting factors are synthesized mainly in the liver. Four clotting factors, prothrombin, Christmas factor, Stuart factor (factor X) and factor VII require vitamin K for their synthesis; vitamin K is a coenzyme for a carboxylase that inserts a second carboxyl group into the gamma carbon of certain glutamic acid residues in the polypeptide chain of these vitamin K-dependent clotting factors.

7. Under appropriate conditions clotted blood may reliquefy. The agent responsible for the dissolution of fibrin is plasmin, a typical serine protease that is the activated form of a plasma proenzyme, plasminogen. Plasminogen can be converted to plasmin by plasma, tissue, and urinary proteases.

8. Platelets are small, disk-shaped anuclear blood cells. Within seconds after the endothelial lining of blood vessels is disrupted, some platelets adhere to the exposed subendothelial structures, such as collagen. The adherent platelets discharge their contents, which then attract other platelets to those already attached to the injured area, and they form a plug that can stop blood flow from small wounds. Among other functions platelets may also provide phopholipids that enhance thrombin formation.

Bibliography

Journal articles

Bennett JS: The molecular biology of platelet membrane proteins, *Semin Hematol* 27:186, 1990.

Davie EW et al: The role of serine proteases in the blood coagulation cascade, *Adv Enzymol* 48:277, 1979.

Esmon CT: The role of protein C and thrombomodulin in the regulation of blood coagulation, *J Biol Chem* 264:4743, 1989.

Fenton JW II: Regulation of thrombin generation and functions, *Semin Thromb Hemostas* 14:234, 1988.

Foster PA, Zimmerman TS: Factor VIII structure and function, *Blood Rev* 3:180, 1989.

Jackson CM, Nemerson Y: Blood coagulation, *Annu Rev Biochem* 49:765, 1980.

Leung LLK, Nachman RL: Complex formation of platelet thrombospondin with fibrinogen, *J Clin Invest* 70:542, 1982.

Mackie IJ, Bull HA: Normal haemostasis and its regulation, *Blood Rev* 3:237, 1989.

Nemerson Y: Tissue factor and hemostasis, *Blood* 71:1, 1988.

Shattil SJ, Bennett JS: Platelets and their membranes in hemostasis: physiology and pathophysiology, *Ann Intern Med* 94:108, 1981.

Suttie JW: Vitamin K-dependent carboxylase, *Ann Rev Biochemistry* 54:459, 1985.

Suttie JW, Jackson CM: Prothrombin structure, activation and biosynthesis, *Physiol Rev* 57:1, 1977.

White JG: Current concepts of platelet structure, *Am J Clin Pathol* 71:373, 1979.

Books and monographs

Colman RW et al: *Hemostasis and thrombosis,* ed 2 Philadelphia, 1987, JB Lippincott.

Esmon JL: *Thrombomodulin.* In Coller BS, editor: Progress in hemostasis and thrombosis, vol 9, Philadelphia, 1989, WB Saunders.

Marcus AJ: *Platelets and their disorders.* In Ratnoff OD, Forbes CD, editors: Disorders of hemostasis, ed 2, Philadelphia, 1991, WB Saunders.

Ogston D: *The physiology of hemostasis,* Cambridge, 1983, Harvard University Press.

Rapaport SI: *Introduction to hematology,* ed 2, Philadelphia, 1987, JB Lippincott.

Ratnoff OD, Forbes CS, editors: Disorders of hemostasis, ed 2, Philadelphia, 1991, WB Saunders.

SECTION

V

THE CARDIOVASCULAR SYSTEM

Robert M. Berne
Matthew N. Levy

CHAPTER

22

The Circuitry

The circulatory, endocrine, and nervous systems constitute the principal coordinating and integrating systems of the body. Whereas the nervous system is primarily concerned with communications and the endocrine glands with regulation of certain body functions, the circulatory system serves to transport and distribute essential substances to the tissues and to remove byproducts of metabolism. The circulatory system also shares in such homeostatic mechanisms as regulation of body temperature, humoral communication throughout the body, and adjustments of oxygen and nutrient supply in different physiological states.

The cardiovascular system that accomplishes these chores is made up of a pump, a series of distributing and collecting tubes, and an extensive system of thin vessels that permit rapid exchange between the tissues and the vascular channels. The primary purpose of this section is to discuss the function of the components of the vascular system and the control mechanisms (with their checks and balances) that are responsible for alteration of blood distribution necessary to meet the changing requirements of different tissues in response to a wide spectrum of physiological and pathological conditions.

Before considering the function of the parts of the circulatory system in detail, it is useful to consider it as a whole in a purely descriptive sense. The heart consists of two pumps in series: one to propel blood through the lungs for exchange of oxygen and carbon dioxide (the **pulmonary circulation**) and the other to propel blood to all other tissues of the body (the **systemic circulation**). Unidirectional flow through the heart is achieved by the appropriate arrangement of effective flap valves. Although the cardiac output is intermittent, continuous flow to the periphery occurs by distension of the aorta and its branches during ventricular contraction (**systole**) and elastic recoil of the walls of the large arteries with forward propulsion of the blood during ventricular relaxation (**diastole**). Blood moves rapidly through the aorta and its arterial branches. The branches become narrower and their walls become thinner and change histologically toward the periphery. From a predominantly elastic structure, the aorta, the peripheral arteries become more muscular until at the arterioles the muscular layer predominates (Fig. 22-1).

In the large arteries frictional resistance is relatively small, and pressures are only slightly less than in the aorta. However, the small arteries offer moderate resistance to blood flow and this resistance reaches a maximum level in the arterioles, sometimes referred to as the stopcocks of the vascular system. Hence the pressure drop is significant in the small arteries and is greatest across the arterioles (Fig. 22-2). Adjustment in the degree of contraction of the circular muscle of these small vessels permits regulation of tissue blood flow and aids in the control of arterial blood pressure.

In addition to a sharp reduction in pressure across the arterioles there is a change from pulsatile to steady flow as pressure continues to decline from the arterial to the venous end of the capillaries. The pulsatile arterial blood flow, caused by the intermittency of cardiac ejection, is damped at the capillary level by the combination of distensibility of the large arteries and frictional resistance in the arterioles. Many capillaries arise from each arteriole so that the total cross-sectional area of the capillary bed is very large, despite the fact that the cross-sectional area of each capillary is less than that of each arteriole. As a result blood flow velocity becomes quite slow in the capillaries, analogous to the decrease in velocity seen at the wide regions of a river. Because the capillaries consist of short tubes whose walls are only one cell thick and because the velocity of blood flow rate is slow, conditions in the capillaries are ideal for the exchange of diffusible substances between blood and tissue.

On its return to the heart from the capillaries blood passes through venules and then through veins of increasing size with a progressive decrease in pressure until the blood reaches the right atrium (see Fig. 22-2). As the heart is approached, the number of veins decreases, the thickness and composition of the vein walls change (see Fig. 22-1), the total cross-sectional area of the venous channels diminishes, and the velocity of blood flow increases (see Fig. 22-2). Also note that

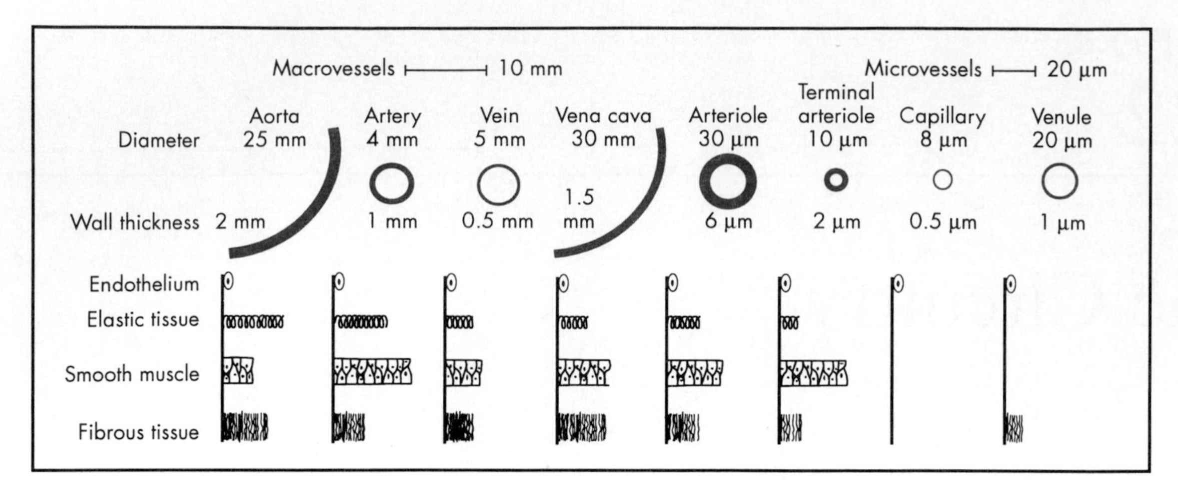

■ **Fig. 22-1** Internal diameter, wall thickness, and relative amounts of the principal components of the vessel walls of the various blood vessels that compose the circulatory system. Cross sections of the vessels are not drawn to scale because of the huge range from aorta and venae cavae to capillaries. (Redrawn from Burton AC: *Physiol Rev* 34:619, 1954.)

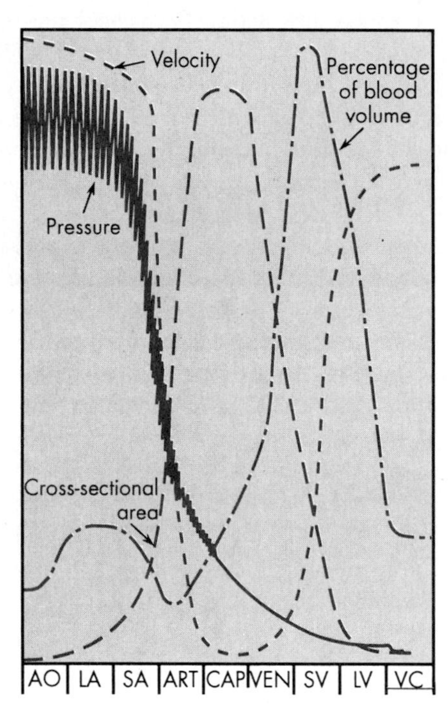

■ **Fig. 22-2** Pressure, velocity of flow, cross-sectional area, and capacity of the blood vessels of the systemic circulation. *The important features are the inverse relationship between velocity and cross-sectional area, the major pressure drop across the arterioles, the maximal cross-sectional area and minimal flow rate in the capillaries, and the large capacity of the venous system.* The small but abrupt drop in pressure in the venae cavae indicates the point of entrance of these vessels into the thoracic cavity and reflects the effect of the negative intrathoracic pressure. To permit schematic representation of velocity and cross-sectional area on a single linear scale, only approximations are possible at the lower values. *AO,* Aorta; *LA,* large arteries; *SA,* small arteries; *ART,* arterioles; *CAP,* capillaries; *VEN,* venules; *SV,* small veins; *LV,* large veins; *VC,* venae cavae.

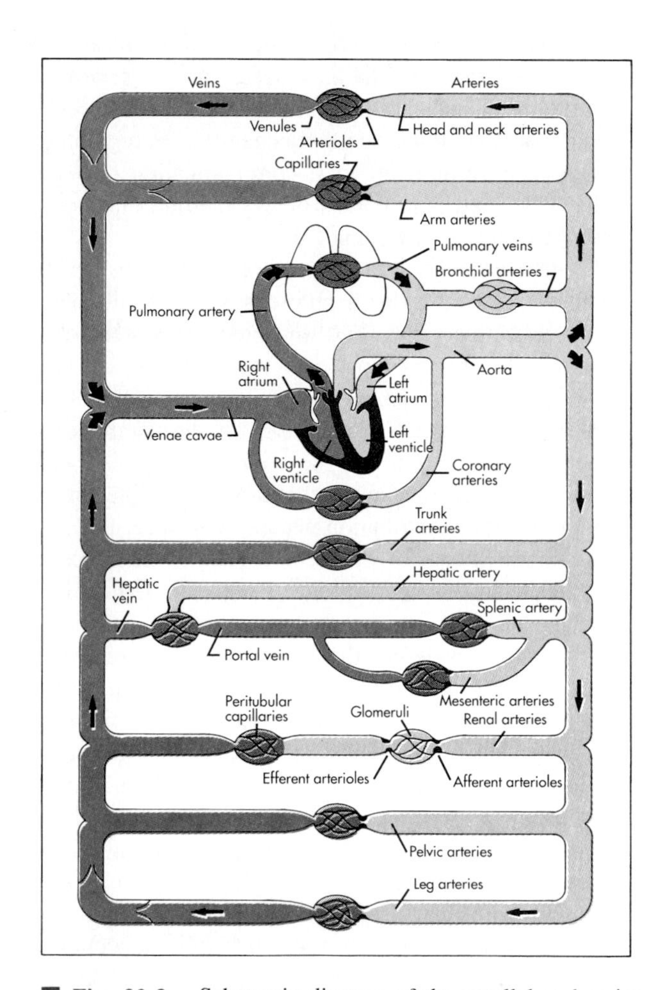

■ **Fig. 22-3** Schematic diagram of the parallel and series arrangement of the vessels composing the circulatory system. The capillary beds are represented by thin lines connecting the arteries (on the right) with the veins (on the left). The crescent-shaped thickenings proximal to the capillary beds represent the arterioles (resistance vessels). (Redrawn from Green HD: In Glasser O, editor: *Medical physics,* vol 1, Chicago, 1944, Year Book Medical Publishers.)

most of the circulating blood is located in the venous vessels (see Fig. 22-2).

Data from a 20-kg dog (Table 22-1) indicate that the number of vessels increases about 3-billion-fold, and the total cross-sectional area increases about 500-fold between the aorta and the capillaries. The volume of blood in the capillaries is only 5% of the total blood volume compared with 11% in the aorta, arteries, and arterioles and 67% in the veins and venules. In contrast blood volume in the pulmonary vascular bed is about equally divided among the arterial, capillary, and venous vessels. The cross-sectional area of the venae cavae is larger than that of the aorta (although not evident from Fig. 22-2 because cross-sectional areas of venae cavae and aorta are so close to zero with a scale that includes the capillaries), and hence the flow is slower than that in the aorta.

Blood entering the right ventricle via the right atrium is pumped through the pulmonary arterial system at a mean pressure about one seventh that in the systemic arteries. The blood then passes through the lung capillaries, where carbon dioxide is released and oxygen taken up. The oxygen-rich blood returns via the pulmonary veins to the left atrium and ventricle to complete the cycle. Thus in the normal intact circulation the total volume of blood is constant, and an increase in the volume of blood in one area must be accompanied by a decrease in another. However, the distribution of the circulating blood to the different regions of the body is determined by the output of the left ventricle and by the contractile state of the arterioles (resistance vessels) of these regions. The circulatory system is composed of conduits arranged in series and in parallel (Fig. 22-3).

Table 22-1 Vascular dimensions in a 20-kg dog

Vessels	*Number*	*Total cross-sectional area (cm^2)*	*Total blood volume (%)*
Systemic			
Aorta	1	2.8	
Arteries	40-110,000	40	11
Arterioles	2.8×10^6	55	
Capillaries	2.7×10^9	1,357	5
Venules	1×10^7	785	
Veins	660,000-110	631	67
Venae cavae	2	3.1	
Pulmonary			
Arteries and arterioles	$1\text{-}1.5 \times 10^6$	137	3
Capillaries	2.7×10^9	1,357	4
Venules and veins	$2 \times 10^6 - 4$	210	5
Heart			
Atria	2		
Ventricles	2		5

(Data from Milnor WR: *Hemodynamics*, Baltimore, 1982, Williams & Wilkins.) The ranges of systemic arteries and veins and of pulmonary venules and veins are given in the direction of blood flow.

CHAPTER

23

Electrical Activity of the Heart

The experiments on "animal electricity" conducted by Galvani and Volta two centuries ago led to the discovery that electrical phenomena were involved in the spontaneous contractions of the heart. In 1855 Kölliker and Müller observed that when they placed the nerve of a nerve-muscle preparation in contact with the surface of a frog's heart, the muscle twitched with each cardiac contraction.

The electrical events that normally take place in the heart initiate its contraction. Disorders in electrical activity can induce serious and sometimes lethal rhythm disturbances.

■ *Transmembrane Potentials*

The electrical behavior of single cardiac muscle cells has been investigated by inserting microelectrodes into the interior of the cell. The potential changes recorded from a typical ventricular muscle fiber are illustrated in Fig. 23-1, *A*. When two electrodes are placed in an electrolyte solution near a strip of quiescent cardiac muscle, no potential difference (point a) is measurable between the two electrodes. At point b one of the electrodes, a microelectrode, was inserted into the interior of a cardiac muscle fiber (see Fig. 23-1). Immediately the galvanometer recorded a potential difference (V_m) across the cell membrane; the potential of the interior of the cell was about 90 mV lower than that of the surrounding medium. Such electronegativity of the interior of the resting cell with respect to the exterior is also characteristic of skeletal and smooth muscle, of nerve, and indeed of most cells within the body (see also Chapter 2).

At point c a propagated action potential was transmitted to the cell impaled with the microelectrode. Very rapidly the cell membrane became depolarized; actually, the potential difference was reversed (positive overshoot), so that the potential of the interior of the cell exceeded that of the exterior by about 20 mV. The rapid upstroke of the action potential is designated phase 0. Immediately after the upstroke there was a brief period of partial repolarization (phase 1), followed by a plateau (phase 2) that persisted for about 0.1 to 0.2 sec. The potential then became progressively more negative (phase 3), until the resting state of polarization was again attained (at point e). Rapid repolarization (phase 3) is a much slower rate of change than is depolarization (phase 0). The interval from the end of repolarization until the beginning of the next action potential is designated phase 4.

The time relationships between the electrical events and the actual mechanical contraction are shown in Fig. 23-2. Rapid depolarization (phase 0) precedes force development and completion of repolarization coincides approximately with peak force. The duration of contraction tends to parallel the duration of the action potential. Also as the frequency of cardiac contraction is increased, the durations of the action potential and the mechanical contraction decrease.

■ *Principal Types of Cardiac Action Potentials*

Two main types of action potentials are observed in the heart, as shown in Fig. 23-1. One type, the fast response, occurs in the normal myocardial fibers in the atria and ventricles and in the specialized conducting fibers (**Purkinje's** fibers). The other type of action potential, the slow response, is found in the sinoatrial (SA) node, the natural pacemaker region of the heart, and in the atrioventricular (AV) node, the specialized tissue involved in conducting the cardiac impulse from atria to ventricles. Furthermore fast responses may be converted to slow responses either spontaneously or under certain experimental conditions. For example, in a myocardial fiber, a gradual shift of the resting membrane potential from its normal level of about −90 mV to a value of about −60 mV converts subsequent action potentials to the slow response. Such conversions may occur spontaneously in those regions of the heart to which the blood supply has been severely curtailed.

As shown in Fig. 23-1, not only is the resting membrane potential of the fast response considerably more

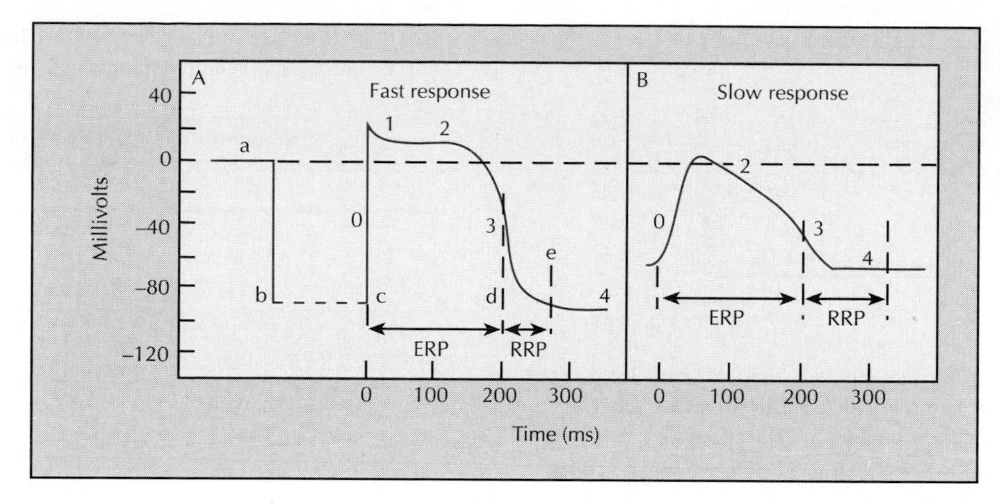

■ **Fig. 23-1** Changes in transmembrane potential recorded from a fast response and a slow response cardiac fiber in isolated cardiac tissue immersed in an electrolyte solution. **A,** At time *a* the microelectrode was in the solution surrounding the cardiac fiber. At time *b* the microelectrode entered the fiber. At time *c* an action potential was initiated in the impaled fiber. Time *c* to *d* represents the effective refractory period (ERP), and time *d* to *e* represents the relative refractory period (RRP). **B,** An action potential recorded from a slow response cardiac fiber. Note that compared to the fast response fiber, the resting potential of the slow fiber is less negative, the upstroke (phase *0*) of the action potential is less steep, the amplitude of the action potential is smaller, phase *1* is absent, and the relative refractory period (RRP) extends well into phase *4* after the fiber has fully repolarized.

negative than that of the slow response, but the slope of the upstroke (phase 0), the amplitude of the action potential, and the extent of the overshoot of the fast response are also greater than in the slow response. The amplitude of the action potential and the steepness of the upstroke are important determinants of propagation velocity. Hence in cardiac tissue characterized by the slow response, conduction velocity is much slower and impulses are more likely to be blocked than in tissues displaying the fast response. Slow conduction and a tendency toward block increase the likelihood of certain rhythm disturbances.

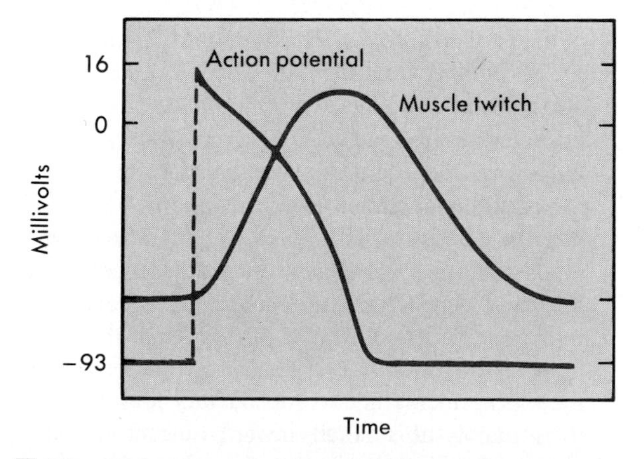

■ **Fig. 23-2** Time relationships between the mechanical tension developed by a thin strip of ventricular muscle and the changes in transmembrane potential. (Redrawn from Kavaler F, Fisher VJ, Stuckey JH: *Bull NY Acad Med* 41:592, 1965.)

■ *Ionic Basis of the Resting Potential*

The various phases of the cardiac action potential are associated with changes in the permeability of the cell membrane, mainly to sodium, potassium, and calcium ions. Changes in cell membrane permeability alter the rate of ion passage across the membrane. The permeability of the membrane to a given ion defines the net quantity of the ion that will diffuse across each unit area of the membrane per unit concentration difference across the membrane. Changes in permeability are accomplished by the opening and closing of ion channels that are specific for the individual ions.

Just as with all other cells in the body (see also Chapter 2), the concentration of potassium ions inside a cardiac muscle cell, $[K^+]_i$, greatly exceeds the concentration outside the cell, $[K^+]_0$, as shown in Fig. 23-3. The reverse concentration gradient exists for Na ions and for unbound Ca ions. Estimates of the extracellular and intracellular concentrations of Na^+, K^+, and Ca^{++}, and of the equilibrium potentials (defined later) for these ions, are compiled in Table 23-1.

The resting cell membrane is relatively permeable to K^+, but much less so to Na^+ and Ca^{++}. Because of the high permeability to K^+ there tends to be a net diffusion of K^+ from the inside to the outside of the cell in

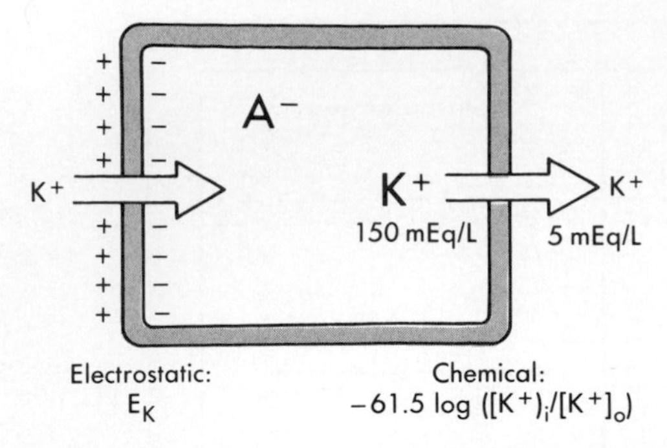

Fig. 23-3 The balance of chemical and electrostatic forces acting on a resting cardiac cell membrane based on a 30:1 ratio of the intracellular to extracellular K^+ concentrations and the existence of a nondiffusible anion (A^-) inside but not outside the cell.

Table 23-1 Intracellular and extracellular ion concentrations and equilibrium potentials in cardiac muscle cells

Ion	*Extracellular concentrations (mM)*	*Intracellular concentrations (mM)**	*Equilibrium potential (mV)*
Na^+	145	10^4	70
K^+	4	135^4	−94
Ca^{++}	2	10^{-4}	132

Modified from Ten Eick RE, Baumgarten CM, Singer DH: *Prog Cardiovasc Dis* 24:157, 1981.
*The intracellular concentrations are estimates of the free concentrations in the cytoplasm.

the direction of the concentration gradient, as shown on the right side of the cell in Fig. 23-3.

Any flux of K^+ that occurs during phase 4 takes place through certain specific K^+ channels. Several types of K^+ channels exist in cardiac cell membranes. Some of these channels are regulated (i.e., opened and closed) by the transmembrane potential, whereas others are regulated by some chemical signal (e.g., the intracellular Ca^{++} concentration). The specific K^+ channel through which K^+ passes during phase 4 is a voltage regulated channel that conducts the K^+ current, called i_{K1}, which is an inwardly rectifying K^+ current, as explained later. Many of the anions (labeled A^-) inside the cell, such as the proteins, are not free to diffuse out with the K^+ (see Fig. 23-3). Therefore as the K^+ diffuses out of the cell and leaves the A^- behind, the deficiency of cations causes the interior of the cell to become electronegative.

Therefore, two opposing forces are involved in the movement of K^+ across the cell membrane. A chemical force, based on the concentration gradient, results in the net outward diffusion of K^+. The counterforce is an electrostatic one; the positively charged K ions are attracted to the interior of the cell by the negative potential that exists there, as shown on the left side of the cell in Fig. 23-3. If the system came into equilibrium, the chemical and the electrostatic forces would be equal.

This equilibrium is expressed by the **Nernst equation** (see Chapter 2) for potassium:

$$E_K = -61.5 \log([K^+]_i/[K^+]_o \quad (1)$$

The right-hand term represents the chemical potential difference at the body temperature of 37° C. The left-hand term, E_K, represents the electrostatic potential difference that would exist across the cell membrane if K^+ were the only diffusible ion. E_K is called the potassium equilibrium potential.

An experimental disturbance in the equilibrium between electrostatic and chemical forces imposed by voltage clamping would cause K^+ to move through the K^+ channels. If the transmembrane potential (V_m) were clamped at a level negative to E_K, the electrostatic forces would exceed the diffusional forces and K^+ would be attracted into the cell; (i.e., the K^+ current would be *inward*). Conversely if V_m were clamped at a level positive to E_K, the diffusional forces would exceed the electrostatic forces and K^+ would leave the cell; (i.e., the K^+ current would be *outward*).

When the measured concentrations of $[K^+]_i$ and $[K^+]_o$ for mammalian myocardial cells are substituted into the Nernst equation, the calculated value of E_K equals about −90 to −100 mV (Table 23-1). This value is close to, but slightly more negative than, the resting potential actually measured in myocardial cells. Therefore a small potential tends to drive K^+ out of the resting cell.

The balance of forces acting on the Na ions is entirely different from that acting on the K ions in resting cardiac cells. The intracellular Na^+ concentration, $[Na^+]_i$, is much lower than the extracellular concentration, $[Na^+]_o$. At 37° C, the sodium equilibrium potential, E_{Na}, expressed by the Nernst equation is:

$$-61.5 \log([Na^+]_i/[Na^+]_o) \quad (2)$$

For cardiac cells E_{Na} is about 40 to 70 mV (Table 23-1). At equilibrium, therefore, an electrostatic force of 40 to 70 mV, oriented with the inside of the cell more positive than the outside, would be necessary to counterbalance the chemical potential for Na^+. However, the actual polarization of the resting cell membrane is just the opposite. The resting membrane potential of myocardial fibers is about −90 mV. Hence both chemical and electrostatic forces act to pull extracellular Na^+ into the cell. The influx of Na^+ through the cell membrane is small, however, because the permeability of the resting membrane to Na^+ is very low. Nevertheless it is mainly this small inward current of positively charged Na ions that causes the potential on the inside of the resting cell membrane to be slightly less negative than the value predicted by the Nernst equation for K^+.

The steady inward leak of Na^+ would gradually depolarize the resting cell membrane were it not for the meta-

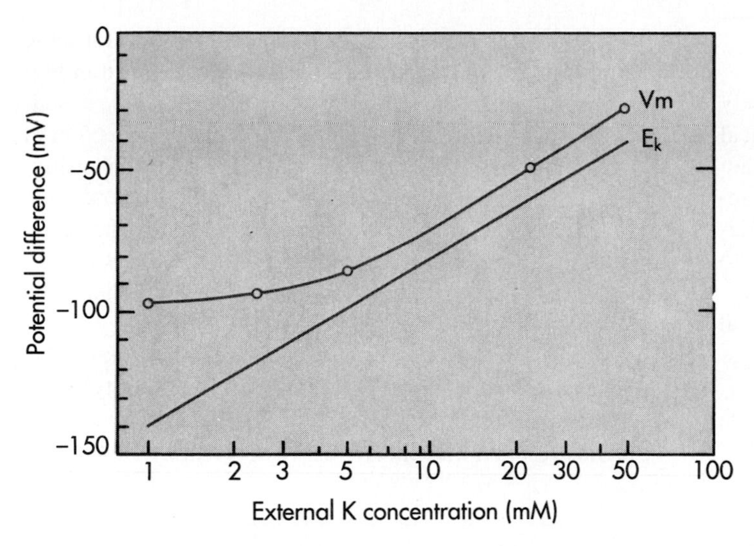

■ **Fig. 23-4** Transmembrane potential of a cardiac muscle fiber varies inversely with the potassium concentration of the external medium *(colored curve)*. The straight black line represents the change in transmembrane potential predicted by the Nernst equation for E_K. (Redrawn from Page E: *Circulation* 26:582, 1962. By permission of the American Heart Association.)

bolic pump that continuously extrudes Na^+ from the cell interior and pumps in K^+. The metabolic pump involves the enzyme Na^+-K^+-activated adenosine triphosphatase (ATPase), which is located in the cell membrane itself (see Chapter 1). Because the pump must move Na^+ against both a chemical and an electrostatic gradient, operation of the pump requires the expenditure of metabolic energy. Increases in $[Na^+]_i$ or in $[K^+]_o$ accelerate the activity of the pump. The quantity of Na^+ extruded by the pump exceeds the quantity of K^+ transferred into the cell by a 3:2 ratio. Therefore the pump itself tends to create a potential difference across the cell membrane, and thus it is termed an electrogenic pump. If the pump is partially inhibited, as by digitalis, the concentration gradients for Na^+ and K^+ are partially dissipated, and the resting membrane potential becomes less negative than normal.

The dependence of the transmembrane potential, V_m, on the intracellular and extracellular concentrations of K^+ and Na^+ and on the conductances (g_K and g_{Na}) of these ions is described by the chord conductance equation:

$$V_m = \frac{g_K}{g_K + g_{Na}} E_K + \frac{g_{Na}}{g_K + g_{Na}} E_{Na} \quad (3)$$

For a given ion (X) the conductance (g_X) is defined as the ratio of the current (i_X) carried by that ion to the difference between the V_m and the Nernst equilibrium potential (E_X) for that ion; that is,

$$g_X = \frac{i_X}{V_m - E_X} \quad (4)$$

The chord conductance equation reveals that the relative, not the absolute, conductances to Na^+ and K^+ determine the resting potential. In the resting cardiac cell g_K is about 100 times greater than g_{Na}. Therefore the chord conductance equation reduces essentially to the Nernst equation for K^+.

When the ratio $[K^+]_i/[K^+]_o$ is decreased experimentally by raising $[K^+]_o$, the measured value of V_m (red line, Fig. 23-4) approximates that predicted by the Nernst equation for K^+ (black line). For extracellular K^+ concentrations of about 5 mM and above the measured values correspond closely with the predicted values. The measured levels are slightly less than those predicted by the Nernst equation because of the small but finite value of g_{Na}. For values of $[K^+]_o$ below about 5 mM g_K decreases as $[K^+]_o$ is diminished. As g_K decreases, the effect of the Na^+ gradient on the transmembrane potential becomes relatively more important, as predicted by equation 3. This change in g_K accounts for the greater deviation of the measured V_m from that predicted by the Nernst equation for K^+ at low levels of $[K^+]_o$ (see Fig. 23-4). Also, in accordance with equation 3, changes in $[Na^+]_o$ have relatively little effect on resting V_m (Fig. 23-5) because of the low value of g_{Na}.

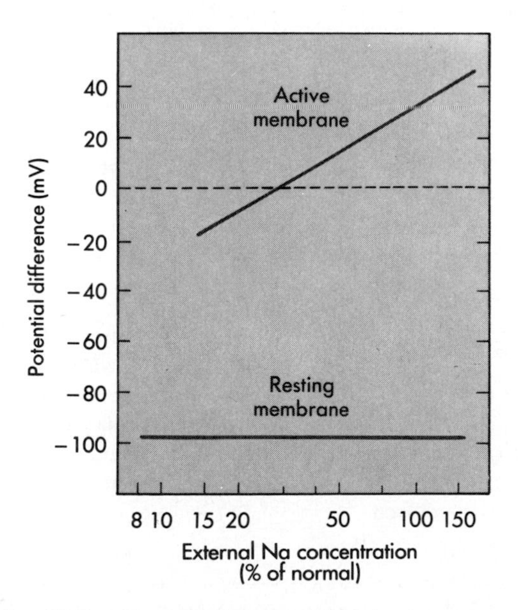

■ **Fig. 23-5** Concentration of sodium in the external medium is a critical determinant of the amplitude of the action potential in cardiac muscle *(upper curve)* but has relatively little influence on the resting potential *(lower curve)*. (Redrawn from Weidmann S: *Elektrophysiologie der Herzmuskelfaser*, Bern, 1956, Verlag Hans Huber.)

Ionic Basis of the Fast Response

Genesis of the upstroke. Any process that abruptly changes the resting membrane potential to a critical value (called the threshold) results in a propagated action potential. The characteristics of fast response action potentials are shown in Fig. 23-1, *A*. The rapid depolarization (phase 0) is related almost exclusively to the inrush of Na^+ by virtue of a sudden increase in g_{Na}. The **amplitude** of the action potential (the potential change during phase 0) varies linearly with the logarithm of $[Na^+]_o$, as shown in Fig. 23-5. When $[Na^+]_o$ is reduced from its normal value of about 140 mM to about 20 mM, the cell is no longer excitable.

The physical and chemical forces responsible for these transmembrane movements of Na^+ are explained in Fig. 23-6. When the resting membrane potential, V_m, is suddenly changed to the threshold level of about −65 mV, the properties of the cell membrane change dramatically. Specific **fast channels** for Na^+ exist in the membrane. These channels can be blocked specifically by the puffer fish toxin, *tetrodotoxin*. Also, many of the drugs used to treat certain cardiac rhythm disturbances act principally on these fast Na^+ channels.

Na^+ moves through these channels in a manner that suggests that the flux is controlled by two types of "gates" in each channel. One of these gates, the m gate, tends to open the channel as V_m becomes less negative

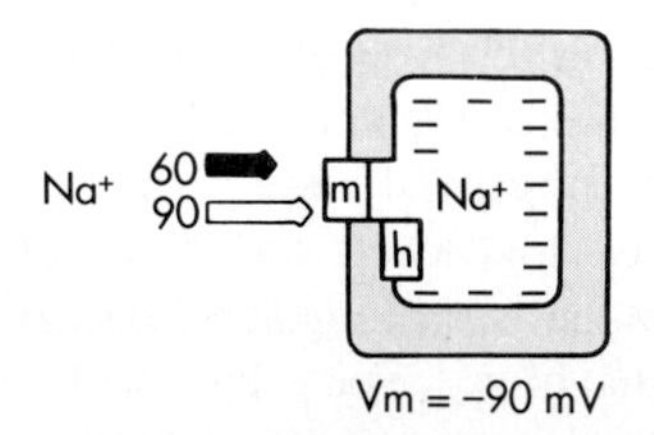

A, During phase 4, the chemical (60 mV) and electrostatic (90 mV) forces favor influx of Na^+ from the extracellular space. Influx is negligible, however, because the activation **(m)** gates are closed.

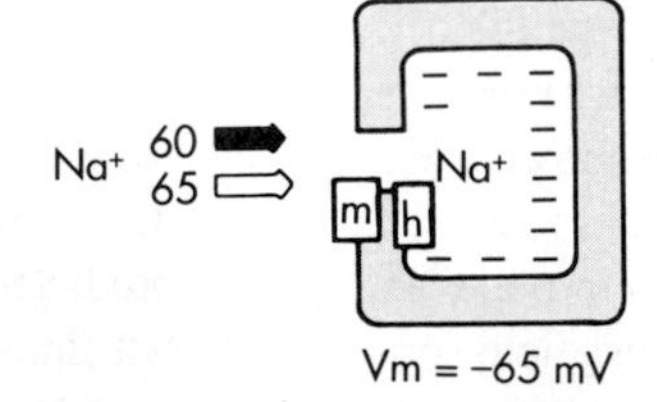

B, If V_m is brought to about −65 V, the **m** gates begin to swing open, and Na^+ begins to enter the cell. This reduces the negative charge inside the cell, and thereby opens still more Na^+ channels, which accelerates the influx of Na^+. The change in V_m also initiates the closure of inactivation **(h)** gates, which operate more slowly than the **m** gates.

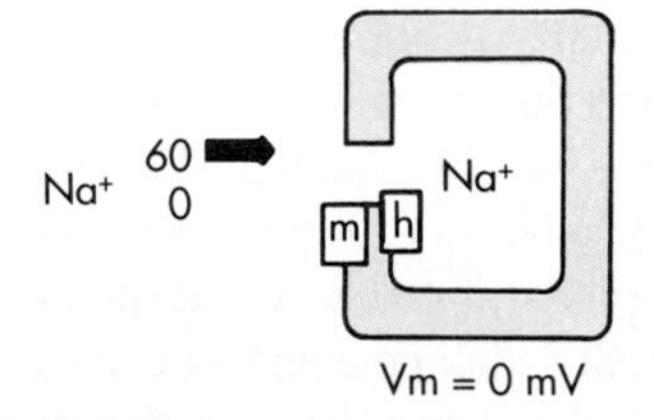

C, The rapid influx of Na^+ rapidly decreases the negativity of V_m. As V_m approaches 0, the electrostatic force attracting Na^+ into the cell is neutralized. Na^+ continues to enter the cell, however, because of the substantial concentration gradient, and V_m begins to become positive.

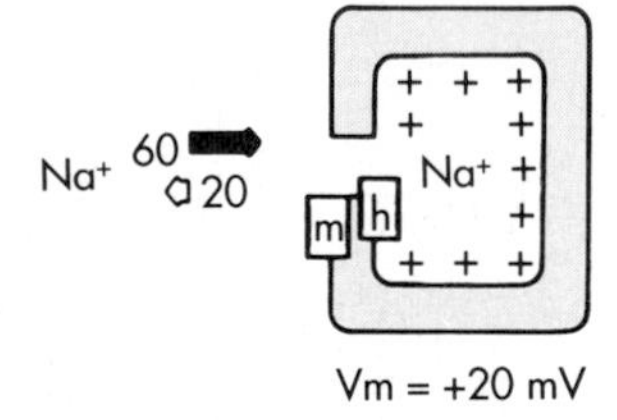

D, When V_m is positive by about 20 mV, Na^+ continues to enter the cell, because the diffusional forces (60 mV) exceed the opposing electrostatic forces (20 mV). The influx of Na^+ is slow, however, because the net driving force is small, and many of the inactivation gates have already closed.

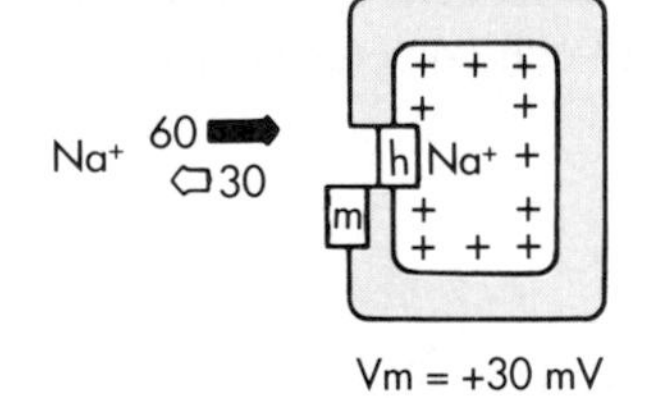

E, When V_m reaches about 30 mV, the **h** gates have now all closed, and Na^+ influx ceases. The **h** gates remain closed until the first half of repolarization, and thus the cell is absolutely refractory during this entire period. During the second half of repolarization, the **m** and **h** gates approach the state represented by panel *A,* and thus the cell is relatively refractory.

Fig. 23-6 The gating of a sodium channel in a cardiac cell membrane during phase 4 *(panel A)* and during various stages of the action potential upstroke *(panels B to E)*. The positions of the **m** and **h** gates in the fast Na^+ channels are shown at the various levels of V_m. The electrostatic forces are represented by the white arrows and the chemical (diffusional) forces by the black arrows.

and is therefore called an activation gate. The other, the h gate, tends to close the channel as V_m becomes less negative and hence is called an inactivation gate. The m and h designations were originally employed by Hodgkin and Huxley in their mathematical model of conduction in nerve fibers.

With the cell at rest, V_m is about −90 mV. At this level the m gates are closed and the **h** gates are wide open, as shown in Fig. 23-6, *A*. The concentration of Na^+ is much greater outside than inside the cell, and the interior of the cell is electrically negative with respect to the exterior. Hence both chemical and electrostatic forces are oriented to draw Na^+ into the cell.

The electrostatic force in Fig. 23-6, *A*, is a potential difference of 90 mV, and it is represented by the white arrow. The chemical force, based on the difference in Na^+ concentration between the outside and inside of the cell, is represented by the black arrow. For a Na^+ concentration difference of about 130 mM, a potential difference of 60 mV (inside more positive than outside) would be necessary to counterbalance the chemical, or diffusional, force, according to the Nernst equation for Na^+ (equation 2). Therefore we may represent the net chemical force favoring the inward movement of Na^+ in Fig. 23-6 (black arrows) as being equivalent to a potential of 60 mV. With the cell at rest the total electrochemical force favoring the inward movement of Na^+ is 150 mV (panel A). The **m** gates are closed, however, and the conductance of the resting cell membrane to Na^+ is very low. Hence virtually no Na^+ moves into the cell: (i.e., the inward Na^+ current is negligible).

Any process that makes V_m less negative tends to open the **m** gates, thereby "activating" the fast Na^+ channels. The activation of the fast channels is therefore called a voltage-dependent phenomenon. The precise potential at which the **m** gates swing open varies somewhat from one Na^+ channel to another in the cell membrane. As V_m becomes progressively less negative, more and more m gates open. As the **m** gates open, Na^+ enters the cell (Fig. 23-6, *B*) by virtue of the chemical and electrostatic forces.

The entry of positively charged Na^+ into the interior of the cell neutralizes some of the negative charges inside the cell and thereby diminishes further the transmembrane potential, V_m. The resultant reduction in V_m, in turn, opens more **m** gates, thereby augmenting the inward Na^+ current. This is called a regenerative process. As V_m approaches about −65 mV, the remaining **m** gates rapidly swing open in the fast Na^+ channels until virtually all of the **m** gates are open (Fig. 23-6, *B*).

The rapid opening of the m gates in the fast Na^+ channels is responsible for the large and abrupt increase in Na^+ conductance, g_{Na}, coincident with phase 0 of the action potential (Fig. 23-7). The rapid influx of Na^+ accounts for the rapid rate of change of V_m during phase 0. The maximum rate of change of V_m (i.e., the maximum dV_m/dt) varies from 100 to 200 V/sec in myocardial cells and from 500 to 1,000 V/sec in Purkinje's fibers. Although the quantity of Na^+ that enters the cell during one action potential alters V_m by more than 100 mV, that quantity of Na^+ is too small to change the intracellular Na^+ concentration measurably. The chemical force remains virtually constant, and only the electrostatic force changes throughout the action potential. Hence the lengths of the black arrows in Fig. 23-6 remain constant at 60 mV, whereas the white arrows change in magnitude and direction.

As Na^+ rushes into the cardiac cell during phase 0, the negative charges inside the cell are neutralized, and V_m becomes progressively less negative. When V_m becomes zero (Fig. 23-6, *C*), an electrostatic force no longer pulls Na^+ into the cell. As long as the fast Na^+ channels are open, however, Na^+ continues to enter the

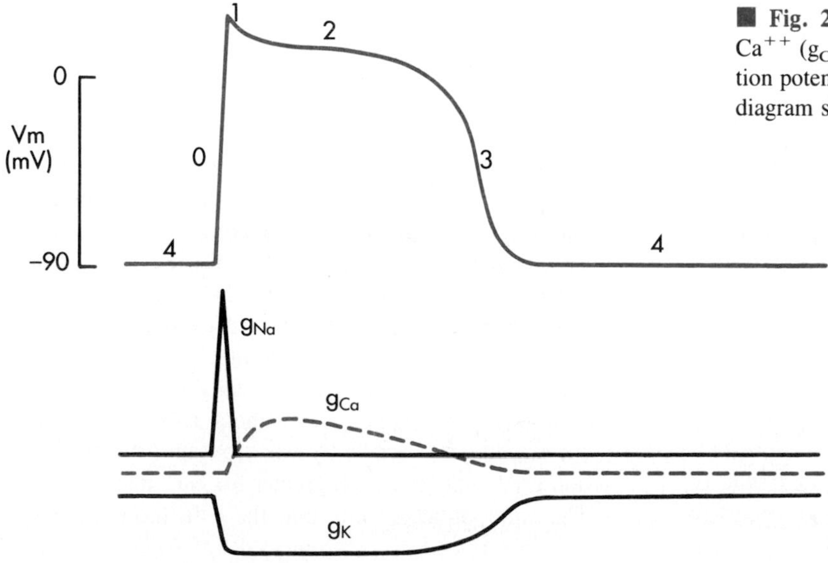

■ **Fig. 23-7** Changes in the conductances of Na^+ (g_{Na}), Ca^{++} (g_{Ca}), and K^+ (g_K) during the various phases of the action potential of a fast-response cardiac cell. The conductance diagram shows directional changes only.

cell because of the large concentration gradient. This continuation of the inward Na^+ current causes the inside of the cell to become positively charged (Fig. 23-6, *D*). This reversal of the membrane polarity is the so-called overshoot of the cardiac action potential. Such a reversal of the electrostatic gradient would, of course, tend to repel the entry of Na^+ (Fig. 23-6, *D*). However as long as the inwardly directed chemical forces exceed these outwardly directed electrostatic forces, the net flux of Na^+ remains inward, although the rate of influx is diminished.

The inward Na^+ current finally ceases when the h (inactivation) gates close (Fig. 23-6, *E*). The activity of the h gates is governed by the value of V_m just as is that of the m gates. However whereas the m gates open as V_m becomes less negative, the h gates close under this same influence. The opening of the m gates occurs very rapidly (in about 0.1 to 0.2 msec), whereas the closure of the h gates requires 1 msec or more. Phase 0 is finally terminated when the h gates have closed and have thereby "inactivated" the fast Na^+ channels. The closure of the h gates so soon after the opening of the m gates accounts for the quick return of g_{Na} to near its resting value (Fig. 23-7).

The h gates then remain closed until the cell has partially repolarized during phase 3 (at about time d in Fig. 23-1, *A*). From time c to time d the cell is in its effective refractory period and will not respond to further excitation. This mechanism prevents a sustained, tetanic contraction of cardiac muscle. Tetanus would of course preclude the normal intermittent pumping action of the heart.

About midway through phase 3 (time d in Fig. 23-1, *A*) the m and h gates in some of the fast Na^+ channels have resumed the states shown in Fig. 23-6, *A*. Such channels are said to have recovered from inactivation. The cell can begin to respond (but weakly at first) to further excitation (see Fig. 23-15). Throughout the remainder of phase 3 the cell completes its recovery from inactivation. By time e in Fig. 23-1, *A*, the h gates have reopened and the m gates have reclosed in the remaining fast Na^+ channels (see Fig. 23-6, *A*).

Statistical characteristics of the "gate" concept. The **patch-clamping technique** has made it possible to measure ionic currents through single membrane channels (see Chapter 3). The individual channels have been observed to open and close repeatedly in a quasirandom sequence. This process is illustrated in Figs. 23-10 and 23-12.

The overall change in ionic conductance of the entire cell membrane at any given time reflects the number of channels that happen to be open at that time. Because the individual channels open and close in an irregular pattern, the overall membrane conductance represents the statistical probability of the open or closed state of the individual channels. The temporal characteristics of the activation process then represent the time course of the increasing probability that the specific channels will be open, rather than the kinetic characteristics of the activation gates in the individual channels. Similarly the temporal characteristics of inactivation reflect the time course of the decreasing probability that the channels will be open and not the kinetic characteristics of the inactivation gates in the individual channels.

Genesis of early repolarization. In many cardiac cells that have a prominent plateau, phase 1 constitutes an early, brief period of limited repolarization between the end of the upstroke and the beginning of the plateau. Phase 1 reflects the activation of a transient outward current, i_{to}, mostly carried by K^+. Activation of these K^+ channels during phase 1 leads to a brief efflux of K^+ from the cell because the interior of the cell is positively charged and because the internal K^+ concentration greatly exceeds the external concentration (Fig. 23-8). This brief efflux of positively charged ions brings about the brief, limited repolarization (phase 1).

Phase 1 is prominent in Purkinje's fibers (see Fig. 23-13) and in epicardial fibers from the ventricular myocardium (Fig. 23-9); it is much less developed in endocardial fibers. When the basic cycle length (BCL) at which the epicardial fibers are driven is increased from 300 to 2,000 msec, phase 1 becomes more pronounced and the action potential duration is increased substantially. The same increase in basic cycle length has no effect on the early portion of the plateau in endocardial fibers, and it has a smaller effect on the action potential duration than it does in epicardial fibers (Fig. 23-9). In the presence of 4-aminopyridine, which blocks the K^+ channels that carry i_{to}, phase 1 disappears in the epicardial action potentials. However this compound does not affect the beginning of the plateau in the endocardial action potentials.

Genesis of the plateau. During the plateau (phase 2) of the action potential Ca^{++} and some Na^+ enter the cell through channels that activate and inactivate much more slowly than do the fast Na^+ channels. During the flat portion of phase 2 (see Fig. 23-9), this influx of positive charge (carried by Ca^{++} and Na^+) is balanced by the efflux of an equal amount of positive charge (carried by K^+). The K^+ exits through channels that conduct mainly the i_{to}, i_K, and i_{K1} currents. The i_{to} current is responsible for phase 1, as described previously, but it is not completely inactivated until after phase 2 has expired. The i_K and i_{K1} currents are described later.

Ca^{++} conductance during the plateau. The slow channels that conduct the inward cationic currents are about 50 to 100 times more permeable to Ca^{++} than to Na^+, and therefore they are usually referred to as calcium channels. Nevertheless about 10% to 20% of the current may be carried by Na^+, because the concentration gradient is so much greater for Na^+ than for Ca^{++}. The entry of these ions into the cell through the slow

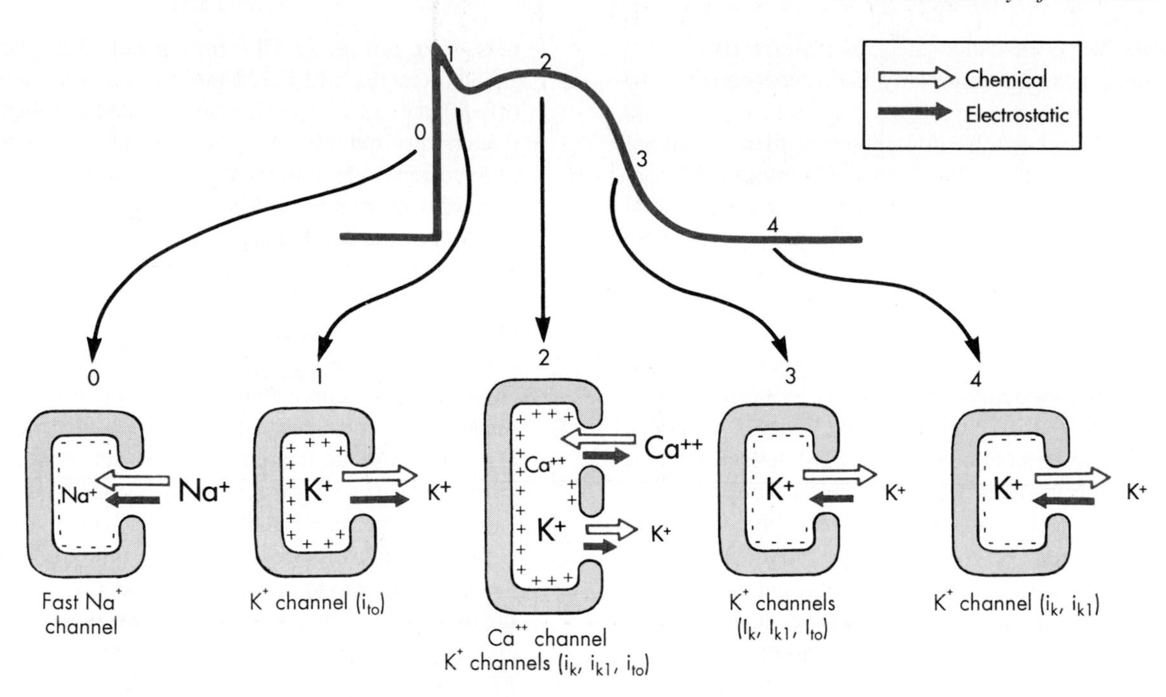

■ **Fig. 23-8** The principal ionic currents and channels that generate the action potential in a cardiac cell.
Phase 0. The chemical and electrostatic forces both favor the entry of Na^+ into the cell through fast Na^+ channels to generate the upstroke.
Phase 1. The chemical and electrostatic forces both favor the efflux of K^+ through i_{to} channels to generate early, partial repolarization.
Phase 2. During the plateau, the net influx of Ca^{++} through Ca^{++} channels is balanced by the efflux of K^+ through i_K, i_{K1}, and i_{to} channels.
Phase 3. The chemical forces that favor the efflux of K^+ through i_K, i_{K1}, and i_{to} channels predominate over the electrostatic forces that favor the influx of K^+ through these same channels.
Phase 4. The chemical forces that favor the efflux of K^+ through i_K and i_{K1} channels exceed very slightly the electrostatic forces that favor the influx of K^+ through these same channels.

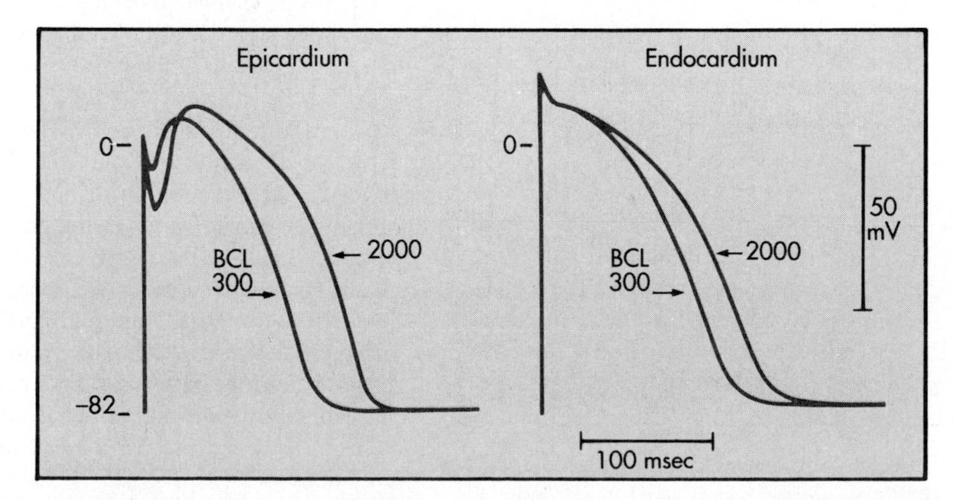

■ **Fig. 23-9** Action potentials recorded from canine epicardial and endocardial strips driven at basic cycle lengths of 300 and 2000 msec. (From Litovsky SH, Antzelevitch C: *J Am Coll Cardiol* 19:1053, 1989.)

channels has been called the slow inward current, i_{si}, but now it is more commonly called the calcium current, i_{Ca}.

The Ca^{++} channels are voltage-regulated channels that are activated as V_m becomes progressively less negative during the upstroke of the action potential. Two types of Ca^{++} channels (L-type and T-type) have been identified in cardiac tissues (see Chapter 3). Some of their important characteristics are illustrated in Fig. 23-10, which displays the Ca^{++} currents generated by voltage-clamping an isolated atrial myocyte. Note that when V_m is suddenly increased to +30 mV from a holding potential of −30 mV (lower panel), an inward Ca^{++} current (denoted by a downward deflection) is activated. Note also that after the inward current reaches its maximum value (in the downward direction), it returns toward zero very gradually (i.e., the channel inactivates very slowly). Thus the current that passes through these channels is long lasting, and they have therefore been designated as **L-type channels.** They are the predominant type of Ca^{++} channels in the heart and are activated during the action potential upstroke when V_m reaches about −10 mV. The L-type channels are blocked by the so-called Ca^{++} channel blocking drugs, such as verapamil, nifedipine, and diltiazem.

The **T-type** (transient) Ca^{++} **channels** are much less abundant in the heart. They are activated at more negative potentials (about −70 mV) than are the L-type channels. Note in Fig. 23-10 (upper panel) that when V_m is suddenly increased to −20 mV from a holding potential of −80 mV, a Ca^{++} current is activated and then is inactivated very quickly. Note how quickly the current returns to zero even though V_m is maintained at −20 mV. The *transient* nature of this current accounts for the designation of the conducting channels as T-type channels. The T-type channels are not blocked by the commonly used Ca^{++} channel blocking drugs, but they can be blocked experimentally by Ni^{++}.

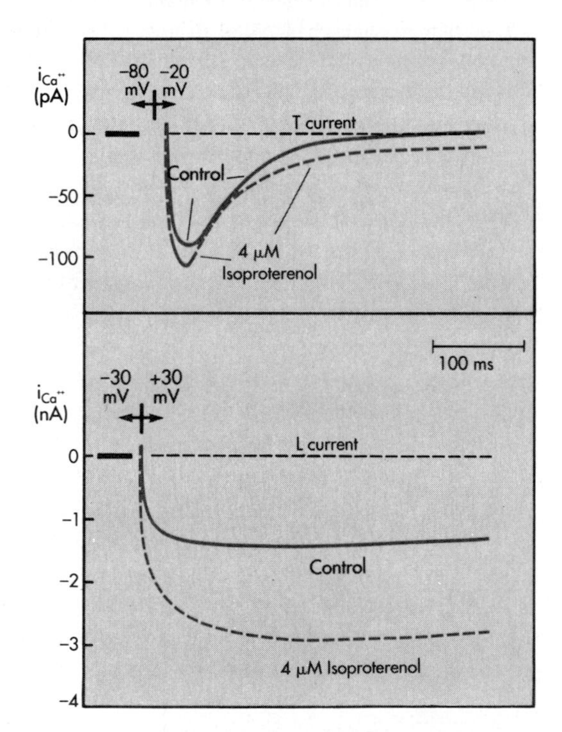

■ **Fig. 23-10** Effects of isoproterenol on the Ca^{++} currents conducted by T-type *(upper panel)* and L-type *(lower panel)* Ca^{++} channels in canine atrial myocytes. *Upper panel,* potential changed from −80 to −20 mV; *lower panel,* potential changed from −30 to +30 mV. (Redrawn from Bean BP: *J Gen Physiol* 86:1, 1985.)

Opening of the Ca^{++} channels is reflected by an increase in Ca^{++} conductance (g_{Ca}), which begins immediately after the upstroke of the action potential (see Fig. 23-7). At the beginning of the action potential the intracellular Ca^{++} concentration is much less than the extracellular concentration (Table 23-1). Consequently with the increase in g_{Ca} there is an influx of Ca^{++} into the cell throughout the plateau. The Ca^{++} that enters the myocardial cell during the plateau is involved in excitation-contraction coupling, as described in Chapters 17 and 24.

Various factors may influence g_{Ca}. This conductance may be increased by catecholamines, such as isoproterenol and norepinephrine. This is probably the principal mechanism by which catecholamines enhance cardiac muscle contractility.

Catecholamines interact with β-adrenergic receptors located in the cardiac cell membranes. This interaction stimulates the membrane bound enzyme adenylyl cyclase, which raises the intracellular concentration of cyclic adenosine monophosphate AMP (see also Chapter 5). This change enhances the activation of the L-type Ca^{++} channels in the cell membrane (see Fig. 23-10, lower panel) and thus augments the influx of Ca^{++} into the cells from the interstitial fluid. However catecholamines do not affect the Ca^{++} current through the T-type channels (upper panel).

The Ca^{++} channel blocking drugs decrease g_{Ca}. By reducing the amount of Ca^{++} that enters the myocardial cells during phase 2, these drugs diminish the strength of the cardiac contraction (Fig. 23-11).

K^+ conductance during the plateau. During the plateau of the action potential, the concentration gradient for K^+ between the inside and outside of the cell is virtually the same as it is during phase 4, but the V_m is positive. Therefore the chemical and electrostatic forces greatly favor the efflux of K^+ from the cell (see Fig. 23-8). If g_K were the same during the plateau as it is during phase 4, the efflux of K^+ during phase 2 would greatly exceed the influx of Ca^{++} and Na^+, and a sustained plateau could not be achieved. However as V_m attains positive values near the end of phase 0, g_K suddenly decreases (see Fig. 23-7).

This reduction in g_K at positive values of V_m is called inward rectification, which is a characteristic of the i_{K1} current. The current-voltage relationship of the K^+ channels that conduct i_{K1} has been determined by

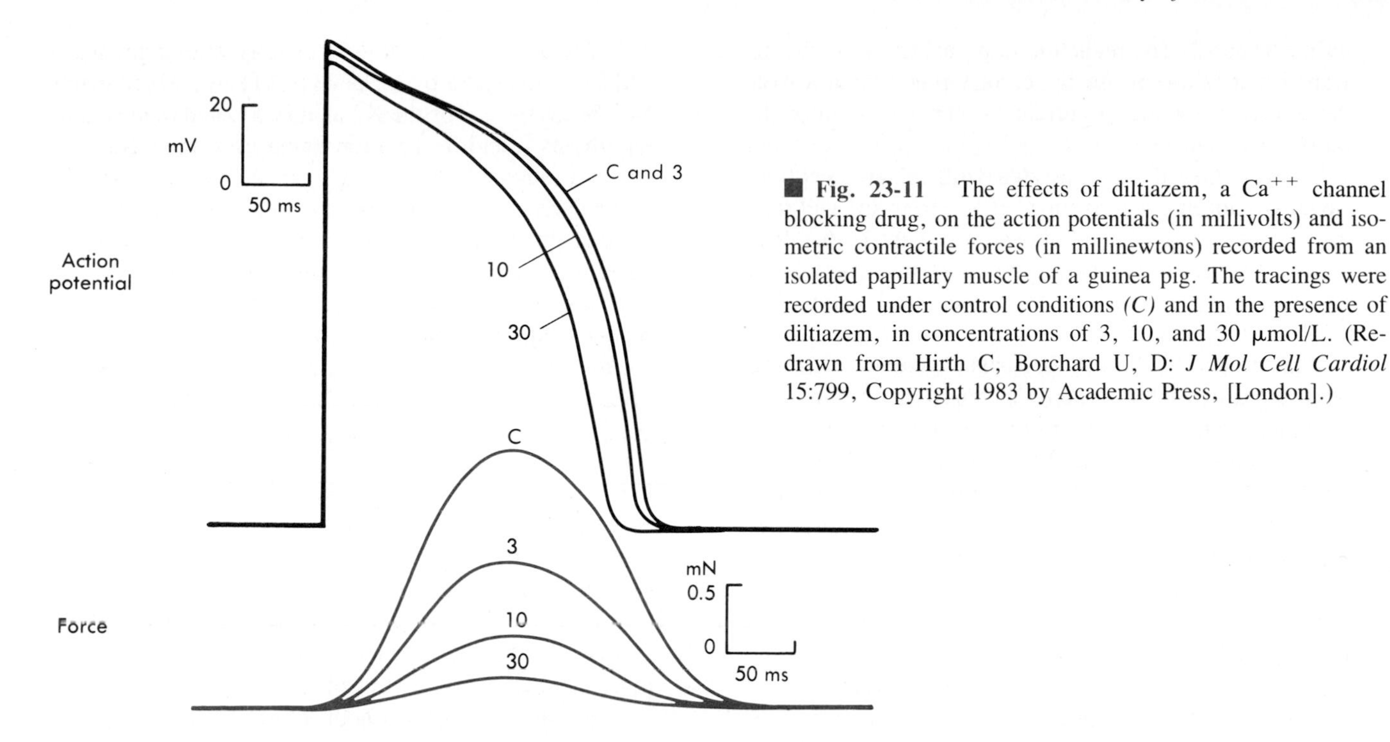

■ **Fig. 23-11** The effects of diltiazem, a Ca^{++} channel blocking drug, on the action potentials (in millivolts) and isometric contractile forces (in millinewtons) recorded from an isolated papillary muscle of a guinea pig. The tracings were recorded under control conditions *(C)* and in the presence of diltiazem, in concentrations of 3, 10, and 30 μmol/L. (Redrawn from Hirth C, Borchard U, D: *J Mol Cell Cardiol* 15:799, Copyright 1983 by Academic Press, [London].)

voltage-clamping cardiac cells (Fig. 23-12). Note that for the cell depicted in the figure, the current-voltage curve intersects the voltage axis at a V_m of −70 mV. The absence of ionic current flow at the point of intersection indicates that the electrostatic forces must have been equal to the chemical (diffusional) forces (see Fig. 23-3) at this potential. Thus in this ventricular cell the Nernst equilibrium potential (E_k) for K^+ must have been −70 mV.

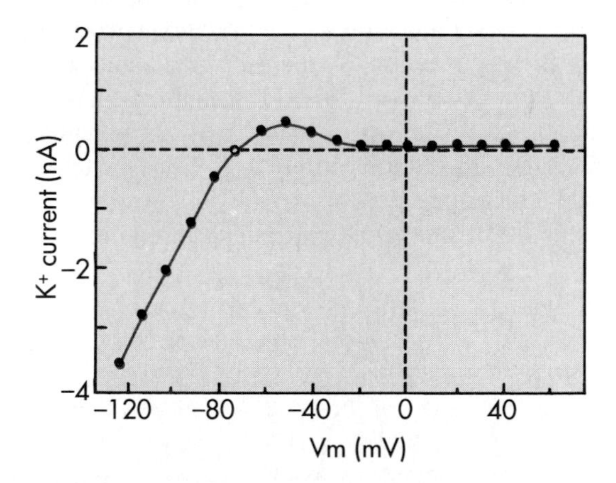

■ **Fig. 23-12** The K^+ currents recorded from a rabbit ventricular myocyte when the potential was changed from a holding potential of −80 mV to various test potentials. Positive values along the vertical axis represent outward currents; negative values represent inward currents. The V_m coordinate of the point of intersection *(open circle)* of the curve with the X axis is the reversal potential; it denotes the equilibrium potential at which the chemical and electrostatic forces are equal. (Redrawn from Giles WR, Imaizumi Y: *J Physiol* (London) 405:123, 1988.)

When the membrane potential was clamped at levels negative to −70 mV in this cardiac cell (see Fig. 23-12), the electrostatic forces exceeded the chemical forces and an *inward* K^+ current was induced (as denoted by the negative values of K^+ current over this range of voltages). Note also that for V_m negative to −70 mV the curve has a steep slope, even at the point of intersection (at which $V_m = E_k$). Thus when V_m equals or is negative to E_K, a small change in V_m induces a large change in K^+ current; that is, g_K is large. During phase 4 the V_m of a myocardial cell is approximately equal to E_k; it is actually slightly less negative (see Fig. 23-4). The substantial g_K that prevails during phase 4 of the cardiac action potential (see Fig. 23-7) is accounted for mainly by the i_{K1} channels.

When the transmembrane potential was clamped at levels positive to −70 mV (see Fig. 23-12), the chemical forces exceeded the electrostatic forces. Therefore the net K^+ currents were *outward* (as denoted by the positive values along the corresponding section of the Y axis). Note that for V_m values positive to −70 mV the curve is relatively flat. Thus a given change in voltage causes only a small change in ionic current; (i.e., g_K is small). Thus g_K is small for outwardly directed K^+ currents, but it is substantial for inwardly directed K^+ currents; i.e., the i_{K1} current is inwardly rectified.

Another factor that contributes to the low g_K during the plateau is **delayed rectification,** which is a characteristic of the i_K channels. These K^+ channels are activated by the voltages that prevail toward the end of phase 0, but activation proceeds very slowly, over several hundreds of milliseconds. Hence activation of these channels tends to increase g_K very slowly and slightly during phase 2. The reduction in g_K achieved by the inward rectification of the i_{K1} current predominates over the tendency for the i_K current to increase g_K throughout the plateau.

The diminished g_K associated with inward rectification prevents an excessive loss of K^+ from the cell during the plateau. The small outward K^+ current that does occur is sufficient to balance the slow inward currents of Ca^{++} and Na^+; hence, V_m remains relatively constant during phase 2.

The effects of altering this balance between the inward currents of Ca^{++} and Na^+ and the outward current of K^+ are demonstrated by the administration of a calcium channel blocking drug. Fig. 23-11 shows that with increasing concentrations of diltiazem the voltage of the plateau becomes less positive, and the duration of the plateau diminishes.

Genesis of final repolarization. The process of final repolarization (phase 3) starts at the end of phase 2, when the efflux of K^+ from the cardiac cell begins to exceed the influx of Ca^{++} and Na^+. At least three outward K^+ currents (i_{to}, i_K, and i_{K1}) reflect the return of g_K to its resting level at the end of the plateau (see Fig. 23-7), and they bring about the rapid repolarization (phase 3) of the cardiac cell (see Fig. 23-8).

The transient outward current (i_{to}) not only accounts for phase 1, as previously described, but also helps determine the duration of the plateau; hence it also helps initiate repolarization. For example, the transient outward current is much more pronounced in atrial than in ventricular myocytes. In atrial cells, therefore, the outward K^+ current exceeds the slow inward Ca^{++} current early in the plateau, whereas the outward and inward currents remain equal for a much longer time in ventricular myocytes. Hence the plateau of the action potential is much less pronounced in atrial than in ventricular myocytes (see Fig. 23-20).

The delayed rectifier **K^+ current** (i_K) is activated near the end of phase 0, but activation is very slow. Hence the outward i_K current tends to increase throughout the plateau. Concurrently the Ca^{++} channels are inactivated after the beginning of the plateau, and therefore the slow inward currents of Ca^{++} and Na^+ are decreasing. As the efflux of K^+ begins to exceed the influx of Ca^{++} and Na^+, V_m becomes progressively less positive, and repolarization is initiated.

The inwardly rectified K^+ **current,** i_{K1}, contributes substantially to the process of repolarization. As the net efflux of cations causes V_m to become more negative during phase 3, the conductance of the channels that carry the i_{K1} current progressively increases. This is reflected by the hump that is evident in the flat portion of the current-voltage curve at V_m values between -20 and -70 mV. Thus as V_m passes through this range of values, the outward K^+ current increases and thereby accelerates repolarization.

Restoration of ionic concentrations. The excess Na^+ that had entered the cell mainly during phases 0 and 2 is eliminated by a Na^+/K^+-ATPase, which ejects Na^+ in exchange for the K^+ that had exited mainly during phases 2 and 3. This ion pump exchanges Na^+ for K^+ in a ratio of 3:2.

Similarly most of the excess Ca^{++} that had entered the cell during phase 2 is eliminated by a Na^+/Ca^{++} exchanger, which exchanges 3 Na^+ for 1 Ca^{++}. However a small fraction of the Ca^{++} is eliminated by an ATP driven Ca^{++} pump (see p. 403).

Ionic Basis of the Slow Response

Fast response action potentials (see Fig. 23-1, *A*) may be considered to consist of three principal components, a spike (phases 0 and 1), a plateau (phase 2), and a period of repolarization (phase 3). In the slow response (see Fig. 23-1, *B*) the first component is absent, and the second and third components account for the entire action potential. In the fast response the spike is produced by the influx of Na^+ through the fast channels. These channels can be blocked by certain compounds, such as tetrodotoxin. When the fast Na^+ channels are blocked, slow responses may be generated in the same fibers under appropriate conditions.

The Purkinje's fiber action potentials shown in Fig. 23-13 clearly exhibit the two response types. In the control tracing (panel A) a prominent notch separates the spike from the plateau. This notch at the beginning of the plateau represents a well-developed phase 1 (see Fig. 23-9) and is ascribable to the transient outward K^+ current, i_{to}. Action potential A is a typical fast response action potential. In action potentials B to E progressively larger quantities of tetrodotoxin were added to the bathing solution to produce a graded blockade of the fast Na^+ channels. It is evident that the spike becomes progressively less prominent in action potentials B to D, and it disappears entirely in E. Thus the tetrodotoxin had a pronounced effect on the spike and only a negligible influence on the plateau. With elimination of the

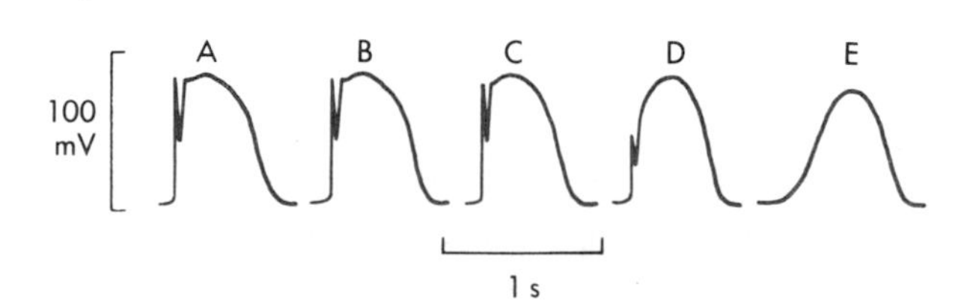

Fig. 23-13 Effect of tetrodotoxin on the action potential recorded in a calf Purkinje's fiber perfused with a solution containing epinephrine and 10.8 mM K^+. The concentration of tetrodotoxin was 0 M in **A,** 3×10^{-8} M in **B,** 3×10^{-7} M in **C,** and 3×10^{-6} M in **D** and E; **E** was recorded later than **D.** (Redrawn from Carmeliet E, Vereecke J: *Pflugers Arch* 313:300, 1969.)

spike (panel E) the action potential resembles a typical slow response.

Certain cells in the heart, notably those in the SA and AV nodes, are normally slow response fibers. In such fibers depolarization is achieved by the slow inward current of Ca^{++} and Na^{+} through the Ca^{++} channels. These ionic events closely resemble those that occur during the plateau of fast response action potentials.

Conduction in Cardiac Fibers

An action potential traveling down a cardiac muscle fiber is propagated by local circuit currents, similar to the process that occurs in nerve and skeletal muscle fibers (see Chapter 3). The characteristics of conduction differ in fast and slow response fibers.

Conduction of the Fast Response

In the fast response the fast Na^{+} channels are activated when the transmembrane potential is suddenly brought to the threshold value of about −70 mV. The inward Na^{+} current then depolarizes the cell very rapidly at that site. This portion of the fiber becomes part of the depolarized zone, and the border is displaced accordingly. The same process then begins at the new border. This process is repeated again and again, and the border moves continuously down the fiber as a wave of depolarization (see Fig. 3-18).

At any given point on the fiber the greater the amplitude of the action potential and the greater the rate of change of potential (dV_m/dt) during phase 0 the more rapid the conduction down the fiber. The amplitude of the action potential equals the difference in potential between the fully depolarized and the fully polarized regions of the cell interior. The magnitude of the local currents is proportional to this potential difference (see Chapter 3). Because these local currents shift the potential of the resting zone toward the threshold value, they are the local stimuli that depolarize the adjacent resting portion of the fiber to its threshold potential. The greater the potential difference between the depolarized and polarized regions (i.e., the greater the amplitude of the action potential) the more efficacious the local stimuli and the more rapidly the wave of depolarization is propagated down the fiber.

The rate of change potential (dV_m/dt) during phase 0 is also an important determinant of the conduction velocity. If the active portion of the fiber depolarizes very gradually, the local currents across the border between the depolarized and polarized regions are very small. Thus the resting region adjacent to the active zone is depolarized very slowly, and consequently each new section of the fiber requires more time to reach threshold.

The level of the resting membrane potential is also an important determinant of conduction velocity. This factor operates through its influence on the amplitude and maximum slope of the action potential. The resting potential may vary for several reasons: (1) it can be altered experimentally by varying $[K^{+}]_o$ (see Fig. 23-4), (2) in cardiac fibers that are intrinsically automatic V_m becomes progressively less negative during phase 4 (see Fig. 23-19, *B*), and (3) during a premature contraction repolarization may not have been completed when the next excitation arrives (see Fig. 23-15). In general the less negative the level of V_m the less the velocity of impulse propagation, regardless of the reason for the change in V_m.

The V_m level affects conduction velocity because the inactivation, or **h,** gates (see Fig. 23-6) in the fast Na^{+} channels are voltage dependent. The less negative the V_m the greater is the number of h gates that tend to close. During the normal process of excitation depolarization proceeds so rapidly during phase 0 that the comparatively slow h gates do not close until the end of that phase. If partial depolarization is produced by a more gradual process, however, such as by elevating the level of external K^{+}, then the gates have ample time to close and thereby inactivate some of the Na^{+} channels. When the cell is partially depolarized, many of the Na^{+} channels are already inactivated, and only a fraction of these channels is available to conduct the inward Na^{+} current during phase 0.

The results of an experiment in which the resting V_m of a bundle of Purkinje's fibers was varied by altering the value of $[K^{+}]_o$ are shown in Fig. 23-14. When $[K^{+}]_o$ was 3 mM (panels A and F), the resting V_m was −82 mV and the slope of phase 0 was steep. At the end of phase 0 the overshoot attained a value of 30 mV. Hence the amplitude of the action potential was 112 mV. The tissue was stimulated at some distance from the impaled cell, and the stimulus artifact (St) appears as a diphasic deflection just before phase 0. The time from this artifact to the beginning of phase 0 is inversely proportional to the conduction velocity.

When $[K^{+}]_o$ was increased to 16 mM (panels B to E), the resting V_m became progressively less negative. Concomitantly the amplitudes and durations of the action potentials and the steepness of the upstrokes all diminished. As a consequence the conduction velocity diminished progressively, as indicated by the distances from the stimulus artifacts to the upstrokes.

At the $[K^{+}]_o$ levels of 14 and 16 mM (panels D and E) the resting V_m had attained levels sufficient to inactivate all the fast Na^{+} channels. The action potentials in panels D and E are characteristic slow responses, presumably mediated by the slow inward current of Ca^{++} and Na^{+}. When the $[K^{+}]_o$ concentration of 3 mM was reestablished (panel F), the action potential was again characteristic of the normal fast response (as in panel A).

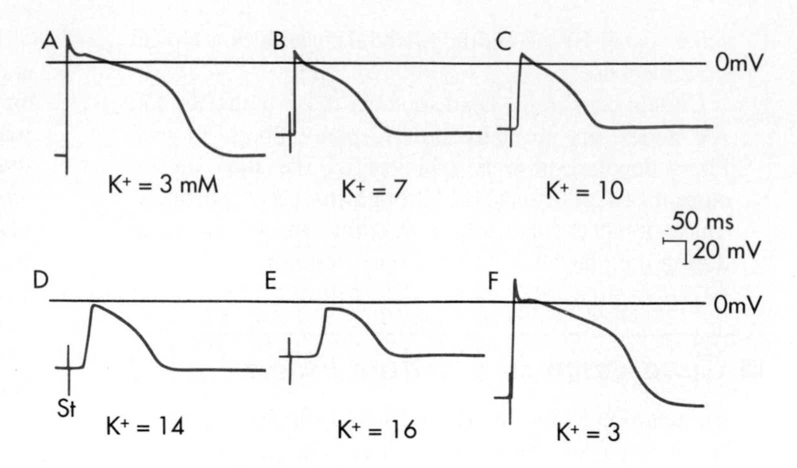

Fig. 23-14 The effect of changes in external potassium concentration on the transmembrane action potentials recorded from a Purkinje's fiber. The stimulus artifact *(St)* appears as a biphasic spike to the left of the upstroke of the action potential. The horizontal lines near the peaks of the action potentials denote 0 mV. (From Myerburg RJ, Lazzara R. In Fisch E, editor: *Complex electrocardiography,* Philadelphia 1973, FA Davis.)

Conduction of the Slow Response

Local circuits (see Fig. 3-18) are also responsible for propagation of the slow response. However the characteristics of the conduction process differ quantitatively from those of the fast response. The threshold potential is about −40 mV for the slow response, and conduction is much slower than for the fast response. The conduction velocities of the slow responses in the SA and AV nodes are about 0.02 to 0.1 m/sec. The fast response conduction velocities are about 0.3 to 1 m/sec for myocardial cells and 1 to 4 m/s for the specialized conducting fibers in the atria and ventricles. Slow responses are more likely to be blocked than are fast responses. Also the former cannot be conducted at such rapid repetition rates.

Cardiac Excitability

Currently more detailed knowledge of cardiac excitability is being acquired because of the rapid development of artificial pacemakers and other electrical devices for correcting serious disturbances of rhythm. The excitability characteristics of cardiac cells differ considerably, depending on whether the action potentials are fast or slow responses.

Fast Response

Once the fast response has been initiated, the depolarized cell is no longer excitable until about the middle of the period of final repolarization (see Fig. 23-1, *A*). The interval from the beginning of the action potential until the fiber is able to conduct another action potential is called the effective refractory period. In the fast response this period extends from the beginning of phase 0 to a point in phase 3 where repolarization has reached about −50 mV (time c to time d in Fig. 23-1, *A*). At about this value of V_m the electrochemical gates (m and h) for many of the fast Na^+ channels have been reset.

Full excitability is not regained until the cardiac fiber has been fully repolarized (time e in Fig. 23-1, *A*). During period d to e in the figure an action potential may be evoked, but only when the stimulus is stronger than that which could elicit a response during phase 4. Period d to e is called the relative refractory period.

When a fast response is evoked during the relative refractory period of a prior excitation, its characteristics vary with the membrane potential that exists at the time of stimulation. The nature of this voltage dependency is illustrated in Fig. 23-15. As the fiber is stimulated later and later in the relative refractory period, the amplitude of the response and the rate of rise of the upstroke increase progressively. Presumably the number of fast Na^+ channels that have recovered from inactivation increases as repolarization proceeds during phase 3. As a consequence of the greater amplitude and upstroke slope of the evoked response the propagation velocity increases as the cell is stimulated later in the relative refractory period. Once the fiber is fully repolarized, the response is constant no matter what time in phase 4 the stimulus is applied. By the end of phase 3 the m and h gates of all channels are in their final positions and no further change in excitability occurs.

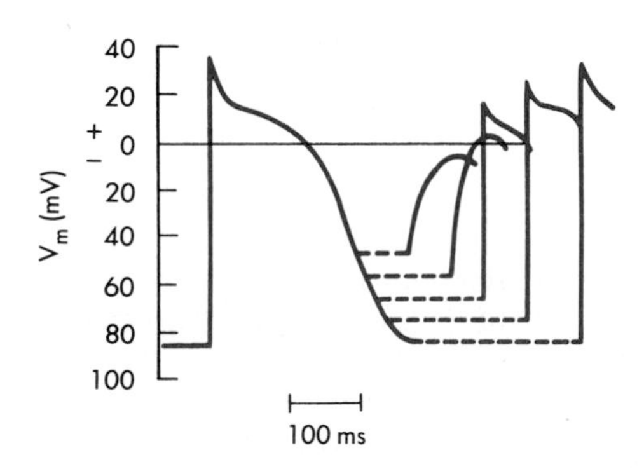

Fig. 23-15 The changes in action potential amplitude and slope of the upstroke as action potentials are initiated at different stages of the relative refractory period of the preceding excitation. (Redrawn from Rosen MR, Wit AL, Hoffman BF: *Am Heart J* 88:380, 1974.)

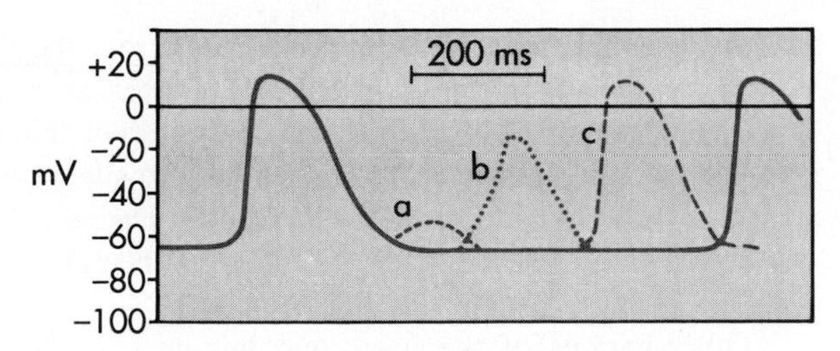

■ **Fig. 23-16** The effects of excitation at various times after the initiation of an action potential in a slow response fiber. In this fiber, excitation very late in phase 3 (or early in phase 4) induces a small, nonpropagated (local) response *(a)*. Later in phase 4, a propagated response *(b)* may be elicited; its amplitude is small and the upstroke is not very steep. This response *b* will be conducted very slowly. Still later in phase 4, full excitability will be regained, and the response *(c)* will display its normal characteristics. (Modified from Singer DH et al: *Prog Cardiovasc Dis* 24:97, 1981.)

■ *Slow Response*

The relative refractory period during the slow response frequently extends well beyond phase 3 (see Fig. 23-1, *B*). Even after the cell has completely repolarized, it may be difficult to evoke a propagated response for some time.

Action potentials evoked early in the relative refractory period are small and the upstokes are not very steep (Fig. 23-16). The amplitudes and upstroke slopes gradually improve as action potentials are elicited later and later in the relative refractory period. The recovery of full excitability is much slower than for the fast response. Impulses that arrive early in its relative refractory period are conducted much more slowly than those that arrive late in that period. The lengthy refractory periods also lead to conduction blocks. Even when slow responses recur at a low repetition rate, the fiber may be able to conduct only a fraction of those impulses.

■ *Effects of Cycle Length*

Changes in cycle length alter the action potential duration of cardiac cells and thus change their refractory periods. Consequently the changes in cycle length are often important factors in the initiation or termination of certain arrhythmias. The changes in action potential durations produced by stepwise reductions in cycle length from 2,000 to 200 msec in a Purkinje's fiber are shown in Fig. 23-17. Note that as the cycle length is diminished, the action potential duration decreases.

This direct correlation between action potential duration and cycle length is mediated by changes in g_K that involve at least two types of K^+ channels, namely,

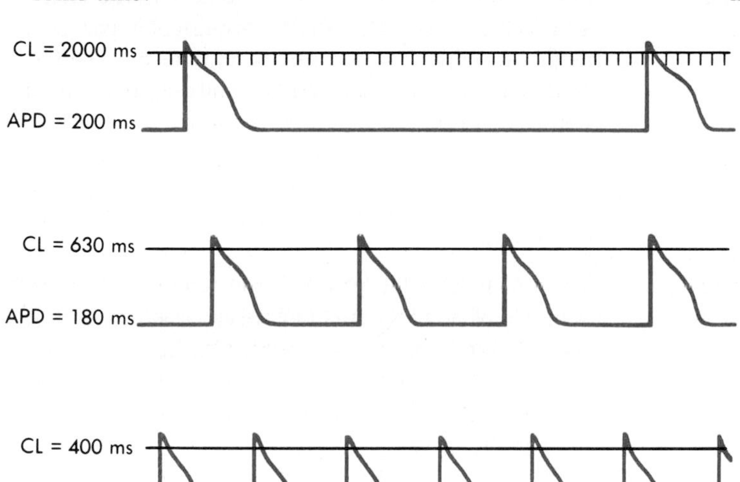

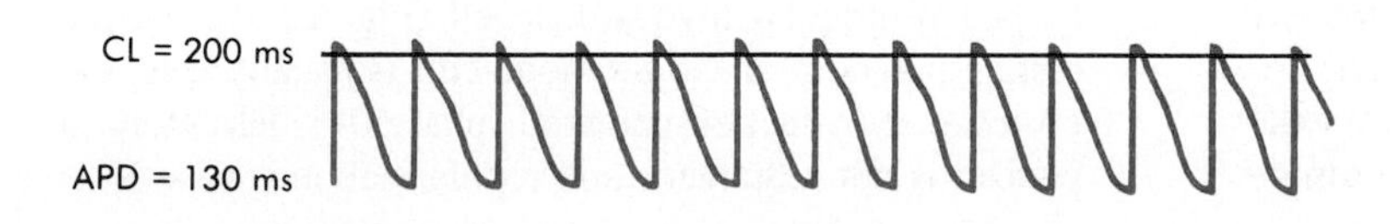

■ **Fig. 23-17** The effect of changes in cycle length *(CL)* on the action potential duration *(APD)* of canine Purkinje's fibers. (Modified from Singer D, Ten Eick RE: *Am J Cardiol* 28:381, 1971.)

those that conduct the delayed rectifier K^+ current, i_K, and those that conduct the transient outward K^+ current, i_{to}.

The i_K current activates slowly and it remains activated for hundreds of milliseconds before it is inactivated, and it also inactivates very slowly. Consequently as the basic cycle length is diminished, each action potential tends to occur earlier in the inactivation period of the i_K current initiated by the preceding action potential. Therefore the shorter the basic cycle length the greater the outward K^+ current during phase 2 and hence the briefer the action potential.

The i_{to} current also influences the relation between cycle length and action potential duration. The i_{to} current varies inversely with the cardiac cycle length. Therefore as the cycle length decreases, the resulting increase in i_{to} shortens the plateau.

This relationship is illustrated by experiments in which the effects of changes in cycle length were compared in epicardial and endocardial ventricular myocytes. Such experiments have demonstrated that a given increase in basic cycle length prolongs the action potential to a greater extent in epicardial than in endocardial fibers (see Fig. 23-9). The i_{to} current is much more prominent in epicardial than in endocardial fibers. Furthermore after the myocyte preparations have been treated with 4-aminopyridine, which blocks i_{to}, the cycle length dependent changes in action potential duration do not differ in epicardial and endocardial myocytes.

Natural Excitation of the Heart

The nervous system controls various aspects of the behavior of the heart, including the frequency at which it beats and the vigor of each contraction. However cardiac function certainly does not require intact nervous pathways. Indeed a patient with a completely denervated heart (a cardiac transplant patient) can function well and can adapt to stressful situations.

The properties of automaticity (the ability to initiate its own beat) and rhythmicity (the regularity of such pacemaking activity) are intrinsic to cardiac tissue. The heart continues to beat even when it is completely removed from the body. If the coronary vasculature is artificially perfused, rhythmic cardiac contraction persists for considerable periods of time. Apparently at least some cells in the walls of all four cardiac chambers are capable of initiating beats; such cells probably reside in the nodal tissues or specialized conducting fibers of the heart.

The region of the mammalian heart that ordinarily generates impulses at the greatest frequency is the SA node; it is called the natural pacemaker of the heart.

Detailed mapping of the electrical potentials on the surface of the right atrium has revealed that two or three sites of automaticity, located 1 or 2 cm from the SA node itself, serve along with the SA node as an atrial pacemaker complex. At times all of these loci initiate impulses simultaneously. At other times the site of earliest excitation shifts from locus to locus, depending on such conditions as the level of autonomic neural activity.

Other regions of the heart that initiate beats under special circumstances are called ectopic foci, or ectopic pacemakers. Ectopic foci may become pacemakers when (1) their own rhythmicity becomes enhanced, (2) the rhythmicity of the higher order pacemakers becomes depressed, or (3) all conduction pathways between the ectopic focus and those regions with greater rhythmicity become blocked.

When the SA node and the other components of the atrial pacemaker complex are excised or destroyed, pacemaker cells in the AV junction usually are the next most rhythmic, and they become the pacemakers for the entire heart. After some time, which may vary from minutes to days, automatic cells in the atria usually become dominant. In the dog the most common site for the ectopic atrial pacemaking region is at the junction between the inferior vena cava and the right atrium. Purkinje's fibers in the specialized conduction system of the ventricles also possess automaticity. Characteristically they fire at a very slow rate. When the AV junction is unable to conduct the cardiac impulse from the atria to the ventricles, suchidioventricular pacemakers in the Purkinje's fiber network initiate the ventricular contractions. Such ventricular contractions occur at a frequency of only 30 to 40 beats/min.

Sinoatrial Node

The SA node is the phylogenetic remnant of the sinus venosus of lower vertebrate hearts. In humans it is about 8 mm long and 2 mm thick. It lies in the groove where the superior vena cava joins the right atrium (Fig. 23-18). The sinus node artery runs lengthwise through the center of the node.

The SA node contains two principal types of cells: (1) small, round cells, which have few organelles and myofibrils, and (2) slender, elongated cells, which are intermediate in appearance between the round and the ordinary atrial myocardial cells. The round cells are probably the pacemaker cells, whereas the transitional cells probably conduct the impulses within the node and to the nodal margins.

A typical transmembrane action potential recorded from a cell in the SA node is depicted in Fig. 23-19, *B*. Compared with the transmembrane potential recorded from a ventricular myocardial cell (Fig. 23-19, *A*), the resting potential of the SA node cell is usually less, the upstroke of the action potential (phase 0) is less steep, a plateau is not sustained, and repolarization (phase 3) is

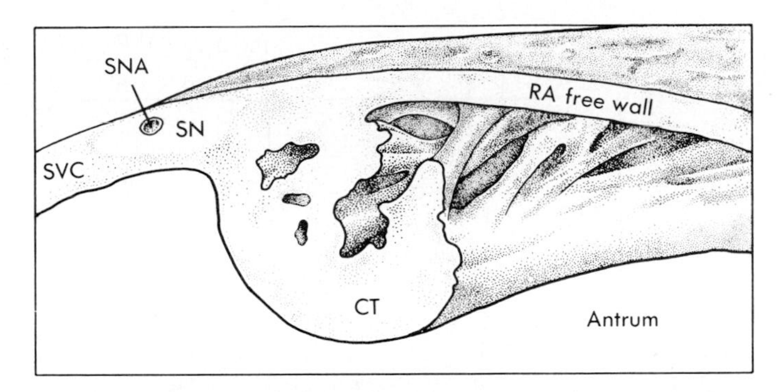

■ **Fig. 23-18** The location of the SA node near the junction between the superior vena cava and the right atrium. *SN,* SA node; *SNA,* sinoatrial artery; *SVC,* superior vena cava; *RA,* right atrium; *CT,* crista terminalis. (Redrawn from James TN: *Am J Cardiol* 40:965, 1977.)

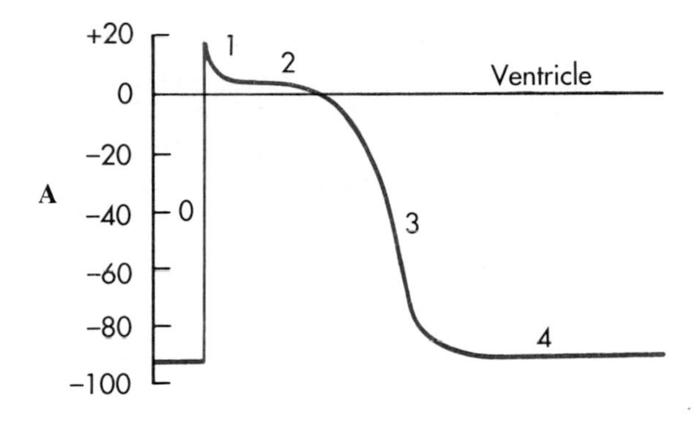

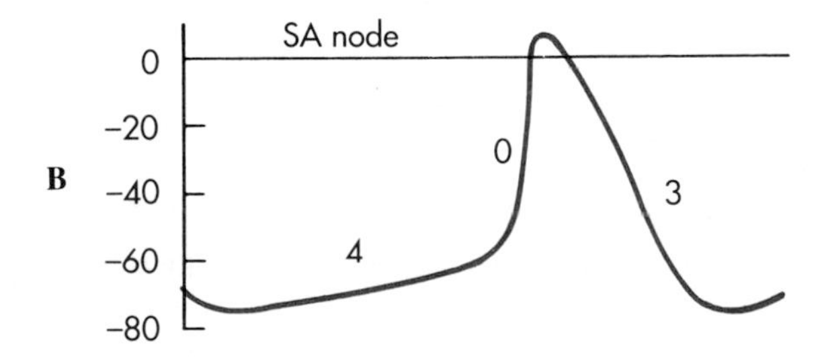

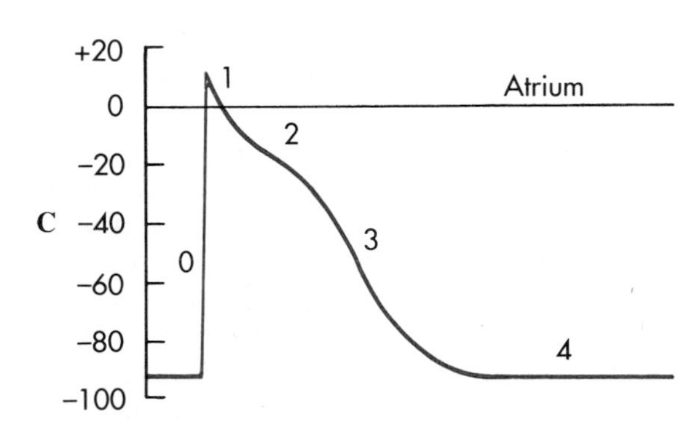

■ **Fig. 23-19** Typical action potentials (in millivolts) recorded from cells in the ventricle, **A,** SA node, **B,** and atrium, **C.** Sweep velocity in **B** is one half that in **A** or **C.** (From Hoffman BF, and Cranefield PF: *Electrophysiology of the heart,* New York, 1960, McGraw-Hill.)

more gradual. These are all characteristic of the slow response. Under ordinary conditions tetrodotoxin has no influence on the SA nodal action potential. This indicates that the upstroke of the action potential is not produced by an inward current of Na^+ through the fast channels.

However the principal distinguishing feature of a pacemaker fiber resides in phase 4. In nonautomatic cells the potential remains constant during this phase, whereas in a pacemaker fiber there is a slow depolarization, called the pacemaker potential, throughout phase 4. Depolarization proceeds at a steady rate until a threshold is attained, and then an action potential is triggered.

The discharge frequency of pacemaker cells may be varied by a change in (1) the rate of depolarization during phase 4, (2) the threshold potential, or (3) the resting potential (Fig. 23-20). With an increase in the rate of depolarization (b to a in Fig. 23-20, *A*) the threshold potential is attained earlier, and the heart rate increases. A rise in the threshold potential (from TP-1 to TP-2 in Fig. 23-20, *B*) delays the onset of phase 0 (from time b to time c), and the heart rate is reduced accordingly. Similarly when the maximum diastolic potential is increased (from a to d), more time is required to reach threshold TP-2 when the slope of phase 4 remains unchanged and the heart rate diminishes.

Ordinarily the frequency of pacemaker firing is controlled by the activity of both divisions of the autonomic nervous system. Increased sympathetic nervous

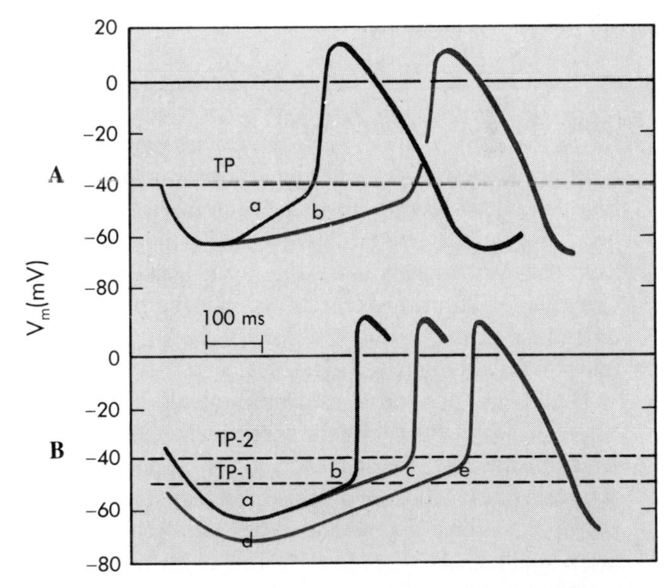

■ **Fig. 23-20** Mechanisms involved in changes of frequency of pacemaker firing. In section **A** a reduction in the slope of the pacemaker potential from *a* to *b* diminishes the frequency. In section **B** an increase in the threshold (from TP-1 to TP-2) or an increase in the magnitude of the resting potential (from *a* to *d*) also diminishes the frequency. (Redrawn from Hoffman BF, Cranefield PF: *Electrophysiology of the heart,* New York, 1960, McGraw-Hill.)

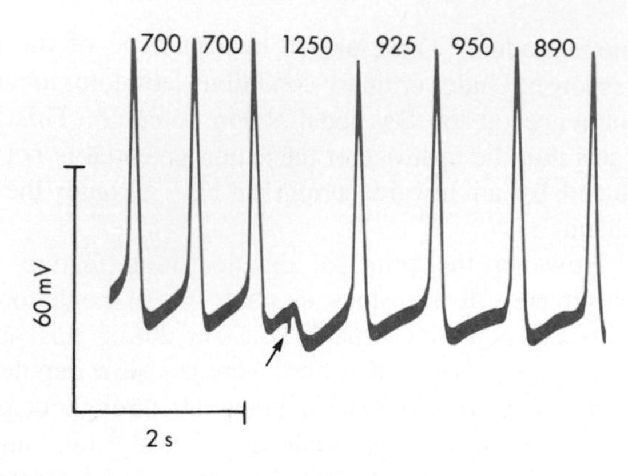

■ **Fig. 23-21** Effect of a brief vagal stimulus *(arrow)* on the transmembrane potential recorded from an SA node pacemaker cell in an isolated cat atrium preparation. The cardiac cycle lengths, in milliseconds, are denoted by the numbers at the top of the figure. (Modified from Jalife J, Moe GK: *Circ Res* 45:595, 1979.)

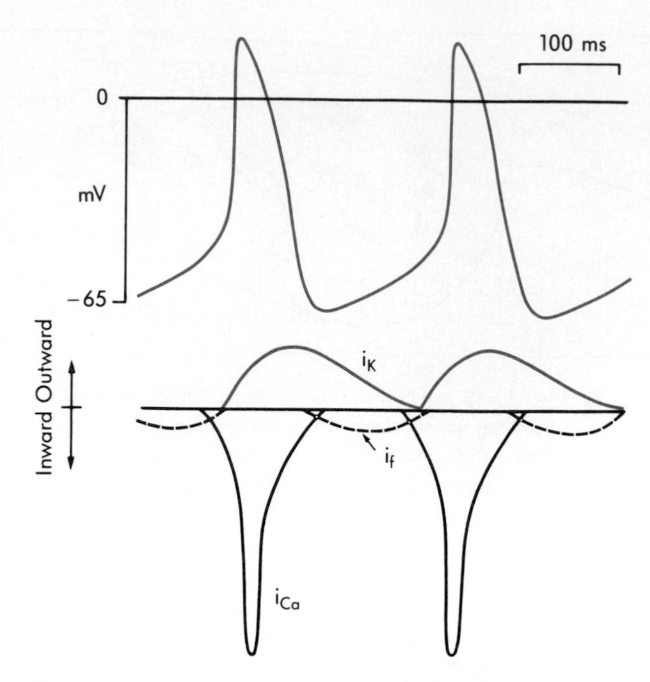

■ **Fig. 23-22** The transmembrane potential changes *(top half)* that occur in SA node cells are produced by three principal currents *(bottom half)*: (1) an inward Ca^{++} current i_{Ca}; (2) a hyperpolarization-induced inward current i_f; and (3) an outward K^+ current, i_K. (Redrawn from Brown HF: *Physiol Rev* 61:644, 1981.)

activity, through the release of norepinephrine, raises the heart rate principally by increasing the slope of the pacemaker potential. Increased vagal activity, through the release of acetylcholine, diminishes the heart rate by hyperpolarizing the pacemaker cell membrane and by reducing the slope of the pacemaker potential (Fig. 23-2).

Changes in autonomic neural activity often also induce a pacemaker shift, where the site of initiation of the cardiac impulse may shift to a different locus within the SA node or to a different component of the atrial pacemaker complex.

■ *Ionic Basis of Automaticity*

Several ionic currents contribute to the slow depolarization that occurs during phase 4 in automatic cells in the heart. In the pacemaker cells of the SA node the diastolic depolarization is ascribable to at least three ionic currents: (1) an inward current, i_f, induced by hyperpolarization; (2) an inward Ca^{++} current, i_{Ca}; and (3) an outward K^+ current, i_K (Fig. 23-22).

The inward current, i_f, is carried mainly by Na^+; the current is conducted through specific channels that differ from the fast Na^+ channels. This current becomes activated during the repolarization phase of the action potential, as the membrane potential becomes more negative than about −50 mV. The more negative the membrane potential becomes at the end of repolarization, the greater is the activation of the i_f current.

The second current responsible for diastolic depolarization is the slow inward current, i_{si}. This current is composed mainly of Ca^{++} and therefore it is usually referred to as the Ca^{++} current, i_{Ca}. This current becomes activated toward the end of phase 4, as the transmembrane potential reaches a value of about −55 mV (see Fig. 23-3).

This Ca^{++} current is carried mainly by T-type Ca^{++} channels, because this channel type activates at such transmembrane potentials (see Fig. 23-10). Once the Ca^{++} channels become activated, the influx of Ca^{++} into the cell increases. The influx of Ca^{++} accelerates the rate of diastolic depolarization, which then leads to the upstroke of the action potential. A decrease in the external Ca^{++} concentration (Fig. 23-23) or the addition of calcium channel blocking agents (Fig. 23-24) diminishes the amplitude of the action potential and the slope of the pacemaker potential in SA node cells.

The progressive diastolic depolarization mediated by the two inward currents, i_f and i_{Ca}, is opposed by a third current, an outward K^+ current, i_K. This efflux of K^+ tends to repolarize the cell after the upstroke of the action potential. The outward K^+ current continues well beyond the time of maximum repolarization, but it diminishes throughout phase 4 (see Fig. 23-22). Hence the opposition of i_K to the depolarizing effects of the two inward currents (i_{Ca} and i_f) gradually decreases.

The ionic basis for automaticity in the AV node pacemaker cells is probably identical to that in the SA node cells. Similar mechanisms probably also account for automaticity in Purkinje's fibers, except that the Ca^{++} current is not involved. Hence the slow diastolic depolarization is mediated principally by the imbalance between the hyperpolarization-induced inward current, i_f, and the outward K^+ current, i_K.

The autonomic neurotransmitters affect automaticity by altering the ionic currents across the cell membranes.

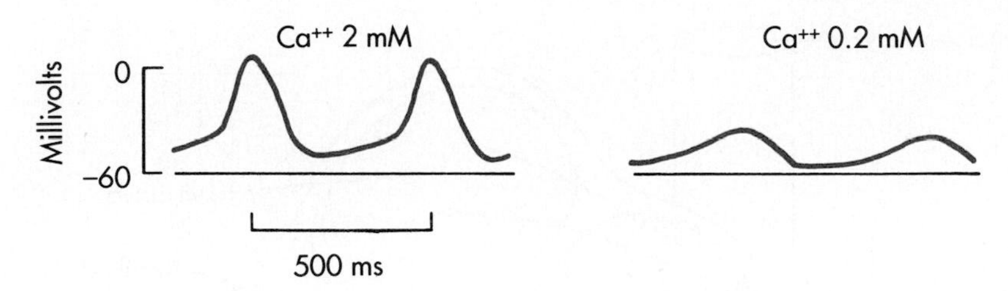

■ **Fig. 23-23** Transmembrane action potentials recorded from an SA node pacemaker cell in an isolated rabbit atrium preparation. The concentration of Ca^{++} in the bath was changed from 2 to 0.2 mM. (Modified from Kohlhardt M, Figulla HR, Tripathi O: *Basic Res Cardiol* 71:17, 1976.)

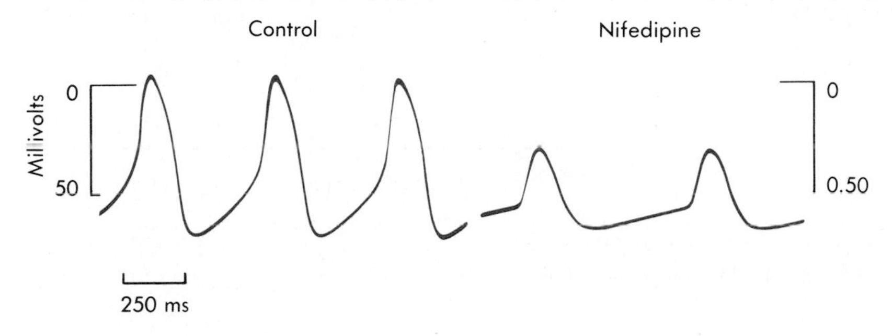

■ **Fig. 23-24** The effects of nifedipine (5.6×10^{-7} M), a Ca^{++} channel blocking drug, on the transmembrane potentials recorded from a rabbit's SA node cell. (From Ning W, Wit, AL: *Am Heart J* 106:345, 1983.)

The adrenergic transmitters increase all three currents involved in SA nodal automaticity. The adrenergically mediated increase in the slope of diastolic depolarization indicates that the augmentations of i_f and i_{Ca} must exceed the enhancement of i_K.

The hyperpolarization (see Fig. 23-21) induced by the acetylcholine released at the vagus endings in the heart is achieved by an increase in g_K. This change in conductance is mediated through activation of specific K^+ channels that are controlled by the cholinergic receptors. Acetylcholine also depresses the i_f and i_{Ca} currents.

■ *Overdrive Suppression*

The automaticity of pacemaker cells becomes depressed after a period of excitation at a high frequency. This phenomenon is known as overdrive suppression. Because of the greater intrinsic rhythmicity of the SA node than of the other latent pacemaking sites in the heart the firing of the SA node tends to suppress the automaticity in the other loci. If an ectopic focus in one of the atria suddenly began to fire at a high rate in an individual with a normal heart rate of 70 beats/min, the ectopic center would become the pacemaker for the entire heart. When that rapid ectopic focus suddenly stopped firing, the SA node might remain quiescent briefly because of overdrive suppression. The interval from the end of the period of overdrive until the SA node resumes firing is called the sinus node recovery time. In patients with the so-called sick sinus syndrome, the sinus node recovery time may be markedly prolonged. The resultant period of asystole might cause syncope.

The mechanism responsible for overdrive suppression appears to be based on the activity of the membrane pump that actively extrudes Na^+ from the cell, in partial exchange for K^+. During each depolarization a certain quantity of Na^+ enters the cell. The more frequently it is depolarized therefore the more Na^+ that enters the cell per minute. At high excitation frequencies the Na^+ pump becomes more active in extruding this larger quantity of Na^+ from the cell interior. The quantity of Na^+ extruded by the pump exceeds the quantity of K^+ that enters the cell; the ratio is 3:2. This enhanced activity of the pump hyperpolarizes the cell because of the net loss of cations from the cell interior. Because of the hyperpolarization the pacemaker potential requires more time to reach the threshold, as shown in Fig. 23-20, *B*. Furthermore when the overdrive suddenly ceases, the Na^+ pump may not decelerate instantaneously but may continue to operate at an accelerated rate for some time. This excessive extrusion of Na^+ opposes the gradual depolarization of the pacemaker cell during phase 4, thereby suppressing its intrinsic automaticity temporarily.

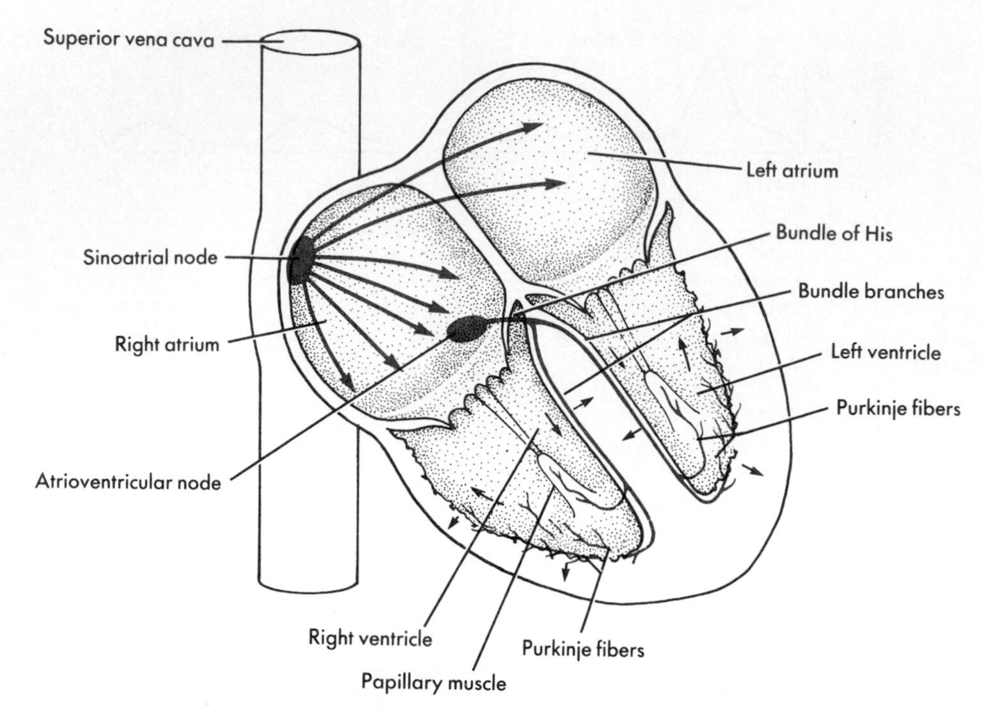

■ **Fig. 23-25** Schematic representation of the conduction system of the heart.

■ *Atrial Conduction*

From the SA node the cardiac impulse spreads radially throughout the right atrium (Fig. 23-25) along ordinary atrial myocardial fibers, at a conduction velocity of approximately 1 m/sec. A special pathway, the anterior interatrial myocardial band (or Bachmann's bundle), conducts the impulse from the SA node directly to the left atrium. Three tracts, the anterior, middle, and posterior internodal pathways, have been described. These tracts consist of a mixture of ordinary myocardial cells and specialized conducting fibers. Some authorities assert that these pathways constitute the principal routes for conduction of the cardiac impulse from the SA to the AV node.

The configuration of the atrial transmembrane potential is depicted in Fig. 23-19, *C*. Compared with the potential recorded from a typical ventricular fiber (see Fig. 23-19, *A*), the plateau (phase 2) is less well developed and repolarization (phase 3) occurs at a slower rate.

■ *Atrioventricular Conduction*

The cardiac action potential proceeds along the internodal pathways in the atrium and ultimately reaches the AV node. This node is approximately 22 mm long, 10 mm wide, and 3 mm thick in adult humans. The node is situated posteriorly on the right side of the interatrial septum near the ostium of the coronary sinus. The AV node contains the same two cell types as the SA node, but the round cells are more sparse and the elongated cells preponderate.

The AV node has been divided into three functional regions: (1) the AN region, the transitional zone between the atrium and the remainder of the node; (2) the N region, the midportion of the AV node; and (3) the NH region, the zone in which nodal fibers gradually merge with the bundle of His, which is the upper portion of the specialized conducting system for the ventricles. Normally the AV node and bundle of His constitute the only pathways for conduction from atria to ventricles. Accessory AV pathways are present in some people, however. Such pathways often serve as a part of a reentry loop (see p. 385), which could lead to serious cardiac rhythm disturbances in these patients.

Several features of AV conduction are of physiological and clinical significance. The principal delay in the passage of the impulse from the atria to the ventricles occurs in the AN and N regions of the AV node. The conduction velocity is actually less in the N region than in the AN region. However, the path length is substantially greater in the AN than in the N region. The conduction times through the AN and N zones account for the delay between the onsets of the P wave (the electrical manifestation of the spread of atrial excitation) and the QRS complex (spread of ventricular excitation) in the electrocardiogram (see Fig. 23-32). Functionally this delay between atrial and ventricular excitation permits optimal ventricular filling during atrial contraction.

In the N region slow response action potentials prevail. The resting potential is about −60 mV, the upstroke velocity is very low (about 5 V/sec), and the conduction velocity is about 0.05 m/sec. Tetrodotoxin, which blocks the fast Na^+ channels, has virtually no effect on the action potentials in this region. Conversely the Ca^{++} channel blocking agents decrease the amplitude and duration of the action potentials (Fig. 23-26) and depress AV conduction. The shapes of the action potentials in the AN region are intermediate between those in the N region and atria. Similarly the action potentials in the NH region are transitional between those in the N region and bundle of His.

The relative refractory period of the cells in the N region extends well beyond the period of complete repolarization (i.e., these cells display postrepolarization refractoriness) (see Fig. 23-16). As the repetition rate of atrial depolarizations is increased, conduction through the AV junction slows. Most of that slowing takes place in the N region. Impulses tend to be blocked at stimulus repetition rates that are easily conducted in other regions of the heart. If the atria are depolarized at a high frequency, only one half or one third of the atrial impulses might be conducted through the AV junction to the ventricles. This protects the ventricles from excessive contraction frequencies, wherein the filling time between contractions might be inadequate. Retrograde conduction can occur through the AV node. However, the propagation time is significantly longer and the impulse is blocked at lower repetition rates during conduction in the retrograde than in the antegrade direction. Finally the AV node is a common site for reentry; the underlying mechanisms are explained on p. 385.

The autonomic nervous system regulates AV conduction. Weak vagal activity may simply prolong the AV conduction time. Stronger vagal activity may cause some or all of the impulses arriving from the atria to be blocked in the node. The delayed conduction or block occurs largely in the N region of the node. The acetylcholine released by the vagal nerve fibers hyperpolarizes the conducting fibers in the N region (Fig. 23-27). The greater the hyperpolarization at the time of arrival of the atrial impulse the more impaired the AV conduction. In the experiment shown in Fig. 23-27 vagus nerve fibers were stimulated intensely (at St) shortly before the second atrial depolarization (A_2). That atrial impulse arrived at the AV node cell when its cell membrane was maximally hyperpolarized. The absence of a corresponding depolarization of the bundle of His shows that the second atrial impulse was not conducted through the AV node. Only a small, nonpropagated response to the second atrial impulse is evident in the recording from the conducting fiber.

The cardiac sympathetic nerves, on the other hand, have a facilitative effect. They decrease the AV conduction time and enhance the rhythmicity of the latent pacemakers in the AV junction. The norepinephrine released at the sympathetic nerve terminals increases the amplitude and slope of the upstroke of the AV nodal action potentials, principally in the AN and N regions of the node.

Ventricular Conduction

The bundle of His passes subendocardially down the right side of the interventricular septum for about 1 cm and then divides into the right and left bundle branches (Figs. 23-25 and 23-28). The right bundle branch is a direct continuation of the bundle of His and it proceeds down the right side of the interventricular septum. The left bundle branch, which is considerably thicker than the right, arises almost perpendicularly from the bundle of His and perforates the interventricular septum. On the subendocardial surface of the left side of the inter-

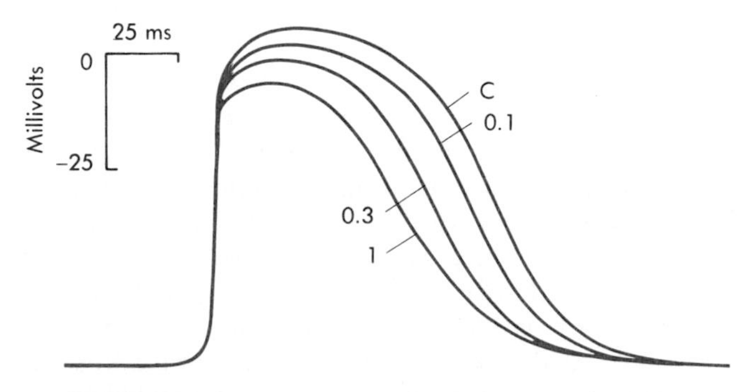

Fig. 23-26 Transmembrane potentials recorded from a rabbit AV node cell under control conditions *(C)* and in the presence of the calcium channel blocking drug, diltiazem, in concentrations of 0.1, 0.3, and 1 μmol/L. (Redrawn from Hirth C, Borchard U, Hafner D: *J Mol Cell Cardiol* 15:799, 1983.)

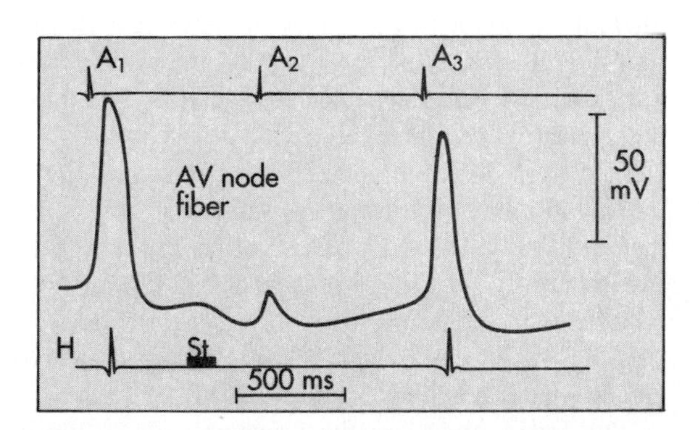

Fig. 23-27 Effects of a brief vagal stimulus *(St)* on the transmembrane potential recorded from an AV nodal fiber from a rabbit. Note that shortly after vagal stimulation, the membrane of the fiber was hyperpolarized. The atrial excitation (A_2) that arrived at the AV node when the cell was hyperpolarized failed to be conducted, as denoted by the absence of a depolarization in the His electrogram *(H)*. The atrial excitations that preceded (A_1) and followed (A_3) excitation A_2 were conducted to the His bundle region. (Redrawn from Mazgalev T et al: *Am J Physiol* 251:H631, 1986.)

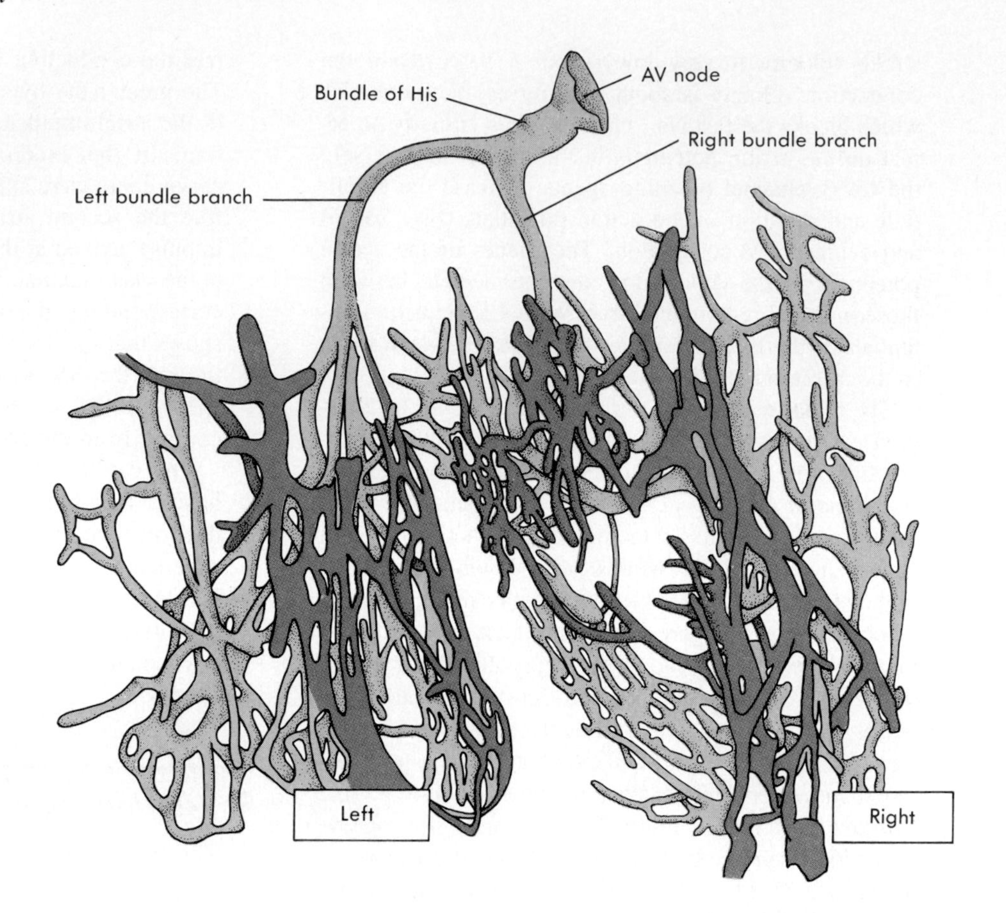

■ **Fig. 23-28** AV and ventricular conduction system of the calf heart. (Redrawn from DeWitt LM: *Anat Rec* 3:475, 1909.)

ventricular septum the main left bundle branch splits into a thin anterior division and a thick posterior division. Clinically, impulse conduction in the right bundle branch, the main left bundle branch, or either division of the left bundle branch may be impaired. Conduction blocks in one or more of these pathways give rise to characteristic electrocardiographic patterns. Block of either of the main bundle branches is known as right or left bundle branch block. Block of either division of the left bundle branch is called left anterior hemiblock or left posterior hemiblock.

The right bundle branch and the two divisions of the left bundle branch ultimately subdivide into a complex network of conducting fibers called Purkinje's fibers, which ramify over the subendocardial surfaces of both ventricles. In certain mammalian species, such as cattle, the Purkinje's fiber network is arranged in discrete, encapsulated bundles (see Fig. 23-28).

Purkinje's fibers are the broadest cells in the heart, 70 to 80 μm in diameter, compared with 10 to 15 μm for ventricular myocardial cells. The large diameter accounts in part for the greater conduction velocity in Purkinje's than in myocardial fibers. Purkinje's cells have abundant, linearly arranged sarcomeres, just as do myocardial cells. However the T-tubular system is absent in the Purkinje's cells of many species, but it is well developed in the myocardial cells.

The conduction of the action potential over the Purkinje's fiber system is the fastest of any tissue within the heart; estimates vary from 1 to 4 m/sec. This permits a rapid activation of the entire endocardial surface of the ventricles.

The action potentials recorded from Purkinje's fibers resembles that of ordinary ventricular myocardial fibers (see Fig. 23-19, *A*). In general phase 1 is more prominent in Purkinje's fiber action potentials (see Fig. 23-13) than in those recorded from ventricular fibers (especially endocardial fibers) and the duration of the plateau (phase 2) is longer.

Because of the long refractory period of the Purkinje's fibers many premature activations of the atria are conducted through the AV junction but are blocked by the Purkinje's fibers. Therefore they fail to evoke a premature contraction of the ventricles. This function of protecting the ventricles against the effects of premature atrial depolarizations is especially pronounced at slow heart rates, because the action potential duration and hence the effective refractory period of the Purkinje's fibers vary inversely with the heart rate (see Fig. 23-17). At slow heart rates the effective refractory period of the Purkinje's fibers is especially prolonged; as the heart rate increases, the refractory period diminishes. Similar directional changes in the refractory period occur in most of the other cells in the heart with changes in rate. However in the AV node the effective refractory period does not change appreciably over the normal range of heart rates, and it actually increases at very rapid heart rates. Therefore at high rates it is the AV node that protects the ventricles when impulses arrive at excessive repetition rates.

The spread of the action potential over the ventricles is of major concern in clinical cardiology. Numerous studies have been conducted to determine the precise course of the wave of excitation under normal and abnormal conditions. Such knowledge serves as a basis for the interpretation of the electrocardiogram. However only the elementary, salient features of ventricular activation are considered here.

The first portions of the ventricles to be excited are the interventricular septum (except the basal portion) and the papillary muscles. The wave of activation spreads into the substance of the septum from both its left and its right endocardial surfaces. Early contraction of the septum tends to make it more rigid and allows it to serve as an anchor point for the contraction of the remaining ventricular myocardium. Also early contraction of the papillary muscles prevents eversion of the AV valves during ventricular systole.

The endocardial surfaces of both ventricles are activated rapidly, but the wave of excitation spreads from endocardium to epicardium at a slower velocity (about 0.3 to 0.4 m/sec). Because the right ventricular wall is appreciably thinner than the left, the epicardial surface of the right ventricle is activated earlier than that of the left ventricle. Also apical and central epicardial regions of both ventricles are activated somewhat earlier than their respective basal regions. The last portions of the ventricles to be excited are the posterior basal epicardial regions and a small zone in the basal portion of the interventricular septum.

■ *Reentry*

Under certain conditions, a cardiac impulse may reexcite some region through which it passed previously. This phenomenon, known as reentry, is responsible for many clinical disturbances of cardiac rhythm. The reentry may be ordered or random. In the ordered variety the impulse traverses a fixed anatomical path, whereas in the random type the path continues to change. The principal example of random reentry is fibrillation (p. 392).

The conditions necessary for reentry are illustrated in Fig. 23-29. In each of the four panels a single bundle (S) of cardiac fibers splits into a left (L) and a right (R) branch. A connecting bundle (C) runs between the two branches. Normally the impulse moving down bundle S is conducted along the L and R branches (panel A). As the impulse reaches connecting link C, it enters from both sides and becomes extinguished at the point of collision. The impulse from the left side cannot proceed farther because the tissue beyond is absolutely refrac-

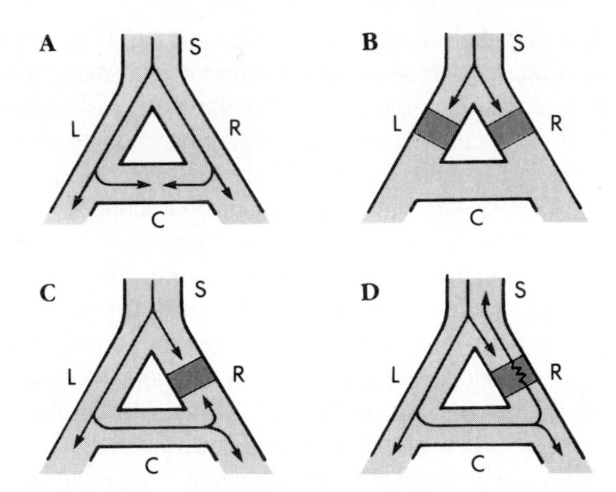

■ **Fig. 23-29** The role of unidirectional block in reentry. In panel **A** an excitation wave traveling down a single bundle *(S)* of fibers continues down the left *(L)* and right *(R)* branches. The depolarization wave enters the connecting branch *(C)* from both ends and is extinguished at the zone of collision. In panel **B** the wave is blocked in the *L* and *R* branches. In panel **C** bidirectional block exists in branch *R*. In panel **D** unidirectional block exists in branch *R*. The antegrade impulse is blocked, but the retrograde impulse is conducted through and reenters bundle *S*.

tory because it has just been depolarized from the other direction. The impulse cannot pass through bundle C from the right for the same reason.

It is obvious from panel B that the impulse cannot make a complete circuit if antegrade block exists in the two branches (L and R) of the fiber bundle. Furthermore if bidirectional block exists at any point in the loop (for example, branch R in panel C), the impulse is not able to reenter.

A necessary condition for reentry is that at some point in the loop the impulse is able to pass in one direction but not in the other. This phenomenon is called unidirectional block. As shown in panel D, the impulse may travel down branch L normally and may be blocked in the antegrade direction in branch R. The impulse that was conducted down branch L and through the connecting branch C may be able to penetrate the depressed region in branch R from the retrograde direction, even though the antegrade impulse was blocked previously at this same site. The antegrade impulse arrives at the depressed region in branch R earlier than the impulse that traverses a longer path and enters branch R from the opposite direction. The antegrade impulse may be blocked simply because it arrives at the depressed region during its effective refractory period. If the retrograde impulse is delayed sufficiently, the refractory period may have ended and the impulse is conducted back into bundle S.

Unidirectional block is a necessary condition for reentry, but not a sufficient one. It is also essential that the effective refractory period of the reentered region be less than the propagation time around the loop. In panel D if the retrograde impulse is conducted through the depressed zone in branch R and if the tissue just beyond is still refractory from the antegrade depolarization, branch S is not reexcited. Therefore the conditions that promote reentry are those that prolong conduction time or shorten effective refractory period.

The functional components of reentry loops responsible for specific arrhythmias in intact hearts are diverse. Some loops are very large and involve entire specialized conduction bundles; others are microscopic. The loop may include myocardial fibers, specialized conducting fibers, nodal cells, and junctional tissues in almost any conceivable arrangement. Also the cardiac cells in the loop may be normal or deranged.

Triggered Activity

Triggered activity is so named because it is always coupled to a preceding action potential. Consequently arrhythmias induced by triggered activity are difficult to distinguish from those induced by reentry. Triggered activity is caused by afterdepolarizations. Two types of afterdepolarizations are recognized: early afterdepolarizations (EADs) and delayed afterdepolarizations (DADs). EADs occur at the end of the plateau (phase 2) or about midway through repolarization (phase 3), whereas DADs occur near the very end of repolarization or just after full repolarization (phase 4).

Early Afterdepolarizations

EADs are more likely to occur when the prevailing heart rate is slow; rapid pacing suppresses EADs. In the experiment shown in Fig. 23-30, EADs were induced by cesium in an isolated Purkinje's fiber preparation. No afterdepolarizations were evident when the preparation was driven at a cycle length of 2 seconds. When the cycle length was increased to 4 seconds. EADs appeared. Most were subthreshold (first two arrows), but one of the EADs did reach threshold and triggered an action potential. When the cycle length was increased to 6 seconds, each driven action potential generated an EAD that triggered a second action potential. Furthermore when the cycle length was increased to 10 seconds, each driven action potential triggered a salvo of four or five additional action potentials.

EADs may be produced experimentally by interventions that prolong the action potential. Because EADs may be initiated at either of two distinct levels of transmembrane potential, namely at the end of the plateau and about midway through repolarization, two different mechanisms may be involved in generating them.

Considerable information has been obtained about the mechanism responsible for those EADs that appear at

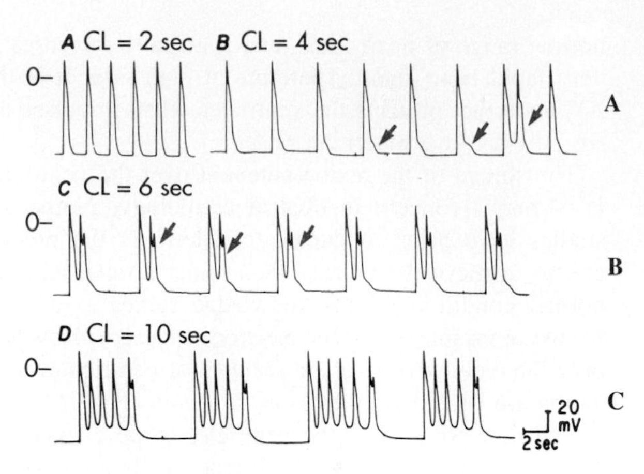

Fig. 23-30 Effect of pacing at different cycle lengths *(CL)* on cesium-induced early afterdepolarizations (EADs) in a canine Purkinje's fiber. (Modified from Damiano BP, Rosen M: *Circulation* 69:1013, 1984, with permission from the American Heart Association.)
A, EADs not evident.
B, EADs first appear *(arrows)*. Third EAD reaches threshold and triggers an action potential *(third arrow)*.
C, EADs that appear after each driven depolarization trigger an action potential.
D, Triggered action potentials occur in salvos.

the end of the plateau. EADs are more likely to occur the more prolonged the action potential. For those action potentials that trigger EADs the plateau appears to be prolonged enough that those Ca^{++} channels that were activated at the beginning of the plateau and then inactivated would have sufficient time to be activated again before the plateau had expired. This secondary activation would trigger an after depolarization.

Less information is available about the cellular mechanisms responsible for those EADs that appear midway through repolarization. The mechanism may be very similar to that for the EADs that occur at plateau potentials, as previously described. The difference may be related to the specific type of Ca^{++} channels involved in the process. Experiments have adduced convincing evidence that the L-type Ca^{++} channels (see p. 372) mediate the EADs that occur at plateau potentials. These channels are activated at levels of V_m that prevail during the plateau. Recent experiments suggest that the EADs that occur at more negative potentials are generated through a similar mechanism, but are mediated instead by T-type Ca^{++} channels. The T-type channels are activated at levels of V_m significantly more negative than the plateau.

Delayed Afterdepolarizations

The salient characteristics of DADs are shown in Fig. 23-31. The transmembrane potentials were recorded from Purkinje's fibers that were exposed to a high con-

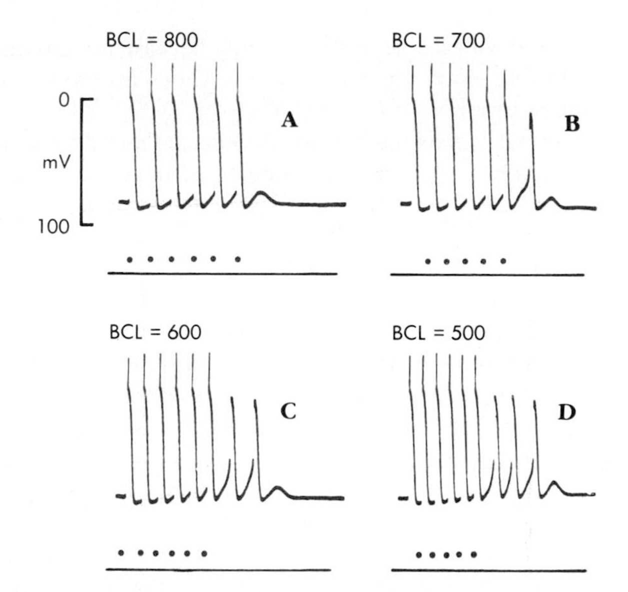

■ **Fig. 23-31** Transmembrane action potentials recorded from isolated canine Purkinje's fibers. Acetylstrophanthidin was added to the bath, and sequences of six driven beats (denoted by the dots) were produced at basic cycle lengths *(BCL)* of 800, 700, 600, and 500 msec. Note that delayed afterpotentials occurred after the driven beats, and that these afterpotentials reached threshold after the last driven beat in panels **B** to **D.** (From Ferrier GR, Saunders JH, Mendez C: *Circ Res* 32:600, 1973, with permission of the American Heart Association.)

centration of acetylstrophanthidin, a digitalis-like substance. In the absence of driving stimuli these fibers were quiescent. In each panel a sequence of six driven depolarizations was induced at various basic cycle lengths.

When the cycle length was 800 msec (panel A), the last driven depolarization was followed by a brief DAD that did not reach threshold. Once that afterdepolarization had subsided, the transmembrane potential remained constant until another driving stimulus was given. The upstroke of a DAD can be detected after each of the first five driven depolarizations.

When the basic cycle length was diminished to 700 msec (panel B), the DAD that followed the last driven beat did reach threshold, and a nondriven depolarization (or extrasystole) ensued. This extrasystole was itself followed by an afterpotential that was subthreshold. Diminution of the basic cycle length to 600 msec (panel C) also evoked an extrasystole after the last driven depolarization. The afterpotential that followed the extrasystole did reach threshold, however, and a second extrasystole occurred. A sequence of three extrasystoles followed the six driven depolarizations that were separated by intervals of 500 msec (panel D). Slightly shorter basic cycle lengths or slightly greater concentrations of acetylstrophanthidin evoked a continuous sequence of nondriven beats, resembling a paroxysmal tachycardia (described on p. 392).

DADs are associated with elevated intracellular Ca^{++} concentrations. The amplitudes of the DADs are increased by interventions that raise intracellular Ca^{++} concentrations. Such interventions include elevated extracellular Ca^{++} concentrations and toxic levels of digitalis glycosides. The elevated levels of intracellular Ca^{++} provoke the oscillatory release of Ca^{++} from the sarcoplasmic reticulum. Hence in myocardial cells the DADs are accompanied by small changes in developed force. The high intracellular Ca^{++} concentrations also activate certain membrane channels that permit the passage of Na^{+} and K^{+}. The net flux of these cations constitutes a transient inward current, i_{ti}, that is at least partly responsible for the afterdepolarization of the cell membrane. The elevated intracellular Ca^{++} may also activate Na^{+}/Ca^{++} exchange (see Chapter 1). This electrogenic exchanger, which takes into the cell three Na^{+} for each Ca^{++} it ejects, also creates a net inward current of cations that would contribute to the DAD.

■ *Electrocardiography*

The electrocardiograph is a valuable instrument because it enables the physician to infer the course of the cardiac impulse simply by recording the variations in electrical potential at various loci on the surface of the body. By analyzing the details of these potential fluctuations, the physician gains valuable insight concerning (1) the anatomical orientation of the heart; (2) the relative sizes of its chambers; (3) a variety of disturbances of rhythm and conduction; (4) the extent, location, and progress of ischemic damage to the myocardium; (5) the effects of altered electrolyte concentrations; and (6) the influence of certain drugs (notably digitalis and its derivatives). The science of electrocardiography is extensive and complex; only its elementary basis is considered here.

■ *Scalar Electrocardiography*

The systems of leads used to record routine electrocardiograms are oriented in certain planes of the body. The diverse electromotive forces that exist in the heart at any moment can be represented by a three-dimensional vector. A system of recording leads oriented in a given plane detects only the projection of the three-dimensional vector on that plane. Furthermore the potential difference between two recording electrodes represents the projection of the vector on the line between the two leads. Components of vectors projected on such lines are not vectors but scalar quantities (having magnitude, but not direction). Hence a recording of the changes with time of the differences of potential between two points on the surface of the skin is called a scalar electrocardiogram.

Configuration of the scalar electrocardiogram. The scalar electrocardiograph detects the changes with time of the electrical potential between some point on the surface of the skin and an indifferent electrode or between pairs of points on the skin surface. The cardiac impulse progresses through the heart in a complex three-dimensional pattern. Hence the precise configuration of the electrocardiogram varies from individual to individual, and in any given individual the pattern varies with the anatomical location of the leads.

In general the pattern consists of P, QRS, and T waves (Fig. 23-32). The PR interval (or more precisely, the PQ interval) is a measure of the time from the onset of atrial activation to the onset of ventricular activation; it normally ranges from 0.12 to 0.20 sec. A considerable fraction of this time involves passage of the impulse through the AV conduction system. Pathological prolongations of this interval are associated with disturbances of AV conduction produced by inflammatory, circulatory, pharmacological, or nervous mechanisms.

The configuration and amplitude of the QRS complex vary considerably among individuals. The duration is usually between 0.06 and 0.10 sec. Abnormal prolongation may indicate a block in the normal conduction pathways through the ventricles (such as a block of the left or right bundle branch). During the ST interval the entire ventricular myocardium is depolarized. Therefore the ST segment lies on the isoelectric line under normal conditions. Any appreciable deviation from the isoelectric line is noteworthy and may indicate ischemic damage of the myocardium. The QT interval is sometimes referred to as the period of "electrical systole" of the ventricles. Its duration is about 0.4 sec, but it varies inversely with the heart rate, mainly because the myocardial cell action potential duration varies inversely with the heart rate (see Fig. 23-17).

In most leads the T wave is deflected in the same direction from the isoelectric line as the major component of the QRS complex, although biphasic or oppositely directed T waves are perfectly normal in certain leads. When the T wave and QRS complex deviate in the same direction from the isoelectric line, it indicates that the repolarization process proceeds in a direction counter to the depolarization process. T waves that are abnormal either in direction or in amplitude may indicate myocardial damage, electrolyte disturbances, or cardiac hypertrophy.

Standard limb leads. The original electrocardiographic lead system was devised by Einthoven. In his lead system the resultant cardiac vector (the vector sum of all electrical activity occurring in the heart at any given moment) was considered to lie in the center of a triangle (assumed to be equilateral) formed by the left and right shoulders and the pubic region (Fig. 23-33). This triangle, called Einthoven's triangle, is oriented in the frontal plane of the body. Hence only the projection

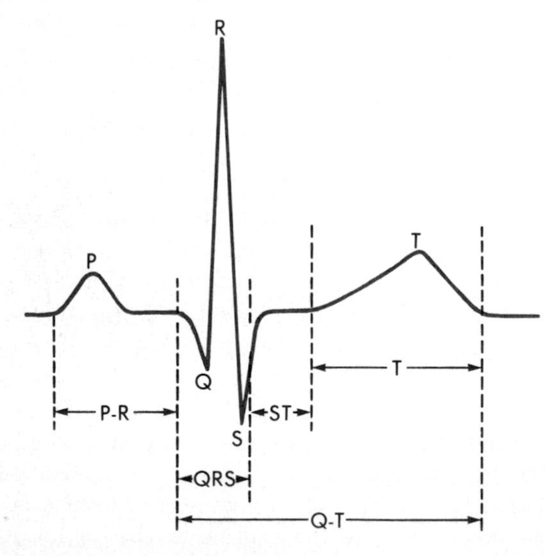

■ **Fig. 23-32** Configuration of a typical scalar electrocardiogram, illustrating the important deflections and intervals.

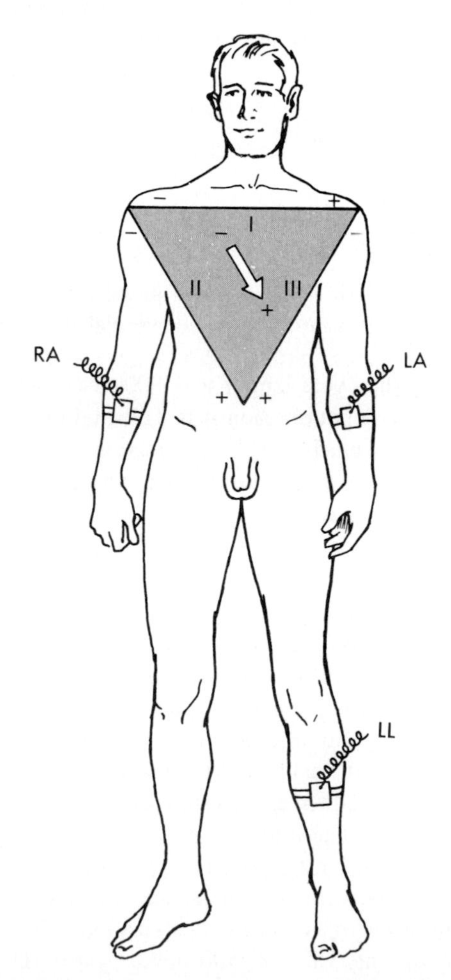

■ **Fig. 23-33** Einthoven triangle, illustrating the galvanometer connections for standard limb leads I, II, and III.

of the resultant cardiac vector on the frontal plane is detected by this system of leads. For convenience the electrodes are connected to the right and left forearms rather than to the corresponding shoulders, because the arms represent simple extensions of the leads from the shoulders. Similarly the leg is taken as an extension of the lead system from the pubis, and the third electrode is connected to the left leg (by convention).

Certain conventions dictate the manner in which these standard limb leads are connected to the galvanometer. Lead I records the potential difference between the left arm (LA) and the right arm (RA). The galvanometer connections are such that when the potential at LA (V_{LA}) exceeds the potential at RA (V_{RA}), the galvanometer is deflected upward from the isoelectric line. In Figs. 23-33 and 23-34 this arrangement of the galvanometer connections for lead I is designated by a (+) at LA and by a (−) at RA. Lead II records the potential difference between RA and LL (left leg) and yields an upward deflection when V_{LL} exceeds V_{RA}. Finally lead III registers the potential difference between LA and LL and yields an upward deflection when V_{LL} exceeds V_{LA}. These galvanometer connections were arbitrarily chosen so that the QRS complexes are upright in all three standard limb leads in most normal individuals.

Let the frontal projection of the resultant cardiac vector at some moment be represented by an arrow (tail negative, head positive), as in Fig. 23-33. Then the potential difference, $V_{LA} - V_{RA}$, recorded in lead I is represented by the component of the vector projected along the horizontal line between LA and RA, as shown in Fig. 23-34. If the vector makes an angle, θ, of 60 degrees with the horizontal (as in the top section of Fig. 23-34), the magnitude of the potential recorded by lead I equals the vector magnitude times cosine 60 degrees. The deflection recorded in lead I is upward because the positive arrowhead lies closer to LA than to RA. The deflection in lead II also is upright because the arrowhead lies closer to LL than to RA. The magnitude of the lead II deflection is greater than that in lead I because in this example the direction of the vector parallels that of lead II; therefore the magnitude of the projection on lead II exceeds that on lead I. Similarly in lead III the deflection is upright, and in this example, where θ = 60 degrees, its magnitude equals that in lead I.

If the vector in the top section of Fig. 23-34 is the result of the electrical events occurring during the peak of the QRS complex, then the orientation of this vector is said to represent the mean electrical axis of the heart in the frontal plane. The positive direction of this axis is taken in the clockwise direction from the horizontal plane (contrary to the usual mathematical convention). For normal individuals the average mean electrical axis is approximately +60 degrees (as in the top section of Fig. 23-34). Therefore the QRS complexes are usually upright in all three leads and largest in lead II.

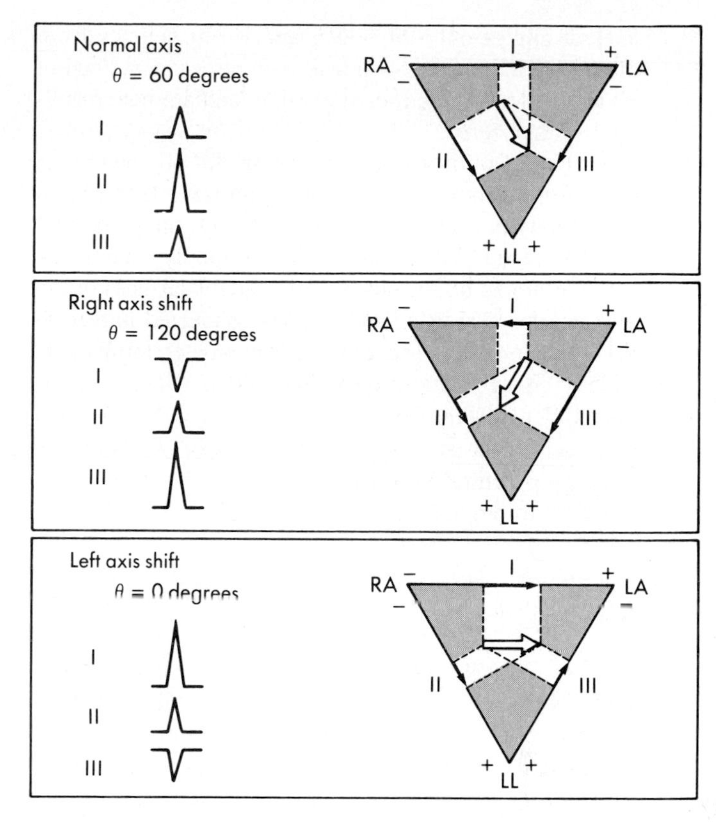

Fig. 23-34 Magnitude and direction of the QRS complexes in limb leads I, II, and III, when the mean electrical axis (θ) is 60 degrees *(top section),* 120 degrees *(middle section),* and 0 degrees *(bottom section).*

Changes in the mean electrical axis may occur with alterations in the anatomical position of the heart or with changes in the relative preponderance of the right and left ventricles. For example, the axis tends to shift toward the left (more horizontal) in short, stocky individuals and toward the right (more vertical) in tall, thin persons. Also with left or right ventricular hypertrophy (increased myocardial mass) the axis shifts toward the hypertrophied side.

With appreciable shift of the mean electrical axis to the right (middle section of Fig. 23-34, where θ = 120 degrees) the displacements of the QRS complexes in the standard leads change considerably. In this case the largest upright deflection is in lead III and the deflection in lead I is inverted because the arrowhead is closer to RA than to LA. With left axis shift (bottom section of Fig. 23-34, where θ = 0 degrees) the largest upright deflection is in lead I, and the QRS complex in lead III is inverted.

As is evident from this discussion, the standard limb leads, I, II, and III, are oriented in the frontal plane at 0, 60, and 120 degrees, respectively, from the horizontal plane. Other limb leads, which are also oriented in the frontal plane, are usually recorded in addition to the standard leads. These "unipolar limb leads" lie along axes at angles of +90, −30, and −150 degrees from the horizontal plane. Such lead systems are described in all textbooks on electrocardiography and are not considered further here.

To obtain information concerning the projections of the cardiac vector on the sagittal and transverse planes of the body in scalar electrocardiography, the precordial leads are usually recorded. Most commonly each of six selected points on the anterior and lateral surfaces of the chest in the vicinity of the heart is connected in turn to the galvanometer. The other galvanometer terminal is usually connected to a central terminal, which is composed of a junction of three leads from LA, RA, and LL, each in series with a 5,000-ohm resistor. The voltage of this central terminal remains at a theoretical zero potential throughout the cardiac cycle.

Arrhythmias

Cardiac arrhythmias reflect disturbances of either impulse propagation or impulse initiation. The principal disturbances of impulse propagation are conduction blocks and reentrant rhythms. Disturbances of impulse initiation include those that arise from the SA node and those that originate from various ectopic foci.

Altered Sinoatrial Rhythms

The frequency of pacemaker discharge varies by the mechanisms described earlier in this chapter (see Fig. 23-20). Changes in SA nodal discharge frequency are usually produced by the cardiac autonomic nerves. Examples of electrocardiograms of **sinus tachycardia** and **sinus bradycardia** are shown in Fig. 23-35. The P, QRS, and T deflections are all normal, but the duration of the cardiac cycle (the PP interval) is altered. Characteristically when sinus bradycardia or tachycardia develops, the cardiac frequency changes gradually and requires several beats to attain its new steady-state value. Electrocardiographic evidence of respiratory cardiac arrhythmia is common and is manifested as a rhythmic variation in the PP interval at the respiratory frequency (p. 422).

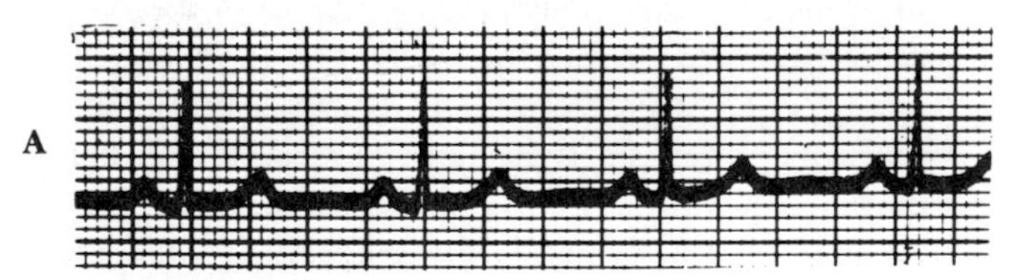

A, Normal sinus rhythm.

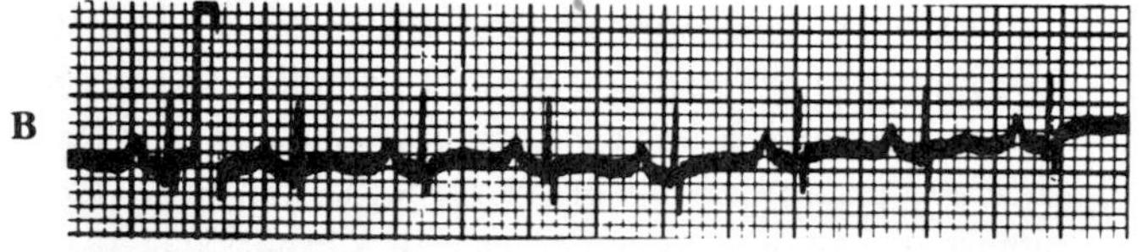

B, Sinus tachycardia.

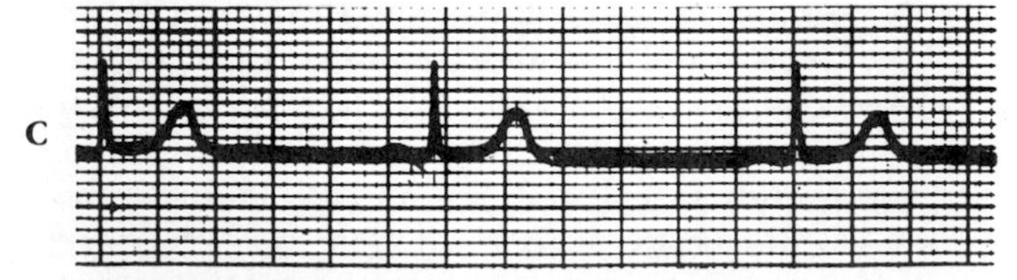

C, Sinus bradycardia.

■ **Fig. 23-35** Sinoatrial rhythms. **A,** Normal sinus rhythm. **B,** Sinus tachycardia. **C,** Sinus bradycardia.

Atrioventricular Transmission Blocks

Various physiological, pharmacological, and pathological processes can impede impulse transmission through the AV conduction tissue. The site of block can be localized more precisely by recording the His bundle electrogram (Fig. 23-36). To obtain such tracings an electrode catheter is introduced into a peripheral vein and is threaded centrally until the tip containing the electrodes lies in the AV junctional region between the right atrium and ventricle. When the electrodes are properly positioned, a distinct deflection (Fig. 23-36, *H*) is registered; it represents the passage of the cardiac impulse down the bundle of His. The time intervals required for propagation from the atrium to the bundle of His (A-H interval) and from the bundle of His to the ventricles (H-V interval) may be measured accurately. Abnormal prolongation of the former or latter interval indicates block above or below the bundle of His, respectively.

Three degrees of AV block can be distinguished, as

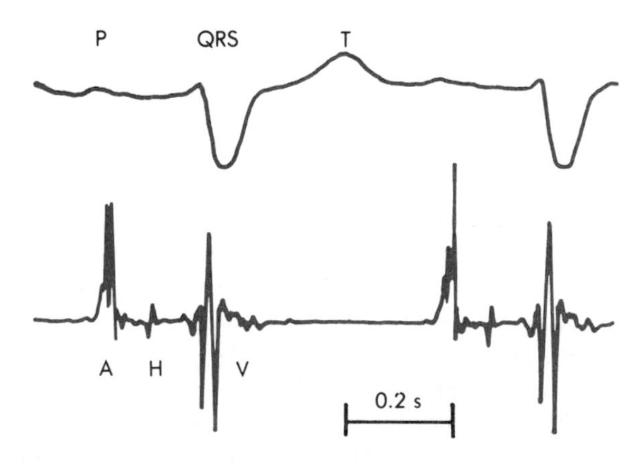

■ **Fig. 23-36** His bundle electrogram (*lower tracing,* retouched) and lead II of the scalar electrocardiogram (*upper tracing*). The deflection, *H,* which represents the impulse conduction over the bundle of His, is clearly visible between the atrial *A,* and the ventricular, *V,* deflections. The conduction time from the atria to the bundle of His is denoted by the A-H interval; that from the bundle of His to the ventricles, by the H-V interval. (Courtesy Dr. J. Edelstein.)

A

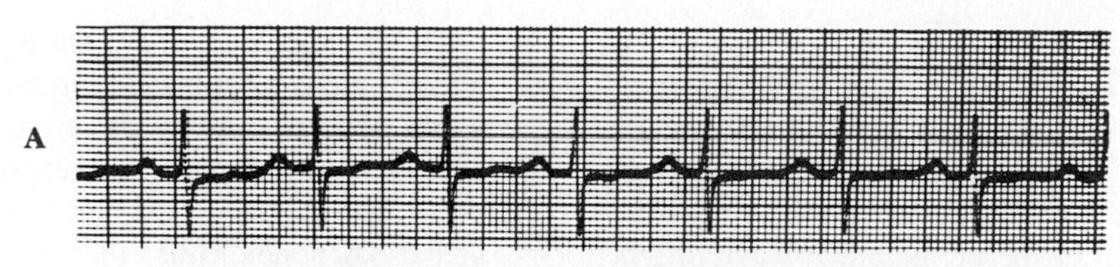

A, First-degree heart block; P-R interval is 0.28 s.

B

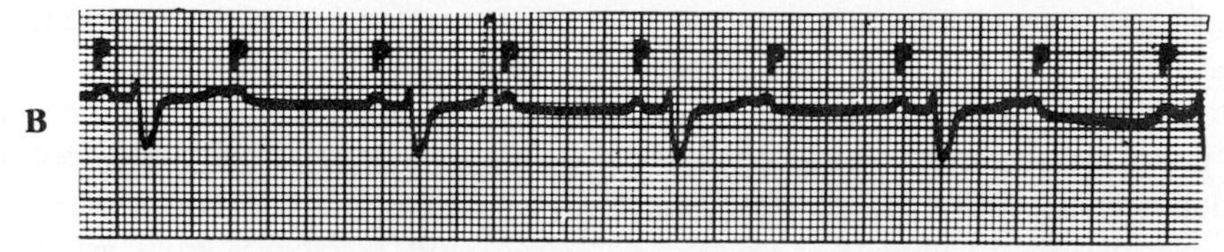

B, Second-degree heart block (2:1).

C

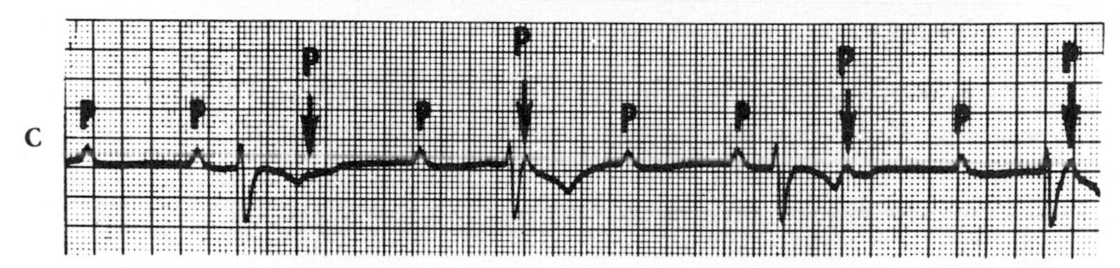

C, Third-degree heart block; note the dissociation between the P waves and the QRS complexes.

■ **Fig. 23-37** AV blocks. **A,** First-degree heart block; PR interval is 0.28 sec. **B,** Second-degree heart block (2:1). **C,** Third-degree heart block; note the dissociation between the P waves and the QRS complexes.

shown in Fig. 23-37. **First-degree AV block** is characterized by a prolonged P-R interval. In Fig. 23-37, *A,* the PR interval is 0.28 second; an interval greater than 0.2 second is abnormal. In most cases of first-degree block the A-H interval of the His bundle electrogram is prolonged, and the H-V interval is normal. Hence the delay is located above the bundle of His (i.e., in the AV node).

In **second-degree AV block** all QRS complexes are preceded by P waves, but not all P waves are followed by QRS complexes. The ratio of P waves to QRS complexes is usually the ratio of two small integers (such as 2:1, 3:1, 3:2). Fig. 23-37, *B* illustrates a typical 2:1 block. Bundle of His electrograms have demonstrated that the site of block may be above or below the bundle of His. A block below the bundle is more serious than one above the bundle because it more often evolves into a third-degree block. Hence an artificial pacemaker is frequently implanted when the block is found to be below the bundle.

Third-degree AV block is often referred to as complete heart block because the impulse is unable to traverse the AV conduction pathway from atria to ventricles. His bundle electrograms reveal that the most common sites of complete block are distal to the bundle of His. In complete heart block the atrial and ventricular rhythms are entirely independent. A typical example is displayed in Fig. 23-37, *C,* where the QRS complexes bear no fixed relationship to the P waves. Because of the slow ventricular rhythm (32 beats/min in this example) circulation is often inadequate, especially during muscular activity. Third-degree block is often associated with syncope (so-called Stokes-Adams attacks) caused principally by insufficient cerebral blood flow. Third-degree block is one of the most common conditions requiring treatment by artificial pacemakers.

■ *Premature Depolarizations*

Premature depolarizations occur at times in most normal individuals but are more common under certain abnormal conditions. They may originate in the atria, AV junction, or ventricles. One type of premature depolarization is coupled to a normally conducted depolarization by a constant coupling interval. If the normal depolarization is suppressed in some way (e.g., by vagal stimulation), the premature depolarization is also abolished. Such premature depolarizations are called coupled extrasystoles, or simply extrasystoles, and they probably reflect a reentry phenomenon (see Fig. 23-29). A second type of premature depolarization occurs as the result of enhanced automaticity in some ectopic focus. This ectopic center may fire regularly and be protected from depolarization by the normal cardiac impulse. If this premature depolarization occurs at a regular interval or at a simple multiple of that interval, the disturbance is called parasystole.

A **premature atrial depolarization** is shown in the electrocardiogram in Fig. 23-38, *A*. The normal interval between beats was 0.89 sec (heart rate, 68 beats/min). The premature atrial depolarization (second P wave in the figure) followed the preceding P wave by only 0.56 sec. The configuration of the premature P wave differs from the configuration of the other, normal P waves because the course of atrial excitation, originating at some ectopic focus in the atrium, is different from the normal spread of excitation originating at the SA node. The QRS complex of the premature depolarization is usually normal in configuration because the spread of ventricular excitation occurs over the usual pathways.

A **premature ventricular depolarization** appears in Fig. 23-38, *B*. Because the premature excitation originated at some ectopic focus in the ventricles, the impulse spread was aberrant and the configurations of the QRS and T waves were entirely different from the normal deflections. The premature QRS complex followed the preceding normal QRS complex by only 0.47 sec. The interval after the premature excitation was 1.28 sec, considerably longer than the normal interval between beats (0.89 sec). The interval (1.75 sec) from the QRS complex just before the premature excitation to the QRS complex just after it was virtually equal to the duration of two normal cardiac cycles (1.78 sec).

A

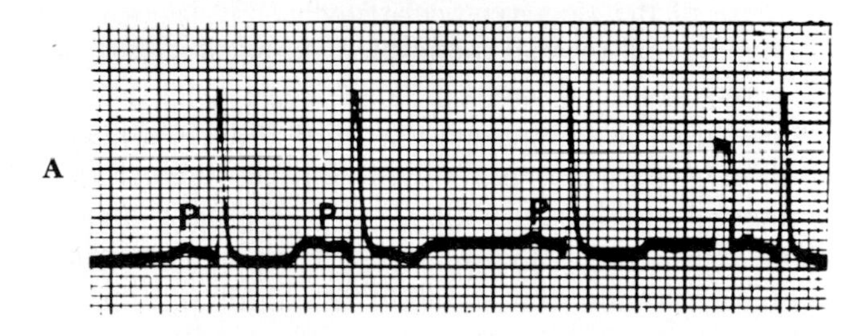

B

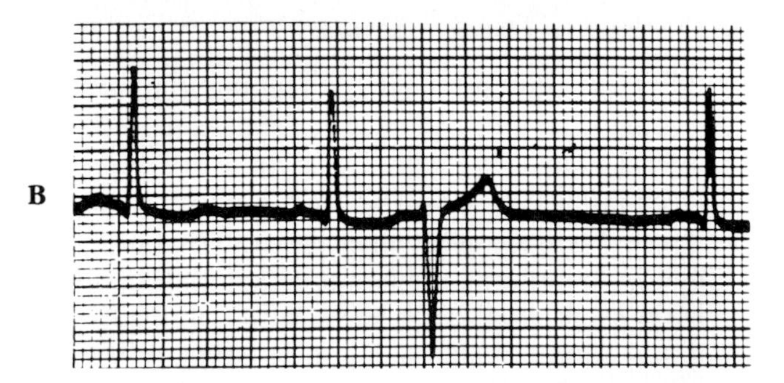

■ **Fig. 23-38** A premature atrial depolarization, **A**, and a premature ventricular depolarization, **B.** The premature atrial depolarization (the second beat in the top tracing) is characterized by an inverted P wave and normal QRS and T waves. The interval following the premature depolarization is not much longer than the usual interval between beats. The brief rectangular deflection just before the last depolarization is a standardization signal. The premature ventricular depolarization, **B,** is characterized by bizarre QRS and T waves and is followed by a compensatory pause.

The prolonged interval that usually follows a premature ventricular depolarization is called a **compensatory pause.** The reason for the compensatory pause after a premature ventricular depolarization is that the ectopic ventricular impulse does not disturb the natural rhythm of the SA node. Either the ectopic ventricular impulse is not conducted retrograde through the AV conduction system or, if it is, the time required is such that the SA node has already fired at its natural interval before the ectopic impulse could have reached it. Likewise the SA nodal impulse usually does not affect the ventricle because the AV junction and perhaps also the ventricles are still refractory from the premature excitation. In Fig. 23-38, *B,* the P wave originating in the SA node at the time of the premature depolarization occurred at the same time as the T wave of the premature cycle and therefore cannot easily be identified in the tracing.

■ *Ectopic Tachycardias*

When a tachycardia originates from some ectopic site in the heart, the onset and termination are typically abrupt, in contrast to the more gradual changes in heart rate in sinus tachycardia. Because of the sudden appearance and abrupt cessation, such ectopic tachycardias are usually called **paroxysmal tachycardias.** Episodes of ectopic tachycardia may persist for only a few beats or for many hours or days, and the episodes often recur. Paroxysmal tachycardias may result from (1) the rapid firing of an ectopic pacemaker, (2) triggered activity secondary to afterpotentials that reach threshold, or (3) an impulse circling a reentry loop repetitively.

Paroxysmal tachycardias originating in the atria or in the AV junctional tissues (Fig. 23-39, *A*) are usually indistinguishable, and therefore both are included in the term **paroxysmal supraventricular tachycardia.** The tachycardia often results from an impulse repetitively circling a reentry loop that includes atrial tissue and the AV junction. The QRS complexes are often normal because ventricular activation proceeds over the normal pathways.

Paroxysmal ventricular tachycardia originates from an ectopic focus in the ventricles. The electrocardiogram is characterized by repeated, bizarre QRS complexes that reflect the aberrant intraventricular impulse conduction (see Fig. 23-39, *B*). Paroxysmal ventricular tachycardia is much more ominous than supraventricular tachycardia because it is frequently a precursor of ventricular fibrillation, a lethal arrhythmia that is described in the next section.

■ *Fibrillation*

Under certain conditions cardiac muscle undergoes an irregular type of contraction that is entirely ineffectual

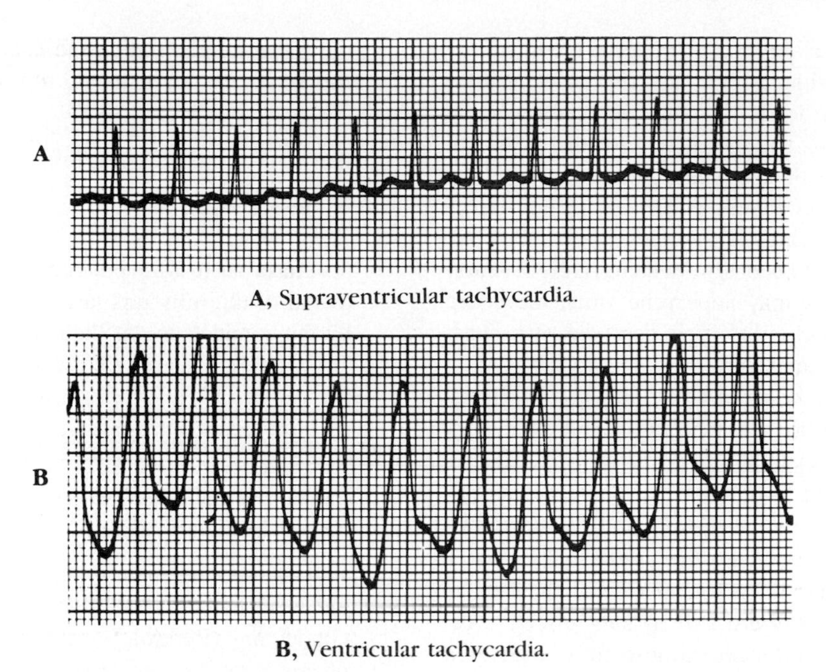

A, Supraventricular tachycardia.

B, Ventricular tachycardia.

■ **Fig. 23-39** Paroxysmal tachycardias. **A,** Supraventricular tachycardia. **B,** Ventricular tachycardia.

in propelling blood. Such an arrhythmia is termed fibrillation and may involve either the atria or the ventricles. Fibrillation probably represents a reentry phenomenon, in which the reentry loop fragments into multiple, irregular circuits.

The tracing in Fig. 23-40, *A,* illustrates the electrocardiographic changes in **atrial fibrillation.** This condition occurs in various types of chronic heart disease. The atria do not contract and relax sequentially during each cardiac cycle and hence do not contribute to ventricular filling. Instead the atria undergo a continuous, uncoordinated, rippling type of activity. In the electrocardiogram there are no P waves; they are replaced by continuous irregular fluctuations of potential, called f waves. The AV node is activated at intervals that may vary considerably from cycle to cycle. Hence there is no constant interval between QRS complexes and therefore between ventricular contractions. Because the strength of ventricular contraction depends on the interval between beats (as explained on p. 430), the volume and the rhythm of the pulse are very irregular. In many patients the atrial reentry loop and the pattern of AV conduction are more regular. The rhythm is then referred to as **atrial flutter.**

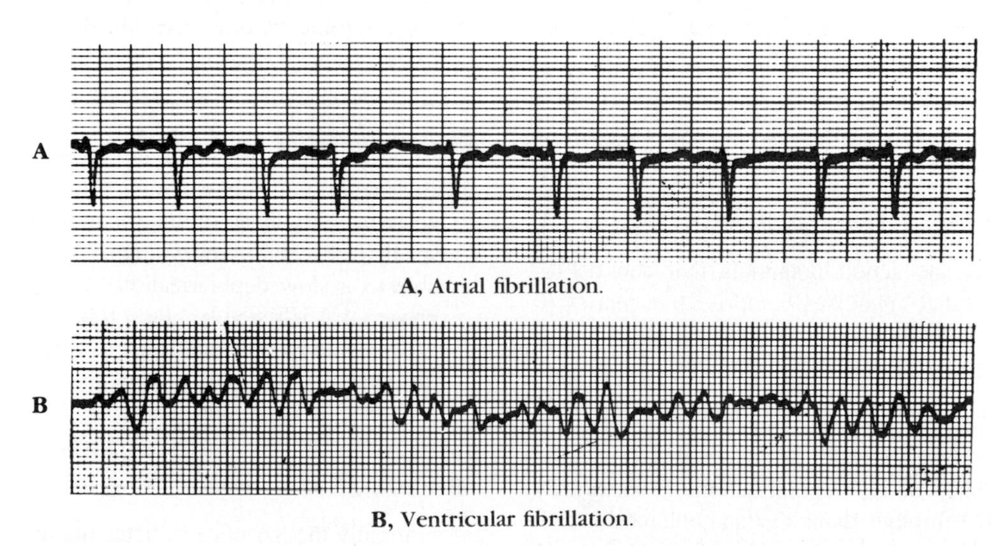

A, Atrial fibrillation.

B, Ventricular fibrillation.

■ **Fig. 23-40** Atrial and ventricular fibrillation. **A,** Atrial fibrillation. **B,** Ventricular fibrillation.

Although atrial fibrillation and flutter are compatible with life and even with full activity, the onset of **ventricular fibrillation** leads to loss of consciousness within a few seconds. The irregular, continuous, uncoordinated twitchings of the ventricular muscle fibers pump no blood. Death ensues unless immediate effective resuscitation is achieved or unless the rhythm reverts to normal spontaneously, which rarely occurs. Ventricular fibrillation may supervene when the entire ventricle, or some portion of it, is deprived of its normal blood supply. It may also occur as a result of electrocution or in response to certain drugs and anesthetics. In the electrocardiogram (see Fig. 23-40, *B*) irregular fluctuations of potential are manifest.

Fibrillation is often initiated when a premature impulse arrives during the vulnerable period. In the ventricles this period coincides with the downslope of the T wave. During this period the excitability of the cardiac cells varies. Some fibers are still in their effective refractory periods, others have almost fully recovered their excitability, and still others are able to conduct impulses but only at very slow conduction velocities. As a consequence the action potentials are propagated over the chambers in several wavelets that travel along circuitous paths and at various conduction velocities. As a region of cardiac cells becomes excitable again, it is ultimately reentered by one of the wave fronts traveling about the chamber. The process is self-sustaining.

Atrial fibrillation may be changed to a normal sinus rhythm by drugs that prolong the refractory period. As the cardiac impulse completes the reentry loop, it may then find the myocardial fibers no longer excitable. However much more dramatic therapy is required in ventricular fibrillation. Conversion to a normal sinus rhythm is accomplished by means of a strong electric current that places the entire myocardium briefly in a refractory state. Techniques have been developed to administer the current through the intact chest wall safely. In successful cases the SA node again takes over as the normal pacemaker for the entire heart.

Summary

Transmembrane Action Potentials

The transmembrane action potentials that can be recorded from cardiac myocytes comprise five phases (0 to 4):

1. Phase 0, upstroke. A suprathreshold stimulus rapidly depolarizes the membrane by activating the fast Na^+ channels.

2. Phase 1, early partial repolarization. Achieved by the efflux of K^+ through channels that conduct the transient outward current, i_{to}.

3. Phase 2, plateau. Achieved by a balance between the influx of Ca^{++} through Ca^{++} channels and the efflux of K^+ through several types of K^+ channels.

4. Phase 3, final repolarization. Initiated when the efflux of K^+ exceeds the influx of Ca^{++}. The resulting partial repolarization rapidly increases the K^+ conductance and rapidly restores full repolarization.

5. Phase 4, resting potential. The transmembrane potential of the fully repolarized cell is determined mainly by the conductance of the cell membrane to K^+.

Types of Action Potentials

Two principal types of action potentials may be recorded from cardiac cells:

1. Fast response action potential. Recorded from atrial and ventricular myocardial fibers and from specialized conducting (Purkinje's) fibers. The action potential is characterized by a large amplitude, steep upstroke, which is produced by the activation of fast Na^+ channels. The effective refractory period begins at the upstroke of the action potential and persists until about midway through phase 3. The fiber is relatively refractory during the remainder of phase 3, but it regains full excitability as soon as it is fully repolarized (phase 4).

2. Slow response action potential. Recorded from normal SA and AV nodal cells and from abnormal myocardial cells that have been partially depolarized. The action potential is characterized by a less negative resting potential, a smaller amplitude, and a less steep upstroke than is the fast response action potential. The upstroke is produced by the activation of Ca^{++} channels. The fiber becomes absolutely refractory at the beginning of the upstroke, but partial excitability may not be regained until very late in phase 3 or after the fiber is fully repolarized. The fiber remains relatively refractory for a significant time after the fiber has fully repolarized.

Automaticity

Automaticity is characteristic of certain cells in the heart, notably those in the SA and AV nodes and in the specialized conducting system. Automaticity is ascribable to a slow depolarization of the membrane during phase 4. Ultimately the transmembrane potential achieves threshold; this leads to the upstroke of the action potential and the firing of the automatic cell.

Cardiac Excitation

Normally the SA node initiates the impulse that induces cardiac contraction. This impulse is propagated from the SA node to the atria, and the wave of excitation ultimately reaches the AV node. Because the cells in the AV node are slow response fibers, the impulse travels very slowly through the AV node. The consequent delay between atrial and ventricular depolarization provides adequate time for atrial contraction to help fill the ventricles.

Disturbances of Impulse Initiation

Impulses may be initiated abnormally (1) by slow diastolic depolarization of automatic cells in ectopic sites or (2) by afterdepolarizations that reach threshold.

1. Ectopic foci. Automatic cells in the atrium, AV node, or His-Purkinje's system may initiate propagated cardiac impulses either because the ordinarily more rhythmic, normal pacemaker cells are suppressed or because the rhythmicity of the ectopic foci is abnormally enhanced.

2. Afterdepolarizations. Under abnormal conditions afterdepolarizations may appear early in phase 3 of a normally initiated beat, or they may be delayed until near the end of phase 3 or the beginning of phase 4. Such afterdepolarizations may themselves trigger propagated impulses.
a. Early afterdepolarization. More likely to occur when the basic cycle length of the initiating beats is very long and when the cardiac action potentials are abnormally prolonged.
b. Delayed afterdepolarizations. More likely to occur when the basic cycle length of the initiating beats is short and when the cardiac cells are overloaded with Ca^{++}.

Disturbances of Impulse Conduction

Disturbances of impulse conduction consist mainly of simple conduction block and reentry.
1. Simple conduction block. Failure of propagation in a cardiac fiber as the result of a disease process (ischemia, inflammation) or a drug.

2. Reentry. A cardiac impulse may traverse a loop of cardiac fibers and reenter previously excited tissue when (1) the impulse is conducted slowly around the loop, and (2) the impulse is blocked unidirectionally in some section of the loop.

Electrocardiogram. The electrocardiogram is recorded from the surface of the body, and it traces the conduction of the cardiac impulse through the heart. The component waves of the electrocardiogram are:

1. P wave. Spread of excitation over the atria.

2. QRS interval. Spread of excitation over the ventricles.

3. T wave. Spread of repolarization over the ventricles.

The electrocardiogram may be used to detect and analyze certain cardiac arrhythmias, such as altered sinoatrial rhythms, atrioventricular conduction blocks, premature depolarizations, ectopic tachycardias, and atrial and ventricular fibrillation.

Bibliography

Journal articles

Armstrong CW: Sodium channels and gating currents, *Physiol Rev* 61:644, 1981.

Bean BP: Multiple types of calcium channels in heart muscle and neurons, *Ann NY Acad Sci* 560:334, 1989.

Bonke FIM et al: Impulse propagation from the SA-node to the ventricles, *Experientia* 43:1044, 1987.

Bouman LN, Jongsma HJ: Structure and function of the sinoatrial node: a review, *Eur Heart J* 7:94, 1986.

Brown HF: Electrophysiology of the sinoatrial node, *Physiol Rev* 62:505, 1982.

DiFrancesco D: Contribution of the "pacemaker" current (i_f) to generation of spontaneous activity in rabbit sino-atrial node myocytes, *J Physiol* 434:23, 1991.

Grant AO: Evolving concepts of cardiac sodium channel function, *J Cardiovasc Electrophysiol* 1:53, 1990.

Horackova M: Transmembrane calcium transport and the activation of cardiac contraction. *Can J Physiol Pharmacol* 62:874, 1984.

January CT, Fozzard HA: Delayed afterdepolarizations in heart muscle: mechanisms and relevance, *Pharmacol Rev* 40:219, 1988.

January CT, Shorofsky S: Early afterdepolarizations: newer insights into cellular mechanisms, *J Cardiovasc Electrophysiol* 1:161, 1990.

Levy MN: Role of calcium in arrhythmogenesis, *Circulation* 80:IV-23, 1989.

Meijler FL, Janse MJ: Morphology and electrophysiology of the mammalian atrioventricular node, *Physiol Rev* 68:608, 1988.

Noble D: The surprising heart: a review of recent progress in cardiac electrophysiology, *J Physiol (Lond)* 353:1, 1984.

Opthof T: Mammalian sinoatrial node, *Cardiovasc Drugs Ther* 1:573, 1988.

Pressler ML, Rardon DP: Molecular basis for arrhythmias: Two nonsarcolemmal ion channels, *J Cardiovasc Electrophysiol* 1:464, 1990.

Reiter M: Calcium mobilization and cardiac inotropic mechanisms *Pharmacol Rev* 40:189, 1988.

Rosen MR: Links between basic and clinical cardiac electrophysiology *Circulation* 77:251, 1988.

Singer DH, Baumgarten CM, Ten Eick RE: Cellular electrophysiology of ventricular and other dysrhythmias: studies on diseased and ischemic heart, *Prog Cardiovasc Dis* 24:97, 1981.

Spach MS, Kootsey JM: The nature of electrical propagation in cardiac muscle. *Am J Physiol* 244:H3, 1983.

Ten Eick RE, Baumgarten CM, Singer DH: Ventricular dysrhythmia: membrane basis, or of currents, channels, gates, and cables, *Prog Cardiovasc Dis* 24:157, 1981.

Tseng GN, Boyden PA: Multiple types of Ca^{2+} currents in single canine Purkinje cells, *Circ Res* 65:1735, 1989.

Zelis R, Moore R: Recent insights into the calcium channels, *Circulation* 80:IV-14, 1989.

Books and monographs

Bouman LN, Jongsma HJ: *Cardiac rate and rhythm*. The Hague, 1982, Martinus Nijhoff.

Cranefield PF, Aronson RS: *Cardiac arrhythmias: The role of triggered activity and other mechanisms,* Mt. Kisco, NY, 1988, Futura Publishing.

Hille B: *Ionic channels of excitable membranes,* ed 2, Sunderland, MA, 1991, Sinauer Associates.

Langer GA, editor: *Calcium and the heart,* New York, 1990, Raven Press.

Levy MN, Vassalle M: *Excitation and neural control of the heart,* Bethesda, Md., 1982, American Physiological Society.

Mazgalev T, Dreifus LS, Michelson EL: *Electrophysiology of the sinoatrial and atrioventricular nodes*. New York, 1988, Alan R. Liss.

Nathan RD: *Cardiac muscle: regulation of excitation and contraction,* Orlando, 1986, Academic Press.

Noble D, Powell T: *Electrophysiology of single cardiac cells,* Orlando, FL, 1987, Academic Press.

Rüegg JC: *Calcium in muscle activation*. New York, 1986, Springer-Verlag.

Sakmann B, Neher E: *Single channel recording,* New York, 1983, Plenum Press.

Sperelakis N: *Physiology and pathophysiology of the heart,* ed 2, Hingham, MA, 1989, Martinus Nijhoff Publishers.

Zipes DP, Jalife J: *Cardiac electrophysiology: from cell to bedside,* Philadelphia, 1990, WB Saunders.

CHAPTER

24

The Cardiac Pump

It is nearly impossible to contemplate the pumping action of the heart without being struck by its simplicity of design, its wide range of activity and functional capacity, and the staggering amount of work it performs relentlessly over the lifetime of an individual. To understand how the heart accomplishes its important task, it is first necessary to consider the relationships between the structure and function of its components.

■ *Structure of the Heart in Relation to Function*

■ *Myocardial Cell*

A number of important morphological and functional differences exist between myocardial and skeletal muscle cells. However, the contractile elements within the two types of cells are quite similar; each skeletal and cardiac muscle cell is made up of **sarcomeres** (from Z line to Z line) containing thick filaments composed of myosin (in the A band) and thin filaments containing actin. The thin filaments extend from the point where they are anchored to the Z line (through the I band) to interdigitate with the thick filaments. As in the case of skeletal muscle, shortening occurs by the sliding filament mechanism (see Chapter 17). Actin filaments slide along adjacent myosin filaments by cycling of the intervening crossbridges, thereby bringing the Z lines closer together.

Skeletal and cardiac muscle show similar length-force relationships. The sarcomere length has been determined with electron microscopy in papillary muscles and intact ventricles rapidly fixed during systole or diastole. The developed force is maximal when the muscle begins its contraction at resting sarcomere lengths of 2 to 2.4 μm for cardiac muscle. At such lengths, there is optimal overlap of thick and thin filaments, and a maximal number of crossbridge attachments. Stretch of the myocardium also increases contractile force by an unexplained increase in sensitivity of the myofilaments to calcium. Developed force of cardiac muscle is less than the maximum value when the sarcomeres are stretched beyond the optimum length, because of less overlap of the actin and myosin filaments, and hence less cycling of the crossbridges. At resting sarcomere lengths shorter than optimum value, the thin filaments overlap each other, which diminishes contractile force.

In general, the fiber length-force relationship for the papillary muscle also holds true for fibers in the intact heart. This relationship may be expressed graphically, as in Fig. 24-1, by substituting ventricular systolic pressure for force and end-diastolic ventricular volume for myocardial resting fiber (and hence sarcomere) length. The lower curve in Fig. 24-1 represents the increment in pressure produced by each increment in volume when the heart is in diastole. The upper curve represents the peak pressure developed by the ventricle during systole at each degree of filling and illustrates the **Frank-Starling relationship** of initial myocardial fiber length (or initial volume) to force (or pressure) development by the ventricle.

Note that the pressure-volume curve in diastole is initially quite flat, indicating that large increases in volume can be accommodated with only small increases in pressure, yet systolic pressure development is considerable at the lower filling pressures. However, the ventricle becomes much less distensible with greater filling, as evidenced by the sharp rise of the diastolic curve at large intraventricular volumes. In the normal intact heart, peak force may be attained at a filling pressure of 12 mm Hg. At this intraventricular diastolic pressure, which is about the upper limit observed in the normal heart, the sarcomere length is 2.2 μm. However, developed force peaks at filling pressures as high as 30 mm Hg in the isolated heart. Even at higher diastolic pressures (>50 mm Hg) the sarcomere length is not greater than 2.6 μm in cardiac muscle. This resistance to stretch of the myocardium at high filling pressures probably resides in the noncontractile constituents of the tissue (connective tissue) and may serve as a safety factor against overloading of the heart in diastole. Usually ventricular diastolic pressure is about 0 to 7 mm Hg, and the average diastolic sarcomere length is about 2.2

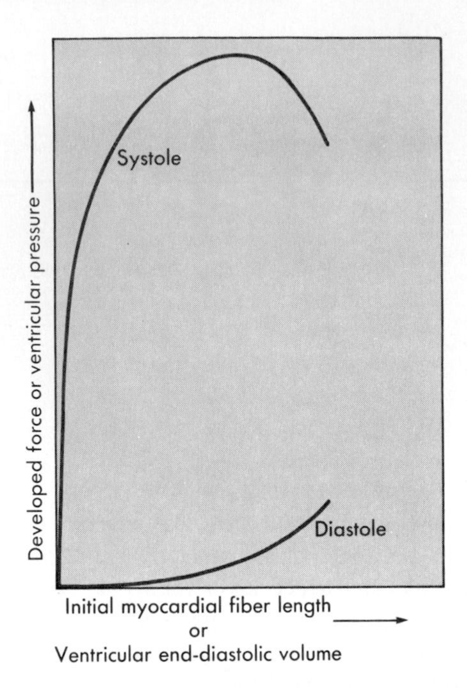

■ **Fig. 24-1** Relationship of myocardial resting fiber length (sarcomere length) or end-diastolic volume to developed force or peak systolic ventricular contraction in the intact dog heart. (Redrawn from Patterson SW, Piper H, Starling EH: *J Physiol* 48:465, 1914.)

μm. Thus the normal heart operates on the ascending portion of the Frank-Starling curve depicted in Fig. 24-1.

If the heart becomes greatly distended with blood during diastole, as may occur in cardiac failure, it is less efficient; more energy is required (greater wall tension) for the distended heart to eject the same volume of blood per beat than for the normal undilated heart. This is an example of Laplace's law (p. 466), which states that the tension in the wall of a vessel (in this case the ventricles) equals the transmural pressure (pressure across the wall, or distending pressure) times the radius of the vessel or chamber. The Laplace relationship applies to infinitely thin-walled vessels but can be applied to the heart if correction is made for wall thickness. The equation is $\tau = Pr/w$ where τ = wall stress, P = transmural pressure, r = radius, and w = wall thickness.

A striking difference in the appearance of cardiac and skeletal muscle is the semblance of a syncytium in cardiac muscle with branching interconnecting fibers (Figs. 24-2 and 24-3). However, the myocardium is not a true anatomical syncytium, because laterally the myocardial fibers are separated from adjacent fibers by their respective sarcolemmas, and the end of each fiber is separated from its neighbor by dense structures, **intercalated disks,** which are continuous with the sarcolemma (Figs. 24-2 to 24-4). Nevertheless, *cardiac muscle functions as a syncytium,* because a wave of depolarization followed by contraction of the entire myocardium (*an all-or-none response*) occurs when a suprathreshold stimulus is applied to any one focus.

As the wave of excitation approaches the end of a cardiac cell, the spread of excitation to the next cell depends on the electrical conductance of the boundary between the two cells. **Gap junctions (nexi)** with high conductances are present in the intercalated disks between adjacent cells (see Figs. 24-2 to 24-4). These gap junctions, which facilitate the conduction of the cardiac impulse from one cell to the next, are made up of **connexons,** which are hexagonal structures that connect the cytosol of adjacent cells. Each connexon consists of six polypeptides surrounding a core channel approximately 1.6 to 2.0 nm wide, which serves as a low resistance pathway for cell-to-cell conductance (see Chapter 4).

Impulse conduction in cardiac tissues progresses more rapidly in a direction parallel to the long axes of the constituent fibers than in a direction perpendicular to the long axes of those fibers. Gap junctions exist in the borders between myocardial fibers that are in contact with each other longitudinally; they are very sparse or absent in the borders between myocardial fibers that lie side by side.

Another difference between cardiac and fast skeletal muscle fibers is in the abundance of mitochondria **(sarcosomes)** in the two tissues. Fast skeletal muscle, which is called on for relatively short periods of repetitive or sustained contraction and which can metabolize anaerobically and build up a substantial oxygen debt, has relatively few mitochondria in the muscle fibers. In contrast, cardiac muscle, which contracts repetitively for a lifetime and requires a continuous supply of oxygen, is very rich in mitochondria (see Figs. 24-2 to 24-4). Rapid oxidation of substrates with the synthesis of adenosine triphosphate (ATP) can keep pace with the myocardial energy requirements because of the large numbers of mitochondria containing the respiratory enzymes necessary for oxidative phosphorylation.

To provide adequate oxygen and substrate for its metabolic machinery, the myocardium is also endowed with a rich capillary supply, about one capillary per fiber. Thus diffusion distances are short; and oxygen, carbon dioxide, substrates, and waste material can move rapidly between myocardial cell and capillary. With respect to exchange of substances between the capillary blood and the myocardial cells, electron micrographs of myocardium show deep invaginations of the sarcolemma into the fiber at the Z lines (see Figs. 24-2 to 24-4). These sarcolemmal invaginations constitute the **transverse-tubular,** or **T-tubular, system.** The lumina of these T-tubules are continuous with the bulk interstitial fluid, and they play a key role in excitation-contraction coupling.

In mammalian ventricular cells, adjacent T-tubules are interconnected by longitudinally running or axial tubules, thus forming an extensively interconnected lattice of "intracellular" tubules (see Fig. 24-4). This T-tu-

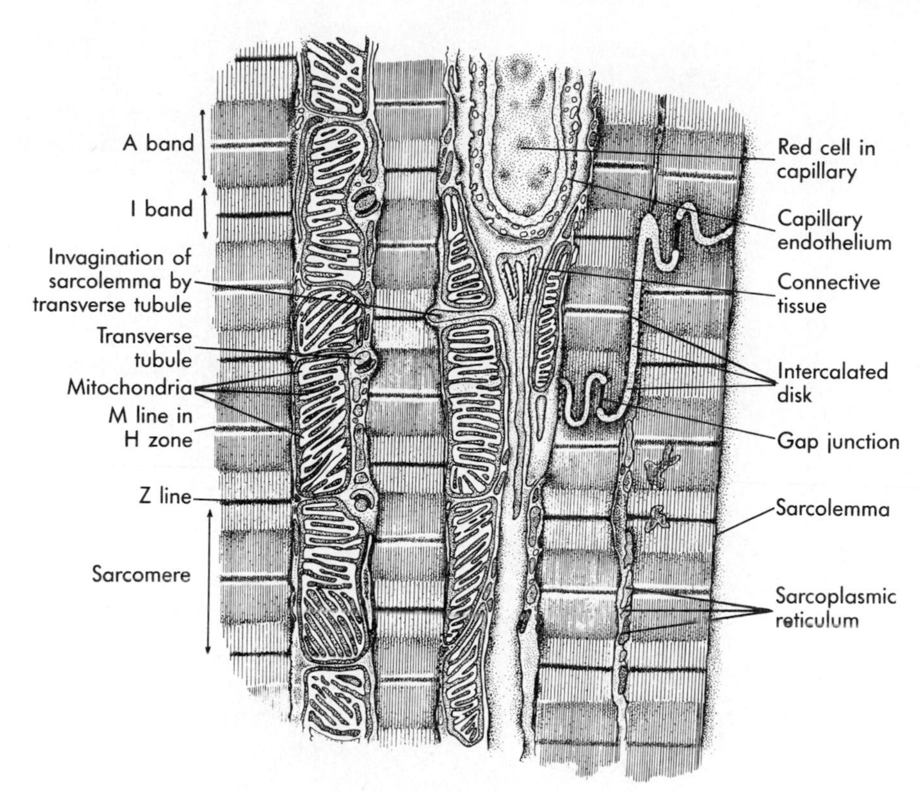

■ **Fig. 24-2** Diagram of an electron micrograph of cardiac muscle showing large numbers of mitochondria and the intercalated disks with nexi (gap junctions), transverse tubules, and longitudinal tubules.

bule system is open to the interstitial fluid, is lined with a basement membrane continuous with that of the surface sarcolemma, and contains micropinocytotic-like vesicles. Thus in ventricular cells the myofibrils and mitochondria have ready access to a space that is continuous with the interstitial fluid. The T-tubular system is absent or poorly developed in atrial cells of many mammalian hearts.

A network of sarcoplasmic reticulum (see Fig. 24-4) consisting of small-diameter sarcotubules is also present surrounding the myofibrils; these sarcotubules are believed to be "closed," because colloidal tracer particles (2 to 10 nm in diameter) do not enter them. They do not contain basement membrane. Flattened elements of the sarcoplasmic reticulum are often found in close proximity to the T-tubular system, as well as to the surface sarcolemma, forming **diads.**

Excitation-contraction coupling. The earliest studies on isolated hearts perfused with isotonic saline solutions indicated the need for optimum concentrations of Na^+, K^+, and Ca^{++}. In the absence of Na^+ the heart is not excitable and will not beat because the action potential depends on extracellular Na ions. In contrast, the resting membrane potential is independent of the Na ion gradient across the membrane (see Fig. 23-5). Under normal conditions the extracellular K^+ concentration is about 4 mM. A reduction in extracellular K^+ has little effect on myocardial excitation and contraction. However, increases in extracellular K^+, if great enough, produce depolarization, loss of excitability of the myocardial cells, and cardiac arrest in diastole. *Ca^{++} is also essential for cardiac contraction;* removal of Ca^{++} from the extracellular fluid results in decreased contractile force and eventually arrest in diastole. Conversely, an increase in extracellular Ca^{++} enhances contractile force, and very high Ca^{++} concentrations induce cardiac arrest in systole (rigor). It is now well documented that free intracellular Ca^{++} is the agent responsible for the contractile state of the myocardium.

Initially a wave of excitation spreads rapidly along the myocardial sarcolemma from cell to cell via gap junctions. Excitation also spreads into the interior of the cells via the T-tubules (see Figs. 24-2 to 24-4), which invaginate the cardiac fibers at the Z lines. (Electrical stimulation at the Z line or the application of ionized Ca to the Z lines in the skinned [sarcolemma removed] cardiac fiber elicits a localized contraction of adjacent myofibrils.) During the plateau (phase 2) of the action po-

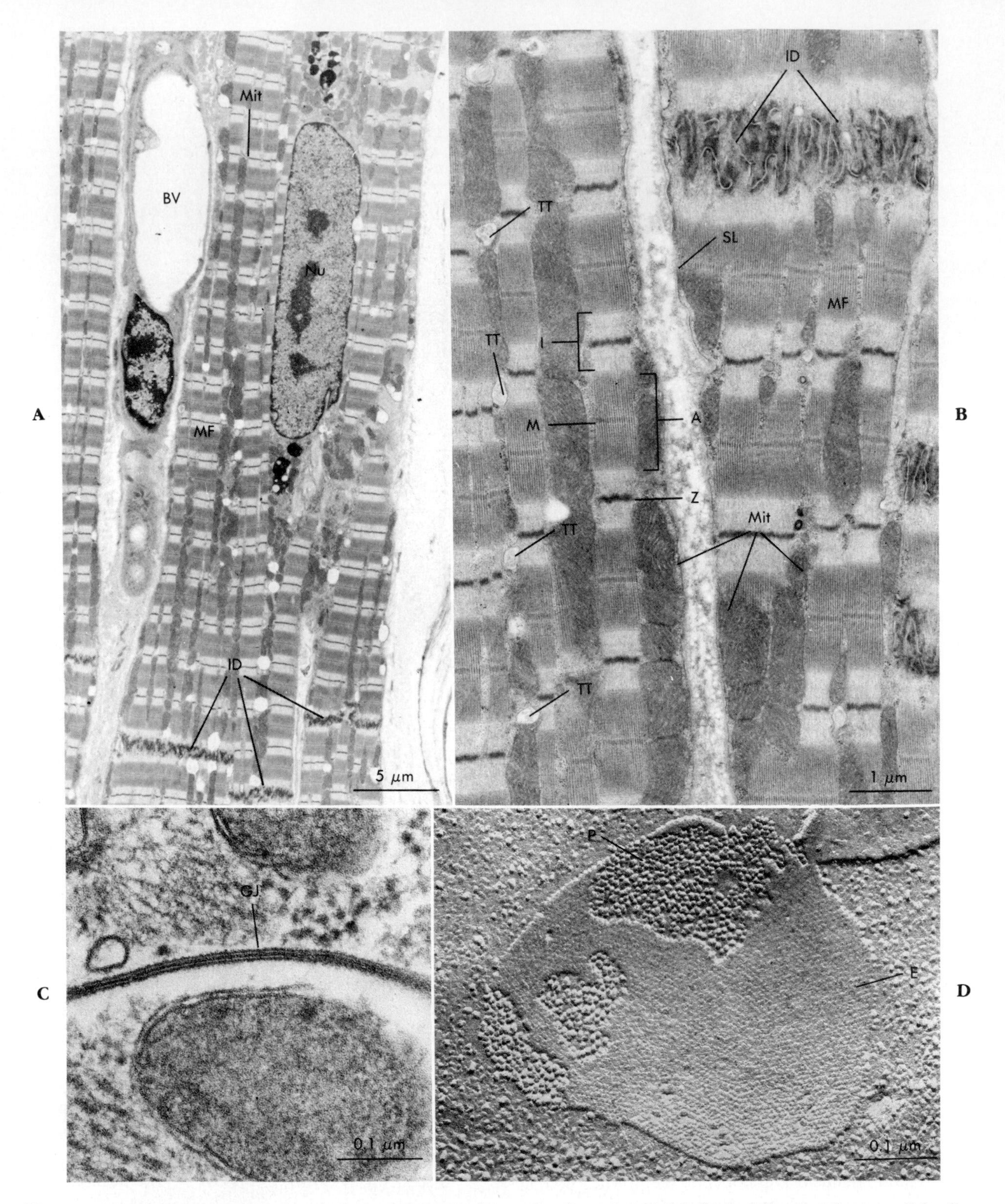

■ **Fig. 24-3** **A,** Low-magnification electron micrograph of a monkey heart (ventricle). Typical features of myocardial cells include the elongated nucleus *(Nu),* striated myofibrils *(MF)* with columns of mitochondria *(Mit)* between the myofibrils, and intercellular junctions (intercalated disks, *ID*). A blood vessel *(BV)* is located between the myocardial cells. **B,** Medium-magnification electron micrograph of monkey ventricular cells, showing details of ultrastructure. The sarcolemma *(SL)* is the boundary of the muscle cells and is thrown into multiple folds where the cells meet at the intercalated disk region *(ID)*. The prominent myofibrils *(MF)* show distinct banding patterns, including the A band *(A),* dark Z lines *(Z),* I band regions *(I),* and M lines *(M)* at the center of each sarcomere unit. Mitochondria *(mit)* occur either in rows between myofibrils or masses just underneath the sarcolemma. Regularly spaced transverse tubules *(TT)* appear at the Z line levels of the myofilbrils. **C,** High-magnification electron micrograph of a specialized intercellular junction between two myocardial cells of the mouse. Called a gap junction *(GJ)* or nexus, this attachment consists of very close apposition of the sarcolemmal membranes of the two cells and appears in thin section to consist of seven layers. **D,** Freeze-fracture replica of mouse myocardial gap junction showing distinct arrays of characteristic intramembraneous particles. Large particles *(P)* belong to the inner half of the sarcolemma of one myocardial cell, whereas the "pitted" membrane face *(E)* is formed by the outer half of the sarcolemma of the cell above.

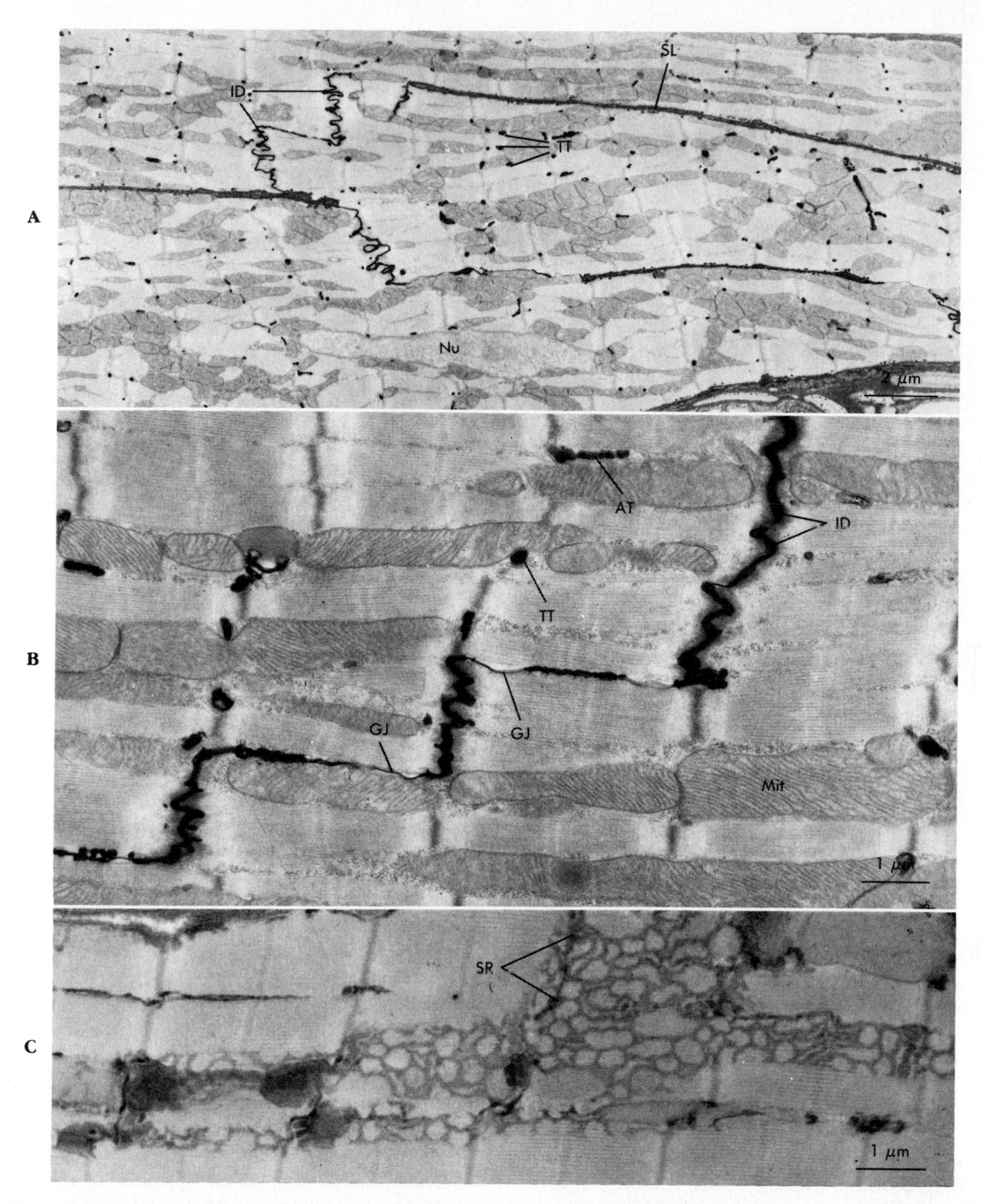

■ **Fig. 24-4** **A,** Low-magnification electron micrograph of the right ventricular wall of a mouse heart. Tissue was fixed in a phosphate-buffered glutaraldehyde solution and postfixed in ferrocyanide-reduced osmium tetroxide. This procedure has resulted in the deposition of electron-opaque precipitate in the extracellular space thus outlining the sarcolemmal borders *(SL)* of the muscle cells and delineating the intercalated disks *(ID)* and transverse tubules *(TT)*. *Nu,* Nucleus of the myocardial cell. **B,** Mouse cardiac muscle in longitudal section, treated as in panel **A.** The path of the extracellular space is traced through the intercalated disk region *(ID),* and sarcolemmal invaginations that are oriented transverse to the cell axis (transverse tubules, *TT*) or parallel to it (axial tubules, *AT*) are clearly identified. Gap junctions *(GJ)* are associated with the intercalated disk. Mitochondria are large and elongated and lie between the myofibrils. **C,** Mouse cardiac muscle. Tissue treated with ferrocyanide-reduced osmium tetroxide so as to identify the internal membrane system (sarcoplasmic reticulum, *SR*). Specific staining of the SR reveals its architecture as a complex network of small-diameter tubules that are closely associated with the myofibrils and mitochondria.

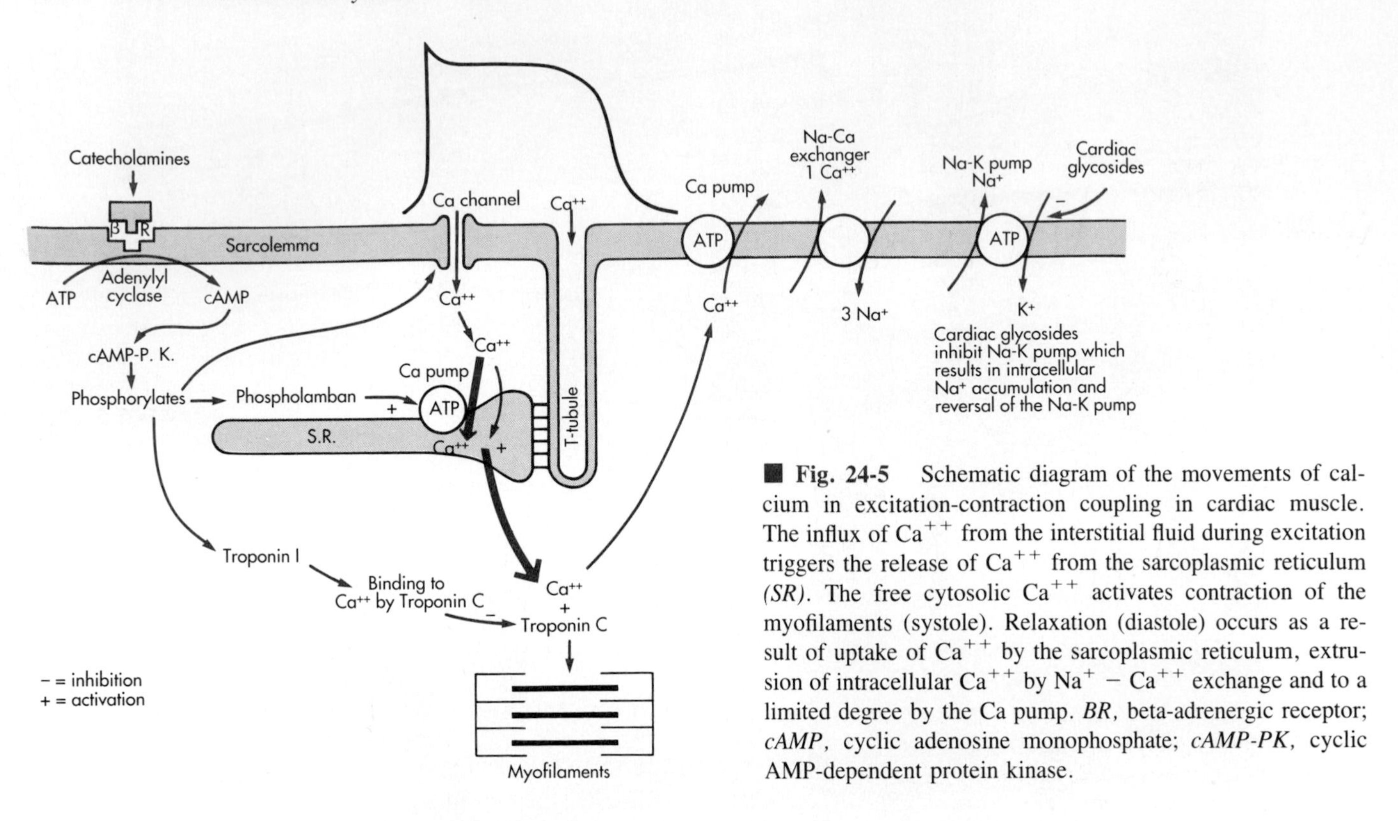

■ **Fig. 24-5** Schematic diagram of the movements of calcium in excitation-contraction coupling in cardiac muscle. The influx of Ca^{++} from the interstitial fluid during excitation triggers the release of Ca^{++} from the sarcoplasmic reticulum *(SR)*. The free cytosolic Ca^{++} activates contraction of the myofilaments (systole). Relaxation (diastole) occurs as a result of uptake of Ca^{++} by the sarcoplasmic reticulum, extrusion of intracellular Ca^{++} by Na^{+} – Ca^{++} exchange and to a limited degree by the Ca pump. *BR*, beta-adrenergic receptor; *cAMP*, cyclic adenosine monophosphate; *cAMP-PK*, cyclic AMP-dependent protein kinase.

tential, Ca^{++} permeability of the sarcolemma increases. Ca^{++} flows down its electrochemical gradient and is largely responsible for the slow inward current (p. 370). Ca^{++} enters the cell through Ca^{++} channels in the sarcolemma and in the invaginations of the sarcolemma, the T-tubules. Opening of the Ca^{++} channels is believed to be caused by phosphorylation of the channel proteins by a cyclic AMP (cAMP)-dependent protein kinase. The primary source of extracellular Ca^{++} is the interstitial fluid (10^{-3}M Ca^{++}). Some Ca^{++} also may be bound to the sarcolemma and to the **glycocalyx,** a mucopolysaccharide that covers the sarcolemma. The amount of calcium entering the cell interior from the extracellular space is not sufficient to induce contraction of the myofibrils, but it serves as a trigger **(trigger Ca^{++})** to release Ca^{++} from the intracellular Ca^{++} stores, the sarcoplasmic reticulum (Fig. 24-5). The cytosolic free Ca^{++} increases from a resting level of about 10^{-7}M to levels of 10^{-6} to 10^{-5}M during excitation, and the Ca^{++} binds to the protein troponin C (see Chapter 18). The Ca^{++}-troponin complex interacts with tropomyosin to unblock active sites between the actin and myosin filaments, which allows crossbridge cycling and hence contraction of the myofibrils (systole).

Mechanisms that raise cytosolic Ca^{++} increase the developed force, and those that lower Ca^{++} decrease the developed force. For example, catecholamines increase the movement of Ca^{++} into the cell by phosphorylation of the Ca^{++} channels by a cAMP-dependent protein kinase. An increase in cytosolic Ca^{++} is also achieved by increasing extracellular Ca^{++} or decreasing the Na^{+} gradient across the sarcolemma.

The sodium gradient can be reduced by increasing intracellular Na^{+} or by decreasing extracellular Na^{+}. Cardiac glycosides increase intracellular Na^{+} by poisoning the Na-K pump, which results in an accumulation of Na^{+} in the cells. The elevated cytosolic Na^{+} reduces the exchange of Na^{+} and Ca^{++} so that less Ca^{++} is removed from the cell. This action elevates cytosolic Ca^{++}, which enhances contractile force.

Developed tension is diminished by a reduction in extracellular Ca^{++}, an increase in the Na^{+} gradient across the sarcolemma, or the administration of Ca^{++} blockers that prevent Ca^{++} from entering the myocardial cell (see Fig. 23-9).

At the end of systole the Ca^{++} influx ceases and the sarcoplasmic reticulum is no longer stimulated to release Ca^{++}. In fact, the sarcoplasmic reticulum avidly takes up Ca^{++} by means of an ATP-energized calcium pump that is stimulated by phospholamban after the phospholamban is phosphorylated by cAMP-dependent protein kinase. Phosphorylation of troponin I inhibits the Ca^{++} binding of troponin C, which permits tropomyosin to again block the sites for interaction between the actin and myosin filaments, and relaxation (diastole) occurs (see Chapter 18). Cardiac contraction and relaxation are both accelerated by catecholamines and adenylyl cyclase activation. The resulting increase in cAMP activates the cAMP-dependent protein kinase, which phosphorylates the Ca channel in the sarcolemma. This allows a greater influx of Ca^{++} into the cell, and thereby accelerates contraction. However, it also accelerates relaxation by phosphorylating phospholamban, which enhances Ca^{++} uptake by the sarcoplasmic reticulum, and by phosphorylating troponin I, which inhibits the Ca^{++} binding of troponin C. Thus the phosphorylations by cAMP-dependent protein kinase serve to increase both the speed of contraction *and* the speed of relaxation.

Mitochondria also take up and release Ca^{++}, but the process is too slow to be involved in excitation-contraction coupling. Only at very high intracellular Ca^{++} levels (pathological states) do the mitochondria take up a significant amount of Ca^{++}.

The Ca^{++} that enters the cell to initiate contraction must be removed during diastole. The removal is primarily accomplished by an electroneutral exchange of 3 Na^+ for 1 Ca^{++} (see Fig. 24-5). Ca^{++} is also removed from the cell by an electrogenic pump that uses energy to transport Ca^{++} across the sarcolemma (see Fig. 24-5).

Myocardial contractile machinery and contractility. Velocity and force of contraction are a function of the intracellular concentration of free Ca ions. *Force and velocity are inversely related; so that with no load, force is negligible and velocity is maximal. In an isometric contraction, where no external shortening occurs, force is maximal and velocity is zero.*

The sequence of events in a ***preloaded*** and ***afterloaded*** isotonic contraction of a papillary muscle is illustrated in Fig. 24-6. Point A represents the resting state in which the preload is responsible for the existing initial stretch. With stimulation the ***contractile element*** begins to shorten; and at point B the ***elastic element*** has been stretched, but the load has not yet been lifted because the overall length of the muscle has not changed; the muscle fibers have shortened at the expense of the elastic element.

This stretch of the elastic element (an expression of the muscle extensibility) consumes a certain amount of energy. To this energy is added the energy of shortening to obtain the total energy expenditure of a single contraction. Stretch of the elastic element is represented in the diagram (see Fig. 24-6) as a progressive rise in force with no external shortening. At point C the force developed by the contractile element has equaled the load (the afterload), and the load has been raised without further stretch of the elastic element. This is represented in the diagram (see Fig. 24-6) as external shortening of the muscle without a further increase in force.

When these observations on papillary muscle are applied to the whole heart, the preload refers to the stretch of the left ventricle just before the onset of contraction (the so-called end-diastolic volume) and the afterload refers to the aortic pressure during the period when the aortic valve is open. The preload can be increased by greater filling of the left ventricle during diastole (see Fig. 24-1). At the lower end-diastolic volumes, increments in filling pressure during diastole elicit a greater systolic pressure during the subsequent contraction until a maximum systolic pressure is reached at the optimum

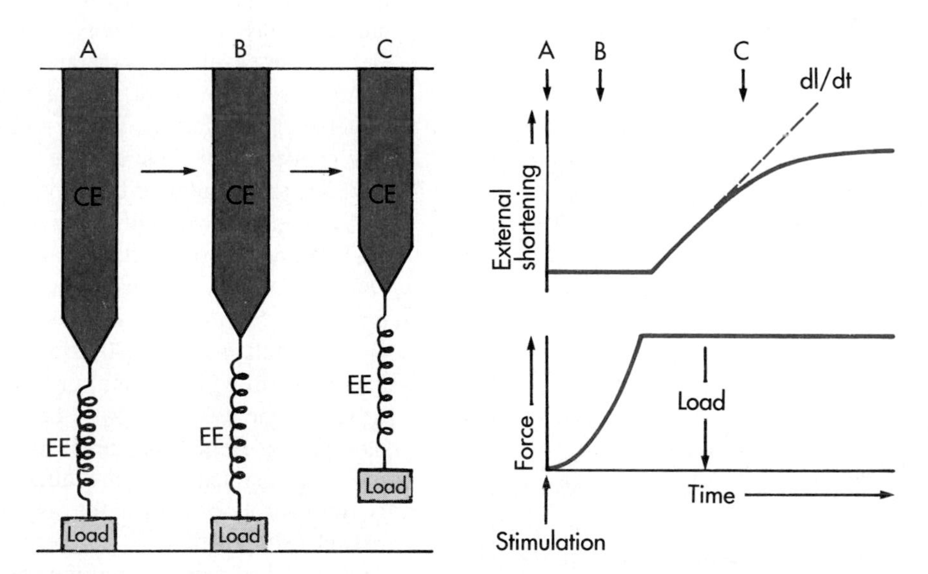

■ **Fig. 24-6** Model for a preloaded (isometric) and afterloaded (isotonic) contraction of papillary muscle. **A,** Muscle at rest. Preload is represented by partial stretch of the elastic element *(EE)*. **B,** Partial contraction of the contractile element *(CE)* with stretch of the EE and no external shortening (the isometric phase of the contraction). **C,** Further contraction of the contractile element with external shortening and lifting of the afterload. The tangent *(dl/dt)* to the initial slope of the shortening curve on the right is the velocity of initial shortening. (Redrawn from Sonneblick EH: *The myocardial cell,* Philadelphia, 1966, University of Pennsylvania Press.)

preload. Further diastolic filling beyond this point results in no further increase in developed pressure; at very high filling pressures, peak pressure development in systole is reduced. At a constant preload, a higher systolic pressure can be reached during ventricular contractions by raising the afterload (e.g., increasing aortic pressure by preventing much of the runoff of blood to the periphery during diastole). Increments in afterload will produce progressively higher peak systolic pressures until the afterload is so great that the ventricle can no longer generate enough force to open the aortic valve. At this point ventricular systole is totally isometric; there is no ejection of blood, and hence no change in volume of the ventricle during systole. The maximal pressure developed by the left ventricle under these conditions is the maximal isometric force of which the ventricle is capable at a given preload. Of course, at preloads below the optimal filling volume an increase in preload can yield a greater maximal isometric force (see Fig. 24-1).

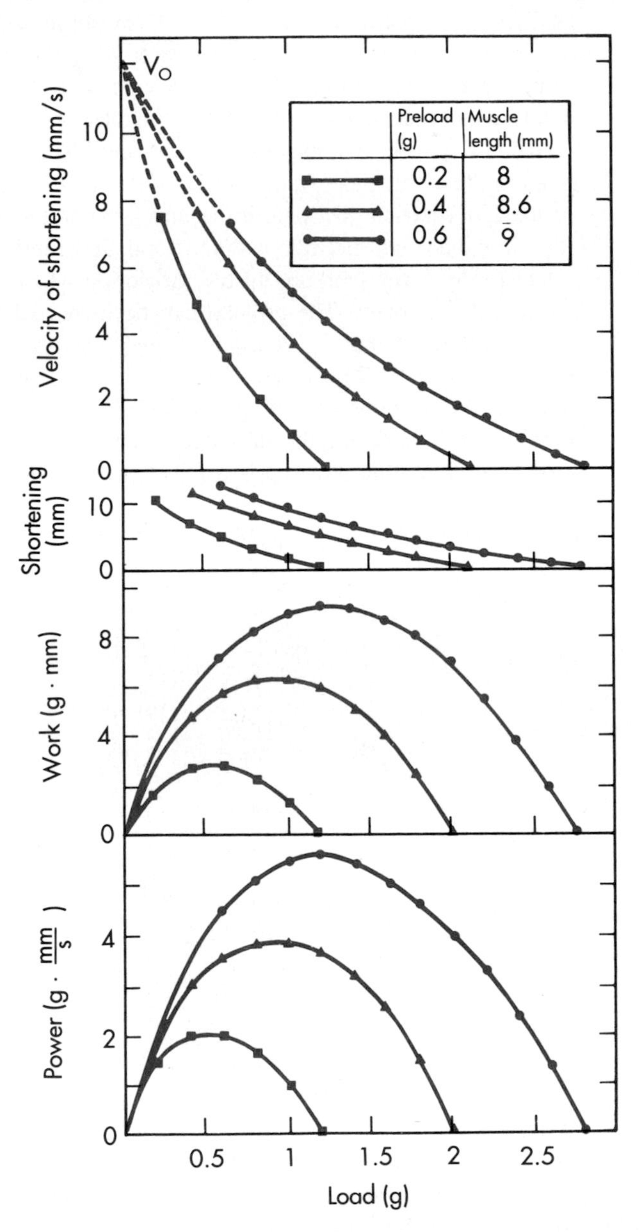

■ **Fig. 24-7** The effect of increasing initial length of a cat papillary muscle on the force-velocity relationship, degree of shortening, muscle work, and muscle power. (Redrawn from *Am J Physiol* 202:931, 1962.)

The preloads and afterloads depend on the characteristics of the vascular system and the behavior of the heart. With respect to the vasculature, the degree of venomotor tone and peripheral resistance influence preload and afterload. With respect to the heart, a change in rate or stroke volume can also alter preload and afterload. Hence, the cardiac and vascular factors are interactive in their effect on preload and afterload (see Chapter 30 for a full explanation).

If the initial velocity of shortening is plotted against the afterload, the force-velocity curves shown in Fig. 24-7 are obtained. The maximum velocity (V_0) may be estimated by extrapolation of the force-velocity curve back to zero load (as indicated by the dotted lines in Fig. 24-7) and represents the maximum rate of cycling of the crossbridges.

Contractility is a measure of the performance of the heart at a given preload and afterload. Contractility is the change in peak isometric force (isovolumic pressure) at a given initial fiber length (end-diastolic volume) and is a function of the rate of cycling of the crossbridges between the actin and myosin filaments. The faster the cycling, the greater the contractility. Augmentation of contractility is observed with certain drugs, such as norepinephrine or digitalis, and with an increase in contraction frequency **(tachycardia).** The increase in contractility **(positive inotropic effect)** produced by any of these interventions is reflected by increments in developed force and V_0.

An increase in initial fiber length produces a more forceful contraction, as shown in Fig. 24-1. However, this greater force development is not associated with any change in contractility, as estimated by V_0 (see Fig. 24-7). Fig. 24-7 also illustrates that at any given preload, the degree of shortening, the work (load × shortening or force × distance), and the power (load × velocity or work/time) all increase with the initial length of the papillary muscle. It is apparent that with an increase in initial fiber length, greater force may be developed, but the estimated V_0 is the same for all three initial lengths. Hence changes in resting length may alter force development but not contractility. This conclusion is, of course, based on the assumption that the displayed extrapolation of the force-velocity curves back to the vertical axis provide the true value for V_0. This assumption has been challenged; some investigators have failed to obtain a hyperbolic force-velocity relationship and have observed changes in V_0 (inotropic state) with changes in initial length (that is, V_0 is **length-dependent**—activation of the contractile system is influenced by muscle length). For this and several other reasons, estimates of V_0 are not a reliable index of contractility.

A reasonable index of myocardial contractility can be obtained from the contour of ventricular pressure curves (Fig. 24-8). A hypodynamic heart is characterized by an elevated end-diastolic pressure, a slowly rising ventricular pressure, and a somewhat reduced ejection phase (curve *C*, Fig. 24-8). A normal ventricle under adrenergic stimulation shows a reduced end-diastolic pressure, a fast-rising ventricular pressure, and a brief ejection phase (curve B, Fig. 24-8). The slope of the ascending limb of the ventricular pressure curve indicates the maximum rate of force development by the ventricle (maximum rate of change in pressure with time; maximum dP/dt, as illustrated by the tangents to the steepest portion of the ascending limbs of the ventricular pressure curves in Fig. 24-8). The slope is maximal during the isovolumic phase of systole (p. 410) and, at any given degree of ventricular filling, provides an index of the initial contraction velocity and hence of contractility. Similarly, one can obtain an indication of the contractile state of the myocardium from the initial velocity of blood flow in the ascending aorta (the initial slope of the aortic flow curve). The **ejection fraction,** which is the ratio of the volume of blood ejected from the left ventricle per beat **(stroke volume)** to the volume of blood in the left ventricle at the end of diastole, is widely used clinically as an index of contractility. Other measurements or combinations of measurements that in general are concerned with the magnitude or velocity of the ventricular contraction also have been used to assess the contractile state of the cardiac muscle. Because no index is entirely satisfactory at present, several indices are currently in use.

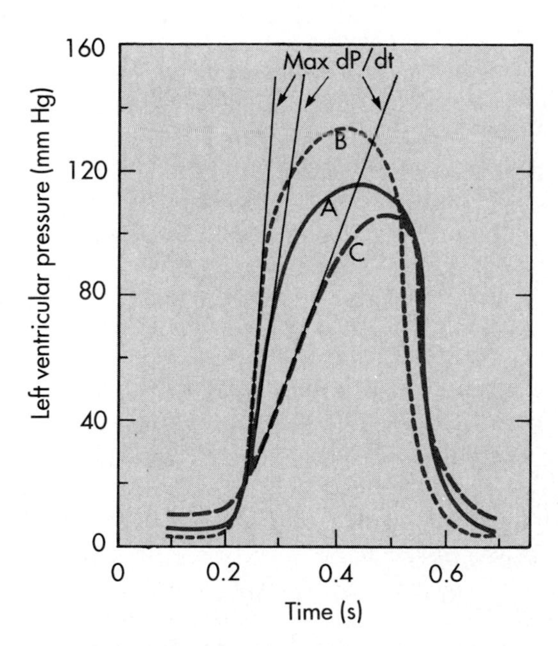

Fig. 24-8 Left ventricular pressure curves with tangents drawn to the steepest portions of the ascending limbs to indicate maximum dP/dt values. *A*, Control; *B*, hyperdynamic heart, as with norepinephrine administration; *C*, hypodynamic heart, as in cardiac failure.

Cardiac Chambers

The atria are thin-walled, low-pressure chambers that function more as large reservoir conduits of blood for their respective ventricles than as important pumps for the forward propulsion of blood. The ventricles were once thought to be made up of bands of muscle. However, it now appears that they are formed by a continuum of muscle fibers that take origin from the fibrous skeleton at the base of the heart (chiefly around the aortic orifice). These fibers sweep toward the apex at the epicardial surface and also pass toward the endocardium as they gradually undergo a 180-degree change in direction to lie parallel to the epicardial fibers and form the endocardium and papillary muscles (Fig. 24-9). At the

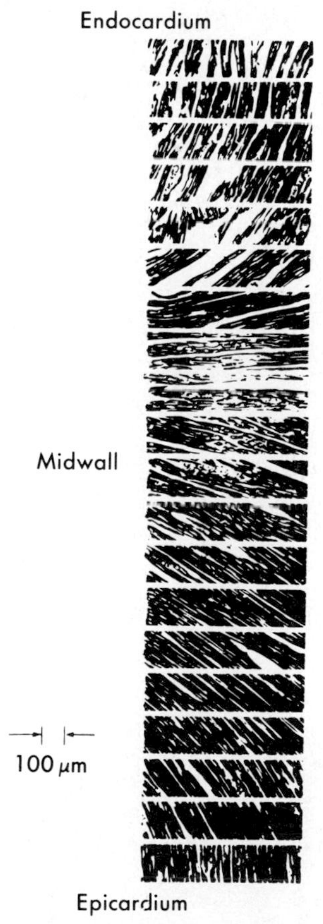

Fig. 24-9 Sequence of photomicrographs showing fiber angles in successive sections taken from the middle of the free wall of the left ventricle from a heart in systole. The sections are parallel to the epicardial plane. Fiber angle is 90 degrees at the endocardium, running through 0 degrees at the midwall to −90 degrees at the epicardium. (From Streeter DD Jr et al: *Circ Res* 24:339, 1969. By permission of the American Heart Association.)

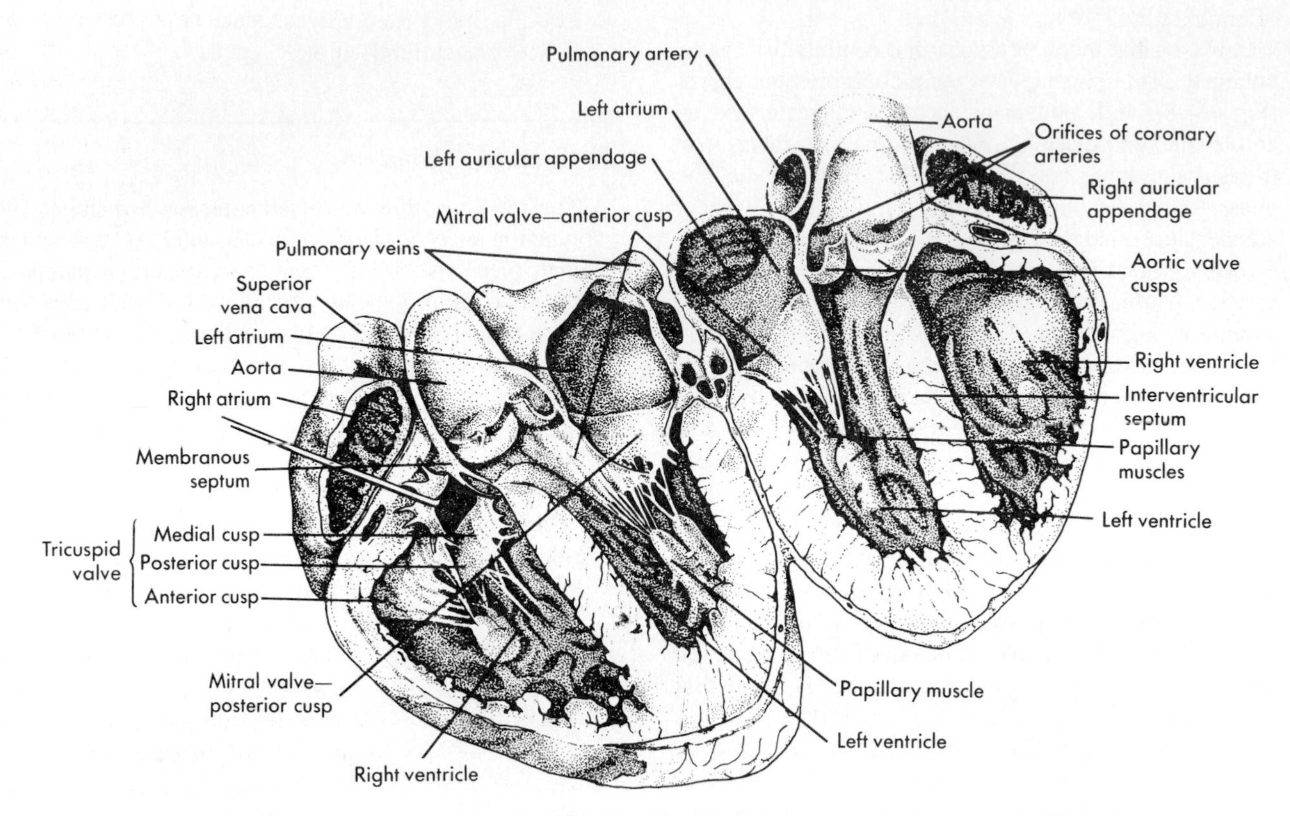

■ **Fig. 24-10** Drawing of a heart split perpendicular to the interventricular septum to illustrate the anatomical relationships of the leaflets of the AV and aortic valves.

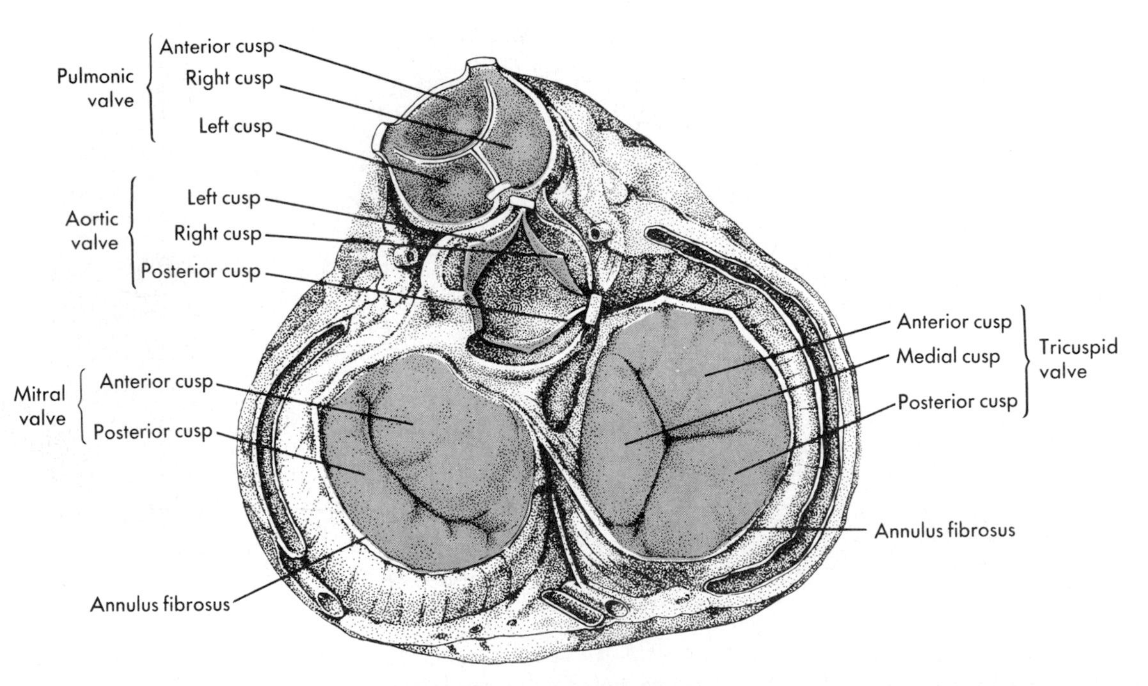

■ **Fig. 24-11** Four cardiac valves as viewed from the base of the heart. Note how the leaflets overlap in the closed valves.

apex of the heart the fibers twist and turn inward to form papillary muscles, whereas at the base and around the valve orifices they form a thick powerful muscle that not only decreases ventricular circumference for ejection of blood but also narrows the AV valve orifices as an aid to valve closure. In addition to a reduction in circumference, ventricular ejection is accomplished by a decrease in the longitudinal axis with descent of the base of the heart. The earlier contraction of the apical part of the ventricles coupled with approximation of the ventricular walls propels the blood toward the outflow tracts. The right ventricle, which develops a mean pressure about one seventh that developed by the left ventricle, is considerably thinner than the left.

Cardiac Valves

The cardiac valves consist of thin flaps of flexible, tough endothelium-covered fibrous tissue firmly attached at the base to the fibrous valve rings. Movements of the valve leaflets are essentially passive, and the orientation of the cardiac valves is responsible for unidirectional flow of blood through the heart. Two types of valves are present in the heart—the **atrioventricular valves,** or **AV valves,** and the **semilunar valves** (Figs. 24-10 and 24-11).

Atrioventricular valves. The valve between the right atrium and right ventricle is made up of three cusps **(tricuspid valve)** whereas that between the left atrium and left ventricle has two cusps **(mitral valve).** The total area of the cusps of each AV valve is approximately twice that of the respective AV orifice so that there is considerable overlap of the leaflets in the closed position (see Figs. 24-10 and 24-11). Attached to the free edges of these valves are fine, strong filaments **(chordae tendineae),** which arise from the powerful papillary muscles of the respective ventricles and prevent eversion of the valves during ventricular systole.

The mechanism of closure of the AV valves has been the subject of considerable investigation, and many factors are thought to play a role in approximating the valve leaflets. In the normal heart the valve leaflets are relatively close during ventricular filling and provide a funnel for the transfer of blood from atrium to ventricle. This partial approximation of the valve surfaces during diastole is believed to be caused by eddy currents behind the leaflets and possibly also by some tension on the free edges of the valves, exerted by the chordae tendineae and papillary muscles that are stretched by the filling ventricle. Movements of the mitral valve leaflets throughout the cardiac cycle are shown in an **echocardiogram** (Fig. 24-12). Echocardiography consists of sending short pulses of high-frequency sound waves (ultrasound) through the chest tissues and the heart and recording the echoes reflected from the various structures. The timing and the pattern of the reflected waves provide such information as the diameter

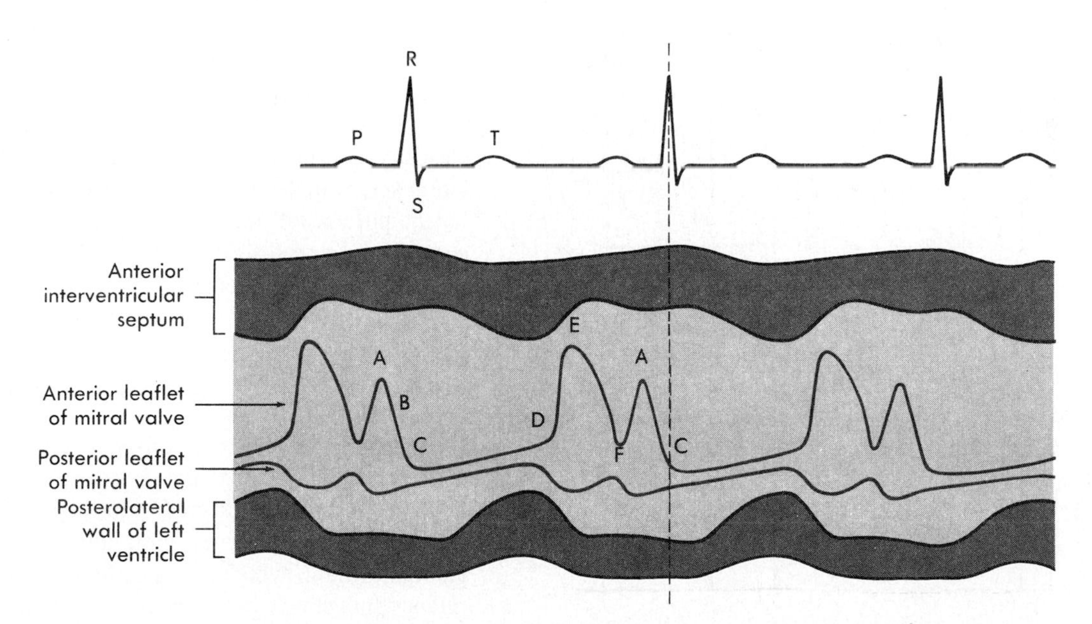

Fig. 24-12 Drawing made from an echocardiogram showing movements of the mitral valve leaflets (particularly the anterior leaflet) and the changes in the diameter of left ventricular cavity and the thickness of the left ventricular walls during cardiac cycles in a normal person. *D* to *C*, Ventricular diastole; *C* to *D*, ventricular systole; *D* to *E*, rapid filling; *E* to *F*, reduced filling (diastasis); *F* to *A*, atrial contraction. Mitral valve closes at *C* and opens *D*. Simultaneously recorded electrocardiogram at top.

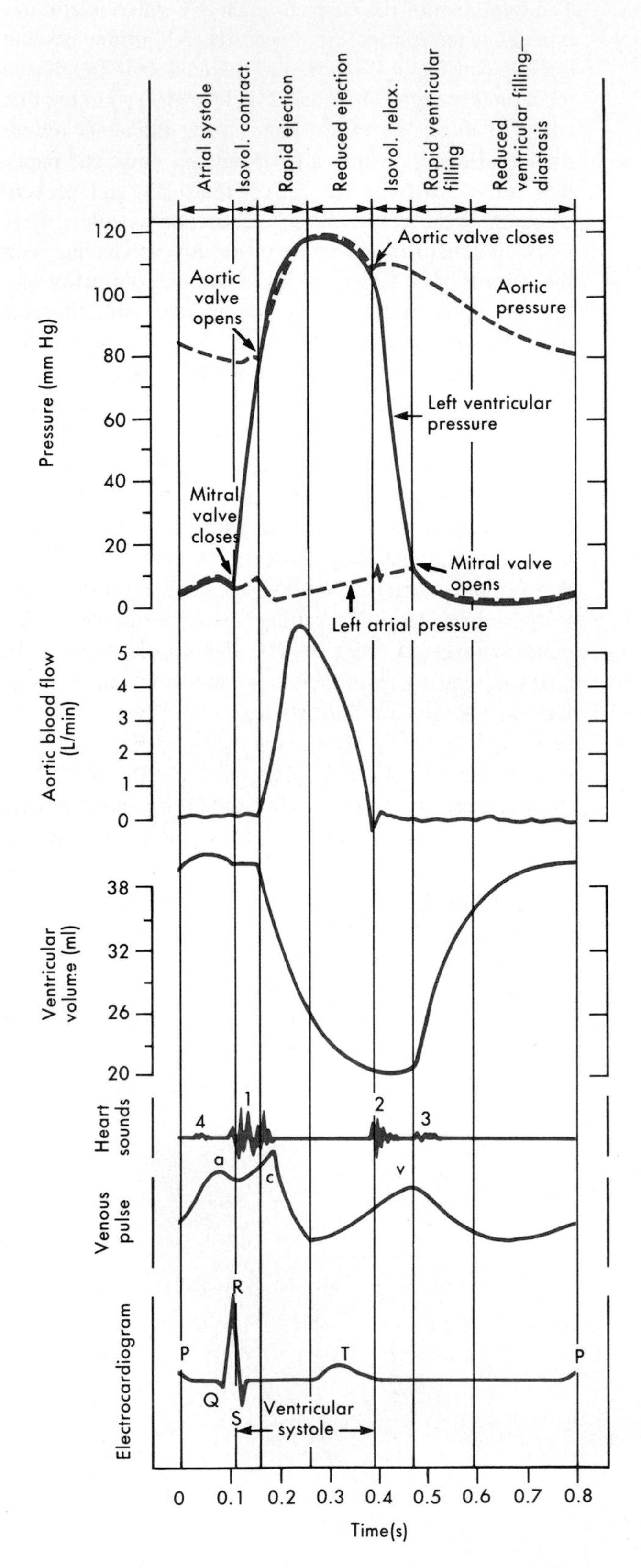

■ **Fig. 24-13** Left atrial, aortic, and left ventricular pressure pulses correlated in time with aortic flow, ventricular volume, heart sounds, venous pulse, and the electrocardiogram for a complete cardiac cycle in the dog.

of the heart, the ventricular wall thickness, and the magnitude and direction of the movements of various components of the heart.

In Fig. 24-12 the echocardiograph is positioned to depict movement of the anterior leaflet of the mitral valve. The posterior leaflet moves in a pattern that is a mirror image of the anterior leaflet, but in the projection shown in Fig. 24-12 the excursions appear much smaller. At point *D* the mitral valve opens, and during rapid filling *(D* to *E)* the anterior leaflet moves toward the ventricular septum. During the reduced filling phase *(E* to *F)* the valve leaflets float toward each other but the valve does not close. The ventricular filling contributed by atrial contraction *(F* to *A)* forces the leaflets apart, and a second approximation of the leaflets follows *(A* to *C)*. At point *C* the valve is closed by ventricular contraction. The valve leaflets, which bulge toward the atrium, stay pressed together during ventricular systole *(C* to *D)*.

Semilunar valves. The valves between the right ventricle and the pulmonary artery and between the left ventricle and the aorta consist of three cuplike cusps attached to the valve rings (see Figs. 24-10 and 24-11). At the end of the reduced ejection phase of ventricular systole, a brief reversal of blood flow occurs toward the ventricles (shown as a negative flow in the phasic aortic flow curve in Fig. 24-13) that snaps the cusps together and prevents regurgitation of blood into the ventricles. During ventricular systole the cusps do not lie back against the walls of the pulmonary artery and aorta but float in the bloodstream approximately midway between the vessel walls and their closed position. Behind the semilunar valves are small outpocketings of the pulmonary artery and aorta **(sinuses of Valsalva),** where eddy currents develop that tend to keep the valve cusps away from the vessel walls. The orifices of the right and left coronary arteries are located behind the right and the left cusps, respectively, of the aortic valve. Were it not for the presence of the sinuses of Valsalva and the eddy currents developed therein, the coronary ostia could be blocked by the valve cusps.

■ *The Pericardium*

The pericardium is an epithelized fibrous sac. It closely invests the entire heart and the cardiac portion of the great vessels and is reflected onto the cardiac surface as the epicardium. The sac normally contains a small amount of fluid, which provides lubrication for the continuous movement of the enclosed heart. The distensibility of the pericardium is small, so that it strongly resists a large, rapid increase in cardiac size. Because of this characteristic, the pericardium plays a role in preventing sudden overdistension of the chambers of the heart. However, in congenital absence of the pericardium or after its surgical removal, cardiac function is within physiological limits. Nevertheless, with the pericardium intact, an increase in diastolic pressure in one ventricle increases the pressure and decreases the compliance of the other ventricle. In contrast to an acute change in intracardiac pressure, progressive and sustained distension of the heart (as occurs in cardiac hypertrophy) or a slow progressive increase in pericardial fluid (as occurs in pericarditis with pericardial effusion) gradually stretches the intact pericardium.

Heart Sounds

Although the heart usually produces four sounds, only two are ordinarily audible through a stethoscope. With electronic amplification the less intense sounds can be detected and recorded graphically as a **phonocardiogram**. This means of registering heart sounds that may be inaudible to the human ear aids in delineating the precise timing of the heart sounds relative to other events in the cardiac cycle.

The first heart sound is initiated at the onset of ventricular systole (see Fig. 24-13) and consists of a series of vibrations of mixed, unrelated, low frequencies (a noise). It is the loudest and longest of the heart sounds, has a crescendo-decrescendo quality, and is heard best over the apical region of the heart. The tricuspid valve sounds are heard best in the fifth intercostal space just to the left of the sternum, and the mitral sounds are heard best in the fifth intercostal space at the cardiac apex.

The first heart sound is chiefly caused by the oscillation of blood in the ventricular chambers and vibration of the chamber walls. The vibrations are engendered in part by the abrupt rise of ventricular pressure with acceleration of blood back toward the atria, but primarily by sudden tension and recoil of the AV valves and adjacent structures with deceleration of the blood by closure of the AV valves. The vibrations of the ventricles and the contained blood are transmitted through surrounding tissues and reach the chest wall where they may be heard or recorded. The intensity of the first sound is a function of the force of ventricular contraction and of the distance between the valve leaflets. When the leaflets are farthest apart, either when the interval between atrial and ventricular systoles is prolonged and AV valve leaflets float apart or when ventricular systole immediately follows atrial systole, the first sound is loudest.

The second heart sound, which occurs with closure of the semilunar valves (see Fig. 24-13), is composed of higher frequency vibrations (higher pitch), is of shorter duration and lower intensity, and has a more snapping quality than the first heart sound. The second sound is caused by abrupt closure of the semilunar valves, which initiates oscillations of the columns of blood and the tensed vessel walls by the stretch and recoil of the closed valve. The second sound caused by closure of the pulmonic valve is heard best in the second thoracic interspace just to the left of the sternum, whereas that caused by closure of the aortic valve is heard best in the same intercostal space but to the right of the sternum. Conditions that bring about a more rapid closure of the semilunar valves, such as increases in pulmonary artery or aortic pressure (e.g., pulmonary or systemic hypertension), will increase the intensity of the second heart sound. In the adult the aortic valve sound is usually louder than the pulmonic, but in cases of pulmonary hypertension the reverse is often true.

A normal phonocardiogram taken simultaneously with an electrocardiogram is illustrated in Fig. 24-14. Note that the first sound, which starts just beyond the peak of the R wave, is composed of irregular waves and is of greater intensity and duration than the second sound, which appears at the end of the T wave. A third and fourth heart sound do not appear on this record.

The third heart sound, which is sometimes heard in children with thin chest walls or in patients with left ventricular failure, consists of a few low-intensity, low-frequency vibrations heard best in the region of the apex. It occurs in early diastole and is believed to be the result of vibrations of the ventricular walls caused by abrupt cessation of ventricular distension and deceleration of blood entering the ventricles. This occurs in overloaded hearts when the ventricular volume is very large and the ventricular walls are stretched to the point where distensibility abruptly decreases. A third heart sound in patients with heart disease is usually a grave sign.

A fourth, or atrial, sound, consisting of a few low-frequency oscillations, is occasionally heard in normal individuals. It is caused by oscillation of blood and cardiac chambers created by atrial contraction (see Fig. 24-13).

Because the onset and termination of right and left

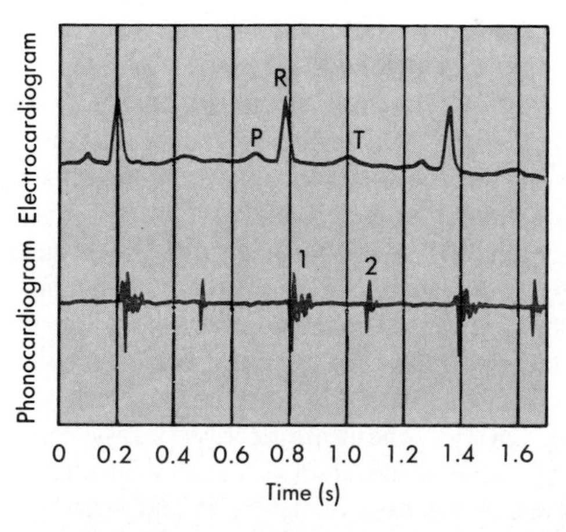

Fig. 24-14 Phonocardiogram illustrating the first and second heart sounds and their relationship to the P, R, and T waves of the electrogram. (Time lines = 0.04 sec.)

ventricular systoles are not precisely synchronous, differences in the time of vibration of the two AV valves or two semilunar valves can sometimes be detected with the stethoscope. Such asynchrony of valve vibrations, which may sometimes indicate abnormal cardiac function, is manifest as a **split sound** over the apex of the heart for the AV valves and over the base for the semilunar valves. The heart sounds also may be altered by deformities of the valves; **murmurs** may be produced, and the character of the murmur serves as an important guide in the diagnosis of valvular disease. When the third and fourth (atrial) sounds are accentuated, as occurs in certain abnormal conditions, triplets of sounds may occur, resembling the sound of a galloping horse. These **gallop rhythms** are essentially of two types—**presystolic gallop** caused by accentuation of the atrial sound, and **protodiastolic gallop** caused by accentuation of the third heart sound.

Cardiac Cycle

Ventricular Systole

Isovolumic contraction. The onset of ventricular contraction coincides with the peak of the R wave of the electrocardiogram and the initial vibration of the first heart sound. It is indicated on the ventricular pressure curve as the earliest rise in ventricular pressure after atrial contraction. The interval of time between the start of ventricular systole and the opening of the semilunar valves (when ventricular pressure rises abruptly) is termed **isovolumic contraction,** because ventricular volume is constant during this brief period (see Fig. 24-13).

The increment in ventricular pressure during isovolumic contraction is transmitted across the closed valves and is evident in Fig. 24-13 as a small oscillation on the aortic pressure curve. Isovolumic contraction also has been referred to as isometric contraction. However, some fibers shorten and others lengthen, as evidenced by changes in ventricular shape; it is therefore not a true isometric contraction.

Ejection. Opening of the semilunar valves marks the onset of the ejection phase, which may be subdivided into an earlier, shorter phase **(rapid ejection)** and a later, longer phase **(reduced ejection).** The rapid ejection phase is distinguished from the reduced ejection phase by (1) the sharp rise in ventricular and aortic pressures that terminates at the peak ventricular and aortic pressures, (2) a more abrupt decrease in ventricular volume, and (3) a greater aortic blood flow (see Fig. 24-13). The sharp decrease in the left atrial pressure curve at the onset of ejection results from the descent of the base of the heart and stretch of the atria. During the reduced ejection period, runoff of blood from the aorta to the periphery exceeds ventricular output and therefore aortic pressure declines. Throughout ventricular systole the blood returning to the atria produces a progressive increase in atrial pressure. Note that during approximately the first third of the ejection period left ventricular pressure slightly exceeds aortic pressure and flow accelerates (continues to increase), whereas during the last two thirds of ventricular ejection the reverse holds true. This reversal of the ventricular/aortic pressure gradient in the presence of continued flow of blood from the left ventricle to the aorta (caused by the momentum of the forward blood flow) is the result of the storage of potential energy in the stretched arterial walls, which produces a deceleration of blood flow into the aorta. The peak of the flow curve coincides in time with the point at which the left ventricular pressure curve intersects the aortic pressure curve during ejection. Thereafter, flow decelerates (continues to decrease) because the pressure gradient has been reversed.

With right ventricular ejection there is shortening of the free wall of the right ventricle (descent of the tricuspid valve ring) in addition to lateral compression of the chamber. However, with left ventricular ejection there is very little shortening of the base-to-apex axis, and ejection is accomplished chiefly by compression of the left ventricular chamber.

The effect of ventricular systole on left ventricular diameter is shown in an echocardiogram (see Fig. 24-12). During ventricular systole (see Fig. 24-12, *C* to *D*) the septum and the free wall of the left ventricle thicken and move closer to each other.

The venous pulse curve shown in Fig. 24-13 has been taken from a jugular vein; and the *c* wave is caused by impact of the adjacent common carotid artery and to some extent by transmission of a pressure wave produced by the abrupt closure of the tricuspid valve in early ventricular systole. Note that except for the *c* wave, the venous pulse closely follows the atrial pressure curve.

At the end of ejection, a volume of blood approximately equal to that ejected during systole remains in the ventricular cavities. This **residual volume** is fairly constant in normal hearts, smaller with increased heart rate or reduced outflow resistance, and larger when the opposite conditions prevail. An increase in myocardial contractility may decrease residual volume and increase stroke volume and ejection fraction, especially in the depressed heart. With severely hypodynamic and dilated hearts, as in **heart failure,** the residual volume can become many times greater than the stroke volume. In addition to serving as a small adjustable blood reservoir, the residual volume to a limited degree can permit transient disparities between the outputs of the two ventricles.

Ventricular Diastole

Isovolumic relaxation. Closure of the aortic valve produces the incisura on the descending limb of the aortic pressure curve and the second heart sound (with some vibrations evident on the atrial pressure curve) and marks the end of ventricular systole. The period between closure of the semilunar valves and opening of the AV valves is termed **isovolumic** (or **isometric**) **relaxation** and is characterized by a precipitous fall in ventricular pressure without a change in ventricular volume.

Rapid filling phase. The major part of the ventricular filling occurs immediately on opening of the AV valves when the blood that had returned to the atria during the previous ventricular systole is abruptly released into the relaxing ventricles. This period of ventricular filling is called the **rapid filling phase.** In Fig. 24-13 the onset of the rapid filling phase is indicated by the decrease in left ventricular pressure below left atrial pressure, resulting in the opening of the mitral valve. The rapid flow of blood from atria to relaxing ventricles produces a decrease in atrial and ventricular pressures and a sharp increase in ventricular volume.

The decrease in pressure from the peak of the *v* wave of the venous pulse is caused by transmission of the pressure decrease incident to the abrupt transfer of blood from the right atrium to the right ventricle with opening of the tricuspid valve. Elastic recoil of the previous ventricular contraction may aid in drawing blood into the relaxing ventricle when residual volume is small, especially when ventricular contractility is enhanced, as produced by catecholamines. However, this mechanism probably does not play a significant role in ventricular filling under most normal conditions.

Diastasis. The rapid filling phase is followed by a phase of slow filling, called **diastasis.** During diastasis, blood returning from the periphery flows into the right ventricle and blood from the lungs into the left ventricle. This small, slow addition to ventricular filling is indicated by a gradual rise in atrial, ventricular, and venous pressures and in ventricular volume (see Fig. 24-13).

Pressure-volume relationship. The changes in left ventricular pressure and volume throughout the cardiac cycle are summarized in Fig. 24-15. The element of time is not considered in this **pressure-volume loop.** Diastolic filling starts at *A* and terminates at *C,* when the mitral valve closes. The initial decrease in left ventricular pressure *(A* to *B),* despite the rapid inflow of blood from the atrium, is caused by progressive ventricular relaxation and distensibility. During the remainder of diastole *(B* to *C)* the increase in ventricular pressure reflects ventricular filling and the passive elastic characteristics of the ventricle. Note that only a small increase in pressure occurs with the increase in ventricular volume during diastole *(B* to *C).* With isovolumic contraction *(C* to *D)* a steep rise in pressure occurs with no change in ventricular volume. At *D* the aortic valve opens and during the first phase of ejection (rapid ejection, *D* to *E*), the large reduction in volume is associated with a continued but less steep increase in ventricular pressure than that which occurred during isovolumic contraction. This volume reduction is followed by reduced ejection *(E* to *F)* and a small decrease in ventricular pressure. The aortic valve closes at *F;* this event is followed by isovolumic relaxation *(F* to *A),* which is characterized by a sharp drop in pressure and no change in volume. The mitral valve opens at *A* to complete one cardiac cycle.

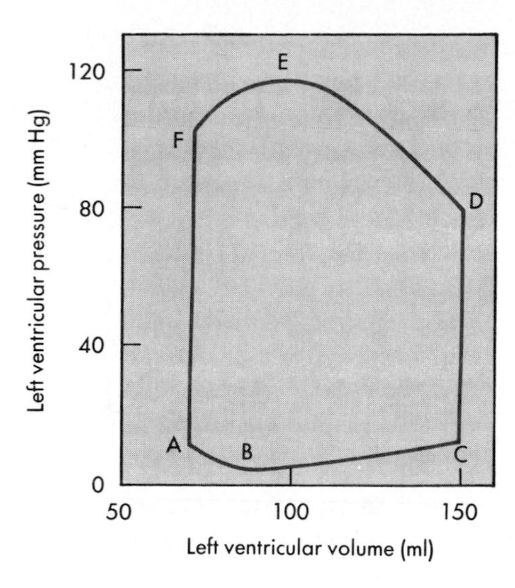

Fig. 24-15 Pressure-volume loop of the left ventricle for a single cardiac cycle *(ABCDEF).*

With changes in preload, afterload, or myocardial contractility, the contours of the pressure volume loops are altered (Fig. 24-16). Increases in preload by greater filling of the ventricles are depicted in Fig. 24-16, *A,* as an extension of the diastolic volume curve upward and to the right (C to C′ and to C″). This greater diastolic filling results in a larger stroke volume (D′ and D″ to F) but peak pressure (E) is unchanged because the afterload (aortic pressure) is unchanged. When the afterload is increased (higher aortic pressure) at a constant preload (Fig. 24-16, *B*), a greater ventricular pressure is reached (D, E, F to D′, E′, F′ and D″, E″, F″). However, less blood is ejected from the ventricle (A to A′ and to A″) (smaller stroke volume and hence a greater residual volume) because of the increased opposition to left ventricular outflow by the high aortic pressure. An increase in contractility without change in preload or afterload, as produced by cardiac sympathetic nerve stimulation or by administration of catecholamines (Fig. 24-16, *C*), results in a greater pressure development and a greater emptying of the ventricle (larger stroke volume and smaller residual volume). This is shown in Fig. 24-16, *C* as a shift from D, E, F, A to D′, E′, F′, A′, and to D″, E″, F″, A″.

Atrial systole. The onset of atrial systole occurs soon after the beginning of the P wave of the electrocardiogram (curve of atrial depolarization), and the transfer of blood from atrium to ventricle made by the peristalsis-like wave of atrial contraction completes the period of ventricular filling. Atrial systole is responsible for the

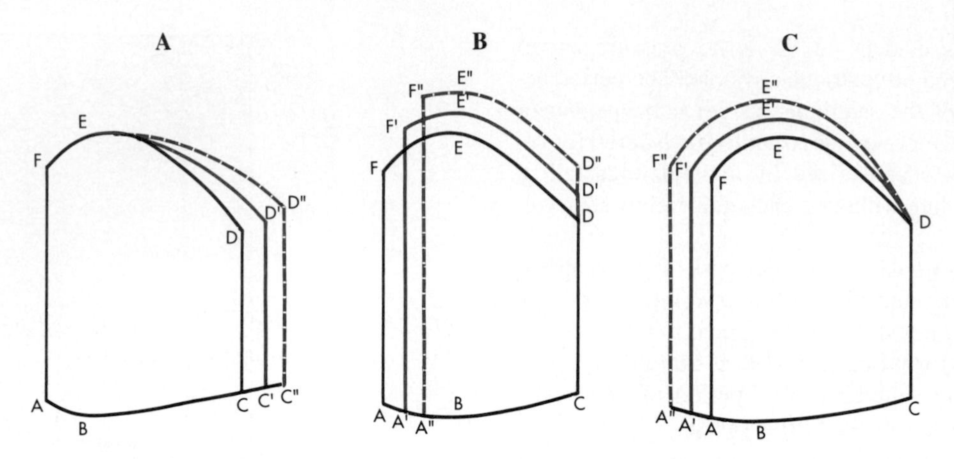

■ **Fig. 24-16** The effects of increases in preload **A,** afterload **B,** and contractility **C,** on the pressure-volume relationships of the left ventricle during a complete cardiac cycle.

small increases in atrial, ventricular, and venous (*a* wave) pressures, as well as in ventricular volume shown in Fig. 24-13. Throughout ventricular diastole, atrial pressure barely exceeds ventricular pressure, indicating a low-resistance pathway across the open AV valves during ventricular filling. A few small vibrations produced by atrial systole constitute the fourth, or atrial, heart sound.

Because there are no valves at the junctions of the venae cavae and right atrium or of the pulmonary veins and left atrium, atrial contraction can force blood in both directions. Actually, little blood is pumped back into the venous tributaries during the brief atrial contraction, mainly because of the inertia of the inflowing blood.

Atrial contraction is not essential for ventricular filling, as can be observed in atrial fibrillation or complete heart block. However, its contribution is governed to a great extent by the heart rate and the structure of the AV valves. At slow heart rates, filling practically ceases toward the end of diastasis, and atrial contraction contributes little additional filling. During tachycardia diastasis is abbreviated and the atrial contribution can become substantial, especially if it occurs immediately after the rapid filling phase when the AV pressure gradient is maximal. Should tachycardia become so great that the rapid filling phase is encroached on, atrial contraction assumes great importance in rapidly propelling blood into the ventricle during this brief period of the cardiac cycle. Of course, if the period of ventricular relaxation is so brief that filling is seriously impaired, even atrial contraction cannot prevent inadequate ventricular filling. The consequent reduction in cardiac output may result in syncope. Obviously, if atrial contraction occurs simultaneously with ventricular contraction, no atrial contribution to ventricular filling can occur. In certain disease states the AV valves may be markedly narrowed **(stenotic).** Under such conditions atrial contraction may play a much more important role in ventricular filling than it does in the normal heart.

Ventricular contraction has been shown to aid indirectly in right ventricular filling by its effect on the right atrium. Descent of the base of the heart stretches the right atrium downward, and pressure measurements indicate a sharp reduction in right atrial pressure associated with acceleration of blood flow in the venae cavae toward the heart. Enhancement of venous return by ventricular systole provides an additional supply of atrial blood for ventricular filling during the subsequent rapid filling phase of diastole. However, this mechanism is probably of little physiological importance, except possibly at rapid heart rates.

■ *Measurement of Cardiac Output*

■ *Fick Principle*

In 1870, the German physiologist, Adolph Fick, contrived the first method for measuring cardiac output in intact animals and people. The basis for this method, called the **Fick principle,** is simply an application of the law of conservation of mass. It is derived from the fact that the quantity of oxygen (O_2) delivered to the pulmonary capillaries via the pulmonary artery plus the quantity of O_2 that enters the pulmonary capillaries from the alveoli must equal the quantity of O_2 that is carried away by the pulmonary veins.

The Fick principle is depicted schematically in Fig. 24-17. The rate, q_1, of O_2 delivery to the lungs equals the O_2 concentration in the pulmonary arterial blood, $[O_2]_{pa}$, times the pulmonary arterial blood flow, Q, which equals the cardiac output; that is,

$$q_1 = Q[O_2]_{pa} \quad (1)$$

Let q_2 be the net rate of O_2 uptake by the pulmonary capillaries from the alveoli. At equilibrium, q_2 equals

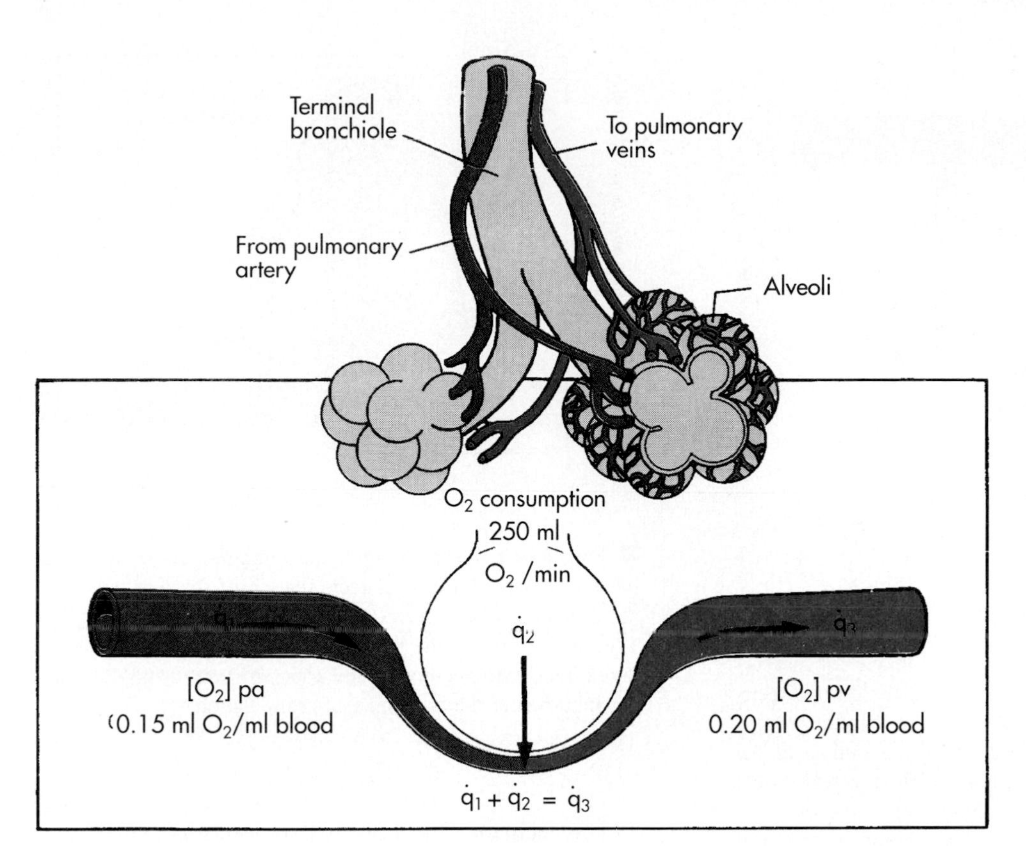

■ **Fig. 24-17** Schema illustrating the Fick principle for measuring cardiac output. The change in color from pulmonary artery to pulmonary vein represents the change in color of the blood as venous blood becomes fully oxygenated.

the **O_2 consumption** of the body. The rate, q_3, at which O_2 is carried away by the pulmonary veins equals the O_2 concentration in the pulmonary venous blood, $[O_2]_{pv}$, times the total pulmonary venous flow, which is virtually equal to the pulmonary arterial blood flow, Q; that is,

$$q_3 = Q[O_2]_{pv} \tag{2}$$

From conservation of mass,

$$q_1 + q_2 = q_3 \tag{3}$$

Therefore

$$Q[O_2]_{pa} + q_2 = Q[O_2]_{pv} \tag{4}$$

Solving for cardiac output,

$$Q = q_2/([O_2]_{pv} - [O_2]_{pa}) \tag{5}$$

Equation 5 is the statement of the Fick principle.

In the clinical determination of cardiac output, O_2 consumption is computed from measurements of the volume and O_2 content of expired air over a given interval of time. Because the O_2 concentration of peripheral arterial blood is essentially identical to that in the pulmonary veins, $[O_2]_{pv}$ is determined on a sample of peripheral arterial blood withdrawn by needle puncture. Pulmonary arterial blood actually represents mixed systemic venous blood. Samples for O_2 analysis are obtained from the pulmonary artery or right ventricle through a catheter. In the past a stiff catheter was used, and it had to be introduced into the pulmonary artery under fluoroscopic guidance. Now, a very flexible catheter with a small balloon near the tip can be inserted into a peripheral vein. As the tube is advanced, it is carried by the flowing blood toward the heart. By following the pressure changes, the physician is able to advance the catheter tip into the pulmonary artery without the aid of fluoroscopy.

An example of the calculation of cardiac output in a normal, resting adult is illustrated in Fig. 24-17. With an O_2 consumption of 250 ml/min, an arterial (pulmonary venous) O_2 content of 0.20 ml O_2/ml blood, and a mixed venous (pulmonary arterial) O_2 content of 0.15 ml O_2/ml blood, the cardiac output would equal 250 ÷ (0.20 − 0.15) = 5000 ml/min.

The Fick principle is also used for estimating the O_2 consumption of organs in situ, when blood flow and the O_2 contents of the arterial and venous blood can be determined. Algebraic rearrangement reveals that O_2 consumption equals the blood flow times the arteriovenous O_2 concentration difference. For example, if the blood flow through one kidney is 700 ml/min, arterial O_2 content is 0.20 ml O_2/ml blood, and renal venous O_2 content is 0.18 ml O_2/ml blood, then the rate of O_2 consumption by that kidney must be 700 (0.20 − 0.18) = 14 ml O_2/min.

■ *Indicator Dilution Techniques*

The indicator dilution technique for measuring cardiac output is also based on the law of conservation of mass and is illustrated by the model in Fig. 24-18. Let a liq-

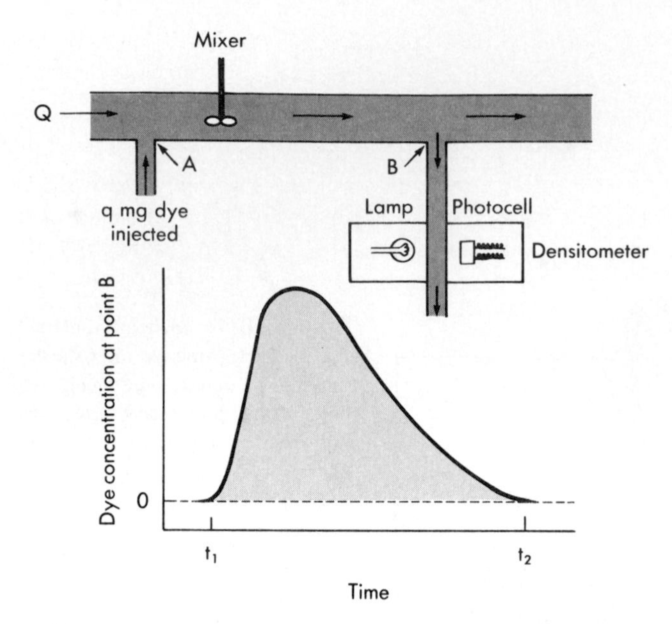

■ **Fig. 24-18** The indicator dilution technique for measuring cardiac output. In this model, in which there is no recirculation, q mg of dye are injected instantaneously at point *A* into a stream flowing at Q ml/min. A mixed sample of the fluid flowing past point *B* is withdrawn at a constant rate through a densitometer. The resulting dye concentration curve at point *B* has the configuration shown in the lower section of the figure.

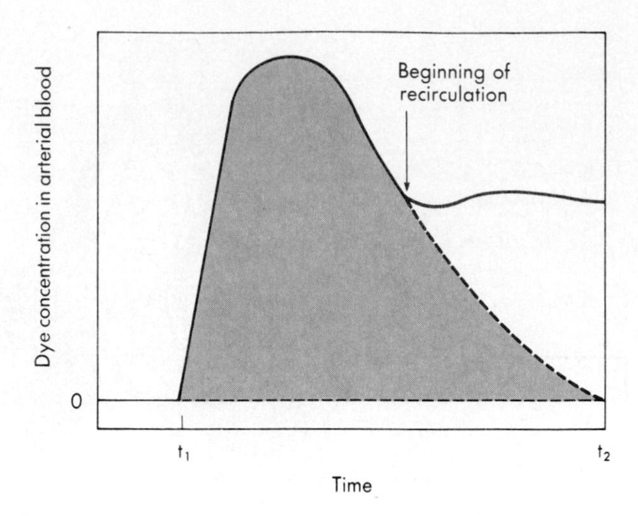

■ **Fig. 24-19** Typical dye concentration curve recorded from a human. Because of recirculation of the dye, the concentration does not return to 0, as in the model in Fig. 24-18. The dashed line on the descending limb represents the semilogarithmic extrapolation of the upper portion of the descending limb before the beginning of recirculation.

uid flow through a tube at a rate of Q ml/sec, and let q mg of dye be injected as a slug into the stream at point *A*. Let mixing occur at some point downstream. If a small sample of liquid is continually withdrawn from point *B* farther downstream and passed through a densitometer, a curve of the dye concentration, c, may be recorded as a function of time, t, as shown in the lower half of the figure.

If no dye is lost between points *A* and *B*, the amount of dye, q, passing point *B* between times t_1 and t_2 will be

$$q = \bar{c}Q(t_2 - t_1) \tag{6}$$

where $\bar{c}$ is the mean concentration of dye. The value of $\bar{c}$ may be computed by dividing the area of the dye concentration by the duration $(t_2 - t_1)$ of that curve; that is

$$\bar{c} = \int_{t_1}^{t_2} c\,dt/(t_2 - t_1) \tag{7}$$

Substituting this value of $\bar{c}$ into equation 6, and solving for Q yields

$$Q = \frac{q}{\int_{t_1}^{t_2} c\,dt} \tag{8}$$

Thus flow may be measured by dividing the amount of indicator injected upstream by the area under the downstream concentration curve.

This technique has been widely used to estimate cardiac output in humans. A measured quantity of some indicator (a dye or isotope that remains within the circulation) is injected rapidly into a large central vein or into the right side of the heart through a catheter. Arterial blood is continuously drawn through a detector (densitometer or isotope rate counter), and a curve of indicator concentration is recorded as a function of time.

Because some of the indicator recirculates and reappears at the site of arterial withdrawal before the entire curve is inscribed, the concentration curve is not as simple as that shown in Fig. 24-18. Instead, on the downstroke of the curve a secondary increase in concentration (Fig. 24-19) appears as the recirculated dye becomes mixed with the last portions of dye still undergoing its primary passage past the site of withdrawal. To compute the area under the concentration curve, the downslope of the curve beyond the beginning of recirculation is extrapolated to zero concentration (dashed line, Fig. 24-19). The extrapolation, of course, introduces some error into the estimation of cardiac output.

Presently the most popular indicator dilution technique is **thermodilution.** The indicator is cold saline. The temperature and volume of the saline are measured accurately before injection. A flexible catheter is introduced into a peripheral vein and advanced so that the tip lies in the pulmonary artery. A small thermistor at the catheter tip records the changes in temperature. The opening in the catheter lies a few inches proximal to the tip. When the tip is in the pulmonary artery, the opening lies in or near the right atrium. The cold saline is injected rapidly into the right atrium through the catheter. The resultant change in temperature downstream is recorded by the thermistor in the pulmonary artery.

The thermodilution technique has the following advantages: (1) an arterial puncture is not necessary; (2) the small volumes of saline used in each determination are innocuous, allowing repeated determinations to be made; and (3) recirculation is negligible. Temperature equilibration takes place as the cooled blood flows through the pulmonary and systemic capillary beds, before it flows by the thermistor in the pulmonary artery the second time. Therefore the curve of temperature change resembles that shown in Fig. 24-18, and the extrapolation errors are averted.

Summary

1. An increase in myocardial fiber length, as occurs with an augmented ventricular filling during diastole (preload), produces a more forceful ventricular contraction. This relation between fiber length and strength of contraction is known as Starling's law of the heart.

2. Although the myocardium is made up of individual cells with discrete membrane boundaries, the cardiac myocytes that comprise the ventricles contract almost in unison, as do those of the atria. The myocardium functions as a syncytium with an all-or-none response to excitation. Cell-to-cell conduction occurs through gap junctions that connect the cytosol of adjacent cells.

3. On excitation, voltage gated calcium channels open to admit extracellular Ca^{++} into the cell. The influx of Ca^{++} triggers the release of Ca^{++} from the sarcoplasmic reticulum. The elevated intracellular Ca^{++} produces contraction of the myofilaments. Relaxation is accomplished by restoration of the resting cytosolic Ca^{++} level by pumping it back into the sarcoplasmic reticulum and by exchanging it for extracellular Na^{+} across the sarcolemma.

4. Velocity and force of contraction are functions of the intracellular concentration of free Ca ions. Force and velocity are inversely related, so that with no load, force is negligible and velocity is maximal. In an isometric contraction, where no external shortening occurs, force is maximal and velocity is zero.

5. In ventricular contraction the preload is the stretch of the fiber by the blood during ventricular filling and the afterload is the aortic pressure against which the left ventricle ejects the blood.

6. Contractility is an expression of cardiac performance at a given preload and afterload. Contractility is increased by stimulation of cardiac sympathetic nerves and decreased by stimulation of the cardiac branches of the vagus nerves.

7. Simultaneous recording of the left atrial, left ventricular and aortic pressures, ventricular volume, heart sounds, and electrocardiogram graphically portray the sequential and related electrical and cardiodynamic events throughout a cardiac cycle.

8. As illustrated by pressure-volume loops of the left ventricle, an increment in preload increases stroke volume, an increase in afterload results in an increase in peak pressure, but a decrease in stroke volume and an increase in contractility produces a greater peak pressure and a greater stroke volume.

9. Cardiac output can be determined, according to the Fick principle, by measuring the oxygen consumption of the body (MVO_2) and by dividing it by the difference between the oxygen content of arterial $[O_2]_{pa}$ and mixed venous $[O_2]_{pv}$ blood: cardiac output $= MVO_2/([O_2]_{pv} - [O_2]_{pa})$. It can also be measured by dye dilution or thermodilution techniques. The greater the cardiac output, the greater the dilution of the injected dye or cold saline by the arterial blood.

Bibliography

Journal articles

Alpert NR, Hamrell BB, Mulieri LA: Heart muscle mechanics, *Ann Rev Physiol* 41:521, 1979.

Bers DM, Lederer WJ, Berlin JR: Intracellular Ca transients in rat cardiac myocytes: role of Na-Ca exchange in excitation-contraction coupling, *Am J Physiol* 258:C944, 1990.

Blaustein MP: Sodium/calcium exchange and the control of contractility in cardiac muscle and vascular smooth muscle, *J Cardiovas Pharmacol* 12:S56, 1988.

Brutsaert DL, Sys SU: Relaxation and diastole of the heart, *Physiol Rev* 69:1228, 1989.

Carafoli E: The homeostasis of calcium in heart cells, *J Mol Cell Cardiol* 17:203, 1985.

Chapman RA: Control of cardiac contractility at the cellular level, *Am J Physiol* 245:H535, 1983.

Fabiato A, Fabiato F: Calcium and cardiac excitation contraction coupling, *Ann Rev Physiol* 41:473, 1979.

Gilbert JC, Glantz SA: Determinants of left ventricular filling and of the diastolic pressure-volume relation, *Circ Res* 64:827, 1989.

Jewell BR: A reexamination of the influence of muscle length on myocardial performance, *Circ Res* 40:221, 1977.

Katz AM: Interplay between inotropic and lusitropic effects of cyclic adenosine monophosphate on the myocardial cell, *Circ* 82:1-7, 1990.

Katz AM: Cyclic adenosine monophosphate effects on the myocardium: a man who blows hot and cold with one breath, *J Am Coll Cardiol* 2:143, 1983.

Sagawa K: The ventricular pressure-volume diagram revisited, *Circ Res* 43:677, 1978.

Sonnenblick E: Force-velocity relations in mammalian heart muscle, *Am J Physiol* 202:931, 1962.

Streeter DD Jr et al: Fiber orientation in the canine left ventricle during diastole and systole, *Circ Res* 24:339, 1969.

Books and monographs

Brady AJ: *Mechanical properties of cardiac fibers*. In *Handbook of physiology*, section 2: *The cardiovascular system—*

the heart, vol 1, Bethesda, Md, 1979, American Physiological Society.

Braunwald E, Ross J Jr, Sonnenblick EH: *Mechanisms of contraction of the normal and failing heart,* ed 2, Boston, 1976, Little, Brown & Co.

Katz AM: *Role of calcium in contraction of cardiac muscle.* In Stone PH, Antman EM, editors: *Calcium channel blocking agents in the treatment of cardiovascular disorders,* Mt. Kisco, New York, 1983, Futura Publishing.

Katz AM, Takenaka H, Watras J: *The sarcoplasmic reticulum.* In Fozzard HA et al, editors: *The heart and cardiovascular system,* New York, 1986, Raven Press.

Parmley WW, Talbot L: *Heart as a pump.* In *Handbook of physiology;* section 2: *The cardiovascular system—the heart, vol 1,* Bethesda, Md, 1979, American Physiological Society.

Ruegg JC: *Calcium in muscle activation,* Springer-Verlag, Heidelberg, 1988.

Sheu SS, Blaustein MP: *Sodium/calcium exchange and control of cell calcium and contractility in cardiac muscle and vascular smooth muscle.* In Fozzard HA, et al, editors: *The heart and cardiovascular system,* New York, 1991, Raven Press.

Sommer JR, Johnson EA: *Ultrastructure of cardiac muscle.* In *Handbook of physiology,* section 2: *The cardiovascular system—the heart,* vol 1, Bethesda, Md, 1979, American Physiological Society.

CHAPTER

25

Regulation of the Heartbeat

The quantity of blood pumped by the heart each minute (i.e., the **cardiac output,** CO) may be varied by changing the frequency of its beats (i.e., the **heart rate,** HR) or the volume ejected per stroke (i.e., the **stroke volume,** SV). Cardiac output is the product of heart rate and stroke volume; i.e.,

$$CO = HR \times SV$$

A discussion of the control of cardiac activity may therefore be subdivided into a consideration of the regulation of pacemaker activity and the regulation of myocardial performance. However, in the intact organism, a change in the behavior of one of these features of cardiac activity almost invariably alters the other.

Certain local factors, such as temperature changes and tissue stretch, can affect the discharge frequency of the SA node. However, the principal control of heart rate is relegated to the autonomic nervous system, and the discussion will be restricted to this aspect of heart rate control. Relative to myocardial performance, intrinsic and extrinsic factors will be considered.

Nervous Control of Heart Rate

In normal adults the average heart rate at rest is approximately 70 beats per minute, but it is significantly greater in children. During sleep the heart rate diminishes by 10 to 20 beats per minute, but during emotional excitement or muscular activity it may accelerate to rates considerably above 100. In well-trained athletes at rest the rate is usually only about 50 beats per minute.

The SA node is usually under the tonic influence of both divisions of the autonomic nervous system. The sympathetic system enhances automaticity, whereas the parasympathetic system inhibits it. Changes in heart rate usually involve a reciprocal action of the two divisions of the autonomic nervous system. Thus an increased heart rate is produced by a diminution of parasympathetic activity and concomitant increase in sympathetic activity; deceleration is usually achieved by the opposite mechanisms. Under certain conditions the heart rate may change by selective action of just one division of the autonomic nervous system, rather than by reciprocal changes in both divisions.

Ordinarily, in healthy, resting individuals parasympathetic tone predominates. Abolition of parasympathetic influences by the administration of atropine usually increases heart rate substantially, whereas abrogation of sympathetic effects by the administration of propranolol usually decreases heart rate only slightly (Fig. 25-1). When both divisions of the autonomic nervous system are blocked, the heart rate of young adults averages about 100 beats per minute. The rate that prevails after complete autonomic blockade is called the **intrinsic heart rate.**

Parasympathetic Pathways

The cardiac parasympathetic fibers originate in the medulla oblongata, in cells that lie in the **dorsal motor nucleus of the vagus** or in the **nucleus ambiguus** (see Chapter 15). The precise location varies from species to species. Centrifugal vagal fibers pass inferiorly through the neck close to the common carotid arteries and then through the mediastinum to synapse with postganglionic cells located on the epicardial surface or within the walls of the heart itself. Most of the cardiac ganglion cells are located near the SA node and AV conduction tissue.

The right and left vagi are distributed differentially to the various cardiac structures. The right vagus nerve affects the SA node predominantly. Stimulation slows SA nodal firing or may even stop it for several seconds. The left vagus nerve mainly inhibits AV conduction tissue to produce various degrees of AV block. However, the efferent vagal fibers overlap, such that left vagal stimulation also depresses the SA node and right vagal stimulation impedes AV conduction.

The SA and AV nodes are rich in **cholinesterase.** Hence the effects of any given vagal impulse are ephemeral because the acetylcholine released at the

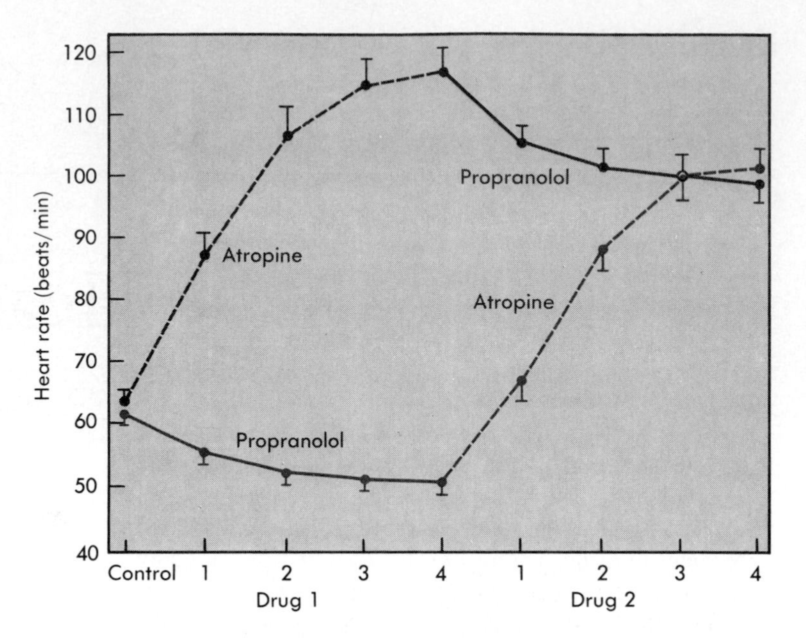

■ **Fig. 25-1** The effects of four equal doses of atropine (0.04 mg/kg total) and of propranolol (0.2 mg/kg total) on the heart rate of 10 healthy young men (mean age, 21.9 years). In half of the trials, atropine was given first *(top curve);* in the other half, propranolol was given first *(bottom curve).* (Redrawn from Katona PG et al: *J Appl Physiol* 52:1652, 1982.)

nerve terminals is rapidly hydrolyzed. Furthermore, the effects of vagal activity on SA and AV nodal function have a very short latency (about 50 to 100 msec), because the released acetylcholine activates special K^+ channels in the cardiac cells. Opening of these channels is so prompt because it does not require the operation of a second messenger system, such as the adenylyl cyclase system. The combination of the brief latency and the rapid decay of the response (because of the abundance of cholinesterase) provides the potential for the vagus nerves to exert a beat-by-beat control of SA and AV nodal function.

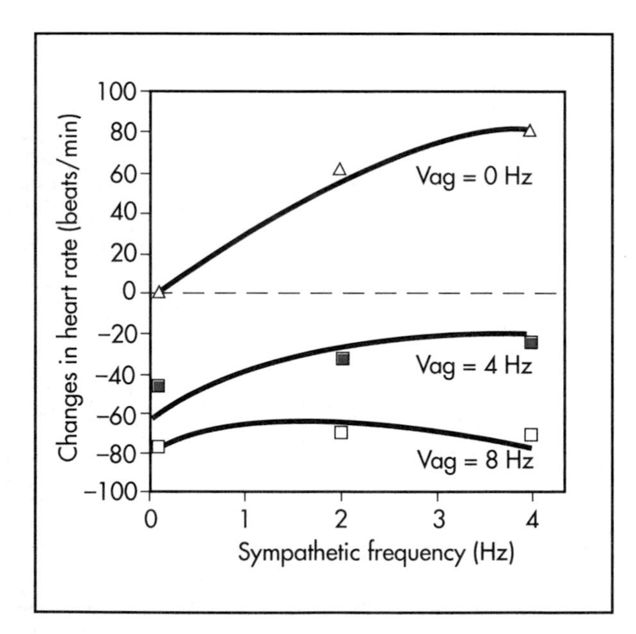

■ **Fig. 25-2** The changes in heart rate in an anesthetized dog when the vagus and cardiac sympathetic nerves were simulated simultaneously. The sympathetic nerves were stimulated at 0, 2, and 4 Hz; the vagus nerves at 0, 4, and 8 Hz. The symbols represent the observed changes in heart rate; the curves were derived from the computed regression equation. (Modified from Levy MN, Zieske H: *J Appl Physiol* 27:465, 1969.)

Parasympathetic influences preponderate over sympathetic effects at the SA node, as shown in Fig. 25-2. As the frequency of sympathetic stimulation in an anesthetized dog was increased from 0 to 4 Hz, the heart rate increased by about 80 beats per minute in the absence of vagal stimulation (Vag = 0 Hz). However, when the vagi were stimulated at 8 Hz, increasing the sympathetic stimulation frequency from 0 to 4 Hz had a negligible influence on heart rate.

■ *Sympathetic Pathways*

The cardiac sympathetic fibers originate in the **intermediolateral columns** of the upper five or six thoracic and lower one or two cervical segments of the spinal cord (see Chapter 15). They emerge from the spinal column through the white communicating branches and enter the paravertebral chains of ganglia. The anatomical details of the sympathetic innervation of the heart vary among mammalian species; the innervation has been elaborated in detail in the dog (Fig. 25-3). The synapses between the preganglionic and postganglionic neurons synapse mainly in the stellate or caudal cervical ganglia, depending on the species. The caudal (or middle) cervical ganglia lie close to the vagus nerves in the superior portion of the mediastinum. Sympathetic and parasympathetic fibers then join to form a complex plexus of mixed efferent nerves to the heart (see Fig. 25-3).

The postganglionic cardiac sympathetic fibers approach the base of the heart along the adventitial sur-

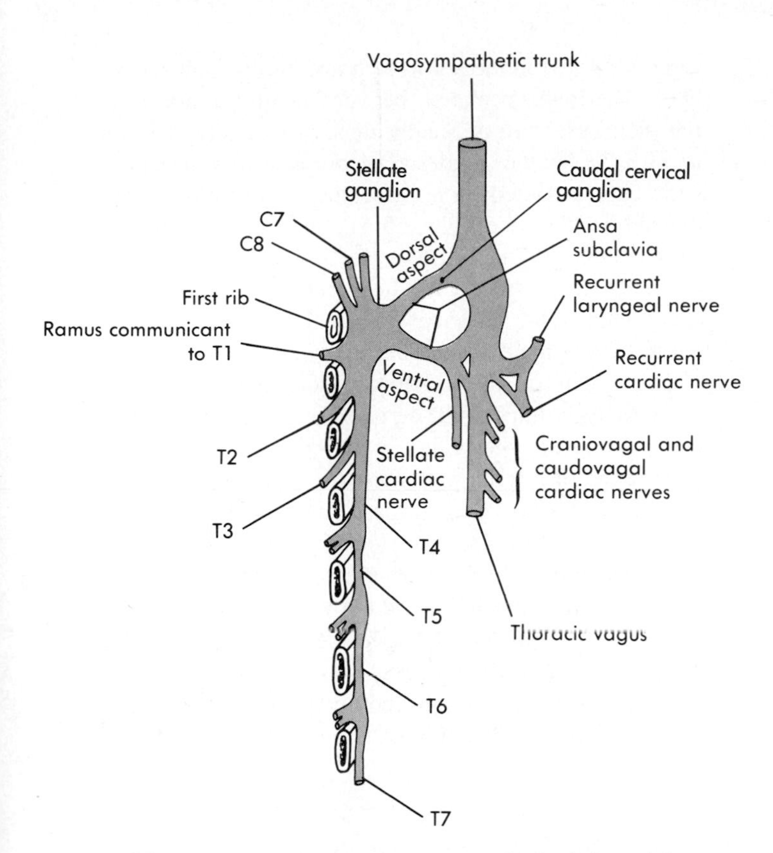

■ **Fig. 25-3** Upper thoracic sympathetic chain and the cardiac autonomic nerves on the right side in the dog. (Modified from Mizeres NJ: *Anat Rec* 132:261, 1958.)

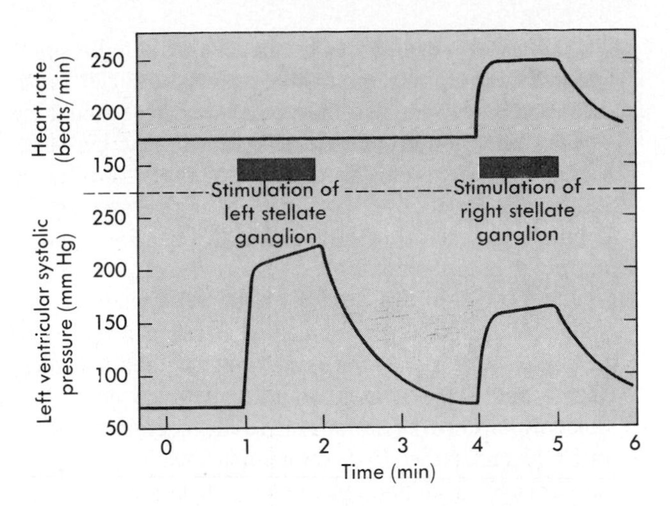

■ **Fig. 25-4** In the dog, stimulation of the left stellate ganglion has a greater effect on ventricular contractility than does right-sided stimulation, but it has a lesser effect on heart rate. In this example, traced from an original record, left stellate ganglion stimulation had no detectable effect at all on heart rate but had a considerable effect on ventricular performance in an isovolumic left ventricle preparation. (From Levy MN: unpublished tracing.)

face of the great vessels. On reaching the base of the heart, these fibers are distributed to the various chambers as an extensive epicardial plexus. They then penetrate the myocardium, usually accompanying the coronary vessels. The adrenergic receptors in the nodal regions and in the myocardium are predominantly of the beta type; (i.e., they are responsive to beta-adrenergic agonists, such as **isoproterenol,** and are inhibited by beta-adrenergic blocking agents, such as **propranolol**).

As with the vagus nerves, there is a differential distribution of the left and right sympathetic fibers. In the dog, for example, the fibers on the left side have more pronounced effects on myocardial contractility than do fibers on the right side, whereas the fibers on the left side have much less effect on heart rate than do the fibers on the right side (Fig. 25-4). In some dogs left cardiac sympathetic nerve stimulation may not affect heart rate at all. This bilateral asymmetry probably also exists in humans. In a group of patients right stellate ganglion blockade caused a mean reduction in heart rate of 14 beats per minute, whereas left-sided blockade decreased heart rate by only 2 beats per minute.

It is evident from Fig. 25-4 that the effects of sympathetic stimulation decay very gradually after the cessation of stimulation, in contrast to the abrupt termination of the response after vagal activity (not shown). Most of the norepinephrine released during sympathetic stimulation is taken up again by the nerve terminals, and much of the remainder is carried away by the bloodstream. These processes are relatively slow. Furthermore, at the beginning of sympathetic stimulation, the facilitatory effects on the heart attain steady state values much more slowly than do the inhibitory effects of vagal stimulation. Recent studies suggest that some of the ionic channels (e.g., the Ca^{++} channels) that regulate cardiac activity may be coupled to the beta-adrenergic receptors *directly* through G proteins, as well as *indirectly* through second messenger systems (mainly the adenylyl cyclase system) (see Chapter 5). Even though the direct coupling may afford a more rapid cardiac response than does the slower second messenger system, sympathetic activity alters heart rate and AV conduction much more slowly than does vagal activity. Therefore, vagal activity can exert beat-by-beat control of cardiac function, whereas sympathetic activity cannot.

■ *Control by Higher Centers*

Dramatic alterations in cardiac rate, rhythm, and contractility have been induced experimentally by stimulation of various regions of the brain (see Chapter 15). In the cerebral cortex the centers regulating cardiac function are located mostly in the anterior half of the brain, principally in the frontal lobe, the orbital cortex, the motor and premotor cortex, the anterior part of the temporal lobe, the insula, and the cingulate gyrus. In the thalamus, tachycardia may be induced by stimulation of the midline, ventral, and medial groups of nuclei. Variations in heart rate also may be evoked by stimulating the posterior and posterolateral regions of the hypothalamus. Stimuli applied to the H_2 fields of Forel in the diencephalon elicit a variety of cardiovascular responses, including tachycardia; such changes simulate closely those observed during muscular exercise. Undoubtedly the cortical and diencephalic centers are responsible for initiating the cardiac reactions that occur during excitement, anxiety, and other emotional states. The hypothalamic centers are also involved in the cardiac response to alterations in environmental temperature. Recent studies have shown that localized temperature changes in the preoptic anterior hypothalamus alter heart rate and peripheral resistance.

Stimulation of the parahypoglossal area of the medulla produces a reciprocal activation of cardiac sympathetic and inhibition of cardiac parasympathetic pathways. In certain dorsal regions of the medulla, distinct cardiac accelerator and augmentor sites have been detected in animals with transected vagi. Stimulation of accelerator sites increases heart rate, whereas stimulation of augmentor sites increases cardiac contractility. The accelerator regions were found to be more abundant on the right and the augmentor sites more prevalent on the left. A similar distribution also exists in the hypothalamus. It appears, therefore, that for the most part the sympathetic fibers descend the brainstem ipsilaterally.

Baroreceptor Reflex

Acute changes in arterial blood pressure reflexly elicit inverse changes in heart rate (Fig. 25-5) via the baroreceptors located in the aortic arch and carotid sinuses (p. 486). The inverse relation between heart rate and arterial blood pressure is usually most pronounced over an intermediate range of arterial blood pressures. In an experiment conducted on a conscious, chronically instrumented monkey (see Fig. 25-5), this range varied between about 70 and 160 mm Hg. Below the intermediate range of pressures, the heart rate maintains a constant, high value, whereas above this pressure range, the heart rate maintains a constant, low value.

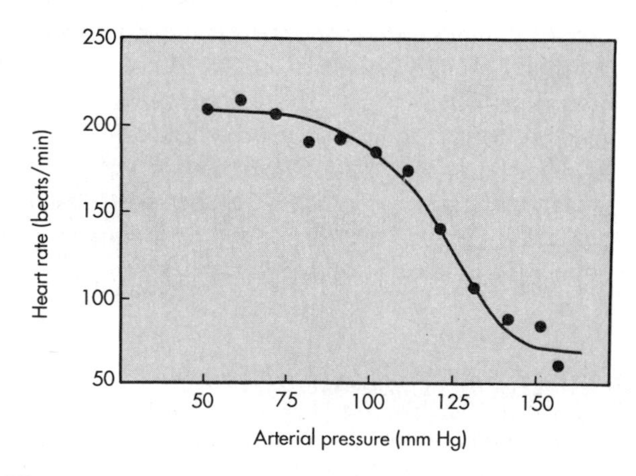

Fig. 25-5 Heart rate as a function of mean arterial pressure in a group of five conscious, chronically instrumented monkeys. The mean control arterial pressure was 114 mm Hg. Pressure was increased above the control value by infusing phenylephrine and was decreased below the control value by infusing nitroprusside. (Adapted from Cornish KG et al: *Am J Physiol* 257:R595, 1989.)

The effects of changes in carotid sinus pressure on the activity in the cardiac autonomic nerves of an anesthetized dog are shown in Fig. 25-6. Over an intermediate range of arterial pressures, the alterations in heart rate are achieved by reciprocal changes in vagal and sympathetic neural activity. Below this range of arterial blood pressures, the high heart rate is achieved by intense sympathetic activity and the virtual absence of vagal activity. Conversely, above the intermediate range of arterial blood pressures, the low heart rate is achieved by intense vagal activity and a constant low level of sympathetic activity.

Bainbridge Reflex and Atrial Receptors

In 1915 Bainbridge reported that infusions of blood or saline accelerated the heart rate in dogs. This increase in heart rate occurred whether arterial blood pressure did or did not rise. Acceleration was observed when-

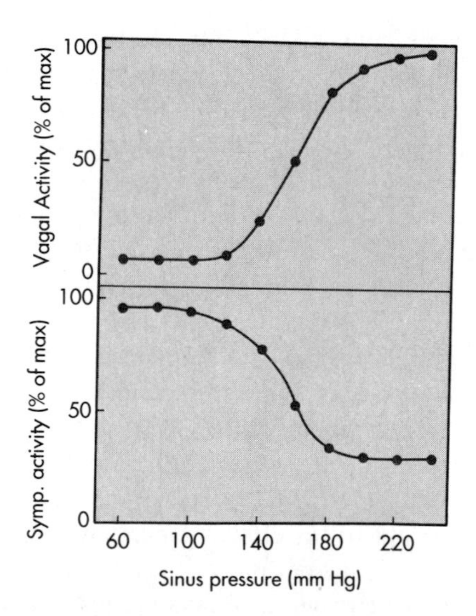

Fig. 25-6 The changes in neural activity in cardiac vagal and sympathetic nerve fibers induced by changes in pressure in the isolated carotid sinuses in an anesthetized dog. Over the pressure range of about 100 to 200 mm Hg, pressure increments increased vagal activity and decreased sympathetic activity. At pressures below about 100 mm Hg, sympathetic activity was maximal (100%), and sympathetic activity remained at a low constant level (about 25% of the maximum value). (Adapted from Kollai M, Koizumi K: *Pflügers Arch Ges Physiol* 413:365, 1989.)

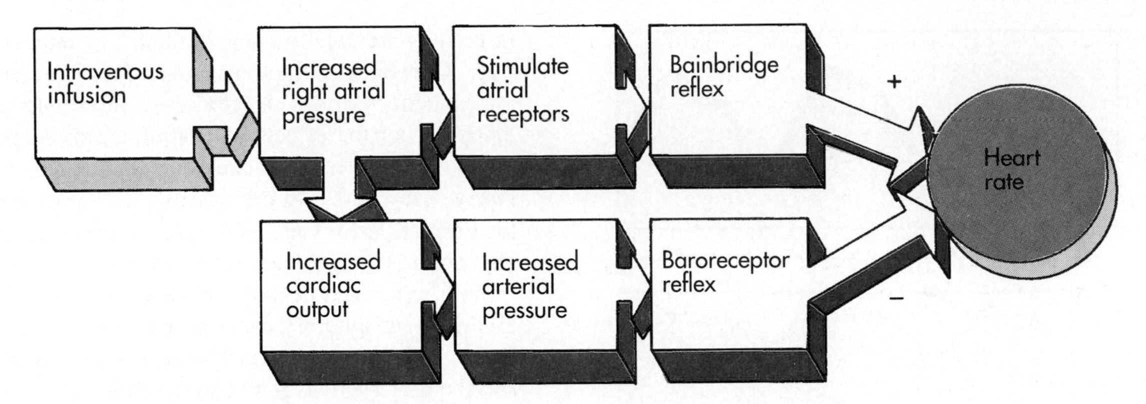

■ **Fig. 25-7** Intravenous infusions of blood or electrolyte solutions tend to increase heart rate via the Bainbridge reflex and to decrease heart rate via the baroreceptor reflex. The actual change in heart rate induced by such infusions is the result of these two opposing effects.

ever central venous pressure rose sufficiently to distend the right side of the heart, and the effect was abolished by bilateral transection of the vagi.

Numerous investigators have confirmed Bainbridge's observations. However, the magnitude and direction of the response depends on the prevailing heart rate. When the heart rate is slow, intravenous infusions usually accelerate the heart. At more rapid heart rates, however, infusions will ordinarily slow the heart. Increases in blood volume not only evoke the **Bainbridge reflex,**

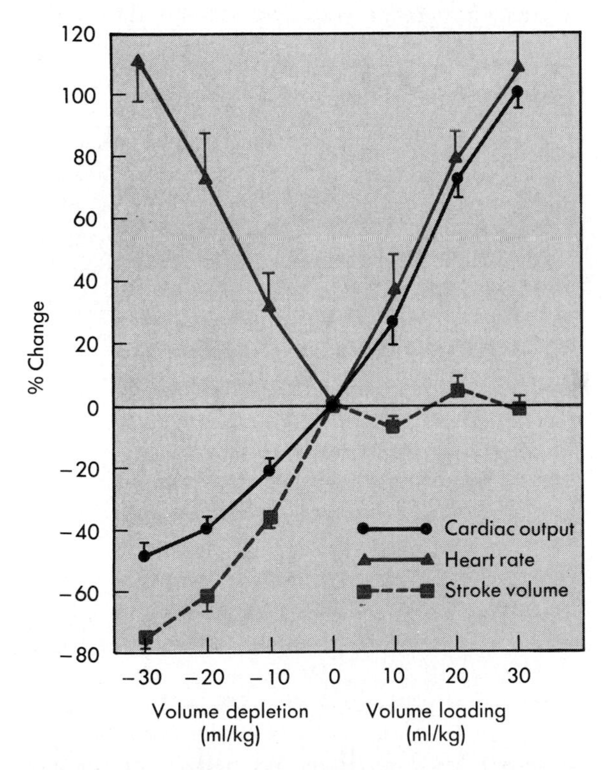

■ **Fig. 25-8** Effects of blood transfusion and of bleeding on cardiac output, heart rate, and stroke volume in unanesthetized dogs. (From Vatner SF, Boettcher DH: *Circ Res* 42:557, 1978, by permission of the American Heart Association.)

but they also activate other reflexes (notably the baroreceptor reflex) that tend to change the heart rate in the opposite direction. The actual change in heart rate evoked by an alteration of blood volume is therefore the resultant of these antagonistic reflex effects (Fig. 25-7).

In unanesthetized dogs, volume loading with blood increased heart rate and cardiac output proportionately (Fig. 25-8). Consequently, stroke volume remained virtually constant. Conversely, reductions in blood volume diminished the cardiac output but increased heart rate. Undoubtedly, the Bainbridge reflex was prepotent over the baroreceptor reflex when the blood volume was raised, but the baroreceptor reflex prevailed over the Bainbridge reflex when the blood volume was diminished.

Receptors that influence heart rate exist in both atria. They are located principally in the venoatrial junctions—in the right atrium at its junctions with the venae cavae and in the left atrium at its junctions with the pulmonary veins. Distension of these atrial receptors sends impulses centripetally in the vagi. The efferent impulses are carried by fibers from both autonomic divisions to the SA node. The cardiac response is highly selective. Even when the reflex increase in heart rate is large, changes in ventricular contractility have been negligible. Furthermore, the increase in heart rate is unattended by an increase of sympathetic activity to the peripheral arterioles.

Stimulation of the atrial receptors also causes an increase in urine volume. Reduced activity in the renal sympathetic nerve fibers might be partially responsible for this diuresis. However, the principal mechanism appears to be a neurally mediated reduction in the secretion of vasopressin (antidiuretic hormone) by the posterior pituitary gland.

A peptide, called **atrial natriuretic peptide** (ANP), is released from atrial tissue in response to increases in blood volume, presumably because of the resultant stretch of the atrial walls. ANP consists of 28 amino ac-

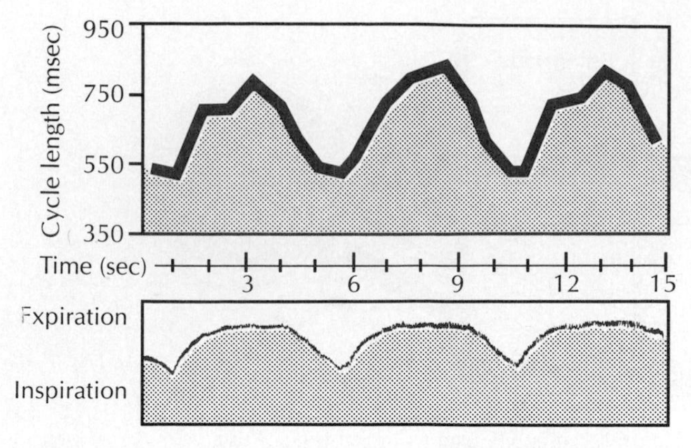

■ **Fig. 25-9** Respiratory sinus arrhythmia in a resting, unanesthetized dog. Note that the cardiac cycle length increases during expiration and decreases during inspiration. (Modified from Warner MR et al: *Am J Physiol* 251:H1134, 1986.)

ids, and it has potent diuretic and natriuretic effects on the kidneys and vasodilator effects on the resistance and capacitance blood vessels. Thus ANP plays an important role in the regulation of blood volume and blood pressure.

■ *Respiratory Sinus Arrhythmia*

Rhythmic variations in heart rate, occurring at the frequency of respiration, are detectable in most individuals and tend to be more pronounced in children. Typically the cardiac rate accelerates during inspiration and decelerates during expiration (Fig. 25-9).

Recordings from the autonomic nerves to the heart reveal that the neural activity increases in the sympathetic fibers during inspiration, whereas the neural activity in the vagal fibers increases during expiration (Fig. 25-10). The acetylcholine released at the vagal endings is removed so rapidly that the rhythmic changes in activity are able to elicit rhythmic variations in heart rate. Conversely, the norepinephrine released at the sympathetic endings is removed more slowly, thus damping out the effects of rhythmic variations in norepinephrine release on heart rate. Hence, rhythmic changes in heart rate are ascribable almost entirely to the oscillations in vagal activity. **Respiratory sinus arrhythmia** is exaggerated when vagal tone is enhanced.

Both reflex and central factors contribute to the genesis of the respiratory cardiac arrhythmia (Fig. 25-11). During inspiration the intrathoracic pressure decreases and therefore venous return to the right side of the heart is accelerated (see p. 507), which elicits the Bainbridge reflex (see Fig. 25-11). After the time delay required for the increased venous return to reach the left side of the heart, left ventricular output increases and raises arterial blood pressure. This in turn reduces heart rate reflexly through baroreceptor stimulation (see Fig. 25-11).

Fluctuations in sympathetic activity to the arterioles cause peripheral resistance to vary at the respiratory frequency. Consequently, arterial blood pressure fluctuates rhythmically, which affects heart rate via the baroreceptor reflex. Stretch receptors in the lungs may also affect heart rate (see Fig. 25-11). Moderate pulmonary inflation may increase heart rate reflexly. The afferent and efferent limbs of this reflex are located in the vagus nerves.

Central factors are also responsible for respiratory cardiac arrhythmia (see Fig. 25-11). The respiratory center in the medulla influences the cardiac autonomic centers. In heart-lung bypass experiments conducted on animals, the chest is open, the lungs are collapsed, venous return is diverted to a pump-oxygenator, and arterial blood pressure is maintained at a constant level. In such experiments rhythmic movements of the rib cage attest to the activity of the medullary respiratory centers, and the movements of the rib cage are often accompanied by rhythmic changes in heart rate at the respiratory frequency. This respiratory cardiac arrhythmia

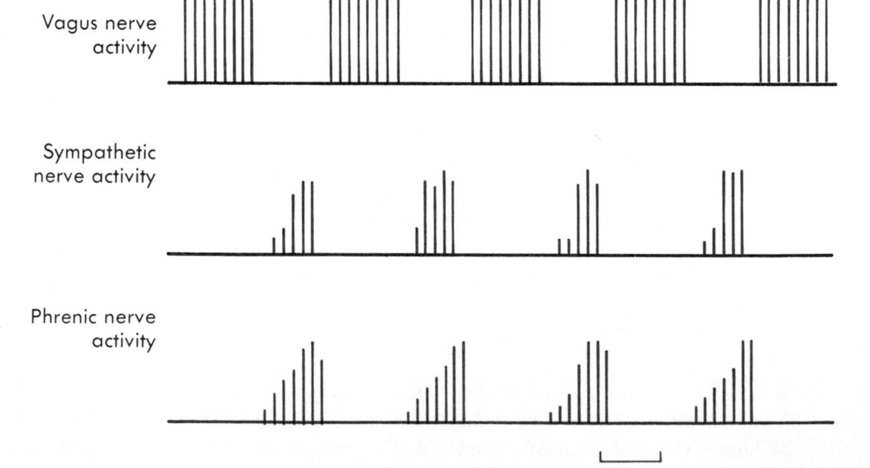

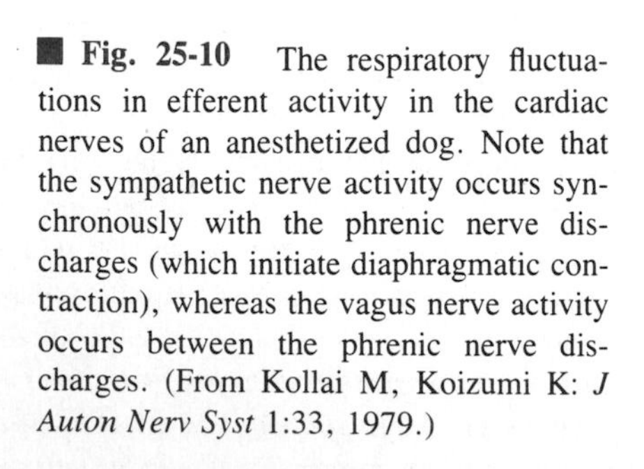

■ **Fig. 25-10** The respiratory fluctuations in efferent activity in the cardiac nerves of an anesthetized dog. Note that the sympathetic nerve activity occurs synchronously with the phrenic nerve discharges (which initiate diaphragmatic contraction), whereas the vagus nerve activity occurs between the phrenic nerve discharges. (From Kollai M, Koizumi K: *J Auton Nerv Syst* 1:33, 1979.)

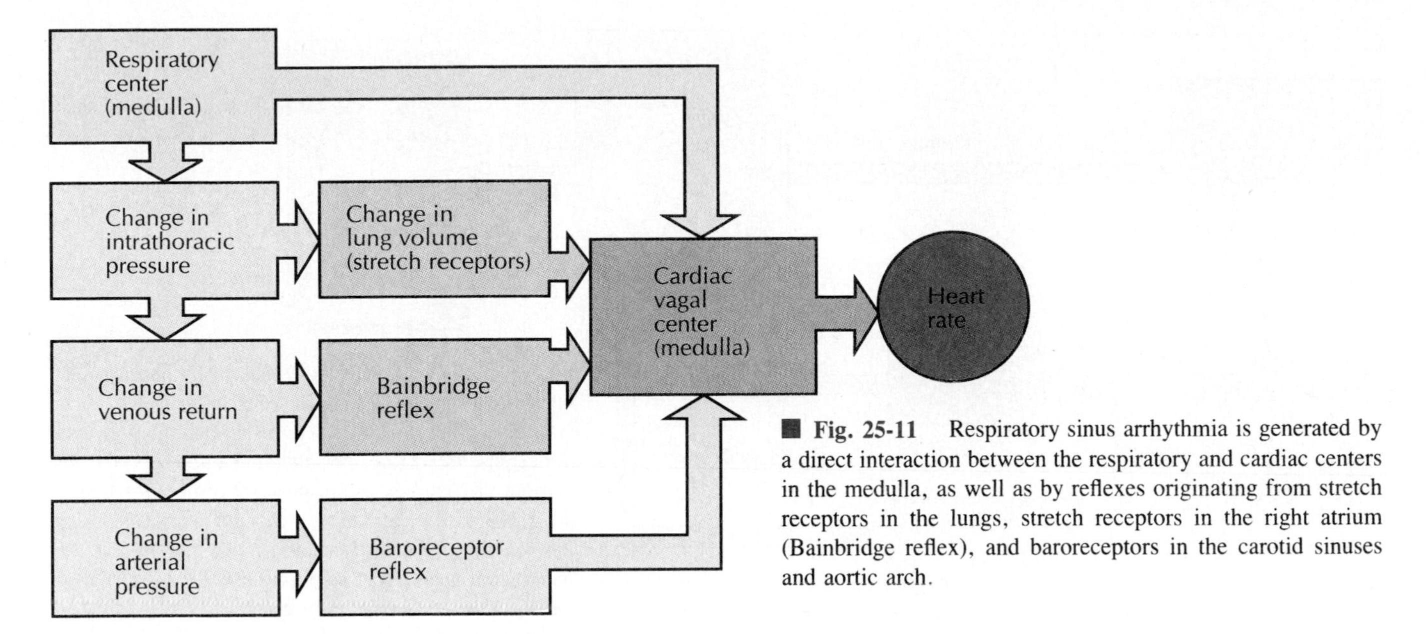

■ **Fig. 25-11** Respiratory sinus arrhythmia is generated by a direct interaction between the respiratory and cardiac centers in the medulla, as well as by reflexes originating from stretch receptors in the lungs, stretch receptors in the right atrium (Bainbridge reflex), and baroreceptors in the carotid sinuses and aortic arch.

is almost certainly induced by an interaction between the respiratory and cardiac centers in the medulla (see Fig. 25-11).

■ *Chemoreceptor Reflex*

The cardiac response to peripheral chemoreceptor stimulation merits special consideration because it illustrates the complexity that may be introduced when one stimulus excites two organ systems simultaneously. In intact

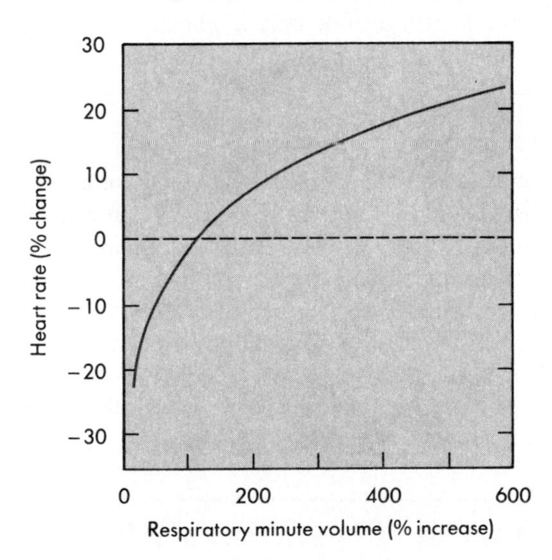

■ **Fig. 25-12** Relationship between the change in heart rate and the change in respiratory minute volume during carotid chemoreceptor stimulation in spontaneously breathing cats and dogs. When respiratory stimulation was relatively slight, heart rate usually diminished; when respiratory stimulation was more pronounced, heart rate usually increased. (Modified from Daly MdeB, Scott MJ: *J Physiol* 144:148, 1958.)

animals, stimulation of the carotid chemoreceptors consistently increases ventilatory rate and depth (see Chapter 37), but ordinarily changes heart rate only slightly.

The directional change in heart rate is related to the enhancement of pulmonary ventilation, as shown in Fig. 25-12. When respiratory stimulation is mild, heart rate usually diminishes; when the increase in pulmonary ventilation is more pronounced, heart rate usually accelerates.

The cardiac response to peripheral chemoreceptor stimulation is the resultant of primary and secondary reflex mechanisms (Fig. 25-13). The primary reflex effect of carotid chemoreceptor excitation is mainly to facilitate the medullary vagal center and thereby to decrease heart rate. Secondary effects are mediated by the respiratory system. The respiratory stimulation by the arterial chemoreceptors tends to inhibit the medullary vagal center. This inhibitory effect varies with the concomitant stimulation of respiration.

An example of the primary inhibitory influence is displayed in Fig. 25-14. In this experiment on an anesthetized dog, the lungs were completely collapsed and blood oxygenation was accomplished by an artificial oxygenator. When the carotid chemoreceptors were stimulated, an intense bradycardia and some degree of AV block ensued. Such effects are mediated primarily by efferent vagal fibers.

The identical primary inhibitory effect also operates in humans. The electrocardiogram in Fig. 25-15 was recorded from a quadriplegic patient who could not breathe spontaneously but required tracheal intubation and artificial respiration. When the tracheal catheter was briefly disconnected to permit nursing care, the patient quickly developed a profound bradycardia. His heart rate was 65 beats per minute just before the tracheal catheter was disconnected. In less than 10 sec af-

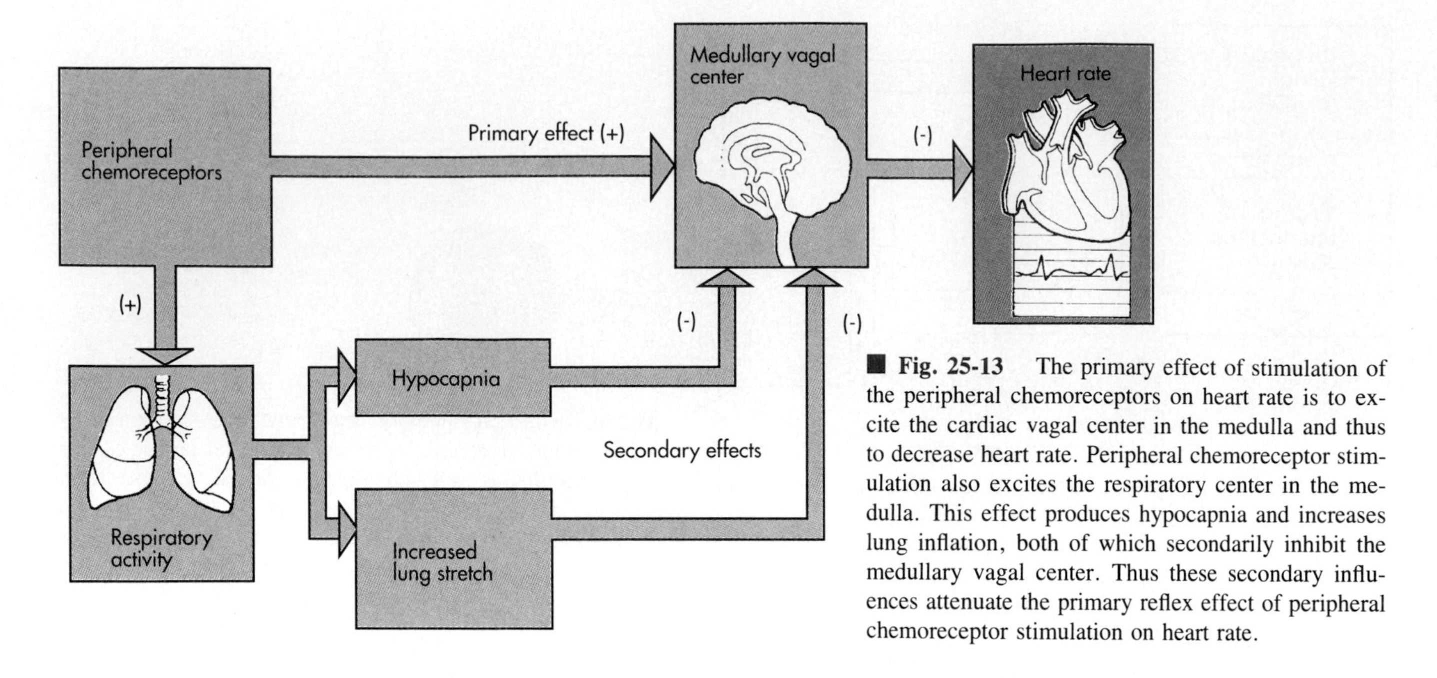

■ **Fig. 25-13** The primary effect of stimulation of the peripheral chemoreceptors on heart rate is to excite the cardiac vagal center in the medulla and thus to decrease heart rate. Peripheral chemoreceptor stimulation also excites the respiratory center in the medulla. This effect produces hypocapnia and increases lung inflation, both of which secondarily inhibit the medullary vagal center. Thus these secondary influences attenuate the primary reflex effect of peripheral chemoreceptor stimulation on heart rate.

ter cessation of artificial respiration, his heart rate dropped to about 20 beats per minute. This bradycardia could be prevented by blocking the effects of efferent vagal activity with atropine, and its onset could be delayed considerably by hyperventilating the patient before disconnecting the tracheal catheter.

The pulmonary hyperventilation that is ordinarily evoked by carotid chemoreceptor stimulation influences heart rate secondarily, both by initiating more pronounced pulmonary inflation reflexes and by producing hypocapnia (see Fig. 25-13). Each of these influences tends to depress the primary cardiac response to chemoreceptor stimulation and thereby to accelerate the heart. Hence when pulmonary hyperventilation is not prevented, the primary and secondary effects tend to neutralize each other, and carotid chemoreceptor stimulation affects heart rate only minimally.

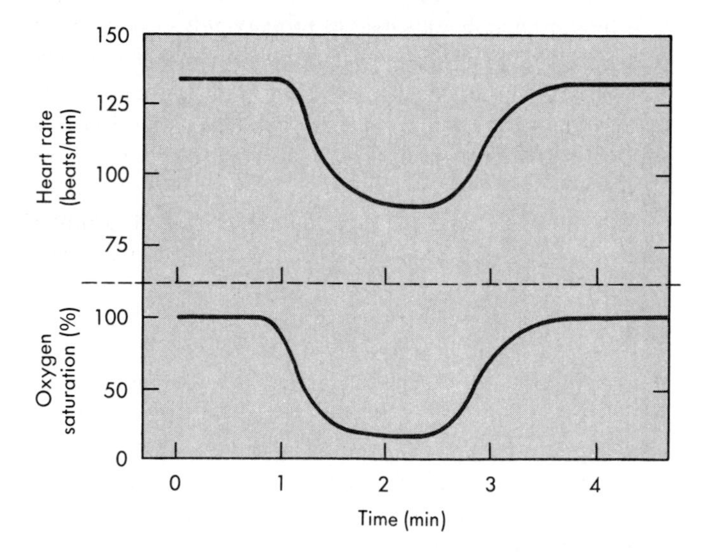

■ **Fig. 25-14** Changes in heart rate during carotid chemoreceptor stimulation in an anesthetized dog on total heart bypass. The lungs remain deflated and respiratory gas exchange is accomplished by an artificial oxygenator. The lower tracing represents the oxygen saturation of the blood perfusing the carotid chemoreceptors. The blood perfusing the remainder of the animal, including the myocardium, was fully saturated with oxygen throughout the experiment. (Modified from Levy MN, DeGeest H, Zieske H: *Circ Res* 18:67, 1966.)

■ *Ventricular Receptor Reflexes*

Sensory receptors located near the endocardial surfaces of the ventricular walls initiate reflex effects similar to those elicited by the arterial baroreceptors. Excitation of these endocardial receptors diminishes the heart rate and peripheral resistance. Other sensory receptors have been identified in the epicardial regions of the ventricles. Ventricular receptors are excited by a variety of mechanical and chemical stimuli, but their physiological functions are not clear.

■ *Intrinsic Regulation of Myocardial Performance*

Just as the heart can initiate its own beat in the absence of any nervous or hormonal control, so also can the myocardium adapt to changing hemodynamic conditions by mechanisms that are intrinsic to cardiac muscle itself. Experiments on denervated hearts reveal that this organ adjusts remarkably well to stress. For example,

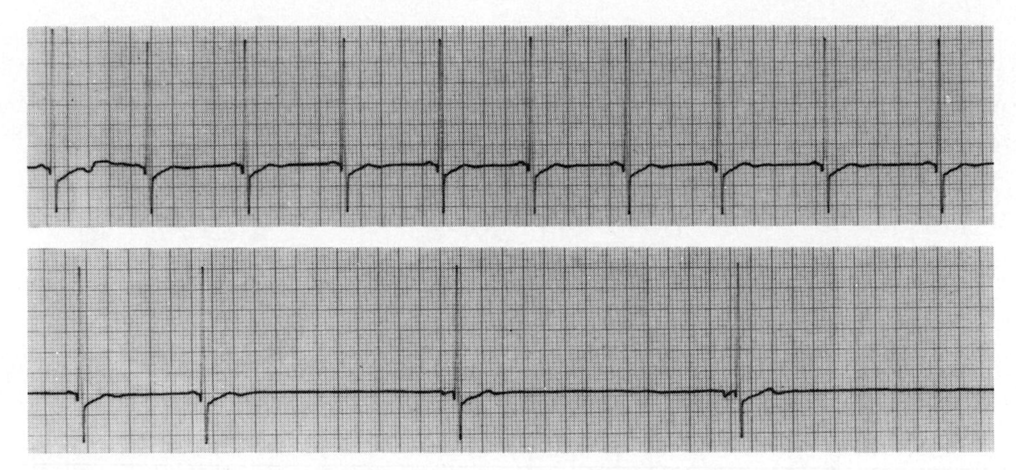

■ **Fig. 25-15** Electrocardiogram of a 30-year-old quadriplegic man who could not breathe spontaneously and required tracheal intubation and artificial respiration. The two strips are continuous. The tracheal catheter was temporarily disconnected from the respirator at the beginning of the top strip, at which time his heart rate was 65 beats per minute. In less than 10 sec, his heart rate decreased to about 20 beats per minute. (Modified from Berk JL, Levy MN. *Eur Surg Res* 9:75, 1977.)

racing greyhounds with denervated hearts perform almost as well as those with intact innervation. Their maximal running speed was found to be only 5% less after complete cardiac denervation. In these dogs the threefold to fourfold increase in cardiac output was achieved principally by an increase in stroke volume. In normal dogs the increase of cardiac output with exercise is accompanied by a proportionate increase of heart rate; stroke volume does not change much (see Chapter 32). It is unlikely that the cardiac adaptation in the denervated animals is achieved entirely by intrinsic mechanisms; circulating catecholamines undoubtedly contribute. If the beta-adrenergic receptors are blocked in greyhounds with denervated hearts, their racing performance is severely impaired.

The heart is partially or completely denervated in a variety of clinical situations: (1) the surgically transplanted heart is totally decentralized, although the intrinsic, postganglionic parasympathetic fibers persist; (2) atropine blocks vagal effects on the heart, and propranolol blocks sympathetic beta-adrenergic influences; (3) certain drugs, such as reserpine, deplete cardiac norepinephrine stores and thereby restrict or abolish sympathetic control; and (4) in chronic congestive heart failure, cardiac norepinephrine stores are often severely diminished, thereby attenuating any sympathetic influences.

The intrinsic cardiac adaptation that has received the greatest attention involves changes in the resting length of the myocardial fibers. This adaptation is designated **Starling's law of the heart,** or the **Frank-Starling mechanism.** The mechanical, ultrastructural, and physiological bases for this mechanism have been explained in Chapter 24.

■ *Frank-Starling Mechanism*

Isolated hearts. In 1895 Frank described the response of the isolated heart of the frog to alterations in the load on the myocardial fibers just before contraction—the **preload.** He observed that as the preload increased the heart responded with a more forceful contraction. Frank recognized that cardiac muscle behavior was similar to that of skeletal muscle when it is stretched to progressively greater initial lengths before contraction.

In 1914 Starling described the intrinsic response of the heart to changes in right atrial and aortic pressure in the canine heart-lung preparation, which is depicted in Fig. 25-16. In this preparation the right atrium is filled with blood from an elevated reservoir. Right atrial pressure is varied either by altering the height of the reservoir or by adjusting a screw clamp on the connecting tube. From the right atrium blood enters the right ventricle, which then pumps it through the pulmonary vessels to the left atrium. The trachea is cannulated and the lungs are artificially ventilated.

The aorta is ligated distal to the arch, and a cannula is inserted into the brachiocephalic artery. Blood is pumped by the left ventricle through this cannula and through rubber tubing that ultimately conducts the blood back through a heating coil to the right atrial reservoir. A pressure limiting system, known as a **Starling resistance,** is installed in the rubber tubing to permit changes in peripheral resistance. The volume of both ventricles is recorded by a special device. Cardiac output is determined by measuring the flow from the rubber tubing back into the venous reservoir connected to the right atrium.

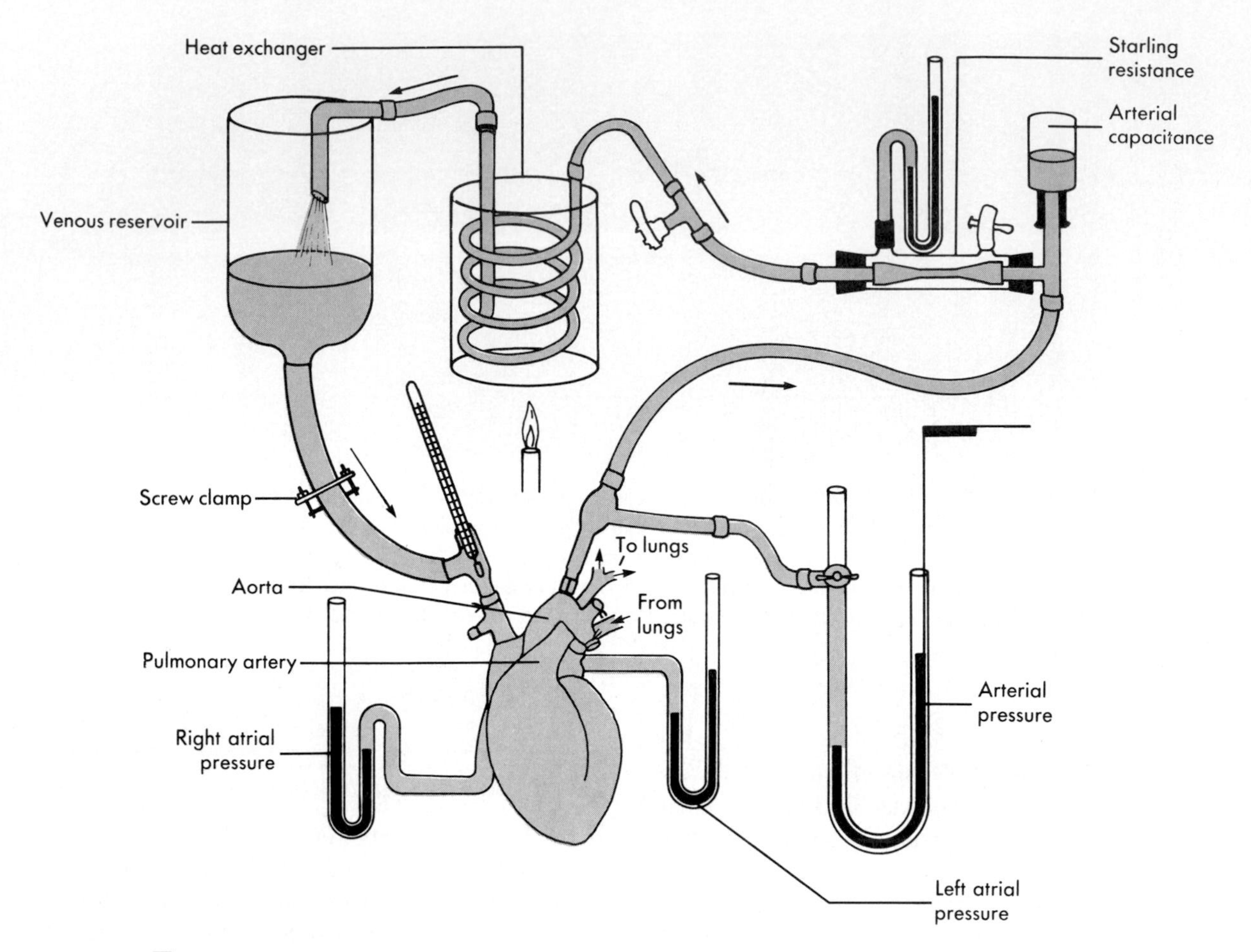

Fig. 25-16 Heart-lung preparation. (Redrawn from Patterson SW, Starling EH: *J Physiol* 48:357, 1914.)

The response of the isolated heart to a sudden augmentation of right atrial pressure (increased preload) is shown in Fig. 25-17. Aortic pressure was permitted to increase only slightly. In the top tracing, an increase of ventricular volume is registered as a downward deflection. Hence the upper border of the tracing represents the systolic volume, the lower border indicates the diastolic volume, and the amplitude of the deflections reflects the stroke volume.

For several beats after the rise in right atrial pressure, the ventricular volume progressively increased. This indicates that during these few beats a disparity must have existed between ventricular inflow during diastole and ventricular output during systole. Thus during this transient period before equilibrium was attained, the stroke volume must have been less than the filling volume. The consequent accumulation of blood dilated the ventricles and lengthened the individual myocardial fibers that comprised the walls of the ventricles.

The increased diastolic fiber length somehow facilitates ventricular contraction and enables the ventricles to pump a greater stroke volume, so that at equilibrium cardiac output exactly matches the augmented venous return. Increased fiber length alters cardiac performance mainly by changing the number of myofilament cross-bridges that can interact and by changing the calcium sensitivity of the myofilaments (see Chapter 24). An optimum fiber length apparently exists, beyond which contraction is actually impaired. Therefore excessively high filling pressures may depress rather than enhance the pumping capacity of the ventricles by overstretching the myocardial fibers.

Changes in diastolic fiber length also permit the isolated heart to compensate for an increase of peripheral resistance. In the experiment depicted in Fig. 25-18 the arterial resistance was abruptly raised in three steps, whereas venous inflow was held constant. Each rise in resistance increased the arterial pressure and ventricular volume. With each abrupt elevation of arterial pressure (increased afterload), the left ventricle was at first unable to pump a normal stroke volume. Because venous return was held constant, the diminution of stroke volume was attended by a rise in ventricular diastolic volume and therefore in the length of the myocardial fibers. This change in end-diastolic fiber length finally enabled the ventricle to pump a given stroke volume against a greater peripheral resistance.

The external work performed per stroke by the left

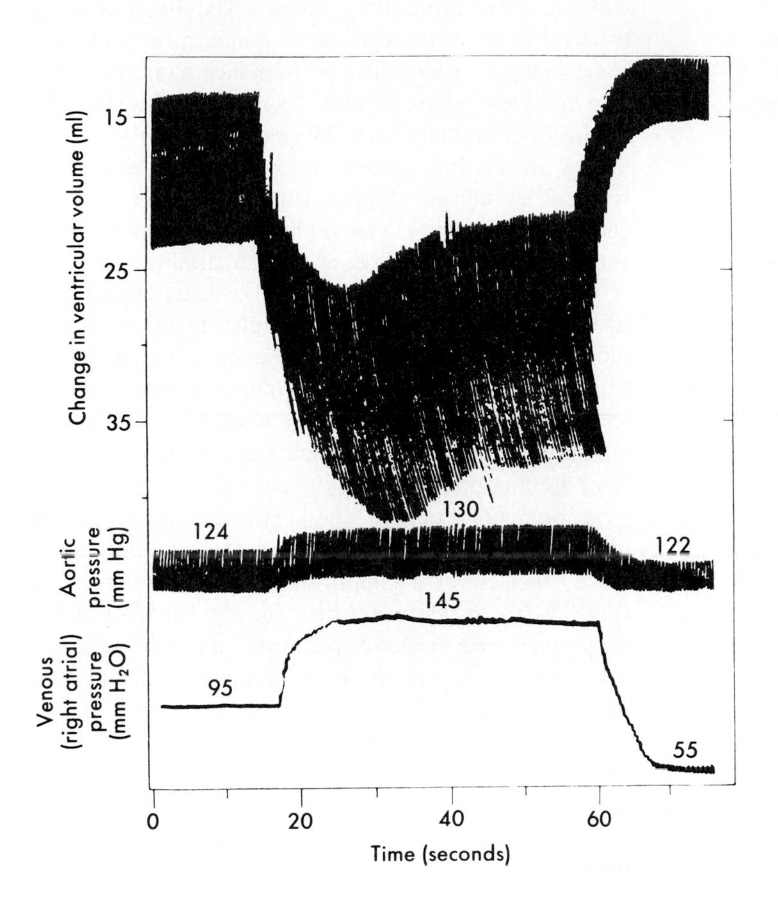

■ **Fig. 25-17** Changes in ventricular volume in a heart-lung preparation when the venous reservoir was suddenly raised (right atrial pressure increased from 95 to 145 mm H_2O) and subsequently lowered (right atrial pressure decreased from 145 to 55 mm H_2O). Note that an increase in ventricular volume is registered as a downward shift in the volume tracing. (Redrawn from Patterson SW, Piper H, Starling EH: *J Physiol* 48:465, 1914.)

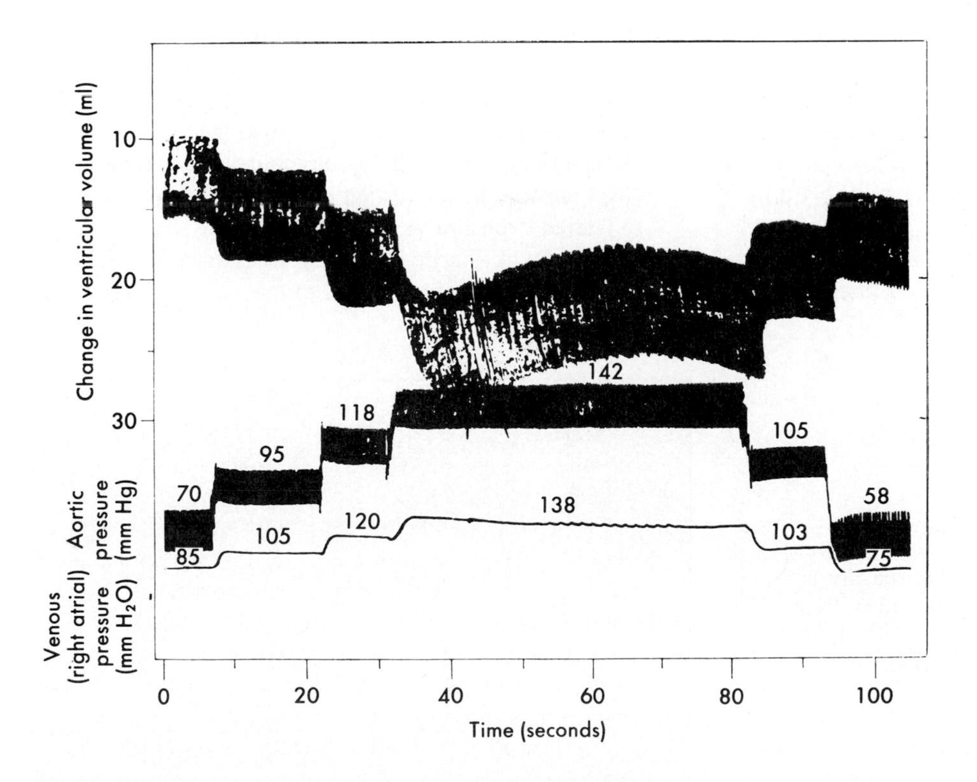

■ **Fig. 25-18** Changes in ventricular volume, aortic pressure, and right atrial pressure in a heart-lung preparation when peripheral resistance was raised and subsequently lowered in several steps. Note that an increase in ventricular volume is registered as a downward shift in the volume tracing. (Redrawn from Patterson SW, Piper H, Starling EH: *J Physiol* 48:465, 1914.)

ventricle is approximately equal to the product of the mean arterial pressure and stroke volume (p. 453). Therefore the increased diastolic length of the cardiac muscle fiber increases the work production by the left ventricle. However, at an excessively high peripheral resistance, further augmentation of resistance will reduce stroke volume and stroke work.

Changes in ventricular volume are also involved in the cardiac adaptation to alterations in heart rate. During bradycardia, for example, the increased duration of diastole permits greater ventricular filling. The consequent augmentation of myocardial fiber length increases stroke volume. Therefore the reduction in heart rate may be fully compensated by the increase in stroke volume, such that cardiac output may remain constant (see Fig. 30-14).

When cardiac compensation involves ventricular dilation, the force required by each myocardial fiber to generate a given intraventricular systolic pressure must be appreciably greater than that developed by the fibers in a ventricle of normal size. The relationship between wall tension and cavity pressure resembles that for cylindrical tubes (p. 466) in that for a constant internal pressure, wall tension varies directly with the radius. As a consequence, the dilated heart requires considerably more oxygen to perform a given amount of external work than does the normal heart.

In the intact animal, the heart is enclosed in the pericardial sac. The relatively rigid pericardium determines the pressure-volume relationship at high levels of pressure and volume. The pericardium exerts this limitation of volume even under normal conditions, when an individual is at rest and the heart rate is slow. In the cardiac dilation and hypertrophy that accompanies chronic heart failure, the pericardium is stretched considerably (Fig. 25-19). The pericardial limitation of cardiac filling is exerted at pressures and volumes that are entirely different from those in normal individuals.

Intact preparations. The major problem of assessing the role of the Frank-Starling mechanism in intact animals and humans is the difficulty of measuring end-diastolic myocardial fiber length. The Frank-Starling mechanism has been represented graphically by plotting some index of ventricular performance along the ordinate and some index of fiber length along the abscissa. The most commonly used indices of ventricular performance are cardiac output, stroke volume, and stroke work. The indices of fiber length include ventricular end-diastolic volume, ventricular end-diastolic pressure, ventricular circumference, and mean atrial pressure.

The Frank-Starling mechanism is better represented by a family of so-called ventricular function curves, rather than by a single curve. To construct a given ventricular function curve, blood volume is altered over a wide range of values, and stroke work and end-diastolic pressure are measured at each step. Similar observations are then made during the desired experimental intervention. For example, the ventricular function curve obtained during a norepinephrine infusion lies above and to the left of a control ventricular function curve (Fig. 25-20). It is evident that, for a given level of left ventricular end-diastolic pressure, the left ventricle performs more work during a norepinephrine infusion than during control conditions. Hence a shift of the ventricular function curve to the left usually signifies an improvement of ventricular contractility (p. 404); a shift to the right usually indicates an impairment of contractility and a consequent tendency toward **cardiac failure.**

A shift in a ventricular function curve does not uniformly indicate a change in contractility, however. **Contractility** is a measure of cardiac performance at a given level of preload and afterload. The end-diastolic pressure is ordinarily a good index of preload, whereas the aortic systolic pressure is a good index of afterload. In assessing myocardial contractility, the cardiac afterload must be held constant as the end-diastolic pressure is varied over a range of values.

The Frank-Starling mechanism is ideally suited for matching the cardiac output to the venous return. Any

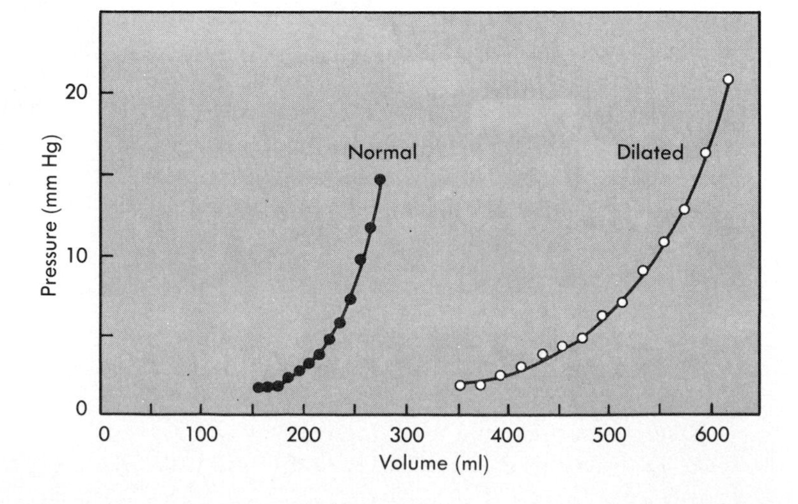

■ **Fig. 25-19** Pericardial pressure-volume relations in a normal dog and in a dog with experimentally induced chronic cardiac hypertrophy. (Modified from Freeman GL, Le Winter MM: *Circ Res* 54:294, 1984, by permission of the American Heart Association.)

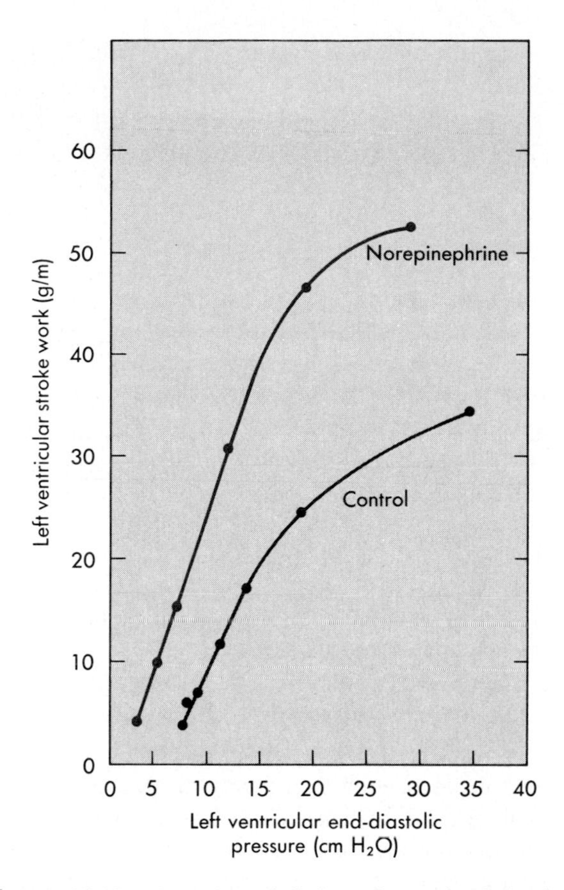

■ **Fig. 25-20** A constant infusion of norepinephrine in a dog shifts the ventricular function curve to the left. This shift signifies an enhancement of ventricular contractility. (Redrawn from Sarnoff SJ et al: *Circ Res* 8:1108, 1960.)

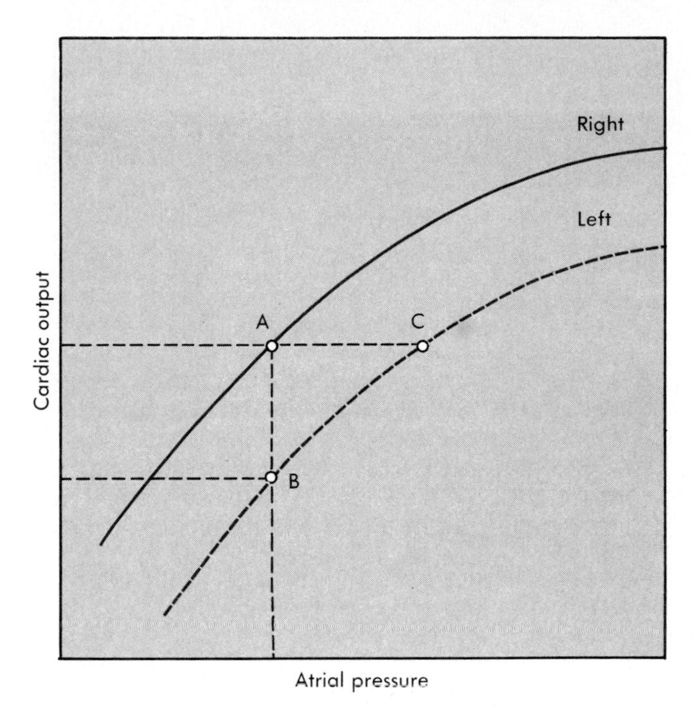

■ **Fig. 25-21** Curves relating the outputs of right and left ventricles to mean right and left atrial pressure, respectively. At a given level of cardiac output, mean left atrial pressure (e.g., point *C*) exceeds mean right atrial pressure (point *A*).

sudden, excessive output by one ventricle soon increases the venous return to the other ventricle. The consequent increase in diastolic fiber length augments the output of the second ventricle to correspond with that of its mate. Therefore it is the Frank-Starling mechanism that maintains a precise balance between the outputs of the right and left ventricles. Because the two ventricles are arranged in series in a closed circuit, even a small, but maintained, imbalance in the outputs of the two ventricles would otherwise be catastrophic.

The curves relating cardiac output to mean atrial pressure for the two ventricles are not coincident; the curve for the left ventricle usually lies below that for the right, as shown in Fig. 25-21. At equal right and left atrial pressures (points *A* and *B*) right ventricular output would exceed left ventricular output. Hence venous return to the left ventricle (a function of right ventricular output) would exceed left ventricular output, and left ventricular diastolic volume and pressure would rise. By the Frank-Starling mechanism, left ventricular output would therefore increase (from *B* toward *C*). Only when the outputs of both ventricles are identical (points *A* and *C*) would the equilibrium be stable. Under such conditions, however, left atrial pressure *(C)* would exceed right atrial pressure *(A)*, and this is precisely the relationship that ordinarily prevails. This difference in atrial pressures accounts for the observation that in **congenital atrial septal defects,** where the two atria communicate, the direction of the shunt flow is usually from left to right.

■ *Rate-Induced Regulation*

The effects of a sustained frequency of contraction on the force developed in an isometrically contracting cat papillary muscle are shown in Fig. 25-22, *B*. Initially the strip of cardiac muscle was stimulated to contract only once every 20 sec. When the muscle was suddenly made to contract once every 0.63 sec, the developed force increased progressively over the next several beats. This progressive increase in developed force induced by a change in contraction frequency is known as the **staircase,** or **Treppe, phenomenon.**

At the new steady state the developed force was more than five times as great as it was at the larger contraction interval. A return to the larger interval (20 sec) had the opposite influence on developed force.

The effect of the interval between contractions on the steady-state level of developed force is shown in panel *A* (see Fig. 25-22) for a wide range of intervals. As the interval is diminished from 300 sec down to about 20 sec, little change occurs in developed force. As the in-

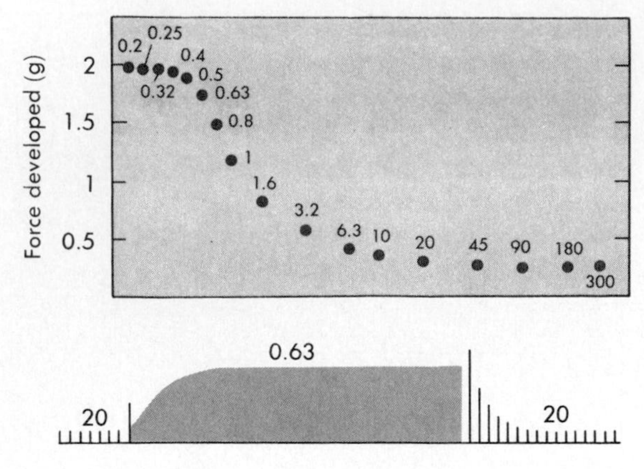

■ **Fig. 25-22** Changes in force development in an isolated papillary muscle from a cat as the interval between contractions is varied. The numbers in both sections of the record denote the interval (in seconds) between beats. In section *A* the points represent the steady-state forces developed at the intervals indicated. (Redrawn from Koch-Weser J, Blinks JR: *Pharmacol Rev* 15:601, 1963.)

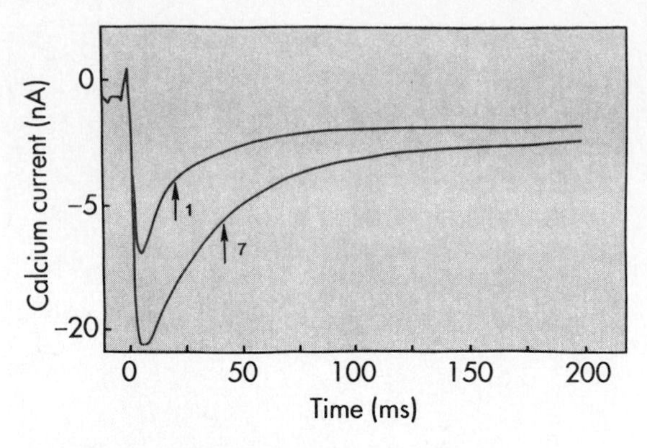

■ **Fig. 25-23** The calcium currents induced in a guinea pig myocyte during the first and seventh depolarizations in a sequence of depolarizations. The arrows indicate the half times of inactivation. Note that during the seventh depolarization, the maximum inward Ca^{++} current and the half time of inactivation were greater than the respective values for the first depolarization. (Modified from Lee KS: *Proc Natl Acad Sci* 84:3941, 1987.)

terval is reduced further, to a value of about 0.5 sec force increases sharply. Further reduction of the interval to 0.2 sec has little additional effect on developed force.

The initial progressive rise in developed force when the interval between beats is suddenly decreased (e.g., from 20 to 0.63 sec in panel *B*) (Fig. 25-22) is mediated by a gradual increase in intracellular Ca^{++} content. Two mechanisms contribute to the rise in Ca^{++} content: (1) an increase in the number of depolarizations per minute, and (2) an increase in the inward Ca^{++} current per depolarization.

With respect to the first mechanism, Ca^{++} enters the myocardial cell during each action potential plateau (see Fig. 23-9). As the interval between beats is diminished, the number of plateaus per minute increases. Even though the duration of each action potential (and of each plateau) decreases as the interval between beats is reduced (see Fig. 23-19), the overriding effect of the increased number of plateaus per minute on the influx of Ca^{++} would prevail, and the intracellular content of Ca^{++} would increase.

With respect to the second mechanism, as the interval between beats is suddenly diminished, the inward Ca^{++} current (i_{Ca}) progressively increases with each successive beat until a new steady state is attained at the new basic cycle length. Fig. 25-23 shows that in an isolated ventricular myocyte that was subjected to repetitive depolarizations, the influx of Ca^{++} into the myocyte increased on successive beats. For example, the maximum i_{Ca} was considerably greater during the seventh depolarization than it was during the first depolarization. Furthermore, the decay of that current (i.e., its inactivation) was substantially slower during the seventh depolarization than during the first depolarization. Both of these characteristics of the i_{Ca} would result in a greater influx of Ca^{++} into the myocyte during the seventh depolarization than during the first depolarization. The greater influx of Ca^{++} would, of course, strengthen the contraction.

Transient changes in the intervals between beats also profoundly affect the strength of contraction. When a premature ventricular systole (Fig. 25-24, beat *A*) occurs, the premature contraction (extrasystole) itself is feeble; whereas the beat after the compensatory pause is very strong. This response is partly ascribable to the Frank-Starling mechanism. Inadequate ventricular filling just before the premature beat accounts partly for the weak premature contraction. Subsequently, the exaggerated degree of filling associated with the compensatory pause explains in part the vigorous postextrasystolic contraction.

Although the Frank-Starling mechanism is certainly involved in the usual ventricular adaptation to a premature beat, it is not the exclusive mechanism. For example, in the ventricular pressure curves recorded from an isovolumic left ventricle preparation (see Fig. 25-24), in which neither filling nor ejection takes place during the cardiac cycle, the premature beat *(A)* is feeble and the succeeding contraction *(B)* is supernormal. Such enhanced contractility in contraction *B* is an example of **postextrasystolic potentiation,** and it may persist for one or more additional beats (e.g., contraction *C*).

The weakness of the premature beat is directly related to the degree of prematurity. Conversely, as the time **(coupling interval)** between the premature beat and the preceding beat is increased, the more nearly normal will be the premature beat. The curve that re-

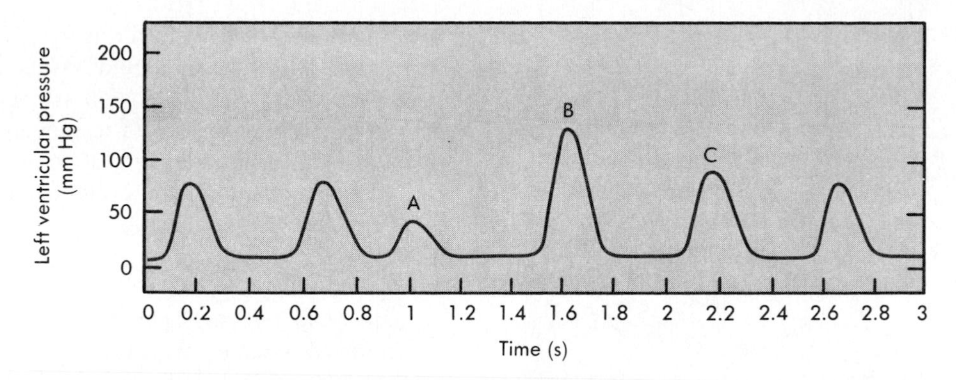

■ **Fig. 25-24** In an isovolumic canine left ventricle preparation a premature ventricular systole (beat *A*) is typically feeble, whereas the postextrasystolic contraction (beat *B*) is characteristically strong, and the enhanced contractility may persist to a diminishing degree over a few beats (e.g., contraction *C*). (From Levy MN: unpublished tracing.)

lates the strength of contraction of a premature beat to the coupling interval is called a **mechanical restitution curve.** Fig. 25-25 shows the restitution curve obtained by varying the coupling intervals of test beats in an isolated ventricular muscle preparation from a guinea pig.

The restitution of contractile strength probably depends on the time course of the intracellular circulation of Ca^{++} during the contraction and relaxation process (see Fig. 24-5). During relaxation the Ca^{++} that dissociates from the contractile proteins is taken up by the sarcoplasmic reticulum for subsequent release. However, about 500 to 800 msec are required before the Ca^{++} that had been taken up becomes available for release in response to the next depolarization.

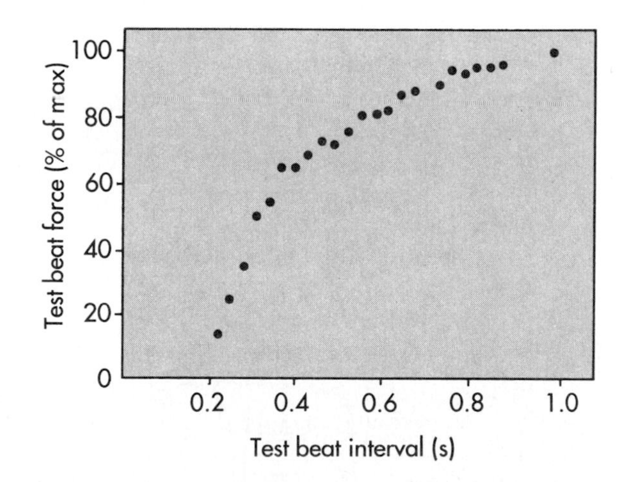

■ **Fig. 25-25** The force generated during premature contractions in a guinea pig isolated ventricular muscle preparation. The muscle was driven to contract once per second. Periodically the muscle was stimulated to contract prematurely. The scale along the X axis denotes the time between the driven and premature beat. The Y axis scale denotes the ratio of the contractile force of the premature beat to that of the driven beat. (Modified from Seed WA, Walker JM: *Cardiovasc Res* 22:303, 1988.)

The premature beat itself (see Fig. 25-24, beat *A*) is feeble probably because not enough time has elapsed to allow much of the Ca^{++} taken up by the sarcoplasmic reticulum during the preceding relaxation to become available for release in response to the premature depolarization. The postextrasystolic beat (see Fig. 25-24, beat *B*), conversely, is considerably stronger than normal. A plausible reason is that after the compensatory pause (p. 392) between beats *A* and *B*, the sarcoplasmic reticulum will have available for release the Ca^{++} that had been taken up during two heartbeats: the extrasystole (beat *A*) and the normal beat that had preceded it.

■ *Extrinsic Regulation of Myocardial Performance*

Although the completely isolated heart can adapt well to changes in preload and afterload, various extrinsic factors also influence the heart in the intact animal. Under many natural conditions these extrinsic regulatory mechanisms may overwhelm the intrinsic mechanisms. These extrinsic regulatory factors may be subdivided into nervous and chemical components.

■ *Nervous Control*

Sympathetic influences. Sympathetic nervous activity enhances atrial and ventricular contractility. Effects of increased cardiac sympathetic activity on the ventricular myocardium are asymmetrical. The cardiac sympathetic nerves on the left side of the body usually have a much greater effect on ventricular contraction than do those on the right side (see Fig. 25-4).

The alterations in ventricular contraction evoked by electrical stimulation of the left stellate ganglion in a canine isovolumic left ventricle preparation are shown in Fig. 25-26. The peak pressure and the maximum rate

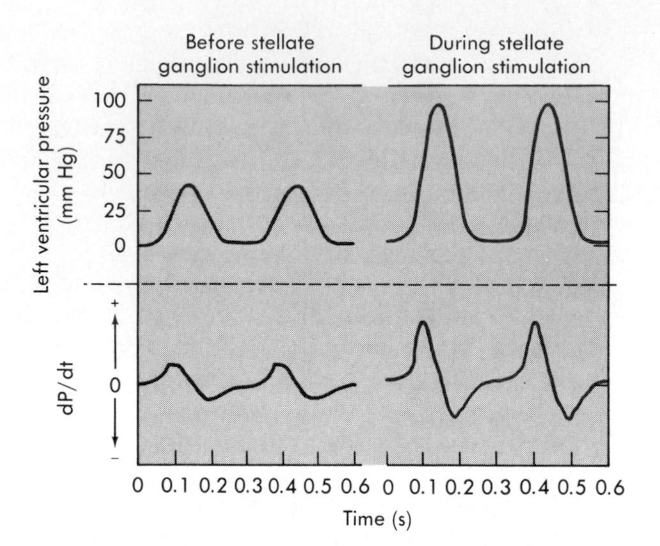

■ **Fig. 25-26** In an isovolumic left ventricle preparation, stimulation of cardiac sympathetic nerves evokes a substantial rise in peak left ventricular pressure and in the maximum rates of intraventricular pressure rise and fall (dP/dt). (From Levy MN: unpublished tracing.)

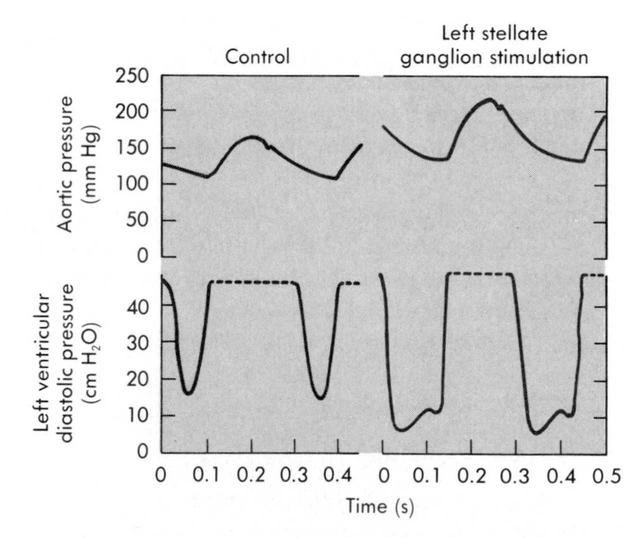

■ **Fig. 25-27** Stimulation of the left stellate ganglion of a dog increases arterial pressure, stroke volume, and stroke work despite a concomitant reduction in ventricular end-diastolic pressure. Note also the abridgement of systole, thereby allowing more time for ventricular filling; the heart was paced at a constant rate. In the ventricular pressure tracings the pen excursion is limited at 45 mm Hg; actual ventricular pressures during systole can be estimated from the aortic pressure tracings. (Redrawn from Mitchell JH, Linden RJ, Sarnoff SJ: *Circ Res* 8:1100, 1960.)

of pressure rise (dP/dt) during systole are markedly increased. Also, the duration of systole is reduced and the rate of ventricular relaxation is increased during the early phases of diastole. The shortening of systole and more rapid ventricular relaxation assist ventricular filling. For a given cardiac cycle length, the abbreviation of systole allows more time for diastole and hence for ventricular filling. In the experiment shown in Fig. 25-27, for example, the animal's heart was paced at a constant rapid rate. Sympathetic stimulation (right panel) shortened systole, which allowed substantially more time for ventricular filling.

Sympathetic nervous activity enhances myocardial performance. Neurally released norepinephrine or circulating catecholamines interact with beta-adrenergic receptors on the cardiac cell membranes (Fig. 25-28). This reaction activates adenylyl cyclase, which raises the intracellular levels of cyclic AMP (cAMP) (see Chapter 5). As a consequence, protein kinases are activated that promote the phosphorylation of various proteins within the myocardial cells. Phosphorylation of specific sarcolemmal proteins activates the calcium channels in the myocardial cell membranes.

Activation of the calcium channels increases the influx of Ca^{++} during the action potential plateau, and more Ca^{++} is released from the sarcoplasmic reticulum in response to each cardiac excitation. The contractile strength of the heart is thereby increased. Fig. 25-29 shows the correlation between the contractile force developed by a thin strip of ventricular muscle and the Ca^{++} concentration (as reflected by the aequorin light signal) in the myocytes as the concentration of isoproterenol (a beta-adrenergic agonist) was increased in the tissue bath.

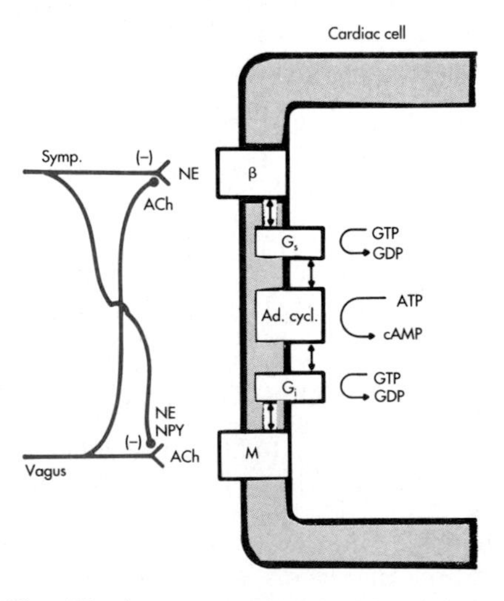

■ **Fig. 25-28** The interneuronal and intracellular mechanisms responsible for the interactions between the sympathetic and parasympathetic systems in the neural control of cardiac function. (From Levy MN. In Kulbertus HE, Franck G, editors: *Neurocardiology,* Mt. Kisco. New York, 1988, Futura Publishing.)

The overall effect of increased cardiac sympathetic activity in intact animals can best be appreciated in terms of families of ventricular function curves. When stepwise increases in the frequency of electrical stimulation are applied to the left stellate ganglion, the ventricular function curves shift progressively to the left. The changes parallel those produced by catecholamine infusions (see Fig. 25-20). Hence, for any given left ventricular end-diastolic pressure, the ventricle is capable of performing more work as the level of sympathetic nervous activity is raised.

During cardiac sympathetic stimulation the increase in work is usually accompanied by a reduction in left ventricular end-diastolic pressure. An example of the response to stellate ganglion stimulation in a heart paced at a constant frequency is shown in Fig. 25-27. In this experiment stroke work increased by about 50%, despite a reduction in the left ventricular end-diastolic pressure. Note also the pronounced shortening of ventricular systole, with the consequent lengthening of the filling period. The reason for the reduction in ventricular end-diastolic pressure is explained on p. 502.

Parasympathetic influences. The vagus nerves inhibit the cardiac pacemaker, atrial myocardium, and AV conduction tissue. The vagus nerves also depress the ventricular myocardium, but the effects are less pronounced. In the isovolumic left ventricle preparation, vagal stimulation decreases the peak left ventricular pressure, maximum rate of pressure development (dP/dt), and maximum rate of pressure decline during diastole (Fig. 25-30). In pumping heart preparations the ventricular function curve shifts to the right during vagal stimulation.

The vagal effects on the ventricular myocardium are achieved by at least two mechanisms, as shown in Fig. 25-28. The acetylcholine (ACh) released from the vagal endings can interact with muscarinic (M) receptors in the cardiac cell membrane. This interaction leads to the inhibition of adenylyl cyclase. The consequent diminution in the intracellular concentration of cyclic AMP leads to a reduction in Ca^{++} conductance of the cell membrane, and hence a decrease in myocardial contractility.

The ACh released from the vagal endings can also inhibit the release of norepinephrine from neighboring sympathetic nerve endings (see Fig. 25-28). The experiment illustrated in Fig. 25-31 demonstrates that stimulation of the cardiac sympathetic nerves *(S)* results in the overflow of substantial amounts of norepinephrine into the coronary sinus blood. Concomitant vagal stimulation (S + V) reduces the overflow of norepinephrine by about 30%. The amount of norepinephrine overflowing into the coronary sinus blood probably parallels the amount released at the sympathetic terminals. Thus, vagal activity can decrease ventricular contractility partly by antagonizing the facilitatory effects of any concomitant sympathetic activity to enhance ventricular contractility. Similarly, sympathetic nerves release norepinephrine and certain neuropeptides, including neuropeptide Y (NPY); norepinephrine and NPY both inhibit the release of acetylcholine from neighboring vagal fibers (see Fig. 25-28).

Baroreceptor reflex. Just as stimulation of the carotid sinus and aortic arch baroreceptors may change heart rate (p. 420), so also may it alter myocardial performance. Evidence of reflex alterations of ventricular contractility is presented in Fig. 25-32. Ventricular function curves were obtained at four levels of carotid sinus pressure. With each successive rise in pressure, the ventricular function curves were displaced farther

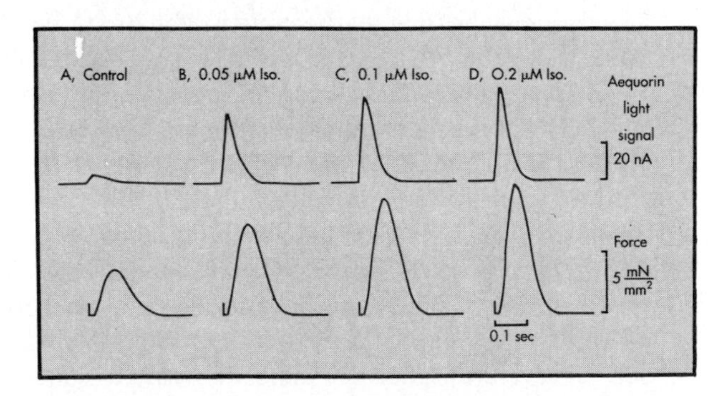

■ **Fig. 25-29** Effects of various concentrations of isoproterenol (Iso) on aequorin light signal (in nA) and contractile force (in mN/mm^2) in a rat ventricular muscle injected with aequorin. The aequorin light signal relects the instantaneous changes in intracellular Ca^{++} concentration. (Modified from Kurihara S, Konishi M: Effects of β-adrenoreceptor stimulation on intracellular Ca transients and tension in rat ventricular muscle, *Pflügers Arch* 409:427, 1987.)

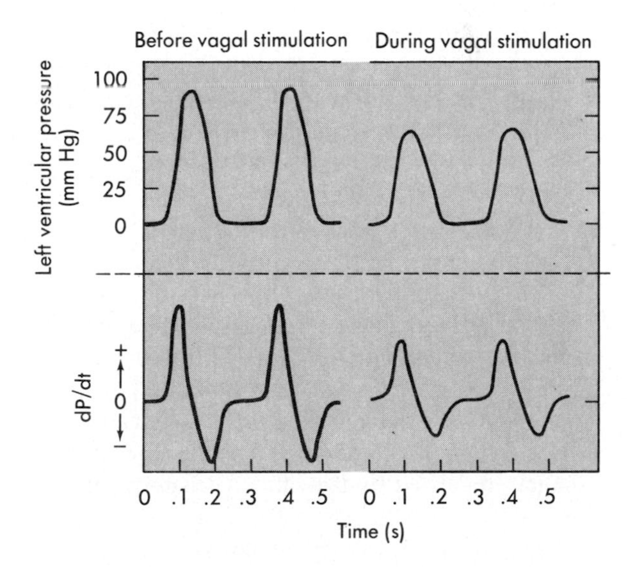

■ **Fig. 25-30** In an isovolumic left ventricle preparation, when the ventricle is paced at a constant frequency, vagal stimulation decreases the peak left ventricular pressure and diminishes the maximum rates of pressure rise and fall (dP/dt). (From Levy MN: unpublished tracing.)

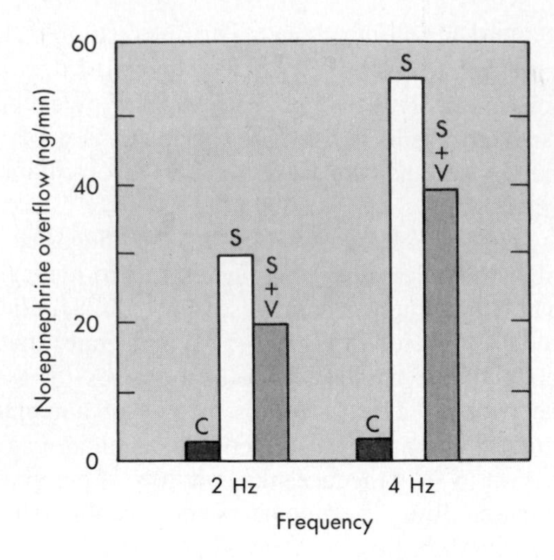

■ **Fig. 25-31** The mean rates of overflow of norepinephrine into the coronary sinus blood in a group of seven dogs under control conditions *(C)*, during cardiac sympathetic stimulation *(S)* at 2 or 4 Hz and during combined sympathetic and vagal stimulation (S + V). The combined stimulus consisted of sympathetic stimulation at 2 or 4 Hz, and vagal stimulation at 15 Hz. (Redrawn from Levy MN, Blattberg B: *Circ Res* 38:81, 1976, by permission of the American Heart Association.)

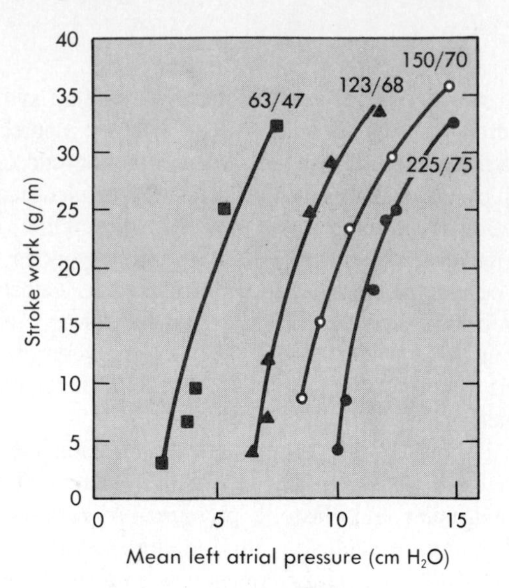

■ **Fig. 25-32** As the pressure in the isolated carotid sinus is progressively raised, the ventricular function curves shift to the right. The numbers at the tops of each curve represent the systolic/diastolic perfusion pressures (in millimeters of mercury) in the carotid sinus regions of the dog. (Redrawn from Sarnoff SJ et al: *Cir Res* 8:1123, 1960.)

and farther to the right, denoting a progressively greater reflex depression of ventricular performance.

In normal, resting individuals and animals the tonic level of sympathetic activity is usually very low. Under such conditions, moderate changes in baroreceptor activity may have little reflex influence on myocardial contractility. In states of augmented sympathetic neural activity, however, the effects of the baroreceptor reflex on contractility may be substantial. In the adaptation to blood loss, for example, a reflex change in myocardial contractility may constitute an important compensation.

■ *Chemical Control*

Hormones

Adrenomedullary hormones. The adrenal medulla is essentially a component of the autonomic nervous system (see also Chapters 15 and 50). The principal hormone secreted by the adrenal medulla is epinephrine, although some norepinephrine is also released. The rate of secretion of catecholamines by the adrenal medulla is largely regulated by the same mechanisms that control the activity of the sympathetic nervous system. The concentrations of catecholamines in the blood rise under the same conditions that activate the sympathoadrenal system. However, the cardiovascular effects of circulating catecholamines are probably minimal under normal conditions. Instead, the cardiac effects of increased sympathoadrenal activity are mainly ascribable to the norepinephrine released at the sympathetic nerve endings in the heart.

The changes in myocardial contractility induced by norepinephrine infusions have been tested in resting, unanesthetized dogs. The maximum rate of rise of left ventricular pressure (dP/dt), an index of myocardial contractility, was found to be proportional to the norepinephrine concentration in the blood (Fig. 25-33). In these same animals, moderate exercise increased the maximum dP/dt by almost 100%, but it raised the circulating catecholamines by only 0.5 ng/ml. Such a rise in blood norepinephrine concentration, by itself, would have had only a negligible effect on left ventricular dP/dt (see Fig. 25-33). Hence, the pronounced change in dP/dt observed during exercise must have been ascribable mainly to the norepinephrine released from the cardiac sympathetic nerve fibers rather than to the catecholamines released from the adrenal medulla.

Adrenocortical hormones. Cardiovascular problems are common in adrenocortical insufficiency **(Addison's disease).** The blood volume tends to fall, which may lead to severe hypotension and cardiovascular collapse, the so-called addisonian crisis (see Chapter 50).

The influence of adrenocortical steroids on the myocardium is controversial. Cardiac muscle removed from adrenalectomized animals and placed in a tissue bath is more likely to fatigue than that obtained from normal animals. In some species the adrenocortical hormones enhance contractility. Furthermore, hydrocortisone potentiates the cardiotonic effects of the catecholamines.

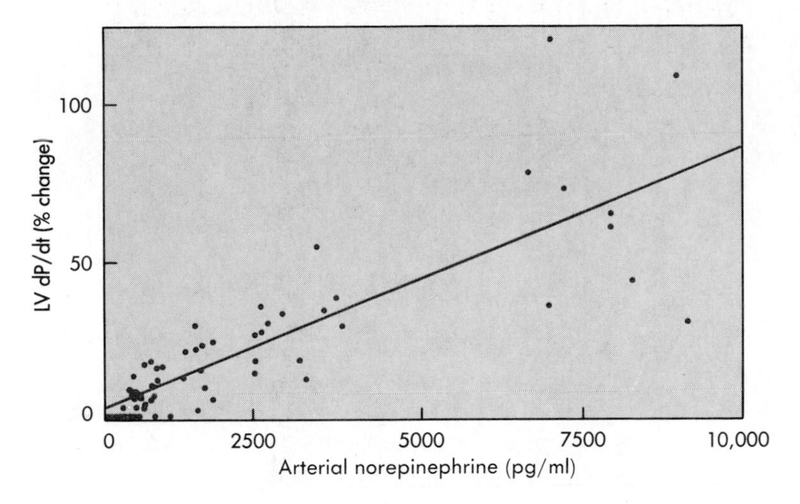

■ **Fig. 25-33** The effect of norepinephrine infusions on ventricular contractility in a group of resting, unanesthetized dogs. The plasma concentrations of norepinephrine *(pg/ml)* plotted along the abscissa are the increments above the control values. The maximum rate of rise of left ventricular pressure *(LV dP/dt)* is plotted along the ordinate as percent change from the control value; it is an index of contractility. (Redrawn from Young MA, Hintze TH, Vatner SF: *Am J Physiol* 248:H82, 1985.)

This potentiation may be mediated in part by an inhibition of the uptake mechanisms for the catecholamines by the adrenocortical steroids.

Thyroid hormones. Cardiac activity is sluggish in patients with inadequate thyroid function **(hypothyroidism);** (i.e., the heart rate is slow and cardiac output is diminished) (see Chapter 49). The converse is true in patients with overactive thyroid glands **(hyperthyroidism).** Characteristically hyperthyroid patients exhibit tachycardia, high cardiac output, palpitations, and arrhythmias (such as atrial fibrillation). In experimental animals the cardiovascular manifestations of hyperthyroidism may be simulated by the administration of thyroxine.

Numerous studies on intact animals and humans have demonstrated that thyroid hormones enhance myocardial contractility. The rates of Ca^{++} uptake and of ATP hydrolysis by the sarcoplasmic reticulum are increased in experimental hyperthyroidism, and the opposite effects occur in hypothyroidism. Thyroid hormones increase protein synthesis in the heart, which leads to cardiac hypertrophy. These hormones also affect the composition of myosin isoenzymes in cardiac muscle. They increase principally those isoenzymes with the greatest ATPase activity, which thereby enhances myocardial contractility.

The cardiovascular changes in thyroid dysfunction also depend on indirect mechanisms. Thyroid hyperactivity increases the body's metabolic rate, and this in turn results in arteriolar vasodilation. The consequent reduction in the total peripheral resistance increases cardiac output, as explained on p. 504. Substantial evidence indicates that with hyperthyroidism either sympathetic neural activity is increased or the sensitivity of the heart to such activity is enhanced. However, other evidence contradicts these conclusions. Studies do agree, however, that thyroid hormone increases the density of beta-adrenergic receptors in cardiac tissue.

Insulin. Insulin has a prominent, direct, positive inotropic effect on the heart (see Chapter 46). The effect of insulin is evident even when hypoglycemia is prevented by glucose infusions and when the beta-adrenergic receptors are blocked. In fact, the positive inotropic effect of insulin is potentiated by beta-adrenergic receptor blockade. The enhancement of contractility cannot be explained satisfactorily by the concomitant augmentation of glucose transport into the myocardial cells.

Glucagon. Glucagon has potent positive inotropic and chronotropic effects on the heart (see Chapter 46). The endogenous hormone probably plays no significant role in the normal regulation of the cardiovascular system, but it has been used to treat a variety of cardiac conditions. The effects of glucagon on the heart closely resemble those of the catecholamines, and certain metabolic effects are similar. Both glucagon and catecholamines activate adenylyl cyclase to increase the myocardial tissue levels of cAMP. The catecholamines activate adenylyl cyclase by interacting with beta-adrenergic receptors, but glucagon activates this enzyme through a different mechanism. Nevertheless, the consequent rise in cAMP increases Ca^{++} influx through the Ca^{++} channels in the sarcolemma and facilitates Ca^{++} release and reuptake by the sarcoplasmic reticulum, just as do the catecholamines.

Anterior pituitary hormones. The cardiovascular derangements in hypopituitarism are related principally to the associated deficiencies in adrenocortical and thyroid function (see Chapter 48). Growth hormone does affect the myocardium, at least in combination with thyroxine. In hypophysectomized animals growth hormone alone has little effect on the depressed heart, whereas thyroxine by itself restores adequate cardiac performance under basal conditions. However, when blood volume or peripheral resistance is increased, thyroxine alone does not restore adequate cardiac function; but the combination of growth hormone and thyroxine does reestablish normal cardiac performance.

Blood gases

Oxygen. Changes in oxygen tension (Pa_{O_2}) of the blood perfusing the brain and the peripheral chemoreceptors affect the heart through nervous mechanisms, as

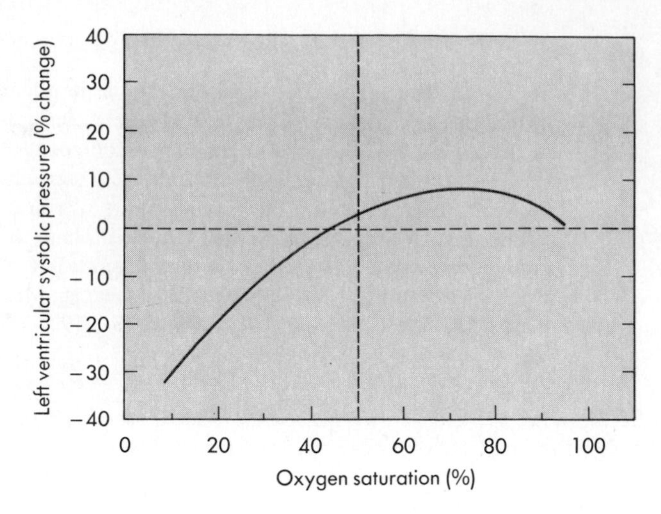

■ **Fig. 25-34** In the isovolumic left ventricle preparation, a reduction in the O_2 saturation of coronary arterial blood to between 45% and 100% stimulates ventricular contractility, whereas an O_2 saturation below 45% depresses ventricular contractility. (Redrawn from Ng ML et al: *Am J Physiol* 211:43, 1966.)

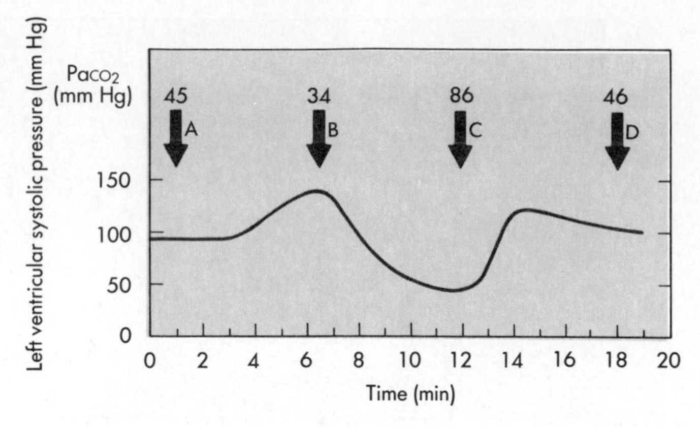

■ **Fig. 25-35** Decrease in $Paco_2$ increases left ventricular systolic pressure (arrow *B*) in an isovolumic left ventricle preparation; a rise in $Paco_2$ (arrow *C*) has the reverse effect. When the $Paco_2$ is returned to the control level (arrow *D*), left ventricular systolic pressure returns to its original value (arrow *A*). (Levy MN: unpublished tracing.)

described earlier in this chapter. These indirect effects of hypoxia are usually prepotent. Moderate degrees of hypoxia characteristically increase heart rate, cardiac output, and myocardial contractility. These changes are largely abolished by beta-adrenergic receptor blockade.

The Pao_2 of the blood perfusing the myocardium also influences myocardial performance directly. The effect of hypoxia is biphasic; moderate degrees are stimulatory and more severe degrees are depressant. As shown in Fig. 25-34, when the O_2 saturation is reduced to levels below 50% in isolated hearts, the peak left ventricular pressures are less than the control levels. However, with less severe degrees of hypoxia (O_2 saturation >50%), the peak pressures exceed the control level.

Carbon dioxide and acidosis. Changes in $Paco_2$ may also affect the myocardium directly and indirectly. The indirect, neurally mediated effects produced by increased $Paco_2$ are similar to those evoked by a decrease in Pao_2.

With respect to the direct effects on the heart, alterations in myocardial performance elicited by changes of $Paco_2$ in the coronary arterial blood are illustrated in Fig. 25-35. In this experiment on an isolated left ventricle preparation, the control $Paco_2$ was 45 mm Hg (arrow *A*). Decreasing the $Paco_2$ to 34 mm Hg (arrow *B*) was stimulatory, whereas increasing $Paco_2$ to 86 mm (arrow *C*) was depressant. In intact animals, systemic hypercapnia activates the sympathoadrenal system, which tends to compensate for the direct depressant effect of the increased $Paco_2$ on the heart.

Neither the $Paco_2$ nor the blood pH are primary determinants of myocardial behavior. The resultant change in intracellular pH is the critical factor. The reduced intracellular pH diminishes the amount of Ca^{++} released from the sarcoplasmic reticulum in response to excitation. The diminished pH also decreases the sensitivity of the myofilaments to Ca^{++}. When they are exposed to a given concentration of Ca^{++}, the lower the prevailing pH, the less force the myofibrils develop.

■ *Summary*

1. Cardiac function is regulated by a number of intrinsic and extrinsic mechanisms.

2. Heart rate is regulated mainly by the autonomic nervous system. Sympathetic nervous activity increases heart rate, whereas parasympathetic (vagal) activity decreases heart rate. When both systems are active, the vagal effects usually dominate.

3. The following reflexes regulate heart rate: baroreceptor, chemoreceptor, pulmonary inflation, atrial receptor (Bainbridge), and ventricular receptor reflexes.

4. The principal intrinsic mechanisms that regulate myocardial contraction are the Frank-Starling mechanism and rate-induced regulation.
a. Frank-Starling mechanism: a change in the resting length of the muscle influences the subsequent contraction by altering the number of interacting cross-bridges between the thick and thin filaments and by altering the affinity of the myofilaments for calcium.
b. Rate-induced regulation: a sustained change in contraction frequency affects the strength of contraction by altering the influx of Ca^{++} into the cell per minute, whereas a transient change in contraction frequency alters contractile strength because an appreciable delay exists between the time that Ca^{++} is taken up by the sarcoplasmic reticulum and the time that it becomes available again for release.

5. The autonomic nervous system regulates myocardial performance mainly by varying the Ca^{++} conductance of the cell membrane via the adenylyl cyclase system.

6. Various hormones, including epinephrine, adrenocortical steroids, thyroid hormones, insulin, glucagon, and anterior pituitary hormones regulate myocardial performance.

7. Changes in the blood concentrations of O_2, CO_2, and H^+ alter cardiac function directly and via the chemoreceptors reflexly.

Bibliography

Journal articles

Bouchard RA, Bose D: Analysis of the interval-force relationship in rat and canine ventricular myocardium, *Am J Physiol* 257:H2036, 1989.

Cantin M, Genest J: The heart as an endocrine gland, *Hypertension* 10(suppl):1-118-121, 1987.

Chernow B, et al: Glucagon: endocrine effects and calcium involvement in cardiovascular actions in dogs, *Circ Shock* 19:393, 1986.

Dampney RAL: Functional organization of central cardiovascular pathways, *Clin Exp Pharmacol Physiol* 8:241, 1981.

Endoh M, Blinks JR: Actions of sympathomimetic amines on the Ca^{2+} transients and contractions of rabbit myocardium: reciprocal changes in myofibrillar responsiveness to Ca^{2+} mediated through α- and β-adrenoceptors, *Circ Res* 62:247, 1988.

Farah AE: Glucagon and the circulation, *Pharmacol Rev* 35:181, 1983.

Hainsworth R: Reflexes from the heart, *Physiol Rev* 71:617, 1991.

Hakumäki MOK: Seventy years of the Bainbridge reflex, *Acta Physiol Scand* 130:177, 1987.

Hathaway DR, March KL: Molecular cardiology: new avenues for the diagnosis and treatment of cardiovascular disease, *J Am Coll Cardiol* 13:265, 1989.

Josephson RA, Spurgeon HA, Lakatta EG: The hyperthyroid heart. An analysis of systolic and diastolic properties in single rat ventricular myocytes, *Circ Res* 66:773, 1990.

Katona PG et al: Sympathetic and parasympathetic cardiac control in athletes and nonathletes at rest, *J Appl Physiol* 52:1652, 1982.

Klein I, Levey GS: New perspectives on thyroid hormone, catecholamines, and the heart, *Am J Med* 76:167, 1984.

Kohmoto O et al: Effects of intracellular acidosis on $(Ca^{2+})_i$ transients, transsarcolemmal Ca^{2+} fluxes, and contraction in ventricular myocytes, *Circ Res* 66:622, 1990.

Kollai M, Koizumi, K: Cardiac vagal and sympathetic nerve responses to baroreceptor stimulation in the dog, *Pflügers Arch* 413:365, 1989.

Kuhn HJ, Bletz C, Rüegg JC: Stretch-induced increase in the Ca^{2+} sensitivity of myofibrillar ATPase activity in skinned fibres from pig ventricles. *Pflügers Arch* 415:741, 1990.

Kurihara S, Konishi M: Effects of β-adrenoceptor stimulation on intracellular Ca transients and tension in rat ventricular muscle, *Pflügers Arch* 409:427, 1987.

Lakatta EG: Starling's law of the heart is explained by an intimate interaction of muscle length and myofilament calcium activation, *J Am Coll Cardiol* 10:1157, 1987.

Levy MN: Autonomic interactions in cardiac control, *Ann NY Acad of Sci* 601:209, 1990.

Löffelholz K, Pappano AJ: The parasympathetic neuroeffector junction of the heart, *Pharmacol Rev* 37:1, 1985.

Reiter M: Calcium mobilization and cardiac inotropic mechanisms, *Pharmacol Rev* 40:189, 1988.

Seed WA, Walker JM: Relation between beat interval and force of the heartbeat and its clinical implications, *Cardiovasc Res* 22:303, 1988.

Winegrad S: Regulation of cardiac contractile proteins, *Circ Res* 55:565, 1984.

Young MA, Hintze TH, Vatner SF: Correlation between cardiac performance and plasma catecholamine levels in conscious dogs, *Am J Physiol* 248:H82, 1985.

Books and monographs

Bishop VS, Malliani A, Thorén P: *Cardiac mechanoreceptors*. In *Handbook of physiology:* section 2: *The cardiovascular system—peripheral circulation and organ blood flow,* vol 3, Bethesda, Md, 1983, American Physiological Society.

Levine HJ, Gaasch WH, editors: *The ventricle: basic and clinical aspects,* Hingham, Mass, 1985, Martinus Nijhoff Publishers.

Levy MN, Martin PJ: *Neural control of the heart.* In *Handbook of physiology:* section 2: *Cardiovascular system—the heart, vol. 1,* Washington, DC, 1979, American Physiological Society.

Opie LH: *Heart: physiology and metabolism,* ed 2, New York, 1991, Raven Press.

Persson PB, Kirchheim HR, editors: *Baroreceptor reflexes: integrative functions and clinical aspects,* Berlin, 1991, Springer-Verlag.

Randall WC, editor: *Nervous control of cardiovascular function,* New York, 1984, Oxford University Press.

Sagawa K et al: *Cardiac contraction and the pressure-volume relationship,* New York, 1988, Oxford University Press.

Sperelakis N, editor: *Physiology and pathophysiology of the heart,* ed 2, Boston, 1989, Kluwer Academic Publishers.

Zucker IH, Gilmore JP: *Reflex control of the circulation,* Boca Raton, 1990, CRC Press.

CHAPTER

26

Hemodynamics

The problem of treating the pulsatile flow of blood through the cardiovascular system in precise mathematical terms is insuperable. The heart is a complicated pump, and its behavior is affected by a variety of physical and chemical factors. The blood vessels are multibranched, elastic conduits of continuously varying dimensions. The blood itself is not a simple, homogeneous solution, but instead is a complex suspension of red and white corpuscles, platelets, and lipid globules dispersed in a colloidal solution of proteins.

Despite these complicating factors, considerable insight may be gained from understanding the elementary principles of fluid mechanics as they pertain to simple physical systems. Such principles will be expounded in this chapter to explain the interrelationships among velocity of blood flow, blood pressure, and the dimensions of the various components of the systemic circulation.

■ *Velocity of the Bloodstream*

In describing the variations in blood flow in different vessels it is first essential to distinguish between the terms **velocity** and **flow.** The former term, sometimes designated as **linear** velocity, refers to the rate of displacement with respect to time and has the dimensions of distance per unit time, for example, cm/sec. The latter term is frequently designated as **volume** flow and has the dimensions of volume per unit time, for example, cm^3/sec. In a conduit of varying cross-sectional dimensions, velocity, v, flow, Q, and cross-sectional area, A, are related by the equation:

$$v = Q/A \qquad (1)$$

The interrelationships among velocity, flow, and area are portrayed in Fig. 26-1. The flow of an incompressible fluid past successive cross sections of a rigid tube must be constant. For a given constant flow the velocity varies inversely as the cross-sectional area (see Fig. 22-2). Thus for the same volume of fluid per second passing from section *a* into section *b*, where the cross-sectional area is five times greater, the velocity diminishes to one fifth of its previous value. Conversely, when the fluid proceeds from section *b* to section *c*, where the cross-sectional area is one tenth as great, the velocity of each particle of fluid must increase tenfold.

The velocity at any point in the system depends not only on area, but also on the flow, Q. This, in turn, depends on the pressure gradient, properties of the fluid, and dimensions of the entire hydraulic system, as discussed in the following section. For any given flow, however, the ratio of the velocity past one cross section relative to that past a second cross section depends only on the inverse ratio of the respective areas; that is,

$$v_1/v_2 = A_2/A_1 \qquad (2)$$

This rule pertains regardless of whether a given cross-sectional area applies to a system that consists of a single large tube or to a system that is composed of several smaller tubes in parallel.

As shown in Fig. 22-2, velocity decreases progressively as the blood traverses the aorta, its larger primary branches, the smaller secondary branches, and the arterioles. Finally a minimum value is reached in the capillaries. As the blood then passes through the venules and continues centrally toward the venae cavae, the velocity progressively increases again. The relative velocities in the various components of the circulatory system are related only to the cross-sectional area. Thus each point on the cross-sectional area curve is inversely proportional to the corresponding point on the velocity curve (see Fig. 22-2).

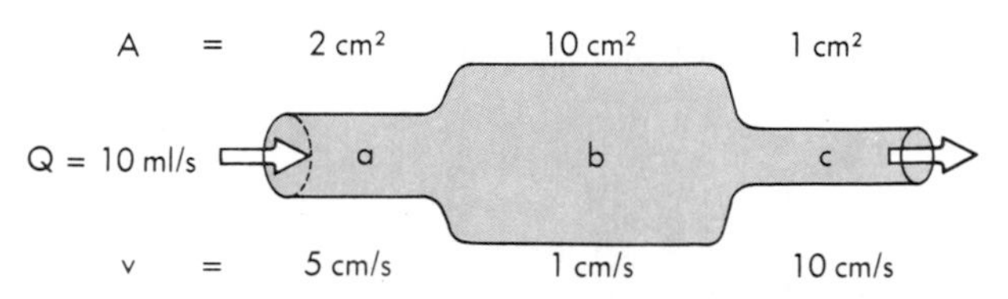

■ **Fig. 26-1** As fluid flows through a tube of variable cross-sectional area, *A*, the linear velocity, *v*, varies inversely as the cross-sectional area.

■ *Relationship Between Velocity and Pressure*

In that portion of a hydraulic system in which the total energy remains virtually constant, changes in velocity may be accompanied by appreciable alterations in the measured pressure. Consider three sections (*A, B,* and *C*) of such a hydraulic system, as depicted in Fig. 26-2. Six pressure probes, or **pitot tubes,** have been inserted. The opening of three of these (*2, 4,* and *6*) are tangential to the direction of flow and hence measure the **lateral,** or **static,** pressure within the tube. The openings of the remaining three pitot tubes (*1, 3,* and *5*) face upstream. Therefore they detect the **total pressure,** which is the lateral pressure plus a dynamic pressure component ascribable to the kinetic energy of the flowing fluid. This dynamic component, P_d, of the total pressure may be calculated from the following equation:

$$P_d = \frac{1}{2}\rho v^2 \qquad (3)$$

where ρ is the density of the fluid, and v is the velocity.

If the midpoints of segments *A, B,* and *C* are at the same hydrostatic level, then the corresponding total pressure, P_1, P_3, and P_5, will be equal, provided that the energy loss from viscosity in these segments is negligible. However, because of the changes in cross-sectional area, the concomitant velocity changes alter the dynamic component.

In sections *A* and *C,* let $\rho = 1$ g/cm^3 and $v = 100$ cm/s. From equation 3

$$P_d = 5000 \text{ dynes/cm}^2$$
$$= 3.8 \text{ mm Hg}$$

because 1330 dynes/cm^2 = 1 mm Hg. In the narrow section, *B,* let the velocity be twice as great as in sections *A* and *C*. Therefore,

$$P_d = 20{,}000 \text{ dynes/cm}^2$$
$$= 15 \text{ mm Hg}$$

Hence, in the wide sections of the conduit the lateral pressures (P_2 and P_6) will be only 3.8 mm Hg less than the respective total pressures (P_1 and P_5), whereas in

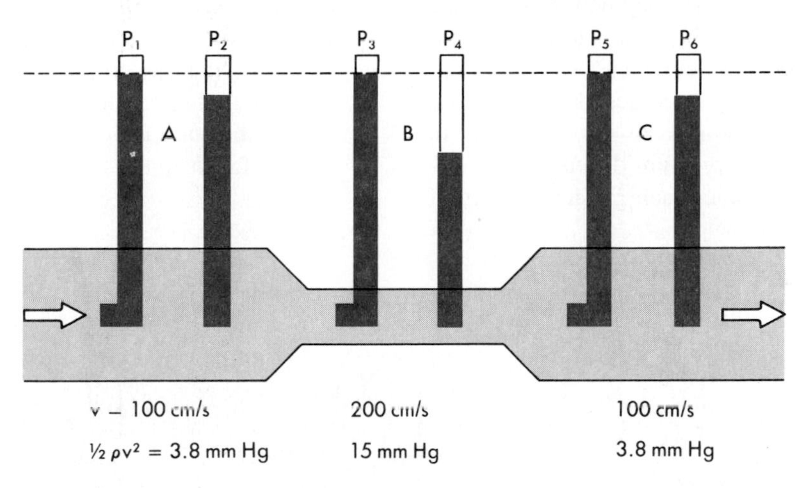

■ **Fig. 26-2** In a narrow section, **B,** of a tube, the linear velocity, *v,* and hence the dynamic component of pressure, $\frac{1}{2}\rho v^2$, are greater than in the wide sections, **A** and **C,** of the same tube. If the total energy is virtually constant throughout the tube (i.e., if the energy loss because of viscosity is negligible), the total pressures (P_1, P_3, and P_5) will not be detectably different, but the lateral pressure, P_4, in the narrow section will be less than the lateral pressures (P_2 and P_6) in the wide sections of the tube.

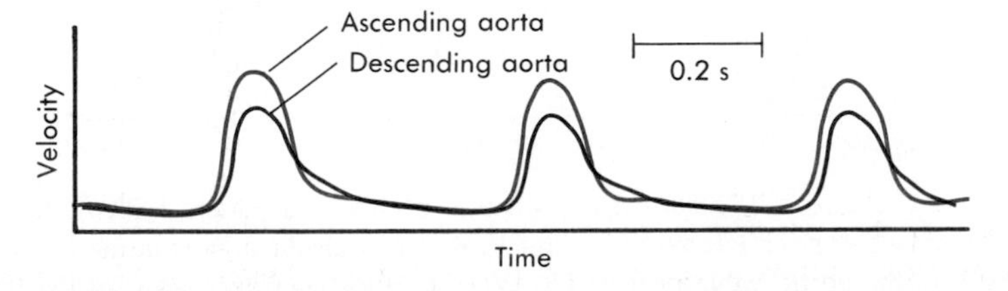

■ **Fig. 26-3** Velocity of the blood in the ascending and descending aorta of a dog. (Redrawn from Falsetti HL et al: *Circ Res* 31:328, 1972, by permission of the American Heart Association.)

the narrow section the lateral pressure (P_4) is 15 mm Hg less than the total pressure (P_3).

The peak velocity of flow in the ascending aorta of normal dogs is about 150 cm/sec. Therefore the measured pressure may vary significantly, depending on the orientation of the pressure probe. In the descending thoracic aorta the peak velocity is substantially less than that in the ascending aorta (Fig. 26-3), and lesser velocities have been recorded in still more distal arterial sites. In most arterial locations the dynamic component will be a negligible fraction of the total pressure, and the orientation of the pressure probe will not materially influence the pressure recorded. At the site of a constriction, however, the dynamic pressure component may attain substantial values. In **aortic stenosis,** for example, the entire output of the left ventricle is ejected through a narrow valve orifice. The high flow velocity is associated with a large kinetic energy, and therefore the lateral pressure is correspondingly reduced.

The pressure tracings shown in Fig. 26-4 were obtained from two pressure transducers inserted into the left ventricle of a patient with aortic stenosis. The transducers were located on the same catheter and were 5 cm apart. When both transducers were well within the left ventricular cavity (Fig. 26-4, *A*), they both recorded the same pressures. However, when the proximal transducer was positioned in the aortic valve orifice (Fig. 26-4, *B*), the lateral pressure recorded during ejection was much less than that recorded by the transducer in the ventricular cavity. This pressure difference was ascribable almost entirely to the much greater velocity of flow in the narrowed valve orifice than in the ventricular cavity. The pressure difference reflects mainly the conversion of some potential energy to kinetic energy. When the catheter was withdrawn still farther, so that the proximal transducer was in the aorta (Fig. 26-4, *C*), the pressure difference was even more pronounced, because substantial energy was lost through friction (viscosity) as blood flowed rapidly through the narrow orifice.

The reduction of lateral pressure in the region of the stenotic valve orifice influences the coronary blood flow in patients with aortic stenosis. The orifices of the right and left coronary arteries are located in the sinuses of Valsalva, just behind the valve leaflets. Hence the initial segments of these vessels are oriented at right angles to the direction of blood flow through the aortic valves. Therefore the lateral pressure is that component of the total pressure that propels the blood through the two major coronary arteries. During the ejection phase of the cardiac cycle, the lateral pressure is diminished by the conversion of potential energy to kinetic energy. This process is grossly exaggerated in aortic stenosis because of the high flow velocities. Angiographic studies in patients with aortic stenosis have revealed that the direction of flow often reverses in the large coronary arteries toward the end of the ejection phase of systole; (i.e., blood flows toward the aorta rather than toward the myocardial capillaries). The decreased lateral pressure in the aorta in aortic stenosis is undoubtedly an important factor in causing this reversal of coronary blood flow.

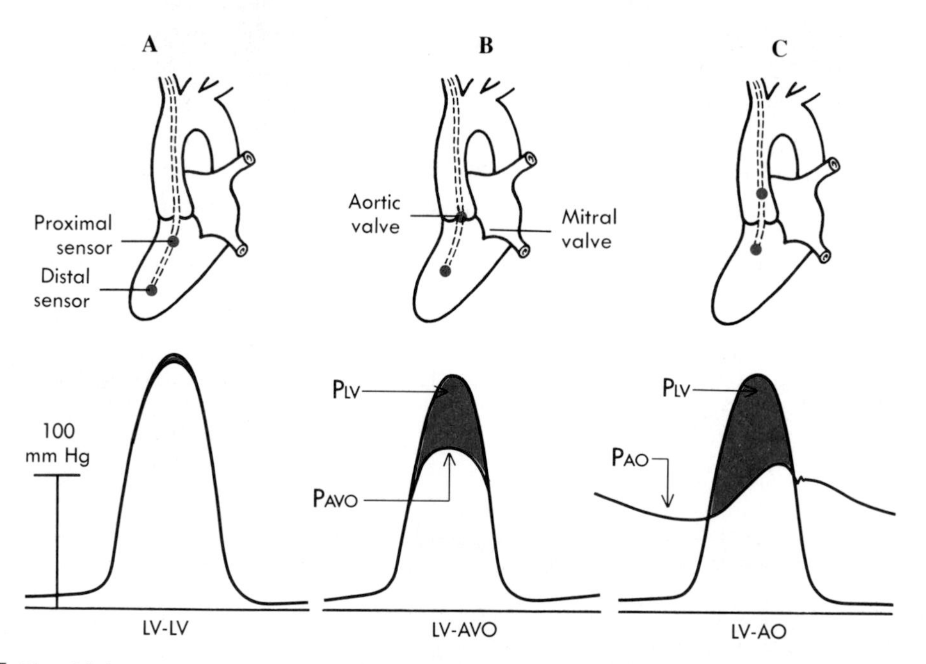

■ **Fig. 26-4** Pressures *(P)* recorded by two transducers in a patient with aortic stenosis. **A,** Both transducers were in the left ventricle *LV-LV*. **B,** One transducer was in the left ventricle, and the other was in the aortic valve orifice *(LV-AVO)*. **C,** One transducer was in the left ventricle, and the other was in the ascending aorta *(LV-AO)*. (Redrawn from Pasipoularides A et al: *Am J Physiol* 246:H542, 1984.)

An important aggravating feature in this condition is that the demands of the heart muscle for oxygen are greatly increased. Therefore *the pronounced drop in lateral pressure during cardiac ejection may contribute to the tendency for patients with severe aortic stenosis to have angina pectoris* (anterior chest pain associated with inadequate blood supply to the heart muscle) and to die suddenly.

■ *Relationship Between Pressure and Flow*

The most fundamental law governing the flow of fluids through cylindrical tubes was derived empirically by Poiseuille. He was primarily interested in the physical determinants of blood flow but substituted simpler liquids for blood in his measurements of flow through glass capillary tubes. His work was so precise and im portant that his observations have been designated **Poiseuille's law.** Subsequently, this same law has been derived theoretically.

Poiseuille's law is applicable to the flow of fluids through cylindrical tubes only under special conditions. It applies to the case of steady, laminar flow of newtonian fluids. The term **steady flow** signifies the absence of variations of flow in time (i.e., a nonpulsatile flow). **Laminar flow** is the type of motion in which the fluid moves as a series of individual layers, with each stratum moving at a different velocity from its neighboring layers (see Fig. 26-12). In the case of flow through a tube, the fluid consists of a series of infinitesimally thin concentric tubes sliding past one another. Laminar flow will be described in greater detail below, where it will be distinguished from turbulent flow. Also a newtonian fluid will be defined more precisely. For the present discussion it may be considered to be a homogeneous fluid, such as water, in contradistinction to a suspension, such as blood.

Pressure is one of the principal determinants of the rate of flow. The pressure, P, in dynes/cm^2, at a distance h centimeters below the surface of a liquid is

$$P = h\rho g \tag{4}$$

where ρ is the density of the liquid in g/cm^3 and g is the acceleration of gravity in cm/sec^2. For convenience, however, pressure is frequently expressed simply in terms of height (h) of the column of liquid above some arbitrary reference point.

Consider the tube connecting reservoirs R_1 and R_2 in Fig. 26-5, *A*. Let reservoir R_1 be filled with liquid to height h_1, and let reservoir R_2 be empty, as in section *1*

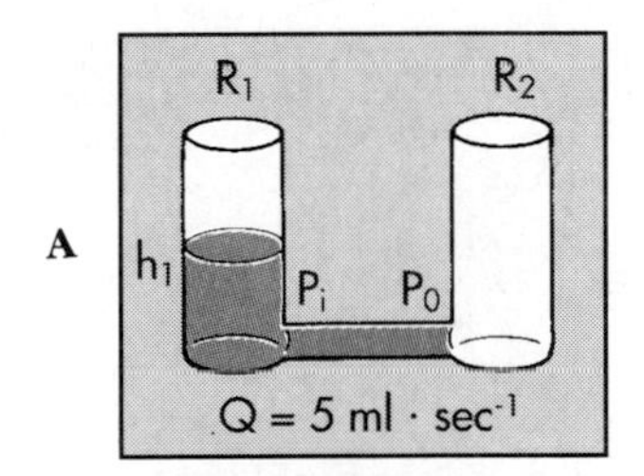

A, When R_1 is empty, fluid flows from R_1 to R_2 at a rate proportional to the pressure in R_1.

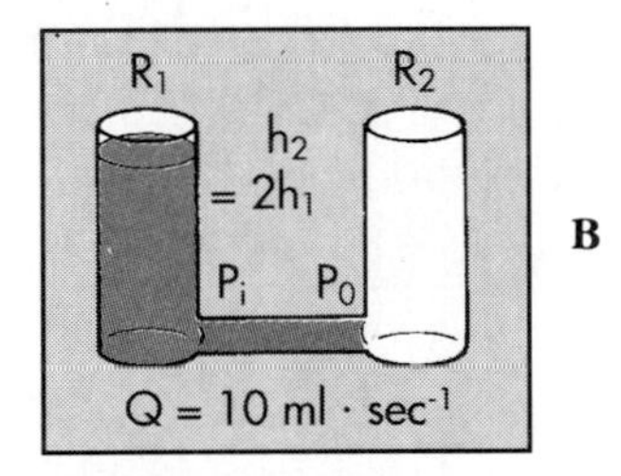

B, When the fluid level in R_1 is increased twofold, the flow increases proportionately.

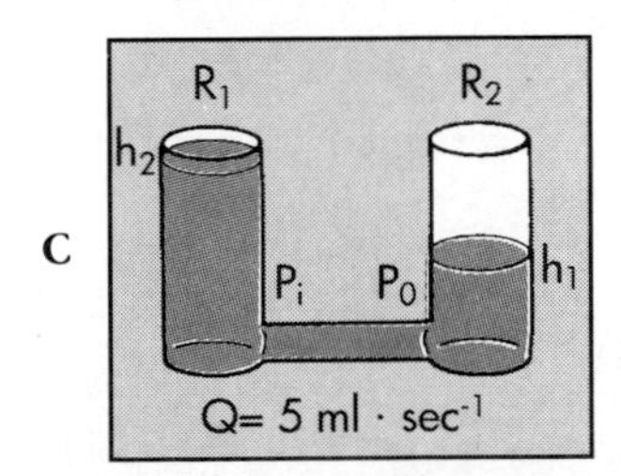

C, Flow from R_1 to R_2 is proportional to the difference between the pressures in R_1 and R_2.

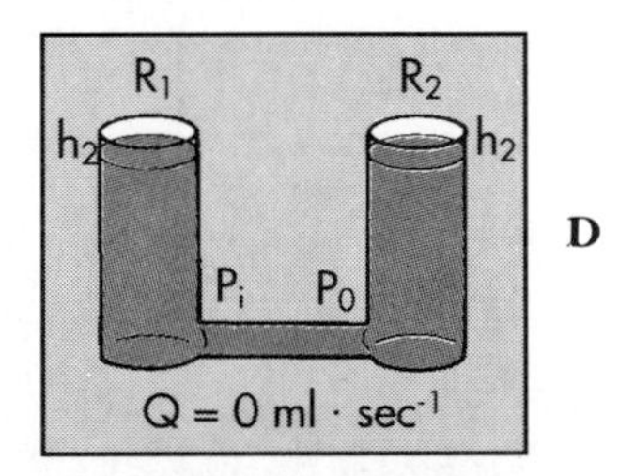

D, When pressure in R_2 rises to equal the pressure in R_1, flow ceases in the connecting tube.

■ **Fig. 26-5** The flow, *Q*, of fluid through a tube connecting two reservoirs, R_1 and R_2, is proportional to the difference between the pressure, P_i, at the inflow end and the pressure, P_o, at the outflow end of the tube. **A,** When R_2 is empty, fluid flows from R_1 to R_2 at a rate proportional to the pressure in R_1. **B,** When the fluid level in R_1 is increased twofold, the flow increases proportionately. **C,** Flow from R_1 to R_2 is proportional to the difference between the pressures in R_1 and R_2. **D,** When pressure in R_2 rises to equal the pressure in R_1, flow ceases in the connecting tube.

of Fig. 26-5. The outflow pressure, P_o, is therefore equal to the atmospheric pressure, which shall be designated as the zero, or reference, level. The inflow pressure, P_1, is then equal to the same reference level plus the height, h_1, of the column of liquid in reservoir R_1. Under these conditions let the flow (Q) through the tube be 5 ml/sec.

If reservoir R_1 is filled to height h_2, which is twice h_1, and reservoir R_2 is again empty (as in panel *B*), the flow will be twice as great (i.e., 10 ml/sec. Thus with reservoir R_2 empty, *the flow will be directly proportional to the inflow pressure,* P_1.

If reservoir R_2 is now allowed to fill to height h_1, and the fluid level in R_1 is maintained at h_2 (as in panel *C*), the flow will again become 5 ml/sec. Thus *flow is directly proportional to the difference between inflow and outflow pressures:*

$$Q \propto P_i - P_o \tag{5}$$

If the fluid level in R_2 attains the same height as in R_1, flow will cease (panel *D*).

For any given pressure difference between the two ends of a tube, the flow will depend on the dimensions of the tube. Consider the tube connected to the reservoir in Fig. 26-6, *A*. With length l_1 and radius r_1, the flow Q_1 is observed to be 10 ml/sec.

The tube connected to the reservoir in panel *B* has the same radius but is twice as long. Under these conditions the flow Q_2 is found to be 5 ml/sec, or only half as great as Q_1. Conversely, for a tube half as long as l_1 the flow would be twice as great as Q_1. In other words, *flow is inversely proportional to the length of the tube:*

$$Q \propto 1/l \tag{6}$$

The tube connected to the reservoir in Fig. 26-6, *C* is the same length as l_1, but the radius is twice as great. Under these conditions the flow Q_3 is found to increase to a value of 160 ml/sec, which is sixteen times greater than Q_1. The precise measurements of Poiseuille revealed that *flow varies directly as the fourth power of the radius:*

$$Q \propto r^4 \tag{7}$$

Thus, in the example above, because $r_3 = 2r_1$, Q_3 will be proportional to $(2r_1)^4$, or $16r_1^4$; therefore, Q_3 will equal $16Q_1$.

Finally, for a given pressure difference and for a cylindrical tube of given dimensions, the flow will vary as a function of the nature of the fluid itself. This flow-determining property of fluids is termed **viscosity,** η, which has been defined by Newton as the ratio of **shear stress** to the **shear rate** of the fluid.

A

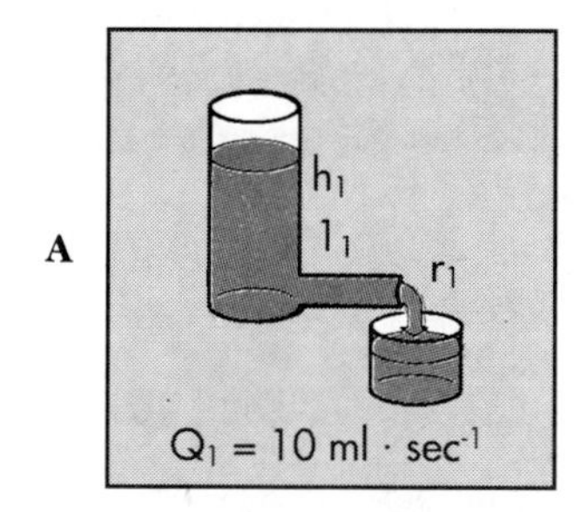

A, Reference condition: for a given pressure, length, radius, and viscosity, let the flow (V_I) equal 10 ml·s^{-1}.

B

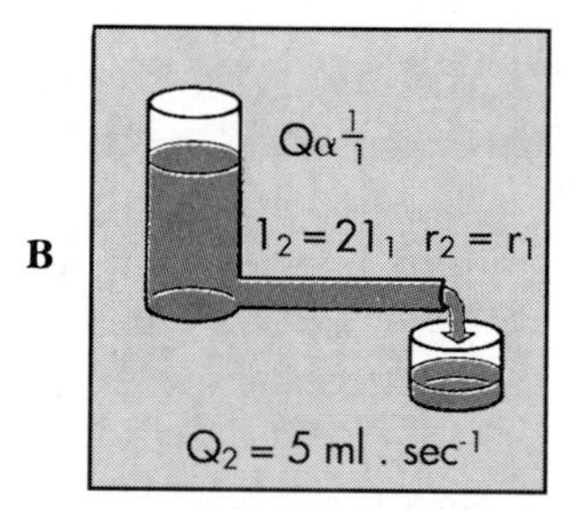

B, If tube length doubles, flow decreases by 50%.

C

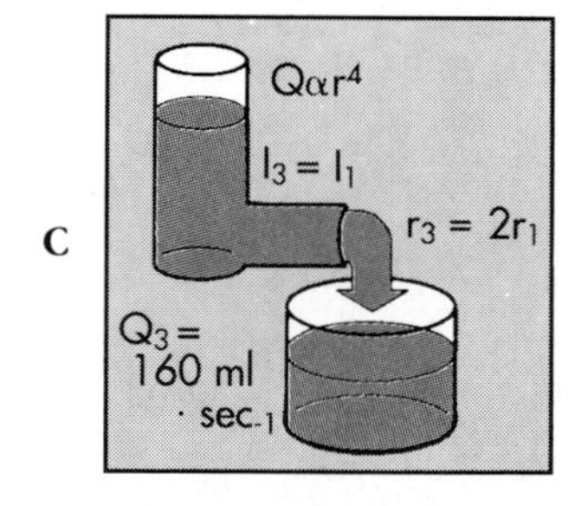

C, If tube radius doubles, flow increases 16-fold.

D

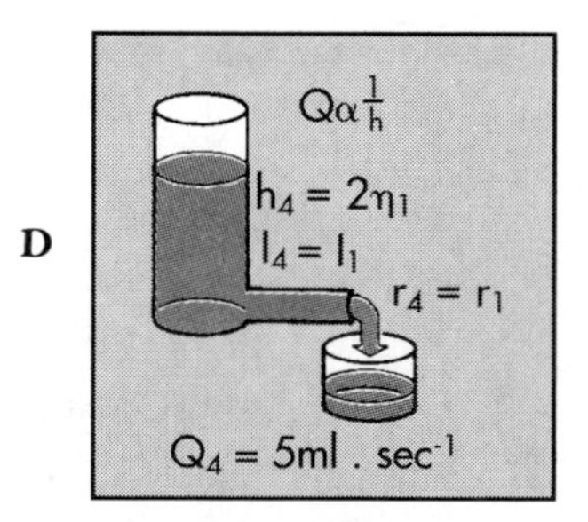

D, If viscosity doubles, flow decreases by 50%.

■ **Fig. 26-6** The flow, Q, of fluid through a tube is inversely proportional to the length, l, and the viscosity, η, and is directly proportional to the fourth power of the radius, r. **A,** Reference condition: for a given pressure, length, radius, and viscosity, let the flow (Q_I) equal 10 ml · sec $^{-1}$. **B,** If tube length doubles, flow decreases by 50%. **C,** If tube radius doubles, flow increases sixteen fold. **D,** If viscosity doubles, flow decreases by 50%.

These terms may be comprehended most clearly by considering the flow of a homogeneous fluid between parallel plates. In Fig. 26-7 let the bottom plate (the bottom of a large basin) be stationary, and let the upper plate move at a constant velocity along the upper surface of the fluid. The **shear stress** τ, is defined as the ratio of F:A, where F is the force applied to the upper plate in the direction of its motion along the upper surface of the fluid and A is the area of the upper plate in contact with the fluid. The shear rate is du/dy, where u is the velocity of a minute fluid element in the direction parallel to the motion of the upper plate and y is the distance of that fluid element above the bottom, stationary plate.

For a movable plate traveling with constant velocity across the surface of a homogeneous fluid, the velocity profile of the fluid will be linear. The fluid layer in contact with the upper plate will adhere to it and therefore will move at the same velocity, U, as the plate. Each minute element of fluid between the plates will move at a velocity, u, proportional to its distance, y, from the lower plate. Therefore, the shear rate will be U/Y where Y is the total distance between the two plates. Because viscosity, η, is defined as the ratio of shear stress, τ, to the shear rate du/dy, in the example illustrated in Fig. 26-7,

$$\eta = (F/A)/(U/Y) \tag{8}$$

Thus the dimensions of viscosity are dynes/cm^2 divided by (cm/sec)/cm, or dyne-sec-cm^{-2}. In honor of Poiseuille, 1 dyne-sec-cm^{-2} has been termed a **poise.** The viscosity of water at 20° C is approximately 0.01 poise, or 1 centipoise. In the case of certain nonhomogeneous fluids, notably suspensions such as blood, the ratio of the shear stress to the shear rate is not constant; such fluids are said to be **nonnewtonian.**

With regard to the flow of newtonian fluids through cylindrical tubes, the flow will vary inversely as the viscosity. Thus in the example of flow from the reservoir in Fig. 26-6, *D,* if the viscosity of the fluid in the reservoir were doubled, then the flow would be halved (5 ml/sec instead of 10 ml/sec).

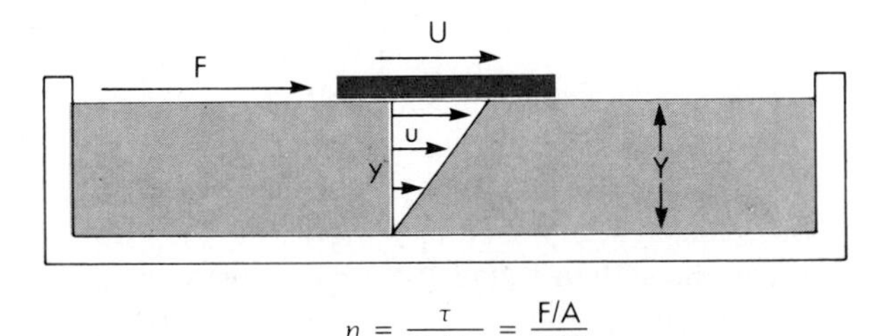

$$\eta = \frac{\tau}{du/dy} = \frac{F/A}{U/Y}$$

■ **Fig. 26-7** For a newtonian fluid, the viscosity, η, is defined as the ratio of shear stress, π, to shear rate, *du/dy*. For a plate of contact area, *A*, moving across the surface of a liquid, π equals the ratio of the force, *F*, applied in the direction of motion to the contact area, *A*, and *du/dy* equals the ratio of the velocity of the plate, *U*, to the depth of the liquid, *Y*.

In summary, *for the steady, laminar flow of a newtonian fluid through a cylindrical tube, the flow, Q, varies directly as the pressure difference,* $P_i - P_o$, *and the fourth power of the radius, r, of the tube, and it varies inversely as the length, l, of the tube and the viscosity,* η, *of the fluid.* The full statement of **Poiseuille's law** is

$$Q = \frac{\pi(P_i - P_o)r^4}{8\eta l} \tag{9}$$

where $\pi/8$ is the constant of proportionality.

■ *Resistance to Flow*

In electrical theory the resistance, R, is defined as the ratio of voltage drop, E, to current flow, I. Similarly, in fluid mechanics the hydraulic resistance, R, may be defined as the ratio of pressure drop, $P_i - P_o$, to flow, Q; P_I and P_o are the pressures at the inflow and outflow ends, respectively, of the hydraulic system. For the steady, laminar flow of a newtonian fluid through a cylindrical tube, the physical components of hydraulic resistance may be appreciated by rearranging Poiseuille's law to give the **hydraulic resistance equation**

$$R = \frac{P_i - P_o}{Q} = \frac{8\eta l}{\pi r^4} \tag{10}$$

Thus, when Poiseuille's law applies, the resistance to flow depends only on the dimensions of the tube and on the characteristics of the fluid.

The principal determinant of the resistance to blood flow through any individual vessel within the circulatory system is its caliber. The resistance to flow through small blood vessels in cat mesentery has been measured, and the resistance per unit length of vessel (R/l) is plotted against the vessel diameter in Fig. 26-8. The resistance is highest in the capillaries (diameter 7 μm), and it diminishes as the vessels increase in diameter on the arterial and venous sides of the capillaries. The values of R/l were found to be virtually proportional to the fourth power of the diameter for the larger vessels on both sides of the capillaries.

Changes in vascular resistance induced by natural stimuli occur by changes in radius. The principal changes are achieved by alterations in the contraction of the circular smooth muscle cells in the vessel wall. However, changes in internal pressure also alter the caliber of the blood vessels, and therefore alter the resistance to blood flow through those vessels. The blood vessels are elastic tubes. Hence, the greater the **transmural pressure** (i.e., the difference between the internal and external pressures), the greater will be the caliber of the vessel, and the less will be its hydraulic resistance.

From Fig. 22-2 it may be noted that the greatest upstream to downstream drop in internal pressure occurs in the arterioles. Because the total flow is the same

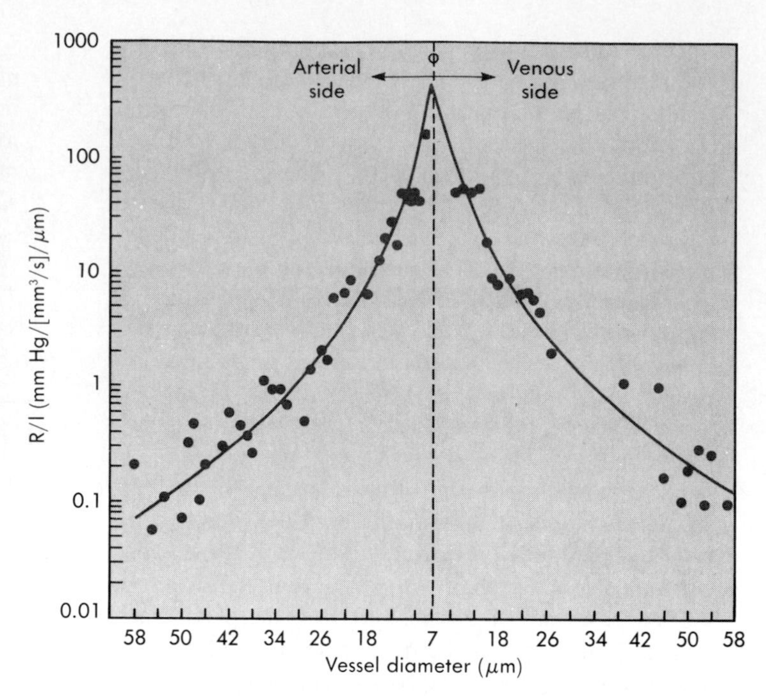

■ **Fig. 26-8** The resistance per unit length *(R/l)* for individual small blood vessels in the cat mesentery. The capillaries, diameter 7 μm, are denoted by the vertical dashed line. Resistances of the arterioles are plotted to the left, and resistances of the venules are plotted to the right of the vertical dashed line. The solid circles represent the actual data. The two curves through the data represent the following regression equations for the arteriole and venule date, respectively: (a) arterioles, $R/l = 1.02 \times 10^6 D^{-4.04}$, and (b) venules, $R/l = 1.07 \times 10^6 D^{-3.94}$. Note that for both types of vessels, the resistance per unit length is inversely proportional to the fourth power (within 1%) of the vessel diameter *(D)*. (Redrawn from Lipowsky HH, Kovalcheck S, Zweifach BW: *Circ Res* 43:738, 1978, by permission of the American Heart Association.)

through the various series components of the circulatory system, it follows that the greatest resistance to flow resides in the arterioles. For example, if R_a represents the resistance of the arterioles, and R_x represents the resistance of any other component of the vascular system in series with the arterioles, then by the definition of hydraulic resistance (equation 10),

$$R_a = (P_i - P_o)_a/Q_a \text{ for the arterioles, and}$$
$$R_x = (P_i - P_o)_x/Q_x \text{ for the other component.}$$

But because the two components are in series, $Q_a = Q_x$, as stated above. Therefore

$$R_a/R_x = (P_i - P_o)_a/(P_i - P_o)_x \qquad (11)$$

That is, *the ratio of the pressure drop across the length of the arterioles to the pressure drop across the length of any other series component of the vascular system is equal to the ratio of the hydraulic resistances of these two vascular components*. The reason why the highest resistance does not reside in the capillaries (as might otherwise be suspected from Fig. 26-8) is related to the relative numbers of parallel capillaries and parallel arterioles, as explained below (p. 445). The arterioles are vested with a thick coat of circularly arranged smooth muscle fibers, by means of which the lumen radius may be varied. From the hydraulic resistance equation, wherein R varies inversely as r^4, it is clear that small changes in radius will alter resistance greatly.

In the cardiovascular system the various types of vessels listed along the horizontal axis in Fig. 22-2 lie in series with one another. Furthermore, the individual members of each category of vessels are ordinarily arranged in parallel with one another (see Fig. 22-3). For example, the capillaries throughout the body are in most instances parallel elements, with the notable exceptions of the renal vasculature (wherein the peritubular capillaries are in series with the glomerular capillaries) and the splanchnic vasculature (wherein the intestinal and hepatic capillaries are aligned in series). Formulas for the total hydraulic resistance of components arranged in series and in parallel have been derived in the same manner as those for similar combinations of electrical resistances.

Three hydraulic resistances, R_1, R_2, and R_3, are arranged in series in the schema depicted in Fig. 26-9. The pressure drop across the entire system—i.e., the difference between inflow pressure, P_i, and outflow pressure, P_o—consists of the sum of the pressure drops across each of the individual resistances (equation *a*). Under steady-state conditions, the flow, Q, through any given cross section must equal the flow through any other cross section. By dividing each component in equation *a* by Q (equation *b*), it becomes evident from the definition of resistance that *for resistances in series, the total resistance,* R_t, *of the entire system equals the sum of the individual resistances, that is,*

$$R_t = R_1 + R_2 + R_3 \qquad (12)$$

For resistances in parallel, as illustrated in Fig. 26-10, the inflow and outflow pressures are the same for all tubes. Under steady-state conditions, the total flow, Q_t, through the system equals the sum of the flows through the individual parallel elements (equation *a*). Because the pressure gradient $(P_i - P_o)$ is identical for all parallel elements, each term in equation *a* may be divided by that pressure gradient to yield equation *b*. From the definition of resistance, equation *c* may be derived. This states that the reciprocal of the total resistance, R_t, equals the sum of the reciprocals of the individual resistances, that is,

$$\frac{1}{R_t} = \frac{1}{R_1} + \frac{1}{R_2} + \frac{1}{R_3} \qquad (13)$$

Stated in another way, if we define hydraulic **conductance** as the reciprocal of resistance, it becomes evident that, *for tubes in parallel, the total conductance is the sum of the individual conductances.*

By considering a few simple illustrations, some of the fundamental properties of parallel hydraulic systems become apparent. For example, if the resistance of the three parallel elements in Fig. 26-10 were all equal, then

$$R_1 = R_2 = R_3$$

Therefore,

$$1/R_t = 3/R_1$$

and

$$R_t = R_1/3$$

Thus the total resistance is less than any of the individual resistances. After further consideration, it becomes evident that *for any parallel arrangement, the total resistance must be less than that of any individual component.* For example, consider a system in which a very high resistance tube is added in parallel to a low-resistance tube. The total resistance must be less than that of the low-resistance component by itself, because the high-resistance component affords an additional pathway, or conductance, for fluid flow.

As a physiological illustration of these principles, consider the relationship between the **total peripheral resistance** (TPR) of the entire systemic vascular bed and the resistance of one of its components, such as the renal vasculature. In an individual with a cardiac output of 5000 ml/min and an arterial pressure of 100 mm Hg, the TPR will be 0.02 mm Hg/ml/min, or 0.02 PRU (peripheral resistance units). Blood flow through one kidney would be approximately 600 ml/min. Renal resistance would therefore be 100 mm Hg/600 ml/min, or 0.17 PRU, which is 8.5 times as great as the TPR.

From Fig. 22-2 it seems paradoxical that the resistance to flow through the arterioles (as manifested by the pressure drop between the arterial and capillary ends of these vessels) is considerably greater than that through certain other vascular components, despite the fact that the total cross-sectional area of the arterioles exceeds that for these same vascular components. For example, the total cross-sectional area of the arterioles greatly exceeds that of the large arteries, yet the resistance to flow through the arterioles exceeds that through the large arteries.

Consideration of simple models of tubes in parallel will help resolve this apparent paradox. In Fig. 26-11 the resistance to flow through one wide tube of cross-sectional area A_w is compared with that through four narrower tubes in parallel, each of area A_n. The total cross-sectional area of the parallel system of four narrow tubes equals the area of the wide tube; that is,

$$A_w = 4A_n \qquad (14)$$

Because resistance, R, is inversely proportional to the fourth power of the radius, r, and since

$$A = \pi r^2 \qquad (15)$$

for cylindrical tubes, it follows that

$$R = k/A^2 \qquad (16)$$

The proportionality constant, k, is related to tube length and fluid viscosity, both of which will be held constant in the example under consideration. From equation 16, the resistances of the wide tube, R_w, and a single narrow tube, R_n, are

$$R_w = k/A_w^2 \qquad (17)$$

$$R_n = k/A_n^2 \qquad (18)$$

From equation 13,

$$\frac{1}{R_t} = \frac{1}{R_n} + \frac{1}{R_n} + \frac{1}{R_n} + \frac{1}{R_n} = \frac{4}{R_n} \qquad (19)$$

Substituting the value of R_n in equation 18,

$$1/R_t = 4A_n^2/k \qquad (20)$$

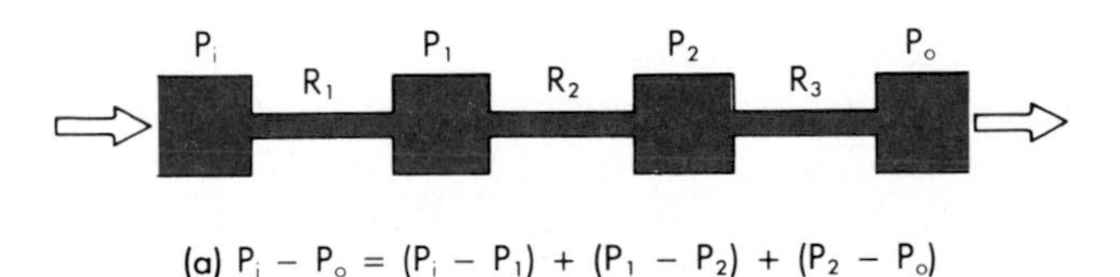

■ **Fig. 26-9** For resistances *(R_1, R_2, and R_3)* arranged in series, the total resistance R_t, equals the sum of the individual resistances.

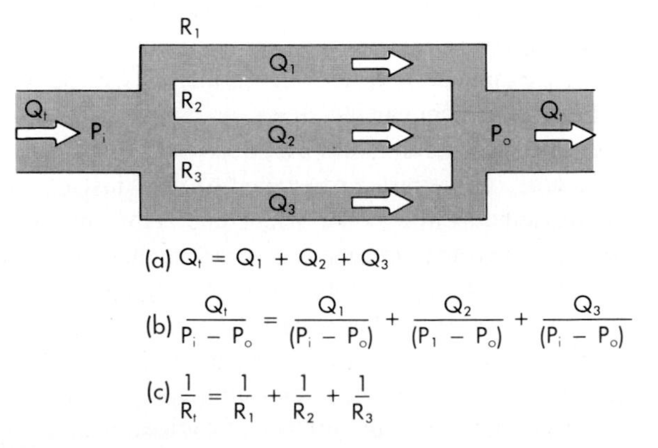

■ **Fig. 26-10** For resistances *(R_1, R_2, and R_3)* arranged in parallel, the reciprocal of the total resistance, R_t, equals the sum of the reciprocals of the individual resistances.

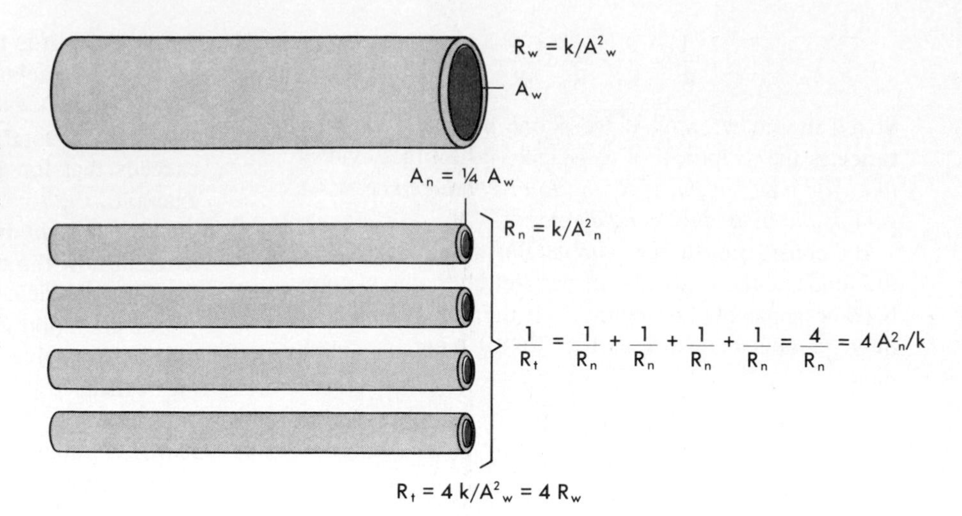

■ **Fig. 26-11** When four narrow tubes, each of area A_n, are connected in parallel, the total cross-sectional area equals the area A_w, of a wide tube of area such that $A_w = 4A_n$. Although the total areas are equal, the total resistance, R_t, to flow through the parallel narrow tubes is four times as great as the resistance, R_w, through the single wide tube.

Rearranging,

$$R_t = k/4A_n^2 \tag{21}$$

From equations 14 and 17,

$$R_t = 4k/A_w^2 = 4R_w \tag{22}$$

Hence the resistance of four such tubes in parallel is four times as great as that of a single tube of equal total cross-sectional area.

If a similar calculation is made for eight such tubes in parallel, with each tube having one fourth the cross-sectional area of the single wide tube, it will be found that the total resistance will equal $2R_w$. In this circumstance the resistance to flow through eight such narrow tubes in parallel will still be twice as great as that through the single tube, despite the fact that the total cross-sectional area for the eight narrow tubes is twice as great as for the single wide tube. This is analogous to the relationship that exists between resistance and area in the circulatory system when comparing the arterioles with the large arteries. Despite the fact that the total cross-sectional area of all the arterioles greatly exceeds that of all the large arteries, the resistance to flow through the arterioles is considerably greater than that through the large arteries.

If the example is carried still further, it will be found that sixteen such narrow tubes in parallel, now with four times the total cross-sectional area of the single wide tube, will exert a resistance to flow just equal to the resistance through the wide tube. Any number of these narrow tubes in excess of sixteen, then, will have a lower resistance than that of the single wide tube. This is analogous to the situation for the arterioles and capillaries. The resistance to flow through a single capillary is much greater than that through a single arteriole (see Fig. 26-8). Yet the number of capillaries so greatly exceeds the number of arterioles, as reflected by the relative difference in total cross-sectional areas, that the pressure drop across the arterioles is considerably greater than the pressure drop across the capillaries (see Fig. 22-2).

■ *Laminar and Turbulent Flow*

Under certain conditions, the flow of a fluid in a cylindrical tube will be **laminar** (sometimes called **streamlined**), as illustrated in Fig. 26-12. At the entrance of the tube all the fluid elements will have the same linear velocities, regardless of their radial positions. In progressing along the tube, however, the thin layer of fluid in contact with the wall of the tube adheres to the wall and hence is motionless. The layer of fluid just central to this external lamina must shear against this motionless layer and therefore moves slowly, but with a finite velocity. Similarly, the adjacent, more central layer travels still more rapidly. Close to the tube inlet the fluid layers near the axis of the tube still move with the same velocity and do not shear against one another. However, at a distance from the tube inlet equal to several tube diameters, laminar flow becomes **fully devel-**

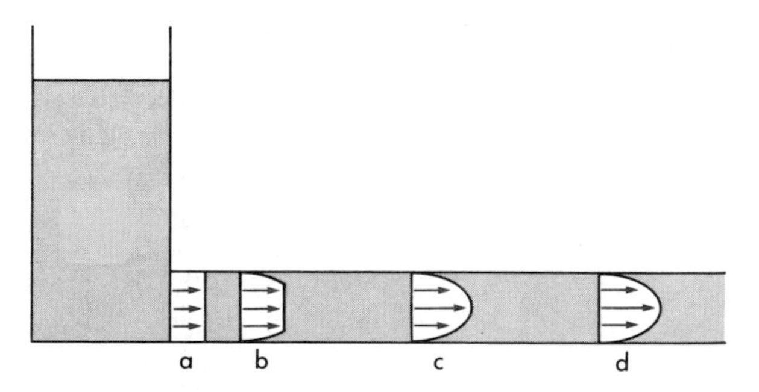

■ **Fig. 26-12** Laminar flow in a cylindrical tube. At the inlet, *a*, the velocities are equal at all radial distances from the center of the tube. Near the inlet, *b*, the velocity profile is flat near the center of the tube, but a velocity gradient is established near the wall. When flow becomes fully developed, *c* and *d*, the velocity profile is parabolic.

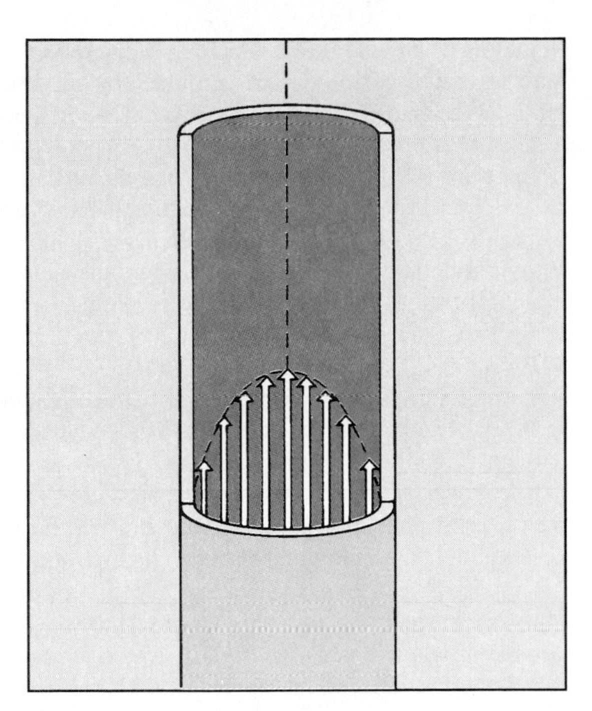

■ **Fig. 26-13** In laminar flow all elements of the fluid move in streamlines that are parallel to the axis of the tube; movement does not occur in a radial or circumferential direction. The layer of fluid in contact with the wall is motionless; the fluid that moves along the axis of the tube has the maximal velocity.

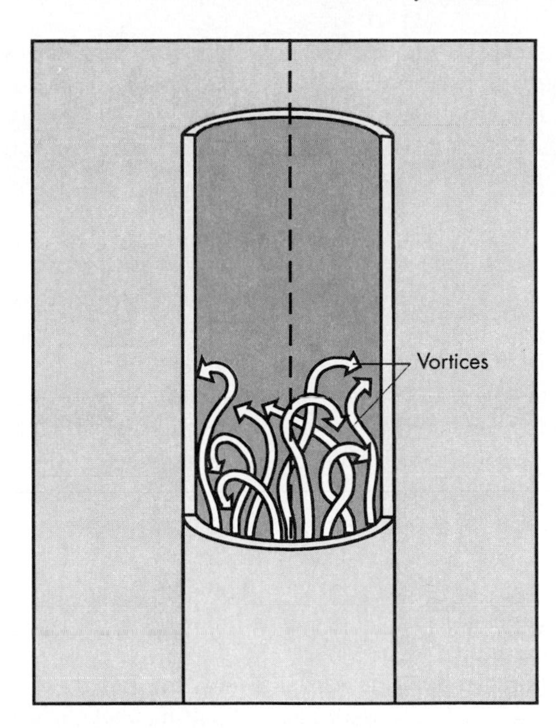

■ **Fig. 26-14** In turbulent flow the elements of the fluid move irregularly in axial, radial, and circumferential directions. Vortices frequently develop.

oped; (i.e., the velocity profiles do not change with longitudinal distance along the tube). In fully developed laminar flow the longitudinal velocity profile is that of a paraboloid (Fig. 26-13). The velocity of the fluid adjacent to the wall is zero, whereas the velocity at the center of the stream is maximum and equal to twice the mean velocity of flow across the entire cross section of the tube. In laminar flow, fluid elements remain in one lamina, or streamline, as the fluid progresses longitudinally along the tube.

Irregular motions of the fluid elements may develop in the flow of fluid through a tube; this flow is called **turbulent.** Under such conditions fluid elements do not remain confined to definite laminae but rapid, radial mixing occurs (Fig. 26-14). A considerably greater pressure is required to force a given flow of fluid through the same tube when the flow is turbulent than when it is laminar. In turbulent flow the pressure drop is approximately proportional to the square of the flow rate, whereas in laminar flow the pressure drop is proportional to the first power of the flow rate. Hence to produce a given flow, a pump such as the heart must do considerably more work if turbulence develops.

Whether turbulent or laminar flow will exist in a tube under given conditions may be predicted on the basis of a dimensionless number called **Reynold's number,** N_R. This number represents the ratio of inertial to viscous forces. For a fluid flowing through a cylindrical tube.

$$N_R = \rho D \bar{v}/\eta \qquad (23)$$

where D is the tube diameter, $\bar{v}$ is the mean velocity, ρ is the density, and η is the viscosity. For $N_R < 2000$, the flow will usually be laminar; for $N_R > 3000$, the flow will be turbulent. Various possible conditions may develop in the transition range of N_R between 2000 and 3000. Because flow tends to be laminar at low N_R and turbulent at high N_R, the definition of N_R indicates that large diameters, high velocities, and low viscosities predispose to turbulence. In addition to these factors, abrupt variations in tube dimensions or irregularities in the tube walls may produce turbulence.

Turbulence is usually accompanied by audible vibrations. When turbulent flow exists within the cardiovascular system, it is usually detected as a **murmur.** The factors listed above that predispose to turbulence may account for murmurs heard clinically. In severe anemia **functional cardiac murmurs** (murmurs not caused by structural abnormalities) are frequently detectable. The physical basis for such murmurs resides in (1) the reduced viscosity of blood in anemia, and (2) the high flow velocities associated with the high cardiac output that usually prevails in anemic patients.

Blood clots, or **thrombi,** are much more likely to develop in turbulent than in laminar flow. One of the problems with the use of artificial valves in the surgical treatment of valvular heart disease is that thrombi may occur in association with the prosthetic valve. The thrombi may be dislodged and occlude a crucial blood vessel. It is thus important to design such valves to avert turbulence.

Shear Stress on the Vessel Wall

In Fig. 26-7 an external force was applied to a plate floating on the surface of a liquid in a large basin. This force, exerted parallel to the surface, caused a shearing stress on the liquid below, thereby producing a differential motion of each layer of liquid relative to the adjacent layers. At the bottom of the basin, the flowing liquid exerted a shearing stress on the surface of the basin in contact with the liquid. By rearranging the equation for viscosity stated in Fig. 26-7, it is apparent that the shear stress, τ, equals η (du/dy); (i.e., the shear stress equals the product of the viscosity and the shear rate). Hence, the greater the rate of flow, the greater the shear stress that the liquid exerts on the walls of the container in which it flows.

For precisely the same reasons, the rapidly flowing blood in a large artery tends to pull the endothelial lining of the artery along with it. This force **(viscous drag)** is proportional to the shear rate (du/dy) of the layers of blood near the wall. For a flow regimen that obeys Poiseuille's law,

$$\tau = 4\eta Q/\pi r^3 \qquad (24)$$

The greater the rate of blood flow (Q) in the artery, the greater will be du/dy near the arterial wall, and the greater will be the viscous drag (τ).

In certain types of arterial disease, particularly in patients with hypertension, the subendothelial layers tend to degenerate locally, and small regions of the endothelium may lose their normal support. The viscous drag on the arterial wall may cause a tear between a normally supported and an unsupported region of the endothelial lining. Blood may then flow from the vessel lumen, through the rift in the lining and dissect between the various layers of the artery. Such a lesion is called a **dissecting aneurysm.** It occurs most commonly in the proximal portions of the aorta and is extremely serious. One reason for its predilection for this site is the high velocity of blood flow, with the associated large values of du/dy at the endothelial wall. The shear stress at the vessel wall also influences many other vascular functions, such as the permeability of the vascular walls to large molecules, the biosynthetic activity of the endothelial cells, the integrity of the formed elements in the blood, and the coagulation of the blood.

Rheological Properties of Blood

The viscosity of a newtonian fluid, such as water, may be determined by measuring the rate of flow of the fluid at a given pressure gradient through a cylindrical tube of known length and radius. As long as the fluid flow is laminar, the viscosity may be computed by substituting these values into Poiseuille's equation. The viscosity of a given newtonian fluid at a specified temperature will be constant over a wide range of tube dimensions and flows. However, for a nonnewtonian fluid the viscosity calculated by substituting into Poiseuille's equation may vary considerably as a function of tube dimensions and flows. Therefore in considering the rheological properties of a suspension such as blood, the term **viscosity** does not have a unique meaning. The terms **anomalous viscosity** and **apparent viscosity** are frequently applied to the value of viscosity obtained for blood under the particular conditions of measurement.

Rheologically, blood is a suspension of formed elements, principally erythrocytes, in a relatively homogeneous liquid, the blood plasma. For this reason the apparent viscosity of blood varies as a function of the **hematocrit ratio** (ratio of volume of red blood cells to volume of whole blood). In Fig. 26-15 the upper curve represents the ratio of the apparent viscosity of whole blood to that of plasma over a range of hematocrit ratios from 0% to 80%, measured in a tube 1 mm in diameter. The viscosity of plasma is 1.2 to 1.3 times that of water. Fig. 26-15 (upper curve) shows that blood, with a normal hematocrit ratio of 45%, has an apparent viscosity 2.4 times that of plasma. In severe anemia blood viscosity is low. With increasing hematocrit ratios the slope of the curve increases progressively; it is especially steep at the upper range of erythrocyte concentrations. A rise in hematocrit ratio from 45% to 70%,

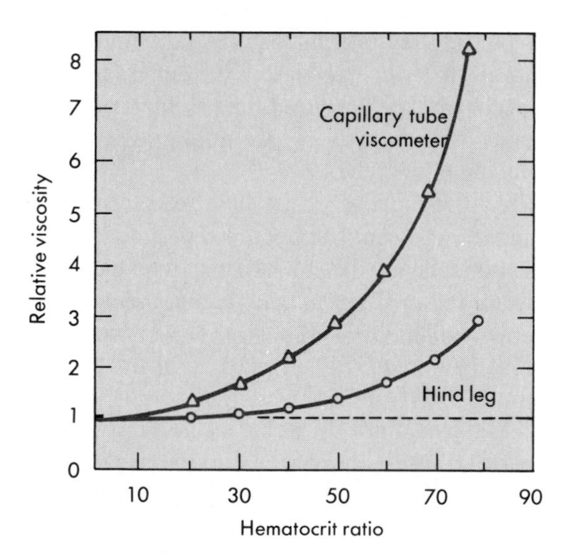

Fig. 26-15 Viscosity of whole blood, relative to that of plasma, increases at a progressively greater rate as the hematocrit ratio increases. For any give hematocrit ratio the apparent viscosity of blood is less when measured in a biological viscometer (such as the hind leg of a dog) than in a conventional capillary tube viscometer. (Redrawn from Levy MN, Share L: *Circ Res* 1:247, 1953.)

which occurs in **polycythemia,** increases the apparent viscosity more than twofold, with a proportionate effect on the resistance to blood flow. The effect of such a change in hematocrit ratio on peripheral resistance may be appreciated when it is recognized that even in the most severe cases of essential hypertension, the total peripheral resistance rarely increases by more than a factor of two. In hypertension the increase in peripheral resistance is achieved by arteriolar vasoconstriction.

For any given hematocrit ratio the apparent viscosity of blood depends on the dimensions of the tube employed in estimating the viscosity. Fig. 26-16 demonstrates that the apparent viscosity of blood diminishes progressively as tube diameter decreases below a value of about 0.3 mm. The diameters of the highest resistance blood vessels, the arterioles, are considerably less than this critical value. This phenomenon therefore reduces the resistance to flow in the blood vessels that possess the greatest resistance.

The apparent viscosity of blood, when measured in living tissues, is considerably less than when measured in a conventional capillary tube viscometer with a diameter greater than 0.3 mm. In the lower curve of Fig. 26-15, the apparent relative viscosity of blood was assessed by using the hind leg of an anesthetized dog as a biological viscometer. Over the entire range of hematocrit ratios, the apparent viscosity was less as measured in the living tissue than in the capillary tube viscometer (upper curve), and the disparity was greater the higher the hematocrit ratio.

The influence of tube diameter on apparent viscosity is ascribable in part to the change in actual composition of the blood as it flows through small tubes. The composition changes because the red blood cells tend to accumulate in the faster axial stream, whereas the blood component that flows in the slower marginal layers is mainly plasma. To illustrate this phenomenon, a reservoir such as R_1 in Fig. 26-5, C has been filled with blood possessing a given hematocrit ratio. The blood in R_1 was constantly agitated to prevent settling and was permitted to flow through a narrow capillary tube into reservoir R_2. As long as the tube diameter was substantially greater than the diameter of the red blood cells, the hematocrit ratio of the blood in R_2 was not detectably different from that in R_1. Surprisingly, however, the hematocrit ratio of the blood contained within the tube was found to be considerably lower than the hematocrit ratio of the blood in either reservoir.

In Fig. 26-17, the relative hematocrit is the ratio of the hematocrit in the tube to that in the reservoir at either end of the tube. For tubes of 500 μm diameter or greater, the relative hematocrit ratio was close to 1. However, as the tube diameter was diminished below 500 μm, the relative hematocrit ratio progressively diminished; for a tube diameter of 30 μm, the relative hematocrit ratio was only 0.6.

That this situation results from a disparity in the relative velocities of the red cells and plasma can be appreciated on the basis of the following analogy. Consider the flow of automobile traffic across a bridge that is 3 miles long. Let the cars move in one lane at a speed of 60 miles per hour and the trucks in another lane at 20 miles per hour, as illustrated in Fig. 26-18. If one car and one truck start out across the bridge each minute, then except for the initial few minutes of traffic flow across the bridge, one car and one truck will arrive at the other end each minute. Yet if one counts the actual number of cars and trucks on the bridge at any moment, there will be three times more of the slower moving trucks than of the more rapidly traveling cars.

Because the axial portions of the bloodstream contain a greater proportion of red cells and move with a greater velocity, the red cells tend to traverse the tube in less time than does the plasma. Therefore the red cells correspond to the rapidly moving cars in the anal-

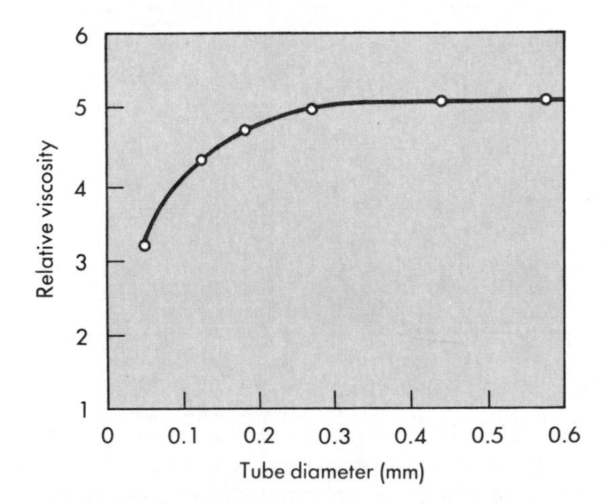

■ **Fig. 26-16** Viscosity of blood, relative to that of water, increases as a function of tube diameter up to a diameter of about 0.3 mm. (Redrawn from Fåhraeus R, Lindqvist T: *Am J Physiol* 96:562, 1931.)

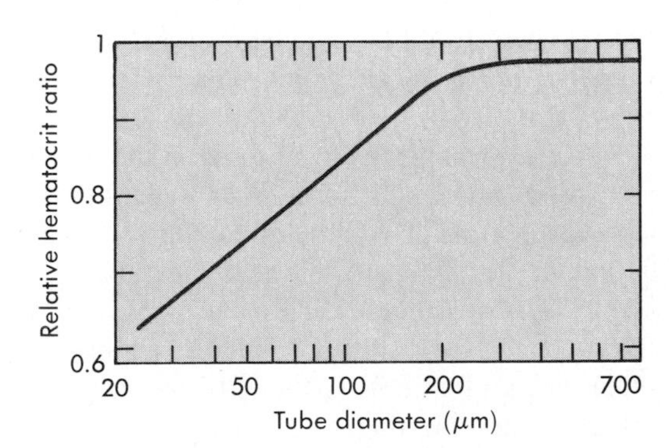

■ **Fig. 26-17** The "relative hematocrit ratio" of blood flowing from a feed reservoir through capillary tubes of various calibers as a function of the tube diameter. The relative hematocrit is the ratio of the hematocrit of the blood in the tubes to that of the blood in the feed reservoir. (Redrawn from Barbee JH, Cokelet GR: *Microvasc Res* 3:6, 1971.)

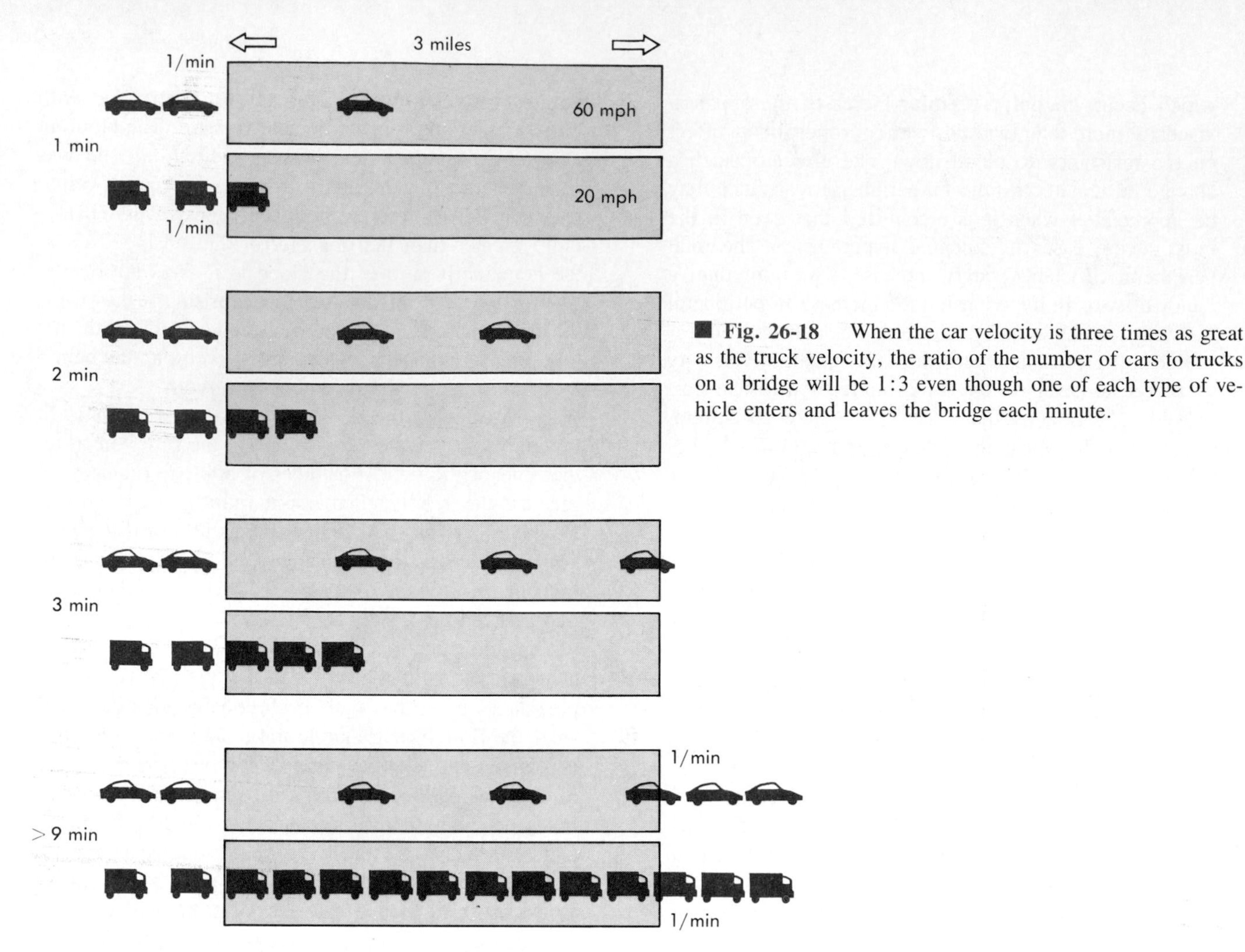

■ **Fig. 26-18** When the car velocity is three times as great as the truck velocity, the ratio of the number of cars to trucks on a bridge will be 1:3 even though one of each type of vehicle enters and leaves the bridge each minute.

ogy, and the plasma corresponds to the slowly moving trucks. Measurement of transit times through various organs has shown that red cells do travel faster than the plasma. Furthermore the hematocrit ratios of the blood contained in various tissues are lower than those in blood samples withdrawn from large arteries or veins in the same animal (Fig. 26-19).

The physical forces responsible for the drift of the erythrocytes toward the axial stream and away from the vessel walls are not fully understood. One factor is the great flexibility of the red blood cells. At low flow (or shear) rates, comparable with those in the microcirculation, rigid particles do not migrate toward the axis of a tube, whereas flexible particles do migrate. The concentration of flexible particles near the tube axis is enhanced by increasing the shear rate.

The apparent viscosity of blood diminishes as the shear rate is increased (Fig. 26-20), a phenomenon called **shear thinning.** The greater tendency of the erythrocytes to accumulate in the axial laminae at higher flow rates is partly responsible for this nonnewtonian behavior. However, a more important factor is that at very slow rates of shear, the suspended cells tend to form aggregates, which would increase viscosity. As the flow is increased, this aggregation would decrease

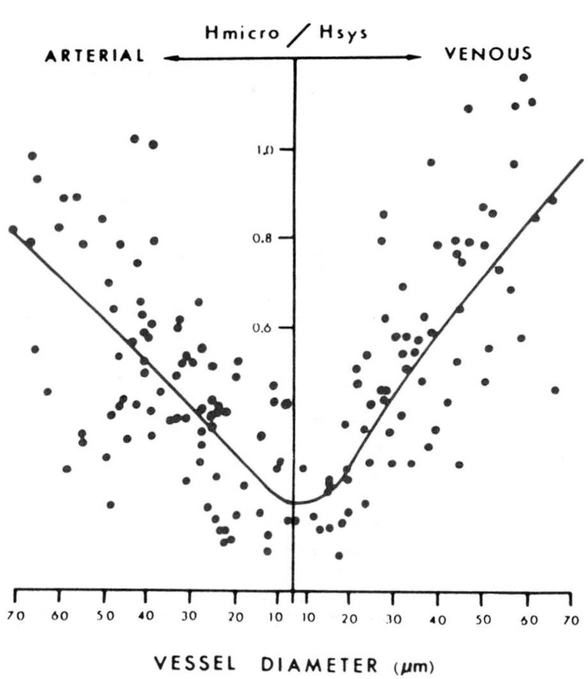

■ **Fig. 26-19** The hematocrit ratio (H_{micro}) of the blood in various sized arterial and venous microvessels in the cat mesentery, relative to the hematocrit ratio (H_{sys}) in the large systemic vessels. The hematocrit ratio is least in the capillaries and tiny venules. (Modified from Lipowsky HH et al: *Microvasc Res* 19:297, 1980.)

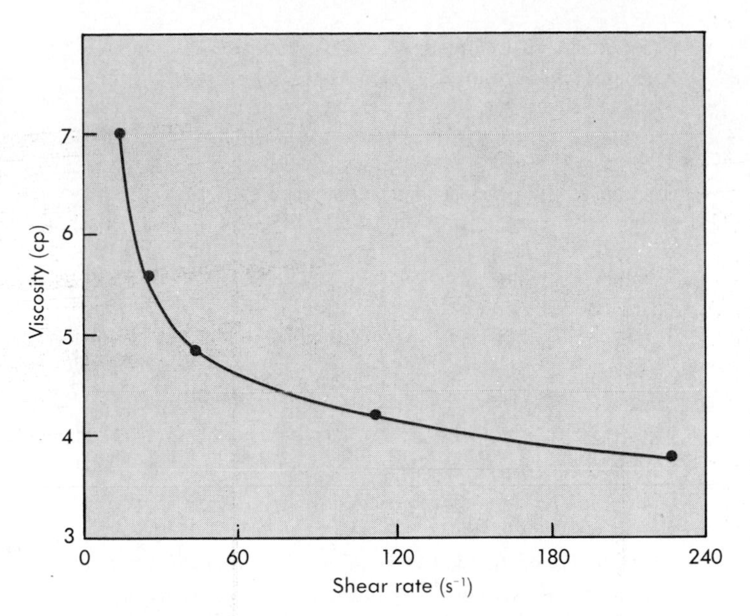

■ **Fig. 26-20** Decrease in the viscosity of blood (centipoise) at increasing rates of shear. The shear rate refers to the velocity of one layer of fluid relative to that of the adjacent layers and is directionally related to the rate of flow. (Redrawn from Amin TM, Sirs JA: *Q J Exp Physiol* 70:37, 1985.)

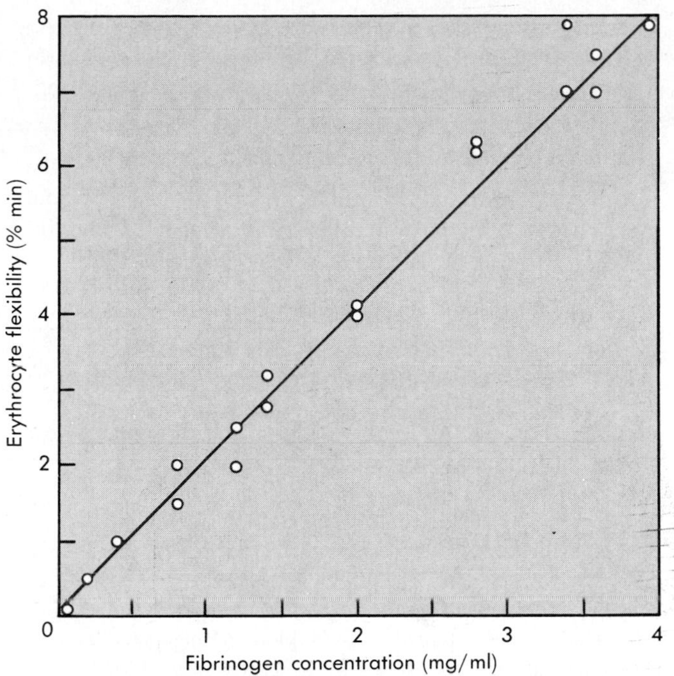

■ **Fig. 26-21** The effect of the plasma fibrinogen concentration on the flexibility of human erythrocytes. (Redrawn from Amin TM, Sirs JA: *Q J Exp Physiol* 70:37, 1985.)

and so also would the apparent viscosity (see Fig. 26-20).

The tendency for the erythrocytes to aggregate at low flows depends on the concentration in the plasma of the larger protein molecules, especially fibrinogen. For this reason the changes in blood viscosity with shear rate are much more pronounced when the concentration of fibrinogen is high. Also, at low flow rates, the leukocytes tend to adhere to the endothelial cells of the microvessels, thereby increasing the apparent viscosity.

The deformability of the erythrocytes is also a factor in shear thinning, especially at high hematocrit ratios. The mean diameter of human red blood cells is about 8 μm, yet they are able to pass through openings with a diameter of only 3 μm. As blood that is densely packed with erythrocytes is caused to flow at progressively greater rates, the erythrocytes become more and more deformed, which diminishes the apparent viscosity of the blood. The flexibility of human erythrocytes is enhanced as the concentration of fibrinogen in the plasma increases (Fig. 26-21). If the red blood cells become hardened, as they are in certain spherocytic anemias, shear thinning may become much less prominent.

■ *Summary*

1. The vascular system is composed of two major subdivisions in series with one another: the systemic circulation and the pulmonary circulation.

2. Each subdivision comprises a number of types of vessels (e.g., arteries, arterioles, capillaries) that are aligned in series with one another. In general, the vessels of a given type are arranged in parallel with each other.

3. The mean velocity (v) of blood flow in a given type of vessel is directly proportional to the total blood flow (Q_t) being pumped by the heart, and it is inversely proportional to the cross-sectional area (A) of all the parallel vessels of that type; (i.e., $v = Q_t/A$).

4. The laterally directed pressure in the bloodstream decreases as the flow velocity increases; the decrement in lateral pressure is proportional to the square of the velocity. The changes are insignificant, however, except when flow is very great.

5. When blood flow is relatively steady and laminar in vessels larger than arterioles, the flow (Q) is proportional to the pressure drop down the vessel ($P_i - P_o$) and to the fourth power of the radius (r), and it is inversely proportional to the length (l) of the vessel and to the viscosity (η) of the fluid; i.e., $Q = \pi(P_i - P_o)r^4/8\eta l$ (Poiseuille's law).

6. For resistances aligned in series, the total resistance equals the sum of the individual resistances.

7. For resistances aligned in parallel, the reciprocal of the total resistance equals the sum of the reciprocals of the individual resistances.

8. Flow tends to become turbulent when (1) flow velocity is high, (2) fluid viscosity is low, (3) tube diameter is larger, or (4) the wall of the vessel is very irregular.

9. Blood flow is nonnewtonian in very small vessels; (i.e., Poiseuille's law is not applicable). The apparent viscosity of the blood diminishes as shear rate (flow) increases and as the tube dimensions decrease.

Bibliography

Journal articles

Amin TM, Sirs JA: The blood rheology of man and various animal species, *Q J Exp Physiol* 70:37, 1985.

Carroll RJ, Falsetti HL: Retrograde coronary artery flow in aortic valve disease, *Circulation* 54:494, 1976.

Chien S: Role of blood cells in microcirculatory regulation, *Microvasc Res* 29:129, 1985.

Cokelet GR, Goldsmith HL: Decreased hydrodynamic resistance in the two-phase flow of blood through small vertical tubes at low flow rates, *Circ Research* 68:1, 1991.

Goldsmith HL: The microrheology of human blood, *Microvasc Res* 31:121, 1986.

Klanchar M, Tarbell JM, Wang DM: In vitro study of the influence of radial wall motion on wall shear stress in an elastic tube model of the aorta, *Circ Res* 66:1624, 1990.

Lipowsky HH, Usami S, Chien S: In vivo measurements of "apparent viscosity" and microvessel hematocrit in the mesentery of the cat, *Microvasc Res* 19:297, 1980.

McKay CB, Meiselman HJ: Osmolality-mediated Fahraeus and Fahraeus-Lindqvist effects for human RBC suspensions, *Am J Physiol* 254:H238, 1988.

Sarelius IH, Duling BR: Direct measurement of microvessel hematocrit, red cell flux, velocity, and transit time, *Am J Physiol* 243:H1018, 1982.

Secomb TW: Flow-dependent rheological properties of blood in capillaries, *Microvasc Res* 34:46, 1987.

Sutera SP et al: Vascular flow resistance in rabbit hearts: "apparent viscosity" of RBC suspensions, *Microvasc Res* 36:305, 1988.

Tangelder GJ et al: Wall shear rate in arterioles in vivo: least estimates from platelet velocity profiles, *Am J Physiol* 254:H1059, 1988.

Thompson TN, La Celle PL, Cokelet GR: Perturbation of red blood cell flow in small tubes by white blood cells, *Pflügers Arch* 413:372, 1989.

Books and monographs

Chien S, Usami S, Skalak R: *Blood flow in small tubes*. In Renkin EM, Michel CC, editors: *Handbook of physiology:* section 2: *The cardiovascular system—microcirculation,* vol 4, Bethesda, Md, 1984, American Physiological Society.

Fung YC: *Biodynamics: circulation,* New York, 1984, Springer-Verlag, New York.

Lowe GDO: *Clinical blood rheology,* vol 1, Boca Raton, 1988, CRC Press.

Milnor WR: *Hemodynamics,* Baltimore, 1982, Williams & Wilkins.

Taylor DEM, Stevens AI, editors: *Blood flow: theory and practice,* New York, 1983, Academic Press.

CHAPTER

27

The Arterial System

■ Hydraulic Filter

The principal function of the systemic and pulmonary arterial systems is to distribute blood to the capillary beds throughout the body. The arterioles, the terminal components of this system, regulate the distribution of flow to the various capillary beds. Between the heart and the arterioles the aorta and pulmonary artery and their major branches constitute a system of conduits of considerable volume and distensibility. An arterial system composed of elastic conduits and high-resistance terminals constitutes a **hydraulic filter** analogous to the resistance-capacitance filters of electrical circuits.

Hydraulic filtering converts the intermittent output of the heart to a steady flow through the capillaries. This important function of the large elastic arteries has been likened to the **Windkessels** of antique fire engines. The Windkessel contained a large volume of trapped air. The compressibility of the trapped air converted the intermittent inflow of water to a steady outflow at the nozzle of the fire hose.

The analogous function of the large elastic arteries is illustrated in Fig. 27-1. The heart is an intermittent pump. The entire stroke volume is discharged into the arterial system during systole, which usually occupies approximately one third of the duration of the cardiac cycle. In fact, as described on p. 410, most of the stroke volume is pumped during the rapid ejection phase, which constitutes about half of the systole. Part of the energy of cardiac contraction is dissipated as forward capillary flow during systole; the remainder is stored as potential energy, in that much of the stroke volume is retained by the distensible arteries (Fig. 27-1, *A, B*). During diastole the elastic recoil of the arterial walls converts this potential energy into capillary blood flow. If the arterial walls were rigid, then capillary flow would cease during diastole (Fig. 27-1, *C, D*).

Hydraulic filtering minimizes the work load of the heart. More work is required to pump a given flow intermittently than steadily; the more effective the filtering, the less the excess work. A simple example will illustrate this point.

Consider first the steady flow of a fluid at a rate of 100 ml/sec through a hydraulic system with a resistance of 1 mm Hg/ml/sec. This combination of flow and resistance would result in a constant pressure of 100 mm Hg, as shown in Fig. 27-2, *A*. Neglecting any inertial effect, hydraulic work, W, may be defined as

$$W = \int_{t_1}^{t_2} P dV \qquad (1)$$

that is, each small increment of volume, dV, pumped is multiplied by the pressure, P, existing at the time, and the products are integrated over the time interval of interest, $t_2 - t_1$, to give the total work, W. For steady flow,

$$W = PV \qquad (2)$$

In the example in Fig. 27-2, *A*, the work done in pumping the fluid for 1 second would be 10,000 mm Hg-ml (or 1.33×10^7 dyne-cm).

Next consider an intermittent pump that puts out the same volume per second but pumps the entire volume at a steady rate over 0.5 second and then pumps nothing during the next 0.5 second. Hence it pumps at the rate of 200 ml/sec for 0.5, as shown in Fig. 27-2, *B* and *C*. In *B* the conduit is rigid and the fluid is incompressible, but the system has the same resistance as in *A*. During the pumping phase of the cycle (systole) the flow of 200 ml/sec through a resistance of 1 mm Hg/ml/sec would produce a pressure of 200 mm Hg. During the filling phase of the pump (diastole) the pressure would be 0 mm Hg in this rigid system. The work done during systole would be 20,000 mm Hg-ml, which is twice that required in the example shown in Fig. 27-2, *A*.

If the system were very distensible, hydraulic filtering would be very effective, and the pressure would remain virtually constant throughout the entire cycle (Fig. 27-2, *C*). Of the 100 ml of fluid pumped during the 0.5 second of systole, only 50 ml would be emitted through

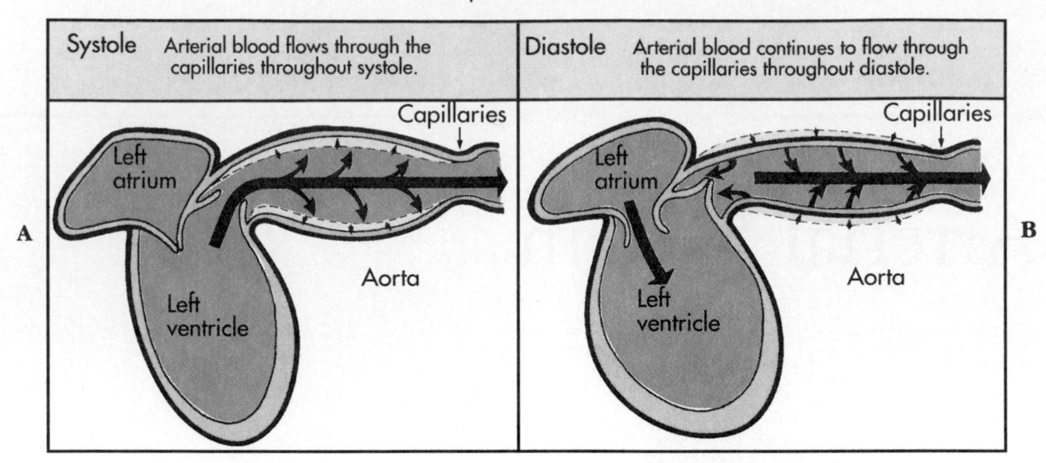

A, When the arteries are normally compliant, a substantial fraction of the stroke volume is stored in the arteries during ventricular systole. The arterial walls are stretched.

B, During ventricular diastole the previously stretched arteries recoil. The volume of blood that is displaced by the recoil furnished continuous capillary flow throughout diastole.

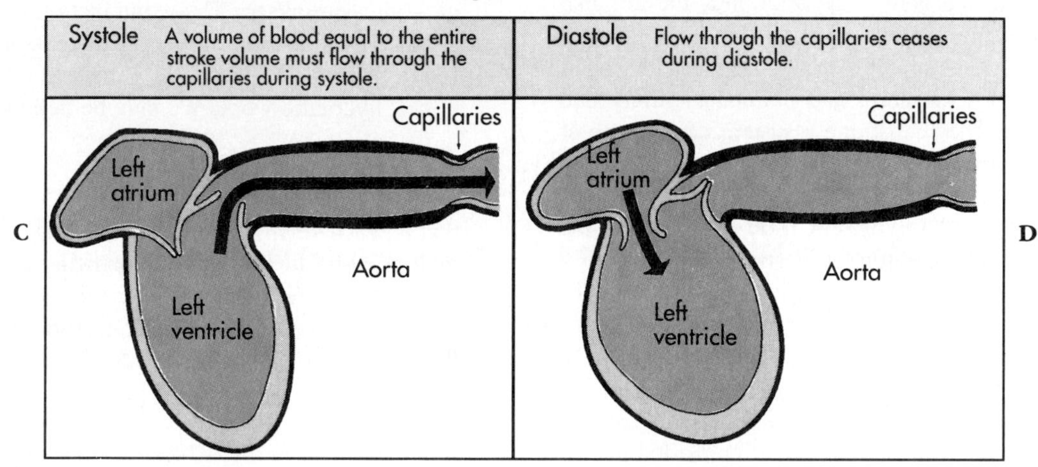

C, When the arteries are rigid, virtually none of the stroke volume can be stored in the arteries.

D, Rigid arteries cannot recoil appreciably during diastole.

■ **Fig. 27-1** When the arteries are normally compliant, blood flows through the capillaries throughout the cardiac cycle. When the arteries are rigid, blood flows through the capillaries during systole, but flow ceases during diastole.

the high-resistance outflow end of the system during systole. The remaining 50 ml would be stored by the distensible conduit during systole and would flow out during diastole. Hence the pressure would be virtually constant at 100 mm Hg throughout the cycle. The fluid pumped during systole would be ejected at only half the pressure that prevailed in Fig. 27-2, *B*, and, therefore, the work would be only half as great. With nearly perfect filtering, as in Fig. 27-2, *C*, the work would be identical to that for steady flow (Fig. 27-2, *A*).

Naturally the filtering accomplished by the systemic and pulmonic arterial systems is intermediate between the examples in Fig. 27-2, *B* and *C*. Ordinarily the additional work imposed by intermittency of pumping, in excess of that for steady flow, is about 35% for the right ventricle and about 10% for the left ventricle. These fractions change, however, with variations in heart rate, peripheral resistance, and arterial distensibility.

■ *Arterial Elasticity*

The elastic properties of the arterial wall may be appreciated by considering first the **static pressure–volume relationship** for the aorta. To obtain the curves shown

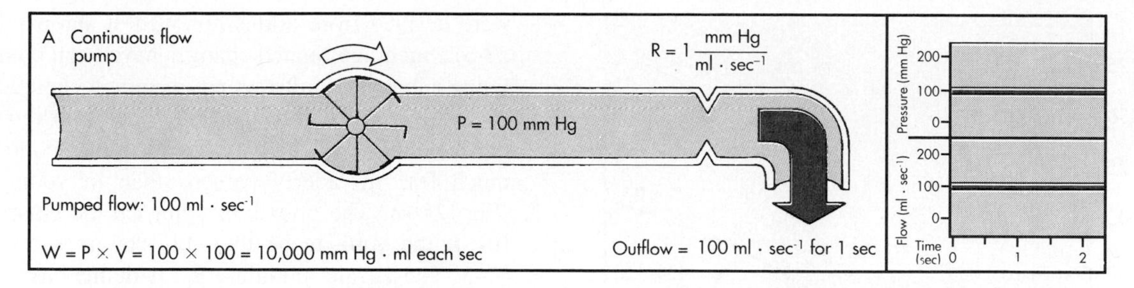

A, The flow is steady, and pressure will remain constant regardless of the distensibility of the conduit.

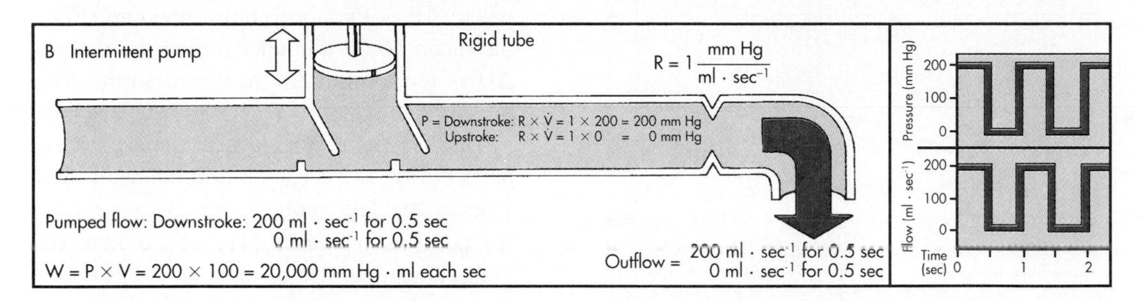

B, The flow produced by the pump is intermittent; it is steady for half the cycle and ceases for the remainder of the cycle. The conduit is rigid, and therefore the flow produced by the pump during its downstroke must exit through the resistance during the same 0.5 second that elapses during the downstroke. The pump must do twice as much work as the pump in **A.**

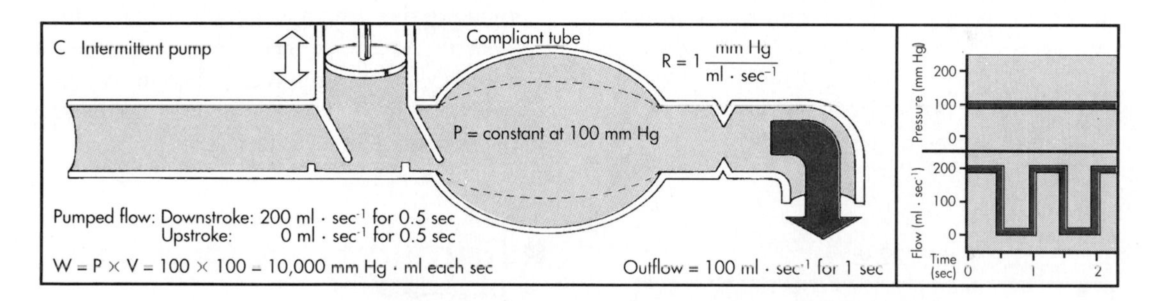

C, The pump operates as in **B,** but the conduit is infinitely distensible. This results in perfect filtering pressure (i.e., the pressure is steady and the flow through the resistance is also steady). The work equals that in **A.**

■ **Fig. 27-2** The relationships between pressure and flow for three hydraulic systems. In each the overall flow is 100 ml $\cdot$ sec^{-1} and the resistance is 1 mm Hg $\cdot$ ml^{-1} $\cdot$ sec.

in Fig. 27-3 aortas were obtained at autopsy from individuals in different age groups. All branches of the aorta were ligated and successive volumes of liquid were injected into this closed elastic system just as successive increments of water might be introduced into a balloon. After each increment of volume the internal pressure was measured. In Fig. 27-3 the curve relating pressure to volume for the youngest age group (curve *a*) is sigmoidal. Although the curve is nearly linear over most of its extent, the slope decreases at the upper and lower ends. At any point the slope (dV/dP) represents the aortic **compliance.** Thus in young individuals the aortic compliance is least at very high and low pressures and greatest over the usual range of pressure variations. This sequence of compliance changes resembles the familiar compliance changes encountered in inflating a balloon. The greatest difficulty in introducing air into the balloon is experienced at the beginning of inflation and again at near-maximum volume, just before rupture of the balloon. At intermediate volumes the balloon is relatively easy to inflate.

It is also apparent from Fig. 27-3 that the curves become displaced downward and the slopes diminish as a function of advancing age. Thus for any pressure above about 80 mm Hg the compliance decreases with age, a manifestation of increased rigidity caused by progressive changes in the collagen and elastin contents of the arterial walls.

The effects of the subject's age on the elastic characteristics of the arterial system that have been described

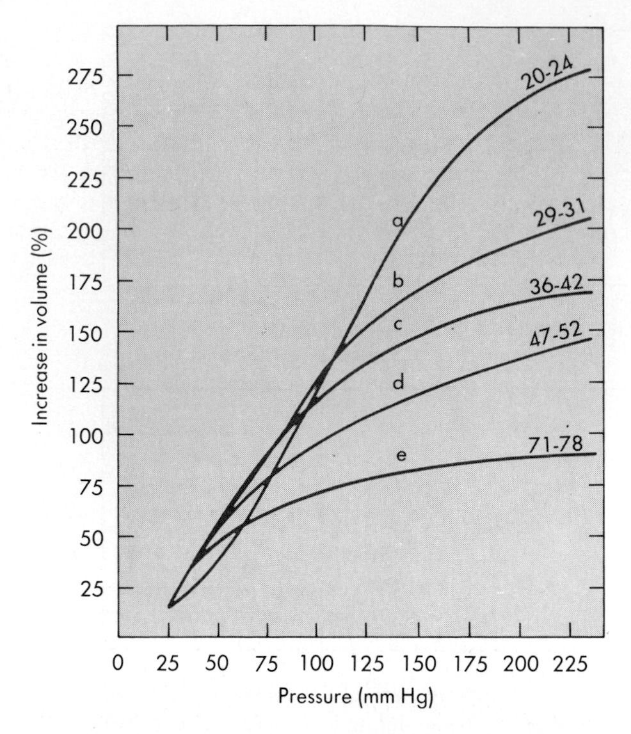

■ **Fig. 27-3** Pressure-volume relationships for aortas obtained at autopsy from humans in different age groups *(denoted by the numbers at the right end of each of the curves)*. (Redrawn from Hallock P, Benson IC: *J Clin Invest* 16:595, 1937.)

were derived from aortas removed at autopsy (see Fig. 27-3). Such age related changes have been confirmed in living subjects by ultrasound imaging techniques. These studies have disclosed that the increase in the diameter of the aorta produced by each cardiac contraction is much less in elderly persons than in young persons (Fig. 27-4). The effects of aging on the **elastic modulus** of the aorta in healthy subjects are shown in Fig. 27-5. The elastic modulus, E_p, is defined as

$$E_p = \Delta P/(\Delta D/D) \quad (3)$$

where ΔP is the aortic pulse pressure (Fig. 27-6), D is the mean aortic diameter during the cardiac cycle, and ΔD is the maximum change in aortic diameter during the cardiac cycle.

The fractional change in diameter ($\Delta D/D$) of the aorta during the cardiac cycle reflects its change in volume as the left ventricle ejects its stroke volume into the aorta each systole. Thus E_p is **inversely** related to compliance, which is the ratio of ΔV to ΔP. Consequently the **increase** in elastic modulus with aging (see Fig. 27-5) and the **decrease** in compliance with aging (see Fig. 27-3) both reflect the stiffening of the arterial walls as individuals age.

The heart is unable to eject its stroke volume into a rigid arterial system as rapidly as into a more compliant system. As compliance diminishes, peak arterial pressure occurs progressively later in systole. Hence the rapid ejection phase of systole is significantly prolonged as aortic compliance decreases.

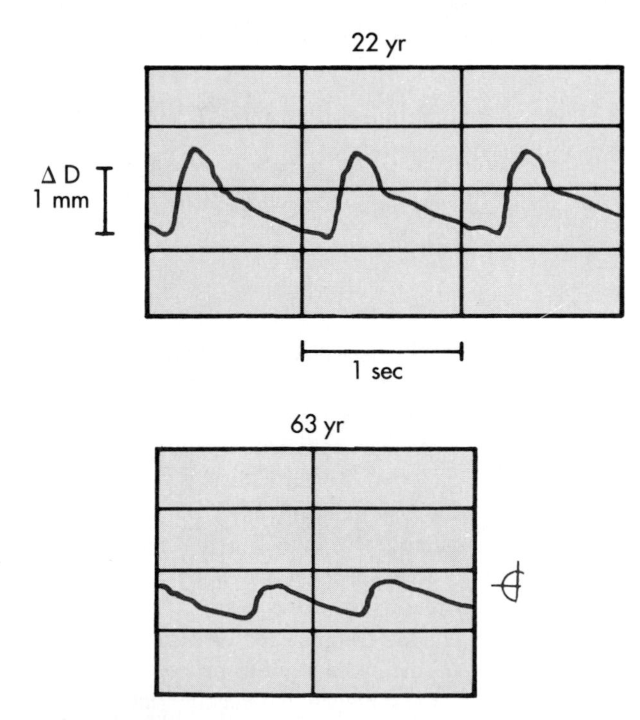

■ **Fig. 27-4** The pulsatile changes in diameter, measured ultrasonically, in a 22-year-old and a 63-year-old man. (Modified from Imura T et al: *Cardiovasc Res* 20:208, 1986.)

■ *Determinants of the Arterial Blood Pressure*

The determinants of the pressure within the arterial system cannot be evaluated precisely. Yet the arterial

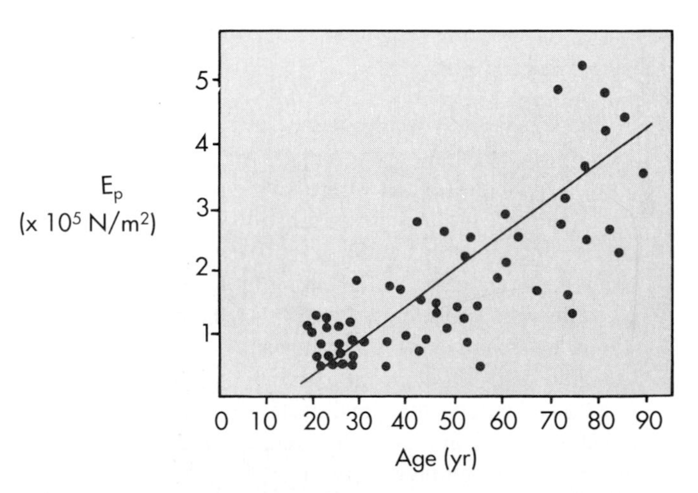

■ **Fig. 27-5** The effects of age on the elastic modulus (E_p) of the abdominal aorta in a group of 61 human subjects. (Modified from Imura T et al: *Cardiovasc Res* 20:208, 1986.)

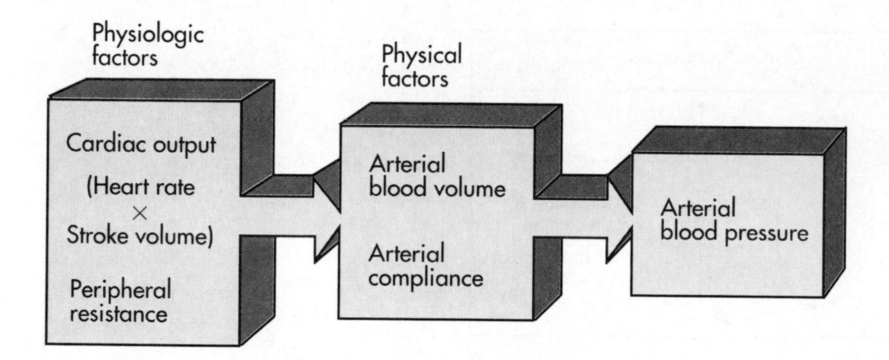

■ **Fig. 27-6** The arterial blood pressure is determined directly by two major physical factors, the arterial blood volume and the arterial compliance. These physical factors are affected in turn by certain physiologic factors, primarily the heart rate, stroke volume, cardiac output (heart rate × stroke volume), and peripheral resistance.

blood pressure is routinely measured in patients, and it provides a useful clue to their cardiovascular status. We will therefore take a simplified approach to explain the principal determinants of the arterial blood pressure. To accomplish this the determinants of the **mean arterial pressure** (defined in the next section) will first be analyzed. The **systolic** and **diastolic arterial pressures** will then be considered as the upper and lower limits of periodic oscillations about this mean pressure. Finally the changes in arterial pressure as the pulse wave progresses from the origin of the aorta toward the capillaries will be discussed.

The determinants of the arterial blood pressure will be arbitrarily subdivided into "physical" and "physiological" factors (Fig. 27-6). The arterial system will be assumed to be a static, elastic system, and the only two "physical" factors considered will be the **blood volume** within the arterial system and the **elastic characteristics** (compliance) of the system. Certain "physiological" factors will be considered, namely, **cardiac output** (which equals **heart rate × stroke volume**) and **peripheral resistance.** Such physiological factors will be shown to operate through one or both of the physical factors however.

■ *Mean Arterial Pressure*

The **mean arterial pressure** is the pressure in the arteries, averaged over time. It may be obtained from an arterial pressure tracing by measuring the area under the curve and dividing this area by the time interval involved, as shown in Fig. 27-7. The mean arterial pressure, $\bar{P}_a$, usually can be approximated satisfactorily from the measured values of the systolic (P_s) and diastolic (P_d) pressures by means of the following formula:

$$\bar{P}_a \cong P_d + \frac{1}{3}(P_s - P_d) \qquad (4)$$

The mean pressure will be considered to depend only on the mean blood volume in the arterial system and on the arterial compliance (see Fig. 27-6). The arterial volume, V_a, in turn, depends on the rate of inflow, Q_h, from the heart into the arteries **(cardiac output)** and the rate of outflow, Q_r, from the arteries through the resistance vessels **(peripheral runoff);** expressed mathematically,

$$dV_a/dt = Q_h - Q_r \qquad (5)$$

If arterial inflow exceeds outflow, then arterial volume

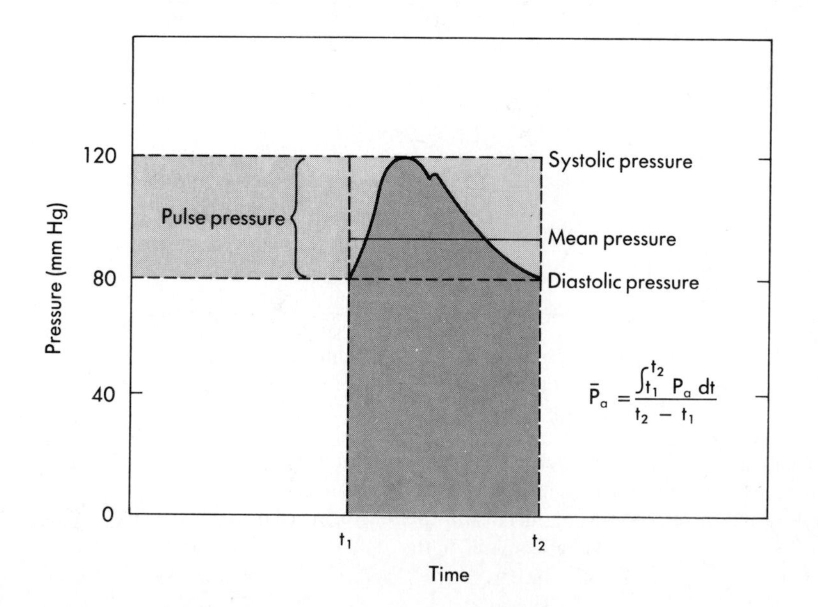

■ **Fig. 27-7** Arterial systolic, diastolic, pulse, and mean pressures. The mean arterial pressure ($\bar{P}_a$) represents the area under the arterial pressure curve *(shaded area)* divided by the cardiac cycle duration *($t_2 - t_1$)*.

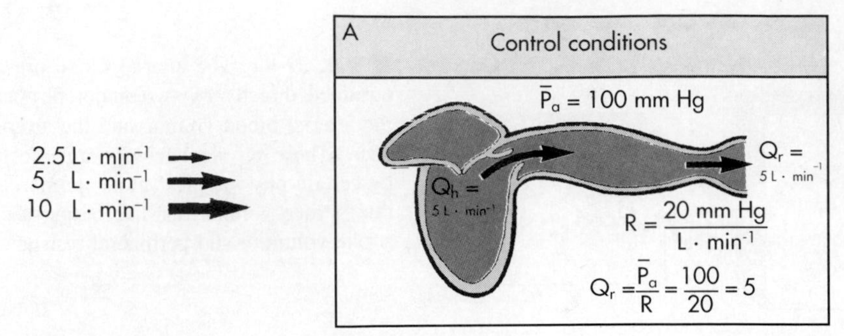

A, Under control conditions $Q_h = 5\ L \cdot min^{-1}$, $\overline{P}_a = 100$ mm Hg $\cdot\ L^{-1}$ min. Q_r must equal Q_h and therefore the mean blood volume *($\overline{V}_a$)* in the arteries will remain constant from heartbeat to heartbeat.

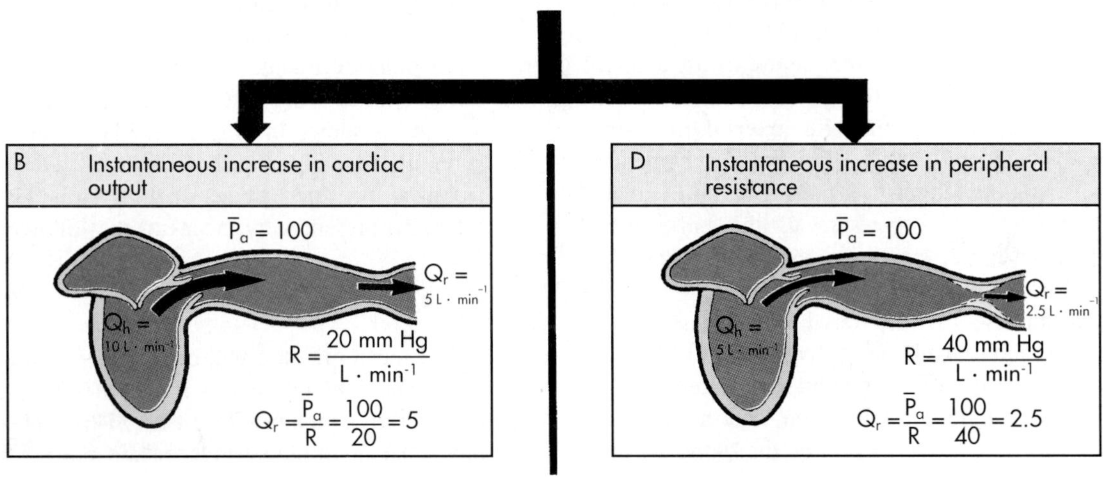

B, If Q_h suddenly increases to 10 $L \cdot min^{-1}$, Q_h will initially exceed Q_r and therefore $\overline{P}_a$ will begin to rise rapidly.

D, If R abruptly increases to 40 mm Hg $\cdot\ L^{-1}$ min, Q_t suddenly decreases and therefore Q_h exceeds $\dot{V}_r$. Thus $\overline{P}_a$ will rise progressively.

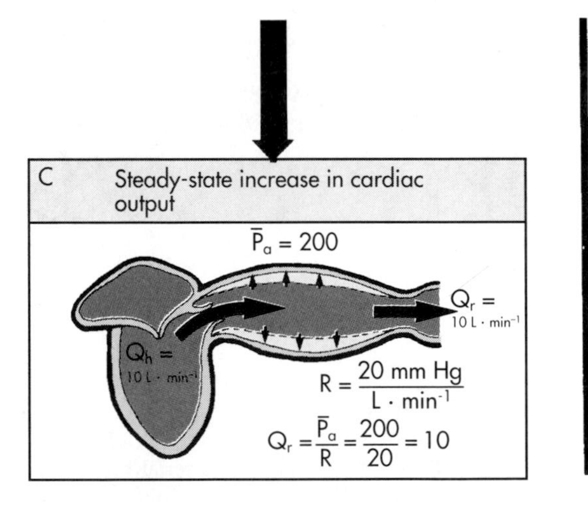

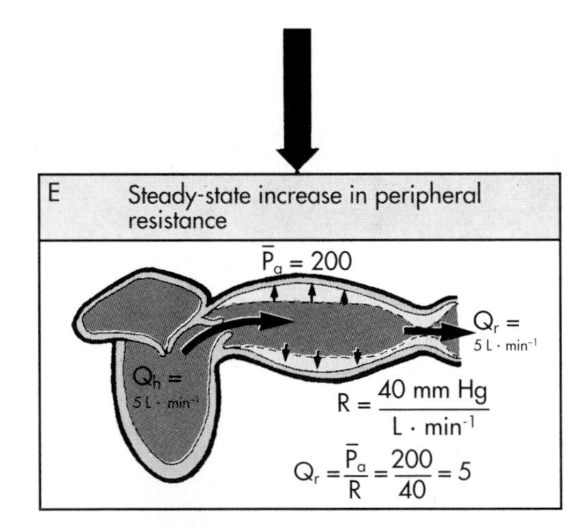

C, The disparity between Q_h and Q_r progressively increases arterial blood volume. The volume continues to increase until $\overline{P}_a$ reaches a level of 200 mm Hg.

E, The excess of Q_h over Q_r accumulates blood in the arteries. Blood continues to accumulate until $\overline{P}_a$ rises to a level of 200 mm Hg.

■ **Fig. 27-8** The relationship of mean arterial blood pressure ($\overline{P}_a$) to cardiac output (Q_h), peripheral runoff (Q_r), and peripheral resistance (R) under control conditions **(A),** in response to an increase in cardiac output **(B & C),** and in response to an increase in peripheral resistance **(D & E).**

increases, arterial walls are stretched more, and pressure rises. The converse happens when arterial outflow exceeds inflow. When inflow equals outflow, arterial pressure remains constant.

Cardiac output. The change in pressure in response to an alteration of cardiac output can be better appreciated by considering some simple examples. Under control conditions let cardiac output be 5 L/min and mean arterial pressure ($\bar{P}_a$) be 100 mm Hg (Fig. 27-8, *A*). From the definition of total peripheral resistance

$$R = (\bar{P}_a - \bar{P}_{ra})/Q_r \tag{6}$$

If $\bar{P}_{ra}$ (mean right atrial pressure) is negligible compared with $\bar{P}_a$,

$$R \cong \bar{P}_a/Q_r \tag{7}$$

Therefore, in the example R is 100/5, or 20 mm Hg/L/min.

Now let cardiac output, Q_h, suddenly increase to 10 L/min (Fig. 27-8, *B*). Instantaneously $\bar{P}_a$ will be unchanged. Because the outflow, Q_r, from the arteries depends on $\bar{P}_a$ and R, Q_r also will remain unchanged at first. Therefore, Q_h, now 10 L/min, will exceed Q_r, still only 5 L/min. This will increase the mean arterial blood volume ($\bar{V}_a$). From equation 5, when $Q_h > Q_r$, then $d\bar{V}_a/dt > 0$; that is, volume is increasing.

Because $\bar{P}_a$ depends on the mean arterial blood volume, $\bar{V}_a$, and the arterial compliance, C_a, an increase in $\bar{V}_a$ will raise the $\bar{P}_a$. By definition

$$C_a = d\bar{V}_a/d\bar{P}_a \tag{8}$$

Therefore,

$$d\bar{V}_a = C_a d\bar{P}_a \tag{9}$$

and

$$\frac{d\bar{V}_a}{dt} = C_a \frac{d\bar{P}_a}{dt} \tag{10}$$

From equation 2,

$$\frac{d\bar{P}_a}{dt} = \frac{Q_h - Q_r}{C_a} \tag{11}$$

Hence P_a will rise when $Q_h > Q_r$, will fall when $Q_h < Q_r$, and will remain constant when $Q_h = Q_r$.

In this example, in which Q_h is suddenly increased to 10 L/min, $\bar{P}_a$ will continue to rise as long as Q_h exceeds Q_r. It is evident from equation 7 that Q_r will not attain a value of 10 L/min until $\bar{P}_a$ reaches a level of 200 mm Hg, as long as R remains constant at 20 mm Hg/L/min. Hence, as $\bar{P}_a$ approaches 200, Q_r will almost equal Q_h and $\bar{P}_a$ will rise very slowly. When Q_h is first raised, however, Q_h is greatly in excess of Q_r, and therefore $\bar{P}_a$ will rise sharply. The pressure-time tracing in Fig. 27-9 indicates that, regardless of the value of C_a, the slope gradually diminishes as pressure rises, to approach a final value asymptotically.

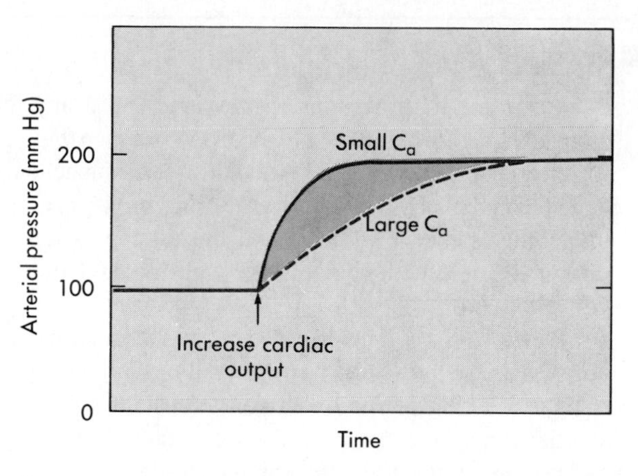

Fig. 27-9 When cardiac output is suddenly increased, the arterial compliance (C_a) determines the rate at which the mean arterial pressure will attain its new, elevated value but does not determine the *magnitude* of the new pressure.

Furthermore the **height** to which $\bar{P}_a$ will rise is independent of the elastic characteristics of the arterial walls. $\bar{P}_a$ must rise to a level such that $Q_r = Q_h$. It is apparent from equation 6 that Q_r depends only on pressure gradient and resistance to flow. Hence C_a determines only the **rate** at which the new equilibrium value of $\bar{P}_a$ will be approached, as illustrated in Fig. 27-9. When C_a is small (rigid vessels), a relatively slight increment in $\bar{V}_a$ (caused by a transient excess of Q_h over Q_r) increases $\bar{P}_a$ greatly. Hence $\bar{P}_a$ attains its new equilibrium level quickly. Conversely when C_a is large, then considerable volumes can be accommodated with relatively small pressure changes. Therefore the new equilibrium value of $\bar{P}_a$ is reached at a slower rate.

Peripheral resistance. Similar reasoning may now be applied to explain the changes in $\bar{P}_a$ that accompany alterations in peripheral resistance. Let the control conditions be identical with those of the preceding example, that is, $Q_h = 5$, $\bar{P}_a = 100$, and $R = 20$ (see Fig. 27-8, *A*). Then let R suddenly be increased to 40 (Fig. 27-8, *D*). Instantaneously $\bar{P}_a$ will be unchanged. With $\bar{P}_a = 100$ and $R = 40$, $Q_r = \bar{P}_a/R = 2.5$ L/min. If Q_h remains constant at 5 L/min, $Q_h \geq Q_r$, and $\bar{V}_a$ will increase; hence $\bar{P}_a$ will rise. $\bar{P}_a$ will continue to rise until it reaches 200 mm Hg (see Fig. 27-8, *E*). At this level $Q_r = 200/40 = 5$ L/min, which equals Q_h. $\bar{P}_a$ will then remain at this new elevated equilibrium level as long as Q_h and R do not change again.

It is clear, therefore, that *the level of the mean arterial pressure depends on cardiac output and peripheral resistance*. It is immaterial whether any change in cardiac output is accomplished by an alteration of heart rate, of stroke volume, or of both. Any change in heart rate that is balanced by a concomitant, oppositely directed change in stroke volume will not alter Q_h. Hence $\bar{P}_a$ will not be affected.

Pulse Pressure

If we assume that the arterial pressure, P_a, at any moment depends on the two physical factors (see Fig. 27-7), arterial blood volume, V_a, and arterial capacitance, C_a, it can be shown that the arterial **pulse pressure** (difference between systolic and diastolic pressures) is principally a function of **stroke volume** and **arterial compliance.**

Stroke volume. The effect of a change in stroke volume on pulse pressure may be analyzed under conditions in which C_a remains virtually constant over a substantial range of pressures. C_a is constant over any linear region of the pressure-volume curve (Fig. 27-10). Volume is plotted along the vertical axis, and pressure is plotted along the horizontal axis; the slope, dV/dP, equals the compliance, C_a.

In an individual with such a linear P_a:V_a curve, the arterial pressure would oscillate about some mean value ($\overline{P}_A$ in Fig. 27-10) that depends entirely on cardiac output and peripheral resistance, as explained previously. This mean pressure corresponds to some mean arterial blood volume, $\overline{V}_A$. The coordinates $\overline{P}_A$, $\overline{V}_A$ define point $\overline{A}$ on the graph. During diastole peripheral runoff from the arterial system occurs in the absence of ventricular ejection of blood, and P_a and V_a diminish to minimum values, P_1 and V_1, just before the next ventricular ejection. P_1 is then, by definition, the **diastolic pressure.**

During the rapid ejection phase of systole the volume of blood introduced into the arterial system exceeds the volume that exits through the arterioles. Arterial pressure and volume therefore rise from point A_1 toward point A_2 in Fig. 27-10. The maximum arterial volume, V_2, is reached at the end of the rapid ejection phase (Fig. 24-13), and this volume corresponds to a peak pressure, P_2, which is the **systolic pressure.**

The **pulse pressure** is the difference between systolic and diastolic pressures ($P_2 - P_1$ in Fig. 27-10), and it corresponds to some **arterial volume increment,** $V_2 - V_1$. *This increment equals the volume of blood discharged by the left ventricle during the rapid ejection phase minus the volume that has run off to the periphery during this same phase of the cardiac cycle.* When a normal heart beats at a normal frequency, the volume increment during the rapid ejection phase is a large fraction of the stroke volume (about 80%). It is this increment that will raise arterial volume rapidly from V_1 to V_2 and hence will cause the arterial pressure to rise from the diastolic to the systolic level (P_1 to P_2 in Fig. 27-10). During the remainder of the cardiac cycle peripheral runoff will greatly exceed cardiac ejection. During diastole, of course, cardiac ejection equals zero. The resultant arterial blood volume decrement will cause volumes and pressures to fall from point A_2 back to point A_1.

If stroke volume is now doubled while heart rate and peripheral resistance remain constant, the mean arterial pressure will be doubled, to $\overline{B}$ in Fig. 27-10. Thus the arterial pressure will now oscillate each heartbeat about this new value of the mean arterial pressure. A normal, vigorous heart will eject this greater stroke volume during a fraction of the cardiac cycle approximately equal to the fraction that prevailed at the lower stroke volume. Therefore the arterial volume increment, $V_4 - V_3$, will be a large fraction of the new stroke volume, and hence it will be approximately twice as great as the previous volume increment ($V_2 - V_1$). With a linear P_a:V_a curve, the greater volume increment will be reflected by a **pulse pressure** ($P_4 - P_3$) that will be approximately **twice as great** as the original pulse pressure ($P_2 - P_1$). With a rise in both mean and pulse

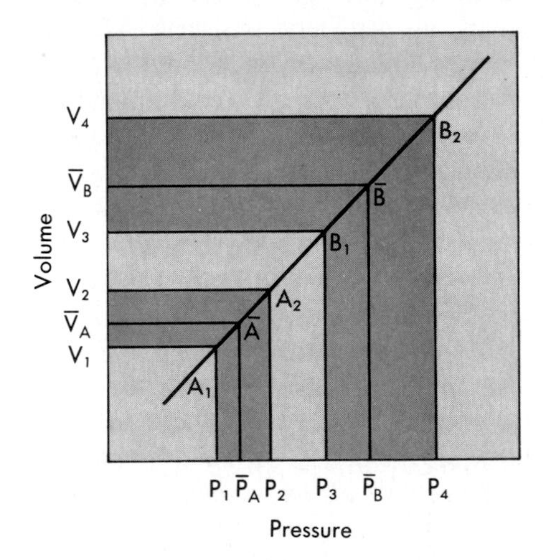

■ **Fig. 27-10** Effect of a change in stroke volume on pulse pressure in a system in which arterial compliance remains constant over the range of pressures and volumes involved. A larger volume increment ($V_4 - V_3$ as compared with $V_2 - V_1$) results in a greater mean pressure ($\overline{P}_B$ as compared with $\overline{P}_A$) and a greater pulse pressure ($P_4 - P_3$ as compared with $P_2 - P_1$).

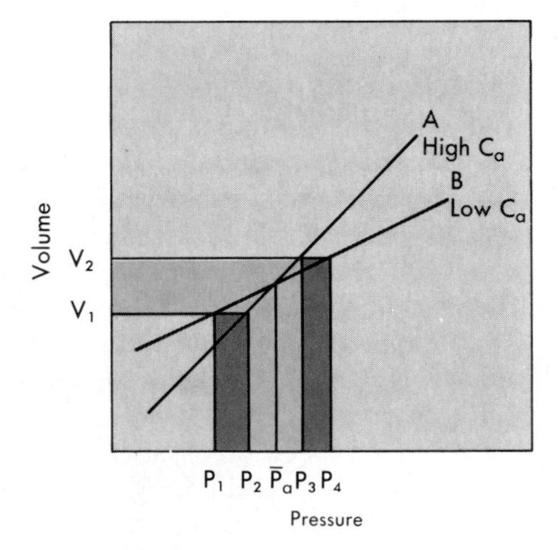

■ **Fig. 27-11** For a given volume increment ($V_2 - V_1$) a reduced arterial compliance (curve *B* as compared with curve *A*) results in an increased pulse pressure ($P_4 - P_1$ as compared with $P_3 - P_2$).

pressures inspection of Fig. 27-10 reveals that the rise in systolic pressure (from P_2 to P_4) exceeds the rise in diastolic pressure (from P_1 to P_3). Thus an increase in stroke volume raises systolic pressure more than it raises diastolic pressure.

Arterial compliance. To assess how arterial compliance affects pulse pressure the relative effects of a given volume increment ($V_2 - V_1$ in Fig. 27-11) in a young person (curve *A*) and in an elderly person (curve *B*) will be compared. Let cardiac output and total peripheral resistance be the same in both people; therefore $\overline{P}_a$ will be the same. It is apparent from Fig. 27-11 that the same volume increment ($V_2 - V_1$) will cause a greater pulse pressure ($P_4 - P_1$) in the less distensible arteries of the elderly individual than in the more compliant arteries of the young one ($P_3 - P_2$). For the reasons enunciated on p. 453 this will impose a greater work load on the left ventricle of the elderly person than on that of the young person, even if the stroke volumes, total peripheral resistances (TPRs), and mean arterial pressures are equivalent.

Fig. 27-12 displays the effects of changes in arterial compliance and in peripheral resistance, R_p, on the arterial pressure in an isolated cat heart preparation. As the compliance was reduced from 43 to 14 to 3.6 units, the pulse pressure increased significantly. In this preparation the stroke volume decreased as the compliance was diminished. This accounts for the failure of the mean arterial pressure to remain constant at the different levels of arterial compliance. The effects of changes in peripheral resistance in this same preparation are described in the next section.

Total peripheral resistance and arterial diastolic pressure. It is often stated that increased TPR mainly affects the level of the diastolic arterial pressure. The validity of such an assertion deserves close scrutiny.

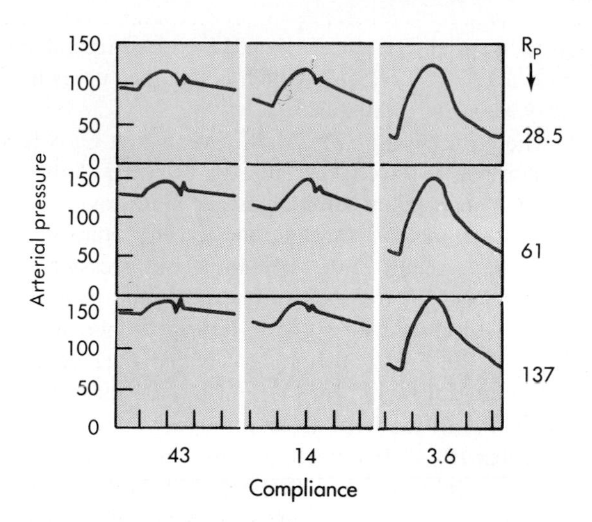

Fig. 27-12 The changes in aortic pressure induced by changes in arterial compliance and peripheral resistance *(R_p)* in an isolated cat heart preparation. (Modified from Elizinga G, Westerhof N: *Circ Res* 32:178, 1973. By permission of the American Heart Association.)

First let TPR be increased in an individual with a linear $P_a:V_a$ curve, as depicted in Fig. 27-13, *A*. If heart rate and stroke volume remain constant, then an increase in TPR will increase $\overline{P}_a$ proportionately (from P_2 to P_5). If the volume increments ($V_2 - V_1$ and $V_4 - V_3$) are equal at both levels of TPR, then the pulse pressures ($P_3 - P_1$ and $P_6 - P_4$) will also be equal. Hence systolic (P_6) and diastolic (P_4) pressures will have been elevated by exactly the same amounts from their respective control levels (P_3 and P_1).

Chronic **hypertension,** a condition characterized by a persistent elevation of TPR, occurs more commonly in older persons than in younger persons. The $P_a:V_a$ curve

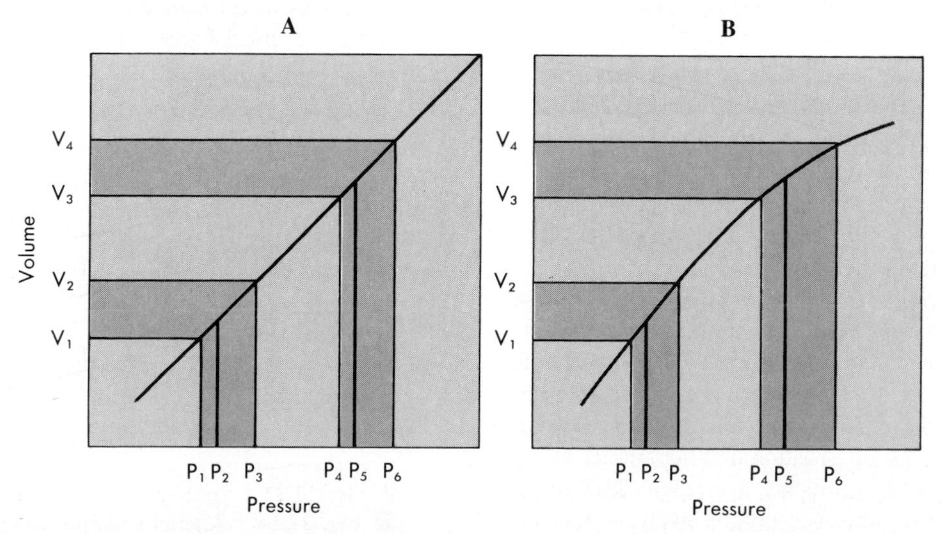

Fig. 27-13 Effect of a change in TPR (volume increment remaining constant) on pulse pressure when the pressure-volume curve for the arterial system is rectilinear, **A**, or curvilinear, **B.**

for a hypertensive patient would therefore resemble that shown in Fig. 27-13, *B*, which is like the curves in Fig. 27-3 for older individuals.

The curve in Fig. 27-13, *B*, reveals that C_a is less at higher than at lower pressures. As before if cardiac output remains contant, an increase in TPR would increase $\bar{P}_a$ proportionately (from P_2 to P_5). For equivalent increases in TPR the elevation of pressure from P_2 to P_5 will be the same in panel *A* as in panel *B*, for reasons discussed on p. 459. If the volume increment ($V_4 - V_3$ in Fig. 27-13, *B*) at elevated TPR were equal to the control increment ($V_2 - V_1$), the pulse pressure ($P_6 - P_4$) in the hypertensive range would greatly exceed that ($P_3 - P_1$) at normal pressure levels. In other words a given volume increment will produce a greater pressure increment (i.e., pulse pressure) when the arteries are more rigid than when they are more compliant. Hence the rise in systolic pressure ($P_6 - P_3$) will exceed the increase in diastolic pressure ($P_4 - P_1$). Thus an increase in peripheral resistance will raise systolic pressure more than it will raise diastolic pressure.

These hypothetical changes in arterial pressure closely resemble those actually seen in patients with hypertension. Diastolic pressure is indeed elevated in such individuals but ordinarily not more than 10 to 40 mm Hg above the average normal level of 80 mm Hg, whereas it is not uncommon for systolic pressures to be elevated by 50 to 150 mm Hg above the average normal level of 120 mm Hg. The combination of increased resistance and diminished arterial compliance would be represented in Fig. 27-12 by a shift in direction from the top left panel to the bottom right panel; that is, both the mean pressure and the pulse pressure would be increased significantly. These results also coincide with the changes predicted by Fig. 27-13, *B*.

Peripheral Arterial Pressure Curves

The radial stretch of the ascending aorta brought about by left ventricular ejection initiates a pressure wave that is propagated down the aorta and its branches. The pressure wave travels much faster than does the blood itself. It is this pressure wave that one perceives by palpating a peripheral artery.

The velocity of the pressure wave varies inversely with the vascular compliance. Accurate measurement of the transmission velocity has provided valuable information about the elastic characteristics of the arterial tree. In general transmission velocity increases with age, confirming the observation that the arteries become less compliant with advancing age (Figs. 27-3 and 27-5). Also velocity increases progressively as the pulse wave travels from the ascending aorta toward the periphery. This indicates that vascular compliance is less in the more distal than in the more proximal portions of the arterial system, a fact that has been confirmed by direct measurement.

The arterial pressure contour becomes distorted as the wave is transmitted down the arterial system; the changes in configuration of the pulse with distance are shown in Fig. 27-14. Aside from the increasing delay in the onset of the initial pressure rise three major changes occur in the arterial pulse contour as the pressure wave travels distally. First the high-frequency components of the pulse, such as the incisura (that is, the notch that appears at the end of ventricular ejection), are damped out and soon disappear. Second the systolic portions of the pressure wave become narrowed and elevated. In the curves shown in Fig. 27-14 the systolic pressure at the level of the knee was 39 mm Hg greater than that recorded in the aortic arch. Third a hump may appear on the diastolic portion of the pressure wave. These changes in contour are pronounced in young individuals, but they diminish with age. In elderly patients the pulse wave may be transmitted virtually unchanged from the ascending aorta to the periphery.

The damping of the high-frequency components of the arterial pulse is largely caused by the viscoelastic properties of the arterial walls. The precise mechanism for the peaking of the pressure wave is controversial. Probably several factors contribute, including (1) reflection, (2) tapering, (3) resonance, and (4) changes in transmission velocity with pressure level. Relative to the first of these mechanisms whenever significant changes in configuration or in dimensions occur (such as at points of branching), pressure waves are reflected backward. Hence pressure at any time and location is determined by the algebraic summation of an antegrade incident wave and retrograde reflected waves. The second factor, tapering, alters the pulse contour because the lumen progressively narrows beyond each successive branch. A pressure wave becomes amplified as it progresses down a tapered tube. With respect to resonance the arterial tree resonates at certain frequencies

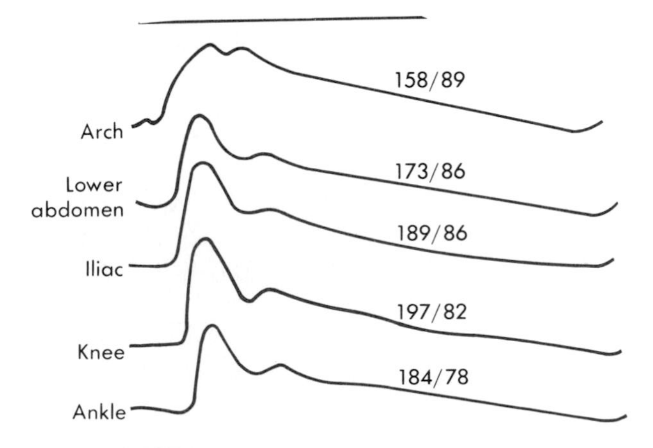

Fig. 27-14 Arterial pressure curves recorded from various sites in an anesthetized dog. (From Remington JW, O'Brien LJ: *Am J Physiol* 218:437, 1970.)

and other frequencies are effectively damped. Finally as stated previously, transmission velocity varies inversely with arterial compliance. Furthermore compliance varies inversely with pressure level (see Fig. 27-3). Hence the points on the pressure curve at the higher pressures tend to travel faster than those at lower pressures. Thus the peak of the arterial pressure curve tends to catch up with the beginning of the same curve. This contributes to the peaking and narrowing of the curve in more distal vessels.

Blood Pressure Measurement in Humans

In hospital intensive care units needles or catheters may be introduced into peripheral arteries of patients, and arterial blood pressure can then be measured **directly** by means of strain gauges. Ordinarily, however, the blood pressure is estimated **indirectly** by means of a **sphygmomanometer.** This instrument consists of an inextensible cuff containing an inflatable bag. The cuff is wrapped around the extremity (usually the arm, occasionally the thigh) so that the inflatable bag lies between the cuff and the skin, directly over the artery to be compressed. The artery is occluded by inflating the bag, by means of a rubber squeeze bulb, to a pressure in excess of arterial systolic pressure. The pressure in the bag is measured by means of a mercury or an aneroid manometer. Pressure is released from the bag at a rate of 2 or 3 mm Hg per heartbeat by means of a needle valve in the inflating bulb (Fig. 27-15).

When blood pressure readings are taken from the arm, the systolic pressure may be estimated by palpating the radial artery at the wrist **(palpatory method).** When pressure in the bag exceeds the systolic level, no pulse will be perceived. As the pressure falls just below the systolic level (Fig. 27-15, *A*), a spurt of blood will pass through the brachial artery under the cuff during the peak of systole and a slight pulse will be felt at the wrist.

The **auscultatory method** is a more sensitive and therefore a more precise method for measuring systolic pressure, and it also permits the estimation of the diastolic pressure level. The practitioner listens with a stethoscope applied to the skin of the antecubital space over the brachial artery. While the pressure in the bag exceeds the systolic pressure, the brachial artery is occluded and no sounds are heard (Fig. 27-15, *B*). When the inflation pressure falls just below the systolic level (120 mm Hg in Fig. 27-15, *A*), a small spurt of blood escapes through the cuff and slight tapping sounds (called **Korotkoff sounds**) are heard with each heartbeat. The pressure at which the first sound is detected represents the **systolic pressure.** It usually corresponds closely with the directly measured systolic pressure.

As inflation pressure continues to fall, more blood es-

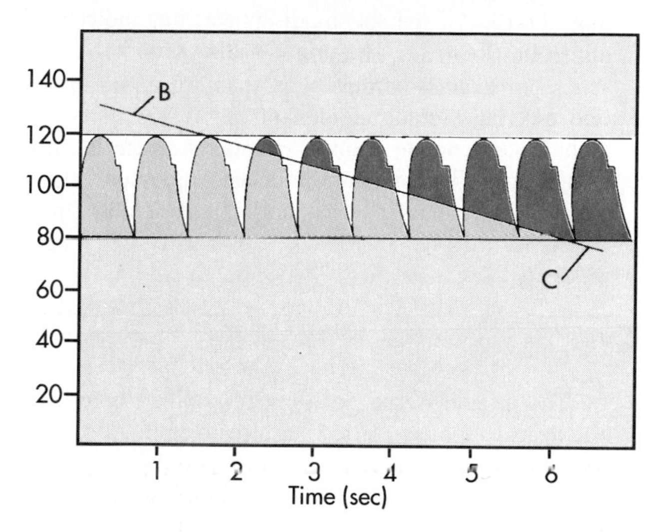

A, Consider that the arterial blood pressure is being measured in a patient whose blood pressure is 120/80 mm Hg. The pressure (represented by the *oblique line*) in a cuff around the patient's arm is allowed to fall from greater than 120 mm Hg (point *B*) to below 80 mm Hg (point *C*) in about 6 seconds.

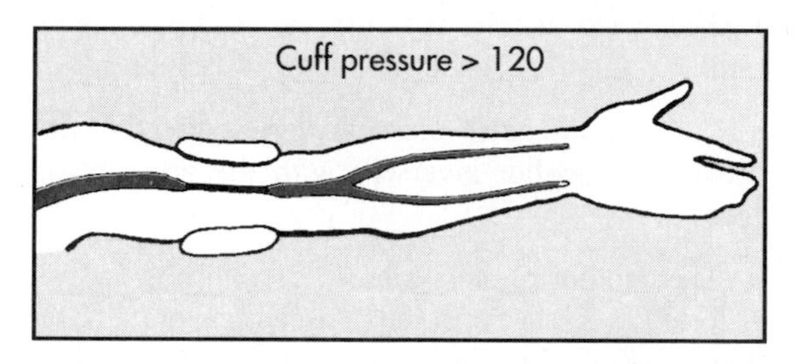

B, When the cuff pressure exceeds the systolic arterial pressure (120 mm Hg), no blood progresses through the arterial segment under the cuff, and no sounds can be detected by a stethoscope bell placed on the arm distal to the cuff.

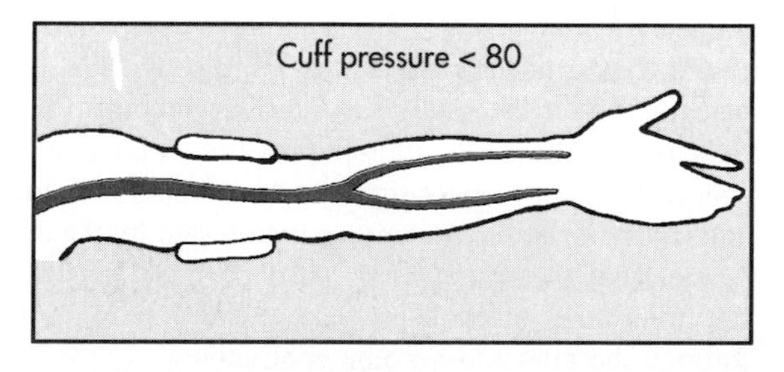

C, When the cuff pressure falls below the diastolic arterial pressure, arterial flow past the region of the cuff is continuous, and no sounds are audible. When the cuff pressure is between 120 and 80 mm Hg, spurts of blood traverse the artery segment under the cuff with each heartbeat, and the Korotkoff sounds are heard through the stethoscope.

Fig. 27-15 Measurement of arterial blood pressure with a sphygmomanometer.

capes under the cuff per beat and the sounds become louder thuds. As the inflation pressure approaches the diastolic level, the Korotkoff sounds become muffled. As they fall just below the diastolic level (80 mm Hg in Fig. 27-15, *A*), the sounds disappear: this indicates the **diastolic pressure.** The origin of the Korotkoff sounds is related to the spurt of blood passing under the cuff and meeting a static column of blood; the impact and turbulence generate audible vibrations. Once the inflation pressure is less than the diastolic pressure, flow is continuous in the brachial artery and sounds are no longer heard (Fig. 27-15, *C*).

Summary

1. The arteries serve not only to conduct blood from the heart to the capillaries but also to store some of the ejected blood during each cardiac systole so that flow can continue through the capillaries during cardiac diastole.

2. The aging process diminishes the compliance of the arteries.

3. The less compliant the arteries the more work the heart must do to pump a given cardiac output.

4. The mean arterial pressure varies directly with the cardiac output and total peripheral resistance.

5. The arterial pulse pressure varies directly with the stroke volume but inversely with the arterial compliance.

6. The contour of the systemic arterial pressure wave is distorted as it travels from the ascending aorta to the periphery. The high-frequency components of the wave are damped, the systolic components are narrowed and elevated, and a hump may appear in the diastolic component of the wave.

7. When blood pressure is measured by a sphygmomanometer in humans, (a) the systolic pressure is manifested by the occurrence of a tapping sound that originates in the artery distal to the cuff as the cuff pressure falls below the peak arterial pressure, which permits spurts of blood to pass through the compressed artery; and (b) the diastolic pressure is manifested by the disappearance of the sound as the cuff pressure falls below the minimum arterial pressure, which permits flow through the artery to become continuous.

Bibliography

Journal articles

Alexander J, Jr et al: Influence of mean pressure on aortic impedance and reflections in the systemic arterial system, *Am J Physiol* 257:H969, 1989.

Blank SG et al: Wideband external pulse recording during cuff deflation: a new technique for evaluation of the arterial pressure pulse and measurement of blood pressure, *Circulation* 77:1297, 1988.

Burattini R et al: Total systemic arterial compliance and aortic characteristic impedance in the dog as a function of pressure: a model based study, *Comput Biomed Res* 20:154, 1987.

Burkhoff D, Alexander J, Jr, and Schipke J: Assessment of Windkessel as a model of aortic input impedance, *Am J Physiol* 255:H742, 1988.

Campbell KB et al: Pulse reflection sites and effective length of the arterial system, *Am J Physiol* 256:H1684, 1989.

Imura T et al: Non-invasive ultrasonic measurement of the elastic properties of the human abdominal aorta, *Cardiovasc Res* 20:208, 1986.

Isnard RN et al: Pulsatile diameter and elastic modulus of the aortic arch in essential hypertension: a noninvasive study, *J Am Coll Cardiol* 13:399, 1989.

Kenner T: Arterial blood pressure and its measurement, *Basic Res Cardiol* 83:107, 1988.

Laskey WK et al: Estimation of total systemic arterial compliance in humans, *J Appl Physiol* 69:112, 1990.

McIlroy MB, Targett RC: Model of the systemic arterial bed showing ventricular-systemic arterial coupling, *Am J Physiol* 254:H609, 1988.

Zuckerman BD, Weisman HF, Yin FCP: Arterial hemodynamics in a rabbit model of atherosclerosis, *Am J Physiol* 257:H891, 1989.

Books and monographs

Dobrin PB: *Vascular mechanics*. In *Handbook of physiology*. Section 2: *The cardiovascular system—peripheral circulation and organ blood flow,* vol III, Bethesda, Md, 1983, American Physiological Society.

Fung YC: *Biodynamics: circulation,* Heidelberg, 1984, Springer-Verlag.

Milnor WR: *Hemodynamics,* Baltimore, 1982, Williams & Wilkins.

Taylor DEM, Stevens AL, editors: *Blood flow: theory and practice,* New York, 1983, Academic Press.

Westerhof N, Gross DR, editors: *Vascular dynamics: physiological perspectives,* New York, 1989, Plenum Press.

CHAPTER

28

The Microcirculation and Lymphatics

The entire circulatory system is geared to supply the body tissues with blood in amounts commensurate with their requirements for oxygen and nutrients. The capillaries, consisting of a single layer of endothelial cells, permit rapid exchange of water and solutes with interstitial fluid. The muscular arterioles, which are the major **resistance vessels,** regulate regional blood flow to the capillary beds, and the venules and veins serve primarily as collecting channels and storage, or **capacitance, vessels.**

The arterioles, which range in diameter from about 5 to 100 μm, have a thick smooth muscle layer, a thin adventitial layer, and an endothelial lining (see Fig. 22-1). The arterioles give rise directly to the capillaries (5 to 10 μm diameter) or in some tissues to **metarterioles** (10 to 20 μm diameter), which then give rise to capillaries (Fig. 28-1). The metarterioles can serve either as thoroughfare channels to the venules, bypassing the capillary bed, or as conduits to supply the capillary bed. There are often cross connections between arterioles and between venules, as well as in the capillary network. Arterioles that give rise directly to capillaries regulate flow through their cognate capillaries by constriction or dilation. The capillaries form an interconnecting network of tubes of different lengths with an average length of 0.5 to 1 mm.

Capillary distribution varies from tissue to tissue. In metabolically active tissues, such as cardiac and skeletal muscle and glandular structures, capillaries are numerous, whereas in less active tissues, such as subcutaneous tissue or cartilage, **capillary density** is low. Also, all capillaries are not of the same diameter, and because some capillaries have diameters less than that of the erythrocytes, it is necessary for the cells to become temporarily deformed in their passage through these capillaries. Fortunately the normal red cells are quite flexible and readily change their shape to conform with that of the small capillaries.

Blood flow in the capillaries is not uniform and depends chiefly on the contractile state of the arterioles. The average velocity of blood flow in the capillaries is approximately 1 mm/sec; however, it can vary from zero to several millimeters per second in the same vessel within a brief period. These changes in capillary blood flow may be of random type or may show rhythmical oscillatory behavior of different frequencies that is caused by contraction and relaxation **(vasomotion)** of the precapillary vessels. This vasomotion is to some extent intrinsic contractile behavior of the vascular smooth muscle and is independent of external input. Furthermore changes in **transmural pressure** (intravascular minus extravascular pressure) influence the contractile state of the precapillary vessels. An increase in transmural pressure, whether produced by increase in venous pressure or by dilation of arterioles, results in contraction of the terminal arterioles at the points of origin of the capillaries, whereas a decrease in transmural pressure elicits precapillary vessel relaxation (see myogenic response, p. 481). In addition, humoral and possibly neural factors also affect vasomotion. For example, when the precapillary vessels contract in response to increased transmural pressure, the contractile response can be overridden and vasomotion abolished. This effect is accomplished by metabolic (humoral) factors (p. 483) when the oxygen supply becomes too low for the requirements of the parenchymal tissue, as occurs in muscle during exercise.

Although reduction of transmural pressure induces relaxation of the terminal arterioles, blood flow through the capillaries obviously cannot increase if the reduction in intravascular pressure is caused by severe constriction of the parent arterioles, metarterioles, or small arteries. Large arterioles and metarterioles also exhibit vasomotion, but in the contraction phase they usually do not completely occlude the lumen of the vessel and arrest blood flow as may occur with contraction of the terminal arterioles. Thus *tissue blood flow may be altered by contraction and relaxation of small arteries, arterioles, and metarterioles.*

Because blood flow through the capillaries provides for exchange of gases and solutes between blood and tissue, it has been termed **nutritional flow,** whereas blood flow that bypasses the capillaries in traveling from the arterial to the venous side of the circulation

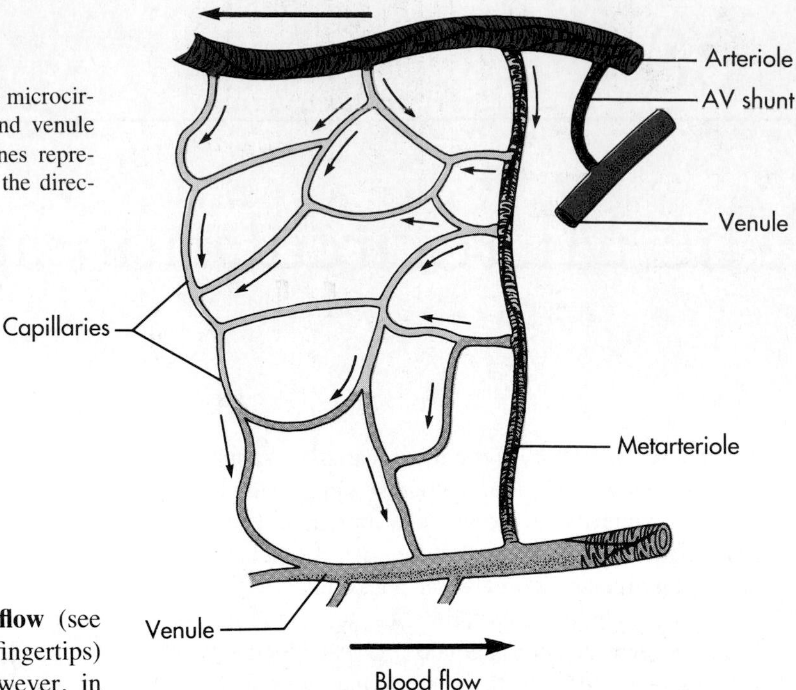

■ **Fig. 28-1** Composite schematic drawing of the microcirculation. The circular structures on the arteriole and venule represent smooth muscle fibers; branching solid lines represent sympathetic nerve fibers. The arrows indicate the direction of blood flow.

has been termed **nonnutritional,** or **shunt flow** (see Fig. 28-1). In some areas of the body (e.g., fingertips) true arteriovenous shunts exist (p. 517). However, in many tissues, such as muscle, evidence of anatomical shunts is lacking. Nevertheless nonnutritional flow can occur and has been termed **physiological shunting.** It is the result of a greater flow of blood through previously open capillaries with either no change or an increase in the number of closed capillaries. In tissues that have metarterioles, shunt flow may be continuous from arteriole to venule during low metabolic activity when many precapillary vessels are closed. When metabolic activity increases in such tissues and more precapillary vessels open, blood passing through the metarterioles is readily available for capillary perfusion.

The true capillaries are devoid of smooth muscle and are therefore incapable of active constriction. Nevertheless the endothelial cells that form the capillary wall contain actin and myosin and can alter their shape in response to certain chemical stimuli. Evidence is lacking, however, that changes in endothelial cell shape regulate blood flow through the capillaries. Hence changes in capillary diameter are passive and are caused by alterations in precapillary and postcapillary resistance.

The thin-walled capillaries can withstand high internal pressures without bursting because of their narrow lumen. This can be explained in terms of the **law of Laplace** and is illustrated in the following comparison of wall tension of a capillary with that of the aorta (Table 28-1). The Laplace equation is

$$T = Pr$$

where

T = Tension in the vessel wall

P = Transmural pressure

r = Radius of the vessel

Wall tension is the force per unit length tangential to the vessel wall that opposes the distending force (Pr) that tends to pull apart a theoretical longitudinal slit in the vessel (Fig. 28-2). Transmural pressure is essentially equal to intraluminal pressure, because extravascular pressure is negligible. The Laplace equation applies to very thin wall vessels, such as capillaries. Wall thickness must be taken into consideration when the

■ **Table 28-1** Vessel wall tension in the aorta and a capillary

	Aorta	*Capillary*
Radius (r)	1.5 cm	5×10^{-4} cm
Height of Hg column (h)	10 cm Hg	2.5 cm Hg
ρ	13.6 g/cm^3	13.6 g/cm^3
g	980 cm/sec^2	980 cm/sec^2
P	$10 \times 13.6 \times 980 = 1.33 \times 10^5$ dyne/cm^2	$2.5 \times 13.6 \times 980 = 3.33 \times 10^4$ dyne/cm^2
w	0.2 cm	1×10^{-4} cm
T = Pr	$(1.33 \times 10^5)(1.5) = 2 \times 10^5$ dyne/cm	$(3.33 \times 10^4)(5 \times 10^{-4}) = 16.7$ dyne/cm
$\sigma = \frac{Pr}{w}$	$\frac{2 \times 10^5}{0.2} = 1 \times 10^6$ dyne/cm^2	$\frac{16.7}{1 \times 10^{-4}} = 1.67 \times 10^5$ dyne/cm^2

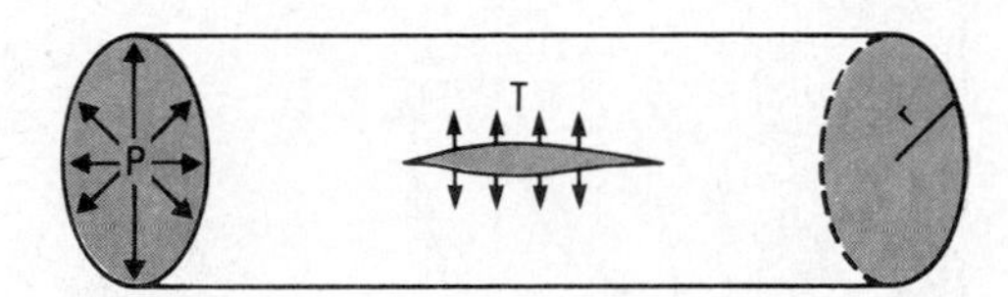

Fig. 28-2 Diagram of a small blood vessel illustrating the law of Laplace: T = Pr, where P = intraluminal pressure, r = radius of vessel, T = wall tension as force per unit length tangential to the vessel wall, tending to pull apart a theoretical longitudinal slit in the vessel.

equation is applied to thick wall vessels, such as the aorta. This is done by dividing Pr (pressure × radius) by wall thickness (w). The equation now becomes

$$\sigma \text{ (wall stress)} = Pr/w$$

To convert pressure in mm Hg (height of Hg column) to dynes per square centimeter, P = hρg, where h = the height of a Hg column in centimeters, ρ = the density of Hg in g/cm^3, g = gravitational acceleration in cm/sec^2, σ = force per unit area, and w = wall thickness.

Thus at normal aortic and capillary pressures the wall tension of the aorta is about 12,000 times greater than that of the capillary (see Table 28-1). When a person is standing quietly capillary pressure in the feet may reach 100 mm Hg. Under such conditions capillary wall tension increases to 66.5 dynes/cm, a value that is still only one three-thousandth that of the wall tension in the aorta at the same internal pressure. However σ (wall stress), which includes wall thickness, is only about 10-fold greater in the aorta than in the capillary.

In addition to providing an explanation for the ability of capillaries to withstand large internal pressures the preceding calculations point out that in dilated vessels, wall tension increases even when internal pressure remains constant and may, under certain circumstances (for example, aneurysm of the aorta), be an important factor in rupture of the vessel. The preceding equation also indicates that as the wall of the vessel becomes thicker, the wall stress decreases. In **hypertension** (high blood pressure) the arterial vessel walls thicken (hypertrophy of the vascular smooth muscle), thereby minimizing the arterial wall stress and hence the possibility of vessel rupture.

The diameter of the resistance vessels is determined by the balance between the contractile force of the vascular smooth muscle and the distending force produced by the intraluminal pressure. The greater the contractile activity of the vascular smooth muscle of an arteriole the smaller its diameter, until a point is reached, in the case of small arterioles, when complete occlusion of the vessel occurs. This is caused by infolding of the endothelium and by trapping of the cells in the vessel. With progressive reduction in the intravascular pressure vessel diameter decreases, as does tension in the vessel wall (law of Laplace). When perfusion pressure is reduced, a point is reached where blood flow ceases even though there is still a positive pressure gradient. This phenomenon has been referred to as the **critical closing pressure,** and its mechanism is still controversial. This critical closing pressure is low when vasomotor activity is reduced by inhibition of sympathetic nerve activity to the vessel, and it is increased when vasomotor tone is enhanced by activation of the vascular sympathetic nerve fibers. It has been suggested that flow stops because of vessel collapse when vascular smooth muscle contractile stress exceeds the stress associated with vessel radius, wall thickness, and intraluminal pressure (law of Laplace).

Vasoactive Role of Endothelium

For many years it was thought that the endothelium was an inert single layer of cells that served solely as a passive filter to permit passage of water and small molecules across the blood vessel wall and to retain blood cells and large molecules (proteins) within the vascular compartment. Recently it has been demonstrated that the endothelium is a source of substances that elicit contraction or relaxation of the vascular smooth muscle.

As shown in Fig. 28-3 prostacyclin can relax vascular smooth muscle via an increase in the cyclic adenosine monophosphate (cAMP) concentration. **Prostacyclin** is formed in the endothelium from arachidonic acid and the process is catalyzed by prostacyclin synthase. To what extent this occurs in vivo is not known, but the prostacyclin may be released by shear stress caused by the pulsatile blood flow. The primary function of prostacyclin is to inhibit platelet adherence to the endothelium and platelet aggregation, thereby to prevent intravascular clot formation.

Of far greater importance in endothelial-mediated vascular dilation are the formation and release of the **endothelial-derived relaxing factor (EDRF)** (see Fig. 28-3). Stimulation of the endothelial cells in vivo, in isolated arteries, or in culture by acetylcholine or several other agents (adenosine triphosphate [ATP], bradykinin, serotonin, substance P, histamine) causes the production and release of EDRF. In blood vessels from which the endothelium has been mechanically removed these agents do not elicit vasodilation. The EDRF (synthesized from L-arginine) activates guanylyl cyclase in the vascular smooth muscle to increase the cyclic guanosine monophosphate (cGMP) concentration, which produces relaxation, possibly by decreasing cytosolic free Ca^{++}. Strong evidence indicates that one form of EDRF is nitric oxide (NO). EDRF release can be stimulated by the shear stress of blood flow on the endothelium, but the physiological role of EDRF in the local regulation of blood flow remains to be elucidated. The drug nitroprusside also increases cGMP, which

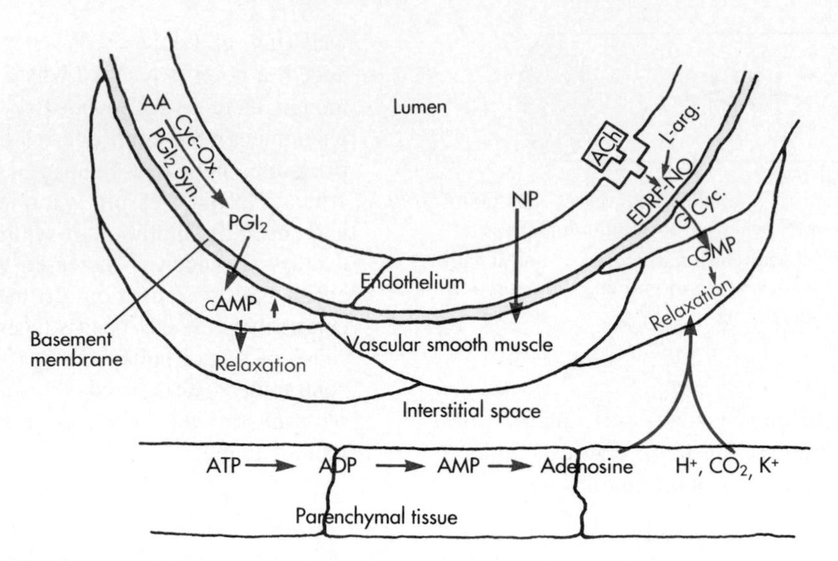

■ **Fig. 28-3** Endothelial and nonendothelial mediated vasodilation. Prostacyclin (PI_2) is formed from arachidonic acid (AA) by the action of cyclooxygenase (Cyc Ox) and prostacyclin synthetase (PGI_2 Syn) in the endothelium and elicits relaxation of the adjacent vascular smooth muscle by increases in cyclic adenosine monophosphate (cAMP). Stimulation of the endothelial cells with acetylcholine (ACh) or other agents *(see text)* results in the formation and release of an endothelial-derived relaxing factor (EDRF), one of which is probably nitric oxide (NO). EDRP stimulates guanylyl cyclase (G Cyc) to increase cyclic guanosine monophosphate (cGMP) in vascular smooth muscle to produce relaxation. The vasodilator agent nitroprusside (NP) acts directly on the vascular smooth muscle. Substances such as adenosine, hydrogen ions (H^+), CO_2, and potassium ions (K^+) can arise in the parenchymal tissue and elicit vasodilation by direct action of the vascular smooth muscle *(see Chapter 29)*.

produces vasodilation, but it acts directly on the vascular smooth muscle and is not endothelial-mediated (see Fig. 28-3). Vasodilator agents such as adenosine, hydrogen ions, CO_2, and potassium may be released from parenchymal tissue and act locally on the resistance vessels (see Fig. 28-3).

The endothelium can also synthesize an **endothelial-derived contracting factor (EDCF),** but little is known about its function. It is apparently not **endothelin,** a small vasoconstrictor peptide that has been isolated from endothelial cells.

■ *Passive Role of Endothelium*

■ *Transcapillary Exchange*

Solvent and solute move across the capillary endothelial wall by three processes: diffusion, filtration, and pinocytosis (by endothelial vesicles).

The permeability of the capillary endothelial membrane is not the same in all body tissues. For example the liver capillaries are quite permeable, and albumin escapes at a rate severalfold greater than from the less permeable muscle capillaries. Also permeability is not uniform along the whole capillary; the venous ends are more permeable than the arterial ends, and permeability is greatest in the venules. The greater permeability at the venous end of the capillaries and in the venules is due to the greater number of pores in these regions of the microvessels.

The sites where filtration occurs have been a controversial subject for a number of years. A little water flows through the capillary endothelial cell membranes, but most of the water flows through apertures **(pores)** in the endothelial wall of the capillaries (Figs. 28-4 and 28-5). Calculations based on the transcapillary movement of solutes of small molecular size led to the prediction of pore diameters of about 4 nm. However electron microscopy failed to reveal pores, and the clefts at the junctions of endothelial cells appeared to be fused at the tight junctions (see Figs. 28-4 and 28-5). However studies on cardiac and skeletal muscle with horseradish peroxidase, a protein with a molecular weight of 40,000, have demonstrated that many of the clefts between adjacent endothelial cells are open. Electron microscopy revealed filling of the clefts with peroxidase from the lumen side of the capillaries with a gap at the narrowest point of about 4 nm, providing morphological support of the physiological evidence for the existence of capillary pores. The clefts (pores) are sparse and represent only about 0.02% of the capillary surface area. In cerebral capillaries, where a blood-brain barrier to many small molecules exists, peroxidase studies do not reveal any pores.

Transcapillary movement of solute (large and small molecules) may occur through channels formed by the

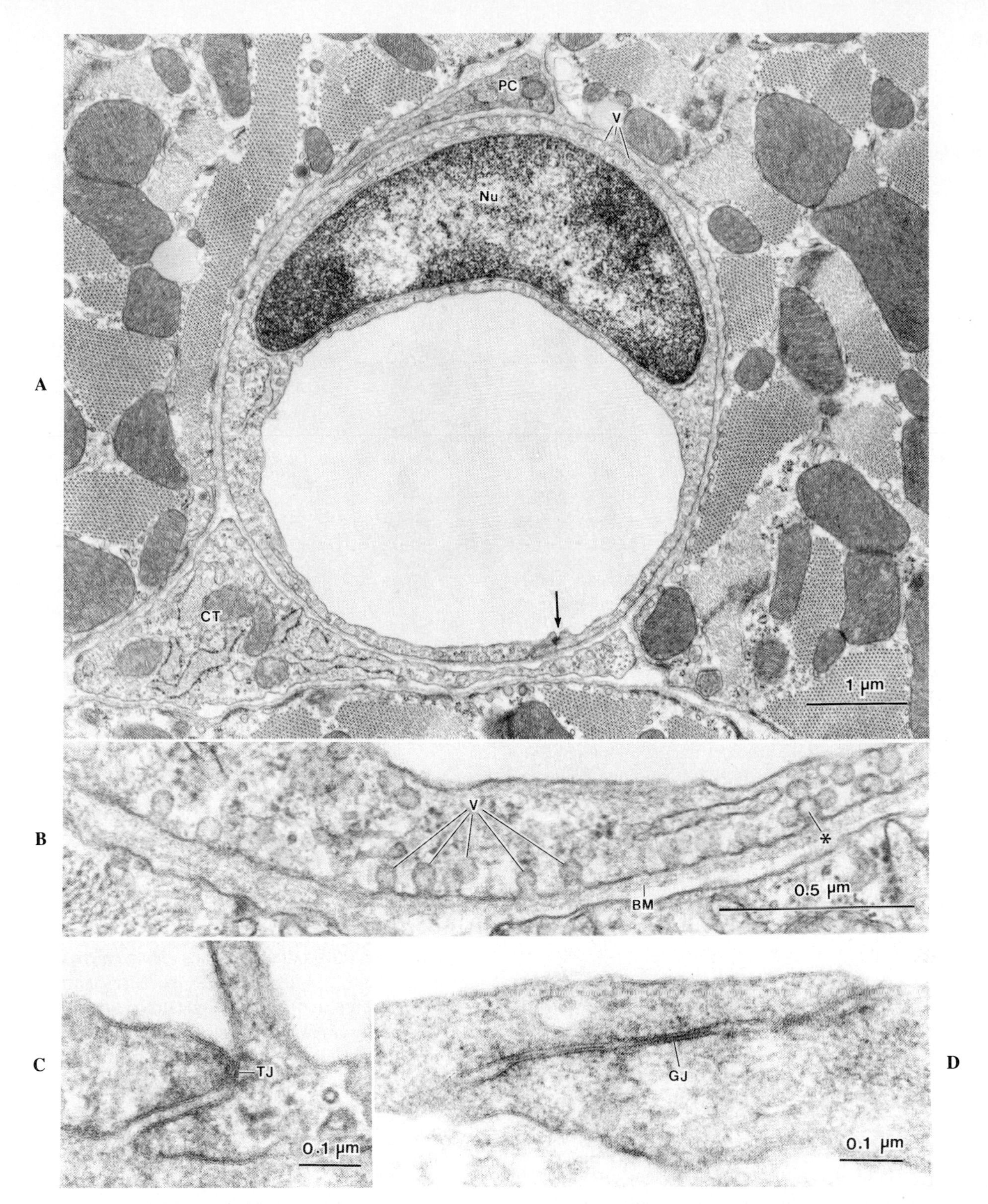

■ **Fig. 28-4** **A,** Cross-sectioned capillary in mouse ventricular wall. Luminal diameter is approximately 4 μm. In this thin section the capillary wall is formed by a single endothelial cell (*Nu,* endothelial nucleus), which forms a junctional complex *(arrow)* with itself. The thin pericapillary space is occupied by a pericyte *(PC)* and a connective tissue *(CT)* cell ("fibroblast"). Note the numerous endothelial vesicles *(V).* **B,** Detail of endothelial cell in panel **A,** showing plasmalemmal vesicles *(V)* that are attached to the endothelial cell surface. These vesicles are especially prominent in vascular endothelium and are involved in transport of substances across the blood vessel wall. Note the complex alveolar vesicle (*). *BM,* Basement membrane. **C,** Junctional complex in a capillary of mouse heart. "Tight" junctions *(TJ)* typically form in these small blood vessels and appear to consist of fusions between apposed endothelial cell surface membranes. **D,** Interendothelial junction in a muscular artery of monkey papillary muscle. Although tight junctions similar to those of capillaries are found in these large blood vessels, extensive junctions that resemble gap junctions in the intercalated disks between myocardial cells often appear in arterial endothelium (example shown at *GJ*).

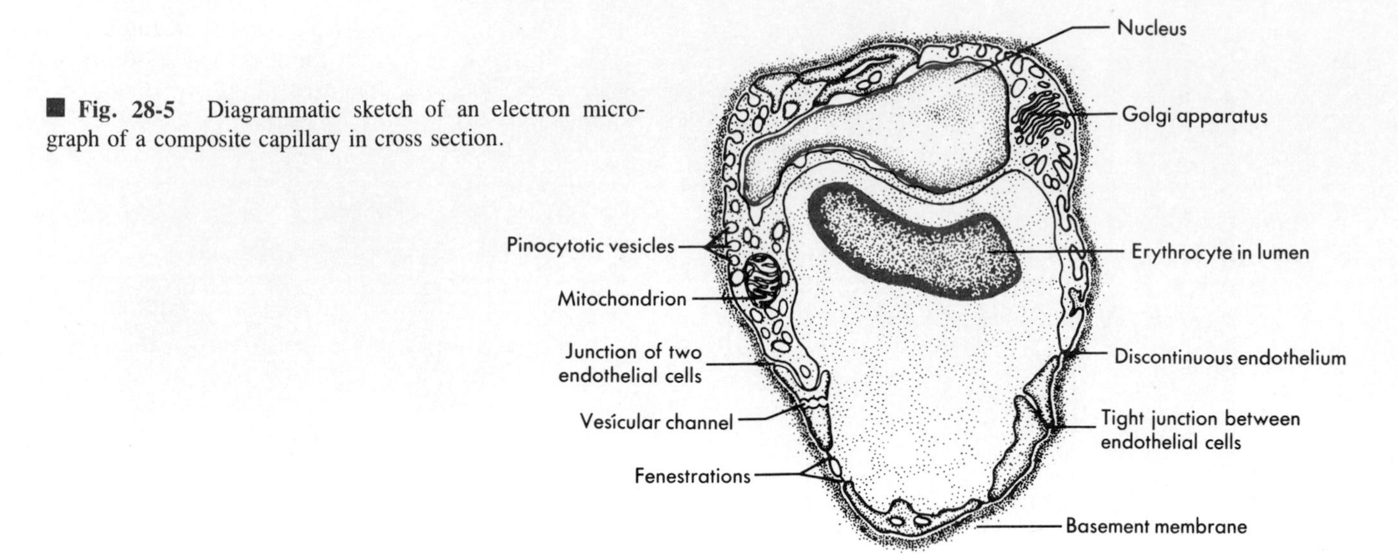

■ **Fig. 28-5** Diagrammatic sketch of an electron micrograph of a composite capillary in cross section.

fusion of vesicles **(vesicular channels)** across the endothelial cells (see Fig. 28-5). However the existence of such channels has been seriously questioned. The basement membrane, a layer of fine fibrillar material around the capillaries, retards the passage of large molecules (greater than 10 nm radius).

In addition to clefts some of the more porous capillaries (for example in kidney and intestine) contain **fenestrations** (see Fig. 28-5) 20 to 100 nm wide, whereas others (such as in the liver) have a **discontinuous endothelium** (see Fig. 28-5). The fenestrations that appear to be sealed by a thin diaphragm are quite permeable to horseradish peroxidase and a number of other tracers. Hence larger molecules can penetrate capillaries with fenestrations or gaps caused by discontinuous endothelium than can pass through the intercellular clefts of the endothelium.

Diffusion. Under normal conditions only about 0.06 ml of water per minute moves back and forth across the capillary wall per 100 g of tissue as a result of filtration and absorption, whereas 300 ml of water per minute per 100 g of tissue does so by diffusion, a 5,000-fold difference.

Relating filtration and diffusion to blood flow, we find that about 2% of the plasma passing through the capillaries is filtered. In contrast the diffusion of water is 40 times greater than the rate that it is brought to the capillaries by blood flow. The transcapillary exchange of solutes is also primarily governed by diffusion. Thus *diffusion is the key factor in providing exchange of gases, substrates, and waste products between the capillaries and the tissue cells.*

The process of diffusion is described by Fick's law:

$$J = -DA\frac{dc}{dx}$$

where

J = Quantity of a substance moved per unit time (t)
D = Free diffusion coefficient for a particular molecule (the value is inversely related to the square root of the molecular weight)
A = Cross-sectional area of the diffusion pathway
$\frac{dc}{dx}$ = Concentration gradient of the solute

Fick's law is also expressed as:

$$J = -PS(C_o - C_i)$$

where

P = Capillary permeability of the substance
S = Capillary surface area
C_i = Concentration of the substance inside the capillary
C_o = Concentration of the substance outside the capillary

Hence the PS product provides a convenient expression of available capillary surface, because permeability is rarely altered under physiological conditions.

In the capillaries diffusion of lipid-insoluble molecules is not free but is restricted to the pores whose mean size can be calculated by measurement of the diffusion rate of an uncharged molecule whose free diffusion coefficient is known. Movement of solutes across the endothelium is quite complex and involves corrections for attractions between solute and solvent molecules, interactions between solute molecules, pore configuration, and charge on the molecules relative to charge on the endothelial cells. It is not simply a question of random thermal movements of molecules down a concentration gradient.

For small molecules, such as water, NaCl, urea, and glucose, the capillary pores offer little restriction to diffusion (low **reflection coefficient**), and diffusion is so rapid that the mean concentration gradient across the capillary endothelium is extremely small. In lipid-insoluble molecules of increasing size diffusion through muscle capillaries becomes progressively more restricted, until diffusion becomes minimal with molecules of a molecular weight above about 60,000. In the case of small molecules the only limitation to net movement across the capillary wall is the rate at which blood flow transports the molecules to the capillary **(flow limited).**

When transport across the capillary is flow limited the concentration of a small molecule solute in the blood reaches equilibrium with its concentration in the interstitial fluid near the origin of the capillary from the cognate arteriole. If an inert small molecule tracer is infused intraarterially, its concentration falls to negligible levels near the arterial end of the capillary (Fig. 28-6, *A*). If the flow is large, the small molecule tracer will be detectable farther downstream in the capillary. A somewhat larger molecule moves farther along the capillary before reaching an insignificant concentration in the blood. The number of still larger molecules that enter the arterial end of the capillary and cannot pass through the capillary pores is the same as the number leaving the venous end of the capillary (see Fig. 28-6, *A*).

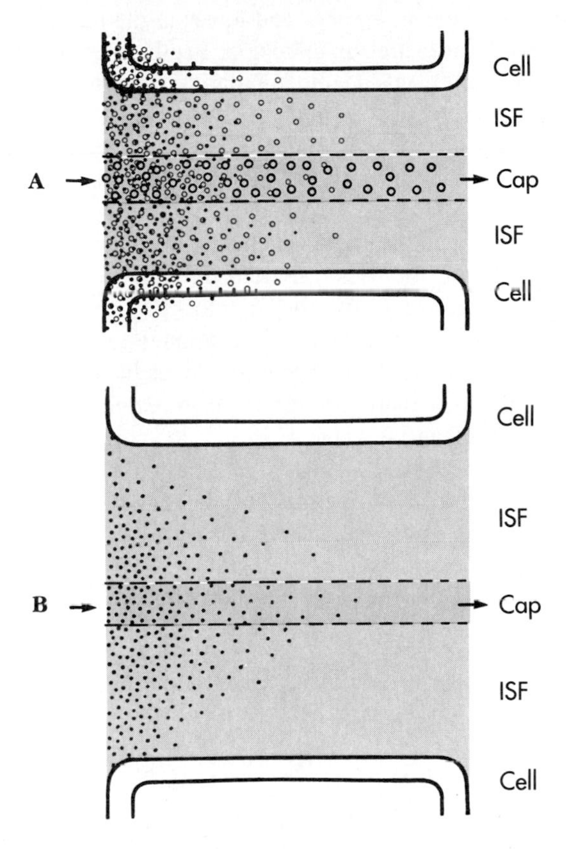

■ **Fig. 28-6** Flow- and diffusion-limited transport from capillaries *(Cap)* to tissue. **A,** Flow-limited transport. The smallest water-soluble inert tracer particles *(black dots)* reach negligible concentrations after passing only a short distance down the capillary. Larger particles *(colored circles)* with similar properties travel farther along the capillary before reaching insignificant intracapillary concentrations. Both substances cross the interstitial fluid *(ISF)* and reach the parenchymal tissue *(cell)*. Because of their size, more of the smaller particles are taken up by the tissue cells. The largest particles *(black circles)* cannot penetrate the capillary pores and hence do not escape from the capillary lumen except by pinocytotic vesicle transport. An increase in the volume of blood flow or capillary density increases tissue supply for the diffusible solutes. Note that capillary permeability is greater at the venous end of the capillary (also in the venule, not shown) because of the larger number of pores in this region. **B,** Diffusion-limited transport. When the distance between the capillaries and the parenchymal tissue is large, as a result of edema or low capillary density, diffusion becomes a limiting factor in the transport of solutes from capillary to tissue even at high rates of capillary blood flow.

In the case of large molecules diffusion across the capillaries becomes the limiting factor **(diffusion limited).** In other words capillary permeability to a large molecule solute limits its transport across the capillary wall (see Fig. 28-6, *A*). The diffusion of small lipid-insoluble molecules is so rapid that diffusion becomes limiting in blood-tissue exchange only when the distances between capillaries and parenchymal cells are large (for example tissue edema or very low capillary density) (see Fig. 28-6, *B*). Furthermore the rate of diffusion is uninfluenced by filtration in the direction opposite to the concentration gradient of the diffusible substance. In fact filtration *or* absorption accelerates the movement of tracer ions from interstitial fluid to blood **(tissue clearance).** The reason for this enhanced tissue clearance is not known, but it may be the result of a stirring effect on the interstitial fluid or to changes in its structure (for example gel to sol transformation or "canals" in a gel matrix).

Movement of lipid-soluble molecules across the capillary wall is not limited to capillary pores (only about 0.02% of the capillary surface), because such molecules can pass directly through the lipid membranes of the entire capillary endothelium. Consequently *lipid-soluble molecules move with great rapidity between blood and tissue. The degree of lipid solubility (oil-to-water partition coefficient) provides a good index of the ease of transfer of lipid molecules through the capillary endothelium.*

Oxygen and carbon dioxide are both lipid soluble and readily pass through the endothelial cells. Calculations based on (1) the diffusion coefficient for O_2, (2) capillary density and diffusion distances, (3) blood flow, and (4) tissue O_2 consumption indicate that the O_2 supply of normal tissue at rest and during activity is not limited by diffusion or the number of open capillaries. Recent measurements of Po_2 and saturation of blood in the microvessels indicate that in many tissues O_2 saturation at the entrance of the capillaries has already decreased to a saturation of about 80% as a result of diffusion of O_2 from arterioles and small arteries. Such studies also have shown that CO_2 loading and the resultant intravascular shifts in the oxyhemoglobin dissociation curve occur in the precapillary vessels. These findings reflect not only the movement of gas to respiring tissue at the precapillary level but also the direct flux of O_2 and CO_2 between adjacent arterioles, venules, and possibly arteries and veins **(countercurrent exchange).** This exchange of gas represents a diffusional shunt of gas around the capillaries, and, at low blood flow rates, it may limit the supply of O_2 to the tissue.

Capillary filtration. *The direction and the magnitude of the movement of water across the capillary wall are determined by the algebraic sum of the hydrostatic and osmotic pressures that exist across the membrane.* An increase in intracapillary hydrostatic pressure favors movement of fluid from the vessel to the interstitial space, whereas an increase in the concentration of osmotically active particles within the vessels favors movement of fluid into the vessels from the interstitial space.

Hydrostatic forces. The hydrostatic pressure (blood pressure) within the capillaries is not constant and depends on the arterial pressure, the venous pressure, and the precapillary (arteriolar) and postcapillary (venule and small vein) resistances. An increase in arterial or venous pressure elevates capillary hydrostatic pressure, whereas a reduction in each has the opposite effect. An increase in arteriolar resistance or closure of arteries reduces capillary pressure, whereas greater venous resistance (venules and veins) increases capillary pressure.

Hydrostatic pressure is the principal force in capillary filtration. However, changes in the venous resistance affect capillary hydrostatic pressure more than do changes in arteriolar resistance. A given change in venous pressure produces a greater effect on capillary hydrostatic pressure than the same change in arterial pressure, and about 80% of an increase in venous pressure is transmitted back to the capillaries.

Despite the fact that capillary hydrostatic pressure (Pc) is variable from tissue to tissue (even within the same tissue) average values, obtained from many direct measurements in human skin, are about 32 mm Hg at the arterial end of the capillaries and 15 mm Hg at the venous end of the capillaries at the level of the heart (Fig. 28-7). When a person stands the hydrostatic pressure is higher in the legs and lower in the head.

Tissue pressure, or more specifically interstitial fluid pressure (P_i) outside the capillaries, opposes capillary filtration, and it is $P_c - P_i$ that constitutes the driving force for filtration. The true value of P_i is still controversial. For years it was assumed to be close to zero in the normal (nonedematous) state. However studies using perforated plastic capsules implanted in the subcutaneous tissue, or wicks inserted through the skin, indicate a negative P_i of from -1 to -7 mm Hg. If the pressures recorded by these techniques are representative of interstitial fluid pressure in undisturbed tissue, then the hydrostatic driving force for capillary filtration is greater than the value of P_c.

Osmotic forces. *The key factor that restrains fluid loss from the capillaries is the osmotic pressure of the plasma proteins*—usually termed the **colloid osmotic pressure** or **oncotic pressure** (π_p). The total osmotic pressure of plasma is about 6,000 mm Hg, whereas the

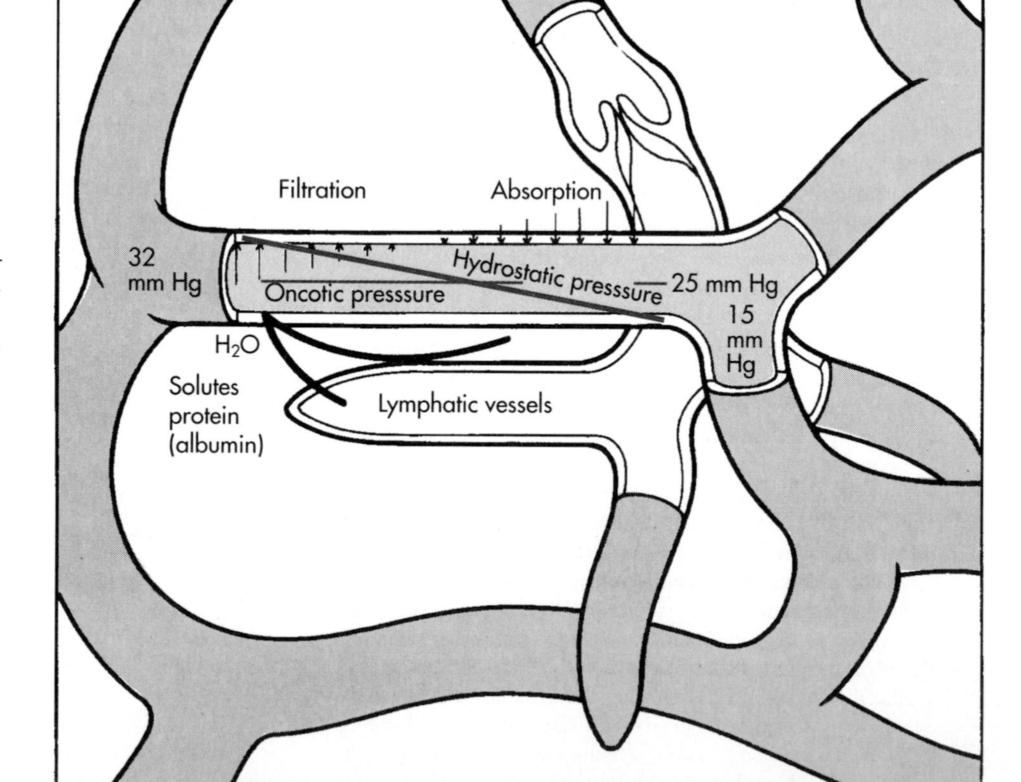

■ Fig. 28-7 Schematic representation of the factors responsible for filtration and absorption across the capillary wall and the formation of lymph.

oncotic pressure is only about 25 mm Hg. However, this low oncotic pressure plays an important role in fluid exchange across the capillary wall, because the plasma proteins are essentially confined to the intravascular space, whereas the electrolytes that are responsible for the major fraction of plasma osmotic pressure are practically equal in concentration on both sides of the capillary endothelium (see Chapter 1). The relative permeability of solute to water influences the actual magnitude of the osmotic pressure. The **reflection coefficient** (σ) is the relative impediment to the passage of a substance through the capillary membrane. The reflection coefficient of water is 0 and that of albumin (to which the endothelium is essentially impermeable) is 1. Filterable solutes have reflection coefficients between 0 and 1. Also different tissues have different reflection coefficients for the same molecule, and therefore movement of a given solute across the endothelial wall varies with the tissue. The true oncotic pressure is defined by

$$\pi = \sigma RT\,(C_i - C_o)$$

where

σ = Reflection coefficient
R = Gas constant
T = Absolute temperature
C_i and C_o = Solute (essentially albumin) concentration inside and outside the capillary

Of the plasma proteins albumin is preponderant in determining oncotic pressure. The average albumin molecule (molecular weight 69,000) is approximately half the size of the average globulin molecule (molecular weight 150,000) and is present in almost twice the concentration of the globulins (4.5 vs. 2.5 g/100 ml of plasma). Albumin also exerts a greater osmotic force than can be accounted for solely on the basis of the number of molecules dissolved in the plasma. Therefore it cannot be completely replaced by inert substances of appropriate molecular size such as dextran. This additional osmotic force becomes disproportionately greater at high concentrations of albumin (as in plasma) and is weak to absent in dilute solutions of albumin (as in interstitial fluid). The reasons for this behavior of albumin are its negative charge at the normal blood pH and the attraction and retention of cations (principally Na^+) in the vascular compartment (the **Gibbs-Donnan effect**) (see Chapter 2). Furthermore albumin binds a small number of chloride ions, thus increasing its negative charge and, hence, its ability to retain more sodium ions inside the capillaries. The small increase in electrolyte concentration of the plasma over that of the interstitial fluid produced by the negatively charged albumin enhances its osmotic force to that of an ideal solution containing a solute of molecular weight of 37,000. If albumin did indeed have a molecular weight of 37,000, it would not be retained by the capillary endothelium because of its small size and obviously could not function as a counterforce to capillary hydrostatic pressure. If, however, albumin did not have an enhanced osmotic force, it would require a concentration of about 12 g of albumin/100 ml of plasma to achieve a plasma oncotic pressure of 25 mm Hg. Such a high albumin concentration would greatly increase blood viscosity and resistance to blood flow through the vascular system. The other factors that contribute to the nonlinearity of the relationship of albumin concentration to osmotic force are not known. About 65% of plasma oncotic pressure is attributable to albumin, about 15% to the globulins, and the remainder to other ill-defined components of the plasma.

Small amounts of albumin escape from the capillaries and enter the interstitial fluid, where they exert a very small osmotic force (0.1 to 5 mm Hg). This force, π_i, is small because of the low concentration of albumin in the interstitial fluid and because at low concentrations the osmotic force of albumin becomes simply a function of the number of albumin molecules per unit volume of interstitial fluid.

Balance of hydrostatic and osmotic forces. The relationship between hydrostatic pressure and oncotic pressure and the role of these forces in regulating fluid passage across the capillary endothelium were expounded by Starling in 1896 and constitute the **Starling hypothesis,** which can be expressed by the equation

$$\text{Fluid movement} = k[(P_c + \pi_i) - (P_i + \pi_p)]$$

where

P_c = Capillary hydrostatic pressure
P_i = Interstitial fluid hydrostatic pressure
π_p = Plasma protein oncotic pressure
π_i = Interstitial fluid oncotic pressure
k = Filtration constant for the capillary membrane

Filtration occurs when the algebraic sum is positive, and absorption occurs when it is negative.

Classically it has been thought that filtration occurs at the arterial end of the capillary and absorption at its venous end because of the gradient of hydrostatic pressure along the capillary. This is true for the idealized capillary as depicted in Fig. 28-7, but direct observations have revealed that many capillaries show filtration for their entire length, whereas others show only absorption. In some vascular beds (for example the renal glomerulus) hydrostatic pressure in the capillary is high enough to result in filtration along the entire length of the capillary. In other vascular beds, such as in the intestinal mucosa, the hydrostatic and oncotic forces are such that absorption occurs along the whole capillary. As discussed earlier in this chapter, capillary pressure is quite variable and depends on several factors, principally the contractile state of the precapillary vessel. In the normal steady state arterial pressure, venous pressure, postcapillary resistance, interstitial fluid hydrostatic and oncotic pressures, and plasma oncotic pressure are relatively constant, and change in precapillary resistance is the determining factor with respect to fluid movement across the wall for any given capillary. Because water moves so quickly across the capillary endothelium, the hydrostatic and osmotic forces are nearly in equilibrium along the entire capillary. Hence filtration and absorption in the normal state occur at very small degrees of imbalance of pressure across the capillary wall. Only a small percentage (2%) of the plasma flowing through the vascular system is filtered, and of this about 85% is absorbed in the capillaries and venules. The remainder returns to the vascular system as lymph fluid along with the albumin that escapes from the capillaries.

In the lungs the mean capillary hydrostatic pressure is only about 8 mm Hg. Because the plasma oncotic pressure is 25 mm Hg and the lung interstitial fluid pressure is approximately 15 mm Hg, the net force slightly favors reabsorption. Nevertheless pulmonary lymph is formed and consists of fluid that is osmotically drawn out of the capillaries by the small amount of plasma protein that escapes through the capillary endothelium. Only in pathological conditions, such as left ventricular failure or stenosis of the mitral valve, does pulmonary capillary hydrostatic pressure exceed plasma oncotic pressure. When this occurs, it may lead to pulmonary edema, a condition that can seriously interfere with gas exchange in the lungs.

Capillary filtration coefficient. The rate of movement of fluid across the capillary membrane (Q_f) depends not only on the algebraic sum of the hydrostatic and osmotic forces across the endothelium (ΔP) but also on the area of the capillary wall available for filtration (A_m), the distance across the capillary wall (Δx), the viscosity of the filtrate (η), and the filtration constant of the membrane (k). These factors may be expressed by the equation

$$Q_f = \frac{kA_m\Delta P}{\eta\Delta x}$$

The dimensions are units of flow per unit of pressure gradient across the capillary wall per unit of capillary surface area. This expression, which describes the flow of fluid through a membrane (pores), is essentially Poiseuille's law for flow through tubes (p. 441).

Because the thickness of the capillary wall and the viscosity of the filtrate are relatively constant, they can be included in the filtration constant, k, and if the area of the capillary membrane is not known, the rate of filtration can be expressed per unit weight of tissue. Hence the equation can be simplified to

$$Q_r = k_t\Delta P$$

where k_t is the capillary filtration coefficient for a given tissue, and the units for Q_f are milliliters per minute per 100 g of tissue per millimeter of mercury pressure.

The rates of filtration and absorption are determined by the rate of change in tissue weight or volume at different mean capillary hydrostatic pressures that are altered by adjustment of arterial and venous pressures. At the isogravimetric or isovolumic point (constant weight or contant volume, respectively, as continuously measured with an appropriate scale or volume recorder) the hydrostatic and osmotic forces are balanced across the capillary wall and there is neither net filtration nor net absorption. An abrupt increase in arterial pressure increases capillary hydrostatic pressure and fluid moves from the capillaries to the interstitial fluid compartment. Because the pressure increment, the weight increase of the tissue per unit time, and the total weight of the tissue are known, the capillary filtration coefficient (or k_t) in milliliters per minute per 100 g of tissue per millimeter of mercury can be calculated. With the isogravimetric and isovolumic techniques it is assumed that 80% of the increments in venous pressure are transmitted back to the capillaries, that precapillary and postcapillary resistances are constant when venous pressure is changed, and that the weight or volume change that occurs immediately after raising venous pressure is the result of vascular distension and not filtration. These assumptions may not always be correct; nevertheless the filtration coefficient constitutes a useful index of capillary permeability and surface area.

In any given tissue the filtration coefficient per unit area of capillary surface, and hence capillary permeability, is not changed by different physiological conditions, such as arteriolar dilation and capillary distension, or by such adverse conditions as hypoxia, hypercapnia, or reduced pH. With capillary injury (toxins, severe burns) capillary permeability increases, as indicated by the filtration coefficient, and significant amounts of fluid and protein leak out of the capillaries into the interstitial space.

Because capillary permeability is constant under normal conditions, the filtration coefficient can be used to determine the relative number of open capillaries (total capillary surface area available for filtration in tissue). For example increased metabolic activity of contracting skeletal muscle induces relaxation of precapillary resistance vessels with opening of more capillaries (**capillary recruitment,** resulting in an increased filtering surface area).

Some protein is apparently required to maintain the integrity of the endothelial membrane. If the plasma proteins are replaced by nonprotein colloids so as to give the same oncotic pressure, the filtration coefficient is doubled and edema occurs. However if as little as 0.2% albumin is added, normal permeability is restored. One reasonable explanation is that the albumin binds to certain sites on the endothelial membrane and alters pore dimensions.

Disturbances in hydrostatic-osmotic balance. Changes in arterial pressure per se may have little effect on filtration, because the change in pressure may be countered by adjustments of the precapillary resistance vessels (autoregulation, p. 481), so that hydrostatic pressure in the open capillaries remains the same. However with severe reduction in arterial pressure, as may occur in hemorrhage, arteriolar constriction mediated by the sympathetic nervous system and fall in venous pressure resulting from the blood loss may occur. These changes lead to a decrease in the capillary hydrostatic pressure. Furthermore the low blood pressure in hemorrhage causes a decrease in blood flow (and hence O_2 supply) to the tissue with the result that vasodilator metabolites accumulate and induce relaxation of arterioles. Precapillary vessel relaxation is also engendered by the reduced transmural pressure. As a consequence of these several factors absorption predominates over filtration and occurs at a larger capillary surface area. This is one of the compensatory mechanisms employed by the body to restore blood volume (p. 539).

An increase in venous pressure alone, as occurs in the feet when one changes from the lying to the standing position, would elevate capillary pressure and enhance filtration. However the increase in transmural pressure causes precapillary vessel closure (myogenic mechanism, p. 481) so that the capillary filtration coefficient actually decreases. This reduction in capillary surface available for filtration protects against the extravasation of large amounts of fluid into the interstitial space (edema). With prolonged standing, particularly when associated with some elevation of venous pressure in the legs (such as that caused by tight garters or pregnancy) or with sustained increases in venous pressure as seen in congestive heart failure, filtration is greatly enhanced and exceeds the capacity of the lymphatic system to remove the capillary filtrate from the interstitial space.

A large amount of fluid can move across the capillary wall in a relatively short time. In a normal individual the filtration coefficient (k_t) for the whole body is about 0.0061 ml/min/100 g of tissue/mm Hg. For a 70-kg man elevation of venous pressure of 10 mm Hg for 10 minutes would increase filtration from capillaries by 342 ml. This would not lead to edema formation, because the fluid is returned to the vascular compartment by the lymphatic vessels. When edema does develop, it usually appears in the dependent parts of the body, where the hydrostatic pressure is greatest, but its location and magnitude are also determined by the type of tissue. Loose tissues, such as the subcutaneous tissue around the eyes or in the scrotum, are more prone to collect larger quantities of interstitial fluid than are firm tissues, such as muscle, or encapsulated structures, such as the kidney.

The concentration of the plasma proteins may also change in different pathological states and, hence, alter the osmotic force and movement of fluid across the capillary membrane. The plasma protein concentration is increased in dehydration (for example water deprivation, prolonged sweating, severe vomiting, and diarrhea), and water moves by osmotic forces from the tissues to the vascular compartment. In contrast the plasma protein concentration is reduced in nephrosis (a renal disease in which there is loss of protein in the urine) and edema may occur. When capillary injury is extensive, as in burns, intravascular fluid and plasma protein leak into the interstitial space. The protein that escapes from the vessel lumen increases the oncotic pressure of the interstitial fluid. This greater osmotic force outside the capillaries leads to additional fluid loss and possibly to severe dehydration of the patient.

Pinocytosis. Some transfer of substances across the capillary wall can occur in tiny pinocytotic vesicles (**pinocytosis**). These vesicles (see Figs. 28-4 and 28-5), formed by a pinching off of the surface membrane, can take up substances on one side of the capillary wall, move by thermal kinetic energy across the cell, and deposit their contents at the other side. The amount of material that can be transported in this way is very small relative to that moved by diffusion. However pinocytosis may be responsible for the movement of large lipid-insoluble molecules (30 nm) between blood and interstitial fluid. The number of pinocytotic vesicles in endothelium varies with the tissue (muscle > lung > brain) and increases from the arterial to the venous end of the capillary.

Lymphatics

The terminal vessels of the lymphatic system consist of a widely distributed closed-end network of highly permeable lymph capillaries that are similar in appearance to blood capillaries. However they are generally lacking in tight junctions between endothelial cells and possess fine filaments that anchor them to the surrounding connective tissue. With muscular contraction these fine strands may distort the lymphatic vessel to open spaces between the endothelial cells and permit the entrance of protein and large particles and cells present in the interstitial fluid. The lymph capillaries drain into larger vessels that finally enter the right and left subclavian veins at their junctions with the respective internal jugular veins. Only cartilage, bone, epithelium, and tissues of the central nervous system are devoid of lymphatic vessels. The plasma capillary filtrate is returned to the circulation by virtue of tissue pressure, facilitated by intermittent skeletal muscle activity, contractions of the lymphatic vessels, and an extensive system of one-way valves. In this respect they resemble the veins, although even the larger lymphatic vessels have thinner walls than the corresponding veins and contain only a small amount of elastic tissue and smooth muscle.

The volume of fluid transported through the lymphatic vessels in 24 hours is about equal to an animal's total plasma volume and the protein returned by the lymphatic vessels to the blood in a day is about one fourth to one half of the circulating plasma proteins. This is the only means whereby protein (albumin) that leaves the vascular compartment can be returned to the blood, because back diffusion into the capillaries cannot occur against the large albumin concentration gradient. Were the protein not removed by the lymph vessels, it would accumulate in the interstitial fluid and act as an oncotic force to draw fluid from the blood capillaries to produce edema. In addition to returning fluid and protein to the vascular bed the lymphatic system filters the lymph at the lymph nodes and removes foreign particles, such as bacteria. The largest lymphatic vessel, the **thoracic duct,** in addition to draining the lower extremities returns protein lost through the permeable liver capillaries and carries substances absorbed from the gastrointestinal tract, principally fat in the form of chylomicrons, to the circulating blood.

Lymph flow varies considerably, being almost nil from resting skeletal muscle and increasing during exercise in proportion to the degree of muscular activity. It is increased by any mechanism that enhances the rate of blood capillary filtration, for example, increased capillary pressure or permeability or decreased plasma oncotic pressure. When either the volume of interstitial fluid exceeds the drainage capacity of the lymphatic vessels or the lymphatic vessels become blocked, as may occur in certain disease states, interstitial fluid accumulates, chiefly in the more compliant tissues (for example subcutaneous tissue), and gives rise to clinical edema.

Summary

1. Blood flow through the capillaries is chiefly regulated by contraction and relaxation of the arterioles (resistance vessels).

2. The capillaries, which consist of a single layer of endothelial cells, can withstand high transmural pressure by virtue of their small diameter. According to the law of Laplace T (wall tension) = P (transmural pressure) × r (radius of the capillary).

3. The endothelium is the source of endothelial-derived relaxing factor (EDRF) and prostacyclin, which relax vascular smooth muscles.

4. Movement of water and small solutes between the vascular and interstitial fluid compartments occurs through capillary pores *mainly by diffusion* but also by filtration and absorption.

5. Because the rate of diffusion is about 40 times greater than the blood flow in the tissue, exchange of small lipid-insoluble molecules is **flow limited.** The larger the molecules the slower the diffusion until with large molecules the lipid insoluble molecules become **diffusion limited.** Molecules larger than about 60 kd are essentially confined to the vascular compartment.

6. Lipid-soluble substances such as CO_2 and O_2 pass directly through the lipid membranes of the capillary, and the ease of transfer is directly proportional to the degree of lipid solubility of the substance.

7. Capillary filtration and absorption are described by the Starling equation: Fluid movement $= k[(P_c + \pi_i) - (P_i + \pi_p)]$ where P_c = capillary hydrostatic pressure, P_i = interstitial fluid hydrostatic pressure, π_i = interstitial fluid oncotic pressure, π_p = plasma protein oncotic pressure. Filtration occurs when the algebraic sum is positive, and absorption occurs when it is negative.

8. Large molecules can move across the capillary wall in vesicles formed from the lipid membrane of the capillaries by a process called **pinocytosis.**

9. Fluid and protein that have escaped from the blood capillaries enter the lymphatic capillaries and are transported via the lymphatic system back to the blood vascular compartment.

Bibliography

Journal articles

Bundgaard M: Transport pathways in capillaries—in search of pores, *Annu Rev Physiol* 42:325, 1980.

Duling BR, Klitzman B: Local control of microvascular function: role in tissue oxygen supply, *Annu Rev Physiol* 42:373, 1980.

Feng Q, Hedner T: Endothelium-derived relaxing factor (EDRF) and nitric oxide. II. Physiology, pharmacology and pathophysiological implications, *Clin Physiol* 10:503, 1990.

Furchgott RF, Vanhoutte PM: Endothelium-derived relaxing and contracting factors, *FASEB J* 3:2007, 1989.

Gore RW, McDonagh PF: Fluid exchange across single capillaries, *Annu Rev Physiol* 42:337, 1980.

Krogh A: The number and distribution of capillaries in muscles with calculation of the oxygen pressure head necessary for supplying the tissue, *J Physiol* 52:409, 1919.

Lewis DH, editor: Symposium on lymph circulation, *Acta Physiol Scand [Suppl]* 463:9, 1979.

Rosell S: Neuronal control of microvessels, *Annu Rev Physiol* 42:359, 1980.

Starling EH: On the absorption of fluids from the connective tissue spaces, *J Physiol* 19:312, 1896.

Books and monographs

Bert JL, Pearce RH: *The interstitium and microvascular exchange*. In *Handbook of physiology*. Section 2. *The cardiovascular system—microcirculation,* vol IV, Bethesda, Md, 1984, American Physiological Society.

Crone C, Levitt DG: *Capillary permeability to small solutes*. In *Handbook of physiology*. Section 2. *The cardiovascular system—microcirculation,* vol IV, Bethesda, Md, 1984, American Physiological Society.

Hudlicka O: Development of microcirculation: *capillary growth and adaptation*. In *Handbook of physiology*. Section 2. *The cardiovascular system—microcirculation,* vol IV, Bethesda, Md, 1984, American Physiological Society.

Krogh A: *The anatomy and physiology of capillaries,* New York, 1959, Hafner.

Luscher TF, Vanhoutte PM: *The endothelium: modulator of cardiovascular function,* Boca Raton, Fla, 1990, CRC Press.

Michel CC: *Fluid movements through capillary walls.* In *Handbook of physiology.* Section 2. *The cardiovascular system—microcirculation,* vol IV, Bethesda, Md, 1984, American Physiological Society.

Mortillaro NA: *Physiology and pharmacology of the microcirculation,* vol 1, New York, 1983, Academic Press.

Renkin EM: *Control microcirculation and blood-tissue exchange.* In *Handbook of physiology.* Section 2. *The cardiovascular system—microcirculation,* vol IV, Bethesda, Md, 1984, American Physiological Society.

Shepro D, D'Amore PA: *Physiology and biochemistry of the vascular wall endothelium.* In *Handbook of physiology.* Section 2. *The cardiovascular system—microcirculation,* vol IV, Bethesda, Md, 1984, American Physiological Society.

Simionescu M, Simionescu N: *Ultrastructure of the microvascular wall: functional correlations.* In *Handbook of physiology.* Section 2. *The cardiovascular system—microcirculation,* vol IV, Bethesda, Md, 1984, American Physiological Society.

Taylor AE, Granger DN: *Exchange of macromolecules across the microcirculation.* In *Handbook of physiology.* Section 2. *The cardiovascular system—microcirculation,* vol IV, Bethesda, Md, 1984, American Physiological Society.

Wiedeman MP: *Architecture.* In *Handbook of physiology.* Section 2. *The cardiovascular system—microcirculation,* vol IV, Bethesda, Md, 1984, American Physiological Society.

Yoffey JM, Courtice FC: *Lymphatics, lymph and the lymphomyeloid complex,* London, 1970, Academic Press.

Zweifach BW, Lipowsky HH: *Pressure-flow relations in blood and lymph microcirculation.* In *Handbook of physiology.* Section 2. *The cardiovascular system—microcirculation,* vol IV, Bethesda, Md, 1984, American Physiological Society.

CHAPTER

29

The Peripheral Circulation and Its Control

The peripheral circulation is essentially under dual control, centrally through the nervous system and locally in the tissues by the environmental conditions in the immediate vicinity of the blood vessels. The relative importance of these two control mechanisms is not the same in all tissues. In some areas of the body, such as the skin and the splanchnic regions, neural regulation of blood flow predominates, whereas in others, such as the heart and brain, this mechanism plays a minor role.

The vessels chiefly involved in regulating the rate of blood flow throughout the body are called the **resistance vessels** (arterioles). These vessels offer the greatest resistance to the flow of blood pumped to the tissues by the heart and thereby are important in the maintenance of arterial blood pressure. Smooth muscle fibers constitute a large percentage of the composition of the walls of the resistance vessels (see Fig. 22-1). Therefore the vessel lumen can be varied from one that is completely obliterated by strong contraction of the smooth muscle, with infolding of the endothelial lining, to one that is maximally dilated as a result of full relaxation of the vascular smooth muscle. Some resistance vessels are closed at any given moment in time and partial contraction **(tone)** of the vascular smooth muscle exists in the arterioles. Were all the resistance vessels in the body to dilate simultaneously, blood pressure would fall precipitously to very low levels.

■ *Vascular Smooth Muscle*

Vascular smooth muscle is the tissue responsible for the control of total peripheral resistance, arterial and venous tone, and distribution of blood flow throughout the body. The smooth muscle cells are small, mononucleate, and spindle shaped. They are generally arranged in helical or circular layers around the larger blood vessels and in a single circular layer around arterioles (Fig. 29-1, *A* and *B*) (also see Chapter 19). Also parts of endothelial cells project into the vascular smooth muscle layer **(myoendothelial junctions)** at various points along the arterioles (Fig. 29-1, *C*). These projections suggest a functional interaction between endothelium and adjacent vascular smooth muscle. In general the close association between action potentials and contraction observed in skeletal and cardiac muscle cells cannot be demonstrated in vascular smooth muscle. Also, vascular smooth muscle lacks transverse tubules.

Graded changes in membrane potential are often associated with increases or decreases in force. Contractile activity is generally elicited by neural or humoral stimuli. The behavior of smooth muscle in different vessels varies. For example, some vessels, particularly in the portal or mesenteric circulation, contain longitudinally oriented smooth muscle that is spontaneously active and that shows action potentials that are correlated with the contractions and the electrical coupling between cells.

The vascular smooth muscle cells contain large numbers of thin, actin filaments and comparatively small numbers of thick, myosin filaments. These filaments are aligned in the long axis of the cell but do not form visible sarcomeres with striations. Nevertheless the sliding filament mechanism is believed to operate in this tissue, and phosphorylation of crossbridges regulates their rate of cycling. Compared to skeletal muscle, the smooth muscle contracts very slowly, develops high forces, maintains force for long periods with low ATP use, and operates over a considerable range of lengths under physiological conditions (see Chapter 19). Cell to cell conduction is via gap junctions as occurs in cardiac muscle (see p. 398).

The interaction of myosin and actin, leading to contraction, is controlled by the myoplasmic Ca^{++} concentration as in other muscles, but the molecular mechanism whereby Ca^{++} regulates contraction differs. For example smooth muscle lacks troponin and fast sodium channels. The increased myoplasmic Ca^{++} that elicits

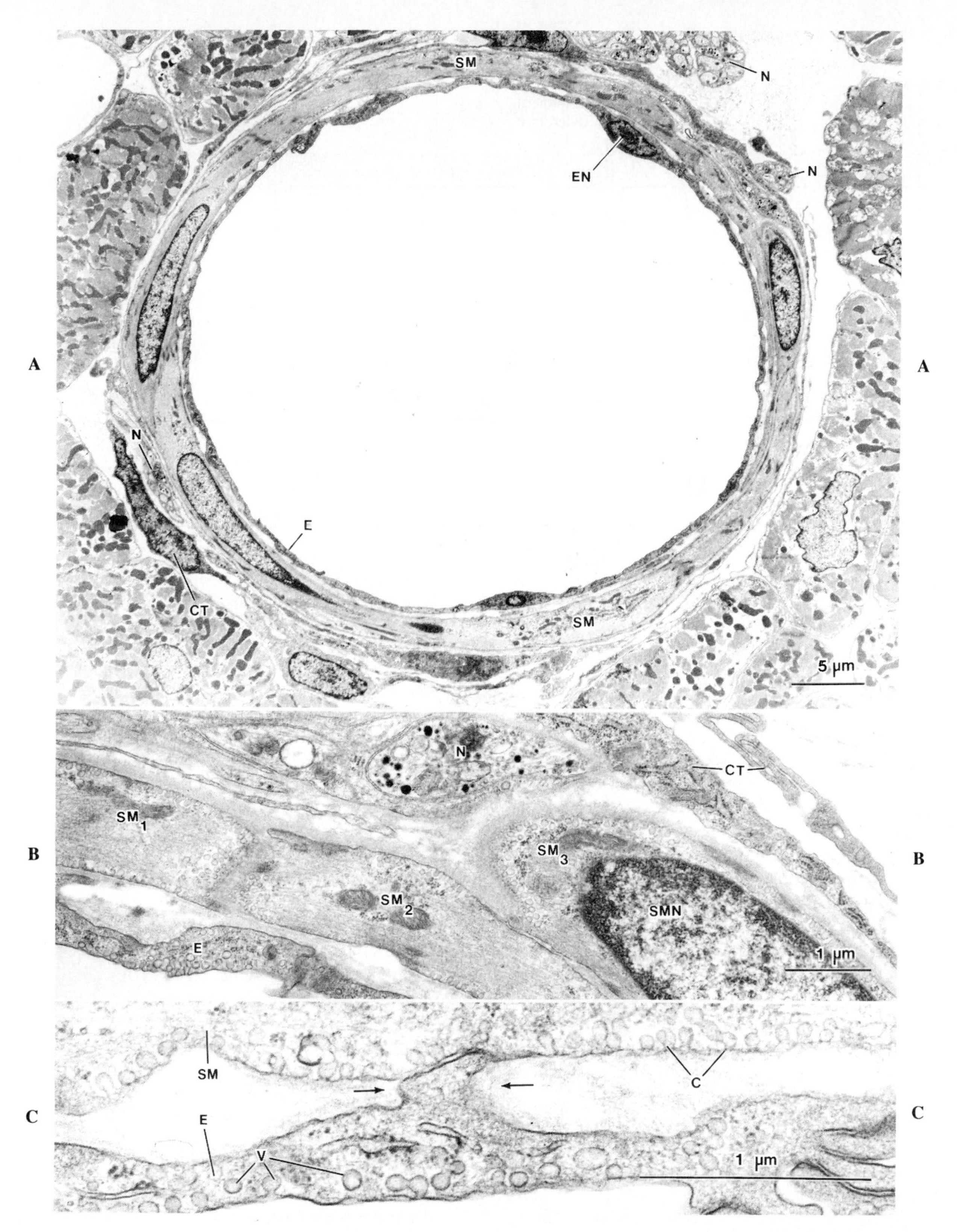

■ **Fig. 29-1** **A,** Low-magnification electron micrograph of an arteriole in cross section (inner diameter of approximately 40 μm) in cat ventricle. The wall of the blood vessel is composed largely of vascular smooth muscle cells *(SM)* whose long axes are directed approximately circularly around the vessel. A single layer of endothelial cells *(E)* forms the innermost portion of the blood vessel. Connective tissue *(CT)* elements such as fibroblasts and collagen make up the adventitial layer at the periphery of the vessel; nerve bundles *(N)* also appear in this layer. *EN,* Endothelial cell nucleus. **B,** Detail of the wall of the blood vessel in panel **A.** This field contains a single endothelial layer *(E),* the medial smooth muscle layer (three smooth muscle cell profiles; SM_1, SM_2, SM_3), and the adventitial layer (containing nerves *[N]* and connective tissue *[CT]*). *SMN,* Smooth muscle nucleus. **C,** Another region of the arteriole, showing the area in which the endothelial *(E)* and smooth muscle *(SM)* layers are apposed. A projection of an endothelial cell *(between arrows)* is closely applied to the surface of the overlying smooth muscle, forming a "myoendothelial junction." Plasmalemmal vesicles are prominent in both the endothelium *(V)* and the smooth muscle cell (where such vesicles are known as "caveolae" *[C]*).

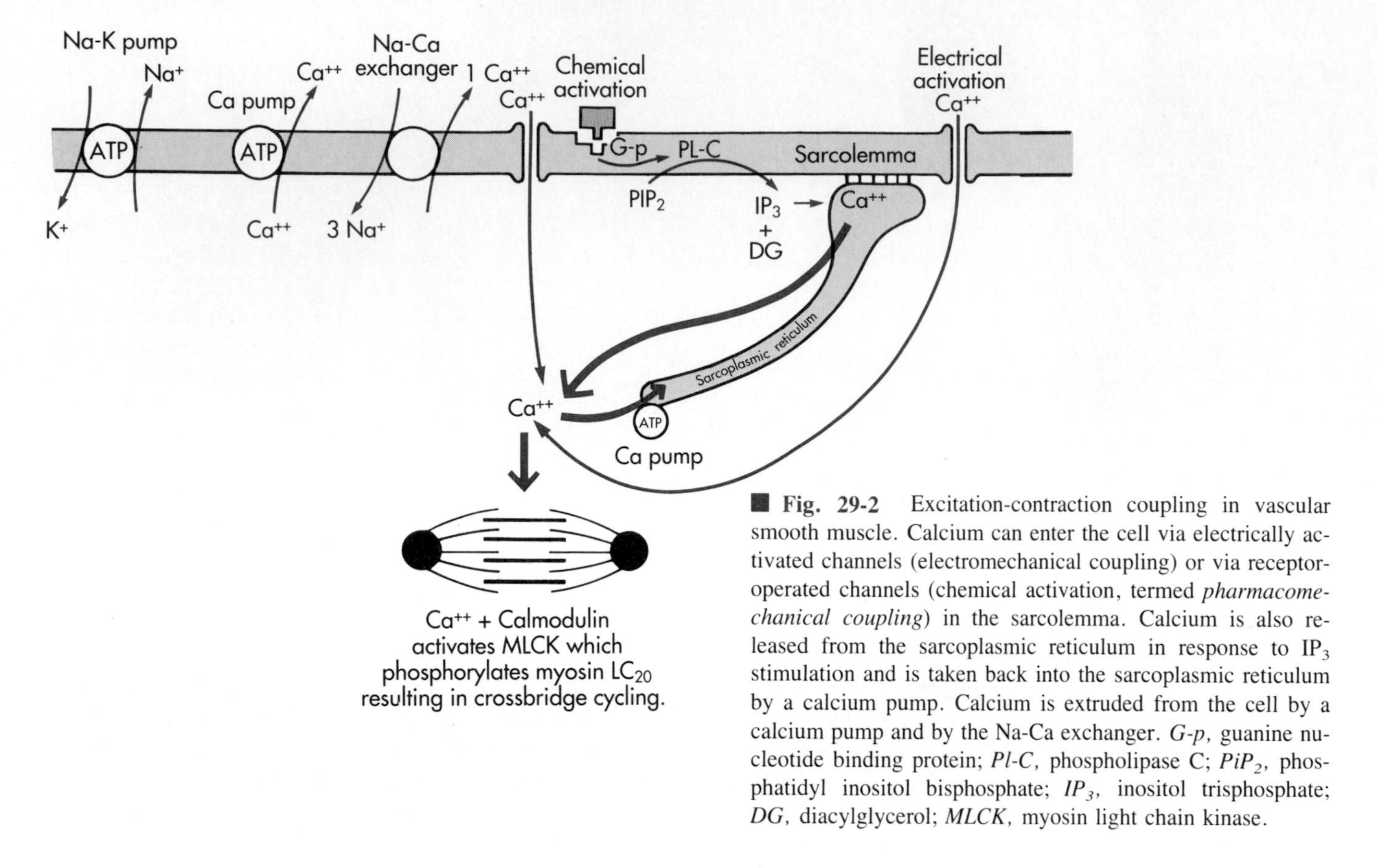

■ **Fig. 29-2** Excitation-contraction coupling in vascular smooth muscle. Calcium can enter the cell via electrically activated channels (electromechanical coupling) or via receptor-operated channels (chemical activation, termed *pharmacomechanical coupling*) in the sarcolemma. Calcium is also released from the sarcoplasmic reticulum in response to IP_3 stimulation and is taken back into the sarcoplasmic reticulum by a calcium pump. Calcium is extruded from the cell by a calcium pump and by the Na-Ca exchanger. *G-p,* guanine nucleotide binding protein; *Pl-C,* phospholipase C; *PiP_2,* phosphatidyl inositol bisphosphate; *IP_3,* inositol trisphosphate; *DG,* diacylglycerol; *MLCK,* myosin light chain kinase.

contraction can come through voltage gated calcium channels **(electromechanical coupling)** and through receptor-operated calcium channels **(pharmacomechanical coupling)** in the sarcolemma, and by release from the sarcoplasmic reticulum (Fig. 29-2). The cells relax when intracellular free Ca^{++} is pumped back into the sarcoplasmic reticulum and is extruded from the cell by the calcium pump in the cell membrane and by the Na-Ca exchanger.

The response to humoral stimuli, which is termed pharmacomechanical coupling, occurs without evidence of electrical excitation and is the predominant mechanism for eliciting contraction of vascular smooth muscle. In the category of pharmacological stimuli are such substances as catecholamines, histamine, acetylcholine, serotonin, angiotensin, adenosine, and prostaglandins.

As illustrated in Fig. 29-2 an agonist activates receptors in the vascular smooth muscle membrane. These receptors in turn activate phospholipase C in a reaction coupled to guanine nucleotide binding proteins (G proteins). The phospholipase C hydrolyzes phosphatidyl inositol bisphosphate in the membrane to yield diacylglycerol and inositol trisphosphate; the latter causes the release of Ca^{++} from the sarcoplasmic reticulum. The Ca^{++} binds to calmodulin, which in turn binds to myosin light chain kinase. This activated Ca^{++}-calmodulin-myosin kinase complex phosphorylates the light chains (20,000 daltons) of myosin. The phosphorylated myosin adenosine triphosphatase (ATPase) is then activated by actin, and the resulting crossbridge cycling initiates contraction.

Finally the sensitivity of the contractile regulatory apparatus to Ca^{++} is increased by agonists. The mechanism for this enhanced sensitivity is still unclear but appears to involve G proteins. Relaxation occurs when the myosin light chain kinase is inactivated by dephosphorylation and the cytosolic Ca^{++} is lowered by sarcoplasmic reticulum uptake and by Ca^{++} extrusion by the Ca pump and the Na-Ca exchanger. Local environmental changes alter the contractile state of vascular smooth muscle, and alterations such as increased temperature or increased carbon dioxide levels induce relaxation of this tissue.

Most of the arteries and veins of the body are supplied to different degrees solely by fibers of the sympathetic nervous system (see Fig. 29-1, *A* and *B*). These nerve fibers exert a tonic effect on the blood vessels, as evidenced by the fact that cutting or freezing the sympathetic nerves to a vascular bed (such as muscle) results in an increase in blood flow. Activation of the sympathetic nerves either directly or reflexly (pp. 484 and 486) enhances vascular resistance. In contrast to the sympathetic nerves the parasympathetic nerves tend to decrease vascular resistance, but they innervate only a small fraction of the blood vessels in the body, mainly in certain viscera and pelvic organs.

■ *Intrinsic or Local Control of Peripheral Blood Flow*

■ *Autoregulation and Myogenic Regulation*

In a number of different tissues the blood flow appears to be adjusted to the existing metabolic activity of the tissue. Furthermore imposed changes in the perfusion pressure (arterial blood pressure) at constant levels of tissue metabolism, as measured by oxygen consumption, are met with vascular resistance changes that tend to maintain a constant blood flow. This mechanism is commonly referred to as **autoregulation** of blood flow and is illustrated graphically in Fig. 29-3. In the skeletal muscle preparation from which these data were gathered the muscle was completely isolated from the rest of the animal and was in a resting state. From a control pressure of 100 mm Hg the pressure was abruptly increased or decreased, and the blood flows observed immediately after changing the perfusion pressure are represented by the closed circles. Maintenance of the altered pressure at each new level was followed within 30 to 60 seconds by a return of flow to or toward the control levels; the open circles represent these steady-state flows. Over the pressure range of 20 to 120 mm Hg, the steady-state flow is relatively constant. Calculation of resistance across the vascular bed (pressure/flow) during steady-state conditions indicates that with elevation of perfusion pressure the resistance vessels constricted, whereas with reduction of perfusion pressure they dilated.

The mechanism responsible for this constancy of blood flow in the presence of altered perfusion pressure is not known, but it appears to be explained best by the **myogenic mechanism.**

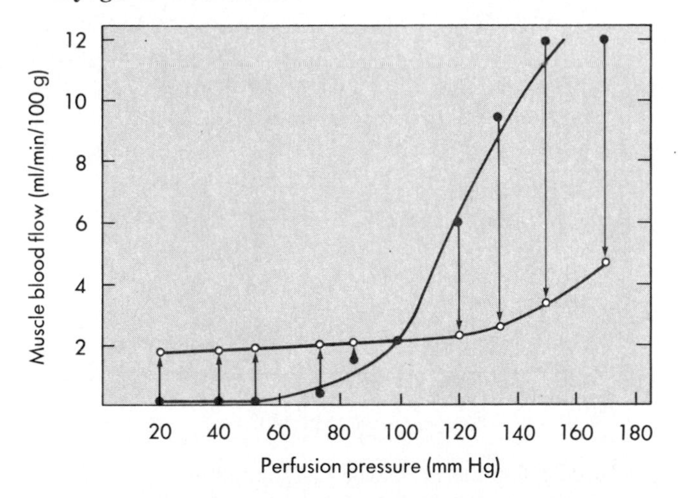

■ **Fig. 29-3** Pressure-flow relationship in the skeletal muscle vascular bed of the dog. The closed circles represent the flows obtained immediately after abrupt changes in perfusion pressure from the control level *(point where lines cross)*. The open circles represent the steady-state flows obtained at the new perfusion pressure. (Redrawn from Jones RD, Berne RM: *Circ Res* 14:126, 1964.)

According to the myogenic mechanism the vascular smooth muscle contracts in response to increased transmural pressure stretch and relaxes with a reduction in transmural pressure. Therefore the initial flow increment produced by an abrupt increase in perfusion pressure that passively distends the blood vessels would be followed by a return of flow to the previous control level by contraction of the smooth muscles of the resistance vessels.

An example of a myogenic response is shown in Fig. 29-4. Arterioles isolated from the heart of young pigs were cannulated at each end, and the transmural pressure (intravascular pressure minus extravascular pressure) and flow through the arteriole could be adjusted to desired levels. With no flow through the arteriole successive increases of transmural pressure elicited progressive decreases in the vessel diameter (Fig. 29-4, *A*). This response was independent of the endothelium because it was identical in intact vessels and in vessels denuded of endothelium (Fig. 29-4, *B*). Arterioles that were relaxed by direct action of nitroprusside on the vascular smooth muscle showed only a passive increase in diameter when transmural pressure was increased. How vessel distension elicits contraction is unsettled, but because stretch of vascular smooth muscle elevates intracellular Ca^{++}, it has been proposed that an increase in transmural pressure activates membrane calcium channels.

Because blood pressure is reflexly maintained at a fairly constant level under normal conditions, operation of a myogenic mechanism would be expected to be minimized. However when one changes position (from lying to standing) a large change in transmural pressure occurs in the lower extremities. The precapillary vessels constrict in response to this imposed stretch, which results in cessation of flow in most capillaries. After flow stops capillary filtration diminishes until the increase in plasma oncotic pressure and the increase in interstitial fluid pressure balance the elevated capillary hydrostatic pressure produced by changing from a horizontal to a vertical position. If arteriolar resistance did not increase with standing, the hydrostatic pressure in the lower parts of the legs would reach such high levels that large volumes of fluid would pass from the capillaries into the interstitial fluid compartment and produce edema.

■ *Endothelial-Mediated Regulation*

As discussed on p. 467, stimulation of the endothelium can elicit a vasoactive response of the vascular smooth muscle. Endothelial-mediated dilation, in the same preparation described for Fig. 29-4, is illustrated in Fig. 29-5. Transmural pressure is kept constant and flow through the isolated arteriole is increased by raising the perfusion fluid reservoir connected to one end of the arteriole and simultaneously lowering the reservoir con-

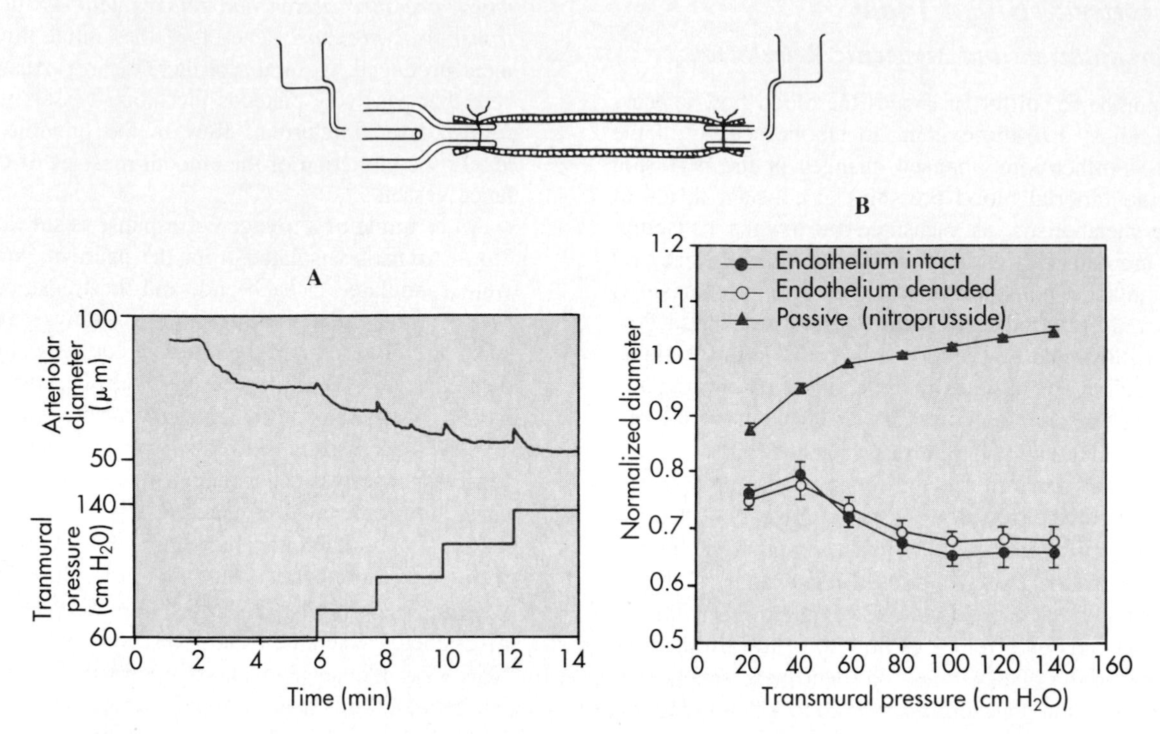

■ **Fig. 29-4** **A,** Constriction of an isolated cardiac arteriole in response to increases in transmural pressure without flow through the blood vessel. **B,** Constrictor response of the arteriole to an increase in transmural pressure is unaffected by removal of its endothelium. When the smooth muscle is relaxed by nitroprusside, the arteriole is passively distended by the increase in transmural pressure. (Redrawn from Kuo L, Davis MJ, Chilian WM: *Am J Physiol* 259:H1063, 1990).

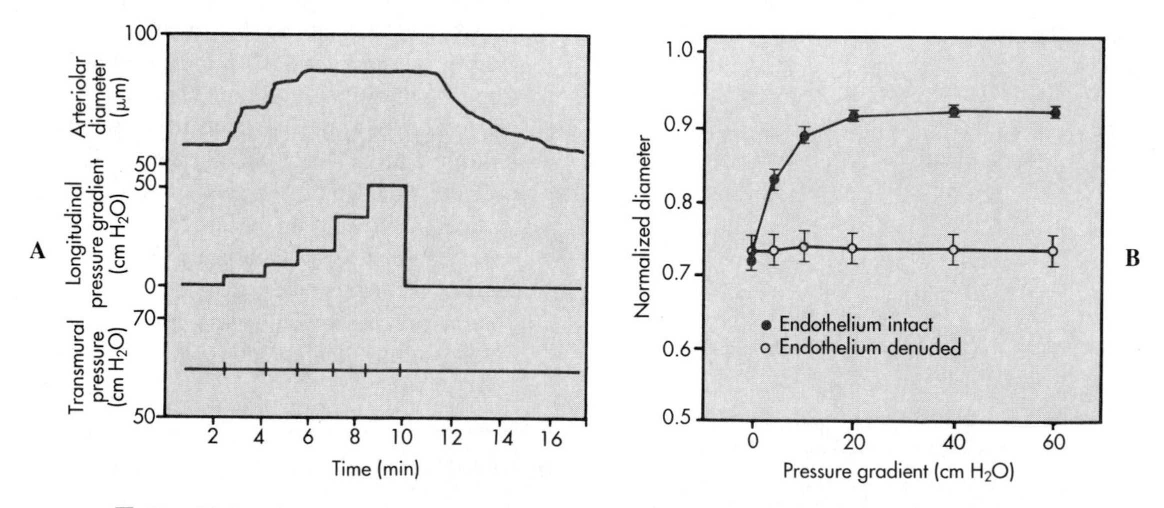

■ **Fig. 29-5** **A,** Flow-induced vasodilation in an isolated cardiac arteriole at constant transmural pressure. Flow was increased progressively by increasing the pressure gradient in the long axis of the arteriole (longitudinal pressure gradient). **B,** Flow-induced vasodilation is abolished by removal of the endothelium of the arteriole. (Redrawn from Kuo L, Davis MJ, Chilian WM: *Am J Physiol* 259:H1063, 1990).

nected to the other end of the arteriole by an equal distance. This elevation of pressure gradient along the vessel axis (longitudinal pressure gradient) increases flow and increases vessel diameter (Fig. 29-5, *A*). If the arteriole is denuded of endothelium, the dilation of the vessel in response to increased flow is abolished (Fig. 29-5, *B*). The vasodilation is presumably caused by the endothelial-derived relaxing factor (EDRF) (see p. 467), which is released from the endothelium in response to the shear stress consequent to the increase in velocity of flow.

Metabolic Regulation

According to the metabolic mechanism blood flow is governed by the metabolic activity of the tissue. Any intervention that results in an O_2 supply that is inadequate for the requirements of the tissue gives rise to the formation of vasodilator metabolites. These metabolites are released from the tissue and act locally to dilate the resistance vessels. When the metabolic rate of the tissue increases or the O_2 delivery to the tissue decreases, more vasodilator substance is released and the metabolite concentration in the tissue increases.

Many substances have been proposed as mediators of metabolic vasodilation. Some of the earliest ones suggested are lactic acid, CO_2, and hydrogen ions. However the decrease in vascular resistance induced by supernormal concentrations of these dilator agents falls considerably short of the dilation observed under physiological conditions of increased metabolic activity.

Changes in O_2 tension can evoke changes in the contractile state of vascular smooth muscle; an increase in Po_2 elicits contraction, and a decrease in Po_2, relaxation. If significant reductions in the intravascular Po_2 occur before the arterial blood reaches the resistance vessels (diffusion through the arterial and arteriolar walls, p. 471), small changes in O_2 supply or consumption could elicit contraction or relaxation of the resistance vessels. However, direct measurements of Po_2 at the resistance vessels indicate that over a wide range of Po_2 (11 to 343 mm Hg) there is no correlation between O_2 tension and arteriolar diameter. Furthermore if Po_2 were directly responsible for vascular smooth muscle tension, one would not expect to find a parallelism between the duration of arterial occlusion and the duration of the reactive hyperemia (flow above control level upon release of an arterial occlusion) (Fig. 29-6). With either short occlusions (5 to 10 seconds) or long occlusions (1 to 3 minutes) the venous blood becomes bright red (well oxygenated) within 1 or 2 seconds after release of the arterial occlusion and hence the smooth muscle of the resistance vessels must be exposed to a high Po_2 in each instance. Nevertheless the longer occlusions result in longer periods of reactive hyperemia. These observations are more compatible with the re-

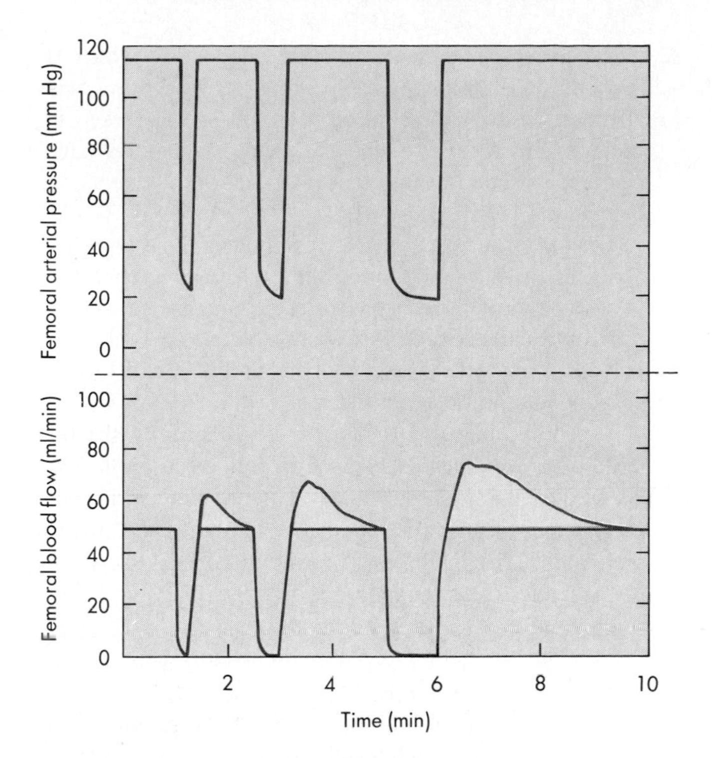

Fig. 29-6 Reactive hyperemia in the hind limb of the dog after 15-, 30-, and 60-second occlusions of the femoral artery.

lease of a vasodilator metabolite from the tissue than with a direct effect of Po_2 on the vascular smooth muscle.

Potassium ions, inorganic phosphate, and interstitial fluid osmolarity can also induce vasodilation. Because K^+ and phosphate are released and osmolarity is increased during skeletal muscle contraction, it has been proposed that these factors contribute to **active hyperemia** (increased blood flow caused by enhanced tissue activity). However significant increases of phosphate concentration and osmolarity are not consistently observed during muscle contraction, and they may produce only transient increases in blood flow. Therefore they are not likely candidates as mediators of the vasodilation observed with muscular activity. Potassium release occurs with the onset of skeletal muscle contraction or an increase in cardiac activity and could be responsible for the initial decrease in vascular resistance observed with exercise or increased cardiac work. However K^+ release is not sustained, despite continued arteriolar dilation throughout the period of enhanced muscle activity. Therefore some other agent must serve as mediator of the vasodilation associated with the greater metabolic activity of the tissue. Reoxygenated venous blood obtained from active cardiac and skeletal muscles under steady-state conditions of exercise does not elicit vasodilation when infused into a test vascular bed. It is difficult to see how oxygenation of the venous blood could alter its K^+ or phosphate content or its osmolarity and thereby destroy its vasodilator effect.

Recent evidence indicates that adenosine, which is involved in the regulation of coronary blood flow, may also participate in the control of the resistance vessels in skeletal muscle; also some of the prostaglandins have been proposed as important vasodilator mediators in certain vascular beds.

Thus there are a number of candidates for the mediator of metabolic vasodilation, and the relative contribution of each of the various factors remains the subject for future investigation. Several factors may be involved in any given vascular bed, and different factors preponderate in different tissues.

Metabolic control of vascular resistance by the release of a vasodilator substance is predicated on the existence of basal vessel tone. This tonic activity, or **basal tone,** of the vascular smooth muscle is readily demonstrable, but in contrast to tone in skeletal muscle it is independent of the nervous system. The factor responsible for basal tone in blood vessels is not known, but one or more of the following factors may be involved: (1) an expression of myogenic activity in response to the stretch imposed by the blood pressure. (2) the high O_2 tension of arterial blood, (3) the presence of calcium ions, or (4) some unknown factor in plasma, because addition of plasma to the bathing solution of isolated vessel segments evokes partial contraction of the smooth muscle.

If arterial inflow to a vascular bed is stopped for a few seconds to several minutes, the blood flow, on release of the occlusion, immediately exceeds the flow before occlusion and only gradually returns to the control level **(reactive hyperemia).** This is illustrated in Fig. 29-6 where blood flow to the leg was stopped by clamping the femoral artery for 15, 30, and 60 seconds. Release of the 60-second occlusion resulted in a peak blood flow 70% greater than the control flow, with a return to control flow within about 110 seconds. When this same experiment is done in humans by inflating a blood pressure cuff on the upper arm, dilation of the resistance vessels of the hand and forearm, immediately after release of the cuff, is evident from the bright red color of the skin and the fullness of the veins. Within limits the peak flow and particularly the duration of the reactive hyperemia are proportional to the duration of the occlusion (see Fig. 29-6). If the extremity is exercised during the occlusion period, reactive hyperemia is increased. These observations and the close relationship that exists between metabolic activity and blood flow in the unoccluded limb are consonant with a metabolic mechanism in the local regulation of tissue blood flow. When the vascular smooth muscle of the arterioles relaxes in response to vasodilator metabolites released by a decrease in the oxygen supply/oxygen demand ratio of the tissue, resistance may diminish in the arteries that feed these arterioles. This results in a greater blood flow than that produced by arteriolar dilation alone. Two possible mechanisms can account for this coordination of arterial and arteriolar dilation. First vasodilation in the microvessels is propagated and when initiated in the arterioles, it can propagate from arterioles back to arteries. Second the metabolite-mediated dilation of the arterioles accelerates blood flow in the feeder arteries. This increases the shear stress on the arterial endothelium and can induce vasodilation by release of EDRF.

Extrinsic Control of Peripheral Blood Flow

Neural Sympathetic Vasoconstriction

A number of regions in the medulla influence cardiovascular activity. Some of the effects of stimulation of the dorsal lateral medulla are vasoconstriction, cardiac acceleration, and enhanced myocardial contractility. Caudal and ventromedial to the **pressor** region is a zone that produces a decrease in blood pressure on stimulation. This **depressor** area exerts its effect by direct spinal inhibition and by inhibition of the medullary pressor region. However the precise mechanism of its depressor actions is still unknown. These areas constitute a center, not in an anatomical sense in that a discrete group of cells is discernible, but in a physiological sense in that stimulation of the pressor region produces the responses mentioned previously. From the vasoconstrictor regions fibers descend in the spinal cord and synapse at different levels of the thoracolumbar region (T1 to L2 or L3). Fibers from the intermediolateral gray matter of the cord emerge with the ventral roots but leave the motor fibers to join the paravertebral sympathetic chains through the white communicating branches. These preganglionic white (myelinated) fibers may pass up or down the sympathetic chains to synapse in the various ganglia within the chains or in certain outlying ganglia. Postganglionic gray branches (unmyelinated) then join the corresponding segmental spinal nerves and accompany them to the periphery to innervate the arteries and veins. Postganglionic sympathetic fibers from the various ganglia join the large arteries and accompany them as an investing network of fibers to the resistance and capacitance vessels (veins) (see Chapter 15).

The vasoconstrictor regions are tonically active, and reflexes or humoral stimuli that enhance this activity result in an increase in frequency of impulses reaching the terminal branches to the vessels, where a constrictor neurohumor **(norepinephrine)** is released and elicits constriction (α-adrenergic effect) of the resistance vessels. Inhibition of the vasoconstrictor areas reduces their tonic activity and hence diminishes the frequency of impulses in the efferent nerve fibers, resulting in vasodilation. In this manner neural regulation of the peripheral circulation is accomplished primarily by alteration of the number of impulses passing down the vasoconstrictor fibers of the sympathetic nerves to the blood vessels. The vasomotor regions may show rhythmic changes in tonic activity manifested as oscillations of arterial pressure. Some occur at the frequency of respiration **(Traube-Hering waves)** and are due to an increase in sympathetic impulses to the resistance vessels coincident with inspiration. Others are independent of and at a lower frequency than respiration **(Mayer waves).**

Sympathetic Constrictor Influence on Resistance and Capacitance Vessels

The vasoconstrictor fibers of the sympathetic nervous system supply the arteries, arterioles, and veins, but neural influence on the larger vessels is of far less functional importance than it is on the microcirculation. Capacitance vessels are apparently more responsive to sympathetic nerve stimulation than are resistance vessels, because they reach maximal constriction at a lower frequency of stimulation than do the resistance vessels. However capacitance vessels do not possess β-adrenergic receptors nor do they respond to vasodilator metabolites. Norepinephrine is the neurotransmitter released at the sympathetic nerve terminals at the blood vessels, and many factors, such as circulating hormones and particularly locally released substances, modify the liberation of norepinephrine from the vesicles of the nerve terminals.

The response of the resistance and capacitance vessels to stimulation of the sympathetic fibers is illustrated in Fig. 29-7. At constant arterial pressure sympathetic fiber stimulation evoked a reduction of blood flow (constriction of the resistance vessels) and a decrease in blood volume of the tissue (constriction of the capacitance vessels). The abrupt decrease in tissue volume is caused by movement of blood out of the capacitance vessels and out of the hindquarters of the cat, whereas the late, slow, progressive decline in volume (to the right of the arrow) is caused by movement of extravascular fluid into the capillaries and hence away from the tissue. The loss of tissue fluid is a consequence of the lowered capillary hydrostatic pressure brought about by constriction of the resistance vessels, with establishment of a new equilibrium of the forces responsible for filtration and absorption across the capillary wall (see p. 473).

In addition to active changes (contraction and relaxation of the vascular smooth muscle) in vessel caliber there are also passive changes caused solely by alteration in intraluminal pressure; an increase in intraluminal pressure produces distension of the vessels, and a decrease produces a reduction in caliber by recoil of the elastic components of the vessel walls.

At basal tone approximately one third of the blood volume of a tissue can be mobilized on stimulation of the sympathetic nerves at physiological frequencies.

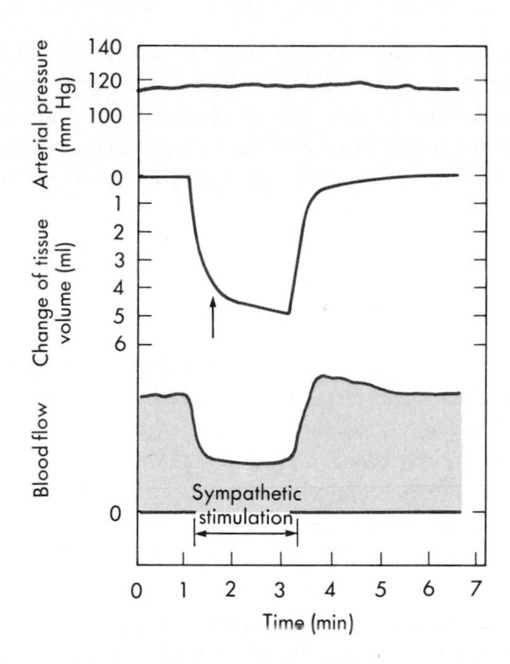

Fig. 29-7 Effect of sympathetic nerve stimulation (2 Hz) on blood flow and tissue volume in the hindquarters of the cat. The arrow denotes the change in slope of the tissue volume curve where the volume decrease caused by emptying of capacitance vessels ceases and loss of extravascular fluid becomes evident. (Redrawn from Mellander S: Acta Physiol Scand 50(suppl 176):1, 1960.)

The basal tone is very low in capacitance vessels; with veins denervated, only small increases in volume are obtained with maximal doses of acetylcholine. Therefore the blood volume at basal tone is close to the maximal blood volume of the tissue. More blood can be mobilized from the skin than from the muscle capacitance vessels, in part because of greater sensitivity of the skin vessels to sympathetic stimulation and because basal tone is lower in skin vessels than in muscle vessels. Therefore in the absence of neural influence the skin capacitance vessels contain more blood than do the muscle capacitance vessels.

Blood is mobilized from capacitance vessels in response to physiological stimuli. In exercise activation of the sympathetic nerve fibers produces constriction of veins and hence augments the cardiac filling pressure. Also in arterial hypotension (as in hemorrhage) the capacitance vessels constrict to aid in overcoming the decreased central venous pressure associated with this condition. In addition the resistance vessels constrict in shock, thereby assisting in the maintenance or restoration of arterial pressure. With arterial hypotension the enhanced arteriolar constriction also leads to a small mobilization of blood from the tissue by virtue of recoil of the postarteriolar vessels when intraluminal pressure is reduced. Furthermore there is mobilization of extravascular fluid because of greater absorption into the capillaries in response to the lowered capillary hydrostatic pressure.

Hence it becomes apparent that neural and humoral stimuli can exert similar or dissimilar effects on different segments of the vascular tree and in so doing can alter blood flow, tissue blood volume, and extravascular volume to meet the physiological requirements of the organism.

Parasympathetic Neural Influence

The efferent fibers of the cranial division of the parasympathetic nervous system supply blood vessels of the head and viscera, whereas fibers of the sacral division supply blood vessels of the genitalia, bladder, and large bowel. Skeletal muscle and skin do not receive parasympathetic innervation. Because only a small proportion of the resistance vessels of the body receives parasympathetic fibers, the effect of these cholinergic fibers on total vascular resistance is small.

Stimulation of the parasympathetic fibers to the salivary glands induces marked vasodilation. A vasodilator polypeptide, **bradykinin,** formed locally from the action of an enzyme on a plasma protein substrate present in the glandular lymphatic vessels has been considered to be the metabolic mediator of the vasodilation produced by chorda tympani stimulation. Whether vasodilation in salivary glands results from the release of a cholinergic neurohumor from nerve endings, from the formation and release of bradykinin, or from both is unsettled. Bradykinin also has been reported to be formed in other exocrine glands, such as the lacrimal glands and the sweat glands. Its presence in sweat is thought to be partly responsible for the dilation of cutaneous blood vessels that occurs with sweating.

Humoral Factors

Epinephrine and norepinephrine exert a profound effect on the peripheral blood vessels. In skeletal muscle epinephrine in low concentrations dilates resistance vessels (**β-adrenergic effect**) and in high concentrations produces constriction (**α-adrenergic effect**). In skin only vasoconstriction is obtained with epinephrine, whereas in all vascular beds the primary effect of norepinephrine is vasoconstriction. When stimulated, the adrenal gland can release epinephrine and norepinephrine into the systemic circulation. However under physiological conditions the effect of catecholamine release from the adrenal medulla is of lesser importance than norepinephrine release produced by sympathetic nerve activation.

Vascular Reflexes

Areas of the medulla that mediate sympathetic and vagal effects are under the influence of neural impulses arising in the baroreceptors, chemoreceptors, hypothalamus, cerebral cortex, and skin. These areas of the medulla are also affected by changes in the blood concentrations of CO_2 and O_2.

Arterial baroreceptors. The **baroreceptors** (or **pressoreceptors**) are stretch receptors located in the carotid sinuses (slightly widened areas of the internal ca-

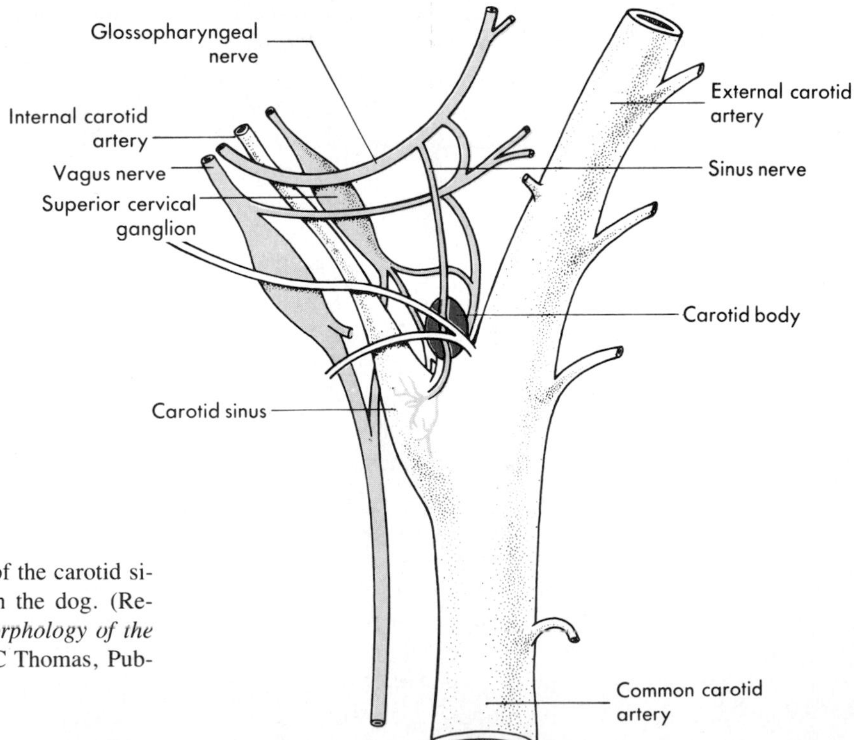

■ **Fig. 29-8** Diagrammatic representation of the carotid sinus and carotid body and their innervation in the dog. (Redrawn from Adams WE: *The comparative morphology of the carotid sinus,* Springfield, Ill, 1958, Charles C Thomas, Publisher.)

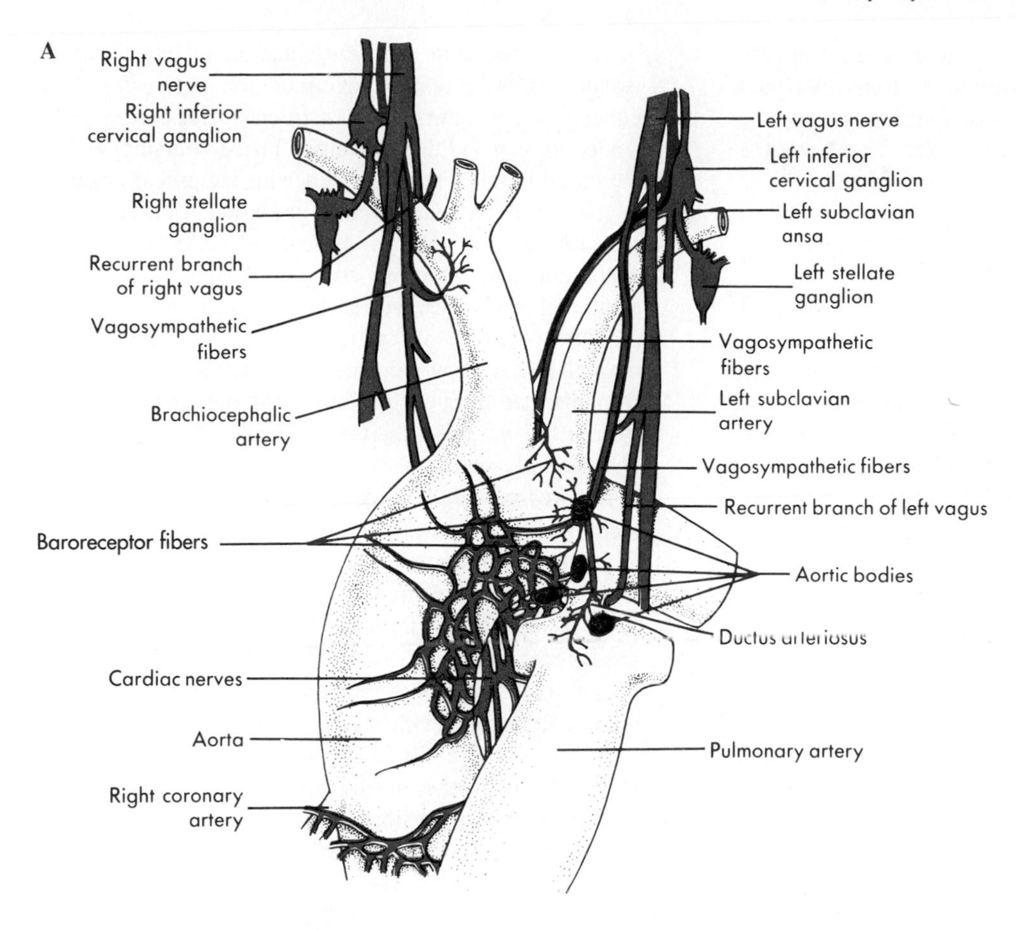

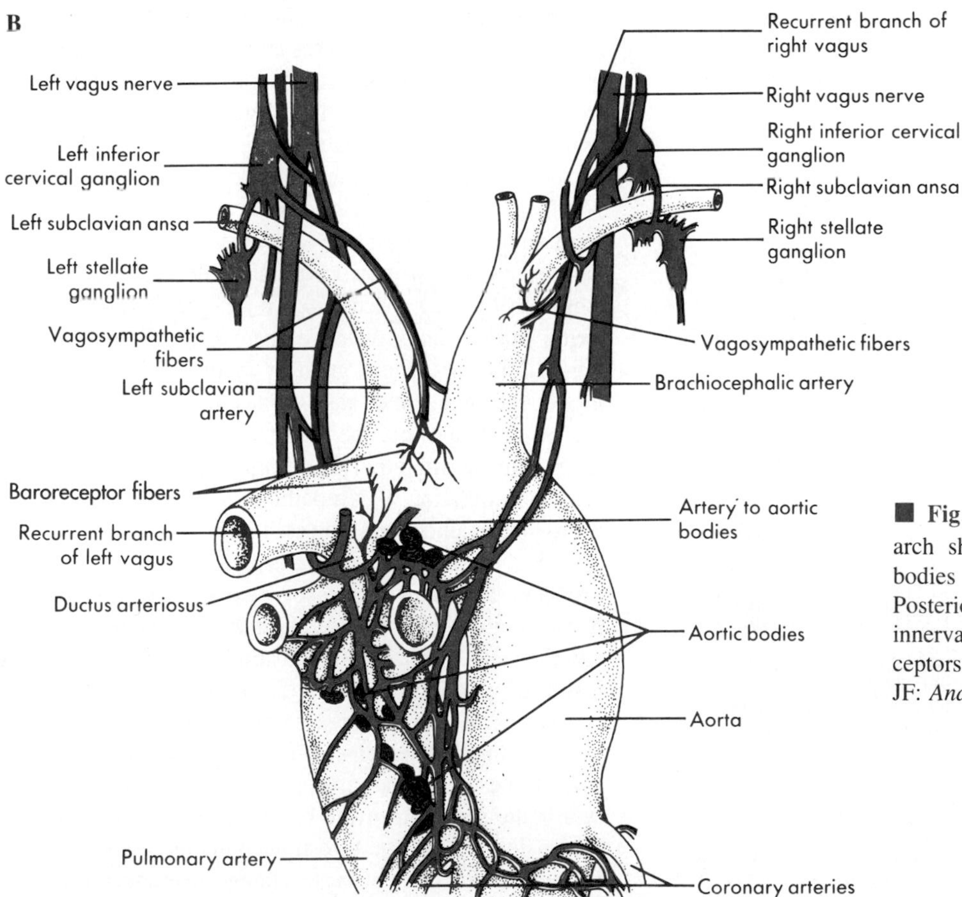

■ **Fig. 29-9** **A,** Anterior view of the aortic arch showing the innervation of the aortic bodies and pressoreceptors in the dog. **B,** Posterior view of the aortic arch showing the innervation of the aortic bodies and pressoreceptors in the dog. (Modified from Nonidez JF: *Anat Rec* 69:299, 1937.)

rotid arteries at their points of origin from the common carotid arteries) and in the aortic arch (Figs. 29-8 and 29-9). Impulses arising in the carotid sinus travel up the sinus nerve to the glossopharyngeal nerve and, via the latter, to the nucleus of the tractus solitarius (NTS) in the medulla. The NTS is the site of central projection of the chemoreceptors and baroreceptors. Stimulation of the NTS inhibits sympathetic nerve impulses to the peripheral blood vessels (depressor), whereas lesions of the NTS produce vasoconstriction (pressor). Impulses arising in the pressoreceptors of the aortic arch reach the NTS via afferent fibers in the vagus nerves. The pressoreceptor nerve terminals in the walls of the carotid sinus and aortic arch respond to the stretch and deformation of the vessel induced by the arterial pressure. The frequency of firing is enhanced by an increase in blood pressure and diminished by a reduction in blood pressure. An increase in impulse frequency, as occurs with a rise in arterial pressure, inhibits the vasoconstrictor regions, resulting in peripheral vasodilation and lowering of blood pressure. Contributing to a lowering of the blood pressure is a bradycardia brought about by stimulation of the vagal regions. The carotid sinus and aortic baroreceptors are not equipotent in their effects on peripheral resistance in response to nonpulsatile alterations in blood pressure. The carotid sinus baroreceptors are more sensitive than those in the aortic arch. Changes in pressure in the carotid sinus evoke greater alterations in systemic arterial pressure than do equivalent changes in aortic arch pressure. However with pulsatile changes in blood pressure the two sets of baroreceptors respond similarly.

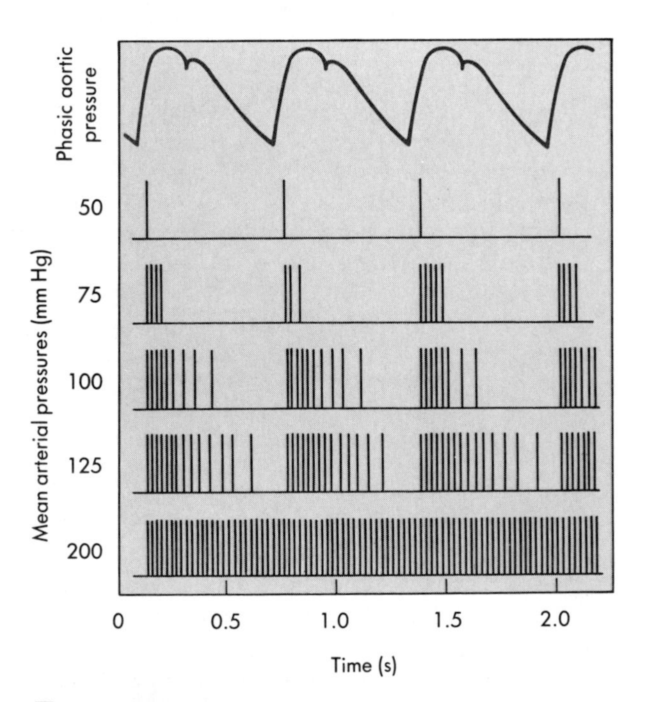

■ **Fig. 29-10** Relationship of phasic aortic blood pressure in the firing of a single afferent nerve fiber from the carotid sinus at different levels of mean arterial pressure.

The carotid sinus with the sinus nerve intact can be isolated from the rest of the circulation and perfused by either a donor animal or an artificial perfusion system. Under these conditions changes in the pressure within the carotid sinus are associated with reciprocal changes in the blood pressure of the experimental animal. The receptors in the walls of the carotid sinus show some adaptation and therefore are more responsive to constantly changing pressures than to sustained constant pressures. This is illustrated in Fig. 29-10, where at normal levels of blood pressure a barrage of impulses from a single fiber of the sinus nerve is initiated in early systole by the pressure rise and only a few spikes are observed during late systole and early diastole. At lower pressures these phasic changes are even more evident, but the overall frequency of discharge is reduced. The blood pressure threshold for eliciting sinus nerve impulses is about 50 mm Hg, and a maximum sustained firing is reached at around 200 mm Hg. Because the pressoreceptors show some degree of adaptation, their response at any level of mean arterial pressure is greater with a large than with a small pulse pressure. This is illustrated in Fig. 29-11, which shows the effects of damping pulsations in the carotid sinus on the frequency of firing in a fiber of the sinus nerve and on the systemic arterial pressure. When the pulse pressure in the carotid sinuses is reduced with an air chamber but mean pressure remains constant, the rate of electrical impulses recorded from a sinus nerve fiber decreases and the systemic arterial pressure increases. Restoration of the pulse pressure in the carotid sinus restores the frequency of sinus nerve discharge and systemic arterial pressure to control levels (see Fig. 29-11).

The resistance increases that occur in the peripheral vascular beds in response to reduced pressure in the carotid sinus vary from one vascular bed to another and thereby produce a redistribution of blood flow. For example in the dog the resistance changes elicited by altering carotid sinus pressure around the normal operating sinus pressure are greatest in the femoral vessels, less in the renal, and least in the mesenteric and celiac vessels. Furthermore the sensitivity of the carotid sinus reflex can be altered. Local application of norepinephrine or stimulation of sympathetic nerve fibers to the carotid sinuses enhances the sensitivity of the receptors in the sinus so that a given increase in intrasinus pressure produces a greater depressor response. A decrease in baroreceptor sensitivity occurs in hypertension when the carotid sinus becomes stiffer and less deformable as a result of the high intraarterial pressure. Under these conditions a given increase in carotid sinus pressure elicits a smaller decrement in systemic arterial pressure than it does at normal levels of blood pressure. In other words the set point of the baroreceptors is raised in hypertension so that the threshold is increased and the receptors are less sensitive to change in transmural pressure.

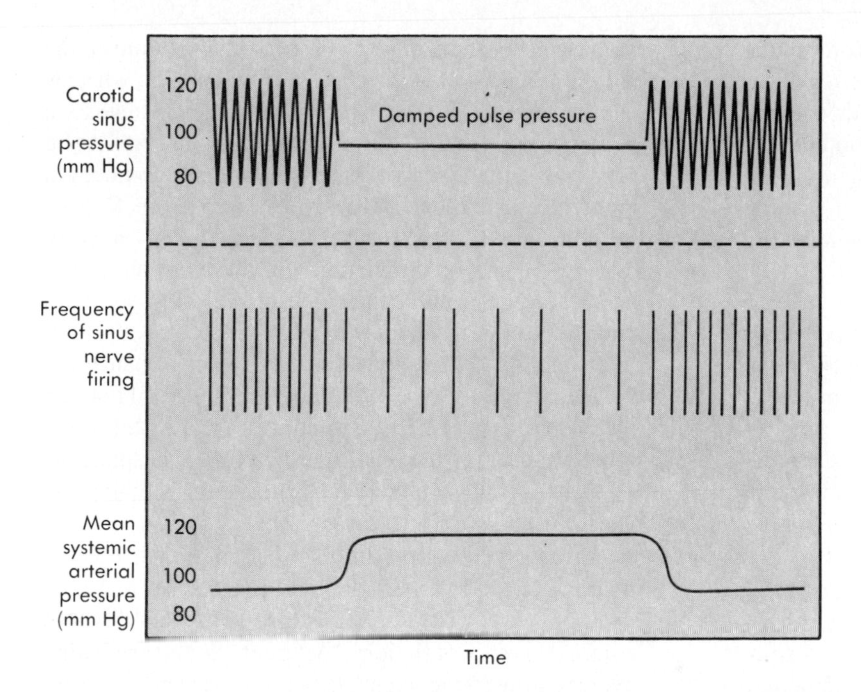

■ **Fig. 29-11** Effect of reducing pulse pressure in the vascularly isolated perfused carotid sinuses *(top record)* on impulses recorded from a fiber of a sinus nerve *(middle record)* and on mean systemic arterial pressure *(bottom record).* Mean pressure in the carotid sinuses *(colored line, top record)* is held constant when pulse pressure is damped.

In some individuals the carotid sinus is quite sensitive to pressure. Hence tight collars or other forms of external pressure over the region of the carotid sinus may elicit marked hypotension and fainting. In some patients with severe coronary artery disease and chest pain **(angina pectoris)** symptoms have been temporarily relieved by stimulation of the sinus nerve by means of a chronically implanted stimulator that can be activated externally. The reduction in blood pressure achieved by sinus nerve stimulation decreases the pressure work of the heart and its oxygen needs. Hence the myocardial ischemia (oxygen deprivation) that is responsible for the pain is alleviated. As would be expected, denervation of the carotid sinus can produce temporary, and in some instances prolonged, hypertension.

The arterial baroreceptors play a key role in short-term adjustments of blood pressure when relatively abrupt changes in blood volume, cardiac output, or peripheral resistance (as in exercise) occur. However long-term control of blood pressure—that is, over days, weeks, and longer—is determined by the fluid balance of the individual, namely, the balance between fluid intake and fluid output. By far the single most important organ in the control of body fluid volume, and hence blood pressure, is the kidney. With overhydration excessive fluid intake is excreted, whereas with dehydration urine output is markedly reduced.

Cardiopulmonary baroreceptors. In addition to the carotid sinus and aortic baroreceptors there are also cardiopulmonary receptors with vagal and sympathetic afferent and efferent nerves. These cardiopulmonary reflexes are tonically active and can alter peripheral resistance with changes in intracardiac, venous, or pulmonary vascular pressures. The receptors are located in the atria, ventricles, and pulmonary vessels.

The atria contain two types of receptors, one that is activated by the tension developed during atrial contraction **(A receptors)** and one that is activated by the stretch of the atria during atrial filling **(B receptors).** Stimulation of these atrial receptors sends impulses up vagal fibers to the vagal center in the medulla. Consequently the sympathetic activity is decreased to the kidney and is increased to the sinus node. These changes in sympathetic activity increase renal blood flow, urine flow, and heart rate.

Activation of the cardiopulmonary receptors can also lower blood pressure reflexly by inhibiting the vasoconstrictor center in the medulla. Stimulation of the receptors inhibits angiotensin, aldosterone, and vasopressin (antidiuretic hormone) release; interruption of the reflex pathway has the opposite effects. Changes in urine volume elicited by changes in cardiopulmonary baroreceptor activation are important in the regulation of blood volume. For example a decrease in blood volume (hypovolemia), as occurs in hemorrhage, enhances sympathetic vasoconstriction in the kidney and increases secretion of renin, angiotensin, aldosterone, and antidiuretic hormone. The renal vasoconstriction (primarily afferent arteriolar) reduces glomerular filtration and increases renin release from the kidney. Renin acts on a plasma substrate to form angiotensin, which increases aldosterone release from the adrenal cortex. The enhanced release of antidiuretic hormone increases water reabsorption. The net results are retention of salt and water by the kidney and a sensation of thirst. Angiotensin II (formed from angiotensin I by converting enzyme) also raises systemic arteriolar tone.

Peripheral chemoreceptors. The chemoreceptors consist of small, highly vascular bodies in the region of the aortic arch and just medial to the carotid sinuses (see Figs. 29-8 and 29-9). They are sensitive to changes in the Po_2, Pco_2, and pH of the blood. Although they are primarily concerned with the regulation of respiration, they reflexly influence the vasomotor regions to a minor degree. A reduction in arterial blood O_2 tension (Pao_2) stimulates the chemoreceptors, and the increase in the number of impulses in the afferent nerve fibers from the carotid and aortic bodies stimulates the vasoconstrictor regions, resulting in increased tone of the resistance and capacitance vessels. The chemoreceptors are also stimulated by increased arterial blood CO_2 tension ($Paco_2$) and reduced pH, but the reflex effect induced is quite small compared to the direct effect of hypercapnia (high $Paco_2$) and hydrogen ions on the vasomotor regions in the medulla. When hypoxia and hypercapnia occur at the same time, the stimulation of the chemoreceptors is greater than the sum of the two stimuli when they act alone. When the chemoreceptors are stimulated simultaneously with a reduction in pressure in the baroreceptors, the chemoreceptors potentiate the vasoconstriction observed in the peripheral vessels. However, when the baroreceptors and chemoreceptors are both stimulated (for example high carotid sinus pressure and low Pao_2), the effects of the baroreceptors predominate.

There are also chemoreceptors with sympathetic afferent fibers in the heart. These cardiac chemoreceptors are activated by ischemia and transmit the precordial pain (angina pectoris) associated with an inadequate blood supply to the myocardium.

Hypothalamus. Optimal function of the cardiovascular reflexes requires the integrity of pontine and hypothalamic structures. Furthermore these structures are responsible for behavioral and emotional control of the cardiovascular system. Stimulation of the anterior hypothalamus produces a fall in blood pressure and bradycardia, whereas stimulation of the posterolateral region of the hypothalamus produces a rise in blood pressure and tachycardia. The hypothalamus also contains a temperature-regulating center that affects the skin vessels. Stimulation by cold applications to the skin or by cooling of the blood perfusing the hypothalamus results in constriction of the skin vessels and heat conservation, whereas warm stimuli result in cutaneous vasodilation and enhanced heat loss (see p. 518).

Cerebrum. The cerebral cortex can also exert a significant effect on blood flow distribution in the body. Stimulation of the motor and premotor areas can affect blood pressure; usually a pressor response is obtained. However vasodilation and depressor responses may be evoked, as in blushing or fainting, in response to an emotional stimulus.

Skin and viscera. Painful stimuli can elicit either pressor or depressor responses, depending on the magnitude and location of the stimulus. Distension of the viscera often evokes a depressor response, whereas painful stimuli on the body surface usually evoke a pressor response. In the anesthetized animal strong electrical stimulation of a sensory nerve produces a strong pressor response. However it is sometimes possible to obtain a depressor response with low-intensity and low-frequency stimulation. Furthermore all vascular beds do not exhibit the same response; in some resistance increases, whereas in others it decreases. In addition muscle contractions can elicit reflex changes in the magnitude of vasoactivity in the muscle. For the most part these reflexes are mediated through the vasomotor areas in the medulla, but there are also spinal areas that can aid in the regulation of peripheral resistance.

Pulmonary reflexes. Inflation of the lungs reflexly induces systemic vasodilation and a decrease in arterial blood pressure. Conversely collapse of the lungs evokes systemic vasoconstriction. Afferent fibers mediating this reflex run in the vagus nerves and possibly to a limited extent in the sympathetic nerves. Their stimulation by stretch of the lungs inhibits the vasomotor areas. The magnitude of the depressor response to lung inflation is directly related to the degree of inflation and to the existing level of vasoconstrictor tone.

Central chemoreceptors. Increases of $Paco_2$ stimulate chemosensitive regions of the medulla and elicit vasoconstriction and increased peripheral resistance. Reduction in $Paco_2$ below normal levels (as with hyperventilation) decreases the degree of tonic activity of these areas in the medulla, thereby decreasing peripheral resistance. The chemosensitive regions are also affected by changes in pH. A lowering of blood pH stimulates and a rise in blood pH inhibits these areas. These effects of changes in $Paco_2$ and blood pH possibly operate through changes in cerebrospinal fluid pH, as appears to be the case for the respiratory center. Whether there are special hydrogen ion chemoreceptors mediating pH-induced vasomotor effects has not been established.

Oxygen tension has relatively little direct effect on the medullary vasomotor region. The primary effect of hypoxia is reflexly mediated via the carotid and aortic chemoreceptors. Moderate reduction of Pao_2 stimulates the vasomotor region, but severe reduction depresses vasomotor activity in the same manner that other areas of the brain are depressed by very low O_2 tensions.

Cerebral ischemia, which may occur because of excessive pressure exerted by an expanding intracranial tumor, results in a marked increase in peripheral vasoconstriction. The stimulation is probably caused by a local accumulation of CO_2 and reduction of O_2 and possibly by excitation of intracranial baroreceptors. With prolonged, severe ischemia central depression eventually supervenes, and the blood pressure falls.

Balance Between Extrinsic and Intrinsic Factors in Regulation of Peripheral Blood Flow

Dual control of the peripheral vessels by intrinsic and extrinsic mechanisms makes possible a number of vascular adjustments that enable the body to direct blood flow to areas where it is needed in greater supply and away from areas whose immediate requirements are less. In some tissues a more or less fixed relative potency of extrinsic and intrinsic mechanisms exists, and in other tissues the ratio is changeable, depending on the state of activity of that tissue.

In the brain and the heart, both vital structures with very limited tolerance for a reduced blood supply, intrinsic flow-regulating mechanisms are dominant. For instance massive discharge of the vasoconstrictor region over the sympathetic nerves, which might occur in severe, acute hemorrhage, has negligible effects on the cerebral and cardiac resistance vessels, whereas skin, renal, and splanchnic blood vessels become greatly constricted.

In the skin the extrinsic vascular control is dominant. Not only do the cutaneous vessels participate strongly in a general vasoconstrictor discharge, but they also respond selectively through hypothalamic pathways to subserve the heat loss and heat conservation function required in body temperature regulation. However intrinsic control can be demonstrated by local changes of temperature that can modify or override the central influence on resistance and capacitance vessels.

In skeletal muscle the interplay and changing balance between extrinsic and intrinsic mechanisms can be clearly seen. In resting skeletal muscle neural control (vasoconstrictor tone) is dominant, as can be demonstrated by the large increment in blood flow that occurs immediately after section of the sympathetic nerves to the tissue. In anticipation of and at the start of exercise, such as running, blood flow increases in the leg muscles. After the onset of exercise the intrinsic flow-regulating mechanism assumes control, and because of the local increase in metabolites, vasodilation occurs in the active muscles. Vasoconstriction occurs in the inactive tissues as a manifestation of the general sympathetic discharge, but constrictor impulses reaching the resistance vessels of the active muscles are overridden by the local metabolic effect. Operation of this dual control mechanism thus provides increased blood where it is re-

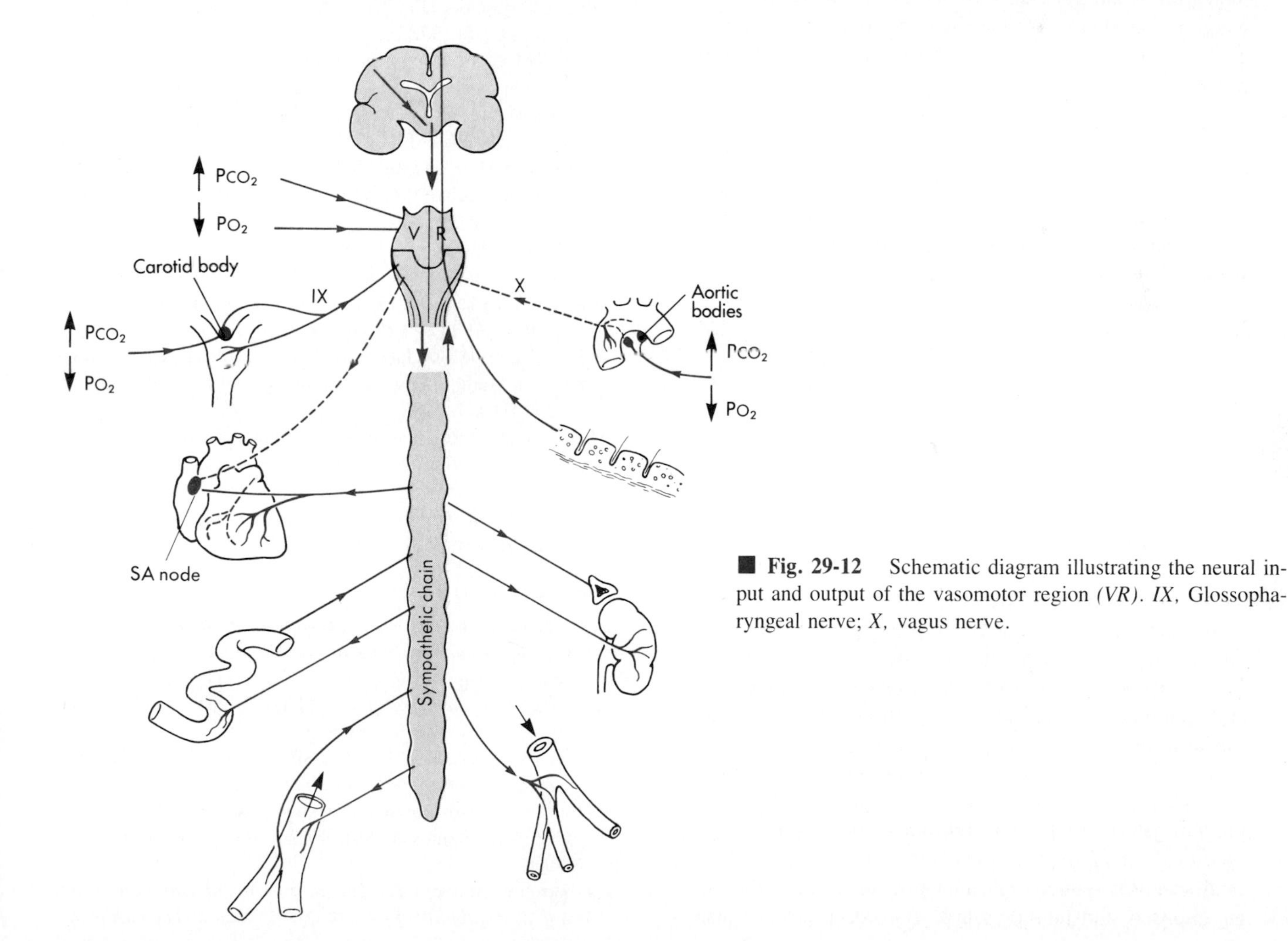

Fig. 29-12 Schematic diagram illustrating the neural input and output of the vasomotor region *(VR)*. *IX*, Glossopharyngeal nerve; *X*, vagus nerve.

quired and shunts it away from relatively inactive areas. Similar effects may be achieved with an increase in $Paco_2$. Normally the hyperventilation associated with exercise keeps $Paco_2$ at normal levels. However, were $Paco_2$ to increase, a generalized vasoconstriction would occur because of stimulation of the medullary vasoconstrictor region by CO_2. In the active muscles, where the CO_2 concentration is highest, the smooth muscle of the arterioles would relax in response to the local Pco_2. Factors affecting and affected by the vasomotor region are summarized in Fig. 29-12.

■ Summary

1. The arterioles, often referred to as the resistance vessels, are important in the regulation of blood flow through their cognate capillaries. The smooth muscle, which comprises a major fraction of the wall of the arterioles, contracts and relaxes in response to neural and humoral stimuli.

2. Most tissues show autoregulation of blood flow, a phenomenon characterized by a constant blood flow in the face of a change in perfusion pressure. A logical explanation of autoregulation is the myogenic mechanism whereby an increase in transmural pressure elicits a contractile response, whereas a decrease in transmural pressure elicits relaxation.

3. The striking parallelism between tissue blood flow and tissue oxygen consumption indicates that *blood flow is largely regulated by a metabolic mechanism*. A decrease in the oxygen supply/oxygen demand ratio of a tissue releases a vasodilator metabolite that dilates arterioles to enhance the oxygen supply.

4. *Neural regulation of blood flow is almost completely accomplished by the sympathetic nervous system*. Sympathetic nerves to blood vessels are tonically active; inhibition of the vasoconstrictor center in the medulla reduces peripheral vascular resistance. Stimulation of the sympathetic nerves constricts resistance and capacitance (veins) vessels. Parasympathetic fibers innervate the head, viscera, and genitalia; they do not innervate skin and muscle.

5. The baroreceptors (pressoreceptors) in the internal carotid arteries and aorta are tonically active and regulate blood pressure on a moment to moment basis. Stretch of these receptors by an increase in arterial pressure reflexly inhibits the vasoconstrictor center in the medulla and induces vasodilation, whereas a decrease in arterial pressure disinhibits the vasoconstrictor center and induces vasoconstriction. The carotid baroreceptors predominate over those in the aorta and respond more vigorously to *changes in pressure* (stretch) than they do to elevated or reduced constant pressures; they adapt to an imposed constant pressure. Baroreceptors are also present in the cardiac chambers and large pulmonary vessels (cardiopulmonary baroreceptors); they have less influence on blood pressure but participate in blood volume regulation.

6. Peripheral chemoreceptors in the carotid bodies and aortic arch and central chemoreceptors in the medulla oblongata are stimulated by a decrease in blood oxygen tension (Pao_2) and an increase in blood carbon dioxide tension ($Paco_2$). Stimulation of these chemoreceptors primarily increases the rate and depth of respiration but also produces peripheral vasoconstriction.

7. Peripheral resistance and hence blood pressure can be affected by stimuli arising in the skin, viscera, lungs, and brain.

8. The combined effect of neural and local metabolic factors is to distribute blood to active tissues and divert it from inactive tissues. In vital structures, such as the heart and brain, and in contracting skeletal muscle the metabolic factors predominate.

■ Bibliography

Journal articles

Belloni FI, Sparks HV: The peripheral circulation, *Annu Rev Physiol* 40:67, 1978.

Berne RM et al: Adenosine in the local regulation of blood flow: a brief overview, *Fed Proc* 42:3136, 1983.

Brown AM: Receptors under pressure—an update on baroreceptors, *Circ Res* 46:1, 1980.

Coleridge HM, Coleridge JCG: Cardiovascular afferents involved in regulation of peripheral vessels, *Annu Rev Physiol* 42:413, 1980.

Donald DE, Shepherd JT: Autonomic regulation of the peripheral circulation, *Annu Rev Physiol* 42:429, 1980.

Hilton SM, Spyer KM: Central nervous regulation of vascular resistance, *Annu Rev Physiol* 42:399, 1980.

Kuo L, Davis JJ, Chilian WM: Endothelium-dependent flow-induced dilation of isolated coronary arterioles, *Am J Physiol* 259:H1063, 1990.

Shen Y-T et al: Relative roles of cardiac receptors and arterial baroreceptors during hemorrhage in conscious dogs, *Circ Res* 66:397, 1990.

Shepherd JT: Reflex control of arterial blood pressure, *Cardiovasc Res* 16:357, 1982.

Books and monographs

Abboud FM, Thames MD: *Interaction of cardiovascular reflexes in circulatory control. In Handbook of physiology.* Section 2: *The cardiovascular system—peripheral circulation and organ blood flow,* vol III, Bethesda, Md, 1983, American Physiological Society.

Bevan JA, Bevan RD, Duckles SP: *Adrenergic regulation of vascular smooth muscle*. In *Handbook of physiology*. Section 2: *The cardiovascular system—vascular smooth muscle,* vol II, Bethesda, Md, 1980, American Physiological Society.

Bishop VS, Malliani A, Thoren P: *Cardiac mechanoreceptors*. In *Handbook of physiology*. Section 2: *The cardiovas-*

cular system, vol III, Bethesda, Md, 1983, American Physiological Society.

Brown AM: *Cardiac reflexes.* In *Handbook of physiology.* Section 2: *The cardiovascular system—the heart,* vol I, Bethesda, Md, 1979, American Physiological Society.

Crass MF, Barnes DC, editors: *Vascular smooth muscle—metabolic, ionic and contractile mechanisms,* New York, 1982, Academic Press.

Eyzaguirre C et al: *Arterial chemoreceptors.* In *Handbook of physiology.* Section 2: *The cardiovascular system—peripheral circulation and organ blood flow,* vol III, Bethesda, Md, 1983, American Physiological Society.

Johnson PC: *The myogenic response.* In *Handbook of physiology.* Section 2: *The cardiovascular system—vascular smooth muscle,* vol II, Bethesda, Md, 1980, American Physiological Society.

Korner PI: *Central nervous control of autonomic cardiovascular function.* In *Handbook of physiology.* Section 2: *The cardiovascular system—the heart,* vol I, Bethesda, Md, 1979, American Physiological Society.

Kovach AGB, Sandos P, and Kollii M, editors: *Cardiovascular physiology: neural control mechanisms,* New York, 1981, Academic Press.

Mancia G, Mark AL: *Arterial baroreflexes in humans.* In *Handbook of physiology.* Section 2: *The cardiovascular system—peripheral circulation and organ blood flow,* vol III, Bethesda, Md, 1983, American Physiological Society.

Mark AL, Mancia G: *Cardiopulmonary baroreflexes in humans.* In *Handbook of physiology.* Section 2: *The cardiovascular system—peripheral circulation and organ blood flow,* vol III, Bethesda, Md, 1983, American Physiological Society.

Mulvany MJ, Strandgaard S, Hammersen F, editors: *Resistance vessels: physiology, pharmacology and hypertensive pathology,* Basel, Switzerland, 1985, S. Karger.

Rothe CF: *Venous system: physiology of the capacitance vessels.* In *Handbook of physiology.* Section 2: *The cardiovascular system—peripheral circulation and organ blood flow,* vol III, Bethesda, Md, 1983, American Physiological Society.

Sagawa K: *Baroreflex control of systemic arterial pressure and vascular bed.* In *Handbook of physiology.* Section 2: *The cardiovascular system—peripheral circulation and organ blood flow,* vol III, Bethesda, Md, 1983, American Physiological Society.

Shepherd JT: *Cardiac mechanoreceptors.* In Fozzard HA et al., editors: *The heart and cardiovascular system, scientific foundations,* ed 2, New York, 1991, Raven Press.

Somlyo AP, Somlyo AV: Smooth muscle structure and function. In Fozzard HA et al, editors: *The heart and cardiovascular system, scientific foundations,* ed 2, New York, 1991, Raven Press.

Sparks HV, Jr: *Effect of local metabolic factors on vascular smooth muscle.* In *Handbook of physiology.* Section 2: *The cardiovascular system—vascular smooth muscle,* vol II, Bethesda, Md, 1980, American Physiological Society.

CHAPTER

30

Control of Cardiac Output: Coupling of Heart and Blood Vessels

Four factors control cardiac output: heart rate, myocardial contractility, preload, and afterload (Fig. 30-1). Heart rate and myocardial contractility are strictly **cardiac factors.** They are characteristics of the cardiac tissues, although they are modulated by various neural and humoral mechanisms. Preload and afterload however depend on the characteristics of both the heart and the vascular system. On the one hand preload and afterload are important **determinants** of cardiac output. On the other hand preload and afterload are themselves *determined by* the cardiac output and certain vascular characteristics. Preload and afterload may be designated **coupling factors,** because they constitute a functional coupling between the heart and blood vessels. The heart pumps the blood around the vascular system. Concomitantly the vessels partly determine the preload and afterload, and hence the vessels regulate the quantity of blood that the heart pumps around the circuit per unit time.

To understand the regulation of cardiac output therefore it is important to appreciate the nature of the coupling between the heart and the vascular system. Guyton and his colleagues have developed graphic techniques that we have modified to analyze the interactions between the cardiac and vascular components of the circulatory system. The graphic analysis involves two simultaneous functional relationships between the **cardiac output** and the **central venous pressure** (that is, the pressure in the right atrium and thoracic venae cavae).

The curve defining one of these relationships will be called the **cardiac function curve.** It is an expression of the well-known Frank-Starling relationship (p. 428) and reflects the dependence of cardiac output on preload (that is, the central venous, or right atrial, pressure). The cardiac function curve is a characteristic of the heart itself and has been studied in hearts completely isolated from the rest of the circulatory system (see Fig. 25-16).

The second curve, which we shall call the **vascular function curve,** defines the dependence of central venous pressure on cardiac output. This relationship depends only on certain vascular system characteristics, namely, the peripheral resistance, the arterial and venous compliances, and the blood volume. The vascular function curve is entirely independent of the characteristics of the heart, and it can be derived even if the heart were replaced by a mechanical pump.

Vascular Function Curve

The vascular function curve defines the changes in central venous pressure evoked by changes in cardiac output; that is, central venous pressure is the **dependent variable** (or **response**), and cardiac output is the **independent variable** (or **stimulus**). This contrasts with the cardiac function curve, for which the central venous pressure (or preload) is the **independent variable** and the cardiac output is the **dependent variable.**

The simplified model of the circulation illustrated in Fig. 30-2 will help explain how the cardiac output determines the level of the central venous pressure. The essential components of the cardiovascular system have been lumped into four elements. The right and left sides of the heart, as well as the pulmonary vascular bed, are considered simply as a **pump,** much like that employed during open heart surgery. The high-resistance microcirculation is designated the **peripheral resistance.** Finally the compliance of the system is subdivided into two components, the **arterial compliance,** C_a, and the **venous compliance,** C_v. As defined on p. 455, compliance (C) is the increment of volume (dV) accommodated per unit change of pressure (dP); that is,

$$C = dV/dP \qquad (1)$$

The venous compliance is about 10 times as great as the arterial compliance. In the example to follow the ratio of C_v to C_a will be set at 19:1 to simplify certain calcu-

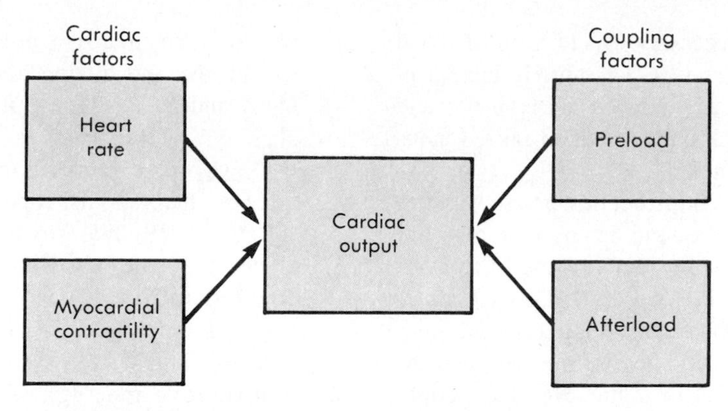

■ **Fig. 30-1** The four factors that determine cardiac output.

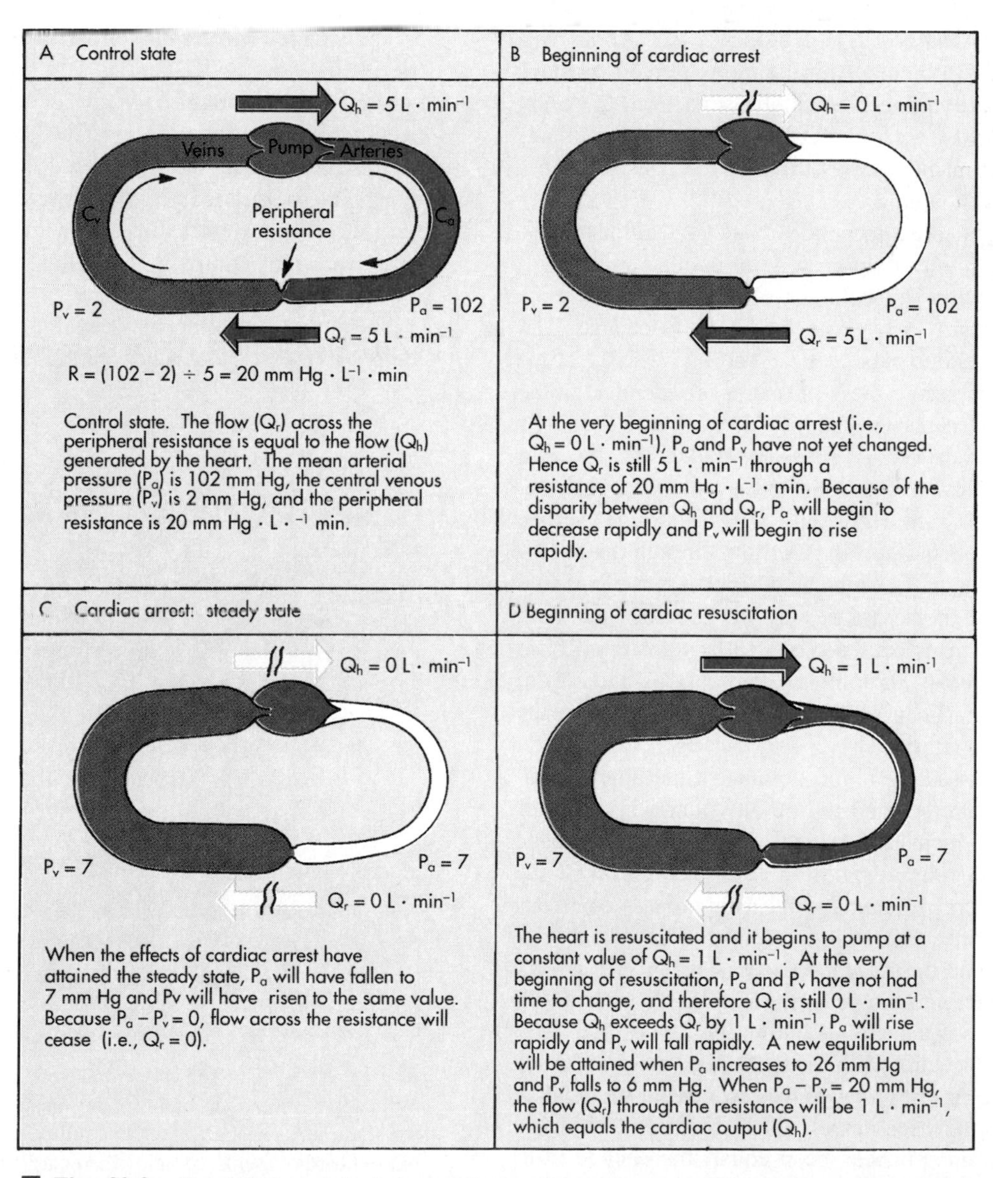

■ **Fig. 30-2** Simplified model of the cardiovascular system, consisting of a pump, arterial compliance (C_a), peripheral resistance, and venous compliance (C_v).

lations. Thus if it were necessary to add x ml of blood to the arterial system to produce a 1 mm Hg increment in arterial pressure, then it would be necessary to add 19x ml of blood to the venous system to raise venous pressure by the same amount.

To illustrate why the central venous pressure varies inversely with the cardiac output let us first give our model the characteristics that resemble those of an average adult person (Fig. 30-2, *A*). Let the flow (Q_h) generated by the heart (i.e., the cardiac output) be 5 L/min, the mean arterial pressure, P_a, be 102 mm Hg, and the central venous pressure, P_v, be 2 mm Hg. The peripheral resistance, R, is the ratio of pressure difference ($P_a - P_v$) to flow (Q_r) through the resistance vessels; this ratio equals 20 mm Hg/L/min. An arteriovenous pressure difference of 100 mm Hg is sufficient to force a flow (Q_r) of 5 L/min through a peripheral resistance of 20 mm Hg/L/min; this flow (Q_r) is precisely equal to the flow (Q_h) generated by the heart. From heartbeat to heartbeat the volume (V_a) of blood in the arteries and the volume (V_v) of blood in the veins remain constant because the volume of blood transferred from the veins to the arteries by the heart equals the volume of blood that flows from the arteries through the resistance vessels and into the veins.

Fig. 30-2, *B*, illustrates the status of the circulation at the very beginning of an episode of cardiac arrest; i.e., $Q_h = 0$. Initially the volumes of blood in the arteries (V_a) and veins (V_v) have not had time to change. The arterial and venous pressures depend on V_a and V_v, respectively. Therefore these pressures are identical to the respective pressures in panel *A* (i.e., $P_a = 102$ and $P_v = 2$). The arteriovenous pressure gradient of 100 mm Hg forces a flow of 5 L/min through the peripheral resistance of 20 mm Hg/L/min. Thus although cardiac output now equals 0 L/min, the flow through the microcirculation equals 5 L/min. In other words the potential energy stored in the arteries by the previous pumping action of the heart causes blood to be transferred from arteries to veins, initially at the control rate, even though the heart can no longer transfer blood from the veins into the arteries.

As time passes the blood volume in the arteries progressively decreases and the blood volume in the veins progressively increases. Because the vessels are elastic structures the arterial pressure falls gradually and the venous pressure rises gradually. This process continues until the arterial and venous pressures become equal (Fig. 30-2, *C*). Once this condition is reached, the flow (Q_r) from the arteries to the veins through the resistance vessels is zero, as is the cardiac output (Q_h).

At zero flow equilibrium (Fig. 30-2, *C*) the pressure attained in the arteries and veins depends on the relative compliances of these vessels. Had the arterial (C_a) and venous (C_v) compliances been equal, the decline in P_a would have been equal to the rise in P_v because the decrement in arterial volume equals the increment in venous volume (principle of conservation of mass). P_a and P_v would have both attained the average of P_a and P_v in panels *A* and *B;* i.e., $P_a = P_v = (102 + 2)/2 = 52$ mm Hg.

However the veins are much more compliant than the arteries; the ratio is approximately equal to the ratio ($C_v{:}C_a = 19$) that we have assumed for the model. Hence the transfer of blood from arteries to veins at equilibrium would induce a fall in arterial pressure 19 times as great as the concomitant rise in venous pressure. As Fig. 30-2, *C*, shows, P_v would increase by 5 mm Hg (to 7 mm Hg), whereas P_a would fall by $19 \times 5 = 95$ mm Hg (to 7 mm Hg). This equilibrium pressure that prevails in the circulatory system in the absence of flow is often referred to as the **mean circulatory pressure,** or the **static pressure.** The pressure in the static system reflects the total volume of blood in the system and the overall compliance ($C_a + C_v$) of the system.

The example of cardiac arrest in Fig. 30-2 provides the basis for understanding the vascular function curves. Two important points on the curve have already been derived, as shown in Fig. 30-3. One point *(A)* represents the normal status (depicted in Fig. 30-2, *A*). At that point, when cardiac output was 5 L/min, P_v was 2 mm Hg. Then, when flow stopped (cardiac output = 0), P_v became 7 mm Hg at equilibrium; this pressure is the mean circulatory pressure, P_{mc}.

The inverse relation between P_v and cardiac output simply expresses that when cardiac output is suddenly

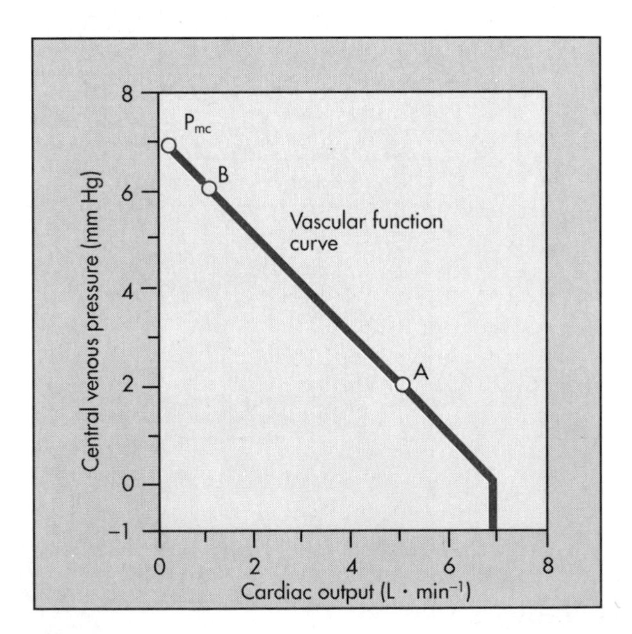

■ **Fig. 30-3** Changes in central venous pressure produced by changes in cardiac output. The mean circulatory pressure (or static pressure), P_{mc}, is the equilibrium pressure throughout the cardiovascular system when cardiac output is 0. Points *B* and *A* represent the values of venous pressure at cardiac outputs of 1 and 5 L/min^{-1}, respectively.

decreased, the rate at which blood flows from arteries to veins through the capillaries is temporarily greater than the rate at which the heart pumps it from the veins back into the arteries. During that transient period a net volume of blood is translocated from arteries to veins; hence P_a falls and P_v rises.

An example of a sudden increase in cardiac output will illustrate how a third point *(B)* on the vascular function curve is derived. Consider that the arrested heart is suddenly restarted, and it immediately begins pumping blood from the veins into the arteries at a rate of 1 L/min (Fig. 30-2, *D*). When the heart first begins to beat, the arteriovenous pressure gradient is zero, and hence no blood is being transferred from the arteries through the capillaries and into the veins. Hence when beating has just resumed, blood is being depleted from the veins at the rate of 1 L/min, and the arterial volume is being repleted at the same rate. Hence P_v begins to fall, and P_a begins to rise. Because of the difference in compliances P_a will rise 19 times more rapidly than P_v will fall.

The resultant pressure gradient will cause blood to flow through the resistance. If the heart maintains a constant output of 1 L/min, P_a will continue to rise and P_v will continue to fall until the pressure gradient becomes 20 mm Hg. This gradient will force a flow of 1 L/min through a resistance of 20 mm Hg/L/min. This gradient will be achieved by a 19 mm Hg rise (to 26 mm Hg) in P_a and a 1 mm Hg fall (to 6 mm Hg) in P_v. This equilibrium value of P_v = 6 mm Hg for a cardiac output of 1 L/min also appears *(B)* on the vascular function curve of Fig. 30-3. It reflects a net transfer of blood from the venous to the arterial side of the circuit and a consequent reduction of P_v.

The reduction of P_v that can be achieved by an increase in cardiac output is limited. At some critical maximum value of cardiac output sufficient fluid will be translocated from the venous to the arterial side of the circuit to reduce P_v below the ambient pressure. In a system of very distensible vessels such as the venous system the vessels will be collapsed by the greater external pressure. This venous collapse constitutes an impediment to venous return to the heart. Hence it limits the maximal value of cardiac output to 7 L/min (see Fig. 30-3), regardless of the capabilities of the pump. For readers interested in the mathematical derivation of these results the basic equations are presented in the next section.

Mathematical Analysis

From the definition of peripheral resistance (p. 445):

$$R = (P_a - P_v)/Q_r, \tag{2}$$

where R is resistance, P_a is arterial pressure, P_v is venous pressure, and Q_r is blood flow through the resistance vessels. At equilibrium, Q_r equals cardiac output, Q_h. Assume that R = 20, and that Q_r had been 0, but that it had then been increased to a constant value of 1 L/min (Fig. 30-4, arrow *1*). If we solve equation 2 for P_a at equilibrium (i.e., $Q_r = Q_h$):

$$P_a = P_v + Q_rR = P_v + (1 \times 20) \tag{3}$$

Thus P_a will increase to a value 20 mm Hg greater than P_v. It will continue to be 20 mm Hg above P_v, as long as the pump output is maintained at 1 L/min and the peripheral resistance remains at 20 mm Hg/L/min.

We can calculate what the actual changes in P_a and P_v will be when Q_h attains a constant value of 1 L/min. The arterial volume increment needed to achieve the required level of P_a depends entirely on the arterial compliance, C_a. For a rigid arterial system (low compliance) this volume will be small; for a distensible system the volume will be large. Whatever the magnitude however the change in volume represents the translocation of some quantity of blood from the venous to the arterial side of the circuit.

For a given total blood volume any increment in arterial volume (ΔV_a) must equal the decrement in venous volume (ΔV_v); that is,

$$\Delta V_a = -\Delta V_v \tag{4}$$

From the definition of compliance

$$C_a = \Delta V_a / \Delta P_a \tag{5}$$

and

$$C_v = \Delta V_v / \Delta P_v \tag{6}$$

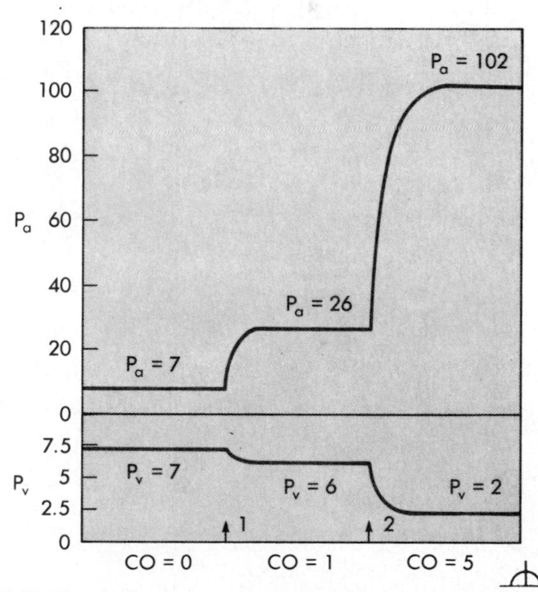

Fig. 30-4 The changes in arterial *(P_a)* and venous *(P_v)* pressures in the circulatory model shown in the preceding figure. The total peripheral resistance is 20 mm Hg/L/min, and the ratio of C_v to C_a is 19:1. The cardiac output *(CO)* is 0 to the left of arrow *1*. It is increased to 1 L/min at arrow *1*, and to 5 L/min at arrow *2*.

By substitution into equation 4

$$\frac{\Delta P_v}{\Delta P_a} = -\frac{C_a}{C_v} \quad (7)$$

Given that C_v is 19 times as great as C_a, then the increment in P_a will be 19 times as great as the decrement in P_v; that is

$$\Delta P_a = -19\Delta P_v \quad (8)$$

To calculate the absolute values of P_a and P_v let ΔP_a represent the difference between the prevailing P_a and the mean circulatory pressure (P_{mc}); that is, let

$$\Delta P_a = P_a - P_{mc} \quad (9)$$

and let ΔP_v represent the difference between the prevailing P_v and the mean circulatory pressure:

$$\Delta P_v = P_v - P_{mc} \quad (10)$$

Substituting these values for ΔP_a and ΔP_v into equation 8

$$P_a - P_{mc} = -19(P_v - P_{mc}) \quad (11)$$

By solving equations 3 and 11 simultaneously:

$$P_a = P_{mc} + 19 \quad (12)$$

and

$$P_v = P_{mc} - 1 \quad (13)$$

Hence for a mean circulatory pressure of 7 mm Hg P_a increases to 26 mm Hg and P_v decreases to 6 mm Hg when Q_h increases from 0 to 1 L/min (see Fig. 30-4). These pressure changes provide the required arteriovenous pressure gradient of 20 mm Hg.

If the pump output is abruptly increased to a constant level of 5 L/min (see Fig. 30-4, arrow *2*) and peripheral resistance remains constant at 20 mm Hg/L/min, an additional volume of blood again will be translocated from the venous to the arterial side of the circuit. It will progressively accumulate in the arteries until P_a reaches a level of 100 mm Hg above P_v, as shown by substitution into equation 3:

$$P_a = P_v + Q_rR = P_v + (5 \times 20) \quad (14)$$

By solving equations 11 and 14 simultaneously we find that P_a rises to a value of 95 mm Hg above P_{mc}, and P_v falls to a value 5 mm Hg below P_{mc}. In Fig. 30-4 therefore P_v declines to 2 mm Hg and P_a rises to 102 mm Hg. The resultant pressure gradient of 100 mm Hg will force a cardiac output of 5 L/min through a constant peripheral resistance of 20 mm Hg/L/min.

Venous Pressure Dependence on Cardiac Output

Experimental and clinical observations have shown that changes in cardiac output do indeed evoke the alterations in P_a and P_v that have been predicted for our simplified model. In an experiment on an anesthesized dog a mechanical pump was substituted for the right ventricle (Fig. 30-5). As the cardiac output, Q, was diminished in a series of small steps P_a fell and P_v rose. Similarly a major coronary artery may suddenly become occluded in a human patient. The resultant **acute myocardial infarction (death of myocardial tissue)** often diminishes cardiac output; this diminution is attended 0by a fall in the arterial pressure and a rise in the central venous pressure.

The changes in P_v evoked by the alterations in blood flow (Q_r) in the experiment illustrated in Fig. 30-5 resemble those derived from our simplified model (see

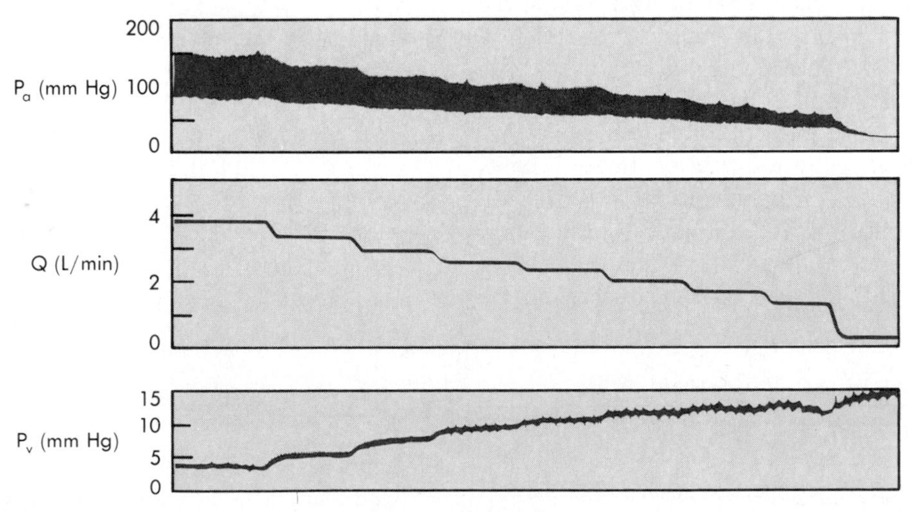

Fig. 30-5 The changes in arterial *(P_a)* and central venous *(P_v)* pressures produced by changes in systemic blood flow *(Q_r)* in a canine right-heart bypass preparation. Stepwise changes in Q_r were produced by altering the rate of a mechanical pump. (From Levy MN: *Circ Res* 44:739, 1979. By permission of the American Heart Association.)

Fig. 30-3). The following equation for P_v as a function of Q_r in the model is derived from equations 2, 7, 9, and 10.

$$P_v = -\frac{RC_a}{C_a + C_v} Q_r + P_{mc} \qquad (15)$$

Note that the slope depends only on R, C_a; and C_v. Note also that when $Q_r = 0$, then $P_v = P_{mc}$; that is, at zero flow P_v equals the mean circulatory pressure.

Blood Volume

The vascular function curve is affected by variations in total blood volume. During circulatory standstill (zero cardiac output) the mean circulatory pressure depends only on total vascular compliance and blood volume, as stated previously. Thus for a given vascular compliance the mean circulatory pressure is be increased when the blood volume is expanded **(hypervolemia)** and decreased when the blood volume is diminished **(hypovolemia).** This is illustrated by the Y-axis intercepts in Fig. 30-6, where the mean circulatory pressure is 5 mm Hg after hemorrhage and 9 mm Hg after transfusion, as compared with the value of 7 mm Hg at the normal blood volume **(normovolemia).**

Furthermore the differences in P_v during hypervolemia, normovolemia, and hypovolemia in the static system are preserved at each level of cardiac output such that the vascular function curves parallel each other (Fig. 30-6). To illustrate consider the example of hypervolemia, in which the mean circulatory pressure is 9 mm Hg. In Fig. 30-6 both P_a and P_v would be 9 mm Hg, instead of 7 mm Hg, when the cardiac output is zero. With a sudden increase in cardiac output to 1 L/min (at arrow *1*, Fig. 30-4) if the peripheral resistance were still 20 mm Hg/L/min, an arteriovenous pressure gradient of 20 mm Hg would still be necessary for 1 L/min to flow through the resistance vessels. This does not differ from the example for normovolemia. Assuming the same ratio of C_v to C_a of 19:1, the pressure gradient would be achieved by a 1 mm Hg decline in P_v and a 19 mm Hg rise in P_a. Hence a change in cardiac output from 0 to 1 L/min would evoke the same 1 mm Hg reduction in P_v irrespective of the blood volume, as long as C_a, C_v, and peripheral resistance were independent of blood volume. Equation 15 also discloses that the slope of the vascular function curve remains constant as long as R, C_v, and C_a do not change.

From Fig. 30-6 it is also apparent that the cardiac output at which $P_v = 0$ varies directly with the blood volume. Therefore the maximal value of cardiac output becomes progressively more limited as the total blood volume is reduced. However the pressure at which the veins collapse (sharp change in slope of the vascular function curve) is not altered appreciably by changes in blood volume. This pressure depends only on the pressure surrounding the central veins.

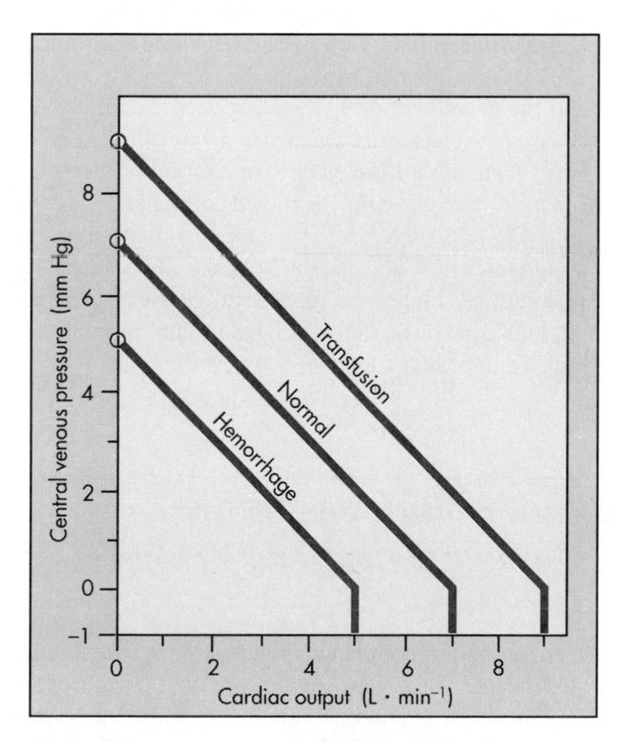

Fig. 30-6 Effects of increased blood volume *(transfusion curve)* and of decreased blood volume *(hemorrhage curve)* on the vascular function curve. Similar shifts in the vascular function curve are produced by increases and decreases, respectively, in venomotor tone.

Venomotor Tone

The effects of changes in venomotor tone on the vascular function curve closely resemble those for changes in blood volume. In Fig. 30-6 for example the transfusion curve could just as well represent increased venomotor tone, whereas the hemorrhage curve could represent decreased tone. During circulatory standstill for a given blood volume the pressure within the vascular system will rise as the tension exerted by the smooth muscle within the vascular walls increases. It is principally the arteriolar and venous smooth muscle that is under notable nervous or humoral control. The fraction of the blood volume located within the arterioles is very small, whereas the blood volume in the veins is large (see Fig. 22-2). Therefore only changes in venous tone can alter the mean circulatory pressure appreciably. Hence mean circulatory pressure rises with increased venomotor tone and falls with diminished tone.

Experimentally the pressure attained shortly after abrupt circulatory standstill is usually above 7 mm Hg, even when blood volume is normal. This is attributable to the generalized venoconstriction elicited by cerebral ischemia, activation of the chemoreceptors, and reduced excitation of the baroreceptors. If resuscitation is not successful, this reflex response subsides as central nervous activity ceases. At normal blood volume the mean circulatory pressure usually approaches a value close to 7 mm Hg.

Blood Reservoirs

Venoconstriction is considerably greater in certain regions of the body than in others. In effect vascular beds that undergo appreciable venoconstriction constitute blood reservoirs. The vascular bed of the skin is one of the major blood reservoirs in people. Blood loss evokes profound subcutaneous venoconstriction, giving rise to the characteristic pale appearance of the skin. The resultant diversion of blood away from the skin liberates several hundred milliliters of blood to be perfused through more vital regions. The vascular beds of the liver, lungs, and spleen are important blood reservoirs. In the dog the spleen is packed with red blood cells and can constrict to a small fraction of its normal size. During hemorrhage this mechanism autotransfuses blood of high erythrocyte content into the general circulation. However in humans the volume changes of the spleen are considerably less extensive.

Peripheral Resistance

The changes in the vascular function curve induced by changes in arteriolar tone are shown in Fig. 30-7. The arterioles contain only about 3% of the total blood volume (see Fig. 22-2). Hence changes in the contractile state of these vessels do not significantly alter the mean circulatory pressure, as stated previously. Thus the family of vascular function curves representing different peripheral resistances converges at a common point on the abscissa.

At any given cardiac output P_v varies inversely with the arteriolar tone, all other factors remaining constant. Arteriolar constriction sufficient to double the peripheral resistance will cause a twofold rise in P_a (p. 459). In the example shown in Fig. 30-4 a change in the cardiac output from 0 to 1 L/min (arrow *1*) caused P_a to rise from 7 to 26 mm Hg, an increment of 19 mm Hg. If peripheral resistance had been twice as great, the same change in cardiac output would have evoked twice as great an increment in P_a.

To achieve this greater rise in P_a, twice as great an increment in blood volume would be required on the arterial side of the circulation, assuming a constant arterial compliance. Given a constant total blood volume this larger arterial volume signifies a corresponding reduction in venous volume. Hence the decrement in venous volume would be twice as great when the peripheral resistance was doubled. If venous compliance remained constant, a twofold reduction in venous volume

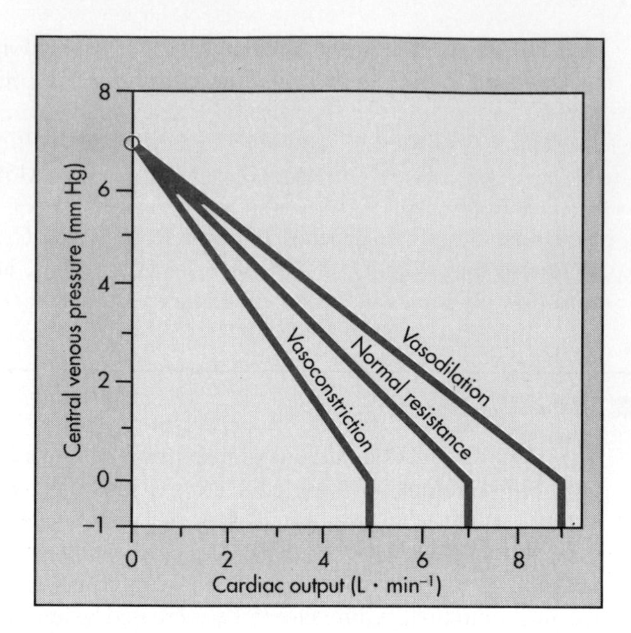

Fig. 30-7 Effects of arteriolar vasodilation and vasoconstriction on the vascular function curve.

would be reflected by a twofold decline in P_v. Therefore in Fig. 30-4 an increase in cardiac output from 0 L/min to 1 L/min (arrow *1*) would have caused a 2 mm Hg decrement in P_v, to a level of 5 mm Hg, instead of the 1 mm Hg decrement that occurred with the normal peripheral resistance. Similarly greater increases in cardiac output would have evoked proportionately greater decrements in P_v under conditions of increased peripheral resistance than with normal levels of resistance.

This relationship between the peripheral resistance and the decrement in P_v, together with the failure of peripheral resistance to affect the mean circulatory pressure, accounts for the clockwise rotation of the vascular function curves with increased peripheral resistance (Fig. 30-7). Conversely arteriolar vasodilation produces a counterclockwise rotation from the same vertical axis intercept. A higher maximal level of cardiac output is attainable with vasodilation than with normal or increased arteriolar tone (see Fig. 30-7).

Interrelationships Between Cardiac Output and Venous Return

Cardiac output and venous return are inextricably interdependent. Clearly except for small, transient disparities the heart is unable to pump any more blood than is delivered to it through the venous system. Similarly because the circulatory system is a closed circuit the venous return must equal the cardiac output over any appreciable interval. The flow around the entire closed circuit depends on the capability of the pump, the characteristics of the circuit, and the total volume of fluid in the system. Cardiac output and venous return are simply two terms for the flow around the closed circuit. Cardiac output is the volume of blood being pumped by the heart per unit time. Venous return is the volume of blood returning to the heart per unit time. At equilibrium these two flows are equal.

The techniques of circuit analysis will be applied in an effort to gain some insight into the control of flow around the circuit. Acute changes in cardiac contractility, peripheral resistance, or blood volume may transiently affect cardiac output and venous return disparately. Except for such brief disparities however such factors simply alter flow around the entire circuit. Whether one thinks of that flow as "cardiac output" or "venous return" is irrelevant. Commonly authors have ascribed the reduction in cardiac output during hemorrhage for example to a decrease in venous return. It will become clear that such an explanation is a blatant example of circular reasoning. Hemorrhage reduces flow around the entire circuit, for reasons to be elucidated. To attribute the reduction in cardiac output to a curtailment of venous return is equivalent to ascribing the decrease in total flow to a decrease in total flow!

Coupling Between the Heart and the Vasculature

In accordance with Starling's law of the heart cardiac output is intimately dependent on the right atrial or central venous pressure. Furthermore the right atrial pressure is approximately equal to the right ventricular end-diastolic pressure, because the normal tricuspid valve constitutes a low-resistance junction between the right atrium and ventricle. In the discussion to follow graphs of cardiac output as a function of central venous pressure (P_v) will be called **cardiac function curves.** Extrinsic regulatory influences may be expressed as shifts in such curves, as indicated previously (p. 428).

A typical cardiac function curve is plotted on the same coordinates as a normal vascular function curve in Fig. 30-8. The cardiac function curve is plotted according to the usual convention; that is, the variable (P_v) plotted along the abscissa is the independent variable (stimulus), and the variable (cardiac output) plotted along the ordinate is the dependent variable (response). In accordance with the Frank-Starling mechanism the cardiac function curve reveals that a rise in P_v causes an increase in cardiac output.

Conversely the vascular function curve describes an inverse relationship between cardiac output and P_v; that is, a rise in cardiac output causes a reduction in P_v. P_v is the dependent variable (or response) and cardiac output is the independent variable (or stimulus) for the vascular function curve. By convention P_v should be scaled along the Y axis and cardiac output should be scaled

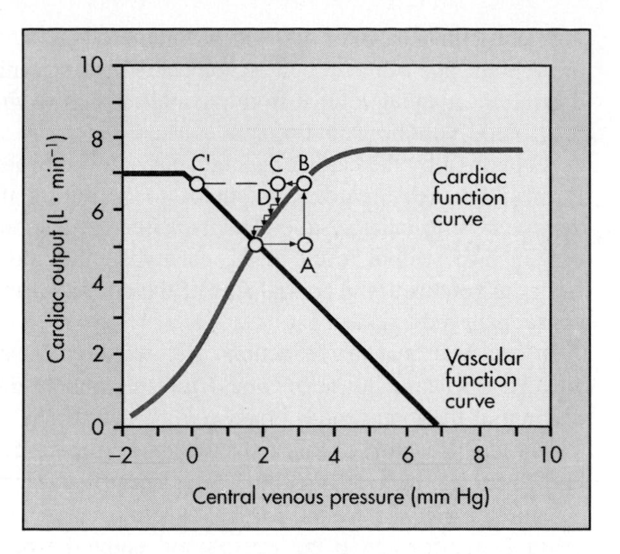

■ **Fig. 30-8** Typical vascular and cardiac function curves plotted on the same coordinate axes. Note that to plot both curves on the same graph it is necessary to reverse the X and Y axes for the vascular function curves shown in Figs. 30-3, 30-6, and 30-7. The coordinates of the equilibrium point, at the intersection of the cardiac and vascular function curves, represent the stable values of cardiac output and central venous pressure at which the system tends to operate. Any perturbation (e.g., when venous pressure is suddenly increased to point *A*) institutes a sequence of changes in cardiac output and venous pressure that restore these variables to their equilibrium values.

along the X axis. Note that this convention is observed for the vascular function curves displayed in Figs. 30-5 to 30-7.

However to include the vascular function curve on the same set of coordinate axes with the cardiac function curve (see Fig. 30-8) it is necessary to violate the plotting convention for one of these curves. We have arbitrarily *violated the convention* for the vascular function curve. Note that the vascular function curve in Fig. 30-8 reflects how P_v (scaled along the X axis) varies in response to a change of cardiac output (scaled along the Y axis).

The **equilibrium point** of a system represented by a given pair of cardiac and vascular function curves is defined by the intersection of these two curves. The coordinates of this equilibrium point represent the values of cardiac output and P_v at which such a system tends to operate. Only transient deviations from such values for cardiac output and P_v are possible, as long as the given cardiac and vascular function curves accurately describe the system.

The tendency of the cardiovascular system to operate about such an equilibrium point may best be illustrated by examining its response to a sudden perturbation. Consider the changes elicited by a sudden rise in P_v from the equilibrium point to point *A* in Fig. 30-8. Such a change might be induced by the rapid injection, during ventricular diastole, of a given volume of blood on the venous side of the circuit, accompanied by the withdrawal of an equal volume from the arterial side so that total blood volume would remain constant.

As defined by the cardiac function curve this elevated P_v would increase cardiac output *(A* to *B)* during the very next ventricular systole. The increased cardiac output, in turn, would result in the net transfer of blood from the venous to the arterial side of the circuit, with a consequent reduction in P_v.

In one heartbeat the reduction in P_v would be small *(B* to *C)* because the heart would transfer only a tiny fraction of the total venous blood volume over to the arterial side. Because of this reduction in P_v the cardiac output during the very next beat diminishes *(C* to *D)* by an amount dictated by the cardiac function curve. Because *D* is still above the intersection point the heart will pump blood from the veins to the arteries at a rate greater than that at which the blood will flow across the peripheral resistance from arteries to veins. Hence P_v will continue to fall. This process will continue in diminishing steps until the point of intersection is reached. Only one specific combination of cardiac output and venous pressure (denoted by the coordinates of the point of intersection) will satisfy simultaneously the requirements of the cardiac and vascular function curves.

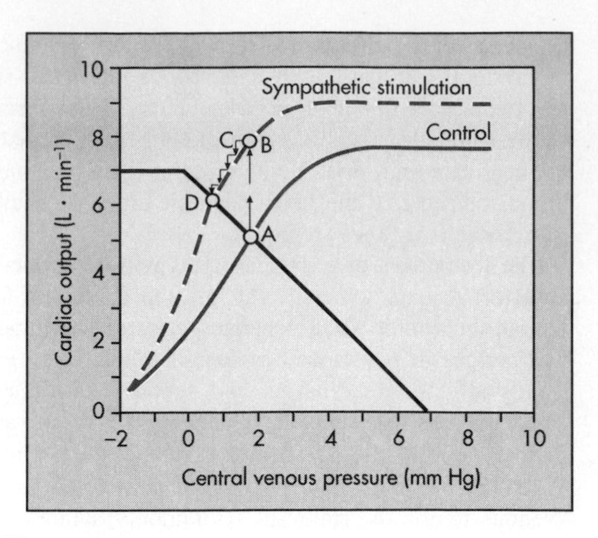

Fig. 30-9 Enhancement of myocardial contractility, as by cardiac sympathetic nerve stimulation, causes the equilibrium values of cardiac output and P_v to shift from the intersection (point *A*) of the control vascular and cardiac function curves *(continuous lines)* to the intersection (point *D*) of the same vascular function curve with the cardiac function curve *(dashed line)* that represents enhanced myocardial contractility.

Myocardial Contractility

Combinations of cardiac and vascular function curves help explain the effects of alterations in ventricular contractility. In Fig. 30-9 the lower cardiac function curve represents the control state, whereas the upper curve reflects improved contractility. This pair of curves is analogous to the family of ventricular function curves described on p. 428. The enhancement of ventricular contractility might be achieved by electrical stimulation of the cardiac sympathetic nerves. If the effects of such stimulation are restricted to the heart, the vascular function curve would be unaffected. Therefore one vascular function curve would suffice, as shown in Fig. 30-9.

During the control state the equilibrium values for cardiac output and P_v are designated by point *A*. Cardiac sympathetic nerve stimulation would abruptly raise cardiac output to point *B* because of the enhanced contractility before P_v would change appreciably. However this high cardiac output would increase the net transfer of blood from the venous to the arterial side of the circuit, and consequently P_v would then begin to fall (point *C*). Cardiac output would continue to fall until a new equilibrium point *(D)*, which is located at the intersection of the vascular function curve with the new cardiac function curve, was reached. The new equilibrium point *(D)* lies above and to the left of the control equilibrium point *(A)*, revealing that sympathetic stimula-

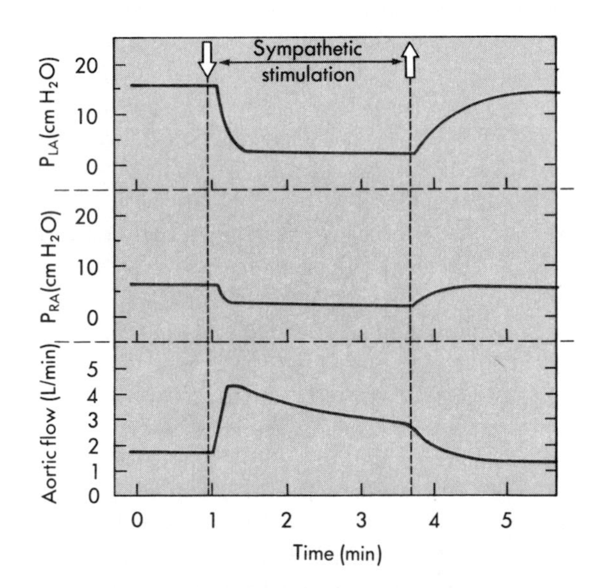

Fig. 30-10 During electrical stimulation of the left stellate ganglion (containing cardiac sympathetic nerve fibers) aortic blood flow increased while pressures in the left atrium *(P_{LA})* and right atrium *(P_{RA})* diminished. These data conform with the conclusions derived from Fig. 30-9, in which the equilibrium values of cardiac output and venous pressure are observed to shift from point *A* to point *D* during cardiac sympathetic nerve stimulation. (Redrawn from Sarnoff SJ et al: *Circ Res* 8:1108, 1960.)

tion evokes a greater cardiac output at a lower level of P_v.

Such a change accurately describes the true response. In the experiment depicted in Fig. 30-10 the left stellate ganglion was stimulated between the two arrows. During stimulation aortic flow (cardiac output) rose quickly to a peak value and then fell gradually to a steady-state value that was significantly greater than the control level. The increased aortic flow was accompanied by reductions in right and left atrial pressures (P_{RA} and P_{LA}).

Blood Volume

Changes in blood volume do not directly affect myocardial contractility, but they do influence the vascular function curve in the manner shown in Fig. 30-6. Therefore to understand the circulatory alterations evoked by a given change in blood volume it is necessary to plot the appropriate cardiac function curve along with the vascular function curves that represent the control and experimental states.

Fig. 30-11 illustrates the response to a blood transfusion. Equilibrium point *B*, which denotes the values for cardiac output and P_v after transfusion, lies above and to the right of the control equilibrium point *A*. Thus transfusion increases both cardiac output and P_v. Hemorrhage has the opposite effect. Pure increases or decreases in venomotor tone elicit responses that are analogous to those evoked by augmentations or reductions, respectively, of the total blood volume, for reasons that were discussed on p. 499.

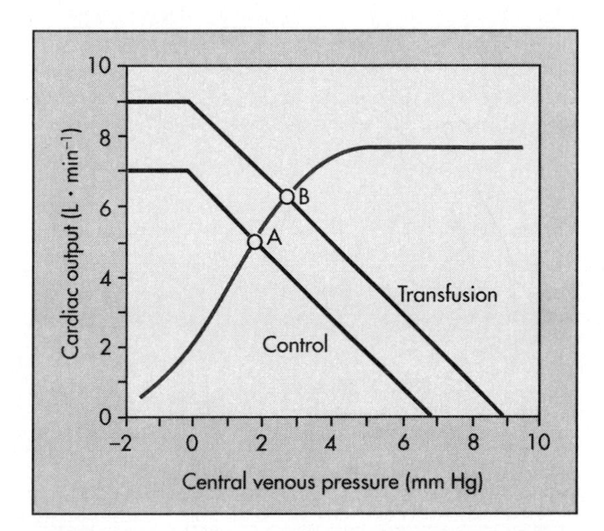

Fig. 30-11 After a blood transfusion the vascular function curve is shifted to the right. Therefore cardiac output and venous pressure are both increased, as denoted by the translocation of the equilibrium point from *A* to *B*.

Heart Failure

Heart failure may be acute or chronic. Acute heart failure may be caused by toxic quantities of drugs and anesthetics or by certain pathological conditions, such as sudden coronary artery occlusion. Chronic heart failure may occur in such conditions as essential hypertension or ischemic heart disease. In these various forms of heart failure myocardial contractility is impaired. Consequently the cardiac function curve is shifted downward and to the right, as depicted in Fig. 30-12.

In acute heart failure blood volume does not change immediately. Therefore the equilibrium point shifts from the intersection of the normal curves (Fig. 30-12, point *A*) to the intersection of the normal vascular function curve with a depressed cardiac function curve (point *B* or *C*).

In chronic heart failure both the cardiac function and the vascular function curves shift. The vascular function curve shifts because of an increase in blood volume caused in part by fluid retention by the kidneys. The fluid retention is related to the concomitant reduction in glomerular filtration rate and to the increased secretion of aldosterone by the adrenal cortex. The resultant hypervolemia is reflected by a rightward shift of the vascular function curve, as shown in Fig. 30-12. Hence with moderate degrees of heart failure P_v is elevated but cardiac output is approximately normal (point *D*). With more severe degrees of heart failure P_v is still higher but cardiac output is subnormal (point *E*).

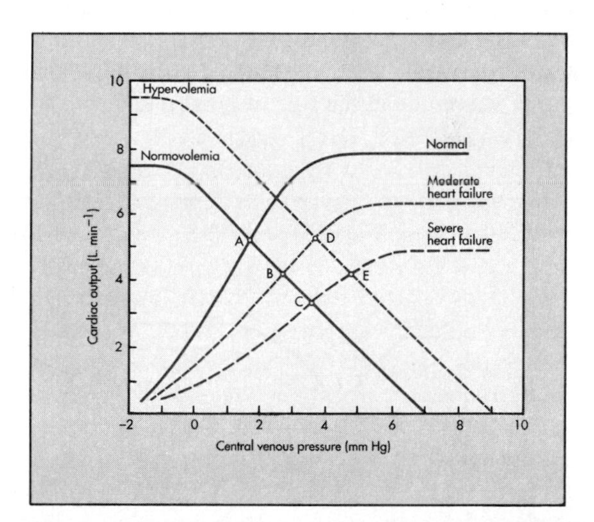

Fig. 30-12 With moderate or severe heart failure the cardiac function curves are shifted to the right. Before any change in blood volume cardiac output decreases and central venous pressure rises (from control equilibrium point *A* to point *B* or point *C*). After the increase in blood volume that usually occurs in heart failure the vascular function curve is shifted to the right. Hence central venous pressure may be elevated with no reduction in cardiac output (point *D*) or (in severe heart failure) with some diminution in cardiac output (point *E*).

Peripheral Resistance

Predictions concerning the effects of changes in peripheral resistance are also complex because both the cardiac and vascular function curves shift. With increased peripheral resistance (Fig. 30-13) the vascular function curve is rotated counterclockwise but it converges to the same P_v-axis intercept as the control curve (see Fig. 30-7); the direction of rotation differs in Figs. 30-7 and 30-13 because the axes were switched for the vascular function curves in the two figures. The cardiac function curve is also shifted downward because at any given P_v the heart is able to pump less blood against a greater afterload. Because both curves are displaced downward the new equilibrium point, *B*, falls below the control point, *A*.

Whether point *B* will fall directly below point *A* or will lie to the right or left of it depends on the magnitude of the shift in each curve. For example if a given increase in peripheral resistance shifts the vascular function curve more than the cardiac function curve, equilibrium point *B* would fall below and to the left of *A;* that is, both cardiac output and P_v would diminish. Conversely if the cardiac function curve is displaced more than the vascular function curve, point *B* will fall below and to the right of point *A;* that is, cardiac output would decrease but P_v would rise.

Role of Heart Rate

Cardiac output is the product of stroke volume and heart rate. The preceding analysis of the control of cardiac output was, in reality, restricted to the control of stroke volume, and the role of heart rate was ignored.

The effect of changes in heart rate on cardiac output will now be considered. The analysis is complex, because a change in heart rate alters the other three factors (preload, afterload, and contractility) that determine stroke volume (see Fig. 30-1). An increase in heart rate for example would decrease the duration of diastole. Hence ventricular filling would be diminished; that is, preload would be reduced. If the proposed increase in heart rate did alter cardiac output, the arterial pressure would change; that is, afterload would be altered. Finally the rise in heart rate would increase the net influx of Ca^{++} per minute into the myocardial cells, and this would enhance myocardial contractility (p. 430).

Heart rate has been varied by artificial pacing in many types of experimental preparations and in humans. The effects on cardiac output have usually resembled the experimental results shown in Fig. 30-14. In that experiment on a dog with third-degree heart block contraction frequency was varied by ventricular pacing. As the ventricular rate was increased from 30 to 60 beats per minute, the cardiac output increased substantially. Presumably at the slowest heart rates within this range the greater filling per cardiac cycle is not adequate to compensate for the decreased number of contractions per minute.

Over the frequency range from 60 to 170 beats per minute however cardiac output did not change very much (Fig. 30-14). Hence as heart rate was increased, the stroke volume was proportionately reduced. The decreased time for filling at the faster rates partly accounts for the observed proportionality between heart rate increase and stroke volume decrease. Also vascular autoregulation tends to hold tissue blood flow constant. This adaptation leads to changes in preload and afterload that maintain cardiac output nearly constant. As the heart

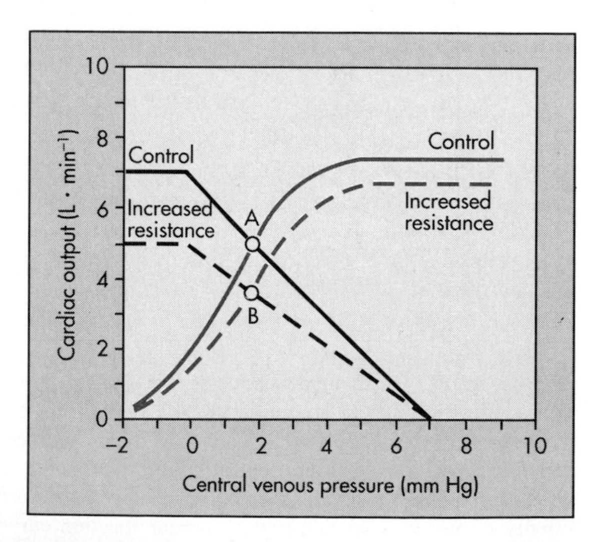

■ **Fig. 30-13** An increase in peripheral resistance shifts the cardiac and the vascular function curves downward. At equilibrium the cardiac output is less *(B)* when the peripheral resistance is high than when it is normal *(A)*.

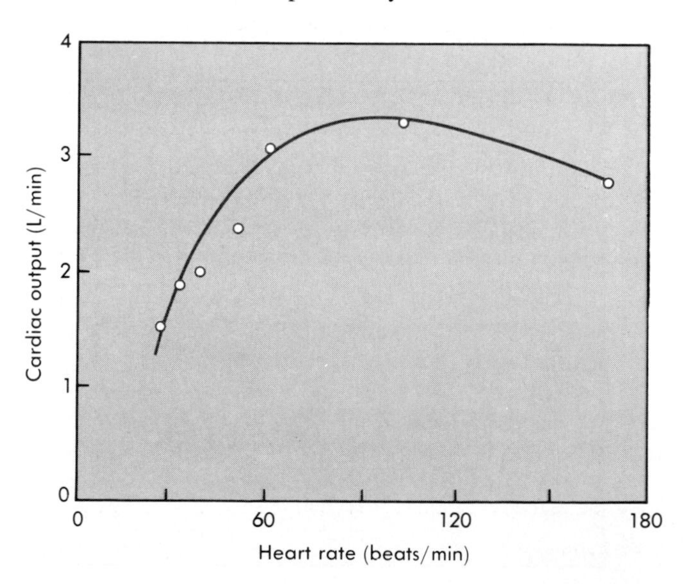

■ **Fig. 30-14** The changes in cardiac output induced by changing the rate of ventricular pacing in a dog with complete heart block. (Redrawn from Miller DE et al: *Circ Res* 10:658, 1962. By permission of the American Heart Association.)

rate is increased to levels in excess of about 150 to 170 beats per minute, the filling time is so severely restricted that compensation is inadequate and cardiac output decreases precipitously (not shown in Fig. 30-14).

During certain activities such as physical exercise the cardiac output increases (see Chapter 32). This change in output is attended predominantly by an increase in heart rate and a slight increase in stroke volume. The temptation is great to conclude that the increase in cardiac output must be caused by the observed increase in heart rate, because of the striking correlation between cardiac output and heart rate. Yet Fig. 30-14 emphasizes that over a wide range of heart rates a change in heart rate has little influence on cardiac output. Several studies on exercising subjects have confirmed that even during exercise changes in pacing frequency do not alter cardiac output very much.

The principal increase in cardiac output during exercise must therefore be ascribed to the pronounced reduction in peripheral vascular resistance. The attendant changes in heart rate are not inconsequential however. If the heart rate cannot increase normally during exercise, the augmentation of cardiac output and the capacity for exercise may be severely limited. The increase in heart rate does play a **permissive role** in augmenting cardiac output, even if it is not proper to assign it a primary, causative role. The mechanisms responsible for raising heart rate in precise proportion to the increase in cardiac output are undoubtedly neural in origin, but the specific components of the reflex arcs remain to be elucidated.

■ *Ancillary Factors that Affect the Venous System and Cardiac Output*

We have oversimplified the interrelationships between central venous pressure and cardiac output. We have described effects evoked by changes in single variables. However because many feedback control loops regulate the cardiovascular system, an isolated change in a single response rarely occurs. A change in blood volume for example reflexly alters cardiac function, peripheral resistance, and venomotor tone. Several auxiliary factors also regulate cardiac output. Such ancillary factors may be considered to modulate the more basic factors that have been considered.

■ *Gravity*

Gravitational forces may affect cardiac output profoundly. It is not unusual for some soldiers standing at attention to faint because of reduced cardiac output. Gravitational effects are exaggerated in airplane pilots during pullouts from dives. The centrifugal force in the footward direction may be several times greater than the force of gravity. Such individuals characteristically black out momentarily during the maneuver, as blood is drained from the cephalic regions and pooled in the lower parts of the body.

The explanation for the reduction in cardiac output under such conditions is often specious. It is argued that when an individual is standing, the forces of gravity impede venous return from the dependent regions of the body. This statement is incomplete however because it ignores the facilitative counterforce on the arterial side of the same circuit.

In this sense the vascular system resembles a U tube. To comprehend the action of gravity on flow through such a system the models depicted in Figs. 30-15 and 30-16 will be analyzed. In Fig. 30-15 all the U tubes represent rigid cylinders of constant diameter. With both limbs of the U tube oriented horizontally *(A)* the flow depends only on the pressures at the inflow and outflow ends of the tube (P_i and P_o, respectively), the viscosity of the fluid, and the length and radius of the tube, in accordance with Poiseuille's equation (p. 443). With a constant cross section the pressure gradient is uniform; hence the pressure midway down the tube (P_m) equals the average of the inflow and outflow pressures.

When the U tube is oriented vertically *(B* to *D)* hydrostatic forces must be taken into consideration. In tube *B* both limbs are open to atmospheric pressure and both ends are located at the same hydrostatic level; hence there is no flow. The pressure at the midpoint of the tube is simply ρhg. It depends on the density of the fluid, ρ; the height of the U tube, h; and the acceleration of gravity, g. In the example the length of the U tube is such that the midpoint pressure is 80 mm Hg.

Now consider tube *C*, where the tube is oriented the same as tube *B*, but where a 100 mm Hg pressure difference is applied across the two ends. The flow precisely equals that in *A*, because the pressure gradient, tube dimensions, and fluid viscosity are all the same. Gravitational forces are precisely equal in magnitude but opposite in direction in the two limbs of the U tube. Because the flow will be the same as that in *A*, there will be a pressure drop of 50 mm Hg at the midpoint because of the viscous losses resulting from flow. Furthermore gravity will tend to increase pressure by 80 mm Hg at the midpoint, just as in tube *B*. The actual pressure at the midpoint of tube *C* will be the resultant of the viscous loss and hydrostatic gain, or 130 mm Hg in this example.

In *D* a pressure gradient of 100 mm Hg is applied to the same U tube, but the tube is oriented in the opposite direction. Gravitational forces will be so directed that the pressure at the midpoint will tend to be 80 mm Hg less than that at the end of the U tube. Viscous losses will still produce a 50 mm Hg pressure drop at the mid-

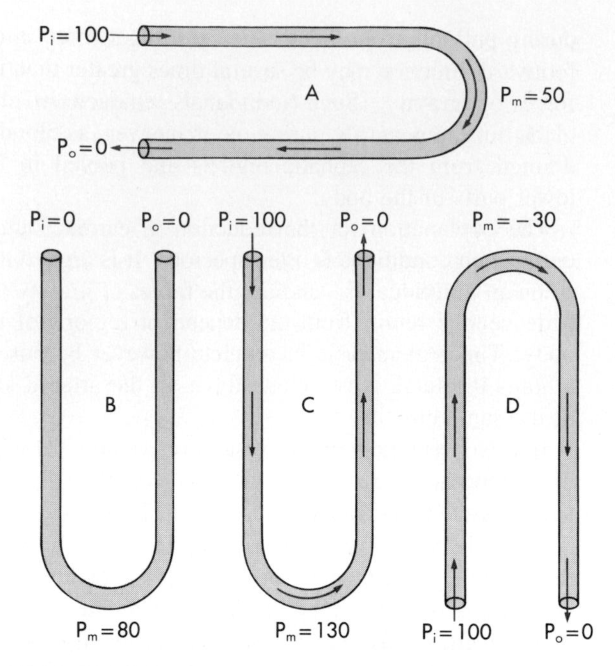

■ **Fig. 30-15** Pressure distributions in *rigid* U tubes with constant internal diameters, all with the same dimensions. For a given inflow pressure (P_i = 100) and outflow pressure (P_o = 0) the pressure at the midpoint (P_m) depends on the orientation of the U tube, but the flow through the tube is independent of the orientation.

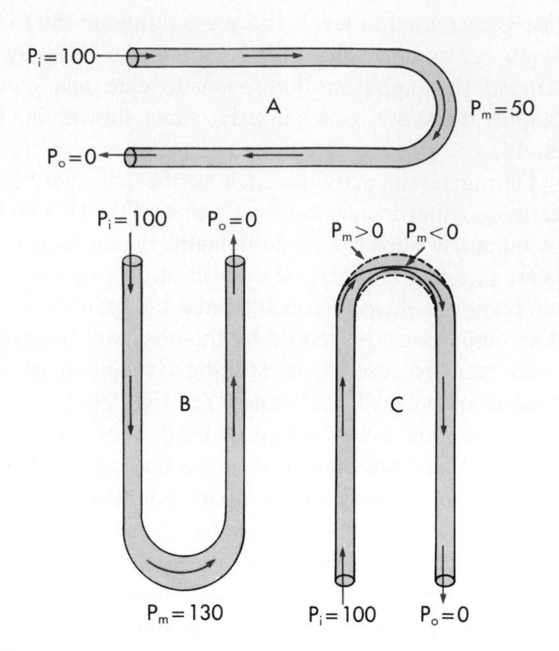

■ **Fig. 30-16** In U tubes with a *distensible section* at the bend, even when inflow (P_i) and outflow (P_o) pressures are the same, the resistance to flow and the fluid volume contained within each tube vary with the orientation of the tube. P_m, pressure at the midpoint of the tube.

point relative to P_i. Hence with this orientation pressure at the midpoint of the U tube will be 30 mm Hg below ambient pressure. Flow will of course be the same as in tubes *A* and *C*, for the reasons stated in relation to *C*.

In a system of rigid U tubes gravitational effects will not alter the rate of fluid flow. However experience shows that gravity does affect the cardiovascular system. The reason is that the vessels are distensible rather than rigid. To explain the gravitational effects the pressures in a set of U tubes with distensible components (at the bends in the tubes of Fig. 30-16) will be examined. In tubes *A* and *B* the pressure distributions will resemble those in tubes *A* and *C*, respectively, of Fig. 30-15. Because the pressure is higher at the bend of tube *B* than at the bend of tube *A* in Fig. 30-16 and because the segments are distensible in this region, the distension at bend *B* will exceed that at bend *A*. The extent of the distension will depend on the compliance of these tube segments. Because flow varies directly with the tube diameter, the flow through *B* will exceed the flow through *A* for a given pressure difference applied at the ends.

Because orienting a U tube with its bend downward actually increases rather than diminishes flow, how then is the observed impairment of cardiovascular function explained when the body is similarly oriented? The explanation is that the cardiovascular system is a closed circuit of constant fluid (blood) volume, whereas the U tube is an open conduit supplied by a fluid source of unlimited volume. In the dependent regions of the cardiovascular system the distension will occur more on the venous than on the arterial side of the circuit, because the venous compliance is so much greater than the arterial compliance. Such venous distension is readily observed on the back of the hands when the arms are allowed to hang down. The hemodynamic effects of such venous distension **(venous pooling)** resemble those caused by the hemorrhage of an equivalent volume of blood from the body. When an adult person shifts from a supine position to a relaxed standing position, from 300 to 800 ml of blood is pooled in the legs. This may reduce cardiac output by about 2 L/min.

The compensatory adjustments to the erect position are similar to the adjustments to blood loss. For example the diminished baroreceptor excitation reflexly speeds the heart, strengthens the cardiac contraction, and constricts the arterioles and veins. The baroreceptor reflex has a greater effect on the resistance than on the capacitance vessels.

Warm ambient temperatures interfere with the compensatory vasomotor reactions, and the absence of muscular activity exaggerates the effects. Many of the drugs used to treat hypertension also interfere with the reflex adaptation to standing. Similarly astronauts exposed to weightlessness lose their adaptations after a few days in space, and they experience difficulties when they first return to earth. When individuals with impaired reflex adaptations stand, their blood pressures may drop substantially. This response is called **orthostatic hypotension,** which may cause lightheadedness or fainting.

When the U tube is rotated so that the bend is directed upward (Fig. 30-16, tube *C*), the effects are opposite to those that take place in tube *B*. The pressure at the bend of tube *C* would tend to be −30 mm Hg, just as in tube *D* of Fig. 30-15. Because the ambient pressure exceeds the internal pressure however the distensible segment of tube *C* will collapse. Flow will then cease, and therefore the decline of pressure associated with viscous flow will cease. In U tube *C* when flow stops the pressure at the top of each limb will be 80 mm Hg less than at the bottom (the hydrostatic pressure difference). Hence in the left, or inflow, limb the pressure will approach 20 mm Hg. As soon as this pressure exceeds ambient pressure (0 mm Hg), the collapsed tubing will be forced open and flow will begin. With the onset of flow however pressure at the bend will again drop below the ambient pressure. Thus the tubing at the bend will flutter; that is, it will fluctuate between the open and closed states.

When an arm is raised, the cutaneous veins in the hand and forearm collapse, for the reasons described previously. Fluttering does not occur here because the deeper veins are protected from collapse by being tethered to surrounding structures. This protection allows these deeper veins to accommodate the flow ordinarily carried by the collapsed superficial veins. The analogy would be to add a rigid tube (representing the deeper veins) in parallel with the collapsible tube (representing the superficial veins) at the bend of tube *C* in Fig. 30-16. The collapsible tube would no longer flutter but would remain closed. All flow would occur through the rigid tube, just as in tube *D* in Fig. 30-15.

The superficial veins in the neck are ordinarily partially collapsed when a normal individual is upright. Venous return from the head is conducted largely through the deeper cervical veins. However when **central venous pressure** is abnormally elevated, the superficial neck veins are distended and they do not collapse even when the subject sits or stands. Such cervical venous distension is an important clinical sign of congestive heart failure.

Muscular Activity and Venous Valves

When a recumbent person stands but remains at rest, the pressure rises in the veins in the dependent regions of the body. The venous pressure in the legs increases gradually and does not reach an equilibrium value until almost 1 minute after standing. The slowness of this rise in P_v is attributable to the venous valves, which permit flow only toward the heart. When the person stands, the valves prevent blood in the veins from actually falling toward the feet. Hence the column of venous blood is supported at numerous levels by these valves; temporarily the venous column consists of many discontinuous segments. However blood continues to enter the column from many venules and small tributary veins, and the pressure continues to rise. As soon as the pressure in one segment exceeds that in the segment just above it, the intervening valve is forced open. Ultimately all the valves are open, and the column is continuous, similar to the state in the outflow limbs of the U tubes shown in Figs. 30-15 and 30-16.

Precise measurement reveals that the final level of P_v in the feet during quiet standing is only slightly greater than that in a static column of blood extending from the right atrium to the feet. This indicates that the pressure drop caused by blood flow from the foot veins to the right atrium is very small. This very low resistance justifies lumping all the veins as a common venous compliance in the circulatory system model illustrated in Fig. 30-2.

When an individual who has been standing quietly begins to walk, the venous pressure in the legs decreases appreciably. Because of the intermittent venous compression exerted by the contracting leg muscles and because of the presence of the venous valves blood is forced from the veins toward the heart. Hence muscular contraction lowers the mean venous pressure in the legs and serves as an **auxiliary pump.** Furthermore it prevents venous pooling and lowers capillary hydrostatic pressure. Thereby it reduces the tendency for edema fluid to collect in the feet during standing.

Respiratory Activity

The normal, periodic activity of the respiratory muscles causes rhythmic variations in vena caval flow. Thus respiration constitutes an auxiliary pump to promote venous return. Coughing, straining at stool, and other activities that require respiratory muscle exertion may affect cardiac output substantially.

The changes in blood flow in the superior vena cava during the normal respiratory cycle are shown in Fig. 30-17. During respiration the reduction in intrathoracic pressure is transmitted to the lumina of the thoracic blood vessels. The reduction in central venous pressure during inspiration increases the pressure gradient between extrathoracic and intrathoracic veins. The consequent acceleration of venous return to the right atrium is displayed in Fig. 30-17 as an increase in superior vena caval blood flow from 5.2 ml/sec during expiration to 11 ml/sec during inspiration.

An exaggerated reduction in intrathoracic pressure achieved by a strong inspiratory effort against a closed glottis (called **Müller's maneuver**) does not increase venous return proportionately. The extrathoracic veins collapse near their entry into the chest when their internal pressures fall below the ambient level. As the veins

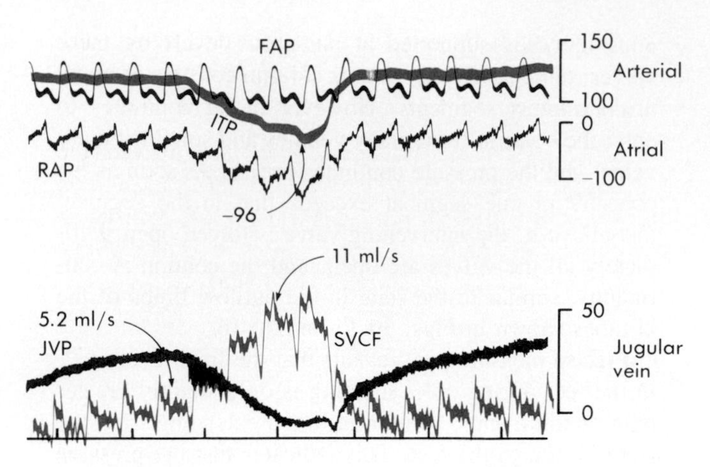

■ **Fig. 30-17** During a normal inspiration intrathoracic *(ITP)*, right atrial *(RAP)*, and jugular venous *(JVP)* pressures decrease, and flow in the superior vena cava (SVCF) increases (from 5.2 to 11 ml/sec). All pressures are in millimeters of water, except for femoral arterial pressure *(FAP)*, which is in millimeters of mercury. (Modified from Brecher GA: *Venous return*, New York, 1956, Grune & Stratton.)

collapse flow into the chest momentarily stops. The cessation of flow raises pressure upstream, forcing the collapsed segment to open again. The process is repetitive; the venous segments adjacent to the chest alternately open and close.

During normal expiration flow into the central veins decelerates. However the mean rate of venous return during normal respiration exceeds the flow during a brief period of **apnea** (cessation of respiration). Hence normal inspiration apparently facilitates venous return more than normal expiration impedes it. In part this must be attributable to the valves in the veins of the extremities and neck. These valves prevent any reversal of flow during expiration. Thus the respiratory muscles and venous valves constitute an **auxiliary pump** for venous return.

Sustained expiratory efforts increase intrathoracic pressure and thereby impede venous return. Straining against a closed glottis (termed **Valsalva's maneuver**) regularly occurs during coughing, defecation, and heavy lifting. Intrathoracic pressures in excess of 100 mm Hg have been recorded in trumpet players and pressures over 400 mm Hg have been observed during paroxysms of coughing. Such pressure increases are transmitted directly to the lumina of the intrathoracic arteries. After cessation of coughing the arterial blood pressure may fall precipitously because of the preceding impediment to venous return.

The dramatic increase in intrathoracic pressure induced by coughing constitutes an **auxiliary pumping mechanism** for the blood, despite its concurrent tendency to impede venous return. During certain diagnostic procedures, such as coronary angiography or electrophysiological testing, patients are at increased risk for ventricular fibrillation. Such patients have been trained to cough rhythmically on command. If ventricular fibrillation does occur, substantial arterial blood pressure increments are generated by each cough, and enough cerebral blood flow may be promoted to sustain consciousness. The cough raises the intravascular pressure equally in intrathoracic arteries and veins. Blood is propelled through the extrathoracic tissues however because the increased pressure is transmitted to the extrathoracic arteries but not to the extrathoracic veins, because of the valves in the extrathoracic veins.

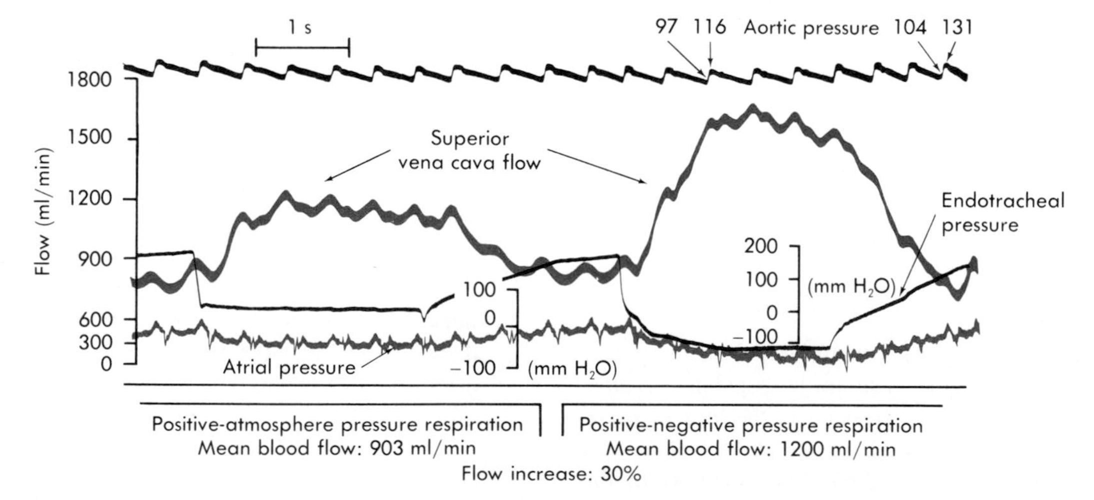

■ **Fig. 30-18** During intermittent positive-pressure respiration the flow in the superior vena cava is approximately 30% greater when the lungs are deflated actively by applying negative endotracheal pressure *(right side)* than when they are allowed to deflate passively against atmospheric pressure *(left side)*. (Modified from Brecher GA: *Venous return*, New York, 1956, Grune & Stratton.)

Artificial Respiration

In most forms of artificial respiration (mouth-to-mouth resuscitation, mechanical respiration) lung inflation is achieved by applying endotracheal pressures above atmospheric pressure, and expiration occurs by passive recoil of the thoracic cage. Thus lung inflation is attended by an appreciable rise in intrathoracic pressure. Vena cava flow decreases sharply during the phase of positive-pressure lung inflation (indicated by the progressive rise in endotracheal pressure in the central portion of Fig. 30-18). When negative endotracheal pressure (indicated by the abrupt decrease in endotracheal pressure in the right half of Fig. 30-18) is used to facilitate deflation, vena cava flow accelerates more than when the lungs are allowed to deflate passively (near the left border of Fig. 30-18).

Summary

1. Two important relationships between cardiac output (CO) and central venous pressure (P_v) prevail in the cardiovascular system. One applies to the heart and the other to the vascular system.

2. With respect to the heart CO varies **directly** with P_v (or preload) over a very wide range of P_v. This relationship is represented by the **cardiac function curve,** and it expresses the Frank-Starling mechanism.

3. With respect to the vascular system P_v varies **inversely** with CO. This relationship is represented by the **vascular function curve,** and it reflects the fact that as CO increases for example a greater fraction of the total blood volume resides in the arteries and a smaller volume resides in the veins.

4. The principal mechanisms that govern the cardiac function curve are the changes in numbers of crossbridges that interact and in the affinity of the contractile proteins for calcium. These mechanisms are evoked by changes in the cardiac filling pressure (preload).

5. The principal factors that govern the vascular function curve are the arterial and venous compliances, the peripheral vascular resistance, and the total blood volume.

6. The equilibrium values of CO and P_v that prevail under a given set of conditions are determined by the **intersection** of the cardiac and vascular function curves.

7. At very low and very high **heart rates** the heart is unable to pump adequate CO. At the very low rates the increment in filling during diastole cannot compensate for the small number of cardiac contractions per minute. At the very high rates the large number of contractions per minute cannot compensate for the inadequate filling time.

8. Gravity influences CO because the veins are so compliant, and substantial quantities of blood tend to pool in the veins of the dependent portions of the body.

9. Respiration changes the pressure gradient between the intrathoracic and extrathoracic veins. Hence respiration serves as an **auxiliary pump,** which may affect the mean level of CO and the transitory changes in stroke volume during the various phases of the respiratory cycle.

Bibliography

Journal articles

Bedford TG, Dormer KJ: Arterial hemodynamics during head-up tilt in conscious dogs. *J Appl Physiol* 65:1556, 1988.

Bromberger-Barnea B: Mechanical effects of inspiration on heart functions: a review. *Fed Proc* 40:2172, 1981.

Freeman GL, Little WC, O'Rourke RA: Influence of heart rate on left ventricular performance in conscious dogs. *Circ Res* 61:455, 1987.

Greenway CV: Mechanisms and quantitative assessment of drug effects on cardiac output with a new model of the circulation, *Pharmacol Rev* 33:213, 1982.

Johnston WE et al: Mechanism of reduced cardiac output during positive end-expiratory pressure in the dog, *Am Rev Respir Dis* 140:1257, 1989.

Levy MN: The cardiac and vascular factors that determine systemic blood flow, *Circ Res* 44:739, 1979.

Miller RR et al: Differential systemic arterial and venous actions and consequent cardiac effects of vasodilator drugs, *Prog Cardiovasc Dis* 24:353, 1982.

Narahara KA, Blettel ML: Effect of rate on left ventricular volumes and ejection fraction during chronic ventricular pacing, *Circulation* 67:323, 1983.

Peters J et al: Negative intrathoracic pressure decreases independently left ventricular filling and emptying, *Am J Physiol* 257:H120, 1989.

Rothe CF: Physiology of venous return: an unappreciated boost to the heart, *Arch Intern Med* 146:977, 1986.

Rothe CF, Gaddis ML: Autoregulation of cardiac output by passive elastic characteristics of the vascular capacitance system, *Circulation* 81:360, 1990.

Shoukas AA, Bohlen HG: Rat venular pressure-diameter relationships are regulated by sympathetic activity, *Am J Physiol* 259:H674, 1990.

Books and monographs

Green JF: *Determinants of systemic blood flow.* In *International review of physiology.* III: *Cardiovascular physiology,* vol 18, Baltimore, 1979, University Park Press.

Guyton AC, Jones CE, Coleman TG: *Circulatory physiology: cardiac output and its regulation,* ed 2, Philadelphia, 1973, WB Saunders.

Rothe CF: *Venous system: physiology of the capacitance vessels.* In *Handbook of physiology.* Section 2. *The cardiovascular system—peripheral circulation and organ blood flow,* vol III, Bethesda, Md, 1983, American Physiological Society.

Smith JJ, editor: Circulatory response to the upright posture, Boca Raton, 1990, CRC Press.

Yin FCP, editor: *Ventricular/vascular coupling,* New York, 1987, Springer-Verlag.

CHAPTER 31

Special Circulations

■ *Coronary Circulation*

■ *Functional Anatomy of Coronary Vessels*

The right and left coronary arteries, which arise at the root of the aorta behind the right and left cusps of the aortic valve, respectively, provide the entire blood supply to the myocardium. The right coronary artery supplies principally the right ventricle and atrium; the left coronary artery, which divides near its origin into the anterior descendens and the circumflex branches, supplies principally the left ventricle and atrium, but there is some overlap. In dogs the left coronary artery supplies about 85% of the myocardium, whereas in humans the right coronary artery is dominant in 50% of individuals, the left coronary artery is dominant in another 20%, and the flow delivered by each main artery is about equal in the remaining 30%.

Coronary blood flow is most commonly measured in humans by thermodilution. Thermodilution is the same procedure used for the measurement of cardiac output (p. 414), except that the indicator (cold saline) is ejected at the tip of a catheter inserted into the coronary sinus via a peripheral vein. The thermal sensor (thermistor) is located on the catheter a few centimeters from the catheter tip. The greater the coronary sinus outflow, the less the temperature decrease produced by the cold saline injection. This method does not measure total coronary blood flow because only about two thirds of coronary arterial inflow returns to the venous circulation through the coronary sinus. However, almost all of the blood flow that does empty into the coronary sinus comes from the left ventricle. Hence the thermodilution method provides a good estimate of left ventricular coronary blood flow. Right and left coronary artery inflow, as well as inflow of the major branches of the left coronary artery, can be measured with reasonable accuracy by injection of a radioactive tracer (e.g., ^{133}Xe) through a catheter threaded into the coronary artery via a peripheral artery. Myocardial clearance of the isotope is monitored with a detector appropriately placed over the precordium.

Recently, newer methods for measurement of coronary blood flow have been developed. In the major coronary arteries, blood flow can be measured by the pulsed Doppler technique. An ultrasound signal is emitted from a crystal at the tip of a cardiac catheter that is inserted into the origin of the artery to be studied. The sound is reflected by the flowing blood, and the frequency shift of the reflected sound is proportional to the velocity of the blood flow. From an estimate of the cross-sectional area of the coronary artery and the measured flow velocity, the blood flow can be calculated. Coronary blood flow can also be estimated by videodensitometry where the movement of a bolus of a radioopaque substance injected into the coronary artery can be monitored by rapid sequential radiographs. Similarly, intracoronary injection of microbubbles and tracking of their movement by echocardiography is used to measure coronary blood flow. Also, ciné computed tomography and magnetic resonance imaging are being developed for determination of total and regional myocardial blood flow.

After passage through the capillary beds, most of the venous blood returns to the right atrium through the coronary sinus, but some reaches the right atrium by way of the anterior coronary veins. There are also vascular communications directly between the vessels of the myocardium and the cardiac chambers; these comprise the **arteriosinusoidal,** the **arterioluminal,** and the **thebesian vessels.** The arteriosinusoidal channels consist of small arteries or arterioles that lose their arterial structure as they penetrate the chamber walls and divide into irregular, endothelium-lined sinuses (50 to 250 μm). These sinuses anastomose with other sinuses and with capillaries and communicate with the cardiac chambers. The arterioluminal vessels are small arteries or arterioles that open directly into the atria and ventricles. The thebesian vessels are small veins that connect capillary beds directly with the cardiac chambers and also communicate with cardiac veins and other thebesian veins. On the basis of anatomical studies, intercommunication appears to exist among all the minute vessels of the myocardium in the form of an extensive plexus of subendocardial vessels. It has been suggested that some myocardial nutrition can be derived from the cardiac cavities through those channels. Isotope-labeled blood in the cardiac chambers does penetrate a short distance into the endocardium, but it does not constitute a significant source of oxygen and nutrients to the myocardium. In the dog a major fraction of the left coronary inflow returns to the right atrium via the coronary sinus, and a small fraction that supplies the interventricular septum returns directly to the right ventricular cavity. Right coronary artery drainage is mainly via the anterior cardiac veins to the right atrium. The epicardial distribution of the coronary arteries and veins is illustrated in Fig. 31-1.

■ *Factors That Influence Coronary Blood Flow*

Physical factors

The primary factor responsible for perfusion of the myocardium is the aortic pressure, which is, of course, generated by the heart itself. Changes in aortic pressure generally evoke parallel directional changes in coronary blood flow. However, alterations of cardiac work, produced by an increase or decrease in aortic pressure, have a considerable effect on coronary resistance. Increased metabolic activity of the heart results in a decrease in coronary resistance, and a reduction in cardiac metabolism produces an increase in coronary resistance. If a cannulated coronary artery is perfused by blood from a pressure-controlled reservoir, perfusion pressure

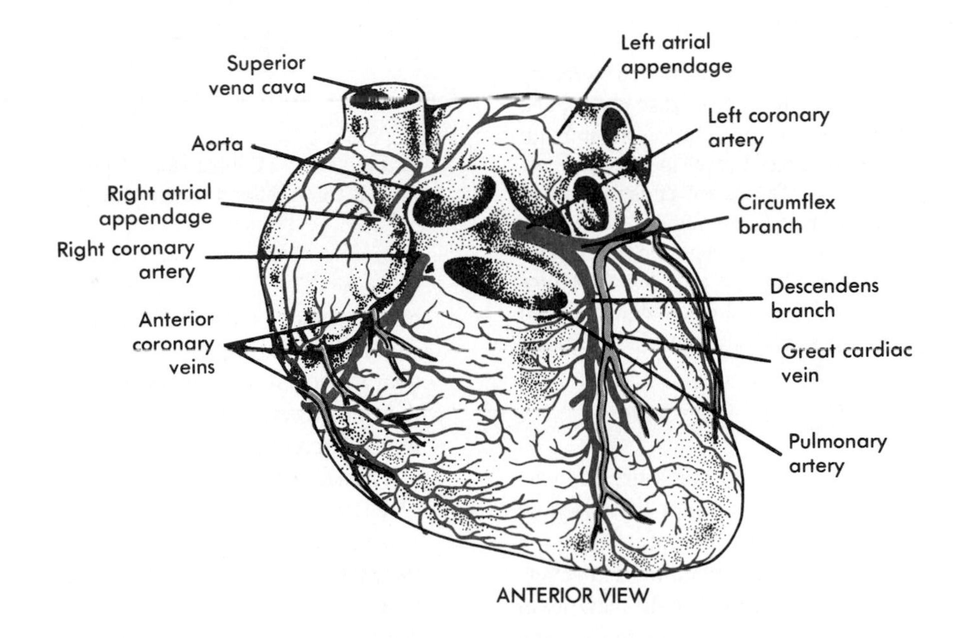

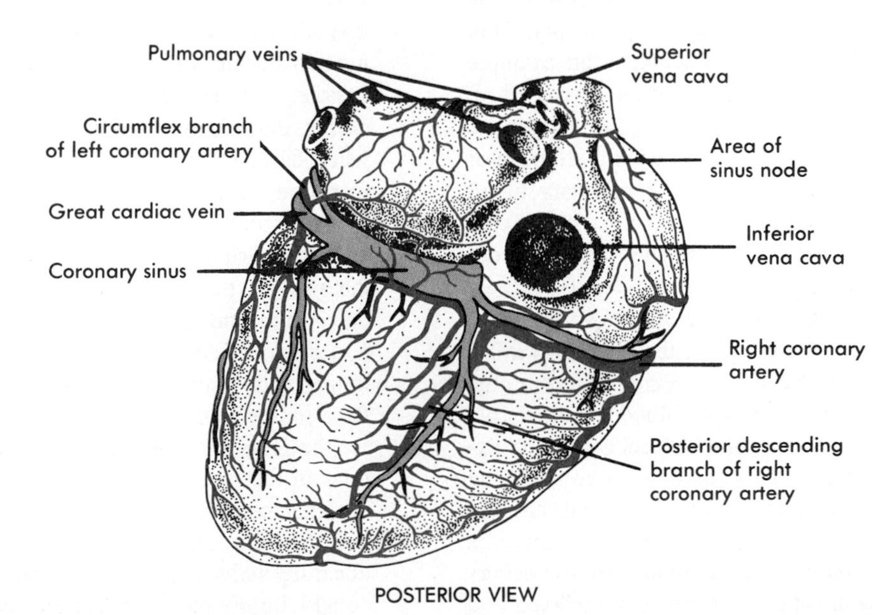

■ **Fig. 31-1** Anterior and posterior surfaces of the heart, illustrating the location and distribution of the principal coronary vessels.

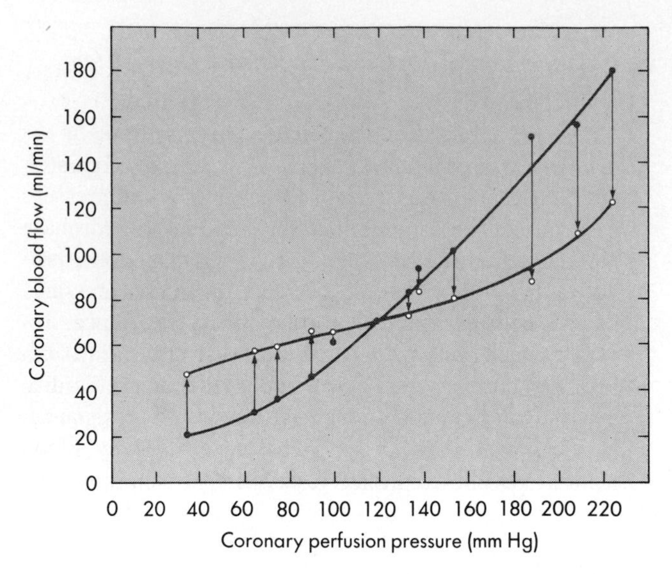

Fig. 31-2 Pressure-flow relationships in the coronary vascular bed. At constant aortic pressure, cardiac output, and heart rate, coronary artery perfusion pressure was abruptly increased or decreased from the control level indicated by the point where the two lines cross. The closed circles represent the flows that were obtained immediately after the change in perfusion pressure, and the open circles represent the steady-state flows at the new pressures. There is a tendency for flow to return toward the control level (autoregulation of blood flow); this is most prominent over the intermediate pressure range (about 60 to 180 mm Hg).

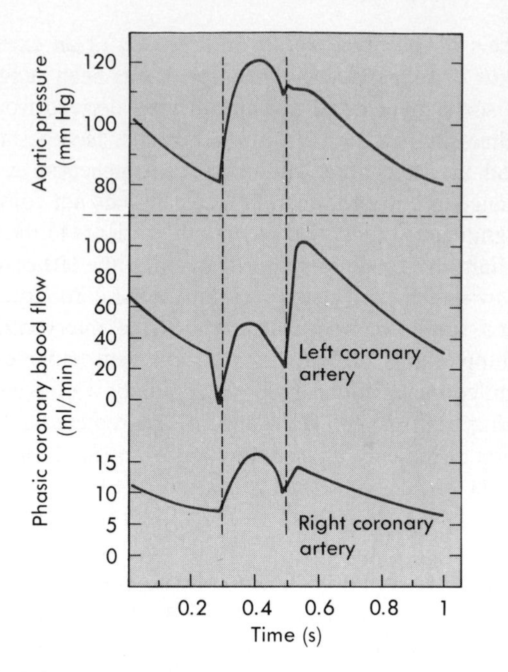

Fig. 31-3 Comparison of phasic coronary blood flow in the left and right coronary arteries.

can be altered without changing aortic pressure and cardiac work. Under these conditions abrupt variations in perfusion pressure produce equally abrupt changes in coronary blood flow in the same direction. However, maintenance of the perfusion pressure at the new level is associated with a return of blood flow toward the level observed before the induced change in perfusion pressure (Fig. 31-2). This phenomenon is an example of autoregulation of blood flow and is discussed in Chapter 29. Under normal conditions blood pressure is kept within relatively narrow limits by the baroreceptor reflex mechanisms so that changes in coronary blood flow are primarily caused by caliber changes of the coronary resistance vessels in response to metabolic demands of the heart.

In addition to providing the head of pressure to drive blood through the coronary vessels, the heart also influences its blood supply by the squeezing effect of the contracting myocardium on the blood vessels that course through it **(extravascular compression** or **extracoronary resistance).** This force is so great during early ventricular systole that blood flow, as measured in a large coronary artery that supplies the left ventricle, is briefly reversed. Maximum left coronary inflow occurs in early diastole, when the ventricles have relaxed and extravascular compression of the coronary vessels is virtually absent. This flow pattern is seen in the phasic coronary flow curve for the left coronary artery (Fig. 31-3). After an initial reversal in early systole, left coronary blood flow follows the aortic pressure until early diastole, when it rises abruptly and then declines slowly as aortic pressure falls during the remainder of diastole.

The minimum extravascular resistance and absence of left ventricular work during diastole are used to advantage clinically to improve myocardial perfusion in patients with damaged myocardium and low blood pressure. The method is called **counterpulsation** and consists of the insertion of an inflatable balloon into the thoracic aorta through a femoral artery. The balloon is inflated during ventricular diastole and deflated during systole. This procedure enhances coronary blood flow during diastole by raising diastolic pressure at a time when coronary extravascular resistance is lowest. Furthermore, it reduces cardiac energy requirements by lowering aortic pressure during ventricular ejection.

Left ventricular myocardial pressure (pressure within the wall of the left ventricle) is greatest near the endocardium and lowest near the epicardium. However, under normal conditions this pressure gradient does not impair endocardial blood flow, because a greater blood flow to the endocardium during diastole compensates for the greater blood flow to the epicardium during systole. In fact, when 10 μm diameter radioactive spheres are injected into the coronary arteries, their distribution indicates that the blood flow to the epicardial and endocardial halves of the left ventricle are approximately equal (slightly higher in the endocardium) under normal conditions. Because extravascular compression is greatest at the endocardial surface of the ventricle, equality of epicardial and endocardial blood flow must mean that the tone of the endocardial resistance vessels is less than that of the epicardial vessels.

Under abnormal conditions, when diastolic pressure in the coronary arteries is low (such as in severe hypotension, partial coronary artery occlusion, or severe aortic stenosis), the ratio of endocardial to epicardial blood flow falls below a value of 1. This change in the ratio indicates that the blood flow to the endocardial regions is more severely impaired than that to the epicardial regions of the ventricle. The redistribution of coronary blood flow is also reflected in an increase in the gradient of myocardial lactic acid and adenosine concentrations from epicardium to endocardium. For this reason the myocardial damage observed in arteriosclerotic heart disease (e.g. following coronary occlusion) is greatest in the inner wall of the left ventricle.

Flow in the right coronary artery shows a similar pattern (see Fig. 31-3); but because of the lower pressure developed during systole by the thin right ventricle, reversal of blood flow does not occur in early systole and systolic blood flow constitutes a much greater proportion of total coronary inflow than it does in the left coronary artery. The extent that extravascular compression restricts coronary inflow can be readily seen when the heart is suddenly arrested in diastole or when ventricular fibrillation is induced. Fig. 31-4 depicts mean left coronary flow when the vessel was perfused with blood at a constant pressure from a reservoir. At the arrow in record *A*, ventricular fibrillation was electrically induced and an immediate and substantial increase in blood flow occurred. Subsequent increase in coronary resistance over a period of many minutes reduced myocardial blood flow to below the level existing before induction of ventricular fibrillation (record *B*, before stellate ganglion stimulation).

Tachycardia and bradycardia have dual effects on coronary blood flow. A change in heart rate is accomplished chiefly by shortening or lengthening of diastole. With tachycardia the proportion of time spent in systole, and consequently the period of restricted inflow, increases. However, this mechanical reduction in mean coronary flow is overridden by the coronary dilation associated with the increased metabolic activity of the more rapidly beating heart. With bradycardia the opposite is true; restriction of coronary inflow is less (more time in diastole) but so are the metabolic (O_2) requirements of the myocardium.

Neural and neurohumoral factors. *Stimulation of the sympathetic nerves to the heart elicits a marked increase in coronary blood flow.* However, the increase in flow is associated with cardiac acceleration and a more forceful systole. The stronger myocardial contractions and the tachycardia (with the consequence that a greater proportion of time is spent in systole) tend to restrict coronary flow, whereas the increase in myocardial metabolic activity, as evidenced by the rate and contractility changes, tends to evoke dilation of the coronary resistance vessels. The increase in coronary blood flow observed with cardiac sympathetic nerve stimula-

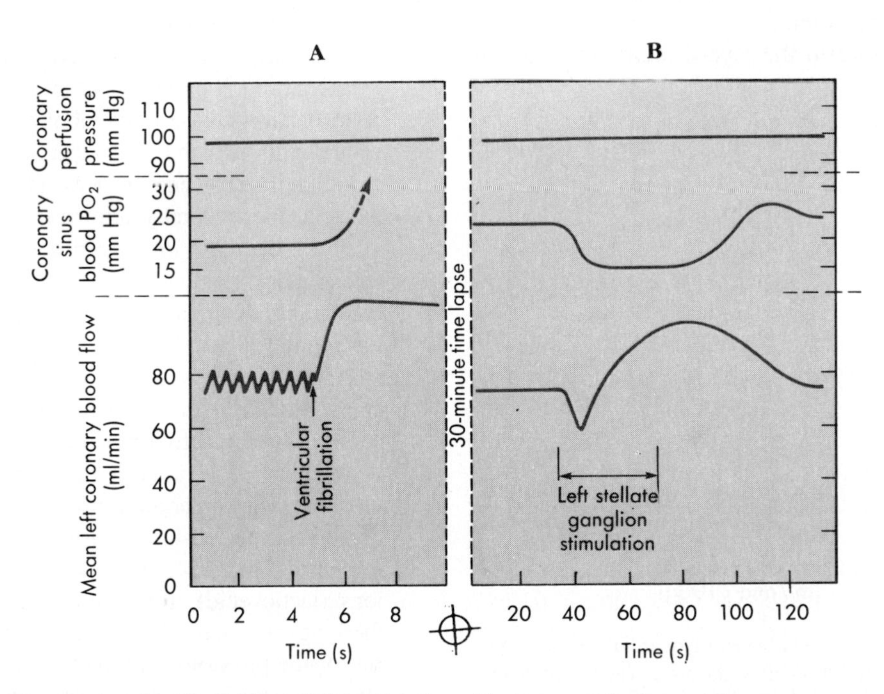

Fig. 31-4 **A,** Unmasking of the restricting effect on ventricular systole on mean coronary blood flow by induction of ventricular fibrillation during constant pressure perfusion of the left coronary artery. **B,** Effect of cardiac sympathetic nerve stimulation on coronary blood flow and coronary sinus blood O_2 tension in the fibrillating heart during constant pressure perfusion of the left coronary artery.

tion is the algebraic sum of these factors. In perfused hearts in which the mechanical effect of extravascular compression is eliminated by cardiac arrest or ventricular fibrillation, an initial coronary vasoconstriction is often observed with cardiac sympathetic nerve stimulation before the vasodilation attributable to the metabolic effect comes into play (Fig. 31-4, *B*).

Furthermore, after the beta receptors are blocked to eliminate the chronotropic and inotropic effects, direct reflex activation of the sympathetic nerves to the heart increases coronary resistance. These observations indicate that *the primary action of the sympathetic nerve fibers on the coronary resistance vessels is vasoconstriction*.

Alpha- and beta-adrenergic drugs and their respective blocking agents reveal the presence of alpha-receptors (constrictors) and beta-receptors (dilators) on the coronary vessels. Furthermore, the coronary resistance vessels participate in the baroreceptor and chemoreceptor reflexes and there is sympathetic constrictor tone of the coronary arterioles that can be reflexly modulated. Nevertheless, coronary resistance is predominantly under local nonneural control.

Vagus nerve stimulation slightly dilates the coronary resistance vessels, and activation of the carotid and aortic chemoreceptors can elicit a small decrease in coronary resistance via the vagus nerves to the heart. The failure of strong vagal stimulation to evoke a large increase in coronary blood flow is not because of insensitivity of the coronary resistance vessels to acetylcholine, because intracoronary administration of this agent elicits marked vasodilation.

Reflexes originating in the myocardium and altering vascular resistance in peripheral systemic vessels, including the coronary vessels, have been conclusively demonstrated. However, the existence of extracardiac reflexes, with the coronary resistance vessels as the effector sites, has not been established.

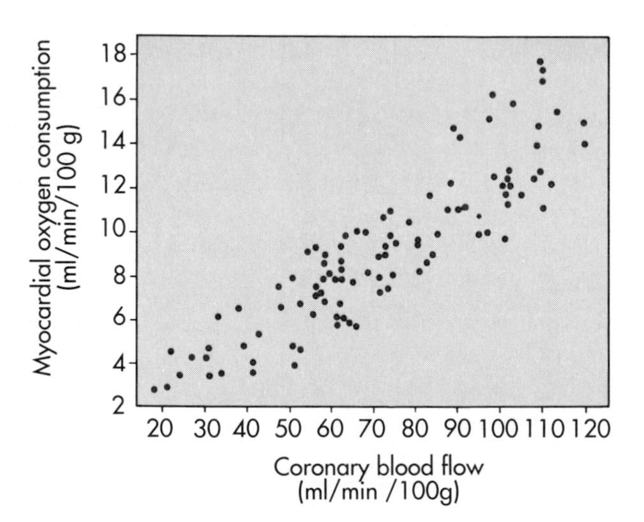

■ **Fig. 31-5** Relationship between myocardial oxygen consumption and coronary blood flow during a variety of interventions that increased or decreased myocardial metabolic rate. (Redrawn from Berne RM, Rubio R: *Coronary circulation*. In *Handbook of physiology, the cardiovascular system—the heart*, vol 1, Bethesda, Md, 1979, American Physiology Society.

Metabolic factors. One of the most striking characteristics of the coronary circulation is the close parallelism between the level of myocardial metabolic activity and the magnitude of the coronary blood flow (Fig. 31-5). This relationship is also found in the denervated heart or the completely isolated heart, whether in the beating or the fibrillating state. The ventricles will continue to fibrillate for many hours when the coronary arteries are perfused with arterial blood from some external source. With the onset of ventricular fibrillation, an abrupt increase in coronary blood flow occurs because of the removal of extravascular compression (see Fig. 31-4). Flow then gradually returns toward, and often falls below, the prefibrillation level. The increase in coronary resistance that occurs despite the elimination of extravascular compression is a manifestation of the heart's ability to adjust its blood flow to meet its energy requirements. The fibrillating heart uses less O_2 than the pumping heart, and blood flow to the myocardium is reduced accordingly.

The link between cardiac metabolic rate and coronary blood flow remains unsettled. However, it appears that *a decrease in the ratio of oxygen supply to oxygen demand (whether produced by a reduction in oxygen supply or by an increment in oxygen demand) releases a vasodilator substance from the myocardium into the interstitial fluid, where it can relax the coronary resistance vessels*. As diagrammed in Fig. 31-6 a decrease in arterial blood oxygen content, coronary blood flow or both, or an increase in metabolic rate decreases the oxygen supply/demand ratio. This causes the release of a vasodilator substance such as adenosine, which dilates the arterioles, thereby adjusting oxygen supply to demand. A decrease in oxygen demand would reduce the vasodilator release and permit greater expression of basal tone. Numerous agents, generally referred to as metabolites, have been suggested as mediators of the vasodilation observed with increased cardiac work. Accumulation of vasoactive metabolites also may be responsible for reactive hyperemia, because the duration of coronary flow after release of the briefly occluded vessel is, within certain limits, proportional to the duration of the period of occlusion. Among the substances implicated are CO_2, O_2 (reduced O_2 tension), hydrogen ions (lactic acid), potassium ions, and adenosine. Of these agents adenosine comes closest to satisfying criteria for the physiological mediator. According to the adenosine hypothesis, a reduction in myocardial O_2 tension produced by low coronary blood flow, hypoxemia, or increased metabolic activity of the heart leads to the myocardial formation of adenosine. This nucleoside crosses the interstitial fluid space to reach the coronary

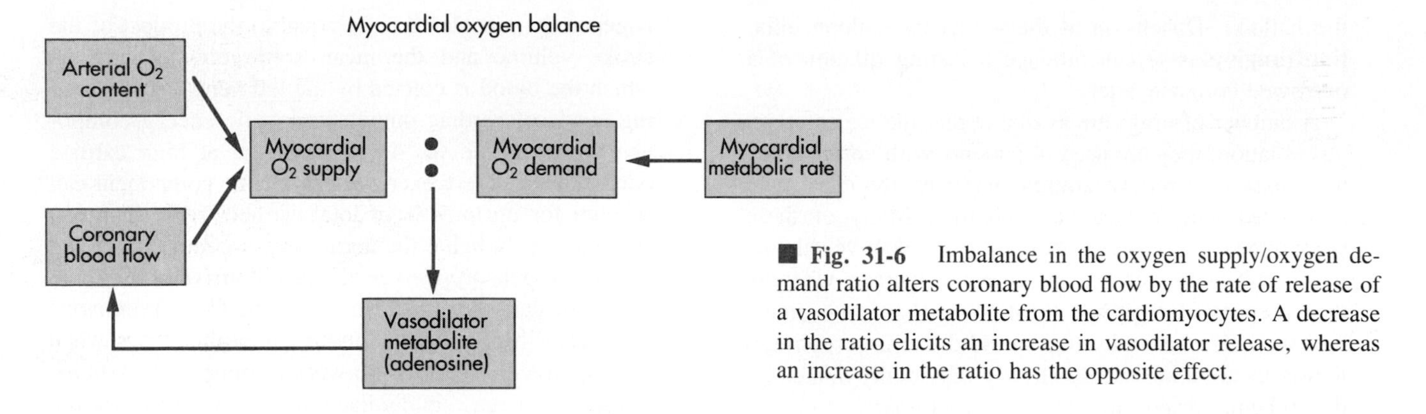

■ **Fig. 31-6** Imbalance in the oxygen supply/oxygen demand ratio alters coronary blood flow by the rate of release of a vasodilator metabolite from the cardiomyocytes. A decrease in the ratio elicits an increase in vasodilator release, whereas an increase in the ratio has the opposite effect.

resistance vessels and induces vasodilation by activating an adenosine receptor.

Although potassium release from the myocardium can account for about half of the initial decrease in coronary resistance, it cannot be responsible for the increased coronary flow observed with prolonged enhancement of cardiac metabolic activity, because its release from the cardiac muscle is transitory. Little evidence supports the concept that CO_2, hydrogen ions, or O_2 play a significant **direct** role in the regulation of coronary blood flow. Factors that alter coronary vascular resistance are schematized in Fig. 31-7.

■ *Coronary Collateral Circulation and Vasodilators*

In the normal human heart virtually no functional intercoronary channels exist, whereas in the dog a few small vessels link branches of the major coronary arteries. Abrupt occlusion of a coronary artery or one of its branches in a human or dog leads to ischemic necrosis and eventual fibrosis of the areas of myocardium supplied by the occluded vessel. However, if narrowing of a coronary artery occurs slowly and progressively over a period of days, weeks, or longer, collateral vessels develop and may furnish sufficient blood to the ischemic myocardium to prevent or reduce the extent of necrosis. The development of collateral coronary vessels has been extensively studied in dogs, and the clinical picture of coronary atherosclerosis, as it occurs in humans, can be simulated by gradual narrowing of the normal dog's coronary arteries. Collateral vessels develop between branches of occluded and nonoccluded arteries. They originate from preexisting small vessels that undergo proliferative changes of the endothelium and smooth muscle, possibly in response to wall stress and chemical agents released by the ischemic tissue.

Numerous surgical attempts have been made to enhance the development of coronary collateral vessels. However, the techniques used do not increase the collateral circulation over and above that produced by coronary artery narrowing alone. When discrete occlusions or severe narrowing occurs in coronary arteries (even vessels as small as 1 mm in diameter), the lesions can be bypassed with a vein graft or the narrow segment can be dilated by inserting a balloon-tipped catheter into the diseased vessel via a peripheral artery and inflating

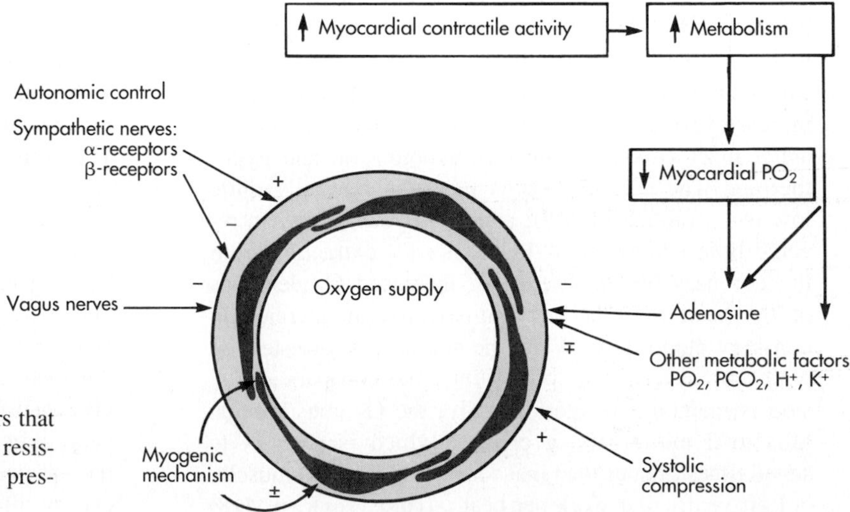

■ **Fig. 31-7** Schematic representation of factors that increase (+) or decrease (−) coronary vascular resistance. The intravascular pressure (arterial blood pressure) stretches the vessel wall.

the balloon. Distension of the vessel by balloon inflation **(angioplasty)** can produce a lasting dilation of a narrowed coronary artery.

A number of drugs are available that induce coronary vasodilation; they are used in patients with coronary artery disease to relieve **angina pectoris,** the chest pain associated with myocardial ischemia. Many of these compounds are nitrites. They are not selective dilators of the coronary vessels, and the mechanism whereby they accomplish their beneficial effects has not been established. The arterioles that would dilate in response to the drugs are undoubtedly already maximally dilated by the ischemia responsible for the symptoms.

In fact, in a patient with marked narrowing of a coronary artery, administration of a vasodilator can fully dilate normal vessel branches proximal to the narrowed segment and reduce the head of pressure to the partially occluded vessel. This will further compromise blood flow to the ischemic myocardium and elicit pain and electrocardiographic changes indicative of tissue injury. This phenomenon, known as **coronary steal,** can be observed with vasodilator drugs such as dipyridamole, which acts by blocking cellular uptake and metabolism of endogenous adenosine.

Nitrites alleviate angina pectoris, at least partly, by reducing cardiac work and myocardial oxygen requirements by relaxing the great veins (decreased preload) and by decreasing blood pressure (decreased afterload). In short, the reduction in pressure work and O_2 requirement must be greater than the reduction in coronary blood flow and O_2 supply consequent to the lowered coronary perfusion pressure. It has also been demonstrated that nitrites dilate large coronary arteries and coronary collateral vessels, thus increasing blood flow to ischemic myocardium and alleviating precordial pain.

Cardiac Oxygen Consumption and Work

The volume of O_2 consumed by the heart is determined by the amount and the type of activity the heart performs. Under basal conditions, myocardial O_2 consumption is about 8 to 10 ml/min/100 g of heart. It can increase severalfold with exercise and decrease moderately under conditions such as hypotension and hypothermia. The cardiac venous blood is normally quite low in O_2 (about 5 ml/dl), and the myocardium can receive little additional O_2 by further O_2 extraction from the coronary blood. Therefore, increased O_2 demands of the heart must be met primarily by an increase in coronary blood flow. When the heartbeat is arrested (as with administration of potassium), but coronary perfusion is maintained experimentally, the O_2 consumption falls to 2 ml/min/100 g or less, which is still six to seven times greater than that for resting skeletal muscle.

Left ventricular work per beat (**stroke work,** p. 428) is generally considered to be equal to the product of the stroke volume and the mean aortic pressure against which the blood is ejected by the left ventricle. At resting levels of cardiac output the kinetic energy component is negligible (p. 439). However, at high cardiac outputs as in severe exercise, the kinetic component can account for up to 50% of total cardiac work. One can simultaneously halve the aortic pressure and double the cardiac output, or vice versa, and still arrive at the same value for cardiac work. However, the O_2 requirements are greater for any given amount of cardiac work when a major fraction is pressure work as opposed to volume work. An increase in cardiac output at a constant aortic pressure (volume work) is accomplished with a small increase in left ventricular O_2 consumption, whereas increased arterial pressure at constant cardiac output (pressure work) is accompanied by a large increment in myocardial O_2 consumption. Thus myocardial O_2 consumption may not correlate well with overall cardiac work. *The magnitude and duration of left ventricular pressure do correlate with left ventricular O_2 consumption.*

The greater energy demand of pressure work over volume work is of great clinical importance, especially in aortic stenosis. In aortic stenosis left ventricular O_2 consumption is increased because of the high intraventricular pressures developed during systole, but coronary perfusion pressure is normal or reduced because of the pressure drop across the narrowed orifice of the diseased aortic valve.

Work of the right ventricle is one seventh that of the left ventricle because pulmonary vascular resistance is much less than systemic vascular resistance.

Cardiac Efficiency

As with an engine, the efficiency of the heart is the ratio of the work accomplished to the total energy used. Assuming an average O_2 consumption of 9 ml/min/100 g for the two ventricles, a 300 g heart consumes 27 ml O_2 min, which is equivalent to 130 small calories at a respiratory quotient of 0.82. Together the two ventricles do about 8 kg/m of work per minute, which is equivalent to 18.7 small calories. Therefore, the gross efficiency is 14%

$$\frac{18.7}{130} \times 100 = 14\% \qquad (1)$$

The net efficiency is slightly higher (18%) and is obtained by subtracting the O_2 consumption of the nonbeating (asystolic) heart (about 2 ml/min/100 g) from the total cardiac O_2 consumption in the calculation of efficiency. It is thus evident that the efficiency of the heart as a pump is relatively low and is comparable to the efficiency of many mechanical devices used in everyday life. With exercise, efficiency improves, because mean blood pressure shows little change, whereas cardiac output and work increase considerably without a proportional increase in myocardial O_2 consumption. The energy expended in cardiac metabolism that does not contribute to the propulsion of blood through the body appears in the form of heat. The energy of the flowing blood is also dissipated as heat, chiefly in passage through the arterioles.

Substrate Utilization

The heart is quite versatile in its use of substrates, and within certain limits the uptake of a particular substrate is directly proportional to its arterial concentration. The utilization of one substrate is also influenced by the presence or absence of other substrates. For example, the addition of lactate to the blood perfusing a heart metabolizing glucose leads to a reduction in glucose uptake, and vice versa. At normal blood concentrations glucose and lactate are consumed at about equal rates, whereas pyruvate uptake is very low but so is its arterial concentration. For glucose the threshold concentration is about 4 mM, and below this blood level no myocardial glucose uptake occurs. Insulin reduces this threshold and increases the rate of glucose uptake by the heart. A very low threshold exists for cardiac utilization of lactate; insulin does not affect its uptake by the myocardium. With hypoxia, glucose utilization is facilitated by an increase in the rate of transport across the myocardial cell wall, whereas lactate cannot be metabolized by the hypoxic heart and is in fact produced by the heart under anaerobic conditions. Associated with lactate production by the hypoxic heart is the breakdown of cardiac glycogen.

Of the total cardiac O_2 consumption, only 35% to 40% can be accounted for by the oxidation of carbohydrate. Thus the heart derives the major part of its energy from oxidation of noncarbohydrate sources. The chief noncarbohydrate fuel used by the heart is esterified and nonesterified fatty acid, which accounts for about 60% of myocardial O_2 consumption in the postabsorptive state. The various fatty acids show different thresholds for myocardial uptake but are generally used in direct proportion to their arterial concentration. Ketone bodies, especially acetoacetate, are readily oxidized by the heart and contribute a major source of energy in diabetic acidosis. As is true of carbohydrate substrates, utilization of a specific noncarbohydrate substrate is influenced by the presence of other substrates, both noncarbohydrate and carbohydrate. Therefore, within certain limits, the heart uses preferentially the substrate that is available in the largest concentration. Most evidence indicates that the contribution to myocardial energy expenditure provided by the oxidation of amino acids is small.

Normally the heart derives its energy by oxidative phosphorylation, in which each mole of glucose yields 36 moles of ATP. However, during hypoxia, glycolysis supervenes and 2 moles of ATP are provided by each mole of glucose. Beta-oxidation of fatty acids is also curtailed. If hypoxia is prolonged, cellular creatine phosphate and eventually ATP are depleted.

In ischemia, lactic acid accumulates (lack of washout) and causes a decrease in intracellular pH. This condition inhibits glycolysis, fatty acid use, and protein synthesis, which results in cellular damage and eventually in necrosis of myocardial cells.

Cutaneous Circulation

The oxygen and nutrient requirements of the skin are relatively small and, in contrast to most other body tissues, the supply of these essential materials is not the chief governing factor in the regulation of cutaneous blood flow. The primary function of the cutaneous circulation is maintenance of a constant body temperature. Consequently, the skin shows wide fluctuations in blood flow, depending on the need for loss or conservation of body heat. Mechanisms responsible for alterations in skin blood flow are primarily activated by changes in ambient and internal body temperatures.

Regulation of Skin Blood Flow

There are essentially two types of resistance vessels in skin: arterioles and **arteriovenous (AV) anastomoses.** The arterioles are similar to those found elsewhere in the body. AV anastomoses shunt blood from arterioles to venules and venous plexuses; hence they bypass the capillary bed. They are found primarily in the fingertips, palms of the hands, toes, soles of the feet, ears, nose, and lips. AV anastomoses differ morphologically from the arterioles in that they are either short, straight, or long coiled vessels about 20 to 40 μm in lumen diameter, with thick muscular walls richly supplied with nerve fibers. These vessels are almost exclusively under sympathetic neural control and become maximally dilated when their nerve supply is interrupted. Conversely, reflex stimulation of the sympathetic fibers to these vessels may produce constriction to the point of complete obliteration of the vascular lumen. Although AV anastomoses do not exhibit **basal tone** (tonic activity of the vascular smooth muscle independent of innervation), they are highly sensitive to vasoconstrictor agents like epinephrine and norepinephrine. Furthermore, AV anastomoses do not appear to be under metabolic control, and they fail to show reactive hyperemia or autoregulation of blood flow. *Thus the regulation of blood flow through these anastomotic channels is governed principally by the nervous system in response to reflex activation by temperature receptors or from higher centers of the central nervous system.*

Most of the skin resistance vessels exhibit some basal tone and are under dual control of the sympathetic nervous system and local regulatory factors, in much the same manner as are resistance vessels in other vascular beds. However, in the case of skin, neural control plays a more important role than local factors. Stimulation of sympathetic nerve fibers to skin blood vessels (arteries and veins, as well as arterioles) induces vasoconstriction, and severance of the sympathetic nerves induces vasodilation. With chronic denervation of the cutaneous blood vessels, the degree of tone that existed before denervation is gradually regained over several weeks. This is accomplished by an enhancement of basal tone that compensates for the degree of tone previously contributed by sympathetic nerve fiber activity. Epinephrine and norepinephrine elicit only vasoconstriction in cutaneous vessels. Whether the increased basal tone following denervation of the skin vessels is the result of their enhanced sensitivity to circulating catecholamines **(denervation hypersensitivity)** has not been established.

Parasympathetic vasodilator nerve fibers do not supply the cutaneous blood vessels. However, stimulation of the sweat glands, which are innervated by cholinergic fibers of the sympathetic nervous system, results in dilation of the skin resistance vessels. Sweat contains an enzyme that acts on a protein moiety in the tissue fluid to produce **bradykinin,** a polypeptide with potent vasodilator properties. Bradykinin formed in the tissue can act locally to dilate the arterioles and increase blood flow to the skin. The skin vessels of certain regions, particularly the head, neck, shoulders, and upper chest, are under the influence of the higher centers of the central nervous system. Blushing, as with embarrassment or anger, and blanching, as with fear or anxiety, are examples of cerebral inhibition and stimulation, respectively, of the sympathetic nerve fibers to the affected regions.

In contrast to AV anastomoses in the skin, the cutaneous resistance vessels show autoregulation of blood flow and reactive hyperemia. If the arterial inflow to a limb is stopped with an inflated blood pressure cuff for a brief period of time, the skin shows a marked reddening below the point of vascular occlusion when the cuff is deflated. This increased cutaneous blood flow (reactive hyperemia) is also manifested by the distension of the superficial veins in the erythematous extremity. Autoregulation of blood flow in the skin is best explained by a myogenic mechanism (see p. 481).

Ambient and body temperature in regulation of skin blood flow. *Because the primary function of the skin is to preserve the internal milieu and protect it from adverse changes in the environment, and because ambient temperature is one of the most important external variables the body must contend with, it is not surprising that the vasculature of the skin is chiefly influenced by environmental temperature.* Exposure to cold elicits a generalized cutaneous vasoconstriction that is most pronounced in the hands and feet. This response is chiefly mediated by the nervous system, because arrest of the circulation to a hand with a pressure cuff and immersion of that hand in cold water results in vasoconstriction in the skin of the other extremities that are exposed to room temperature. With the circulation to the chilled hand unoccluded, the reflex vasoconstriction is caused in part by the cooled blood returning to the general circulation and stimulating the temperature-regulating center in the anterior hypothalamus. Direct application of cold to this region of the brain produces cutaneous vasoconstriction.

The skin vessels of the cooled hand also show a direct response to cold. Moderate cooling or exposure for brief periods to severe cold (0° to 15° C) results in constriction of the resistance and capacitance vessels, including AV anastomoses. However, prolonged exposure of the hand to severe cold has a secondary vasodilator effect. Prompt vasoconstriction and severe pain are elicited by immersion of the hand in water near 0° C but are soon followed by dilation of the skin vessels with reddening of the immersed part and alleviation of the pain. With continued immersion of the hand, alternating periods of constriction and dilation occur, but the skin temperature rarely drops to as low a degree as it did with the initial vasoconstriction. Prolonged severe cold, of course, results in tissue damage. The rosy faces of people working or playing in a cold environment are examples of cold vasodilation. However, the blood flow through the skin of the face may be greatly reduced despite the flushed appearance. The red color of the slowly flowing blood is in large measure the result of the reduced oxygen uptake by the cold skin and the cold-induced shift to the left of the oxyhemoglobin dissociation curve.

Direct application of heat produces not only local vasodilation of resistance and capacitance vessels and AV anastomoses but also reflex dilation in other parts of the body. The local effect is independent of the vascular nerve supply, whereas the reflex vasodilation is a combination of anterior hypothalamic stimulation by the returning warmed blood and of stimulation of receptors in the heated part. However, evidence for a reflex from peripheral temperature receptors is not as definitive for warm stimulation as it is for cold stimulation.

The proximity of the major arteries and veins to each other permits considerable heat exchange (countercurrent) between artery and vein. Cold blood that flows in veins from a cooled hand toward the heart takes up heat from adjacent arteries, resulting in warming of the venous blood and cooling of the arterial blood. Heat exchange is, of course, in the opposite direction with exposure of the extremity to heat. Thus heat conservation is enhanced and heat gain is minimized during exposure of extremities to cold and warm environments, respectively.

■ *Skin Color and Special Reactions of the Skin Vessels*

The color of the skin is, of course, caused in large part by pigment; however, in all but very dark skin, the degree of pallor or ruddiness is mainly a function of the amount of blood in the skin. With little blood in the venous plexus the skin appears pale, whereas with moderate to large quantities of blood in the venous plexus, the skin shows color. Whether this color is bright red, blue, or some shade between is determined by the degree of oxygenation of the blood in the subcutaneous vessels. For example, a combination of vasoconstriction and reduced hemoglobin can produce an ashen gray color of the skin, whereas a combination of venous engorgement and reduced hemoglobin can result in a dark purple hue. Skin color provides little information about the rate of cutaneous blood flow. There may coexist rapid blood flow and pale skin when the AV anastomoses are open, and slow blood flow and red skin when the extremity is exposed to cold.

White reaction and triple response. If the skin of the forearm of many individuals is lightly stroked with a blunt instrument, a white line appears at the site of the stroke within 20 seconds. The blanching becomes maximum in about 30 to 40 seconds and then gradually disappears within 3 to 5 minutes. This response is known as a **white reaction** and has been attributed to capillary contraction, because it occurs in the denervated limb and is unaffected by arrest of the limb circulation. Because all direct evidence indicates that capillaries do not contract and because skin color is mainly a result of blood content of venous plexuses, venules, and small veins, it seems logical to attribute the white reaction to venous constriction induced by mechanical stimulation.

If the skin is stroked more strongly with a sharp pointed instrument, a **triple response** is elicited. Within 3 to 15 seconds a thin **red line** appears at the site of the stroke, followed in about 15 to 30 seconds by a red blush, or **flare,** extending out 1 to 2 cm from either side of the red line. This in turn is followed in 3 to 5 minutes by an elevation of the skin along the red line; with gradual fading of the red line as the elevation, a **wheal,** becomes more prominent. The red line is probably caused by dilation of the vessels because of mechanical stimulation. The flare, however, is the result of dilation of neighboring arterioles caused to relax by an **axon reflex** originating at the site of mechanical stimulation. In an axon reflex the nerve impulse travels centripetally in the cutaneous sensory nerve fiber and then antidromically down the small branches of the afferent nerve to adjacent arterioles to elicit vasodilation (Fig. 31-8). The flare is not affected by acute section or anesthetic block of the sensory nerve central to the point of branching, whereas it is abolished when the nerve degenerates after section. The wheal is caused by increased capillary permeability induced by the trauma. Fluid containing protein leaks out of the capillaries locally and produces edema at the site of injury. Because the triple response can be elicited by an intradermal injection of histamine, the triple response has been attributed to histamine, which can be released from mast cells by action of a peptide that originates in the sensory nerve endings. In addition vasoactive agents such as substance P, ATP, and calcitonin gene-related peptide (CGRP) can be released from sensory nerve endings and may interact in a complex manner to elicit the triple response.

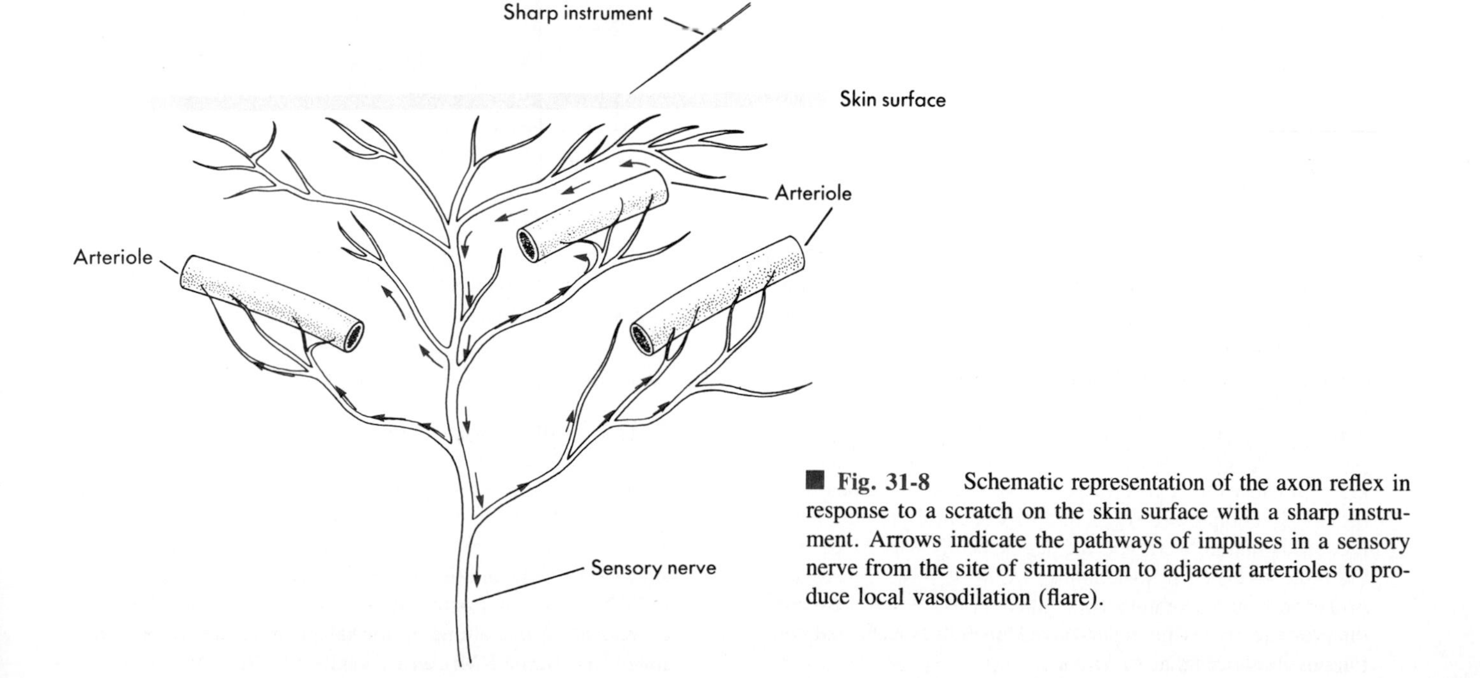

■ **Fig. 31-8** Schematic representation of the axon reflex in response to a scratch on the skin surface with a sharp instrument. Arrows indicate the pathways of impulses in a sensory nerve from the site of stimulation to adjacent arterioles to produce local vasodilation (flare).

Skeletal Muscle Circulation

The rate of blood flow in skeletal muscle varies directly with the contractile activity of the tissue and the type of muscle. Blood flow and capillary density in red (slow-twitch, high-oxidative) muscle are greater than in white (fast-twitch, low-oxidative) muscle. In resting muscle the precapillary arterioles exhibit asynchronous intermittent contractions and relaxations, so that at any given moment, a very large percentage of the capillary bed is not perfused. Consequently, total blood flow through quiescent skeletal muscle is low (1.4 to 4.5 ml/min/100 g). With exercise the resistance vessels relax and the muscle blood flow may increase manyfold (up to fifteen to twenty times the resting level), the magnitude of the increase depending largely on the severity of the exercise.

Regulation of Skeletal Muscle Blood Flow

Control of muscle circulation is achieved by neural and local factors. As with all tissues, physical factors such as arterial pressure, tissue pressure, and blood viscosity influence muscle blood flow. However, another physical factor comes into play during exercise—the squeezing effect of the active muscle on the vessels. With intermittent contractions, inflow is restricted and venous outflow is enhanced during each brief contraction. The presence of the venous valves prevents backflow of blood in the veins between contractions, thereby aiding in the forward propulsion of the blood (Fig. 31-9). With strong sustained contractions the vascular bed can be compressed to the point where blood flow actually ceases temporarily.

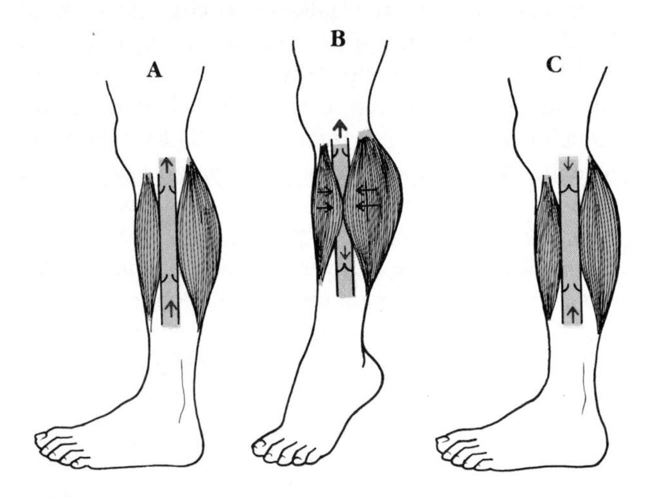

Fig. 31-9 Action of the muscle pump in venous return from the legs. **A,** Standing at rest the venous valves are open and blood flows upward toward the heart by virtue of the pressure generated by the heart and is transmitted through the capillaries to the veins from the arterial side of the vascular system **(vis a tergo). B,** Contraction of the muscle compresses the vein so that the increased pressure in the vein drives blood toward the thorax through the upper valve and closes the lower valve in the uncompressed segment of the vein just below the point of muscular compression. **C,** Immediately after muscle relaxation the pressure in the previously compressed venous segment falls, and the reversed pressure gradient causes the upper valve to close. The valve below the previously compressed segment opens because pressure below it exceeds that above it and the segment fills with blood from the foot. As blood flow continues from the foot, the pressure in the previously compressed segment rises. When it exceeds the pressure above the upper valve, this valve opens and continuous flow occurs as in part *A*.

Neural factors. Although the resistance vessels of muscle possess a high degree of basal tone, they also display tone attributable to continuous low frequency activity in the sympathetic vasoconstrictor nerve fibers. The basal frequency of firing in the sympathetic vasoconstrictor fibers is quite low (about 1 to 2 per second), and maximum vasoconstriction is observed at frequencies as low as 8 to 10 per second. Stimulation of the sympathetic nerve fibers to skeletal muscle elicits vasoconstriction that is caused by the release of norepinephrine at the fiber endings. Intraarterial injection of norepinephrine elicits only vasoconstriction, whereas low doses of epinephrine produce vasodilation in muscle and large doses cause vasoconstriction.

The tonic activity of the sympathetic nerves is greatly influenced by reflexes from the baroreceptors. An increase in carotid sinus pressure results in dilation of the vascular bed of the muscle, and a decrease in carotid si-

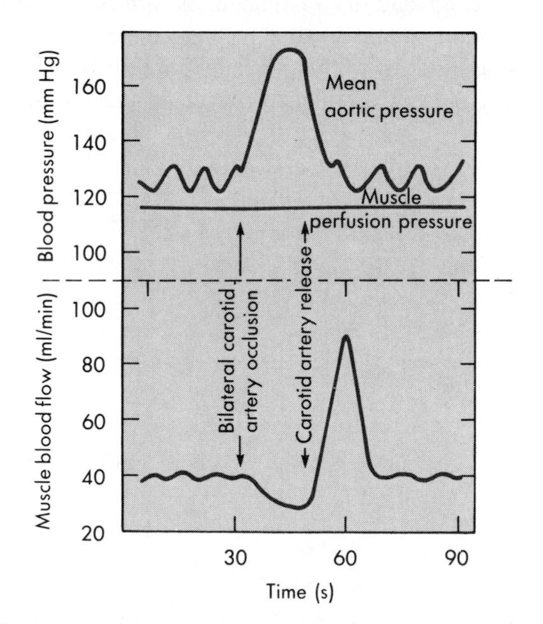

Fig. 31-10 Evidence for participation of the muscle vascular bed in vasoconstriction and vasodilation mediated by the carotid sinus baroreceptors after common carotid artery occlusion and release. In this preparation the sciatic and femoral nerves constituted the only direct connection between the hind leg muscle mass and the rest of the dog. The muscle was perfused by blood at a constant pressure that was completely independent of the animal's arterial pressure. (Redrawn from Jones RD, Berne RM: *Am J Physiol* 204:461, 1963.)

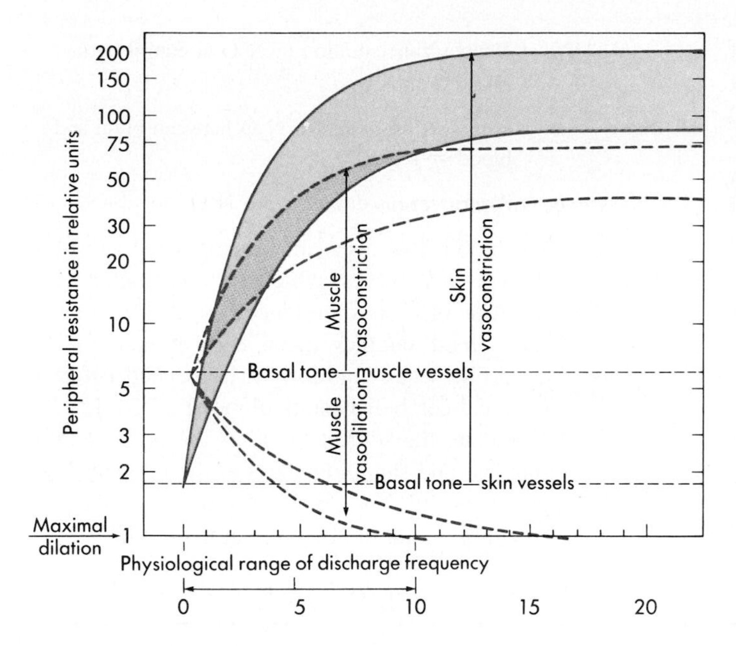

■ **Fig. 31-11** Basal tone and the range of response of the resistance vessels in muscle *(dashed line)* and skin *(shaded area)* to sympathetic nerve stimulation. Peripheral resistance plotted on a logarithmic scale. (Redrawn from Celander O, Folkow B: *Acta Physiol Scand* 29:241, 1953.)

nus pressure elicits vasoconstriction (Fig. 31-10). When the existing sympathetic constrictor tone is high, as in the experiment illustrated in Fig. 31-10, the decrease in blood flow associated with common carotid artery occlusion is small; but the increase following the release of occlusion is large. The vasodilation produced by baroreceptor stimulation is caused by inhibition of sympathetic vasoconstrictor activity. Because muscle is the major body component on the basis of mass and thereby represents the largest vascular bed, the participation of its resistance vessels in vascular reflexes plays an important role in maintaining a constant arterial blood pressure.

A comparison of the vasoconstrictor and vasodilator effects of the sympathetic nerves to blood vessels of muscle and skin is summarized in Fig. 31-11. Note the lower basal tone of the skin vessels, their greater constrictor response, and the absence of active cutaneous vasodilation.

Local factors. Whether neural or local factors predominate in the regulation of skeletal muscle blood flow depends on the activity of the muscle. In resting muscle the neural factors predominate and superimpose neurogenic tone on the nonneural basal tone (see Fig. 31-11). Section of the sympathetic nerves to muscle abolishes the neural component of vascular tone and unmasks the intrinsic basal tone of the blood vessels. *Neural and local blood flow regulating mechanisms oppose each other, and during muscle contraction the local vasodilator mechanism supervenes.* However, during exercise strong sympathetic nerve stimulation slightly reduces the vasodilation induced by locally released metabolites. The identity of the local mediator(s) of the vasodilation that occurs with exercise has not been determined; several mediators are probably involved. Adenosine is released from ischemic muscle and in some cases from contracting muscle during unimpaired blood flow, but the nucleoside accounts for part of the exercise-induced vasodilation only in fast oxidative muscles. Skeletal muscle effectively autoregulates blood flow, particularly during active contractions. As in other cases of autoregulation the mechanism is probably myogenic (see p. 481).

■ Cerebral Circulation

Blood reaches the brain through the internal carotid and vertebral arteries. The latter join to form the basilar artery, which, in conjunction with branches of the internal carotid arteries, forms the **circle of Willis.** A unique feature of the cerebral circulation is that it all lies within a rigid structure, the cranium. Because intracranial contents are incompressible, any increase in arterial inflow, as with arteriolar dilation, must be associated with a comparable increase in venous outflow. The volume of blood and of extravascular fluid can vary considerably in most tissues. In brain the volume of blood and extravascular fluid is relatively constant; changes in either of these fluid volumes must be accompanied by a reciprocal change in the other. In contrast to most other organs, the rate of total cerebral blood flow is held within a relatively narrow range and in humans averages 55 ml/min/100 g of brain.

■ Estimation of Cerebral Blood Flow

Total cerebral blood flow can be measured in humans by the nitrous oxide (N_2O) method, which is based on the Fick principle (p. 412). The subject breathes a gas mixture of 15% N_2O, 21% O_2, and 64% N_2 for 10 minutes, which is sufficient time to permit equilibration of the N_2O between the brain tissue and the blood leaving the brain. Simultaneous samples of arterial (any artery) blood and mixed cerebral venous (internal jugular vein) blood are taken at the start of N_2O inhalation and at 1-minute intervals throughout the 10-minute period of N_2O administration. From these data, the cerebral blood flow can be calculated by the Fick equation:

$$\text{Cerebral blood flow} = \frac{\text{Amount of } N_2O \text{ taken up by brain during time } (t_2 - t_1)}{\text{AV difference of } N_2O \text{ across brain during time } (t_2 - t_1)} \quad (2)$$

Because the arterial and venous concentrations are continuously changing with time, it is necessary to get the true AV difference during the period of N_2O inhalation by integration of the AV difference over the 10-minute period. This is represented in Fig. 31-12, *A*, by the shaded area between the arterial and venous N_2O concentration curves constructed from the blood concentrations observed at successive 1-minute intervals during N_2O administration. The amount of N_2O removed by the brain, as well as the concentration of N_2O in the brain tissue, are unknown. Because the partition coefficient between brain and blood is about 1 and because equilibrium of N_2O between the brain and the blood leaving the brain is reached by the end of 10 minutes, the concentration of N_2O in the brain tissue closely approximates that of the cerebral venous blood in the 10-minute sample. The total weight of the brain is not known, and so for convenience the concentration in brain tissue is multiplied by 100 to express the cerebral blood flow (CBF) in ml/min/100 g of brain tissue. The equation is:

$$\text{CBF} = \frac{V_{10} \cdot S \cdot 100}{\int_0^{10} (A - V)dt} \quad (3)$$

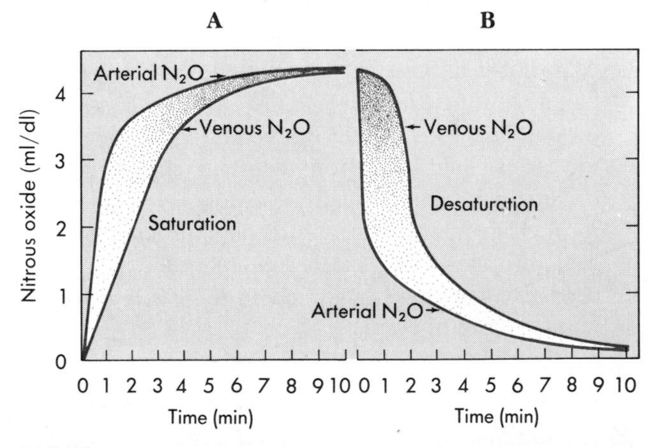

■ **Fig. 31-12** Concentrations of N_2O in arterial and cerebral venous blood during saturation, **A**, and desaturation, **B**. The shaded areas represent the arteriovenous differences of N_2O during the 10-minute period of N_2O inhalation and the 10 minutes after discontinuing the N_2O administration.

V_{10} = Venous concentration of N_2O at equilibrium (at 10 min)

S = Partition coefficient of N_2O between brain and blood = 1

A − V = Arteriovenous difference of N_2O

Cerebral blood flow also can be calculated from the desaturation A − V curves, which are constructed from N_2O concentrations of simultaneously drawn arterial and venous blood samples taken each minute for 10 minutes, starting when equilibrium is reached between brain tissue and cerebral venous blood (Fig. 31-12, *B*). In this procedure the subject breathes the N_2O mixture for 10 minutes, and sampling starts at the moment N_2O inhalation is stopped. The only difference from the preceding equation is that the denominator becomes

$$\int_{10}^{20} (V - A)dt \quad (4)$$

Cerebral blood flow and its distribution to different areas of the brain can be measured in animals by injection into the internal carotid artery of microspheres (about 15 μm) labeled with radioactive substances. The microspheres become lodged in the arterioles and capillaries, the brain tissue is sampled, and the radioactivity of the tissue is determined. Blood flow to each tissue sample is proportional to the radioactivity in that sample. By the use of microspheres labeled with different radioactive isotopes, several measurements of cerebral blood flow can be made. A gamma counter is used to measure each isotope independently of the other isotopes in the sample. This method is also used for measurements of blood flow in other tissues, such as the myocardium, and is the standard technique for **regional blood flow** measurements. One can also measure cerebral blood flow in animals with the use of ^{14}C-antipyrine, which is taken up by the brain in proportion to the blood flow. The brain is then sliced, and the radioactivity of the slice is determined by radioautography. The advantage of these methods over the N_2O method or the direct measurement of venous outflow from the brain is that blood flow to different regions of the brain can be determined. The obvious disadvantage is that the animals must be sacrificed to obtain the necessary samples of brain tissue.

Recently the development of multiple collimated scintillation detectors built into a helmet that fits over the cranium has made possible the measurement of regional blood flow (cortical blood flow) in animals and humans. An inert radioactive gas (such as ^{133}Xe) is injected into an internal carotid artery; and from its rate of washout from the brain, regional cerebral blood flow can be determined. The radioactive gas also may be given by inhalation, but more sophisticated techniques are required to eliminate noncerebral blood flow and to distinguish between blood flow to cortical (gray matter) and deep cerebral (white matter) tissue.

■ *Regulation of Cerebral Blood Flow*

Of the various body tissues, the brain is the least tolerant of ischemia. Interruption of cerebral blood flow for as little as 5 seconds results in loss of consciousness, and ischemia lasting just a few minutes results in irreversible tissue damage. Fortunately regulation of the cerebral circulation is primarily under direction of the brain itself. Local regulatory mechanisms and reflexes originating in the brain tend to maintain cerebral circulation relatively constant in the presence of possible adverse extrinsic effects such as sympathetic vasomotor nerve activity, circulating humoral vasoactive agents, and changes in arterial blood pressure. Under certain conditions the brain also regulates its blood flow by initiating changes in systemic blood pressure. For example, elevation of intracranial pressure results in an increase in systemic blood pressure. This response, called **Cushing's phenomenon** is apparently caused by ischemic stimulation of vasomotor regions of the medulla. It aids in maintaining cerebral blood flow in such conditions as expanding intracranial tumors.

Neural factors. The cerebral vessels receive innervation from the cervical sympathetic nerve fibers that accompany the internal carotid and vertebral arteries into the cranial cavity. The importance of neural regulation of the cerebral circulation is controversial, but the prevalent belief is that relative to other vascular beds sympathetic control of the cerebral vessels is weak and that the contractile state of the cerebral vascular smooth muscle depends mainly on local metabolic factors. There are no known sympathetic vasodilator nerves to the cerebral vessels; but the vessels do receive parasympathetic fibers from the facial nerve, which produce a slight vasodilation on stimulation.

Local factors. *Generally total cerebral blood flow is constant. However, regional cortical blood flow is associated with regional neural activity.* For example, movement of one hand results in increased blood flow only in the hand area of the contralateral sensory-motor and premotor cortex. Also, talking, reading, and other stimuli to the cerebral center are associated with increased blood flow in the appropriate regions of the contralateral cortex (Fig. 31-13). The vasodilation in the cortex is not associated with enhanced local cerebral metabolism nor another adenosine release. It appears to be mediated by the local release of nitric oxide. Stimulation of the retina with flashes of light increases blood flow only in the visual cortex. Glucose uptake also corresponds with regional cortical neuronal activity. For example, when the retina is stimulated by light, uptake of ^{14}C-2-deoxyglucose is enhanced in the visual cortex. This analogue of glucose is taken up and phosphorylated by cerebral neurons but cannot be metabolized further. The magnitude of its uptake is determined from radioautographs of slices of the brain. The mediator of the link between cerebral activity and blood flow has not been established, but three possible candidates are pH, potassium, and adenosine.

It is well known that the cerebral vessels are very sensitive to carbon dioxide tension. Increases in arterial blood CO_2 tension ($Paco_2$) elicit marked cerebral vasodilation; inhalation of 7% CO_2 results in a twofold increment in cerebral blood flow. By the same token de-

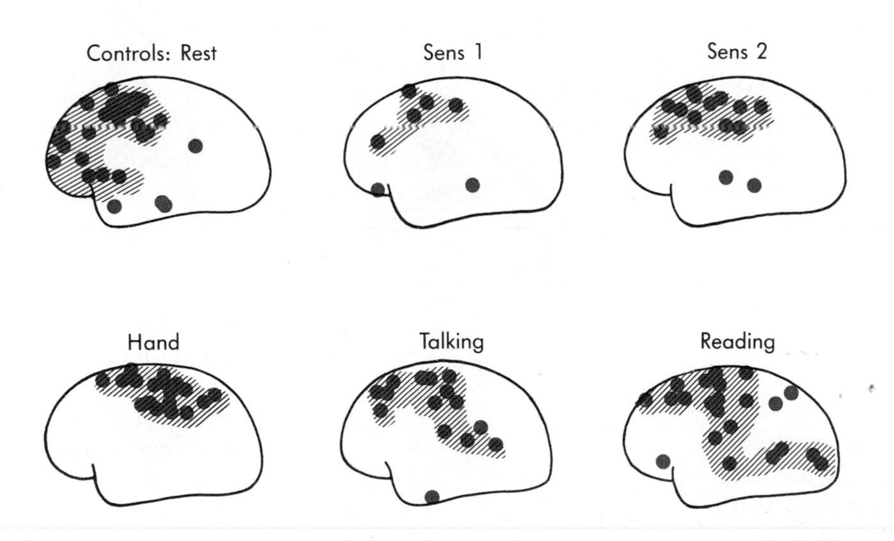

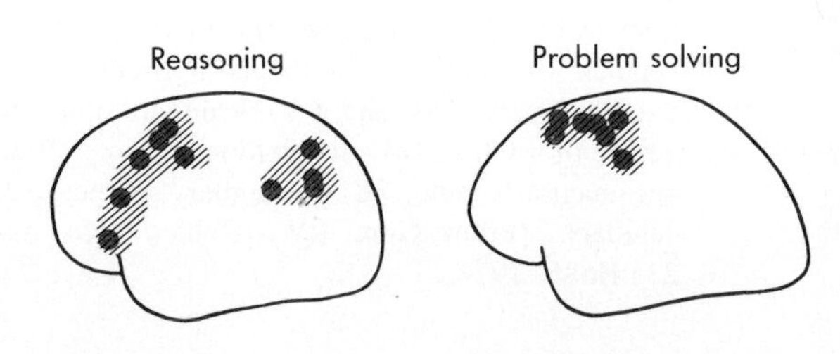

■ **Fig. 31-13** Effects of different stimuli on regional blood flow in the contralateral human cerebral cortex. Sens 1, low intensity electrical stimulation of the hand; Sens 2, high intensity electrical stimulation of the hand (pain). (Redrawn from Ingvar DH: *Brain Res* 107:181, 1976.)

creases in $Paco_2$, such as elicited by hyperventilation, produce a decrease in cerebral blood flow. CO_2 produces changes in arteriolar resistance by altering perivascular (and probably intracellular vascular smooth muscle) pH. By independently changing Pco_2 and bicarbonate concentration, it has been demonstrated that pial vessel diameter (and presumably blood flow) and pH are inversely related, regardless of the level of the Pco_2.

Carbon dioxide can diffuse to the vascular smooth muscle from the brain tissue or from the lumen of the vessels, whereas hydrogen ions in the blood are prevented from reaching the arteriolar smooth muscle by the **blood-brain barrier.** Hence, the cerebral vessels dilate when the hydrogen ion concentration of the cerebrospinal fluid is increased but show only minimal dilation in response to an increase in the hydrogen ion concentration of the arterial blood. Despite the responsiveness of the cerebral vessels to pH changes, the precise role of hydrogen ions in the regulation of cerebral blood flow remains obscure. The initiation of increases in cerebral blood flow produced by seizures has been reported to be associated with transient increases rather than decreases in perivascular pH. Also, the intracellular and extracellular decreases in pH that occur with electrical stimulation of the brain or hypoxia often occur after cerebral blood flow has increased in response to the stimulus. Furthermore, with prolonged hypocapnia, cerebrospinal fluid pH may return to control levels in the face of a persistent reduction in cerebral blood flow. Therefore, pH probably plays no significant role in the normal regulation of cerebral blood flow.

With respect to K^+, such stimuli as hypoxia, electrical stimulation of the brain, and seizures elicit rapid increases in cerebral blood flow and are associated with increases in perivascular K^+. The increments in K^+ are similar to those which produce pial arteriolar dilation when K^+ is applied topically to these vessels. However, the increase in K^+ is not sustained throughout the period of stimulation. Hence, only the initial increment in cerebral blood flow can be attributed to the release of K^+.

Adenosine levels of the brain increase with ischemia, hypoxemia, hypotension, hypocapnia, electrical stimulation of the brain, or induced seizures. When it is applied topically, adenosine is a potent dilator of the pial arterioles. In short, any intervention that either reduces the O_2 supply to the brain or increases the O_2 need of the brain results in rapid (within 5 seconds) formation of adenosine in the cerebral tissue. Unlike pH or K^+, the adenosine concentration of the brain increases with initiation of the stimulus and remains elevated throughout the period of O_2 imbalance. The adenosine released into the cerebrospinal fluid during conditions associated with inadequate brain O_2 supply is available to the brain tissue for reincorporation into cerebral tissue adenine nucleotides.

All three factors, pH, K^+, and adenosine may act in concert to adjust the cerebral blood flow to the metabolic activity of the brain, but how these factors interact to accomplish this regulation of cerebral blood flow remains to be elucidated.

The cerebral circulation shows reactive hyperemia and excellent autoregulation between pressures of about 60 and 160 mm Hg. Mean arterial pressures below 60 mm Hg result in reduced cerebral blood flow and syncope, whereas mean pressures above 160 may lead to increased permeability of the blood-brain barrier and cerebral edema. Autoregulation of cerebral blood flow is abolished by hypercapnia or any other potent vasodilator, and none of the candidates for metabolic regulation of cerebral blood flow has been shown to be responsible for this phenomenon. Hence, autoregulation of cerebral blood flow is probably attributable to a myogenic mechanism, although experimental proof is still lacking.

Splanchnic Circulation

The splanchnic circulation consists of the blood supply to the gastrointestinal tract, liver, spleen, and pancreas. Several features distinguish the splanchnic circulation,

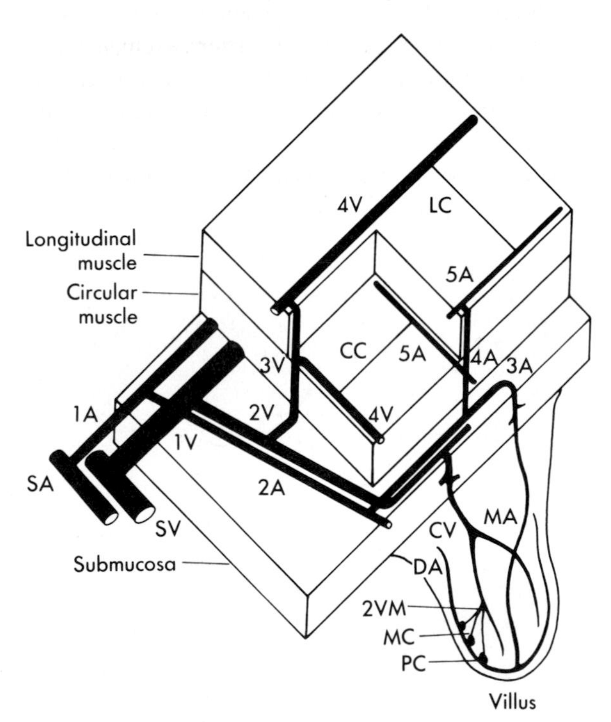

Fig. 31-14 The distribution of small blood vessels to the rat intestinal wall. *SA,* Small artery; *SV,* small vein; *1A* to *5A,* first- to fifth-order arterioles; *1V* to *4V,* first- to fourth-order venules; *CC* and *LC,* capillaries in circular and longitudinal muscle layers; *MA* and *CV,* main arteriole and collecting venule of a villus; *DA,* distribution arteriole; *2VM,* second-order mucosal venule; *PC,* precapillary sphincter; *MC,* mucosal capillary. (From Gore RW, Bohlen HG: *Am J Physiol* 233:H685, 1977.)

the most noteworthy of which is that two large capillary beds are partially in series with one another. The small splanchnic arterial branches supply the capillary beds in the gastrointestinal tract, spleen, and pancreas. From these capillary beds, the venous blood ultimately flows into the portal vein, which normally provides most of the blood supply to the liver. However, the hepatic artery also supplies blood to the liver.

■ *Intestinal Circulation*

Anatomy. The gastrointestinal tract is supplied by the celiac, superior mesenteric, and inferior mesenteric arteries. The superior mesenteric artery is the largest of all the aortic branches and carries over 10% of the cardiac output. Small mesenteric arteries form an extensive vascular network in the submucosa (Fig. 31-14). Their branches penetrate the longitudinal and circular muscle layers and give rise to third- and fourth-order arterioles. Some third-order arterioles in the submucosa become the main arterioles to the tips of the villi.

The direction of the blood flow in the capillaries and venules in a villus is opposite to that in the main arteriole (Fig. 31-15). This arrangement constitutes a countercurrent exchange system. An effective countercurrent multiplier in the villus facilitates the absorption of sodium and water. The countercurrent exchange also permits diffusion of O_2 from arterioles to venules. At low flow rates, a substantial fraction of the O_2 may be shunted from arterioles to venules near the base of the villus, thereby curtailing the supply of O_2 to the mucosal cells at the tip of the villus. When intestinal blood flow is reduced, the shunting of O_2 is exaggerated, which may cause extensive necrosis of the intestinal villi.

Neural regulation. The neural control of the mesenteric circulation is almost exclusively sympathetic. Increased sympathetic activity constricts the mesenteric arterioles, precapillary sphincters, and capacitance vessels. These responses are mediated by alpha-receptors, which are prepotent in the mesenteric circulation; however, beta-receptors are also present. Infusion of a beta-receptor agonist, such as isoproterenol, causes vasodilation.

During fighting or in response to artificial stimulation of the hypothalamic "defense" area, pronounced vasoconstriction occurs in the mesenteric vascular bed. This shifts blood flow from the temporarily less important intestinal circulation to the more crucial skeletal muscles, heart, and brain.

Autoregulation. Autoregulation in the intestinal circulation is not as well developed as it is in certain other vascular beds, such as those in the brain and kidney. The principal mechanism responsible for autoregulation is metabolic, although a myogenic mechanism probably also participates. The adenosine concentration in the mesenteric venous blood rises fourfold after brief arterial occlusion. Adenosine is a potent vasodilator in the mesenteric vascular bed and may be the principal metabolic mediator of autoregulation. However, potassium and altered osmolality may also contribute to the overall response.

The O_2 consumption of the small intestine is more rigorously controlled than is the blood flow. In one series of experiments, the O_2 uptake of the small intestine

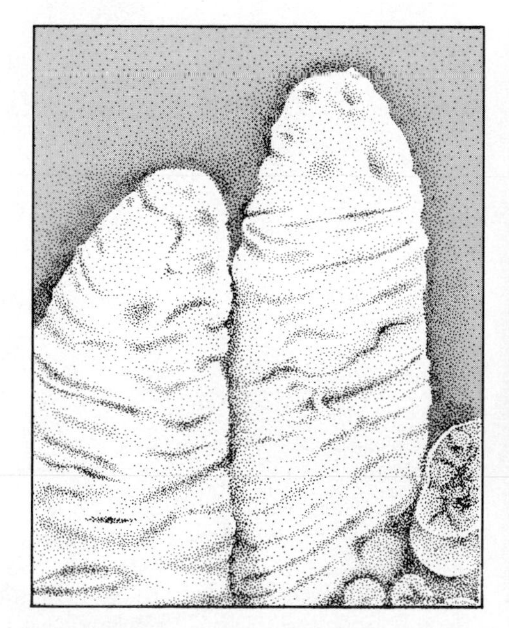

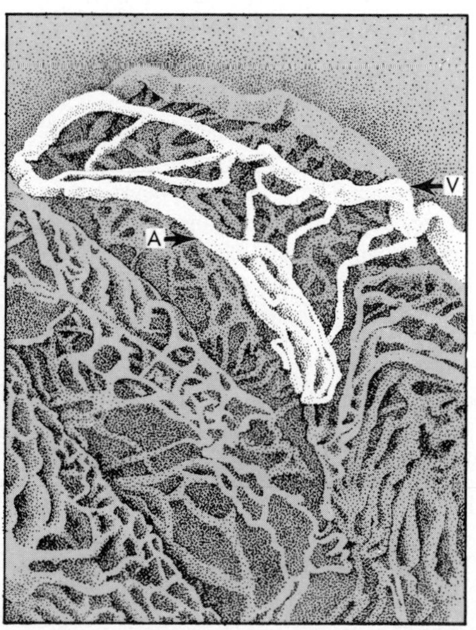

■ **Fig. 31-15** Scanning electron micrographs of rabbit intestinal villi *(left panel)* and corrosion cast of the microcirculation in the villus *(right panel)*. *A,* Arteriole; *V,* venule. (From Gannon BJ, Gore RW, Rogers PAW: *Biomed Res* 2[Suppl.]:235, 1981.)

remained constant when the arterial perfusion pressure was varied between 30 and 125 mm Hg.

Functional hyperemia. Food ingestion increases intestinal blood flow. The secretion of certain gastrointestinal hormones contributes to this hyperemia. Gastrin and cholecystokinin augment intestinal blood flow, and they are secreted when food is ingested. The absorption of food affects the intestinal blood flow. Undigested food has no vasoactive influence, whereas several products of digestion are potent vasodilators. Among the various constituents of chyme, the principal mediators of mesenteric hyperemia are glucose and fatty acids.

■ *Hepatic Circulation*

Anatomy. The blood flow to the liver normally is about 25% of the cardiac output. The flow is derived from two sources, the portal vein and the hepatic artery. Ordinarily the portal vein provides about three fourths of the blood flow. The portal venous blood already has passed through the gastrointestinal capillary bed, and therefore much of the O_2 already has been extracted. The hepatic artery delivers the remaining one fourth of the blood, which is fully saturated with O_2. Hence, about three fourths of the O_2 used by the liver is derived from the hepatic arterial blood.

The small branches of the portal vein and hepatic artery give rise to terminal portal venules and hepatic arterioles (Fig. 31-16). These terminal vessels enter the hepatic acinus (the functional unit of the liver) at its center. Blood flows from these terminal vessels into the sinusoids, which constitute the capillary network of the liver. The sinusoids radiate toward the periphery of the acinus, where they connect with the terminal hepatic venules. Blood from these terminal venules drains into progressively larger branches of the hepatic veins, which are tributaries of the inferior vena cava.

Hemodynamics. The mean blood pressure in the portal vein is about 10 mm Hg and that in the hepatic artery is about 90 mm Hg. The resistance of the vessels upstream to the hepatic sinusoids is considerably greater than that of the downstream vessels. Consequently, the pressure in the sinusoids is only 2 or 3 mm Hg greater than that in the hepatic veins and inferior vena cava. The ratio of presinusoidal to postsinusoidal resistance in the liver is much greater than is the ratio of precapillary to postcapillary resistance for almost any other vascular bed. Hence, drugs and other interventions that alter the presinusoidal resistance usually affect the pressure in

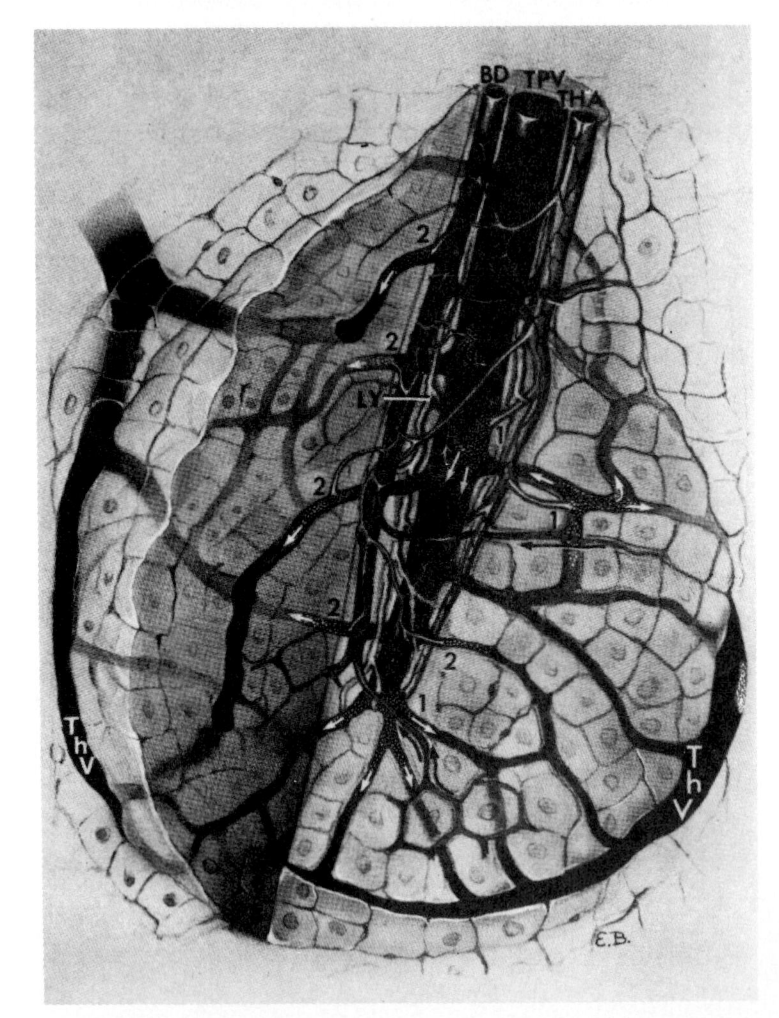

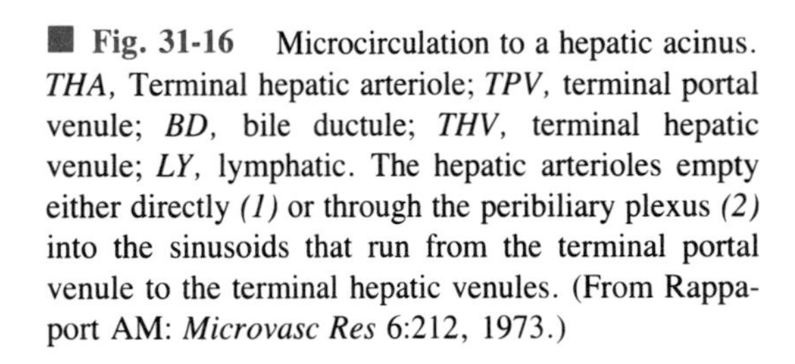
■ **Fig. 31-16** Microcirculation to a hepatic acinus. *THA,* Terminal hepatic arteriole; *TPV,* terminal portal venule; *BD,* bile ductule; *THV,* terminal hepatic venule; *LY,* lymphatic. The hepatic arterioles empty either directly *(1)* or through the peribiliary plexus *(2)* into the sinusoids that run from the terminal portal venule to the terminal hepatic venules. (From Rappaport AM: *Microvasc Res* 6:212, 1973.)

the sinusoids only slightly. Such changes in presinusoidal resistance have little effect on the fluid exchange across the sinusoidal wall. Conversely, changes in hepatic venous (and in central venous) pressure are transmitted almost quantitatively to the hepatic sinusoids and profoundly affect the transsinusoidal exchange of fluids. When central venous pressure is elevated, as in congestive heart failure, plasma water transudes from the liver into the peritoneal cavity, leading to **ascites.**

Regulation of flow. Blood flows in the portal venous and hepatic arterial systems vary reciprocally. When blood flow is curtailed in one system, the flow increases in the other system. However, the resultant increase in flow in one system usually does not fully compensate for the initiating reduction in flow in the other system.

The portal venous system does not autoregulate. As the portal venous pressure and flow are raised, resistance either remains constant or it decreases. The hepatic arterial system does autoregulate, however.

The liver tends to maintain a constant O_2 consumption, because the extraction of O_2 from the hepatic blood is very efficient. As the rate of O_2 delivery to the liver is varied, the liver compensates by an appropriate change in the fraction of O_2 extracted from the blood. This extraction is facilitated by the distinct separation of the presinusoidal vessels at the acinar center from the postsinusoidal vessels at the periphery of the acinus (see Fig. 31-16). The substantial distance between these types of vessels prevents the countercurrent exchange of O_2, contrary to the condition that exists in an intestinal villus (see Fig. 31-15).

The sympathetic nerves constrict the presinusoidal resistance vessels in the portal venous and hepatic arterial systems. Neural effects on the capacitance vessels are more important, however. The effects are mediated mainly by alpha-receptors.

Capacitance vessels. The liver contains about 15% of the total blood volume of the body. Under appropriate conditions, such as in response to hemorrhage, about half of the hepatic blood volume can be rapidly expelled. Hence, the liver constitutes an important blood reservoir in humans. In certain other species, such as the dog, the spleen is a more important blood reservoir. Smooth muscle in the capsule and trabeculae of the spleen contract in response to increased sympathetic neural activity, such as occurs during exercise or hemorrhage. However, this mechanism does not exist in humans.

Fetal Circulation

The circulation of the fetus shows a number of differences from that of the postnatal infant. The fetal lungs are functionally inactive, and the fetus depends completely on the placenta for O_2 and nutrient supply. Oxygenated fetal blood from the placenta passes through the umbilical vein to the liver. A major fraction passes through the liver, and a small fraction bypasses the liver to the inferior vena cava through the **ductus venosus** (Fig. 31-17). In the inferior vena cava, blood from the ductus venosus joins blood returning from the lower trunk and extremities; this combined stream is in turn joined by blood from the liver through the hepatic veins. The streams of blood tend to maintain their identity in the inferior vena cava and are divided into two streams of unequal size by the edge of the interatrial septum **(crista dividens).** The larger stream, which is mainly blood from the umbilical vein, is shunted to the left atrium through the **foramen ovale,** which lies between the inferior vena cava and the left atrium (inset, Fig. 31-17). The other stream passes into the right atrium, where it is joined by superior vena caval blood returning from the upper parts of the body and by blood from the myocardium. In contrast to the adult, in whom the right and left ventricles pump in series, in the fetus the ventricles operate essentially in parallel. Because of the large pulmonary resistance, less than one third of the right ventricular output goes through the lungs. The remainder passes through the **ductus arteriosus** from the pulmonary artery to the aorta at a point distal to the origins of the arteries to the head and upper extremities. Flow from pulmonary artery to aorta occurs because pulmonary artery pressure is about 5 mm Hg higher than aortic pressure in the fetus. The large volume of blood coming through the foramen ovale into the left atrium is joined by blood returning from the lungs and is pumped out by the left ventricle into the aorta. About one third of the aortic blood goes to the head, upper thorax, and arms and the remaining two thirds go to the rest of the body and the placenta. The amount of blood pumped by the left ventricle is about 20% greater than that pumped by the right ventricle, and the major fraction of the blood that passes down the descending aorta flows by way of the two umbilical arteries to the placenta.

In Fig. 31-17 the distribution of fetal blood flow is given in a percentage of the combined right and left ventricular outputs. Note that over half of the combined cardiac output is returned directly to the placenta without passing through any capillary bed. Also indicated in Fig. 31-17 are the O_2 saturations of the blood (numbers in parentheses) at various points of the fetal circulation. Fetal blood leaving the placenta is 80% saturated, but the saturation of the blood passing through the foramen ovale is reduced to 67% by mixing with desaturated blood returning from the lower part of the body and the liver. Addition of the desaturated blood from the lungs reduces the O_2 saturation of left ventricular blood to 62%, which is the level of saturation of the blood reaching the head and upper extremities. The blood in the right ventricle, a mixture of desaturated superior vena caval blood, coronary venous blood, and inferior vena caval blood, is only 52% saturated with O_2. When

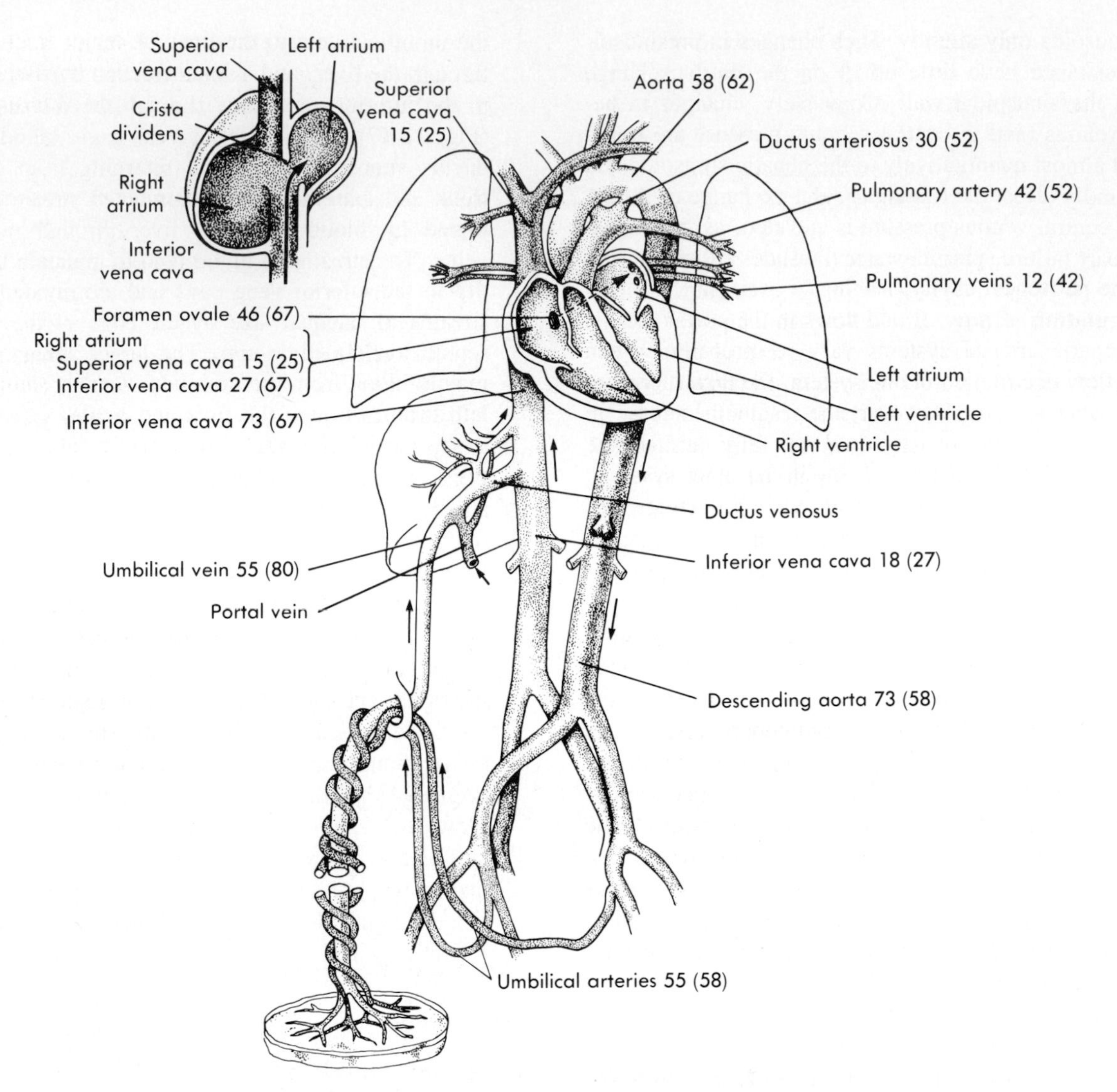

■ **Fig. 31-17** Schematic diagram of the fetal circulation. The numbers without parentheses represent the distribution of cardiac output as a percentage of the sum of the right and left ventricular outputs, and the numbers within parentheses represent the percentage of O_2 saturation of the blood flowing in the indicated blood vessel. The insert at the upper left illustrates the direction of flow of a major portion of the inferior vena cava blood through the foramen ovale to the left atrium. (Values for percentage distribution of blood flow and O_2 saturations are from Dawes GS, Mott JC, Widdicombe JG: *J Physiol* 126:563, 1954.)

the major portion of this blood traverses the ductus arteriosus and joins that pumped out by the left ventricle, the resultant O_2 saturation of blood traveling to the lower part of the body and back to the placenta is 58% saturated. Thus it is apparent that the tissues receiving blood of the highest O_2 saturation are the liver, heart, and upper parts of the body, including the head.

At the placenta the chorionic villi dip into the maternal sinuses, and O_2, CO_2, nutrients, and metabolic waste products exchange across the membranes. The barrier to exchange is quite large, and the equilibrium of O_2 tension between the two circulations is not reached at normal rates of blood flow. Therefore, the O_2 tension of the fetal blood leaving the placenta is very low. Were it not for the fact that fetal hemoglobin has a greater affinity for O_2 than does adult hemoglobin, the fetus would not receive an adequate O_2 supply. The fetal oxyhemoglobin dissociation curve is shifted to the left so that at equal pressures of O_2 fetal blood will carry significantly more O_2 than will maternal blood. If the mother is subjected to hypoxia, the reduced blood O_2 tension is reflected in the fetus by tachycardia and an increase in blood flow through the umbilical vessels. If the hypoxia persists or if flow through the umbilical vessels is impaired, fetal distress occurs and is first manifested as bradycardia. In early fetal life the high cardiac glycogen levels that prevail (which gradually decrease to adult levels by term) may protect the heart from acute periods of hypoxia.

Circulatory Changes That Occur at Birth

The umbilical vessels have thick, muscular walls that are very reactive to trauma, tension, sympathomimetic amines, bradykinin, angiotensin, and changes in O_2 tension. In animals in which the umbilical cord is not tied, hemorrhage of the newborn is prevented by constriction of these large vessels in response to one or more of these stimuli. Closure of the umbilical vessels produces an increase in total peripheral resistance and of blood pressure. When blood flow through the umbilical vein ceases, the ductus venosus, a thick-walled vessel with a muscular sphincter, closes. What initiates closure of the ductus venosus is still unknown. The asphyxia, which starts with constriction or clamping of the umbilical vessels, plus the cooling of the body activate the respiratory center of the newborn infant. With the filling of the lungs with air, pulmonary vascular resistance decreases to about one tenth of the value existing before lung expansion. This resistance change is not caused by the presence of O_2 in the lungs, because the change is just as great if the lungs are filled with nitrogen. However, filling the lungs with liquid does not reduce pulmonary vascular resistance (see Chapter 34).

The left atrial pressure is raised above that in the inferior vena cava and right atrium by (1) the decrease in pulmonary resistance, with the resulting large flow of blood through the lungs to the left atrium, (2) the reduction of flow to the right atrium caused by occlusion of the umbilical vein, and (3) the increased resistance to left ventricular output produced by occlusion of the umbilical arteries. This reversal of the pressure gradient across the atria abruptly closes the valve over the foramen ovale, and fusion of the septal leaflets occurs over a period of several days.

With the decrease in pulmonary vascular resistance, the pressure in the pulmonary artery falls to about one half its previous level (to about 35 mm Hg). This change in pressure, coupled with a slight increase in aortic pressure, reverses the flow of blood through the ductus arteriosus. However, within several minutes the large ductus arteriosus begins to constrict, producing turbulent flow, which is manifest as a murmur in the newborn. Constriction of the ductus arteriosus is progressive and usually complete within 1 to 2 days after birth. Closure of the ductus arteriosus appears to be initiated by the high O_2 tension of the arterial blood passing through it, because pulmonary ventilation with O_2 or with air low in O_2 induces, respectively, closure and opening of this shunt vessel. Whether O_2 acts directly on the ductus or through the release of a vasoconstrictor substance is not known. Similarly, in a heart-lung preparation made from a newborn lamb, the ductus arteriosus may be made to close with high Pao_2 and to open with low Pao_2. The mechanism whereby increases in Pao_2 produce closure of the ductus arteriosus is unknown, but changes in the concentrations of bradykinin, prostaglandins, and adenosine in blood or ductal tissue may be involved.

At birth the walls of the two ventricles are approximately of the same thickness, with a possibly slight preponderance of the right ventricle. Also present in the newborn is thickening of the muscular levels of the pulmonary arterioles, which is apparently responsible in part for the high pulmonary vascular resistance of the fetus. After birth the thickness of the walls of the right ventricle diminishes, as does the muscle layer of the pulmonary arterioles; the left ventricular walls increase in thickness. These changes are progressive over a period of weeks after birth.

Failure of the foramen ovale or ductus arteriosus to close after birth is occasionally observed and constitutes two of the more common congenital cardiac abnormalities that are now amenable to surgical correction.

Summary

Coronary Circulation

1. The physical factors that influence coronary blood flow are the viscosity of the blood, the frictional resistance of the vessel walls, the aortic pressure, and the extravascular compression of the vessels within the walls of the left ventricle. Left coronary blood flow is restricted during ventricular systole as a result of extravascular compression, and it is greatest during diastole when the intramyocardial vessels are not compressed.

2. Neural regulation of coronary blood flow is much less important than is metabolic regulation. Activation of the cardiac sympathetic nerves directly constricts the coronary resistance vessels. However, the enhanced myocardial metabolism caused by the associated increase in heart rate and contractile force produces vasodilation, which overrides the direct constrictor effect of sympathetic nerve stimulation. Stimulation of the cardiac branches of the vagus nerves slightly dilates the coronary arterioles.

3. A striking parallelism exists between metabolic activity of the heart and coronary blood flow. A decrease in oxygen supply or an increase in oxygen demand apparently releases a vasodilator that decreases coronary resistance. Of the known factors (CO_2, O_2, H^+, K^+, adenosine) that can mediate this response, adenosine appears to be the most likely candidate.

4. In response to gradual occlusion of a coronary artery, collateral vessels from adjacent unoccluded arter-

ies develop and supply blood to the compromised myocardium distal to the point of occlusion.

5. The myocardium functions only aerobically and in general uses substrates in proportion to their arterial concentration.

Skin Circulation

1. Most of the resistance vessels in the skin are under dual control of the sympathetic nervous system and local vasodilator metabolites, but the arteriovenous anastomoses found in the hands, feet, and face are solely under neural control.

2. The main function of skin blood vessels is to aid in the regulation of body temperature by constricting to conserve heat and dilating to lose heat.

3. Skin blood vessels dilate directly and reflexly in response to heat and constrict directly and reflexly in response to cold.

Skeletal Muscle Circulation

1. Skeletal muscle blood flow is regulated centrally by the sympathetic nerves and locally by the release of vasodilator metabolites.

2. At rest, neural regulation of blood flow is paramount, but it yields to metabolic regulation during muscle contractions.

Cerebral Circulation

1. Cerebral blood flow is predominantly regulated by metabolic factors, especially CO_2, K^+, and adenosine.

2. Increased regional cerebral activity produced by stimuli such as touch, pain, hand motion, talking, reading, reasoning, and problem solving are associated with enhanced blood flow in the activated areas of the contralateral cerebral cortex.

Intestinal Circulation

1. The microcirculation in the intestinal villi constitutes a countercurrent exchange system for O_2. This places the villi in jeopardy in states of low blood flow.

2. The splanchnic resistance and capacitance vessels are very responsive to changes in sympathetic neural activity.

Hepatic Circulation

1. The liver receives about 25% of the cardiac output; about three fourths of this comes via the portal vein and about one fourth via the hepatic artery. When flow is diminished in either the portal or hepatic systems, flow in the other system usually increases, but not proportionately.

2. The liver tends to maintain a constant O_2 consumption, in part because its mechanism for extracting O_2 from the blood is so efficient.

3. The liver normally contains about 15% of the total blood volume. It serves as an important blood reservoir for the body.

Fetal Circulation

1. In the fetus a large percentage of the right atrial blood passes through the foramen ovale to the left atrium, and a large percentage of the pulmonary artery blood passes through the ductus arteriosus to the aorta.

2. At birth the umbilical vessels, ductus venosus, and ductus arteriosus close by contraction of their muscle layers. The reduction in the pulmonary vascular resistance caused by lung inflation is the main factor that reverses the pressure gradient between the atria, thereby closing the foramen ovale.

Bibliography

Journal articles

Abboud FM, editor: Regulation of the cerebral circulation (symposium), *Fed Proc* 40:2296, 1981.

Baron JF et al: Independent role of arterial O_2 tension in local control of coronary blood flow, *Am J Physiol* 258:H1388, 1990.

Belardinelli L, Linden J, Berne RM: The cardiac effects of adenosine, *Prog Cardiovasc Dis* 32:73, 1989.

Belloni FL: The local control of coronary blood flow, *Cardiovasc Res* 13:63, 1979.

Berne RM: Role of adenosine in the regulation of coronary blood flow, *Circ Res* 47:807, 1980.

Berne RM, Winn HR, Rubio R: The local regulation of cerebral blood flow. *Prog Cardiovasc Dis* 24:243, 1981.

Brunner JJ et al: Carotid sinus baroreceptor control of splanchnic resistance and capacity, *Am J Physiol* 255:H1305, 1988.

Buckberg GD et al: Variable effects of heart rate on phasic and regional left ventricular muscle blood flow in anesthetized dogs, *Cardiovasc Res* 9:1, 1975.

Feigl EO: Coronary physiology, *Physiol Rev* 63:1, 1983.

Feigl EO: Parasympathetic control of coronary blood flow in dogs, *Circ Res* 25:509, 1969.

Feigl EO: Sympathetic control of coronary circulation, *Circ Res* 20:262, 1967.

Greenway CV, Lautt WW: Distensibility of hepatic venous

resistance sites and consequences on portal pressure, *Am J Physiol* 254:H452, 1988.

Gregg DE: The natural history of coronary collateral development, *Circ Res* 35:335, 1974.

Heymann MA, Iwamoto HS, Rudolf AM: Factors affecting changes in the neonatal systemic circulation, *Ann Rev Physiol* 43:371, 1981.

Heymann MA, Rudolph AM: Control of the ductus arteriosus, *Physiol Rev* 55:62, 1975.

Klocke FJ, Ellis AK: Control of coronary blood flow, *Ann Rev Med* 31:489, 1980.

Klocke FJ et al: Coronary pressure-flow relationships—controversial issues and probable implications, *Circ Res* 56:310, 1985.

Kontos HA: Regulation of the cerebral circulation, *Ann Rev Physiol* 43:397, 1981.

Laughlin MH: Skeletal muscle blood flow capacity: role of muscle pump in exercise hyperemia, *Am J Physiol* 253:H993, 1987.

Lautt WW: Hepatic nerves: a review of their functions and effects, *Can J Physiol Pharmacol* 58:105, 1980.

Murray PA, Belloni FL, Sparks HV: The role of potassium in the metabolic control of coronary vascular resistance in the dog, *Circ Res* 44:767, 1979.

Olsson RA: Local factors regulating cardiac and skeletal muscle blood flow, *Ann Rev Physiol* 43:385, 1981.

Olsson RA, Bunger R: Metabolic control of coronary blood flow, *Prog Cardiovasc Dis* 29:369, 1987.

Olsson RA, Pearson JD: Cardiovascular purinoceptors, *Physiol Rev* 70:761, 1990.

Rubio R, Berne RM: Regulation of coronary blood flow, *Prog Cardiovasc Dis* 18:105, 1975.

Schwartz LM, McKenzie JE: Adenosine and active hyperemia in soleus and gracilis muscle of cats, *Am J Physiol* 259:H1295, 1990.

Shepherd AP, Riedel GL: Intramural distribution of intestinal blood flow during sympathetic stimulation, *Am J Physiol* 255:H1091, 1988.

Wearn JT et al: The nature of the vascular communications between the coronary arteries and the chambers of the heart, *Am Heart J* 9:143, 1933.

White CW, Wilson RF, Marcus ML: Methods of measuring myocardial blood flow in humans, *Prog Cardiovasc Dis* 31:79, 1988.

Books and monographs

Berne RM, Rubio R: *Coronary circulation*. In *Handbook of physiology:* section 2: *The cardiovascular system—the heart,* vol 1, Bethesda, Md, 1979, American Physiological Society.

Berne RM, Winn HR, Rubio R: *Metabolic regulation of cerebral blood flow*. In *Mechanisms of vasodilation*—Second Symposium, New York, 1981, Raven Press.

Donald DE: Splanchnic circulation. In *Handbook of physiology:* section 2: *The cardiovascular system—peripheral circulation and organ blood flow,* vol 3, Bethesda, Md, 1983, American Physiological Society.

Faber JJ, Thornburg KL: *Placental physiology,* New York, 1983, Raven Press.

Gootman N, Gootman PM, editors: *Perinatal cardiovascular function,* New York, 1983, Marcel Dekker.

Granger DN et al: *Microcirculation of the intestinal mucosa*. In *Handbook of physiology, gastrointestinal system,* vol 1, Bethesda, Md, 1989, American Physiological Society.

Greenway CV, Lautt WW: *Hepatic circulation*. In *Handbook of physiology, gastrointestinal system; motility and circulation,* vol 1, Bethesda, Md, 1989, American Physiological Society.

Gregg DE: *Coronary circulation in health and disease,* Philadelphia, 1950, Lea & Febiger.

Grover RF et al: *Pulmonary circulation*. In *Handbook of physiology:* section 2: *The cardiovascular system—peripheral circulation and organ blood flow,* vol 3, Bethesda, Md, 1983, American Physiological Society.

Guth PH, Leung FW, Kauffman GL Jr: *Physiology of gastric circulation*. In *Handbook of physiology, gastrointestinal system,* vol 1, Bethesda, Md, 1989, American Physiological Society.

Heistad DD, Kontos HA: *Cerebral circulation*. In *Handbook of physiology:* section 2: *The cardiovascular system—peripheral circulation and organ blood flow,* vol 3, Bethesda, Md, 1983, American Physiological Society.

Hellon R: *Thermoreceptors*. In *Handbook of physiology:* section 2: *The cardiovascular system—peripheral circulation and organ blood flow,* vol 3, Bethesda, Md, 1983, American Physiological Society.

Lewis T: *Blood vessels of the human skin and their responses,* London, 1927, Shaw & Son.

Longo LD, Reneau DD, editors: *Fetal and newborn cardiovascular physiology,* vol 1, *Developmental aspects,* New York, 1978, Garland STPM Press.

Marcus ML: *The coronary circulation in health and disease,* New York, 1983, McGraw-Hill.

Morgan HE, Rannels DE, McKee EE: *Protein metabolism of the heart*. In *Handbook of physiology:* section 2: *The cardiovascular system—the heart,* vol 1, Bethesda, Md, 1979, American Physiological Society.

Mott JC, Walker DW: *Neural and endocrine regulation of circulation in the fetus and newborn*. In *Handbook of physiology:* section 2: *The cardiovascular system—peripheral circulation and organ blood flow,* vol 3, Bethesda, Md, 1983, American Physiological Society.

Olsson RA, Bunger R, Spaan JAE.: *The coronary circulation*. In Fozzard HA et al, editors: *The heart and cardiovascular system,* ed 2, New York, 1991, Raven Press.

Owman C, Edvinsson L, editors: *Neurogenic control of the brain circulation,* Oxford, 1977, Pergamon Press.

Randle PJ, Tubbs PK: *Carbohydrate and fatty acid metabolism*. In *Handbook of physiology:* section 2: *The cardiovascular system—the heart,* vol 1, Bethesda, Md, 1979, American Physiological Society.

Roddie EC: *Circulation to skin and adipose tissue*. In *Handbook of physiology:* section 2: *The cardiovascular system—peripheral circulation and organ blood flow,* vol 3, Bethesda, Md, 1983, American Physiological Society.

Schaper W: The collateral circulation of the heart, New York, 1971, North-Holland Publishing.

Shepherd JT: *Circulation to skeletal muscle*. In *Handbook of physiology:* section 2: *The cardiovascular system—peripheral circulation and organ blood flow,* vol 3, Bethesda, Md, 1983, American Physiological Society.

Sparks HV Jr, Wangler RD, Gorman MW: *Control of the coronary circulation*. In Sperelakis N, editor: *Physiology and pathophysiology of the heart,* ed 2, Boston, 1989, Wolters-Kluwer Publishers.

CHAPTER

32

Interplay of Central and Peripheral Factors in the Control of the Circulation

The primary function of the circulatory system is to deliver the supplies needed for tissue metabolism and growth and to remove the products of metabolism. To explain how the heart and blood vessels serve this function, it has been necessary to analyze the system morphologically and functionally and to discuss the mechanisms of action of the component parts in their contribution to maintaining adequate tissue perfusion under different physiological conditions. Once the functions of the various components are understood, it is essential that their interrelationships in the overall role of the circulatory system be considered. Tissue perfusion depends on arterial pressure and local vascular resistance; and arterial pressure, in turn, depends on cardiac output and total peripheral resistance (TPR). Arterial pressure is maintained within a relatively narrow range in the normal individual, a feat that is accomplished by reciprocal changes in cardiac output and TPR. However, cardiac output and peripheral resistance are each influenced by a number of factors, and it is the interplay among these factors that determines the level of these two variables. The autonomic nervous system and the baroreceptors play the key role in regulating blood pressure. However, from the long-range point of view the control of fluid balance by the kidney, adrenal cortex, and central nervous system, with maintenance of a constant blood volume, is of the greatest importance.

In a well-regulated system one way to study the extent and sensitivity of the regulatory mechanism is to disturb the system and observe its response to restore the preexisting steady state. Disturbances in the form of physical exercise and hemorrhage will be used to illustrate the effects of the various factors that go into regulation of the circulatory system.

■ *Exercise*

The cardiovascular adjustments in exercise comprise a combination and integration of neural and local (chemical) factors. The neural factors consist of (1) **central command,** (2) reflexes originating in the contracting muscle, and (3) the baroreceptor reflex. Central command is the cerebrocortical activation of the sympathetic nervous system that produces cardiac acceleration, increased myocardial contractile force, and peripheral vasoconstriction. Reflexes can be activated intramuscularly by stimulation of mechanoreceptors (stretch, tension) and "chemoreceptors" (products of metabolism) in response to muscle contraction. Impulses from these receptors travel centrally via small myelinated (group III) and unmyelinated (group IV) afferent nerve fibers. The group IV unmyelinated fibers may represent the muscle chemoreceptors; no morphological chemoreceptor has been identified. The central connections of this reflex are unknown, but the efferent limb is the sympathetic nerve fibers to the heart and peripheral blood vessels. The baroreceptor reflex has been described on p. 486, and the local factors that influence skeletal muscle blood flow (metabolic vasodilators) are described on pp. 483 and 521. Vascular chemoreceptors do not play a significant role in regulation of the cardiovascular system in exercise. The pH, P_{CO_2}, and P_{O_2} of arterial blood are normal during exercise, and the vascular chemoreceptors are located on the arterial side of the circulatory system.

■ *Mild to Moderate Exercise*

In humans or in trained animals, anticipation of physical activity inhibits the vagal nerve impulses to the heart and increases sympathetic discharge. The concerted inhibition of parasympathetic areas and activation of sympathetic areas of the medulla on the heart result in an increase in heart rate and myocardial contractility. The tachycardia and the enhanced contractility increase cardiac output.

Peripheral resistance. At the same time that cardiac stimulation occurs, the sympathetic nervous system also elicits vascular resistance changes in the periphery. In skin, kidneys, splanchnic regions, and inactive muscle, sympathetic-mediated vasoconstriction increases vascular resistance, which diverts blood away from these areas (Fig. 32-1). This increased resistance in vascular beds of inactive tissues persists throughout the period of exercise.

As cardiac output and blood flow to active muscles increase with progressive increments in the intensity of exercise, blood flow to the splanchnic and renal vasculatures decreases. Blood flow to the myocardium increases, whereas that to the brain is unchanged. Skin blood flow initially decreases during exercise and then increases as body temperature rises with increments in duration and intensity of exercise. Skin blood flow finally decreases when the skin vessels constrict as the total body O_2 consumption nears maximum (see Fig. 32-1).

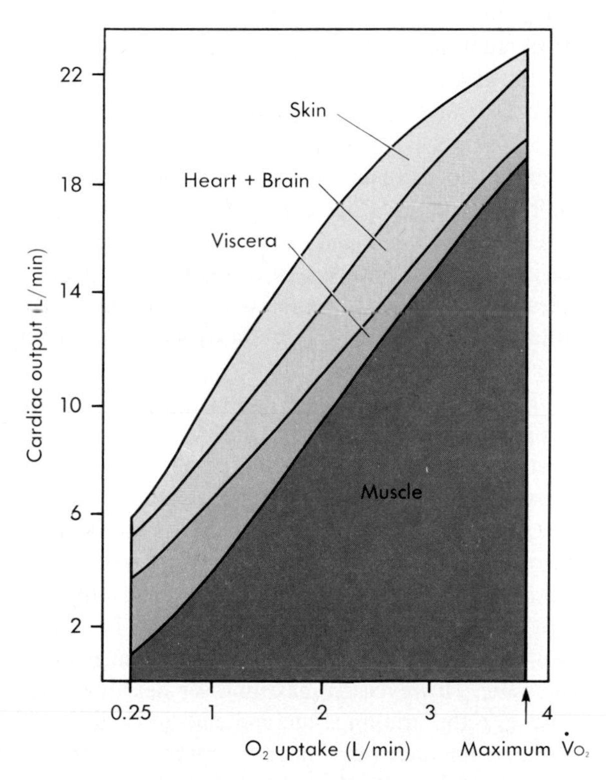

■ **Fig. 32-1** Approximate distribution of cardiac output at rest and at different levels of exercise up to the maximum O_2 consumption ($\dot{V}O_{2max}$) in a normal young man. Viscera refers to splanchnic and renal blood flow. (Redrawn from Ruch HP, Patton TC: *Physiology and biophysics,* ed 12, 1974, WB Saunders.)

The major circulatory adjustment to prolonged exercise involves the vasculature of the active muscles. Local formation of vasoactive metabolites induces marked dilation of the resistance vessels, which progresses with increases in the intensity level of exercise. Potassium is one of the vasodilator substances released by contracting muscle, and it may be in part responsible for the initial decrease in vascular resistance in the active muscles. Other contributing factors may be the release of adenosine and a decrease in pH during sustained exercise. The local accumulation of metabolites relaxes the terminal arterioles. Blood flow through the muscle may increase fifteen to twenty times above the resting level. This metabolic vasodilation of the precapillary vessels in active muscles occurs very soon after the onset of exercise, and the decrease in TPR enables the heart to pump more blood at a lesser load and more efficiently (less pressure work, p. 516) than if TPR were unchanged. Only a small percentage of the capillaries is perfused at rest, whereas in actively contracting muscle all or nearly all of the capillaries contain flowing blood **(capillary recruitment).** The surface available for exchange of gases, water, and solutes is increased manyfold. Furthermore, the hydrostatic pressure in the capillaries is increased because of the relaxation of the resistance vessels. Hence, there is a net movement of water and solutes into the muscle tissue. Tissue pressure rises and remains elevated during exercise as fluid continues to move out of the capillaries and is carried away by the lymphatics. Lymph flow is increased as a result of the increase in capillary hydrostatic pressure and the massaging effect of the contracting muscles on the valve-containing lymphatic vessels.

The contracting muscle avidly extracts O_2 from the perfusing blood (increased AV-O_2 difference, Fig. 32-2), and the release of O_2 from the blood is facilitated by the nature of oxyhemoglobin dissociation. The reduction in pH caused by the high concentration of CO_2 and the formation of lactic acid, and the increase in temperature in the contracting muscle contribute to shifting the oxyhemoglobin dissociation curve to the right. At any given partial pressure of O_2, less O_2 is held by the hemoglobin in the red cells; consequently O_2 removal from the blood is more effective. Oxygen consumption may increase as much as sixtyfold with only a fifteenfold increase in muscle blood flow. Muscle myoglobin may serve as a limited O_2 store in exercise and can release attached O_2 at very low partial pressures. However, it can facilitate O_2 transport from capillaries to mitochondria by serving as an O_2 carrier.

Cardiac output. The enhanced sympathetic drive and the reduced parasympathetic inhibition of the sinoatrial node continue during exercise; consequently, tachycardia persists. If the work load is moderate and constant, the heart rate will reach a certain level and remain there throughout the period of exercise. However, if the work load increases, a concomitant increase in

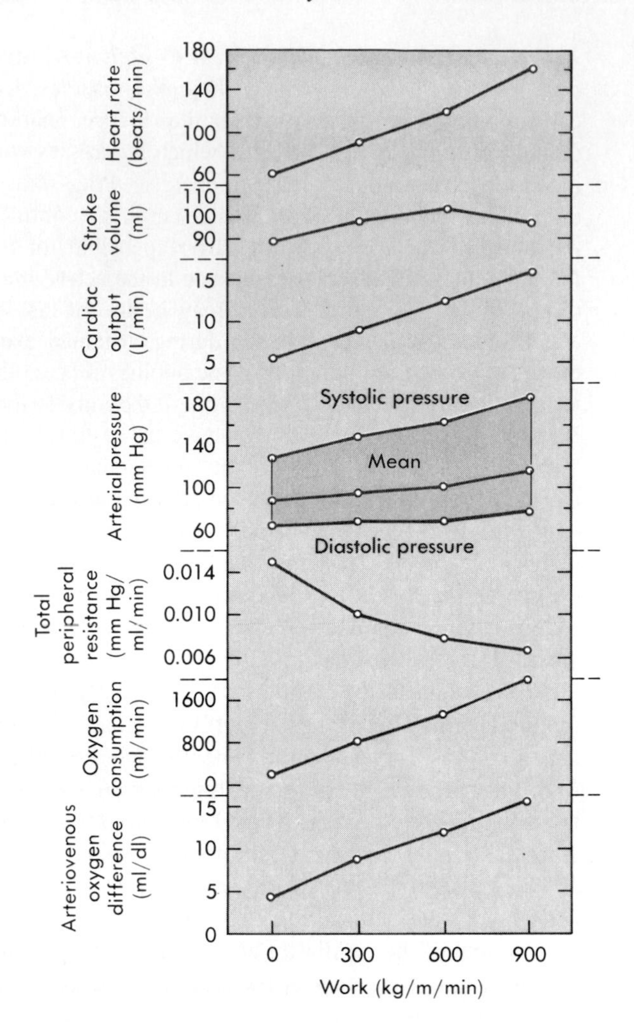

Fig. 32-2 Effect of different levels of exercise on several cardiovascular variables. (Date from Carlsten A, Grimby G: *The circulatory response to muscular exercise in man,* Springfield, Ill, 1966, Charles C Thomas, Publisher.)

heart rate occurs until a plateau is reached in severe exercise at about 180 beats per minute (see Fig. 32-2). In contrast to the large increment in heart rate, the increase in stroke volume is only about 10% to 35% (see Fig. 32-2), the larger values occurring in trained individuals. (In very well-trained distance runners, whose cardiac outputs can reach six to seven times the resting level, stroke volume reaches about twice the resting value.)

Thus it is apparent that *the increase in cardiac output observed with exercise is achieved principally by an increase in heart rate.* If the baroreceptors are denervated, the cardiac output and heart rate responses to exercise are sluggish when compared to the changes in animals with normally innervated baroreceptors. However, in the absence of autonomic innervation of the heart, as produced experimentally in dogs by total cardiac denervation, exercise still elicits an increment in cardiac output comparable to that observed in normal animals, but chiefly by means of an elevated stroke volume. However, if a beta-adrenergic receptor blocking agent is given to dogs with denervated hearts, exercise performance is impaired. The beta-adrenergic receptor blocker apparently prevents the cardiac acceleration and enhanced contractility caused by increased amounts of circulating catecholamines and hence limits the increase in cardiac output necessary for maximum exercise performance.

Venous return. In addition to the contribution made by sympathetically mediated constriction of the capacitance vessels in both exercising and nonexercising parts of the body, venous return is aided by the working skeletal muscles and the muscles of respiration. The intermittently contracting muscles compress the vessels that course through them, and in the case of veins with their valves oriented toward the heart, pump blood back toward the right atrium. The flow of venous blood to the heart is also aided by the increase in the pressure gradient developed by the more negative intrathoracic pressure produced by deeper and more frequent respirations. In humans, little evidence exists that blood reservoirs contribute much to the circulating blood volume, with the exception of the skin, lungs, and liver. In fact, blood volume is usually reduced slightly during exercise, as evidenced by a rise in the hematocrit ratio, because of water loss externally by sweating and enhanced ventilation, and by fluid movement into the contracting muscle. The fluid loss from the vascular compartment into contracting muscle reaches a plateau as interstitial fluid pressure rises and opposes the increased hydrostatic pressure in the capillaries of the active muscle. The fluid loss is partially offset by movement of fluid from the splanchnic regions and inactive muscle into the bloodstream. This influx of fluid occurs as a result of a decrease of hydrostatic pressure in the capillaries of these tissues and of an increase in plasma osmolarity because of movement of osmotically active particles into the blood from the contracting muscle. In addition, reduced urine formation by the kidneys helps to conserve body water.

The large volume of blood returning to the heart is so rapidly pumped through the lungs and out into the aorta that central venous pressure remains essentially constant. Thus the Frank-Starling mechanism of a greater initial fiber length does not account for the greater stroke volume in moderate exercise. Chest x-ray films of individuals at rest and during exercise reveal a decrease in heart size in exercise, which is in harmony with the observations of a constant ventricular filling pressure. However, in maximum or near-maximum exercise, right atrial pressure and end-diastolic ventricular volume do increase. Thus, the Frank-Starling mechanism contributes to the enhanced stroke volume in very vigorous exercise.

Arterial pressure. If the exercise involves a large proportion of the body musculature, such as in running or swimming, the reduction in total vascular resistance can be considerable (see Fig. 32-2). Nevertheless, arterial pressure starts to rise with the onset of exercise, and the increase in blood pressure roughly parallels the severity of the exercise performed (see Fig. 32-2). Therefore, the increase in cardiac output is proportionally greater than the decrease in TPR. The vasoconstriction produced in the inactive tissues by the sympathetic nervous system (and to some extent by the release of catecholamines from the adrenal medulla) is important for maintaining normal or increased blood pressure, because sympathectomy or drug-induced block of the adrenergic sympathetic nerve fibers results in a decrease in arterial pressure **(hypotension)** during exercise.

Sympathetic-mediated vasoconstriction also occurs in active muscle when additional muscles are activated after about half of the total skeletal musculature is contracting. In experiments in which one leg is working at maximum levels and then the other leg starts to work, blood flow decreases in the first working leg. Furthermore, blood levels of norepinephrine rise significantly in exercise, and most of it comes from sympathetic nerves in the active muscles.

As body temperature rises during exercise, the skin vessels dilate in response to thermal stimulation of the heat-regulating center in the hypothalamus, and TPR decreases further. This would result in a decline in blood pressure were it not for the increasing cardiac output and constriction of arterioles in the renal, splanchnic, and other tissues.

In general, mean arterial pressure rises during exercise as a result of the increase in cardiac output. However, the effect of enhanced cardiac output is offset by the overall decrease in TPR so that the mean blood pressure increase is relatively small. Vasoconstriction in the inactive vascular beds contributes to the maintenance of a normal arterial blood pressure for adequate perfusion of the active tissues. The actual pressure attained represents a balance between cardiac output and TPR (p. 459). Systolic pressure usually increases more than diastolic pressure, which results in an increase in pulse pressure (see Fig. 32-2). The larger pulse pressure is primarily attributable to a greater stroke volume and to a lesser degree to a more rapid ejection of blood by the left ventricle with less peripheral runoff during the brief ventricular ejection period.

Severe Exercise

In severe exercise taken to the point of exhaustion, the compensatory mechanisms begin to fail. Heart rate attains a maximum level of about 180 beats per minute, and stroke volume reaches a plateau and often decreases, resulting in a fall in blood pressure. Dehydration occurs. Sympathetic vasoconstrictor activity supersedes the vasodilator influence on the cutaneous vessels and has the hemodynamic effect of a slight increase in effective blood volume. However, cutaneous vasoconstriction also decreases the rate of heat loss. Body temperature is normally elevated in exercise, and reduction in heat loss through cutaneous vasoconstriction can, under these conditions, lead to very high body temperatures with associated feelings of acute distress. The tissue and blood pH decrease, as a result of increased lactic acid and CO_2 production, and the reduced pH is probably the key factor that determines the maximum amount of exercise a given individual can tolerate because of muscle pain, subjective feeling of exhaustion, and inability or loss of will to continue. A summary of the neural and local effects of exercise on the cardiovascular system is schematized in Fig. 32-3.

Postexercise Recovery

When exercise stops, an abrupt decrease in heart rate and cardiac output occurs—the sympathetic drive to the heart is essentially removed. In contrast, TPR remains low for some time after the exercise is ended, presumably because of the accumulation of vasodilator metabolites in the muscles during the exercise period. As a result of the reduced cardiac output and persistence of vasodilation in the muscles, arterial pressure falls, often below preexercise levels, for brief periods. Blood pressure is then stabilized at normal levels by the baroreceptor reflexes.

Limits of Exercise Performance

The two main forces that could limit skeletal muscle performance in the human body are the rate of O_2 utilization by the muscles and the O_2 supply to the muscles. Muscle O_2 usage is probably not critical, because during exercise maximum O_2 consumption ($\dot{V}O_{2max}$) by a large percentage of the body muscle mass is not increased when additional muscles are activated. If muscle O_2 utilization were limiting, recruitment of more contracting muscle would use additional O_2 to meet the enhanced O_2 requirements and would thereby increase total body O_2 consumption. Limitation of O_2 supply could be caused by inadequate oxygenation of blood in the lungs or limitation of the supply of O_2 -laden blood to the muscles. Failure to fully oxygenate blood by the lungs can be excluded because even with the most strenuous exercise at sea level arterial blood is fully saturated with O_2. Therefore, O_2 delivery (blood flow, because arterial blood O_2 content is normal) to the active muscles appears to be the limiting factor in muscle performance. This limitation could be caused by the inability to increase cardiac output beyond a certain level as a result of a limitation of stroke volume, because heart rate reaches maximum levels before $\dot{V}O_{2max}$ is reached. However, blood pressure provides the energy for muscle perfusion, and blood pressure depends on peripheral

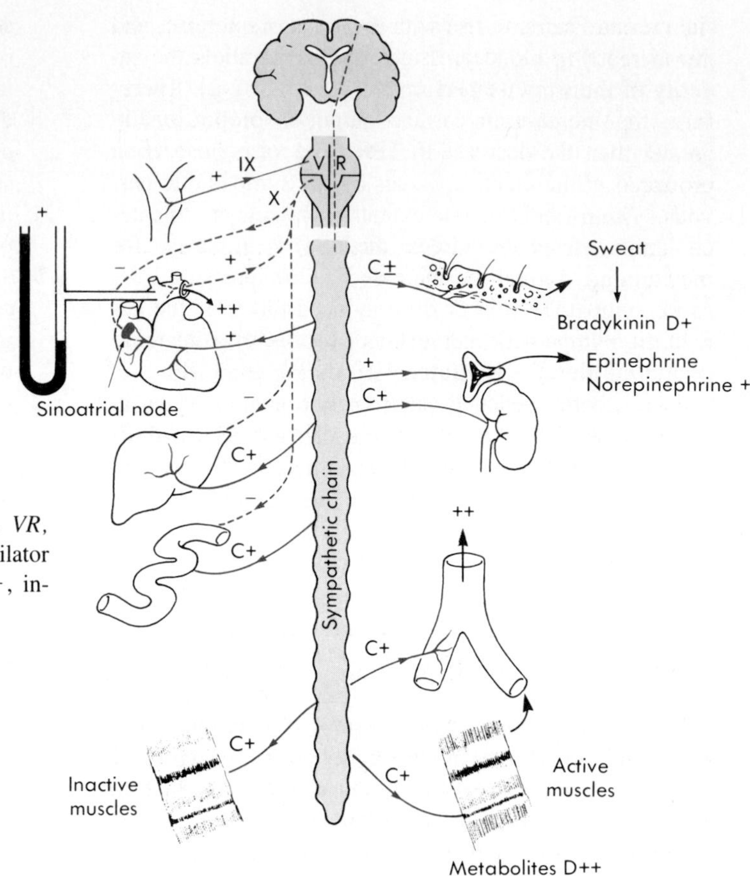

■ **Fig. 32-3** Cardiovascular adjustments in exercise. *VR*, Vasomotor region; *C*, vasoconstrictor activity; *D*, vasodilator activity; *IX*, glossopharyngeal nerve; *X*, vagus nerve; +, increased activity; −, decreased activity.

resistance, as well as on cardiac output. With increasing levels of exercise at peak $\dot{V}O_{2max}$ and peak cardiac output, blood pressure would fall as more muscle vascular beds dilate in response to locally released vasodilator metabolites if some degree of centrally mediated vasoconstriction (baroreceptor reflex) did not occur in the resistance vessels of the active muscles. Hence, the adjustment of resistance in the active muscles appears to be a contributing factor in the limitation of whole body exercise. *The major factor is the pumping capacity of the heart.* With exercise of a small group of muscles, such as the hand, when the cardiovascular system is not severely taxed, the limiting factor is unknown but lies within the muscle.

■ *Physical Training and Conditioning*

The response of the cardiovascular system to regular exercise is to increase its capacity to deliver O_2 to the active muscles and to improve the ability of the muscle to use O_2. The $\dot{V}O_{2max}$ is quite reproducible in a given individual and varies with the level of physical conditioning. Training progressively increases the $\dot{V}O_{2max}$, which reaches a plateau at the highest level of conditioning. Highly trained athletes have a lower resting heart rate, greater stroke volume, and lower peripheral resistance than they had before training or after deconditioning (becoming sedentary). The low resting heart rate is caused by a higher vagal tone and a lower sympathetic tone. With exercise, the maximum heart rate of the trained individual is the same as that in the untrained but is attained at a higher level of exercise. The trained person also exhibits a low vascular resistance that is inherent in the muscle. For example, if an individual exercises one leg regularly over an extended period and does not exercise the other leg, the vascular resistance is lower and the $\dot{V}O_{2max}$ is higher in the "trained" leg than in the "untrained" leg. Also, the well-trained athlete has a lower resting sympathetic outflow to the viscera than does a sedentary counterpart.

Physical conditioning is also associated with greater extraction of O_2 from the blood (greater AV-O_2 difference) by the muscles but not with an improvement in cardiac contractility. With long-term training capillary density in skeletal muscle increases. One can speculate that the number of arterioles also increases which can account for the decrease in muscle vascular resistance. The number of mitochondria increases, as do the oxidative enzymes in the mitochondria. Also, it appears that ATPase activity, myoglobin, and enzymes involved in lipid metabolism increase with physical conditioning. Endurance training, such as running or swimming, produces an increase in left ventricular volume without an increase in left ventricular wall thickness. In contrast, strength exercises, such as weight lifting, appear to produce some increase in left ventricular wall thickness (hypertrophy) with little effect on ventricular volume. However, this increase in wall thickness is small relative to that observed in hypertension in which the elevation of afterload persists because of the high peripheral resistance.

Hemorrhage

In an individual who has lost a large quantity of blood, the principal findings are related to the cardiovascular system. The arterial systolic, diastolic, and pulse pressures diminish and the pulse is rapid and feeble. The cutaneous veins are collapsed and fill slowly when compressed centrally. The skin is pale, moist, and slightly cyanotic. Respiration is rapid, but the depth of respiration may be shallow or deep.

Course of Arterial Blood Pressure Changes

Cardiac output decreases as a result of blood loss (p. 503). The changes in mean arterial pressure evoked by an acute hemorrhage in experimental animals are illustrated in Fig. 32-4. If sufficient blood is withdrawn rapidly to bring mean arterial pressure to 50 mm Hg, the pressure tends to rise spontaneously toward control over the subsequent 20 or 30 minutes. In some animals (curve *A*, Fig. 32-4) this trend continues, and normal pressures are regained within a few hours. In other animals (curve *B*), after an initial pressure rise, the pressure begins to decline and it continues to fall at an accelerating rate until death ensues.

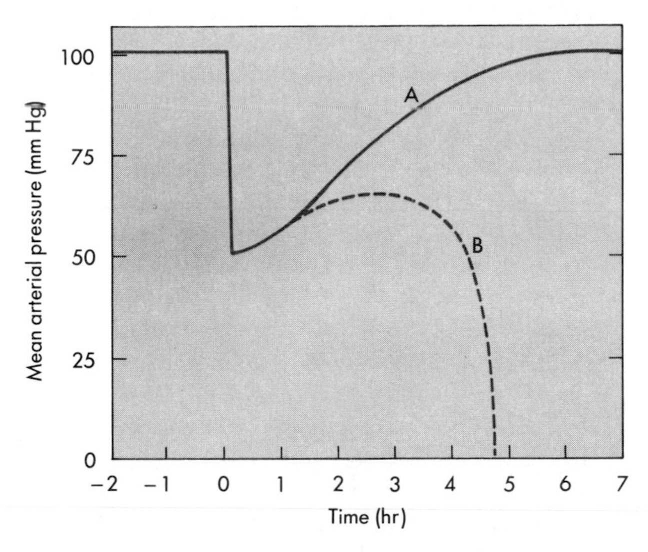

■ **Fig. 32-4** The changes in mean arterial pressure after a rapid hemorrhage. At time zero the animal is bled rapidly to a mean arterial pressure of 50 mm Hg. After a period in which the pressure returns toward the control level, some animals will continue to improve until the control pressure is attained (curve *A*). However, in other animals the pressure will begin to decline until death ensues (curve *B*).

In other experiments, animals are bled to a given hypotensive level, for example, to 35 mm Hg, by connecting a peripheral artery to a reservoir elevated to an appropriate height (Fig. 32-5). The arterial blood runs rapidly into the reservoir until the pressures become equilibrated and then continues to flow into the reservoir at a progressively slower rate for about 2 hours. This gradual increase in shed blood volume reflects the same compensatory mechanisms that produced the rise in arterial blood pressure after transitory hemorrhage in the experiment depicted in Fig. 32-4. However, in the experiment represented in Fig. 32-5, as the arterial pressure tends to rise higher than the pressure established by the reservoir, blood flows from the cannulated vessel into the reservoir.

About 2 hours after the beginning of hemorrhage, the arterial pressure tends to fall below the established level; and blood begins to flow from the reservoir to the animal (Fig. 32-5). Once about 50% of the maximum shed volume has returned to the animal, rapid reinfusion of the remaining shed blood improves arterial pressure only transiently. The arterial pressure then falls progressively until the animal dies. This progressive deterioration of cardiovascular function is termed **shock.** At some point the deterioration becomes irreversible; a lethal outcome can be retarded only temporarily by any known therapy, including massive transfusions of donor blood.

Compensatory Mechanisms

The changes in arterial pressure immediately after an acute blood loss (see Fig. 32-4) and in shed blood volume during the initial stages of sustained hemorrhage

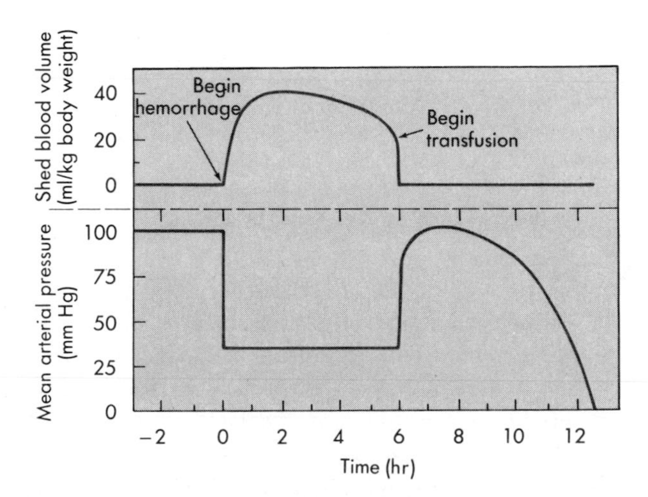

■ **Fig. 32-5** Changes in shed blood volume and mean arterial pressure during and after a 6-hour period of hemorrhage sufficient to hold mean arterial pressure at 35 mm Hg. The shed blood was reinfused after 6 hours.

(see Fig. 32-5) indicate that certain compensatory mechanisms must be operating. Any mechanism that raises the arterial pressure toward normal in response to the reduction in pressure may be designated a **negative feedback mechanism.** It is termed "negative" because the secondary change in pressure is opposite to the initiating change. The following negative feedback responses are evoked: (1) the baroreceptor reflexes, (2) the chemoreceptor reflexes, (3) cerebral ischemia responses, (4) reabsorption of tissue fluids, (5) release of endogenous vasoconstrictor substances, and (6) renal conservation of salt and water.

Baroreceptor reflexes. The reduction in mean arterial pressure and in pulse pressure during hemorrhage decreases the stimulation of the baroreceptors in the carotid sinuses and aortic arch (see Chapter 29). Several cardiovascular responses are thus evoked, all of which tend to return the arterial pressure toward normal. Reduction of vagal tone and enhancement of sympathetic tone increase heart rate and enhance myocardial contractility.

The increased sympathetic discharge also produces generalized venoconstriction, which has the same hemodynamic consequences as a transfusion of blood (p. 499). Sympathetic activation constricts certain blood reservoirs, which provides an autotransfusion of blood into the circulating bloodstream. In the dog considerable quantities of blood are mobilized by contraction of the spleen. In humans the spleen is not an important blood reservoir. Instead, the cutaneous, pulmonary, and hepatic vasculatures probably constitute the principal blood reservoirs.

Generalized arteriolar vasoconstriction is a prominent response to the diminished baroreceptor stimulation during hemorrhage. The reflex increase in peripheral resistance minimizes the fall in arterial pressure resulting from the reduction of cardiac output. Fig. 32-6 shows the effect of an 8% blood loss on mean aortic pressure in a group of dogs. If both vagi are cut to eliminate the influence of the aortic arch baroreceptors and only the carotid sinus baroreceptors are operative (left panel), this hemorrhage decreases mean aortic pressure by 14%. This pressure change did not differ significantly from the pressure decline (12%) evoked by the same hemorrhage before vagotomy (not shown). When the carotid sinuses are denervated and the aortic baroreceptor reflexes are intact, the 8% blood loss decreases mean aortic pressure by 38% (middle panel). Hence, the carotid sinus baroreceptors are more effective than the aortic baroreceptors in attenuating the fall in pressure. The aortic baroreceptors must also be operative, however, because when both sets of afferent baroreceptor pathways are interrupted, an 8% blood loss reduces arterial pressure by 48%.

Although the arteriolar vasoconstriction is widespread during hemorrhage, it is by no means uniform. Vasoconstriction is most severe in the cutaneous, skeletal muscle, and splanchnic vascular beds and is slight or absent in the cerebral and coronary circulations. In many instances the cerebral and coronary vascular resistances are diminished. Thus the reduced cardiac output is redistributed to favor flow through the brain and the heart.

The severe cutaneous vasoconstriction accounts for the characteristic pale, cold skin of patients suffering from blood loss. Warming the skin of such patients improves their appearance considerably, much to the satisfaction of well-meaning individuals rendering first aid. However, it also inactivates an effective, natural compensatory mechanism—to the possible detriment of the patient.

In the early stages of mild to moderate hemorrhage,

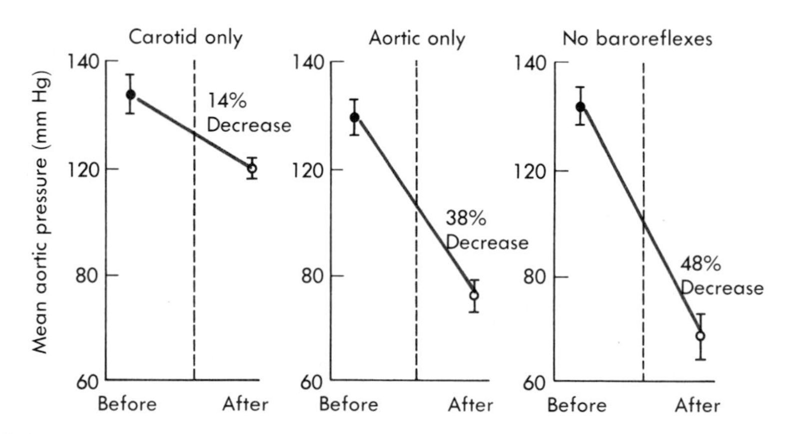

■ **Fig. 32-6** The changes in mean aortic pressure in response to an 8% blood loss in a group of eight dogs. *Left panel,* The carotid sinus baroreceptor reflexes were intact, and the aortic reflexes were interrupted. *Middle panel,* The aortic reflexes were intact, and the carotid sinus reflexes were interrupted. *Right panel,* All sinoaortic reflexes were abrogated. (From Shepherd JT: *Circulation* 50:418, 1974. By permission of the American Heart Association; derived from the data of Edis AJ: *Am J Physiol* 221:1352, 1971.)

the changes in renal resistance are usually slight. The tendency for increased sympathetic activity to constrict the renal vessels is counteracted by autoregulatory mechanisms (p. 481). With more-prolonged and severe hemorrhages, however, renal vasoconstriction becomes intense. The reductions in renal circulation are most severe in the outer layers of the renal cortex. The inner zones of the cortex and outer zones of the medulla are spared.

The severe renal and splanchnic vasoconstriction during hemorrhage favors the heart and brain. However, if such constriction persists too long, it may be harmful. Frequently patients survive the acute hypotensive period only to die several days later from kidney failure resulting from renal ischemia. Intestinal ischemia also may have dire effects. In the dog, for example, intestinal bleeding and extensive sloughing of the mucosa occur after only a few hours of hemorrhagic hypotension. Furthermore, the low splanchnic flow swells the centrilobular cells in the liver. The resultant obstruction of the hepatic sinusoids raises the portal venous pressure, which intensifies the intestinal blood loss. Fortunately the pathological changes in the liver and intestine are usually much less severe in humans than in dogs.

Chemoreceptor reflexes. Reductions in arterial pressure below about 60 mm Hg do not evoke any additional responses through the baroreceptor reflexes, because this pressure level constitutes the threshold for stimulation (see Chapter 29). However, low arterial pressure may stimulate peripheral chemoreceptors because of hypoxia in the chemoreceptor tissue consequent to inadequate local blood flow. Chemoreceptor excitation enhances the already existent peripheral vasoconstriction evoked by the baroreceptor reflexes. Also, respiratory stimulation assists venous return by the auxiliary pumping mechanism described on pp. 507 and 520.

Cerebral ischemia. When the arterial pressure is below about 40 mm Hg, the resulting cerebral ischemia activates the sympathoadrenal system. The sympathetic nervous discharge is several times greater than the maximum activity that occurs when the baroreceptors cease to be stimulated. Therefore, the vasoconstriction and facilitation of myocardial contractility may be pronounced. With more severe degrees of cerebral ischemia, however, the vagal centers also become activated. The resulting bradycardia may aggravate the hypotension that initiated the cerebral ischemia.

Reabsorption of tissue fluids. The arterial hypotension, arteriolar constriction, and reduced venous pressure during hemorrhagic hypotension lower the hydrostatic pressure in the capillaries. The balance of these forces promotes the net reabsorption of interstitial fluid into the vascular compartment. The rapidity of this response is displayed in Fig. 32-7. In a group of cats, 45% of the estimated blood volume was removed over a 30-minute period. The mean arterial blood pressure declined rapidly to about 45 mm Hg. The pressure then

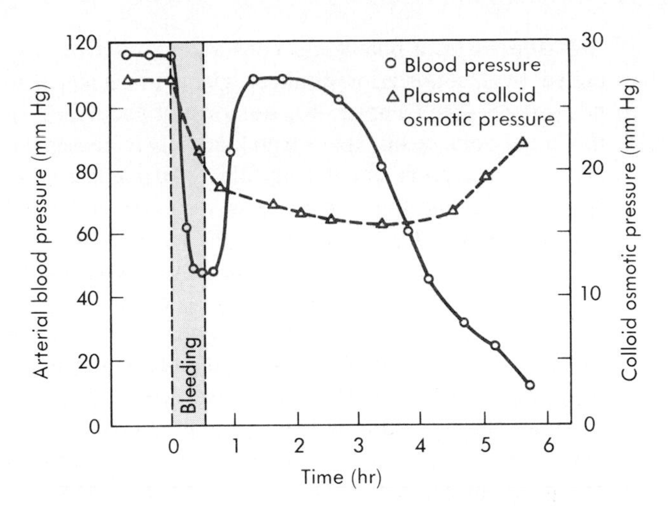

■ **Fig. 32-7** The changes in arterial blood pressure and plasma colloid osmotic pressure in response to withdrawal of 45% of the estimated blood volume over a 30-minute period, beginning at time zero. The data are the average values for 23 cats. (Redrawn from Zweifach BW: *Anesthesiology* 41:157, 1974.)

returned rapidly, but only temporarily, to near the control level. The plasma colloid osmotic pressure declined markedly during the bleeding and continued to decrease more gradually for several hours. The reduction in colloid osmotic pressure reflects the dilution of the blood by tissue fluids.

Considerable quantities of fluid thus may be drawn into the circulation during hemorrhage. About 0.25 ml of fluid per minute per kilogram of body weight may be reabsorbed. Approximately 1 liter of fluid per hour might be autoinfused into the circulatory system of an average individual from the interstitial spaces after an acute blood loss.

Considerable quantities of fluid also may be slowly shifted from intracellular to extracellular spaces. This fluid exchange is probably mediated by the secretion of cortisol from the adrenal cortex in response to hemorrhage. Cortisol appears to be essential for a full restoration of the plasma volume after hemorrhage.

Endogenous vasoconstrictors. The **catecholamines,** epinephrine and norepinephrine, are released from the adrenal medulla in response to the same stimuli that evoke widespread sympathetic nervous discharge (see Chapter 50). Blood levels of catecholamines are high during and after hemorrhage. When animals are bled to an arterial pressure level of 40 mm Hg, the catecholamines increase as much as fifty times.

Epinephrine comes almost exclusively from the adrenal medulla, whereas norepinephrine is derived both from the adrenal medulla and the peripheral sympathetic nerve endings. These humoral substances reinforce the effects of sympathetic nervous activity listed previously.

Vasopressin, a potent vasoconstrictor, is actively secreted by the posterior pituitary gland in response to hemorrhage (see Chapter 48). Removal of about 20% of the blood volume in experimental animals increases vasopressin secretion about fortyfold. The receptors responsible for the augmented release are the sinoaortic baroreceptors and the receptors in the left atrium.

The diminished renal perfusion during hemorrhagic hypotension leads to the secretion of **renin** from the juxtaglomerular apparatus (see Chapter 42). This enzyme acts on a plasma protein, **angiotensinogen,** to form **angiotensin,** a very powerful vasoconstrictor.

Renal conservation of water. Fluid and electrolytes are conserved by the kidneys during hemorrhage in response to various stimuli, including the increased secretion of vasopressin (antidiuretic hormone) noted previously. The lower arterial pressure decreases the glomerular filtration rate, and thus curtails the excretion of water and electrolytes (see Chapter 42). Also the diminished renal blood flow raises the blood levels of angiotensin, as described previously. This polypeptide accelerates the release of **aldosterone** from the adrenal cortex. Aldosterone, in turn, stimulates sodium reabsorption by the renal tubules, and water accompanies the sodium that is actively reabsorbed.

Decompensatory Mechanisms

In contrast to the negative feedback mechanisms just described, latent **positive feedback mechanisms** are also evoked by hemorrhage. Such mechanisms exaggerate any primary change initiated by the blood loss. Specifically, positive feedback mechanisms aggravate the hypotension induced by blood loss and tend to initiate **vicious cycles,** which may lead to death. The operation of positive feedback mechanisms is manifest in curve *B* of Fig. 32-4.

Whether a positive feedback mechanism will lead to a vicious cycle depends on the **gain** of that mechanism. Gain is defined as the ratio of the secondary change evoked by a given mechanism to the initiating change itself. A gain greater than 1 induces a vicious cycle; a gain less than 1 does not. For example, consider a positive feedback mechanism with a gain of 2. If, for any reason, mean arterial pressure decreases by 10 mm Hg, the positive feedback mechanism would then evoke a secondary pressure reduction of 20 mm Hg, which in turn would cause a further decrement of 40 mm Hg; (i.e., each change would induce a subsequent change that was twice as great). Hence mean arterial pressure would decline at an ever-increasing rate until death supervened, much as is depicted by curve *B* in Fig. 32-4.

Conversely, a positive feedback mechanism with a gain of 0.5 would indeed exaggerate any change in mean arterial pressure but would not necessarily lead to death. For example, if arterial pressure suddenly decreased by 10 mm Hg, the positive feedback mechanism would initiate a secondary, additional fall of 5 mm Hg. This, in turn, would provoke a further decrease of 2.5 mm Hg. The process would continue in ever-diminishing steps, with the arterial pressure approaching an equilibrium value asymptotically.

Some of the more important positive feedback mechanisms include (1) cardiac failure, (2) acidosis, (3) inadequate cerebral blood flow, (4) aberrations of blood clotting, and (5) depression of the reticuloendothelial system.

Cardiac failure. The role of cardiac failure in the progression of shock during hemorrhage is controversial. All investigators agree that the heart fails terminally, but opinions differ concerning the importance of cardiac failure during earlier stages of hemorrhagic hypotension. Shifts to the right in ventricular function curves (Fig. 32-8) constitute experimental evidence of a progressive depression of myocardial contractility during hemorrhage.

The hypotension induced by hemorrhage reduces the coronary blood flow and therefore depresses ventricular function. The consequent reduction in cardiac output leads to a further decline in arterial pressure, a classical example of a positive feedback mechanism. Further-

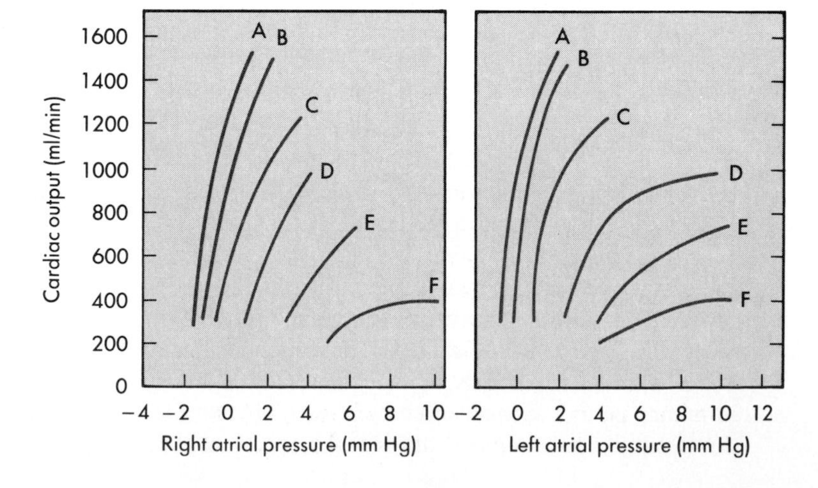

Fig. 32-8 Ventricular function curves for the right and left ventricles during the course of hemorrhagic shock. Curves *A* represent the control function curve; curves *B*, 117 min; curves *C*, 247 min; curves *D*, 380 min; curves *E*, 295 min; and curves *F*, 310 min after the initial hemorrhage. (Redrawn from Crowell JW, Guyton AC: *Am J Physiol* 203:248, 1962.)

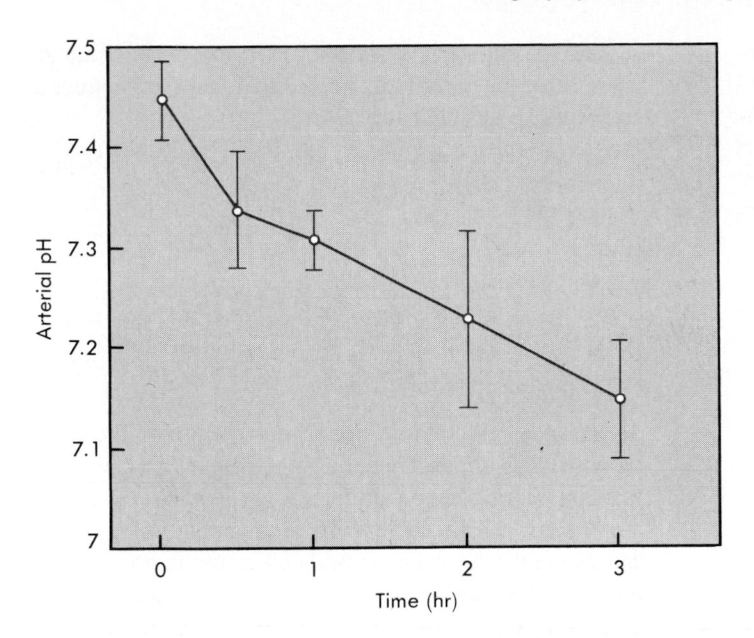

■ **Fig. 32-9** The reduction in arterial blood pH (mean ± SD) in a group of 11 dogs whose blood pressure had been held at a level of 35 mm Hg by bleeding into a reservoir, beginning at time zero. (Modified from Markov AK et al: *Circ Shock* 8:9, 1981.)

more, the reduced tissue blood flow leads to an accumulation of vasodilator metabolites, which decreases peripheral resistance and therefore aggravates the fall in arterial pressure.

Acidosis. The inadequate blood flow during hemorrhage affects the metabolism of all cells in the body. The resultant stagnant anoxia accelerates the production of lactic acid and other acid metabolites by the tissues. Furthermore, impaired kidney function prevents adequate excretion of the excess H^+, and generalized metabolic acidosis ensues (Fig. 32-9). The resulting depressant effect of acidosis on the heart further reduces tissue perfusion and thus aggravates the metabolic acidosis. Acidosis also diminishes the reactivity of the heart and resistance vessels to neurally released and circulating catecholamines, thereby intensifying the hypotension.

Central nervous system depression. The hypotension in shock reduces cerebral blood flow. Moderate degrees of cerebral ischemia induce a pronounced sympathetic nervous stimulation of the heart, arterioles, and veins, as stated previously. With severe degrees of hypotension, however, the cardiovascular centers in the brainstem eventually become depressed because of inadequate blood flow to the brain. The resultant loss of sympathetic tone then reduces cardiac output and peripheral resistance. The resulting reduction in mean arterial pressure intensifies the inadequate cerebral perfusion.

Various endogenous **opioids,** such as **enkephalins** and **beta-endorphin,** may be released into the brain substance or into the circulation in response to the same stresses that provoke circulatory shock. Enkephalins exist along with catecholamines in secretory granules in the adrenal medulla, and they are released together in response to stress. Similar stimuli release beta-endorphin and adrenocorticotrophic hormone (ACTH) from the anterior pituitary gland. These opioids depress the centers in the brainstem that mediate some of the compensatory autonomic adaptations to blood loss, endotoxemia, and other shock-provoking stresses. Conversely, the opioid antagonist, **naloxone,** improves cardiovascular function and survival in various forms of shock.

Aberrations of blood clotting. The alterations of blood clotting after hemorrhage are typically biphasic—an initial phase of hypercoagulability followed by a secondary phase of hypocoagulability and fibrinolysis (see Chapter 21). In the initial phase intravascular clots, or **thrombi,** develop within a few minutes of the onset of severe hemorrhage, and coagulation may be extensive throughout the minute blood vessels.

Thromboxane A_2 may be released from various ischemic tissues. It aggregates platelets, and more thromboxane A_2 is released from the trapped platelets, which serves to trap additional platelets. This form of positive feedback intensifies and prolongs the clotting tendency. The mortality from certain standard shock-provoking procedures has been reduced considerably by anticoagulants such as heparin.

In the later stages of hemorrhagic hypotension, the clotting time is prolonged and fibrinolysis is prominent. It was mentioned previously that in the dog, hemorrhage into the intestinal lumen is common after several hours of hemorrhagic hypotension. Blood loss into the intestinal lumen would, of course, aggravate the hemodynamic effects of the original hemorrhage.

Reticuloendothelial system. During the course of hemorrhagic hypotension, reticuloendothelial system (RES) function becomes depressed. The phagocytic activity of the RES is modulated by an opsonic protein. The opsonic activity in plasma diminishes during shock, which may account in part for the depression of RES function. As a consequence, the antibacterial and antitoxic defense mechanisms are impaired. Endotoxins from the normal bacterial flora of the intestine constantly enter the circulation. Ordinarily they are inactivated by the RES, principally in the liver. When the RES is depressed, these endotoxins invade the general circulation. Endotoxins produce a form of shock that resembles in many respects that produced by hemorrhage. Therefore, depression of the RES aggravates the hemodynamic changes caused by blood loss.

In addition to their role in inactivating endotoxin, the macrophages release many of the mediators that are associated with shock: acid hydrolases, neutral proteases, certain coagulation factors, and the arachidonic acid derivatives—prostaglandins, thromboxanes, and leukotrienes. Macrophages also release certain regulatory proteins, called **monokines,** that modulate temperature regulation, intermediary metabolism, hormone secretion, and the immune system.

Interactions of Positive and Negative Feedback Mechanisms

Hemorrhage provokes a multitude of circulatory and metabolic derangements. Some of these changes are compensatory; others are decompensatory. Some of these feedback mechanisms possess a high gain; others, a low gain. Furthermore, the gain of any specific mechanism varies with the severity of the hemorrhage. For example, with only a slight loss of blood, mean arterial pressure is within the range of normal and the gain of the baroreceptor reflexes is high. With greater losses of blood, when mean arterial pressure is below about 60 mm Hg (that is, below the threshold for the baroreceptors), further reductions of pressure will have no additional influence through the baroreceptor reflexes. Hence below this critical pressure the baroreceptor reflex gain will be zero or near zero.

As a general rule, with minor degrees of blood loss, the gains of the negative feedback mechanisms are high, whereas those of the positive feedback mechanisms are low. The converse is true with more severe hemorrhages. The gains of the various mechanisms are additive algebraically. Therefore, whether a vicious cycle develops depends on whether the sum of the various gains exceeds 1. Total gains in excess of 1 are, of course, more likely with severe losses of blood. Therefore, to avert a vicious cycle, serious hemorrhages must be treated quickly and intensively, preferably by whole blood transfusions, before the process becomes irreversible.

Summary

Exercise

1. In anticipation of exercise the vagus nerve impulses to the heart are inhibited and the sympathetic nervous system is activated by central command. The result is an increase in heart rate, myocardial contractile force, and regional vascular resistance.

2. With exercise, vascular resistance increases in skin, kidneys, splanchnic regions, and inactive muscles and decreases in active muscles. The increase in cardiac output is mainly caused by the increase in heart rate. Stroke volume increases only slightly. Total peripheral resistance decreases, oxygen consumption and blood oxygen extraction increase, and systolic and mean blood pressure increase slightly.

3. As body temperature rises during exercise, the skin blood vessels dilate. However, when heart rate becomes maximal during severe exercise, the skin vessels constrict. This increases the effective blood volume but causes greater increases in body temperature and a feeling of exhaustion.

4. The limiting factor in exercise performance is the delivery of blood to the active muscles.

Hemorrhage

1. Acute blood loss induces the following hemodynamic changes: tachycardia, hypotension, generalized arteriolar vasoconstriction, and generalized venoconstriction.

2. Acute blood loss invokes a number of negative feedback (compensatory) mechanisms, such as baroreceptor and chemoreceptor reflexes, responses to moderate cerebral ischemia, reabsorption of tissue fluids, release of endogenous vasoconstrictors, and renal conservation of water and electrolytes.

3. Acute blood loss also induces a number of positive feedback (decompensatory) mechanisms, such as cardiac failure, acidosis, central nervous system depression, aberrations of blood coagulation, and depression of the reticuloendothelial system.

4. The outcome of an acute blood loss depends on the gains of the various feedback mechanisms and on the interactions between the positive and negative feedback mechanisms.

Bibliography

Journal articles

Abel FI: Myocardial function in sepsis and endotoxin shock, *Am J Physiol* 257:R1265, 1989.

Averill DB, Scher AM, Feigl EO: Angiotensin causes vasoconstriction during hemorrhage in baroreceptor-denervated dogs, *Am J Physiol* 245:H667, 1983.

Bernton EW, Long JB, Holaday JW: Opioids and neuropeptides: mechanisms in circulatory shock, *Fed Proc* 44:290, 1985.

Bevegåard BS, Shepherd JT: Regulation of the circulation during exercise in man, *Physiol Rev* 47:178, 1967.

Blomqvist CG, Saltin B: Cardiovascular adaptations to physical training, *Annu Rev Physiol* 45:169, 1983.

Bond RF, Johnson G III: Vascular adrenergic interactions during hemorrhagic shock, *Fed Proc* 44:281, 1985.

Brengelmann GL: Circulatory adjustments to exercise and heat stress, *Annu Rev Physiol* 45:191, 1983.

Briand R, Yamaguchi N, Gagne J: Plasma catecholamine and glucose concentrations during hemorrhagic hypotension in anesthetized dogs, *Am J Physiol* 257:R317, 1989.

Christensen NJ, Galbo H: Sympathetic nervous activity during exercise, *Annu Rev Physiol* 45:139, 1983.

Clausen JP: Effect of physical training on cardiovascular adjustments to exercise in man, *Physiol Rev* 57:779, 1977.

Connett RJ, Pearce FJ, Drucker WR: Scaling of physiological responses: a new approach for hemorrhagic shock, *Am J Physiol* 250:R951, 1986.

Cornish KG, Gilmore JP, McCulloch T: Central blood volume and blood pressure in conscious primates, *Am J Physiol* 254:H693, 1988.

Eldridge FL et al: Stimulation by central command of locomotion, respiration, and circulation during exercise, *Resp Physiol* 59:313, 1985.

Filkins JP: Monokines and the metabolic pathophysiology of septic shock, *Fed Proc* 44:300, 1985.

Hosomi H et al: Interactions among reflex compensatory systems for posthemorrhage hypotension, *Am J Physiol* 250:H944, 1986.

Laughlin MH, Armstrong RB: Muscle blood flow during locomotory exercise, *Exerc Sport Sci Rev* 13:95, 1985.

Lefer AM: Eicosanoids as mediators of ischemia and shock, *Fed Proc* 44:275, 1985.

Liard JF: Vasopressin in cardiovascular control: role of circulating vasopressin, *Clin Sci* 67:473, 1984.

Ludbrook J: Reflex control of blood pressure during exercise, *Annu Rev Physiol* 45:155, 1983.

Ludbrook J, Graham WF: The role of cardiac receptor and arterial baroreceptor reflexes in control of the circulation during acute change of blood volume in the conscious rabbit, *Circ Res* 54:424, 1984.

Mitchell JH, Kaufman MP, Iwamoto GA: The exercise pressor reflex: its cardiovascular effects, afferent mechanisms, and central pathways, *Annu Rev Physiol* 45:229, 1983.

Saltin B, Rowell LB: Functional adaptations to physical activity and inactivity, *Fed Proc* 39:1506, 1980.

Sanders JS, Mark AL, Ferguson DW: Importance of aortic baroreflex in regulation of sympathetic responses during hypotension, *Circulation* 79:83, 1989.

Schadt JC, Ludbrook J: Hemodynamic and neurohumoral responses to acute hypovolemia in conscious mammals, *Am J Physiol* 260:H305, 1991.

Share L: Role of vasopressin in cardiovascular regulation, *Physiol Rev* 68:1248, 1988.

Shen Y-T et al: Relative roles of cardiac receptors and arterial baroreceptors during hemorrhage in conscious dogs, *Circ Res* 66:397, 1990.

Vatner SF, Pagani M: Cardiovascular adjustments to exercise: hemodynamics and mechanisms, *Prog Cardiovasc Dis* 19:91, 1976.

Books and monographs

Altura BM, Lefer AM, Schumer W: *Handbook of shock and trauma,* vol 1, *Basic science,* New York, 1983, Raven Press.

Bond RF, Adams HR, Chaudry IH, associate editors: *Perspectives in shock research,* New York, 1988, Alan R Liss.

Brooks GA, Fahey TD: *Exercise physiology—human bioenergetics and its applications,* New York, 1984, John Wiley & Sons.

Carlsten A, Grimby G: *The circulatory response to muscular exercise in man,* Springfield, Ill, 1966, Charles C Thomas, Publisher.

Janssen HF, Barnes CD, editors: *Circulatory shock: basic and clinical implications,* New York, 1985, Academic Press.

Lind AR: *Cardiovascular adjustments to isometric contractions: static effort.* In *Handbook of physiology:* section 2: *The cardiovascular system—peripheral circulation and organ blood flow,* vol 3, Bethesda, Md, 1983, American Physiological Society.

Mitchell JH, Schmidt RF: *Cardiovascular reflex control by afferent fibers from skeletal muscle receptors.* In *Handbook of physiology;* section 2: *The cardiovascular system—peripheral circulation and organ blood flow,* vol 3, Bethesda, Md, 1983, American Physiological Society.

Roth BL, Nielsen TB, McKee AE, editors: *Molecular and cellular mechanisms of septic shock.* In: *Progress in clinical and biological research,* vol 286, New York, 1988, Alan R Liss.

Rowell LB: *Integration of body systems in exercise.* In Berne RM, Levy MN, editors: *Principles of physiology,* St Louis, 1990, Mosby–Year Book.

Rowell LB: Human circulation: regulation during physical stress, New York, 1986, Oxford University Press.

SECTION
VI

THE RESPIRATORY SYSTEM

Norman C. Staub, Sr.

CHAPTER

33

Structure and Function of the Respiratory System

Breathing is an automatic, rhythmic, and centrally regulated mechanical process by which the contraction and relaxation of the skeletal muscles of the diaphragm, abdomen, and rib cage cause gas to move into and out of the terminal respiratory units (functional alveoli) of the lung. **Respiration** is the overall process of controlled oxidation of metabolites for the production of useful energy by living organisms. It includes breathing.

For successful steady-state respiration to occur in multicellular organisms, **oxygen** (O_2) and other nutrients must be brought to the vicinity of the respiring cells, whereas **carbon dioxide** (CO_2) and other waste products must be taken away. Large animals, including humans, make use of two systems for this process: a **circulatory system,** which carries whatever is needed to and from the tissue cells, and a **breathing system** (a gas exchanger), which carries oxygen and carbon dioxide between the environment (ambient atmosphere) and the alveoli of the lung.

■ *Functions and Definitions*

■ *Principal Function of the Lung*

The functional design of the lung (Fig. 33-1) is to provide an adequate distribution of inspired air and pulmonary blood flow, such that the exchange of O_2 and CO_2 between the alveolar gas and the pulmonary capillary blood is accomplished with a minimal expenditure of **energy** (work of breathing and of the right ventricle).

■ *Ventilation and Perfusion*

The process of breathing is measured by **ventilation** (frequency × depth of breathing). The exchange of O_2 and of CO_2 in the lung is measured by their concentration differences between inspired air and expired gas.

The process of **perfusion** is measured by the cardiac output (heart rate × stroke volume of the right ventricle). The exchange of O_2 and of CO_2 in pulmonary cap-

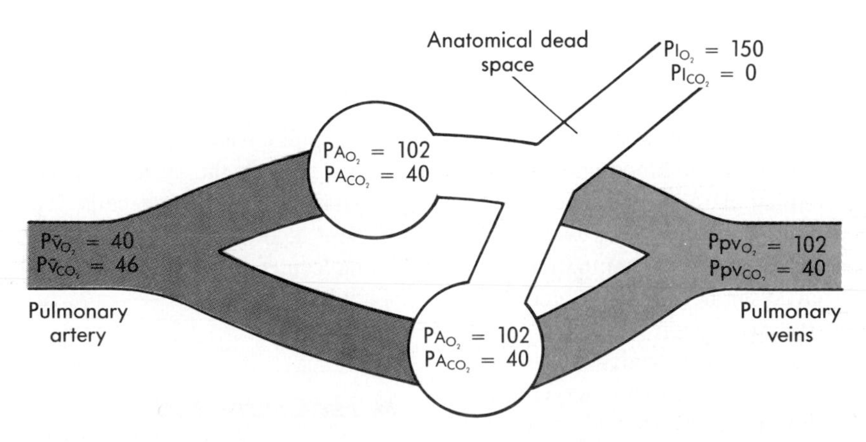

■ **Fig. 33-1** Simple lung model showing two normal parallel lung units. Both units receive equal quantities of fresh air and blood flow for their size. The blood and alveolar gas partial pressures, P, are normal.

illary blood is measured by their concentration differences between blood in the pulmonary artery (mixed venous blood) and in the pulmonary veins, left atrium, or any systemic artery.

The differences between O_2 and CO_2 **partial pressures** (fraction of total gas pressure attributed to a particular molecular species) and between alveolar gas and systemic arterial blood are useful in determining overall lung function efficiency. Efficiency depends on the adequacy of the matching of alveolar ventilation (volume flow rate of fresh air to each part of the lung) to perfusion (volume flow rate of blood to each part of the lung).

Ideally the **ventilation/perfusion ratios** of all parts of the lungs are identical.

Blood Gas Transport

Oxygenated blood, leaving the lungs via the pulmonary veins, is pumped by the left ventricle through the systemic arteries to the capillaries associated with all of the respiring cells of the body. Likewise the principal waste product of metabolism, CO_2, is transported away from the respiring cells via the systemic veins to the lung for elimination. The process of blood O_2 and CO_2 transport is an important part of respiration.

For (convective) blood O_2 transport, concentration is the important variable. Because oxygen is rather insoluble in water, it is critical to know something about the special protein, hemoglobin, within the erythrocytes. The remarkable property of hemoglobin is its ability to combine rapidly and reversibly with oxygen so as to effectively increase its solubility in blood manyfold. The **oxygen-hemoglobin equilibrium curve** (Fig. 33-2) is the empirical (experimental) relationship between the partial pressure of oxygen in blood and the relative amount (percent saturation) bound to normal hemoglobin. Because the normal blood hemoglobin concentration is 150 g/L (15 g/100 ml blood), it is very effective in increasing blood oxygen concentration. Arterial blood contains nearly 200 ml O_2/L at the normal oxygen partial pressure of 100 mm Hg.

The cardiac output in a resting human adult is about 5 L/min. Only about 25% of the oxygen bound to hemoglobin is exchanged between the blood and systemic tissues during each circulation (arterial − venous O_2 difference = 50 ml/L). This difference has two beneficial effects: (1) it maintains a high oxygen partial pressure difference between systemic capillary blood and tissue cells, which is the driving pressure for oxygen diffusion into the cells, and (2) it provides a reserve of oxygen molecules for use by cells in an emergency (as when capillary blood flow is briefly slowed or stopped by muscle contraction).

Even in the heaviest steady-state exercise sustainable by average normal humans, cardiac output is unlikely to increase to more than 3 times the resting level (15 L/min). Because oxygen usage by the body **(oxygen consumption)** in steady state exercise may increase sixfold from the resting value (from 250 ml O_2/min to 1500 ml O_2/min), an additional 25% of the O_2 bound to hemoglobin in systemic arterial blood must be unloaded in the capillaries. The threefold increase in blood flow multiplied by the doubling of O_2 removal from blood accounts for the sixfold increase in O_2 consumption. Some highly trained athletes can increase their exercise cardiac output, ventilation, or oxygen consumption to more than twice the values just stated, but such people are not "average normal."

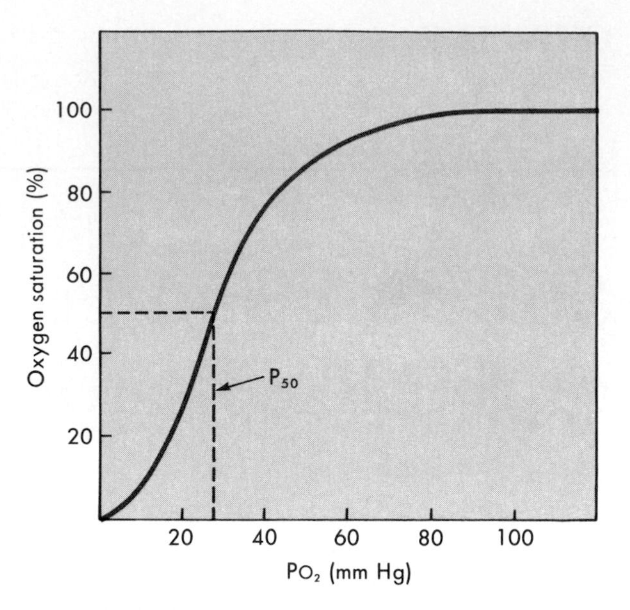

Fig. 33-2 The standard hemoglobin-oxygen equilibrium curve (HbO_2). It is used to convert between oxygen pressure (abscissa) and fractional saturation of the hemoglobin in erythrocytes (ordinate). The P_{50} is used to compare different hemoglobins.

Gas transport from the environment to cells is largely by the two bulk (convective) transport processes (ventilation and blood flow) already mentioned. However, this situation differs at two key interfaces; namely between gas in the alveoli and the blood flowing through the pulmonary capillaries and between the blood flowing through the systemic capillaries and the mitochondria of the respiring cells. The only process available here is **diffusion,** the passive thermodynamic flow of molecules between regions with different chemical activities.

The Control System

The overall efficiency of breathing depends on the regulation of ventilation, blood flow, and ventilation/perfusion matching by various external mechanisms (nervous system, humoral substances), and internal mechanisms (terminal respiratory unit distensibility, resistance to air flow, and resistance to blood flow) (Fig. 33-3). Respiration, i.e., cellular demand for oxygen, is not normally regulated in the sense that steady-state oxygen consumption is not affected by changes in breathing, inspired gas composition, cardiac output, or blood composition. In the larger sense, however, respiration of cells is regulated on demand through very complex local and central events (Fig. 33-4). Further details are presented in Chapter 37.

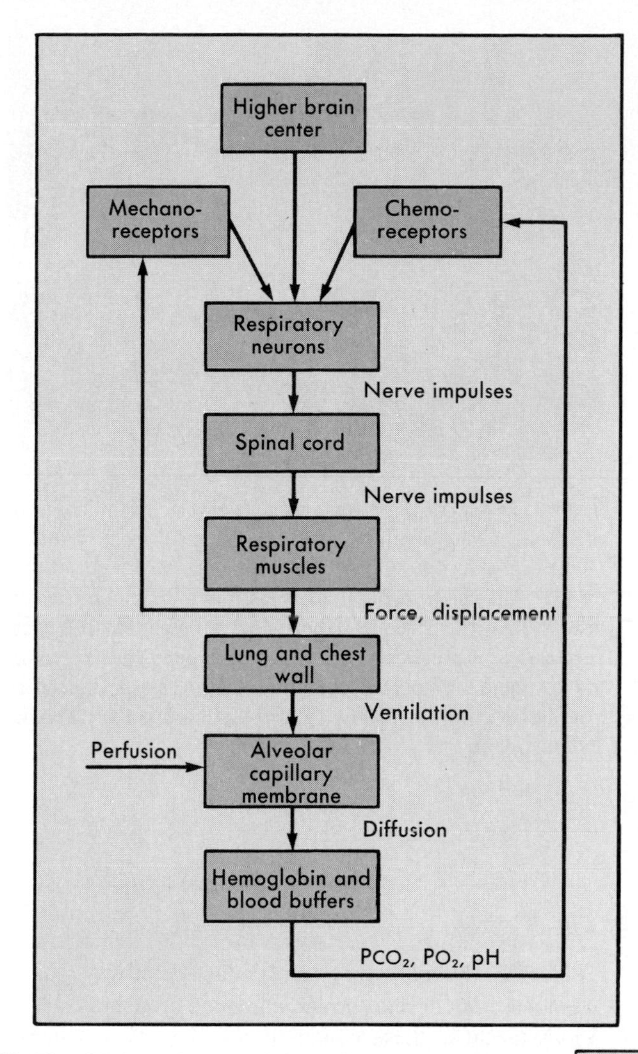

■ **Fig. 33-3** Diagram of the respiratory control system. Ventilation and perfusion join near the bottom, and their output sets carbon dioxide and oxygen partial pressures, and in part arterial hydrogen ion concentration, pH. These outputs feed back to the controllers via chemoreceptors located in strategic places.

■ *Structural Basis of Breathing*

The normal human adult lungs weigh 900 to 1000 g; blood content is 40% to 50%. At end-expiration **(functional residual capacity, FRC)** the gas volume is about 2.4 L, whereas at maximal inspiration **(total lung capacity, TLC)** it may be greater than 6 L.

Inspiration, caused by coordinated neural influences from the respiratory control centers in the brainstem, is the active phase of breathing. It causes the diaphragm and external intercostal muscles to contract. The muscle contraction causes the thoracic cavity to expand, which lowers the pressure in the pleural space surrounding the lungs. As the pressure falls in the pleural space, the distensible lungs expand passively, which causes the pressure in the terminal air spaces (alveolar ducts and alveoli) to decrease. As the pressure decreases, fresh air flows down the branching airways into the terminal air

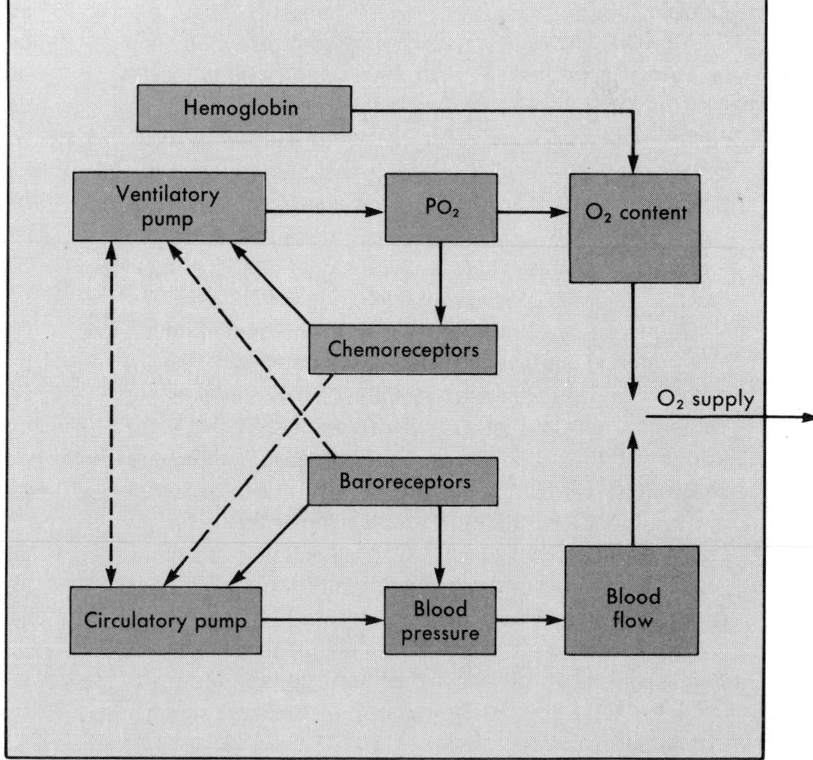

■ **Fig. 33-4** Because the main purpose of breathing is to supply O_2 to all cells of the body, the interactions between the respiratory and cardiovascular system require tight control.

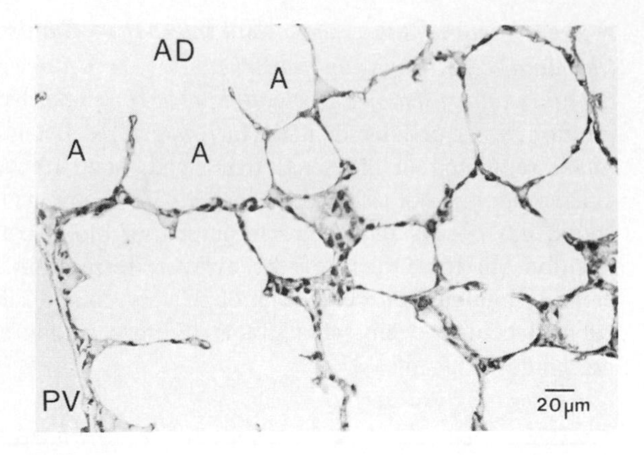

■ **Fig. 33-5** A low power microscopic view from a normal inflated human lung. The space occupied by gas is very large, and the alveolar to capillary tissue pathway for diffusion is minimized. *A* = alveoli; *AD* = alveolar duct. (Courtesy Albertine KH, Thomas Jefferson University, Philadelphia.)

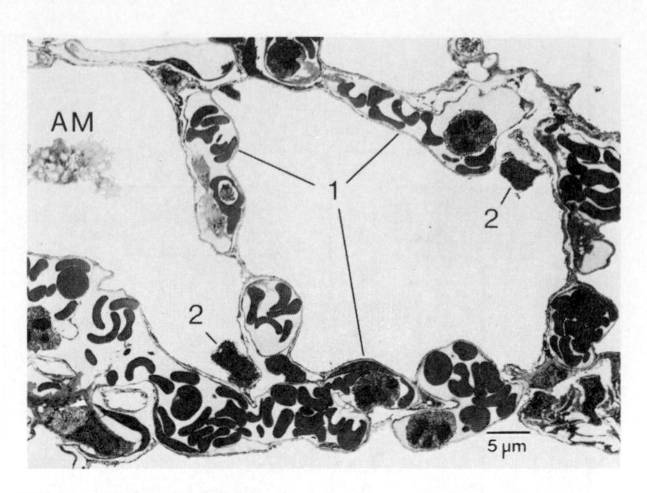

■ **Fig. 33-6** Low power transmission electron micrograph of a human lung alveolus. The capillaries fill most of the alveolar walls so that O_2 and CO_2 exchange will proceed efficiently. The main, type 1, alveolar lining epithelium covers nearly all of the alveolus. The type 2 cells are small but are very active in producing, packaging, secreting, and recycling the air-liquid surfactant. At the left margin is a portion of an alveolar macrophage, *AM*. (Courtesy Albertine KH, Thomas Jefferson University, Philadelphia.)

spaces until the pressures are equalized, which marks the end of inspiration. During **expiration** (the mostly passive phase of breathing) the process is reversed, pleural and alveolar pressures rise, and gas flows out of the lung.

During growth and development the lung conforms to the shape of the pleural cavities in such a manner as to minimize structural stress (mechanical forces tend to distort the lung tissue elements). With breathing, the lung expands in all directions as the chest cavity expands.

Fig. 33-5 is a light microscopic picture of a thin slice of an expanded normal lung. The alveolar tissue occupies only a tiny fraction of the total area. Fig. 33-6 is an electron micrograph of several alveolar walls. The average distance between the alveolar gas and the hemoglobin in the red blood cells is 1.5 μm. Diffusion, which is very rapid over short distances, efficiently exchanges the respiratory gases across the alveolar-capillary interface.

The myriad anatomical alveoli give the human lung a total internal surface area of approximately 1 m^2/kg body weight at FRC. This vast area (70 m^2 total) is not necessary for the distribution of gas within the lung. Rather it fulfills the need for distributing the pulmonary blood flow (cardiac output) into a very thin film (about one red blood cell thick) in such a manner that even under stressful circumstances, such as exercise, the time spent by each erythrocyte flowing along the capillaries is sufficient to permit equilibration of O_2 and CO_2 between blood and gas; the tissue phase of gaseous diffusion (air to blood distance) is minimized; and the resistance to blood flow is low because of the large number of parallel pathways.

■ *The Airways*

There are two main types of conducting airways: cartilaginous bronchi and the membranous bronchioles (Fig. 33-7). Because these conducting airways do not generally participate in gas exchange, their portion of each breath is wasted (anatomical dead space). Their combined length and cross-sectional area is such that the dead space is about 30% of each normal breath, whereas resistance to airflow is low.

In addition to their supporting cartilage plates, the **bronchi** are lined by a pseudostratified columnar epithelium, which rests on spiral bands of smooth muscle. The bronchi are able to dilate or constrict independently of lung volume. Among the numerous cell types in the epithelium are a large number of cells with cilia, whose rhythmic beating in a thin liquid layer effectively transports the surface film of mucus and particles out of the lung by way of the trachea.

The bronchi decrease in diameter and length with each successive branching but the sum of the cross-sectional areas of the two "child" bronchi is greater than that of the "parent." The cartilage support also gradually disappears. In airways about 1 mm diameter the cartilage disappears completely; by convention all subsequent airways are called **bronchioles.**

In addition to being small and having no cartilage support, the bronchioles have a simple cuboidal epithelium. An important functional difference exists between

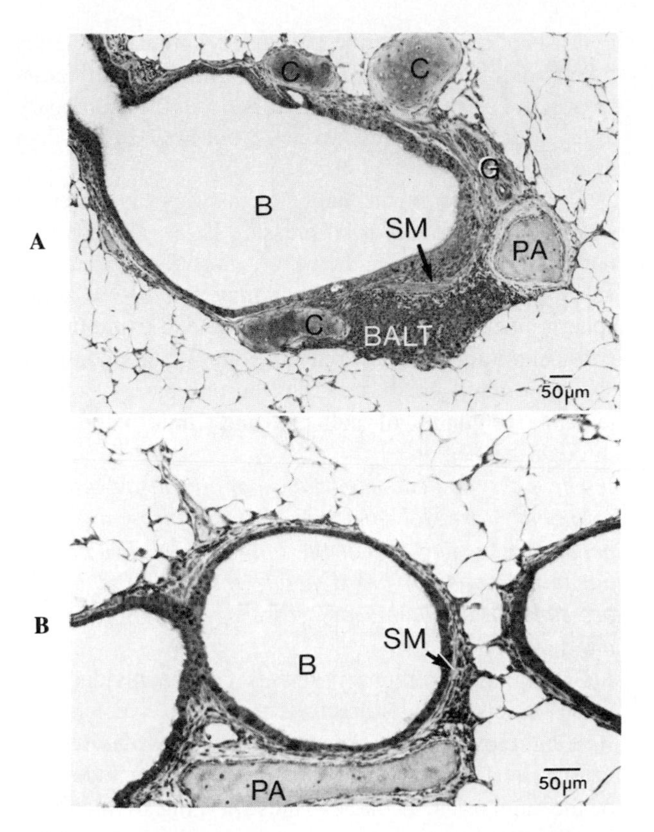

■ **Fig. 33-7** Two human airways. **A,** 2 mm diameter cartilaginous bronchus, *BR*. Bits of cartilage, *C*, and submucosal glands, *G* are indicated. BALT refers to bronchus associated lymphoid tissue. No further mention of it will be made. **B,** Membranous bronchiole about 0.2 mm diameter. There is no cartilage, and the lining epithelium is simple cuboidal. Both sections show the accompanying pulmonary artery, *PA*, and the submucosal smooth muscle, *SM*. (Courtesy Albertine KH, Thomas Jefferson University, Philadelphia.)

bronchioles and bronchi; i.e., the bronchioles are embedded directly into the connective tissue framework of the lung so that their diameter depends on lung volume.

The blood supply of the airways, from the trachea to the terminal bronchioles, is via the bronchial arteries, a systemic source. Bronchial blood flow is normally about 1.0% of cardiac output. The functions of the bronchial circulation are to provide nutrition to the airways and larger pulmonary blood vessels, to warm and add water vapor to condition inspired air, and to provide substrate for airway cellular metabolism and secretions.

Not seen in the usual histological sections are the motor and sensory nerves of the airways, which participate in the reflex regulation of breathing, airway caliber, glandular secretion, and in bronchial vasomotor control. The subepithelial smooth muscle bands receive their motor innervation from the parasympathetic branch of the autonomic nervous system (vagus nerve). When the airway smooth muscle contracts, the lumen is narrowed.

Sensory fibers are located beneath and within the intercellular junctions of the epithelial cells. The best understood sensory receptors, located in the bronchi (chiefly, the trachea and main stem bronchi), are sensitive to physical distortion (stretch) and to chemical substances (irritants). All along the airways, particularly in the bronchioles and in the alveolar walls, are small, nonmyelinated, slowly conducting C-fibers. Although these fibers are nonspecific, they occur in great numbers and can be stimulated by a variety of chemical mediators by which various pulmonary-cardiac reflex responses are elicited. Their role in normal regulation of the lungs and heart, however, is not well understood.

Eventually the bronchioles develop outpouchings, which are the primitive alveoli. The first bronchioles with alveoli are called **respiratory bronchioles,** to convey the concept that they participate in gas exchange. With successive branching (generations) of the respiratory bronchioles, the number and size of the anatomical alveoli increase until the walls of the bronchioles are almost completely replaced by the mouths of the alveoli. These final airway branches are called **alveolar ducts.**

■ *Pulmonary Circulation*

The pulmonary artery (perfusion) accompanies and branches in close relationship with the airways (ventilation). Thus the physiological theme of *ventilation/perfusion matching for efficient lung function* is reflected in the anatomical pattern of the bronchovascular relations. The pulmonary veins, on the other hand, lie within the interlobular and interlobar connective tissue septae, where they receive blood from many terminal respiratory units.

Pulmonary arteries and veins larger than about 50 μm in diameter in the normal adult human contain smooth muscle and are capable of active regulation of their diameter, which alters resistance to blood flow.

The pulmonary vasculature is richly innervated. The motor nerve to the smooth muscle comes from the sympathetic branch of the autonomic nervous system. In contrast to the systemic circulation, however, the normal pulmonary circulation shows little evidence of active external regulation. The extensive sensory innervation, located in the adventitia surrounding the blood vessels, can be stimulated by vascular pressure changes (stretch) and various chemical substances, but its role in regulation of pulmonary and cardiac events is unclear. Most active regulation is caused by local metabolic influences.

The **bronchial circulation** is the nutritive supply to all lung support structures (airways, vessels, connective tissue septae, and pleura). The pulmonary circulation is the nutritive supply to the alveolar walls.

The **pulmonary capillaries** form an extensive interdigitating network within the alveolar walls. Most of the surface area (70% to 80%) of the alveolar wall overlies red cells when the capillaries are well filled with blood. The total capillary surface area is nearly as great as the alveolar surface area.

Anatomically, the maximal capillary volume of the human adult lung is about 200 ml, whereas at rest the effective volume is about 70 ml (1 ml/kg) at FRC. Capillary volume can be increased by opening of closed or compressed segments. This is called **recruitment** and occurs with increased cardiac output, as in exercise. Capillary volume can also be increased by enlarging open capillaries as their internal pressure rises. This is called **distension** and occurs when the lungs become congested by rising left atrial pressure, as in left heart failure.

Another feature of the alveolar capillary network is that the capillaries are continuous over several alveoli. The average path traveled by red blood cells is 600 to 800 μm before they enter the venous drainage system. The pulmonary capillary blood volume at any instant is about equal to the stroke volume of the right ventricle. That means that, at an average heart rate of 75 beats/min, red blood cells will remain within the capillaries for one cardiac cycle (about 0.8 sec). This time is more than adequate for the diffusion of O_2 and CO_2, which occurs rapidly (<0.25 sec to equilibration) through the very thin barrier mentioned previously.

From the top to the bottom of the lung (Fig. 33-8) the hydrostatic pressure in the pulmonary circulation varies by approximately 1 cm H_2O/cm height (in the direction of gravity). Thus at FRC the pressure in the pulmonary artery near the bottom is about 25 cm H_2O greater than near the top.

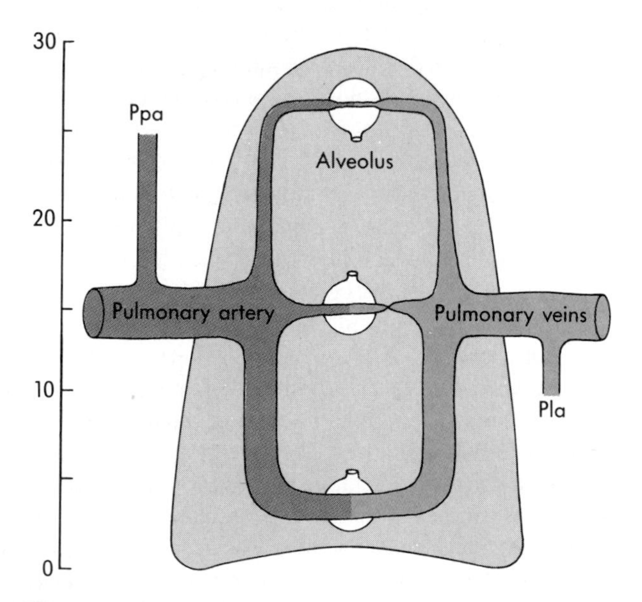

■ **Fig. 33-8** The distribution of pulmonary blood flow in the lung is sensitive to both the pulmonary arterial and left atrial pressures because pulmonary vascular pressures are typically low compared with systemic ones. The three main blood flow zones are shown.

The pressures in the lung's veins vary in the same manner. Because left atrial pressure is less than pulmonary arterial pressure, however, the pressure in the veins near the top of the lung may fall below atmospheric pressure (alveolar pressure) at end-expiration or end-inspiration. The anatomical effect of this compressive transmural pressure is that the pulmonary capillaries and venules will collapse and limit blood flow through that region.

The physiological importance of the distribution of pulmonary arterial and venous pressures over the height of the lung is that the lung can be divided into functional zones of blood flow, which depend on the pressures in the pulmonary vessels relative to alveolar pressure.

Finally, the pulmonary vessels can be divided into those vessels that are directly affected by alveolar pressure and those that are not: **alveolar** and **extra-alveolar vessels,** respectively. The alveolar vessels include most of the capillaries. If one expands the lung by high positive alveolar gas pressure, it is possible to squeeze most of the alveolar capillaries so that they contain no blood, but it is not possible to squeeze the blood out of the arteries and veins because the surrounding lung tissue elements are effectively pulling them open as lung volume increases.

■ *The Terminal Respiratory Unit*

The functional unit of the lung, in terms of O_2 and CO_2 exchange, is called the **terminal respiratory unit.** The human adult lung has about 60,000 terminal respiratory units. Each unit contains approximately 5000 anatomical alveoli and 250 alveolar ducts. Fig. 33-9 is a model that shows the relationships among the various structural elements of the lung.

■ *The Chest Wall*

The functional chest wall includes not only the rib cage and diaphragm but also the abdominal cavity and anterior abdominal muscles. Because the lungs are passive during breathing, it is up to the muscles of the chest wall, principally the diaphragm, to expand the thoracic cavity and thereby cause inspiration to occur.

The shape of the lungs must be the same as that of the thoracic cavity. The visceral and parietal pleuras cover the surfaces of the lungs and thoracic cavity, respectively. The lung and the chest wall pleuras are coupled together by a thin layer (about 20 μm thick) of liq-

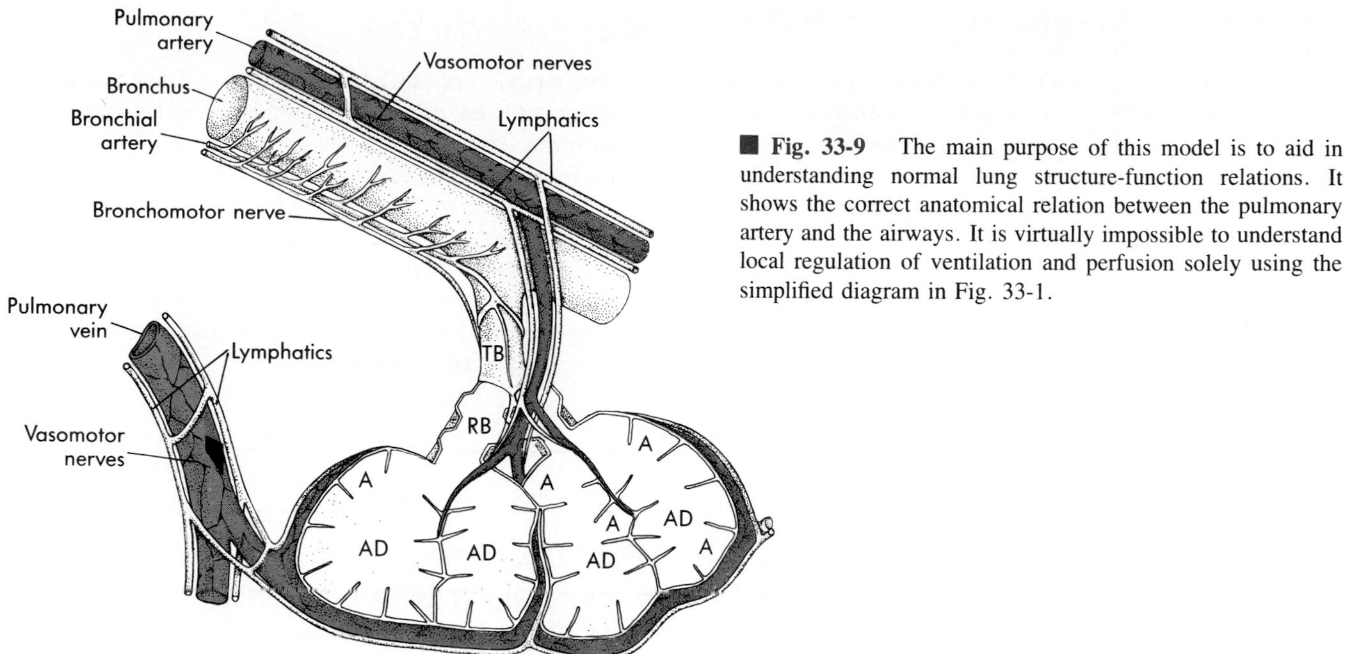

■ **Fig. 33-9** The main purpose of this model is to aid in understanding normal lung structure-function relations. It shows the correct anatomical relation between the pulmonary artery and the airways. It is virtually impossible to understand local regulation of ventilation and perfusion solely using the simplified diagram in Fig. 33-1.

uid. The liquid coupling allows the lung to move relative to the chest wall during breathing and thereby to accommodate itself with minimum stress to changes in thoracic configuration.

The thoracic cavity enlarges in all dimensions as the rib cage rotates upward and outward (pail-handle motion), and the diaphragm descends into the abdomen (Fig. 33-10).

The diaphragm is the main muscle of the chest wall. In the chest roentgenogram it appears as a dome-shaped

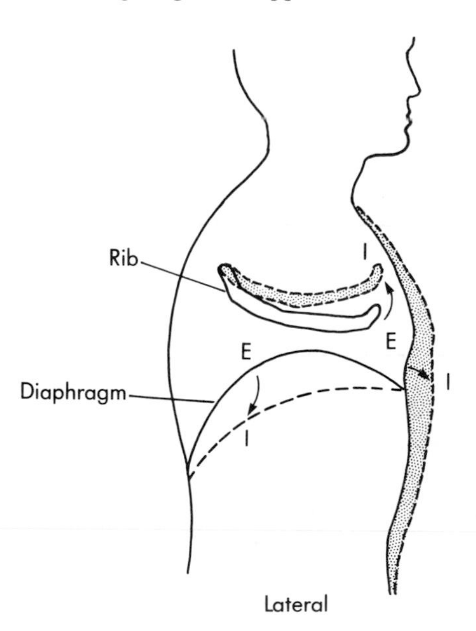

■ **Fig. 33-10** Chest and rib movements both contribute to the expansion of the chest cavity during breathing. Note that the anterior abdominal wall is included in the functional chest wall because it influences FRC and also diaphragm movement.

structure that separates the thoracic and abdominal cavities. The blood supply to the diaphragm is from branches of the intercostal arteries. The veins drain centrally into the inferior vena cava. The diaphragm is innervated by the two phrenic nerves, which have their origins at the 3rd to 5th cervical (spinal cord) segments and which descend laterally in the mediastinum to the right and left leaves of the diaphragm.

In humans the 12 ribs on each side articulate with the thoracic vertebrae. The only motion permitted for the ribs is to rotate upward toward the horizontal plane (pail-handle motion). This motion increases the cross-sectional area of the thorax when viewed along its cephalo-caudal axis.

The principal inspiratory muscles of the rib cage are the external intercostals. When they contract the ribs rotate upward. The internal intercostals are expiratory muscles. When they contract the ribs rotate downward. The accessory muscles (sternocleidomastoids and the scalenes) pull up on the rib cage and thereby assist inspiration. The rib cage muscles are supplied by intercostal arteries and veins and are innervated by motor and sensory intercostal nerves.

An important characteristic of the rib cage is its stiffness. In normal quiet breathing the rib cage may contribute up to half of the active inspiratory volume change. Diaphragmatic contraction causes the remainder. The rib cage also contributes passively because it prevents any inward (paradoxical) movement of the thoracic wall as pleural pressure becomes more subatmospheric during inspiration.

At low lung volume **(residual volume, RV)** it is possible to generate very negative intrathoracic pressure (−80 mm Hg) by attempting to inspire against a closed glottis (Müeller's maneuver). Near TLC, the maximal lung volume that one can voluntarily obtain decreases rapidly to zero. This effect limits maximal inspiratory volume.

In many diseases FRC is increased. Examples include asthma with marked expiratory airway resistance and emphysema characterized by destruction of lung elastic tissue. Under these conditions the muscles of inspiration operate at a disadvantage. Their ability to generate lower pleural pressures that cause lung expansion is reduced.

The work of taking a breath is done on the lung by the inspiratory muscles. Work is low under normal conditions so that the respiratory muscles have a large reserve capacity. In certain diseases fatigue of the chest wall muscles, especially the diaphragm, may occur and be the immediate cause of respiratory failure.

Review of Some Basic Physical Gas Laws

Table 33-1 lists many of the commonly used respiratory variables together with their units for a normal adult for resting and steady-state exercise conditions.

Universal Gas Law

The equation of state for ideal gases relates three variables, pressure (P), temperature (T), and volume (V), to the number (n) of moles of gas by a proportionality factor, R, the gas constant. Thus

$$nR = PV/T \qquad (1)$$

This law is mainly used by respiratory physiologists to convert volume between different conditions by setting two PV/T conditions (1 and 2) equal to each other (nR in equation 1 is constant):

$$V_2 = V_1 \times \frac{P_1}{P_2} \times \frac{T_2}{T_1} \qquad (2)$$

Partial Pressure

The concept of the partial pressure or tension of a gas is important. *In any volume the total gas pressure of all molecular species is the sum of the individual pressures that would exist if each species was alone in the same volume*. The law of partial pressures is based on the as-

Table 33-1 Values of cardiopulmonary physiological variables for a normal adult, 70 kg body weight, at rest and during steady state exercise

	Units	*Rest*	*Exercise*
Constants			
Hemoglobin concentration	g/L	150	
O_2 capacity of hemoglobin	ml O_2/g Hb	1.34	
Atmospheric pressure (sea level)	mm Hg	760	
Water vapor pressure (37° C)	mm Hg	47	
Standard conditions (STPD)	°K, mm Hg, mm Hg	273, 760, 0	
Body conditions (BTPS)	°K, mm Hg, mm Hg	310, 760, 47	
Cardiovascular			
Cardiac output	L/min	5	15
Heart rate	per min	60	180
Systemic arterial pressure, mean	mm Hg	90	100
Right atrial pressure, mean	cm H_2O	2	4
Pulmonary vascular pressures (left atrial level)			
Pulmonary arterial, mean	cm H_2O	19	30
Left atrial, mean	cm H_2O	11	15
Lung variables, BTPS			
Functional residual capacity	L	2.4	2.4
Total lung capacity	L	6	6
Tidal volume	L	0.5	2
Frequency	per min	12	15
Metabolism, STPD			
Carbon dioxide production	ml/min	200	1200
Oxygen consumption	ml/min	250	1500
Respiratory exchange ratio		.80	.80
Mechanics			
Pleural pressure, mean, at FRC	cm H_2O	−5	−3.5
Chest wall compliance at FRC	L/cm H_2O	0.2	0.2
Lung compliance at FRC	L/cm H_2O	0.2	0.3
Airway resistance	cm H_2O × L/sec	2.0	1.5

sumption that the gas molecules do not interact, which is nearly so for the respiratory gases (O_2, CO_2, N_2, H_2O). The law of partial pressures is a direct consequence of the universal gas law, which can be readily seen by keeping volume and temperature constant and relating n (moles of gas) to pressure (P) using equation 1.

Dry room air at atmospheric pressure, P_B, = 760 mm Hg contains 21% O_2, 79% N_2 and 0% CO_2. If we remove all of the nitrogen and let the oxygen fill the room at constant temperature, the universal gas law requires that the pressure of oxygen (Po_2) equal $0.21 \times P_B$ = 0.21×760 mm Hg = 160 mm Hg; similarly, P_{N_2} = 600 mm Hg.

The partial pressures of gases dissolved in water (blood plasma, interstitial, or intracellular liquids) are equal to the partial pressures in the gas phase that is in equilibrium with the liquid. In the normal resting human, blood that leaves the pulmonary capillaries has come into equilibrium with all of the alveolar gases. However, small quantities of venous blood from the bronchial venules and the thebesian vessels (heart) contaminate pulmonary venous outflow so that the partial pressure of O_2 (Pa_{O_2}) in systemic arterial blood (a) is normally 5 to 10 mm Hg less than that in alveolar gas. Thus, Pa_{O_2} = 90 − 100 mm Hg. Table 33-2 lists the normal partial pressures of the respiratory gases at various physiologically important locations in the body.

Water Vapor Pressure

When air is inspired, it is warmed to 37° C in the nose, throat, and trachea and becomes saturated with water vapor at 37° C. This is an obligatory event and occurs rapidly. The heat and water vapor are supplied by the pulmonary and bronchial blood flows. The water vapor exerts a mandatory partial pressure, P_{H_2O} = 47 mm Hg.

The total quantity of water in expired gas is not trivial. It is about 40 mg/L, which over the course of 24 hours accounts for nearly half of the insensible (obligatory) water loss from the body.

The pulmonary physiologist is often interested only in dry gas volumes and partial pressures. The water vapor partial pressure can be eliminated by subtracting it from total gas pressure. Thus inspired gas (I) in the trachea still has its normal concentration (fraction, F) of the respiratory gases in room air, $F_{I_{O_2}}$ = 0.21. Thus the partial pressures of oxygen in dry inspired air is reduced to $0.21 \times (760 - 47)$ mm Hg = 150 mm Hg.

In the alveoli some oxygen diffuses into the pulmonary capillary blood. This diffusion further reduces the partial pressure of the oxygen in alveolar gas. At a normal respiratory exchange ratio, R = 0.80, the fraction, F, of O_2 in alveolar (A) gas (F_{AO_2}) = 0.143. Thus normally $P_{A_{O_2}}$ (alveolar Po_2) = $0.143 \times (760 - 47)$ = 102 mm Hg. The deficit in O_2 between inspired air and alveolar gas is made up mainly by the CO_2 that enters from the blood (F_{ACO_2} = 0.056). $P_{A_{CO_2}}$ = $0.056 \times (760 - 47)$ mm Hg = 40 mm Hg. Normally $P_{A_{CO_2}}$ (alveolar Pco_2) is closely regulated. It is the main controlled variable in breathing (see Chapter 37).

Conditions

The universal gas law is used to correct respiratory gas volumes under three conditions (see equation 2). These are ambient temperature and pressure saturated (ATPS), body temperature pressure saturated (BTPS), and standard temperature pressure dry (STPD). Lung volumes should be measured under the existing conditions, BTPS. But if the volumes are recorded in a spirometer, ATPS, corrections are required. Standard conditions (0 ° C, 760 mm Hg, dry) must be used when referring to oxygen and carbon dioxide consumptions.

Blood Oxygen Concentration

The hemoglobin-oxygen (HbO_2) equilibrium curve (see Fig. 33-2) is a plot of the HbO_2 saturation (So_2) as a function of the partial pressure of O_2 (Po_2). Clearly the relationship between So_2 and Po_2 is a complex function. The main use of the graph is to determine one of the variables using a knowledge of the other.

What the curve does not show is that, even at equilibrium, rapid dissociation and association chemical reactions are occurring. In other words a given point on the equilibrium curve corresponds to the situation where $O_2 + Hb \rightarrow HbO_2$ exactly equals $HbO_2 \rightarrow Hb + O_2$. One needs to know the speed (kinetics) of these two si-

Table 33-2 Total and partial pressures of respiratory gases in ideal alveolar gas and blood at sea level barometric pressure (760 mm Hg)

	Ambient air (dry)	*Moist tracheal air*	*Alveolar gas (R = 0.80)*	*Systemic arterial blood*	*Mixed venous blood*
Po_2	160	150	102	102	40
Pco_2	0	0	40	40	46
P_{H_2O}, 37° C	0	47	47	47	47
P_{N_2}	600	563	571*	571	571
P_{total}	760	760	760	760	704†

*P_{N_2} is increased in alveolar gas by 1% because R < 1 normally.
†P_T in venous blood is reduced because Po_2 has decreased more than Pco_2 has increased.

multaneous reactions to fully understand the physiological value of the hemoglobin-oxygen equilibrium curve.

From the equilibrium curve one can read that at Po_2 = 100 mm Hg the saturation of blood (So_2) = 97.4%. However, some oxygen is also dissolved (in physical solution), even though oxygen is not very soluble in blood. The dissolved oxygen concentration is determined by the solubility coefficient, $\alpha = 22.8$ ml/(L × 760 mm Hg). Thus 1 L of blood dissolves 3.0 ml O_2 for a Po_2 = 100 mm Hg.

Because hemoglobin is able to chemically bind 1.34 ml O_2/g and the normal hemoglobin concentration (Hb) in blood is 150 g/L, the **O_2 capacity of hemoglobin** = 1.34 ml O_2/g Hb × 150 g Hb/L = 200 ml O_2/L.

Normal blood in the aorta has an So_2 = 97.0% because the Pa_{O_2} = 90 mm Hg. This means that the bound oxygen content is 97.0% of capacity. To obtain the total O_2 concentration in blood add the chemically bound O_2 to that in physical solution at the appropriate Po_2. For systemic arterial blood the total concentration of oxygen, Cao_2 = (0.97 × 200 + 0.03 × 100) ml O_2/L = 197 ml O_2/L. The correction for dissolved O_2 is usually small, but when a person breathes 100% O_2 it may be substantial (18 to 20 ml/L of blood).

Conservation of Mass

A physical principle widely used in physiology (but often disguised by special names) is the conservation of mass law. *The law states that in a closed system (such as the whole body) the total number of atoms remains constant.* Because oxygen is entering the body at the rate of 250 ml/min, the same quantity of oxygen must leave the body, else the conservation of mass law will be violated. The oxygen that leaves the body does so either as carbon dioxide (CO_2) or water vapor (H_2O).

The ratio of carbon dioxide production ($\dot{V}co_2$) to oxygen consumption ($\dot{V}o_2$) is called the **respiratory exchange ratio (R)** = $\dot{V}co_2/\dot{V}o_2$. R in the steady state ranges between 0.7 and 1.0. The variability in the respiratory exchange ratio is due to the metabolic fuel being burned by the body. The fuel mixture determines the number of CO_2 molecules produced for a given number of O_2 molecules consumed. In Table 33-1 the average respiratory quotient is listed as 0.80, which is reasonable for a mixed diet of carbohydrates, protein, and fat. Thus for a resting $\dot{V}o_2$ = 250 ml/min, $\dot{V}co_2$ = 0.80 × 250 ml/min = 200 ml/min.

Because $\dot{V}o_2$ is the quantity of oxygen disappearing into the body, the oxygen must be removed by the cardiac output, $\dot{Q}$, as it flows through the pulmonary capillaries. Applying the conservation of mass law, one can equate the oxygen consumption with the blood flow and the oxygen concentration difference between blood entering and leaving the pulmonary capillaries to obtain the important Fick equation for measuring cardiac output (see also Chapter 24):

$$\dot{Q} = \frac{\dot{V}o_2}{(Cao_2 - C\bar{v}o_2)} \tag{3}$$

One can also measure the pulmonary blood flow using carbon dioxide production and its mixed venous to arterial concentration difference.

Normally at rest $C\bar{v}o_2$ = 150 ml/L and Cao_2 = 200 ml/L whole blood. Therefore the normal resting pulmonary blood flow is:

$$\dot{Q} = \frac{250 \text{ ml } O_2/\text{min}}{(200 - 150) \text{ ml } O_2/\text{L}} = 5 \text{ L/min} \tag{4}$$

Lung Volumes and Ventilation

Because tissues engaged in aerobic metabolism use O_2 and produce CO_2, they remove O_2 from systemic capillary blood and add CO_2 to it. This lowers the partial pressure of oxygen in mixed venous blood ($P\bar{v}_{O_2}$) to below that of alveolar gas and raises the partial pressure of carbon dioxide in mixed venous blood ($P\bar{v}_{CO_2}$) to above that of alveolar gas. The normal partial pressures of the common respiratory gases in systemic arterial and mixed venous (pulmonary arterial) blood are listed in Table 33-2.

Half of the process of ventilation/perfusion matching involves ventilation of the alveoli in such a way as to increase the Pa_{O_2} well above that of mixed venous blood. This causes oxygen to diffuse along its partial pressure gradient and loads oxygen into the pulmonary capillary blood. Ventilation lowers the Pa_{CO_2} below that in mixed venous blood. This causes CO_2 to diffuse along its partial pressure gradient and reduces the CO_2 content of the pulmonary capillary blood.

When body metabolism increases, as in exercise, $P\bar{v}_{O_2}$ will decrease and $P\bar{v}_{CO_2}$ will increase as the body consumes more oxygen and produces more carbon dioxide. To keep the arterial blood partial pressures (Pa_{O_2}, Pa_{CO_2}) at steady levels, the lungs—by the process of alveolar ventilation—must supply to the alveoli an amount of O_2 equal to the amount removed from the blood by the metabolizing tissues and must remove from the alveoli the amount of CO_2 that was added to venous blood by the systemic tissues.

The main purpose of ventilation is to maintain an optimal composition of alveolar gas. Think of alveolar gas as a buffer (stabilizing) compartment of gas lying between the environment (ambient air) and pulmonary capillary blood. Oxygen is continuously removed from alveolar gas, and CO_2 is continuously added to it by blood flowing through the alveolar wall capillary network. Oxygen is supplied to the alveolar gas, and CO_2 is removed from it by the cyclic process of ventilation, the inspiration of fresh air followed by the expiration of alveolar gas. The cyclic nature of ventilation suggests the importance of the buffering effect of a large alveolar gas volume.

Total Ventilation and Alveolar Ventilation

Total ventilation is the volume of air entering or leaving the nose or mouth during each breath or each minute. It can be measured breath-to-breath by volume recorders, such as the spirometer in Fig. 33-11.

The volume of each breath is called the **tidal volume, VT,** which varies with age, sex, body position, and metabolic activity. The average normal value for a resting adult is 0.5 L (500 ml). The largest possible tidal volume is the **vital capacity, VC.**

Alveolar ventilation, $\dot{V}$A, is the volume of fresh air that enters the alveoli in each breath or in each minute. Alveolar ventilation is always less than total ventilation; how much less depends on the anatomical dead space and tidal volume.

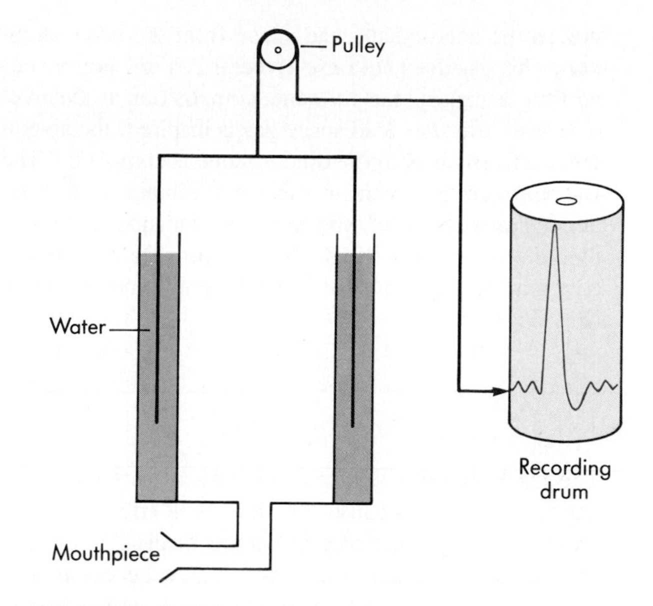

Fig. 33-11 The spirometer is a useful device for measuring all lung volumes (except residual volume) and airflow ratio. The subject breathes in and out of the water-sealed chamber, which moves the delicately balanced float that moves a pen on a rotating drum.

Anatomical Dead Space and Tidal Volume

Fresh air does not go directly to the terminal respiratory units. It first flows through the conducting airways (nose, mouth, pharynx, larynx, trachea, bronchi, and bronchioles). In the conducting airways O_2 and CO_2 do not significantly exchange between gas and blood. Therefore that portion of the fresh inspired air is called the **anatomical dead space, VD.** Although the anatomical dead space can be measured, it is generally assumed to be approximately 2 ml/kg of ideal body weight.

At the end of a normal expiration (just before the next inspiration begins), the conducting airways are filled with alveolar gas, which has an ideal PO_2 of 102 mm Hg and a PCO_2 of 40 mm Hg. Thus as a tidal inspiration begins, the alveoli must first receive the gas that

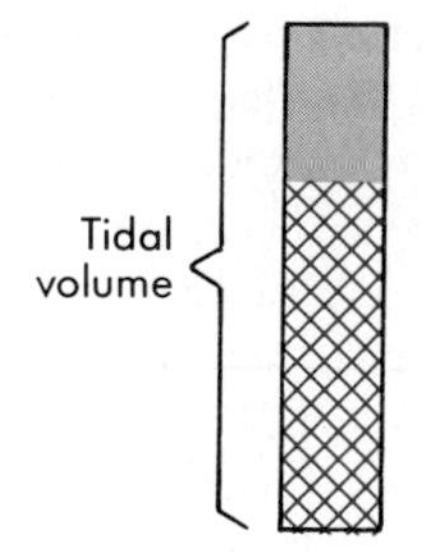

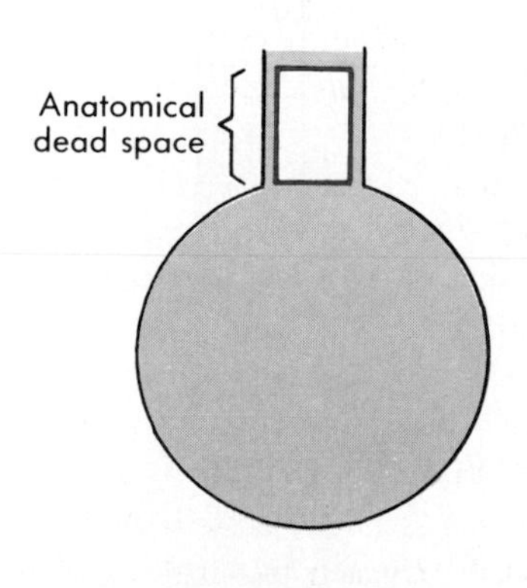

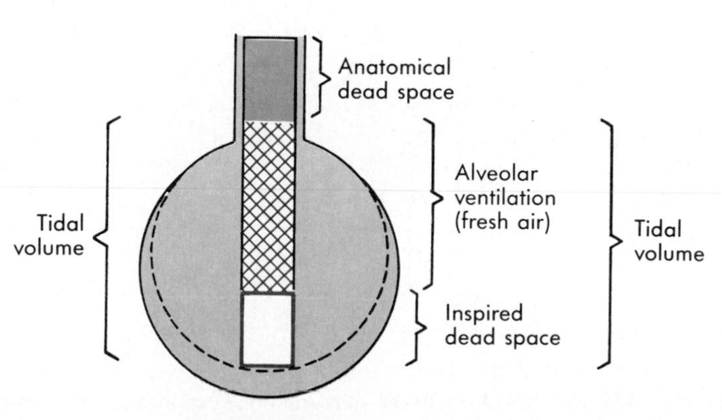

Fig. 33-12 Tidal volume and alveolar ventilation are equal, but the air entering the alveoli contains the deadspace gas from the previous breath plus some of the new air. The fresh air in the anatomical deadspace is wasted (i.e., it does not contribute to gas exchange).

was in the anatomical dead space from the last exhalation. This gas does not raise alveolar Po_2 or lower alveolar Pco_2 because it has the same composition as the alveolar gas. After the dead space gas is inspired, the alveoli receive fresh air until the tidal volume is completed. The last portion of the fresh air, of course, remains in the conducting airways. The volume of the dead space, V_D, and the volume of each breath, V_T, are important factors in determining the amount of alveolar ventilation per breath (Fig. 33-12).

Lung Volumes and Capacities

The lungs do not collapse to the airless state with each expiration partly because the chest wall (ribs, intercostal muscles, and diaphragm) becomes stiffer at the end of expiration. Indeed, the lungs cannot be emptied of gas, even by the most forceful expiration; some gas still remains, i.e., the **residual volume (RV)** (Fig. 33-13).

The FRC is the volume of air that remains at the end of a normal expiration. The FRC is not actively regulated; it is determined by the passive mechanical relationship between the chest wall and the lungs, although airflow dynamics may contribute in some conditions.

The FRC acts as a buffer against large changes in alveolar Po_2 with each breath. If FRC was very small, Pa_{O_2} would fluctuate markedly with each breath decreasing toward that of mixed venous blood in the pulmonary capillaries at end-expiration ($P\bar{v}_{O_2}$ = 40 mm Hg) and rising toward that of moist tracheal air (Po_2 = 150 mm Hg) with inspiration. The advantage of cyclic ventilation of a relatively large space (FRC = 2.4 L; 40% of TLC, normally) by a small volume of alveolar ventilation per breath ($V_T - V_D$) is that fluctuations of PA_{O_2} and PA_{CO_2} and, consequently, of Pa_{O_2} and Pa_{CO_2} are minimal (about 4 mm Hg for the former and 3 mm Hg for the latter). For practical purposes physiologists and physicians tend to ignore breath-to-breath fluctuations in alveolar and blood gas tensions in normal people. However in some circumstances the fluctuation may be important for the control of breathing.

The various static lung volumes and capacities are shown in Fig. 33-13, which also includes a tracing in real time of a normal spirogram. There are 4 nonoverlapping volumes together with several capacities, each of which includes 2 or more volumes. Of the 4 named volumes, all, except residual volume, can be measured directly by volume recorders, such as the spirometer (see Fig. 33-11).

Diseases of the lungs or chest wall affect lung volumes and capacities in various ways. The most frequent change is in the vital capacity, VC, which may be greatly reduced. The reduction may be caused by limited expansion (restrictive disease) or to an abnormally large residual volume (obstructive airway disease). In strenuous exercise tidal volume may have to increase to one half the vital capacity to ensure adequate alveolar ventilation. Thus limitation of exercise capacity is often an early sign of lung disease that limits VC.

Changes in the residual volume (RV) are important to the assessment of lung function; hence it is frequently measured in standard pulmonary function tests. One common method called the single breath volume of dilution is another application of the widely used conservation of mass law.

One inspires a known volume of gas of markedly different composition from normal alveolar gas, measures the dilution of the gas, then uses the conservation of mass law to calculate the residual volume. Most commonly used are the insoluble inert gases, helium or neon. For example, a normal person inspires 4.4 L, V_I, of a mixture containing 10% helium from a spirometer at 21° C. On expiration a sample of alveolar gas con-

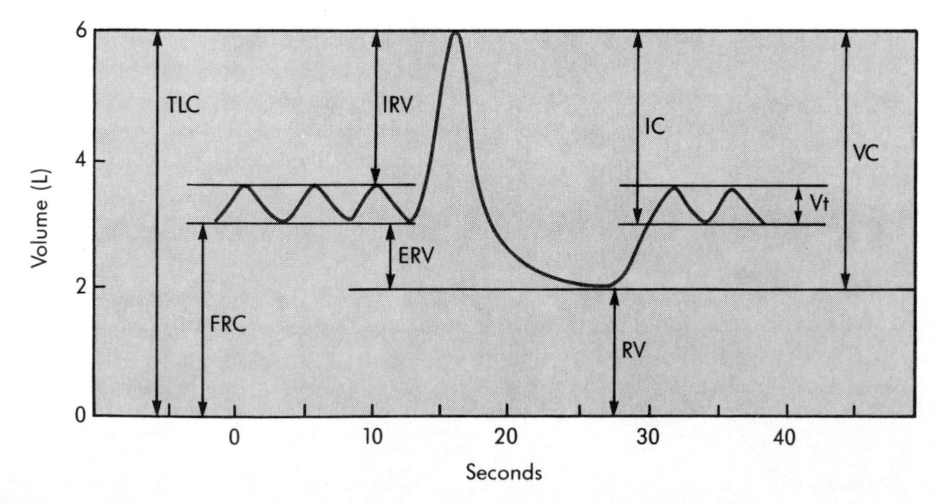

Fig. 33-13 On the spirogram one can measure the forced vital capacity, arguably the single most useful test of lung function. The named lung volumes and capacities are marked.

tains 8% helium. The inspired volume was diluted by the gas already in the lungs. To compute RV we set up the mixing equation using conservation of mass:

$$(C_1 \times V_1) + (C_2 \times V_2) = (C_3 \times V_3) \qquad (5)$$

If V_1 = RV, then C_1 is zero because helium is not normally present in air, so the first term drops out, which is convenient and explains why physiologists use foreign inert gases. Let V_3 be the total volume of gas (RV + V_I). Substitute the necessary helium concentrations to obtain residual volume. There is one small problem: the inspired gas in the spirometer is ATPS, therefore the inspired volume has to be corrected up to 37° C to achieve the correct RV (BTPS).

As alveolar O_2 tension, $P_{A_{O_2}}$, is the main determinant of the rate of diffusion of O_2 from alveolar gas into pulmonary capillary blood, the amount of alveolar ventilation is more important in determining $P_{A_{O_2}}$ than is the size of the FRC.

Alveolar Ventilation

The alveolar ventilation/breath, when multiplied by the frequency of breathing (respiratory rate), f, gives the alveolar ventilation per minute, $\dot{V}_A$, just as tidal volume × frequency gives the total expired ventilation per minute, $\dot{V}_E$. If V_T = 500 ml and the normal respiratory frequency is 12/min, then $\dot{V}_E = V_T \times f$ = 500 ml × 12/min = 6,000 ml/min.

To obtain alveolar ventilation subtract the anatomical dead space from the tidal volume and multiply by the frequency of breathing; $\dot{V}_A = f \times (V_T - V_D)$. Actually the anatomical dead space is not often measured. The volume of dilution principle is used to calculate the alveolar ventilation in a manner similar to the previous example about residual volume. In this application, however, instead of using a foreign gas, it is more convenient to use CO_2 because by definition no CO_2 exchange occurs in the anatomical dead space. Therefore ventilation of the dead space contributes nothing to the expired CO_2, although it does contribute to expired volume. Usually all gas concentrations are expressed as fractions, F, of total dry gas pressure. In terms of F_{CO_2}, the volume of dilution equation is as follows:

$$(\dot{V}_A \times F_{ACO_2}) + (\dot{V}_D \times F_{DCO_2}) = \dot{V}_E \times F_{ECO_2} \qquad (6)$$

where the second term on the left represents dead space CO_2 production, which is normally zero, as explained previously. Normal alveolar gas contains 5.6% CO_2 [F_{ACO_2} = 40 mm Hg/(760 − 47 mm Hg) = 0.056]. If the expired ventilation had been collected in a spirometer, ATPS, then a correction to BTPS would have been required because $\dot{V}_A$ is always reported under body conditions.

Because the main purpose of ventilation is to maintain an optimal concentration of alveolar gases, alveolar ventilation is in balance (steady state) for oxygen when it matches O_2 use with O_2 supply. For example if we use the normal values of tidal volume of 500 ml and dead space volume of 150 ml, the alveolar ventilation per breath is 350 ml. If a normal person breathes 12 times/min, the alveoli is supplied with an alveolar ventilation, $\dot{V}_A$, of 4200 ml/min.

As will be described later in this chapter, it is not the oxygen supply to the alveoli but the partial pressure of CO_2 in arterial blood, Pa_{CO_2}, that is closely regulated. If the Pa_{CO_2} is held at 40 mm Hg, which can be equated with the average $P_{A_{CO_2}}$ = 40 mm Hg, then while the subject breathes air at sea level, the $P_{A_{O_2}}$ is maintained at 102 mm Hg.

The adequacy of ventilation is described in terms of Pa_{CO_2} because it is much more common for the inspired oxygen concentration to be decreased, as when one goes to higher altitude, or increased, as when one breathes enriched oxygen mixtures. Normal alveolar ventilation means that Pa_{CO_2} = 40 mm Hg. **Hyperventilation** (overventilation) for a particular metabolic state means that Pa_{CO_2} is <40 mm Hg. **Hypoventilation** (underventilation), which is the more common condition encountered in patients with severe lung diseases, means that Pa_{CO_2} >40 mm Hg.

Wasted Ventilation (Physiological Dead Space)

In a perfect lung all alveoli receive ventilation ($\dot{V}_A$) and blood flow ($\dot{Q}$) in the same proportion. This is referred to as uniform ventilation/perfusion ratios ($\dot{V}_A/\dot{Q}$). The ideal condition does not exist even in the healthiest individuals and may be markedly abnormal in diseased lungs. The concept of wasted ventilation or physiological dead space is used clinically to describe the deviation of ventilation relative to blood flow from ideal (Fig. 33-14).

Clearly ventilation of the anatomical dead space is necessary, but it is wasted because no useful gas exchange occurs. But in addition suppose that some terminal lung units do not receive any pulmonary blood flow. For example, on a gross scale imagine that one of the branch pulmonary arteries is blocked by a blood clot **(embolus).** The alveolar ventilation to that lung is wasted because it does not participate in any useful exchange. Thus the physiological dead space is the sum of anatomical dead space, V_D, and the portion of $\dot{V}_A$/breath that ventilates lung units, which receive no blood flow. Because the expired alveolar volume contains gas from unperfused and normally perfused alveoli, the true $P_{A_{CO_2}}$ is underestimated. The systemic arterial carbon dioxide partial pressure, Pa_{CO_2}, is generally accepted as being equal to the carbon dioxide partial pressure, $P_{A_{CO_2}}$, for the properly ventilated and perfused lung.

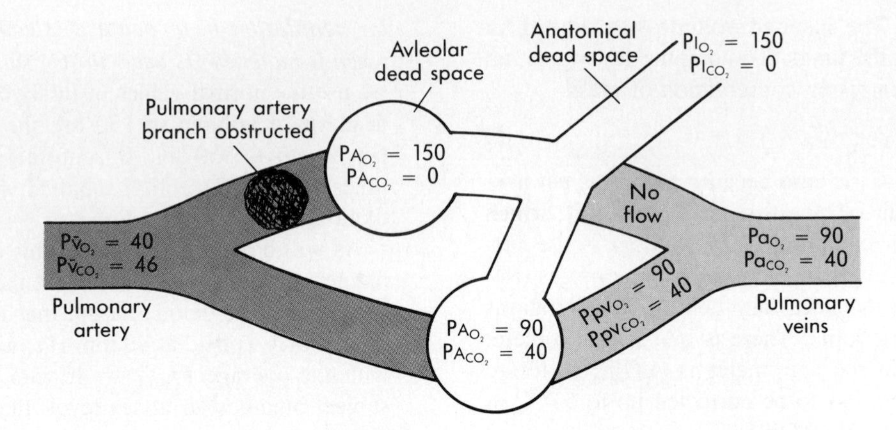

■ **Fig. 33-14** Wasted ventilation includes the anatomical dead space plus any portion of the alveolar ventilation that does not exchange O_2 or CO_2 with the pulmonary blood flow. This simple sketch shows one pulmonary artery completely obstructed by a blood clot, but in most patients the anatomical definition of the wasted ventilation is not easily made.

■ *Alveolar Ventilation Equation*

The alveolar ventilation equation, the central equation of pulmonary physiology, describes the reciprocal (hyperbolic) relationship between alveolar ventilation and PA_{CO_2}. All discussions of the adequacy of ventilation and of alveolar oxygen or carbon dioxide partial pressures come directly from this equation.

$$\dot{V}A,\ L/min = \frac{\dot{V}CO_2,\ ml/min}{PA_{CO_2},\ mm\ Hg} \times K \qquad (7)$$

If the barometric pressure is 760 mm Hg and body temperature is 33° C, then the correction factor, K, is 0.863 mm Hg × L/ml, when $\dot{V}A$ is given in L/min, BTPS, and the CO_2 production, $\dot{V}CO_2$, is in ml/min, STPD (Fig. 33-15).

The alveolar ventilation equation is correct for any CO_2 production, even during heavy exercise, where $\dot{V}CO_2$ may be 6 times the resting value (see dashed line in Fig. 33-15). The equation carries with it the correct implication that PCO_2 is the controlled variable for the regulation of alveolar ventilation.

Using equation 7 one can compute the wasted ventilation as an index of the efficiency of ventilation. Normally it is less than 35% of $\dot{V}E$. The wasted ventilation, of course, is total ventilation minus the effective alveolar ventilation ($\dot{V}E - \dot{V}A$).

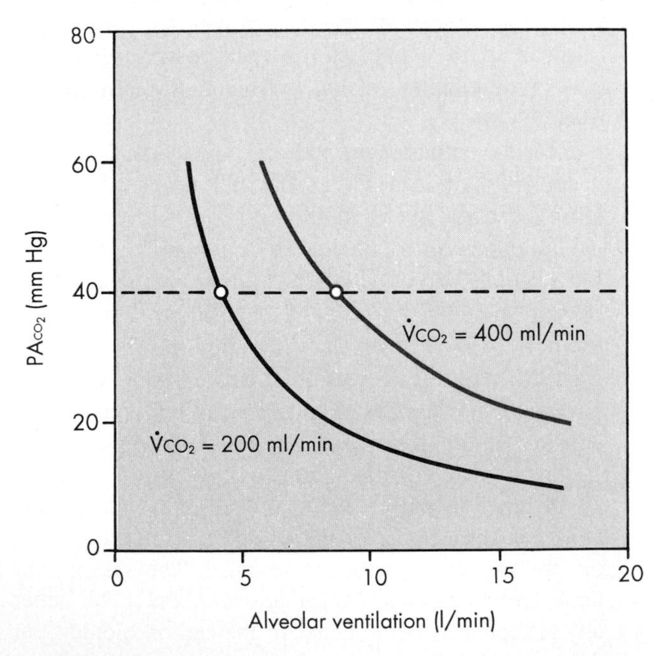

■ **Fig. 33-15** The relationship between alveolar ventilation and alveolar PCO_2 is a hyperbola. The other variable required is the CO_2 production. Two examples are shown.

■ *Alveolar Gas Equation*

If PA_{CO_2} is known, one can calculate PA_{O_2} by the alveolar gas equation:

$$PA_{O_2},\ mm\ Hg = \qquad (8)$$

$$F\times (PB - PH_2O) - PA_{CO_2} \times I_{O_2} \left[FI_{O_2} + \frac{1 - FI_{O_2}}{R} \right]$$

$PB - PH_2O$ is the total dry gas pressure as previously defined. If R = 1 or if one breathes 100% oxygen (FI_{O_2}= 1.0), the complex term in the square brackets equals 1. Normally, the term in brackets evaluates to 1.2 (FI = 0.21; R = 0.80). Because we used Pa_{CO_2} (not PA_{CO_2}) in the alveolar gas equation, we obtain the PA_{O_2} for the effectively ventilated lung.

■ *O_2 - CO_2 Diagram*

Pa_{CO_2}, R, and PA_{O_2} are tightly related when one breathes air. The relationship is not linear because of

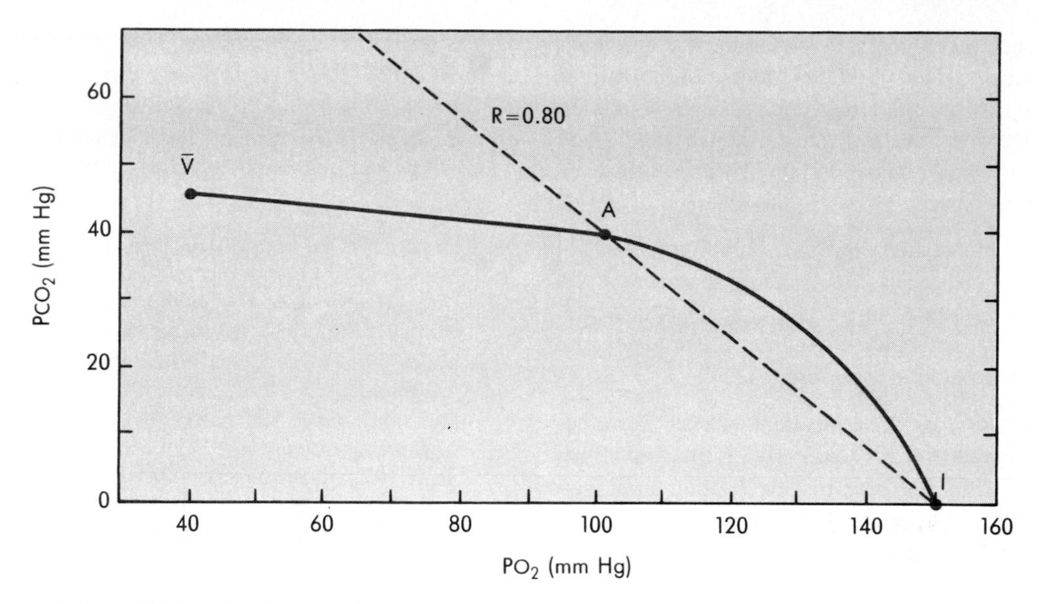

Fig. 33-16 The O_2-CO_2 diagram in its simplest form. All pairs of alveolar partial pressures for a given condition are shown. The inspired point, *I*, for normal sea level dry tracheal air, $\bar{V}$, is mixed venous blood. Point *A* (mean alveolar gas) depends on the respiratory quotient, *R*.

the hyperbolic relation in the alveolar ventilation equation (see Fig. 33-15) and the correction factor in the alveolar gas equation. But the O_2 and CO_2 relations have been carefully worked out and presented in graphic form as the O_2-CO_2 diagram. The O_2-CO_2 diagram in its basic form (Fig. 33-16) is useful and easy to read. The line begins on the x-axis at a point labeled I, which indicates inspired air (moist tracheal air at sea level) with PI_{O_2} = 150 mm Hg and $PICO_2$ = O. The point labeled A represents normal alveolar gas when the respiratory exchange ratio, R, = 0.80. The PA_{CO_2} = 40 mm Hg and Pa_{O_2} = 102 mm Hg. The line continues to the left with a fairly flat trajectory, and it ends at a point labeled, $\bar{v}$, which represents the normal mixed venous blood gas tensions; $P\bar{v}_{O_2}$ = 40 mm Hg and $P\bar{v}_{CO_2}$ = 46 mm Hg. If some alveoli receive no ventilation, the blood passing through their capillaries exits into the pulmonary veins with the same gas tensions as mixed venous blood (physiological right-left shunt). Every possible combination of PA_{O_2} and PA_{CO_2}, under the conditions described, lies on the line in the figure. All points on the line to the left of A represent hypoventilation, whereas all points to the right represent hyperventilation.

Summary

1. The main function of the lung is to bring fresh air (ventilation) into close contact with blood flowing in the pulmonary capillaries (perfusion) so that the exchange of oxygen and carbon dioxide will take place efficiently.

2. The lung is constructed for efficient matching of ventilation to perfusion, as reflected by the coordinated branching pattern of the airways and the pulmonary arteries. The airways are of two types; cartilaginous bronchi and membranous bronchioles. One pulmonary artery branch accompanies each airway and branches with it. The lung has the most extensive capillary network surface area of any organ. The erythrocytes remain in the capillaries long enough for gas exchange to reach equilibrium, even in heavy exercise. Groups of alveolar ducts and their alveoli, together with their supplying arteries, are combined into small functional elements called terminal respiratory units. These units are arranged in parallel, which ensures the efficient distribution of inspired air and mixed venous blood.

3. Functionally the chest wall includes not only the diaphragm and rib cage but also the abdomen. Coordinated muscle contraction in the chest wall enlarges the thoracic cavity during inspiration, while the lungs expand passively in all directions to fill the cavity.

4. The universal gas law is used to convert gas volumes between differing pressures and temperatures. Each kind of gas in a mixture, including water vapor, exerts a partial pressure in both gas and liquid phases proportional to its fractional concentration in the gas phase.

5. The red protein hemoglobin in erythrocytes binds O_2 rapidly and reversibly. The oxygen capacity (100% saturation) of normal blood is 200 ml O_2/L.

6. The conservation of mass law is widely used to measure cardiac output and various components of

breathing, such as alveolar ventilation, oxygen consumption, and carbon dioxide production. Alveolar ventilation is the useful (fresh air) portion of minute ventilation that reaches the gas exchange units. The portion of each breath that fills the airway is called the anatomical dead space.

7. TLC is composed of several separate volumes and overlapping capacities, of which the most important are residual volume, functional residual capacity, and vital capacity. Except for residual volume, all volumes and capacities can be directly measured.

8. Alveolar P_{CO_2} or its equivalent arterial P_{CO_2} describes the adequacy of alveolar ventilation. The alveolar ventilation equation describes the relation among alveolar ventilation, arterial P_{CO_2}, and carbon dioxide production (metabolism).

9. The alveolar gas equation permits the calculation of mean alveolar P_{O_2} if P_{CO_2} is known. The O_2 -CO_2 diagram shows all possible alveolar oxygen and carbon dioxide partial pressures under given physiological conditions.

Bibliography

Journal articles

Staub NC: The interdependence of pulmonary structure and function, *Anesthesiology* 24:831, 1963.

Books and monographs

Hayek HV: *The human lung,* New York, 1960, Hafner.

Macklem PT: *Symbols and abbreviations.* In Fishman A, editor: Handbook of physiology, section 3, *Respiration,* Bethesda, Md, 1985, American Physiological Society.

Miller WS: *The lung,* ed 2, Springfield, 1947, Charles C Thomas.

Staub NC: *Basic respiratory physiology,* New York, 1991, Churchill-Livingstone.

Staub NC, Albertine, KH: *The structure of the lung relative to its principal function.* In Murray JF, Nadel JA, editors: *Textbook of respiratory medicine,* Philadelphia, 1988, WB Saunders.

Weibel ER: *Morphometry of the human lung,* New York, 1963, Academic Press.

CHAPTER

34

Mechanical Properties in Breathing

Statics

The static mechanical properties of the lung (L) and chest wall (w) encompass a major part of modern respiratory physiology and have important manifestations in such diverse diseases as emphysema, pulmonary fibrosis, and respiratory distress syndromes. When examining respiratory mechanics one must treat the lungs and chest wall as passive. That is, the muscles of the chest wall (diaphragm and intercostal muscles) must not contract. This condition occurs normally only at functional residual capacity (FRC) (end expiration) but can be achieved clinically when the lungs and chest wall are moved by a mechanical ventilator in a relaxed subject or in a patient who is paralyzed.

Lung Distensibility

Elastance, E, describes the resistance of an object to deformation by an external force. **Compliance, C,** has the opposite connotation. Thus, compliance is the reciprocal of elastance: $C = 1/E$.

Elastic Recoil of the Lung

The elastance or compliance of the lung is determined from the pressure-volume curve. The lung is not a perfectly elastic body because of its complex anatomical structure and because of a major additional force, the alveolar air-liquid surface tension. A normal pressure-volume curve of the lung is shown in Fig. 34-1. The main points to note are that the curve must be read in the direction of the arrows, the end points show sharp discontinuities, and inflation and deflation follow different paths: that is the pressure-volume cycle exhibits hysteresis, which must be explained.

The structures of the lung are not uniform springs; they consist of collagen and elastic fibers, giant glyco protein molecules in the interstitial matrix, and various cells, such as the alveolar epithelium and capillary endothelium. Some of the lung tissues are compliant (easily stretched), such as cells, interstitial ground substance molecules, and elastic fibers. Collagen fibers however are not easily stretched. As lung volume increases, they restrict lung expansion and deformation more and more. This causes the lung to become stiffer and the slope of the pressure-volume curve to decrease.

The Measure of Lung Distensibility

The compliance of the lung, C_L, is the measure of its distensibility. It is the slope of the line between any two points on the deflation pressure-volume curve; that is,

$$C_L = \Delta V_L/\Delta P_L, \quad (1)$$

where ΔV_L is the change in volume and ΔP_L is the change in translung pressure. The units of lung compliance are liters per centimeter of water (L/cm H_2O). Lung elastance, E_L, is the reciprocal of compliance: that is it is the measure of the lung's resistance to distension:

$$E_L = \Delta P_L/\Delta V_L. \quad (2)$$

With reference to transorgan pressures the pressure difference across the wall is always measured from the inside to the outside. For the lung transorgan pressure, P_L, is defined as: $P_L = P_A - Ppl$, where P_A is total alveolar gas pressure and Ppl is the pressure in the pleural space.

The Deflation Pressure-Volume Curve

The static deflation limb of the pressure-volume curve of a normal, air-filled human lung is shown in Fig. 34-2. Zero applied translung pressure is indicated by the thin vertical line. Pressures to the right are positive and act to distend the lung; pressures to the left are negative and act to compress it.

The ordinate shows relative lung volume from zero to

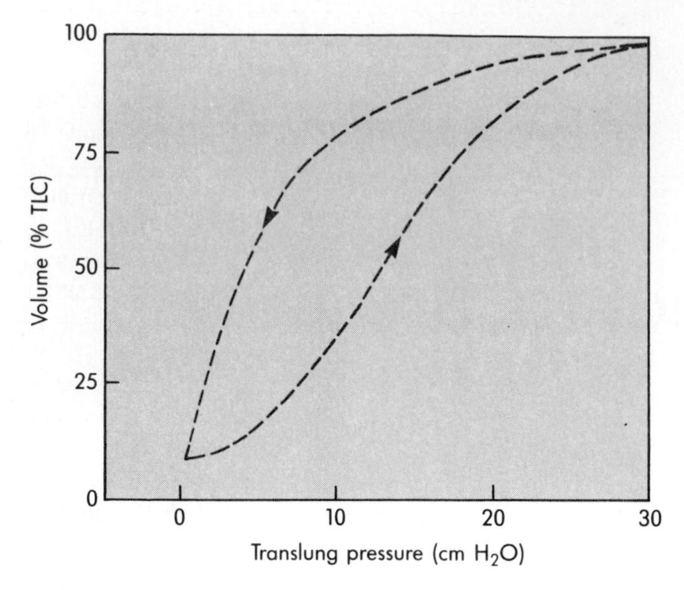

■ **Fig. 34-1** The air pressure-volume curve is rich in information about the mechanical properties of the lung. The curve can only be traced in the direction shown by the arrows. Inflation and deflation lines are different mainly because the phenomenon of air-liquid surface tension.

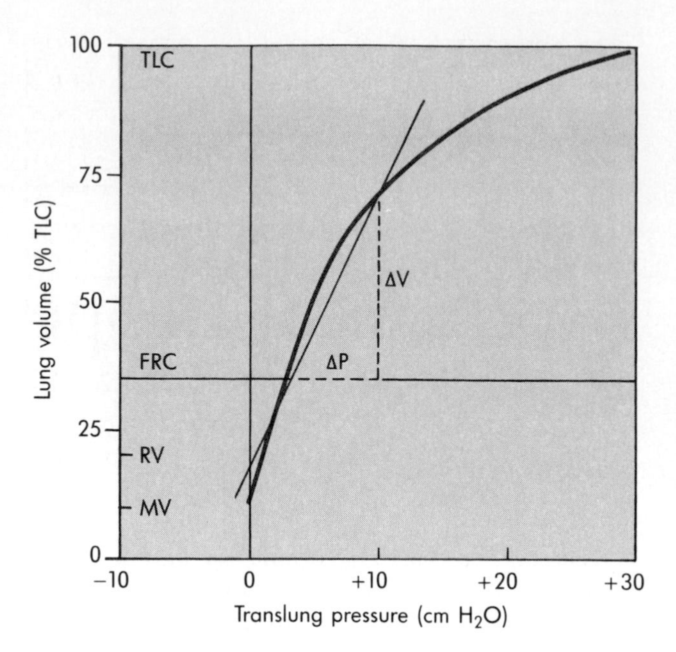

■ **Fig. 34-2** Deflation pressure-volume curve of a normal human lung demonstrating how compliance is obtained.

total lung capacity, TLC. Also marked are **residual volume (RV), (FRC)**, and **minimal volume (MV)**. The last represents the unstressed volume. When the chest is opened, as at thoracic surgery, and the lungs are allowed to recoil until the translung pressure equals zero, the lung does not collapse to the airless condition but retains approximately 10% of its total gas capacity: that is, its minimal volume. Reexamine Fig. 34-1.

Minimal volume for the lung as a whole does not occur in life because the chest wall prevents it. But some regions of the lung may reach minimal volume during normal breathing, especially in older people. One of the major chronic lung diseases **(emphysema)** is characterized by degeneration of the elastic tissue. The lung loses its elastance (becomes more compliant), so that minimal volume occurs within the normal FRC range. This causes residual volume, which cannot be less than minimal volume, to increase markedly (Fig. 34-3).

The deflation pressure-volume curve shown in Fig. 34-2 is for a normal adult human lung that has been inflated with air to total lung capacity and then allowed to deflate very slowly. The purpose of this standard procedure is to ensure reproducible initial conditions. The ordinate is correct for any lung but does not indicate what the absolute volumes are in liters.

The deflation pressure-volume curve is curvilinear. The slope is less (low compliance or high elastance) over its upper third but is steep (high compliance or low elastance) over its lower portion (between 0 and 10 cm H_2O). Thus the lung can be easily inflated when the transpulmonary pressure is low. This is beneficial because normally, even in exercise, we breathe over the lower 70% of the pressure-volume curve. Tidal volume in exercise seldom exceeds half of vital capacity. The work of breathing, $P \times \Delta V$, is less than if one breathed near the top of the curve, as may occur when a person is suffering from asthma, where the narrowed airways cause airflow obstruction and lead to increased FRC (Fig. 34-3).

The normal operating point of the pressure-volume curve is FRC, represented by a thin horizontal line in Fig. 34-2. The elastic recoil pressure of the lung at FRC averages 3.5 cm H_2O at a standard volume of 2,400 ml (40% TLC; 25% VC).

To calculate the compliance of the lung between any two points on the deflation P-V curve, one divides the change in volume, ΔV, by the change in the translung pressure, ΔP_L. Because only the *pressure difference* counts, it does not matter whether pleural pressure is -3.5 cm H_2O and alveolar pressure is zero (natural or "negative pressure" breathing) or whether pleural pressure is $+6.5$ and alveolar pressure is $+10$ cm H_2O (positive pressure ventilation).

To calculate the compliance of the lung from Fig. 34-2 for any volume change, one must know the absolute volume change. Suppose it is 1.8 L. By reading down to the X-axis from the two volume points, the change in translung pressure is 6.5 cm H_2O (10.0 − 3.5); $C_L = \Delta V/\Delta P_L = 1.8$ L/6.5 cm $H_2O = 0.28$ L/cm H_2O. Similarly lung elastance, E_L, $= 1/C_L = 3.6$ cm H_2O/L. Lung compliance or its reciprocal, elastance, is always a positive number.

What happens to lung volume when transmural pressure becomes negative, that is, when pleural pressure exceeds the alveolar pressure? This is undefined in Fig.

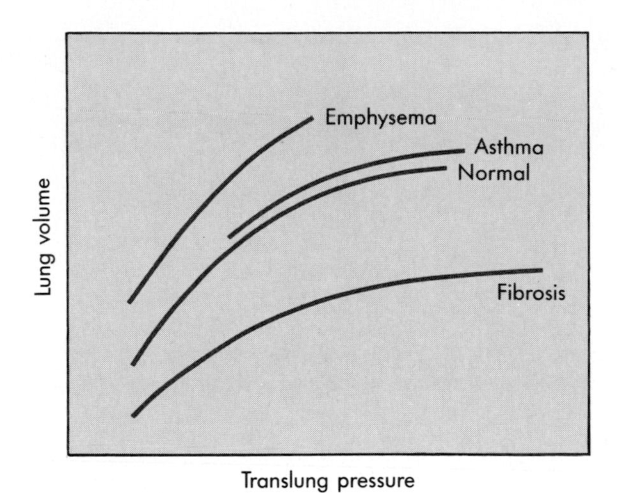

■ **Fig. 34-3** Deflation pressure volume curves of a normal adult and in three common chronic lung diseases.

34-2 because ordinarily one cannot decrease lung volume below minimal volume because the compressive pressure causes the airways to collapse.

When interpreting lung compliance or elastance, due regard should be given to lung volume. Clearly, compliance is greater in large lungs than in small ones simply because they are larger. Think of the difference between a child and an adult. The usual correction for size is to calculate compliance per unit lung volume (C_L/vol).This is called **specific compliance** = C_L/lung volume.

■ *Surface Forces and Lung Recoil*

The adult human lung contains about 300 million tiny anatomical alveoli. This arrangement has the beneficial effect of vastly increasing the surface area for O_2 and CO_2 exchange but has the disadvantage that each anatomical alveolus has a very small radius, r (average r in the adult human is 110 μm at FRC). This geometrical condition plus the presence of an interface between air in the alveoli and the watery alveolar tissue introduces the complication of **surface tension** (see below).

The significance of the air-liquid surface tension is best demonstrated by determining the complete inflation and deflation pressure-volume curve of the lungs, when they are filled with liquid (for example 0.9% NaCl solution) and by comparing the result with the air P-V loop. This comparison is shown in Fig. 34-4. The air P-V loop is the same as in Fig. 34-1.

The pressure volume curve of the liquid-filled lung is displaced to the left (lower distending pressure) at any given volume, especially the inflation limb. *The significance of the shift is that it requires much less pressure to maintain lung volume when the lung is filled with liquid (saline) than when filled with gas.* In some disease conditions (notably respiratory distress syndrome of the newborn) the deflation limb of the air pressure-volume curve is displaced to the right (closer to the inflation limb). That means that higher pressures are necessary to hold the lungs inflated.

The difference between the air and liquid P-V loops is not caused by tissue elastic forces (collagen, elastin, etc), because these are not affected differently by filling

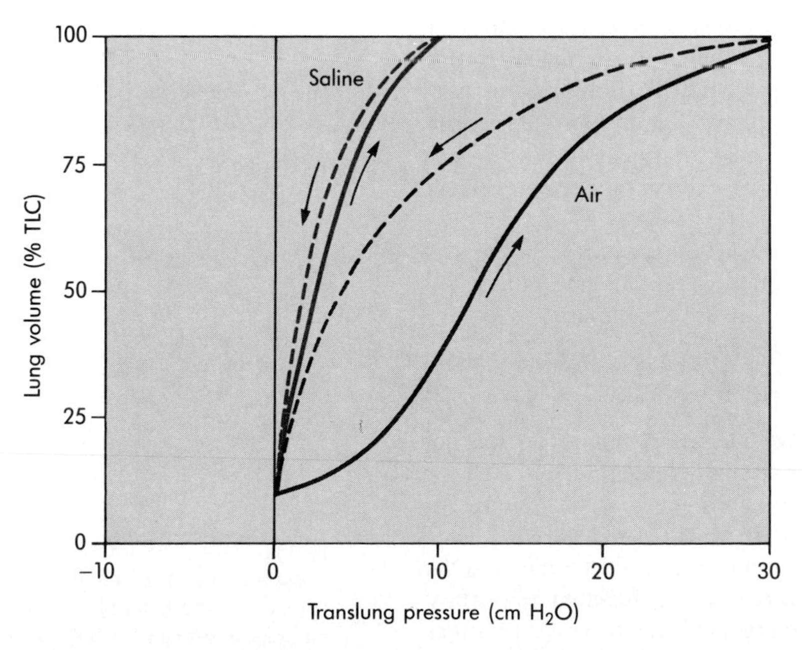

■ **Fig. 34-4** Note the marked difference between the air pressure-volume curve from Fig. 34-1 and the saline-filled lung. It is much easier to inflate the lung without air.

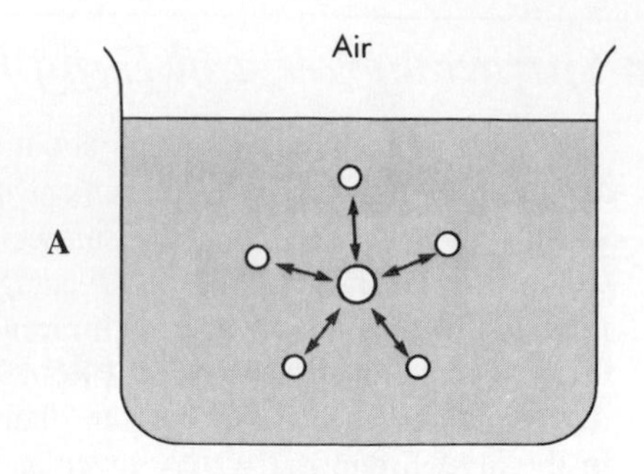

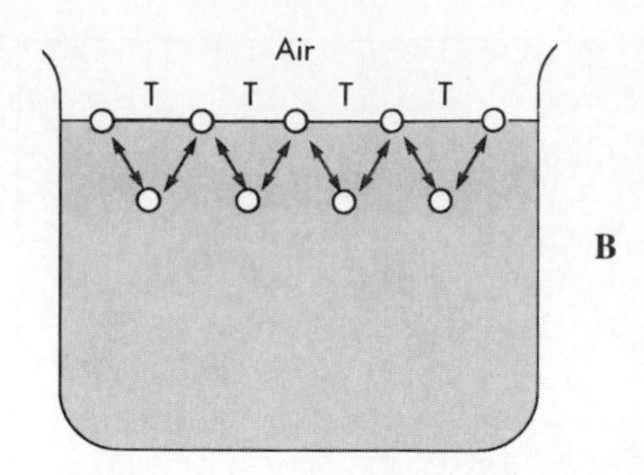

Fig. 34-5 Intermolecular forces that generate surface tension. **A,** Force is relatively uniform on molecules in the interior. **B,** At the surface the molecules are unbalanced, pulled toward the interior, and generate a tension, *T,* in the plane of the surface.

the lung with liquid vs. air. What has changed is that the alveolar air-liquid interface has disappeared. The pressure necessary to maintain a given lung volume is the sum of the pressures necessary to overcome the elastic recoil of tissue elements and of the elastic "skin" (air-liquid interface) at the alveolar surface.

What is Surface Tension?

Within the bulk phase of a liquid, such as water (Fig. 34-5, *A*), intermolecular forces of attraction exist among the water molecules. On the average these forces are equal in all directions. But at the surface (Fig. 34-5, *B*) the water molecules do not have equal attraction forces in all directions. The lateral attraction at the surface among the water molecules exerts a force within the plane of the surface, which acts as an elastic tension.

Surface tension is determined in terms of units of force per unit length. In a hemisphere, which for this purpose is analogous to an anatomical alveolus, the surface tension acts along the two major radii of curvature in the curved surface to hold the sphere together in opposition to the transmural distending pressure, as in Fig. 34-6.

The relationship between the tension, T, in the surface and the transmural distending pressure, Ptm, is described by the law of LaPlace (see Chapter 28). For hemispherical alveoli the law is:

$$Ptm = 2 \times T/r \qquad (3)$$

Ptm is the portion of the translung pressure, P_L, that accounts for the difference between the liquid and air pressure-volume curves at any given volume during inflation or deflation in Fig. 34-4.

The surface tension in the alveoli of the lung has the special property of varying as a function of surface area. In Fig. 34-4 the inflation limb of the air pressure-volume curve requires a higher translung pressure at any volume than does the deflation limb. This can be explained by the difference in surface tension between inflation and deflation: higher as the lung surface expands and lower as the lung surface contracts. The variable surface tension accounts for most of the hysteresis seen in the air pressure-volume curve. Note that hysteresis disappears in the saline solution filled lung.

The amazing feature of the lung's surface is the presence of special phospholipid molecules. These molecules allow the surface tension to vary in a cyclic manner as alveolar surface area changes during breathing. In the ordinary range of breathing the variation in surface tension is small, and it cycles around the equilibrium value of 28 dynes/cm at 37° C. With a large breath however as the surface expands the surface tension rises to 50 dynes/cm. This rising tension opposes expansion of the lung and requires increased transpulmonary pressure to enlarge the lung. However it has the beneficial effect of allowing more molecules of the surface-tension-lowering material to enter the air-liquid interface.

When the lung next begins to deflate, the surface molecules are compressed together. This acts to lower surface tension to less than 10 dynes/cm. Consequently

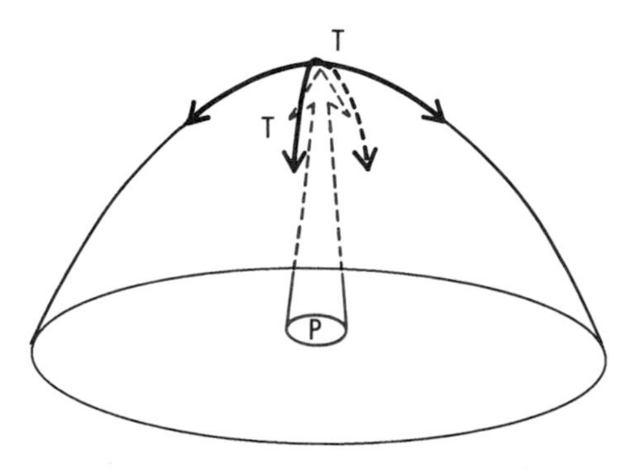

Fig. 34-6 As translung pressure inflates an alveolus, it is opposed by air-liquid surface tension along the two principal axes of the hemisphere. The pressure is also opposed by the tensile strength of the alveolar wall structures. However the tensile strength is weak except near TLC, as shown in the liquid-filled lung in Fig. 34-4.

transpulmonary pressure decreases substantially before lung volume changes very much, as the deflation limb of the air P-V curve in Fig. 34-4 illustrates. The decreasing surface tension during deflation also stabilizes the smallest alveoli, so that they do not shrink faster than the largest alveoli.

In Fig. 34-4 the pressure difference between the air and liquid deflation curves at 50% VC is 2 cm H_2O (1 cm H_2O = 960 dyne/cm^2). If the average radius of the anatomical alveoli in the human lung is 120 μm (120 × 10^{-4} cm) at 50% then the surface tension is P = 2 × T/r; 1,960 = 2T/(120 × 10^{-4}); T = 12 dynes/cm.

Many substances, called surfactants (wetting agents or detergents), lower the surface tension of water. Even the proteins in plasma reduce the air-liquid surface tension from 70 dynes/cm (pure water) to 50 dynes/cm.

The alveolar surfactant molecules are less attracted by the water molecules in the bulk phase because a portion of the molecule is hydrophobic. When the surface area of the alveoli changes the surfactant molecules at the air-liquid interface are compacted (deflation) or spread apart (inflation). It is this effect that causes the alveolar surface tension to vary with lung volume.

The physiological advantages of lung surfactant are the following: (1) it reduces the work of breathing; that is, it reduces the muscular effort needed to expand the lungs; (2) it lowers the elastic recoil at low lung volume (FRC) thus helping to prevent the alveoli from collapsing at the end of each expiration (surfactant deficiency is the basic defect in the ***respiratory distress syndrome*** *that may occur in the lungs of underdeveloped newborns); (3) it stabilizes alveoli that tend to deflate at different rates because the terminal respiratory unit that deflates more rapidly generates a lower surface tension.* This slows its rate of volume decrease and allows a more slowly deflating unit with a higher surface tension to "catch up."

Origin, Composition, Turnover, and Regulation of the Surface-Active Material in the Lung

Lung surfactant consists of a complex phospholipid-protein material, which is produced by the type 2 alveolar epithelial cells (see Fig. 33-7) and is secreted onto the alveolar surface. The main surface-tension-lowering substance in lung surfactant is **dipalmitoyl phosphatidylcholine (DPPC),** which contains two 16-carbon, saturated fatty acid chains. The lipid chains are hydrophobic, whereas the phosphotidylcholine is hydrophilic. The molecules orient themselves at the air-liquid interface with the fatty acid residues arranged vertically, projecting out from the bulk phase. Because this arrangement is mechanically stable the lung surface film resists compression during lung deflation and is able to lower surface tension.

Lung surfactant is continually replaced, as some old molecules leave the surface film and some new ones enter it. The stability of the alveolar air-liquid interfacial film depends on the sustained metabolism of the type 2 alveolar epithelial cells, which produce and also recycle the surfactant.

Beginning in late fetal development (third trimester) the pulmonary surface-active material is produced as the lung matures and becomes ready for air breathing. Immediately after birth babies have to fill their liquid-filled lungs with air. The first cry of the newborn indicates that its lungs have been successfully inflated with air. Although the surfactant does not make the initial air inflation easier, it does keep the alveoli inflated at end expiration. This inflation makes subsequent breathing easier to achieve.

Chest Wall Distensibility

The functional chest wall includes the rib cage, diaphragm, and abdomen. These components work together, under central nervous system control, to cause the cyclic process of breathing. Diseases that affect the chest wall can be very serious and may be the immediate cause of death (for example multiple broken ribs, which cause a collapsible rib cage; inspiratory muscle paralysis in poliomyelitis). Understanding of the function of the chest wall is vital to complete understanding of pulmonary mechanics.

Elastic Recoil of the Chest Wall

Consider for a moment the condition of the chest wall relative to that of the lungs at end-expiration (FRC). In Fig. 34-2 the translung static recoil pressure at FRC is 3.5 cm H_2O. At that point alveolar pressure, PA, is zero (equal to atmospheric pressure). Thus pleural pressure, Ppl, must be −3.5 cm H_2O. The static recoil pressure across the chest wall, Pw, must be: Pw = Ppl − Pbs. Pleural pressure is now the pressure on the inside of the chest wall and Pbs is body surface pressure (normally atmospheric). Thus at FRC chest wall transorgan pressure is Pw = −3.5 cm H_2O. The negative sign indicates that the chest wall must be under compression; that is, it is below its unstressed position.

At FRC the lungs are above their unstressed volume and the chest wall is below its unstressed volume. *It is these equal but opposite elastic recoil forces that determine the position of the respiratory system (lungs and chest wall) at end-expiration.* If air is injected between the parietal and visceral pleuras (creating a pneumothorax) to break the liquid seal, which couples the lungs to the chest wall, the lungs became smaller and the thoracic cavity becomes larger.

The passive pressure-volume curve of the chest wall

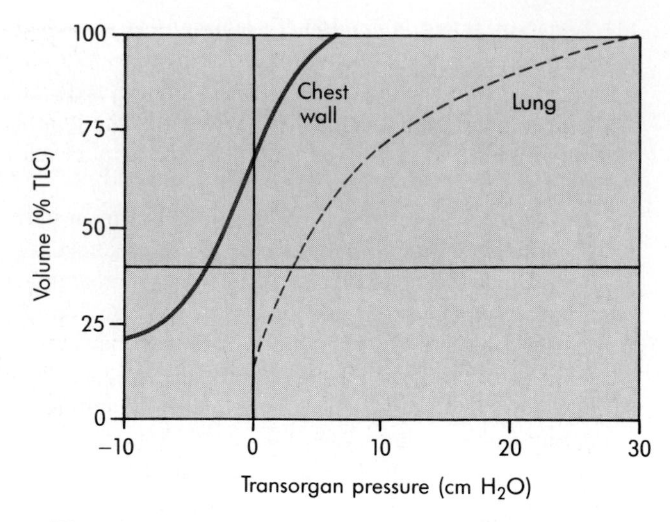

■ **Fig. 34-7** The pressure-volume curve of the chest wall. The pressure difference is pleural pressure minus body surface pressure. At FRC the chest wall and lung are recoiling equally but in opposite directions. At zero transorgan pressure the chest is expanded, while the lung collapses to minimal volume.

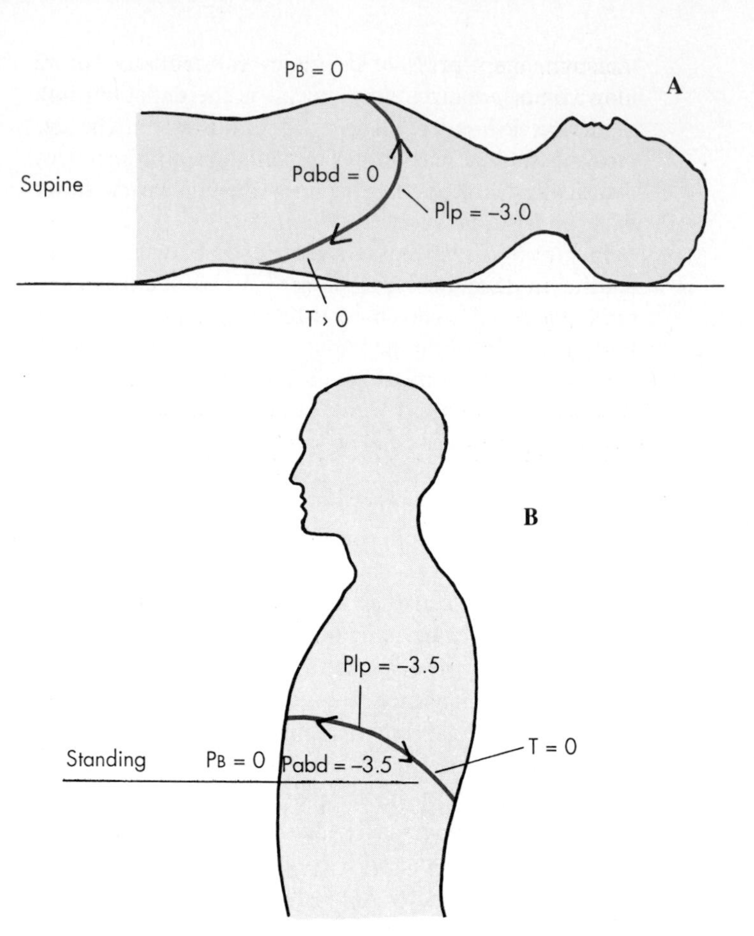

■ **Fig. 34-8** The role of the diaphragm and abdominal wall as part of the functional chest wall. **A,** Supine, the weight of abdominal contents pushes against the diaphragm until appropriate tension develops. **B,** Standing, the liver and other abdominal contents tend to sink, and the abdominal wall protrudes until adequate tension is created.

is added to that of the lung in Fig. 34-7. The pressure-volume curve of the chest wall is always to the left of that of the lung. Below FRC the chest wall becomes progressively stiffer until at residual volume Pw is about −20 cm H_2O. Above FRC the P-V curve of the chest wall is steep (high compliance, low elastance) and nearly linear. At TLC trans–chest wall pressure is only about +7 cm H_2O.

Near TLC the stiff lung limits thoracic expansion. This explains why a person with degenerative lung tissue disease (emphysema) may have a TLC above normal, as shown Fig. 34-3. Near RV it is the stiff chest wall that becomes the factor that limits collapse of the lungs to minimal volume.

Even though the chest wall is normally under compression, the compliance of the chest wall, Cw, is a positive number, because the slope of the P-V curve is positive. Over the range of normal breathing chest wall compliance is similar to lung compliance, as can be seen in Fig. 34-7.

■ *Transdiaphragmatic Pressure*

One of the least intuitive aspects of chest wall mechanics is that the diaphragm and abdomen are important components of the chest wall (reexamine Fig. 33-10). At FRC if trans–chest wall pressure, Pw, is equal everywhere, then the pressure across the diaphragm must be equal to Pw and therefore the intraabdominal pressure must be zero. In supine humans with the diaphragm and abdominal muscles completely relaxed abdominal pressure under the diaphragm is indeed equal to atmospheric pressure (Fig. 34-8, *A*). Thus FRC and pleural pressure are less when the person is in the supine position because the diaphragm is pushed into the thoracic cavity by the higher abdominal pressure until the diaphragm develops sufficient passive tension to oppose the transdiaphragmatic pressure. In the intensive care unit this may be a serious problem in patients with lungs that tend to collapse as a result of increased alveolar surface tension. These patients, generally supine, are mechanically ventilated. Sometimes high pressures are required to inflate the lungs adequately for gas exchange. The high inflation pressure imposes the attendant danger of barotrauma (rupture of the lung caused by the high pressure). If the patient is turned to the prone position, abdominal pressure on the diaphragm is reduced and ventilating the lungs becomes easier.

When a person stands however abdominal pressure decreases as the weight of the abdominal contents pushes down and out against the anterior abdominal wall (Fig. 34-8, *B*). At equilibrium there is no tension in the diaphragm. The abdominal pressure beneath the diaphragm is the same as pleural pressure, and transdiaphragm pressure is zero. Thus the diaphragm moves downward into the abdominal cavity; this explains why FRC increases upon standing. In the third trimester of pregnancy the enlarged uterus pushes upward and outward so that abdominal wall tension is high. In the standing position the diaphragm does not descend and therefore FRC remains reduced.

The key point is that *the pressure difference across all components of the chest wall at equilibrium must be the same*.

■ *Elastic Recoil of the Entire Respiratory System*

Normally the lungs and chest wall move together. The volume changes are identical because the lungs are coupled to the chest wall by molecular cohesion within the pleural space liquid layer. The total pressure across the respiratory system (Prs) is the sum of translung (P_L) and trans–chest wall (P_W) pressures:

$$Prs = P_L + P_W = P_A - Pbs. \quad (4)$$

The pressure across the lung–chest wall system only has meaning in terms of lung and chest wall distensibility, when all the muscles of breathing, including the abdominal muscles, are completely relaxed (passive pressure-volume curve). At FRC Prs = 0, because P_A = Pbs. At every volume the passive pressure across the respiratory system must be the sum of the trans–chest wall pressure and the translung pressure. Fig. 34-9 shows the pressure-volume curve for the respiratory system in addition to those of the lung and chest wall.

The slope of the P-V curve for the respiratory system is less at every point than that for the either the lung or the chest wall. This means that the compliance of the respiratory system (Crs) must be less than that of either the lung or the chest wall.

Calculation of compliance is complicated by the fact that compliances in series must be added as reciprocals: $1/Crs = 1/C_L + 1/C_W$. Thus it is preferable to use elastance instead of compliance. The elastance of the respiratory system (Ers) is readily measured by adding the elastances of the lung (E_L) and chest wall (E_W) together:

$$Ers = E_L + E_W. \quad (5)$$

At FRC $E_L = E_W = 3.0$ cm H_2O/L, Ers = 6.0 cm H_2O/L, Crs = 1/Ers = 0.17 L/cm H_2O, which is half of either lung or chest wall compliance.

In disease the distensibility of the various components of the respiratory system, individually or together, may be profoundly changed. For example the lungs may become very stiff as a result of increased deposition of collagen (fibrosis; Fig. 34-3), or they may become very flaccid as a result of destruction of their elastic fibers (emphysema; Fig. 34-3). The chest wall may be distorted by congenital malformations or by diseases that produce stiffening (kyphoscoliosis, abdominal distension, extreme obesity).

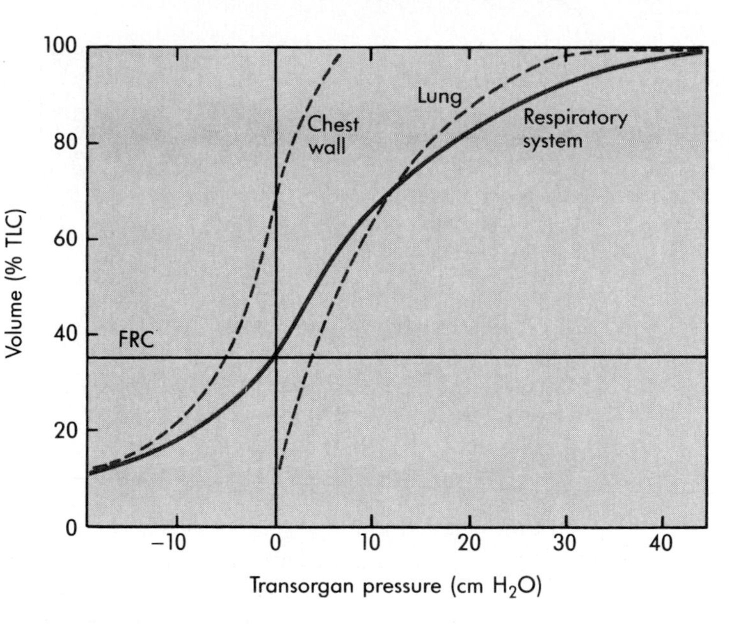

■ **Fig. 34-9** Pressure-volume curve of the entire respiratory system. Only at FRC does the passively inflated lung-thorax system have no pressure difference between the alveoli and the body surface.

■ *Dynamics*

The normal resting breathing rate of 12/min allows only 2 seconds for inspiration and 3 seconds for expiration, scarcely a static condition. Indeed breathing adds a set of dynamic mechanical properties that affect the pressures within the lung-thorax system.

■ *Dynamic Lung Compliance*

Breathing over the normal range of FRC to FRC + V_T produces a dynamic pressure-volume loop as shown in Fig. 34-10, which also includes the static curve in Fig. 34-1. The tidal loop is located approximately in the center of the static P-V loop. The important differences are that (1) FRC is slightly reduced, causing the translung static pressure to be 5 cm H_2O, not 3.5 cm H_2O; (2) a normal 0.5 L tidal volume increases lung volume to 50% of TLC, where the translung pressure is 7.5 cm H_2O; and (3) the inspiratory and expiratory curves show hysteresis, although this is caused mainly by frictional energy losses (resistance) during airflow, not to changes in surface tension, because the surface area change is small.

The mean **dynamic compliance** of the lung (dynC_L) during breathing can be calculated as the slope of the

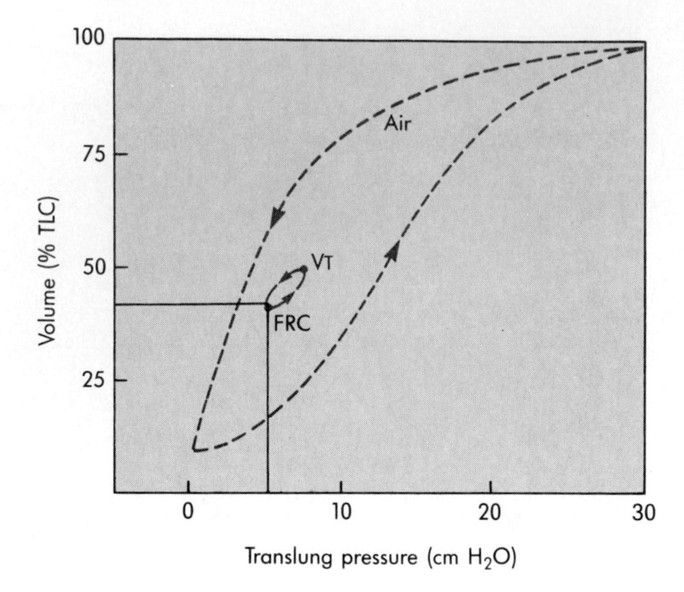

■ **Fig. 34-10** Dynamic pressure-volume curve for tidal breathing. All possible air pressure-volume curves lie within the boundaries of the large pressure-volume curve.

line that joins the end-inspiratory and end-expiratory points of no flow. Thus $dynC_L = \Delta V/\Delta P_L = 0.5/2.5 = 0.2$ L/cm H_2O. The average dynamic compliance is less than the static compliance. Table 33-2 lists a dynamic compliance of 0.2 L/cm H_2O at rest, but of 0.3 L/cm H_2O in exercise. The latter is higher because the tidal volume in exercise is much larger (up to half of vital capacity).

Dynamic compliance is less than static compliance because the minimal change in alveolar surface area associated with tidal breathing is inadequate to draw new surfactant molecules into the air-liquid interface. The recoil of the lung (elastance) increases with time as surface tension rises. The slowly decreasing FRC is detected by lung deflation receptors that drive the central (brain) breathing controller occasionally to increase tidal volume. Sighing is a single large lung inflation that occurs every few minutes. It restores the normal surfactant layer and the FRC. Yawning accomplishes the same result.

■ *Dynamic Chest Wall Compliance*

The dynamic compliance of the chest wall is not different from its static compliance. However the trans–chest wall pressure at FRC during tidal breathing must be −5 cm H_2O to oppose the translung pressure. Because the lungs are stiffer during quiet breathing FRC decreases a little until the trans–chest wall pressure, Pw, equals −5 cm H_2O, which is sufficient to balance the translung pressure. Remember that at FRC the lungs and chest wall are recoiling equally but in opposite directions. The pressure across the respiratory system (Prs) must equal zero.

The active pressure-volume curve of the chest wall. One further complication in dynamic breathing is that the chest wall has an active pressure-volume curve caused by the contraction of the muscles of breathing in the chest wall. The active pressure-volume curve of the chest wall during normal breathing is the mirror image of the passive pressure-volume curve of the lung, as shown in Fig. 34-11. At any point of no air flow, such as end-inspiration, the pressure in the alveoli and the pressure around the chest wall equal zero: that is the respiratory system appears to be infinitely compliant, as shown by the vertical line at Prs = 0 in Fig. 34-11.

■ *Resistance to Airflow*

The hysteresis loop of the tidal breathing pressure-volume curve means that some of the work ($P \times \Delta V$) of breathing done by the chest wall muscles during inspiration is not recovered during expiration. This is chiefly caused by frictional losses (resistance) during airflow.

During breathing the total pressure difference across the lung is the sum of the pressure required to overcome elastic recoil and the pressure caused by airflow resistance.

$$P_L = (P_A - Ppl) + (Pao - P_A) = Pao - Ppl, \quad (6)$$

where Pao represents pressure at the airway opening (nose or mouth). At end-inspiration and end-expiration, when airflow stops momentarily, $Pao = P_A$. Dynamic compliance can thus be calculated, as already described. During airflow $Pao - P_A \neq 0$. The difference may be positive or negative, depending on whether air is flowing into or out of the lung.

■ *Physical Factors that Determine Resistance to Airflow in Rigid Tubes*

Resistance in the airways (Raw) is calculated as driving pressure divided by flow ($\dot{V}$):

$$Raw = \frac{Pao - P_A}{\dot{V}} \quad (7)$$

The factors that affect laminar airflow resistance are described by Poiseuille's equation (see also Chapter 26):

$$Raw = \frac{8}{\pi} \times \eta \times \frac{1}{r^4}, \quad (8)$$

consisting of a constant ($8/\pi$), viscosity (η), and geometry (length/radius^4). Resistance is inversely proportional to the fourth power of the airway radius, r, which is the main factor that affects airway resistance. Small changes greatly affect airway resistance. For example if mean airway diameter decreases by only 10%, resistance increases by nearly 50%.

Airflow resistance is also affected by the fact that even in quiet breathing flow is turbulent (see also Chap-

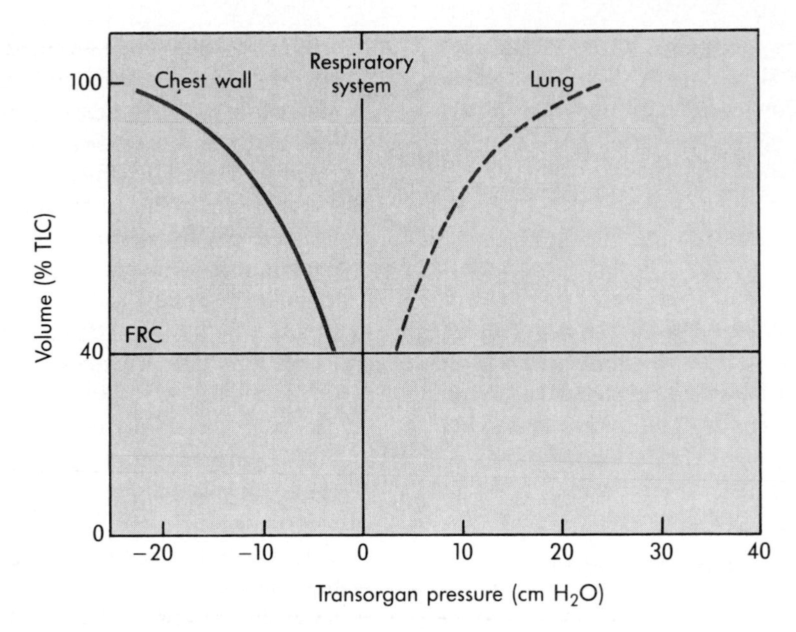

Fig. 34-11 In normal active breathing the chest wall follows a curve that is the mirror image of that of the lung. At any point of no flow the pressure across the lung-chest wall system is zero.

ter 26) in the upper airways (nose, mouth, glottis, and bronchi). Thus an additional term must be added to the driving pressure in the resistance equation (35),

$$P = 2.4 \times V + 0.03 \times \dot{V}^2, \quad (9)$$

where the first term describes the pressure drop caused by laminar flow and the second term (containing velocity squared) describes the pressure drop caused by turbulence (see Fig. 26 -14). Turbulence causes little extra pressure until $\dot{V}^2$ becomes large, as in exercise. Turbulence accounts for the sounds one hears over the chest during breathing. Laminar flow is silent.

Airflow resistance is sometimes divided between the upper airways (airways > 2 mm diameter) and lower airways (airways < 2 mm diameter); Raw = $R_{\text{large airways}} + R_{\text{small airways}}$. Normally 70% to 80% of total airway resistance is in the large airways. Small airway resistance is low because airflow velocity is very low as a result of the large effective cross-sectional area of many small tubes in parallel. Resistances of tubes in parallel are added reciprocally (see Chapter 26).

$$\frac{1}{R} = \frac{1}{R_1} + \frac{1}{R_2} + \ldots + \frac{1}{R_n}, \quad (10)$$

where n represents the number of parallel airways of a given radius. For example the huge number, n = 30,000, of terminal bronchioles makes their equivalent resistance very low, even though r is only 0.03 cm (300 μm).

In disease the longitudinal distribution of airway resistance may vary markedly. In infants a life-threatening viral disease known as **bronchiolitis** (inflamed bronchioles) causes edema (swelling) of the lining mucosa of the bronchioles, The edema may drastically reduce the radii of the bronchioles. A reduction by half increases their contribution to resistance 16-fold because of the fourth power factor (equation 8).

Measurement of Airway Resistance

To measure total airways resistance one needs to measure alveolar pressure, P_A, and air flow velocity, $\dot{V}$. An easy way to measure airflow velocity is with a device known as a pneumotachograph, which detects the very small pressure drop caused by flow across a low fixed resistance. The more difficult problem is to measure the alveolar pressure, which is the downstream pressure during inspiration and the upstream pressure during expiration. Because alveolar pressure is not the same as ambient pressure during airflow, the measurement is made by briefly occluding the airway opening during breathing. At the instant after airflow stops, Pao = P_A.

During normal tidal breathing airflow reaches a peak velocity during inspiration of about +0.5 L/sec and during expiration of −0.5 L/sec (the minus sign indicates flow out of the lung). Alveolar pressure decreases below the pressure at the airway opening by 0.8 cm H_2O at peak airflow during inspiration, and it exceeds the pressure at the airway opening by 1.2 cm H_2O during expiration. The difference occurs because the airways are narrower during expiration, and therefore resistance is increased. Thus at peak velocity during inspiration Raw = $\Delta P/\dot{V}$ = (Pao − P_A)/$\dot{V}$ = 1.6 cm H_2O × sec/L. Likewise during expiration Raw = 2.4 cm H_2O × sec/L.

The normal range for airway resistance in adult humans is 1.5 to 2.0 cm H_2O × sec/L (see Table 33-2). As the signs of the airflow and the alveolar pressure difference are always the same, resistance is always indicated by a positive number.

Total Translung Pressure During Breathing

The total pressure difference from ambient to pleural space during airflow is the pressure difference caused by lung distensibility (dynamic compliance) plus the pressure difference caused by airflow resistance. The sum of these pressures gives P_L in equation 4.

Physical Factors that Influence Airway Resistance

In rigid cylindrical tubes resistance during laminar flow is fixed because the physical dimensions of the tube never change, regardless of the transmural pressure across the tube walls. But if the tube is distensible and collapsible the transmural pressure may have important nonlinear effects.

Fig. 34-12 is a model of flow in a nonrigid tube. In panel A the pressure in the tube always exceeds the external pressure. Therefore the transmural pressure is positive. The tube expands and flow resistance decreases. In panel B the external pressure is raised above the outflow or downstream pressure. Thus as flow resistance dissipates the driving pressure, the pressure inside the tube falls. When it falls below the external pressure, the tube begins to narrow. As transmural pressure becomes more negative, the tube is compressed more and more until a new stable condition in which the collapsed zone (always at the outflow end of the tube) limits the flow rate is reached.

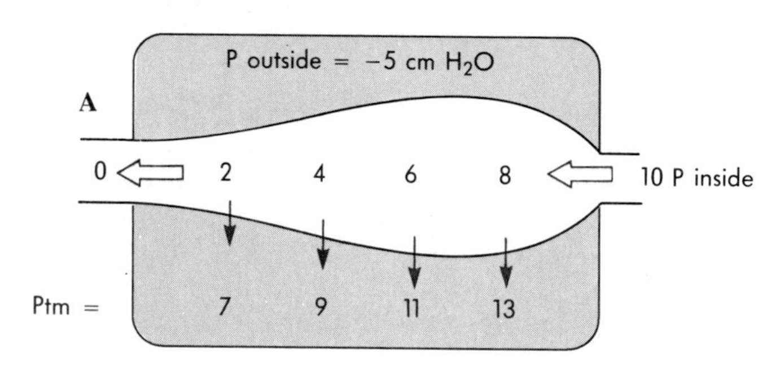

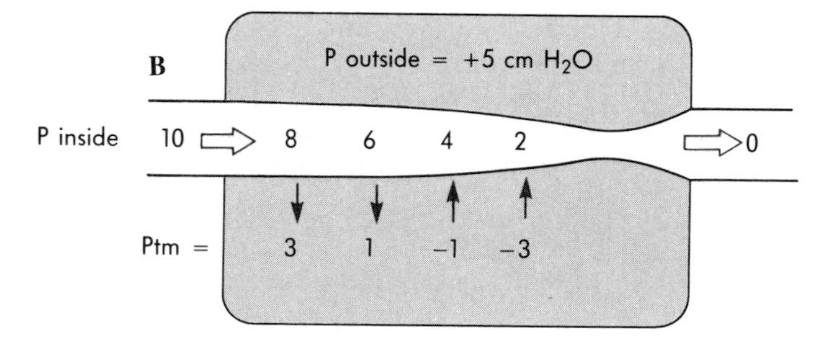

Fig. 34-12 Flow in a distensible and collapsible tube is affected by the lateral transmural pressure as well as the longitudinal pressure. **A,** Analogy to lung inflation, where the pressure outside the airways is subatmospheric, which distends the tubes. **B,** Analogy to lung deflation where the pressure outside the airways may become positive and compress them near the outlet.

The regulation of airflow in this manner is called **dynamic airway compression.** Under such conditions airflow is independent of the total driving pressure. During expiration pleural pressure rises and increases alveolar pressure above the pressure at the airway opening. Because the cartilagenous airways narrow during expiration, airway resistance is higher during expiration. The bronchioles however are much less sensitive to the phases of breathing because they depend on lung volume, which in normal tidal breathing does not change much (<20%).

When a healthy person takes a very large inspiration from FRC to TLC and then breathes out as hard and fast as possible (forced vital capacity) pleural pressure rises to a high level and generates the large driving pressure necessary. This driving pressure raises pleural pressure above the pressure at the airway opening, as shown in Fig. 34-12. The large airways are compressed near the thoracic outlet (main stem bronchi or trachea). As the airways are compressed, flow velocity is increased, reducing the lateral pressure, according to the Bernoulli principle (see Chapter 26), and leading to further compression. This limits expiratory airflow. The dynamic compression quickly reaches a stable condition, such that no matter how hard one tries, one cannot increase the rate of expiratory flow. The reason is that any further rise in pleural pressure causes more airway compression, which prevents any increase in flow. This generates a maximum expiratory flow rate that is *independent of effort* (see the spirogram tracing of a forced vital capacity maneuver [Fig. 33-13]).

Coughing is a naturally occurring example of dynamic airway compression. In the degenerative lung disease called emphysema cartilagenous airway support is lost. As the transpulmonary pressure at any lung volume is reduced, pleural pressure at the end of a normal inspiration is closer to atmospheric pressure than it is in a normal individual. In addition the airways are further weakened by degeneration of the cartilage and elastic fibers. During expiration in order to generate sufficient airflow in a reasonable period to provide adequate minute alveolar ventilation the individual tries to increase expiratory airflow by increasing pleural pressure. The elevated pleural pressure further compresses the large airways, thus increasing expiratory resistance further. Emphysema is often called **chronic obstructive**

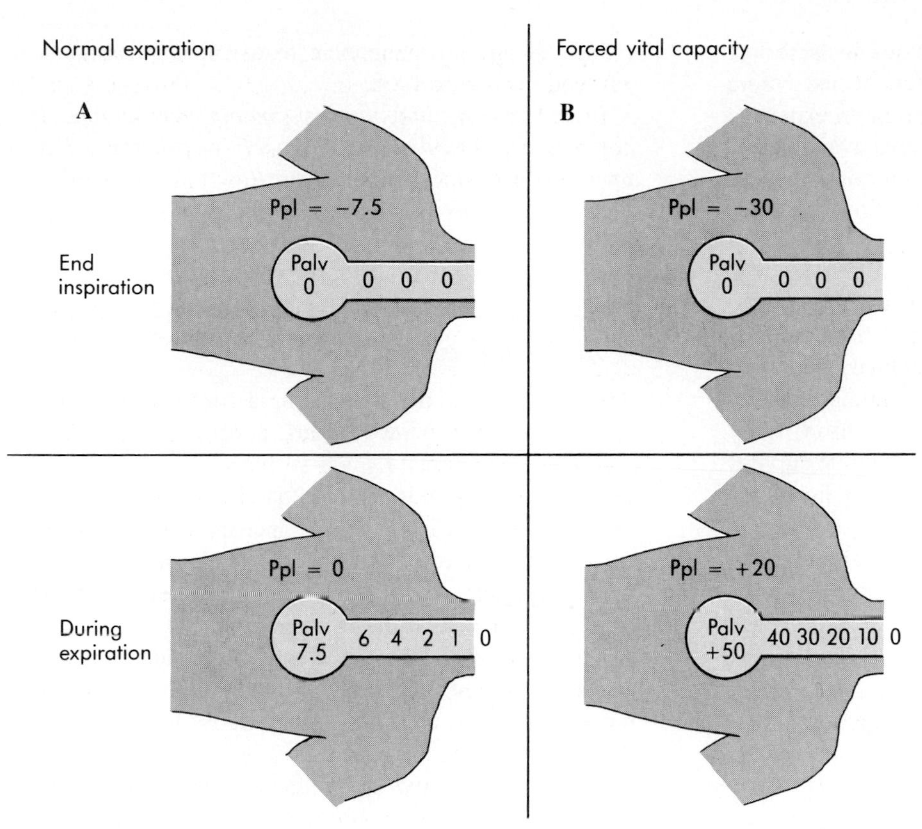

■ **Fig. 34-13** **A,** No dynamic compression during normal expiration. **B,** Generation of dynamic airway compression during a forced vital capacity maneuver.

pulmonary disease. Emphysematous patients learn to inspire quickly and breathe out slowly to prevent airway collapse. As chronic obstructive pulmonary disease becomes more severe, the afflicted patient can generate less and less maximal expiratory flow.

On the other hand a person who has hyperirritable airway disease **(asthma)** has a different problem. His or her airways are narrowed by smooth muscle contraction, mucosal edema, or excessive secretions. Thus both inspiratory and expiratory flow resistance are increased. However the patient with asthma usually has normal lung elastance. He or she breathes at an increased FRC (see Fig. 34-3), which acts to distend the airways during inspiration and limits compression during expiration.

■ *Flow-Volume Relationship*

If one plots flow velocity against the lung volume at each point along the breathing loop, one generates a flow-volume curve. Fig. 34-14 shows two stylized flow-volume curves. The small loop is the flow-volume curve for a normal tidal breath. The large loop is a maximal flow-volume curve. Inspiratory flow is only limited by effort, but the expiratory flow curve is the maximal flow that can be achieved no matter how much effort is expended. The maximal flow-volume curve allows one to assess dynamic airway compression.

One reason that inspiration cannot produce high flow

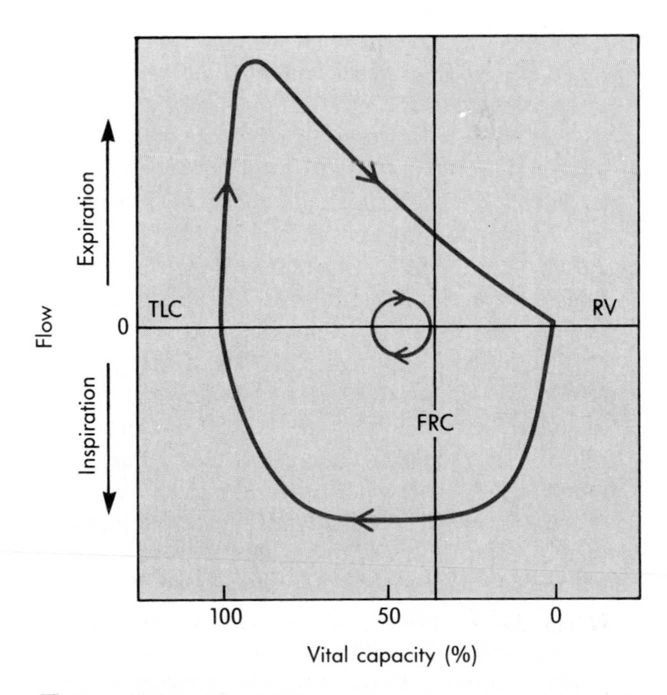

■ **Fig. 34-14** Normal flow-volume loops for tidal breathing (small) and forced vital capacity. Expiration in the forced vital capacity maneuver shows dynamic compression, which becomes more limiting as lung volume decreases.

velocity is that inspiratory effort decreases as the thorax enlarges: the strongest inspiratory effort occurs before the diaphragm and chest wall muscles have expanded the thoracic cavity. Even though the force available decreases, inspiratory flow tends to remain constant until near total lung capacity, at which point flow falls rapidly to zero.

During expiration the curve is entirely different. Contracting the powerful anterior abdominal muscles and the intercostal muscles generates a high flow velocity when thoracic volume is near TLC. Velocity decreases rapidly (approximately linearly) as volume decreases, but not as a result of decreasing muscular effort. Once the linear portion of the expiratory flow-volume curve is reached, *it is effort-independent*. No matter how hard the subject tries he or she cannot exceed the maximal flow at the given volume. Dynamic compression of the airways limits expiratory flow and this limitation increases as lung volume decreases.

Neurohumoral Regulation of Airway Resistance

The vagus nerve is the motor nerve to the smooth muscle of all the cartilagenous airways and probably to the membranous bronchioles and alveolar ducts. Stimulation of the efferent vagal fibers, either reflexly or electrically, leads to airway constriction, a decrease in anatomical dead space volume, VD, and an increase in airway resistance, Raw.

Stimulation of the sympathetic nerves has the opposite effect. In humans sympathetic nerves to the lungs innervate vascular smooth muscle, submucosal glands, and parasympathetic ganglia, but probably not airway smooth muscle. However the airway smooth muscle has adrenergic receptors, so that local diffusion of the sympathetic postganglionic neurotransmitter, norepinephrine, inhibits airway constriction.

In another class of inhibitory innervation the main transmitter is similar or identical to vasoactive intestinal peptide. Its physiological role is uncertain.

A number of agents act reflexly via the vagus nerve to constrict the airways or to induce cough. Inhalation of smoke, dust, cold air, and irritant substances has this effect. For example exercise-induced asthma is common among people who have hyperirritable airways.

Some substances affect airway smooth muscle directly. Agents that tend to constrict airways in this manner include decreased PA_{CO_2}, histamine, acetylcholine, adrenergic β-receptor antagonists, thromboxane A_2, prostaglandin F2α, and leukotrienes C_4 and D_4. Histaminelike compounds are sometimes used as provocative agents in testing patients who may have hypersensitive airways (asthma).

Agents that tend to dilate airways include increased PA_{CO_2}, adrenergic α- and β-receptor agonists, atropine (which blocks postganglionic parasympathetic impulses), and prostaglandin E_2.

In the normal lung airway smooth muscle tone is continuously modulated, presumably to provide a balance between airway resistance (Raw) and anatomical dead space (VD) to maintain minimal work for each breath.

The Work of Breathing

The work (W) involved in taking a single breath is defined as $W = P \times \Delta V$. Work is measured in joules (pressure × volume; 1 joule = 10 L × cm H_2O), and the rate of doing work (power) is measured in watts (1 W = 1 J/sec). Clearly if one increases tidal volume, one does more work to inflate the lung, and if one increases minute ventilation (increased volume flow/time), one increases power utilization.

The work of breathing is commonly divided into two portions: namely the work of moving the lungs and the work of moving the chest wall. Many diseases affect the work of breathing.

The work of moving the chest wall cannot be assessed during active breathing for the same reason that the compliance of the chest wall cannot be measured during active breathing; there is no simple way to assess the mechanical work done by the muscles of breathing on themselves. However if a subject is paralyzed either through disease or through a neuromuscular blocking drug, his or her ventilation can be maintained by a mechanical device. Under these conditions the lungs and thorax are moved passively by the ventilator, so that the total work of breathing can be determined. From such measurements the total work of a single breath at normal tidal volume is about 0.25 J, of which half is expended in moving the lung. The power requirement of normal breathing is approximately 50 mW (0.25 Je/breath × 1 breath/5 sec = 0.05 W).

For the lungs alone one can equate the work done on the lung (only) during a single breath with the shaded area bounded by FRC and FRC + VT and the pressure axis, as Fig. 34-15 shows. During inspiration (Fig. 34-15, *A*) work is done to move the lung, whereas during expiration (Fig. 34-15, *B*) work is done by the potential energy stored in the elastic recoil of the lung to restore FRC. The difference between the work (shaded areas) is due to the energy used to overcome frictional (resistive) losses. Normally more potential energy is available to do work in expiration (because of the change in surface tension) than is necessary to overcome airway resistance. Because in the steady state no energy can be left over, the extra energy is used to pull the chest wall back to its resting configuration. Remember that during tidal breathing the chest wall is compressed below its unstressed position and therefore loses potential energy during inspiration. In addition some of the excess en-

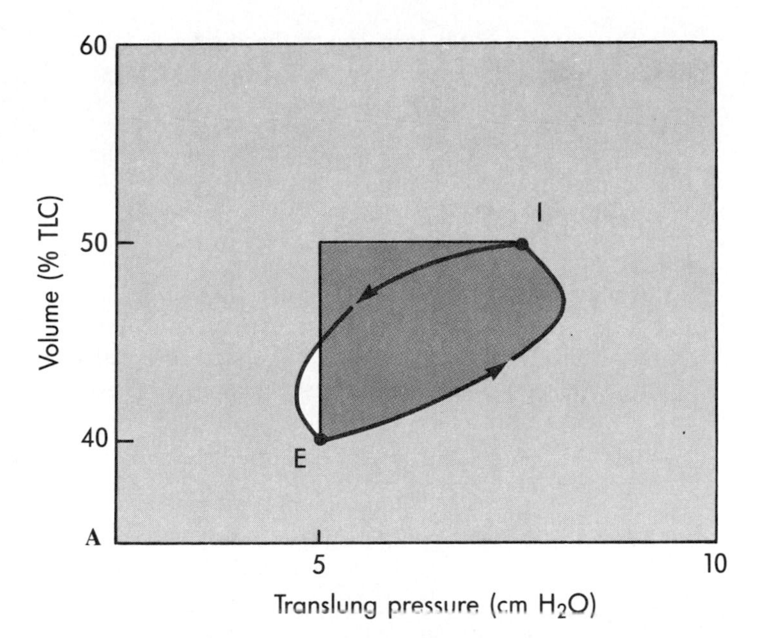

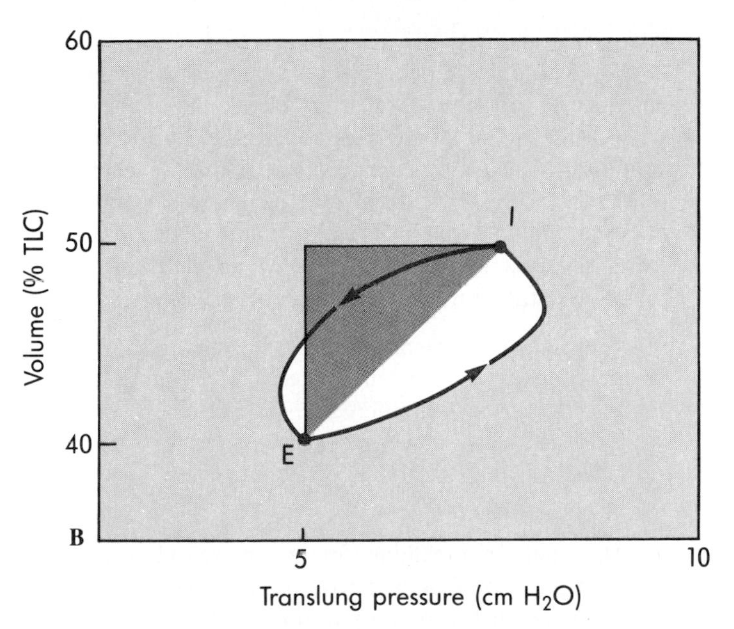

■ **Fig. 34-15** Work done on the lung during a normal inspiration **A,** includes frictional losses not recovered during expiration, **B.**

ergy is used to impart velocity to the air leaving the airway opening.

During exercise the work of each breath increases proportionally more than does tidal volume because airflow velocity is greater during exercise and this increases turbulence. A 10-fold increase in airflow rate increases the turbulence factor by 100 ($\dot{V}^2$ in equation 9).

During the pulmonary function test known as maximum voluntary ventilation an average person may move 160 L/min. Reasonable values are 4 L/breath × 40 breaths/min. The resistive driving pressure across the respiratory system may reach 20 cm H_2O. The power requirement is 5 W (100 times normal), even though ventilation is increased only thirtyfold.

■ *Oxygen Cost and Efficiency of Breathing*

During resting tidal breathing the oxygen cost of quiet breathing ($\dot{V}O_2$ of the respiratory muscles) cannot be reliably measured. The increase in oxygen consumption associated with forced ventilation however can be measured and it amounts to an O_2 cost of breathing of 1% to 2% (3 to 5 ml O_2/min) of the resting oxygen consumption of the whole body (250 ml/min). The mechanical efficiency of breathing is defined as the work required to inflate the lungs divided by the energy consumed by the muscles of breathing:

$$\text{Efficiency} = \frac{P \times \Delta V}{O_2 \text{ cost of breathing}}. \quad (11)$$

The efficiency of the breathing apparatus is low (less than 10%) compared with that of skeletal muscle in general; muscle efficiency is about 15% to 20%.

In patients who have diseases that increase lung elastance (reduce compliance) or increase airway resistance the work of breathing may be markedly increased. Ordinarily a large safety factor prevails for the muscles of breathing, such that they are adequately supplied with oxygen and metabolic substrates to do their work indefinitely (aerobic metabolism). But muscle fatigue can occur when the elastic or resistive load is too great for the available blood supply. Under such conditions **respiratory fatigue** occurs. This means the muscles of breathing are unable to ventilate the lungs adequately to supply the body's oxygen requirements. Thus $P_{A_{CO_2}}$ rises. Respiratory muscle fatigue can be corrected by reducing the workload (for example by dilating the airways in asthma) or by mechanically assisting ventilation (for example in emphysema complicated by bronchitis).

Although the normal oxygen cost of breathing is less than 2% of the resting oxygen consumption of the whole body, during voluntary hyperventilation one may use up to 30% of total oxygen consumption to move the lungs and chest wall. Obviously in patients who are already using more O_2 than normal to move air at rest the oxygen cost of breathing may limit exercise capability.

Sufficient potential energy is stored in the lungs and thorax during inspiration to restore the breathing apparatus to its FRC position. If the work of inspiration is increased by increased resistance to airflow, as in asthma, the potential energy stored in the lungs and chest wall is not increased, because the extra work is dissipated as frictional (heat) loss during inspiration. Because expiratory resistance is also increased elastic potential energy may not be sufficient to move air fast enough for the lungs to reach their original FRC position within the expiratory time available. If expiratory airflow continues right up to the beginning of the next inspiration alveolar pressure is greater than the pressure at the airway opening at the end of expiration. This is referred to as "intrinsic PEEP," meaning a built-in form of **positive end-expiratory pressure** that is not caused by an external device. Consequently FRC increases until sufficient elastic energy is stored by the lungs to achieve a new steady state. This explains why patients have hyperinflated lungs during periodic attacks of asthma (see Fig. 34-3). Intrinsic PEEP is the state in which the usual potential energy stored in the breathing apparatus is insufficient to return the lungs to their normal FRC within the time available. Breathing is a dynamic process, and therefore time is an important variable.

One of the difficulties associated with increased work of breathing in pulmonary disease is that as breathing becomes more labored (requires more muscular effort), the low efficiency of the breathing apparatus increases the oxygen cost of breathing exorbitantly. If all of the increase in oxygen consumption is required just to meet the demands of the breathing apparatus then additional oxygen is not available for other metabolic requirements. This lack can be extremely limiting to patients with severe lung disease, such as those who have advanced emphysema or infants who have marked airway narrowing (bronchiolitis or severe asthma).

Whether a control mechanism senses the work of breathing is not clear. In disease the tendency is to breathe rapidly and shallowly when lung compliance is increased and to breathe slowly and deeply when airflow resistance is increased. These two factors, resistance and compliance (R and C), affect breathing in opposite ways. Their product, R × C (the time constant), is another important variable, which is discussed further in Chapter 36.

■ *Summary*

1. The normal lungs are compliant (easy to distend) at low volume but become much less compliant (stiffer) near total lung capacity. Lung compliance is determined from the slope of the pressure-volume curve during deflation.

2. Lung distensibility during inflation is low compared to its distensibility during deflation. The main reason is the variable air-liquid surface tension in the alveoli, which have a very small radius. The cause of the variable surface tension is a special surface-active material called surfactant (secreted by type 2 alveolar epithelial cells), whose main component is dipalmitoyl phosphotidylcholine.

3. At end-expiration the lungs recoil to a smaller volume, but the chest wall recoils in the opposite direction toward a larger volume. Thus the chest wall is normally under compression in the range of tidal breathing.

4. The pressure difference across the diaphragm-abdominal wall (an important functional component of the chest wall) is the same as that across the rib cage.

5. Because the lungs and chest wall move together the total compliance of the respiratory system is less than that of either part alone.

6. The dynamic compliance of the lung is less than its static compliance mainly as the result of a rise in air-liquid surface tension toward its equilibrium value during tidal breathing. An occasional large sighing breath restores the surfactant and increases compliance. Chest wall dynamic compliance is the same as its static compliance.

7. Because of resistance to flow in the airways alveolar pressure equals ambient pressure only at points of no airflow. During breathing the pressure difference across the lung is the sum of the elastic and resistive components. Breath sounds are caused by turbulence in the upper airways.

8. An important physical factor affecting airway resistance is the transmural pressure across the bronchi, whose walls are distensible and collapsible. In a forced expiratory maneuver pleural pressure may rise above large airway pressure, dynamically compress the airways, and limit peak expiratory flow velocity. The flow-volume relationship clearly shows expiratory flow limitation during a forced vital capacity maneuver.

9. The vagus nerve (motor nerve to the airway smooth muscle) reflexly regulates airway resistance. Many substances (histamine, thromboxane, leukotrienes, substance P) can cause constriction and increased airflow resistance, whereas high CO_2, catecholamines, and atropine cause dilation.

10. The work of breathing for one breath at rest is small. In pulmonary disease work may be increased to such an extent that respiratory respiratory failure occurs (increased Pa_{CO_2}).

■ *Bibliography*

Journal articles

Clements JA: Surface phenomena in relation to pulmonary function, *Physiologist* 5:11, 1962.

Dayman H: Mechanics of air flow in health and emphysema, *J Clin Invest* 30:1175, 1951.

Derenne JPH, Macklem PT, CS Roussos: The respiratory muscles: mechanics, control and pathophysiology, *Am Rev Respir Dis* 48:119, 373, 581, 1978.

Fry DL, Hyatt RE: Pulmonary mechanics: a unified analysis of the relationship between pressure, volume and gas flow in the lungs of normal and diseased human subjects, *Am J Med* 29:672, 1960.

Macklem PT, Mead J: Resistance of central and peripheral airways measured by a retrograde catheter, *J Appl Physiol* 22:395, 1962.

Mead J: Mechanical properties of lungs, *Physiol Rev* 41:281, 1961.

Otis AB: The work of breathing, *Physiol Rev* 34:449, 1954.

Pattle RE: Properties, function and origin of the alveolar lining layer, *Nature* 175:1125, 1955.

Rahn H et al: The pressure volume diagram of the thorax and lung, *J Appl Physiol* 146:161, 1946.

Books and monographs

Bates DV, Macklem PT, Christie RV: *Respiratory function in disease,* ed 2, Philadelphia, 1971, WB Saunders.

Forgars P: *Lung sounds,* London, 1978, Bailliere Tindall.

Forster RE et al: *The lung,* ed 3, Chicago, 1984, Year Book.

Murray JF: *The normal lung,* ed 2, Philadelphia, 1986, WB Saunders.

Nunn JF: *Applied respiratory physiology,* ed 3, London, 1987, Butterworths.

CHAPTER

35

Pulmonary and Bronchial Circulations: Ventilation/Perfusion Ratios

Pulmonary blood flow is the denominator of the ventilation-perfusion ratio, $\dot{V}_A/\dot{Q}$, and is of importance equal to breathing in determining the overall efficiency of gas exchange. The right ventricle pumps the mixed venous blood through the pulmonary arterial distribution system, then through the alveolar wall capillaries (where O_2 is added and CO_2 is removed), and on to the left atrium. A major problem of pulmonary hemodynamics is to understand how all of the cardiac output flows through the pulmonary circulation at a much lower pressure than through the systemic circulation. The difference is caused in part by the enormous number of resistance vessels (small muscular pulmonary arteries) and in part by the normally dilated state of these resistance vessels. That does not mean that the pulmonary vascular bed has little active vasomotor control. The pulmonary circulation has potent local mechanisms to balance perfusion with ventilation and thereby maintain the arterial P_{O_2} and P_{CO_2} at or near normal levels.

The small muscular arteries and arterioles of the pulmonary circulation have much less smooth muscle than do those of the systemic circulation; this lack of smooth muscle is part of the reason for the lower resistance in the pulmonary circuit. The importance of having low resistance in the pulmonary circulation is that 100% of the cardiac output can flow through the lung without expenditure of a large amount of metabolic energy.

The alveolar wall capillary network is the main exchange system of the pulmonary circulation. The pulmonary capillary bed is a remarkable structure. In resting adults it contains about 75 ml of blood, spread out in a vast array of thin-walled interconnecting vessels. During exercise the capillary blood volume increases, approaching the maximal anatomical capillary volume, which is about 200 ml. The average thickness of the capillary walls, which is only 0.1 μm, helps to optimize oxygen diffusion between the alveolar gas and the hemoglobin in the red blood cells. The total capillary surface area for gas exchange is about 70 m^2 (40 times body surface area).

Lung capillary blood volume is approximately equal to the stroke volume of the right ventricle under most conditions. At rest the erythrocytes entering the pulmonary capillaries remain there for 0.75 seconds, more than adequate time for O_2 and CO_2 equilibration between alveolar gas and blood.

The total blood volume of the pulmonary circulation (main pulmonary artery to left atrium) is 500 ml, which is 10% of the total circulating blood volume. In life the lung is 40% blood by weight. The vascular volume, except the blood in the capillaries, is distributed approximately equally between the arterial and venous vessels. The large pulmonary vascular volume serves as a capacitance reservoir (buffer) for the left atrium. If venous return to the right ventricle suddenly changes the left ventricle diastolic filling does not change appreciably for two to three cardiac cycles. During normal breathing the right and left ventricles are out of phase. As pleural pressure falls during inspiration the right ventricle receives more blood because the pressure gradient for flow into the heart (venous return) is increased, whereas the left ventricle ejects less blood because the left ventricle is in a lower pressure chamber (thorax) relative to the systemic arteries, which are surrounded by atmospheric pressure. During expiration the opposite occurs.

■ *Pressure and Resistance*

Fig. 35-1 compares the normal pressures (in millimeters of mercury [mm Hg]) in the human pulmonary circulation with those in the systemic circulation. The data are for an adult at rest lying supine. However pulmonary physiologists usually refer to pulmonary vascular pressure in terms of centimeters of water [cm H_2O]). In Ta-

Table 35-1 Pressures in the pulmonary circulation of normal, resting supine adult humans

	mm Hg	*cm H_2O*
Pulmonary artery*		
Systolic/diastolic	24/9	33/11
Mean	14	19
Arterioles		
Mean	12	16
Capillaries		
Mean	10.5	14
Venules		
Mean	9	12
Left atrium		
Mean	8	11

*The pulmonary arterial and left atrial pressures are measured at cardiac catheterization; the former directly, the latter by wedging the arterial catheter into a branch of the pulmonary artery. The capillary pressure is computed by a standard equation.

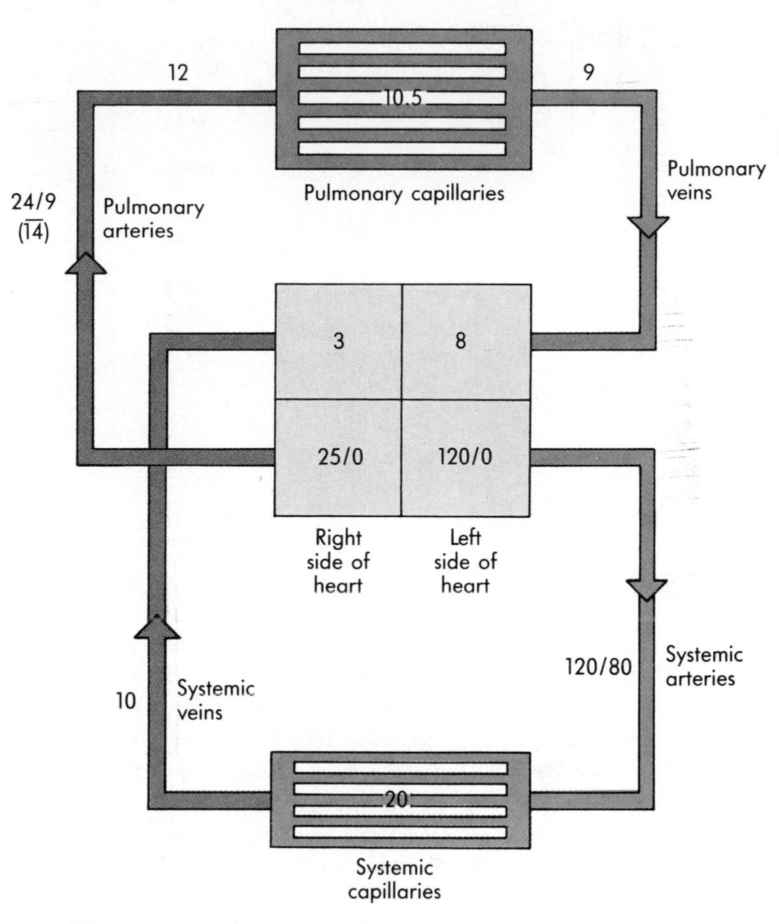

Fig. 35-1 Pressures (mm Hg) in the pulmonary and systemic circulations in a normal, resting supine human adult.

ble 35-1 are listed the normal pulmonary vascular pressures in both units. The pressure in the pulmonary artery is about one fifth that in the aorta. Left atrial pressure is higher than right atrial pressure.

Pulmonary vascular resistance (PVR) at the normal resting cardiac output ($\dot{Q}$) of 5 L/min and at mean pulmonary arterial pressure (Ppa) = 19 cm H_2O and mean left atrial pressure (Pla) = 11 cm H_2O is.

$$PVR = \frac{Ppa - Pla}{\dot{Q}} = \frac{(19 - 11)\ \text{cm}\ H_2O}{5\ \text{L/min}} = \frac{1.6\ \text{cm}\ H_2O \times \text{min}}{\text{L}} \quad (1)$$

Pulmonary vascular resistance is less than 10% of that in the systemic vascular bed.

Pressure-Flow Curves

A useful view of pulmonary hemodynamics is obtained by measuring the changes in driving pressure (Ppa − Pla) as cardiac output varies. The resulting graphs are called pressure-flow curves. Two representative curves in different physiological conditions are shown schematically in Fig. 35-2. Vascular resistance is represented on a pressure-flow curve by the slope of the line from the origin to a given point on the curve. The shapes of the pressure-flow curves are due to the greater distensibility (higher compliance) of the pulmonary vessels at low than at high distending pressures. On curve 1 (normal) the resistance during exercise, E, is less than at rest, R. Alveolar hypoxia (line 2) causes vasoconstriction; therefore a greater driving pressure (and consequently resistance) exists at any given flow. When one exercises while hypoxic the resistance still may decrease, as shown by point HE.

Pulmonary Blood Flow

Although the Fick equation is still the primary reference for measuring cardiac output, there are several quicker methods in common use, all based on what is known as the indicator dilution principle (see Chapter 24). In the 1980s the thermodilution (caloric) method began to supersede all others. It is widely used in cardiac catheterization laboratories, intensive care units, and recovery rooms. The procedure and the calculations are automated and easy to use. The only limitation to its wide application is the need to have a pulmonary arterial (or alternately a systemic arterial) catheter equipped with a sensitive heat detector (thermistor) to measure the small temperature changes.

Shunts Between the Right and Left Sides of the Heart

Right-to-left shunt. An important functional test of the adequacy of the pulmonary circulation is the determination of the fraction of the cardiac output that effec-

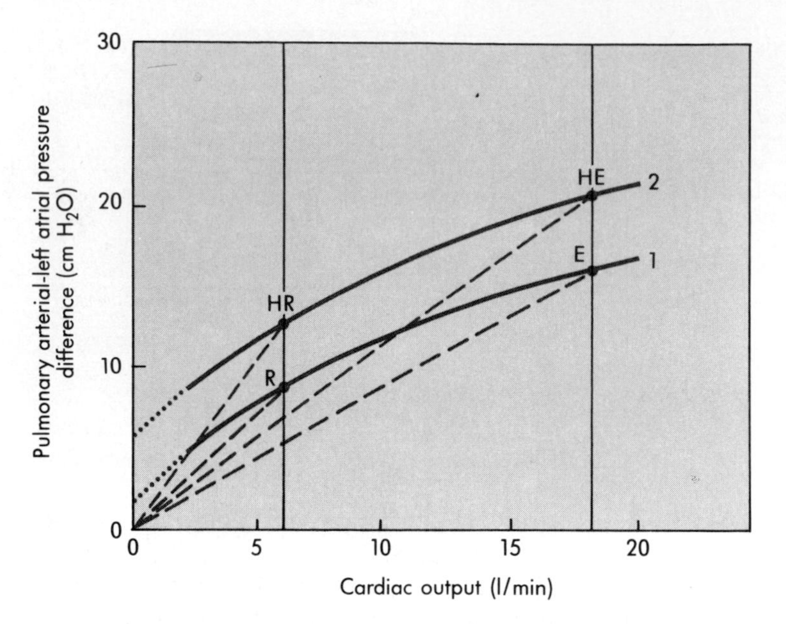

■ **Fig. 35-2** Representative pressure-flow curves that might be obtained in the human pulmonary circulation. Line 1 is normal. R is the normal resting condition, and E is maximum steady-state exercise. The slopes of the dashed lines from the origin to the points on the curve represent pulmonary vascular resistance ($\Delta P/\dot{Q}$). Curve 2 shows hypoxic pulmonary vasoconstriction. It lies above line 1, meaning there is increased resistance at any given flow, (e.g., point HR). Notice the positive Y-axis intercept; (i.e., flow stops even though the driving pressure in not zero).

tively flows through pulmonary capillaries (exchanges with alveolar gas) and the fraction that bypasses the lungs to enter the systemic arteries without becoming oxygenated. The latter fraction, called **venous admixture,** includes true anatomical shunts. A small amount of venous admixture is normal because some venous blood enters the left atrium and ventricle by way of the bronchopulmonary venous anatomoses and the intracardiac Thebesian veins. This amounts to 1% to 2% of the cardiac output normally, but in some diseases bronchial anastamotic blood flow may rise to 10% to 20% of cardiac output. In certain congenital anomalies the right-to-left anatomical shunt may be 50% of cardiac output. Fig. 35-3 shows a large right-to-left shunt through an unventilated lung and the effect of the shunt on the arterial blood gases. Compare it with Fig. 33-1. Right-to-left shunts always reduce systemic arterial oxygen tension and concentration.

There is a simple clinical method for estimating small shunt fractions, if the person is breathing 100% oxygen: The shunt is *1% of cardiac output for each 20 mm Hg difference* between alveolar and arterial oxygen tensions.

If one thinks of venous admixture as a stream of blood being pumped around and around the body without delivering any O_2 or picking up any CO_2, then venous admixture reduces the efficiency of blood gas exchange. Thus venous admixture is the blood flow equivalent of wasted ventilation.

Left-to-right shunt. Left-to-right shunts on the other hand do not affect the systemic arterial oxygen tension. A left-to-right shunt always exists in addition to the systemic flow; that is pulmonary blood flow exceeds systemic blood flow. The characteristic finding at cardiac catheterization in a person who has a left-to-right intrathoracic shunt is a steep increase in the oxygen concentration of blood somewhere on the right side of the heart. When one knows which chamber has the increase, one knows the anatomy of the shunt and can estimate its size from the shunt flow. This information is vital to the cardiac surgeon before he or she attempts to repair the defect.

■ *Occurrence of Shunts in Lung Disease*

Right-to-left shunts (venous admixture) occur commonly in pulmonary diseases and in some forms of con-

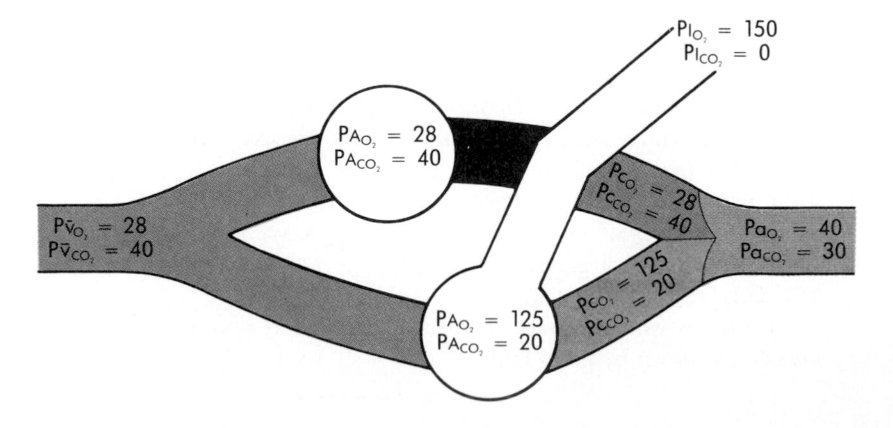

■ **Fig. 35-3** Schema of venous admixture (right-to-left shunt). Notice the marked decrease in the arterial P_{O_2}.

genital heart disease. Left-to-right shunts are not common and are usually due to congenital heart disease. They occur, albeit rarely, in congenital pulmonary disease and indicate a bizarre rearrangement of lung vascular anatomy.

Distribution of Blood Flow

The normal pressure in the aorta (120/80 mm Hg) is the pressure at the level of the heart (see Fig. 35-1). In the upright position the pressure varies by 0.74 mm Hg/cm body height. In a standing adult human (175 cm tall) systemic arterial pressure is 130 mm Hg higher in the feet than in the head.

The effects of gravity are relatively greater in the pulmonary circulation because the pressures in the pulmonary artery and left atrium are much lower than systemic vascular pressures. The normal upright adult lung at TLC is about 30 cm high. The average pressures shown in Table 35-1 are only valid at the level of the heart (left atrium). When lung volume is at TLC the bottom of the lung is about 15 cm below the left atrium. That means that for each 1 cm below the left atrium the pulmonary arterial pressure increases by 1 cm H_2O. At the bottom of the lung in the costodiaphragmatic recess pulmonary arterial pressure is 34 cm H_2O and left atrial pressure is 26 cm H_2O. Although the higher pressures toward the bottom do not change the driving pressure (8 cm H_2O), they do change the transmural distending pressure. One consequence is that the higher transmural pressure toward the bottom of the lung distends the tubes and decreases the resistance to flow. Thus flow is greater toward the bottom of the lung because resistance is less, not because driving pressure is greater.

The opposite effect occurs as one goes up the lung. One must subtract 1 cm H_2O/cm height. When the lung is at TLC mean pulmonary arterial pressure at the top is 4 cm H_2O. During systole the pressure is higher and during diastole lower. Late in diastole, when the pulmonary arterial pressure is at its lowest value, the pressure in the pulmonary artery at the top of the lung may drop to zero or below. The effective left atrial pressure equals zero at 11 cm above the reference level. Above that the venous outflow pressure is less than alveolar (atmospheric) pressure.

The pressure pattern is important because changing pulmonary arterial pressure affects the distribution of blood flow over the height of the lung. As arterial pressure rises relatively more flow is shifted upward. The effect is readily predicted on the basis of the distensibility of the lung vessels. The distribution of flow is illustrated in Fig. 35-4, which is the graphic realization of Fig. 33-8. Although the higher pressures toward the bottom do not change the driving pressure, they do change the transmural (distending) pressure.

The pressure in the microvessels near the bottom of the lung is much higher than at the top; this relationship increases the hydrostatic pressure favoring liquid filtration. In congestive heart failure (high left atrial pressure) interstitial liquid (edema) tends to accumulate first at the bottom of the lung.

Alveolar and Extraalveolar Vessels

In Fig. 33-8 the extraalveolar vessels are represented by the main arterial and venous tubes, whereas the alveolar vessels are mainly the alveolar wall capillaries, represented as very thin-walled soft tubes. The pressure outside the alveolar vessels is alveolar gas pressure, which is ordinarily atmospheric. The pressure outside the extraalveolar vessels however is below alveolar pressure and is similar to pleural pressure (-5 cm H_2O at FRC). When the lung expands and pleural pressure falls the pressure outside the extraalveolar vessels decreases rel-

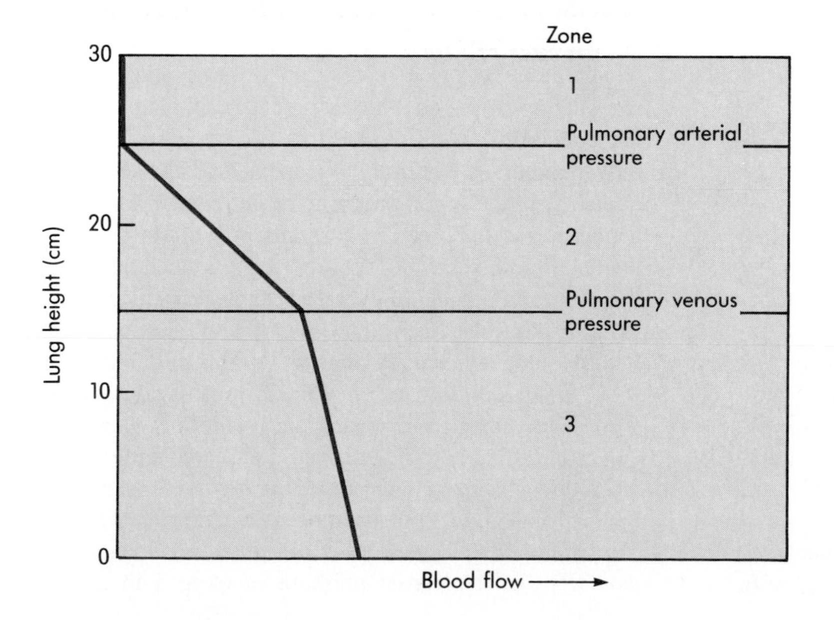

Fig. 35-4 The distribution of flow in the three main lung zones. In zone 1, where alveolar pressure exceeds both venous and arterial pressures, there is no flow. In zone 2, where pulmonary arterial pressure exceeds alveolar pressure, flow is regulated by alveolar pressure compressing the outflow from the alveolar capillaries. In zone 3, the driving pressure is constant because arterial and venous pressures exceed alveolar pressure, but flow continues to increase going down zone 3 because of the passive distention of the vessels by the increasing transmural pressure.

ative to alveolar pressure. This causes the extraalveolar vessels to enlarge during inspiration, thus increasing pulmonary blood volume.

On the other hand alveolar pressure does not vary much with breathing. Near the top of the lung, where the effective venous pressure falls below alveolar pressure, the transmural pressure across the small veins becomes a compressive pressure at the point where the veins leave the alveolar wall network. As the blood vessels are compressed, vascular resistance increases.

The uneven distribution of blood flow caused by gravity is divided into three zones, depending on the relative values of pulmonary arterial, venous, and alveolar pressures. The effect of gravity on blood flow distribution is shown in Fig. 35-4.

In zone 1 flow is zero because pulmonary arterial pressure in less than alveolar pressure. In zone 3 flow is high and changes with distance down the lung only because of the increasing transmural pressure. In zone 2 the alveolar pressure exceeds the venous outflow pressure. The transmural pressure of these vessels is negative: that is $(Pla - P_A) < 0$. Any small veins exposed to alveolar pressure are compressed at the outflow end of the alveolar compartment.

Although all three zones can exist in the human lung, pulmonary arterial pressure is normally high enough that the zone 1 condition of no flow does not occur, and the condition described for zone 2 is limited to the upper third of the lung because left atrial pressure is normally well above alveolar pressure.

During mechanical ventilation however alveolar pressure is artifically increased, because positive pressure is applied to the airway opening. This pressure increases the amount of lung in zone 2, because alveolar pressure rises relative to pulmonary venous pressure, increasing resistance to blood flow in that zone.

When pulmonary vascular pressures are increased, the distribution of blood flow over the height of the lung is more uniform. Thus in heavy exercise blood flow distribution is nearly equal throughout the lung.

Regulation of Pulmonary Blood Flow

Longitudinal Distribution of Resistance

The resistance to blood flow in the pulmonary arteries and arterioles is much less than that in the systemic circulation. In a person breathing normally the alveolar wall capillaries contribute 35% to 40% of the total resistance, the arteries about 50%, and the veins 10% to 15%. In the systemic circulation the arterioles contribute 75% of the resistance, whereas the capillary resistance is trivial (<5%).

What was trivial in the systemic capillaries, namely a pressure fall of several centimeters of water (cm H_2O), is a substantial quantity in the pulmonary circulation. In Table 35-1 the average pressure difference (Ppa − Pla) across the lung is 19 − 11 = 8 cm H_2O. An approximately 3 cm H_2O pressure drop occurs in the alveolar wall capillaries. This pressure drop depends on the dimensions and distensibility of the alveolar wall capillary network, because there is no active regulation of capillary dimensions. The fraction of resistance within the alveolar wall vessels is sensitive to lung volume (less at low volume), blood flow rate (less at high flow), and vascular distending pressure (greater at low pressure).

Passive Regulation

When humans perform submaximal sustained exercise cardiac output may rise threefold. This increase in blood flow is accommodated by the pulmonary circulation without an equivalent rise in the pulmonary vascular driving pressure. This occurs because of recruitment and distension of microvessels by the increasing transmural pressure in the very small vessels (arterioles, capillaries, and venules). As already noted in humans during exercise the volume of blood in the pulmonary capillaries may double, substantially decreasing the contribution of the capillaries to resistance. At a threefold increase in flow the driving pressure across the lungs may increase only about 50%, indicating that the calculated pulmonary vascular resistance must have decreased by 50%.

Active Regulation

Although the passive effects of pressure on the distensible pulmonary vascular bed are prominent, active regulation occurs under physiological and pathological conditions. Smooth muscle associated with the small muscular pulmonary arteries, arterioles, and veins is adequate to alter pulmonary vascular resistance substantially. The smooth muscle may hypertrophy remarkably in pathological conditions.

A number of naturally occurring substances selectively affect the vasomotor tone of pulmonary arterial or venous vessels. Some are normally constrictors: for example thromboxane A_2, α-adrenergic catecholamines, histamine, angiotensin, several prostaglandins, neuropeptides, leukotrienes, serotonin, endothelin, and increased P_{CO_2}. Some are normally dilators: for example increased $P_{A_{O_2}}$, β-adrenergic catecholamines, prostacyclin, endothelial-derived relaxing factor, acetylcholine when vascular tone is increased, bradykinin, and dopamine.

Alveolar oxygen tension. *The factor of overriding importance in governing minute-to-minute regulation of the pulmonary circulation is the alveolar oxygen tension ($P_{A_{O_2}}$).* The partial pressure of oxygen in the air spaces is far more important than the oxygen tension within the mixed venous blood. PA_{O_2} is so important because the small muscular pulmonary arteries are surrounded by the alveolar gas of the terminal respiratory units they subserve (Fig. 33-9). Ordinarily the PA_{O_2} is high (100 mm Hg), so the smooth muscle cells of microvessels in or near the alveolar walls are bathed in the highest oxygen tensions of any organ. Oxygen diffuses through the thin alveolar walls into the smooth muscle cells.

The response of pulmonary vascular smooth muscle to low Po_2 is vastly different from that of the systemic circulation. In the lung low alveolar oxygen tension constricts the nearby arterioles. Conversely in the systemic circulation low oxygen tension relaxes the local resistance vessels, permitting passive dilation to occur. The low Po_2 appears to act directly on the vascular smooth muscle cells. As alveolar oxygen tension falls in any region of the lung the adjacent arterioles constrict. *This constriction decreases local blood flow and shifts it to other regions of the lung.* Very small changes in local vascular resistance cause marked shifts of blood flow without any significant effect on overall pulmonary vascular resistance, provided that the volume of lung involved is not greater than 20% of lung mass.

On the other hand a global reduction in alveolar oxygen tension, such as occurs at high altitude or in breathing of low oxygen mixtures, increases pulmonary vascular resistance. Pulmonary vascular resistance in acute alveolar hypoxia may be more than twice normal. This effect is sustained in humans and in most normal mammals. The effect of alveolar oxygen tension on pulmonary vascular resistance is chiefly on the arterioles and small muscular arteries. In many types of lung disease hypoxic vasoconstriction is a major compensatory factor. On the other hand if a disease process (such as lobar pneumonia) inhibits hypoxic vasoconstriction, arterial oxygen partial pressure may fall because blood flow cannot be redistributed away from the poorly ventilated region. Thus an important clinical test of the vasoconstrictor component of pulmonary hypertension is the determination of the effect of oxygen inhalation on pulmonary vascular resistance.

Thromboxane and prostacyclin. The most important of the vasoconstrictors is **thromboxane A_2,** a product of arachidonic acid metabolism. It is produced within the pulmonary circulation in several types of acute lung injury. Many cells, including leucocytes, macrophages, and endothelial cells, can produce and release thromboxane. Thromboxane A_2 is one of the most powerful constrictors of pulmonary arterial and venous smooth muscle. Thromboxane constricts both arteries and veins. As with oxygen the effect is mainly localized to the region where the thromboxane is released, although recirculation and diffusion through the lung interstitium sometimes spread its effect to airways and other vessels.

Prostacyclin (Prostaglandin I_2), a potent vasodilator, is another product of arachidonic acid metabolism. Endothelial cells are probably the chief source of this substance. However little thromboxane or prostacyclin is normally produced. Thus the normal smooth muscle tone of the pulmonary vascular bed is low.

Bronchial Circulation

The bronchial arteries supply water and other nutrients to the mucosal cells and glands of the airways down to and including the terminal bronchioles. They also nourish the pleura, interlobular septal supporting tissues, and pulmonary arteries and veins. The bronchial circulation does not normally supply the terminal respiratory unit, which receives its nutrition via the alveolar wall capillaries.

The pressure in the main bronchial arteries is nearly the same as in the aorta, so that there is a high driving pressure regardless of body position. Although bronchial blood flow is less than 1% of cardiac output, it is appropriate for the portion of the lung tissue it serves.

About half of the bronchial blood flow returns to the right side of the heart via the bronchial veins. The remainder flows through small bronchopulmonary anastomoses (<100 μm diameter) into the pulmonary veins and thereby contributes to the normal venous admixture (right-to-left shunt). In certain inflammatory diseases of the airways (for example bronchitis, bronchiectasis, bronchogenic carcinoma) the bronchial circulation expands dramatically and may contribute as much as 10% to 20% venous admixture.

When the pulmonary circulation is obstructed (thrombosis or embolism), the bronchial arteries dilate and develop connections with precapillary vessels of the pulmonary circulation. When this occurs the blood flowing through the alveolar capillaries is systemic blood, which takes up little oxygen from the alveolar gas but can still give up carbon dioxide. Thus the bronchial circulation keeps the lung alive when the pulmonary blood flow is stopped.

The bronchial circulation participates in conditioning the inspired air, especially under circumstances in which air bypasses the upper air passages, such as when breathing through the mouth during exercise. The bronchial circulation also warms the incoming air. The inhaled air is completely warmed and humidified in the upper air passages. Hence there is no evaporation from the alveolar surfaces.

Matching Ventilation to Perfusion

Ventilation and Perfusion

The concept of matching gas (ventilation) and blood (perfusion) for successful oxygen and carbon dioxide exchange is central to an understanding of gas exchange. Inequalities of the distribution of ventilation/perfusion ratios are the most common causes of inefficient O_2 and CO_2 exchange.

In the ideal lung the ventilation/perfusion ratio ($\dot{V}A/\dot{Q}$) is uniformly distributed. Because resting normal alveolar ventilation, $\dot{V}A$, is 4.2 L/min and normal resting pulmonary blood flow, $\dot{Q}$, is 5 L/min, normal $\dot{V}A/\dot{Q}$ = 4.2/5.0 = 0.84 (no units). These values yield the normal Pa_{O_2} = 100 mm Hg and Pa_{CO_2} = 40 mm Hg at R = 0.80 on the O_2-CO_2 diagram (Fig. 33-16).

In exercise $\dot{V}A/\dot{Q}$ rises. For example during maximum steady-state exercise $\dot{V}O_2$ = 1,500 ml/min (6 times resting), $\dot{V}A$ = 26 liters/min and $\dot{Q}$ = 15 L/min, so $\dot{V}A/\dot{Q}$ = 1.7. But the arterial blood gases are normal. Thus to interpret what the overall lung $\dot{V}A/\dot{Q}$ ratio means one needs to know respiratory exchange ratio, R, and steady-state oxygen consumption, $\dot{V}O_2$, as well as ventilation and blood flow.

In the real lung neither ventilation nor blood flow is uniformly distributed, even in a healthy normal young adult. In cardiopulmonary disease patients the most frequent cause of systemic arterial hypoxemia is not hypoventilation or venous admixture (right-to-left shunt): it is uneven alveolar ventilation in relation to alveolar blood flow.

Ventilation-perfusion mismatching is generally expressed in terms of its effect on the alveolar-arterial PO_2 difference. There is a parallel but much smaller effect on CO_2, which is generally ignored because the main cause of CO_2 retention is hypoventilation.

Two extreme instances of ventilation-perfusion mismatching have already been introduced: wasted ventilation (Chapter 33) and venous admixture (discussed earlier in this chapter).

Wasted Ventilation

To extend what was introduced in Chapter 33 (Fig. 33-14) wasted ventilation may occur clinically, when a large blood clot (pulmonary embolism) obstructs a branch of the pulmonary artery. In Fig. 33-14 lung A has its blood flow cut off. If compensatory changes in ventilation or blood flow do not occur, and if half of the blood flow and alveolar ventilation went to each lung before the obstruction, then after the occlusion all of the blood flow is diverted to lung B, although half of the ventilation still goes to lung A. Clearly the ventilation to lung A is wasted, because it fails to oxygenate any of the mixed venous blood. The overall efficiency of lung ventilation decreases, because more than half the power used in breathing moves air that serves no useful purpose.

The ventilation/perfusion ratio of lung A is infinite because the denominator (blood flow) equals zero (2.1/0). But what is the ventilation/perfusion ratio of lung B? It cannot be the normal value of 0.84. Lung B receives its normal portion of alveolar ventilation but *all* of the cardiac output. Its ventilation/perfusion ratio is 2.1/5.0 = 0.42.

Fortunately compensatory changes begin almost immediately, and various mechanisms shift ventilation from the useless lung to the functional lung. This shift moves the $\dot{V}A/\dot{Q}$ back toward its normal value. If this did not occur, the Pa_{CO_2} would increase and Pa_{O_2} would decrease (hypoventilation).

Venous Admixture

Venous admixture is shown schematically in Fig. 35-3. In this example lung A receives no ventilation but still has its normal blood flow. Clearly hypoxemia will result because 50% of cardiac output is shunted. The $\dot{V}A/\dot{Q}$ of lung A = 0/2.5 = 0. A right-to-left shunt is equivalent to a ventilation/perfusion ratio equal to zero. On the other hand ventilation of lung B has doubled relative to its blood flow. The $\dot{V}A/\dot{Q}$ of lung B = 4.2/2.5 = 1.68. This is clearly hyperventilation, as defined in Chapter 33. Thus the arterial PCO_2 decreases, although not very much. The oxygen tension of blood in the pulmonary veins of the ventilated lung rises. However because of the flat slope of the HbO_2 equilibrium curve when the PO_2 is near 100 mm Hg the blood flow through lung B cannot compensate for the low PO_2 of the shunt through lung A. Therefore Pa_{O_2} decreases markedly.

Other Ventilation/Perfusion Distributions

If alveolar dead space ($\dot{Q}$ = 0; $\dot{V}A/\dot{Q} = \infty$) and right-to-left shunt ($\dot{V}A$ = 0; $\dot{V}A/\dot{Q}$ = 0) represent the extremes of ventilation-perfusion mismatching, then all other possible ventilation/perfusion ratios must lie between them.

The $\dot{V}A/\dot{Q}$ distribution throughout the normal lung is not homogeneous, although the average value is 0.84. Some lung units are overventilated and some are underventilated. Fig. 35-5 shows the distribution of $\dot{V}A/\dot{Q}$ in a normal human adult. The average normal value (0.84) is shown as the thin vertical line. The ordinate shows absolute units of ventilation or perfusion (liters/minute). The ventilation/perfusion ratios are shown on the X-axis. A logarithmic X-axis scale is used to show the wide range of $\dot{V}A/\dot{Q}$ ratios that may be encountered in diseased lungs.

An elegant way to measure the $\dot{V}A/\dot{Q}$ distribution in the lung is the **multiple inert gas procedure,** in which six inert gases of differing gas/blood solubility ratios are infused intravenously until a steady state of gas elimination and infusion is established. From the infused and expired partial pressures of each gas the most likely pattern of $\dot{V}A/\dot{Q}$ distribution is determined.

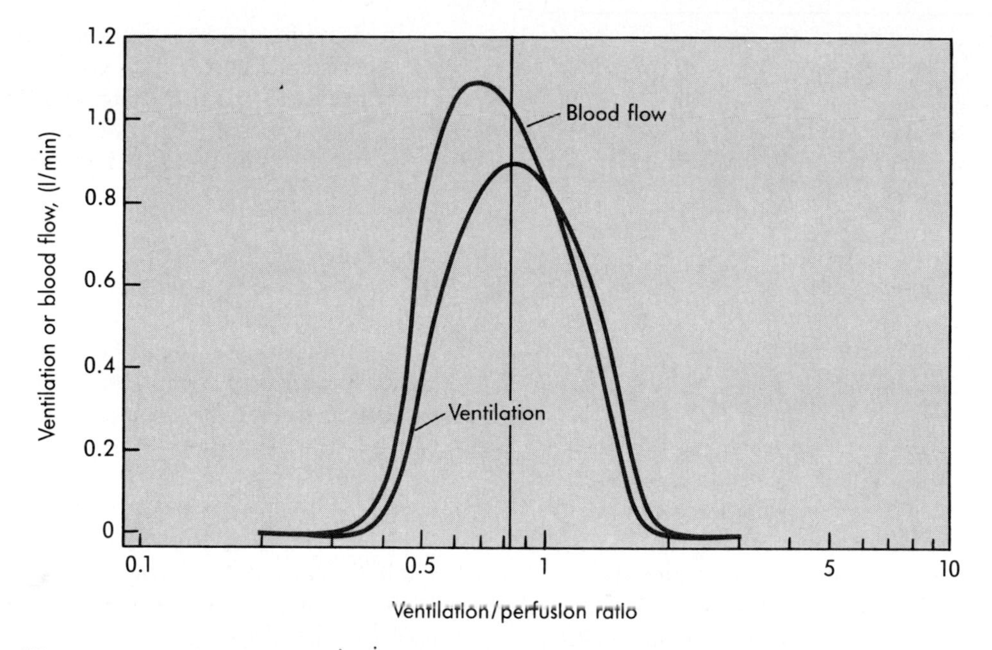

Fig. 35-5 Normal human $\dot{V}A/\dot{Q}$ distribution curves. The ordinate refers to blood flow or ventilation (L/min). Their quotient is the ventilation/perfusion ratio shown on the abcissa on a logarithmic scale. The overall normal $\dot{V}A/\dot{Q} = 0.84$ is shown by the thin vertical line.

O_2-CO_2 diagram. Another way to look at the $\dot{V}A/\dot{Q}$ distribution is shown is Fig. 35-6, which is a modification of the O_2-CO_2 diagram shown in Fig. 33-16. As before point A represents the normal Pa_{CO_2} and Pa_{O_2} for an ideal $\dot{V}A/\dot{Q} = 0.84$. Some representative $\dot{V}A/\dot{Q}$ ratios from Fig. 35-5 are included. Notice that these are not arranged linearly along the line. All $\dot{V}A/\dot{Q}$ ratios greater than the normal value lie to the right of A and represent overventilation. The extreme value of dead space ventilation is the inspired point, I. All $\dot{V}A/\dot{Q}$ ratios less than 0.84 (overperfused) lie to the left of point A. The mixed venous point, $\bar{v}$, represents a $\dot{V}A/\dot{Q} = 0$ (venous admixture).

For the range of normal $\dot{V}A/\dot{Q}$ shown PCO_2 does not change much. At low $\dot{V}A/\dot{Q}$ however the oxygen tension falls steeply but the PCO_2 cannot rise above that of mixed venous blood.

Any deviation of $\dot{V}A/\dot{Q}$ from the ideal value impairs the efficiency of O_2 and CO_2 transfer: that is it increases the differences between alveolar to arterial gas tension, especially that for oxygen.

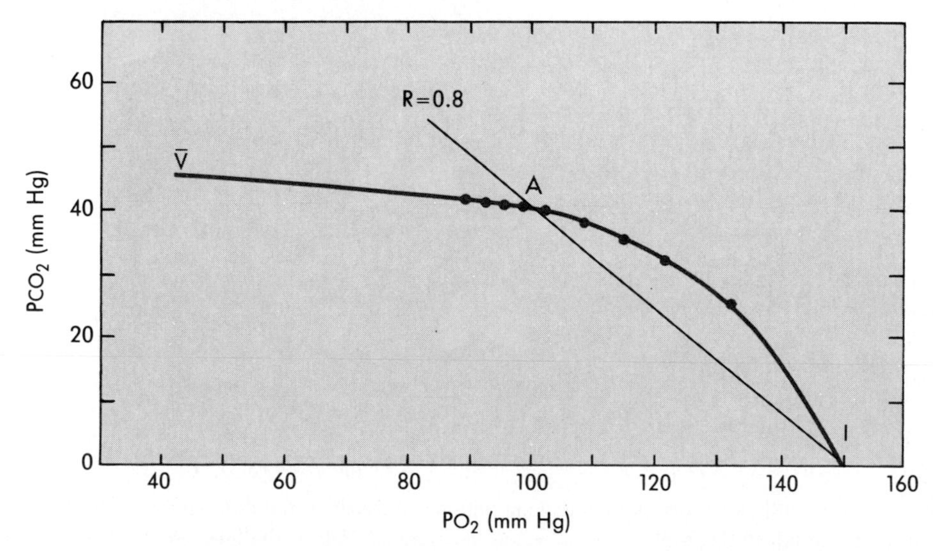

Fig. 35-6 The O_2-CO_2 diagram showing the range in the normal lung. I is the inspired gas ($\dot{V}A/\dot{Q} = \infty$) and ($\bar{v}$) is venous admixture ($\dot{V}A/\dot{Q} = 0$). Ideal $\dot{V}A/\dot{Q}$ is shown at *A*.

Distribution of Ventilation

The normal lung is not uniformly ventilated. *Regional* nonuniform ventilation is mainly due to gravitational effects, such as those moving from top to bottom in the upright human lung. *Local* nonuniform ventilation among terminal respiratory units is due to variable airway resistance, R, or compliance, C, and is described by the time constant ($\tau = R \times C$).

Regional Ventilation Distribution

Each level of lung is suspended from and supported by the lung above it (see Chapter 33). Ultimately the force that maintains lung support is transmitted through the layer of pleural liquid between the lung and the chest wall. Consequently as one moves up the lung the mass of lung that must be supported below each level increases. This means that the weight of the lung pulling down or away from the chest wall increases. Thus pleural pressure decreases. If pleural pressure decreases, then static translung pressure ($P_L = P_A - Ppl$) must increase.

The result is that the alveoli are expanded more near the top at all lung volumes below total lung capacity. Indeed the relative alveolar volume follows the compliance curve of the lung. Fig. 35-7 shows the end-expiratory volume of typical alveoli at three places along the pressure-volume curve.

However the data in Fig. 35-7 are somewhat misleading in the sense that the curve shown is the deflation pressure-volume curve, not the dynamic breathing curve. Over the normal tidal volume range the differences in regional ventilation are small because the slope of the P-V curve is nearly constant. That means that the regional distribution of ventilation on a gravitational basis is less than expected.

Nevertheless the lower portion of the lung tends to be ventilated more because the end-expiratory volume (FRC) is slightly less and the compliance is slightly greater than at the top of the lung (Fig. 35-7, middle).

Furthermore the regional ventilation distribution is markedly affected by body position: less when the body is prone than when it is supine. This effect is caused in part by abdominal pressure on the diaphragm that affects FRC. Indeed an effective treatment of some patients whose lungs are difficult to ventilate mechanically is simply turning them to the prone position.

After a person takes a single maximal inspiration of 100% O_2 the regional distribution of ventilation in large part determines the slope of what is called the expired alveolar plateau (Fig. 35-8), because the better ventilated alveoli tend to empty first. This is the basis of a simple, useful pulmonary function test for the uniformity of ventilation distribution, known as the **single breath nitrogen test.**

Local Ventilation Distribution

Uneven distribution of ventilation, which is not dependent on gravity, is due to varying time constants, $\tau = R \times C$. Recent evidence suggests that local variation in

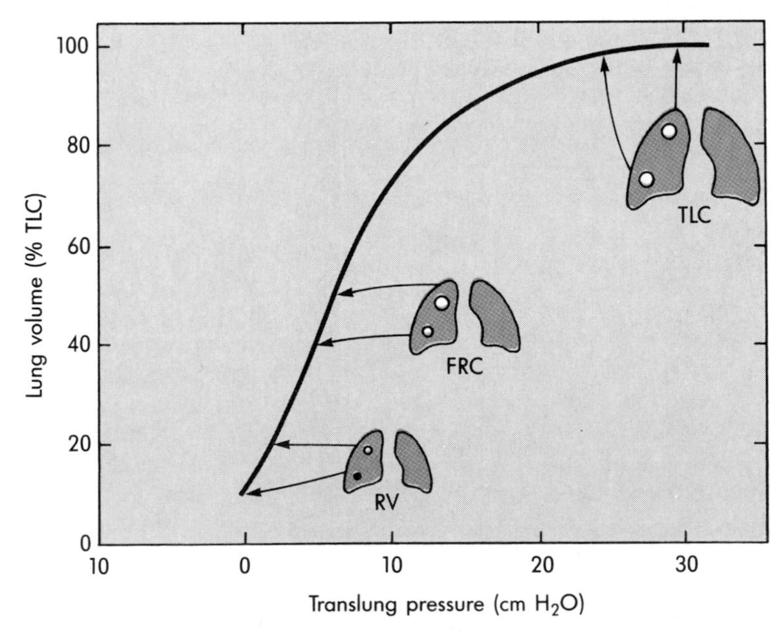

Fig. 35-7 Regional distribution of lung volume. Because of the differences in pleural pressure from the top to the bottom, the translung pressure of units at the lung apex will be greater than those at the base. The effect is very marked at residual volume, less at FRC, and disappears at total lung capacity.

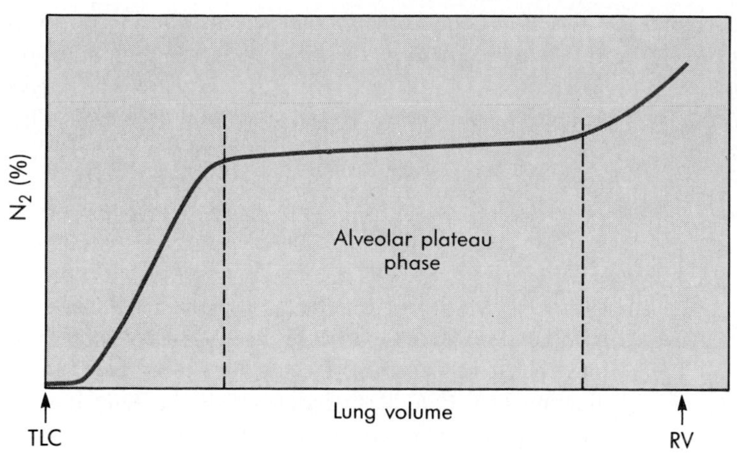

■ **Fig. 35-8** The single breath nitrogen washout curve is a simple useful test of regional ventilation distribution because well-ventilated units (short time constant) empty faster than less well ventilated units (long time constant). The early part of the curve represents dead space washout. The long alveolar plateau rises slowly if ventilation distribution is relatively uniform, as shown here. The final phase shows very late, slowly emptying alveoli. This phase is accentuated with age.

ventilation distribution is more important, even in the normal lung, than previously supposed. In other words resistance and compliance differences among terminal respiratory units vary substantially in functional and anatomical terms. Fig. 35-9 illustrates the time constant concept. A longer time constant means slower ventilation; thus a unit with increased resistance, increased compliance, or both takes longer to fill, when other factors are constant. In addition a decrease in compliance of a unit decreases its absolute volume at any given translung pressure.

Because breathing is a dynamic process time is important even at rest. At the normal breathing rate of 12/min inspiratory time, T_I, is 2 seconds and expiratory time, T_E, is 3 seconds. These times may not be long enough for every terminal respiratory unit to achieve a new steady state. In diseases such as chronic obstructive pulmonary disease (emphysema) some lung units may fill and empty very slowly. When chest physicians refer to "fast" and "slow" alveoli they are referring to the time constants of those units.

Fig. 35-9 shows three units, one is normal, one has twice the airway resistance of the others, and one has reduced compliance. The graph shows the time course of filling the units during a normal 2 second inspiration. The normal unit, N, and the low compliance unit reach a new steady state, although the volume of the stiff unit is markedly reduced. The unit with increased R fills only 80% but is actually better ventilated than the low compliance unit. If inspiration were prolonged, the high resistance unit would eventually fill.

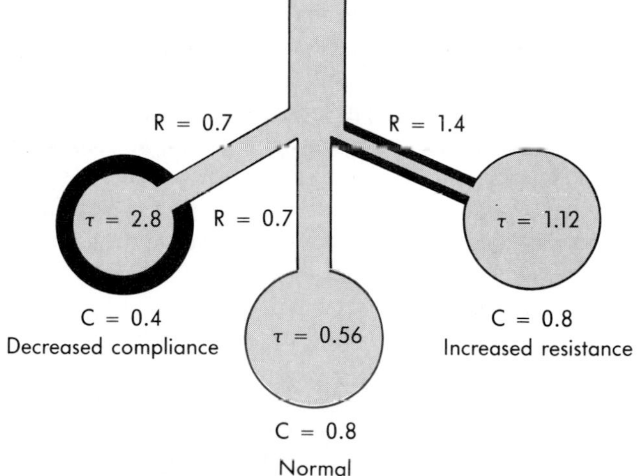

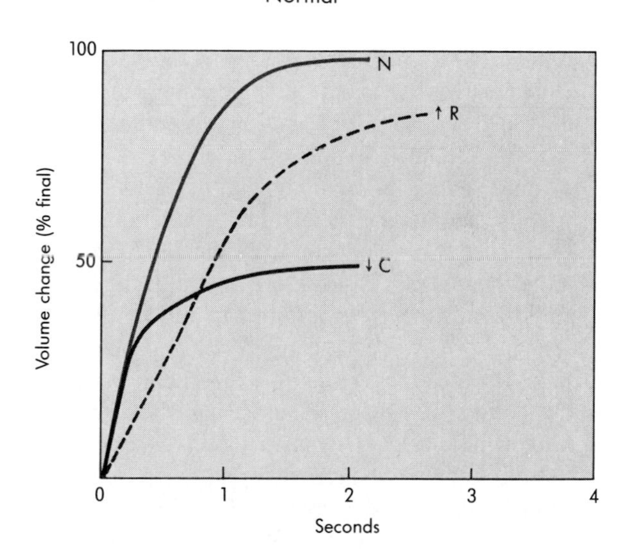

■ **Fig. 35-9** Examples of local regulation of ventilation because of variation of resistance *(R)* or compliance *(C)* of individual lung units. The normal lung is represented by unit labeled N, which has a normal time constant, τ, of 0.56 sec. This unit reaches 97% of final equilibrium in 2 sec, the normal inspiratory time, as shown in the inset graph. One unit has a twofold increase in resistance. Hence its time constant is doubled. That unit fills more slowly and reaches only 80% equilibrium during a normal breath. The unit is underventilated. One unit has reduced compliance (stiff), which acts to reduce its time constant. This unit fills faster than the normal unit but only receives half the ventilation of a normal unit.

■ *Distribution of Perfusion*

■ *Regional Perfusion Distribution*

Gravity affects the regional distribution of blood flow in the lung by determining the transmural distending pressure of the blood vessels and the relative arterial, venous, and alveolar pressures (lung zones) (see Fig. 33-8). However the gravitational effect on the blood flow gradient from the top to the bottom of the lung is probably a substantial overestimate. Much of the original data was obtained from lungs at total lung capacity, not at FRC or over the tidal range. For example zone 1 is small in the normal human in the upright position, except possibly during a brief portion of each cardiac cycle (late in diastole). Likewise regional hypoxic pulmonary vasoconstriction may affect the lower lung more.

Local Perfusion Distribution

Nongravitational uneven distribution of blood flow to terminal respiratory units may be important. There appear to be two components: One is anatomical (geometrical): that is the diameter, length, and branching angles of the vessels to each unit. This congenital factor is not affected by large changes of flow. Such apparently random anatomical variations also occur in cardiac and skeletal muscle.

The second component is the alveolar oxygen tension, which affects the arterial resistance to the various lung units by altering vasomotor tone. Normally there is not much hypoxic vasoconstrictor effect, but what there is probably acts on the lower $\dot{V}A/\dot{Q}$ regions to divert flow upward.

The Effect of $\dot{V}A/\dot{Q}$ Mismatching

Local $\dot{V}A/\dot{Q}$

Unfortunately no simple graphic approach can identify the nongravitational distribution of ventilation/perfusion ratios in the lung. However one can examine the effect of mismatching of ventilation to perfusion in a general way by determining the alveolar to arterial difference in oxygen tension. Notice that PO_2 in Fig. 35-6 varies by more than 40 mm Hg over the height of the lung. Blood leaving units near the bottom has a lower PO_2 and consequently a lower SO_2 than blood leaving units near the top. Also one must remember that there is more flow to the bottom of the lung. Increasing alveolar PO_2 above normal does not have much effect on arterial oxygen saturation or concentration, because the HbO_2 equilibrium curve is so flat above $PO_2 = 70$ mm Hg. When any $\dot{V}A/\dot{Q}$ mismatching occurs the arterial PO_2 is always less than the ideal value. In order to make a precise calculation one would need to know the blood flow through each lung region.

Although venous admixture caused by venous blood that bypasses the normal lung (anatomical shunt) is only 1% to 2% of cardiac output, the $\dot{V}A/\dot{Q}$ mismatch contribution to venous admixture may be 4% to 5%. Thus the normal systemic arterial PO_2 in humans is between 85 and 90 mm Hg (an alveolar to arterial PO_2 difference of 10 to 15 mm Hg). Any alveolar to arterial PO_2 difference of less than 20 mm Hg is considered normal.

The consequence of the normal distribution of $\dot{V}A/\dot{Q}$ is a slight decrease in arterial oxygen saturation to about 96.9% ($Pa_{O_2} = 90$ mm Hg), that is about 0.5% less than the ideal arterial oxygen saturation of 97.4%.

Ventilation/Perfusion Mismatching and Arterial PCO_2

Although $\dot{V}A/\dot{Q}$ mismatching is generally stated in terms of the decrease in arterial oxygen tension, there must be a comparable effect on Pa_{CO_2}, as the O_2-CO_2 diagram (see Fig. 35-6) indicates. If the lung has reduced efficiency in oxygen uptake, it must also have reduced efficiency in CO_2 output. However examination of the O_2-CO_2 diagram shows that the effects on Pa_{CO_2} are much less than those on Pa_{O_2}. Indeed many people who have fairly severe lung disease and considerable $\dot{V}A/\dot{Q}$ mismatching may have nearly normal Pa_{CO_2}. The explanation is twofold: the CO_2 equilibrium curve of blood is almost linear (see Chapter 36), and the arterial PCO_2 has an overriding effect on ventilation. Contrary to what one might think the regulatory mechanisms of ventilation are far more sensitive to Pa_{CO_2} than to Pa_{O_2}.

Compensation for $\dot{V}A/\dot{Q}$ Mismatching

Obviously the extreme conditions represented in Fig. 33-14 and 35-3 are unstable. There is an enormous wasted ventilation in the former and wasted blood flow in the latter. As a result Pa_{O_2} and Pa_{CO_2} are adversely affected.

Fortunately a number of compensatory mechanisms come into play to alleviate the $\dot{V}A/\dot{Q}$ mismatch. Initially these are centrally mediated (brain controller) effects on ventilation, which in both examples shown rise as a result of the increased Pa_{CO_2}. But increasing total ventilation is only a stopgap measure because the wasted ventilation fraction is not affected (Fig. 33-14).

Local factors also come into play, and these are very important. When ventilation is wasted the local PA_{CO_2} falls. This fall lowers the hydrogen ion concentration, $[H^+]$, in and around the associated airway smooth muscle. The fall in $[H^+]$ leads to airway constriction and a shift of ventilation away from the terminal respiratory units with high $\dot{V}A/\dot{Q}$ ratios. The compensation can be very effective, depending, of course, on how big the affected lung unit is.

If blood flow is reduced to a terminal respiratory unit, the local alveolar cell metabolism is affected; notably surfactant production or release is decreased. An increase in the air-liquid interfacial tension reduces the unit's compliance and its FRC volume decreases. This outcome provides an effective compensatory mechanism for severely underperfused units. It is a slow mechanism (hours to days) however compared with the very fast changes in total ventilation (seconds) and local airway tone (seconds to minutes) caused by changes in P_{CO_2}.

The effectiveness of hypoxic vasoconstriction in shifting flow away from an underventilated unit depends on the fraction of lung involved. If the vessels are constricted to only a small fraction (<20%) of the lung mass, then shifting flow away from the underventilated units is very effective because the vasoconstriction has little impact on overall pulmonary hemodynamics: that is pulmonary arterial pressure is not increased.

When a large fraction (>20%) of the lung is involved then hypoxic vasoconstriction increases pulmonary arterial pressure. The extreme case is global alveolar hypoxia (high altitude), in which pulmonary arterial pressure may be doubled. This actually improves the $\dot{V}_A/\dot{Q}$ distribution because the rise in pulmonary arterial pressure tends to even out the normal gravitational maldistribution of blood flow.

■ Summary

1. In normal humans pulmonary vascular resistance is much less than in the systemic circulation. Comparing pressure-flow curves is an excellent way to assess changes in pulmonary hemodynamics under various conditions.

2. Blood flow from the systemic venous to the systemic arterial system without being fully oxygenated (right-to-left shunt) decreases systemic arterial oxygen tension and concentration. In reverse or left-to-right shunts pulmonary blood flow is increased, but systemic arterial oxygen tension and concentration, and systemic blood flow are normal.

3. The distribution of pulmonary blood flow is sensitive to gravity over the height of the air-filled lung because the pulmonary arterial and left atrial pressures are normally low and the vessels are distensible or collapsible. This relationship gives rise to three possible flow conditions: zone 1, no flow; zone 2, flow regulated by compression of microvessels at the outflow from the alveolar walls; zone 3, flow dependent on driving pressure, vascular geometry, and smooth muscle tone.

4. Although normally passive regulation of pulmonary blood flow distribution predominates, active regulation may become very important. The main regulator is alveolar oxygen tension. Alveolar hypoxia leads to an immediate and sustained local increase in vascular resistance by causing the small pulmonary arteries to constrict. Thromboxane and prostacyclin are powerful constrictors and dilators, respectively.

5. The bronchial circulation nourishes the walls of the airways and blood vessels, warms and humidifies incoming air, and supplies substrate to airway glands, mucosa, and smooth muscle.

6. The distribution of ventilation/perfusion ratios among terminal respiratory units is not uniform, even in the normal lung, and may show marked deviations from normalcy in disease. The limiting ratios are wasted ventilation and venous admixture.

7. Ventilation is distributed nonuniformly in the lung on regional (gravitational) and local (nongravitational) bases. Regional distribution predominates in the normal upright human. Local factors affecting ventilation distribution to each terminal respiratory unit are its resistance and compliance, whose product is the time constant. The time constant determines the rate at which the unit is ventilated.

8. Blood flow is distributed nonuniformly in the lung on regional (gravitational) and local (nongravitational) bases. Regional distribution predominates in the normal upright human. The main local factor affecting perfusion distribution is the alveolar oxygen tension. Hypoxic vasoconstriction is very effective provided the amount of lung tissue affected is small. The overall effect of ventilation/perfusion distribution can be judged by determining the alveolar-arterial oxygen tension difference, which is normally 10 to 15 mm Hg.

■ Bibliography

Journal articles

Bhattacharya J, Staub NC: Direct measurement of microvascular pressures in the isolated perfused dog lung, *Science* 210:327, 1980.

Deffenbach ME et al: The bronchial circulation: small, but a vital attribute of the lung, *Am Rev Respir Dis* 135:463, 1987.

Hyman AL, Spannhake EW, Kadowitz PJ: Prostaglandins and the lung, *Am Rev Respir Dis* 117:111, 1978.

Lenfant C: Measurement of ventilation/perfusion distribution with alveolar arterial differences, *J Appl Physiol* 18:1090, 1963.

Milic-Emili J et al: Regional distribution of inspired gas in the lung, *J Appl Physiol* 21:749, 1966.

Mitzner W: Resistance of the pulmonary circulation, *Clin Chest Med* 4:127, 1983.

West JB, Dollery CT, Naimarh A: Distribution of blood flow in isolated lung: relation to vascular and alveolar pressures, *J Appl Physiol* 19:713, 1964.

West JF: Ventilation-perfusion relationships, *Am Rev Respir Dis* 116:919, 1977.

Books and monographs

Forster RE, II et al: *The lung: physiologic basis of pulmonary function tests,* ed 3, Chicago, 1986, Yearbook.

Grover RF et al: *Pulmonary circulation.* In Shepherd JT, Abboud FM, *Handbook of physiology: the cardiovascular system III,* Bethesda, Md, 1983, Am Physiol Society.

Harris P, Heath D: *The human pulmonary circulation,* ed 2, Edinburgh, 1978, *Churchill Livingstone.*

Nunn JF: *Applied respiratory physiology,* ed 3, London, 1987, Butterworths.

CHAPTER

36

Transport of Oxygen and Carbon Dioxide: Tissue Oxygenation

■ *Oxygen Transport*

One liter of plasma holds only 3 ml O_2 in physical solution at the ideal arterial Po_2 of 100 mm Hg. If O_2 transport in blood depended on dissolved O_2 to sustain normal resting oxygen consumption of 250 ml/min, cardiac output would have to be more than 80 L/min, even if all of the oxygen could be extracted from the plasma. Fortunately, blood carries the iron-containing red protein, hemoglobin, packed into erythrocytes. Hemoglobin, at its normal concentration of 150 g/L, permits whole blood to carry 65 times more oxygen than does plasma at a Po_2 of 100 mm Hg.

■ *The Hemoglobin-Oxygen Equilibrium Curve*

Fig. 36-1 shows the HbO_2 equilibrium curve of normal human blood at 37° C and at a hydrogen ion concentration (H^+) of 40 nmol/L (40×10^{-9} mol/L; pH = 7.40). The curve shows fractional saturation as a function of Po_2, but it gives little information about oxygen transport. One needs to know more about the rates of the chemical reactions between O_2 and Hb within erythrocytes and the concentration of hemoglobin in blood to make any physiological use of the equilibrium curve.

■ *Hemoglobin, the Oxygen Carrier Protein*

Briefly, hemoglobin consists of four O_2-binding heme molecules (iron-containing porphyrin rings) and globin (protein) chains; one heme group per chain. The molecular weight of hemoglobin is 66,500. There are three important physiological properties of the chemical binding between hemoglobin and O_2:

1. Hemoglobin combines *reversibly* with O_2. The oxygen-containing form is called **oxyhemoglobin,** HbO_2, and the unoxygenated form is called **hemoglobin,** Hb (also reduced, or deoxyhemoglobin). The iron in the four heme pigments is in the ferrous (Fe^{++}; reduced) state.

Obviously, HbO_2 saturation is not limited to multiples of 25%, 50%, 75%, or 100%. Thus, when arterial blood is 90% saturated, it follows that some of the Hb molecules have four oxygens bound, some have three, and a few have two or one. The statistical average of all oxygen bound

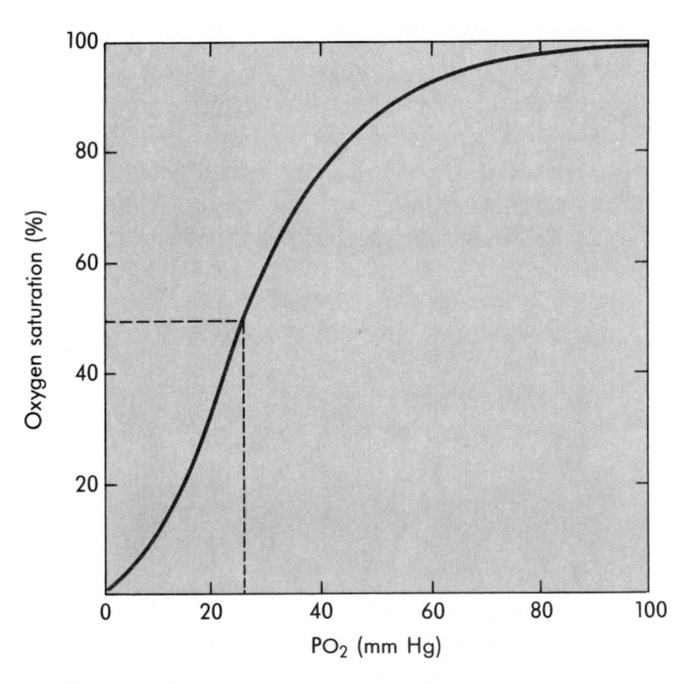

■ **Fig. 36-1** The standard hemoglobin-oxygen (HbO_2) equilibrium curve. It relates oxygen partial pressure (abscissa) and fractional saturation of hemoglobin in erythrocytes (ordinate). The P_{50} is used to compare the curves under different conditions or among different hemoglobins. The curve, as it stands, gives no information about O_2 transport.

to hemoglobin molecules relative to the total amount that can be bound is its **oxygen saturation** (So_2).
2. Molecular oxygen associates with or dissociates from hemoglobin very rapidly (in milliseconds), even when the hemoglobin is densely packed in erythrocytes. This fast reaction is critical for O_2 transport because blood remains in the exchange capillaries less than 1 second.
3. The sigmoid (S-shaped) HbO_2 equilibrium curve is caused by *molecular interaction* among the four heme groups. It affects hemoglobin's affinity for oxygen, such that it is well suited both for the loading of O_2 in the lungs (PA_{O_2} = 100 mm Hg) and the unloading of O_2 in the tissue capillaries ($P\overline{cap}o_2$= 55 mm Hg). The shape of the curve maximizes the quantity of O_2 transported from the lungs and the quantity released at a reasonably high Po_2 in the systemic capillaries. The functional importance of the flat upper portion of the curve is that the arterial oxygen saturation (Sao_2) does not change much when PA_{O_2} falls as a result of ventilation/perfusion mismatching, global hypoventilation or low ambient Po_2 (high altitude). For example, at Po_2 = 70 mm Hg the So_2 = 94.1%, which is a trivial decrease from the ideal 97.4% at Po_2 = 100 mm Hg.

The Hbo_2 equilibrium curve may be modified by a number of physiological or pathological factors. The curve can be affected in two ways; a change in its position (called a right or left shift); or a change in its shape (resulting from a loss of heme group interaction). The latter interferes much more with O_2 transport than does the former.

The maximum amount of O_2 that can be bound to Hb (So_2 = 100%) per unit of blood is called its **oxygen capacity.** Because 1 g of functional Hb combines with 1.34 ml O_2, the O_2 capacity of normal blood is 150 g Hb/L × 1.34 ml O_2 g Hb = 200 ml O_2/L. Although the normal hemoglobin concentration in blood is 150 g/L, individuals with anemia have a lower concentration. In chronic anemia the circulating hemoglobin concentration may be <50 g/L. Even then the patient may only complain of a little tiredness or the inability to do much exercise.

Oxygen Transport in Blood

A **hematocrit** (fraction of blood that is red cells) less than 30% is usually associated with a disease process that affects erythrocyte turnover. Such processes include iron deficiency, decreased erythrocyte production, or shortened circulating red cell lifetime (normally, 120 days). The erythrocyte concentration (hematocrit) affects the viscosity of blood as it flows through small vessels (see Chapter 26). One extreme is the complete absence of red blood cells, a condition incompatible with human life. Another extreme is polycythemia (e.g., in chronic high altitude sickness), in which hematocrit levels approaching the theoretical maximum of 80% may occur. The blood is so viscous that the heart cannot pump enough of it through the capillaries. The power (work per unit time) requirement for the heart to transport the normal quantity of oxygen is at a minimum when the hematocrit equals about 40%. When the hematocrit is less than 30% or greater than 55%, the heart has to do much more work to pump adequate quantities of blood to meet the oxygen demands of the tissues.

Factors Affecting the HbO_2 Equilibrium Curve

The standard HbO_2 equilibrium curve shown in Fig. 36-1 applies exactly only under the following conditions: human hemoglobin type A, hydrogen ion concentration (H^+) = 40 nmol/L (pH = 7.40), Pco_2 = 40 mm Hg, temperature = 37° C and 2,3-diphosphoglycerate concentration (2,3-DPG) = 15 μmol/g Hb.

When the concentration of any of the last four factors (H^+, Pco_2, temperature, 2,3-DPG) increases, the affinity of hemoglobin for oxygen decreases. It is customary to describe the effect in terms of the P_{50} (Po_2 for 50% saturation). The P_{50} increases when any one of the factors increases, as shown by the long dashed curve in Fig. 36-2. The entire HbO_2 curve is shifted proportionally to the right of the standard curve. The change is in the position of the curve, not its shape. When Hb affinity for O_2 changes, the P_{50} changes in the opposite direction.

Conversely, when the value of any of the last four factors falls (see Fig. 36-2, short dashed line), the affinity of hemoglobin for oxygen increases and the P_{50} decreases. The entire HbO_2 curve is shifted to the left. Again, the change is in the position of the curve, not its shape.

The effect of H^+ is caused by a greater affinity of hydrogen ions for deoxygenated hemoglobin than for oxyhemoglobin. CO_2, which forms carbonic acid in plasma, has its principal effect by increasing the H^+ concentration.

The temperature effect is the usual one for chemical equilibria: increased temperature favors dissociation and decreased temperature favors association of O_2 with Hb.

The compound 2,3-diphosphoglycerate (2,3-DPG) is present in erythrocytes in high concentration relative to that in other cells, because mature red blood cells, having no mitochondria, respire by anaerobic metabolism (glycolysis), which produces 2,3-DPG as a side reaction. 2,3-DPG binds to reduced hemoglobin more strongly than to Hbo_2 and reduces its affinity for oxy-

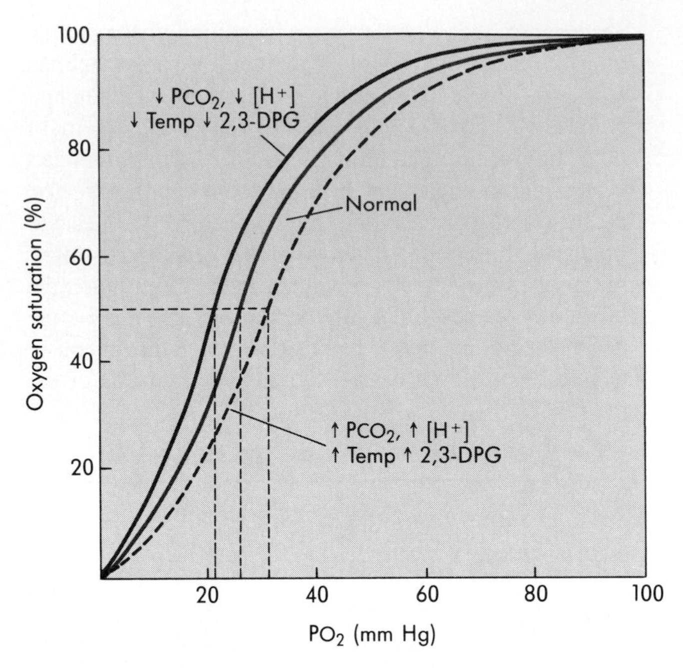

Fig. 36-2 Changes within the erythrocytes of the hydrogen ion concentration $[H^+]$, CO_2, temperature, and 2,3 diphophoglycerate concentration affect the affinity of O_2 for hemoglobin in the directions indicated by the long dashed line (decreased affinity; shift to the right) and the short dashed (increased affinity; shift to the left) line. A line at So_2 50% is also shown. Affinity changes are quantified by their effect on P_{50}. The curves are all of the same shape, only their positions have shifted.

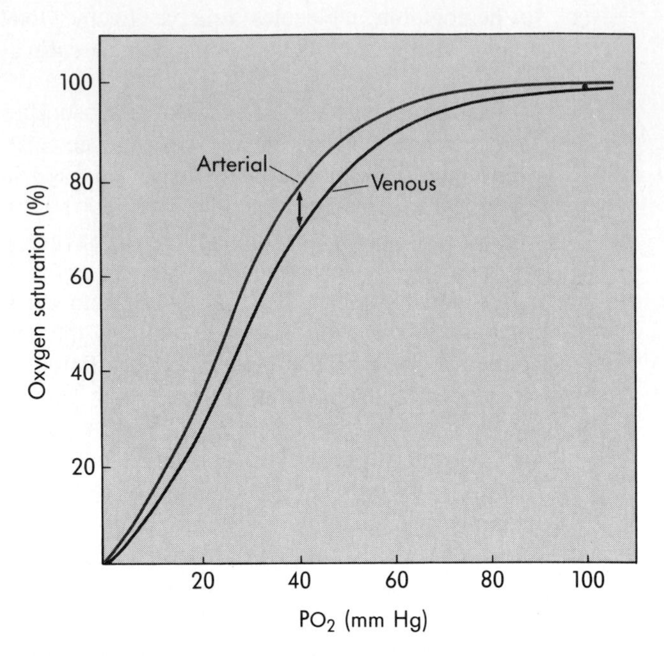

Fig. 36-3 Normal arterial and venous HbO_2 equilibrium curves. The effect in the lung of the shift to the left caused by decreased hydrogen ion concentration speeds up the oxygen uptake process. In the systemic capillaries significant O_2 unloading begins about Po_2 = 70 and continues until the red cells leave the capillaries at Po_2 = 36 mm Hg. The shift between the arterial and venous curves is more apparent at low Po_2, but is just as important in the lung. The P_{50} of the venous curve is about 29 mm Hg compared with the P_{50} of the arterial curve of 26 mm Hg.

gen. Increases in 2,3-DPG occur in chronic hypoxemia (decreased Pa_{O_2}) and when blood (H^+) decreases (increased pH), whereas decreases in 2,3-DPG occur in blood stored for transfusions.

In everyday life shifts in H^+, Pco_2, temperature, and 2,3-DPG are small. For example, venous blood has a higher Pco_2 (46 mm Hg) and H^+ concentration (42 nmol/L; pH = 7.38) than does arterial blood (Pco_2 = 40 mm Hg and H^+ = 40 nmol/L [pH = 7.40]). Therefore the Hbo_2 curve of mixed venous blood lies slightly to the right of the curve of arterial blood. The P_{50} increases from 26 to 29 mm Hg. Fig. 36-3 compares the Hbo_2 equilibrium curves of arterial and mixed venous blood. Some diseases and physiological conditions produce large shifts in the position of the curve.

Myoglobin and Carbon Monoxide Hemoglobin

Separately, each of the four globin chains with its heme group combines with one molecule of O_2 in a manner similar to that of myoglobin, which is the single chain heme pigment in skeletal muscle cells (molecular weight 16,500 Daltons). The myoglobin-oxygen equilibrium curve (MbO_2) lies to the left of the HbO_2 equilibrium curve; it has a different shape (hyperbola) because each molecule has only one heme group, and therefore there cannot be any molecular interaction among the heme groups. The two curves are compared in Fig. 36-4. Because of the very low Po_2 at which myoglobin binds O_2, it is unsuited for O_2 transport. However, it is useful for storing O_2 temporarily in skeletal muscle cells, where the Po_2 is normally low and in the range over which myoglobin is only partly saturated. The half saturation partial pressure (P_{50}) of oxymyoglobin (5 mm Hg) is a reasonable estimate of the Po_2 at the mitochondria of resting skeletal muscle cells.

Fig. 36-4 also shows the equilibrium curve for the pathological carbon monoxide hemoglobin (HbCO) has more than 250 times the *affinity* (attraction) for Hb than does oxygen because carbon monoxide combines readily with Hb, but it does not dissociate unless the Pco is very low.

How the Position of the HbO_2 Curve Affects O_2 Transport

A shift to the right (decreased affinity) reduces the quantity of oxygen that can be taken up at any $P_{A_{O_2}}$ as blood flows through the lung capillaries. But as the HbO_2 equilibrium curve is nearly flat above a Po_2 of 70 mm Hg, positional shifts in the curve have little effect

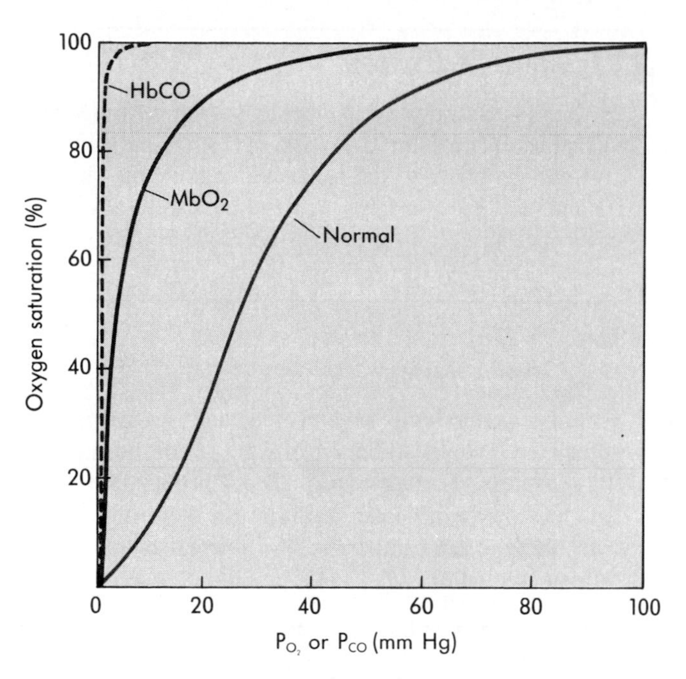

Fig. 36-4 Comparison among the saturation curves of hemoglobin for oxygen and the deadly gas carbon monoxide, and the normal skeletal muscle oxygen carrier protein myoglobin. The P_{50} of myoglobin is about 5 mm Hg, which is close to the normal intracellular Po_2 of muscle. Myoglobin serves as a temporary intracellular O_2 storage mechanism, useful when blood flow is interrupted briefly by phasic muscle contraction. Carbon monoxide hemoglobin has a P_{50} of 0.1 mm Hg.

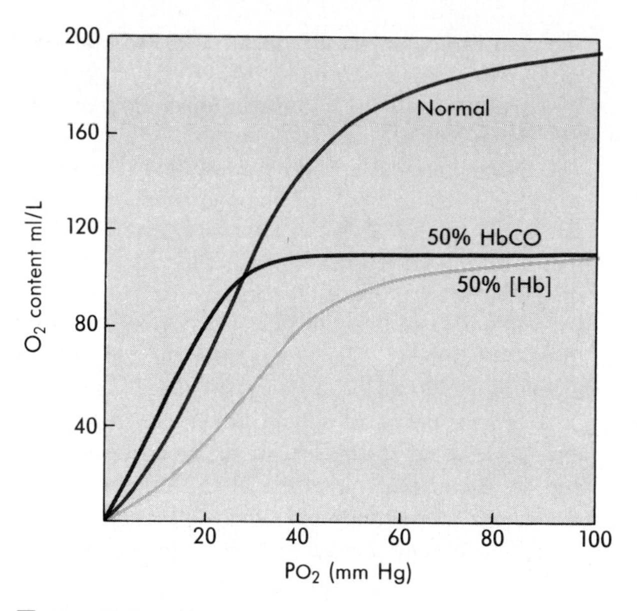

Fig. 36-5 Comparison of oxygen content curves under various conditions: Normal, 50% reduction in circulating hemoglobin (anemia); 50% HbCO. The anemia and HbCO curves both show the same decreased oxygen content in arterial blood. However, the HbCO curve has a much more profound effect in lowering the venous Po_2.

on the arterial oxygen concentration. A shift to the right of the HbO_2 equilibrium curve, however, enhances the quantity of oxygen that can be dissociated (given up) by blood as it flows through the systemic tissue capillaries. *This has the beneficial effect of increasing the delivery of oxygen at a given partial pressure.* The rising Pco_2 and H^+ in systemic capillary blood causes a slight shift to the right (P_{50} increases by 3 mm Hg). Note the small vertical arrows between the arterial and venous blood curves in Fig. 36-3.

Conversely, a shift to the left (increased affinity) increases the binding of oxygen by hemoglobin at any specified partial pressure. Such a slight shift (P_{50} decreases by 3 mm Hg) occurs in the lung capillaries. It hastens the uptake of oxygen, which is beneficial when blood flows rapidly, as in exercise. In the systemic capillaries, however, a shift toward the left decreases the amount of oxygen that can be unloaded at any given partial pressure. This shift could lead to tissue hypoxia (insufficient O_2 for aerobic metabolism) unless it is compensated (e.g., by an increased blood flow).

Oxygen Capacity of Blood

Breathing affects only the Pa_{O_2}. Other factors (dietary iron, vitamin B_{12}, the hormone erythropoietin) regulate blood hemoglobin concentration. Unfortunately the usual HbO_2 equilibrium curve (see Fig. 36-1) does not show changes in blood O_2 capacity, because the ordinate is saturation (a relative value). If one changes the Y-axis to reveal the absolute concentration of oxygen (ml O_2/L), then one can appreciate the effect of changes in O_2 capacity (Fig. 36-5).

Fig. 36-5 shows the effect of moderately severe anemia (Hb decreased by half). The uppermost line is the normal HbO_2 equilibrium curve (Hb = 150 g/L, HbCO = 0%) in terms of O_2 concentration at the normal oxygen capacity of blood = 200 ml O_2/L (100% So_2).

A 50% reduction in hemoglobin concentration (75 g Hb/L) is represented by the lowermost line. The curve is changed neither in position nor in shape; on a saturation-Po_2 graph it appears normal. On the O_2 content graph, however, it is moved *down* to reflect the reduction in the O_2 capacity to 100 ml O_2/L. The arterial Po_2 and P_{50} are not affected.

The mixed venous Po_2 cannot be accurately shown because people with severe anemias compensate for the low blood O_2 capacity by increasing cardiac output by various amounts. The heart accomplishes this increased output without undue work because the blood viscosity is low (reduced hematocrit).

How the Shape of the HbO_2 Curve Affects O_2 Transport

Changes in the shape of the HbO_2 curve almost always have deleterious effects on oxygen transport because *the shape changes usually increase hemoglobin affinity for O_2. This means that the unloading of oxygen in systemic capillaries must occur at a lower Po_2, and therefore tissue Po_2 decreases.*

Carbon monoxide (CO) poisoning is the most important clinical example of this phenomenon; it is not a rare event. Fig. 36-5 shows the curve for 50% HbCO in blood with a normal hemoglobin concentration (middle line). The arterial Po_2 is the same as for the anemic person, but the mixed venous Po_2 is moved far to the left at Po_2 = 13 mm Hg. The reasons for the very low $P\bar{v}_{O_2}$ are twofold. The HbCO decreases the O_2 capacity by 50% (moves the curve downward). More important, the curve is shifted to the left of the normal and anemia curves, because of the effect of CO binding tightly to two of the four heme groups in each hemoglobin molecule. This binding markedly disturbs the chemical equilibrium between the association and dissociation of oxygen (increased affinity for O_2) and destroys the heme group interaction. The affinity of CO for Hb is so great relative to that of O_2 (250-fold) that, even at very low concentrations of CO in alveolar gas, a substantial quantity of HbCO will be in the blood in the steady state.

There are a number of genetic variants of hemoglobin, nearly all of which adversely affect the HbO_2 equilibrium curve. Normal adult human hemoglobin is called type A. The most common variant is fetal hemoglobin (HbF), which is the major hemoglobin in the fetus. It is normally replaced soon after birth by hemoglobin A. HbF has slightly less affinity for O_2 (shift to the right). However, it is less affected by 2,3-DPG, so that functionally HbF has a greater affinity for O_2 (shift to the left), which is beneficial to the fetus whose arterial Po_2 is low (<40 mm Hg).

Most genetic variants of hemoglobin shift the curve to the left and change its shape. Some hemoglobin variants destroy the interaction among the globin chains and cause hemoglobin to revert toward its ancestral myoglobin behavior, thus rendering it useless for O_2 transport. Such Hb forms are lethal, if they constitute a substantial fraction of the circulating hemoglobin.

One special form called hemoglobin S crystallizes into long rods when Po_2 is low and H^+ is increased. The rods distort the shape of the erythrocytes into sickle cells. Although purified HbS binds O_2 as does HbA, erythrocytes containing it have more 2,3-DPG; consequently, the Hbo_2 curve is shifted to the right.

Finally, another factor that affects both the position, shape, and O_2 capacity is oxidation of the iron to its ferric state (Fe^{+++}) from the normal ferrous state (Fe^{++}). Fortunately, reducing enzymes present in erythrocytes keeps the Hb in its ferrous (functional) form. However, certain chemicals (nitrates and sulphates) can increase the oxidized hemoglobin content and cause methemoglobinemia. Oxidized hemoglobin does not bind or transport oxygen.

Oxygen Diffusion

Oxygen is transported from ambient air to the tissue mitochondria in two critical steps: (1) diffusion from the alveolar gas phase to the red blood cells in the capillaries and (2) diffusion from the systemic capillaries to the mitochondria in the tissue cells.

Oxygen Diffusion Across the Alveolocapillary Barrier

Fortunately, the lungs have evolved so that O_2 diffusion across the alveolocapillary barrier, a seriously rate-limiting process, is maximized. (See Chapter 33 for the structure of the alveolar wall and the flow of erythrocytes through the capillaries.) For respiratory gases, the diffusion equation is:

$$\dot{V}o_2 = D \times A \times (P_{A_{O_2}} - \overline{Pcapo_2}) \quad (1)$$

where $\dot{V}o_2$ is the volume flow of gas by diffusion (oxygen consumption); D is the diffusion coefficient (a number dependent on the gas's molecular size and solubility); A is the effective alveolar surface area across which the diffusion occurs; and L is the path length along which the diffusion occurs. The O_2 partial pressure difference is the driving force. The mean oxygen pressure in the pulmonary capillary blood ($\overline{Pcapo_2}$) is a complex function, not only of the rate of diffusion across the air-blood barrier but also of the process of diffusion and chemical reaction that occurs when oxygen penetrates the erythrocytes and reacts chemically with hemoglobin.

Under normal resting conditions and even under stressful conditions, such as heavy exercise or cardiac and pulmonary diseases, the time of red cell transit through the capillaries is nearly always adequate for the hemoglobin in the red cells to come into equilibrium with the alveolar gas; *that is, diffusion is not rate limiting*. The capillary blood volume in the lungs is about the same as the stroke volume of the right ventricle, so that the erythrocytes have approximately the duration of one cardiac cycle for gas to exchange.

Inert gases (e.g., nitrogen, anesthetic gases) equilibrate between alveolar gas and pulmonary capillary blood extremely rapidly (several milliseconds). The reason for the slow transfer rate of oxygen is the enormous unfilled O_2 capacity of mixed venous blood (50 ml/L). In other words, the "effective" solubility of oxygen in blood is markedly increased by hemoglobin. Although the chemical reaction, $Hb + O_2 \rightarrow HbO_2$, is rapid (milliseconds), equilibration is significantly delayed because of the low solubility of O_2 in the water within the alveolar wall tissue and plasma.

One example of a diffusion limitation for oxygen arises when someone exercises heavily (increased oxygen demand) while breathing low inspired oxygen (alveolar hypoxia). Under those conditions the driving pressure for oxygen diffusion ($P_{A_{O_2}} - P\overline{cap}o_2$) across the alveolar-capillary barrier is reduced at the same time that the oxygen demand ($\dot{V}o_2$) and the cardiac output have increased.

Some clinical physiologists believe that a limitation of O_2 diffusion can be demonstrated in certain lung diseases where the alveolar walls are thickened or scarred. Unfortunately, it is not easy to separate the diffusion component from the ventilation/perfusion mismatching or capillary obliteration that the disease has caused. A conservative view is not to invoke inadequate diffusion as a cause of decreased $P_{A_{O_2}}$ until other causes have been excluded.

Oxygen Diffusion to the Mitochondria

At the second critical point in the O_2 transport chain, in the peripheral tissues, the diffusion conditions are less favorable. Tissue diffusion really is the rate-limiting process. The diffusion equation is:

$$\dot{V}o_2 = \frac{D \times A \times (P\overline{cap}o_2 - P\overline{t}o_2)}{L} \qquad (2)$$

A is the systemic capillary surface area and L is the path length from the capillaries to the mitochondria. $P\overline{cap}o_2$ and $P\overline{t}o_2$ are the mean systemic capillary and the mean tissue Po_2's, respectively.

Arterial blood enters the systemic capillaries at a Po_2 of 100 mm Hg. However, according to Fig. 36-1, $Pcapo_2$ must fall substantially before any significant quantity of O_2 dissociates from hemoglobin. Thus, the mean capillary Po_2 is closer to 40 mm Hg than to 100 mm Hg (about 55 mm Hg). The force driving oxygen diffusion to the tissue cells depends on the partial pressure difference between the capillary blood and the most distant mitochondria. Fortunately, mitochondria are able to carry out oxidative metabolism at Pto_2 as low as 1 to 2 mm Hg.

Under normal resting conditions it is unusual to find $Pto_2 < 5$ mm Hg, which means that all mitochondria are receiving adequate oxygen. A reasonable estimate of mean resting Pto_2 is about 10 mm Hg. Pto_2 falls not only along the capillary but also in the radial direction away from the capillary.

The most important factor that affects tissue O_2 diffusion is the path length over which oxygen must diffuse to reach the cells (Fig. 36-6). In the myocardium, which is a critical tissue that requires a large oxygen supply per unit mass, the capillaries are about 25 μm apart, the width of one left ventricular muscle fiber. Thus, from each capillary oxygen must diffuse outward within a tissue cylinder of about 13 μm radius. That seems little enough but it is ten times farther than across the alveolar-capillary barrier in the lung. As the diffusion rate is affected by the path length squared (L^2), oxygen molecules require about 100 times longer to diffuse the 13 μm from the capillary to the farthest heart muscle mitochondria than from the alveolar gas to the

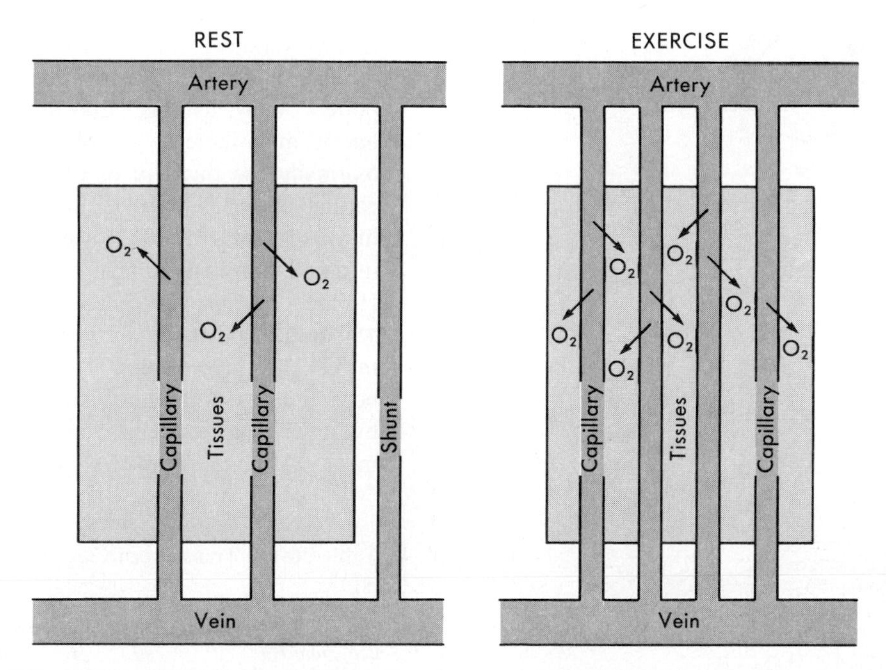

Fig. 36-6 Tissue oxygen transport is sensitive not only to capillary Po_2 but also to the distance over which oxygen must diffuse to reach all mitochondria. In skeletal muscle there are about three times as many capillaries as required for adequate oxygenation under baseline conditions. During exercise these capillaries are perfused, thereby markedly decreasing the oxygen diffusion distance.

red cells in the pulmonary capillaries. In addition mitochondria along the way are extracting oxygen, so that tissue Po_2 falls in a complex manner. In brain cortex the capillaries are some 40 μm apart, and in resting skeletal muscle they are about 80 μm apart.

The most effective way for the body to improve oxygen delivery to tissue cells is to decrease the diffusion path length by recruitment of more capillaries. This process also increases the surface area of the capillaries across which oxygen diffusion occurs. This process occurs in skeletal muscle, where functional capillary density increases threefold in heavy exercise (see Fig. 36-6).

Although the rate of oxygen dissociation from hemoglobin is slower than the rate of association, under most conditions the Po_2 of venous blood leaving an organ is very close to the Po_2 of the end-capillary blood.

Tissue Oxygen Extraction

The normal arterial-venous oxygen concentration difference is 50 ml O_2/L. More important, net O_2 transport occurs only in arterial blood—from the alveolar capillaries to the systemic capillaries (Fig. 36-7). The remaining HbO_2 (150 ml O_2/L) serves as a reserve of O_2 for use in emergencies (transient states of inadequate blood flow) and to maintain the Po_2 of systemic capillary blood at a level sufficiently high to drive adequate O_2 diffusion to the tissues.

Under maximum steady-state exercise conditions, the (a-$\bar{v}$) O_2 concentration difference increases to about 100 ml O_2/L of blood. When this is multiplied by the threefold increase in cardiac output, it accounts for a sixfold increment in oxygen consumption. It is uncommon for steady-state mixed venous oxygen saturation to fall below 50% ($P\bar{v}_{O_2}$ = 29 mm Hg) at $P\bar{v}_{CO_2}$ = 46 mm Hg, although the venous blood from individual organs may do so.

The normal myocardium extracts the maximum quantity of oxygen per volume of blood even at rest; that is, the coronary sinus (venous) oxygen saturation is about 50% (see also Chapter 31). Increases in oxygen demand by the heart are met almost entirely by an increase in blood flow.

Clearly a number of clinically relevant O_2 transport deficiencies may occur. These include hypoxic hypoxia (inadequate O_2 uptake into blood in the lung), stagnant hypoxia (inadequate blood flow either to the entire body or to a particular organ), anemic hypoxia (inadequate carrying capacity of blood oxygen as in CO poisoning), and histotoxic hypoxia (biochemical blockade of mitochondrial respiration (as in cyanide poisoning).

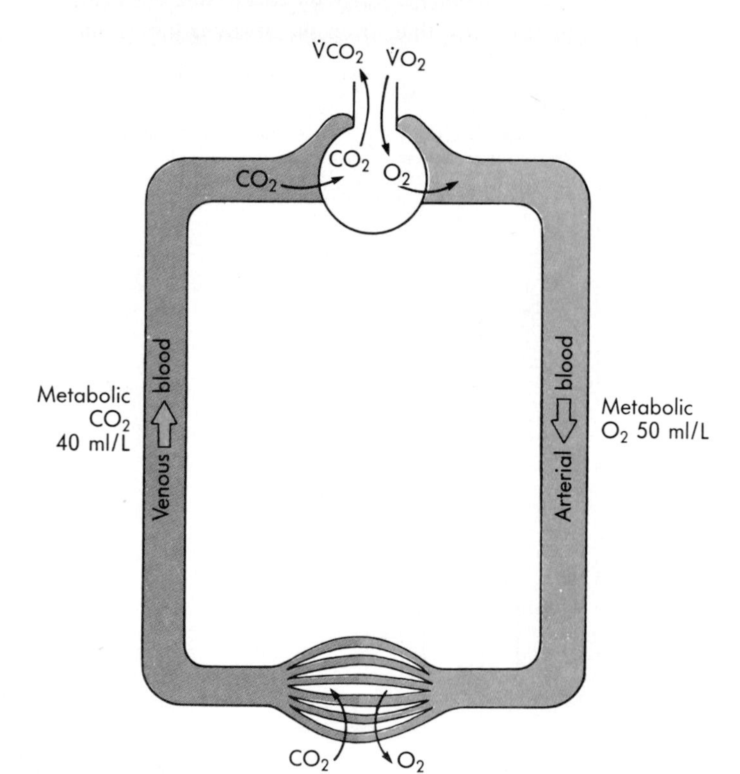

Fig. 36-7 Scheme of oxygen and carbon dioxide transport in blood. Metabolic oxygen transport occurs only in arterial blood; metabolic CO_2 transport occurs only in venous blood.

Carbon Dioxide Transport

CO_2 Concentration in Blood

One of the chief products of cellular metabolism is carbon dioxide (CO_2). As Fig. 36-7 illustrates, CO_2 is carried by venous blood to the lung, where it is eliminated in the expired gas. The total amount of CO_2 thus transported, $\dot{V}co_2$, averages 200 ml/min in a resting adult, but it may increase sixfold in steady-state exercise. Normally the quantity of CO_2 transported per liter of cardiac output is 40 ml (5 L/min × 40 ml/L = 200 ml/min) (see Table 36-1). However, there is far more CO_2 in blood than that. In fact, at a normal Pa_{CO_2} of 40 mm Hg the total CO_2 concentration in arterial blood is about 480 ml/L, and in mixed venous blood it is about 520 ml/L. The huge excess of CO_2 in blood and in other body liquids is required for the maintenance of the hydrogen ion concentration of the internal environment. Although the two processes (metabolic CO_2

Table 36-1 Transport of CO_2 per liter of normal human blood

Pco_2, mm Hg	*Arterial* 40	*Mixed venous* 46	*a-$\bar{v}$ difference* 6
Dissolved (ml/L)	25	29	4
Carbamino (ml/L)	24	38	14
HCO_3^- (ml/L)	433	455	22
TOTAL (ml/L)	482	522	40

transport and hydrogen ion regulation) occur simultaneously and interact in venous blood, it is sometimes useful to think of them as two independent physiological processes.

Mechanisms of CO_2 Transport by Blood

As CO_2 is formed in cells, it increases the tissue P_{CO_2} to above that of the arterial blood entering the capillaries. The normal mean resting P_{tCO_2} is 50 mm Hg. The CO_2, which is 20 times more soluble in water than oxygen (α_{CO_2}, 37° C = 0.7 ml/(L × mm Hg), diffuses from the cells along its partial pressure gradient into the capillary blood. As Table 36-1 and Fig. 36-8 show, CO_2 in blood is carried in three forms: **dissolved CO_2** (a tiny fraction is converted to carbonic acid (H_2CO_3)), **bicarbonate ion** (HCO_3^-), and a form bound to hemoglobin and plasma proteins **(carbamino-CO_2).** Although the mechanism is somewhat complex, it is important to understand.

Dissolved CO_2 is hydrated to form carbonic acid:

$$CO_2 + H_2O \leftrightarrows H_2CO_3 \quad (3)$$

By physiological standards this reaction is rather slow (several seconds) in tissue liquid or plasma. However, it is markedly accelerated (milliseconds) inside the red blood cell because of the enzyme **carbonic anhydrase.** This enzyme is present in erythrocytes but is virtually absent in plasma and interstitial liquid (Fig. 36-8). The equilibrium shown in Equation 3 is far to the left; that is, in the direction of dissolved CO_2. Thus, there is very little carbonic acid.

As carbonic acid is formed, however, it rapidly dissociates into ions,

$$H_2CO_3 \rightarrow H^+ + HCO_3^- \quad (4)$$

This reaction is spontaneous and very fast (microseconds). It would not proceed far, however, unless the H^+ ions produced were removed from the reaction site. The H^+ ions are removed from solution by combining chemically with the enormous amount of hemoglobin and, to a lesser extent, with the plasma proteins.

The newly formed bicarbonate ions diffuse out of the red blood cells into plasma in exchange for Cl^-, because the erythrocyte cell membrane is permeable to both molecules. Thus, the total CO_2 reaction may be written as:

$$CO_2 + H_2O \rightarrow H_2CO_3 \rightarrow H^+ + HCO_3^- + Hb \rightarrow HHb \quad (5)$$

The reaction proceeds rapidly to the right in the direction of HCO_3^- formation, when CO_2 is added to blood.

Fig. 36-9 shows the CO_2 equilibrium curve of blood. Over the range of P_{CO_2} that is generally encountered in humans, the relationship is nearly linear; it is very different from the HbO_2 equilibrium curve. Two lines are shown, because the CO_2 equilibrium curve is significantly affected by the oxygen saturation of hemoglobin; deoxygenated hemoglobin is a weaker acid (less dissociated into H^+ and Hb^- ions) than is oxygenated hemoglobin. As hemoglobin becomes deoxygenated in the systemic capillaries, its ability to bind with CO_2 (to form carbamino-CO_2) is increased.

The fact that the CO_2 equilibrium curve is nearly lin-

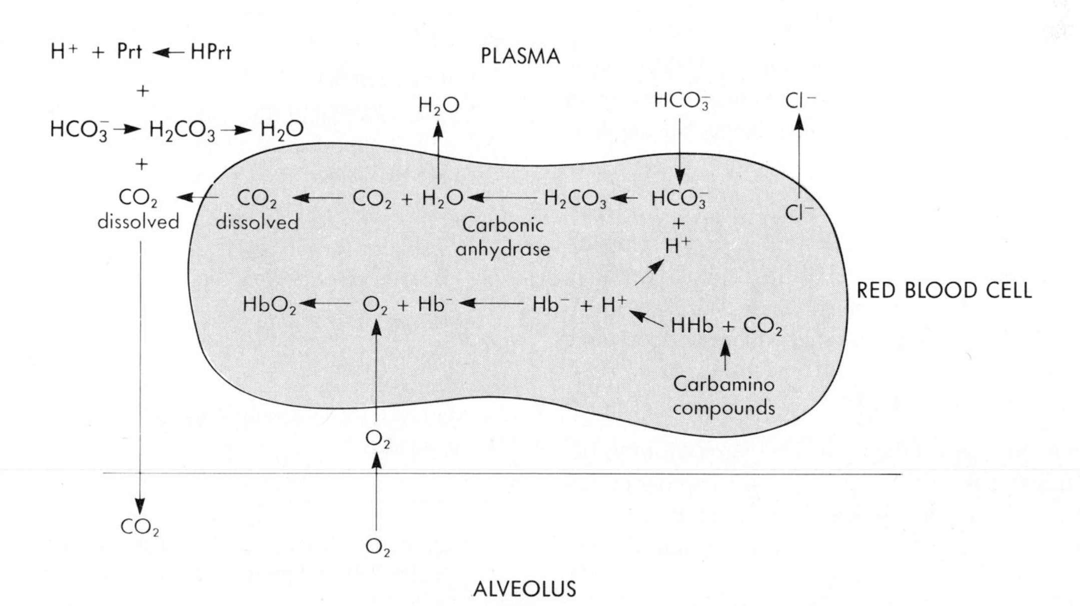

Fig. 36-8 CO_2 exchange within blood. The chemical reactions among O_2, CO_2, Hb, Cl^- and H_2O are shown for O_2 uptake and CO_2 excretion in the pulmonary capillaries. In the systemic tissues the same processes occur in reverse.

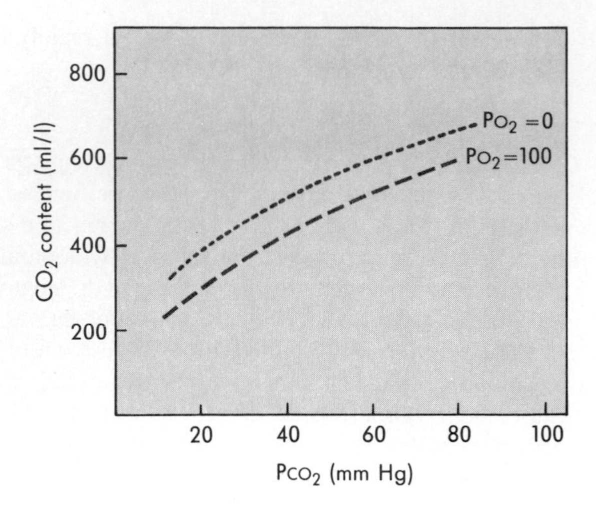

■ **Fig. 36-9** The blood CO_2 equilibrium curves (arterial and venous). Compared with those of oxygen, the CO_2 curves are essentially straight lines over the usual range of blood Pco_2.

ear, whereas the HbO_2 equilibrium is markedly curvilinear over the useful ranges of Pco_2 and Po_2, respectively, partially explains why the alveolar to arterial partial pressure difference for oxygen is much greater than that for carbon dioxide, when ventilation and perfusion are mismatched (Chapter 35).

■ *Summary*

1. Hemoglobin quickly and reversibly binds with oxygen. The oxygen carrying capacity of normal human blood (150 g Hb/L) is 200 ml O_2/L.

2. The position of the hemoglobin-oxygen (HbO_2) equilibrium curve is well suited for the loading of oxygen in the lungs and for unloading it in systemic tissue capillaries.

3. Normal physiological factors that affect the HbO_2 curve are hydrogen ion concentration, Pco_2, temperature, and the concentration of 2,3-diphosphoglycerate in erythrocytes. Increases in any of these shifts the position of the HbO_2 curve to the right, thereby decreasing hemoglobin affinity for oxygen (increased partial pressure for half saturation [P_{50}]).

4. A 3 mm Hg shift to the right (decreased affinity) occurs as blood passes through systemic capillaries because Pco_2 and hydrogen ion concentration increase. This shift increases O_2 unloading in the systemic capillaries; a reverse shift occurs in the lung favoring O_2 uptake.

5. The shape of the HbO_2 equilibrium curve is such that in the systemic capillaries hemoglobin releases large quantities of oxygen as Po_2 falls below 70 mm Hg.

6. The shape of the curve is affected by carbon monoxide, which has 250 times greater affinity for hemoglobin than does oxygen. Even in very low concentrations, carbon monoxide combines avidly with hemoglobin, interferes with oxygen binding, and destroys heme group interactions.

7. The transport of oxygen between body compartments is governed by diffusion at a rate dependent on the Po_2 gradient. The principal diffusion limitation for oxygen is from the systemic capillaries through several microns of interstitial liquid to the mitochondria of the respiring cells.

8. Carbon dioxide is produced by aerobic metabolism in mitochondria, diffuses into systemic capillary blood, and is transported to the lungs, where it diffuses into alveolar gas and is exhaled.

9. CO_2 combines chemically with water to produce carbonic acid slowly. In erythrocytes the enzyme carbonic anhydrase catalyzes this reaction. Once formed, carbonic acid dissociates instantly into hydrogen ions and bicarbonate ions. The hydrogen ions are removed by combining chemically with the enormous amount of hemoglobin within the red cells. The biocarbonate ions diffuse out of the red cells in exchange for chloride ions.

■ *Bibliography*

Journal articles

Crowell JW, Smith EE: Determinants of the optimal hematocrit, *J Appl Physiol* 22:501-504, 1967.

Staub NC: Alveolar arterial oxygen tension due to diffusion, *J Appl Physiol* 18:673-680, 1963.

Wasserman K, Whipp BJ: Exercise physiology in health and disease, *Am Rev Resp Dis* 112:219-249, 1975.

Books & monographs

Comroe JH Jr: *Physiology of respiration,* ed 2, 1974, Chicago, Year Book.

Murray JF: *The normal lung,* ed 2, Philadelphia, 1986, WB Saunders.

Nunn JF: *Applied respiratory physiology,* ed 3, London, 1987, Butterworths.

Schmidt RF, Thews G: *Human physiology* (English edition), Berlin, 1983, Springer.

CHAPTER

37

Control of Breathing

Both the rate and depth of breathing are regulated so that Pa_{CO_2} is maintained close to 40 mm Hg. Remember that in Chapter 33 (*see* wasted ventilation) we assumed that arterial (a) and alveolar (A) carbon dioxide tensions were equal ($Pa_{CO_2} = PA_{CO_2}$), so that they will be used interchangeably. However, it is PCO_2 that is controlled — just as the alveolar ventilation equation (Equation 33-6) implied. When PA_{CO_2} is regulated, PA_{O_2} is automatically set (alveolar gas equation; equation 33-7) to an appropriate value, depending, of course, on the ambient partial pressure of oxygen.

The Pa_{CO_2}-sensitive mechanism is the main controller. It operates on a breath-by-breath basis as we go about our daily business, scarcely ever thinking about our breathing patterns. But the Pa_{CO_2} controller can be overridden in systemic arterial hypoxemia (e.g., acclimatization to living at high altitude) by a Pa_{O_2}-sensitive controller. If arterial oxygen tension were to decrease below 60 mm Hg, the supply of oxygen to the systemic tissue mitochondria might be impaired because the oxygen diffusion gradient between the capillaries and the tissue cells is reduced (Equation 36-2).

■ *Central Organization of Breathing*

■ *Breathing is Mainly Controlled at the Brainstem Level*

Two separate but overlapping patterns are involved in breathing: the metabolic (automatic) control pattern and the behavioral (voluntary) control pattern. **Metabolic breathing** is concerned with oxygen delivery to the mitochondria and with acid-base balance because of the effect of breathing on PA_{CO_2}, one of the independent hydrogen ion control variables. Experience teaches that metabolic breathing can be overridden briefly. However, within a minute or so the metabolic control system reasserts its authority.

The metabolic controller (respiratory neuronal groups in Fig. 37-1) resides within the brainstem. It was once believed that the neurons responsible for inspiration and expiration were located in specific brainstem centers, but it is now known that the organization is much looser.

Surrounding and interdigitating throughout the brainstem is a loose network of interneurons known as the **reticular activating system.** This network influences the brainstem controller by affecting the state of alertness (wakefulness) of the brain.

Less is known about the **behavioral control system,** except that higher brain center controllers in the thalamus and cerebral cortex are concerned. These are necessary to coordinate breathing in relation to the many complex but volitional motor activities that use the lungs and chest wall (e.g., talking, singing, suckling, swallowing, coughing, sneezing, defecation, partuition, anxiety, and fear). Indeed, this is one area in which

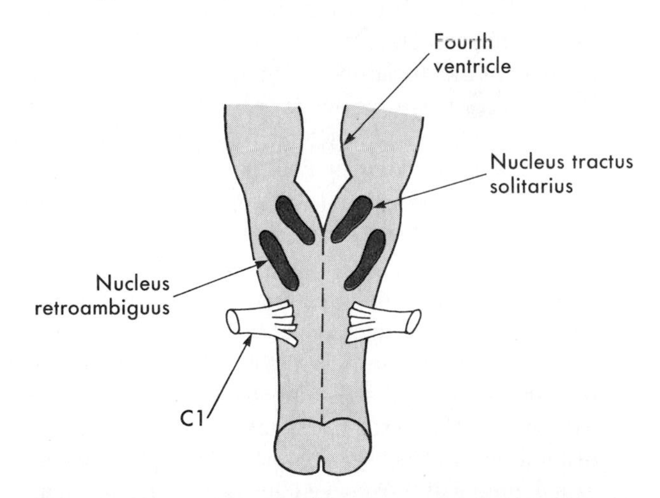

■ **Fig. 37-1** The metabolic controller of breathing is located in the medulla (most primitive portion of the brain). The neurons are located mainly in two ill-defined areas called the *nucleus tractus solitarius* and the *nucleus retoambiguus.* These colorful, descriptive names were given by 19th century anatomists before function was established. These nuclei subserve other basic cardiovascular functions in addition to breathing control. The reticular activating system is not shown but surrounds the nuclei, connecting with them and with higher brain centers to control the state of alertness.

"natural experiments" (patients with neurological disorders) often give important insights. A number of pathways carry the cortical and thalamic descending axons to the primary brainstem controllers.

Various ablation, stimulation, and recording techniques have been used to explore the brain and to locate breathing control areas. At least two regions in the brainstem function as intrinsic breathing controllers: the medullary respiratory area and the pneumotaxic center in the pons (anterior brainstem). The brainstem provides nearly complete basic regulation because the pattern of breathing is essentially normal even when the medulla is separated from the rest of the brain.

The precise anatomical and functional organization of the respiratory medullary neurons is still under investigation. Substantial evidence indicates that two separate neuronal networks in the medulla are crucial (see Fig. 37-1). These networks include neuronal groups in the nucleus tractus solitarius, located dorsally near the exit of the ninth cranial nerve, and a ventrally located group of neurons (nucleus retroambiguus) that extends rostrally from the first cervical spinal segment to the caudal border of the pons.

The dorsal group of neurons discharges mainly in inspiration. The ventral group contains both inspiratory and expiratory neurons. Neurons from both the ventral and dorsal groups direct the respiratory activity of the muscles of the chest wall and abdomen.

Efferent fibers from the dorsal and ventral group cross over and travel contralaterally in the ventrolateral portion of the spinal cord to reach the spinal motor neurons. The fiber tracts for voluntary control are separate. Afferent nerve fibers from the peripheral chemoreceptors, baroreceptors, and pulmonary mechanoreceptors synapse in the dorsal motor group.

The mechanism by which these networks of medullary neurons cause switching between inspiration and expiration is not completely clear. Most of the evidence supports the concept that rhythmic breathing depends on a continuous (tonic) inspiratory drive from the dorsal motor group with intermittent (phasic) excitatory inputs. This means that breathing results from the reciprocal inhibition of interconnected neuronal networks.

A network of neurons in the pons **(pneumotaxic center)** influences the switching between inspiration and expiration. When the pneumotaxic center is inactivated, inspiration becomes greatly prolonged. The pattern is called **apneusis** (prolonged inspiration lasting tens of seconds).

Thoracic **mechanoreceptors**, chiefly stretch receptors in the walls of the airways, are critical in setting the breathing frequency. The stretch receptors are stimulated by lung inflation. The afferent impulses travel up the vagus nerves and affect the duration of both inspiration and expiration. Neurons whose firing is excited by lung inflation have been found in the ventral motor group. Stretch receptor input therefore mainly influences inspiration. Interruption of the stretch receptor traffic by cutting or cooling the vagus nerves markedly prolongs the duration of inspiration (Fig. 37-2).

By artificially inflating the lung at different times in the breathing cycle, the effect of lung volume changes on the timing of inspiration and expiration can be demonstrated. For example, inflation of the lung during phrenic nerve activity can terminate inspiration. The effect of lung inflation on the respiratory pattern depends on the volume of gas introduced; more volume is required early in inspiration, less volume is required late in inspiration. After vagotomy, lung inflation has no effect on the duration of inspiration.

One complication to this explanation, however, is that the input from the pulmonary stretch receptors has less influence when the subject is conscious, so that additional factors are involved in the control of breathing frequency.

The regulation of expiratory time is less well understood, although it is closely correlated with changes in inspiratory time. These reflex effects on respiratory timing have important benefits. During spontaneous breathing, termination of inspiration prevents overinflation. Shortening inspiratory time also facilitates the higher breathing frequency needed for adequate increases in ventilation when breathing is stimulated (breathing CO_2 enriched gases, exercise). When the resistance to inspiratory airflow is increased, the decreased rate of lung inflation prolongs inspiration. Hence more time is available

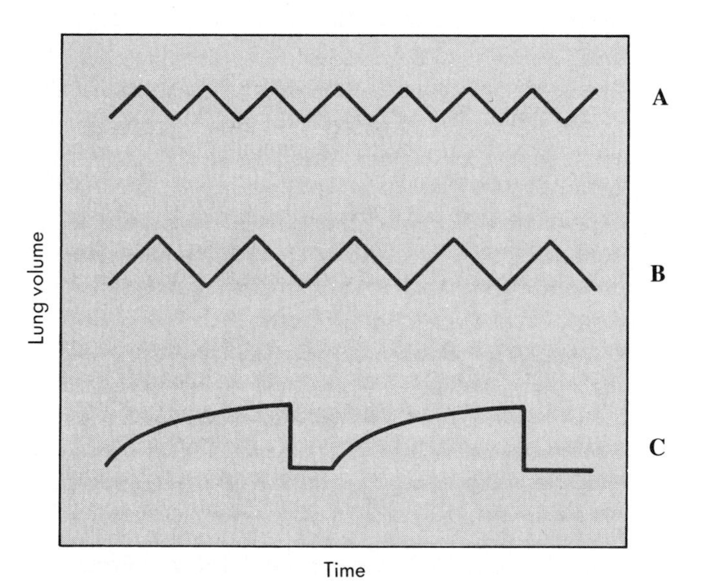

■ **Fig. 37-2** Some basic patterns of breathing. **A,** Normal breathing at about 15/min in man; **B,** the effect of removing sensory input from various lung receptors (mainly stretch) is to lengthen each breathing cycle and to increase tidal volume so that alveolar ventilation is little affected; **C,** when input from the cerebral cortex and thalamus are eliminated (nonspecific environmental stimuli) and added to vagal blockade, the result is prolonged inspiratory activity broken after several seconds by brief expirations (apneusis).

for gas to enter the lung and a constant tidal volume can be maintained. On the other hand, increased lung volume or obstruction to airflow during expiration increases the time available for lung emptying by forestalling the next inspiratory effort.

These fundamental observations can be synthesized by an operational model of the brainstem mechanisms that generate the respiratory rhythm (Fig. 37-3). Signals from the central and peripheral chemoreceptors impinge on a pool of inspiratory neurons (pool A; dorsal motor group). These neurons project to the spinal motor neurons involved with breathing and increase their activity (induce inspiratory muscle contraction). Indeed, the time course of the increase in activity resembles that of the changes in tidal volume. Increased input from various chemoreceptors (see later in this chapter) also increases the neural activity.

The central inspiratory activity stimulates another pool of neurons (B pool; see ventral motor group in Fig. 37-3), probably also in the nucleus tractus solitarius. In addition, pool B receives signals from pulmonary stretch receptors. As the lung expands, the stretch receptor signals increase and are added to the pool input.

Pool B, in turn, activates pool C, called the "inspiratory cutoff switch" because its output inhibits the main inspiratory neurons in pool A. When a critical level of excitation occurs in pool C, the activity of pool A is extinguished and expiration occurs. An increase in chemoreceptor activity raises the "cutoff switch" threshold and enhances central inspiratory activity. Ablation of the pneumotaxic center in the pons also raises the "cutoff switch" threshold.

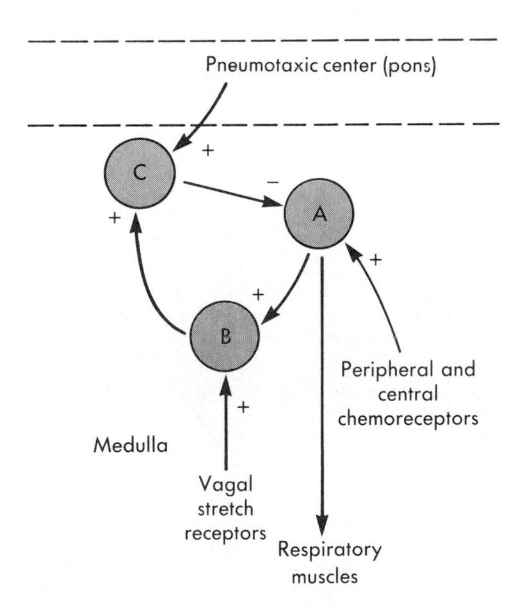

■ **Fig. 37-3** The basic wiring diagram of the brainstem ventilatory controller. The signs on the main outputs (arrows) of the neuron pools indicate whether the outputs are excitatory (+) or inhibitory (−). Pool *A* is tonically active providing inspiratory stimuli to the muscles of breathing (apneusis). Other centers feed into pool *C* (inspiratory cut-off switch), which sends inhibitory impulses to pool *A*. Pool *B* is stimulated by pool *A* and provides additional stimulation to the muscles of breathing. But pool *B* also stimulates pool *C*. Afferent information (feedback) from various sensors acts at different locations. A pneumotaxic center exists in the anterior pons, probably receiving input from the cerebral cortex. It also feeds into the pool *C* group.

This scheme of breathing pattern generation (i.e., tonic inspiratory activity inhibited only when sufficient sensory signals are received) is attractive because no basic cyclic neuronal activity has been found. The scheme also integrates the effects of CO_2 and O_2 (chemoreceptor activity) and certain reflex stimuli into various normal and abnormal breathing patterns. If this scheme seems unnecessarily complex and vague, one must remember that evolution does not work by devising the most efficient or simplest system. Often the first system that works is used. Furthermore, man's understanding of brain function is still primitive, so that the control mechanism employed may actually be the best one.

Higher brain centers modulate the function of the respiratory controller. As mentioned previously, projections from the cerebral motor cortex that subserve behavioral (volitional) control descend to the respiratory neurons in the brainstem via the corticobulbar tracts and to the spinal motor neurons via the corticospinal tracts, which are located in the dorsolateral columns of the cord. Strange as it seems, discrete lesions in the spinal cord may eliminate metabolic (rhythmic) activity in the muscles of breathing but leave voluntary breathing intact. This works only as long as the person does not fall asleep.

■ *Spinal Integration*

Descending neural impulses from the brain reach various segments of the spinal cord, where they are integrated with intrasegmental and intersegmental activity to modulate the membrane potential of the motor cells. This modulation leads to rhythmic increases and decreases of their excitability (see Fig. 37-3).

Excitation and inhibition of respiratory muscles involve segmental interneuronal networks and descending influences. For example, inhibition of antagonist muscles takes place through interneurons that connect the motor neurons of inspiratory and expiratory muscles. Stretch of intercostal muscles or electrical stimulation of dorsal roots in thoracic segments T9 to T12 excites intercostal and phrenic motor neurons to enlarge the thoracic cavity. In contrast, stimuli to segments T1 to T8 inhibit phrenic motor neuron activity and terminate inspiration.

Spinal reflexes are important because they augment muscle force when respiratory resistance is increased or compliance is decreased. Intercostal muscle to spinal motor neuron reflexes may be especially beneficial to the newborn whose cartilagenous rib cage is very compliant. The baby's rib cage needs to be stabilized during inspiration so that the subatmospheric pleural pressure does not suck the rib cage inward. Intercostal muscle stretch receptors sense the inward movement of the rib cage as pleural pressure decreases during inspiration. These receptors reflexly stimulate the intercostal muscle motor neurons, which cause contraction to oppose the distortion.

Chemoreceptor Control of Breathing

Carbon Dioxide

In conscious humans the central chemoreceptors, located at or near the ventrolateral surface of the medulla (between the origins of the seventh and tenth cranial nerves (Fig. 37-4)), account for 70% to 80% of the CO_2 induced increases in ventilation. These cells respond to changes in the hydrogen ion (H^+) concentration of the surrounding brainstem interstitial fluid. The peripheral chemoreceptors (which lie outside the central nervous system chiefly in the carotid bodies) account for the remainder. Both central and peripheral chemoreceptors respond in proportion to the level of Pa_{CO_2}.

Elevations of Pa_{CO_2} up to 100 mm Hg cause a linear increase in ventilation (see Fig 37-5). In normal awake man, hyperventilation, which may drastically lower Pa_{CO_2}, rarely causes apnea (no breathing). On the other hand, in anesthetized animals and man, artificial hyperventilation may produce apnea. This difference is believed to be caused by nonspecific environmental stimuli (e.g., noise, light, or touch) that maintain the reticular activating system in a condition of alertness.

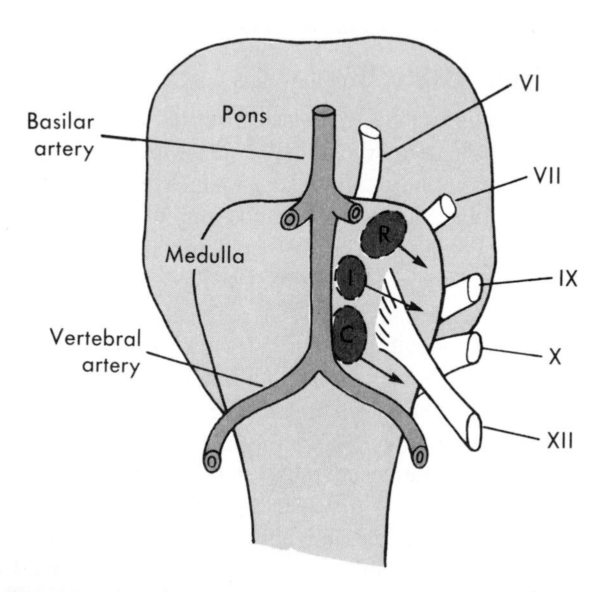

Fig. 37-4 The locations of the central CO_2 ($[H^+]$)-sensitive areas on the ventrolateral medulla. The receptor cells are not actually at the surface but are close to it. *R, I,* and *C* stand for rostral, intermediate, and caudal receptor areas.

Sensitivity to CO_2 is determined from the slope of the line that relates ventilation to Pa_{CO_2} (Figs. 37-5, 37-6). Although inspired CO_2 quickly equilibrates with the brain, considerable time is required for ventilation to reach its final steady state. Ventilation and the cerebral venous P_{CO_2} reach steady-state values at about the same time. Arterial P_{CO_2} reaches its steady-state level long before the brain venous P_{CO_2} does.

In normal individuals the average ventilatory response to inspired CO_2 is about 2.5 L/(mm Hg × min). The CO_2 response varies considerably among individuals because of differences in body size, age, sex, genetic makeup, or personality.

The effects of chronic metabolic acidosis and alkalosis on the ventilatory response to CO_2 are shown in Fig. 37-6. At any level of P_{CO_2} ventilation is greater in acidosis and less in alkalosis.

Acid injected directly into the blood immediately stimulates the peripheral chemoreceptors. Hence, ventilation increases and the Pa_{CO_2} falls. The diffusion of CO_2 out of the brain interstitial fluid is much faster than any compensatory mechanisms. Thus the brain interstitial fluid initially becomes alkaline (H^+ falls), and central chemoreceptor activity diminishes. Over many hours as the brain HCO_3^- slowly decreases, the H^+ rises and consequently ventilation steadily rises towards a new steady state.

The carotid bodies (peripheral chemoreceptors) con-

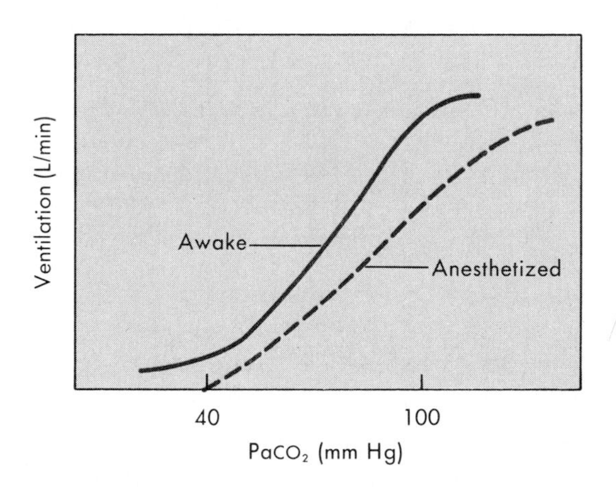

Fig. 37-5 The CO_2-ventilation dose-response curve. Ventilation is sensitive to Pa_{CO_2} (PA_{CO_2}) as shown by the solid line (awake). The curve is defined by the slope of the straight portion and the extrapolated x-axis intercept. The normal operating point is at 40 mm Hg. Note that in the awake (alert) state ventilation is relatively insensitive to decreases in CO_2 below 40 mm Hg threshold. When the reticular activating system is turned off as during anesthesia, the CO_2 response is shifted to the right and the slope is decreased; (i.e., the central controller's threshhold is increased and its sensitivity is decreased). This is a rapid response and also occurs in sleep.

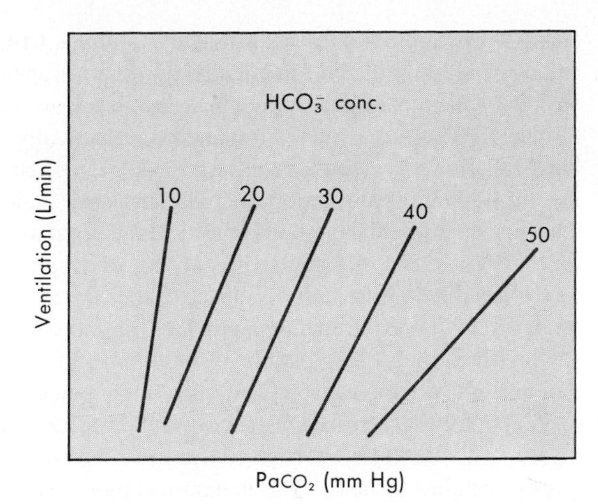

■ **Fig. 37-6** The ventilatory response to P_{CO_2} is affected by the bicarbonate concentration ($[HCO_3^-]$) of the brainstem interstitial fluid. As $[HCO_3^-]$ decreases (acidosis) below its normal level of about 25 mEq/L, the ventilatory response becomes more sensitive. When $[HCO_3^-]$ rises, (alkalosis) the CO_2 response decreases. These effects are due to the buffering effect of bicarbonate on hydrogen ion concentration.

tribute about 25% to the total ventilatory response to CO_2. Because the peripheral chemoreceptors react rapidly to changes in inspired CO_2, their response to hypercapnia can be evaluated by measuring the immediate change in ventilation that occurs in the first few breaths after an abrupt change in inspired CO_2 concentration.

■ *Brain Blood Flow Influences Breathing*

Brain interstitial fluid P_{CO_2} also depends on cerebral blood flow, because the brain cells produce CO_2 at a nearly constant rate. As cerebral blood flow rises, the interstitial fluid P_{CO_2} falls. Because hypercapnia increases cerebral blood flow, the effects of CO_2 on the cerebral vessels influences the relationship between ventilation and Pa_{CO_2}.

■ *Oxygen*

Reduced arterial P_{O_2} increases ventilation primarily by exciting sensors in the carotid body (innervated by the ninth cranial nerve). The relationship between Pa_{O_2} and ventilation is hyperbolic, as shown in Fig. 37-7, *A*. However, if the carotid and aortic bodies are removed, hypoxia depresses breathing, because a reduction in brain P_{O_2} depresses the brainstem neuronal activity. In addition, hypoxia increases brain blood flow, which tends to lower brain P_{CO_2}. In fact the interaction between O_2 and CO_2 is very important in assessing the chemical control of breathing.

Hypoxia accentuates the effects of hypercapnia and

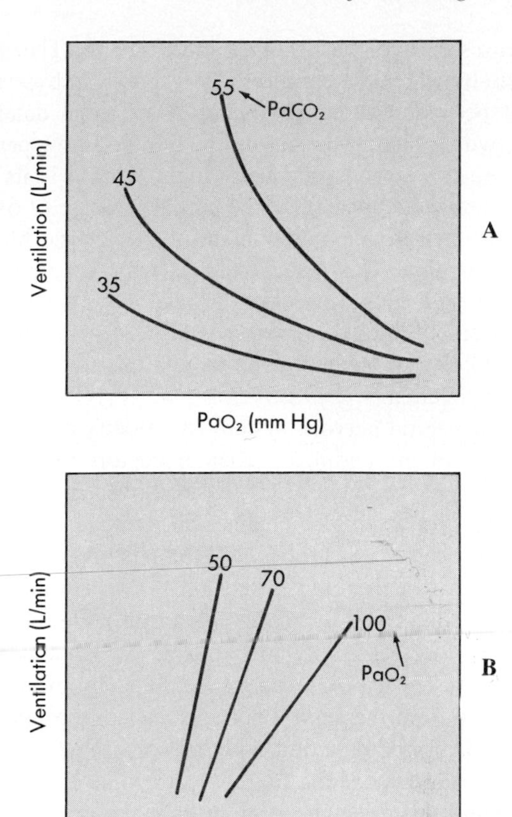

■ **Fig. 37-7** The effects of hypoxia *(A)* and hypercapnia *(B)* on ventilation as the other respiratory gas partial pressure is varied. **A,** At given P_{CO_2}, ventilation increases more and more as P_{O_2} falls. In the normal P_{CO_2} range, between 35 and 45 mm Hg, there is little stimulation of breathing until P_{O_2} falls below 60 mm Hg. The response is mediated through the carotid body chemoreceptors. **B,** The ventilatory response to carbon dioxide is enhanced by hypoxia but the effects are less than in **A.**

acidosis on peripheral chemoreceptor activity (Fig. 37-7, *B*), which is especially important in mediating the immediate increase in ventilation that occurs when blood (H^+) is increased quickly. Similarly, increases in Pa_{CO_2} enhance the ventilatory response to hypoxia (see Fig. 37-7, *A*). If constant P_{CO_2} is not maintained when gases low in oxygen are inhaled, the response to hypoxia is attenuated because the increasing ventilation leads to a decrease in the PA_{CO_2}, as shown by the alveolar ventilation equation.

Arterial P_{O_2} and P_{CO_2} rise and fall by 2 to 3 mm Hg during each breathing cycle as the alveolar volume and the pulmonary capillary blood flow change, but only the changing CO_2 affects the carotid body. The absolute level of CO_2 partial pressure and its rate of change are both involved in the stimulation of these chemoreceptors. Thus, carotid body discharges are enhanced by increasing tidal volume, even when the mean level of P_{CO_2} is unchanged.

Neural activity in afferent chemoreceptor fibers increases as hypoxia becomes more severe. The mechanism by which the carotid body responds to hypoxia has not been elucidated completely, but some details are known. Although the carotid body blood flow per gram of organ weight is unusually high, so also is its metabolic rate. It appears that the partial pressure of oxygen, rather than the oxygen content of the arterial blood, is the stimulus. This detail explains why in carbon monoxide poisoning (in which arterial O_2 content is decreased but Pa_{O_2} is normal; see Fig. 36-4) there is little stimulation to breathe, even though one might be dying of hypoxemia.

The central nervous system can modify both the sensitivity of the peripheral chemoreceptors to alterations in Po_2 and the range of O_2 tensions over which they respond. For example, changes in sympathetic nervous activity can alter carotid body blood flow, which then would affect the oxygen tension sensed by the carotid body cells. Thus, decreases in carotid body blood flow increase its sensitivity to hypoxia. Also, carotid body responses to hypoxia can be inhibited by efferent discharges from the central nervous system through axons that travel in the carotid body nerve.

Chronic hypoxemia (e.g., as in people living at high altitude) depresses the ventilatory response to hypoxia, presumably because of adaptation of the chemoreceptors or of the neurons in the central nervous system. This depressant effect is especially pronounced if the chronic hypoxemia begins at birth.

Mechanical Control of Breathing

Three Kinds of Receptors in the Lungs

Sensory receptors in the lungs and airways, as in other hollow viscera, are stimulated by irritation of the lining layers and changes in distending forces. The afferent fibers of these receptors travel to the brain in the vagus nerves.

There are three types of pulmonary receptors: stretch receptors, which are located within the smooth muscle layer of the extrapulmonary airways; irritant receptors, which ramify among airway epithelial cells and have a distribution similar to the stretch receptors; and unmyelinated C fibers, which are situated in the lung interstitium and alveolar walls.

The stretch receptors are excited by an increase in bronchial transmural pressure, and they adapt slowly to a sustained stimulus. As the lung is inflated, they inhibit inspiration and promote expiration. They are responsible for the **Hering-Breuer reflex,** which produces apnea in response to large lung inflations and augments expiratory muscle contraction. This reflex is weak in adult humans, but it may be prominent in newborns.

Irritant receptor and C fibers receptor neurons rapidly adapt when subjected to a sustained stimulus. Irritant receptors are stimulated chemically by noxious agents, such as sulfur dioxide, ammonia, or inhaled antigens (pollens). They are also stimulated mechanically by lung inflation, by increases in airflow, by particulate matter impinging on the bronchial surfaces, and by changes in bronchial smooth muscle tone. Irritant receptor stimulation augments the activity of inspiratory motor neurons. Stimulation constricts the airways and interacts with the stretch receptors to promote rapid, shallow breathing. This pattern of breathing, in combination with airway constriction, may limit penetration of potentially harmful substances into the lung and thereby prevent these substances from reacting with the gas exchanging surfaces. The irritant receptors may also help to maximize lung compliance by initiating periodic sighs (large breaths) that occur during normal breathing. These sighs expand the alveolar surface area and replenish the surface active molecules. The chemical mediators released in the lung during allergic reactions (histamine, leukotriene C_4, or bradykinin) also stimulate irritant receptors. The augmentation of inspiratory activity and increases in breathing frequency produced by irritant receptor excitation may enhance ventilation during asthmatic attacks when the work of breathing is greatly increased.

The unmyelinated C fibers are excited by distortion of the lung's interstitium, such as by liquid accumulation (edema), and by several chemicals, including histamine and capsaicin (the active irritant in pepper, which is widely used to stimulate C fibers). Activation of the C fibers causes laryngeal closure and apnea, followed by rapid, shallow breathing. Stimulation of C fibers, together with the irritant receptors, may be responsible for the tachypnea (rapid breathing) seen in patients with multiple pulmonary emboli, lung edema, or pneumonia.

Several Kinds of Receptors in the Chest Wall

Like all skeletal muscles, the muscles of breathing (diaphragm, intercostals, abdominal wall) develop tension that depends on their initial length (preload) and their afterload. The preload varies with posture, and the afterload varies with chest wall expansion and the resistance to airflow. Receptors in the chest wall reflexly modify motor nerve discharge to the breathing muscles in such a manner that ventilation changes are minimized, despite varying conditions.

Receptors in the chest wall include the joint, tendon, and muscle spindle receptors. Joint receptor activity varies with the extent and speed of rib movement. Tendon organs in the intercostal muscles and the diaphragm monitor the force of muscle contraction and tend to inhibit inspiration. Muscle spindles are abundant in the intercostal and abdominal wall muscles, but scarce in the diaphragm. The spindles help coordinate breathing during changes in posture and speech. They also help stabilize the rib cage when breathing is impeded by increases in airway resistance or decreases in lung compliance.

Fig. 37-8 shows the operation of the intercostal muscle spindle and its neural connections. Spindles are located on intrafusal muscle fibers, which are aligned in parallel with the main muscle bundles that elevate the ribs. Motor innervation of the intercostal muscles originates in ventral horn motor neurons. The intrafusal fibers, on the other hand, are innervated by gamma motor neurons.

Passive stretch of an intercostal spindle, as occurs during lateral flexion of the trunk, increases afferent spindle activity. Through a monosynaptic segmental reflex such spindle excitation causes contraction of the adjacent muscle fibers, which helps restore the upright position. The spindles can also be stretched by efferent gamma neuron discharge, which causes contraction of the intrafusal fiber itself. Some fusimotor fibers fire phasically, so that their discharge rises during inspiration and falls during expiration; other fusimotor fibers are tonically active. Without phasic fusimotor activity, spindle discharge would decrease when the extrafusal fibers in the external intercostal muscles contract during inspiration. Simultaneous activation of fusimotor and motor neurons causes the spindles to be under continuous stretch during inspiration. This enhances the contribution made by the intercostal muscles to breathing. If inspiratory movements are impeded, afferent activity heightens in a spindle innervated by a phasically active fusimotor fiber. This increases inspiratory muscle force and helps to preserve tidal volume.

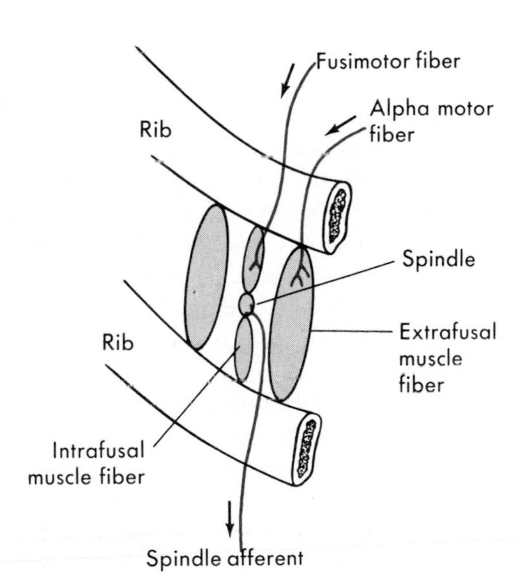

■ **Fig. 37-8** Intercostal muscle spindle and its innervation. The motor nerve (fusiform fiber) has its origin in the small gamma motor neurons of the ventral horns in the spinal cord. The afferent neurons travel to the cord and participate in segmental, intersegmental, and central reflex control of posture and breathing.

Spindle afferent fibers project to the cerebral cortex and provide the information that allows respiratory movements to be perceived consciously. The sensation of breathlessness during exertion or in various lung diseases may result from an imbalance in the demand for muscle shortening and the actual degree of shortening, as reflected in afferent spindle activity.

■ *Respiratory Failure*

Lung disease frequently depresses the ventilatory responses to CO_2 and O_2. The depression may be a result of increased work of breathing, decreased efficiency of gas exchange, impaired chest wall muscle function, or reduced responses to chemical or mechanical stimuli.

Some patients with severe impairment of chest wall muscle function develop respiratory failure, which means that ventilation cannot keep up with O_2 demand. According to the alveolar ventilation equation, a rise in P_{CO_2} must occur. Those patients who have the poorest chemosensitivity seem to be the most likely to develop elevated P_{CO_2} and CO_2 retention when the performance of the chest bellows is reduced.

■ *Abnormal Breathing Patterns*

Breathing is normally a smoothly recurring cyclic process, but in some diseases of the central nervous system episodes of apnea recur.

Cheyne-Stokes breathing is a manifestation of instability of ventilatory control in which tidal volume waxes and wanes cyclically in association with recurrent periods of apnea (Fig. 37-9). This breathing pattern, during which the blood concentrations of oxygen and carbon dioxide fluctuate markedly, may appear during hypoxia or sleep and immediately after voluntary hyperventilation.

Delays in information transfer within the respiratory system occur when the transit time around the circulation is prolonged; that is, when cardiac output is reduced, as in congestive heart failure. When the time required for the blood to circulate from the lungs to the central and peripheral chemoreceptors is prolonged, the controller may react inappropriately (see Fig. 37-9).

Apneustic breathing with its prolonged inspiratory pauses has previously been discussed on p. 600. In **Biot's breathing** periods of normal breathing are interrupted suddenly by periods of apnea. The mechanism for this pattern is unclear; it may be a variant of Cheyne-Stokes breathing. It occurs in patients with central nervous system diseases, especially meningitis.

Grossly irregular breathing occurs in some patients with medullary lesions and occasionally in persons with absent responses to chemical stimuli (primary alveolar hypoventilation).

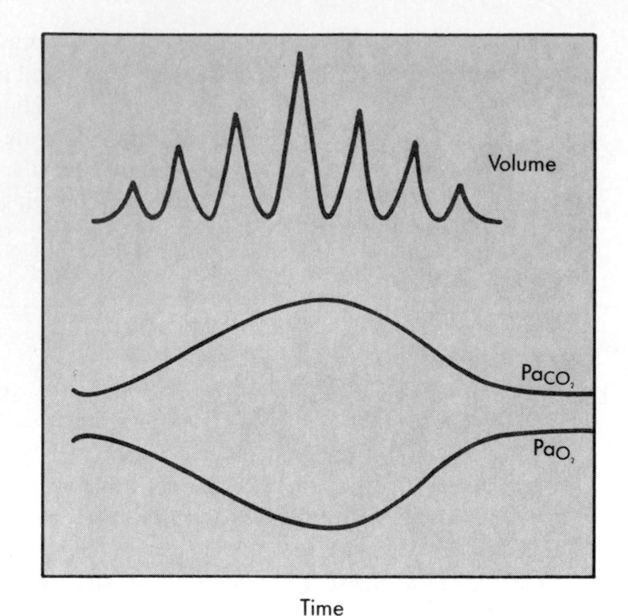

■ **Fig. 37-9** In Cheyne-Stokes breathing tidal volume and consequently arterial blood gases wax and wane. Generally Cheyne-Stokes breathing is a sign of vasomotor instability, particularly low cardiac output.

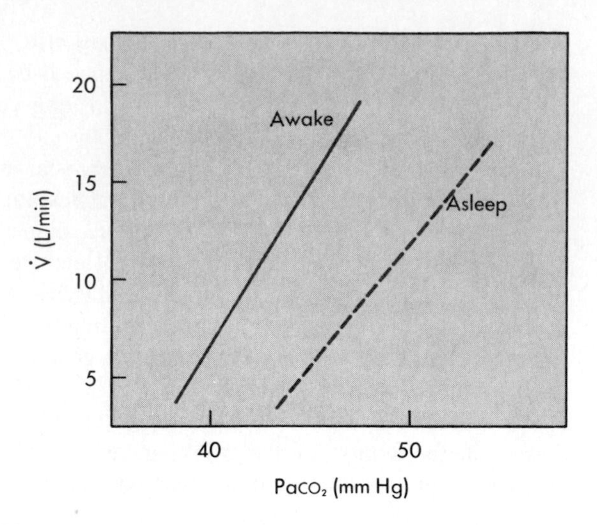

■ **Fig. 37-10** The ventilatory response to CO_2 during wakefulness and during sleep. The threshold of the response is higher (shift to the right) and the sensitivity is less (decreased slope).

Increased breathing with hypocapnia occurs in diseases that excite irritant receptors or lung C fibers (e.g., asthma and pulmonary embolism) or that cause metabolic acidosis. In diabetic coma this form of hyperventilation is called **Kussmaul's breathing.**

■ *Sleep*

In alert, conscious humans stimuli from the environment act reflexly via brain centers to excite breathing. Such stimuli help to sustain breathing even at low levels of chemical drive caused by their effect on alertness (reticular activating system).

Fluctuations in the state of alertness occur, even in awake subjects. However, they are more pronounced during sleep and may markedly affect breathing. The neural and biochemical mechanisms that produce sleep involve an excitatory and inhibitory interplay among various areas of the brain. The reticular activating system is shut down during sleep.

Sleep is not a homogeneous phenomenon. It can be divided into two main stages, a slow wave stage and a rapid eye movement (REM) stage, during which dreaming occurs (see Chapter 16). Each stage is associated with fairly characteristic changes in muscular, cardiovascular, and respiratory activity.

■ *Regulation of Breathing in Sleep*

A reduction in the level of environmental stimulation and withdrawal of cerebral influences on the medullary controllers decrease ventilation and thereby increase the arterial P_{CO_2} during slow wave sleep. Systemic blood pressure and heart rate are reduced. Hypercapnic ventilatory responses are attenuated, and the CO_2 response curve is shifted to the right (Fig. 37-10). REM sleep is divided into phasic and tonic stages. In tonic REM sleep breathing maintains its regularity, but tidal volume may decrease. In addition, the ventilatory responses to inspired CO_2 are further reduced. The ventilatory responses to hypoxia, however, are maintained. External stimuli and changes in blood gas tensions are less effective in producing arousal in REM sleep than in slow wave sleep. Phasic REM sleep is associated with irregular breathing patterns, because the intrinsic activity of higher brain centers dominates respiratory neuron activity.

Ventilation and the adequacy of gas exchange also depend on the caliber of the upper airways. The upper airways (nose, pharynx, and larynx) are convoluted and semirigid but they include movable structures. During inspiration the pressure within the upper airways and the extrathoracic trachea becomes slightly subatmospheric. This tends to compress the upper airway and displaces structures such as the tongue. Consequently, upper airway caliber is reduced, and resistance to air flow increases. These effects are counterbalanced by the actions of the upper airway muscles, which enlarge and stiffen the various structures. In awake subjects all of these muscles are tonically active. In sleep, however, the activity of the upper airway muscles is markedly reduced.

A variety of neural reflexes modulates breathing during sleep. Mechanical stimulation of the airways in animals in either slow wave or REM sleep elicits reflex responses that differ from those observed in the awake state. For example, laryngeal stimulation during wakefulness produces coughing, but it causes apnea during REM sleep. Irritant and stretch reflexes not only affect the activity of chest wall muscles but also influence the muscles of the upper airway. The excitation of pulmonary stretch receptors by lung inflation can markedly reduce the level of activity of the vocal cord abductor and tongue protrusor muscles.

The phasic and tonic activity of skeletal muscle decreases during sleep, particularly in the REM stage. The loss of activity in upper airway muscles during sleep is far greater than that of the diaphragm. Because of the relative loss of tone in the upper airway, the negative airway pressure created by the diaphragm during inspiration may be sufficient to occlude the upper airway. Brief periods of upper airway obstruction occur even in normal people during sleep.

Two Kinds of Sleep Apnea

Apneic periods occur during sleep in one third of normal individuals and are particularly frequent among older men. Apnea may last for more than 10 seconds and may be associated with reductions in arterial oxygen saturation to 75% or less. These apneas appear during all sleep stages but are more common in the lighter stages of slow wave and REM sleep.

Sleep apneas have been classified into two distinct categories: central and obstructive (Fig. 37-11). **Central apnea** is characterized by a cessation of all breathing efforts. In **obstructive apnea,** despite persistent breathing efforts, air flow ceases because of total obstruction of the upper airway. Snoring is a manifestation of partial obstruction of the upper airway. Arousal may be an important element in terminating sleep apnea; it may result from chemoreceptor excitation by hypoxia and hypercapnia. Breathing disturbances that occur during sleep may be primary factors in certain diseases. In patients with such disturbances, prolonged and frequent obstructive apneas occur. The recurrent periods of hypoxia and hypercapnia may lead to polycythemia, rightsided heart failure, and pulmonary hypertension.

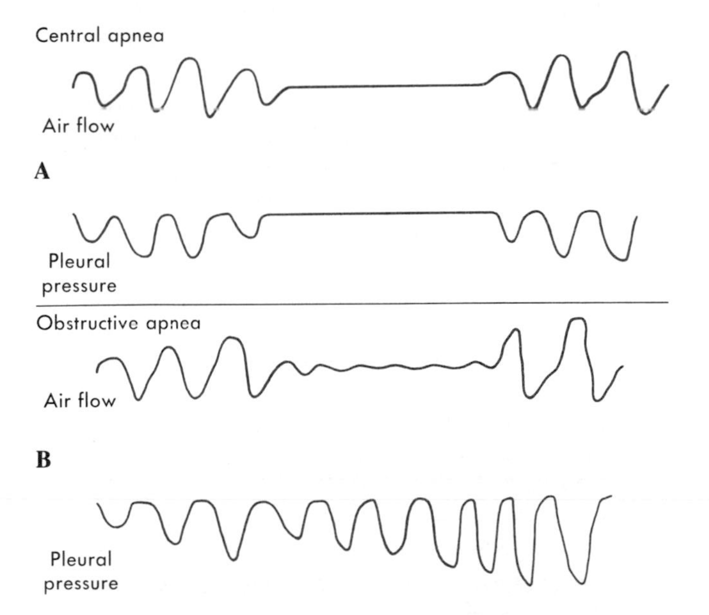

Fig. 37-11 The two main types of sleep apnea. **A,** Central apnea is characterized by no attempt to breathe, as demonstrated by no change in pleural pressure. **B,** In obstructive sleep apnea, pleural pressure swings are augmented, indicating increased airflow resistance because of upper airway obstruction.

Exercise

The ability to exercise depends on the capacity of the cardiovascular and respiratory systems, acting in concert, to increase O_2 delivery to the tissues and to remove the excess CO_2. The circulatory adjustments are covered in Chapter 32. Only the respiratory adaptations are considered here.

Because of the low resistance and great distensibility of the lung's vascular bed, the increase in blood flow during exercise is accompanied by only a moderate rise in pulmonary vascular pressure. More capillaries are perfused, and the area available for gas diffusion increases; although the time spent by erythrocytes in the capillaries is somewhat decreased. The alveolar-arterial Po_2 difference decreases slightly from the resting value in moderate exercise. This reduction reflects the more even distribution of ventilation/perfusion ratios. However, as exercise intensifies, the Po_2 difference begins to widen as the nonsteady state (exhaustion) develops.

The airways distend slightly during exercise and increase the anatomical dead space. However, the physiologic dead space tends to decrease because of the improvement in ventilation/perfusion matching.

The ventilatory adjustments that take place are geared to the intensity of exercise and to its duration. In very brief, intense exercise, as in the 100 m dash, the breath is frequently held until the end of the exercise. With more prolonged exercise, however, ventilation is elevated above the resting level; and it increases even more as exercise becomes more strenuous.

The highest level of work that can be performed without inducing a sustained metabolic acidosis is called the anaerobic threshold (Fig. 37-12). Acid-base balance is normal during steady-state exercise (up to about a sixfold increase in O_2 consumption), when O_2 delivery to the mitochondria is adequate to meet all energy requirements. However, further increases in the level of exercise cross the anaerobic threshold, after which energy requirements can be satisfied only by a combination of aerobic metabolism and anaerobic glycolysis. The lactic acid formed during glycolysis diffuses into the blood and increases the H^+ ion concentration. The level of the anaerobic threshold is higher in athletes than in untrained subjects. When the level of exercise is below the anaerobic threshold, ventilation is

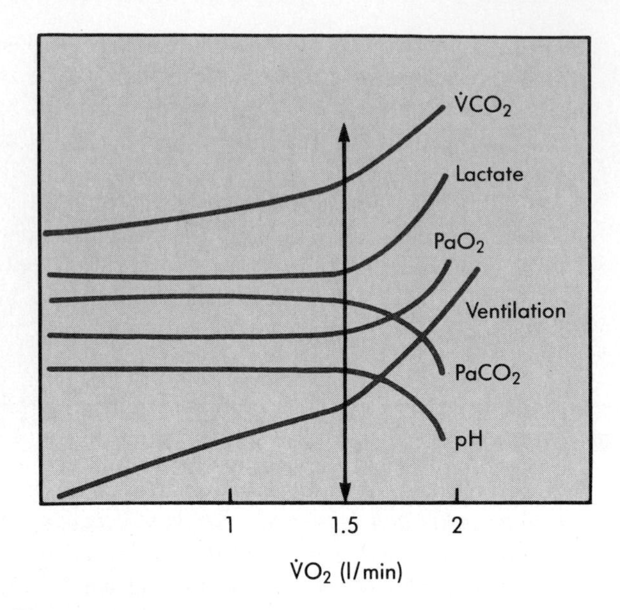

■ **Fig. 37-12** Metabolic changes in exercise. The anaerobic threshhold is marked by the sudden change in the measured variables mainly because of developing lactic acidosis as gycolysis takes over more and more of the muscle energy supply as a result of the relative failure of the body to supply oxygen to the muscles at the rate demanded by the level of exercise.

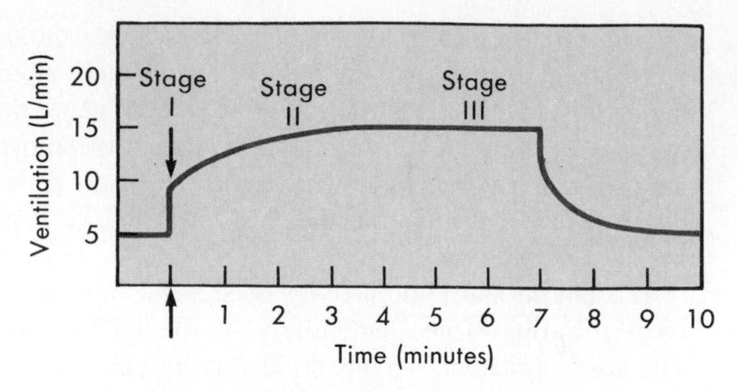

■ **Fig. 37-13** The three stages of exercise. I) onset; II) transient period of adjustment; III) the steady state.

linearly related to both CO_2 production and O_2 consumption (Fig. 37-12). Arterial Po_2, Pco_2, and pH are virtually unchanged, although venous values are markedly altered. The rise in arterial H^+ that occurs when the level of exercise exceeds the anaerobic threshold stimulates the carotid body. Ventilation increases out of proportion to the rise in O_2 consumption. Thus, because of the hyperventilation that results from the increasing blood lactic acid at very high work intensities, Pa_{CO_2} falls and Pa_{O_2} rises.

■ *Mechanism of Exercise Hyperpnea*

The increase in ventilation is the most obvious and important respiratory adjustment to exercise. It takes 4 to 6 minutes after the beginning of moderate exercise before a steady rate of ventilation is reached; with severe exercise it may take even longer. Exercise increases ventilation in fairly distinct stages (Fig. 37-13). In stage I ventilation increases abruptly; in stage II ventilation increases more gradually; and in stage III ventilation remains constant. Both neural and chemical factors regulate ventilation during exercise.

Some of the mechanisms whereby ventilation can be increased during exercise are by activation of the following: (1) cardiovascular mechanoreceptors in the systemic or the pulmonary circulation, (2) temperature receptors, central or peripheral chemoreceptors; (3) mechanoreceptors in muscle, (4) receptors that monitor metabolic activity in muscle or blood gas concentrations in the mixed venous blood. Ventilation may also be increased by general environmental stimulation of the central nervous system. It is not obvious which of the above-mentioned signals mediates the tight coupling of ventilation to metabolic rate. Despite very large changes in CO_2 production and O_2 consumption, the steady-state arterial Po_2 and Pco_2 levels remain remarkably constant. Hence stimulation of chemoreceptors alone by changes in arterial blood gases cannot account for exercise hyperpnea. Some other factor must regulate ventilation during exercise.

Fig. 37-14 illustrates the response of the central controller to increasing Pa_{CO_2} at rest ($\dot{V}co_2$ = 200 ml/min) and moderate exercise ($\dot{V}co_2$ = 800 ml/min). If the exercise-induced change in ventilation were caused by the ventilatory drive from the newly produced CO_2 accumulating in the bloodstream, the response line would intersect the curve for the higher CO_2 at point Y. This point defines the elevation in arterial Pco_2 that would occur during exercise, if CO_2 were the only factor responsible for the hyperpnea.

The fact that Pa_{CO_2} is unchanged during exercise (below the anaerobic threshold) indicates that some factor sensitive to metabolic rate has increased breathing. This factor shifts the CO_2 response line by some unknown mechanism. As mentioned earlier, the Pco_2 (H^+) in arterial blood fluctuates with breathing and produces corresponding fluctuations in peripheral chemoreceptor discharge. The oscillations of carotid body activity are greater during exercise because of the increased level of metabolic activity and the larger tidal volume. This may represent the additional drive that maintains ventilation in proportion to metabolic rate.

■ *Respiratory Effects of Varying Inspired Oxygen*

Barometric pressure decreases with increasing altitude (Fig. 37-15); the relation is approximately exponential. Thus, inspired Po_2 falls as a person ascends from sea

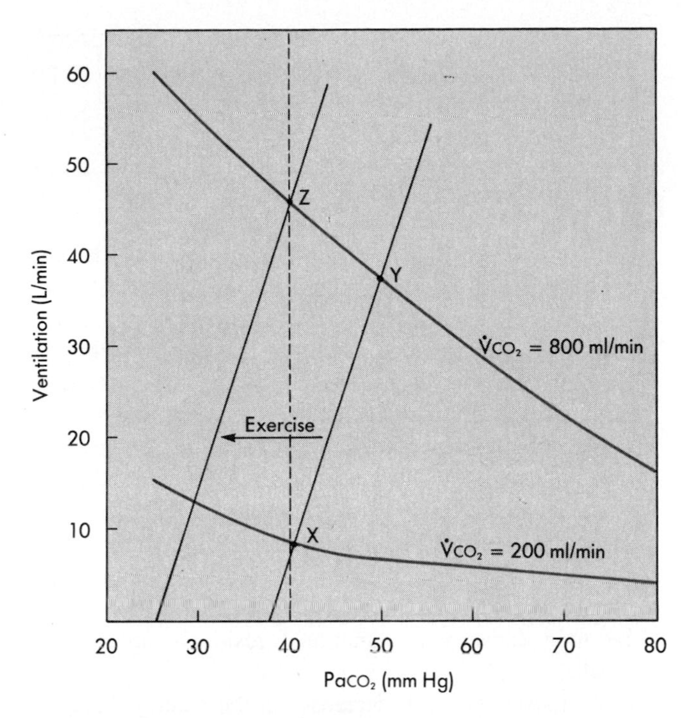

■ **Fig. 37-14** Changes in the CO_2 ventilation line between rest and exercise. Based on the CO_2 response line at rest *(X)*, P_{CO_2} ought to intercept the exercise CO_2 production curve at Y, if ventilation were driven by CO_2. In fact, however, the response is shifted toward more sensitivity *(Z)* maintaining $P_{A_{CO_2}}$ at 40 mm Hg. Thus ventilation has increased without any apparent rise in Pa_{CO_2}.

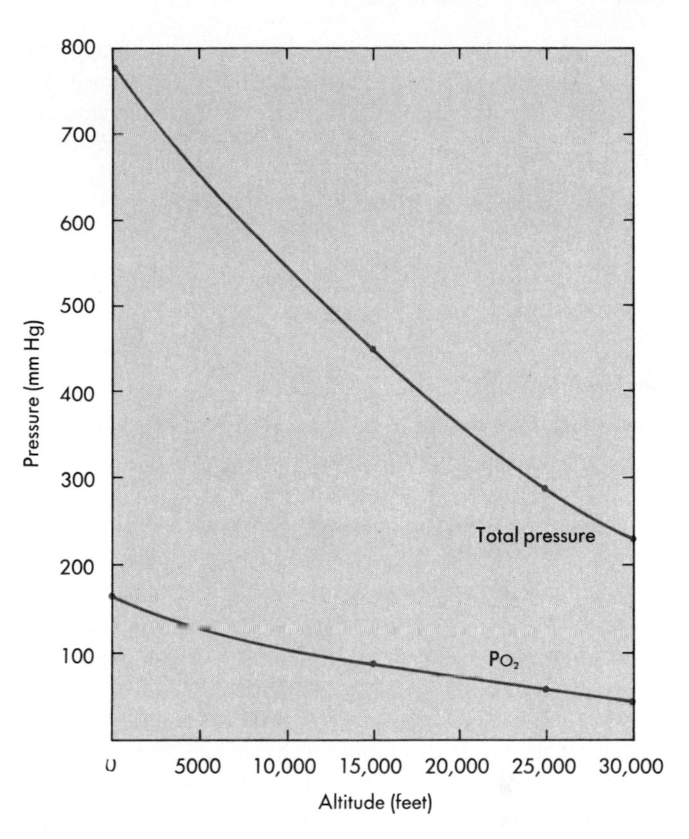

■ **Fig. 37-15** Barometric pressure and P_{O_2} fall exponentially as one acends to high altitude (x-axis).

level to high altitudes, just as the alveolar gas equation requires. The resulting hypoxemia elicits a variety of compensatory responses. Some of these occur quickly, while others develop gradually during prolonged exposure.

The initial hyperventilation that occurs at high altitude is caused by stimulation of the peripheral chemoreceptors, particularly the carotid bodies. The increase in ventilation reduces the arterial P_{CO_2} and H^+. This decreases the excitation of central chemoreceptors and thus limits the increase in ventilation.

After 2 or 3 days at high altitude ventilation increases further. This increase is part of the process of acclimatization. This secondary increase in ventilation occurs, in part, by the following mechanisms: (1) the renal excretion of sodium bicarbonate reduces plasma HCO_3^- concentration and returns blood hydrogen ion concentration toward normal; (2) the HCO_3^- concentration is decreased in brain interstitial fluid, probably by a metabolic process that moves sodium ions from the brain interstitial fluid into the blood; (3) anaerobic metabolism in the hypoxic brain allows lactate to replace bicarbonate ions. Another important alteration that occurs is a shift to the left of the ventilatory response to CO_2 and a steepening of the slope of the response curve. Consequently, the threshold for a stimulatory effect occurs at a lower P_{CO_2}. Persons who live at high altitudes and who are chronically hypoxic have a diminished ventilatory response to hypoxia. The response to hypercapnia, on the other hand, is the same in people who live at high and low altitudes. Hypoxic desensitization develops over years and is more likely to occur when chronic exposure to hypoxia begins in infancy.

A similar blunted response to hypoxia is found in patients who are hypoxemic because they have a type of congenital heart disease in which blood is shunted from the right to the left side of the heart. It is not clear whether the blunted ventilatory response is caused by some depressant effect of chronic hypoxia on the central nervous system or whether it originates at the peripheral chemoreceptors.

Exposure to high altitude also increases the hemoglobin concentration in the blood and consequently augments its oxygen-carrying capacity. The increase in hemoglobin is caused by the release of erythropoietin from the kidney. This hormone accelerates red cell production in the bone marrow. During exposure to high altitude the concentration of 2,3-diphosphoglycerate also rises in the red cell. This decreases the affinity of hemoglobin for O_2 (Chapter 36). The P_{50} of the blood is shifted left, which improves delivery of oxygen in the systemic capillaries.

Occasionally, tolerance of high altitude disappears and serious symptoms develop, such as ventilatory depression, polycythemia, or heart failure. This intolerance to altitude (chronic mountain sickness) is relieved by descent to a lower altitude or by administration of oxygen.

Effect of Age on Breathing

The Newborn

One of the first changes that must occur after birth is to transform the fluid-filled fetal lung to one containing air. High distending forces are needed in the first few breaths to overcome the high surface tension that opposes alveolar expansion with air. *With successful air inflation, the alveolar surfactant is recruited into the air-liquid interface so that functional residual capacity stabilizes and the work of breathing is diminished.*

The ability of the newborn to maintain air in the lungs also depends on how well the rib cage can resist the collapsing forces produced by contraction of the diaphragm. In the premature infant the chest is very pliable and surfactant production may be poor, thereby increasing the danger of lung collapse. Resistance to airflow is higher, but the specific lung compliance is nearly the same in the infant as in the adult.

Pulmonary vascular resistance is very high in the fetus because the vessels are constricted and have as much smooth muscle as systemic vessels (see Chapter 31).

Breathing may be irregular at birth, particularly in the premature infant. The patterns of breathing range from regular to frequent apneic episodes. In preterm infants apnea is predominantly central; that is, breathing movements are absent. Obstructive apnea may occur after the first month of life, and it may predispose to the sudden infant death syndrome. Responsiveness to CO_2 is less well developed the more immature the infant. Preterm infants respond to a reduction in inspired O_2 concentrations with a transient increase in ventilation for approximately 1 minute, followed by a sustained depression. In normal infants the response is similar during the first week of life. This biphasic response to hypoxemia in the neonate is explained by initial stimulation of the peripheral chemoreceptors, followed by an overriding depression of the brain stem respiratory center.

Changes in lung volume reflexly alter the timing of breathing much more in the newborn than in the adult. A small but sustained increase in lung volume in the newborn significantly prolongs the expiratory time and decreases the breathing frequency, whereas lung deflation reflexly increases the respiratory rate. The increase in breathing rate is caused primarily by a shortening of expiratory time and may help the infant maintain an adequate functional residual capacity. Pulmonary irritant reflexes have been elicited in the neonate; direct stimulation of the lining of the tracheal wall augments respiratory efforts.

The Elderly

Pulmonary performance tends to decline above the age of 30. The changes proceed at a variable rate that is dependent both on the "aging process" and on the extent of exposure to noxious agents in the environment. As lung elasticity decreases, the transpulmonary pressure at a given lung volume decreases. Thus, the bronchi, particularly in the dependent lung, close at higher lung volumes (Chapter 34). This accounts for the elevated functional residual capacity in the elderly. With age the chest wall stiffens because of structural changes in the rib cage. The compliances in lung and chest wall change in opposite directions; therefore, changes in total lung capacity or functional residual capacity are small.

As muscle strength decreases in the elderly, vital capacity and forced expiratory flow rates decrease. Also the internal surface area of the lung decreases, because the alveoli become wider and shallower because of loss of elastic fibers in the alveolar walls and alveolar ducts.

Ventilation/perfusion ratios become more variable with advancing age, and as a result the arterial Po_2 falls about 3 mm Hg/decade. The arterial Pco_2, however, is still regulated at 40 mm Hg, but the ventilatory responses to both hypercapnia and hypoxia are reduced.

Summary

1. There are two main schemes for the regulation of breathing. *Metabolic* (automatic) control is concerned with oxygen delivery and with acid-base balance (Pa_{CO_2}). *Behavioral* (voluntary) control is related to coordinated activities in which breathing may be temporarily suspended or altered.

2. The control system consists of a *central controller* (driver) located in the brainstem (medulla and pons); an *effector* (mainly the muscles of the chest wall but also the smooth muscle of the airways); and various sensors, which report back to the central controller the results of the intended action.

3. The modern view of the brainstem controller is that it contains a tonically active inspiratory neuron pool that receives input from a wide variety of sensors. The summed sensory input generally inhibits inspiratory activity. Higher centers regulate behavioral breathing by temporarily overriding the brainstem pattern generator.

4. The sensory component includes *central chemoreceptors* (on or near the surface of the medulla); *peripheral chemoreceptors* (carotid bodies); and *propriocep-*

tors (lung stretch, irritant and C fiber receptors, plus diaphragm, intercostal and abdominal muscle spindles, tendon and joint organs).

5. The medullary receptors are most sensitive to Pa_{CO_2}. The initial ventilatory response to CO_2 is large and occurs rapidly.

6. The peripheral chemoreceptors in the carotid bodies are mainly sensitive to reduced arterial oxygen tension. The response to hypoxia comes into play only at low Pa_{O_2} (<60 mm Hg).

7. Irritant receptors in the large airways protect the delicate alveolar surfaces from particles, chemical vapors, and physical factors.

8. C fiber receptors in the terminal respiratory units are stimulated by distortion of the alveolar walls (lung congestion or edema).

9. Many factors influence ventilation in exercise in a manner not completely understood, but there is no doubt that the controlled variable is Pa_{CO_2}.

10. Sleep is a complex phenomenon consisting of several phases during which breathing control varies. Both the sensitivity to CO_2 and O_2 are diminished, possibly because the reticular activating system is depressed.

11. Acute and chronic hypoxia affect breathing somewhat differently because of slow adjustments in cerebral spinal fluid (H^+), which alters CO_2 sensitivity.

Bibliography

Journal articles

Bennett FM et al: Dynamics of ventilatory response to exercise in humans, *J Appl Physiol* 51:194, 1981.

Cherniack NS: Sleep apnea and its causes, *J Clin Invest* 73:1501, 1984.

Coleridge JCG, Coleridge HM: Afferent vagal fibre innervation of the lung and airways and its functional significance, *Rev Physiol Biochem Pharmacol* 99:2-110, 1984.

Dempsey JA, Vidruk EH, Mitchell GS: Pulmonary control systems in exercise: update, *Fed Proc 44* 2260-2270, 1985.

Euler C von: On the central pattern generator for the basic breathing rhythmicity, *J Appl Physiol* 55:1647, 1983.

Eyzaguirre C, Zapata P: Perspectives in carotid body research, *J Appl Physiol* 57:931, 1984.

Fisher JT et al: Respiration in newborns: development of the control of breathing, *Am Rev Resp Dis* 125:650, 1982.

Long S, Duffin J: The neuronal determinants of respiratory rhythm, *Prog Neurobiol* 27:101, 1986.

Lydic R: State-dependent aspects of regulatory physiology, *Faseb J* 1:6-15, 1987.

Mitchell RA, Burger AJ: Neural regulation of respiration, *Am Rev Resp Dis* 111:206-224, 1975.

Pack AI: Sensory inputs to the medulla, *Annu Rev Physiol* 43:73, 1981.

Phillipson EA: Control of breathing during sleep, *Am Rev Resp Dis* 118:909-939, 1978.

Richter DW: Generation and maintenance of the respiratory rhythm, *J Exp Biol* 100:93, 1982.

Schiaefke ME: Central chemosensitivity: a respiratory drive, *Rev Physiol Biochem Pharmacol* 90:171, 1981.

Wasserman K, Whipp BJ: Exercise physiology in health and disease, *Am Rev Res Dis* 112:219-249, 1975.

Whipp BJ, Ward SA: Cardiopulmonary coupling during exercise, *J Exp Biol* 100:175, 1982.

Books and monographs

Comroe JH Jr: *Physiology of respiration,* ed 2, Chicago, 1974, Year Book.

Nunn FJ: *Applied respiratory physiology,* ed 3, London, 1987, Butterworths.

SECTION

VII

THE GASTROINTESTINAL SYSTEM

Howard C. Kutchai

CHAPTER

38

Gastrointestinal Motility

■ *Structure and Innervation of the Gastrointestinal Tract*

■ *Structure of the Wall of the Gastrointestinal Tract*

The structure of the gastrointestinal tract varies greatly from region to region, but there are common features in the overall organization of the tissue. Fig. 38-1 depicts the general layered structure of the wall of the gastrointestinal tract.

The **mucosa** consists of an epithelium, the lamina propria, and the muscularis mucosae. The nature of the epithelium varies greatly from one part of the digestive tract to another. The **lamina propria** consists largely of loose connective tissue containing collagen and elastin fibrils. The lamina propria is rich in several types of glands and contains lymph nodules and capillaries. The **muscularis mucosae** is the thin innermost layer of intestinal smooth muscle. In some parts of the gastrointestinal tract, the muscularis mucosae has an inner circular layer and an outer longitudinal layer. Contractions of the muscularis mucosae throw the mucosa into folds and ridges.

The **submucosa** consists largely of loose connective tissue with collagen and elastin fibrils. In some regions submucosal glands are present. The larger blood vessels of the intestinal wall travel in the submucosa.

The **muscularis externa** characteristically consists of two substantial layers of smooth muscle cells: an inner circular layer and an outer longitudinal layer. Contractions of the muscularis externa mix the contents in the lumen and propel them in a controlled fashion toward the anus. The circular layer is three to five times thicker than the longitudinal layer, and it is mainly responsible for mixing and propulsion in the gastrointestinal tract. In man and most mammals, the circular layer may be divided into an **inner dense circular layer,** composed of smaller more closely-packed smooth muscle cells, and a thicker **outer circular layer.**

The wall of the gastrointestinal tract contains a great many neurons that are highly interconnected. A dense network of nerve cells in the submucosa is called the **submucosal plexus** (Meissner's plexus). The prominent **myenteric plexus** (Auerbach's plexus) is located between the circular and longitudinal smooth muscle layers. The myenteric and submucosal plexuses consist of ganglia that contain neuronal cell bodies and bundles of unmyelinated fibers that interconnect the ganglia. In addition, other nonganglionated plexuses that consist of networks of bundles of unmyelinated fibers originate in the myenteric or submucosal plexuses. The submucosal and myenteric plexuses (intramural plexuses), together with the other neurons and plexuses of the gastrointestinal tract, constitute the **enteric nervous system,** which helps integrate the motor and secretory activities of the gastrointestinal system. If the sympathetic and parasympathetic nerves to the gut are cut, many of the motor and secretory activities continue to occur because of control by the enteric nervous system.

The **serosa,** or adventitia, is the outermost layer and consists mainly of connective tissue covered with a layer of squamous mesothelial cells.

■ *Innervation of the Gastrointestinal Tract*

Sympathetic innervation. Sympathetic innervation of the gastrointestinal tract is primarily via postganglionic adrenergic fibers whose cell bodies are in prevertebral and paravertebral plexuses. The celiac, superior and inferior mesenteric, and hypogastric plexuses provide postganglionic sympathetic innervation to various segments of the gastrointestinal tract. Most of the sympathetic fibers terminate in the submucosal and myenteric plexuses. Activation of the sympathetic nerves usually has an inhibitory effect on synaptic transmission in the enteric plexuses. Some sympathetic fibers innervate blood vessels of the gastrointestinal tract, causing vasoconstriction. Other sympathetic fibers innervate glandular structures in the wall of the gut. Relatively few of the sympathetic fibers terminate in the muscu-

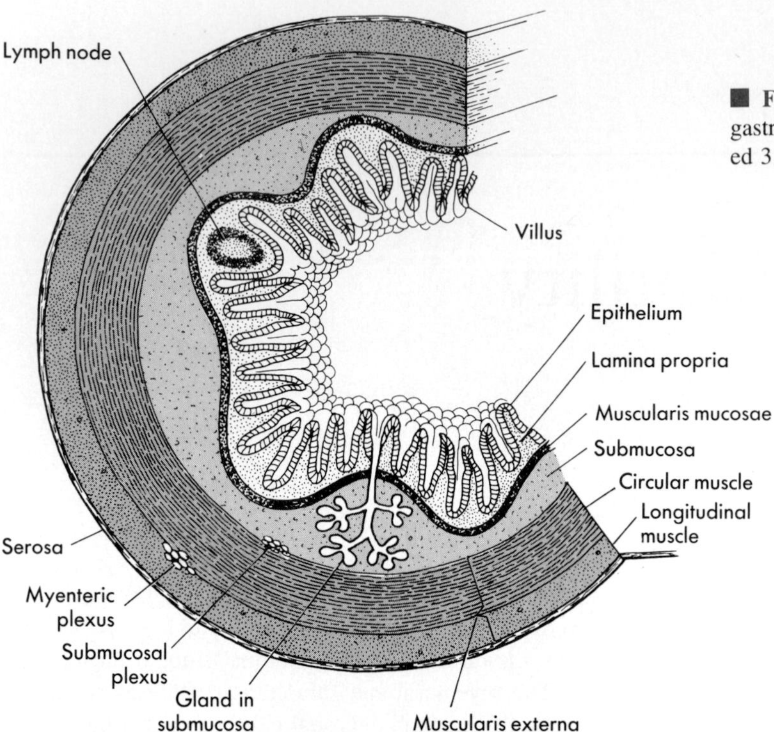

■ **Fig. 38-1** The general organization of the layers of the gastrointestinal tract. (Modified from Ham, AW: *Histology,* ed 3, Philadelphia, 1957, JB Lippincott Co.)

laris externa. Stimulation of the sympathetic input to the gastrointestinal tract inhibits motor activity of the muscularis externa but stimulates contraction of the muscularis mucosae and certain sphincters. The inhibitory effect of the sympathetic nerves on the muscularis externa is not a direct action on the smooth muscle cells, because there are few sympathetic nerve endings in the muscularis externa. Rather the sympathetic nerves act to influence neural circuits in the intrinsic plexuses that provide input to the smooth muscle cells. This effect may be reinforced by the action of the sympathetic nerves in reducing blood flow to the muscularis externa. Other fibers that travel with the sympathetic nerves are cholinergic; still others release neurotransmitters that remain to be identified. Fig. 38-2 summarizes the sympathetic innervation of the gastrointestinal tract.

Parasympathetic innervation. Parasympathetic innervation of the gastrointestinal tract down to the level of the transverse colon is provided by branches of the vagus nerve. The remainder of the colon receives parasympathetic fibers from the pelvic nerves by way of the hypogastric plexus. These parasympathetic fibers are predominantly cholinergic. Other fibers that travel in the vagus and its branches have other transmitters that have not been identified. Vagal fibers terminate predominantly on the ganglion cells in the intramural plexuses. The ganglion cells innervate the smooth muscle and secretory cells of the gastrointestinal tract. Parasympathetic input usually stimulates the motor and secretory activity of the gut. Fig. 38-3 illustrates the parasympathetic innervation of the gastrointestinal tract.

Enteric nervous system. The myenteric and submucosal plexuses are the most well-defined plexuses in the wall of the gastrointestinal tract. Other nonganglionated plexuses have been identified. The myenteric and submucosal plexuses are networks of nerve fibers and ganglion cell bodies. Some of the incoming axons are parasympathetic fibers, and others are sympathetic fibers. Interneurons in the plexuses connect intrinsic afferent sensory fibers with efferent neurons to smooth muscle and secretory cells. Consequently, the myenteric and submucosal plexuses can control a good deal of coordinated activity in the absence of extrinsic innervation to the gastrointestinal tract. Axons of plexus neurons innervate gland cells in the mucosa and submucosa, smooth muscle cells in the muscularis externa and muscularis mucosae, and intramural endocrine and exocrine cells. Afferent fibers from mechanoreceptors and chemoreceptors in the mucosa or deeper in the wall of the gastrointestinal tract synapse in the plexuses, so that local reflex activity occurs (Fig. 38-4). The functions of the intramural plexuses are discussed in more detail later in this chapter.

Afferent fibers. Intrinsic and extrinsic afferent fibers in the gut provide the afferent limbs of reflex arcs that are both local and central (see Fig. 38-4). Chemoreceptor and mechanoreceptor endings are present in the mucosa and in the muscularis externa. The cell bodies of some of these sensory receptors are located in the my-

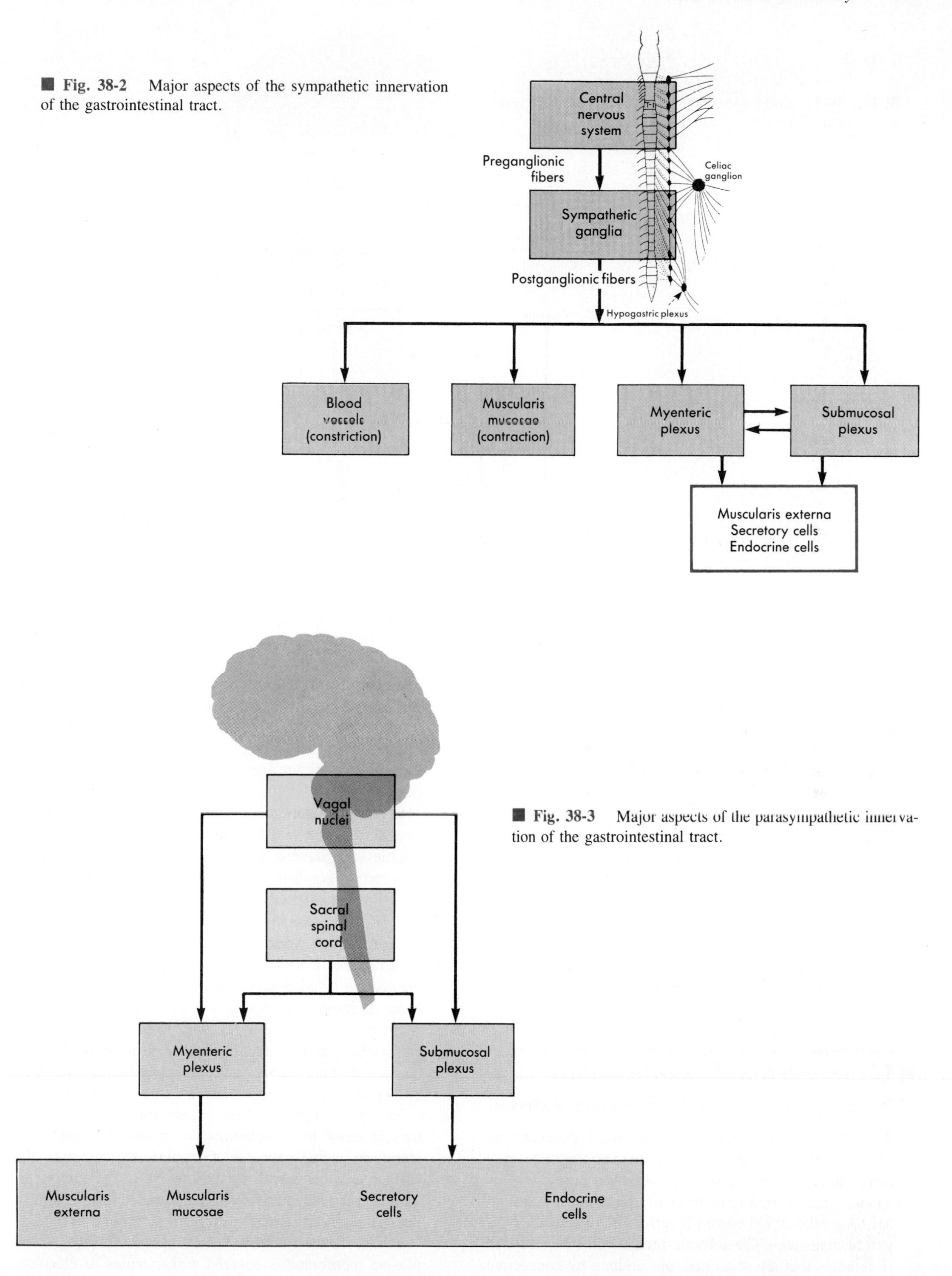

■ Fig. 38-2 Major aspects of the sympathetic innervation of the gastrointestinal tract.

■ Fig. 38-3 Major aspects of the parasympathetic innervation of the gastrointestinal tract.

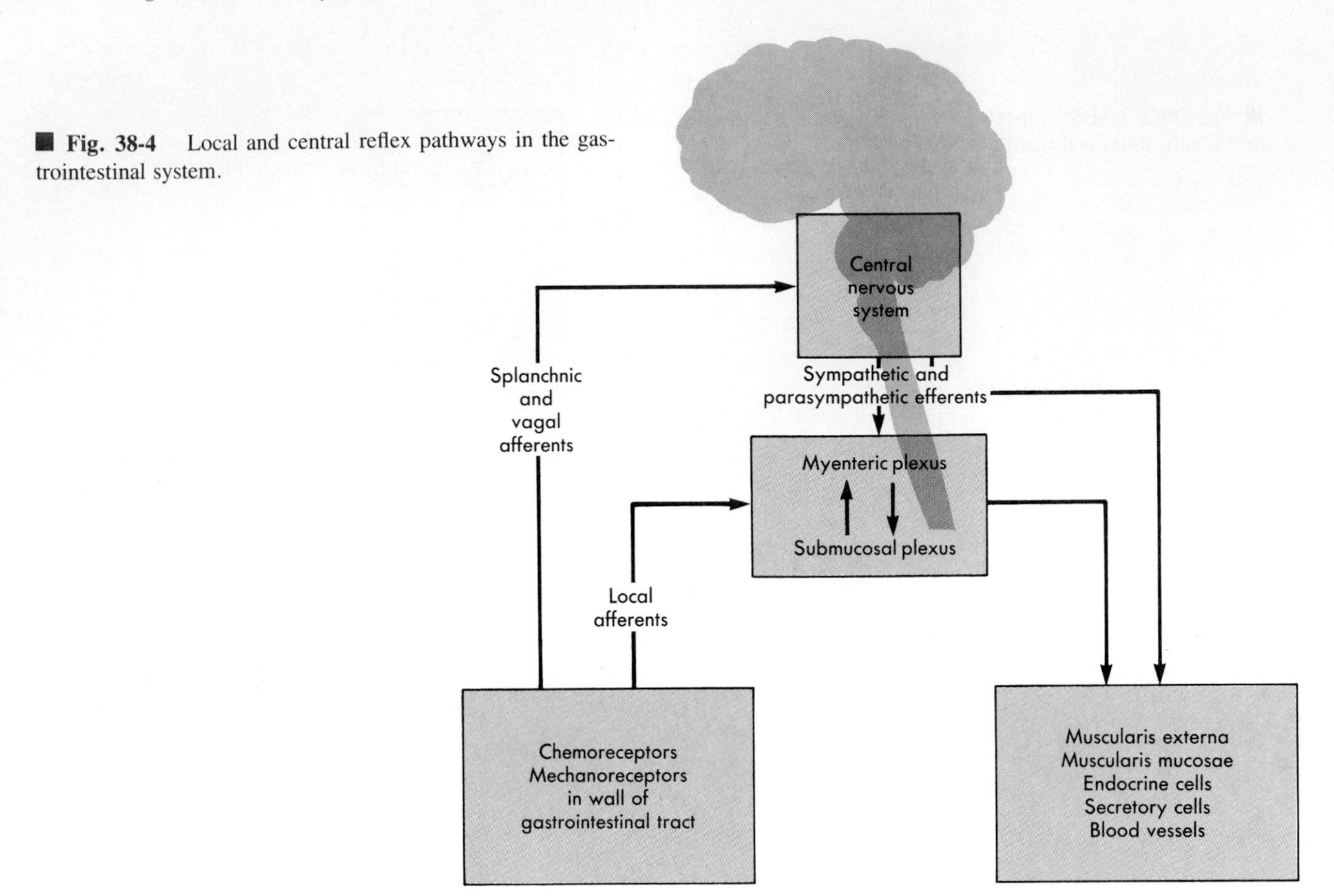

■ **Fig. 38-4** Local and central reflex pathways in the gastrointestinal system.

enteric and submucosal plexuses. The axons of some of these receptor cells synapse with other cells in the plexuses to mediate local reflex activity. Others of these receptors send their axons back to the central nervous system. The cell bodies of these sensory neurons are located more centrally. Afferent fibers in the vagus, which greatly outnumber vagal motor fibers, have cell bodies principally in the nodose ganglion, whereas many sensory fibers that travel via the sympathetic nerves have their cell bodies in dorsal root ganglia. *The number of sensory afferent fibers from the gastrointestinal tract is large, and the complex afferent and efferent innervation of the gastrointestinal tract allows for fine control of secretory and motor activities by intrinsic and central pathways.*

■ Gastrointestinal Smooth Muscle

■ Smooth Muscle Cells of the Muscularis Externa

The properties of smooth muscle were discussed in Chapter 19. Some of the properties pertinent to gastrointestinal smooth muscle are reviewed here.

The smooth muscle cells of the gastrointestinal tract are long (about 500 μm in length) and slender (5 to 20 μm in diameter). The smooth muscle cells are arranged in bundles that are separated and defined by connective tissue. The bundles branch and frequently anastomose with other bundles. The bundles are about 200 μm across, and a cross-section of a bundle contains several hundred cells. Because of the high degree of electrical coupling among the smooth muscle cells, the functional contractile unit is a bundle rather than a single cell.

Gastrointestinal smooth muscle cells are much smaller than skeletal muscle cells. As a result, smooth muscle cells have a much larger surface/volume ratio. The resting conductances to the major ions (Na^+, K^+, Cl^-) are considerably less than those in skeletal muscle. Therefore, in spite of the increased surface/volume ratio in smooth muscle, the energy cost of maintaining ion gradients across the plasma membrane is comparable to that of skeletal muscle. The plasma membrane contains both the Na^+, K^+-ATPase and a Ca^{++}-ATPase to accumulate K^+ and to extrude Na^+ and Ca^{++}. Cl^- is accumulated against a gradient, but the nature of the active transport responsible for Cl^- uptake remains to be elucidated.

The plasma membranes of gastrointestinal smooth muscle cells have numerous invaginations that form structures called caveolae. Caveolae may increase the surface area of the cell by 50% to 70%. The interior of the caveolae is accessible to macromolecules present in the extracellular fluid.

A significant fraction of the interior surface of the plasma membrane is covered with plaques of electron-dense material called **dense bands.** Frequently the extracellular surface of the plasma membrane opposite a dense band appears to be attached via microfibrils to collagen fibers in the extracellular matrix. These attachments may function as intramuscular "microtendons" that allow the force of contraction of individual cells to be transferred to the muscle as a whole. Occasionally a dense band of one cell may be aligned with a dense band of a neighboring cell. In such cases the neighboring plasma membranes are frequently closely apposed between the dense bands, with electron-dense material bridging the gap between the cells. This structure is known as an **intermediate junction.** Intermediate junctions are believed to represent mechanical connections that allow transmission of contractile force between neighboring cells.

Excitation-Contraction Coupling

As in skeletal and cardiac muscle, the level of intracellular Ca^{++} plays a central role in regulating the contraction of smooth muscle. In smooth muscle, contraction is initiated by the Ca^{++} that crosses the plasma membrane during the action potential, as well as by Ca^{++} released from the sarcoplasmic reticulum. The less well developed the sarcoplasmic reticulum, the more the smooth muscle depends on extracellular Ca^{++} for contraction. As discussed in Chapter 19, regulation of contraction by Ca^{++} in smooth muscle occurs by a different mechanism than in skeletal muscle.

Action potentials. Action potentials in gastrointestinal smooth muscle are more prolonged (10 to 20 msec duration) than those of skeletal muscle and have little or no overshoot. The rising phase of the action potential is caused by ion flow through channels that are relatively slow to open and conduct both Ca^{++} and Na^{+}. The Ca^{++} that enters the cell during the action potential plays a significant role in initiating contraction. Repolarization is aided by a delayed increase in K^{+} conductance. In gastrointestinal smooth muscle the resting membrane potential does not remain constant in time. Typically, slow oscillations of the resting potential, called **slow waves,** occur (Fig. 38-5). The depolarizing phase of the slow waves may trigger trains of action potentials.

Electrical Coupling Between Smooth Muscle Cells

Neighboring cells are said to be well coupled electrically if a perturbation of the membrane potential of one cell spreads rapidly, and with little decrement, to the

Longitudinal muscle of human lower jejunum

V L.m O.c.m. I.c.m. F 420 mg 0 −16 mV −50 10 sec

Outer circular muscle of dog jejunum

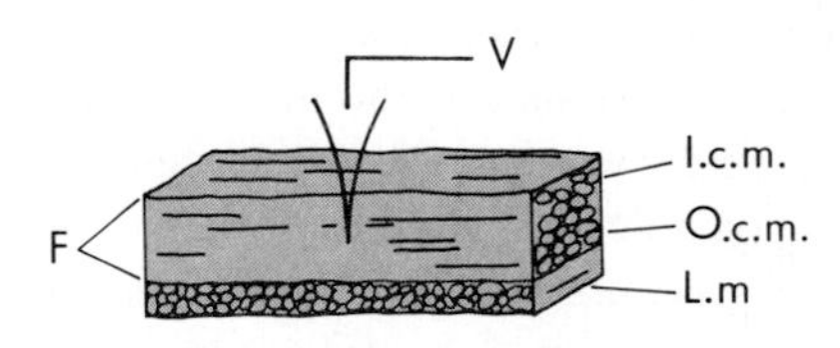

−12 mV −60 0.2 g 10 sec

Inner circular muscle of dog jejunum

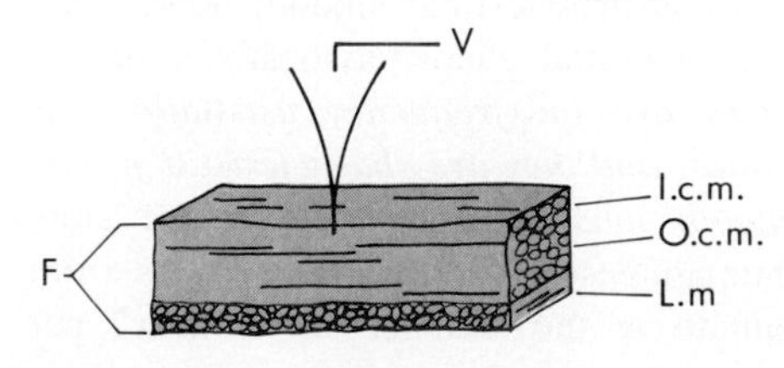

1.8 g −13 mV −50 10 sec

Fig. 38-5 Electrical and mechanical activity in the small intestine. Intracellular electrical activity (colored traces) and contractile activity (black traces) from the whole thickness preparations indicated. *F,* Force transducer; *V,* intracellular microelectrode; *Lm,* longitudinal muscle layer; *Ocm,* outer circular muscle layer; *Icm,* inner circular muscle layer. (Modified from Hara Y et al: *J Physiol,* [London] 372:501, 1986.)

other cell. The smooth muscle cells of the muscularis externa are well coupled.

The smooth muscle cells of the circular layer are somewhat better coupled than those of the longitudinal layer. The cells of the circular layer are joined by frequent gap junctions that provide low-resistance pathways that allow the spread of electrical current from one cell to another.

Because the electrical resistance of membranes is much higher than the resistance of the cytoplasm, the resistance along the long axis of smooth muscle cells is less than that in the transverse direction. Thus an electrical depolarization in the circular muscle is readily conducted circumferentially and quickly depolarizes a ring of circular muscle near the original site of depolarization. The ring of depolarization then spreads more slowly along the long axis of the gut.

In the intact gastrointestinal tract the longitudinal smooth muscle cells of the muscularis externa are also coupled electrically. Because of the low electrical resistance along the length of the smooth muscle cells, an electrical disturbance is conducted along the long axis of the gut in the longitudinal layer.

Contractility of Gastrointestinal Smooth Muscle

The length/tension curve. The length/tension curve of gastrointestinal smooth muscle is similar in shape to that of skeletal muscle, but it has a much broader maximum. This gives gastrointestinal smooth muscle the ability to develop force effectively over a greater range of muscle length. This property may reflect the organization of muscle cells within the tissue and the organization of the contractile elements within the cells more than it reflects the intrinsic length/tension properties of the contractile elements themselves.

Relationship between membrane potential and tension. Gastrointestinal smooth muscle cells contract phasically in response to slow waves and action potentials. The action potentials usually occur in bursts at the peak of the slow waves and enhance phasic contractions that are superimposed on the baseline level of contraction (see Fig 38-5). Because smooth muscle cells contract rather slowly (about 10 times slower than skeletal muscle), the individual contractions caused by each action potential in a burst are not visible as distinct twitches but rather sum temporally to produce a smoothly increasing level of tension. The increase in tension in response to a burst of action potentials is proportional to the number of action potentials in the burst.

Between bursts of action potentials the tension developed by gastrointestinal smooth muscle falls, but not to zero. This nonzero "resting," or baseline, tension developed by the smooth muscle is called **tone.** The tone of gastrointestinal smooth muscle may be altered by neurotransmitters, hormones, or drugs.

Responses to stretch. Gastrointestinal smooth muscle may respond to stretch or release of stretch. In many cases rapidly stretching gastrointestinal smooth muscle will lead to an immediate increase in the frequency of action potentials and an increase in contractile tension—a phenomenon known as **stress activation.** This may be followed by a decrease in tension back toward the original level—a phenomenon known as **stress relaxation.** Both these responses to stretch play roles in the motility of the various segments of the gastrointestinal tract.

Electrophysiology of Gastrointestinal Smooth Muscle Cells

The resting membrane potential. The resting membrane potential of gastrointestinal smooth muscle cells ranges from about -40 mV to around -80 mV. As discussed in Chapter 2, the relative membrane conductances of K^+, Na^+, and Cl^- are important in determining the resting membrane potential. Compared with skeletal muscle, gastrointestinal smooth muscle cells have a higher ratio of Na^+ conductance to K^+ conductance. This contributes to the somewhat lower resting membrane potential of gastrointestinal smooth muscle.

If the relative conductances and the equilibrium potentials for K^+, Na^+, and Cl^- in guinea pig teniae coli (a well-studied preparation) are used in the chord conductance equation (Chapter 2), a predicted resting membrane potential of about -40 mV for the teniae coli is computed. When the resting membrane potential of teniae coli is measured with microelectrodes, a value of about -60 mV is obtained. The potential difference across the plasma membrane of the resting teniae coli is about 20 mV larger than can be accounted for by diffusion of ions.

What is responsible for the extra 20 to 25 mV of polarization across the plasma membrane? The current view is that the Na^+, K^+ pump is **electrogenic,** and it contributes about 20 mV to the resting membrane potential. Because 3 Na^+ are extruded for every 2 K^+ taken up, the pump produces a net outward flow of positive charge, which contributes to the membrane potential. Experiments indicate that the electrogenic Na^+, K^+ pump is responsible for a significant part of the resting potential in gastrointestinal smooth muscle (Fig. 38-6). If the Na^+, K^+ pump of guinea pig teniae coli is inhibited with ouabain, the resting membrane potential changes from about -60 to near -40 mV. When ouabain is washed out, the resting potential returns to near -60 mV.

Slow waves. In gastrointestinal smooth muscle the resting membrane potential characteristically varies in time. *Slow waves are low-frequency oscillations of membrane potential, and they are characteristic of gastrointestinal smooth muscle.* The frequency of slow waves in the human gastrointestinal tract varies from about 3 per minute in the stomach to about 12 per

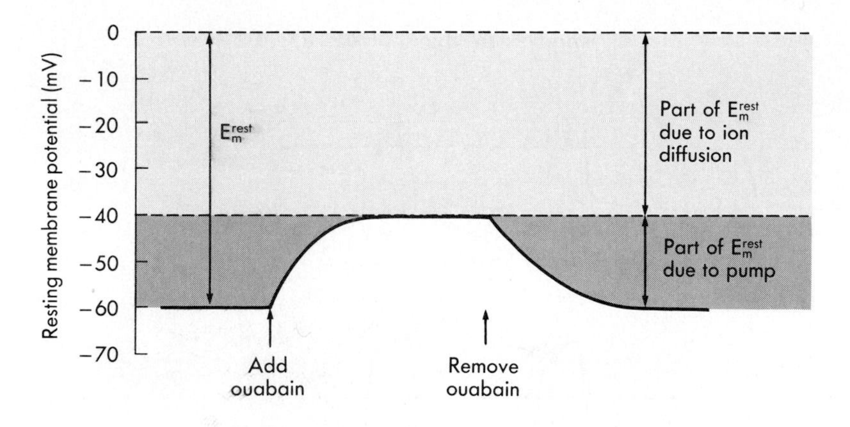

■ **Fig. 38-6** Idealized representation of an experiment suggesting that a significant part of the resting membrane potential of gastrointestinal smooth muscle is caused by the electrogenic Na+, K+ pump.

minute in the duodenum. Fig. 38-5 illustrates slow waves and associated contractions in the different smooth muscle layers of dog and human jejunum. Slow waves have also been called **the basic electrical rhythm** or **the electrical control activity** of the gastrointestinal tract.

As shown in Fig. 38-5, slow waves differ in the various smooth muscle layers of the gastrointestinal tract. Slow waves also vary from one segment of the gastrointestinal tract to another. In some segments of the gastrointestinal tract the slow waves appear to be sinusoidal (Fig. 38-5, *B*). In other locations the form of the slow wave resembles that of the action potential of cardiac ventricular muscle (but the slow wave is longer). The rising phase is caused by entry of Ca^{++} or of Na^+ and Ca^{++} through voltage-gated channels. The plateau phase represents a balance between influx of Na^+ and Ca^{++} and the efflux of K^+ via Ca^{++}-activated K^+ channels. The repolarizing phase occurs when the Na^+/Ca^{++} channels close. The Ca^{++}-activated K^+ and delayed K^+ channels contribute to the repolarization.

Interstitial cells: the generators of slow waves. Slow waves are generated by specialized **interstitial cells,** which have morphological and biochemical characteristics of both fibroblasts and smooth muscle cells. A thin layer of interstitial cells, three to five cells thick, is present at the border between the circular and longitudinal smooth muscle layers. In humans and other species that have gastrointestinal tracts with an inner dense circular muscle layer, another thin layer of interstitial cells is present between the inner dense circular and the outer circular smooth muscle. The interstitial cells put out processes that make gap junctions with the intestinal smooth muscle cells in the layers on either side of the interstitial layer. In this way the slow wave that is generated in the interstitial cells is rapidly conducted through all the cells of the muscularis externa via the gap junctions that interconnect the smooth muscle cells. The interstitial cells may be compared with the cells of the sinoatrial node of the heart, in that they serve as the pacemakers for the electrical and contractile activity of the smooth muscle of the muscularis externa. In some cases the depolarizing phase of the slow wave is preceded by a ramplike depolarization called a **prepotential.**

Modulation of slow waves by enteric neurons. Whereas the generation of slow waves does not require the action of neurons, the enteric nervous system modulates the amplitude and the frequency of slow waves. Acetylcholine and substance P, released from excitatory motor nerve endings, increase the amplitude of slow waves. Such excitatory agonists may also increase the frequency of slow waves or the duration of the plateau phase of the slow wave. Excitatory agonists can increase the frequency of action potentials during the plateau phase of the slow waves. Action potentials are generated when the depolarization during the rising phase is sufficient to activate the voltage-gated Ca^{++} channels that are responsible for the action potentials—that is, there is an effective **electrical threshold** for generating action potentials (Fig. 38-7). By increasing the amplitude of the slow wave, excitatory agonists cause more of the slow wave to be above this threshold, so that more action potentials are fired.

Inhibitory agonists, such as VIP and nitric oxide (NO) released from enteric neurons, and circulating epinephrine have the opposite effect. By decreasing the amplitude of slow waves, inhibitory agonists decrease the extent to which the plateau phase of the slow wave is above the electrical threshold. In this way inhibitory agonists decrease the frequency of action potentials or may even abolish them.

Contraction of smooth muscle evoked by slow waves. In general the contraction of gastrointestinal smooth muscle can be evoked by the depolarizing phase of the slow wave in the absence of action potentials. Such responses are typically seen in the outer circular muscle layer of the small intestine (Fig. 38-5, *B*). When the depolarization reaches a **mechanical threshold**, contraction begins, even in the absence of action potentials. An inhibitory agonist may decrease the amplitude of the slow wave to the extent that the depolarizing phase does not reach the mechanical threshold, so that contraction fails to occur in response to the damped slow wave. An excitatory agonist, by increasing the amplitude and duration of the slow wave, increases the extent to which

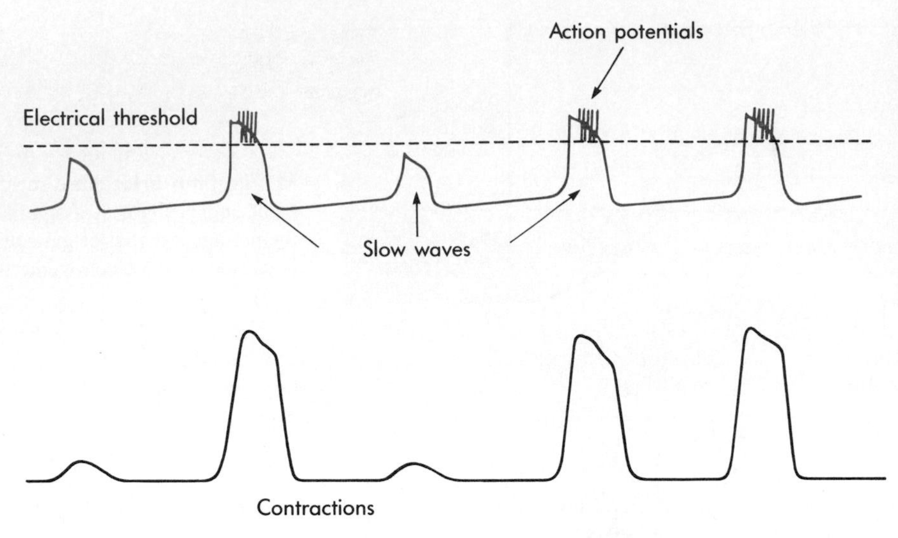

■ **Fig. 38-7** Electrical threshold for generation of action potentials. When the slow wave reaches the electrical threshold, trains of action potential spikes are generated. There are no action potentials on those slow waves that fail to reach the threshold. The black trace shows contractile activity. Contraction is much stronger when action potentials occur. (Modified from Sarna, SK: In vivo myoelectric activity: methods, analysis, and interpretation. In Wood, JD, editor: *Handbook of physiology,* section 6: The gastrointestinal system, vol I, p 2, Bethesda, 1989, American Physiological Society.)

the depolarization exceeds the mechanical threshold and enhances contraction. Excitatory agonists may cause action potentials to appear on the crests of the slow waves in circular muscle. Ca^{++} entry during the action potentials elicits a stronger contraction of the circular muscle than would occur in the absence of action potentials. The greater the number and frequency of action potentials, the stronger is the contraction.

In longitudinal small intestinal smooth muscle (Fig. 38-5, *A*) and in the inner dense circular muscle layer (Fig. 38-5, *C*), contractions are usually associated with the presence of action potentials on the crests of the slow waves. Excitatory agonists increase the frequency of action potentials and the strength of contractions. Inhibitory agonists diminish the frequency of action potentials or may even abolish action potentials.

■ *Neuromuscular Interactions in the Gastrointestinal Tract*

Neuromuscular interactions in the gastrointestinal tract do not involve true neuromuscular junctions with specialization of the postjunctional membrane, as occurs at the motor endplate. The neurons of the myenteric plexus send axons to the smooth muscle layers, and each axon branches extensively to innervate many smooth muscle cells.

The circular smooth muscle layer of the muscularis externa is heavily innervated by nerve terminals that form close associations with the plasma membranes of the smooth muscle cells. Neuromuscular gaps of about 20 nm are typical in the circular layer. Innervation is by fibers that are both excitatory and inhibitory to the electrical and contractile activity of the smooth muscle cells.

The longitudinal smooth muscle cells are much less richly innervated by the neurons of the myenteric plexus, and the neuromuscular contacts are not as intimate. Gaps of about 80 nm separate nerve terminals from the plasma membrane of the smooth muscle cells they innervate.

■ *Integration and Control of Gastrointestinal Motor Activities*

Control of the contractile activities of gastrointestinal smooth muscle involves the central nervous system, the intrinsic plexuses of the gut, humoral factors, and electrical coupling among the smooth muscle cells. The motor behavior of particular segments of the gastrointestinal tract is discussed in the succeeding sections of this chapter.

This section outlines the structural and functional properties of the enteric nervous system that underlie control and integration of contractile function. A major conclusion is that the gastrointestinal tract displays a great deal of intrinsic control. The enteric nervous system can mediate control and integration of most of the contractile behavior of the gut without intervention by the autonomic nervous system. The autonomic nervous system modulates patterns of muscular and secretory activity that are controlled by the enteric nervous system.

The Enteric Nervous System

The plexuses of the wall of the gastrointestinal tract constitute a semiautonomous nervous system that controls the motor and secretory activities of the digestive system. This enteric nervous system contains about 10^8 neurons (about as many neurons as in the spinal cord). Fig. 38-8 depicts the major components of the enteric nervous system and their locations in the wall of the intestine. The **myenteric plexus** and the **submucosal plexus** consist of ganglia, which are interconnected by tracts of fine, unmyelinated nerve fibers. The neurons in

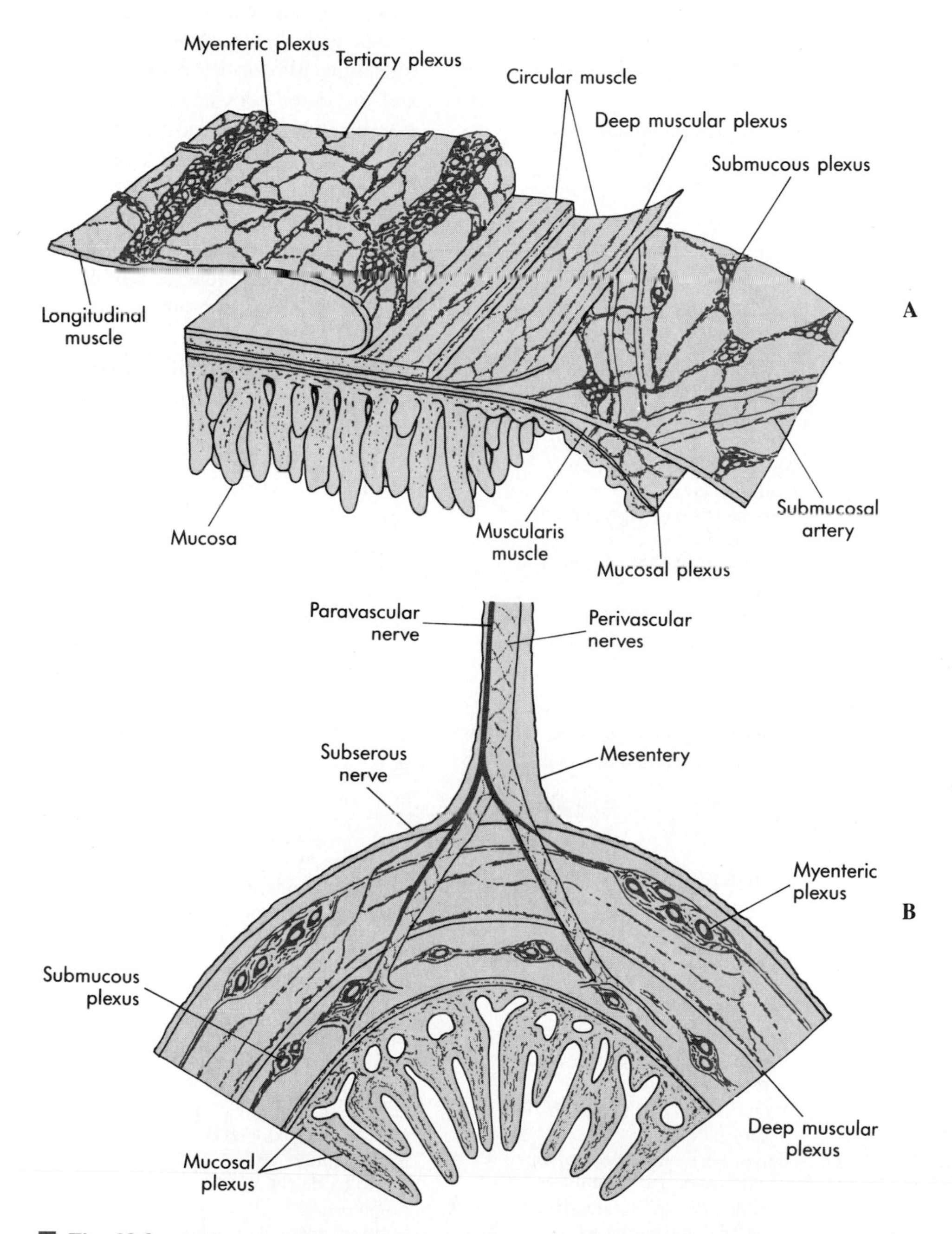

Fig. 38-8 Major neural plexuses of the small intestine. (**A,** Seen in whole mounts; **B,** seen in transverse section.) The two ganglionated plexuses are the myenteric and submucosal plexuses. Fibers originating in the myenteric and submucosal plexuses form the nonganglionated plexuses: the tertiary plexus (which innervates the longitudinal layer of muscularis externa), the deep muscular plexus (which supplies the inner dense circular muscle), and the mucosal plexus. Neurons and neuronal processes are shown in color (Redrawn from Furness, JB and Costa, M: *Neuroscience* 5:1, 1980.)

the ganglia include sensory neurons, with their sensory endings in the wall of the gastrointestinal tract. Neurons sensitive to mechanical deformation, to particular chemical stimuli, and to temperature have been identified. Some of the neurons in the enteric ganglia are effector neurons that send axons to smooth muscle cells of the circular or longitudinal layers, to secretory cells of the gastrointestinal tract, or to gastrointestinal blood vessels. Some of the neurons in the enteric ganglia are interneurons that are part of a network of neurons that integrates the sensory input to the ganglia and formulates the response of the effector neurons. The myenteric and submucosal plexuses give rise to bundles of nerve fibers that form nonganglionated plexuses in the longitudinal and circular muscle, in the muscularis mucosae, and in the mucosa. Particularly prominent among these nonganglionated plexuses are the mucosal plexus and the deep muscular plexus. The **mucosal plexus** is particularly evident at the bases of mucosal glands and in the cores of intestinal villi. The **deep muscular plexus** lies between the inner dense circular and the outer circular muscle layers of the muscularis externa and supplies most of the motor nerves to the inner dense circular muscle layer.

The extrinsic innervation of the gastrointestinal tract, via sympathetic and parasympathetic nerves, projects primarily onto the neurons of myenteric and submucosal plexuses to excite or inhibit particular plexus neurons. In this way the autonomic nervous system influences the motor and secretory functions of the gastrointestinal tract. Significant control of gastrointestinal activities is exercised by the enteric neurons independent of sympathetic or parasympathetic input. The major types of neurons in the gastrointestinal tract, both intrinsic and extrinsic, are listed in Table 38-1.

■ **Table 38-1** Neuronal components of the enteric nervous system

Intrinsic neurons (Cell bodies in myenteric or submucosal ganglia)

Type of neuron	Function
Motoneurons	
To muscle	Enteric excitatory motoneurons
	Enteric inhibitory motoneurons
To arterioles	Enteric vasodilator neurons
To epithelia	Enteric secretomotor neurons (cholinergic and noncholinergic)
	Motoneurons to gastric parietal cells
	Motoneurons to gastrointestinal endocrine cells
Sensory neurons	Distension (stretch)-sensitive neurons
	Chemoceptive neurons
Associative neurons	Interneurons in motility vasomotor and secretomotor pathways
Intestinofugal neurons	Neurons with cell bodies in enteric ganglia and terminals in prevertebral ganglia

Extrinsic neurons (Cell bodies in vagal nuclei, sympathetic ganglia, dorsal root ganglia, or spinal cord)

Neural pathway	Function
In motor pathways	Sympathetic motility-inhibiting neurons
	Sympathetic vasoconstrictor neurons
	Sympathetic secretomotor-inhibiting neurons
	Vagal inputs to enteric excitatory pathways
	Vagal inputs to enteric inhibitory pathways
	Pelvic nerve inputs to enteric excitatory and inhibitory pathways and to enteric vasodilator pathways
	Vagal inputs promoting gastrin and acid secretion
In sensory pathways	Mechanoreceptor neurons
	Chemoceptive neurons
	Nociceptive neurons

Modified from Costa M, Furness JB: In Makhlouf GM, editor: *Handbook of physiology, section 6: the gastrointestinal system, vol 2, Neural and endocrine biology,* Bethesda, Md 1989, American Physiological Society.

■ *Neuromodulatory Substances of the Enteric Nervous System*

The number of neuromodulatory substances present in the wall of the gastrointestinal tract approximates the

■ **Table 38-2** Neurochemically-defined nerve cell bodies in myenteric ganglia guinea pig small intestine

Neurochemical	*Proportion, %*
Aromatic amine handling	0.5
Acetylcholinesterase	High
Calcium-binding protein	30
Cholecystokinin	6
Calcitonin gene–related peptide	2
Choline acetyltransferase	High
Dynorphin	49
Enkephalin	51
Gastrin-releasing peptide	19
Monoamine oxidase B	10
Neuropeptide Y	28
Nitric oxide (NO)	10
Somatostatin	6
Substance P	37
Vasoactive intestinal peptide	39

Modified from Costa M, Furness JB: In Makhlouf GM, editor: *Handbook of physiology, section 6: the gastrointestinal system, vol 2, Neural and endocrine biology,* Bethesda, Md, 1989, American Physiological Society.

There are approximately 10,000 neurons in the myenteric ganglia in a 1 cm long segment of guinea pig ileum. Typically a given neuron contains more than one of these markers of putative transmitter or neuroactive compounds.

number of such compounds in the brain. Most of the known neuromodulators are present in both gut and brain. The functions of some of these neuroactive substances in the central nervous system were discussed in Chapter 4.

Tables 38-2 and 38-3 show the neuroactive substances (or markers of these substances) that are present in the myenteric and submucosal ganglia of guinea pig small intestine. Also shown are the proportions of neurons that contain certain neuroactive substances. *In most cases the enteric neurons contain more than one putative transmitter or neuromodulator, and in many instances corelease of more than one substance at enteric nerve endings may occur. The combination of neuroactive substances present in a particular neuron correlates with the morphology and function of that neuron and with its projections.* The functional roles of putative neurotransmitters and neuromodulators of enteric neurons are presented in Table 38-4.

Functions of Enteric Neurons

Myenteric neurons. *Myenteric ganglia contain motoneurons, sensory neurons, and interneurons.* The majority of myenteric neurons are motor neurons to the longitudinal and circular muscle layers of muscularis externa. Excitatory motoneurons release acetylcholine onto muscarinic receptors on smooth muscle cells; they also release substance P. Inhibitory motoneurons release VIP and probably NO. Other myenteric neurons synapse on neurons in the submucosal plexus. About a third of the neurons in myenteric ganglia are sensory neurons; their receptive endings may be located in any of the layers of the wall of the gastrointestinal tract. Most myenteric interneurons release acetylcholine onto nicotinic receptors on myenteric motoneurons or on other interneurons.

Submucosal neurons. *Most neurons in submucosal ganglia regulate the secretory activities of glandular, endocrine, and epithelial cells.* Stimulatory secretomotor neurons release acetylcholine and VIP onto gland cells and epithelial cells. There are sensory neurons in submucosal ganglia; these neurons respond to chemical stimuli or to mechanical deformation of the mucosa. They mediate secretomotor reflexes. Other submucosal neurons are interneurons to other submucosal neurons or to myenteric neurons (ACh onto nicotinic receptors). Submucosal ganglia contain vasodilator neurons that release ACh or VIP onto the smooth muscle cells of mucosal blood vessels.

Table 38-3 Neurochemically-defined nerve cell bodies in submucosal ganglia of guinea pig small intestine

Neurochemical	*Proportion, %*
Dynorphin/galanin/vasoactive intestinal peptide	45
Choline acetyltransferase/cholecystokinin/calcitonin gene–related peptide/(galanin)/neuropeptide Y/somatostatin	20
Choline acetyltransferase/substance P	11
Choline acetyltransferase	14
NO synthase	10
Aromatic amine handling	11

Modified from Costa M, Furness JB: In Makhlouf GM, editor: *Handbook of physiology, section 6: the gastrointestinal system, vol 2, Neural and endocrine biology,* Bethesda, Md, 1989, American Physiological Society.

There are about 7,000 neurons in the submucosal ganglia of a 1 cm length of guinea pig ileum.

Morphology and Electrophysiology of Enteric Neurons

Enteric neurons are either unipolar or multipolar in their form. **Unipolar enteric neurons** have from several to many short, branching dendritic processes and a single axon. The axons of unipolar neurons vary greatly in length, and the single axon may branch extensively. **Multipolar enteric neurons** have from 3 to 12 long (1 to 3 mm), smooth processes. Enteric interneurons and motoneurons are unipolar. Enteric sensory neurons are multipolar. Fig. 38-9 illustrates the morphology of motor and sensory neurons of myenteric ganglia.

The electrophysiological characteristics of unipolar enteric neurons differ from those of multipolar enteric neurons (Fig. 38-10). The unipolar neurons are classified as **S-type cells,** because they receive fast synaptic input. The multipolar cells, classified as **AH-type cells,** do not receive fast synaptic input and, when stimulated, tend to fire only a few action potentials followed by a prolonged afterhyperpolarization.

The fast excitatory postsynaptic potentials (fast EPSPs) of S-type neurons are caused by the release of acetylcholine onto nicotinic receptors (Fig. 38-10, *A,* trace *b*). S cells tend to be tonic neurons that respond to stimulation with a prolonged burst of action potentials that continues as long as the cell is stimulated. Slow EPSPs can also be observed in S neurons. S cells can fire action potentials at rates up to 150/second. The action potentials of S cells are blocked by tetrodotoxin; thus fast Na^+ channels appear to be responsible for the depolarizing phase of S cell action potentials.

AH-type neurons do not display fast EPSPs; they do not appear to receive cholinergic input. When AH cells are stimulated they fire a small number of action potentials (one to four). Then, even when stimulation is maintained, they enter a phase of prolonged afterhyperpolarization that lasts for 5 to 20 seconds (Fig. 38-10, *B,* traces *b* and *c*). The AH cells are phasic neurons. The action potentials generated in the cell bodies of AH

Table 38-4 Substances that may be neurotransmitters or neuromodulators in the enteric nervous system

Substance	*Location and role*
Acetylcholine (ACh)	Primary excitatory transmitter to muscle, to intestinal epithelium, to parietal cells, to some gut endocrine cells, and at neuro-neuronal synapses
Adenosine triphosphate (ATP)	Probably contributes to transmission from enteric inhibitory muscle motoneurons
γ-Aminobutyric acid (GABA)	Present in different populations of neurons, depending on species and region Does not appear to be a primary neurotransmitter
Calcitonin gene-related peptide (CGRP)	Present in some secretomotoneurons and interneurons Role unknown
Cholecystokinin (CCK)	Present in some secretomotoneurons and in some interneurons May contribute to excitatory transmission Generally excites muscle
Dynorphin (DYN) and dynorphin-related peptides	Present in secretomotoneurons, interneurons, and motoneurons to muscle Does not appear to be a primary transmitter
Enkephalin (ENK) and enkephalin-related peptides	Present in interneurons and muscle motoneurons In most regions these substances probably provide feedback inhibition of transmitter release
Galanin	Present in secretomotoneurons, descending interneurons, and inhibitory motoneurons in human intestine Role unknown
Gastrin-releasing peptide (GRP) (mammalian bombesin)	Excitatory transmitter to gastrin cells Also found in nerve fibers to muscle and in interneurons, where its roles are not known
Neuropeptide Y	Present in secretomotoneurons, where it appears to inhibit secretion of water and electrolytes Also present in interneurons and inhibitory muscle motoneurons
Nitric oxide (NO)	A cotransmitter from enteric inhibitory muscle motoneurons Possible transmitter at neuro-neuronal synapses
Noradrenaline	Noradrenergic nerve fibers in the intestine are not strictly enteric: They are of sympathetic origin Major roles are to inhibit motility in nonsphincter regions, to contract the muscle of the sphincters, to inhibit secretomotor reflexes, and to act as vasoconstrictor neurons to enteric arterioles
Serotonin (5-HT)	Appears to participate in excitatory neuro-neuronal transmission
Somatostatin	Despite its widespread distribution in enteric neurons, no clearly defined roles have been established
Tachykinins (substance P, neurokinin A, neuropeptide K and neuropeptide γ)	Excitatory transmitters to muscle; and are cotransmitters with ACh May contribute to excitatory neuro-neuronal transmission
Vasoactive intestinal peptide (VIP) (and peptide histidine isoleucine [PHI])	Excitatory transmitter from secretomotoneurons Possibly a transmitter of enteric vasodilator neurons Contributes to transmission from enteric inhibitory muscle motoneurons

Modified from Furness JB et al: *Trends Neurosci* 15:66, 1992.

neurons are only partially blocked by tetrodotoxin or by removal of extracellular Ca^{++}; both Na^{+} and Ca^{++} channels appear to be involved. The afterhyperpolarization of AH neurons is caused by opening of Ca^{++}-activated K^{+} channels. The multiple, long (1 to 3 mm) processes of AH cells fire action potentials, which are blocked by tetrodotoxin (unlike action potentials in the soma). These processes thus appear to be axons, not dendrites.

AH neurons receive excitatory input from neurons that cause slow EPSPs in the AH cells. The transmitter that mediates slow EPSPs is unknown. VIP, substance P, somatostatin, and serotonin are candidates. The slow EPSP is caused by closing of K^{+} channels that are open in the unstimulated cell. Slow EPSPs evoked by maintained stimulation of these inputs can counteract the effects of opening of the Ca^{++}-activated K^{+} channels. In this way, they prevent the hyperpolarization that normally occurs in AH cells. Slow EPSPs transiently convert AH neurons from phasic to tonic behavior (Fig. 38-11).

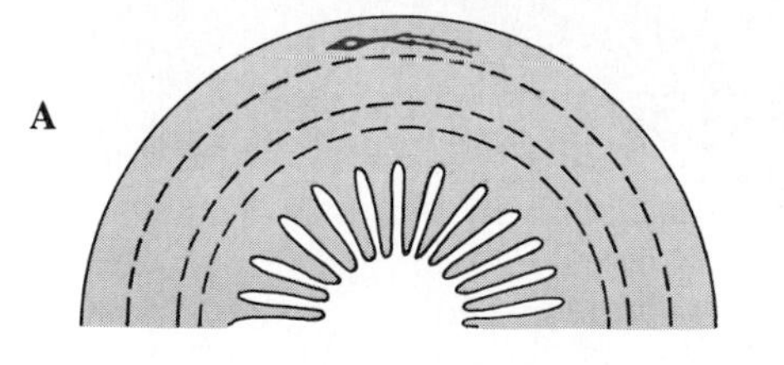

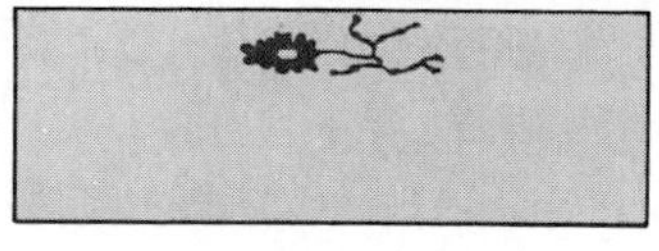

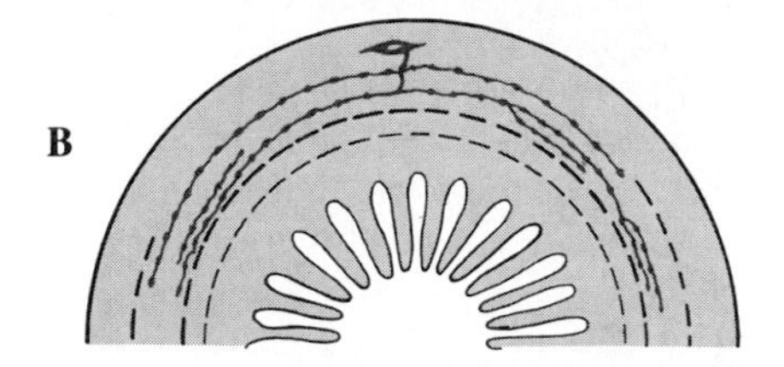

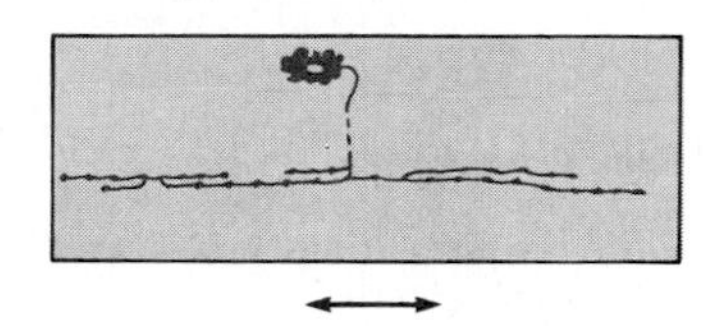

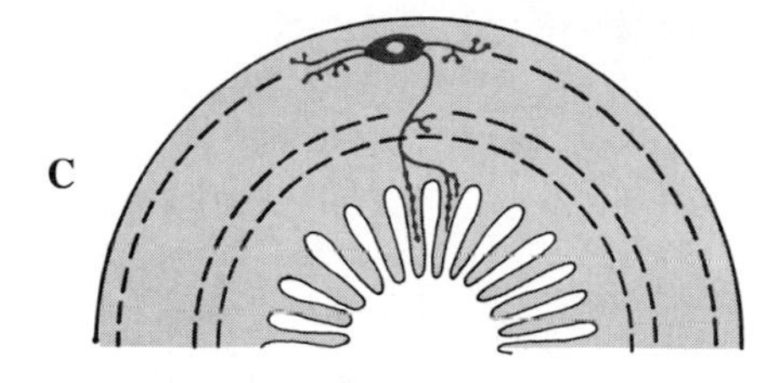

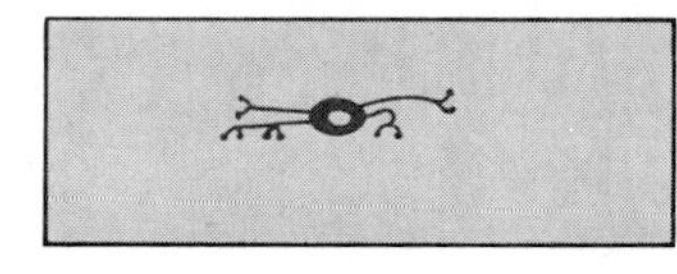

■ **Fig. 38-9** Morphology of neurons of the myenteric plexus of the small intestine. **A,** Motoneuron that projects to the tertiary plexus that supplies the longitudinal layer of muscularis externa. **B,** Motoneuron that innervates the circular muscle layer. **C,** Sensory neuron with receptor endings in the mucosa. The drawings on the left represent the neurons as seen in a transverse section of the small intestine. On the right are depicted the neurons as viewed from the serosal surface of the intestine slit open and laid flat; the arrows indicate the circumferential direction. *L,* Longitudinal muscle layer; *c,* circular muscle layer; *s,* submucosa; *m,* mucosa. (Modified from Furness, JB et al.: *Arch Histol Cytol* 52(suppl.):161, 1989.)

■ *Reflexes in the Enteric Nervous System*

Enteric reflexes play major roles in controlling the motility of the gastrointestinal tract and its secretory and absorptive activities. About one-third of the neurons in the myenteric plexus are sensory. These neurons respond to distension of the gut wall, mechanical deformation of the mucosa, or chemical stimuli (such as acidity). Input from these sensory neurons is integrated in myenteric ganglia, and efferent fibers are activated to muscle, gland, and epithelial cells. One of the most common patterns of enteric reflex is illustrated in Fig. 38-12. Activation of a sensory neuron, shown in the center of the figure, typically stimulates motility oral to the sensory stimulus and inhibits motility anal to the stimulus. Other sensory neurons with cell bodies in dorsal root ganglia and in vagal nuclei elicit central reflexes that provide another level of control that modulates the effects of the enteric reflexes (Fig. 38-12). The sensory neurons for different sensory modalities often converge onto the same interneurons and motoneurons. *Distending a segment of intestine characteristically elicits ascending excitation and descending inhibition. Mechanical stimulation of the mucosal surface in that segment elicits a similar response. The same motoneurons are activated in response to either stimulus and mucosal deformation potentiates the response to distension.*

■ *Chewing (Mastication)*

Chewing can be carried out voluntarily, but it is more frequently a reflex behavior. Chewing serves to lubricate the food by mixing it with salivary mucus, to mix starch-containing food with the salivary α-amylase, and to subdivide the food so that it can be mixed more readily with the digestive secretions of the stomach and duodenum.

■ *Swallowing*

Swallowing can be initiated voluntarily, but thereafter it is almost entirely under reflex control. The **swallowing reflex** is a rigidly ordered sequence of events that results in the propulsion of food from the mouth to the stomach, at the same time inhibiting respiration and preventing the entrance of food into the trachea (Fig. 38-13). The afferent limb of the swallowing reflex begins with tactile receptors, most notably those near the opening of the pharynx. Sensory impulses from these receptors are transmitted to certain areas in the medulla. The central integrating areas for swallowing lie in the medulla and lower pons; they are collectively called the "swallowing center." Motor impulses travel from the swallowing center to the musculature of the pharynx and upper esophagus via various cranial nerves.

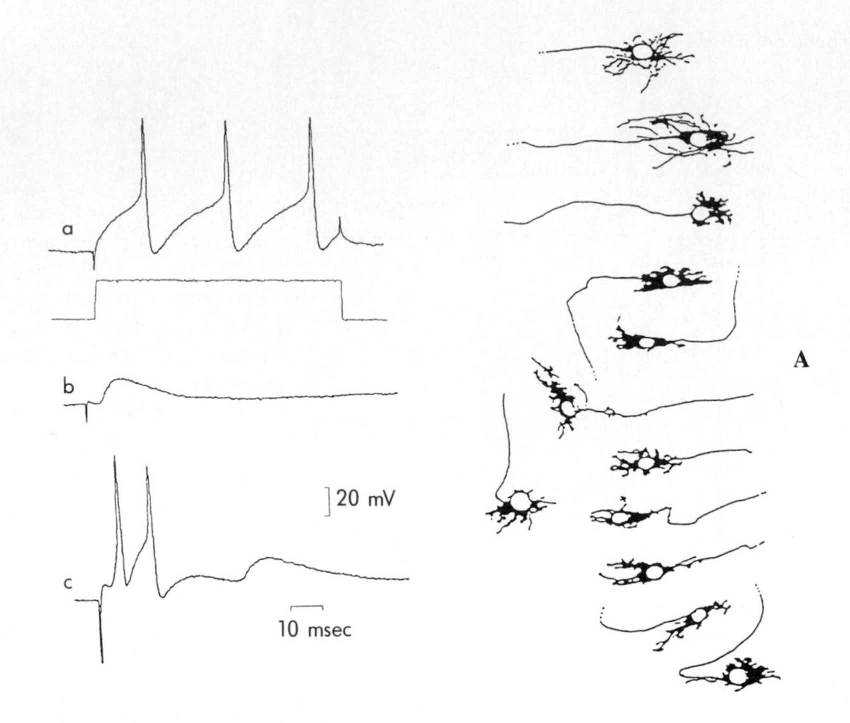

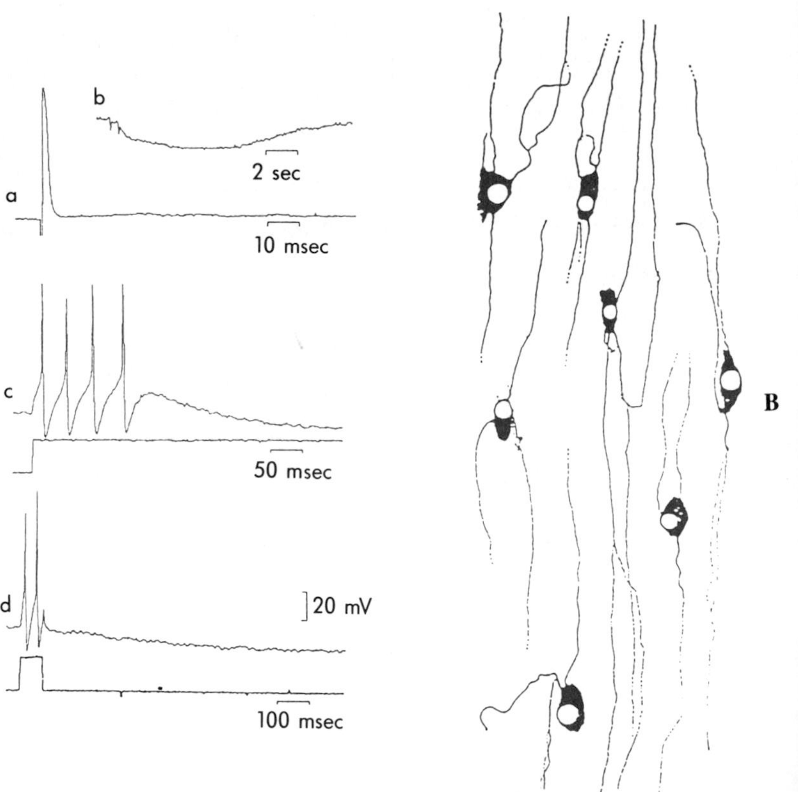

■ **Fig. 38-10** Electrophysiology and morphology of neurons of myenteric ganglia. **A, S-type neurons.** These are motoneurons with characteristic unipolar morphology. S-neurons often fire a train of action potentials in response to maintained current injection (trace a). S-neurons receive fast synaptic input: a single fast EPSP elicited by a minimal stimulus to an internodal strand is shown in trace b. A single larger stimulus to the internodal strand elicits two antidromic action potentials followed by two EPSPs (trace c). **B, AH-type neurons** have a multipolar morphology. A single action potential is fired in response to an antidromic stimulus applied to an internodal strand (trace a). Trace b shows a similar response to that in trace a but on a much slower time base. The action potential spike is not resolved in trace b and the slow time base shows the slow time course of the afterhyperpolarization that characterizes these neurons. In response to maintained intracellular current injection the AH neuron fires only four action potential spikes (trace c). Afterhyperpolarization inhibits further action potentials. In trace d, two action potentials are fired in response to a 75 msec intracellular current injection. Electrophysiological recordings from myenteric neurons of proximal guinea pig colon are unpublished results of Dr. Terence K. Smith (provided by Dr. Smith). (Morphological representations of enteric neurons are redrawn from Furness, JB, Bornstein, JC, Trussell, DC: *Cell Tissue Res* 254:561, 1988.)

The oral or voluntary phase of swallowing is initiated by separating a bolus of food from the mass in the mouth with the tip of the tongue. The bolus to be swallowed is moved upward and backward in the mouth by pressing first the tip of the tongue and later also the more posterior portions of the tongue as well against the hard palate. This forces the bolus into the pharynx, where it stimulates the tactile receptors that initiate the swallowing reflex.

The pharyngeal stage of swallowing involves the following sequence of events, which occur in less than 1 second:

1. The soft palate is pulled upward, and the palatopharyngeal folds move inward toward one

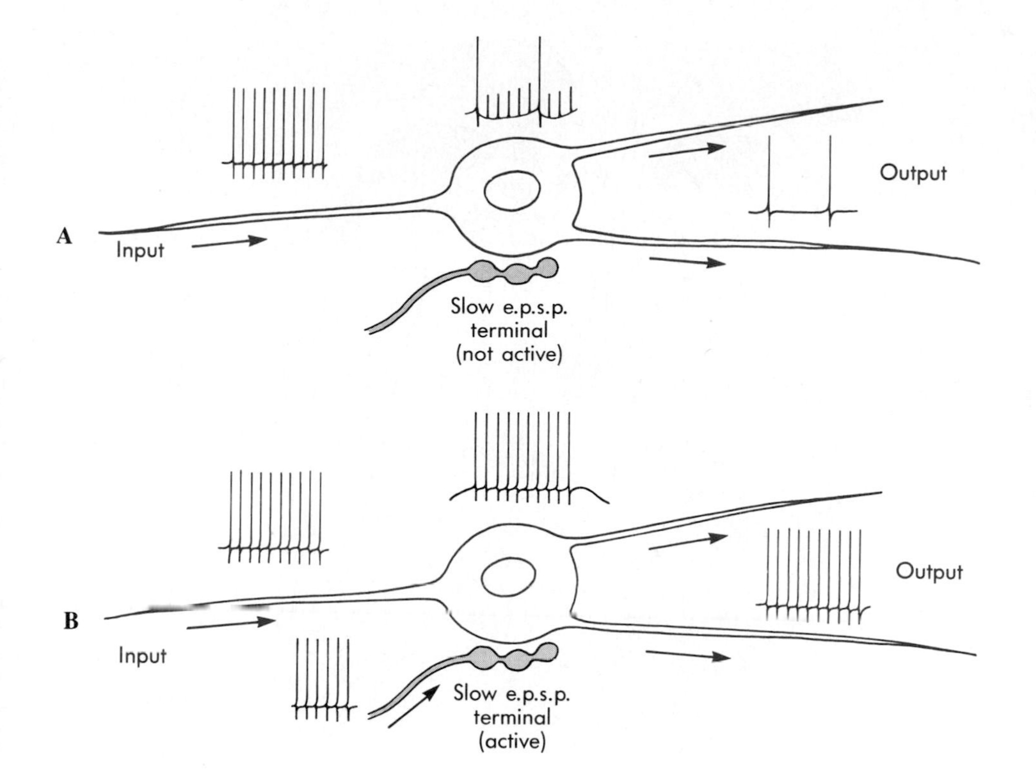

■ **Fig. 38-11** Control of the firing pattern of enteric AH (sensory) neurons. **A,** Response of an AH neuron to a train of excitatory stimuli. Because of the afterhyperpolarization that follows each action potential, the AH cell does not fire an action potential in response to each incoming EPSP. **B,** Modulatory effect of slow EPSPs on the behavior of the AH neuron. The slow EPSP results from closing of K^+ channels and thus counteracts the effect of the open K^+ channels that cause afterhyperpolarization. As a result, during slow EPSP activity, the AH cell is converted to a rapidly firing neuron that can respond to each excitatory stimulus with an action potential. (Modified from Furness, JB and Costa, M: *The enteric nervous system,* Edinburgh, 1987, Churchill Livingstone.)

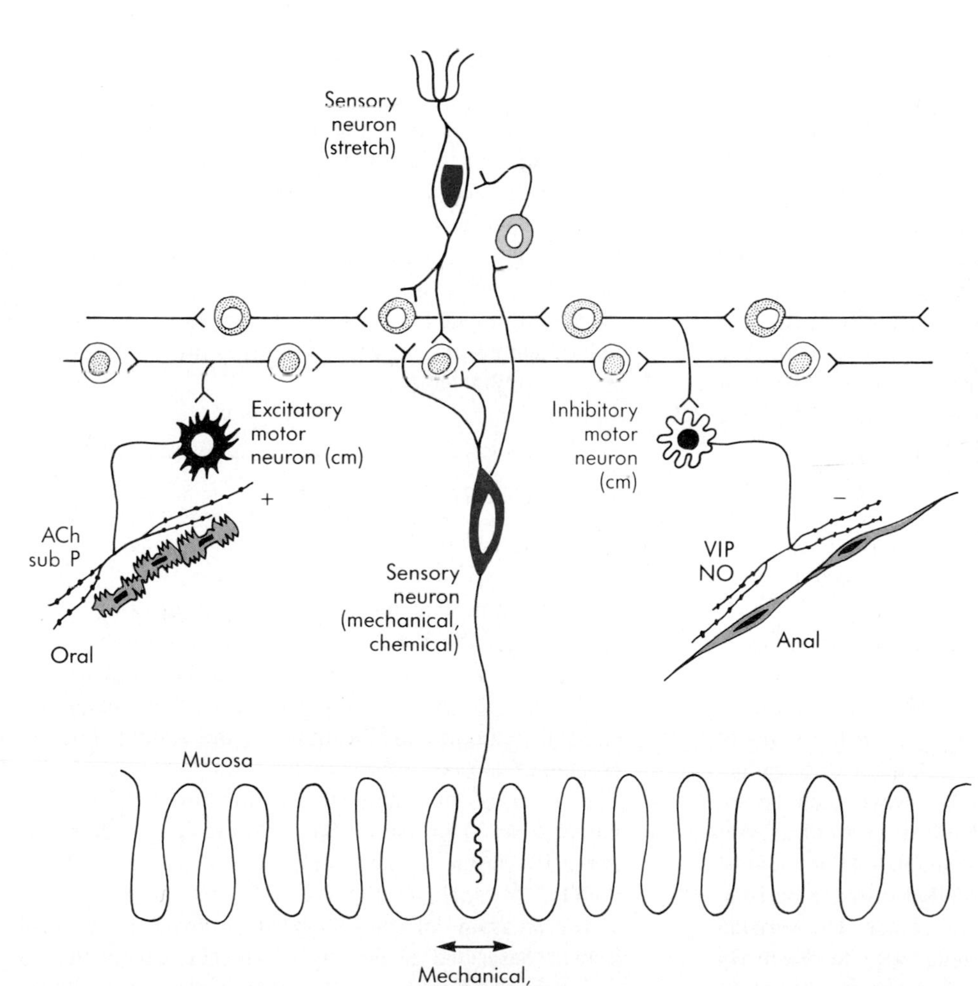

■ **Fig. 38-12** Diagram of circuit of enteric reflexes that control motoneurons to circular muscle (cm) of muscularis externa. In the center of the figure are two sensory neurons: a stretch-sensitive neuron (white cytoplasm, colored nucleus) and mechanico- or chemo-sensitive neuron with its ending in the mucosa (colored cytoplasm, white nucleus). Stimulation of either of these sensory neurons results in activation of ascending (oral) excitatory pathways and descending inhibitory (anal) pathways to circular muscle. Note that the output of the two sensory neurons converges onto the same interneurons and motoneurons. (Courtesy Dr. TK Smith.)

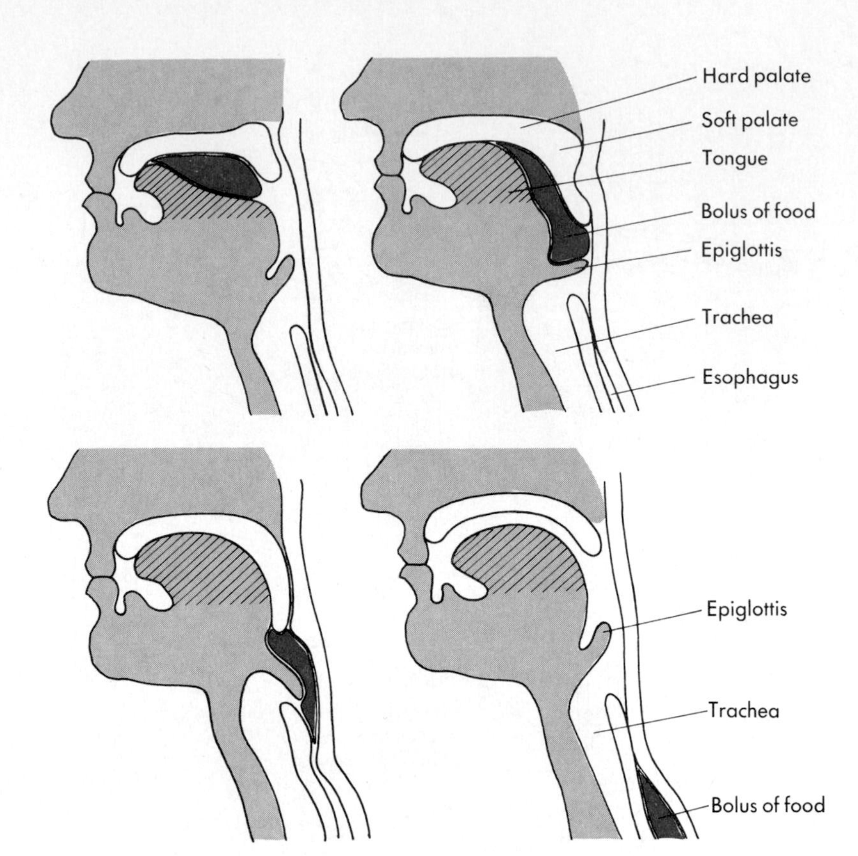

■ **Fig. 38-13** Major events involved in the swallowing reflex. (Modified from Johnson, LR: *Gastrointestinal physiology,* ed 2, St Louis, 1981, Mosby–Year Book.

another. This prevents reflux of food into the nasopharynx and provides a narrow passage through which the food moves into the pharynx.

2. The vocal cords are pulled together, and the epiglottis covers the opening to the larynx. The larynx is moved upward against the epiglottis. These actions prevent food from entering the trachea.
3. The upper esophageal sphincter relaxes to receive the bolus of food. Then the superior constrictor muscles of the pharynx contract strongly to force the bolus deeply into the pharynx.
4. A peristaltic wave is initiated with the contraction of the superior constrictor muscles of the pharynx, and it moves toward the esophagus. This forces the bolus of food through the relaxed upper esophageal sphincter.

During the pharyngeal stage of swallowing, respiration is reflexly inhibited.

The esophageal phase of swallowing also is partially controlled by the swallowing center. After the bolus of food passes the upper esophageal sphincter, the sphincter reflexly constricts. A peristaltic wave then begins just below the upper esophageal sphincter and traverses the entire esophagus in about 10 seconds (Fig. 38-14). This initial wave of peristalsis, called primary peristalsis, is controlled by the swallowing center. The peristaltic wave travels down the esophagus at 3 to 5 cm/second and takes 5 to 10 seconds to reach the stomach. Should the primary peristalsis be insufficient to clear the esophagus of food, the distension of the esophagus would initiate another peristaltic wave that begins at the site of distension and moves downward. This latter type of peristalsis, termed secondary peristalsis, is partially mediated by the enteric nervous system, because it occurs (albeit more weakly) in the extrinsically denervated esophagus. Input from esophageal sensory fibers to the central and enteric nervous systems is involved in modulating esophageal peristalsis.

■ *Esophageal Function*

After food is swallowed, the esophagus functions as a conduit to move the food from the pharynx to the stomach. It is important to prevent air from entering at the upper end of the esophagus and to keep corrosive gastric contents from refluxing back into the esophagus at its lower end. Reflux is particularly problematic because the pressure in the body of the resting thoracic esophagus closely approximates intrathoracic pressure. Thus it is less than atmospheric pressure and less than intraabdominal pressure. The pressure in the short abdominal section of the esophagus mirrors the intraabdominal pressure.

The structure of the esophagus follows the general scheme described earlier in this chapter, except that in the upper third of the esophagus both the inner circular and outer longitudinal muscle layers are striated. In the

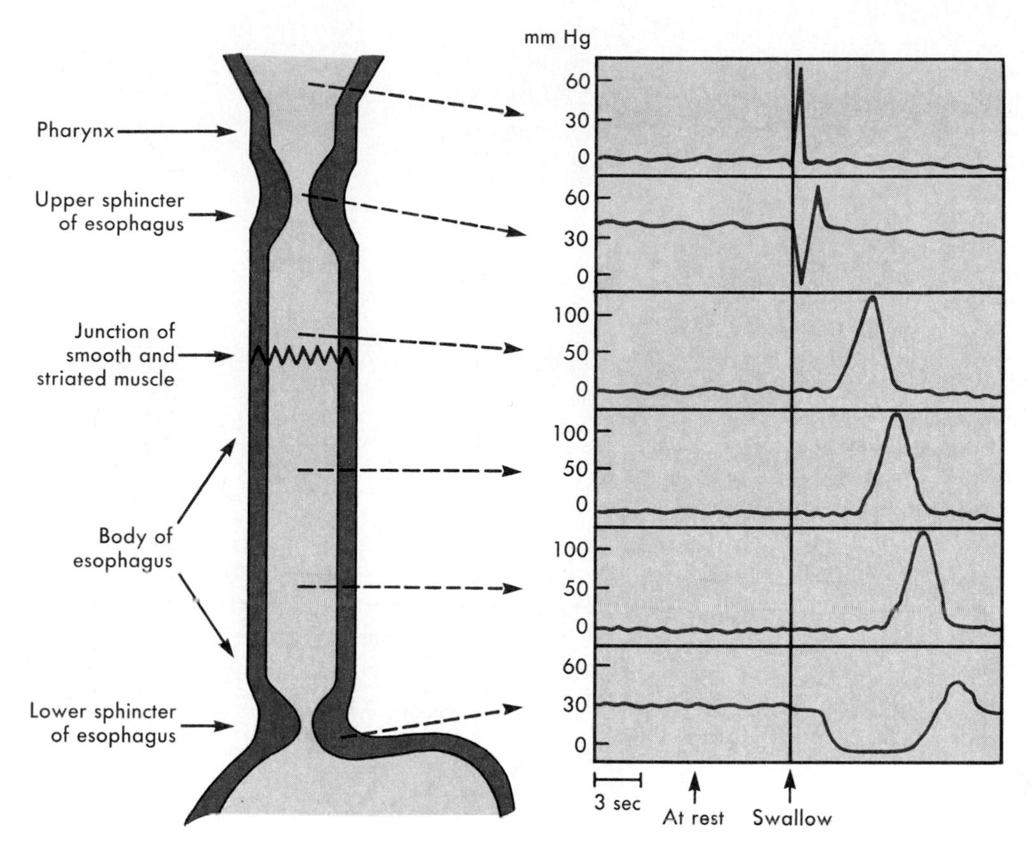

■ **Fig. 38-14** Pressures in the pharynx, esophagus, and esophageal sphincters during swallowing. Note the reflex relaxation of the upper and lower esophageal sphincters and the timing of the relaxation. (From Christensen, JL: In Christensen, J, and Wingate, DL, editors: *A guide to gastrointestinal motility,* Bristol, UK, 1983, John Wright & Sons.)

lower third of the esophagus the muscle layers are composed entirely of smooth muscle cells. In the middle third, skeletal and smooth muscle coexist, there being a gradient from all skeletal above to all smooth below.

The esophageal musculature, both striated and smooth, is primarily innervated by branches of the vagus nerve. Somatic motor fibers of the vagus nerve that originate in the nucleus ambiguus form motor endplates on striated muscle fibers. Visceral motor nerves, primarily from the dorsal motor nucleus of the vagus nerve, are preganglionic parasympathetic fibers. They synapse primarily on the nerve cells of the myenteric plexus. Neurons of the myenteric plexus innervate the smooth muscle cells of the esophagus and communicate with one another. The neural circuits that control the esophagus are schematized in Fig. 38-15.

The upper and lower ends of the esophagus function as sphincters to prevent the entry of air and gastric contents, respectively, into the esophagus. The sphincters are known as the **upper esophageal** (or pharyngeoesophageal) **sphincter** and the **lower esophageal** (or gastroesophageal) **sphincter. The upper esophageal sphincter** is formed by a thickening of the circular layer of striated muscle at the upper end of the esophagus; the sphincter consists of the cricopharyngeus muscle and lower fibers of the inferior pharyngeal constrictor. The pressure in the upper esophageal sphincter is about 40 mm Hg at rest. The lower esophageal sphincter is difficult to identify anatomically, but the lower 1 to 2 cm of the esophagus functions as a sphincter. In normal individuals the pressure at the lower esophageal sphincter is always greater than that in the stomach.

The **lower esophageal sphincter** opens when a wave of esophageal peristalsis reaches it. The opening is vagally mediated. In the absence of esophageal peristalsis the sphincter must remain tightly closed to prevent reflux of gastric contents, which would cause esophagitis and the sensation of heartburn.

■ *Control of the Lower Esophageal Sphincter*

Control of LES tone. The resting pressure in the lower esophageal sphincter (LES) is about 30 mm Hg. The tonic contraction of the circular musculature of the LES is regulated by nerves, both intrinsic and extrinsic, and by hormones and neuromodulators. A significant fraction of basal tone in the LES is mediated by vagal cholinergic nerves. Stimulating sympathetic nerves to the LES also causes contraction. This effect is mediated

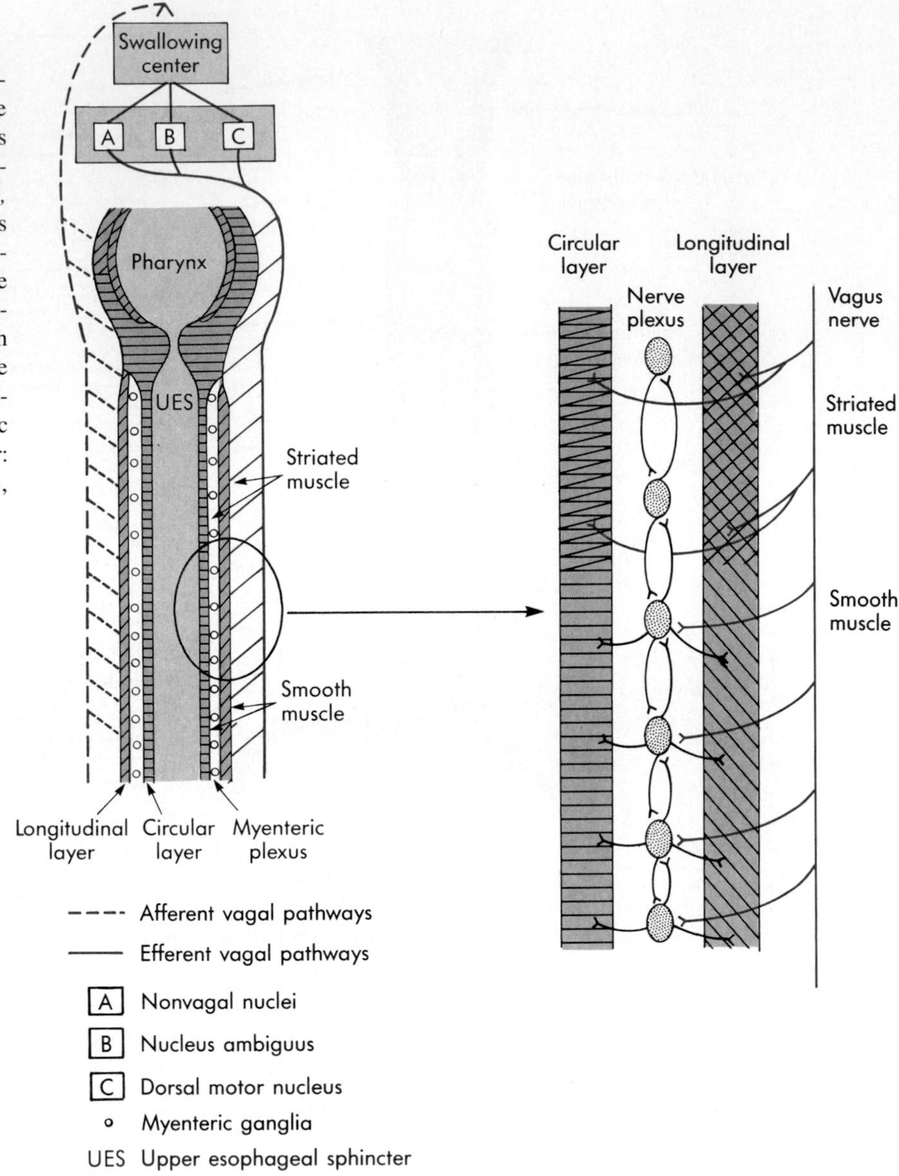

■ **Fig. 38-15** Neural control of esophageal motility. *Left,* Motor neurons reach the esophagus in branches of the vagus nerves and sensory feedback to the swallowing center is carried by vagal afferent fibers. *Right,* Enlarged view of circled region of esophagus to show vagal somatic motor neurons that innervate the striated skeletal muscle of the pharynx and upper esophagus and vagal visceral motoneurons that innervate the smooth muscle of the lower esophagus. Note that the vagal visceral motoneurons terminate predominantly on neurons of the myenteric plexus. (Modified from Johnson, LR, editor: *Gastrointestinal physiology,* ed 3, St Louis, 1985, Mosby–Year Book.)

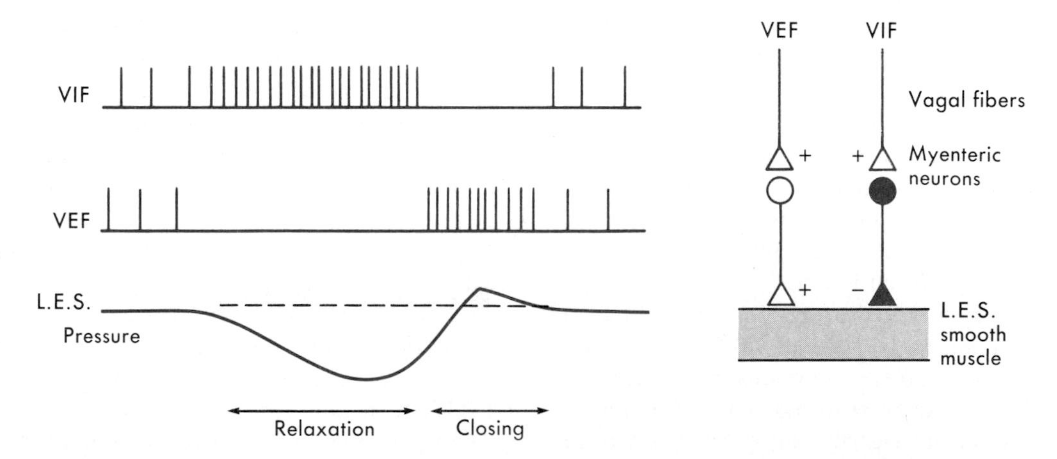

■ **Fig. 38-16** Schematic representation of vagal control of the lower esophageal sphincter *(LES)*. Note that relaxation of the *LES* is associated with an increase in the firing rate in vagal inhibitory fibers *(VIF)* and a decreased frequency of action potentials in vagal excitatory fibers *(VEF)*. Reciprocal changes occur when the sphincter regains its resting tone. (From Miolan, JP, and Roman, C: *J Physiol* [Paris] 74:709, 1978.)

by norepinephrine acting on α-receptors. Norepinephrine may also act presynaptically to inhibit release of acetylcholine from vagal cholinergic fibers. Significant LES tone and regulation of that tone persist when the extrinsic nerves to the LES are sectioned. Hence the enteric nervous system plays a major role in regulating the function of the LES. A number of the gut-brain peptides can influence the tone of the LES, but the physiological roles of these substances remain to be determined.

Relaxation of the LES. The innervation, both intrinsic and extrinsic, of the LES is both excitatory and inhibitory (Fig. 38-16). A major component of the relaxation of the LES that occurs in response to primary peristalsis in the esophagus is mediated by vagal fibers that are inhibitory to the circular muscle and that use VIP and/or NO as their transmitter. A decrease in the impulse traffic in vagal excitatory nerves, which are mostly cholinergic, promotes relaxation of the sphincter. The enteric nervous system helps to control relaxation of the LES, but the mechanisms involved remain to be elucidated.

During swallowing in some individuals the LES fails to relax sufficiently to allow food to enter the stomach, a condition known as **achalasia.** Therapy for achalasia may involve surgically weakening the lower esophageal sphincter. Individuals with **diffuse esophageal spasm** have prolonged and painful contraction of the lower part of the esophagus after swallowing, instead of the normal esophageal peristaltic wave. Achalasia, incompetence of the lower sphincter, and diffuse esophageal spasm are the most common disorders of esophageal function.

Gastric Motility

The motility of the stomach serves the following major functions: (1) to allow the stomach to serve as a reservoir for the large volume of food that may be ingested at a single meal, (2) to fragment food into smaller particles and to mix chyme with gastric secretions so that digestion can begin, and (3) to empty gastric contents into the duodenum at a controlled rate. Fig. 38-17 shows the major anatomical divisions of the stomach.

The stomach has features that allow it to carry out each of these functions. The fundus and the body of the stomach can accommodate volume increases as large as 1.5 L without a marked increase in intragastric pressure. Contractions of the fundus and body are normally weak, so that much of the gastric contents remains relatively unmixed for long periods. Thus the fundus and the body serve the reservoir functions of the stomach. In the antrum, however, contractions are vigorous and thoroughly mix antral chyme with gastric juice and subdivide food into smaller particles. The antral contractions serve to empty the gastric contents in small squirts into the duodenal bulb. *The rate of gastric emptying is adjusted by a number of mechanisms so that chyme is not delivered to the duodenum too rapidly.* The physiological mechanisms that underlie this behavior are discussed later.

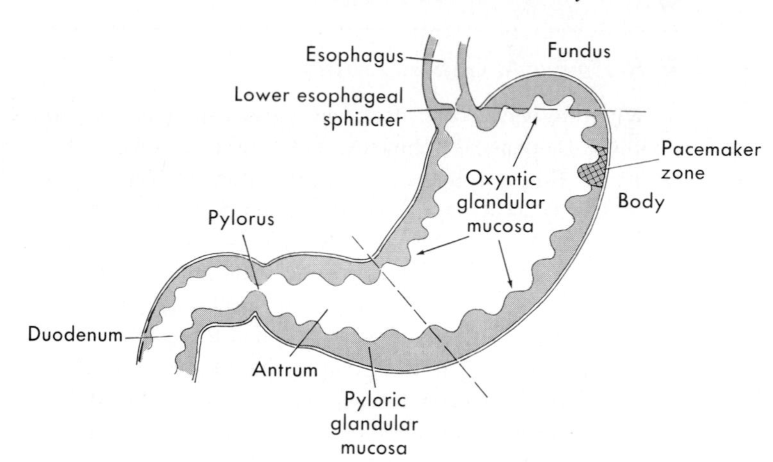

Fig. 38-17 The major anatomical divisions of the stomach.

Structure and Innervation of the Stomach

The basic structure of the gastric wall follows the scheme presented in Fig. 38-1. The circular muscle layer of muscularis externa is more prominent than the longitudinal layer. An incomplete inner layer of obliquely oriented muscle cells is present only on the anterior and posterior sides of the stomach. The muscularis externa of the fundus and the body is relatively thin, but that of the antrum is considerably thicker, and it increases in thickness toward the pylorus.

The stomach is richly innervated by extrinsic nerves and by the neurons of the submucosal and myenteric plexuses. Axons from the cells of the intramural plexuses innervate smooth muscle and secretory cells.

Extrinsic innervation comes via the vagus nerves and from the celiac plexus. In general, cholinergic nerves stimulate gastric smooth muscle motility and gastric secretions, whereas adrenergic fibers inhibit these functions. A number of other gastric transmitter and modulator substances have been identified.

Numerous sensory afferent fibers leave the stomach in the vagus nerves, and some travel with sympathetic nerves. Other sensory fibers are the afferent arms of intrinsic reflex arcs via the intramural plexuses of the stomach. These sensory fibers respond to intragastric pressure, gastric distension, intragastric pH, or pain. The afferent fibers signaling gastric distension and intragastric pressure have been implicated in satiety.

Responses to Gastric Filling

When a wave of esophageal peristalsis reaches the lower esophageal sphincter, the sphincter reflexly relaxes. This is followed by a relaxation, known as **receptive relaxation,** of the fundus and body of the stomach. The stomach also will relax if it is directly filled with gas or liquid. Both of these responses are greatly diminished if the vagus nerves are sectioned; the vagi are a major efferent pathway for reflex relaxation of the stomach. The vagal fibers that mediate this response release VIP. When the stomach relaxes in response to filling, the sensory afferent fibers that report gastric distension constitute the afferent limb of the response.

The smooth muscle of the fundus and body facilitates receptive relaxation and the reservoir function of the stomach. The smooth muscle cells in the fundus have a particularly low resting membrane potential, about −50 mV. At this resting membrane potential the fundic smooth muscle cells are partly contracted. In response to hyperpolarization induced by impulses in inhibitory nerve fibers, the smooth muscle cells relax. The smooth muscle of the body of the stomach has a lower resistance to stretch than the smooth muscle of the antrum. Both of these properties enhance the ability of the fundus and body to accommodate a large increase in volume with little increase in intragastric pressure.

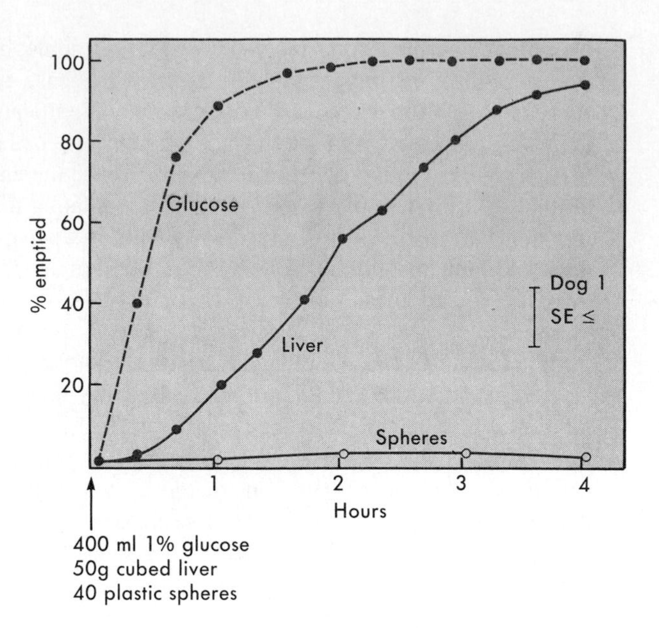

Fig. 38-18 Rates of emptying of different materials from dog stomach. A solution (1% glucose) is emptied faster than a digestible solid (cubed liver). An indigestible solid (plastic spheres, 7 mm in diameter) remains in the stomach under these conditions. (From Hinder, RA, and Kelly, KA: *Am J Physiol* 233:E335, 1977.)

Mixing and Emptying of Gastric Contents

The muscle layers in the fundus and body are thin, and weak contractions are characteristic of these parts of the stomach. As a result, the contents of the fundus and the body tend to form layers based on their density. The gastric contents may remain unmixed for as long as 1 hour after eating. Fats tend to form an oily layer on top of the other gastric contents. Fats thus are emptied later than other gastric contents. Liquids can flow around the mass of food contained in the body of the stomach and are emptied more rapidly into the duodenum. Large or indigestible particles are retained in the stomach for a longer period of time (Fig. 38-18).

Gastric contractions usually begin in the middle of the body of the stomach and travel toward the pylorus. The contractions increase in force and velocity as they approach the gastroduodenal junction. As a result, the major mixing activity occurs in the antrum, the contents of which are mixed rapidly and thoroughly with gastric secretions. Immediately after eating, the antral contractions are relatively weak, but as digestion proceeds, they become more forceful. The frequency of gastric contraction is about three per minute.

Because of the acceleration of the peristaltic wave, the terminal end of the antrum and the pylorus contract almost simultaneously. This is known as **systolic contraction of the antrum.** The peristaltic wave pushes the antral contents ahead of the ring of contraction. Most waves are strong enough to squirt a small fraction of antral contents into the duodenal bulb. The squirt is terminated by the abrupt closure of the pyloric sphincter. Because of its small diameter, the sphincter closes early in the systolic contraction of the antrum. The forceful contraction of the terminal end of the antrum rapidly forces antral contents back into the more proximal part of the antrum. The vigorous backward motion of antral content, called **retropulsion,** effectively mixes the antral contents with gastric secretions and aids in mechanical disruption of food particles.

After an animal feeds, regular contractions of the antrum occur at the frequency of the gastric slow waves. As discussed later, the rate of gastric emptying is regulated by mechanisms that feed back to diminish the force of antral contractions. Contractions of the antrum after ingestion of food vary from moderately forceful to weak.

In a fasted animal a different pattern of antral contractions occurs. The antrum is quiescent for 1 to 2 hours; then a short period of intense electrical and motor activity occurs that lasts for 10 to 20 minutes. This activity is characterized by strong contractions of the antrum with a relaxed pylorus. During this period even large chunks of material that remain from the previous meal are emptied from the stomach (Fig. 38-19). The period of intense contractions is followed by another 1- to 2-hour period of quiescence. This cyclical pattern of contractions in the stomach is part of a pattern of con-

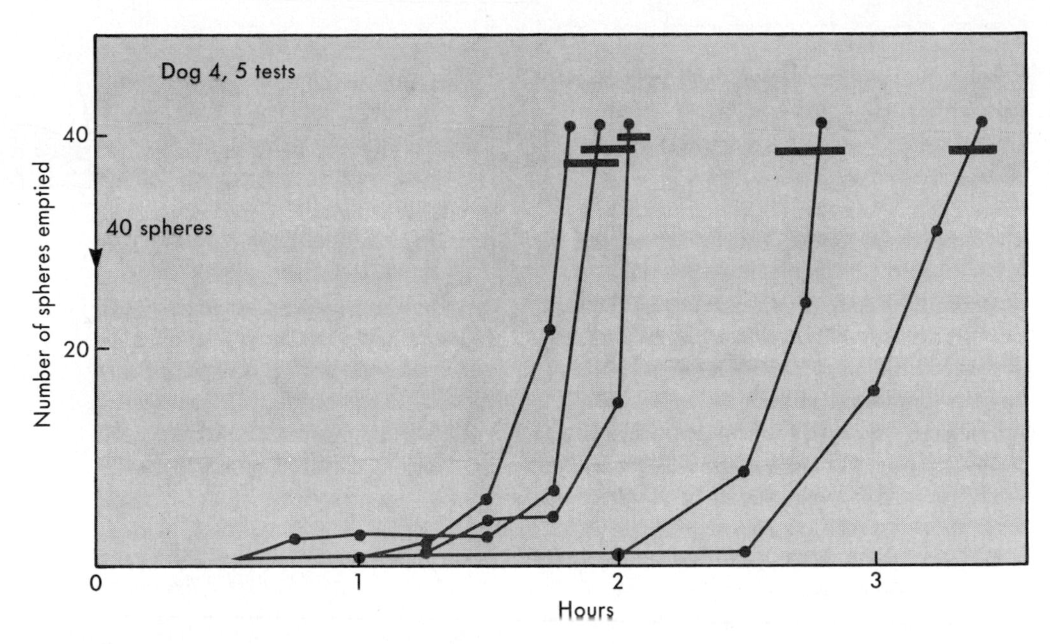

■ **Fig. 38-19** Emptying of plastic spheres (7 mm in diameter) from dog stomach during fasting. Vigorous gastric contractions that are associated with phase III of the migrating myoelectric complex (solid bars) empty the plastic spheres into the duodenum. (From Mroz, CT, and Kelly, KA: *Surgery* 145:369, 1977.)

tractile activity that periodically sweeps from the stomach to the terminal ileum. This pattern of cyclical contractile activity is known as the **migrating myoelectric complex** and is discussed later.

■ *Electrical Activity that Underlies Gastric Contractions*

The gastric peristaltic waves occur at the frequency of the gastric slow waves that are generated by a "pace maker zone" (see Fig. 38-17) on the greater curvature near the middle of the body of the stomach. These waves are conducted toward the pylorus. Isolated strips of gastric smooth muscle have intrinsic rhythmicity and generate slow waves at a characteristic frequency. The pacemaker zone has the highest intrinsic frequency, and for that reason acts as the pacemaker. In the human stomach the frequency of slow waves is about three per minute. In an individual at rest the frequency of slow waves is relatively constant. Injections of gastrin can increase the frequency of slow waves to about four per minute. Secretin can slow the frequency of the slow waves or, in high doses, can even abolish them. The physiological roles of these responses to gastrin and secretin have not been established.

The gastric slow wave is triphasic (Fig. 38-20). Its shape is reminiscent of action potentials in cardiac muscle. However, the gastric slow wave lasts about 10 times longer than the cardiac action potential and does not overshoot. The initial, rapid depolarizing phase of the slow wave appears to be caused by entry of Ca^{++} via voltage-gated Ca^{++} channels. The plateau of the slow wave is believed to be caused by the entry of Ca^{++} and Na^{+} through slower voltage-gated channels. Repolarization of the membrane is associated with a delayed increase in K^{+} conductance.

Gastric smooth muscle contracts when the depolarization during the slow wave exceeds the threshold for contraction (see Fig. 38-20). The greater the extent of depolarization and the longer the cell remains depolarized above the mechanical threshold, the greater the

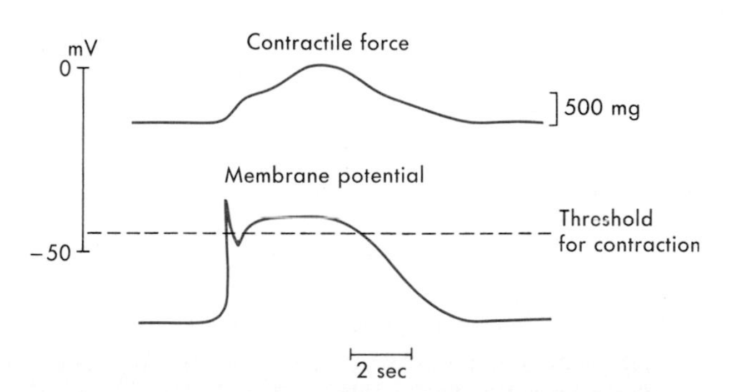

■ **Fig. 38-20** Relationship between contraction of smooth muscle of dog stomach *(upper tracing)* and intracellularly recorded slow wave *(lower tracing)*. Note the triphasic shape of the slow wave in gastric smooth muscle. Contraction occurs when the depolarizing phase of the slow wave exceeds the threshold for contraction, even though there are no action potential spikes on the plateau of the slow wave. (From Szurszewski, J.: In Johnson, LR, editor: *Physiology of the gastrointestinal tract,* New York, 1981, Raven Press.)

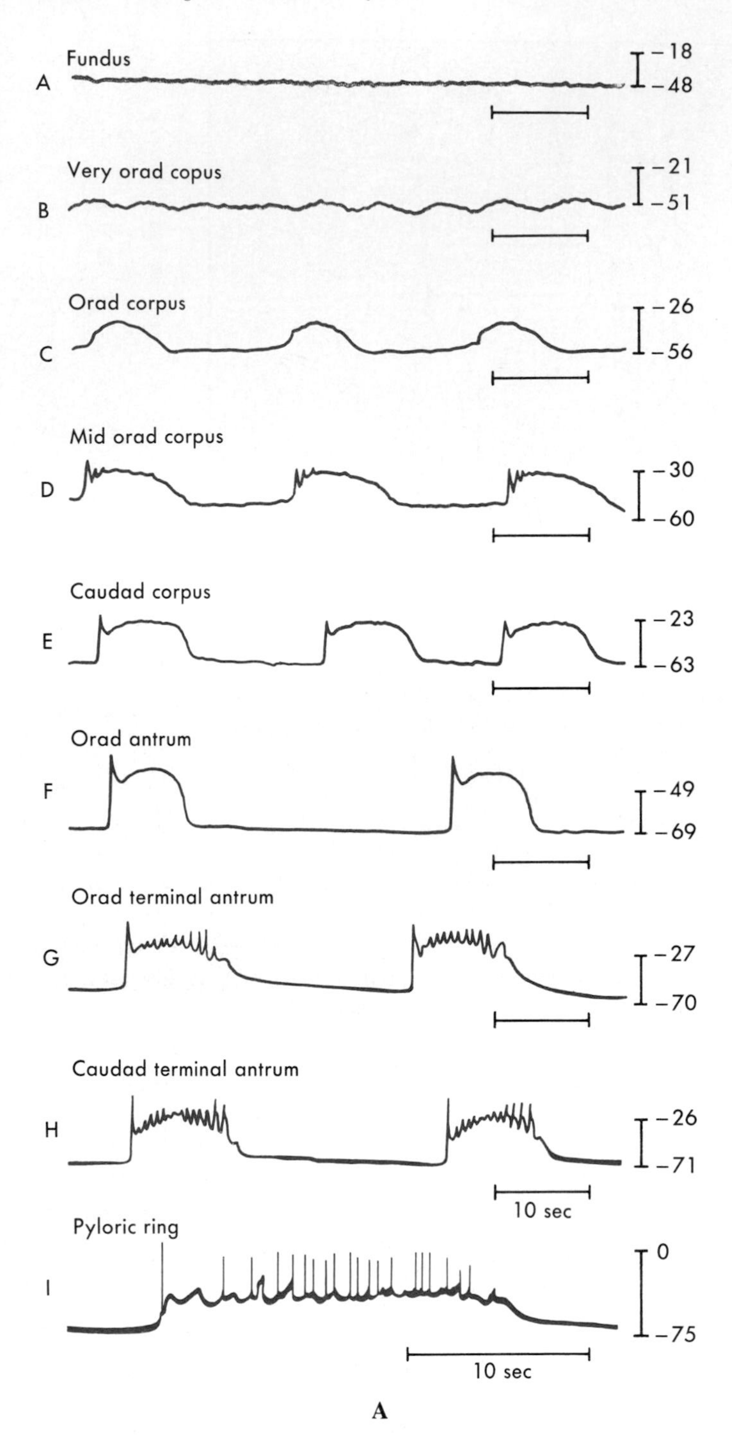

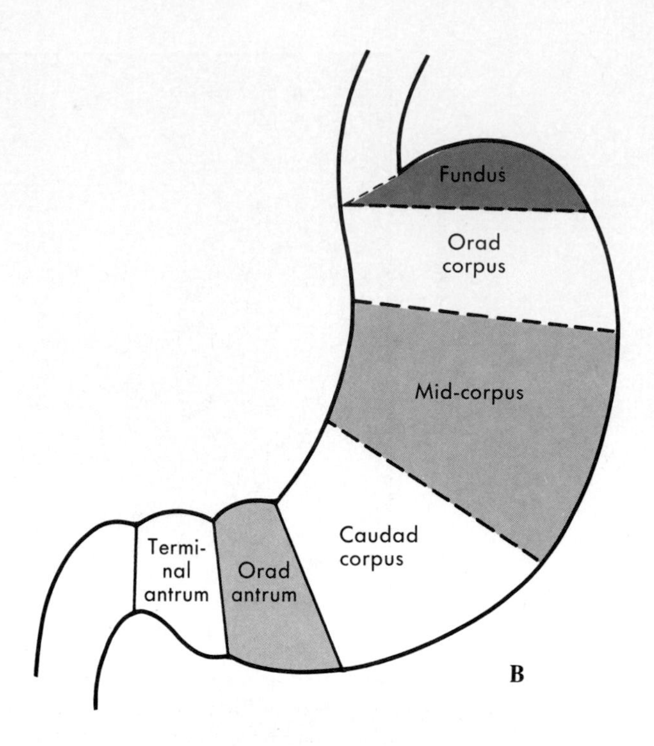

■ **Fig. 38-21** Intracellular recordings, **A,** of the electrical activity in smooth muscle cells of isolated strips of various regions, **B,** of a dog's stomach. Note that the slow waves are absent in the fundus and weak in the oral corpus—and gain in strength and definition toward the antrum. Only in the terminal antrum do action potential spikes occur on the plateaus of the slow waves. The action potential spikes are associated with more vigorous contractions of the terminal antrum. In the intact stomach the slow waves in the different parts of the stomach have the same frequency because they are driven by the same pacemaker. In these records the intrinsic slow wave frequency differs in the isolated strips of muscle. (From Szurszewski, J.: In Johnson, LR, editor: *Physiology of the gastrointestinal tract,* New York, 1981, Raven Press.)

force of contraction. Acetylcholine and gastrin stimulate gastric contractility by increasing the amplitude and duration of the plateau phase of the gastric slow wave. Norepinephrine and neurotensin, which reduce the force of gastric contractions, decrease the extent of depolarization and the duration of the plateau phase.

In the terminal antrum and pylorus, wavelike depolarizations and action potential spikes are generated during the plateau phase of the slow wave (Fig. 38-21). The action potential spikes are associated with increased force of contraction of the antral muscle. The force of antral contraction varies greatly from moment to moment, with periods of moderate peristaltic mixing movements being interspersed among periods of intense propulsive contractions. *The more vigorous contractions occur in response to trains of action potential spikes during the plateaus of the slow waves.*

■ *The Gastroduodenal Junction*

The pylorus separates the gastric antrum from the first part of the duodenum, the duodenal bulb (or cap). It is debatable whether the pylorus constitutes a true anatomical sphincter, but it functions physiologically as a sphincter in many respects. The circular smooth muscle of the pylorus forms two ring-like thickenings that are followed by a connective tissue ring that separates pylorus from duodenum (Fig. 38-22). The mucosa, submucosa, and muscle layers of the pylorus and duodenal bulb are separate, with the exception of a few longitudinal muscle fibers that cross over the junction. How-

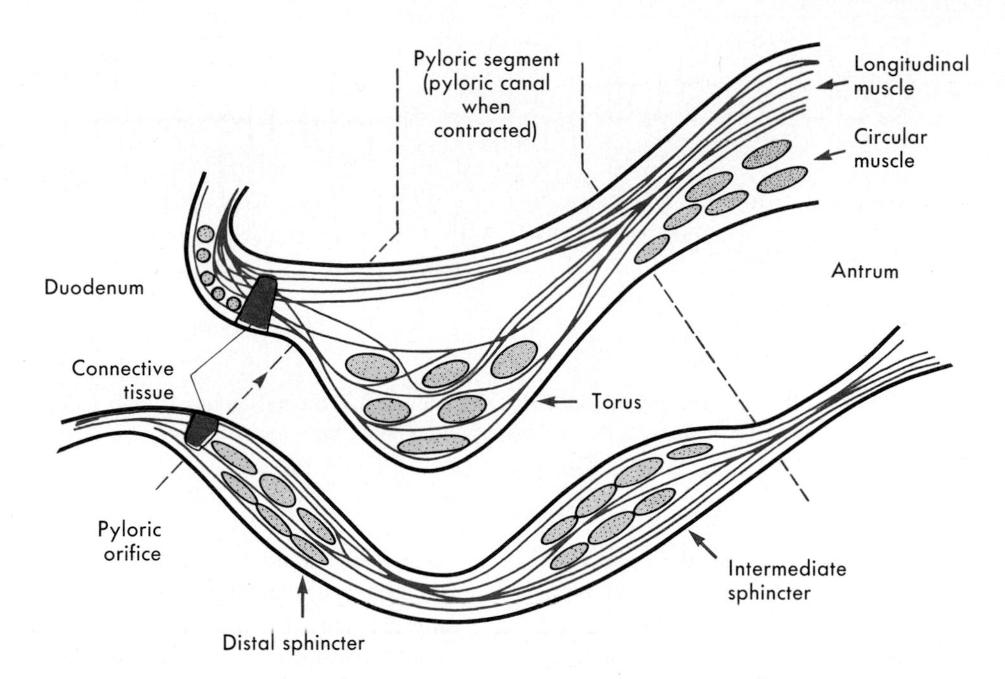

■ **Fig. 38-22** Diagrammatic representation of a longitudinal section through the pyloric sphincter. Note the connective tissue ring that separates the pyloric sphincter from the duodenum. (Modified from Schulze-Delrier, K, et al.: In Roman, C, editor: *Gastrointestinal motility,* 1984, Lancaster, UK, MTP Press.)

ever, the myenteric plexuses of the pylorus and duodenal bulb are continuous.

The duodenum has a basic electrical rhythm of 10 to 12 per minute, far faster than the 3 per minute of the stomach. The duodenal bulb is influenced by the basic electrical rhythms of both the stomach and the postbulbar duodenum. It thus contracts somewhat irregularly. The antrum and duodenum are coordinated; when the antrum contracts, the duodenal bulb is relaxed.

The essential functions of the gastroduodenal junction are (1) to allow the carefully regulated emptying of gastric contents at a rate commensurate with the ability of the duodenum to process the chyme, and (2) to prevent regurgitation of duodenal contents back into the stomach. The gastric mucosa is highly resistant to acid, but it may be damaged by bile. The duodenal mucosa has the opposite properties. Thus too rapid gastric emptying may lead to duodenal ulcers, whereas regurgitation of duodenal contents often contributes to gastric ulcers.

The pylorus is densely innervated by both vagal and sympathetic nerve fibers. Sympathetic postganglionic fibers release norepinephrine, which acts on α-adrenergic receptors to increase constriction of the sphincter. Vagal fibers are both excitatory and inhibitory to pyloric smooth muscle. Cholinergic vagal fibers stimulate constriction of the sphincter. Inhibitory vagal fibers release another transmitter, possibly VIP, that mediates relaxation of the sphincter.

Cholecystokinin, gastrin, gastric inhibitory peptide, and secretin all cause constriction of the pyloric sphincter. Whether these hormones physiologically modulate the function of the pyloric sphincter is not known. Cholecystokinin is a good candidate for a physiological regulator because it constricts the pyloric sphincter at concentrations that activate gallbladder contraction.

■ *Regulation of the Rate of Gastric Emptying*

The emptying of gastric contents is regulated by both neural and humoral mechanisms. The duodenal and jejunal mucosa have receptors that sense acidity, osmotic pressure, and fat content.

The presence of fatty acids or monoglycerides in the duodenum dramatically decreases the rate of gastric emptying. The rate of the gastric slow waves remains essentially unchanged, but the contractility of the pyloric sphincter is greatly increased so that the rate of gastric emptying is much less.

The chyme that leaves the stomach is usually hypertonic, and it becomes more hypertonic because of the action of the digestive enzymes in the duodenum. Hypertonic solutions in the duodenum slow gastric emptying.

Duodenal contents with a pH less than about 3.5 retard gastric emptying. The presence of amino acids and peptides in the duodenum also may slow gastric emptying.

As a result of these mechanisms, (1) fat is not emptied into the duodenum at a rate greater than that at which it can be emulsified by the bile acids and lecithin

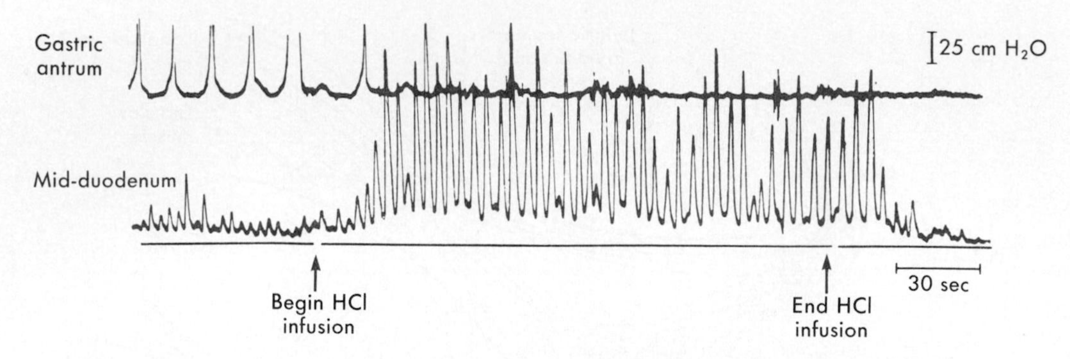

■ **Fig. 38-23** Effect of instilling 100 mM HCl at 6 ml/min into the duodenum of a dog. Contractile activities of gastric antrum *(upper tracing)* and mid duodenum *(lower tracing)* are shown. (From Brink, BM, et al.: *Gut* 6:163, 1965.)

of the bile; (2) acid is not dumped into the duodenum more rapidly than it can be neutralized by pancreatic and duodenal secretions and by other mechanisms; and (3) in general, the rates at which the other components of chyme are presented to the small intestine correspond to the rates at which the small intestine can process those components.

Mechanisms that regulate gastric emptying. The slowing of gastric emptying in response to fatty acids, low pH, or hypertonicity of duodenal contents is mediated by neural and hormonal mechanisms. The interplay between the neural and hormonal mechanisms and their relative importance are not well understood. As an example, consider the response of gastric emptying to the presence of acid in the duodenum.

Acid in the duodenum. Fig. 38-23 shows that when acid is instilled into the duodenum of a dog, the force of contractions of the gastric antrum promptly decreases and duodenal motility promptly increases. This response has a neural component, because cutting the branches of the vagus to the stomach and duodenum markedly decreases the response. The presence of acid in the duodenum also releases **secretin,** which diminishes the rate of gastric emptying by inhibiting antral contractions and stimulating contraction of the pyloric sphincter. Secretin also stimulates the output of the bicarbonate-rich secretions of the pancreas and liver. These secretions buffer the duodenal pH. Partly for this reason, the duodenal pH that elicits the response shown in Fig. 38-23 is much lower than the pH that exists in the duodenum under physiological conditions. Neural and hormonal mechanisms that appear to control the rate of gastric emptying are schematically depicted in Fig. 38-24.

Fat-digestion products. The presence of fatty acids—as well as monoglycerides and diglycerides—in the duodenum and jejunum activates neural and hormonal mechanisms that decrease the rate of gastric emptying. Unsaturated fatty acids elicit this response more effectively than saturated fatty acids. Fatty acids of 14 carbons or more are more effective than short-chain and medium-chain fatty acids. Although the decreased gastric emptying may have a neural component, this response is probably caused mainly by release of **cholecystokinin** from the duodenum and jejunum. Cholecystokinin stimulates contractions of the gastric antrum and constriction of the pyloric sphincter. The net effect of cholecystokinin is to decrease the rate of gastric emptying. Physiological levels of cholecystokinin have been shown to diminish the rate of gastric emptying. The presence of fatty acids in the duodenum and jejunum also elicits the release of another hormone, **gastric inhibitory peptide,** that decreases the rate of gastric emptying. A similar decrease in the rate of gastric emptying is elicited by the presence of fatty acids in the ileum.

Osmolarity of duodenal contents. When hyperosmolar solutions are present in the duodenum and jejunum, the rate of gastric emptying is slowed. Osmoreceptors in the mucosa or submucosa of the duodenum and jejunum may be responsible for this reaction. The decreased gastric emptying in response to hyperosmolarity may have a neural component. Hyperosmolarity in the duodenum releases an unidentified hormone that diminishes the rate of gastric emptying.

Peptides and amino acids in the duodenum. Peptides and amino acids release **gastrin** from G cells located in the antrum of the stomach and the duodenum. Gastrin acts, probably at physiological levels, to increase the strength of antral contractions and to increase constriction of the pyloric sphincter. Under most circumstances the net effect of gastrin is to diminish the rate of gastric emptying. The presence of tryptophan in the duodenum strongly inhibits the rate of gastric emptying. Gastrin is involved in the response to tryptophan.

■ *Vomiting*

Vomiting is the expulsion of gastric (and sometimes duodenal) contents from the gastrointestinal tract via the mouth. Vomiting often is preceded by a feeling of nau-

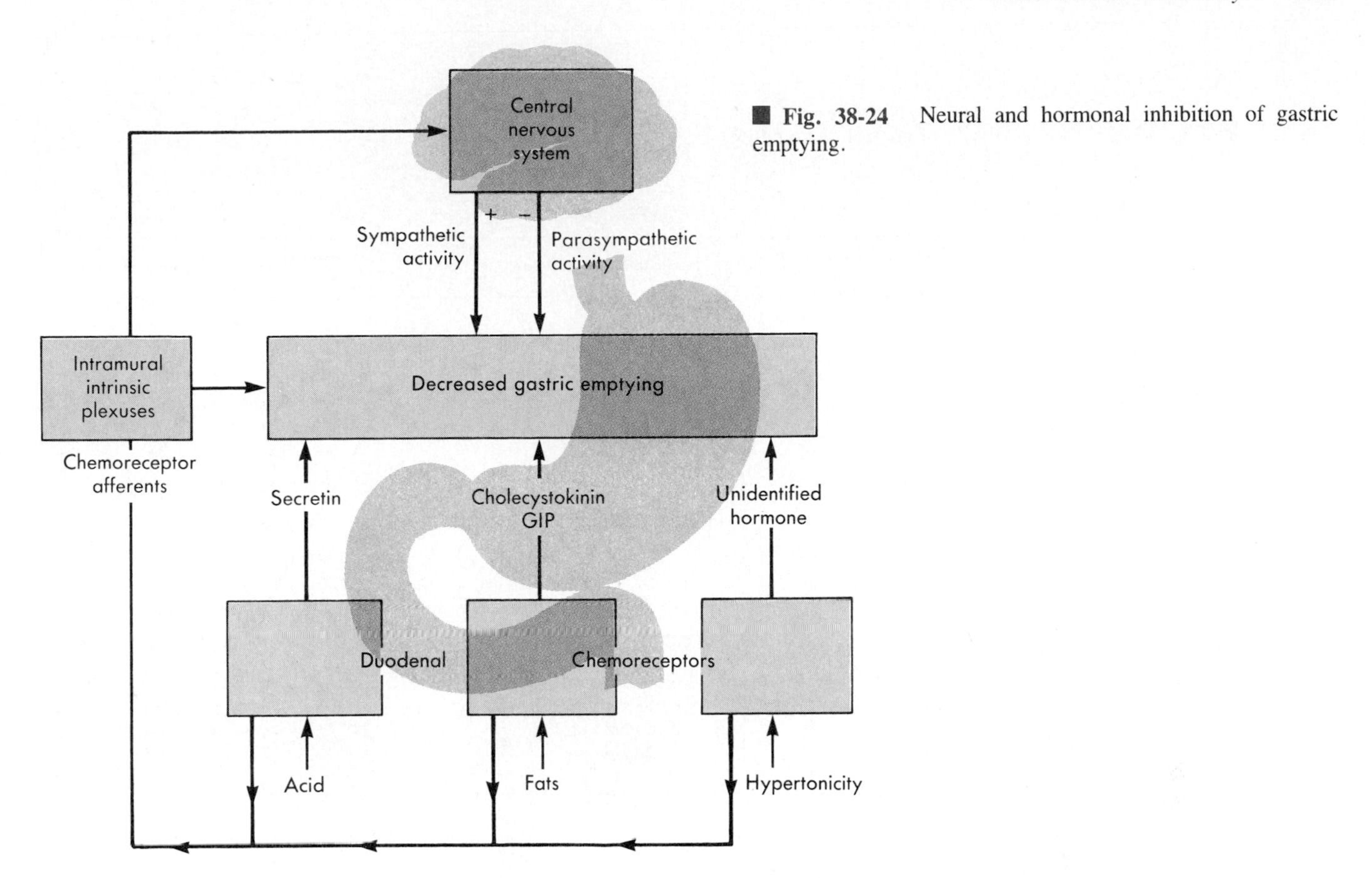

■ **Fig. 38-24** Neural and hormonal inhibition of gastric emptying.

sea, a rapid or irregular heart beat, dizziness, sweating, pallor, and pupillary dilation. Vomiting usually is also preceded by **retching,** in which gastric contents are forced up into the esophagus but do not enter the pharynx. A series of retches of increasing strength often precedes vomiting.

Vomiting is a reflex behavior controlled and coordinated by certain medullary centers (Fig. 38-25). Electrical stimulation of an area of the medulla known as the **vomiting center** elicits prompt vomiting without preliminary retching. Stimulation of another medullary locus leads to retching without subsequent vomiting. It appears that important interactions between these two areas in the medulla occur in typical vomiting. A large number of different areas in the body have receptors that provide afferent input to the vomiting center. Distension of the stomach and duodenum is a strong stimulus that elicits vomiting. Tickling the back of the throat, painful injury to the genitourinary system, dizziness, and certain other stimuli are among the diverse events that can bring about vomiting.

Certain chemicals, called **emetics** can elicit vomiting. Some emetics do this by stimulating receptors in the stomach or more commonly in the duodenum. The widely used emetic ipecac works via duodenal receptors. Certain other emetics act at the level of the central nervous system on receptors in the floor of the fourth ventricle in an area known as the **chemoreceptor trigger zone.** The chemoreceptor trigger zone is in or near the area postrema, which lies on the blood side of the blood-brain barrier, and thus can be reached by most blood-borne substances.

When the vomiting reflex is initiated, the sequence of events is the same regardless of the stimulus that initiates the reflex. Early events in the vomiting reflex include a wave of reverse peristalsis that sweeps from the middle of the small intestine to the duodenum. The pyloric sphincter and the stomach relax to receive intestinal contents. The gastric relaxation is mediated by vagal inhibitory fibers. Then a forced inspiration occurs against a closed glottis. This results in a decreased intrathoracic pressure, whereas the lowering of the diaphragm causes an increase in intraabdominal pressure. Then a forceful contraction of abdominal muscles sharply elevates intraabdominal pressure, driving gastric contents into the esophagus. The lower esophageal sphincter relaxes reflexly to receive gastric contents, and the pylorus and antrum contract reflexly, which prevents orthograde flow of gastric contents.

When a person retches, the upper esophageal sphincter remains closed, preventing vomiting. When the respiratory and abdominal muscles relax, the esophagus is emptied by secondary peristalsis into the stomach. Often a series of stronger and stronger retches precedes vomiting.

When a person vomits, the rapid propulsion of gastric contents into the esophagus is accompanied by a reflex relaxation of the upper esophageal sphincter. Re-

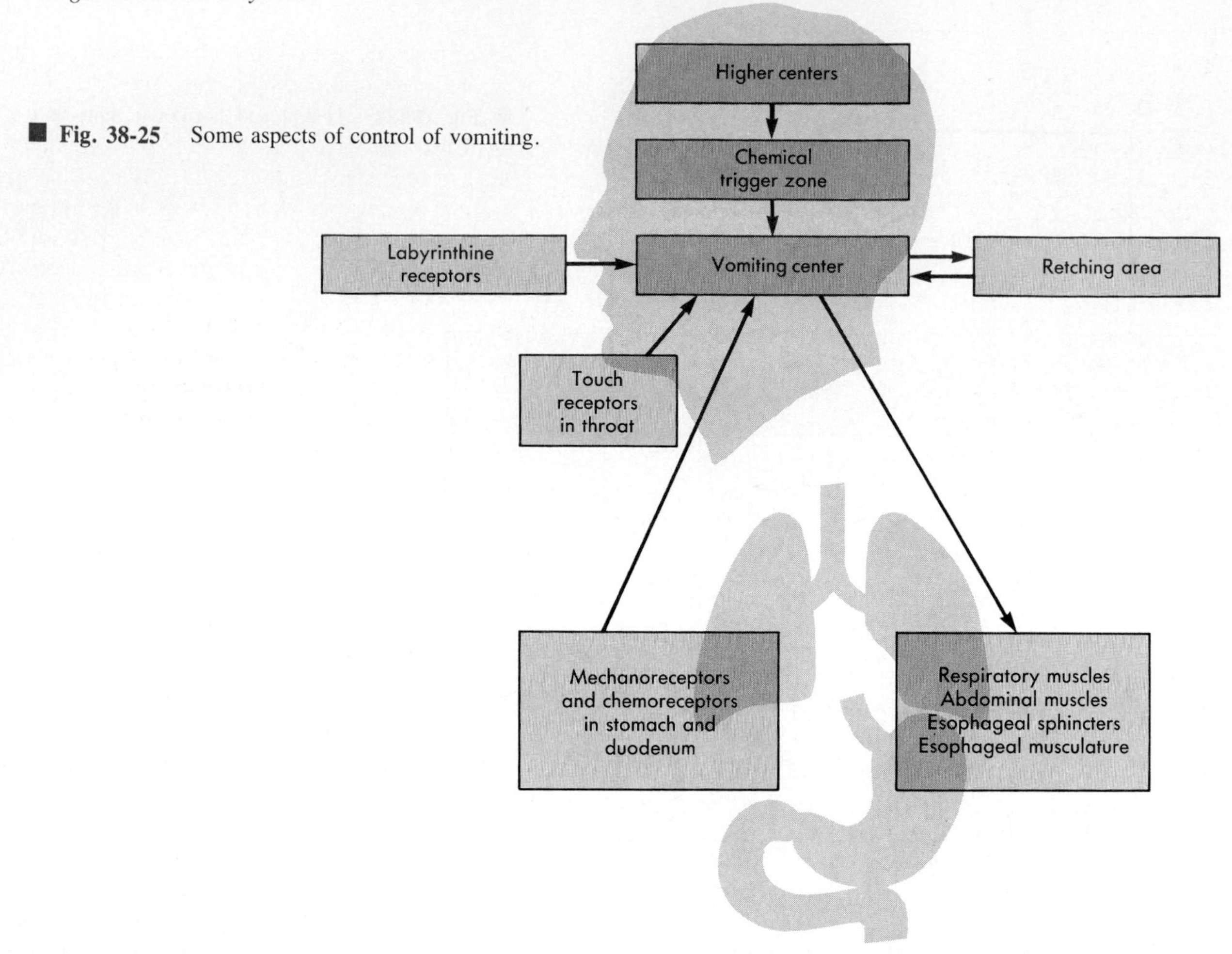

■ **Fig. 38-25** Some aspects of control of vomiting.

laxation is accomplished by a forward movement of the hyoid bone and the larynx. This allows vomitus to be projected into the pharynx and mouth. Entry of vomitus into the trachea is prevented by approximation of the vocal chords, closure of the glottis, and inhibition of inspiration.

■ *Motility of the Small Intestine*

The small intestine, the largest segment of the gastrointestinal system, accounts for about three fourths of the length of the human gastrointestinal tract. The small intestine is about 5 m in length, and chyme typically takes 2 to 4 hours to traverse the small intestine. The first 25 cm or so of the small intestine is the duodenum, which has no mesentery and can be distinguished from the rest of the small intestine histologically. The remaining small intestine is divided into the jejunum and the ileum. The jejunum is more proximal and makes up about 40% of the length of the small bowel. The ileum is the distal part of the small intestine, and it accounts for approximately 60% of the length of the small intestine.

The small intestine, particularly the duodenum and jejunum, is the site of most digestion and absorption. The movements of the small intestine mix chyme with digestive secretions, bring fresh chyme into contact with the absorptive surface of the microvilli, and propel chyme toward the colon.

The most frequent type of movement of the small intestine is termed **segmentation.** Segmentation is characterized by closely spaced contractions of the circular muscle layer. The contractions divide the small intestine into small neighboring segments. In rhythmic segmentation the sites of the circular contractions alternate, so that a given segment of gut contracts and then relaxes. Segmentation is effective at mixing chyme with digestive secretions and bringing fresh chyme into contact with the mucosal surface. **Peristalsis** is the progressive contraction of successive sections of circular smooth muscle. The contractions move along the gastrointestinal tract in an aboral direction. Short peristaltic waves do occur in the small intestine, but they usually involve only a small length of intestine. Peristaltic rushes that travel along much of the length of the small bowel are rare.

As in other parts of the digestive tract, the slow waves of the smooth muscle cells determine the timing of intestinal contractions. The intramural plexuses can control segmentation and short peristaltic movements in the absence of extrinsic innervation. However, extrinsic nerves mediate certain long-range reflexes and modulate the excitability of small intestinal smooth muscle.

Electrical Activity of Small Intestinal Smooth Muscle

Regular slow waves occur all along the small intestine (see Fig. 38-5). The frequency is highest (11 to 13 per minute in humans) in the duodenum, and it declines along the length of the small bowel (to a minimum 8 or 9 per minute in humans in the terminal part of the ileum). As in the stomach, the slow waves may or may not be accompanied by bursts of action potential spikes during the depolarizing part of the slow waves. The slow wave frequency determines the maximum possible frequency of intestinal contractions. The time interval between intestinal contractions tends to be some multiple of the time interval between slow waves. Action potential bursts are localized to short segments of the intestine.

From the duodenum toward the ileum, there tends to be a phase lag in the small intestinal slow waves. Thus it appears that the slow waves are propagated in the orthograde direction. When peristaltic contractions do occur, they are propagated along the intestine at the apparent velocity of the slow waves.

The basic electrical rhythm of the small intestine is independent of extrinsic innervation; the extrinsically denervated intestine can carry out segmentation and short peristaltic movements. The frequency of the action potential spike bursts that enhance contractile force depends on the excitability of the smooth muscle cells of the small intestine. The excitability of the smooth muscle (and thus the frequency of spikes) is influenced by circulating hormones, by the autonomic nervous system, and by enteric neurons. Excitability is enhanced by parasympathetic nerves and is inhibited by sympathetic nerves, both acting primarily on the intramural plexuses. Even though much of the direct control of intestinal motility resides in the intramural plexuses, the extrinsic innervation of the small intestine (parasympathetic via the vagus nerves and sympathetic nerves from the celiac and superior mesenteric plexuses) are important in modulating contractile activity. The extrinsic neural circuits are essential for certain long-range intestinal reflexes discussed later.

Contractile Behavior of the Small Intestine

Contractions of the duodenal bulb mix chyme with pancreatic and biliary secretions, and the contractions propel the chyme along the duodenum. Contractions of the duodenal bulb follow antral systolic contractions. This helps prevent regurgitation of duodenal contents back into the stomach. The basic electrical rhythm of the stomach is about 3 per minute, and that in the duodenal bulb is about 11 per minute. Via the longitudinal muscle fibers that cross from stomach to duodenum, and

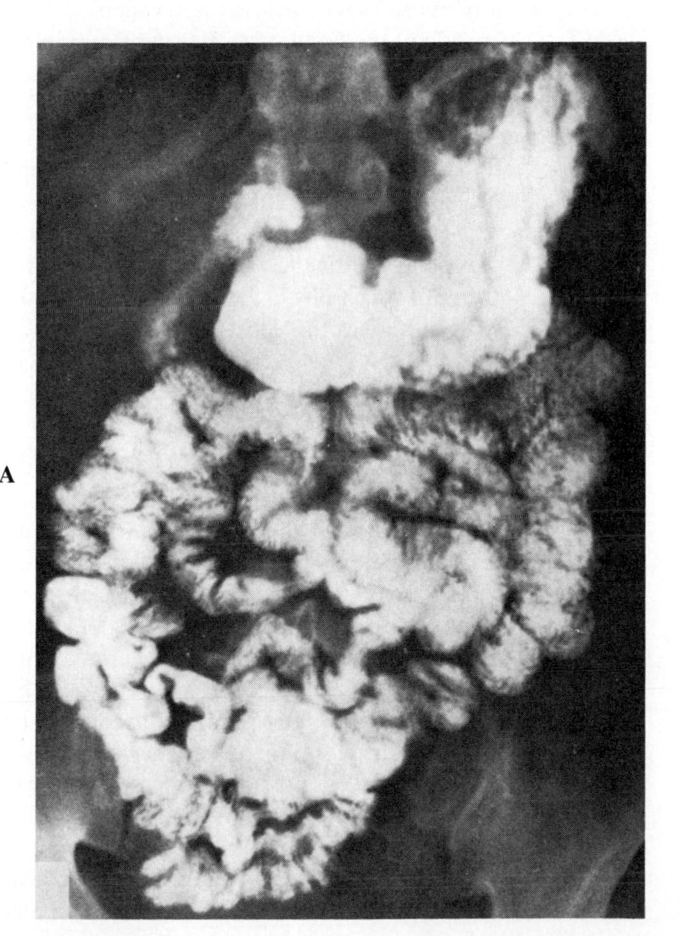

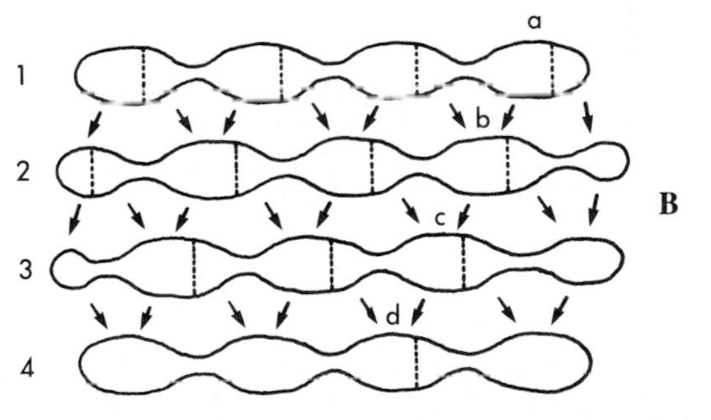

Fig. 38-26 **A,** X-ray view showing the stomach and small intestine filled with barium contrast medium in a normal individual. Note that segmentation of the small intestine divides its contents into ovoid segments. **B,** The sequence of segmental contractions in a portion of a cat's small intestine. Lines 1 through 4 indicate successive patterns in time (note the return in line 4 of the same pattern that existed in line 1). The dotted lines indicate where contractions will occur next. The arrows show the direction of chyme movement. In this case segmentation occurred from 18 to 21 times per minute. (**A** from Gardner, EM, et al.: *Anatomy: a regional study of human structure,* ed 4, Philadelphia, 1975, WB Saunders Co. **B** redrawn from Cannon, WB: *Am J Physiol* 6:251, 1902.)

through enteric neurons, the gastric slow waves can influence the duodenal smooth muscle. Thus about every fifth duodenal slow wave is often somewhat increased in amplitude because of the propagated gastric slow wave.

Segmentation is the most frequent type of movement by the small intestine. Segmentation is characterized by localized contractions of rings of the circular smooth muscle. These contractions divide the small intestinal contents into oval segments (Fig. 38-26). When the contracted segments relax, neighboring segments contract. Segmentation occurs at a frequency similar to that of the small intestinal slow waves: about 11 or 12 contractions per minute in the duodenum and 8 or 9 contractions per minute in the ileum. The frequency of segmental contractions is not as constant as that of the slow waves. A group of sequential contractions tends to be followed by a period of rest. The decrease in segmental contraction frequency along the small intestine is illustrated in Fig. 38-27.

In the jejunum, contractile activity usually occurs in bursts, separated by an interval in which contractions are weak or absent. Because the bursts are approximately 1 minute apart, the pattern has been termed the **minute rhythm** of the jejunum (Fig. 38-28).

Segmentation causes a good deal of back-and-forth movement of intestinal contents. It mixes intestinal contents with digestive secretions and circulates the chyme across the surface of the mucosal epithelium. Because of the phase relationship among small intestinal slow waves, contraction in one segment is followed by contraction of a neighboring segment farther down the intestine. Such a sequence of contractions propels chyme in an aboral direction. The decreasing frequency of segmentation as the chyme moves down the small intestine also contributes to aboral propulsion.

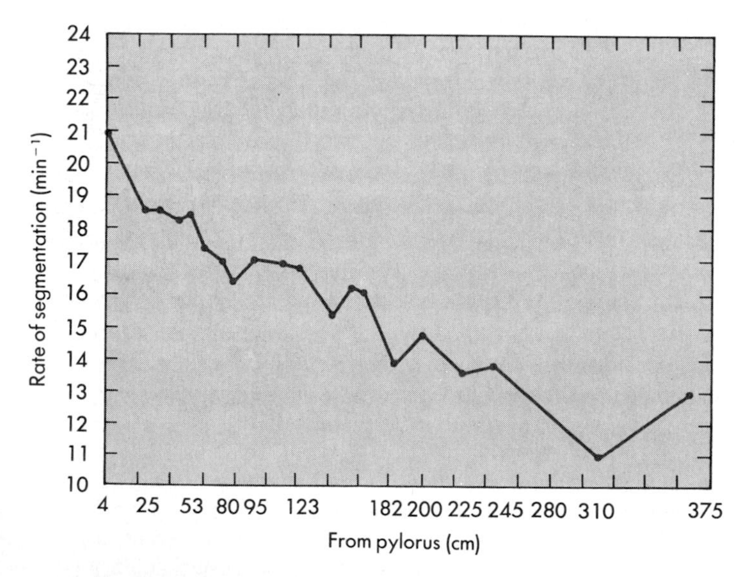

■ **Fig. 38-27** The rate of segmentation decreases along the length of the small intestine of the rabbit. The data were collected from nearly 30 rabbits. (Redrawn from Alvarez, WC: *Am J Physiol* 37:267, 1915.)

Short-range peristalsis also occurs in the small intestine, although much less frequently than segmentation. A peristaltic wave rarely traverses a large part of the small intestine. Typically a peristaltic contraction will die out after traveling about 10 cm. The relatively low rate of net propulsion of chyme in the small intestine allows time for digestion and absorption. Whereas intestinal motility is increased after eating, the net rate of chyme movement is actually decreased. This response depends on the extrinsic nerves to the gut.

■ *Intestinal Reflexes*

Intestinal reflexes can occur along a considerable length of the gastrointestinal tract. These depend to some extent on the function of both intrinsic and extrinsic nerves.

When a bolus of material is placed in the small intestine, the intestine may contract oral to the bolus and relax aboral to the bolus; this response was originally called the **law of the intestine.** This may propel the bolus in an aboral direction, like a peristaltic wave.

Overdistension of one segment of the intestine relaxes the smooth muscle in the rest of the intestine. This response is known as the **intestinointestinal reflex,** and it requires intact extrinsic innervation.

The stomach and the terminal part of the ileum interact reflexly. Distension of the ileum decreases gastric motility, a response called the **ileogastric reflex.** Elevated secretory and motor functions of the stomach increase the motility of the terminal part of the ileum and accelerate the movement of material through the ileocecal sphincter. This response is called the **gastroileal reflex.** The hormone gastrin may play a role in this response, since gastrin, at physiological levels, increases ileal motility and relaxes the ileocecal sphincter.

■ *The Migrating Myoelectric Complex*

The contractile behavior of the small intestine (already discussed) is characteristic of the period after ingestion of a meal. In a *fasted* individual or after the processing of the previous meal, small intestinal motility follows a different pattern that is characterized by bursts of intense electrical and contractile activity separated by longer quiescent periods. This pattern appears to be propagated from the stomach to the terminal ileum (Fig. 38-29) and is known as the **migrating myoelectric complex,** the migrating motor complex, or **MMC.** The MMC in the stomach was discussed previously.

The MMC repeats every 75 to 90 minutes in humans, and it occurs in four phases. Phase I, the quiescent

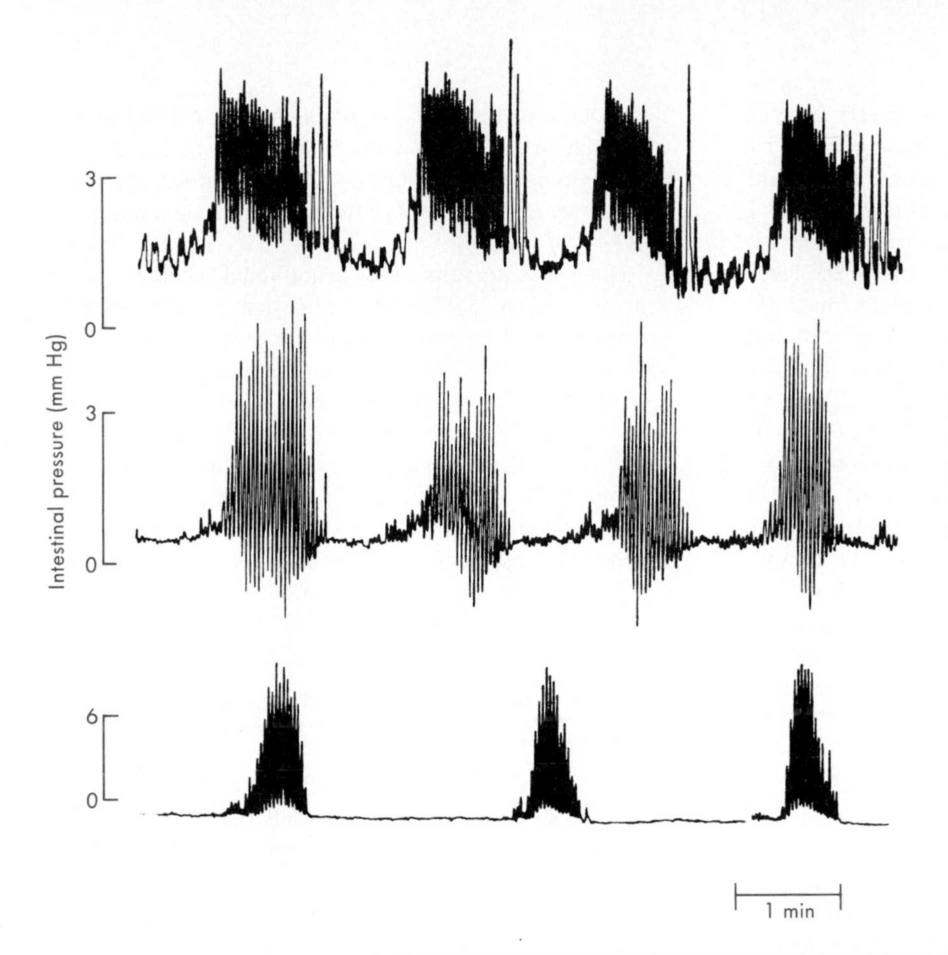

■ **Fig. 38-28** The "minute rhythm" recorded from jejunums of three different ferrets in the fed state. Similar minute rhythms have been described for the human jejunum. (From Collman, PI, et al.: *J Physiol* [London] 345:65, 1983.)

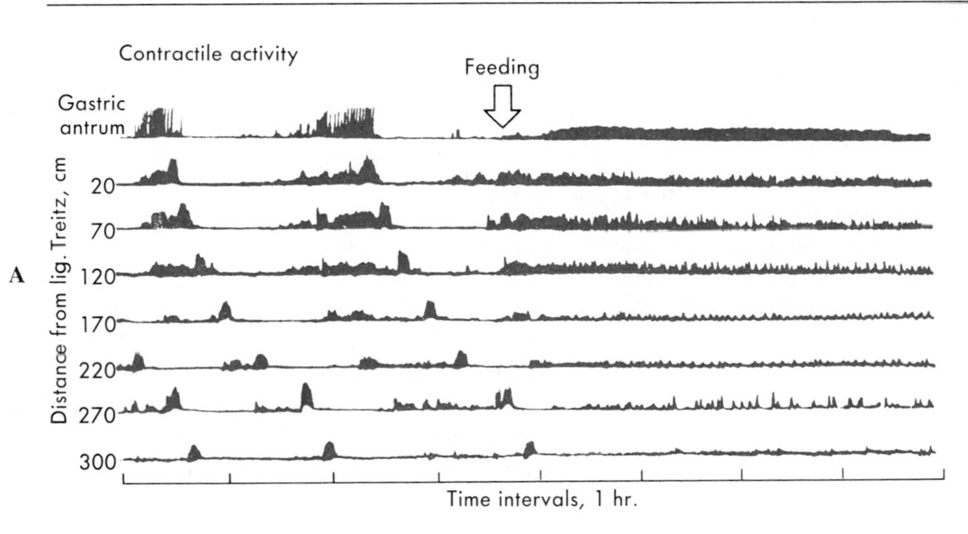

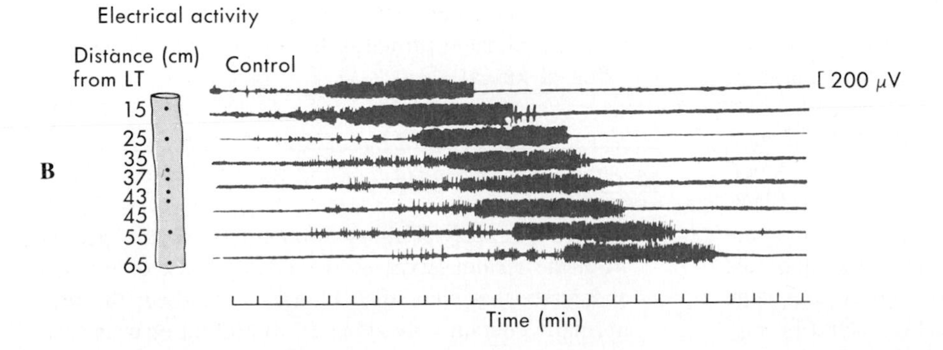

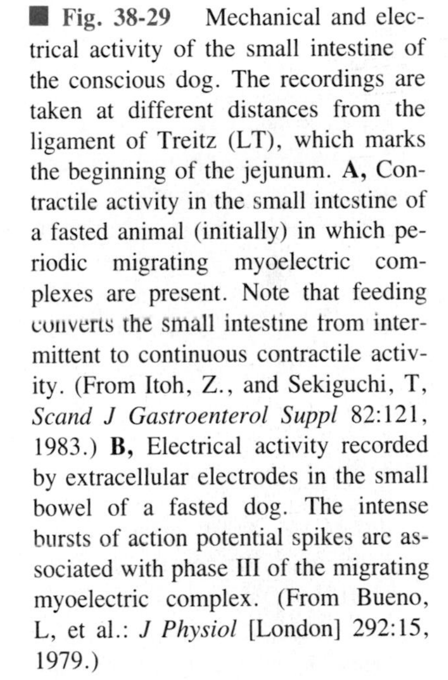
■ **Fig. 38-29** Mechanical and electrical activity of the small intestine of the conscious dog. The recordings are taken at different distances from the ligament of Treitz (LT), which marks the beginning of the jejunum. **A,** Contractile activity in the small intestine of a fasted animal (initially) in which periodic migrating myoelectric complexes are present. Note that feeding converts the small intestine from intermittent to continuous contractile activity. (From Itoh, Z., and Sekiguchi, T, *Scand J Gastroenterol Suppl* 82:121, 1983.) **B,** Electrical activity recorded by extracellular electrodes in the small bowel of a fasted dog. The intense bursts of action potential spikes are associated with phase III of the migrating myoelectric complex. (From Bueno, L, et al.: *J Physiol* [London] 292:15, 1979.)

phase, is characterized by slow waves with very few action potentials and very few contractions. In phase II irregular action potentials and contractions occur and gradually increase in intensity and frequency. Phase III is a period of intense electrical and contractile activity lasting from 3 to 6 minutes. During phase IV the electrical and contractile activity rapidly decline. Phase IV varies in duration and blends almost imperceptibly into phase I. About the time that one MMC reaches the distal ileum, the next MMC begins in the stomach.

The contractions of phase III of the MMC, both in the stomach and in the small intestine, are more vigorous and more propulsive than the contractions that occur in the fed individual. Phase III sweeps the small bowel clean and empties its contents into the cecum. Thus the MMC has been termed the "housekeeper" of the small intestine. By periodically sweeping the contents of the small intestine into the cecum, the MMC inhibits the migration of colonic bacteria into the distal ileum.

Propagation of the MMC

The mechanisms by which the MMC propagates are incompletely understood. The MMC is not a rigid reflex sequence that can only originate in the stomach. Occasionally an MMC originates in the upper small bowel, and then it propagates normally. The enteric nervous system plays a major role in propagation of the MMC. If the enteric neurons are blocked at a particular level of the small intestine, the MMC is not propagated past that point. The small bowel can be cut into several segments and then reanastomosed. This procedure interrupts the enteric nervous system but disturbs the extrinsic nerves only slightly. The small intestine then behaves like a series of independent segments. Each segment generates MMCs that propagate along that segment but not to the next segment.

Generation of the MMC

The MMC usually originates in the stomach, sometimes as high as the lower esophageal sphincter. However, MMCs occasionally begin in the duodenum or jejunum and sometimes even in the ileum. The initiation of the MMC is not fully understood. Both neural and hormonal mechanisms are probably involved in initiating the MMC.

Some investigators contend that the MMC is normally initiated by vagal impulses to the stomach. A corollary of this hypothesis is that the timing of the MMC is determined by pattern-generating circuitry in the central nervous system. This viewpoint is partly supported by experiments in which the cervical vagi of dogs were exteriorized in loops of skin so that they could be rapidly and reversibly inactivated by cooling. Cooling the cervical vagi to 4° C caused all MMC activity in the stomach to cease. However, the MMC in the duodenum, jejunum, and ileum continued at about the same frequency, but phase II activity did not precede phase III.

Other investigators favor a hormonal mechanism for initiation of the MMC. They note that chronic extrinsic denervation of the stomach and upper small bowel does not abolish MMC activity. Motilin is currently the favored candidate as the hormone responsible for initiation of the MMC. Intravenous administration of motilin initiates MMCs that strongly resemble the naturally occuring MMC. In addition, plasma motilin levels oscillate in phase with the MMCs (Fig. 38-30). Other hormones that may influence the MMC include substance P, somatostatin, and neurotensin.

Contractile Activity of the Muscularis Mucosae

Sections of the muscularis mucosae contract irregularly at a rate of about three contractions per minute. These contractions alter the pattern of ridges and folds of the mucosa, mix the luminal contents, and bring different parts of the mucosal surface into contact with freshly mixed chyme. Especially in the proximal part of the small intestine, the villi themselves contract irregularly. This may aid in emptying the central lacteals of the villi and thus enhance intestinal lymph flow.

Emptying the Ileum

The **ileocecal sphincter** separates the terminal end of the ileum from the cecum, the first part of the colon. Normally the sphincter is closed, but short-range peristalsis in the terminal part of the ileum causes relaxation of the sphincter and allows a small amount of chyme to squirt into the cecum. Distension of the terminal ileum relaxes the sphincter reflexly. Distension of the cecum causes the sphincter to contract and prevents additional emptying of the ileum. The gastroileal reflex enhances ileal emptying after eating. Under normal conditions the ileocecal sphincter allows ileal chyme to enter the colon at a slow enough rate that the colon can absorb most of the salts and water of the chyme. The ileocecal sphincter is coordinated primarily by the neurons of the intramural plexuses.

Colonic Motility

The colon receives about 1500 ml of chyme per day from the terminal part of the ileum. Most of the salts and water that enter the colon are absorbed; the feces normally contain only about 50 to 100 ml of water each day. Colonic contractions mix the chyme and circulate

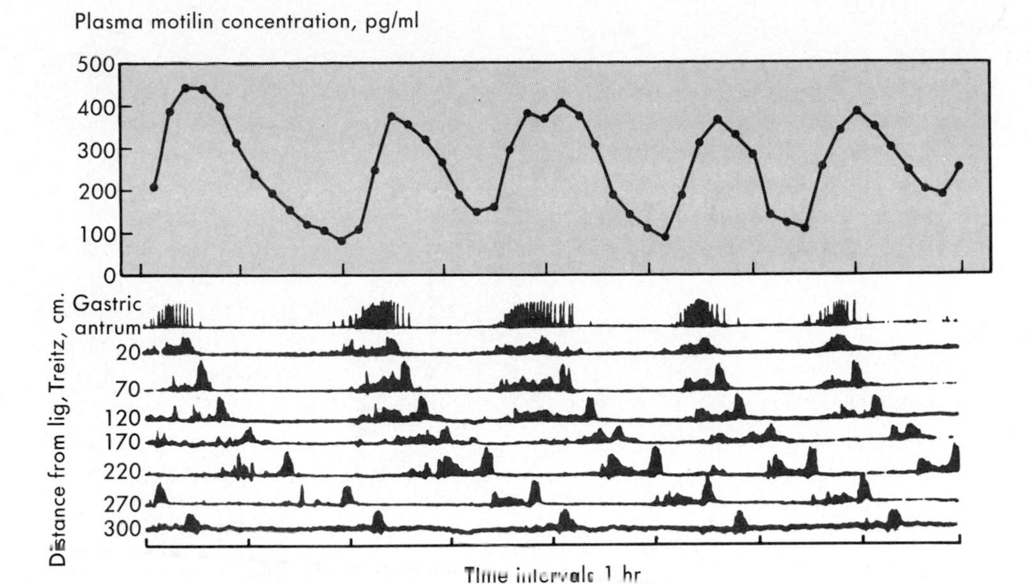

■ **Fig. 38-30** The relationship between plasma levels of motilin *(upper tracing)* and the occurrence of migrating myoelectric complexes in the dog stomach and small intestine *(lower records)*. (From Itoh, Z, and Sekiguchi, T: *Scand J Gastroenterol Suppl* 82:121, 1983.)

it across the mucosal surface of the colon. As the chyme becomes semisolid, this mixing resembles a kneading process. The progress of colonic contents is slow, about 5 to 10 cm/hour at most. One to three times daily a wave of contraction, called a **mass movement,** occurs. A mass movement resembles a peristaltic wave in which the contracted segments remain contracted for some time. Mass movements serve to push the contents of a significant length of colon in an aboral direction.

■ *Structure and Innervation of the Large Intestine*

As shown in Fig. 38-31, the major subdivisions of the large intestine are the cecum, the ascending colon, the transverse colon, the descending colon, the sigmoid colon, the rectum, and the anal canal. The structure of the wall of the large bowel follows the general plan presented earlier in this chapter, but the longitudinal muscle layer of the muscularis externa is concentrated into three bands called the **teniae coli.** In between the teniae coli the longitudinal layer is quite thin. The myenteric plexus is more dense under the teniae coli. The longitudinal muscle of the rectum and anal canal is substantial and continuous.

The extrinsic innervation of the large intestine is predominantly autonomic. Parasympathetic innervation of the cecum and the ascending and transverse colon is via branches of the vagus nerve; that of the descending and sigmoid colon, the rectum, and the anal canal is via the pelvic nerves from the sacral spinal cord. The parasympathetic nerves end primarily on neurons of the intramural plexuses. Sympathetic innervation to the large intestine comes to the proximal part of the bowel via the superior mesenteric plexus, to the distal part of the large intestine from the inferior mesenteric and superior hypogastric plexuses, and to the rectum and anal canal from the inferior hypogastric plexus.

Stimulation of sympathetic nerves stops colonic movements. Vagal stimulation causes segmental contractions of the proximal part of the colon. Stimulation of the pelvic nerves brings about expulsive movements of the colon and sustained contraction of some segments.

The anal canal usually is kept closed by the internal and external anal sphincters. The internal anal sphincter is a thickening of the circular smooth muscle of the anal canal. The external anal sphincter is more distal, and it consists entirely of striated muscle. The external anal sphincter is innervated by somatic motor fibers via the pudendal nerves, which allow it to be controlled both reflexly and voluntarily.

At most times the tonic contraction of the puborectal muscle pulls the upper anal canal forward to cause a sharp angle between the rectum and the anal canal and prevent feces from entering the anal canal. The innervation of the puborectal muscle is the same as that of the external anal sphincter.

■ *Motility of the Cecum and Proximal Part of the Colon*

Most contractions of the cecum and proximal part of the large bowel are segmental, and they are more effective

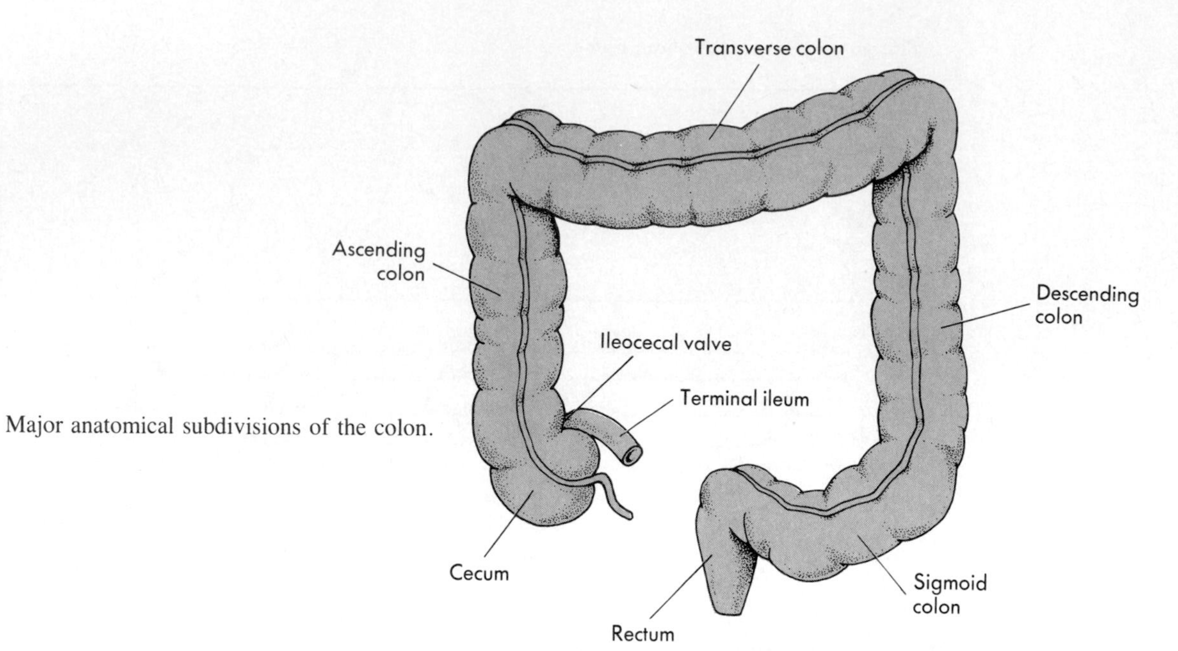

■ **Fig. 38-31** Major anatomical subdivisions of the colon.

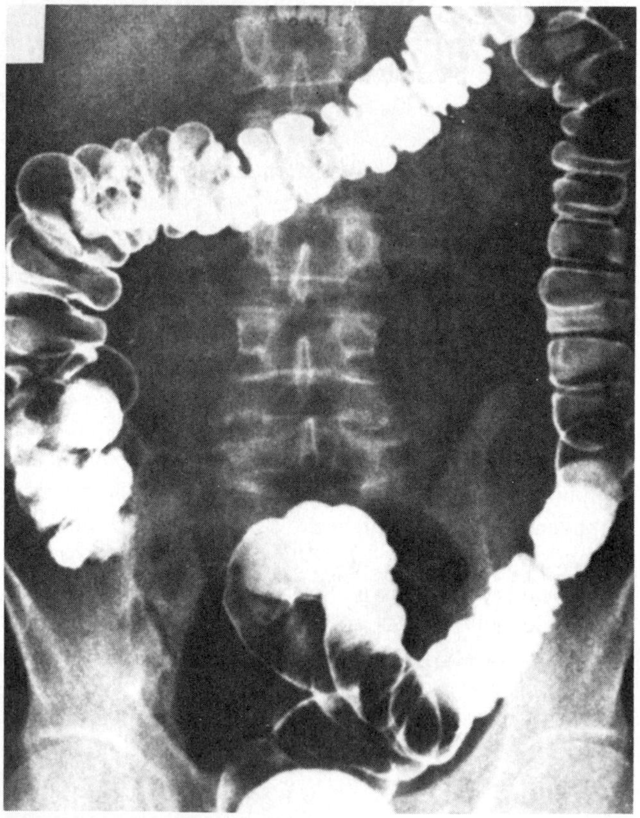

at mixing and circulating the contents than at propelling them. The mixing action facilitates absorption of salts and water by the mucosal epithelium.

Localized segmental contractions divide the colon into neighboring ovoid segments called **haustra** (Fig. 38-32). Hence segmentation in the colon is known as **haustration.** The most dramatic difference between haustration and the segmentation that occurs in the small intestine is the regularity of the segments (haustra) produced by haustration and the large length of bowel involved in haustration (see Fig. 38-32). Segmental thickenings of the circular smooth muscle are probably an anatomical component of haustration. The pattern of haustra is not fixed, however. It fluctuates as segments of the large bowel relax and contract. This results in back-and-forth mixing of luminal contents, known as haustral shuttling. When a few neighboring haustra are emptied in a proximal to distal direction, net propulsion results; this is called **segmental propulsion.** Less frequently a few neighboring haustra will empty because of a concerted contraction of their smooth muscle; this is termed **systolic multihaustral propulsion.**

In the proximal colon, "antipropulsive" patterns predominate. Reverse peristalsis and segmental propulsion toward the cecum both occur. These antipropulsive movements tend to retain chyme in the proximal colon and thus facilitate the absorption of salts and water from the proximal colon.

The net rate of chyme flow in the proximal part of the colon is only about 5 cm/hour in a fasting individual. Colonic propulsive motility increases after eating, and the net rate of flow increases to about 10 cm/hour. About one to three times daily a mass movement occurs that empties a large portion of the proximal part of the colon in an aboral direction. A mass movement resembles a peristaltic wave in which the contracted segments

■ **Fig. 38-32** X-ray image showing a prominent haustral pattern in the colon in a normal individual. (From Keats, TE: *An atlas of normal roentgen variants*, ed 2, Chicago, 1979, Year Book.)

of the gut remain contracted for a longer period than in peristalsis. Just before a mass movement takes place in a segment of the colon, haustral contractions relax. Then a ring of contraction of the circular muscle begins at the proximal end of the segment and rapidly elongates until one entire segment is contracted. Then that segment of bowel relaxes, haustra reappear, and the dominant motility pattern resumes.

Motility of the Central and Distal Parts of the Colon

Typically the flow of material from ascending to transverse colon takes place very slowly. The exception is the occasional (1 to 3 per day) mass movements in the proximal colon; they rapidly fill the transverse colon. The movements of the transverse and descending colons are not antipropulsive. Rather, localized haustral contractions predominate, with some short-range peristalsis. These movements result in kneading of the bowel contents and in slow orthograde transport of the forming feces, which become steadily more solid as they progress down the colon. Only during a mass movement is there rapid movement of feces along a considerable length of bowel.

Control of Colonic Movements

As in other segments of the gastrointestinal tract, the intramural plexuses control the contractile behavior of the colon, and the extrinsic autonomic nerves to the colon serve a modulatory function. The defecation reflex, discussed later, is an exception to this rule; it requires the function of the spinal cord via the pelvic nerves. Enteric neurons are both excitatory and inhibitory to colonic smooth muscle. The net effect is inhibitory, because blocking enteric neurons with drugs leads to sustained contractions of the circular muscle of the colon. In *Hirschsprung's disease,* in which enteric neurons are congenitally absent, tonic contractions of the colonic circular muscle lead to obstruction of the colon.

Control by extrinsic nerves. Branches of the vagus nerves supply the proximal colon. Experiments that involve cooling the vagi or injecting cholinergic antagonists support the interpretation that the vagus nerves tonically stimulate the motility of the proximal colon.

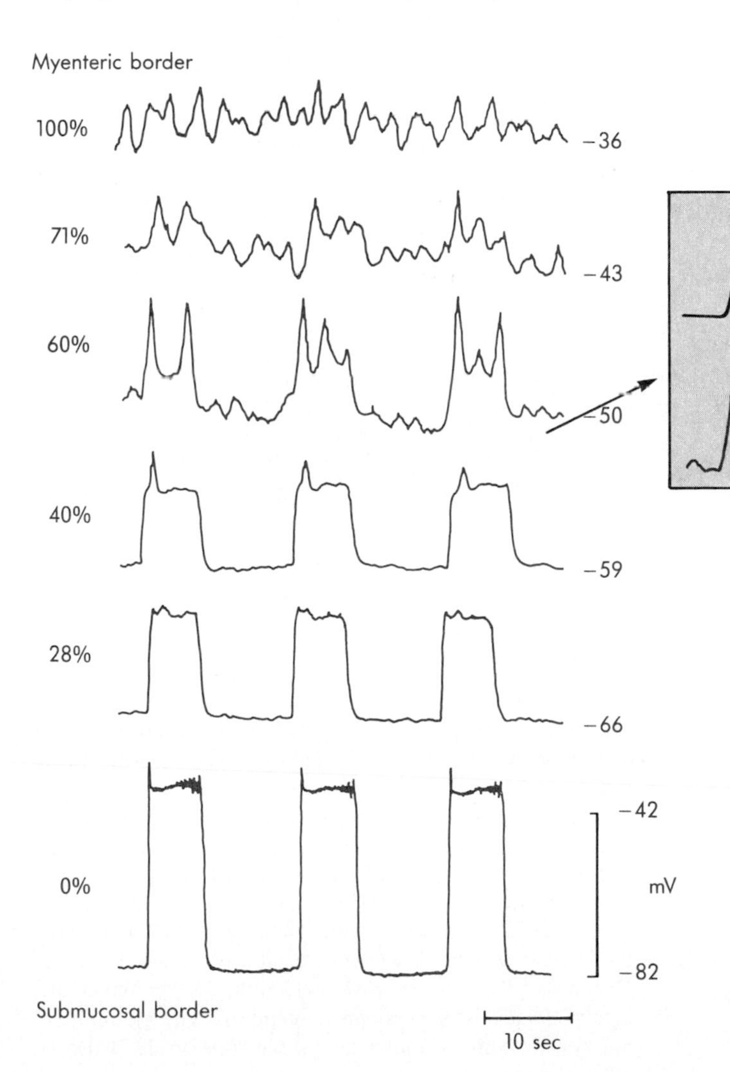

Fig. 38-33 Slow waves and myenteric potential oscillations in the colon. Shown are six intracellular microelectrode recordings of membrane potentials in smooth muscle cells of the circular layer of canine colon. The records are from different depths in the circular muscle layer with 0% representing the submucosal border and 100% the myenteric border. Near the myenteric border only myenteric potential oscillations are seen. Near the submucosal border only slow waves influence the potential. In the middle of the circular layer the two types of activity are summed. The box to the right shows a single slow wave and the resulting contraction *(black trace)* at the depth indicated. (Modified from Smith, TK, Reed, JB, and Sanders, KM: *Am J Physiol* 252:C290, 1987.)

Some of the tonic vagal stimulation is mediated by cholinergic fibers and some by noncholinergic fibers.

Cutting the sympathetic nerves to the colon enhances contractile activity. Sympathetic nerves tonically inhibit colonic motility. This effect is mediated by norepinephrine acting on neurons of the intramural plexuses.

Electrophysiology of the colon and control of colonic motility. Slow waves and myenteric potential oscillations occur in the colon The colonic muscle of cats and dogs has been more extensively studied than human colonic muscle. This muscle contains two classes of rhythm-generating cells. Cells near the submucosal border of the circular muscle generate classical slow waves at a frequency of about 6 per minute. The slow waves resemble those of the stomach in waveform, with a rapid depolarizing phase, a plateau phase, and a repolarizing phase. The upstroke of the colonic slow wave is caused by opening of channels that conduct both Na^+ and Ca^{++}, and the plateau phase is caused by a balance of inward current through L-type Ca^{++} and outward current through Ca^{++}-activated K^+ channels.

Another class of pacemaker cells, located between the longitudinal and circular muscle layers, generates **myenteric potential oscillations** of small, irregular amplitude. The frequency of these oscillations is about 20 per minute (Fig. 38-33).

The cells that generate the slow waves and myenteric potential oscillations in the colon have not been identified unequivocally, but they are probably similar to the interstitial cells that generate slow waves elsewhere in the gastrointestinal tract. The interstitial cells are not neurons and have characteristics of both smooth muscle cells and fibroblasts.

Both slow waves and myenteric potential oscillations are conducted decrementally through the circular muscle. The circular muscle near the submucosal border exhibits only the slow waves. The circular muscle near the myenteric plexus is influenced solely by the myenteric potential oscillations. Near the center of the circular muscle layer these two influences are summed (Fig. 38-33).

Longitudinal muscle layer. The membrane potential of longitudinal muscle displays the myenteric potential oscillations (Fig. 38-34). Single action potential spikes are triggered irregularly at the peaks of the myenteric potential oscillations. A contraction of the longitudinal muscle is elicited by each action potential spike. The longitudinal muscle cells are poorly coupled, and thus the occurrence of action potential spikes is often highly localized (see Fig. 38-34).

Circular muscle layer. In contrast to longitudinal muscle, the circular muscle cells (at least in canine and feline colon) do not usually fire action potential spikes. The amplitude of the slow waves is highest at the submucosal border and decreases toward the myenteric border of the circular muscle. However, the baseline resting membrane potential becomes progressively more depolarized toward the myenteric border. Hence the extent of depolarization reached at the peak of the slow wave or myenteric potential oscillation does not vary greatly with depth in the circular muscle (see Fig. 38-33). Contractions of colonic circular muscle are irregu-

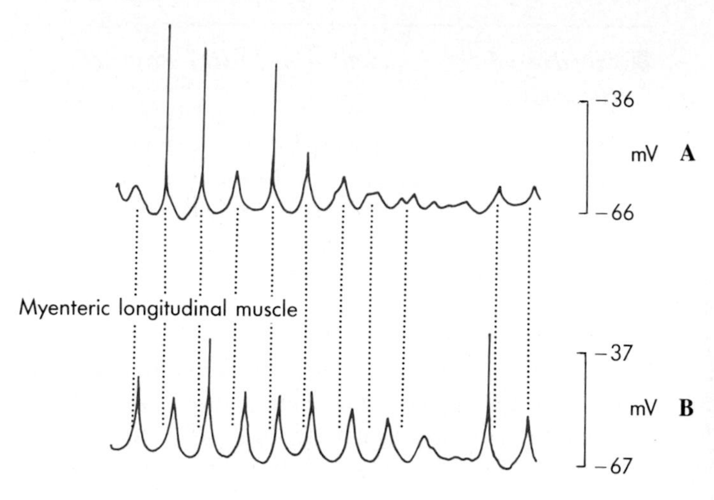

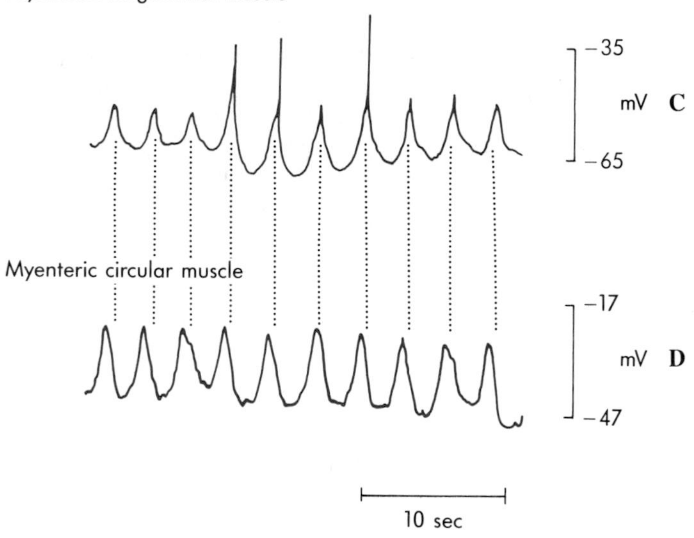

■ **Fig. 38-34** Myenteric potential oscillations and action potentials in longitudinal muscle of colon. Intracellular microelectrode recordings from individual smooth muscle cells of dog colon. **A** and **B** show recordings from smooth muscle cells near the serosal and myenteric edges of the longitudinal muscle, respectively. The more serosal cells fire more frequent action potentials that are larger in amplitude. Traces **C** and **D** compare longitudinal and circular cells on either side of the myenteric border. Even though these cells display myenteric potential oscillations at the same frequency and phase, the circular smooth muscle cells exhibit no action potentials. (Modified from Sanders, KM and Smith, TK. In Wood, JD, editor: *Handbook of physiology,* section 6: The gastrointestinal system, vol I. Motility and circulation, p 1, Bethesda, 1989, American Physiological Society.

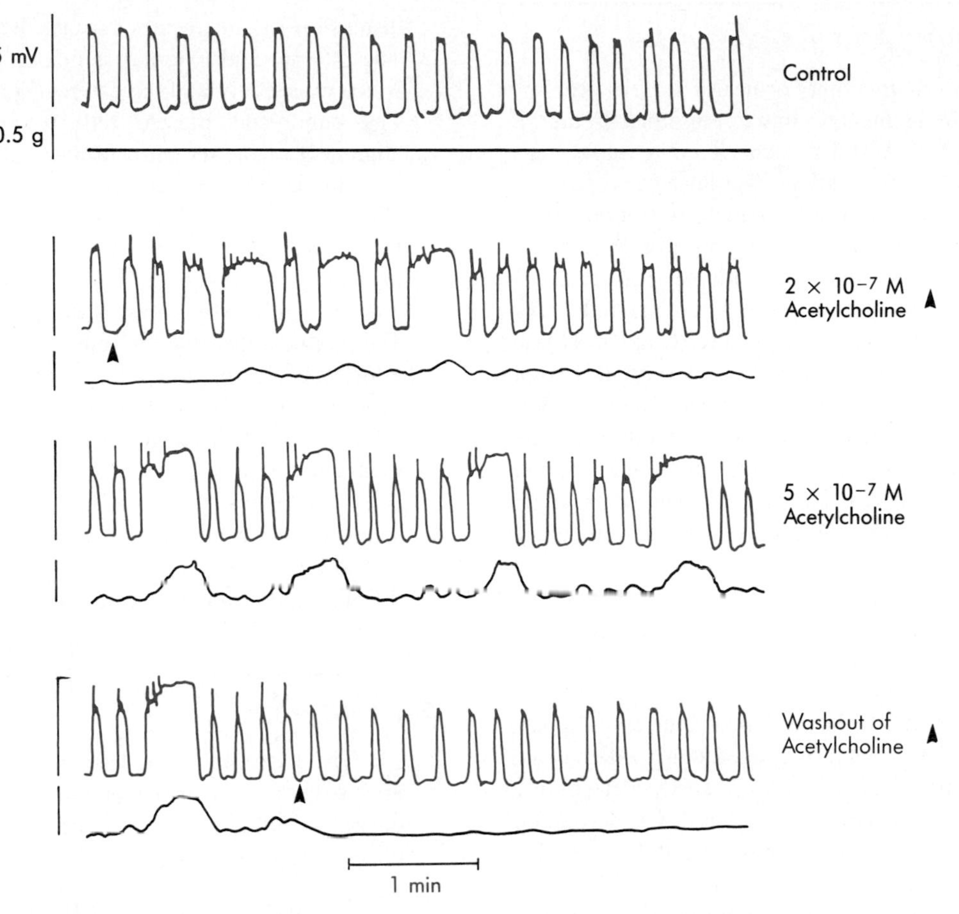

■ **Fig. 38-35** Effects of acetylcholine on electrophysiological and contractile behavior of circular muscle cells of canine colon. In each panel the top trace *(color)* is the membrane potential recording and the bottom trace *(black)* is the contractile response. Superfusion of the preparation with acetylcholine causes a somewhat irregular lengthening of slow waves. The longer slow waves elicit phasic contractions. (Modified from Huizinga, JD, Chang, NE, Diamant, NE, and El-Sharkawy, TY: *J Pharm Exp Ther* 231:692, 1984.)

lar and occur in response to slow waves of increased duration (Fig. 38-35).

Control of colonic contractions. As is the case elsewhere in the gastrointestinal tract, the smooth muscle cells of the colon are influenced by enteric neurons and by endocrine and paracrine modulators. As in the small intestine, intrinsic cholinergic neurons are excitatory to smooth muscle cells of longitudinal and circular muscle. When acetylcholine or other cholinergic agonists are applied to longitudinal muscle, the frequency of action potential spikes and contractions is increased. Cholinergic stimulation of longitudinal muscle can result in each myenteric potential oscillation triggering an action potential spike and a contraction.

In circular colonic muscle cholinergic agonists do not usually cause action potentials, but they elicit somewhat irregular prolonged slow waves (see Fig. 38-35). The prolonged slow waves are associated with contractions of the circular muscle. Increased entry of Ca^{++} during the prolonged plateau phase of the slow wave elicits the contractile response.

Reflex control. Distension of one part of the colon elicits reflex relaxation of other parts of the colon. This **colonocolonic reflex** is mediated partly by the sympathetic fibers that supply the colon.

The motility of proximal and distal colon and the frequency of mass movements increase reflexly after a meal enters the stomach. This so-called **gastrocolic reflex** is ill-defined; its afferent and efferent limbs are uncertain. The gastrocolic reflex has a rapid neural component, which is elicited by stretching the stomach. The efferent limb is probably carried by parasympathetic nerve fibers to the proximal (via vagus nerves) and distal (via pelvic nerves) segments of the colon. The gastrocolic reflex may also have a slower, hormonal component, for which cholecystokinin and gastrin are the leading mediator candidates. The blood levels of both cholecystokinin and gastrin increase after a meal, and the concentrations that these hormones attain in the blood are high enough to stimulate colonic motility in vitro.

The Rectum and Anal Canal

The rectum is usually empty or nearly so. The rectum is more active in segmental contractions than is the sigmoid colon. Thus rectal contents tend to move retrograde into the sigmoid colon. The anal canal is tightly closed by the anal sphincters. Before defecation the rectum is filled as a result of a mass movement in the sigmoid colon. Filling the rectum brings about reflex relaxation of the internal anal sphincter and reflex constriction of the external anal sphincter (Fig. 38-36) and causes the urge to defecate. Persons who lack functional motor nerves to the external anal sphincter defecate involuntarily when the rectum is filled. The reflex reactions of the sphincters to rectal distension are transient. If defecation is postponed, the sphincters regain their normal tone, and the urge to defecate temporarily subsides.

Defecation

When an individual feels the circumstances are appropriate, he or she voluntarily relaxes the external anal sphincter to allow defecation to proceed. Defecation is a complex behavior involving both reflex and voluntary actions. The integrating center for the reflex actions is in the sacral spinal cord but is modulated by higher centers. The efferent pathways are cholinergic parasympathetic fibers in the pelvic nerves. The sympathetic nervous system does not play a significant role in normal defecation.

Before defecation the smooth muscle layers of the descending colon and the sigmoid colon contract in a mass movement to force feces toward the anus. The distension of the rectum signals the urge to defecate. The internal and external sphincters relax, the internal sphincter reflexly and the external sphincter voluntarily. The puborectal muscle reflexly relaxes and allows alignment of the rectum and the anal canal.

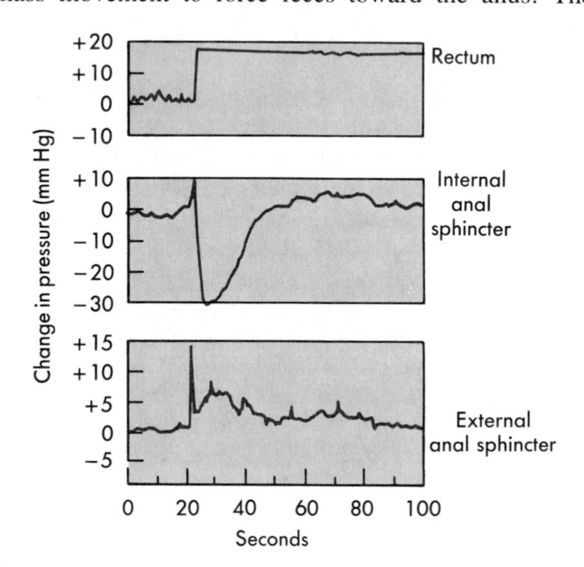

■ **Fig. 38-36** Responses of internal and external anal sphincters to a prolonged distension of the rectum. Note that the responses of the sphincters are transient. (Redrawn from Schuster, MM, et al.: *Bull Johns Hopkins Hosp* 116:79, 1965.)

Voluntary actions are also important in defecation. The external anal sphincter is voluntarily held in the relaxed state. Intraabdominal pressure is elevated to aid in expulsion of feces. Evacuation is normally preceded by a deep breath, which moves the diaphragm downward. The glottis is then closed, and contractions of the respiratory muscles on full lungs elevate both the intrathoracic and the intraabdominal pressure. Contractions of the muscles of the abdominal wall further increase intraabdominal pressure, which may be as high as 200 cm H_2O and helps to force feces through the relaxed sphincters. The muscles of the pelvic floor are relaxed to allow the floor to drop. This helps to straighten out the rectum and prevent rectal prolapse.

Summary

1. The gastrointestinal tract has a characteristic layered structure consisting of mucosa, submucosa, muscularis externa, and serosa; this structure varies from one segment to another.

2. The gastrointestinal tract receives both sympathetic and parasympathetic innervation. Autonomic nerves influence the motor and secretory activities of the gastrointestinal tract and regulate the caliber of blood vessels of the gastrointestinal tract.

3. The smooth muscle of the muscularis external mixes and propels the contents of the gastrointestinal tract. Gastrointestinal smooth muscle cells are electrically coupled, and their membrane potential oscillates with a rhythm characteristic of each segment of the gastrointestinal tract. The membrane potential oscillations, called slow waves, control the timing and the force of contractions of gastrointestinal smooth muscle.

4. The nerve plexuses of the gastrointestinal tract, the enteric nervous system, contain about 10^8 neurons, as many as in the spinal cord. The enteric nerve plexuses contain motoneurons, sensory neurons, and interneurons. Enteric sensory neurons function as the afferent arms of enteric reflex arcs by which the enteric nervous system controls much of the motor and secretory activities of the gastrointestinal tract. The autonomic nervous system modulates the activities of the enteric nervous system.

5. Swallowing is a reflex coordinated by a swallowing center in the medulla oblongata. The swallowing reflex is initiated by touch receptors in the pharynx. This reflex involves a series of ordered and coordinated motor impulses to muscles of the pharynx, upper esophageal

sphincter, esophageal striated muscle, esophageal smooth muscle, and lower esophageal sphincter.

6. The motor activity of the stomach mixes food with gastric juice and mechanically subdivides the food. Gastric emptying is closely regulated, which ensures that gastric contents are not emptied into the duodenum at a rate faster than the duodenum and jejunum can neutralize the gastric acid and process the chyme. Hormonal and neural mechanisms initiated by the presence of acid, fats, peptides, and hypertonicity in the duodenum regulate gastric emptying.

7. Segmentation is the major contractile activity in the small intestine. Segmental contractions mix and circulate intestinal contents but are not very propulsive. The slow rate of transport of small intestinal contents allows adequate time for digestion and absorption.

8. In a fasted animal a different pattern of motility called the migrating myoelectric complex (MMC) occurs. The MMC is characterized by 75 to 90-minute periods of quiescence interrupted by periods of vigorous and intensely propulsive contractions that last 3 to 6 minutes. The contractile phase of the MMC begins in the stomach and appears to propagate along the entire length of the small intestine. The MMC sweeps the stomach and the small intestine clear of any debris that remains from the previous meal.

9. In the proximal colon, antipropulsive contractions predominate, which allow time for absorption of salts and water. In the transverse and descending colon haustral contractions mix and knead colonic contents to facilitate extraction of salts and water. Mass movements that occur in the colon 1 to 3 times daily sweep colonic contents in the anal direction.

10. Filling of the rectum with feces initiates the defecation reflex. The integrating center for the defecation reflex is in the sacral spinal cord, and the pelvic nerves are the principal motor pathway that regulates the actions of the distal colon, the rectum, the anal canal, and the internal and external anal sphincters in defecation. Both reflex and voluntary activities are involved in defecation.

Bibliography

Journal articles

Bornstein JC, Furness JB: Correlated electrophysiological and histochemical studies of submucous neurons and their contribution to understanding enteric neural circuits, *J Auton Nerv Syst* 25:1, 1988.

Bornstein JC et al: Synaptic responses evoked by mechanical stimulation of the mucosa in morphologically characterized myenteric neurons of the guinea-pig ileum, *J Neurosci* 11:505, 1991.

Bywater RAB, et al: The enteric nervous system in the control of motility and secretion, *Digest Dis* 5:193, 1987.

Costa M et al: Chemical coding of neurons in the gastrointestinal tract, *Adv Exp Med Biol* 298:17, 1991.

Furness JB et al: Correlated functional and structural analysis of enteric neural circuits, *Arch Histol Cytol* 52(suppl):161, 1989.

Furness JB et al: Roles of peptides in the enteric nervous system, *Trends Neurosci* 15:66, 1992.

Furness JB et al: Shapes of nerve cells in the myenteric plexus of the guinea-pig small intestine revealed by the intracellular injection of dye, *Cell Tissue Res* 254:561, 1988.

Hara Y et al: Electrophysiology of smooth muscle of the small intestine of some mammals, *J Physiol* 372:501, 1986.

Langton P, et al: Spontaneous electrical activity of interstitial cells of Cajal isolated from canine proximal colon, *Proc Nat Acad Sci USA* 86:7280, 1989.

Smith TK, et al: Interactions between reflexes evoked by distension and mucosal stimulation: electrophysiological studies of guinea-pig ileum, *J Auton Nerv Syst* 34:69, 1991.

Smith TK et al: Distension-evoked ascending and descending reflexes in the circular muscle of guinea pig ilium: an intracellular study, *J Auton Nerv Syst* 29:203, 1990.

Smith TK et al: Interaction of two electrical pacemakers in muscularis of canine proximal colon, *Am J Physiol* 252:C290, 1987.

Wood JD: Enteric neurophysiology, *Am J Physiol* 247:G585, 1984.

Books and monographs

Christensen J, Wingate DL, editors: *A guide to gastrointestinal motility,* Bristol, UK, 1983, John Wright and Sons.

Davenport HW: *Physiology of the digestive tract,* ed 5, Chicago, 1985, Year Book.

Furness JB, Bornstein JC: *The enteric nervous system and its extrinsic connections.* In Yamada T, editor: *Textbook of gastroenterology,* vol 1, Philadelphia, 1991, JB Lippincott.

Furness JB, Costa M: *The enteric nervous system,* Edinburgh, 1987, Churchill Livingstone.

Grundy D: *Gastrointestinal motility: the integration of physiological mechanisms,* Lancaster, UK, 1985, MTP Press.

Johnson LR, editor: *Physiology of the gastrointestinal tract,* ed 2, vols 1 and 2, New York, 1987, Raven Press.

Kamm MA, Lennard-Jones JE, editors: *Gastrointestinal transit,* Petersfield, U K, 1991, Wrightson Biomedical Publishing.

Makhlouf GM, editor: *Handbook of physiology, section 6: the gastrointestinal system,* vol 2. *Neural and endocrine biology,* Bethesda, Md, 1989, American Physiological Society.

Sanders KM, Smith TK: *Electrophysiology of colonic smooth muscle.* In Wood JD, editor: *Handbook of physiology,* section 6: *the gastrointestinal system,* vol 1.*Motility and circulation,* part 1, Bethesda, Md, 1989, American Physiological Society.

Wood JD, editor: *Handbook of physiology,* section 6*: the gastrointestinal system,* vol 1. *Motility and circulation* (parts 1 and 2), Bethesda, Md, 1989, American Physiological Society.

Wood JD: *Electrical and synaptic behavior of enteric neurons.* In Wood JD, editor: *Handbook of physiology,* section 6*: the gastrointestinal system,* vol 1. *Motility and circulation, part 1,* Bethesda, Md 1989, American Physiological Society.

CHAPTER

39

Gastrointestinal Secretions

■ *Regulation of Secretion—General Aspects*

This chapter deals with glandular secretion of fluids and compounds that have important functions in the digestive tract. In particular, the secretions of salivary glands, gastric glands, the exocrine pancreas, and the liver are considered. In each case the nature of the secretions and their functions in digestion are discussed, and the regulation of the secretory processes is emphasized.

The secretions mentioned above are elicited by the action of specific effector substances on the secretory cells. These substances may be classified as neurocrine, endocrine, or paracrine. **Neurocrine** substances are released from the endings of neurons that innervate the secretory cells. **Endocrine** modulators are produced by specific cells located some distance from their target cell, and they reach the target cell via the circulation. **Paracrine** regulatory substances are released in the neighborhood of the target secretory cell and reach the target cell by diffusion.

Most of the substances that elicit secretion (and other cellular responses as well) do so by a few basic intracellular mechanisms. For example, the secretion of α-amylase by salivary glands can be elicited by β-adrenergic agents. This process has much in common with the cellular events by which cholecystokinin elicits secretion of pancreatic enzymes.

A substance that stimulates a particular cell to secrete is called a **secretagogue.** The number of secretagogues is large, but the cellular mechanisms of action of secretagogues are few. The action of secretagogues occurs by the signal transduction pathways that were discussed in Chapter 5 (Figs. 5-1 through 5-5). Most secretagogues increase the intracellular level of cyclic adenosine monophosphate (cAMP) or increase the turnover of certain inositol-containing phospholipids in the plasma membrane of the secretory cell. The phospholipid turnover may result in an increase in the concentration of calcium ions in the cytosol. Ca^{++} and cAMP are **second messengers** that initiate chains of events that culminate in increased secretion (or in other responses in nonsecretory cells).

■ *Secretion of Saliva*

In humans the salivary glands produce about 1 L of saliva each day. Saliva lubricates food for greater ease of swallowing. It also facilitates speaking. Saliva contains an **α-amylase,** also known as ptyalin, which begins the digestion of starch. In people lacking functional salivary glands a condition called xerostomia (dry mouth), dental caries, and infections of the buccal mucosa are prevalent. Secretion of saliva is an active process. The ionic composition of the fluid produced by the acinar cells is similar to that of plasma. The epithelial cells that line the ducts of the salivary glands modify the ionic composition as the saliva flows by. The functions of the salivary glands are primarily controlled by the autonomic nervous system; both sympathetic and parasympathetic stimulation enhance the overall rate of salivary secretion. Physiologically the parasympathetic system is more important.

■ *Functions of Saliva*

Mucins (glycoproteins) produced mainly by the submaxillary and sublingual glands lubricate food so that it may be more readily swallowed. The major digestive function of saliva results from the action of salivary amylase on starch. Ptyalin is an enzyme that has the same specificity as the α-amylase of pancreatic juice. Salivary amylase cleaves the internal α-1,4 glycosidic linkages in starch, but it cannot hydrolyze the terminal α-1,4 linkages or the α-1,6 linkages at the branch points. The major products of salivary amylase action are maltose, maltotriose, and oligosaccharides containing an α-1,6 branch point (called α-limit dextrins). The pH optimum of salivary amylase is about 7, but it has activity between pH levels of 4 and 11. Amylase action continues in the mass of food in the stomach, and it is terminated only when the contents of the antrum are mixed with enough gastric acid to lower the pH of the antral contents to below 4. More than half the starch in a well-chewed meal may be reduced to small oligosaccharides by amylase action. However, because of the large capacity of the pancreatic α-amylase to digest starch in the small intestine, the absence of salivary amylase causes no malabsorption of starch. Other components of saliva, present in smaller amounts, include RNAase, DNAase, lysozyme, peroxidase, lingual lipase, kallikrein, and secretory IgA.

The salivary glands have an extremely high blood flow. The capillaries that supply the ducts typically coalesce into venules, which in turn break up into another set of capillaries that supply the acini. The acini are thus partly supplied by a portal circulation.

The Major Salivary Glands and Their Structure

In humans, the **parotid glands,** the largest glands, are entirely serous glands. Their watery secretion lacks mucins. The **submaxillary** and **sublingual glands** are mixed mucous and serous glands, and they secrete a more viscous saliva containing mucins. Many smaller salivary glands are present in the oral cavity. The microscopic structure of mixed salivary glands is depicted in Fig. 39-1. The salivary glands structurally resemble the exocrine pancreas in many respects. The serous acinar cells, located in the end-pieces, have apical zymogen granules that contain amylase and certain other salivary proteins as well (Fig. 39-2). Mucous acinar cells secrete glycoprotein mucins into the saliva. The intercalated ducts that drain the acini are lined with columnar epithelial cells. The intercalated ducts drain the acini into somewhat larger ducts, the striated ducts, which empty into still larger excretory ducts, and so forth. A single large duct brings the secretions of each major gland into the mouth.

The current concept is that a primary secretion is elaborated in the secretory end-pieces. The cells that line the intercalated ducts may augment the secretions of the end-piece cells. The basal membranes of serous acinar cells often have many thin, fingerlike projections that greatly increase the surface area of the basal membrane. The cells that line the duct system modify the primary secretion. The striated ducts that drain the intercalated ducts are so named because of striations in the basal cytoplasm of their epithelial cells. The striations are due to numerous mitochondria aligned with regular infoldings of the basal plasma membrane. This structure suggests that the striated ducts may modify the primary salivary secretion. Such a role for the larger excretory ducts has been well established.

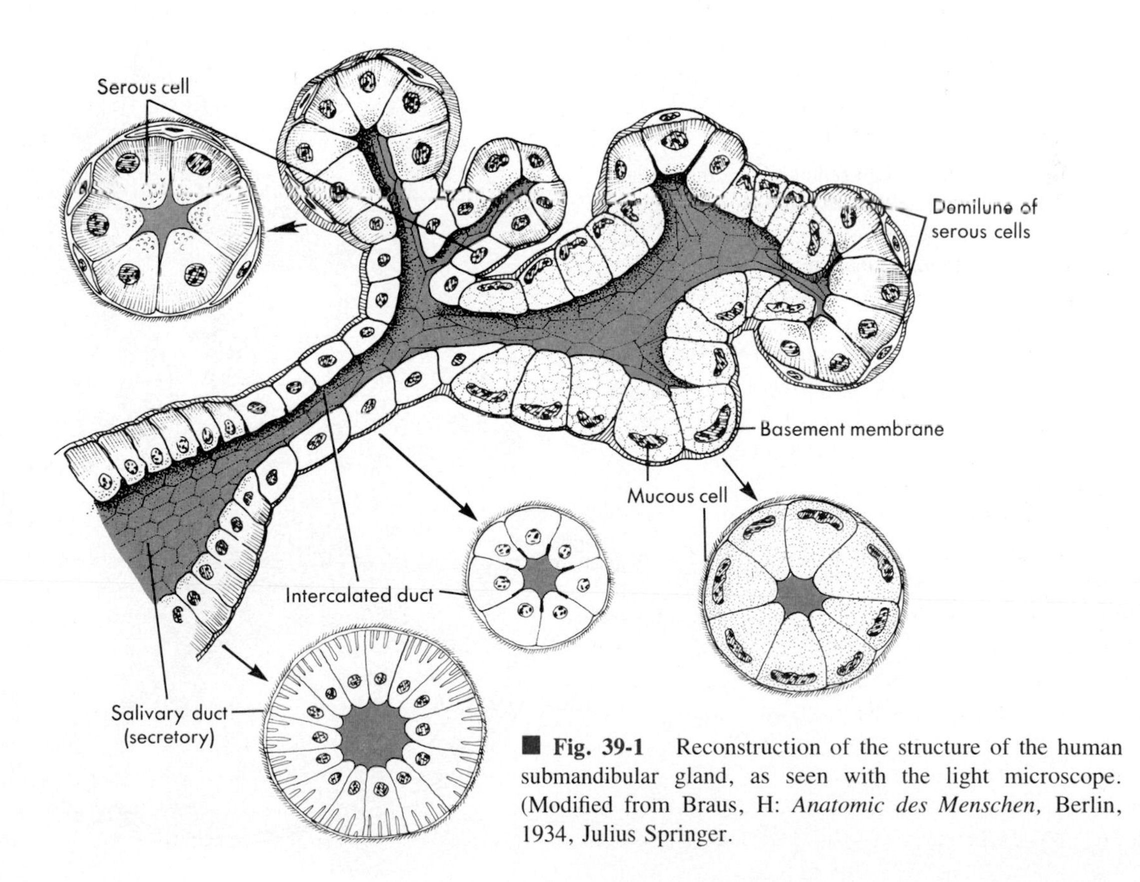

Fig. 39-1 Reconstruction of the structure of the human submandibular gland, as seen with the light microscope. (Modified from Braus, H: *Anatomic des Menschen,* Berlin, 1934, Julius Springer.

Metabolism and Blood Flow of Salivary Glands

For their size the salivary glands produce a prodigious volume flow of saliva: The maximal rate in humans is about 1 ml per minute per gram of gland. Salivary glands have a high rate of metabolism and a high blood flow, both being proportional to the rate of saliva formation. The blood flow to maximally secreting salivary glands is approximately 10 times that of an equal mass of actively contracting skeletal muscle. Stimulation of the parasympathetic nerves to salivary glands increases blood flow by dilation of the vasculature of the glands. The hormone **vasoactive intestinal polypeptide (VIP)** coexists with acetylcholine in parasympathetic nerve terminals in the salivary glands. VIP is released along with acetylcholine when the parasympathetic nerves are stimulated and contributes to vasodilation during secretory activity.

The Ionic Composition of Saliva and Secretion of Water and Electrolytes

In humans, saliva is always hypotonic to plasma. As shown in Fig. 39-3 the Na^+ and Cl^- concentrations are less than those of plasma, but K^+ and HCO_3^- concentrations are greater than those of plasma. The tonicity of saliva and its ionic composition vary from species to species and from one salivary gland to another in the same species. The greater the rate of secretory flow, the higher is the tonicity of the saliva; at maximal flow rates the tonicity of saliva in humans is about 70% of that of plasma. The pH of saliva from resting glands is slightly acidic. During active secretion, however, the saliva becomes basic, with pH approaching 8. The increase in pH with secretory flow rate is partly caused by the increase in salivary HCO_3^- concentration. The concentration of K^+ in saliva is almost independent of salivary flow rate over a wide range of flows. However, at very low salivary flow rates the concentration of K^+ in saliva increases steeply with decreasing flow rate. K^+ concentration can reach levels of 100 mmol/L or more.

Studies in which small samples of fluid are taken from the intercalated ducts of rat salivary glands show that the intercalated ducts contain a fluid that resembles plasma in its concentration of Na^+, K^+, Cl^-, and probably also HCO_3^-. Fluid in the intercalated ducts is either isotonic or slightly hypertonic to plasma.

Available evidence is consistent with a **two-stage model** of salivary secretion (Fig. 39-4). The two-stage model postulates the following:

1. The end-pieces, perhaps with the participation of intercalated ducts, produce an isotonic primary secretion. The amylase concentration and the rate of fluid secretion vary with the level and type of stimulation. However, the electrolyte composition of the secretion is fairly constant, and the levels of Na^+, K^+, and Cl^- are close to plasma levels.

2. The execretory ducts, and probably the striated ducts too, modify the primary secretion by extracting Na^+ and Cl^- from and adding K^+ and HCO_3^- to the sa-

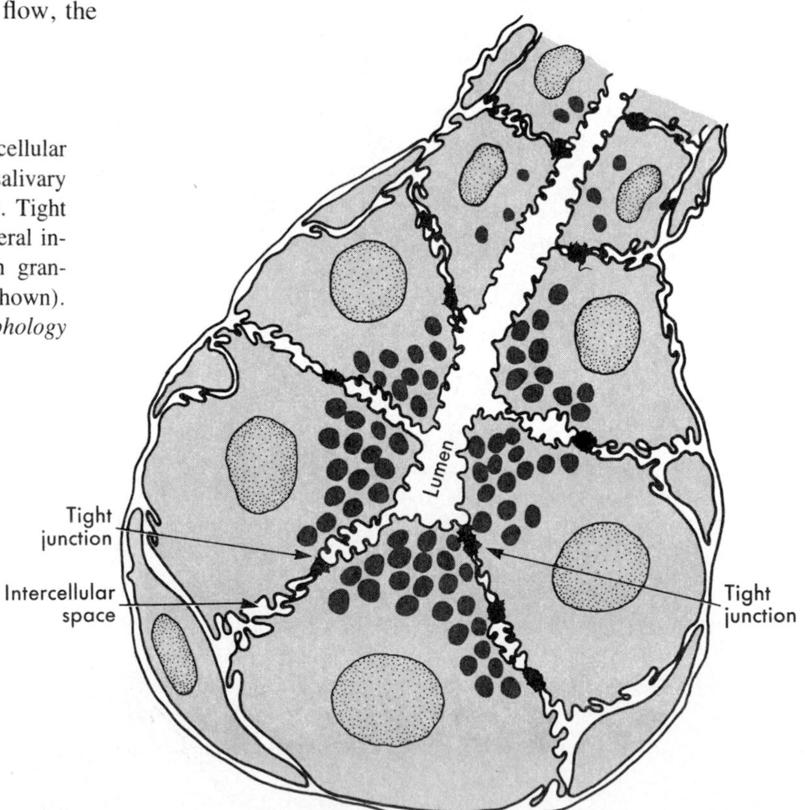

Fig. 39-2 A schematic representation of the cellular morphology of a secretory end-piece of a serous salivary gland. Secretory canaliculi drain into the acinar lumen. Tight junctions separate the secretory canaliculi from the lateral intercellular spaces. Colored circles represent zymogen granules. Acinar cells are coupled by gap junctions (not shown). (Redrawn from Young JA and Van Lennep EW: *Morphology of salivary glands*, London, 1978, Academic Press.)

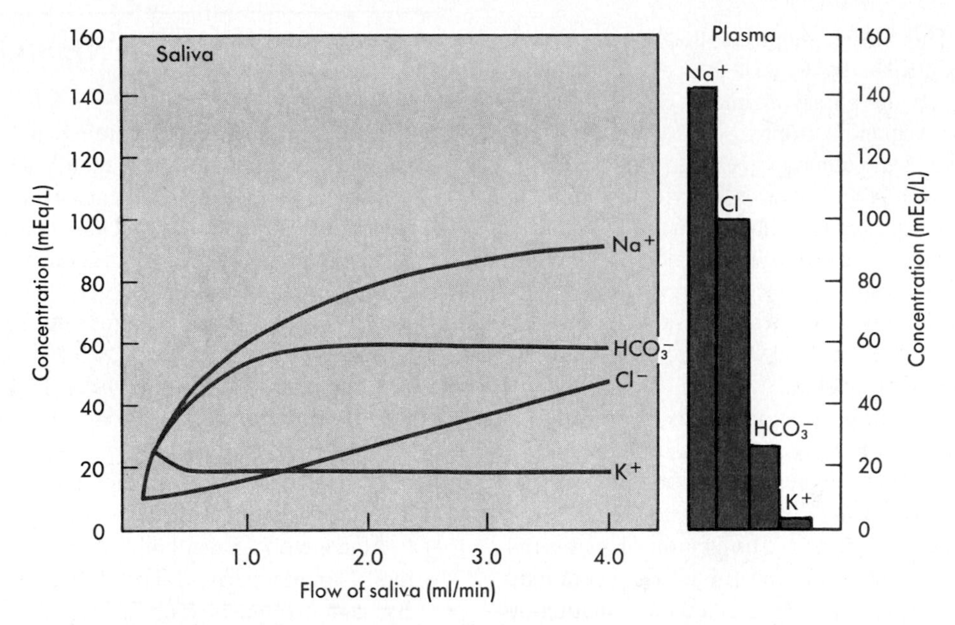

■ **Fig. 39-3** Average composition of the parotid saliva as a function of the rate of salivary flow. (From Thaysen JH et al: *Am J Physiol* 178:155, 1954.)

liva. The amylase concentration of saliva is at least as great as that in the primary secretion. Thus the ducts may reabsorb water, and they do not add to the volume of saliva.

As saliva flows down the ducts it becomes progressively more hypotonic. Thus the ducts remove more ions from saliva than they contribute to it.

The faster the flow rate of the saliva down the striated and excretory ducts, the closer to isotonicity is the saliva. In spite of the gradient for osmotic water flow out of the saliva, relatively little water is absorbed in the ducts, primarily because of the low permeability of the ductular epithelium to water.

■ *Secretion of Salivary Amylase*

In their apical cytoplasm, serous acinar cells have zymogen granules (see Fig. 39-2) that contain salivary amylase. Formation of zymogen granules proceeds by the following pathway. Salivary amylase is synthesized on the ribosomes of rough endoplasmic reticulum and enters the cisternae of the endoplasmic reticulum. Smooth vesicles containing the newly synthesized enzyme molecules move toward the Golgi apparatus, where the enzymes become encapsulated in membrane-bound vacuoles. The contents of the vacuoles are condensed, and the resulting vesicles take up residence in the apical cytoplasm of the cell, where they are recognizable as zymogen granules. When the gland is stimulated to secrete, the zymogen granules fuse with the plasma membrane. Their contents are released into the lumen of an acinus by exocytosis. Salivary amylase also may be secreted by a pathway that is independent of the zymogen granules.

■ *Neural Control of Salivary Gland Function*

The primary physiological control of the salivary glands is by the autonomic nervous system. By contrast, control of secretion of the other major gastrointestinal accessory glands is primarily hormonal. Stimulation of either sympathetic or parasympathetic nerves to the salivary glands stimulates salivary secretion, but the effects of the parasympathetic nerves are stronger and more long lasting. Interruption of the sympathetic nerves causes no major defect in the function of the sal-

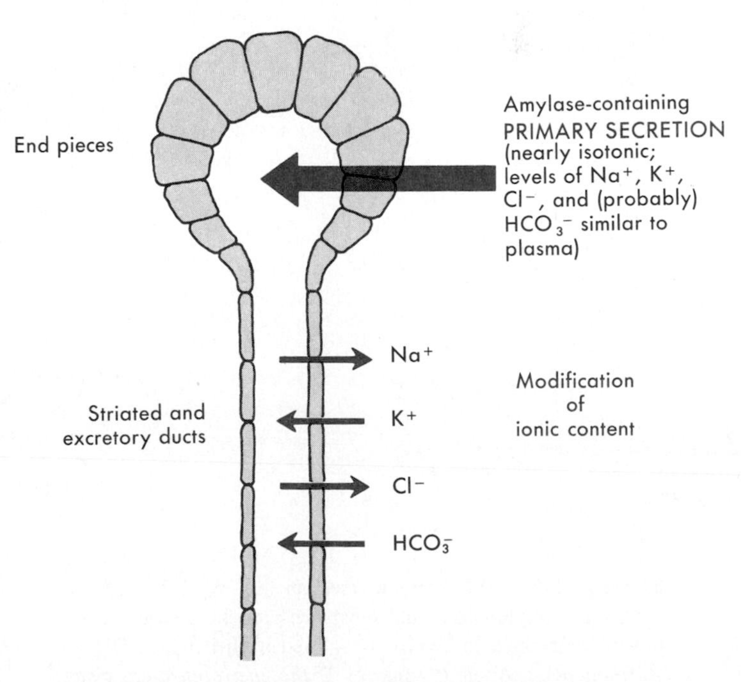

■ **Fig. 39-4** Schematic representation of the two-stage model of salivary secretion.

ivary glands. This effect suggests that the essential physiological control is by way of the parasympathetic nervous system. If the parasympathetic supply is interrupted the salivary glands atrophy.

Postganglionic sympathetic fibers to the salivary glands come from the superior cervical ganglion. Preganglionic parasympathetic fibers come via branches of the facial and glossopharyngeal nerves (cranial nerves VII and IX, respectively), and they synapse with postganglionic neurons in or near the salivary glands. The cells of the end-pieces and ducts are supplied with parasympathetic nerve endings.

Parasympathetic stimulation increases the synthesis and secretion of amylase and mucins, influences the transport activities of the ductular epithelium, greatly increases blood flow to the glands, and stimulates glandular metabolism and growth. Stimulation of parasympathetic nerves increases the rate of HCO_3^- secretion, but inhibits the reabsorption of Na^+ and the secretion of K^+ by the ductular epithelial cells.

Sympathetic stimulation and circulating catecholamines stimulate secretion of saliva rich in ptyalin, K^+, and HCO_3^-, primarily via β-adrenergic receptors. Sympathetic stimulation causes contraction of myoepithelial cells around the acini and ducts and constriction of blood vessels, with consequent reductions in salivary gland blood flow. The increased salivary secretion that results from stimulation of sympathetic nerves is transient.

■ *Cellular Mechanisms of Neural Regulation of Salivary Secretion*

Acinar cells. Understanding of secretory mechanisms of acinar cells, particularly serous cells, has increased dramatically in recent years. The neuroeffector substances that stimulate acinar cell secretions act primarily by one of the two general mechanisms outlined at the beginning of this chapter; that is, by elevating intracellular cAMP or by increasing the level of Ca^{++} in the cytosol.

Acetylcholine, norepinephrine, substance P, and vasoactive intestinal polypeptide are released in salivary glands by specific nerve terminals (Fig. 39-5). Each of these neuroeffectors may increase the secretion of amylase and the flow of saliva.

Acinar cells have both α- and β-adrenergic receptors. Norepinephrine binds to both classes of receptors. By using agonists that are specific for one receptor class it has been found that activation of β-receptors (the β_1-subtype) in salivary glands elevates intracellular cAMP. Vasoactive intestinal polypeptide also increases cAMP. By contrast, acetylcholine, substance P, and norepinephrine acting on α-receptors (α_1-subtype) act by mobilizing cellular Ca^{++} via hydrolysis of PIP_2 in the plasma membrane.

In general, effectors that increase cellular cAMP elicit a primary secretion that is richer in amylase than the secretion evoked by agents that increase intracellu-

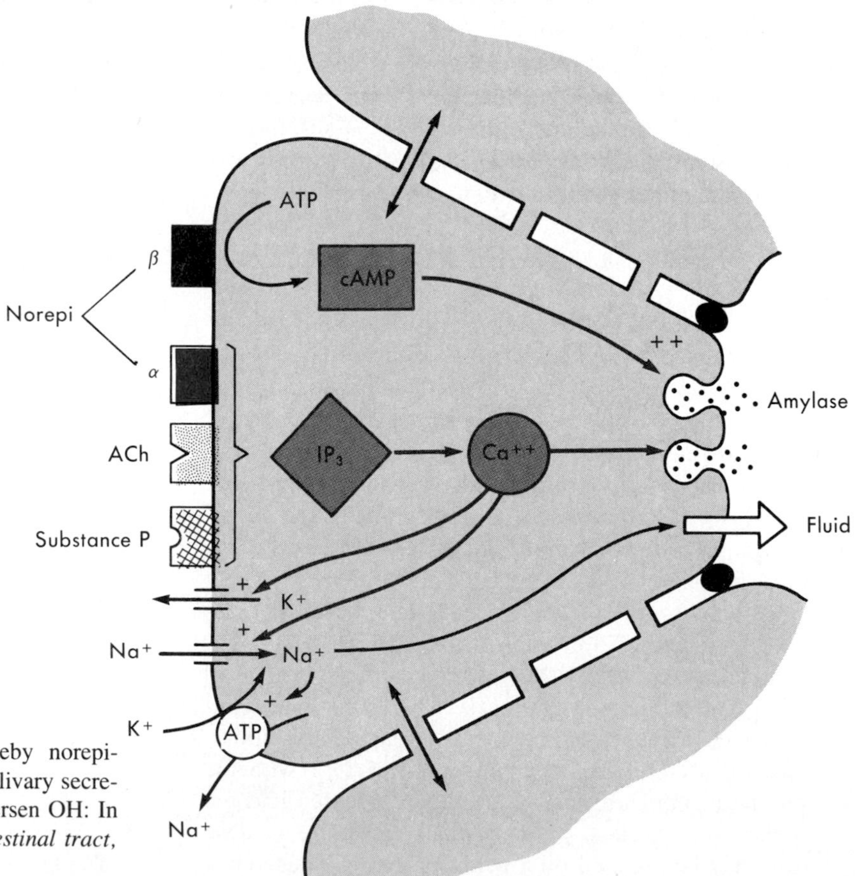

■ **Fig. 39-5** The cellular mechanisms whereby norepinephrine, acetylcholine, and substance P evoke salivary secretion by salivary acinar cells. (Modified from Petersen OH: In Johnson, RL, editor: *Physiology of the gastrointestinal tract,* New York, 1981, Raven Press.)

lar Ca^{++}. Substances that increase intracellular Ca^{++} increase the volume of acinar cell secretion to a greater extent than compounds that increase cellular cAMP. The ionic composition of acinar cell secretions is relatively independent of the mode of stimulation.

Effectors that act by elevating cAMP. β-Adrenergic agonists such as isoproterenol elevate cAMP. The elevation of cAMP correlates with the increase of amylase secretion and the volume of the primary secretion. The mechanisms whereby elevated cAMP enhances secretion are not well understood. In rat parotid gland, β-adrenergic agonists stimulate phosphorylation of three specific proteins of the plasma membrane or endoplasmic reticulum. The identity of these proteins and a relationship between their phosphorylation and enhanced secretion remain to be demonstrated.

Effectors that mobilize cellular Ca^{++}. Cholinergic agents, α-adrenergic agonists, and substance P each bind to distinct classes of receptors. However, they all act by mobilizing a common pool of intracellular Ca^{++}, probably located in the endoplasmic reticulum of the acinar cell. When the Ca^{++} pool has been discharged by one of the classes of effector in the absence of extracellular Ca^{++}, the cell cannot then respond to the other types of effectors. Refilling of the intracellular Ca^{++} pool depends on extracellular Ca^{++}.

All agonists that mobilize intracellular Ca^{++} cause electrophysiological changes in acinar cells by altering the fluxes of particular ions across the plasma membrane. Elevation of intracellular Ca^{++} opens membrane channels that conduct both Na^+ and K^+. The conductance of these Ca^{++}-gated channels for K^+ exceeds that for Na^+; the reversal potential for the channel is about −50 mV. The resting membrane potential of the acinar cell is usually more negative than −50 mV, so that opening these channels results in depolarization. However, when the resting potential is more positive than −50 mV, Ca^{++}-mobilizing agonists hyperpolarize the acinar cell. The fluxes of Na^+ and K^+ through these channels cause a large increase in intracellular Na^+ and a decrease in K^+. The increase in intracellular Na^+ concentration stimulates the Na^+, K^+-ATPase to take up K^+ and extrude Na^+. Because the Na^+, K^+-pump is electrogenic its increased activity hyperpolarizes the acinar cell. Thus stimulation of acinar cells by Ca^{++}-mobilizing agonists causes a characteristic two-phase change in membrane potential, which is called the secretory potential (Fig. 39-6). This first phase is due to the opening of Ca^{++}-gated cation channels, and the second phase is due to electrogenic ion pumping. The second phase can be blocked by ouabain and other inhibitors of the Na^+, K^+-ATPase.

The secretory potential appears to be a consequence rather than a cause of the events that elicit acinar cell secretion. The cellular mechanisms whereby elevated intracellular Ca^{++} enhances secretion by acinar cells are poorly understood. Exocytotic release of amylase from zymogen granules is a major mechanism of amylase secretion. Increased intracellular Ca^{++} is the trigger for exocytosis in acinar cells, as it is in nerve terminals.

Potential role of cyclic guanosine monophosphate. Stimulation of acinar cells by Ca^{++}-mobilizing effectors may also increase the intracellular levels of cyclic guanosine monophosphate (cGMP), which may correlate with the activation of secretion. By contrast, substance P, a powerful secretagogue, has no effect on cGMP levels. Thus cGMP is probably not directly involved in evoking acinar cell secretions. cGMP may play other roles, perhaps in regulation of growth or development of acinar cells.

Ionic Mechanisms of Salivary Secretion

Ionic mechanisms of acinar cell secretion. Fig. 39-7 shows a current simplified view of ionic mechanisms whereby Na^+, Cl^-, and HCO_3^- are secreted into the lumen of the secretory end-piece. The basolateral membranes of the acinar cells contain the Na^+, K^+-ATPase and the electroneutral Na,K,2Cl cotransporter that is present in many types of epithelial cells. The Na,K,2Cl transporter uses the energy of the electrochemical potential gradient of Na^+ created by the Na^+, K^+-ATPase to actively transport K^+ and Cl^- into the cell. As a result, the intracellular electrochemical potential of Cl^- is greater than in extracellular fluid. Cl^- then flows down its electrochemical potential gradient into the lumen of the acinus via an electrogenic anion channel in the apical membrane of the acinar cell. The electrogenic transport of Cl^- causes the lumen to become electrically negative with respect to the basolateral side of the cell; this provides an electrical force that drives Na^+ into the lumen via the tight junctions, which are leaky, especially to cations. A Na^+/H^+ exchanger in the basolateral membrane promotes alkalization of the cy-

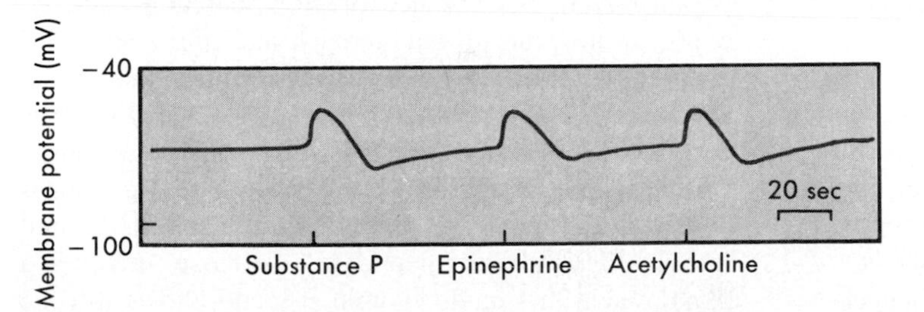

Fig. 39-6 Membrane potential responses of isolated acinar cells of rat parotid gland to application of secretagogues: substance P, epinephrine, and acetylcholine. (Redrawn from Gallacher, DV and Petersen, OH: *Nature* 83:393, 1980.)

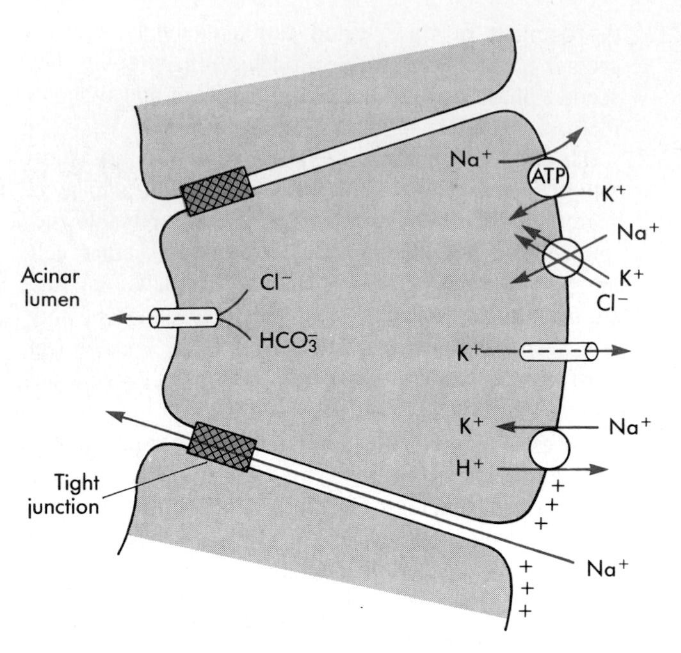

■ **Fig. 39-7** A simplified model of ionic transport processes involved in the secretion of Na^+, Cl^-, and HCO_3^- by salivary acinar cells. The operation of these ion transport processes is discussed in the text. (Young, JA et al: In Johnson, LR, editor, *Physiology of the gastrointestinal tract,* vol 1, ed 2, New York, 1987, Raven Press.)

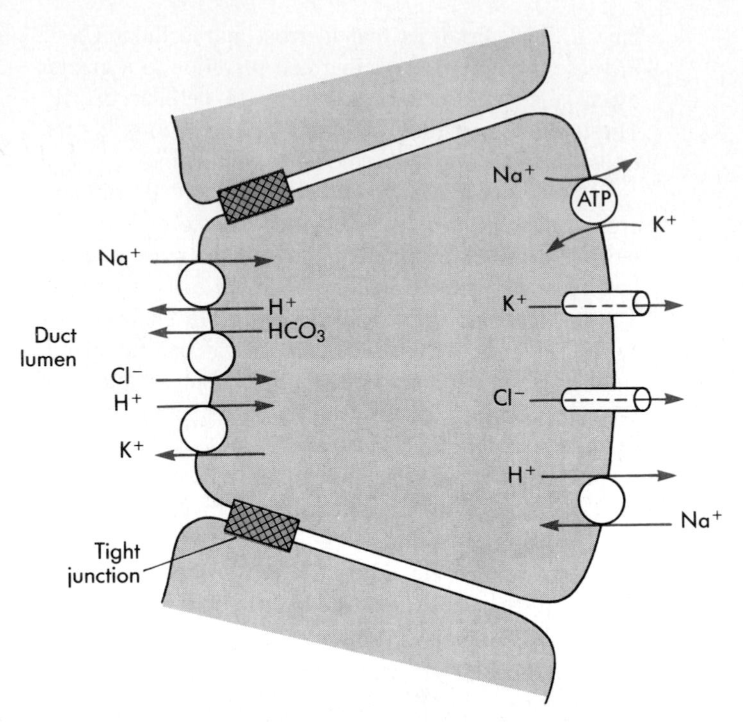

■ **Fig. 39-8** A simplified model of the ionic transport processes involved in the reabsorption of Na^+ and Cl^- and the secretion of K^+ and HCO_3^- by the ductular epithelial cells of salivary glands. See the text for discussion of this model. (Young, JA et al: In Johnson, LR, editor, *Physiology of the gastrointestinal tract,* vol 1, ed 2, New York, 1987, Raven Press.)

tosol, which raises the intracellular HCO_3^- concentration. HCO_3^- is secreted into the lumen via the electrogenic anion channel in the apical membrane. A basolateral Ca^{++}-activated K^+ channel allows electrogenic efflux of K^+ that helps to maintain the electronegativity of the cytosol. This negativity provides part of the driving force for electrogenic Cl^- and HCO_3^- transport across the apical membrane. Secretory agonists, both those that mobilize Ca^{++} and those that elevate cyclic AMP, stimulate *all* of these ionic transport processes.

Ionic mechanisms of salivary ductular epithelial cells. Fig. 39-8 shows a simplified model of the ionic transport processes that operate in the epithelial cells of the excretory ducts and probably in the striated ducts as well. The basolateral Na^+, K^+-ATPase maintains the electrochemical potential gradients of Na^+ and K^+. These gradients power the other ionic transport processes involved in the absorption of Na^+ and Cl^- and the secretion of K^+ and HCO_3^-. The parallel operation of Na^+/H^+ and Cl^-/HCO_3^- exchangers in the apical membrane results in the electroneutral uptake of NaCl. Cl^- leaves the cell at the basolateral surface via an electrogenic Cl^- channel, while Na^+ is pumped out by the Na^+, K^+-ATPase. A basolateral K^+ channel enables K^+ to flow as a counter ion to balance the electrogenic Cl^- efflux. The secretion of HCO_3^- into the lumen across the apical membrane via an anion exchanger is promoted by alkalinization of the cytosol caused by Na^+/H^+ exchange at both apical and basolateral membranes. K^+ secretion into the lumen is accomplished by a K^+/H^+ exchanger in the apical membrane. In contrast to the leaky tight junctions present in secretory endpieces, the tight junctions between ductular epithelial cells are relatively impermeable to water and ions.

■ *Gastric Secretion*

The stomach performs a number of functions. It serves as a reservoir, allowing the ingestion of large meals. It empties its contents into the duodenum at a rate consistent with the ability of the duodenum and small intestine to deal with them. The major secretions of the stomach are HCl, pepsinogens, intrinsic factor, and mucus. HCl keeps gastric contents relatively free of microorganisms. In the absence of HCl secretion bacterial overgrowth occurs in the stomach and upper small intestine. HCl catalyzes the cleavage of pepsinogens to active pepsins and provides a low pH, at which **pepsins** can begin digestion of proteins. Pepsin is not required for normal digestion. In its absence, ingested protein is completely digested by pancreatic proteases and small intestinal brush border peptidases. **Intrinsic factor** is a glycoprotein that binds vitamin B_{12} and allows it to be absorbed by the ileal epithelium. Intrinsic factor is the only gastric secretion required for life. The hormone **gastrin** is secreted by G cells in the antrum and helps regulate gastric acid secretion. **Mucous secretions** protect the gastric mucosa from mechanical and chemical destruction.

Structure of the Gastric Mucosa

The surface of the gastric mucosa (Fig. 39-9, *A*) is covered by columnar epithelial cells that secrete mucus and an alkaline fluid that protects the epithelium from mechanical injury and from gastric acid. The surface is studded with gastric pits; each pit is the opening of a duct into which one or more gastric glands empty. The gastric pits are so numerous that they account for a significant fraction of the total surface area.

The gastric mucosa can be divided into three regions, based on the structures of the glands present. The region just below the lower esophageal sphincter is called the **cardiac glandular region.** In humans the cardiac glandular region is, at most, a few centimeters wide. The glands of this region are tortuous and contain primarily mucus-secreting cells. The remainder of the gastric mucosa is divided into the oxyntic **(acid-secreting) glandular region,** above the notch, and the **pyloric glandular region,** below the notch.

The structure of a gastric gland from the oxyntic glandular region is illustrated in Fig. 39-9, *B*. The surface epithelial cells extend a bit into the duct opening. In the narrow neck of the gland are the **mucous neck cells,** which secrete mucus. Deeper in the gland are **parietal** or **oxyntic cells,** which secrete HCl and intrinsic factor, and **chief** or **peptic cells,** which secrete pepsinogen. Oxyntic cells are particularly numerous in glands in the fundus.

The glands of the pyloric glandular region contain few oxyntic and peptic cells; mucus-secreting cells predominate there. The pyloric glands also contain **G cells** which secrete gastrin.

Surface epithelial cells are exfoliated into the lumen at a considerable rate during normal gastric function. They are replaced by mucous neck cells, which differentiate into columnar surface epithelial cells and migrate up out of the necks of the glands. The capacity of the stomach to repair damage to its epithelial surface in this way is impressive.

Gastric Acid Secretion

The fluid secreted into the stomach is called gastric juice. Gastric juice is a mixture of the secretions of the surface epithelial cells and the secretions of gastric glands. Among the important components of gastric

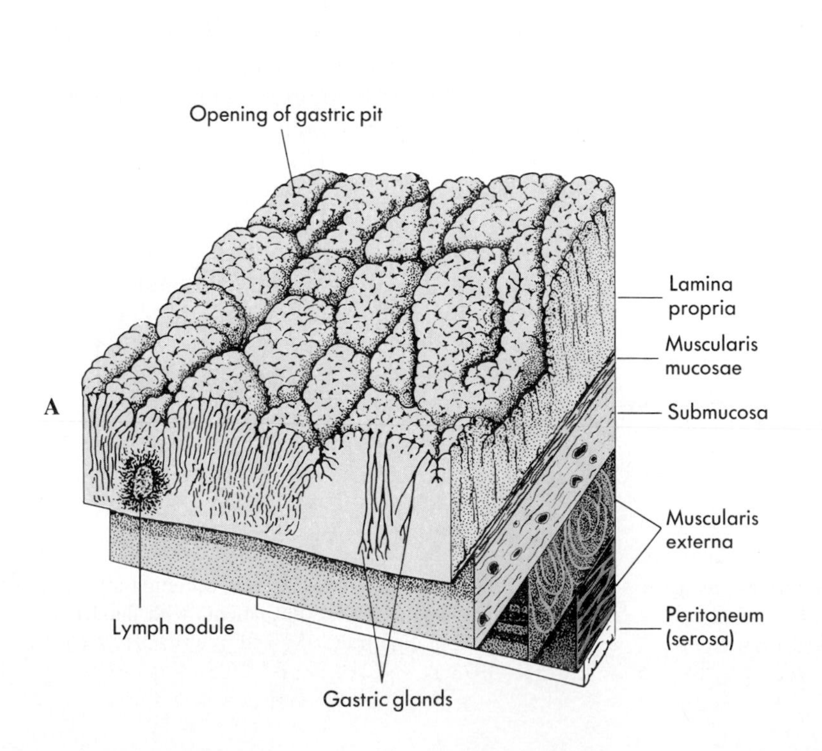

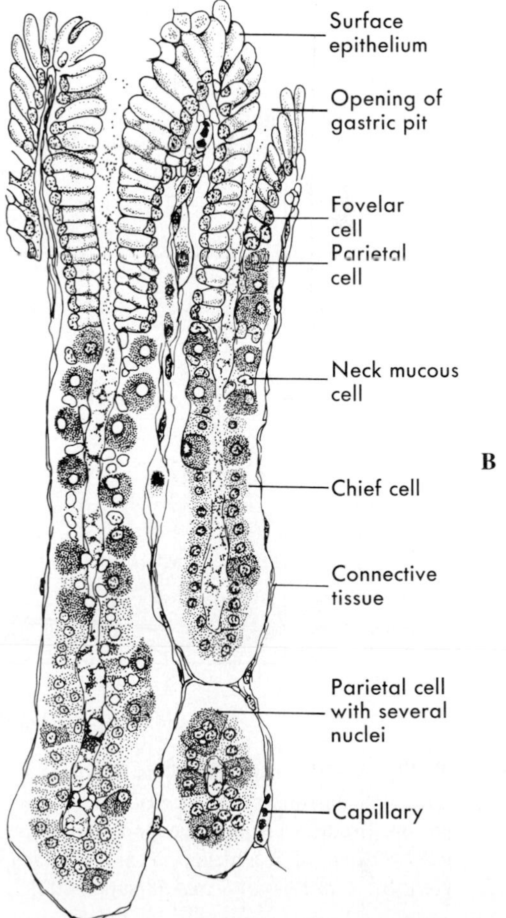

Fig. 39-9 Structure of the gastric mucosa. **A,** Reconstruction of part of the gastric wall. (Redrawn from Braus, H.: *Anatomie des menschen,* Berlin, 1934, Julius Springer.) **B,** Two gastric glands from a human stomach. (Redrawn from Weiss, L., editor: *Histology: cell and tissue biology,* ed 5, New York, 1981, Elsevier.)

juice are salts and water, HCl, pepsins, intrinsic factor, and mucus. Secretion of all of these components increases after a meal.

Ionic Composition of Gastric Juice

The ionic composition of gastric juice depends on the rate of secretion. Fig. 39-10 shows that the higher the secretory rate, the higher the concentration of hydrogen ion. At lower secretory rates $[H^+]$ diminishes, and $[Na^+]$ increases. $[K^+]$ in gastric juice is always higher than in plasma, and consequently prolonged vomiting may lead to hypokalemia. At all rates of secretion Cl^- is the major anion of gastric juice. At high rates of secretion the composition of gastric juice resembles that of an isotonic solution of HCl. At low rates of secretion gastric juice is hypotonic to plasma. When secretion is stimulated the gastric juice approaches isotonicity. The cellular mechanisms by which the ionic composition of gastric juice varies are not well understood. Gastric HCl converts pepsinogen to pepsin, provides an acid pH at which pepsin is active, and kills most ingested bacteria.

Rate of Secretion of Gastric Acid

The rate of gastric acid secretion varies considerably among individuals, partly because of variations in the number of parietal cells. Basal (unstimulated) rates of gastric acid production typically range from about 1 to 5 mEq/hour in humans. On maximal stimulation with histamine or pentagastrin, production rises to 6 to 40 mEq/hour. Patients with gastric ulcers secrete less HCl on the average and patients with duodenal ulcers secrete more HCl on the average than do normal individuals (Fig. 39-11).

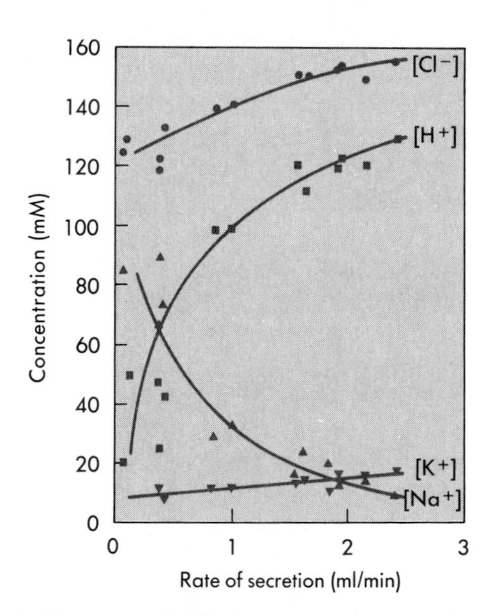

Fig. 39-10 Concentrations of major ions in the gastric juice as a function of the rate of secretion in a normal young person. (Redrawn from Davenport, HW: *Physiology of the digestive tract,* ed 5, Chicago, Year Book; adapted from Nordgren, B.: *Acta Physiol Scand* 58[suppl. 202]:1, 1963.

Morphological Changes That Accompany Gastric Acid Secretion

Parietal cells have a distinctive ultrastructure (Fig. 39-12). They have an elaborate system of branching **secretory canaliculi,** which course through the cytoplasm and are connected by a common outlet to the luminal surface of the cell. Microvilli line the surfaces of the canaliculi. In addition the cytoplasm of the parietal cells contains extensive tubules and vesicles—the **tubulovesicular system.**

When parietal cells are stimulated to secrete, a pronounced morphological change occurs (see Fig. 39-12). Tubules and vesicles of the tubulovesicular system fuse with the plasma membrane of the secretory canaliculi, greatly diminishing the content of tubulovesicles and greatly increasing the surface area of the secretory canaliculi. The tubulovesicles contain the HCl secretory apparatus. The extensive membrane fusion that occurs on stimulation greatly increases the number of HCl pumping sites available at the surface of the secretory canaliculi.

The Cellular Mechanism of Gastric Acid Secretion

The mucosal surface of the stomach is always electrically negative with respect to the serosal surface. In the resting stomach the mucosa is −60 to −80 mV (negative with respect to the serosa). When acid secretion is stimulated the potential difference falls to −30 to −50 mV. Thus Cl^-, the ultimate source of which is the plasma, is transported from the extracellular fluid into

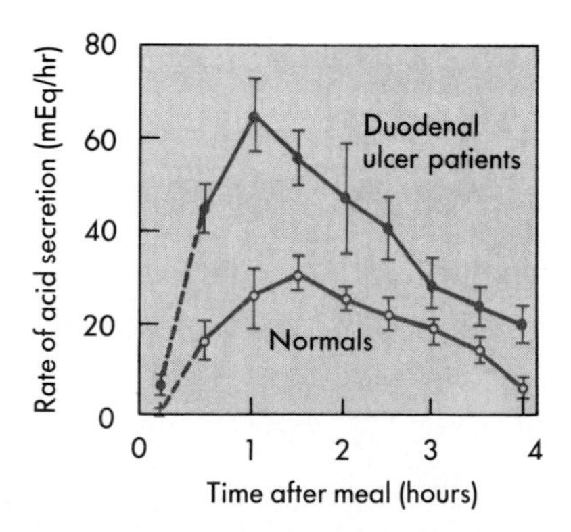

Fig. 39-11 Rate of gastric acid secretion after a meal in six normal subjects and seven patients with duodenal ulcers. (Redrawn from Fordtran, JS, and Walsh, JH: *J Clin Invest* 52:645, 1973.)

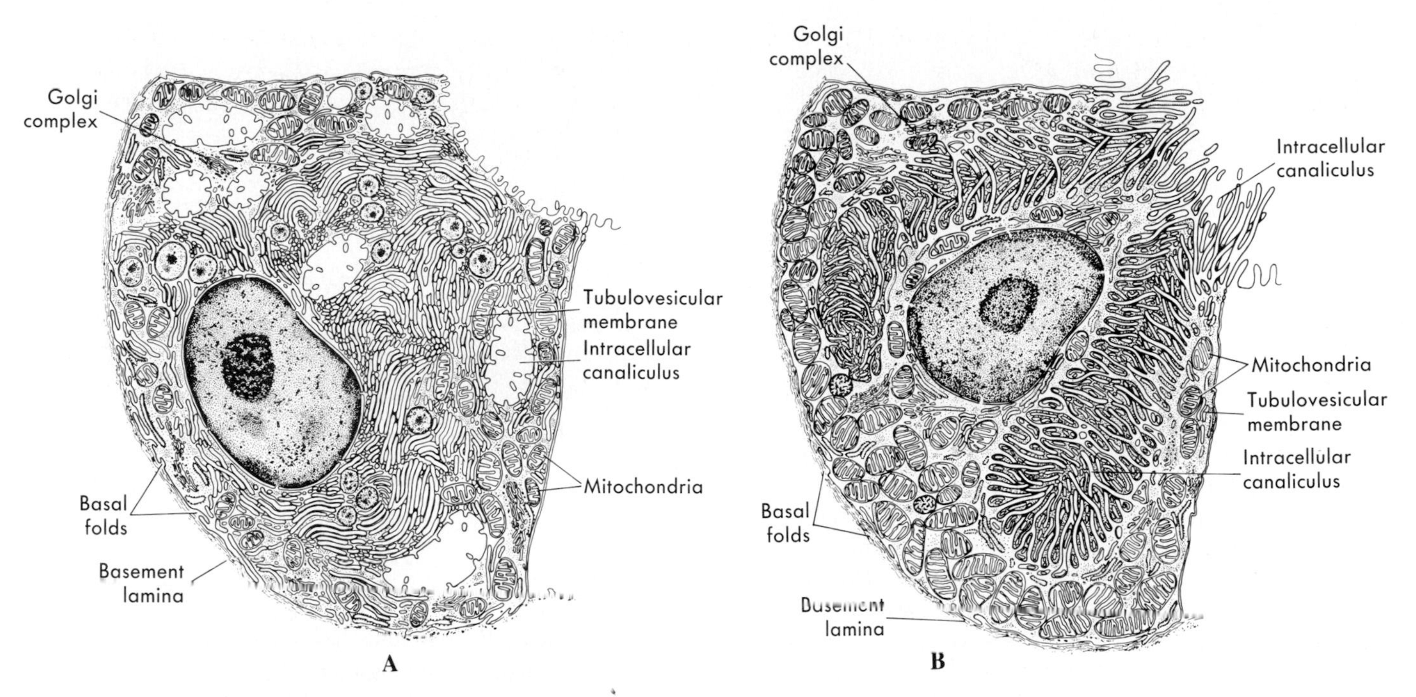

■ **Fig. 39-12** **A,** Drawing of a resting parietal cell with cytoplasm full of tubulovesicles and an internalized intracellular canaliculus. **B,** An acid-secreting parietal cell. Tubulovesicles have fused with the membrane of the intracellular canaliculus, which is now open to the lumen of the gland and lined with abundant long microvilli. (Redrawn from Ito, S: In Johnson, LR, editor: *Physiology of the gastrointestinal tract,* New York, 1981, Raven Press.)

the lumen of the stomach against both electrical and concentration gradients. Hydrogen ion moves down an electrical gradient into the lumen of the stomach but against a much larger chemical concentration gradient. At maximal rates of secretion H^+ is pumped against a concentration gradient that is more than one million to one. Thus energy is required for transport of both H^+ and Cl^-.

The H^+, K^+-ATPase. The H^+, K^+-ATPase that secretes H^+ ions into the secretory canaliculus of the parietal cell is a member of the family of P-type ion-transporting ATPases. This family includes the Na^+, K^+-ATPase and the Ca^{++}-ATPases of sarcoplasmic reticulum and plasma membrane. The amino acid sequence of the H^+, K^+-ATPase is more than 60% homologous to that of the Na^+, K^+-ATPase. Homology is particularly high in the ATP-binding domain and near the aspartate residue that is phosphorylated as part of the transport cycle (see Chapter 1). Fig. 39-13 is a model of the secondary structure of the H^+, K^+-ATPase. The actively pumping form of the H^+, K^+-ATPase is probably a dimer. The H^+, K^+-ATPase has a β-subunit that is about 40% homologous in amino acid sequence to the β-subunit of the Na^+, K^+-ATPase. The function of the β-subunit of the H^+, K^+-ATPase is not yet known.

Drugs that inhibit the H^+, K^+-ATPase have recently been discovered. Substituted benzimidazoles, such as omeprazole, are unreactive at neutral pH, but are converted by H^+ to a form that is reactive with sulfhydryl groups on the H^+, K^+-ATPase. Omeprazole inactivates the H^+, K^+-ATPase; the reaction is irreversible under physiological conditions. Omeprazole is used for treatment of Zollinger-Ellison syndrome. In this disease a gastrin-secreting tumor results in persistent stimulation of gastric HCl secretion; this disorder was difficult to treat before the availability of omeprazole.

Cellular mechanisms of H^+ and Cl^- secretion. A simplified model of the ionic mechanisms involved in secretion of H^+ and Cl^- by the parietal cell is shown in Fig. 39-14. H^+ and HCO_3^- are derived from CO_2 produced by the metabolism of the parietal cell. H^+ is secreted into the lumen of the secretory canaliculus by the H^+, K^+-ATPase in exchange for K^+. Cl^- flows from the cytosol to the lumen of the canaliculus through an electrogenic Cl^- channel.

The secretion of H^+ into the secretory canaliculus causes the pH of the parietal cell to become alkaline. If the alkalinity could not be dissipated, H^+ secretion would cease. To dissipate the excess base, HCO_3^- and OH^- leave the cell at the basolateral membrane via the anion exchanger (see Chapter 1) in exchange for Cl^-. HCO_3^- and OH^- leave the cell down their electrochemical potential gradients; this exit of HCO_3^- and OH^- powers the active uptake of Cl^- against its electrochemical potential gradient. The active uptake of Cl^- at the basolateral membrane is essential for its passive flow through the Cl^- channel in the membrane of the secretory canaliculus. The basolateral membrane also contains the Na^+, K^+-ATPase, a Na^+, H^+ exchanger, and two K^+ channels that are important in stimulation of HCl secretion by agonists.

Stimulation of the cell to secrete H^+ and Cl^-.

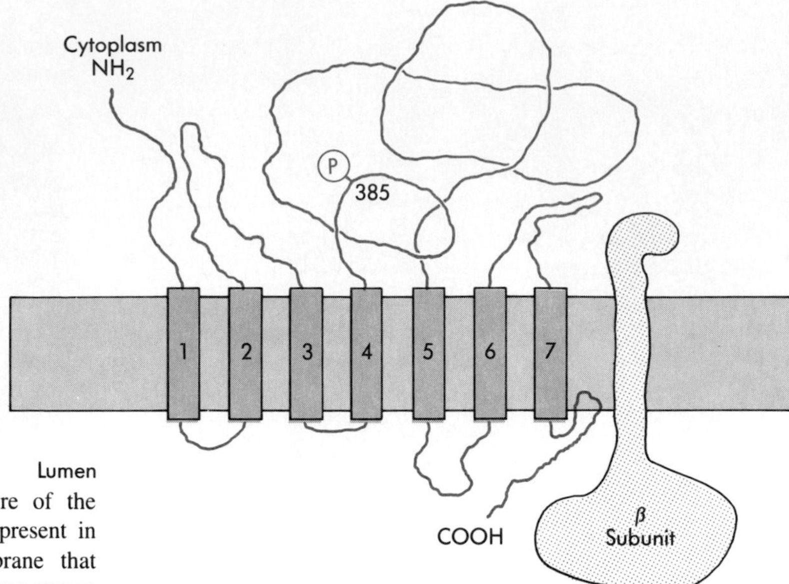

■ **Fig. 39-13** Model of the secondary structure of the H^+,K^+-ATPase. This ion-transporting ATPase is present in the tubovesicular membranes and in the membrane that bounds the secretory canaliculus. The H^+,K^+-ATPase pumps H^+ ions into the lumen of the secretory canaliculus in exchange for K^+ taken into the parietal cell. The phosphorylation site, aspartate 385, is identified by *P*. The β-subunit is a glycoprotein that may function similarly to the β-subunit of the Na^+,K^+-ATPase. (Modified from Rabon, EC and Reuben, MA: *Ann Rev Physiol* 52:321, 1990.)

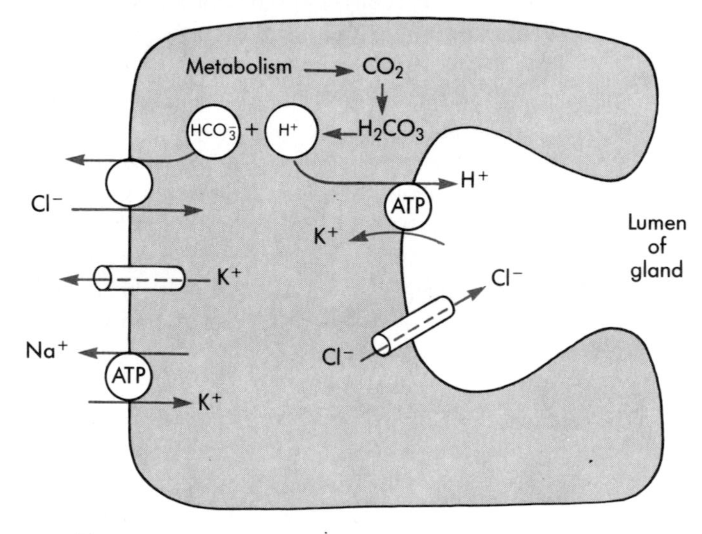

■ **Fig. 39-14** Simplified view of the major ion transport processes involved in secretion of H^+ and Cl^- by parietal cells. H^+ and HCO_3^- are derived from CO_2 produced by the metabolism of the parietal cell. H^+ is secreted into the lumen of the secretory canaliculus by the H^+,K^+-ATPase in exchange for K^+. Cl^- flows from the cytosol to the lumen of the canaliculus through an electrogenic Cl^- channel. The cytosolic alkalinity generated by H^+ secretion is dissipated by HCO_3^- and OH^- leaving the cell at the basolateral membrane via the anion exchanger in exchange for Cl^-. HCO_3^- and OH^- leave the cell down their electrochemical potential gradients; this exit of HCO_3^- and OH^- powers the active uptake of Cl^- against its electrochemical potential gradient.

Recent patch clamp studies have shown that the dominant conductance of the basolateral membrane is to K^+. At least two types of K^+ channels exist: One type is activated by elevated cyclic AMP, and the other is activated by increased intracellular Ca^{++}.

The membrane that bounds the secretory canaliculus, the apical membrane, has a single class of Cl^- channels. At rest few K^+ channels are present in the apical membrane. When the cell is stimulated to secrete, the number of K^+ channels may increase. The Cl^- conductance of the apical membrane is dramatically increased by cyclic AMP and by membrane hyperpolarization caused by a dramatic increase in the average channel open time. In addition cyclic AMP and Ca^{++} cause the insertion of more Cl^- channels and H^+, K^+-ATPase molecules into the apical membrane (see Fig. 39-12).

The three physiological agonists of HCl secretion are ACh, histamine, and gastrin. ACh, acting on M_3 muscarinic receptors, increases the Ca^{++} conductance of the basolateral membrane and mobilizes intracellular Ca^{++}. Histamine, acting on H_2 receptors, activates adenylyl cyclase and elevates cellular cyclic AMP. Gastrin elevates intracellular Ca^{++} via the inositol lipid cascade. Elevation of either cyclic AMP or Ca^{++} activates K^+ channels in the basolateral membrane and thereby hyperpolarizes the cell. Hyperpolarization and cyclic AMP both markedly elevate the Cl^- conductance of the apical membrane. The secretory agonists also cause the fusion of tubovesicles with the apical membrane, and thereby increase the number of H^+, K^+-ATPase molecules and Cl^- channels that participate in H^+ and Cl^- secretion.

■ *Pepsins*

The pepsins are a group of proteases secreted primarily by the chief cells of the gastric glands and often collectively referred to as pepsin. The pepsins fall into two electrophoretic classes: group I, secreted mostly in the oxyntic glandular mucosa, and group II, secreted throughout the stomach and by Brunner's glands of the duodenum.

Pepsins are secreted as inactive proenzymes known as **pepsinogens.** Because of gastric acidity, cleavage of acid-labile linkages converts pepsinogens to pepsins; the lower the pH, the more rapid the conversion. Pepsins also act proteolytically on pepsinogens to form more pepsins.

The pepsins have unusually low pH optima; they have their highest proteolytic activity at pH 3 and below. Pepsins may digest as much as 20% of the protein in a typical meal. When the duodenal contents are neutralized, pepsins are inactivated irreversibly.

Pepsinogens are contained in membrane-bound zymogen granules in the chief cells. The contents of the zymogen granules are released by exocytosis when the chief cells are stimulated to secrete. Chief cells without zymogen granules also can secrete pepsinogens.

Intrinsic Factor

Intrinsic factor is a glycoprotein that has a molecular weight of about 55,000 and is secreted by the parietal cells of the stomach. Intrinsic factor is required for normal intestinal absorption of vitamin B_{12}. Vitamin B_{12} (in all its physiological forms) binds to intrinsic factor. The intrinsic factor—B_{12} complex is highly resistant to digestion. Receptors in the mucosa of the ileum bind the complex, and the B_{12} is taken up by the ileal mucosal epithelial cells. Intrinsic factor is released in response to the same stimuli that evoke secretion of gastric acid from the parietal cells.

The Gastric Mucosal Barrier: Mucus and Bicarbonate

Secretions that contain glycoprotein mucins are viscous and sticky and are collectively termed **mucus.** Mucus adheres to the gastric mucosa and protects the mucosal surface from abrasion by lumps of food and from chemical damage caused by the acid and digestive enzymes.

Mucous neck cells in the necks of gastric glands secrete a clear mucus sometimes called **soluble mucus.** Soluble mucus is not present in the resting stomach. Secretion of soluble mucus is stimulated by sham feeding and by some of the same stimuli that enhance acid and pepsinogen secretion, especially by acetylcholine released from parasympathetic nerve endings near the gastric glands.

The surface epithelial cells secrete a mucus containing different mucins. This mucus is cloudy in appearance and thus is termed **visible mucus.** The surface epithelial cells also secrete watery fluid with Na^+ and Cl^- concentrations similar to plasma but with higher K^+ (four times) and HCO_3^- (two times) concentrations than in plasma. The high $[HCO_3^-]$ makes the visible mucus alkaline. Visible mucus is secreted by the resting mucosa and lines the stomach with a sticky, viscous, alkaline coat. When food is eaten the rates of secretion of visible mucus and of HCO_3^- by the surface epithelial cells increase. Among the stimuli that enhance secretion are mechanical stimulation of the mucosa and stimulation of either sympathetic or parasympathetic nerves to the stomach.

The mucus forms a gel on the luminal surface of the mucosa. The gel protects the mucosa from mechanical damage from chunks of food. The alkaline fluid it entraps protects the mucosa against damage by HCl and pepsin. The mucus and alkaline secretions are part of the **gastric mucosal barrier** that prevents damage to the mucosa by gastric contents.

The mucus layer prevents the bicarbonate-rich secretions of the surface epithelial cells from rapidly mixing with the contents of the gastric lumen. Consequently, the surfaces of gastric epithelial cells are bathed in their own bicarbonate-rich secretions. HCO_3^- buffers H^+ ions that diffuse from the lumen to the epithelial surface (Fig. 39-15). Experiments with pH microelectrodes showed that the surface of the epithelial cells can be maintained at a slightly alkaline pH, despite a luminal pH of about 2. The protection depends on both mucus and HCO_3^- secretion; either mucus alone or HCO_3^- alone could not hold the pH at the epithelial cell surface near neutral. The unstirred layer provided by the mucus retards convective mixing of epithelial cell secretions with luminal contents and slows diffusion of H^+ to the surface and HCO_3^- into the lumen. When the unstirred layer is 1 mm thick the diffusion times for H^+ and HCO_3^- are about 10 minutes, but this time delay would not prevent a rise in the concentration of H^+ at the epithelial surface without a continuous secretion of HCO_3^- to neutralize the H^+ as it arrives.

Mucus is stored in large granules in the apical cytoplasm of mucous neck cells and surface epithelial cells. It is released by exocytosis, by dissolution of the apical membrane of the cell, or by exfoliation of an entire epithelial cell into the mucous coat. The normal epithelium can replace exfoliated surface cells effectively even when extensive cell loss occurs.

The mechanism of HCO_3^- secretion by surface epithelial cells is not well understood. Carbonic anhydrase in the epithelial cells supports HCO_3^- secretion, which suggests that the source of HCO_3^- is CO_2 produced by metabolism. Secretion of HCO_3^- may involve exchange of HCO_3^- for Cl^- across the luminal plasma membrane of the epithelial cell. The maximal rate of HCO_3^- secretion is about 10% of the highest rate of gastric acid secretion. Elevated Ca^{++} in the serosal fluid and cholinergic agonists stimulate secretion of HCO_3^-, but hista-

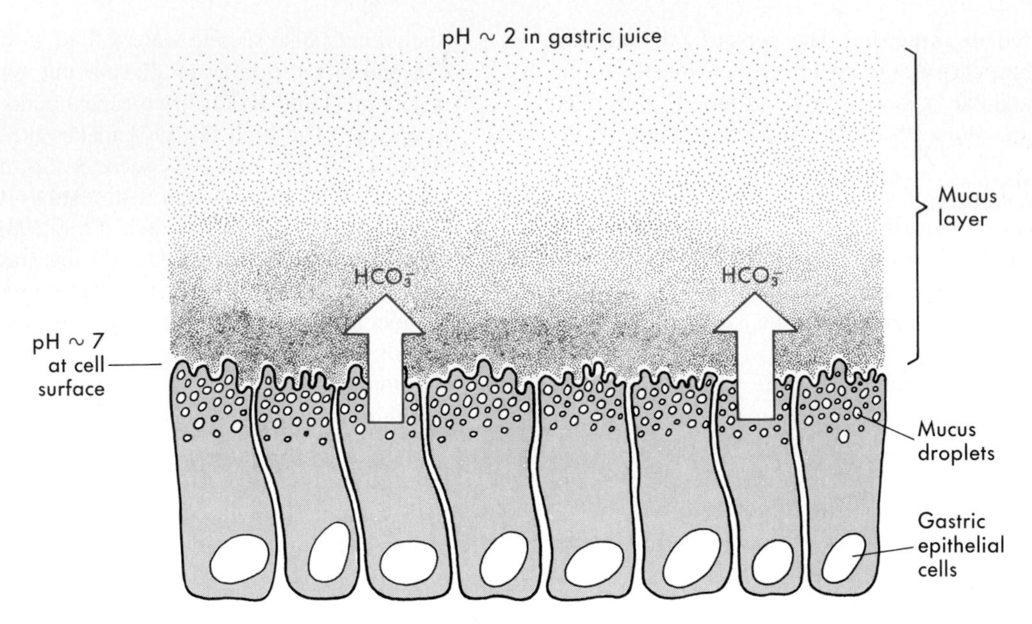

■ **Fig. 39-15** The protection provided the mucosal surface of the stomach by the mucus layer. Buffering by the bicarbonate-rich secretions of the surface epithelial cells and the restraint to convective mixing owing to the high viscosity of the mucus layer allow the pH at the cell surface to remain near 7, whereas the pH in the gastric juice is 1 to 2.

mine and gastrin have little effect. α-Adrenergic agonists decrease bicarbonate secretion. This effect may play a role in pathogenesis of stress ulcers: a chronically elevated level of circulating epinephrine may suppress secretion of HCO_3^- sufficiently to diminish protection of the epithelial cell surface. Aspirin and other nonsteroidal antiinflammatory agents inhibit secretion of both mucus and bicarbonate; prolonged use of these drugs may lead to damage of the mucosal surface. Certain prostaglandins enhance mucus and HCO_3^- secretion and may protect an individual who is susceptible to gastric ulcers.

The structure of mucus. Mucins are the major components of mucus. The mucins produced by surface epithelial cells of the stomach are glycoproteins that are about 80% carbohydrate. About 65% of the sugars present in the carbohydrate side chains of gastric mucins are galactose and *N*-acetylglucosamine. The protein backbone is rich in threonine, serine, and proline, and the carbohydrate side chains are linked to the hydroxyl groups of these amino acid residues. Intact mucins have molecular weights near 2 million. They consist of four similar monomers with molecular weights of 500,000, linked together by disulfide cross-links (Fig. 39-16). Each monomer is largely covered by carbohydrate side chains that protect it from proteolytic degradation. The portion of the monomer that participates in the disulfide cross-links is rich in cystine and free of carbohydrate; thus it is vulnerable to proteolysis. Pepsins cleave bonds in the central region of the tetramer and release fragments roughly as large as monomers. The tetrameric mucins form a gel when their concentration exceeds about 50 mg/ml. The mucin monomers do not form a gel. Proteolysis of mucins by pepsin thus dissolves the gel. Maintenance of the protective mucus layer requires that new tetrameric mucins be secreted to replace mucins that are cleaved by pepsins.

■ *Control of Acid Secretion*

■ *Control of HCl Secretion at the Level of the Parietal Cell*

Parietal cell agonists. Acetylcholine, histamine, and gastrin each binds to a distinct class of receptors on the plasma membrane of the parietal cell and directly stimulates the parietal cell to secrete HCl (Fig. 39-17). Acetylcholine is released near parietal cells by cholinergic nerve terminals. Gastrin, a hormone, is produced by G cells in the mucosa of the gastric antrum and the duodenum and reaches parietal cells via the bloodstream. Histamine is released from **ECL (enterochromaffin-like)** cells in the gastric mucosa and diffuses to the parietal cells (hence histamine is a paracrine regulator).

Acetylcholine, histamine, and gastrin are important in the regulation of HCl secretion. Specific antagonists for acetylcholine (e.g., atropine) and histamine (e.g., cimetidine) not only block the effects of acetylcholine and histamine, respectively, but also inhibit acid secretion in response to any effective stimulus. A physiological role for gastrin has been established by correlating

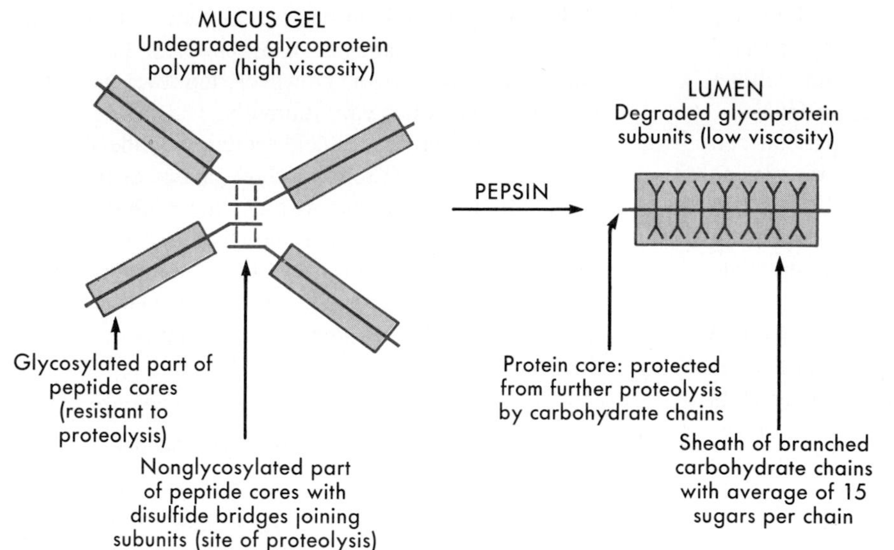

■ **Fig. 39-16** Schematic representation of the structure of gastric mucus glycoprotein before and after hydrolysis by pepsin. (Redrawn from Allen, A: *Br Med Bull* 34:28, 1978.)

the rate of gastric acid secretion with the level of gastrin in the blood. Under many circumstances each of the primary agonists (acetylcholine, histamine, and gastrin) is able to potentiate the acid secretion elicited by the other two mediators. The mutual potentiation of the primary mediators appears to explain the lack of specificity in vivo of atropine and cimetidine, substances that are believed to be specific receptor antagonists.

Cellular mechanisms of parietal cell agonists. *Acetylcholine, histamine, and gastrin are the three natural secretagogues that act directly on parietal cells to enhance the rate of acid secretion.*

Acetylcholine stimulates acid secretion by binding to M_3 muscarinic receptors on the basal membrane of the parietal cell (see Fig. 39-17). Binding of acetylcholine to its receptors opens Ca^{++} channels and allows Ca^{++}

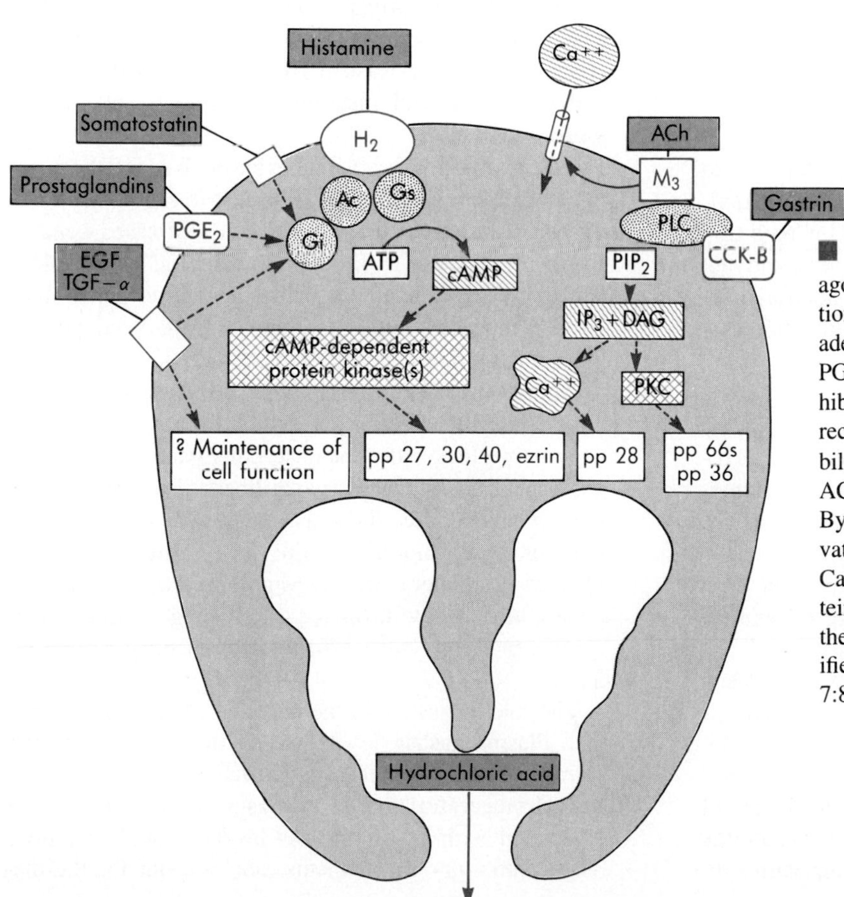

■ **Fig. 39-17** Signal transduction mechanisms of agonists and antagonists of parietal cell HCl secretion. Histamine, acting on H_2 receptors, activates adenylyl cyclase by activating G_s. Somatostatin and PGE_2, as well as EGF and TGF-α, activate G_i and inhibit adenylyl cyclase. Acetylcholine binding to M_3 receptors and gastrin binding to CCK-B receptors mobilize intracellular Ca^{++} by the inositol lipid cascade. ACh also increases Ca^{++} entry via Ca^{++} channels. By these signal transduction pathways agonists activate protein kinases (protein kinases A and C and Ca/CaM kinases) that phosphorylate intracellular proteins, some of which are indicated. The functions of these phosphoproteins remain to be elucidated. (Modified from Chew, CS: *Curr Opinion Gastroenterol* 7:856, 1991.)

to enter the cell, increasing the cytosolic level of free Ca^{++}. Acetylcholine also causes Ca^{++} to be released from intracellular stores. The increased intracellular Ca^{++} concentration results in the secretion of HCl.

Histamine binds to H_2 receptors on the parietal cell membrane and activates adenylate cyclase in the plasma membrane, thereby increasing cytosolic level of cAMP. The increased cAMP stimulates a cAMP-dependent protein kinase that phosphorylates proteins that regulate acid secretion.

Gastrin is not as potent a direct stimulant of parietal cells as acetylcholine and histamine. The direct actions of gastrin are not blocked by muscarinic antagonists (e.g., atropine) or by H_2 receptor blockers (e.g., cimetidine). Gastrin elevates Ca^{++} in the parietal cell, via the inositol phosphate pathway. **Proglumide** is a gastrin receptor antagonist. The physiological response to elevated levels of gastrin in the blood is markedly attenuated by cimetidine. Thus a major component of the physiological response to gastrin may be caused by gastrin-stimulated release of histamine.

Histamine is certainly a major physiological mediator of HCl secretion. Cimetidine blocks a large portion of the acid secretion elicited by any known secretagogue. How histamine levels in the vicinity of parietal cells are regulated is not known. ECL cells are present in the gastric mucosa, and these cells synthesize and store histamine. Upon stimulation by acetylcholine or gastrin the ECL cells release histamine, which diffuses to nearby parietal cells to stimulate HCl secretion. Parietal cells actively take up histamine and inactivate it by methylation.

Endogenous antagonists of acid secretion. **Prostaglandins** of the E and I series, **somatostatin,** and **EGF (epidermal growth factor)** act on parietal cells to inhibit the secretion of HCl. Their mechanisms of inhibition are shown in Fig. 39-17. Prostaglandins bind to a PGE_2 receptor on the parietal cell and thereby activate G_i, the GTP-binding protein that inhibits adenylyl cyclase. The prostaglandin analogue **misoprostal** suppresses HCl secretion in this way.

Somatostatin, released from D cells in the gastric and intestinal mucosae, also inhibits HCl secretion by activating G_i. Somatostatin also may inhibit acid secretion evoked by the Ca^{++}-mobilizing agonists. Somatostatin might also inhibit release of histamine by ECL cells.

EGF and certain other growth factors (such as TGF-α) inhibit acid secretion by activating G_i and by inhibiting the effects of ACh and gastrin.

In vivo control of the rate of acid secretion. When the stomach has been empty for several hours, HCl is secreted at a significant basal rate, which is about 10% of the maximal rate. The regulation of the basal secretory rate is not well understood. Background levels of acetylcholine and histamine may tonically stimulate the parietal cells. The basal rate of HCl secretion varies diurnally, being highest in the evening and lowest in the morning. The mechanisms responsible for the diurnal variation are not known. However, the level of gastrin in the blood does not vary diurnally.

After a meal the rate of acid secretion by the stomach increases promptly. There are three phases of increased acid secretion in response to food: the **cephalic phase** (elicited before food reaches the stomach), the **gastric phase** (elicited by the presence of food in the stomach), and the **intestinal phase** (elicited by mechanisms originating in the duodenum and upper jejunum).

The Cephalic Phase of Gastric Secretion

The cephalic phase of gastric secretion is normally elicited by the sight, smell, and taste of food. The cephalic phase may be studied in isolation by means of **sham feeding.** In sham feeding, food is chewed but not swallowed or is swallowed and then diverted to the outside of the body by an esophageal fistula. The acid secretion rate during the cephalic phase can be as much as 40% of the maximum rate.

The cephalic phase of gastric acid secretion is mediated entirely by impulses in the vagus nerves; bilateral truncal vagotomy completely abolishes the cephalic phase. Cholinergic vagal fibers and cholinergic neurons of the intramural plexuses are the principal mediators of the cephalic phase. Acetylcholine released from these neurons directly stimulates parietal cells to secrete HCl. The acetylcholine also stimulates acid secretion indirectly by releasing gastrin from G cells in the antrum and duodenum and histamine from ECL cells in the gastric mucosa.

Other stimuli sensed in the brain, besides those related to the presence of food, may evoke acid secretion via vagal impulses. For example, a decreased concentration of glucose in the cerebral arterial blood elicits acid secretion. Inhibition of brain glucose metabolism by 2-deoxyglucose (a nonmetabolizable analogue of glucose) augments gastric acid secretion.

Atropine blocks most but not all of the acid secretion in response to cephalic-phase stimuli. However, truncal vagotomy blocks the response completely, which suggests that noncholinergic vagal fibers also participate in the cephalic phase. The responses to cephalic stimuli that increase blood gastrin levels are completely blocked by bilateral truncal vagotomy but are not completely blocked by atropine, which suggests that a noncholinergic vagal neuroeffector evokes some gastrin release.

The role of gastrin in the cephalic phase is controversial. Plasma gastrin levels do not increase consistently in response to sham feeding in man. In certain experimental preparations large increases in blood gastrin are associated with relatively low levels of acid secretion. Vagal inhibitory mechanisms may account for the discrepancy between gastrin levels and acid secretion in these experiments.

The presence of a low pH in the antrum diminishes the amount of HCl secreted during the cephalic phase. In the absence of food in the stomach to buffer the acid secreted, the pH of the antral contents falls rapidly during the cephalic phase. The low pH limits the amount of acid secreted by a direct effect on parietal cells.

Regulation of gastric acid secretion by the central nervous system. At the present time knowledge of cerebral control of gastric acid secretion is fragmentary. The hypothalamus participates in the regulation of gastric acid secretion. Electrical stimulation of the lateral hypothalamus increases gastric acid secretion, whereas stimulation of the ventromedial hypothalamus inhibits acid secretion. Injection of 2-deoxyglucose into the lateral hypothalamus increases gastric acid secretion, suggesting that this area enhances gastric acid secretion when cerebral glucose metabolism is impeded. The amygdala and the limbic system may also regulate acid secretion.

A complex interplay between norepinephrine, γ-aminobutyric acid (GABA), and certain neuropeptides may occur in the brain in the control of gastric acid secretion. Norepinephrine appears to be the neurotransmitter in some central neural pathways that inhibit gastric acid secretion, whereas GABA is the transmitter in certain pathways that stimulate acid secretion.

A number of brain peptides may regulate gastric acid secretion. Thyrotropin-releasing hormone and gastrin may affect pathways that enhance acid secretion. Neurotensin, calcitonin, somatostatin, and certain opioid peptides (β-endorphin, Met-enkephalin, and dynorphin) may act centrally to inhibit acid secretion. Dopaminergic neural pathways, both in the brain and in the enteric nervous system, inhibit gastric acid secretion. Neuropeptide Y also has central inhibitory effects on acid secretion. The interrelationships among the various central nervous system pathways that influence gastric acid secretion and their physiological significance remain to be elucidated.

Chronic exposure to nicotine increases gastric acid secretion. Nicotine appears to exert this effect on the central nervous system.

■ *The Gastric Phase of Gastric Secretion*

The gastric phase of gastric secretion is brought about by the presence of food in the stomach. The principal stimuli are distension of the stomach and the presence of amino acids and peptides resulting from the actions of pepsins. Most of the acid secreted in response to a meal is secreted during the gastric phase.

When either the body or the antrum of the stomach is distended, mechanoreceptors are stimulated. These mechanoreceptors serve as the afferent arms of local and central reflexes that bring about secretion. Both local and central responses are largely cholinergic. Afferent and efferent pathways of the central reflexes are in the vagus nerve (i.e., they are **vagovagal reflexes**).

When the oxyntic glandular area is distended, local and central reflexes bring about release of acetylcholine near parietal cells and stimulate HCl secretion directly. Distension of the body (oxyntic gland area) of the stomach brings about gastrin release from the antral mucosa via a vagal reflex. Distension of the pyloric glandular area (antrum) enhances gastrin release and, via a vagal reflex, increases acid secretion by the oxyntic glandular mucosa.

All the responses elicited by gastric distension can be blocked effectively by bathing the mucosal surface with an acid solution with pH 2 or less. Once the buffering capacity of the gastric contents is saturated, gastric pH falls rapidly and greatly inhibits further acid release. In this way the acidity of gastric contents regulates itself. In patients with duodenal ulcers, acid secretion is less inhibited by the presence of acid in the antrum.

The presence of amino acids or peptides in the stomach elicits secretion of gastric acid. Intact proteins do not have this effect. The various amino acids differ greatly in their abilities to stimulate acid secretion; tryptophan and phenylalanine are particularly potent stimuli of acid secretion.

Amino acids and peptides may directly stimulate parietal cells, but this action is controversial. Amino acids and peptides do act directly on G cells in the gastric antrum to release gastrin. This appears to be the principal way in which amino acids and peptides stimulate acid secretion during the gastric phase.

Other frequently ingested substances that elicit gastric acid secretion include calcium ions, caffeine, and alcohol. Pure caffeine stimulates acid secretion, perhaps by inhibiting the phosphodiesterase that breaks down cAMP. Acid secretion is enhanced by coffee, but caffeine is not the mediator of this effect, because decaffeinated coffee is equally effective. Ethanol in high concentrations stimulates gastric acid secretion; dilute ethanol is not an effective stimulus.

Gastric distension potentiates the stimulatory effects of chemical stimuli on acid secretion during the gastric phase.

■ *The Intestinal Phase of Gastric Secretion*

The presence of chyme in the duodenum brings about neural and endocrine responses that first stimulate and later inhibit secretion of acid by the stomach. Early in gastric emptying, when the pH of gastric chyme is above 3, the stimulatory influences predominate. Later, when the buffer capacity of gastric chyme is exhausted and the pH of chyme emptied into the duodenum falls below pH 2, inhibitory influences prevail. Tables 39-1

Table 39-1 Major mechanisms for stimulation of gastric acid secretion

Phase	*Stimulus*	*Pathway*	*Stimulus to parietal cell*
Cephalic	Chewing, swallowing, etc.	Vagus nerve to:	
		1. Parietal cells	Acetylcholine
		2. G cells	Gastrin
Gastric	Gastric distension	Local and vagovagal reflexes to:	
		1. Parietal cells	Acetylcholine
		2. G cells	Gastrin
Intestinal	Protein digestion products in duodenum	1. Intestinal G cells	Gastrin
		2. Intestinal endocrine cells	Enterooxyntin

Modified from Johnson LR, editor: *Gastrointestinal physiology*, ed 3, St Louis, 1985, CV Mosby; adapted from M.I. Grossman.

Table 39-2 Major mechanisms for inhibition of gastric acid secretion

Region	*Stimulus*	*Mediator*	*Inhibit gastrin release*	*Inhibit acid secretion*
Antrum	Acid (pH < 3.0)	None, direct	+	
Duodenum	Acid	Secretin	+	+
		Bulbogastrone	+	+
		Nervous reflex		+
Duodenum and jejunum	Hyperosmotic solutions	Unidentified enterogastrone		+
	Fatty acids, monoglycerides	Gastric inhibitory peptide	+	+
		Cholecystokinin		+
		Unidentified enterogastrone		+

Modified from Johnson LR, editor: *Gastrointestinal physiology*, ed 3, St. Louis, 1985, CV Mosby; adapted from M.I. Grossman.

and 39-2 summarize the major mechanisms that control gastric acid secretion.

Stimulation of gastric secretion. During the intestinal phase, gastric secretion is brought about by distension of the duodenum and by the presence of products of protein digestion (peptides and amino acids) in the duodenum, primarily via endocrine mechanisms. The duodenum and proximal jejunum contain G cells that release gastrin when stimulated by peptides and amino acids. The gastrin is carried in the blood to the parietal cells and stimulates them to secrete acid. A second hormone is also released from the duodenum in response to the presence of chyme, and it also acts directly on the parietal cells to stimulate acid secretion and to potentiate the effect of gastrin. This hormone has been named **enterooxyntin,** and its chemical identity and its physiological role in humans remain to be defined. Amino acids in the blood can stimulate gastric acid secretion; thus absorbed amino acids may help to stimulate acid secretion during the intestinal phase.

Inhibition of gastric acid secretion. Several different mechanisms that operate during the intestinal phase inhibit gastric secretion (Table 39-2). The stimuli for these mechanisms are the presence of acid, fat digestion products, and hypertonicity in the duodenum and proximal part of the jejunum.

Acid solutions in the duodenum cause the release of **secretin** into the bloodstream. Secretin inhibits gastric acid secretion in two ways: it inhibits gastrin release by G cells, and it inhibits the response of parietal cells to gastrin. Acid in the duodenum also inhibits gastric acid secretion via a local nervous reflex.

Acid in the duodenal bulb releases the hormone **bulbogastrone,** which has not been chemically characterized. Bulbogastrone, like secretin, inhibits gastrin-stimulated acid secretion by the parietal cells.

Fatty acids with 10 or more carbons and monoglycerides, the major products of triglyceride digestion, in the duodenum and proximal part of the jejunum release two hormones: **gastric inhibitory peptide** and **cholecystokinin.** Gastric inhibitory peptide inhibits acid secretion by suppressing gastrin release and by directly inhibiting secretion of acid from the parietal cells. Cholecystokinin also inhibits acid secretion by parietal cells, but this effect may not be physiologically significant.

Another hormone, as yet unidentified, may also inhibit gastric acid secretion in response to fats in the duodenum. Hyperosmotic solutions in the duodenum release another unidentified hormone that inhibits gastric acid secretion.

Pepsinogen Secretion

Most of the agents that stimulate parietal cells to secrete acid also elicit release of pepsinogens from chief cells. Hence the rates of release of acid and pepsinogens from the gastric glands are highly correlated. Acetylcholine is a potent stimulus for the chief cells to release pepsinogens. When cholinergic fibers that enhance acid secre-

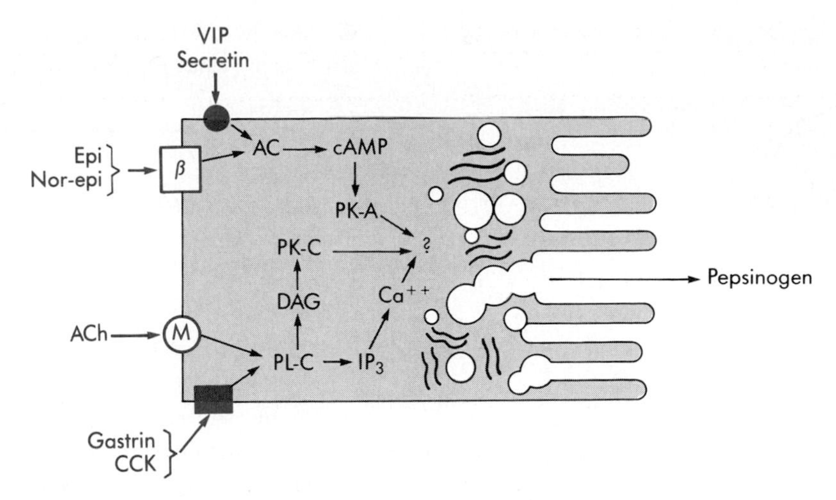

■ **Fig. 39-18** Cellular mechanisms of secretagogues that elicit pepsinogen secretion by chief cells. Secretin, VIP, and β-adrenergic agonists act via receptors that increase the activity of adenylyl cyclase *(AC)* and increase the level of cyclic AMP, resulting in activation of cyclic AMP-dependent protein kinase *(PK-A)*. Cholinergic agonists, CCK, and gastrin act by the inositol phosphate pathway resulting in increased intracellular Ca^{++} and activation of protein kinase C *(PK-C)*. *PL-C,* phospholipase C; *DAG,* diacylglycerol; IP_3, inositol trisphosphate. (Modified from Hersey, SJ: In *Handbook of physiology,* Sec 6, Vol III, Washington, DC, 1989, American Physiological Society.)

tion are active, so are cholinergic fibers that release acetylcholine near chief cells. Gastrin also stimulates chief cells to secrete pepsinogens. Acid in contact with the gastric mucosa enhances the output of pepsinogens by a local reflex. Secretin and cholecystokinin, hormones released by the duodenal mucosa in response to acid and fat digestion products, respectively, *stimulate* chief cells to secrete pepsinogens. The relative importance of these various secretagogues remains to be determined.

Studies with isolated gastric glands have provided information about the cellular mechanisms of action of the compounds that mediate release of pepsinogens (Fig. 39-18). β-Adrenergic agonists release pepsinogens by increasing the level of cAMP in chief cells. Secretin shares this mechanism of eliciting secretion. Acetylcholine, cholecystokinin, and gastrin stimulate release of pepsinogens by increasing the intracellular level of Ca^{++}. How elevated cAMP and Ca^{++} levels increase the release of pepsinogens is not well understood.

■ *Pancreatic Secretion*

The human pancreas weighs less than 100 g, and each day it elaborates 1 L, 10 times its mass, of pancreatic juice. The pancreas is unusual in having both endocrine and exocrine secretory functions. Its principal endocrine secretions are insulin and glucagon, whose functions are discussed in Chapter 46. The exocrine secretions of the pancreas are important in digestion. Pancreatic juice is composed of an aqueous component, rich in bicarbonate, that helps to neutralize duodenal contents, and an enzyme component that contains enzymes for digesting carbohydrates, proteins, and fats. Pancreatic exocrine secretion is controlled by both neural and hormonal signals, elicited primarily by the presence of acid and digestion products in the duodenum. Secretin plays the major role in eliciting secretion of the aqueous component, and cholecystokinin stimulates the secretion of pancreatic enzymes.

■ *Structure and Innervation of the Pancreas*

The structure of the exocrine pancreas resembles that of the salivary glands. Microscopic blind-ended tubules are surrounded by polygonal acinar cells whose primary function is to secrete the enzyme component of pancreatic juice. The acini are organized into lobules. The tiny ducts that drain the acini are called intercalated ducts (Fig. 39-19). The intercalated ducts empty into somewhat larger intralobular ducts. The intralobular ducts of a particular lobule drain into a single extralobular duct that empties that lobule into larger ducts. The larger ducts converge into a still larger main collecting duct that drains the pancreas and enters the duodenum along with the common bile duct.

The cells of both acini and ducts are joined by junctional complexes that consist of tight junctions, zonulae adherens, and desmosomes. The tight junctions constitute a permeability barrier between the luminal fluid and the extracellular fluid that bathes the basolateral sur-

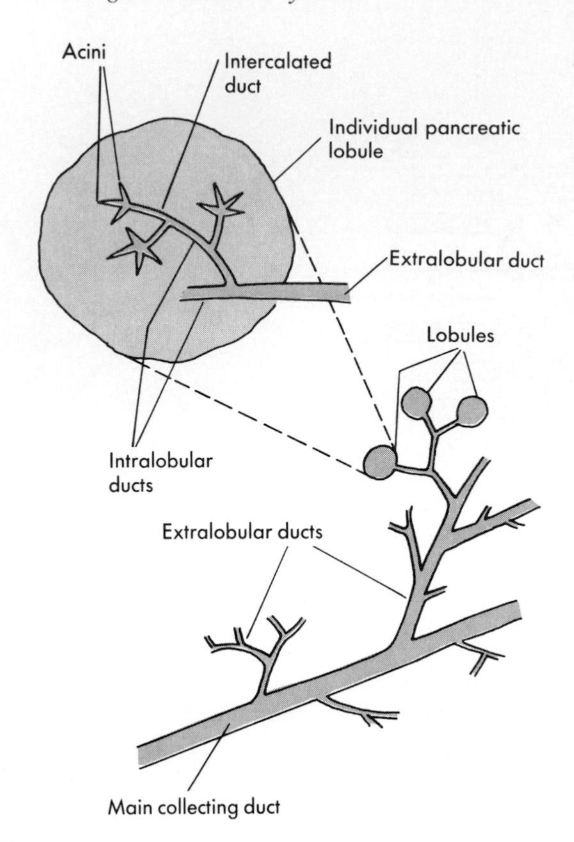

■ **Fig. 39-19** The duct system of the pancreas. (Redrawn from Swanson, CH and Solomon, AK: *J Gen Physiol* 62:407, 1973.)

■ **Table 39-3** Volumes occupied by various cellular elements in guinea pig pancreas

Cell type	*Volume occupied (%)*
Acinar	82.0
Duct	3.9
Endocrine	1.8
Blood vessels	3.7
Extracellular space	9.4

Data of Bolender RP: *J Cell Biol* 61:269, 1974.

faces of the cells. The tight junctions are impermeable to macromolecules but relatively leaky to water and ions. Acinar and duct cells make gap junctions with their neighbors. These junctions allow rapid communication of changes in membrane potential and permit exchange of molecules with molecular weights of 1400 or less.

The pancreas is supplied by branches of the celiac and superior mesenteric arteries. The portal vein is the exit pathway for pancreatic blood flow. The acini and islets are supplied by separate capillary nets. Some of those capillaries that supply the islets converge into venules, which then break up into second capillary networks around the acini.

The endocrine cells of the pancreas reside in the **islets of Langerhans.** Although islet cells account for less than 2% of the volume of the pancreas (Table 39-3), their hormones are essential in regulating metabolism. Insulin, glucagon, somatostatin, and pancreatic polypeptide are hormones released from cells of the islets of Langerhans. Each of these hormones, when administered intravenously, influences the exocrine secretion of the pancreas. However, the physiological roles of these hormones in the regulation of acinar and duct cells remain to be elucidated.

The pancreas is innervated by preganglionic parasympathetic branches of the vagus. Vagal fibers synapse with cholinergic neurons that are within the pancreas and that innervate both acinar and islet cells. Postganglionic sympathetic nerves from the celiac and superior mesenteric plexuses innervate pancreatic blood vessels. In general, secretion of pancreatic juice is stimulated by parasympathetic activity and inhibited by sympathetic activity.

■ *The Aqueous Component of Pancreatic Juice*

The aqueous component of pancreatic juice is elaborated principally by the columnar epithelial cells that line the ducts. The Na^+ and K^+ concentrations of pancreatic juice are similar to those in plasma. HCO_3^- and Cl^- are its major anions. The HCO_3^- concentration varies from about 30 mEq/L at low rates of secretion to over 100 mEq/L at high secretory rates (Fig. 39-20). HCO_3^- and Cl^- concentrations vary reciprocally. As secreted by the duct cells the aqueous component is slightly hypertonic and has a high HCO_3^- concentration. As it flows down the ducts, water equilibrates across the epithelium to make the pancreatic juice isotonic, and some HCO_3^- exchanges for Cl^- (Fig. 39-21). The faster the flow rate, the less is the time available for HCO_3^--Cl^- exchange, and the higher is the HCO_3^- concentration in the juice.

Under resting conditions the aqueous component is produced primarily by the intercalated and other intralobular ducts. When secretion is stimulated by secretin, however, the additional flow comes mostly from the extralobular ducts (see Fig. 39-21). Secretin is the major physiological stimulus for secretion of the aqueous component. The secretin-stimulated juice secreted by the extralobular ducts resembles the resting secretion produced by the intralobular ducts, but the extralobular secretion has a slightly higher HCO_3^- concentration.

When acinar cells are stimulated by cholecystokinin to secrete proteins, a small volume of fluid is also secreted in the lumen of the acinus. This fluid is isotonic and resembles plasma in its ionic composition. It is not clear whether the acinar cells themselves are the source of the acinar fluid. The centroacinar cells that reside near the junction between the acini and the intercalated ducts, or the epithelial cells of the intercalated ducts

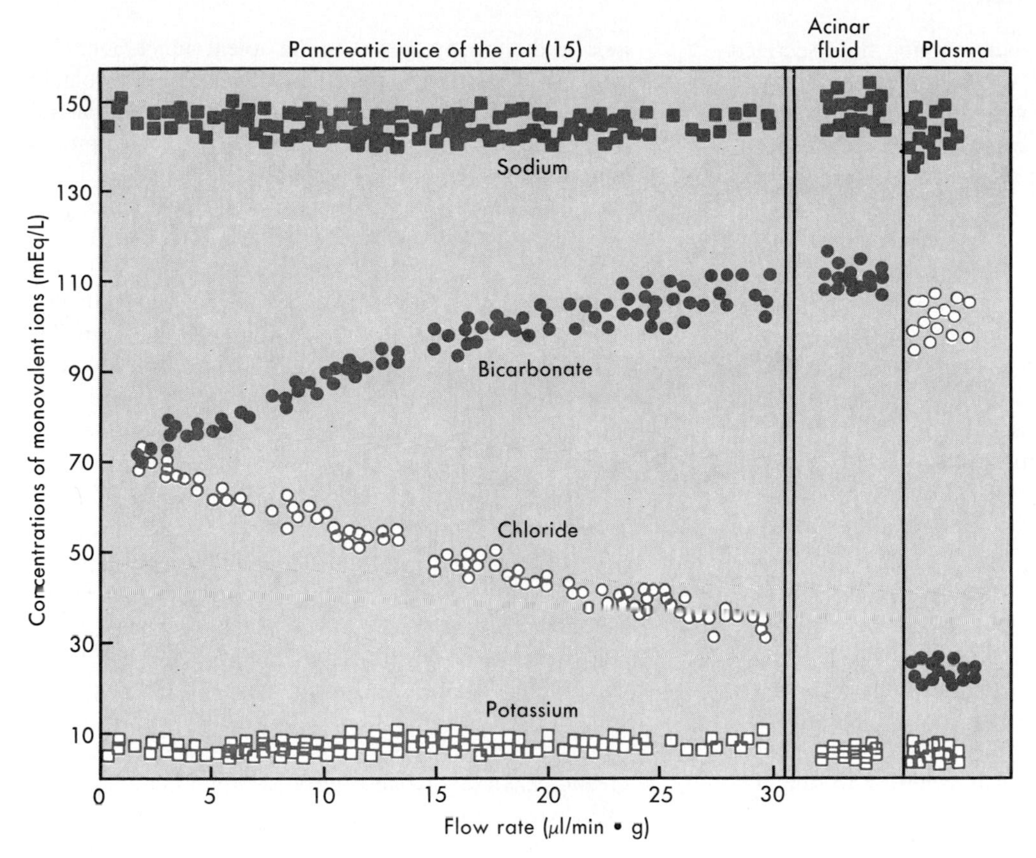

Fig. 39-20 Concentrations of the major ions in pancreatic juice as functions of the secretory flow rate. The concentrations of the ions in acinar fluid and plasma are shown for reference. Secretion was stimulated by intravenous injection of secretin. (Redrawn from Mangos, JA, and McSherry, NR: *Am J Physiol* 221:496, 1971.)

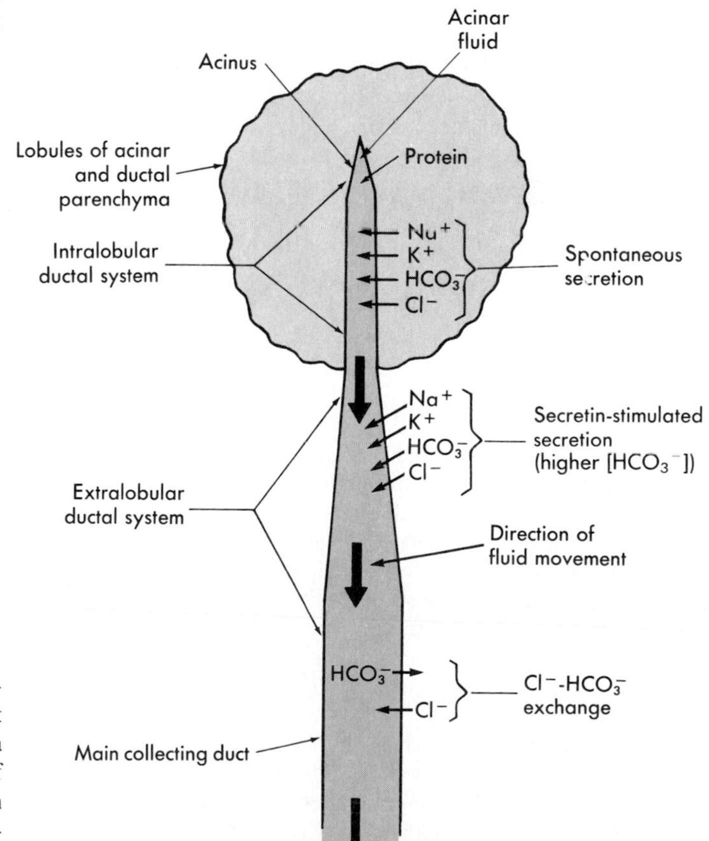

Fig. 39-21 The locations of some of the transport processes involved in the elaboration of the aqueous component of pancreatic juice. Acinar fluid: isotonic; resembles plasma in concentrations of Na^+,K^+, Cl^-, and HCO_3^-; high level of pancreatic enzymes; secretion stimulated by cholecystokinin and acetylcholine. (Redrawn from Swanson, CH and Solomon, AK: *J Gen Physiol* 62:407, 1973.)

cannot be excluded as the source of the fluid secreted into the acini.

When secretion of the aqueous component of pancreatic juice is stimulated by secretin, the composition of the acinar fluid and the fluid secreted by the intralobular ducts does not change. However, the extralobular ducts respond to secretin stimulation by elaborating a secretion of greater volume and with higher bicarbonate levels than are found in the resting secretion.

The cellular events that result in secretion of bicarbonate-rich fluid by the cells of the intralobular and extralobular ducts are not well understood. Bicarbonate secretion depends on the presence of bicarbonate in the plasma: Both the concentration of bicarbonate in pancreatic juice and its rate of production are proportional to the plasma level of bicarbonate. Thus the source of bicarbonate in pancreatic juice appears to be plasma bicarbonate, rather than CO_2 produced by duct cell metabolism.

A postulated mechanism for bicarbonate-rich secretion by epithelial cells of the extralobular ducts of the pancreas is shown in Fig. 39-22. Blood is the principal source of the bicarbonate secreted into the duct lumen. The blood perfusing the ducts is acidified by the H^+, K^+-ATPase and the Na^+/H^+ exchanger in the basolateral membrane of the ductular epithelial cells, and thus CO_2 is formed in the blood. CO_2 diffuses back into the epithelial cell and across the epithelium into the pancreatic juice; CO_2 is hydrated to form carbonic acid, which dissociates into H^+ and HCO_3^-. Elimination of H^+ at the basolateral membrane elevates the HCO_3^- concentration in the cytosol. Bicarbonate is then secreted into the lumen by the Cl^-/HCO_3^- exchanger and by an electrogenic anion channel in the apical membrane of the epithelial cell. Secretin elevates cyclic AMP in the cytosol of the duct cell. Cyclic AMP increases the open time of the anion channel in the apical membrane. Cyclic AMP also causes cytosolic tubovesicles that contain H^+, K^+-ATPase molecules to fuse with the basolateral membrane and thereby increases the number of proton pumps in the membrane. Not shown in Fig. 39-22 is the Na^+, K^+-ATPase of the basolateral membrane; this ATPase provides much of the driving force for H^+ extrusion by the Na^+/H^+ exchanger. The basolateral membrane also contains a K^+ channel that is activated by Ca^{++} and depolarization. Ca^{++} agonists open this K^+ channel and thus hyperpolarize the basolateral membrane and increase the driving force for Na^+/H^+ exchange. In this way Ca^{++}-mobilizing agonists, such as CCK, enhance the action of secretin.

The Enzyme Component of Pancreatic Juice

Table 39-4 lists proteins secreted by pancreatic acinar cells. The secretions of the acinar cells comprise the enzyme component of pancreatic juice. This component contains enzymes important for the digestion of all the major classes of foodstuffs. In the total absence of pancreatic enzymes, malabsorption of lipids, proteins, and carbohydrates occurs.

The proteases of pancreatic juice are secreted in inactive zymogen form. The major pancreatic proteases are **trypsin, chymotrypsin,** and **carboxypeptidase.** They are secreted as trypsinogen, chymotrypsinogen, and procarboxypeptidase, respectively. Trypsinogen is specifically activated by **enteropeptidase,** formerly known as enterokinase, which is secreted by the duodenal mucosa. Trypsin then activates trypsinogen, chymotrypsinogen, and procarboxypeptidase. Trypsin and chymotrypsin cleave certain peptide bonds to reduce polypeptide chains to smaller peptides. Carboxypeptidase spe-

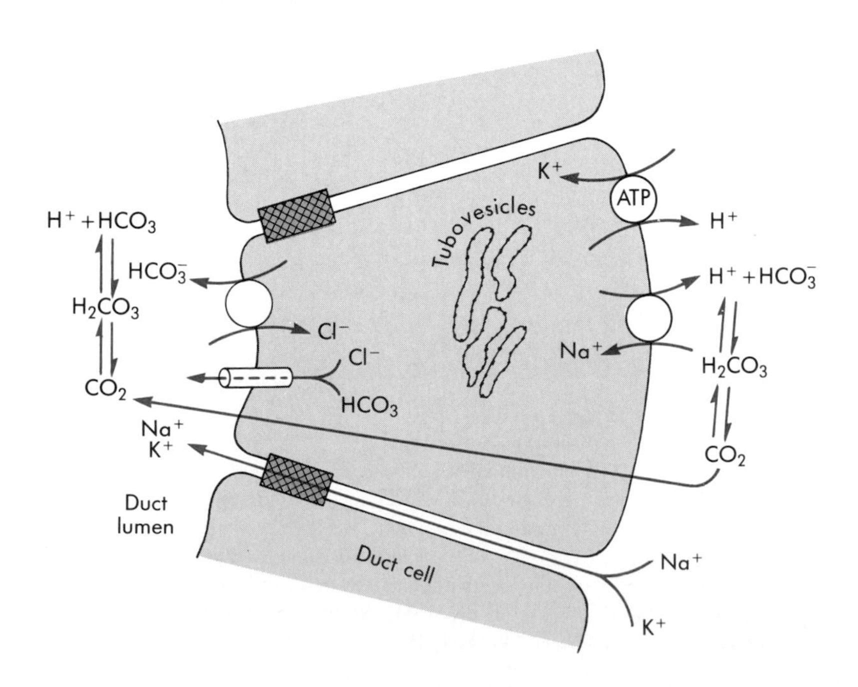

Fig. 39-22 Postulated mechanism for secretion of bicarbonate-rich secretion by epithelial cells of the extralobular ducts of the pancreas. Blood is the principal source of the bicarbonate secreted into the duct lumen. The blood perfusing the ducts is acidified by the H^+,K^+-ATPase and the Na^+/H^+ exchanger in the basolateral membrane of the ductular epithelial cells, resulting in the formation of CO_2 in the blood. CO_2 diffuses across the epithelium into the pancreatic juice. Elimination of H^+ at the basolateral membrane elevates the HCO_3^- concentration in the cytosol. Bicarbonate is then secreted into the lumen by the Cl^-/HCO_3^- exchanger and by an electrogenic anion channel in the apical membrane of the epithelial cell. Secretin elevates cyclic AMP in the duct cell. This increases the open time of the anion channel in the apical membrane and causes the cytosolic tubovesicles that contain H^+,K^+-ATPase molecules to fuse with the basolateral membrane, increasing the number of proton pumps there.

■ **Table 39-4** Some of the proteins of human pancreatic juice

Protein	*Molecular weight*	*Mass proportion (%)*
α-Amylase	54,800	5.3
Triacylglycerol hydrolase	50,500	0.7
Phospholipase A_2	17,500	—
Colipase 1	—	—
Colipase 2	—	—
Procarboxypeptidase A1	46,000	16.8
Procarboxypeptidase A2	47,000	8.1
Procarboxypeptidase B1	47,000	4.4
Procarboxypeptidase B2	47,000	2.9
Trypsinogen 1	28,000	23.1
Trypsinogen 2	26,000	—
Trypsinogen 3	26,700	16.0
Chymotrypsinogen	29,000	1.7
Proelastase 1	30,500	3.1
Proelastase 2	30,500	1.2

Data from Scheele G et al: *Gastroenterol* 80:461, 1981.

cifically removes amino acids from the C-terminal ends of peptide chains.

Pancreatic juice contains an α-amylase that is secreted in active form. Pancreatic amylase has the same substrate specificity as salivary amylase, and it cleaves only interior α-1,4 links in starch to yield maltose, maltotriose, and various α-limit dextrins.

Pancreatice juice also contains a number of lipid-digesting enzymes, or **lipases.** Among the major pancreatic lipases are triacylglycerol hydrolase, cholesterol ester hydrolase, and phospholipase A_2. Triacylglycerol hydrolase (sometimes called pancreatic lipase) acts on triglycerides to cleave specifically the ester bonds at the 1 and 1′ positions to release two free fatty acids and leave a 2-monoglyceride. Cholesterol ester hydrolase acts on cholesterol esters to produce cholesterol and free fatty acids. Phospholipase A_2 specifically cleaves the fatty acyl ester bond at the 2 carbon of a phosphoglyceride to produce a free fatty acid and a 1-lysophosphatide.

Among the other enzymes contained in pancreatic juice are ribonuclease and deoxyribonuclease, which reduce RNA and DNA, respectively, to their constituent nucleotides. The nucleotides are absorbed to some extent and are also reduced to nucleosides and free purine and pyrimidine bases by brush border enzymes. A trypsin inhibitor present in pancreatic juice prevents premature activation of hydrolytic enzymes within the pancreatic ducts.

■ *Secretion of the Enzyme Component*

Pancreatic acinar cells contain numerous membrane-bound zymogen granules in their apical cytoplasm. These granules contain most of the enzymes of pancreatic juice. Typically the number of zymogen granules increases between meals. After a meal the zymogen granules release their contents by exocytosis into the duct lumen, and the density of zymogen granules in the acinar cells decreases. Pancreatic enzymes and zymogens are synthesized on membrane-bound ribosomes, cross the membrane of the rough endoplasmic reticulum to enter the cisternae, and are concentrated somewhat in the cisternae of the rough endoplasmic reticulum. Smooth vesicles bud off from the rough endoplasmic reticulum. These vesicles carry the enzymes and zymogens to the Golgi apparatus, where they are packaged into zymogen granules (secretory vesicles). In response to stimulation of the acinar cell to secrete, the zymogen granules fuse with the plasma membrane to dump their contents into the lumen by exocytosis. The processes of synthesis, storage, and release of proteins by acinar cells are summarized in Fig. 39-23.

The zymogen granules are not a necessary route of secretion of pancreatic enzymes and zymogens. Acinar cells that have been completely depleted of zymogen granules by prolonged stimulation nonetheless can secrete pancreatic enzymes at high rates. We do not yet understand the way in which the zymogen granule secretory pathway and the granule-free pathway interact under physiological conditions.

In addition to digestive enzymes and other proteins, the acinar cells, perhaps with the participation of the intercalated ducts, elaborate a significant volume of fluid. Na^+ and Cl^- are the major ionic constituents of this fluid. Fig. 39-24 shows a postulated mechanism for secretion of the ionic solution that accompanies the enzyme component of pancreatic secretion.

■ *Regulation of Pancreatic Exocrine Secretion*

The secretory activities of duct and acinar cells of the pancreas are controlled by hormones and by substances released from nerve terminals. Stimulation of the vagal branches to the pancreas enhances the rate of secretion of the enzyme and aqueous components of pancreatic juice. Activation of sympathetic fibers inhibits pancreatic secretion, perhaps principally by decreasing blood flow to the pancreas. Secretin and cholecystokinin, hormones released from the duodenal mucosa in response to particular constituents of duodenal contents, stimulate secretion of the aqueous and enzyme components, respectively. Because the aqueous and enzyme components of pancreatic juice are separately controlled, the composition of the juice varies from less than 1% to as much as 10% protein. Substances other than secretin and cholecystokinin also modulate pancreatic exocrine function.

The cephalic phase of pancreatic secretion. Sham feeding induces the secretion of a low volume of pancreatic juice with a high protein content. Gastrin re-

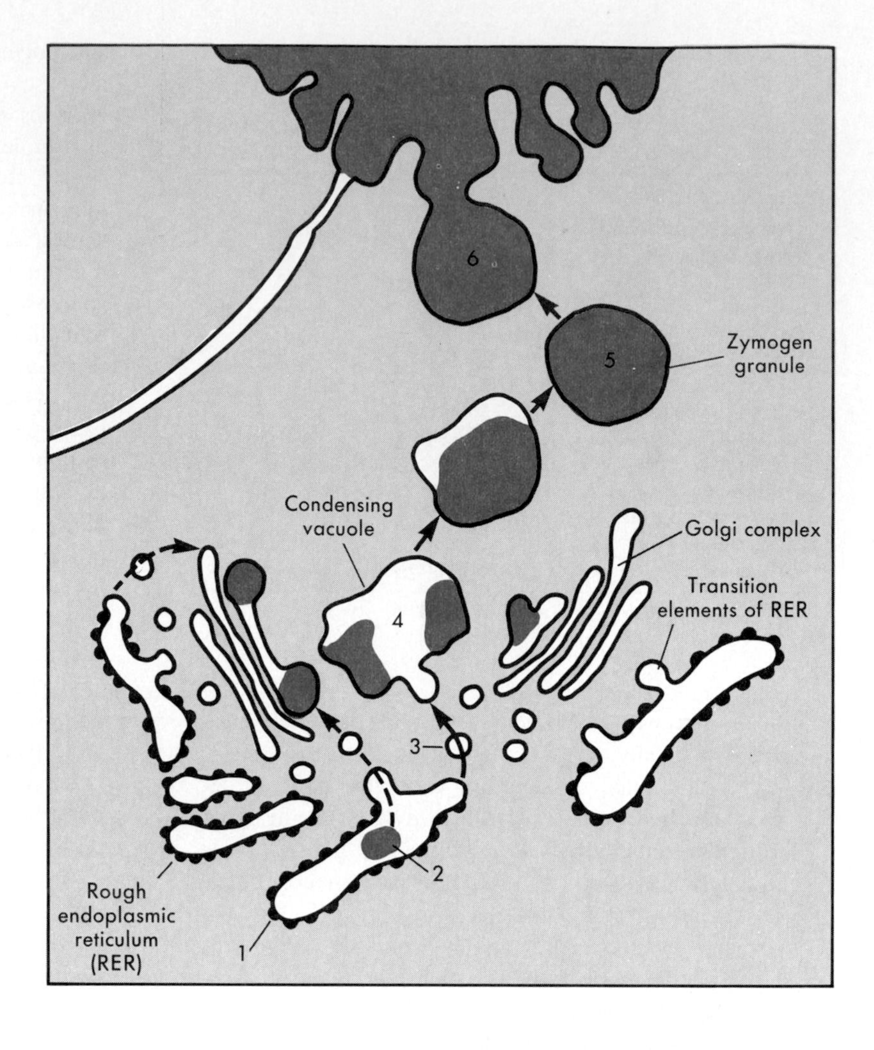

Fig. 39-23 Steps involved in processing of secretory proteins by pancreatic acinar cells: *1*, synthesis in rough endoplasmic reticulum; *2*, posttranslational modification in lumen of rough endoplasmic reticulum; *3*, transfer of proteins to the Golgi complex; *4*, modification in the Golgi complex and concentration of the protein in condensing vacuoles; *5*, storage in zymogen granules; *6*, exocytosis. (Redrawn from Gorelick, FS, and Jamieson, JD: In Johnson, RL, editor: *Physiology of the gastrointestinal tract*, New York, 1981, Raven Press.)

leased from the mucosa of the gastric antrum in response to vagal impulses is a major mediator of pancreatic secretion during the cephalic phase. Gastrin is a member of the same class of peptides as cholecystokinin but is markedly less potent as a pancreatic secretagogue than is cholecystokinin. However, levels of gastrin that elicit acid secretion by the parietal cells also evoke enzyme secretion by pancreatic acinar cells. Acidification of antral contents, which blocks release of gastrin during the cephalic phase, also blocks the cephalic phase of pancreatic secretion.

The gastric phase of pancreatic secretion. During the gastric phase of secretion gastrin is released in response to gastric distension and to the presence of amino acids and peptides in the antrum of the stomach. The gastrin released during the gastric phase continues to stimulate secretion by the pancreas. In addition, vagovagal reflexes elicited by stretching either the fundus or the antrum of the stomach evoke secretion of small volumes of pancreatic juice with high enzyme content.

The intestinal phase of pancreatic secretion. In the intestinal phase of secretion certain components of the chyme present in the duodenum and upper jejunum evoke pancreatic secretion. Acid in the duodenum and upper jejunum elicits the secretion of a large volume of pancreatic juice rich in bicarbonate but poor in pancreatic enzymes. The hormone secretin is a major mediator of this response to acid. Secretin is released by certain cells in the mucosa of the duodenum and upper jejunum in response to acid in the lumen. Secretin is released when the pH of duodenal contents is 4.5 or below; when the pH of duodenal chyme is below about 3, the quantity of titratable acid present, rather than the pH, is the major determinant of the amount of secretin released. *Secretin directly stimulates the cells of the pancreatic ductular epithelium to secrete the bicarbonate-rich aqueous component of the pancreatic juice.*

The presence of peptides and certain amino acids, especially tryptophan and phenylalanine, in the duodenum brings about the secretion of pancreatic juice that is rich in protein components. In the duodenum fatty acids of chain length longer than eight carbon atoms and monoglycerides of these fatty acids also elicit secretion of protein-rich pancreatic juice. *Cholecystokinin is the most important physiological mediator of this response to the digestion products of proteins and lipids.* Cholecystokinin is a hormone released by particular cells in the duodenum and upper jejunum in response to these digestion products. This hormone directly stimulates the

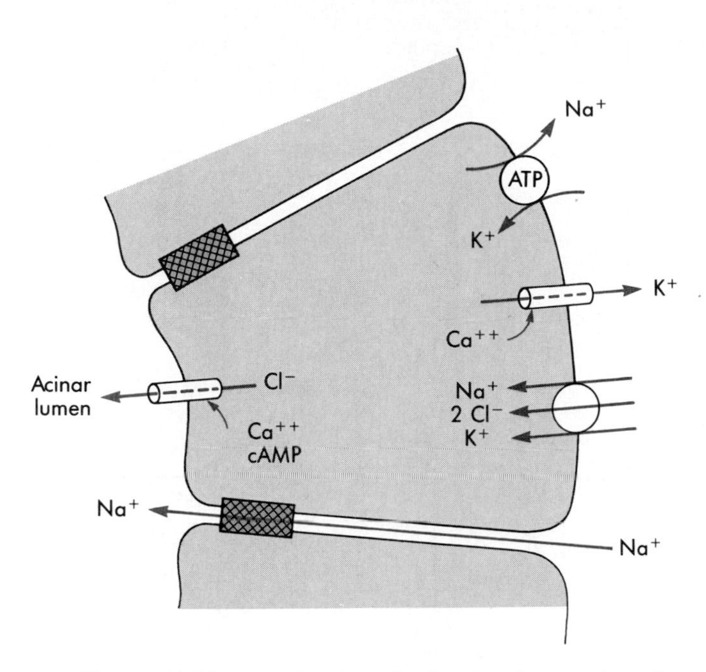

■ **Fig. 39-24** Postulated mechanism for the secretion of a NaCl-rich fluid by the pancreatic acinar cells, perhaps with participation of the cells that line the intercalated ducts. The basolateral membrane has an electroneutral transporter that uses the energy of the Na^+ gradient to take up one Na^+, one K^+, and two Cl^-; with the latter two ions being transported against their electrochemical potential gradients into the cell. Chloride leaves the cell at the apical membrane via a chloride channel that is activated by Ca^{++} and cyclic AMP. Na^+ enters the lumen of the acinus via the leaky tight junctions between the acinar cells. The basolateral membrane has a Ca^{++}-activated K^+ channel. Ca^{++}-mobilizing agonists thus hyperpolarize the cell and enhance the electrical driving forces for basolateral Na^+-driven Cl^- entry and for apical Cl^- efflux.

acinar cells to release the contents of their zymogen granules.

Cholecystokinin has little direct effect on the ductular epithelium of the pancreas, but it potentiates the stimulatory effect of secretin on the ducts. Secretin is a weak agonist of acinar cells, but it potentiates the effect of cholecystokinin on acinar cells.

The neural component of the intestinal phase of pancreatic secretion is not well-characterized. Vagotomy decreases the pancreatic response to the presence of chyme in the duodenum by about 50%. The pancreas responds more rapidly to the presence of acids, peptides, or fats in the duodenum than can be accounted for by the responses to secretin and cholecystokinin. The rapid component of pancreatic secretion in response to chyme in the duodenum is greatly diminished by vagotomy. The initial rapid increase in pancreatic secretion during the intestinal phase is probably mediated mainly by enteropancreatic vagovagal reflexes.

■ *Regulatory Molecules That Influence Pancreatic Exocrine Secretion*

Cholecystokinin and secretin are not the only endogenous substances that modulate pancreatic secretion. Table 39-5 lists some of the compounds that can stimulate or inhibit pancreatic secretion. The physiological significance of the actions of these substances is not yet understood. The second messenger of some of these regulatory molecules is cAMP or Ca^{++}. The second messengers of some of the regulators are not known. Specific receptors for some of the regulatory molecules in Table 39-5 are present in the pancreas.

■ *Cellular Mechanisms of the Mediators of Pancreatic Exocrine Secretion*

Acinar cells. Fig. 39-25 shows six classes of receptors that mediate the responses of pancreatic acinar cells

■ **Table 39-5** Substances that influence pancreatic exocrine function

Cellular mechanism	*Substances*	*Actions on pancreas*
Enhanced turnover of inositol phospholipids and elevation of intracellular Ca^{++}	CCK, gastrin, ACh, gastrin-releasing peptide substance P	Enhanced secretion of digestive enzymes and Cl^--rich fluid by acinar cells, trophic effects, increased cellular metabolism
Activation of adenylyl cyclase and elevation of cAMP	Secretin, VIP, calcitonin gene-related peptide, peptide histidine isoleucine	Secretion of HCO_3^--rich fluid by duct cells, potentiation of effects of Ca^{++}-mobilizing agonists
Inhibition of adenylyl cyclase and decrease of cAMP	Somatostatin	Inhibition of secretions of acinar and duct cells
Activation of receptor-associated tyrosine kinase	Insulin, insulin-like growth factors, EGF	Potentiation of enzyme synthesis and secretion, trophic effects, increased cell metabolism, maintenance of differentiated functions

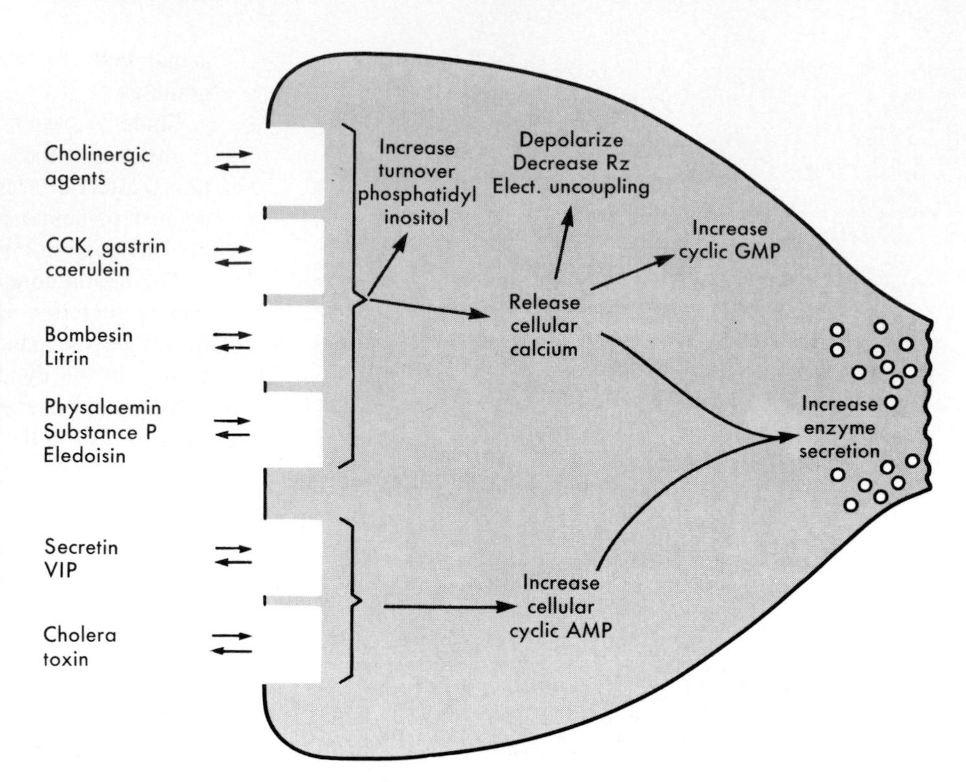

■ **Fig. 39-25** Representation of the cellular mechanisms of action of secretagogues on pancreatic acinar cells. Six discrete classes of receptors are proposed for secretagogues. Two receptor types are linked to adenylate cyclase and to increased intracellular cAMP. The other four receptor classes are coupled to turnover of inositol phosphatides and to increased intracellular free Ca^{++}. (Redrawn from Jensen RT, and Gardner, JD: *Adv Cyclic Nucleotide Res* 17:375, 1984.)

to secretagogues. Secretin and vasoactive intestinal polypeptide are related peptides that compete for receptor binding. At least two kinds of receptors for secretin and vasoactive intestinal polypeptide exist: one kind prefers secretin, and the other type prefers vasoactive intestinal polypeptide. Binding of secretin and vasoactive intestinal polypeptide to these receptors elevates intracellular cAMP, the second messenger for these secretagogues.

Four classes of secretagogues exert their effects by increasing the level of intracellular calcium by the inositol phosphate pathway (see Fig. 39-25). Among the endogenous secretagogues that act in this way are acetylcholine, cholecystokinin, gastrin, and substance P. Gastrin and cholecystokinin compete for the same receptor.

Whereas the second messengers of the secretagogues shown in Fig. 39-25 have been identified, further details of the mechanisms of action are unclear. Phosphorylation of specific proteins by protein kinases activated by cAMP, Ca^{++}, or diglyceride may play a role in the stimulation of secretion brought about by these secretagogues.

Many of the secretagogues that elevate intracellular Ca^{++} also increase cytosolic levels of cGMP in the acinar cell. The increased cGMP does not directly elicit pancreatic exocrine secretion, but cGMP may play a regulatory role in the acinar cells.

Secretagogues that act via cAMP potentiate the effect of secretagogues that use Ca^{++} as second messenger, and vice versa. The maximal enzyme secretion in response to the two secretagogues acting together is greater than the maximal response to either agonist alone. Secretagogues that act via a common second messenger do not potentiate one another's effects.

Extralobular duct epithelial cells. Secretin is the major physiological agonist for secretion of bicarbonate-rich fluid by the epithelial cells that line the extralobular ducts (see Fig. 39-22). Occupation of the secretin receptor in these cells enhances the activity of adenylyl cyclase and elevates intracellular cAMP. VIP also elevates cyclic AMP and enhances secretion by the ducts. There are VIP-containing neurons in the pancreas, but their physiological role in regulating duct secretion remains to be elucidated. CCK alone is not an agonist, but as discussed previously, it enhances the effect of secretin on duct epithelial cells. β-Adrenergic agonists enhance, but α-adrenergic agonist inhibit, duct secretions. Acetylcholine is not an effective agonist at the level of the duct cells.

Somatostatin, glucagon, and pancreatic polypeptide —three substances released from the endocrine pancreas—all inhibit the secretion of fluid and electrolytes by the extralobular duct epithelium. The mechanisms of these effects remain to be determined.

■ *Functions of Liver and Gallbladder*

■ *Structure of the Liver*

One view of hepatic histology is shown in Fig. 39-26. Each liver lobule is organized around a central vein. At the periphery of the lobule blood enters the sinusoids from branches of the portal vein and the hepatic artery. In the sinusoids blood flows toward the center of the lobule between plates of hepatic cells that are one or two hepatocytes thick. Each hepatocyte is thus in direct contact with sinusoidal blood because of the large fenestrations between the endothelial cells that line the sinusoids. The intimate contact of a large fraction of the hepatocyte surface with blood contributes to the ability of the liver to clear the blood effectively of certain classes of compounds. Biliary canaliculi lie between adjacent hepatocytes, and the canaliculi drain into bile ducts at the periphery of the lobule.

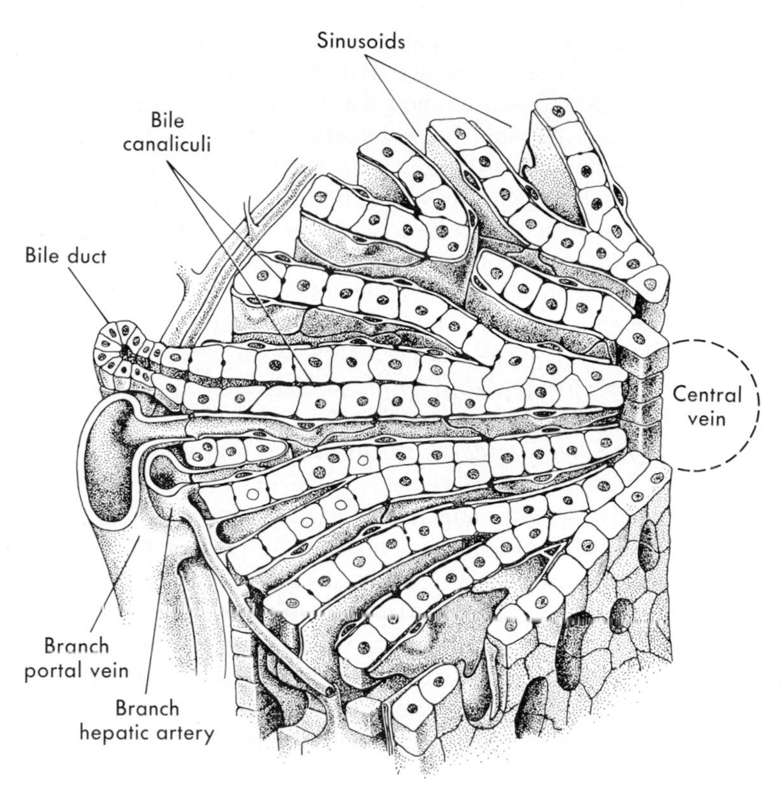

Fig. 39-26 Diagrammatic representation of a hepatic lobule. A central vein is located in the center of the lobule with plates of hepatic cells disposed radially. Branches of the portal vein and hepatic artery are located on the periphery of the lobule, and blood from both perfuses the sinusoids. Peripherally located bile ducts drain the bile canaliculi that run between the hepatocytes. (Redrawn from Bloom, W, and Fawcett, DW: *A textbook of histology,* ed 10, Philadelphia, 1975, WB Saunders.)

Functions of the Liver

Bile secretion is the principal digestive function of the liver. A discussion of the composition of bile, the mechanisms of bile formation, and regulation of bile synthesis and secretion follows. The liver performs a large number of other functions that are vital to the health of the organism. This book does not have a separate chapter on the liver, but each function of the liver is discussed in the context of the individual organ system most affected. Nevertheless, it may be useful at this point to mention some of the most important activities of the liver. The liver is essential in regulating metabolism, in synthesizing certain proteins, in serving as a storage site for certain vitamins and iron, in degrading certain hormones, and in inactivating and excreting certain drugs and toxins. Bile is the only route of excretion of most heavy metals.

The liver regulates the metabolism of carbohydrates, lipids, and proteins. Liver and skeletal muscle are the two major sites of glycogen storage in the body. When the level of glucose in the blood is high, glycogen is deposited in the liver. When blood glucose is low, liver glycogen is broken down to glucose (the process is called **glycogenolysis**), and the glucose is then released into the blood. In this way the liver helps to maintain a relatively constant blood glucose level. The liver is also the major site of **gluconeogenesis,** the conversion of amino acids, lipids, or simple carbohydrate substances (lactate, for example) into glucose. Carbohydrate metabolism by the liver is regulated by several hormones (Chapters 45, 46, and 50).

The liver is also involved in lipid metabolism. As described in Chapter 40, lipids absorbed from the intestine leave the intestine in chylomicrons in the lymph. Lipoprotein lipase on the endothelial cell surface of blood vessels hydrolyzes some of the triglyceride in the **chylomicrons,** thereby allowing glycerol and fatty acids to be taken up by adipocytes. This results in formation of **chylomicron remnants** that are rich in cholesterol. Chylomicron remnants are taken up by hepatocytes and degraded. Hepatocytes synthesize and secrete very-low-density lipoproteins. Very-low-density lipoproteins are then converted to the other types of serum lipoproteins: low-density, intermediate-density, and high-density lipoproteins. These lipoproteins are the major sources of cholesterol and triglycerides for most other tissues of the body. Cholesterol present in bile represents the only route of excretion of cholesterol. Hepatocytes are thus a principal source of cholesterol in the body and are the major site of excretion of cholesterol. These cells play an important role in regulation of serum cholesterol levels.

In certain physiological and pathological conditions, β-oxidation of fatty acids provides the major source of energy for the body. In the liver, acetyl coenzyme A liberated from fatty acids condenses to form acetoacetate. Acetoacetate is converted to β-hydroxybutyrate and acetone. These three compounds are called **ketone bodies.** Ketone bodies are released from hepatocytes and carried in the circulation to other tissues, where they are metabolized. The hepatic functions that regulate lipid metabolism are also subject to endocrine control.

The liver is centrally involved in protein metabolism.

When proteins are catabolized, amino acids are deaminated to form ammonia (NH_3). Ammonia cannot be further metabolized by most tissues and becomes toxic at levels achievable by metabolism. Ammonia is dissipated by conversion to urea, mainly in the liver. The liver also synthesizes all the nonessential amino acids.

The liver synthesizes certain proteins, among them the plasma lipoproteins discussed previously. Plasma contains numerous other proteins; among them are albumins, globulins, fibrinogens, and other proteins involved in blood clotting. With the exception of the γ-globulins produced by plasma cells, the liver synthesizes all the plasma proteins.

The liver stores certain substances important in metabolism. Next to hemoglobin in red blood cells the liver is the most important storage site for iron. Certain vitamins, most notably vitamins A, D, and B_{12}, are stored in the liver. Hepatic storage protects the body from limited dietary deficiencies of these vitamins.

The liver is a major site for the degradation and excretion of hormones. Epinephrine and norepinephrine are inactivated by oxidation (catalyzed by monoamine oxidase) and methylation (catalyzed by catechol-*O*-methyltransferase). Both of these enzymes are abundant in hepatocytes. Certain polypeptide hormones are degraded by liver cells. The liver inactivates and excretes steroid hormones. For example, cortisol, a principal glucocorticoid, is reduced in the liver to inactive tetrahydrocortisol and then conjugated to glucuronic acid. The conjugated derivative is water-soluble and enters the circulation to be excreted in the urine.

The liver transforms and excretes a large number of drugs and toxins. Drugs and toxins are frequently converted to inactive forms by reactions that occur in hepatocytes. However, some drugs and toxins are activated by hepatic transformation, and some drugs are converted to toxic products. The smooth endoplasmic reticulum of hepatocytes contains systems of enzymes and cofactors, known as mixed function oxidases, that are responsible for oxidative transformations of many drugs. Certain other enzymes in the endoplasmic reticulum catalyze the conjugation of many compounds with glucuronic acid, glycine, or glutathione. Other drug transformations that occur in hepatocytes include acetylation, methylation, reduction, and hydrolysis. The transformations that occur in the liver render many drugs more water soluble, and thus they are more readily excreted by the kidneys. Some drug metabolites are secreted into bile. Many organic anions and cations are actively secreted into bile by Na^+-dependent mechanisms. Steroids and related molecules are secreted into bile by a carrier-mediated mechanism, perhaps by active transport.

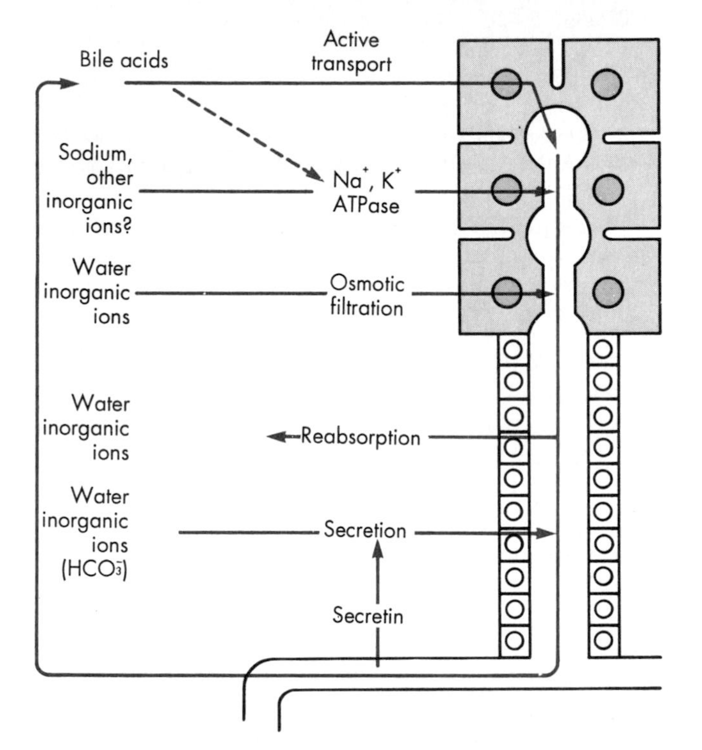

■ **Fig. 39-27** Schematic view of bile secretion by hepatocytes and bile duct epithelium. The hepatocytes secrete bile acids, bilirubin, and protein into the bile canaliculi in an isotonic fluid with Na^+, Cl, and HCO_3^- at concentrations that resemble those in plasma. The bile duct epithelial cells contribute an aqueous secretion with a bicarbonate concentration higher and a chloride concentration lower than those of plasma. (Modified from Erlinger, S: In Schiff, L, and Schiff, LR, editors: *Diseases of the liver*, ed 5, Philadelphia, 1982, JB Lippincott.)

■ *Bile Secretion*

The foregoing discussion briefly described various vital functions of the liver. The hepatic function most important to the digestive tract is the secretion of bile. The mechanisms of bile secretion and the control of bile secretion are the major topics of this section.

Bile, elaborated by hepatocytes, contains bile acids, cholesterol, lecithin, and bile pigments. These constituents are all synthesized and secreted by hepatocytes into the bile canaliculi along with an isotonic fluid that resembles plasma in its concentrations of Na^+, K^+, Cl^-, and HCO_3^-. The bile canaliculi merge into ever larger ducts and finally into a single large bile duct. The epithelial cells that line the bile ducts secrete a watery fluid that is rich in bicarbonate and contributes to the volume of bile that leaves the liver. Contractions of the bile canaliculi and small bile ducts enhance the flow of bile toward the larger bile ducts.

The secretory function of the liver resembles that of the exocrine pancreas. In both organs the major parenchymal cell type elaborates a primary secretion containing the substances responsible for the major digestive function of the organ. In both pancreas and liver the primary secretion is isotonic and contains Na^+, K^+, and Cl^- at concentrations near plasma levels, and the primary secretion is stimulated by *cholecystokinin*. In both pancreas and liver the epithelial cells that line the duct system modify the primary secretion. When stimulated by *secretin* these epithelial cells contribute an aqueous secretion that is characterized by high bicarbonate concentrations (Fig. 39-27).

In the periods between meals, bile is diverted into the gallbladder. The gallbladder epithelium extracts salts and water from the stored bile, and the bile acids are thereby concentrated five- to twentyfold. After an individual has eaten, the gallbladder contracts and empties its concentrated bile into the duodenum. The most potent stimulus for emptying of the gallbladder is cholecystokinin, which is released by the duodenal mucosa primarily in response to the presence of fats and their digestion products. From 250 to 1500 ml of bile enters the duodenum each day.

Bile acids emulsify lipids, thereby increasing the surface area available to lipolytic enzymes. Bile acids then form mixed micelles with the products of lipid digestion. This process increases the transport of lipid digestion products to the brush border surface and in this way enhances absorption of lipids by the epithelial cells. Bile acids are actively absorbed, chiefly in the terminal part of the ileum. A small fraction of bile acids escapes absorption and is excreted. The returning bile acids are avidly taken up by the liver and are rapidly resecreted during the course of digestion. The entire bile acid pool (approximately 3 g on average) is recirculated twice in response to a typical meal. The recirculation of the bile is known as the **enterohepatic circulation.** About 10% to 20% of the bile acid pool is excreted in the feces each day and is replenished by hepatic synthesis of new bile acids. Fig. 39-28 summarizes some major aspects of the enterohepatic circulation.

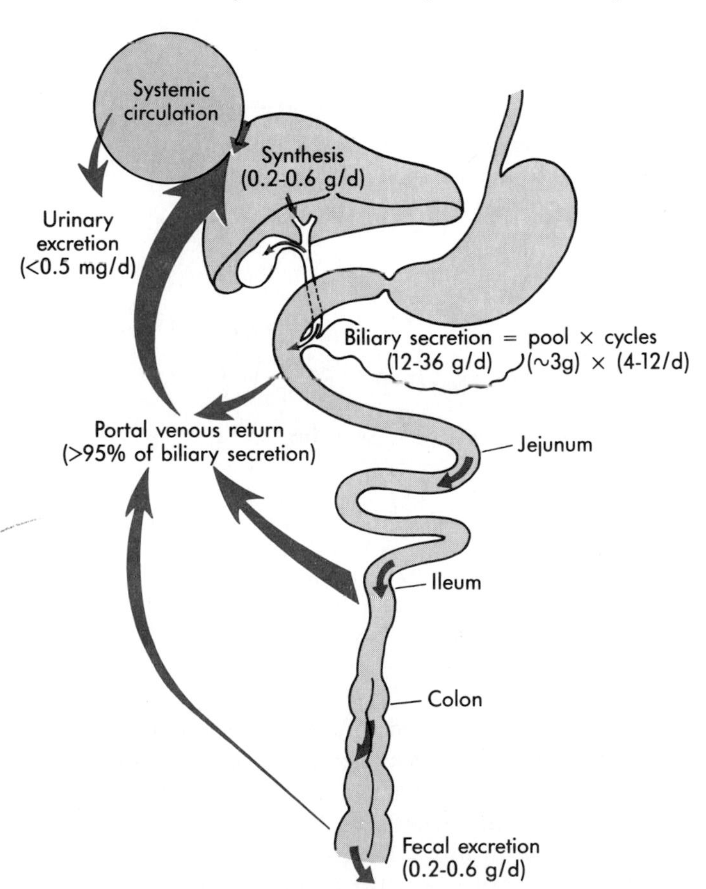

Fig. 39-28 Representation of key components of the enterohepatic circulation of bile acids in normal humans. (Redrawn from Carey, MC and Cahalane, MJ: In Arias, IM et al: *The liver: biology and pathobiology,* ed 2, 1988, New York, Raven Press.).

The Fraction of Bile Secreted by Hepatocytes

The bile acids. Bile acids comprise about 65% of the dry weight of bile. Other important compounds secreted by the hepatocytes into the bile include lecithin, cholesterol, bile pigments, and proteins (Fig. 39-29).

Bile acids have a steroid nucleus and are synthesized by the hepatocytes from cholesterol. The major bile acids synthesized by the liver are called **primary bile acids** (Fig. 39-30). These are cholic acid (3-hydroxyl groups) and chenodeoxycholic acid (2 hydroxyl groups). The presence of the carboxyl and hydroxyl groups makes the bile acids much more water soluble than the cholesterol from which they are synthesized.

Bacteria in the digestive tract dehydroxylate bile acids to form **secondary bile acids.** The major secondary bile acids (see Fig. 39-30) are deoxycholic acid (from dehydroxylation of cholic acid) and lithocholic acid (from dehydroxylation of chenodeoxycholic acid). Bile contains both primary and secondary bile acids.

Bile acids normally are secreted conjugated with glycine or taurine (see Fig. 39-30). Conjugated bile acids

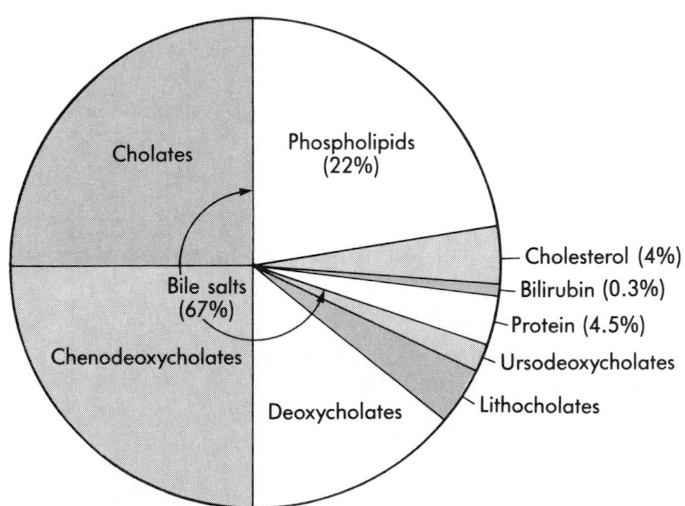

Fig. 39-29 Major components (percentage of dry weight) of bile in healthy humans. (Redrawn from Carey, MC and Cahalane, MJ: In Arias, IM et al.: *The liver: biology and pathobiology,* ed 2, 1988, New York, Raven Press.)

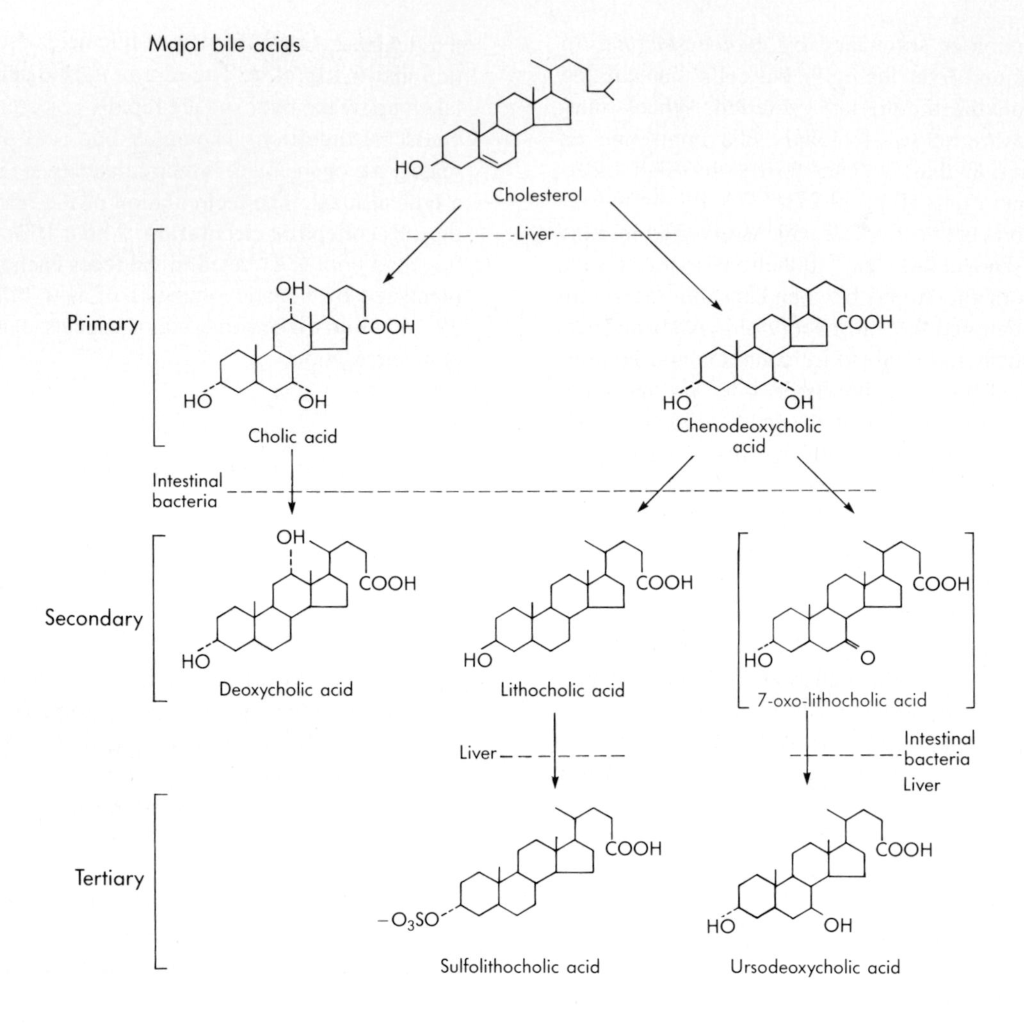

■ **Fig. 39-30** Structures and sites of conversion of major primary, secondary, and tertiary bile acids in human bile. At the bottom of the figure the conjugation of cholic acid with glycine or taurine is illustrated. (Redrawn from Carey, MC and Cahalane, MJ: In Arias, IM et al: *The liver: biology and pathobiology*, ed 2, 1988, New York, Raven Press.)

contain glycine or taurine linked by a peptide bond between the carboxyl group of an unconjugated bile acid and the amino group of glycine or taurine. The peptide bond of conjugated bile acids is extremely resistant to hydrolysis by the pancreatic carboxypeptidases. The pK_a's of the carboxyl groups of unconjugated bile acids are near neutral pH, but the pK_a's of conjugated bile acids are considerably lower. Thus at the near-neutral pH of the gastrointestinal tract the conjugated bile acids are more completely ionized, and thus more water soluble, than the unconjugated bile acids. Conjugated bile acids therefore exist almost entirely as salts of various cations (mostly Na^+) and hence often are called **bile salts.**

The steroid nucleus of bile acids is roughly planar. In solution bile acids have their polar (hydrophilic) groups—the hydroxyl groups, the carboxyl moiety of glycine or taurine, and the peptide bond—all on one surface of the molecule. The other surface is quite hydrophobic (Fig. 39-31, *A*). This makes the bile acid molecule amphipathic, that is, having both hydrophilic and hydrophobic domains. Conjugated bile acids are more amphipathic than unconjugated ones. Because they are amphipathic, bile acids tend to form molecular aggregates, called micelles, by turning their hydrophobic faces inside and away from water and their hydrophilic surfaces toward the water (Fig. 39-31, *B*). Whenever bile acids are present above a certain concentration, called the critical micelle concentration, bile acid micelles will form. Above this concentration any additional bile acid will go into the micelles exclusively and not into molecular solution. In bile, the bile acids are normally present at a concentration well above the critical micelle concentration.

Phospholipids and cholesterol in bile. Hepatocytes secrete phospholipids, predominantly lecithins, and cholesterol into bile. Bile is the major route for excretion of cholesterol from the body. Lecithin and cholesterol are secreted into bile in the form of bilayer lipid vesicles. The vesicles fuse transiently with the canalicular plasma membrane and then are extruded into the lumen of the canaliculus. Because the lecithin-cholesterol vesicles are greatly outnumbered by micelles of bile acids, the lecithin and cholesterol partition into the hydrophobic interior of the micelles to form mixed micelles of lecithin, cholesterol and bile acid, and the lecithin-cholesterol vesicles slowly disappear.

Cholesterol, being apolar, partitions into the center of the micelle. Lecithin, because it is amphipathic, buries its fatty acyl chains in the micelle interior and leaves its polar head group near the micelle surface (Fig. 39-31, *B*). The lecithin increases the amount of cholesterol that can be solubilized in the micelles. If more cholesterol is present in the bile than can be solubilized in the micelles, crystals of cholesterol may form in the bile. These crystals are important in formation of cholesterol gallstones (the most common kind of gallstones) in the duct system of the liver or more commonly in the gallbladder.

Bile pigments. When senescent red blood cells are degraded in reticuloendothelial cells, the porphyrin moiety of hemoglobin is converted to bilirubin. Bilirubin is released into the plasma, where it is bound to albumin. Hepatocytes efficiently remove bilirubin from blood in the sinusoids via a protein-mediated transport mechanism in the hepatocyte plasma membrane that faces the sinusoids. In the hepatocytes bilirubin is conjugated with one or two glucuronic acid molecules, and the bilirubin glucuronides are secreted into the bile, probably by an active transport mechanism. Unconjugated bilirubin is not secreted into the bile. Bilirubin is yellow and contributes to the yellow color of bile. Colonic bacteria convert bilirubin to mesobilirubinogen and then to urobilinogen. Some of the urobilinogen is absorbed in the colon. A fraction of the absorbed urobilinogen is excreted in the urine, and the remainder is resecreted into bile.

Proteins in bile. After being concentrated in the gallbladder, bile contains protein at 5 to 50 mg/ml. Proteins in the bile include most of the proteins present in plasma, including immunoglobulins and the apoproteins of serum lipoproteins. Secretory IgG in bile may defend against ingested antigens. A number of peptide hormones are present in bile. The hepatocytes themselves are the source of other biliary proteins, including lysosomal hydrolases and enzymes associated with other hepatocyte organelles. The mechanisms of protein secretion into bile are not well characterized, and the possible physiological functions of proteins in bile are not clear.

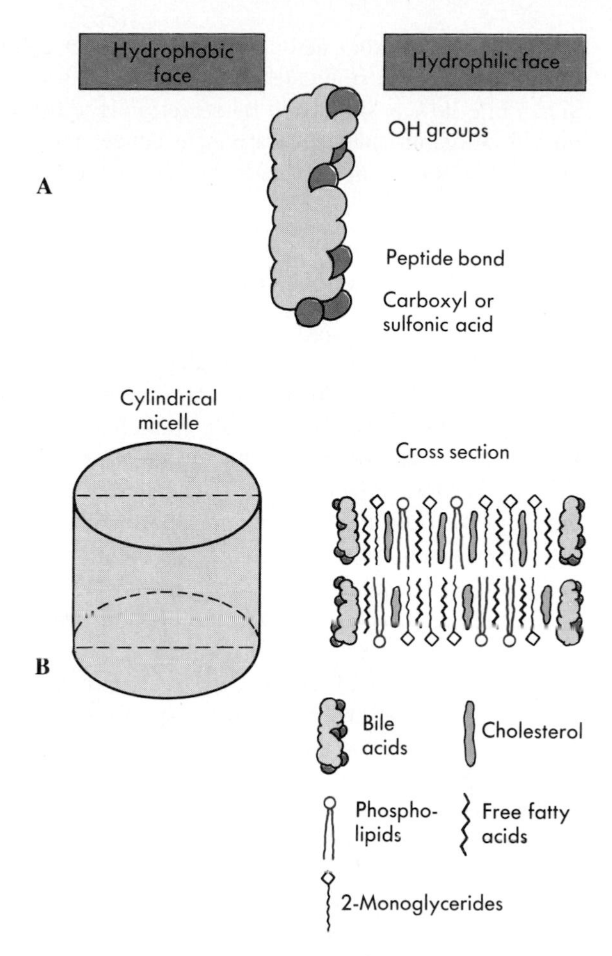

■ **Fig. 39-31** Structure of bile acids and micelles. **A,** A bile acid molecule in solution. The molecule is amphipathic in that it has a hydrophilic face and a hydrophobic face. The amphipathic structure is key in the ability of the bile acids to emulsify lipids and to form micelles. **B,** A model of the structure of a bile acid-lipid mixed micelle.

■ *Secretion by Bile Duct Epithelium*

The epithelial cells that line the bile ducts contribute an aqueous secretion that can account for about 50% of the total volume of bile. The secretion of the bile duct epithelium is isotonic and contains Na^+ and K^+ at levels similar to plasma. However, the concentration of HCO_3^- is greater and the concentration of Cl^- is less than in plasma. The secretory activity of the bile duct epithelium is specifically stimulated by the hormone secretin. When bile flow is stimulated by secretin alone the volume flow of bile and its bicarbonate concentration increase, but content of bile acids does not increase.

Cellular Mechanism of Bile Formation

Secretion of bile acids. Fig. 39-32 illustrates the current understanding of the cellular mechanisms responsible for secretion of bile by hepatocytes into bile canaliculi. As is the case for all epithelial cells, the plasma membrane of the hepatocyte is polarized such that the membrane facing the bile canaliculus is different from the basolateral membrane (the membrane facing the sinusoid plus the lateral cell membrane that faces adjacent hepatocytes).

Bile acids are taken up from the portal blood by hepatocytes by at least four different transport mechanisms (see Fig. 39-32). Hepatocytes take up conjugated bile acids from the blood by two transport mechanisms: one dependent on Na^+, the other independent of Na^+. The Na^+-dependent mechanism, which has the highest affinity for trihydroxy conjugated bile acids, uses the electrochemical potential gradient of Na^+ to actively transport bile acids into the hepatocyte. The Na^+-independent transporter catalyzes facilitated transport of conjugated bile acids into the hepatocyte. Unconjugated bile acids are taken up by an anion exchanger that takes up bile acid in exchange for hydroxide or bicarbonate ion. Unconjugated bile acids are also taken up by simple diffusion; monohydroxy bile acids, being nonpolar, are most able to use this route of entry.

In the cytosol of the hepatocyte, most bile acids are bound to bile-acid binding proteins. Bile acids are present in hepatocytes at high levels. If bile acids were not largely bound to intracellular binding proteins, the bile acids might disrupt the membranes of cellular organelles. The protein-mediated transport mechanism(s) by which bile acids are secreted across the canalicular membrane into bile remain to be characterized. The prevailing view is that bile acids cross the canalicular membrane by protein-mediated facilitated transport, which is driven partly by the electrical potential difference across the canalicular membrane (about 35 mV, cytosol negative).

The tight junctions joining hepatocytes that surround the canaliculi are leaky to water-soluble molecules of less than macromolecular dimensions. The leakage of bile acids from the canalicular lumen back into the sinusoidal blood is limited by micelle formation in the lumen (micelles are too large to back diffuse) and by the relative impermeability of the tight junctions to anions.

Secretion of water and electrolytes into bile. Water and electrolytes are secreted into the bile canaliculi. The major ions are present in canalicular fluid at the same concentrations as they are in plasma. The osmotic pressure of bile acids in the canalicular lumen is believed to draw water into the canaliculi via the leaky tight junctions that join hepatocytes. The flow of water is thought to carry with it the ions of the extracellular

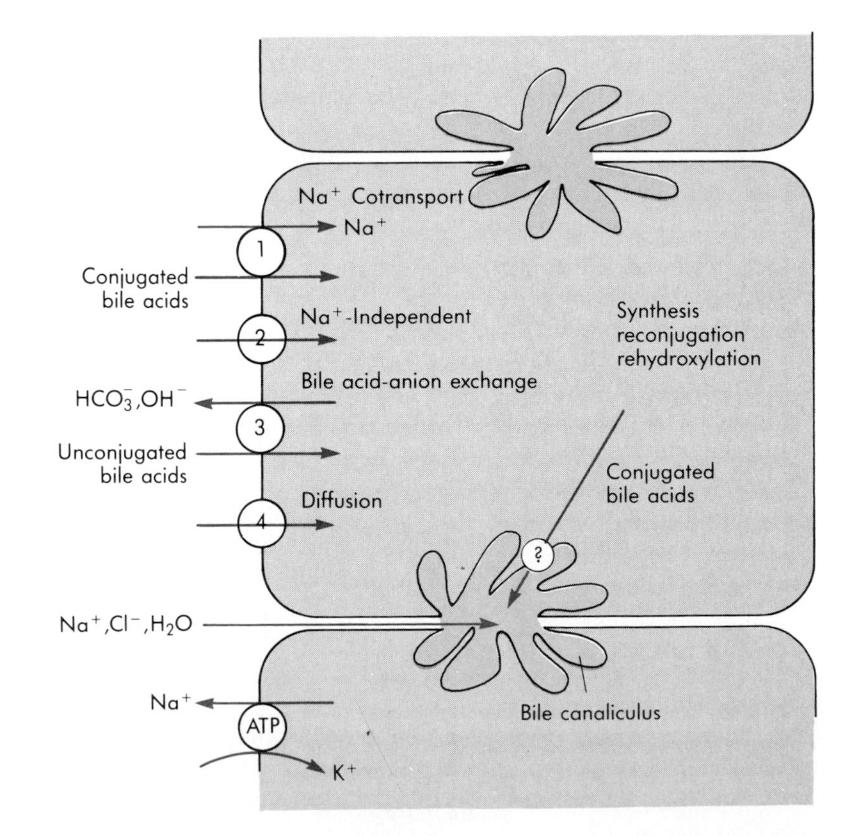

Fig. 39-32 Postulated mechanisms for uptake and secretion of bile acids by hepatocytes. Hepatocytes take up conjugated bile acids from the blood by two transport mechanisms—one dependent on Na^+, the other independent of Na^+. Unconjugated bile acids are taken up by an anion exchange mechanism and by simple diffusion. The protein-mediated transport mechanism(s) whereby bile acids are secreted across the canalicular membrane into bile remain to be characterized. Na^+, Cl^-, and water enter the bile via the leaky tight junctions that join the hepatocytes.

fluid by solvent drag (see Fig. 39-32). It is also clear that the canalicular and basolateral plasma membranes have ion transport proteins that aid in the transport of ions into canalicular bile. The ion transport mechanisms of the hepatocyte have not been completely defined, but they resemble the ion transport pathways of the pancreatic acinar cell (see Fig. 39-24).

The bile ducts secrete a fluid rich in bicarbonate. As for the pancreatic duct secretions, the major physiological stimulus for the secretion of water and electrolytes by the bile duct epithelial cells is the hormone secretin. It is likely that the ion transport mechanisms are involved, and the ways that they are regulated by secretin are similar to those in the pancreatic ductular epithelial cells (see Fig. 39-22).

Bile Concentration and Storage in the Gallbladder

Between meals the tone of the sphincter of Oddi, which guards the entrance of the common bile duct into the duodenum, is high. Thus most bile flow is diverted into the gallbladder. The gallbladder is a small organ, having a capacity of 15 to 60 ml (average about 35 ml) in humans. Many times this volume of bile may be secreted by the liver between meals. The gallbladder concentrates the bile by absorbing Na^+, Cl^-, HCO_3^-, and water from the bile so that the bile acids can be concentrated from 5 to 20 times in the gallbladder. K^+ is concentrated in the bile when water is absorbed, and then K^+ is absorbed by simple diffusion.

The active transport of Na^+ is the primary active process in the concentrating action of the gallbladder; Cl^- and HCO_3^- are absorbed to preserve electroneutrality.

Bile contains significant concentrations of Ca^{++} and $CO_3^=$. The gallbladder epithelium secretes H^+ into bile and carbonate is thereby converted to bicarbonate. Both calcium and bicarbonate are absorbed by the gallbladder epithelial cells. These absorptive processes prevent the precipitation of calcium carbonate in the gallbladder lumen as bile is concentrated.

Because of its high rate of water absorption the gallbladder serves as a model for water and electrolyte transport by tight-junctioned epithelia. The **standing gradient mechanism** for fluid absorption was first proposed for the gallbladder. A key initial observation was that, when fluid was being reabsorbed by the gallbladder, the lateral intercellular spaces between the epithelial cells were large and swollen. When fluid transport was blocked, for example, by poisoning the Na^+ pumps with ouabain, the intercellular spaces almost disappeared. These observations strongly suggested that the intercellular spaces are a major route of fluid flow during absorption. A current view is that Na^+ is actively pumped into the lateral intercellular spaces. The Na^+ pumps are believed to be especially dense near the

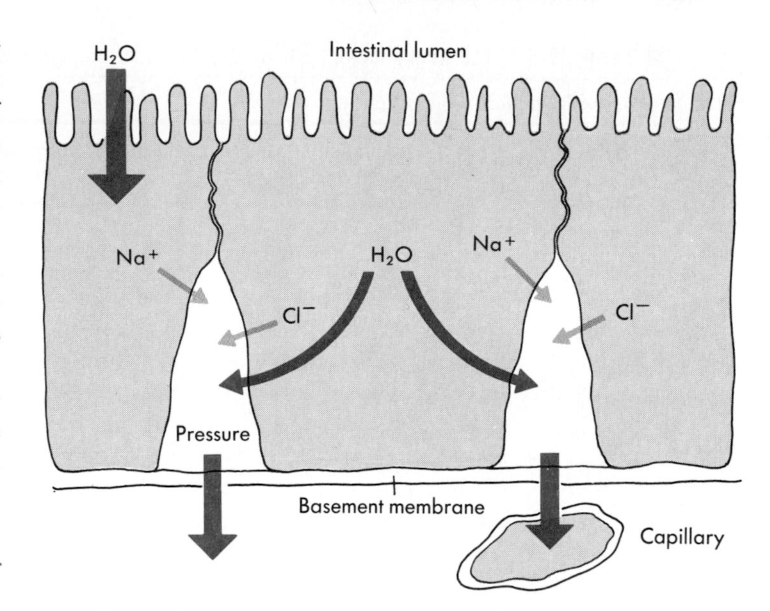

Fig. 39-33 Water absorption from the gallbladder by the mechanism of the standing osmotic gradient. Na^+ is actively pumped into the lateral intercellular spaces; Cl^- follows. Water is drawn by osmosis to enter the intercellular spaces, elevating the hydrostatic pressure. Water, Na^+, and Cl^- are filtered across the porous basement membrane and enter the capillaries.

mucosal (apical) end of the channel (Fig. 39-33). Cl^- and HCO_3^- also are transported into the intercellular space, probably because of an electrical potential created by Na^+ transport. The high ion concentration near the apical end of the intercellular space causes the fluid there to be hypertonic. This produces an osmotic flow of water from the lumen via adjacent cells into the intercellular space. Water distends the intercellular channel because of increased hydrostatic pressure. Because of water flow from adjacent cells the fluid becomes less hypertonic as it flows down the intercellular channel, and it is essentially isotonic when it reaches the serosal (basal) end of the channel. Ions and water flow across the basement membrane of the epithelium and are carried away by the capillaries.

Emptying of the Gallbladder

Emptying of the gallbladder contents into the duodenum begins several minutes after the start of a meal. Intermittent contractions of the gallbladder force bile through the partially relaxed sphincter of Oddi. During the cephalic and gastric phases of digestion gallbladder contraction and relaxation of the sphincter are mediated by cholinergic fibers in branches of the vagus nerve (Table 39-6). Stimulation of sympathetic nerves to the gallbladder and duodenum inhibits emptying of the gallbladder. Vasoactive intestinal polypeptide (VIP) inhib-

Table 39-6 The major factors affecting gallbladder emptying and bile synthesis and secretion

Phase of digestion	*Stimulus*	*Mediating factor*	*Response*
Cephalic	Taste and smell of food; food in mouth and pharynx	Impulses in branches of vagus nerve Gastrin?	Increased rate of gallbladder emptying
Gastric	Gastric distension	Impulses in branches of vagus nerve; Gastrin?	Increased rate of gallbladder emptying
Intestinal	Fat digestion products in duodenum	Cholecystokinin	Increased rate of gallbladder emptying; increased rate of bile acid secretion
	Acid in duodenum	Secretin	Increased rate of secretion of bicarbonate-rich fluid by the bile duct epithelium (this effect strongly potentiated by cholecystokinin)
	Absorption of bile acids in the distal part of ileum	High concentration of bile acids in portal blood	Stimulation of bile acid secretion; inhibition of bile acid synthesis
Interdigestive period	Low rate of release of bile to duodenum	Low concentration of bile acids in portal blood	Stimulation of bile acid synthesis; inhibition of bile acid secretion

its gallbladder contractions; VIP-containing nerve terminals are present in the wall of the gallbladder.

The highest rate of gallbladder emptying occurs during the intestinal phase of digestion. The strongest stimulus for the emptying is the hormone ***cholecystokinin.*** Cholecystokinin is released by the duodenal mucosa most strongly in response to the presence of fat digestion products and essential amino acids in the duodenum. Cholecystokinin reaches the gallbladder via the circulation, and it causes strong contractions of the gallbladder and relaxation of the sphincter of Oddi. Substances that mimic the actions of cholecystokinin in promoting gallbladder emptying are called ***cholecystagogues.*** Gastrin and cholecystokinin share a common sequence of five amino acids at their C-terminals. Gastrin is not nearly as potent a cholecystagogue as cholecystokinin, but gastrin may play a role in eliciting gallbladder contractions during the cephalic and gastric phases.

Under normal circumstances the rate of gallbladder emptying is sufficient to keep the concentration of bile acids in the duodenum above the critical micelle concentration.

Intestinal Absorption of Bile Acids and Their Enterohepatic Circulation

The functions of bile acids in emulsifying dietary lipid and in forming mixed micelles with the products of lipid digestion are discussed in Chapter 40.

Normally, by the time chyme reaches the terminal part of the ileum, dietary fat is almost completely absorbed. Bile acids are then absorbed. The epithelial cells of the distal part of the ileum actively take up bile acids against a large concentration gradient. The active transport system has a higher affinity for conjugated bile acids.

Bile acids also have a fair degree of lipid solubility. Thus they also can be taken up by simple diffusion. Bacteria in the terminal part of the ileum and colon deconjugate bile acids and also dehydroxylate them to produce secondary bile acids. Both deconjugation and dehydroxylation lessen the polarity of bile acids, enhancing their lipid solubility and their absorption by simple diffusion.

Typically about 0.5 g of bile acids escapes absorption and is excreted in the feces each day. This quantity is 10% to 20% of the total bile acid pool, and normally it is replenished by synthesis of new bile acids by the liver.

Bile acids, whether absorbed by active transport or simple diffusion, are transported away from the intestine in the portal blood, mostly bound to plasma proteins. In the liver, hepatocytes avidly extract the bile acids from the portal blood. In a single pass through the liver the portal blood is essentially cleared of bile acids. Bile acids in all forms, primary and secondary, both conjugated and deconjugated, are taken up by the hepatocytes. The hepatocytes reconjugate almost all the deconjugated bile acids and rehydroxylate some of the secondary bile acids. These bile acids are secreted into the bile along with newly synthesized bile acids.

Control of Bile Acid Synthesis and Secretion

The rate of return of bile acids to the liver is a major influence on the rate of synthesis and secretion of bile acids. Bile acids in the portal blood stimulate the secretion of bile acids by the hepatocytes but inhibit the synthesis of bile acids (Fig. 39-34). This is called the choleretic effect of bile acids. Substances that act to enhance bile acid secretion are known as **choleretics.**

Long after a meal, when bile acids already have been returned to the liver, the level of bile acids in the portal

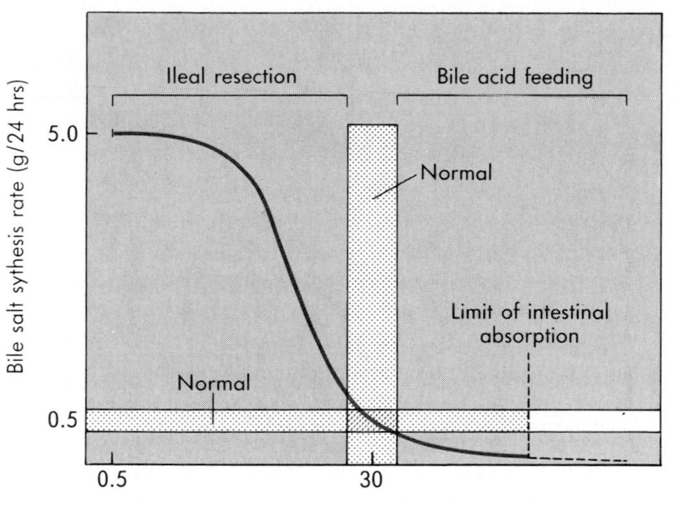

■ **Fig 39-34** A reciprocal relationship exists between the rate of de novo synthesis of bile acids by hepatocytes and the rate of secretion of bile acids. When bile acids return to the liver during digestion, synthesis is inhibited and the energy of the hepatocyte is used for reprocessing and secretion of the returning bile acids. Removal of part of the distal ileum decreases the rate of return of bile acids in the portal blood, resulting in a greater rate of synthesis of bile acids. Feeding bile acids results in increased levels of bile acids in portal blood and in greater inhibition of bile acid synthesis. (Redrawn from Carey, MC and Cahalane, MJ: In Arias, IM et al.: *The liver: biology and pathobiology*, ed 2, 1988, New York, Raven Press.)

blood is low. Hence synthesis of new bile acids is not inhibited, and synthesis proceeds at near maximal rates. The rate of secretion of bile acids is low because secretion is not being stimulated by bile acids in the portal blood, and the gallbladder fills rather slowly with bile.

After a meal the presence of fat digestion products in the duodenum causes release of cholecystokinin. This brings about emptying of gallbladder contents into the duodenum and strongly stimulates secretion of bile acids by the hepatocytes. Secretin, released by acidic chyme in the duodenum, increases the secretion of bicarbonate-rich fluid by the bile duct epithelium. At this time the liver continues to synthesize bile acids at a high rate and to secrete them in response to cholecystokinin. This continues until bile acids are absorbed from the terminal aspect of the ileum and return to the liver in the portal blood.

Bile acids returning to the liver inhibit the synthesis of new bile acids but stimulate high rates of secretion. Bile acids that are taken up are rapidly reconjugated (some secondary bile acids are rehydroxylated) and resecreted almost immediately. So powerful is the stimulus to resecrete the returning bile acids that the entire pool of bile acids (1.5 to 31.5 g) recirculates twice in response to a typical meal. In response to a single meal with a very high fat content, the bile acid pool may recirculate five or more times. Table 39-6 summarizes major aspects of control of gallbladder emptying and bile synthesis and secretion.

Gallstones

The most common type of gallstones contains cholesterol as its major component. Cholesterol is essentially insoluble in water. When bile contains more cholesterol than can be solubilized in the bile acid–lecithin micelles, crystals of cholesterol form in the bile. Such bile is said to be **supersaturated** with cholesterol. The greater the concentration of bile acids and lecithin in bile, the greater is the amount of cholesterol that can be contained in the mixed micelles. Lecithin is important in this regard because lecithin-cholesterol mixed micelles can solubilize more cholesterol than can micelles of bile acids alone.

At night most normal individuals secrete bile that is supersaturated with cholesterol. This occurs because the rate of bile acid secretion is particularly low as a result of the absence of stimulation by bile acids returning in the portal blood. During the day, however, the average rate of bile acid secretion is higher, and bile is unsaturated with cholesterol in normal persons. Individuals with cholesterol gallstones tend to secrete bile that is supersaturated with cholesterol both night and day. Most cholesterol gallstones contain crystallized bilirubin at their cores. This suggests that bilirubin crystals serve as nucleation sites for the crystallization of cholesterol.

Bile pigment stones are the other major class of gallstones. Their major constituent is the calcium salt of unconjugated bilirubin. Conjugated bilirubin is quite soluble and does not form insoluble calcium salts in bile. Bile may contain elevated levels of unconjugated bilirubin because hepatocytes are deficient in forming the glucuronide or because of excessive deconjugation by glucuronidase.

Intestinal Secretions

The mucosa of the intestine, from the duodenum through the rectum, elaborates secretions that contain mucus, electrolytes, and water. The total volume of intestinal secretions is about 1500 ml/day. The mucus in the secretions protects the mucosa from mechanical damage. The nature of the secretions and the mechanisms that control secretion vary from one segment of the intestine to another.

Duodenal Secretions

The duodenal submucosa contains branching glands that elaborate a secretion rich in mucus. The glands have ducts that empty into the crypts of Lieberkühn. The duodenal epithelial cells probably also contribute to duodenal secretions, but most of the secretions are produced by the glands. The duodenal secretion contains mucus and an aqueous component that does not differ greatly from plasma in its concentrations of the major ions. Gastrin, secretin, and cholecystokinin stimulate duodenal secretion, but a physiological role for any of these hormones has not yet been established.

Secretions of the Small Intestine

Goblet cells, which lie among the columnar epithelial cells of the small intestine, secrete mucus that lubricates the mucosal surface and protects it against mechanical damage. During the course of normal digestion an aqueous secretion is elaborated by the epithelial cells at a rate only slightly less than the rate of fluid absorption by these cells. So the net absorption of fluid that normally occurs is the result of much larger unidirectional absorptive and secretory flows. As discussed on p. 703, cholera toxin greatly increases the rate of aqueous secretion by the small intestine, particularly by the jejunum. This process leads to secretory diarrhea.

Secretions of the Colon

The secretions of the colon are smaller in volume but richer in mucus than small intestinal secretions. The mucus is produced by the numerous goblet cells of the colonic mucosa. The aqueous component of colonic secretions is alkaline because of secretion of HCO_3^- in exchange for Cl^-, and the secretion is rich in K^+. The production of colonic secretions is stimulated by mechanical irritation of the mucosa and by activation of cholinergic pathways to the colon. Stimulation of sympathetic nerves to the colon decreases the rate of colonic secretion.

Summary

1. The epithelial cells that line the gastrointestinal tract and the cells of various glands associated with the gastrointestinal tract produce secretions that contain water, electrolytes, and proteins. These secretions have important functions. The physiological regulation of gastrointestinal secretions is effected by intrinsic and extrinsic neurons, hormones, and paracrine mediators.

2. Salivary glands produce a hypotonic fluid with bicarbonate and potassium concentrations in excess of plasma levels. Saliva contains an α-amylase that begins the digestion of starch. Mucus in saliva lubricates food. Parasympathetic nerves are the key regulators of salivary secretion.

3. The stomach serves as a reservoir for ingested food that empties gastric contents into the duodenum in a regulated fashion. Parietal cells secrete HCl and instrinsic factor into the stomach. Chief cells secrete pepsinogens. The regulation of HCl secretion involves extrinsic and intrinsic nerves with acetylcholine as the major stimulatory neurotransmitter. Gastrin, a hormone released by G cells in glands in the gastric antrum and in the duodenum, and histamine, a paracrine agonist released by ECL (enterochromaffin-like) cells in the stomach, are important physiological agonists of HCl secretion. HCl catalyzes the conversion of pepsinogens to active pepsins. Pepsins convert a significant fraction of ingested protein to oligopeptides. Mucus and bicarbonate secretions form the "gastric mucosal barrier" that protects the epithelial cells of the stomach from the effects of HCl and pepsins.

4. The pancreas produces a bicarbonate-rich fluid that contains enzymes that are essential for the digestion of carbohydrates, proteins, and fats. Pancreatic acinar cells produce the enzyme component of pancreatic juice; the intralobular and extralobular ducts secrete much of the aqueous component (water and electrolytes) of pancreatic juice. Cholecystokinin (CCK) is the major physiological agonist of acinar cell secretion of the enzyme component. Secretin is the major stimulus for secretion of bicarbonate-rich fluid by the extralobular ducts. CCK and secretin are hormones released by cells in the duodenum and jejunum in response to the presence of fat digestion products and acid, respectively.

5. The liver produces and the gallbladder concentrates a secretion called bile. Bile is a bicarbonate-rich fluid containing bile acids, bile pigments, lecithin, cholesterol, and numerous other components. Bile acids play a vital role in the digestion and absorption of lipids. Bile acids first emulsify lipids, thus increasing the surface area available for the action of lipid-digesting enzymes. Then bile acids form micelles with the products of lipid digestion. Micelles are small enough to diffuse among the intestinal microvilli, thus allowing the entire brush border membrane to participate in lipid absorption. The parenchymal cells of the liver, the hepatocytes, are responsible for secreting the organic components of bile. The cells of the bile ducts secrete a bicarbonate-rich fluid. CCK is a major secretagogue for secretion by the hepatocytes. Secretin stimulates the bile ducts to produce their bicarbonate-rich fluid. Bile acids are absorbed in the terminal ileum and return to the liver in the portal vein. Hepatocytes rapidly clear the blood of bile acids and resecrete them. Bile acids in the portal blood are a powerful stimulus to the hepatocytes to resecrete bile acids. The entire bile acid pool may be recirculated two to five times in response to a single meal. The secretion, return, and resecretion of bile acids is known as the enterohepatic circulation of bile acids.

Bibliography

Journal articles

Allen A, Garner A: Mucus and bicarbonate secretion in the stomach and their possible role in mucosal protection, *Gut* 21:249, 1980.

Blitzer BL, Boyer JL: Cellular mechanisms of bile formation, *Gastroenterology* 82:346, 1982.

Chew CS: Intracellular mechanisms in control of acid secretion, *Curr Opinion Gastroenterol* 7:856, 1991.

Coleman R: Biochemistry of bile secretion, *Biochem J* 244:249, 1987.

Crawford JM, Berken CA, Gollan JL: Role of the hepatocyte microtubular system in the excretion of bile salts and biliary lipid: implications for intracellular vesicular transport, *J Lipid Res* 29:144, 1988.

Demarest JR, Loo DDF: Electrophysiology of the parietal cell, *Annu Rev Physiol* 52:307, 1990.

Grotmol T, et al: Secretin-dependent HCO_3^- from pancreas and liver, *J Intern Med* 228(suppl 1):47, 1990.

Hoffman AF: Current concepts of biliary secretion, *Dig Dis Sci* 34(suppl):16S, 1989.

Klaasen CD, Watkins JB III: Mechanisms of bile formation, hepatic uptake, and biliary excretion, *Pharmacol Rev* 36:1, 1984.

LaRusso NF: Proteins in bile: how they get there and what they do, *Am J Physiol* 247:G199, 1984.

Liedtke CM: Regulation of chloride transport in epithelia, *Annu Rev Physiol* 51:143, 1989.

Petersen OH, Gallacher DV: Electrophysiology of pancreatic and salivary acinar cells, *Annu Rev Physiol* 50:65, 1988.

Putney JW Jr: Identification of cellular activation mechanisms associated with salivary secretion, *Annu Rev Physiol* 48:75, 1986.

Rabon EC, Reuben MA: The mechanism and structure of the gastric H,K-ATPase, *Annu Rev Physiol* 52:321, 1990.

Reuss L: Ion transport across gallbladder epithelium, *Physiol Rev* 69:503, 1989.

Scharschmidt BF, Lake JR: Hepatocellular bile acid transport and ursodeoxycholic acid hypercholeresis, *Dig Dis Sci* 34(suppl):5S, 1989.

Tache Y: CNS peptides and regulation of gastric acid secretion, *Annu Rev Physiol* 50:17, 1988.

Walsh JH: Peptides as regulators of gastric acid secretion, *Annu Rev Physiol* 50:41, 1988.

Watanabe S, Phillips MJ: Ca^{++} causes active contraction of bile canaliculi: direct evidence from microinjection studies, *Proc Nat Acad Sci USA* 81:6164, 1984.

Books and monographs

Allen A, editor: *Mechanisms of mucosal protection in the upper gastrointestinal tract,* 1984, New York, Raven Press.

Arias IM, et al: *The liver: biology and pathobiology,* ed 2, New York, 1988, Raven Press.

Flemstrom G: Gastric and duodenal mucosal bicarbonate secretion. In Johnson LR, editor: *Physiology of the gastrointestinal tract,* vol 2, ed 2, New York, 1987, Raven Press.

Forte JG, Wolosin M: HCl secretion by the gastric oxyntic cell. In Johnson LR, editor: *Physiology of the gastrointestinal tract,* vol 1, ed 2, New York, 1987, Raven Press.

Forte JG, Soll AH: Cell biology of hydrochloric acid secretion. In: *Handbook of physiology,* section 6, vol 3, Washington, DC, 1989, American Physiological Society.

Gardner JD, Jensen RT: Secretagogue receptors on pancreatic acinar cells. In Johnson LR, editor: *Physiology of the gastrointestinal tract,* vol 2, ed 2, New York, 1987, Raven Press.

Hernandez DE, Glavin GB, editors: *Neurobiology of stress ulcers,* Ann NY Acad Sci, vol 299, New York, 1990, New York Academy of Sciences.

Hersey SJ: Cellular basis of pepsinogen secretion. In: *Handbook of physiology,* section 6, vol 3, Washington, DC, 1989, American Physiological Society.

Hootman SR, Williams JA: Stimulus-secretion coupling in the pancreatic acinus. In Johnson LR, editor: *Physiology of the gastrointestinal tract,* vol 2, ed 2, New York, 1987, Raven Press.

Sachs G: The gastric proton pump: The H^+, K^+-ATPase. In Johnson LR, editor: *Physiology of the gastrointestinal tract,* vol 1, ed 2, New York, 1987, Raven Press.

Schultz I: Electrolyte and fluid secretion in the exocrine pancreas. In Johnson LR, editor: *Physiology of the gastrointestinal tract,* vol 2, ed 2, New York, 1987, Raven Press.

Soll AH, Berglindh T: Physiology of isolated gastric glands and parietal cells: receptors and effectors regulating function. In Johnson LR, editor: *Physiology of the gastrointestinal tract,* vol 1, ed 2, New York, 1987, Raven Press.

Young JA, et al: Secretion by the major salivary glands. In Johnson LR, editor: *Physiology of the gastrointestinal tract,* vol 1, ed 2, New York, 1987, Raven Press.

CHAPTER

40

Digestion and Absorption

Digestion and Absorption of Carbohydrates

Carbohydrates in the Diet

Plant starch, amylopectin, is the major source of carbohydrate in most human diets. *There is no nutritional requirement for carbohydrate per se, but it is usually the principal source of calories.* Amylopectin is a high molecular weight ($>10^6$), branched molecule of glucose monomers. A smaller proportion of dietary starch is amylose, a smaller molecular weight ($<10^5$), linear α-1,4 linked polymer of glucose. Cellulose is a β-1,4 linked glucose polymer. Intestinal enzymes cannot hydrolyze β-glycosidic linkages; thus cellulose and other molecules with β-glycosidic linkages remain undigested. Cellulose is a major component of dietary "fiber." The amount of the animal starch, glycogen, typically ingested varies widely among cultures and among individuals within a given culture. Sucrose and lactose are the principal dietary disaccharides, and glucose and fructose are the major monosaccharides.

Digestion of Carbohydrates

The structure of a branched starch molecule is depicted in Fig. 40-1. Starch is a polymer of glucose, and it consists of chains of glucose units linked by α-1,4 glycosidic bonds. The α-1,4 chains have branch points formed by α-1,6 linkages, and the starch molecule is highly branched. The digestion of starch begins in the mouth with the action of the α-amylase, formerly known as ptyalin, contained in salivary secretions. This enzyme catalyzes the hydrolysis of the *internal* α-1,4 links of starch but cannot hydrolyze the α-1,6 branching links. (The α-amylase secreted by the pancreas has the same specificity.) As shown in Fig. 40-1, the principal products of α-amylase digestion of starch are maltose, maltotriose, and branched oligosaccharides known as α-limit dextrins. The action of the salivary α-amylase continues until the food in the stomach is mixed with gastric acid. Considerable digestion of starch by the salivary α-amylase may occur normally, but this enzyme is not required for the complete digestion and absorption of the starch ingested. After the salivary

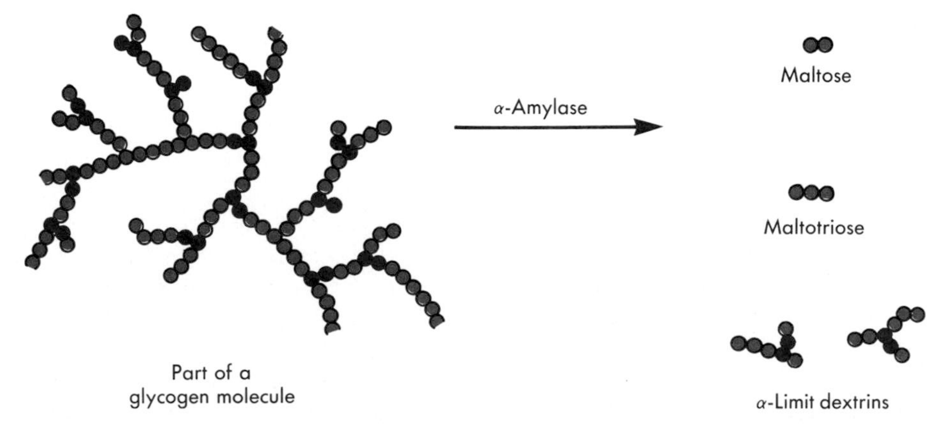

Fig. 40-1 Structure of a branched starch molecule and the action of α-amylase. The circles represent glucose monomers. The black circles show glucose units linked by α-1,6 linkages at the branch points. The α-1,6 linkages and terminal α-1,4 bonds cannot be cleaved by α-amylase. The colored glucose monomers are linked by α-1,4 linkages.

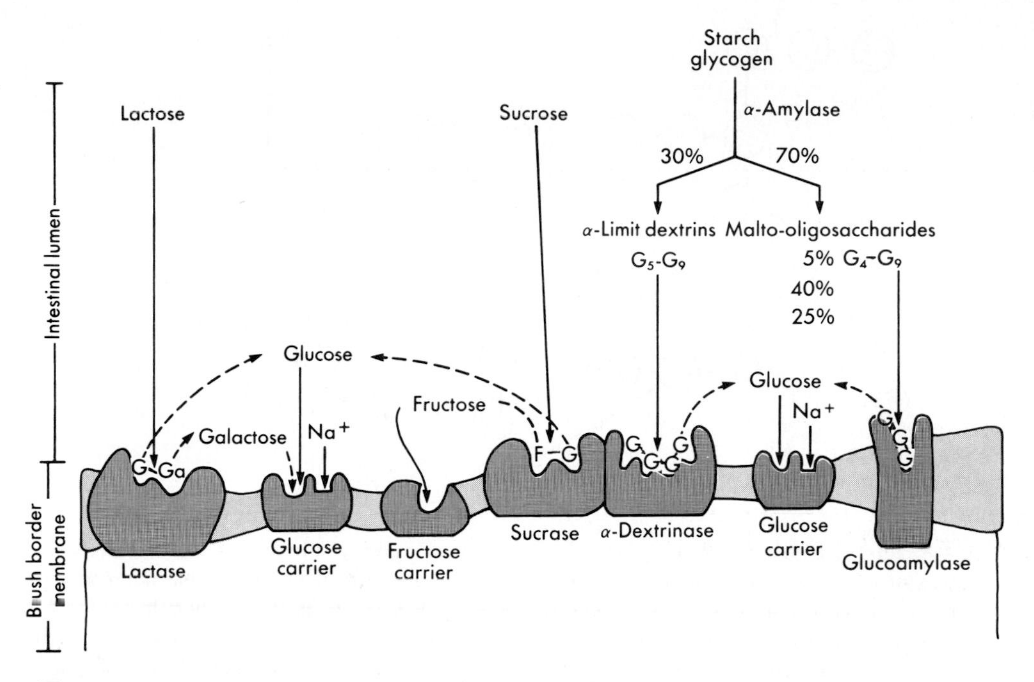

■ **Fig. 40-2** Functions of the major brush border oligosaccharidases. The glucose, galactose, and fructose molecules released by enzymatic hydrolysis are then transported into the epithelial cell by specific transport carrier proteins. *G,* Glucose; *Ga,* galactose; *F,* fructose. (From Gray, GM: *N Engl J Med* 292:1225, 1975.)

α-amylase is inactivated by gastric acid, no further processing of carbohydrate occurs in the stomach.

The pancreatic secretions contain a highly active α-amylase. The products of starch digestion by this enzyme are the same as for the salivary α-amylase (see Fig. 40-1), but the total activity of the pancreatic enzyme is considerably greater than the salivary amylase. The pancreatic α-amylase is most concentrated in the duodenum. Within 10 minutes after entering the duodenum, starch is largely converted to the following small oligosaccharides: maltose, maltotriose, α-1,4 linked maltooligosaccharides (from four to nine glucose units long), and α-limit dextrins containing from five to nine glucose monomers.

The further digestion of these oligosaccharides is accomplished by enzymes that reside in the brush border membrane of the epithelium of the duodenum and jejunum (Fig. 40-2). The major brush border oligosaccharidases are **lactase,** which splits lactose into glucose and galactose, **sucrase,** which splits sucrose into fructose and glucose, **α-dextrinase** (which is also called isomaltase and which "debranches" the α-limit dextrins by cleaving the α-1,6 linkages at the branch points), and **glucoamylase,** which breaks maltooligosaccharides down to glucose units.

Both α-dextrinase and sucrase also cleave α-1,4 glucose-glycosidic linkages. The digestion of α-limit dextrins proceeds by the sequential removal of glucose monomers from the nonreducing ends. When branch points are encountered, they are cleaved by α-dextrinase (Fig. 40-3). Sucrase and isomaltase are noncovalently associated subunits of a single protein. The two enzymes are synthesized as a single polypeptide chain that is cleaved into two polypeptides after insertion into the brush border membrane (Fig. 40-4). The two enzyme activities are not affected by their molecular interactions, so the function for this association remains obscure. The activities of these four enzymes are highest in the brush border of the upper jejunum, and they gradually decline through the rest of the small intestine.

■ *Absorption of Carbohydrates*

The duodenum and upper jejunum have the highest capacity to absorb sugars. The capacities of the lower jejunum and ileum are progressively less. The only dietary monosaccharides that are well absorbed are glucose, galactose, and fructose. Glucose and galactose are actively taken up by the brush border epithelial cells through a fairly well-characterized transport system (Chapter 1). Glucose and galactose compete for entry; other sugars are less effective competitors. Transport is inhibited by phlorhizin and by metabolic inhibitors.

The active entry of glucose and galactose into the intestinal epithelial cells is stimulated by the presence of Na^+ in the lumen. Similarly, the entry of Na^+ into the epithelial cell across the brush border membrane is stimulated by glucose or galactose in the lumen. Na^+ and glucose or galactose are transported into the cell by the same membrane protein, which has two Na^+-binding sites and one sugar-binding site. Na^+ enters the cell

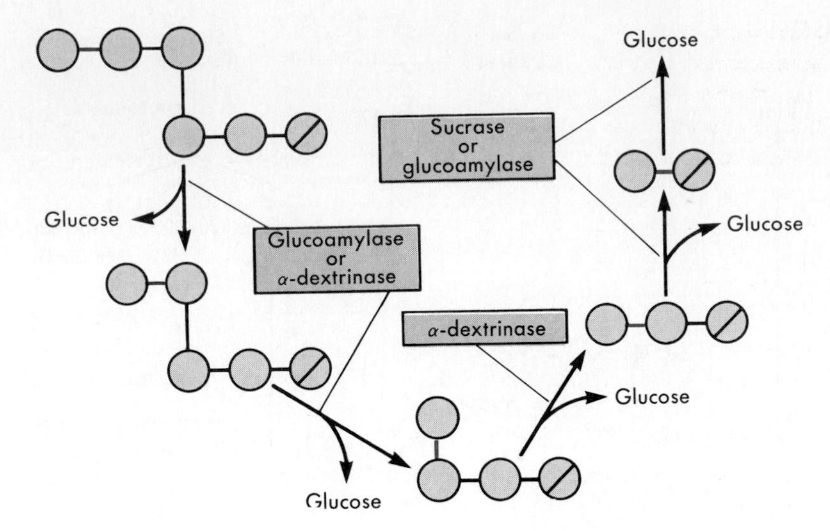

■ **Fig. 40-3** Cleavage of an α-limit dextrin by the oligosaccharidases of the brush border plasma membrane. Glucose monomers are removed in sequence, beginning at the nonreducing end of the molecule. Note the overlapping specificities of glucoamylase and α-dextrinase and of sucrase and glucoamylase. α-Dextrinase (isomaltase) is the only enzyme that cleaves the α-1,6 linkages at the branch points of the α-limit dextrins. (From Gray, GM: Carbohydrate absorption and malabsorption. In Johnson, RL, editor: *Physiology of the gastrointestinal tract,* New York, 1981, Raven Press.)

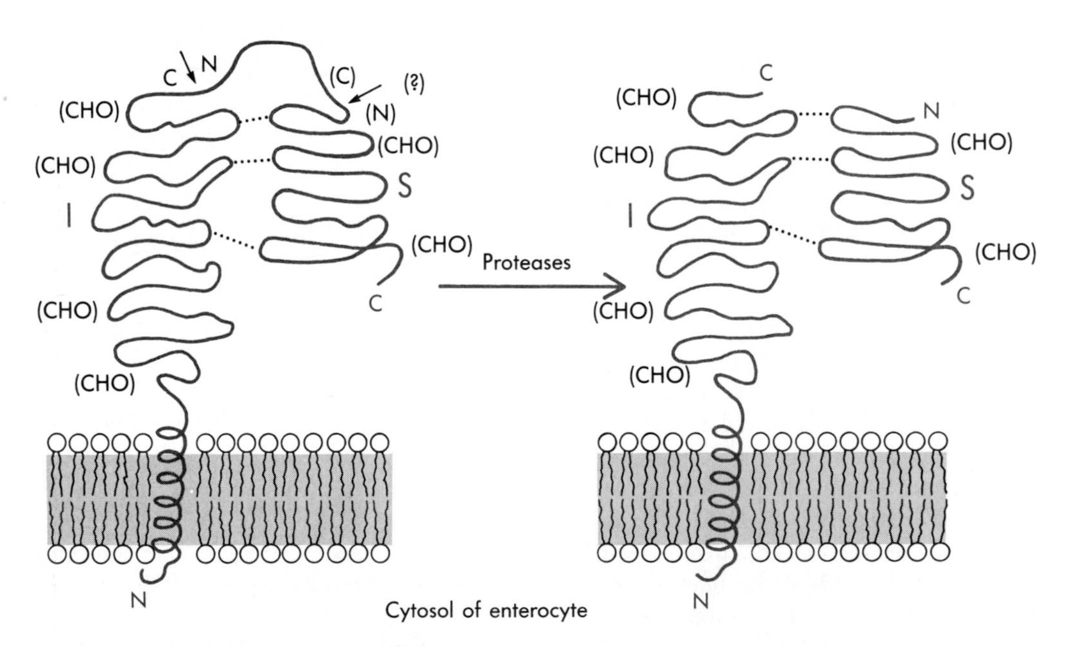

■ **Fig. 40-4** The sucrase-isomaltase complex is inserted into the endoplasmic reticulum membrane as a single polypeptide chain. When endoplasmic reticulum vesicles fuse with the brush border membrane of small intestinal epithelial cells, pancreatic proteases cleave the protein into its sucrase *(S)* and isomaltase *(I)* polypeptides by acting at the sites shown by the arrows. *CHO* indicates the presence of carbohydrate side chains and the dotted lines represent noncovalent interactions between the S and I subunits. (Redrawn from Dahlqvist, A and Semenza, G: *J Pediatr Gastroenterol Nutr* 4:857, 1985.)

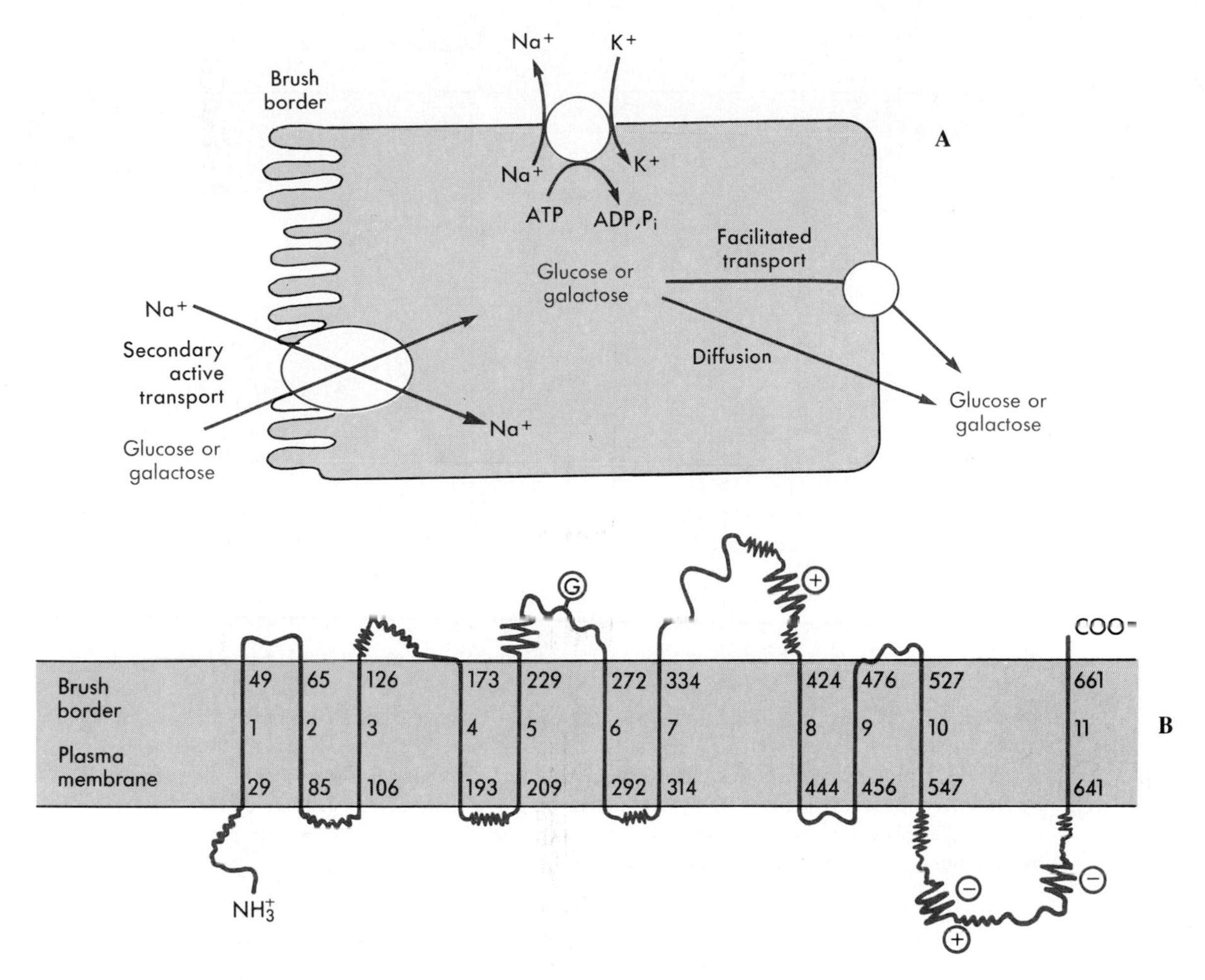

■ **Fig. 40-5** Glucose and galactose absorption in the small intestine. **A,** Glucose and galactose enter the epithelial cell at the brush border against a concentration gradient. The Na^+ gradient provides the energy for glucose entry. Glucose and galactose leave the cell at the basolateral membrane by facilitated transport and simple diffusion. **B,** Hypothetical secondary structure of the Na^+-monosaccharide co-transporter of small intestine. *G* indicates a potential N-linked glycosylation site. + and − indicate clusters of basic and acidic amino acid residues, respectively. (From Hediger, MA, et al: *Nature* 330:379, 1987).

down a large electrochemical potential gradient; both concentration and electrical forces drive it into the cell. The energy released by the flow of Na^+ down its electrochemical potential is harnessed to force glucose or galactose into the cell *against* a concentration gradient for the sugar. The electrochemical potential gradient for Na^+ is created by Na^+, K^+-ATPase molecules in the basal and lateral plasma membranes of the intestinal epithelial cells. Na^+ is transported by primary active transport by the sodium-potassium pump. The sugars are transported by secondary active transport, because their active transport depends on the electrochemical potential gradient of another species (Na^+). Glucose and galactose leave the intestinal epithelial cell at the basal and lateral plasma membranes via facilitated transport and by simple diffusion to some extent, and they diffuse into the mucosal capillaries. Fig. 40-5 summarizes some major features of glucose and galactose absorption.

The absorption of fructose by the intestine is less well understood. Fructose does not compete for the glucose-galactose system, and fructose transport is not linked to Na^+ absorption. Yet it is transported almost as rapidly as glucose and galactose and much more rapidly than other monosaccharides. Hence it is presumed that fructose uptake is mediated by a membrane protein. Active transport of fructose has been observed in rat intestine but has not been demonstrated in the human intestine.

Absorption of carbohydrate from different food sources. Monosaccharides and disaccharides are completely absorbed in the small intestine. The rate and extent of absorption of starch may vary with the starch-containing foodstuff (Fig. 40-6, *A;* Table 40-1). Starch-

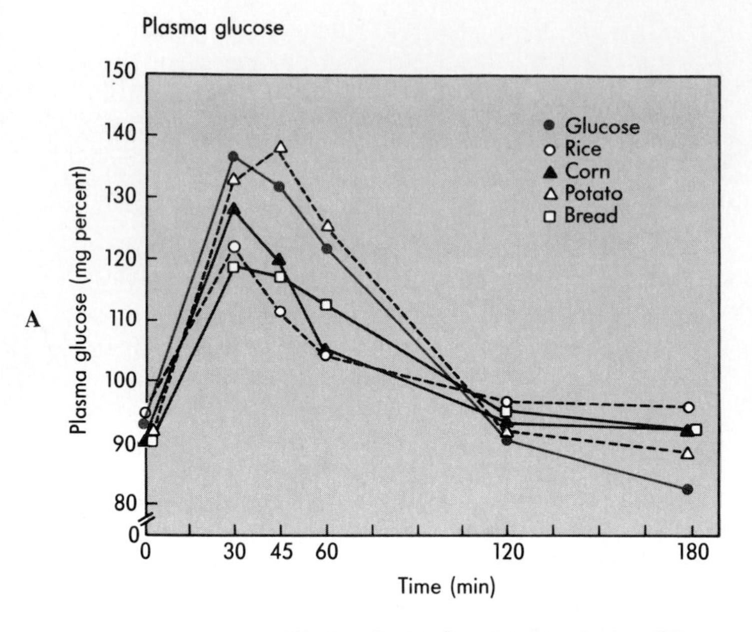

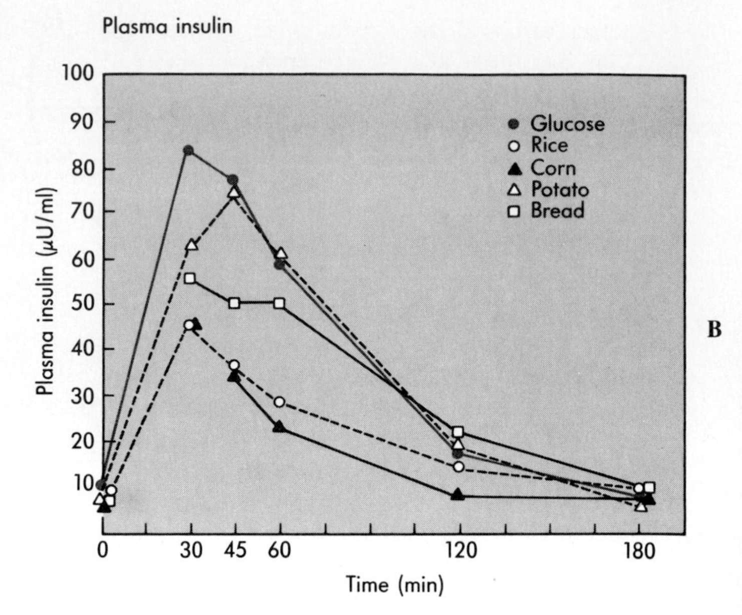

■ **Fig. 40-6** Starch is absorbed at different rates from different foodstuffs. Healthy human volunteers were fed 50 g of carbohydrate as the different foods indicated. Plasma glucose, **A,** and insulin levels, **B,** were monitored for 3 hours thereafter. The rate of plasma glucose increase and the maximum plasma level attained depends on the foodstuff. There is a positive correlation between the maximum level of blood glucose attained after eating a particular foodstuff and the peak level of plasma insulin attained. Note that relatively small differences in peak blood glucose levels for the different carbohydrates produced greater increments in the peak insulin levels and in the total amounts of insulin released. (Redrawn from Crapo, PA et al: *Diabetes* 26:1178, 1977.)

■ **Table 40-1** Absorption of glucose from various carbohydrate-containing foods

Food	*Glycemic index*
Glucose	100
Carrots	92
Corn flakes	80
Rice, white	72
Potatoes, raw	70
Bread, white	69
Shredded wheat	67
Bananas	62
Corn	59
Pear	51
All-Bran	51
Spaghetti	50
Potatoes, sweet	48
Orange	40
Apple	39
Beans, navy	31
Beans, kidney	29
Sausages	28

Healthy adult men and women were fed amounts of the various food listed sufficient to provide 50 g of carbohydrate. The response of blood glucose levels was determined as illustrated in Fig. 40-6, *A*. The glycemic index is defined as the area under the blood glucose vs. time curve from the time of eating to two hours after eating expressed as a percentage of the area under the curve after ingesting 50 g of glucose. Data are the means of 5 to 10 individuals. (Data from Jenkins DJA, et al.: *Am J Clin Nutr* 34:362, 1981.)

containing foods that are more rapidly absorbed result in a greater rate and extent of insulin release from the pancreas (Fig. 40-6, *B*).

Extent of absorption of carbohydrates. In healthy humans fed a test meal containing 20 or 60 grams of starch, about 6% to 10% escapes absorption in the small intestine and is passed on to the colon, where it serves as an excellent carbon source for colonic bacteria. This failure to absorb carbohydrate completely is not associated with any untoward symptoms.

■ *Carbohydrate Malabsorption Syndromes*

Malabsorption of carbohydrates is usually caused by a deficiency in one of the digestive enzymes of the brush border.

Lactose malabsorption syndrome. This common disorder is caused by a deficiency of lactase in the brush border of the duodenum and jejunum. Undigested lactose cannot be absorbed. Thus it is passed on to the colonic bacteria, which avidly metabolize the lactose, producing gas and metabolic products that enhance colonic motility.

Individuals with this disorder are said to be **lactose intolerant.** The symptoms of this disorder, and the other carbohydrate malabsorption syndromes as well, are intestinal distension, borborygmi, and diarrhea. **Borborygmi** are the gurgling noises made by the intestine as it mixes gas and liquid.

More than 50% of the adults in the world are lactose intolerant. This condition seems to be genetically determined. In Oriental societies lactose intolerance among adults is almost universal. The majority of northern European adults, on the other hand, are lactose tolerant. Many black American adults are lactose intolerant. Many lactose-intolerant adults simply do not drink milk or eat certain milk products and thereby avoid the symptoms without being aware that they have the disorder. The presence of lactose in the diet may induce a higher level of intestinal lactase activity than would be present in the absence of dietary lactose.

Congenital lactose intolerance. This is rare. Infants who lack lactase have diarrhea when they are fed breast milk or formula containing lactose. The resulting dehydration and electrolyte imbalance is life threatening. Such infants do well when fed a formula containing sucrose or fructose instead of lactose.

Sucrase-isomaltase deficiency. A deficiency of sucrase and isomaltase activities in the small intestinal mucosa is an autosomal recessive, inherited disorder that results in intolerance to ingested sucrose. About 10% of Greenland's Eskimos and as many as 0.2% of North Americans have sucrase-isomaltase deficiency. In this disorder either the synthesis of sucrase and isomaltase is suppressed or the enzymes are destroyed by antibodies. Individuals with sucrase-isomaltase deficiency do well on diets low in sucrose.

Glucose-galactose malabsorption syndrome. This is a rare hereditary disorder caused by a defect in the brush border active transport system for glucose and galactose. Ingestion of glucose, galactose, or starch leads to flatulence and severe diarrhea. Fructose is well tolerated and can be fed to infants with this disorder. The brush border saccharidases are normal in this disease.

Some diagnostic procedures for carbohydrate intolerance. The most common diagnostic test is the **oral sugar tolerance test.** The patient is given an oral dose of the sugar in question, and the levels of that sugar in the patient's blood and feces are monitored. If the patient is intolerant of the administered sugar, diarrhea will ensue and that sugar will fail to appear in the blood, but it will appear in the feces. In suspected saccharidase deficiency the most definitive test is to take a biopsy sample of the jejunal mucosa and assay it for the deficient enzyme.

Digestion and Absorption of Proteins

The amount of dietary protein varies greatly among cultures and among individuals within a culture. In poor societies it is difficult for an adult to obtain the amount of protein (0.5 to 0.7 g/day/kg of body weight) required to balance normal catabolism of proteins, and it is still more difficult for children to get the relatively greater amounts of protein required to sustain normal growth. In wealthier societies, chiefly in industrially developed countries, a typical individual may ingest protein far in excess of the nutritional requirement.

The gastrointestinal tract must deal with the 10 to 30 g of protein per day contained in digestive secretions and a similar amount of protein in desquamated epithelial cells, in addition to the protein ingested.

In normal humans essentially all ingested protein is digested and absorbed. Most of the protein in digestive secretions and desquamated cells is also digested and absorbed. The small amount of protein in the feces is derived principally from colonic bacteria, desquamated cells, and proteins in mucous secretions of the colon. In humans, ingested protein is almost completely absorbed by the time the meal has traversed the jejunum.

Digestion of Proteins

Digestion in the stomach. Pepsinogen is secreted by the chief cells of the stomach and is converted by hydrogen ions to the active enzyme **pepsin.** The extent to which pepsin hydrolyzes dietary protein is significant but highly variable. At most, about 15% of dietary protein may be reduced to amino acids and small peptides by pepsin. The duodenum and small intestine have such a high capacity to process protein that the total absence of pepsin does not impair the digestion and absorption of dietary protein.

Digestion in the duodenum and small intestine. Proteases secreted by the pancreas play a major role in protein digestion. The most important of these proteases are **trypsin, chymotrypsin, carboxypeptidases A** and **B,** and **elastase.** The pancreatic juice contains these enzymes in inactive, proenzyme forms. The enzyme **enteropeptidase** (also called enterokinase), secreted by the mucosa of the duodenum and jejunum, converts trypsinogen to trypsin. Trypsin acts autocatalytically to activate trypsinogen and also converts the other proenzymes to the active enzymes (Fig. 40-7). The pancreatic proteases are present at high activities in the duodenum and rapidly convert dietary protein to small peptides. About 50% of the ingested protein is digested and absorbed in the duodenum.

The brush border of the duodenum and the small intestine contains a number of peptidases. These peptidases are integral membrane proteins whose active sites face the intestinal lumen. The brush border enzymes are richest in the proximal jejunum. They reduce the peptides produced by pancreatic proteases to oligopeptides and amino acids. The brush border peptidases include aminopeptidases (that cleave single amino acids from

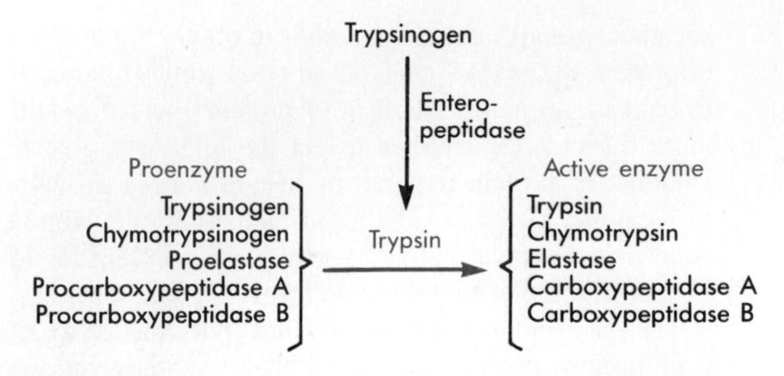

■ **Fig. 40-7** Conversion of inactive proenzymes of pancreatic juice to active enzymes by the action of trypsin. Trypsinogen of pancreatic juice is proteolytically converted to active trypsin by enteropeptidase (also known as enterokinase) secreted by the epithelial cells of duodenum and jejunum. Trypsin then activates the other proenzymes of pancreatic juice as shown.

the N-terminals of peptides), dipeptidases (that cleave dipeptides to amino acids), and dipeptidyl aminopeptidases (that cleave a dipeptide from the N-terminal end of a peptide). Fig. 40-8 illustrates some major proteases in the small intestine.

The principal products of protein digestion by pancreatic proteases and brush border peptidases are small peptides and amino acids. The small peptides (primarily dipeptides, tripeptides, and tetrapeptides) are about three or four times more concentrated than the single amino acids. As discussed below, small peptides and amino acids are transported across the brush border plasma membrane into intestinal epithelial cells. Small peptides are then hydrolyzed by peptidases in the cytosol of the epithelial cells; consequently, single amino acids and a few dipeptides appear in the portal blood. The cytosolic peptidases are more abundant than the brush border peptidases, and the cytosolic peptidases are particularly active against dipeptides and tripeptides (which are transported with high efficiency across the

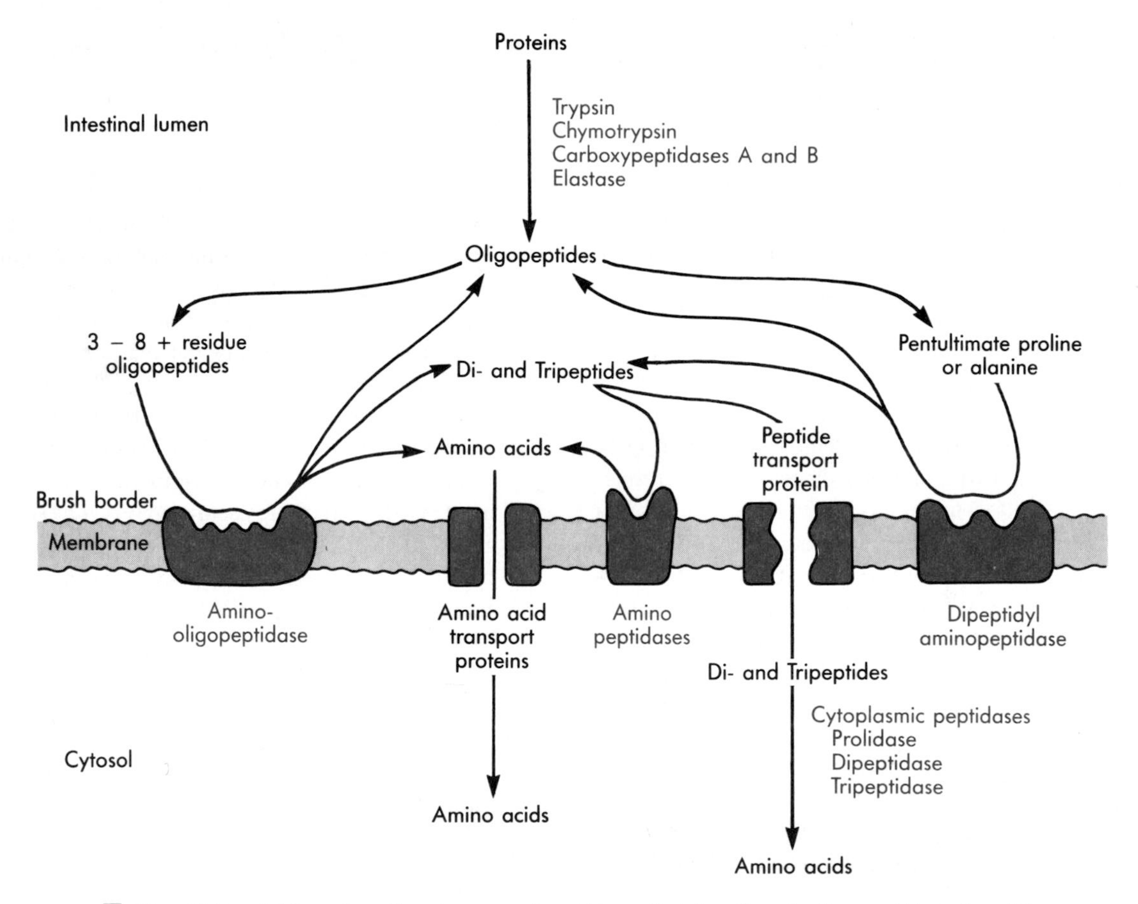

■ **Fig. 40-8** A hierarchy of proteases and peptidases that functions in the small intestine. The pancreatic proteases convert dietary proteins to oligopeptides. Brush border peptidases then convert the oligopeptides to amino acids (about 70%) and di- and tri-peptides (about 30%). The amino acids are taken up across the brush border membrane by amino acid transporters and the small peptides by a peptide transporter. In the cytosol of the enterocyte di- and tripeptides are cleaved to single amino acids. (Modified from Van Dyke, RW: Mechanisms of digestion and absorption of food. In Sleisenger, MH and Fordtran, JS, editors: *Gastrointestinal disease,* ed 4, Philadelphia, 1989, WB Saunders Co.)

brush border membrane). The brush border peptidases, on the other hand, are mainly active against peptides of four or more amino acids.

Absorption of the Products of Protein Digestion

Intact proteins and large peptides. These are not absorbed by humans to an extent that is nutritionally significant, but amounts sufficient to trigger an immunological response can be absorbed. M cells of the mucosal immune system take up small amounts of luminal proteins. In ruminants and rodents, but not in humans, the neonatal intestine has a high capacity for the specific absorption of immune globulins present in colostrum. This is vital in the development of normal immune competence in ruminants and rodents. Absorption takes place by receptor-mediated endocytosis.

Absorption of small peptides. Dipeptides and tripeptides are transported across the brush border membrane. The rate of transport of dipeptides or tripeptides usually exceeds the rate of transport of individual amino acids. In the experiments shown in Fig. 40-9, glycine was absorbed by the human jejunum less rapidly as the amino acid than it was from glycylglycine or from glycylglycylglycine. In this case the transport capacity of the jejunal mucosa was greater for the dipeptide and the tripeptide than for the amino acid; the apparent affinities for glycine and glycylglycine were similar.

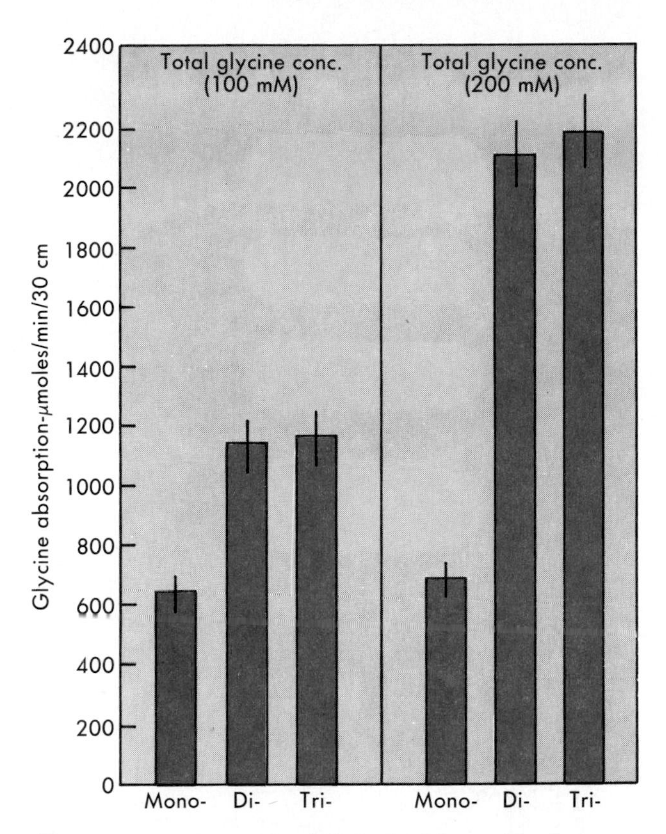

Fig. 40-9 Glycine is absorbed more rapidly by the human small intestine in the form of diglycine or triglycine than as free glycine. (From Adibi, SA and Morse, EL: *J Clin Invest* 60:1008, 1977.

A single membrane transport system with broad specificity is probably responsible for absorption of small peptides. The transport system has high affinity for dipeptides and tripeptides, but very low affinities for peptides of four or more amino acid residues. The transport system is stereospecific and prefers peptides of the physiological L-amino acids. The affinity is higher for peptides of amino acids with bulky side chains. Transport of dipeptides and tripeptides across the brush border plasma membrane is a secondary active transport process powered by the electrochemical potential difference of Na^+ across the membrane. The total amount of each amino acid that enters intestinal epithelial cells in the form of dipeptides or tripeptides is considerably greater than the amount that enters as the single amino acid.

Absorption of amino acids. With respect to amino acid transport properties, the brush border plasma membrane of small intestinal epithelial cells differs considerably from the basolateral plasma membrane. Normally amino acids are transported across the brush border plasma membrane into the enterocyte by way of certain specific amino acid transport systems. Transport of amino acids out of the epithelial cell across the basolateral membrane occurs by other transporters. Some of the transporters depend on the Na^+ gradient (as previously described for glucose and galactose absorption), whereas other transport systems are independent of Na^+. The Na^+-dependent brush border transport systems are found only in epithelial cells. Brush border membranes of the small intestine and proximal renal tubule are similar with respect to the nature of Na^+-dependent amino acid transport processes. The amino acid transporters in the brush border plasma membrane that do not depend on the Na^+ gradient are similar to amino acid transport systems that have been found in nonepithelial cells.

Other transport systems are responsible for transporting amino acids across the basolateral plasma membrane of the intestinal epithelial cell. As is true for the brush border membrane, some of the amino acid transporters present in the basolateral membrane depend on Na^+ and others do not. The basolateral membrane, however, is less highly differentiated than the brush border membrane, in that all of the amino acid transporters present in the basolateral membrane occur in certain nonepithelial cells. For most amino acids simple diffusion is a significant pathway across both brush border and basolateral membranes. The more hydrophobic the amino acid and the larger its concentration gradient across the membrane, the greater the importance of diffusion. Fig. 40-10 summarizes the processes responsi-

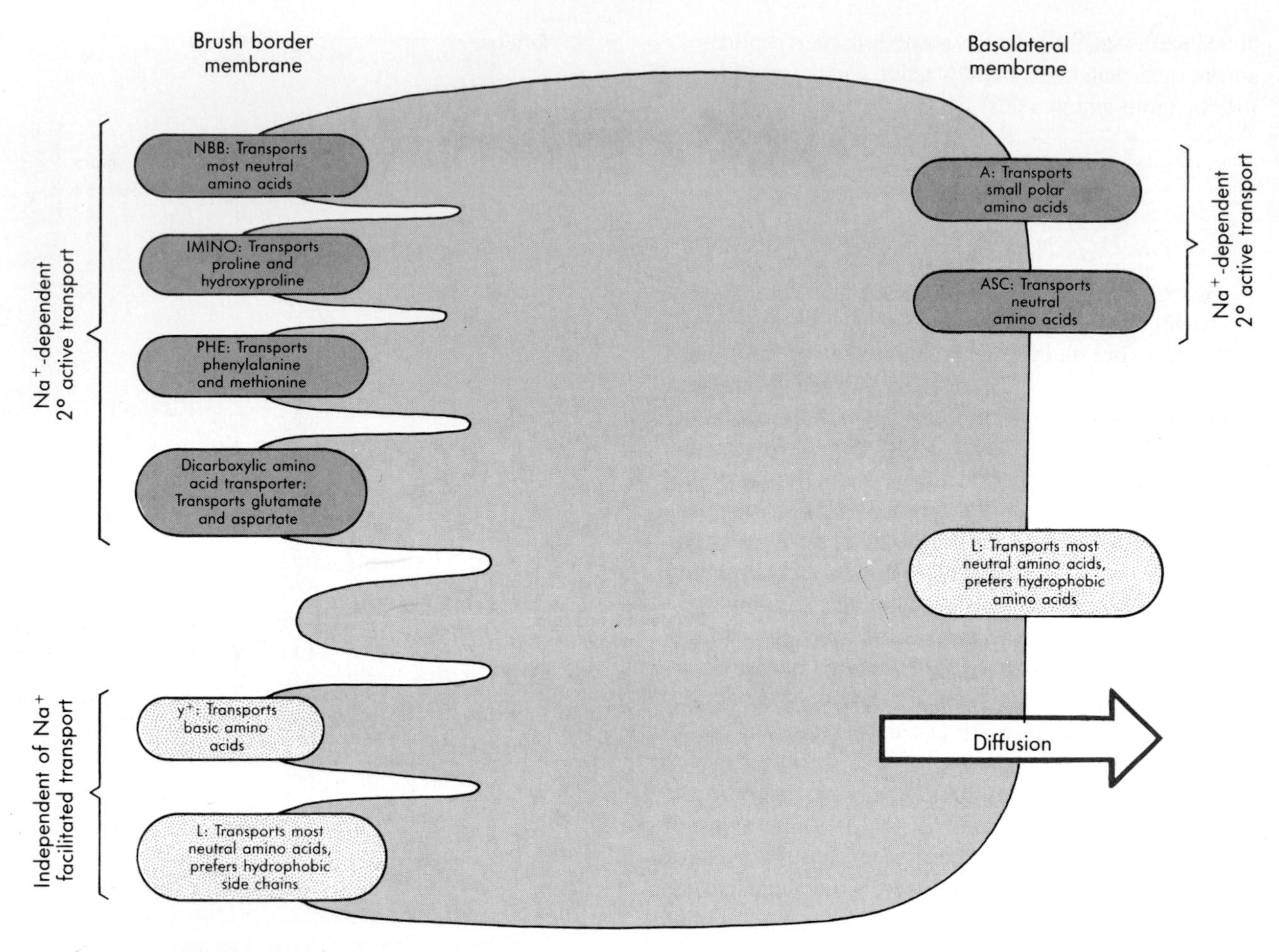

■ **Fig. 40-10** Amino acid transport proteins present in the epithelial cells of the small intestine. Transporters for basic and acidic amino acids and multiple transporters for neutral amino acids are present in the plasma membrane of the enterocyte. The transporters in the brush border membrane differ from those of the basolateral membrane. In addition, amino acids can diffuse at appreciable rates across both membranes; the more hydrophobic the amino acid, the more rapidly it can be transported by diffusion.

ble for amino acid transport in epithelial cells of the small intestine.

Brush border membranes. Three Na^+-dependent active transport systems for amino acids occur in brush border membranes. All three systems are present only in epithelial cell types, such as the epithelial cells of the proximal renal tubule. The system called **neutral brush border (NBB)** transports most of the neutral amino acids, both hydrophobic and hydrophilic. The system called PHE transports primarily phenylalanine and methionine. The IMINO acid transport system handles proline and hydroxyproline.

The L and y^+ systems present in the brush border do not depend on the Na^+. These systems have been described in nonepithelial cells. The L system is present in all eukaryotic cell types and transports neutral amino acids, with preference for those with hydrophobic side chains. The y^+ system transports basic amino acids such as lysine and arginine.

The basolateral membrane. Two Na^+-dependent systems transport neutral amino acids across the basolateral plasma membrane. The A system prefers small, hydrophilic amino acids, and the ASC system also prefers small neutral amino acids. The Na^+-independent L system (also present in brush border membranes) transports neutral amino acids across the basolateral membrane and has high affinity for hydrophobic amino acids. All of the transport systems present in the basolateral membrane also occur in nonepithelial cells. The basolateral membrane is more permeable to amino acids than is the brush border membrane. Therefore diffusion is a more important pathway for transport across the basolateral plasma membrane, especially for amino acids with hydrophobic side chains.

■ *Defects of Amino Acid Absorption*

Hartnup's disease. This rare hereditary disease involves defective renal transport of neutral amino acids.

Intestinal absorption of neutral amino acids is also decreased. Hartnup's disease may involve defects in the NBB system of the epithelial cells of the jejunum and the proximal renal tubule.

Prolinuria. This condition is rare and involves defective renal and intestinal reabsorption of proline. Prolinuria appears to be due to a defect in the IMINO system.

Because of the effective absorption of dipeptides and tripeptides by the intestinal epithelial cells, neither Hartnup's disease nor prolinuria results in malnutrition.

■ *Intestinal Absorption of Salts and Water*

Under normal circumstances humans absorb almost 99% of the water and ions presented to them in ingested food and in gastrointestinal secretions. Thus net fluxes of water and ions are normally from the lumen to the blood. In most cases the net fluxes of water and ions are the differences between much larger unidirectional fluxes from lumen to blood and from blood to lumen.

■ *Absorption of Water*

Typically, about 2 L of water is ingested each day and approximately 7 L/day is contained in gastrointestinal secretions. Only about 50 to 100 ml of water per day is lost in the feces. Thus the gastrointestinal tract typically absorbs almost 9 L/day.

Very little net absorption occurs in the duodenum, but the chyme is brought to isotonicity here. The chyme that is delivered from the stomach is often hypertonic. The action of digestive enzymes creates still more osmotic activity. The duodenum is highly water permeable, and very large fluxes of water occur from lumen to blood and from blood to lumen. Usually the net flux is from blood to lumen because of the hypertonicity of the chyme. *Large net water absorption occurs in the small intestine;* the *jejunum is more active than the ileum in absorbing water. The net absorption that occurs in the colon is relatively small, about 400 ml/day. However, the colon can absorb water against a larger osmotic pressure difference than can the rest of the gastrointestinal tract.* Fig. 40-11 summarizes the handling of water by the gastrointestinal tract.

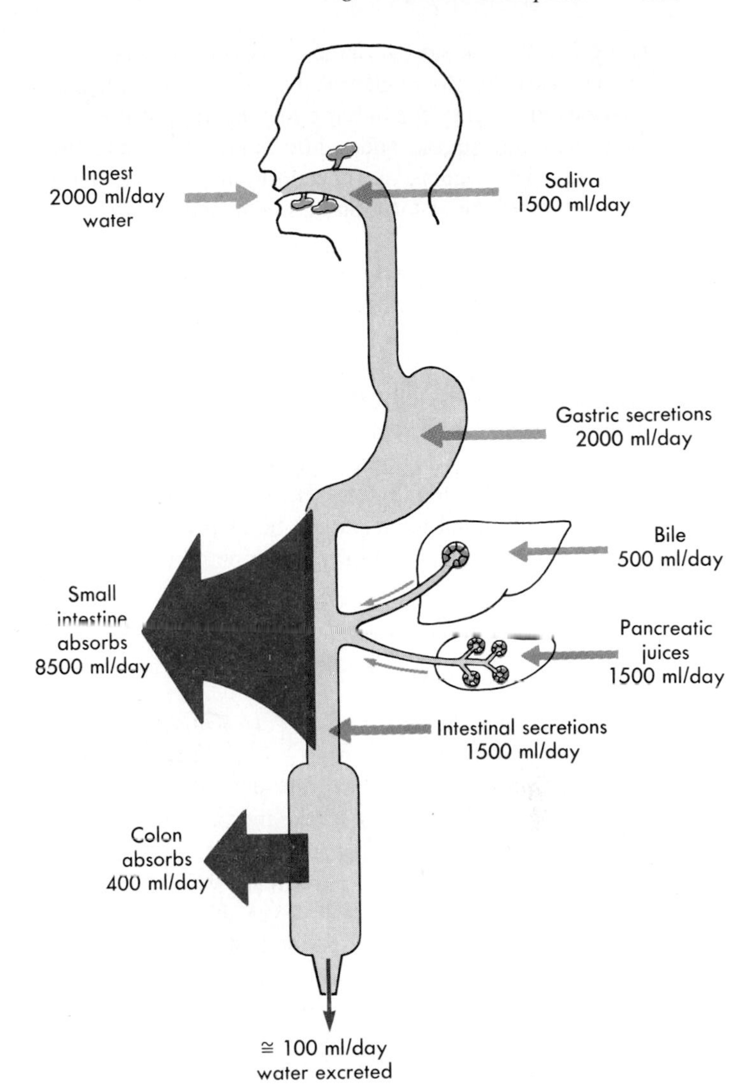

■ **Fig. 40-11** Overall fluid balance in the human gastrointestinal tract. About 2 L of water is ingested each day, and 7 L of various secretions enters the gastrointestinal tract. Of this total of 9 L, 8.5 is absorbed in the small intestine. About 500 ml is passed on to the colon, which normally absorbs 80% to 90% of the water presented to it.

■ *Absorption of Na^+*

Na^+ is absorbed along the entire length of the intestine. As is the case with water, net absorption is the result of large, unidirectional fluxes from blood to lumen and from lumen to blood. The unidirectional fluxes are greater in the proximal gut than in the distal intestine. The fluxes correlate with the greater brush border surface area per unit length in the jejunum, and this ratio diminishes toward the ileum and is still smaller in the colon. Na^+ crosses the brush border membrane down an electrochemical gradient, and it is actively extruded from the epithelial cells by the Na^+, K^+-ATPase in the basal and lateral plasma membrane. Normally the contents of the small bowel are isotonic to plasma. Luminal contents have about the same Na^+ concentration as plasma, so that Na^+ absorption normally takes place in the absence of a significant concentration gradient. Na^+ absorption is active, however, and can occur against a small electrochemical potential difference for Na^+.

In the jejunum the net rate of absorption of Na^+ is highest. Here Na^+ absorption is enhanced by the presence in the lumen of glucose, galactose, and neutral amino acids. These substances and Na^+ cross the brush border membrane on the same transport proteins. Na^+ moves down its electrochemical potential gradient and provides the energy for moving the sugars (glucose and galactose) and neutral and acidic amino acids into the epithelial cells against a concentration gradient. Thus Na^+ enhances the absorption of sugars and amino acids and vice versa.

In the ileum the net rate of Na^+ absorption is smaller. Na^+ absorption is only slightly stimulated by sugars and amino acids because the sugar and amino acid transport proteins are less concentrated in the ileum. The ileum can absorb Na^+ against a larger electrochemical potential than can the jejunum.

In the colon Na^+ is normally absorbed against a large electrochemical potential difference. Sodium concentrations in the luminal contents can be as low as 25 mM, compared with about 120 mM in the plasma.

Absorption of Cl^- and HCO_3^-

In the proximal duodenum HCO_3^- is secreted into the lumen. In the jejunum both Cl^- and HCO_3^- are absorbed in large amounts. By the end of the jejunum most of the HCO_3^- of the hepatic and pancreatic secretions has been absorbed. *In the ileum Cl^- is absorbed, but HCO_3^- is normally secreted.* If the HCO_3^- concentration in the lumen of the ileum exceeds about 45 mM, then the flux from lumen to blood exceeds that from blood to lumen, and net absorption occurs. *In the colon the transport of these ions is qualitatively similar to that in the ileum, in that Cl^- is absorbed and bicarbonate is usually secreted.*

Absorption of K^+

As with the other ions, the net movement of potassium across the intestinal epithelium is the difference between large unidirectional fluxes from lumen to blood and from blood to lumen. *In the jejunum and in the ileum the net flux is from lumen to blood.* As the volume of intestinal contents is reduced through the absorption of water, K^+ is concentrated, providing a driving force for the movement of K^+ across the intestinal mucosa and into the blood. Evidence for active transport of K^+ in the small intestine is lacking. *In the colon K^+ may be secreted or absorbed.* Net secretion occurs when the luminal concentration is less than about 25 mM, above 25 mM, net absorption occurs. Under most circumstances there is net secretion of K^+ in the colon; the secretory process is active.

Because most absorption of K^+ is a consequence of its enhanced concentration in the lumen caused by the absorption of water, significant K^+ loss may occur in diarrhea. If diarrhea is prolonged, the K^+ level in the extracellular fluid compartment of the body falls. Because of the importance of maintaining the normal K^+ level in the extracellular fluid compartment, especially for the heart and other muscles, life-threatening consequences such as cardiac dysrhythmias may ensue. Table 40-2 summarizes the transport of Na^+, K^+, Cl^-, and HCO_3^- in the small and large intestines.

Mechanisms of Salt and Water Absorption by the Intestine

The tight junctions. The epithelial cells that line the intestine are connected to their neighbors by tight junctions near their luminal surfaces. The tight junctions are leakiest in the duodenum, a bit tighter in the jejunum, still tighter in the ileum, and tightest in the colon.

Transcellular versus paracellular transport. Because the tight junctions are leaky, some fraction of the water and ions that traverse the intestinal epithelium passes between the epithelial cells, rather than passing through them. Transmucosal movement by passing through the tight junctions and the lateral intercellular spaces is called **paracellular transport.** Passage through the epithelial cells is termed **transcellular transport.**

Because the tight junctions in the duodenum are very leaky, significant proportions of the large unidirec-

Table 40-2 Transport of Na^+, K^+, Cl^-, and HCO_3^- in the small and large intestines

Segment of intestine	*Na^+*	*K^+*	*Cl^-*	*HCO_3^-*
Jejunum	Actively absorbed; absorption enhanced by sugars, neutral amino acids	Passively absorbed when concentration rises because of absorption of water	Absorbed	Absorbed
Ileum	Actively absorbed	Passively absorbed	Absorbed, some in exchange for HCO_3^-	Secreted, partly in exchange for Cl^-
Colon	Actively absorbed	Net secretion occurs when (K^+) concentration in lumen <25 mM	Absorbed, some in exchange for HCO_3^-	Secreted, partly in exchange for Cl^-

tional fluxes of water and ions that take place in the duodenum occur via the paracellular pathway. The proportions of water or a particular ion that pass through the transcellular and paracellular routes are determined by the relative permeabilities of the two pathways for the substance in question. Even in the ileum, where the junctions are much tighter than in the duodenum, the paracellular pathway contributes more to the total ionic conductance of the mucosa than does the transcellular pathway.

Villous versus crypt cells. The highly differentiated epithelial cells near the tips of the villi are specialized for absorption of water and ions, whereas the less differentiated cells in the crypts produce net secretion of water and ions.

Ion Transport by Intestinal Epithelial Cells

The movement of water across the intestinal epithelium is secondary to the movement of ions. Significant progress has recently been made in characterizing the ionic transport processes that occur in intestinal epithelial cells. Nevertheless, important issues remain unresolved.

Ion transport in the jejunum. The basolateral plasma membrane contains the Na^+, K^+-ATPase (Fig. 40-12). As a result of the active extrusion of Na^+ ions from the cytoplasm by the Na^+, K^+-ATPase, the electrochemical potential of Na^+ in the cytoplasm is much less than in the luminal fluid. Na^+ enters the epithelial cell by flowing across the brush border plasma membrane, moving down its large electrochemical potential gradient. Some of the Na^+ enters on the same transport proteins with sugars or amino acids, as described previously.

The luminal plasma membrane also contains a Na^+/H^+ exchange protein. Some of the energy released by Na^+ moving down its electrochemical potential gradient is harnessed to actively extrude H^+ into the lumen. Some of the H^+ in the lumen then reacts with HCO_3^- from bile and pancreatic juice to produce H_2CO_3, some of which then forms CO_2 and H_2O. The CO_2 diffuses readily across the epithelial cells and is carried away in the blood. Acidification of the luminal fluid appears to be the major mechanism for absorption of HCO_3^- in the jejunum.

The extrusion of H^+ from the cytoplasm creates a high cytoplasmic concentration of HCO_3^- in the epithelial cells. HCO_3^- leaves the cell by crossing the basolateral plasma membrane by facilitated transport.

The electrogenic effect of the Na^+, K^+-ATPase in the basolateral membrane and the electrogenic entry of Na^+ with sugars and amino acids at the luminal surface both tend to produce an electrical potential difference (lumen negative) across the epithelium. The electrical potential difference causes Cl^- to flow through the tight junctions into the lateral intercellular spaces. The tight junctions in the jejunum are leaky, so large fluxes of Cl^- occur by this route. The large paracellular flux of Cl^- almost short-circuits the electrogenic effects of Na^+ transport, and as a result the potential difference across the jejunum is only a few millivolts (lumen negative). The flow of Cl^- via the tight junctions appears to be a major route of Cl^- absorption in the jejunum. Some Cl^- flows through the cells, but the transport processes involved are not well characterized.

Note that the only primary active transport is by the

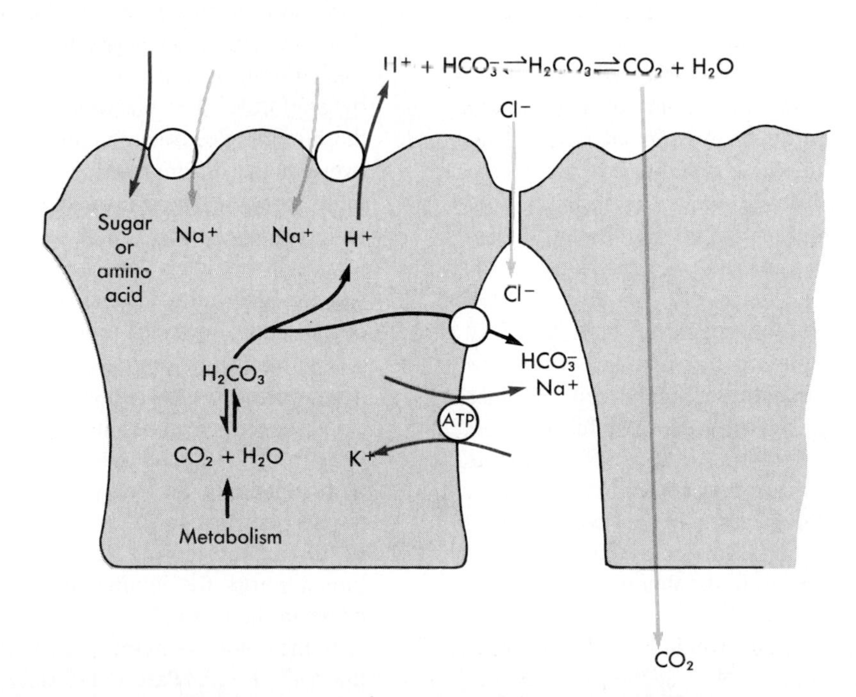

■ **Fig. 40-12** A summary of major ion transport processes that occur in the jejunum.

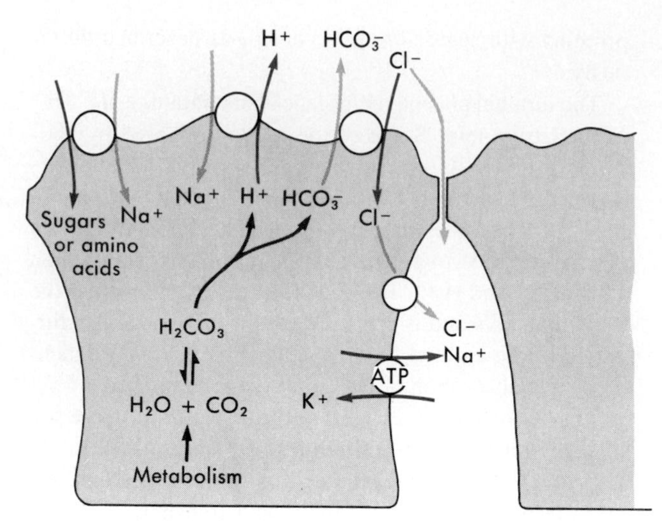

■ **Fig. 40-13** A summary of major ion transport processes that occur in the ileum.

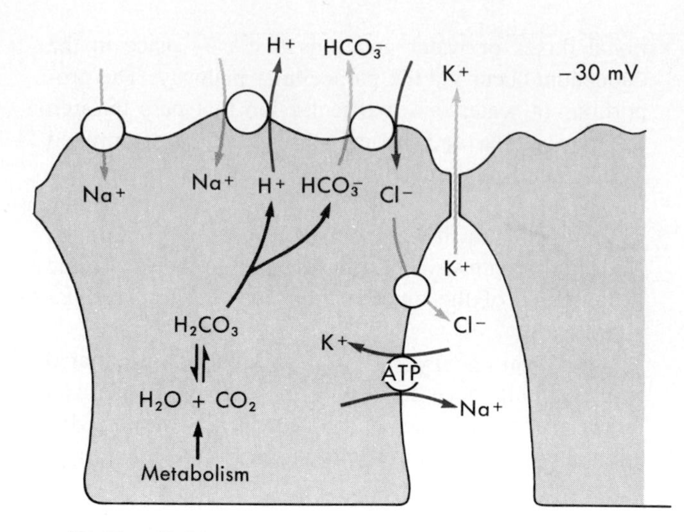

■ **Fig. 40-14** A summary of major ion transport processes that occur in the colon.

Na^+, K^+-ATPase. The resulting Na^+ gradient is responsible for the active extrusion of H^+. The extrusion of H^+ brings about absorption of HCO_3^-. The electrogenic effects of Na^+ transport create an electrical potential difference that powers Cl^- absorption in the jejunum.

Ion transport in the ileum. The ionic transport processes that occur in the ileum (Fig. 40-13) resemble those in the jejunum. Basolateral Na^+, K^+-ATPase, electrogenic Na^+ entry with sugars and amino acids (to a lesser extent than in the jejunum), and luminal Na^+/H^+ exchange are all present in the ileum.

The brush border plasma membrane of the ileum contains a Cl^-/HCO_3^- exchange protein that is not present in the jejunum. As in the jejunum, the active extrusion of H^+ into the lumen is driven by Na^+ entry. The extrusion of H^+ elevates the intracellular concentration of HCO_3^-. The HCO_3^- then flows down its electrochemical potential gradient into the lumen, and part of the energy liberated is used to bring Cl^- into the epithelial cell against its electrochemical potential gradient. Cl^- then leaves the cell at the basolateral membrane by facilitated transport.

The net effect of Na^+/H^+ exchange and the Cl^-/HCO_3^- exchange occurring together is the electroneutral entry of Na^+ and Cl^- into the epithelial cell. Because the extrusion of H^+ is driven by the downhill entry of Na^+, and the resulting downhill efflux of HCO_3^- drives the uphill entry of Cl^-, the energy for Cl^- entry comes, albeit indirectly, from the electrochemical potential gradient of Na^+. Thus the actual linkage of the ionic transport processes to metabolism in the ileum occurs via the Na^+, K^+-ATPase.

As in the jejunum, the electrogenic Na^+, K^+-ATPase and the electrogenic entry of Na^+ at the brush border cause a small electrical potential difference across the ileal mucosa (lumen negative). The electrical potential difference causes Cl^- to flow through the tight junctions into the intercellular spaces. The electrically driven flux of Cl^- in the ileum is smaller than it is in the jejunum because the ileal junctions are tighter than those in the jejunum.

Ion transport in the colon. Ion transport in the colon (Fig. 40-14) is characterized by luminal Na^+/H^+ and Cl^-/HCO_3^- exchange pumps and by basolateral Na^+, K^+-ATPase and facilitated Cl^- transport, as is the case in the ileum.

The Na^+/sugar and Na^+/amino acid cotransport systems are not present in the colon, but another process that mediates the electrogenic entry of Na^+ into the epithelial cells is present. The latter process is inhibited by amiloride. The electrogenic Na^+, K^+-ATPase and the luminal electrogenic entry of Na^+ cause an electrical potential difference (lumen negative) across the colonic mucosa. The transmucosal potential difference in the colon is about -30 mV, much larger than in the jejunum or the ileum. The greater electrical potential difference across the colonic mucosa is due partly to the decreased leakiness of tight junctions in the colon.

The tight junctions in all regions of the intestine are more permeable to cations than to anions. The large electronegativity in the lumen of the colon causes K^+ to flow from the intercellular spaces to the lumen via the tight junctions. This may be the major mechanism for the net secretion of K^+ that usually occurs in the colon. Facilitated transport of K^+ from cytoplasm to intestinal lumen across the luminal plasma membrane may also occur in the colon.

In the colon, as in the other segments of the intestine, the Na^+, K^+-ATPase is the only transport process that is directly linked to metabolic energy.

The Mechanism of Water Absorption

The absorption of water depends on the absorption of ions, principally Na^+ and Cl^-. Under normal circumstances, water absorption in the small intestine occurs in the absence of an osmotic pressure difference between the luminal contents and the blood in the intestinal capillaries. Water absorption by the colon typically proceeds against an osmotic pressure gradient. For a long time students of the gastrointestinal tract were puzzled by the absorption of water in the apparent absence of an osmotic pressure gradient to power the absorption.

Our current understanding is that water absorption occurs by a mechanism known as **standing gradient osmosis** (see Fig. 39-33). The major features of the standing gradient osmotic mechanism follow:

1. Active pumping of Na^+ into the lateral intercellular spaces by the Na^+, K^+-ATPase
2. Entry of Cl^- into the lateral intercellular spaces by flow from the lumen via the tight junctions or from the adjacent epithelial cells by facilitated transport
3. Presence of slightly hypertonic fluid near the luminal ends of the lateral intercellular spaces
4. Entry of water by osmosis into the lateral intercellular spaces
5. Hydrostatic flow of water and ions down the lateral intercellular spaces and across the epithelial basement membrane

Na^+, K^+-ATPase molecules are particularly concentrated in the basolateral plasma membrane that surrounds the luminal ends of the intercellular spaces. Because of the high rate of Na^+ pumping and the narrowness of the luminal ends of the intercellular spaces, the Na^+ concentration near the luminal ends of the intercellular spaces tends to reach levels above the luminal concentration. Chloride enters the intercellular spaces from the lumen via the tight junctions (driven by the transmucosal electrical potential difference) and/or from the adjacent epithelial cells by facilitated transport across the basolateral membrane.

The NaCl concentration in the luminal ends of the lateral intercellular spaces is high enough that the fluid there is hypertonic to the luminal contents and to the cytoplasm of the adjacent epithelial cells. Because of the hypertonicity in the intercellular space, water flows by osmosis into the intercellular space from the adjacent epithelial cells and from the lumen via the tight junctions. The inflow of water raises the hydrostatic pressure and dilates the intercellular channels. Fluid flows down the intercellular space because of the hydrostatic pressure gradient, and water and ions flow across the basement membrane of the epithelium, which offers little resistance to their passage. As the hypertonic fluid flows down the intercellular channels, water continues to enter the channels by osmosis from the adjacent epithelial cells. By the time the fluid reaches the basement membrane, it is essentially isotonic to the cytoplasm of the epithelial cells. In this way isotonic fluid is taken up at the brush border and is discharged through the basement membrane, to be carried away by the intestinal capillaries. The slight gradient of tonicity in the lateral intercellular space, hypertonic at the luminal end of the intercellular space and isotonic at the serosal end, gives the standing gradient mechanism of fluid absorption its name.

Because most water absorption takes place in the absence of a transmucosal osmotic pressure difference, the absorption of the end products of digestion, particularly sugars and amino acids, plays an important role in water absorption. The absorption of sugars and amino acids allows more water to be absorbed.

Secretion of Electrolytes and Water by Cells in Lieberkühn's Crypts

The normal net absorption of electrolytes and water by the small intestine is the result of large unidirectional fluxes from lumen to blood and from blood to lumen. *Mature epithelial cells near the tips of the villi are active in net absorption, whereas more immature cells in Lieberkühn's crypts function as net secretors of electrolytes and water.*

A current view of the ionic transport mechanisms that function in the crypt cells is shown in Fig. 40-15. Cl^- is actively taken up at the basolateral plasma membrane by the Na/K/2Cl cotransporter (see Chapter 1). This transporter uses the electrochemical potential difference of Na^+ to actively transport Cl^- and K^+ into the cell.

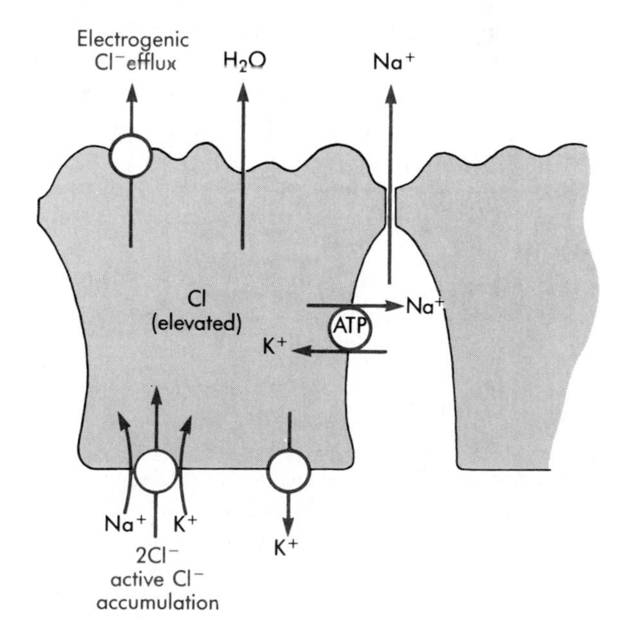

Fig. 40-15 The ion transport pathways involved in secretion of Cl^-, Na^+, and water by the epithelial cells in Lieberkühn's crypts in the small intestine.

Cl^- leaves the cell at the luminal membrane via an electrogenic Cl^- channel. Na^+ is transported into the lumen; the transport is driven by the net luminal electronegativity produced by the electrogenic Cl^- secretion into the lumen. Efflux of K^+ via a K^+ channel in the basolateral membrane prevents K^+ from accumulating in the cytosol of the crypt cell. This process maintains an electrical potential (cytosol negative) across the basolateral membrane that maintains the electrochemical driving force for entry of Na^+ (and thus for Cl^- entry as well).

The open time of the luminal Cl^- channel is enhanced by cyclic AMP. The basolateral K^+ channel is activated by Ca^{++}. Thus net secretion by the crypt cells is enhanced by agonists that elevate intracellular cyclic AMP (e.g., prostaglandins and vasoactive intestinal peptide and by Ca^{++}-mobilizing agonists (e.g., acetylcholine). The effects of agonists that elevate cyclic AMP are potentiated by agonists that increase cytosolic Ca^{++}, and vice versa.

Control of Intestinal Absorption of Electrolytes and Water

The ion transport activities of cells at the villous tip and in the crypts are subject to complex regulation by physiological and pharmacological substances. Hormones, paracrine agonists, and substances released from neurons in the gastrointestinal wall control the intestinal absorption and secretion of electrolytes and water. Some of the substances that influence intestinal transport of electrolytes and water are listed in the box below. In most cases the cellular mechanisms of action and the physiological significance of these regulatory substances remain to be fully elucidated.

The autonomic nervous system. Stimulation of sympathetic nerves to the small intestine or elevated levels of epinephrine in the circulation enhances the absorption of NaCl and water. Stimulation of parasympathetic nerves decreases net salt and water absorption.

The enteric nervous system. Agonists released by enteric neurons are important regulators of intestinal salt and water transport. Somatostatin and enkephalins are potent stimulators of net salt and water absorption by the small intestine. The physiological significance of regulation by somatostatin and enkephalins is not fully understood, but the large number of enteric neurons in the small intestine that contain these agonists (Table 38-2) supports the view that they are important in regulating salt and water transport.

The enteric immune system. Histamine and other substances released from mast cells in the wall of the intestine are potent stimulators of salt and water secretion in the small intestine. These agonists probably contribute to the secretory diarrhea that is characteristic of inflammatory diseases of the small intestine.

Adrenal hormones. Glucocorticoids stimulate the net absorption of salt and water in the small intestine. Aldosterone strongly enhances absorption of NaCl and water, and secretion of K^+, by the colon (also by the ileum to a lesser extent). Aldosterone increases the number of electrogenic Na^+ channels in the luminal plasma membrane of the colon (Fig. 40-14) and increases the density of Na^+, K^+-ATPase molecules in the basolateral membrane. Aldosterone has similar effects on the epithelial cells of the distal tubule of the kidney. The enhanced absorption of NaCl and water in the colon and the kidney induced by aldosterone is an important mechanism in the body's compensatory response to dehydration. Glucocorticoids also enhance salt and water absorption by increasing the density of Na^+, K^+-ATPases in the basolateral plasma membrane in the small intestine and the colon.

Endogenous substances that influence intestinal transport of electrolytes and water

Substances that stimulate absorption of salts and water	Substances that stimulate secretion of salts and water
α-Adrenergic agonists	Acetylcholine
Dopamine	Histamine
Enkephalins and other opioids	Prostaglandins
Somatostatin	Adenosine
Angiotensin	Substance P
Glucocorticoids	Serotonin
Neuropeptide Y	Neurotensin
Prolactin	VIP
	Secretin
	Motilin
	Gastrin
	GIP
	CCK
	Vasopressin

Pathophysiological Alterations of Salt and Water Absorption

The general causes of abnormalities in the absorption of salts and water include (1) deficiency of a normal ion transport system, (2) failure to absorb a nonelectrolyte normally with resultant osmotic diarrhea, (3) hypermotility of the intestine leading to abnormally rapid flow of intestinal contents past the absorptive epithelium; and (4) an enhanced rate of net secretion of water and electrolytes by the intestinal mucosa. Examples of each of these classes of abnormalities follow.

Failure to absorb an electrolyte normally. In **congenital chloride diarrhea,** ion transport and water absorption occur normally in the duodenum and jejunum, which normally lack the Cl^-/HCO_3^- exchanger. However, the Cl^-/HCO_3^- exchange transport system in the brush border plasma membrane of the ileum and colon is missing or grossly deficient. As a result, chloride absorption is severely impaired. This leads to diarrhea in which the stools contain an unusually high chloride concentration. In this disease the concentration of Cl^- in the stool exceeds the sum of the concentrations of Na^+ and K^+. The Na^+/H^+ exchange pump continues to operate, so that H^+ is eliminated in the feces without HCO_3^- to neutralize it. The net loss of H^+, with retention of HCO_3^-, contributes to a metabolic alkalosis.

Failure to absorb a nutrient normally. In any of the carbohydrate malabsorption syndromes the sugar that is retained in the lumen of the small intestine increases the osmotic pressure of the luminal contents. Water is retained as a result, and an increased volume of chyme is passed on to the colon. The increased volume flow may overwhelm the ability of the colon to absorb electrolytes and water, resulting in pronounced diarrhea. In addition, the high level of carbohydrates provides a medium that supports increased growth and metabolism of colonic bacteria. The increased production of CO_2 by colonic bacteria contributes to gassiness and borborygmi, and certain products of bacterial metabolism may inhibit absorption of electrolytes by the colonic epithelium.

Hypermotility of the intestine. The causes of hypermotility of the intestine are not well understood. Hypermotility of the small intestine may deliver electrolytes and water to the colon at faster rates than they can be absorbed by the colonic epithelial cells. Consequently, hypermotility of the colon may result in the elimination of feces before the maximum amount of salts and water can be extracted from them. Hypermotility may add to other factors that cause diarrhea. In cases of fat malabsorption, colonic bacteria metabolize lipids and produce certain waste products, such as hydroxylated fatty acids, that enhance the motility of the colon and inhibit salt and water absorption by the colonic epithelium.

Enhanced secretion of water and electrolytes. An increased secretion of water and electrolytes is probably the most important mechanism in serious diarrheal diseases. As mentioned previously, immature epithelial cells in Lieberkühn's crypts normally function to secrete Na^+, Cl^-, and H_2O. When the secretory activities of the crypt cells are elevated, the unidirectional secretory flux may exceed the unidirectional absorptive flux, so that net secretion prevails.

Cholera is perhaps the best understood type of **secretory diarrhea.** The toxin released by *Vibrio cholerae* in the intestine binds to receptors on the luminal surface of the crypt cells of the small intestine. The binding of cholera toxin to its receptors leads to irreversible activation of adenylyl cyclase in the plasma membranes of these cells. The resulting increase in cAMP greatly increases the secretion of Na^+, Cl^-, and water by the cells, by increasing the conductance of the luminal membrane to Cl^-, as discussed previously.

Other agents that elevate cAMP in intestinal epithelial cells also lead to secretion of water and electrolytes. Vasoactive intestinal polypeptide, which is present in certain enteric neurons and also circulates as a hormone, acts in this way. Certain individuals with islet cell tumors of the pancreas suffer from a watery diarrhea known as **pancreatic cholera.** In this disorder plasma levels of vasoactive intestinal polypeptide are elevated, and this may be the cause of the secretory diarrhea. Certain prostaglandins augment secretion by elevating cAMP in crypt cells.

The epithelial cells of the crypts of the small intestine are also stimulated to secrete electrolytes and water by elevated intracellular Ca^{++} levels. Acetylcholine, serotonin, substance P, and neurotensin may elicit intestinal secretion of water and electrolytes by increasing intracellular Ca^{++}. All of these agents are present in intrinsic neurons of the intestinal wall, and thus they may contribute to diarrhea in certain pathological situations.

Absorption of Calcium

Calcium ions are actively absorbed by all segments of the intestine. The duodenum and jejunum are especially active and can concentrate Ca^{++} against a greater than tenfold concentration gradient. The rate of absorption of Ca^{++} is much greater than that of any other divalent ion, but still 50 times slower than Na^+ absorption.

The ability of the intestine to absorb Ca^{++} is regulated. Animals receiving a calcium-deficient diet increase their ability to absorb Ca^{++}. Animals receiving high-calcium diets are less able to absorb Ca^{++} Intestinal absorption of Ca^{++} is stimulated by vitamin D and slightly stimulated by parathyroid hormone by a mechanism that is not yet fully understood.

Cellular mechanism of calcium absorption

The brush border membrane. A current view of the cellular mechanism of Ca^{++} absorption by the epithelial cells of the small intestine is shown in Fig. 40-16.

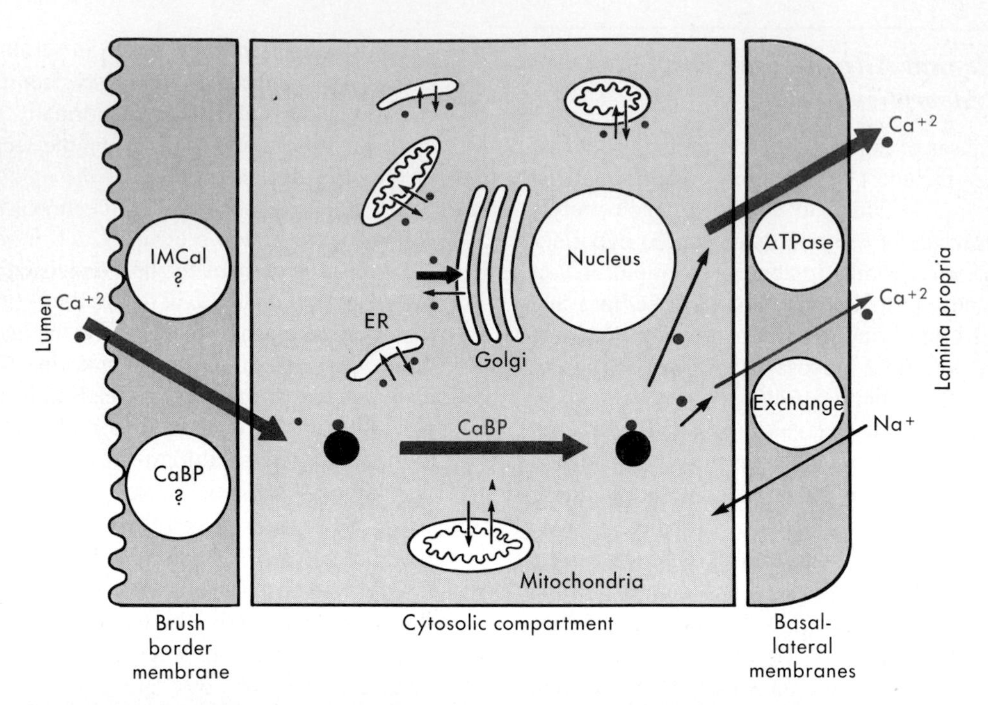

■ **Fig. 40-16** Cellular mechanisms of Ca^{++} absorption in the small intestine. Ca^{++} crosses the brush border plasma membrane by pathways that are poorly characterized, but may involve intestinal membrane calcium-binding protein (IMCal) or calcium-binding protein (CaBP). In the cytosol of the enterocyte Ca^{++} is bound to a soluble CaBP. Ca^{++} is extruded across the basolateral membrane by a Ca^{++}-ATPase and a Na^{+}/Ca^{++} exchange mechanism. Colored dots represent Ca^{++}; large black dots represent cytosolic Ca^{++}-binding protein. (From Wasserman, RH, and Fullmer, CS: *Ann Rev Physiol* 45:375, 1983.)

Ca^{++} moves down its electrochemical potential gradient across the brush border membrane into the cytosol. An integral protein of the brush border plasma membrane is called the intestinal membrane calcium-binding protein (IMCal), and it may function as a membrane transporter for Ca^{++}. Some of the cytosolic calcium-binding protein (CaBP, also called **calbindin**) is associated with the brush border membrane; its function in the brush border membrane has not been established.

Epithelial cell cytosol. The cytosol of the intestinal epithelial cells contains CaBP. In mammals the CaBP has a molecular weight of about 10,000 and binds two calcium ions with high affinity. As discussed below, the level of CaBP in the epithelial cells correlates well with the capacity to absorb Ca^{++}. It has been proposed that CaBP allows large amounts of Ca^{++} to traverse the cytosol, while avoiding concentrations of free Ca^{++} high enough to form insoluble salts with intracellular anions. CaBP appears to be an essential component of Ca^{++} absorption. Free Ca^{++} and Ca^{++} bound to CaBP are in dynamic exchange with Ca^{++} in mitochondria and endoplasmic reticulum, but the role of these organelles in Ca^{++} absorption is not clear.

The basolateral membrane. The basolateral plasma membrane contains two transport proteins capable of ejecting Ca^{++} from the cell against its electrochemical potential gradient. A Ca^{++}-ATPase in the basolateral membrane is a primary active transport protein that splits ATP and uses the energy to transport Ca^{++}. The Na^{+}/Ca^{++} exchanger present in the basolateral membrane uses the energy of the Na^{+} gradient to extrude Ca^{++} by secondary active transport. The Na^{+}/Ca^{++} exchanger is more effective at high levels of free intracellular Ca^{++}, whereas at low levels of free intracellular Ca^{++} the Ca^{++}-ATPase is the major mechanism for Ca^{++} extrusion. Ca^{++} itself stimulates the activity of the Ca^{++}-ATPase by binding to calmodulin. The calcium-calmodulin complex then stimulates a protein kinase, which phosphorylates the Ca^{++}-ATPase, thereby enhancing its enzymatic and transport activities.

Actions of vitamin D. Vitamin D is essential for normal levels of calcium absorption by the intestine. In rickets, a disease caused by vitamin D deficiency, the rate of absorption of Ca^{++} is very low. Fig. 40-17 illustrates the effects of administration of vitamin D to chicks with rickets.

The actions of vitamin D are discussed in Chapter 47. Vitamin D may stimulate each phase of absorption of Ca^{++} by the epithelium of the small intestine: passage across the brush border membrane, traversal of the cytosol, and active extrusion across the basolateral membrane. Vitamin D, like other steroid hormones, exerts its major effects by binding to nuclear receptors and

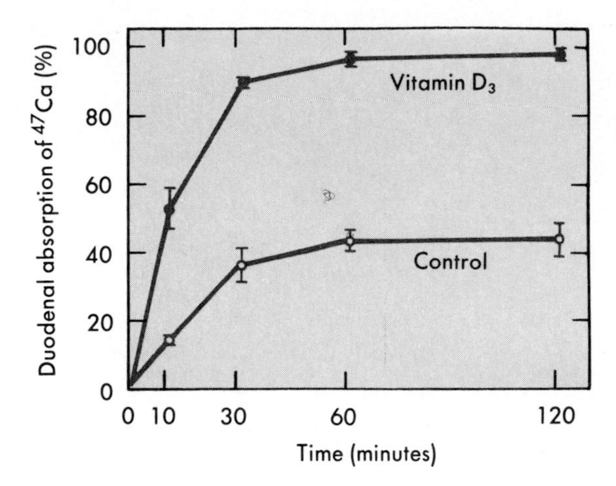

■ **Fig. 40-17** Effects of vitamin D_3 on absorption of Ca^{++} by chick duodenum. Control animals were fed a diet deficient in vitamin D (lower curve). Upper curve, Ca^{++} absorption by animals fed the same diet, but administered vitamin D_3 24 hours before the experiment. (Redrawn from Wasserman, RH: J *Nutrition* 77:69, 1962.)

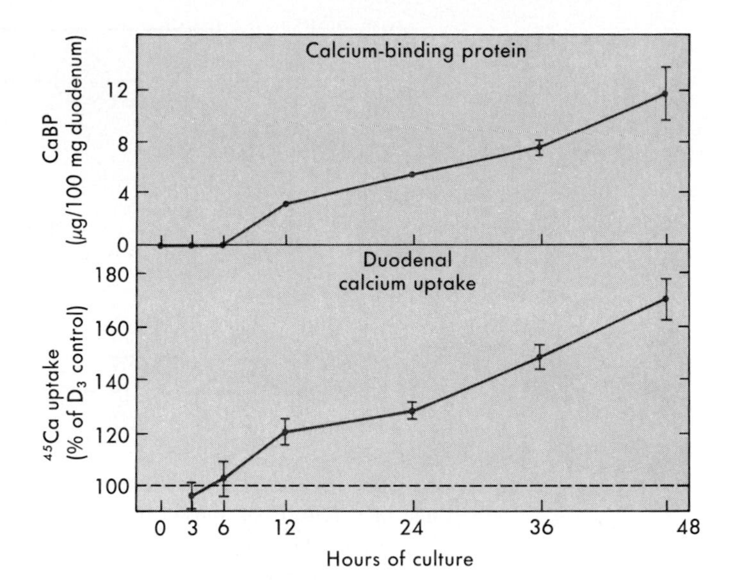

■ **Fig. 40-18** The correlation between the levels of CaBP and the rate of calcium uptake by chick duodenum in organ culture. At time zero, vitamin D_3 was added to the culture medium. (Redrawn from Corradino, RA: *Endocrinology* 94:1607, 1974.)

stimulating the synthesis of messenger RNA that codes for particular proteins. Vitamin D_3 induces the synthesis of the cytosolic CaBP (Fig. 40-18). The CaBP level correlates well with the capacity of the small intestine to absorb Ca^{++}. Vitamin D also increases the level of the intestinal membrane CaBP that may be involved in transporting Ca^{++} across the brush border plasma membrane. In addition, vitamin D increases the level of the basolateral Ca^{++}-ATPase that actively pumps calcium out of the enterocyte.

■ *Absorption of Iron*

A typical adult in Western societies ingests about 15 to 20 mg of iron daily. Of the 15 to 20 mg ingested, only 0.5 to 1 mg is absorbed by normal adult men and 1 to 1.5 mg is absorbed by premenopausal adult women. Iron depletion, caused by hemorrhage for example, results in increased iron absorption. Growing children and pregnant women also absorb increased amounts of iron.

Iron absorption is limited because iron tends to form insoluble salts, such as hydroxide, phosphate, and bicarbonate, with anions that are present in intestinal secretions. Iron also tends to form insoluble complexes with other substances commonly present in food, such as phytate, tannins, and the fiber of cereal grains. These iron complexes are more soluble at low pH. Therefore HCl secreted by the stomach enhances iron absorption, whereas iron absorption is commonly low in individuals deficient in acid secretion. Ascorbate effectively promotes iron absorption. Ascorbate forms a soluble complex with iron, preventing it from forming insoluble complexes, and ascorbate reduces Fe^{3+} to Fe^{2+}. Fe^{2+} has much less tendency to form insoluble complexes than Fe^{3+} and, for this reason, Fe^{2+} is absorbed much better.

Heme iron is relatively well absorbed; about 20% of the heme ingested is absorbed. Proteolytic enzymes release heme groups from proteins in the intestinal lumen. Heme is probably taken up by facilitated transport by the epithelial cells that line the upper small intestine. In the epithelial cell, iron is split from the heme by reactions involving xanthine oxidase. No intact heme is transported into the portal blood.

Cellular mechanism of iron absorption. One current view of iron absorption is depicted in Fig. 40-19. The epithelial cells of the duodenum and jejunum release an iron-binding protein into the lumen. The protein is called transferrin, and it is very similar, but not identical, to the transferrin that is the principal iron-binding protein of plasma. In the lumen of the duodenum and jejunum, transferrin binds iron; two iron ions can be bound by each transferrin molecule. Receptors on the brush border surface of the duodenum and jejunum bind the transferrin-iron complex, and the complex is taken up into the epithelial cell, probably by receptor-mediated endocytosis. Endocytic uptake of iron-transferrin by hepatocytes and reticulocytes also has been demonstrated. In the cytosol of the intestinal epithelial cell, transferrin apparently acts as a soluble iron carrier. Much of the transferrin, after it releases its bound iron, is resecreted into the lumen. Much of the brush border receptor for the iron-transferrin complex is recycled back into the brush border membrane.

Iron ultimately appears in plasma bound to plasma transferrin. Postulated steps that occur between uptake

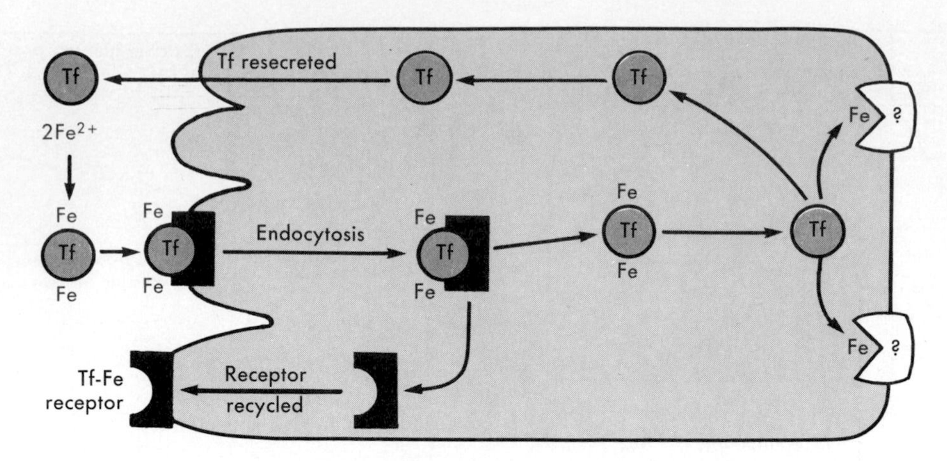

Fig. 40-19 A current view of the mechanism of iron absorption by the epithelial cells of the small intestine. A form of transferrin (Tf) is secreted into the lumen, where it binds Fe^{++}. The Fe_2-Tf complex is taken up by receptor-mediated endocytosis and the receptor and the Tf are recycled. Iron is transported across the basolateral membrane by a poorly understood process and appears in portal blood bound to the plasma form of transferrin.

by the enterocyte of the transferrin-iron complex and the appearance of iron in the plasma include release of iron from transferrin in the enterocyte, transport of iron across the basolateral membrane, and binding of iron to plasma transferrin. These processes are poorly understood. Lysosomes may play a role in release of iron from the secreted form of transferrin. The transport of iron across the basolateral membrane requires metabolic energy but is otherwise undefined.

Regulation of iron absorption. Iron absorption is regulated in accordance with the body's need for iron. In chronic iron deficiency or after hemorrhage, the capacity of the duodenum and jejunum to absorb iron is elevated. The intestine also protects the body from the consequences of absorbing too much iron. The excretion of iron is limited, and thus the absorption of more iron than is needed may lead to iron overload. Iron overload can result from chronic ingestion of large amounts of absorbable iron. This condition is common in certain African tribes that regularly consume a home-brewed beer high in iron content. In the genetic disease called **idiopathic hemochromatosis,** an excessive amount of iron is absorbed from a diet that is normal in iron content.

An important mechanism for preventing excess absorption of iron is the almost irreversible binding of

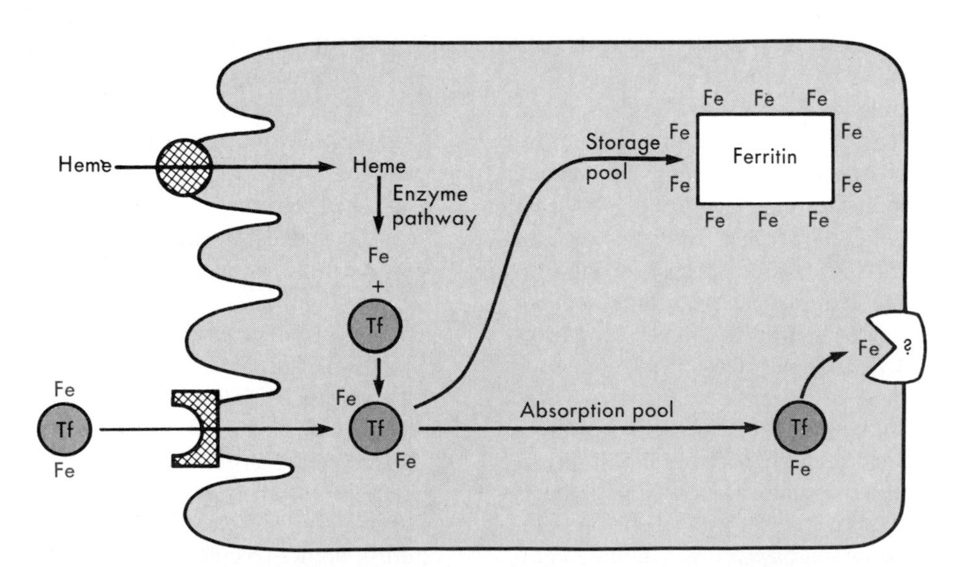

Fig. 40-20 In the enterocytes of the small intestine iron bound to transferrin (Tf) is available for transport across the basolateral membrane. Iron bound to ferritin is, however, unavailable for absorption and is lost into the lumen when the cell desquamates.

iron to ferritin in the intestinal epithelial cell. Iron bound to ferritin is not available for transport into the plasma (Fig. 40-20), but it is lost into the intestinal lumen and excreted in the feces when the intestinal epithelial cell desquamates. The amount of apoferritin present in the intestinal epithelial cells may determine how much iron can be trapped in this nonabsorbable pool. The synthesis of apoferritin is stimulated at the translational level by iron, and this protects against absorption of excessive amounts of iron.

The capacity of the duodenum and jejunum to absorb iron increases after a hemorrhage, with a time lag of 3 to 4 days. The intestinal epithelial cells require this time to migrate from their sites of formation in Lieberkühn's crypts to the tips of the villi, where they are most involved in absorptive activities. The iron-absorbing capacity of the epithelial cells appears to be programmed when the cells are in Lieberkühn's crypts. The mechanisms that alter the capacity of the epithelial cells to absorb iron are poorly characterized. Iron absorption by duodenal and jejunal epithelial cells is subject to multiple controls. The brush border membranes of the duodenum and jejunum of an iron-deficient animal have an increased number of receptors for the complex of iron with transferrin and thus absorb the iron-transferrin complex from the lumen more rapidly. Other regulatory factors are probably also involved.

Absorption of Other Ions

Magnesium. Magnesium is absorbed along the entire length of the small intestine, with about half the normal dietary intake being absorbed. The regulation of the rate of Mg^{++} absorption is poorly understood.

Phosphate. Phosphate is also absorbed all along the small intestine. Some phosphate may be absorbed by active transport. Little is known about the regulation of phosphate absorption.

Copper. Copper is absorbed in the jejunum, with approximately 50% of the ingested load being absorbed. Copper is secreted in the bile bound to certain bile acids, and this copper is lost in the feces. In individuals who fail to secrete sufficient amounts of copper in the bile, the body's copper pool grows and copper accumulates in certain tissues. The mechanisms by which absorption and excretion of copper are regulated are poorly characterized.

Absorption of Water-Soluble Vitamins

Most water-soluble vitamins can be absorbed by simple diffusion if they are taken in sufficiently high doses. Nevertheless, specific transport mechanisms play important roles in the normal absorption of most water-soluble vitamins. Table 40-3 summarizes current knowledge of these transport mechanisms.

Ascorbic acid (vitamin C). Vitamin C is absorbed in the proximal ileum by active transport. The active transport of ascorbate depends on the presence of Na^+ in the lumen. Na^+ and ascorbate are probably cotransported into the cell and the energy for ascorbate transport comes from the energy in the electrochemical potential gradient of Na^+.

Biotin. Biotin is actively taken up by the epithelial cells of the upper small intestine by a mechanism that depends on Na^+ in the lumen.

Folic acid (pteroylglutamic acid). Folic acid is absorbed by carrier-mediated transport (perhaps by active transport) in the jejunum. Pteroylpolyglutamates apparently share this transport pathway. The polyglutamates are cleaved intracellularly to the monoglutamate derivative, which is transported into the blood. Another dietary source of folic acid, 5-methyltetrahydrofolate, is absorbed by simple diffusion.

Nicotinic acid. Nicotinic acid is absorbed by the jejunum partly by an Na^+-dependent, saturable mechanism. The details of nicotinic acid absorption remain to be elucidated.

Pyridoxine (vitamin B_6). Evidence favors simple diffusion as the mechanism of pyridoxine absorption.

Riboflavin (vitamin B_2). Riboflavin is absorbed in the proximal small intestine, probably by facilitated transport. The presence of bile acids enhances the absorption of riboflavin by an unknown mechanism.

Thiamin (vitamin B_1). Thiamin is absorbed by an Na^+-dependent active transport mechanism in the jejunum. Some thiamin is phosphorylated in the jejunal epithelial cells, but the thiamin that appears in the blood is primarily free thiamin.

Vitamin B_{12}. A specific active transport process also has been implicated in the absorption of vitamin B_{12}. The four physiologically important forms of B_{12} are cyanocobalamin, hydroxycobalamin, deoxyadenosylcobalamin, and methylcobalamin. In the absence of vitamin B_{12} the maturation of red blood cells is retarded, and **pernicious anemia** ensues. Because of its medical importance, a good deal of attention has been paid to the absorption of vitamin B_{12}. The dietary requirement for B_{12} is fairly close to the maximal absorption capacity for the vitamin. Enteric bacteria synthesize vitamin B_{12} and other B vitamins, but the colonic epithelium lacks specific mechanisms for their absorption.

Storage in liver. The liver contains a large store of vitamin B_{12} (2 to 5 mg). Vitamin B_{12} is normally present in the bile (0.5 to 5 μg daily), but about 70% of this is normally reabsorbed. Because only about 0.1% of the store is lost daily, even if absorption totally ceases the store will last for 3 to 6 years.

Gastric phase. Most of the cobalamins present in food are bound to proteins. The low pH in the stomach and the digestion of proteins by pepsin release free co-

Table 40-3 Properties of transport of water-soluble vitamins in small intestine

Vitamin	MW	Species	Site of absorption in intestine	Process involved	K_m	Other properties*
L-Ascorbic acid	176	Human	Ileum	Active transport		1
		Guinea pig	Ileum	Active transport	0.3 mM	1
Dehydroascorbic acid	174	Guinea pig	Ileum	Facilitated diffusion and intracellular reduction	0.2 mM	4
Biotin	244	Hamster, rat, mouse, squirrel, chipmunk	Proximal	Facilitated diffusion	1-10 μM	1
Choline	121	Guinea pig	Jejunum	Facilitated diffusion	0.6 mM	4
		Chick	?	Facilitated diffusion	0.1 mM	1
		Hamster	Ileum	Facilitated diffusion	0.2 mM	
Folic acid (pteroylglutamic acid)	441	Rat	Jejunum	Facilitated diffusion (active transport?)	6.1 μM	1, 4
	Rabbit	Jejunum	Folate-OH^- exchange	0.2 μM		
Inositol	180	Hamster	Small intestine	Active transport	0.1 mM	1
		Chicken	Small intestine	Facilitated diffusion	0.1 mM	1
Nicotinic acid	123	Bullfrog	Proximal	Active transport		1
Nicotinamide	122	Chicken	Proximal	Intracellular metabolism	†	4
Pantothenic acid	219	Rat, chicken	Small intestine	Active transport	17 μM	1
Riboflavin	376	Human	Proximal	Facilitated diffusion		2
		Rat	Jejunum	Active transport	0.4 μM	1, 3
Thiamine	337	Rat	Jejunum	Active transport	0.2-0.6 μM	1, 3
Vitamin B_6	206	Rat, hamster, human	Proximal	Intracellular metabolism		3, 4

From Rose RC: *Intestinal transport of water-soluble vitamins.* In *Handbook of Physiology, section 6: the gastrointestinal system,* volume IV, Bethesda, Md, 1991, American Physiological Society.
MW, molecular weight
*1, Na^+ dependent; 2, bile salt dependent; 3, phosphorylated; 4, metabolized other than by phosphorylation.
†None demonstrated.

balamins. The free cobalamins are rapidly bound to a number of cobalamin-binding glycoproteins known as R proteins. R proteins are present in saliva and in gastric juice and bind cobalamins tightly over a wide pH range. The R proteins of saliva and gastric juice have molecular weights near 60,000 and are closely related to transcobalamins I, II, and III.

Intrinsic factor (IF) is a cobalamin-binding protein that is secreted by the gastric parietal cells. IF is a glycoprotein with 15% carbohydrate and a molecular weight of about 45,000. The rate of IF secretion usually parallels the rate of HCl secretion. IF binds cobalamins with less affinity than the R proteins, so in the stomach most of the cobalamin present in food is bound to R proteins.

Intestinal phase. Pancreatic proteases begin degradation of the complexes between R proteins and cobalamins. This degradation greatly lowers the affinities of the R proteins for cobalamins, so that cobalamins are transferred to IF. IF and IF-cobalamin complexes are very resistant to digestion by pancreatic proteases. As described later, the normal mechanism for absorption involves brush border receptors for IF-cobalamin complexes. These receptors do not recognize the R protein–cobalamin complexes. Thus in pancreatic insufficiency, when R proteins are not degraded, cobalamins remain bound to R proteins. These complexes are not available for absorption, and therefore vitamin B_{12} deficiency may ensue.

Absorption of vitamin B_{12}. Fig. 40-21 summarizes a current view of the mechanism of vitamin B_{12} absorption. The normal absorption of cobalamins depends on the presence of IF. When IF binds vitamin B_{12} the IF undergoes a conformational change that favors the formation of dimers; each dimer binds two vitamin B_{12} molecules. The brush border plasma membranes of the epithelial cells of the ileum contain a receptor protein that recognizes and binds the IF-B_{12} dimer. Free IF does not compete for binding and the receptor does not recognize free cobalamins.

Binding to the receptor is required for uptake of vitamin B_{12} into the cell. It is not known whether the IF-B_{12} complex or B_{12} alone enters the ileal epithelial cell. The most convincing evidence favors the hypothesis that B_{12} is split from IF extracellularly and that free B_{12} enters the cell by an active transport mechanism.

After the binding of the IF-B_{12} complex to the ileal receptors, B_{12} is slowly transported through the epithelial cell and into the blood. Vitamin B_{12} does not appear in the blood until 4 hours after it is fed, and the

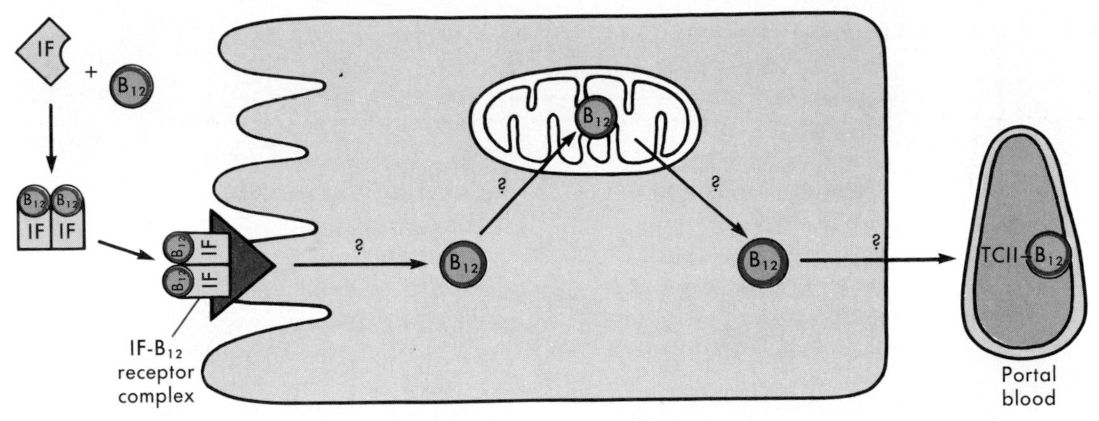

■ **Fig. 40-21** Mechanism of vitamin B_{12} absorption by epithelial cells of the ileum.

peak B_{12} level in plasma occurs 6 to 8 hours after feeding. The reason for this delay is not well understood, but some evidence suggests that for much of the lag period B_{12} is located predominantly in the mitochondria of the epithelial cells and that some conversion of cobalamin to deoxyadenosylcobalamin occurs there.

The exit of vitamin B_{12} from the cells of the ileal epithelium is even less understood than its entry. Facilitated or active transport is presumably involved. Most of the B_{12} absorbed appears in the portal blood bound to transcobalamin II, a globulin. Transcobalamin II is synthesized in the liver, but the ileal epithelium also makes this protein. The transcobalamin II–B_{12} complex is rapidly cleared from the portal blood by the liver by receptor-mediated endocytosis. Other tissues that take up the transcobalamin-B_{12} complex include the kidney, spleen, heart, and placenta. Reticulocytes and fibroblasts also take up the complex.

Absorption in the absence of IF. In the complete absence of IF, about 1% to 2% of an ingested load of B_{12} will be absorbed. If massive doses of B_{12} are taken (about 1 mg/day), enough can be absorbed to treat pernicious anemia. The IF-independent mechanism shows no maximal absorptive capacity, does not appear to be limited to the ileum, and shows a much shorter lag time (about 1 hour) than IF-dependent absorption.

Pernicious anemia and other diseases involving the malabsorption of vitamin B_{12}. In the absence of sufficient levels of B_{12} the maturation of red cells is retarded, and anemia results. Pernicious anemia is due to atrophy of the gastric mucosa, with almost complete inability to secrete HCl, pepsin, and IF. Most patients with pernicious anemia have serum antibodies against parietal cells. However, it is not clear whether the antibodies cause the disease or are the response to gastric damage from some other cause.

Pernicious anemia in childhood is rare and has three forms: (1) an autoimmune type of pernicious anemia that has the characteristics just described; (2) congenital IF deficiency, in which pepsin and acid secretion are normal but IF secretion is deficient; and (3) congenital B_{12} malabsorption syndrome, in which there is normal gastric function and normal levels of IF are secreted, but B_{12} absorption is deficient because of a defect in the ileal IF-B_{12} receptors.

■ *Digestion and Absorption of Lipids*

The primary lipids of a normal diet are triglycerides. The diet contains smaller amounts of sterols, sterol esters, and phospholipids (Fig. 40-22). Because lipids are only slightly soluble in water, they pose special problems to the gastrointestinal tract at every stage of their processing. In the stomach, lipids tend to separate out into an oily phase. In the duodenum and small intestine, lipids are emulsified with the aid of bile acids. The large surface area of the emulsion droplets allows access of the water-soluble lipolytic enzymes to their substrates. The digestion products of lipids form small molecular aggregates, known as **micelles**, with the bile acids. The micelles are small enough to diffuse among the microvilli and allow absorption of the lipids from molecular solution at the intestinal brush border. The digestion and absorption of lipids are more complex than for any other class of nutrients and are thus more frequently subject to malfunction.

■ *In the Stomach*

Because fats tend to separate out into an oily phase, they tend to be emptied from the stomach later than the other gastric contents. Despite the presence of gastric lipase, little digestion of lipids occurs in the stomach. Any tendency to form emulsions with phospholipids or other natural emulsifying agents is inhibited by the high acidity. Fat in the duodenum strongly inhibits gastric emptying. This ensures that the fat is not emptied from the stomach more rapidly than it can be accommodated

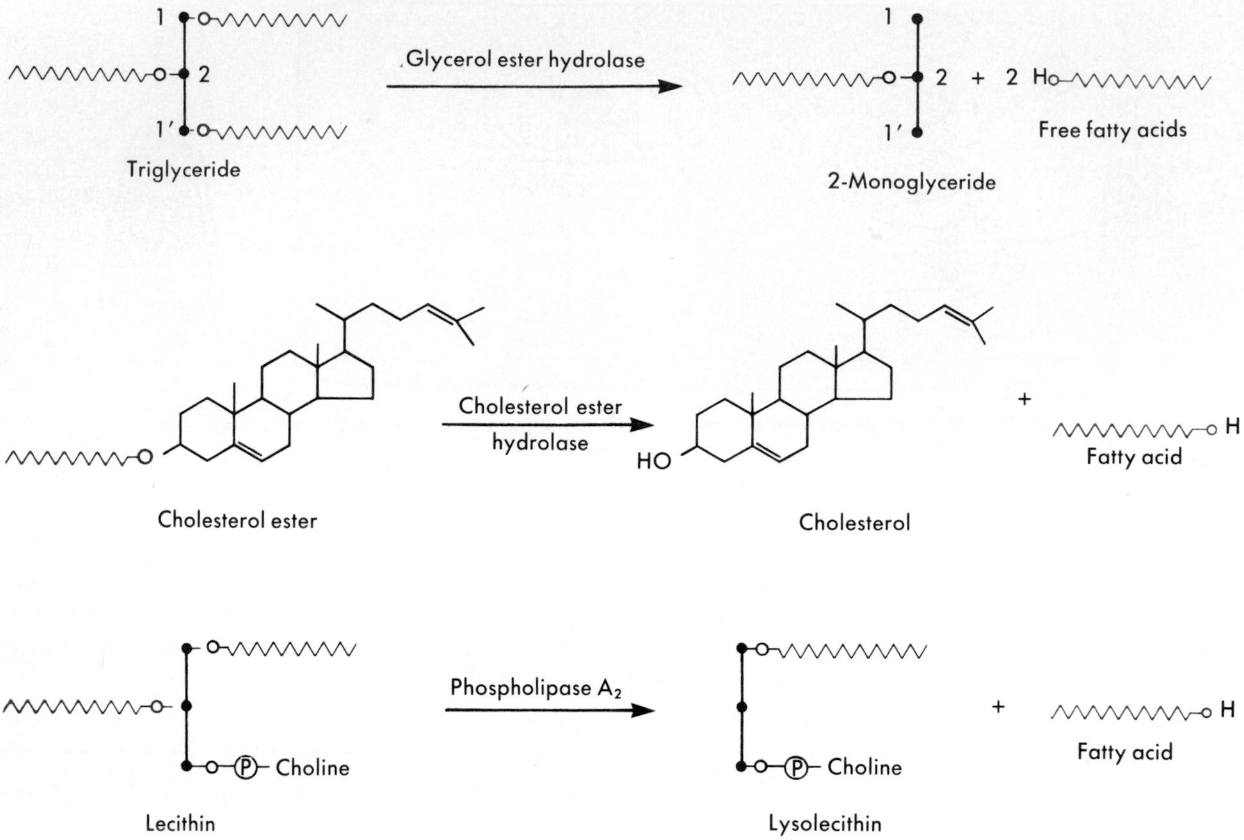

Glycerol ester hydrolase
Triglyceride
2-Monoglyceride
Free fatty acids

Cholesterol ester hydrolase
Cholesterol ester
Cholesterol
Fatty acid

Phospholipase A_2
Lecithin
Lysolecithin
Fatty acid

■ **Fig. 40-22** Action of major pancreatic lipases. The cleavage of lipids by glycerol ester hydrolase (pancreatic lipase), cholesterol ester hydrolase, and phospholipase A_2 is illustrated.

by the duodenal mechanisms that provide for emulsification and digestion.

Significant hydrolysis of triglycerides occurs in the stomach. The enzymes responsible are known as **preduodenal lipases.** These enzymes have acid pH optima. In rats the principal preduodenal lipase is **lingual lipase,** which is produced by von Ebner glands under the circumvallate papillae. In humans lingual lipase is probably a very minor component of preduodenal lipase, the major component being **gastric lipase** produced by gland cells in the fundus of the stomach. Under normal circumstances the amount of pancreatic lipase is so great that the absence of preduodenal lipase would not cause malabsorption of triglycerides. However, when pancreatic lipase is grossly deficient (in marked pancreatic insufficiency) or pancreatic lipase is inactive because of the high acidity in the upper small intestine (in Zollinger-Ellison syndrome), the hydrolysis of triglycerides by gastric lipase may be essential for digestion and absorption of triglycerides and other lipids as well (because of the effect of 2-monoglycerides and fatty acids in promoting formation of mixed micelles).

■ *Digestion of Lipids and Micelle Formation*

The lipolytic enzymes of the pancreatic juice are water-soluble molecules and thus have access to the lipids only at the surfaces of the fat droplets. The surface available for digestion is increased many thousand times by emulsification of the lipids. Bile acids themselves are rather poor emulsifying agents. However, with the aid of lecithin, which is present in high concentration in the bile, the bile acids emulsify dietary fats. The **emulsion droplets** are around 1 μm in diameter, and they have a large surface area on which the digestive enzymes can work.

The pancreatic secretions. Pancreatic juice contains the major lipolytic enzymes responsible for digestion of lipids (see Fig. 40-22). The most important digestive enzymes are as follows:

1. **Glycerol ester hydrolase** (also called simply pancreatic lipase), which cleaves the 1 and 1′ fatty acids preferentially off a triglyceride to produce two free fatty acids and one 2-monoglyceride
2. **Cholesterol esterase,** which cleaves the ester

bond in a cholesterol ester to give one fatty acid and free cholesterol

3. **Phospholipase A_2,** which cleaves the ester bond at the 2 position of a glycerophosphatide to yield, in the case of lecithin, one fatty acid and one lysolecithin

The formation of micelles. Bile acids form micelles with the products of fat digestion, especially 2-monoglycerides. The micelles are multimolecular aggregates (about 5 nm in diameter) containing about 20 to 30 molecules. 2-Monoglycerides and lysophosphatides tend to have their hydrophobic acyl chains in the interior of the micelle and their more polar portions facing the surrounding water. Bile acids are flat molecules that have a polar face and a nonpolar face (Fig. 39-31). Much of the surface of the micelles is covered with bile acids, with the nonpolar face toward the lipid interior of the micelle and the polar face toward the outside. Extremely hydrophobic molecules, such as long-chain fatty acids, cholesterol, and certain fat-soluble vitamins, tend to partition into the interior of the micelle. Phospholipids and monoglycerides tend to have their more polar ends facing the outside aqueous phase. Micelles contain almost no intact triglyceride.

Bile acids must be present at a certain minimum concentration, called the **critical micelle concentration,** before micelles will form. Conjugated bile acids have a much lower critical micelle concentration than the unconjugated forms. In the normal state bile acids are always present in the duodenum at greater than the critical micelle concentration.

Lipids and lipid digestion products in the micelles are in rapid exchange with lipid digestion products in the aqueous solution surrounding the micelle. In this way the micelles keep the aqueous solution surrounding them saturated with 2-monoglycerides, various fatty acids, cholesterol, and lysophosphatides. These lipids are present in the aqueous solution at a low concentration because of their limited water solubility.

Primary Enzymes of Lipid Digestion

Glycerol ester hydrolase (pancreatic lipase). The glycerol ester hydrolase of pancreatic juice has a molecular weight of about 50,000, and it is rather specific for triglycerides. Pancreatic lipase has very low activity against triglycerides in molecular solution. It is active on droplets or emulsions of triglycerides. The enzyme operates at the interface between the aqueous phase and the triglyceride-containing oil phase. Its activity is proportional to the surface area of the oil phase. There is a huge excess of pancreatic lipase. Based on samples of duodenal contents, sufficient pancreatic lipase is present during lipid digestion to hydrolyze in 1 minute an amount of trigyceride almost as large as the average daily intake of triglyceride.

Colipase. Pancreatic lipase is essentially completely inactivated by bile salts at physiological concentrations. A protein of 10,000 molecular weight, known as colipase, is present in pancreatic juice. Colipase is able to relieve the inactivation of lipase by bile salts. Bile salts inhibit the activity of pancreatic lipase by binding to the surface of triglyceride-containing oil droplets, thereby preventing the pancreatic lipase from binding. Colipase displaces bile salts from the surface of oil droplets. One pancreatic lipase molecule then binds to each colipase molecule. Lipase binds to colipase more tightly in the presence of droplets of triglyceride. The lipase-colipase complex is very active in cleaving the 1 and 1′ fatty acids from triglycerides at the surface of the droplet. Colipase also has a site that binds to bile salt micelles. This suggests that a lipase-colipase complex may simultaneously bind to a fat droplet and to a bile salt micelle. This binding might allow a direct transfer of the products of lipase action, and other contents of the fat droplet, to the bile salt micelle.

Cholesterol esterase. Cholesterol esterase is probably identical to a nonspecific enzyme, known as nonspecific lipase, that cleaves fatty acid ester linkages in a variety of lipid substrates. In humans nonspecific lipase has a molecular weight of about 100,000. The enzyme forms dimers in the presence of bile salts, and in this form it is protected from proteolytic digestion. The dimeric enzyme is active in hydrolyzing fatty acids from cholesterol esters, lysophospholipids, triglycerides, 2-monoglycerides, and fatty acyl esters of vitamins A, D, and E. The total activity of nonspecific lipase in pancreatic juice is small compared with the activity of pancreatic lipase.

Phospholipase A_2. Phospholipase A_2 is secreted as a proenzyme by the pancreas. Tryptic cleavage activates phospholipase A_2 against phospholipids emulsified by bile salts. Phospholipase A_2 is a highly stable protein with a molecular weight of 14,000. The enzyme requires calcium ions for activity. Phospholipase A_2 hydrolyzes the fatty acid at the number 2 position of phosphatidylcholine, phosphatidylethanolamine, phosphatidylserine, phosphatidylglycerol, and cardiolipin to yield free fatty acids and lysophosphatides. Sphingolipids and glycosphingolipids are not good substrates for phospholipase A_2. The fatty acyl ester linkage at the number 1 position is not cleaved by this enzyme and is hydrolyzed poorly by pancreatic lipase; therefore lysophosphatides are the principal form in which phospholipids are absorbed. Phospholipase A_2 is much more active against phospholipids in bile acid–mixed micelles than against phospholipid bilayers.

Absorption of the Products of Lipid Digestion

Transport into the intestinal epithelial cell. micelles are important in the absorption of the products of lipid digestion and in the absorption of most other fat-soluble molecules (such as the fat-soluble vitamins). The micelles diffuse among the microvilli that form the brush border. The presence of micelles tends to keep the aqueous solution in contact with the brush border saturated with fatty acids, 2-monoglycerides, cholesterol, and other micellar contents. In this way the huge surface area of the brush border is made available for the absorption of the micellar contents (Fig. 40-23).

Because of their high lipid solubility, the fatty acids, 2-monoglycerides, cholesterol, and lysolecithin can readily diffuse across the brush border membrane. A membrane-associated fatty acid binding protein in the brush border plasma membrane appears to play a role in the uptake of long-chain fatty acids. Free fatty acids, 2-monoglycerides, and the other products of lipid digestion can diffuse across the brush border plasma membrane so rapidly that this step does not limit the rate of their uptake. The main limitation to the rate of lipid uptake by the epithelial cells of the upper small intestine is the diffusion of the mixed micelles through an unstirred layer (or diffusion boundary layer) on the luminal surface of the brush border plasma membrane. Partly because of the convoluted surface of the intestinal mucosa, the fluid in immediate contact with the epithelial cell surface is not readily mixed with the bulk of the luminal contents. The effective thickness of the unstirred layer ranges from 200 to 500 μm. Nutrients present in the well-mixed contents of the intestinal lumen must diffuse through the unstirred layer to reach the brush border plasma membrane (see Fig. 40-23). A concentration gradient exists across the unstirred layer, with micelles and lipid digestion products in lower concentration at the brush border surface than in the well-mixed contents of the lumen. A pH gradient also exists across the unstirred layer, such that the fluid in immediate contact with the brush border plasma membrane is about 1 pH unit more acidic than the bulk luminal contents. The lower pH at the brush border surface may enhance absorption of fatty acids, because protonated fatty acids are more lipid soluble than ionized ones.

Cholesterol is absorbed more slowly than most of the other constituents of the micelles. Therefore, as the micelles progress down the small intestine, they become more concentrated in cholesterol. The duodenum and jejunum are most active in fat absorption, and most ingested fat is absorbed by the midjejunum. The fat present in normal stools is not ingested fat (which is completely absorbed), but fat from colonic bacteria and from desquamated intestinal epithelial cells.

Inside the intestinal epithelial cell. The products of lipid digestion make their way to the smooth endoplasmic reticulum. A recently discovered cytoplasmic fatty acid–binding protein may play a role in transporting fatty acids to the smooth endoplasmic reticulum, or it may simply function to prevent fatty acids from forming fat droplets prematurely. The fatty acid–binding protein has a molecular weight of 12,000. It has higher

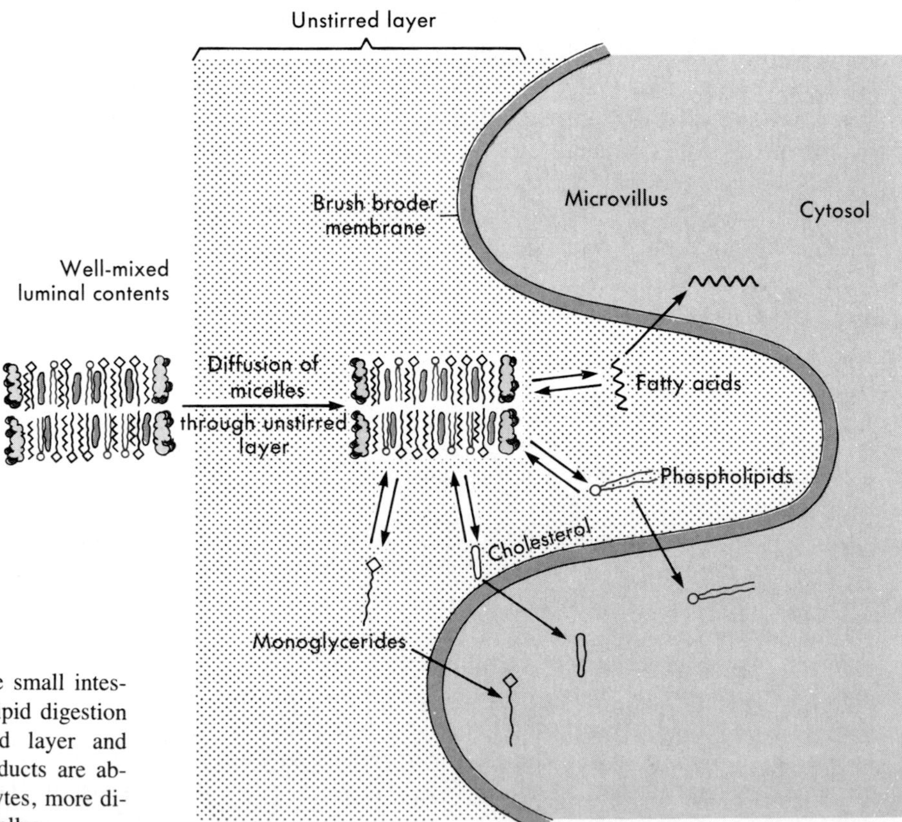

■ **Fig. 40-23** Lipid absorption in the small intestine. Mixed micelles of bile acids and lipid digestion products diffuse through the unstirred layer and among the microvilli. As digestion products are absorbed from free solution by the enterocytes, more digestion products partition out of the micelles.

affinity for unsaturated fatty acids than for saturated ones and binds long-chain fatty acids more tightly than short- or medium-chain fatty acids.

In the smooth endoplasmic reticulum, which is engorged with lipid after a meal, considerable chemical reprocessing goes on (Fig. 40-24). The 2-monoglycerides are reesterified with fatty acids at the 1 and 1′ carbons to re-form triglycerides. Lysophospholipids are reconverted to phospholipids. Cholesterol is reesterified to a considerable extent, although some free cholesterol remains. The processing of 2-monoglycerides and lysophospholipids is essentially complete, since the blood levels of these components are negligible. The intestinal epithelial cells are also capable of de novo synthesis of lipids to some extent.

Chylomicron formation and transport. The reprocessed lipids, along with those that are synthesized de novo, accumulate in the vesicles of the smooth endoplasmic reticulum. Phospholipids tend to cover the external surfaces of these lipid droplets, with their hydrophobic acyl chains in the fatty interior and their polar head groups toward the aqueous exterior. The lipid droplets, of the order of 1 nm in diameter at this point, are known as chylomicrons. About 10% of their surface is covered by β-lipoprotein, which is synthesized in the intestinal epithelial cells.

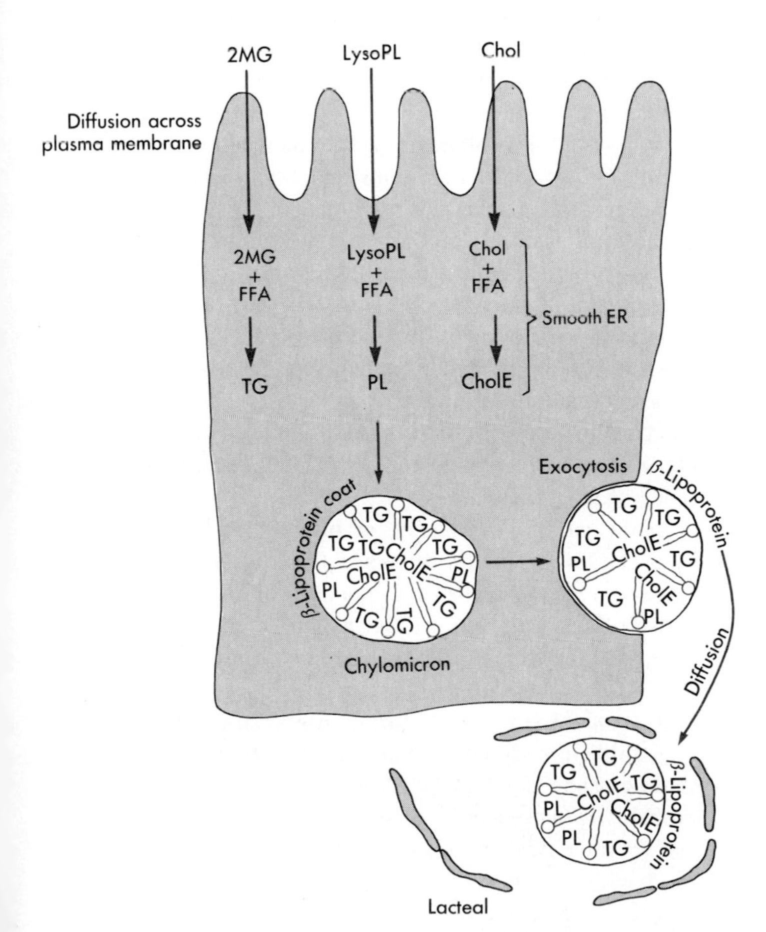

■ **Fig. 40-24** Lipid resynthesis in the epithelial cells of the small intestine, chylomicron formation, and subsequent transport of chylomicrons. *FFA,* Free fatty acid; *2MG,* 2-monoglyceride; *TG,* triglyceride; *lysoPL,* lysophospholipid; *PL,* phospholipid; *Chol,* cholesterol; *CholE,* cholesterol ester.

Chylomicrons are ejected from the cell by exocytosis (see Fig. 40-24). The β-lipoprotein is important in this process; when it is absent, the intestinal epithelial cells become engorged with lipid, and lipid absorption is severely impaired. The chylomicrons leave the cells at the level of the nuclei and enter the lateral intercellular spaces. Chylomicrons are too large to pass through the basement membrane that invests the mucosal capillaries. However, they do enter the lacteals, which have sufficiently large fenestrations for the chylomicrons to pass through. The chylomicrons leave the intestine with the lymph, primarily via the thoracic duct, and are dumped into the venous circulation.

Chylomicrons are approximately spherical and vary greatly in size (60 to 750 nm). When large amounts of lipid are being absorbed, large chylomicrons are formed; when little lipid is being absorbed the chylomicrons formed tend to be small. Triglyceride is the predominant component, making up 85% to 92% of the chylomicrons. Phospholipids cover about 80% of the surface of the chylomicrons and account for 6% to 8% of their mass. Biliary phospholipids are used preferentially for this purpose. Apolipoproteins cover the remaining 20% of the chylomicron surface. Cholesterol and cholesterol esters are present in the triglyceride-rich core of the particles, each making up about 1% of the chylomicron mass. Apolipoproteins of the C class constitute about 50% of the protein present in chylomicrons. There are smaller proportions of apo A-I (15% to 35%), apo A-IV (10%), apo B (10%), and apo E (5%).

Synthesis by enterocytes of apolipoproteins is required for normal absorption of chylomicrons. A graphic demonstration of the role of apolipoproteins occurs in the disease known as **abetalipoproteinemia,** in which enterocytes are deficient in their ability to synthesize apo B. In this disease the transport of chylomicrons from enterocytes to intestinal lymph is greatly diminished, with the result that the enterocytes become engorged with nascent chylomicrons.

Many, but not all, of the apoproteins associated with chylomicrons present in intestinal lymph are synthesized by intestinal epithelial cells. Hepatocytes are the other major source of apolipoproteins. The major source of certain of these apoproteins is controversial. It is becoming clear that the intestine is an important source of certain apoproteins of plasma lipoproteins. Recent evidence suggests that even when fat is absent from the intestine, significant amounts of apo A-I and apo B may be secreted into intestinal lymph. The view is emerging that the intestine plays a major role in the body's synthesis and metabolism of serum lipoproteins and that these intestinal processes are subject to multiple levels of regulation.

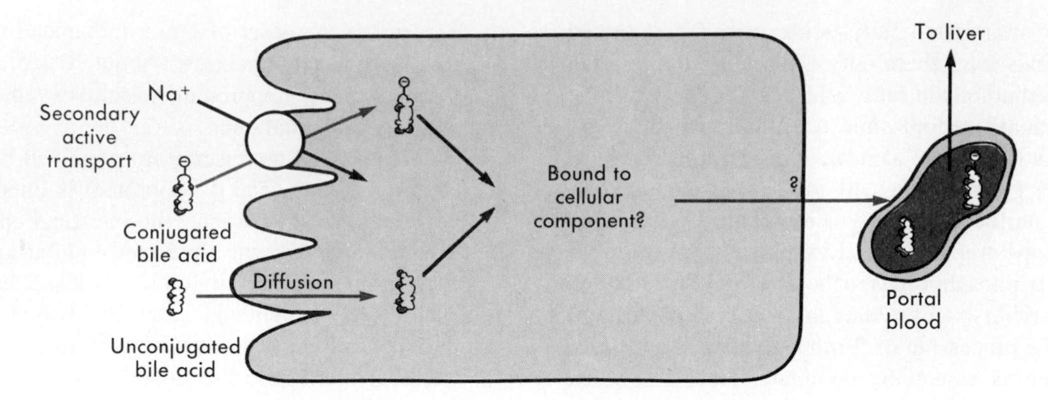

Fig. 40-25 Absorption of bile acids by epithelial cells of the terminal ileum. Bile acids are absorbed both by simple diffusion and by Na^+-powered secondary active transport. Conjugated bile acids are absorbed avidly by active transport. Unconjugated bile acids are absorbed chiefly by simple diffusion.

Absorption of Bile Acids

The absorption of dietary lipids is typically complete by the midjejunum. Bile acids, by contrast, are absorbed largely in the terminal part of the ileum. As for other fat-soluble substances, the unstirred layer is an important barrier to bile acid absorption. Bile acids cross the brush border plasma membrane by two routes: by an active transport process and by simple diffusion (Fig. 40-25). The active process is secondary active transport, powered by the Na^+ gradient across the brush border membrane. In this process one Na^+ ion is cotransported across the brush border plasma membrane with one bile acid molecule. Conjugated bile acids are the principal substrates for active absorption; deconjugated bile acids have poor affinity for the transporter. Deconjugated bile acids are less polar than conjugated bile acids and are thus better absorbed by simple diffusion. The fewer hydroxyl groups on a bile acid, the poorer substrate it is for active absorption. For this reason, dehydroxylation of bile acids by enteric bacteria to form secondary bile acids enhances absorption of bile acids by diffusion.

Other aspects of the absorption of bile acids are less understood. It is suspected that bile acids are bound to cytosolic structures or macromolecules, which remain to be identified, in intestinal epithelial cells. The process by which bile acids traverse the basolateral plasma membrane of the enterocyte has not been characterized.

Absorbed bile acids are carried away from the intestine in the portal blood. Hepatocytes avidly extract bile acids, essentially clearing them from the blood in a single pass through the liver. In the hepatocytes most deconjugated bile acids are reconjugated, and some secondary bile acids are rehydroxylated. The reprocessed bile acids, together with newly synthesized bile acids, are secreted into bile.

Malabsorption of Lipids

Malabsorption of lipids occurs more frequently than malabsorption of proteins or carbohydrates. Among the general causes of lipid malabsorption are bile deficiency, pancreatic insufficiency, and the intestinal mucosal atrophy that occurs in some disease states.

In cases of **bile deficiency** and **pancreatic insufficiency** the levels of bile acids and lipolytic enzymes, respectively, must be severely reduced before serious malabsorption occurs. In both cases the quantity of fecal fat is roughly proportional to the quantity ingested.

Even in the **complete absence of bile acids,** significant hydrolysis of triglyceride occurs. The rate of absorption of fatty acids from triglycerides may be 50% of normal. Cholesterol, cholesterol esters, and fat-soluble vitamins are much less water soluble than fatty acids, and their absorption is grossly deficient in the absence of bile acids.

In the **complete absence of pancreatic lipases** all lipid classes are poorly absorbed. This is probably because of the necessity of 2-monoglycerides and lysophosphatides (products of the action of pancreatic lipases) for the formation of mixed micelles with bile acids.

In **tropical sprue** and **gluten enteropathy** the intestinal epithelium is flattened and the density of microvilli is decreased. Lipid malabsorption in these diseases is probably a consequence of the marked decrease in the surface area available for lipid absorption.

Absorption of Fat-Soluble Vitamins

Because of their solubility in nonpolar solvents, the fat-soluble vitamins (A, D, E, and K) partition into the mixed micelles formed by the bile acids and lipid digestion products. As is the case for other lipids, the fat-soluble vitamins enter the intestinal epithelial cell by diffusing across the brush border plasma membrane. There is no evidence that the passage of fat-soluble vitamins across the brush border plasma membrane is mediated by membrane proteins. In general, the presence of bile acids and lipid digestion products enhances the absorption of fat-soluble vitamins. In the intestinal epithelial cell the fat-soluble vitamins enter the chylomicrons and leave the intestine in the lymph. In the absence of bile acids a significant fraction of the ingested load of a fat-soluble vitamin may be absorbed and leave the intestine in the portal blood.

Vitamin A. Vitamin A (retinol) is better absorbed than is provitamin A (β-carotene) and vitamin A aldehyde (retinal). In the epithelial cells of the intestine, β-carotene and retinal are converted to retinol. Vitamin A is present in thoracic duct lymph primarily as fatty acid esters of retinol. Vitamin A acid (retinoic acid) is absorbed independently of the bile acid–mixed micelles and leaves the intestine in the portal blood.

Vitamin D. Vitamin D is absorbed principally in the jejunum, primarily as the free vitamin. Most fatty acid esters of vitamin D are hydrolyzed in the intestinal lumen before absorption. With normal oral loads, 55% to 99% of the vitamin D ingested is absorbed.

Vitamin E. The absorption of vitamin E (α-tocopherol) requires the presence of bile acid–mixed micelles. Small oral doses of vitamin E are almost completely absorbed. With larger doses, a significant fraction escapes absorption and appears in the feces. In the intestinal epithelial cell, vitamin E enters the chylomicrons and leaves the intestine in the lymph.

Vitamin K. Vitamins K_1 and K_2, which have hydrophobic side chains, partition into bile acid–mixed micelles, are absorbed from free solution, enter the chylomicrons, and leave the intestine in the lymph. Vitamin K_3 (2-methylnaphthoquinone) lacks a side chain, is absorbed independently of the mixed micelles, and leaves the intestine primarily in the portal blood.

■ *Summary*

1. The α-amylases of saliva and pancreatic juice cleave branched starch into maltose, maltotriose, and α-limit dextrins. These digestion products are then reduced to glucose molecules by glucoamylase and isomaltase, carbohydrate-digesting enzymes on the brush border plasma membrane. The brush border also contains the disaccharidases sucrase and lactase that cleave sucrose and lactose into monosaccharides that can be transported into the enterocyte by the monosaccharide transport proteins of the brush border membrane (i.e., the glucose-galactose transporter and the fructose transporter).

2. Protein digestion begins in the stomach by the action of pepsins. The pancreatic proteases rapidly cleave proteins in the duodenum and jejunum to oligopeptides. Peptide-cleaving enzymes on the brush border membrane reduce oligopeptides to single amino acids and to dipeptides and tripeptides. Amino acids are taken into the enterocyte by an array of amino acid–transporting proteins in the brush border membrane. Dipeptides and tripeptides are taken up by a brush border peptide transport protein with broad specificity.

3. A typical human ingests 2 L of water per day, and about 7 L enter the gastrointestinal tract in gastrointestinal secretions. About 99% of the water presented to the gastrointestinal tract is absorbed; approximately 100 ml of water escape into feces each day. The absorption of water is powered by the absorption of ions and nutrients, predominantly in the small intestine. The mature epithelial cells at the tips of small intestinal villi are active in absorption of water and electrolytes. Cells in Lieberkühn's crypts are net secretors of water and ions. The net absorption that usually occurs in the small intestine is the result of much larger absorptive and secretory fluxes. In secretory diarrheal disease, such as cholera, the secretory fluxes in the crypt cells increase and the absorptive fluxes in cells at the villous tips are inhibited.

4. Calcium is actively absorbed in the small intestine. Vitamin D stimulates the absorption of Ca^{++} by enhancing the synthesis of cytosolic calbindin, a Ca^{++}-binding protein. Ca^{++} is transported across the basolateral membrane by the Ca^{++}-ATPase and the Na^{+}/Ca^{++} exchange protein. The capacity of the intestinal epithelial cells to absorb Ca^{++} is regulated in accordance with the body's need for Ca^{++}.

5. About 5% of inorganic iron ingested is absorbed by the small intestine; approximately 20% of heme iron is absorbed. The small intestinal epithelial cells secrete a transferrin into the lumen. The complex of iron with transferrin is taken up at the small intestinal brush border by receptor-mediated endocytosis. In the enterocyte some iron is bound to ferritin and is unavailable for absorption. The capacity to absorb iron increases in response to hemorrhage. Iron appears in the portal blood bound to transferrin.

6. Most water-soluble vitamins are taken up by specific transporters in the small intestinal brush border membrane. Vitamin B_{12} is bound to R proteins in saliva and gastric juice. When R proteins are digested, vitamin B_{12} is bound by intrinsic factor (IF). Receptors on the ileal brush border membrane recognize the IF-B_{12} complex and allow vitamin B_{12} to be absorbed by the ileal enterocyte. Pernicious anemia is caused by a deficiency of IF. Vitamin B_{12} appears in the plasma bound to transcobalamin II.

7. Triglyceride is the principal dietary lipid. Lipids form droplets in the stomach and are emulsified in the duodenum by bile acids. Emulsification greatly in-

creases the surface area available for the action of lipid-digesting enzymes of the pancreatic juice. The products of triglyceride digestion, 2-monoglycerides and fatty acids, form mixed micelles with bile acids. Cholesterol, fat-soluble vitamins, and other lipids partition into the micelles. Mixed micelles are small enough to diffuse among the microvilli and thus greatly enhance the brush border surface area available for lipid absorption. In the enterocyte triglycerides and phospholipids are resynthesized and packaged along with other lipids into chylomicrons. Chylomicrons are coated with apolipoproteins and released at the basolateral membrane by exocytosis. Chylomicrons leave the intestine in the lymphatic vessels and the thoracic duct.

■ Bibliography

Journal articles

Binder HJ: The pathophysiology of diarrhea, *Hosp Pract* 19(Oct):107, 1984.

Bisgaier CL, Glickman RM: Intestinal synthesis, secretion, and transport of lipoproteins, *Annu Rev Physiol* 45:625, 1983.

Charlton RW, Bothwell TH: Iron absorption, *Annu Rev Med* 34:55, 1983.

Crapo PA, Reaven G, Olefsky J: Postprandial plasma-glucose and -insulin responses to different complex carbohydrates, *Diabetes* 26:1178, 1977.

Hediger MA et al: Expression cloning and cDNA sequencing of the Na^+/glucose co-transporter, *Nature* 330:379, 1987.

Jenkins DJA et al: Glycemic index of foods: a physiological basis for carbohydrate exchange, *Am J Clin Nutr* 34:362, 1981.

Seetharam B, Alpers DH: Absorption and transport of cobalamin (vitamin B_{12}), *Annu Rev Nutr* 2:343, 1982.

Stephen AM, Haddad AC, Phillips SF: Passage of carbohydrate into the colon: direct measurements in humans, *Gastroenterology* 85:589, 1983.

Stevens BR, et al: Intestinal transport of amino acids and sugars: advances using membrane vesicles, *Annu Rev Physiol* 46:417, 1984.

Stremmel W: Uptake of fatty acids by jejunal mucosal cells is mediated by a fatty acid binding membrane protein, *J Clin Invest* 82:2001, 1988.

Wasserman RH, Fullmer CS: Calcium transport proteins, calcium absorption, and vitamin D, *Annu Rev Physiol* 45:375, 1983.

Young S, Bomford A: Transferrin and cellular iron exchange, *Clin Sci* 67:273, 1984.

Books and monographs

Alpers DH: *Digestion and absorption of carbohydrates and proteins*. In Johnson LR, editor: *Physiology of the gastrointestinal tract,* ed 2, vol 2, New York, 1987, Raven Press.

Barrett KE, Dharmsathaphorn K: *Secretion and absorption: small intestine and colon*. In Yamada T, editor: *Textbook of gastroenterology,* vol 1, Philadelphia, 1991, JB Lippincott.

Cooke HJ: *Neural and humoral regulation of small intestinal electrolyte transport*. In Johnson LR, editor: *Physiology of the gastrointestinal tract,* ed 2, vol 2, New York, 1987, Raven Press.

Davenport HW: *Physiology of the digestive tract,* ed 5, Chicago, 1982, Year Book.

Davidson NO, Magun AM, Glickman RM: *Enterocyte lipid absorption and secretion*. In Field M and Frizzell RA, editors: *Handbook of Physiology,* section 6, vol 4, Bethesda, Md, 1991, The American Physiological Society.

Donowitz M, Welsh MJ: *Regulation of mammalian small intestinal electrolyte secretion*. In Johnson LR, editor: *Physiology of the gastrointestinal tract,* ed 2, vol 2, New York, 1987, Raven Press.

Gray GM: *Dietary protein processing: intraluminal and enterocyte surface events*. In Field M and Frizzell RA, editors: *Handbook of Physiology,* section 6, vol 4, Bethesda, Md, 1991, The American Physiological Society.

Greenberger NJ: *Gastrointestinal disorders: a pathophysiologic approach,* ed 3, Chicago, 1986, Year Book.

Johnson LR, editor: *Gastrointestinal physiology,* ed 4, St Louis, 1991, Mosby–Year Book.

Rose RC: *Intestinal transport of water-soluble vitamins*. In Field M and Frizzell RA, editors: *Handbook of physiology,* section 6, vol 4, Bethesda, Md, 1991, The American Physiological Society.

Shiau Y-F: *Lipid digestion and absorption*. In Johnson LR, editor: *Physiology of the gastrointestinal tract,* ed 2, vol 2, New York, 1987, Raven Press.

Sullivan SK, Field M: *Ion transport across mammalian small intestine*. In Field M and Frizzell RA, editors: *Handbook of physiology,* section 6, vol 4, Bethesda, Md, 1991, The American Physiological Society.

Van Dyke RW: *Mechanisms of digestion and absorption of food*. In Sleisenger MH, Fordtran JS, editors: *Gastrointestinal disease,* ed 4, Philadelphia, 1989, WB Saunders.

SECTION
VIII

THE KIDNEY

Bruce A. Stanton
Bruce M. Koeppen

CHAPTER

41

Elements of Renal Function

■ *Overview of the Kidney*

The kidneys are both regulatory and excretory organs. By their excretion of water and solutes the kidneys are able to regulate the volume and composition of the body fluids within a very narrow range despite wide variations in the intake of food and water. As a consequence of the kidneys' homeostatic role the tissues and cells of the body are able to carry out their normal functions in a relatively constant environment.

The kidneys have several major functions, including the following:

1. Regulation of body fluid osmolality and volume
2. Regulation of electrolyte balance
3. Regulation of acid-base balance
4. Excretion of metabolic products and foreign substances
5. Production and secretion of hormones

The control of body fluid osmolality is important for the maintenance of normal cell volume in all tissues of the body. Control of the volume of the body fluids is necessary for normal function of the cardiovascular system. The kidneys, working in an integrated fashion with components of the cardiovascular, endocrine, and central nervous systems, accomplish these tasks by regulating the excretion of water and NaCl.

The kidneys play an essential role in regulating the amount of several important inorganic ions in the body, including Na^+, K^+, Cl^-, HCO_3^-, H^+, Ca^{++}, Mg^{++}, and $PO_4^{\equiv}$. The kidneys also contribute to the maintenance of organic ion balance. For example the excretion of many of the intermediates of the Krebs cycle (e.g., citrate, succinate) is controlled by the kidneys. To maintain an appropriate balance the excretion of any one of these electrolytes must be in balance with the daily intake. If intake exceeds excretion the amount of a particular electrolyte in the body increases. Conversely if excretion exceeds intake the amount decreases. For many of these electrolytes the kidneys are the sole or primary route of excretion. Thus electrolyte balance is achieved by carefully matching daily excretion by the kidneys with daily intake.

Many of the body's metabolic functions are exquisitely sensitive to pH. Thus the pH of the body fluids must be maintained within very narrow limits. This regulation is accomplished by buffers within the body fluids and by the coordinated action of the lungs and kidneys.

The kidneys excrete a number of end products of metabolism that are no longer needed by the body. These so-called waste products include urea (from amino acids), uric acid (from nucleic acids), creatinine (from muscle creatine), end-products of hemoglobin metabolism, and metabolites of hormones. The kidneys eliminate these substances from the body at a rate that matches their production. Thus the kidneys regulate their concentrations within the body fluids. The kidneys also represent an important route for elimination of foreign substances from the body, such as drugs, pesticides, and other chemicals ingested in the food.

The kidneys are important endocrine organs that produce and secrete renin, prostaglandins, kinins, 1,25-dihydroxyvitamin D_3, and erythropoietin. Renin activates the renin-angiotensin-aldosterone system, which is important in regulating blood pressure as well as sodium and potassium balance. Prostaglandins and kinins (e.g., bradykinin) are vasoactive and are important in the regulation and modulation of renal blood flow. Together with angiotensin II, prostaglandins and kinins influence systemic blood pressure. 1,25-Dihydroxyvitamin D_3 is necessary for normal reabsorption of Ca^{++} by the gastrointestinal tract and for its deposition in bone. Erythropoietin stimulates red blood cell formation by the bone marrow.

■ *Functional Anatomy of the Kidney*

Structure and function are closely linked in the kidneys. Consequently an appreciation of the gross anatomical

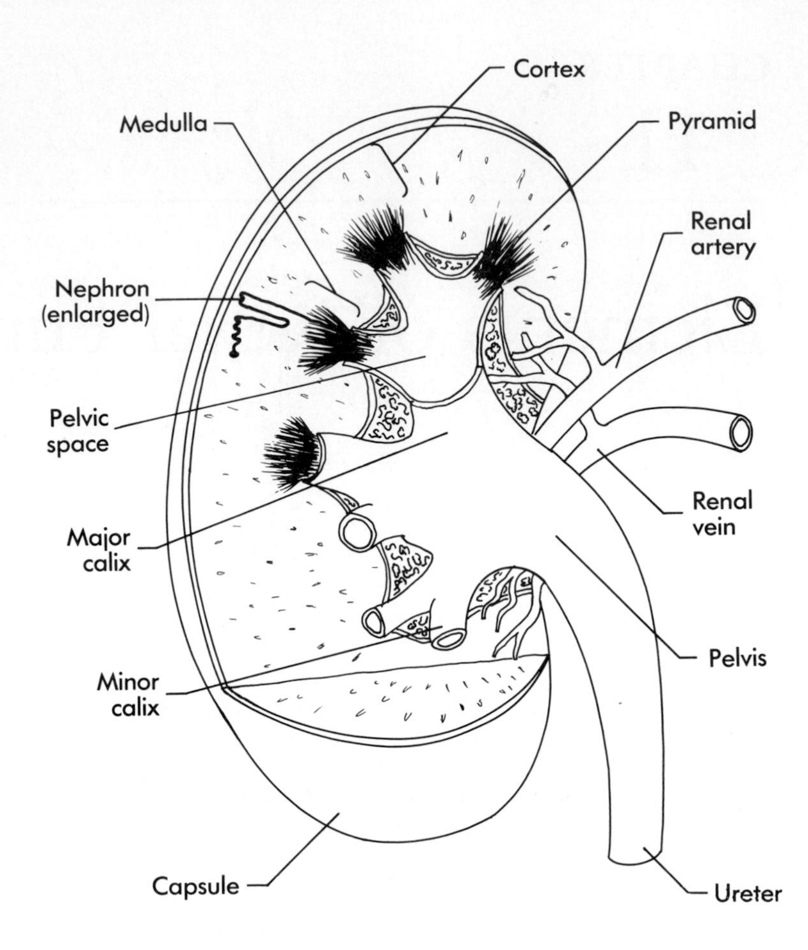

■ **Fig. 41-1** Structure of the human kidney, cut open to show internal structures. (Modified from Marsh DJ: *Renal physiology,* New York, 1983, Raven Press.)

and histological features of the kidneys is a prerequisite for an understanding of their function.

■ *Gross Anatomy*

The gross anatomical features of the mammalian kidney are illustrated in Fig. 41-1. The medial side of each kidney contains an indentation through which pass the renal artery and vein, nerves, and pelvis. On the cut surface of a bisected kidney two regions are evident: an outer region called the **cortex** and an inner region called the **medulla.** The cortex and medulla are composed of **nephrons** (the functional units of the kidney), blood vessels, lymphatic vessels, and nerves. The medulla in the human kidney is divided into 8 to 18 conical masses, the **renal pyramids.** The base of each pyramid originates at the corticomedullary border and the apex terminates in the papilla, which lies within the **pelvic space.** The **pelvis** represents the upper expanded region of the **ureter,** which carries urine from the pelvic space to the urinary bladder. In the human kidney the pelvis divides into two or three open-ended pouches, the **major calices,** which extend outward from the dilated end of the pelvis. Each major calix divides into **minor calices,** which collect the urine from each papilla. The walls of the calices, pelvis, and ureters contain smooth muscle, which contracts to propel the urine toward the **bladder** for storage until elimination.

The blood flow to the two kidneys is equal to 25% (1.25 L/min) of the cardiac output in resting subjects. However the kidneys constitute less than 0.5% of the total body weight. As illustrated in Fig. 41-2 the **renal artery** enters the kidney alongside the ureter, and it branches to form progressively the **interlobar artery,** the **arcuate artery,** the **interlobular artery** (cortical radial artery), and the **afferent arteriole,** which leads into the **glomerular capillaries.** The glomerular capillaries coalesce to form the **efferent arteriole,** which leads into a second capillary network, the **peritubular capillaries,** which supply blood to the nephron. The vessels of the venous system run parallel to the arterial vessels and progressively form the **interlobular vein** (cortical radial vein), the **arcuate vein,** the **interlobar vein,** and the **renal vein,** which courses beside the ureter.

■ *Ultrastructure of the Nephron*

The functional unit of the kidney is the nephron. Each human kidney contains approximately 1.2 million nephrons, which are hollow tubes composed of a single cell layer. The nephron consists of a renal corpuscle (glomerulus), a proximal tubule, a loop of Henle, a distal

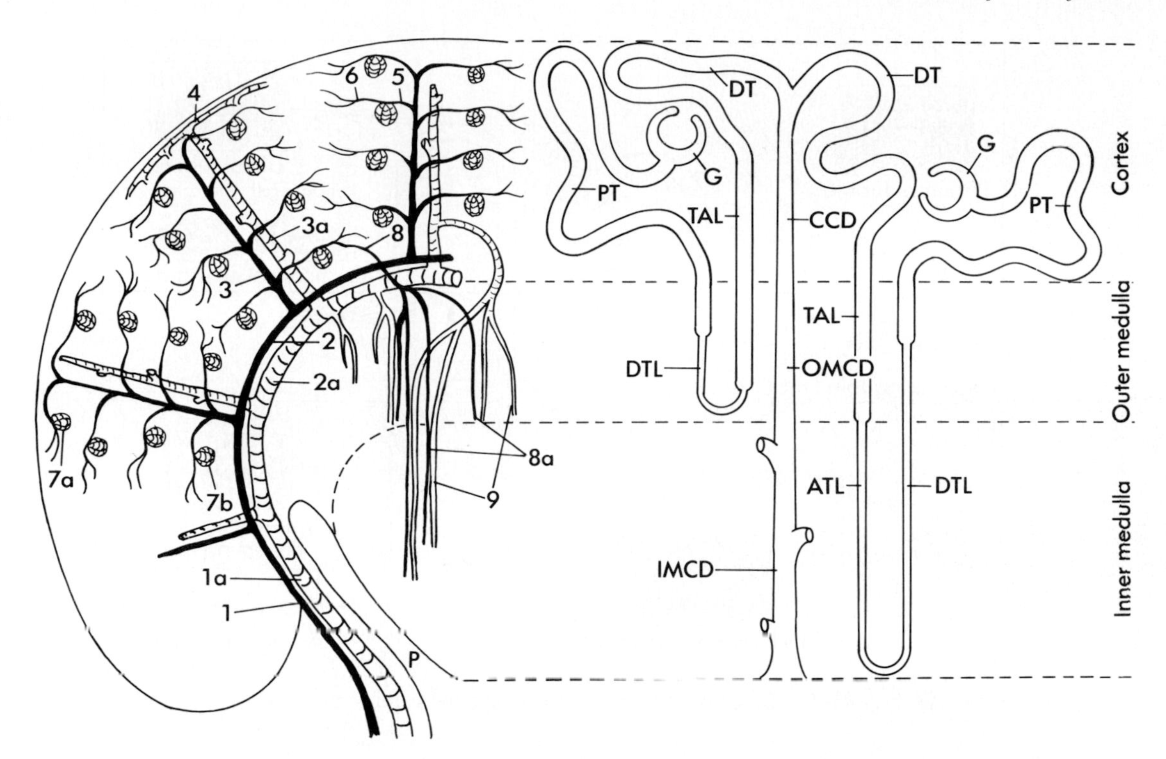

■ **Fig. 41-2** **(Left panel)** Organization of the vascular system of the human kidney. This scheme depicts the course and distribution of the intrarenal blood vessels; peritubular capillaries are not shown. Not drawn to scale. The renal artery branches to form interlobar arteries *(1)*, which give rise to arcuate arteries *(2)*. Arcuate arteries lead to interlobular arteries *(3)* that ascend toward the renal capsule and branch to form afferent arterioles *(5)*. Afferent arterioles branch to form glomerular capillary networks *(7a, 7b)*, which then coalesce to form efferent arterioles *(6)*. The efferent arterioles of the outer cortical nephrons form capillary networks *(not shown)* that suffuse the cells in the cortex. The efferent arterioles of the juxtamedullary nephrons divide into descending vasa recta *(8)*, which form capillary networks that supply blood to the outer and inner medulla *(8a)*. Blood from the peritubular capillaries enters, consecutively, the stellate vein *(4)*, interlobular vein *(3a)*, arcuate vein *(2a)*, and interlobar vein *(1a)*. Blood from the ascending vasa recta *(9)* enters the arcuate and interlobular veins. P, pelvis. **(Right panel)** Organization of the human nephron. A superficial nephron is illustrated on the left and a juxtamedullary nephron is illustrated on the right. Not drawn to scale. *DT*, distal tubule; *PT*, proximal tubule; *G*, glomerulus; *CCD*, cortical collecting duct; *TAL*, thick ascending limb; *DTL*, descending thin limb; *OMCD*, outer medullary collecting duct; *ATL*, ascending thin limb; *IMCD*, inner medullary collecting duct. The loop of Henle includes the straight portion of the PT, and the DTL, ATL, and TAL. (Modified from Kriz W, Bankir L: *Am J Physiol* 254:F1, 1988.)

tubule, and a collecting duct system.* The renal **glomerulus** consists of glomerular capillaries and **Bowman's capsule.** The **proximal tubule** initially forms several coils followed by a straight piece that descends toward the medulla. The next segment is **Henle's loop,** which is composed of the straight part of the proximal tubule, the descending thin limb (which ends at a hairpin turn), the ascending thin limb (only in nephrons with long loops of Henle), and the thick ascending limb. Near the end of the thick ascending limb the nephron passes between the afferent and efferent arterioles that supply the renal glomerulus of the same nephron. This short segment of the thick ascending limb is called the **macula densa.** The **distal tubule** begins a short distance beyond the macula densa and extends to the point in the cortex where two or more nephrons join to form a cortical **collecting duct.** The cortical collecting duct enters the medulla and becomes the outer medullary collecting duct and then the inner medullary collecting duct.

*The organization of the nephron is actually much more complicated than presented here; however for simplicity and clarity of presentation in subsequent chapters the nephron is divided into four segments. For details on the subdivisions of the four nephron segments, consult the references by Kriz and Bankir, Kriz and Kaissling, and Tisher and Madsen. The collecting duct system is not actually part of the nephron. However for simplicity we consider the collecting duct system as part of the nephron.

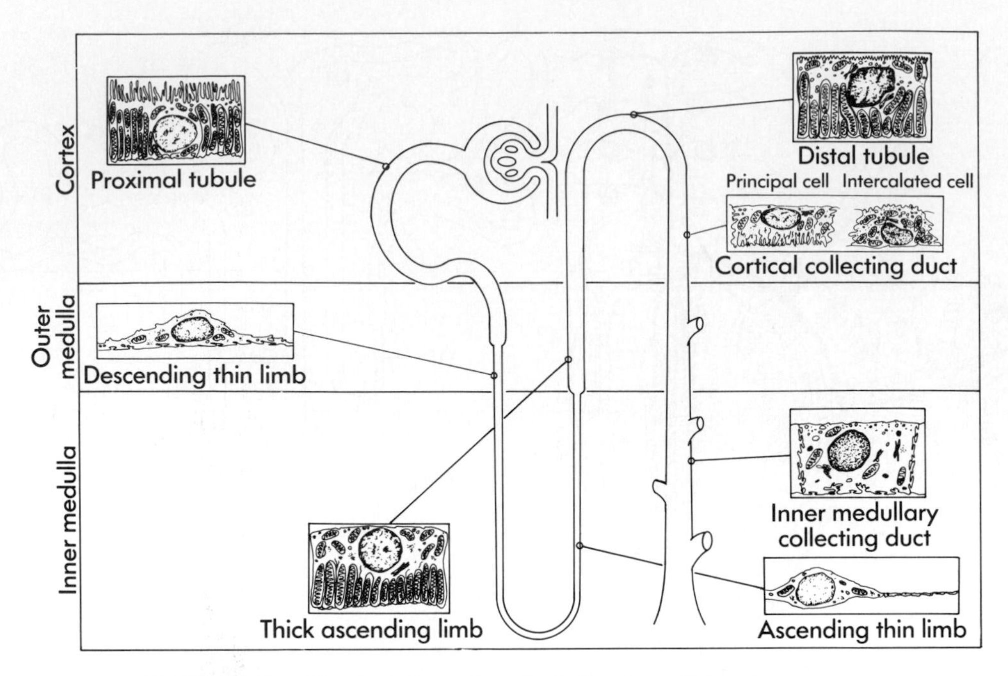

Fig. 41-3 Diagram of a nephron including the cellular ultrastructure.

Each nephron segment is composed of cells that are uniquely suited to perform specific transport functions (Fig. 41-3). Proximal tubule cells have an extensively amplified apical membrane (the urine side of the cell) called the brush border, which is present only in the proximal tubule. The basolateral membrane (the blood side of the cell) is highly invaginated. These invaginations contain many mitochondria. In contrast the descending thin limb and ascending thin limb of Henle's loop have poorly developed apical and basolateral surfaces and few mitochondria. The cells of the thick ascending limb and the distal tubule have abundant mitochondria and extensive infoldings of the basolateral membrane. The collecting duct is composed of two cell types: **principal cells** and **intercalated cells**. Principal cells have a moderately invaginated basolateral membrane and contain few mitochondria. Intercalated cells have a high density of mitochondria. The final segment of the nephron, the inner medullary collecting duct, is composed of inner medullary collecting duct cells.

Nephrons may be subdivided into superficial and juxtamedullary types (Fig. 41-2, *right panel*). The glomerulus of each **superficial nephron** is located in the outer region of the cortex (Fig. 41-2, *right panel*). Its loop of Henle is short, and its efferent arteriole branches into peritubular capillaries that surround the tubular segments of its own and adjacent nephrons. This capillary network conveys oxygen and important nutrients to the tubular segments, delivers substances to the tubules for secretion (i.e., the movement of a substance from the blood into the tubular fluid), and serves as a pathway for the return of reabsorbed water and solutes to the circulatory system. A few species, including humans, also possess very short superficial nephrons whose loops of Henle never enter the medulla.

The glomerulus of each **juxtamedullary nephron** is located in the region of the cortex adjacent to the medulla (Fig. 41-2, *left panel, 7b*). The juxtamedullary nephrons differ anatomically from the superficial nephrons in three important ways: (1) the glomerulus is larger, (2) the loop of Henle is longer and extends deeper into the medulla, and (3) the efferent arteriole forms not only a network of peritubular capillaries but also a series of vascular loops called the **vasa recta.** As illustrated in Fig. 41-2 *(left panel),* the vasa recta descend into the medulla, where they form capillary networks that surround the collecting ducts and ascending limbs of Henle's loop. The blood returns to the cortex in the ascending vasa recta. Although less than 0.7 percent of the renal blood flow enters the vasa recta, these vessels subserve important functions, including conveying oxygen and important nutrients to the tubular segments, delivering substances to the tubules for secretion, serving as a pathway for the return of reabsorbed water and solutes to the circulatory system, and concentrating and diluting the urine (see Chapter 42).

Ultrastructure of the Glomerulus

The first step in urine formation begins with the ultrafiltration of plasma across the glomerular capillaries. The term **ultrafiltration** refers to the passive movement of an essentially protein-free fluid from the glomerular capillaries into Bowman's space. To understand the process of ultrafiltration one must appreciate the anat-

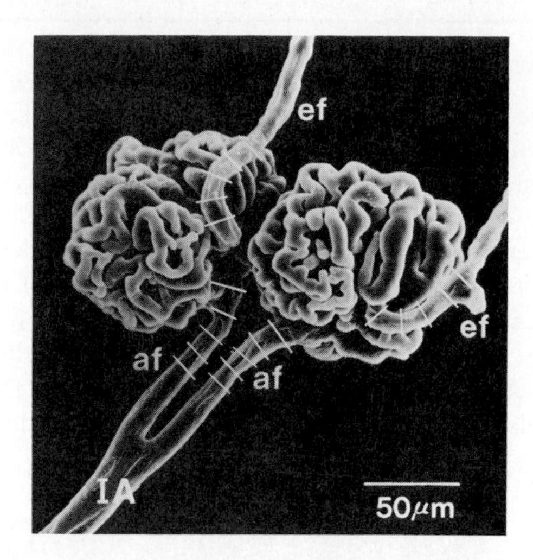

■ **Fig. 41-4** Scanning electron micrograph of interlobular artery (IA) afferent arterioles (af), glomerular capillaries, and efferent arterioles (ef). The renal vessels were filled with an acryl resin by arterial injection. The white bars on the afferent and efferent arterioles indicate that the arterioles are 15 to 20 μm in diameter. (From Kimura K et al: *Am J Physiol* 259:F936, 1990.)

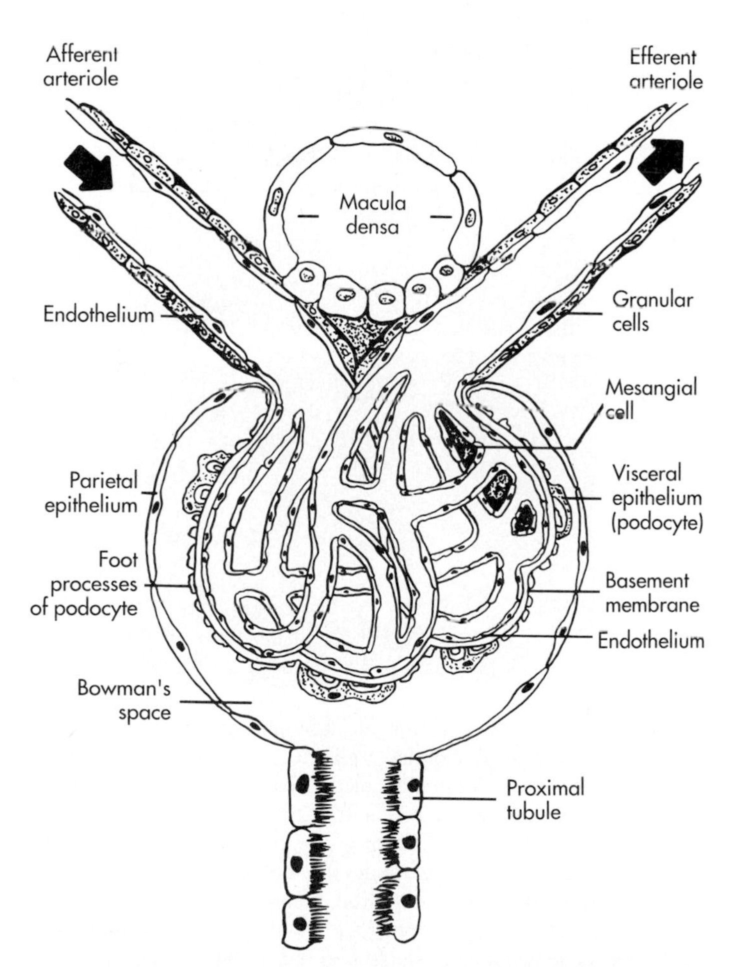

■ **Fig. 41-5** Anatomy of the glomerulus and the juxtaglomerular apparatus. (Modified from Koushanpour E, Kriz W: *Renal physiology-principles, structure and function,* ed 2, Berlin, 1986, Springer-Verlag.)

omy of the glomerulus. The glomerulus consists of a network of capillaries supplied by the afferent arteriole and drained by the efferent arteriole (Fig. 41-4). During development the glomerular capillaries press into the closed end of the proximal tubule. The capillaries are covered by epithelial cells called **podocytes,** which form the **visceral layer** of Bowman's capsule (Fig. 41-5). The visceral cells are reflected at the vascular pole to form the **parietal layer** of Bowman's capsule. The space between the visceral layer and the parietal layer is called **Bowman's space,** which becomes the lumen of the proximal tubule at the urinary pole of the glomerulus.

The endothelial cells of glomerular capillaries are covered by a **basement membrane,** which is surrounded by podocytes (Figs. 41-5 and 41-6). The capillary endothelium, basement membrane, and foot processes of podocytes form the so-called **filtration barrier** (Fig. 41-6). The endothelium is fenestrated (i.e., contains 700 Å holes) and is freely permeable to water; to small solutes such as sodium, urea, and glucose; and even to small protein molecules. Because the fenestrations are relatively large (700 Å) the endothelium acts as a filtration barrier only to cells. The basement membrane consists of three layers (lamina rara interna, lamina densa, and lamina rara externa) and is an important filtration barrier to plasma proteins. The podocytes, which are endocytic, have long fingerlike processes that completely encircle the outer surface of the capillaries (Fig. 41-7). The processes interdigitate to cover the basement membrane and are separated by gaps called **filtration slits.** Each filtration slit is bridged by a thin diaphragm, which contains pores with dimensions of 40 Å × 140 Å. Therefore the filtration slits retard the filtration of some proteins and macromolecules that pass through the endothelium and basement membrane. Because the endothelial cells, basement membrane, and the filtration slits contain negatively charged glycoproteins some molecules are held back because of size and charge. For molecules with an effective molecular radius between 18 and 36 Å cationic molecules are filtered more readily than anionic molecules (see p. 735).

Another important component of the glomerulus is the **mesangium,** which consists of **mesangial cells** and the **mesangial matrix** (Figs. 41-5 and 41-8). Mesangial cells surround glomerular capillaries, provide structural support for the glomerular capillaries, secrete the extracellular matrix, exhibit phagocytic activity, and secrete prostaglandins. Because mesangial cells exhibit contractile activity and are adjacent to glomerular capillaries they may influence the glomerular filtration rate by regulating blood flow through glomerular capillaries. Mesangial cells located outside the glomerulus (between the afferent and efferent arterioles) are called extraglomerular mesangial cells (or lacis cells or Goormaghtigh cells). Lacis cells, like mesangial cells, exhibit phagocytic activity.

The Juxtaglomerular Apparatus

The structures that compose the juxtaglomerular apparatus include (1) the macula densa of the thick ascending limb, (2) the extraglomerular mesangial cells, and (3) the renin-producing **granular cells** of the afferent and efferent arterioles (see Fig. 41-5). Macula densa cells represent a morphologically distinct region of the thick ascending limb that passes through the angle formed by the afferent and efferent arterioles. The cells of the macula densa contact the extraglomerular mesangial cells and the granular cells of the afferent and efferent

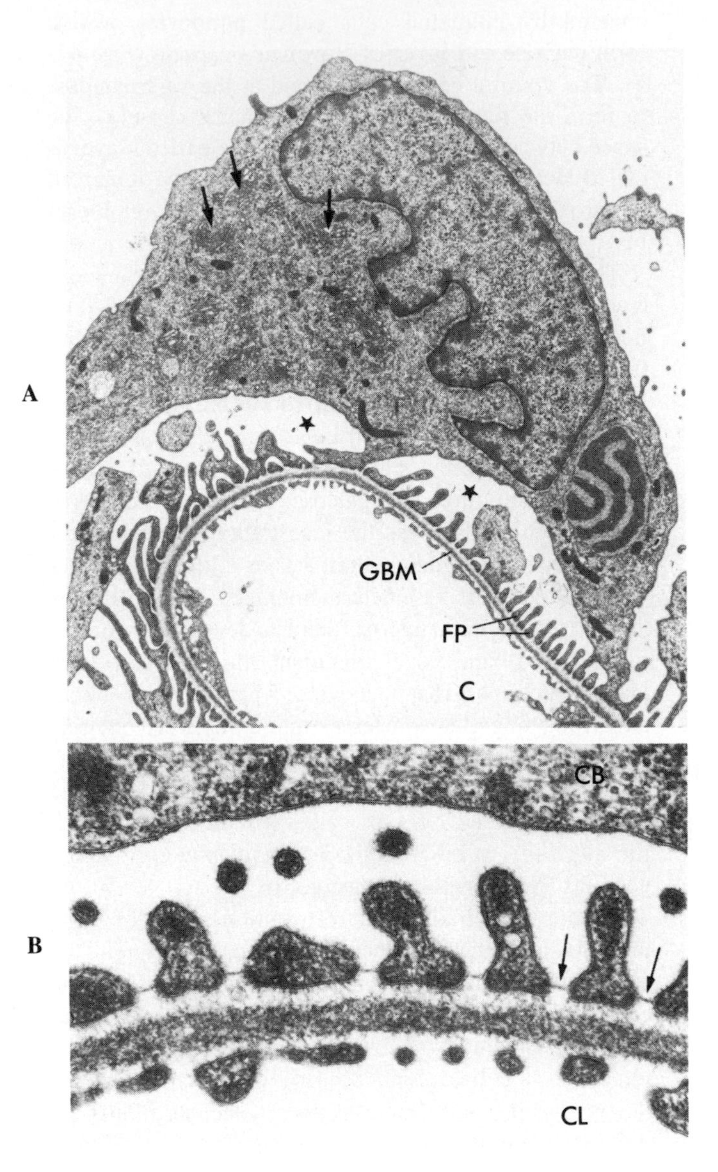

■ **Fig. 41-6** **A,** Electron micrograph of a podocyte surrounding a glomerular capillary. The cell body of the podocyte contains a large nucleus with indentations. Cell processes of the podocyte form the interdigitating foot processes (FP). The arrows in the cytoplasm of the podocyte indicate the well-developed Golgi apparatus. *C,* capillary; *GMB,* glomerular basement membrane. Stars indicate Bowman's space. (Magnification ~5,700×.) **B,** Electron micrograph of the filtration barrier of a glomerular capillary. *CL,* capillary lumen; *CB,* cell body of a podocyte. The filtration barrier is composed of three layers: the endothelium with large pores, the basement membrane, and the foot processes. Note the diaphragm bridging the floor of the filtration slits *(arrows).* (Magnification ~42,700×.) (From Kriz W, Kaissling B: *Structural organization of the mammalian kidney.* In Seldin DW, Giebisch G, editors: *The kidney: physiology and pathophysiology,* ed 2, New York, 1992, Raven Press.)

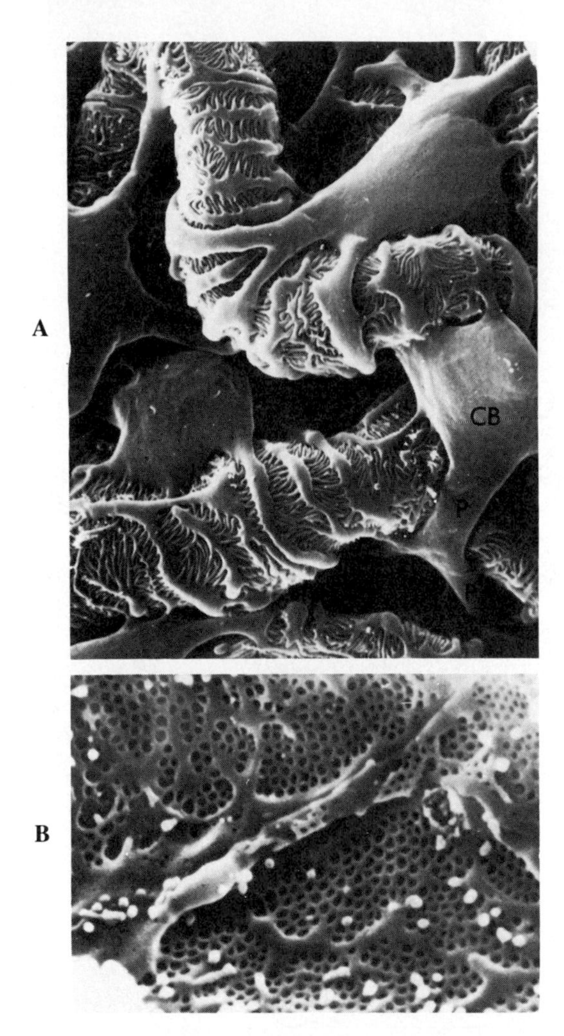

■ **Fig. 41-7** **A,** Scanning electron micrograph showing the outer surface of glomerular capillaries. This is the view that would be seen from Bowman's space. Processes (P) of podocytes run from the cell body (CB) toward the capillaries, where they ultimately split into foot processes. Interdigitation of the foot processes creates the filtration slits. (Magnification ~2,500×.) **B,** Scanning electron micrograph of the inner surface (blood side) of a glomerular capillary. The fenestrations of the endothelial cells are seen as small 700 Å holes. (Magnification ~12,000×.) (From Kriz W, Kaissling B: *Structural organization of the mammalian kidney.* In Seldin DW, Giebisch G, editors: *The kidney: physiology and pathophysiology,* ed 2, New York, 1992, Raven Press.)

arterioles. Granular cells of the afferent and efferent arterioles are modified smooth muscle cells that manufacture, store, and release **renin.** Renin is involved in the formation of angiotensin II and ultimately in the secretion of aldosterone (see Chapter 42). The juxtaglomerular apparatus is one component of an important feedback mechanism (i.e., tubuloglomerular feedback mechanism) that is involved in the autoregulation of renal blood flow and glomerular filtration rate. Details of this feedback mechanism are discussed below.

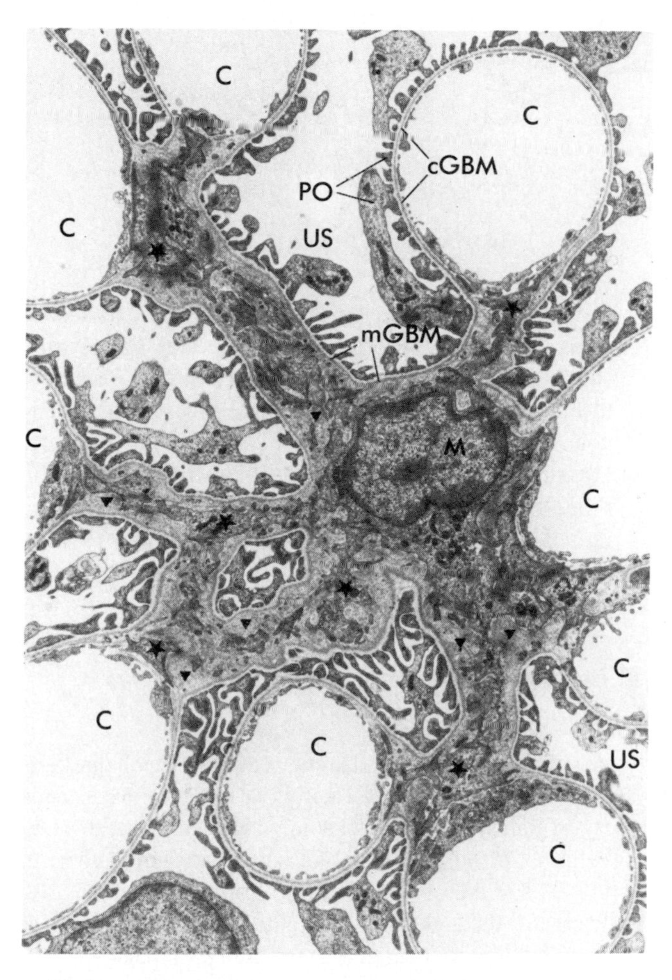

■ **Fig. 41-8** Electron micrograph of the central region in the glomerulus showing glomerular capillaries and mesangial cells. The area between capillaries containing mesangial cells is called the mesangium. *C,* glomerular capillaries; *cGBM,* capillary glomerular basement membrane surrounded by foot processes of podocytes and endothelial cells; *mGBM,* mesangial glomerular basement membrane surrounded by foot processes of podocytes (PO) and mesangial cells; *M,* mesangial cell body that gives rise to several processes, some marked by stars; *US,* urinary space. Note the extensive extracellular matrix surrounding mesangial cells *(marked by triangles).* (Magnification ~4,100×.) (From Kriz W, Kaissling B: *Structural organization of the mammalian kidney.* In Seldin DW, Giebisch G, editors: *The kidney: physiology and pathophysiology,* ed 2, New York, 1992, Raven Press.)

■ *Innervation of the Kidney*

Renal nerves help regulate renal blood flow, glomerular filtration rate, and salt and water reabsorption by the nephron. The nerve supply to the kidneys consists of sympathetic nerve fibers that originate mainly in the celiac plexus. The kidney receives no parasympathetic innervation. **Adrenergic fibers** innervating the kidneys release norepinephrine and dopamine. The adrenergic fibers lie adjacent to the smooth muscle cells of the major branches of the renal artery (interlobar, arcuate, and interlobular arteries) and the afferent and efferent arterioles. Moreover the renin-producing granular cells of the afferent and efferent arterioles are innervated by sympathetic nerves. Renin secretion is elicited by increased sympathetic activity. Nerve fibers also innervate the proximal tubule, loop of Henle, distal tubule, and collecting duct; activation of these nerves enhances sodium reabsorption by these nephron segments.

■ *Anatomy and Physiology of the Lower Urinary Tract*

■ *Gross Anatomy and Histology*

Once urine leaves the renal calices and pelvis it flows through the **ureters** and enters the **bladder,** where urine is stored (Fig. 41-9). The ureters are muscular tubes 30 cm long. They enter the bladder on its posterior aspect near the base, above the bladder neck. The triangular region of the posterior bladder wall, called the **trigone**, lies above the entrance to the posterior urethra and below the point where the ureters enter the bladder. The bladder is composed of two parts: the **fundus,** or body, which stores urine, and the **neck,** which is funnel-shaped and connects with the **urethra.** The bladder neck, which is 2 to 3 cm long, is also called the posterior urethra. In females the posterior urethra is the end of the urinary tract and the point of exit of urine from the body. In males urine flows through the posterior urethra into the anterior urethra, which extends through the penis. Urine leaves the urethra through the **external meatus.**

The renal calices, pelvis, ureter, and bladder are lined with a transitional epithelium composed of several layers of cells: basal columnar cells, intermediate cuboidal cells, and superficial squamous cells. This epithelium is surrounded by a mixture of spiral and longitudinal smooth muscle fibers that are not arranged in discrete layers. The bladder is also lined with a transitional epithelium that is surrounded by a mixture of smooth muscle fibers, called the **detrusor muscle.** Detrusor muscle fibers are arranged at random. They do not form layers except close to the bladder neck, where the fibers form three layers: inner longitudinal, middle circular, and outer longitudinal. Muscle fibers in the

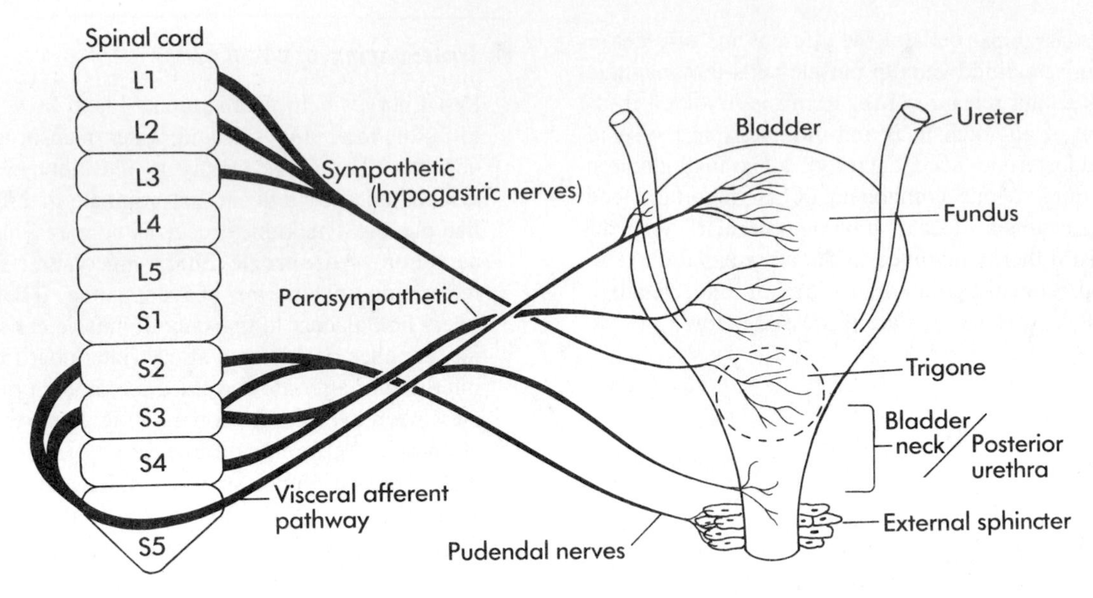

■ **Fig. 41-9** The anatomy of the lower urinary tract and its innervation. See text for details.

bladder neck form the **internal sphincter,** which is not a true sphincter but a thickening of the bladder wall formed by converging muscle fibers. The internal sphincter is not under conscious control. Its inherent tone prevents emptying of the bladder until appropriate stimuli initiate urination. The urethra passes through the urogenital diaphragm, which contains a layer of skeletal muscle called the **external sphincter.** This muscle is under voluntary control and can be used to prevent or interrupt urination, especially in males. In females the external sphincter is poorly developed; thus it is less important in voluntary bladder control. The smooth muscle cells in the lower urinary tract are electrically coupled, exhibit action potentials, contract when stretched, and respond to parasympathetic neurotransmitters.

The walls of the ureters, bladder, and urethra are highly folded and thereby very distensible. In the bladder and urethra these folds are called **rugae.** As the bladder fills with urine the rugae flatten out and the volume of the bladder increases with very little change in intravesical pressure. To illustrate the highly compliant nature of the bladder the volume of this structure can increase from a minimal volume of 10 ml after urination to 400 ml with a pressure change of only 5 cm H_2O.

■ *Innervation of the Bladder*

Innervation of the bladder and urethra is important in controlling urination. The smooth muscle of the bladder neck receives sympathetic innervation from the hypogastric nerves. **α-adrenergic receptors,** located mainly in the bladder neck and the urethra, cause contraction. Stimulation of these receptors facilitates storage of urine by inducing closure of the urethra. **Parasympathetic fibers** via pelvic nerves (muscarinic) innervate the body of the bladder and cause a sustained bladder contraction. Sensory fibers of the pelvic nerves (visceral afferent pathway) also innervate the fundus. The **pudendal nerves** innervate the skeletal muscle fibers of the external sphincter and excitatory impulses cause contraction.

■ *Passage of Urine from the Kidney to the Bladder*

As urine collects in the renal calices stretch promotes their inherent pacemaker activity. The pacemaker activity initiates a peristaltic contraction that begins in the calices, spreads to the pelvis and along the length of the ureter, and thereby forces urine from the renal pelvis toward the bladder. The peristaltic wave, caused by action potentials generated by the pacemaker, passes along the smooth muscle syncytium. The ureters are innervated with sensory nerve fibers (pelvic nerves). When the ureter is blocked with a renal stone reflex constriction of the ureter around the stone elicits severe pain.

■ *Micturition*

Micturition is the process of emptying the urinary bladder. Two processes are involved: (1) progressive filling of the bladder until the pressure rises to a critical value and (2) a neuronal reflex called the **micturition reflex,** which empties the bladder. *The micturition re-*

flex is an automatic spinal cord reflex. However it can be inhibited or facilitated by centers in the brain stem and the cerebral cortex.

Filling of the bladder stretches the bladder wall and causes it to contract. Contractions are the result of a reflex initiated by stretch receptors in the bladder. Sensory signals from the bladder fundus enter the spinal cord via pelvic nerves and return directly to the bladder through parasympathetic fibers in the same nerves. Stimulation of parasympathetic fibers causes intense stimulation of the detrusor muscle. Because the smooth muscle in the bladder is a **syncytium** (i.e., electrically coupled) stimulation of the detrusor also causes the muscle cells in the neck of the bladder to contract. Because the muscle fibers of the bladder outlet are oriented both longitudinally and radially, contraction opens the bladder neck and allows urine to flow through the posterior urethra. Voluntary relaxation of the external sphincter by cortical inhibition of the pudendal nerve permits the flow of urine through the external meatus. *Voluntary relaxation of the external sphincter is required and may be the event that initiates micturition*. Interruption of the hypogastric sympathetic nerves and the pudendal nerves to the lower urinary tract does not alter the micturition reflex. However destruction of the parasympathetic nerves results in complete bladder dysfunction.

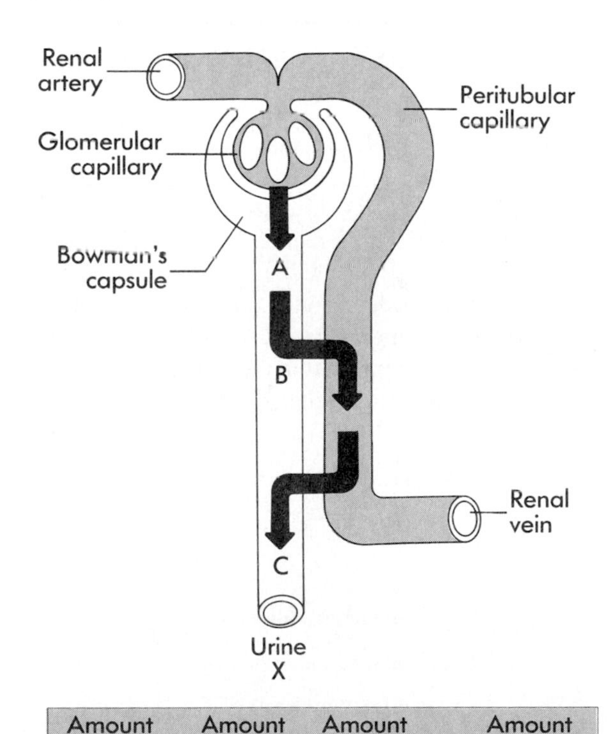

Fig. 41-10 Schematic representation of the entire nephron population of both kidneys depicting the three general processes that determine and modify the composition of the urine: glomerular filtration *(A)*, tubular reabsorption *(B)*, and tubular secretion *(C)*.

Assessment of Renal Function

The amount of a substance that appears in the urine reflects the coordinated action of the nephron's various segments and represents three general processes:

1. Glomerular filtration
2. Reabsorption of the substance from the tubular fluid back into the blood
3. Secretion of the substance from the blood into the tubular fluid

These three processes are illustrated in Fig. 41-10, in which the entire nephron population of both kidneys is represented by a single nephron.

This section develops the renal clearance concept, which can be used to quantitate these processes. In addition the use of clearance to measure the rates of glomerular filtration and renal blood flow are discussed.

Renal Clearance

The concept of renal clearance is based on the principle of mass balance (Fig. 41-11). The renal artery is the single input source to the kidney, whereas the renal vein and ureter constitute the two output routes. Maintaining mass balance the following relationship can be derived:

$$P_x^a \cdot RPF^a = (P_x^v \cdot RPF^v) + (U_x \cdot \dot{V}) \qquad (1)$$

where

P_x^a and P_x^v = concentrations of substance x in the renal artery and renal vein plasma

RPF^a and RPF^v = renal plasma flow rates in the artery and vein

U_x = concentration of x in the urine

$\dot{V}$ = urine flow rate per minute.

Using this relationship one can quantitate the amount of x excreted in the urine versus the amount returned to the systemic circulation in the renal venous blood.

The principle of renal clearance (C_x) emphasizes the excretory function of the kidney; it considers only the rate at which a substance is excreted into the urine, and not the rate at which it is returned to the systemic circulation in the renal vein. Therefore in terms of mass balance (equation 1), the urinary excretion rate of x ($U_x \cdot \dot{V}$) is proportional to the plasma concentration of x (P_x^a).

$$P_x^a \sim U_x \cdot \dot{V} \qquad (2)$$

To equate the urinary excretion rate of x to its renal arterial plasma concentration one must determine the rate at which x is removed from the plasma by the kidneys. This removal rate is the clearance (C_x).

$$P_x^a \cdot C_x = U_x \cdot \dot{V} \qquad (3)$$

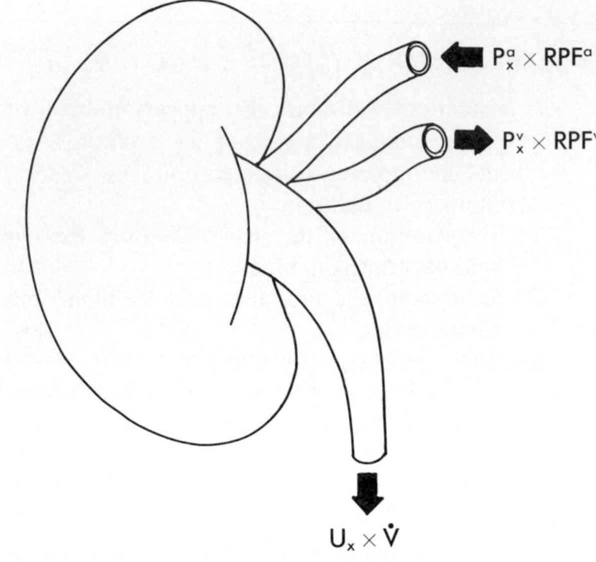

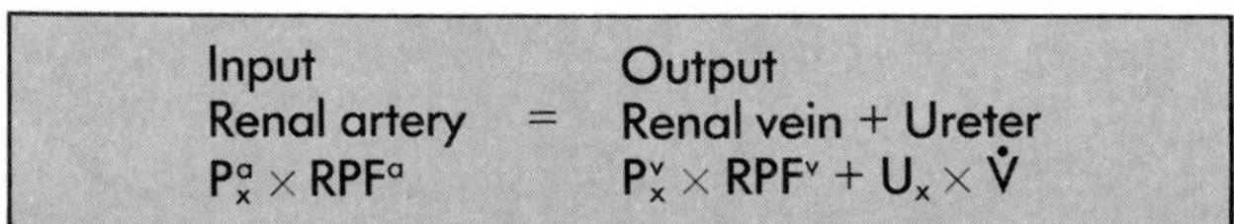

■ **Fig. 41-11** Mass balance relationships for the kidney. See text for definition of symbols.

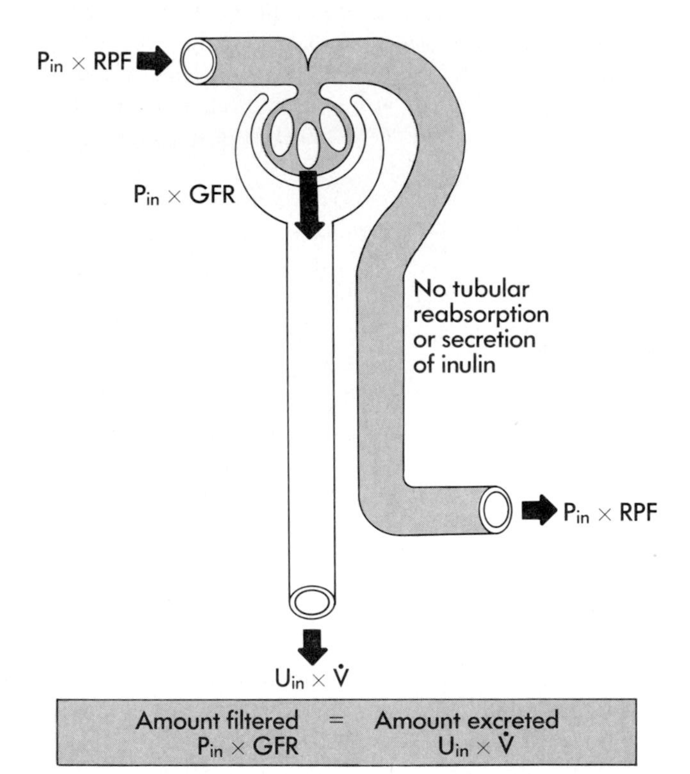

■ **Fig. 41-12** Renal handling of inulin. Inulin is freely filtered at the glomerulus and is neither reabsorbed, secreted, nor metabolized by the nephron. P_{in}, plasma inulin concentration; *RPF*, renal plasma flow; *GFR*, glomerular filtration rate; U_{in}, urinary concentration of inulin; $\dot{V}$, urine flow rate. Note that all the inulin entering the kidney in the renal artery is not filtered at the glomerulus (normally 15% to 20% of plasma and therefore inulin is filtered). The portion that is not filtered is returned to the systemic circulation in the renal vein.

Rearranging equation 3 and assuming that the concentration of x in the renal artery plasma is identical to its concentration in a plasma sample from any peripheral blood vessel (e.g., arm vein blood sample: P_x) the following relationship is obtained:

$$C_x = \frac{U_x \cdot \dot{V}}{P_x} \tag{4}$$

Clearance *has the dimensions of volume/time, and it represents the volume of plasma from which all the substance has been removed and excreted into the urine per unit time*. This last point is best illustrated by considering the following example:

If a substance is present in the urine at a concentration of 100 mg/ml and the urine flow rate is 1 ml/min, then the excretion rate for this substance is calculated as

$$\text{Excretion rate} = U_x \cdot \dot{V} = (100 \text{ mg/ml}) \times (1 \text{ ml/min}) = 100 \text{ mg/min} \tag{5}$$

If this substance is present in the plasma at a concentration of 1 mg/ml, then its clearance according to equation 4 is:

$$C_x = \frac{U_x \cdot \dot{V}}{P_x} = \frac{100 \text{ mg/min}}{1 \text{ mg/ml}} = 100 \text{ ml/min} \tag{6}$$

That is, 100 ml of plasma is completely cleared of substance X each minute.

■ *Glomerular Filtration Rate and Clearance of Inulin*

Inulin is a polyfructose molecule (molecular weight, ca. 5,000) that can be used to measure the **glomerular filtration rate (GFR).** It is not produced endogenously by the body and therefore must be administered intravenously. Inulin is freely filtered at the glomerulus and is neither reabsorbed, secreted, nor metabolized by the cells of the nephron. As illustrated in Fig. 41-12 the amount of inulin excreted in the urine per minute equals the amount of inulin filtered at the glomerulus each minute:

$$\text{Amount filtered} = \text{Amount excreted} \tag{7}$$
$$GFR \cdot P_{in} = U_{in} \cdot \dot{V}$$

where

GFR = glomerular filtration rate

P_{in} and U_{in} = plasma and urine concentrations of inulin

$\dot{V}$ = urine flow rate.

If equation 7 is solved for the GFR,

$$GFR = \frac{U_{in} \cdot \dot{V}}{P_{in}} \tag{8}$$

This equation is the same form as that for clearance (see equation 4). Thus *the clearance of inulin provides a means for determining the GFR*.

Inulin is not the only substance that can be used to

measure the GFR. Any substance that meets the following criteria will serve as an appropriate marker for the measurement of GFR:

1. The substance must be freely filtered by the glomerulus.
2. The substance must not be reabsorbed or secreted by the nephron.
3. The substance must not be metabolized or produced by the kidney.
4. The substance must not alter GFR.

Whereas inulin is used extensively in experimental studies the fact that it must be infused intravenously limits its use in clinical settings. Consequently **creatinine** is used to estimate the GFR in clinical practice. Creatinine is a by-product of skeletal muscle creatine metabolism. It is produced at a relatively constant rate, and the amount produced is proportional to the muscle mass. With regard to the measurement of GFR creatinine has an advantage over inulin in that it is produced endogenously, thus obviating the need for intravenous infusion. However creatinine is not a perfect substance to measure GFR, because it is secreted to a small extent by the organic cation secretory system in the proximal tubule (see p. 743). The error introduced by this secretory component is approximately 10%. Thus the amount of creatinine excreted in the urine exceeds the amount expected from filtration alone by 10%. However the method used to quantitate the plasma creatinine concentration overestimates the true value by 10%. Consequently the two errors cancel, and *the creatinine clearance provides a reasonably accurate measure of the GFR.*

As illustrated in Fig. 41-12 not all the inulin (or any substance used to measure GFR) entering the kidney in the renal arterial plasma is filtered at the glomerulus. Likewise not all of the plasma entering the kidney and therefore the glomerulus is filtered.* The portion of plasma that is filtered is termed the **filtration fraction;** it is determined as

$$\text{Filtration fraction} = \frac{\text{GFR}}{\text{RPF}} \qquad (9)$$

where, again, RPF is renal plasma flow. Under normal conditions the filtration fraction averages 0.15 to 0.20. This means that only 15% to 20% of the plasma that enters the glomerulus is actually filtered. The remaining 80% to 85% continues on through the glomerular capillaries into the peritubular capillaries and finally is returned to the systemic circulation in the renal vein.

Filtration Plus Tubular Reabsorption and Glucose Clearance

Renal glucose excretion is determined by the amount filtered at the glomerulus minus the amount subsequently reabsorbed from the tubular fluid by the cells of the nephron. The amount of glucose filtered at the glomerulus, termed the **filtered load,** is determined by the magnitude of the GFR and the plasma glucose concentration (P_G):

$$\text{Filtered load (glucose)} = \text{GFR} \cdot P_G \qquad (10)$$

Glucose is reabsorbed from tubular fluid by the cells of the proximal tubule. The transport mechanism for glucose reabsorption is described in detail below. For the purposes of this discussion the glucose transport system can be viewed quantitatively as if it has a maximum transport rate. This maximum transport rate is termed the **tubular transport maximum (T_m).** The T_m for glucose varies from individual to individual and on average has a value of 375 mg/min. Thus *when the filtered load of glucose is less than 375 mg/min all the glucose is reabsorbed and returned to the body via the renal vein; no glucose appears in the urine, and the clearance of glucose is zero.* When the filtered load exceeds 375 mg/min however the maximal amount will be reabsorbed and returned to the body via the renal vein; the remainder is excreted in the urine, and some glucose is thereby cleared from the body. Fig. 41-13 illustrates these two cases.

The two cases illustrated show that the renal handling of glucose depends on the plasma glucose concentration. Fig. 41-14 depicts the relationships of the plasma glucose concentration to its filtered load, excretion rate, and tubular reabsorption rate. As indicated by equation 10 the filtered load for glucose increases linearly as the plasma glucose concentration is increased (GFR is constant). If the filtered load is below the T_m glucose reabsorption is complete, and the reabsorptive rate increases linearly with the filtered load. As a consequence at these low plasma glucose concentrations the excretion rate is zero. When the T_m is reached, the reabsorptive rate is constant, and in the face of an increasing filtered load glucose appears in the urine. The plasma glucose concentration at which glucose first appears in the urine is called the **plasma threshold.** Beyond this point the excretion rate increases linearly and parallels the filtered load.

Careful examination of Fig. 41-14 shows that the reabsorption and excretion curves display a nonlinear transition at the plasma threshold concentration. This "rounding" of the curves is termed splay. **Splay** likely represents heterogeneity in the T_m value for individual nephrons. Thus the T_m for any given nephron may be slightly higher or slightly lower than the mean T_m for all the nephrons.

Filtration Plus Tubular Secretion and PAH Clearance

p-Aminohippuric acid (PAH) is an organic anion excreted into the urine by the processes of glomerular fil-

*Nearly all (90%) of the plasma that enters the kidney in the renal artery passes through the glomerulus.

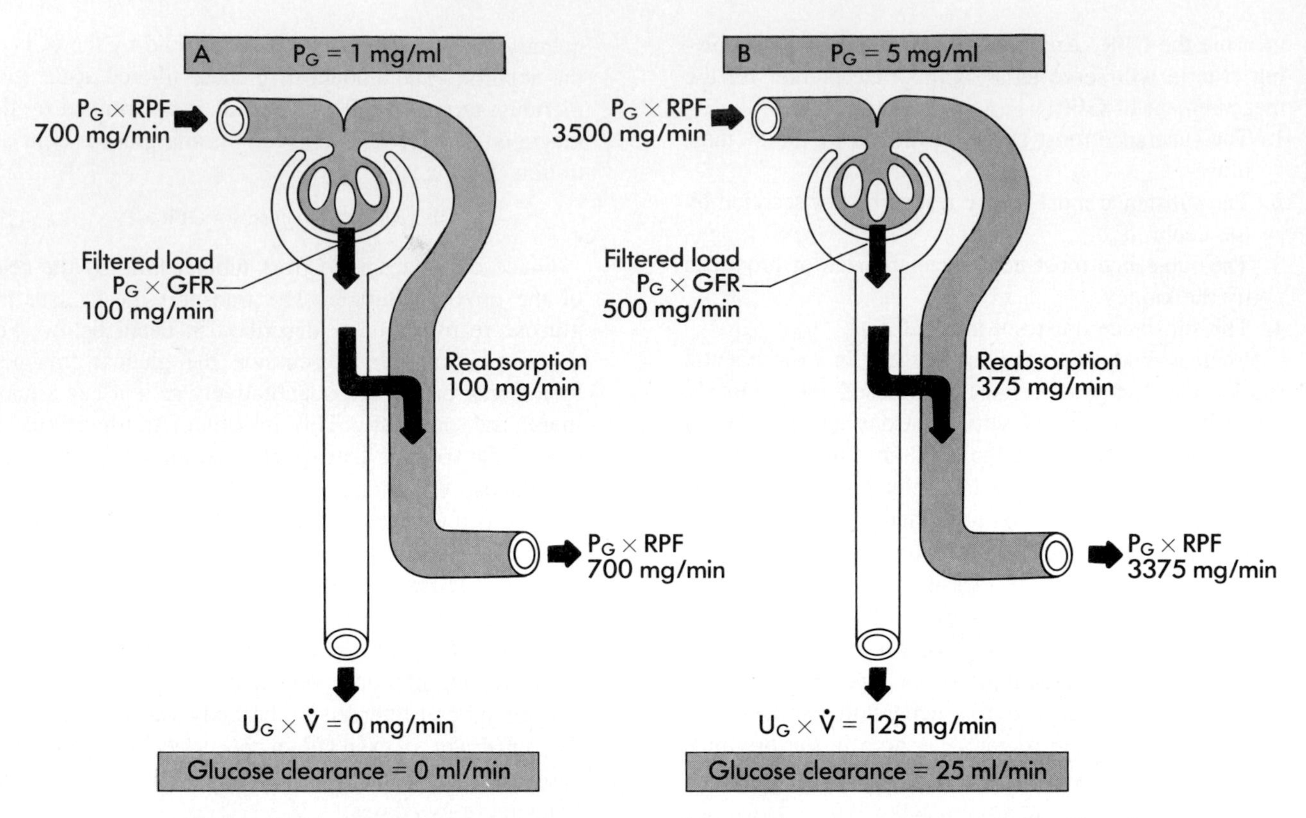

■ **Fig. 41-13** Renal handling of glucose at two different plasma concentrations. **A,** The filtered load is less than the tubular maximum (T_m) for glucose. **B,** The filtered load exceeds the T_m for glucose. For both cases the renal plasma flow (RPF) is 700 ml/min, the glomerular filtration rate (GFR) is 100 ml/min, and the glucose T_m is 375 mg/min. P_G, plasma glucose concentration; U_G, urine glucose concentration; $\dot{V}$, urine flow rate. The clearance of glucose is calculated from equation 4.

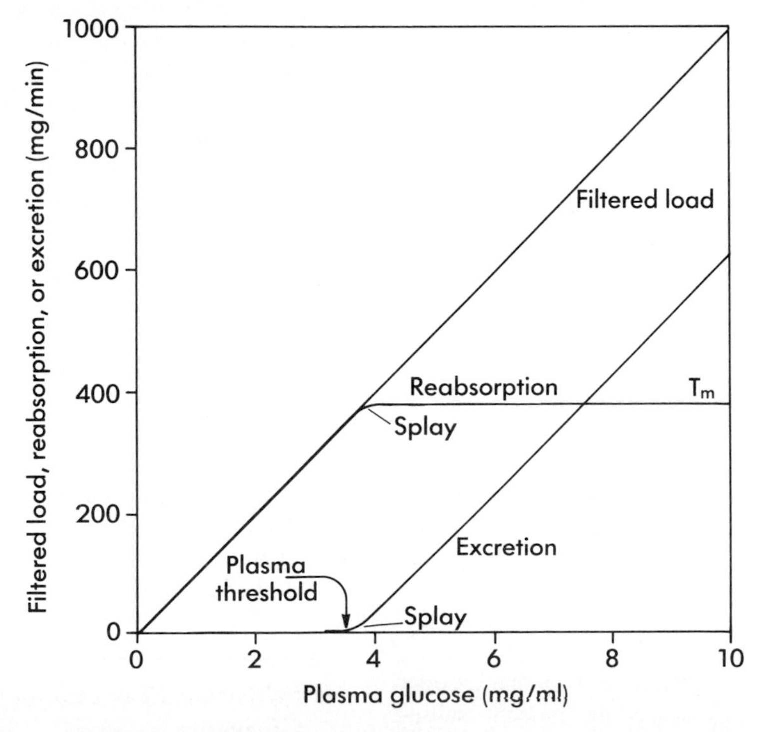

■ **Fig. 41-14** Dependence of the glucose filtered load, excretion rate, and reabsorptive transport rate on the plasma glucose concentration. See text for details.

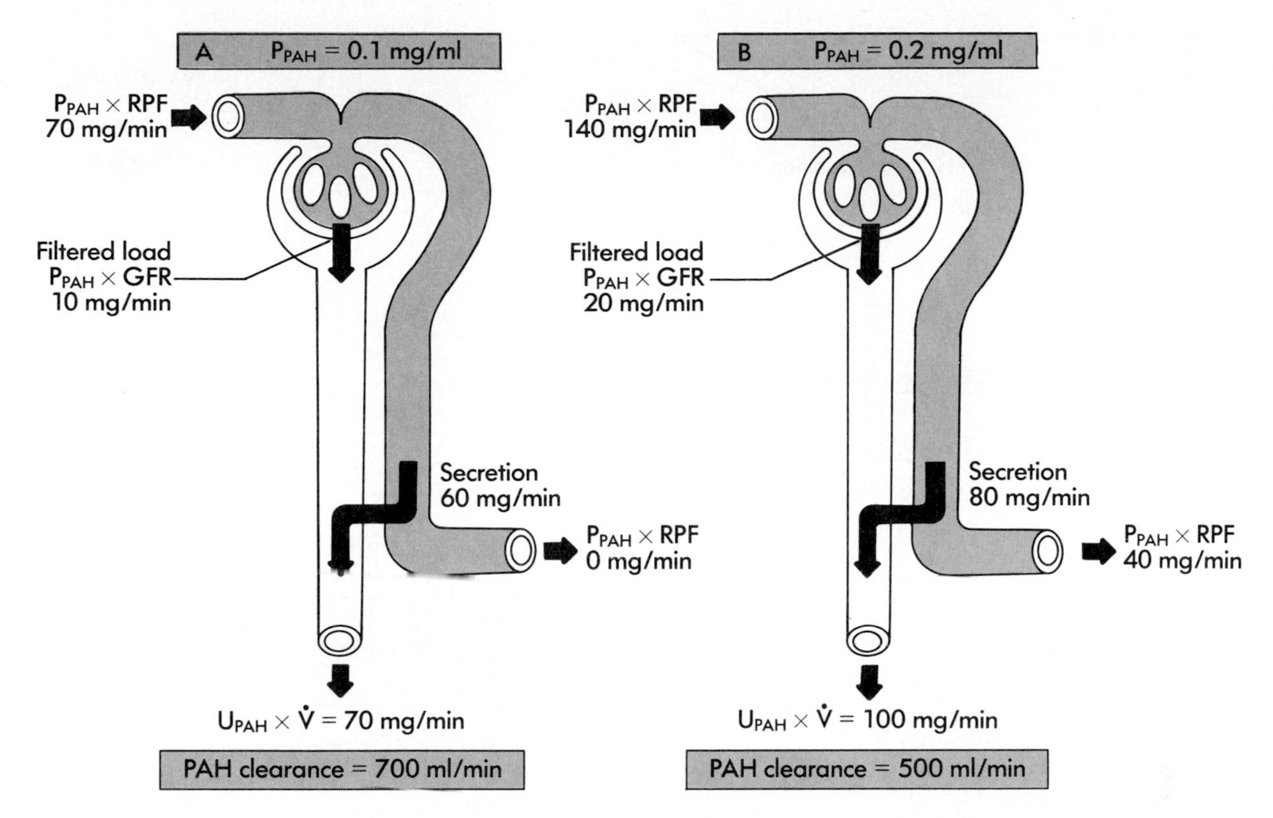

■ **Fig. 41-15** Renal handling of p-aminohippuric acid (PAH) at two different plasma concentrations (P_{PAH}). **A,** The P_{PAH} is less than the value that would lead to saturation of the PAH secretory mechanism. **B,** The elevated P_{PAH} results in delivery of PAH to the secretory mechanism exceeding the transport maximum (T_m). For both cases the renal plasma flow (RPF) is 700 ml/min, the glomerular filtration rate (GFR) is 100 ml/min, and the T_m for PAH is 80 mg/min. U_{PAH}, urine PAH concentration; $\dot{V}$, urine flow rate. The clearance of PAH is calculated from equation 4.

tration and tubular secretion. Thus the amount excreted will be the sum of the filtered load plus the secreted component. As with inulin PAH is not produced in the body and therefore must be infused. PAH is transported by the organic anion secretory system of the proximal tubule; this system is discussed in detail later in this chapter. For this discussion it is sufficient to recognize that, like glucose, the PAH transport mechanism has a maximum value. Because secretion of PAH occurs from the peritubular capillary into the tubular lumen it is not the filtered load of PAH that determines whether or not the secretory mechanism is saturated but rather the amount of PAH delivered to the peritubular capillaries. The T_m for PAH varies from individual to individual and has an average value of 80 mg/min. Delivery of PAH to the peritubular capillaries at a lower rate causes virtually all of the PAH to be secreted into the tubular fluid, and thus little PAH remains in the renal vein plasma. When the plasma PAH concentration is increased (generally above 0.12 mg/ml) the delivery of PAH to the peritubular capillaries exceeds 80 mg/min. As a result 80 mg/min is secreted into the tubule lumen, and the remainder is returned to the systemic circulation via the renal vein. These two cases are summarized in Fig. 41-15.

The renal handling of PAH, like that of glucose, varies as a function of the plasma PAH concentration. Fig. 41-16 summarizes the relationship of the plasma PAH concentration to its filtered load, secretory rate, and excretion rate. At a constant GFR the filtered load of PAH increases linearly with the increase in plasma PAH concentration. The secretion process becomes saturated between plasma PAH concentrations of 0.1 and 0.2 mg/ml. Below the T_m virtually all of the PAH that enters the kidney is excreted. When the Tm is exceeded the secretory component is constant and the excretion rate for PAH increases in parallel with the filtered load. Splay is seen in both the secretion and excretion curves. The splay again reflects the heterogeneity in the T_m value for the PAH secretory mechanism among different nephrons.

■ *Renal Plasma Flow and Renal Blood Flow*

When the plasma PAH concentration is low so that the T_m of the secretory mechanism is not exceeded (gener-

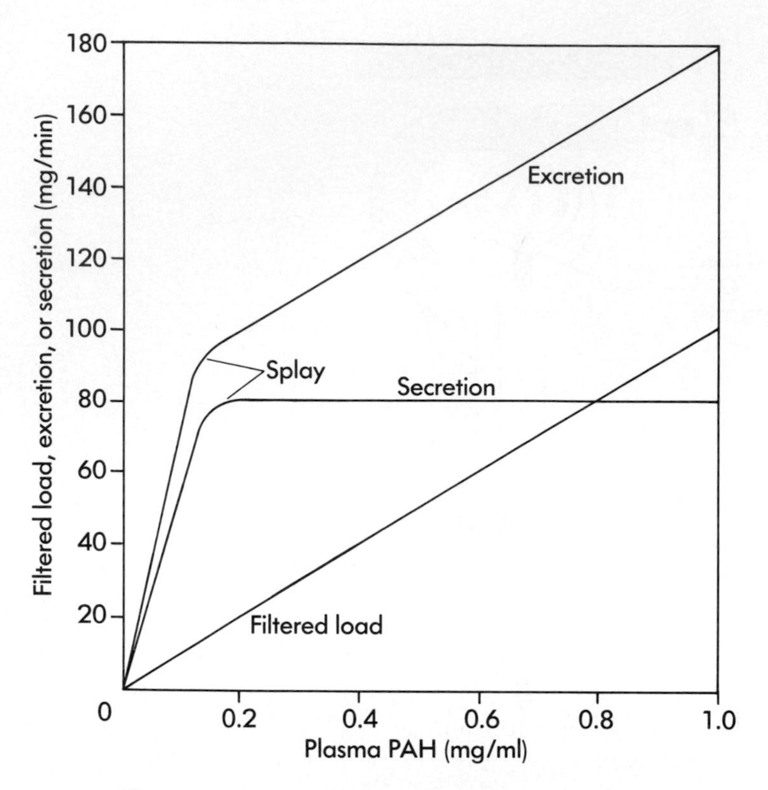

■ **Fig. 41-16** Dependence of PAH filtered load, excretion, and secretory transport rate on the plasma PAH concentration. See text for details.

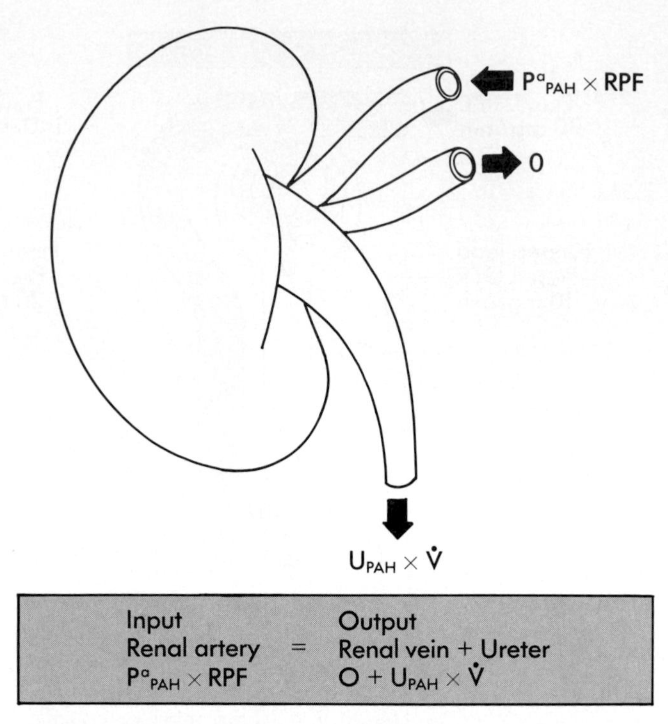

■ **Fig. 41-17** Mass balance relationships for the use of PAH clearance to measure the renal plasma flow (RPF). P^a_{PAH}, plasma PAH concentration; U_{PAH}, urine PAH concentration; $\dot{V}$, urine flow rate. P^a_{PAH} must be low so that the T_m for PAH is not exceeded.

ally at plasma [PAH] below 0.12 mg/ml), its clearance can be used to measure the **renal plasma flow.**

Fig. 41-17 depicts the renal handling of PAH in terms of whole kidney mass balance and illustrates why when nonsaturating (i.e., T_m for PAH is not exceeded) concentrations of PAH are used its clearance provides a measure of the renal plasma flow (RPF). The amount of PAH that arrives at the kidneys per minute is simply the product of the plasma PAH concentration (P^a_{PAH}) and the RPF. Because all of the PAH is excreted into the urine and none is returned to the systemic circulation via the renal vein the following mass balance relationship holds true:

$$RPF \cdot P^a_{PAH} = U_{PAH} \cdot \dot{V} \qquad (11)$$

where

U_{PAH} = urine PAH concentration
$\dot{V}$ = urine flow rate

Rearranging and solving for the RPF results in the following equation:

$$RPF = \frac{U_{PAH} \cdot \dot{V}}{P^a_{PAH}} \qquad (12)$$

This equation conforms to the general clearance equation (equation 4). Thus at low plasma PAH concentrations PAH clearance is equal to renal plasma flow. At high plasma PAH concentrations however the PAH secretory mechanism is saturated and a significant amount of PAH appears in the renal venous blood. Under this condition, equations 11 and 12 do not hold and the clearance of PAH does not equal the renal plasma flow.

The relationship between PAH clearance and renal plasma flow described here is idealized. Even at plasma PAH concentrations that do not exceed the T_m some PAH still appears in the renal venous blood. The reason for this is related to the anatomy of the nephron and of the renal blood vessels. The PAH secretory mechanism is located in the proximal tubule. Consequently if all of the PAH entering the renal artery were to be secreted into the tubular fluid all of the plasma would have to flow through the peritubular capillaries surrounding the proximal tubule. Approximately 90% of plasma does in fact flow through peritubular capillaries that surround the proximal tubules. However 10% does not; this plasma perfuses some of the medullary structures, the renal capsule, and parts of the renal hilum. Thus the PAH in this plasma cannot be secreted, and this portion of PAH is returned to the systemic circulation in the renal vein plasma. In recognition of the fact that the clearance of PAH does not provide a fully accurate measure of the renal plasma flow (i.e., it underestimates the true value by approximately 10%) it is more appropriate to refer to the clearance of PAH as providing a measure of the **effective renal plasma flow (ERPF)**—effective in the sense that this value represents plasma flow past portions of the nephron that can effectively secrete PAH.

The clearance of PAH also can be used to estimate the **renal blood flow (RBF).** Whole blood consists of a cellular fraction and a plasma fraction. Normally the plasma accounts for 50% to 60% of the blood volume and the cells account for the remainder. The fraction of a blood sample that is composed of cells is termed the **hematocrit** (HCT). Normally the hematocrit is in the range of 40% to 50%. Once the hematocrit is known renal blood flow can be calculated as

$$\text{RBF} = \frac{\text{RPF}}{1 - \text{HCT}} \tag{13}$$

Using Clearance to Estimate Transport Mechanisms

As depicted in Fig. 41-10 the excretion rate for any substance can be determined as

$$\text{Excretion rate} = \text{Filtered load} - \text{Reabsorption rate} + \text{Secretion rate} \tag{14}$$

$$U_x \cdot \dot{V} = \text{GFR} \cdot P_x - R + S$$

Most substances are filtered and either reabsorbed or secreted. The important exceptions to this general rule are K^+ and urea, which undergo filtration as well as both reabsorption and secretion. If a substance is known to be filtered freely at the glomerulus comparison of its clearance with that of inulin (measure of GFR) indicates the net handling of the substance by the kidney. Thus:

1. If its clearance is less than that of inulin, the substance is reabsorbed by the nephron (e.g., glucose).
2. If its clearance is greater than that of inulin, the substance is secreted (e.g., PAH).
3. If its clearance equals that of inulin, the substance is only filtered.

For those substances that are both reabsorbed and secreted the clearance reflects the dominant transport system.

The conclusions obtained about transport mechanism from the analysis of clearance values must be considered carefully. For example suppose the renal handling of a substance (x) occurs solely by glomerular filtration. If substance x were filtered freely, its clearance would be equal to that of inulin. However consider what happens if 50% of x is bound to plasma protein. Because only the unbound portion can be filtered and thus excreted the clearance of substance x will be less than that of inulin by 50%. If we did not know in advance that x was partially protein-bound, we would conclude erroneously that substance x was reabsorbed by the nephron.*

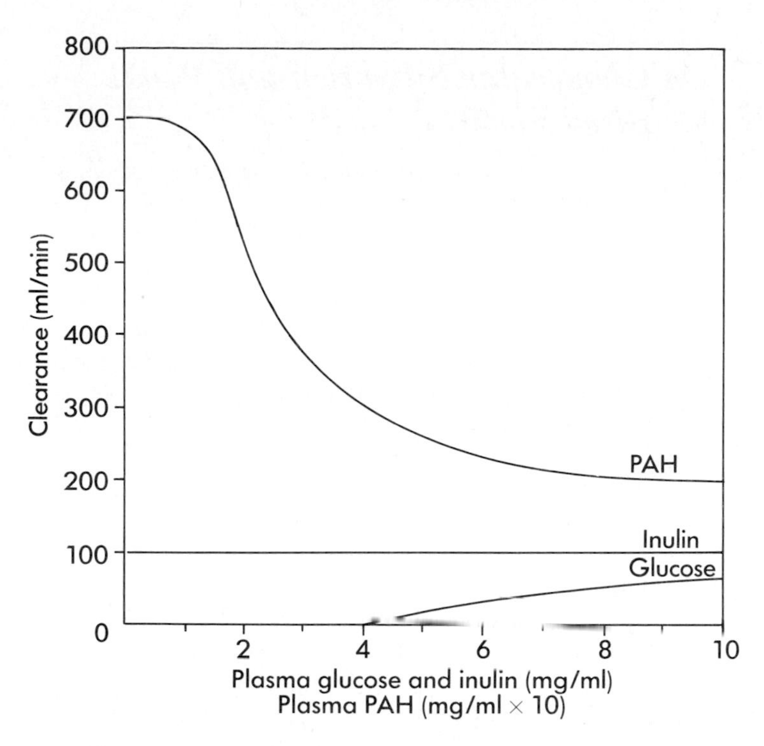

Fig. 41-18 The dependence of clearance of PAH, inulin, and glucose on their plasma concentrations.

One final point regarding the use of clearance to assess renal function relates to the dependency of clearance on the plasma concentration of the substance. Fig. 41-18 depicts the clearances for inulin, glucose, and PAH as a function of their plasma concentrations (for all substances and at all plasma concentrations the GFR is constant at 100 ml/min). Because the GFR is constant the clearance of inulin is also constant regardless of its plasma concentration. In contrast the clearances of glucose and PAH vary with their plasma concentrations. At low plasma concentrations the clearance of PAH exceeds that of inulin whereas the clearance of glucose is zero. As the plasma concentrations of PAH and glucose increase their clearances approach that of inulin. The reason for this convergence of the PAH and glucose clearances is that at high plasma concentrations the secretory (PAH) and reabsorptive (glucose) mechanisms are saturated (T_m is exceeded) and the filtered load becomes a much larger fraction of the total amount of the substance excreted in the urine.

*When the renal clearance of a protein-bound substance is calculated the total plasma concentration (bound plus unbound) is used. However only the unbound portion can be filtered and only this portion is used to calculate the filtered load.

Glomerular Filtration and Renal Blood Flow

The first step in the formation of urine is the production of an ultrafiltrate of the plasma by the glomerulus. The ultrafiltrate is devoid of cellular elements and is essentially protein-free. The concentrations of salts and organic molecules, such as glucose and amino acids, are similar in the plasma and ultrafiltrate. Ultrafiltration is driven by Starling forces (see p. 473) across the glomerular capillaries, and changes in these forces and in renal plasma flow alter the glomerular filtration rate. Glomerular filtration rate and renal plasma flow normally are held within very narrow ranges by a phenomenon called autoregulation (see pp. 481 and 737). This section will review the composition of the glomerular filtrate, the dynamics of its formation, and the relationship between renal plasma flow and glomerular filtration rate. In addition the factors that contribute to the autoregulation of glomerular filtration rate and renal blood flow will be discussed.

Determinants of Ultrafiltrate Composition

The unique structure of the glomerular filtration barrier (capillary endothelium, basement membrane, and filtration slits of the podocytes) determines the composition of the ultrafiltrate of plasma. *The glomerular filtration barrier restricts the filtration of molecules on the basis of size and electrical charge.* Table 41-1 illustrates the effect of size on filtration. In general molecules with a radius less than 18 Å are filtered freely, molecules larger than 36 Å are not filtered, and molecules between 18 and 36 Å are filtered to various degrees. For example albumin, an anionic protein that has an effective molecular radius of 35 Å, is filtered poorly: approximately 7 grams of albumin is filtered each day.* Because albumin is reabsorbed avidly by the proximal tubule however almost none appears in the urine.

Fig. 41-19 illustrates how electrical charge affects the filtration of dextrans by the glomerulus. Dextrans are a family of exogenous polysaccharides that are manufactured in various molecular weights, as well as in an electrically neutral form or with negative charges (polyanionic) or positive charges (polycationic). At constant charge as the size (i.e., effective molecular radius) increases filtration decreases. For any given molecular radius cationic molecules are more readily filtered than are anionic molecules. The restriction of anionic molecules is explained by the presence of negatively charged glycoproteins on the surface of all components of the glomerular filtration barrier. These charged glycoproteins repel similarly charged molecules. Because most plasma proteins are negatively charged the negative charge on the filtration barrier restricts the filtration of proteins that have a molecular radius of 18 to 36 Å.

*Approximately 70,000 g/day of albumin passes through the glomeruli. Therefore the filtration of 7 g/day represents only 0.01%. This is well below the filtration fraction for substances that are freely filtered (15% to 20%).

Table 41-1 Relationship between molecular radius and glomerular filterability

Substance	*Molecular wt (g)*	*Molecular radius (Å)*	*Filterability*
Water	18	1.0	1.0
Sodium	23	1.4	1.0
Urea	60	1.6	1.0
Glucose	180	3.6	1.0
Sucrose	342	4.4	1.0
Inulin	5,500	14.8	0.98
Myoglobin	17,000	19.5	0.75
Egg albumin	43,500	28.5	0.22
Hemoglobin	68,000	32.5	0.03
Serum albumin	69,000	35.5	<0.01

A value of 1.0 for filterability indicates that it is filtered freely.

The importance of the negative charges on the filtration barrier in restricting the filtration of plasma proteins is illustrated in Fig. 41-20. Removal of negative charges from the filtration barrier causes proteins to be filtered solely on the basis of their effective molecular radius. Hence at any molecular radius between approximately 18 and 36 Å the filtration of polyanionic proteins will exceed the filtration that prevails in the normal state, in which the filtration barrier has anionic charges. In a number of glomerular diseases the negative charge on the filtration barrier is lost. As a result filtration of proteins is increased and proteins appear in the urine **(proteinuria).**

Dynamics of Ultrafiltration

The forces responsible for the glomerular filtration of plasma are the same as those involved in fluid exchange across all capillary beds. *Ultrafiltration occurs because* ***Starling forces*** *drive fluid from the lumen of glomerular capillaries, across the filtration barrier, into Bowman's space.* As shown in Fig. 41-21 the Starling forces across glomerular capillaries are similar to the forces that promote filtration across other capillary beds and include hydrostatic and oncotic pressures (see p. 473). The hydrostatic pressure in the glomerular capillary (P_{GC}) is oriented so as to promote the movement of fluid from the glomerular capillary into Bowman's space. Because the glomerular ultrafiltrate is essentially protein-free the oncotic pressure in Bowman's space (π_{BS}) is near zero. Therefore P_{GC} is the only force that favors filtration, and it is opposed by the hydrostatic pressure in Bowman's space (P_{BS}) and the oncotic pressure in the glomerular capillary (π_{GC}).

As illustrated in Figs. 41-21 and 41-22 a net ultrafiltration pressure (P_{UF}) of 10 mm Hg exists at the affer-

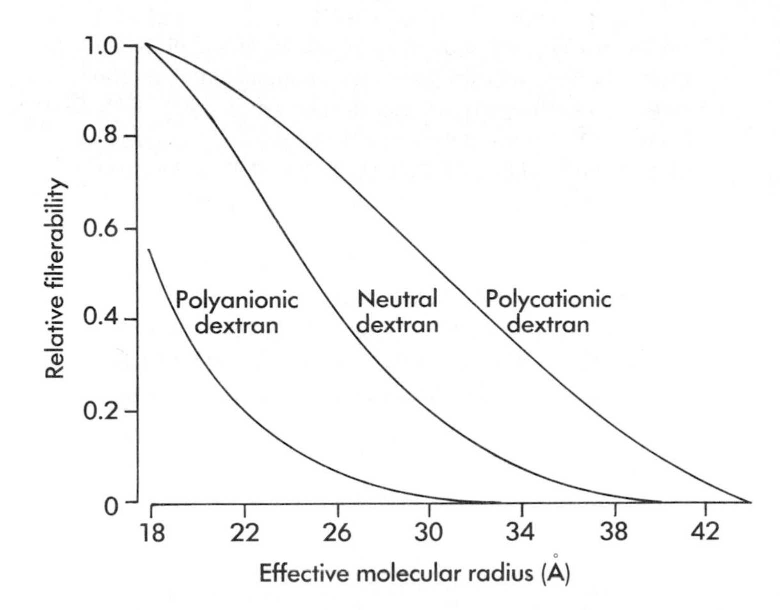

■ **Fig. 41-19** Influence of size and electrical charge of dextran on its filterability. A value of 1 indicates that it is filtered freely, whereas a value of 0 indicates that it is not filtered. The filterability of neutral dextrans between approximately 18 and 36 Å depends on charge.

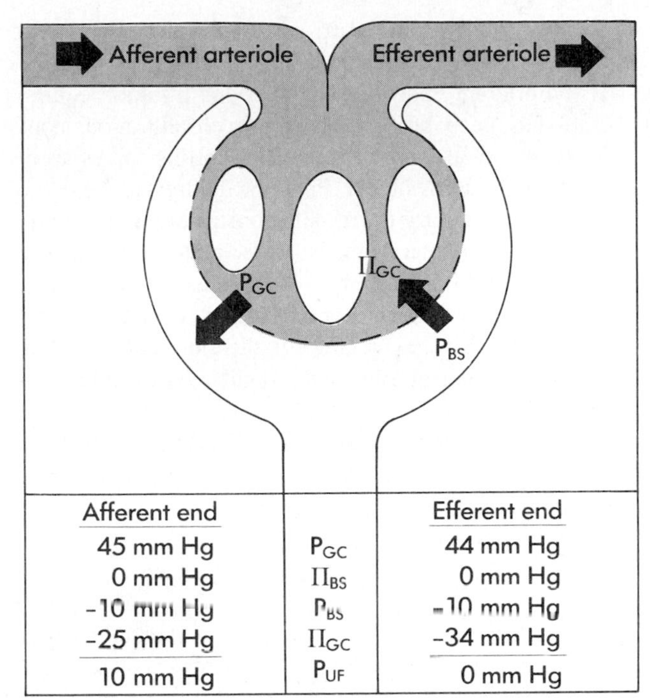

■ **Fig. 41-21** Schematic representation of an idealized glomerular capillary and the Starling forces across the filtration barrier. P_{UF}, net ultrafiltration pressure; P_{GC}, glomerular capillary hydrostatic pressure; P_{BS}, Bowman's space hydrostatic pressure; π_{GC}, glomerular capillary oncotic pressure; π_{BS}, Bowman's space oncotic pressure.

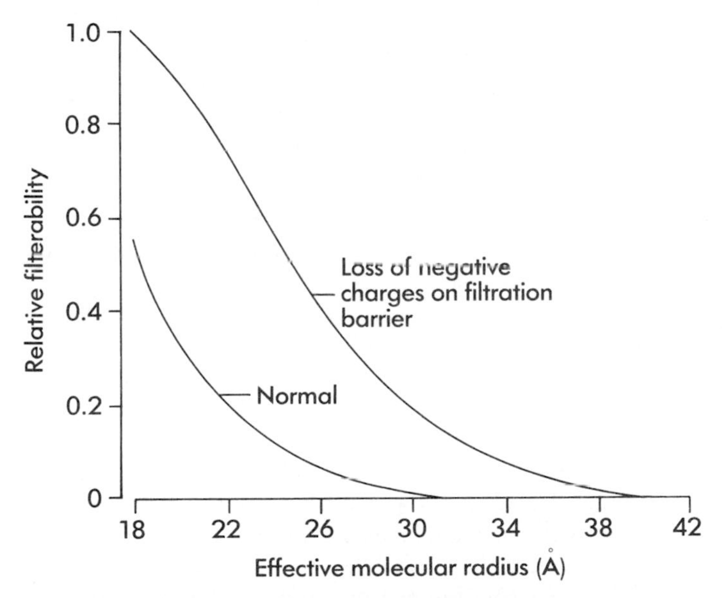

■ **Fig. 41-20** Reduction of the negative charges of the glomerular wall results in filtration of proteins on the basis of size only. Injection of antiglomerular basement membrane antibodies into experimental animals reduces the number of fixed anionic charges on the glomerular wall. In this condition, which is known as nephrotoxic serum nephritis, the relative filterability of proteins depends only on the molecular radius. In this pathophysiological condition the excretion of polyanionic proteins (effective molecular radius of 18 to 36 Å) in the urine increases because more proteins of this size are filtered.

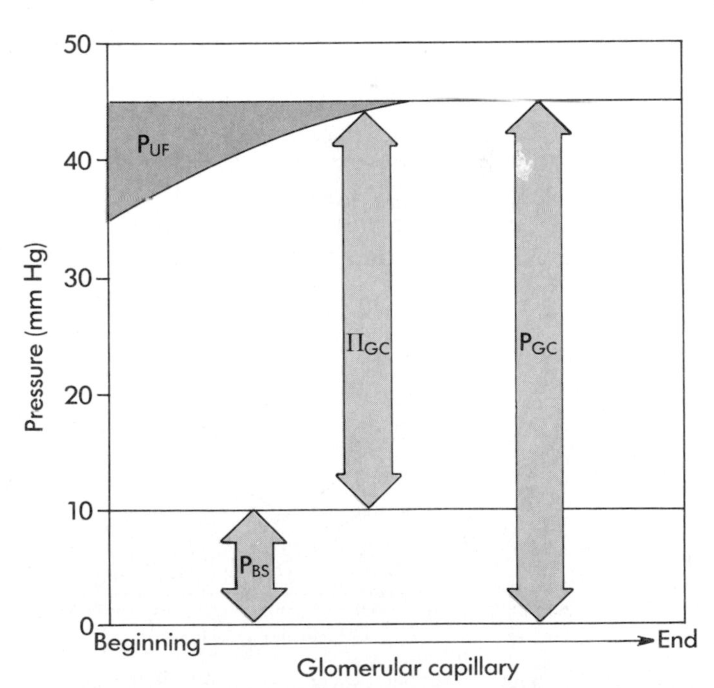

■ **Fig. 41-22** Relationship between the hydrostatic pressure in the glomerular capillary (P_{GC}) and Bowman's space (P_{BS}) and the oncotic pressure in the glomerular capillary (π_{GC}) along the length of an idealized glomerular capillary. P_{UF}, net ultrafiltration pressure.

ent end of the glomerulus, whereas at the efferent end the P_{UF} is zero (where $P_{UF} = P_{GC} - P_{BS} - \pi_{GC}$). Thus at the efferent end of the glomerulus filtration equilibrium has been achieved and net ultrafiltration stops. Two additional points concerning Starling forces are illustrated in Fig. 41-22. First the hydrostatic pressure within the capillary is virtually constant along its length. The only force that changes appreciably during the process of ultrafiltration is π_{GC}. The increase in π_{GC} results from the filtration of water and solute. Because water is filtered and protein is retained in the glomerular capillary the protein concentration in the capillary rises and π_{GC} increases.

The glomerular filtration rate (GFR) is proportional to the sum of the Starling forces that exists across the capillaries $[(P_{GC} - P_{BS}) - (\pi_{GC} - \pi_{BS})]$ times the ultrafiltration coefficient, K_f:

$$\text{GFR} = K_f\,[(P_{GC} - P_{BS}) - (\pi_{GC} - \pi_{BS})] \qquad (15)$$

The K_f is the product of the intrinsic permeability of the glomerular capillary and the glomerular surface area available for filtration. Although the P_{UF} in glomerular capillaries is similar to that in other capillary beds, the rate of glomerular filtration is considerably greater in glomerular capillaries, mainly because the K_f is approximately 100 times higher (GFR is 180 L/day, whereas net filtration across all the other capillaries in the body is 2 L/day).

The second point concerning Starling forces is that plasma flow rate is an important determinant of GFR. Fig. 41-23 shows that P_{UF}, and thereby GFR, depend on the plasma flow through the glomerular capillaries.

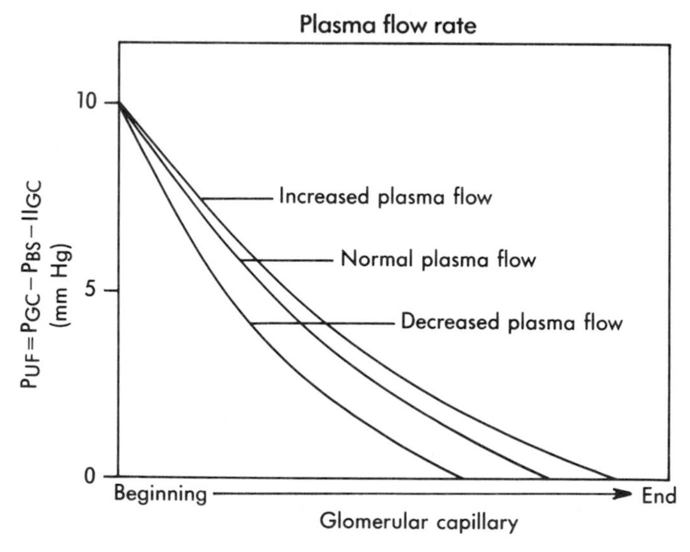

■ **Fig. 41-23** Relationship between plasma flow rate along an idealized glomerular capillary and the net ultrafiltration pressure, $P_{UF} = (P_{GC} - P_{BS} - \pi_{GC})$. Increased plasma flow increases P_{UF} primarily by reducing the increase in π_{GC} along the glomerular capillaries, and decreased plasma flow decreases P_{UF} primarily by increasing the rise in π_{GC}.

As plasma flow increases π_{GC} increases less rapidly and more filtrate is formed before the capillary oncotic pressure rises sufficiently to stop filtration (i.e., $P_{UF} = 0$). Conversely if plasma flow rate decreases, π_{GC} increases more rapidly and less filtrate is formed before the capillary π_{GC} rises sufficiently to stop filtration.

The GFR can be altered by changing K_f or by changing any of the Starling forces. Physiologically however GFR is affected in two main ways.

1. An increase in P_{GC} enhances GFR, and a decrease in P_{GC} depresses GFR. Changes in arterial blood pressure are the most frequent cause of variations in P_{GC}.
2. Variations in renal blood flow: As afferent arteriolar plasma flow increases GFR rises. As plasma flow in the capillaries increases π_{GC} rises more slowly and the net ultrafiltration pressure increases. A fall in plasma flow decreases GFR.

Pathological conditions and drugs may also affect GFR, mainly by changing π_{GC}, P_{BS}, and K_f. Thus GFR may change by three additional mechanisms:

1. Changes in π_{GC}: An inverse relationship exists between π_{GC} and GFR. Alterations in π_{GC} result from changes in protein metabolism outside the kidney.
2. Changes in K_f: Increased K_f enhances GFR, whereas decreased K_f reduces GFR. Some kidney diseases reduce K_f by reducing the number of filtering glomeruli. Some drugs and hormones that vasodilate the glomerular arterioles increase K_f and some that vasoconstrict them decrease K_f.
3. Changes in P_{BS}: Increased P_{BS} reduces GFR, whereas decreased P_{BS} facilitates GFR. Acute obstruction of the urinary tract (e.g., occlusion of the ureter by a kidney stone) increases P_{BS}.

■ *Renal Blood Flow*

In resting subjects the blood flow to the kidneys (1.25 L/min) is equal to 25% of the cardiac output. However the kidneys constitute less than 0.5% of the total body weight. Blood flow through the kidneys serves several important functions, including the following:

1. Determining the GFR
2. Modifying the rate of solute and water reabsorption by the proximal tubule
3. Participating in the concentration and dilution of the urine
4. Delivering oxygen, nutrients, and hormones to the cells of the nephron and returning carbon dioxide and reabsorbed fluid and solutes to the general circulation.

The relationship between GFR and renal blood flow will be discussed later in this chapter; the other functions of renal blood flow will be discussed later.

The equation for blood flow through any organ is

$$Q = \Delta P/R \quad (16)$$

where

Q = blood flow

ΔP = mean arterial pressure minus venous pressure for that organ

R = resistance to flow through that organ (see p. 443).

Accordingly RBF is equal to the pressure difference between the renal artery and the renal vein divided by the renal vascular resistance:

$$RBF = \frac{\text{Aortic pressure} - \text{renal venous pressure}}{\text{Renal vascular resistance}} \quad (17)$$

The afferent arteriole, the efferent arteriole, and the interlobular artery are the major resistance vessels in the kidney; therefore they determine renal vascular resistance. The kidneys, like most other organs, regulate their blood flow by adjusting the vascular resistance in response to changes in arterial pressure (Fig. 41-24). This adjustment in resistance is so precise that blood flow remains constant as arterial blood pressure changes between 90 and 180 mm Hg. GFR is also regulated over the same range of arterial pressures. The phenomenon whereby renal blood flow (RBF) and GFR are maintained constant is called **autoregulation** (see p. 481). As the term indicates autoregulation is achieved by changes in vascular resistance exclusively within the kidney. Because both GFR and RBF are regulated over the same range of pressures and because renal plasma flow (RPF) is an important determinant of GFR it is not surprising that the same mechanisms regulate both flows.

Two mechanisms are responsible for autoregulation of RBF and GFR: one responds to changes in arterial pressure, and another responds to changes in renal tubular flow. The pressure-sensitive mechanism, the so-called **myogenic mechanism,** is related to an intrinsic property of vascular smooth muscle: the tendency to contract when it is stretched (see p. 481). Accordingly when arterial pressure rises and the renal afferent arteriole is stretched, the smooth muscle contracts. Because the increase in the resistance of the arteriole offsets the increase in pressure, blood flow and therefore GFR remain constant (i.e., RBF is constant if the ratio of ΔP/ΔR is kept constant).

The second mechanism responsible for autoregulation of GFR and RBF, the flow-dependent mechanism, is known as **tubuloglomerular feedback** (Fig. 41-25). This mechanism involves a feedback loop in which the flow of tubular fluid (or some other factor, such as the rate of NaCl reabsorption, which increases in direct proportion to flow) is sensed by the macula densa of the juxtaglomerular apparatus (JGA) and converted into a signal that affects GFR. When GFR increases and causes the flow rate of tubular fluid at the macula densa to rise, the JGA sends a signal that causes RBF and GFR to return to normal levels. In contrast when GFR and tubular flow rate past the macula densa decrease, the JGA sends a signal causing RBF and GFR to increase to normal levels. The signal affects RBF and GFR mainly by changing the resistance of the afferent arteriole. The major unknowns about tubuloglomerular feedback concern the variable that is sensed at the JGA and the effector substance that alters the resistance of the afferent arteriole. Some have suggested that flow-dependent changes in NaCl reabsorption are sensed by the macula densa. The effector mechanism may involve the renin-angiotensin system (see Chapter 42) or other vasoactive substances, such as prostaglandins, catecholamines, adenosine, angiotensin II, or kinins.

Because animals engage in many activities that can change arterial blood pressure, having mechanisms that maintain RBF and GFR constant despite changes in arterial pressure is highly desirable. If RBF and GFR were to rise or fall suddenly in proportion to changes in blood pressure, urinary excretion of fluid and solutes would also change suddenly because alterations in GFR influence water and solute excretion. Such changes in water and solute excretion without comparable alterations in intake would alter fluid and water balance. Accordingly *autoregulation of GFR and RBF provides an effective means of uncoupling renal function from arterial pressure and ensures that fluid and solute excretion remain constant.*

Two points concerning autoregulation should be made: (1) autoregulation is absent below arterial pressures of 90 mm Hg and above arterial pressures of 180 mm Hg and (2) despite autoregulation GFR and RBF can be changed, under appropriate conditions, by several hormones (see later in chapter).

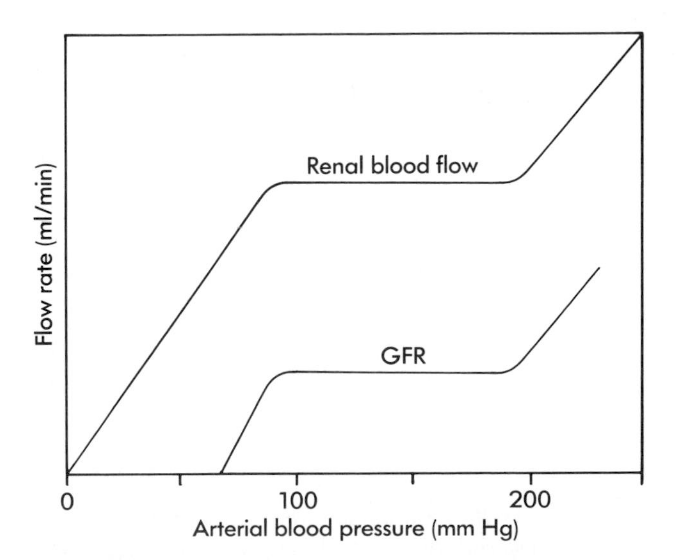

Fig. 41-24 Relationships of arterial blood pressure, renal blood flow, and glomerular filtration rate (GFR). Autoregulation of blood flow and GFR is maintained as blood pressure changes from 90 to 180 mm Hg.

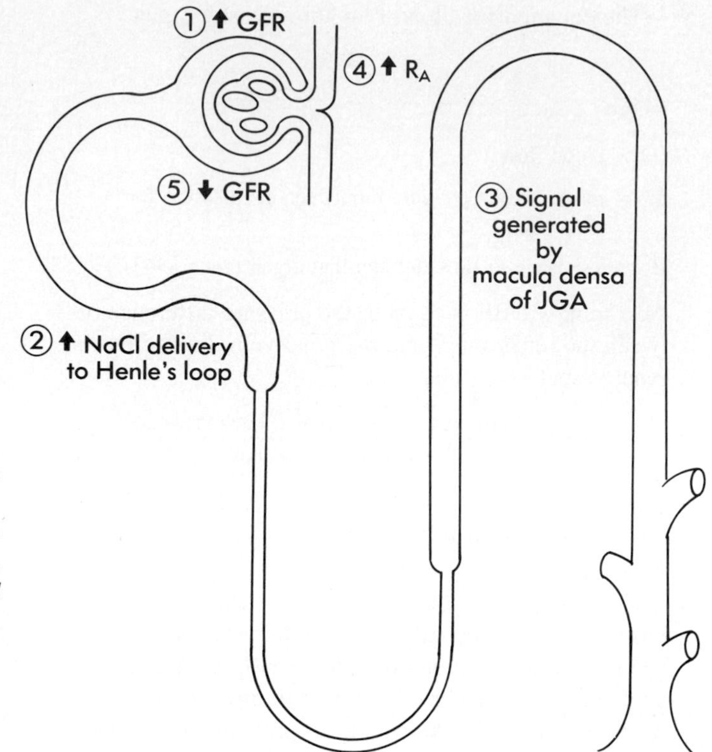

■ **Fig. 41-25** Tubuloglomerular feedback. An increase in GFR *(1)* increases NaCl delivery into the loop of Henle *(2)*, which is sensed by the macula densa and converted into a signal *(3)* that increases afferent arteriolar resistance (R_A) (4), which decreases GFR *(5)*. *(Adapted from Cogan MG: Fluid and electrolytes: physiology and pathophysiology,* Norwalk, Conn, 1991, Appleton & Lange.)

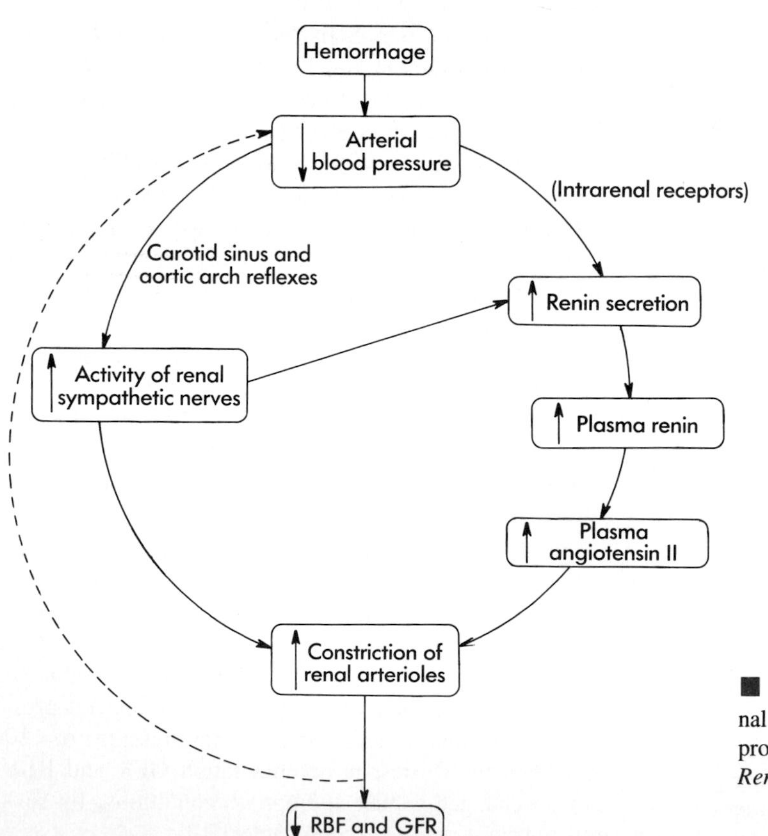

■ **Fig. 41-26** Pathway by which hemorrhage activates renal sympathetic nerve activity and stimulates angiotensin II production. See text for details. (Modified from Vander AJ: *Renal physiology*. ed 2, New York, 1980, McGraw-Hill.)

Table 41-2 Filtration, excretion, and reabsorption of water, electrolytes, and solutes

Substance	*Measure*	*Filtered*	*Excreted*	*Reabsorbed*	*Filtered load reabsorbed (%)*
Water	L/day	180	1.5	178.5	99.2
Na^+	mEq/day	25,200	150	25,050	99.4
K^+	mEq/day	720	100	620	86.1
Ca^{++}	mEq/day	540	10	530	98.2
HCO_3^-	mEq/day	4320	2	4,318	99.9+
Cl^-	mEq/day	18,000	150	17,850	99.2
Glucose	mmol/day	800	0.5	799.5	99.9+
Urea	g/day	56	28	28	50.0

From Stanton BA, Koeppen BM. Berne RM, Levy MN, editors. *Principles of physiology,* St Louis, 1990 Mosby-Year Book.)
The filtered amount of any substance is calculated by multiplying the concentration of that substance in the ultrafiltrate by the glomerular filtration rate (e.g., the filtered load of Na^+ is calculated as $[Na^+]_{ultrafiltrate}$ (140 mEq/L) × glomerular filtration rate (180 L/day) = 25,200 mEq/day).

Regulation of Renal Blood Flow

Several hormones have a major effect on RBF. Hormones that cause vasoconstriction and thereby decrease RBF and GFR include epinephrine, norepinephrine, angiotensin II, and adenosine. Hormones that cause vasodilation and thereby increase RBF and GFR include the prostaglandins PGE_2 and PGI_2.

Sympathetic control. The afferent and efferent arterioles are innervated by sympathetic neurons that release **norepinephrine.** Norepinephrine and circulating **epinephrine,** secreted by the adrenal medulla, cause vasoconstriction by binding to α_1-adrenoceptors on the arterioles and thereby decrease RBF and GFR.

Angiotensin II. Another major hormone that regulates RBF is **angiotensin II** (see Chapter 42 for details on the renin-angiotensin system). Angiotensin II constricts both the afferent and efferent arterioles* and decreases RBF. Fig. 41-26 illustrates how norepinephrine, epinephrine, and angiotensin II decrease RBF and GFR during hemorrhage. Hemorrhage decreases arterial blood pressure thus activating the sympathetic nerves to the kidneys. Norepinephrine elicits an intense vasoconstriction of the afferent and efferent arterioles thereby decreasing RBF and GFR. The rise in sympathetic activity also increases the production of epinephrine and angiotensin II, causing further vasoconstriction and a fall in RBF. The rise in the resistance of the kidney and other vascular beds increases total peripheral resistance; this resistance, by increasing blood pressure (BP = cardiac output × total peripheral resistance), offsets the fall in mean arterial blood pressure elicited by hemorrhage. Hence this system works to preserve arterial pressure at the expense of maintaining normal RBF and GFR. This example illustrates the important point that although autoregulatory mechanisms can prevent the effects of changes in arterial pressure on RBF and GFR, when needed, sympathetic nerves and angiotensin II have important salutary effects on RBF and GFR.

Prostaglandins. Prostaglandins do not regulate RBF or GFR in subjects who are in the basal state. However during pathophysiological conditions, such as hemorrhage, prostaglandins (notably PGE_2 and PGI_2) are produced locally within the kidneys. These substances vasodilate the afferent and efferent arterioles and thereby dampen the vasoconstrictor effects of sympathetic nerves and angiotensin II. This effect of prostaglandins is important because it prevents severe and potentially harmful renal vasoconstriction and renal ischemia. Prostaglandin synthesis is stimulated by sympathetic nerve activity and angiotensin II.

*The efferent arteriole is more sensitive to angiotensin II than is the afferent arteriole. Therefore with low concentrations of angiotensin II, constriction of the efferent arteriole predominates. However with high concentrations of angiotensin II constriction of both afferent and efferent arterioles occurs.

Tubular Function

Formation of urine involves three basic processes: ultrafiltration of plasma by the glomerulus, reabsorption of water and solutes from the ultrafiltrate, and secretion of selected solutes into the tubular fluid. Although 180 L of essentially protein-free fluid are filtered by the human glomeruli each day, only 1% to 2% of the water, less than 1% of the filtered Na^+, and variable amounts of the other solutes are excreted in the urine (Table 41-2). By the processes of reabsorption and secretion the renal tubules modulate the volume and composition of the urine (Table 41-3). Consequently the tubules precisely control the volume, osmolality, composition, and pH of the intracellular and extracellular fluid compartments.

Because of the importance of tubular reabsorption and secretion the first part of this section will define some basic transport mechanisms kidney cells use to reabsorb and secrete solutes. Then NaCl and water reabsorption and some of the factors and hormones that regulate reabsorption will be discussed. Details on acid-base transport, K^+, Ca^{++}, Mg^{++}, and $PO_4^{\equiv}$ transport, and their regulation are provided in Chapter 43.

■ **Table 41-3** Composition of the urine

Substance	*Concentration*
Na^+	50-130 mEq/L
K^+	20-70 mEq/L
NH_4^+	30-50 mEq/L
Ca^{++}	5-12 mEq/L
Mg^{++}	2-18 mEq/L
Cl^-	50-130 mEq/L
$PO_4^=$	20-40 mEq/L
Urea	200-400 mM
Creatinine	6-20 mM
pH	5.0-7.0 units
Osmolality	500-800 mOsm/kg H_2O
Glucose*	0
Amino acids*	0
Protein*	0
Blood*	0
Ketones*	0
Leukocytes*	0
Bilirubin*	0

*These values represent average ranges. Asterisks indicate that the presence of these substances in freshly voided urine are measured with dipstick reagent strips. These small strips of plastic contain reagents that change color in a semiquantitative manner in the presence of specific compounds. Water excretion ranges between 0.5 and 1.5 L/day. (Table modified from Valtin HV: *Renal physiology*, ed 2, Boston, 1983, Little, Brown & Co.)

■ *General Principles of Transepithelial Solute and Water Transport*

Solutes may be transported across cell membranes by passive or active mechanisms. As defined in Chapter 1 passive movement of solutes occurs down an electrochemical gradient. Active transport results in the movement of a solute against an electrochemical gradient and requires energy derived from metabolic processes. Transport that is directly coupled to an energy source (hydrolysis of ATP) is termed **primary active transport.** As detailed later the primary active transport of Na^+ by the **Na^+-K^+-ATPase** pump is central to the functioning of the kidney. Transport coupled indirectly to an energy source (e.g., energy stored in an ion gradient) is termed **secondary active transport.** The reabsorption of amino acids by the kidney is an example of such a secondary active process (see later in this chapter).

In contrast to solutes, which are transported by both passive and active mechanisms in the kidney, water is always transported by passive processes. The driving force for water movement is an osmotic gradient. Water

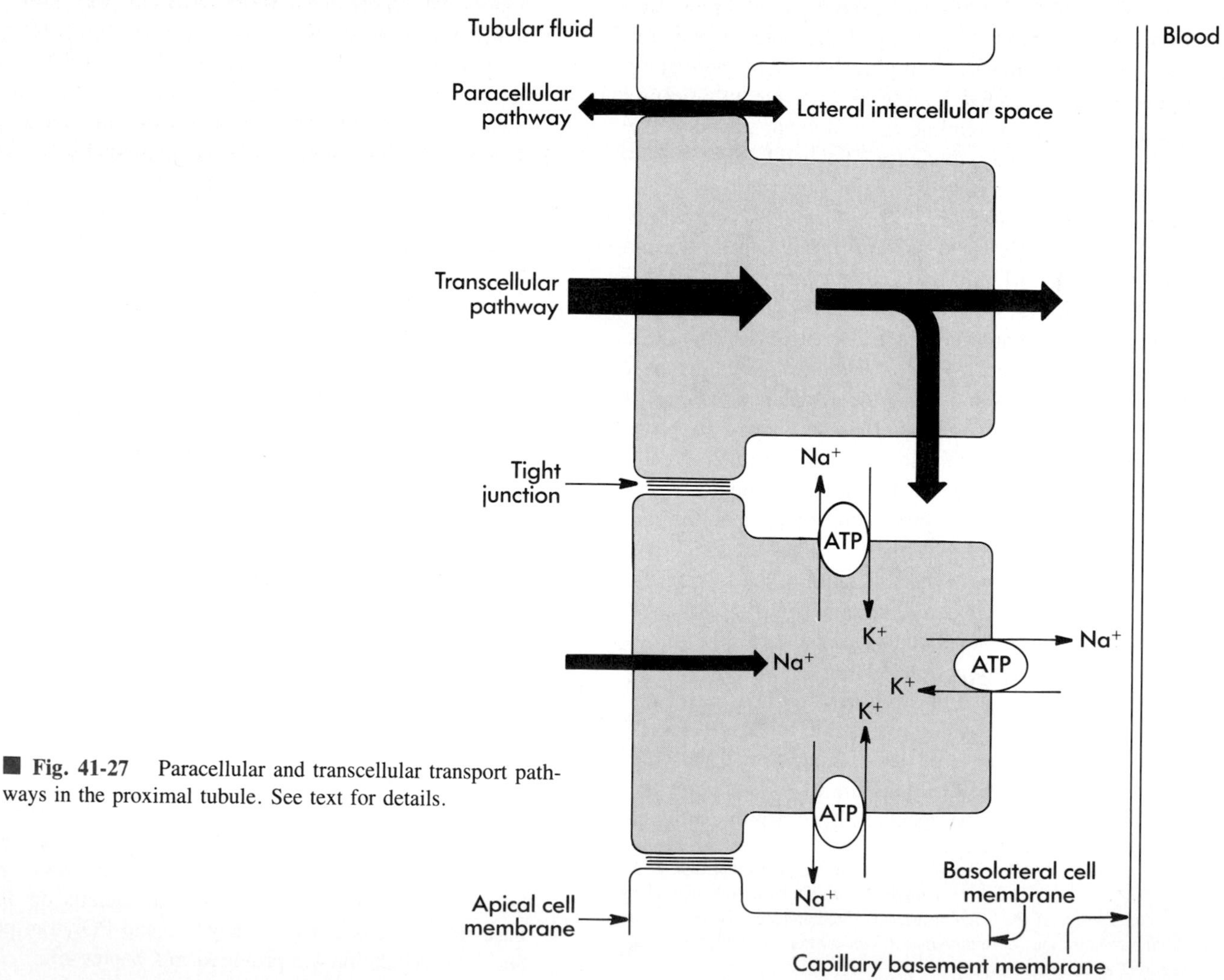

■ **Fig. 41-27** Paracellular and transcellular transport pathways in the proximal tubule. See text for details.

moves from an area of low osmolality to an area of high osmolality.

The nephron, like other epithelia such as the intestine, can transport solutes and water from one side of the tubule to the other. **Reabsorption** *is the net transport of a substance from the tubular lumen into the blood, whereas* **secretion** *is the net transport in the opposite direction.*

As illustrated in Fig. 41-27 renal cells are held together by **tight junctions.** Below the tight junctions the cells are separated by **lateral intercellular spaces.** The tight junctions separate the apical membranes from the basolateral membranes. An epithelium can be compared with a six-pack of soda in which the cans are the cells and the plastic holder represents the tight junctions.

In the nephron a substance can be reabsorbed or secreted across cells, the so-called **transcellular pathway,** or between cells, the so-called **paracellular pathway** (Fig. 41-27). Na^+ reabsorption by the proximal tubule is a good example of transport by the transcellular pathway. Na^+ reabsorption in this nephron segment depends on the operation of the Na^+-K^+-ATPase pump (see Fig. 41-27). The Na^+-K^+-ATPase pump, which is located exclusively in the basolateral membrane, moves Na^+ out of the cell into the blood and moves K^+ into the cell. Thus the operation of the Na^+-K^+-ATPase pump lowers intracellular Na^+ concentration and increases intracellular K^+ concentration. Because intracellular Na^+ is low (12 mEq/L) and the Na^+ concentration in tubular fluid is high (140 mEq/L) Na^+ moves across the apical cell membrane down a chemical concentration gradient from the tubular lumen into the cell. The Na^+-K^+-ATPase pump senses the addition of Na^+ to the cell and is stimulated to increase its rate of Na^+ extrusion into the blood and thereby returns intracellular Na^+ to normal levels (see p. 15). Thus transcellular Na^+ reabsorption by the proximal tubule is a two-step process: (1) movement across the apical membrane into the cell down a chemical concentration gradient established by the Na^+-K^+-ATPase pump and (2) movement across the basolateral membrane against an electrochemical gradient via the Na^+-K^+-ATPase pump.

The reabsorption of Ca^{++}, Mg^{++}, and K^+ across the proximal tubule is a good example of paracellular transport. Some of the water reabsorbed across the proximal tubule traverses the paracellular pathway. Some solutes dissolved in this water, in particular Ca^{++}, Mg^{++}, and K^+, are entrained in the reabsorbed fluid and are thereby reabsorbed by the process of **solvent drag** (for additional details of transcellular and paracellular reabsorption see later in this chapter).

Proximal Tubule

The proximal tubule reabsorbs approximately 67% of the filtered water, Na^+, Cl^-, K^+, and other solutes (Fig. 41-28). In addition virtually all of the glucose and amino acids filtered by the glomerulus are reabsorbed. The key element in proximal tubule reabsorption is the Na^+-K^+-ATPase pump in the basolateral membrane. *The reabsorption of every substance, including water, is linked to the operation of the Na^+-K^+-ATPase pump.*

NaCl and water reabsorption. NaCl and water reabsorption by the proximal tubule is commonly divided into two phases: (1) reabsorption of Na^+ with glucose, amino acids, $PO_4^{\equiv}$, lactate, and HCO_3^- in the first half of the proximal tubule and (2) reabsorption of Na^+, mainly with Cl^-, in the second half of the proximal tubule. This distinction is based on the types of solute

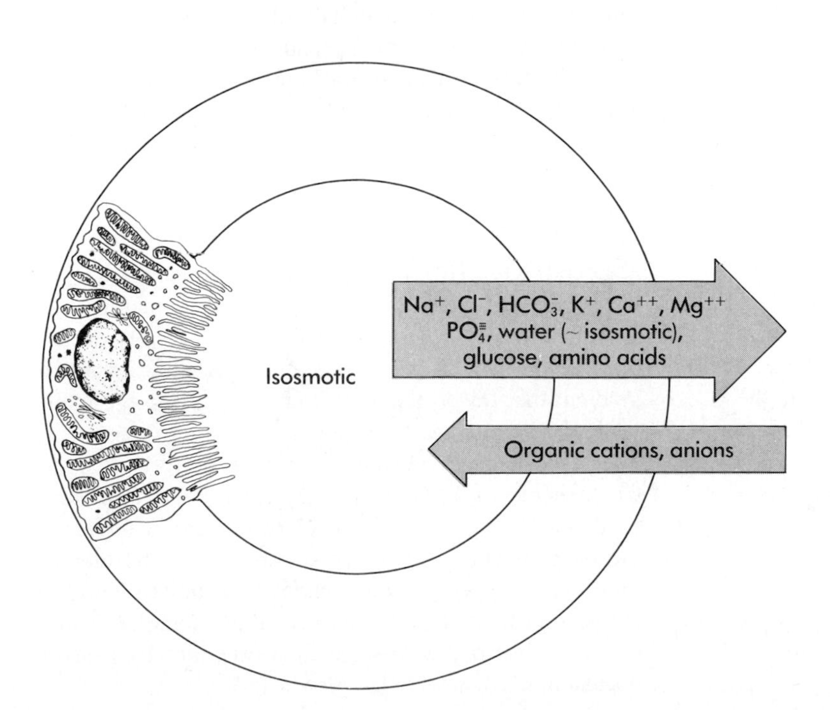

Fig. 41-28 Tubular profile of the proximal tubule illustrating the cellular ultrastructure and the primary transport characteristics. The proximal tubule reabsorbs 67% of the filtered Na^+, Cl^-, HCO_3^-, K^+, Ca^{++}, $PO_4^{\equiv}$, and water and all of the filtered glucose and amino acids. This segment also secretes organic cations and anions. (Modified from Burg MB: *Renal handling of Na^+, Cl^-, water, amino acids, and glucose.* In Brenner BM, Rector FC Jr, editors: *The kidney,* ed 2, Philadelphia, 1987, WB Saunders.)

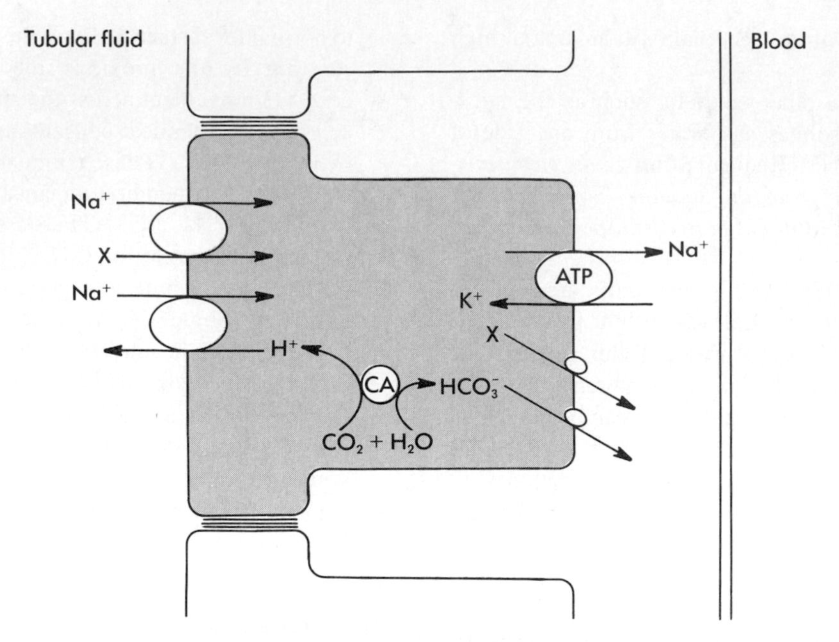

■ **Fig. 41-29** Transport processes in the first half of the proximal tubule. Na^+-X-cotransport protein indicates the presence of five unique symporters. X represents either glucose, amino acids, phosphate, Cl^-, or lactate. CO_2 and H_2O combine inside the cells to form H^+ and HCO_3^- in a reaction facilitated by the enzyme carbonic anhydrase (CA).

transport systems present in the early and late portions of the proximal tubule, as well as the composition of tubular fluid at these sites.

During the *first phase* of proximal tubule reabsorption Na^+ uptake into the cell is coupled with organic solutes and anions and H^+ (Fig. 41-29). Na^+ entry into the cell across the apical membrane is mediated by specific **symporter** and **antiporter** proteins and not by simple diffusion.* Na^+ enters proximal cells by Na^+-glucose, Na^+-amino acid, Na^+-$PO_4^{\equiv}$, and Na^+-lactate symporters. Na^+ entry is also coupled with H^+ extrusion from the cell by the Na^+-H^+ antiporter (Fig. 41-29). H^+ secretion by the Na^+-H^+ antiporter results in HCO_3^- reabsorption (see Chapter 43 for details). The Na^+ that enters the cell across the apical membrane, by either a symport or an antiport mechanism, enters the blood via the Na^+-K^+-ATPase. The solutes and anions that enter the cell with Na^+ (e.g., glucose, amino acids, $PO_4^{\equiv}$, and lactate) exit across the basolateral membrane by passive mechanisms. In summary in the first phase of proximal reabsorption, which occurs in the first half of the proximal tubule, reabsorption of Na^+ is coupled to that of HCO_3^-, glucose, amino acids, $PO_4^{\equiv}$, and lactate. Reabsorption of glucose, amino acids, and lactate is so avid that these solutes are completely removed from the tubular fluid in the first half of the proximal tubule.

In the *second phase* of proximal tubular reabsorption, which occurs in the second half of the proximal tubule, NaCl is reabsorbed via both the transcellular and paracellular pathways (Fig. 41-30). Na^+ is reabsorbed with Cl^- rather than with organic anions or HCO_3^- as the accompanying anion. This occurs because the tubular fluid entering the second half of the proximal tubule contains very little glucose and amino acids and because the tubular fluid has a high concentration of Cl^- (140 mEq/L versus 105 mEq/L in the first half of the proximal tubule) and a low concentration of HCO_3^- (5 mEq/L versus 25 mEq/L in the first half of the proximal tubule). The Cl^- concentration is high because in the first half of the proximal tubule Na^+ is preferentially reabsorbed with HCO_3^-, glucose, and organic anions, leaving behind a solution that becomes enriched in Cl^-.

Paracellular NaCl reabsorption (Fig. 41-30) occurs because the rise in [Cl^-] in the tubular fluid creates a gradient that favors the diffusion of Cl^- from the tubular lumen across the tight junctions and into the lateral intercellular space. Movement of the negatively charged Cl^- generates a positive transepithelial voltage (tubular fluid positive relative to the blood), which causes the diffusion of positively charged Na^+ out of the tubular fluid across the tight junction into the blood. Thus in the second half of the proximal tubule some Na^+ and Cl^- is reabsorbed across the tight junctions by passive diffusion.

*As described in Chapter 1 symporters and antiporters are examples of coupled transport proteins. Symporters couple the movement of two or more molecules in the same direction across the membrane. Antiporters couple the movement of two or more molecules in opposite directions across the membrane.

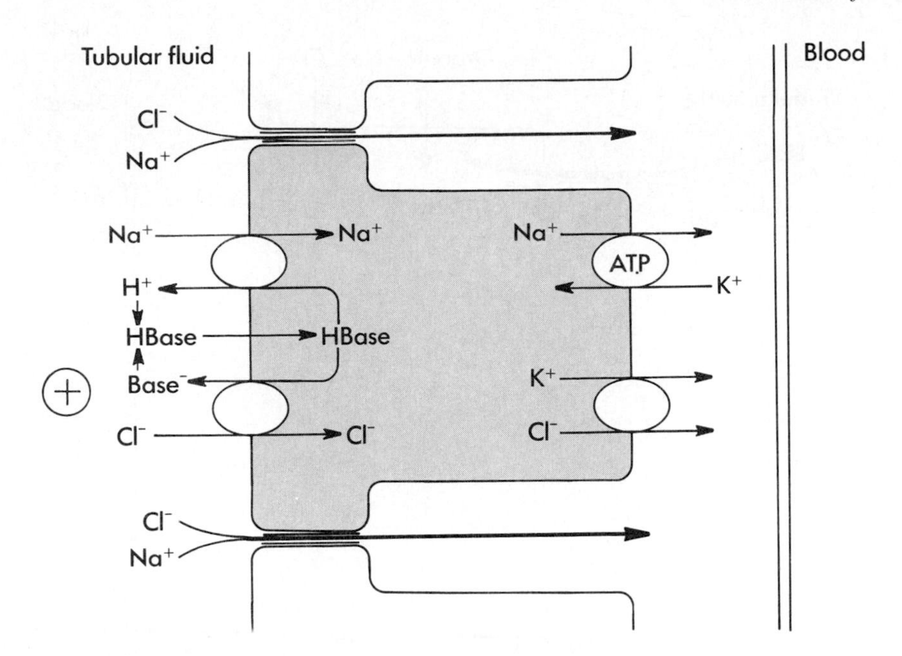

■ **Fig. 41-30** Transport processes in the second half of the proximal tubule. Na^+ and Cl^- enter the cell across the apical membrane by the operation of parallel Na^+-H^+ and Cl^--$Base^-$ antiporters. Because the secreted H^+ and $Base^-$ combine in the tubular fluid to form HBase (which reenters the cell) the net result is NaCl uptake. Base may be OH^-, formate (HCO_2^-), oxalate$^-$, or HCO_3^-. The lumen positive transepithelial voltage is generated by the diffusion of Cl^- (lumen-to-blood) across the tight junction.

NaCl reabsorption by the second half of the proximal tubule also occurs by a transcellular route (see Fig. 41-30). NaCl enters the cell across the luminal membrane by the parallel operation of Na^+-H^+ and Cl^--$Base^-$ antiporters. Because the secreted H^+ and $Base^-$ combine in the tubular fluid and reenter the cell by passive diffusion the operation of the Na^+/H^+ and $Cl^-/Base^-$ antiporters is equivalent to NaCl uptake from tubular fluid into the cell. Na^+ leaves the cell by the Na^+-K^+-ATPase pump, and Cl^- leaves the cell and enters the blood by a KCl symporter protein in the basolateral membrane.

In summary proximal tubule reabsorption of Na^+ and Cl^- occurs across the paracellular pathway and across the transcellular pathway. Approximately 17,000 mEq of the 25,200 mEq of sodium filtered each day is reabsorbed in the proximal tubule (~67% of the filtered load). Of this two thirds moves across the transcellular pathway, while the remaining one third moves across the paracellular pathway.

Fig. 41-31 illustrates the mechanism of water reabsorption by the proximal tubule. The reabsorption of Na^+ with organic solutes, HCO_3^-, and Cl^- increases the osmolality of the lateral intercellular space. This occurs because some Na^+-K^+-ATPase pumps and organic solute, HCO_3^-, and Cl^- transporters are located on the lateral cell membranes and deposit these solutes in this space. Furthermore some NaCl also enters the lateral intercellular space by diffusion across the tight junction. Because the lateral intercellular space becomes slightly hyperosmotic (3 to 5 mOsm/kg H_2O) with respect to tubular fluid and because the proximal tubule is highly permeable to water, water flows by osmosis across both the tight junctions and the proximal tubular cells into this hyperosmotic compartment. Accumulation of fluid and solutes within the lateral intercellular space increases the hydrostatic pressure in this compartment; the pressure in turn forces fluid and these solutes to move into the capillaries. Thus *water reabsorption follows solute reabsorption in the proximal tubule*. The reabsorbed fluid is essentially *isoosmotic* to plasma. (The osmolality of the reabsorbed fluid is actually slightly hyperosmotic to plasma. However for simplicity renal physiologists usually consider the fluid to be isoosmotic, and we shall follow this policy). An important consequence of osmotic water flow across the proximal tubule is that some solutes, especially K^+, Ca^{++}, and Mg^{++}, are entrained in the reabsorbed fluid and are thereby reabsorbed by the process of solvent drag (Fig. 41-31).

Because the reabsorption of virtually all organic solutes, Cl^-, other ions, and water is coupled to Na^+ reabsorption, changes in Na^+ reabsorption influence the reabsorption of water and the other solutes by the proximal tubule.

Organic anion and organic cation secretion. In addition to reabsorbing solutes and water the proximal tubule also secretes organic cations and organic anions

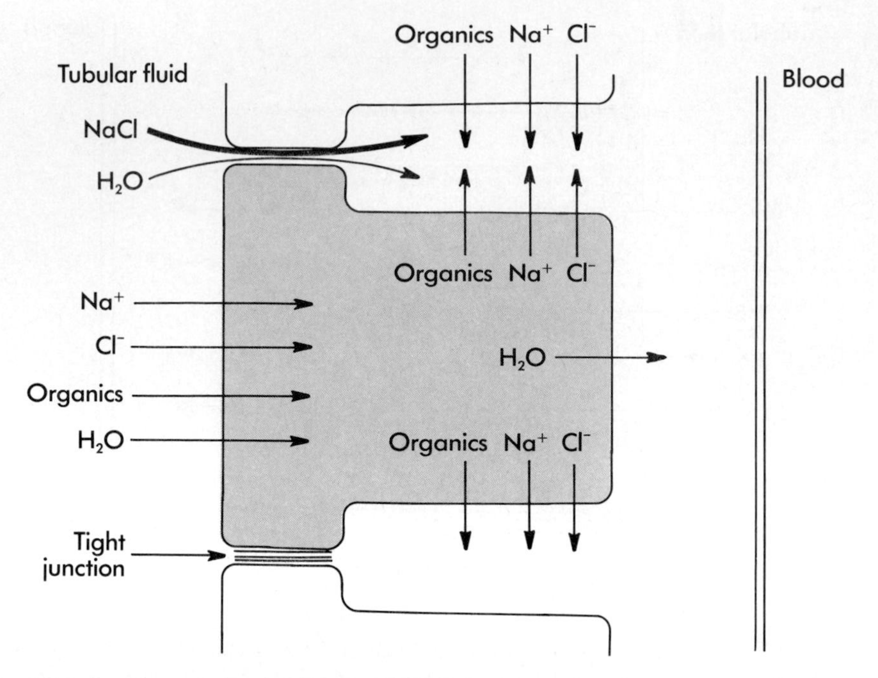

Fig. 41-31 Routes of water reabsorption across the proximal tubule. Transport of Na^+, Cl^-, and organic solutes into the lateral intercellular space increases the osmolality of this compartment, which establishes the driving force for osmotic water reabsorption across the proximal tubule. Water flows across the paracellular and cellular pathways.

Table 41-4 Some organic anions secreted by the proximal tubule

Endogenous anions	Drugs
cAMP	Acetazolamide
Bile salts	Chlorothiazide
Hippurates	Furosemide
Oxalate	Penicillin
Prostaglandins	Probenecid
Urate	Salicylate (aspirin)
	Hydrochlorothiazide
	Bumetanide

Table 41-5 Some organic cations secreted by the proximal tubule

Endogenous cations	Drugs
Creatinine	Atropine
Dopamine	Isoproterenol
Epinephrine	Cimetidine
Norepinephrine	Morphine
	Quinine
	Amiloride

(see Tables 41-4 and 41-5 for a partial listing). Many of these substances are end-products of metabolism, and they circulate in the plasma. The proximal tubule also secretes numerous exogenous organic compounds, including p-aminohippurate (PAH) and drugs such as penicillin. Because many of these organic compounds can be bound to plasma proteins they are not readily filtered. Therefore excretion by filtration alone eliminates only a small portion of these potentially toxic substances from the body. Excretion rates of these substances are high because they are secreted from the peritubular capillaries into the tubular fluid. Because the kidney removes virtually all organic ions and drugs from the plasma entering the kidney, it is evident that these secretory mechanisms are very powerful and that they serve a vital function by clearing these substances from the plasma.

Fig. 41-32 illustrates the mechanism of PAH transport across the proximal tubule as an example of organic anion secretion. This secretory pathway has a T_m, has a low specificity, and is responsible for the secretion of all organic anions listed in Table 41-5. The organic anion PAH, which can be used to measure RPF, has been used to unravel the details of this pathway. PAH is taken up into the cell, across the basolateral membrane, and against its chemical gradient by a PAH-dicarboxylate and tricarboxylate antiport mechanism The dicarboxylates and tricarboxylates accumu-

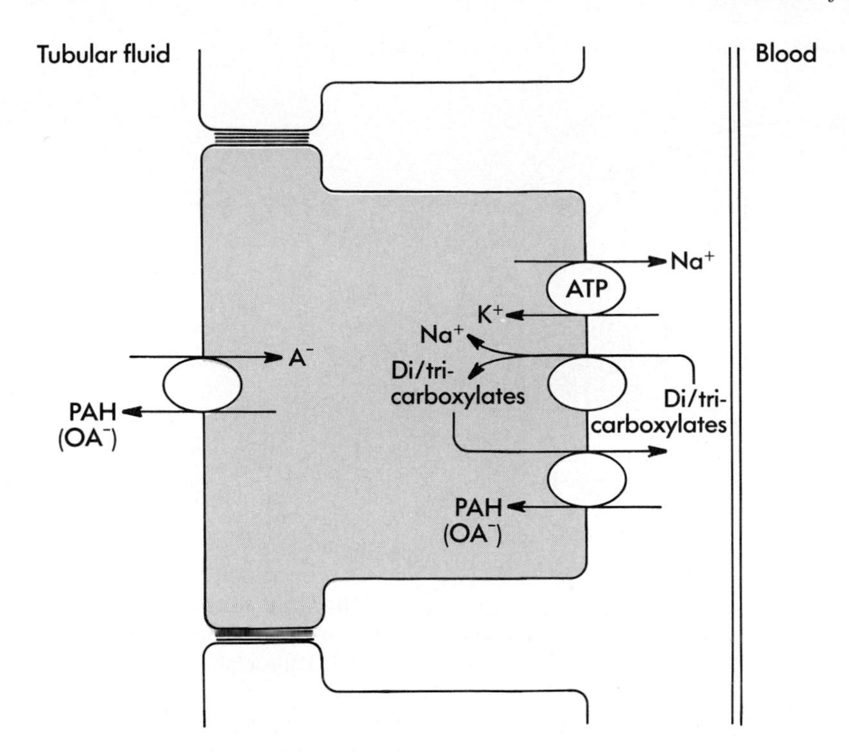

■ **Fig. 41-32** Organic anion secretion (PAH) across the proximal tubule. PAH or another organic anion (OA^-) enters the cell across the basolateral membrane by a PAH-dicarboxylate and tricarboxylate antiport mechanism. The uptake of dicarboxylate and tricarboxylates into the cell, against their chemical gradients, is driven by the movement of Na^+ into the cell. The dicarboxylates and tricarboxylates recycle across the basolateral membrane. PAH leaves the cell across the apical membrane, down its chemical gradient, by a PAH-organic anion (OA^-) antiport mechanism. The OA^- indicates one of several possible anions (e.g., urate).

late inside the cell by a Na^+-dicarboxylate and tricarboxylate symporter, also present in the basolateral membrane. Thus PAH uptake into the cell against its chemical gradient is coupled to the exit of dicarboxylates and tricarboxylates out of the cell down their chemical gradients by the antiport mechanism. The resulting high intracellular concentration of PAH provides the driving force for PAH exit across the luminal membrane into the tubular fluid via a PAH-anion antiporter (see Fig. 41-32). Because all organic anions compete for transport, elevated plasma levels of one anion inhibit the secretion of the others (e.g., a reduction of penicillin secretion can be produced by infusing PAH).

The active secretory pathway for organic cations in the proximal tubule is analogous to that of organic anions, although the precise details of the transport mechanisms have not been elucidated. The pathway for organic cations is nonspecific, several organic cations compete for the transport pathway, and the pathway has a T_m.

Protein reabsorption. The proximal tubule also reabsorbs proteins. As mentioned previously peptide hormones, small proteins, and even small amounts of larger proteins, such as albumin, are filtered by the glomerulus. Although filtration of proteins is small (the concentration of proteins in the ultrafiltrate is only 40 mg/L), the amount of protein filtered per day is significant because the GFR is so high (filtered protein = GFR × [protein] in the ultrafiltrate: thus filtered protein = 180 L/day × 40 mg/L = 7.2 g/day). These proteins are partially degraded by enzymes on the surface of the proximal tubule cells, and they are taken up into the cell by endocytosis. Once inside the cell endocytic vesicles fuse with lysosomes, which contain enzymes that digest the proteins and peptides into their constituent amino acids. The amino acids leave the cell across the basolateral membrane and are returned to the blood. Normally this mechanism reabsorbs virtually all of the protein filtered, and hence the urine is essentially protein-free. However because the mechanism is easily saturated, protein appears in the urine if the amount of protein filtered increases. Disruption of the glomerular barrier to proteins increases the filtration of proteins and results in proteinuria (appearance of protein in the urine).

■ *Henle's Loop*

Henle's loop reabsorbs approximately 20% of the filtered Na^+, Cl^-, and K^+ (Fig. 41-33). Ca^{++}, HCO_3^-,

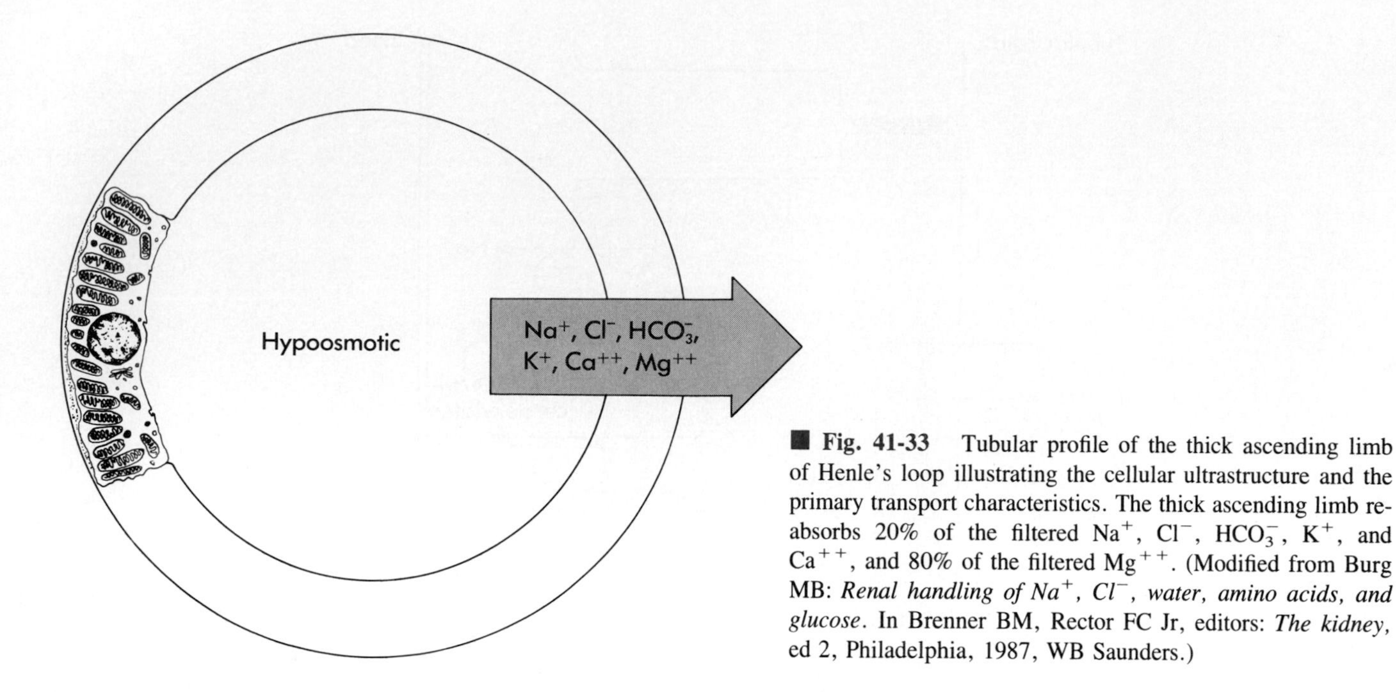

■ **Fig. 41-33** Tubular profile of the thick ascending limb of Henle's loop illustrating the cellular ultrastructure and the primary transport characteristics. The thick ascending limb reabsorbs 20% of the filtered Na^+, Cl^-, HCO_3^-, K^+, and Ca^{++}, and 80% of the filtered Mg^{++}. (Modified from Burg MB: *Renal handling of Na$^+$, Cl$^-$, water, amino acids, and glucose*. In Brenner BM, Rector FC Jr, editors: *The kidney*, ed 2, Philadelphia, 1987, WB Saunders.)

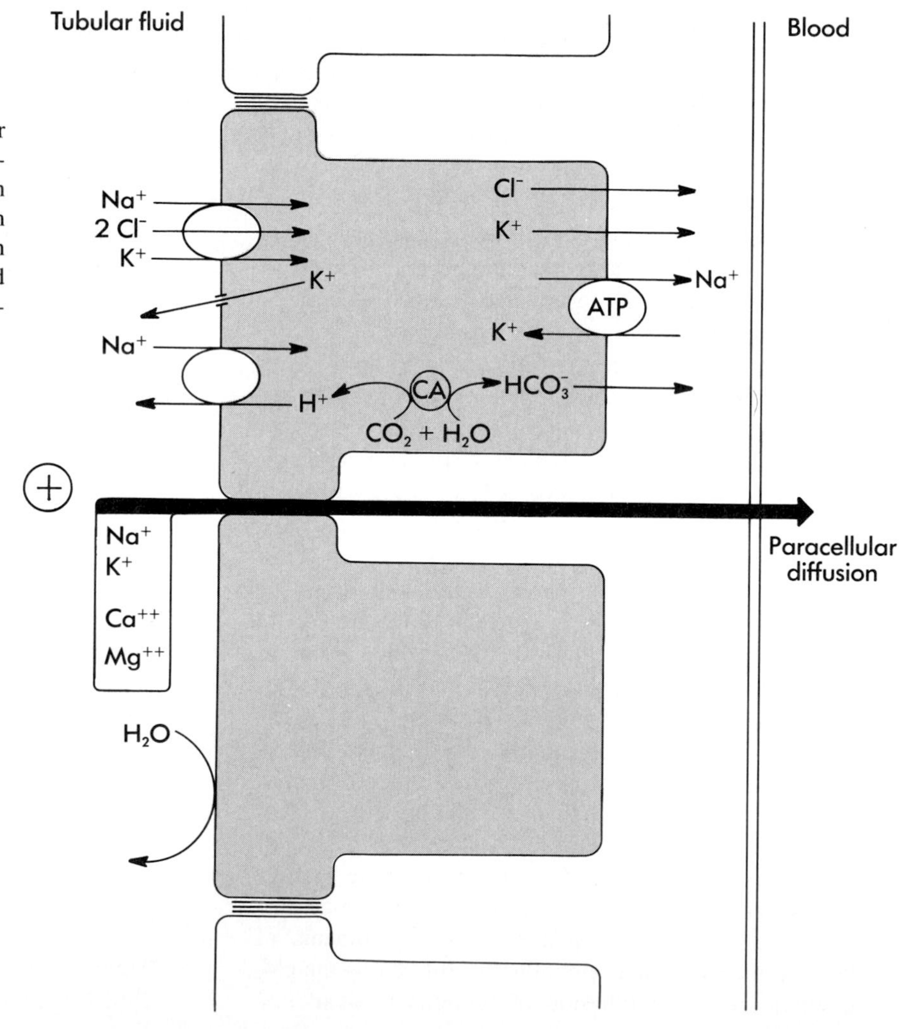

■ **Fig. 41-34** Transport mechanisms for Na^+ and Cl^- reabsorption in the thick ascending limb of Henle's loop. The lumen positive transepithelial voltage results from the unique location of transport proteins in the apical and basolateral membrane and plays a major role in driving passive paracellular reabsorption of cations.

and Mg^{++} are also reabsorbed in the loop of Henle (see Chapter 43). This reabsorption occurs almost exclusively in the thick ascending limb. By comparison the ascending thin limb has a much lower reabsorptive capacity, and the descending thin limb does not reabsorb significant amounts of solutes (see Chapter 42).

The loop of Henle reabsorbs approximately 20% of the filtered water. This reabsorption, however, occurs exclusively in the descending thin limb. *The ascending limb is impermeable to water.*

The key element in solute reabsorption by the thick ascending limb is the Na^+-K^+-ATPase pump in the basolateral membrane (Fig. 41-34). As with reabsorption in the proximal tubule the reabsorption of every substance by the thick ascending limb is linked to the Na^+-K^+-ATPase pump. The operation of the Na^+-K^+-ATPase pump maintains a low cell [Na^+]. This low [Na^+] provides a favorable chemical gradient for the movement of Na^+ from the tubular fluid into the cell. The movement of Na^+ across the apical membrane into the cell is mediated by the $1Na^+$-$2Cl^-$-$1K^+$ symporter, which couples the movement of $1Na^+$ with $2Cl^-$ and $1K^+$. This symporter protein uses the potential energy released by the downhill movement of Na^+ and Cl^- to drive the uphill movement of K^+ into the cell. Inhibition of this symport protein by loop diuretics, such as furosemide, completely inhibits Na^+ and Cl^- reabsorption by the thick ascending limb.

A Na^+-H^+ antiporter in the apical cell membrane also mediates Na^+ reabsorption, as well as H^+ secretion (HCO_3^- reabsorption) in the thick ascending limb (see Fig. 41-34). Na^+ leaves the cell across the basolateral membrane via the Na^+-K^+-ATPase pump, and K^+, Cl^-, and HCO_3^- leave the cell across the basolateral membrane by separate pathways.

The voltage across the thick ascending limb is positive in the tubular fluid relative to the blood because of the unique location of transport proteins in the apical and basolateral membranes. The important points to recognize are that increased salt transport by the thick ascending limb increases the magnitude of the positive voltage in the lumen and that this voltage is an important driving force for the reabsorption of several cations, including Na^+, K^+, Ca^{++}, and Mg^{++} across the paracellular pathway (see Fig. 41-34). Thus salt reabsorption across the thick ascending limb occurs by transcellular and paracellular pathways. Fifty percent of transport is transcellular and 50% is paracellular.

Because the thick ascending limb is very impermeable to water, reabsorption of Na^+, Cl^-, and other solutes reduces the osmolality of tubular fluid to less than 150 mOsm/kg H_2O.

Distal Tubule and Collecting Duct

The distal tubule and the collecting duct reabsorb approximately 12% of the filtered Na^+ and Cl^-, secrete variable amounts of K^+ and H^+, and reabsorb a variable amount of water. The initial segment of the distal tubule (early distal tubule) reabsorbs Na^+, Cl^-, and Ca^{++} and, like the thick ascending limb, is impermeable to water (Fig. 41-35). Na^+ and Cl^- entry into the cell across the apical membrane is mediated by a NaCl symporter (Fig. 41-36). Na^+ leaves the cell via the Na^+-K^+-ATPase pump and Cl^- leaves the cell by diffusion via channels. NaCl reabsorption is inhibited by thiazide diuretics that inhibit the Na^+-Cl^- symporter. Thus *the active dilution of the tubular fluid begins in the thick ascending limb and continues in the early segment of the distal tubule.*

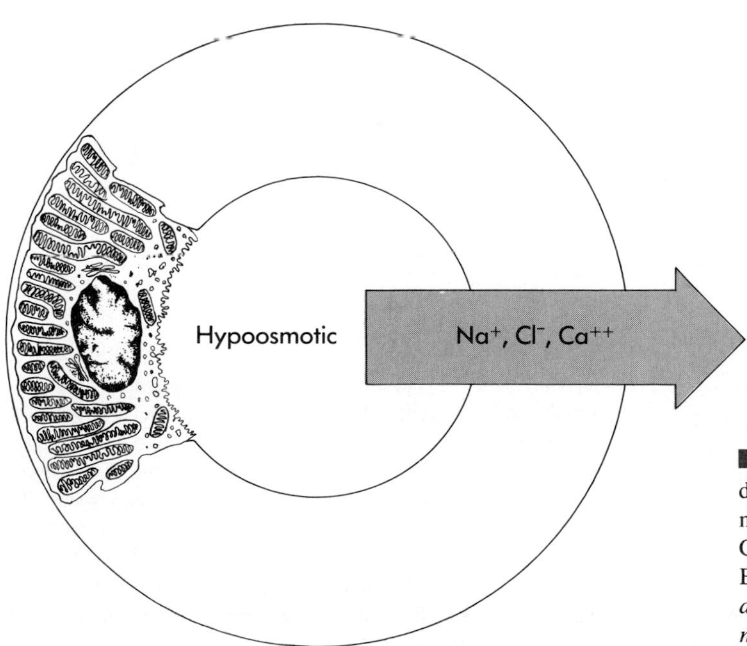

Fig. 41-35 Tubular profile of the early segment of the distal tubule illustrating the cellular ultrastructure and the primary transport characteristics. This segment reabsorbs Na^+, Cl^-, and Ca^{++} but is impermeable to water. (Modified from Burg MB: *Renal handling of Na^+, Cl^-, water, amino acids, and glucose*. In Brenner BM, Rector FC Jr, editors: *The kidney,* ed 2, Philadelphia, 1987, WB Saunders.)

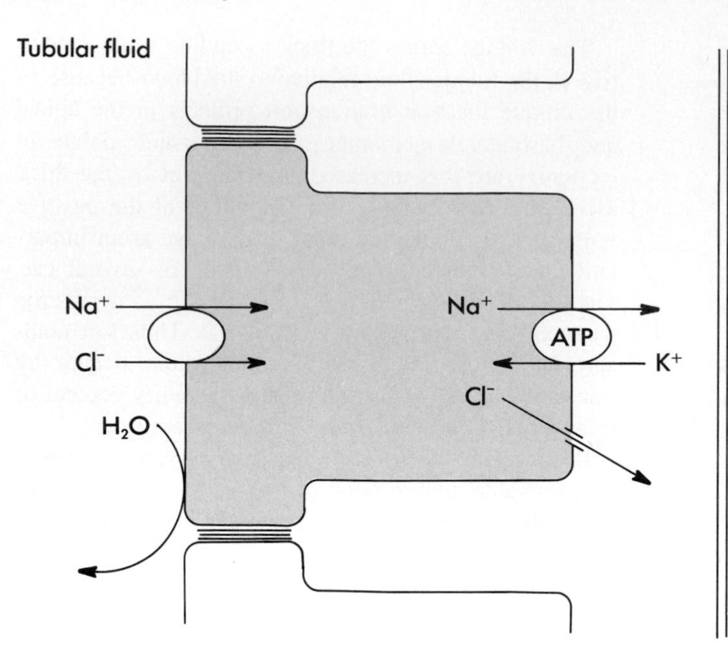

■ **Fig. 41-36** Transport mechanism for Na^+ and Cl^- reabsorption in the early segment of the distal tubule. See the text for details.

The last segment of the distal tubule (late distal tubule) and the collecting duct are composed of two cell types, **principal cells** and **intercalated cells** (Fig. 41-37). Principal cells reabsorb Na^+ and water and secrete K^+. Intercalated cells secrete H^+ (reabsorb HCO_3^-) and thus are important in regulating acid-base balance (see Chapter 43). Intercalated cells also reabsorb K^+. Both Na^+ reabsorption and K^+ secretion by principal cells depend on the activity of the Na^+-K^+-ATPase pump in the basolateral membrane (Fig. 41-38). This enzyme maintains a low cell [Na^+], which provides a favorable chemical gradient for the movement of Na^+ from the tubular fluid into the cell. Because Na^+ enters the cell across the apical membrane by diffusion through channels in the membrane, the negative potential inside the cell facilitates Na^+ entry. Na^+ leaves the cell across the basolateral membrane and enters the blood via the Na^+-K^+-ATPase pump. Although the collecting duct reabsorbs significant amounts of Cl^- the mechanism and route of transport are not completely understood.

K^+ is secreted by principal cells from the blood into the tubular fluid in two steps (see Fig. 41-38). K^+ uptake across the basolateral membrane is mediated by the Na^+-K^+-ATPase pump. Because the [K^+] inside the cells is high (140 mEq/L) and the [K^+] in tubular fluid is low (~10 mEq/L), this ion diffuses down its concentration gradient across the apical cell membrane into the tubular fluid. Although the negative potential inside the cells tends to retain K^+ within the cell, the combined electrochemical gradient across the apical membrane favors K^+ secretion from the cell into the tubular fluid. Additional details of K^+ secretion and its regulation are considered in Chapter 43.

Intercalated cells secrete H^+ and reabsorb HCO_3^- and K^+. H^+ is secreted across the apical membrane against an electrochemical gradient. Secretion is mediated by an

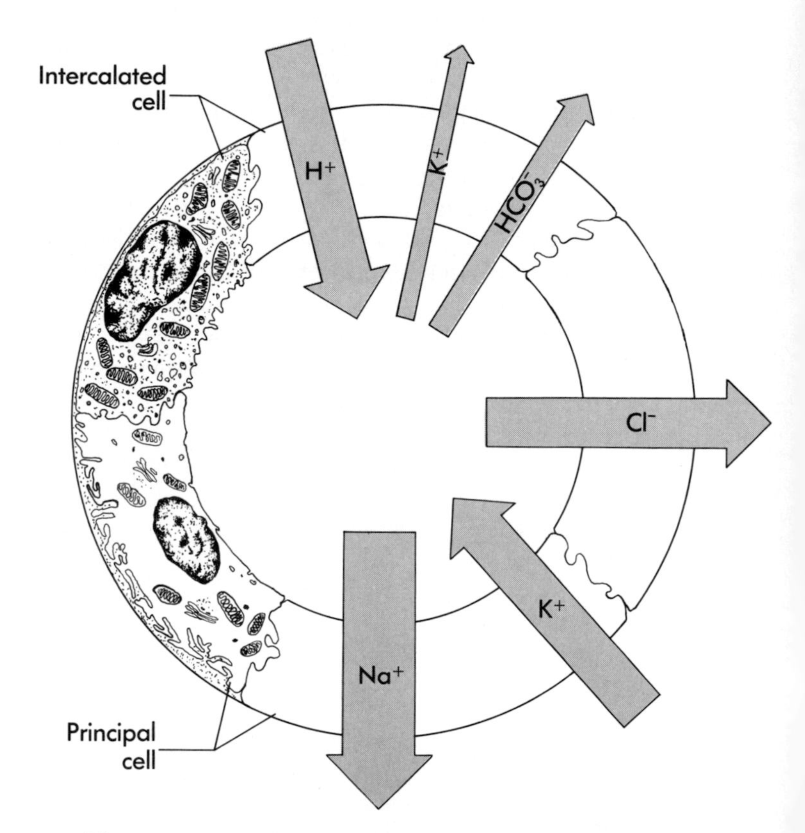

■ **Fig. 41-37** Tubular profile of last segment of the distal tubule and the collecting duct illustrating the cellular ultrastructure and the primary transport characteristics. Intercalated cells secrete H^+ and reabsorb HCO_3^- and K^+, whereas principal cells secrete K^+ and reabsorb Na^+ and water. The mechanism of Cl^- reabsorption and the pathway are not completely understood. (Modified from Burg MB: *Renal handling of Na⁺, Cl⁻, water, amino acids, and glucose*. In Brenner BM, Rector FC Jr, editors: *The kidney*, ed 2, Philadelphia, 1987, WB Saunders.)

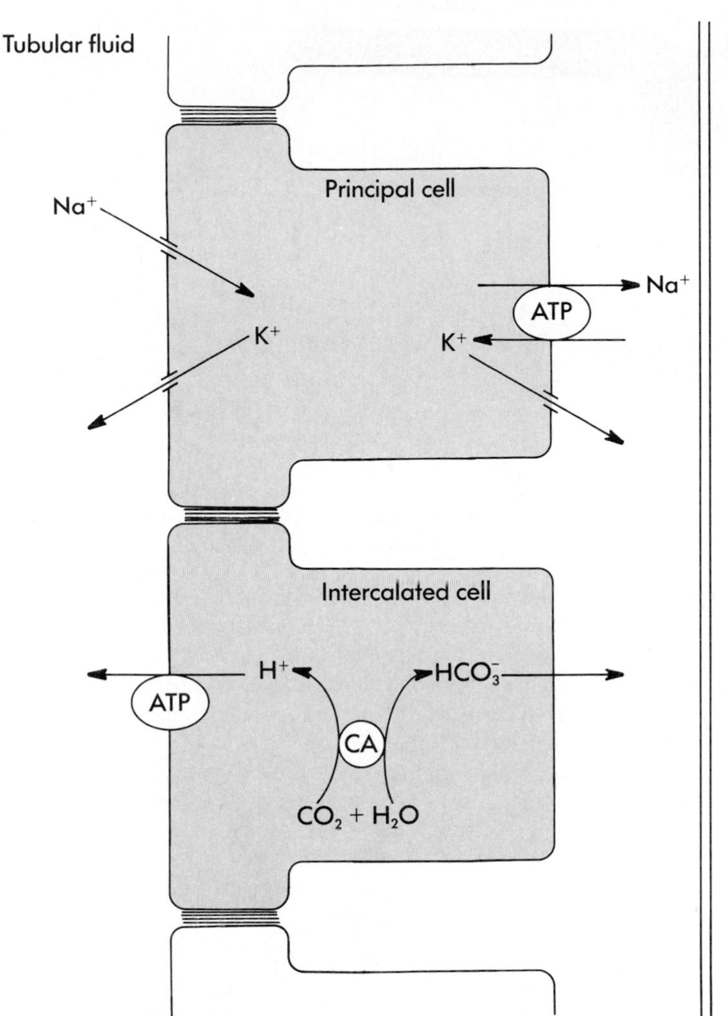

■ **Fig. 41-38** Transport pathways in principal cells and intercalated cells of the distal tubule and collecting duct. See the text for details.

H^+-ATPase transport mechanism (see Fig. 41-38). The generation of H^+ inside the cell is facilitated by the enzyme carbonic anhydrase, and it results in the production of HCO_3^-. For each H^+ secreted into the tubular fluid one HCO_3^- leaves the cell across the basolateral membrane. Additional details of H^+ secretion and bicarbonate reabsorption by the collecting duct are considered in Chapter 42. The mechanism of K^+ reabsorption by intercalated cells is not completely understood.

■ *Regulation of NaCl and Water Reabsorption*

Starling forces. Some of the most important factors that regulate the reabsorption of solutes and water across the proximal tubule and the loop of Henle are Starling forces* (Fig. 41-39). As described previously Na^+, Cl^-, HCO_3^-, amino acids, glucose, and water are transported into the intercellular space of the proximal tubule. Starling forces between this space and the peritubular capillaries facilitate the movement of the reabsorbate into the capillaries. Starling forces that favor movement from the interstitium into the peritubular capillaries are the capillary oncotic pressure (π_{cap}) and the hydrostatic pressure in the intercellular space(P_{IS}). The opposing Starling forces are the interstitial oncotic pressure (π_{IS}) and the capillary hydrostatic pressure (P_{cap}). Normally the sum of the Starling forces favors movement of solute and water from the interstitium into the capillary. However some of the solutes and fluid that enter the lateral intercellular space leak back into the proximal tubular fluid (see Fig. 41-39). Starling forces do not affect transport by the distal tubule and collecting duct because these segments are less permeable to H_2O than is the proximal tubule.

Starling forces across the peritubular capillaries sur-

*Starling forces across the wall of the peritubular capillaries are the hydrostatic pressure in the peritubular capillary (P_{cap}), lateral intercellular space (P_{IS}), oncotic pressure in the peritubular capillary (π_{cap}), and lateral intercellular space (π_{IS}). Thus the reabsorption of water that results from sodium transport from tubular fluid into the lateral intercellular space is modified by the Starling forces. Thus:

$$Q = K_f(P_{IS} - P_{cap}) - (\pi_{cap} - \pi_{IS})$$

where Q equals flow (positive numbers indicate flow from the intercellular space into blood).

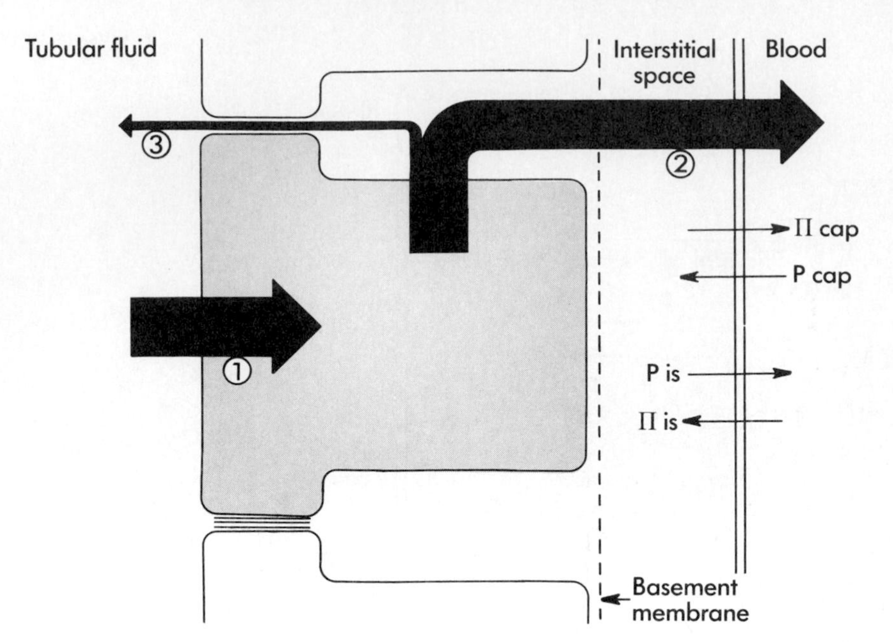

■ **Fig. 41-39** Routes of solute and water transport across the proximal tubule and the Starling forces that modify reabsorption. *(1)* The amount of solute and water moving across the apical membrane. This solute and water then crosses the lateral cell membrane, where some reenters the tubule fluid *(indicated by arrow 3),* and the remainder enters the interstitial space and then flows into the capillary *(indicated by arrow 2).* Starling forces across the capillary wall determine the amount of fluid flowing through pathway *2* versus *3*. Transport mechanisms in the apical cell membranes determine the amount of solute and water entering the cell *(pathway 1).* π_{cap}, capillary oncotic pressure; P_{cap}, capillary hydrostatic pressure; π_{is}, interstitial fluid oncotic pressure; P_{is}, interstitial hydrostatic pressure. Arrows indicate direction of water movement in response to each force.

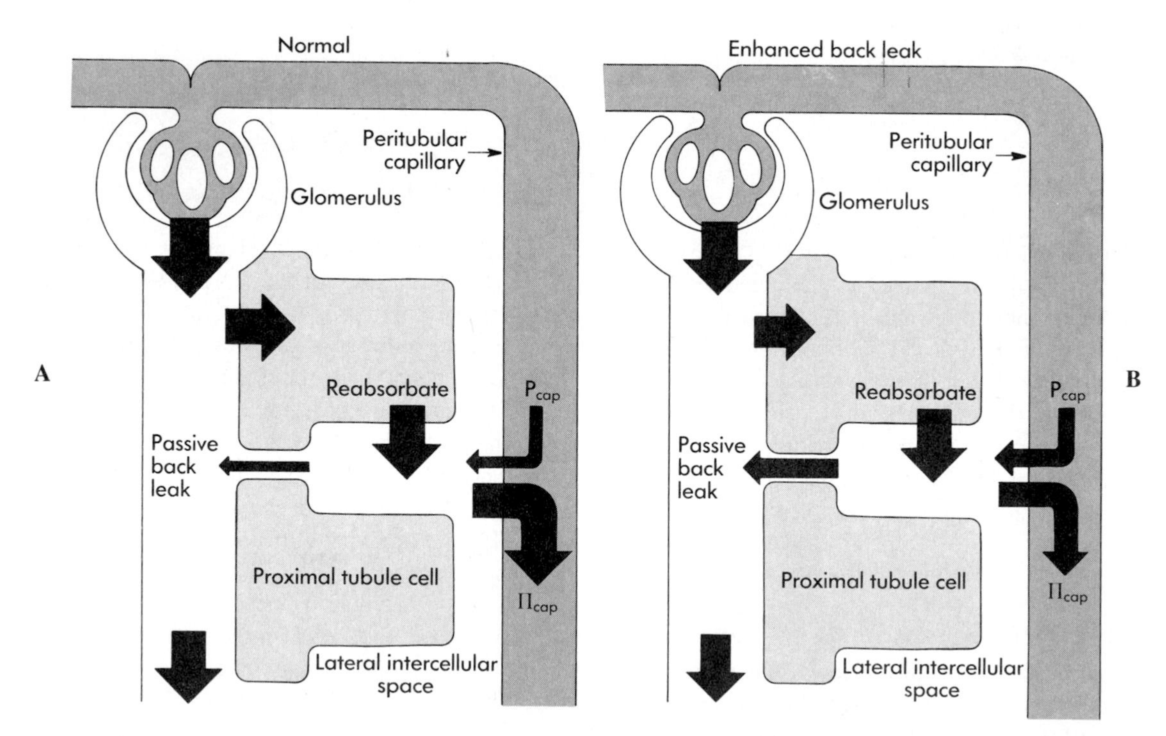

■ **Fig. 41-40** Effects of a decrease in GFR, caused by an increased diameter of the efferent arteriole, on the Starling forces across the peritubular capillary and the ensuing effects on the passive backleak of solutes and water across the tight junctions of the proximal tubule. **A,** Normal. **B,** Enhanced backleak. Dilation of the efferent arteriole decreases hydrostatic pressure in the glomerular capillary, resulting in a fall in GFR. Efferent arteriole dilation also causes an increase in the hydrostatic pressure of the peritubular capillary (P_{cap}) and a fall in the oncotic pressure of the peritubular capillary (π_{cap}). This increases the passive backleak of solutes and water across the tight junction thereby reducing net solute and water reabsorption. Width of arrows is proportional to the amount of solute and water movement.

rounding the proximal tubule are readily altered. Dilation of the efferent arteriole increases P_{cap}, whereas constriction of the efferent arteriole decreases P_{cap}. An increase in P_{cap} inhibits solute and water reabsorption by increasing the backleak across the tight junction, whereas a decrease in P_{cap} stimulates reabsorption by decreasing backleak across the tight junction (see Fig. 41-39).

The oncotic pressure in the peritubular capillary is determined in part by the rate of formation of the glomerular ultrafiltrate (Fig. 41-40). Assuming a constant plasma flow in the afferent arteriole as less ultrafiltrate is formed (i.e., as GFR decreases) the plasma proteins become less concentrated in the plasma that enters the efferent arteriole and peritubular capillary. Hence the peritubular oncotic pressure decreases. Thus the peritubular oncotic pressure is directly related to the filtration fraction (FF = GFR/RPF). A fall in the FF, in the present example resulting from a decrease in GFR at constant RPF, decreases the peritubular capillary oncotic pressure. This decrease in turn increases backflux of solutes and water from the lateral intercellular space into the tubular fluid and thereby decreases net solute and water reabsorption across the proximal tubule. An increase in the FF has the opposite effects.

Glomerulotubular balance. The importance of Starling forces in regulating solute and water reabsorption by the proximal tubule is underscored by the phenomenon of glomerulotubular balance **(GT balance)**. Spontaneous changes in GFR markedly alter the filtered load of sodium (filtered load = GFR × $[P_{Na^+}]$). Unless such changes were rapidly accompanied by adjustments in Na^+ reabsorption, urine Na^+ excretion would fluctuate widely and disturb the Na^+ balance of the whole body. However spontaneous changes in GFR do not alter Na^+ balance because of the phenomenon of GT balance. In GT balance when body Na^+ balance is normal, Na^+ and water reabsorption increase in parallel with an increase in GFR and filtered load of Na^+. Thus a constant fraction of the filtered Na^+ and water is reabsorbed in the proximal tubule despite variations in GFR. The net result of GT balance is to reduce the impact of GFR changes on the amount of Na^+ and water excreted in the urine.

Two mechanisms are responsible for GT balance. One is related to the oncotic and hydrostatic pressures between the peritubular capillaries and the lateral intercellular space (i.e., Starling forces), and the other is related to the filtered load of glucose and amino acids. As an example of the first mechanism an increase in GFR (at constant RPF) raises the protein concentration above normal in the glomerular capillary plasma. This protein-rich plasma leaves the glomerular capillaries, flows through the efferent arteriole, and enters the peritubular capillaries. The increased oncotic pressure in the peritubular capillaries augments the movement of solute and fluid from the lateral intercellular space into the peritubular capillaries, thereby increasing net solute and water reabsorption.

The second mechanism responsible for GT balance is initiated by an increase in the filtered load of glucose and amino acids. As discussed earlier in this chapter the reabsorption of Na^+ is coupled to that of glucose and amino acids. The rate of Na^+ reabsorption therefore depends in part on the filtered load of glucose and amino acids. As GFR and the filtered load of glucose and amino acids increase, Na^+ and water reabsorption also rises.

In addition to GT balance another physiological mechanism operates to minimize changes in the filtered load of Na^+. As described earlier in this chapter an increase in GFR and thus in the amount of Na^+ filtered by the glomerulus, activates the tubuloglomerular feedback mechanism, which returns GFR and the filtration of Na^+ to normal values. Thus spontaneous changes in GFR (e.g., those caused by changes in posture), increase the amount of Na^+ filtered for only a few minutes. Until GFR returns to normal values the mechanisms that underlie GT balance maintain urinary sodium excretion constant and thereby maintain Na^+ homeostasis (see Chapter 42).

Hormones. Table 41-6 summarizes the effects of several hormones on Na^+, Cl^-, and water reabsorption by the segments of the nephron. All of the hormones listed in Table 41-6 act within minutes except aldosterone, which exerts its action on Na^+ reabsorption with a delay of 1 hour. The regulation by these hormones of K^+, Ca^{++}, $PO_4^{\equiv}$, and Mg^{++} reabsorption will be described in Chapter 43.

Sympathetic nervous system. The sympathetic nervous system also regulates Na^+, Cl^-, and water reabsorption by the proximal tubule and loop of Henle. Activation of sympathetic nerves (e.g., after blood volume depletion) stimulates reabsorption, whereas inhibition of these nerves has the opposite effect.

Table 41-6 Hormones that regulate NaCl and water reabsorption

Segment	*Hormone*	*Effects on NaCl and water reabsorption*
Proximal tubule		
	Angiotensin II	↑ NaCl ↑ H_2O
	Glucocorticoids	↑ NaCl ↑ H_2O
Thick ascending limb		
	Aldosterone	↑ NaCl
	Vasopressin	↑ NaCl
Distal tubule/collecting duct		
	Aldosterone	↑ NaCl
	Atrial natriuretic peptide	↓ NaCl ↓ H_2O
	Prostaglandins	↓ NaCl ↓ H_2O
	Bradykinin	↓ NaCl
	Vasopressin	↑ NaCl ↑ H_2O

Summary

1. The functional unit of the kidney is the nephron, which consists of a glomerulus, proximal tubule, loop of Henle, distal tubule, and collecting duct system.

2. The glomerulus is composed of glomerular capillaries and Bowman's capsule.

3. The juxtaglomerular apparatus is one component of an important feedback mechanism that regulates renal blood flow and glomerular filtration rate. The structures that compose the juxtaglomerular apparatus include the macula densa, the extraglomerular mesangial cells, and the renin-producing granular cells.

4. The lower urinary tract is composed of the ureters, bladder, and urethra. Micturition is the process of emptying the urinary bladder. The micturition reflex is an automatic spinal cord reflex; however it can be inhibited or facilitated by centers in the brain stem and cerebral cortex.

5. Urine is formed by three general processes: glomerular filtration, reabsorption of solutes and water from the ultrafiltrate into the peritubular capillaries, and secretion of selected solutes from the peritubular capillaries into the tubular fluid.

6. The rate of glomerular filtration is calculated by measuring the clearance of inulin or creatinine.

7. Effective renal plasma flow is determined by the clearance of PAH.

8. Renal clearance equations can be used to determine whether a substance undergoes either net reabsorption or secretion by the nephron.

9. The first step in the production of urine is the formation of an ultrafiltrate of plasma by the glomerulus. Starling forces across the glomerular capillaries drive an ultrafiltrate of plasma from the glomerular capillaries into Bowman's space. The ultrafiltrate is devoid of cellular elements, contains very little protein, but otherwise is identical to plasma. Proteins with a molecular radius smaller than 18 Å are readily filtered; proteins between 18 and 36 Å are filtered to varying degrees, depending on size and charge (cationic proteins are more readily filtered than anionic proteins); and proteins with molecular radii greater than 36 Å are not filtered.

10. Renal blood flow (1.25 L/min) is 25% of the cardiac output, yet the kidneys constitute less than 0.5% of the body weight. Renal blood flow serves several important functions, including determining the glomerular filtration rate; modifying solute and water reabsorption by the proximal tubule; participating in concentration and dilution of the urine and delivery of oxygen, nutrients, and hormones to the cells of the nephron; and return of carbon dioxide and reabsorbed fluid and solutes to the general circulation.

11. Renal blood flow and glomerular filtration rate are maintained constant, despite changes in arterial blood pressure between 90 and 180 mm Hg, by the phenomenon of autoregulation. Autoregulation is achieved by the myogenic reflex and tubuloglomerular feedback.

12. The four major segments of the nephron determine the composition and volume of the urine by the processes of selective reabsorption of solutes and water and secretion of solutes.

13. Reabsorption allows the kidneys to retain those substances that are essential and to thereby regulate their levels in the plasma. The reabsorption of Na^+, Cl^-, other anions, and organic solutes together with water constitutes the major function of the nephron. Approximately 25,200 mEq of Na^+ and 178 L of water are reabsorbed each day. The proximal tubule reabsorbs 67% of the glomerular ultrafiltrate and the loop of Henle reabsorbs about 20%. The distal segments of the nephron (distal tubule and collecting duct system) have a more limited reabsorptive capacity. However in the distal segments the final adjustments in the composition and volume of the urine are made and most of the regulation by hormones and other factors is effected.

14. Secretion of substances into tubular fluid is a means of excreting various by-products of metabolism and serves to eliminate exogenous organic anions and bases (e.g., drugs) from the body. Many organic compounds are bound to plasma proteins and therefore are unavailable for ultrafiltration. Thus secretion is their major route of excretion in the urine.

Bibliography

Journal articles

Baylis C, Blantz RC: Glomerular hemodynamics, *News Physiol Sci* 1:86, 1986.

Giebisch G, Boulpaep EL, editors: Symposium on cotransport mechanisms in renal tubules, *Kidney Int* 36:333, 1989.

Gottschalk CW, editor: Renal and electrolyte physiology section: tubuloglomerular feedback mechanisms, *Annu Rev Physiol* 49:249, 1987.

Kriz W, Bankir L: A standard nomenclature for structures of the kidney, *Am J Physiol* 254:F1, 1988.

Takabatake T, Thurau K, editors: Tubuloglomerular feedback system, *Kidney Int Suppl* 32:1, 1991.

Books and monographs

Berry CA, Rector FC, Jr: *Renal transport of glucose, amino acids, sodium, chloride and water*. In Brenner BM, Rector FC Jr, editors: *The kidney*, ed 4, Philadelphia, 1991, WB Saunders.

Bradley WE: *Physiology of the urinary bladder*. In Walsh PC et al, editors: *Cambell's urology*, ed 5, Philadelphia, 1986, WB Saunders.

Byrne JH, Schultz SG: *An introduction to membrane transport and bioelectricity*, New York, 1988, Raven Press.

Cogan MG: *Fluid and electrolytes: physiology and pathophysiology,* Norwalk, Conn, 1991, Appleton & Lange.

Dworkin LD, Brenner BM: *Biophysical basis of glomerular filtration.* In Seldin DW, Giebisch G, editors: *The kidney: physiology and pathophysiology,* ed 2, New York, 1992, Raven Press.

Dworkin LD, Brenner BM: *The renal circulations.* In Brenner BM, Rector FC Jr, editors: *The kidney,* ed 4, Philadelphia, 1991, WB Saunders.

Kassier JP, Harrington JT: *Laboratory evaluation of renal function.* In Schrier RW, Gottschalk CW, editors: *Diseases of the kidney,* ed 4, Boston, 1988, Little, Brown & Co.

Koeppen BM, Stanton BA: *Renal physiology,* St. Louis, 1992, Mosby–Year Book, Inc.

Koushanpour E, Kriz W: *Renal physiology,* ed 2, Berlin, 1986, Springer-Verlag.

Kriz W, Kaissling B: *Structural organization of the mammalian kidney.* In Seldin DW, Giebisch G, editors: *The kidney: physiology and pathophysiology,* ed 2, New York, 1992, Raven Press.

Maddox DA, Brenner BM: *Glomerular ultrafiltration.* In Brenner BM, Rector FC Jr, editors: *The kidney,* ed 4, Philadelphia, 1991, WB Saunders.

Rose BD: *Clinical physiology of acid-base and electrolyte disorders,* ed 3, New York, 1989, McGraw-Hill.

Tanagho EA: *Anatomy of the lower urinary tract.* In Walsh PC et al, editors: Cambell's urology, ed 5, Philadelphia, 1986, WB Saunders.

Tanagho EA: *Anatomy of the genitourinary tract.* In Tanagho EA, McAnich JW, editors: *Smith's general urology,* ed 12, Norwalk, Conn, 1988, Appleton & Lange.

Tisher CC, Madsen KM: *Anatomy of the kidney.* In Brenner BM, Rector FC Jr, editors: *The kidney,* ed 4, Philadelphia, 1991, WB Saunders.

Ulfendal H, Wolgast M: *Renal circulation and lymphatics.* In Seldin DW, Giebisch G, editors: *The kidney: physiology and pathophysiology,* ed 2, New York, 1992, Raven Press.

CHAPTER

42

Control of Body Fluid Osmolality and Volume

One of the major functions of the kidneys is to maintain the volume and composition of the body fluids constant despite wide variation in the daily intake of water and solutes. In this chapter the regulation of body fluid osmolality and volume is discussed. As an introduction to this material the normal volume and compositions of the various body fluid compartments are reviewed.

The Body Fluid Compartments

Volumes of Body Fluid Compartments

Water accounts for approximately 60% of the body weight. The water content of the body varies with the amount of adipose tissue. Because the water content of adipose tissue is lower than that of other tissue, increasing amounts of adipose tissue reduce the fraction of total body weight due to water. The percentage of body weight attributed to water also varies with age. In the newborn infant water constitutes about 75% of the body weight. This decreases to the adult value of 60% by the age of 1 year.

The **total body water** is distributed between two major compartments, which are divided by the cell membrane (Fig. 42-1). The **intracellular fluid** (ICF) compartment is the larger compartment; it contains approximately two thirds of the total body water. The remaining one third is contained in the **extracellular fluid** (ECF) compartment.* Expressed as percentages of body weight the volumes of total body water, ICF, and ECF can be estimated as

$$\text{Total body water} = 0.6 \times (\text{Body weight})$$

$$\text{ICF} = 0.4 \times (\text{Body weight})$$

$$\text{ECF} = 0.2 \times (\text{Body weight})$$

*In these and all subsequent calculations it is assumed that 1 L of fluid (e.g., ICF and ECF) has a mass of 1 kg. This allows interconversion between units of osmolality and volume.

The extracellular fluid compartment is further subdivided into **interstitial fluid** and **plasma;** these fluids are separated by the capillary endothelium. The interstitial fluid, which represents the fluid surrounding the cells in the various tissues of the body, comprises three fourths of the extracellular fluid volume. Included in this compartment is water contained within the bone and dense connective tissue. The plasma volume represents the remaining one fourth of the extracellular fluid.

Measurement of Body Fluid Volumes

The volume of a body fluid compartment can be measured by adding a marker to the compartment and measuring its concentration once equilibrium has been reached. For example if 5 g of a marker is added to a beaker of water of unknown volume and the concentration of this marker in the beaker is 10 g/L after complete equilibration, the volume of water in the beaker is

$$\text{Volume} = \frac{\text{Amount}}{\text{Concentration}} = \frac{5 \text{ g}}{10 \text{ g/L}} = 0.5 \text{ L} \quad (1)$$

For a substance to be a good marker of a particular body fluid compartment it must be confined to and evenly distributed throughout that compartment. In addition the marker should be physiologically inert, not metabolized, and easily measured. Table 42-1 lists

Table 42-1 Substances used as markers for body fluid compartments

Body fluid compartment	*Marker*
Total body water	Tritiated water
Extracellular fluid	Inulin, mannitol
Plasma	^{125}I-albumin, Evans blue

No marker exists for the intracellular fluid compartment; its volume is calculated as the difference between the total body water and extracellular fluid volume.

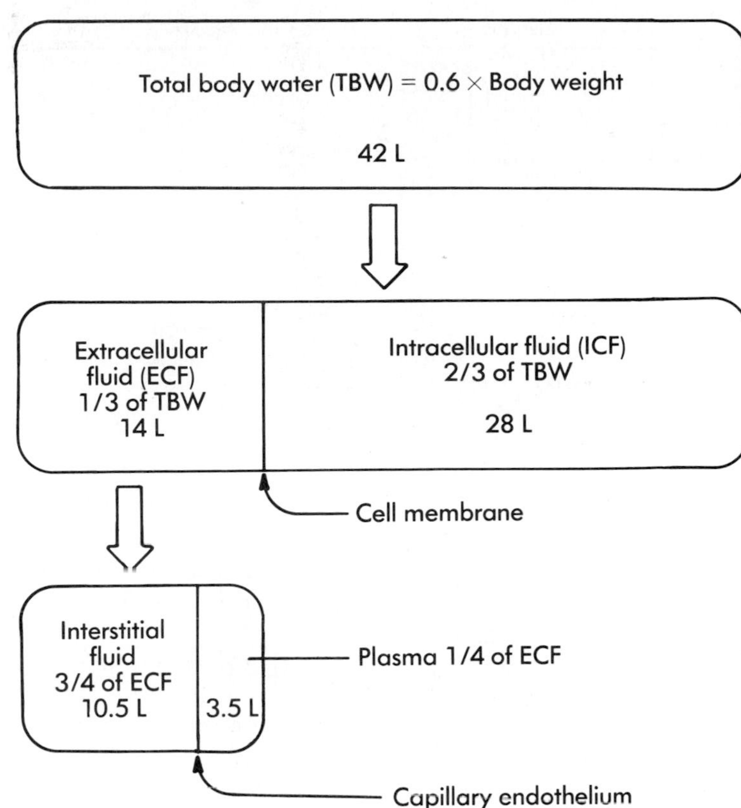

■ **Fig. 42-1** Relationship between the volumes of the major body fluid compartments. The actual values shown are calculated for a 70 kg individual.

some markers that have been used to measure the volumes of the various body fluid components. Note that no marker exists for measuring the intracellular fluid volume. This volume is therefore calculated as the difference between the total body water and the extracellular fluid.

In practice the measurement of the volumes of body fluid compartments is more complicated because the markers used are excreted from the body while they are also distributing throughout the compartments. However by quantitating the amount excreted one can make appropriate corrections.

■ *Composition of Body Fluid Compartments*

The concentrations of the major cations and anions in the ECF (interstitial fluid and plasma) are illustrated in Fig. 42-2. Na^+ is the major cation of the ECF, and Cl^- and HCO_3^- are the major anions. The ionic compositions of the two compartments of the ECF are very similar because these compartments are separated only by the capillary endothelium, and this barrier is freely permeable to small ions. The major difference between the interstitial fluid and plasma is that the plasma contains significantly more protein. The differential concentrations of protein in the interstitial fluid and plasma can affect the distribution of cations and anions between these compartments; plasma proteins have a net negative charge and tend to increase the cation concentrations and reduce the anion concentrations in the plasma compartment. However this effect is small, and the ionic compositions of the interstitial fluid and plasma can be considered to be identical. Because of its abundance Na^+ (and its attendant anions, primarily Cl^- and HCO_3^-) is the major determinant of the osmolality of the ECF. Accordingly a rough estimate of the ECF osmolality can be obtained by simply doubling the sodium concentration $[Na^+]$. For example if the plasma $[Na^+]$ is 145 mEq/L, the osmolality of plasma and ECF can be estimated as

$$\text{Plasma osmolality} = 2(\text{plasma } [Na^+]) = 290 \text{ mOsm/kg } H_2O \quad (2)$$

Because water is in osmotic equilibrium across the capillary endothelium and across the plasma membrane of cells, measurement of the plasma osmolality also provides a measure of the osmolality of the ECF and ICF.

The composition of the ICF is more difficult to measure and can vary significantly from one tissue to another (Fig. 42-2). In contrast to the ECF the $[Na^+]$ of the ICF is extremely low. K^+ is the predominant cation of the ICF. This asymmetrical distribution of Na^+ and K^+ across the plasma membrane is maintained by the activity of the ubiquitous Na^+-K^+-ATPase pump. By its action Na^+ is extruded from the cell in exchange for K^+.

The anion composition of the ICF also differs markedly from that of the ECF. The $[Cl^-]$ of the ICF is low compared with that of the ECF. The major ICF anions are phosphates, organic anions, and protein.

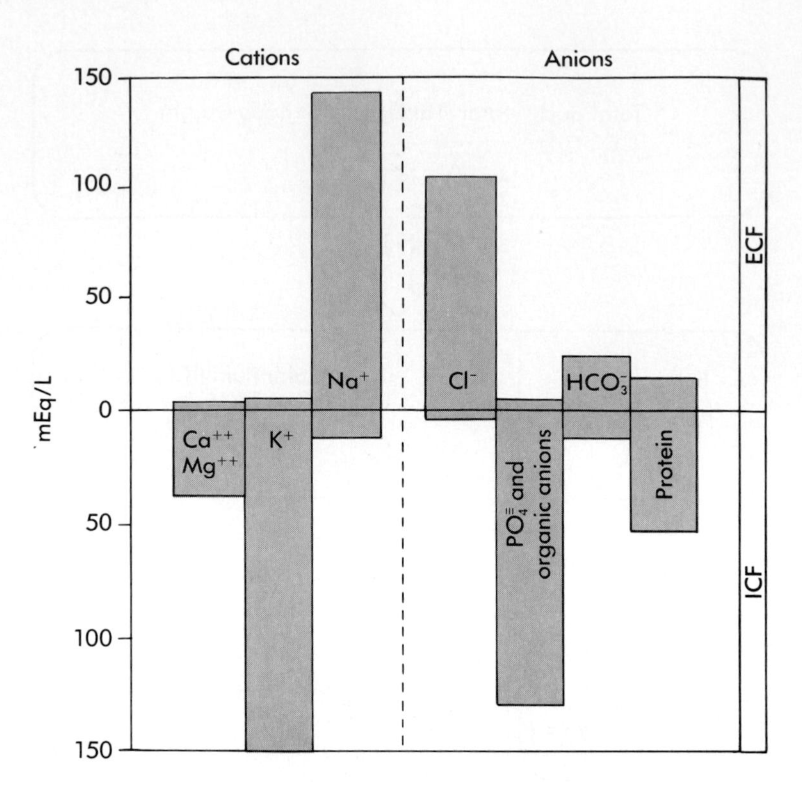

■ **Fig. 42-2** Concentrations of the major cations and anions in extracellular (ECF) and intracellular (ICF) fluids. The concentrations of Ca^{++} and Mg^{++} are the sum of these two ions. Concentrations represent the total of free and complexed ions. The total concentration of cations equals that of anions in both compartments.

■ *Fluid Exchange between Body Fluid Compartments*

Water moves freely between the various body fluid compartments. Two forces determine this movement: hydrostatic pressure and osmotic pressure. Hydrostatic pressure (generated by the pumping of the heart) and the osmotic pressure of the plasma proteins (oncotic pressure) are important determinants of fluid movement across the capillary endothelium (see p. 472), whereas only osmotic pressure differences between the ICF and ECF are responsible for fluid movement across cell membranes. Because the plasma membranes of cells are highly permeable to water, a change in the osmolality of either the ICF or ECF moves water rapidly between these compartments. Thus, *except for transient changes, the ICF and ECF compartments are in osmotic equilibrium*.

In contrast to the movement of water the movement of ions across cell membranes is more variable and depends on the presence of specific membrane transporters. Consequently as a first approximation fluid exchange between the ICF and ECF under pathophysiological conditions can be analyzed by assuming that appreciable shifts of ions between the compartments do not occur. This can best be illustrated by considering the consequences of adding water, NaCl, or isotonic saline to the ECF (Fig. 42-3).

Example 1—addition of 2 L of water to the ECF. When water is added to the ECF, the osmolality of this compartment is reduced. If no fluid shifts were to occur, the osmolality of the ECF compartment would decrease from 290 to 254 mOsm/kg H_2O. However the cell membranes are freely permeable to water, and the osmotic gradient resulting from the addition of water to the ECF compartment causes water to move into the cells. This shift of water increases the volume of both the ICF and the ECF above their initial values. The osmolality of the body fluids is also reduced, but to a lesser degree than would occur in the absence of such a shift. The volumes and osmolalities of the ICF and ECF after equilibration (see Fig. 42-3) are calculated as follows:

Initial Conditions

Initial total body water	= 0.6 × (70 kg) = 42 L
Initial ICF volume	= 0.4 × (70 kg) = 28 L
Initial ECF volume	= 0.2 × (70 kg) = 14 L
Initial total body osmoles	= (Total body water)(Body fluid osmolality) = (42 L)(290 mOsm/kg H_2O) = 12,180 mOsm
Initial ICF osmoles	= (ICF Volume)(Body fluid osmolality) = (28 L)(290 mOsm/kg H_2O) = 8,120 mOsm
Initial ECF osmoles	= Total body osmoles − ICF osmoles = 12,180 mOsm − 8,120 mOsm = 4,060 mOsm

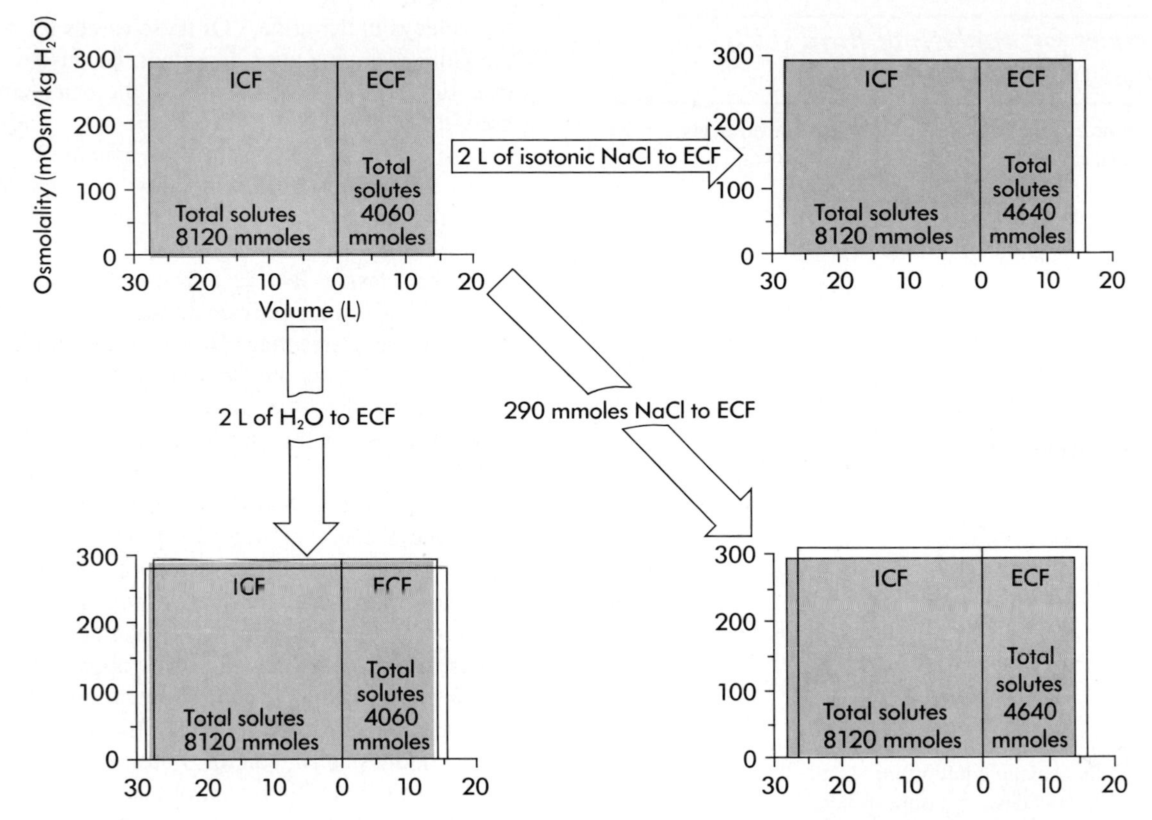

Fig. 42-3 Alterations in the volumes and osmolality of the body fluid compartments resulting from the addition to the ECF of 2 L of water, 290 mmoles of NaCl (580 mOsm of solute), and 2 L of isotonic NaCl. The original conditions are indicated by shading.

Final conditions

$$\text{Final osmolality} = \frac{\text{Total body osmoles}}{\text{New total body water}}$$

$$= \frac{12{,}180 \text{ mOsm}}{44 \text{ L}} = 277 \text{ mOsm/kg } H_2O$$

$$\text{Final ICF volume} = \frac{\text{ICF osmoles}}{\text{New osmolality}}$$

$$= \frac{8{,}120 \text{ mOsm}}{277 \text{ mOsm/kg } H_2O} = 29.3 \text{ L}$$

$$\text{Final ECF volume} = \text{New body water} - \text{Final ICF volume}$$

$$= 44 \text{ L} - 29.3 \text{ L} = 14.7 \text{ L}$$

Example 2—addition of 290 mmoles NaCl to the ECF. Because of the Na^+-K^+-ATPase in the membrane of cells NaCl is effectively restricted to the ECF space. Because NaCl dissociates into two particles, this adds 580 mOsm to the ECF compartment (e.g, 290 mmol × 2 = 580 mOsm). If no fluid shifts occur the osmolality transiently increases from 290 to 331 mOsm/kg H_2O. However such an increase in ECF osmolality moves water out of the ICF. As a consequence of this fluid shift the volume of the ICF decreases and the volume of the ECF increases above their initial values. The volumes and osmolalities of the ICF and ECF after equilibration (see Fig. 42-3) are calculated as follows:

$$\text{Final osmolality} = \frac{\text{New total body osmoles}}{\text{Total body water}}$$

$$= \frac{12{,}180 \text{ mOsm} + 580 \text{ mOsm}}{42 \text{ L}}$$

$$= 304 \text{ mOsm/kg } H_2O$$

$$\text{Final ICF volume} = \frac{\text{ICF osmoles}}{\text{New osmolality}}$$

$$= \frac{8{,}120 \text{ mOsm}}{304 \text{ mOsm/kg } H_2O} = 26.7 \text{ L}$$

$$\text{Final ECF volume} = \text{Total body water} - \text{Final ICF volume}$$

$$= 42 \text{ L} - 26.7 \text{ L} = 15.3 \text{ L}$$

Example 3—addition of 2 L of isotonic saline to the ECF. Addition of isotonic saline (osmolality = 290 mOsm/kg H_2O) to the ECF does not change the osmolality of this compartment. Consequently there is no osmotic driving force for a shift of fluid between the ICF and ECF. The entire 2 L of isotonic saline solution remains in the ECF. The volumes and osmolalities of the

Principles for analysis of fluid shifts between ICF and ECF

The volumes of the various body fluid compartments can be estimated in the normal adult by the following:

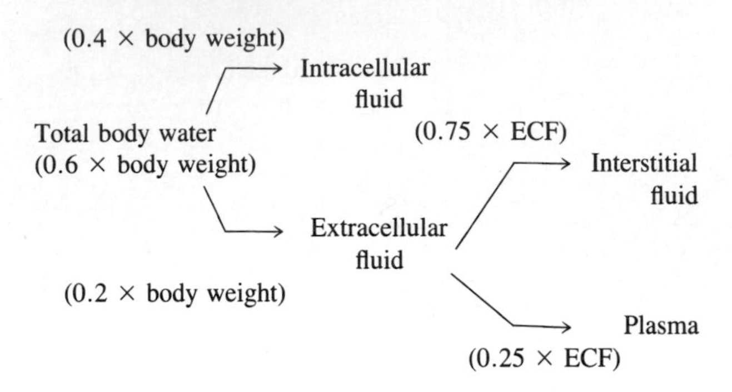

All exchanges of water and solutes with the external environment occur through the ECF (e.g., intravenous infusion, intake, or loss via the gastrointestinal tract). Changes in the ICF are secondary to fluid shifts between the ECF and ICF. Fluid shifts occur only if the perturbation of the ECF alters its osmolality.

Except for brief periods (seconds to minutes) the ICF and ECF are in osmotic equilibrium. A measurement of plasma osmolality provides a measure of both ECF and ICF osmolality.

A simplifying assumption is that equilibration between the ICF and ECF occurs only by movement of water, and not by movement of osmotically active solutes.

Conservation of mass must be maintained, especially when considering either addition of water and/or solutes to the body or their excretion from the body.

(From Koeppen BM, Stanton BA: *Renal physiology,* St Louis, 1992, Mosby–Year Book.

ICF and ECF after equilibration (see Fig. 42-3) are calculated as follows:

Final osmolality = 290 mOsm/kg H_2O (unchanged from initial value)

Final ECF volume = Initial ECF volume + 2 L isotonic saline = 14 L + 2 L = 16 L

Final ICF volume = 28 L (unchanged from initial value)

The box above summarizes the basic principles and approaches that can be used to examine problems related to the exchange of fluid between ICF and ECF.

■ *Control of Body Fluid Osmolality: Urine Concentration and Dilution*

Water is lost from the body by several routes, including (1) the lungs during respiration, (2) the skin by perspiration, (3) the gastrointestinal tract in the feces, and (4) the kidneys in the urine.* Of these routes for water loss the kidneys are the most important, because renal water excretion is regulated to maintain the osmolality of the body fluids constant.

Disorders of water balance are manifested by alterations in the body fluid osmolality (e.g., plasma). Because the major determinant of the plasma osmolality is Na^+ (with its anions Cl^- and HCO_3^-) these disorders alter the plasma [Na^+]. When evaluating abnormal plasma [Na^+] in an individual, suspecting a problem in Na^+ balance is tempting. However the problem relates not to Na^+ balance but to water balance. Changes in Na^+ balance alter the volume of the extracellular fluid and not its osmolality (see below).

When water intake is low or when water is lost from the body by other routes (e.g., perspiration, diarrhea), the kidneys conserve water by producing a small volume of urine that is hyperosmotic with respect to plasma. When water intake is high a large volume of hypoosmotic urine is produced. In a normal individual the urine osmolality can vary from approximately 50 to 1,200 mOsm/kg H_2O, and the urine volume can vary from as much as 20 L/day to as little as 0.5 L/day.

The kidneys can control water excretion independently of their ability to control the excretion of a number of other physiologically important substances (e.g., Na^+, K^+, H^+, urea). Indeed this ability is necessary for survival because it allows water balance to be achieved without perturbing the other homeostatic functions of the kidneys.

In this section the mechanism by which the kidneys excrete either **hypoosmotic** (dilute) or **hyperosmotic** (concentrated) urine is discussed. In addition the control of vasopressin secretion is considered, as is its important role in regulating the excretion of water by the kidneys.

■ *Antidiuretic Hormone*

Antidiuretic hormone (ADH), or **vasopressin,** acts on the kidneys to regulate the volume and osmolality of the urine. When plasma ADH levels are low, a large volume of urine is excreted **(diuresis),** and the urine is dilute.† When plasma ADH levels are elevated, a small volume of urine is excreted **(antidiuresis)** and the urine is concentrated. Fig. 42-4 illustrates the effect of ADH on the urine flow rate and the osmolality of this urine. Also illustrated is the excretion of total solute (e.g., Na^+, K^+, H^+, urea) by the kidneys. Note that ADH does not appreciably alter the excretion of solute. This underscores the fact that ADH controls water excretion

*The loss of water via the lungs, skin, and gastrointestinal tract is collectively termed insensible water loss because of the difficulty associated with its measurement. In a normal individual insensible water loss is approximately 1 to 1.2 L/day.

†*Diuresis* refers simply to a large urine output. When the urine primarily contains water, the condition is referred to as a water diuresis.

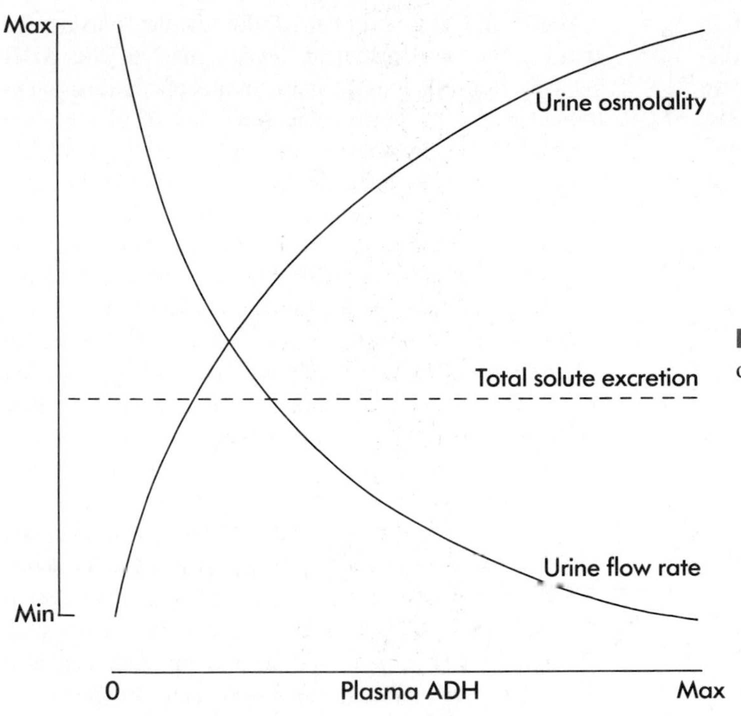

■ **Fig. 42-4** Relationship of plasma ADH levels to urine osmolality, urine flow rate, and total solute excretion.

and maintains water balance without altering the excretion and homeostatic control of other substances.

ADH is a small peptide nine amino acids in length. It is synthesized in neuroendocrine cells located within the supraoptic and paraventricular nuclei of the hypothalamus (Fig. 42-5). The synthesized hormone is packaged in granules, which are transported down the axon of the cell and stored in the nerve terminals located in the **neurohypophysis (posterior pituitary).**

The secretion of ADH by the posterior pituitary can be influenced by several factors. The two primary physiological regulators of ADH secretion are osmotic (plasma osmolality) and hemodynamic (blood volume and pressure). Other factors that can alter ADH secretion include nausea (stimulates), atrial natriuretic peptide (inhibits), and angiotensin II (stimulates). A number of drugs also affect ADH secretion. For example nicotine stimulates secretion, whereas ethanol inhibits its secretion.

Osmotic control of ADH secretion. Changes in the osmolality of the body fluids play the most important role in regulating ADH secretion. Changes in osmolal-

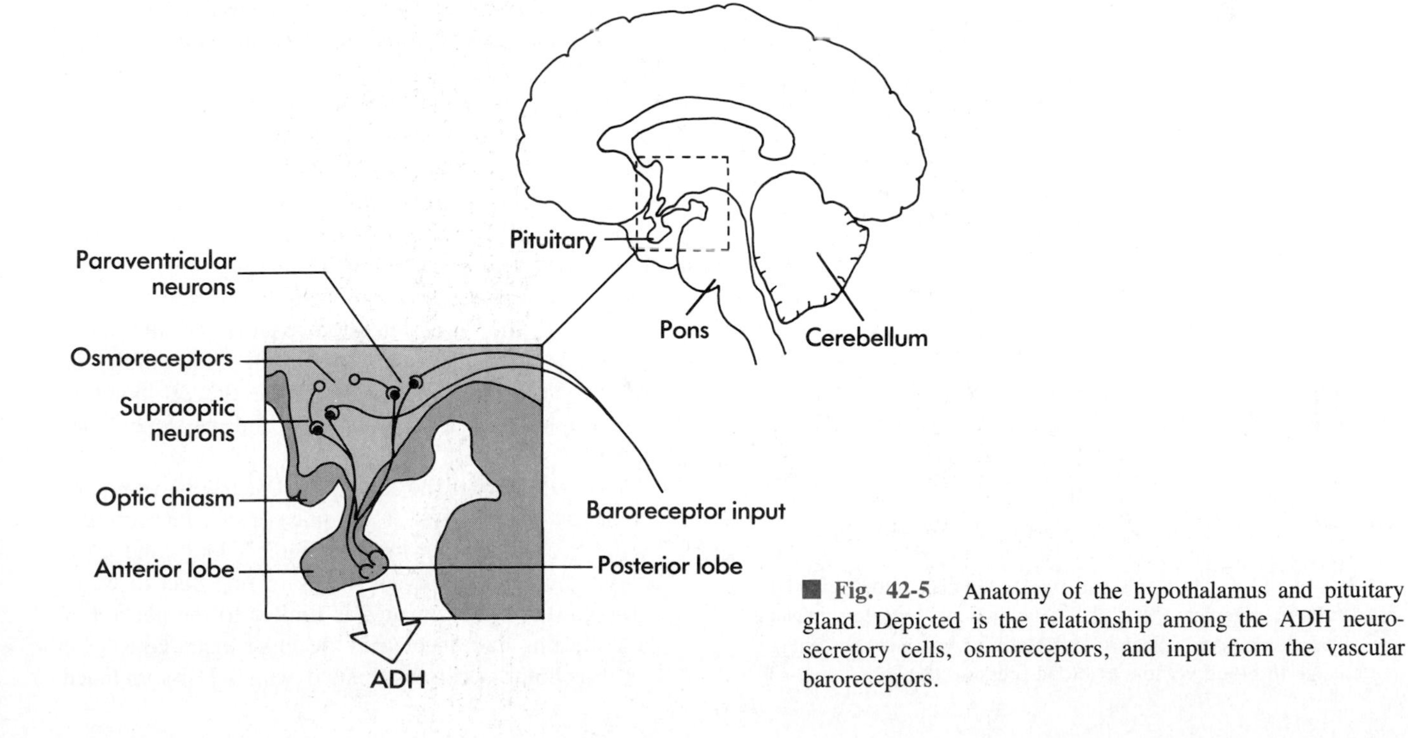

■ **Fig. 42-5** Anatomy of the hypothalamus and pituitary gland. Depicted is the relationship among the ADH neurosecretory cells, osmoreceptors, and input from the vascular baroreceptors.

ity as small as 1% are sufficient to alter ADH secretion significantly. Cells located in the hypothalamus but distinct from those that synthesize ADH are involved in sensing changes in body fluid osmolality. These cells, termed **osmoreceptors,** appear to behave as osmometers and sense changes in body fluid osmolality by either shrinking or swelling. Osmoreceptors respond *only* to plasma solutes that are *effective osmoles* (see Chapter 48). For example urea is an ineffective osmole when the function of the osmoreceptors is considered. Thus elevation of only the plasma urea concentration has little effect on ADH secretion.

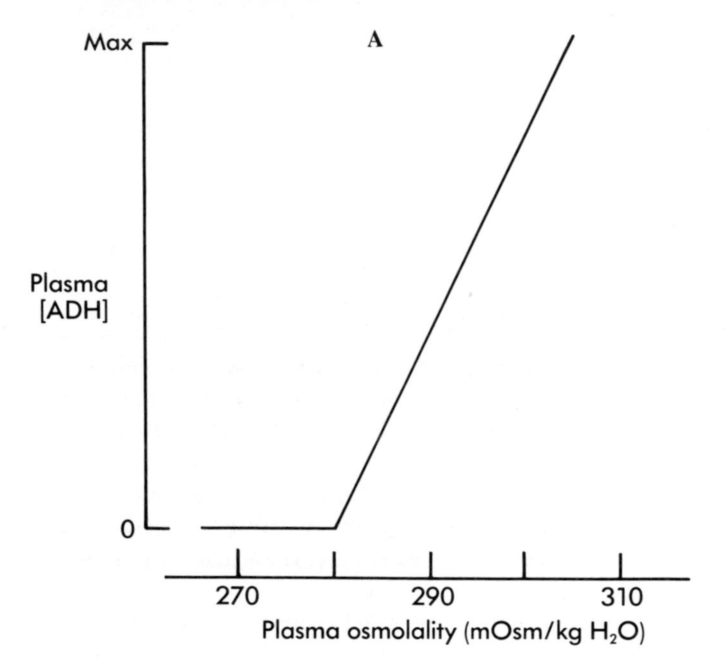

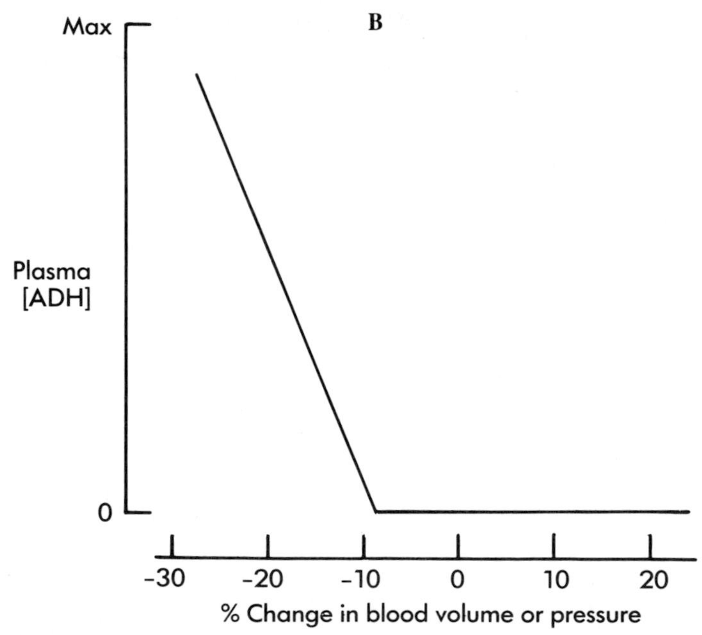

■ **Fig. 42-6** Osmotic and hemodynamic control of ADH secretion. Depicted are the relationships between plasma ADH levels and plasma osmolality **(A),** and percentage change in blood volume or blood pressure **(B).**

When the effective osmolality of the plasma increases, the osmoreceptors send signals to the ADH synthesizing cells located in the supraoptic and paraventricular nuclei of the hypothalamus and ADH secretion is stimulated. Conversely when the effective osmolality of the plasma is reduced, secretion is inhibited. Because ADH is rapidly degraded in the plasma, circulating levels can be reduced to zero within minutes after secretion is inhibited. As a result the ADH system can respond rapidly to fluctuations in plasma osmolality.

Fig. 42-6, *A* illustrates the effect of changes in plasma osmolality on circulating ADH levels. The **set point** of the system is defined as the plasma osmolality value at which ADH secretion begins to increase. Below this set point virtually no ADH is released. Above the set point the slope of the relationship is quite steep; this slope reflects the sensitivity of this system. The set point varies among individuals and is genetically determined. In healthy adults the set point varies from 280 to 290 mOsm/kg H_2O. Several physiological factors such as alterations in blood volume and pressure can also change the set point, as discussed later. Pregnancy is also associated with a decrease in the set point.

Hemodynamic control of ADH secretion. A decrease in blood volume or pressure also stimulates ADH secretion. The receptors activated by this response are located in both the low-pressure (left atrium and pulmonary vessels) and the high-pressure (aortic arch and carotid sinus) sides of the circulatory system. These receptors respond to stretch and are termed **baroreceptors** (see later and p. 486). Signals from these receptors are relayed to the ADH secretory cells of the supraoptic and paraventricular hypothalamic nuclei via afferent fibers in the vagus and glossopharyngeal nerves. The sensitivity of the baroreceptor system is less than that of the osmoreceptors; a 5% to 10% decrease in blood volume or pressure is required to stimulate ADH secretion (Fig. 42-6, *B*).

Alterations in blood volume and pressure also effect the response to changes in body fluid osmolality (Fig. 42-7). With a decrease in blood volume or pressure the set point is shifted to lower osmolality values and the slope of the relationship is steeper. In terms of survival of an individual faced with circulatory collapse this means that the kidney continues to conserve water, even though the water retention reduces the osmolality of the body fluids. With an increase in blood volume or pressure the opposite action occurs, the set point is shifted to higher osmolality values, and the slope is decreased.

ADH actions on the kidney. ADH has two primary actions on the kidneys. It stimulates NaCl reabsorption by the thick ascending limb of Henle's loop, and it increases the permeability of the collecting duct to water and urea (the effect on urea is limited to the portion of the collecting duct located in the inner medulla).

The cellular events associated with ADH-stimulated

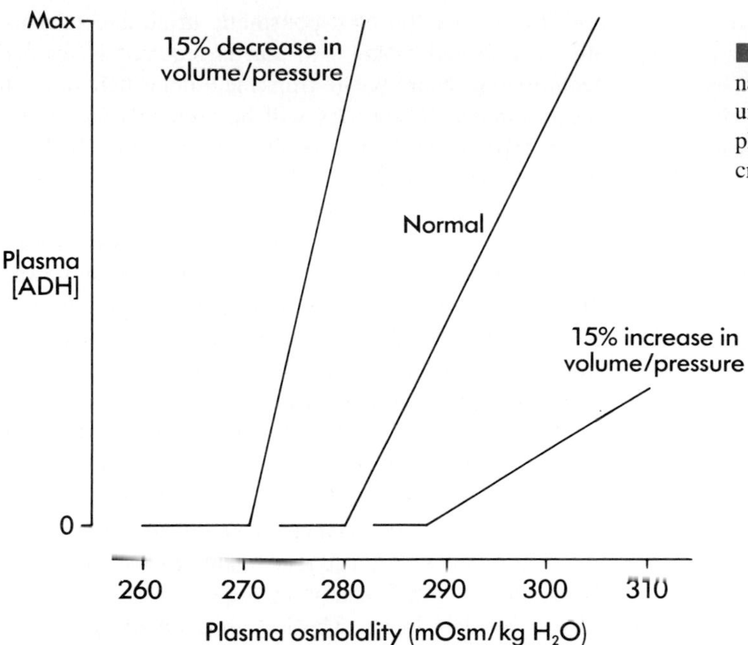

■ **Fig. 42-7** Interaction between osmotic and hemodynamic stimuli for ADH secretion. With decreased blood volume and pressure the osmotic set point is shifted to lower plasma osmolality values, and the slope is increased. An increase in blood volume and pressure has the opposite effect.

transport by the thick ascending limb have not been completely elucidated. However it is known that net reabsorption of NaCl is increased as a result of the enhanced activity of the $1Na^+$, $2Cl^-$, $1K^+$ symporter located in the apical membrane of the cell. As discussed later, this symporter augments the ability of the kidney to concentrate the urine.

ADH increases the permeability of the collecting duct to water (Fig. 42-8). ADH binds to a receptor on the basolateral membrane of the cell. Binding to this receptor, which is coupled to adenylyl cyclase, increases the intracellular levels of **cyclic adenosine monophosphate** (cAMP). The rise in intracellular cAMP activates one or more protein kinases; activation in turn results in the insertion of vesicles containing **water channels** into the apical membrane. These water channels are preformed and reside in vesicles located beneath the apical membrane of the cell. With the removal of ADH these water channels are reinternalized into the cell, and the apical membrane once again becomes impermeable to water. This shuttling of water channels into and out of the apical membrane provides a rapid mechanism for controlling membrane permeability to water. The basolateral membrane is freely permeable to water. Thus any water that enters the cell through the apical membrane water channels exits across the basolateral membrane. These processes result in the net reabsorption of water from the tubule fluid into the peritubular capillaries.

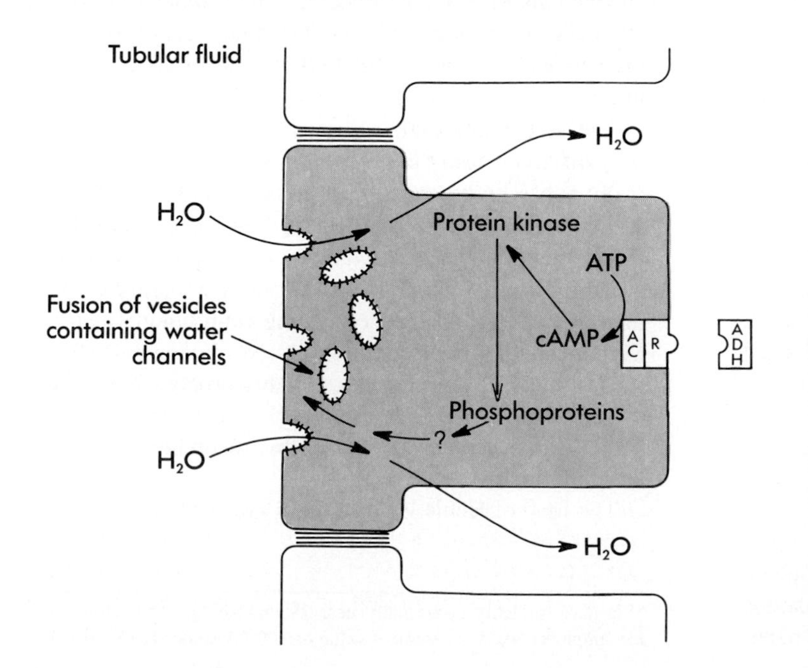

■ **Fig. 42-8** Cellular events associated with action of ADH on the cells of the collecting duct. *R*, ADH receptor; *AC*, adenylyl cyclase. See text for details.

ADH also increases the permeability of the inner medullary portion of the collecting duct to urea. The increase in permeability to urea is also mediated by cAMP. When ADH binds to its membrane receptor, the cAMP levels rise within the inner medullary collecting duct cells. Ultimately and by mechanisms not yet defined this rise in intracellular cAMP activates specific urea transporters in the membrane, and thereby urea permeability is increased.

Thirst

In addition to affecting the secretion of ADH changes in plasma osmolality and in blood volume or pressure lead to alterations in the perception of *thirst.* When body fluid osmolality is increased or when the blood volume or pressure is reduced, the individual perceives thirst. Of these stimuli hypertonicity is the more potent. Increases in plasma osmolality of only 2% to 3% produce a strong desire to drink, whereas decreases in blood volume and pressure in the range of 10% to 15% are required to produce the same response.

The neural centers involved with the thirst response have not been completely defined. It appears that osmoreceptors similar to, but distinct from, those involved in ADH release respond to changes in plasma osmolality. Like the ADH osmoreceptors those involved in thirst respond only to effective osmoles. Even less is known about the pathways involved in the thirst response to decreased blood volume or pressure; the pathways may be the same as those involved in the ADH system (see above).

The sensation of thirst is satisfied by the act of drinking, even before sufficient water is absorbed from the gastrointestinal tract to correct the plasma osmolality. Oropharyngeal and upper gastrointestinal receptors appear to be involved in this response. However relief of the thirst sensation via these receptors is short-lived. The desire to drink is completely satisfied only when the plasma osmolality or blood volume and pressure are corrected.

It should be apparent that *both the ADH and thirst systems work in concert to maintain water balance*. An increase in the plasma osmolality invokes drinking, and via ADH action on the kidneys water is conserved. Conversely when the plasma osmolality is decreased, thirst is suppressed, and in the absence of ADH renal water excretion is enhanced.

Countercurrent Multiplication by the Loop of Henle

The production of urine that is either hypoosmotic or hyperosmotic with respect to plasma requires that solute be separated from water at some point along the nephron. The production of hypoosmotic urine is conceptually easy to understand. All that is required is for the nephron to reabsorb solute from the tubular fluid but not allow water to follow. As will be seen this can occur under appropriate conditions in the ascending limb of Henle's loop, the distal tubule, and the collecting duct.

The excretion of hyperosmotic urine is more difficult to envision. This process requires removing water from the tubular fluid and leaving solute behind. Because water can move only passively, driven by an osmotic gradient, the kidneys must be able to generate a hyperosmotic environment that can then be used to remove water from the tubular fluid. Indeed such a hyperosmotic environment is generated in the renal medulla. The importance of the loop of Henle in the production of hyperosmotic urine is underscored by the fact that whereas the kidneys of all vertebrates can produce dilute urine, only the kidneys of birds and mammals can elaborate hyperosmotic urine, and it is only their kidneys that possess Henle's loops. The key to understanding how the loop of Henle functions in this regard is to recognize its ability to operate as a **countercurrent multiplier** (Fig. 42-9).

The loop of Henle consists of two parallel limbs with tubular fluid flowing in opposite directions **(countercurrent flow).** Fluid flows into the medulla in the descending limb and out of the medulla in the ascending limb. In the idealized situation shown in Fig. 42-9 (panel 1) the fluid within both the descending and ascending limbs as well as the surrounding interstitial fluid has an initial osmolality equal to that of plasma (300 mOsm/kg H_2O).* The ascending limb is impermeable to water and reabsorbs solute from the tubular fluid. Thus fluid within the ascending limb becomes diluted (Fig. 42-9, panel 2). This separation of solute and water by the ascending limb is termed the **single effect** of the countercurrent multiplication process. The solute removed from the ascending limb tubular fluid accumulates in the surrounding interstitial fluid and raises its osmolality.

The descending limb has permeability characteristics very different from those of the ascending limb. Specifically it has high permeability to water but low permeability to solute; consequently the hyperosmotic medullary interstitium causes water to move out of the descending limb. With equilibration the osmolality of the tubular fluid within the descending limb equals that of the surrounding interstitial fluid. As depicted in Fig. 42-9, (panel 3) the net effect of this process is the establishment of a 200 mOsm/kg H_2O osmotic gradient between the tubular fluids in the ascending and descending limbs.

The loop of Henle is not a static system because fresh

*Plasma osmolality is normally near 290 mOsm/kg H_2O. However for simplicity we will assume a value of 300 mOsm/kg H_2O.

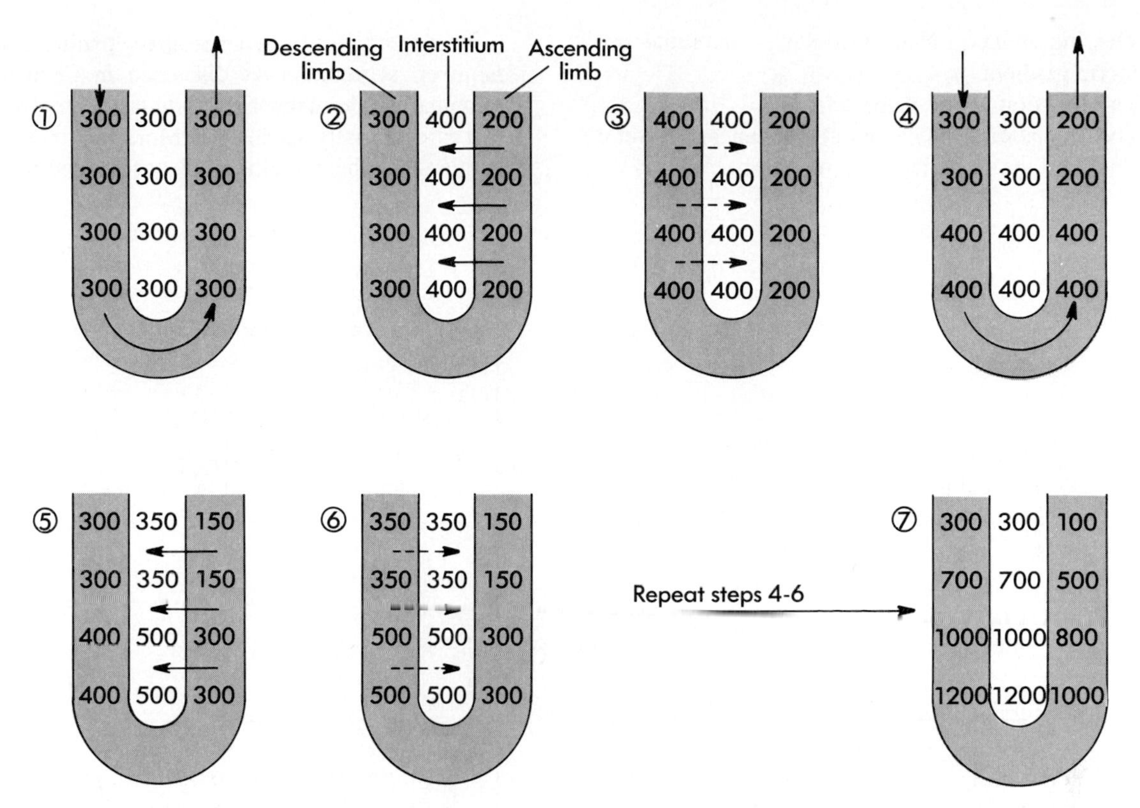

■ **Fig. 42-9** The process of countercurrent multiplication by the loop of Henle. Initially *(1)* fluid in the loop of Henle and interstitium has an osmolality equal to that of plasma (300 mOsm/kg H_2O). The transport of solute out of the ascending limb into the interstitium represents the single effect of separating solute from water *(2 and 5)*. The osmotic pressure gradient between the interstitium and the descending limb results in passive movement of water out of the descending limb *(3 and 6)*. In the steady state with continuous tubular fluid flow *(4)* the single effect is multiplied along the length of the loop, establishing a standing osmotic gradient *(7)*.

tubular fluid constantly enters the descending limb from the proximal tubule. Because proximal tubular reabsorption is essentially an isosmotic process (see Chapter 41), the osmolality of the fluid entering Henle's loop is equal to that of plasma. As this new fluid flows into the descending limb the hyperosmotic fluid previously formed in the descending limb flows into the ascending limb (Fig. 42-9, panel 4). Separation of solute from water again occurs in the ascending limb, which in turn leads to the reestablishment of the 200 mOsm/kg H_2O osmotic gradient between the two limbs (Fig. 42-9, panels 5 and 6). With sufficient time and in the steady state the situation depicted in panel 7 of Fig. 42-9 results. The net effect of this process is that the single effect of separating solute from water is multiplied by the countercurrent flow of fluid in the loop, such that the fluid at the bend of the loop is hyperosmotic to plasma. Also a standing interstitial fluid osmotic gradient is established from the junction of the medulla and cortex into the inner medulla.

As illustrated in panel 7 of Fig. 42-9 fluid having an osmolality of 300 mOsm/kg H_2O enters the medulla, in the descending limb of Henle's loop, and fluid having an osmolality of 100 mOsm/kg H_2O exits the loop in the ascending limb. For mass balance to exist one or more other routes must be available for solute to leave the medulla. As described in more detail below this excess solute leaves the medulla by way of the collecting duct (hyperosmotic urine) and the ascending vasa recta.

It should be apparent that countercurrent multiplication is an energy-efficient process. By its operation a considerable osmotic gradient is generated between fluid in the tip of the loop of Henle and the systemic plasma (1,200 − 300 = 900 mOsm/kg H_2O). The energy cost associated with the establishment of this gradient is quite modest and represents only the energy expended in generating the 200 mOsm/kg H_2O gradient that exists between adjacent segments of the descending and ascending limbs.

To understand how the process of countercurrent multiplication results in the excretion of either dilute or concentrated urine the following points must be considered in some detail:

1. The transport and permeability properties of the various portions of Henle's loop, distal tubule, and collecting duct

2. The importance of the medullary interstitial osmotic gradient
3. The function of the vasa recta
4. The transition from a diuretic to an antidiuretic state and the important role of ADH

Transport and permeability properties of the nephron segments. As discussed in Chapter 41, the proximal tubule reabsorbs solute and water by a process that is essentially isosmotic. Moreover this applies regardless of whether dilute or concentrated urine is pro-

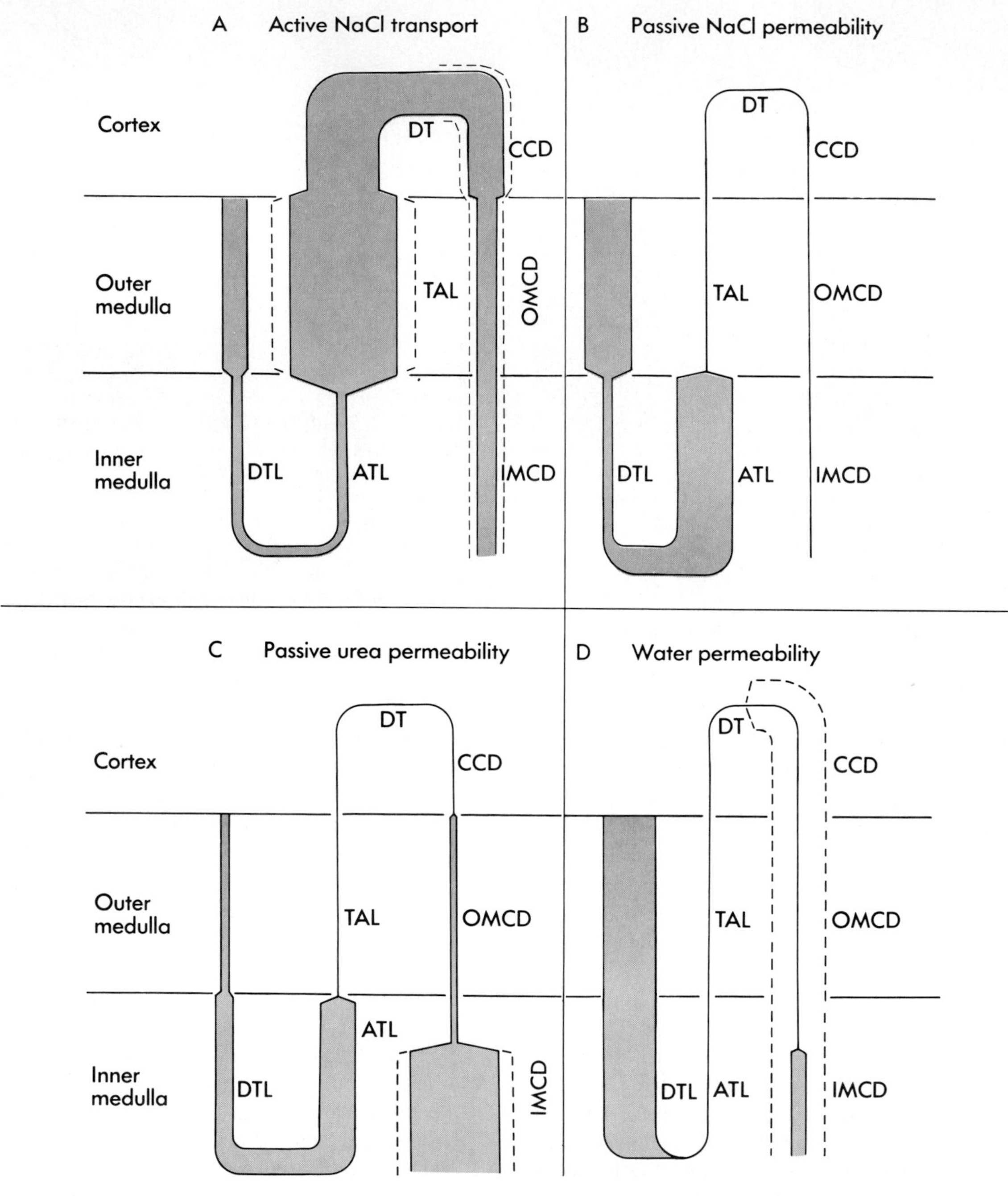

■ **Fig. 42-10** Transport and passive permeability properties of the nephron segments involved in the dilution and concentration of urine. The width of the tubular segment is proportional to the magnitude of the parameter. The solid lines depict the situation in the absence of ADH. The dashed lines show the effect of ADH. *DTL,* descending thin limb; *ATL,* ascending thin limb; *TAL,* thick ascending limb; *DT,* distal tubule; *CCD,* cortical collecting duct; *OMCD,* outer medullary collecting duct; *IMCD,* inner medullary collecting duct. (Adapted from Knepper MA, Rector FC, Jr: *The kidney,* ed 4, Philadelphia, 1991, WB Saunders).

duced. Thus solute and water are not separated in the proximal tubule. Consequently to understand the processes involved in the dilution and concentration of the urine we need to focus on all nephron segments distal to the proximal tubule.

As already noted a dilute urine is produced when plasma ADH levels are low. Under this condition (Fig. 42-10, *D*, solid lines) all tubule segments, except the descending thin limb of Henle's loop, have very low permeability to water. Reabsorption of solute (NaCl) by these segments (Fig. 42-10, *A*) therefore dilutes the luminal fluid. Because of the large transport capacity of the thick ascending limb of Henle's loop it is the major site of dilution of the tubule fluid. Indeed the thick ascending limb of Henle's loop is often referred to as the **diluting segment** of the kidney.

Fig. 42-10 points out some other important features of these nephron segments. First with the exception of the ascending thin limb of Henle's loop and outer medullary portion of the descending limb of Henle's loop the passive permeability of these nephron segments to NaCl is quite low (see Fig. 42-10, *B*). The relatively high permeability of the ascending thin limb, coupled with the high [NaCl] in the tubular fluid (see later in chapter), allows passive efflux of NaCl out of the tubule lumen. Because this segment is also impermeable to water the efflux of NaCl results in separation of solute and water and thus dilution of the luminal fluid. Second only the nephron segments deep within the medulla have significant passive permeabilities to urea (see Fig. 42-10, *C*). As discussed in the following section this allows urea to be accumulated in the inner medulla; urea accumulation is a critical requirement of maximal urine concentrating ability.

Medullary interstitial osmotic gradient. The medullary interstitium plays a critical role in the urine concentrating and diluting process. The osmolality of this fluid provides the driving force for the absorption of water from the descending thin limb of Henle's loop and the terminal portions of the collecting duct.

The principal constituents of the medullary interstitial fluid are NaCl and urea,* but the distribution of these solutes is not uniform throughout the medulla. At the junction of the cortex and outer portion of the medulla the interstitial fluid has an osmolality similar to that of plasma (300 mOsm/kg H_2O), with virtually all osmoles attributable to NaCl. The osmolality of the interstitial fluid rises progressively with increasing depth into the medulla and attains a value of approximately 1,200 mOsm/kg H_2O at the deepest portion of the inner medulla (i.e., papilla). At this point the distribution of osmoles is roughly 600 mOsm/kg H_2O attributable to NaCl and 600 mOsm/kg H_2O attributable to urea.

*Urea can be an ineffective osmole. This is the case for those cells that have high urea permeability (e.g., red blood cells). However urea does behave as an effective osmole in most of the segments of the distal nephron, and in these segments it causes osmotic water flow.

This **medullary interstitial osmotic gradient** for NaCl and urea is generated by the processes of countercurrent multiplication and urea trapping. In the previous section the role of countercurrent multiplication in producing the medullary interstitial osmotic gradient was illustrated. This process applies to the accumulation of NaCl, which is deposited in the interstitium as a result of transport out of the lumen of the thin ascending and thick ascending limbs of Henle's loop. Understanding the process of medullary urea accumulation is more complex; it requires consideration of the urea permeability properties of the various nephron segments (see Fig. 42-10, *C*).

Urea enters the tubule by glomerular filtration. A portion of the filtered urea is reabsorbed back into the blood by the proximal tubule. However this urea reabsorption is accompanied by water such that the [urea] of the tubular fluid that enters the medulla in the descending limb of Henle's loop is nearly equal to that of the plasma. Of the remaining nephron segments only the inner medullary collecting duct has high permeability to urea (see Fig. 42-10, *C*). The lower but significant permeabilities of the descending thin and ascending thin limbs of Henle's loop will be considered shortly (see later in this chapter). Consequently the major portion of urea that enters the descending thin limb of Henle's loop from the proximal tubule remains trapped in the tubule lumen until it reaches the inner medullary collecting duct. At this point some of the urea can leave the lumen of the inner medullary collecting duct and enter the medullary interstitium. *The movement of urea is passive,* and it occurs in the direction of a favorable urea concentration gradient.

In the presence of ADH the reabsorption of water from the tubular fluid across the cortical and outer medullary portions of the collecting duct causes the [urea] in the tubular fluid to rise. When this fluid reaches the inner medullary collecting duct urea diffuses out of the tubule into the interstitium. In the absence of ADH fluid reaching the inner medullary collecting duct is dilute and has low [urea]. As a result urea diffuses from the medullary interstitium into the lumen of the collecting duct. This urea secretion however is limited by the reduced permeability of the inner medullary collecting duct to urea in the absence of ADH (see Fig. 42-10, *C*).

Both the descending thin and ascending thin limbs of Henle's loop are permeable to urea, although this permeability is less than that of the inner medullary collecting duct (see Fig. 42-10, *C*). Normally the [urea] in both of these segments is less than that in the surrounding interstitial fluid. Consequently urea enters the tubular fluid at these sites. This recycling of urea from the inner medullary collecting duct into the thin limbs of Henle's loop aids in the trapping and accumulation of urea in the inner medulla interstitium.

As already described the hyperosmotic medullary interstitium is important for the process of countercurrent multiplication, and specifically for the reabsorption of water from the descending thin limb of Henle's loop. The hyperosmotic medullary interstitium is also essential for concentrating the tubular fluid within the collecting duct. Because of the hyperosmotic medullary interstitium and in the presence of ADH (see following section) water is reabsorbed from the collecting duct. Reabsorption of water is driven by the osmotic gradient between the luminal fluid and the interstitium. Water movement out of the collecting duct is passive; therefore the osmolality of the medullary interstitium defines maximal urine osmolality. For example if the osmolality of the interstitial fluid at the papilla is 1,200 mOsm/kg H_2O then the osmolality of the urine can equal but not exceed this value. The urine osmolality can be less than that of the interstitium; this occurs when ADH levels are less than maximal.

Because a hyperosmotic medullary interstitium is essential for concentrating the urine any process that reduces this osmolality impairs the ability to concentrate the urine maximally. For example if the medullary interstitium osmolality were reduced to 800 mOsm/kg H_2O the maximal osmolality of the urine would also be 800 mOsm/kg H_2O.

Vasa recta function—countercurrent exchange. As described in Chapter 41 the blood supply to the renal medulla consists of **vasa recta,** which run in parallel to Henle's loops. The vasa recta serve two important functions: they provide nutrients and oxygen to the tubules in the medulla, and, more importantly, they help maintain the medullary interstitial osmotic gradient by acting as **countercurrent exchangers** (Fig. 42-11). The vasa recta remove the excess water and solute that are added to the medullary interstitium by the various nephron segments in the medulla. The excess water and solute would otherwise dissipate the osmotic gradient.

The osmolality of plasma entering the medulla via the descending vasa recta equals that of the systemic plasma (300 mOsm/kg H_2O). As the plasma flows deeper into the medulla it equilibrates with the surrounding interstitial fluid by gaining solute and losing water (the vasa recta are very permeable to both solute and water) and at the papilla has an osmolality equal to that of the surrounding interstitium (e.g., 1,200 mOsm/kg H_2O). If the vasa recta were to exit the medulla at this point, they would reduce the osmolality of the interstitium by removing the accumulated solutes. Because of the countercurrent flow arrangement however the blood reequilibrates as it ascends back through the medulla. As a result, the plasma exiting the medulla in the ascending vasa recta is only slightly hyperosmotic to systemic plasma and the hyperosmotic medullary interstitium is preserved. As noted above the vasa recta also remove the excess water and solutes that are added to the interstitial fluid by the various nephron segments in the medulla.

Diuresis versus antidiuresis and the role of ADH. During the diuretic state ADH is absent, and the nephron segments beginning with the ascending thin limb of Henle's loop and continuing on through the collecting duct have low permeability to water (see Fig. 42-10, *D*). Consequently the separation of solute from water

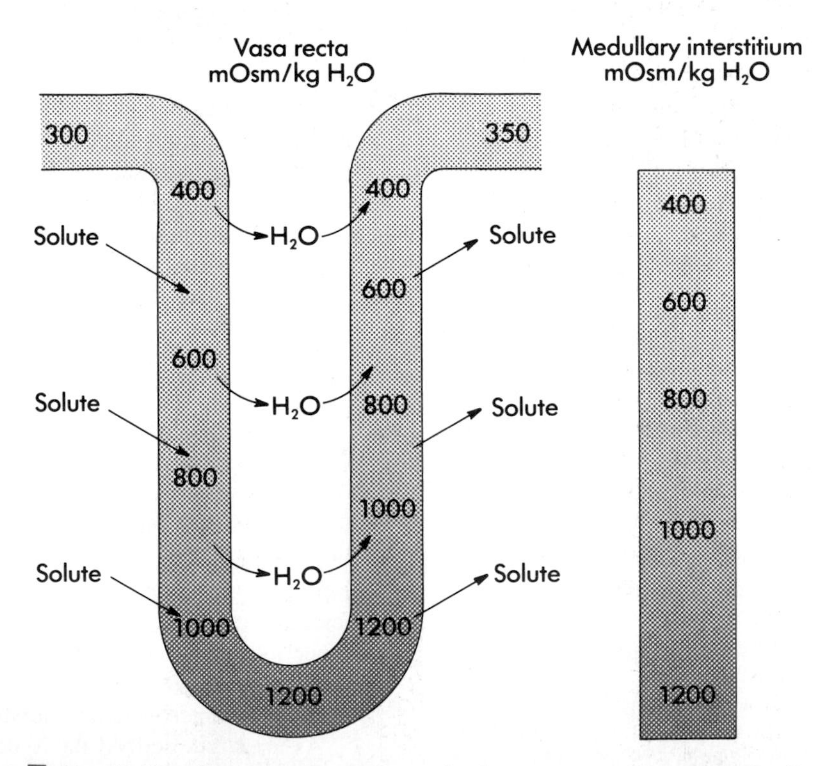

■ **Fig. 42-11** Countercurrent exchange by the vasa recta. See text for details.

by these nephron segments results in the elaboration of dilute urine. The only change necessary to allow a transition to the antidiuretic state is that ADH be present and effective. As described above the primary actions of ADH on the kidneys are to increase the permeability of the collecting duct to water and urea and to stimulate NaCl transport by the thick ascending limb of Henle's loop.

With regard to water permeability ADH exerts its effect on the nephron segment beginning with the terminal one third of the distal tubule and continuing throughout the entire collecting duct. Because the ascending thin and thick ascending limbs of Henle's loop are impermeable to water both in the absence and in the presence of ADH fluid that enters the distal tubule is always hypoosmotic to plasma (approximately 100 mOsm/kg H_2O). Therefore when ADH is present water exits the tubule lumen by osmosis into the surrounding interstitial tissue because of the existing osmotic gradient. With maximal levels of ADH the tubular fluid reaches osmotic equilibrium with the interstitial fluid as it flows down the collecting duct. Thus in the cortex the tubular fluid is concentrated to a value equal to that of plasma (300 mOsm/kg H_2O), and at the papilla it has an osmolality equal to that of the surrounding medullary interstitial fluid (1,200 mOsm/kg H_2O). Because individual collecting ducts join as they descend through the medulla many more cortical collecting ducts than medullary collecting ducts exist. Consequently more water is reabsorbed in the cortex than in the medulla.

The action of ADH on the thick ascending limb of Henle's loop to enhance NaCl reabsorption and on the inner medullary collecting duct to increase permeability to urea augments the accumulation of these solutes in the medullary interstitium. This action serves to maintain the osmotic gradient between interstitium and collecting duct lumen at a time when water reabsorption would dissipate the osmotic gradient in the medullary interstitium.

Integrated View of the Urine Concentrating Process

Fig. 42-12 summarizes the essential features of the urine concentrating process. As depicted ADH levels are maximal and concentrated urine is excreted. For the production of dilute urine the mechanisms are the same, except that ADH is not present, and therefore water is not reabsorbed from the lumen of the collecting duct.

The following steps describe the urine concentrating process. The numbers correspond to those in Fig. 42-12.

1. Fluid that enters the descending thin limb of Henle's loop from the proximal tubule is isoosmotic with respect to plasma. This reflects the isoosmotic nature of solute and fluid reabsorption in the proximal tubule (see Chapter 41). The osmolality of this solution is approximately 300 mOsm/kg H_2O; the majority of the osmoles are due to NaCl.
2. The descending thin limb is highly permeable to water and much less so to NaCl and urea. Consequently as fluid flows deeper into the hyperosmotic medulla water is reabsorbed. By this process fluid at the bend of the loop has an osmolality equal to that of the surrounding interstitial fluid (1,200 mOsm/kg H_2O). Most of the osmoles are due to NaCl, although the other components of the tubular fluid are concentrated as well. The urea concentration of the tubular fluid is also beginning to increase, partly because water is reabsorbed from the tubular fluid, and also because urea diffuses into the descending thin limb from the medullary interstitium **(urea recycling).**
3. The ascending thin limb is essentially impermeable to water but highly permeable to NaCl. As the NaCl-rich tubular fluid moves up the ascending thin limb NaCl diffuses passively into the medullary interstitium (recall that NaCl accounts for 600 mOsm/kg H_2O of the interstitial fluid osmolality, whereas nearly 1,000 mOsm/kg H_2O are attributable to NaCl in the tubular fluid that enters the ascending thin limb). This passive reabsorption of NaCl without concomitant water reabsorption begins to dilute the tubular fluid. Because the ascending thin limb is permeable to urea, urea also diffuses from the medullary interstitium into the tubule lumen (urea recycling).
4. The thick ascending limb of Henle's loop is also essentially impermeable to water. Active reabsorption of NaCl by this nephron segment further dilutes the tubular fluid. As noted in the previous section ADH also stimulates the active reabsorption of NaCl by the thick ascending limb of Henle's loop. Dilution occurs to such a degree that fluid leaving the thick ascending limb is hypoosmotic with respect to plasma (approximately 100 mOsm/kg H_2O).
5. Fluid reaching the collecting duct is hypoosmotic with respect to the surrounding interstitial fluid. Therefore in the presence of ADH water diffuses out of the tubule lumen; this begins the process of urine concentration. The maximum osmolality that the fluid in the cortical collecting duct can attain is approximately 300 mOsm/kg H_2O, which is the osmolality of the surrounding interstitial fluid and plasma. Although the fluid at this point is the same osmolality as that which entered the descending thin limb its composition has been altered dramatically. Because of reabsorption by the preceding nephron segments NaCl accounts for a much smaller portion of the total tubular fluid osmolality. Instead the urine osmolality reflects the

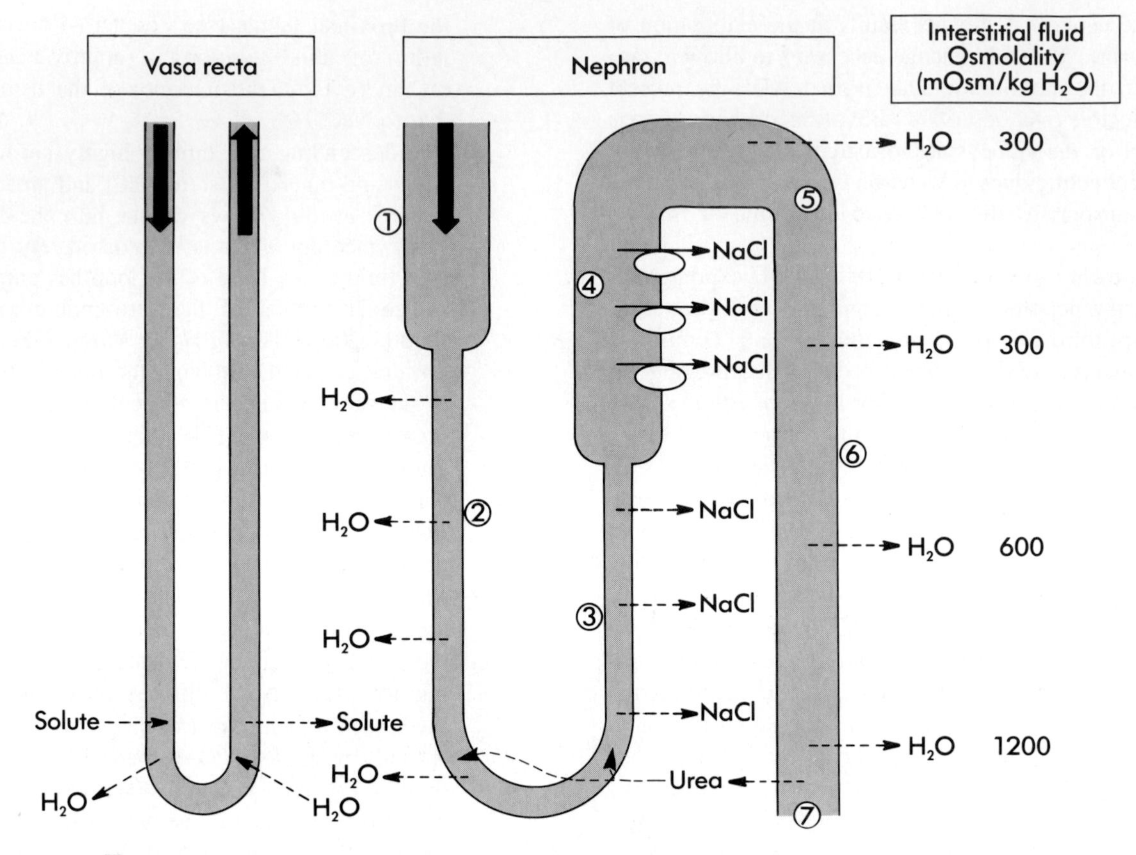

■ **Fig. 42-12** Mechanism for the excretion of concentrated urine. ADH levels are maximal. Passive movements of water and solutes are indicated by dashed lines. Numbers refer to description of process in text. NaCl is actively absorbed in the thick ascending limb of Henle's loop.

presence of urea (filtered urea, plus urea added in the descending thin and ascending thin limbs of Henle's loop) and other nonreabsorbed solutes (e.g., creatinine).

6. As fluid in the collecting duct continues through the medulla water is reabsorbed. This process increases the osmolality of the tubular fluid; this increase is caused mainly by urea. In the inner medulla the collecting duct is permeable to urea; moreover ADH increases this permeability. Because the urea concentration of the tubular fluid is increased by water reabsorption, some urea diffuses out of the tubule lumen and into the medullary interstitium.
7. The urine has an osmolality of 1,200 mOsm/kg H_2O and contains high concentrations of urea and other nonreabsorbed solutes. Because urea in the tubular fluid tends to equilibrate with the interstitial urea its concentration in the urine does not exceed that of the interstitium (approximately 600 mmol/L). The high NaCl concentration of the interstitium drives additional water reabsorption, concentrating the other nonreabsorbed solutes in the tubular fluid.

■ *Quantitating Renal Diluting and Concentrating Ability*

The central process of the dilution and concentration of the urine is the single effect of separating solute from water. Through this separation the kidneys in a sense generate a volume of water that is free of all solute. When the urine is dilute this **solute-free water** is excreted from the body; when the urine is concentrated it is returned to the systemic circulation. The concept of **free-water clearance** provides a means for quantitating the ability of the kidneys to generate solute-free water. This concept follows directly from renal clearance as described in Chapter 41.

The clearance of total solute from plasma by the kidneys can be calculated as

$$C_{osm} = \frac{U_{osm} \cdot \dot{V}}{P_{osm}} \qquad (3)$$

where

C_{osm} = the **osmolar clearance**

U_{osm} = urine osmolality

$\dot{V}$ = urine flow rate

P_{osm} = plasma osmolality

C_{osm} has units of volume/unit time.

Consider the following hypothetical case the kidneys must excrete 300 mOsm of solute. This is accomplished by producing urine that is either isoosmotic ($U_{osm}/P_{osm} = 1$), hypoosmotic ($U_{osm}/P_{osm} < 1$), or hyperosmotic ($U_{osm}/P_{osm} > 1$) with respect to plasma. We will examine each of these conditions separately (Fig. 42-13).

Example 1—isoosmotic urine. If the urine flow rate under this condition is 2 ml/min, C_{osm} is calculated as

$$C_{osm} = \frac{300 \text{ mOsm/kg } H_2O \cdot 2 \text{ ml/min}}{300 \text{ mOsm/kg } H_2O} = 2 \text{ ml/min} \quad (4)$$

Note that the urine flow rate ($\dot{V}$) is equal to C_{osm}. We will consider the significance of this after considering the situation in which the 300 mOsm of solute is excreted in a dilute urine.

Example 2—dilute urine. In this situation the 300 mOsm of solute are excreted in twice the volume. Thus U_{osm} = 150 mOsm/kg H_2O and $\dot{V}$ = 4 ml/min. Accordingly under this condition C_{osm} is calculated as

$$C_{osm} = \frac{150 \text{ mOsm/kg } H_2O \cdot 4 \text{ ml/min}}{300 \text{ mOsm/kg } H_2O} = 2 \text{ ml/min} \quad (5)$$

Note that C_{osm} is unchanged; however $\dot{V}$ now exceeds C_{osm}. The difference between these two values represents the solute-free water excreted by the kidneys and is termed the **free-water clearance (C_{H_2O})**. C_{H_2O} is calculated by the following formula:

$$C_{H_2O} = \dot{V} - C_{osm} \quad (6)$$

Equation 6 can be expressed in the following way: The total urine output ($\dot{V}$) consists of two hypothetical components. The first contains all the solute in a volume that has an osmolality equal to that of the plasma (i.e., $U_{osm}/P_{osm} = 1$) and represents a volume from which no separation of solute and water has occurred. The second is a volume of solute-free water and defined as C_{H_2O}.

For the conditions described above:

$$C_{H_2O} = 4 \text{ ml/min} - 2 \text{ ml/min} = 2 \text{ ml/min} \quad (7)$$

For the conditions described in example 1 (isoosmotic urine) the C_{H_2O} is calculated to be

$$C_{H_2O} = 2 \text{ ml/min} - 2 \text{ ml/min} = 0 \quad (8)$$

Thus when urine is isoosmotic no excretion of solute-free water occurs.

Example 3—concentrated urine. In concentrated urine the 300 mOsm of solute is excreted in half the urine volume. Thus U_{osm} = 600 mOsm/kg H_2O and $\dot{V}$ = 1 ml/min. C_{osm} is calculated as

$$C_{osm} = \frac{600 \text{ mOsm/kg } H_2O \cdot 1 \text{ ml/min}}{300 \text{ mOsm/kg } H_2O} = 2 \text{ ml/min} \quad (9)$$

Note that under this condition C_{osm} again is unchanged; however $\dot{V}$ is now less than C_{osm}. The difference (1 ml/min) between these values represents the solute-free water reabsorbed by the kidneys and returned to the systemic circulation.

$$C_{H_2O} = 1 \text{ ml/min} - 2 \text{ ml/min} = -1 \text{ ml/min} \quad (10)$$

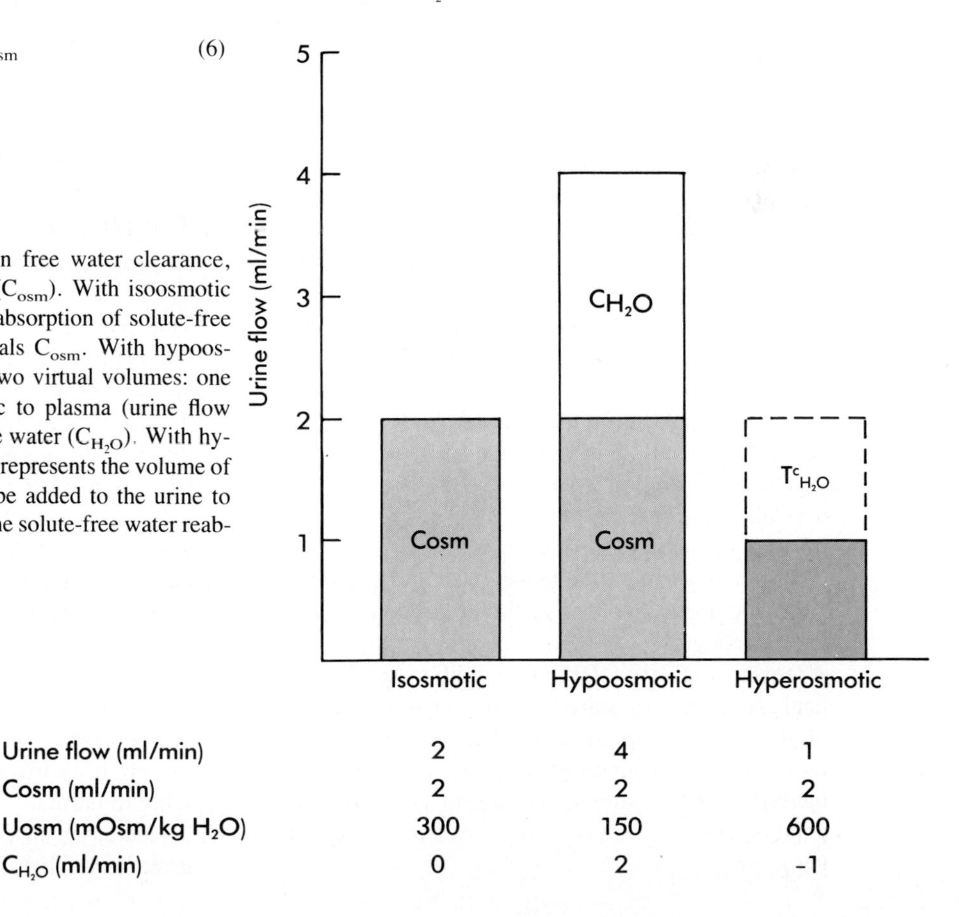

■ **Fig. 42-13** Relationship between free water clearance, urine volume and osmolar clearance (C_{osm}). With isoosmotic urine there is neither excretion nor reabsorption of solute-free water (C_{H_2O}) and urine flow rate equals C_{osm}. With hypoosmotic urine the urine is divided into two virtual volumes: one that contains solute and is isoosmotic to plasma (urine flow rate = C_{osm}) and one that is solute-free water (C_{H_2O}). With hyperosmotic urine T_{H_2O} (negative C_{H_2O}) represents the volume of solute-free water that would have to be added to the urine to make it isoosmotic to plasma. This is the solute-free water reabsorbed by the kidneys.

By convention negative values of C_{H_2O} are expressed as $T^c_{H_2O}$ **(tubular conservation of water).** By analogy for dilute urine the excretion of concentrated urine can be expressed in the following way: $T^c_{H_2O}$ represents the volume of water that would have to be added to the urine to reduce its osmolality to a value equal to that of the plasma. This represents the volume of solute-free water added to the systemic circulation.

Discussion. Two points regarding the above examples require emphasis: First changes in free-water excretion and/or reabsorption occur without changes in solute excretion (C_{osm} is unchanged). This underscores what was stated at the beginning of this chapter: The control of water balance by the kidneys is independent of the control of excretion of other solutes. Second C_{H_2O} and $T^c_{H_2O}$ reflect net processes within the kidneys. For example if C_{H_2O} is equal to zero no net separation of solute and water by the kidneys occurs. In actuality solute and water are separated by the ascending limb of Henle's loop. However this solute-free water is reabsorbed by the collecting duct (i.e., solute and water recombined), instead of being excreted (i.e., as dilute urine). The net effect therefore is that C_{H_2O} is zero.

The determination of C_{H_2O} and $T^c_{H_2O}$ can provide important information about the function of those portions of the nephron involved in the production of dilute and concentrated urine. Whether the kidneys excrete or reabsorb free water depends on the absence or presence of ADH. When no ADH is present or levels are low solute-free water is excreted. When ADH levels are high solute-free water is reabsorbed. The magnitude of C_{H_2O} and $T^c_{H_2O}$ varies from individual to individual and depends on a number of factors.

The following factors are required to enable the kidneys to excrete maximally solute-free water (C_{H_2O}):

1. ADH must be absent. This prevents water reabsorption by the collecting duct.
2. The tubular structures that can separate solute from water (i.e., dilute the luminal fluid) must function normally. In the absence of ADH the following nephron segments can function to dilute the luminal fluid:
   a. Ascending thin limb of Henle's loop.
   b. Thick ascending limb of Henle's loop.
   c. Distal tubule.
   d. Collecting duct.
   Because of its high transport rate the thick ascending limb is quantitatively the most important of these segments involved in the separation of solute and water.
3. Adequate delivery of tubular fluid to the above nephron sites is required for maximal separation of solute and water. Factors that reduce delivery (e.g., decreased glomerular filtration rate or enhanced proximal tubule reabsorption) impair the ability of the kidneys to excrete maximal C_{H_2O}.

In the normal adult the maximal value of C_{H_2O} is about 10% of the GFR. Thus if the GFR is 180 L/day the maximum C_{H_2O} is 18 L/day. This means that under appropriate conditions an individual could ingest 18 L of water over a 24 hour period and the kidneys would be able to excrete this as solute-free water; thus the osmolality of the body fluids would be maintained at a normal level. If however water intake exceeded this capacity body fluid osmolality would fall.

The following conditions are required for the kidneys to reabsorb solute-free water maximally ($T^c_{H_2O}$):

1. ADH must be present to allow water reabsorption by the collecting duct. Also, the collecting duct must be responsive to ADH.
2. The nephron segments important in establishing the hyperosmotic medullary interstitium must function normally:
   a. Thin ascending limb of Henle's loop.
   b. Thick ascending limb of Henle's loop.
   Because of its high transport capacity the thick ascending limb is quantitatively the more important segment.
3. A hyperosmotic medullary interstitium is necessary to allow water reabsorption by the collecting duct.
4. Adequate delivery of fluid to the ascending limb of Henle's loop is needed. The separation of solute and water at this site is needed to maintain the hyperosmotic medullary interstitium.

In a normal adult $T^c_{H_2O}$ has a maximal value of approximately 8 L/day. Thus if water loss from other sources (e.g., through respiration and perspiration and in the stool) is 1.2 L/day an individual can live for approximately 6 to 7 days without water.

■ *Control of Extracellular Fluid Volume*

The major solutes of the extracellular fluid (ECF) are the salts of Na^+. Of these NaCl is the most abundant and therefore the most important. Because Na^+ (with its salts) is the major determinant of the osmolality of the ECF it is commonly assumed that alterations in Na^+ balance disturb ECF osmolality. However this is not usually the case because of the ADH secretory mechanism and thirst and their roles in regulating water balance. For example if NaCl were added to the ECF without any H_2O both $[Na^+]$ and osmolality of the ECF would increase. The increase in osmolality would stimulate thirst and release of ADH from the posterior pituitary. The increased ingestion of water in response to thirst, together with the ADH-induced decrease in water excretion by the kidney, would restore the ECF osmolality to normal. However the volume of the ECF would be increased in proportion to the amount of water ingested, which in turn depends on the amount of NaCl added to the ECF. Conversely a decrease in the NaCl content of the ECF decreases the volume of this compartment. Thus provided the ADH and thirst systems function normally alterations in the NaCl content of the ECF result in parallel changes in ECF volume.

The kidneys are the major route for excretion of NaCl from the body. Consequently they are important in regulating the volume of the ECF. Under normal conditions the kidneys maintain constant ECF volume by adjusting the excretion of NaCl to match the amount ingested in the diet. If ingestion exceeds excretion volume of the ECF increases above normal, whereas the opposite occurs if excretion exceeds ingestion. To defend itself against changes in ECF volume the body relies on a system that monitors the volume of this compartment and then sends signals to the kidneys so that NaCl excretion can be adjusted appropriately.

This section reviews the physiological aspects of the volume receptors, considers the various signals that act on the kidneys to regulate NaCl excretion, and discusses the response of the various portions of the nephron to these signals.

■ *Concept of Effective Circulating Volume*

To understand the role of the kidneys in regulating the volume of the ECF one must consider the concept of the **effective circulating volume** (ECV). The ECV is not a measurable and distinct body fluid compartment; rather it reflects the adequacy of tissue perfusion. Thus ECV is related to the "fullness" of and "pressure" within the vascular tree. In the normal individual the ECV varies in parallel with the volume of the ECF.

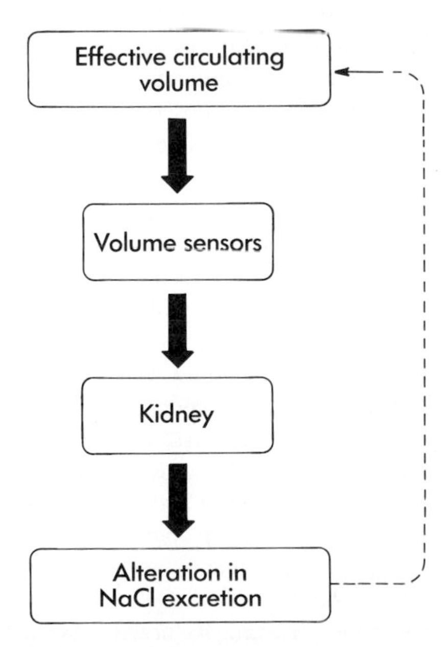

■ **Fig. 42-14** General scheme for monitoring and controlling the effective circulating volume.

However in some pathological conditions this relationship is not maintained. For example in patients with heart disease the ECV may be low because of reduced cardiac output, yet the volume of the ECF may be greater than normal.

The important point to consider regarding the ECV is that the kidneys alter their excretion of NaCl in response to changes in this effective volume. When the ECV is decreased renal NaCl excretion is reduced. This adaptive response, as explained later in this chapter, serves to restore the ECV to its normal value and maintains adequate tissue perfusion. Conversely an increase in the ECV enhances renal NaCl excretion (termed **natriuresis**). Fig. 42-14 summarizes the components of the ECV regulatory system, each of which is considered in more detail in the next section.

In the normal individual the terms *ECV* and *ECF* can be, and quite often are, interchanged. However it is the ECV, especially under certain pathological conditions, that determines renal Na^+ excretion. Consequently to provide a framework for the eventual understanding of the pathophysiological basis of some of these clinically important conditions the remaining sections of this chapter refer primarily to the ECV.

■ *ECV Sensors*

Table 42-2 lists the various volume sensors in the body. Sensors have been identified in the vascular tree, which monitor both fullness and pressure. These **vascular volume receptors** appear to be the primary sensors for the ECV. Receptors related to the control of NaCl excretion by the kidneys have also been postulated to exist within the brain and the liver. Less is known about these latter two groups of receptors. The receptors in the vascular system respond to stretch (baroreceptors) and are considered in more detail in the following sections.

Vascular low-pressure volume receptors. Baroreceptors are located within the walls of the cardiac atria and the pulmonary vessels and respond to distension of these structures (see p. 489). Because of the low pres-

■ **Table 42-2** ECV sensors

Vascular
Low pressure
Cardiac atria
Pulmonary vasculature
High pressure
Carotid sinus
Aortic arch
Juxtaglomerular apparatus of kidney (afferent arteriole)
Central nervous system
Hepatic

sure within the atria and pulmonary vessels these receptors respond primarily to the fullness of the vascular tree. These baroreceptors send signals to the brain (hypothalamic and medullary regions) via afferent fibers in the vagus nerve; distension increases the number of impulses/second. This vagal input alters sympathetic outflow from the central nervous system (CNS) and secretion of ADH by the posterior pituitary. The mechanisms whereby sympathetic nerves and ADH alter renal handling of NaCl and water are considered in detail later in this chapter.

The cardiac atria possess an additional mechanism related to control of renal NaCl excretion and thus ECV. The myocytes of the atria synthesize and store a peptide hormone. This hormone, termed atrial natriuretic peptide (ANP), is released when the atria are distended (i.e., expansion of the ECV). By the mechanisms outlined in subsequent sections ANP rapidly increases the excretion of NaCl and water by the kidneys.

Vascular high-pressure volume receptors. Baroreceptors are also present in the arterial side of the vascular tree (see p. 486). These receptors are located in the wall of the aortic arch, the carotid sinus, and the afferent arteriole of the kidney, and they respond primarily to changes in blood pressure. The aortic arch and carotid baroreceptors, like the low-pressure receptors, send input to the hypothalamic and medullary centers of the brain via the vagus and glossopharyngeal nerves; increased arterial pressure leads to increased number of impulses/second. This input also induces alterations in sympathetic outflow from the CNS and in ADH secretion by the posterior pituitary (see later in this chapter).

The juxtaglomerular apparatus (see Chapter 41), and in particular the afferent arteriole of the glomerulus, acts as a high-pressure baroreceptor. Changes in perfusion pressure at this site lead to alterations in the secretion of renin. Renin in turn determines levels of angiotensin II and aldosterone, both of which regulate renal Na^+ excretion (see later in this chapter).

Volume Receptor Signals

Once the various volume receptors have detected a change in the ECV, signals are sent to the kidneys, which adjust NaCl and water excretion appropriately. Both neural and hormonal signals have been identified in Table 42-3.

Table 42-3 Signals involved in the control of renal Na^+ and water excretion

Renal sympathetic nerves (↑ activity: ↓ Na^+ excretion)
↓ Glomerular filtration rate
↑ Renin secretion
↑ Proximal tubule Na^+ reabsorption
Renin-angiotensin-aldosterone (↑ secretion: ↓ Na^+ excretion)
↑ Angiotensin II levels stimulate proximal tubule Na^+ reabsorption
↑ Aldosterone levels stimulate collecting duct Na^+ reabsorption
↑ ADH secretion
Atrial natriuretic peptide (↑ secretion: ↑ Na^+ excretion)
↑ Glomerular filtration rate
↓ Renin secretion
↓ Aldosterone secretion
↓ Na^+ reabsorption by the collecting duct
↓ ADH secretion
ADH (↑ secretion: ↓ H_2O and Na^+ excretion)
↑ H_2O absorption by the collecting duct
↑ NaCl reabsorption by the thick ascending limb of Henle's loop
↑ Na^+ reabsorption by the collecting duct

Renal sympathetic nerves. As described in Chapter 41 sympathetic nerve fibers innervate the afferent and efferent arterioles of the glomerulus, as well as cells of the nephron. When the ECV is reduced the diminished afferent input to the CNS from the low- and high-pressure baroreceptors increases renal sympathetic nerve activity. This increased renal sympathetic nerve activity has the following effects:

1. Constriction of the afferent and efferent arterioles: This vasoconstriction (the effect appears to be greater on the afferent arteriole) decreases the volume of plasma flowing through the glomerulus and decreases the hydrostatic pressure within the lumen of the glomerular capillary. The net effect of these changes on glomerular hemodynamics is that the GFR is reduced. With this decrease in GFR the filtered load of Na^+ is reduced (Filtered load = GFR × plasma [Na^+]).
2. Stimulation of renin secretion by the cells of the afferent and efferent arterioles: As described in detail below renin subsequently increases the circulating levels of angiotensin II and aldosterone.
3. Direct stimulation of NaCl reabsorption by the proximal tubule and Henle's loop.

As will be described in subsequent sections the combined effect of these actions contributes to an overall decrease in the excretion of NaCl; this is an adaptive response that restores the ECV to its normal value.

When the volume of the ECV is increased renal sympathetic nerve activity is reduced. This action generally reverses the effects previously described.

Renin-angiotensin-aldosterone system. Smooth muscle cells in the afferent and efferent arterioles of the glomerulus are the site of synthesis, storage, and release of **renin.** Three factors are important in stimulating renin secretion:

1. Perfusion pressure: As noted previously the afferent arteriole behaves as a high-pressure baroreceptor. When perfusion pressure to the kidney is re-

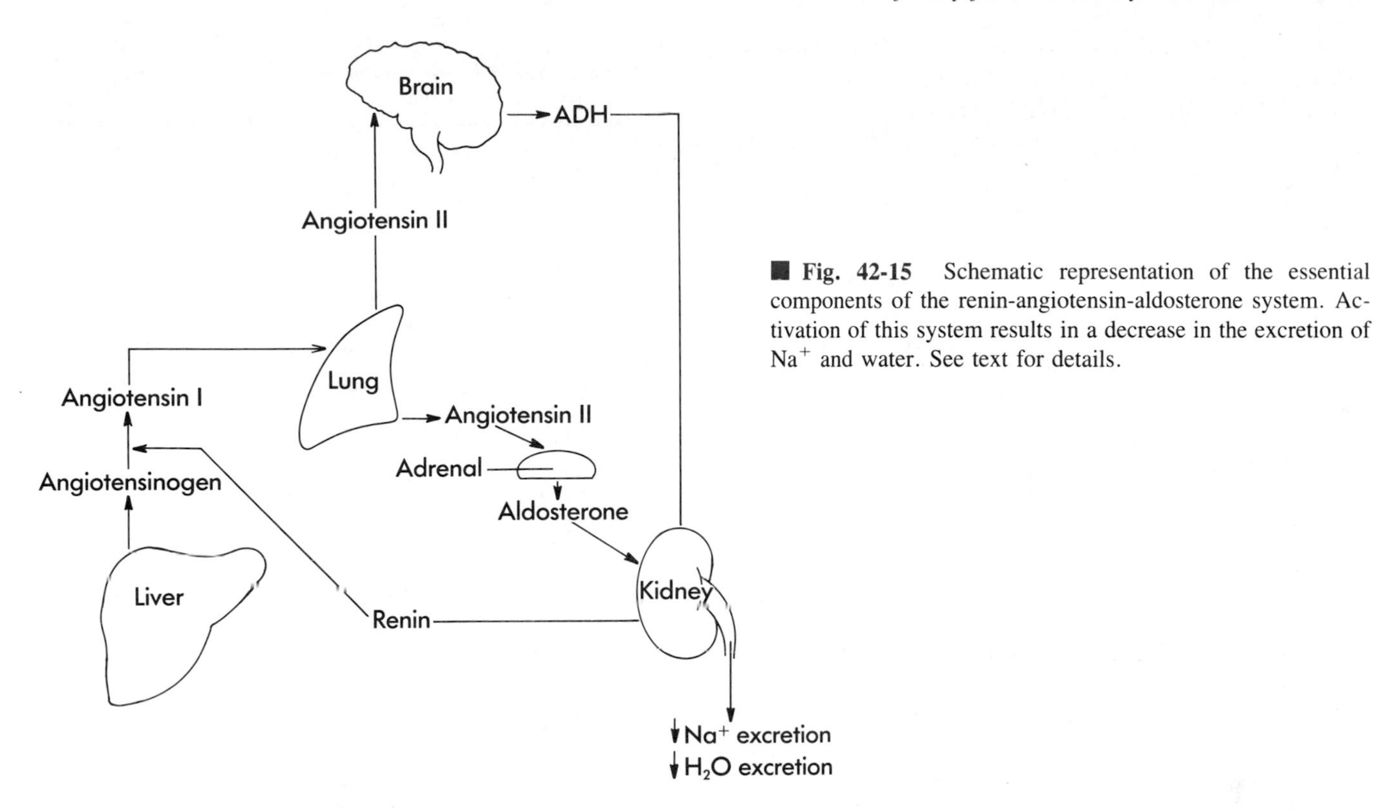

■ **Fig. 42-15** Schematic representation of the essential components of the renin-angiotensin-aldosterone system. Activation of this system results in a decrease in the excretion of Na^+ and water. See text for details.

duced, renin secretion is stimulated. Conversely an increase in perfusion pressure inhibits renin release.

2. Sympathetic nerve activity: Activation of the sympathetic nerve fibers that innervate the afferent and efferent arterioles increases renin secretion. Renin secretion decreases as renal sympathetic nerve activity decreases.
3. Delivery of NaCl to the macula densa: As discussed previously (see Chapter 41) delivery of NaCl to the macula densa regulates the GFR. With increased NaCl delivery GFR decreases, whereas decreased NaCl delivery increases GFR. Renin secretion is also altered by NaCl delivery to the macula densa. When NaCl delivery is decreased renin secretion is enhanced; conversely when NaCl delivery is increased renin secretion is inhibited.

Fig. 42-15 summarizes the essential components of the renin-angiotensin-aldosterone system. Renin alone does not directly alter renal function. Renin is a proteolytic enzyme, and its substrate is a circulating peptide (angiotensinogen), which is produced by the liver. Angiotensinogen is cleaved by renin to a 10 amino acid peptide (angiotensin I). Angiotensin I, like renin, does not directly affect the kidneys. It is cleaved to an 8 amino acid peptide **(angiotensin II)** by a converting enzyme. This converting enzyme is found in high concentration in the lung, and virtually all angiotensin I is converted to angiotensin II in a single pass through the pulmonary circulation. Angiotensin II acts as a negative feedback regulator of renin secretion (i.e., inhibits renin secretion). In addition angiotensin II has the following important physiologic functions:

1. Stimulation of aldosterone secretion by glomerulosa cells of the adrenal cortex
2. Vasoconstriction of systemic and renal arterioles, which increases blood pressure
3. Stimulation of ADH secretion from the posterior pituitary and stimulation of the thirst center within the hypothalamus
4. Enhancement of NaCl reabsorption by the proximal tubule

Angiotensin II is an important secretagogue for **aldosterone.** Aldosterone is a steroid hormone produced by the glomerulosa cells of the adrenal cortex, and has a number of important actions on the kidneys (see Chapters 41 and 43). With regard to the regulation of the ECV aldosterone acts to reduce NaCl excretion by stimulating its reabsorption by the thick ascending limb of Henle's loop, distal tubule, and collecting duct.

Aldosterone stimulates Na^+ reabsorption by the principal cells in the late portion of the distal tubule and collecting duct (Fig. 42-16). Aldosterone enters the cell and binds to a cytoplasmic receptor. The hormone-receptor complex interacts with specific binding sites on the deoxyribonucleic acid (DNA) and thereby regulates the transcription of messenger ribonucleic acid (mRNA). Translation of these messages increases the levels of a number of proteins important in the process of Na^+ reabsorption by the cell. These aldosterone-induced proteins may include new apical membrane Na^+

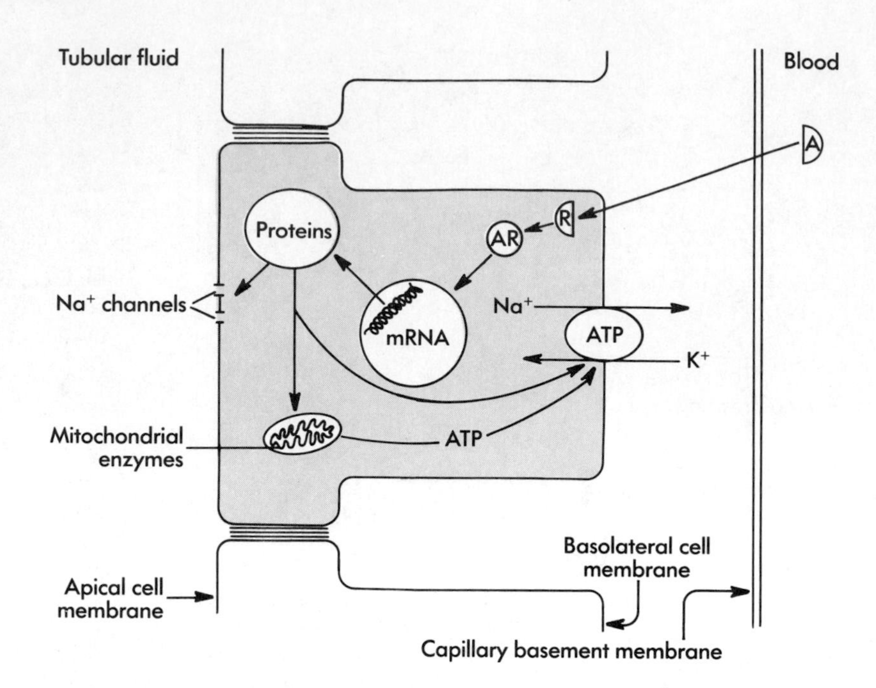

■ **Fig. 42-16** Cellular actions of aldosterone on the principal cell of the collecting duct. Aldosterone *(A)* binds to an intracellular receptor *(R)*. The receptor-aldosterone complex *(AR)* regulates the transcription of specific mRNA and ultimately the synthesis of proteins. These aldosterone-induced proteins include apical membrane Na^+ channels, mitochondrial enzymes, and Na^+-K^+-ATPase. As a result of aldosterone's actions reabsorption of Na^+ from the tubular fluid is increased.

channels, enzymes needed for the synthesis of adenosine triphosphate (ATP), and Na^+-K^+-ATPase pump. By these actions Na^+ entry into the cell across the apical membrane is enhanced, as is its extrusion from the cell across the basolateral membrane. Because Na^+ reabsorption by the distal tubule and collecting duct generates a lumen negative transepithelial voltage (see Chapter 41) this enhanced reabsorption of Na^+ from the luminal fluid increases the magnitude of this voltage. The passive movement of Cl^- from lumen to blood across the paracellular pathways is enhanced by this voltage change. Thus aldosterone increases the reabsorption of NaCl from the tubular fluid. Reduced levels of aldosterone result in a decrease in the amount of NaCl reabsorbed by the principal cells.

As noted previously aldosterone also enhances NaCl reabsorption by cells of the thick ascending limb of Henle's loop. However the precise cellular mechanisms have not been elucidated.

In general activation of the **renin-angiotensin-aldosterone system** decreases excretion of NaCl by the kidneys. This system is suppressed when the ECV is expanded, and renal NaCl excretion is therefore enhanced.

Atrial natriuretic peptide. As noted previously the atrial myocytes produce and store a peptide hormone (**atrial natriuretic peptide** [**ANP**]) that promotes NaCl and water excretion by the kidneys. ANP is released with atrial stretch as would occur with expansion of the ECV. The circulating form of ANP is 28 amino acids in length, and in general its actions antagonize those of the renin-angiotensin-aldosterone system. Actions of ANP include the following:

1. Vasodilation of the afferent and efferent arterioles of the glomerulus; this increases plasma flow and the GFR.* With the increase in GFR the filtered load of Na^+ is increased.
2. Inhibition of renin secretion by the afferent and efferent arterioles.
3. Inhibition of aldosterone secretion by the glomerulosa cells of the adrenal cortex. ANP reduces aldosterone secretion by two mechanisms. First angiotensin II-induced aldosterone secretion is reduced secondary to ANP inhibition of renin secretion. Second ANP acts directly on the glomerulosa cells of the adrenal cortex to inhibit aldosterone secretion.
4. Inhibition of NaCl reabsorption by the collecting duct. This inhibition is partly caused by reduced levels of aldosterone, but ANP also acts directly on the cells of the collecting duct. ANP acts through its second-messenger cyclic guanosine monophosphate (cGMP) to inhibit NaCl reabsorption. This effect occurs predominantly in the medullary portion of the collecting duct.
5. Inhibition of ADH secretion by the posterior pituitary and inhibition of ADH action on the collecting duct. Inhibition of ADH secretion and action reduces water reabsorption by the collecting duct and thereby increases excretion of water in the urine.

ANP is also an important vasodilator. This effect accounts for the reduction in blood pressure induced by ANP.

Taken together these effects of ANP increase excretion of NaCl and water. Hypothetically a reduction in

*Because GFR is increased, the effect of ANP on the afferent arteriole must predominate. Thus capillary hydrostatic pressure and renal plasma flow are increased.

circulating levels of ANP would be expected to decrease NaCl and water excretion. However the evidence that a reduction in circulating levels of ANP is important in this regard is not convincing.

Antidiuretic hormone. When the ECV is reduced and in response to decreased baroreceptor input ADH secretion by the posterior pituitary is stimulated. As discussed previously ADH increases the permeability of the collecting duct to water, and it enhances NaCl reabsorption by the thick ascending limb of Henle's loop and to a lesser degree the collecting duct. Thus NaCl and water excretion is reduced; this effect tends to restore the ECV to normal. An increase in the ECV reduces ADH secretion in response to increased baroreceptor input and contributes to the associated increase in NaCl and water excretion.

Control of Na^+ Excretion With Normal ECV

The maintenance of normal ECV,* which is termed **euvolemia,** requires precise balance of the amount of NaCl ingested and the amount excreted from the body. Because the kidneys are the major route for NaCl excretion the amount of NaCl in the urine reflects dietary intake. Thus *in a euvolemic individual daily urine NaCl excretion equals daily NaCl intake.*

The kidneys can vary the amount of NaCl they excrete over a wide range. Under conditions of NaCl restriction (i.e., low NaCl diet) virtually no NaCl appears in the urine. Conversely in individuals who ingest large quantities of NaCl renal excretion can exceed 1,000 mEq/day. The response of the kidneys to variations in dietary NaCl usually takes several days, especially when large changes occur. During the transition period excretion does not match intake and the individual is in either **positive** (intake > excretion) or **negative** (intake < excretion) **NaCl balance.** In the example shown in Fig. 42-17 NaCl intake is abruptly increased. During the transition period required for the kidney to increase its excretion a state of positive balance exists. The retained NaCl is added to the extracellular fluid, which increases its volume; water is retained in parallel by the ADH system to maintain plasma osmolality constant. The individual detects this increase in ECF volume by a gain in body weight (1 L of ECF ≅ 1 kg). Note that after several days NaCl excretion again equals intake but at a higher level. When the intake is abruptly returned to its original level a period of negative balance exists, but after a few days excretion again equals intake. The decrease in ECF volume causes the body weight to return to its original level.

To understand how NaCl excretion is regulated by the kidney the general features of Na^+ handling along the nephron must be understood. Fig. 42-18 schematically summarizes the contribution of each nephron seg-

*During euvolemia the ECV varies in direct proportion to the volume of the ECF. Thus the two terms can be interchanged.

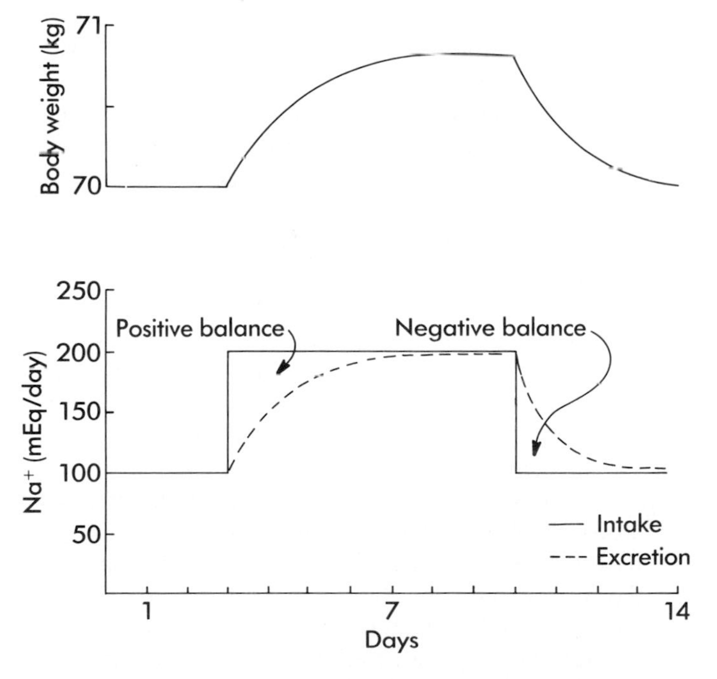

■ **Fig. 42-17** Response to a step increase and decrease in NaCl intake. NaCl excretion lags behind abrupt changes in NaCl intake. The transient periods of positive and negative balance result in alterations in the volume of the extracellular fluid, which are detected as changes in body weight.

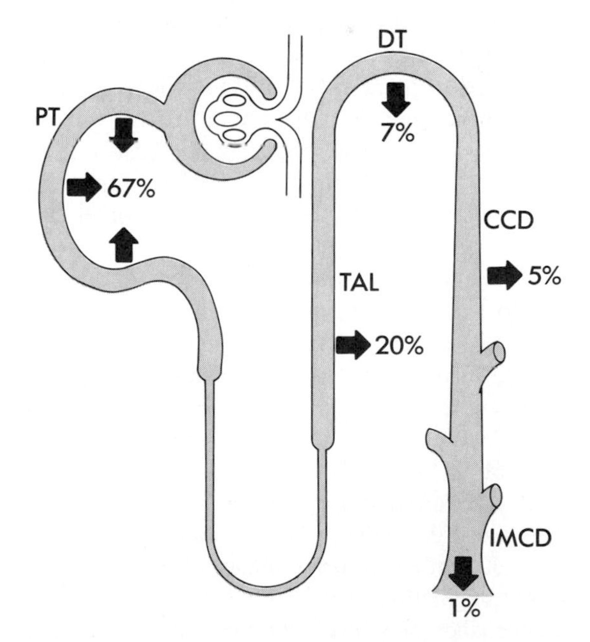

■ **Fig. 42-18** Segmental Na^+ reabsorption. The percentage of the filtered load of Na^+ reabsorbed by each nephron segment is indicated. *PT,* proximal tubule; *TAL,* thick ascending limb; *DT,* distal tubule; *CCD,* cortical collecting duct; *IMCD,* inner medullary collecting duct.

ment to the reabsorption of the filtered load of Na^+ under euvolemic conditions (the specific cellular mechanisms of Na^+ transport have been discussed in Chapter 41). In the following discussion only Na^+ reabsorption is considered. Although not specifically addressed Cl^- reabsorption occurs in parallel.

In a normal adult the filtered load of Na^+ can be calculated as

$$\text{Filtered load of } Na^+ = (\text{GFR})(\text{Plasma } [Na^+]) \quad (11)$$
$$= (180 \text{ L/day})(140 \text{ mEq/L})$$
$$= 25{,}200 \text{ mEq/day}$$

With a typical diet only 1% or less of the filtered Na is excreted in the urine (<250 mEq/day). Because of the large filtered load of Na^+ small changes in Na^+ reabsorption by the nephron can have a large effect on Na^+ balance and thus on the volume of the ECF. For example an increase in the excretion of Na^+ from 1% to 3% of the filtered load would represent an additional loss of approximately 500 mEq/day. Because the ECF $[Na^+]$ is 140 mEq/L Na^+ loss of this magnitude would decrease the volume of the ECF by more than 3 L (500 mEq/day ÷ 140 mEq/L = 3.6 L).

Regulation of renal Na^+ excretion requires the integrated action of all segments of the nephron. During euvolemia Na^+ handling by the nephron can be divided into two general processes:

1. Na^+ reabsorption by the proximal tubule, Henle's loop, and distal tubule is regulated such that a relatively constant portion of the filtered load of Na^+ is delivered to the collecting duct. The combined function of these nephron segments delivers 6% of the filtered load to the beginning of the collecting duct (see Fig. 42-18).
2. Reabsorption of Na^+ by the collecting duct is regulated such that the amount of Na^+ excreted in the urine matches the amount ingested in the diet. Thus the collecting duct is the site where final adjustments in Na^+ excretion that maintain the individual in the euvolemic state are made.

Mechanisms for maintaining constant Na^+ delivery to the collecting duct. A number of mechanisms serve to maintain delivery of a constant fraction of the filtered load of Na^+ to the beginning of the collecting duct. These mechanisms include:

1. Autoregulation of GFR and thus the filtered load of Na^+
2. Glomerulotubular balance
3. Load dependency of Na^+ reabsorption by the loop of Henle and distal tubule

Autoregulation of the GFR (see Chapter 41) permits maintenance of a relatively constant filtration rate over a wide range of perfusion pressures. Because of this constant filtration rate the filtered load of Na^+ is also held constant.

Despite the autoregulatory control of GFR small variations do occur. If these changes were not compensated for by an appropriate adjustment in Na^+ reabsorption by the nephron marked changes in Na^+ excretion would result. However Na^+ reabsorption, especially by the proximal tubule, changes in parallel with changes in GFR. This phenomenon is termed glomerulotubular balance (GT balance). By the process of GT balance reabsorption of Na^+ mainly by the proximal tubule is adjusted to match the GFR. Thus if GFR increases the amount of Na^+ reabsorbed by this nephron segment increases as well. The opposite occurs if GFR decreases. Details of the phenomenon of GT balance are described in Chapter 41.

The final mechanism that contributes to the maintenance of constant delivery of Na^+ to the beginning of the collecting duct is the ability of both Henle's loop and the distal tubule to increase their Na^+ reabsorptive rates in response to an increase in Na^+ delivery. Of these two segments Henle's loop, and in particular the thick ascending limb, has the greater capacity to increase Na^+ reabsorption in response to increased delivery. The mechanism involved in this phenomenon probably reflects that under normal conditions the transport capacity of these segments is not saturated. Thus an increase in delivery is simply dealt with by an increase in reabsorption.

Regulation of collecting duct Na^+ reabsorption. With a constant delivery of Na^+ small adjustments in collecting duct reabsorption are sufficient to balance excretion with intake (recall that a 2% change in the fractional excretion of Na^+ represents more than 3 L of ECF). Aldosterone is the main regulator of collecting duct Na^+ reabsorption and thus Na^+ excretion (see earlier discussion). When aldosterone levels are elevated Na^+ reabsorption by the collecting duct is increased (excretion decreased), whereas Na^+ reabsorption is decreased (excretion increased) when aldosterone levels are suppressed.

A number of other hormones also alter Na^+ reabsorption by the collecting duct. For example ANP, prostaglandins, and bradykinin inhibit Na^+ reabsorption. ADH, in addition to enhancing water reabsorption by the collecting duct, stimulates Na^+ reabsorption. However at present the relative roles of these other hormones on the regulation of collecting duct Na^+ reabsorption during euvolemia are not resolved.

As long as variations in the dietary intake of NaCl are small, the mechanisms just described can regulate renal Na^+ excretion appropriately and thereby maintain the volume of the ECV at a normal level. Large changes in NaCl intake however cannot be handled effectively by these mechanisms. Consequently large alterations in NaCl intake alter ECV. When this occurs additional factors that act on the kidneys to adjust Na^+ reabsorption to reestablish the euvolemic state are called into play.

Maura K. Carroll

Control of Na^+ Excretion With Increased ECV

An increase in ECV above normal is detected by the volume sensors, and signals that increase the excretion of Na^+ are sent to the kidneys (Fig. 42-19). The signals that act on the kidneys include the following:

1. Decreased activity of the renal sympathetic nerves
2. Release of ANP from atrial myocytes
3. Inhibition of ADH secretion from the posterior pituitary
4. Decreased renin secretion and thus decreased production of angiotensin II and decreased secretion of aldosterone by the adrenal cortex

The important difference between an increase in the ECV above normal and the euvolemic state is that the renal response is not limited to the collecting duct; rather, it involves the entire nephron. Three general responses to an increase in the ECV occur (see Fig. 42-19).

1. The GFR increases. The GFR increases mainly as a result of the decrease in sympathetic nerve activity. Sympathetic fibers innervate the afferent and efferent arterioles of the glomerulus and control their diameter. Decreased sympathetic nerve activity leads to arteriolar dilation. Because the effect appears to be greater on the afferent arteriole the hydrostatic pressure within the glomerular capillaries increases. Renal plasma flow also increases as a result of dilation of the afferent and efferent arterioles. As a result of these changes in glomerular hemodynamics GFR increases. ANP also increases GFR by dilating the afferent and efferent arterioles. With the increase in GFR the filtered load of Na^+ increases in parallel.
2. Reabsorption of Na^+ decreases in the proximal tubule. Several mechanisms appear to be involved in reducing Na^+ reabsorption by the proximal tubule, but the precise role of each is controversial. The sympathetic nerve fibers that innervate this region of the nephron stimulate Na^+ reabsorption. Hence the decreased sympathetic nerve activity that results from the increased ECV may contribute to the decreased Na^+ reabsorption. In addition angiotensin II directly stimulates Na^+ reabsorption by the proximal tubule. Because angiotensin II levels are also reduced under this condition proximal tubule Na^+ reabsorption may be decreased as a result. Because of the dilation of the afferent and efferent arterioles (see above) the hydrostatic pressure within the peritubular capillaries is increased. This alteration in the Starling forces reduces the absorption of solute (e.g., Na^+) and water (see Chapter 41 for mechanism).

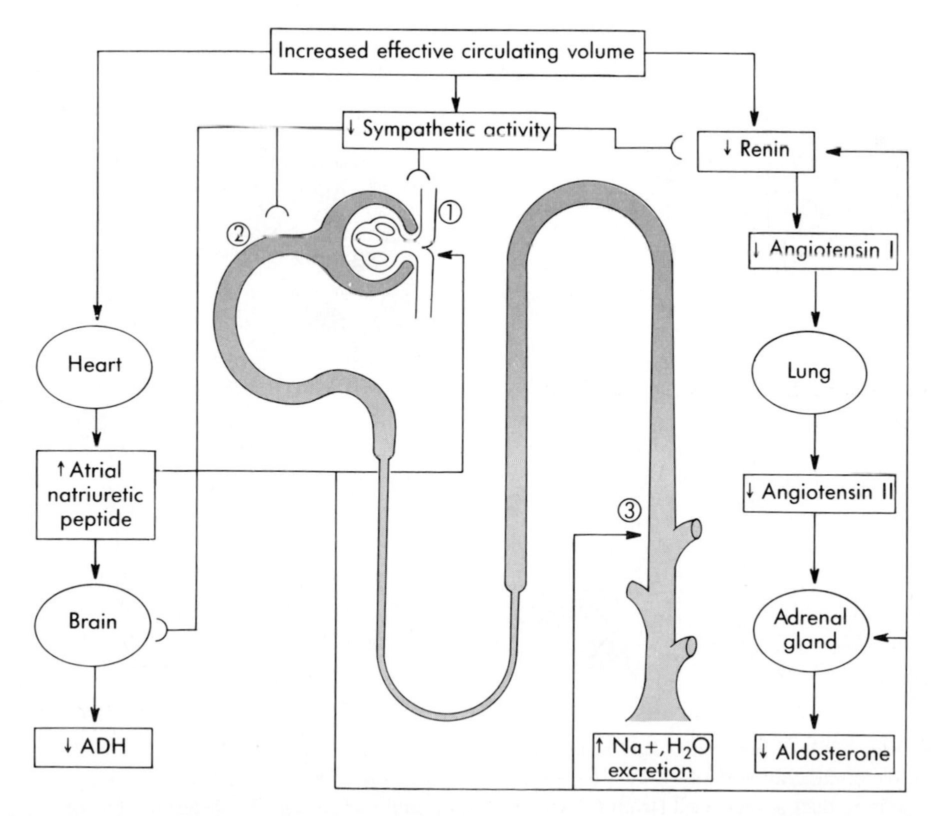

Fig. 42-19 Integrated response to expansion of the effective circulating volume. Numbers refer to description of the response in the text.

3. Na^+ reabsorption decreases in the collecting duct. Both the increase in filtered load of Na^+ and the decrease in proximal tubule reabsorption result in the delivery of large amounts of NaCl to Henle's loop and the distal tubule. The sympathetic nerves and aldosterone both stimulate NaCl reabsorption by Henle's loop. Therefore the reduced nerve activity and low aldosterone levels that occur in the presence of expanded ECV could in theory reduce NaCl reabsorption in this region of the nephron. However because reabsorption by the thick ascending limb is load-dependent, these effects are offset, and the fraction of the filtered load of Na^+ reabsorbed by the loop of Henle is actually increased. Nevertheless the amount of Na^+ delivered to the beginning of the collecting duct is greater than it is in the euvolemic state (Fig. 42-20).

The amount of Na^+ delivered to the beginning of the collecting duct varies in proportion to the degree of ECV expansion. A large load of Na^+ can overwhelm the reabsorptive capacity of the collecting duct, and the reabsorptive capacity can be even further reduced by the actions of ANP and by the decrease in the circulating levels of aldosterone.

One final component to be considered in the response to ECV expansion is the excretion of water. As Na^+ excretion is increased, plasma osmolality begins to fall and decreases the secretion of ADH. ADH secretion is also decreased by activity of the baroreceptors (see above) and by the action of ANP. In addition ANP inhibits the action of ADH on the collecting duct. Together these effects decrease water reabsorption by the collecting duct and thereby increase water excretion by the kidneys. Thus the excretion of NaCl and water occurs in concert, the ECV is restored to normal, and body fluid osmolality remains constant.

In summary the response of the nephron to expansion of the ECV involves the integrated action of all its component parts. The filtered load is increased, proximal tubule reabsorption is reduced (note: GFR is increased, whereas proximal reabsorption is decreased; thus GT balance does not occur under this condition), and the delivery of Na^+ to the beginning of the collecting duct is increased. This combination results in the excretion of a larger fraction of the filtered load of Na^+. The excretion of Na^+ exceeds intake (negative balance), and the ECV is restored to its normal value. The time course of this restoration (hours to days) depends on the degree to which the ECV is increased above normal and on the magnitude of the difference between the intake and excretion of Na^+.

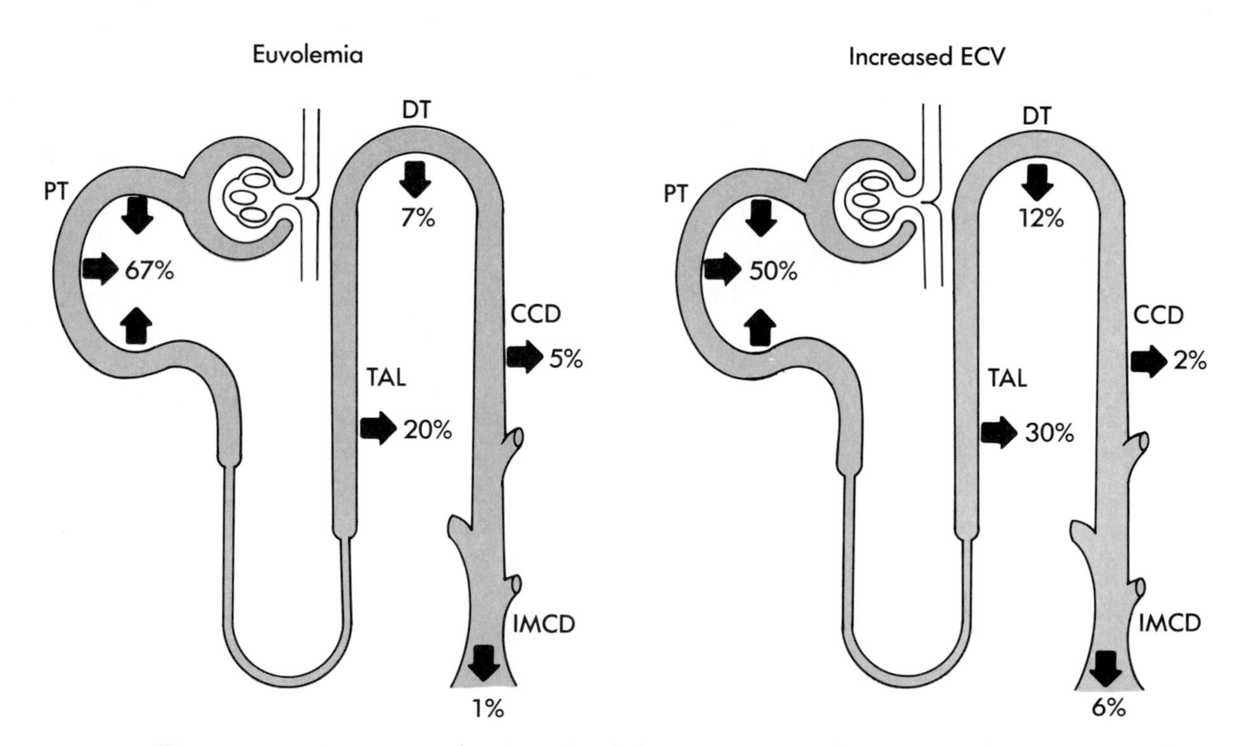

■ **Fig. 42-20** Segmental Na^+ reabsorption during euvolemia and after expansion of the effective circulating volume (ECV). Note that with expansion of the ECV delivery of Na^+ to the collecting duct is increased (from 6% to 8%). With inhibition of Na^+ reabsorption by the collecting duct Na^+ excretion is increased (from 1% to 6%). *PT,* proximal tubule; *TAL,* thick ascending limb; *DT,* distal tubule; *CCD,* cortical collecting duct; *IMCD,* inner medullary collecting duct.

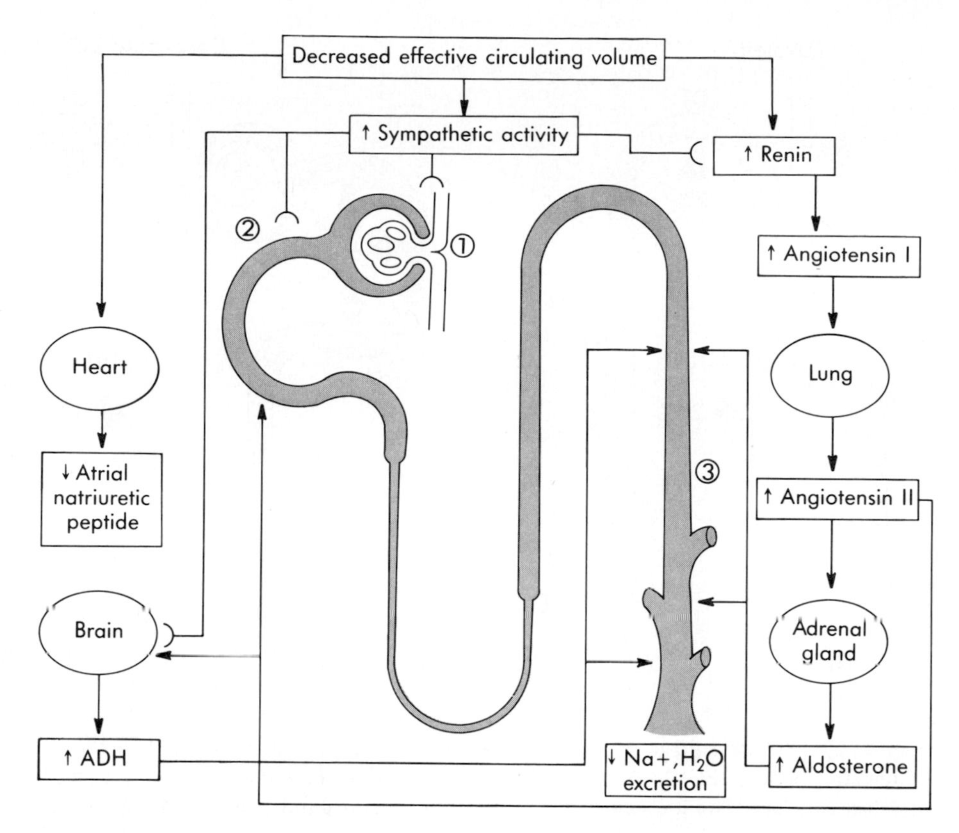

■ **Fig. 42-21** Integrated response to a decrease in the effective circulating volume. Numbers refer to description of the response in text.

■ *Control of Na$^+$ Excretion With Decreased ECV*

A decrease in ECV is detected by the volume sensors. Signals are sent to the kidneys, and Na$^+$ and water excretion are reduced. The signals involved are essentially opposite of those described above for the response to expansion of the ECV (Fig. 42-21). They include the following:

1. Increased renal sympathetic nerve activity
2. Increased secretion of renin, which results in increased levels of angiotensin II and thus increased secretion of aldosterone by the adrenal cortex
3. Inhibited ANP secretion by the atrial myocytes
4. Stimulated ADH secretion by the posterior pituitary

The response of the nephron to a contraction of the ECV involves all nephron segments. The general response is as follows:

1. The GFR decreases. Afferent and efferent arterioles constrict as a result of increased renal sympathetic nerve activity. The effect appears to be greater on the afferent arteriole, with the result that the hydrostatic pressure in the glomerular capillary falls. Renal plasma flow also declines. The net effect of these changes in glomerular hemodynamics is that GFR decreases. This in turn reduces the filtered load of Na$^+$.
2. Na$^+$ reabsorption by the proximal tubule is increased. Several mechanisms may augment Na$^+$ reabsorption in this segment. For example increased sympathetic nerve activity directly stimulates Na$^+$ reabsorption. Angiotensin II also directly stimulates proximal tubule reabsorption. As a result of constriction of the afferent and efferent arterioles the hydrostatic pressure within the peritubular capillaries is reduced. This action facilitates the movement of fluid from the lateral intercellular space into the capillary and thereby stimulates proximal tubule reabsorption of solute and water (see Chapter 41 for a discussion on the details of this mechanism).
3. Na$^+$ reabsorption by the collecting duct is enhanced. Both the reduction in filtered load and the enhanced proximal tubule reabsorption decrease the delivery of Na$^+$ to the loop of Henle and the distal tubule. Increased sympathetic nerve activity, ADH, and aldosterone stimulate Na$^+$ reabsorption by the thick ascending limb. Aldosterone and ADH also increase Na$^+$ reabsorption in the distal tubule. Because sympathetic nerve activity is increased and levels of ADH and aldosterone

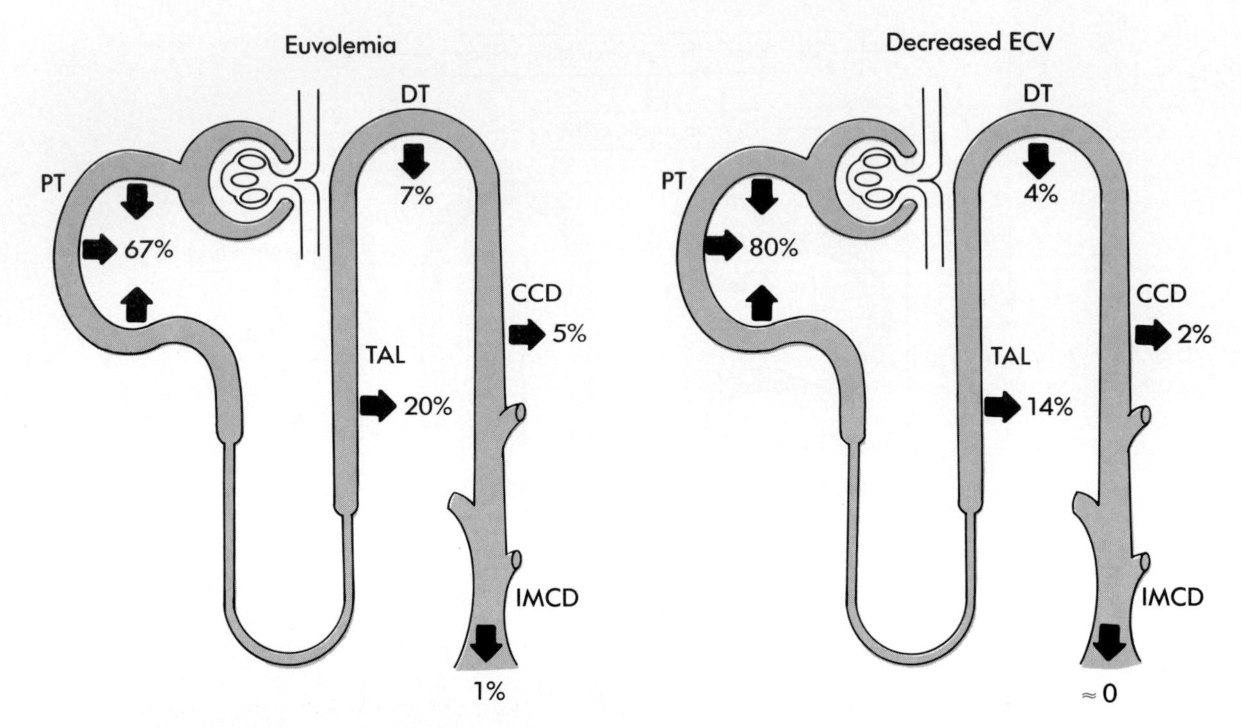

Fig. 42-22 Segmental Na^+ reabsorption during euvolemia and after a decrease in the effective circulating volume (ECV). Note that with contraction of the ECV delivery of Na^+ to the collecting duct is reduced (from 6% to 2%). The collecting duct reabsorbs virtually all of the Na^+ it receives, and Na^+ excretion is reduced to near zero. *PT*, proximal tubule; *TAL*, thick ascending limb; *DT*, distal tubule; *CCD*, collecting duct; *IMCD*, inner medullary collecting duct.

are elevated with decreased ECV, Na^+ reabsorption by these segments may increase. However Na^+ transport by both the thick ascending limb and the distal tubule is load-dependent. This factor offsets the stimulatory effects of increased sympathetic nerve activity, ADH, and aldosterone; the fraction of the filtered load of Na^+ reabsorbed by these segments is actually less than that seen in the euvolemic state. Nevertheless the net result of the decrease in GFR, together with enhanced proximal tubule reabsorption and reabsorption (albeit at reduced rates), by Henle's loop and the distal tubule, is that less Na^+ is delivered to the beginning of the collecting duct (Fig. 42-22).

The small amount of Na^+ delivered to the collecting duct is virtually all reabsorbed because transport in this segment is enhanced. This stimulation of Na^+ reabsorption by the collecting duct is caused mainly by the increased levels of aldosterone. Additionally ANP, which inhibits collecting duct reabsorption, is not present, and ADH, which can stimulate Na^+ reabsorption, is present at high levels.

Finally water reabsorption by the collecting duct is enhanced by the action of ADH, whose levels are elevated by activation of the baroreceptors. As a result water excretion is reduced. This, together with the NaCl retained by the kidneys, restores the ECV while it maintains body fluid osmolality constant.

In summary the response of the nephron to contraction of the ECV, like that seen with an expanded ECV, involves the integrated action of all its component segments. The filtered load of Na^+ is decreased, proximal tubule reabsorption is enhanced (note: GFR is decreased, whereas proximal reabsorption is increased; thus GT balance does not occur under this condition), and the delivery of Na^+ to the beginning of the collecting duct is reduced. This decreased Na^+ delivery together with the enhanced capacity of the collecting duct to reabsorb Na^+ virtually eliminates all of the Na^+ from the urine. Thus excretion of Na^+ is less than intake (positive balance) and ECV is reexpanded. The time course of this reexpansion (hours to days) depends on the degree to which the ECV is reduced from normal and on the dietary intake of Na^+.

Edema and the Role of the Kidneys

Edema is the accumulation of excess fluid within the interstitial space. The Starling forces across the capillaries determine the movement of fluid out of the vascular compartment into the interstitial space. Alterations of these forces under pathological conditions can lead to increased movement of fluid from the vascular space into the interstitium, resulting in edema formation. However for edema to be detected clinically (e.g., swelling of ankles), NaCl and water must be retained by the kidneys.

The role of the kidneys in the formation of edema can be appreciated by recognizing that the interstitial compartment must contain 2 to 3 L of excess fluid before edema is detectable. Because the source of this fluid is the vascular compartment (i.e., plasma), which has a volume of 3 to 4 L in the normal individual, the movement of 2 to 3 L out of the plasma compartment into the interstitial compartment would result in a marked decrease in blood pressure. As described below this decrease in blood pressure would prevent further movement of fluid from the vascular compartment into the interstitial compartment. However retention of NaCl and water by the kidneys replenishes the plasma volume and thereby maintains the blood pressure. As a result accumulation of fluid in the interstitial compartment continues and edema develops.

Alterations in Starling forces. In Chapters 28 and 41, the Starling forces and the way they determine the movement of fluid across the wall of the capillary were considered. Under normal conditions and considering all capillary beds (exclusive of the glomerulus), approximately 20 L/day move out of the capillary at the arteriole end, 18 L/day are absorbed back into the capillary at the venous end, and 2 L/day return to the circulation via the lymphatic vessels. Edema results from a change in the Starling forces that alters these fluid dynamics.

Capillary hydrostatic pressure (P_{cap}). Increasing the P_{cap} favors the movement of fluid out of the capillary or retards its movement into the capillary and thereby promotes the formation of edema. Normally the resistance of the precapillary arteriole is well regulated such that changes in systemic blood pressure do not result in marked alterations in the P_{cap}. However alterations in the pressure within the venous side of the circulation do have an important effect on P_{cap}. Consequently an increase in the venous pressure elevates P_{cap}. This action reduces the amount of fluid reabsorbed from the interstitium back into the capillary lumen, resulting in the accumulation of edema fluid.

One of the common conditions that produce an increase in P_{cap} is heart failure. Because of poor cardiac performance the pressure in the venous circulation is elevated. P_{cap} can also be elevated secondary to venous thrombosis.

Plasma oncotic pressure (π_{cap}). A decrease in π_{cap} favors movement of fluid out of the capillary lumen and inhibits its reabsorption from the interstitium. Alterations in π_{cap} result primarily from changes in the plasma [albumin]. In some renal diseases the permeability of the glomerular capillary is abnormally high. As a result large quantities of albumin are filtered and lost in the urine. If the rate of loss exceeds the rate at which albumin is synthesized by the liver, the plasma [albumin] falls and edema can form.

Lymphatic obstruction. Obstruction of the lymphatic vessels interferes with and reduces the volume of interstitial fluid that is returned to the circulation via this route. As a result edema can form. The most common cause of lymphatic obstruction is malignancy. When malignant cells spread to lymph nodes obstruction can occur.

Capillary permeability. Increase in the permeability of the capillary favors increased movement of fluid across the capillary wall. This favors both capillary-to-interstitium movement at the arteriolar end and interstitium-to-capillary movement at the venous end. Because permeability is increased, albumin can accumulate in the interstitium. The oncotic pressure of this albumin results in the accumulation of excess fluid in the interstitial compartment, and edema formation ensues.

The role of the kidneys. In each of the conditions described previously excess fluid accumulates in the interstitium at the expense of the plasma. As a result if the plasma volume is not maintained near a normal level, this accumulation of fluid is self-limiting. For example consider the situation that exists in heart failure. Because of decreased cardiac performance the venous pressure is elevated. This elevation raises capillary hydrostatic pressure, and net accumulation of fluid in the interstitium occurs. Because the source of this fluid is the plasma and assuming plasma volume is not maintained by some mechanism (see below), plasma volume decreases. This decrease in turn decreases both venous pressure and capillary hydrostatic pressure, and movement of fluid into the interstitium ceases.

Fig. 42-23 illustrates what actually occurs when an alteration in the capillary Starling forces causes fluid accumulation in the interstitium. As fluid moves from the plasma into the interstitium plasma volume is decreased. This is sensed by the volume receptors as a decrease in the ECV. Appropriate signals, which result in a decrease in the excretion of NaCl and water, are sent to the kidneys and plasma volume is restored. With restoration of the plasma volume and assuming the condition altering the Starling forces still exists (e.g., untreated heart failure) additional fluid moves into the interstitium and edema forms. This cycle continues until a new steady state is reached at the level of the capillary such that the amount of fluid moving out of the capillary lumen into the interstitium is again balanced by the amount moving in the opposite direction. The new steady state can be attained even when the underlying cause is uncorrected, because as edema fluid accumulates in the interstitial compartment, the hydrostatic pressure in this compartment increases. This increase in pressure occurs as a result of the limited compliance of

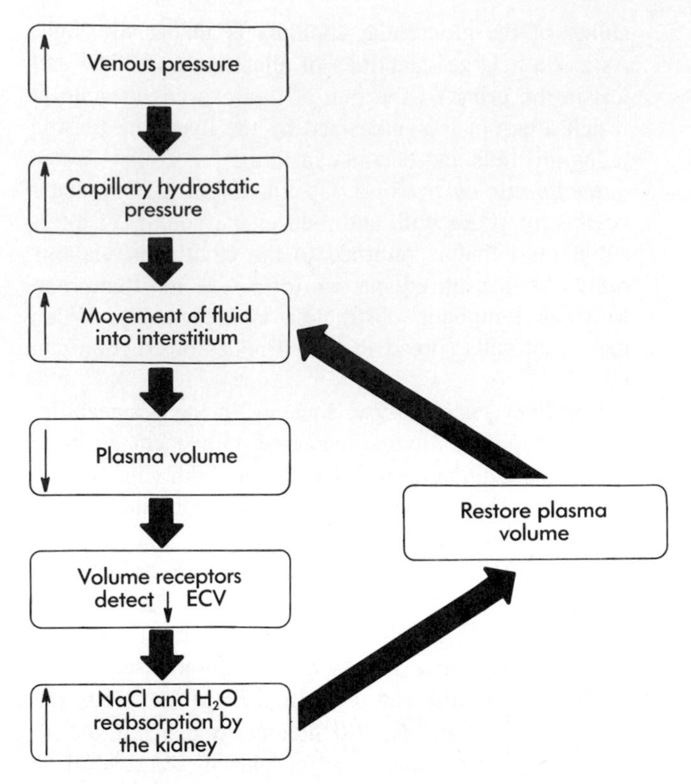

■ **Fig. 42-23** Steps involved in the development of edema resulting from an increase in venous pressure (e.g., during heart failure). NaCl and water retention by the kidneys maintains plasma volume, allowing the accumulation of fluid in the interstitium. See text for details.

the surrounding tissue (e.g., skin).* Eventually this pressure rises to a level where net accumulation in the interstitial compartment ceases.

The importance of NaCl retention by the kidneys in the formation of edema provides two approaches for treatment. The first involves dietary manipulation. The ultimate source of NaCl is the diet; thus if dietary intake of NaCl is restricted, the amount that can be retained by the kidneys is reduced, with the result that edema formation is limited. The second approach involves inhibition of the kidneys' ability to retain NaCl. This is accomplished by the use of **diuretics,** which act by inhibiting Na^+ transport mechanisms in the nephron. Thus by their action Na^+ excretion is increased and ECF volume reduced.

■ *Summary*

1. The osmolality and volume of the body fluids are maintained within a narrow range despite wide variation in water and solute intake. The kidneys play the central role in this regulatory process by virtue of their ability to vary the excretion of water and solutes.

2. Regulation of body fluid osmolality requires that water intake and loss from the body be equal. This involves the integrated interaction of the ADH secretory and thirst centers of the hypothalamus and the ability of the kidneys to excrete urine that is either hypoosmotic or hyperosmotic with respect to the body fluids. When body fluid osmolality increases, ADH secretion and thirst are stimulated. ADH acts on the kidneys to increase the permeability of the collecting duct to water. Hence water is reabsorbed from the lumen of the collecting duct, and a small volume of hyperosmotic urine is excreted. This renal conservation of water, together with increased water intake, restores body fluid osmolality to normal. When body fluid osmolality decreases, ADH secretion and thirst are suppressed. In the absence of ADH the collecting duct is impermeable to water, and a large volume of hypoosmotic urine is excreted. With this increased excretion of water and with a decreased intake of water caused by suppression of thirst the osmolality of the body fluids is restored to normal.

3. Central to the process of concentrating and diluting the urine is the loop of Henle. The transport of NaCl by the loop allows the separation of solute and water, which is essential for the elaboration of hypoosmotic urine. By this same mechanism the interstitial fluid in the medullary portion of the kidney is rendered hyperosmotic. This hyperosmotic medullary interstitial fluid in turn provides the osmotic driving force for the reabsorption of water from the lumen of the collecting duct when ADH is present.

4. The volume of the extracellular fluid is determined by Na^+ balance. When intake of Na^+ exceeds excretion the extracellular fluid volume increases (positive Na^+ balance). Conversely when excretion of Na^+ exceeds intake extracellular fluid volume decreases (negative Na^+ balance). The kidneys are the primary route for Na^+ excretion. The coordination of Na^+ intake and excretion and thus the maintenance of a normal extracellular fluid volume requires the integrated action of the kidneys with the cardiovascular and sympathetic nervous systems. Cardiovascular volume receptors detect changes in extracellular fluid volume and, by sympathetic and hormonal signals, effect appropriate adjustments in Na^+ excretion by the kidneys.

5. Under normal conditions (euvolemia) Na^+ excretion by the kidneys is matched to the amount of Na^+ ingested in the diet. The kidneys accomplish this by reabsorbing virtually all of the filtered load of Na^+ (typically less than 1% of the filtered load is excreted). During euvolemia the collecting duct is responsible for making small adjustments in urinary Na^+ excretion to effect

*The situation is analogous to the increase in pressure that occurs as a balloon is inflated. Initially the volume of the balloon increases without a large increase in pressure. However as the balloon reaches its limit of distensibility the pressure increases.

Na^+ balance. The major factor regulating collecting duct Na^+ reabsorption is aldosterone, which acts to stimulate Na^+ reabsorption.

6. When the volume of the extracellular fluid is increased, the volume receptors initiate a response that ultimately leads to increased excretion of Na^+ by the kidneys and the return of the extracellular fluid to its normal volume. The components of this response include a decrease in sympathetic outflow to the kidney, suppression of the renin-angiotensin-aldosterone system, and release from the cardiac atria of atrial natriuretic peptide. The actions of these effectors enhance the glomerular filtration rate, which increases the filtered load of Na^+, and Na^+ reabsorption by the proximal tubule and collecting duct is reduced. Together these changes in renal Na^+ handling enhance Na^+ excretion.

7. When the volume of the extracellular fluid is decreased, the above sequence of events is reversed (increased sympathetic outflow to the kidney, activation of the renin-angiotension-aldosterone system, and suppression of atrial natriuretic peptide secretion). This action decreases the glomerular filtration rate, enhances reabsorption of Na^+ by the proximal tubule and collecting duct, and thus reduces Na^+ excretion.

Bibliography

Journal articles

Bayliss PH: Osmoregulation and control of vasopressin secretion in healthy humans, *Am J Physiol* 253:R671, 1987.

Brenner BM et al: Diverse biological actions of atrial natriuretic peptide, *Physiol Rev* 70:665, 1990.

De Wardner HE: The control of sodium excretion, *Am J Physiol* 235:F163, 1978.

DiBona GF: Role of renal nerves in volume homeostasis, *Acta Physiol Scand* 139:18, 1990.

Dzau VJ et al: Molecular biology of the renin-angiotensin system, *Am J Physiol* 255:F563, 1988.

Gonzalez-Campoy JM et al: Escape from the sodium-retaining effects of mineralocorticoids: role of ANF and intrarenal hormone systems, *Kidney Int* 35:767, 1989.

Greenwald L, Stetson D: Urine concentration and the length of the renal papilla, *News Physiol Sci* 3:46, 1988.

Gross P, Ritz E, editors: Water metabolism, *Kidney Int* 32(suppl 21):August 1987.

Jamison RL, Maffly RH: The urinary concentrating mechanism, *N Engl J Med* 295:1059, 1976.

Knepper MA, Star RA: The vasopressin-regulated urea transporter in renal inner medullary collecting duct, *Am J Physiol* 259:F393, 1990.

Marver D, editor: Corticosteroids and the kidney. *Semin Nephrol* 10:July 1990.

O'Neil RG: Aldosterone regulation of sodium and potassium transport in the cortical collecting duct, *Semin Nephrol* 10:365, 1990.

Sands JM et al: Vasopressin effects on urea and H_2O transport in inner medullary collecting duct subsegments, *Am J Physiol* 253:F823, 1987.

Sands JM et al: Hormone effects on NaCl permeability of rat inner medullary collecting duct, *Am J Physiol* 255:F421, 1988.

Schlatter E: Antidiuretic hormone regulation of electrolyte transport in the distal nephron, *Renal Physiol Biochem* 12:65, 1989.

Schrier RW: Pathogenesis of sodium and water retention in high-output and low-output cardiac failure, nephrotic syndrome, cirrhosis, and pregnancy, *N Engl J Med* 319:1065 and 1127, 1988.

Books and monographs

Badr K Ichikawa I: *Physical and biological properties of body fluid electrolytes.* In Ichikawa I, editor: *Pediatric textbook of fluids and electrolytes,* Baltimore, 1990, Williams & Wilkins.

Fanestil DD: *Compartmentation of body water.* In Maxwell MH, Kleeman CR, Narins RG, editors: *Clinical disorders of fluid and electrolyte metabolism,* ed 4, New York, 1987, McGraw-Hill.

Hays RM: *Cell biology of vasopressin.* In Brenner BM, Rector FC Jr, editors: *The kidney,* ed 4, Philadelphia, 1991, WB Saunders.

Knepper MA, Rector FC Jr: *Urinary concentration and dilution.* In Brenner BM, Rector FC Jr, editors: *The kidney,* ed 4, Philadelphia, 1991, WB Saunders.

Robertson GL, Berl T: *Pathophysiology of water metabolism.* In Brenner BM, Rector FC Jr, editors: *The kidney,* ed 4, Philadelphia, 1991, WB Saunders.

Moe GW, Legault L, Skorecki KL: *Control of extracellular fluid volume and pathophysiology of edema formation.* In Brenner BM, Rector FC Jr, editors: *The kidney,* ed 4, Philadelphia, 1991, WB Saunders.

Rose BD: *Clinical physiology of acid-base and electrolyte disorders,* ed 3, New York, 1989, McGraw-Hill.

Seldin DW, Giebisch G, editors: *The regulation of sodium and chloride balance,* New York, 1990, Raven Press.

Yoshioka T, Iitaka K, Ichikawa I: *Body fluid compartments.* In Ichikawa I, editor: *Pediatric textbook of fluids and electrolytes,* Baltimore, 1990, Williams & Wilkins.

CHAPTER

43

Regulation of Potassium, Calcium, Magnesium, Phosphate, and Acid-Base Balance

■ *Potassium*

K^+ is one of the most abundant cations in the body. It is critical for many cell functions, and its concentration in cells and extracellular fluid remains constant despite wide fluctuations in dietary K^+ intake. Two sets of regulatory mechanisms safeguard K^+ homeostasis. First, several mechanisms regulate internal K^+ balance and control the $[K^+]$ in the intracellular and extracellular fluid. Another set of mechanisms holds external K^+ balance constant by adjusting renal K^+ excretion to match dietary K^+ intake. The kidneys maintain external K^+ balance. This section will focus on the hormones and factors that influence the distribution of K^+ between the intracellular and extracellular fluid compartments **(internal K^+ balance)** and the hormones and factors that regulate the amount of K^+ in the body **(external K^+ balance).**

■ *Overview of K^+ Homeostasis*

Total body K^+ has been estimated at 50 mEq/kg of body weight, or 3,500 mEq for a 70 kg individual. Ninety eight percent of the K^+ in the body is within cells where its average concentration is 150 mEq/L. A high intracellular $[K^+]$ is required for many cell functions, including growth and division, the function of enzymes, acid-base balance, volume regulation, excitability, and contraction. Only 2% of total body K^+ is located in the extracellular fluid, where its normal concentration is 4 mEq/L. The large concentration difference of K^+ across cell membranes (146 mEq/L) is maintained by the operation of the Na^+-K^+-ATPase pump. This gradient is important for maintaining the potential difference across cell membranes. Thus K^+ is critical for the excitability of nerve and muscle cells, as well as for the contractility of cardiac, skeletal, and smooth muscle cells.

When the $[K^+]$ in the extracellular fluid exceeds 5.5 mEq/L, an individual is **hyperkalemic.** Hyperkalemia reduces the resting membrane potential (the voltage becomes less negative) and increases the excitability of neurons, cardiac cells, and muscle cells (see Chapter 2). Severe, rapid increases in plasma $[K^+]$ can lead to cardiac arrest and death. In contrast when the $[K^+]$ of the extracellular fluid is less than 3.5 mEq/L, an individual is **hypokalemic.** A decline in extracellular $[K^+]$ hyperpolarizes the resting cell membrane potential (the voltage becomes more negative) and reduces the excitability of neurons, cardiac cells, and muscle cells. Severe hypokalemia can lead to paralysis, cardiac arrhythmias, a decreased ability to concentrate the urine, metabolic alkalosis, increased production of NH_4^+ by the kidneys, and death.

■ *Internal K^+ Distribution*

After a meal the K^+ absorbed by the gastrointestinal tract rapidly enters the extracellular fluid. If all of the K^+ ingested during a normal meal (~50 mEq) were to remain in the extracellular fluid compartment, plasma $[K^+]$ would increase by a potentially lethal 3.6 mEq/L (50 mEq of K^+ added to 14 L of extracellular fluid). This rise in plasma $[K^+]$ is attenuated by the rapid (minutes) uptake of K^+ into cells. Because the excretion of K^+ by the kidneys after a meal is relatively slow (hours), the uptake of K^+ by cells is essential to prevent life-threatening hyperkalemia. To maintain external K^+ balance, all of the K^+ absorbed by the gastro-

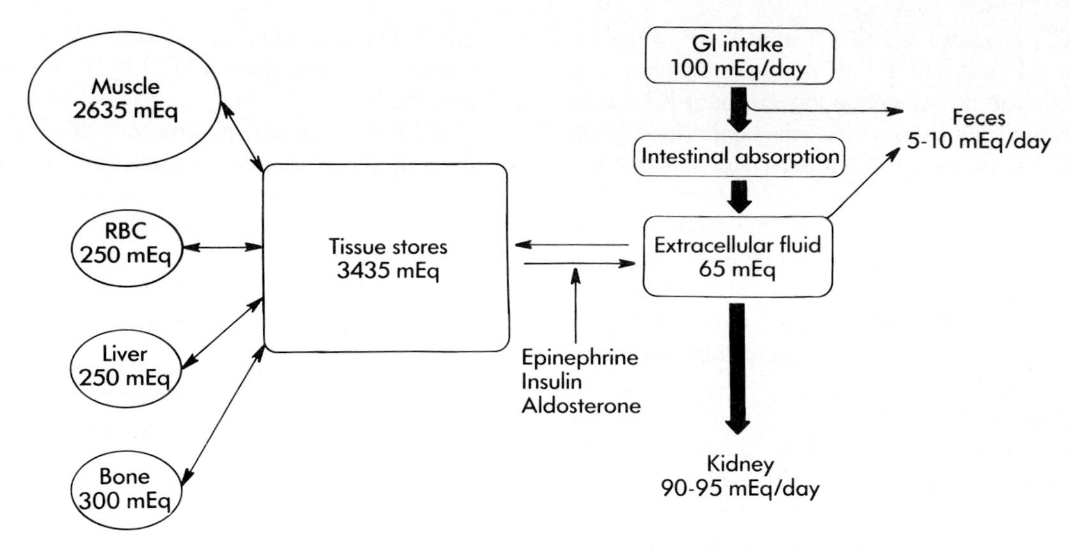

■ **Fig. 43-1** Overview of potassium homeostasis. See text for details.

intestinal tract must eventually be excreted by the kidneys. This K^+ is slowly excreted such that after 6 hours it is eliminated from the body, thereby maintaining external K^+ homeostasis.

Several hormones promote the uptake of K^+ into cells after a rise in plasma [K^+] and thereby prevent dangerous **hyperkalemia** (Fig. 43-1 and Table 43-1). These hormones include epinephrine, insulin, and aldosterone. They increase K^+ uptake into skeletal muscle, liver, bone, and red blood cells by stimulating the Na^+-K^+-ATPase pump. Hyperkalemia, subsequent to K^+ absorption by the gastrointestinal tract, stimulates insulin secretion from the pancreas, aldosterone release from the adrenal cortex, and epinephrine secretion from the adrenal medulla. In contrast **hypokalemia** inhibits release of these hormones.

Epinephrine. Catecholamines affect the distribution of K^+ across cell membranes by activating α and β_2-adrenergic receptors. Stimulation of α receptors increases plasma [K^+], whereas stimulation of β_2 receptors decreases plasma [K^+]. Alpha receptor activation post-exercise is important in preventing subsequent hypokalemia.The importance of β_2 receptors is illustrated by two observations. First the rise in plasma [K^+] after a K^+-rich meal is greater if the person has been pretreated with propranolol, a β-adrenergic blocker. Second the release of epinephrine during stress (e.g., coronary ischemia) can rapidly lower plasma [K^+].

Insulin. Insulin also stimulates K^+ uptake into cells. The importance of insulin is illustrated by two observations. First, the rise in plasma [K^+] after a K^+-rich meal is greater in patients with diabetes mellitus (i.e., insulin deficiency). Second, infusions of insulin or of glucose (to increase insulin levels) are two forms of acute therapy to treat hyperkalemia. *Insulin is the most important hormone that shifts K^+ into cells after ingestion of K^+ in a meal.*

Aldosterone. Aldosterone, like catecholamines and insulin, also promotes K^+ uptake into cells (see Chapter 41). A rise in aldosterone levels (e.g., primary aldosteronism) causes hypokalemia, and a fall in aldosterone levels (e.g., Addison's disease) causes hyperkalemia.

Thus far the discussion has focused on hormones that regulate the distribution of K^+ across cell membranes. Other factors, however, influence K^+ movements across the cell membrane but they are not homeostatic mechanisms because they displace plasma [K^+] from normal levels (see Table 43-1).

Acid-base balance. Changes in acid-base balance have important effects on plasma [K^+] (see end of chapter). Metabolic acidosis increases, and metabolic alkalosis decreases plasma [K^+], whereas respiratory acid-base disorders do not appreciably affect plasma [K^+]. For example, a metabolic acidosis produced by the addition of inorganic acids (e.g., HCl, H_2SO_4), but not by organic acids (e.g., lactic acid, acetic acid, keto acids), to the extracellular fluid increases plasma [K^+].

■ **Table 43-1** Factors influencing the distribution of K^+ between the intracellular and extracellular fluid

Physiological: maintain plasma [K^+] constant
Epinephrine
Insulin
Aldosterone
Pathophysiological: displace plasma [K^+] from normal
Acid-base balance
Plasma osmolality
Cell lysis
Exercise

Plasma [K^+] increases 0.2 to 1.7 mEq/L for every 0.1 unit fall in pH. The reduced pH promotes movement of H^+ into cells and the reciprocal movement of K^+ out of cells. Metabolic alkalosis has the opposite effect; plasma [K^+] decreases as K^+ moves into cells and H^+ leaves cells. The mechanism responsible for this shift is not fully understood. Some have proposed that the movement of H^+ occurs as the cells buffer changes in the [H^+] of the extracellular fluid (see below). As H^+ moves across the cell membranes, K^+ moves in the opposite direction, and thus cations are neither gained nor lost across the cell membranes.*

Plasma osmolality. The osmolality of the plasma also influences the distribution of K^+ across cell membranes. An increase in the osmolality of the extracellular fluid enhances K^+ release by cells and thus increases extracellular [K^+]. The plasma K^+ level may increase by 0.4 to 0.8 mEq/L for a 10 mOsm/kg H_2O elevation in plasma osmolality. Hypoosmolality has the opposite action.

As plasma osmolality increases, water will leave cells because of the osmotic gradient across the plasma membrane. Water leaves cells until the intracellular osmolality becomes equal to that of the extracellular fluid. This loss of water shrinks cells and causes cell [K^+] to rise. The rise in intracellular [K^+] provides a driving force for K^+ efflux from the cells. This sequence increases plasma [K^+]. A fall in plasma osmolality has the opposite effect.

Cell lysis. Cell lysis causes hyperkalemia. Severe trauma (e.g., burns), some diseases such as tumor lysis syndrome, and **rhabdomyolysis** (i.e., destruction of skeletal muscle) cause cell destruction and release of K^+ (and other cell solutes) into the extracellular fluid. In addition gastric ulcers may cause seepage of red blood cells into the gastrointestinal tract. The blood cells are digested, and the K^+ released from the cells is absorbed and causes hyperkalemia.

Exercise. K^+ is released from skeletal muscle cells during exercise. Release of K^+ and the ensuing hyperkalemia depend on the degree of exercise. Plasma [K^+] increases by 0.3 mEq/L with slow walking and may increase up to 2.0 mEq/L with severe exercise. These changes in plasma [K^+] usually do not produce symptoms and are reversed after several minutes of rest. However, for individuals (1) who have disorders in internal K^+ balance (certain endocrine disorders), (2) whose ability to excrete K^+ is impaired (renal failure) or (3) who are on certain medications, exercise can lead to potentially life-threatening hyperkalemia. For example, individuals taking β-adrenergic blockers for hypertension may raise plasma [K^+] by 2 to 4 mEq/L during exercise.

Acid-base balance, plasma osmolality, cell lysis, and exercise do not maintain plasma [K^+] at a normal value and therefore do not contribute to K^+ homeostasis. The extent to which these pathophysiological states alter plasma [K^+] depends on the integrity of the homeostatic mechanisms that regulate plasma [K^+] (e.g., secretion of epinephrine, insulin, and aldosterone).

*Although organic acids produce a metabolic acidosis, they do not cause hyperkalemia. Two possible explanations have been suggested for the inability of organic acids to cause hyperkalemia. First, the organic anion may enter the cell with H^+, thereby eliminating K^+ for H^+ exchange across the membrane. Second, organic anions may stimulate insulin secretion, which drives K^+ into cells, counteracting the direct effect of the acidosis that moves K^+ out of cells.

External K^+ Balance: Excretion by the Kidneys

The kidneys play the major role in maintaining external body K^+ balance (see Fig. 43-1). The kidneys excrete 90% to 95% of the K^+ ingested in the diet. Excretion equals intake even when intake increases by as much as tenfold. This equality between urinary excretion and dietary intake underscores the importance of the kidneys in maintaining K^+ homeostasis. Although small amounts of K^+ are lost each day in the stool and sweat (~5% to 10% of the K^+ ingested in the diet), this amount is essentially constant, is not regulated, and therefore is relatively much less important than is the K^+ excreted by the kidneys.* *The primary event in determining urinary K^+ excretion is K^+ secretion from the blood into the tubular fluid by the cells of the distal tubule and collecting duct system.* The transport pattern of K^+ by the major nephron segments is illustrated in Fig. 43-2.

Because K^+ is not bound to plasma proteins, it is freely filtered by the glomerulus. Normally, urinary K^+ excretion is 15% of the amount filtered. Accordingly, K^+ must be reabsorbed by the nephron. When dietary K^+ intake is augmented, however, K^+ excretion can, under some experimental conditions, exceed the amount filtered; thus K^+ is also secreted.

The proximal tubule reabsorbs 67%, and the loop of Henle reabsorbs 20% of the filtered K^+. In both segments reabsorption is a constant fraction of the amount filtered. In contrast to these segments, which are capable of only reabsorbing K^+, *the distal tubule and the collecting duct have the dual capacity to reabsorb and secrete K^+*. The rate of K^+ reabsorption or secretion by the distal tubule and collecting duct is variable and depends on several hormones and other factors. When K^+ intake is normal (100 mEq/day), K^+ is secreted by these segments. A rise in dietary K^+ intake increases K^+ secretion such that the amount of K^+ that appears in the urine may equal 80% of the amount filtered (see Fig. 43-2). In contrast, a low K^+ diet activates K^+ reabsorption along the distal tubule and collecting duct such that urinary excretion falls to 1% of the K^+ fil-

*Loss of K^+ in the feces can become significant during diarrhea.

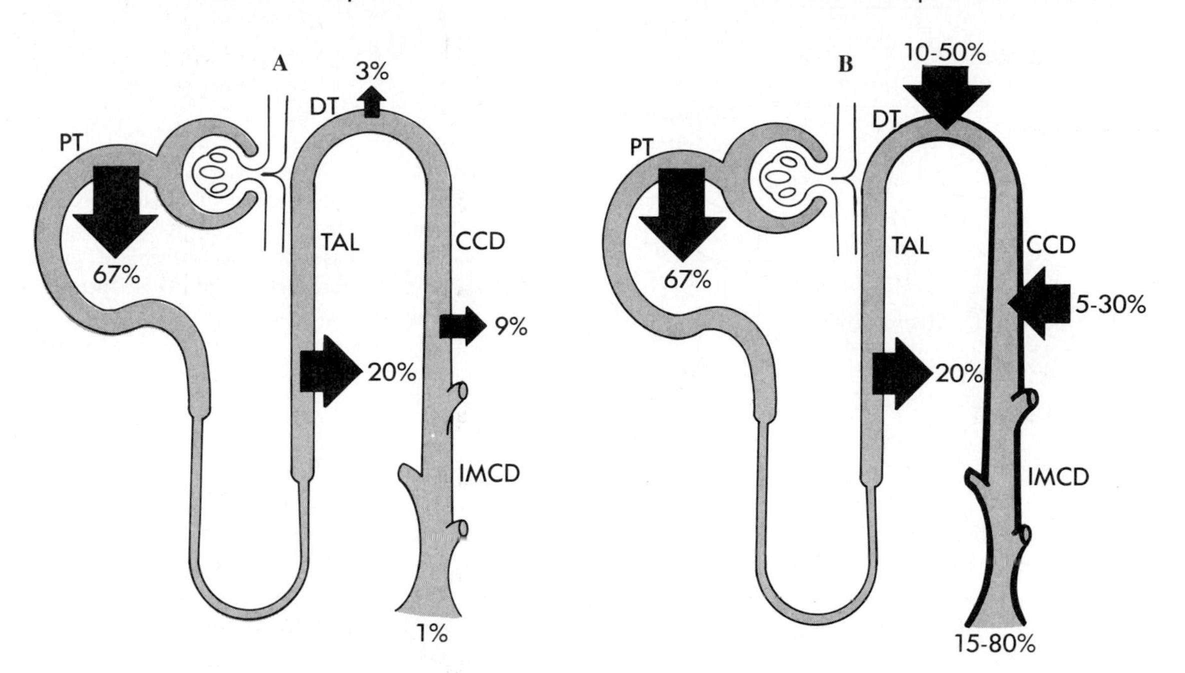

■ **Fig. 43-2** K^+ transport along the nephron. K^+ excretion depends on the rate and direction of K^+ secretion by the distal tubule and the collecting duct. Percentages refer to the amount of filtered K^+ reabsorbed or secreted by each nephron segment. **A,** Dietary K^+ depletion. An amount of K^+ equal to 1% of the filtered load of K^+ is excreted. **B,** Normal and increased dietary K^+ intake. An amount of K^+ equal to 15% to 80% of the filtered load is excreted. *PT,* proximal tubule; *TAL,* thick ascending limb; *DT,* distal tubule; *CCD,* cortical collecting duct; *IMCD,* inner medullary collecting duct.

tered by the glomerulus (see Fig. 43-2). The kidneys are not able to reduce K^+ excretion to the same low levels as they can for Na^+. Therefore hypokalemia can develop in individuals placed on a K^+-deficient diet.

Because the magnitude and direction of K^+ transport by the distal tubule and collecting duct are variable, the overall rate of urinary K^+ excretion is determined by these tubular segments. The remainder of this section will focus on the mechanisms of K^+ transport by the distal tubule and the collecting duct and on the factors and hormones that regulate K^+ transport by these segments.

■ *Cellular Mechanisms of K^+ Transport by the Distal Tubule and Collecting Duct*

Fig. 43-3 illustrates the cellular mechanism of K^+ secretion by principal cells in the distal tubule and collecting duct. Secretion from blood into tubular fluid is a two-step process involving (1) uptake across the basolateral membrane by the Na^+-K^+-ATPase pump and (2) diffusion of K^+ from the cell into the tubular fluid. The operation of the Na^+-K^+-ATPase pump creates a high intracellular $[K^+]$, which provides the chemical driving force for K^+ exit across the apical membrane through K^+ channels. Although K^+ channels are also present in the basolateral membrane, K^+ preferentially leaves the cell across the apical membrane and enters the tubular fluid for two reasons. First the electrochemical gradient of K^+ across the apical membrane favors the downhill movement into the tubular fluid. Second the permeability of the apical membrane to K^+ is greater than that of the basolateral membrane.

The three major factors that control the rate of K^+ secretion by the distal tubule and the collecting duct are (see Fig. 43-3):

1. The activity of the Na^+-K^+-ATPase pump
2. The driving force (electrochemical gradient) for K^+ movement across the apical membrane
3. The permeability of the apical membrane to K^+.

Every change in K^+ secretion results from an alteration in one or more of these parameters.

In contrast, the cellular pathways and mechanisms of K^+ reabsorption in the distal tubule and collecting duct system are not completely understood. The intercalated cells may reabsorb K^+ by an H^+-K^+-ATPase transport mechanism located in the apical membrane. This transporter mediates K^+ uptake in exchange for H^+. However, the pathway of K^+ exit from intercalated cells into the blood is unknown. Reabsorption of K^+ is activated by a low K^+ diet.

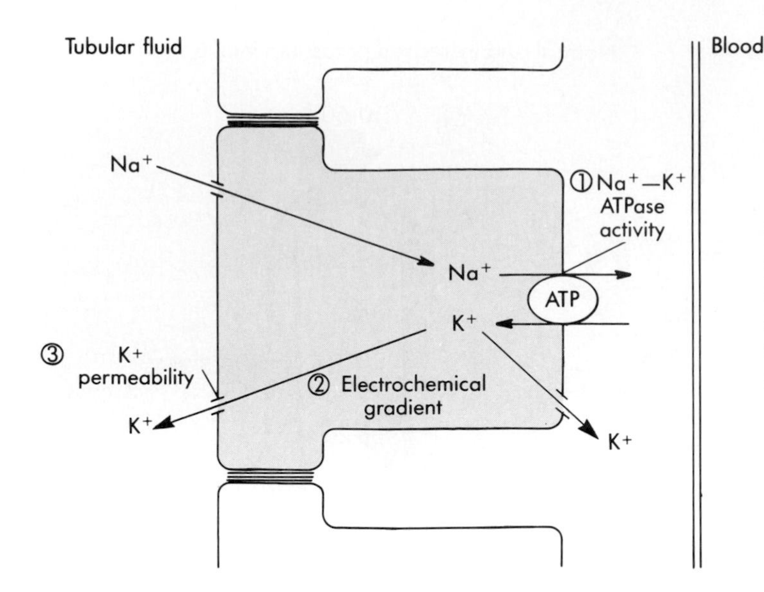

■ **Fig. 43-3** Cellular mechanism of K^+ secretion by the principal cell in the distal tubule and collecting duct. The numbers indicate the sites where K^+ secretion is regulated: (1) Na^+-K^+-ATPase; (2) electrochemical gradient of K^+ across the apical membrane; and (3) K^+ permeability of the apical membrane.

■ *Regulation of K^+ Excretion*

Regulation of K^+ excretion occurs mainly as a result of alterations in K^+ secretion by the principal cells of the distal tubule and collecting duct. Table 43-2 summarizes the major factors and hormones that modulate K^+ secretion by these cells. Plasma [K^+] and aldosterone are the major physiological regulators of K^+ secretion. ADH, the flow rate of tubular fluid, acid-base balance, and the [Na^+] of tubular fluid also modify K^+ secretion. However, they are less important than plasma [K^+] and aldosterone.

Plasma [K^+]. Plasma [K^+] is an important determinant of K^+ secretion by the distal tubule and collecting duct (Fig. 43-4). Hyperkalemia (e.g., high K^+ diet or rhabdomyolysis) rapidly (in minutes) stimulates secretion. Several mechanisms are involved. First hyperkalemia stimulates the Na^+-K^+-ATPase pump and thereby increases K^+uptake across the basolateral membrane. This raises intracellular [K^+] and increases the driving force for K^+ exit across the apical membrane. Second hyperkalemia also increases the permeability of the apical membrane to K^+. Third hyperkalemia stimulates aldosterone secretion by the adrenal cortex, which, as discussed below, acts synergistically with K^+ to stimulate K^+ secretion.

■ **Table 43-2** Major factors and hormones regulating K^+ secretion by the distal tubule and collecting duct

Plasma [K^+]
Aldosterone
Flow rate of tubular fluid
Antidiuretic hormone
Acid-base balance
[Na^+] of tubular fluid

Hypokalemia (e.g., low K^+ diet or diarrhea) decreases K^+ secretion by actions opposite to those described for hyperkalemia. Hence hypokalemia inhibits the Na^+-K^+-ATPase pump, decreases the driving force for K^+ efflux across the apical membrane, reduces the permeability of the apical membrane to K^+, and causes a reduction in plasma [aldosterone].

Aldosterone. Fig. 43-5 illustrates the effect of aldosterone on K^+ secretion by the distal tubule and collecting duct. Aldosterone, in addition to stimulating Na^+ reabsorption, enhances K^+ secretion by increasing the amount of Na^+-K^+-ATPase in principal cells. This elevates cell [K^+]. Aldosterone also increases the driving force for K^+ exit across the apical membrane and increases the permeability of the apical membrane to K^+. Aldosterone secretion is increased by hyperkalemia and angiotensin II (following activation of the renin-angiotensin system) and decreased by hypokalemia and ANP. Stimulation of K^+ secretion by aldosterone occurs after a 1-hour lag period and attains its highest level after 1 day.

Flow rate of tubular fluid. A rise in the flow of tubular fluid (e.g., diuretic treatment, extracellular fluid volume expansion) rapidly stimulates K^+ secretion, whereas a fall in flow (e.g., extracellular fluid volume contraction) reduces K^+ secretion (Fig. 43-6). Increments in tubular fluid flow are more effective in stimulating K^+ secretion as dietary K^+ intake is increased from a low K^+ diet to a high K^+ diet (see Fig. 43-6). Alterations in tubular fluid flow influence K^+ secretion by changing the driving force for K^+ exit across the apical membrane. As potassium is secreted into the tubular fluid, the [K^+] of the fluid increases. This will decrease the driving force for K^+ exit across the apical membrane and thereby reduce the rate of secretion. An

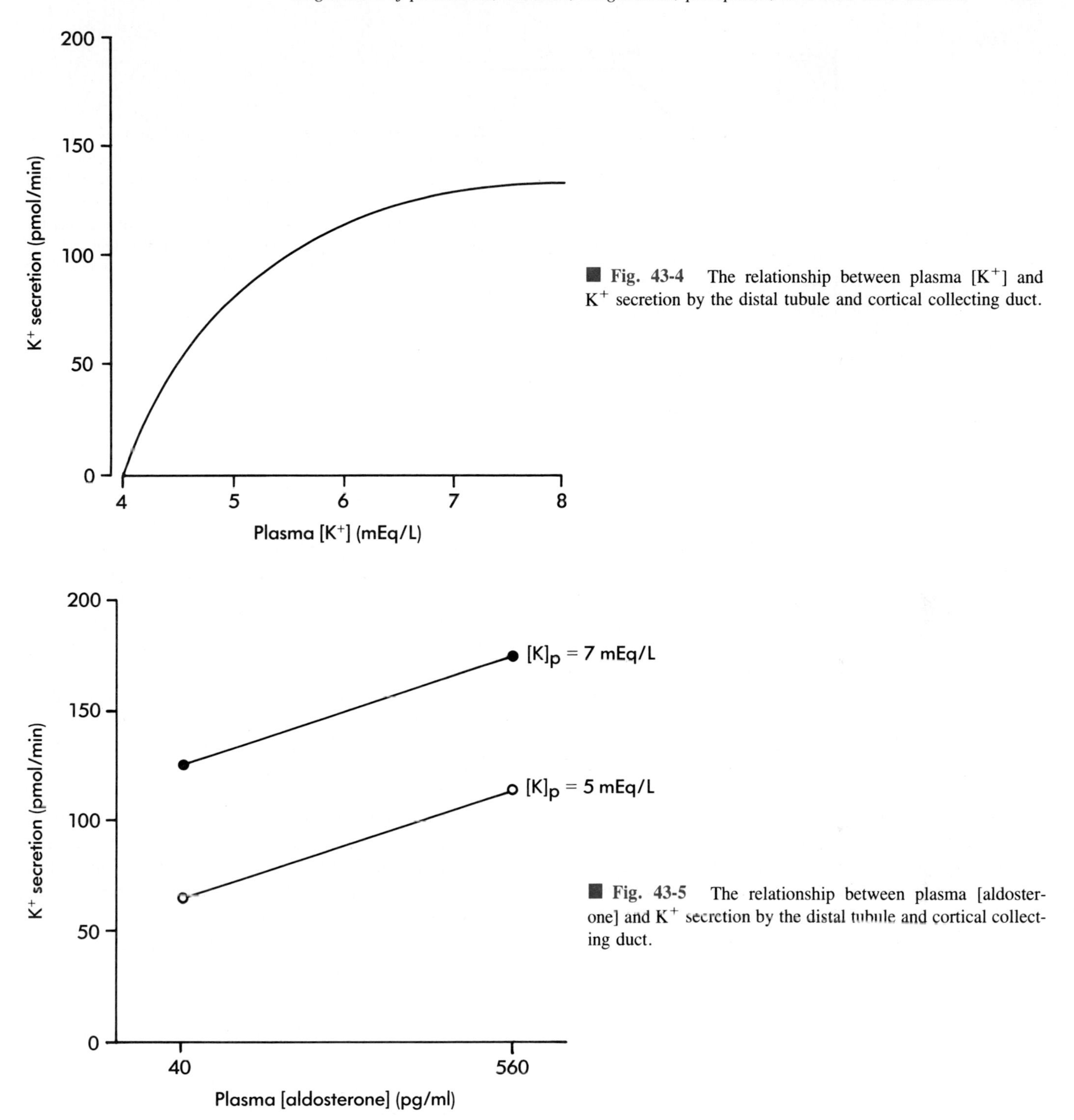

Fig. 43-4 The relationship between plasma $[K^+]$ and K^+ secretion by the distal tubule and cortical collecting duct.

Fig. 43-5 The relationship between plasma [aldosterone] and K^+ secretion by the distal tubule and cortical collecting duct.

increase in tubular fluid flow minimizes the rise in tubular fluid $[K^+]$ as the secreted K^+ is washed downstream. As a result, K^+ secretion is stimulated by an increase in the flow of tubular fluid. *Because diuretic drugs increase the flow of tubular fluid through the distal tubule and collecting duct, they also enhance urinary K^+ excretion.* In contrast a decline in tubular fluid flow inhibits K^+ secretion. A decline in the tubular fluid flow facilitates the rise in tubular fluid $[K^+]$ and thereby reduces secretion.

Antidiuretic hormone (ADH). ADH increases the driving force for K^+ exit across the apical membrane of principal cells in the distal tubule and collecting duct. However, because ADH also decreases tubular fluid flow, which reduces K^+ secretion, it does not change K^+ secretion by the distal tubule and collecting duct. Changes in ADH levels do not alter urinary K^+ excretion because the stimulatory effect of ADH on K^+ secretion is offset by the inhibitory effect of decreased tu-

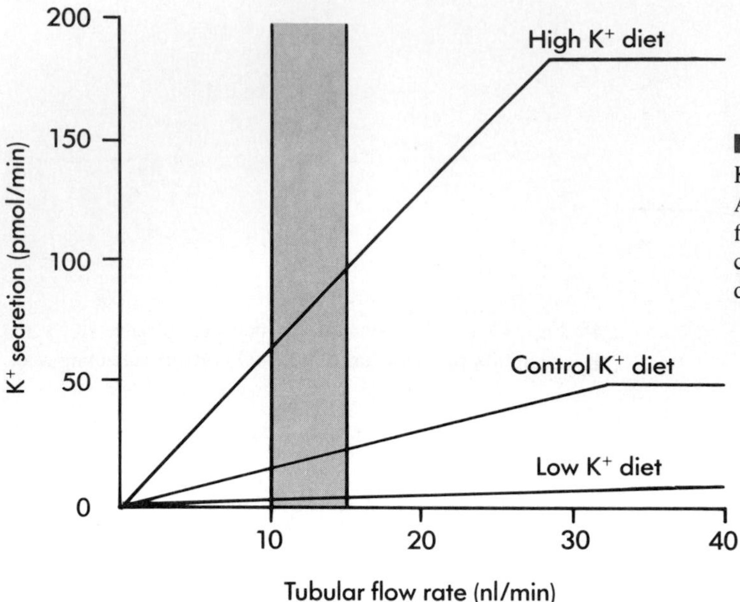

■ **Fig. 43-6** Relationship between tubular flow rate and K^+ secretion by the distal tubule and cortical collecting duct. A high K^+ diet increases the slope of the relationship between flow rate and secretion and increases the maximum rate of secretion. A low K^+ diet has the opposite effect. Shaded bar indicates the flow rate under most physiological conditions.

bular fluid flow (Fig. 43-7). If ADH did not increase the driving force favoring K^+ secretion, urinary K^+ excretion would fall as ADH levels increase and K^+ balance would be modulated by alterations in water balance. Thus these opposing effects of ADH on K^+ secretion enable urinary K^+ excretion to be maintained constant despite wide fluctuations in water excretion.

Acid-base balance. Another factor that modulates K^+ secretion is the $[H^+]$ of the extracellular fluid. Acute alterations (over a period of minutes to hours) in the pH of the plasma influence K^+ secretion by the distal tubule and collecting duct. Alkalosis increases secretion, and acidosis decreases secretion (Fig. 43-8; see end of chapter). Acidosis reduces K^+ secretion by two mechanisms: (1) it inhibits the Na^+-K^+-ATPase pump, reduces cell $[K^+]$, and thereby reduces the driving force for K^+ exit across the apical membrane, and (2) it reduces the permeability of the apical membrane to K^+ (see Fig. 43-3). Alkalosis has the opposite effect.

The effect of metabolic acidosis on K^+ excretion is time-dependent. When a metabolic acidosis is prolonged for several days, urinary K^+ excretion is stimulated. This occurs because chronic metabolic acidosis inhibits water and NaCl reabsorption by the proximal tubule and thereby increases the flow of tubular fluid through the distal tubule and collecting duct. This rise in tubular fluid flow offsets the effects of acidosis on cell $[K^+]$ and apical membrane permeability such that K^+ secretion rises. Thus metabolic acidosis may either inhibit or stimulate potassium excretion depending on the duration of the disturbance.

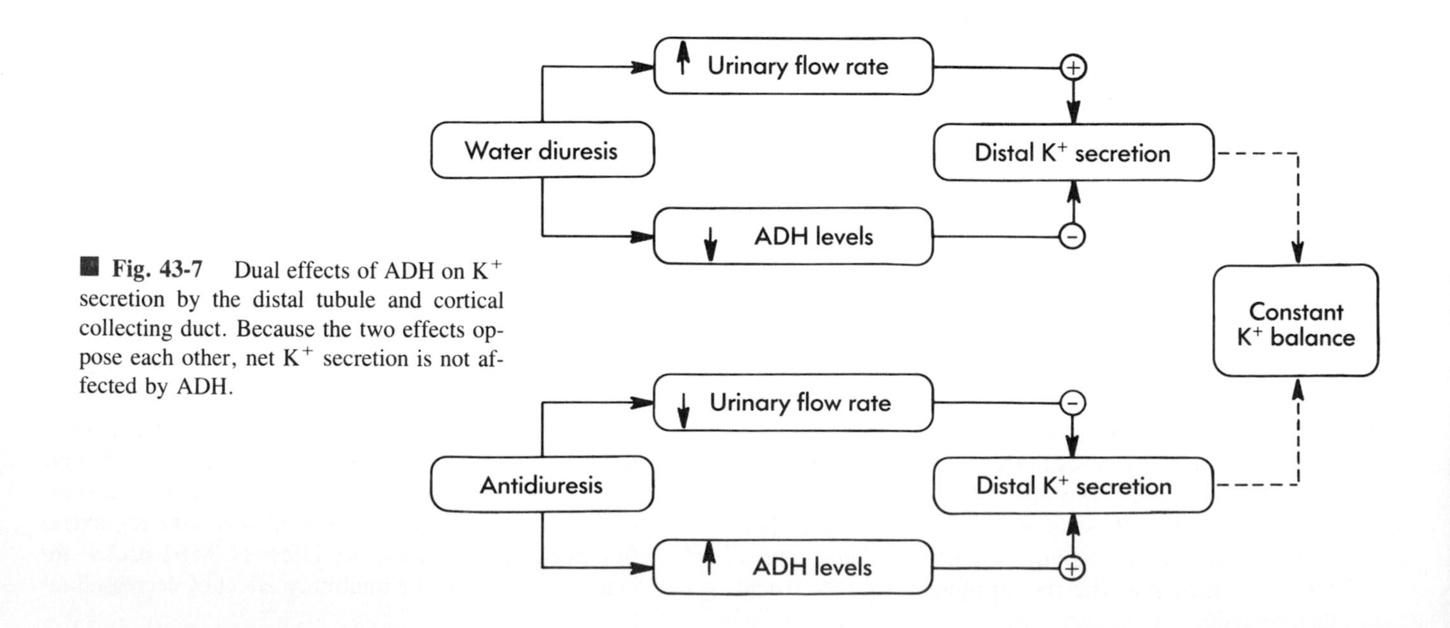

■ **Fig. 43-7** Dual effects of ADH on K^+ secretion by the distal tubule and cortical collecting duct. Because the two effects oppose each other, net K^+ secretion is not affected by ADH.

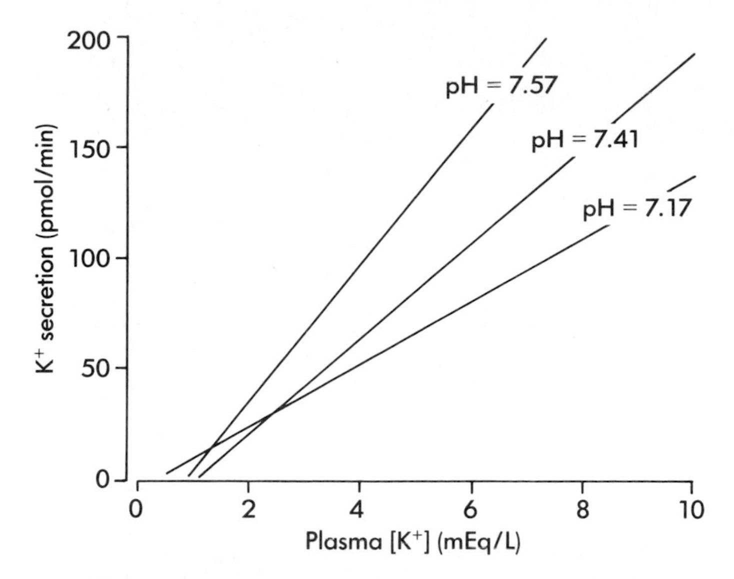

Fig. 43-8 Effect of plasma pH on the relationship between plasma [K^+] and K^+ secretion by the distal tubule and cortical collecting duct.

[Na^+] of tubular fluid. The [Na^+] of tubular fluid can, under some conditions, be an important determinant of K^+ secretion. A rise in [Na^+] stimulates secretion, whereas a fall in concentration has the opposite effect. An increase in the [Na^+] of the tubular fluid enhances Na^+ movement across the apical membrane of principal cells and increases intracellular [Na^+]. This stimulates the Na^+-K^+-ATPase pump and accelerates K^+ uptake across the basolateral membrane. The rise in intracellular [K^+] in turn increases the favorable driving force for K^+ exit across the apical membrane and thereby enhances K^+ secretion. The opposite occurs with a decrease in tubular fluid [Na^+].

Multivalent Ions

Ca^{++}, Mg^{++}, and inorganic phosphate ($PO_4^{\equiv}$)*, are multivalent ions that subserve many complex and vital functions. In a normal adult the renal excretion of these ions is balanced by gastrointestinal absorption. If body stores decline substantially, gastrointestinal absorption, bone resorption, and renal tubular reabsorption increase and return body stores to normal levels. During growth and during pregnancy, intestinal absorption exceeds urinary excretion, and these ions accumulate in newly formed fetal tissue and bone. In contrast bone disease (e.g., **osteoporosis,** the decreased mineralization of bone) or a decline in lean body mass increases urinary mineral loss without a change in intestinal absorption. In these conditions a net loss of Ca^{++}, Mg^{++}, and $PO_4^{\equiv}$ from the body occurs.

*At physiological pH phosphate exists as $HPO_4^{=}$ and $H_2PO_4^-$ (pK = 6.8). For simplicity we collectively refer to these ion species as $PO_4^{\equiv}$.

Table 43-3 Body content and distribution of Ca^{++}, Mg^{++}, and $PO_4^{\equiv}$

		Compartment		
Ion	*Body content*	*Bone*	*Intracellular*	*Extracellular*
Ca^{++}	1,300 gm	99%	1%	0.10%
Mg^{++}	26 gm	54%	45%	1.00%
$PO_4^{\equiv}$	700 gm	86%	14%	0.03%

From this brief introduction it is evident that the kidneys, in conjunction with the gastrointestinal tract and bone, play a major role in maintaining Ca^{++}, Mg^{++}, and $PO_4^{\equiv}$ homeostasis. Accordingly, this section will discuss Ca^{++}, Mg^{++}, and $PO_4^{\equiv}$ handling by the kidney with an emphasis on the hormones and factors that regulate urinary excretion.

Ca^{++}

Calcium ions play a major role in many processes, including bone formation, cell division and growth, blood coagulation, hormone-response coupling, and electrical stimulus-response coupling (such as muscle contraction and neurotransmitter release). Ninety-nine percent of Ca^{++} is stored in bone, 1% is found in the intracellular fluid, and 0.1% is located in the extracellular fluid (Table 43-3). The total [Ca^{++}] in plasma is about 2.5 mM mEq/L, and its concentration is normally maintained within very narrow limits. A low [Ca^{++}] increases the excitability of nerve and muscle cells. Diseases that lower plasma [Ca^{++}] cause hypocalcemic tetany, which is characterized by spasms of skeletal muscle. Hypercalcemia causes cardiac arrhythmias and decreased neuromuscular excitability.

Overview of Ca^{++} homeostasis. The maintenance of Ca^{++} homeostasis is a function of two variables: (1) the total amount of Ca^{++} in the body, and (2) the distribution of Ca^{++} between the intracellular and extracellular fluid compartments. Total body Ca^{++} is determined by the relative amounts of Ca^{++} absorbed by the gastrointestinal tract and excreted by the kidneys (Fig. 43-9). Ca^{++} absorption by the gastrointestinal tract is mediated by an active, carrier-mediated transport mechanism that is stimulated by 1,25-dihydroxyvitamin D_3 (1,25[OH]$_2D_3$) (see Chapter 47). Net Ca^{++} absorption is normally 175 mg/day, but it can increase to 600 mg/day when 1,25(OH)$_2D_3$ levels rise. Ca^{++} excretion by the kidneys is equal to the amount absorbed by the gastrointestinal tract (175 mg/day), and changes in parallel with reabsorption of Ca^{++} by the gastrointestinal tract. Thus Ca^{++} balance is maintained because the amount of Ca^{++} ingested in an average diet (1,000 mg/day) is equal to the amount lost in the feces (825 mg/day: the fraction that escapes reabsorption by the gastrointestinal

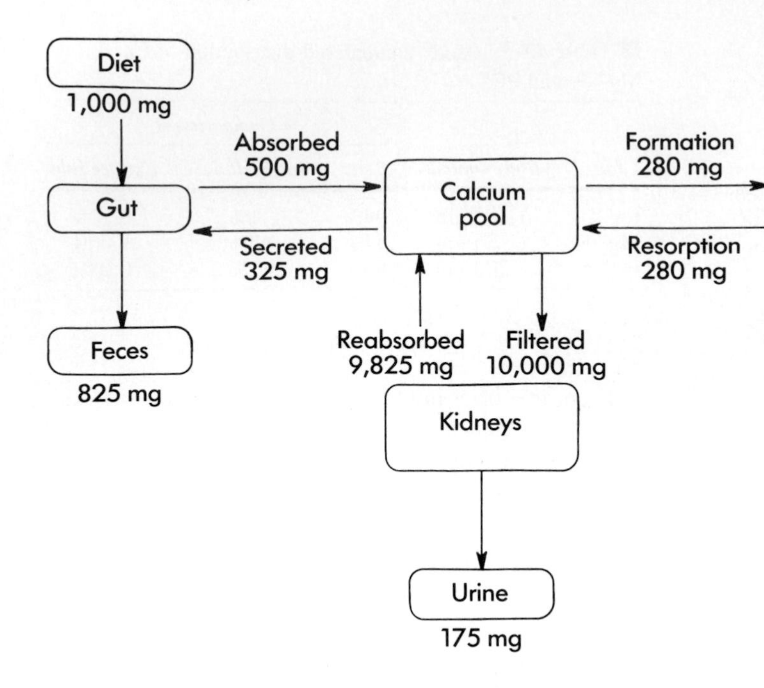

■ **Fig. 43-9** Overview of Ca^{++} homeostasis. See text for details.

tract) plus the amount excreted in the urine (175 mg/day).

The other variable that controls Ca^{++} homeostasis is the distribution of Ca^{++} between bone and the extracellular fluid. Two hormones, parathyroid hormone (PTH) and $1,25(OH)_2D_3$, are the most important hormones controlling this variable and thereby regulate plasma $[Ca^{++}]$ (see Chapter 47). PTH is secreted by the parathyroid glands, and its secretion is stimulated by a decline in plasma $[Ca^{++}]$ (i.e., hypocalcemia). PTH increases plasma $[Ca^{++}]$ by:

1. Stimulating bone resorption.
2. Increasing Ca^{++} reabsorption by the kidneys.
3. Stimulating the production of $1,25(OH)_2D_3$ which in turn increases Ca^{++} absorption by the gastrointestinal tract and stimulates bone resorption.

Hypercalcemia reduces PTH secretion, which leads to actions opposite to those described previously.

Ca^{++} transport along the nephron. Approximately 50% of the Ca^{++} in plasma is ionized, 45% is bound to plasma proteins (mainly albumin), and 5% is complexed to several anions, including HCO_3^-, citrate, $PO_4^{\equiv}$ and $SO_4^=$ (Table 43-4). The pH of the plasma influences this distribution. Acidosis increases the percentage of ionized calcium at the expense of Ca^{++} bound to proteins, whereas alkalosis decreases the percentage of ionized calcium, again by altering Ca^{++} bound to proteins. Ca^{++} available for filtration consists of the ionized fraction and that complexed with anions. Thus about 55% of the Ca^{++} in the plasma is available for glomerular filtration.

Normally, 99% of the filtered Ca^{++} is reabsorbed by the nephron (Fig. 43-10). The proximal tubule reabsorbs 70% of the filtered Ca^{++}. Another 20% is reabsorbed in the loop of Henle (mainly the thick ascending

■ **Table 43-4** Forms of Ca^{++}, Mg^{++}, and $PO_4^{\equiv}$ in plasma

		Percentage of total		
Ion	*mM*	*Ionized (%)*	*Protein bound (%)*	*Complexed (%)*
Ca^{++}	2.5	50	45	5
Mg^{++}	1.0	55	30	15
$PO_4^{\equiv}$	1.3	84	10	6

Ca^{++} and Mg^{++} are bound (i.e., complexed) to various anions in the plasma, including HCO_3^-, citrate, $PO_4^{\equiv}$ and $SO_4^=$. $PO_4^{\equiv}$ is complexed to various cations including Na^+ and K^+.

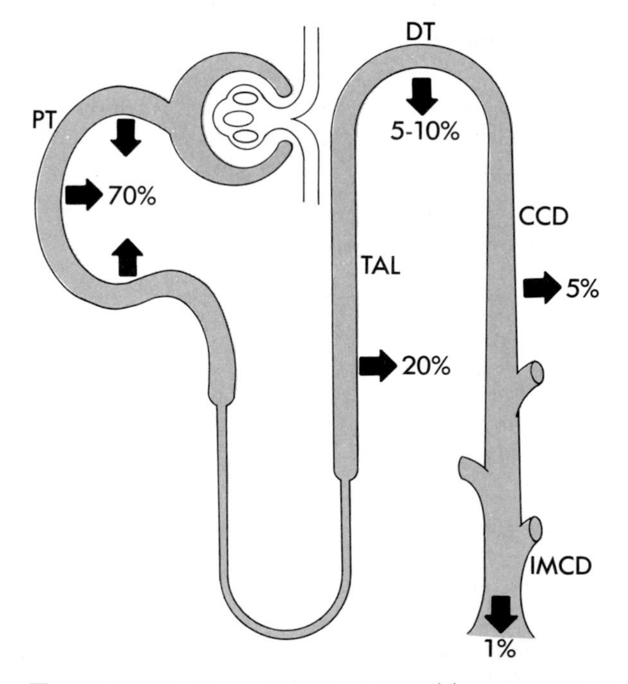

■ **Fig. 43-10** Transport pattern of Ca^{++} along the nephron. Percentages refer to the amount of the filtered Ca^{++} reabsorbed by each nephron segment. Approximately 1% of the filtered Ca^{++} is excreted. *PT,* proximal tubule; *TAL,* thick ascending limb; *DT,* distal tubule; *CCD,* cortical collecting duct; *IMCD,* inner medullary collecting duct.

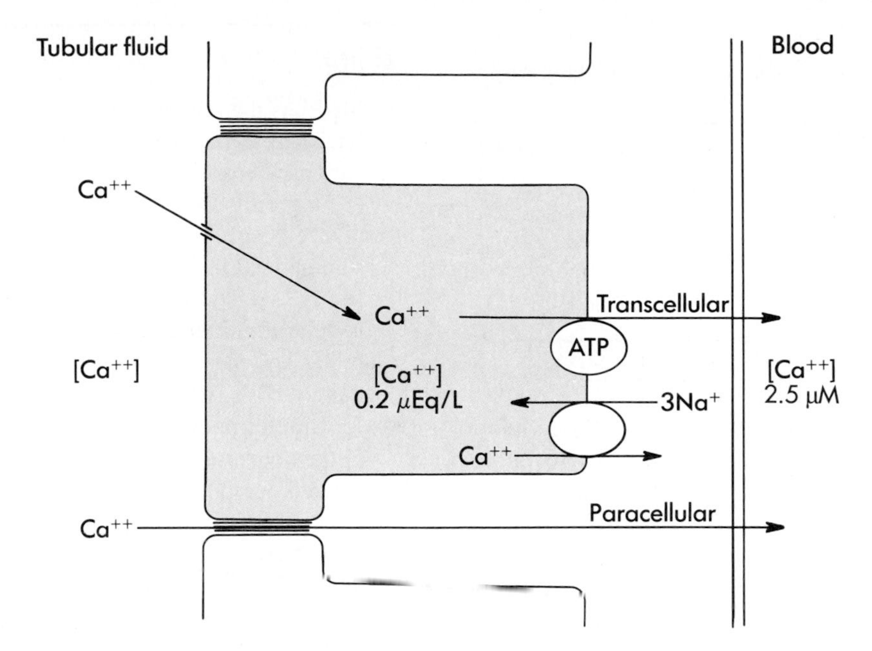

■ **Fig. 43-11** Cellular scheme of Ca^{++} reabsorption by the proximal tubule. Ca^{++} is reabsorbed by transcellular and paracellular routes. The mechanism of Ca^{++} diffusion into the cell across the apical membrane has not been characterized but is likely to occur via Ca^{++} channels.

limb), 5% to 10% is reabsorbed by the distal tubule, and <5% is reabsorbed by the collecting duct. About 1% (8.6 mEq/day or 175 mg/day) is excreted in the urine. This fraction is equal to the net amount absorbed daily by the gastrointestinal tract.

Cellular mechanisms of Ca^{++} reabsorption. Ca^{++} reabsorption by the proximal tubule occurs by two pathways: transcellular and paracellular (Fig. 43-11). Ca^{++} reabsorption across the cellular pathway (i.e., transcellular) accounts for one-third of proximal reabsorption. Ca^{++} reabsorption through the cell is an active process, and it occurs in two steps. Ca^{++} diffuses across the apical membrane into the cell down its electrochemical gradient. This gradient is exceptionally steep because the cell Ca^{++} concentration is only 0.1μM/L, about 10,000-fold less than that in the tubular fluid (1.5mM). The cell interior is electrically negative with respect to the luminal side of the apical membrane, which also favors Ca^{++} entry into the cell. Ca^{++} is extruded across the basolateral membrane against its electrochemical gradient. The mechanism for the active extrusion of Ca^{++} has not been identified, but it could occur by either a Ca^{++}-ATPase or a $3Na^{+}$-Ca^{++} antiporter (see Chapter 24). Two-thirds of Ca^{++} is reabsorbed between cells across the tight junctions (i.e., paracellular pathway). This passive paracellular reabsorption of Ca^{++} occurs by solvent drag along the entire length of the proximal tubule (see p. 741), and is also driven by the positive luminal voltage in the second half of the proximal tubule (see p. 742).

Ca^{++} reabsorption by Henle's loop is restricted to the thick ascending limb. Ca^{++} is reabsorbed via a cellular and a paracellular route by mechanisms similar to those described for the proximal tubule, except that Ca^{++} is not reabsorbed by solvent drag in this segment (recall that the thick ascending limb is impermeable to water). In both the proximal tubule and the thick ascending limb, Ca^{++} and Na^{+} reabsorption parallel each other because of the significant component of Ca^{++} reabsorption that occurs by passive, paracellular mechanisms secondary to Na^{+} reabsorption and by the generation of the lumen-positive transepithelial voltages in these segments. Therefore *changes in Na^{+} reabsorption will also result in parallel changes in Ca^{++} reabsorption by the proximal tubule and thick ascending limb.*

In the distal tubule and collecting duct, where the voltage in the tubule lumen is electrically negative with respect to the blood, Ca^{++} reabsorption is entirely active because Ca^{++} is reabsorbed against its electrochemical gradient. The mechanisms of Ca^{++} reabsorption by these segments have not been elucidated. *Because the reabsorption of Ca^{++} and Na^{+} by the distal tubule and the collecting duct are independent and are differentially regulated, urinary Ca^{++} and Na^{+} excretion do not always change in parallel.*

Regulation of Ca^{++} excretion. The following box summarizes the hormones and factors regulating urinary Ca^{++} excretion. **PTH** exerts the most powerful control on renal Ca^{++} excretion. This hormone stimulates overall Ca^{++} reabsorption. Although PTH inhibits reabsorption of fluid, and therefore of Ca^{++} by the proximal tubule, it dramatically stimulates Ca^{++} reabsorp-

■ Table 43-5 Hormones and factors influencing urinary Ca^{++} excretion

Increase excretion	Decrease excretion
Decrease in [PTH]	Increase in [PTH]
ECF expansion	ECF contraction
$PO_4^{=}$ depletion	$PO_4^{=}$ loading
Metabolic acidosis	Metabolic alkalosis
	$1,25(OH)_3$

tion by the thick ascending limb of Henle's loop and the distal tubule. As a result, urinary Ca^{++} excretion declines. An increase in plasma [$PO_4^{=}$] (e.g., increased dietary intake of $PO_4^{=}$ or ingestion of large amounts of $PO_4^{=}$ -containing antacids) elevates PTH levels and thereby decreases Ca^{++} excretion, whereas a decline in plasma [$PO_4^{=}$] (e.g., dietary $PO_4^{=}$ depletion) has the opposite effect.

Changes in the extracellular fluid volume alter Ca^{++} excretion mainly by affecting Na^+ and fluid reabsorption in the proximal tubule. Contraction of the extracellular fluid volume increases Na^+ and water reabsorption by the proximal tubule and thereby enhances Ca^{++} reabsorption. Accordingly, urinary Ca^{++} excretion declines. Expansion of the extracellular fluid volume has the opposite effect.

Acidosis increases Ca^{++} excretion, whereas alkalosis decreases excretion. The regulation of Ca^{++} reabsorption by pH occurs in the distal tubule by an unknown mechanism.

Finally, $1,25(OH)_2D_3$ increases Ca^{++} reabsorption by the distal tubule and thereby decreases Ca^{++} excretion.

■ *Mg^{++}*

Mg^{++} is the second most abundant intracellular electrolyte, and it has many biochemical roles, including activation of enzymes and regulation of protein synthesis. It is also important for bone formation. The distribution of Mg^{++} in the body and its forms in the plasma are summarized in Tables 43-3 and 43-4. Fifty-four percent of Mg^{++} is located in bone, 45% is located in the intracellular fluid, and 1% is located in the extracellular fluid. The plasma [Mg^{++}] is about 1 mM/L. Approximately 30% is protein-bound and therefore unavailable for ultrafiltration by the glomerulus. The Mg^{++} that is filtered consists of an ionized fraction (55%) and a nonionized component (15%) that is complexed to HCO_3^-, citrate, $HPO_4^{=}$, and $SO_4^{=}$. Accordingly, the Mg^{++} concentration in the glomerular ultrafiltrate is 30% less than the Mg^{++} concentration in the plasma.

Overview of Mg^{++} homeostasis. A general scheme of Mg^{++} homeostasis is illustrated in Fig. 43-12. As with Ca^{++}, the maintenance of [Mg^{++}] in body fluids is a function of two variables: (1) the total amount of Mg^{++} in the body and (2) the distribution of Mg^{++} between the intracellular and extracellular fluid compartments. Total body Mg^{++} is determined by the relative amount of net Mg^{++} absorption by the gastrointestinal tract and excretion by the kidneys. Body Mg^{++} balance is maintained by the kidneys because of their ability to excrete in the urine an amount of Mg^{++} equal to the amount absorbed by the gastrointestinal tract (i.e., 75 mg/day). Although the gastrointestinal absorption of Mg^{++} is not regulated as closely as that of Ca^{++}, when dietary intake is restricted, intestinal absorption rises. The hormones and factors responsible for the regulation of intestinal Mg^{++} absorption are unknown.

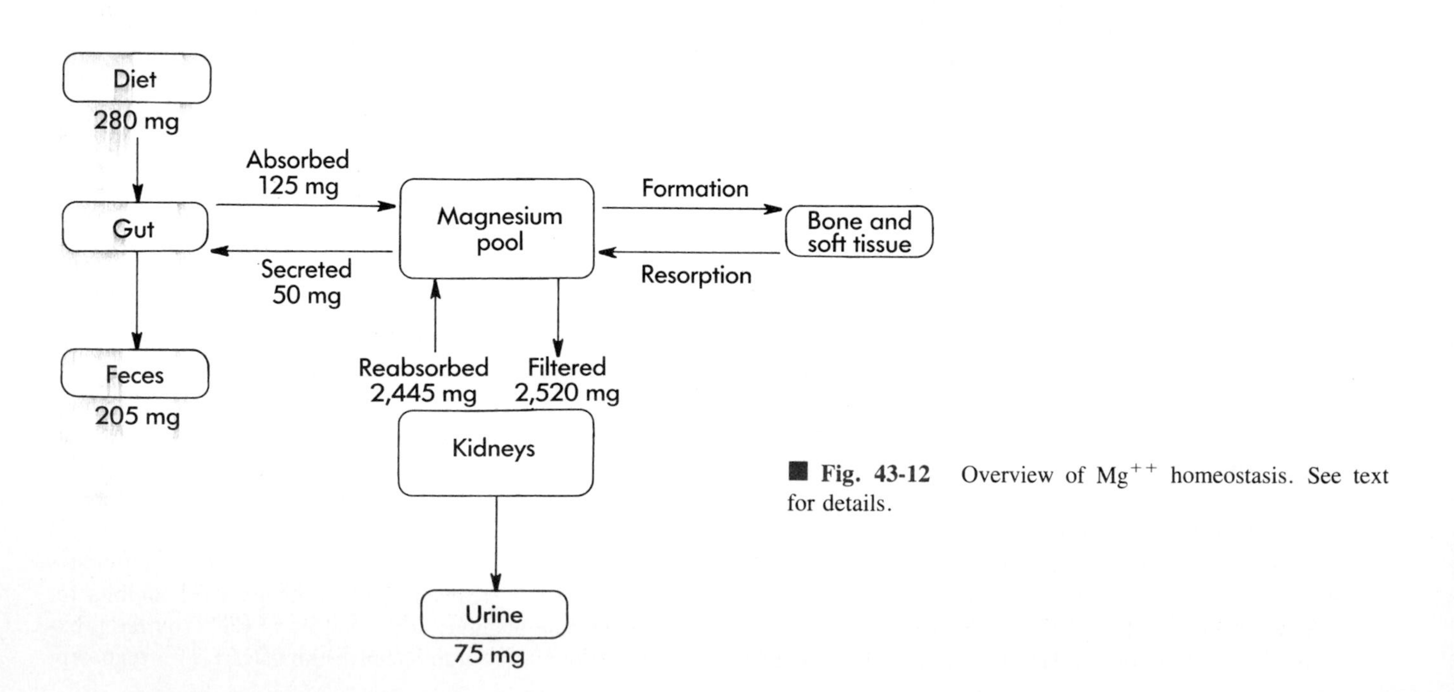

■ Fig. 43-12 Overview of Mg^{++} homeostasis. See text for details.

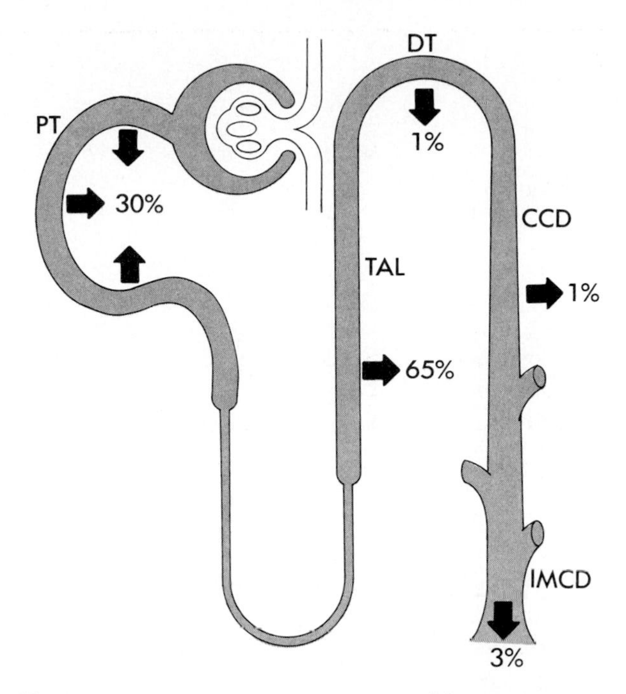

Fig. 43-13 Transport pattern of Mg^{++} along the nephron. Mg^{++} is reabsorbed mainly by the proximal tubule and the thick ascending limb of Henle's loop. Percentages refer to the amount of the filtered Mg^{++} reabsorbed by each nephron segment. Approximately 3% of the filtered Mg^{++} is excreted. *PT,* proximal tubule; *TAL,* thick ascending limb; *DT,* distal tubule; *CCD,* cortical collecting duct; *IMCD,* inner medullary collecting duct.

The other variable that regulates Mg^{++} homeostasis is the distribution of Mg^{++} between the intracellular fluid and the extracellular fluid. The hormones and factors that regulate this distribution have not been identified.

Mg^{++} transport along the nephron. The kidneys play a vital role in Mg^{++} homeostasis, and they excrete 3% of the filtered Mg^{++}. Mg^{++} reabsorption by the nephron exhibits a T_m that is increased by PTH. The pattern of Mg^{++} transport along the nephron is shown in Fig. 43-13. Approximately 30% of the filtered Mg^{++} is reabsorbed by the proximal tubule. The thick ascending limb of Henle's loop is the major site of Mg^{++} transport; it reabsorbs 65% of the filtered Mg^{++}. Reabsorption is passive and is driven by the lumen-positive transepithelial voltage across this nephron segment. *Alterations in urinary Mg^{++} excretion usually arise from changes in Mg^{++} reabsorption by the thick ascending limb.* Very little Mg^{++} is reabsorbed by the distal tubule and collecting duct (<2%). The mechanisms of Mg^{++} transport along the nephron are poorly understood. However, reabsorption is thought to be passive and to be driven by the lumen-positive transepithelial voltage in the second half of the proximal tubule and the thick ascending limb.

Regulation of Mg^{++} excretion. Table 43-6 summarizes the major hormones and factors that regulate

Table 43-6 Hormones and factors influencing urinary Mg^{++} excretion

Increase excretion	Decrease excretion
Hypercalcemia	Hypocalcemia
Hypermagnesemia	Hypomagnesemia
ECF expansion	ECF contraction
Decrease of [PTH]	Increase of [PTH]
Acidosis	Alkalosis

Mg^{++} excretion. An increase in excretion is caused by hypercalcemia, hypermagnesemia, extracellular fluid volume expansion, acidosis, and a decrease in PTH levels. A decrease in excretion is caused by hypocalcemia, hypomagnesemia, extracellular volume contraction, alkalosis, and an increase in PTH levels. All of these regulatory influences have a direct effect on Na^+ reabsorption and thus the transepithelial voltage. By virtue of changing the transepithelial voltage, these regulatory influences modulate Mg^{++} reabsorption across the paracellular pathway. Also, alterations in extracellular fluid volume modulate Na^+ and water reabsorption and hence influence Mg^{++} reabsorption by affecting solvent drag and paracellular diffusion of Mg^{++}.

$PO_4^{\equiv}$

$PO_4^{\equiv}$ is an important component of many organic molecules, including DNA, RNA, ATP, and intermediates of metabolic pathways. It is also a major constituent of bone. Its concentration in plasma is an important determinant in bone formation and resorption. In addition urinary $PO_4^{\equiv}$ is an important buffer (titratable acid) for the maintenance of acid-base balance. Eighty-six percent of $PO_4^{\equiv}$ is located in bone, 14% is located in the intracellular fluid, and 0.03% is located in the extracellular fluid (see Table 43-3). The plasma $[PO_4^{\equiv}]$ is 1 mM. The forms of $PO_4^{\equiv}$ in the plasma are summarized in Table 43-4. Approximately 10% of the $PO_4^{\equiv}$ in the plasma is protein-bound and therefore unavailable for ultrafiltration by the glomerulus. Accordingly, the $[PO_4^{\equiv}]$ in the ultrafiltrate is 10% less than that in plasma.

Overview of $PO_4^{\equiv}$ homeostasis. A general scheme of $PO_4^{\equiv}$ homeostasis is shown in Fig. 43-14. The maintenance of $PO_4^{\equiv}$ homeostasis is a function of two variables: (1) the amount of $PO_4^{\equiv}$ in the body and (2) the distribution of $PO_4^{\equiv}$ between the intracellular fluid and the extracellular fluid compartments. Total body $PO_4^{\equiv}$ is determined by the relative amount of $PO_4^{\equiv}$ absorbed by the gastrointestinal tract versus the amount excreted by the kidneys. $PO_4^{\equiv}$ absorption by the gastrointestinal tract occurs by active and passive mechanisms; it increases as dietary $PO_4^{\equiv}$ rises, and it is stimulated by $1,25(OH)_2D_3$.

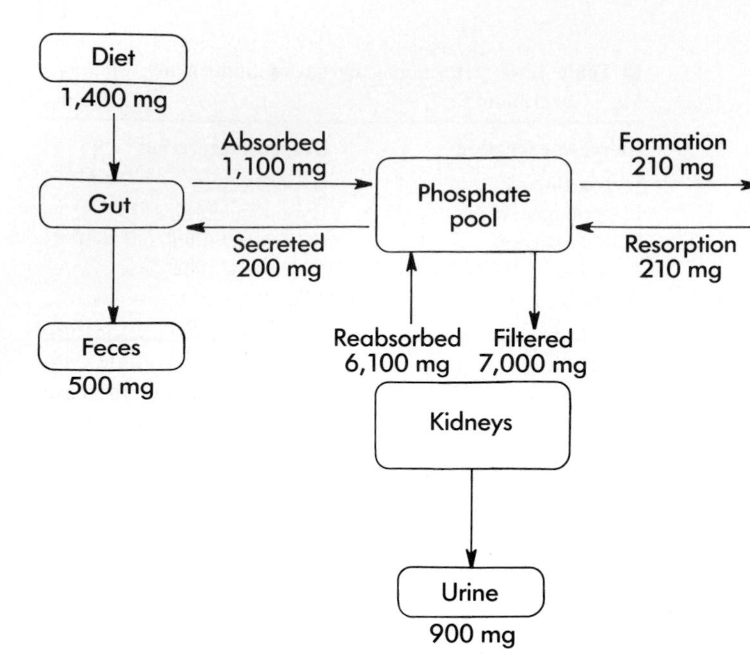

Fig. 43-14 Overview of $PO_4^{=}$ homeostasis. See text for details.

Despite changes in $PO_4^{=}$ intake between 800 and 1,500 mg/day, total body $PO_4^{=}$ balance is maintained by the kidneys, which excrete an amount of $PO_4^{=}$ in the urine equal to the amount absorbed by the gastrointestinal tract. Thus, the kidneys play a vital role in maintaining $PO_4^{=}$ homeostasis.

The second variable that maintains $PO_4^{=}$ homeostasis is the distribution of $PO_4^{=}$ between bone and the intracellular and extracellular fluid compartments. The release of $PO_4^{=}$ from intracellular stores is stimulated by the same hormones (PTH and $1,25[OH]_2D_3$) that release Ca^{++} from this pool. Thus the release of $PO_4^{=}$ is always accompanied by a release of Ca^{++}. The kidneys also contribute importantly to the regulation of plasma $[PO_4^{=}]$. This can be illustrated by considering the $PO_4^{=}$ titration curve in (Fig. 43-15). Recalling the glucose titration curve presented in Chapter 41, it should be evident that the tubular mechanisms for $PO_4^{=}$ reabsorption share many properties with glucose transport. These properties include a T_m, a splay, and a threshold. However, the titration curves for glucose and $PO_4^{=}$ differ significantly in several respects. First the T_m for $PO_4^{=}$ is only slightly above the normal filtered load. Hence a small increase in plasma $[PO_4^{=}]$ increases the filtered load such that the T_m is exceeded and urinary $PO_4^{=}$ excretion rises (see Fig. 43-15). This in turn causes plasma $[PO_4^{=}]$ to fall. Accordingly the kidneys regulate plasma $[PO_4^{=}]$. A second unique aspect of the $PO_4^{=}$ titration curve is that the T_m for $PO_4^{=}$ is variable and is regulated by dietary $PO_4^{=}$ intake. A high $PO_4^{=}$ diet decreases the T_m, and a low $PO_4^{=}$ diet increases the T_m. This effect of dietary $PO_4^{=}$ intake on the T_m is independent of changes in PTH levels.

$PO_4^{=}$ transport along the nephron. Fig. 43-16 illustrates the transport pattern of $PO_4^{=}$ along the nephron. The proximal tubule reabsorbs 80% of the $PO_4^{=}$ filtered by the glomerulus, and the distal tubule reabsorbs 10%. In contrast the loop of Henle and the collecting duct reabsorb negligible amounts of $PO_4^{=}$. Therefore 10% of the filtered load of $PO_4^{=}$ is excreted.

$PO_4^{=}$ reabsorption by the proximal tubule occurs mainly, if not exclusively by a transcellular route (Fig. 43-17). $PO_4^{=}$ uptake across the apical membrane occurs by a $2Na^+$-$PO_4^{=}$ symport mechanism. $PO_4^{=}$ exits across the basolateral membrane by a $PO_4^{=}$-anion antiporter. The cellular mechanisms of $PO_4^{=}$ reabsorption by the distal tubule and collecting duct have not been characterized.

Regulation of $PO_4^{=}$ excretion. Table 43-7 summarizes the major hormones and factors that regulate $PO_4^{=}$ excretion. All act on the proximal tubule and either stimulate or inhibit $PO_4^{=}$ reabsorption. PTH is the most

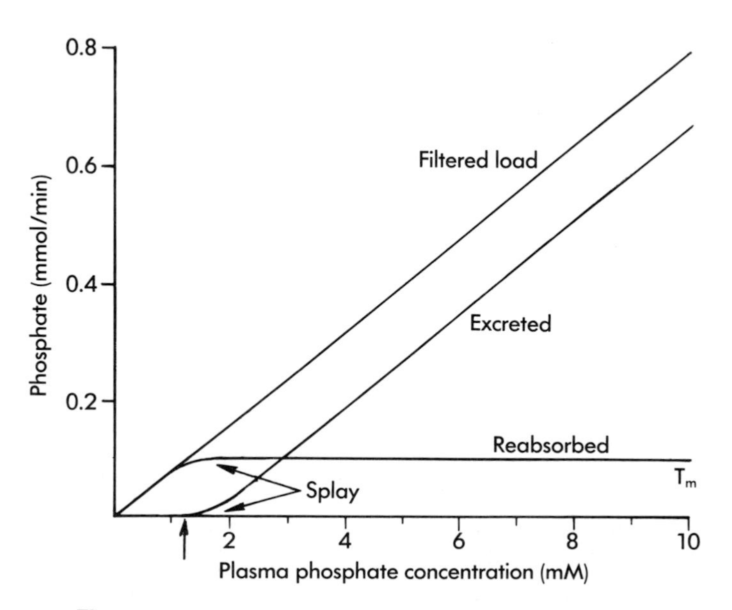

Fig. 43-15 Titration curve for $PO_4^{=}$. The amount of filtered $PO_4^{=}$ that was reabsorbed is calculated as the difference between the amount filtered minus the amount excreted. The arrow pointing to the X-axis indicates the normal plasma $[PO_4^{=}]$. T_m, transport maximum.

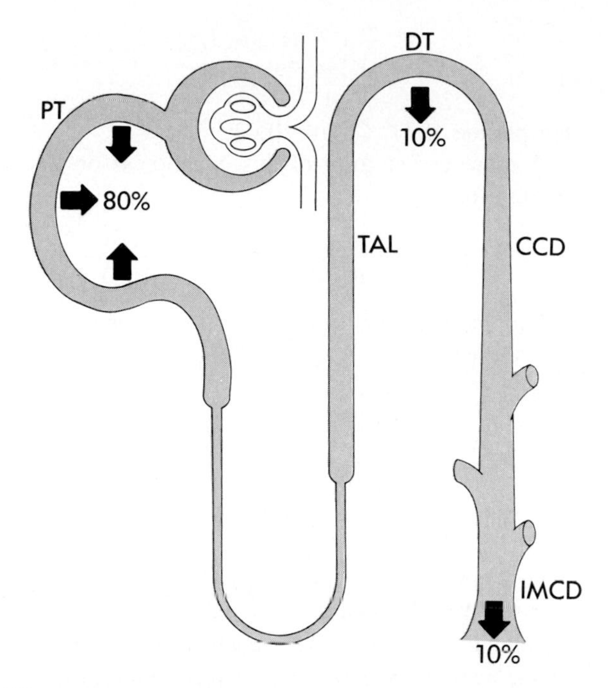

■ **Fig. 43-16** Transport pattern of $PO_4^=$ along the nephron. $PO_4^=$ is reabsorbed primarily by the proximal tubule and distal tubule. Percentages refer to the amount of the filtered $PO_4^=$ reabsorbed by each nephron segment. Approximately 10% of the filtered $PO_4^=$ is excreted. *PT*, proximal tubule; *TAL*, thick ascending limb; *DT*, distal tubule; *CCD*, cortical collecting duct; IMCD, inner medullary collecting duct.

important hormone that controls $PO_4^=$ excretion. PTH increases cAMP production and inhibits $PO_4^=$ reabsorption by the proximal tubule. Dietary $PO_4^=$ intake also regulates $PO_4^=$ excretion by mechanisms unrelated to changes in PTH levels. $PO_4^=$ loading increases excretion, whereas $PO_4^=$ depletion decreases excretion. Changes in dietary $PO_4^=$ intake modulate $PO_4^=$ transport by altering the transport rate of each $2Na^+$-$PO_4^=$ symporter without changing the number of transporters.

Extracellular fluid volume also affects $PO_4^=$ excretion: volume expansion increases excretion, and volume contraction decreases excretion. Glucocorticoids increase the excretion of $PO_4^=$. Glucocorticoids increase the delivery of $PO_4^=$ to the distal tubule and collecting duct by inhibiting proximal tubular $PO_4^=$ reabsorption. This inhibition enables the distal tubule and collecting duct to secrete more H^+ and to generate more HCO_3^-. $PO_4^=$ is an important urinary buffer, and as described in the next section, is termed titratable acid. For example, in the absence of glucocorticoids (e.g., Addison's disease), $PO_4^=$ excretion is depressed, as is the ability of the kidneys to excrete titratable acid and to generate new HCO_3^-. Acid-base balance also influences $PO_4^=$ excretion; acidosis increases $PO_4^=$ excretion, and alkalosis decreases $PO_4^=$ excretion.

■ *Renal Regulation of Acid-Base Balance*

The concentration of H^+ in the body fluids is low compared with that of other ions. For example, Na^+ is present at a concentration some three million times

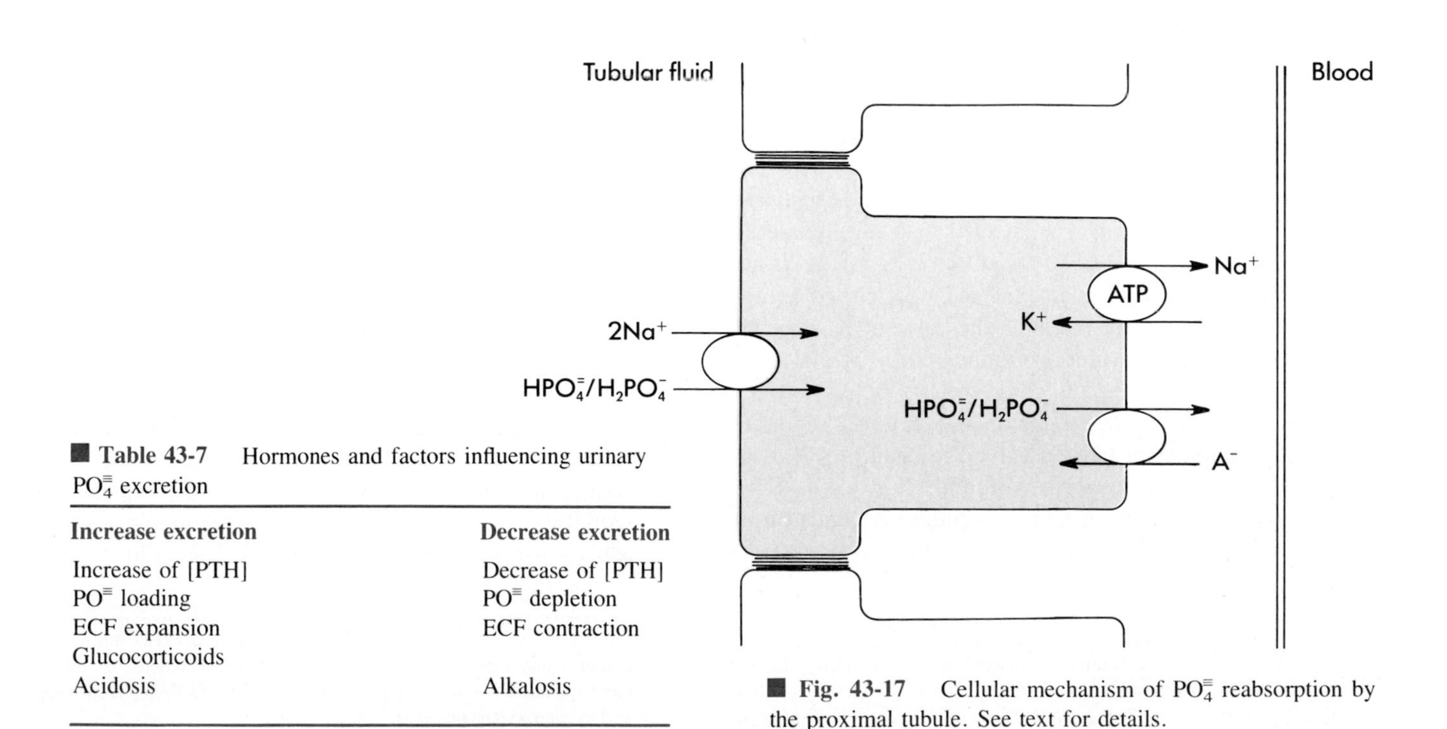

■ **Table 43-7** Hormones and factors influencing urinary $PO_4^=$ excretion

Increase excretion	Decrease excretion
Increase of [PTH]	Decrease of [PTH]
$PO^=$ loading	$PO^=$ depletion
ECF expansion	ECF contraction
Glucocorticoids	
Acidosis	Alkalosis

■ **Fig. 43-17** Cellular mechanism of $PO_4^=$ reabsorption by the proximal tubule. See text for details.

greater than that of H^+ ($[Na^+]$ = 140 mEq/L; $[H^+]$ = 40 nEq/L). Because of the low $[H^+]$ of the body fluids, it is commonly expressed as the negative logarithm, or pH.

Many of the metabolic functions of the body are exquisitely sensitive to pH, and normal function can occur over only a very narrow range. The pH range that is generally compatible with life is 6.8 to 7.8 (160 to 16 nEq/L of H^+). This section will examine the mechanisms that the body uses to maintain the pH of the body fluids within this normal range. Special emphasis is placed on the role of the kidneys (see Chapter 37 for a more detailed discussion of the role of the lungs in acid-base balance).

The CO_2/HCO_3^- Buffer System

Quantitatively HCO_3^- is the most important buffer in the extracellular fluid (normal plasma $[HCO_3^-] \cong 24$ mEq/L). The **CO_2/HCO_3^- buffer system** differs from other buffer systems of the body (e.g., phosphate) because it is under the dual regulation of the lungs and kidneys. This can best be understood by considering the reactions of CO_2 in water.

$$CO_2 + H_2O \xleftrightarrow{\text{Carbonic anhydrase}} H_2CO_3 \leftrightarrow H^+ + HCO_3^- \quad (1)$$

The first reaction (hydration/dehydration of CO_2) is the rate-limiting step. This reaction, which is normally slow, is greatly accelerated in the presence of the enzyme carbonic anhydrase (CA). The second reaction is the ionization of H_2CO_3, and it is virtually instantaneous.

To quantitate these reactions one can consider H^+ and HCO_3^- as products and CO_2 and H_2CO_3 as reactants. Thus:

$$K' = \frac{[H^+]\,[HCO_3^-]}{[CO_2]\,[H_2CO_3]} \quad (2)$$

Because this simplification combines the dissociation reaction ($H_2CO_3 \rightleftharpoons H^+ + HCO_3^-$) with the hydration/dehydration reaction ($CO_2 + H_2O \rightleftharpoons H_2CO_3$), K′ is not a true dissociation constant. Instead it is termed an **apparent dissociation constant.** The value of K′ depends on temperature and solution composition. For plasma at 37° C, K′ has a value of $10^{-6.1}$ (pK′ = 6.1).

The terms in the denominator of equation 2 represent the total amount of CO_2 dissolved in solution. Most of this CO_2 is in the gas form, with only 0.3% as H_2CO_3. Because the amount of CO_2 in solution depends on its partial pressure (P_{CO_2}) and its solubility (α)*, equation 2 can be rewritten as:

$$K' = \frac{[H^+][HCO_3^-]}{\alpha P_{CO_2}} \quad (3)$$

For plasma at 37° C, α = 0.03.

A more useful form of this equation is obtained by solving for $[H^+]$:

$$[H^+] = \frac{K'\alpha P_{CO_2}}{[HCO_3^-]} \text{ or } [H^+] = 24\frac{P_{CO_2}}{[HCO_3^-]} \quad (4)$$

Taking the negative logarithm of both sides of the equation yields:

$$-\log[H^+] = \frac{-\log[K'] + -\log\alpha P_{CO_2}}{-\log[HCO_3^-]} \quad (5)$$

$$pH = pK' + \log\frac{[HCO_3^-]}{\alpha P_{CO_2}} \text{ or } pH = 6.1 + \log\frac{[HCO_3^-]}{0.03 P_{CO_2}} \quad (6)$$

Equation 6 is the **Henderson-Hasselbalch equation.** Inspection of this equation shows that the pH of the extracellular fluid (ECF) varies when either the $[HCO_3^-]$ or the P_{CO_2} is altered. Disturbances of acid-base balance that result from a change in the ECF $[HCO_3^-]$ are termed **metabolic acid-base disorders,** whereas those that result from a change in the P_{CO_2} are termed **respiratory acid-base disorders.** These disorders are considered in more detail in a subsequent section. The kidneys are mainly responsible for regulating the $[HCO_3^-]$, whereas the lungs control the P_{CO_2}.

Metabolic Production of Acid and Alkali

In the normal individual the metabolism of the dietary food stuffs produces a number of substances that affect the acid-base status of the individual. When insulin is present and the tissues are adequately perfused, cellular metabolism of carbohydrates and fats produces large quantities of CO_2 (15 to 20 mole/day)*. This CO_2, which is a potential acid in the body fluids (as H_2CO_3), is termed **volatile acid,** and is excreted from the body by the lungs. In addition the normal diet contains a number of constituents whose metabolism produces acids other than CO_2. These are termed **nonvolatile acids.**

Metabolites of amino acids constitute a major portion of the nonvolatile acid production. The sulfur-containing amino acids cysteine and methionine yield sulphuric acid when metabolized, whereas hydrochloric acid results from the metabolism of the cationic amino acids lysine, arginine, and histidine. This acid production is partially offset by the metabolism of the anionic amino acids aspartate and glutamate, which results in the pro-

*α is not strictly a gas solubility constant, rather it is a constant that relates the P_{CO_2} to the total concentration of H_2CO_3 and the dissolved CO_2.

*In the absence of insulin, or when tissue hypoxia exists, the carbohydrates and fats are incompletely metabolized. When this occurs large quantities of nonvolatile acids are produced (e.g., lactic acid and β-hydroxybutyric acid).

Table 43-8 Metabolic production of nonvolatile acid and alkali from the diet

Food source	*Acid/alkali produced*	*Quantity (mmoles/day)*
Carbohydrates	Normally none	0
Fats	Normally none	0
Amino acids		
Sulfur-containing*	H_2SO_4	
Cationic†	HCl	100
Anionic‡	HCO_3^-	
Organic anions	HCO_3^-	−60
Phosphate	$H_2PO_4^-$	30
Total		**70**

*Cysteine, methionine.
†Lysine, arginine, histidine.
‡Aspartate, glutamate.

duction of HCO_3^-. Considering all amino acid metabolism, approximately 100 mmole/day of nonvolatile acid is produced (assuming a 100 g/day intake of protein).

Phosphate (as $H_2PO_4^-$) constitutes another nonvolatile acid load to the body. Typically the acid load associated with the ingestion of phosphates is in the range of 30 mmole/day.

Finally, the diet contains a number of organic anions (e.g., citrate), which, when metabolized, produce HCO_3^-. On average the metabolism of these anions produces approximately 60 mmole/day of HCO_3^-.

Table 43-8 summarizes the production of nonvolatile acids and HCO_3^- in an individual on a typical diet. On balance, acid production exceeds HCO_3^- production, with the net effect being the addition of approximately 70 mmole/day of nonvolatile acid to the body. When expressed in terms of body weight, nonvolatile acid production equals 1 mmole/day/kg body weight. The production of nonvolatile acids is highly dependent on the diet. For example, acid production can be less on a vegetarian diet.

The nonvolatile acids produced during metabolism do not circulate as free acids but are immediately buffered in the ECF.

$$H_2SO_4 + 2NaHCO_3 \leftrightarrow Na_2SO_4 + 2CO_2 + 2H_2O \quad (7)$$

$$HCl + NaHCO_3 \leftrightarrow NaCl + CO_2 + H_2O \quad (8)$$

This titration process yields the Na^+ salts of the strong acids and removes HCO_3^- from the ECF. *To maintain acid-base balance the kidney must excrete these Na^+ salts and replenish the HCO_3^- lost by titration.*

Overview of Renal Acid Excretion

To maintain acid-base balance the kidneys must excrete an amount of acid equal to the nonvolatile acid production. In addition they must prevent the loss of HCO_3^- in the urine. This latter task is quantitatively more important because the filtered load of HCO_3^- is approximately 4,320 mEq/day (24 mEq/L × 180 L/day = 4,320 mEq/day) compared with only 70 mEq/day of nonvolatile acid excretion.

Both the reabsorption of filtered HCO_3^- and the excretion of acid are accomplished through the process of H^+ secretion by the nephrons. Thus in a single day the nephrons must secrete approximately 4,390 mEq of H^+ into the tubular fluid. Most of the H^+ does not leave the body in the urine but serves to reabsorb the filtered load of HCO_3^-. Only 70 mEq are excreted. As a result of this acid excretion, the urine is normally acidic.

Theoretically the kidneys could excrete the nonvolatile acids and replenish the HCO_3^- lost during their titration by reversing the reactions shown in equations 7 and 8. Because the pKs of these acids are so low, however, this process would require a urine pH of 1.0. However, the minimum urine pH attainable by the kidneys is only 4.0 to 4.5. Consequently, the kidneys cannot excrete the free acids but must excrete their Na^+ salts, while at the same time excreting the H^+ with other urinary buffers. The two major urinary buffers are ammonia (NH_3/NH_4^+) and phosphate ($HPO_4^=/H_2PO_4^-$). Although both ammonia and phosphate are titrated during the process of urinary acidification, only phosphate, and to a lesser degree other buffer species (e.g., creatinine), are termed **titratable acid.** This distinction reflects the methods traditionally used to quantitate urinary acidification. Urinary ammonia is quantitated by direct measurement, whereas phosphate is estimated by measuring the number of OH^- equivalents needed to titrate an acidic urine to a value of 7.4, which is normally the pH of the plasma and glomerular filtrate. Because of this titration assay the nonammonia urinary buffers (primarily phosphate) are termed collectively titratable acid.

The overall process of **net acid excretion** (NAE) by the kidneys can be quantitated as follows:

$$NAE = [(U_{NH_4^+} \cdot \dot{V}) + (U_{TA} \cdot \dot{V})] - (U_{HCO_3^-} \cdot \dot{V}) \quad (9)$$

where: $U_{NH_4^+} \cdot \dot{V}$ and $U_{TA} \cdot \dot{V}$ are the rates of H^+ excretion (mEq/day) as NH_4^+ and titratable acid (TA), and $U_{HCO_3^-} \cdot \dot{V}$ is the amount of HCO_3^- lost in the urine (equivalent to adding H^+ to the body). *To maintain acid-base balance, net acid excretion must equal nonvolatile acid production.*

HCO_3^- Reabsorption Along the Nephron

Fig. 43-18 summarizes HCO_3^- reabsorption along the length of the nephron. Because this process prevents HCO_3^- loss in the urine, it is frequently referred to as **reabsorption of filtered HCO_3^-.**

About 85% of the filtered load of HCO_3^- is reabsorbed in the proximal tubule (Fig. 43-19). The apical membrane of the proximal tubule cell contains a Na^+-H^+ an-

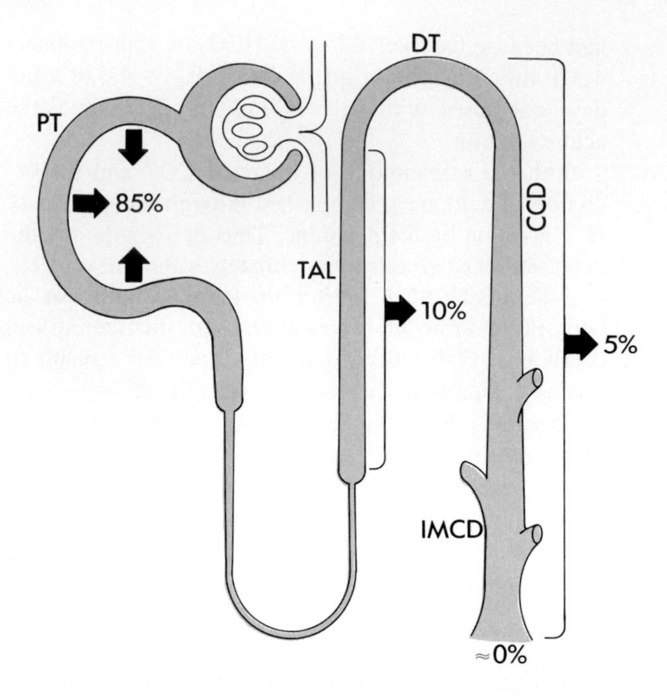

■ **Fig. 43-18** Segmental reabsorption of HCO_3^-. The fraction of the filtered load of HCO_3^- reabsorbed by the various segments of the nephron is shown. Normally, the entire filtered load of HCO_3^- is reabsorbed. *PT,* proximal tubule; *TAL,* thick ascending limb; *DT,* distal tubule; *CCD,* cortical collecting duct; *IMCD,* inner medullary collecting duct.

tiporter. This antiporter secretes H^+ into the tubular fluid. The energy for this secretion is derived from the lumen-to-cell Na^+ gradient. Recent evidence indicates that a portion of H^+ secretion is also mediated by a H^+-ATPase. Within the cell H^+ and HCO_3^- are produced in a reaction catalyzed by carbonic anhydrase (see equation 1). The H^+ is secreted into the tubular fluid, whereas the HCO_3^- exits the cell across the basolateral membrane and is returned to the peritubular blood. HCO_3^- does not simply diffuse from the cell across the basolateral membrane. Instead its movement is coupled to other ions. The majority of the HCO_3^- exits via a symporter that couples the efflux of one Na^+ with 3 HCO_3^-. In addition some of the HCO_3^- exits in exchange for Cl^- (Cl^--HCO_3^- antiporter). Within the tubular fluid the secreted H^+ combines with the filtered HCO_3^- to form H_2CO_3. This is rapidly converted to CO_2 and H_2O by carbonic anhydrase, which is present in the apical membrane of proximal tubule cells and exposed to the tubular fluid contents. Because the tubule is highly permeable to both CO_2 and H_2O, these are rapidly reabsorbed. The net effect of this process is that for each HCO_3^- removed from the tubular fluid, one HCO_3^- appears in the peritubular blood.

An additional 10% of the filtered load of HCO_3^- is reabsorbed by the loop of Henle. The majority of this HCO_3^- is reabsorbed by the cells of the thick ascending limb. The mechanism for HCO_3^- reabsorption by the thick ascending limb cells appears to be very similar to that described previously for the proximal tubule. In particular, H^+ is secreted into the tubular fluid by a Na^+-H^+ antiporter in the apical membrane. HCO_3^-, coupled to Na^+ (Na^+-3HCO_3^- symporter), exits the cell across the basolateral membrane and is returned to the peritubular blood.

The distal tubule and collecting duct reabsorb the small amount of HCO_3^- that escapes reabsorption by the proximal tubule and loop of Henle (5% of the filtered load). The reabsorption mechanism (Fig. 43-20) does not depend on Na^+ (i.e., apical membrane Na^+-H^+ antiporter) contrary to the mechanism in the earlier nephron segments. In the collecting duct H^+ is secreted by the intercalated cell (see Chapter 41). Within the inter-

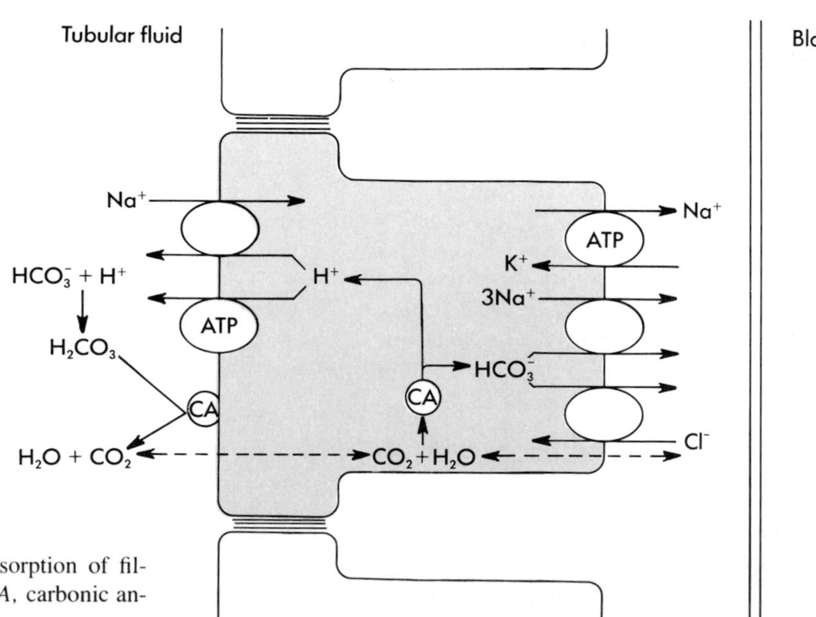

■ **Fig. 43-19** Cellular mechanism for reabsorption of filtered HCO_3^- by cells of the proximal tubule. *CA,* carbonic anhydrase. See text for details.

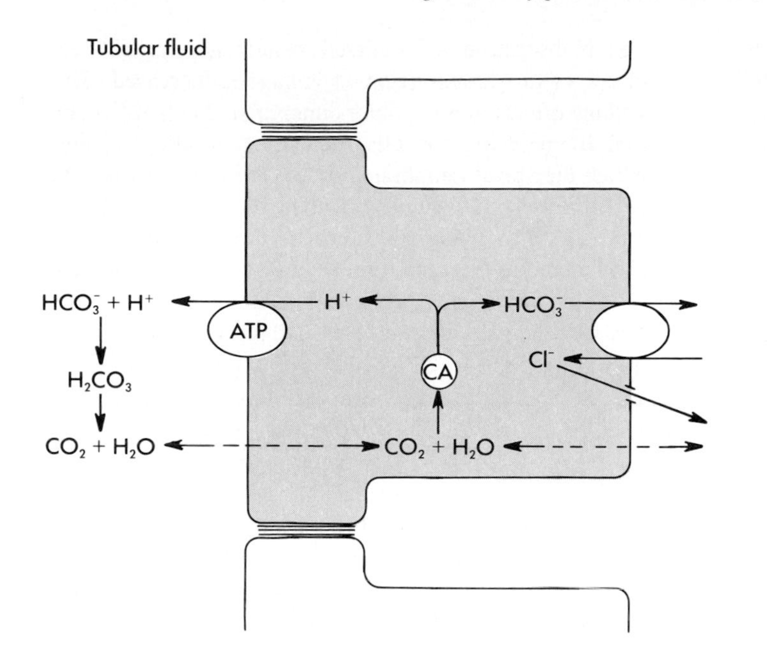

■ **Fig. 43-20** Cellular mechanism for reabsorption of filtered HCO_3^- by the intercalated cell of the collecting duct. *CA*, carbonic anhydrase. See text for details.

calated cell H^+ and HCO_3^- are produced by the hydration of CO_2. This reaction is catalyzed by carbonic anhydrase. The H^+ is secreted into the tubular fluid by an apical membrane H^+-ATPase. The HCO_3^- exits the cell across the basolateral membrane in exchange for Cl^- (Cl^--HCO_3^- antiporter) and enters the peritubular capillary blood. The apical membrane of the cells of the collecting duct are only slightly permeable to H^+, and the pH of the tubular fluid can be rendered quite acidic. Indeed, the most acidic tubular fluid along the nephron (pH = 4.0 to 4.5) is produced at this site. By comparison, the proximal tubule is much more permeable to H^+ and HCO_3^- and the tubular fluid pH falls to only 6.5 in this segment.

■ *Regulation of HCO_3^- Reabsorption*

HCO_3^- reabsorption is regulated by several factors, which act at the proximal tubule, thick ascending limb of Henle's loop, and collecting duct (box).

Because of glomerulotubular balance, any change in the filtered load of HCO_3^-, as a result of alterations in the GFR, is matched by an appropriate change in HCO_3^- reabsorption by the proximal tubule. Thus proximal tubule HCO_3^- reabsorption increases with an increase in filtered load and decreases when the filtered load is reduced.

The major fraction of proximal tubule HCO_3^- reabsorption occurs by the Na^+-H^+ antiporter located in the

■ **Table 43-9** Factors regulating H^+ secretion (HCO_3^- reabsorption) by the nephron

Factor	Nephron site of action
Increasing H^+ secretion	
Increase in filtered load of HCO_3^-	Proximal tubule
Decrease in ECF volume	Proximal tubule
Decrease in plasma [HCO_3^-] (↓ pH)	Proximal tubule, thick ascending limb, and collecting duct
Increase in blood P_{CO_2}	Proximal tubule, thick ascending limb, and collecting duct
Aldosterone	Collecting duct
Decreasing H^+ secretion	
Decrease in filtered load of HCO_3^-	Proximal tubule
Increase in ECF volume	Proximal tubule
Increase in plasma [HCO_3^-] (↑ pH)	Proximal tubule, thick ascending limb, and collecting duct
Decrease in blood P_{CO_2}	Proximal tubule, thick ascending limb, and collecting duct

From Stanton BA, Koeppen BM: Regulation of potassium, calcium magnesium, phosphate, and acid base balance, In Berne RM, Levy MN, editors: *Principles of physiology*. St Louis, 1990, Mosby–Year Book.

apical membrane of the cells. Consequently, factors that regulate Na^+ homeostasis (see Chapter 42) will alter HCO_3^- reabsorption secondarily. Thus expansion of the ECF, which inhibits proximal tubule Na^+ reabsorption, will also decrease the reabsorption of HCO_3^-. Conversely, HCO_3^- reabsorption is enhanced when the ECF is decreased.

As might be expected, changes in systemic acid-base balance also affect HCO_3^- reabsorption. Systemic acidosis, whether produced by a decrease in the plasma $[HCO_3^-]$ (metabolic) or by an increase in the P_{CO_2} (respiratory), stimulates HCO_3^- reabsorption in the proximal tubule, loop of Henle, and collecting duct. This stimulation probably occurs as a result of acidification of the intracellular fluid, which in turn produces a more favorable cell-to-lumen H^+ gradient and favors H^+ secretion. Also, acidification of the intracellular fluid in the intercalated cells of the collecting duct results in the insertion of more H^+-ATPase into the apical membrane. With more H^+-ATPase, H^+ secretion and thus HCO_3^- reabsorption is enhanced. Conversely, metabolic and respiratory alkalosis inhibit HCO_3^- reabsorption in the proximal tubule, loop of Henle, and collecting duct. The mechanisms involved are opposite of those described previously for stimulation of HCO_3^- reabsorption with acidosis.

The last major regulatory factor for HCO_3^- reabsorption is aldosterone. Aldosterone stimulates H^+ secretion by the intercalated cells of the collecting duct. This effect reflects both a direct action of the hormone on the intercalated cell, as well as an indirect effect by aldosterone stimulation of Na^+ reabsorption by the principal cell. With regard to the indirect effect of aldosterone, Na^+ reabsorption by the principal cells of the collecting duct generates a lumen-negative transepithelial voltage. When Na^+ reabsorption is stimulated by aldosterone, the magnitude of this lumen negative voltage is increased. This voltage effect, in turn, favors intercalated cell H^+ secretion by reducing the electrochemical gradient against which the apical membrane H^+-ATPase must pump; the H^+-ATPase is affected by a voltage gradient. The cell mechanism by which aldosterone directly stimulates intercalated cell H^+ secretion has not yet been elucidated. As would be expected, collecting duct H^+ secretion is decreased when aldosterone levels are reduced.

Formation of New HCO_3^-: The Role of Ammonia

As discussed previously, the reabsorption of HCO_3^- is important for the maintenance of acid-base balance. HCO_3^- loss in the urine decreases the plasma $[HCO_3^-]$ and is equivalent to the addition of H^+ to the body. However, HCO_3^- reabsorption alone does not replenish the HCO_3^- that was lost during the titration of the nonvolatile acids produced by metabolism. For the maintenance of acid-base balance, the kidney must replace this lost HCO_3^- with new HCO_3^-. The production of new HCO_3^- is critically dependent on the availability of urinary buffers. Fig. 43-21 illustrates how the titration of urinary buffers results in the formation of new HCO_3^-. Secretion of H^+ results in the excretion of the H^+. HCO_3^- is produced in the cell from the hydration of CO_2, and it is added to the blood.

As noted previously, the two major urinary buffers are ammonia (NH_3/NH_4^+) and phosphate ($HPO_4^=/H_2PO_4^-$). Phosphate is derived solely from the diet. The amount excreted in the urine as titratable acid therefore depends on the filtered load minus the amount reabsorbed by the nephron. Ammonia is produced by the kidneys, and its

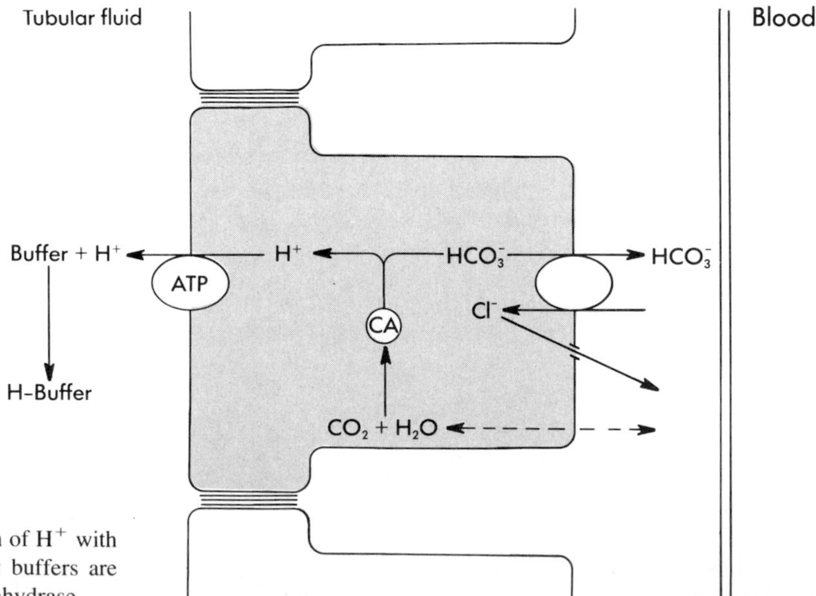

Fig. 43-21 General scheme for the excretion of H^+ with non HCO_3^- urinary buffers. The primary urinary buffers are NH_3 and $HPO_4^=$ (titratable acid). *CA*, carbonic anhydrase.

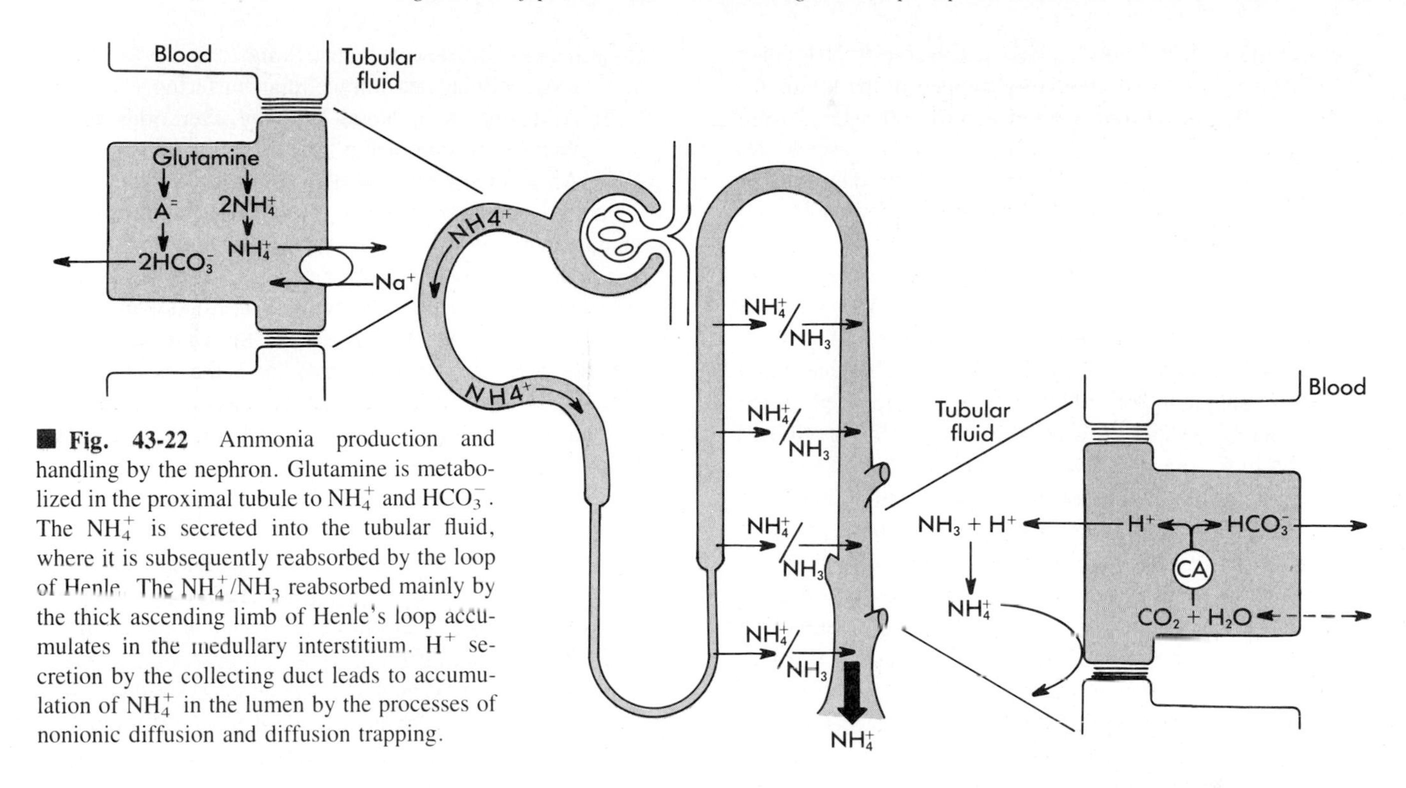

■ **Fig. 43-22** Ammonia production and handling by the nephron. Glutamine is metabolized in the proximal tubule to NH_4^+ and HCO_3^-. The NH_4^+ is secreted into the tubular fluid, where it is subsequently reabsorbed by the loop of Henle. The NH_4^+/NH_3 reabsorbed mainly by the thick ascending limb of Henle's loop accumulates in the medullary interstitium. H^+ secretion by the collecting duct leads to accumulation of NH_4^+ in the lumen by the processes of nonionic diffusion and diffusion trapping.

synthesis and subsequent excretion can be regulated in response to the acid-base requirements of the body. Therefore ammonia is the more important urinary buffer.

The production of new HCO_3^- and the importance of the NH_3/NH_4^+ buffer pair can be summarized for the kidneys as a whole by the following general scheme:

$$\text{Glutamine} \rightarrow 2NH_4^+ + 2HCO_3^- \qquad (10)$$

Glutamine is metabolized by the kidneys, the NH_4^+ is excreted into the urine, whereas the HCO_3 is returned to the systemic circulation to replenish the HCO_3^- lost in the titration of nonvolatile acids. The formation of new HCO_3^- by this reaction depends on the ability of the kidney to excrete the NH_4^+ in the urine. If the NH_4^+ is not excreted in the urine, but instead enters the systemic circulation, it will titrate plasma HCO_3^-, and thus it will negate the process of new HCO_3^- generation.* The process by which the kidneys excrete NH_4^+ is complex (Fig. 43-22).

NH_4^+ is produced in the proximal tubule cells from glutamine. Each molecule of glutamine produces two molecules of NH_4^+ and a divalent anion. Metabolism of this divalent anion ultimately provides two molecules of HCO_3^-.

*The mechanism by which NH_4^+ titrates HCO_3^- is indirect and occurs by the following mechanism:

$$2NH_4^+ \rightarrow \text{urea} + 2H^+ \text{ and } 2H^+ + 2HCO_3^- \rightarrow 2H_2O + 2CO_2$$

The metabolism of NH_4^+ to urea occurs in the liver.

$$\text{Glutamine} \leftrightarrow 2NH_4^+ + \text{Anion}^= \leftrightarrow 2HCO_3^- + 2NH_4^+ \qquad (11)$$

The HCO_3^- exits the cell across the basolateral membrane and enters the peritubular blood as new HCO_3^-. NH_4^+ exits the cell across the apical membrane and enters the tubular fluid. A major mechanism for the secretion of NH_4^+ into the tubular fluid involves the Na^+-H^+ antiporter, with NH_4^+ substituting for H^+. A large portion of the NH_4^+ secreted by the proximal tubule is reabsorbed by the thick ascending limb of Henle's loop. The NH_4^+ accumulates in the medullary interstitium and is then secreted into the tubular fluid by the cells of the collecting duct. Secretion of NH_4^+ by the collecting duct involves nonionic diffusion and diffusion trapping. Ammonia reabsorbed by the thick ascending limb accumulates in the medullary interstitial fluid where it exists as both NH_4^+ and NH_3^+.* The collecting duct does not have a specific transport mechanism for the secretion of NH_4^+ nor are the cells passively permeable to NH_4^+. However, the cells of the collecting duct are permeable to NH_3, and NH_3 will diffuse from the medullary interstitium into the lumen of the collecting duct. As described previously, H^+ secretion by the collecting duct intercalated cells acidifies the luminal fluid. Consequently NH_3 diffusing from the medullary interstitium into the collecting duct lumen **(nonionic diffusion)** will be protonated to NH_4^+ by the acidic tubular fluid. Be-

*Ammonia is a weak base and will be present as both NH_4^+ and NH_3 with the relative amounts of each species determined by the pKa. (pKa = 9.0).

cause the collecting duct is less permeable to NH_4^+ than to NH_3, the NH_4^+ is effectively trapped in the tubule lumen **(diffusion trapping)** and thereby eliminated from the body in the urine.

Note that H^+ secretion by the collecting duct is critical for the excretion of NH_4^+. If collecting duct H^+ secretion is inhibited, the NH_4^+ reabsorbed by the thick ascending limb will not be excreted in the urine. Instead it will be returned to the systemic circulation, where it will titrate HCO_3^-, as described previously. If this occurs, net acid excretion by the kidneys is reduced, and insufficient quantities of new HCO_3^- are added to the systemic circulation to replenish that which was titrated by the buffering of nonvolatile acids.

An important feature of the NH_4^+ system is that it can be regulated. Alterations in extracellular fluid pH, presumably by affecting intracellular pH, change NH_4^+ production. During systemic acidosis the enzymes in the proximal tubule cell responsible for the metabolism of glutamine are stimulated. This stimulation involves the synthesis of new enzyme and requires several days for complete adaptation. Elevated levels of this enzyme increase NH_4^+ production and allow more H^+ excretion and enhanced production of new HCO_3^-. Conversely, alkalosis reduces NH_4^+ production.

Plasma $[K^+]$ also alters NH_4^+ production. Hyperkalemia inhibits NH_4^+ production, whereas hypokalemia stimulates it. The mechanism whereby plasma K^+ alters NH_4^+ production is not completely understood. Alterations in plasma $[K^+]$ may change intracellular $[H^+]$ by exchanging H^+ for K^+ (see above), and this change in intracellular pH may control NH_4^+ production. During hyperkalemia, exchange of extracellular K^+ for intracellular H^+ by this mechanism would raise intracellular pH and thereby inhibit NH_4^+ production. The opposite would occur during hypokalemia.

Response to Acid-Base Disorders

The pH of the body fluids is maintained within a very narrow range (pH = 7.40 ± 0.02). **Acidosis** exists when the pH of plasma falls below 7.40, whereas **alkalosis** exists when the pH is greater that 7.40*. When the acid-base disorder is characterized by a primary change in the $[HCO_3^-]$, it is termed a metabolic disorder. When the disorder is characterized by a primary change in the P_{CO_2}, it is termed a respiratory disorder.

The body has three general mechanisms to defend itself against changes in body fluid pH with acid-base disturbances. These mechanisms are:

1. Extracellular and intracellular buffering.
2. Adjustments in blood P_{CO_2} by alterations in the ventilatory rate of the lungs.
3. Adjustments in renal acid excretion.

These mechanisms all work in concert to minimize the change in blood pH. However, by themselves they do not return the blood pH to its normal value. Restoration of the blood pH to its normal value requires correction of the underlying process(es) that produced the acid-base disorder. For example, metabolism of fats in the absence of insulin leads to the accumulation of keto-acids (a nonvolatile acid) in the blood and the development of a metabolic acidosis. The acid-base defense mechanisms minimize the fall in pH that occurs in this condition, but normal acid-base balance is not restored until insulin is administered and keto-acid production ceases.

Extracellular and intracellular buffering. The first line of defense to acid-base disorders is extracellular and intracellular buffering. The response of the extracellular buffers is virtually instantaneous. Cellular buffering is somewhat slower; it can take several minutes.

Metabolic disorders that result from the addition of nonvolatile acids or alkali to the body fluids are buffered in the extracellular fluid and in the intracellular fluid (see below). *The CO_2/HCO_3^- buffer system is the most important extracellular buffer.* Buffering of acid and alkali by the CO_2/HCO_3^- buffer system involves the following reaction:

$$H^+ + HCO_3^- \leftrightarrow H_2CO_3 \leftrightarrow H_2O + CO_2 \qquad (12)$$

When nonvolatile acid is added to the body fluids (or alkali is lost from the body), the above reaction is driven to the right, HCO_3^- is consumed during the process of buffering the acid load, and the plasma $[HCO_3^-]$ is reduced. Thus **metabolic acidosis** is characterized by having a low pH and $[HCO_3^-]$. Conversely when nonvolatile alkali is added to the body fluids (or acid is lost from the body), the reaction is driven to the left, and as a consequence the $[HCO_3^-]$ increases. Thus **metabolic alkalosis** is characterized by an increased pH and $[HCO_3^-]$.

Although the CO_2/HCO_3^- buffer system is the main extracellular buffer, additional extracellular buffering occurs with phosphate and plasma protein.

$$\begin{aligned} H^+ + HPO_4^{=} &\leftrightarrow H_2PO_4^- \\ H^+ + \text{Protein}^- &\leftrightarrow \text{H-Protein} \end{aligned} \qquad (13)$$

The combined action of the CO_2/HCO_3^-, $HPO_4^{=}$ and plasma protein buffering processes accounts for 40% to 50% of the buffering of a nonvolatile acid load and 60% to 70% of a nonvolatile alkali load. The remainder of the buffering under these two conditions occurs intracellularly.

Intracellular buffering involves the movement of H^+ into cells (during buffering of nonvolatile acid) or the movement of H^+ out of cells (during buffering of non-

*In clinical practice plasma pH values in the range of 7.35 to 7.45 are considered normal. For simplicity, we will consider the single value of 7.40 to represent the normal plasma pH, and deviations from this value are deemed abnormal. Similarly the normal range for P_{CO_2} is 35 to 45 mm Hg. However, a P_{CO_2} = 40 mm Hg is used as the normal reference value. Lastly we use a value of 24 mEq/L for the normal $[HCO_3^-]$, even though the normal range is from 23 to 25 mEq/L.

volatile alkali).* H^+ is titrated inside the cell by both $HPO_4^=/H_2PO_4^-$ and protein.

With respiratory acid-base disorders body fluid pH changes are affected by alterations in $[H_2CO_3]$, which in turn are determined directly by the P_{CO_2} (see equation 1). Virtually all buffering in respiratory acid-base disorders occurs inside cells. When P_{CO_2} rises **(respiratory acidosis),** CO_2 moves into the cell, where it combines with H_2O to form H_2CO_3, which then dissociates to H^+ and HCO_3^-. The H^+ is buffered by cellular proteins, and the HCO_3^- exits the cell and raises the plasma $[HCO_3^-]$. This process is reversed when the P_{CO_2} is reduced **(respiratory alkalosis).** Under this condition the hydration reaction ($H_2O + CO_2 \leftrightarrow H_2CO_3$) is shifted to the left by the decrease in P_{CO_2}. This in turn shifts the dissociation reaction ($H_2CO_3 \leftrightarrow H^+ + HCO_3^-$) to the left and thereby reduces the plasma $[HCO_3^-]$.

Respiratory defense. The lungs represent the second line of defense (see Chapter 37). As indicated by the Henderson-Hasselbalch equation, changes in the P_{CO_2} will alter the blood pH: an increase in P_{CO_2} decreases pH, and a decrease in P_{CO_2} increases pH.

The ventilatory rate is the main determinant of the P_{CO_2}. Increased ventilation decreases the P_{CO_2}, which normally is 40 mm Hg. With maximal hyperventilation the P_{CO_2} can be reduced to approximately 10 mm Hg. Conversely the P_{CO_2} increases with decreased ventilation. Because hypoventilation induces hypoxia and hypoxia is a potent stimulator of ventilation, the degree to which the P_{CO_2} can be increased is limited. In a normal individual the P_{CO_2} generally does not exceed 60 mm Hg. Both the blood P_{CO_2} and pH are important regulators of the ventilatory rate. Chemoreceptors located in the brain (ventral surface of medulla) and in the periphery (carotid and aortic bodies) sense changes in P_{CO_2} and $[H^+]$ and cause changes in the ventilatory rate (see Chapter 37). In metabolic acidosis the increase in $[H^+]$ (decrease in pH) increases the ventilatory rate. Conversely in metabolic alkalosis the decrease in $[H^+]$ (increase in pH) decreases the ventilatory rate. The respiratory response to metabolic acid-base disturbances may require several hours to complete.

Renal defense. The third and final line of defense is the kidneys. In response to an alteration in the plasma pH and P_{CO_2} the kidneys make appropriate adjustments in the excretion of HCO_3^- and net acid (Table 43-4). Completion of the renal response requires several days.

In the case of acidosis (increase in $[H^+]$ or P_{CO_2}), secretion of H^+ by the nephron is stimulated and the entire filtered load of HCO_3^- is reabsorbed. Also the production and excretion of NH_4^+ is stimulated thereby increasing net acid excretion by the kidney (see equation 9). The new HCO_3^- generated during the process of net acid excretion is returned to the body, and the plasma $[HCO_3^-]$ increases.

In alkalosis (decrease in $[H^+]$ or P_{CO_2}), secretion of H^+ by the nephron is inhibited. Consequently net acid excretion and HCO_3^- reabsorption are reduced. HCO_3^- will appear in the urine and thereby reduces the plasma $[HCO_3^-]$.

*The movement of H^+ into cells is associated in some instances with the release of cellular K^+. Thus with acidosis, hyperkalemia may result, whereas with alkalosis, hypokalemia may result (see above). However, the degree of cellular H^+-K^+ exchange depends on the nature of the acid and/or alkali. In general, metabolic acid-base disorders are associated with alterations in K^+ but respiratory disorders are not. Also metabolic disorders that involve organic anions (e.g., lactic acid) have less effect on plasma K^+ than do mineral acids (e.g., HCl).

Simple Acid-Base Disorders

Table 43-10 summarizes the primary alteration and the subsequent defense mechanisms for the various simple acid-base disorders. The respiratory and renal defense mechanisms in simple acid-base disorders are commonly referred to as **compensatory responses.** Accordingly the lungs compensate for metabolic disorders and the kidneys compensate for respiratory disorders.

Metabolic acidosis. *Metabolic acidosis is characterized by a low plasma $[HCO_3^-]$ and a low pH.* This condition can develop either by the addition of nonvolatile acid to the body (e.g., with diabetic ketoacidosis), by the loss of nonvolatile alkali (e.g., with diarrhea), or by the failure of the kidneys to excrete enough net acid to replenish the HCO_3^- used to titrate nonvolatile acids (e.g., with renal failure). As described above, H^+ will be buffered in both the extracellular and intracellular fluid. The fall in pH will stimulate the respiratory centers, and the ventilatory rate will increase **(respiratory compensation).** This reduces the P_{CO_2}, which further helps to minimize the fall in plasma pH. In general the P_{CO_2} decreases by 1.2 mm Hg for every 1 mEq/L fall in

Table 43-10 Characteristics of simple acid-base disorders

Disorder	*Plasma pH*	*Primary alteration*	*Defense mechanisms*
Metabolic acidosis	↓	↓ Plasma $[HCO_3^-]$	ICF and ECF buffers; hyperventilation (↓ P_{CO_2})
Metabolic alkalosis	↑	↑ Plasma $[HCO_3^-]$	ICF and ECF buffers; hypoventilation (↑ P_{CO_2})
Respiratory acidosis	↓	↑ P_{CO_2}	ICF buffers; ↑ renal H^+ excretion
Respiratory alkalosis	↑	↓ P_{CO_2}	ICF buffers; ↓ renal H^+ excretion

the plasma [HCO_3^-]. Thus if the plasma [HCO_3^-] decreases to 14 mEq/L from a normal value of 24 mEq/L, the expected decrease in Pco_2 would be 12 mm Hg and the measured Pco_2 would be 28 mm Hg (normal Pco_2 = 40 mm Hg).

Renal excretion of net acid increases. This occurs by eliminating all HCO_3^- from the urine (enhanced reabsorption of filtered HCO_3^-) and by increasing ammonia excretion (enhanced production of new HCO_3^-). If the process that initiated the acid-base disturbance is corrected, the enhanced excretion of acid by the kidney will ultimately return the pH and [HCO_3^-] to normal. Correction of the pH will also return the ventilatory rate to normal.

Metabolic alkalosis. *Metabolic alkalosis is characterized by an elevated plasma [HCO_3^-] and pH.* This can occur either by the addition of nonvolatile alkali to the body (e.g., ingestion of antacids), or more commonly from loss of nonvolatile acid (e.g., loss of gastric HCl with vomiting). Buffering occurs in the extracellular and intracellular fluid compartments. The increase in pH inhibits the respiratory centers, the ventilatory rate is reduced, and the Pco_2 rises (respiratory compensation). With appropriate respiratory compensation Pco_2 increases 0.7 mm Hg for every 1 mEq/L rise in plasma [HCO_3^-]. The renal compensatory response to metabolic alkalosis is to increase the excretion of HCO_3^- by reducing its reabsorption along the nephron. Normally this occurs quite rapidly and effectively. However, when the alkalosis occurs in the setting of reduced effective circulating volume (e.g., vomiting), Na^+ reabsorption (and therefore HCO_3^- reabsorption) is enhanced in the proximal tubule (see Chapter 42). Renal excretion of HCO_3^- and correction of the alkalosis will occur with restoration of a normal effective circulating volume. As is the case with metabolic acidosis, renal excretion of HCO_3^- will eventually return the pH and [HCO_3^-] to their normal values, provided that the underlying cause of the initial disturbance is corrected. Correction of the pH also returns the ventilatory rate to normal.

Respiratory acidosis. *Respiratory acidosis is characterized by an elevated Pco_2 and reduced plasma pH* (see also Chapter 37). It results from decreased gas exchange across the alveoli, either as a result of inadequate ventilation (e.g., drug-induced depression of the respiratory centers) or impaired gas diffusion (e.g., pulmonary edema). In contrast to the metabolic disorders, buffering during respiratory acidosis occurs almost entirely in the intracellular compartment (see above). Both the increase in Pco_2 and the decrease in pH stimulate HCO_3^- reabsorption by the kidney and stimulate ammonium excretion **(renal compensation).** Together these responses increase net acid excretion and generate new HCO_3^-. However, the renal compensatory response takes several days. Consequently respiratory acid-base disorders are commonly divided into acute and chronic phases. In the acute phase, the renal compensatory response has not had sufficient time to occur and the body relies on intracellular buffering to minimize the change in pH. In this phase of buffering, plasma [HCO_3^-] increases 1 mEq/L for every 10 mm Hg rise in Pco_2. In the chronic phase, renal compensation occurs and the plasma [HCO_3^-] increases 3.5 mEq/L for each 10 mm Hg rise in Pco_2. Correction of the underlying disorder returns the Pco_2 to normal; secondarily the renal excretion of acid decreases to its initial level.

Respiratory alkalosis. *Respiratory alkalosis is characterized by a reduced Pco_2 and elevated plasma pH.* It results from increased gas exchange in the lungs, usually caused by increased ventilation elicited by stimulation of the respiratory centers (e.g., by drugs or CNS disorders). Hyperventilation can also occur as a response to anxiety or fear. Buffering is mainly intracellular, and in the acute phase of respiratory alkalosis it accounts for a 2 mEq/L decrease in the plasma [HCO_3^-] for every 10 mm Hg fall in Pco_2. The elevated pH and reduced Pco_2 inhibits HCO_3^- reabsorption by the nephron and reduces ammonium excretion (renal compensation). These two effects reduce net acid excretion. The response takes several days and results in a 5 mEq/L decrease in plasma [HCO_3^-] for every 10 mm Hg reduction in Pco_2. Correction of the underlying disorder returns the Pco_2 to normal; secondarily the renal excretion of acid increases to its initial level.

Analysis of Acid-Base Disorders

Analysis of a clinical acid-base disorder is directed at identification of the underlying cause so that appropriate therapy can be initiated. Often the patient's medical history and associated physical findings provide valuable clues about the nature and origin of an acid-base disorder. An analysis of an arterial blood sample is frequently required. For example, consider the following data:

pH = 7.35 [HCO_3^-] = 16 mEq/L Pco_2 = 30 mm Hg

The acid-base disorder represented by values such as these can be determined by the following three-step approach (Fig. 43-23).

1. *Examination of the pH:* By first examining the pH, the underlying disorder can be classified as either an acidosis (pH $<$ 7.40) or an alkalosis (pH $>$ 7.40). Note that the defense mechanisms of the body, by themselves, cannot correct an acid-base disorder. Thus even if the defense mechanisms are completely operative, the pH will still indicate the origin of the initial disorder. In the example shown, the pH of 7.35 indicates an acidosis.
2. *Determination of metabolic versus respiratory disorder:* Simple acid-base disorders are either metabolic or respiratory. To determine which dis-

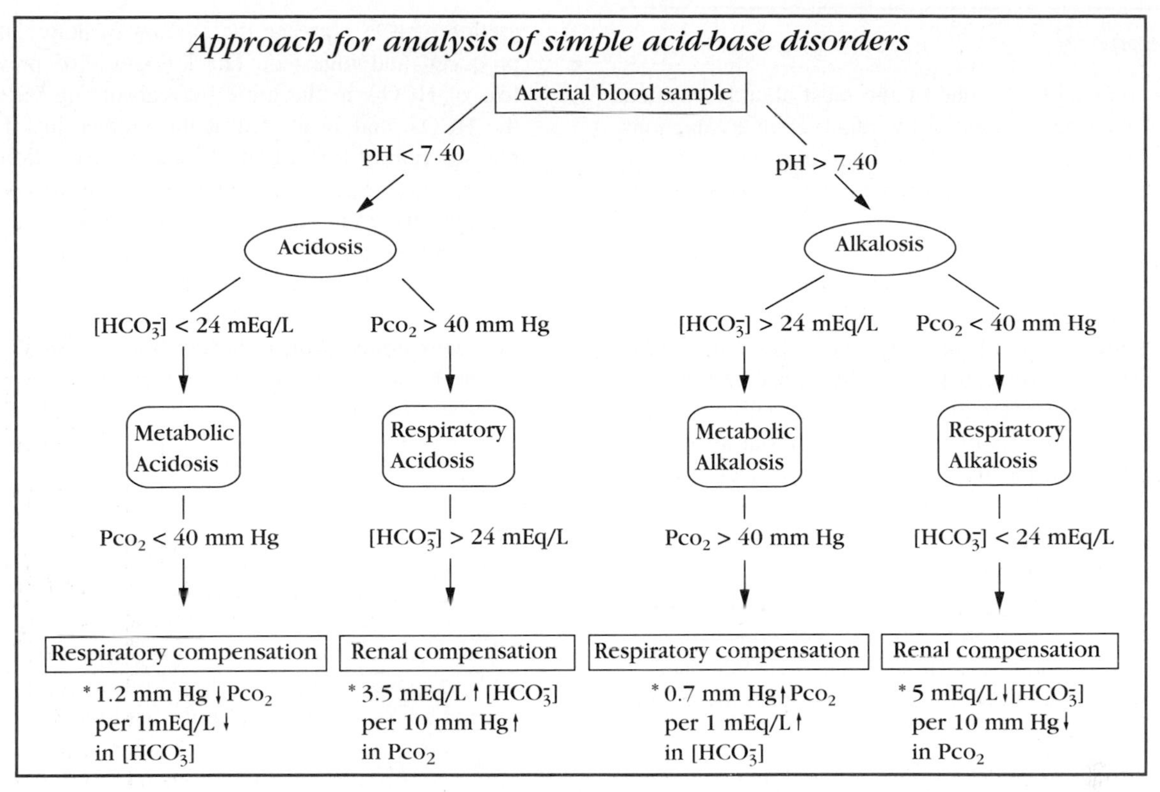

Fig. 43-23 Approach for analysis of simple acid base disorders.

order is present, the $[HCO_3^-]$ and Pco_2 must be examined. As indicated by the Henderson-Hasselbalch equation, a pH value below 7.40 (i.e., acidosis) could be the result of a decrease in the $[HCO_3^-]$ (metabolic) or an increase in the Pco_2 (respiratory). Alternatively a pH value above 7.40 (alkalosis) could be the result of an increase in the $[HCO_3^-]$ (metabolic) or a decrease in the Pco_2 (respiratory). For the previous example, the $[HCO_3^-]$ is reduced from normal (normal = 23 to 25 mEq/L), as is the Pco_2 (normal = 40 mm Hg). The disorder must therefore be a metabolic acidosis; it cannot be a respiratory acidosis because the Pco_2 is reduced.

3. *Analysis of compensatory response.* Metabolic disorders result in compensatory changes in ventilation and thus in the Pco_2. Respiratory disorders result in compensatory changes in renal acid excretion and thus in the plasma $[HCO_3^-]$. In an appropriately compensated metabolic acidosis, the Pco_2 will be decreased, whereas with a compensated metabolic alkalosis the Pco_2 will be elevated. With respiratory acidosis, complete compensation elevates the $[HCO_3^-]$, whereas with respiratory alkalosis complete compensation reduces the $[HCO_3^-]$. In the example the Pco_2 is reduced from normal, and the magnitude of this reduction (10 mm Hg decrease for a 8 mEq/L increase in $[HCO_3^-]$) is as expected. Therefore the acid-base disorder is a simple metabolic acidosis with appropriate respiratory compensation.

If the compensatory response is not appropriate, a **mixed acid-base disorder** should be suspected (see below). A mixed acid-base disorder simply reflects the presence of two or more underlying causes for the acid-base disturbance. A mixed acid-base disorder is suspected when analysis of the arterial blood gas indicates that compensation has not been appropriate. For example, consider the following data:

$$pH = 6.96 \qquad [HCO_3^-] = 12\ mEq/L \qquad Pco_2 = 55\ mm\ Hg$$

Following this three-step approach, it indicates that the disturbance is an acidosis that has both a metabolic component ($[HCO_3^-] < 24$ mEq/L) and a respiratory component ($Pco_2 > 40$ mm Hg). Thus this disorder is mixed. Such a disorder would be manifest in an individual with a history of chronic pulmonary disease such as emphysema (chronic respiratory acidosis), and the patient also develops an acute gastrointestinal illness with diarrhea. Because diarrhea fluid contains large quantities of HCO_3^-, the loss of this fluid from the body induces a metabolic acidosis. Note that the finding of a normal pH with abnormal Pco_2 and $[HCO_3^-]$ also indicates that the disorder is mixed.

Summary

1. *Potassium.* K^+ is one of the most abundant cations in the body and is crucial for many cellular functions, including cell growth and division and the excitability of nerve and muscle. K^+ homeostasis is maintained by (1) hormones that regulate internal K^+ balance (the distribution of K^+ between the intracellular and extracellular fluid), and (2) by the kidneys (which adjust K^+ excretion to match dietary K^+ intake). Internal K^+ balance is maintained by insulin, epinephrine, and aldosterone. In contrast, cell lysis and exercise and changes in acid-base balance and in plasma osmolality often disrupt K^+ homeostasis by altering the normal distribution of K^+ between the intracellular and extracellular fluid compartments. K^+ excretion by the kidneys is determined by the rate of K^+ secretion by the distal tubule and collecting duct. K^+ secretion by these tubular segments is stimulated by increases in plasma $[K^+]$, aldosterone, tubular flow rate, ADH, metabolic alkalosis, and the $[Na^+]$ of tubular fluid. Acute metabolic acidosis inhibits K^+ excretion; chronic metabolic acidosis stimulates K^+ excretion.

2. *Multivalent ions.* Ca^{++}, Mg^{++}, and inorganic phosphate ($PO_4^{\equiv}$) are multivalent ions that subserve many vital functions. The kidneys, in conjunction with the gastrointestinal tract and bone, play a vital role in regulating Ca^{++}, Mg^{++}, and $PO_4^{\equiv}$ homeostasis. Plasma Ca^{++} is regulated by PTH and $1,25(OH)_2D_3$. Ca^{++} excretion by the kidneys is determined by the rate of Ca^{++} reabsorption by the thick ascending limb of Henle's loop. Ca^{++} reabsorption by the thick ascending limb is regulated mainly by PTH, which stimulates Ca^{++} reabsorption. Although Mg^{++} has many important biological roles, relatively little is known about the factors that regulate plasma $[Mg^{++}]$ and urinary Mg^{++} excretion. Plasma $[PO_4^{\equiv}]$ is regulated by PTH and $1,25(OH)_2D_3$. A fall in $[PO_4^{\equiv}]$ stimulates release of PTH and $1,25(OH)_2 D_3$, which cause the release of Ca^{++} and $PO_4^{\equiv}$ from bone into the ECF. $PO_4^{\equiv}$ excretion by the kidneys is determined mainly by the rate of reabsorption in the proximal tubule. PTH inhibits $PO_4^{\equiv}$ reabsorption by the proximal tubule and enhances urinary $PO_4^{\equiv}$ excretion.

3. *Acid-base balance.* The pH of the body fluids is maintained within a narrow range by the coordinated function of the lungs and kidneys. Volatile (CO_2 derived) and nonvolatile acids, together with any acid or alkali ingested in the diet, must be excreted for acid-base balance to be maintained. The lungs are the excretory route for the volatile acid, whereas the kidneys are the route for excretion of the nonvolatile acid. The body uses buffer systems to minimize changes in body fluid pH; the CO_2/HCO_3^- buffer system of the extracellular fluid is the most important.

The kidneys maintain acid-base balance by their excretion of acid equal to the amount of nonvolatile acid produced and ingested. The kidneys also prevent the loss of HCO_3^- in the urine by reabsorbing virtually all the HCO_3^- that is filtered at the glomerulus. Both the reabsorption of filtered HCO_3^- and the excretion of acid are accomplished by secretion of H^+ by the nephrons. Because the minimum pH of the urine is only 4.0 to 4.5, urinary buffers are necessary for effective excretion of acid. The two major urinary buffers are ammonia and phosphate (titratable acid). Ammonia is the more important urinary buffer because its production by the kidneys and excretion is regulated in response to acid-base disturbances.

4. *Acid-base disturbances.* Acid-base disturbances are of two general types: respiratory and metabolic. Respiratory acid-base disorders result from primary alterations in the blood P_{CO_2}. Elevation of P_{CO_2} produces acidosis, and the kidneys respond by an increase in excretion of acid. Conversely, reduction of P_{CO_2} produces alkalosis and renal acid excretion is reduced. The kidneys respond to respiratory acid-base disorders over several hours to days. Metabolic acid-base disorders result from primary alterations in the plasma $[HCO_3^-]$, which in turn results from addition of acid to or loss of alkali from the body. In response to metabolic acidosis, pulmonary ventilation is increased, which decreases the P_{CO_2}. An increase in the $[HCO_3^-]$ causes alkalosis. This decreases pulmonary ventilation, which elevates the P_{CO_2}. The pulmonary response to metabolic acid-base disorders occurs in a matter of minutes.

Bibliography

Journal articles

DuBose TD Jr: Reclamation of filtered bicarbonate, *Kidney Int* 38:584, 1990.

DuBose TD Jr et al: Ammonium transport in the kidney: new physiological concepts and their clinical implications, *J Am Soc Nephrol* 1:1193, 1991.

Friedman PA: Renal calcium transport: sites and insights, *News Physiol Sci* 3:17, 1988.

Gluck SL: Cellular and molecular aspects of renal H^+ transport, *Hosp Prac* 24:149, 1989.

Knepper MA, Packer R, Good DW: Ammonia transport in the kidney, *Physiol Rev* 69:179, 1989.

Kurtz I, editor: Renal ammoniagenesis, *Miner Electrolyte Metab* 16:241, 1990.

Books and monographs

Alpern RJ, Stone DK, Rector FC Jr: Renal acidification mechanisms. In Brenner BM, Rector FC Jr, editors: *The kidney*, ed 4, Philadelphia, 1991, WB Saunders.

Berndt TJ, Knox FG: Renal regulation of phosphate excretion. In Seldin DW, Giebisch G, editors: *The kidney: physiology and pathophysiology*, ed 2, New York, 1992, Raven Press.

Cogan MG, Rector FC Jr: Acid-base disorders. In Brenner BM, Rector FC Jr, editors: *The kidney*, ed 4, Philadelphia, 1991, WB Saunders.

Giebisch G, Malnic G, Berliner RW: Renal transport and control of potassium excretion. In Brenner BM, Rector FC Jr, editors: *The kidney,* ed 4, Philadelphia, 1991, WB Saunders.

Seldin DW, Giebisch G, editors: *The regulation of acid-base balance,* New York, 1990, Raven Press.

Stanton BA, Giebisch G: Renal potassium transport. In Windhager EE, editor: *Handbook of physiology: renal physiology,* ed 2, New York, 1992, Oxford University Press.

Suki WN, Rouse D: Renal transport of calcium, magnesium. In Brenner BM, Rector FC Jr, editors: *The kidney,* ed 4, Philadelphia, 1991, WB Saunders.

Valtin H, Gennari FJ: *Acid-base disorders: basic concepts and clinical management,* Boston, 1987, Little, Brown & Co.

SECTION
IX

THE ENDOCRINE SYSTEM

Saul M. Genuth

CHAPTER

44

General Principles of Endocrine Physiology

Endocrinology was classically defined as a discipline concerned with the "internal secretions of the body." The original concept—that a chemical substance called a hormone, liberated by one kind of a cell, is carried by the bloodstream to act on a distant target cell—represented a major advance in physiological understanding. It suggested a basic mechanism for maintaining the stability of the internal milieu in the face of irregular nutrient, mineral, and water fluxes, as well as physical alterations in the environment. Secretion of the hormone was evoked by a specific change in that milieu; as a result of the hormone's action on its target cells, the change was counteracted and chemical or physical homeostasis (the desired status quo) was restored. However, this basic homeostatic notion has grown increasingly complex, and the overall mission of the endocrine system is now understood also to include regulation of growth, maturation, body mass, reproduction, senescence, and behavior.

A diversity of endocrine cell types, locations, and organization has been found; moreover the number of known hormones and their molecular variety has multiplied. Target cells of hormone action now include other hormone-producing cells, and some hormones now are known to require chemical modification at intermediate sites between the gland of origin and the target cells before their mission can be accomplished. A complementary group of target cell substances—hormone receptors—has been found to play an essential role in mediating hormone action, which is known to involve multiple intracellular mechanisms. Much of the control of hormone secretion has been found to depend on the feedback principle described on p. 817. In addition, an important relationship between neural and endocrine function has been established. Because of this complexity and diversity, students are in danger of being overwhelmed unless they learn to relate the functional characteristics of each component of the endocrine system to a set of unifying principles. The major purpose of this chapter is to provide such a framework for the subsequent chapters.

Relationship Between Endocrine and Neural Physiology

In a conceptual sense the nervous system and the endocrine system have important functional similarities. Each is basically a system for signaling. Each operates in a stimulus-response manner. Each transmits signals that in some cases are highly localized, narrow, and unitary in purpose, and in other cases are widespread, broad, and diverse in purpose. Each system is crucial to the cooperative physiological functioning of the multiplicity of highly differentiated cells, tissues, and organs that make up the human organism. Therefore *the nervous system and the endocrine system together must integrate incoming stimuli so as to integrate the organism's response to changes in its external and internal environment.* Recent advances in immunohistochemistry, cellular physiology, and molecular biology reveal increasingly intimate relationships between neural function and hormone secretion. The very distinction between a hormone and a neurotransmitter has grown blurred. It may not be fanciful either to think of a circulating hormone as a signal molecule freed from the confines of a single axon to reach all responsive cells, or to think of a neurotransmitter as a signal molecule captured from the circulation to provide a restricted chemical connection between two specific cells separated in space.

This close relationship between the nervous and endocrine systems is illustrated by several common characteristics:

1. Neurons and endocrine cells are capable of secreting into the blood stream.
2. Endocrine cells and neurons generate electrical potentials and can be depolarized.

3. Peptides originally discovered as products of endocrine cells are now found to have neurotransmitter functions as well. Likewise, molecules nominally considered to be neurotransmitters can act as hormones.
4. A single cell can produce biogenic amine neurotransmitters and peptide hormone molecules.
5. A single gene can be transcribed and translated to yield either or both a peptide neurotransmitter and a peptide hormone.

Two examples illustrate the coordinated function of the endocrine and nervous systems. (1) A fall in the plasma glucose to a dangerously low level is sensed in the brain and liver. The sympathetic nervous system, neurohormones from the hypothalamus, and hormones from the anterior pituitary gland, the adrenal cortex, the adrenal medulla, and the pancreatic islets all act on target cells in liver, muscle, and adipose tissue to restore plasma glucose to normal. (2) A significant decrease in the circulating blood volume is sensed by baroreceptors, the cardiac atria, the kidney, and the brain. The sympathetic nervous system, a neurohormone from the posterior pituitary gland, and hormones from the cardiac atria, the adrenal medulla, the adrenal cortex, and the kidney act on target cells in blood vessels and kidney to restore blood volume.

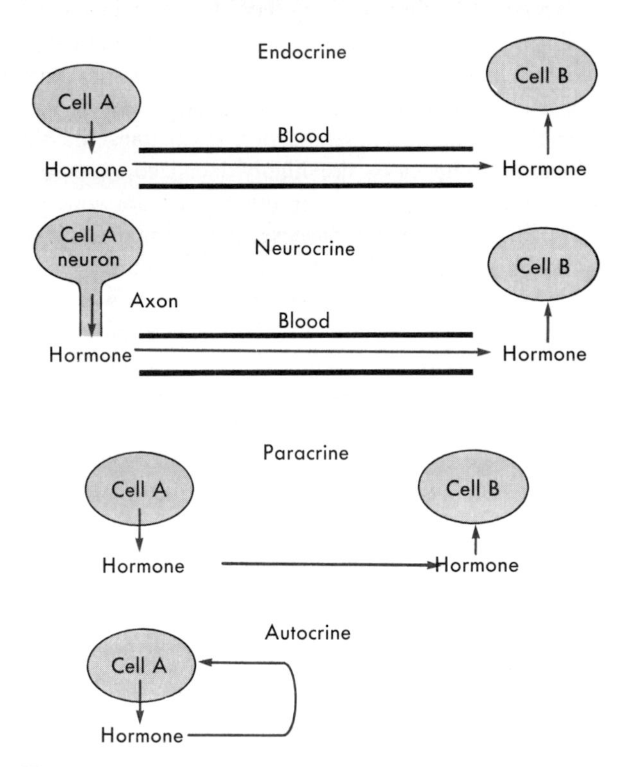

■ **Fig. 44-1** Schematic representation of mechanisms for cell-to-cell signaling via hormone molecules. In *endocrine* function the signal is carried to a distant target via the blood stream. In *neurocrine* function the hormone signal originates in a neuron, and after axonal transport to the blood stream it is carried to a distant target cell. In *paracrine* function the hormone signal is carried to an adjacent target cell over short distances via the interstitial fluid. In *autocrine* function the hormone signal acts back on the cell of origin or adjacent identical cells.

As our understanding has grown, the spectrum of hormonal signaling has expanded and we now define endocrine, neurocrine, paracrine, and autocrine effects (Fig. 44-1). **Endocrine** function is transmission of a molecular signal from an endocrine cell through the blood stream to a distant target cell. **Neurocrine** function is transmission of a molecular signal from a neuron first down its axon and then into the blood stream to a relatively distant target cell. **Paracrine** function is transmission of a molecular signal from one cell type to a neighboring different cell type by diffusion through intercellular fluid channels. **Autocrine** function is transmission of a molecular signal released into the intercellular fluid back to the cell of origin or to neighboring identical cells.

According to the route by which it is transmitted, the same messenger molecule may function as an endocrine hormone (blood stream conveyance), as a neurotransmitter (axonal conveyance), as a neurohormone (combined axonal and blood stream conveyance), or as a paracrine or autocrine hormone (local conveyance). The effect produced by the signaling molecule then depends on the target cell and the intracellular mechanisms that are activated. For example, a hypothalamic cell (see Chapter 48) may secrete the neurohormone somatostatin into blood, whereby it reaches the pituitary gland and inhibits the release of a hormone that stimulates growth; a brain cell may transmit somatostatin to a second brain cell so as to alter behavior; and one cell in the pancreatic islets may release somatostatin into the fluid bathing adjacent cells and inhibit their release of insulin (see Chapter 46).

Our knowledge of the relationship between neural and endocrine function also helps to illuminate the pathogenesis of some endocrine diseases. For example, in certain families, separate neoplasms tend to arise within seemingly disparate neural and endocrine tissue in the same individual; the connection between such events is becoming clearer. In other such instances, a single neoplasm may be associated with manifestations that can be traced to simultaneous production of a peptide hormone and a neurotransmitter molecule.

■ *Types of Hormones*

Hormone molecules fall into three general chemical classes. The first to be discovered—the amines—includes thyroid hormones and catecholamines. Both originate from the amino acid tyrosine and retain the aliphatic α-amino group. Introduction of a second hydroxyl group in the ortho- position on the benzene ring is characteristic of the catecholamines, whereas iodination of the benzene ring distinguishes the thyroid hormones. The second group is composed of proteins and peptides. In some instances, protein hormones with similar structures but dissimilar missions have originated from a common ancestral gene during evolution. In other instances a single progenitor protein gives rise to several hormone offspring of different sizes, with some sharing amino acid sequences and some having overlapping action. The third chemical group is the steroids, which include adrenal cortical and reproductive gland hormones and the active metabolites of vitamin D. Cholesterol is the common precursor in this class. Modification of side chains, hydroxylation at various sites, and ring aromatization confer individual biological activities on the various steroid hormones.

Hormone Synthesis

A protein or peptide hormone is synthesized on the rough endoplasmic reticulum in the same biochemical manner as other proteins. The appropriate amino acid sequence is dictated by a specific messenger RNA that results from transcription of the hormone gene. In general, a single gene determines the structure and synthesis of a single protein or peptide hormone. However, multiple genes containing the same sequence or only slightly varied sequences directing a single peptide hormone have also been described. Alternatively, a unique gene may give rise to more than one primary RNA message by inclusion or exclusion of particular exons in the processes of excision and splicing. Thus one gene may direct the synthesis of different peptides in various cells.

The DNA molecules for many protein hormones have been cloned and structured, allowing deduction of the structure of the primary gene product. This has led to discovery of previously unknown additional peptide products whose function remains to be ascertained. Recombinant DNA technology also now permits synthesis of authentic human protein hormones such as insulin and growth hormone for therapeutic purposes.

The process of peptide hormone synthesis is illustrated in Fig. 44-2. Translation of the mature RNA message begins with an N-terminal signal peptide sequence. When this is complete, translation temporarily ceases, while the signal peptide attaches the message to the endoplasmic reticulum receptors via "docking proteins." Translation then resumes until the entire encoded peptide sequence, known at this stage as a **preprohormone,** is formed. The signal peptide is then cleaved, resulting in a **prohormone,** which is simultaneously directed into the cisternal space for transport to the Golgi apparatus. The prohormone contains the hormone along with other peptide sequences, some of which may function to ensure proper folding of the hormone peptide chain so as to permit formation of intramolecular linkages. Other peptide sequences within

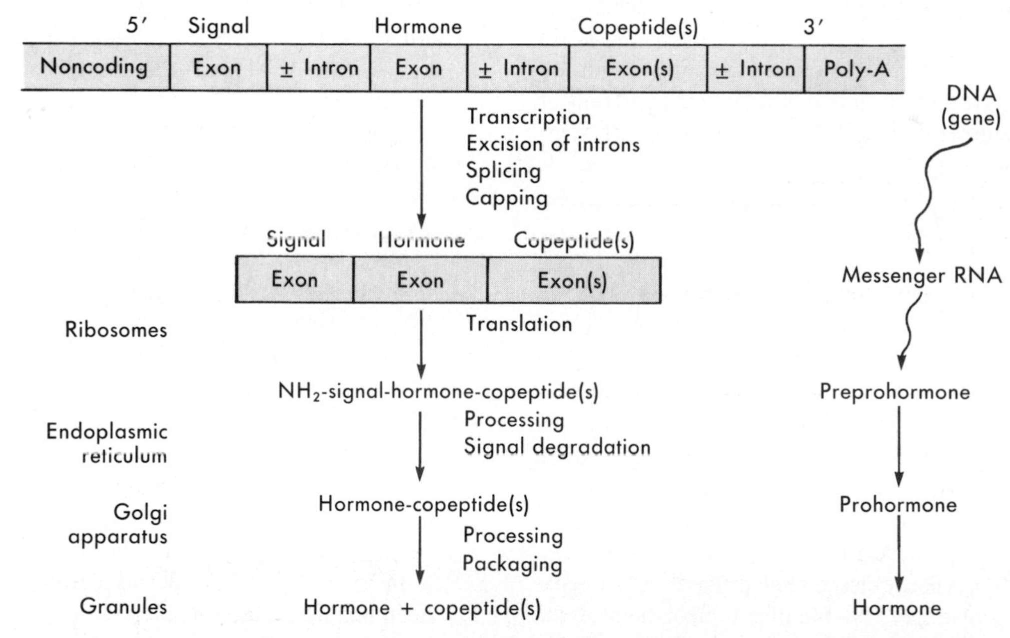

Fig. 44-2 Schematic representation of peptide hormone synthesis. In the nucleus the primary gene transcript undergoes excision of introns, splicing of the exons, capping of the 5′ end, and addition of poly-A at the 3′ end. The resultant mature messenger RNA enters the cytoplasm, where it directs the synthesis of a preprohormone peptide sequence on ribosomes. In this process the N-terminal signal is removed, and the resultant prohormone is transferred vectorially into the endoplasmic reticulum. The prohormone undergoes further processing and packaging in the Golgi apparatus. After final cleavage of the prohormone within the granules, they contain the hormone and copeptides ready for secretion by exocytosis.

the prohormone may have related or independent functions of their own; they are cosecreted with the hormone. Within the Golgi apparatus, the prohormone is packaged for storage in secretory granules. The latter may contain proteolytic enzymes necessary for subsequent conversion of prohormone to hormone or for elimination of copeptides. Golgi processing of prohormones to hormones may also involve glycosylation and phosphorylation. In addition, many secretory granules of peptide hormones contain a soluble acidic protein of unknown function—**chromogranin.**

The possibility of genetic errors in protein and peptide hormone synthesis is appreciable. Single amino acid substitutions or deletions in active sites can alter the degree or specificity of biological action. If such errors occur in accessory peptides of preprohormones, they may prevent normal processing to the active hormone. In a number of instances, such mutations are rare causes of hormone deficiency states.

The amine and steroid hormones are synthesized from tyrosine and cholesterol, respectively, through a sequential series of discrete enzymatic reactions. The intermediate products of the pathway may have hormonal properties of their own. This multiplicity of distinctive steps leads to the prediction that a variety of defects in hormone synthesis (and their clinical consequences) could arise from single gene–single enzyme mutations and deletions or from selective drug-induced enzyme inhibition. In fact, numerous examples of congenital enzyme deficiencies in adrenal and gonadal steroid hormone and in thyroid hormone biosynthesis do occur. The resultant clinical states may reflect both the deficiency of product hormone and the accumulation of hormone precursors.

■ *Hormone Release*

Both protein and catecholamine hormones are stored in secretory granules. Hormone release is then accomplished by the process of exocytosis (Fig. 44-3). The stimulus to secretion is usually followed by an immediate rise in cytosolic calcium taken up from the extracellular fluid and mobilized from intracellular bound stores. This initiates movement of secretory vesicle organelles to the plasma membrane, a process involving participation of microtubule and microfilament elements. Specific vesicle-associated proteins and a guanosine triphosphate (GTP) binding protein help attach the secretory granules to appropriate sites. After fusion of the granule and plasma membranes, gap junctions and small pores form and the hormone is released into the extracellular fluid along with copeptides, cleavage enzymes, chromogranin, and any other granule contents. The membrane material from the empty core may then be reprocessed. In addition to the stimulated release of protein hormones from storage granules by exocytosis, a low direct basal rate of constitutive secretion of newly synthesized prohormones, partially processed prohormones, or the hormone itself also occurs.

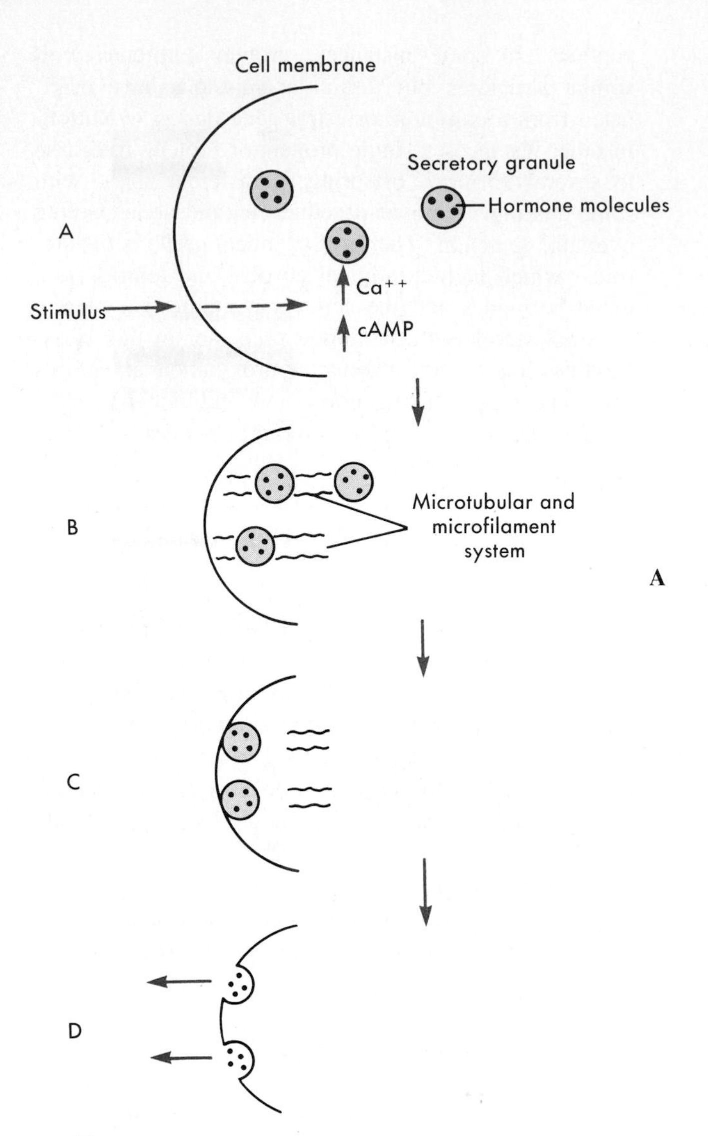

■ **Fig. 44-3** **A,** Secretion of peptide hormones via exocytosis is initiated *(A)* by application of a stimulus that raises intracellular Ca^{++} levels and also usually raises cytosolic cAMP. The secretory granules are lined up and translocated to the plasma membrane *(B)* via activation of a microtubular and microfilament system. The membrane of the secretory granule fuses with that of the cell *(C)*. The common membrane is lysed *(D)*, releasing the hormone into the interstitial space.

In the case of thyroid and steroid hormones storage does not take place in discrete granules, although the hormones may be compartmentalized in the cell. Once the hormones have appeared in free form within the cytoplasm, they apparently leave the cell by simple transfer through the plasma membrane.

These modes of hormone synthesis and release are essentially unicellular. However, more complicated patterns of hormone production also are encountered. Two adjacent cell types in a single gland may interact so that

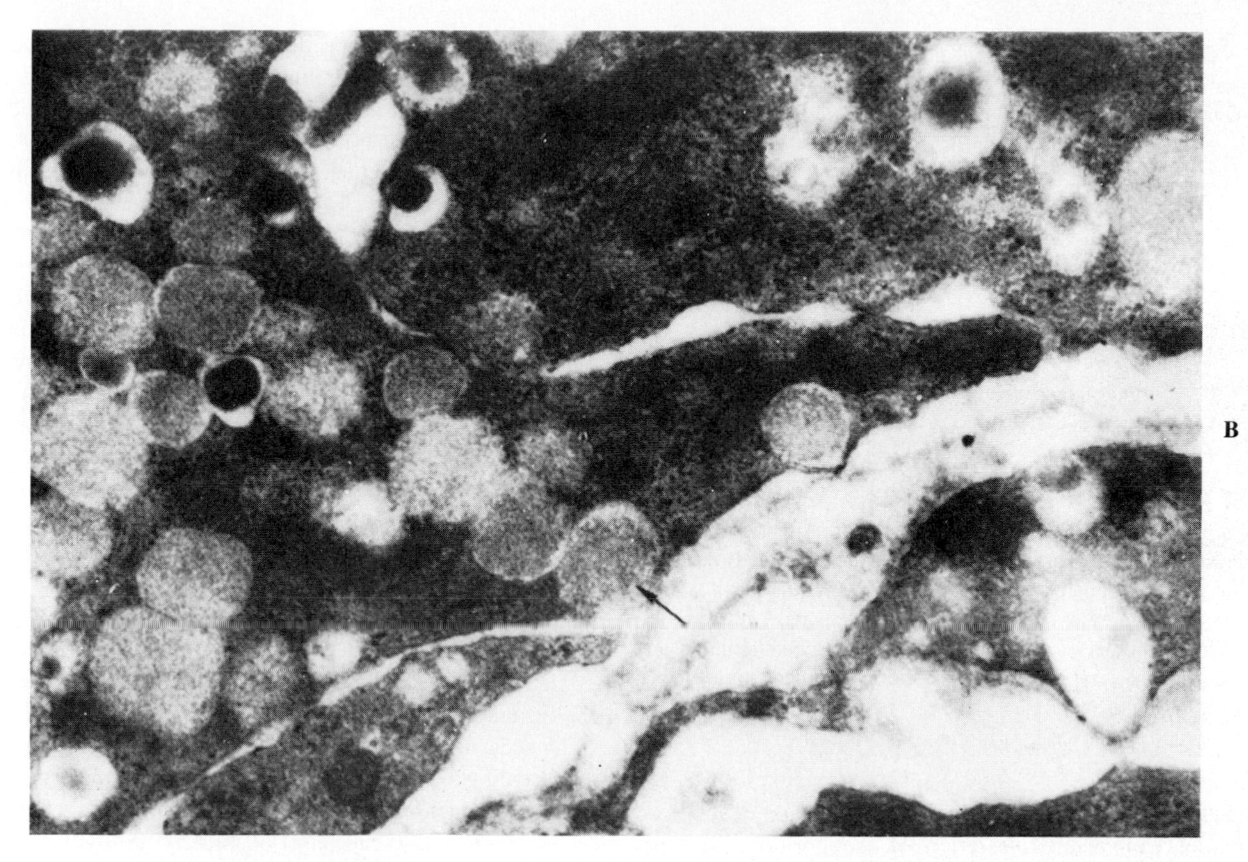

■ **Fig. 44-3, cont'd** **B,** Insulin secretory granules in a β-cell being stimulated by glucose. The arrow indicates a granule undergoing exocytosis. (From Lacy PE: Beta cell secretion—from the standpoint of a pathobiologist, *Diabetes* 19:895, 1970. From the American Diabetes Association.)

hormone A from cell A is modified in cell B to produce hormone B with an entirely different spectrum of biological effects. In this manner, for example, estrogens are produced from androgens in the gonads. In a further extension of this principle, peripheral tissues that are ordinarily considered to be nonendocrine may carry out similar conversions. A second mode of hormone production involves modification of a precursor molecule of low activity to one of higher activity by successive steps in several tissues. For example, a sterol synthesized in the skin requires actions by the liver and kidney to produce the most potent vitamin D hormone. Lastly, production of peptide hormones can even occur in the circulation itself from a protein precursor. A prototype for this pattern is the synthesis of angiotensin—a peptide hormone—from a protein secreted by the liver and acted on sequentially by enzymes from the kidney and the lung.

■ *Regulation of Hormone Secretion*

A number of general mechanisms that govern the secretion of hormones are the following:

Feedback control
Hormone-hormone
Substrate-hormone
Mineral-hormone

Neural control	*Chronotropic control*
Adrenergic	Oscillating
Cholinergic	Pulsatile
Dopaminergic	Diurnal rhythm
Serotoninergic	Sleep-wake cycle
Endorphinergic-enkephalinergic	Menstrual rhythm
	Seasonal rhythm
Gabaergic	Developmental rhythm

The feedback principle is universally operative. **Negative feedback** *is most common and acts to limit the excursions in output of each partner in the pair* (Fig. 44-4). In the simplest instance hormone A, which stimulates secretion of hormone B, in turn will be inhibited by an excess of hormone B. This straightforward mechanism dominates the relationship between hormones of the pituitary gland and its target glands. Secretion of a hormone that either accelerates the production or retards the utilization of a particular substrate will increase its concentration in plasma; the resultant circulating excess

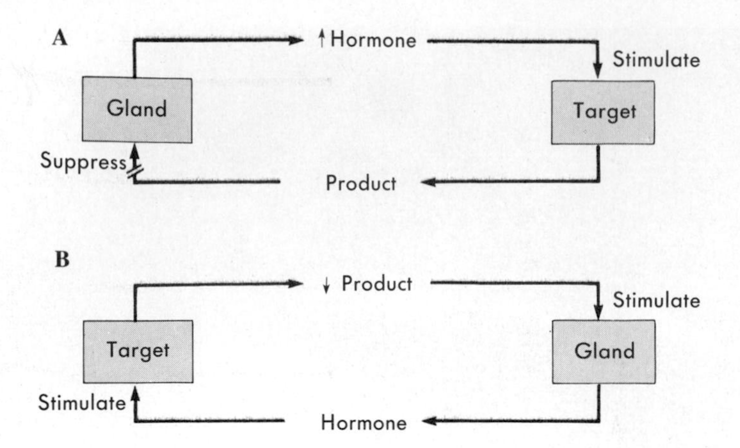

■ **Fig. 44-4** Negative feedback principle. **A,** A primary increase in hormone secretion stimulates a greater output of product from the target cell. The product then feeds back on the gland to suppress further hormone secretion. In this fashion hormone excess is limited or prevented. **B,** A primary decrease in output of product from the target cell stimulates the gland to secrete hormone. The hormone then stimulates a greater output of product from the target cell. In this fashion the product deficiency is limited or corrected.

of that substrate will inhibit further secretion of the hormone. Conversely, secretion of the hormone will be stimulated by a circulating deficit of that substrate. On the other hand, the secretion of a hormone that impedes the production or accelerates the utilization of a particular substrate, thus decreasing its plasma concentration, will be stimulated by a circulating excess but inhibited by a circulating deficit of the substrate.

Positive feedback, which is less common, *acts to **amplify** the initial biological effect of the hormone.* Thus hormone A, which stimulates secretion of hormone B, in turn may be initially stimulated to greater secretion rates by hormone B, but only through a limited dose response range. Once sufficient biological momentum for secretion of hormone B has been obtained, other influences, including negative feedback, will reduce the response of hormone A to fit the final biological purpose.

Neural control acts to evoke or suppress hormone secretion in response to both external and internal stimuli. These may arise from visual, auditory, olfactory, gustatory, tactile, or pressure sensations and may be perceived consciously or unconsciously. Pain, emotion, sexual excitement, fright, injury, and stress all can modulate hormone secretion through neural mechanisms. Examples include the release of oxytocin, which fills the milk ducts in response to the stimulus of suckling, or the release of aldosterone, which augments the volume of the circulatory system in response to upright posture.

Many hormones are secreted in distinct pulses. Furthermore, certain patterns are dictated by rhythms that may be genetically encoded or acquired. Circadian or daily rhythms can be demonstrated even within certain individual cells in culture. For example, pineal gland cells show regular and coordinated 24-hour variation in synthesis of the hormone melatonin, in the enzyme N-acetyltransferase necessary for melatonin synthesis, and in the stimulatory molecule cyclic-adenosine-monophosphate (cAMP). A similar intrinsic 24-hour cycle can be demonstrated for a number of whole organism endocrine and neural physiological processes, even in a timeless environment lacking day or night. However, such intrinsic endocrine oscillations often become entrained with independent light-dark, day-night, or sleep-wake cycles and can be shifted forward or backward with such cycles. The suprachiasmatic nucleus of the hypothalamus may be the center for intrinsic oscillations or cycles; modifications of these cycles occur via input from the retina, thalamus, midbrain, hippocampus, and pineal gland.

Seasonal variation in hormone secretion also occurs, which may reflect the influence of temperature, sunlight, or tides. Some of these rhythms appear atavistic in humans, having probably served to fulfill biological needs of evolutionary ancestors. Perhaps the most intriguing of all are those patterns of hormone secretion that coincide with and are unique to developmental stages, such as the onset of puberty. A diverse and still growing number of neurotransmitter molecules carries these signals to the endocrine cells.

■ *Hormone Action*

Three major sequential steps are involved in eliciting the response of a target cell to a hormonal stimulus:

1. The hormone must be recognized by a specific cell receptor.
2. The hormone receptor complex must be coupled to a signal-generating mechanism.
3. The generated signal (second messenger) then causes a quantitative change in intracellular processes by altering the activity or concentration of enzymes, carrier proteins, and so on. Only if all three steps are intact can a particular hormone effectively stimulate a particular cell (Fig. 44-5.)

There are two basic schemes for accomplishing these steps. In the first, mainly employed by peptide/protein and catecholamine hormones, the receptor for the hormone and the signal-generating system is located within or adjacent to the plasma membrane of the cell. In this case, the essential information for triggering the response actually lies in the receptor molecule. By its occupancy, the hormone changes the receptor's conformation and allows transmission of the information the receptor contains. The hormone is essentially only an extracellular signal. This type of hormone response is elicited within seconds to minutes. In the second case, largely employed by steroid and thyroid hormones, the

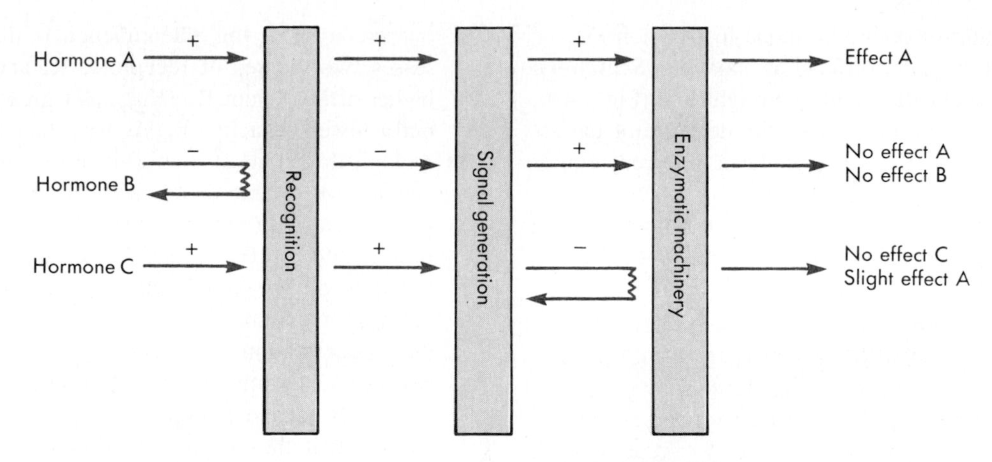

■ **Fig. 44-5** Hormone-cell interaction. Hormone A is recognized by this cell through binding with its specific receptor. An intracellular signal is generated that stimulates the appropriate enzymatic machinery, and effect A is produced. Hormone B is not recognized because the cell lacks a receptor for it. Thus hormone B cannot produce effect A in this cell even though it might operate through an identical enzymatic machinery in its own target cell. Hormone C may be slightly recognized by this cell through individual overlap with the receptor to hormone A. Although a weak intracellular signal may be generated, no effect of C results because the cell lacks the appropriate enzymatic machinery. To a minor extent, however, hormone C may produce effect A.

hormone must enter the cell, occupy the receptor, and in combination with the receptor, interact with DNA molecules in the nucleus to generate a message. In this situation, the essential information for triggering a response lies in the hormone and the receptor, coupled together. The hormone is a true intracellular signal. This type of hormonal action requires minutes to hours or even days for its expression.

■ *Receptor Kinetics*

A given hormone may act through different receptor molecules in different target cells. However, multiple actions of a given hormone in the same target cell are initiated by binding to a sole receptor. There are generally 2,000 to 100,000 receptor molecules per cell. This number often guarantees that receptor availability will not be rate limiting for hormone action.

The association of a hormone with its receptor is a reversible reaction that appears to obey the following molecular chemical kinetics:

$$[H] + [R] \rightleftharpoons [HR] \tag{1}$$

$$K_{assoc} = \frac{[HR]}{[H]\,[R]} \tag{2}$$

$$\frac{[HR]}{[H]} = K_{assoc} \times [R] \tag{3}$$

where

[H] = free hormone in solution

[R] = unoccupied receptor

[HR] = bound hormone = occupied receptor

R_0 = initial receptor capacity = [R] + [HR]

K_{assoc} = affinity constant

If a constant amount of receptor is incubated in vitro with increasing concentrations of hormone, the amount of bound hormone increases until receptor occupancy reaches 100%. At this point the number of bound hormone molecules equals the total number of originally available receptor molecules, i.e., the receptor capacity (R_o). At the same time, as the free hormone concentration approaches infinity and the receptor occupancy approaches 100%, the ratio of bound hormone to free hormone progressively decreases and approaches zero. That is, as

$$\begin{gathered}[H] \rightarrow \text{Infinity} \\ [HR] \rightarrow R_o \\ \frac{[HR]}{[H]} \rightarrow 0\end{gathered} \tag{4}$$

The data obtained by incubating receptors with increasing amounts of hormone can be plotted in a meaningful way by simple substitution in equation 3. Since

$$[R] = R_o - [HR]$$

$$\frac{[HR]}{[H]} = K_{assoc} \times (R_o - [HR]) \tag{5}$$

$$\frac{[HR]}{[H]} = -K_{assoc}\,[HR] + K_{assoc}\,R_o \tag{6}$$

$$\frac{\text{Bound hormone}}{\text{Free hormone}} = -K_{assoc} \times \text{Bound hormone} + K_{assoc} \times \text{Receptor capacity}$$

Plotting the ratio of bound hormone to free hormone as a function of bound hormone is called a **Scatchard plot.** This theoretically yields a straight line (Fig. 44-6, *A*). The slope of the line equals the negative of the association constant (K_{assoc}), and the x intercept equals R_o (the receptor capacity).

In practice many Scatchard plots of hormone binding to receptor yield exponential curves (Fig. 44-6, *B*). One interpretation of this phenomenon is that the cell possesses two classes of receptors, R_1 and R_2. R_1 has a higher affinity than R_2 (K_{assoc1} is greater than K_{assoc2}) but a lower capacity (R_{o1} is less than R_{o2}). Often full biological action of a hormone is quantitatively accounted for by interaction with the apparent high affinity, low-capacity receptor, and requires occupancy of as little as 5% to 10% of the total receptor capacity. The function of the remaining "spare receptors" is not known for certain. However, they can act to increase the concentration of receptor-bound hormone (HR) when either hormone concentration or receptor affinity is low. A second interpretation of nonlinear Scatchard plots is that the affinity of unoccupied receptor molecules for hormone is decreased by the presence of adjacent occupied receptor molecules. This phenomenon is termed **negative cooperativity.** It has the effect of moderating the increase in receptor-bound hormone and thus moderating the increase in hormone action should hormone concentration rise too precipitously. Although Scatchard plots provide a conceptual framework for visualizing the kinetics of hormone-receptor interaction, the complete accuracy of the derived receptor capacities and association constants is open to question. Therefore they are best used to assess the effects of physiological manipulations in a defined system rather than to compare values obtained in different systems.

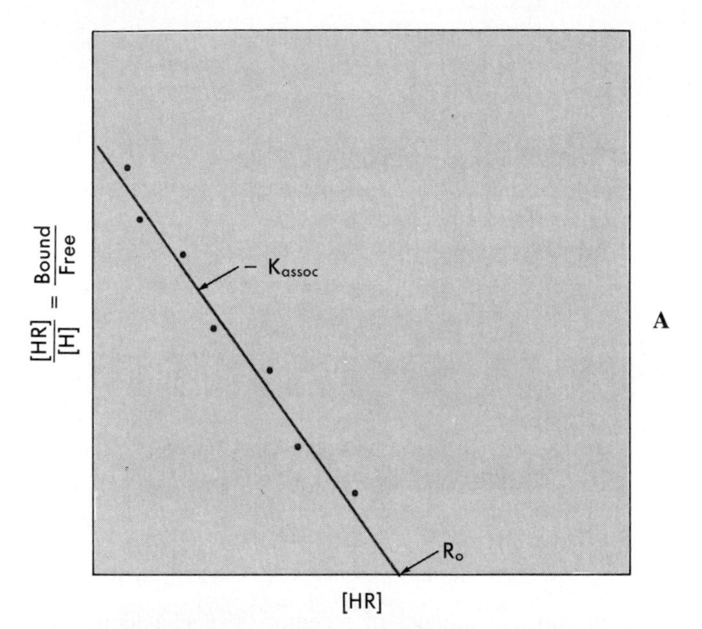

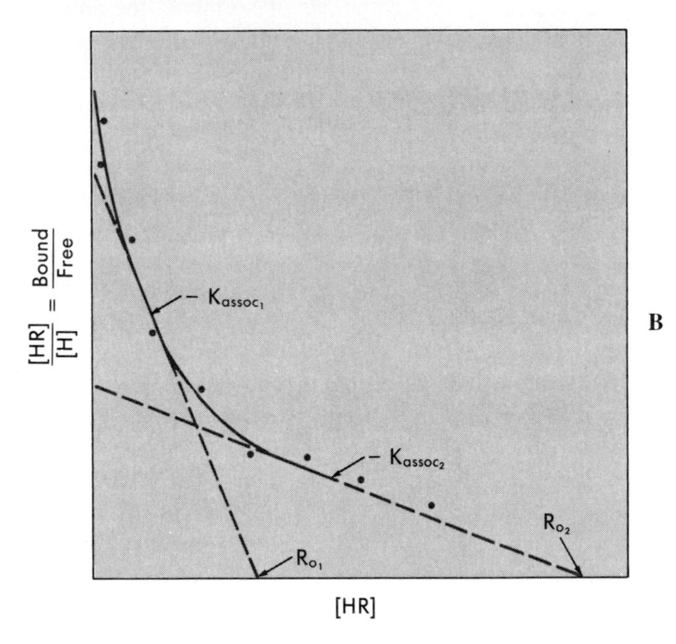

■ **Fig. 44-6** Scatchard plot. **A,** Linear plot results when the hormone reacts with a single receptor class and no cooperativity is present. **B,** An exponential plot results when the hormone reacts with multiple classes of receptors in the same cell or when the reaction with a single class of receptor is influenced by cooperativity. The negative of the association constant, K_{assoc}, equals the slope of the line. The receptor number, R_o, equals the intercept with the x axis.

Regulation of the hormone receptor provides another mechanism for regulating hormone action. Rearrangement of equation 5 to provide an expression for HR yields

$$[HR] = R_o \times \frac{K_{assoc}\,[H]}{K_{assoc}\,[H] + I} \qquad (7)$$

Thus it is seen that HR is directly proportional to R_o, the initial receptor number. An increase in receptor number (R_o) raises the maximal [HR] obtainable at saturating concentrations of hormone. This would raise the maximal responsiveness of the cell for those hormone effects in which receptor binding, rather than a later step in hormone action, is rate limiting. This may be seen where there are no spare receptors, as is generally the case for steroid and thyroid hormones. At submaximal hormone concentrations the increase in [HR] produced by an increase in R_o would enhance the sensitivity of the cell. This is commonly seen with peptide hormones.

Receptor capacity often is regulated by its own hormone. In many instances the regulation is inverse, i.e., a sustained excess of hormone decreases the number of its receptors per cell, which is **down regulation.** This acts to lessen the effect of chronic exposure to excess hormone. However, in some instances, at low concentrations of hormone, the relationship is direct, i.e., the hormone appears to recruit its own receptors—**up regulation**—after a period of exposure. This amplifies the cell's response to the hormone. An increase in receptor affinity (K_{assoc}) also will increase [HR] and the sensitivity of the cell to hormone stimulation. Receptor affinity can be altered by factors such as pH, osmolality, ion concentrations, and substrate levels. In a growing number of diseases, abnormalities of the hormone receptor are involved in pathogenesis. Therefore interest in receptor structure, biosynthesis, degradation, function, and regulation is currently intense.

Plasma Membrane Receptor Systems

Peptide and catecholamine hormone receptors are large complexes, often composed of subunits, and of molecular weight 50,000 to 200,000. They all appear to be glycoproteins, and in several instances their structure resembles that of immunoglobulins.

A specific extracellular portion of the receptor binds the hormone. The intramembranous portion may simply span the plasma membrane once or wind in and out of it a number of times, thereby anchoring the receptor and/or allowing interaction with more than one signal-generating mechanism within the plasma membrane. The intracellular tail of the receptor may contain a separate signal-generating mechanism. Some evidence suggests that the amino acid sequence within the receptor that binds a peptide hormone may be an internal image of a complementary amino acid sequence within the hormone. In other words, the DNA molecule directing synthesis of the receptor and that directing synthesis of the peptide hormone may have certain sequences with complementary base pairs.

The receptors tend to be concentrated in cellular microvilli. Hormone binding to these receptors may change their conformation and their distribution within the plasma membrane. This leads to clustering of the hormone-receptor complexes. After receptor activation is completed, internalization of the complexes occurs at these sites by endocytosis, sometimes via coated pits or vesicles. Within the cell, lysozomal degradation of the complex occurs, with either destruction of the individual hormone and receptor molecules or recycling of the receptor molecules back into the plasma membrane. It remains unclear whether an internalized hormone-receptor complex can also mediate certain intracellular actions of the hormone before its disruption.

Coupling by G-Proteins

For many hormones, occupancy of the plasma membrane receptor initiates a sequence of reactions within the membrane bilayer. A family of coupling molecules known as G-proteins functionally links various receptors to nearby effector molecules. The latter, in turn, generate second messengers that mediate the hormones' intracellular actions (Fig. 44-7). All G-proteins are trimers; each has a unique α-subunit but a β-γ dimer subunit that is similar among the family members. The α-subunits bind to receptors, to effector molecules, and to guanosine diphosphate (GDP) and guanosine triphosphate (GTP). The β-γ-dimeric subunit may serve to attach the G-protein to the plasma membrane.

In its inactive state the G-protein is bound to GDP. Creation of a hormone receptor complex causes the appropriate adjacent G-protein alpha-subunit to bind to the occupied receptor, then to release its bound GDP and to bind instead GTP. The displacement of GDP by GTP activates the G-protein by causing the α-subunit to dissociate from both the hormone receptor and from its own β-γ-subunit. The α-subunit-GTP then moves to and binds to a nearby membrane effector molecule such as adenylyl cyclase, an ion channel carrier protein, or phospholipase-C. The activity of the effector molecule is either stimulated or inhibited by the α-subunit-GTP complex. In this process, the α-subunit then hydrolyzes its bound GTP to GDP and inorganic phosphate. The α-subunit-GDP then reassociates with its β-γ-dimer; this process reconstitutes the original inactive G-protein. The latter is again free to interact with another hormone-occupied receptor molecule and start another cycle. This coupling cycle is repeated with the G-protein α-subunit shuttling between receptor and effector molecules until the hormone dissociates from its receptor or is internalized and degraded. Mg^{++} is required for the G-protein cycle, which is driven by the energy derived from the hydrolysis of GTP to GDP. Because the effector generates many intracellular second messenger molecules, the G-protein mechanism of transduction amplifies greatly the original extracellular hormone signal (see Chapter 5).

Second Messengers

G-proteins couple hormone-receptor complexes to at least three different effector systems: the adenylyl cyclase-cAMP system, the calcium-calmodulin system, and the membrane phospholipase-phospholipid system (see Chapter 5).

The **adenylyl cyclase-cAMP** system was the first to be described and initiated the concept of a second messenger. The plasma membrane enzyme adenylyl cyclase catalyzes formation of cAMP from adenosine triphosphate (ATP) with Mg^{++} as cofactor. A stimulatory G-protein (Gs) therefore increases intracellular cAMP levels, whereas an inhibitory G-protein (G_1) decreases cAMP levels.

An increase in cAMP stimulates the activation of **protein kinase A** (Fig. 44-8, p. 823). This, in turn, activates a number of enzymes in numerous metabolic pathways by phosphorylating their kinases. Alternatively, cAMP-stimulated phosphorylation may deactivate other enzymes. Thus after hormone binding to re-

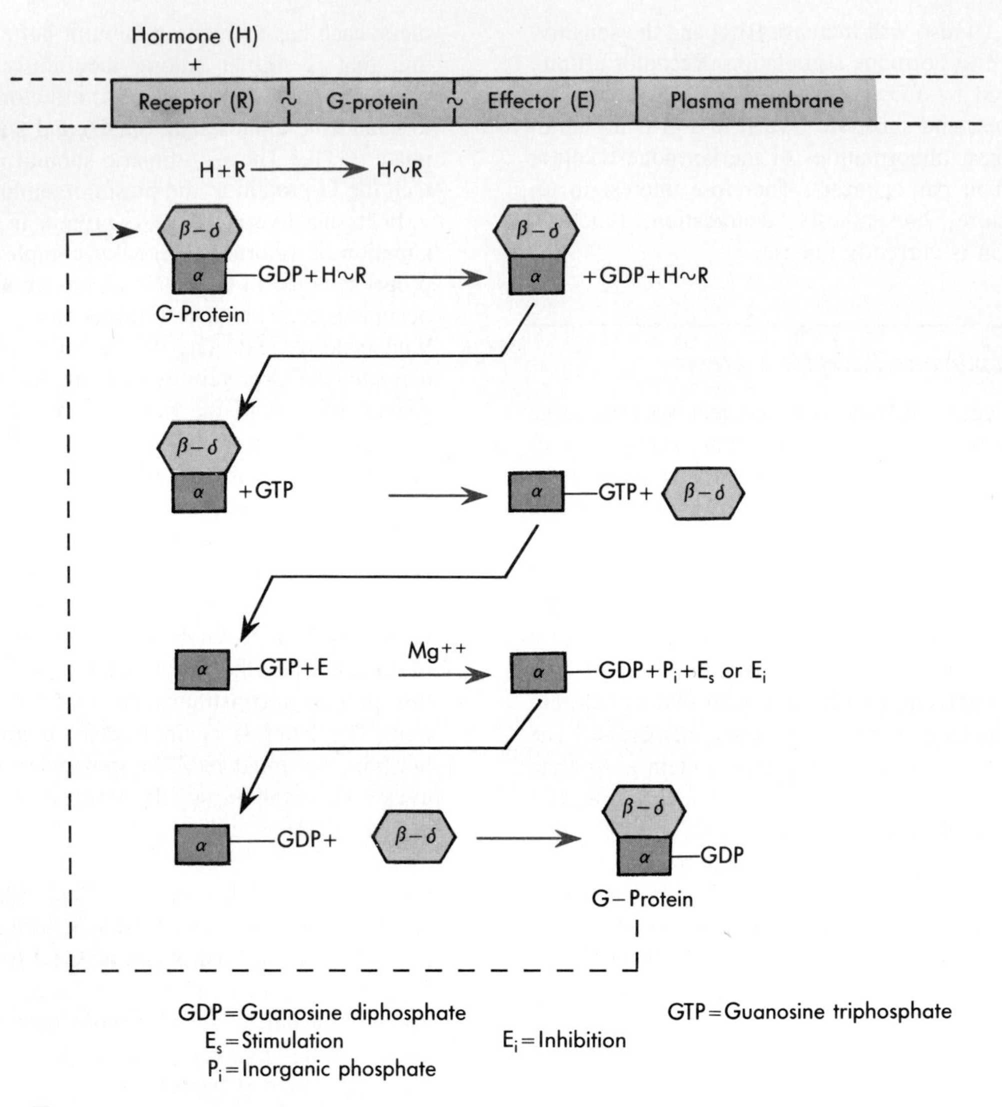

■ **Fig. 44-7** Mechanism of G-protein mediated hormone actions. G-protein transduces the hormone-receptor complex message to either stimulate or inhibit the effector in the plasma membrane. A GDP-GTP cycle fuels the sequence. See text for detailed description.

ceptor, this system generates a cascade of effects that ultimately changes the flux of metabolites in the cell. In the end either the storage or the release of an important metabolite may be facilitated. Activation or deactivation of reciprocal pathways by the same hormone can augment the result, for example, by simultaneously inhibiting the release pathway while stimulating the storage pathway. Adenylyl cyclase and cAMP are ubiquitously distributed. Therefore the specificity of enzyme responses to a particular hormone also may depend on the compartmentalization of cAMP increases within the cell or on the proximity of the activated adenylyl cyclase to target enzyme(s).

cAMP may also act as a hormone second messenger by altering gene expression. Target DNA molecules have a **cAMP regulatory element (CRE)** that binds a protein transcription factor known as **cAMP response element binding protein (CREB).** cAMP activates protein kinase A, which then phosphorylates CREB. The phosphorylated CREB is then capable of binding to CRE in a complex with another transcription protein. The result is to stimulate or inhibit RNA polymerase and transcription of the gene and thereby to stimulate or inhibit synthesis of a specific protein.

The actions of cAMP are terminated by its hydrolysis; this reaction is catalyzed by the enzyme phosphodiesterase. Because the activity of phosphodiesterase is also modulated by hormones via a G-protein, the level of cAMP is under dual regulation. Two hormones can function antagonistically if one stimulates adenylyl cyclase and the other stimulates phosphodiesterase.

A second system for transduction of the hormone signal is the **calcium-calmodulin system** (Fig. 44-9). As a result of hormone occupancy of its receptor, a specific

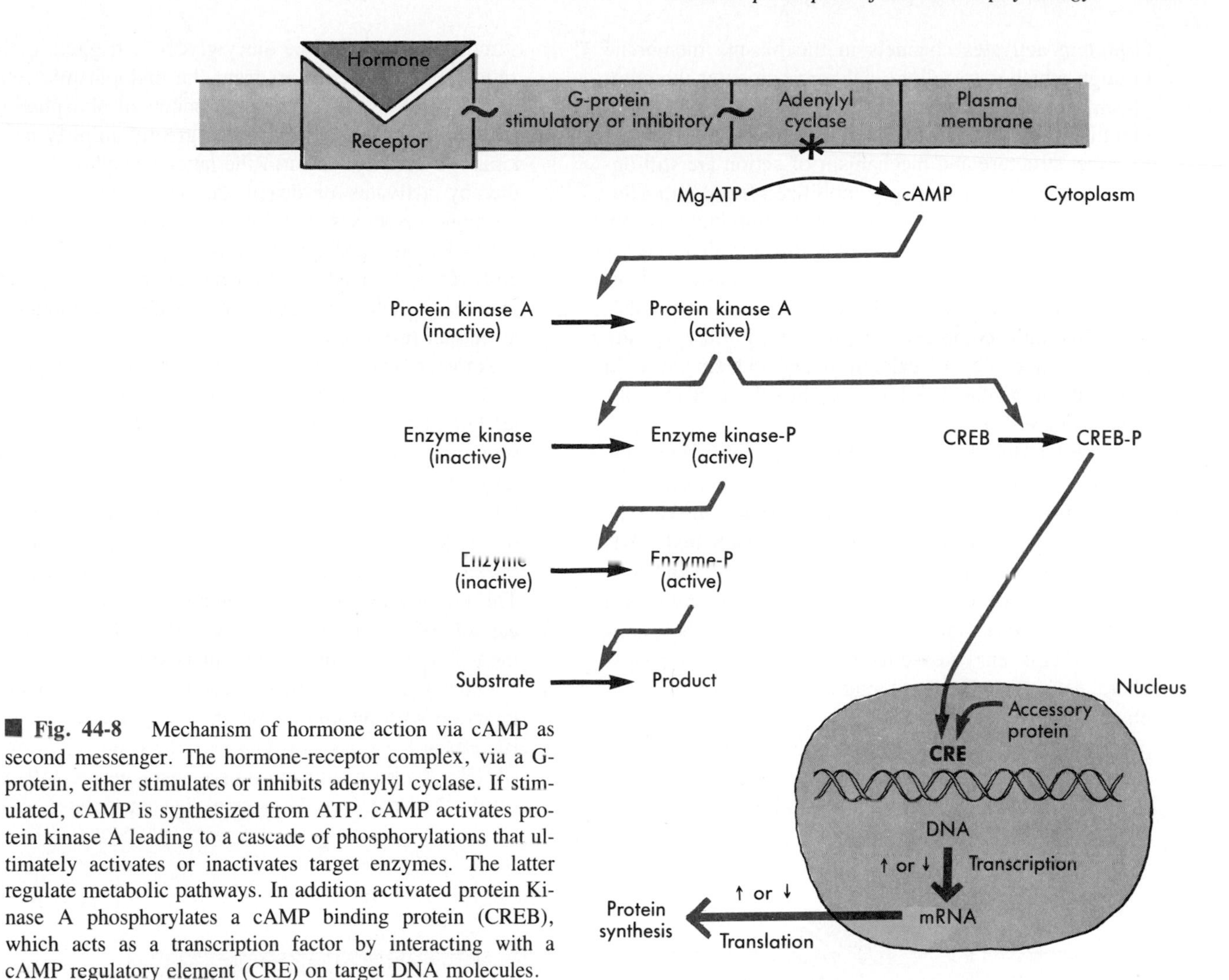

■ **Fig. 44-8** Mechanism of hormone action via cAMP as second messenger. The hormone-receptor complex, via a G-protein, either stimulates or inhibits adenylyl cyclase. If stimulated, cAMP is synthesized from ATP. cAMP activates protein kinase A leading to a cascade of phosphorylations that ultimately activates or inactivates target enzymes. The latter regulate metabolic pathways. In addition activated protein Kinase A phosphorylates a cAMP binding protein (CREB), which acts as a transcription factor by interacting with a cAMP regulatory element (CRE) on target DNA molecules.

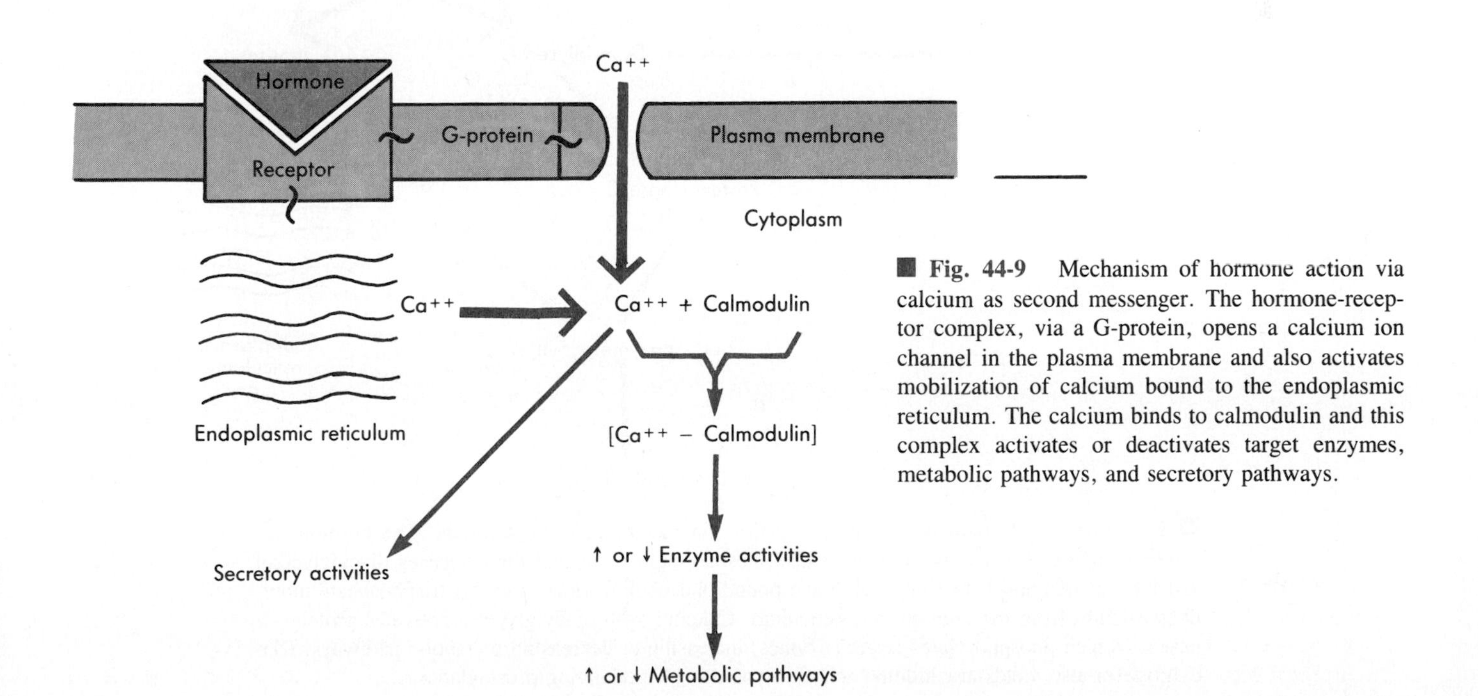

■ **Fig. 44-9** Mechanism of hormone action via calcium as second messenger. The hormone-receptor complex, via a G-protein, opens a calcium ion channel in the plasma membrane and also activates mobilization of calcium bound to the endoplasmic reticulum. The calcium binds to calmodulin and this complex activates or deactivates target enzymes, metabolic pathways, and secretory pathways.

G-protein activates channels in the plasma membrane through which extracellular calcium can enter the cytoplasm. Although a plasma membrane protein has been identified as essential to creating the calcium channel, its exact structure and mechanism of action are still unknown. Calcium may also be mobilized from intracellular reservoirs in the endoplasmic reticulum and possibly the mitochondria. The increased cytosolic calcium combines with its ubiquitous and specific binding protein, calmodulin, in various proportions. The different calcium-calmodulin complexes amplify or diminish the activities of a variety of calcium-dependent enzymes. In this fashion, metabolite levels within the cell are ultimately altered.

A third system for transduction of the initial hormone receptor signal is through intermediates generated from **plasma membrane phospholipids** (Fig. 44-10). The key phospholipid is phosphatidylinositol, which undergoes phosphorylation to phosphatidylinositol-4,5-bisphosphate. Hormone occupancy of its receptor initiates a G-protein activation of membrane-bound phospholipase-C. This enzyme splits phosphatidyl-4,5-bisphosphate to a diacylglycerol and inositol-1,4,5-trisphosphate. The diacylglycerols are potent activators of protein kinase-C. The latter is a calcium-dependent enzyme, and the diacylglycerol markedly increases its affinity for calcium. **Inositol trisphosphate** (IP_3) released simultaneously with the diacylglycerol, triggers mobilization of bound calcium from the endoplasmic reticulum. Therefore the combined products of phosphatidylinositol bisphosphate breakdown greatly amplify protein kinase-C activity. In turn, the latter phosphorylates and thereby activates or deactivates enzymes involved in hormone responses. Finally, arachidonic acid, which is derived from subsequent hydrolysis of the diacylglycerol, serves as a substrate for rapid synthesis of prostaglandins, which are themselves capable of modulating hormonal responses.

Another type of signal generation that is independent of direct G-protein transductions arises in the intracellular tail of plasma membrane receptors. Binding of the protein hormone changes the conformation of the receptor, exposing an intracellular site capable of receptor autophosphorylation. As a result, the receptor becomes a tyrosine kinase capable of phosphorylating tyrosine residues on other enzymes such as kinases or phosphatases. This increases or decreases their enzymatic activity, and a cascade of effects similar to that initiated by cAMP can then alter specific intracellular metabolic processes.

Other second messengers resulting from hormone interaction with plasma membrane receptors may also exist. For at least one hormone, atrial natriuretic hormone (ANH), cyclic guanosine monophosphate (cGMP) may fulfill this role. A G-protein that modulates cGMP

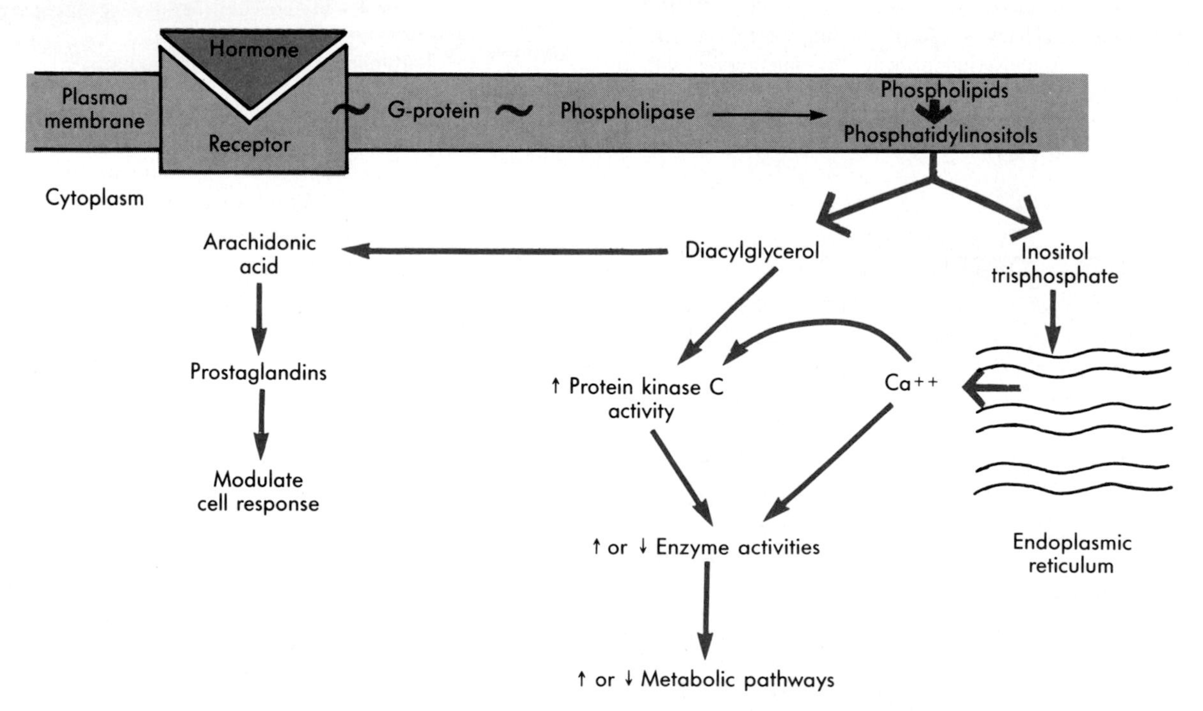

■ **Fig. 44-10** Mechanism of hormone action via membrane phospholipids. The hormone-receptor complex, via a G-protein, activates phospholipase C, which then releases diacylglycerol and inositol trisphosphate from membrane bound phosphoinositides. Inositol trisphosphate mobilizes calcium from the endoplasmic reticulum. Calcium with diacylglycerol activates protein kinase C, which phosphorylates target enzymes, increasing or decreasing metabolic pathways. Diacylglycerol also yields arachidonic acid for synthesis of modulating prostaglandins.

phosphodiesterase regulates cGMP levels. Finally, peptide and protein hormones can also cause increases or decreases in the concentration, as opposed to the activity, of enzymes and their messenger RNAs. This implies modification of gene expression. Because the cAMP mechanism described earlier cannot explain all these effects, other mechanisms for transmitting signals from the hormone-occupied plasma membrane receptor to the nucleus must exist.

As complex as each of these systems seems individually, it must be appreciated that a single hormone may operate through one or more of them simultaneously or in sequence. Each messenger may subserve a different function of the hormone. Moreover, the systems can interact with each other in important ways. For example, calcium-calmodulin can stimulate not only adenylyl cyclase activity itself, but also the activity of phosphodiesterase, the enzyme that hydrolyzes cAMP. Thus cAMP levels, initially increased by a rapid hormone activation of adenylyl cyclase, may be subsequently dampened by a later response to the same hormone involving phosphodiesterase activity. Likewise, in some situations, cAMP and protein kinase A products can inhibit the activity of the phospholipid messenger system by decreasing the generation of diacylglycerols. Thus an early stimulation of cell processes by hormone action through protein kinase C may be later dampened by hormone action through protein kinase A. Finally, cAMP-activated phosphorylation of plasma membrane receptors themselves may decrease their affinity for hormones. This constitutes still another negative feedback mechanism for limiting hormone action or its duration.

■ *Intracellular Receptor Systems*

In contrast to peptide and catecholamine hormones, steroid and thyroid hormones enter the cell and bind within minutes to predominantly nuclear receptors (Fig 44-11). These are oligomeric phosphorylated proteins having a molecular weight of 50,000 to 200,000 and three functional domains. They are all coded for by a superfamily of genes related to the cis-oncogenes. The C-terminus domain is unique for each of these receptors and binds the ligand hormone. The middle domain exhibits up to 50% homology among the various receptors and contains the DNA-binding site. The remaining N-terminus domain varies and has no assignable function as yet. The native receptors exist in inactive form, probably as oligomers and possibly in association with blocking proteins. Binding of the hormone to the C-terminus domain transforms or **activates** the receptor, a step that involves a change in size and conformation. The blocking protein is displaced, and this frees the middle domain for interaction with DNA. Phosphorylation of the receptor may also play a role in its activation. Within the nucleus, **acceptor proteins** are recruited and combined with the hormone receptor complex, the latter as dimers. This permits the DNA-binding fingers (Fig. 44-11) of the receptor to interact with a steroid or thyroid hormone regulatory element (HRE) on a target DNA molecule. It also fixes the DNA to the nuclear matrix. HREs are generally 8 to 15 base pairs long. Occupancy of the HRE activates various promotor elements on the DNA molecule, usually but not necessarily upstream from the capsite at the 5′ end. Constitutive or basal protein transcription factors may also be recruited for the hormone to act.

The hormone-activated promoter nucleotide sequences initiate transcription at the start site of the specific gene message by RNA polymerase II. The resultant "immature" RNA then undergoes maturation to messenger RNA by capping, excision of untranslated nucleotide sequences, and splicing. Translation of the RNA message in the cytoplasm results in synthesis of specific target proteins of the hormone. This mechanism of hormone induction may include both enhancement of the activity of positive regulatory elements and relief of repression exerted by negative regulatory elements in the gene. On the other hand, suppression of gene transcription can also result from steroid or thyroid hormone receptor effects. In these cases, binding of the hormone receptor complex to the HRE probably displaces a positive transcription factor from the DNA molecule. Since the gene is present in all cells, the expression of hormone action in specific cells depends primarily on the presence of its receptor. However, within the nucleus of non–target cells, other proteins may also prevent hormone action by masking the acceptor protein. Finally, very early in development, otherwise hormone-responsive genes may be "closed" in non–target cells by processes such as methylation.

Proteins whose synthesis is regulated up or down by thyroid and steroid hormones may be enzymes, structural proteins, receptor proteins, transcriptional proteins that will regulate the expression of other genes, or proteins that are exported by the cell. By this hormone mechanism, enzymes are either induced or repressed rather than activated or inactivated as they are by the plasma membrane messenger systems. Response of metabolic pathways can either be accelerated or retarded. Other consequences of hormone action include alterations in the processing of the primary RNA product, in the turnover of messenger RNA molecules, or in posttranslational modification of proteins. The transcription mechanism of steroid and thyroid hormones explains the fact that hours are usually required for many of their biological effects to be evident. The rate and magnitude of their action may be further limited by specific factors such as the number of hormone receptor molecules or of acceptor protein molecules per cell or by nonspecific factors such as the concentration of RNA polymerase II, of the enzymes of protein synthesis or processing, of transfer RNA, and of amino acid substrates.

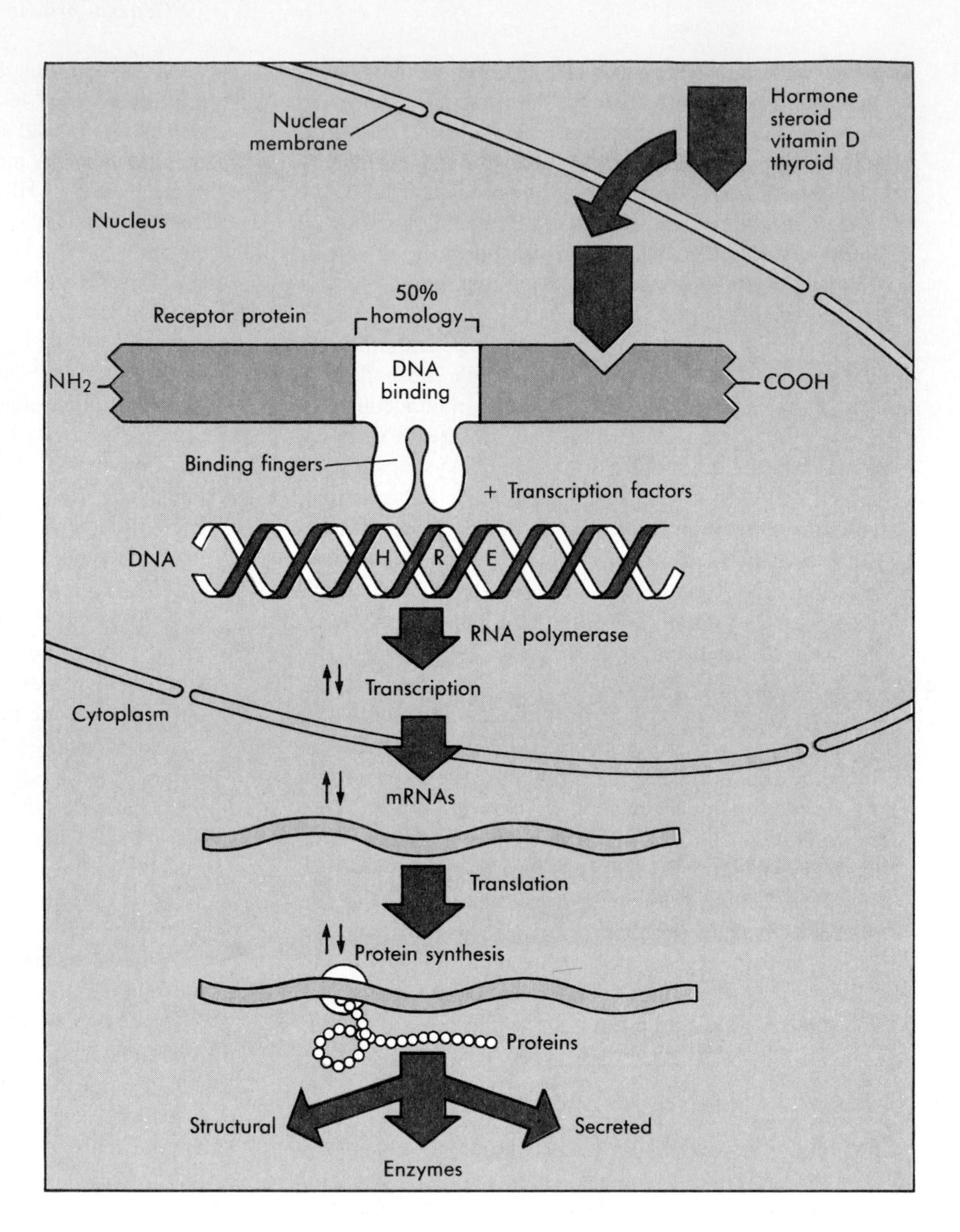

■ **Fig. 44-11** Mechanism of action of vitamin D, steroid, and thyroid hormones. The hormone combines with the C-terminus of a specific intracellular receptor protein. The DNA binding midportion of the receptor protein changes conformation permitting it to interact with a hormone regulatory element in target DNA molecules. Gene transcription and synthesis of protein products are thereby stimulated or repressed.

■ *Responsivity to Hormones*

The final outcome of the interaction of a hormone with its target cell depends on a number of factors. These include hormone concentration, receptor number, duration of exposure, intervals between consecutive exposures, intracellular conditions such as concentrations of rate-limiting enzymes, cofactors or substrates, and the concurrent effects of antagonistic or synergistic hormones. Hormonal effects are not "all or none" phenomena. The dose-response curve for the action of a hormone is generally complex and often exhibits a sigmoidal shape (Fig. 44-12, *A*). An intrinsic basal level of activity may be observed independent of added hormone and long after any previous exposure. A certain minimal threshold concentration of hormone then is required to elicit a measurable response. The effect that is obtained at saturating doses of hormone defines the maximal responsiveness of the target cell. The concentration of hormone required to elicit a half-maximal response is an index of the sensitivity of the target cell. Alterations in the dose-response curve in vivo can take two general forms (Fig. 44-12, *B*). (1) A decrease in maximal responsiveness could be caused by a decrease in the number of functional target cells, in the total number of receptors per cell, in the concentration of an enzyme being activated by the hormone, or in the concentration of a precursor essential to the final product of hormone action; it also could be caused by an increase in the concentration of a noncompetitive inhibitor. (2) A decrease in hormone sensitivity could be caused by a decrease in the number or affinity of hormone receptors, alterations in the concentration of modulating cofactors, an increase in the rate of hormone degradation, or increases in antagonistic hormones.

The normal range of responsiveness to a hormone is

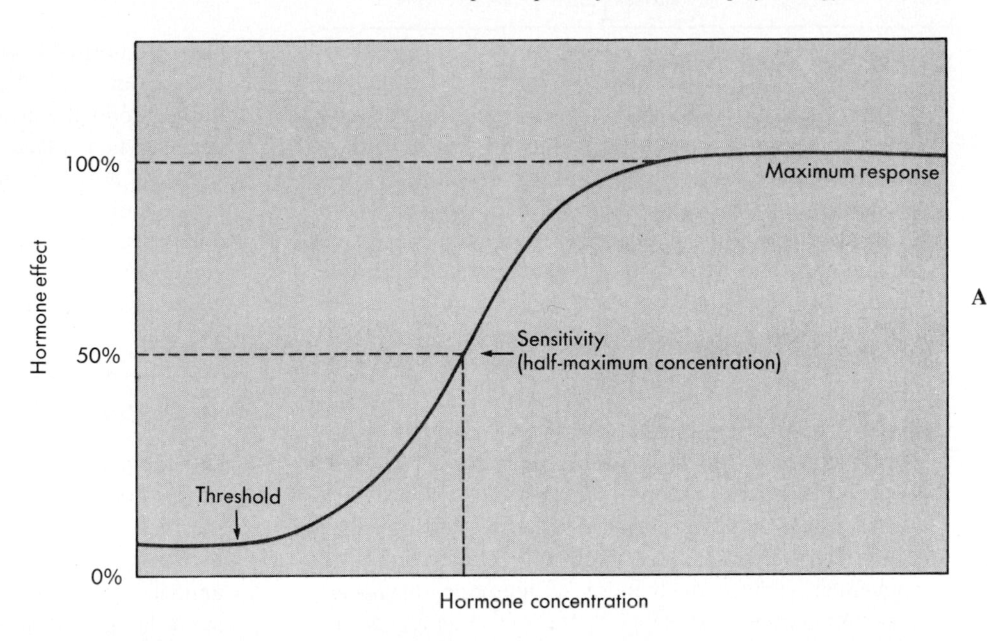

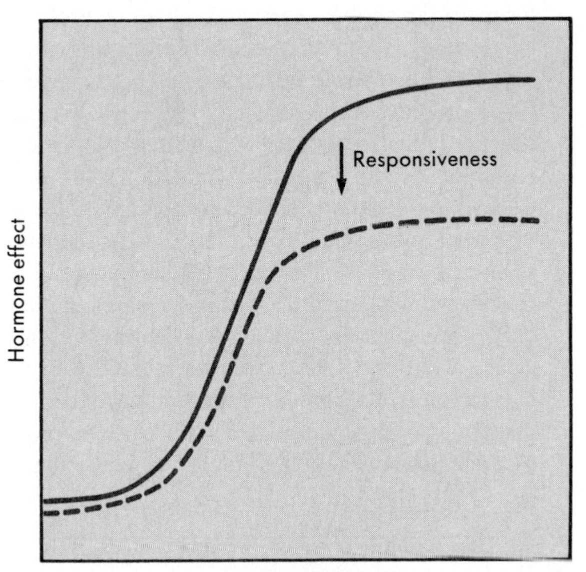

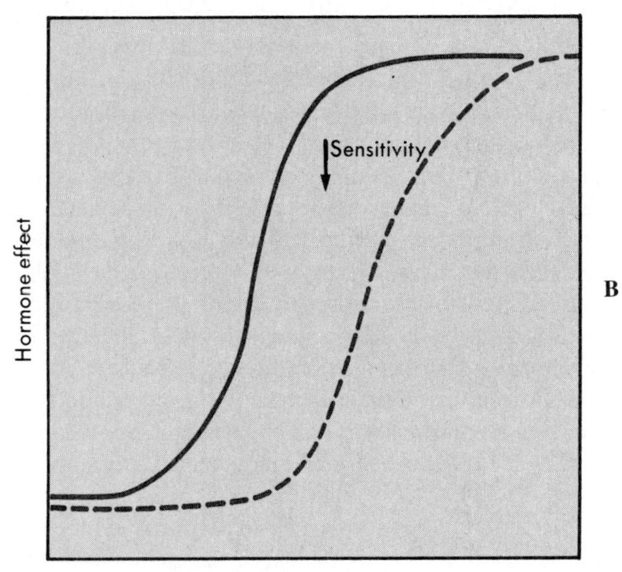

■ **Fig. 44-12** **A,** The general shape of a hormone dose-response curve. Sensitivity is most often expressed as the concentration of the hormone that produces a half-maximal response. **B,** Alterations in the dose-response curve can take the form of a change in maximal responsiveness *(left panel),* a change in sensitivity *(right panel),* or both.

usually rather broad and is in part a result of the physiological variability created by the aforementioned factors within and among normal individuals. However, the very exquisiteness with which hormonal effects can be modulated is an important component in achieving one major objective of hormonal regulation: metabolic stability.

■ *Hormone Transport*

After secretion, hormones enter pools characterized by widely varying size, volume of distribution, degree of compartmentalization, and fractional turnover rates. Initially all hormones enter the plasma pool, where they may circulate either as free molecules or bound to specific carrier proteins. Catecholamine and most peptide and protein hormones circulate unbound, although exceptions are noted. In contrast, steroid and thyroid hormones circulate bound to specific globulins that are synthesized in the liver. The extent of protein binding markedly influences the exit rates of hormones from plasma into interstitial fluid and hence to target cells. As noted in Table 44-1 the plasma half-life of a hormone is directly correlated with the percentage of protein binding. For example, thyroxine is 99.95% protein bound and has a plasma half-life of 6 days, whereas aldosterone is only 15% bound and has a plasma half-life of 25 minutes. Larger and more complex protein hormones tend to have longer half-lives than do smaller proteins and peptides. Hormone exit from plasma does not have to be entirely irreversible. In some instances hormone molecules may return to plasma from other compartments, possibly after dissociation from cell membrane receptors. The return to plasma can occur by way of the lymphatic channels.

Hormone Disposal

Irreversible removal of hormone is a result of target cell uptake, metabolic degradation, and urinary or biliary excretion. The sum of all removal processes is expressed in the term metabolic clearance rate (MCR). In a steady state this is defined as volume of plasma cleared/unit time, which equals mass removed/unit time divided by circulating mass/unit volume, i.e.,

$$\text{MCR} = \frac{\text{mg/minute removed}}{\text{mg/ml of plasma}} = \frac{\text{ml cleared}}{\text{minute}} \qquad (8)$$

MCR is one expression of the efficiency with which a hormone is removed from plasma. The ratio of MCR to the volume of distribution of a hormone is a measure of its fractional turnover rate (K). The plasma half-life, which is inversely related to K, is a cruder but more conveniently determined index of hormone disappearance. As shown in Table 44-1, MCR is inversely correlated with plasma half-life and usually also inversely correlated with the percentage of protein binding.

The kidney and liver are the major sites of hormone extraction and degradation. Renal clearance of hormone is reduced greatly by protein binding to globulins in the plasma; e.g., less than 1% of secreted cortisol appears unchanged in the urine because only the small, free fraction of plasma cortisol is filtered by the glomerulus. On the other hand, about 30% of cortisol metabolites are excreted in the urine, since they are generally unbound or only loosely bound to protein. Peptide and smaller protein hormones are filtered to some degree by the glomerulus. Usually, however, they subsequently undergo tubular reabsorption and degradation within the kidney, so that only a small amount appears in the final urine.

Metabolic degradation occurs by enzymatic processes that include proteolysis, oxidation, reduction, hydroxylation, decarboxylation, and methylation. Virtually all hormones are extracted from the plasma and degraded to some extent by the liver. In addition, glucuronidation and sulfation of hormones or their metabolites may be carried out, with the conjugates subsequently excreted in the bile or the urine. Some hormonal degradation appears to take place during interaction with target tissues. As noted previously, internalization of a portion of the hormone–plasma membrane receptor complex does occur, and hormone may be degraded within the cell. Hormone action and hormone degradation may be quantitatively linked in these situations.

Hormone Measurement

In large part the history of endocrinology is the history of methodology for hormone measurement. Initially hormones all were measured by biological assay, i.e., by measuring an effect that they produced. When whole animals served as test subjects, sensitivity was necessarily low, and precision was greatly influenced by many variables that were difficult to control. As more knowledge of hormone action developed, biological assays could be carried out on organs in vivo and later on tissues in vitro. The sensitivity improved, precision was enhanced, and specificity was increased. Nevertheless, tissue assays still had major limitations. Physicochemical methods, such as spectrophotometry and fluorometry, were subsequently developed with sensitivity that permitted their use for catecholamine, thyroid and steroid hormones, or their metabolites. Even then often only the relatively higher concentrations in tissue, urine, or effluent blood from the gland of origin could be accurately determined. Furthermore, laborious chromatographic procedures sometimes were required to separate hormones with similar chemical groupings.

Radioimmunoassay

A dramatic advance in the acquisition of information and understanding in endocrinology followed the devel-

Table 44-1 Correlation of plasma half-life and metabolic clearance of hormones with their structure and degree of protein binding

Hormone	*Protein binding (percent)*	*Plasma half-life (days)*	*Metabolic clearance (ml/minute)*
Thyroid			
Thyroxine	99.97	6	0.7
Triiodothyronine	99.7	1	18
Steroids			
Cortisol	94	0.07	140
Testosterone	89	0.04	860
Aldosterone	15	0.016	1100
Proteins			
Thyrotropin (molecular weight 28,000)	Nil	0.034	50
Insulin (molecular weight 6,000)	Nil	0.006	800

opment of radioimmunoassay and related methods in 1957. In one stroke, sensitivity was brought to the level of normal plasma hormone concentrations, precision to the level where changes of 10% to 20% could be reliably determined, and ease, cost, and sample volume to the point where as many as 50 values could be obtained from an individual in a single day. Assay specificity also was enhanced to the point where it has become possible to reliably discriminate between a hormone and its precursor (e.g., insulin and proinsulin) or a hormone and its slightly modified metabolites (e.g., thyroxine, triiodothyronine, and reverse triiodothyronine).

The principle of radioimmunoassay is shown in Fig. 44-13. The sample, which may be plasma, urine, cerebrospinal fluid, or a tissue extract, is incubated with a fixed amount of radioactively labeled hormone (the tracer) and a fixed amount of a specific hormone-binding protein, most often an antibody. The nonradioactive hormone molecules in the sample compete with the tracer hormone molecules for the binding sites on the protein. The concentration of binding sites is fixed and limiting. Therefore progressive increases in the number of nonradioactive hormone molecules in the sample will displace more and more of the tracer hormone molecules from the binding sites. At the end of the incubation the bound tracer hormone molecules are separated from those that are free. The radioactivity in the individual bound and free tracer fractions is counted. If the hormone concentration in the sample is relatively high, the percentage of radioactivity remaining in the bound fraction will be relatively low, and vice versa (Fig. 44-13, *A*). The absolute amount of hormone in the sample is calculated by comparison with a standard curve generated by incubating varying amounts of authentic hormone with the tracer and binding proteins (Fig. 44-13, *B*).

The specificity of radioimmunoassay rests on the reaction of a hormone with a binding site that uniquely recognizes the hormone molecule. The introduction of monoclonal antibodies and the technique of using two

Hormone + Binding protein —Incubate→ Bound + Free | Bound/Free

3.0

1.0

A

0.33

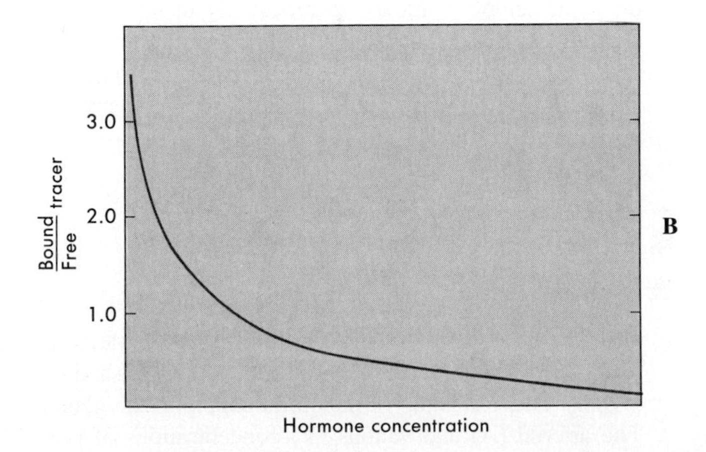

■ **Fig. 44-13** **A,** The principle of radioimmunoassay. Note that concentrations of tracer hormone and binding protein are fixed. The addition of increasing amounts of unlabeled hormone in sample or standard in effect dilutes the tracer hormone molecules and displaces them from the binding protein. This results in a decreasing ratio of bound/free tracer hormone. **B,** Typical radioimmunoassay standard curve is exponential. The curve, however, often can be made linear by reciprocal or logarithmic transformation of the data.

● Tracer hormone

● Hormone in sample or standard

) Binding protein

antibodies directed against different epitopes on the hormone molecule (immunoradiometric assays) has greatly improved specificity. However, subtle alterations in structure that render a hormone devoid of biological activity may still permit its easy recognition by induced antibodies whose determinants happen to be removed from the biologically active sites. Therefore elevated levels of such a hormone molecule may be measured by a radioimmunoassay even though clinical signs of hormone deficiency exist. This problem may sometimes be obviated by using a normal tissue receptor protein instead of an induced antibody as the binding protein in the assay.

Still radioimmunoassays may not always distinguish completely or sufficiently between similar hormones secreted by the same gland (e.g., two adrenal steroid hormones), between a peptide hormone and its prohormone, or between a hormone and its metabolic products. This may give rise to experimental or clinical misinterpretations unless the specificity of each assay is carefully documented. Preliminary separation procedures can eliminate this problem. High pressure liquid chromatography has become especially useful.

The great sensitivity of radioimmunoassay rests on the high association constants of antigen-antibody reactions, which can be around 10^{10} to 10^{12} L/mole. This permits use of very low concentrations of reactants. High sensitivity also derives from the ability to measure very small amounts of radioactivity or more recently from the use of chemical luminescence or conjugated enzymatic activity as indications of the hormone-antibody reaction. In clinical assays, however, it is still sometimes difficult to differentiate pathologically low levels from normal basal levels of hormone in plasma. In such instances the conclusion that glandular hypofunction exists may have to be made from the inference that the hormone level is not elevated, despite the presence of a strong physiological or pharmacological stimulus.

The precision of radioimmunoassay stems from the simplicity of handling and its suitability for automation. When samples are analyzed in replicate in a single assay, the coefficient of variation is 5% to 10%. When the same sample is measured repeatedly in separate assays, a slightly higher coefficient of variation is observed. Samples with high concentration of hormone should be diluted to bring them into the most precise portion of the standard curve.

Estimates of Hormone Secretion

Secretion Rate

Absolute quantitation of the output of a single hormone by an individual gland can only be accomplished in vivo by catheterization of the blood supply to the gland. The arterial (A) and venous (V) concentrations of hormone, in mass per unit volume, and the blood flow (BF) across the gland, in volume per unit time, must be measured. The rate of secretion in mass per unit time is then $(V - A) \times BF$. This method for assessing secretion rate is suitable for animal studies but is not applicable to clinical circumstances. However, sampling of veins that drain glands can be used diagnostically to localize the source of excess hormone production. This is true for multiple glands, such as the parathyroid glands, or for hormones, such as steroids, that may be secreted from either the adrenal glands or the gonads.

Blood Production Rate

A less direct but frequently satisfactory method to estimate hormone secretion is the measurement of the production rate (PR) in mass per unit time. This is the total amount of the hormone entering the peripheral circulation per unit time; in a steady state it will equal the total amount of hormone leaving the circulation. Therefore production rate is determined by measuring plasma concentration (P) and the MCR (p. 828):

$$PR = P \times MCR \tag{9}$$

Although this measurement primarily is carried out under research conditions, it can be used in select circumstances for clinical diagnosis.

Plasma Levels

A simple plasma concentration provides a valid index of hormone production rate when the metabolic clearance of the hormone is within normal limits and can be taken as a constant; since $PR = P \times MCR$, PR is proportional to P if MCR is a constant. This is the theoretical basis for employing plasma hormone measurements alone as an index of activity of the gland of origin. However, the release of many hormones is characterized by episodic spurts and diurnal variation. It therefore may be hazardous to conclude too much from a single plasma value. Multiple measurements or measurements taken at different times of the day may be needed. To reduce the number of laboratory analyses, multiple samples of equal volume can be pooled, and a single careful measurement of this pool then yields the average plasma concentration over the interval in which the samples were collected.

Urinary Excretion

Measurement of urinary hormone excretion is cumbersome because it requires accurately timed collections. However, it offers the advantage of, in effect, averaging plasma fluctuations over the collection period. Furthermore, in some instances the quantity of a hormone metabolite in the urine far exceeds the plasma or urinary level of the hormone itself and is therefore more easily measurable. Urinary excretion (UV) is equal to the urinary concentration (U) times the urine flow (V). Urinary excretion of the hormone is related to production rate and the renal clearance (C) of the hormone (Chapter 41), as follows:

$$C = \frac{UV}{P}; \; P = \frac{UV}{C} \qquad (10)$$

By substitution into equation 9:

$$PR = UV \times \frac{MCR}{C} \qquad (11)$$

Hence PR is proportional to UV when MCR/C can be taken as a constant. Thus urinary excretion of a hormone is a valid index of production rate (and hence secretion rate) when the following two conditions are fulfilled: (1) the kidney must be contributing its usual fraction of the total metabolic clearance; and (2) within the kidney itself the usual proportioning of hormone between intrarenal degradation and urinary excretion must be maintained. The chief sources of error in employing urinary excretion as a reflection of hormone secretion are general impairment in renal function, a change in the pattern of degradation to a metabolic product excreted differently by the kidney, and incomplete collections of urine. When there is little diurnal variation in hormone secretion and degradation, incomplete collections can be partially compensated for by indexing hormone excretion to simultaneously determined creatinine excretion.

Summary

1. The function of the endocrine system is to regulate metabolism, fluid status, growth, and sexual development. The endocrine and nervous systems work conjointly to maintain homeostasis.

2. Hormones are signaling molecules that are conveyed by the blood stream (endocrine), by neural axons and the blood stream (neurocrine), or by local diffusion (paracrine, autocrine).

3. Hormone molecules may be proteins, peptides, catecholamines, steroids, or iodinated tyrosine derivatives.

4. Protein and peptide hormone synthesis involves processing of a primary gene transcript called a prohormone. Such processing includes proteolytic cleavage, glycosylation, and phosphorylation. Thyroid hormone and catecholamines are synthesized from tyrosine and steroid hormones from cholesterol by multiple enzyme reactions.

5. Peptide and protein hormones and catecholamines are stored in granules and secreted by exocytosis. Thyroid hormone is stored within protein molecules in large quantities; steroid hormones are not stored at all. Both are released by diffusion.

6. Protein and peptide and catecholamine hormones act on target cells via specific protein receptors located in the plasma membranes; the hormone-receptor complexes transduce signals through second messengers. Stimulatory or inhibitory G-proteins link the receptor to membrane mechanisms that generate cAMP, Ca^{++}, diacylglycerols, and inositol triphosphate. These molecules act intracellularly to increase or decrease enzyme activities.

7. Thyroid and steroid hormones act by means of specific protein receptors located in the nucleus. The hormone-receptor complex interacts with promoter, enhancer, or negative enhancer elements in DNA molecules to induce or repress expression of target genes. In turn, this leads to increases or decreases in the concentration of enzymes and other proteins.

8. The sensitivity of an organism to hormone action is expressed as the hormone concentration that produces half-maximum activity. The sensitivity can be influenced by changes in receptor number, affinity, hormone degradation rate, or competitive antagonists. The maximum effect produced by saturating concentrations of hormone can be influenced by the number of target cells, receptor number, concentration of target enzymes, or noncompetitive antagonists.

9. Hormone secretion is measured directly by arterial-venous concentration gradients and flow rates across the gland. Clinically, plasma levels and urinary excretion rates are used as indirect indices of hormonal secretion rates. These are valid as long as metabolic or renal clearance of the hormone is normal.

10. The principal method of hormone measurement is radioimmunoassay or its analogs. The method makes use of the reaction between a hormone and an induced antibody. Technically, this is a competitive binding assay with high sensitivity and specificity, particularly when monoclonal antibodies are employed.

Bibliography

Journal articles

Birnbaumer L et al: Molecular basis of regulation of ionic channels by G proteins, *Rec Prog Horm Res* 45:121, 1989.

Bost KL et al: Similarity between the corticotrophin ACTH receptor and a peptide encoded by an RNA that is complementary to ACTH mRNA, *Proc Natl Acad Sci* 82:1372, 1985.

Carson-Jurica MA et al: Steroid receptor family: structure and function, *Endocr Rev* 11:201, 1990.

Chambon P et al: Promoter elements of genes coding for proteins and modulation of transcription by estrogens and progesterone, *Recent Prog Horm Res* 40:1, 1984.

Fradkin JE et al: Specificity spillover at the hormone receptor: exploring its role in human disease, *N Engl J Med* 320:640, 1989.

Gordon P et al: Internalization of polypeptide hormones; mechanism, intracellular localization and significance, *Diabetologia* 18:263, 1980.

Lacy PE: Beta cell secretion—from the standpoint of a pathobiologist, *Diabetes* 19:895, 1970.

Lefkowitz R et al: Mechanisms of membrane-receptor regulation: biochemical, physiological, and clinical insights derived from studies of the adrenergic receptors, *N Engl J Med* 310:1570, 1984.

Reiter RJ: Pineal melatonin: cell biology of its synthesis and of its physiological interactions, *Endocr Rev* 12:151, 1991.

Sherwin RS et al: A model of the kinetics of insulin in man, *J Clin Invest* 53:1481, 1974.

Spiegel AM: Guanine nucleotide binding protein and signal transduction, *Vitam Horm* 44:47, 1988.

Tait JF: The use of isotopic steroids for the measurement of production rates in vivo, *J Clin Endocrinol* 23:1285, 1963.

Turek FW: Circadian neural rhythms in mammals, *Annu Rev Physiol* 47:49, 1985.

Wilkin TJ: Mechanisms of disease: receptor autoimmunity in endocrine disorders, *N Engl J Med* 323:1318, 1990.

Zor U: Role of cytoskeletal organization in the regulation of adenylate cyclase-cyclic adenosine monophosphate by hormones, *Endocrine Rev* 4:1, 1984.

Books and monographs

Clark JH, Schrader WT, O'Malley BW: *Mechanisms of action of steroid hormones*. In Wilson JD, Foster DF, editors: *Textbook of endocrinology,* ed 8, Philadelphia, 1992, WB Saunders.

Exton JH, Blackmore PF: *Calcium-mediated hormonal responses*. In DeGroot LJ, editor: *Endocrinology,* ed 2, Philadelphia, 1989, WB Saunders.

Habener JF: *Genetic control of hormone formation*. In Wilson FJ, Foster DF, editors: *Textbook of endocrinology,* ed 8, Philadelphia, 1992, WB Saunders.

Kahn CR, Smith RJ, Chin WW: *Mechanism of action of hormones that act at the cell surface*. In Wilson JD, Foster DF, editors: *Textbook of endocrinology,* ed 8, Philadelphia, 1992, WB Saunders.

Nishizuka Y: *Membrane phospholipids and the mechanism of action of hormones*. In LaBrie F, Proulx L, editors: *Endocrinology* Proceedings of the seventh International Congress of Endocrinology, Quebec City, 1984, Amsterdam, 1984, Elsevier Science Publishers.

Yalow RS et al: *Introduction and general considerations*. In Odell WD, Daughaday WH, editors: *Principles of competitive protein-binding assays,* Philadelphia, 1971, JB Lippincott.

CHAPTER
45

Whole Body Metabolism

Metabolism may be broadly defined as the sum of all the chemical (and physical) processes involved (1) in producing energy from exogenous and endogenous sources, (2) in synthesizing and degrading structural and functional tissue components, and (3) in disposing of resultant waste products. Regulating the rate and direction of many basic aspects of metabolism is one of the major functions of the endocrine system. Therefore a firm grasp of the fundamentals of metabolism is essential if one is to understand the important influence of hormones on body functions.

■ *Energy Metabolism*

■ *Balance*

The laws of thermodynamics require that energy balance be constantly maintained in living organisms. However, energy may be obtained in various forms, may be stored in other forms, and may be expended in many different ways. Therefore numerous interconversions of chemical, mechanical, and thermal energy are possible within the basic rule that *in the steady state, energy input must always equal energy output.* Fig. 45-1 illustrates this overall flow of energy through the human organism.

■ *Energy Input*

Energy input consists of foodstuffs, which are classified into three major chemical categories: carbohydrate, fat, and protein. The complete combustion of each chemical type yields characteristic amounts of energy, expressed as joules or kilocalories per gram (1 kcal = 4,184 joules). Combustion also requires characteristic amounts of oxygen, depending on the proportions of carbon, hydrogen, and oxygen in the substance. However, for each class of foodstuff, the energy yield per liter of oxygen used is quite similar because the ratio of carbon to hydrogen atoms is also similar in each class (Table 45-1). Within the body the carbon skeletons of carbohydrate and protein can be converted to fat, and their potential energy can be stored more efficiently in that manner. The carbon skeletons of protein can be converted to carbohydrate when that energy source is

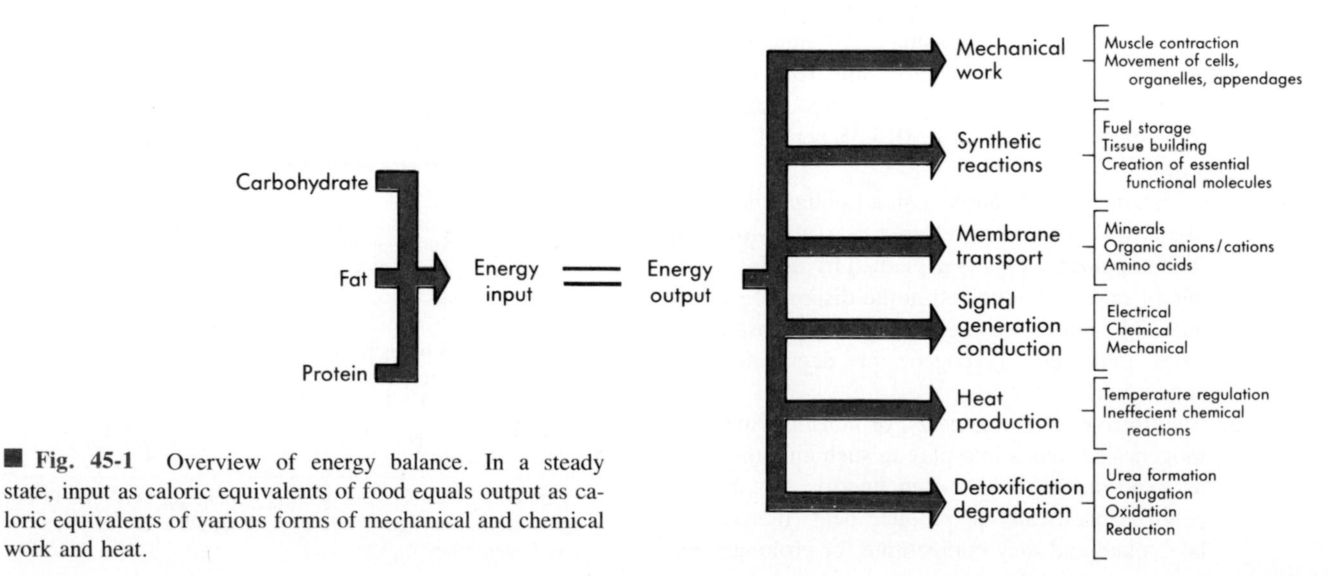

■ **Fig. 45-1** Overview of energy balance. In a steady state, input as caloric equivalents of food equals output as caloric equivalents of various forms of mechanical and chemical work and heat.

Table 45-1 Energy equivalents of foodstuffs

	Kilocalories produced per gram	O_2 *used (L/g)*	*Kilocalories produced per liter of* O_2	*Respiratory quotient*
Carbohydrate	4.2	0.84	5.0	1.00
Fat	9.4	2.00	4.7	0.70
Protein	4.3*	0.96*	4.5	0.80
Typical fuel mix	—	—	4.8	0.85

*Each gram of protein oxidized yields 0.16 g of urinary nitrogen. Thus these values should be multiplied by 6.25 to express them per gram of urinary nitrogen.

specifically needed. However, there is no significant conversion of carbon atoms from fat to carbohydrate.

Energy Output

Energy output can be divided into several distinct and measurable components.

1. At rest, energy is expended in a myriad of synthetic and degradative chemical reactions; in generating and maintaining gradients of ions and other molecules across cell and organelle membranes; in the creation and conduction of signals, particularly in the nervous system; in the mechanical work of respiration and circulation of the blood; and in obligate heat loss to the environment. This absolute minimal energy expenditure is called the **basal** or **resting metabolic rate** (**BMR** or RMR). In the adult human BMR amounts to an average daily expenditure of 20 to 25 kcal (84 to 105 kjoules)/kg body weight (or 1.0 to 1.2 kcal/minute) and requires the use of approximately 200 to 250 ml oxygen/minute. About 40% of the BMR is accounted for by the central nervous system and 20% to 30% by the skeletal muscle mass.

   The BMR is linearly related to lean body mass and body surface area. It declines in the elderly, partly because lean body mass declines with age. BMR is increased by raising environmental temperature. During sleep BMR falls 10% to 15%. Studies in identical twins and families suggest that some of the variation in BMR is genetically determined.
2. Ingestion of food causes a small obligate increase in energy expenditure referred to as **diet-induced thermogenesis.** This is explained by the increased rate of reactions involved in the disposition of the ingested calories, such as storage of glucose in the large molecule, glycogen, or degradation of amino acids to urea.
3. **Facultative thermogenesis,** or nonshivering thermogenesis, comes into play in such circumstances as exposure to cold, when energy may be expended specifically to produce heat. It may also be evoked and may compensate for prolonged exposure to caloric excess as a means of limiting weight gain.
4. Energy is also expended by sedentary individuals in spontaneous physical activity such as "fidgeting," at least some of which is unconscious and seemingly purposeless.
5. The additional energy expended in occupational labor and purposeful exercise (Table 45-2) varies greatly among individuals, as well as from day to day and from season to season. This component generates the greatest need for variation in daily caloric intake and underscores the importance of energy stores to buffer temporary discrepancies between energy output and intake.

Of a total average daily expenditure of 2,300 kcal (9,700 kjoules) in a sedentary adult, basal metabolism accounts for 75%, dietary thermogenesis for 7%, and spontaneous physical activity for 18%. Up to an additional 4,000 kcal may be used in daily physical work. During short periods of occupational or recreational exercise, energy expenditure can increase more than tenfold over basal levels.

Energy Generation

The basic chemical currency of energy in all living cells are the two high-energy phosphate bonds contained in

Table 45-2 Estimates of energy expenditure in adults

Activity	*Calorie expenditure (kcal/minute)*
Basal	1.1
Sitting	1.8
Walking, 2.5 miles/hour	4.3
Walking, 4.0 miles/hour	8.2
Climbing stairs	9.0
Swimming	10.9
Bicycling, 13 miles/hour	11.1
Household domestic work	2 to 4.5
Factory work	2 to 6
Farming	4 to 6
Building trades	4 to 9

Data from Kottke FJ: Animal energy exchange. In Altman PL, editor: *Metabolism,* Bethesda, Md, 1968, Federation of American Society for Experimental Biology.

adenosine triphosphate (ATP). To a much lesser extent, other purine and pyrimidine nucleotides (guanosine triphosphate, cytosine triphosphate, uridine triphosphate, inosine triphosphate) also serve as energy sources after the energy from ATP is transferred to them. In muscle, creatine phosphate is a high-energy molecule of particular importance.

The two terminal P-O bonds of ATP each contain about 12 kcal of potential energy per mole under physiological conditions. These bonds are in constant flux. They are generated by oxidative reactions and are consumed as the energy is either (1) transferred into other high-energy bonds involved in synthetic reactions (e.g., amino acid + ATP → amino acyl AMP), (2) expended in creating lower-energy phosphorylated metabolic intermediates (e.g., glucose + ATP → glucose-6-phosphate), or (3) converted to mechanical work (e.g., propulsion of spermatozoa). Because the production and transfer of energy is only 65% efficient, about 18 kcal of substrate are required to generate each terminal P-O bond of ATP. In a normal day when 2,300 kcal are turned over, about 128 moles, or 63 kg, of ATP (a mass approximating body weight) are generated and expended. An overview of energy production with generation of ATP from the major substrates is shown in Fig. 45-2.

The combustion of carbohydrates, chiefly glucose with lesser amounts of fructose and galactose, includes two major phases:

1. During the cytoplasmic anaerobic phase known as glycolysis (Embden-Meyerhof pathway), each partially oxidized glucose molecule has yielded two molecules of pyruvate but only 8% of its energy content. Glycolysis can serve as a sole source of energy only briefly because (a) the supply of glucose is limited, and (b) the accumulated

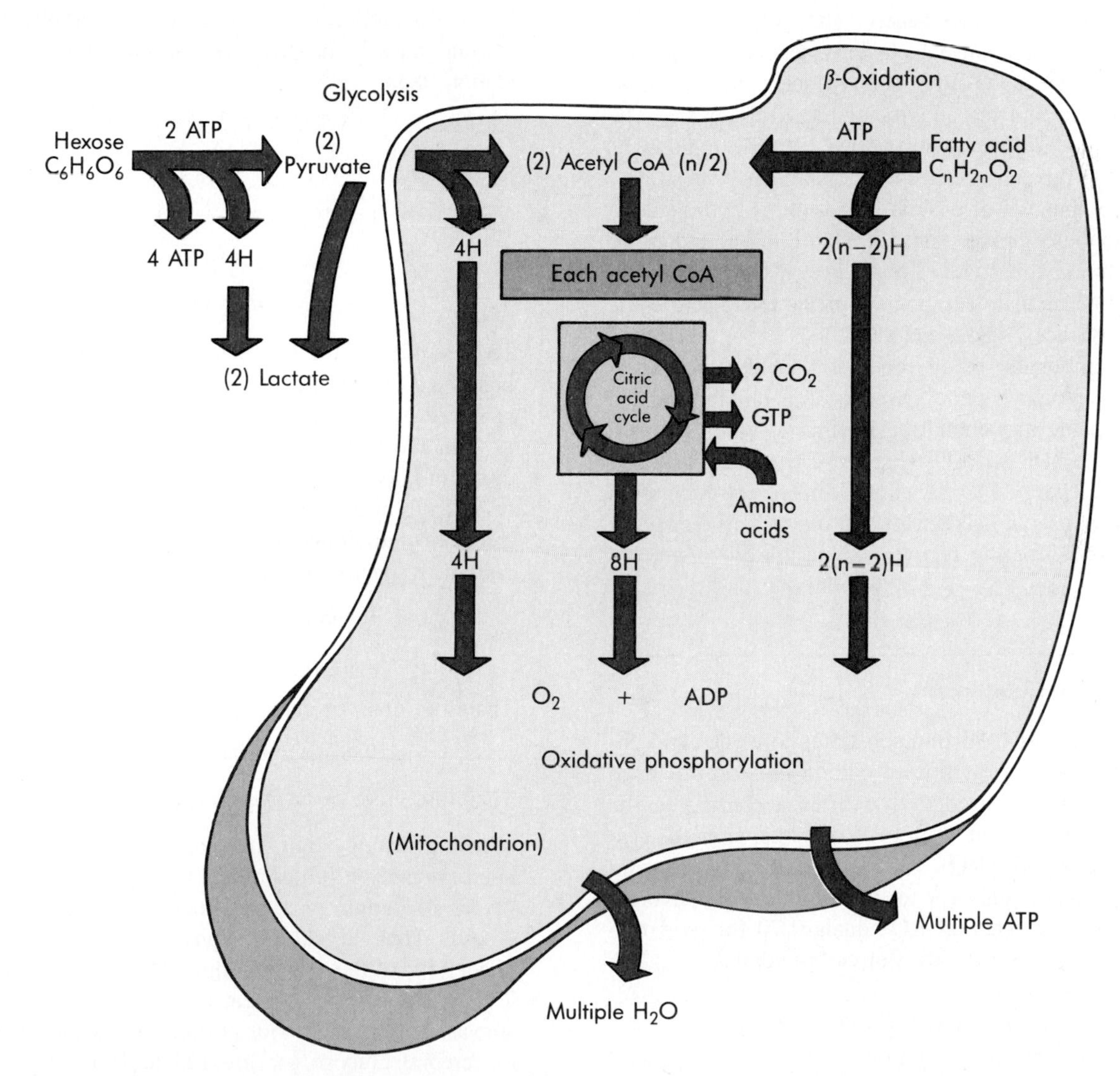

■ **Fig. 45-2** An overview of energy production. Glycolysis and each turn of the citric acid cycle supply only 2 ATP equivalents apiece. ATP is generated mainly when hydrogens removed from carbohydrate, fat, or protein substrates are oxidized in the mitochondria. *ATP*, Adenosine triphosphate; *ADP*, adenosine diphosphate; *GTP*, guanosine triphosphate.

pyruvate must be syphoned off by reduction to lactate, a metabolite that is ultimately noxious.

2. During the mitochondrial aerobic phase, the two pyruvate molecules are oxidized to CO_2 by the citric acid cycle (Krebs cycle), and the remaining energy is liberated. In this pathway acetyl coenzyme A (acetyl CoA), initially formed by oxidative decarboxylation of pyruvate, is condensed with oxaloacetate to form citrate. Through a cyclic series of reactions the carbons of acetyl CoA appear as CO_2, and oxaloacetate is regenerated.

The combustion of fatty acids, the major energy component of fats, proceeds in the mitochondria through a repetitive biochemical sequence known as β-oxidation. This process releases two carbons at a time as acetyl CoA until the entire fatty acid molecule is broken down. The acetyl CoA is disposed of by way of the citric acid cycle, as already described. A variable portion of fatty acid oxidation in the liver stops at the last four carbons and yields acetoacetic and β-hydroxybutyric acids. The production of ketoacids results from an imbalance between the flow of fatty acids into the liver mitochondria and the activity of the Krebs cycle. The water-soluble ketoacids are released by the liver to be oxidized in other tissues as additional energy substrates.

The combustion of protein first requires hydrolysis to its component amino acids. Each of these undergoes degradation by individual pathways, which ultimately lead to intermediate compounds of the citric acid cycle and then to acetyl CoA and CO_2.

The combustion of all foodstuffs yields large numbers of hydrogen atoms. These hydrogens are oxidized to H_2O in the mitochondrion in linkage with phosphorylation of ADP to ATP (Fig. 45-2). In this process, three high-energy P-O bonds are formed for each atom of oxygen used. This yields an overall efficiency of 60% to 65% for the recovery of usable chemical energy.

Respiratory Quotient

In the process of oxidizing substrates to meet basal energy needs, the proportion of carbon dioxide produced (V_{CO_2}) to oxygen used (V_{O_2}) varies according to the fuel mix. The ratio of V_{CO_2} to V_{O_2} is known as the **respiratory quotient (RQ).** As indicated by the following equations, RQ equals 1.0 for oxidation of carbohydrate (e.g., glucose), whereas RQ equals 0.70 for oxidation of fat (e.g., palmitic acid). For carbohydrates:

$$\underset{\text{Glucose}}{C_6H_{12}O_6} + 6\ O_2 \rightarrow 6\ CO_2 + 6\ H_2O$$

$$RQ = \frac{6\ CO_2}{6\ O_2} = 1.0 \quad (1)$$

For fats:

$$\underset{\text{Palmitic acid}}{C_{15}H_{31}COOH} + 23\ O_2 \rightarrow 16\ CO_2 + 16\ H_2O$$

$$RQ = \frac{16\ CO_2}{23\ O_2} = 0.70 \quad (2)$$

The RQ for protein reflects that of the individual RQs of the amino acids and averages 0.80. Ordinarily, protein is a minor energy source. The small contribution of protein oxidation to the overall RQ can be corrected for by measuring the urinary excretion of the nitrogen that results from the metabolism of amino acids.

By determining the RQ corrected for protein, one can then calculate the proportion of carbohydrate and fat in the fuel mix being oxidized.

For example, when the O_2 consumed = 210 ml/minute and the CO_2 produced = 174 ml/minute:

$$RQ = \frac{174}{210} = 0.83$$

If C is the proportion of energy due to carbohydrate oxidation and F is the proportion of energy due to fat oxidation, then:

$$1 \times C + 0.7 \times F = 0.83$$

Since C + F = 1:

$$1 \times C + 0.7 \times (1 - C) = 0.83$$

$$C = 0.43$$

$$F = 0.57$$

That is, 43% of the calories are being provided by carbohydrate oxidation, and 57% of the calories are being provided by fat oxidation.

With a knowledge of the actual rate of energy expenditure in kilocalories per minute and the standard values of 4 kcal/g of carbohydrate and 9 kcal/g of fat, one can calculate the actual rates of oxidation of carbohydrate and fat, respectively, in grams per minute:

$$210 \text{ ml of } O_2 \text{ consumed per minute} = 1.05 \text{ kcal/minute}$$

$$4 \times \text{gram of carbohydrate} = 0.43 \times 1.05$$

Therefore gram of carbohydrate = 0.113/minute

$$9 \times \text{gram of fat} = 0.57 \times 1.05$$

Therefore gram of fat = 0.067/minute

In the resting adult the rate of glucose oxidation is approximately 120 mg/minute (170 g/day), whereas that of fat oxidation is approximately 60 mg/minute (90 g/day). Thus in caloric equivalents glucose supplies about 45% of the energy required for basal metabolism. The central nervous system is an obligate glucose consumer and uses the major portion of this fuel. Estimates of cerebral glucose use are 125 to 150 g/day. In contrast, the large muscle mass oxidizes primarily fatty acids in resting individuals. However, like all tissues, the muscle mass uses at least some glucose, thereby maintaining sufficient concentrations of Krebs cycle intermediates for efficient disposal of acetyl CoA and completion of fatty acid oxidation.

■ *Energy Storage and Transfers*

The intake of energy in the form of food is periodic. Its time course does not match either the constant rate of energy expenditure in the basal state or that expended during intermittent muscle work. Therefore the organism must have mechanisms for storing ingested energy for future use. The greatest part of these energy reserves (75%) is in the form of fat as triglycerides, stored in adipose tissue (Fig. 45-3). In normal-weight humans fat constitutes 10% to 30% of body weight, but it can reach 80% in very obese individuals. Fat is a particularly efficient storage fuel because of its high caloric density (i.e., 9 kcal/g) and because it engenders little additional weight as intracellular water. Fat stores can supply energy needs for up to 2 months in totally fasted individuals of normal weight. Triglycerides are formed by esterification of free fatty acids, largely derived from the diet, with α-glycerol phosphate. However, free fatty acids can also be synthesized from acetyl CoA derived from oxidation of glucose; thus carbohydrate can be converted to fat in liver and adipose tissue and its energy stored in that more efficient form (Fig. 45-4).

Protein (4 kcal/g) constitutes almost 25% of the potential energy reserves (see Fig. 45-3), and the component amino acids can contribute to the glucose supply. However, virtually all proteins serve some vital structural and functional role. Therefore their use as a major source of energy is deleterious and only arises as a last resort before death from fasting.

Carbohydrate (4 kcal/g) in the form of a glucose polymer, glycogen, forms less than 1% of total energy reserves (see Fig. 45-3). However, this portion is critical for support of central nervous system metabolism and for short bursts of intense muscle work. Approximately one fourth of the glycogen stores (75 to 100 g) is in the liver, and about three fourths (300 to 400 g) is in the muscle mass. Liver glycogen can be made available to other tissues by the process of glycogenolysis and glucose release. Muscle glycogen can be used only by muscle, since this tissue lacks the enzyme glucose-6-phosphatase, which is required for release of glucose into the bloodstream.

Glycogen can be formed from all three major dietary sugars. In addition, in the liver (and to a much lesser extent in the kidney) glucose itself can also be synthesized de novo from the three carbon precursors pyruvate, lactate, and glycerol and from parts of the carbon skeleton of all 20 amino acids in protein except leucine and lysine. This process, known as **gluconeogenesis,** converts two pyruvate molecules to glucose, but it is not a simple reversal of all the reactions of glycolysis (see Fig. 45-4). The chemical free energy change is too large to permit efficient backward flow of the glycolytic reactions at three steps: (1) pyruvate to phosphoenolpyruvate, (2) fructose-1,6-diphosphate to fructose-6-phosphate, and (3) glucose-6-phosphate to glucose. Substitution of simple phosphatase reactions reverses the last two steps. However, the first step requires energy input in the form of ATP and guanosine triphosphate (GTP). It is also important to realize that *net glucose synthesis cannot occur from acetyl CoA,* even though carbon atoms from acetyl CoA can become part of oxaloacetate and then part of glucose molecules by

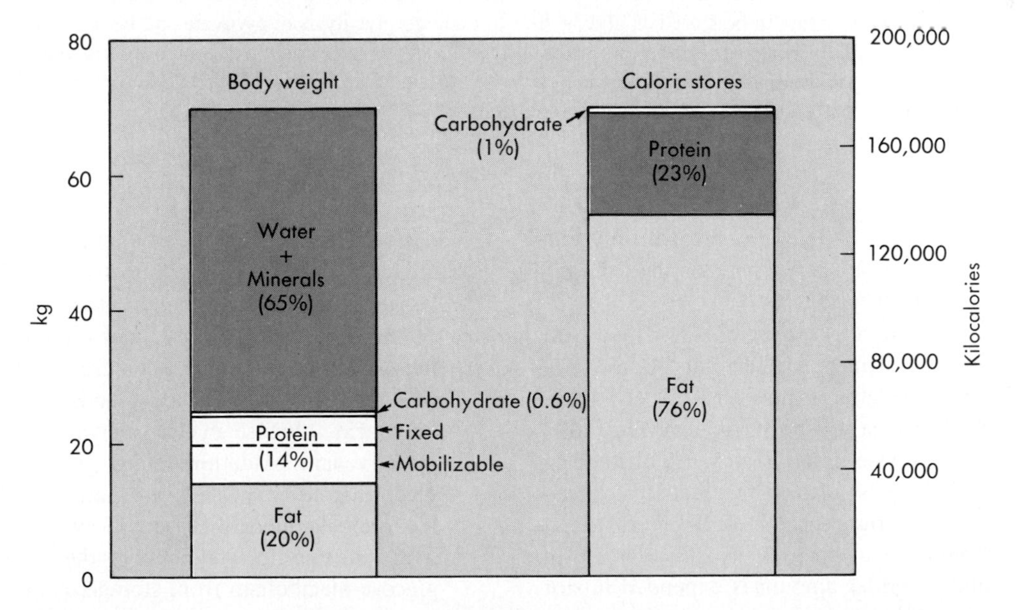

■ **Fig. 45-3** The composition of an average 70 kg human shown in terms of weight (left) and caloric stores (right). Note the trivial proportion of carbohydrate stores relative to fat stores.

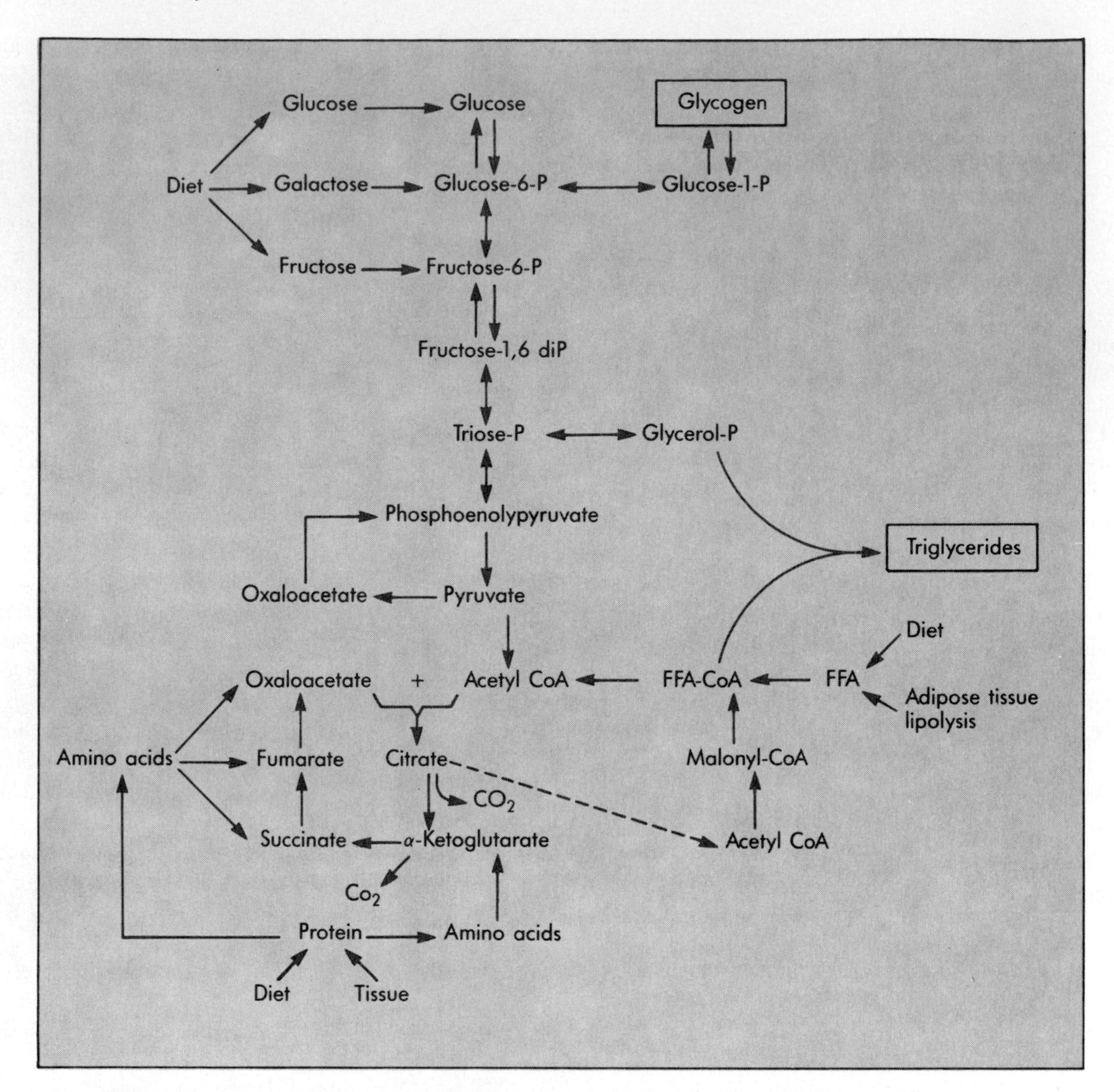

■ **Fig. 45-4** Chemical pathways of energy transfer and storage. Carbohydrates funnel through glucose-6-phosphate to be stored as glycogen or to undergo glycolysis to pryuvate and be used for synthesis of fatty acids. The latter are esterified with glycerol-phosphate and stored as triglycerides. Amino acids arising from protein are converted to glucose via citric acid cycle intermediates and pyruvate in the process known as gluconeogenesis.

means of the citric acid cycle. Thus fat can only contribute to carbohydrate stores by way of the 3-carbon glycerol moiety of triglycerides.

The processes of energy storage and transfer themselves expend energy. This partly accounts for the stimulation of oxygen use after a meal (i.e., diet-induced thermogenesis). The cost of storing dietary fatty acids as triglycerides in adipose tissue is only 3% of the original calories, and the cost of storing glucose as glycogen is only 7% of the original calories. In contrast, conversion of carbohydrate to fat uses up 23% of the original calories, and a similar amount is expended in storing dietary amino acids as protein or in converting them to glycogen.

Because glucose and fatty acids are alternative and in effect competing energy substrates, some relationship between their use and their synthesis and storage within cells could be expected. Under circumstances in which fatty acid supply and plasma free fatty acid levels are increased, glucose uptake by cells is decreased (Randle fatty acid–glucose cycle). Furthermore, intermediates of fatty acid oxidation retard glycolysis and promote gluconeogenesis. Thus increased use of fatty acids as fuel shifts glucose metabolism away from oxidation in liver and muscle, and in liver the increased use shifts glucose metabolism from storage as glycogen to release into the blood stream. Conversely, when dietary glucose is plentiful, glycolysis is augmented, more acetyl

CoA is generated from pyruvate, and more citrate is formed in the mitochondria (see Fig. 45-3). The citrate diffuses back into the cytoplasm, where it is a potent activator of the first step in the synthesis of fatty acids (acetyl CoA → malonyl CoA). Citrate is also split back into oxaloacetate and acetyl CoA; the latter provides the substrate for fatty acid synthesis. Further, other glucose molecules are oxidized via the pentose shunt pathway, which generates NADPH, also needed for fatty acid synthesis. Thus when the supply of glucose is plentiful, there is net conversion of glucose carbon to fatty acid carbon with the following stoichiometry:

$$4\frac{1}{2} \text{ glucose} + 4\ O_2 \rightarrow \text{Palmitic acid} + 11\ CO_2$$

The resultant RQ is $11 \div 4 = 2.75$. Therefore, whenever the RQ in a human is greater than 1.0, lipogenesis from glucose is likely occurring. In addition, glycolysis will produce more glycerol phosphate from triose phosphates. The combination of increased fatty acid synthesis and glycerol phosphate availability results in accentuated synthesis of triglycerides and reduced oxidation of fat. Thus increased carbohydrate utilization shifts fat metabolism from oxidation to storage. Many of these intrinsic chemical checks and balances are also reinforced by hormonal signals.

In addition to these intracellular relationships, transfer of energy between organs is another important metabolic process. The stored energy contained within adipose tissue triglycerides is transported as free fatty acids to the liver. There part of the energy (not the carbon atoms) is effectively transferred to glucose molecules, because as fatty acids are oxidized, gluconeogenesis is stimulated concurrently, as previously described. The newly synthesized glucose molecules in turn can then be transported to muscle tissue, where their energy is released during glycolysis and applied to muscle contraction. Furthermore, if the lactate produced exceeds the ability of the muscle to oxidize it rapidly enough in the citric acid cycle, the lactate can be returned to the liver, where it may again be built back up into glucose molecules. From this viewpoint the liver is a flexible and versatile organ that can transmute and transfer energy from fuel depots to working tissues.

■ *Carbohydrate Metabolism*

Dietary carbohydrates give rise to various sugars (hexoses), the most important of which are glucose, fructose, and galactose. In addition to serving as energy sources, sugars are also components of glycoproteins, glycopeptides, and glycolipids. These have structural and functional roles as basement membrane collagen, mucopolysaccharides, nerve cell myelin, hormones, and hormone receptors.

Glucose is the central molecule in carbohydrate metabolism. Other sugars are metabolized through the glucose pathways. The initial step in glucose metabolism is transport of the hexose across the cell membrane down a normally large concentration gradient from extracellular fluid to cytoplasm. This process is called facilitated diffusion, as opposed to active transport against a concentration gradient (see Chapter 1). Facilitated diffusion is carried out by a family of at least five glucose transport proteins that are coded for by closely related genes. Each transport protein has approximately 450 amino acids and a molecular weight of about 50,000. The transport protein spans the plasma membrane twelve times and has both its amino and carboxy terminias within the cytoplasm. The Glut-1 transporter is expressed constitutively and is responsible for the low level of basal glucose uptake required to sustain the energy generation process by all cells. This transporter is increased by fasting and decreased by an excess of glucose. The Glut-4 transporter is expressed exclusively in cardiac and skeletal muscle and adipose tissue and is specifically responsible for the glucose utilization that is stimulated by the hormone insulin (see later discussion). Glut-2 is expressed by hepatic and renal tubular cells that transfer glucose out of the cell and into the extracellular fluid and plasma. $Glut_2$ is also present in insulin-secreting cells of the pancreatic islets where it maintains virtual equilibrium between the glucose concentrations in the extracellular fluid and cytoplasm.

Basal plasma glucose levels are tightly regulated around an average concentration of 80 mg/dl^{-1} (4.5 $\mu mol/L^{-1}$), with a range of 60 to 115 mg/dl^{-1}. When the plasma level falls below 60 mg/dl^{-1}, brain uptake of the sugar and utilization of oxygen decrease in parallel. Central nervous system function becomes progressively impaired, and death may ensue. The major products of glycolysis, lactate and pyruvate, circulate at average concentrations of 0.7 and 0.07 mM, respectively. This 10:1 ratio of lactate to pyruvate prevails even when the glycolytic rate changes. However, when tissues are deprived of oxygen, the equilibrium between the two shifts toward lactate, the reduced molecule; plasma concentration ratios as high as 30:1 may then be observed. Very high concentrations of lactate produce metabolic acidosis.

In the basal state glucose turnover is about 2 $mg/kg^{-1}/minute^{-1}$ (11 $\mu mol/kg^{-1}/minute^{-1}$), which is equivalent to about 9 $g/hour^{-1}$ or 225 g/day^{-1} in adults. Approximately 55% of glucose use results from terminal oxidation, of which the brain accounts for the greatest part (Fig. 45-5). Another 20% results from glycolysis; the resulting lactate then returns to the liver for resynthesis into glucose (Cori cycle). Reuptake by the liver and other splanchnic tissues accounts for the remaining 25% of glucose use. Most of glucose use (about 70%) in the basal state is independent of insulin, a hormone with otherwise important regulatory effects on glucose metabolism.

The circulating pool of glucose is only slightly larger

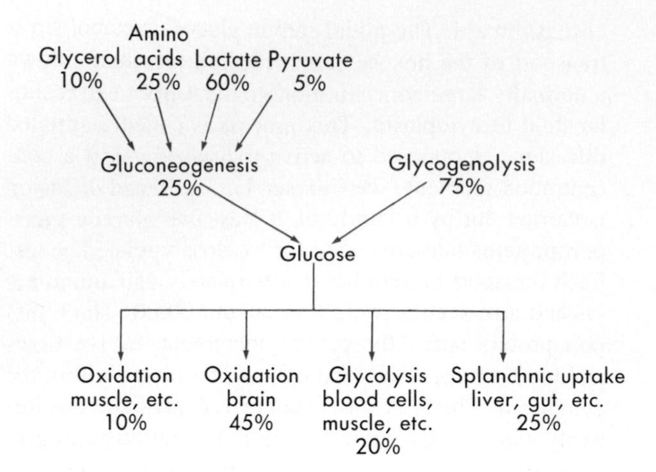

■ **Fig. 45-5** Quantitative overview of glucose turnover in overnight fasted humans. The disposal of circulating glucose *(lower portion)* is largely by oxidation and glycolysis, with the brain acting as the single largest consumer. The production of glucose in the circulation *(upper portion)* is largely from glycogenolysis. The proportionate contribution of glucose precursors to gluconeogenesis, with lactate predominating, is also shown. The average overall rate of glucose turnover is 2.0 mg/kg/minute or 11 μmoles/kg/minute.

than the liver output in 1 hour. This pool is only sufficient to maintain brain oxidation for 3 hours, even if all other glucose use ceased. This emphasizes the crucial importance of continuous hepatic production of glucose in the fasting state. About 75% of this production results from glycogenolysis and 25% from gluconeogenesis. Hepatic uptake and use of circulating lactate accounts for more than half the glucose supplied by gluconeogenesis. The remainder is largely accounted for by amino acids, especially alanine. The supply of lactate comes from glycolysis in muscle, red blood cells, white blood cells, and a few other tissues. The amino acid precursors come from proteolysis of muscle. Despite their importance, however, simply increasing the supply of precursors does not increase the rate of gluconeogenesis. The necessary enzymes must be upregulated by hormonal modulation or by hepatic autoregulatory responses to the prevailing glucose concentrations.

When an individual ingests glucose after overnight fasting, approximately 70% of the load is assimilated by peripheral tissues, mainly muscle, and about 30% by splanchnic tissues, mainly liver. Only 20% to 30% of a glucose load is oxidized during the 3 to 5 hours required for its absorption from the gastrointestinal tract. The remainder is stored as glycogen, partly in muscle and partly in liver. Glucose initially stored as muscle glycogen can later be transferred to the liver by undergoing glycolysis to lactate, which is released into the circulation; the lactate is then taken up by the liver, rebuilt into glucose, and stored as glycogen in that organ. During the period of peak absorption of exogenous glucose, hepatic output of the sugar is largely unnecessary and is greatly reduced from basal levels. These metabolic adaptations are facilitated by coordinated secretion of the pancreatic islet hormones, insulin and glucagon.

■ *Protein Metabolism*

The average adult body contains 10 kg of protein, of which about 6 kg is metabolically active. This pool turns over continuously; the degradative and synthetic reactions involved are estimated to account for 20% of the BMR. Approximately 50 g of amino acids are released daily by proteolysis from muscle, the main endogenous repository, and degraded. Therefore daily dietary intake of 50 g of protein or 0.8 g/kg is ordinarily sufficient. When accretion of lean body mass is taking place (e.g., in growing children, pregnant women, persons recovering from prior weight loss), daily protein requirements increase to 1.5 to 2.0 g/kg^{-1}.

All proteins are composed of the same 20 amino acids. Half of these are called **essential amino acids** because their carbon skeletons, the corresponding α-ketoacids, cannot be synthesized by humans. (The essential amino acids are threonine, methionine, valine, leucine, isoleucine, phenylalanine, tyrosine, tryptophane, lysine, and in infants, histidine.) Once present, however, they can be converted to the essential amino acids by transamination. The other half, the **nonessential amino acids,** can be synthesized endogenously because the appropriate carbon skeletons can be built from glucose metabolites in the citric acid cycle. The essential amino acids must be supplied in the diet, with individual minimal requirements ranging from 0.5 to 1.5 g/day^{-1}. All 20 amino acids are required for normal protein synthesis; therefore a deficiency of even one essential amino acid disrupts this process.

Protein sources vary greatly in their biologic effectiveness, depending in part on the ratio of essential to nonessential amino acids. Milk and egg proteins are of the highest quality in this regard. During infancy and childhood about 40% of the protein intake should consist of essential amino acids in order to support growth. In adults this requirement falls to 20%. In addition to their incorporation into proteins, many of the amino acids, including some essential ones, are precursors for important molecules, such as purines, pyrimidines, polyamines, phospholipids, creatine, carnitine, methyl donors, thyroid and catecholamine hormones, and neurotransmitters.

The plasma concentrations of the individual amino acids range widely from 20 $\mu mol/L^{-1}$ to 500 $\mu mol/L^{-1}$. All 20 amino acids are completely oxidized to CO_2 and H_2O after removal of the amino group. Each traverses a specific degradative pathway. (Refer to standard biochemistry textbooks for details.) However, all these pathways converge into three general metabolic processes: gluconeogenesis, ketogenesis, and ureagene-

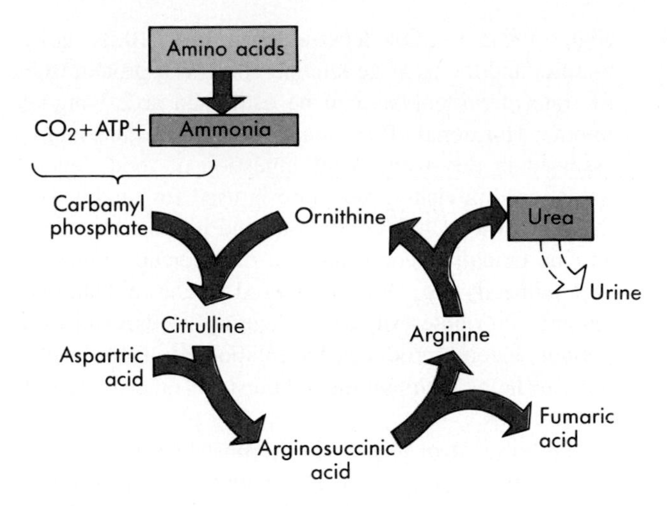

■ **Fig. 45-6** The Krebs-Henseleit urea cycle for disposal of amino acid ammonia.

sis. Except for leucine and lysine, all the amino acids can contribute carbon atoms for the synthesis of glucose. Five ketogenic amino acids give rise either to acetoacetate or its CoA precursors. In the degradation of all amino acids, ammonia is released. Ammonia, incorporated mainly into glutamine and alanine molecules, is then transported to the liver. In the liver ammonia is "detoxified" by incorporation into urea, a metabolically inert molecule. The synthesis of urea by the Krebs-Henseleit cycle is depicted in Fig. 45-6. The urea resulting from protein degradation is excreted by the kidney (see Chapter 42).

In the healthy adult under steady-state conditions, the total daily nitrogen excreted in the urine as urea plus ammonia, along with minor losses of nitrogen in the feces (0.4 g/day^{-1}) and skin (0.3 g/day^{-1}), is equal to the nitrogen released during metabolism of exogenous and endogenous protein. Such an individual is said to be in **nitrogen balance.** When there is no dietary protein intake, the sum of urea plus ammonia nitrogen in the urine reflects almost quantitatively the rate of endogenous protein degradation. When protein breakdown is greatly accelerated by tissue trauma or disease, urinary urea plus ammonia nitrogen may exceed protein nitrogen intake. In these two cases the individual is said to be in **negative nitrogen balance.** In a growing child or in a previously malnourished individual undergoing protein repletion with gain in body mass, urinary urea plus ammonia nitrogen excretion is less than the intake of protein nitrogen. This individual is said to be in **positive nitrogen balance.**

Measurements of external nitrogen balance do not themselves provide insight into the dynamic internal equilibrium between protein synthesis and protein degradation. The latter must be estimated by labeling body protein with an isotopic amino acid tracer, such as ^{15}N-glycine, and determining the flux of protein from measurements of ^{15}N-specific activity (Fig. 45-7). Such studies show that healthy adults, receiving isocaloric diets containing adequate protein, degrade and synthesize protein at a rate of 3 to 4 g/kg/day. Individual synthesis rates for most proteins are not known, but the hepatic synthesis of albumin accounts for about 5% of the

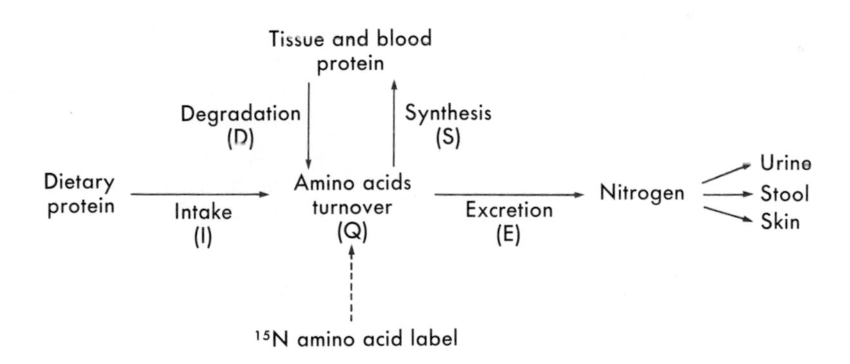

1. Label body pool
2. At isotopic steady state
3. Q = D+I = S+E
4. D = Q−I
5. S = Q−E

$$Q = \frac{\text{Tracer inflow}}{\text{Pool specific activity}} = \frac{\text{CPM}}{\text{Day}} \div \frac{\text{CPM}}{\text{GM}} = \frac{\text{GM}}{\text{Day}}$$

■ **Fig. 45-7** Schematic overview of body protein turnover. Body protein turnover is assessed by labeling the amino acid pool with a stable isotope tracer, ^{15}N, and simultaneously determining nitrogen balance. Input to the amino acid pool is from dietary protein *(I)* and from tissue degradation *(D)*; output from the amino acid pool is via tissue protein synthesis *(S)* and via oxidation and excretion as nitrogen *(E)*. At an isotopic steady state, the turnover or flux *(Q)* equals input *(I + D)*, which equals output *(S + E)*. When *Q* and *I* are measured, *D* can be calculated. Similarly, when *Q* and *E* are measured, *S* can be calculated. When *I* = *E*, the individual is said to be in nitrogen balance. In this state, *D* = *S*.

above total. The rate of total body protein synthesis is diminished when the diet is severely deficient in energy, in total protein, or in one of the essential amino acids. In such situations, the rate of protein degradation usually also diminishes, but not to the same extent as synthesis, so that net loss of body protein results.

In addition to the flux of total body protein, each individual amino acid undergoes its own unique turnover. Two that have been well studied are leucine, an essential amino acid, and alanine, a nonessential amino acid. In the postabsorptive state the only source of leucine is endogenous protein degradation (Fig. 45-8, *A*). Oxidation accounts for 20% of leucine disappearance from plasma (approximately 4 g/day), whereas 80% is reincorporated into new protein molecules. From a knowledge of the rate of leucine production (0.2 mg/kg/minute) and the average leucine content of protein (8%) the rate of proteolysis can be estimated at 2.5 mg/kg/minute. Hormonal effects on protein breakdown can be assessed in this way. After ingestion of food, leucine oxidation diminishes and more is used for protein synthesis. However, the continuous and irreversible rate of leucine oxidation necessitates a daily dietary intake of approximately 1 g. With prolonged fasting, oxidation of leucine contributes slightly to energy needs. More important, leucine produced by catabolism of dispensible proteins helps maintain the synthesis of other more critical proteins.

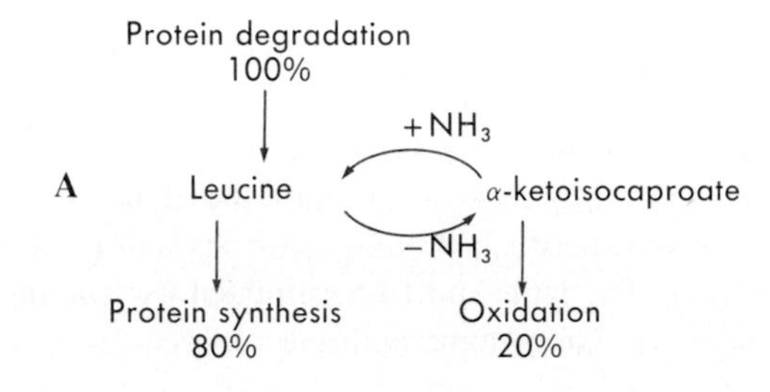

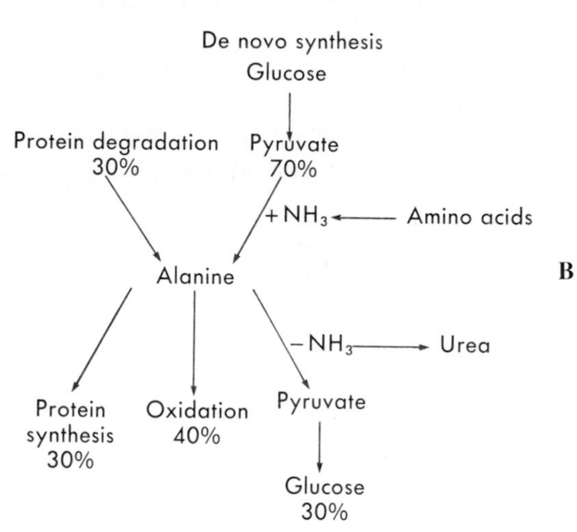

■ **Fig. 45-8** **A,** Quantitative turnover of leucine, an essential amino acid, in overnight fasted humans. Protein degradation is the only source for this amino acid whose carbon skeleton (α-ketoisocaproate) cannot be synthesized. The majority of leucine disposal is via reincorporation into protein synthesis. The leucine that is lost daily by oxidation must be replaced by dietary intake. The average overall rate of leucine turnover is 0.2 mg/kg/minute or 1.5 μmoles/kg/minute. **B,** Quantitative turnover of alanine, a nonessential amino acid, in overnight fasted humans. Alanine production is mostly by de novo synthesis from pyruvate, with a lesser contribution from protein degradation. Alanine disposal takes place via reincorporation into protein, oxidation, and gluconeogenesis. Alanine molecules, which arise from pyruvate and are reconverted to pyruvate, act as carriers for ammonia transferred from other amino acids. (Data in **A** from Matthews D et al: *Am J Physiol* 238:473, 1980; data in **B** from Chochinov R et al: *Diabetes* 27:287, 1978.)

The turnover of alanine in the postabsorptive state is more complex (Fig. 45-8, *B*). About 30% of alanine is produced from protein degradation. The other 70% of alanine arises from de novo synthesis: the carbon skeleton from glucose by way of pyruvate and the nitrogen by transamination from other amino acids. The disposal of plasma alanine reflects three major routes and processes; about 40% is oxidized after reconversion to pyruvate, about 30% reappears as glucose via gluconeogenesis, and the remaining 30% is reused for protein synthesis. Fig. 45-8, *B*, shows that a glucose-alanine cycle, similar to the glucose-lactate (Cori cycle), can be envisioned. In this cycle alanine functions as a carrier of amino groups. Except for glutamine, alanine is the predominant amino acid released by muscle. It is formed from pyruvate by transamination from other amino acids released during muscle proteolysis. Alanine is then extracted by the liver where the amino group is removed for urea synthesis, thereby regenerating pyruvate for gluconeogenesis. It should be clear from Fig. 45-8, *B*, that although alanine is a gluconeogenic amino acid, no net synthesis of glucose from protein results from this particular pathway. This is so even though fully 10% of the glucose molecules can be traced back to alanine molecules. Carbon atoms are merely shuttling from one glucose molecule to another.

■ *Fat Metabolism*

Fat represents almost half the total daily substrate for oxidation (about 100 g, or 900 kcal). The usual daily intake in the United States is also approximately 100 g, or 40% of total calories. The major component of both dietary and storage fat is triglycerides. Exogenous triglycerides from the diet are absorbed as chylomicrons, whereas endogenous triglycerides are synthesized largely in the liver. Both consist of long-chain saturated and monounsaturated fatty acids (chiefly palmitic, stearic, and oleic acids) esterified to glycerol. Because these fatty acids can also be synthesized in the liver and adipose tissue, in an overall sense, no strict dietary requirement exists for fat. However, about 3% to 5% of fatty acids are polyunsaturated and cannot be synthe-

■ **Table 45-3** Average lipid concentrations in postabsorptive plasma

	mg/dl	*μmole/L*
Ketoacids	10	0.1
Free fatty acids	10	0.4
Triglycerides	100	1.2
Cholesterol (total)	185	4.8
Low density	120	
High density	50	
Very low density	15	

sized in the body. These are termed **essential dietary fatty acids** (linoleic, linolenic, and arachidonic) because they are required as precursors for certain membrane phospholipid and glycolipid substances, as well as for important intracellular mediators known as prostaglandins. Another component of fat is the steroid molecule cholesterol, which serves a variety of specific functions in membranes and is the precursor for bile acids and steroid hormones. Cholesterol is both ingested and synthesized by most cells.

Table 45-3 presents the average basal concentrations of the most important plasma lipids. Although free fatty acids circulate at the lowest concentration, their plasma half life is by far the shortest (2 minutes), and their rate of turnover is the greatest (up to 200 g/day). Thirty percent to 40% of plasma FFA molecules are oxidized, largely by muscle, where they are the major fuel; 50% to 70% are reesterified to triglycerides. The rate of oxidation is largely proportional to the plasma FFA concentration. Plasma FFA is the direct source of 50% of total lipid oxidation; the remainder represents oxidation of intracellular lipids prestored in heart, muscle, and liver, along with a small contribution from plasma triglycerides. The recycling of plasma FFA by means of lipolysis and reesterification ordinarily accounts for only a trivial amount of daily energy expenditure. De novo synthesis of FFA from acetyl CoA in adipose tissue is also quantitatively insignificant in humans. Plasma ketoacids, derived from beta oxidation, can also be oxidized by muscle but become an important fuel only there and in the central nervous system during prolonged fasting.

The production, transport, and fate of plasma lipoproteins is very complex. Triglycerides and cholesterol, which form the major components of plasma lipids (see Table 45-3), circulate as complex lipoprotein particles ranging in size from 75 to 1500 μm. The nonpolar, hydrophobic core of these particles contains triglyceride and cholesterol esters, whereas their polar surfaces consist of phospholipid, cholesterol, and numerous apoproteins. The plasma lipoproteins are classified according to their physical densities (Table 45-4). As would be expected, the lowest density particles contain primarily triglycerides. As the particles increase in density, the triglyceride proportion decreases while that of phospholipid and protein increases. Numerous apoproteins varying in size and charge are found among the lipoprotein classes. Their varied functions include facilitation of triglyceride transport out of the intestine and liver, activation of enzymes of lipoprotein metabolism, and attachment of the lipoprotein particles to specific cell surface receptors. Apoproteins also can be exchanged by lipoprotein particles, a process that facilitates the normal traffic of lipids through the blood.

Chylomicrons circulate after a meal. They are formed from dietary fat and transported across the intestinal wall into the blood, as detailed in Chapter 40. They are rapidly cleared from plasma, with a half-life of 5 minutes. On the capillary endothelial surfaces of adipose tissue, muscle, and heart, chylomicron triglyceride is partly hydrolyzed by the enzyme lipoprotein lipase (Fig. 45-9). The latter is activated by apoprotein CII. The liberated free fatty acids are transported across the endothelial cells and taken up for resynthesis and storage as intracellular triglycerides. The remaining particles, known as chylomicron remnants, now contain less triglyceride and are enriched with cholesterol esters by interaction with high-density lipoprotein (HDL) particles. These remnants have also acquired apoprotein E, which with apoprotein B48 directs their uptake by the liver. There the remnants are degraded to free fatty acids, glycerol, free cholesterol, and the protein portion to amino acids. Thus the net result of chylomicron metabolism is to transfer most of the dietary triglycerides to adipose tissue and the dietary cholesterol to the liver.

VLDL particles are the major source of plasma triglycerides in the postabsorptive state. VLDL particles are synthesized and secreted largely by the liver, with a smaller contribution from the intestine. The liver uses free fatty acids, which are produced by de novo synthe-

■ **Table 45-4** Major lipoprotein classes

			Cholesterol			
	Density	*Triglyceride (%)*	*Free (%)*	*Esters (%)*	*Phospholipid (%)*	*Protein (%)*
Chylomicrons	<0.94	85	2	4	8	2
VLDL	0.94-1.006	60	6	16	18	10
IDL	1.006-1.019	30	8	22	22	18
LDL	1.019-1.063	7	10	40	20	25
HDL	1.063-1.21	5	4	15	30	50

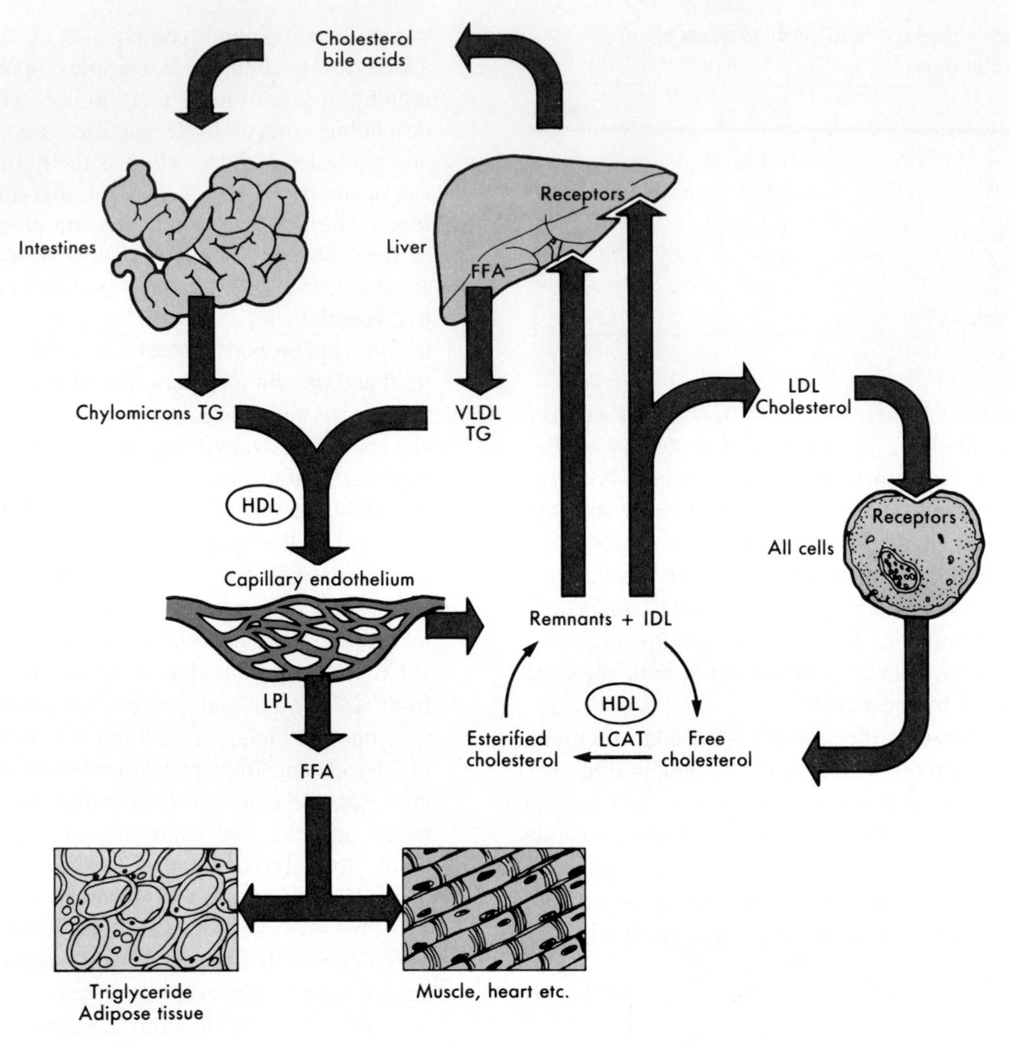

■ **Fig. 45-9** A schematic overview of lipoprotein metabolism and major aspects of lipid turnover in humans. Exogenous triglycerides (TG, chylomicrons absorbed from the intestine) and endogenous triglycerides (very-low-density lipoproteins, VLDL, produced in the liver) both give rise to free fatty acids (FFA) for storage in adipose tissue and oxidation in muscle. High-density lipoprotein particles (HDL) facilitate, and the enzyme lipoprotein lipase (LPL) directly catalyzes liberation of the free fatty acids from triglycerides. The resultant particles, called remnants, from chylomicrons and intermediate-density lipoproteins (IDL) from VLDL undergo further change in the circulation, which is also facilitated by HDL. The ratio of esterified cholesterol to free cholesterol is increased in the remnant and IDL particles by the enzyme lecithin-cholesterol acyltransferase (LCAT). The remnant particles are then taken up by the liver for further metabolism. The IDL particles are partly taken up by the liver and partly converted to cholesterol-rich low-density lipoprotein (LDL) particles. The latter are then taken up by virtually all cells after interaction with specific LDL receptors. Cholesterol, either synthesized in the liver or extracted from remnant and IDL particles, is also excreted into the intestine, partly as bile acids.

sis, derived from intrahepatic hydrolysis of triglycerides, or taken up directly from the plasma. Under normal dietary circumstances about 15 g of VLDL are produced per day. However, this can increase threefold to sixfold to accommodate high-calorie or high-carbohydrate diets. VLDL particles have a plasma half-life of about 2 hours. The initial phase of VLDL metabolism follows the same route as chylomicrons (see Fig. 45-9). After lipolysis by lipoprotein lipase and interaction with HDL particles, the partially triglyceride-depleted and cholesterol-enriched particle is called intermediate density lipoprotein (IDL) (Table 45-4). About half of the IDL particles are taken up by the liver, directed by apoprotein E and apoprotein B100. However, the other 50% is converted to low-density lipoprotein (LDL) particles in the blood and liver by an unknown mechanism. These particles transfer exogenous and endogenous cholesterol to other tissues.

LDL is the major cholesterol compound in plasma (Table 45-3), and it turns over with a half-life of 24 hours. The average daily amount of cholesterol absorbed from the diet (300 mg) plus the average amount synthesized (600 mg) is balanced by the daily excretion in the form of bile salts and neutral steroids. Hepatic synthesis of cholesterol varies inversely with the dietary intake, thereby maintaining a constant input. The cholesterol esters in LDL particles, generated as described earlier, are taken up by numerous tissues after interaction with specific LDL receptors that recognize their apoprotein E and apoprotein B100 (see Fig. 45-9). The internalized cholesterol esters are hydrolyzed in lysosomes to free cholesterol, which is then used by different cells for various purposes. It also may be reesterified (a reaction catalyzed by acyl-CoA-cholesterol acyltransferase) and then stored. The free cholesterol level within the cell modulates itself. Cholesterol diminishes its own further uptake into the cell by down regulating the LDL receptor, and it reduces its own intracellular de novo synthesis by suppressing the rate-limiting enzyme, hydroxymethylglutaryl-CoA (HMG-CoA) reductase.

HDL particles (Table 45-4) perform crucial functions in transferring lipid components between other lipoprotein particles and ultimately between organs. They also facilitate enzyme activities in the metabolism of lipoproteins. HDL particles synthesized in the liver contain primarily apoprotein C and apoprotein E; those that are synthesized in the intestine contain primarily apoprotein A. Nascent HDL is released as small, relatively dense spherical particles. Shortly afterward, they acquire phospholipids and a bilayer discoid shape and equilibrate their apoprotein contents. The plasma half-life of HDL is 5 to 6 days.

HDL assists in the previously described hydrolysis of chylomicron and VLDL triglycerides to FFA (see Fig. 45-9) by providing apoprotein-C for the activation of lipoprotein lipase. HDL also facilitates the flow of excess plasma triglycerides back to the liver and the flow of cholesterol to peripheral cells and to the liver (see Figs. 45-9 and 45-10). Thus the smaller and denser subfraction HDL_3 accepts free cholesterol from peripheral cells, from chylomicrons, and from VLDL, remnant, and IDL particles. Esterification of the cholesterol by the plasma enzyme lecithin-cholesterol acyl transferase (LCAT) is activated by HDL apoprotein A_1. The cholesterol ester is then exchanged for triglycerides in VLDL and chylomicrons by the cholesterol-ester trans-

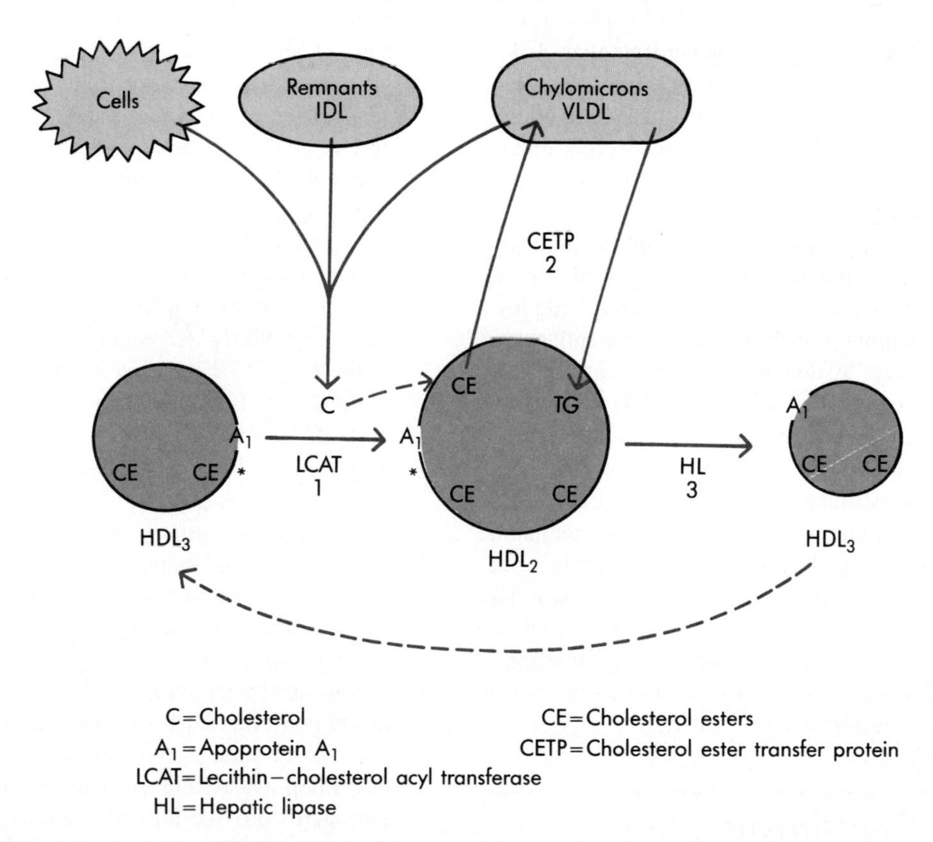

■ **Fig. 45-10** Functioning of high density lipoprotein (HDL). HDL_3 particles accept and esterify cholesterol from cells and other lipoprotein particles. Apoprotein A_1 activates the key enzyme LCAT. The resultant particle exchanges cholesterol esters for triglycerides with chylomicrons and VLDL, becoming more buoyant HDL_2 particles. These in turn are relieved of the triglycerides by the action of hepatic lipase and cycle back to HDL_3 particles.

fer protein; in the process, the HDL changes to the larger and less dense HDL_2 subfraction. The cholesterol esters then can either be taken up by the liver as part of remnant, IDL, and LDL particles, or they can be delivered to the peripheral cells as LDL particles. The HDL_2 can itself cycle back to HDL_3, as the previously acquired triglycerides are hydrolyzed by hepatic lipase (see Fig. 45-10). The ratio of HDL_3/HDL_2 thus fluctuates; it is higher during alimentary lipemia and lower during fasting. HDL levels in plasma correlate positively with lipoprotein lipase levels and negatively with hepatic lipase levels. They also correlate negatively with tissue cholesterol content.

The factors that influence the relative proportions of HDL and LDL particles are of great public health importance. HDL levels are increased in females and by estrogens, exercise conditioning, moderate alcohol intake, and several drugs. They are decreased in males and by androgens, smoking, obesity, a high polyunsaturated fat diet, and certain drugs. LDL levels are increased in males and by smoking, obesity, a sedentary life-style, and certain drugs. These important effects partly explain the fact that male gender, a lack of physical fitness, obesity, and smoking are strong risk factors for cardiovascular disease and death. These, as well as certain genetic diseases (e.g., LDL receptor deficiency or mutation), elevate the ratio of LDL cholesterol to HDL cholesterol in plasma. A ratio greater than 4 is a high risk factor for cardiovascular disease. Strenuous public health efforts are underway to lower this ratio in the U.S. population through diet change, weight reduction, cessation of smoking, and increased exercise.

Major abnormalities in lipoprotein levels are caused by specific defects in the pathways of lipid metabolism. A deficiency in lipoprotein lipase activity leads to excessive and prolonged chylomicron and triglyceride levels after fat-containing meals; this can cause inflammation of the pancreas. A deficiency of cell-surface LDL receptors leads to very high plasma LDL cholesterol levels and premature coronary artery disease. Synthesis of an abnormal apoprotein E, which cannot efficiently direct lipoprotein particles to interact with cell receptors, leads to accumulation of chylomicron remnants and IDL in plasma. In this situation, both triglyceride and cholesterol levels are elevated with increased risk of coronary artery disease. Conversely, a deficiency of apoprotein B leads to very low levels of chylomicrons, VLDL, and LDL in plasma, with accumulation of triglyceride in the intestines and the liver.

■ *Metabolic Adaptations*

■ *Fasting*

In the fasting state the individual totally depends on endogenous substrates for energy (Fig. 45-11). Mobilization of glucose provides essential fuel for the central nervous system; release of free fatty acids provides for the oxidative needs of the other tissues. An increase in protein degradation to amino acids is also a fundamental feature of this response. The fasting individual is said to be in a state of **catabolism** because carbohydrate, fat, and protein stores are all decreasing.

The liver supplies glucose to the circulation initially by augmenting glycogenolysis. After 12 to 15 hours of fasting, however, hepatic glycogen stores are greatly depleted, and a rapid enhancement of gluconeogenesis fills the void. To supply glucose precursors, 75 to 100 g of muscle protein are broken down daily during the first few days. This is reflected in a rising excretion of nitrogen in the urine. Gluconeogenesis is also supported by the provision of 15 to 20 g of glycerol daily, which is released during the accelerated lipolysis of triglycerides in adipose tissue. Glucose oxidation in muscle and liver is spared as increasing quantities of free fatty acids become available. Muscle lipoprotein lipase activity increases to facilitate uptake of triglycerides for oxidation; adipose lipoprotein lipase activity decreases to dampen uptake of triglyceride for storage.

A portion of fatty acid oxidation in liver yields the ketoacids β-hydroxybutyrate and acetoacetate. These can also be oxidized by muscle cells, further sparing use of glucose. The net shift away from glucose and toward fatty acid oxidation lowers the respiratory quotient. These adaptations are also reflected in changing plasma concentrations of substrates (Fig. 45-12). The concentrations of glucose and the major gluconeogenic amino acid alanine decrease, whereas the concentrations of free fatty acids, glycerol, and branch-chain amino acids such as leucine increase. High levels of the strong ketoacids produce a tendency to metabolic acidosis and a slight reduction in blood pH. Much of this pattern of adaptation is mediated by hormonal modification, particularly decreasing insulin and increasing glucagon (Fig. 45-12).

As fasting is prolonged beyond a few days, other important adaptations occur. Total energy expenditure, reflected in the BMR, decreases 10% to 20%, limiting the drain on energy stores. The central nervous system no longer depends entirely on glucose as an energy source, and two thirds of its needs are eventually met by the ketoacids. As less glucose is needed for oxidation, gluconeogenesis diminishes and protein breakdown declines to 25 to 30 g/day^{-1}. In long-term fasting body weight diminishes by an average of 300 g/day^{-1}, of which two thirds is accounted for by fat and one third by lean tissue. Of the latter, 25% constitutes protein and 75% intracellular water and electrolytes. In long-term fasting fatty acids provide 90% of the total energy expenditure. As long as sufficient fluid is ingested, an individual of normal weight can survive up to 60 days. About that time fat stores are almost exhausted, protein degradation suddenly accelerates, and death follows. The flow of substrates shown in Fig. 45-11 is reversed after feed-

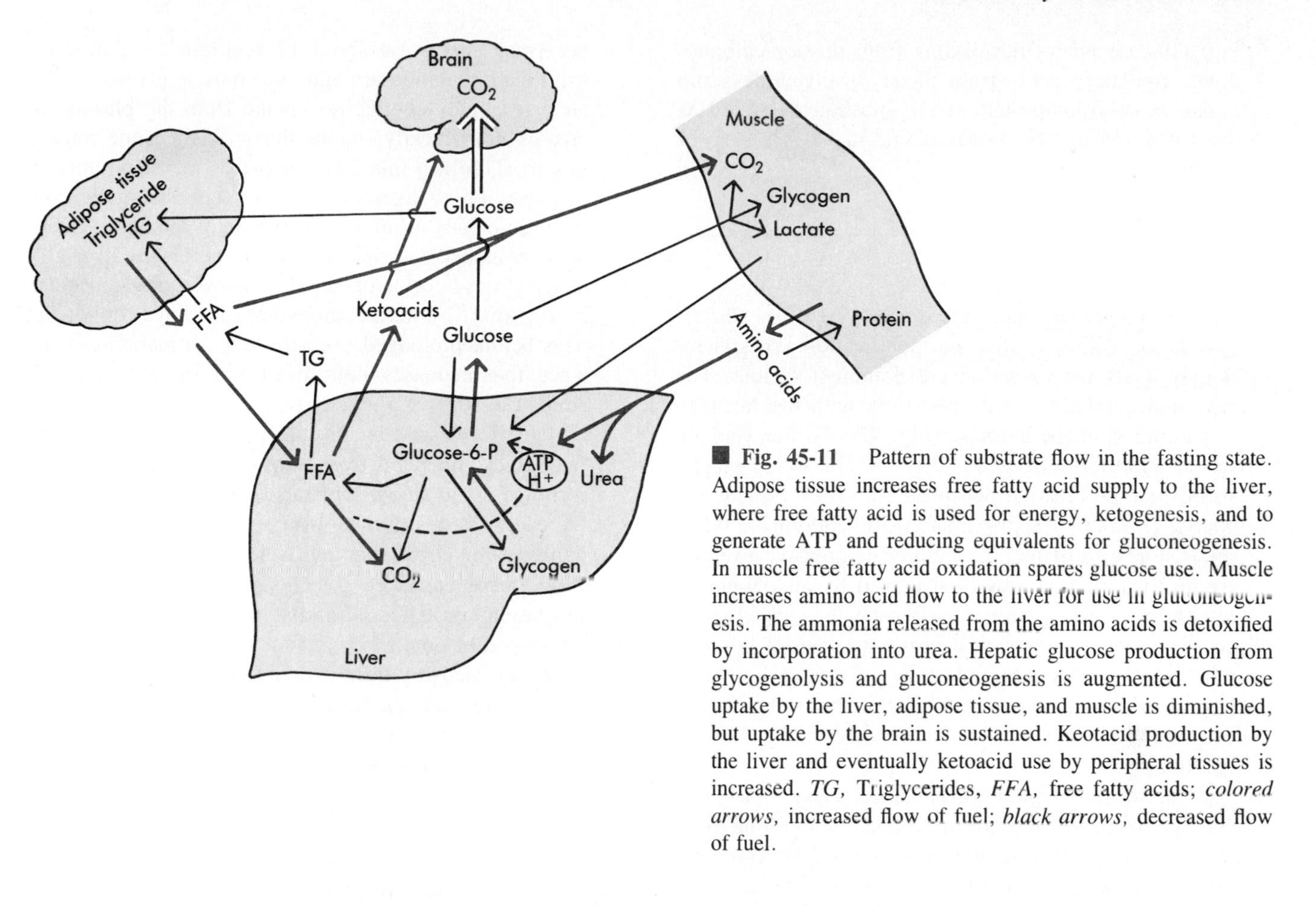

■ **Fig. 45-11** Pattern of substrate flow in the fasting state. Adipose tissue increases free fatty acid supply to the liver, where free fatty acid is used for energy, ketogenesis, and to generate ATP and reducing equivalents for gluconeogenesis. In muscle free fatty acid oxidation spares glucose use. Muscle increases amino acid flow to the liver for use in gluconeogenesis. The ammonia released from the amino acids is detoxified by incorporation into urea. Hepatic glucose production from glycogenolysis and gluconeogenesis is augmented. Glucose uptake by the liver, adipose tissue, and muscle is diminished, but uptake by the brain is sustained. Keotacid production by the liver and eventually ketoacid use by peripheral tissues is increased. *TG,* Triglycerides, *FFA,* free fatty acids; *colored arrows,* increased flow of fuel; *black arrows,* decreased flow of fuel.

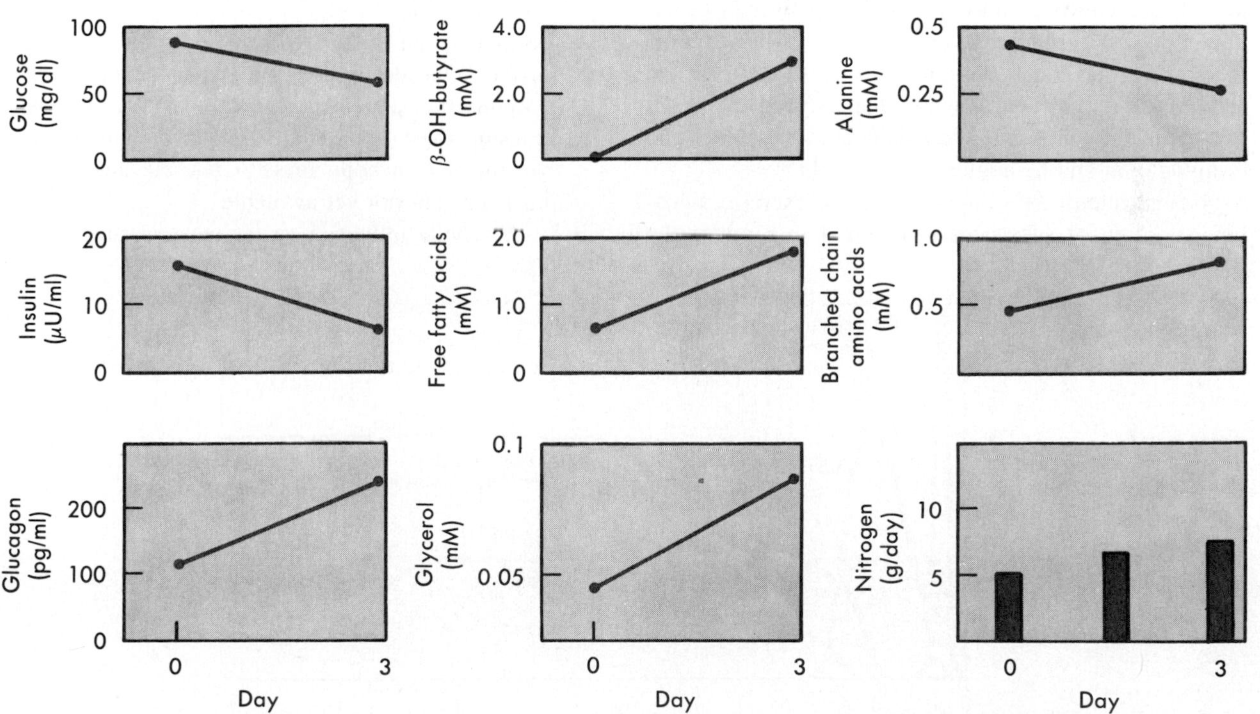

■ **Fig. 45-12** Changes in plasma substrate levels, urine nitrogen excretion, and plasma insulin and glucagon during 3 days of fasting in humans. Note the rise in lipid derived fuels and products of proteolysis. These changes are mediated in part by a fall in insulin and a rise in glucagon (hormones from the pancreatic islets). (Redrawn from Felig P et al: *J Clin Invest* 48:584, 1969. From the American Society for Clinical Investigation.)

ing. Glucose enters the plasma from dietary carbohydrate, free fatty acids from dietary triglycerides, and amino acids from protein. Each substrate is stored as described and the individual is said to be in a state of **anabolism.**

Exercise

The metabolic response to exercise resembles the response to fasting, in that the mobilization and generation of fuels for oxidation are dominant factors. The type and amounts of substrate vary with the intensity and duration of the exercise (Fig. 45-13). For very intense, short-term exercise (e.g., a 10- to 15-second sprint), stored creatine phosphate and ATP provide the energy at a rate of approximately 50 $kcal/min^{-1}$. When these stores are depleted, additional intensive exercise for up to 2 minutes can be sustained by breakdown of muscle glycogen to glucose-6-phosphate, with glycolysis yielding the necessary energy (at a rate of 30 $kcal/min^{-1}$). This **anaerobic** phase is not limited by depletion of muscle glycogen at this point, but rather by the accumulation of lactic acid in the exercising muscles and the circulation.

After several minutes of exhaustive anaerobic exercise, an oxygen debt of 10 to 12 L can be built up. This must be repaid before the exercise can be repeated. From 6 to 8 L are required either to rebuild the accumulated lactic acid back into glucose in the liver or to oxidize it to CO_2. About 2 L are required to replenish normal muscle ATP and creatine phosphate content. An additional 2 L will replenish the oxygen normally present in the lungs and body fluids and oxygen bound to myoglobin and hemoglobin.

For less intense but longer periods of exercise, **aerobic** oxidation of substrates is required to produce the necessary energy (at about 12 $kcal/min^{-1}$). Substrates from the circulation are added to muscle glycogen. After a few minutes glucose uptake from the plasma increases dramatically, up to thirtyfold in some muscle groups. To offset this drain, hepatic glucose production increases up to fivefold. Initially this is largely from glycogenolysis. With exercise of long duration, gluconeogenesis becomes increasingly important as liver glycogen stores become depleted. However, endurance can be improved by high-carbohydrate feedings for several days before prolonged exercise (e.g., a marathon run), since this increases both liver and muscle glycogen stores. To support gluconeogenesis, amino acids are increasingly released by muscle proteolysis. Eventually, fatty acids, liberated from adipose tissue triglycerides, form the predominant substrate, supplying two thirds of the energy needs during sustained exercise. Except for increases in circulating pyruvate and lactate resulting from greatly enhanced glycolysis, the pattern of change in plasma substrates is similar to that of fasting, only telescoped in time.

During recovery from exercise, muscle and liver glycogen stores must be rebuilt; this requires energy input. A significant amount of energy is also expended during this period as futile recycling of FFA back into triglycerides.

Regulation of Energy Stores

As stated, the preponderance of stored energy consists of fat. What determines the proper quantity of this energy reserve, and what regulates it? Does an ideal relationship exist between fat mass and either total body weight or lean body mass? Clear-cut answers to these questions are not yet available.

A genetic influence on fat mass is suggested by (1)

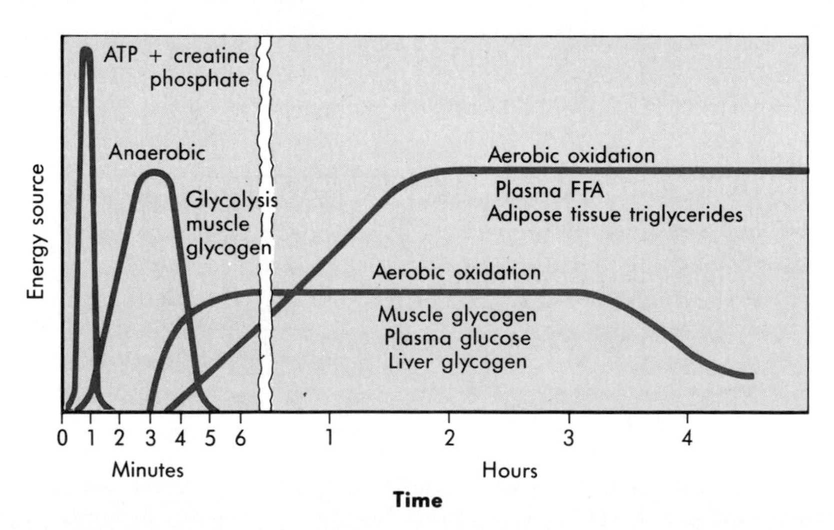

Fig. 45-13 Energy sources during exercise. Note the sequential use of stored high-energy phosphate bonds, glycogen, circulating glucose, and circulating free fatty acids (FFA). The latter dominate in sustained exercise.

the greater similarity of adipose stores in identical twins than in fraternal twins and (2) the tendency for body mass of adopted children to resemble that of their biological parents more than that of their adoptive parents. Environmental influences, specifically the quality and quantity of the food available, are suggested by the excessive weight gain of certain laboratory animals in response to presentation of high-fat or "junk food" diets, as well as by the much greater prevalence of obesity in affluent westernized societies than in other populations. Also, the human species has more energy-storage adipose cells per unit body mass than any other species except whales, which may contribute to this propensity for obesity.

Some data suggest the existence of a particular set point for energy stores in each individual. Once adult weight is reached, it tends to be constant, at least until middle age, at which point most humans incur at least a modest weight gain to a new, higher constant level with a higher proportion of body fat. Normal laboratory animals subjected to overfeeding or underfeeding experiments will return to their original weight and degree of fatness when again allowed free access to food. They will do this not only by adjusting food intake, but also by adjusting energy expenditure in the appropriate direction.

Decreases in energy stores caused by excessive expenditure are compensated by increased caloric intake. Control of appetite appears to reside in the hypothalamus. In rodents evidence exists for both a hunger center in the lateral hypothalamus and a satiety center in the ventromedial hypothalamus.

A large variety of afferent signals to the hypothalamus contribute to stimulating the appetite. These include the sight, smell, taste, sweetness, and palatability of food; a 15% reduction in plasma glucose; and a decrease in FFA oxidation in the liver. Eating is inhibited by glucose in the duodenum or portal vein. A variety of peptides of gastrointestinal origin, some found also in the brain, and a variety of neurotransmitters carry appetite regulatory signals to and within the central nervous system (see box). Some appear nutrient-specific, such as serotonin (glucose) and enterostatin (fat). At present, it is difficult to synthesize a coherent schema of overall control of caloric intake.

Modulators of feeding behavior

Stimulate	Inhibit
Norepinephrine	Epinephrine/norepinephrine
γ amino butyric acid	Serotonin
Endorphins	Corticotropin releasing hormone
Neuropeptide Y	Cholecystokinin
Galanin	Glucagon
	Calcitonin
	Bombesin
	Enterostatin
	? Insulin

Increases in energy stores resulting from excessive food intake can be compensated for by increased energy expenditure. This may occur via thermogenic processes, such as "futile cycles" that are wasteful of ATP (e.g., glucose → glucose-6-phosphate → glucose), ion pumping via the membrane enzyme Na^+, K^+ -ATPase, or uncoupling of ATP formation from mitochondrial oxidation. Various hormones, as well as the sympathetic nervous system, can regulate energy expenditure.

Pathologic accumulation of energy stores as fat (i.e., obesity) is a major health problem in many countries. A body weight 20% above average increases the risk of disorders such as diabetes and hypertension, and a body weight 50% above average greatly increases the risk of death. The cause of human obesity is not known. Some obese individuals behave as if they have an elevated set point for energy stores; they defend this set point tenaciously by decreasing energy expenditure when caloric intake is reduced by dieting. In some studies individuals with lower BMRs are at greater risk for future weight gain. However, once obesity is present, BMR, diet-induced thermogenesis, and the energy cost of exercise are generally equal to or greater than those of normal weight individuals.

Other obese individuals behave as though appetite and caloric intake are uncoupled from any perception of energy stores; these individuals may regulate appetite abnormally as opposed to having an elevated set point. In obese humans the profile of hormones involved in regulation of fat metabolism also generally favors deposition rather than mobilization of fat. In addition, adipose tissue of obese humans contains elevated levels of lipoprotein lipase, the key enzyme that transfers circulating triglycerides into cells. Although abnormalities of this sort are intriguing, none has been proved to be the primary cause of human obesity.

Summary

1. Energy input as carbohydrate, fat, and protein calories must equal energy expenditure. The latter is composed of basal, diet-induced thermogenesis and sedentary activity components plus exercise and heavy labor needs. Basal metabolic rate (75% of total) is proportional to body mass, is in part genetically determined, and declines with aging.

2. Fatty acids are the major fuel in most tissues except for the central nervous system and red blood cells, where glucose is the major and obligatory oxidative substrate. Depending on availability, fatty acids and glucose are competitive substrates in muscle mass and liver.

3. Anaerobic glycolysis, beta oxidation of fatty acids,

and disposal of acetyl CoA by the Krebs cycle are the biochemical mechanisms that generate ATP by oxidative phosphorylation. The overall efficiency of energy yield is 65%.

4. Energy is primarily stored as adipose tissue triglycerides with lesser amounts as protein. Carbohydrate stores are trivial, hence the need for efficient glucose production by the liver (gluconeogenesis).

5. During long-term fasting, gluconeogenesis from amino acids, glycerol, and lactate is required to sustain central nervous system metabolism and other critical functions dependent on glucose. The pathway of gluconeogenesis is partly a reversal of glycolysis but requires special steps from pyruvate. Increased use of fatty acids during fasting greatly increases production of the ketoacids β-hydroxybutyrate and acetoacetate.

6. Endogenous protein turnover obligates a daily ingestion of protein, in particular, essential amino acids such as leucine. Such amino acids are irreversibly degraded, and their carbon skeletons cannot be synthesized. The carbon skeletons of nonessential amino acids, such as alanine, are degraded and resynthesized daily.

7. Fat metabolism involves a variety of circulating lipoprotein particles that transfer triglycerides and cholesterol, either originating in the diet or from hepatic synthesis, to and from various tissues. Low density lipoprotein particles, rich in cholesterol, play a role in the development of atherosclerosis, whereas high density lipoprotein particles have a protective effect.

8. Energy needs during exercise are met in sequence by stored muscle creatine phosphate plus ATP, stored muscle glycogen, anaerobic glycolysis, and, finally, aerobic oxidation of glucose and then fatty acids taken up from the plasma. These substrates are supplied by hepatic glycogenolysis and gluconeogenesis and adipose tissue lipolysis, respectively.

9. Correlation of energy intake and expenditure with energy stores is a complex process, probably controlled in the hypothalamus and involving numerous amine and peptide neurotransmitters and neuromodulators. Obesity can result from an altered set point of energy stores, from unregulated caloric intake, or from decreased energy use.

Bibliography

Journal articles

Bogardus C et al: Familial dependence of the resting metabolic rate, *N Engl J Med* 315:96, 1986.

Bray GA, York DA, Fisler JS: Experimental obesity, *Vitam Horm* 45:1, 1988.

Cahill GF: Starvation in man, *N Engl J Med* 282:668, 1970.

Eckel RH: Lipoprotein lipase: a multifunctional enzyme relevant to metabolic diseases, *N Engl J Med* 320:1060, 1989.

Exton JH: Gluconeogenesis, *Metabolism* 21:945, 1972.

Felig P et al: Amino acid metabolism during prolonged starvation, *J Clin Invest* 48:584, 1969.

Ferrannini E et al: The disposal of an oral glucose load in health subjects: a quantitative study, *Diabetes* 34:580, 1985.

Foster D: From glycogen to ketones—and back (Banting Lecture 1984), *Diabetes* 33:1188, 1984.

Giesecke K et al: Protein and amino acid metabolism during early starvation as reflected by excretion of urea and methylhistidines, *Metabolism* 38:1196, 1989.

Groop LC et al: Role of free fatty acids and insulin in determining free fatty acid and lipid oxidation in man, *J Clin Invest* 87:83, 1991.

Harris RBS: Role of set-point theory in regulation of body weight, *FASEB J* 4:3310, 1990.

Haymond M, Miles J: Branched chain amino acids as a major source of alanine nitrogen in man, *Diabetes* 31:86, 1982.

Katz L et al: Splanchnic and peripheral disposal of oral glucose in man, *Diabetes* 32:675, 1983.

Morley JE, Levine AS: Nutrition: the changing scene—the central control of appetite, *Lancet* 1:398, 1983.

Ravussin E et al: Determinants of 24-hour energy expenditure in man: methods and results using a respiratory chamber, *J Clin Invest* 78:1568, 1986.

Reeds P, James W: Nutrition: the changing scene—protein turnover, *Lancet* 1:571, 1983.

Schaefer E et al: Pathogenesis and management of lipoprotein disorders, *N Engl J Med* 312:1300, 1985.

Sims EAH: Energy balance in human beings: the problems of plentitude, *Vitam Horm* 43:1, 1986.

Sims EAH, Danforth E Jr: Expenditure and storage of energy in man, *J Clin Invest* 79:1019, 1987.

Tall AR: Plasma high density lipoproteins, *J Clin Invest* 86:379, 1990.

Wasserman DH, Cherrington AD: Hepatic fuel metabolism during muscular work: role and regulation, *Am J Physiol* 260:E811, 1991.

Wolfe BM et al: Effect of elevated free fatty acids on glucose oxidation in normal humans, *Metabolism* 37:323, 1988.

Monograph and books

Flatt JP: *The biochemistry of energy expenditure*. In Bray G, editor: *Recent advances in obesity research, II*. Proceedings of the Second International Congress on Obesity, Los Angeles, 1978, Newman Publishing.

Leibowitz SF: *Brain neurotransmitters and hormones in relation to eating behavior and its disorders*. In Björntorp P, Brodoff BN, editors: *Obesity*, Philadelphia, 1992, JP Lippincott.

CHAPTER

46

Hormones of the Pancreatic Islets

■ *Anatomy of the Pancreatic Islets*

The islets of the pancreas secrete two major hormones, **insulin** and **glucagon;** these hormones are rapid and powerful regulators of metabolism. Together they coordinate the flow of endogenous glucose, free fatty acids, amino acids, and other substrate molecules to ensure that energy needs are met in the basal state and during exercise. In addition they coordinate the efficient disposition of the nutrient input from meals. They accomplish these functions primarily by actions on the liver, muscle mass, and adipose tissue. Other hormonelike products of islet cells (including amylin, pancreastatin, somatostatin, and pancreatic polypeptide) may play subsidiary roles in the regulation of metabolism.

The strategic location of the islets reflects these functions. Insulin and glucagon are released in response to nutrient inflow from the gut and to gastrointestinal secretogogues, as are the products of the exocrine pancreas (Chapter 39). The proximity of the islets to the pancreatic acini may permit them to have local effects on exocrine pancreatic function. Islet hormones are secreted into the pancreatic vein and then into the portal vein. This arrangement permits these hormones to join the nutrient stream after meals, and it preferentially exposes the liver, which is the central organ in substrate traffic, to higher hormone concentrations than the peripheral tissues receive. This vascular arrangement also allows the liver to extract variable amounts of insulin and glucagon on first pass and thereby to modulate their availability to other tissues. Because insulin and glucagon are often secreted and act in reciprocal fashion—when one is needed, the other usually is not—the ratio of their concentrations may be more critical than their actual levels.

There are approximately one million islets and they constitute 1% to 1.5% of the human pancreas mass. Each islet contains on average 2,500 cells composed of four types. β-cells, the unique source of insulin, compose 60% to 70%; α-cells, the source of glucagon, compose 20% to 25%; δ-cells, the source of somatostatin, compose 10%. The remainder are PP cells, the source of pancreatic polypeptide.

The intimate structure, vascular supply, and distribution of the islets are also important to their function. Each islet consists of a core of β-cells (Fig. 46-1, *A*) with either a mantle of α- or δ-cells (in the dorsal body and tail of the pancreas; Fig. 46-1, *A*) or a mantle of δ- and PP cells in the ventral head of the organ. If islets are disaggregated experimentally and the individual cells are dispersed, these cells spontaneously reaggregate into islets if brought back into contact in culture. Gap junctions exist between neighboring cells (Fig. 46-1, *B*) and permit the flow of molecules less than 1,000 molecular weight and of electrical currents between them.

The endocrine cells of the islets arise from common endodermal ancestors in the pancreatic ducts. However, they also show some characteristics of neuroectodermal cells, such as polarization/depolarization and synthesis of neurotransmitter amines such as gamma-aminobutyric acid (GABA). The developmental factors that lead to selective expression of different hormone genes and to differential functions are unknown. The islets are identifiable by the fourth week of human gestation and are capable of hormone secretion by the tenth week.

The islets are exceedingly well vascularized; they receive 10% of the pancreatic blood supply. Small arterioles enter the core of the islet and break up into a network of capillaries with fenestrated endothelium. These come together into venules, which then carry the blood to the mantle of the islet. This portal arrangement allows high concentrations of insulin from the β-cell core to bathe the α-, δ-, and PP cells of the respective mantles. Such a vascular pattern suggests local paracrine regulation. Each β- and α-cell has a basal (arterial) and an apical (venous) face. Between the lateral surfaces of neighboring β-cells run canaliculi that span the distance between the arteriolar and venous ends of the cell. The canaliculi carry interstitial fluid in a venous direction and permit selective exposure of the lateral cell surfaces to regulatory molecules, such as glucose (Fig. 46-2).

A

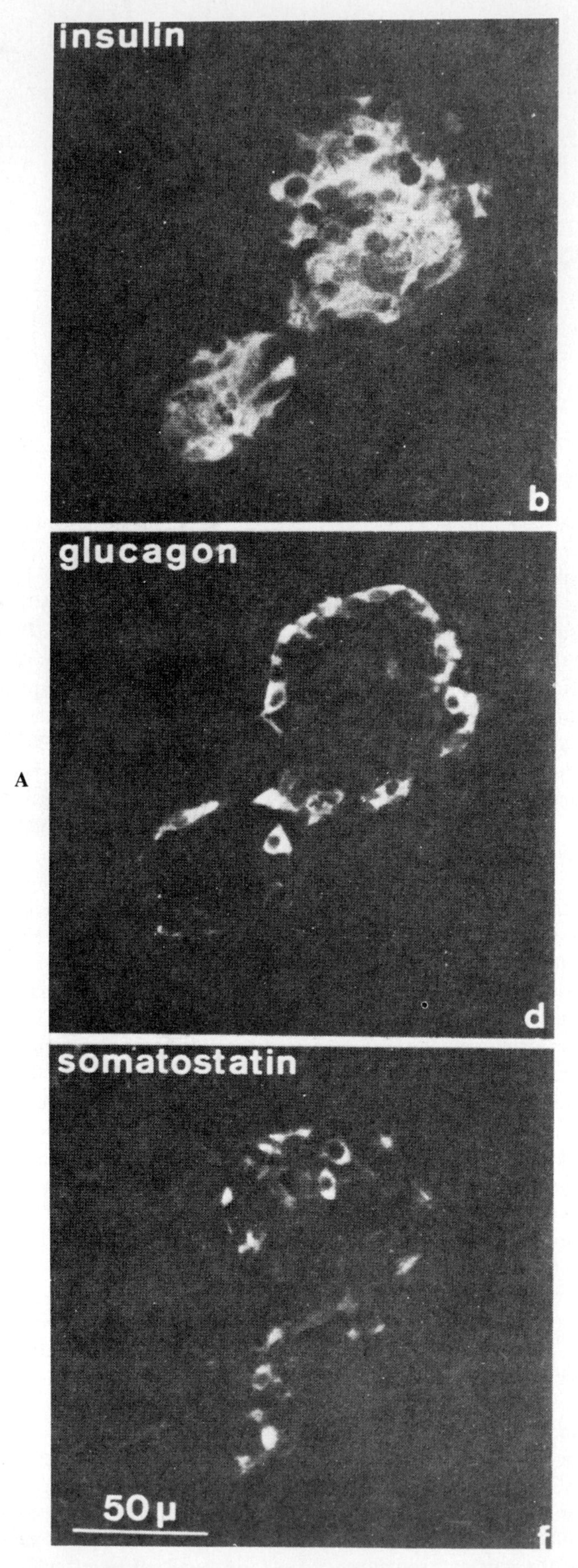

■ **Fig. 46-1, A** For legend see opposite page.

■ **Fig. 46-1** **A,** Human islet stained by immunohistochemical methods shows the predominance of β cells and the peripheral distribution of α-cells and δ-cells (from Unger RH et al: *Annu Rev Phys,* 40:307, 1978. **B,** Electron micrograph of adjacent β- and α-cells in an islet showing the presence of gap junctions *(GJ)* and tight junctions *(TJ)* between them. (From Orci L et al: Cell contacts in human islets of Langerhaus, *J Clin Endocrinol Metab* 41:841, 1975)

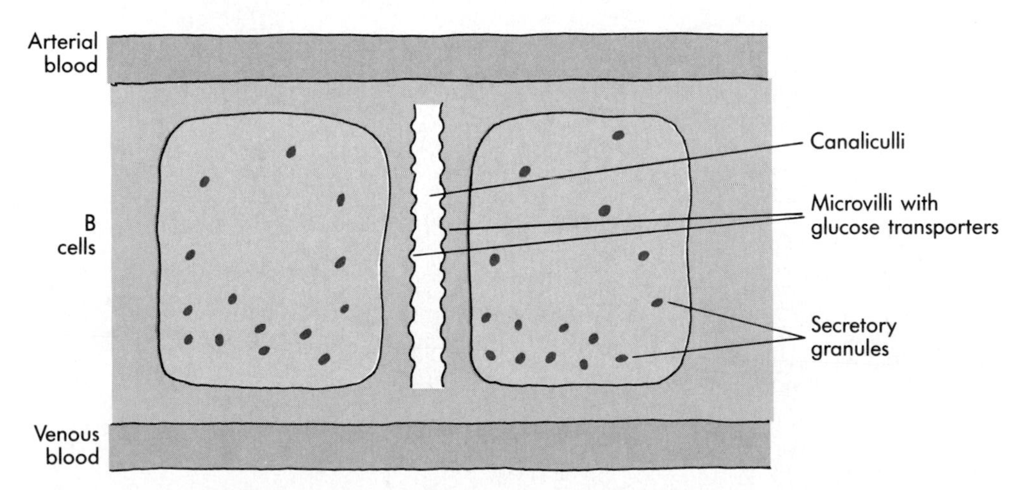

■ **Fig. 46-2** Orientation of islet β-cells to their circulatory supply. Secretory granules are concentrated on the venous face of the cell. Glucose transporters are concentrated in microvilli on the lateral surfaces in contact with channels (canaliculi) that contain intercellular fluid and connect arterial and venous vessels.

The islets are innervated by parasympathetic, sympathetic, and peptidergic nerves. The δ-cells in the mantle are dendritic in shape and send granule-containing processes into the β-cell core; these features suggest neurocrine or paracrine pathways of intraislet regulation.

Islet cells contain secretory granules with smooth membranes within which hormone is stored. These granules are more heavily distributed in the apical venous side of the cell (see Fig. 46-2). A system of microtubules is present, often lying in parallel bundles that separate linear rows of secretory granules. In addition microfilaments containing myosin and actin form a web adjacent to the plasma membrane and in association with the microtubules. Agents that destroy microtubules or prevent their function inhibit hormone release, whereas agents that cause hypercontraction of microfilaments enhance it. These structures may facilitate active movement of the secretory granules to the plasma membrane. Movement of insulin-containing granules through the cytoplasm of β-cells occurs at a rate of 1.5 μm/second.

Insulin

Structure and Synthesis

Among the peptide hormones, insulin has been of preeminent historical, physiological, and clinical importance. It was the first hormone to be isolated from animal sources in pure enough form to be administered therapeutically—and with dramatic life-saving effects. It was the first to have its amino acid sequence and tertiary structure elucidated and the first peptide hormone for which a mechanism of action on the cell membrane was demonstrated. It was also the first to be measured by radioimmunoassay; indeed, the entire concept of this landmark analytical method developed from studies of the metabolism of radiolabeled insulin in humans. The biosynthesis of peptide hormones from larger precursor molecules was observed initially in insulin. Finally, it was the first mammalian peptide hormone whose biosynthesis by recombinant DNA technology was achieved.

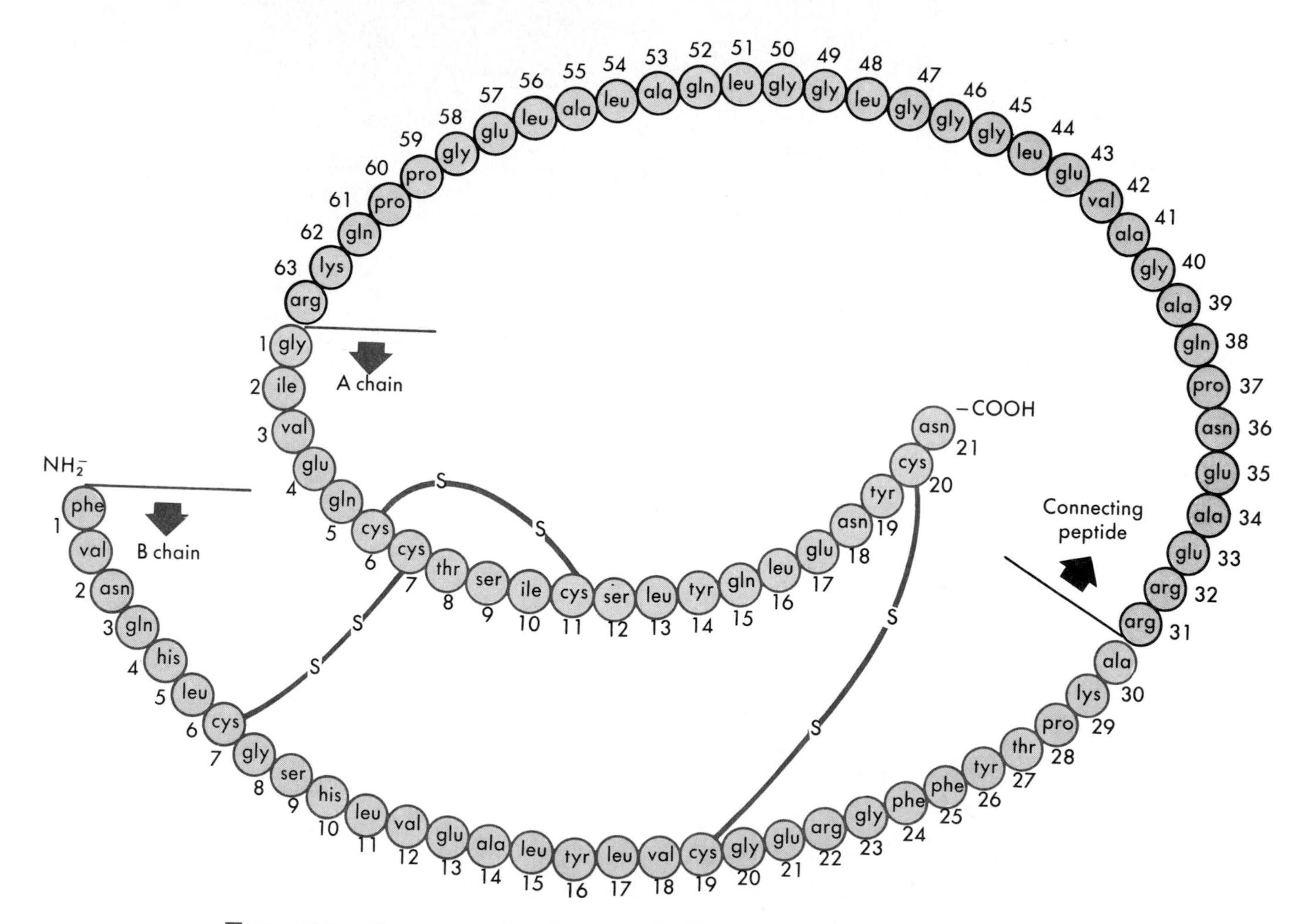

■ **Fig. 46-3** The structure of porcine proinsulin. The solid area is the insulin molecule released by cleavage of the connecting peptide. (Redrawn from Shaw WN, Chance RE: Effect of porcine proinsulin in vitro on adipose tissue and diaphragm of the normal rat, *Diabetes* 17:737, 1968. From the American Diabetes Association.)

Insulin is a peptide consisting of two straight chains (Fig. 46-3). Its molecular weight is 6,000. The A chain, containing 21 amino acids, and the B chain, containing 30 amino acids, are linked by two disulfide bridges. In addition, the A chain contains an intrachain disulfide ring (see Fig. 46-3). Insulins from various species are virtually equipotent in humans. The features most conserved in vertebrate evolution are the positions of the three disulfide bonds, the N-terminal and C-terminal amino acids of the A chain, and the hydrophobic character of the amino acids at the C-terminal of the B chain. These areas determine the tertiary structure, critical to biological activity, that resides within the B chain. Insulin monomers readily form dimers, three of which in the presence of zinc form a crystalline hexameric unit with a threefold axis passing through two zinc atoms. Crystalline zinc insulin is the basic pharmaceutical preparation of greatest importance in therapy.

Synthesis of insulin occurs as shown in Fig. 46-4. The insulin gene on human chromosome 11 is the ancestral member of a superfamily coding for a variety of related growth factors. It is composed of four exons and two introns. The gene directs the synthesis of preproinsulin, a precursor of molecular weight 11,500, containing four sequential peptides: an N-terminal signal peptide, the B chain of insulin, a connecting peptide, and the A chain of insulin (see Fig. 46-3). The N-terminal 23 amino acid signal is rapidly cleaved at the site of synthesis while the proinsulin chain is being completed. During guided passage of proinsulin to the Golgi apparatus, establishment of the disulfide linkages yields the "folded" proinsulin molecule with a molecular weight

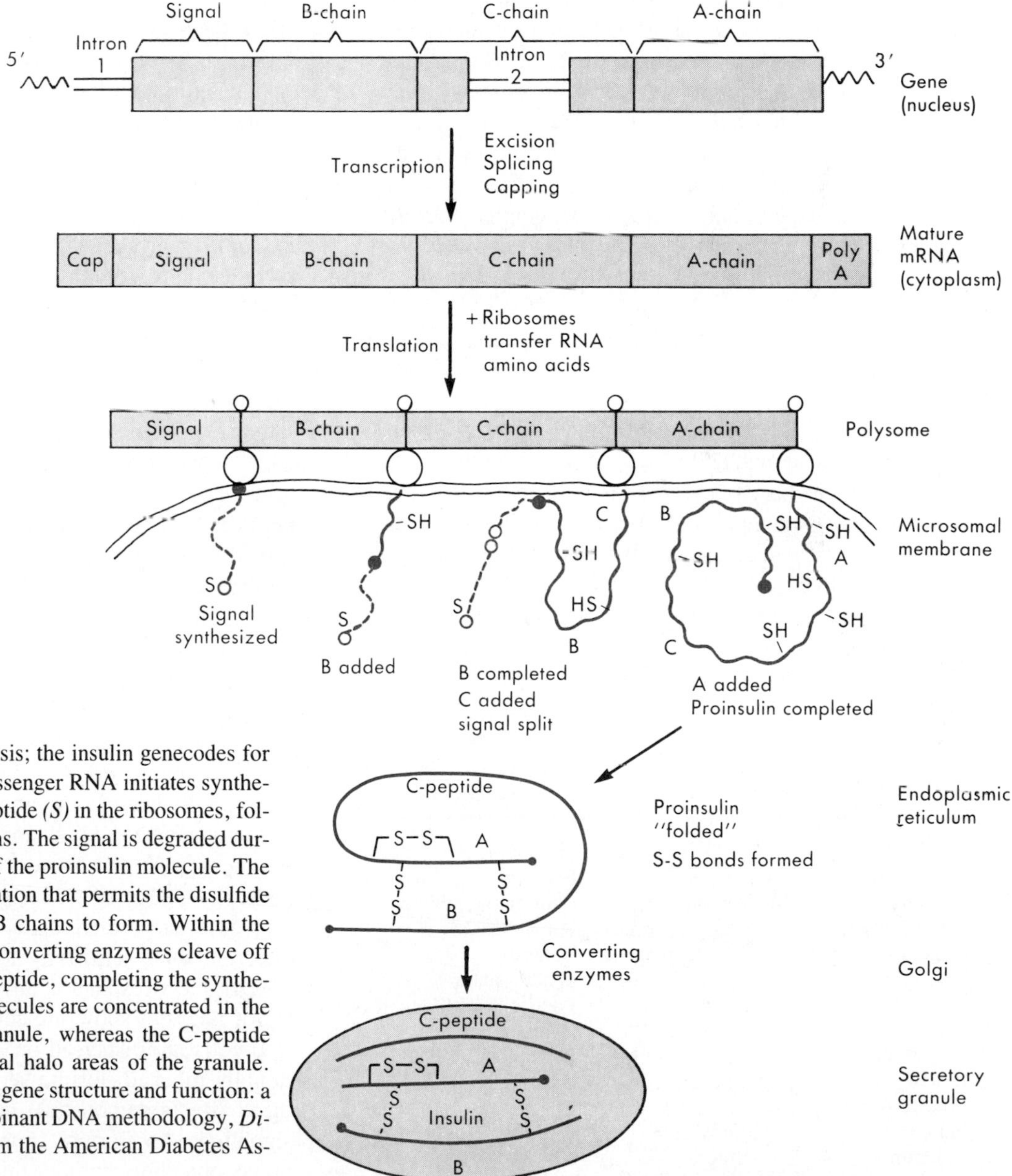

■ **Fig. 46-4** Insulin synthesis; the insulin genecodes for preproinsulin. The mature messenger RNA initiates synthesis of the N-terminal signal peptide *(S)* in the ribosomes, followed by the B, C, and A chains. The signal is degraded during the course of completion of the proinsulin molecule. The latter is folded into a conformation that permits the disulfide linkages between the A and B chains to form. Within the Golgi and secretory granule, converting enzymes cleave off the C chain, known as the C-peptide, completing the synthesis of insulin. The insulin molecules are concentrated in the electron-dense core of the granule, whereas the C-peptide molecules are in the peripheral halo areas of the granule. (From Permutt M et al: Insulin gene structure and function: a review of studies using recombinant DNA methodology, *Diabetes Care* 7:386, 1984. From the American Diabetes Association.)

of 9,000. The disulfide bonded A and B chains of insulin are linked to a connecting peptide (C-peptide) through two basic residues each at the C-terminal of the B chain and the N-terminal of the A chain (see Fig. 46-3). During its packaging into granules by the Golgi apparatus, proinsulin is slowly cleaved by specific trypsin-like and carboxypeptidase-like enzymes that split off the arg-arg and lys-arg residues, respectively. The generated insulin and C-peptide molecules are each retained in the granules and released in equimolar amounts. The association of insulin with zinc takes place as the secretory granules mature. The zinc insulin crystals then form the dense central core of the granule, whereas C-peptide is present in the clear space between the membrane and the core.

In the transgolgi, 99% of proinsulin is sorted toward processing in the fashion that produces the storage depot for regulated insulin release. One percent of proinsulin is sorted toward nongranule processing and is the source of a low rate of constitutive insulin secretion. Insulin synthesis is stimulated by glucose or feeding and decreased by fasting. Glucose rapidly increases the rate of initiation of translation of the insulin mRNA and more slowly increases transcription of the insulin gene. A cAMP regulatory element and a possible separate glucose regulatory element have been identified in the insulin gene. *In general, synthesis and secretion of insulin are closely linked.*

Insulin Secretion

A large number of factors can stimulate or inhibit insulin release (Table 46-1). Because of the preeminent importance of glucose, the mechanism whereby this substrate acutely stimulates insulin secretion has been intensively investigated. The following sequence seems to occur (Fig. 46-5):

1. A specific transporter (Glut-2), concentrated in the microvilli of the canaliculi between β-cells (Fig. 46-2) facilitates diffusion of glucose into the β-cell and maintains its concentration essentially equal to that of the interstitial fluid.
2. Considerable evidence supports the proposal that the enzyme glucokinase acts as the fundamental glucose sensor that controls the subsequent β-cell response. The enzyme has a K_m for glucose of 5 mM, which is in the physiological range. Inhibition of glucokinase activity blocks insulin release. Phosphorylation of glucose by glucokinase is rate-limiting for islet glucose use, and the rate of insulin secretion parallels that of glucose oxidation. However, an insulin-releasing signal must be generated downstream from glucose-6-phosphate, because 3-carbon substrates (such as glyceraldehyde) are equipotent to glucose.
3. Very rapid rises in ATP, the ATP/ADP ratio, NADH and NADPH, and H^+ all follow glucose uptake.
4. An ATP-sensitive K^+ channel closes, K^+ efflux is suppressed, and the β-cell depolarizes. This opens a voltage-regulated Ca^{++} channel, and intracellular Ca^{++} rapidly increases. The elevated Ca^{++} activates the mechanism of secretory granule movement, perhaps via a myosin light chain kinase, and exocytosis of insulin follows.

All of the above occur within 1 minute of exposure to glucose. A similar sequence with specific transporters and enzymatic steps may underlie the stimulatory action of other fuels, such as amino acids and ketoacids (Fig. 46-5). A rise in cAMP levels in β-cells also follows exposure to glucose; a cAMP-dependent protein kinase stimulates insulin release, possibly by phosphorylating the proteins involved in exocytosis (Fig. 46-5). Stimulatory G proteins mediate the insulin-releasing effects of such peptides as glucagon, whereas inhibitory G proteins mediate the insulin suppressive effects of such peptides as somatostatin. A stimulatory G protein linked to phospholipase C mediates the ability of acetylcholine to stimulate insulin release via generation of phosphatidylinositol second messengers and increased protein kinase C activity. A class of sulfonylurea drugs stimulate insulin release by directly closing the K^+ channel; these drugs are useful in the treatment of type II noninsulin-dependent diabetes. In contrast, the drug diazoxide opens the K^+ channel and inhibits insulin release and is useful in treatment of hyperinsulinism.

Many nuances to glucose-stimulated insulin release may prove important to future understanding of the pathophysiology of diabetes. Newly synthesized insulin molecules are preferentially secreted. Individual β-cells differ in their sensitivity to glucose and only 50% to 60% of them respond at any one time. Those at the center of the islet show greater and faster responsiveness. If β-cells are dispersed or their gap junctions are functionally blocked, insulin secretion is markedly reduced; these effects demonstrate the importance of cell-to-cell signaling. An intrinsic nonneural oscillation of membrane potential, cytoplasmic Ca^{++} concentration, and

Table 46-1 Insulin secretion

Increased by		Decreased by
D-Glucose	Glucagon	Fasting
Galactose	Glucagon-like peptide 1	Exercise
Mannose	Gastric inhibitory polypeptide	Endurance training
Glyceraldehyde	Secretin	Somatostatin
Protein	Cholecystokinin	Galanin
Arginine	Vagal activity	Pancreastatin
Lysine	Acetylcholine	Interleukin-1
Leucine	β-adrenergic activity	α-adrenergic activity
Alanine	Sulfonylurea drugs	Prostaglandin E_2
Ketoacids		Diazoxide
Free fatty acids		
Potassium		
Calcium		

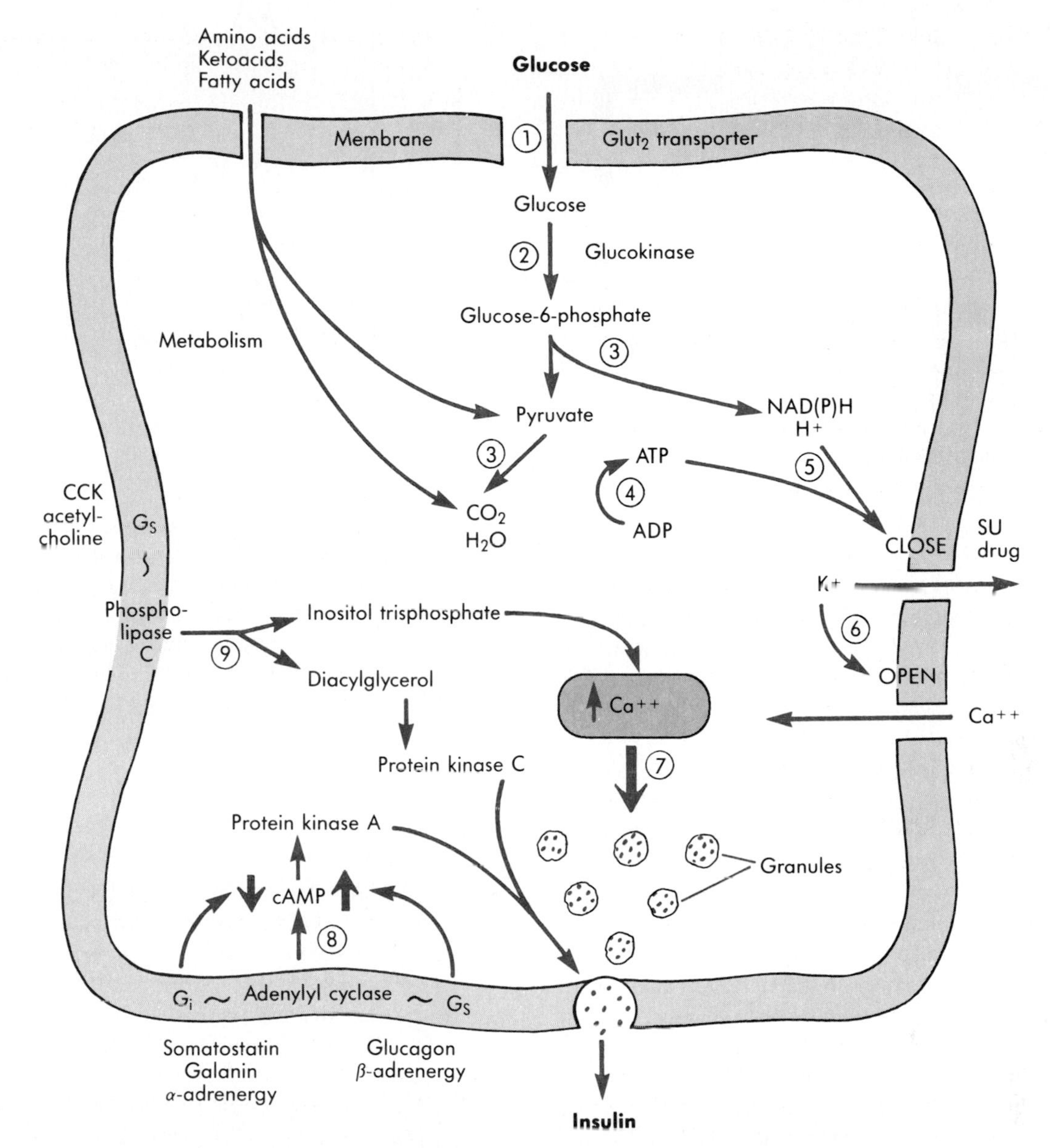

■ **Fig. 46-5** Current concepts of regulation of insulin secretion by the β cell. Glucose transport *(1)* and *glucokinase* catalyzed phosphorylation *(2)* raise glucose-6-phosphate levels. Metabolism *(3)* subsequently leads to increased ATP levels *(4)* and NAD(P)H levels *(5)* that inhibit or close a potassium channel *(6)* and open a calcium channel *(6)*. Increased calcium levels then trigger exocytosis of insulin granules. Other modulators of secretion act via the adenylyl cyclase-cAMP-protein kinase pathway and the phospholipase-phosphoinositide pathway.

insulin release exists with a cycle of 13 to 15 minutes. Because plasma insulin levels follow a similar cycle, a pacemaker must coordinate islet function throughout the pancreas.

■ *Regulation of Insulin Secretion*

In the broadest sense, insulin secretion is governed by a feedback relationship with exogenous nutrient supply (Fig. 46-6). When substrate supply is abundant, insulin is secreted in response; the hormone then stimulates use of these same incoming nutrients and simultaneously inhibits the mobilization of the analogous endogenous substrates. When nutrient supply is low or absent, insulin secretion is dampened and mobilization of endogenous fuels is enhanced.

Glucose is the stimulant of greatest importance in the human being. Because insulin in turn stimulates the use of glucose, this substrate-hormone pair forms a feedback system for close regulation of plasma glucose levels. The relationship between plasma insulin and

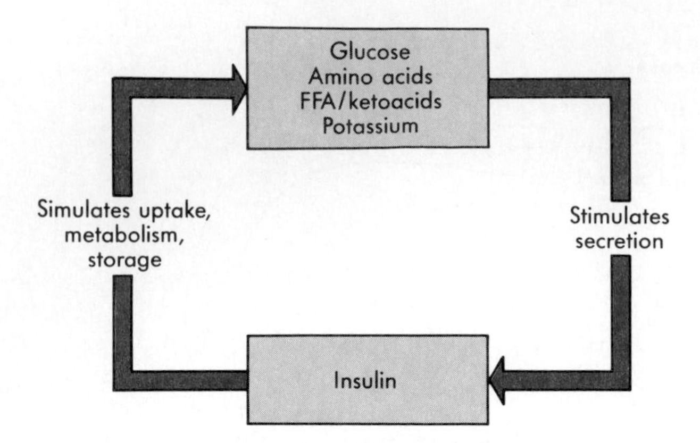

■ **Fig. 46-6** Feedback relationship between insulin and nutrients. Those nutrients that stimulate insulin secretion are the same nutrients whose disposal is facilitated by insulin. *FFA,* Free fatty acids.

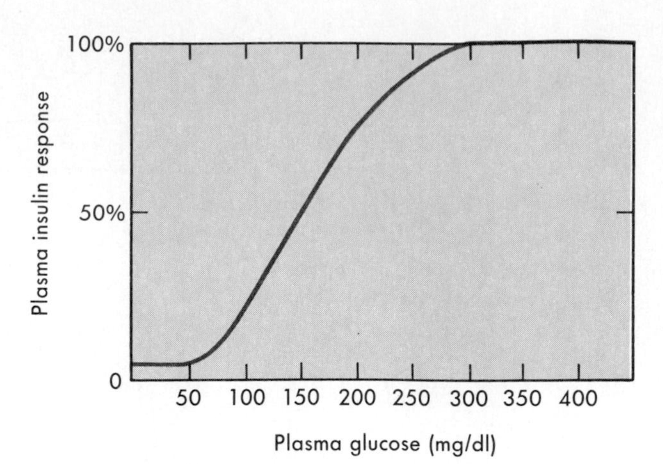

■ **Fig. 46-7** Approximate in vivo relationship between plasma glucose and insulin secretion, the latter being assessed by the plasma insulin response to stepwise infusion of glucose in humans. No insulin is secreted below a plasma glucose level of 50 mg/dl. Half maximum secretion occurs at 125 to 150 mg/dl. (Redrawn from Karam JH et al: "Staircase" glucose stimulation of insulin secretion in obesity; measurement of beta cell sensitivity and capacity, *Diabetes* 23:763, 1974. From the American Diabetes Association.)

plasma glucose is sigmoidal (Fig. 46-7). Virtually no insulin is secreted below a plasma glucose threshold of about 50 mg/dl. A half-maximum insulin secretory response occurs at a plasma glucose level of about 150 mg/dl and a maximum insulin response at levels of 300 mg/dl^{-1}.

Both in vitro and in vivo insulin secretion exhibits a biphasic response to a continuous glucose stimulus (Fig. 46-8). Within seconds of exposure to glucose, there is an immediate pulse of insulin release, peaking at 1 minute and then returning toward baseline. After 10 minutes of continuous stimulation, a second phase of secretion begins with a slower rise of insulin to a second plateau, which can be maintained for many hours in normal individuals. A number of explanations for the biphasic response have been suggested: (1) labile and stable storage compartments exist that contain granules with different sensitivities to glucose; (2) the second phase represents newly synthesized insulin and results from glucose stimulation of the synthetic process; (3) a feedback inhibitor of insulin release is rapidly generated after glucose stimulation and then slowly removed.

When glucose is given orally, a greater insulin response is elicited than when plasma glucose is comparably elevated by intravenous administration. This is accounted for by one or several gastrointestinal hormones released in response to meals and capable of potentiating glucose-stimulated insulin secretion (see Table 46-1). Gastric inhibitory polypeptide and glucagon-like peptide 1 are probably the most important of these insulinogogues. This prompt gastrointestinal mechanism of insulinogenesis moderates the early rise in plasma glucose that follows the ingestion and absorption of a carbohydrate meal. In contrast, somatostatin of islet origin **(neurocrine)** and of intestinal origin **(endocrine)** dampens insulin responses to meals.

Insulin secretion also is stimulated by oral protein; this is mediated by the amino acids resulting from digestion of the protein. The basic amino acids, arginine

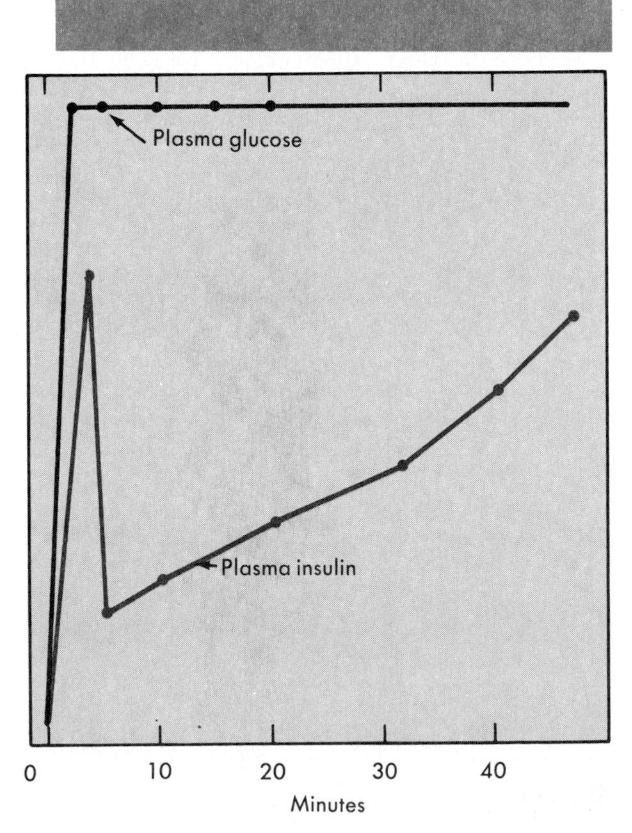

■ **Fig. 46-8** Insulin response to glucose infusion shows a rapid first phase of release followed by a fall and a later slower second phase.

and lysine, are the most potent stimulants; leucine, alanine, and others contribute modestly to this effect. Glucose and amino acids are synergistic in their actions, so that the insulin rise that follows a meal represents more than the additive effect of its carbohydrate and protein content. Triglycerides and fatty acids have only a small stimulatory effect in humans and these may be mostly indirect by means of gastric inhibitory polypeptide. Ketoacids at concentrations that prevail during prolonged fasting stimulate insulin secretion; although modest, this effect may help sustain a critical low level of insulin when β-cell stimulation by glucose and amino acids is reduced.

Both potassium and calcium are essential for normal insulin responses to glucose. Thus relative insulin deficiency occurs in subjects depleted of potassium or calcium. Magnesium exerts at least a modulatory effect. Sympathetic nerves and epinephrine stimulate insulin secretion via their β-adrenergic receptors but inhibit insulin secretion via their α-adrenergic receptors. Galanin may transmit the inhibitory impulses of some sympathetic nerves. Parasympathetic activity via the vagus nerve increases insulin release, and there is a cephalic phase of insulin secretion, which in animals has been demonstrated to precede the entrance of food into the gastrointestinal tract.

A large number of other hormones directly or indirectly produce hyperplasia of the β-cells and lead to chronic increases in insulin secretion. The list includes cortisol, growth hormone, estrogen-progesterone, human placental lactogen, and thyroid hormones. These hormones increase insulin secretion largely by antagonizing its action and increasing its need by peripheral tissues. A feedback effect of insulin on its own secretion has been demonstrated that is independent of its hypoglycemic action. During an insulin infusion plasma C-peptide levels (representing β-cell activity) decrease even if plasma glucose is held constant.

The net result of these many physiological influences is to maintain an average basal peripheral plasma insulin level of 10 μU/ml (6×10^{-11} M) in normal humans. After several days of fasting this value declines over 50%. A similar decrease occurs during prolonged exercise. Plasma insulin increases threefold to tenfold after a typical meal, usually peaking 30 to 60 minutes after eating is initiated (Fig. 46-9). Plasma C-peptide fluctuates similarly (see Fig. 46-9). In addition to the 15-minute cycles and bursts of secretion with meals, there is a higher amplitude cycle with a 2-hour frequency that is entrained by glucose and that represents feedback regulation. Obesity—a large increase in adipose mass—markedly increases insulin secretion (see Fig. 46-9). Physical conditioning by endurance training has the opposite effect. The liver is regularly exposed to insulin concentrations two or three times higher than those of other organs in the basal state and transiently five to ten times higher after β-cell stimulation.

Insulin circulates unbound to any carrier protein. Its half-life in plasma is 5 to 8 minutes, its volume of distribution is about 20% of body weight, and its metabolic clearance rate is 800 ml/minute. In humans estimates of basal insulin delivery rate to the peripheral circulation are about 0.5 to 1 units/hour (20 to 40 μg/hour). During meals the delivery rate increases up to tenfold, and the total daily peripheral delivery of insulin is about 30 units. Since the liver removes 50% of portal vein insulin on the first pass, the actual β-cell secretory rate is approximately 60 units/day. The initial hepatic extraction is decreased by glucose and meals, permitting a greater escape of insulin to the periphery for stimulation of glucose uptake.

Insulin is metabolized largely in the kidney and liver by specific enzymes that split the disulfide bonds, producing separate A and B chains. Very little insulin is excreted unchanged in the urine. Degradation of insulin also occurs in association with its plasma membrane receptor after internalization by target cells. Circulating antibodies produced by therapeutic injections of animal insulins slow the rate of insulin disposal and affect the timing of insulin availability to its target cells.

Although C-peptide is secreted in amounts that are equimolar to insulin, its basal peripheral plasma levels are approximately fivefold higher, averaging 1 ng/ml (3×10^{-10} M). This difference is caused by a lower rate of metabolic clearance for C-peptide. In contrast to insulin, C-peptide undergoes no significant hepatic extraction. Therefore plasma C-peptide measurements actually provide more direct information about β-cell function. 24-hour urine C-peptide excretion also may be used as an index of daily insulin secretion. Proinsulin is released by the β-cell and forms up to 20% of total insulin immunoreactivity in the basal state. During stimulation absolute proinsulin levels increase more slowly and to a lesser degree than does insulin. Although proinsulin can exert insulin effects in vitro and in vivo, it has only 5% to 10% the activity of insulin. At present, no biological function can be ascribed to either proinsulin or C-peptide.

Insulin Action on Cells

The first and overall rate-limiting step in insulin action is transport of the hormone through the capillary wall. Once at the target cell, insulin combines with a plasma membrane glycoprotein receptor (Fig. 46-10, p. 861). This tetramer contains an extracellular α-subunit that is disulfide-bonded to a β-subunit that transverses the plasma membrane and largely resides within the cytoplasm. Two such identical α-β dimers are joined extracellularly by another disulfide (Fig. 46-10). Each α-β dimer contains approximately 1,500 amino acids (α = 730, molecular weight 135,000; β =620, molecular weight 95,000). The β-subunit has 194 extracellular

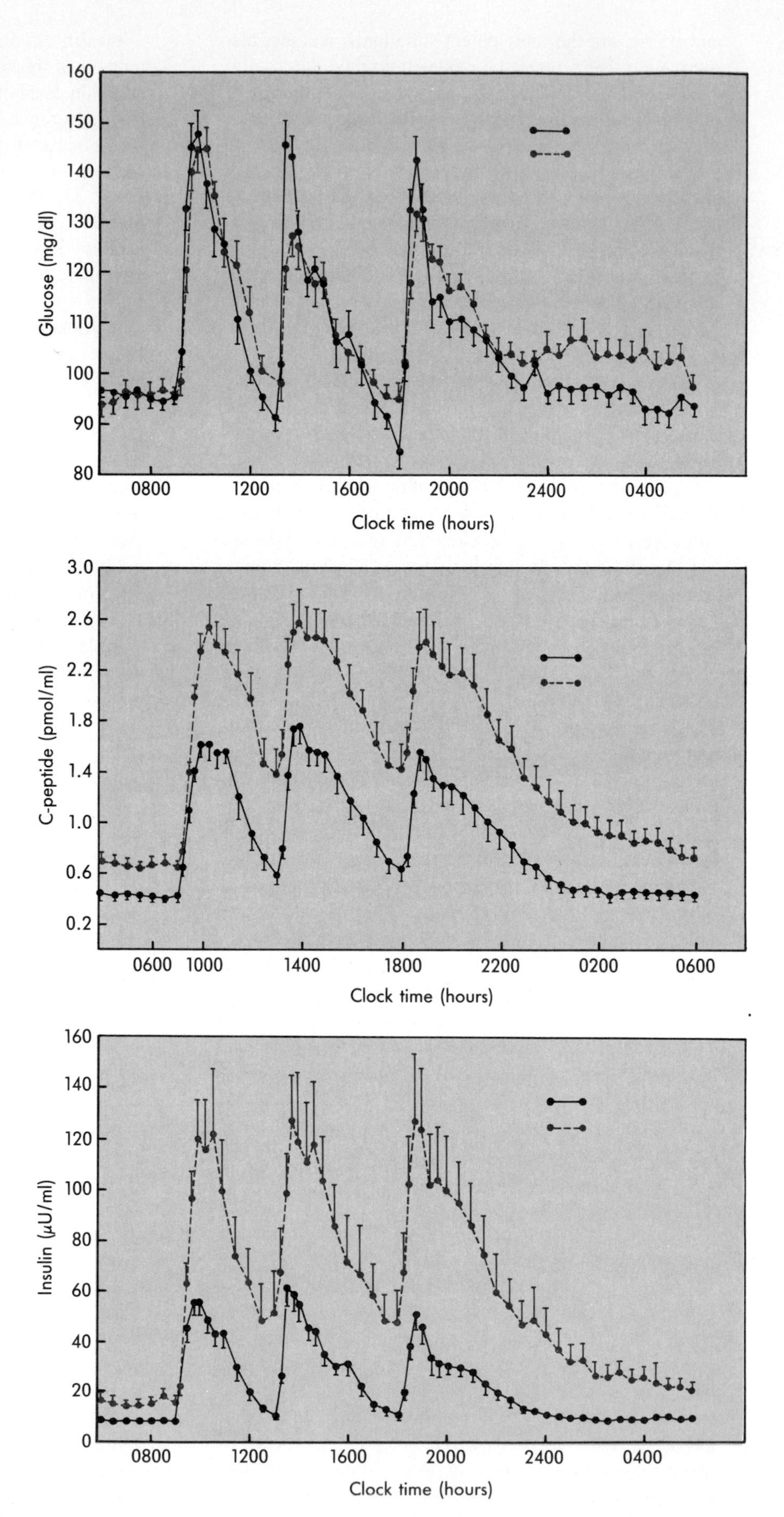

■ **Fig. 46-9** Twenty-four hour profiles of plasma glucose, C-peptide, and insulin in normal weight (solid lines) and obese (dotted lines) humans. Note the increases with each meal, the rapid return toward baseline, and the exaggerated β cell responses in obesity. (Redrawn from Polonsky K et al: β *J Clin Invest* 81:442, 1988. From the American Society for Clinical Investigation.)

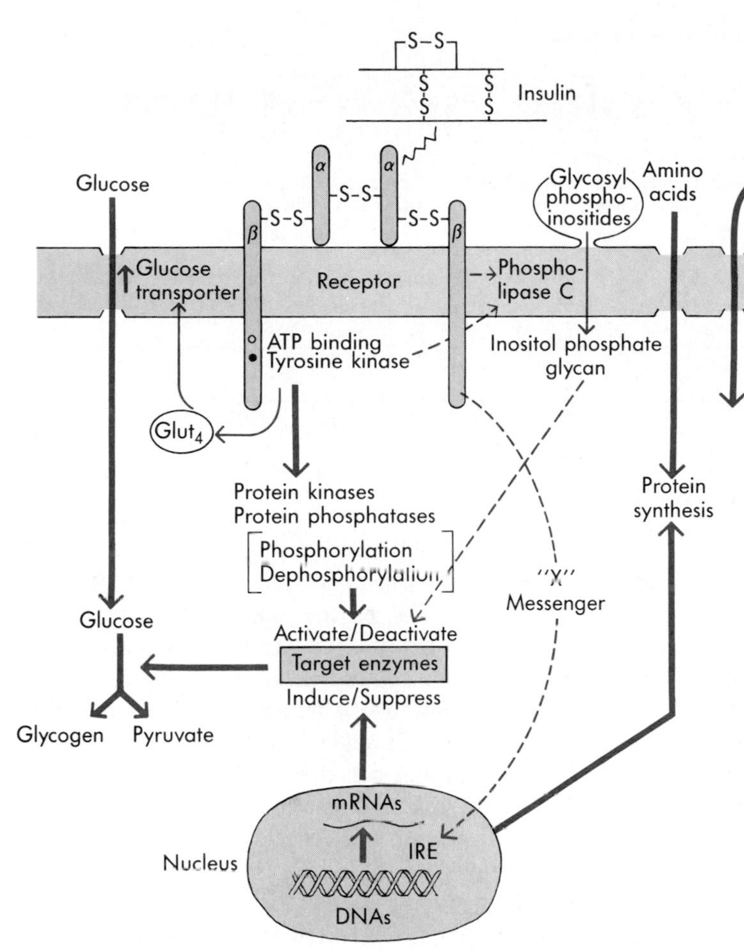

■ **Fig. 46-10** Mechanisms of insulin action on cells. Insulin binding to α-subunit of its receptor causes autophosphorylation by ATP of an intracellular β-subunit receptor site that generates tyrosine kinase activity. Glucose transporters are then moved to the plasma membrane and facilitate glucose entry. The receptor tyrosine kinase is presumed to phosphorylate protein kinases and phosphatases, which in turn activate or deactivate target enzymes of glucose metabolism by phosphorylation or dephosphorylation. (This may be aided by generation of a separate inositol phosphate glycan second messenger.) By still undefined mechanisms, a transcription factor(s) is generated that stimulates or represses gene transcription through an insulin response element (IRE) in DNA molecules. Also independently potassium, magnesium, and phosphate entries into the cell are facilitated.

residues, a 23–amino acid transmembrane anchor, and an intracytoplasmic component with 403 residues. The insulin receptor gene on chromosome 17 is a member of a superfamily that codes for other growth factor receptors. It contains 22 exons (Fig. 46-11) that code for a proreceptor molecule containing in sequence a signal peptide, the α subunit, a proreceptor processing site, and the β subunit. After translation and removal of the signal peptide, two proreceptor molecules associate into a disulfide dimer, after which a basic amino acid cleavage site on each dimer is removed. This yields the discrete α- and β-subunits that are also linked by disulfide bonds. The functional anatomy of the insulin receptor gene is shown in Fig. 46-11. Of great importance are exons 17 through 21, which code for the receptor tyrosine kinase activity.

After insulin binding, a conformational change leads to aggregation of receptors with clustering in coated pits. The hormone receptor complex is subsequently internalized by endocytosis; the hormone is degraded; and the receptor is either degraded, stored, or recycled back to the plasma membrane. Insulin down regulates the insulin receptor by increasing its rate of degradation and suppressing its synthesis. This phenomenon plays a role in the decreased sensitivity to insulin seen in obese subjects (see Fig. 46-9). For some actions, full biological activity of insulin is expressed when only 5% of its receptor sites are occupied; therefore spare receptors exist.

Insulin action results in multiple metabolic events in several loci including the plasma membrane itself, the cytoplasm, the mitochondria, and the nucleus. Whether one receptor signal mechanism initiates all of these events is not certain. At present the most certain mechanism of signal transduction is via the receptor tyrosine kinase activity in the beta subunit. Most or all of the metabolic and mitogenic effects of insulin are lost if the tyrosine kinase domain of the receptor is deleted or altered by mutagenesis. The activity of this kinase is constitutively repressed by the unoccupied extracellular α-subunit. When a single molecule of insulin binds to its α-subunit site, this repression is relieved, possibly by a conformational change transmitted through the receptor molecule. The tyrosine kinase site is stimulated and autophosphorylates the β-subunit at three key tyrosines (amino acids 1146, 1150, and 1151), with ATP as substrate. In turn, the now fully active tyrosine kinase is then presumed to phosphorylate serine, threonine, and tyrosine residues on a variety of intracellular protein substrates. These may include protein kinases, protein phosphatases, membrane phospholipases, and G proteins. Thus this single receptor mechanism can account for multiple cascades of events in which enzymes

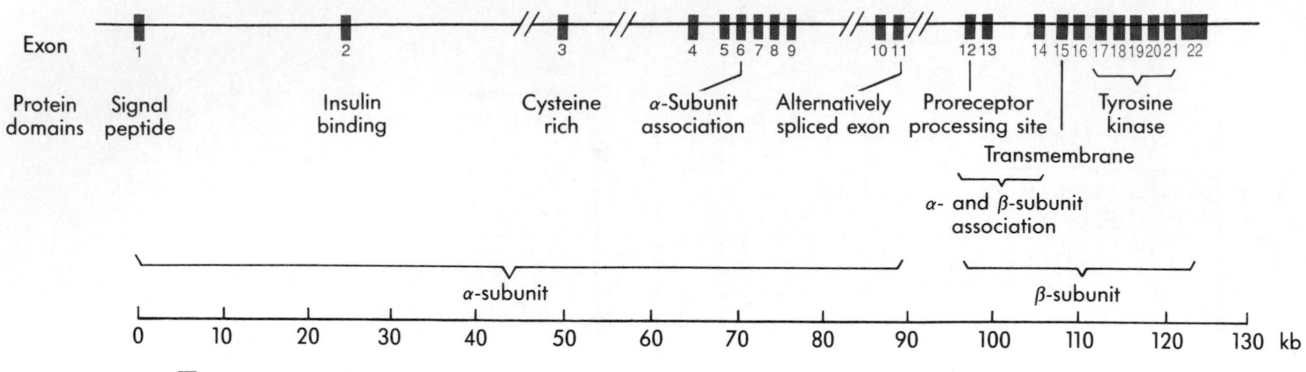

■ **Fig. 46-11** The anatomy of the insulin receptor gene. Modified from Seino S, Seino M, Bell GI: *Diabetes* 39(2):129, 1990. From the American Diabetes Association.)

are either activated or inactivated by covalent modification via phosphorylation or dephosphorylation. However, the precise substrates of metabolic importance for the receptor tyrosine kinase remain to be identified.

Other receptor domains participate in insulin actions, even those mediated by the tyrosine kinase site. Tyrosines 1316 and 1322 in the C terminal tail are also phosphorylated, but they are not required for metabolic signaling. However, if they are mutated, mitogenic activity of insulin is greatly increased. If the C terminal tail of the β-subunit is truncated, leaving intact the tyrosine kinase site, metabolic activity of insulin is markedly reduced, but again, mitogenic actions are enhanced.

An entirely unique intracellular mediator arising from membrane phospholipids has also been proposed (see Fig. 46-10). Binding of insulin to its receptor generates an inositol phosphate glycan molecule from glycosyl phosphoinositides anchored to a plasma membrane protein. This molecule can transduce specific hormone actions. In addition some insulin actions are mimicked by diacylglycerols and protein kinase C activation. In some insulin-stimulated cells an increase in phosphodiesterase activity reduces cAMP levels. In addition insulin inhibits cAMP binding to protein kinase A and therefore its activity.

Insulin also has effects that result from stimulation or inhibition of the expression of numerous genes. A specific transacting DNA-binding protein that is induced by insulin has been described. However, no specific messenger and no consensus sequence of nucleotides corresponding to a unique "insulin responsive element" has yet been found. Because insulin binding to the nucleus, the Golgi, and the endoplasmic reticulum has been described, it is possible that internalized hormone may have direct actions as well.

Following transduction of the insulin-receptor signal(s), a number of processes are rapidly stimulated by the hormone (see Fig. 46-10). Within 1 minute, glucose transport into muscle and adipose cells is increased fivefold to twentyfold by activation of a glucose carrier system in the plasma membrane. A glucose transporter (Glut-4), specifically expressed in muscle and adipose tissue, facilitates diffusion of glucose down its concentration gradient. This process does not require energy. Insulin acutely recruits Glut-4 from a cytoplasmic pool of vesicles to the plasma membrane; insulin also increases the activity of the transporter. Transcription of the Glut-4 gene is stimulated more slowly. The importance of this insulin action is underscored by the fact that at low physiological insulin concentrations, glucose transport is the rate-limiting step in glucose use. At high physiological insulin concentrations, such as after a meal, the rate-limiting step shifts to an unknown point in intracellular glucose metabolism. In an independent manner, insulin also facilitates the cellular uptake of amino acids, potassium, magnesium, and phosphate (see Fig. 46-10).

In addition to making glucose and amino acids available, insulin directs their disposition once these compounds are within various cells. Conversion of glucose to glycogen, to pyruvate and lactate, and in some cells to fatty acids are all stimulated to varying degrees. From the amino acids, synthesis of specific proteins such as albumin, casein, and certain enzymes is selectively enhanced. Some of these effects are mediated rapidly by altering target enzyme activities through covalent modification. Examples include activation of acetyl-CoA-carboxylase by phosphorylation and activation of pyruvate dehydrogenase by dephosphorylation. Other effects are slowly mediated by altering the number of target enzyme molecules through transcriptional or translational effects. More than 50 mRNA levels are now known to be regulated by insulin. Key examples include glucokinase and pyruvate kinase, which are increased by the hormone, and phosphoenolpyruvate carboxykinase and phosphofructose-2-kinase, which are decreased by the hormone.

■ *Actions on Flow of Fuels*

Insulin is the hormone of abundance. When the influx of nutrients exceeds concurrent energy needs and rates

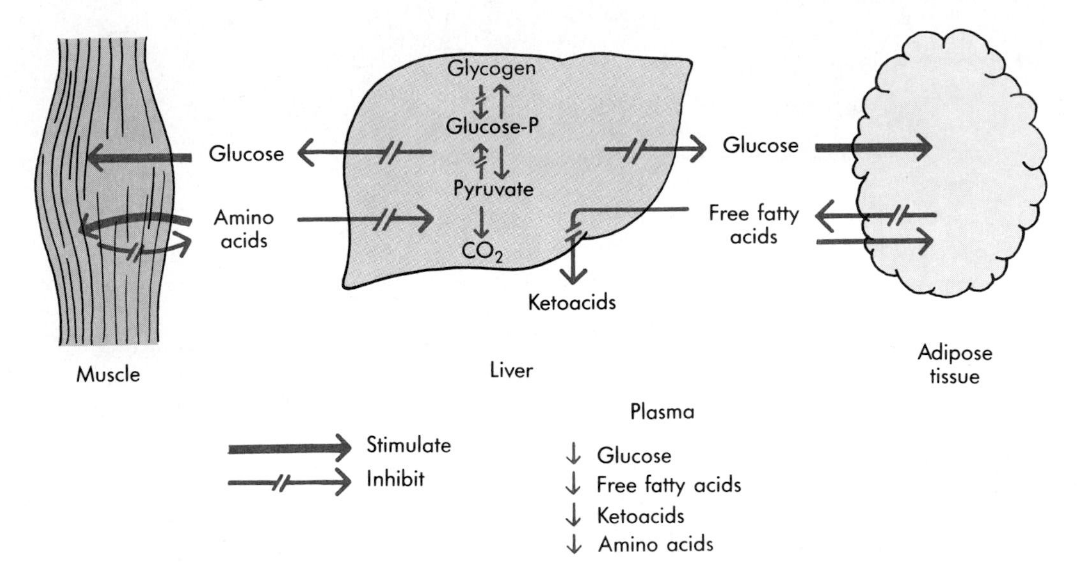

■ **Fig. 46-12** Effect of insulin on the overall flow of fuels results in tissue uptake and sequestration of glucose, fatty acids, and amino acids, with a resultant decrease in their plasma levels.

of anabolism, the secreted insulin induces efficient storage of the excess nutrients while suppressing mobilization of endogenous substrates. The stored nutrients then can be made available during subsequent fasting periods to maintain glucose delivery to the central nervous system and free fatty acid delivery to the muscle mass and viscera. The major targets for insulin action are the liver, the adipose tissue, and the muscle mass. Fig. 46-12 displays the general flow of substrates produced by insulin, and Fig. 46-13 shows some of the important metabolic control points where insulin acts directly or indirectly in the liver.

Carbohydrate metabolism. Insulin stimulates glucose oxidation and storage and simultaneously inhibits glucose production. Therefore insulin either lowers the basal circulating glucose concentration or limits the rise in plasma glucose that results from a dietary carbohydrate load. This is accomplished by a number of insulin effects.

Liver. In the liver, extracellular glucose levels equilibrate rapidly with intracellular levels by means of the $Glut_2$ transporter. Insulin enhances inward movement of glucose by inducing hepatic glucokinase and stimulating phosphorylation of glucose to glucose-6-phosphate. Insulin then promotes storage of the glucose by activating the glycogen synthase enzyme complex. At the same time, insulin stimulates the flow of glucose to pyruvate and lactate (glycolysis) by increasing the activities of the committed enzymes phosphofructokinase and pyruvate kinase. Insulin also reduces hepatic glucose output rapidly by inhibiting glycogenolysis through decreasing glycogen phosphorylase activity and, gradually, by decreasing glucose-6-phosphatase levels (see Fig. 46-13). In addition, insulin inhibits gluconeogenesis. This is accomplished by decreasing hepatic uptake of precursor amino acids and their availability from muscle (see Fig. 46-12). Insulin also decreases the levels or activities of the committed gluconeogenic enzymes pyruvate carboxylase, phosphoenolypyruvate carboxykinase, and fructose-1, 6-diphosphatase. Finally, insulin diminishes the supply of phosphoenolpyruvate for gluconeogenesis by increasing the activities of pyruvate kinase and pyruvate dehydrogenase (see Fig. 46-13).

Many of these hepatic effects of insulin require concurrent administration of glucose and represent amplification of biochemical events that are induced by glucose itself. Some also may be augmented by other consequences of insulin action. For example insulin decreases free fatty acid delivery to the liver, β-oxidation and generation of acetyl CoA; insulin stimulates the conversion of acetyl CoA to malonyl CoA through activation of acetyl-CoA carboxylase, further decreasing hepatic levels of acetyl CoA. Because acetyl CoA is an important allosteric activator of the gluconeogenic enzyme, pyruvate carboxylase, insulin decreases the rate of gluconeogenesis in this fashion, too. Insulin also suppresses gluconeogenesis by inhibiting the secretion of its antagonistic hormone, glucagon.

If insulin is infused continuously in the basal state, plasma glucose declines; however, hepatic glucose production recovers from its insulin-induced nadir after a short time, reflecting three counterregulatory processes: (a) an autoregulatory response of the liver to a lowered plasma glucose concentration, probably mediated by the direct effects of glucose on phosphorylase and other enzymes; (b) a sensing of hypoglycemia by the hypothalamus, leading to stimulation of glucose output via activation of the sympathetic nervous system; and (c) a sensing of hypoglycemia by the α-cells of the pancreatic islets, which then secrete the insulin antagonist, glucagon. These mechanisms are essential to protection

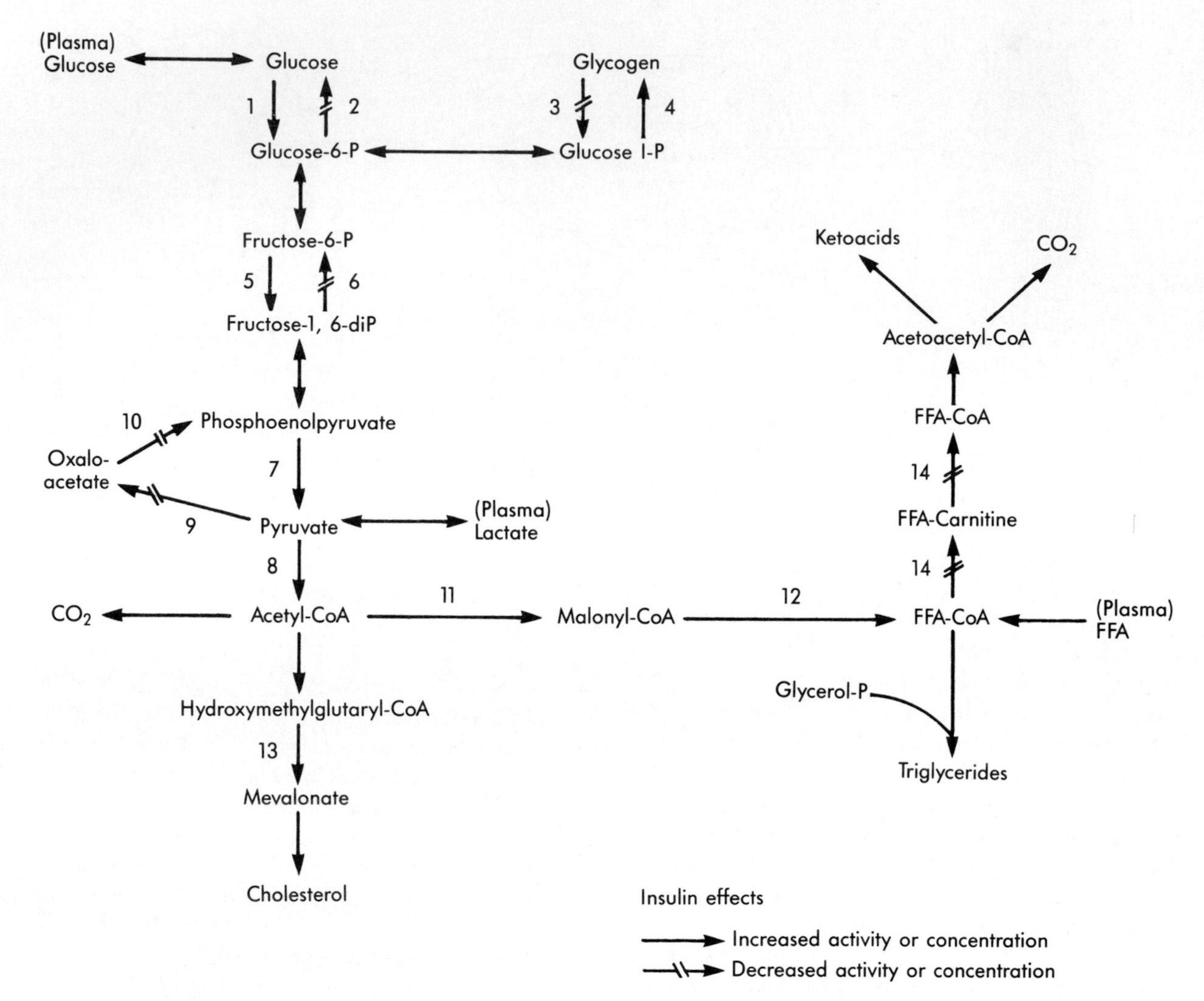

Fig. 46-13 Insulin actions on glucose and fatty acid metabolism in the liver. The pathway from pyruvate to glucose via oxaloacetate constitutes gluconeogenesis. The key enzymes involved that are influenced by insulin are as follows: 1, glucokinase; 2, glucose-6-phosphatase; 3, phosphorylase; 4, glycogen synthase; 5, phosphofructokinase; 6, fructose-1,6-diphosphatase; 7, pyruvate kinase; 8, pyruvate dehydrogenase; 9, pyruvate carboxylase; 10, phosphoenolpyruvate carboxykinase; 11, acetyl-CoA carboxylase; 12, fatty acid synthase; 13, hydroxy-methylglutaryl-CoA reductase; and 14, carnitine acyltransferase. The antagonistic hormone glucagon has the opposite effects on virtually all of these enzymes.

of the central nervous system from glucose deprivation because of the unchecked action of insulin.

Muscle. In muscle, insulin stimulates the transport of glucose into the cell, as just described. Depending on insulin concentration, 20% to 50% of the translocated glucose undergoes glycolysis and oxidation mainly caused by activation of pyruvate dehydrogenase. The remainder is specifically directed into storage as glycogen by insulin activation of glycogen synthase. It is of interest that the slow twitch "high oxidative" type I fibers, which are more dependent on fatty acid fuel, are more sensitive to insulin action on glucose uptake than are the fast twitch "glycolytic" more glucose-dependent type II muscle fibers. Muscle blood flow also increases as a consequence of insulin action.

Adipose tissues. In adipose tissue, as in muscle, insulin stimulates the transport of glucose into the cells. This glucose is then metabolized to α-glycerophosphate, which is used in the esterification of fatty acids and permits their storage as triglycerides. To a minor extent glucose can also be converted to fatty acids (see Fig. 46-13).

Fat metabolism. The metabolism of both endogenous and exogenous fat is profoundly influenced by insulin. The net overall effect is to enhance storage and to block mobilization and oxidation of fatty acids (see Fig. 46-12). Insulin rapidly lowers the circulating levels of free fatty acids and ketoacids and may eventually reduce the level of triglycerides.

Adipose tissues. In adipose tissue the deposition of

fat is stimulated by insulin in several ways. Most important, insulin profoundly inhibits hormone-sensitive adipose tissue lipase activity. It may do this by decreasing the levels of cAMP and inhibiting protein kinase-A. By suppressing lipolysis and the release of stored fatty acids, insulin lowers their rate of delivery to the liver and peripheral tissues. A major consequence is a marked reduction in the generation of ketoacids. In addition insulin stimulates the use of ketoacids by the peripheral tissues. Thus insulin is the major and perhaps the sole antiketogenic hormone.

Insulin also actively promotes deposition of circulating fat into adipose tissue. The enzyme adipose tissue lipoprotein lipase, which catalyzes hydrolysis of VLDL and chylomicron triglycerides to FFA (see Chapter 45) is induced by insulin. This action makes available FFA for transfer into adipose cells (see Fig. 46-12). α-Glycerophosphate, needed for esterification of the FFA, is generated from glyceraldehyde phosphate; the necessary enzyme, glyceraldehyde phosphate dehydrogenase, is also induced by insulin.

Muscle. Within muscle, insulin suppresses the enzyme lipoprotein lipase in inverse proportion to its stimulation of glucose uptake. Thus in muscle FFA uptake and oxidation are inhibited by insulin, especially in the type I oxidative fibers that are most dependent on FFA. These overall actions of insulin within muscle reinforce the principle that glucose and FFA are competitive energy substrates.

Liver. Within the liver, insulin is also antiketogenic and lipogenic. Free fatty acids entering from the circulation are shunted away from β-oxidation and from ketogenesis (see Figs. 46-12 and 46-13). They are instead reesterified with glycerophosphate, derived either from insulin-stimulated glycolysis or from glycerol via the enzyme glycerophosphate kinase. Fatty acids are also synthesized from glucose under the influence of insulin. Mitochondrial acetyl-CoA generated from glucose-derived pyruvate by the action of pyruvate dehydrogenase is transferred to the cytoplasm, where it is converted to malonyl-CoA by the action of acetyl-CoA carboxylase. This rate-limiting step in fatty acid synthesis is activated by insulin, which also induces the final enzyme involved, fatty acid synthase. Moreover, insulin increases the activity of the hexose monophosphate shunt by inducing the enzyme glucose-6-phosphate dehydrogenase. This generates the supply of reduced triphosphopyridine nucleotide, which is also needed for fatty acid synthesis.

The intrahepatic antiketogenic action of insulin may also be mediated by stimulation of malonyl-CoA formation because malonyl CoA inhibits the enzyme carnitine acyltransferase (see Fig. 46-13). The latter enzyme is responsible for transferring free fatty acids from the cytoplasm into the mitochondria for oxidation and conversion to ketoacids.

Insulin also favors hepatic synthesis of cholesterol from acetyl-CoA by activating the rate-limiting enzyme, hydroxymethylglutaryl-CoA reductase. In parallel with increasing hepatic triglyceride storage, insulin decreases apolipoprotein B synthesis; the net result is an acute suppression of VLDL release. However, under conditions of continuous, long-term hyperinsulinemia, the high rate of VLDL synthesis ultimately elevates circulating triglyceride levels.

Protein metabolism. Insulin enhances protein and amino acid sequestration in all target tissues (see Fig. 46-12). *Thus insulin is an anabolic hormone.* If administered in the basal state, insulin lowers the plasma levels of many amino acids. During the assimilation of a protein meal, the increase in insulin secretion limits the rise of amino acids, especially of the plasma branch chain amino acids leucine, valine, and isoleucine. In muscle, insulin stimulates the sodium-dependent transport of amino acids across the cell membrane. Insulin also stimulates the general rate of protein synthesis in vitro. However, in vivo this effect is evident mostly when amino acids are in abundance after a meal. The mechanisms include increases in gene transcription for numerous proteins (e.g., albumin), as well as increases in rates of mRNA translation. General RNA synthesis is increased, and RNA degradation is decreased by insulin.

Even more dramatically, insulin inhibits proteolysis. This is manifest by suppression of the release of the branch chain and aromatic amino acids from muscle and inhibition of their oxidation. In contrast, insulin has little effect on the flux of alanine.

As mentioned earlier, the genes for insulin and its receptor are closely related to those for a variety of tissue growth factors. These include somatomedins (insulin-like growth factors 1 and 2), epidermal growth factor, nerve growth factor, colony-stimulating growth factor, platelet-derived growth factor, and relaxin. Insulin is not only an anabolic hormone, but it also stimulates the synthesis of macromolecules in such tissues as cartilage and bone and thereby contributes to body growth. It also stimulates the transcription of related gene growth factors such as insulin growth factor I. Hence the insulin-deprived young animal or human has a reduced lean body mass and is retarded in height and maturation.

Other actions. Both protein anabolism and the storage of glucose as glycogen require concomitant cellular uptake of potassium, phosphate, and magnesium. Insulin stimulates translocation of all three minerals into muscle cells and of potassium and phosphate into the liver. These effects are separable from the hormone's effect on glucose transport. Insulin secreted in response to a carbohydrate load lowers serum potassium, phosphate, and magnesium levels, and insulin is considered to be one of the normal regulators of potassium balance. Another effect of insulin on electrolyte balance is to increase reabsorption of potassium, phosphate, and sodium by the tubules of the kidney. These renal effects contribute to anabolism by conserving vital intracellular electrolytes. Sodium is necessary for formation of additional extracellular fluid required when lean body mass is being expanded.

The overall consumption of glucose by the central nervous system is independent of insulin. However, selected areas of the brain, particularly the hypothalamus and its adjacent capillary endothelium contain insulin receptors and are insulin responsive. Insulin administration initially increases the firing rate of single neurons from the "satiety" area of the hypothalamus before the suppressive effect of insulin-induced hypoglycemia appears. Insulin also reaches the cerebral spinal fluid from plasma and, when it is injected into the cerebral ventricular system of baboons, it decreases food intake. Thus a role for insulin in appetite regulation seems likely, either directly or by affecting hypothalamic glucose metabolism.

Correlation of Insulin Secretion and Action

The insulin sensitivity of tissues relates well to prevailing plasma insulin levels in various physiological states (Fig. 46-14). Suppression of lipolysis and FFA mobilization and consequent ketogenesis are the most sensitive insulin actions. These are followed by inhibition of hepatic glucose production and of muscle proteolysis as exemplified by release of branch chain amino acids (see Fig. 46-14). Stimulation of muscle glucose uptake requires considerably higher insulin concentrations.

The normal postabsorptive plasma insulin concentration of about 10 μU/ml permits a finely regulated flow of free fatty acid substrates for energy with minimal ketogenesis during the daily nocturnal fasting period. In addition sufficient glycogenolysis is permitted so as to sustain the plasma glucose level. Shortly after eating, insulin concentrations rise to 30 μU/ml and hepatic glucose output becomes greatly suppressed. With a further increase to postprandial levels of 50 to 100 μU/ml, muscle proteolysis is inhibited and glucose uptake (and amino acid uptake) by peripheral tissues is strongly stimulated, facilitating use of substrate at a time of abundance. Under maximum insulin stimulation at approximately 200 μU/ml, peripheral glucose use increases from 2 mg/kg/minute to 12 mg/kg/minute. Of this total increase, two thirds is accounted for by storage as glycogen. Thus the β-cell responds to the physiological need of the moment with insulin delivery rates that provide appropriate hormone concentrations for regulating substrate fluxes.

A further subtlety is appreciated when the response of each individual's β-cells to glucose is compared with the peripheral glucose uptake in response to insulin (Fig. 46-15). It is apparent that the least insulin-sensitive individuals secrete the greatest amount of insulin.

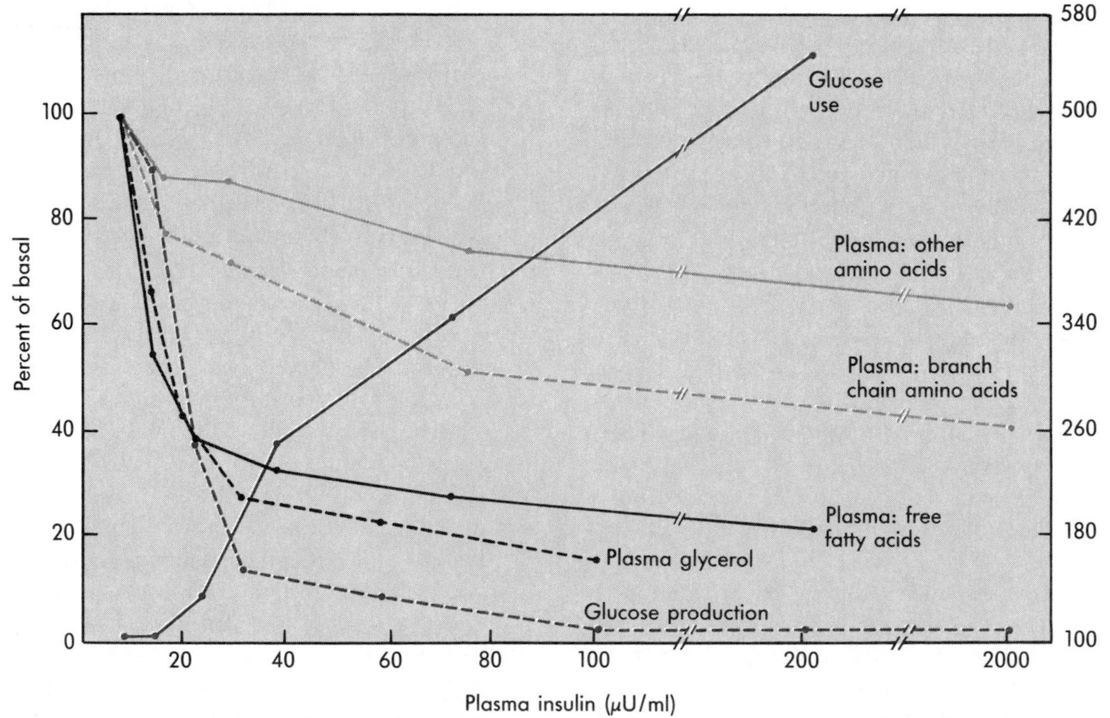

Fig. 46-14 Plasma dose response curves for insulin actions in humans. Note marked sensitivity of inhibition of lipolysis and glucose production with lesser sensitivity for stimulation of glucose production. Effects on amino acid metabolism are intermediate.

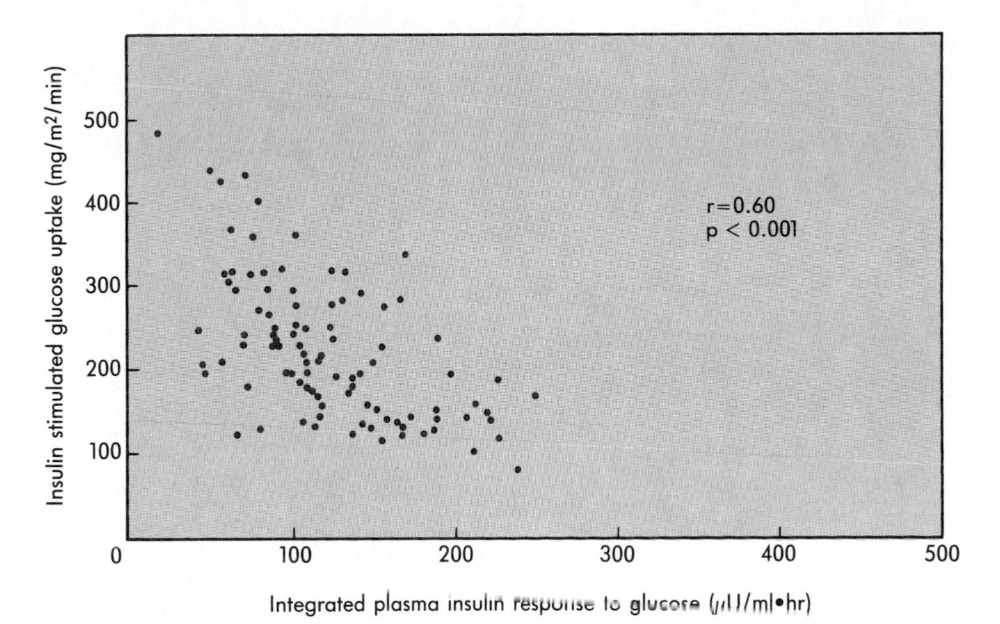

■ **Fig. 46-15** Relationship between responsiveness of peripheral tissue glucose uptake to insulin and responsiveness of β-cells to glucose. Note that the less responsive tissues are to insulin, the greater is the amount of insulin secreted when glucose is provided orally to humans. This feedback system helps to maintain plasma glucose in the physiological range. (Modified from Hollenbeck C, Reaven GM: Variations in insulin-stimulated glucose uptake in health individuals with normal glucose tolerance, *J Clin Endocrinol Metab* 64(6):1169, 1987. From the Endocrine Society.)

This fine coordination between insulin availability (β-cell function) and insulin need (tissue responsiveness) maintains plasma glucose in a narrow range. This feedback operates in obese and in elderly individuals, in whom insulin secretion rates rise to accommodate decreased tissue sensitivity. This feedback occurs also in physically trained individuals, in whom insulin secretion falls as tissue sensitivity to the hormone increases. During puberty a selective resistance to insulin action on glucose metabolism leads to an increase in insulin secretion that facilitates amino acid uptake for tissue growth.

■ *Other β-Cell Products*

The β-cell granule contains peptides (structurally unrelated to insulin) that are synthesized by direction of other genes, but are packaged with insulin and coreleased during exocytosis. Amylin is a 37-amino acid peptide synthesized from a precursor that is expressed by a calcitonin-related gene (see Chapter 47). The granule content and the plasma levels of amylin, both in the basal state and after stimulation, are approximately 1% to 2% those of insulin. Amylin has a noncompetitive antagonistic action to insulin, especially decreasing glucose uptake and metabolism in muscle. Amylin also tends to polymerize, and fibrils containing this peptide accumulate extracellularly within the islets of individuals with noninsulin-dependent diabetes mellitus. The physiological and pathophysiological roles of amylin remain to be determined. Pancreastatin is a 49-amino acid peptide that is structurally related to and may be a product of the processing of chromogranin A (see Chapter 44). This peptide inhibits insulin secretion. Because it is coreleased with insulin from the β-cell granule, pancreastatin may participate in an autofeedback regulation of insulin secretion.

■ *Glucagon*

■ *Structure and Synthesis*

Glucagon is a single straight-chain peptide hormone of 29 amino acids and molecular weight 3,500. The amino acid sequences are identical in mammalian glucagons, and both the structure and function of glucagon appear highly conserved in evolution. The N-terminal residues 1 to 6 are essential for receptor binding and biological activity.

Glucagon is synthesized via a preproglucagon by islet α-cells under the direction of a gene located on human chromosome 2 (Fig. 46-16). The gene is a member of a superfamily that codes for vasoactive intestinal peptide, gastric inhibitory peptide, secretin, and growth hormone-releasing hormone (Chapters 39 and 48). The functions of the copeptides produced by the α-cells (see Fig. 46-16) are unknown. The same gene in intestinal L

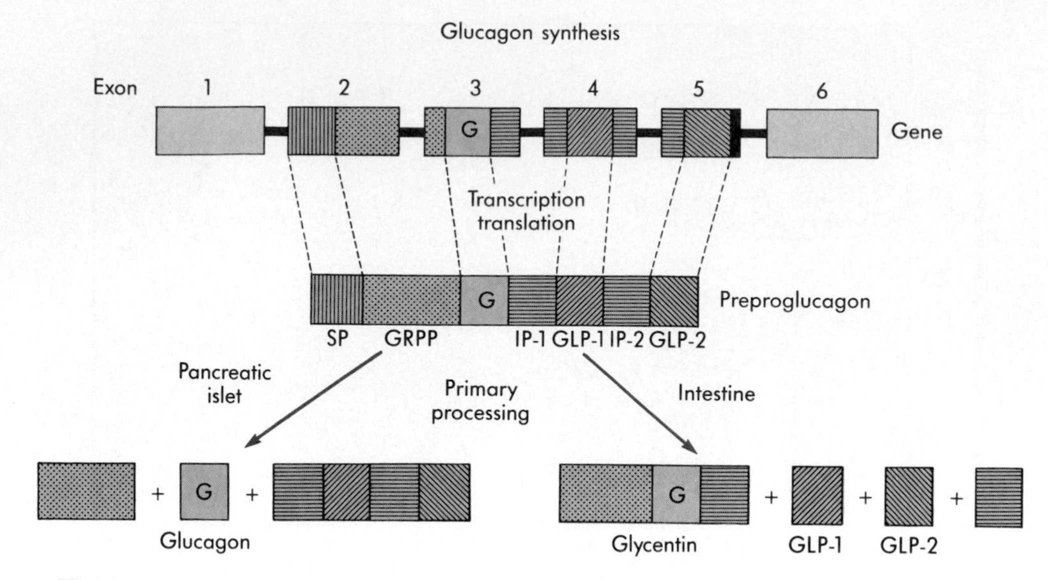

Fig. 46-16 Glucagon synthesis. The gene is composed of 6 exons that yield preproglucagon. After elimination of the signal peptide (SP), the α-cells of the pancreas process to glucagon, a glucagon related polypeptide (GRPP), and a C-terminal peptide. In the intestinal L-cells, glucagon itself is not produced but glucagon-like peptides 1 and 2 (GLP-1 and GLP-2) are secreted along with a larger glucagon containing peptide called glycentin (Modified from Phillipe J: Structure and pancreatic expression of the insulin and glucagon gene, Endocr Rev 12(3):252, 1991. From the Endocrine Society.)

cells expresses preproglucagon, but processing yields glucagon-like peptides of several sorts, instead of glucagon. Both glucose and insulin decrease α-cell glucagon synthesis by repressing gene transcription. Both a cAMP responsive element and an insulin responsive element are present in the gene. The directional blood flow in the islets promotes the bathing of mantle α-cells with high concentrations of inhibitory insulin from the core β-cells. Glucagon is stored in dense granules and is released by exocytosis. This process is inhibited if α-cell Ca^{++} levels are decreased.

Regulation of Secretion

The most important principle governing glucagon secretion is the maintenance of normoglycemia. Exactly opposite to insulin, glucagon is secreted in response to glucose deficiency and acts to increase circulating glucose levels (Fig. 46-17). Hypoglycemia causes a twofold to fourfold increase in plasma levels, whereas hyperglycemia lowers them approximately 50%. Although glucose directly regulates glucagon secretion in vitro, its effect is strongly modulated by insulin. Glucagon secretion is stimulated much more by low glucose levels if insulin is absent. Conversely, the presence of insulin greatly potentiates the suppressive effect of high glucose levels on the α cell. In general the specificity of α-cell responses to glucose analogues, metabolites, and blockers of intracellular glucose metabolism is similar to that exhibited by β-cells.

Glucagon secretion also is stimulated by a protein meal and, most powerfully, by amino acids such as arginine and alanine. However, the α-cell response to protein is greatly dampened if glucose is administered concurrently. This interaction is partly via insulin; positive glucagon responses to amino acids are restrained by insulin excess and augmented by insulin deficiency. Free fatty acids—like glucose—exert a suppressive effect. Glucagon responses to orally ingested nutrients (as opposed to their intravenous delivery) may also be reinforced by the release of gastrointestinal secretagogues that augment insulin secretion. Exceptions are GLP-1 (a proglucagon product of the intestine) (see Fig. 46-16) and secretin, both of which inhibit α-cell glucagon release. The sum of all these individual influences is that the ingestion of ordinary meals produces much less variation in plasma glucagon than in plasma insulin levels. This is partly explained by the offsetting effects of the carbohydrate and protein portions of the meal on the α-cell, as compared with the synergistic effects of these nutrients on the β-cell.

Fasting for 3 days increases plasma glucagon twofold. Exercise of sufficient intensity and duration also increases plasma glucagon. Neural mechanisms may mediate some of these responses. In particular, vagal stimulation and acetylcholine acutely increase glucagon secretion. Of importance to the physician is that a variety of stresses, including infection, burns, tissue infarction, and major surgery, all increase glucagon secretion promptly. This phenomenon is probably mediated by the sympathetic nervous system via outflow from the

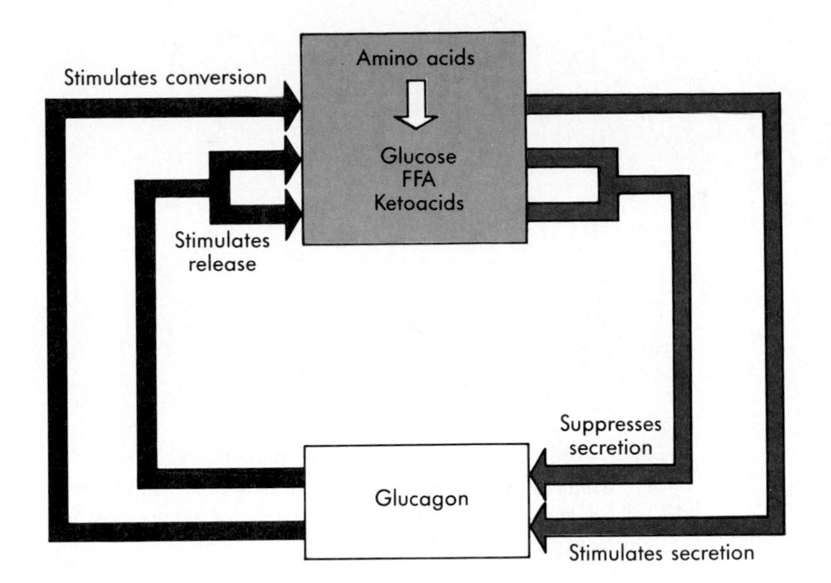

■ **Fig. 46-17** Feedback relationship between glucagon and nutrients. Glucagon stimulates production and release of glucose, free fatty acids *(FFA),* and ketoacids, which in turn suppress glucagon secretion. Amino acids stimulate glucagon secretion, and glucagon in turn stimulates amino acids' conversion to glucose.

ventromedial hypothalamus to α-adrenergic receptors in the α-cells. The excess of glucagon often leads to clinically significant hyperglycemia. The neurohormone somatostatin inhibits the secretion of glucagon, probably by paracrine or neurocrine effects made possible by the islet microarchitecture (see Fig. 46-1).

Monomeric α-cell glucagon circulates unbound in plasma at a basal concentration of 50 to 100 pg/ml (2×10^{-11} M). It has a half-life of 6 minutes and a metabolic clearance rate of about 600 ml/minute. The daily secretion rate of glucagon is estimated to be 100 to 150 μg. A portion of glucagon immunoreactivity is caused by related α-cell and intestinal L-cell peptides. The ratio of portal vein to peripheral vein glucagon concentrations is about 1.5 in the basal state, and about 50% of glucagon is extracted by the liver on a single passage. The kidney is the other major locus of glucagon degradation. Less than 1% of the glucagon filtered by the glomerulus is excreted in the urine.

■ *Hormone Actions*

In almost all respects the actions of glucagon are exactly opposite to those of insulin. Glucagon promotes mobilization rather than storage of fuels, especially glucose (Fig. 46-18). Both hormones act at numerous similar control points in the liver (see Fig. 46-13). Indeed, some investigators suggest that glucagon is the primary hormone regulating hepatic glucose production and ketogenesis, with insulin's role being that of glucagon antagonist.

Glucagon binds in the plasma membrane to a hepatic glycoprotein receptor of 63,000 molecular weight. Signal transduction is via a stimulatory G-protein, adenylyl cyclase, and cAMP as second messenger (see Chapter 44). cAMP-activated protein kinase A initiates a cascade of phosphorylations that activate or deactivate a number of enzyme kinases or phosphatases by covalent modification. The first and best-studied example is phosphorylase. Activated protein kinase A converts inactive phosphorylase kinase to active phosphorylase kinase. The latter then converts inactive phosphorylase to active phosphorylase, and glycogenolysis increases. In contrast phosphorylation of the enzymes phosphofructokinase and pyruvate kinase decreases their activity so that glycolysis is inhibited.

The dominant effect of glucagon is in the liver. Its actions on adipose tissue and muscle are minor unless insulin is virtually absent. Glucagon exerts an immediate and profound glycogenolytic effect through activation of glycogen phosphorylase. The glucose-1-phosphate released is prevented from undergoing resynthesis to glycogen by a simultaneous inhibition of glycogen synthase. Glucagon also stimulates gluconeogenesis by several mechanisms. The hepatic extraction of amino acids, especially alanine, is increased. The activities of the gluconeogenic enzymes pyruvate carboxylase and phosphoenolpyruvate carboxykinase are increased, whereas that of the glycolytic enzyme pyruvate kinase is decreased. The enzyme pair phosphofructokinase/fructose-1,6-diphosphatase determines the flow between fructose-6-phosphate and fructose-1,6-disposphate and therefore the relative rates of glycolysis and gluconeogenesis (see Fig. 51-13). The activities of these two enzymes, in turn, are reciprocally related by the level of fructose-2,6-biphosphate. This important metabolite is very sensitive hormonally (Fig. 46-19).

Glucagon phosphorylates the unique bidirectional enzyme 6-phosphofructo-2-kinase/fructose-2,6-biphosphatase via cAMP-dependent protein kinase A. When phosphorylated, this enzyme is a phosphatase that decreases fructose-2,6-biphosphate levels; when dephosphorylated, the enzyme is a kinase that increases fructose-2,6-biphosphate levels. Thus glucagon action lowers fructose-2,6-biphosphate levels; this, in turn, de-

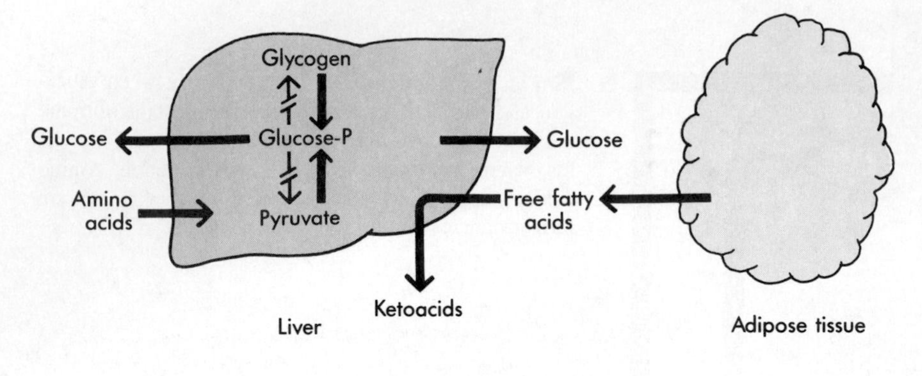

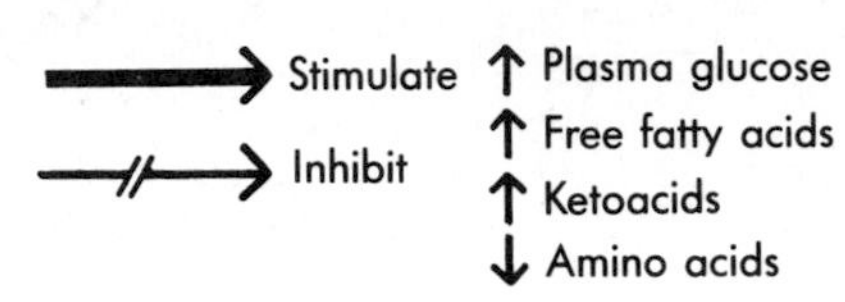

■ **Fig. 46-18** Effect of glucagon on the overall flow of fuels results in tissue release of glucose, fatty acids, and ketoacids into the circulation and hepatic uptake of amino acids for gluconeogenesis.

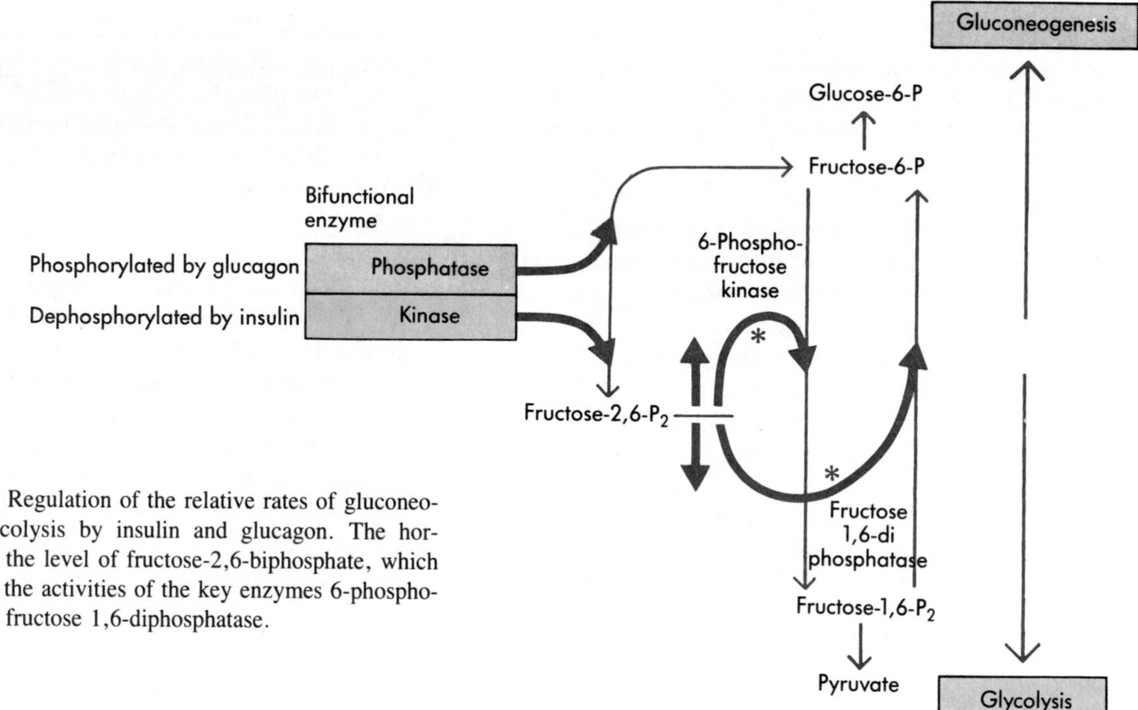

■ **Fig. 46-19** Regulation of the relative rates of gluconeogenesis and glycolysis by insulin and glucagon. The hormones modulate the level of fructose-2,6-biphosphate, which in turn regulates the activities of the key enzymes 6-phosphofructokinase and fructose 1,6-diphosphatase.

creases phosphofructokinase and increases phospho-1,6-biphosphatase activities (see Fig. 46-19). The result is an increase in gluconeogenesis and a decrease in glycolysis. Insulin has exactly the opposite effects on fructose-2, 6-biphosphate levels (see Fig. 46-19).

The crucial importance of glucagon to the maintenance of basal hepatic glucose output is shown by the marked decline of 75% that follows the selective inhibition of glucagon secretion by somatostatin. The powerful glycogenolytic and hyperglycemic action of glucagon is exhibited at plasma hormone concentrations of 150 to 500 pg/ml and occurs even in the presence of insulin levels somewhat above basal levels (20 to 30 μU/ml). However, this action is transient; the acute initial stimulation of hepatic glucose output, which occurs during continuous glucagon administration, wanes after about 30 minutes because of hepatic autoregulation by hyperglycemia and stimulation of insulin release. However, if glucagon is given in a more physiological, fluctuating pattern, each increment in the hormone causes a response. Glucagon has little or no influence on glucose use by peripheral tissues. Thus hyperglucagonemia has no effect on the plasma glucose levels that are generated by an exogenous glucose load, as long as the insulin response is normal. The gluconeogenic action of glucagon is also reflected in the ability of the hormone to increase the rate of disposal of infused amino acids and their degradation to urea. However, glucagon has no particular influence on branch chain amino acid levels; this suggests that it has little or no effect on muscle proteolysis.

Another intrahepatic action of glucagon is to direct incoming free fatty acids away from triglyceride synthesis and toward β-oxidation. Thus glucagon is a ketogenic and a hyperglycemic hormone. Glucagon inactivates acetyl-CoA carboxylase, the rate-limiting step in free fatty acid synthesis from cytoplasmic acetyl CoA. This results in lower levels of malonyl CoA, an allosteric inhibitor of carnitine acyltransferase. In turn, this allows a faster rate of influx of fatty acyl CoA into the mitochondrion for conversion to ketoacids. Simultaneously the greater rate of β-oxidation of free fatty acids increases the generation of acetyl CoA, which activates pyruvate carboxylase, helping to increase gluconeogenesis. Glucagon also activates adipose tissue lipase, thereby increasing lipolysis, the delivery of free fatty acids to the liver, and ketogenesis. Although glucagon's ketogenic actions are physiologically relevant, they are easily nullified by rather small amounts of insulin, particularly at the adipose tissue locus. Finally, hepatic hydroxy-methylglutaryl-CoA reductase activity is inhibited by glucagon, thereby decreasing hepatic cholesterol synthesis.

Other actions of glucagon include inhibition of renal tubular sodium resorption, causing natriuresis, and activation of myocardial adenylyl cyclase, causing a moderate increase in cardiac output. The latter effect has been of rare therapeutic value in refractory heart failure. The possibility that glucagon may act locally as a central nervous system hormone in the regulation of appetite is suggested by the presence of mRNA for the hormone in the brain.

Insulin/Glucagon Ratio

It should now be apparent that substrate fluxes are very sensitive to the relative availability of insulin and glucagon. The usual molar ratio of insulin to glucagon in plasma is about 2.0. Under circumstances that require mobilization and increased use of endogenous substrates, the insulin/glucagon ratio drops to 0.5 or less. This is seen in fasting, in prolonged exercise, and in the neonatal period when the infant is abruptly cut off from maternal fuel supplies but is not yet able to assimilate exogenous fuel efficiently. The ratio drops because of both decreased insulin secretion and increased glucagon secretion (see Fig. 45-12). A low insulin/glucagon ratio facilitates increased glycogenolysis, amino acid mobilization, and gluconeogenesis to maintain glucose supply to the central nervous system. Lipolysis is also enhanced, increasing free fatty acid flow to muscle and liver for oxidation. During exercise the low insulin/glucagon ratio still permits muscle glucose uptake because this is stimulated by an exercise-specific mechanism that is at least partly independent of insulin.

Conversely, under circumstances in which substrate storage is advantageous, as after a pure carbohydrate load or a mixed meal, this ratio rises to 10 or more, mainly because of increased insulin secretion. The high ratio enhances glucose uptake, oxidation, and conversion to liver and muscle glycogen while suppressing unneeded proteolysis and lipolysis. An interesting example of only a small and insignificant change in the insulin/glucagon ratio occurs after ingesting a pure protein meal. In this situation insulin secretion increases, facilitating muscle uptake of amino acids and their synthesis into proteins. At the same time glucagon secretion also increases. This prevents the decrease in hepatic glucose output and hypoglycemia that would ensue if the extra insulin action were completely unopposed. Ingestion of pure fat has little direct influence on the insulin/glucagon ratio. During a mixed meal the increase in the ratio facilitates clearance of chylomicrons by activation of adipose tissue lipoprotein lipase.

Somatostatin Secretion and Action

The neurohormone somatostatin was originally discovered in hypothalamic extracts as an inhibitor of growth hormone secretion (see Chapter 48). However, subsequent studies showed that somatostatin is synthesized by the δ cells of the pancreatic islets and by intestinal cells. In the δ-cells the processing of the primary somatostatin gene product, prosomatostatin, yields mostly a 14-amino acid peptide (SS-14). In intestinal cells a 28-amino acid peptide with an N-terminal extension is probably the major product (SS-28). Many of their biological actions are similar. Somatostatin secretion is stimulated by glucose, amino acids, free fatty acids, a variety of gastrointestinal hormones, glucagon, and β-adrenergic and cholinergic neurotransmitters. It is inhibited by insulin and α-adrenergic neurotransmitters. Exocytosis of somatostatin granules is stimulated by cAMP. Plasma levels of SS-28, but not of SS-14, increase with mixed meals. The plasma half-life of somatostatin is only a few minutes.

Somatostatin is a profound inhibitor of both insulin and glucagon secretion. The administration of somatostatin also decreases the assimilation rate of all nutrients from the gastrointestinal tract. This is accomplished by inhibitory actions on gastric, duodenal, and gallbladder motility; on the secretion of hydrochloric acid, pepsin, gastrin, secretin, and intestinal juices; and on pancreatic exocrine function. Somatostatin also inhibits the absorption of glucose, xylose, and triglycerides across the mucosal membrane. SS-14 and SS-28 may participate in a feedback arrangement, whereby entrance of food into the gut stimulates their release so as to prevent rapid nutrient overload. SS-14 may be released locally and may act in neurocrine and paracrine fashion to limit insulin and glucagon responses to meals. SS-28, which is released in particular after ingestion of fat, may function as a true

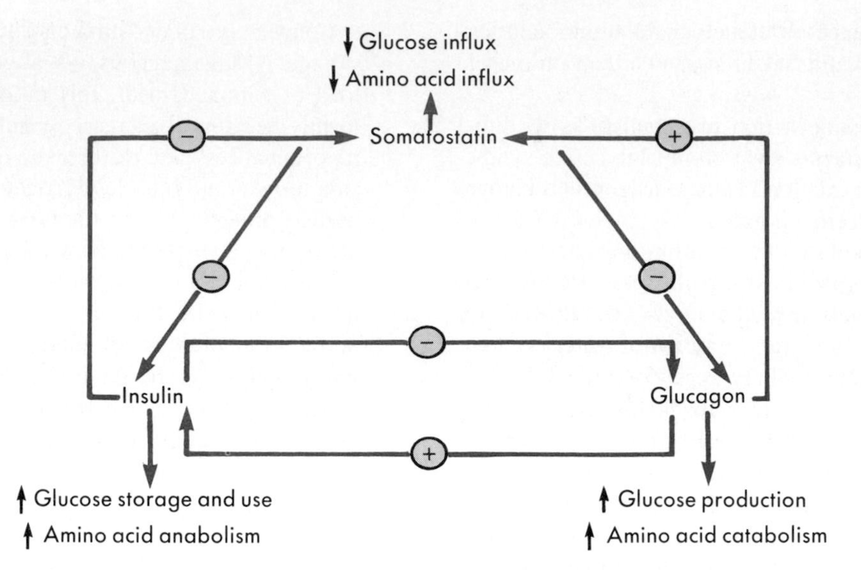

■ **Fig. 46-20** Schema interrelating the effects of somatostatin, insulin, and glucagon on each other's secretion with their effects on glucose and amino acid metabolism. (Modified from Unger RH et al. *Annu Rev Physiol* 40:307, 1978.)

hormone, i.e., it may reach the pancreatic islets via the blood stream. The interaction among the α-cells, β-cells, δ-cells, and intestinal cell products may coordinate the rates of bulk movement, digestion, and absorption of nutrients with the rates of nutrient uptake by the liver and peripheral tissues (Fig. 46-20).

■ *Pancreatic Polypeptide*

Pancreatic polypeptide is a specific product of the PP-cells. It has 36 amino acids and a distinctive C-terminal tyrosine-amide residue. It belongs to a family of such molecules as neuropeptide-Y in the hypothalamus and peptide Y-Y in the gastrointestinal tract. Pancreatic polypeptide is secreted in response to food ingestion via gastrointestinal secretogogues and cholinergic stimulation. Pancreatic polypeptide is also stimulated by hypoglycemia and inhibited by glucose administration. Its most well-defined action is to inhibit exocrine pancreatic secretion, partly by inhibiting the uptake of precursor amino acids by the acinar cells. Its true physiological importance is unclear, but elevated plasma levels of pancreatic polypeptide serve as markers for the presence of islet cell tumors and their response to treatment.

■ *Clinical Syndromes of Islet Cell Dysfunction*

■ *Insulin*

Insulin excess is usually caused by tumors of the β-cells. The cardinal manifestation is a low plasma glucose level (<50 mg/dl) in the fasting state. Because the sympathetic nervous system is actuated quickly by abrupt hypoglycemia, bursts of insulin hypersecretion produce episodes of rapid heart rate, nervousness, sweating, and hunger. With chronic insulin excess and persistent hypoglycemia, disturbed central nervous system function results in bizarre behavior, defects in cerebration, loss of consciousness, or convulsions. The need to ingest large amounts of carbohydrate combined with the lipogenic effect of insulin produces weight gain. The restraining effect of insulin on glucagon secretion aggravates the tendency to hypoglycemia. The diagnosis is established by demonstrating fasting plasma insulin and C-peptide levels inappropriately high for the prevailing glucose levels. Removal of the tumor cures the condition. Failing that, drugs that inhibit insulin secretion palliate the hypoglycemia.

Primary deficiency of insulin is the consequence of β-cell destruction. This disorder, known as insulin dependent or type 1 diabetes mellitus, results from a genetically conferred vulnerability to a currently unspecified insult. An autoimmune process contributes to β-cell damage. As a consequence of insulin lack (and glucagon excess), glucose production is augmented, and the efficiency of peripheral glucose use is reduced until a new equilibrium between these processes is reached at a very high plasma glucose level (300 to 1,000 mg/dl). An increased rate of gluconeogenesis is supported by uninhibited proteolysis. Muscle anabolism is correspondingly reduced; thus nitrogen balance becomes negative. Uninhibited lipolysis elevates plasma free fatty acids and body fat decreases. Ketogenesis is markedly stimulated, whereas peripheral ketoacid use is inhibited. This can greatly elevate plasma levels of the strong carboxylic acids, acetoacetic and β OH-butyric. As they are neutralized by sodium bicarbonate, carbonic acid is formed, which dissociates to carbon dioxide and water. Compensatory pulmonary hyperventilation is stimulated, which lowers P_{CO_2}. Despite this, the blood pH may finally fall to less than 6.8 and death from diabetic ketoacidosis ensues.

Before this terminal point a classic constellation of symptoms is observed. Because of the high plasma glucose levels, the filtered load of glucose exceeds the renal tubular capacity for reabsorption. Glucose therefore is excreted in the urine in large quantities, causing, by its osmotic effect, increased excretion of water and salts and frequent urination. Thirst is stimulated by the hyperosmolality of the plasma and by the hypovolemia. The loss of glucose is also a caloric drain, which the patient attempts to recoup by eating more, adding to hyperglycemia, as the insulin-deficient individual is unable to store carbohydrate efficiently. This catabolic state is manifested by loss of lean body mass, adipose tissue, and body fluids. Deficits of nitrogen, potassium, phosphate, magnesium, and other intracellular components develop as these are excreted in the urine. Exercise capacity may be impaired because of the reduction in muscle and liver glycogen stores.

Osmotic fluid shifts secondary to the high plasma glucose, and its conversion to other sugars, such as sorbitol, may cause the lens of the eye to swell, blurred vision, and even formation of cataracts. Other malfunctions whose pathogenesis is less clear include initially increased glomerular filtration, slowed transmission of nerve impulses and decreased resistance to infectious agents. Numerous proteins, including hemoglobin, albumin, and collagen, are nonenzymatically glucosylated by formation of an adduct between the aldehyde group of glucose and free amino groups. This and other slow processes may contribute to the long-term tissue damage in the retina, kidneys, nerves, and cardiovascular system.

Insulin treatment systematically lowers plasma levels of glucose, free fatty acids, and ketoacids to normal and reduces urine nitrogen losses (Fig. 46-21). This result is achieved by direct actions of insulin and also by diminishing the secretion of the insulin antagonist glucagon (see Fig. 46-21).

Another more common form of diabetes mellitus, noninsulin-dependent or type 2, often is associated with obesity. In this disease the major target tissues are resistant to the action of insulin. The locus of resistance is distal to the insulin receptor binding site, but defects in receptor tyrosine kinase activity, glucose transport, and activities of insulin-sensitive enzymes have been found. In addition there is a derangement in β-cell recognition of glucose as a stimulus, so that first phase insulin secretion is lost, though a delayed release does occur. Fasting plasma glucose levels are elevated because of excessive hepatic glucose production; postprandially, glucose levels are further elevated, particularly after carbohydrate intake. However, accelerated lipolysis and ketogenesis are rarely seen in the absence of intercurrent illness.

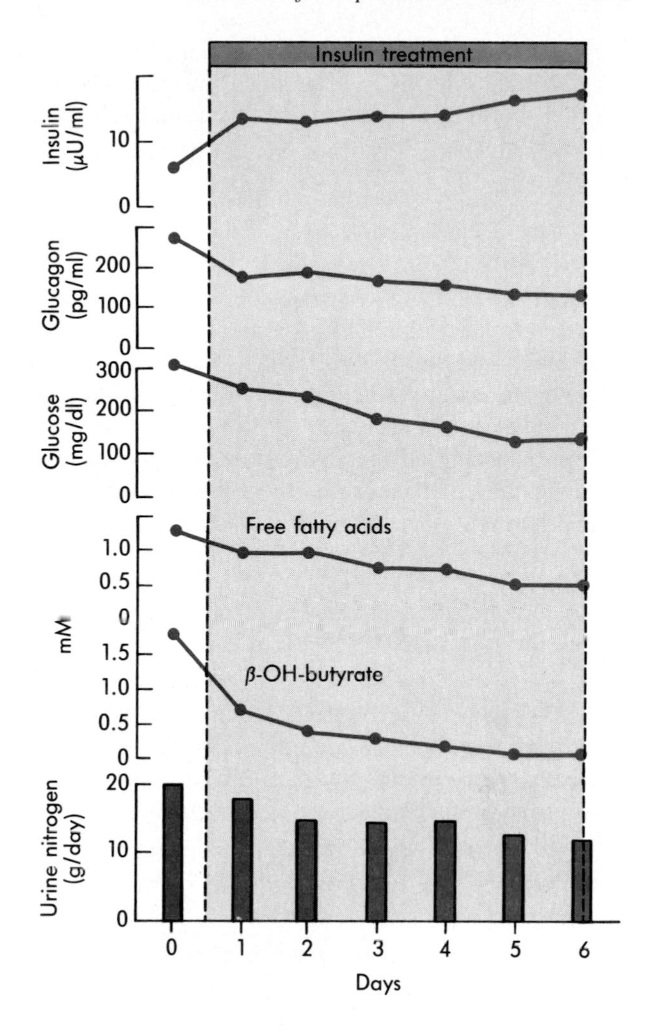

■ **Fig. 46-21** The effects of insulin replacement on plasma hormone and substrate levels and on urine losses of glucose and nitrogen in insulin-deficient diabetic humans.

Treatment of type 2 diabetes does not ordinarily require insulin administration. Caloric regulation, weight reduction if obesity is present, and use of sulfonylurea drugs simultaneously improve tissue responsiveness to endogenous insulin and β-cell responsiveness to glucose. In late stages insulin administration is required.

■ *Glucagon*

Primary glucagon excess is produced by α-cell tumors. The catabolic action of the hormone is exhibited by loss of weight and by a peculiar destructive skin lesion. Plasma levels of glucose and of ketoacids are elevated. The marked increase in gluconeogenesis causes a generalized reduction of plasma amino acids and an increase in urinary nitrogen. Severe diabetes mellitus is uncommon because of the compensatory increase in insulin secretion. The predictable consequence of glucagon deficiency would be fasting hypoglycemia. In fact, however, primary cases have been documented only rarely. Other hormonal adjustments would probably sustain plasma glucose if glucagon were missing.

Somatostatin

Tumors of the δ-cells produce somatostatin excess. This is manifest by inhibition of nutrient absorption, excess stool fat, and weight loss. Plasma insulin and glucagon levels are low, and the low insulin may lead to modest hyperglycemia.

Somatostatin analogues are now used therapeutically to alleviate diarrhea caused by unregulated gastrointestinal hormone secretion and to inhibit excessive release of various peptide and protein hormones by neoplasms.

Summary

1. The pancreatic islets are composed of insulin-secreting β-cells (the majority), glucagon-secreting α-cells, somatostatin-secreting δ-cells, and pancreatic polypeptide-secreting cells. The microarchitecture and blood flow arrangements permit paracrine and neurocrine functioning, as well as direct cell-to-cell communication through gap junctions.

2. Insulin is a major glucoregulatory, antilipolytic, antiketogenic, and anabolic hormone. It consists of two straight chain peptides held together by disulfide bonds and synthesized from a single chain precursor (proinsulin).

3. Insulin secretion is stimulated by meals, glucose, and other fuels, gastrointestinal peptides, and cholingeric and β-adrenergic stimuli. Release is inhibited by fasting and by exercise, both circumstances where fuel mobilization is required.

4. Insulin promotes fuel storage. Its activities in order of decreasing sensitivity follow: inhibition of adipose tissue lipolysis and of ketogenesis; inhibition of hepatic glycogenolysis, gluconeogenesis, and glucose release; inhibition of muscle proteolysis; and stimulation of muscle glucose uptake and storage as glycogen. Insulin also stimulates uptake of amino acids, phosphate, and magnesium, as well as protein synthesis.

5. Insulin acts through a plasma membrane receptor that possesses tyrosine kinase activity. This leads to covalent modulation of the activities of enzymes involved in glucose and fatty acid metabolism. Insulin also affects gene expression of numerous enzymes and other proteins by unknown messengers.

6. Insulin decreases plasma levels of glucose, free fatty acids, ketoacids, glycerol, and branch chain and other amino acids. Insulin deficiency leads to hyperglycemia, loss of lean body and adipose tissue mass, growth retardation, and ultimately to metabolic ketoacidosis.

7. Glucagon is a straight chain peptide released in response to hypoglycemia and to amino acids. Its secretion is suppressed by glucose, free fatty acids, and insulin. Glucagon secretion increases during prolonged fasting and exercise.

8. Glucagon promotes mobilization of glucose. It acts primarily on the liver to stimulate glycogenolysis and gluconeogenesis, as well as fatty acid oxidation and ketogenesis. cAMP is its second messenger, and covalent modification of enzyme activities by phosphorylation is the main mechanism of action. Glucagon increases the plasma levels of glucose, free fatty acids, and ketoacids, but it decreases amino acid levels.

9. The insulin/glucagon ratio controls the relative rates of glycolysis and gluconeogenesis by altering hepatic fructose-2,6-biphosphate levels. This metabolite, in turn, regulates the rate and direction of flow between fructose-1-phosphate and fructose-1,6-biphosphate. The two hormones have antagonistic effects at numerous other liver enzyme steps in glucose and fatty acid metabolism.

10. Somatostatin is a neuropeptide of islet and intestinal cell origin. It decreases the motility of the gastrointestinal tract, gastrointestinal secretions, digestion and absorption of nutrients, and the secretion of both insulin and glucagon. Somatostatin is secreted in response to meals; its actions, along with those of insulin and glucagon, coordinate nutrient input with substrate disposal.

Bibliography

Journal articles

Amiel SA et al: Insulin resistance of puberty: a defect restricted to peripheral glucose metabolism, *J Clin Endocrinol Metab* 72:277, 1991.

Boden G et al: Role of glucagon in disposal of an amino acid load, *Am J Physiol* 259:E225, 1990.

Bonadonna RC et al: Dose-dependent effect of insulin on plasma free fatty acid turnover and oxidation in humans, *Am J Physiol* 259:E726, 1990.

Cahill GF: Starvation in man, *N Engl J Med* 282:668, 1970.

Exton JH: Some thoughts on the mechanism of action of insulin, *Diabetes* 40:521, 1991.

Frank HJL et al: A direct in vitro demonstration of insulin binding to isolated brain microvessels, *Diabetes* 30:757, 1981.

Fukagawa NK et al: Insulin dose-dependent reductions in plasma amino acids in man, *Am J Physiol* 250:E13, 1986.

Gottesman I et al: Insulin increases the maximum velocity for glucose uptake without altering the Michaelis constant in man: evidence that insulin increases glucose uptake merely by providing additional transport sites, *J Clin Invest* 70:1310, 1982

Granner DK, Andreone TL: Insulin modulation of gene expression, *Diabetes/Metab Rev* 1:139, 1985.

Hollenbeck C, Reaven GM: Variations in insulin-stimulated glucose uptake in healthy individuals with normal glucose tolerance, *J Clin Endocrinol Metab* 64:1169, 1987.

Itoh M et al: Antisomatostatin gamma globulin augments secretion of both insulin and glucagon in vitro: evidence for a physiologic role for endogenous somatostatin in the regulation of pancreatic alpha and beta cell function, *Diabetes* 29:693, 1980.

Jackson RA et al: Influence of aging on glucose homeostasis, *J Clin Endocrinol Metab* 55:840, 1982.

Jefferson LS: Role of insulin in the regulation of protein synthesis, *Diabetes* 29:487, 1980.

Kahn SE et al: Evidence of cosecretion of islet amyloid polypeptide and insulin by beta cells, *Diabetes* 39:634, 1990.

King DS et al: Insulin secretory capacity in endurance-trained and untrained young men, *Am J Physiol* 259:E155, 1990.

Krasinski SD et al: Pancreatic polypeptide and peptide YY gene expression, *Ann NY Acad Sci* 611:73, 1990.

Liljenquist JE et al: Evidence for an important role of glucagon in the regulation of hepatic glucose production in normal man, *J Clin Invest* 59:369, 1977.

Magnuson MA: Glucokinase gene structure: functional implications of molecular genetic studies, *Diabetes* 39:523, 1990.

Matschinsky FM: Glucokinase as glucose sensor and metabolic signal generator in pancreatic beta cells and hepatocytes, *Diabetes* 39:647, 1990.

Meda P et al: Rapid and reversible secretion change during uncoupling of rat insulin-producing cells, *J Clin Invest* 86:759, 1990.

Mueckler M: Family of glucose-transporter genes: implications for glucose homeostasis and diabetes, *Diabetes* 39:6, 1990.

Nurjhan N et al: Insulin dose-response characteristics for suppression of glycerol release and conversion to glucose in humans, *Diabetes* 35:1326, 1986.

Olefsky JM: The insulin receptor: a multifunctional protein, *Diabetes* 39:1009, 1991.

Orskov C et al: Proglucagon products in plasma of noninsulin-dependent diabetics and nondiabetic controls in the fasting state and after oral glucose and intravenous arginine, *J Clin Invest* 87:415, 1991.

Philippe J: Structure and pancreatic expression of the insulin and glucagon genes, *Endocr Rev* 12:252, 1991.

Polonsky KS, Given BD, Van Cauter E: Twenty-four-hour profiles and pulsatile patterns of insulin secretion in normal and obese subjects, *J Clin Invest* 81:442, 1988.

Reichlin S: Somatostatin, *N Engl J Med* 309:1495, 1983

Robertson RP, Seaquist ER, Walseth TF: G proteins and modulation of insulin secretion, *Diabetes* 40:1, 1991.

Saltiel AR: Second messengers of insulin action, *Diabetes Care* 13:244, 1990.

Schwartz TW: Pancreatic polypeptide: a hormone under vagal control, *Gastroenterology* 85:1411, 1983.

Seino S, Seino M, Bell GI: Human insulin receptor gene, *Diabetes* 39:129, 1990.

Sturis J et al: Entrainment of pulsatile insulin secretion by oscillatory glucose infusion, *J Clin Invest* 87:439, 1991.

Thiebaud D et al: The effect of graded doses of insulin on total glucose uptake, glucose oxidation, and glucose storage in man, *Diabetes* 31:957, 1982.

Unger RH et al: Insulin, glucagon, and somatostatin secretion in the regulation of metabolism, *Annu Rev Physiol* 40:307, 1978.

Weigle DS: Pulsatile secretion of fuel regulatory hormones, *Diabetes* 36:764, 1987.

Weir GC, Bonner-Weir S: Islets of Langerhans: the puzzle of intraislet interactions and their relevance to diabetes, *J Clin Invest* 85:983, 1990.

CHAPTER

47

Endocrine Regulation of Calcium and Phosphate Metabolism

■ *Calcium and Phosphate Pools and Turnover*

The maintenance of calcium and phosphate homeostasis is dependent on major contributions from three organ systems, the intestinal tract, the skeleton, and the kidneys, with facilitatory but essential contributions from the skin and the liver. Before detailing the hormonal mechanisms of regulation it is necessary to present an overview of calcium and phosphate metabolism. The structural and functional aspects of the bone mass relevant to regulation of calcium and phosphate must also be appreciated. Without such an initial overview it is difficult to tie together the individual endocrine components of the system.

The calcium ion is of such fundamental importance to all biological systems that its concentration must be kept within specific limits of physiological tolerance in several compartments. The resting intracellular cytosolic concentration of free calcium is only 10^{-7}M. This can transiently increase 10- to 100-fold in the course of calcium ion involvement in creation or maintenance of action potentials, contraction and motility, cytoskeletal rearrangements, cell division, secretion, and modulation of enzyme activities. The total pool of intracellular free calcium is estimated to be only 0.2 mg. An additional 9 g of intracellular calcium is present in bound form in the endoplasmic reticulum, the mitochondria, and the plasma membrane. This constitutes an immediately accessible storage pool and contributes to the structural integrity of the cell. The extracellular concentration of free calcium is approximately 10^{-3}M, or four orders of magnitude higher than the intracellular concentration. This huge gradient is maintained by specialized membranes and calcium pumps. (see Chapter 1) The proper extracellular concentration is essential for generation of normal membrane potentials, for calcium uptake into the cells in the course of contraction and exocytosis, for normal blood clotting, and for modulation of plasma enzyme activities. This total extracellular pool constitutes about 1 g. The skeleton and teeth contain 1 to 2 kg of calcium, depending on body size, or 99% of the total. This pool of bound calcium functions in skeletal structure, protection of internal organs, and locomotion.

The normal range of calcium in the plasma is 8.6 to 10.6 mg/dl (2.15 to 2.65 mmole/L = 4.3 to 5.3 mE/L). For any individual however the variation from day to day is generally less than 10%. Approximately 50% of plasma calcium is in ionized, biologically active form; 10% is complexed in nonionic but ultrafilterable forms, such as calcium bicarbonate; and 40% is bound to proteins, mainly albumin. The equilibrium between ionized and protein-bound calcium is shifted toward increased binding as the blood pH increases. Thus alkalosis decreases and acidosis increases the plasma ionized calcium concentration.

Fig. 47-1 details the normal turnover of calcium in the body. Daily dietary calcium intake can range from 200 to 2,000 mg. The percentage of dietary calcium that is absorbed from the gut is inversely related to intake but the relation is curvilinear. An adaptive increase in fractional absorption is one important mechanism for maintaining normal body calcium stores in the face of dietary deprivation, whereas an adaptive decrease prevents overload in the face of dietary surfeit. At an average daily intake of 1,000 mg about 35% is absorbed. The same amount of calcium, 350 mg, ultimately must be excreted to maintain balance. About 150 mg is secreted back into the intestine and excreted in the stools, along with the unabsorbed fraction from the diet. The remaining 200 mg is excreted in the urine. The kidney filters about 10,000 mg of calcium per day (non-protein-bound calcium concentration × glomerular filtration rate = 60 mg/L × 170 L/day). However, approximately 98% is reabsorbed in the tubules. Alteration in

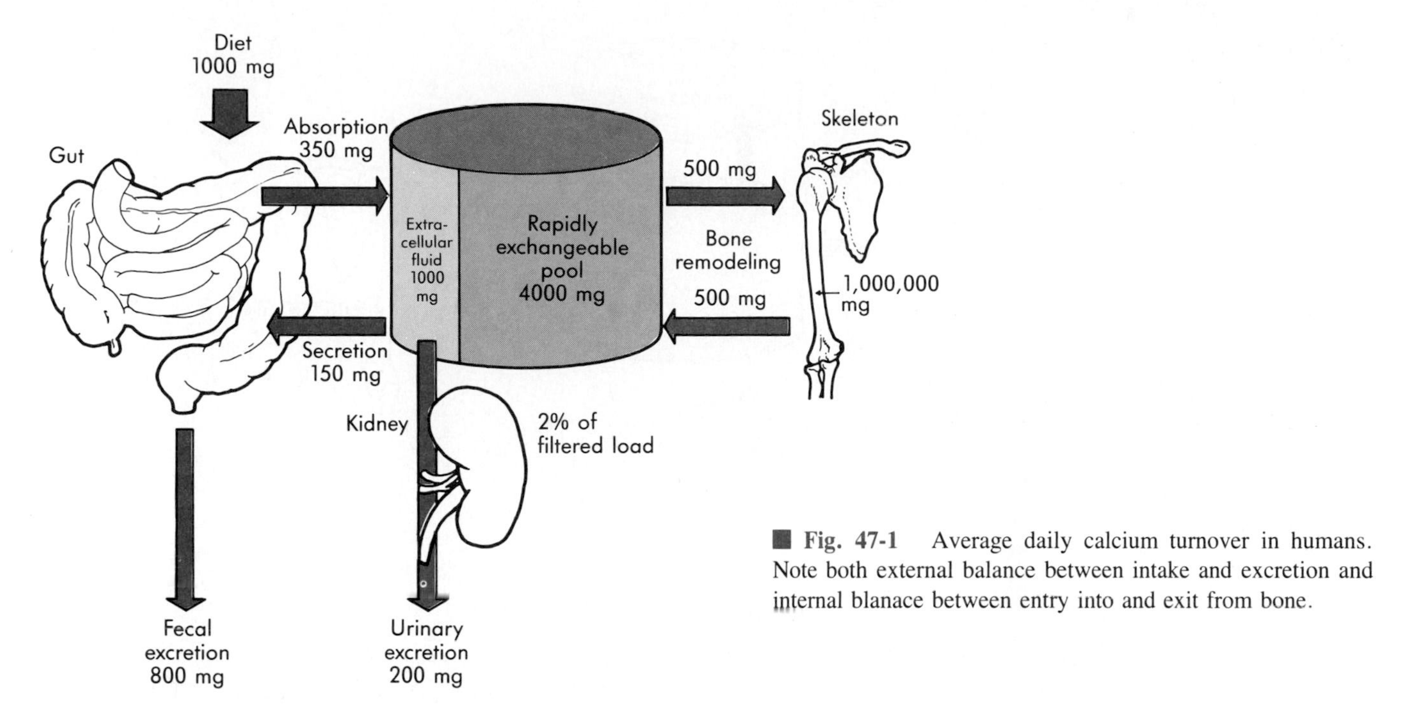

■ **Fig. 47-1** Average daily calcium turnover in humans. Note both external balance between intake and excretion and internal blanace between entry into and exit from bone.

the small fraction of filtered calcium that is finally excreted provides a sensitive means of maintaining calcium balance. Both dietary calcium intake and absorption from the gut are diminished in the senescent period of life. This contributes to declining bone mass and risk of fracture.

Calcium enters and exits from an extracellular pool of 1,000 mg. This in turn is in equilibrium with a rapidly exchanging pool of several times that size. This pool probably represents the surface of recently or partially mineralized bone. A total of 500 mg is "irreversibly" removed from the extracellular space by bone formation, and 500 mg is returned to it by bone resorption in the process that constitutes normal bone remodeling.

The phosphate ion is also of critical importance to all biological systems. Phosphate is an integral component of all glycolytic compounds from glucose-l-phosphate to phosphoenolpyruvate. It is part of the structure of high-energy transfer compounds, such as adenosine triphosphate (ATP) and creatine phosphate; of cofactors such as NAD, NADP, and thiamine pyrophosphate; of DNA and RNA; and of lipids such as phosphatidylcholine. Phosphate functions as a covalent modifier of numerous enzymes. It is the major intracellular anion and a major anion in the crystalline structure of bone and teeth.

The normal concentration of phosphate in the plasma is 2.5 to 4.5 mg/dl (0.81 to 1.45 mmole/L). Because the valence of phosphate changes with pH, it is less meaningful to express phosphate concentration in milliequivalents per liter. The turnover of phosphate is shown in Fig. 47-2. In contrast to that of calcium the percentage of phosphate absorbed from the diet is relatively constant. Thus net absorption of phosphate from the gut is linearly related to intake and adaptive regulation at this site is of less importance. Therefore urinary excretion provides the major mechanism for preserving phosphate balance. Of the daily filtered load of approximately 6,000 mg (plasma concentration × glomerular filtration rate = 36 mg/L × 170 L/day) renal tubular reabsorption can vary from 70% to 100%, with an average of 90%. This provides the needed flexibility to compensate for large swings in dietary intake.

Soft tissue stores of phosphate, such as in the muscle mass, are large. Rapid transfer between these stores and the extracellular fluid is an important factor in minute-to-minute regulation of plasma phosphate concentration. About 250 mg, or half the total extracellular fluid pool of 500 mg, enters and leaves the bone mass daily in the process of remodeling.

The divalent cation magnesium (Mg^{++}) is related in some metabolic respects to calcium and phosphate. Magnesium is essential in neuromuscular transmission and serves as a cofactor in numerous enzyme reactions, most notably those involving energy transfers via ATP and those concerned with ribosomal protein synthesis. The normal range of magnesium in plasma is 1.8 to 2.4 mg/dl (1.5 to 2.0 mEq/L). One third of plasma magnesium is bound to protein. The body content is about 25,000 mg, of which 50% is present in the skeleton. Virtually all the rest is present in the intracellular fluid, where magnesium functions with potassium as a major

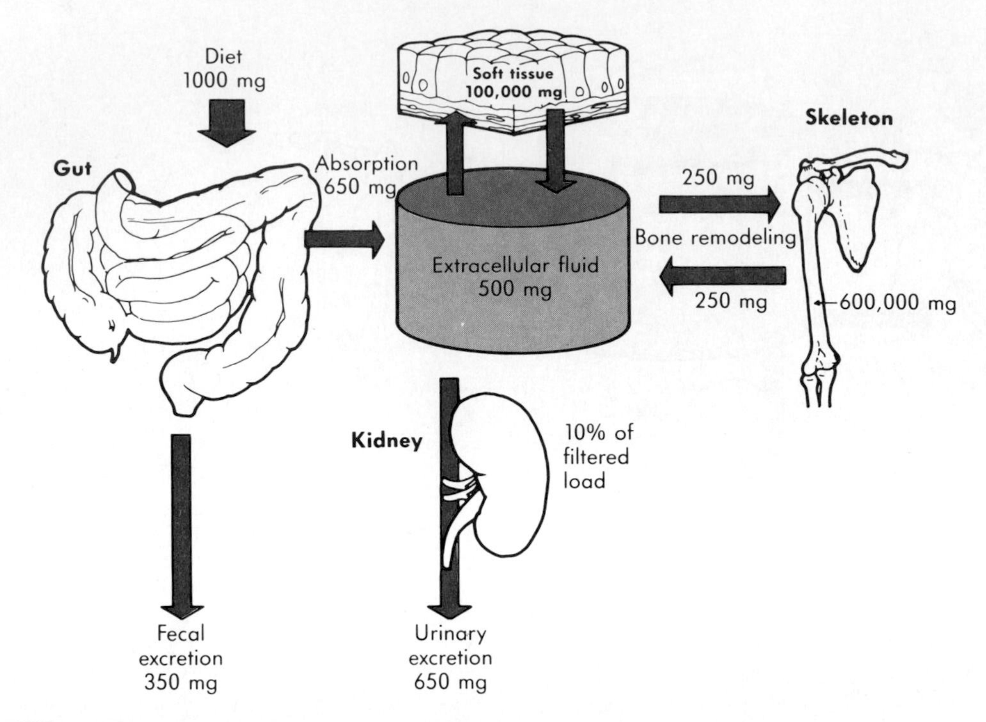

■ **Fig. 47-2** Average daily phosphate turnover in humans. Note both external balance between intake and excretion and internal balance between entry into and exit from bone.

cation. The daily intake ranges from 300 to 500 mg. Forty percent of this is absorbed, and in a steady state the same amount, 120 to 200 mg, is excreted in the urine.

■ *Bone Dynamics*

A detailed account of the development of the skeleton, its structural properties, and its mechanical function is beyond the purview of endocrine physiology. However, a brief review of the organization of bone in relation to its function as a mineral reservoir is essential to understanding hormonal regulation of calcium and phosphate metabolism.

About 80% of the bone mass consists of cortical bone comprising chiefly the dense, concentric outer layers of the appendicular skeleton (long bones) and the thinner outer layer of the flat bones. About 20% consists of trabecular bone, bridges of bone spicules comprising chiefly the larger inner parts of the axial skeleton (skull, ribs, vertebrae, and pelvis), and the smaller interior of the shafts of long bones (Fig. 47-3). Though of lesser mass trabecular bone has five times as much total surface area as cortical bone, making the former disproportionately important in calcium dynamics.

Throughout life the bone mass is continuously turning over by a well-regulated coupling of the processes of bone formation and resorption. This coupling is achieved in all types of bone within individual microscopic units, called osteons or bone modeling units. The chemical or mechanical signals that coordinate local rates of formation and resorption are incompletely understood. During the growth years formation exceeds resorption, and skeletal mass increases. Linear growth occurs at the ends of long bones by replacement of cartilage in specialized areas known as epiphyseal plates. These close off at the end of puberty when adult height is reached. Increase in bone width occurs by apposition to the outer surfaces of cortical bone under the connective tissue covering, the periosteum. A peak in bone mass is reached between ages 20 and 30 years. Thereafter equal rates of formation and resorption stabilize the bone mass until age 35 to 40 years, at which time resorption begins to exceed formation, and the total mass slowly decreases.

This process of bone turnover in the adult is known as **remodeling;** it is one of the major mechanisms for maintaining calcium homeostasis. As much as 15% of the total bone mass normally turns over each year in the remodeling process. Endocrine diseases that disrupt the coupling of formation and resorption are more florid in their manifestations when they are superimposed on the natural disequilibrium that exists during the early phase of growth and the late phase of senescence. Women are affected by such diseases more than men because women have a 25% smaller peak bone mass and a more rapid physiological rate of early senescent loss.

Three major cell types are recognized in histological sections of bone: **osteoblasts, osteocytes,** and **osteo-**

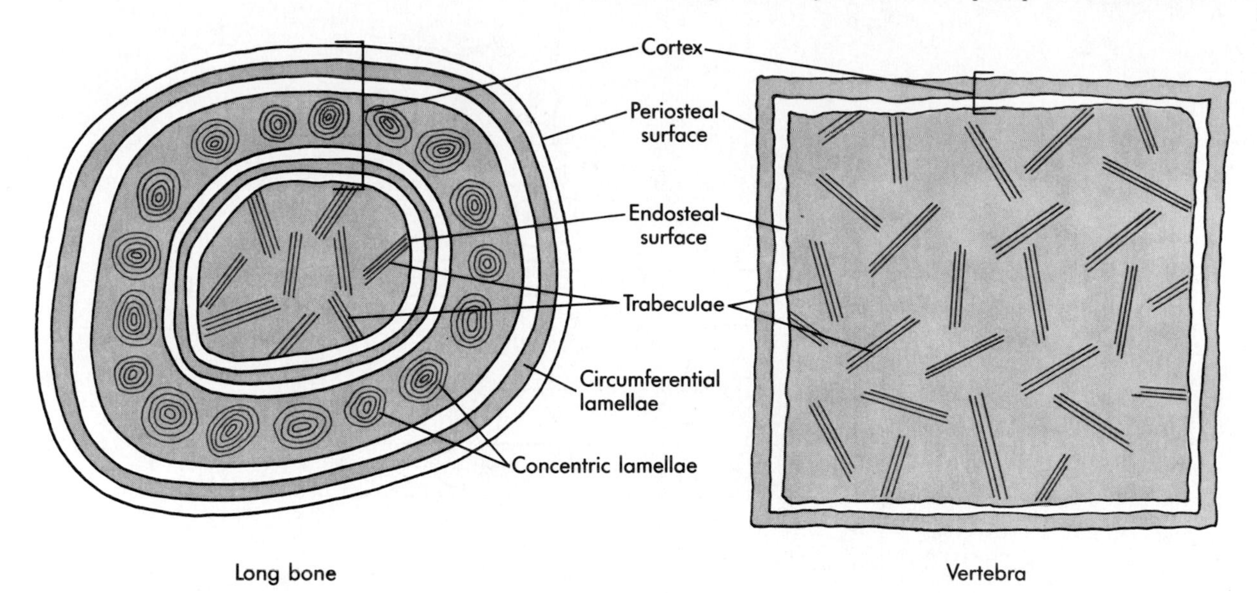

■ **Fig. 47-3** Schematic cross-sectional representations of a long bone, primarily cortical, and a vertebra, primarily trabecular. The long bone is distinguished by a thick outer cortex containing circumferential rings (lamellae) within which concentric lamellae known as haversian canals are present. These contain nutrient vessels. The inner portions of long bones and vertebrae consist of bridging spicules of bone or trabeculae in organized lamellar arrays. Between the trabeculae are bone marrow elements and connective tissue cells.

clasts (Fig. 47-4). The first two arise from primitive cells, called osteoprogenitor cells, within the connective tissue of the mesenchyme. These cells are similar to fibroblast precursors. A variety of proteins in the bone, known as skeletal growth factors, attract osteoprogenitor cells, direct their differentiation into osteoblasts, and stimulate their growth. The osteoclasts arise from the same precursors that give rise to circulating monocytes and tissue macrophages (i.e., promonocytes or monoblasts).

Bone formation is carried out by active osteoblasts, which synthesize and extrude collagen into the adjacent extracellular space. The collagen fibrils line up in regular arrays and produce an organic matrix known as **osteoid** within which calcium is then deposited as amorphous masses of calcium phosphate. About ten days elapse between osteoid formation and mineralization. Once initiated, however, most of the calcium phosphate is deposited within a period of 6 to 12 hours. Thereafter hydroxide and bicarbonate ions are gradually added to the mineral phase and mature hydroxyapatite crystals are slowly formed. These have a calcium/phosphate ratio of 2.2 by weight and 1.7 by moles. As this completely mineralized bone accumulates and surrounds the osteoblast, that cell decreases its synthetic activity and becomes an interior osteocyte (Fig. 47-4). Osteoblastic activity, therefore, is observed only along the surfaces of bone, whether they are the concentric lamaellae of cortical bone or the linear lamellae of the interior bridging trabeculae (Fig. 47-3). Along these surfaces are resting osteocytes or osteoblasts that may initiate the cycle of remodeling in each osteon.

The mineralization process requires normal plasma concentrations of calcium and phosphate and is dependent on vitamin D. The enzyme **alkaline phosphatase** and possibly other macromolecules from the osteoblast also participate in this process. Of note are **osteonectin,** a protein of 32,000 molecular weight, and **osteocalcin,** a protein of 6,000 molecular weight. These two proteins form 1% to 2% of all bone protein. Osteonectin binds to collagen, and the complex, in turn, binds hydroxyapatite crystals. Osteocalcin is distinguished by the presence of γ-carboxyglutamate residues, which have an affinity for calcium and a strong avidity for uncrystalized hydroxyapatite. Although their precise roles in bone formation are unclear, both alkaline phosphatase and osteocalcin circulate in plasma and their levels are markers of osteoblastic activity.

Within each bone unit minute fluid-containing channels, called **canaliculi,** traverse the mineralized bone; through these channels the interior osteocytes remain connected with surface cells via syncytial cell processes (Fig. 47-4). This arrangement permits transfer of calcium from the enormous surface area of the interior to the exterior of the bone units and thence into the extracellular fluid. This transfer process, which is carried out by the osteocytes, is known as **osteocytic osteolysis.** It probably does not actually decrease bone mass but simply removes calcium from the most recently formed crystals.

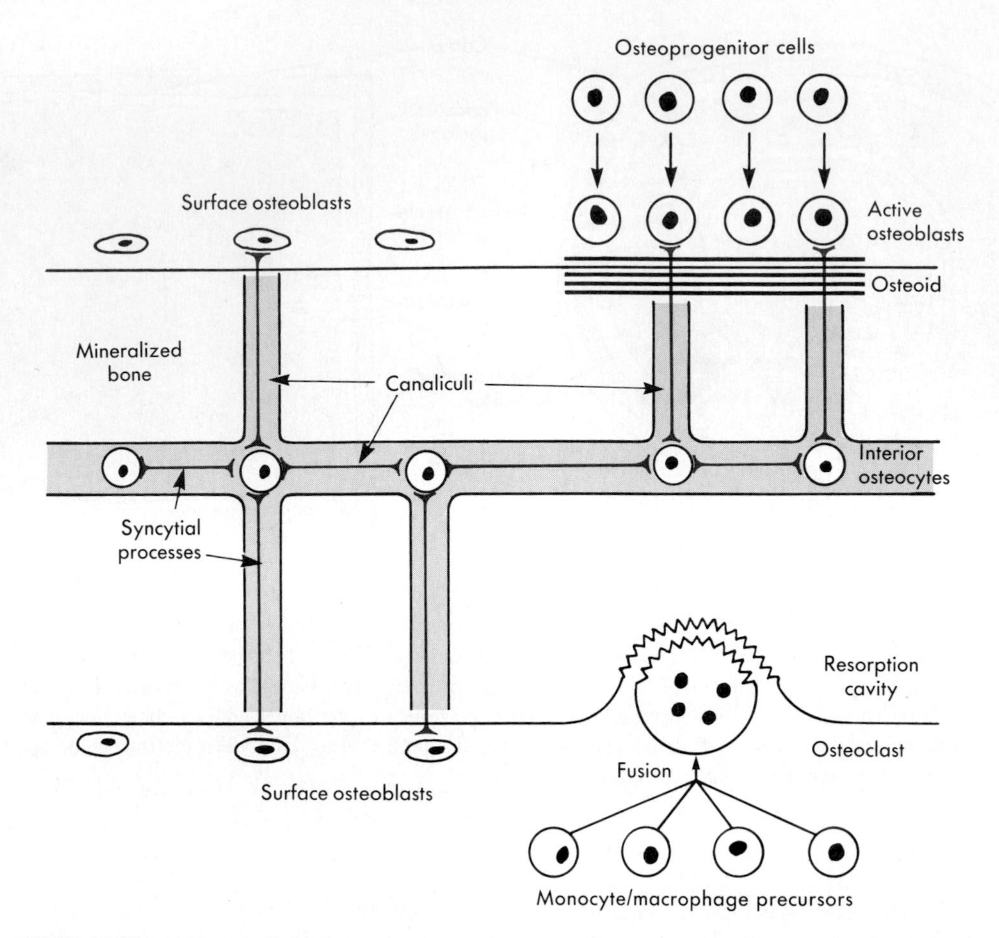

■ **Fig. 47-4** The relationships between bone cells and bone surfaces. The canaliculi provide a huge interface between the interior surfaces of mineralized bone and intercellular fluid. This permits efficient osteolysis with transfer of calcium and phosphate to the exterior via syncytial processes connecting interior and surface osteocytes. (Redrawn from Avioli LV et al: *Bone metabolism and disease*. In Bondy PK, Rosenberg LE: *Metabolic control and disease*, Philadelphia, 1980, WB Saunders.)

In contrast the process of resorption of bone does not merely extract calcium; it also destroys the entire matrix, thereby diminishing the bone mass. The cell responsible for this is the osteoclast, which is a giant multinucleated cell formed by fusion of several precursor cells (Fig. 47-4). The osteoclast contains large numbers of mitochondria and lysosomes. It attaches to the surface of the bone modeling unit. At the point of attachment a ruffled border is created by infolding of the plasma membrane. Within this zone the process of bone dissolution is carried out by collagenase, lysosomal enzymes, and phosphatase. Thus the osteoclast literally tunnels its way into the mineralized bone. Calcium, phosphate, the amino acids hydroxyproline and hydroxylysine that are unique to collagen, and fluorescent products of collagen crosslinkages known as pyridinolines are all released into the extracellular fluid. Urinary excretion rates of the organic products provide quantitative indices of bone resorption.

As noted previously resorption and formation are closely coordinated within each bone modeling unit. It is now believed that the remodeling process begins with a chemical signal(s) from resting osteocytes or osteoblasts (Fig. 47-5). The signal stimulates recruitment and differentiation of osteoclast precursors and their activation. The osteoclasts resorb a segment of bone as described, after which macrophages "clean up" the residue of the resorption process. Osteoblasts are then recruited to the same site and fill in the newly created resorption cavity. Thus the first phase of the remodeling cycle is actually resorption, which lasts about 10 days. The second phase, formation, takes about 3 months to complete. The remodeling cycle is influenced by a large array of hormones (Table 47-1). These can affect any one or more of the various steps either positively or negatively, or even both, depending on the concentration of hormone and the duration of exposure. As a general principle whether the primary effect of a hormone is on formation or resorption of bone, the phenomenon of coupling will secondarily alter the other process in the same direction. Therefore the net effect of any endocrine abnormality will depend on the degree

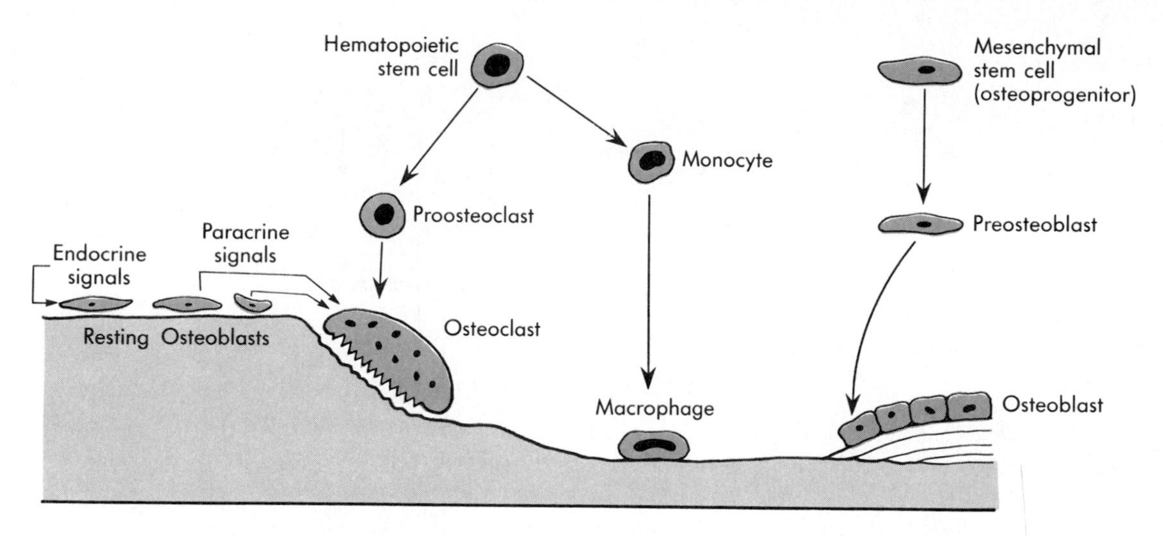

Fig. 47-5 The process of bone remodelling. Endocrine signals to resting osteoblasts generate local paracrine signals to nearby osteoclasts and osteoclast precursors. The osteoclasts resorb an area of mineralized bone, and local macrophages complete the cleanup of dissolved elements. The process then reverses to formation as osteoblast precursors are recruited to the site and differentiate into active osteoblasts. These lay down new organic matrix and mineralize it. Thus new bone replaces the previously resorbed mature bone. (Modified from Raisz LG: *N Engl J Med* 318:820, 1988.)

to which the total bone mass is defended by the compensatory process of coupling.

Vitamin D

Vitamin D is one of the two major regulators of calcium metabolism. Through a hormonally active metabolite vitamin D increases calcium absorption from the gut and also calcium resorption from bone. Both actions raise or sustain plasma calcium. Vitamin D has similar effects on phosphate.

Table 47-1 Major effects of various hormones on bone

Bone formation	*Bone resorption*
Stimulated by	**Stimulated by**
Growth hormone (somatomedins)	Parathyroid hormone
Insulin	Vitamin D
Estrogen	Cortisol
Androgen	Thyroid hormone
Vitamin D	Prostaglandins
Transforming growth factor β	Interleukin-1
Skeletal growth factor	Tumor necrosis factor
Bone derived growth factor	
Platelet derived growth factor	**Inhibited by**
Calcitonin	Estrogen
	Androgen
Inhibited by	Calcitonin
Parathyroid hormone	Transforming growth factor β
Cortisol	γ-Interferon

Vitamin D Production

There are two sources of vitamin D in humans: that produced in the skin by ultraviolet irradiation (D_3) and that ingested in the diet (D_3 and D_2). In this sense vitamin D is not a classic hormone of endocrine gland origin. But its pathway of molecular modification to yield active metabolites and its mechanism of action similar to that of secreted steroid hormones justify classifying it as a hormone.

The structures of vitamin D_3, its precursors, and its metabolites are shown in Fig. 47-6. Vitamin D_2, which differs only in having an additional double bond at the 21 to 22 position, is derived from the plant sterol, ergosterol, by ultraviolet radiation and is a major dietary source in many countries. It has identical biological actions to vitamin D_3, and henceforth the term vitamin D is used to indicate both forms.

Vitamin D_3 synthesis occurs primarily in **keratinocytes,** cells in the epidermis of the skin. Under the influence of summer sunlight (ultraviolet radiation [UV] at 290 to 315 nm) 7-dehydrocholesterol is photoconverted to previtamin D_3 (Fig. 47-6). A minimum of 20 mJ radiation energy/cm^2 skin is required. The amount of previtamin D_3 formed is quantitatively related in exponential fashion to the UV input. Previtamin D_3 is then spontaneously converted over 3 days to vitamin D_3 in a reaction driven by thermal energy from sunshine. Continuous exposure to sunlight also causes photodegradation of previtamin D_3 to inactive products in reactions catalyzed by UV radiation at 315 to 330 nm. Thus

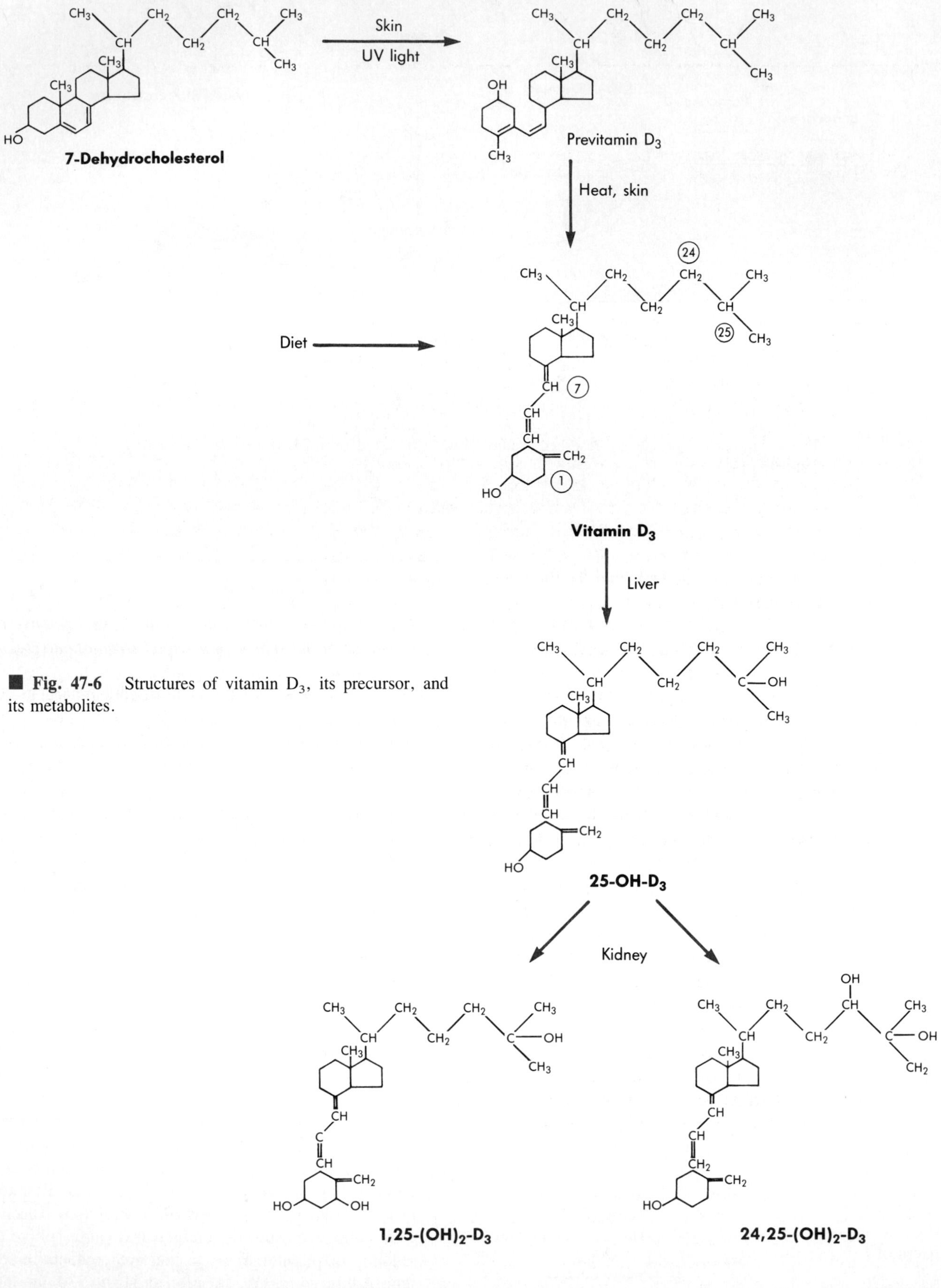

7-Dehydrocholesterol

Skin
UV light

Previtamin D_3

Heat, skin

Diet

Vitamin D_3

Liver

25-OH-D_3

Kidney

1,25-$(OH)_2$-D_3

24,25-$(OH)_2$-D_3

Fig. 47-6 Structures of vitamin D_3, its precursor, and its metabolites.

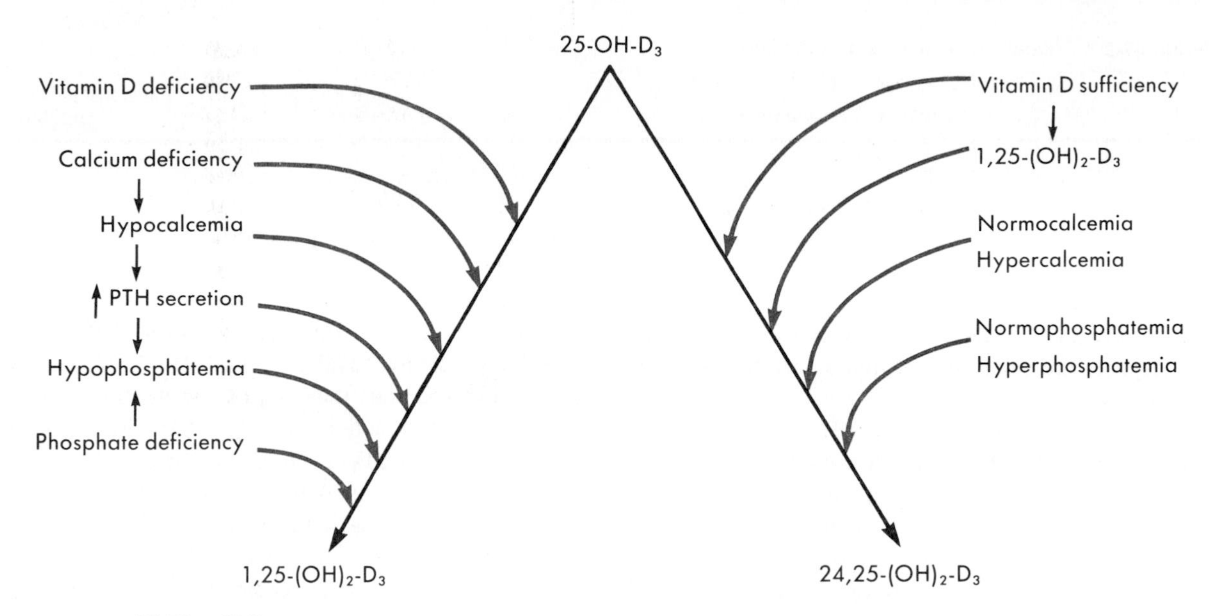

Fig. 47-7 Factors that regulate conversion of 25-OH-D_3 to either 1,25-$(OH)_2$-D_3 or 24,25-$(OH)_2$-D_3. The former is increased by either calcium or phosphate lack.

sunlight both stimulates vitamin D_3 production and restrains overproduction.

In winter or in sunlight-poor climates dietary vitamin D may be essential for health. The most important sources are fish, liver, and irradiated milk. The minimum daily dietary requirement is approximately 2.5 μg (100 units), and the recommended daily intake is 10 μg. Because of its fat solubility absorption of vitamin D from the gut is mediated by bile salts and occurs via the lymphatic glands. The normal body pool of Vitamin D is about 1000 μg. Excesses are efficiently stored in adipose tissue and liver and can take several months to be dissipated.

Vitamin D has very little intrinsic biological activity. It must undergo successive hydroxylations in order to act as a hormone (Fig. 47-6). In the liver it is hydroxylated by a microsomal and mitochondrial enzyme to 25-OH-D_3 in a reaction dependent on NADPH and O_2. Although 25-OH-D_3 is two to five times more potent than vitamin D_3 in vivo and can be shown to have intrinsic activity in vitro, its physiological function in calcium regulation also appears to be largely that of a precursor. From the liver 25-OH-D_3 is transported to the kidney (and possibly to other sites), where it undergoes alternative fates (Fig. 47-6).

Hydroxylation in the 1 position to 1,25-$(OH)_2$-D_3 occurs in the mitochondria of the proximal convoluted and straight tubules. The 1-α-hydroxylase is a mixed function P_{450} steroid hydroxylase that requires NADPH, O_2, and a flavoprotein known as renoredoxin or ferrodoxin. 1,25-$(OH)_2$-D_3 is at least 10 times as potent as vitamin D in vivo. It is unquestionably the metabolite that expresses most, if not all, the activity of the vitamin. Alternatively 25-OH-D_3 may be hydroxylated in the 24 position to 24,25-$(OH)_2$-D_3 in a mitochondrial reaction that again requires NADPH and O_2. The biological activity of 24,25-$(OH)_2$-D_3 remains controversial. It is only one twentieth as potent as 1,25-$(OH)_2$-D_3 in most assay systems and may only represent a means of inactivating vitamin D_3. However some evidence suggests that 24,25-$(OH)_2$-D_3 could have a separate role in expressing certain vitamin D actions.

Regulation of overall 1,25-dihydroxy-D_3 production largely occurs in the kidney. There is no known feedback by calcium or related hormones on vitamin D_3 synthesis. 25-OH-D_3 production in the liver is decreased by 1,25-$(OH)_2$-D_3, and the latter also increases the clearance of 25-OH-D_3 from plasma. But the major locus of regulation is at the alternate 1- or 24-hydroxylase step (Fig. 47-7). 25-OH-D_3 is preferentially directed toward the active metabolite 1,25-$(OH)_2$-D_3 whenever there is a lack of vitamin D, a lack of calcium, or a lack of phosphate. In vitamin D–deficient states it may be partly the lack of 1,25-$(OH)_2$-D_3 itself that enhances 1-hydroxylation, because 1,25-$(OH)_2$-D_3 is a suppressor of 1-hydroxylase activity. When vitamin D is sufficient the 1-hydroxylase enzyme also is still subject to regulation by calcium and phosphate. Calcium deprivation leads to hypocalcemia, which in turn stimulates parathyroid hormone (PTH) hypersecretion. The lowered plasma calcium and the elevated PTH concentration both independently stimulate 1-hydroxylase activity. In addition the PTH excess leads to a phosphate diuresis (discussed later); the resultant hypophosphatemia and lowered renal cortical phosphate content also stimulate 1-hydroxylase activity. Finally phosphate deprivation itself leads to hypophosphatemia and directly to an increase in 1-hydroxylase activity. Thus the supply of active 1,25-$(OH)_2$-D_3 is augmented whenever the mobilization of calcium or phosphate from intestine and bone into the extracellular fluid is needed. Conversely augmentation of the supply of rela-

Table 47-2 Vitamin D metabolism in humans

	Plasma concentration (μg/L)	*Plasma half-life (days)*	*Estimated production rate (μg/day)*
1,25-$(OH)_2$-D_3	0.03	0.25	1
24,25-$(OH)_2$-D_3	2	15 to 40	1
25-OH-D_3	30	15	10

tively inactive 24,25-$(OH)_2$-D_3 is favored when calcium and phosphate are plentiful and bone accretion can be sustained.

Vitamin D_3, 25-OH-D_3, 1,25-$(OH)_2$-D_3, and 24,25-$(OH)_2$-D_3 all circulate bound to an α-globulin. The concentrations, approximate half-lives, and estimated daily production rates for the key metabolites in humans are shown in Table 47-2. It is evident that 1,25-$(OH)_2$-D_3 has by far the lowest concentration and the shortest half-life of the three, befitting its potency. Certain relationships among the plasma concentrations are also of physiological significance. 1,25-$(OH)_2$-D_3 concentration is ordinarily independent of 25-OH-D_3 concentration. Except in severe vitamin D–deficiency or toxicity states the regulatory factors previously outlined maintain the appropriate concentration of active metabolite irrespective of the supply of precursor. In contrast 24,25-$(OH)_2$-D_3 concentration is ordinarily directly proportional to 25-OH-D_3 concentration. This suggests that 24-hydroxylation is the key sluice for disposing of excess precursor. 1,25-$(OH)_2$-D_3 can be further hydroxylated to 1,24,25-$(OH)_3$-D_3, a compound of little activity. Thus if an excess of 1,25-$(OH)_2$-D_3 should develop, it can still be inactivated by 24-hydroxylation.

Some 30 other hydroxylated vitamin D_3 metabolites without apparent function have been described. Both active and inactive vitamin D_3 metabolites undergo biliary excretion and enterohepatic recycling. A loss of recycling caused by intestinal disease may contribute to the development of vitamin D_3 deficiency.

Vitamin D Actions

1,25-$(OH)_2$-D_3 acts through the general nuclear mechanism outlined for steroid hormones. After combination with its receptor the hormone receptor complex stimulates or inhibits transcription of target DNA molecules to messenger RNA molecules. These actions require up to 24 hours to become manifest. One major product is a series of calcium binding proteins, called **calbindins,** of various molecular weights. In intestinal cells calbindin 28-KD_a, with 250 amino acids, binds four calcium ions per mole and calbindin-9-KD_a, with 80 amino acids, binds two calcium ions per mole. These calbindins have significant homology with calmodulin and myosin-light chain. In addition via its receptor (but not through a transcriptional mechanism) 1,25-$(OH)_2$-D_3 rapidly stimulates an increase in cGMP and intracellular calcium levels. 1,25-$(OH)_2$-D_3 also upregulates its own receptor, thus amplifying the hormones effects.

The major action of 1,25-$(OH)_2$-D_3 is to stimulate absorption of calcium from the intestinal lumen against a concentration gradient. The hormone localizes to the nuclei of intestinal villus and crypt cells but not to goblet or submucosa cells. It initially acts on the brush order, possibly to change the lipid concentration of the membrane. Calcium entry from the intestinal lumen is rapidly stimulated. Hours later the calbindin concentration rises and these molecules may act to ferry calcium across the intestinal cell or to buffer the high calcium concentrations resulting from initial entry of the ion. Calbindin's distribution in the intestinal tract and at the tip of the villi correlates with a role in vitamin D_3–dependent calcium transport. Finally calcium is extruded from the opposite side of the cell and enters the bathing capillary blood. This action of 1,25-$(OH)_2$-D_3 is responsible for the adaptation, described previously, whereby intestinal calcium absorption adjusts to alteration in dietary calcium. By independent mechanisms active absorption of phosphate and magnesium across the intestinal cell membrane is also stimulated by 1,25-$(OH)_2$-D_3.

Another major target organ of 1,25-$(OH)_2$-D_3 is bone, where complex effects are seen. Osteoblasts, but not osteoclasts, have 1,25-$(OH)_2$-D_3 receptors. Nonetheless 1,25-$(OH)_2$-D_3 in high concentrations directly stimulates bone resorption. This probably occurs by stimulating a paracrine signal originating in the osteoblast cell. Thereby 1,25-$(OH)_2$-D_3 increases the recruitment, differentiation, and fusion of precursors to active osteoclasts, which carry out their resorptive missions. Exceedingly low concentrations of 1,25-$(OH)_2$-D_3 also can stimulate osteocytic osteolysis and mobilization of bone calcium independently of osteoclast action.

On the other hand the normal mineralization of newly formed osteoid along a calcification front is also critically dependent on vitamin D. In its absence, excess osteoid accumulates from osteoblastic activity, and the bone so formed is weakened. It remains enigmatic whether this action of vitamin D is entirely accounted for by augmentation of the supply of calcium and phosphate in the fluid bathing the osteoblast, or whether 1,25-$(OH)_2$-D_3 acts in a direct manner on the bone matrix or cells to hasten mineralization. A direct effect on bone formation is supported by the fact that 1,25-

$(OH)_2$-D_3 induces the synthesis of osteocalcin and of fibronectin; the hormone also suppresses the synthesis of type I collagen by osteoblasts. In any case the first observable effect of vitamin D replacement on the bone from vitamin D–deficient animals and humans is the reappearance of a normal mineralized front. Vitamin D, possibly through its 24,25-$(OH)_2$-D_3 metabolite, may also stimulate cartilage development at some stage.

An important feedback action of 1,25-$(OH)_2$-D is to repress the gene directing the synthesis of PH. This action is facilitated by 1,25-$(OH)_2$-D_3 induction of its own receptor as well as a calbindin in parathyroid cells.

1,25-$(OH)_2$-D_3 has a weak stimulatory effect on renal calcium reabsorption. It also stimulates calcium transport into skeletal and cardiac muscle; muscle weakness and cardiac dysfunction can result from vitamin D deficiency in animals. Other diverse actions of uncertain physiological significance include enhancement of the secretion of insulin and of the pituitary hormone, prolactin.

An interesting new role for vitamin D in immunomodulation has emerged. Macrophages, monocytes, and transformed lymphocytes possess the ability to synthesize 1,25-$(OH)_2$-D_3 from 25-OH-D. Activated T-lymphocytes also express the 1,25-$(OH)_2$-D_3 receptor. In turn the hormone can decrease production of interleukin-2, γ-interferon, and other lymphokines. Ultimately the proliferation of T- and B-lymphocytes and even immunoglobulin synthesis by the latter are decreased by 1,25-$(OH)_2$-D_3. Thus this molecule may participate in autocrine and/or paracrine actions important to regulating immune responses. Macrophages and osteoclasts have common ancestors. Therefore, 1,25-$(OH)_2$-D_3 could mediate an autocatalytic system whereby tissue macrophages are recruited when they are needed for tissue response to injury, or osteoclasts are recruited when they are needed for bone resorption. Finally keratinocytes synthesize 1,25-$(OH)_2$-D_3 and also respond to the hormone. Hence, the steroid may play a possible autocrine or paracrine role in skin turnover.

Parathyroid Hormone

The parathyroid glands secrete **parathyroid hormone (PTH),** the major regulator of calcium metabolism. Its paramount effect is to increase plasma calcium levels by stimulation of bone resorption, renal tubular calcium reabsorption, and 1,25-$(OH)_2$-D_3 synthesis. PTH decreases plasma phosphate concentration at the same time by inhibition of renal tubular phosphate reabsorption.

There are four parathyroid glands that develop at 5 to 14 weeks of gestation from the third and fourth branchial pouches. They descend to lie posterior to the thyroid gland. The lower pair actually is derived from the third branchial pouch and traverses a longer embryological descent. Ectopic locations in the neck and mediastinum occur. The total weight of adult parathyroid tissue is about 130 mg, and that of any one gland is 30 to 50 mg. Their blood supply is from the thyroid arteries. Samples drawn from the left and right thyroid veins can localize the site of unilateral parathyroid gland hyperfunction.

The histological appearance of the parathyroid glands changes with age. The predominant cell, known as the **chief cell,** is present throughout life and is the normal source of PTH. Resting chief cells have abundant glycogen, an involuted Golgi apparatus, and a few clusters of secretory granules. Active cells have little glycogen, a large convoluted Golgi apparatus with vacuoles and vesicles, and a granular endoplasmic reticulum. During hormone secretion numerous granules may be seen undergoing exocytosis. A second cell type—the **oxyphil cell**—is distinguished by an eosinophilic cytoplasm and first appears at puberty. Its normal function is not known, but its ultrastructural appearance suggests energy production and storage rather than hormone synthesis. Under some circumstances it does appear to be capable of hormone release. Either of these cell types may give rise to hypersecretion of PTH.

Synthesis and Release of PTH

PTH is a single-chain protein of 9,000 molecular weight, containing 84 amino acids (Fig. 47-8). The biological activity resides in the N-terminal portion of the molecule within amino acids 1 to 27. The first two amino acids are essential; alteration, elimination, or backward extension of the N terminus markedly reduces hormonal activity. The larger C terminus portion of the molecule may also subserve other functions that are unrelated to calcium metabolism.

The synthesis of PTH begins with prepro-PTH (Fig. 47-8), a 115-amino-acid protein that contains all the information encoded in the gene for PTH. As the peptide chain of prepro-PTH grows to its complete length on the ribosomes, first 2 and then 23 amino acids of the N-terminal signal sequence are enzymatically removed, leaving pro-PTH. This 90-amino-acid molecule is then transported to the Golgi apparatus, within which another 6 amino acids are removed by a tryptic enzyme. Processing of pro-PTH to PTH is normally very efficient, as only 7% of the glandular content of PTH immunoreactivity is accounted for by the precursor. Some of the resulting PTH is packaged for storage in mature secretory granules and release by exocytosis. In addition some newly synthesized PTH may be transported, still in Golgi vesicles, directly through the cell for immediate release. Cleavage of PTH between amino acids 33 and 40 (Fig. 47-8) also occurs within the gland, and

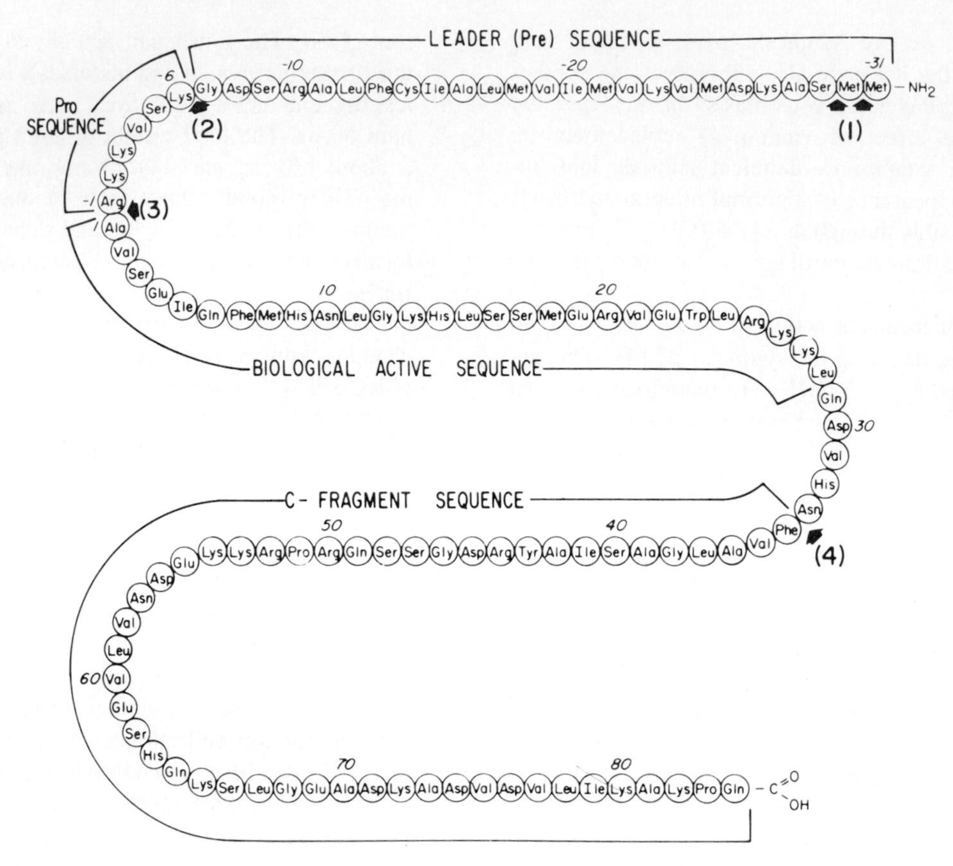

■ **Fig. 47-8** Structure and processing of preproparathyroid hormone. The N-terminal leader (signal) sequence is cleaved in two steps *(steps 1* and *2)* leaving proparathyroid hormone. Removal of 6 more amino acids *(step 3)* yields the 1-84 parathyroid hormone structure. Further cleavage within the gland as well as in peripheral tissues *(step 4)* yields as the major metabolic product of parathyroid hormone, a C-terminal fragment that is biologically inactive. (Redrawn from Habener J et al: *Physiol Rev* 64:985, 1984.)

therefore not all of the synthesized molecules reach the circulation.

The dominant regulator of parathyroid gland activity is the plasma calcium level. The hormone and the ion form a negative feedback pair. Secretion of PTH is inversely related to the plasma calcium concentration in a sigmoidal fashion (Fig. 47-9). Maximum secretory rates are achieved below a total calcium concentration of 7 mg/dl (ionized calcium, 3.5 mg/dl). As total calcium concentration increases to 11 mg/dl (ionized calcium, 5.5 mg/dl) PTH secretion is progressively diminished to a persistent low basal rate that is not suppressible by further elevation of plasma calcium. It is actually the ionized fraction of plasma calcium that regulates PTH secretion, which responds to alterations within seconds even if total calcium concentration is kept constant. The set point for half-maximum secretion is an ionized calcium level of about 4.5 mg/dl. Because of the steep dose-response slope a 2- to 3-mg/dl decrease in ionized calcium stimulates a 400% increase in PTH secretion. The greater the *rate of fall* in ionized calcium the larger is the PTH response. This relation indicates that the parathyroid cells possess an "anticipatory" capability.

Suppression of PTH secretion by calcium influx into the parathyroid chief cells represents a notable exception to the usual rule that calcium influx into the cytoplasm of endocrine cells stimulates hormonal secretion. The mechanism underlying this paradoxical effect has been extensively studied. Current evidence favors the existence of a plasma membrane protein receptor for ionized calcium that responds to changes in extracellular concentration of the ion. A low ionized calcium signal is transduced via a stimulatory G-protein to adenylyl cyclase. The consequent increase in cAMP stimulates exocytosis of PTH-containing secretory granules. During this process intracellular calcium concentration may rise transiently, but it is not certain whether this plays a role in stimulating exocytosis. Of interest, trivalent ions such as lanthanum can act similarly to an increase in extracellular calcium concentration; they decrease cAMP levels and inhibit PTH secretion, even though they are unable to enter the cell. This supports a mecha-

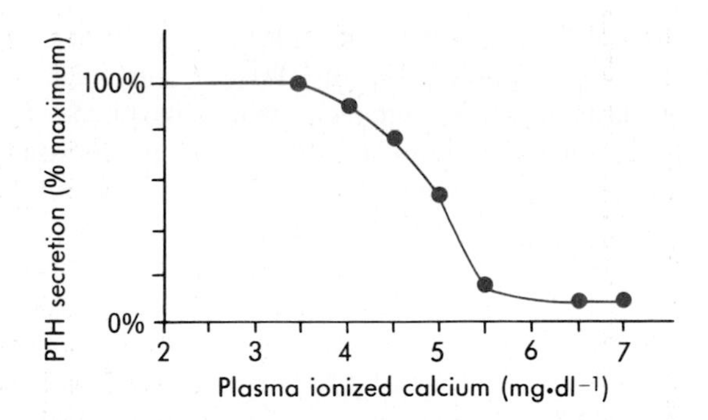

Fig. 47-9 The inverse relationship between parathyroid hormone secretion and plasma ionized calcium concentration in humans. (Redrawn from Brent GA et al: *J Clin Endocrinol Metab* 67:944, 1988. From The Endocrine Society.)

nism that regulates PTH release via a plasma membrane receptor. However the intracellular calcium level can also transiently increase on exposure to these ions. Finally parathyroid cells whose plasma membranes have been made permeable can be directly stimulated to secrete PTH by exposure to guanosine triphosphate (GTP).

In addition to rapidly regulating secretion calcium also modulates PTH synthesis and degradation within the glands. Exposure to a high calcium concentration for hours to days depresses the rate of PTH synthesis by decreasing gene transcription. Processing of preproPTH is not affected. Exposure to a high calcium concentration for days to weeks decreases proliferation of parathyroid cells. Thus elevated calcium levels decrease PTH synthesis, stores, and release, as well as, ultimately, parathyroid cell mass. Conversely hypocalcemia increases PTH synthesis, stores, and secretory rates and ultimately stimulates growth of the glands.

The divalent cation Mg^{++} modulates PTH secretion in a manner analogous to that of Ca^{++}, although Mg^{++} is less effective on a molar basis. Therefore Mg^{++} is of much less importance in its normal physiological range (1.5 to 2.5 mEq/L). However chronic hypomagnesemia strongly inhibits PTH synthesis, and in severely magnesium-depleted individuals there is a reduced rate of PTH release. In addition hypomagnesemia impairs the response of target tissues to PTH.

Despite the close physiological relation of phosphate to calcium, no direct effects of phosphate on the parathyroid glands have been demonstrated. However a rise in plasma phosphate does cause an immediate fall in ionized calcium levels, which in turn stimulates PTH secretion. As is elaborated later this resultant increase in PTH helps moderate hyperphosphatemia by increasing renal phosphate excretion. 1,25-$(OH)_2$-D_3 inhibits transcription of the PTH gene and decreases PTH secretion. In addition the sterol inhibits proliferation of parathyroid cells. This constitutes another negative feedback loop that regulates calcium metabolism.

Phosphodiesterase inhibitors (by increasing cAMP), epinephrine (via its β-adrenergic receptor), dopamine, and histamine (via H_2 receptors) all stimulate PTH secretion. α-Adrenergic agonists and prostaglandins inhibit PTH secretion by decreasing cAMP levels. Co-secreted chromogranin and related products feed back negatively on PTH secretion; this relation may indicate autocrine regulation. An excess of aluminum, commonly present in antacids, is also inhibitory. PTH secretion is inherently pulsatile. In addition there is a circadian rhythm with a nocturnal peak; this rhythm is independent of plasma calcium levels. The physiological significance of many of these observations remains to be determined.

Under normal circumstances neither prepro-PTH nor pro-PTH is secreted from the glands. Furthermore neither of these molecules appears to have intrinsic biological activity. One or more products of the intraglandular degradation of PTH, however, are released. In addition PTH undergoes rapid metabolism in the peripheral tissues. The hormone is predominantly split in the liver, producing as the major species a circulating 6,000 molecular weight carboxy-terminus fragment that is probably acted upon further in the kidney. Amino terminal fragments generated during cleavage of PTH also circulate but are not biologically active. The plasma concentration of PTH, measured by radioimmunoassays specific for the intact molecule, is 10 to 50 pg/ml (approximately 3×10^{-12} M).

There is no evidence for protein binding of circulating PTH or its metabolites. The plasma half-life of the intact hormone is 20 to 30 minutes. The carboxy terminal fragment has a much longer plasma half-life (6 to 12 hours). It can also be measured by immunoassay, and in the absence of renal failure it is an index of chronic hypersecretion.

PTH Actions

PTH action is initiated by binding to a plasma cell membrane receptor. This is a glycoprotein composed of multiple subunits of 85,000 molecular weight. The 14 to 34 region of PTH contains the binding sequence. In all target cells activation of adenylyl cyclase raises intracellular cAMP levels, and most of the effects of PTH can be mimicked by cAMP. The subsequent intracellular events mediated by cAMP are not known in detail. Presumably cAMP triggers a protein kinase cascade that leads to phosphorylation of proteins necessary for enhanced transport of calcium and other ions. Phosphatidylinositol products may also have second messenger roles.

Independent of cAMP, PTH also stimulates the uptake of calcium by the cytosol from the bathing bone fluid. Whether this calcium is itself another intracellular second messenger or whether it only acts to modulate the adenylyl cyclase response by counterregulatory inhibition remains moot. The initial uptake of calcium is reflected in a slight transient hypocalcemia, which follows PTH administration and precedes the classic hypercalcemic response. The presence of 1,25-$(OH)_2$-D_3 is required for the exhibition of the full spectrum of PTH actions. It remains to be determined whether the sterol stimulates synthesis of an essential intracellular protein mediator, or whether it only augments the pool of extracellular calcium from which the rapid early uptake occurs. A sufficient intracellular concentration of magnesium is also necessary for maximum PTH responsiveness.

The overall effect of PTH is to increase plasma calcium and decrease plasma phosphate by acting on three major target organs: bone, kidney, and gastrointestinal tract. All three actions ultimately increase calcium influx into the plasma, raising the concentration. In contrast, the actions on bone and gut, which increase phosphate influx, are overwhelmed by the action on the kidney, which increases phosphate efflux, so that plasma phosphate concentration falls (Fig. 47-10).

Bone. PTH accelerates removal of calcium from bone by two processes. Its initial effect is to stimulate osteolysis. This process causes transfer of calcium from the bone canalicular fluid into the osteocyte and thence out the opposite side of the cell into the extracellular fluid. Replenishment of calcium in the canalicular fluid probably then occurs from the surface of partially mineralized bone. Phosphate does not appear to be mobilized with calcium in this process. A second, more slowly developing effect of PTH is to stimulate the osteoclasts to resorb completely mineralized bone. In this process both calcium and phosphate are released for transfer into the extracellular fluid and the organic bone matrix is hydrolyzed by increased activity of collagenase and of lysosomal enzymes. PTH initially increases the ruffled border of the osteoclast and the clear resorptive zone that develops between it and the mineralized bone. This is followed by PTH stimulation of osteoclast size, RNA synthesis, and number of nuclei. Ultimately proliferation of osteoclasts is increased by PTH. The giant osteoclasts create large resorption cavities in both cortical and trabecular bone. PTH also induces increases in acid phosphatase and carbonic anhydrase and accumulation of lactic acid and citric acid. The resultant lowering of ambient pH contributes to the resorptive process. As a consequence of bone destruction hydroxyproline and other products are released into the plasma and then excreted into the urine.

As in the case of 1,25-$(OH)_2$-D_3 there are no PTH receptors on osteoclasts, and the resorptive effects of PTH can only be demonstrated in vitro if osteoblasts are also present as intermediary cells. Osteoblasts, and particularly a larger variety termed **proosteoblasts,** do have PTH receptors and respond to the hormone in a number of ways. This includes early alteration in their shape and cytoskeletal arrangement, probably by cAMP-induced phosphorylation of myosin light chains.

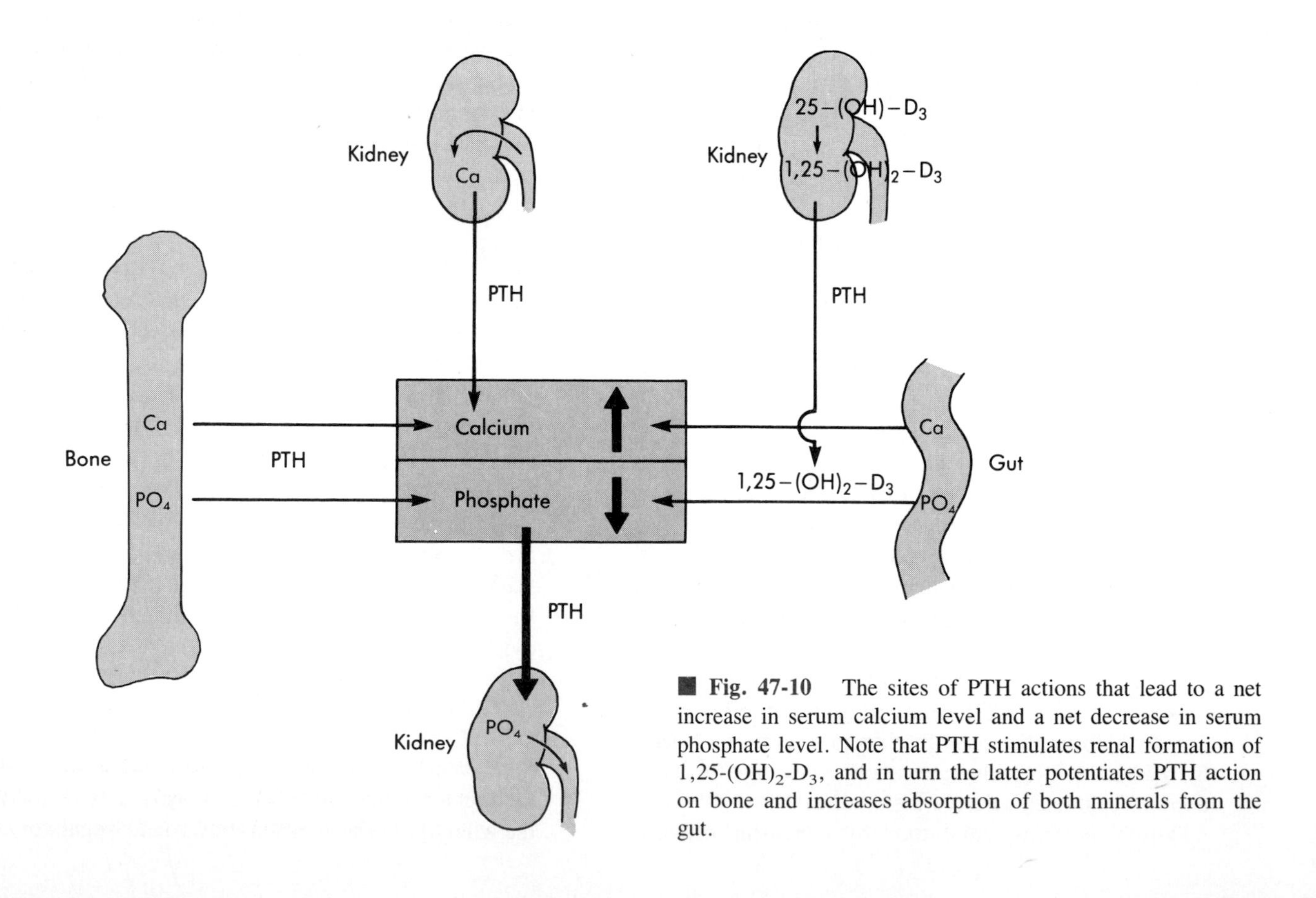

■ **Fig. 47-10** The sites of PTH actions that lead to a net increase in serum calcium level and a net decrease in serum phosphate level. Note that PTH stimulates renal formation of 1,25-$(OH)_2$-D_3, and in turn the latter potentiates PTH action on bone and increases absorption of both minerals from the gut.

PTH also inhibits the synthesis of collagen by osteoblasts, probably at the level of transcription. The proosteoblast has long syncytial processes that extend through the bone matrix and intertwine with other cells and vascular structures. Thus they could mediate the resorptive effects of PTH by stimulating secretion of osteoblast products with paracrine effects on neighboring osteoclasts and their precursors. The resorptive effects of PTH on bone are achieved by the elevated concentrations of hormone that result from stimulation of the parathyroid glands by hypocalcemia; thus they serve the purpose of restoring the plasma calcium level to normal.

However PTH also has an anabolic action on bone. In cultures this results in an increase in the number of osteoblasts and in collagen synthesis. When given in low doses and intermittently PTH stimulates bone formation in humans. In bone biopsy specimens there may be seen an increase in woven (as opposed to lamellar) bone and a proliferation of fibroblasts adjacent to the areas of increased resorptive activity. The plasma level of alkaline phosphatase, an osteoblastic enzyme whose activity parallels bone formation, is often increased by PTH. These anabolic actions may be mediated by IGF-I, the messenger ribonucleic acid (mRNA) levels of which are increased in osteoblasts by PTH. The net effect of sustained increases in PTH may be either a decrease or an increase in total skeletal mass. This undoubtedly depends on concomitant factors that affect bone remodeling, such as the availability of calcium, phosphate, and vitamin D.

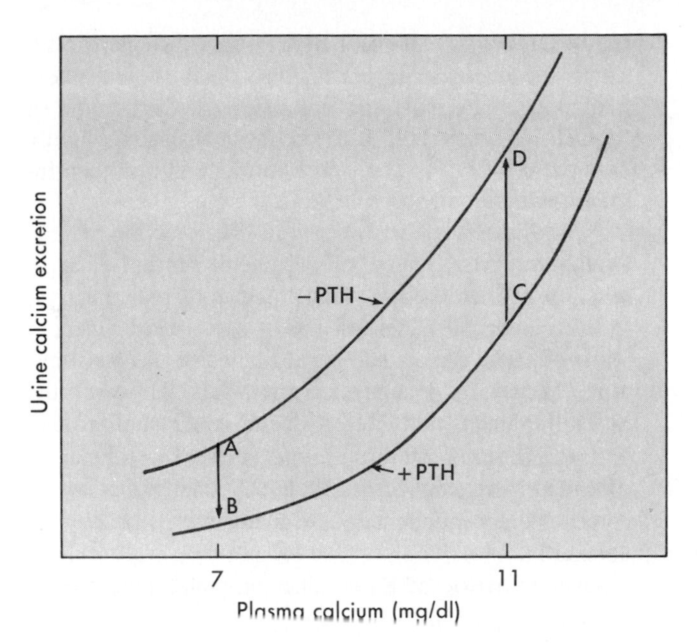

■ **Fig. 47-11** Effect of PTH on the relationship between urine calcium excretion and plasma calcium level. At a low plasma calcium level, PTH secretion is stimulated, and the hormone shifts urine calcium excretion from point *A* to point *B,* thus conserving calcium. At a high plasma calcium level, PTH secretion is suppressed, and urine calcium excretion is shifted from point *C* to point *D,* thus disposing of excess calcium. (Redrawn from Nordin BEC et al: *Lancet* 2:1280, 1969).

Kidney. PTH increases the reabsorption of calcium from the distal tubule of the kidney. By this mechanism the PTH that is secreted in response to hypocalcemia helps raise a depressed plasma calcium concentration. This effect is mediated by PTH stimulation of cAMP production at the capillary surface of the renal tubular cell. The cAMP is transported to the luminal surface where it activates unidentified protein kinases that are located in the brush border and are involved in calcium reabsorption. The relationship between urinary calcium excretion and plasma calcium concentration is shifted to the right by PTH (Fig. 47-11). Therefore acute stimulation of PTH secretion by calcium deficiency helps prevent hypocalcemia by causing a greater fraction of the filtered calcium to be reabsorbed. However by virtue of its bone effects prolonged excess of PTH elevates plasma calcium levels and, with it, the filtered load of calcium. This has a much greater influence on calcium excretion than does the effect of PTH on renal tubular reabsorption. Therefore the absolute amount of calcium excreted in the urine is elevated in PTH excess states, despite an increase in fractional reabsorption in the tubules.

The most dramatic effect of PTH on the kidney is to inhibit the reabsorption of phosphate in the proximal tubule. Within minutes of its administration PTH increases urinary phosphate excretion. As befits its second messenger role in this action cAMP excretion into the urine also increases just before that of phosphate. The phosphaturic effect of PTH allows disposition of the extra phosphate released by PTH-stimulated bone resorption. Otherwise simultaneous plasma elevation of calcium and phosphate would occur, with the potential danger of precipitating calcium-phosphate complexes in critical tissues. In contrast under circumstances of primary phosphate deprivation plasma calcium tends to rise, thereby suppressing PTH secretion. This in turn increases tubular phosphate reabsorption, conserving this essential mineral.

PTH also inhibits the reabsorption of sodium and bicarbonate in the proximal tubule in a manner parallel to that of phosphate reabsorption inhibition. This action may prevent the occurrence of metabolic alkalosis, which could result from the release of bicarbonate during the dissolution of hydroxyapatite crystals in bone. In addition PTH stimulates reabsorption of magnesium by the renal tubules; this response helps to conserve this important cation.

A most important action of PTH in the kidney is to stimulate the synthesis of 1,25-$(OH)_2$-D_3. This occurs via increased levels of cAMP; protein kinase A phosphorylates and activates a protein phosphatase, which then dephosphorylates the ferroprotein renoredoxin. In its dephosphorylated active form, renoredoxin is essential to the activity of the 1-hydroxylase enzyme. In addition the decrease in plasma and renal cortical phosphate caused by PTH also enhances 1-hydroxylation of 25-OH-D_3. Thus 1,25-$(OH)_2$-D_3 synthesis is critically dependent on PTH. The sterol hormone in turn then increases calcium absorption from the gut.

Miscellaneous. Alterations in the function of the central nervous system, of peripheral nerves, of muscles, of other endocrine glands, and of the vascular system are seen in states of PTH excess or deficiency. Most of these effects can be attributed to the concomitant changes in plasma calcium levels. However the possibility exists that PTH or one of its metabolites may have direct action on these tissues, either by stimulation of calcium transfer across their cell membranes or by some calcium-independent mechanism still to be discovered.

Overall action of PTH. The summated biochemical effects of PTH are illustrated in Fig. 47-12, which demonstrates the results of administering the hormone to a hypoparathyroid patient. There are a prompt increase in plasma calcium and a decrease in plasma phosphate. The renal tubular reabsorption of phosphate falls. Urinary excretion of calcium initially declines as its tubular reabsorption increases. However as plasma calcium continues to increase, the filtered load goes up and urinary calcium excretion also rises. Urinary excretion of hydroxyproline increases as a result of PTH-stimulated bone resorption.

Parathyroid Hormone-Related Protein

Recently a new parathyroid hormone-related peptide or protein (PTH_{rp}) has been found. It was originally discovered as a product of human cancers that were of

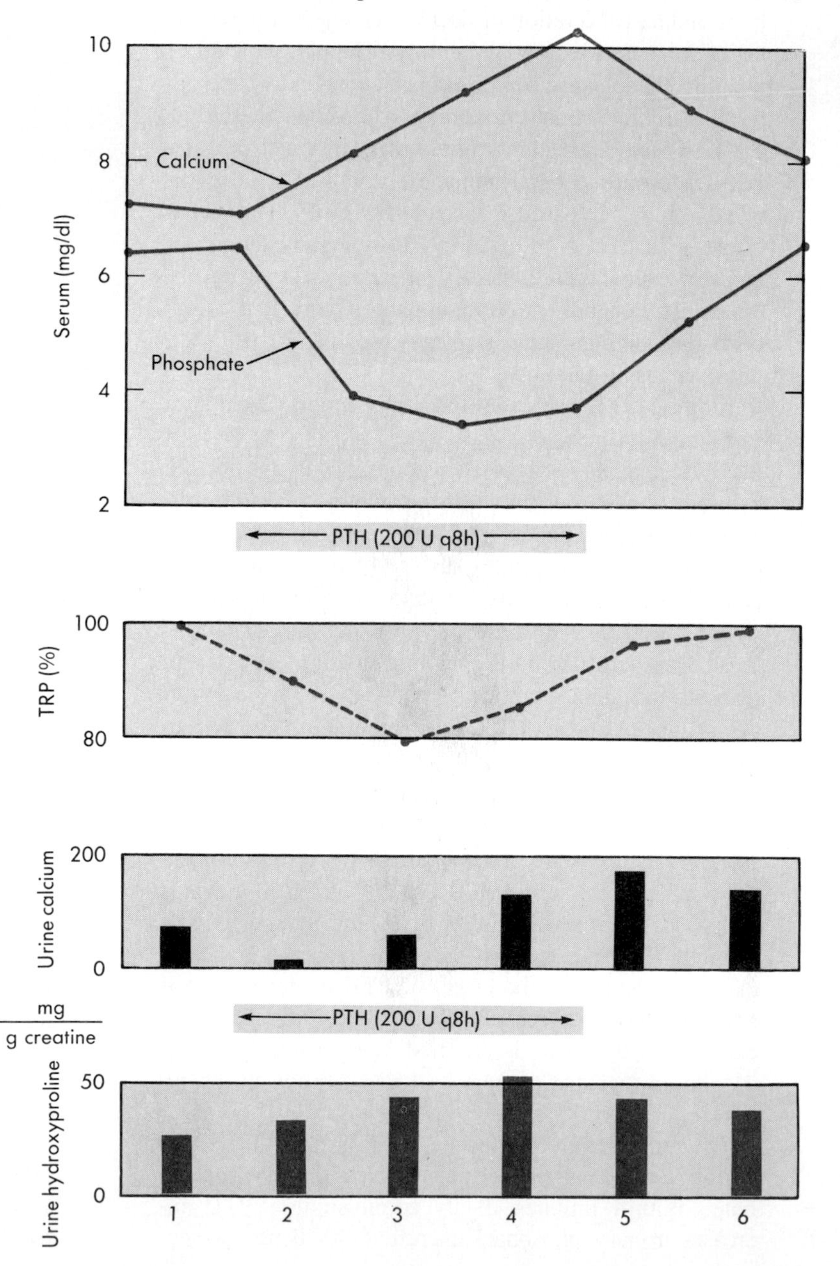

Fig. 47-12 Effect of PTH administration to a PTH-deficient human. Plasma calcium level increases and plasma phosphate level decreases. Urine calcium initially declines because of PTH action on the renal tubule; urine calcium excretion subsequently increases as plasma calcium level rises and the filtered load of calcium rises in parallel. Urine hydroxyproline increases because of enhanced bone resorption. Tubular resorption of phosphate (TRP) falls so that urine phosphate (not shown) excretion increases. Numbers at bottom refer to days.

squamous cell origin and that were associated with hypercalcemia. This molecule is now known to be expressed as well by at least some normal tissues, such as skin keratinocytes, lactating mammary epithelium, and probably fetal parathyroid glands.

The gene for PTH_{rp} and the gene for PTH are on paired chromosomes and evolve from a common ancestor. Three products of amino acid length 139 to 173 are expressed by alternate splicing of the PTH_{rp} gene primary transcript. As a result of striking homology between the N-terminal amino acids of PTH_{rp} and region 1 to 13 of PTH, PTH_{rp} exhibits all the actions of PTH on bone and kidney. It does so by binding to the PTH receptor, although the amino acids in the binding sequences of the two molecules are not homologous.

In addition a completely different segment of PTH_{rp}, between amino acids 75 to 84, uniquely stimulates placental calcium transport. Thus one function for PTH_{rp} derived from the fetus may be to maintain the 30% to 40% increased ionized calcium concentration gradient that exists between fetal and maternal plasma. In addition PTH_{rp} is present in milk; it may function in the gastrointestinal tract of the infant or conceivably escape degradation and play a systemic role in calcium homeostasis.

Calcitonin

Synthesis and Release of Calcitonin

The parafollicular, or C, cells of the thyroid gland secrete another protein hormone, **calcitonin,** which influences calcium metabolism. Whereas PTH acts to increase plasma calcium, calcitonin acts to lower it. Although uncertainty exists about the significance of its physiological role in humans, calcitonin is likely to be an important regulator of plasma calcium in lower animals that live in an aquatic environment high in calcium. The parafollicular cells are of neural crest origin. In lower animals they form a discrete ultimobranchial gland from the sixth branchial pouch. In humans they are concentrated in the central portion of the lateral lobes of the thyroid gland, forming 0.1% of the epithelial cells. They are relatively large cells with a pale cytoplasm that contains small secretory granules enclosed in membranes. Pathological secretion of calcitonin occurs with parafollicular cell tumors and others that are of neural crest origin and that secrete amine or peptide hormones.

Calcitonin is a straight-chain peptide composed of 32 amino acids. It has a molecular weight of 3,400. The hormone contains a seven-membered disulfide ring at the N terminus and prolineamide at the C terminus. Both fish and animal calcitonins are active in humans. The biologically active core of the molecule probably resides in its central region, although some evidence suggests that the entire peptide sequence is required. The synthesis of calcitonin proceeds from a preprohormone of 17,000 molecular weight. The hormone is packaged in granules along with N-terminal and C-terminal copeptides. Their physiological roles are being explored.

The gene for calcitonin illustrates the significant relationship between the endocrine and nervous systems. In some cells the primary RNA transcript is processed to a messenger RNA that encodes for preprocalcitonin and direct synthesis of calcitonin. However, in other cells (both in thyroid and nervous tissue) the same primary RNA transcript is processed to a different messenger RNA that encodes the precursor and directs the synthesis of an entirely different peptide. This molecule, known as **calcitonin gene related peptide (CGRP),** also circulates in human plasma, and it probably arises from perivascular nerves. CGRP is a potent vasodilator and cardiac inotropic agent. The evolutionary path by which a hormone and a neuropeptide arose from *alternate* expression of the same gene and the functional significance of this relationship remain to be determined.

The major stimulus to secretion of calcitonin is a rise in plasma calcium. However, the degree of response seen in various species is related to their need to prevent hypercalcemia. Vertebrates that originated in fresh water (of low calcium concentration) but migrated into the sea (with a calcium concentration of 40 mg/dl) were the first to require and to develop a calcium-lowering hormone. When vertebrates moved to land, the emphasis in calcium economy shifted toward defense against hypocalcemia rather than against hypercalcemia. PTH was developed, and the importance of calcitonin in plasma calcium regulation probably declined. Calcitonin circulates in humans at concentrations of 10 to 20 pg/ml (5×10^{-12}) and increases 2- to 10-fold after an acute increase in plasma calcium level of as little as 1 mg/dl. Much larger responses of the hormone to calcium infusion are elicited in patients with calcitonin-secreting tumors. Likewise in such patients a sharp reduction in ionized calcium lowers plasma calcitonin levels. The stimulating effect of calcium on calcitonin secretion involves an increase in intracellular cAMP levels. Ingestion of food stimulates calcitonin secretion without elevating plasma calcium. This is mediated by several gastrointestinal hormones, of which gastrin is the most potent. The physiological significance of this phenomenon is not yet clear, but gastrin responsiveness provides a useful diagnostic test for states of calcitonin hypersecretion.

Circulating calcitonin is heterogeneous; immunoreactive molecules both larger and smaller than the native hormone have been found in the plasma of patients with tumors. Whether precursors are secreted normally is not known. Calcitonin appears to be largely degraded and cleared by the kidney. Its plasma half-life is less than 1 hour.

Calcitonin Actions

The immediate target cell of calcitonin is the osteoclast. Binding of calcitonin to plasma membrane receptors is followed by an elevation of the intracellular cAMP level. This second messenger initiates calcitonin actions in all target cells. The subsequent intracellular events are obscure. Sequestration of calcium in membrane bound pools, thus lowering cytosol calcium concentration, has been advanced as a mechanism for reducing the efflux of calcium from bone cells into the extracellular fluid. The affected osteoclasts lose their ruffled borders, undergo cytoskeletal rearrangment, exhibit reduced motility, and detach from bone surfaces.

The major effect of calcitonin administration is a rapid fall in plasma calcium levels. The magnitude of this decrease is directly proportional to the baseline rate of bone turnover. Thus young growing animals are most affected, whereas in adults, who have more stable skeletons, only a minimal response is seen. Inhibition of bone resorption by osteoclasts, in the basal state and when these are stimulated by PTH, is the immediate mechanism of the hypocalcemic action. An escape phenomenon is noted however, possibly caused by down regulation of calcitonin receptors. Continued provision of calcitonin eventually decreases the number of osteoclasts as well as their activity. More dense bone with fewer resorption cavities eventually results.

Calcitonin is clearly a physiological antagonist to PTH with respect to calcium. However with respect to phosphate, it has the same net effect as PTH; i.e., it causes a decrease in plasma phosphate level. This is due to inhibition of bone resorption as well as promotion of phosphate entry into bone. There is also a small increase in urinary phosphate excretion after calcitonin administration. The hypophosphatemic effect is independent of the hypocalcemic effect.

Calcitonin arose in evolution at a time when animals were exposed continually to high concentrations of calcium in sea water, and a hormone that could prevent hypercalcemia was useful. The importance of calcitonin to normal human calcium economy is unclear. Under ordinary circumstances the absorption of dietary calcium loads produces little, if any, elevation of the plasma calcium level. Whether the increase in calcitonin provoked by eating helps prevent the development of postprandial hypercalcemia is not clear. Calcitonin deficiency resulting from complete removal of the thyroid gland does not lead to hypercalcemia, although disposition of an acute calcium load may be retarded. A chronic excess of calcitonin generated either by tumor secretion or by exogenous administration, does not produce hypocalcemia. At present it may be most reasonable to conclude that any effects of calcitonin deficiency or excess are easily offset by appropriate adjustment of PTH and vitamin D levels.

On the other hand the possibility that calcitonin participates significantly in the regulation of bone remodeling cannot be easily excluded. It could participate in fetal skeletal development. The fact that plasma calcitonin is lower in women than in men and also declines with aging may imply a functional role for the hormone in the common development of accelerated bone loss after the menopause. In lower vertebrates and mammals a role for calcitonin appears likely in diverse calcium-related processes, such as lactation, production of egg shells, and protection of the skeleton from the calcium drain of pregnancy. Calcitonin has found clinical uses in the acute treatment of hypercalcemia and in certain bone diseases where a sustained reduction in osteoclastic resorption is therapeutically beneficial. Finally, the discovery of calcitonin in a number of locations throughout the body—in the pituitary gland and hypothalamus and within cells of neural crest origin—has raised the possibility that calcitonin also may have paracrine and neurotransmitter functions. In this regard calcitonin has analgesic properties independent of the opioid system.

Integrated Hormonal Regulation of Calcium and Phosphate

From all the foregoing it should be clear that a complex interplay of several hormones acting on a number of tissues is responsible for maintenance of normal concentrations of calcium and phosphate in body fluids. This integrated system is best visualized by tracing the compensatory responses to deprivation of calcium and of phosphate (Figs. 47-13 and 47-14).

Calcium deprivation, with hypocalcemia as the signal, primarily stimulates PTH secretion (Fig. 47-13). PTH increases urinary phosphate excretion, thereby decreasing the plasma phosphate level and renal cortical phosphate content. All three factors—hypocalcemia, excess PTH, and hypophosphatemia—act to stimulate the production of 1,25-$(OH)_2$-D_3. The latter raises plasma calcium levels toward normal by increasing absorption of calcium from the gastrointestinal tract and, in concert with PTH, by increasing osteocytic and osteoclastic bone resorption. PTH further acts to return calcium to the plasma by increasing its reabsorption from the renal tubular urine. Thus this beautifully integrated response to calcium deprivation increases the flux of calcium into the extracellular fluid. Simultaneously the extra phosphate that enters with the calcium from the bone and gut is disposed of by excretion in the urine. The recovery of plasma calcium concentration to normal will shut off PTH hypersecretion by negative feedback. 1,25-$(OH)_2$-D_3 synthesis will then decline. 24,25-$(OH)_2$-D_3 synthesis will increase, and the whole sequence will diminish. As a further safety valve should the compensatory rise in plasma calcium concentration

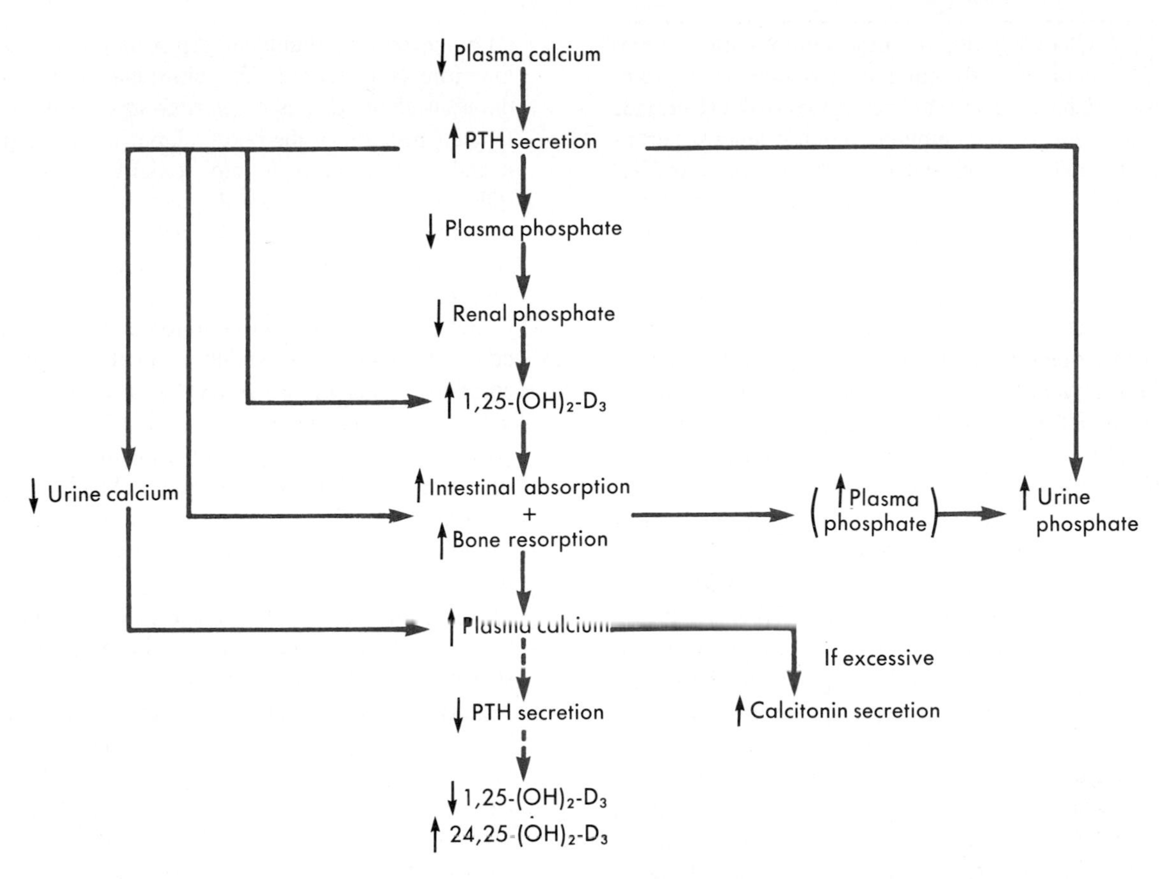

■ **Fig. 47-13** The compensatory response to calcium deprivation. See text for explication.

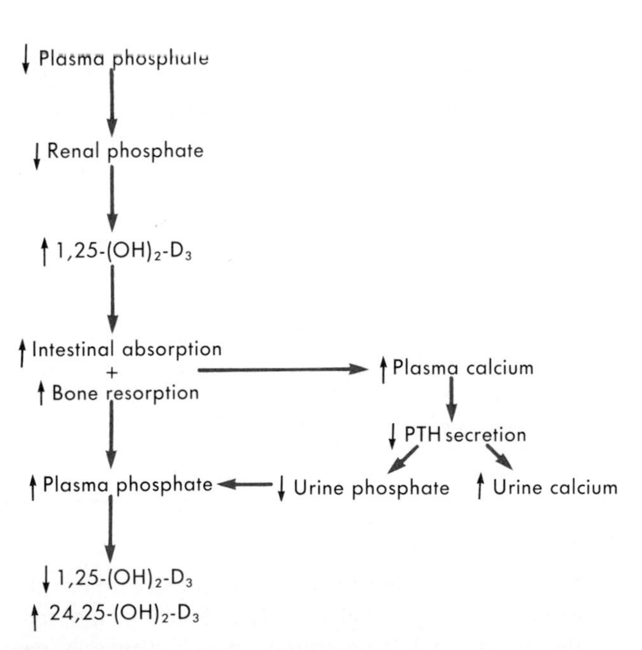

■ **Fig. 47-14** The compensatory response to phosphate deprivation. See text for explication.

exceed normal levels, stimulation of calcitonin secretion would act to moderate it as well.

In a contrasting sequence phosphate deprivation via hypophosphatemia will directly stimulate 1,25-$(OH)_2$-D_3 production (Fig. 47-14). The latter increases the flux of phosphate into the extracellular fluid by stimulating its absorption from the gut and by stimulating bone resorption. The extra calcium that simultaneously enters raises plasma calcium and the increase in plasma calcium suppresses PTH secretion. The absence of PTH causes tubular reabsorption of phosphate to increase, conserving urinary phosphate and aiding in the restoration of plasma phosphate levels to normal. At the same time the lack of PTH permits easier disposal of the extra calcium that was mobilized by diminishing its renal tubular reabsorption and increasing its excretion in the urine. As plasma phosphate returns to normal 24,25-$(OH)_2$-D_3 production is favored, 1,25-$(OH)_2$-D_3 levels decline, and the whole process is reversed.

The combined arrangement of dual hormone regulation and dual hormone action permits selective defense of either plasma calcium or plasma phosphate level, without creating a circulatory excess of the other. The same principles apply in reverse to imposition of excess calcium or excess phosphate loads of either endogenous or exogenous sources.

Certain characteristics of these homeostatic systems for adjustment of body calcium and phosphate stores deserve emphasis. The renal responses of PTH provide the most rapid (within minutes) defense against perturbations of both calcium and phosphate stores. As PTH secretion ranges from very high to very low levels the rate of urinary calcium excretion can rise 25-fold from approximately 0.05 to 1.2 mg/min, and that of phosphate can fall from 2 to 0 mg/min. A sudden 2 to 3 mg/dl increase in either calcium or phosphate concentration in the extracellular fluid can be corrected within 24 hours by the kidney, acting under the appropriate alteration in PTH levels. In the face of complete phosphate and calcium deprivation renal conservation of phosphate is complete, whereas that of calcium is not quite so. The gastrointestinal component of this homeostatic system is both slower and narrower in range. As a result of variations in 1,25-$(OH)_2$-D_3 the absorption of dietary calcium increases from 20% to 70% as calcium intake decreases from 2,000 to 200 mg/day. Thus absorbed calcium can effectively range from 140 to 400 mg/day. Hormonal effects on phosphate absorption are even less striking, since the latter is virtually a linear function of dietary intake. Bone responses to regulatory fluctuations in both PTH and 1,25-$(OH)_2$-D_3 are rapid when produced by osteocytic osteolysis and relatively slow when caused by osteoclastic resorption. However the capacity for compensatory calcium and phosphate uptake and release is enormous. In humans 10-fold variations in calcium turnover have been observed. Finally an important major difference between the renal and gastrointestinal mechanisms on the one hand and the bone mechanisms on the other hand must be borne in mind. The compensatory responses of the kidney and the gut have the virtue of defending total body *and bone* stores of calcium and phosphate against erosion or inundation. In contrast the skeletal mechanisms of defense against perturbations of plasma calcium and phosphate have the disadvantage that their long-term employment eventually sacrifices the chemical and structural integrity of the bone mass.

Clinical Syndromes of Hormone Dysfunction

Vitamin D

Toxicity results from excessive administration of vitamin D or, in a few diseases, from overproduction of 1,25-$(OH)_2$-D_3 by immune system cells. Absorption of calcium from the gut and resorption of bone are enhanced. This elevates plasma and urinary calcium levels with the same clinical consequences noted below for hyperparathyroidism. In contradistinction to the latter however, vitamin D excess raises, rather than lowers, serum phosphate because the hypercalcemia shuts off PTH secretion. Tubular reabsorption of phosphate therefore is increased, and phosphate is retained. The duration of toxicity is greatest from exogenous vitamin D itself because of the body's large storage capacity for it and its low turnover rate; toxicity is least for 1,25-$(OH)_2$-D_3, which is rapidly removed from the body. Treatment consists of blocking the effects of vitamin D with cortisol analogues or calcitonin while waiting for the excess to be cleared.

Deficiency of vitamin D action can result from inadequate sunlight, lack of dietary intake, diminished absorption, or defective hydroxylation in the liver or kidney, or defective D receptors. This leads to decreased gastrointestinal absorption of calcium and phosphate. The resultant fall in plasma calcium level is buffered by stimulation of PTH secretion. This in turn strongly accentuates the fall in plasma phosphate level by decreasing its tubular resorption in the kidney and increasing its excretion in the urine. Urinary calcium excretion on the other hand is reduced by the PTH excess. The strongly negative phosphate balance and more modestly negative calcium balance together decrease the rate of bone mineralization. An excess of osteoid accumulates on bone surfaces, and no regular line of calcification can be seen.

Two clinical pictures result from vitamin D deficiency. In children the centers of endochondral ossification at the epiphyseal plates are most critically affected, producing growth failure and the characteristic deformities and x-ray films of **rickets.** In adults softening and bending of long bones, with fracture lines along nutrient arteries, create the condition known as ***osteomalacia.*** If lack of either sunlight or dietary vitamin D is the cause, plasma 25-(OH)-D_3 levels (and in severe cases 1,25-$(OH)_2$-D_3 levels) are low. If resistance to vitamin D action is the cause, 25-(OH)-D_3 and 1,25-$(OH)_2$-D_3 levels are elevated. If kidney failure is the cause plasma 25-(OH)-D_3 is high, whereas plasma 1,25-$(OH)_2$-D_3 levels are low. In general plasma calcium is reduced, as are plasma phosphate levels (except in cases of kidney failure), and plasma PTH levels are elevated. The levels of alkaline phosphatase and osteocalcin are also high in the plasma because of overactivity of the osteoblasts. Treatment of simple deficiency consists of replacement doses of vitamin D itself, with supplemental calcium. In cases caused by biosynthetic defects (for example hepatic or renal disease) or target tissue unresponsiveness large doses of vitamin D, 25-OH-D_3, or 1,25-$(OH)_2$-D_3 may be needed.

PTH

Hyperparathyroidism results from enlargement of one or more parathyroid glands. The clinical picture reflects the effects of hypercalcemia, which depresses neuromuscular excitability. This produces dulled mentation, lethargy, anorexia, constipation, and muscle weakness. If urinary excretion of calcium is excessive, renal stones may form. With marked chronic hypersecretion of PTH massive bone resorption causes weakening of the bones, pain, fractures, and deformities. Persistent hypercalcemia also can lead to reduced renal function secondary to deposition of calcium in the kidney cells. The diagnosis is established by demonstrating elevated plasma calcium levels, decreased plasma phosphate levels, and elevated PTH levels. Treatment is surgical removal of the hyperfunctioning tissue.

Hypoparathyroidism most often is caused by autoimmune atrophy or inadvertent surgical removal of the glands. In rare cases the problem is end-organ resistance to PTH action attributable to a mutant G-protein and deficient generation of cAMP in target cells. Calcium absorption from the gut is diminished because of defective synthesis of 1,25-$(OH)_2$-D_3 and there is also decreased bone resorption. The resultant hypocalcemia increases neuromuscular excitability. This produces hyperactive reflexes, spontaneous muscle contractions, convulsions, and laryngeal spasm with airway obstruction. Low plasma calcium levels and elevated plasma phosphate levels are essential to the diagnosis. Plasma PTH levels are low when hypoparathyroidism is caused by destruction of the glands. Plasma PTH level is elevated when target tissue unresponsiveness is the problem in which case administration of exogenous PTH does not increase urinary excretion of cAMP and phosphate. The usual treatment of hypoparathyroidism is replacement with calcium supplements and 1,25-$(OH)_2$-D_3.

■ *Summary*

1. Calcium participates critically in a myriad of biological functions, including neurotransmission, hormone secretion and action, enzyme activities, muscle contraction, and blood clotting. It is also the chief mineral contributing to the structural integrity of the skeleton and teeth.

2. Extracellular Ca^{++} concentration (approximately 10^{-3} M) is controlled closely, in part to help regulate the wide transient swings in the much lower intracellular concentration (approximately 10^{-7} M).

3. Phosphate ion is critical to all major enzymatic pathways involved in energy generation, substrate disposition, and protein and other macromolecule synthesis. Phosphate is also the anion partner of calcium in bone structure.

4. Calcium balance depends on dietary intake, fractional gastrointestinal absorption, renal excretion, and internal movement of calcium into and out of skeletal reservoirs.

5. Phosphate balance reflects dietary intake, renal excretion, and internal shifts among extracellular fluid, large soft tissue contents, and the skeletal reservoir.

6. Bone is a complex organ with cells specifically devoted to a continuous process of remodeling. In this process mineralized bone is resorbed by osteoclasts (releasing calcium and phosphate) and is formed by osteoblasts (assimilating calcium and phosphate). This process is augmented during growth periods and slows with aging.

7. Vitamin D is a steroid molecule either synthesized from cholesterol in the skin by UV light or absorbed from the diet. It undergoes successive modifications in the liver and kidney to 1,25-$(OH)_2$-D_3, the active metabolite.

8. 1,25-$(OH)_2$-D acts via its osteoblast nuclear receptor to increase calcium (and phosphate) absorption from the gastrointestinal tract. The hormone is therefore critical to supplying calcium for bone formation and growth, as well as other calcium-dependent processes. It also enhances bone resorption. Overall, 1,25-$(OH)_2$-D_3 increases plasma calcium and plasma phosphate levels.

9. Parathyroid hormone (PTH) is a straight chain peptide synthesized from a prohormone in four parathyroid glands. PTH is released by exocytosis in response to a decrease in plasma calcium. Its synthesis and secretion are suppressed by calcium and 1,25-$(OH)_2$-D_3.

10. PTH acts via a plasma membrane receptor and cAMP (a) to increase osteoclastic bone resorption, (b) to increase renal tubular reabsorption of calcium, (c) to increase 1,25-$(OH)_2$-D_3 synthesis in the kidney, (d) to decrease renal tubular reabsorption and increase urinary excretion of phosphate. Overall PTH increases plasma calcium and decreases plasma phosphate.

11. Calcium deficiency evokes a synergistic sequence that increases PTH and 1,25-$(OH)_2$-D_3 secretion. Their combined actions increase the inflow of calcium and restore plasma levels to normal, and they simultaneously dispose of the inflow of extra phosphate by enhancing its renal excretion.

12. In contrast phosphate deprivation evokes a synergistic sequence that increases 1,25-$(OH)_2$-D_3 secretion but suppresses PTH secretion. The result is to restore plasma phosphate toward normal while disposing of the inflow of extra calcium by increasing its renal excretion.

13. Calcitonin is a peptide hormone synthesized in C cells within the thyroid gland. It is a PTH antagonist in bone and is secreted in response to hypercalcemia. Thus it acts to lower the plasma level of calcium.

Bibliography

Journal articles

Austin LA et al: Calcitonin: physiology and pathophysiology, *N Engl J Med* 304:269, 1981.

Bickle DD et al: 1,25-Dihydroxyvitamin D_3 production by human keratinocytes: kinetics and regulation, *J Clin Invest* 78:557, 1986.

Bordier P et al: Vitamin D metabolites and bone mineralization in man, *J Clin Endocrinol Metab* 46:284, 1978.

Brent GA, et al: Relationship between the concentration and rate of change of calcium and serum intact parathyroid hormone levels in normal humans, *J Clin Endocrinol Metab* 67:994, 1988.

Brown EM: Extracellular Ca^{2+} sensing, regulation of parathyroid cell function, and role of Ca^{2+} and other ions as extracellular (first) messengers, *Physiol Rev* 71:371, 1991.

Burger E et al: In vitro formation of osteoclasts from long-term cultures of bone marrow mononuclear phagocytes, *J Exp Med* 156:1604, 1982.

Canalis E: The hormonal and local regulation of bone formation, *Endocr Rev* 4:62, 1983.

DeLuca H, Schnoes H: Vitamin D: recent advances, *Annu Rev Biochem* 52:411, 1983.

Epstein S: Serum and urinary markers of bone remodeling: assessment of bone turnover, *Endocr Rev* 9:437, 1988.

Fitzpatrick LA: Differences in the actions of calcium versus lanthanum to influence parathyroid hormone release, *Endocrinology* 127:711, 1990.

Fukayama S et al: Human parathyroid hormone (PTH)-related protein and human PTH: comparative biological activities on human bone cells and bone resorption, *Endocrinology* 123:2841, 1988.

Gross M, Kumar R: Physiology and biochemistry of vitamin D–dependent calcium binding proteins, *Am J Physiol* 259:195, 1990.

Holick MF: Skin: site of the synthesis of vitamin D and a target tissue for the active form, 1,25-dihydroxyvitamin D_3, *Ann NY Acad Sci* 548:14, 1988.

Ishimi Y et al: Regulation by calcium and 1,25-$(OH)_2$-D_3 of cell proliferation and function of bovine parathyroid cells in culture, *J Bone Miner Res* 5:755, 1990.

Kumar R: Metabolism of 1,25-dihydroxyvitamin D_3, *Physiol Rev* 64:478, 1984.

Mahonen A et al: Effect of 1,25-$(OH)_2$-D_3 on its receptor mRNA levels and osteocalcin synthesis in human osteosarcoma cells, *Biochem Biophys Acta* 30:1048, 1990.

Mallette LE: The parathyroid polyhormones: new concepts in the spectrum of peptide hormone action, *EndocrRev* 12:110, 1991.

Mawer EB: Clinical implications of measurements of circulating vitamin D metabolites, *Clin Endocrinol Metab* 9:63, 1980.

Nemeth EF, Scarpa A: Are changes in intracellular free calcium necessary for regulating secretion in parathyroid cells? *Ann NY Acad Sci* 493:542, 1987.

Nijweide PJ et al: Cells of bone: proliferation, differentiation and hormonal regulation, *Physiol Rev* 66:885, 1986.

Nissenson RA et al: Synthetic peptides comprising the amino-terminal sequence of a parathyroid hormone-like protein from human malignancies: binding to parathyroid hormone receptors and activation of adenylate cyclase in bone cells and kidney, *J Biol Chem* 263:12866, 1988.

Norman A et al: The vitamin D endocrine system: steroid metabolism, hormone receptors, and biological response (calcium binding proteins), *Endocr Rev* 3:331, 1982.

Oetting M et al: Guanine nucleotides are potent secretagogues in permeabilized parathyroid cells, *FEBS Lett* 208:99, 1986.

Orloff JJ et al: Parathyroid hormone-like proteins: biochemical responses and receptor interactions, *Endocr Rev* 10:476, 1989.

Parfitt AM: The actions of parathyroid hormone on bone: relation to bone remodeling and turnover, calcium homeostasis, and metabolic bone disease. I. Mechanisms of calcium transfer between blood and bone and their cellular basis: morphologic and kinetic approaches to bone turnover, *Metabolism* 25:809, 1976.

Parfitt AM: The actions of parathyroid hormone on bone: relation to bone remodeling and turnover, calcium homeostasis, and metabolic bone disease. II. PTH and bone cells: bone turnover and plasma calcium regulation, *Metabolism* 25:909, 1976.

Parthemore JG et al: Calcitonin secretion in normal human subjects, *J Clin Endocrinol Metab* 47:184, 1978.

Raisz LG: Direct effects of vitamin D and its metabolites on skeletal tissue, *Clin Endocrinol Metab* 9:27, 1980.

Raisz LG: Local and systemic factors in the pathogenesis of osteoporosis, *N Engl J Med* 318:818, 1988.

Reichel H et al: The role of the vitamin D endocrine system in health and disease, *N Engl J Med* 320:980, 1989.

Rouleau MF et al: Characterization of the major parathyroid hormone target cell in the endosteal metaphysis of rat long bones, *J Bone Miner Res* 10:1043, 1990.

Shigeno C et al: Parathyroid hormone receptors are plasma membrane glycoproteins with asparagine-linked oligosaccharides, *J Biol Chem* 263:3872, 1988.

Silver J et al: Regulation by vitamin D metabolites of parathyroid hormone gene transcription in vivo in the rat, *J Clin Invest* 78:1296, 1986.

Stern PH: Vitamin D and bone, *Kidney Int* 29:S17, 1990.

Webb AR et al: Sunlight regulates the cutaneous production of vitamin D_3 by causing its photodegradation, *J Clin Endocrinol Metab* 68:882, 1989.

Yamamoto M et al: Hypocalcemia increases and hypercalcemia decreases the steady-state level of parathyroid hormone messenger RNA in the rat, *J Clin Invest* 83:1053, 1989.

CHAPTER

48

The Hypothalamus and Pituitary Gland

The hypothalamus-pituitary consortium forms the most complex and dominant portion of the entire endocrine system. Its internal anatomical and functional relationships are elaborate and subtle. The output of the hypothalamus-pituitary unit regulates the function of the thyroid, adrenal, and reproductive glands and also controls somatic growth, lactation, milk secretion, and water metabolism. Two hormones, antidiuretic hormone (ADH or vasopressin) and oxytocin, are synthesized by neurons in the hypothalamus but are stored and secreted by the posterior pituitary gland, or **neurohypophysis.** A group of tropic hormones, adrenocorticotropic hormone (ACTH), thyroid-stimulating hormone (TSH), luteinizing hormone (LH), follicle-stimulating hormone (FSH), growth hormone (GH), and prolactin, are synthesized, stored, and secreted by endocrine cells in the anterior pituitary gland, or **adenohypophysis.** However a set of releasing and inhibiting hormones that are produced in the hypothalamus and travel to the adenohypophysis regulates the synthesis and secretion of these adenohypophyseal tropic hormones. All of this emanates from a mass of only 500 mg of pituitary tissue in association with 10 g of adjacent hypothalamus.

■ *Anatomy*

A knowledge of the embryological development of the pituitary gland is crucial to understanding its anatomy and function. *The fully developed gland is really an amalgam of hormone-producing glandular cells (the adenohypophysis) and neural cells with secretory function (the neurohypophysis).* The anterior endocrine portion of the pituitary develops from an upward outpouching of ectodermal cells from the roof of the oral cavity (Rathke's pouch). This pouch eventually pinches off and becomes separated from the oral cavity by the sphenoid bone of the skull. The lumen of the pouch is reduced to a small cleft. The posterior neural portion of the pituitary develops from a downward outpouching of ectoderm from the brain in the floor of the third ventricle. The lumen of this pouch is obliterated inferiorly as the sides fuse into the infundibular process. Superiorly the lumen remains contiguous with and forms a recess in the adult third ventricle. The upper portion of this neural stalk expands to invest the lowest portion of the hypothalamus and is called the **median eminence.** The cleftlike remnant of Rathke's pouch demarcates the interwoven anterior and posterior portions of the pituitary. In some animals, but not in humans, cells in the area of Rathke's pouch and adjacent to the neurohypophysis form a distinct intermediate lobe. The entire pituitary gland sits in a socket of sphenoid bone called the sella turcica. A reflection of the dura mater, called the diaphragm, extends across the top of the sella turcica and separates the bulk of the pituitary gland from the brain. However the neural stalk penetrates the diaphragm, maintaining its continuity with the brain. These anatomical relationships are shown in Fig. 48-1, *B*. The human pituitary gland can be visualized by computerized axial tomography (CAT scanning) and nuclear magnetic resonance imaging (Fig. 48-1, *A*).

The blood supply to this consortium of neural and endocrine tissue is complex. In the posterior pituitary the neural tissue of the infundibular process derives its blood mostly from the inferior hypophyseal artery, and the capillary plexus thereof drains into the dural sinus. The neural tissue of the upper stalk and of the median eminence is supplied largely by the superior hypophyseal artery. After investing the axons in these areas, the capillary plexus emanating from this artery forms a set of long portal veins that carry the blood downward into the anterior pituitary. There these portal veins give rise to a second capillary plexus that supplies the endocrine cells with the majority of their blood, which is then drained off into the dural sinus. The anterior pituitary receives its remaining blood via a set of short portal veins originating in the capillary plexus of the inferior

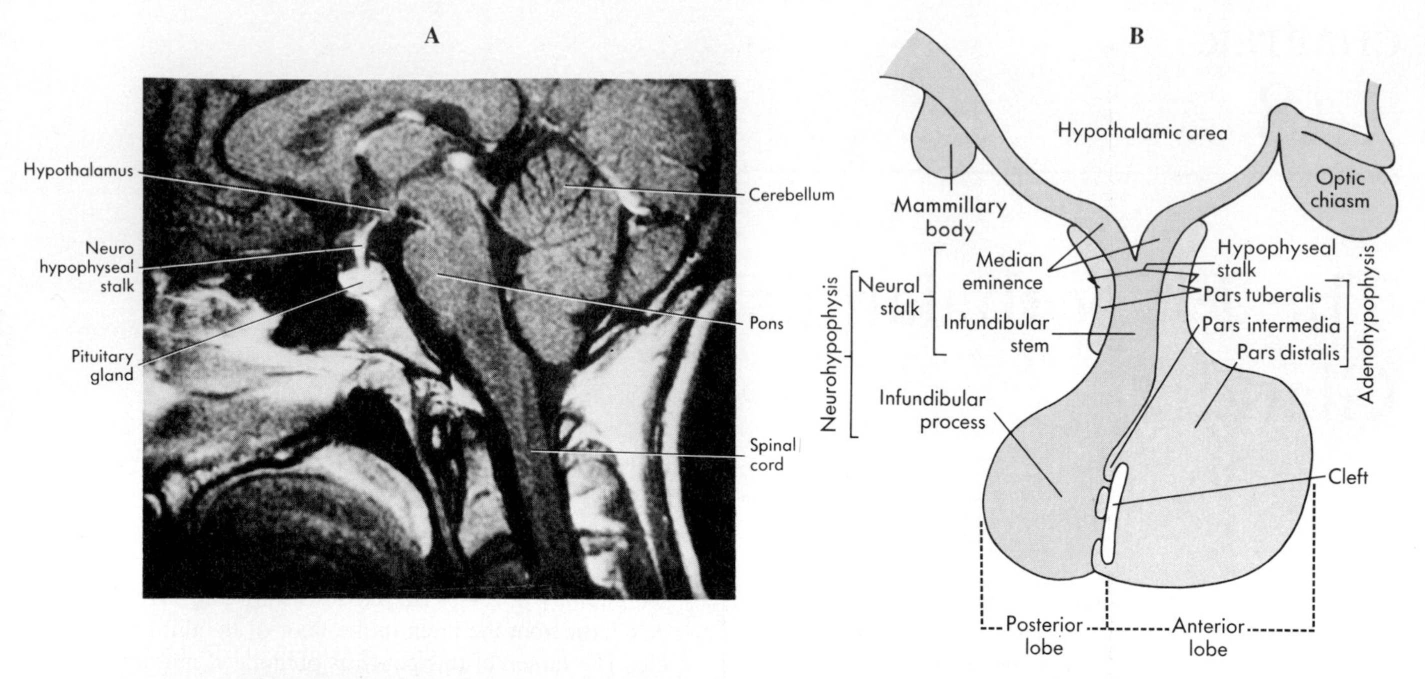

■ **Fig. 48-1** **A,** Magnetic resonance image of the head shows the proximity of the hypothalamus and pituitary gland (Courtesy Steven Wiener, M.D.). **B,** Diagram of the pituitary gland shows its division into the adenohypophysis and neurohypophysis. (**B,** adapted from an original painting by Frank H. Netter, M.D., from The CIBA Collection of Medical Illustrations, Division of CIBA-Geigy Corporation.)

hypophyseal artery within the neural stalk. Thus there is little or no direct arterial blood supply to the adenohypophyseal cells. Furthermore it should be noted that the anterior pituitary gland lies outside the blood-brain barrier.

The implications of this involved anatomical arrangement become apparent when the functional relationships are examined in Fig. 48-2. The neurohypophysis represents a collection of axons whose cell bodies lie in the hypothalamus. Peptide hormones synthesized in the cell bodies of these hypothalamic neurons travel down their axons in neurosecretory granules to be stored in the nerve terminals lying in the posterior pituitary gland. These terminals consist of neurosecretory vesicles that are invested by modified astroglial cells known as pituicytes. Upon stimulation of the cell bodies the granules are released from the axonal terminals by exocytosis; the peptide hormones then enter the peripheral circulation via the capillary plexuses of the inferior hypophyseal artery. Thus a single cell performs the entire process of hormone synthesis, storage, and release in the classic example of neurocrine function.

In contrast the adenohypophysis is a collection of endocrine cells that are regulated by bloodborne stimuli originating in neural tissue. Cell bodies of particular hypothalamic neurons synthesize releasing hormones and inhibiting hormones, which travel in packets down their axons only as far as the median eminence. Here they are stored as neurosecretory granules in the nerve terminals. After stimulation of these hypothalamic cells the releasing or inhibiting hormones are discharged into the median eminence and enter the capillary plexus of the superior hypophyseal artery. They are transported down the long portal veins and exit from the secondary capillary plexus to reach their specific endocrine target cells in the adenohypophysis. These cells respond to the releasing or inhibiting hormones by increasing or decreasing their output of tropic hormones. The latter enter the same second capillary plexus through which they ultimately reach the peripheral circulation. Thus two cells, one neural and one endocrine, participate in the processes leading to synthesis and release of the anterior pituitary tropic hormones, a combination of neurocrine and endocrine function. Key evidence in support of this functional arrangement includes the following observations:

1. Neural tracts containing hypothalamic peptides can be traced by immunohistochemical techniques down to the median eminence where they end in proximity to capillaries (Fig. 48-3).
2. Direct measurement of hypothalamic peptides in pituitary portal venous blood shows concentrations 10-fold to 20-fold higher than in peripheral blood.
3. Exposure of anterior pituitary tissue in perfusion systems or in tissue culture to individual hypothalamic peptides causes specific patterns of stimulation or inhibition of the release of corresponding tropic hormones.

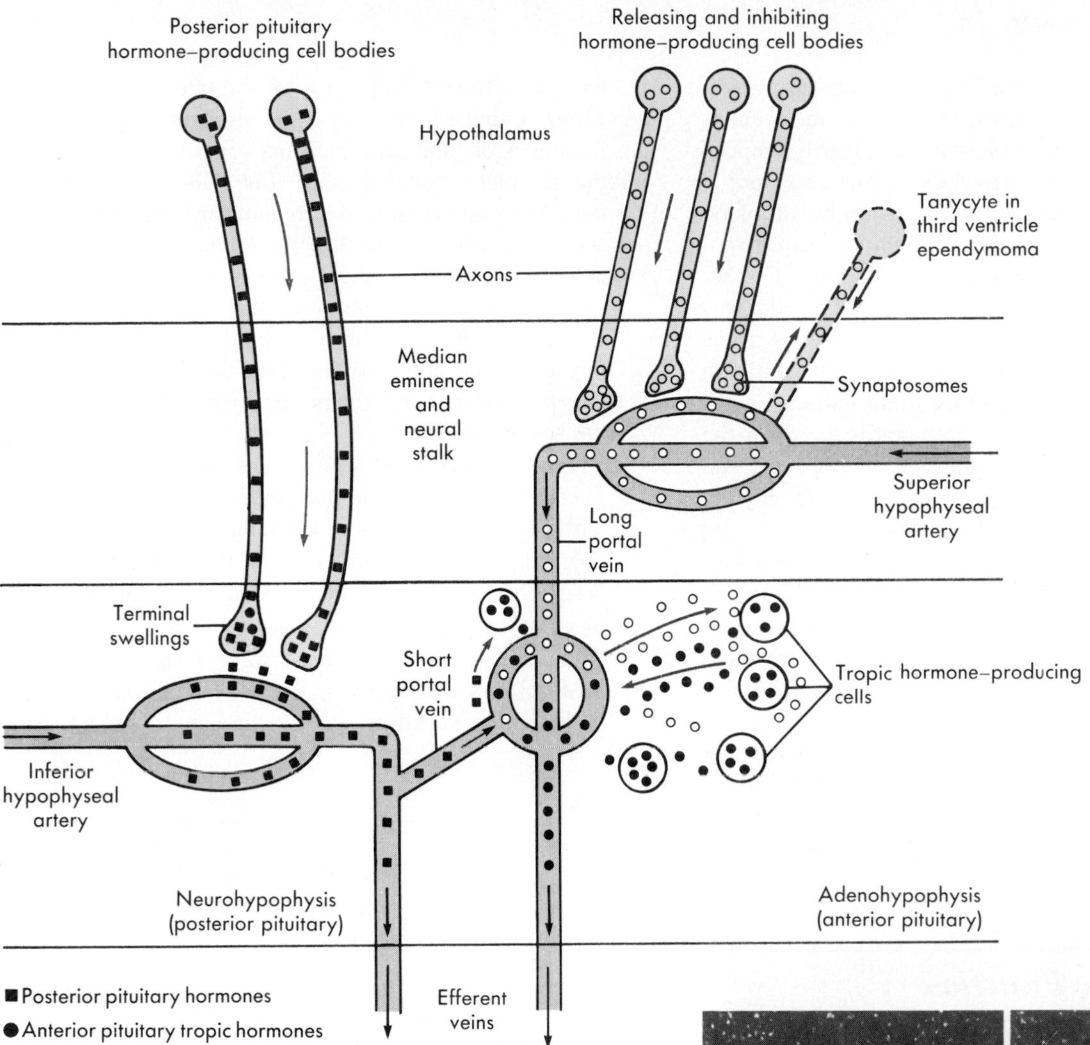

■ **Fig. 48-2** Anatomical and functional relationships between the hypothalamus, the pituitary gland, and its blood supply. Arrows indicate direction of movement of hormone molecules. Note that the adenohypophysis has no direct arterial supply but receives blood from the median eminence, which contains hypothalamic releasing and inhibiting hormones.

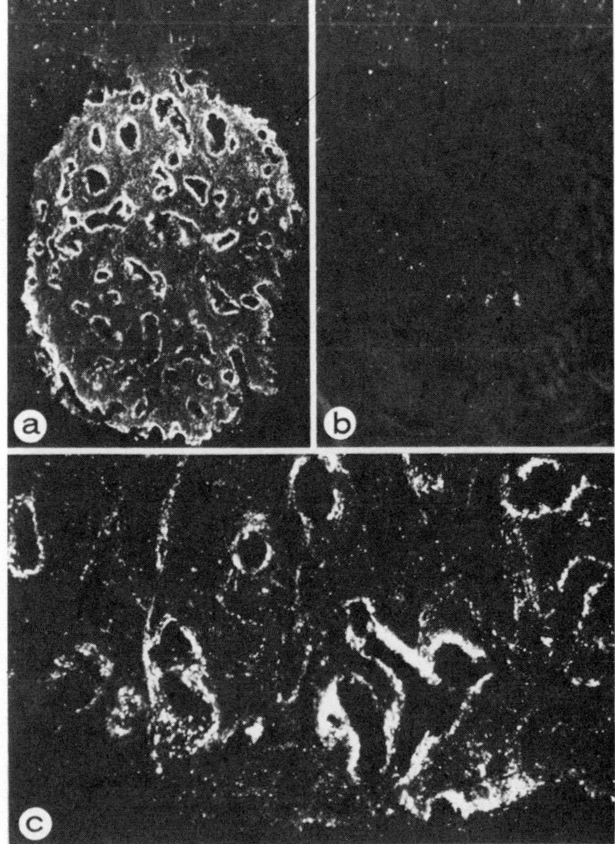

■ **Fig. 48-3** Immunohistochemical localization of growth hormone-releasing hormone (GHRH) in the median eminence of the squirrel monkey. **A,** Section stained with fluorescent-labeled antibody to GHRH shows localization of GHRH around capillaries of the median eminence. **B,** Section stained with fluorescent-labeled control serum shows little reaction, demonstrating specificity of the antibody to GHRH. **C,** Higher power of **A** showing GHRH in axonal tracts ending in the vicinity of the capillaries. (From Bloch B et al: *Nature* 301:607, 1983. From Macmillan Journals.)

The preceding description implies an entirely unidirectional arrangement. However not all the venous drainage from the anterior pituitary necessarily empties directly into the systemic circulation. The short portal veins may act as conduits for reverse flow of blood from the anterior pituitary cells through the neurohypophyseal capillary plexus back up to the neurons in either the median eminence or the hypothalamus itself. This direction of flow would permit anterior pituitary tropic hormones to bathe the neurons in high concentration without impedance from the blood-brain barrier.

It is also possible that two-way traffic between the cerebrospinal fluid and both the neurohypophysis and adenohypophysis may exist. Specialized ependymal cells in the interior recess of the third ventricle send long processes down that interdigitate with blood vessels in the median eminence and infundibular stalk. (Fig. 48-2). These cells, known as **pituitary tanycytes,** could facilitate transfer of regulatory substances from the cerebrospinal fluid to the pituitary. They could also allow posterior pituitary peptide hormones, hypothalamic releasing or inhibiting hormones, or even anterior pituitary tropic hormones to have access to the brain via the cerebrospinal fluid. Further definition of the physiological significance of these possible relationships is awaited.

■ *Hypothalamic Function*

The hypothalamus clearly plays a key role in regulating pituitary function. It can be considered a central relay station for collecting and integrating signals from diverse sources and funneling them to the pituitary. The hypothalamus receives afferent nerve tracts from the thalamus, the reticular activating substance, the limbic system (amygdala, olfactory bulb, hippocampus, and habenula), the eyes, and remotely from the neocortex. Some of the connections are multisynaptic. Through this input pituitary function can be influenced by pain, sleep or wakefulness, emotion, fright, rage, olfactory sensations, light, and possibly even thought. It can be coordinated with patterned behavior and mating responses. The proximity of other hypothalamic nuclei that govern thirst, appetite, temperature regulation, and autonomic nervous system function also allows coordination between the output of pituitary hormones and a wide variety of basic functions.

Hypothalamohypothalamic tracts may help to integrate multiple simultaneous pituitary responses with each other, as well as to regulate pituitary function in accordance with change in temperature, energy needs, or fluid balance. The neurotransmitters involved in afferent impulses to the hypothalamus are largely norepinephrine, acetylcholine, and serotonin. Dopamine, acetylcholine, and γ-aminobutyric acid (GABA), and the opioid peptide, β-endorphin, act as neurotransmitters for efferent impulses to the median eminence. These impulses regulate the discharge of releasing hormones or inhibiting hormones into the capillaries of the median eminence (see Fig. 48-2). In addition neurotransmitters from the hypothalamus may reach the portal vein blood and directly influence the output of anterior pituitary tropic hormones. The presence of neurotransmitter receptors in adenohypophyseal cells supports such an additional regulatory route. Finally dopamine and β-endorphin also modulate efferent outflow by transmitting signals between different areas of the hypothalamus.

The hypothalamus-pituitary axis is also under the influence of blood-borne substances from the periphery. Virtually all of the tropic hormones from the adenohypophysis cause changes either in the concentrations of peripheral gland hormones (thyroid, adrenal, gonadal) or of substrates, such as glucose or free fatty acids. Conditions exist for at least three levels of humoral feedback, as illustrated in Fig. 48-4. Peripheral gland hormones or substrates arising from tissue metabolism

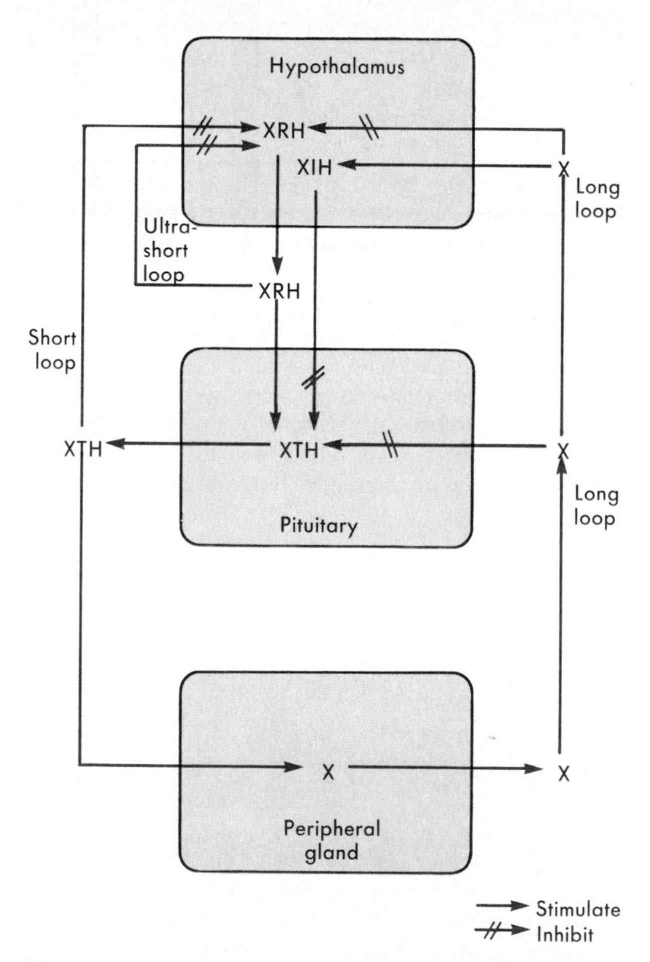

■ **Fig. 48-4** Negative feedback loops regulating hormone secretion in a typical hypothalamus-pituitary-peripheral gland axis. *X,* Peripheral gland hormone; *XTH,* pituitary tropic hormone; *XRH,* hypothalamic releasing hormone; *XIH,* hypothalamic inhibiting hormone.

can exert feedback control on both the hypothalamus and the anterior pituitary gland. This is known as **long-loop feedback** and is usually negative, although it can occasionally be positive. Negative feedback can also be exerted by the tropic hormones themselves through effects on the synthesis or discharge of the related hypothalamic releasing or inhibiting hormones. This is known as **short-loop feedback.** Since tropic hormones do not ordinarily cross the blood-brain barrier, short-loop feedback may occur either by specialized transport across fenestrated endothelial cells of the capillaries that bathe hypothalamic neurons or by retrograde flow through the short portal veins, as just described. Finally hypothalamic releasing hormones may even inhibit their own synthesis and discharge or stimulate the discharge of a paired hypothalamic inhibiting hormone. This is called **ultrashort-loop feedback.** It could occur by neurotransmission between two hypothalamic cells, or it could be mediated by transport of the releasing hormone via the pituitary tanycytes to the cerebrospinal fluid and thence back to the hypothalamus.

The anterior pituitary gland is the central point of the hypothalamic pituitary-peripheral gland axis. At this level hypothalamic releasing hormones and peripheral target gland hormones can usually be viewed as antagonists, one accelerating and one braking anterior pituitary hormone secretion. The short- and ultrashort-loop feedback mechanisms help to set the balance finally achieved.

Table 48-1 lists the currently known or suspected hypothalamic releasing or inhibiting hormones.

To some extent the hypothalamus can be subdivided into endocrinologically distinct functional areas. The lateral hypothalamus largely accepts afferent impulses and relays them to the neurosecretory nuclei of the anterior and medial basal portions. The anterior segment of the hypothalamus contains two well-defined collections of large neurons, the supraoptic and paraventricular nuclei, which are responsible for the synthesis of the two posterior pituitary peptide hormones. Their axons project primarily to the posterior pituitary, though some fibers also project to the median eminence and to other neurons in the floor of the third ventricle and the brain stem. Immediately beneath the third ventricle in the arcuate nucleus and the periventricular nucleus of the medial basal hypothalamus, small neurons responsible for synthesis of the various hypothalamic releasing and inhibiting hormones tend to be clustered. Some are also located in the paraventricular nucleus. The axons of these small neurons project to the median eminence.

■ **Table 48-1** Hypothalamic hormones and factors

Hormone	*Predominant hypothalamic localization*	*Structure*	*Target pituitary hormones*
Thyrotropin-releasing hormone (TRH)	Paraventricular	pGLU-HIS-PRO-NH_2	Thyrotropin Prolactin Growth hormone (pathological)
Gonadotropin–releasing hormone (GnRH)	Arcuate	pGLU-HIS-TRP-SER-TYR-GLY-LEU-ARG-PRO-GLY-NH_2	Luteinizing hormone Follicle-stimulating hormone Growth hormone (pathological)
Corticotropin-releasing hormone (CRH)	Paraventricular	SER-GLN-GLU-PRO-PRO-ILE-SER-LEU-ASP-LEU-THR-PHE-HIS-LEU-LEUARG-GLU-VAL-LEU-GLU-MET-THR-LYS-ALA-ASP-GLN-LEU-ALA-GLN-GLN-ALA-HIS-SER-ASN-ARG-LYS-LEU-LEU-ASP-ILE-ALA-NH_2	Adrenocorticotropin β- and γ-Lipotropin β-Endorphins
Growth hormone–releasing hormone (GHRH)	Arcuate	TYR-ALA-ASP-ALA-ILE-PHE-THR-ASN-SER-TYR-ARG-LYS-VAL-LEU-GLY-GLN-LEU-SER-ALA-ARG-LYS-LEU-LEU-GLN-ASP-ILE-MET-SER-ARG-GLN-GLN-GLY-GLU-SER-ASN-GLN-GLU-ARG-GLY-ALA-ARG-ALA-ARG-LEU-NH_2	Growth hormone
Growth hormone–inhibiting hormone (somatostatin)	Anterior periventricular	ALA-GLY-CYS-LYS-ASN-PHE-PHE-TRP-LYS-THR-PHE-THR-SER-CYS	Growth hormone Prolactin Thyrotropin Adrenocorticotropin (pathological)
Prolactin-inhibiting factor (PIF)	Arcuate	Dopamine	Prolactin Growth hormone (pathological)
Prolactin-releasing factor (PRF)	Not known	Not established	Prolactin

However a scattering of cells, also containing hypothalamic releasing hormones, is present in numerous other areas of the hypothalamus. In these neurons the peptides may have particular neurotransmitter roles related to or distinct from their known endocrine functions. By immunohistochemical mapping the anatomical separation between dense collections of the hypothalamic peptides is sufficient to indicate that, in general, only one cell type produces each neurohormone. However in at least one instance two peptides (corticotropin-releasing hormone and antidiuretic hormone) are co-localized within certain hypothalamic neurons.

All hypothalamic peptides have been named on the basis of the anterior pituitary hormone whose secretion they were originally discovered to influence. Although it was initially presumed that each tropic hormone might be under the control of a unique hypothalamic releasing or inhibiting hormone and that each hypothalamic hormone would have only one target anterior pituitary cell, the actual physiology is more complex. The tripeptide conventionally known as **thyrotropin-releasing hormone (TRH)** can also stimulate secretion of prolactin. Somatostatin, discovered as a growth hormone–inhibiting factor, can also inhibit the secretion of thyrotropin. Furthermore, pathologically functioning adenohypophysial cells of a given type may express receptors for hypothalamic peptides that the normal cell line of this type does not. In addition the hypothalamic peptide hormones have also been found outside the hypothalamus, in such diverse areas as the cerebral cortex, limbic area, spinal cord, autonomic ganglia, sensory neurons, and pancreatic islets and throughout the gastrointestinal tract. In these areas they serve neuromodulatory roles related to or independent of their endocrine function.

The hypothalamic peptides are synthesized via preprohormones as described in Chapter 44. Many common features characterize their functional behavior.

Characteristics of hypothalamic-releasing hormones

1. Secretion in pulses
2. Action on specific plasma membrane receptors
3. Transduction of signals through calcium, membrane phospholipid products, and cAMP as second messengers
4. Stimulation of release of stored target anterior pituitary hormones
5. Stimulation of synthesis of target anterior pituitary hormones at the transcriptional level
6. Modification of biological activity of target anterior pituitary hormones by posttranslational effects
7. Stimulation of hyperplasia and hypertrophy of target cells
8. Modulation of effects by regulation of own receptors

The secretion of hypothalamic releasing hormones into the pituitary portal veins is pulsatile, a pattern apparently dependent on an intrinsic neural oscillator within the cells of origin. This pulsatility is critical for maintaining the appropriate pattern and level of secretion of their target anterior pituitary hormones. The releasing hormones react with plasma membrane receptors in the anterior pituitary cells, following which cytosolic calcium and then cAMP levels increase. In addition diacylglycerols, inositol phosphates, and arachidonic acid from membrane phospholipids help mediate the intracellular effects that follow. Specific proteins are presumably phosphorylated by activated protein kinase A or C. Granule exocytosis is rapidly stimulated, with release of stored tropic hormones. In addition tropic hormone synthesis is stimulated by increasing the levels of specific messenger RNAs. In some instances the biological activity of the target pituitary hormones may also be increased after translation by modifying their content of sugars or sialic acid or by phosphorylation.

The anterior pituitary contains at least five endocrine cell types (Table 48-2). These cannot be completely distinguished by conventional histological staining and are not localized to exclusive areas. However the development of immunohistochemical techniques employing hormone-specific antisera has permitted each type to be identified specifically. In addition "null cells" containing no known hormones are present. The distribution of cell types within the gland is not random, and evidence suggests a strong likelihood of paracrine interactions among them. In rat pituitary glands corticotrophs, usually found in the center of the gland, may associate with particular somatotrophs. Gonadotrophs and mammotrophs intertwine and form junctional complexes. Studies with cells sorted in vitro suggest that a GnRH stimulated product of the gonadotroph in turn stimulates the mammotroph to release its hormone prolactin.

Mathematical modeling of the plasma profiles of human anterior pituitary hormones suggests that there is little or no tonic secretion of these hormones. Rather secretion is episodic, prompted by pulses of hypothalamic releasing hormones. Secretion bursts probably last only 4 to 16 minutes. The longer duration of the resultant plasma peaks (90 to 140 minutes) reflects relatively slow metabolic clearance rates.

Anterior Pituitary Hormones

Thyrotropic Hormone

TSH is a glycoprotein hormone whose function is to regulate the growth and metabolism of the thyroid gland and the secretion of its hormones, **thyroxine (T_4)** and **triiodothyronine (T_3)**. The TSH-producing cells normally form 3% to 5% of the adult human anterior pituitary population, and they are found predominantly in the anteromedial area of the gland. These cells develop at about 13 weeks of gestation at the same time that the fetal thyroid gland is beginning to secrete thyroid hormone. In the adult pituitary TSH is stored in small secretory granules.

■ **Table 48-2** Anterior pituitary cells and hormones

Cell	*% Pituitary population*	*Products/molecular weight*	*Targets*
Corticotroph	15-20	Adrenocorticotropin (ACTH), 4,500 β-Lipotropin, 11,000	Adrenal gland Adipose tissue Melanocytes
Thyrotroph	3-5	Thyrotropin (TSH), 28,000	Thyroid gland
Gonadotroph	10-15	Luteinizing hormone (LH), 28,000 Follicle-stimulating hormone (FSH), 33,000	Gonads
Somatotroph	40-50	Somatotropin, growth hormone (HGH), 22,000	All tissues
Mammotroph	10-25	Prolactin, 23,000	Breasts Gonads

TSH has a molecular weight of 28,000 and contains 15% carbohydrate bound covalently to the peptide chains. The hormone is made of two subunits tightly associated by noncovalent forces. The α subunit of 96 amino acids is nonspecific, being a component also of two other anterior pituitary hormones (FSH and LH), as well as of a placental hormone (human chorionic gonadotropin). The β-subunit of 110 amino acids confers the specific biological activity on the TSH molecule. However both α and β-subunits are required for receptor binding and subsequent hormone action.

Separate genes, located on different chromosomes, code for the individual messenger RNAs of the α- and β subunits. In each instance during translation a signal N-terminal peptide is eliminated from the primary translation product (termed a **prehormone**). Subsequently the N-glycosidic-linked sugar moieties that are rich in mannose and protect the nascent molecule from premature proteolysis are added. During transport from the rough endoplasmic reticulum and packaging in the Golgi apparatus the carbohydrate units are further modified, sialic acid and sulfate are added, and intramolecular disulfide bonds are formed. These changes assure the proper conformation that permits the two individual subunits to combine in the mature TSH molecule. Expression of the α and β subunit genes is separately regulated, though there must be coordination between them. Ordinarily an excess of the nonspecific α subunit is produced, but selective addition of an extra O-linked oligosaccharide renders the unneeded α subunits incapable of combining with β subunits. Transcription of both TSH subunit genes is stimulated by the hypothalamic thyrotropin-releasing hormone (TRH) and is suppressed by thyroid hormone. In addition TRH and thyroid hormone modulate the glycosylation process so as to increase or decrease biological activity, respectively. Transcription of the α-subunit gene is also regulated by cAMP.

The secretion of TSH is reciprocally regulated by two major factors. TRH increases the rate of secretion, whereas thyroid hormone decreases it in negative feedback fashion (Fig. 48-5). As a result of this balance TSH is secreted in a relatively steady even if somewhat pulsatile fashion. This befits the role of TSH, which is to stimulate a target gland whose own output is meant to be steady because the actions of its hormones (thyroxine and triiodothyronine) wax and wane slowly. In other words although the hypothalamic-pituitary-thyroid gland axis is set with great precision, it moves to higher or lower levels in hours or days rather than in minutes. After intravenous administration of TRH plasma TSH levels rise as much as 10-fold and return toward baseline levels by 60 minutes (Fig. 48-6). With repeated TRH injections the TSH response diminishes over time

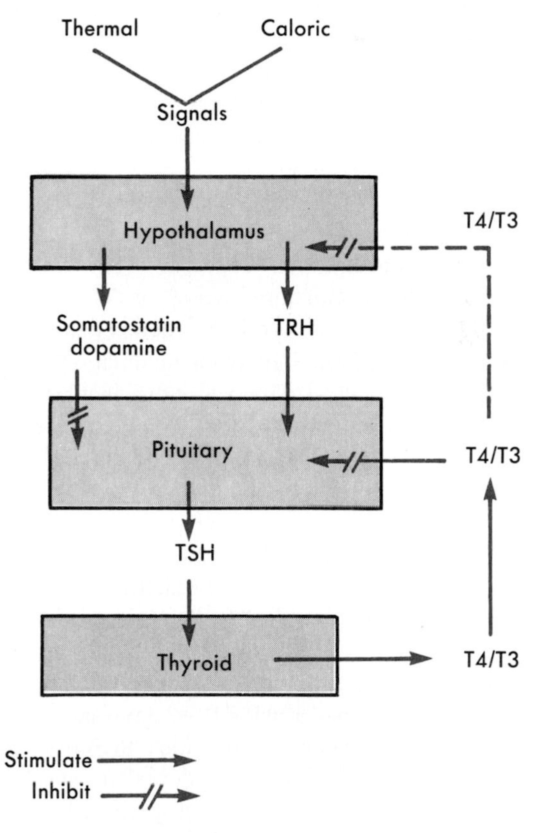

■ **Fig. 48-5** Regulation of TSH secretion. Thyroxine (T_4) and triiodothyronine (T_3) exert negative feedback on the pituitary by blocking the action of TRH. Negative feedback of T_4 and T_3 at the level of the hypothalamus is less well established. Somatostatin and dopamine each inhibit TSH secretion tonically.

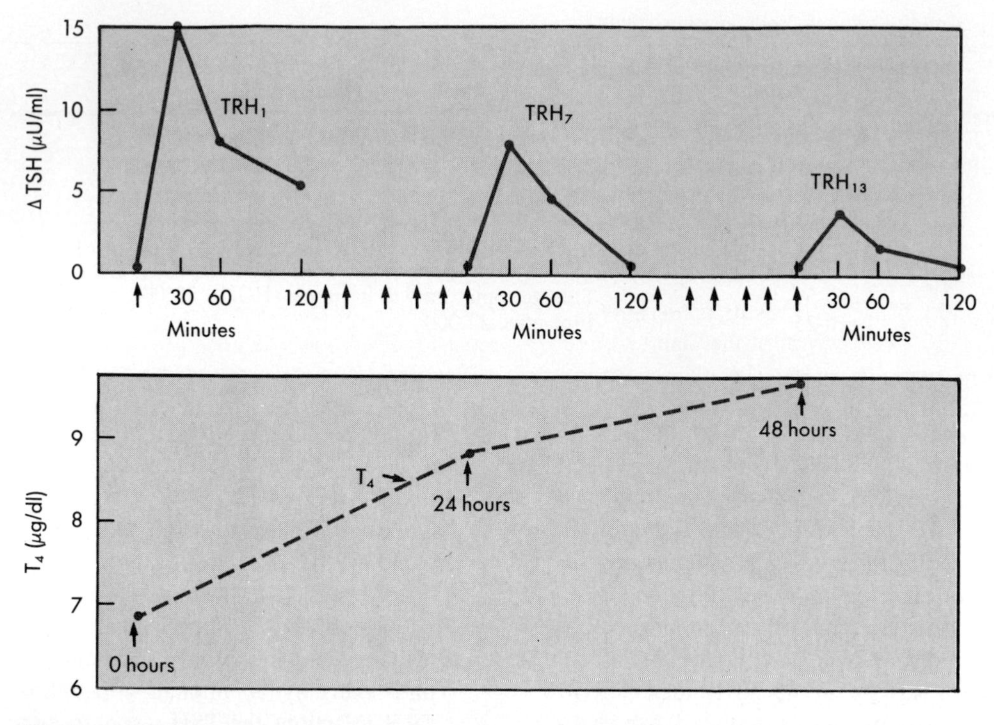

■ **Fig. 48-6** Pituitary and thyroid gland responses to repetitive injections of TRH every 4 hours for 48 hours in humans. Note that as plasma thyroxine (T_4) increases as a result of stimulation by TSH, the TSH responses to TRH are progressively blunted. (Redrawn from Snyder PJ: *J Clin Invest* 52:2305, 1973. From The American Society for Clinical Investigation.)

primarily because the secondarily stimulated thyroid gland increases its output of T_4 and T_3 (Fig. 48-6). This demonstrates vividly the negative feedback regulation of TSH secretion. These small increments of thyroid hormone concentration suppress TSH secretion by blocking the stimulatory action of TRH; conversely small decrements of thyroid hormone augment TSH responsivity to TRH. Significant modulation of TSH secretion is associated with variations in plasma thyroid hormone concentrations of only 10% to 30% above or below the individual's baseline level. The thyrotroph's response to continuous TRH stimulation also is limited by down regulation of the TRH receptor.

The intracellular mediator of thyroid hormone's effect on TSH is likely T_3. Furthermore T_3 generated within the pituitary cell from T_4 is more effective and important in this regard than is T_3 that enters from the circulation. The suppressive effect of thyroid hormone on TSH release has a half-life of days and is prevented if protein synthesis by the pituitary thyrotroph is inhibited. This suggests that T_3 induces the synthesis of a protein with TSH-suppressing properties. In addition T_3 decreases the number of TRH receptors. Although thyroid hormone clearly inhibits TSH secretion and synthesis at the pituitary level, evidence from microinjection directly into the hypothalamus indicates that T_3 may also reduce the synthesis or release of TRH. Because of negative feedback individuals who have thyroid diseases that result in chronic deficiency of thyroid hormone **(hypothyroidism)** have very high plasma TSH levels and hyperplasia of the thyrotrophs.

Physiological modulation of TSH secretion (and consequently of thyroid hormone output) occurs in at least two circumstances, namely, fasting and exposure to cold. TSH responsiveness to TRH and possibly TRH release itself are diminished during total fasting. This coincides with a decrease in metabolic rate and appears teleologically useful. In animals TSH secretion is augmented by exposure to cold, but this has been demonstrated only infrequently in humans. Since TSH increases thermogenesis via stimulation of the thyroid gland this too is a logical response. Other hormonal and neural influences have been noted. There is a slight diurnal variation in TSH secretion, with the highest levels occurring at night. A tonic inhibitory effect on TSH secretion is exerted by the hypothalamic peptide somatostatin and the neurotransmitter dopamine. Furthermore thyroid hormone may suppress TSH secretion partly by stimulating the release of these hypothalamic inhibitors. Cortisol (a hormone from the adrenal cortex) decreases both TRH and TSH secretion; growth hormone also reduces TSH secretion.

TSH normally circulates in plasma at a concentration of 0.5 to 5 μU/ml, approximating a concentration of 10^{-11} M. Daily TSH production is about 165,000 μU, which is equivalent to the entire content of one normal pituitary gland. The metabolic clearance rate of TSH is 50 L/day. This is inversely related to the degree of glycosylation. In normal individuals the α subunit is also individually secreted and circulates at low levels. When TSH secretion is chronically hyperstimulated in response to deficient function of the thyroid gland, both β and α subunits circulate individually in elevated amounts.

The only TSH actions of importance are those exerted on the thyroid gland. TSH, as its name implies, is tropic: that is, it promotes growth of the gland and stimulates all aspects of its function. The glandular uptake of iodide, its organification, the completion of thyroid hormone synthesis, and the subsequent release of thyroid gland products are all stimulated by TSH. These effects are described in detail in Chapter 49. TSH binds to a plasma membrane receptor, and cAMP is the second messenger for many of the hormone's effects on thyroid hormone synthesis and release and on thyroid cell growth and differentiation.

The sole pathological effects of an excess or deficiency of TSH are those of increased or decreased thyroid gland function, as described in Chapter 49.

Adrenocorticotropic Hormone

Adrenocorticotropic hormone (ACTH) is an anterior pituitary polypeptide hormone whose function is to regulate the growth and secretion of the adrenal cortex. Its most important target gland hormone is **cortisol.** The corticotrophs form 20% of the anterior pituitary population. Although these are largely localized to the pars distalis of the anterior lobe (see Fig. 48-1, *B*), ACTH producing cells may exist in the intermediate lobe in animals and in pathological human situations. Corticotrophs are distinguished ultrastructurally by the presence of large numbers of microfilaments. The hormone is stored in secretory granules that have a clear space between the contents and the membrane. In the human fetus ACTH synthesis and secretion begin at 10 to 12 weeks of gestation, just before the development of the adrenal cortex.

ACTH is a straight chain peptide with 39 amino acids and a molecular weight of 4,500. The N-terminal 1 to 24 sequence contains full biological activity, and sequence 5 to 10 is critical for stimulating the adrenal cortex. The remaining C-terminal portion probably only prolongs the hormone's action by protecting it against enzymatic degradation.

Synthesis of ACTH. As shown in Fig. 48-7, a single gene controls the transcription of a mature messenger RNA that directs the synthesis of a 31,000 molecular weight protein known as preproopiomelanocortin. Sequential processing of the latter in the human gives rise to ACTH along with several other products that are cosecreted into the plasma. These include β-lipotropin, γ-lipotropin, β-endorphin, and the N-terminal peptide. Melanocyte-stimulating hormone (MSH) activity is also contained within several of these peptides. Some ACTH

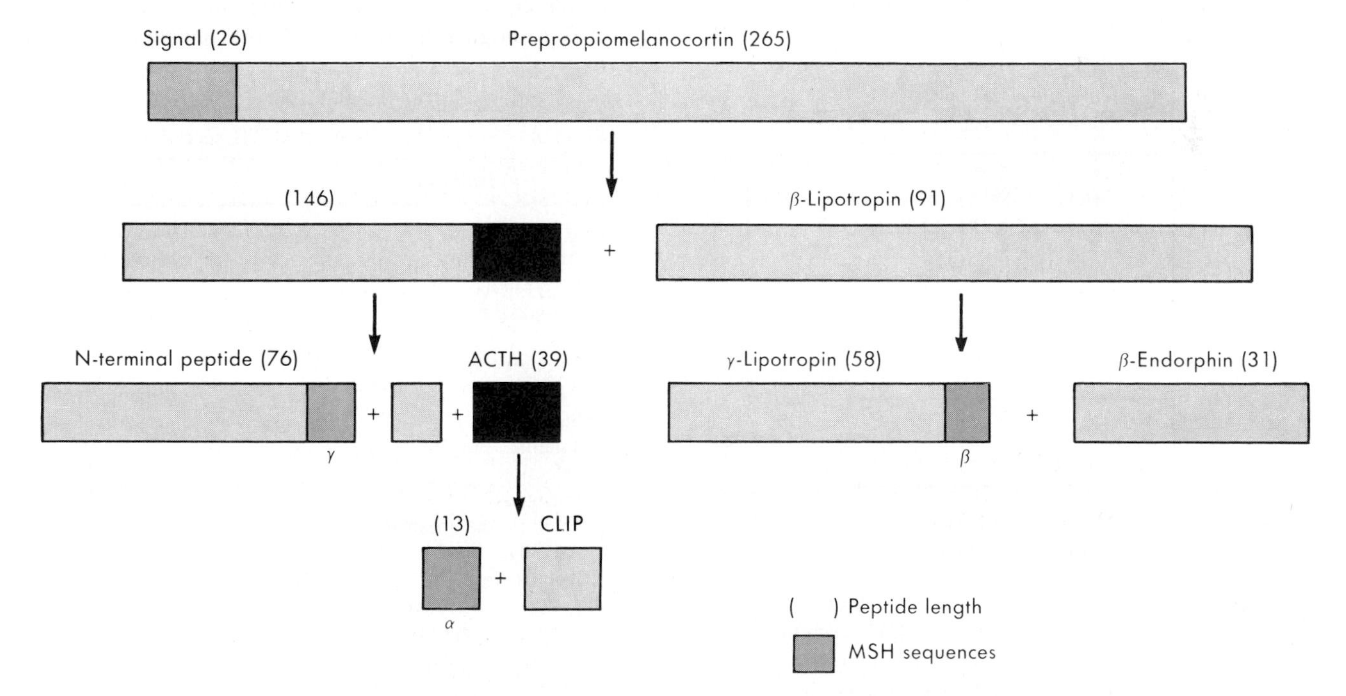

■ **Fig. 48-7** The processing of preproopiomelanocortin. In the anterior lobe of the human pituitary, ACTH, β-lipotropin, γ-lipotropin, β-endorphin, and a 76-amino acid N-terminal fragment are end products that are released. In other species, ACTH is further cleaved to α-MSH and corticotropin-like intermediate peptide (CLIP) in the neural intermediate lobe.

molecules may undergo posttranslational phosphorylation or glycosylation.

In certain animal species but not in humans, the cells of the intermediate neural lobe further cleave ACTH to α-MSH and a second product known as corticotropin-like intermediate peptide (CLIP). In addition these cells may cleave γ-MSH from the N-terminal peptide and β-MSH from γ-lipotropin (see Fig. 48-7). A similar pattern of processing from the same precursor molecule does take place in human extrapituitary tissue (brain, hypothalamus, gastrointestinal tract, pancreatic islets, adrenal medulla) where ACTH and these related peptides may subserve completely different functions. Because there is little evidence for the independent existence of α-, β-, or γ-MSH in human plasma, the expression of melanocyte-stimulating activity may depend on the presence of these amino acid sequences within the parent molecules.

Finally the N-terminal pentapeptide of β-endorphin is identical to metenkephalin, with which it shares analgesic and mood-modifying effects of opioids. The brain enkephalins however do not arise by cleavage of β-endorphin but are synthesized from an entirely different precursor directed by a separate gene.

Secretion of ACTH. The regulation of ACTH secretion is among the most complex of all the pituitary hormones (Fig. 48-8). The hormone exhibits circadian rhythms, cyclic bursts, feedback control, and responses to a wide variety of stimuli (Table 48-3). Although the mechanisms for each form of control are not completely clear, the hypothalamic **corticotropin-releasing hormone (CRH)** is the important mediator. CRH is a peptide with 41 amino acids originating in small cells of the paraventricular nucleus (see Table 48-1). It stimulates the release of ACTH and its proopiomelanocortin coproducts via cAMP as second messenger. Antidiuretic hormone (ADH) also exhibits corticotropin-releasing activity and under particular physiological circumstances it augments the primary effect of CRH. The gene that directs the synthesis of prepro-CRH has considerable homology with the genes for prepro-ADH and preproopiomelanocortin itself, suggesting a common evolutionary starting point for these functionally related molecules.

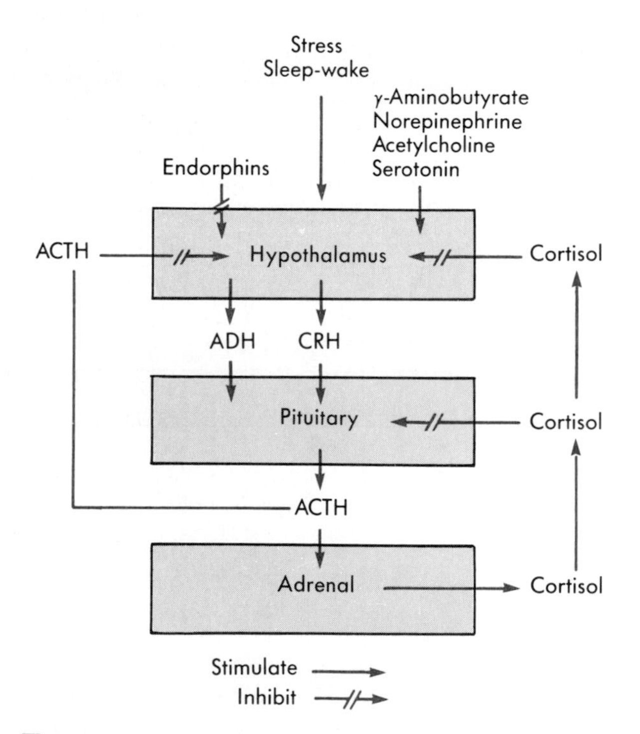

■ **Fig. 48-8** Regulation of ACTH secretion. CRH and ADH (antidiuretic hormone) stimulate ACTH secretion. Cortisol exerts negative feedback (1) at the pituitary level by blocking CRH action and (2) at the hypothalamus level by inhibiting CRH release. Gamma aminobutyric acid, norepinephrine, acetylcholine, and serotonin are positive modulators, whereas endorphins and ACTH itself are negative modulators of CRH release.

ACTH secretion has a markedly diurnal pattern. As shown in Fig. 48-9 a large peak occurs 2 to 4 hours before awakening. Thereafter the average level decreases to a nadir just before or after falling asleep. A rise and fall in the major adrenocortical hormone, cortisol, is entrained in this ACTH pattern. The clock time of the diurnal pattern can be shifted by systematically altering the sleep-wake cycle for a number of days; however the ACTH peak is not entrained with a specific stage of sleep. The circadian rhythm is diminished or abolished by loss of consciousness, blindness, or constant exposure to either dark or light. The nocturnal ACTH peak is primarily generated in the hypothalamus by CRH release and does not depend on negative feedback from its target, the adrenal gland. Nevertheless the nocturnal peak is augmented by prior cortisol deficiency and con-

■ **Table 48-3** Regulation of ACTH secretion

Stimulation	*Inhibition*
Cortisol decrease	Cortisol increase
Adrenalectomy	Enkephalins
Metyrapone	Opioids
Sleep-wake transition	ACTH
Stress	Somatostatin
Hypoglycemia	
Anesthesia	
Surgery	
Trauma	
Infection	
Pyrogens	
Psychiatric disturbance	
Anxiety	
Depression	
Antidiuretic hormone	
α-Adrenergic agonists	
β-Adrenergic antagonists	
Serotonin	
γ-Aminobutyric acid (GABA)	
Acetylcholine	
Interleukins	
GI peptides	

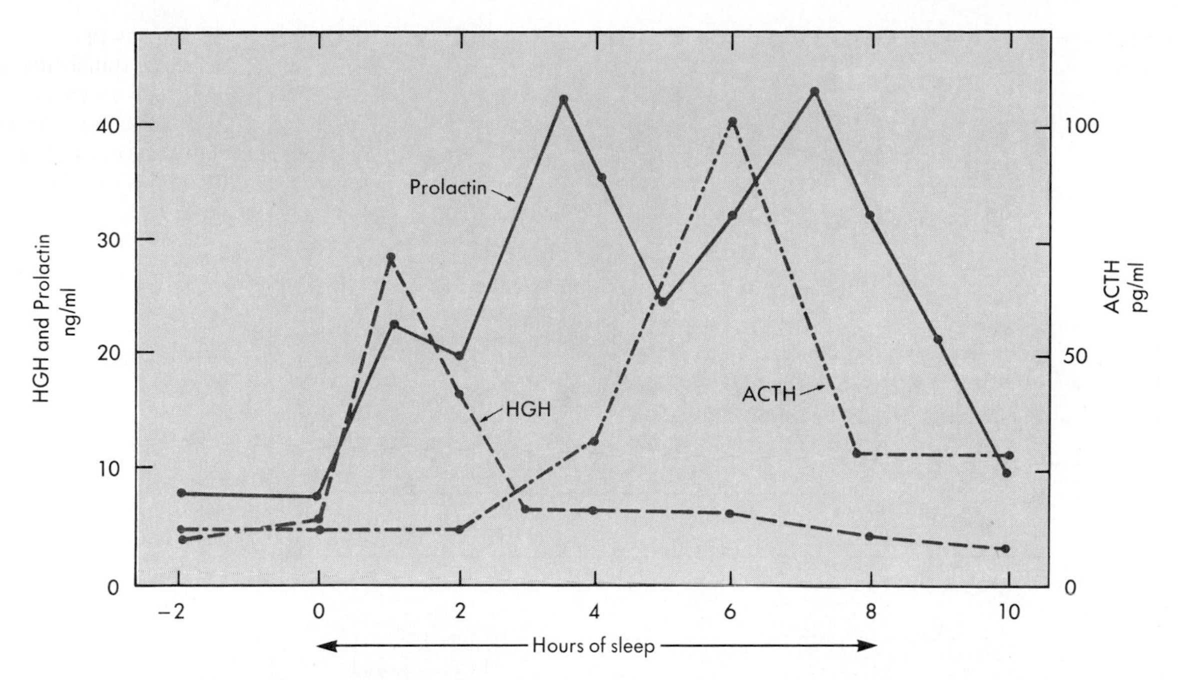

■ **Fig. 48-9** Nocturnal release of ACTH, HGH, and prolactin. Note the distinctive pattern for each hormone. (Redrawn from Takahashi Y et al: *J Clin Invest* 47:2079, 1968; Berson SA et al: *J Clin Invest* 47:2725, 1968; from The American Society for Clinical Investigation; and Sassin JF et al: *Science* 177:1205, 1972. From the American Association for the Advancement of Science.)

versely can be completely suppressed by excess cortisol. The diurnal pattern is composed of pulses of ACTH release with little or no tonic secretion. Very frequent plasma sampling suggests up to three pulses per hour, each pulse lasting about 20 minutes. Major ACTH peaks appear to be caused by increased amplitude rather than increased frequency of secretory bursts. As expected β-endorphin pulses occur simultaneously with those of ACTH, whereas cortisol pulses follow 10 minutes later. Men exhibit both a greater frequency and a greater amplitude of ACTH pulses than do women.

Feedback inhibition of ACTH secretion is effected by its peripheral target hormone, cortisol, or by any synthetic analogue with a potency proportional to its other cortisol-like activity (see Fig. 48-8). The suppressive action may outlive the duration of cortisol exposure. Conversely when cortisol action is blocked by an antagonist, cortisol secretion is reduced by disease, or cortisol release is pharmacologically inhibited (see Fig. 48-10), ACTH secretion is stimulated. Cortisol suppresses ACTH secretion at the pituitary level by blocking the stimulatory action of CRH (Fig. 48-11). Cortisol also decreases the synthesis of ACTH by inhibiting transcription of preproopiomelanocortin. In addition cortisol blocks hypothalamic release of CRH. The negative feedback effects of cortisol on diurnal and stress-induced ACTH release are also indirectly mediated by neural input from the hippocampus to the CRH neurons of the hypothalamus. Two distinct cortisol receptors in the hippocampus provide a range of affinities to accommodate the usual range of plasma cortisol levels.

ACTH may also inhibit its own secretion by decreasing CRH release, an example of short-loop feedback. Chronic deficiency of cortisol leads to persistent elevation of plasma ACTH, but the diurnal and pulsatile patterns are preserved, indicating their basic nonfeedback origin. Chronic autonomous hypersecretion of cortisol or long-term therapeutic administration of cortisol analogues for various diseases leads to functional atrophy of the CRH-ACTH axis. Several months may be required for recovery of this axis after the suppressive influence has been removed.

ACTH secretion responds most strikingly to stressful stimuli, a response that is critical to survival. Numerous factors that elicit the stress reaction in humans are noted in Table 48-3. The response to insulin-induced hypoglycemia is illustrated in Fig. 48-12, p. 909. In some instances such as major abdominal surgery or severe psychiatric disturbance the stress-induced hypersecretion of ACTH completely overrides negative feedback, and it cannot be suppressed by even the maximum level of cortisol secretion of which the adrenal cortex is capable. Stress also often obliterates the diurnal variation of ACTH levels, though pulsatility persists. The pathways by which each particular stress is signaled, sensed, and then stimulates CRH (and ADH) secretion vary. For example hypothalamic sensitivity to glucose levels per se, augmented by norepinephrine (via α-adrenergic receptors) and serotonin input, may induce

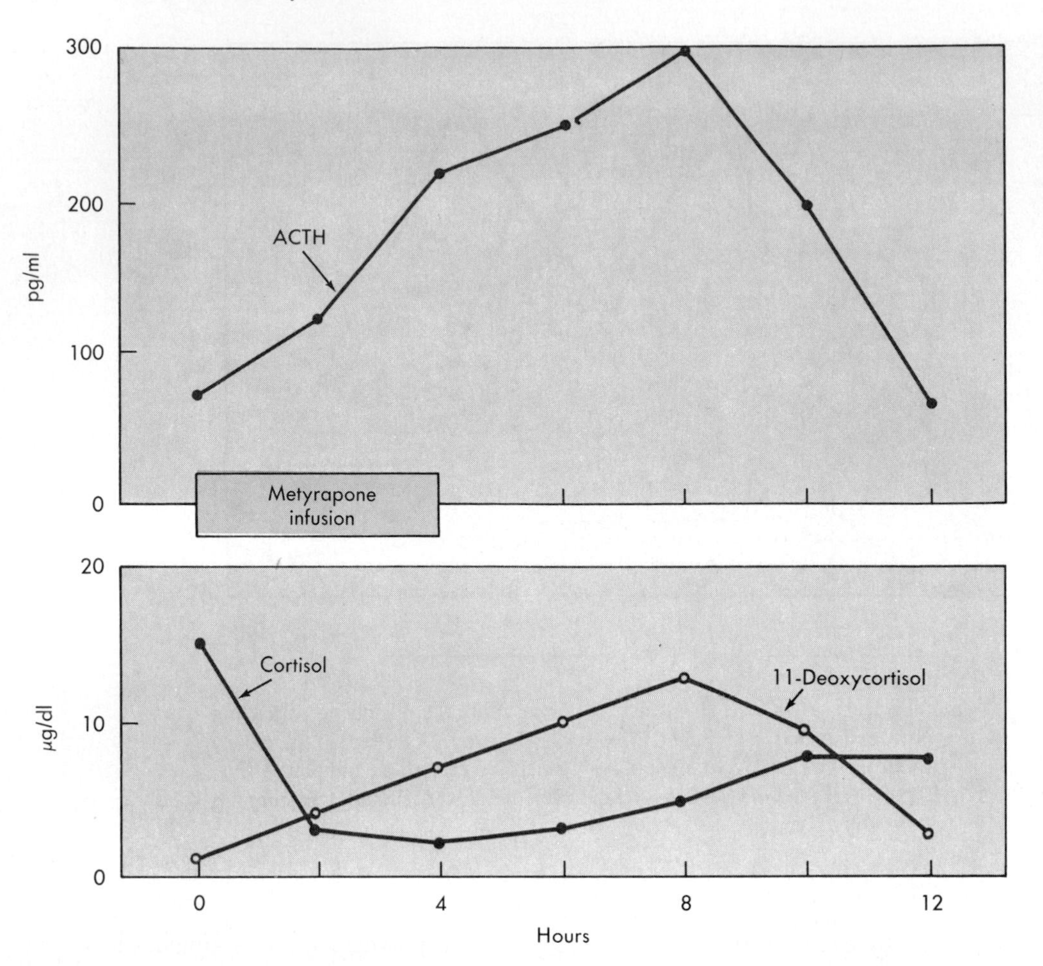

■ **Fig. 48-10** Negative feedback stimulation of ACTH release by metyrapone, a drug that blocks the conversion of 11-deoxycortisol to cortisol. (Redrawn from Jubiz W et al: *Arch Intern Med* 125:468, 1970. From the American Medical Association.)

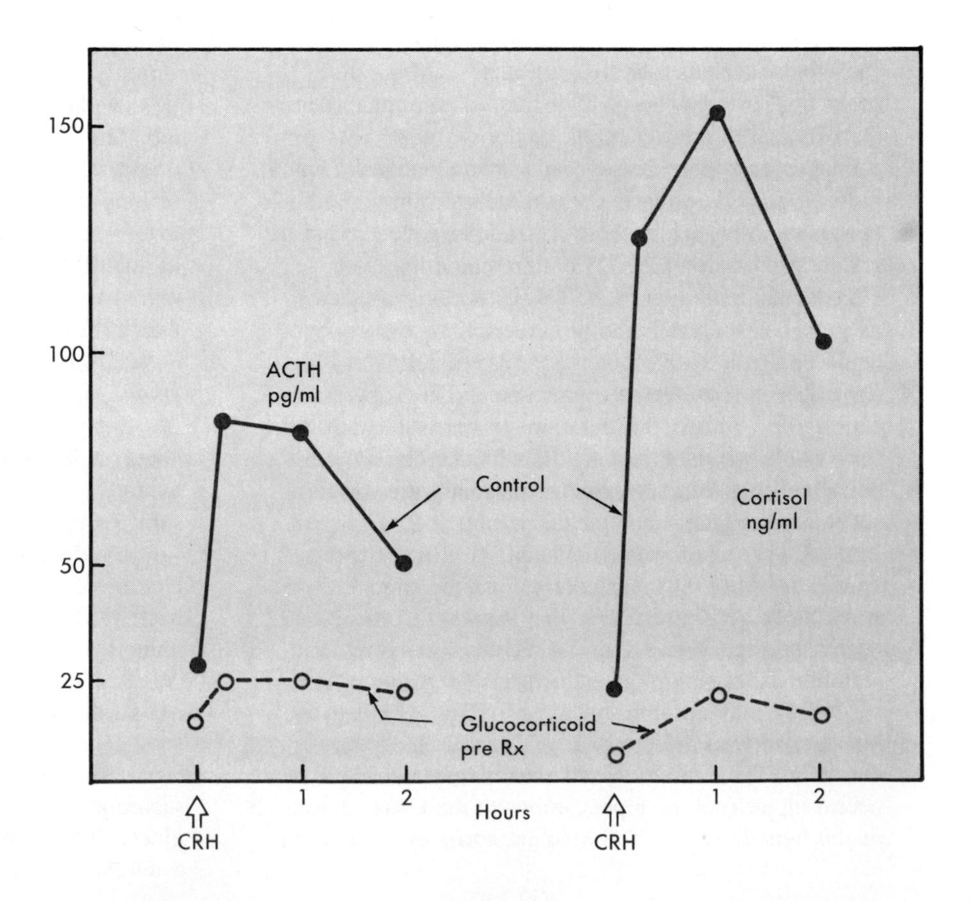

■ **Fig. 48-11** Plasma ACTH and cortisol responses to administration of CRH. Pretreatment with a synthetic glucocorticoid (analog of cortisol) suppresses the action of CRH on the pituitary. The diminished ACTH response leads secondarily to a diminished secretion of cortisol by the adrenal glands. (Redrawn from Copinschi G et al: *J Clin Endocrinol Metab* 57:1287, 1983.)

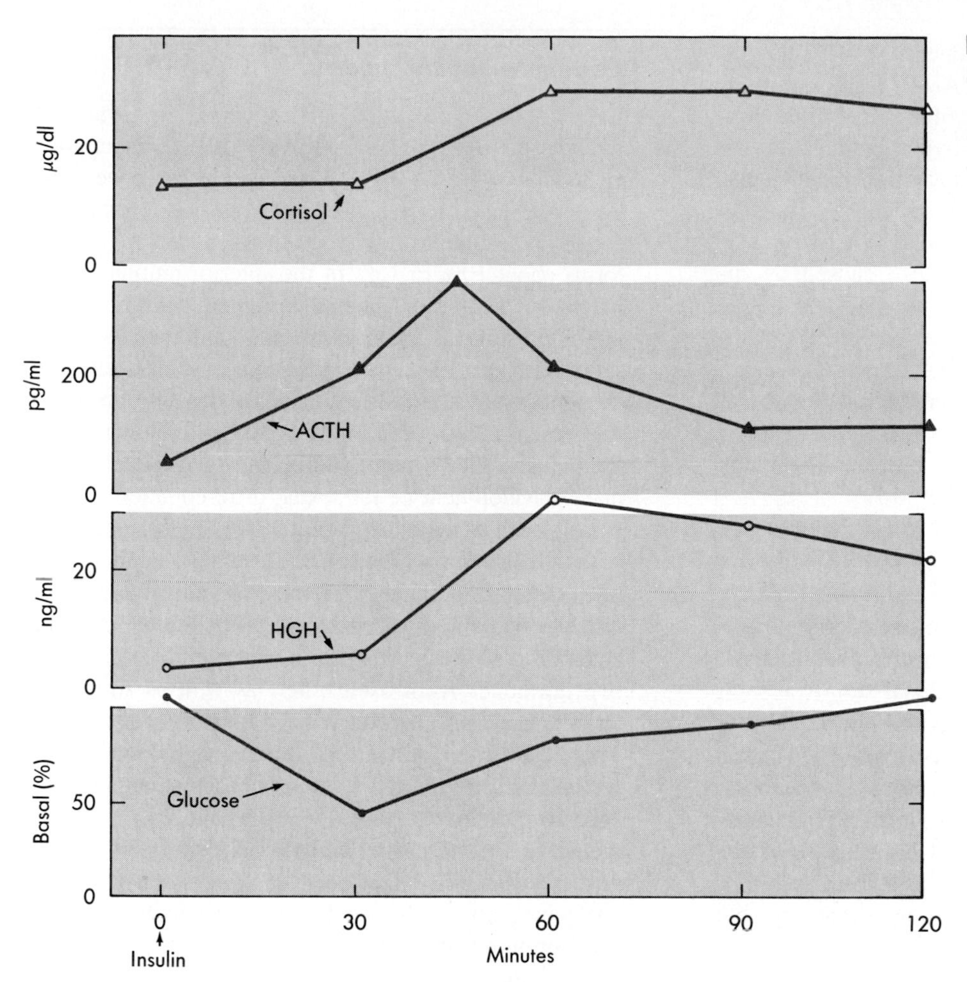

■ **Fig. 48-12** Stimulation of ACTH, cortisol, and *HGH* secretion by insulin-induced hypoglycemia in humans. (Redrawn from Ichikawa Y et al: *J Clin Endocrinol Metab* 34:895, 1972.)

the ACTH response to hypoglycemia. Other stress responses may be modulated by acetylcholine or GABA.

However in its most general sense stress is a life-threatening situation that usually evokes both CRH secretion and activation of the sympathetic nervous system. As detailed in Chapter 50 these have mutually reinforcing actions. CRH and norepinephrine stimulate each other's neuronal release. CRH itself, independent of ACTH, increases arousal, motor activity, vocalization, and sensitivity to auditory stimuli, and it reduces growth hormone secretion, sexual activity, and gonadotropin secretion. This pattern of response appears useful to the endangered individual.

ACTH circulates unbound in plasma with a half-life of 15 minutes. Basal concentrations at 6 AM range from 20 to 100 pg/ml (average 50 pg/ml or 10^{-11} M). Daily production is probably less than 100 µg, compared with an average adult human pituitary content of 250 to 500 µg. The bulk of the ACTH is secreted in a limited period of each day, which allows time for recovery of stores.

Action of ACTH. *ACTH stimulates the growth of those specific zones of the adrenal cortex concerned with secretion of cortisol and other steroid hormones.* In this respect the effect is to increase the size rather than the number of adrenal cells. In the absence of ACTH profound atrophy of the relevant adrenal zones occurs. ACTH action follows binding to a specific plasma membrane receptor and a rise in cAMP. A number of steps in the synthesis of adrenal steroids are thereby stimulated by ACTH, and these are detailed in Chapter 50. Because of the rapidity of synthesis ACTH promptly causes secretion of adrenocortical hormones. Adrenal responsiveness to ACTH is attenuated and delayed by prior chronic underexposure to the tropic hormone; conversely responsiveness is accentuated by prior chronic overexposure.

In animals extraadrenal actions of ACTH, such as stimulation of lipolysis, have been described. In addition ACTH synthesis and receptors for ACTH are present in the brain and gastrointestinal tract, where the peptide may have neuromodulatory or paracrine functions. A feedback relationship between ACTH and the immune system has emerged. ACTH receptors and ACTH secretion occur in lymphocytes; in turn interleukin products from lymphocytes stimulate ACTH release by corticotrophs.

ACTH, because of its MSH sequences, increases skin pigmentation. In amphibians MSH acts on melanocytes, causing the dispersal of melanin pigment granules within these cells and their dendrites. This action, which is probably mediated by cAMP, results in darkening of the skin. In humans it is more likely that peptides with MSH activity cause hyperpigmentation by stimulating melanin synthesis and the transfer of melanin from the melanocytes to epidermal cells.

The pathological effects of a primary excess or deficiency of ACTH are essentially those caused by increased or decreased secretion of adrenocortical hormones. These are described in Chapter 50. Hyperpigmentation of the skin characterizes those diseases in which large increases in ACTH secretion occur.

Secretion and Actions of Other Proopiomelanocortin Peptides

As already noted the remaining peptides in proopiomelanocortin are under the same transcriptional and translational control as ACTH. The functional significance of this fact is still not well understood, though it could reflect a coordinated physiological response. The plasma level of each peptide rises and falls in parallel with the ACTH level in feedback and stress situations. The molar ratios of these levels differ from 1.0 because of differences in metabolic clearance rates. The lipotropins were so named because of their lipolytic activity, but their physiological role in mobilizing fatty acids from human adipose tissue is unknown. Similarly the low plasma levels—or even all of the sources—of circulating β-endorphin are of uncertain significance. When administered to humans β-endorphins (like opioids) inhibit ACTH secretion, as well as gonadotropin secretion. They also stimulate prolactin, insulin, and glucagon secretion. These observations may simply point to neurocrine actions of β-endorphin generated in the hypothalamus or paracrine actions of the peptide generated in the pancreatic islets. Preliminary studies suggest that the N-terminal peptide of proopiomelanocortin may promote growth of the adrenal cortex or contain specific androgenic steroid stimulating sequences.

Gonadotropic Hormones

LH and **FSH** are glycoproteins whose function is to regulate the development, growth, pubertal maturation, reproductive processes, and sex steroid hormone secretion of the gonads of either sex. Both hormones are secreted by a single cell type, the gonadotroph, which forms about 10% to 15% of the anterior pituitary population and which is scattered throughout the gland. By immunohistological processes, the same cell generally stains for both FSH and LH, but occasional cells contain only one or the other gonadotropin. Both hormones are present by 10 to 12 weeks of fetal life; however neither is absolutely required for initial intrauterine gonadal development or for sexual differentiation.

LH, with a molecular weight of 28,000, and FSH, with a molecular weight of 33,000, have similar structures. Each is composed of the common pituitary hormone α subunit (molecular weight, 14,000; 92 amino acids) and a unique β subunit, which differentiates the two hormones from each other, as well as from TSH and human chorionic gonadotropin (HCG) (Fig. 48-13). The α and β subunits are held together by noncovalent forces. Disulfide bridges create tertiary structures. The carbohydrate moieties are 15% (LH) and 25% (FSH) by weight and contain oligosaccharides composed of mannose, galactose, fucose, galactosamine, acetylglucosamine, and sialic acid. The carbohydrate groups function in receptor binding and postreceptor responses, whereas the sialic acid residues decrease the rate of hormone degradation. Neither the β subunit of LH nor that of FSH is biologically active by itself. The α subunit is required for binding of gonadotropins to their receptors.

The details of LH and FSH biosynthesis are similar to those already described for TSH. Individual genes code for the two subunits; transcription of the β subunit gene is rate-limiting for gonadotropin synthesis. Ribosomal assembly of the two peptide chains is followed by addition and later modification of the carbohydrate moieties. As a result of this the bioactivity of secreted LH and FSH molecules can vary considerably in different physiological circumstances. In women the stores of

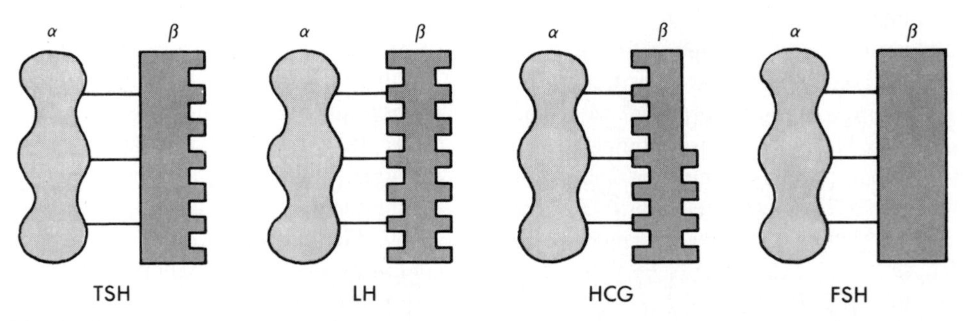

■ **Fig. 48-13** Structural similarities among TSH, LH, HCG, and FSH are depicted schematically. Note all share the same α-subunit.

both LH and FSH in their secretory granules fluctuate throughout the menstrual cycle and are highest just before ovulation. The patterns of release suggest more than one intracellular pool of LH.

Secretion of LH and FSH. The regulation of LH and FSH secretion is highly complex, embodying pulsatile, periodic, diurnal, and cyclic, and stage of life elements. It is also different in women and men. The main factors controlling gonadotropin secretion are discussed in this chapter; their reproductive function in both genders will be reiterated and amplified in Chapter 51. Both the secretion of LH and that of FSH are stimulated primarily by a single hypothalamic hormone, known either as **gonadotropin releasing hormone (GnRH)** or **luteinizing hormone–releasing hormone (LHRH).** As the latter name implies, it causes a much greater increase in LH than in FSH secretion. Whether a separate hypothalamic releasing hormone with a greater specificity for FSH exists remains uncertain. GnRH is a decapeptide (see Table 48-1) that is synthesized from a large prohormone that also gives rise to other products. The cells of origin of GnRH are predominantly in the arcuate nucleus and the preoptic area of the hypothalamus. Whether these two clusters have different functional roles is not yet known. After transport to the median eminence GnRH is stored in small granules.

GnRH neurons are under dopaminergic, serotonergic, noradrenergic and endorphinergic influence. In particular GnRH neurons are closely associated with dopamine neurons within the arcuate nucleus of the hypothalamus. Dopamine inhibits LH secretion directly and by decreasing GnRH release. Endorphins also inhibit GnRH release and LH secretion. Norepinephrine input to GnRH neurons is probably stimulatory in humans. Neural input from the retina to the hypothalamus accounts for the influence of light/dark cycles on GnRH release. In some species melatonin from the pineal gland may mediate seasonal variations in gonadotropin secretion and in reproductive activity; these variations are related to daylight length. Melatonin, which inhibits gonadotropin release, is itself suppressed by light. In this way the hormone may mediate environmental influences on reproduction. However the role of melatonin in regulating human gonadotropin secretion is not established. Connections from the olfactory bulb probably transfer reproductive signals received from another individual via pheromones, which are airborne or waterborne chemical exciters or inhibitors. Menstrual function in women and sperm production in men are commonly lost during prolonged physical or psychic stress. This may be mediated by CRH, which inhibits GnRH release by augmenting endorphinergic tone.

GnRH binds to a gonadotroph plasma membrane receptor, causing it to microaggregate. Calcium-calmodulin and phosphatidylinositol products are generated as principal second messengers. The exocytosis of gonadotropin secretory granules is rapidly stimulated, followed by an increase in transcription of the LH and

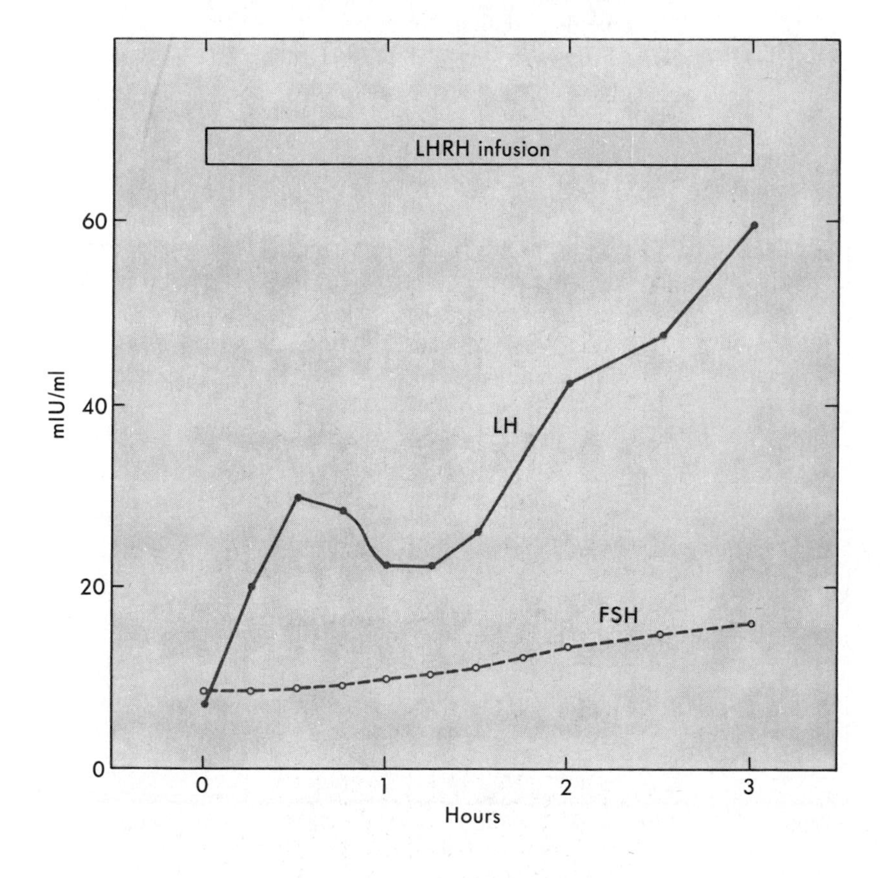

■ **Fig. 48-14** Stimulation of gonadotropin release by GnRH. Note biphasic response of LH and uniphasic response of FSH. (Redrawn from Wang CF et al: *J Clin Endocrinol Metab* 42:718, 1976.)

FSH β subunit genes. The bioactivity of LH and FSH is also increased by modification of their carbohydrate composition.

GnRH infusion causes a biphasic response in plasma LH; the initial peak is reached at 30 minutes, followed by a secondary rise beginning at 90 minutes and continuing for hours thereafter (Fig. 48-14). In contrast GnRH causes only a uniphasic progressive rise in FSH (see Fig. 48-14). In normal women LH is secreted in pulses characterized by a 15-minute upsurge and a falloff with a half-life of 60 minutes. These peaks of plasma LH have a periodicity varying from 1 to 7 hours, depending on the phase of the menstrual cycle. The amplitude of the pulses can be equivalent to 100% changes in the plasma LH level, except at the time of ovulation, when it is much greater. Normal men also exhibit 8 to 10 secretory bursts of LH per day (Fig. 48-15).

Much evidence indicates that the pulsatile secretion of LH is due primarily to pulsatile secretion of GnRH rather than to a rapidly fluctuating sensitivity of the gonadotrophs to the releasing hormone (Fig. 48-16). Moreover pulsatility does not depend on the presence of sex steroid hormones from target glands, because agonadal individuals and postmenopausal women exhibit even sharper spikes of the plasma LH level. Pulsatile secretion of LH is minimal in young children but increases sharply as puberty approaches, at first occurring only at night. During the initial stages of puberty this produces a nocturnal peak of LH. Although this diurnal

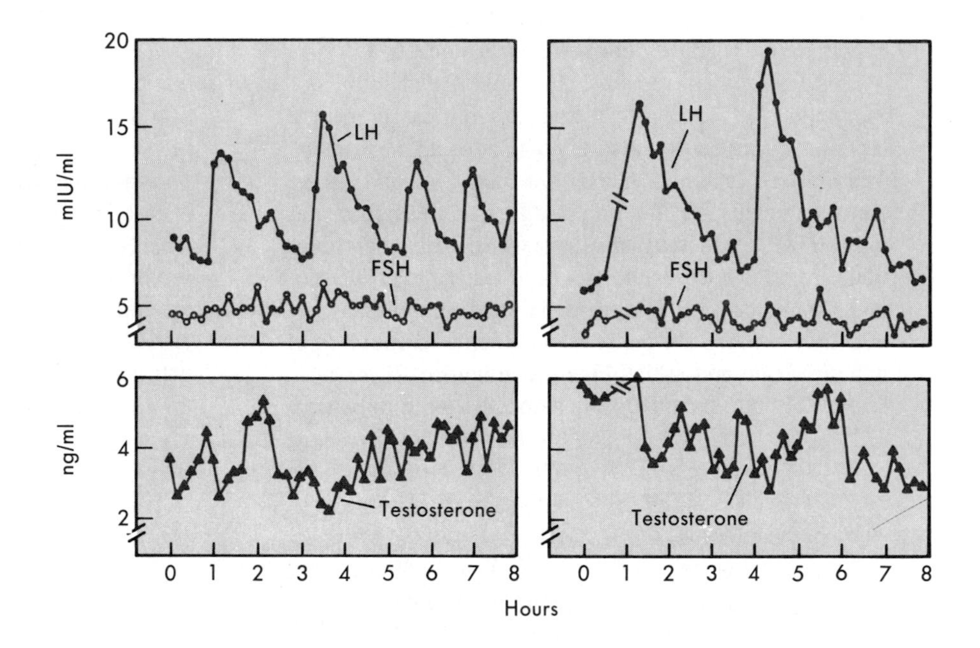

■ **Fig. 48-15** Pulsatile fluctuations in plasma LH levels and its target hormone, testosterone, in men. (Redrawn from Naftolin F et al: *J Clin Endocrinol Metab* 36:285, 1973.)

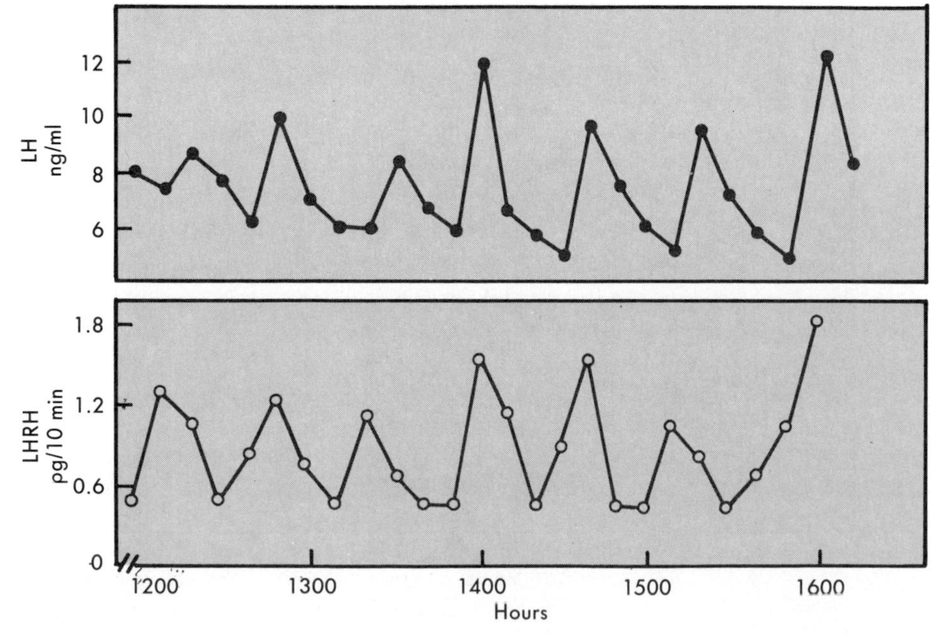

■ **Fig. 48-16** Fluctuation of peripheral vein plasma LH levels and portal vein plasma GnRH levels in unanesthetized ovariectomized female sheep. Each pulse of LH is coordinated with a pulse of GnRH. This supports the view that pulsatility of LH release is dependent on pulsatile stimulation of the pituitary by GnRH. (From Levine J et al: *Endocrinology* 111:1449, 1982.)

pattern lasts only 1 or 2 years, disappearing as puberty is completed, the pulsatility of LH secretion becomes fixed. The most striking feature of LH secretion in women, as opposed to men, is its monthly cyclicity. This results from a complex interaction between the GnRH neuron-gonadotroph unit and sequential changes in ovarian steroid secretion.

FSH secretion also exhibits a pulsatile pattern, usually synchronized with LH, but of lesser magnitude (see Fig. 48-15). The possibility of a separate hypothalamic releasing hormone for FSH has been suggested by the fact that the ratio of FSH to LH levels in plasma can fluctuate considerably. However the frequency of LHRH pulses may be a factor, because a decreased frequency is associated with an increased FSH/LH ratio. In addition the sensitivity of the individual FSH and LH secreting gonadotrophs may be differentially affected by feedback from gonadal sex steroid and protein hormones.

The critical importance of pulsatile GnRH release is shown by the fact that GnRH agonists with long biological half-lives (superagonists) eventually suppress LH and FSH secretions by down-regulating GnRH receptors. These superagonists are used therapeutically in clinical circumstances where it is important to eliminate endogenous gonodotrophic cell funcion or even gonadal function.

Feedback regulation of gonadotropins. The secretion of both LH and FSH is regulated by gonadal products. However the patterns and mechanisms are more complex than those that have thus far been described for TSH and ACTH. It is best to consider first the basic framework, which is of the classic negative feedback type. Plasma levels of FSH and LH are both elevated by surgical or functional absence of the gonads. FSH however is usually increased proportionally more than LH. A number of gonadal products from at least two gonadal cell types normally act to restrain the secretion of each gonadotropin. The basic schema is depicted in Fig. 48-17.

The major androgen, **testosterone,** from the Leydig cells of the testis and the interstitial cells of the ovary, inhibits the release of LH. The major estrogen, **estradiol,** which arises from the granulosa cells of the ovary and the Sertoli cells of the testis (as well as by conversion from testosterone in peripheral tissues), also inhibits the release of LH. Both the amplitude and the frequency of the LH pulses are affected; such changes indicate pituitary and hypothalamic sites of feedback.

Estradiol and testosterone administration blunt the response of the gonadotroph to a single pulse of GnRH. Conversely in estradiol-deficient women and in testosterone-deficient men LH responses to GnRH are exaggerated. In addition to inhibiting release of LH (and FSH) in some circumstances estradiol decreases their synthesis by lowering the levels of messenger RNA for the common α subunit and for the specific β subunits of the gonadotropins. Some of these actions of estradiol

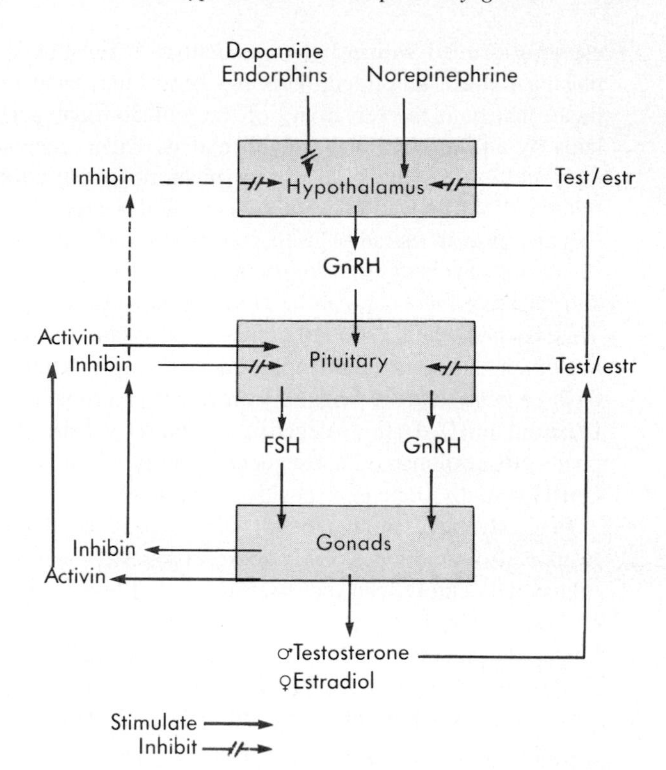

Fig. 48-17 Regulation of gonadotropin secretion. GnRH stimulates LH and FSH release. The gonadal steroids, estradiol in women and testosterone in men, exert negative feedback (1) at the pituitary level by blocking GnRH action and (2) at the hypothalamus level by inhibiting GnRH release. Negative modulation by endorphins and dopamine may mediate some of the steroid hormone feedback. Separate gonadal protein products selectively suppress (inhibin) or stimulate (activin) FSH release.

may be mediated by altering the number of GnRH receptors on the gonadotroph. In addition to these pituitary effects estradiol and testosterone also act on the hypothalamus to decrease GnRH secretion, probably via interaction with endorphin neurons in the hypothalamus. The latter then complete this pathway of negative feedback by suppressing discharge of GnRH from the median eminence into the portal blood.

FSH secretion is also inhibited by estradiol and testosterone, as they block the pituitary response to GnRH. However feedback inhibition of FSH secretion is specifically carried out by another gonadal product, a glycoprotein called **inhibin.** Secreted by ovarian granulosa cells and testicular Sertoli cells, inhibin suppresses FSH β subunit synthesis, GnRH stimulated FSH release, and possibly GnRH secretion. In contrast inhibin has much less if any effect on LH secretion.

Engrafted on this negative feedback framework are striking positive feedback effects of estradiol in women. When estradiol is administered in an appropriate dose range and for a sufficient number of days, LH response to GnRH is *augmented* rather than reduced. Furthermore if GnRH is administered repetitively to properly estradiol-primed women, the cumulative increments in plasma LH are amplified. This has been interpreted to mean that both the sensitivity of the gonadotroph (perhaps by an increase in the number of its GnRH receptors) and its LH stores have been enhanced by estradiol treatment. Moreover in women so treated a rapid further increase in estradiol itself causes a significant rise in plasma levels of LH. Aspects of positive feedback and negative feedback can be observed simultaneously. This occurs when estradiol-deficient agonadal women are given initial estradiol replacement therapy. After 7 days of treatment the originally elevated basal levels of LH (and FSH) decline (negative feedback) yet the capacity to respond to subsequent repetitive doses of GnRH actually increases (positive feedback).

Progesterone, another major steroid product of the ovary, also modulates LH release. Progesterone, administered acutely, can increase plasma LH levels 24 to 48 hours later. Progesterone can also either enhance or blunt the positive feedback effects of estradiol on GnRH responsiveness, depending on the timing with which the two hormones are administered. However continuous administration of progesterone (or analogues) inhibits gonadotropin secretion. This is one basic mechanism of action of the oral contraceptives. Still other protein products of the gonads influence FSH secretion. **Activin,** related structurally to inhibin, stimulates FSH synthesis and release. **Follistatin** inhibits FSH secretion. Thus complex interactions between gonadal products and gonadotropin secretion help to shape the hormonal pattern of the menstrual cycle.

Prolactin, a mammotropic hormone from the anterior pituitary, also inhibits GnRH release and lowers basal secretion of LH and FSH. Although the physiological significance of the effect is not well understood, it explains many cases of infertility and pathological loss of menses. Finally LH can inhibit secretion of its own releasing hormone via short-loop negative feedback. The route of access of LH to the hypothalamic GnRH neuron may be retrograde flow in the pituitary portal veins or via the tanycytes.

LH and FSH both circulate unbound to plasma proteins. The concentrations of both are in the range of 5 to 20 mIU/ml in men and in reproductive-age women. In the latter the levels of both hormones are higher in the first half of the menstrual cycle than in the second half; in addition both hormones show sharp, single-day peaks at the time of ovulation. Basal plasma concentrations of each hormone are of the order of 10^{-11} M. The metabolic clearance rates of LH and FSH, respectively, are 36 and 20 L/day, and their half-lives in plasma are approximately 1 and 3 hours. The slower rate of degradation of FSH reflects a high sialic acid content. In contrast to the trivial excretion of other peptide hormones 10% of the daily production of LH and FSH appears in the urine. This permits employment of urinary gonadotropin measurements as a reflection of integrated plasma concentrations. Such measurements are particularly useful when plasma levels are low, as in children. Home measurements of urinary gonadotropins by women can help them anticipate ovulation, assist in conception, and detect early pregnancy. The common α subunit is secreted to a small extent by normal gonadotrophs. In contrast the individual β subunits of LH and FSH are secreted only by hyperstimulated or neoplastic gonadotrophs.

Actions of gonadotropins. LH and FSH bind to specific plasma membrane receptors; in each case cAMP is generated as the second messenger. FSH stimulates ovarian granulosa cells and testicular Sertoli cells to synthesize and secrete estradiol and a variety of protein products essential to oogenesis and spermatogenesis, respectively. LH stimulates ovarian interstitial (thecal) cells and testicular Leydig cells to secrete testosterone and other products with vital roles in reproduction. The actions of gonadotropins are discussed in detail in Chapter 51.

Abnormalities in secretion. Disorders (usually neoplasms) of the hypothalamus or pituitary gland may produce deficiency of one or both gonadotropins. With rare exceptions this leads to a loss of reproductive capacity in adults and can cause some regression of already established secondary sexual characteristics. If gonadotropin deficiency occurs before puberty, the development of secondary sexual characteristics, the expected rapid growth spurt, and skeletal maturation are all prevented. Slow growth may continue for a long time, producing the eunuchoidal habitus, that is, a tall, juvenile-appearing adult. A primary excess of FSH or LH secretion is exceedingly rare, and a unique clinical picture cannot be described. A persistently elevated LH/FSH ratio is associated with disordered reproductive function in women (polycystic ovary syndrome).

Growth Hormone (Somatotropin)

Growth hormone stimulates postnatal somatic growth and development. In addition it has numerous actions on protein, carbohydrate, and fat metabolism. The hormone originates in anterior pituitary cells that make up 40% to 50% of the adult gland. It is stored in large dense granules. Typically somatotrophs stain with acidophilic dyes. In humans these cells can form tumors that secrete growth hormone and produce a highly distinctive disease called *acromegaly*.

Growth hormone (GH) is a single-chain polypeptide with a molecular weight of 22,000. It contains 191 amino acids and 2 disulfide bridges (Fig. 48-18). The three-dimensional structure is not fully known, but sites at each end of the molecule participate in binding to receptors. Only primate growth hormones are active in humans. Until recently this required that GH-deficient individuals be treated with hormone extracted from human pituitaries. However GH is now produced by re-

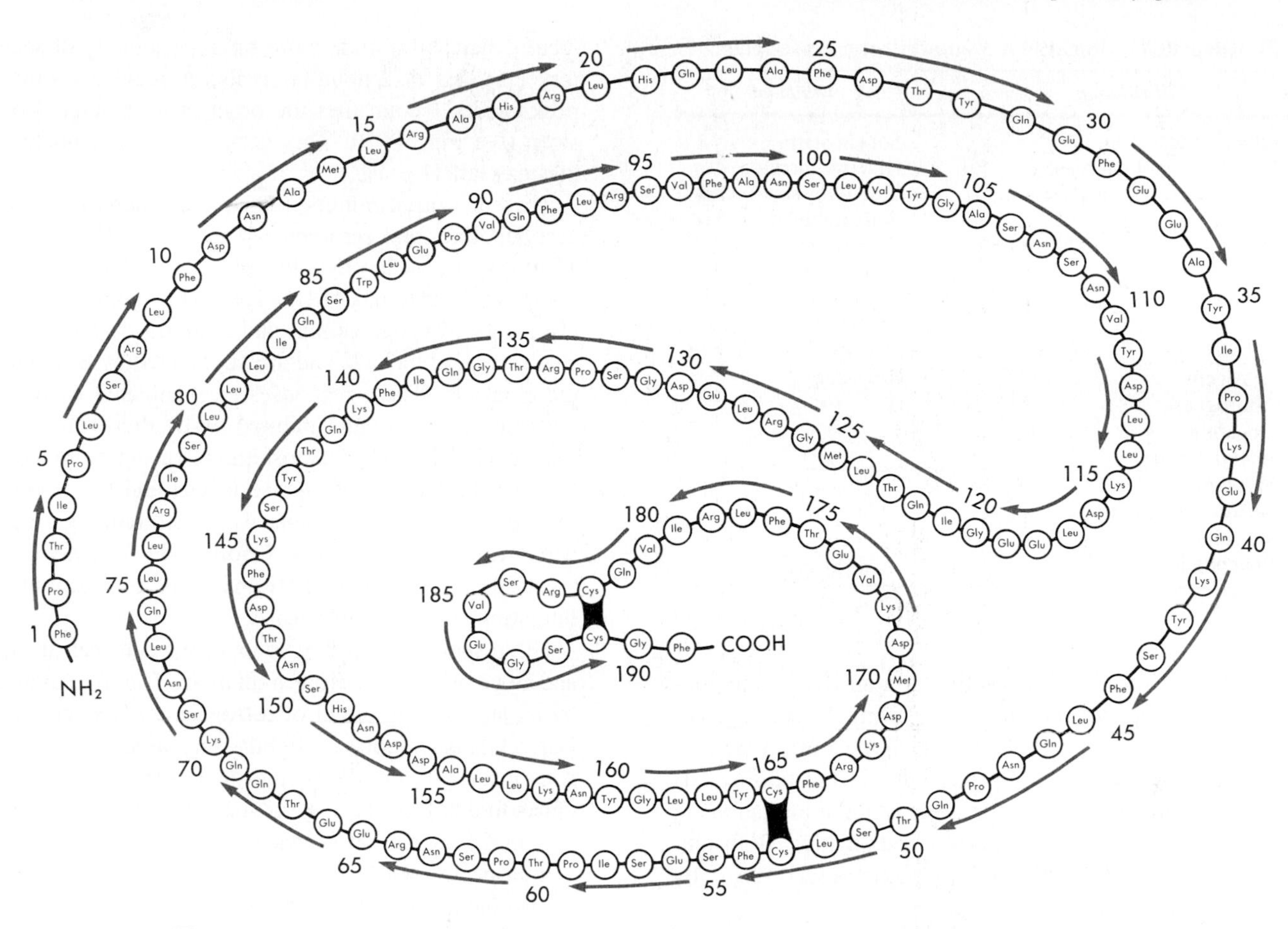

Fig. 48-18 Structure of human growth hormone. Though consisting of nearly 200 amino acids and containing two disulfide bridges, growth hormone is a single polypeptide chain (From Li C et al: *Proc Natl Acad Sci* 74:1016, 1977.)

combinant DNA technology; this method assures a dependable therapeutic supply.

Synthesis and release of GH. The normal pituitary GH gene is a member of a gene family that also directs the synthesis of the structurally related hormones prolactin, human placental lactogen, and a growth hormone variant produced exclusively in the placenta. Four additional genes capable of directing synthesis of slightly larger GH molecules are present but are not normally expressed. A constitutive protein transcription factor that binds to the promotor region of the GH gene appears essential to its selective expression in the pituitary gland. The gene transcribes a messenger RNA that directs synthesis of a prehormone. Subsequently a signal peptide is removed and the hormone in final form is stored in granules. The synthesis of GH is increased by its specific hypothalamic releasing hormone, GHRH (see Table 48-1), which causes rapid stimulation of gene transcription. Thyroid hormone and cortisol synergistically induce GH synthesis by transcriptional mechanisms.

The release of GH requires activation of a microtubular system that carries the granules to the plasma membrane before exocytosis. Release is stimulated by GHRH, after the latter binds to a plasma membrane receptor. cAMP and Ca^{++} are primary mediators, and phosphatidylinositol products are secondary mediators of GHRH action. Prostaglandins are also potent stimulators of GH release in vitro.

Another hypothalamic peptide, **somatostatin** (see Table 48-1) is a powerful inhibitor of GH release. Somatostatin blocks GHRH stimulation in a noncompetitive manner. An N-terminal extended form, known as somatostatin-28, is also physiologically active, but it exhibits different dose-response characteristics. Somatostatin acts through its own plasma membrane receptor, in part by decreasing intracellular cAMP and calcium levels. GH is secreted in pulses, which are due to pulsatile release of GHRH into the portal blood. Somatostatin diminishes the degree to which the somatotroph is able to respond to the GHRH pulses and may also reduce their frequency.

Secretion of GH. As seen in Table 48-4 GH secretion is under many different influences. An acute fall in plasma levels of either of the major energy-yielding substrates, glucose or free fatty acids (FFA), produces an increase in GH. For example when insulin is administered intravenously the plasma level of GH rises

Table 48-4 Regulation of growth hormone secretion

Stimulation	*Inhibition*
Glucose decrease	Somatostatin
Free fatty acid decrease	Glucose increase
Amino acid increase (arginine)	Free fatty acid increase
Fasting	Somatomedins
Prolonged caloric deprivation	Growth hormone
Stage IV sleep	β-Adrenergic agonists
Exercise	Cortisol
Stress (Table 48-3)	Senescence
Puberty	Obesity
Estrogens	Pregnancy
Androgens	
Dopamine	
Acetylcholine	
Serotonin	
α-Adrenergic agonists	
γ-Amino butyric acid	
Enkephalins	

2-fold to 10-fold 30 to 60 minutes after the plasma glucose level has declined to below 50 mg/dl (see Fig. 48-10). Conversely a carbohydrate-rich meal or a pure glucose load causes a prompt decrease in the plasma GH level of at least 50%. Responses to similar alterations in free fatty acid levels are slower and smaller. Both glucose and free fatty acids mainly suppress GH release by increasing somatostatin.

A high protein meal or the infusion of a mixture of amino acids raises the plasma GH level; arginine is the most consistent amino acid stimulator. However prolonged protein calorie deprivation or total fasting also stimulate GH secretion, probably by decreasing negative feedback (see below). Exercise and various stresses, including blood drawing, anesthesia, fever, trauma, and major surgery, increase GH secretion quickly. In addition to GH spikes produced by these factors there is an underlying basal periodicity of secretory episodes at 2-hour intervals. A regular nocturnal peak occurs 1 hour after the onset of deep stage 3 or 4 sleep (see Fig. 48-9). This is preceded by a nocturnal plasma GHRH peak.

The neurotransmitters dopamine, norepinephrine, acetylcholine, and serotonin all increase GH secretion. They may act at the hypothalamic or median eminence level by stimulating GHRH. The GH responses to exercise, stress, hypoglycemia, and arginine are reduced by α-adrenergic blockade and are augmented by β-adrenergic blockade. These responses are facilitated by α_2-adrenergic receptors and inhibited by β-adrenergic receptors in GHRH and somatostatin releasing neurons. In contrast the sleep-induced rise in GH and the response to hypoglycemia are enhanced by serotoninergic pathways from the brain stem. Cholinergic pathways augment GH response to GHRH and to all other stimuli by inhibiting somatostatin release.

GH responsiveness is greater in women than in men and is greatest just before ovulation. This is explained by an augmenting effect of estrogens on GH secretion. Daily GH secretion is slightly increased in children, rises further during the period of puberty, and then declines to adult levels (Fig. 48-19). A further late reduction in GH secretion in response to GHRH and other stimuli occurs with aging. This may be partly responsible for the decline in lean body mass and metabolic rate as well as the increase in adipose mass that characterizes human senescence. During puberty the increase in GH pulses and daily secretion correlates with the rate of increase in height. Furthermore especially tall adults demonstrate greater responses to GHRH than do adults of average height. Thus the final height of humans may be partly determined by their inherent GH secretory capacity.

Feedback regulation of GH is also complex (Fig. 48-

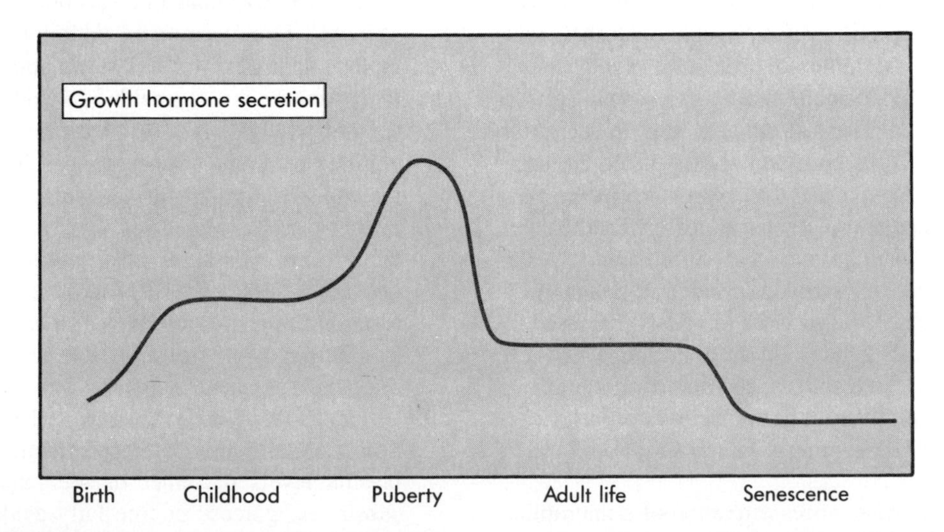

Fig. 48-19 Lifetime pattern of growth hormone (GH) secretion. GH levels are higher in children than adults with a peak period during puberty. GH secretion declines with aging.

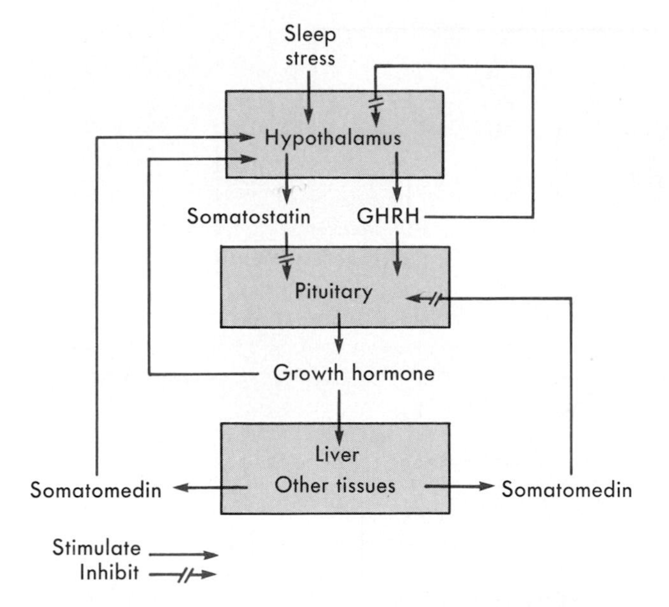

■ **Fig. 48-20** Regulation of growth hormone (GH) secretion. The hypothalamic peptide (GHRH) stimulates growth hormone release, whereas the hypothalamic peptide somatostatin inhibits it. Negative feedback is by the peripheral mediator of HGH action: somatomedin. Negative feedback occurs both via somatomedin inhibition of GHRH action and by somatomedin stimulation of somatostatin release. HGH inhibits its own secretion by short-loop feedback. In addition GHRH inhibits its own release via ultra short-loop feedback. In both of these cases the negative feedback is probably via increasing somatostatin release.

20). For example GH inhibits its own secretion. Administration of exogenous GH dampens subsequent endogenous GH responsiveness to a number of stimuli, including hypoglycemia and stage IV sleep. The mechanisms may involve short-loop feedback because GH stimulates the synthesis and release of somatostatin (see Fig. 48-20). However somatostatin synthesis and release are also increased by **somatomedins,** which are peptide mediators of GH action that are generated in the periphery. Somatomedins also act at the pituitary level to decrease responsiveness to GHRH (see Fig. 48-20). Finally GHRH itself, in doses too small to stimulate the somatotrophs but administered in a manner that allows access to the hypothalamus, may decrease rather than increase GH secretion. This paradoxical effect likely results from a stimulation of somatostatin release and may reflect the existence of axonal connections between GHRH neurons and somatostatin neurons.

Other negative regulatory influences are also noted. Although basal cortisol levels stimulate GH gene expression, an excess of this steroid decreases GH responses to GHRH. Insulin represses GH gene expression. A decline in pituitary GH secretion occurs during the latter part of pregnancy, perhaps in response to production of placental GH and placental lactogen. Obese individuals exhibit dampened GH responses to all stimuli, including GHRH itself. These responses are restored to normal by weight reduction.

The normal basal plasma GH concentration is 1 to 5 ng/ml (about 10^{-10} M). This may increase 10-fold to 50-fold under various stimuli. The plasma half-life of GH is 20 minutes and the metabolic clearance rate is 350 L/day. Daily secretion in normal adults is 300 to 500 μg, only 5% of the large pituitary store. A minor portion of the circulating immunoreactive GH consists of larger variants whose biological activity is unknown. Circulating GH is bound to a plasma protein whose structure is virtually identical to the extracellular portion of the hepatic plasma membrane GH receptor. Although only 0.002% of secreted GH is excreted unchanged by the kidney, daily urinary GH excretion correlates well with the integrated 24-hour plasma GH profile.

Actions of GH. *GH is a hormone with profound anabolic action. In its absence animals and humans show stunted growth. When it is administered to GH-deficient individuals it causes prompt nitrogen retention, hypoaminoacidemia, and decreased urea production because the amino acids are diverted from oxidation to protein synthesis as growth ensues.*

The multiplicity of GH targets and effects is indicated in Fig. 48-21. The most striking and specific effect is the stimulation of linear growth that results from GH action on the epiphyseal cartilage or growth plates of long bones. All aspects of the metabolism of the cartilage-forming cells, the chondrocytes, are stimulated. This includes the incorporation of proline into collagen and its conversion to hydroxyproline and the incorporation of sulfate into the proteoglycan chondroitin. Together chondroitin and collagen form the resilient extracellular matrix of cartilage. GH also stimulates the proliferation of chondrocytes, as well as their synthesis of DNA, RNA, and proteins. In support of protein synthesis GH also stimulates cellular uptake of amino acids.

Other tissues share in the anabolic response. GH increases the activity and probably the number of bone modeling units. After GH administration initially urinary hydroxyproline and calcium excretion increase, indicating activation of osteoclastic resorption. However subsequently plasma osteocalcin levels rise, indicating osteoblastic responses, and ultimately total bone mass and mineral content are increased by GH. Visceral organs (liver, kidney, pancreas, intestines), endocrine glands (adrenals, parathyroids, pancreatic islets), skeletal muscle, heart, skin, and connective tissue all undergo hypertrophy and hyperplasia in response to GH. In most instances this is reflected in an enhanced functional capacity of the enlarged organ. For example renal plasma flow, glomerular filtration, cardiac output, and hepatic clearance of test substances are all increased by GH.

GH has several actions on carbohydrate and lipid metabolism. Normal levels of GH are required to sustain

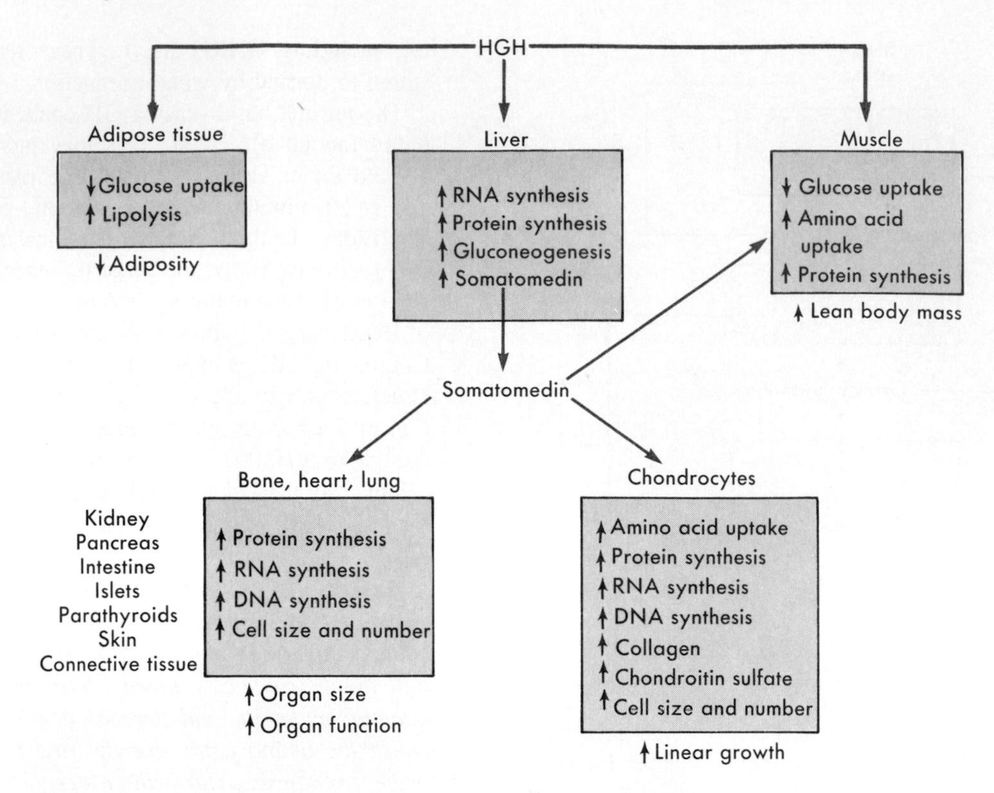

■ **Fig. 48-21** Biological actions of GH. The effects on linear growth, organ size, and lean body mass are mediated by somatomedin produced in the liver.

normal pancreatic islet function; in the absence of GH insulin secretion declines. However an excess of GH causes decreased glucose uptake in insulin-sensitive tissues, such as muscle, and increased hepatic glucose output. Insulin secretion then becomes elevated to compensate for the GH-induced insulin resistance. GH is also lipolytic; this characteristic leads to increases in plasma FFA and ketoacids, especially when insulin secretion cannot compensate adequately. The increased FFA levels may themselves contribute to GH-induced insulin resistance. Increased oxidation of fat and decreased oxidation of glucose is reflected in a decreased respiratory quotient; total metabolic rate is also usually increased. On balance GH is a diabetogenic hormone.

GH increases extracellular fluid by stimulating the renin-angiotensin-aldosterone axis (Chapter 49) and by suppressing atrial natriuretic peptide (ANP). Proximal tubular phosphate reabsorption and plasma phosphate are increased by GH. Calcium absorption from the gut is enhanced, probably by GH stimulation of 1,25-(OH) $_2$-D_3 production.

Mechanisms of GH action. A family of plasma membrane GH receptors of sizes varying from 40,000 to 110,000 molecular weight are present in target cells ranging from the liver to adipose tissue. All the receptors are composed of glycoprotein disulfide linked subunits that span the plasma membrane. It is the length of the intracellular cytoplasmic receptor tail that varies most from tissue to tissue. Synthesis of GH receptors requires GH itself to be present, but an excess of GH downregulates its receptors. The GH receptor is also induced by insulin and estrogens and is repressed by fasting. The mechanism of signal transduction after GH binding to its receptor is still unclear, but none of the known second messengers seems to mediate GH actions.

Earlier observations demonstrated that 12 hours elapse after administration of GH before its anabolic, growth-promoting effects become evident. It is now apparent that many if not all of these effects require the generation of a family of peptide hormone intermediaries. These are known as **somatomedins** or **insulin-like growth factors (IGFs).**

These compounds have a molecular weight of about 7,000 and are structurally related to proinsulin. IGF-1 is a 70-amino-acid straight chain peptide with 50% homology to the A-chain and B-chain domains of proinsulin (see Fig. 46-3). IGF-2 has 70% homology with IGF-1 in these same domains. Both somatomedins however have distinctive C-chain domains. Circulating somatomedins originate primarily as a result of GH stimulated production in the liver. In contrast to the sharp and rapid fluctuations of plasma GH, somatomedin concentrations are relatively stable. For example they do not increase after hypoglycemia as GH does. In general plasma IGF-2 is threefold to fourfold higher in concentration than IGF-1. The plasma half-life of somatomedins is also longer than that of GH, because they circulate bound to several carrier proteins. A large binding protein is synthesized in the liver in response to GH stimulation. A smaller binding protein is independent of GH but its production is suppressed by insulin. The binding proteins probably control access of circulating somatomedins to target cells under different physiological circumstances.

The growth-promoting effects of GH can be largely accounted for by the somatomedins. These have been shown to stimulate typical GH responses in cartilage, muscle, adipose tissue, fibroblasts, and tumor cells in vitro. Somatomedins also stimulate nitrogen retention and enhance renal function in vivo. Most evidence suggests that somatomedin molecules generated locally within clones of GH target cells (for example osteoblasts) and acting in autocrine or paracrine fashion may be more important than somatomedin molecules derived from the plasma. Somatomedins bind to specific plasma membrane receptors. The receptor for IGF-1 is a dimer, structurally similar to the insulin receptor, and it exhibits autophosphorylation and tyrosine kinase activity. It binds both insulin and IGF-2, though with lower affinities. The receptor for IGF-2 is a monomer that is dissimilar to those of IGF-1 and insulin. It binds IGF-1 with lower affinity and insulin not at all. The exact roles of cAMP and other second messengers in IGF-1 and IGF-2 actions remain undefined.

The cross-reactivities described may assume biological importance when very high concentrations of either somatomedins or of insulin exist. For example some patients with tumors that secrete somatomedins develop spontaneous hypoglycemia, because of activation of the insulin receptor. In other patients with insulin receptor deficiency who have (in compensation) extremely high plasma insulin levels excessive soft tissue growth develops because of activation of the somatomedin receptors.

Plasma somatomedins are increased by administration of GH, with a time lag of 12 to 18 hours, and they disappear from GH-deficient individuals. IGF-1 is clearly GH-dependent, and its plasma levels are very sensitive to changes in GH availability; IGF-2 levels do not reflect GH status nearly as well. During adolescence the increased secretion of GHRH and of GH produces an increase in plasma levels of IGF-1. The levels of IGF-1 correlate well with the progression of pubertal growth. Although GH itself is not necessary for fetal growth, one or more of the somatomedins produced in the placenta may be. Somatomedin production is reduced by factors that can override GH. Fasting, protein deprivation, and insulin deficiency all lead to diminished liver production of somatomedins and to a decrease in their plasma levels, despite increases in GH secretion. Indeed in these pathophysiological states the lack of somatomedin may be the cause of the elevated GH levels through negative feedback. Estrogens also decrease somatomedin production; this may account for their antagonism to GH action despite their stimulation of growth hormone secretion. Cortisol also diminishes somatomedin levels.

Overall role of GH in substrate flow. It is useful to review the interactions between GH and insulin in common physiological circumstances, as presented in Fig. 48-22. When protein and energy intake are both ample, the absorbed amino acids are used for protein synthesis and to stimulate growth. Hence both GH and insulin secretion are stimulated by amino acids, and together they augment the production of somatomedins. The latter in turn stimulate accretion of lean body mass. (These actions are probably directly enhanced by insulin as well.) The insulin antagonistic effect of the GH molecule itself on carbohydrate metabolism is also useful at this time; it helps to prevent hypoglycemia, which might result from insulin stimulation in the absence of carbohydrate.

On the other hand when a carbohydrate load is ingested and insulin secretion is correspondingly increased, GH secretion is suppressed. In this circumstance accelerated generation of somatomedins is not needed because protein anabolism is not advantageous in the absence of amino acid inflow. Neither is insulin antagonism necessary; on the contrary unrestrained expression of insulin action permits efficient storage of the excess carbohydrate calories. Finally when an individual is fasting insulin secretion falls, partly because of a fall in plasma glucose levels. Although this increases GH secretion the calorie deficit and significant deficiency of insulin lead to a decrease in somatomedin production. Again this is appropriate in a situation where an increase in protein anabolism is disadvantageous and protein catabolism is essential. However the increase in GH may still be beneficial during fasting, because it enhances lipolysis, decreases peripheral glucose use, and increases glucose production.

Clinical syndromes of GH dysfunction. Deficiency of GH in children can result from hypothalamic dysfunction, pituitary tumors, a biologically incompetent GH molecule, failure to generate somatomedins normally, or receptor deficiency. Short stature and correspondingly delayed bone maturation are the consequences. Mild obesity is common and puberty is usually delayed. In the adult no obvious signs are evident. The diagnosis of GH deficiency is established by demonstrating low plasma GH levels, which fail to rise after stimulation. This is confirmed by low levels of somatomedins. Replacement treatment with GH causes nitrogen retention, increased lean body mass, and decreased adipose mass. In children growth velocity increases. Pubescence occurs and fertility is established.

Sustained hypersecretion of GH results from pituitary tumors and produces a unique syndrome called **acromegaly.** If hypersecretion of GH begins before puberty is completed, the individual grows very tall and has long arms and legs. In adults only periosteal bone growth can be increased by GH, leading to widened fingers, toes, hands, and feet; prominent bony ridges

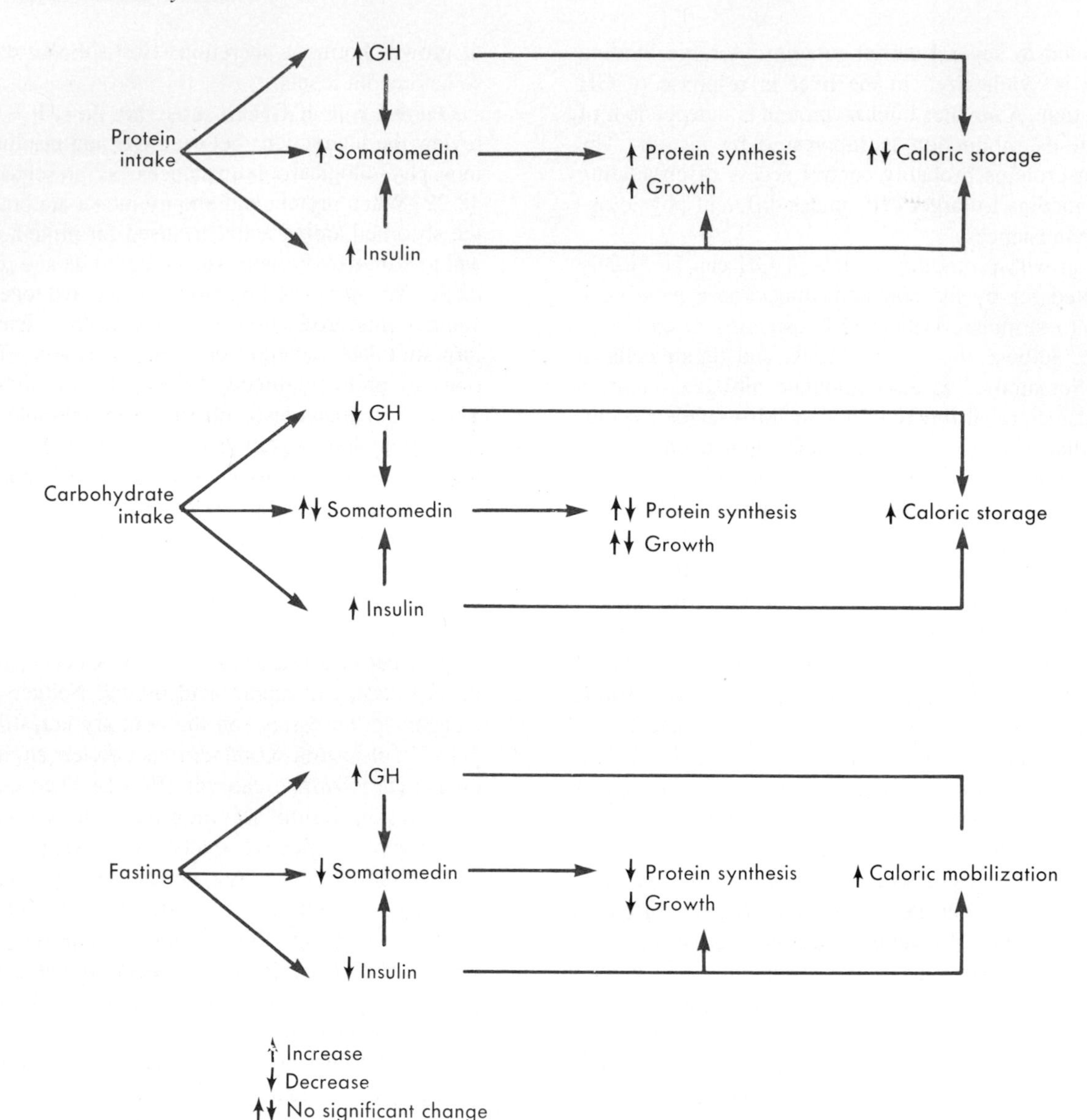

■ **Fig. 48-22** Complementary regulation of GH and insulin secretion coordinate nutrient availability with anabolism and caloric flux. Note that both hormones are increased by protein, and both stimulate protein synthesis.

above the eyes; and a prominent lower jaw. Facial features are coarsened by accumulation of excess soft tissue and a bulbous nose. The tongue is enlarged, and the skin is thick, whereas subcutaneous fat is sparse. Virtually all organ sizes are increased. Enlargement of the heart and accelerated atherosclerosis often lead to a shortened life span. The insulin antagonistic effect of GH produces an abnormal tolerance to carbohydrate or even frank diabetes mellitus requiring treatment with insulin. The diagnosis is confirmed by demonstrating elevated plasma GH levels, which are not suppressed when glucose is administered. Plasma levels of somatomedins are also high. Definitive treatment requires surgical removal of the tumor. Somatostatin analogues and dopaminergic agents that diminish HGH hypersecretion are also useful.

■ *Prolactin*

In humans prolactin is a protein hormone principally concerned with stimulating breast development and milk production. In addition it exerts an influence on reproductive function. Prolactin originates in specific anterior pituitary cells that comprise 10% to 25% of the pituitary population. They increase in number during pregnancy and lactation and with estrogen treatment.

Prolactin is a single-chain protein of molecular weight 23,000 with 198 amino acids and three disulfide bridges. Its gene and structure are homologous to those of GH (see Fig. 48-18) but the molecule has a major midportion loop. Synthesis of prolactin proceeds from a prehormone. The N-terminal signal peptide is cleaved

Table 48-5 Regulation of prolactin secretion

Stimulation	*Inhibition*
Pregnancy	Dopamine
Estrogen	Dopaminergic agonists
Nursing—breast manipulation	Somatostatin
Sleep	GnRH associated–peptide
Stress (Table 48-3)	Prolactin
TRH	γ-Aminobutyric acid (GABA)
Dopaminergic antagonists	
Opioids	
Serotonin	
Histamine antagonists (H_2)	
Adrenergic antagonists	
Candidate prolactin-releasing factors	
Vasoactive intestinal peptide	
Peptide histidine-isoleucine	
Oxytocin	
Angiotensin-2	
Neurotensin	
Galanin	
Neurophysin-2	
Intermediate lobe product	

and transient N-glycosylation takes place before arrival in the Golgi apparatus. There the hormone molecules subsequently destined to be stored in granules and released by acute stimuli (or secreted during pregnancy) are deglycosylated. However some of the N-glycosylated molecules escape complete processing and are secreted constitutively. These molecules form a major part of circulating prolactin in nonpregnant women, and they have lower biological activity. Transcription of the prolactin gene is regulated by factors that also regulate secretion of the hormone. Thus TRH increases prolactin messenger RNA, whereas dopamine decreases it.

Secretion of prolactin. Table 48-5 lists the most important influences on prolactin secretion. Consistent with its essential role in lactation, prolactin secretion increases steadily during pregnancy. This is probably mediated by the large increase in estrogen, which stimulates hyperplasia of prolactin-producing cells and synthesis of the hormone by inducing transcription of the gene. In addition although estrogen does not itself stimulate the release of prolactin, it enhances responsiveness to other stimuli. If a new mother does not nurse her child, the plasma level of prolactin declines 3 to 6 weeks after delivery to the normal range of nonpregnant women. However suckling (or any other form of nipple stimulation) maintains elevated levels of prolactin secretion, especially for the first 8 to 12 weeks (Fig. 48-23).

Like other tropic hormones prolactin secretion rises at night (see Fig. 48-9). The first peak appears 60 to 90 minutes after the onset of slow-wave sleep and subsequent peaks occur later, after cycles of REM (rapid eye movement) sleep. Stresses of various sorts, including anesthesia, surgery, insulin-induced hypoglycemia, fear, and mental tension, all cause prolactin release. The function of sleep or stress-induced prolactin release is unknown.

The pathways for regulating prolactin release in each physiological circumstance remain to be worked out. However uniquely among the pituitary hormones, pro-

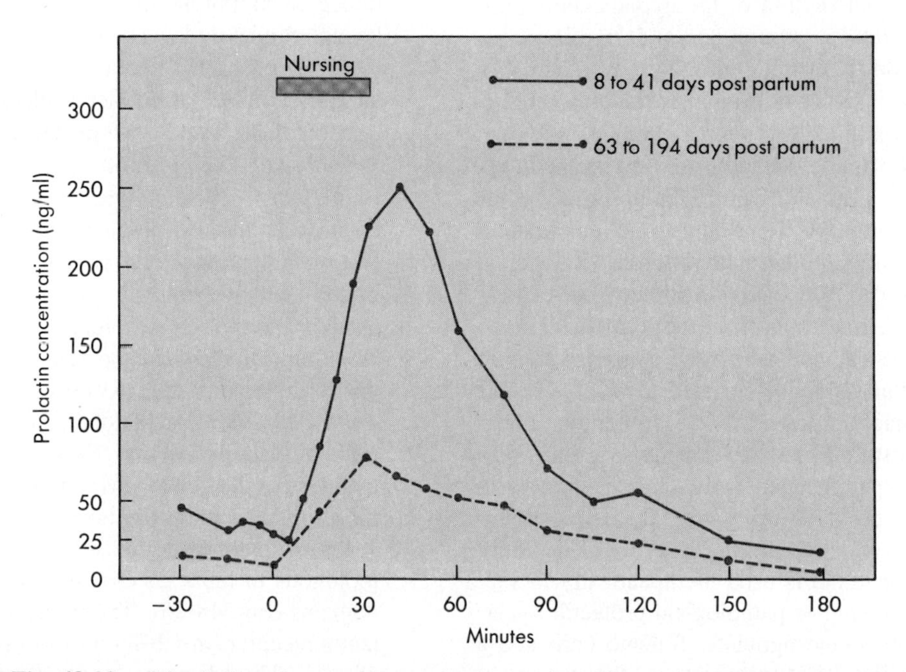

Fig. 48-23 Stimulation of prolactin secretion by nursing. Note decreased responses with increasing interval of time from delivery. (Redrawn from Noel GL et al: *J Clin Endocrinol Metab* 38:413, 1974.)

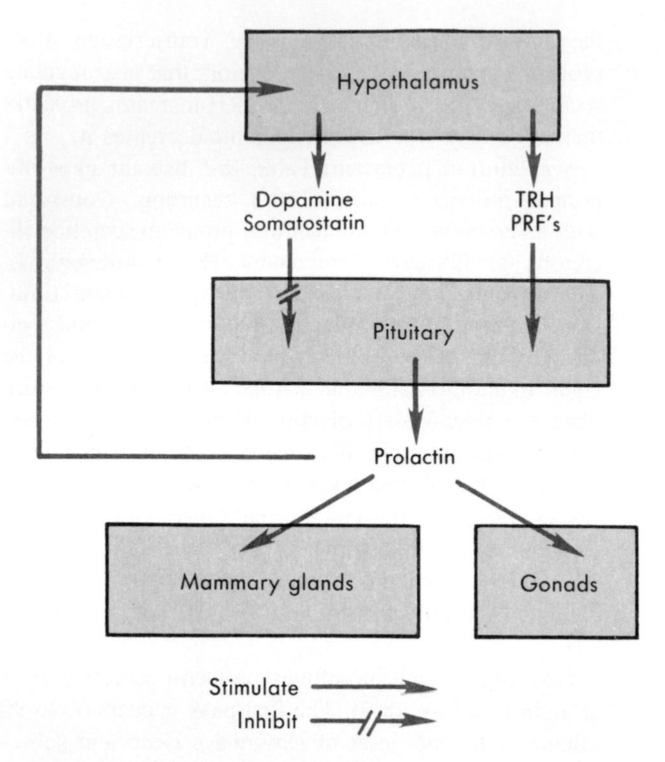

■ **Fig. 48-24** Regulation of prolactin secretion. The predominant mode of hypothalamic regulation is tonic inhibition by dopamine with a contribution from somatostatin. TRH and a number of prolactin-releasing factor (PRF) candidates stimulate prolactin release but their relative roles are uncertain. Prolactin itself exacts negative feedback by stimulating secretion of dopamine.

lactin secretion is tonically *inhibited* by the hypothalamus (Fig. 48-24). Disruption of the hypothalamic-pituitary connection induces prompt and enduring increases in plasma prolactin. Dopamine has many characteristics that qualify it for the role of primary **prolactin-inhibiting factor (PIF),** although it is not a hypothalamic peptide. This catecholamine strongly inhibits prolactin release, either when generated within the brain in vivo or when applied to pituitary tissue in vitro. After binding to dopamine receptors on the mammotrophs its action is mediated by lowering the levels of calcium and cAMP.

A dopaminergic tract runs from the hypothalamus to the median eminence, and dopamine concentrations in the pituitary portal veins are elevated to levels capable of inhibiting prolactin release in vitro. In addition dopamine from the posterior pituitary (arriving via short portal veins) and dopamine generated or concentrated in adjacent anterior pituitary cells (by paracrine or autocrine action) contribute to chronic inhibition. The inhibitory effect of dopamine is used therapeutically to suppress undesired nursing or pathological prolactin hypersecretion with dopamine agonists. Somatostatin and a peptide resulting from the processing of the gene transcript of GnRH are other hypothalamic inhibitors of prolactin secretion.

Prolactin inhibits its own secretion via a short-loop feedback. It does so by directly increasing the synthesis and release of dopamine, its hypothalamic inhibitor. The hypothalamus has positive as well as negative effects on prolactin. There are a number of candidates for the role of prolactin-releasing hormone (PRH) (Table 48-5). TRH, for example stimulates prolactin synthesis and release, acting through specific membrane receptors in the mammotroph and phosphatidylinositol products. When TSH secretion is chronically increased as a result of negative feedback from the thyroid gland, prolactin also tends to increase modestly, probably because of increased endogenous TRH. It is doubtful however that TRH is the sole or the most important stimulator of prolactin release. Suckling for example does not produce an acute rise in plasma levels of TSH, as would be expected if TRH were the mediator of the plasma prolactin increase. Some evidence links oxytocin to this particular prolactin-stimulating function. A number of other peptides that are found in the hypothalamus or median eminence and that have receptors in the anterior pituitary have prolactin-releasing activity (Table 48-5), but their functional roles are unclear. Similarly the physiological significance of the positive effects of opioid peptides and serotonin and of the negative effects of norepinephrine on prolactin release remains to be established.

Normal basal plasma concentrations of prolactin are about 10 ng/ml (5×10^{-10}) and are similar in women and men. The half-life of the hormone is 20 minutes. Its metabolic clearance rate is 110 L/day, and the daily production is estimated at 350 μg. The kidney is one likely organ of degradation, because patients who experience renal failure often have high plasma prolactin levels. Prolactin is also present in amniotic fluid; its source is pregnancy-modified cells of the uterus stimulated by a placental prolactin-releasing factor.

Biological effects of prolactin. *Prolactin participates in stimulating the original development of breast tissue and its further hyperplasia during pregnancy. It is the principal hormone responsible for lactogenesis.* During prepubertal and postpubertal life prolactin, together with estrogens, progesterone, cortisol, and growth hormone, stimulates the proliferation and branching of ducts in the female breast. During pregnancy prolactin, along with estrogen and progesterone, causes development of lobules of alveoli within which milk production occurs. Finally, after parturition prolactin, together with insulin and cortisol, stimulates milk synthesis and secretion.

Prolactin binds to plasma membrane receptors homologous to those of GH in their extracellular binding domains (Fig. 48-25). The intracytoplasmic tails of prolactin receptors are different from and shorter than those of GH. This disparity suggests different intracellular forms of signal transduction. No second messengers have yet been identified. Consequent to receptor bind-

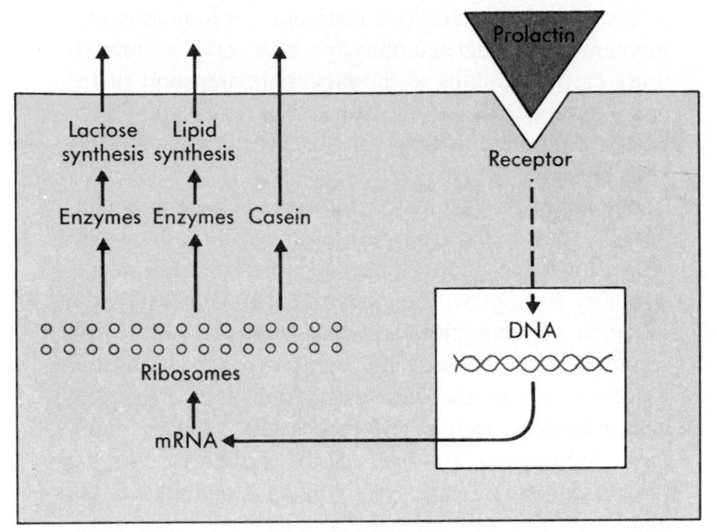

■ **Fig. 48-25** Mechanism of prolactin action. Following binding to a plasma membrane receptor, an unknown second messenger(s) stimulates expression of genes directing synthesis of enzymes essential for production of the milk components: lactose, casein, and lipids.

ing prolactin induces transcription of genes for the milk proteins casein, lactalbumin, and β-lactoglobulin and also stabilizes their messenger RNAs. Galactosyl transferase and N-acetyllactosamine synthetase are concurrently induced. These enzymes are necessary for synthesis of lactose, the major sugar in milk. The synthesis of fatty acids and phospholipids is also stimulated by prolactin specifically in breast tissue, but not in adipose tissue. Prolactin upregulates the number of its own receptors. Estrogen also increases the number of prolactin receptors. However estrogen and progesterone directly antagonize the stimulatory effect of prolactin on milk synthesis.

Prolactin has both stimulating and inhibiting effects on reproduction, depending in part on the phase of the reproductive process during which it acts. Excess prolactin blocks the synthesis and release of GnRH; this action causes the loss of normal GnRH pulses and prevents ovulation in females and normal sperm production in males. Prolactin can both induce and repress gene transcription for certain enzymes essential to gonadal steroid hormone production, depending on which cell type is affected and, in females, on the stage of the menstrual cycle. Certain reproductive behavioral effects of prolactin have been described, such as inhibition of libido in humans and stimulation of parental protective behavior toward the newborn in animals.

Interest is increasing in certain effects of prolactin on cell growth and proliferation. These effects are demonstrable in cultures of mammary, epithelial and lymphoma cell lines. Although they resemble GH effects, they are expressed through prolactin receptors. In addition prolactin, like GH, may induce an intermediary growth molecule synthesized and released by the liver, in analogy to the somatomedins. Finally prolactin has biphasic effects on immune responses, which suggest a role for the hormone in the balance between acceptance of fetal tissues by the mother and protection of maternal tissues from fetal invasion.

Clinical syndromes of prolactin dysfunction. In women prolactin deficiency caused by destruction of the anterior pituitary results in inability to lactate. Prolactin excess results from hypothalamic dysfunction or from pituitary tumors. In women this causes infertility and even complete loss of menses. Less often lactation unassociated with pregnancy (galactorrhea) occurs. In men decreased testosterone secretion and sperm production result from prolactin excess. Stimulation of breast development is uncommon and lactation is rare. In both sexes decreased libido is noted. The diagnosis is established by demonstrating a high plasma prolactin level. Therapy may consist of surgical ablation of tumor tissue. However treatment with dopaminergic drugs also reduces prolactin secretion to normal, reversing adverse effects on reproduction and galactorrhea.

■ *Posterior Pituitary Hormones*

Two nonapeptides of homologous structure (Fig. 48-26), **antidiuretic hormone (ADH),** also known as arginine vasopressin (AVP), and **oxytocin,** are secreted from the posterior pituitary gland. *The primary role of ADH is to conserve body water and regulate the tonicity of body fluids. The primary role of oxytocin is to eject milk from the lactating mammary gland.* Although their functions are different, the synthesis, storage, and mode of secretion of the two hormones are similar and will be discussed together.

Both hormones are synthesized in the cell bodies of hypothalamic neurons. ADH largely originates in the supraoptic nucleus and oxytocin largely in the paraventricular nucleus of the hypothalamus, although each hormone is synthesized in the alternate site. The genes directing synthesis of the respective preprohormones are remarkably similar and are located close together on the same chromosome (Fig. 48-27). The hormone sequences are contained within exon 1.

In addition to ADH or oxytocin the gene products include distinctive proteins, known as **neurophysins,** of molecular weight 10,000. Neurophysin-1 for oxytocin and neurophysin-2 for ADH are virtually identical in their large central cores (corresponding to exon-2 in each gene). The neurophysins differ in their N-terminal portions, which are coded for by exon-1 (see Fig. 48-27) and in their C-terminal portions, which are coded for by exon-3. In the case of ADH an additional glycopeptide is coded for by exon-3. After processing of the preprohormones ADH and oxytocin are packaged to-

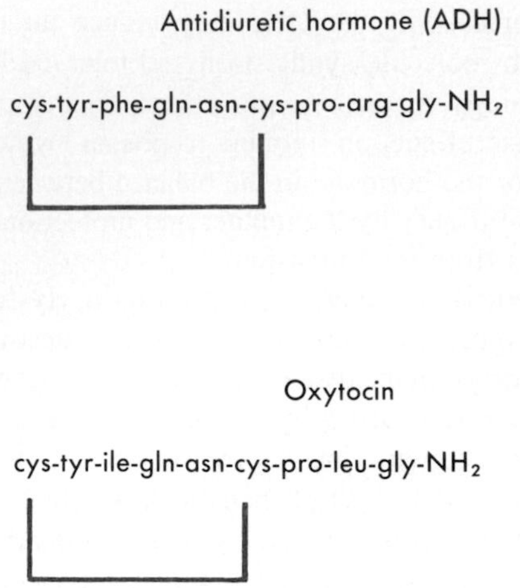

Antidiuretic hormone (ADH)

cys-tyr-phe-gln-asn-cys-pro-arg-gly-NH_2

Oxytocin

cys-tyr-ile-gln-asn-cys-pro-leu-gly-NH_2

■ **Fig. 48-26** Structures of posterior pituitary peptides. The alternate term for ADH is arginine vasopressin (AVP).

gether with their respective neurophysins in neurosecretory granules. The neurophysins may serve as low-affinity carrier proteins in the process of transport of the neurohormones down the axons. The latter end in the posterior pituitary as terminal swellings, known as **Herring bodies.**

Release of ADH or oxytocin occurs when a nerve impulse is transmitted from the cell body in the hypothalamus down the axon, where it depolarizes the neurosecretory vesicles within the terminal Herring body. An influx of calcium into the neurosecretory vesicle then results in hormone secretion by exocytosis. During this process the hormone dissociates from its neurophysin, and separately each enters the closely adjacent capillary. Subsequent passage of the hormone into the bloodstream is by endocytosis into the endothelial cell and by diffusion through pores in the fenestrated capillary endothelium.

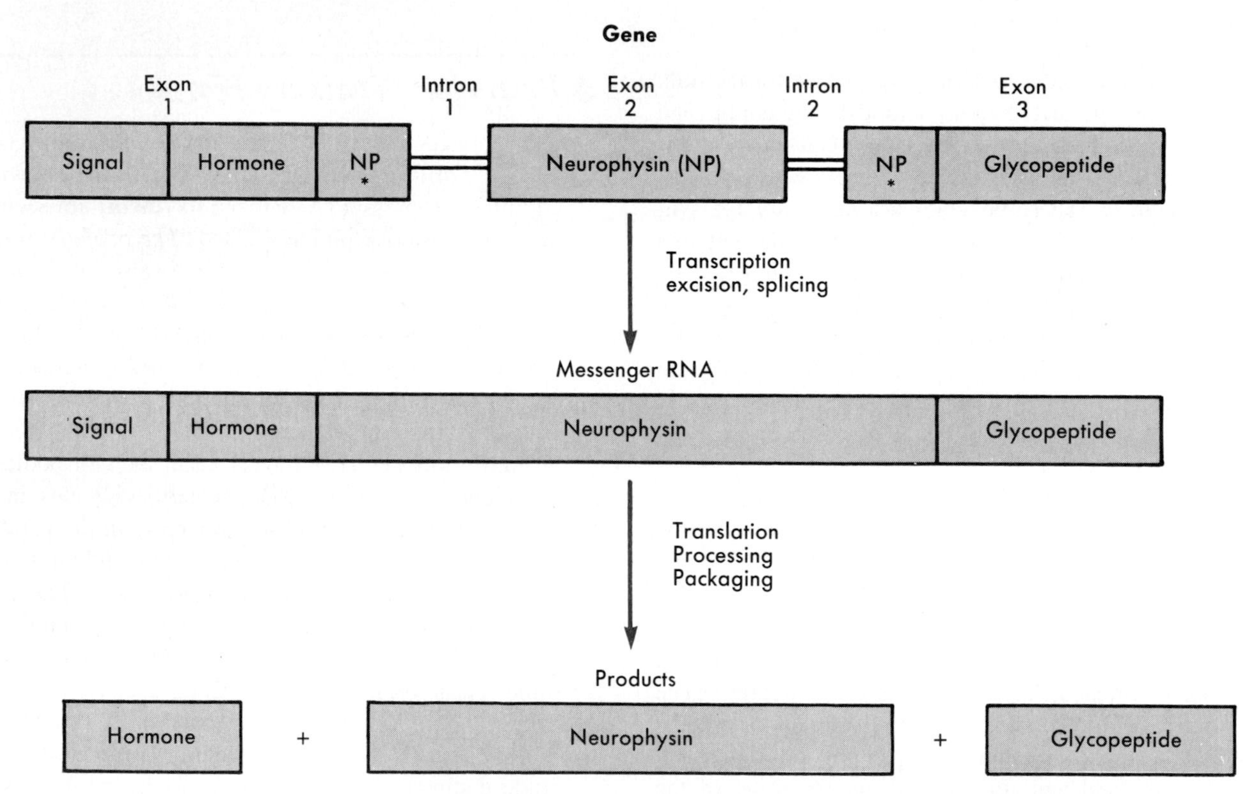

■ **Fig. 48-27** Schematic representation of the synthesis of the two posterior pituitary peptides ADH and oxytocin. The two gene structures are similar. In each case, exon-1 codes for the signal peptide, the hormone, and a variable portion of its corresponding neurophysin. Exon-2 is virtually identical in the two genes and codes for the homologous large central core of each neurophysin. Exon-3 codes for the remaining variable portion of each neurophysin. In the case of ADH only, exon-3 contains a base sequence extension that codes for a C-terminal glycopeptide, which is coreleased with ADH. (Redrawn from Richter D, Ivell R: Gene organization, biosynthesis, and chemistry of neurohypophyseal hormones. In Imura H: *The pituitary gland,* New York, 1985, Raven Press.)

Secretion of ADH

Consonant with its role in water metabolism secretion of ADH is primarily regulated by osmotic and volume stimuli (Table 48-6). Water deprivation produces an increase in the osmolality of plasma and hence of the fluids bathing the brain. This causes the loss of intracellular water from osmoreceptor neurons in the hypothalamus. Although these could be identical with the magnocellular neurons that secrete ADH, current evidence favors the presence of a distinct population of osmoreceptor neurons with connections to the ADH neurons. In either case the shrinkage of cell volume or increase in intracellular osmolality causes ADH to be released. Conversely water ingestion suppresses osmoreceptor firing and consequently shuts off ADH release. This begins by reflex neural stimulation shortly after water is swallowed; the plasma ADH level then declines further as the water is absorbed from the gut and plasma osmolality falls.

If plasma osmolality is directly increased by administration of solutes, only those that do not freely or rapidly penetrate cell membranes, such as sodium, cause ADH release. Substances, such as urea, that enter cells rapidly do not stimulate ADH secretion, because they do not produce osmotic disequilibrium between extracellular and intracellular fluids. A selective increase in sodium concentration of the cerebrospinal fluid also increases ADH secretion. The hypothalamic osmoreceptors are extraordinarily sensitive, being responsive to changes in osmolality of only 1% to 2% (Fig. 48-28). An increase in plasma osmolality of 1 mOsm/kg increases the ADH level 0.2 to 0.3 pg/ml. If water deprivation is prolonged, ADH synthesis is also increased. In response to plasma hyperosmolarity osmoreceptor neurons also stimulate thirst. In humans the threshold for this action is close to or somewhat higher than the threshold for ADH release of around 282 mOsm/kg (see Fig. 48-28). Therefore ADH secretion is at least as important as thirst in maintaining normal body water content.

ADH release is also stimulated by a decrease of 5% to 10% in total circulating blood volume, central blood volume, cardiac output, or blood pressure (Fig. 48-29). Hemorrhage is a potent stimulus to ADH release. Quiet standing or tilting and positive pressure breathing, which reduce central blood volume, also increase ADH secretion, particularly if blood pressure falls. Conversely administration of blood or isotonic saline solution, which increases total circulating blood volume, or immersion to the neck in water, which increases central blood volume, suppresses ADH release. Hypovolemia is perceived by a number of pressure (rather than volume) sensors (see Chapters 25 and 29). These include carotid and aortic baroreceptors, stretch receptors in the walls of the left atrium and pulmonary veins, and pos-

Table 48-6 Regulation of ADH secretion

Stimulation	*Inhibition*
Extracellular fluid osmolality increase	Extracellular fluid osmolality decrease
Volume decrease	Volume increase
Pressure decrease	Temperature decrease
Cerebrospinal fluid sodium increase	α-Adrenergic agonists
	γ-Aminobutyric acid (GABA)
Pain	Ethanol
Nausea and vomiting	Cortisol
Stress (Table 48-3)	Thyroid hormone(?)
Hypoglycemia	Atrial natriuretic peptide
Temperature increase	
Senescence	
Drugs	
Nicotine	
Opiates	
Barbiturates	
Sulfonylureas	
Antineoplastic agents	

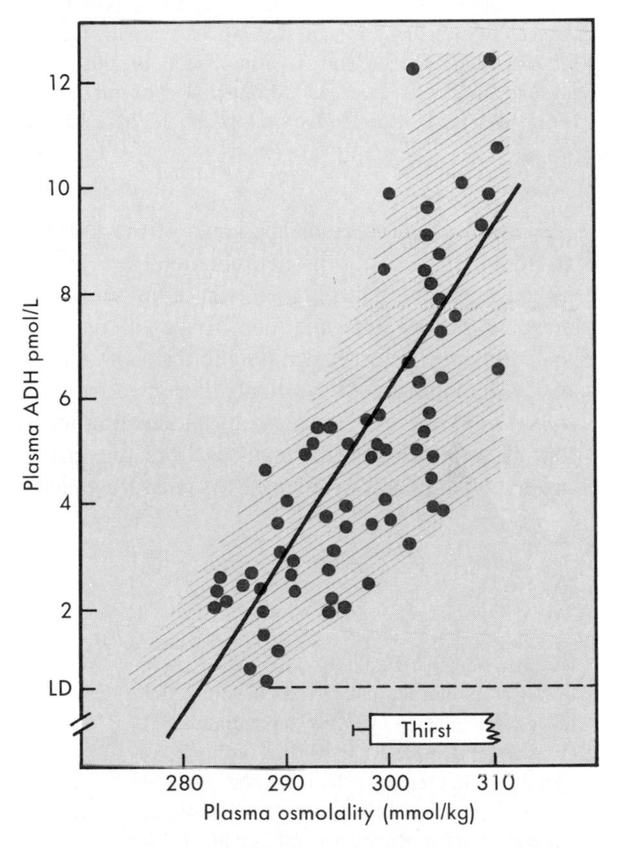

Fig. 48-28 Correlation between plasma ADH and plasma osmolality in humans. As plasma osmolality is increased by infusing hypertonic sodium chloride, ADH secretion is stimulated and plasma ADH rises over a linear concentration range. This response to hyperosmolality just precedes the response of thirst. Both ADH release and thirst lead to increases in body water that limit further increments in the osmolality of body fluids. (From Baylis P: *Clin Endocrinol Metab* 12:747, 1983.)

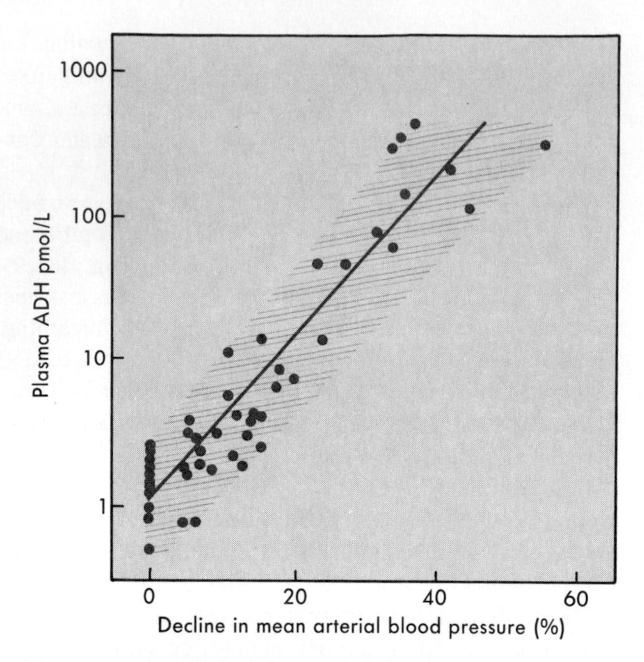

■ **Fig. 48-29** Correlation between plasma ADH and declining blood pressure in humans. As blood pressure is *decreased* by infusing an agent that blocks sympathetic ganglion function, ADH secretion is stimulated. In this response, plasma ADH rises over an exponential concentration range. (From Bayliss P: *Clin Endocrinol Metab* 12:747, 1983.)

sibly the juxtaglomerular apparatus of the kidney. The afferent impulses of this neurohumoral arc are carried by the ninth and tenth cranial nerves to their respective nuclei in the medulla and then by way of the midbrain via adrenergic neurotransmitters to the supraoptic nuclei of the hypothalamus. Normally the pressure receptors *tonically inhibit* ADH release by modulating an inhibitory flow of adrenergic impulses from the medulla to the hypothalamus. A decrease in pressure reduces the flow of impulses from the baroreceptors to the brainstem; this relieves the inhibition on the hypothalamus, resulting in increased ADH secretion.

Hypovolemia also stimulates the generation of renin and angiotensin directly within the brain; this local angiotensin, in addition to stimulating thirst, enhances the release of ADH. Furthermore, volume regulation of ADH is partly mediated or reinforced by atrial natriuretic hormone (ANH). When circulating volume is increased, ANH is released by atrial myocytes and ANH generated locally in supraoptic hypothalamic neurons acts to inhibit ADH release. Plasma ADH rises to much higher levels in response to hypotension than in response to hyperosmolarity (compare Fig. 48-29 with Fig. 48-28). This reflects the fact that the vascular system is less sensitive to the hormone than is the kidney.

The interaction between the two major stimuli of ADH release is shown in Fig. 48-30. Increases or decreases in circulating volume reinforce the osmolar responses by raising or lowering, respectively, the threshold for osmotic release of ADH. Thus hypovolemia sensitizes the system to hyperosmolarity. However under circumstances of marked hypovolemia baroregulation overrides osmotic regulation and ADH secretion is stimulated, even though plasma osmolality may be below 270 mOsm/kg.

Secretion of ADH is also influenced by a number of other conditions (see Table 48-6). Pain, emotional stress, heat, and a variety of drugs are stimulators; nausea and vomiting are especially potent. Aged humans secrete increased ADH, probably in compensation for a lesser ability of their kidneys to concentrate urine. Ethanol is a commonly encountered inhibitor; as little as 30 to 90 ml of whiskey is sufficient to suppress ADH secretion. Cortisol and thyroid hormones restrain ADH release; in their absence ADH may be secreted even though plasma osmolality is low.

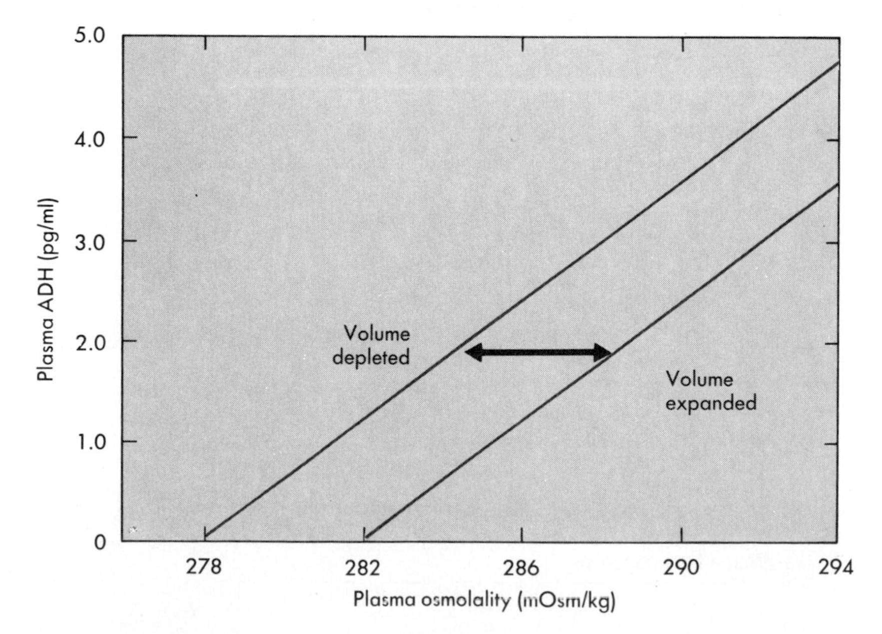

■ **Fig. 48-30** Regulation of ADH secretion by the interaction between plasma osmolality and plasma volume in humans. Increases or decreases in plasma volume, respectively, increase or decrease the threshold for ADH release in response to osmolality. (Redrawn from Robertson GL et al: *J Clin Endocrinol Metab* 42:613, 1976. From The Endocrine Society.)

ADH circulates at an average basal concentration of 1 pg/ml (10^{-12} M). The plasma half-life is 8 minutes, although the half-life of biological action may be up to 20 minutes. Metabolic clearance of ADH increases with its plasma level and averages 600 ml/min at 10 pg/ml. Urinary clearance of ADH consistently averages 5% of total metabolic clearance and 50% of glomerular filtration rate. Therefore urinary excretion rates are a valid index of ADH secretion. The latter is normally about 1 mg/day.

During water deprivation secretion increases threefold to fivefold. Transient 50-fold increases can occur with hemorrhage (Chapter 32), severe pain, or nausea. Neurophysin-2 also circulates in plasma and its levels rise and fall parallel with ADH. The C-terminal glycopeptide of prepro-ADH origin is also present in plasma. No functional role for these peptides in peripheral tissues has yet been identified.

Actions of ADH

The major action of ADH is on renal cells that are responsible for reabsorbing free (i.e., osmotically unencumbered) water from the glomerular filtrate (Chapter 42). These ADH-responsive cells line the distal convoluted tubules and collecting ducts of the renal medulla. ADH binds to a specific plasma membrane receptor (known as the V_2 receptor) on the capillary side of the cell, where it activates adenylyl cyclase. The increase in intracellular cAMP activates a protein kinase on the opposite luminal side of the cell. This phosphorylates currently unidentified membrane proteins after which the permeability of the cell membrane to water is enhanced. Microtubular and microfilamentous elements of the cell cytoskeleton participate in this process. As a result of ADH action, cytoplasmic tubular structures containing water-conducting particles are carried to and fused with the luminal plasma membrane.

The resultant increase in membrane permeability permits back diffusion of water along an osmotic gradient from the hypotonic tubular urine that emerges from the loop of Henle to the hypertonic interstitial fluid of the renal medulla. The mechanisms for establishing this gradient are discussed in Chapter 42. Recent evidence suggests that ADH also acts on the ascending limb of the loop of Henle to enhance sodium transport into the medullary interstitium. This helps create the osmotic gradient for water reaborption. The net result of ADH action is to increase the osmolality of urine to a maximum that is fourfold greater than that of the glomerular filtrate or plasma. In other words ADH significantly reduces free water clearance by the kidney.

Water deprivation stimulates ADH secretion, resulting in decreased free water clearance and enhanced conservation of water. A water load decreases ADH secretion, resulting in increased free water clearance and more efficient excretion of the load. Thus ADH and water form a negative feedback loop. The sigmoidal dose-response relation between plasma levels of ADH and urine osmolality is shown in Fig. 48-31. Most of the renal effect occurs at plasma ADH levels between 2 and 5 pg/ml. In this range urine osmolality correlates directly with plasma ADH concentrations. An increase in plasma ADH concentration of only 0.3 pg/ml increases urine osmolality from 60 to 300 mOsm/kg in water-loaded humans. At 5 pg/ml a maximum ADH effect of 900 mOsm/kg is achieved.

A number of factors blunt the action of ADH on the tubular cell: solute diuresis, chronic water loading (which reduces medullary hyperosmolarity), prostaglandin E (which interferes with ADH activation of adenylyl cyclase), ANH, cortisol, potassium deficiency, calcium excess, and lithium. Certain sulfonylureas, used in the treatment of diabetes mellitus, and other drugs potentiate ADH action.

ADH may subserve other functions in addition to its

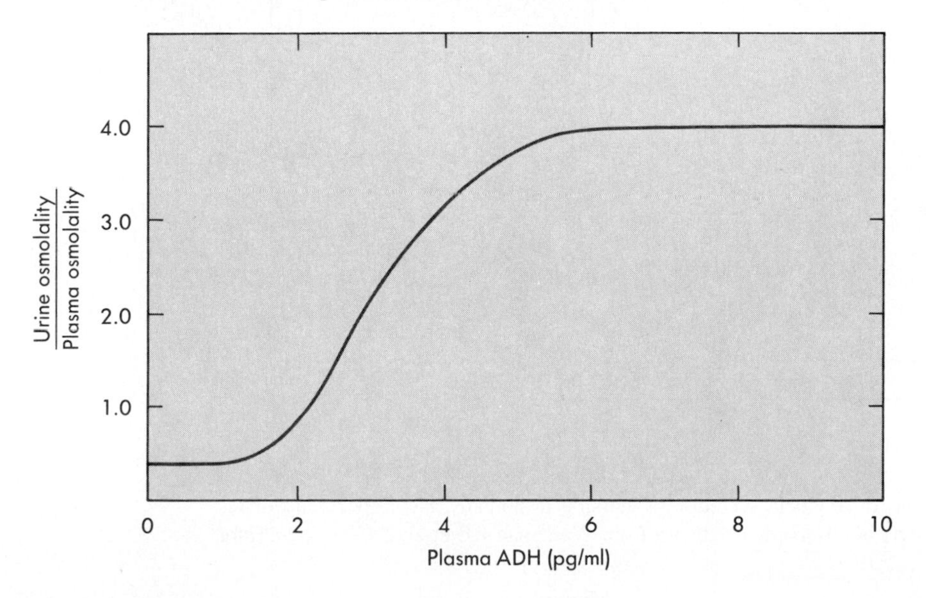

Fig. 48-31 Dose-response curve for the effect of ADH to increase renal tubular reabsorption of free water, expressed as the ratio of urine to plasma osmolality. (Data from Moore WW: *Fed Proc* 30:1387, 1971.)

primary role in water metabolism. It contributes to increasing vascular tone in response to hemorrhage by binding to arteriolar smooth muscles via the V_1 receptor and causing them to constrict. This action is mediated by Ca^{++} and phospholipase-C-generated second messengers. In contrast a V_2 receptor-mediated vasodilator effect of ADH may prevent much change of blood pressure in the usual physiological range of ADH secretion. However, when ADH is administered systemically in large doses it elevates the blood pressure and constricts the coronary and splanchnic beds. The last effect has been exploited therapeutically in controlling serious gastrointestinal bleeding.

ADH also functions as a corticotropin-releasing factor via axons that transmit the peptide to the median eminence. From there via the portal veins it reaches the anterior pituitary, where it interacts with a V_1-like receptor. In addition ADH may serve as a neurotransmitter in the brain to facilitate memory. Finally ADH, present in local high concentrations and acting via a V_1 receptor in a paracrine manner, stimulates smooth muscle contraction in the human spermatic cord. This may facilitate ejaculation of sperm.

Oxytocin Secretion and Actions

Suckling is the major stimulus for oxytocin release. Afferent neural impulses are carried from sensory receptors in the nipple to the spinal cord, where they ascend in the spinothalamic tract. From relays in the brainstem and midbrain they reach the paraventricular nuclei of the hypothalamus; via a cholinergic synapse they trigger oxytocin release within seconds from the neurosecretory vesicles in the posterior pituitary. As suckling is continued, oxytocin synthesis and transfer down the hypothalamic axon are also stimulated. As shown in Fig. 48-32 the stimulus of suckling is specific for oxytocin, because no release of ADH is noted. Correspondingly the various stimuli for ADH secretion stimulate little or no oxytocin release in humans. Opioid (endorphinergic) input to the hypothalamus inhibits oxytocin responses to stimuli. Oxytocin circulates unbound and exhibits a plasma half-life of 3 to 5 minutes. It is degraded by the kidneys and liver. Oxytocin binds to a plasma membrane receptor distinct from those of ADH. Increases in calcium levels and in phosphatidylinositol products mediate oxytocin actions.

The unique effect of oxytocin is to cause contraction of the myoepithelial cells of the alveoli of the mammary glands. This forces milk from the alveoli into the ducts, from where it is evacuated by the infant. No other hormone has this action. Estrogens augment and catecholamines block the action of oxytocin. Men have about the same basal plasma oxytocin levels as do women. Although the levels rise acutely during orgasm, the function of oxytocin in men is not clear.

Oxytocin also has a powerful action on smooth muscle in the uterus. Rhythmic contractions of the myometrium are stimulated by very small doses, which act by slightly lowering the threshold for membrane depolarization. Large doses lower the threshold still further, prevent repolarization and spiking discharges, and induce a sustained tetanic contraction. Neither maternal

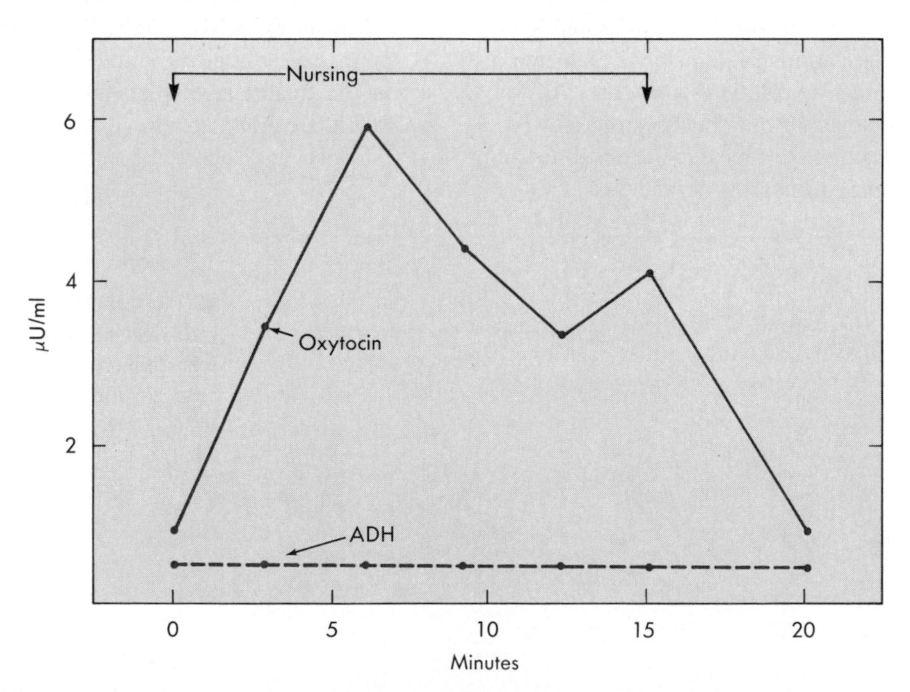

Fig. 48-32 Stimulation of oxytocin secretion by nursing in humans. Note specificity of response as no release of ADH is observed. (Redrawn from Weitzman RE et al: *J Clin Endocrinol Metab* 41:836, 1980.)

plasma oxytocin levels nor fetal oxytocin availability bears a consistent relationship to the progress of human labor. Therefore despite the ability to stimulate rhythmic uterine contractions oxytocin seems more a contributing factor than an essential hormone of human parturition. However after delivery it may play an important role in the sustained contractions that help to maintain hemostasis after evacuation of the placenta. Oxytocin is used to induce labor in women who are physiologically ready, and it is also used therapeutically to decrease immediate postpartum bleeding. Oxytocin and its receptor are also present in the ovary. An endocrine or a paracrine role for the hormone in terminating the corpus luteum at the end of the menstrual cycle (Chapter 51) has been suggested.

Clinical Syndromes of Posterior Pituitary Hormone Dysfunction

ADH deficiency is caused by destruction or dysfunction of the supraoptic and paraventricular nuclei of the hypothalamus. The posterior pituitary gland can be selectively removed without seriously affecting the availability of the hormone because the sectioned axons can still secrete ADH. The inability to produce concentrated urine is the hallmark of ADH deficiency, a condition called **diabetes insipidus.** In normal humans water deprivation can be compensated for by an increase in urine osmolality to 1,000 to 1,400 mOsm/kg. Individuals who totally lack ADH cannot achieve osmolalities higher than that of plasma (290 mOsm/kg) and often not higher than 50 to 200 mOsm/kg. Because a typical diet generates up to 900 mOsm of solute per day for obligatory excretion by the kidney, urine volumes as high as 12 L may be required in diabetes insipidus, in contrast to the usual 1 to 3 L. The patient therefore urinates frequently both day and night and must also drink excess fluids to replace the loss of water. Chronic elevation of serum osmolality (greater than 290 mOsm/kg) and of serum sodium (greater than 145 mEq/L) are common. If for any reason the patient loses access to fluids, hyperosmolarity and life-threatening dehydration may ensue. Replacement of ADH prevents this sequence and relieves the frequent urination and thirst.

ADH excess, inappropriate to either the osmolarity or volume of the body fluids, can result from a variety of conditions: (1) increased secretion caused by central nervous system disease, trauma, or psychosis; (2) ectopic production of the hormone by tumors; or (3) potentiation of hormone secretion or action by drugs. The reduction in free water clearance caused by ADH, combined with voluntary or involuntary fluid intake, leads to water retention. Plasma sodium concentration and osmolality are significantly lowered, whereas urine osmolality is increased. Characteristically sodium excretion in the urine is also increased, despite the hyponatremia, as a result of a compensatory increase in ANH secretion. Both intracellular and extracellular fluid volumes are expanded. The swelling of brain cells and hypoosmolality cause headache, nausea, lethargy, somnolence, convulsions, and coma when plasma osmolality declines to below 250 mOsm/kg and plasma sodium to below 125 mEq/L. Water restriction is a logical and effective acute treatment. It may have to be supplemented with drugs that induce a hypotonic diuresis or with hypertonic sodium chloride solutions to raise osmolality more rapidly.

Oxytocin deficiency leads to difficulty in nursing because of poor milk ejection. Oxytocin excess is unknown as a clinical syndrome.

Summary

1. The hypothalamic-pituitary unit regulates growth, lactation, water metabolism, and the functions of the thyroid gland, adrenal glands, and gonads.

2. Peptide hormones synthesized in some hypothalamic neurons pass down their axons to be stored in and released into the circulation from the posterior pituitary gland. Other hypothalamic peptides travel down axons to the median eminence, from which they are released into a portal venous circulation that carries them to the anterior pituitary gland. There they stimulate or inhibit release of target hormones.

3. Hypothalamic releasing and inhibiting peptides are secreted in pulses and induce effects via Ca^{++}, cAMP, and phosphatidyl inositol products as messengers. They stimulate or inhibit transcription, modulate translation, and stimulate or inhibit secretion of the target anterior pituitary hormones.

4. The anterior pituitary gland contains five functional cell types in close proximity, suggesting paracrine interactions. These are thyrotrophs, adrenocorticotrophs, gonadotrophs, somatotrophs, and mammotrophs. Each secretes a hormone(s) in response to hypothalamic stimulation. Peripheral target gland hormones or other peripheral products feed back negatively to inhibit their respective anterior pituitary hormones.

5. Thyrotropin (TSH) is a glycoprotein that contains α and β subunits. TSH stimulates secretion and growth of the thyroid gland. It is released in response to thyrotropin-releasing hormone and to thyroid hormone deficiency.

6. Adrenocorticotrophin (ACTH) is a peptide that stimulates the secretion and growth of the adrenal cortex. It is synthesized from a multifunctional precursor and is secreted in bursts in response to corticotropin-releasing hormone, to cortisol deficiency, and to various stresses, such as trauma and hypoglycemia.

7. Follicle-stimulating hormone (FSH) and luteinizing hormone (LH) are glycoprotein products of a single gonadotrophic cell. They share common α subunits with each other and with TSH but have distinctive β subunits.

8. FSH stimulates development of ovarian follicles in women and spermatogenesis in men.

9. LH stimulates steroid hormone secretion in both genders, primarily estradiol and its precursors in women, testosterone in men. LH also stimulates ovulation. Estradiol and testosterone as well as specific gonadal protein hormones feed back on FSH and LH secretion in complex ways.

10. Growth hormone (GH) is a protein hormone that stimulates cartilage development and growth, bone growth, and accretion of lean body mass. It acts via a peptide mediator (somatomedin) produced in the liver and many other cells. GH also has insulin antagonistic actions and stimulates lipolysis. It is secreted in response to growth hormone–releasing hormone, hypoglycemia, amino acids, and stress. Its secretion is inhibited by somatostatin from the hypothalamus, somatomedin, and glucose.

11. Prolactin is structurally similar to GH but specifically stimulates growth of the mammary glands and production of milk. Prolactin is normally tonically inhibited by dopamine from the hypothalamus. Its synthesis is markedly increased during pregnancy and is augmented by estrogens. Its release is stimulated by suckling.

12. Antidiuretic hormone (ADH) is a small peptide that acts on the renal tubules via cAMP as messenger. ADH increases the reabsorption of free water and the final urine osmolality. It also has vasoconstrictor actions via a separate plasma membrane receptor. ADH is secreted from the posterior pituitary in response to an increase in plasma osmolality or to a decrease in plasma volume or blood pressure.

13. Oxytocin (OCT) is structurally very similar to ADH but acts specifically on the mammary gland to cause release of milk. It is secreted in response to suckling. OCT also causes contraction of the uterus and plays a role in the overall process of parturition.

Bibliography

Journal articles

Argente J et al: Relationship of plasma growth hormone–releasing hormone levels to pubertal changes, *J Clin Endocrinol Metab* 63:680, 1986.

Brixen K et al: A short course of recombinant human growth hormone treatment stimulates osteoblasts and activates bone remodeling in normal human volunteers, *J Bone Miner Res* 5:609, 1990.

Charlton JA, Baylis PH: Mechanisms responsible for mediating the antidiuretic action of vasopressin (editorial), *J Endocrinol* 118:3, 1988.

Chin WW: Hormonal regulation of thyrotropin and gonadotropin gene expression *Clin Res* 36:484, 1988.

Chou JL et al: In vitro and in vivo growth and casein gene expression of mouse mammary tumor epithelial cells in response to hormones, *Exp Cell Res* 186:250, 1990.

Clemmons D, Van Wyk J: Factors controlling blood concentration of somatomedin C, *Clin Endocrinol Metab* 13:113, 1984.

Conn PM, Crowley WF, Jr: Gonadotropin-releasing hormone and its analogues, *N Engl J Med* 324:93, 1991.

Daughaday WH, Rotwein P: Insulin-like growth factors I and II: peptide, messenger, ribonucleic acid and gene structures, serum, and tissue concentrations, *Endocrinol Rev* 10:68, 1989.

Denef C: Paracrine interactions in the anterior pituitary, *Clin Endocrinol Metab* 15:1, 1986.

English DE et al: Evidence for a role of the liver in the mammotrophic action of prolactin, *Endocrinology* 126:2252, 1990.

Gharib SD et al: Molecular biology of the pituitary gonadotropins, *Endocr Rev* 11:177, 1990.

Giraldi A, et al: Oxytocin and the initiation of parturition, a review, *Dan Med Bull* 37:377, 1990.

Gitay-Goren H et al: Effects of prolactin on steroidogenesis and cAMP accumulation in rat luteal cell cultures, *Mol Cell Endocrinol* 65:195, 1989.

Hall K, Sara V: Somatomedin levels in childhood, adolescence and adult life, *Clin Endocrinol Metab* 13:91, 1984.

Hirsch AT et al: Vasopressin-mediated forearm vasodilation in normal humans, evidence for a vascular vasopressin V_2 receptor, *J Clin Invest* 84:418, 1989.

Holl RW et al: Thirty-second sampling of plasma growth hormone in man: correlation with sleep stages, *J Clin Endocrinol Metab* 72:854, 1991.

Iannotti JP: Growth plate physiology and pathology, *Orthop Clin North Am* 21:1, 1990.

Iranmanesh A et al: Intensive venous sampling paradigms disclose high frequency adrenocorticotropin release episodes in normal men, *J Clin Endocrinol Metab* 71:1276, 1990.

Jacobson L, Sapolsky R: The role of the hippocampus in feedback regulation of the hypothalamic-pituitary-adrenocortical axis, *Endocr Rev* 12:118, 1991.

Jones MT, Gillham B: Factors involved in the regulation of adrenocorticotropic hormone/beta lipotropic hormone, *Physiol Rev* 68:743, 1988.

Kelly PA et al: The prolactin/growth hormone receptor family, *Endocr Rev* 12:235, 1991.

LaBarbera AR, Rebar RW: Reproductive peptide hormones, generation, degradation, reception, and action, *Clin Obstet Gynecol* 33:576, 1990.

Lamberts SW, Macleod RM: Regulation of prolactin secretion at the level of the lactotroph, *Physiol Rev* 70:279, 1990.

Magner JA: Thyroid stimulating hormone: biosynthesis, cell biology and bioactivity, *Endocr Rev* 11:354, 1990.

Matsumoto A, Bremner W: Modulation of pulsatile gonadotropin secretion by testosterone in man, *J Clin Endocrinol Metab* 58:609, 1984.

Miller N et al: Short-term effects of growth hormone on fuel oxidation and regional substrate metabolism in normal man, *J Clin Endocrinol Metab* 70:1179, 1990.

Moller J et al: Expansion of extracellular volume and suppression of atrial natriuretic peptide after growth hormone administration in normal man, *J Clin Endocrinol Metab* 72:768, 1991.

Moses AM, Steciak E: Urinary and metabolic clearances of arginine vasopressin in normal subjects, *Am J Physiol* 251:R365, 1986.

Nicoll CS et al: Structural features of prolactins and growth hormones that can be related to their biological properties, *Endocr Rev* 7:169, 1986.

Norsk P, Epstein M: Effects of water immersion on arginine vasopressin release in humans, *J Appl Physiol* 64:1, 1988.

Pelletier G et al: Identification of human anterior pituitary cells by immunoelectron microscopy, *J Clin Endocrinol Metab* 46:534, 1978.

Richter D: Molecular events in expression of vasopressin and oxytocin and their cognate receptors, *Am J Physiol* 255:F207, 1988.

Salomon F et al: The effects of treatment with recombinant human growth hormone on body composition and metabolism in adults with growth hormone deficiency, *N Engl J Med* 321:1797, 1989.

Samuels MH et al: Pathophysiology of pulsatile and copulsatile release of thyroid-stimulating hormone, luteinizing hormone, follicle stimulating hormone, and alpha-subunit, *J Clin Endocrinol Metab* 71:425, 1990.

Shurmeyer T et al: Human corticotropin-releasing factor in man: pharmacokinetic properties and dose-response of plasma adrenocorticotropin and cortisol secretion, *J Clin Endocrinol Metab* 59:1103, 1984.

Snyder SH: Brain peptides as neurotransmitters, Science 209:976, 1980.

Southworth MB et al: The importance of signal pattern in the transmission of endocrine information: pituitary gonadotropin responses to continuous and pulsatile gonadotropin-releasing hormone, *J Clin Endocrinol Metab* 72:1286, 1991.

Spencer SA et al: Growth hormone receptor and binding protein, *Rec Prog Horm Res* 46:165, 1990.

Thompson CJ et al: Reproducibility of osmotic and nonosmotic tests of vasopressin secretion in men, *Am J Physiol* 260:R533, 1991.

Veldhuis J et al: Endogenous opiates modulate the pulsatile secretion of biologically active luteinizing hormone in man, *J Clin Invest* 72:2031, 1983.

Veldhuis J et al: Twenty-four-hour rhythms in plasma concentrations of adenohypophyseal hormones are generated by distinct amplitude and/or frequency modulation of underlying pituitary secretory bursts, *J Clin Endocrinol Metab* 71:1616, 1990.

Weitzman RE et al: The effect of nursing on neurohypophyseal hormone and prolactin secretion in human subjects, *J Clin Endocrinol Metab* 41:836, 1980.

Wennink JM et al: Growth hormone secretion patterns in relation to LH and testosterone secretion throughout normal male puberty, *Acta Endocrinol (Copenh)* 123:263, 1990.

Williams TD et al: Atrial natriuretic peptide inhibits postural release of renin and vasopressin in humans, *Am J Physiol* 255:R368, 1988.

Winer LM et al: Basal plasma growth hormone levels in man: new evidence for rhythmicity of growth hormone secretion, *J Clin Endocrinol Metab* 70:1678, 1990.

Books and monographs

deKretzer D et al: *The pituitary and tests, clinical and experimental studies,* Berlin, 1983, Springer-Verlag.

Frohman LA et al: The physiological and pharmacological control of anterior pituitary hormone secretion. In Dunn A, Nemeroff C, editors: *Behavioral neuroendocrinology,* New York, 1983, Spectrum Publications.

Guillemin R: Neuroendocrine interrelations. In Bondy P, Rosenberg LE, editors: *Metabolic control and disease,* Philadelphia, 1980, WB Saunders.

Kato Y et al: Regulation of prolactin secretion. In Imura H, editor: *The pituitary gland,* New York, 1985, Raven Press.

Keith LD, Kendall JW: Regulation of ACTH secretion. In Imura H, editor: *The pituitary gland,* New York, 1985, Raven Press.

Numa S, Imura H: ACTH and related peptides: gene structure and biosynthesis. In Imura H, editor: *The pituitary gland,* New York, 1985, Raven Press.

Reichlin S: Neuroendocrinology. In Foster D, Wilson J, editors: *Williams textbook of endocrinology,* ed 8, Philadelphia, 1992, WB Saunders.

Riskind PN, Martin JB: Functional anatomy of the hypothalamic-anterior pituitary complex. In Degroot LJ, editor: *Endocrinology,* Philadelphia, 1989, WB Saunders.

Thorner MO et al: The anterior pituitary. In Wilson JD, Foster DF, editors: *Williams textbook of endocrinology,* Philadelphia, 1992, WB Saunders.

Seo H: Growth hormone and prolactin: chemistry, gene organization, biosynthesis, and regulation of gene expression. In Imura H, editor: *The pituitary gland,* New York, 1985, Raven Press.

Verbalis J, Robinson A: Neurophysin and vasopressin: newer concepts of secretion and regulation. In Imura H, editor: *The pituitary gland,* New York, 1985, Raven Press.

CHAPTER

49

The Thyroid Gland

The thyroid gland was the first endocrine gland to be recognized as such by those symptoms associated with excess or deficient function. This recognition occurred in the mid-nineteenth century. The early speculation that the deficiency of an internal secretion caused the clinical state associated with atrophy of the gland was borne out when replacement with crude thyroid extracts became the first example of successful hormonal therapy in 1891. Soon thereafter the most important mission of the thyroid gland was discovered to be regulation of the overall rate of body metabolism. In addition, the gland was found to be critical for normal growth and development.

The thyroid gland develops from endoderm associated with the pharyngeal gut. It descends to the anterior part of the neck, where half lies on either side of the trachea. Abnormalities in its developmental descent may lead to final locations anywhere from the base of the tongue to the anterior mediastinum. By 11 to 12 weeks of gestational age the gland is capable of synthesizing and secreting its own thyroid hormones under the stimulus of fetal thyroid-stimulating hormone (TSH). Both are absolutely required for subsequent normal intrauterine development of the central nervous system and skeleton (although not for body growth), because only small amounts of maternal thyroid hormone can reach the fetus.

The two lobes of the adult thyroid gland together weigh approximately 20 g. They receive a rich blood supply from the thyrocervical arteries and innervation from the autonomic nervous system. The histological structure is shown in Fig. 49-1, *A*. The hormone-producing, cuboidal epithelial cells form circular follicles 200 to 300 μm in diameter. Within their lumens newly synthesized hormone is stored in the form of a colloid material. The base of each cell is covered by a basement membrane, and tight junctions connect adjacent cells at both their basal and apical (luminal) portions. When the gland is under intensive stimulation, the endocrine cells enlarge and assume a more columnar shape with their nuclei at the base (Fig. 49-1, *B*). The lumens of the follicles then appear scalloped because of endocytic resorption of the hormone-containing colloid (see Fig. 49-1, *B*). Evidence suggests that the epithelial cells are polyclonal and quite heterogeneous in their capacity to perform various steps in hormone synthesis in response to stimulation.

Scattered within the gland in close association with the epithelial cells is a separate line of parafollicular cells, called **C-cells.** These are the source of the polypeptide hormone, calcitonin, which is discussed in the section dealing with calcium metabolism in Chapter 47.

■ *Synthesis and Release of Thyroid Hormones*

The secretory products of the thyroid gland are **iodothyronines,** a series of compounds resulting from the coupling of two iodinated tyrosine molecules. Approximately 90% of the output is 3,5,3′,5′-tetraiodothyronine **(thyroxine, or T_4),** 10% is 3,5,3′-triiodothyronine **(T_3)**. and less than 1% is 3,3′,5′-triiodothyronine **(reverse T_3, or rT_3).** Normally these three compounds are secreted in the same proportions as they are stored in the gland. However, T_3 is the molecule most responsible for the actions of thyroid hormone.

Because of the unique role of iodide in thyroid physiology, a description of thyroid hormone synthesis properly begins with a consideration of iodide turnover (Fig. 49-2). An average of 400 μg of iodide per person is ingested daily in the United States. In a steady state virtually the same amount is excreted in the urine. Iodide is actively concentrated in the thyroid gland, the salivary glands, and gastric glands. About 70 to 80 μg of iodide is taken up daily by the thyroid gland from a circulating pool that ranges from 250 to 750 μg. If this extrathyroidal iodide pool is labeled with a small dose of radioactive iodine (^{123}I or ^{131}I), the percentage of uptake of this tracer in 24 hours (8% to 35%) gives a dynamic index of thyroid gland activity. The total iodide content of the thyroid gland averages 7500 μg, virtually all of which is in the form of iodothyronines. In a

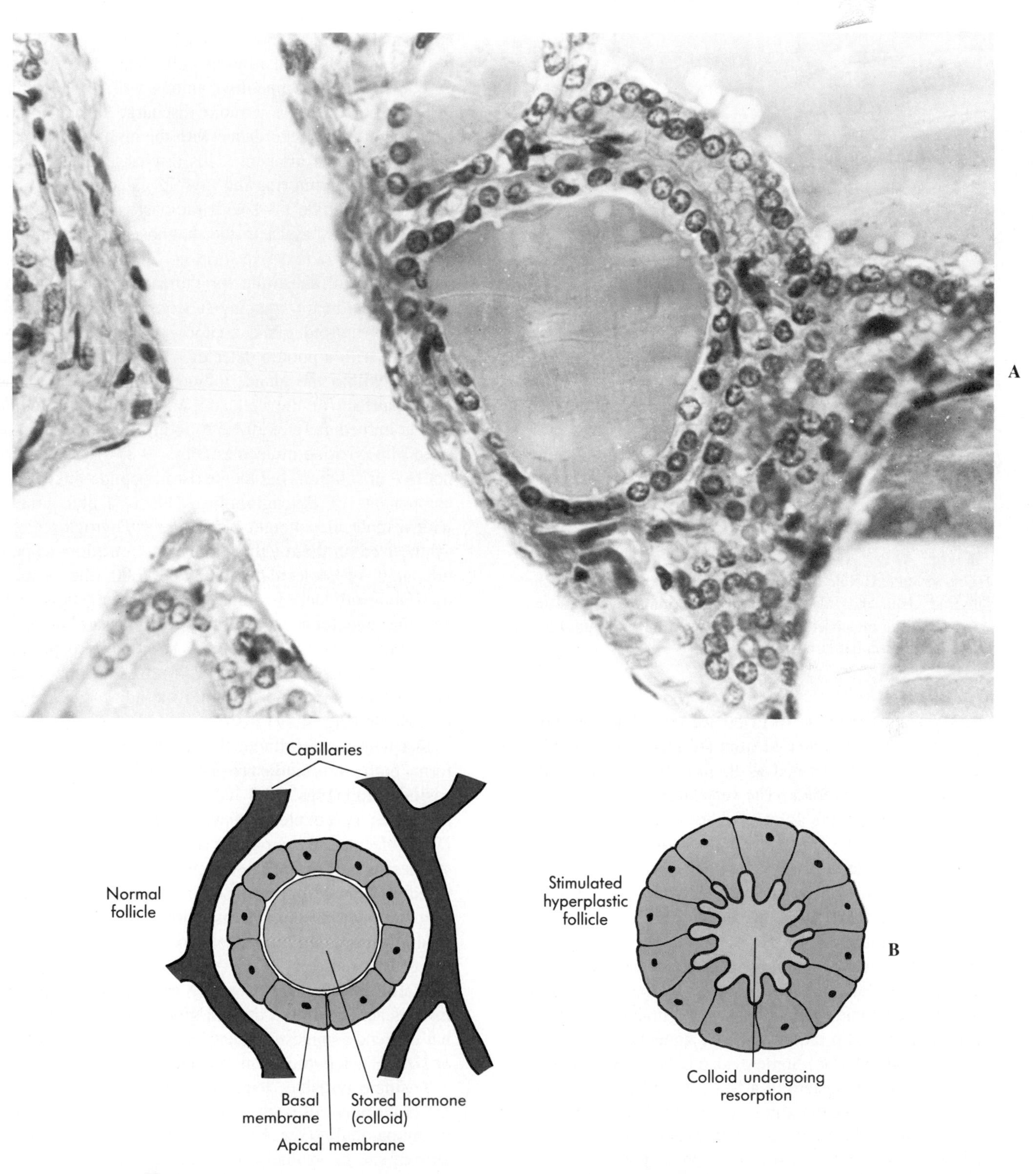

■ **Fig. 49-1** **A,** Photomicrograph of thyroid gland follicle. **B,** Schematic drawing of normal thyroid gland follicle and a follicle stimulated by thyrotropin. Note change in shapes from cuboidal to columnar, relocation of nuclei to base of cells, and scalloped appearance of follicle lumen.

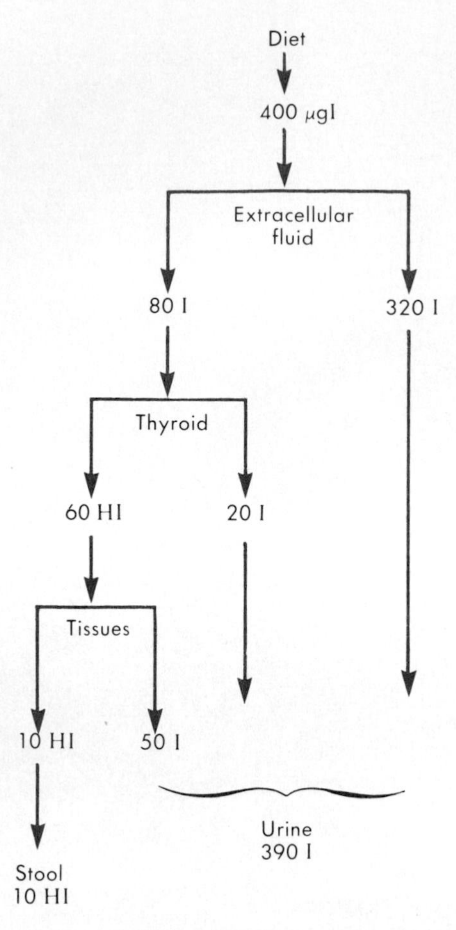

■ **Fig. 49-2** Average daily iodide turnover in humans (United States). Note that 20% of the intake is taken up by the thyroid gland and 15% turns over in hormone synthesis and disposal. The unneeded excess is excreted in the urine. *I*, Iodide; *HI*, hormonal iodide.

steady-state condition 70 to 80 μg of iodide, or about 1% of the total, is released from the gland daily. Of this amount, 75% is secreted as thyroid hormone, and the remainder is free iodide. The very large ratio (100:1) of iodide stored in the form of hormone to the amount turned over daily, can protect the individual from the effects of iodide deficiency for about 2 months. Further conservation of iodide then is affected by a marked reduction in its renal excretion as the circulating concentration and filtered load fall.

Iodide is actively transported into the gland against chemical and electrical gradients; normally a thyroid/plasma-free iodide ratio of 30 is maintained. This so-called **iodide trap** requires energy generation by oxidative phosphorylation. Some evidence links the operation of this transport system to a Na^+, K^+-ATPase. The iodide trap is markedly stimulated by TSH, probably mediated through cAMP. This stimulation may also require the synthesis of a specific protein, possibly the iodide carrier itself. The trap displays saturation kinetics. A primary reduction in dietary iodide intake depletes the circulating iodide pool and greatly enhances the activity of the iodide trap. Under these circumstances the percentage thyroidal uptake of iodide can reach 80% to 90%. If the lack of iodide is severe enough, thyroid hormone deficiency results.

A number of anions, such as thiocyanate (CNS^-), perchlorate ($HClO_4^-$), and pertechnetate ($T_cO_4^-$), act as competitive inhibitors of active iodide transport. If for any reason iodide cannot be rapidly incorporated into tyrosine after its uptake by the cell, then administration of one of these competitive anions will, by blocking further uptake, cause a rapid discharge of the iodide from the gland in accordance with the high thyroid/plasma concentration gradient. This discharge can be demonstrated by monitoring the thyroid gland in vivo after labeling the iodide pool with radioactive tracer, a maneuver that may assist in the diagnosis of biosynthetic defects. In its radioactive form as $^{99m}TcO_4$, pertechnetate is a useful substitute for radioactive iodide in the measurement of the trapping function and in the visualization of thyroid gland anatomy by external isotope scanning with a photon detector.

Once within the gland, iodide rapidly moves to the apical surface of the cell and into the lumen. Iodide (I^-) is immediately oxidized to iodine (I^o) and incorporated into tyrosine molecules (Fig. 49-3). The latter are not free in solution, but they exist in peptide linkages as components of **thyroglobulin.** This is a glycoprotein with a molecular weight of 670,000. Thyroglobulin is synthesized on the rough endoplasmic reticulum as peptide units of molecular weight 330,000 (the primary translation product of its messenger RNA). These units combine and then carbohydrate moieties are added in transit to the Golgi apparatus. The completed protein, incorporated in small vesicles, moves to the apical plasma membrane and then into the adjacent lumen of the follicle (Fig. 49-4).

Just within the follicle, thyroglobulin is iodinated to form both **monoiodotyrosine (MIT)** and **diiodotyrosine (DIT)** (Figs. 49-3 and 49-4). Thereafter two DIT molecules are coupled to form T_4, or one MIT and one DIT molecule are coupled to form T_3. Very little rT_3 is synthesized. This entire sequence of reactions is catalyzed by **thyroid peroxidase,** an enzyme complex adjacent to or within the cell membrane bordering the follicular lumen. The immediate oxidant (electron acceptor) for the reaction iodide → iodine is hydrogen peroxide. The mechanism whereby hydrogen peroxide is itself generated in the thyroid gland remains debatable, but evidence suggests reduction of oxygen by NADPH or NADH via cytochrome reductases.

A single tyrosine, located at the fifth position from the N-terminus of each of the two constituent peptides of thyroglobulin (site A), is a preferential but not an exclusive site of synthesis of T_4 or T_3. Both iodide and this tyrosine may be initially oxidized to free radicals that are complexed to the peroxidase enzyme and then combined to form MIT or DIT. Coupling may subse-

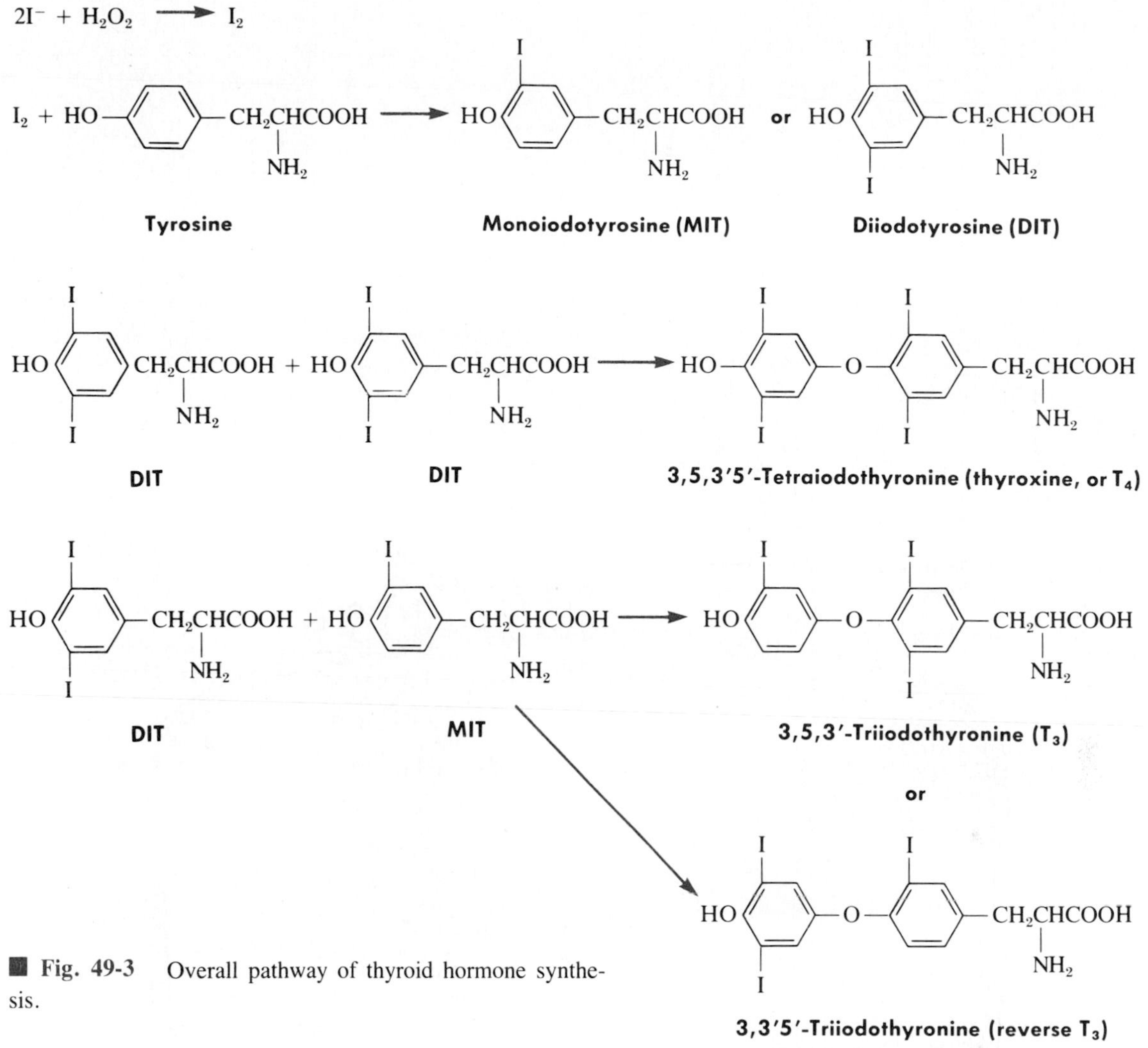

$$2I^- + H_2O_2 \longrightarrow I_2$$

I_2 + HO–(ring)–$CH_2CHCOOH(NH_2)$ ⟶ HO–(ring, I)–$CH_2CHCOOH(NH_2)$ **or** HO–(ring, I, I)–$CH_2CHCOOH(NH_2)$

Tyrosine **Monoiodotyrosine (MIT)** **Diiodotyrosine (DIT)**

HO–(ring, I, I)–$CH_2CHCOOH(NH_2)$ + HO–(ring, I, I)–$CH_2CHCOOH(NH_2)$ ⟶ HO–(ring, I, I)–O–(ring, I, I)–$CH_2CHCOOH(NH_2)$

DIT **DIT** **3,5,3′5′-Tetraiodothyronine (thyroxine, or T_4)**

HO–(ring, I, I)–$CH_2CHCOOH(NH_2)$ + HO–(ring, I)–$CH_2CHCOOH(NH_2)$ ⟶ HO–(ring, I)–O–(ring, I, I)–$CH_2CHCOOH(NH_2)$

DIT **MIT** **3,5,3′-Triiodothyronine (T_3)**

or

HO–(ring, I, I)–O–(ring, I)–$CH_2CHCOOH(NH_2)$

3,3′5′-Triiodothyronine (reverse T_3)

■ **Fig. 49-3** Overall pathway of thyroid hormone synthesis.

quently be facilitated by orienting MIT or DIT at site A into juxtaposition with another MIT or DIT molecule buried deeper within thyroglobulin. The latter MIT or DIT donates its iodinated phenolic ring to the MIT or DIT at site A. This produces T_4 or T_3 at site A and leaves dehydroalanine in peptide linkage within the deeper portion of thyroglobulin.

The entire sequence of thyroglobulin iodination is carried out very rapidly; labeled iodide appears in hormone molecules 1 minute after in vivo administration, and by 1 hour 90% to 95% of iodide is organically bound. The usual distribution of iodoaminoacids, as residues per molecule of thyroglobulin, is MIT, 7; DIT, 6; T_4, 2; and T_3, 0.2. Approximately one third of the iodine in thyroglobulin is in the form of calorigenic hormone (T_4 and T_3). Certain factors regulate the proportion of T_3 to T_4 that is synthesized. When iodide availability is restricted, the formation of T_3 is favored. Because T_3 is three times as potent as T_4, this response provides more active hormone per molecule of organified iodide. The proportion of T_3 also is increased when the gland is hyperstimulated by TSH or other activators.

Once thyroglobulin has been iodinated, it is stored in the lumen of the follicle as **colloid**. Release of the peptide-linked T_4 and T_3 into the bloodstream requires proteolysis of the thyroglobulin. Histochemical and radiographic studies have demonstrated that the colloid is retrieved from the lumen of the follicle by the epithelial cell through the process of endocytosis. The plasma cell membrane forms pseudopods that engulf a pocket of colloid. After this portion of the luminal content has been pinched off by the plasma cell membrane, it appears as a colloid droplet within the cytoplasm (Fig. 49-4). The droplet moves through the cytoplasm in a basal direction, probably as a result of microtubule and microfilament function. At the same time lysosomes move from the base toward the apex of the cell and fuse with the colloid droplets. The action of the lysosomal proteases then releases free T_4 and T_3, which leave the cell through the plasma membrane at the basal end and enter the bloodstream via the adjacent rich capillary plexus.

The MIT and DIT molecules, which also are released on proteolysis of thyroglobulin, are rapidly deiodinated

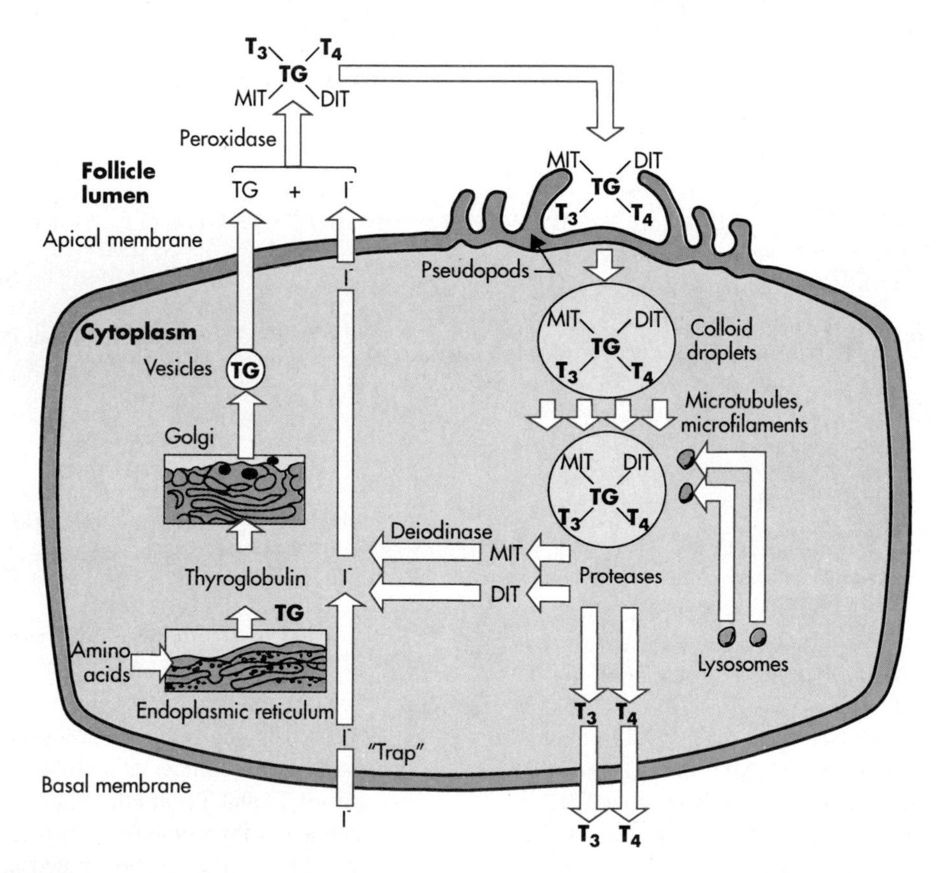

■ **Fig. 49-4** Histological demonstration of the process of resorption of colloid. **A,** Unstimulated follicles, **B,** Within minutes of TSH administration colloid droplets are seen inside the follicular cells. (From Wollman SH et al: *J Cell Biol* 21:191, 1964. From The Rockefeller University Press.)

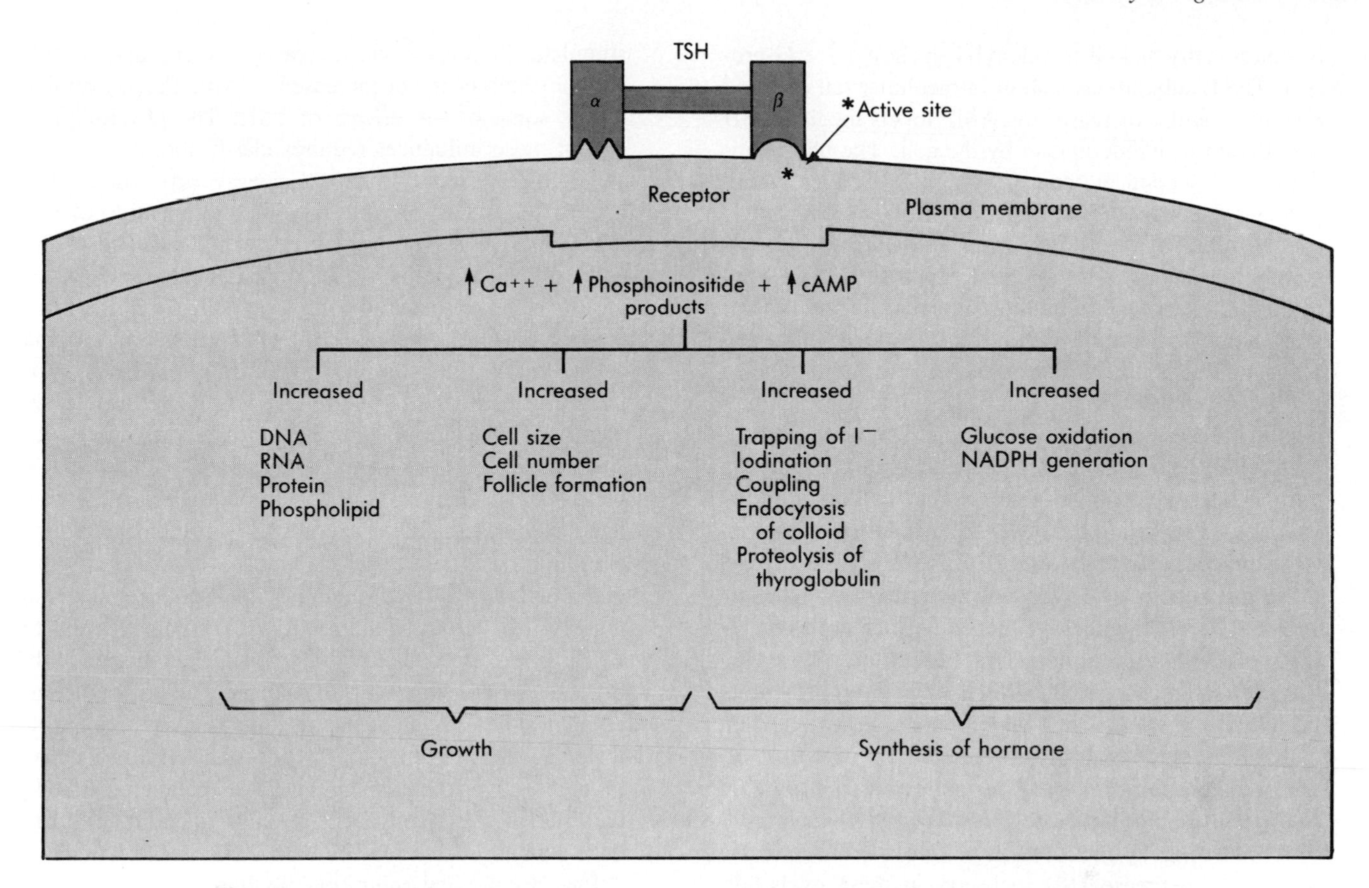

■ **Fig. 49-5** TSH actions on the thyroid cell. Cyclic adenosine monophosphate (cAMP) along with calcium ions (Ca^{++}) and phosphoinositol products act as second messengers generated by TSH binding to its receptor. All steps in thyroid hormone production, as well as many aspects of thyroid cell metabolism and growth, are stimulated by TSH.

within the follicular cell by the enzyme **deiodinase**. (Fig. 49-4.) Since these compounds are metabolically useless and would be lost in the urine if secreted, their deiodination retrieves the iodide for recycling into T_4 and T_3 synthesis. Only minor amounts of intact thyroglobulin leave the follicular cell under normal circumstances. However, in certain pathological states thyroglobulin escapes at a greater rate and circulating concentrations of the storage protein increase. Each of the preceding steps in the sequence of thyroid hormone synthesis and release is discrete, so that specific congenital biosynthetic defects occur that result in thyroid hormone deficiency.

■ *Regulation of Thyroid Gland Activity*

The most important regulator of thyroid gland function and growth is the hypothalamic-pituitary TRH-TSH axis (see Chapter 48 and Figs. 48-5 and 48-6). Because TSH secretion shows only minimal diurnal or day-to-day variation, thyroid hormone secretion and plasma concentrations are also relatively constant. Small nocturnal increases in secretion of TSH and release of T_4 have been described; they are presently of unknown physiological significance. TSH stimulates the process of iodide trapping and of each step in T_4 and T_3 synthesis. It also stimulates the endocytosis of colloid, the proteolysis of thyroglobulin, and the release of T_4 and T_3 from the gland. Sustained TSH stimulation leads to hypertrophy and hyperplasia of the follicular cells. The enlarged cells show an increased volume of endoplasmic reticulum, increased numbers of ribosomes, a larger and more complex Golgi apparatus, and an increase in DNA synthesis. Proliferation of capillaries also is observed, and thyroid blood flow increases. In the absence of TSH, marked atrophy of the gland occurs. However, a low basal level of thyroid hormone production and release continues in humans, seemingly independent of TSH.

The effects of TSH are exerted through a multiplicity of actions (see Fig. 49-5). The initial step is binding to a plasma membrane receptor of approximately 750 amino acids and molecular weight 85,000. The extracellular α subunit of the receptor binds the α and β subunits of TSH. The β subunit of the receptor probably winds through the plasma membrane several times and is functionally linked to adenylyl cyclase by a G-protein. The β subunit has a short intracellular tail of about 80 amino acids. Increases in cAMP levels mediate TSH stimulation of iodide uptake by the cell. The phosphatidylinositol second messenger system as well as cAMP probably mediate TSH stimulation of other steps in thyroid hormone synthesis.

Within minutes of thyroid cell exposure to TSH, thyroglobulin, stored within the follicular lumen, undergoes endocytosis, and colloid droplets appear in the cytoplasm. Shortly thereafter an increase in iodide uptake and in peroxidase activity can be demonstrated. Coincident with these actions on hormone synthesis and release, TSH also stimulates glucose oxidation, especially via the hexose monophosphate shunt. This may be the means for generating the NADPH needed for the peroxidase reaction. Further effects of TSH on the thyroid gland are seen after a delay of hours to days. Nucleic acid and protein synthesis is generally increased, via effects on both transcription and translation. These actions probably relate to the effects of TSH on promoting growth of the gland. They may involve local production of IGF and epidermal growth factor.

The regulation of thyroid hormone secretion by TSH is under exquisite feedback control (Figs. 48-5 and 48-6). Circulating T_4 and T_3 each produce feedback to the pituitary to decrease TSH secretion; as their levels fall, TSH secretion increases. It is free T_4 and T_3, not the protein-bound portions, that regulate pituitary TSH output. The pituitary gland is capable of deiodinating T_4 to T_3, and the latter is the final effector molecule in turning off TSH. In situations in which the diseased thyroid gland produces thyroid hormones autonomously or in which exogenous thyroid hormone is administered chronically, plasma TSH is reduced greatly.

Another important regulator of thyroid gland function is iodide itself, which has a biphasic action. At relatively low levels of iodide intake, thyroid hormone synthesis is positively correlated with iodide availability. However, as the intake of iodide exceeds 2 mg/day, an intraglandular concentration of iodide or of some organic iodine product is reached that is inhibitory to the iodide trap and to the biosynthetic mechanism; thus hormone production declines back to normal. This autoregulatory phenomenon is known as the ***Wolff-Chaikoff effect.*** In unusual instances the inhibition of hormone synthesis by iodide can be great enough to induce thyroid hormone deficiency. In addition, probably by an independent mechanism, an excess of iodide eventually also inhibits its own binding by the follicular cells

Other modes of autoregulation may play a role in preventing excessive responses to TSH stimulation. Thyroglobulin inhibits binding of TSH to its receptors, as well as the response of adenylyl cyclase to the tropic hormone. In addition, T_4 and T_3 also directly inhibit the thyroid gland in vitro. Adrenergic, VIPergic, and cholinergic innervation is present. Epinephrine and VIP stimulate T_4 release via increased cAMP, and acetylcholine inhibits it via increased cGMP. Prostaglandins mimic some of the effects of TSH. The physiological role of these influences requires clarification.

Thyroid hormones increase energy expenditure and heat production. Therefore it would be logical to expect that thyroid hormone availability would respond to the body's changing caloric and thermal status. In fact, ingestion of an excess of calories, particularly in the form of carbohydrate, increases T_3 production and plasma concentration and the metabolic rate, whereas prolonged fasting leads to corresponding decreases. However, similar fluctuations in T_4 do not occur. Therefore, since most T_3 arises from circulating T_4 (Table 49-1), peripheral mechanisms are more important in mediating these changes than are alterations in thyroid gland secretion. In animals exposure to cold increases thyroid gland activity. Humans living in cold polar zones also increase T_3 production and TSH responsiveness to TRH. In the neonatal period when the infant suddenly becomes responsible for self-maintenance of body temperature, there is an acute rise in TSH secretion followed by a rise in plasma T_4 to levels well above those of adults. Over the ensuing weeks or months plasma T_4 then subsides to a range that remains stable in adult life until a small decline occurs with senescence.

Pharmacological inhibition of thyroid gland activity is of considerable therapeutic importance. A class of drugs known as **thiouracils** blocks the synthesis of T_4 and T_3 by acting on the peroxidase complex. Because of the inhibition of organification, iodide taken up by activity of the trap is rapidly discharged again, as shown by studies with radioactive iodine. After continuous administration for weeks the stores of thyroid hormone (and of iodide) become depleted. The thiouracils are very effective in the treatment of ***hyperthyroidism.*** Lithium inhibits the release of thyroid hormones and, secondarily, their synthesis, probably by blocking adenylyl cyclase and cAMP accumulation. Finally, a large excess of iodide, in addition to the effects previously noted, can also promptly inhibit thyroid hormone re-

Table 49-1 Thyroid hormone turnover

	T_4	T_3	rT_3
Daily production (μg)	90	35	35
From thyroid (%)	100	25	5
From T_4 (%)	—	75	95
Extracellular pool (μg)	850	40	40
Plasma concentration			
Total (μg/dl)	8.0	0.12	0.04
Free (ng/dl)	2.0	0.28	0.20
Half-life (days)	7	1	0.8
Metabolic clearance (L/day)	1	26	77
Fractional turnover per day (%)	10	75	90

lease. Although this is a transient action, the administration of iodide may benefit individuals with severe hyperthyroidism.

Metabolism of Thyroid Hormones

Table 49-1 shows the daily production rates, pool sizes, plasma concentrations, half-lives, metabolic clearances, and fractional turnovers of T_4, T_3, and rT_3. T_4 is clearly the dominant secreted and circulating form of thyroid hormone. In contrast, the major portion of T_3 and virtually all of rT_3 come secondarily from circulating T_4, rather than primarily from thyroid gland secretion. Thus T_4 serves primarily as a prohormone for T_3, in addition to providing some intracellular action of its own. This "storage" function of plasma T_4 also is reflected in its much lower metabolic clearance and fractional turnover rates, compared with those of T_3 or rT_3. A small amount of intact thyroglobulin also is secreted, and it circulates at an average plasma concentration of 5 ng/ml.

Most conversion of T_4 to T_3 occurs in tissues, such as the liver and possibly the kidneys, with high blood flows and rapid exchanges with plasma. This process supplies circulating T_3 for uptake by other tissues in which local T_3 generation is too restricted to provide sufficient thyroid hormone action.

Secreted T_4 and T_3 circulate in the bloodstream almost entirely bound to proteins. Normally only 0.03% of total plasma T_4 and 0.3% of total plasma T_3 are in the free state (see Table 49-1). However, these are the critical fractions that are *biologically active*, not only in exerting thyroid hormone effects on peripheral tissues but in pituitary feedback as well. The major binding protein is **thyroxine-binding globulin (TBG).** This is a glycoprotein α-globulin with a molecular weight of 63,000 which is synthesized in the liver. Each TBG molecule binds one molecule of T_4; at the normal TBG concentration of 1.5 ng/dl, 20 μg of T_4 can be bound per deciliter.

About 70% to 80% of circulating T_4 and T_3 is bound to TBG; most of the remainder is bound to albumin and to a thyroid-binding prealbumin (TBPA) and 3% to lipoproteins. Compared with TBG, albumin and TBPA have much lower affinities but much higher capacities for T_4 and T_3. Ordinarily, however, only alterations in TBG concentration significantly alter total plasma T_4 and T_3 levels.

Two biological functions have been ascribed to TBG. First, it maintains a large circulating reservoir of T_4, which buffers against acute changes in thyroid gland function. Even the instant addition to the plasma of the calorigenic hormone needed for an entire day would cause only a 10% increase in the total T_4 concentration (see Fig. 49-6). Conversely, after removal of the thyroid gland it would take 1 week for the plasma T_4 concentration to fall as much as 50%. *Second, the binding of plasma T_4 and T_3 to large proteins prevents the loss of these relatively small hormone molecules into the urine and thereby helps conserve iodide.*

The reservoir function of TBG is best understood by examining the chemical equilibrium between T_4 and TBG. This equilibrium governs the distribution of the hormone between the free (T_4) and bound (T_4·TBG) forms.

$$T_4 + TBG \leftrightarrows T_4 \cdot TBG \qquad (1)$$

$$Keq = \frac{[T_4 \cdot TBG]}{[T_4]\,[TBG]} \qquad (2)$$

$$\frac{[T_4]}{[T_4 \cdot TBG]} = \frac{\text{Free } T_4}{\text{Bound } T_4} = \frac{1}{Keq\,[TBG]} \qquad (3)$$

$$[T_4] = [T_4 \cdot TBG] \times \frac{1}{Keq\,[TBG]} \qquad (4)$$

A temporary decrease in free T_4, caused by a decrease in thyroid gland output or accelerated uptake by target cells, can be rapidly compensated for by a dissociation of bound T_4 (T_4·TBG), until the new ratio of T_4/(T_4·TBG) returns to that required by Keq (equation 3). A temporary increase in free T_4, caused by endogenous secretion or exogenous administration, can be rapidly compensated for by association of the excess with TBG, because normally only 30% of the available T_4 binding sites on TBG are occupied. Of course sustained decreases or increases in T_4 supply, resulting from thyroid disease, must eventually lead to sustained decreases or increases in free T_4, because the latter is directly proportional to T_4·TBG (equation 4).

It is important to note that a primary change in TBG concentration will also disturb the ratio of free to bound T_4 (equation 3). In this circumstance the normal thyroid gland must increase or decrease its rate of hormone secretion appropriately, until the new equilibrium state restores the free T_4 level to normal. TBG concentration can decrease because of reduced hepatic synthesis (liver disease) or excessive loss in the urine (kidney disease). Free T_4 then will increase temporarily. In compensation, pituitary TSH secretion will be suppressed by negative feedback. T_4 output by the thyroid gland then will decrease until the new, lower steady-state level of T_4·TBG yields a normal level of free T_4 (equation 4). Hepatic synthesis of TBG also can be stimulated (e.g., by estrogen administration or pregnancy), leading to an increase in TBG concentration. Free T_4 will decrease temporarily; this will stimulate pituitary secretion of TSH, and consequently T_4 output by the thyroid gland will increase. This will continue until the elevated level of T_4·TBG is sufficient to restore the free T_4 level to normal in a new steady-state condition.

Identical qualitative considerations govern the circulating levels of free and bound T_3. However, the buffering action of TBG is less effective, because the Keq for T_3 is an order of magnitude lower than that for T_4 (2

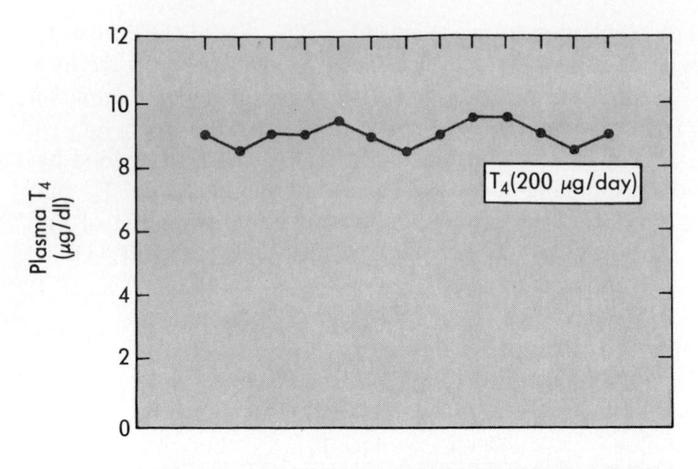

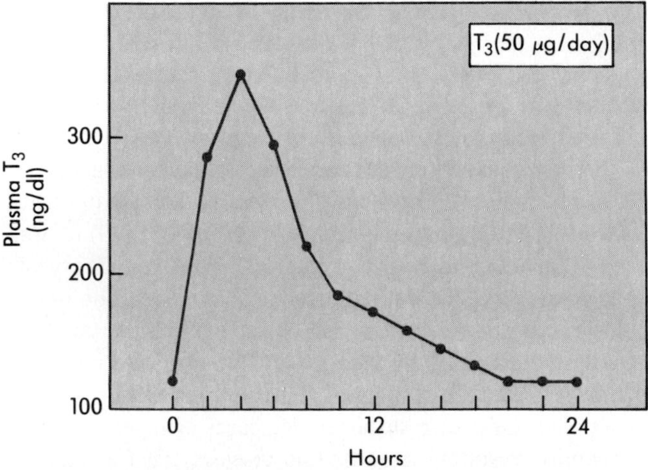

■ **Fig. 49-6** Effect of administering a single day's supply of calorigenic hormone by mouth to hypothyroid individuals. Note that 50 μg of T_3 elevates plasma T_3 levels for many hours, whereas 200 μg of T_4 causes no significant change in either plasma T_4 or T_3. This occurs because of the much larger pool size and tighter protein binding of T_4 than of T_3. (Redrawn from Saberi M et al. Serum thyroid hormone and thyrotropin concentrations during thyroxine and triiodothyronine therapy, *J Clin Endocrinol Metab* 39:923, 1974.)

$\times 10^9$ versus 2×10^{10}, respectively) and because the total extrathyroidal pool of T_3 is only a twentieth that of T_4 (see Table 49-1). Thus rapid addition of T_3 equivalent to the calorigenic hormone needed for an entire day produces greater swings in the concentrations of total and free T_3 (Fig. 49-6). Although alterations in TBG are not usually caused by thyroid gland disease, they must be taken into account when plasma thyroid hormone levels are measured for diagnostic purposes.

The peripheral metabolism of circulating thyroid hormones is outlined in Fig. 49-7. The majority of the 80 to 90 μg of T_4 that is released by the gland daily undergoes deiodination. However, approximately 15% is irreversibly excreted in the bile as the various iodothyronines in glucuronide or sulfate conjugates. Tetraiodoacetic acid and triiodoacetic acid are less important metabolites. The liver, kidney, and skeletal muscle are the major sites of degradation. The overall rate of disposal of T_4 is directly related to the free T_4 concentration in the plasma and to the intracellular T_4 content, particularly of the liver. Thus T_4 increases its own degradative metabolism. The entire cascade of products—T_3, T_2, and T_1—of the sequential deiodination steps is regularly increased in the plasma in *hyperthyroidism* (T_4 excess) and is usually decreased in *hypothyroidism* (T_4 deficiency).

■ *Relationship between Hormone Metabolism and Hormone Action*

The initial step in T_4 metabolism, the intracellular conversion of T_4 to either T_3 or rT_3, is of critical importance to thyroid hormone action. T_3 is the hormone of greatest biological activity, whereas rT_3 has almost no apparent calorigenic action. Therefore factors that regulate the relative rates of outer ring versus inner ring monodeiodination (Fig. 49-7) also determine the final biological effect of secreted T_4. The T_3 generating activity, ***5′-monodeiodination,*** is supplied by several tissue-specific types of the enzyme 5′ monodeiodinase. In the pituitary gland and certain areas of the brain, the isoenzyme has high affinity for T_4 and favors local T_3 generation. In the liver and kidney a different isoenzyme has a lower affinity but a higher capacity for T_4 and also has a high affinity and capacity for rT_3. In these organs the adaptation favors regulation of general T_3 supply as well as disposition of T_4 excess. This deiodinase isoenzyme is distinguished by the presence of the rare amino acid selenocysteine in its composition. The essential trace element selenium therefore plays a role in thyroid physiology.

In humans the normal distribution of T_4 products is 45% T_3 and 55% rT_3. An increase in T_4 concentration leads to a decrease in its conversion to T_3. Thus the biological effects of T_4 excess or deficiency are automatically mitigated to a slight extent by accelerated or retarded metabolic inactivation, respectively.

Other states and factors are associated with reduced conversion of T_4 to T_3 and often with a reciprocally enhanced conversion of T_4 to rT_3. These include the gestational period, fasting, stressful states, catabolic diseases, hepatic disease, renal failure, thiouracil drugs, and β-adrenergic blockade. In many instances inhibition of the enzyme 5′-monodeiodinase appears to explain this switch. As seen in Fig. 49-7, such inhibition would decrease production of T_3 from T_4 (reducing plasma T_3) and simultaneously decrease degradation of rT_3 to $3,3'T_2$ (increasing plasma rT_3). The reduction of 5′ monodeiodinase activity may result from decreased glucose metabolism, increased FFA metabolism, and excess secretion of the stress hormone cortisol.

The biological effects of T_4 are largely a result of its intracellular conversion to T_3. When administered exoge-

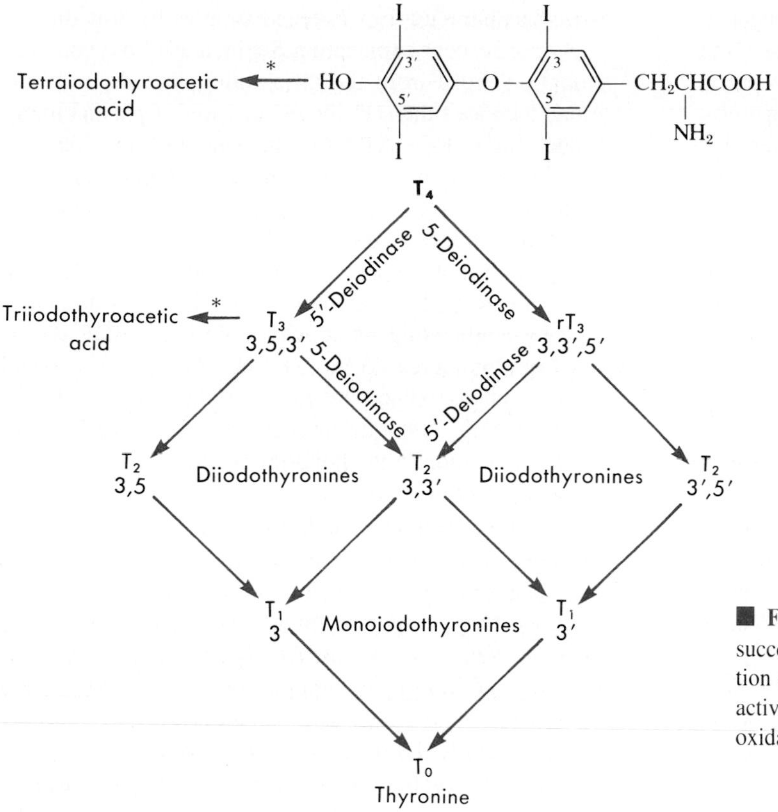

■ **Fig. 49-7** Peripheral metabolism of thyroxine (T_4) by successive deiodinations. A key regulatory step is the proportion of T_4 undergoing the initial deiodination to metabolically active T_3 versus metabolically inactive rT_3. Asterisks signify oxidative deamination and decarboxylation.

nously, T_3 is three to four times more potent than T_4 in humans. However, the intrinsic biological activity of T_4 is still debatable. Evidence favoring activity of T_4 is found in situations in which a low plasma T_4 level is accompanied by a state of biological thyroid deficiency despite a normal plasma T_3, and in situations in which a clinically normal state exists with a normal plasma T_4 concentration despite a low plasma T_3. Furthermore, in the absence of endogenous thyroid gland function the maintenance of a euthyroid state requires exogenous doses of T_3 that sustain supranormal plasma T_3 levels, whereas only doses of T_4 that sustain normal plasma T_4 levels are required. The T_3 that has been generated from T_4 intracellularly might be more efficient than the T_3 reaching its intracellular sites of action from the circulation.

■ *Intracellular Actions of Thyroid Hormone*

Free T_4 and T_3 enter cells by a carrier-mediated, energy-dependent process. Within the cell most if not all of the T_4 undergoes conversion to T_3 (or rT_3). T_3 and T_4 bind to a nuclear receptor protein that has a molecular weight of 50,000 and that is associated with chromatin, usually template-inactive. Multiple forms of the thyroid receptor exist and are expressed in a tissue-specific manner by separate genes. The structures are related to c-erbA$_s$ (avian erythroblastosis virus) oncogene products and to the steroid receptor family. T_3 has 10 times the affinity for the receptor as T_4 and 100 times the affinity as rT_3. The T_3-receptor complex binds to a cis-acting thyroid regulatory element (TRE) on target DNA molecules where it stimulates or inhibits gene transcription. One endogenous T_3 receptor analogous molecule does not bind T_3, but it does bind to the TRE and inhibits T_3 action. A host of T_3 target genes exist including those for growth hormone, osteocalcin, myosin chains, malic enzyme, TSH, and the T_3 receptor itself. A cell-specific, constitutive basal element is necessary to permit T_3-receptor complex action on certain genes in some cells and at some stages of development. Subsequent to DNA binding, a large number of messenger RNA levels are increased or decreased, and the synthesis of the related proteins are altered accordingly.

In addition to nuclear receptors, thyroid hormone-binding sites have been identified in ribosomes, mitochondria, and the plasma membrane. These may mediate posttranscriptional and pretranslational events, such as association of messenger RNA with ribosomes, or posttranslational processes, such as membrane transport.

The responsiveness of tissues to T_3 correlates well with their nuclear receptor capacity and with the degree of receptor saturation. In humans normally about half the available T_3 nuclear receptor sites are occupied. Several indices of T_3 action have been shown to correlate linearly with the percentage occupancy of receptors; occupancy was calculated from serum T_3 concentration and the known affinities of human liver and kidney T_3 receptors. In some tissues T_3 may downregulate its own receptor by inhibiting its synthesis, providing still another means for receptor modulation of T_3 action. Since thyroid hormone appears to act largely through influencing transcription, many of its effects can be blocked by inhibitors of protein synthesis. This mechanism accounts for the 12- to 48-hour delay before most of the hormone's effects become evident in vivo. Indeed several weeks of T_4 replacement are required before all the consequences of the hypothyroid state are eliminated.

A complete explanation of the multitude of thyroid hormone actions on an intracellular basis is still difficult. A large catalogue of thyroid-induced enzyme and substrate changes can be listed, but no final common pathway can be offered incontrovertibly as a unifying theory of hormone action. More likely, actions exist at multiple loci, which vary in different tissues. General effects in the nucleus include stimulation of RNA polymerase and phosphoprotein kinases and the synthesis of other nuclear proteins. These nuclear effects are followed or paralleled by an increase in the biogenesis of mitochondria and their rate of respiration. The number, size, inner membrane components and areas, and protein synthesis and RNA synthesis of mitochondria are all increased by thyroid hormone. These effects may also involve, at least secondarily, mitochondrial DNA. Key respiratory enzyme activities, such as NADPH cytochrome C reductase and cytochrome oxidase, are increased. α-Glycerophosphate dehydrogenase and pyridine nucleotide transhydrogenases, important in regulating the levels of pyridine nucleotide cofactors, are likewise increased.

The level of malic enzyme, involved in providing NADPH for fatty acid synthesis, and its messenger RNA, is also greatly augmented by T_3. In this interesting example, T_3 acts synergistically with high carbohydrate feeding to provide an enzyme important in converting an excess of glucose to triglycerides. Here T_3 may be acting as a multiplier of a primary carbohydrate-generated signal. The activities of a host of other enzymes concerned with glucose oxidation and gluconeogenesis also are augmented by thyroid hormone. Induction of another thyroid specific protein (labeled S_4) in the liver precedes an increase in fatty acid synthase activity.

The clinical observation that thyroid excess appears to increase the rate of oxygen use without increasing useful work output has suggested a decrease in the efficiency with which high-energy phosphate bonds are formed during aerobic respiration. Early studies with supraphysiological doses of thyroid hormone supported such a theory. However, this hypothesis has not proved tenable. The normal P/O ratios of approximately 3 subsequently observed in muscle from hyperthyroid humans has particularly weakened support for this theory.

A more recent explanation for increased oxygen consumption arose from the observation that thyroid hormone increases the activity and amount of plasma membrane Na^+, K^+–ATPase, an enzyme essential for membrane cation transport (see also Chapter 1). Ouabain, an inhibitor of this enzyme, also blocks the action of thyroid hormone on respiration. Since the sodium pump is responsible for up to 80% of energy turnover in various tissues, large amounts of ADP would be generated by augmenting its activity. The extra ADP resulting from the increased Na^+, K^+-ATPase activity would then stimulate oxygen utilization in the mitochondria.

In keeping with this mechanism is a report that T_3 binds to the inner mitochondrial membrane enzyme, adenine nucleotide translocase, which is responsible for transporting ADP into and ATP out of the mitochondria. However, in some thyroid sensitive tissues Na^+, K^+-ATPase accounts for only 15% of total oxygen utilization. Hence, such a unitary mechanism of thyroid hormone action is dubious. Another recent suggestion is that thyroid hormone simultaneously stimulates fatty acid synthesis and fatty acid oxidation; in effect, it may operate a futile thermogenic energy cycle.

In tissues, such as brain, in which oxygen consumption is not stimulated, thyroid hormones increase the synthesis of specific structural or functional proteins. In brain and other tissues thyroid hormone also stimulates the transport of amino acids across the cell membrane, thereby facilitating protein synthesis. On the other hand, in muscle proteolytic and lysosomal enzyme activities are also increased by thyroid hormone.

Whole Body Actions of Thyroid Hormone on Metabolism

The most obvious in vivo effect of thyroid hormone is to increase the basal rate of oxygen consumption and heat production (Fig. 49-8). This action is demonstrable in all tissues except the brain, gonads, and spleen. Resting oxygen use in humans ranges from about 150 ml/min in the hypothyroid state to about 400 ml/min in the hyperthyroid state (normal 250 ml/min). When standardized to body surface area, the basal metabolic rate ranges from −40% to +80% of normal at the clinical extremes of thyroid function. The respiratory quotient (RQ) reflecting fuel mix is not affected. Glucose and fatty acid oxidation are both increased, as are glucose recycling and fatty acid-triglyceride recycling. Thyroid hormone does not augment specifically either diet-induced or facultative oxygen utilization, and it may not change the efficiency of energy use with exercise.

Of necessity, thermogenesis increases concomitantly with oxygen use. Thus changes in body temperature parallel fluctuations in thyroid hormone availability. The potential increase in body temperature, however, is

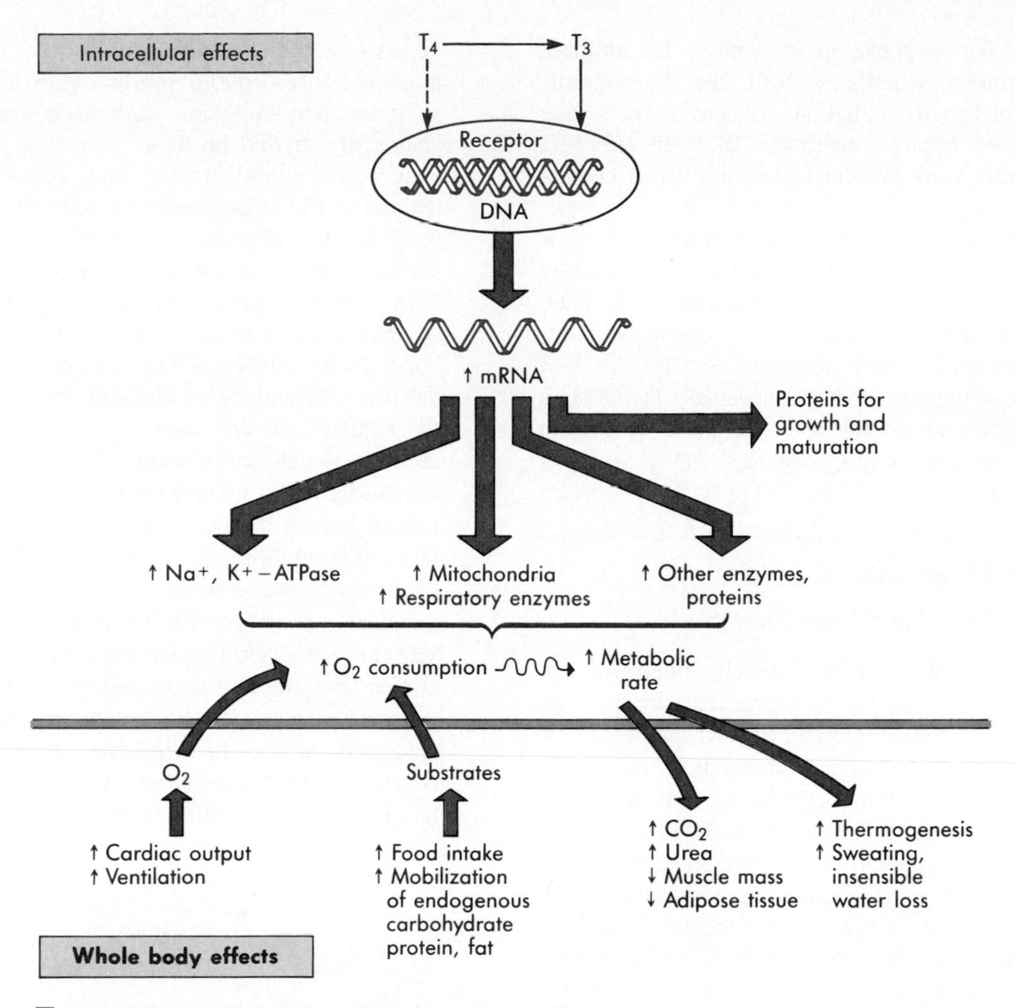

Fig. 49-8 Overall schema of thyroid hormone effects. The upper portion represents intracellular actions; the lower portion represents whole body effects.

moderated by a compensatory increase in heat loss through appropriate thyroid hormone-mediated increases in blood flow, sweating, and ventilation.

Thyroid hormone could not stimulate oxygen utilization for long without also enhancing oxygen supply. Thus T_4 and T_3 increase the resting rate of ventilation sufficiently to maintain a normal arterial P_{O_2} in the face of increased oxygen utilization and a normal P_{CO_2} in the face of increased carbon dioxide production. Additionally, a small increase in red blood cell mass is effected—enhancing the oxygen-carrying capacity—probably through stimulation of erythropoietin production.

Most important, thyroid hormone increases cardiac output, ensuring sufficient oxygen delivery to the tissues. The resting heart rate and the stroke volume are both increased. The speed and force of myocardial contractions are enhanced. Systolic blood pressure is augmented and diastolic blood pressure is decreased, so that the net result is a widened pulse pressure. This reflects the combined effects of the increased stroke volume and a substantial reduction in peripheral vascular resistance resulting from blood vessel dilation in skin, muscle, and heart. This in turn is secondary to the increase in tissue metabolism that thyroid hormone induces (Chapters 29 and 30).

The cardiac inotropic effects are partly indirect, via adrenergic stimulation, and partly direct. Myocardial calcium uptake and adenylyl cyclase activity are increased and enhance contractile force. Thyroid hormone induces the myosin heavy chain alpha gene and represses the beta gene, and thereby increases the velocity of myocardial contraction. The calcium-ATPase of the sarcolemmal reticulum is increased; this facilitates sequestration of calcium during diastole and shortens the relaxation time.

Stimulation of oxygen use must ultimately depend on the provision of necessary substrates for oxidation. T_4 and T_3 augment glucose absorption from the gastrointestinal tract and potentiate the respective stimulatory effects of epinephrine, norepinephrine, glucagon, cortisol, and growth hormone on gluconeogenesis, on lipolysis, on ketogenesis, and on proteolysis of the labile protein pool. The overall metabolic effect of thyroid hormone therefore may be aptly described as accelerating the response to starvation. In addition, thyroid hormone stimulates both the biosynthesis of cholesterol and its oxidation, its conversion to bile acids, and its biliary secretion. The net effect is to decrease the body pool and plasma level of cholesterol.

The metabolic disposal of steroid hormones, of B vitamins, and of many administered drugs is increased by thyroid hormone. Therefore the endogenous secretion rates of hormones, such as cortisol, the dietary requirements of vitamins, such as thiamine, and the doses of drugs, such as digoxin, that are necessary to maintain normal or effective plasma levels of these substances are all increased by thyroid hormone.

■ *Thyroid Hormone and the Sympathetic Nervous System*

One of the prominent but incompletely understood features of thyroid hormone is its interaction with the sympathetic nervous system. Certain effects, such as the increases in metabolic rate, heat production, heart rate, motor activity, and central nervous system excitation, also are produced by the adrenergic catecholamines epinephrine and norepinephrine. An indisputable explanation for this striking similarity remains to be found. Thyroid hormone does not increase the levels of catecholamine hormones or their metabolites in blood, urine, or tissues. But increased levels of cAMP, a beta–adrenergic second messenger, are found in plasma, urine, and muscle from hyperthyroid persons. Furthermore the cAMP response to epinephrine in cultured myocardial cells is augmented by T_3, which also increases the number of β-adrenergic receptors in heart muscle. Synergism between catecholamines and thyroid hormones also may be required for maximum thermogenesis, lipolysis, glycogenolysis, and gluconeogenesis to occur. The importance of this physiological issue is underscored by the fact that adrenergic blockade, particularly of the β-receptors, attenuates some of the cardiovascular and central nervous system manifestations of hyperthyroidism.

■ *Thyroid Hormone Effects on Growth and Development*

Another major effect of thyroid hormone is on growth and maturation. Perhaps the most spectacular example is seen in the process of metamorphosis. Endogenous thyroid hormone levels are very low in amphibians until just before the major stage of metamorphosis. At this point, the hormone levels increase sharply, paralleling the rapid change from the larval to the adult form, after which they again decline. Addition of exogenous thyroid hormone to the fluid bathing tadpoles accelerates all aspects of their metamorphosis, including limb growth, tail resorption, shortening of the gastrointestinal tract, and induction of hepatic ureagenesis. Biochemically thyroid hormone increases protein and nucleic acid synthesis in the limb buds, increases proteolytic and hydrolytic enzyme activities in the tail, and increases the hepatic content of carbamyl phosphate synthase, the rate-limiting enzyme in the urea cycle.

In humans and other mammals, thyroid hormone stimulates endochondral ossification, linear growth of bone, and maturation of the epiphyseal bone centers. T_3 enhances the maturation and activity of chondrocytes in the cartilage growth plate (Fig. 49-9), in part by increasing somatomedin action. In addition, T_3 may accelerate growth by facilitating the synthesis and secretion of growth hormone, as previously noted. Interestingly, thyroid hormone is not required for linear growth until after birth, whereas it is already essential for normal maturation of growth centers in the bones of the developing fetus. The regular progression of tooth development and eruption is dependent on thyroid hormone, as is the normal cycle of growth and maturation of the epidermis and its hair follicles. Because the normal degradative processes in these structural and integumentary tissues also are stimulated by thyroid hormone, ex-

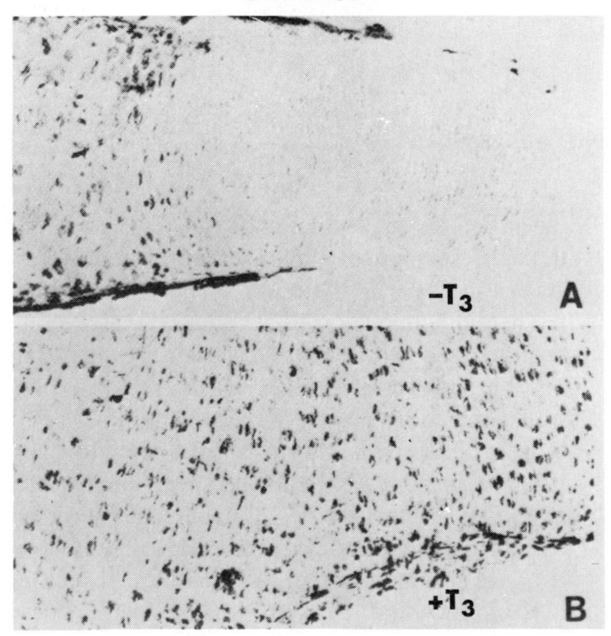

■ **Fig. 49-9** Effect of thyroid hormone on cartilage growth. Cartilage from the growth plates of pigs was incubated in vitro with or without T_3 for 3 days. Chondrocytes were then identified by staining sections for the enzyme alkaline phosphatase. As seen in **B,** addition of T_3 caused a marked increase in the rate of proliferation of new chondrocytes as compared to **A.** (From Burch WM, Lebovitz HE: Triiodothyronine stimulates maturation of porcine growth-plate cartilage in vitro, *J Clin Invest* 70:496, 1982.)

cess exposure to T_4 and T_3 also can cause desquamation of skin and hair loss.

Thyroid hormone alters the characteristics of subcutaneous tissue by inhibiting the synthesis of mucopolysaccharides (glycosoaminoglycans) in the intercellular ground substance as well as that of **fibronectin,** an adherence-producing molecule elaborated by fibroblasts.

T_3 receptors are present in osteoblasts, and thyroid hormone stimulates bone remodeling. This is manifested by increases in plasma levels of alkaline phosphatase and osteocalcin and in urinary excretion of hydroxyproline and pyridinium crosslink compounds. The net result of continued thyroid hormone excess is a decrease in total bone mass.

A critical set of actions of thyroid hormone is on the development of the central nervous system. If thyroid hormone is deficient in utero and early infancy, growth of the cerebral and cerebellar cortex, proliferation of axons and branching of dendrites, and the process of myelinization are all decreased. Irreversible brain damage can result when the deficiency of thyroid hormone is not recognized and treated promptly after birth. The anatomical defects are paralleled by biochemical abnormalities. In various areas of the brain of hypothyroid embryos the cell size, the RNA and protein content, the amount of tubulin and microtubule associated protein, the protein and lipid content of myelin and the rates of protein synthesis are all decreased. Such enzymes as succinic dehydrogenase, essential for energy generation, and as galactosyl sialyl transferase, essential for myelin formation, as well as receptors for neurotransmitters are also diminished.

The crucial role of thyroid hormone in central nervous system development is underscored by a number of adaptive phenomena seen specifically in the neonatal brain that increase the biological effectiveness of thyroid hormone at this critical time. T_3 receptors in the cortex and other areas are increased. The activity of brain 5′-monodeiodinase is augmented, enhancing local conversion of T_4 to T_3, whereas the activity of brain 5-monodeiodinase is diminished, reducing the local degradation of T_3 to $3,3'T_2$ (see Fig. 49-7). Thyroid hormone also enhances wakefulness, alertness, responsiveness to various stimuli, auditory sense, awareness of hunger, memory, and learning capacity. Normal emotional tone also depends on proper thyroid hormone availability. Furthermore, the speed and amplitude of peripheral nerve reflexes are increased by thyroid hormone, as is motility of the gastrointestinal tract.

In both women and men thyroid hormone plays an important permissive role in the regulation of reproductive function. The normal ovarian cycle of follicular development, maturation, and ovulation, the homologous testicular process of spermatogenesis, and the maintenance of the healthy pregnant state all are disrupted by significant deviations of thyroid hormone from the normal range. In part these may be caused by alterations in the metabolism or availability of steroid hormones; e.g., thyroid hormone stimulates hepatic synthesis and release of sex steroid-binding globulin.

Normal function of skeletal muscles also requires normal amounts of thyroid hormone. This may well be related to the regulation of energy production and storage in this tissue. Concentrations of creatine phosphate are reduced by an excess of T_4 and T_3; the inability of muscle to take up and phosphorylate creatine leads to its increased urinary excretion.

Kidney size, renal plasma flow, glomerular filtration rate, and tubular transport maximums for a number of substances also are increased by thyroid hormone.

■ *Clinical Syndromes of Thyroid Dysfunction*

Hyperthyroidism results from a number of causes. Most commonly the entire gland undergoes hyperplasia as a result of autoimmune stimulation **(Graves' disease),** in which antibodies formed against the TSH receptor bind to it and mimic TSH action. Next most commonly, one or more areas of the thyroid form benign neoplasms, which escape from the normal hypothalamic-pituitary regulation. Least common causes are inflammation of the thyroid, excessive pituitary secretion of TSH, or ingestion of exogenous T_4 or T_3.

The patient suffering from an excess of thyroid hormone presents one of the most striking pictures in clinical medicine. *The large increase in metabolic rate causes the highly characteristic combination of weight loss concomitant with an increased intake of food.* The increased heat production causes discomfort in warm environments, excessive sweating, and a greater intake of water. The increase in adrenergic activity is manifested by a rapid heart rate, hyperkinesis, tremor, nervousness, and a wide-eyed stare. Weakness is caused by a loss of muscle mass as well as a specific thyrotoxic myopathy. Other symptoms include emotional lability, breathlessness during exercise, and difficulty swallowing or breathing because of compression of the esophagus or trachea by the enlarged thyroid gland **(goiter)**.

The diagnosis is established by demonstrating an elevated serum T_4 or T_3 level (appropriately corrected for any abnormalities in TBG concentrations). In most instances the thyroid uptake of iodine (labeled with ^{131}I or ^{123}I) is excessive. Serum TSH levels are low, since the pituitary is inhibited by the high levels of T_4 and T_3. The most definitive treatment is ablation of thyroid tissue, either by radiation effects of ^{131}I or by surgery. Alternatively, thiouracil drugs, which block hormone synthesis, are given for 18 months.

Hypothyroidism most often results from idiopathic atrophy of the gland—thought to be preceded by a chronic autoimmune inflammatory reaction. This form of **lymphocytic thyroiditis** is also an autoimmune con-

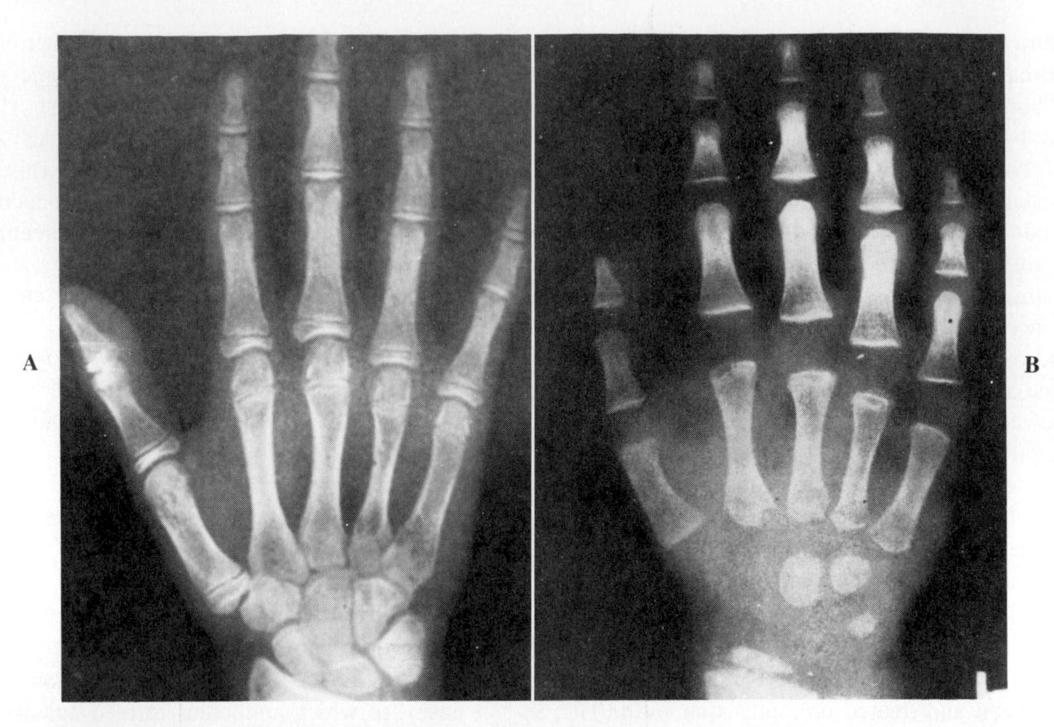

■ **Fig. 49-10** Hand x-ray films of a 13-year-old normal child (**A**), and a 13-year-old hypothyroid child (**B**). Note that the hypothyroid child has a marked delay in development of the small bones of the hands, in growth centers at either end of the fingers, and in the growth center of the distal end of the radius. (**A,** from Tanner JM et al: *Assessment of skeletal maturity and prediction of adult height (TW2 method),* New York, 1975, Academic Press; **B,** from Andersen HJ: Nongoitrous hypothyroidism. In Gardner LI, editor: *Endocrine and genetic diseases of childhood and adolescence,* Philadelphia, 1975, WB Saunders.)

dition. In this case the antibodies produced may block hormone synthesis or have cytotoxic properties. Other causes include radiation damage, surgical removal, nodular goiters, hypothalamic or pituitary destruction, and, in certain areas of the world, iodide deficiency. Rarest of all is resistance to the action of thyroid hormones, possibly because of deficiency of receptors.

The clinical picture is, in many respects, the exact opposite of that seen in hyperthyroidism. *The lower than normal metabolic rate leads to weight gain without appreciable increase in caloric intake*. The decreased thermogenesis lowers body temperature and causes intolerance to cold, decreased sweating, and dry skin. There is decreased adrenergic activity, with bradycardia; a generalized slowing of movement, speech, and thought; lethargy; sleepiness; and a lowering of the upper eyelids (ptosis). An accumulation of mucopolysaccharides—ground substance—in the tissues causes an accumulation of fluid, as well. This myxedema produces puffy features, an enlarged tongue, hoarseness, joint stiffness, effusions in the pleural, pericardial, and peritoneal spaces, and entrapment of peripheral and cranial nerves with consequent dysfunction. Constipation, loss of hair, menstrual dysfunction, and anemia are other signs. Notably hypothyroidism in infancy or childhood causes marked retardation of growth and even greater slowing in the maturation of the epiphyseal growth centers of the bone (Fig. 49-10). If hypothyroidism is present at birth and remains untreated for even 2 to 4 weeks, the central nervous system will not undergo its normal maturation process in the first year of life. Developmental milestones, such as sitting, standing, and walking, will be late, and severe irreversible mental retardation can result. Such individuals are known as **cretins.**

The diagnosis of hypothyroidism is made by finding a low serum T_4 level. Serum TSH will be elevated because of negative feedback unless the hypothyroidism is caused by hypothalamic or pituitary disease. If the pituitary is at fault, TSH levels will be low and will not respond to administration of TRH. Replacement therapy with T_4 is curative. T_3 is not needed, since it will be generated intracellularly from the administered T_4.

The stepwise nature of thyroid hormone synthesis offers multiple possibilities for congenital hypothyroidism caused by specific enzyme deficiencies. These syndromes are characterized by the symptoms of childhood hypothyroidism noted previously, plus thyroid gland enlargement (congenital goiter) resulting from persistent hypersecretion of TSH. Table 49-2 lists the best understood of these rare syndromes, with the biochemical findings that point to the lesion. Replacement therapy

Table 49-2 Some congenital defects in thyroid hormone synthesis

Defect	*Diagnostic pattern*
Iodide trap	Decreased uptake of radioactive iodine; decreased salivary/blood ratio of radioactive iodine
Peroxidase	Increased early uptake of radioactive iodine*; rapid discharge by perchlorate
Deiodinase	Increased uptake of radioactive iodine*; increased MIT and DIT in urine
Coupling	Increased uptake of radioactive iodine*; increased MIT and DIT and decreased T_4 and T_3 in thyroid tissue

*Radioactive iodine uptake is increased because of increased TSH secretion, which stimulates the iodide trap.

with T_4 corrects the hormone deficiency in each instance and reduces the size of the goiter. Large amounts of iodine alone can be used successfully if the defect is in the iodide trap.

Summary

1. The thyroid gland is the source of tetraiodothyronine (thyroxine, T_4) and triiodothyronine (T_3). The basic endocrine unit in the gland is a follicle consisting of a single circular layer of epithelial cells surrounding a central lumen that contains colloid or stored hormone.

2. T_4 and T_3 are synthesized from tyrosine and iodine by the enzyme complex, peroxidase. Tyrosine is incorporated in peptide linkages within the protein thyroglobulin. After iodination, two iodotyrosine molecules are coupled to yield the iodothyronines.

3. Secretion of stored T_4 and T_3 requires retrieval of thyroglobulin from the follicle lumen by endocytosis. To support hormone synthesis, iodide is both actively concentrated by the gland and conserved within it by recovery from the iodotyrosine that escapes secretion.

4. Thyrotropin (TSH) acts on the thyroid gland via its plasma membrane receptor and cAMP to stimulate all steps in the production of T_4 and T_3. These steps include iodide uptake, iodination and coupling, and retrieval from thyroglobulin. TSH also stimulates glucose oxidation, protein synthesis, and growth of the epithelial cells.

5. More than 99.5% of the T_4 and T_3 circulate bound to thyroid-binding globulin (TBG). Only the free fractions of T_4 and T_3 are biologically active. Changes in TBG levels require corresponding changes in thyroid hormone secretion to maintain normal concentrations of free T_4 and T_3.

6. T_4 functions largely as a prohormone. Monodeiodination of the outer ring yields 75% of the daily production of T_3, which is the principal active hormone. Alternatively, monodeiodination of the inner ring yields reverse T_3, which is biologically inactive. Proportioning of T_4 between T_3 and reverse T_3 regulates the availability of active thyroid hormone.

7. Thyroid hormone is a major regulator of and increases the basal metabolic rate. T_3 combines with its receptor in target cell nuclei; the T_3-receptor complex interacts with many target DNA molecules to induce or suppress synthesis of a variety of enzymes and other proteins. The result is to increase oxygen utilization by mechanisms that include increases in the size and number of mitochondria, in Na^+, K^+-ATPase activity, and in the rates of glucose and fatty acid oxidation and synthesis.

8. Additional important actions of thyroid hormone are to increase heart rate, cardiac output, and ventilation and to decrease peripheral resistance. These subserve the increased tissue oxygen demand. The corresponding increase in heat production leads to increased sweating. Substrate mobilization and disposal of metabolic products is enhanced.

9. Other thyroid hormone effects on the central nervous system and skeleton are crucial to normal growth and development. In the absence of the hormone, brain development is retarded and cretinism results. The stature shortens and the bones fail to mature. In adults thyroid hormone increases the rates of bone resorption and of degradation of skin and hair.

Bibliography

Journal articles

Acheson K et al: Thyroid hormones and thermogenesis: the metabolic cost of food and exercise, *Metabolism* 33:262, 1984.

Bantle JP et al: Common clinical indices of thyroid hormone action: relationships to serum free 3,5,3′-triiodothyronine concentration and estimated nuclear occupancy, *J Clin Endocrinol Metab* 50:286, 1980.

Danforth E Jr: The role of thyroid hormone and insulin in the regulation of energy metabolism, *Am J Clin Nutr* 38:1006, 1983.

Dillmann WH: Biochemical basis of thyroid hormone action in the heart, *Am J Med* 88:626, 1990.

Gavaret JM et al: Formation of dehydroalanine residues during thyroid hormone synthesis in thyroglobulin, *J Biol Chem* 255:5281, 1980.

Izumo S et al: All members of the MHC multigene family respond to thyroid hormone in a highly tissue-specific manner, *Science* 231:597, 1986.

Klein I: Thyroid hormone and the cardiovascular system, *Am J Med* 88:631, 1988.

Larsen PR et al: Inhibition of intrapituitary thyroxine to 3,5,3′-triiodothyronine conversion prevents the acute suppression of thyrotropin release by throxine in hypothyroid rats, *J Clin Invest* 64:117, 1979.

Larsen PR et al: Relationships between circulating and intracellular thyroid hormones, physiological and clinical implications, *Endocr Rev* 2:87, 1981.

Lazar MA et al: Nuclear thyroid hormone receptors, *J Clin Invest* 86:1777, 1990.

Martial JA et al: Regulation of growth hormone gene expression; synergistic effects of thyroid and glucocorticoid hormones, *Proc Natl Acad Sci USA* 74:4293, 1977.

Misrahi M et al: Cloning, sequencing and expression of human TSH receptor, *Biochem Biophys Res Commun* 166:394, 1990.

Murata Y et al: Thyroid hormone inhibits fibronectin synthesis by cultured human skin fibroblasts, *J Clin Endocrinol Metab* 64:334, 1987.

Mutvei A et al: Thyroid hormone and not growth hormone is the principle regulator of mammalian mitochondrial biogenesis, *Acta Endocrinol (Copenh)* 121:223, 1989.

Nelson BD: Thyroid hormone regulation of mitochondrial function. Comments on the mechanism of signal transduction, *Biochem Biophys Acta* 1018:275, 1990.

Oppenheimer JH et al: Stimulation of hepatic mitochondrial alpha-glycerophosphate dehydrogenase and malic enzyme by L-triiodothyronine; characteristics of the response with specific nuclear thyroid hormone binding sites fully saturated, *J Clin Invest* 59:517, 1977.

Oppenheimer JH: Thyroid hormone action at the nuclear level, *Ann Intern Med* 102:374, 1985.

Piolino V et al: Thermogenic effect of thyroid hormones: interactions with epinephrine and insulin, *Am J Physiol* 259:E305, 1990.

Reed HL et al: Changes in serum triiodothyronine (T3) kinetics after prolonged Antarctic residence: the polar T3 syndrome, *J Clin Endocrinol Metab* 70:965, 1990.

Roger PO et al: Thyrotropin is a potent growth factor for normal human thyroid cells in primary culture, *Biochem Biophys Res Commun* 149:707, 1987.

Samuels HH et al: Regulation of gene expression by thyroid hormone, *Annu Rev Physiol* 51:623, 1989.

Schimmel M et al: Thyroidal and peripheral production of thyroid hormones, *Ann Intern Med* 87:760, 1977.

Smith TJ et al: Regulation of glycosaminoglycan synthesis by thyroid hormone in vitro, *J Clin Invest* 70:1066, 1982.

Sterling K: Direct thyroid hormone activation of mitochondria: identification of adenine nucleotide translocase (AdNT) as the hormone receptor, *Trans Assoc Am Physicians* 100:284

Stocker WW et al: Coupled oxidative phosphorylation in muscle of thyrotoxic patients, *Am J Med* 44:900, 1968.

Vagenakis AG et al: Effect of starvation on the production and metabolism of thyroxine and triiodothyronine in euthyroid obese patients, *J Clin Endocrinol Metab* 45:1305, 1977.

Vulsma T et al: Maternal-fetal transfer of thyroxine in congenital hypothyroidism due to a total organification defect or thyroid agenesis, *N Eng J Med* 321:13, 1989.

Books and monographs

Galton VA: Thyroid hormone action in amphibian metamorphosis. In Oppenheimer JH, Samuels HH, editors: *Molecular basis of thyroid hormone action,* New York, 1983, Academic Press.

Greer MA et al: Thyroid secretion. In *Handbook of physiology,* section 7: Endocrinology, Thyroid, vol 3, Baltimore, 1974, American Physiological Society.

Reichlin S: Neuroendocrine control of thyrotropin secretion. In Ingebar SH, Braverman LE, editors: *Werner's the thyroid,* Philadelphia, 1986, JB Lippincott.

Taurog A: Hormone synthesis: thyroid iodine metabolism. In Ingbar, SH, Braverman LE, editors: *Werner's the thyroid,* Philadelphia, 1986, JB Lippincott.

CHAPTER

50

The Adrenal Glands

The adrenal glands are complex, multifunctional endocrine organs, which are essential for life. Severe illness results from their atrophy, and death follows their complete removal.

Each adrenal gland is really a combination of two separate functional entities (Fig. 50-1). The outer zone, or **cortex,** which comprises 80% to 90% of the gland, is derived from mesodermal tissue and is the source of corticosteroid hormones. The inner zone, or **medulla,** comprising the other 10% to 20%, is derived from neuroectodermal cells of the sympathetic ganglia and is the source of catecholamine hormones.

The adrenal glands are located in the retroperitoneum just above each kidney. Their total weight is 6 to 10 g. They receive arterial blood from branches of the aorta, the renal arteries, and the phrenic arteries; they have one of the body's highest rates of blood flow per gram of tissue. Arterial blood enters sinusoidal capillaries in the cortex. The blood then drains into medullary venules, an arrangement that exposes the medulla to relatively high concentrations of corticosteroids from the cortex. The right adrenal vein drains directly into the inferior vena cava, whereas the left drains into the renal vein on that side. Catheterization of each adrenal vein for blood sampling is possible, but catheterization is more difficult on the right. The human adrenals can be visualized radiographically by computed tomography (CT) scanning or magnetic resonance imaging.

■ *The Adrenal Cortex*

The major hormones of the cortex are (1) the **glucocorticoids, cortisol** and **corticosterone,** which are critical to life by virtue of their effects on carbohydrate and protein metabolism; (2) a **mineralocorticoid, aldosterone,** which is vital to maintaining sodium and potassium balance; and (3) precursors to the **sex steroids, androgens** and **estrogens** which contribute to establishing and maintaining secondary sexual characteristics. Particular medical interest in cortisol and in glucocorticoids in general was heightened greatly by the discovery of their potent antiinflammatory effects. In supraphysiological doses, these have been used to treat a wide variety of diseases.

The cortical portion of the adrenal gland differentiates by 8 weeks of intrauterine life and is initially much larger than the adjacent kidney. At this time the cortex contains two zones. The **peripheral neocortex,** comprising 15% of the cortex, is undifferentiated and relatively inactive. The inner 85%, known as the **fetal cortex,** is highly active and is responsible for fetal adrenal steroid production throughout almost all intrauterine life. Shortly after birth, the fetal cortex begins to involute; hemorrhage and necrosis occur, and this zone disappears completely in 3 to 12 months. Concurrently, the thin outer zone enlarges, differentiates, and becomes the permanent three-layered adrenal cortex of the normal human. Each layer or zone mainly secretes one of the three types of corticosteroids (see Fig. 50-1).

The histological appearance of the three mature cortical zones differs. The outermost **zona glomerulosa** is very thin and consists of small cells with numerous elongated mitochondria that possess lamellar cristae.

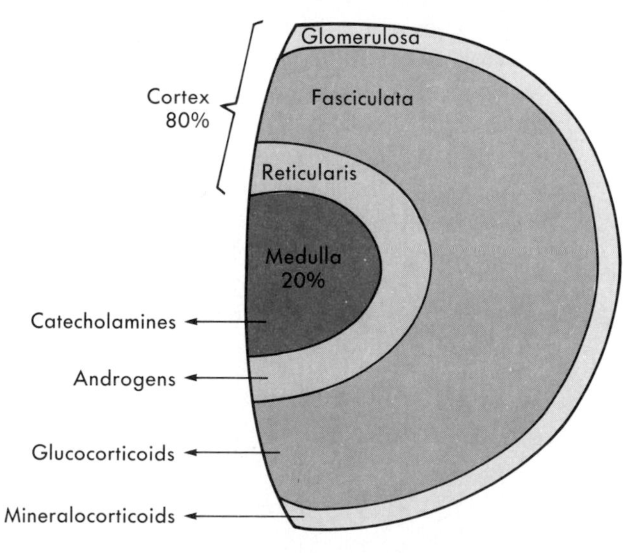

■ **Fig. 50-1** Schematic representation of the zones of the adrenal gland and their main secretory products.

The middle **zona fasciculata** is the widest zone and consists of columnar cells that form long cords. The cytoplasm is highly vacuolated and contains lipid droplets. The mitochondria are distinguished by their large size and numerous vesicular cristae within their membranes. The innermost **zona reticularis** contains networks of interconnecting cells with fewer lipid droplets but mitochondria similar to those of fasciculata cells. Under stimulation by adrenocorticotropic hormone (ACTH), the size and the number of cells in the fasciculata and reticularis increase. Their mitochondria become larger and more numerous and develop central ribosomes, vesicular cristae, and polylamellar membranes that extend to nearby cholesterol-containing vacuoles. In addition, the endoplasmic reticulum increases. These changes relate to ACTH effects on steroid hormone synthesis as detailed later.

Synthesis of Adrenocortical Hormones

All hormones of the adrenal cortex represent chemical modifications of the steroid nucleus shown in Fig. 50-2. Potent glucocorticoids require the presence of a ketone at the 3 position and hydroxyl groups at the 11 and 21 positions. Potent mineralocorticoids require an oxygenated carbon at the 18 position. Potent androgens are characterized by the elimination of the $C_{20\text{-}21}$ side chain and the presence of an oxygenated carbon at the 17 position. Estrogens are characterized by aromatization of the A ring. During adrenal development cells migrate inward from the outermost layer; as they do, they gain 17-hydroxylating activity and lose 18-hydroxylating activity. Those that migrate furthest then lose 11-hydroxylating activity.

The precursor for all adrenocortical hormones is **cholesterol,** which is actively taken up from the plasma by adrenal cells. Specific adrenal plasma membrane receptors bind circulating low-density lipoproteins rich in cholesterol. After transfer into the cell by endocytosis, the cholesterol is largely esterified and stored in cytoplasmic vacuoles. Cholesterol is also synthesized to a minor degree in adrenal cells from acetylcoenzyme A (acetyl CoA) by the usual pathway. Under basal circumstances free cholesterol from plasma is the major source used for hormone synthesis. When production of corticosteroids is stimulated, however, stored cholesterol becomes an increasingly important precursor.

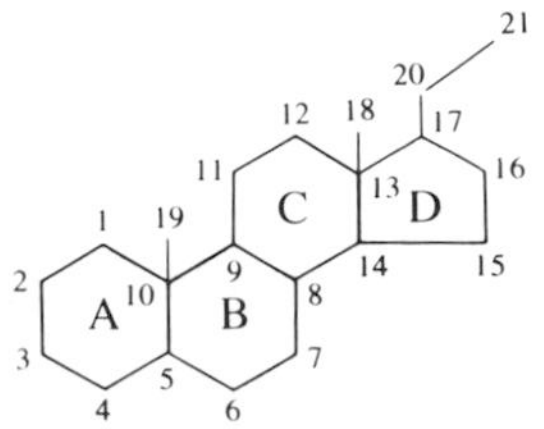

Fig. 50-2 The adrenocorticosteroid nucleus.

Most of the synthetic reactions from cholesterol to active hormones involve **cytochrome P-450 enzymes**, which are mixed oxygenases that catalyze steroid hydroxylations. These hydrophobic hemoproteins are localized in the lipophilic membranes of the endoplasmic reticulum and mitochondrial cristae. Molecular oxygen is split so that one oxygen atom interposes between the carbon and hydrogen of the steroid site; the other oxygen atom is reduced by hydrogen to H_2O. NADPH and, to some extent, NADH, which are generated by oxidation of a variety of substrates, are the ultimate donors of the hydrogen. A flavoprotein enzyme, adrenoxin reductase, and an iron-containing protein, adrenoxin, are intermediates in the transfer of hydrogen from NADPH to the P-450 enzymes.

Glucocorticoids. The synthesis of glucocorticoids occurs largely in the zona fasciculata with a smaller contribution from the zona reticularis. **Cortisol** is the dominant glucocorticoid for humans. However, if cortisol synthesis is blocked but the pathway to **corticosterone** is open, increased synthesis of the latter can provide the necessary glucocorticoid activity. The sequence of reactions for glucocorticoid synthesis is shown in Fig. 50-3. The intracellular localization of the various steps is illustrated in Fig. 50-4. Cholesterol is first transported from its storage vacuoles to the mitochondria by an active process (Fig. 50-5). The initial reaction converting cholesterol to Δ^5-pregnenolone is catalyzed by a side chain cleavage enzyme ($P_{450_{scc}}$) also known as 20,22-desmolase. This intramitochondrial complex carries out successive hydroxylations followed by cleavage of the cholesterol side chain. This step limits the rate for adrenal steroidogenesis in general. The Δ^5-product, pregnenolone, is then converted to 11-deoxycortisol by successive steps within the endoplasmic reticulum. The latter steroid, after transfer back to the mitochondria, is hydroxylated in the 11 position. The endproduct, cortisol, rapidly diffuses out of the cell. The last and critical step for glucocorticoid synthesis, 11-hydroxylation, is very efficient in humans; 95% of the 11-deoxycortisol formed is converted to cortisol.

The order of hydroxylations from Δ^5-pregnenolone to 11-deoxycortisol is not invariant. Even the 3B-ol-dehydrogenase and $\Delta^{4,5}$-isomerase reactions can occur after all the hydroxylations, rather than before. However, the sequences presented in Fig. 50-3 (17-21-11) are compatible with the usual pattern of precursor accumulation that occurs when the various hydroxylases are either chemically blocked or congenitally deficient. A limited degree of 18-hydroxylation in the zona fasciculata produces small amounts of 18-OH-cortisol, 18-OH-deoxycorticosterone (18-OH-DOC), and 18-OH-corticosterone.

It is important to note that the *critical hormone, cortisol is not stored in the adrenocortical cell to any significant extent.* Hence, an acute need for increased amounts of circulating cortisol requires rapid activation of the entire synthetic sequence from cholesterol and

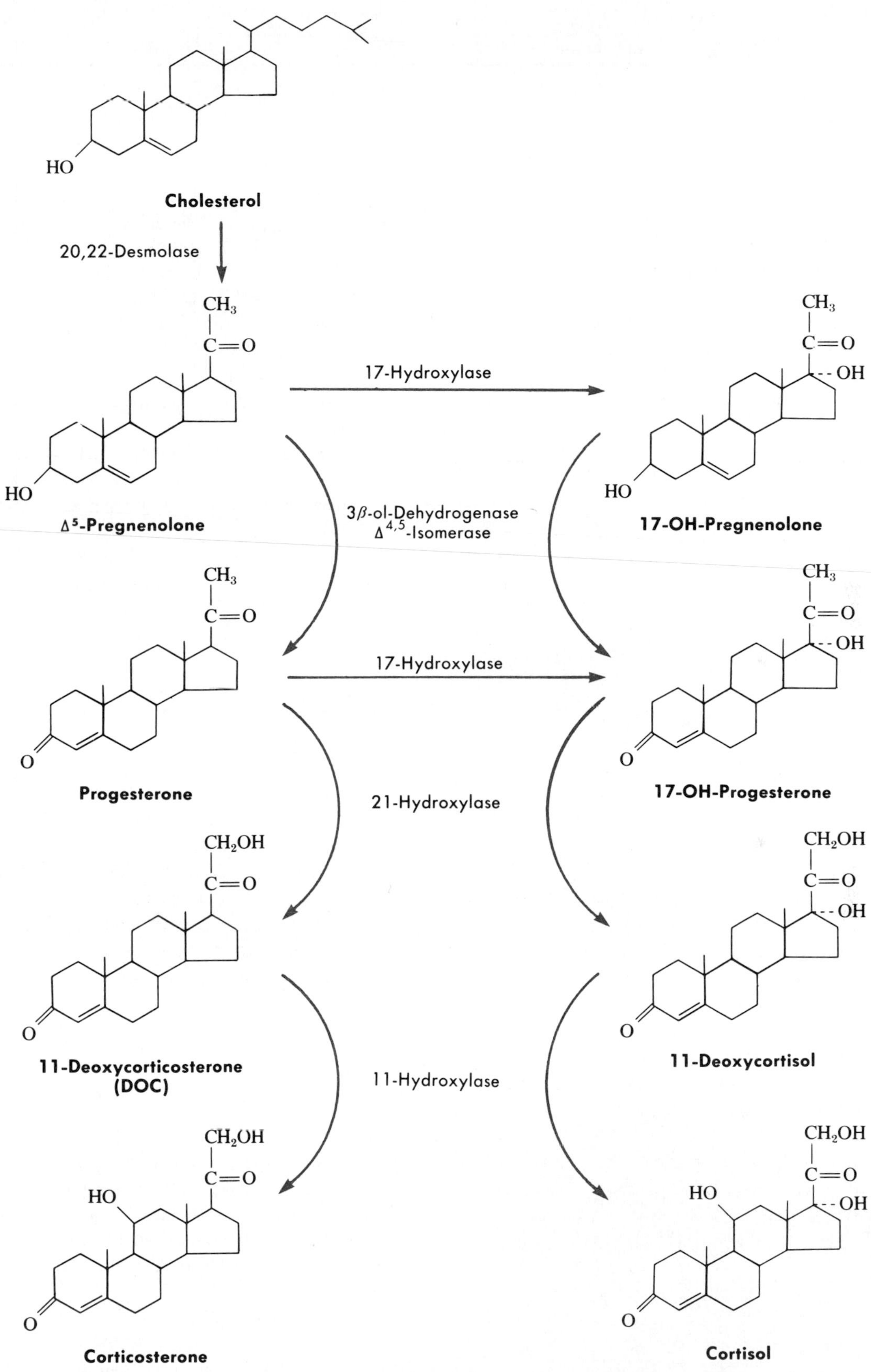

HO

Cholesterol

20,22-Desmolase

CH_3
C=O

17-Hydroxylase

CH_3
C=O
OH

HO

Δ^5-Pregnenolone

3β-ol-Dehydrogenase
$\Delta^{4,5}$-Isomerase

HO

17-OH-Pregnenolone

CH_3
C=O

17-Hydroxylase

CH_3
C=O
OH

O

Progesterone

21-Hydroxylase

O

17-OH-Progesterone

CH_2OH
C=O

CH_2OH
C=O
OH

O

11-Deoxycorticosterone (DOC)

11-Hydroxylase

O

11-Deoxycortisol

CH_2OH
C=O
HO

CH_2OH
C=O
HO
OH

O

Corticosterone

O

Cortisol

■ **Fig. 50-3** Synthesis of glucocorticoids in the zona fasciculata. Cortisol is the major product in humans.

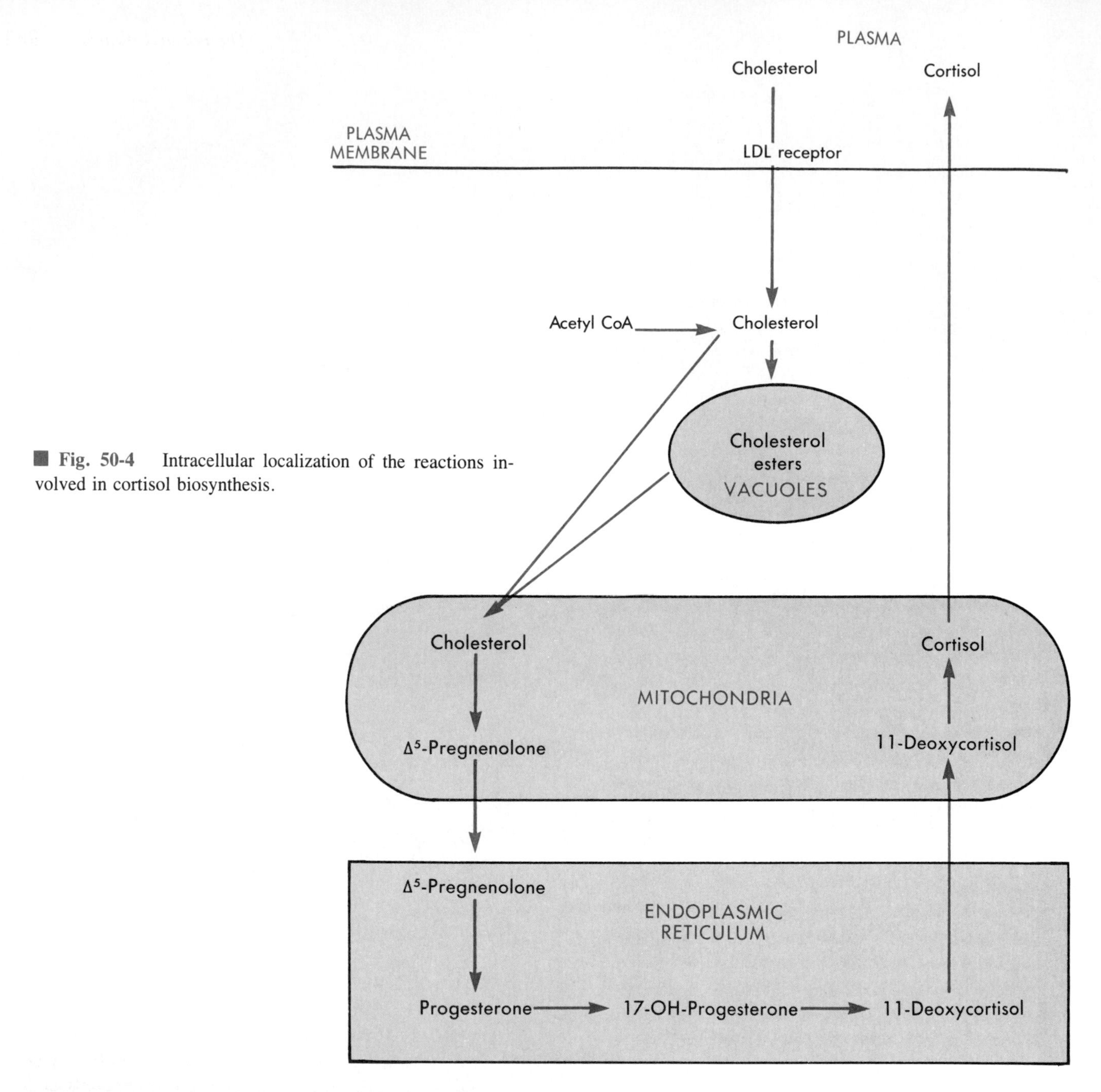

■ **Fig. 50-4** Intracellular localization of the reactions involved in cortisol biosynthesis.

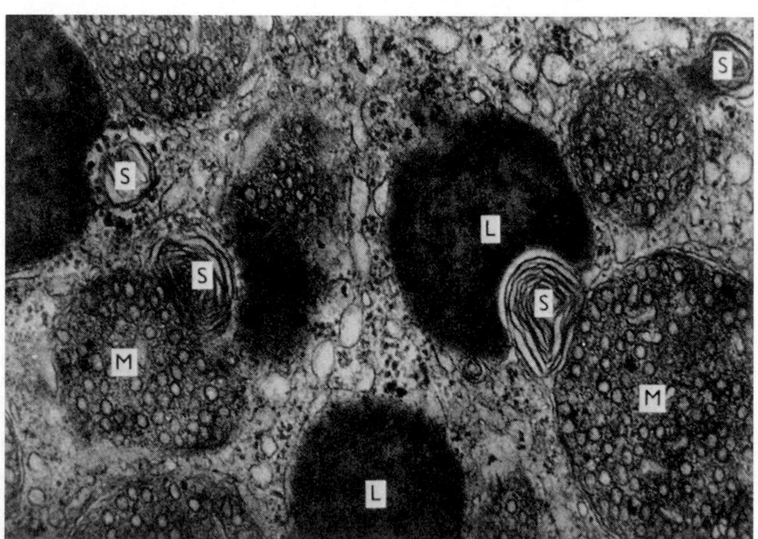

■ **Fig. 50-5** Electron micrograph of rat cell from adrenal cortex. Note how lipid vacuoles (L) containing cholesterol attach to mitochondria (M) via syncitial membranes (S) to accomplish transfer of the cholesterol into the mitochondria for the first step in steroid hormone synthesis. (From Merry BJ: *J Anat* 119:611, 1975, Cambridge University Press.)

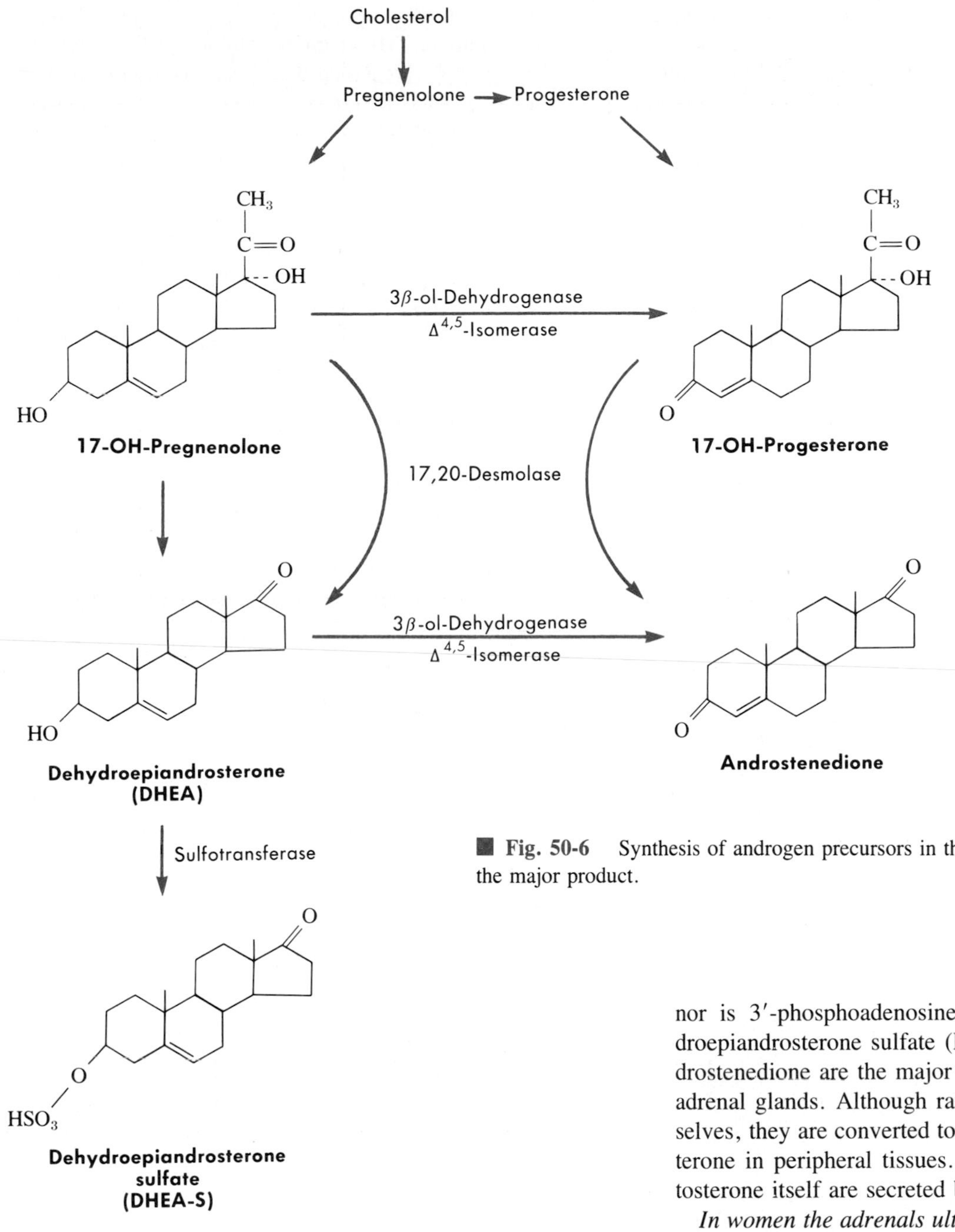

Cholesterol
Pregnenolone → Progesterone
17-OH-Pregnenolone
17-OH-Progesterone
3β-ol-Dehydrogenase
$\Delta^{4,5}$-Isomerase
17,20-Desmolase
Dehydroepiandrosterone (DHEA)
Androstenedione
Sulfotransferase
Dehydroepiandrosterone sulfate (DHEA-S)

■ **Fig. 50-6** Synthesis of androgen precursors in the zona reticularis. DHEA-S is the major product.

most particularly that of the rate-limiting $P_{450_{scc}}$ desmolase reaction.

Androgens and estrogens. The synthesis of sex steroid precursors occurs largely in the zona reticularis. The 17-hydroxylated derivatives of Δ^5-pregnenolone and progesterone are the starting points for androgen and estrogen synthesis. Removal of the $C_{20\text{-}21}$ side chain by a microsomal desmolase-like reaction is the key step, yielding dehydroepiandrosterone (DHEA) and androstenedione, respectively, as shown in Fig. 50-6. This reaction is catalyzed by the same enzyme, $P_{450_{c17}}$, that catalyzes the previous 17-hydroxylation step. DHEA is sulfated by a specific enzyme; the sulfate donor is 3′-phosphoadenosine 5′-phosphosulfate. Dehydroepiandrosterone sulfate (DHEA-S), DHEA, and androstenedione are the major androgenic products of the adrenal glands. Although rather weak androgens themselves, they are converted to the potent androgen testosterone in peripheral tissues. Only tiny amounts of testosterone itself are secreted by the zona reticularis.

In women the adrenals ultimately supply 50% to 60% of the androgenic hormone requirements. Adrenal androgen precursors are of little biological importance to men because the testes produce a large quantity of testosterone. The further conversion within the adrenal cortex of testosterone and androstenedione to estradiol and estrone, respectively, is of little quantitative significance in women until the ovaries cease to function after menopause. Then estrogens secreted directly from the adrenal glands or arising in the periphery from adrenal precursors become the only source for this biological activity. It is very important to note that 17-hydroxylation is the last reaction common to the synthesis of cortisol and the adrenal androgens (compare Fig. 50-3 with Fig. 50-6). When cortisol synthesis is impaired at any point beyond this step, the accumulation of 17-hydroxypregnenolone and 17-hydroxyprogesterone leads to greatly increased androgen synthesis.

Mineralocorticoids. The synthesis of **aldosterone**, the major mineralocorticoid, is carried out exclusively by the zona glomerulosa (Fig. 50-7). The sequence from cholesterol to corticosterone is identical to that in the zona fasciculata. The C_{18} methyl group of corticosterone is then hydroxylated and converted to an aldehyde by a mitochondrial P_{450} mixed oxygenase to yield aldosterone, which is rapidly released. 18-OH corticosterone is not a direct intermediate but is a byproduct of the enzymatic reaction. The same gene that directs the synthesis of the 11-hydroxylation enzyme is also responsible for the 18-oxidation step. The factors that regulate the relative degree of 11- and 18-hydroxylation in a given cell are not completely understood. DOC and 18-OH DOC also have some mineralocorticoid activity. However, only rarely are they secreted in physiologically significant amounts by the zona fasciculata under ACTH stimulation.

Inhibitors of adrenocortical hormone synthesis. A number of drugs that are capable of blocking steroid synthesis at various steps have diagnostic and therapeutic usefulness. **Metyrapone** inhibits 11-hydroxylation, the last and critical step in cortisol synthesis. Administration of this drug creates acute cortisol deficiency, thereby stimulating ACTH secretion via negative feedback. As a result of the increase in ACTH, adrenal production of the immediate precursor to cortisol, 11-deoxycortisol (see Fig. 50-3), increases markedly, demonstrating the reserve capacity of the normal hypothalamic-pituitary ACTH axis (Fig. 48-10). Failure to respond indicates a hypothalamic-pituitary disease that has ablated this function. Aldosterone secretion is also decreased by metyrapone by virtue of its inhibitory effect on 18-oxygenation.

Aminoglutethimide is a potent inhibitor of the desmolase reaction, thereby decreasing all adrenal steroid synthesis. This drug has been useful in treating patients with excessive cortisol output, as well as in women with breast cancer to diminish estrogen production. **Ketoconazole**, an antifungal agent, also inhibits several steps in adrenocorticosteroid synthesis with therapeutic efficacy.

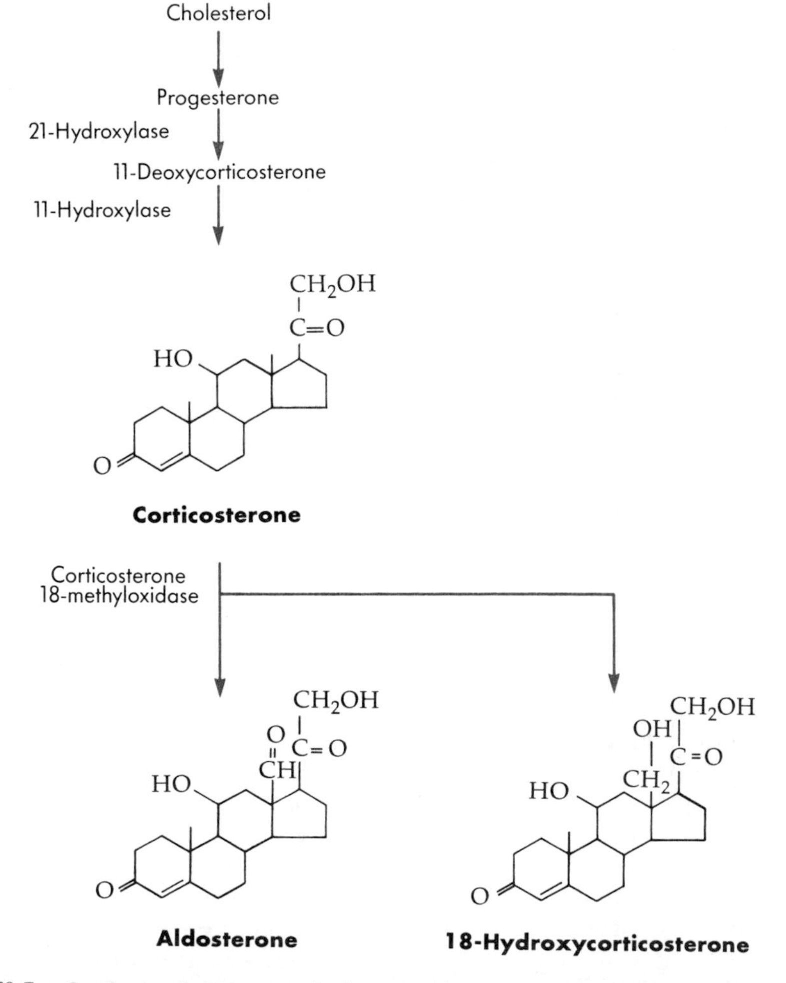

■ **Fig. 50-7** Synthesis of aldosterone in the zona glomerulosa. Note that 18-hydroxycorticosterone is a byproduct of the 18 methyl oxidation step.

Metabolism of Adrenocorticosteroids

Cortisol circulates in plasma 75% to 80% bound to a specific corticosteroid-binding α_2-globulin called **transcortin.** This glycoprotein has a molecular weight of 52,000 and binds a single molecule of cortisol. The normal concentration of transcortin is 3 mg/dl, with a binding capacity of 20 μg cortisol/dl. An additional 15% of plasma cortisol is bound to albumin, leaving only 5% to 10% free. The concentration of transcortin, and therefore of total cortisol, is increased during pregnancy and by estrogen administration; the physiological effects are determined by principles similar to those discussed with regard to thyroxine binding (Chapter 49).

Transcortin may have functions of its own when bound to cortisol, because cellular receptors for this complex have been found that activate adenylyl cyclase. Transcortin is also cleaved by a leukocyte enzyme, a process that could increase free cortisol concentration at sites of inflammation. The plasma half-life of cortisol is about 70 minutes, and the metabolic clearance rate averages 200 L/day. The free fraction of cortisol in plasma is filtered by the kidney, but only about 0.3% of the total daily secretion, or approximately 50 μg, is excreted in the urine.

Cortisol is in equilibrium with its 11-keto analogue cortisone. This interconversion is catalyzed by 11β-ol-dehydrogenase, which is present in many tissues. The glucocorticoid activity of cortisone is dependent on its conversion to cortisol. The vast majority of cortisol and cortisone is metabolized in the liver; the reduced metabolites are conjugated and excreted in the urine as glucuronides. About half of these excretory products are normally derived from cortisol and half from cortisone. (Fig. 50-8).

The measurement of urinary metabolites provides a reliable index of cortisol secretion as long as hepatic and renal functions are intact. Of particular use is the 17,21-dihydroxy-20-ketone configuration of tetrahydrocortisol and tetrahydrocortisone (Fig. 50-8). The so-

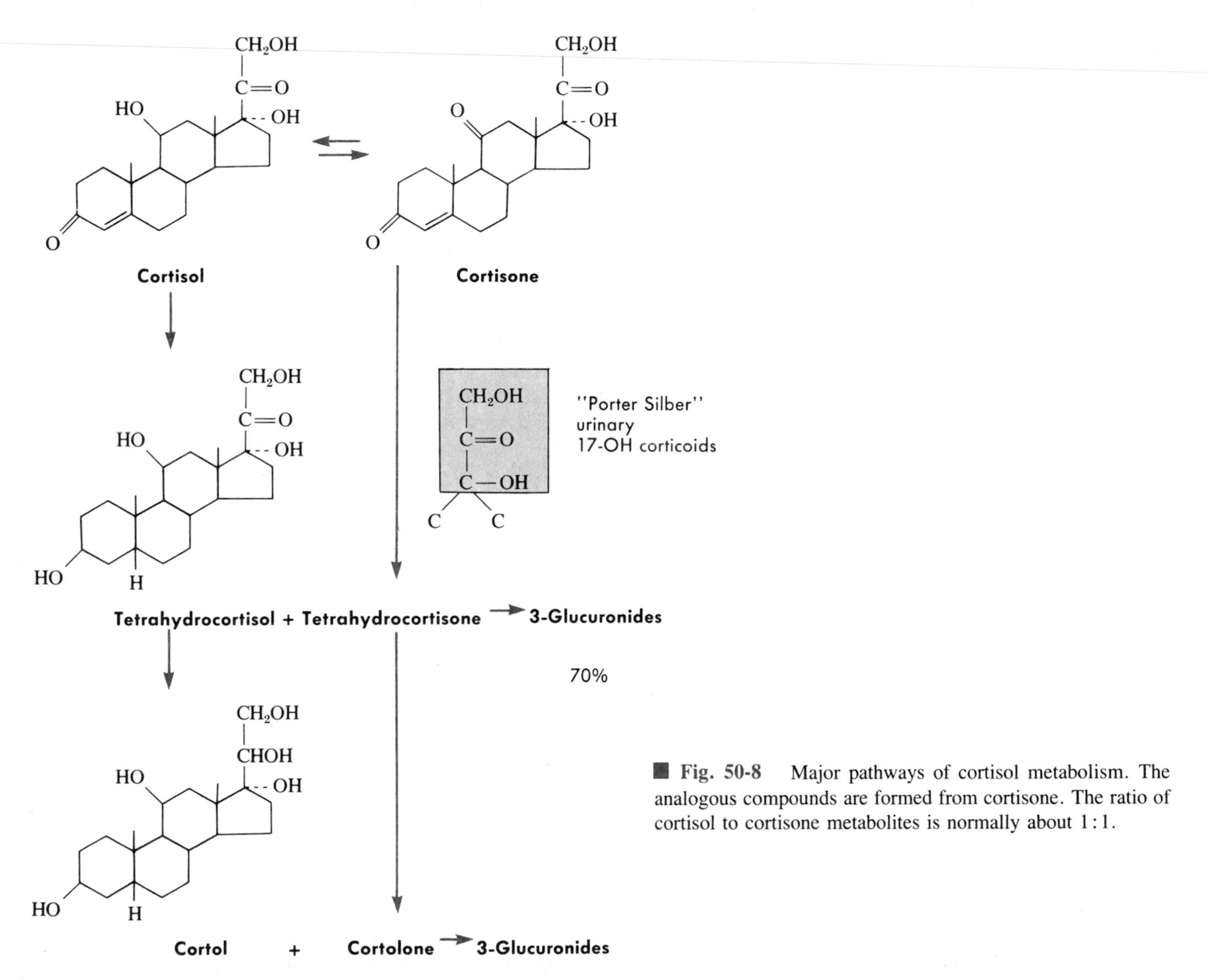

Fig. 50-8 Major pathways of cortisol metabolism. The analogous compounds are formed from cortisone. The ratio of cortisol to cortisone metabolites is normally about 1:1.

called "**17-hydroxycorticoids**" represent up to 50% of the total daily cortisol secretion. Normal values of 17-hydroxycorticoids range from 2 to 12 mg/day, and are slightly higher in men than in women, as are cortisol secretion rates. Daily urinary 17-hydroxycorticoid excretion may be corrected for differences in lean body mass by expressing the result as a ratio to the urinary creatinine excretion. Adrenocortical responsiveness to ACTH (or metyrapone) or suppressibility by exogenous synthetic glucocorticoids can be assessed by measuring urinary 17-hydroxycorticoids daily.

The cortisol precursors, progesterone and 17-hydroxyprogesterone, are metabolized to cortols known as pregnanediol and pregnanetriol, respectively. In adult females these urinary metabolites reflect both adrenal and ovarian secretion. In prepubertal children, however, elevation of urinary pregnanetriol specifically indicates increased secretion of adrenal 17-hydroxy-progesterone and is, therefore, a valuable marker for congenital blocks in cortisol secretion.

Aldosterone circulates in plasma bound to a specific aldosterone-binding globulin, to transcortin, and to albumin. Overall binding is weaker than for cortisol. Hence, the plasma half-life is only 20 minutes, and the metabolic clearance rate is 1,600 L/day. Ninety percent of aldosterone is cleared by the liver in a single passage. In the liver aldosterone is reduced to tetrahydroaldosterone, the major metabolite that is excreted in the urine as the 3-glucuronide. A smaller portion of aldosterone is excreted simply as the 18-glucuronide. The latter, however, is the urinary form that is commonly measured. The values in subjects with a normal sodium diet range from 5 to 20 μg/day.

The metabolism of androgens, in general, involves reduction of the 3-ketone group and the A ring in the liver. The two isomers formed, androsterone and etiocholanolone are then excreted in the urine. These metabolites, however, are not specific for the adrenal gland because they arise from gonadal androgens as well. DHEA-S is entirely excreted directly in the urine and is adrenal-specific.

Androsterone, etiocholanolone, and DHEA-S together make up the major part of a urinary fraction called 17-ketosteroids. Normal values range from 5 to 14 mg/day in women and 8 to 20 mg/day in men. Two thirds of this fraction is normally derived from adrenal and one third from gonadal androgen secretions. A large increase in urinary 17-ketosteroid excretion almost always indicates an adrenal abnormality. Measurement of urinary 17-ketosteroids is especially useful diagnostically in women or children who show evidence of virilization. Similar information can be obtained by measurement of plasma or urinary DHEA-S.

Regulation of Zona Fasciculata and Zona Reticularis Functions

The secretion of cortisol by the zona fasciculata is under the exclusive control of the hypothalamic-pituitary

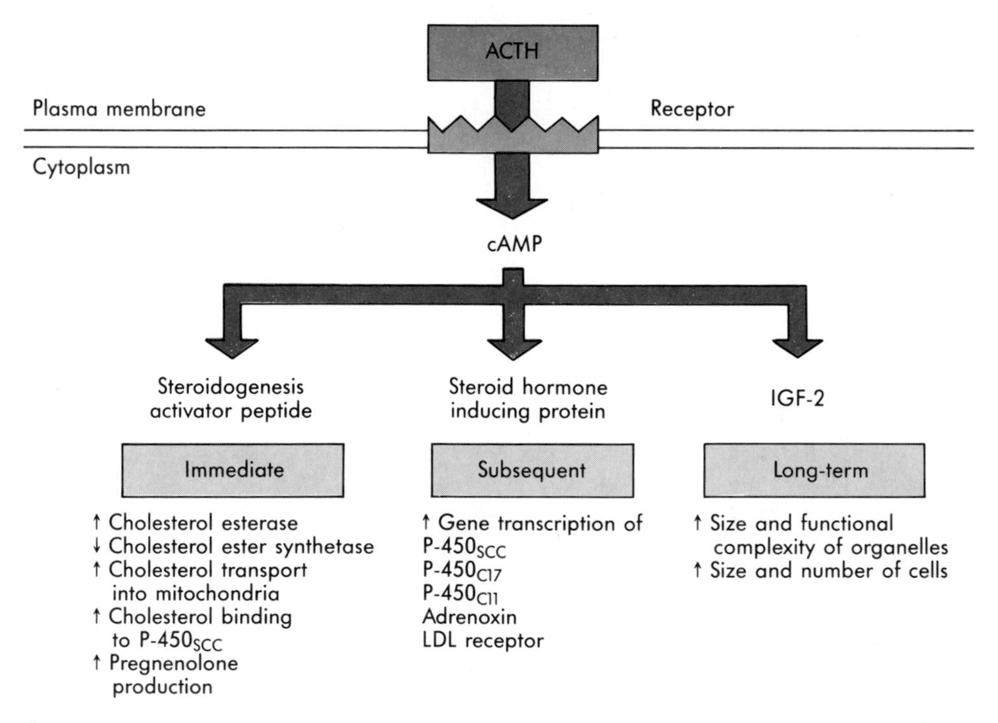

Fig. 50-9 An overview of ACTH actions on target adrenocortical cells. See text for details. *cAMP,* Cyclic adenosine monophosphate; *IGF-2,* insulin growth factor 2; *LDL,* low-density lipoprotein.

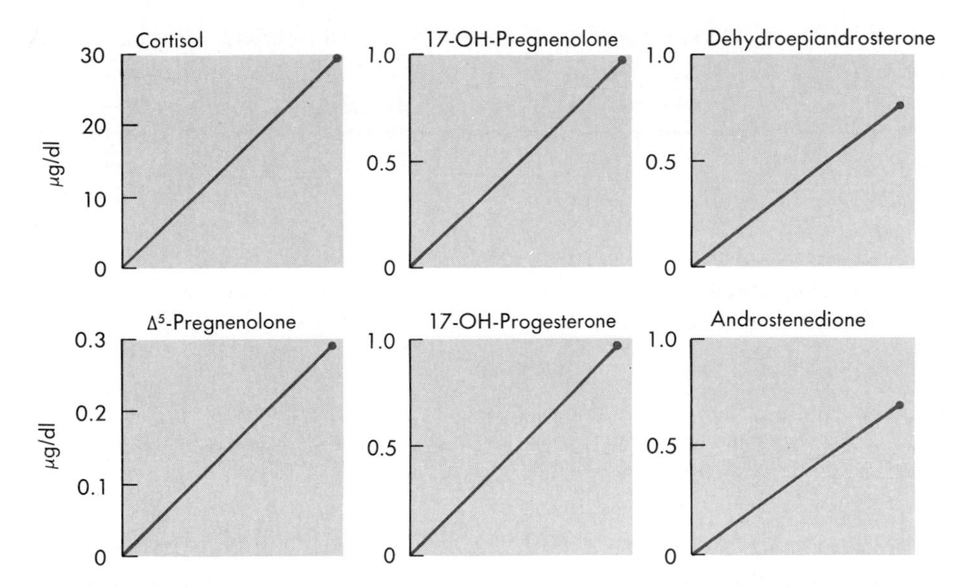

■ **Fig. 50-10** Plasma adrenocortical hormone responses to a 1-hour infusion of ACTH in humans. Note increase of precursors and hormonally active products. These changes indicate that ACTH activates the entire biosynthetic sequence. (Redrawn from Lachelin, GCL et al: *J Clin Endocrinol Metab* 49:892, 1979.)

CRH-ACTH axis. (See Chapter 48 and Figs. 48-8 through 48-12.) The secretion of adrenal androgens is likewise regulated by ACTH, but evidence suggests the possible existence of a separate tropic hormone product of the proopiomelanocortin gene that acts specifically on the zona reticularis. In the absence of ACTH, adrenocortical secretion virtually ceases.

ACTH initiates its action by binding to its plasma membrane receptor, a step that requires calcium (Fig. 50-9). This binding is followed by activation of adenylyl cyclase and a rise in cAMP levels as the principal second messenger. Phosphatidylinositol products play adjunctive second messenger roles. Protein kinase A and C probably then phosphorylate various protein mediators of ACTH action. A **steroidogenesis activator protein** mediates immediate hydrolysis of stored cholesterol esters, transfer of the released cholesterol into the mitochondria (Fig. 50-5) by a sterol carrier protein, binding of the cholesterol to $P_{450_{scc}}$ and activation of the rate-limiting desmolase reaction. Later, a **steroid hormone-inducing protein** increases transcription of the genes for $P_{450_{scc}}$, $P_{450_{c17}}$, $P_{450_{c11}}$, adrenoxin, and the LDL receptor. Thus all steps in corticosteroid hormone synthesis from cholesterol entry to final products are activated by ACTH. ACTH also acts on the cytoskeleton to bring cholesterol-containing vacuoles into intimate association with the mitochondria. Chronically, perhaps by stimulating local IGF generation, ACTH increases the size and number of adrenal cells and their mitochondria. This trophic effect is enhanced by virtue of the fact that ACTH upregulates its own receptor and that both ACTH and IGF increase each other's receptor number.

As seen in Fig. 50-10, plasma levels of cortisol, adrenal androgens, and their precursors rise within minutes of intravenous ACTH administration to humans. A sustained increase in cortisol secretion of two- to fivefold is obtained when the fully active 1 to 24 sequence of ACTH is infused. Plasma ACTH concentrations of about 300 pg/ml, or 6 times the basal level, are maximally effective in the short term. However, when the adrenal gland is chronically hyperstimulated by ACTH, it undergoes hyperplasia and the capacity for cortisol secretion rises up to twentyfold.

As detailed in Chapter 48, all of the factors that influence ACTH secretion likewise affect cortisol secretion; plasma levels of the latter generally follow those of the former by 15 to 30 minutes. Thus cortisol secretion, like that of ACTH, exhibits distinct diurnal variation with a peak just before the subject awakes in the morning and a nadir at or near zero just after the subject

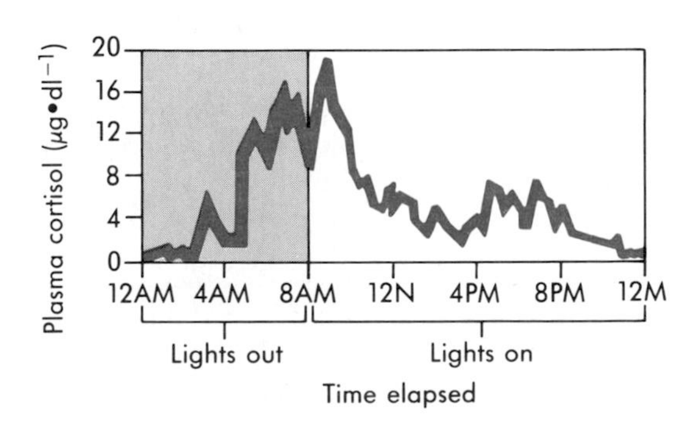

■ **Fig. 50-11** Pulsatile and diurnal nature of cortisol secretion. (Redrawn from Weitzman ED et al: *J Clin Endocrinol Metab* 33:14, 1971, The Endocrine Society.)

Table 50-1 Relative glucocorticoid and mineralocorticoid potency of natural corticosteroids and some synthetic analogues in clinical use*

	Glucocorticoid	*Mineralocorticoid*
Cortisol	1.0	1.0
Cortisone (11-keto)	0.8	0.8
Corticosterone	0.5	1.5
Prednisone (1.2 double bond)	4	<0.1
6α-Methylprednisone (Medrol)	5	<0.1
9α-Fluoro-16α-hydroxyprednisolone (triamcinolone)	5	<0.1
9α-Fluoro-16α-methylprednisolone (dexamethasone)	30	<0.1
Aldosterone	0.25	500
Deoxycorticosterone	0.01	30
9α-Fluorocortisol	10	500

*All values are relative to the glucocorticoid and mineralocorticoid potencies of cortisol, which have each been set at 1.0 arbitrarily. Cortisol actually has only 1/500 the potency of the natural mineralocorticoid aldosterone.

falls asleep (Fig. 50-11). The diurnal curve of total and free plasma cortisol includes 7 to 13 pulses or episodes of secretion per day. Half of the total daily cortisol is secreted within the major predawn burst.

Detailed analysis of plasma cortisol curves obtained by ultrafrequent sampling indicates that the peaks are determined by the frequency and duration of secretory bursts rather than by changes in the absolute rate of cortisol secretion in milligrams per minute. Thus the basal unstimulated rate of secretion is near zero and ACTH produces an all-or-none adrenal response. The reason for each of the daytime bursts of cortisol is unknown. A consistent burst after lunch suggests that cortisol secretion may be entrained with feeding patterns, though not necessarily in response to plasma substrate fluctuations. Other adrenal steroids, such as DHEA, show profiles parallel to that of cortisol; any disparities reflect differences in their metabolic clearance rates.

Plasma cortisol levels are increased by the stress of surgery, burns, infection, fever, psychosis, electroconvulsive therapy, acute anxiety, prolonged and strenuous exercise, and hypoglycemia. If attendant pain is prevented by disruption of the sensory input to the hypothalamus or by opioid analgesia, the cortisol response is blocked. Plasma cortisol levels are decreased promptly by the administration of synthetic glucocorticoids, such as dexamethasone (Table 50-1), which suppresses ACTH secretion by negative feedback (see Fig. 48-11). A single dose of dexamethasone that is biologically equivalent to twice the daily secretion rate of cortisol is sufficient to block completely the nocturnal ACTH peak and the morning rise in plasma cortisol level. Under circumstances of stress, the normal diurnal pattern of cortisol secretion may be lost, and feedback suppressibility may be impaired.

The average normal 8 AM plasma levels of cortisol and other adrenal steroids in humans, as well as their estimated secretion rates, are given in Table 50-2. The dominance of cortisol over corticosterone as a glucocorticoid is evident. Under severe stress, the maximal rate of cortisol secretion is 300 to 400 mg/day. Therefore this amount is usually provided to patients who lack adrenal function and are either acutely ill or must undergo surgery; without the steroid they may not survive. Plasma cortisol concentration varies little with age, but secretion rates are correlated with lean body mass. In contrast, DHEA and DHEA-S increase during puberty from childhood to adult levels. These adrenal steroids then decline significantly concommitant with the aging process. Some evidence suggests a correlation between low levels of DHEA and increased risk of cardiovascular disease later in life.

Table 50-2 Average 8 AM plasma concentration and secretion rates of adrenocortical steroids in adult humans

	Plasma concentration (μg/dl)	*Secretion rate (mg/day)*
Cortisol	13	15
Corticosterone	1	3
11-Deoxycortisol	0.16	0.40
Deoxycorticosterone	0.07	0.20
Aldosterone	0.009	0.15
18-OH Corticosterone	0.009	0.10
Dehydroepiandrosterone sulfate	115	15

Actions of Glucocorticoids

Cortisol is essential for life. Although provision of carbohydrate and of a pure mineralocorticoid or sodium chloride can postpone death, human beings cannot survive total adrenalectomy for long without glucocorticoid replacement. The hormone maintains glucose production from protein, facilitates fat metabolism, supports vascular responsiveness, and modulates central nervous system function. In addition, cortisol affects skeletal turnover, muscle function, immune responses, and renal function. The net effect of its metabolic actions is catabolic or antianabolic. The term *permissive* has been used to describe many of cortisol's actions, implying that the hormone may not directly *initiate* so much as *allow* certain processes to occur. Several illustrations may serve to better define this permissive role:

1. Cortisol may amplify the effect of another hormone on a process that it does not affect directly. For example, cortisol does not itself stimulate glycogenolysis but augments the stimulation of glycogenolysis by glucagon.
2. Cortisol and glucagon individually increase the activity of the enzyme phosphoenolpyruvate carboxykinase. However, their combined effect on this important regulatory step of gluconeogenesis is synergistic rather than additive.
3. The enzyme tyrosine transaminase is inducible by cortisol but not normally by its substrate tyrosine. However, in the presence of submaximal doses of cortisol, tyrosine administration will now induce the enzyme.

In vitro and in vivo, the effects of cortisol may be evident within minutes (inhibition of ACTH release), but they usually require hours (increase in plasma glucose) or days (induction of glucose-6-phosphatase to be expressed). Cortisol enters target cells freely and is then bound to a glucocorticoid receptor that can be found in both the cytoplasm and the nucleus. The glucocorticoid receptor (GR: also known as the type II glucocorticoid receptor) consists of four identical subunits. It belongs to the superfamily of steroid thyroid and retinoid receptors (Chapter 44). GR has a widespread tissue distribution and appears to be the same molecule in all cells. However, its concentration varies with cell type, degree of cellular differentiation, and phase of the cell cycle (lowest in G-I and M, highest in S and G-II). The GR concentration is also downregulated by glucocorticoids.

Cortisol combines with its receptor noncovalently but strongly in a manner that transforms the receptor into a molecule capable of binding to a specific glucocorticoid regulatory element (GRE) on target DNA molecules (Fig. 50-12). Some evidence suggests a mechanism whereby binding of cortisol to the C-terminal portion of GR displaces an inhibitory heat shock protein from a nearby site on the receptor; this changes the receptor conformation in its midportion so that it can interact with DNA. For example, a C-terminal truncated version of GR can bind to a GRE even without cortisol. GREs are usually upstream from the gene promotor but may be downstream or even within the gene; there may be multiple regulatory elements. Constitutive protein transcription factors may facilitate binding or in the blocking of the GR to the GRE or block it.

Once the cortisol-receptor complex has bound to a GRE, initiation or repression of gene transcription takes place. The latter may occur by displacement of adjacent positive factors from the DNA molecule. In any one cell type only a limited number of genes are affected. For example, the growth hormone gene is induced by cortisol in the anterior pituitary gland but not in the liver. Conversely, the gene for the enzyme tyrosine aminotransferase is induced by cortisol in the liver but not in the anterior pituitary. GR can bind other steroids, for example, progesterone; GREs can bind other steroid receptors, for example, the progesterone receptor. However, without the simultaneous combination of glucocorticoid plus GR plus GRE, there is no glucocorticoid effect on gene expression.

The mineralocorticoid receptor (MR) has strong homology to the GR in its C-terminus and mid-molecule domains and is also known as the type I glucocorticoid receptor. This receptor molecule binds mainly mineralocorticoids, yet it also actually binds cortisol ten times more strongly than does the GR or Type II glucocorticoid receptor. The MR may mediate some cortisol actions at low basal concentrations of the glucocorticoid. The selective CNS distribution of the MR also suggests that it mediates certain specific actions of cortisol in the brain.

Other intracellular mechanisms of cortisol action also probably exist. Cortisol does not generally alter intracellular cAMP levels. It does, however, synergize with the nucleotide in a number of situations, and some actions of cortisol can be mimicked by cAMP. Cortisol may also act by altering cGMP levels and the phospholipid component of various intracellular membranes. Even in a single cell type various enzyme changes produced by cortisol are not necessarily linked. This observation bespeaks multiple mechanisms of action.

Effects on metabolism. *The most important overall action of cortisol is to facilitate the conversion of protein to glycogen* (Fig. 50-13). Cortisol enhances the mobilization of muscle protein for gluconeogenesis by accelerating protein degradation and inhibiting protein synthesis. In humans the blood production rates of essential and nonessential amino acids are increased by cortisol, indicating protein breakdown (Fig. 50-14). The plasma concentrations of the branch chain amino acids increase; however, alanine levels do not rise because the conversion of alanine to glucose is also markedly increased by cortisol (Fig. 50-14).

Although the combined catabolic and antianabolic action of cortisol in normal amounts is physiologically beneficial, a continuous excess of glucocorticoid action produces a continuous drain on body protein stores, most notably in muscle, bone, connective tissue, and skin. This drain cannot be compensated for by dietary protein because of the inhibition of protein synthesis. Cortisol further stimulates the transformation of the proteolytically derived amino acids into glucose precursors and thence into glucose (Fig. 50-13). Table 50-3 lists enzymes that are induced by cortisol and that are important to gluconeogenesis and to the disposition of ammonia released from gluconeogenic amino acids.

Glucocorticoids are critical for the survival of a fasting animal or human. Without them there is little in-

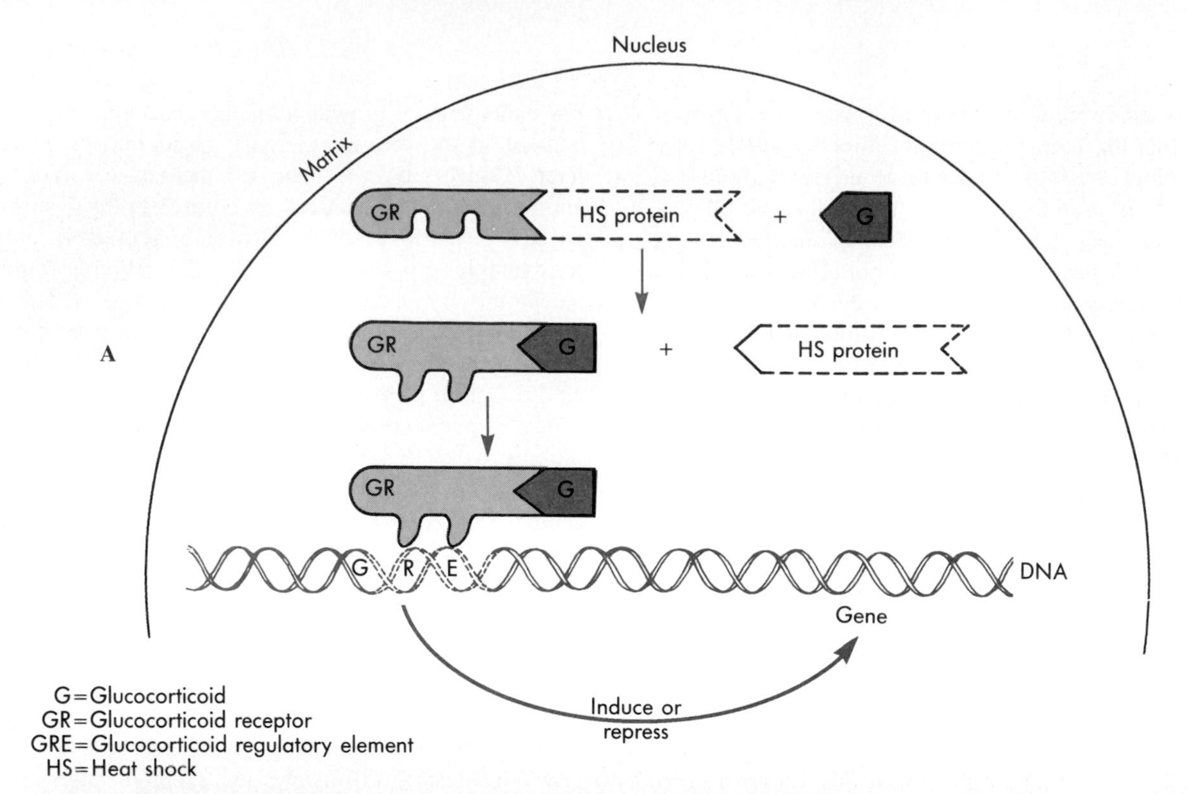

B

■ **Fig. 50-12** **A,** Mechanism of glucocorticoid (cortisol) action. The hormone (G) displaces a heat shock protein from the ligand binding protion of the glucocorticoid receptor (GR), changing the latter's conformation. The hormone-receptor complex then interacts with the glucocorticoid regulatory element (GRE) in target DNA molecules. **B,** Electron micrographs of DNA molecules from murine mammary tumor virus. The DNA molecules were incubated with purified glucocorticoid receptor complexes that appear as dark areas along the DNA molecules. Note that the glucocorticoid receptor complexes bind to the same site near one end of each of the DNA molecules. Bar at lower left represents 100 nm. (From Payvar F et al: *Cell* 35:381, 1983.)

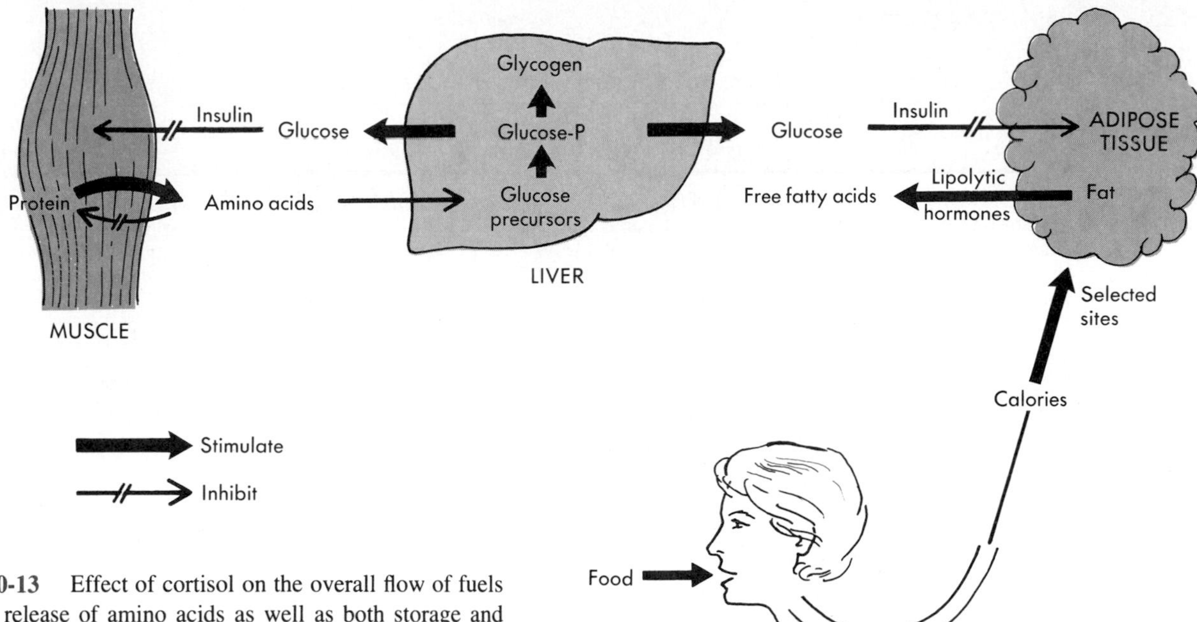

■ **Fig. 50-13** Effect of cortisol on the overall flow of fuels facilitates release of amino acids as well as both storage and release of glucose and fatty acids.

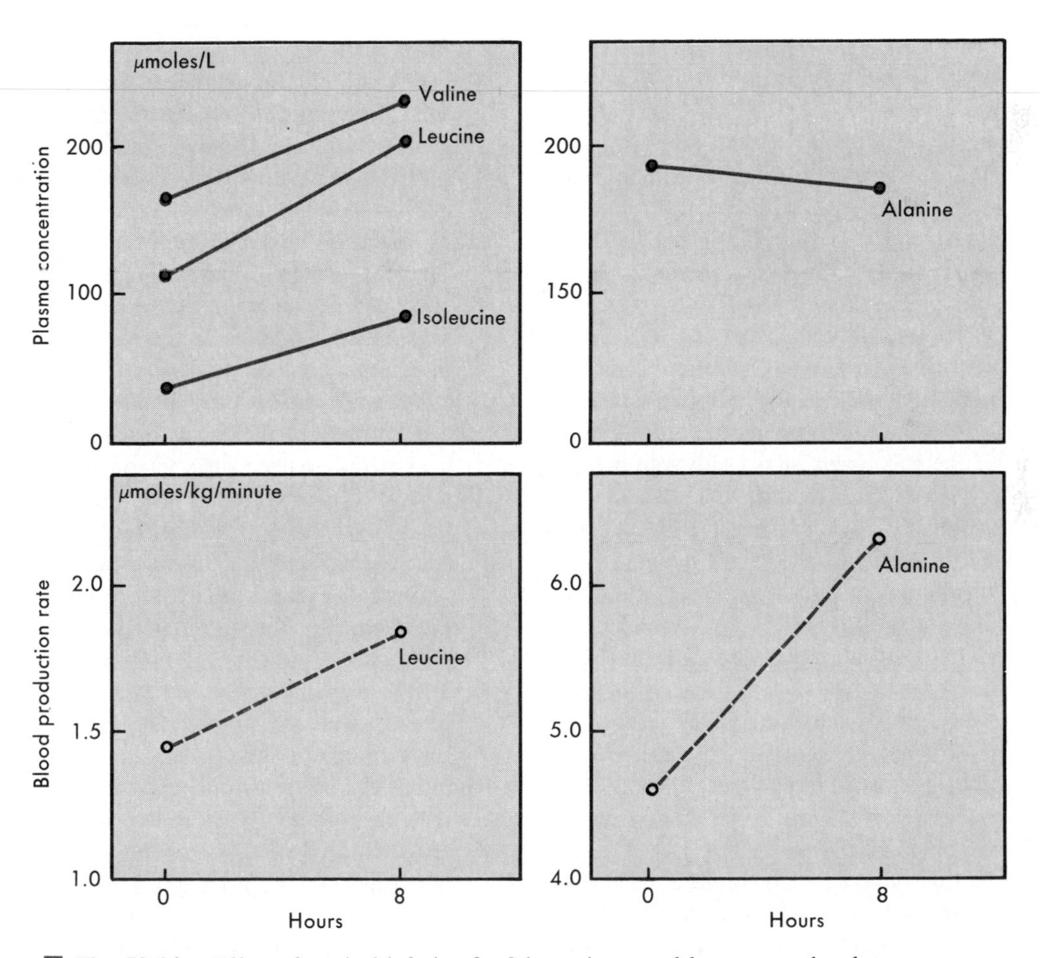

■ **Fig. 50-14** Effect of cortisol infusion for 8 hours in normal humans on the plasma concentrations and blood production rates of amino acids. Production rates were determined isotopically as outlined in Chapter 45. The plasma concentrations of leucine and two other branch chain amino acids increase. The increase of leucine is accomplished by an increase in its production rate and suggests that cortisol stimulates proteolysis. Although the blood production rate of alanine also increases, its plasma concentration does not, because cortisol simultaneously stimulates alanine use by augmenting its conversion to glucose. (Redrawn from Simmons PS et al: *J Clin Invest* 73:412, 1984.)

Table 50-3 Enzymes whose activities are increased by cortisol*

Provide carbon precursors	*Convert pyruvate to glycogen*	*Release glucose*	*Dispose of ammonia liberated from amino acids in urea cycle*
Alanine transaminase	Pyruvate carboxylase	Glucose 6-phosphatase	Ariginine synthetase
Tyrosine transaminase	Phosphoenolpyruvate carboxykinase		Arginosuccinase
Tryptophan pyrrolase	Phosphoglyceraldehyde dehydroge		Ariginasenase
Threonine dehydrase	Aldolase		
Serine dehydrase	Fructose 1,6-diphosphatase		
	Phosphohexoisomerase		
	Glycogen synthase		

*Increase varies from 130% for glycogen synthetase to 1,000% for alan pyrrolase to 4 days for aldolase.

crease in proteolysis, as evidenced by lack of increase in urinary nitrogen excretion. Therefore when liver glycogen stores are depleted, gluconeogenesis from protein is deficient, and death from hypoglycemia may ensue. The secretion of cortisol is increased only modestly, if at all, by fasting. It is the *previous* exposure to normal levels of cortisol and the maintenance of those levels during fasting that "permit" amino acid mobilization to be augmented.

Cortisol plays a similar role in the defense against hypoglycemia that is evoked by insulin. Although the rapid release of the glycogenolytic hormones, glucagon and epinephrine, is primarily responsible for the rapid recovery of plasma glucose levels, the *prior* action of cortisol leads to the buildup of sufficient glycogen stores on which the other hormones can act. During the later phase of recovery from hypoglycemia, cortisol itself increases hepatic glucose production and decreases glucose utilization. Cortisol also enhances glucagon release by cells of the pancreatic islets. Although the major impact of cortisol is on liver glycogen, an excess of the hormone eventually increases plasma glucose levels. This occurs because cortisol powerfully antagonizes the actions of insulin on glucose metabolism, inhibiting insulin-stimulated glucose uptake in muscle and adipose tissue and reversing insulin suppression of hepatic glucose production. As shown in Fig. 50-15, cortisol decreases tissue sensitivity but not maximal responsiveness to insulin. This antagonism takes place largely at postreceptor steps; e.g., insulin represses whereas cortisol induces transcription of the gene for phosphoenolpyruvate carboxykinase.

In part, cortisol plays an analogous role in fat metabolism (Fig. 50-13). Although only weakly lipolytic itself, the presence of cortisol is necessary for maximal stimulation of fat mobilization by epinephrine, growth hormone, and other lipolytic peptides. Thus during fasting, cortisol permits accelerated release of energy stores as fatty acids and of glycerol for gluconeogenesis. However, cortisol actions on body fat are quite complex. The hormone increases appetite and caloric intake. It also increases differentiation of preadipocytes to adipocytes and stimulates lipogenesis by increasing adipocyte lipoprotein lipase and glucose-6-phosphate dehydrogenase activity. These actions are especially prominent in certain areas of the body. Therefore an excess of cortisol finally results in obesity with a peculiar distribution of fat that favors the abdomen, trunk, and face but spares the extremities.

Overall, cortisol is an important diabetogenic, antiinsulin hormone. Its hyperglycemic, lipolytic, and ketogenic actions are usually exhibited only when its secretion is stimulated by stress. Then cortisol potentiates and extends the duration of the hyperglycemia evoked by glucagon, epinephrine, and growth hormone, and accentuates loss of body protein. These diabetogenic and catabolic actions are markedly amplified when insulin secretion is deficient and are commonly encountered in that clinical setting.

Other effects. Cortisol has actions on the structure and function of numerous organs. Many of these have been deduced from observations in patients with disorders of cortisol secretion or in experimental animals that have been rendered either cortisol-deficient or have been treated with excess hormone.

Cortisol exerts a dual action on muscle function. In the absence of the hormone, the contractility and work performance of skeletal and cardiac muscle decline. The inotropic action of cortisol on skeletal muscle may be exerted at the myoneural junction via an increase in acetylcholine synthesis. In addition, cortisol may improve cardiac function by increasing myocardial β-adrenergic receptors. Yet an excess of cortisol causes decreased muscle protein synthesis, increased muscle catabolism, protein wastage, a consequent reduction of muscle mass, and muscle weakness. The ratio of the insulin-sensitive, slow oxidative type I muscle fibers to the fast glycolytic type II-B muscle fibers is decreased by cortisol. This effect adds to the insulin resistance.

Bone. Cortisol inhibits bone formation by several mechanisms. First, the synthesis of type I collagen, the fundamental component of bone matrix, is reduced by cortisol. Second, the rate of differentiation of osteoprogenitor cells to active osteoblasts is decreased (Chapter

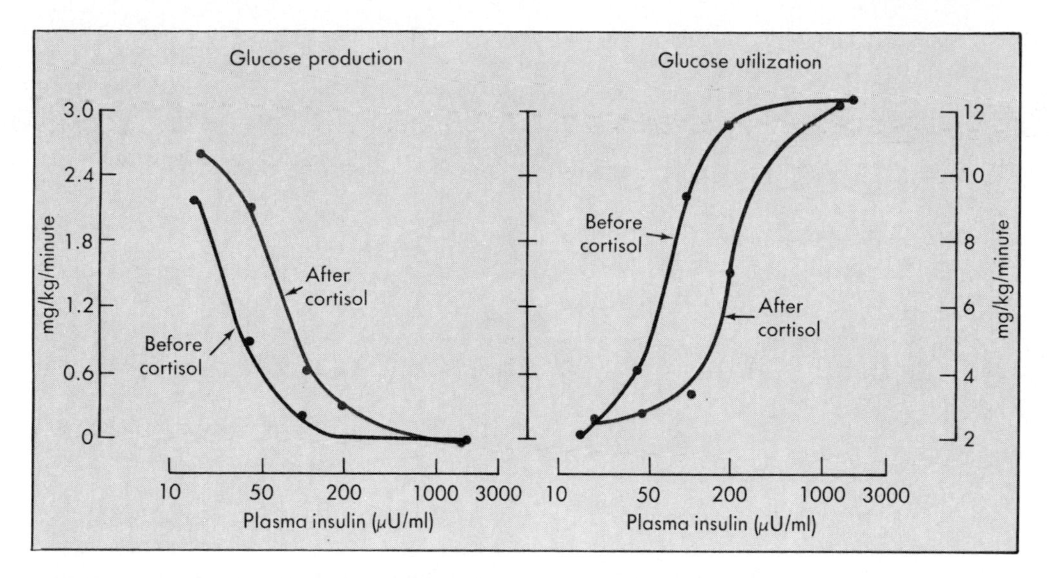

Fig. 50-15 The effect of cortisol on glucose turnover in response to increasing levels of insulin in a 24-year-old man. Cortisol decreases the sensitivity to insulin (the dose response curve is shifted to the right) both with regard to insulin inhibition of glucose production and insulin stimulation of glucose use. (Redrawn from Rizza RA et al: *Am J Med* 70:169, 1981).

47). Third, cortisol decreases the absorption of calcium from the intestinal tract by antagonizing the action of 1,25-$(OH)_2$-D_3 and possibly also by diminishing the synthesis of this active vitamin D metabolite. This results in reduced availability of calcium for bone mineralization. Finally, cortisol increases the rate of bone resorption. Thus one major consequence of a cortisol excess is an overall reduction in bone mass **(osteoporosis).**

Connective tissue. Inhibition of collagen synthesis by cortisol produces thinning of the skin and of the walls of capillaries. The consequent fragility of the capillaries leads to their easy rupture and intracutaneous hemorrhage.

Vascular system. Cortisol is required for the maintenance of normal blood pressure. Besides sustaining myocardial performance, the hormone permits normal responsiveness of arterioles to the constrictive action of catecholamines and may decrease production of vasodilator prostaglandins. Cortisol also helps to maintain blood volume by decreasing the permeability of the vascular endothelium.

Kidney. Cortisol increases the rate of glomerular filtration by decreasing preglomerular resistance and increasing glomerular plasma flow. The hormone is also essential for the rapid excretion of a water load. In the absence of cortisol, the secretion of antidiuretic hormone and its action on the renal tubule are enhanced. Therefore free water clearance is diminished, and maximal dilution of the urine cannot occur (Fig. 50-16). Cortisol is also required for generation of ammonium ion from glutamate in response to acid loads. Glucocorticoids also increase phosphate excretion by decreasing its reabsorption in the proximal tubules.

Central nervous system. Both type I and type II glucocorticoid receptors are present in various areas of the brain, particularly in the limbic system. Cortisol decreases REM sleep but increases both slow wave sleep and time awake. The nocturnal rise in ACTH and cortisol generally precedes S_1, or light sleep. Cortisol specifically decreases the ability to detect a salty taste; glucocorticoids in general dampen acuity with regard to gustatory, olfactory, auditory, and visual stimuli. On the other hand, they appear to improve the ability to integrate those sensations that are perceived and to organize appropriate responses. In excess, cortisol can cause

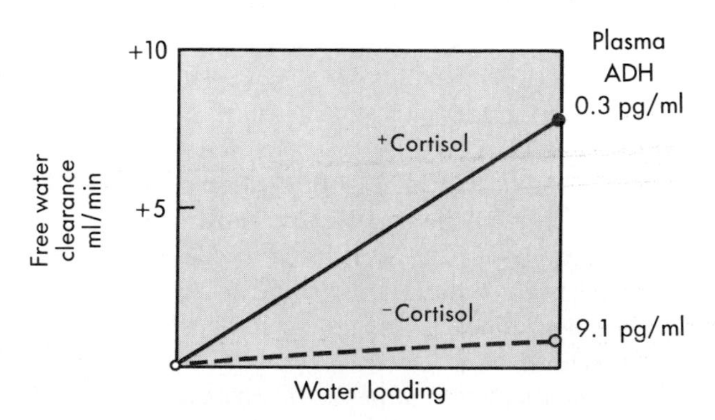

Fig. 50-16 Effect of cortisol on the response of adrenalectomized dogs to water loading. In the absence of the hormone, there is little increase in free water clearance and antidiuretic hormone (ADH) levels remain high. Cortisol replacement allows a normal suppression of ADH levels and a sharp increase in free water clearance. (Modified from Boykin J et al: *J Clin Invest* 62:738, 1978. From the American Society for Clinical Investigation.)

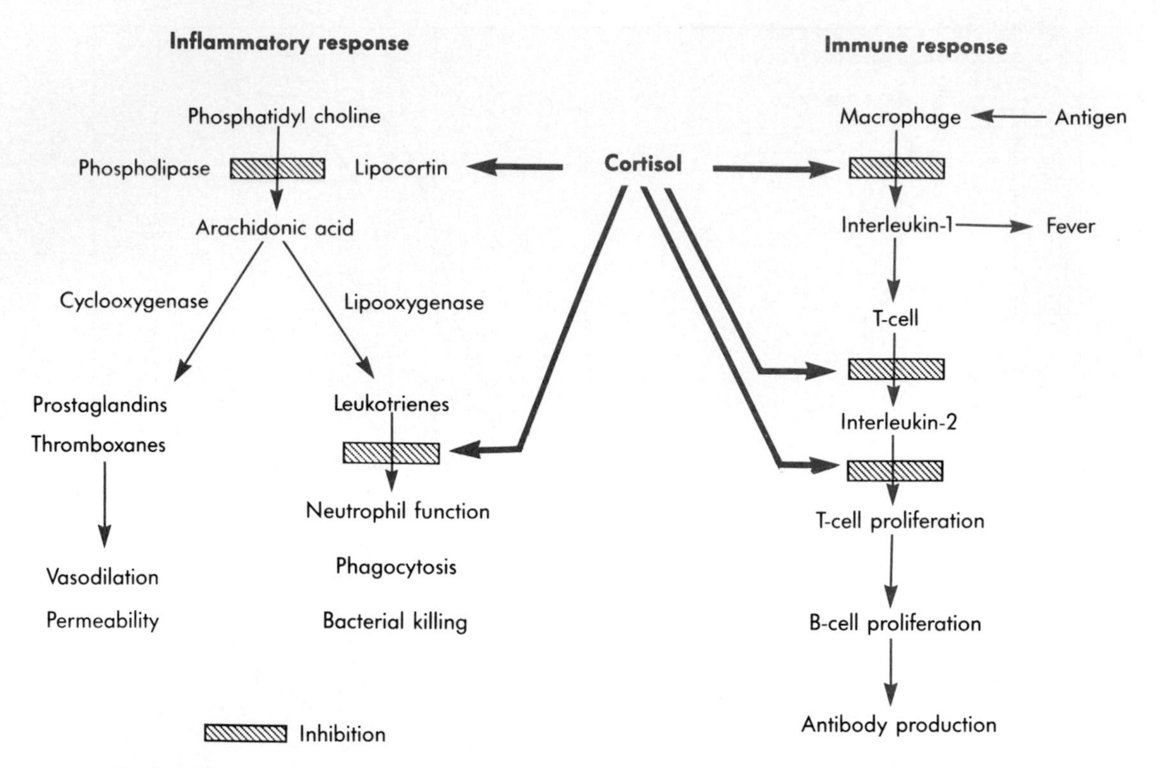

■ **Fig. 50-17** Mechanisms of cortisol inhibition of inflammatory and immune responses. Cortisol suppresses production of mediators of inflammation derived from arachidonic acid, neutrophil functions, and T lymphocyte cell mediated immune responses.

insomnia, can strikingly elevate or depress moods, and can lower the threshold for seizure activity.

Fetus. Cortisol facilitates in utero maturation of the central nervous system, retina, skin, gastrointestinal tract, and lungs. The latter two have been best studied. The digestive enzyme capacity of the intestinal mucosa changes from a fetal pattern to a mature adult pattern under the influence of cortisol. This permits the newborn to use disaccharides present in milk. Timely preparation of the fetal lung to permit satisfactory breathing immediately after birth is facilitated by cortisol. The rate of development of the alveoli, flattening of the lining cells, and thinning of the lung septa are increased by the hormone. Most important, during the last weeks of gestation the synthesis of surfactant, a phospholipid vital for maintaining alveolar surface tension, is increased. The effect is mediated by increasing the activity of key enzymes in the surfactant biosynthetic pathway, including phosphatidyl acid phosphatase and choline phosphotransferase.

Inflammatory and immune responses. *Cortisol has a profound influence on the complex set of reactions evoked by tissue trauma, chemical irritants, foreign proteins, or infection* (Fig. 50-17). The immediate local reaction to injury consists of dilation of capillaries and changes in the endothelial cell membranes that increase microvascular permeability and enhance the trapping of circulating leukocytes at the site of injury. These reactions, mediated by prostaglandins, thromboxanes, and leukotrienes, are profoundly inhibited by all glucocorticoids through suppression of the synthesis and release of these mediators. This results because cortisol induces lipocortin, a phosphoprotein that inhibits the activity of the enzyme phospholipase A_2. This enzyme releases arachidonic acid from its linkage to phosphatidyl choline. Arachidonic acid is the immediate precursor to prostaglandins, thromboxanes, and leukotrienes, and its production, the rate-limiting step in the synthesis of these mediators of inflammation, is thus inhibited by cortisol. Glucocorticoids also stabilize lysosomes and thereby reduce the local release of proteolytic enzymes and hyaluronidase that contribute to tissue swelling. The differentiation and proliferation of local inflammatory mast cells (but not their release of histamine) is inhibited by cortisol.

The recruitment of circulating leukocytes to the site of trauma or infection is inhibited by cortisol. The hormone decreases margination of leukocytes from blood vessels and their adherence to capillary endothelium. This process requires interaction between chemotactic peptides that attract the leukocytes and specific endothelial cell surface receptors. Cortisol inhibits the production and binding of these peptides to their receptors. Cortisol also decreases the phagocytic and bacteriocidal activity of neutrophils and the leukotriene-stimulated respiratory burst that accompanies these activities. Be-

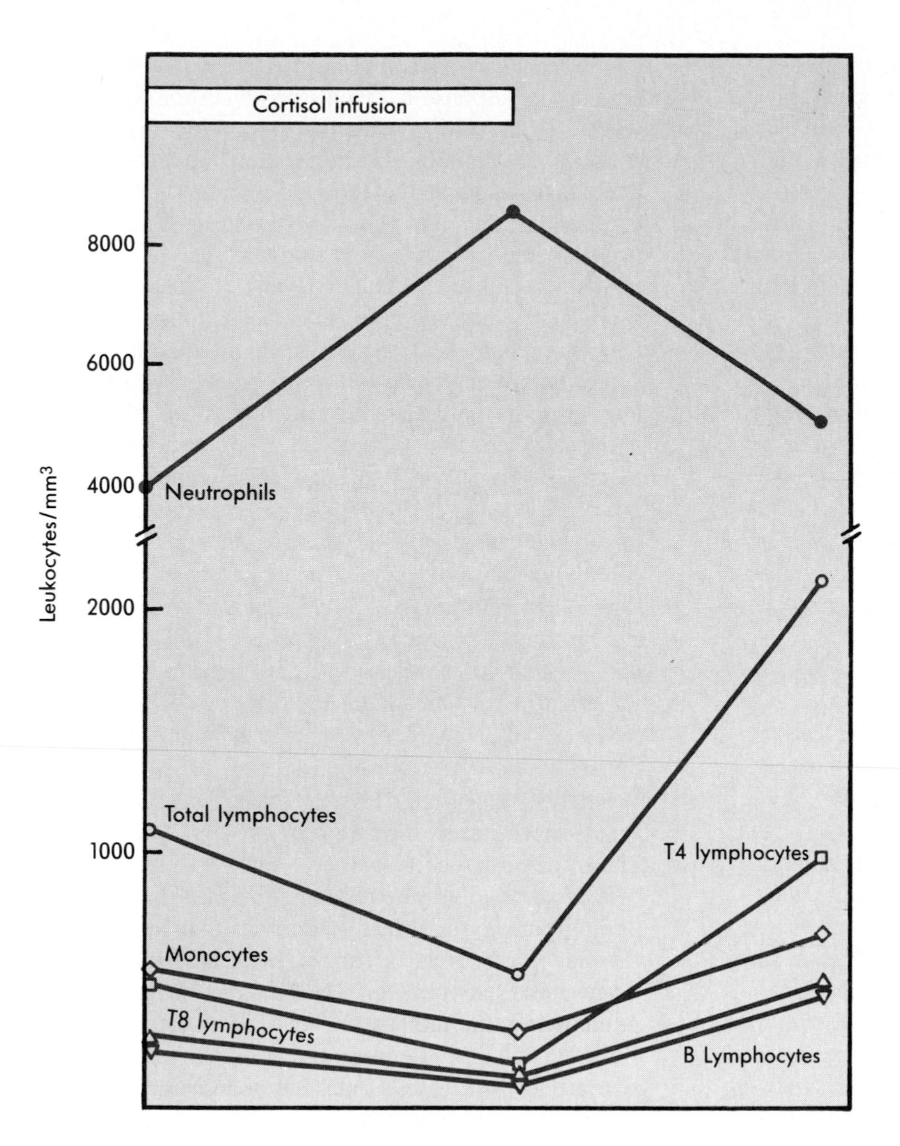

■ **Fig. 50-18** Effects of cortisol on circulatory leukocytes. Note increase in neutrophils and decrease in monocytes and lymphocytes of all types. T_4 helper lymphocytes were disproportionately reduced. Eosinophils (not shown) also decrease. (Data from Calvano SE et al: *Surg Gynecol Obstet* 164:509, 1987.)

cause cortisol increases the release of neutrophils from bone marrow, their circulating number actually increases (Fig. 50-18), although their effectiveness decreases. In contrast, cortisol decreases the number of circulating eosinophils. Cortisol also decreases the proliferation of fibroblasts, their synthesis, and the deposition of fibrils, which form the basis for the chronic inflammatory response to injury. The net result of these actions is to impede the ability to deal locally in an effective manner with irritants or organisms and to prevent the walling off of infection.

Cortisol also suppresses immune responses to foreign substances. The hormone decreases the number of circulating thymus-derived lymphocytes, (T cells), especially the proportion of helper T_4 lymphocytes (see Fig. 50-18), as well as their transport to the site of antigenic stimulation, and their function. Thus cell-mediated immunity, as manifested by the rejection of transplanted tissue, is markedly inhibited by the hormone.

The mechanism of inhibition is multifactorial (see Fig. 50-17). When a foreign protein, or **antigen,** intrudes into the body, it is picked up by a monocyte/macrophage. This cell presents the antigen to T cells and simultaneously elaborates interleukin-1, a peptide lymphokine that activates a subset of T-cells with helper or inducer function. In turn, the helper T-cell secretes interleukin-2, a peptide that stimulates the proliferation of still more T-cells. Cortisol inhibits the production of interleukin-1 and interleukin-2, as well as that of gamma interferon and other macrophage and lymphocyte products. Lymphocyte proliferation is arrested in cell stages G_0 and G_1 and the differentiation of monocytes to macrophages is inhibited by cortisol. T-cells act in part by recruiting and activating B-lymphocytes which produce antibodies directed against the antigen. However, the proliferation and differentiation of B-lymphocytes and their production of antibodies are probably not influenced directly by cortisol. Neither is the degradation of antibodies nor their specific reaction with antigen molecules affected by glucocorticoids.

The antiinflammatory action of glucocorticoids also includes suppression of the usual febrile response to infections or tissue injury. This probably occurs from decreased production of interleukin-1, which acts as an endogenous pyrogen.

Immune reactivity demonstrates a diurnal variation that reflects inversely the pattern of cortisol secretion. Moreover, interleukin-1 and other lymphocyte products stimulate ACTH secretion. Thus a negative feedback relationship between the endocrine and immune systems is becoming evident.

The metabolic actions of cortisol are essential to the survival of the severely stressed, traumatized, or infected individual. On the other hand, many of the defense mechanisms incorporated in the response to injury are inhibited by elevated levels of glucocorticoids. To explain this paradox, it has been suggested that permissive lower levels of cortisol may be required for the initial metabolic (and possibly immunological) responses to stress. The evoked higher levels of cortisol may later serve to limit cellular and tissue reactions so that they do not themselves seriously damage the organism, for example, by autoimmune reactions.

The antiinflammatory and immunosuppressive actions of glucocorticoids are used in treating nonendocrine diseases. Administered in high doses, they represent a two-edged sword. When the symptoms of tissue injury resulting from disease are functionally disabling or life threatening or when the rejection of transplanted organs or tissues must be prevented, glucocorticoids are dramatically beneficial. However, if glucocorticoids are administered therapeutically for very long, they may increase the susceptibility to infections or allow their dissemination, and they may prevent normal wound healing after injury. These serious adverse effects, along with diabetes, osteoporosis, and psychiatric disorders, enjoin physicians to prescribe glucocorticoids only when no safer form of treatment can succeed. This does not apply to the use of cortisol as replacement therapy in individuals who have lost adrenocortical function.

Action of Adrenal Androgens

The adrenal steroids DHEA-S, DHEA, and androstenedione are relatively weak androgens. Their physiological function is largely expressed by their peripheral conversion to the potent androgen testosterone. In females, testosterone of ultimate adrenal origin sustains normal pubic and axillary hair. It may possibly also contribute to red blood cell production. In males, testosterone of testicular origin far exceeds that of adrenal origin, rendering the latter physiologically unimportant.

Regulation of Zona Glomerulosa Function

The principal function of **aldosterone,** *the major mineralocorticoid, is to sustain extracellular fluid volume by conserving body sodium.* Hence, aldosterone is largely secreted in response to signals arising from the kidney when a reduction in circulating fluid volume is sensed. As shown in Fig. 50-19, when sodium depletion is produced, (for example, by dietary restriction), the fall in extracellular fluid and plasma volume causes a decrease in renal arterial blood flow and pressure. The juxtaglomerular cells of the kidney respond to this change by secreting the enzyme renin into the peripheral circulation. As detailed in Chapter 42, renin acts on its substrate, **angiotensinogen** (an α_2-globulin of hepatic origin), to form the decapeptide **angiotensin I.** The latter is then further cleaved by a converting enzyme of pulmonary origin to the octapeptide **angiotensin II.** This potent vasoconstrictor binds to specific adrenal plasma membrane receptors, after which calcium and phosphatidylinositol products are generated as second messengers. Protein kinase-C is translocated to the plasma membrane and activated. Subsequently, the desmolase and 18-methylcorticosteroid oxidase steps in the synthesis of aldosterone arc stimulated (Fig. 50-7). Minute increases in plasma angiotensin are sufficient to stimulate maximal aldosterone release.

After 5 days of only 10 mEq sodium intake, aldosterone secretion rates increase four- to eightfold. The renin and aldosterone response to hypovolemia can also be evoked rapidly by hemorrhage, by assumption of the upright posture for several hours, or by an acute diuresis. Such maneuvers increase plasma aldosterone two- to fourfold. Conversely, when excess sodium is ingested and extracellular fluid volume expands, renin release, angiotensin II generation, and aldosterone secretion are all suppressed. Thus the juxtaglomerular cells and the zona glomerulosa form a physiological feedback system. Sodium deprivation induces aldosterone hypersecretion via renin and angiotensin. When the additional aldosterone has caused sufficient sodium retention to restore extracellular fluid and plasma volume to normal, renin release is dampened, and aldosterone hypersecretion ceases. In this manner daily aldosterone secretion ranges from 50 μg with a dietary sodium intake of 150 mEq to 250 μg with a dietary sodium intake of 10 mEq.

The release of renin is enhanced by increased sympathetic neural activity, which is induced by hypovolemia, via norepinephrine and β-adrenergic receptors in the kidney. Hence, β-adrenergic antagonists depress renin and aldosterone responses to sodium depletion. The release of renin is also stimulated by certain local prostaglandins; therefore prostaglandin synthesis inhibitors (nonsteroidal antiinflammatory agents) also reduce aldosterone responses. Short-loop feedback inhibition of renin release is exerted by angiotensin II, but there is no direct feedback on the juxtaglomerular cells by aldosterone. Atrial natriuretic peptide (ANP) reinforces the effects of the renin-angiotensin system on aldosterone secretion. In response to volume expansion, atrial myo-

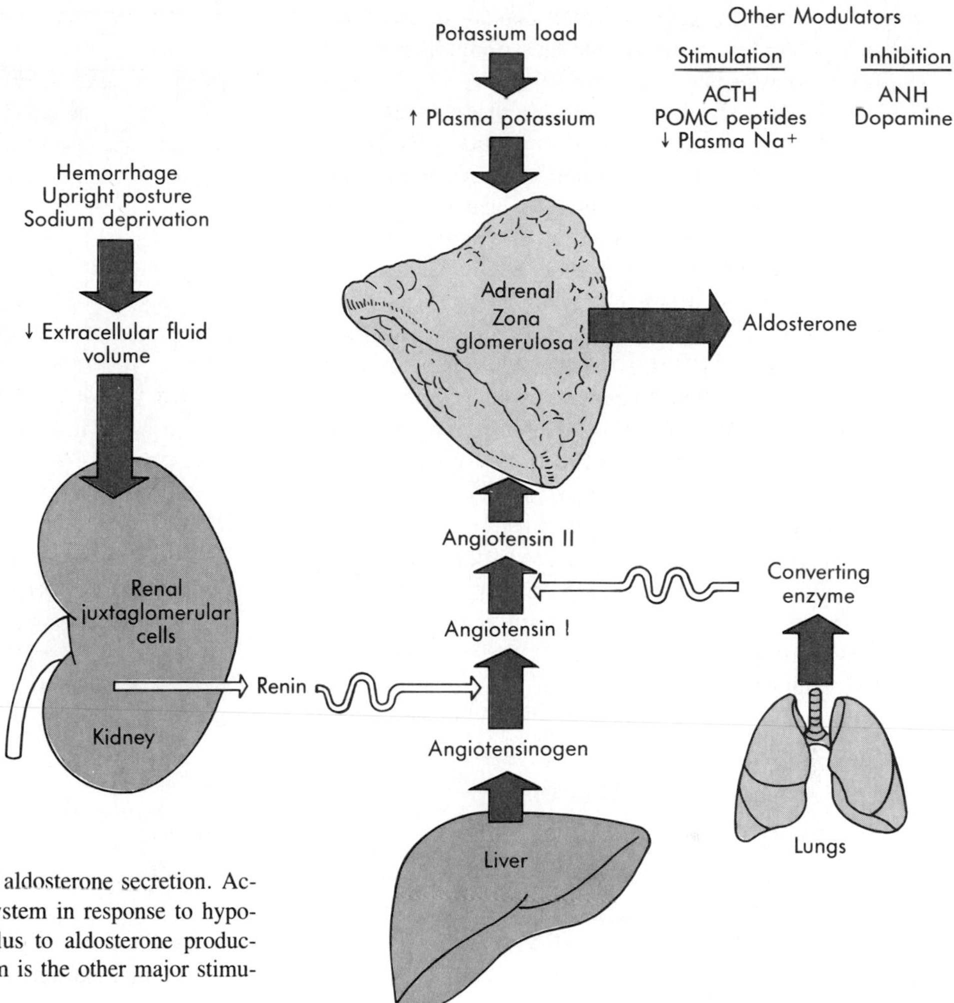

■ **Fig. 50-19** The regulation of aldosterone secretion. Activation of the renin-angiotensin system in response to hypovolemia is the predominant stimulus to aldosterone production. Elevation of plasma potassium is the other major stimulus.

cytes release ANP, which binds to specific receptors in the zona glomerulosa and inhibits the synthesis and release of aldosterone. This direct inhibitory effect is mediated by decreased cAMP and increased cGMP levels. ANP also reduces aldosterone secretion indirectly by decreasing renin release.

Aldosterone also participates in a vital physiological feedback relationship with potassium (Fig. 50-19). Another major function of aldosterone is to facilitate the clearance of potassium from the extracellular fluid, and concordantly potassium acts as an important stimulator of aldosterone secretion. In humans an acute infusion of potassium that raises plasma levels only 0.5 mEq/L immediately increases plasma aldosterone threefold. An increase in dietary potassium from 40 to 200 mEq/day increases plasma aldosterone sixfold in supine resting subjects. Conversely, potassium depletion lowers aldosterone secretion. Potassium stimulates aldosterone release by depolarizing the adrenal cell membrane, opening voltage-dependent calcium channels, and increasing intracellular calcium concentration. Accordingly, calcium channel blockers inhibit aldosterone release.

ACTH, in doses and in a manner similar to those that increase corticol secretion from the zona fasciculata, also stimulates aldosterone secretion (see Fig. 50-19). However, this stimulatory effect of ACTH in vivo wanes after several days. As sodium is retained and extracellular fluid volume rises secondary to the increased action of aldosterone, the release of renin and angiotensin is suppressed, ANP is stimulated, and together this returns aldosterone levels to baseline.

The physiological role of ACTH in regulating aldosterone output appears limited to a tonic one; that is, when ACTH is deficient, the zona glomerulosa is less able to respond to the primary stimulus of sodium depletion. This debility is seldom critical in patients with hypopituitarism. ACTH also stimulates the secretion of DOC and 18-OH-DOC from the zona fasciculata. Under rare circumstances these steroids can generate clinical syndromes of mineralocorticoid excess.

Stimulation of aldosterone secretion by the three major factors noted earlier is interrelated. A low sodium intake or a low plasma sodium level potentiates aldosterone responsiveness to angiotensin, potassium, and ACTH. The increased sensitivity to angiotensin that results from sodium depletion is explained partly by increased binding of the peptide hormone to its receptors and partly by enhanced activity of the biosynthetic pathway. Conversely, if the potassium content of the adrenal cell is depleted, the responses to angiotensin and ACTH are diminished.

There is evidence for still other regulatory factors of physiological significance. A glycopeptide of anterior pituitary origin and likely a product of proopiomelanocortin other than ACTH (Chapter 48) can stimulate aldosterone secretion, but its normal role requires clarification. The neural transmitters acetylcholine and serotonin are stimulatory, whereas dopamine decreases aldosterone secretion via an inhibitory G-protein and lowered levels of cAMP. Aldosterone responses to other stimuli are enhanced by dopamine antagonists.

In humans the plasma aldosterone level shows a definite diurnal fluctuation, with the highest concentration occurring at 8 AM and the lowest at 11 PM. Although this profile correlates with similar directional changes in plasma renin and plasma cortisol levels, the diurnal pattern of aldosterone arises independently because it is not affected by variation in sodium intake, posture, ACTH suppression by exogenous glucocorticoids, or plasma potassium.

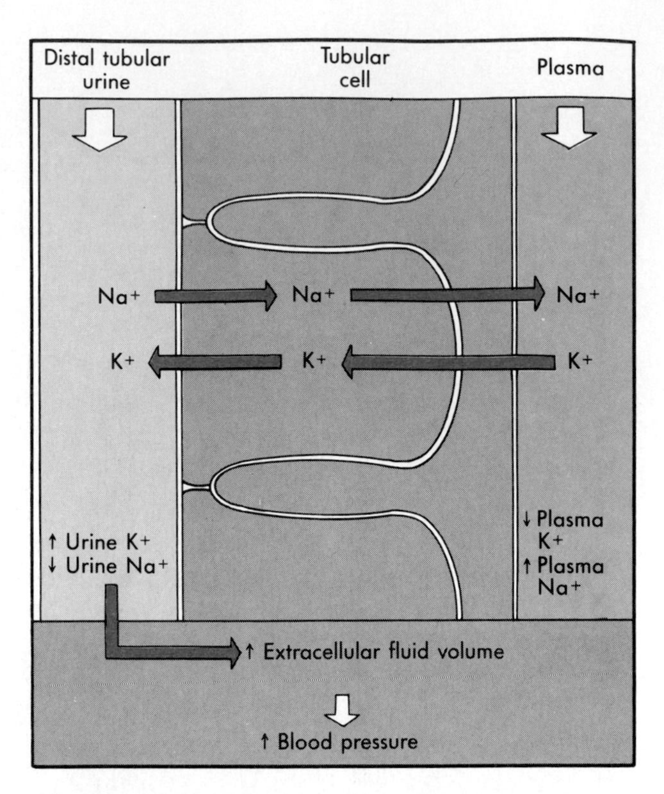

■ **Fig. 50-20** The action of aldosterone on the renal tubule. Sodium reabsorption from tubular urine is stimulated. Simultaneously, potassium secretion into the tubular urine is increased.

■ *Actions of Aldosterone and Other Mineralocorticoids*

Aldosterone binds to the mineralocorticoid receptor in target cells, and the complex effects transcriptional changes. A variety of messenger RNAs and proteins are induced, which mediate the hormone's effects. A lag of 1 to 2 hours is required between exposure to aldosterone and its onset of action.

The kidney is the major site of mineralocorticoid activity. Aldosterone stimulates the active reabsorption of sodium from the tubular urine by the cells of the collecting ducts and late distal convoluted tubules. The sodium is transported through the tubular cell and back into the capillary blood. Thus net urinary sodium excretion is diminished, and the vital extracellular cation is conserved (Fig. 50-20). Because water is passively reabsorbed with the sodium, there is little increase in plasma sodium concentration; and extracellular fluid volume expands in an isotonic fashion. Although only 3% of total sodium reabsorption is regulated by aldosterone, its deficiency produces a significant negative sodium balance.

Aldosterone acts at the following loci: (1) at the apical (luminal) surface of renal tubular cells to increase the number of membrane channels through which sodium enters the cell along an electrochemical gradient; (2) at the basal (capillary) surface of the cell to increase Na^+,K^+-ATPase, which pumps the sodium out; (3) in the mitochondria, stimulating Krebs cycle reactions, such as citrate synthase, that help generate the needed energy for extrusion of sodium into the interstitial fluid and capillary blood; and (4) in the cytosol to increase phospholipase activity and fatty acid synthesis, possibly for membrane generation.

Aldosterone stimulates the active secretion of potassium out of the tubular cell and into the urine (see Fig. 50-20) concurrently with sodium reabsorption. This does not constitute a stoichiometric exchange of potassium for sodium. Nonetheless, the active reabsorption of sodium is thought to create an electronegative condition in the tubular lumen, which facilitates the passive transfer of potassium into the tubular urine. The ability to excrete a potassium load depends on distal nephron flow and sodium delivery. Aldosterone allows a small amount of potassium secretion to occur, even when sodium is restricted. However, the extent of kaliuresis increases in parallel with the rate of delivery of sodium to the distal tubule. Thus a high sodium intake will greatly exacerbate urinary potassium losses caused by aldosterone.

Most of the potassium that is excreted daily results from distal tubular secretion. Hence, the presence of aldosterone is critical for disposal of the daily dietary potassium load at a normal plasma potassium concentration. Potassium flux, unlike that of sodium, does not entrain the movement of water; therefore, in the absence of aldosterone, potassium retention can cause a dangerous rise in plasma potassium. An excess of the hormone causes a significant decrease in plasma potassium.

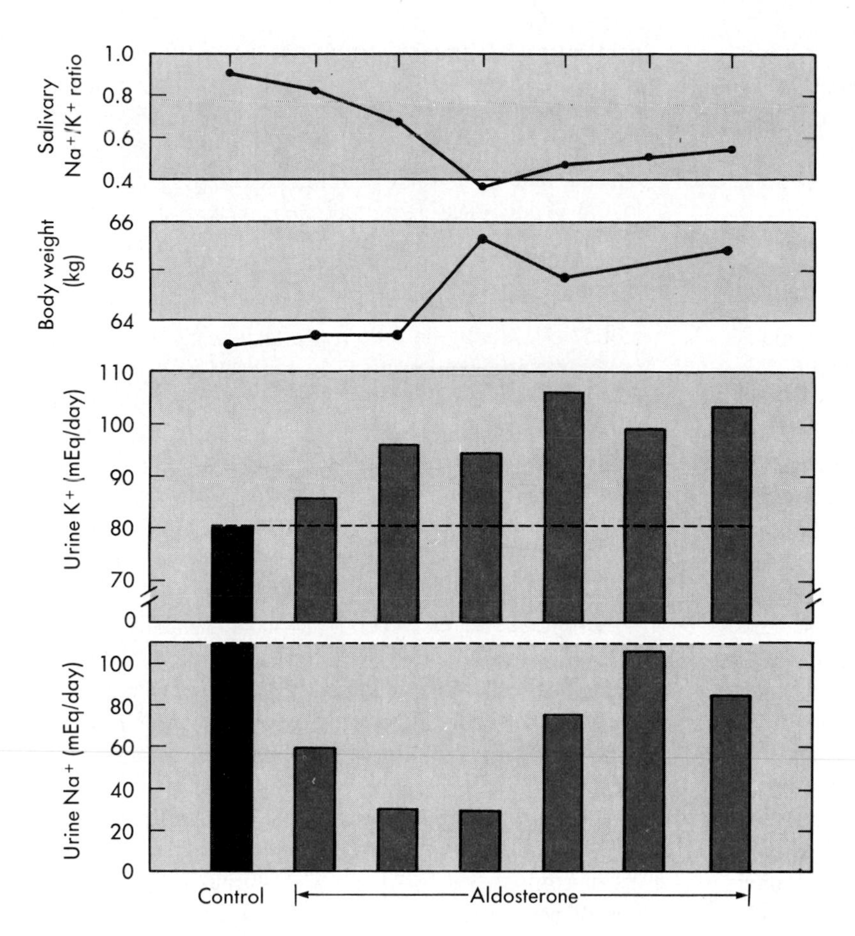

■ **Fig. 50-21** The effects of aldosterone administration in a normal human. Note eventual escape from sodium retention with stabilization of body weight yet continuing loss of potassium in the urine. Dashed lines represent levels of sodium and potassium intake. (Redrawn from August JT et al: *J Clin Invest* 37:1549, 1958. From The American Society for Clinical Investigation.)

Continued administration of aldosterone in the face of a normal sodium intake produces sodium retention, weight gain (Fig. 50-21), and an increase in blood pressure as a result of the expanded extracellular fluid volume. However, after several days and an accumulation of 200 to 300 mEq of sodium, retention ceases, balance is achieved, and body weight stabilizes. This escape is thought to be caused by a depression in proximal tubule sodium reabsorption as a result of expansion of the extracellular fluid and subsequent release of ANP. Nonetheless, aldosterone-induced potassium loss continues because sodium delivery to the distal tubule is maintained.

In addition to its effects on potassium secretion, aldosterone enhances tubular secretion of hydrogen ion as sodium is reabsorbed. Therefore aldosterone excess leads to the development of a mild systemic metabolic alkalosis, which can be further aggravated by the depletion of potassium. Ammonium excretion is also increased; however, the final urine pH is usually alkaline because the expansion of extracellular fluid volume inhibits bicarbonate reabsorption. In contrast, a deficiency of aldosterone produces a metabolic acidosis. Finally, aldosterone also stimulates the excretion of magnesium.

Aldosterone additionally affects mineral transport in other organs. The hormone stimulates sodium reabsorption from the colon while enhancing potassium excretion in the feces. Similarly, the hormone decreases the ratio of sodium to potassium in perspiration and in saliva (see Fig. 50-21). These actions, however, have little importance in overall cation balance. Aldosterone significantly affects sodium and potassium exchange between the extracellular fluid and the intracellular fluid. The net result is to increase potassium content of the intracellular space.

An effect of clinical importance is the increased blood pressure that results from an excess of aldosterone. In part, this is an indirect consequence of the retention of sodium, expansion of the extracellular fluid volume, and a slight increase in cardiac output. In addition, the sodium and water content of the arteriolar cells may increase; the resultant swelling narrows the arteriolar lumen and increases peripheral resistance. Finally, arterial smooth muscles possess mineralocorticoid receptors, and aldosterone antagonists lower peripheral resistance. Thus, aldosterone may have a direct vasoconstrictor action.

Although cortisol binds well to the mineralocorticoid receptor, it does not ordinarily contribute significantly to mineralocorticoid action because the kidney has very high levels of 11-beta-hydroxysteroid dehydrogenase. This enzyme inactivates cortisol locally by converting it to cortisone. Aldosterone actions in the kidney are blocked by high concentrations of progesterone and 17-

hydroxyprogesterone. An important inhibitor in clinical usage as a diuretic and antihypertensive agent is spironolactone. These inhibitors are competitive antagonists that bind to the mineralocorticoid receptor.

Clinical Syndromes of Adrenocortical Dysfunction

Hypofunction. Destruction of the adrenal cortex, **Addison's disease,** is ultimately incompatible with life. Except in cases of surgical removal, Addison's disease usually progresses slowly and leads to the gradual development of glucocorticoid, adrenal androgen, and mineralocorticoid deficiencies. A lack of cortisol leads to loss of appetite with weight loss, malaise, fatigue, lethargy, muscle weakness, nausea, vomiting and abdominal pain, fever, poor tolerance of minor medical or surgical stress, fasting hypoglycemia, and an increase in circulating lymphocytes and eosinophils with a reduction in neutrophils. A loss of adrenal androgens may contribute to anemia and, in females, a reduction in pubic and axillary hair. Because of negative feedback, the secretion of ACTH and all proopiomelanocortin products increases as the cortisol levels decline; and their melanocyte-stimulating activity produces striking hyperpigmentation of the skin. The diagnosis is confirmed by demonstrating low plasma cortisol levels and decreased urinary excretion of 17-hydroxycorticoids. Plasma ACTH levels are elevated; and if exogenous ACTH is administered, the levels of cortisol and its urinary metabolites fail to increase. Deficiency of aldosterone is marked by polyuria (which is caused by natriuresis), dehydration, hypotension, hyperkalemia, hyponatremia, and metabolic acidosis. Plasma and urinary aldosterone levels are low. Plasma renin and angiotensin levels are elevated consequent to the stimulus of sodium depletion.

Adrenal insufficiency may be caused by ACTH deficiency, resulting from disease of the hypothalamus or pituitary. The clinical picture, then, is that described earlier for loss of cortisol and adrenal androgen, but aldosterone and plasma potassium remain normal. Hyperpigmentation does not occur because ACTH and copeptides are not present.

Treatment of acute adrenal insufficiency (adrenal crisis) consists of large doses of intravenous cortisol and sufficient isotonic sodium chloride infusion to restore normal extracellular fluid volume and lower plasma potassium levels. For maintenance patients require oral cortisol and a synthetic mineralocorticoid such as 9α-fluorocortisol (Table 50-1).

Hyperfunction. The most common cause of endogenous adrenocortical hormone excess is bilateral hyperplasia of the adrenal cortex resulting from hypersecretion of ACTH. Tumors that give rise to autonomous hormone secretion also occur in the various adrenocortical zones. The major manifestations of increased glucocorticoids include (1) obesity with a peculiar distribution of fat involving the cheeks, the supraclavicular areas, the posterior cervicothoracic junction, the trunk, and the abdomen—the extremities being spared; (2) a loss of bone mass (osteoporosis), vertebral fractures, and necrosis of the hips; (3) a loss of connective tissue integrity, associated with fragile capillaries, easy bruisability, and thin skin through which the underlying blood vessels may be seen (purple striae); (4) increased protein catabolism, resulting in atrophy and weakness of the muscles of the trunk and extremities, poor wound healing, and stunted growth in children; (5) abnormal carbohydrate metabolism, or frank diabetes; (6) impaired response to infections; and (7) insomnia and psychosis. All these pathological consequences can also be produced by large the therapeutic doses of synthetic glucocorticoids.

Adrenal androgen hypersecretion is clinically detectable in females by signs of masculinization. This includes loss of regular menses, regression of breast tissue, increased body hair, acne, deepening of the voice, enlargement of the clitoris, increased muscularity, and heightened libido.

The diagnosis of endogenous glucocorticoid excess is made by demonstrating elevated plasma or urinary levels of cortisol, loss of its normal diurnal variation, and loss of normal suppressibility by exogenous glucocorticoids. Adrenal androgen excess is marked by elevated urinary excretion of 17-ketosteroids and elevated plasma levels of DHEA-S, androstenedione, and, in women, testosterone. If the pituitary gland dictates cortisol hypersecretion, plasma ACTH is elevated. If autonomous adrenal tissue has developed, then negative feedback will cause low levels of plasma ACTH.

A primary excess of mineralocorticoid produces a clinical syndrome characterized by hypertension, a slightly expanded extracellular fluid volume, hypokalemia with metabolic alkalosis, and slight hypernatremia. The diagnosis is established by demonstrating that plasma and urinary aldosterone (or rarely, DOC or 18-OH-DOC) are elevated even when the patient has a high sodium intake. If the secretion of aldosterone is autonomous, plasma renin levels are low, being suppressed by the expanded extracellular fluid. Obstructive lesions of the renal arteries, which reduce perfusion pressure, stimulate excess renin and, secondarily, hypersecretion of aldosterone.

Treatment of excess cortisol or aldosterone secretion by adrenal tumors is best accomplished by removal of the neoplasm. If the basic lesion is within the pituitary gland, modern microsurgical techniques permit removal of the small ACTH-producing tumors. Drugs that inhibit corticosteroid hormone biosynthesis or that antagonize cortisol or aldosterone actions are also useful.

Biosynthetic defects. A number of congenital enzyme deficiencies in the pathways of adrenocortical hormone synthesis occur. These include deficiencies of desmolase, 3β-ol-dehydrogenase, 21-hydroxylase, 11-hydroxylase, 17-hydroxylase, and 18-hydroxylase. *Adrenogenital syndrome* is the term given to this group of disorders. The consequences of each biosynthetic defect can be predicted (see Figs. 50-3 and 50-6); the product of the reaction will be deficient, and the precursors will be increased enormously as a result of negative feedback to the pituitary gland or, occasionally, to the juxtaglomerular cells as well.

The most common biochemical lesion encountered, 21-hydroxylase deficiency, serves as a good example. Plasma levels of cortisol and 11-deoxycortisol tend to be low, whereas plasma levels of the immediate precursor, 17-hydroxyprogesterone, are greatly elevated (see Fig. 50-3). The corresponding urinary metabolites show the same pattern. The block in cortisol synthesis leads to increased ACTH secretion, which stimulates all open pathways and causes immense hyperplasia of the zona fasciculata and zona reticularis. As seen in Fig. 50-6, 17-hydroxyprogesterone serves as the major precursor for adrenal androgens. Therefore in 21-hydroxylase–deficient patients, androgens are secreted in excess as a result of hyperstimulation by ACTH. Plasma levels of DHEA, DHEA-S, androstenedione, and testosterone are high, as is excretion of the urinary 17-ketosteroids.

The clinical consequences of this block and its overflow are dramatic. The high adrenal androgen levels in female fetuses cause a masculinized pattern of development of the external genitalia. Thus they have penilelike clitorides and scrotumlike labia. The ambiguous genitalia can lead to incorrect gender assignment at birth. If not treated promptly, the androgen excess causes early acceleration of linear growth and early appearance of pubic and axillary hair but suppression of gonadal function and reproduction in both genders. Ultimately, the individual is short as a result of premature closure of the growth centers in the bones. If the 21-hydroxylase block is severe, cortisol and aldosterone secretion may be so impaired as to cause episodes of adrenal crisis. This entire sequence can be reversed by supplying appropriate amounts of both cortisol and an aldosterone substitute.

The Adrenal Medulla

The adrenal medulla is the source of the circulating catecholamine hormone **epinephrine.** It also secretes small amounts of **norepinephrine**—nominally a neurotransmitter—which in select circumstances may also function as a hormone. These compounds have diverse effects on metabolism as well as on virtually all organ systems in the body. The adrenal medulla represents essentially an enlarged and specialized sympathetic ganglion. However, the neuronal cell bodies of the medulla do not have axons; instead they discharge their catecholamine hormones directly into the bloodstream, thus functioning as endocrine rather than nerve cells. The adrenal medulla is formed in parallel with the peripheral sympathetic nervous system. At about 7 weeks of gestation, neuroectodermal cells from the neural crest invade the anlage of the primitive adrenal cortex. There they develop into the medulla, which begins to secrete during gestation and by birth is completely functional. The development of sympathetic nervous tissue and induction of neural hormone synthesis is stimulated by **nerve growth factor**.

Adrenal medullary tissue in the adult weighs about 1 g and consists of chromaffin cells (so named for their affinity for chromium stains). These are organized in cords and clumps in intimate relationship with venules that drain the adrenal cortex and with nerve endings from cholinergic preganglionic fibers of the sympathetic nervous system. Within the chromaffin cells are numerous granules of 100 to 300 nm diameter, similar to those found in postganglionic sympathetic nerve terminals. These granules consist of the catecholamine hormones epinephrine and norepinephrine (20% by weight), adenosine triphosphate and other nucleotides (15%), protein (35%), and lipid (20%). They also contain enkephalins, β-endorphin, other proopiomelanocortin peptides, and chromagranin.

The adrenal medulla is usually activated in association with the rest of the sympathetic nervous system and acts in concert with it. Some actions of the neurotransmitter norepinephrine (which is released locally at the effector site of the postganglionic sympathetic nerve endings) are duplicated and amplified by the hormone epinephrine, which reaches similar sites via the circulation. However, epinephrine has unique effects of its own, some of which modulate those of norepinephrine. Furthermore, under certain circumstances, e.g., during hypoglycemia, the adrenal medulla is probably activated selectively.

Synthesis and Storage of Catecholamine Hormones

The catecholamine hormones are synthesized within the chromaffin cell by a series of reactions shown in Fig. 50-22. The first, catalyzed by the enzyme tyrosine hydroxylase, is the rate-limiting step in the sequence and occurs in the cytoplasm. The conversion of tyrosine to dihydroxyphenylalanine (dopa) requires molecular oxygen, a tetrahydropteridine, and NADPH. Norepinephrine inhibits this reaction by negative feedback. The conversion of dopa to dopamine is catalyzed by a nonspecific aromatic L-amino acid decarboxylase that employs pyridoxal phosphate as a cofactor. The dopamine thus formed in the cytoplasm must be taken up by the chromaffin granule before it can be acted on further.

The next enzyme in the sequence, **dopamine β-hydroxylase,** is present exclusively within the granule. In

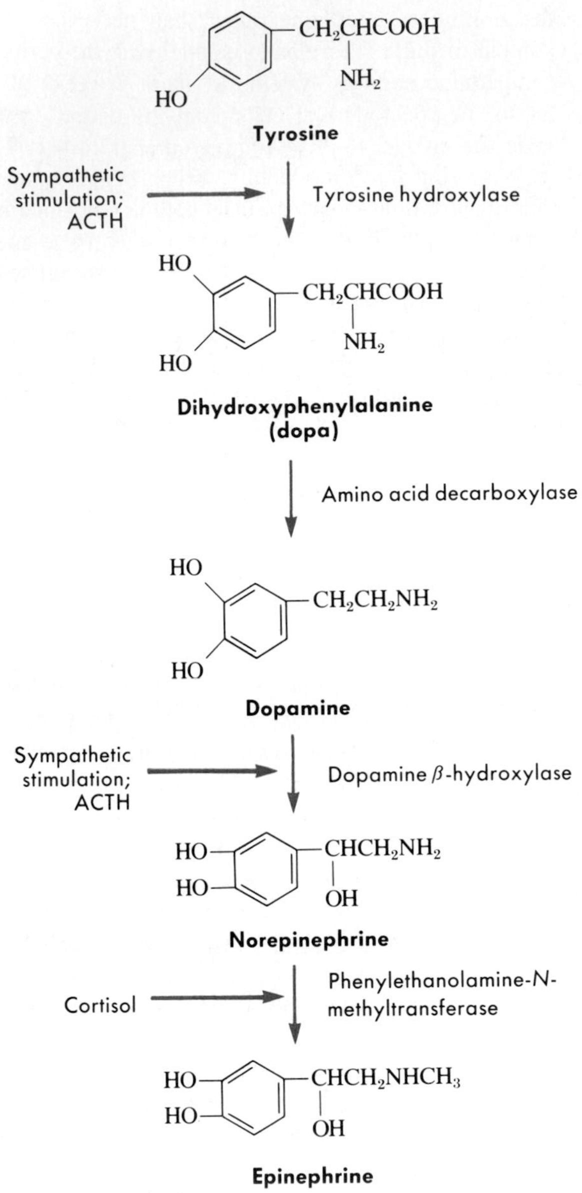

$CH_2CHCOOH$ / NH_2 / HO

Tyrosine

Sympathetic stimulation; ACTH → Tyrosine hydroxylase

HO / HO / $CH_2CHCOOH$ / NH_2

Dihydroxyphenylalanine (dopa)

Amino acid decarboxylase

HO / HO / $CH_2CH_2NH_2$

Dopamine

Sympathetic stimulation; ACTH → Dopamine β-hydroxylase

HO / HO / $CHCH_2NH_2$ / OH

Norepinephrine

Cortisol → Phenylethanolamine-*N*-methyltransferase

HO / HO / $CHCH_2NHCH_3$ / OH

Epinephrine

■ **Fig. 50-22** Pathway of catecholamine hormone synthesis in the adrenal medulla. The dopamine β-hydroxylase reaction occurs within the secretory granule. Note stimulatory effects of ACTH, cortisol, and sympathetic nerve impulses at various points.

the presence of molecular oxygen and a hydrogen donor, it catalyzes the formation of norepinephrine from dopamine. In approximately 15% of the granules, the sequence ends here, and the norepinephrine is stored. In the rest, norepinephrine diffuses back into the cytoplasm. There it is N-methylated by phenylethanolamine N-methyltransferase using S-adenosylmethionine as the methyl donor. The resultant epinephrine is then taken back up into the chromaffin granule where it is stored as the predominant adrenal medullary hormone. The uptake of dopamine, norepinephrine, and epinephrine by the secretory granules is an active process that requires adenosine triphosphate (ATP) and magnesium. The storage of the catecholamine hormones at such high intragranular concentrations also requires energy in the form of ATP. One mole of the nucleotide is present in a complex with 4 moles of catecholamine and chromagranin.

The synthesis of epinephrine and norepinephrine is regulated by several factors. Acute sympathetic stimulation activates tyrosine hydroxylase, possibly by decreasing cytoplasmic catecholamine levels and relieving the feedback inhibition. Chronic stimulation of the preganglionic fibers induces increased concentrations of both tyrosine hydroxylase and dopamine β-hydroxylase, thus helping to ensure maintenance of catecholamine output in the face of continuous demand. The mechanism of induction may involve a cAMP-dependent protein kinase. ACTH, acting directly, helps to sustain the levels of the same two enzymes under stressful conditions. In contrast, cortisol specifically induces the N-methyltransferase and therefore selectively stimulates epinephrine synthesis. The anatomical relationship between the medulla and the cortex subserves this action, since blood from the cortex with a high concentration of cortisol directly perfuses the chromaffin cells.

■ *Metabolism of Cathecholamines*

Essentially all the circulating epinephrine is derived from adrenal medullary secretion. In contrast, most of the circulating norepinephrine is derived from sympathetic nerve terminals and from the brain, having escaped immediate local re-uptake from synaptic clefts. However, the metabolic fate of epinephrine and norepinephrine merges into one or two major excretory products.

Epinephrine and norepinephrine have extremely short life spans in the circulation, allowing rapid turnoff of their dramatic effects. Half-lives are in the range of 1 to 3 minutes. The metabolic clearance rate of epinephrine is reported to be 3.5 to 6.0 L/minute, and that of norepinephrine is 2.0 to 4.0 L/minute. Both hormones increase their clearance rates still further by activating β-adrenergic receptors, another mechanism that helps to limit their actions. Only 2% to 3% of catecholamines are disposed of unchanged in the urine. The normal total daily excretion is about 50 μg, of which 20% is epinephrine and 80% is norepinephrine. Another 100 μg is excreted as sulfate or glucuronide conjugates. The vast majority of epinephrine is metabolized within the adrenal medullary chromaffin cell when synthesis exceeds the capacity for storage. Circulating epinephrine and norepinephrine are metabolized in many tissues, but predominantly in the liver and kidney.

The catecholamine hormones are metabolized by the reaction sequences shown in Fig. 50-23. The key enzymes are **catecholamine O-methyltransferase** and the

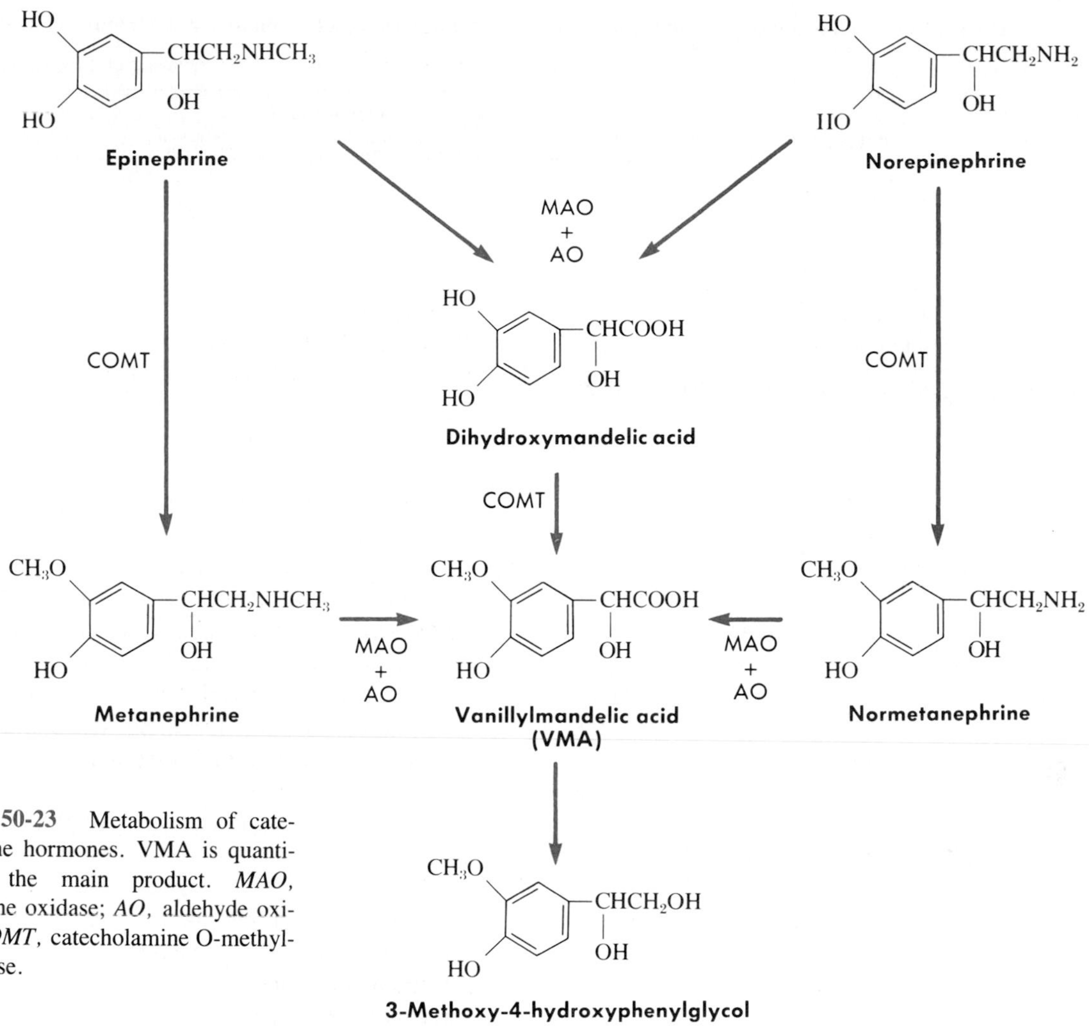

Epinephrine

Norepinephrine

MAO + AO

COMT

Dihydroxymandelic acid

COMT

COMT

Metanephrine

MAO + AO

Vanillylmandelic acid (VMA)

MAO + AO

Normetanephrine

3-Methoxy-4-hydroxyphenylglycol

■ **Fig. 50-23** Metabolism of catecholamine hormones. VMA is quantitatively the main product. *MAO,* Monamine oxidase; *AO,* aldehyde oxidase; *COMT,* catecholamine O-methyltransferase.

combination of **monamine oxidase** and **aldehyde oxidase.** O-Methylation and oxidative deamination can be carried out in either order, giving rise to several products that are then excreted in the urine. O-Methylation alone yields an average daily excretion of metanephrine (from epinephrine) plus normetanephrine (from norepinephrine) of 300 μg. In contrast, the excretion of the common deaminated products, vanillylmandelic acid (VMA) and methoxyhydroxyphenylglycol (MOPG), averages 4000 μg and 2000 μg, respectively. Under normal circumstances epinephrine accounts for only a very minor proportion of urinary VMA and MOPG; the majority, derived from norepinephrine, largely reflects activity of the sympathetic nervous system. Activity of the adrenal medulla can only be assessed *specifically* by measurement of urinary free epinephrine or plasma epinephrine levels.

■ *Regulation of Adrenal Medullary Secretion*

Secretion from the adrenal medulla forms an integral part of the "fight-or-flight" reaction evoked by stimulation of the sympathetic nervous system. Thus perception or even anticipation of danger or harm (anxiety), trauma, pain, hypovolemia from hemorrhage or fluid loss, hypotension, anoxia, extremes of temperature, hypoglycemia, and severe exercise cause rapid secretion of epinephrine (and probably norepinephrine) from the adrenal medulla. These stimuli are sensed at various higher levels in the sympathetic nervous system and responses are initiated in the hypothalamus and brainstem (Chapter 15). Usually activation of the adrenal medulla follows activation of the sympathetic nervous system and is evoked when the stimuli are more intense.

The final common effector pathway activating the adrenal medulla consists of cholinergic preganglionic fibers in the greater splanchnic nerve. On stimulation, acetylcholine is released from the nerve terminals. This neurotransmitter depolarizes the chromaffin cell membrane by increasing its permeability to sodium. This, in turn, induces an influx of calcium ions, which stimulate exocytosis of the secretory granules. Epinephrine, norepinephrine, ATP, the enzyme dopamine β-hydroxylase, opioid peptides, and chromagranin are all released

Table 50-4 Comparison of circulating concentrations of catecholamine hormones with biologically effective concentrations

		Plasma epinephrine (pg/ml)		*Plasma norepinephrine (pg/ml)*	
Physiological state	*Relevant biological action*	*Observed*	*Effective range for relevant biological action*	*Observed*	*Effective range for relevant biological action*
Basal	—	34	—	228	—
Upright position	↑ Heart rate and blood pressure	73	50-125	526	1800
↓ Plasma glucose	↑ Plasma glucose	230	150-200	262	1800
Severe hypoglycemia	—	1500	—	770	—
Diabetic ketoacidosis	↑ Lipolysis and ketosis ↓ Insulin	510	100-400	1270	1800

Data from Clutter WE et al: *J Clin Invest* 66:94, 1980; Silverberg AB et al: *Am J Physiol* 234:E252, 1978; and Christensen NJ: *Diabetes* 23:1, Jan 1974.

into the circulation. The membranous material of the granule is retained and probably recycled.

Basal plasma epinephrine levels are 25 to 50 pg/ml (6×10^{-10}M). A daily basal delivery rate of 150 μg can be estimated, and this rate of epinephrine release can increase greatly with physiological stimuli (Table 50-4). For example, under the stimulation of a modest fall in plasma glucose to 60 mg/dl, epinephrine concentrations rise to 230 pg/ml. If epinephrine is infused exogenously at a rate sufficient to maintain this approximate level, a rise in plasma glucose results. Hence, the adrenal medulla is quite capable of secreting enough epinephrine to contribute to glucose homeostasis.

The same is true of cardiovascular responses to the hormone. An increase in heart rate and systolic blood pressure can be produced by the concentrations of epinephrine that are generated endogenously by assumption of the upright position (Table 50-4). In addition, the high concentrations of epinephrine that occur in such illnesses as diabetic ketoacidosis are quite capable of contributing to the pathological state by stimulation of lipolysis and ketosis. Thus epinephrine functions as a true hormone in all these situations.

In contrast, circulating norepinephrine levels do not generally increase to levels sufficient to produce relevant biological actions (Table 50-4). Therefore, norepinephrine does not usually function in an endocrine fashion, although it may do so in severe, stressful illnesses such as myocardial infarction. Instead, norepinephrine's effects on metabolic processes, such as glucose production or lipolysis, result from its role as a neurotransmitter, wherein the necessary high concentrations are generated locally at the effector site.

Actions of Catecholamines

Epinephrine and norepinephrine exert their effects on a group of plasma membrane receptors designated β-1, β-2, α-1, and α-2. Epinephrine reacts well with β-1 and β-2 receptors, but its potency is less with α-receptors. Norepinephrine reacts predominantly with α receptors, less strongly with β-1 receptors, and only weakly with β-2 receptors. β-1, β-2, and α-2 receptors are structurally similar. They are single-unit glycoproteins with molecular weights around 64,000. α-1 Receptors differ from the others and have molecular weights of 80,000. β-1 and β-2 Receptors are coupled to and stimulate adenylyl cyclase, so that cAMP is the second messenger for these biological effects. Protein kinase A is activated and a cascade of changes in enzyme activities follows.

The α-2 receptor, in contrast, couples with an inhibitory G-protein so that hormone binding decreases cAMP levels and protein kinase A activity. The α-1 receptor is coupled to the phosphatidylinositol membrane system; protein kinase C, along with calcium, mediates the hormone effects.

Continuous stimulation of catecholamine release or exposure to catecholamine agonists downregulates adrenergic receptor numbers and induces partial refractoriness to hormone action. Conversely, sympathectomy increases receptor number and enhances sensitivity to catecholamines. Acute exposure to catecholamine hormones produces rapid desensitization to subsequent doses. This effect is caused by phosphorylation of the various receptors by protein kinase A or C. This renders them inaccessible to further hormone binding. This receptor desensitization process is a form of rapid intracellular negative feedback, which limits hormone actions.

A listing of epinephrine and norepinephrine actions is presented in Table 50-5. Both catecholamine hormones increase glucose production. They stimulate glycogenolysis in the liver via β-receptors by activating phosphorylase through the same cAMP-initiated cascade as is produced by glucagon. Glycogen synthase activity is concurrently restrained. The adrenal medullary epinephrine response to hypoglycemia is not needed as long as glucagon secretion is intact. However, in the absence of glucagon epinephrine becomes essential for recovery from hypoglycemia.

Epinephrine and norepinephrine also stimulate gluconeogenesis, by activation of α-receptors on the liver

■ Table 50-5 Some actions of catecholamine hormones

β	α
Epinephrine > norepinephrine	*Norepinephrine > epinephrine*
↑ Glycogenolysis	↑ Gluconeogenesis (α1)
↑ Lipolysis and ketosis (β1)	
↑ Calorigenesis (β1)	
↓ Glucose utilization	
↑ Insulin secretion (β2)	↓ Insulin secretion (α2)
↑ Glucagon secretion (β2)	
↑ Muscle K^+ uptake (β2)	
↑ Cardiac contractility (β1)	
↑ Heart rate (β1)	
↑ Conduction velocity (β1)	
↑ Arteriolar dilation: ↓ BP (β2) (muscle)	↑ Arteriolar vasoconstriction; ↑ BP (α1) (Splanchnic, renal, cutaneous, genital)
↑ Muscle relaxation (β2)	↑ Sphincter contraction (α1)
Gastrointestinal	Gastrointestinal
Urinary	Urinary
Bronchial	Platelet aggregation (α2)
	Sweating ("adrenergic")
	Dilation of pupils

cells. In addition, they stimulate muscle glycogenolysis, which increases plasma lactate levels and provides additional substrate to the liver. Simultaneously epinephrine inhibits insulin-mediated glucose uptake by muscle and adipose tissue. The catecholamines also stimulate glucagon secretion while inhibiting insulin secretion. All these actions help restore plasma glucose and its delivery to the central nervous system. At the same time epinephrine activates adipose tissue lipase, thereby increasing plasma free fatty acids, their B-oxidation in muscle and liver, and ketogenesis. These effects on flow of fuels are shown in Fig. 50-24.

When the catecholamine hormones are secreted during exercise, they promote (1) use of muscle glycogen stores by stimulating phosphorylase, (2) efficient hepatic reutilization of lactate released by the exercising muscle, and (3) provision of free fatty acids as alternate fuels. When epinephrine secretion is stimulated by "stress," such as during illness or surgery, its actions on fuel turnover contribute significantly to induction of hyperglycemia and ketosis; i.e., it is a diabetogenic hormone.

Epinephrine also increases the basal metabolic rate by 7% to 15%. This action increases nonshivering thermogenesis as well as diet-induced thermogenesis. It is, therefore, an important part of the response to cold exposure and helps to regulate overall energy balance and stores. In neonates of many species, brown adipose tissue is an important site where catecholamines increase heat production. Here they stimulate proton conductance into the mitochondria, thereby uncoupling ATP synthesis from oxygen utilization.* In most metabolic effects epinephrine is much more potent than norepinephrine. The latter, nonetheless, contributes importantly to metabolism via the activity of the sympathetic nervous system. For example, sympathetic nervous system activity, with norepinephrine as the mediator, decreases with fasting and increases after feeding. In this way norepinephrine helps to adapt total energy utilization to its availability by modulating thermogenesis. In contrast, epinephrine secretion increases modestly during prolonged fasting and also 4 to 5 hours after a meal, in both cases in response to a declining plasma glucose level. This serves the purpose of helping to sustain glucose production for use by the central nervous system.

*The role of this specialized adipose tissue in human physiology is controversial.

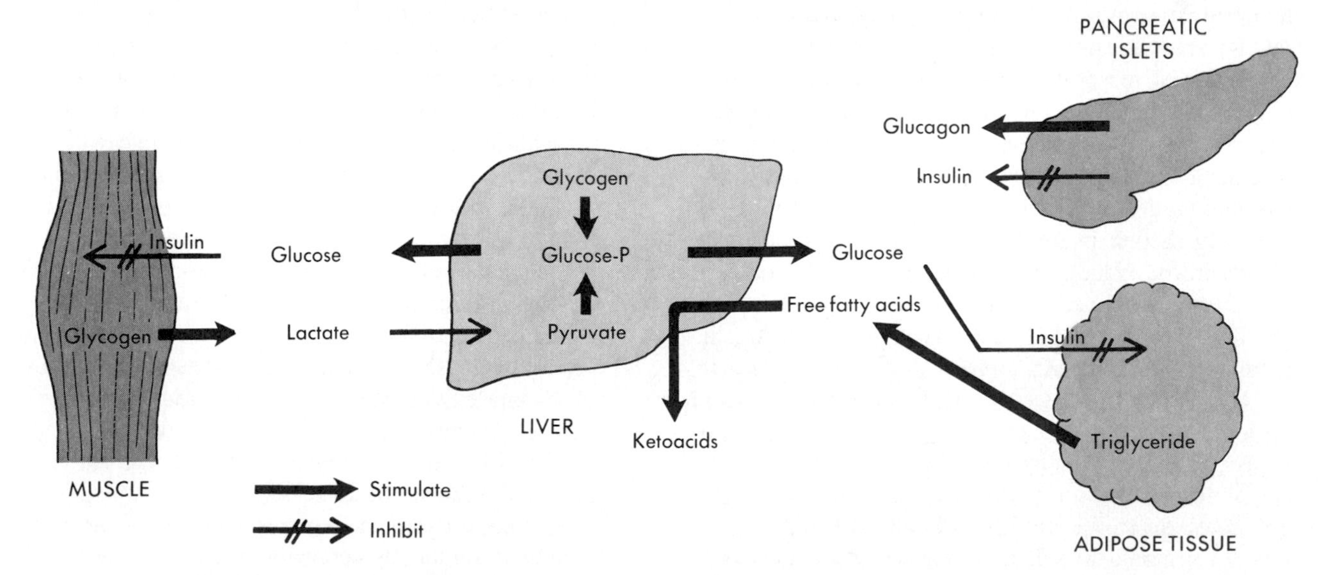

■ Fig. 50-24 Metabolic actions of epinephrine. The hormone stimulates glucose production and inhibits glucose use. It also stimulates lipolysis and ketogenesis. Insulin secretion is inhibited. The net effect is a rise in plasma glucose, free fatty acids, and ketoacids.

The cardiovascular effects of epinephrine reinforce its metabolic actions. Cardiac output rises as a result of an increase in heart rate and contractile force; on the other hand, arteriolar constriction is selectively produced in the renal, splanchnic, and cutaneous beds, whereas the muscle arterioles are dilated. Systolic blood pressure increases, whereas diastolic blood pressure remains unchanged or decreases slightly. The net effect of these changes is to shunt blood toward exercising muscles while maintaining coronary and cerebral blood flow (Chapters 29 and 32). This guarantees delivery of substrate for energy production to the critical organs in the fight-or-flight situation.

During exposure to cold constriction of cutaneous vessels helps to conserve heat, reinforcing epinephrine's thermogenic action. The inhibition of gastrointestinal and genitourinary motor activity, the relaxation of bronchioles to prevent expiratory airway obstruction and improve gas exchange, and the dilation of pupils to permit better distant vision are also of benefit to the endangered individual.

The cardiovascular consequences of catecholamines also initially benefit the individual who has suffered major trauma, circulatory failure, or hypoxia. Without them, death may rapidly ensue. However, prolonged secretion of catecholamines eventually becomes deleterious. Reduced blood flow to the kidneys leads to renal failure; reduced blood flow to the splanchnic bed leads to hepatic failure as well as intestinal paralysis and necrosis; reduced blood flow to many tissues leads to decreased oxygenation with increased lactate production, which, in the face of decreased lactate utilization in the liver, produces metabolic (lactate) acidosis (Chapter 32).

Catecholamines exhibit a number of other diverse actions. By interacting with β-receptors in the vascular tree and lowering blood pressure, they may stimulate ADH release and water retention; conversely, by interacting with α receptors and raising blood pressure, they may inhibit ADH release and cause water loss (Chapter 48). Catecholamines increase renin release by stimulation of β-receptors in the kidney. This increases aldosterone secretion, which in turn enhances sodium retention. This action is augmented by local catecholamine effects in the kidney on the distribution of blood flow and on tubular function. Epinephrine stimulates influx of potassium into muscle cells via β-2 receptors. This helps to prevent hyperkalemia.

The interaction between catecholamine and thyroid hormone function remains an important area of investigation. Thyroid hormone secretion is enhanced by catecholamines under some circumstances and the peripheral conversion of T_4 to T_3 is stimulated via β-2 receptors. In general, thyroid hormone sensitizes animals to catecholamine actions by complex mechanisms. Epinephrine increases parathyroid hormone, calcitonin, growth hormone, and gastrin secretion. Whether these effects are produced in physiological circumstances remains to be determined.

■ *Therapeutic Usage*

Catecholamine agonists and antagonists are in widespread use in medicine. A group of agonists called **amphetamines** are used as nasal decongestants, appetite suppressants, and general stimulants. All of these may be prescribed or sold over the counter, but their illicit availability has become a public health problem as well. They may cause hypertension; exacerbate tachycardia, palpitations, and nervousness in hyperthyroid patients; or, rarely, increase plasma glucose in diabetic patients. In large doses they can produce life-threatening "highs." Both β-adrenergic antagonists and α-adrenergic antagonists are used in the treatment of hypertension, and the former are also of value in the treatment of coronary artery disease. β-antagonists are also used effectively to counteract the hyperactive adrenergic state in hyperthyroid patients.

■ *Pathological Secretion of Catecholamines*

Spontaneous deficiency of epinephrine is unknown as an adult disease, and adrenalectomized patients do not require epinephrine replacement. Hypersecretion of epinephrine and norepinephrine from tumors of the chromaffin cells **(pheochromocytomas)** results in a well-defined syndrome. Dramatic clinical episodes are produced by sudden spurts of catecholamines. These can result from any stress or from a rapid change in posture.

Sudden severe headache, palpitations, chest pain, extreme anxiety with a sense of impending death, cold perspiration, pallor of the skin caused by vasoconstriction, and blurred vision may occur. Blood pressure may be extremely high, for example, 250/150. If primarily epinephrine is being secreted, the heart rate will be increased; if norepinephrine is the predominant hormone, the heart rate will be decreased reflexly in response to the marked hypertension. In addition to these episodes, chronic catecholamine excess may produce weight loss, as a result of an increased metabolic rate, and hyperglycemia caused by inhibition of insulin secretion.

The diagnosis is established by detecting high levels of plasma epinephrine or norepinephrine when the patient is recumbent and at rest. In addition, urinary excretion of free catecholamines, metanephrines, and VMA will usually be increased.

Definitive treatment requires removal of the adrenal medullary tumor. Symptomatic treatment is provided by α-adrenergic antagonists, which lower the elevated blood pressure and β-adrenergic antagonists, which reduce tachycardia.

Integration of the Response to Stress

The adrenal medulla and adrenal cortex are both major participants in the adaptation to stress. Their intimate anatomic juxtaposition mirrors a fundamental functional relationship between the adrenergic nervous system and the corticotropin-releasing hormone—adrenocorticotropic hormone-cortisol axis. Recent advances justify presenting an integrated overview of the neuroendocrine adaptation to stress (Fig. 50-25).

Stress is perceived by many areas of the brain, from the cortex down to the brainstem. Major stresses activate both CRH neurons and adrenergic neurons in the hypothalamus. The activation is mutually reinforcing because norepinephrine increases CRH release and CRH increases adrenergic discharge (see Fig. 50-25). CRH release ultimately elevates plasma cortisol levels; adrenergic stimulation elevates plasma catecholamine levels. Together these hormones increase glucose production; epinephrine does so rapidly by activating glycogenolysis, and cortisol more slowly by providing amino acid substrate for gluconeogenesis. Together they shift glucose utilization toward the central nervous system and away from peripheral tissues. Epinephrine also rapidly augments free fatty acid supply to the heart and to muscles, and cortisol facilitates the lipolytic response. Both hormones raise blood pressure and cardiac output and improve delivery of substrates to tissues that are critical to the immediate defense of the organism. If the stress involves tissue trauma or invasion, high cortisol levels eventually act to restrain the initial inflammatory and immune responses so that they do not lead to irreparable damage.

The neurotransmitter norepinephrine and the neuropeptide CRH can produce other adaptive responses to stress. A general state of arousal and vigilance, an activation of defensively useful behavior, and appropriate aggressiveness result from adrenergic stimuli to the pertinent brain centers. At the same time CRH input to other hypothalamic neurons inhibits growth hormone and gonadotropin release, presumably because growth and reproduction are not useful functions during stress. This is reinforced by the excess of cortisol, which also suppresses growth and ovulation. In addition, CRH inhibits sexual activity and feeding, again inappropriate activities when the organism perceives itself to be in immediate serious danger. Thus the adaptation to stress represents a prime example of the integration of the nervous system and the endocrine system.

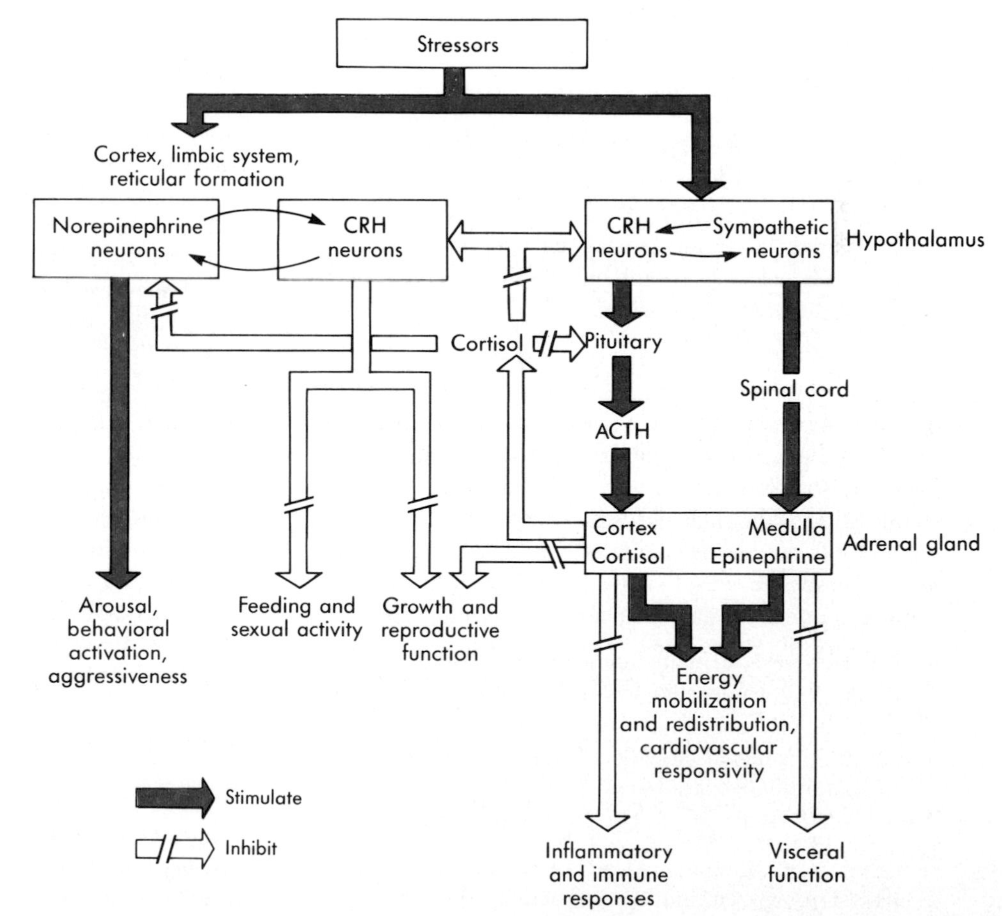

Fig. 50-25 Integrated responses to stress mediated by the sympathetic nervous system and the hypothalamic-pituitary-adrenocortical axis. The responses are mutually reinforcing, both at the central and peripheral levels. Negative feedback by cortisol also can limit an overresponse that might be harmful to the individual. *Colored arrows,* stimulation; *open arrows,* inhibition; *CRH,* corticotropin-releasing hormone; *ACTH,* adrenocorticotropic hormone.

Summary

1. The two adrenal glands consist of an outer three-layered cortex and an inner medulla. The cortex secretes three types of steroid hormones: cortisol, a glucocorticoid; aldosterone, a mineralocorticoid; and androgen precursors. The medulla secretes the catecholamine hormones, epinephrine and norepinephrine. The adrenal glands are richly vascularized and essential to survival because of the cortisol they produce.

2. All adrenocorticosteroids are synthesized from cholesterol by sequential enzymatic steps consisting of side chain cleavage and hydroxylation of key sites in the steroid nucleus. Cortisol specifically requires an 11-hydroxyl group, aldosterone, an 18-hydroxyl group, and androgens, a 17-hydroxyl group for respective activities. The mitochondrial and microsomal enzymes involved are P_{450} mixed oxygenases. Steroid hormones are not stored directly; increased secretory demands require rapid synthesis from stored cholesterol.

3. Cortisol and androgen secretion are regulated by adrenocorticotropin (ACTH). The pituitary hormone acts through a plasma membrane receptor with cAMP as the main second messenger. ACTH stimulates cellular uptake of cholesterol, its movement from storage vacuoles into mitochondria, and all subsequent biosynthetic steps.

4. Cortisol has major effects on protein, glucose, and fat metabolism. It acts via a nuclear receptor and modulation of gene expression of numerous enzymes and proteins. Cortisol increases muscle proteolysis and stimulates hepatic conversion of the liberated amino acids into glucose and storage as glycogen. Cortisol also inhibits insulin-stimulated glucose uptake by muscle. Cortisol stimulates caloric intake and favors deposition of fat in selected sites. By inhibiting collagen synthesis, cortisol reduces bone formation and impairs the integrity of capillaries and skin.

5. Cortisol strongly inhibits the entire process of inflammation, including the recruitment of neutrophils and the release of prostaglandin and leukotriene mediators. It also inhibits the immune system and prevents proliferation of thymus-derived lymphocytes and production of lymphokines. These actions underlie the broad therapeutic use of synthetic analogues of cortisol as antiinflammatory and immune suppressant agents.

6. Aldosterone is a major regulator of sodium, potassium, and fluid balance. It acts on the renal tubule via a specific nuclear receptor and gene expression. Sodium reabsorption is increased with concomitant expansion of extracellular fluid. Potassium excretion is simultaneously increased and plasma potassium concentration is lowered.

7. Aldosterone secretion is regulated by the renin-angiotensin system. In response to sodium deprivation, production of angiotensin-2 is increased. This hormone stimulates aldosterone secretion via Ca^{++} and phosphatidylinositol second messengers. Aldosterone secretion is also directly stimulated by potassium.

8. The adrenal medulla is an enlarged, specialized sympathetic ganglion. It synthesizes epinephrine and norepinephrine from tyrosine and stores these catecholamine hormones in granules. They are released in response to stimulation of preganglionic cholinergic sympathetic nervous system fibers. Hypoglycemia, hypovolemia, hypotension, stress, and pain are major stimuli.

9. Epinephrine acts as a true hormone to increase glycogenolysis in liver and muscle and lipolysis in adipose tissue. Epinephrine also decreases insulin-stimulated glucose uptake and increases the metabolic rate. cAMP and Ca^{++} are second messengers. Epinephrine increases plasma glucose, free fatty acids, and ketoacids. Numerous vascular and visceral actions of the sympathetic nervous system are reinforced by circulating epinephrine.

Bibliography

Journal articles

Allison AC, Lee SW: The mode of action of anti-rheumatic drugs. I. Anti-inflammatory and immunosuppressive effects of glucocorticoids, *Prog Drug Res* 33:63, 1989.

Barbarino A et al: Corticotropin-releasing hormone inhibition of gonadotropin release and the effect of opioid blockade, *J Clin Endocrinol Metab* 68:523, 1989.

Barrett PQ et al: Role of calcium in angiotensin II-mediated aldosterone secretion, *Endocr Rev* 10:496, 1989.

Born J et al: Influences of cortisol on auditory evoked potentials (AEPs) and mood in humans, *Neuropsychobiology* 20:145, 1989.

Burnstein KC, Cidlowski JA: Regulation of gene expression by glucocorticoids, *Annu Rev Physiol* 51:603, 1989.

Calvano SE et al: Comparison of numerical and phenotypic leukocyte changes during constant hydrocortisone infusion in normal humans with those in thermally injured patients, *Surg Gynecol Obstet* 164:509, 1987.

Clutter W et al: Epinephrine plasma metabolic clearance rates and physiologic thresholds for metabolic and hemodynamic actions in man, *J Clin Invest* 66:94, 1980.

Cryer PE: Physiology and pathophysiology of the human sympathoadrenal neuroendocrine system, *N Engl J Med* 303:436, 1980.

Darmaun D et al: Physiological hypercortisolemia increases proteolysis, glutamine, and alanine production, *Am J Physiol* 255:E366, 1988.

DeFeo P et al: Contribution of cortisol to glucose counterregulation in humans, *Am J Physiol* 257:E35, 1989.

Estaban NV et al: Daily cortisol production rate in man determined by stable isotope dilution/mass spectrometry, *J Clin Endocrinol Metab* 72:39, 1991.

Fauci AS et al: Glucocorticosteroid therapy: mechanisms of action and clinical considerations, *Ann Intern Med* 84:304, 1976.

Garcia-Robles R, Ruilope LM: Pharmacological influences on aldosterone secretion, *J Steroid Biochem* 27:941, 1987.

Goldberg AL et al: Hormonal regulation of protein degradation and synthesis in skeletal muscle, *Fed Prac* 39:31, 1980.

Gustafsson J et al: Biochemistry, molecular biology and physiology of the glucocorticoid receptor, *Endocrine Rev* 8:185, 1987.

Hauner H et al: Glucocorticoids and insulin promote the differentiation of human adipocyte precursor cells into fat cells, *J Clin Endocrinol Metab* 64:832, 1987.

Horber FF et al: Differential effects of prednisone and growth hormone on fuel metabolism and insulin antagonism in humans, *Diabetes* 40:141, 1991.

Horrocks PM et al: Patterns of ACTH and cortisol pulsatility over twenty-four hours in normal males and females, *Clin Endocrinol (Oxf)* 32:127, 1990.

Jackson R et al: Synthetic ovine corticotropin-releasing hormone: simultaneous release of propiolipomelanocortin peptides in man, *J Clin Endocrinol Metab* 58:740, 1984.

Kirkham BW, Panayi GS: Diurnal periodicity of cortisol secretion, immune reactivity and disease activity in rheumatoid arthritis: implications for steroid treatment, *Br J Rheumatol* 28:154, 1989.

Landsberg L, Young JB: The role of the sympathetic nervous system and catecholamines in the regulation of energy metabolism, *Am J Clin Nutr* 38:1018, 1983.

Lefkowitz RJ, Caron MG: Adrenergic receptors: molecular mechanisms of clinically relevant recognition, *Clin Res* 33e:395, 1985.

Lundgren JD et al: Mechanisms by which glucocorticosteroids inhibit secretion of mucus in asthmatic airways, *Am Rev Respir Dis* 141(2, pt 2):S52, 1990.

Matthews DE et al: Effect of epinephrine on amino acid and energy metabolism in humans, *Amer Jour Physiol* 258:E948, 1990.

Miller WL: Molecular biology of steroid hormone synthesis, *Endocr Rev* 9:295, 1988.

Munck A et al: Physiological functions of glucocorticoids in stress and their relation to pharmacological actions, *Endocr Rev* 5:25, 1984.

Peers SH, Flower RJ: The role of lipocortin in corticosteroid actions, *Am Rev Respir Dis* 141:S18, 1990.

Penhoat A et al: Synergistic effects of corticotropin and insulin-like growth factor I on corticotropin receptors and corticotropin responsiveness in cultured bovine adrenocortical cells, *Biochem Biophys Res Commun* 165:355, 1989.

Prummel MF et al: The course of biochemical parameter of bone turnover during treatment with corticosteroids, *J Clin Endocrinol Metab* 72:382, 1991.

Quinn SJ, Williams GH: Regulation of aldosterone secretion, *Annu Rev Physiol* 50:409, 1988.

Ramachandran J et al: Corticotropin receptors, *Ann NY Acad Sci* 512:415, 1987.

Rebuffe-Scrive M et al: Muscle and adipose tissue morphology and metabolism in Cushing's syndrome, *J Clin Endocrinol Metab* 67:1122, 1988.

Rizza RA et al: Cortisol-induced insulin resistance in man: impaired suppression of glucose production and stimulation of glucose utilization due to a postreceptor defect of insulin action, *J Clin Endocrinol Metab* 54:131, 1982.

Rosner W: The functions of corticosteroid-binding globulin and sex hormone-binding globulin: recent advances, *Endocr Rev* 11:80, 1990.

Santiago JV et al: Epinephrine, norepinephrine, glucagon, and growth hormone release in association with physiological decrements in the plasma glucose concentration in normal and diabetic man, *J Clin Endocrinol Metab* 51:877, 1980.

Schenker Y: Atrial natriuretic hormone and aldosterone regulation in salt-depleted state, *Am J Physiol* 257:E583, 1989.

Schleimer RP: Effects of glucocorticosteroids on inflammatory cells relevant to their therapeutic applications in asthma, *Am Rev Respir Dis* 141:S59, 1990.

Silverberg A et al: Norepinephrine: hormone and neurotransmitter in man, *Am J Physiol* 234:E252, 1978.

Simpson ER, Waterman MR: Regulation of the synthesis of steroidogenic enzymes in adrenal cortical cells by ACTH, *Annu Rev Physiol* 50:427, 1988.

Suda T et al: Immunoreactive corticotropin-releasing factor in human plasma, *J Clin Invest* 76:2026, 1985.

Taylor AL, Fishman LM: Corticotropin-releasing hormone, *N Engl J Med* 319:213, 1988.

Umeki S, Soejima R: Hydrocortisone inhibits the respiratory burst oxidase from human neutrophils in whole-cell and cell-free systems, *Biochem Biophys Acta* 1052:211, 1990.

Veldhuis JD et al: Amplitude modulation of a burstlike mode of cortisol secretion subserves the circadian glucorticoid rhythm, *Am J Physiol* 257:E6, 1989.

Wong MM et al: Long-term effects of physiologic concentrations of dexamethasone on human bone-derived cells, *J Bone Miner Res* 5:803, 1990.

Wortsman J et al: Adrenomedullary response to maximal stress in humans, *Am J Med* 77:779, 1984.

Yamakado M et al: Extrarenal role of aldosterone in the regulation of blood pressure, *Am J Hypertens* 1:276, 1988.

Young DB: Quantitative analysis of aldosterone's role in potassium regulation, *Am J Physiol* 255:F811, 1988.

Books and monographs

Crabbe J: Mechanism of action of aldosterone. In Degroot LJ, editor: *Endocrinology,* Philadelphia, 1989, WB Saunders.

Meikle AW: Secretion and metabolism of the corticosteroids and adrenal function and testing. In Degroot LJ, editor: *Endocrinology,* Philadelphia, 1989, WB Saunders.

Nelson DH: The secretion of the adrenal cortex and steroid biosynthesis. In Smith LH Jr, editor: *The adrenal cortex: physiological function and disease,* vol 18, Major problems in internal medicine, Philadelphia, 1980, WB Saunders.

Numa S, Imura H: ACTH and related peptides: gene structure and biosynthesis. In Imura H: *The pituitary gland,* New York, 1985, Raven Press.

CHAPTER

51

The Reproductive Glands

The endocrine glands that have been discussed thus far are essential to the maintenance of the life and well-being of the individual. In contrast, the endocrine function of the gonads is primarily concerned with maintaining the life and well-being of the species. The evolution of sexual reproduction has required the development of highly complex patterns of gonadal function. These are concerned with the development, maturation, and nurturance of the individual male and female germ cells, their successful union, and the subsequent growth and development of the newly created individual within the body of the mother. There are many obvious differences between the functioning of the testes and the ovaries, but there are also important basic conceptual similarities and operational homologies. Therefore the subject of human gonadal endocrinology is presented as a single unit in the following sequence: (1) sexual differentiation, (2) homologous aspects of gonadal structure and function, (3) testicular function, (4) ovarian function, and (5) endocrine aspects of pregnancy.

■ *Sexual Differentiation*

The process of sexual differentiation, i.e., the pattern of development of the gonads, genital ducts, and external genitalia, produces the most fundamental and obvious differences between the genders. However, during the first 5 weeks of gestation, the gonads of males and females are indistinguishable and their genital tracts are unformed. From this stage of the "indifferent gonad" to that of the completed normal individual of either gender lies the process of sexual differentiation (Figs. 51-1, 51-2, and 51-3). Before this process is described, the gonadal cell lines and their functions that are common to both genders need to be considered (Fig. 51-1).

The primordial germ cells generate the **oogonia** and **spermatogonia** that undergo eventual reductional division and maturation into large numbers of ova and sperm, respectively. Only a few of each will eventually unite with each other to reproduce the species, in a manner that guarantees an almost infinite variety of individual characteristics. One cell line of the indifferent gonad becomes the granulosa cells of the ovarian follicle and the Sertoli cells of the testicular seminiferous tubules. The function of these cells is homologous, to sustain or "nurse" the germ cells, foster their maturation, and guide their movement into the genital duct system. This cell line is also the main source of estrogenic hormones in females. Another gonadal cell line, the interstitial cells, gives rise to theca cells in the ovary and Leydig cells in the testis. The primary function of this cell line is to secrete androgenic hormones. These are essential for masculine development and sperm production and, in females, as precursors for estrogen synthesis.

The final maleness or femaleness of an individual is best characterized in terms of differences in genetic sex, in gonadal sex, and in genital (phenotypic) sex.

■ *Genetic Sex*

The normal male chromosome complement is 44 autosomes and 2 sex chromosomes, X and Y. The presence of the Y chromosome is a positive and the single most constant determinant of maleness. Without either a Y chromosome or, in rare cases, material from the Y chromosome translocated to an X chromosome, neither testes nor a masculine genital pattern can develop. A testis-determining factor (TDF) has been localized to the distal part of the Y_P, the short arm of the human Y chromosome. It codes for a protein with a "zinc finger" configuration capable of binding to DNA molecules. Either identical to the TDF or close by and linked to it is a gene that codes for a minor histocompatibility antigen known as H-Y. This is one of two antigens involved in rejection of male tissue by female recipients in certain inbred animal strains. This H-Y antigen has a molecular weight of 18,000 and is present on the surface of all male cells except diploid germ cells. A receptor for this protein has been localized to gonadal tissue. In the presence of this H-Y antigen, indifferent gonads from female embryos virilize in culture and disassociated ovar-

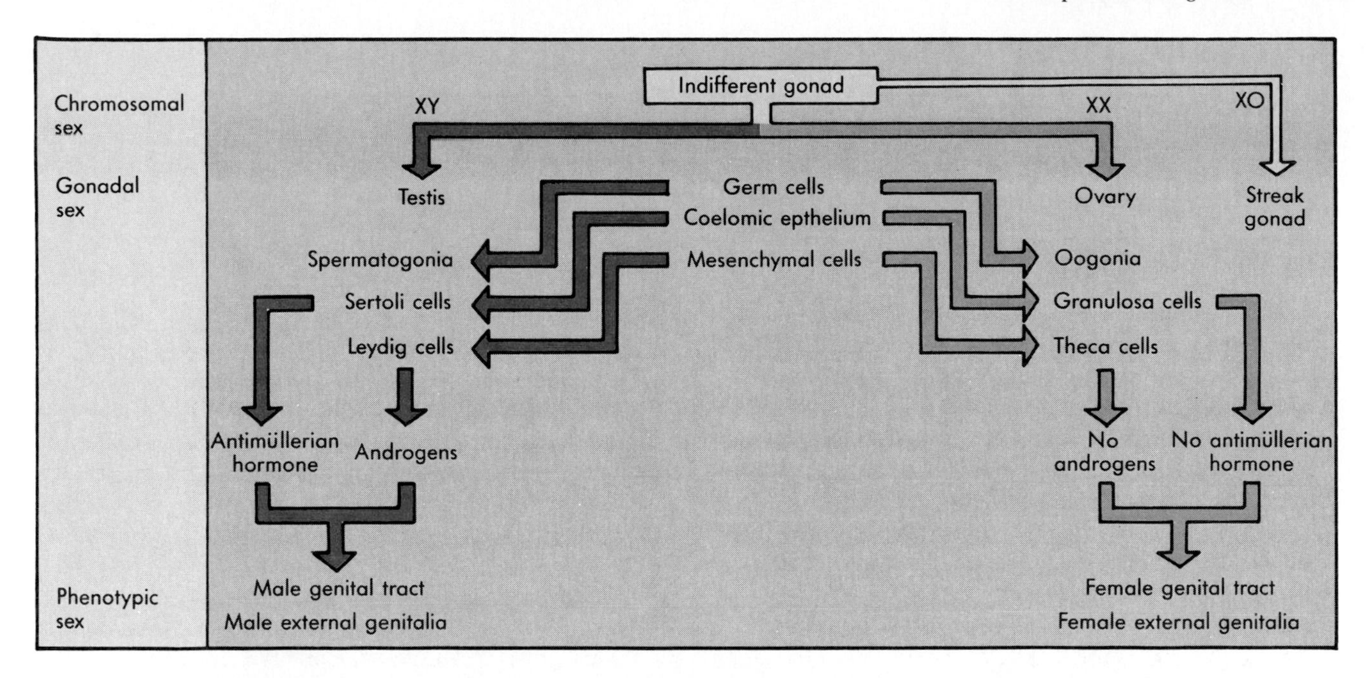

■ **Fig. 51-1** An overview of the development of the cells of the ovary and testis from the primitive indifferent gonad. The hormonal products from the testis and the absence of these products from the ovary determine the gender differences in the internal genital tracts and the external genitalia.

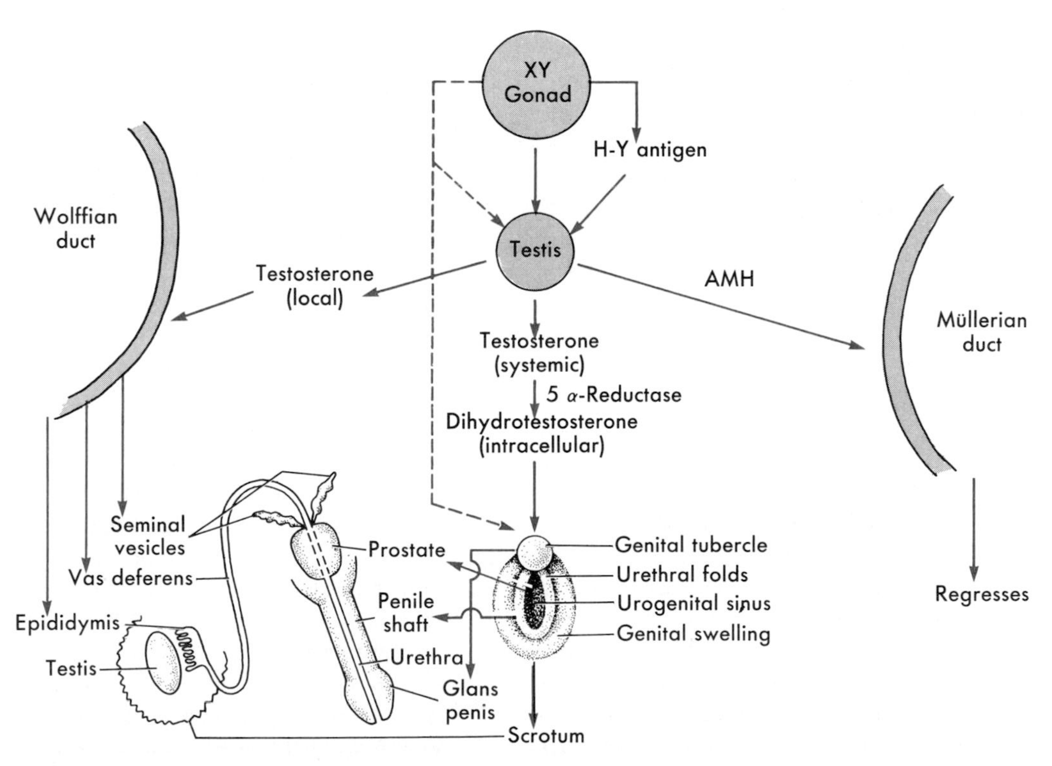

■ **Fig. 51-2** Development of the human male reproductive organs and tract. Note the dependence on hormone products of the gonad. *AMH,* Anti-Müllerian hormone.

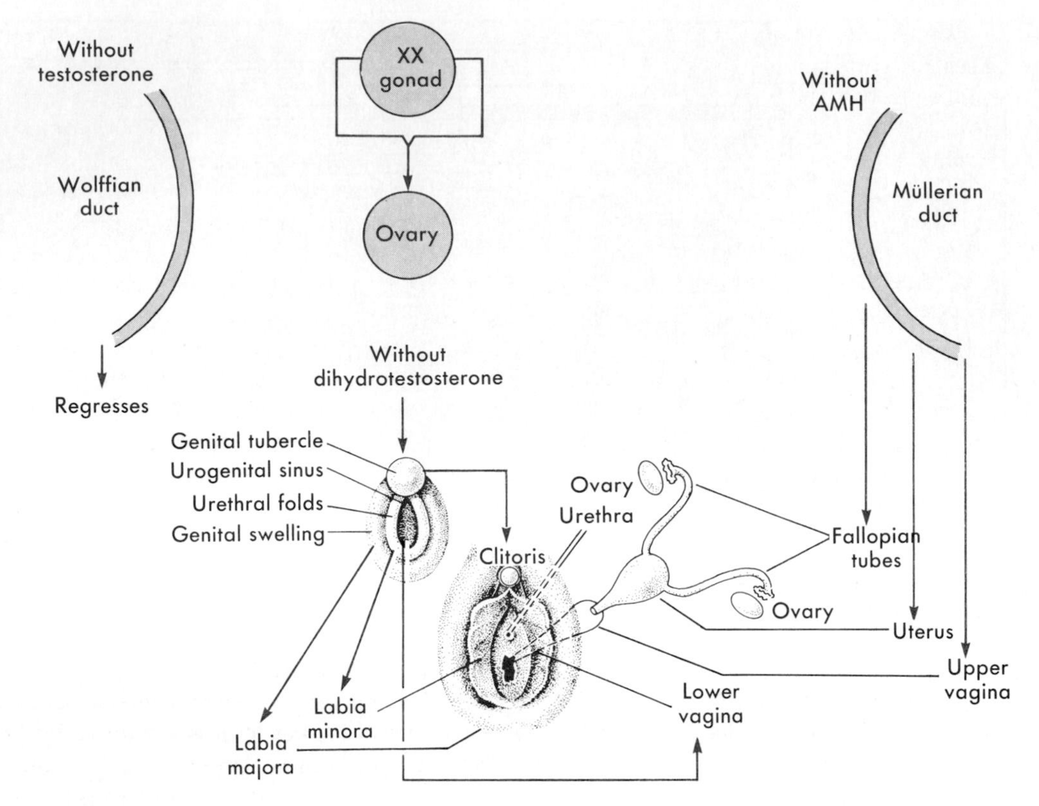

■ **Fig. 51-3** Development of the human female reproductive organs and tract. Note the independence from hormonal products of the gonad. In the absence of any gonads the female format results. *AMH,* Anti-Müllerian hormone.

ian cells reaggregate in a testicular format in vitro. A second H-Y antigen coded for by DNA located on the Y_q, the long arm of the Y chromosome, appears to be involved in the development of spermatogenesis.

Although Y chromosomal DNA is essential, it is not sufficient for complete maleness. Virilization of the genital ducts and external genitalia requires the presence of an androgen hormone receptor, which is only coded for by the X chromosome. In addition, some evidence suggests that a TDF allele may be present on the X chromosome. This has led to the proposal that a dose effect of TDF rather than a qualitative effect may determine maleness. Finally, autosomes may also participate in directing the organization of the primitive gonad into a normally functioning testis.

The normal female chromosome complement is 44 autosomes and two sex chromosomes, XX. Both of the X chromosomes are active in germ cells and are essential for the genesis of a normal ovary (see Fig. 51-1). In addition, evidence suggests that there is participation of autosomes in ovarian development, since rare individuals with a normal complement of XX sex chromosomes inherit defective gonads as an autosomal recessive trait. In contrast, differentiation of the genital ducts and external genitalia along normal female lines requires that only a single X chromosome be active in directing transcription within the cell. The second X chromosome of a normal XX female is inactivated early in all extragonadal tissues. Therefore, if an abnormality of meiosis or mitosis produces an individual with only a single X chromosome (XO karotype), the phenotype will still be female even though the gonads will be defective (see Fig. 51-1).

Gonadal sex. During the initial 5 weeks of fetal life, the gonads develop along sexually indistinguishable lines. Primordial germ cells differentiate in the 5-day-old blastocyst. At 22 to 24 days, the germ cells are present within the yolk sack endoderm. They then migrate to the genital ridge where they associate with mesonephric tissue. This indifferent gonad lasts for 7 to 10 days and consists of coelomic epithelium, the precursors of granulosa (female) and Sertoli (male) cells; mesenchymal stromal cells, the precursors of theca and Leydig cells; and the germ cells (see Fig. 51-1). The assembly is organized as an outer cortex and inner medulla.

In a normal male at 6 to 7 weeks of age, the seminiferous tubules begin to form as the Sertoli cells enclose the germ cells. The Leydig cells appear at 8 to 9 weeks. At that point, a recognizable testis is present and testosterone secretion is established. The medulla of the testis dominates anatomically, whereas the cortex regresses.

In the normal female, differentiation of the indifferent gonad into an ovary does not start until 9 weeks of age. At this time *both X chromosomes* within the germ cells become activated (see Fig. 51-1), an absolute requirement for further development and survival. The germ cells begin to undergo mitosis, giving rise to oogonia, which continue to proliferate. Shortly thereafter meiosis is initiated in some oogonia, and they become surrounded by granulosa cells and stroma, from the latter of which interstitial cells subsequently appear. The germ cells are now known as primary oocytes and remain in the diplotene or late prophase stage of meiosis (prophase) until possible ovulation many years later. In contrast to the male gonad, the cortex predominates in the developed ovary while the medulla regresses. The capacity of the primitive ovary to synthesize estrogenic hormones develops at about the same time that testosterone synthesis begins in the testis.

Genital (Phenotypic) Sex

Up to this point of fetal development sexual differentiation does not require any known hormonal products. However, differentiation of the genital ducts and of the external genitalia does require hormones. *The guiding principle is that positive hormonal influences, normally arising from the gonad, are required to produce the masculine format. In the absence of any gonadal hormonal input the feminine format will result.*

During the sexually indifferent stage, from 3 to 7 weeks, two genital ducts develop on each side. In the male at about 9 to 10 weeks, the wolffian, or mesonephric, ducts begin to grow and eventually give rise to the epididymis, the vas deferens, the seminal vesicles, and the ejaculatory duct (see Fig. 51-2). This constitutes the system for delivering sperm from the testis to the penis. The differentiation of the wolffian ducts is preceded by the appearance of testosterone-secreting Leydig cells in the testis. It is the androgenic hormone **testosterone** that stimulates the growth and differentiation of the wolffian ducts in the male. Furthermore, the testosterone produced by each testis acts unilaterally on its own wolffian duct (see Fig. 51-2), as shown by gonadal transplantation experiments or by testosterone implantations. Testosterone does not have to be converted to its hormonally active product, dihydrotestosterone, to act within the wolffian duct cells, as it does in some other tissues (described later). Indeed these cells do not develop the 5α-reductase activity necessary for this conversion until after they have fully differentiated. In the female the wolffian ducts begin to regress at 10 to 11 weeks because the ovary does not secrete testosterone.

The müllerian ducts arise parallel to and in part from the wolffian ducts on each side. In the male these ducts begin to regress at 7 to 8 weeks, about the same time that the Sertoli cells of the testis appear. These cells produce a glycoprotein hormone called **müllerian-inhibiting factor (MIF),** or **antimüllerian hormone (AMH),** which causes atrophy of the müllerian ducts. AMH is a glycoprotein composed of two identical subunits, each of molecular weight 70,000. It is coded for by a gene from a superfamily of growth-regulating factors, including transforming growth factors α and β, epidermal growth factor, and inhibin. AMH probably acts by blocking the stimulatory effect of epidermal growth factor on the müllerian system. In addition, AMH may participate in organizing the testis into seminiferous tubules and may initiate descent of the testis into the inguinal area. Although AMH is found in the postnatal testis, it has no known role after birth.

AMH is not produced by granulosa cells until late in gestation or after birth. Therefore, in the female who lacks AMH, the müllerian ducts are allowed to grow and differentiate into fallopian tubes at the upper ends, whereas at the lower ends they join to form the uterus, cervix, and upper vagina (see Fig. 51-3). This process is completed at 18 to 20 weeks of age. It does not require any ovarian hormone.

The external genitalia of both sexes begin to differentiate at 9 to 10 weeks. They are derived from the same anlage: the genital tubercle, the genital swelling, the urethral, or genital, folds, and the urogenital sinus. In the male, testosterone must be secreted into the circulation of the fetus and subsequently must be converted to **dihydrotestosterone** within the cells of the anlage tissues for normal differentiation of the external genitalia to occur. As a result of dihydrotestosterone stimulation, the genital tubercle grows into the glans penis, the genital swellings fold and fuse into the scrotum, the urethral folds enlarge and enclose the penile urethra and corpora spongiosa, and the urogenital sinus gives rise to the prostate gland (see Fig. 51-2). In addition to androgen hormone activity, the presence of the androgen receptor is required in these target tissues.

In the normal female or in the absence of any gonads the anlage tissues develop without apparent hormonal influences into the clitoris, labia majora, labia minora, and lower vagina, respectively (see Fig. 51-3). The possibility that hormones derived from the placenta might be involved in shaping the normal female phenotype cannot be excluded at present. If the normal female fetus is exposed to an excess of testosterone or other androgens (e.g., from the adrenal glands) during the period of differentiation of the external genitalia, a male pattern can result. However, once the female pattern of differentiation has been achieved, such exposure cannot change it to the male pattern, although it can cause enlargement of the clitoris.

The androgen production necessary for sexual differentiation does not depend on fetal pituitary gonadotropins. An LH-like hormone, chorionic gonadotropin from the placenta, stimulates testosterone production by the Leydig cells. Placental steroid hormone precursors, such as pregnenolone, might serve as a source of fetal androgens, obviating the necessity for gonadotropin stimulation of the reactions from cholesterol to pregnenolone (see later discussions). The growth of the external genitalia in the latter part of gestation requires fetal pituitary LH to stimulate the necessary quantity of androgen (or possibly estrogen in the female).

Other aspects of phenotypic sexual differentiation are not evident until long after birth. These include differences between the constant pattern of gonadotropin secretion in the male versus the cyclic pattern in the female, the different degree of breast development, and the psychological identification with a unique gender. It is difficult at present to be certain what factors imprint or regulate these traits in humans. Evidence from rodent studies suggests that circulating androgens induce the fetal hypothalamus to set a constant pattern of gonadotropin secretion in the postpubertal male. To do so, they require metabolism to estrogens within the target neurons. In the absence of androgens the cyclic pattern of the female ultimately results. This would constitute another instance in which the female pattern was the "neutral pattern," whereas the male pattern required an action ultimately derived from the Y chromosome. It is not certain whether the same mechanism operates in humans.

Mammary gland development in the rodent embryo also is clearly under androgen regulation. In its absence a normal female breast develops; in its presence the ductal system is suppressed. However, in the human, male-female differences in breast development are not apparent before puberty. At that time the hormonal milieu in the female induces growth and differentiation of breast tissue, whereas that in the male suppresses it.

A large body of clinical evidence suggests that psychological gender identification is independent of hormonal regulation or even of the phenotype of the genitalia; instead it appears to depend on rearing cues and possibly other factors. However, exceptions to this are noted in certain cases of male pseudohermaphrodites raised as girls and possibly other factors. In such individuals significant growth of the penis under pubertal testosterone stimulation may cause a reversal of psychosocial gender from female to male.

Abnormalities of Sexual Differentiation

Anatomical aberrations that result from certain genetic errors are listed in Table 51-1. Sexual differentiation can be distorted by abnormalities in either sex chromosomes or autosomes.

The XO chromosomal karyotype produces individuals with only a vestigial gonadal streak, because they do not have either the ovarian organizational input of two active X chromosomes or the testicular organizational input of the Y chromosome. The absence of testicular function in turn leads to müllerian duct development, female external genitalia, and wolffian duct regression.

XY individuals who completely lack the capacity to respond to androgenic hormones because of receptor deficiency (the X-linked testicular feminization syndrome) develop testes because of the presence of the Y chromosome. They demonstrate müllerian duct regression, caused by the presence of AMH. However, they show no growth or development of the wolffian ducts nor masculinization of their external genitalia, since without receptors there is no effective testosterone or dihydrotestosterone action. The external genitalia are feminine.

XY individuals who have any one of five known defects in testosterone biosynthesis will develop testes because of the presence of the Y chromosome, and will show müllerian duct regression because of the presence of AMH. However, depending on the degree of the testosterone deficiency, the wolffian duct structures are variably underdeveloped, and the external genitalia may show simple failure of the urethral folds to fuse completely or an entirely female pattern.

XY individuals who are deficient only in the conversion of testosterone to dihydrotestosterone have a normal testis because of the presence of the Y chromosome. They demonstrate müllerian duct regression because of the presence of AMH. The development of the epididymis, vas deferens, and seminal vesicles is normal because of the presence of testosterone. However, they have external genitalia that vary from a partial to a

Table 51-1 Examples of abnormal development of the reproductive system

Genetic state	*Gonad*	*Müllerian duct*	*Wolffian duct*	*External genitalia*
XY, normal ♂	Testis	Regressed	Developed	♂
XX, normal ♀	Ovary	Developed	Regressed	♀
XO, Turner's syndrome	Streak*	Developed	Regressed	♀
XY, loss of X-linked gene for androgen receptor	Testis	Regressed	Regressed	♀
XY, deficient testosterone synthesis	Testis	Regressed	Regressed to variably developed	♀/♂
XY, deficient 5α-reductase	Testis	Regressed	Developed	♀/♂
XXY, Klinefelter's syndrome	Dysgenetic testis	Regressed	Developed	♂
XX, adrenal 21- or 11-hydroxylase deficiency	Ovary	Developed	Regressed	♀/♂

*A fibrous streak essentially devoid of germ cells.

complete female pattern because of the lack of dihydrotestosterone.

XX individuals with a deficiency of adrenal 21- or 11-hydroxylase enzymes overproduce androgens in utero (see Chapter 50, p. 953). They have ovaries because of the presence of two X chromosomes and normal müllerian duct development because of the absence of AMH. Their wolffian structures regress because of the absence of local gonadal testosterone and the relatively late exposure to adrenal androgen excess. However, depending on the severity of the enzymatic block and the degree of resultant androgen hypersecretion, their external genitalia show variable degrees of the male pattern. This ranges from mild enlargement of the clitoris to complete scrotal fusion of the labia and a persistent urogenital sinus.

Individuals with supernumerary X chromosomes develop testes if a Y chromosome is also present and ovaries if it is not. Their genital ducts and external genitalia develop normally. However, spermatogenesis and seminiferous tubule development are markedly deficient in male individuals with XXY chromosomes **(Klinefelter's syndrome).** Female individuals with XXX chromosomes may have shortened reproductive lives. The mechanisms whereby extra X chromosomes damage germ cell function are unknown. The XYY karyotype is associated with defective spermatogenesis.

Common Aspects of Gonadal Function

Pathway of Gonadal Steroid Synthesis

Both genders use a common pathway of steroid hormone biosynthesis in gonadal tissue. It is essentially identical to that of the adrenal cortex. The enzyme characteristics, localizations, and cofactor requirements are also those previously described for the adrenal glands in Chapter 50. As shown in Fig. 51-4, cholesterol, either generated by de novo synthesis from acetyl CoA or taken up from the circulating plasma pool, is the starting compound. In the gonads, in situ cholesterol synthesis may be quantitatively more important than in the adrenal glands. Cholesterol side chain cleavage (the 20, 22-desmolase step) appears to be rate limiting for synthesis of progesterone, of the androgens testosterone, dihydrotestosterone, and androstanediol and of the estrogens estradiol and estrone. The 17-hydroxylase and 17, 20 desmolase steps are catalyzed by a single molecule.

Two parallel routes to testosterone are evident, but the Δ_5 pathway from pregnenolone is favored (see Fig. 51-4). Oxidation of the A ring by the 3-β-ol-dehydrogenase-isomerase complex can take place at any level from pregnenolone to androstenediol. Only a small quantity of testosterone undergoes 5α-reduction to dihydrotestosterone and a further α-reduction of the 3-ketone position to 5α-androstanediol within the testis (see Fig. 51-4). Androgens are the obligate precursors of estrogens. The key step is aromatization of the A ring, a reaction heavily favored in the ovary and placenta. The aromatase enzyme complex is a cytochrome P_{450} localized in the endoplasmic reticulum. It sequentially hydroxylates the 19-methyl group, oxidizes it to the aldehyde, hydroxylates the 2 position, and then creates a 1-2 double bond by reduction. Following these steps, the 19-carbon is removed by decarboxylation, and the characteristic benzene ring is formed. Estradiol and estrone result from testosterone and androstenedione, respectively. The two estrogens also may be interconverted by 17-hydroxysteroid dehydrogenase.

Other Gonadal Products

Testicular Sertoli cells and ovarian granulosa cells synthesize and secrete numerous peptide and protein products that act in endocrine, paracrine, and even autocrine fashion to modulate the process of gametogenesis. Inhibins and activins are members of the same superfamily of growth regulating factors as AMH. They are constructed by combining three basic subunits in various combinations as shown in Fig. 51-5. **Inhibin** is a glycoprotein that circulates in plasma and inhibits GnRH-stimulated FSH secretion by the pituitary gland. **Activin** has the opposite action and stimulates FSH secretion. Each likely also has intragonadal actions, which are described later. **Follistatin** is another FSH-suppressing protein of entirely unrelated structure. It may act by binding activin. Insulin-like growth factor-1 (IGF-1 or somatomedin-C) and transforming growth factors α and β are also synthesized by this cell line and modulate cell growth and hormonal responses within the gonads. Leydig cells synthesize and secrete proopiomelanocortin products (see Chapter 48) and oxytocin. A peptide functionally resembling GnRH but structurally dissimilar to it is also present in the gonads. In addition, a variety of trace metal–binding proteins, steroid-binding proteins, proteases, prostaglandins, lymphokines, and extracellular matrix molecules, such as laminin, collagen types I and IV, and proteoglycans, are also produced. These have local functions in the nurturence and development of the germ cells and the later exodus of ova and sperm from their respective gonadal enclaves.

Gonadotropin Actions in the Gonads

The general framework for the hypothalamic-pituitary-gonadal axis was presented in Chapter 48(see Fig. 48-17). LH and FSH are the coordinate pituitary regulators of gonadal function. Through negative feedback, their synthesis and secretion are increased by decreases in gonadal function. Luteinizing hormone (LH) stimulates the interstitial cell line of male and female gonads pri-

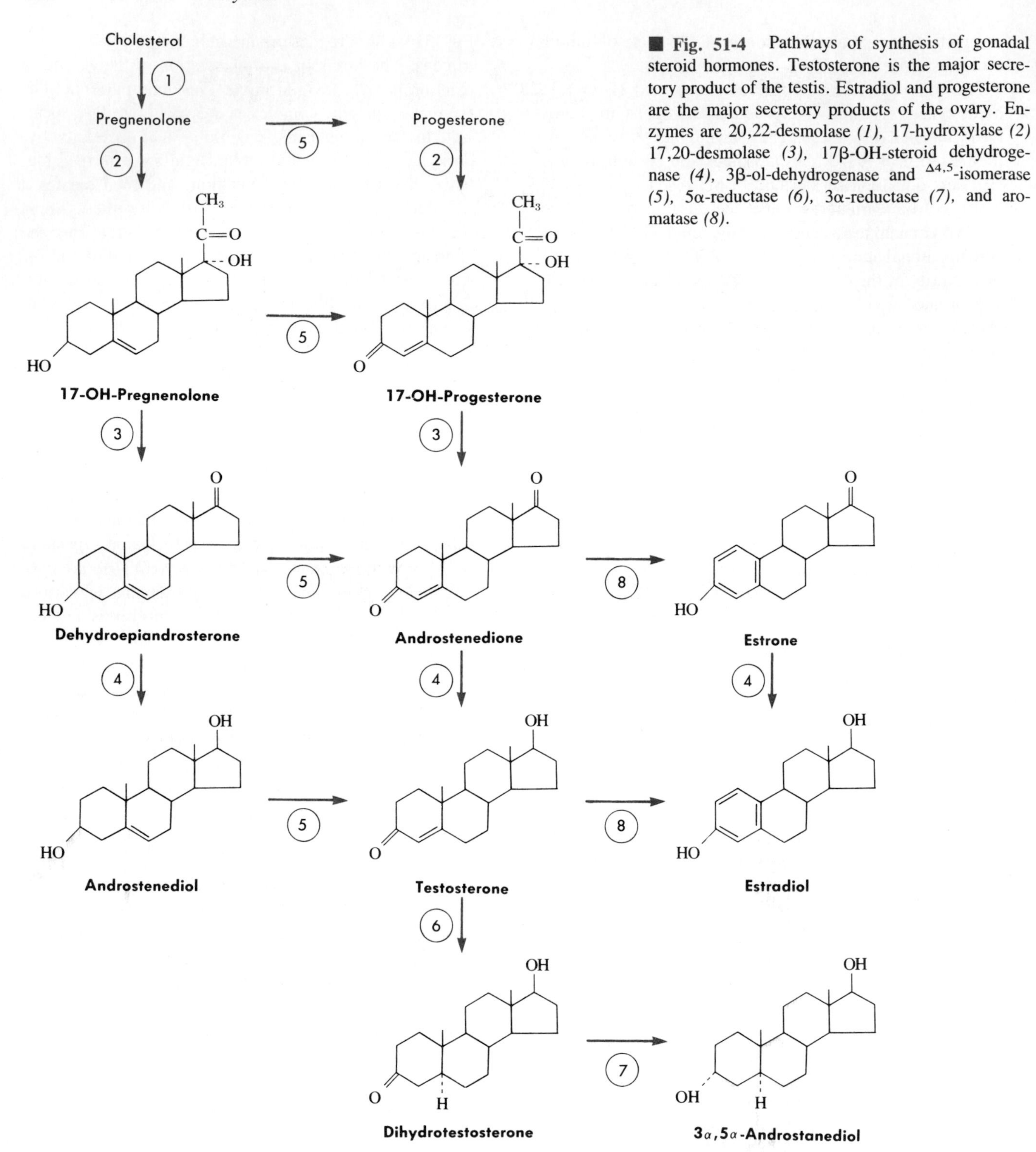

■ **Fig. 51-4** Pathways of synthesis of gonadal steroid hormones. Testosterone is the major secretory product of the testis. Estradiol and progesterone are the major secretory products of the ovary. Enzymes are 20,22-desmolase *(1)*, 17-hydroxylase *(2)* 17,20-desmolase *(3)*, 17β-OH-steroid dehydrogenase *(4)*, 3β-ol-dehydrogenase and $^{\Delta 4,5}$-isomerase *(5)*, 5α-reductase *(6)*, 3α-reductase *(7)*, and aromatase *(8)*.

marily to secrete androgens. LH binds to a plasma membrane receptor that appears to be a single polypeptide of 80,000 molecular weight and that associates in oligomers. It has a large extracellular portion, spans the plasma membrane seven times, and terminates in an intracellular carboxy tail. It acts by means of a G-protein, adenylyl cyclase, and cAMP as a second messenger.

The interaction of LH with its receptors is exquisitely sensitive. As little as 1% receptor occupancy may be sufficient for stimulation, and 5% to 10% occupancy may produce maximal cellular responsiveness to the hormone. Continued stimulation of gonadal cells by LH leads to downregulation of its receptors and reduced responsivity to the hormone. Prostaglandins have been implicated as additional intermediaries in LH action, possibly potentiating cAMP effects.

LH increases uptake and mobilization of cholesterol and its conversion to pregnenolone by stimulating the

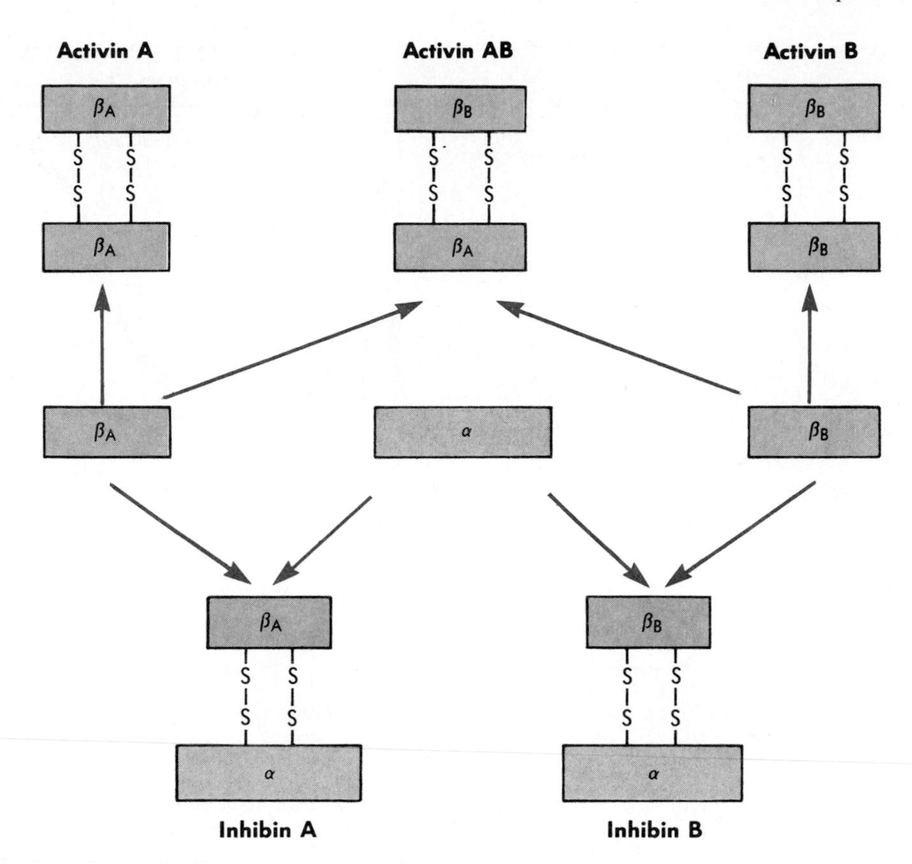

■ **Fig. 51-5** Synthesis of inhibins from a common α-subunit and two distinct β-subunits. Activins are synthesized by combining the β-subunits only.

17, 20-desmolase reaction. It also stimulates transcription of the gene for the enzyme 17-hydroxylase-17,20 desmolase and the cofactor adrenoxin.

FSH acts on ovarian granulosa cells and testicular Sertoli cells by means of a plasma membrane receptor with partial homology to the LH receptor. The subsequent increase in cAMP concentration is followed by an increase in transcription of the aromatase gene and a marked stimulation of estrogen synthesis. FSH also stimulates synthesis of inhibin and numerous other protein products of Sertoli and granulosa cells. A very important effect of FSH is to increase the number of LH receptors in granulosa cells and thereby to amplify their sensitivity to LH.

In addition to their actions on steroidogenesis, LH and FSH produce diverse metabolic effects on their target gonadal cells. Glucose oxidation and lactic acid production are increased, and this may lead to local vasodilation. The long-term tropic effects of the two hormones depend on stimulation of amino acid transport, RNA synthesis, and general protein synthesis.

■ *Age-Related Changes in Gonadotropin Secretion*

The hypothalamic-pituitary-gonadal axis is unique in that it undergoes extreme changes throughout the human life span. Although the patterns of females and males differ, certain common aspects are noteworthy.

■ *Intrauterine and Childhood Pattern*

In humans GnRH is present in the hypothalamus at 4 weeks, and FSH and LH are present in the pituitary gland by 10 to 12 weeks of gestation. A broad peak of gonadotropin concentrations occurs in fetal plasma at midgestation (Fig. 51-6). After the concentrations drop to low levels before birth, they transiently increase (more prolonged in females) again at about 2 months of age. For the rest of childhood both gonadotropins are secreted at very low levels. These changes are mirrored by fluctuations of plasma testosterone in males and estradiol in females.

■ *Puberty*

The transition from a nonreproductive to a reproductive state during puberty requires maturation of the entire hypothalamic-pituitary-gonadal axis. Before the child reaches age 10 years, plasma LH and FSH levels are low despite low concentrations of gonadal hormones and inhibin. Blockade of normally inhibitory opioid receptors does not increase plasma LH and FSH. There-

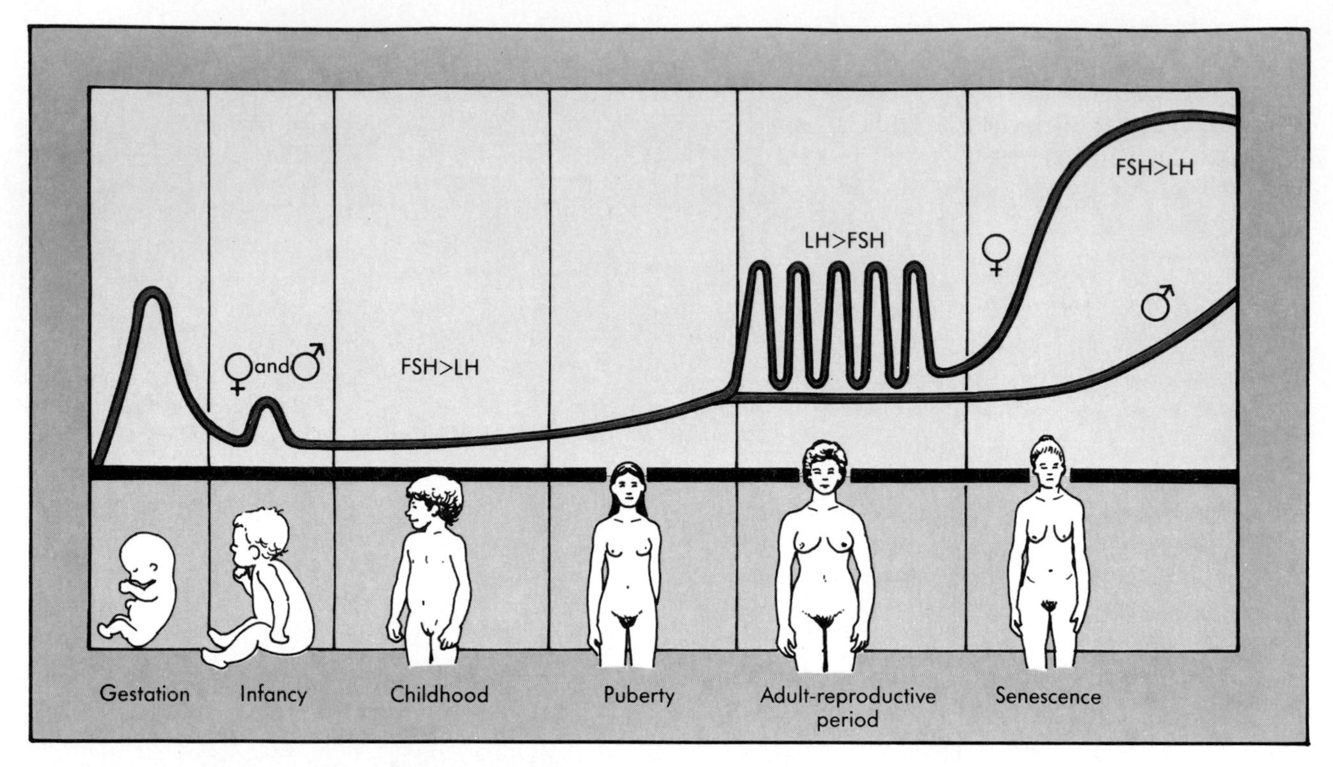

■ **Fig. 51-6** The pattern of gonadotropin secretion throughout life. Note transient peaks during gestation and early infancy and low levels thereafter in childhood. Women subsequently develop monthly cyclic bursts, with luteinizing hormone *(LH)* exceeding follicle-stimulating hormone *(FSH);* men do not. Both genders show increased gonadotropin production after age 50 years, with FSH exceeding LH.

fore either the negative feedback system is inoperative or the hypothalamus and pituitary gland are exquisitely sensitive to testosterone, estradiol, and inhibin. One factor in puberty may thus be the gradual maturing of hypothalamic neurons that leads to an increased synthesis and release of GnRH. The time and rate of onset of this maturational process may well be genetically preprogrammed because familial patterns are apparent. Other central nervous system components may influence this process. Nocturnal secretion of melatonin from the pineal gland declines from childhood to adult life, and destruction of the gland may cause premature puberty. However, it has not been established that in humans, melatonin is a normal supressor of gonadotropins during childhood or the rate of decline of melatonin is a normal regulator of the onset of puberty.

As puberty approaches, a pulsatile pattern of LH and FSH secretion appears. The ratio of plasma LH to FSH rises as the pulse frequency increases. Furthermore, during early and middle puberty, but at no other time of life, a nocturnal peak in LH secretion is observed. This then disappears as adult status is reached (Fig. 51-7). The gonad itself is not necessary for these changes in GnRH and gonadotropins to occur.

During early puberty the responsiveness of the pituitary gland to GnRH changes so that LH exceeds FSH output. This may result from increased synthesis and storage of LH in response to pulsatile GnRH secretion because the latter allows better maintenance of GnRH receptors. Although the gonadal target cells respond to LH in childhood, their responsiveness is augmented during puberty. Therefore testosterone levels in males and estradiol levels in females increase sharply. In addition, FSH stimulates a pubertal rise of inhibin levels in both genes. Thus this period can be viewed as a cascade of increasing maturation from the hypothalamic to the pituitary to the gonadal level. Once the adult pattern of gonadotropin secretion is established, the basal plasma concentrations of LH and FSH (approximately 10^{-11} molar) are similar in men and women. An important distinguishing feature between the genders is the additional establishment of a dramatic monthly gonadotropin cycle in females only, with the LH bursts greatly exceeding the FSH bursts (see Fig. 51-6).

■ *Climacteric*

In both genders a loss of gonadal responsiveness to gonadotropin stimulation occurs after the fifth decade of life. In males this is gradual, and some reproductive capacity usually persists into the eighth decade. In fe-

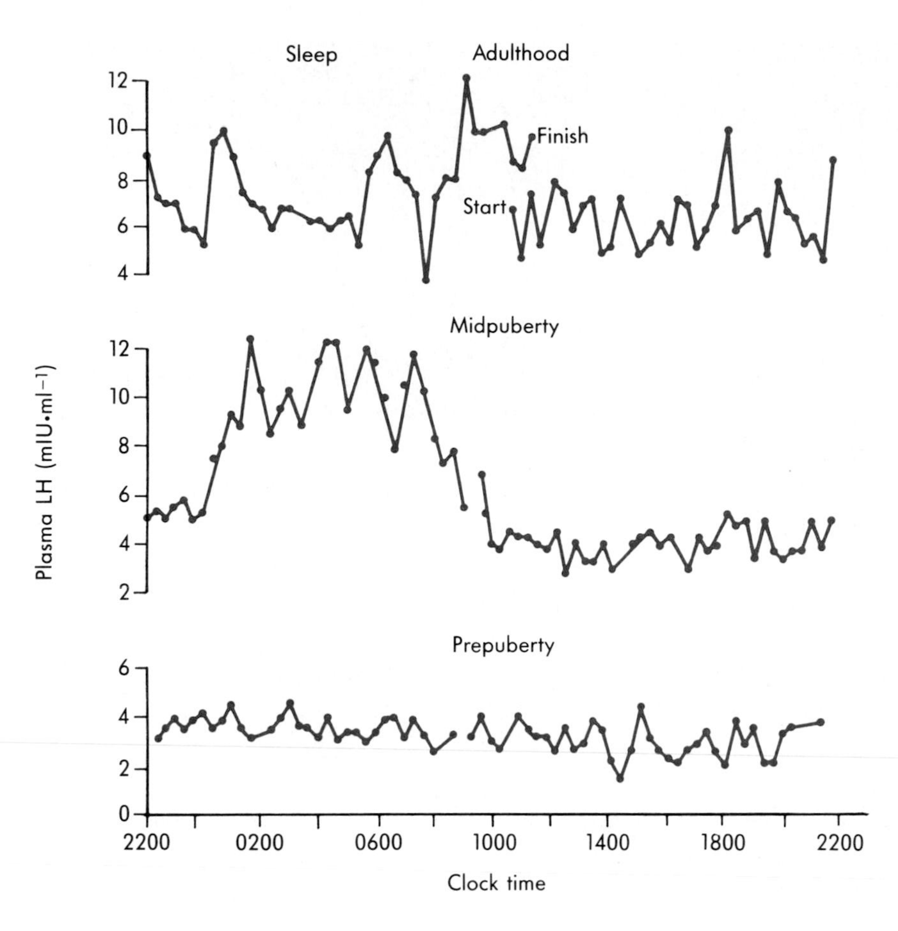

■ **Fig. 51-7** Pubertal changes in LH secretion. The pattern of secretion becomes much more pulsatile. In addition a nocturnal peak in LH appears early in puberty and then disappears when puberty is completed. Males and females both show these changes. (Redrawn from Boyar RM et al: *N Engl J Med* 287:582, 1972.)

males reproductive capacity is lost completely, and menopause occurs. In both genders, however, negative feedback leads to elevated plasma gonadotropin levels. The FSH level rises more than the LH level, and the increase is more distinct in females (see Fig. 51-6).

■ The Testes

■ Anatomy

The human testes are normally situated in the scrotum, where they are maintained at a temperature 1 to 2 ° C below that of the body core temperature. This lower temperature is essential to normal sperm production and is partly maintained by intertwined coiling of arteries and veins to facilitate heat exchange between them. Each testis weighs about 40 g and has a long diameter of 4.5 cm. The testes receive blood from the spermatic arteries, which arise directly from the aorta. The right spermatic vein drains into the inferior vena cava, whereas the left drains into the ipsilateral renal vein. Eighty percent of the adult testis is made up of the seminiferous tubules; the remaining 20% is composed of supportive connective tissue, throughout which the Leydig cells are scattered (Fig. 51–8). The seminiferous tubules are a coiled mass of loops; each loop begins and ends in a single duct, the tubulus rectus. The tubuli recti, in turn, anastomose in the rete testis and eventually drain via the ductuli efferentes into the epididymis. The latter constitutes a storage and maturation depot for spermatozoa. From the epididymis the spermatozoa are carried via the vas deferens and ejaculatory duct into the penis, to be emitted during copulation.

The structure of the adult seminiferous tubule is complex (Fig. 51-8). Each seminiferous tubule is bounded by a basement membrane, separating it from the Leydig cells, the peritubular (myoid) cells, and the surrounding connective tissue. Immediately beneath the basement membrane are spermatogonia (germ cells) and Sertoli cells. As the spermatogonia divide and develop successively into spermatocytes and spermatids, a column of cells is formed that reaches from the basement membrane to the lumen of the tubule and culminates in the spermatozoa. In contrast, the cytoplasm of each Sertoli cells extends all the way from the basement membrane to the lumen. This cytoplasm invests the spermatogonia and its germ cell line successors (see Fig. 51-8).

Special processes of the Sertoli cell cytoplasm fuse into tight junctions, which create two compartments of intercellular space between the basement membrane and the lumen of the tubule. The spermatogonia and early primary spermatocytes lie within the proximal basal compartment, whereas the later spermatocytes and subsequent stages in spermatozoon development lie in the distal adluminal compartment. This separation continues a barrier to the blood, which is partly established by

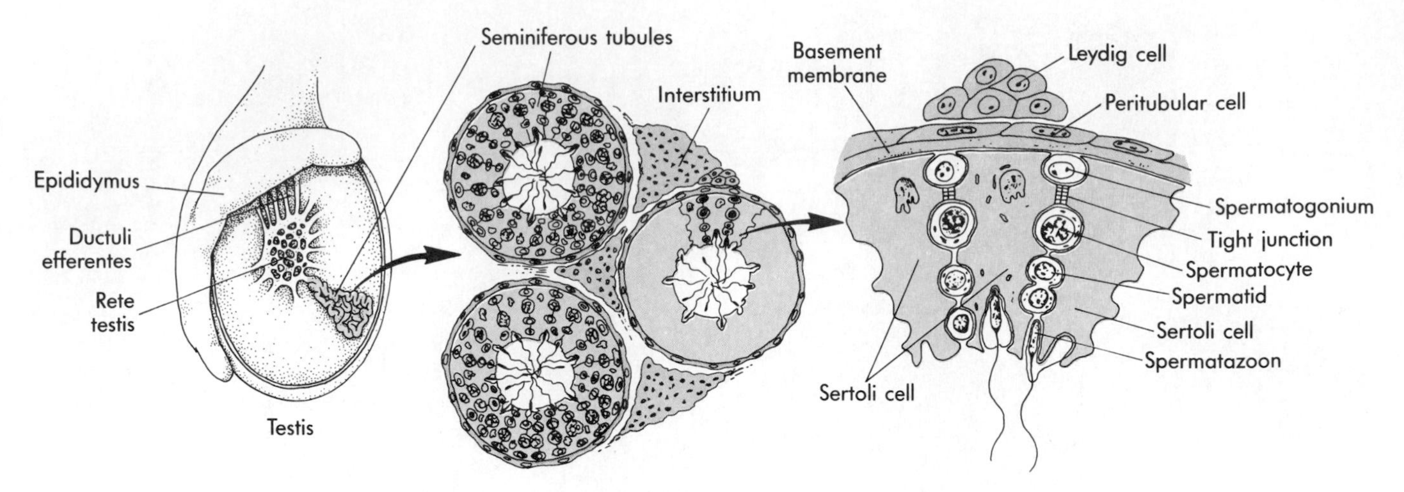

■ **Fig. 51-8** Schematic representation of the architecture of the testis. Note that the Leydig cells and peritubular cells are separated from the spermatogenic tubules. Within the latter the germ cell line is completely invested by cytoplasm of the surrounding Sertoli cells. In addition tight junctions between adjacent Sertoli cells separate the ancestral spermatogonia from their descendent spermatocytes, spermatids, and spermatozoa. Thus a blood testis barrier effectively filters plasma permitting only selected substances to reach the developing germ cells from Sertoli cell cytoplasm. (Redrawn from Skinner MK: Cell-cell interactions in the testis, *Endocr Rev* 12:45-77, 1991. From The Endocrine Society.)

overlapping peritubular cells and by the basement membrane. In even more discriminating fashion, the cytoplasm of the adjacent Sertoli cells excludes a variety of circulating substances from the intercellular fluid that bathes the maturing germ cells and also from the seminiferous tubular fluid in which the spermatozoa will begin their outward journey. Another consequence of this barrier is to prevent late spermatogenic products from reaching the blood stream, where, if recognized as foreign, they could evoke rejection mechanisms.

In short, the testis consists of separate but interacting functional elements (Figs. 51-8 and 51-9). The first element is the Leydig cells, which are pure steroid-secreting cells whose major product, testosterone, has both vital local effects on germ cell replication and actions on distant target cells. The second element consists of peritubular myoid cells (see Fig. 51-8), which secrete local regulatory products and may produce physical effects on the tubules and the vasculature. The third element is the seminiferous tubules, which carry out the process of spermatogenesis while bathed in locally generated testosterone and Sertoli cell products.

■ *The Biology of Spermatogenesis*

The production of sperm is an ongoing process throughout the reproductive life of the male. Approximately 100 to 200 million sperm are produced daily. In generating this large number of sperm, the spermatogonia must renew themselves by cell division. This situation differs fundamentally from that in the female, who at birth has a fixed number of oocytes, which decreases throughout her life.

Each spermatogonium can give rise to 64 spermatozoa. The extraordinary metamorphosis from spermatogonium to spermatozoon is depicted in Fig. 51-10. The first two mitotic divisions of a spermatogonium give rise to four cells: a single resting cell *(Ad)* that will eventually serve as the ancestor of a later generation of sperm and three active cells *(Ap)*. The latter divide by further mitoses to yield type B spermatogonia, which then give rise to a number of primary spermatocytes. These cells enter the prophase of meiosis, the first reduction division, in which they remain for about 20 days. This process occurs within the basal compartment.

The complex process of chromosomal reduplication, synapsis, crossover, division, and separation is reflected histologically in the changing appearance of the primary spermatocytes (see Fig. 51-10). After completion of meiosis (late pachytene spermatocyte, see Fig. 51-10), their daughter cells, the secondary spermatocytes, immediately divide again in the adluminal compartment. The products, called spermatids, each contain 22 autosomes and either an X or a Y sex chromosome. The spermatids lie near the lumen of the seminiferous tubule. They are attached to the abutting Sertoli cells by specialized junctions. The spermatocytes and spermatids of each generation are connected with each other through intercellular bridges. In a process termed **spermiogenesis,** spermatids undergo nuclear condensation, shrinkage of cytoplasm, formation of an acrosome, and development of a tail to emerge as flagellated spermatozoa. The spermatozoa are then extruded into the lumen

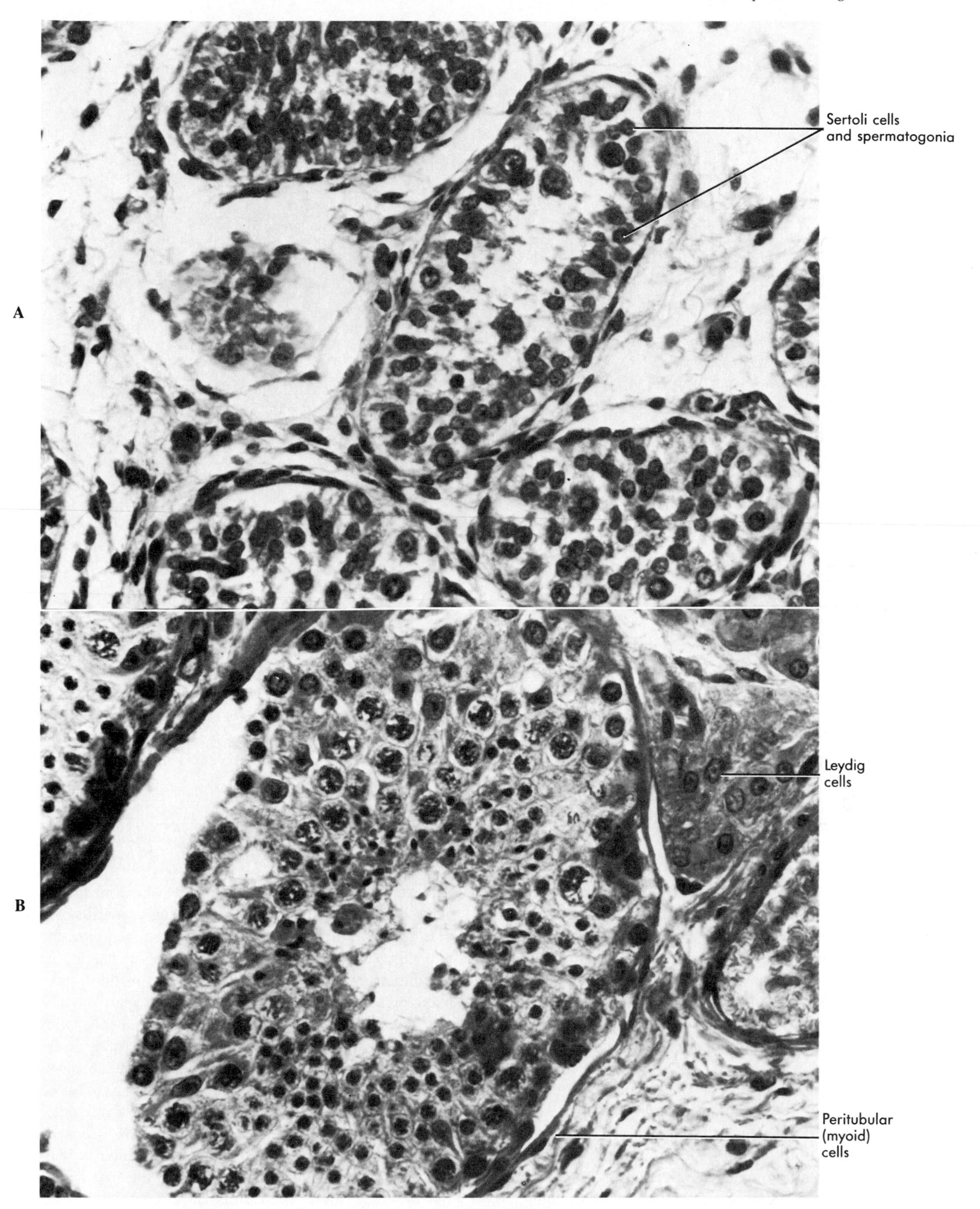

■ **Fig. 51-9** Histological sections of the testis from prepubertal **(A)** and postpubertal **(B)**, males. Note absence of Leydig cells and of active spermatogenesis prior to puberty. (Courtesy Dr. Howard Levin.)

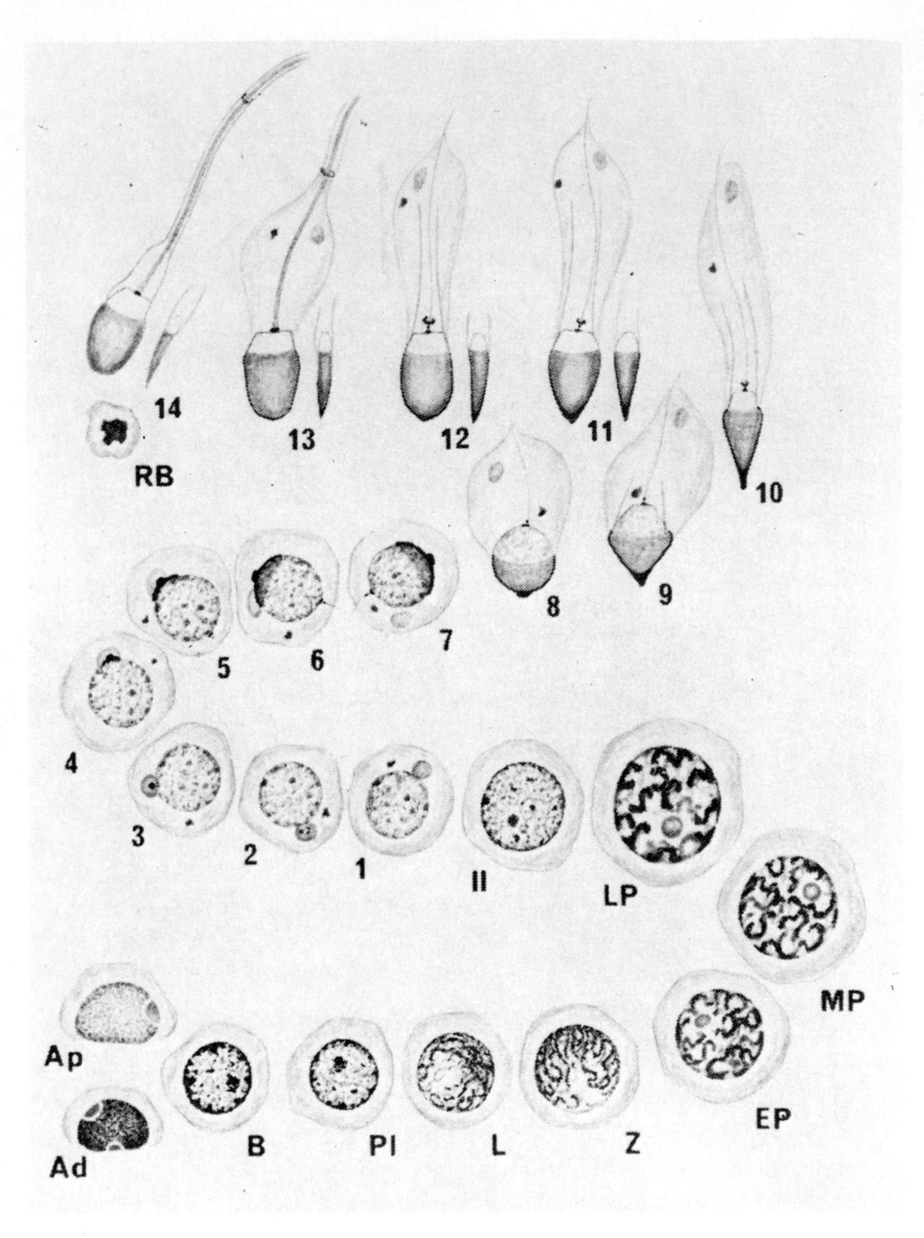

■ **Fig. 51-10** Development of sperma-tozoa from spermatogonia in the subhuman primate. *Ad,* Dark spermatogonium; *Ap,* pale spermatogonium; *B,* type B spermatogonium; *Pl,* preleptotene primary spermatocyte; *L,* Leptotene spermatocyte; *Z,* Zygotene spermatocyte; *EP, MP, LP,* early, middle, and late pachytene spermatocytes; *II,* secondary spermatocyte; *1-7,* early spermatids; *8-13,* late spermatids with progressive formation of flagellum; *14,* spermatozoon; *RB,* residual body. (Redrawn from Clermont Y: *Physiol Rev* 52:198, 1972).

of the tubule by a process called **spermiation,** during which most of the cytoplasm of the spermatozoa remains embedded in the cytoplasm of a Sertoli cell.

Once the spermatozoa have entered the seminiferous tubules, they consist of linear structures with several components (Fig. 51-11). The head contains the nucleus and an acrosomal cap in which are concentrated hydrolytic and proteolytic enzymes, which facilitate penetration of the ovum and possibly also the mucous plug of the female cervix. The middle piece, or body, contains mitochondria, which generate the motile energy of the spermatozoon. The chief piece of the tail contains stored ATP and pairs of contractile microtubules down its entire length, one pair in the center and nine pairs around the circumference. Crossbridging arms contain **dynein,** a magnesium-dependent ATPase, which catalyzes the conversion of ATP energy into a sliding movement between the microtubules. This imparts flagellar motion to the spermatozoa. Both cAMP and Ca^{++} are involved in regulating sperm motility.

In a human approximately 70 days are required for the entire sequence of development from spermatogonia to spermatozoa. However, individual resting spermatogonia do not start into the process of spermatogenesis randomly. Cycles of spermatogenesis exist with distinct cycle times. Groups of adjacent resting spermatogonia initiate a new cycle about every 16 days, thus constituting one "generation." At about the same time that the primary spermatocytes of one cycle enter prophase, a second cycle of spermatogonia is activated. A third cycle begins approximately synchronously with the appearance of the spermatids from the first cycle. By the time these spermatids have completed their transformation into spermatozoa, a fourth cycle of spermatogonia has been started.

Around the circumference of any individual seminif-

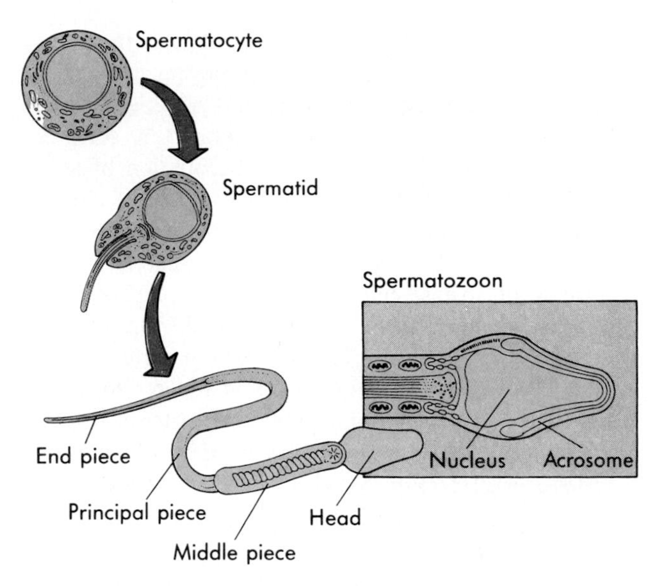

■ **Fig. 51-11** Schematic representation of the morphological alterations in the development of the spermatozoon from the spermatocyte.

erous tubule several spermatogenic cycles may be in process simultaneously. Within each cycle approximately six stages of cellular development can be identified histologically. This gives rise to several specific cellular constellations that exist side by side. In some mammals, but not in humans, spermatogenic cycles are repeated in a defined topographical relationship to each other along the length of each seminiferous tubule. This has been termed the **wave of spermatogenesis.**

There may not be total separation of the individual germ cells that constitute the successive descendents of type B spermatogonia lying within the adluminal compartment of the tubule. Continuity of cytoplasm and possibly cell-to-cell intercommunication may exist. Because of these possibilities and because of the regular topographical association of particular stages of spermatogenesis in neighboring cycles, products of germ cells in one stage of spermatogenesis may initiate or regulate events in other stages.

■ *Sexual Functioning*

After spermiation, the spermatozoa reach the epididymis, which they traverse in a variable period of up to 24 days. During this time, they undergo further maturation and gain motility. By the time they reach the vas deferens, they have lost all of their cytoplasm. It is still unclear to what extent this process is preprogrammed within the spermatozoa or to what extent it depends on critical functions of the epididymal cells.

Spermatozoa are initially drawn into the epididymis by seminiferous tubular fluid currents generated by the peritubular myoid cells or by contraction of the testicular capsule. The epididymis is lined by specialized epithelial cells and is surrounded by contractile muscle cells. The growth and differentiation of the epididymis, as well as the motility and fertility of the migrating sperm, depend on androgens. Marked changes in fluid osmolality, in electrolytes, and in the concentrations of many small molecules occur progressively within the length of the epididymis. This suggests homologies with the function of the renal tubules. A number of proteins provided by epididymal and seminiferous tubular fluid bind to the membranes of sperm and enhance their ultimate function. These proteins include a forward-mobility protein, an acrosomal stabilizing or inhibiting factor, and a zona pellucida binding protein. All told, the total amount of sperm stored in the epididymis is about the equivalent of a single ejaculate or a single day's production.

Delivery of spermatozoa into the female genital tract occurs by ejaculation from the vas deferens. To the contents of the vas deferens are added successive fluids. The initial secretions from the prostate gland contain citrate, calcium, zinc, and acid phosphatase. The alkalinity of prostatic fluid helps neutralize the acid pH of the semen and the vaginal and cervical secretions. The terminal portion of the ejaculate is composed largely of secretions from the seminal vesicles. These contain fructose, an important oxidative substrate for the spermatozoa. Seminal vesicle secretions also contain prostaglandins, which may stimulate contractions of the uterus and fallopian tubes—helping propel the spermatozoa toward the ovum.

Seminal fluid also contains LH, FSH, prolactin, testosterone, estradiol, inhibins, endorphins, oxytocin, kallikreins, relaxin (see later discussion), proteases, plasminogen activator, and sperm-coating proteins. The concentrations of these substances are usually higher than those in plasma; this suggests that they have been secreted by cells of the genital tract.

Delivery of the sperm into the vagina requires penile erection, which is caused by filling of the venous sinuses with blood. This converts the penis into a firm organ for penetration. The venous sinuses are filled by the coupling of arteriolar dilation with venous constriction: the erection is under parasympathetic control. Ejaculation is effected by sympathetic nervous system impulses. A typical emission contains 200 million to 400 million spermatozoa in a volume of 3 to 4 ml. Once within the vagina, the spermatozoa's rate of flagellated movement is up to 44 mm/minute. The life span of sperm in the female genital tract is approximately 2 days.

Ejaculated sperm cannot immediately fertilize an ovum. In vivo fertilization can take place only after the sperm has been in the milieu of the female reproductive tract for 4 to 6 hours, a process termed **capacitation.** In vitro human fertilization can take place after the spermatozoa have been washed free of seminal fluid. This suggests that materials in the female genital tract either remove or neutralize substances that are located on the sperm surface and that would otherwise prevent union with an ovum. In the course of capacitation, cholesterol is withdrawn from the sperm membrane, within which the surface proteins redistribute; in addition, calcium influx occurs and motility becomes more whiplike.

Although the process of capacitation is poorly understood, it results in unique patterns of sperm motility that may enhance the penetration of the ovum. In addition, and most important, capacitation results in the acrosomal reaction, which consists of fusion of the acrosomal membrane with that of the outer sperm membrane. The membrane fusion creates pores through which the acrosomal hydrolytic and proteolytic enzymes can escape. These then create a path for penetration of the sperm through the protective investments of the ovum.

Hormonal Regulation of Spermatogenesis

Knowledge of the endocrine mechanisms that govern human spermatogenesis is incomplete. Clearly, adult functioning of the GnRH-LH/FSH-testicular axis, as illustrated in Fig. 48-17, is essential for spermatogenesis to occur at all. *Of critical importance is the pulsatile release of GnRH and consequent arrival of LH and FSH at their target cells.* Studies in men with congenital GnRH deficiency show that administration of the releasing hormone will only induce fertility if it is given in appropriately sized pulses and timed intervals, but not if it is given continuously.

The prepubertal testis contains essentially only resting spermatogonia; neither Leydig cells nor peritubular cells of the adult myoid character are present (see Fig. 51-9). The Sertoli cells are quiescent and do not exhibit cyclic structural or biochemical alterations. Pubertal activation of gonadotropin secretion leads to dramatic and very complex changes. The proximity of several stimulated endocrine cell types (with multiple secretory products) to each other and to the germinal cell line creates many potential and observed paracrine and possibly autocrine effects. These effects are in addition to central feedback actions. The specific or critical nature of each such effect and the point in spermatogenesis at which it operates is still difficult to determine.

During fetal life the transient midgestation surge of pituitary FSH, LH, and testosterone release may stimulate the transformation of the primordial germ cells into resting spermatogonia. Withdrawal of fetal gonadotropins then leaves the spermatogonia in suspended development throughout childhood, possibly through operation of a local meiosis inhibitor. Shortly after FSH secretion begins to rise at the onset of puberty, the spermatogonia are activated. *FSH stimulates the Sertoli cells whose functions are, in turn, required for initial germ cell mitotic and early meiotic activity.* These will be described in detail later. LH stimulates the Leydig cells to secrete testosterone, which presumably diffuses across the basement membrane and into Sertoli cells. The latter contain androgen receptors. In some unknown manner, the high local concentration of testosterone (100-fold greater than in plasma) is essential for completion of the later stages of spermatogenesis. In men who lack LH, for example, testosterone replacement to normal systemic plasma levels cannot sustain spermatogenesis. It is possible though unproven that testosterone also enters and acts within germ cells; the latter have recently been shown in rodents to contain both the active metabolite dihydrotestosterone and the 5-α-reductase enzyme necessary for its production from testosterone (see Fig. 51-4). Conversion of testosterone to estradiol within the Sertoli cell makes estradiol available as another possible local spermatogenesis modulator that is ultimately derived from LH. LH may also act to stimulate the final step of spermiation.

Once regular spermatogenesis has been established during puberty, it can continue to a slight extent in adults with very low or absent levels of both FSH and LH (Fig. 51-12), provided testosterone is present in high amounts. Under these circumstances, the number of sperm is markedly reduced, but the sperm are normal in appearance. Selective restoration of either FSH alone or LH alone can increase sperm numbers (see Fig. 51-12), but both seem to be required to reach normal levels. Still other studies have shown that in some men, after a suitable period of exposure to FSH, LH alone can maintain sufficient sperm production (15 to 60 million per ejaculate) to permit fertility. Neither FSH nor LH appears to act directly on germ cells, i.e., they act only on the intervening Sertoli and Leydig cells, respectively.

Other pituitary hormones may play subsidiary roles. Prolactin receptors are present on Leydig cells. This hormone increases LH receptors and synergizes with LH to stimulate androgen production. Growth hormone deficiency clearly delays the onset of reproductive function. The role of growth hormone may be to stimulate local production of insulin-like growth factors by Sertoli cells.

Despite the cyclic nature of spermatogenesis locally, the testis as a whole is continuously releasing spermatozoa. Furthermore, even though gonadotropin release is pulsatile, the mean daily plasma levels of FSH and LH are essentially constant in adult men. It is conjectural whether local differences in gonadotropin receptors or differences in peritubular myoid cell control of local capillary blood flow that restricts gonadotropin availability are responsible for the cyclic and topographical nature of the spermatogenic process. The situation is clearly different from that of the ovary, in which a phasic pattern of gonadotropin secretion, highlighted by a single distinct burst, produces the single monthly release of an ovum.

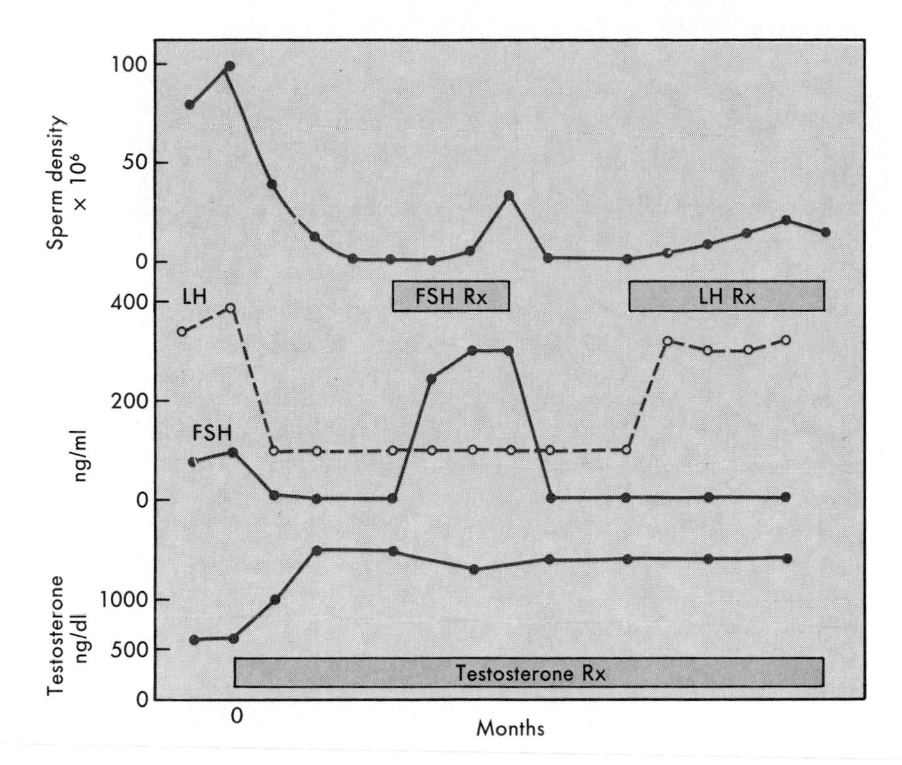

■ **Fig. 51-12** The individual effects of FSH and LH on human sperm production. Normal men were given sufficient testosterone to suppress endogenous FSH and LH secretion by negative feedback. As a result sperm density declined to very low levels. Selective restoration of either FSH or LH individually raised sperm levels. However, neither gonadotropin alone could return sperm production to normal. (Redrawn from Matsumoto AM et al: *J Clin Invest* 72:1005, 1983, and from Matsumoto AM et al: *J Clin Endocrinol Metab* 59:882, 1984.)

The Sertoli cells and their responses to FSH and testosterone are crucial to spermatogenesis. Before puberty, their function is unknown, other than to secrete AMH during sexual differentiation. After puberty the Sertoli cells do not undergo any further cell divisions. Each is in contact with up to five other Sertoli cells and with an estimated 47 germ cells in different stages of development. Ectoplasmic processes from Sertoli and germ cells invaginate into each other's plasma membranes. In close association with the cycle of spermatogenesis, Sertoli cells undergo regular changes in the activity and shape of the nucleus; in the size, shape, and branching of the cytoplasmic processes; in the concentrations of lipid and glycogen; in mitochondrial function; and in enzyme content. These changes relate to the processing of the germ cells in such an intimate and regular manner as to suggest that they respond in part to signals from the germ cells.

The cytoplasmic processes of the Sertoli cells, extending from the basement membrane to the lumen of the seminiferous tubule, act as conduits, between which the various stages of germ cells move in their passage to the lumen (see Fig. 51-8). As spermatocytes mature, new tight junctions between the investing Sertoli cells develop behind them and the old tight junctions ahead of them "unzip." In this way, spermatocytes pass from the basal to the adluminal compartment without breaking the integrity of the blood-testis barrier formed by Sertoli cell cytoplasm. The Sertoli cell cytoplasm acts as a filter, permitting only certain substances, presumably those advantageous to spermatogenesis, to reach the spermatocytes.

A wide variety of products are synthesized and secreted by the Sertoli cells in response to FSH. Some of these secretions are directed vectorially into the lumen of the seminiferous tubule (see Fig. 51-8). FSH induces the enzyme aromatase and stimulates estradiol production by Sertoli cells from androgen precursors of Leydig cell origin. In response to FSH and testosterone acting synergistically, an **androgen-binding glycoprotein (ABP)** with a molecular weight of 90,000 and an amino acid sequence identical to that of the circulating sex steroid–binding globulin (but with a different carbohydrate composition) is synthesized. ABP binds testosterone, dihydrotestosterone, and estradiol with high affinity and thereby regulates their availability to germ cells in the seminiferous tubular fluid and to sperm within the epididymis (Figs. 51-8 and 51-13). ABP also may regulate the inhibitory effect of estradiol on Leydig cell testosterone synthesis (see Fig. 51-13).

Inhibin, activin, and various growth factors are also synthesized under the influence of FSH and testosterone. The loss of the spermatogenic cells specifically leads to the loss of inhibin secretion and suggests that a signal from spermatogenic cells also modulates synthesis of inhibin by the Sertoli cells (see Fig. 51-13). In addition to their central feedback roles of inhibiting and stimulating FSH secretion, respectively, inhibin and activin may also have reciprocal local actions on the neighboring cells (see Fig. 51-13). For example, inhibin increases whereas activin decreases testosterone secretion by the Leydig cells; thus indirectly FSH can also influence Leydig cell function by modulating production of inhibin and activin. FSH promotes the availability of iron, copper, vitamin A, and crucial sphingolipids to the germ cells by stimulating synthesis of

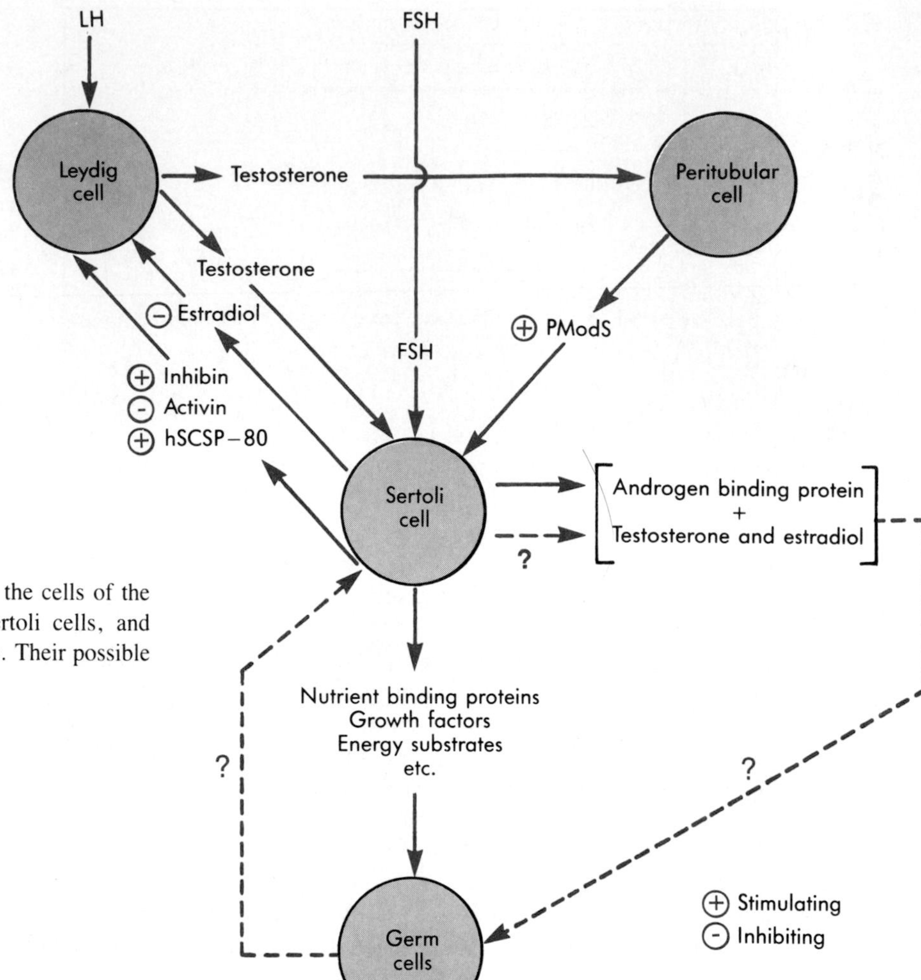

■ **Fig. 51-13** Potential interactions among the cells of the testis. hSCSP-80 is a protein product of Sertoli cells, and PMods is a protein product of peritubular cells. Their possible effects are described in the text.

their binding proteins. The latter then extract their respective ligands from plasma and transfer them to the germ cells. FSH also increases the glucose metabolism of Sertoli cells, and the resulting pyruvate and lactate are efficient energy substrates for the spermatogenic cells. Finally, FSH-stimulated proteases and plasminogen activator likely play roles in spermiation.

Either by mechanical or chemical means, the Sertoli cell also is responsible for the ejection of the spermatozoa into the lumen. In this process the nucleus of the spermatozoon is oriented toward the base of the tubule. The bulk of the cytoplasm then is squeezed out past the nucleus and shed as the residual body, while the spermatozoon is cast free. The residual body and other fragments are then phagocytosed by the Sertoli cells and subsequently degraded.

Other paracrine interactions may be important in maintaining the proper testicular environment in support of spermatogenesis (see Fig. 51-13). Thus testosterone from the Leydig cells stimulates differentiation and proliferation of the peritubular myoid cells. The latter secrete a protein, currently known as PModS, that stimulates Sertoli cell function. The Sertoli cells secrete several products that modulate Leydig cell growth and steroid secretion. Each of these pathways may vary in functional activity and significance at different points in the cycle of spermatogenesis.

■ *Secretion and Metabolism of Androgens*

Testosterone, the major androgenic hormone, is synthesized as described previously (see Fig. 51-4). Its synthesis and release by the Leydig cells are regulated by LH; therefore plasma testosterone levels show small pulses throughout the day (see Fig. 48-15). There is, in addition, a superimposed diurnal trend, such that plasma testosterone is about 25% lower at 8:00 PM than at 8:00 AM. The response to prolonged stimulation with exogenous LH is a rapid rise in plasma testosterone, followed by a brief decline and then a second plateau. The temporary decrease may be caused by downregulation of LH receptors by the peak concentrations of gonadotropin.

Testosterone also gives rise to two other potent androgens: dihydrotestosterone and 5α-androstanediol

Table 51-2 Turnover of gonadal steroids in adult men

Steroid	*Plasma concentration (ng/dl)*	*Blood production rate (μg/day)*	*Metabolic clearance rate (L/day)*
Testosterone	650	7000	1100
Dihydrotestosterone	45	300*	600
5α-Androstanediol	12	200*	1800
Androstenedione	120	2400	2000
Estradiol	3.0	50†	1700
Estrone	2.5	60†	2500

*About 60% to 80% produced peripherally from testosterone.
†About 80% to 90% produced peripherally from testosterone and androstenedione, respectively.

(see Fig. 51-4). The major portion of circulating dihydrotestosterone and androstanediol is derived from the reduction of testosterone in peripheral tissues. The plasma levels, blood production rates, and metabolic clearances of these androgens are shown in Table 51-2. The testosterone precursor, androstenedione, also is secreted in major amounts by the Leydig cells (see Fig. 51-4), but it contributes little per se to androgen action. The two estrogens—estradiol and estrone—are produced in significant amounts in men. However, only a trivial fraction of the daily production is by direct testicular secretion in response to LH. The majority is derived from circulating testosterone and androstenedione by aromatization, largely in the adipose tissue and liver.

Leydig cell function varies distinctively during the life span of the individual. As shown in Fig. 51-14, plasma testosterone rises to levels of 400 ng/dl in the fetus at the time that the external genitalia are undergoing differentiation to the masculine pattern. By birth, however, these levels have declined to less than 50 ng/dl. Very soon thereafter plasma testosterone begins to rise, reaching a peak of 150 to 200 ng/dl at 4 to 8 weeks of age. Its physiological significance is not known. Plasma testosterone again falls to low levels and remains so throughout childhood, corresponding to the absence of Leydig cells (see Fig. 51-9).

At about age 11 plasma testosterone begins a steep rise, reaching an adult plateau at about age 17 (see Fig. 51-14). This is sustained for about 50 years. During the seventh and eighth decades of life plasma testosterone gradually declines because of loss of Leydig cell responsiveness to stimulation. Although decreasing testosterone levels may be associated with a decline in libido and a decrease in bone and muscle mass, spermatogenesis itself is remarkably well preserved in most octogenarians.

Only 1% to 2% of circulating testosterone is in the free form. Sixty-five percent of testosterone is bound to a liver-derived sex steroid–binding globulin (SSBG), also known as testosterone-estradiol-binding globulin. This is a β-globulin glycoprotein with a molecular weight of 88,000. Most of the remainder is bound to albumin and other proteins. SSBG also binds dihydrotestosterone and 5α-androstanediol. Only the free and the loosely bound albumin fractions of testosterone and the other androgens are biologically active. The SSBG-bound fractions serve

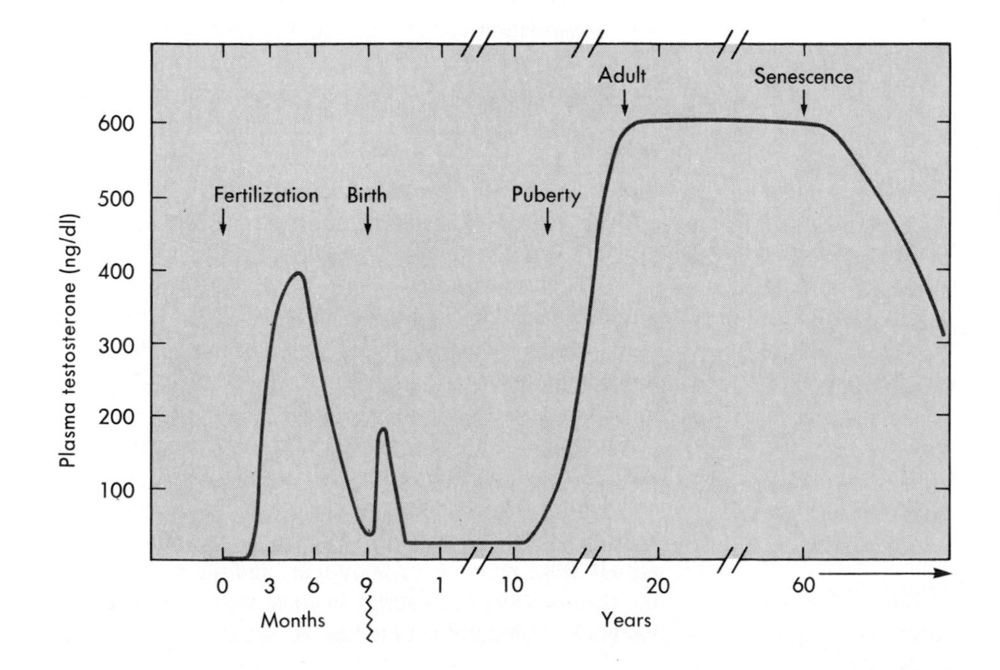

Fig. 51-14 Plasma testosterone profile during the life span of a normal male. (Redrawn from Griffin JE et al: *The testis.* In Bondy PK, Rosenberg LE: *Metabolic control and disease,* Philadelphia, 1980, WB Saunders: and Winter, JSD et al: Pituitary-gonadal relations in infancy, *J Clin Endocrinol Metab* 42:679, 1976.)

as circulating reservoirs, similar to those of thyroid hormone and cortisol.

The concentration of SSBG is increased by estrogens and is decreased by androgens. Reciprocally, then, estrogen reduces the percentage of free testosterone, whereas androgen increases it. About 1% of the daily production of testosterone (70 μg) is excreted daily in the urine as glucuronide. Most of the remainder is metabolized to two 17-ketosteroid products that are excreted in the urine (Fig. 51-15). In men only 30% of the total urinary 17-ketosteroids derives from testosterone; most arises from adrenal androgen precursors. Therefore measurement of plasma testosterone (and occasionally urine testosterone) is the mainstay for assessing Leydig cell function.

The extratesticular effects of testosterone and related androgens can be divided into two major categories: those pertaining specifically to reproductive function and secondary sexual characteristics and those pertaining more generally to stimulation of tissue growth and maturation. Similar intracellular mechanisms are involved in both categories. In general, the model for steroid hormone effects is applicable (see Fig. 44-11).

Testosterone diffuses freely into cells. In many but not all target cells it rapidly undergoes reduction to dihydrotestosterone (DHT) and, in some, to 5α-androstanediol (see Fig. 51-4). The relevant hydroxysteroid dehydrogenases are microsomal in location and employ NADPH as the reductant. DHT is much more potent than testosterone in some biological actions. However, all three steroids bind to a single nuclear protein of the steroid receptor superfamily. The absolute requirement for this androgen receptor is best illustrated by the syndrome of **testicular feminization** in humans and mice. Lacking the gene for androgen receptor synthesis, XY individuals with testes show complete failure to masculinize their genital ducts or external genitalia.

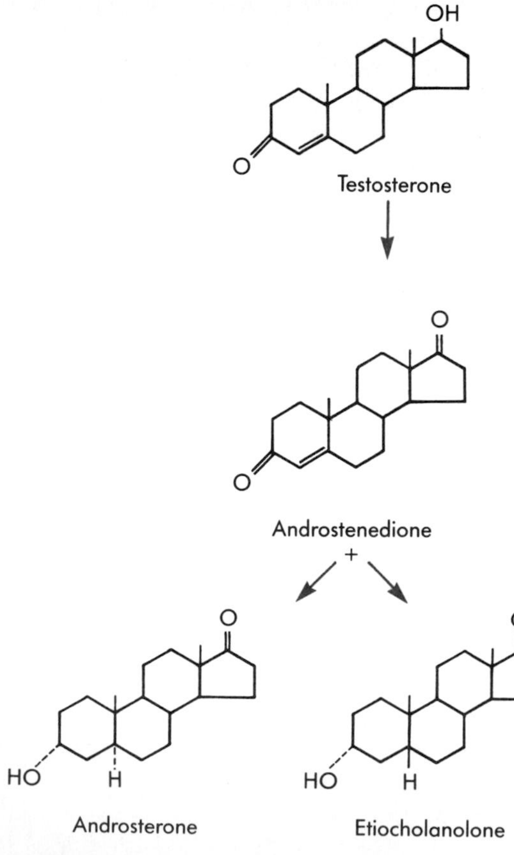

Testosterone

Androstenedione

Androsterone

Etiocholanolone

■ **Fig. 51-15** Metabolism of testosterone.

The androgen hormone–receptor complex moves into the nucleus, where it interacts with chromosomal DNA and nuclear proteins. The result is a marked stimulation of RNA polymerase, of various messenger RNAs, and of the synthesis of proteins. In addition, the activities of enzymes concerned with DNA synthesis, such as thymidine kinase and DNA polymerase, are increased. Virtually all actions of androgens are blocked by inhibitors of RNA or protein synthesis. Therefore they require induction of new enzyme molecules, as opposed to allosteric or covalent activation of existing enzyme molecules.

In target tissues, such as the prostate gland and seminal vesicles, polyamine (e.g., spermine and putrescine) synthesis is stimulated by androgens, and these compounds in turn enhance RNA synthesis. Androgens also stimulate remarkable growth of these accessory organs of reproduction, characterized by hypertrophy and hyperplasia of the epithelial cells, stromal components, and blood vessels.

By far the major circulating androgen is testosterone (Table 51-2), which can in part be considered a prohormone for dihydrotestosterone and 5α-androstanediol, much as thyroxine is a prohormone for triiodothyronine. However, testosterone definitely has intrinsic hormonal activity of its own in fetal and adult tissues that lack the enzyme 5α-reductase. An interesting source of such information comes from studies of XY individuals with testes who have congenital 5α-reductase deficiency and cannot produce DHT. These individuals have feminized external genitalia at birth, but during puberty they undergo selective masculinization in response to the rising testosterone secretion and produce sperm.

A classification of androgen effects, according to the probable actual effector molecule is shown in Fig. 51-16. DHT is specifically required in the fetus for the differentiation of the genital tubercle, genital swellings, genital folds, and urogenital sinus into the penis, scrotum, penile urethra, and prostate, respectively. It is required again during puberty for growth of the scrotum and prostate and stimulation of prostatic secretions. DHT or 5α-androstanediol stimulates the hair follicles and produces the typical male pattern. This consists of beard growth, a diamond-shaped pubic escutcheon, relatively large amounts of body hair, and the recession of the temporal hairline, which in some men culminates in baldness. Increased production of sebum by the seba-

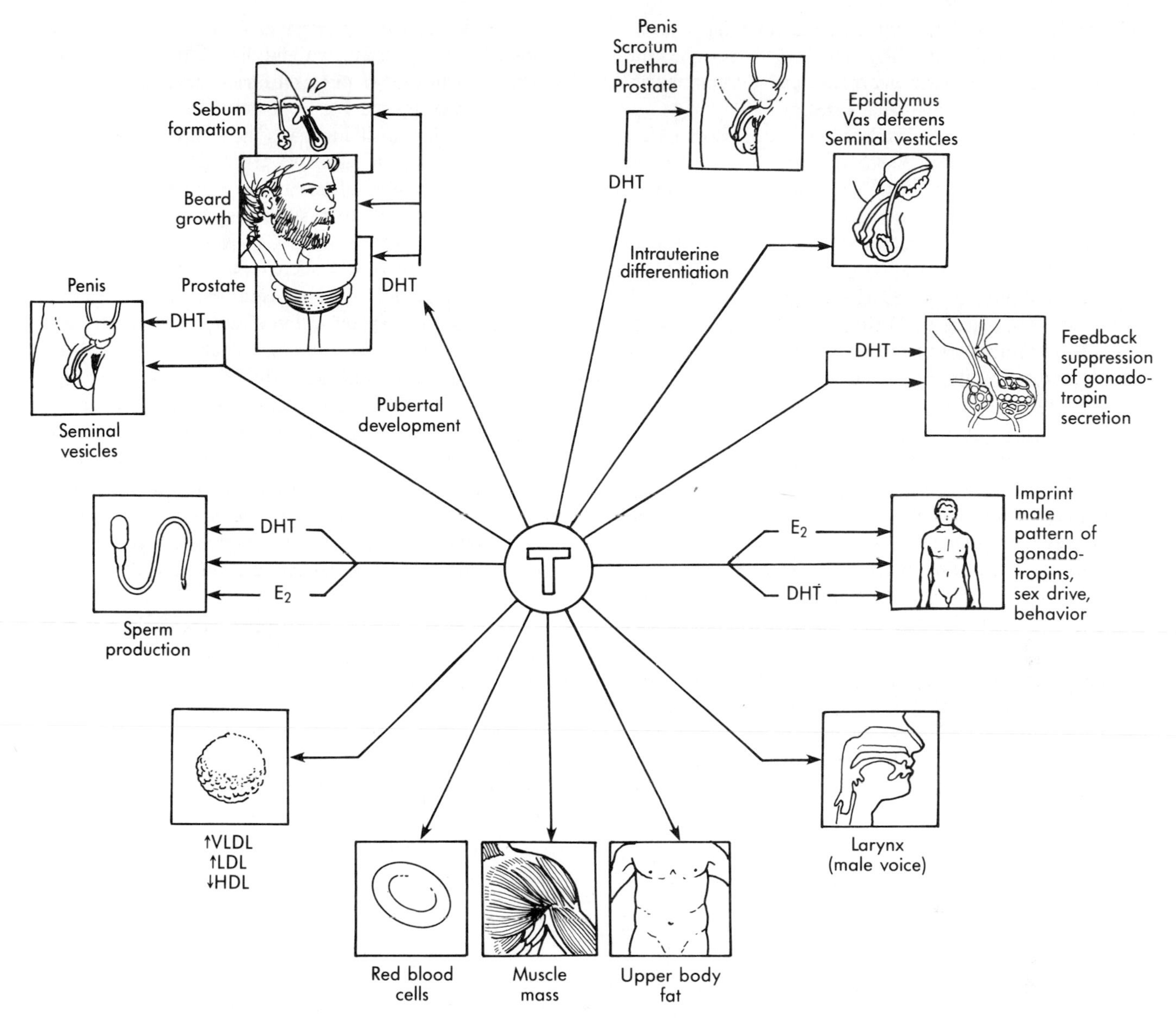

■ **Fig. 51-16** The spectrum of androgen effects. Note some effects result from the action of testosterone *(T)* itself, whereas others are mediated by dihydrotestosterone *(DHT)* and possibly estradiol *(E_2)* after they are produced from testosterone. *VLDL, LDL,* and *DHL* are very-low-density, low-density, and high-density lipoproteins.

ceous glands also is brought about by DHT or 5α-androstanediol.

Testosterone, on the other hand, specifically stimulates the differentiation of the wolffian ducts into the epididymis, vas deferens, and seminal vesicles. During puberty testosterone with or without DHT causes enlargement of the penis and of the seminal vesicles. It also causes enlargement of the larynx and thickening of the vocal cords, resulting in a deeper voice. Although testosterone is the major local hormone required for initiation and maintenance of spermatogenesis, DHT may participate in sperm production as well.

Testosterone itself also first stimulates the pubertal growth spurt and then causes cessation of linear growth by closure of the epiphyseal growth centers. Androgen receptors in osteoblasts transduce testosterone-stimulated increases in levels of transforming growth factors. Testosterone causes enlargement of the muscle mass in boys during puberty. In subsequent adult life administration of testosterone causes nitrogen retention in both sexes, reflecting protein anabolism. It is noteworthy that the hypothalamus lacks significant 5α-reductase activity. Thus suppression of gonadotropin secretion by negative feedback is largely a direct function of testosterone, with a possible small additional effect from circulating dihydrotestosterone.

Testosterone has important effects on lipid metabolism. It increases circulating LDL cholesterol and decreases circulating HDL cholesterol and favors accumulation of upper body, abdominal, and visceral fat. These lipid abnormalities are associated with an increased risk of cardiovascular disease.

Certain other diverse androgenic actions can be ascribed to testosterone. These include (1) initiation of sexual drive (libido) and the ability to achieve a physiologically complete erection (potency), (2) suppression of mammary gland growth, (3) stimulation of hematopoiesis and maintenance of a normal red blood cell mass, (4) stimulation of renal sodium reabsorption, (5) stimulation of aggressive behavior, and (6) suppression of hepatic synthesis of SSBG, cortisol-binding globulin, and thyroxine-binding globulin.

Male Puberty

Beginning at an average age of 10 to 11 and ending at an average age of 15 to 17, males develop full reproductive function, Leydig cell proliferation, and adult levels of androgenic hormones (see Figs. 51-9 and 51-14). Secondary to activation of the testis, males acquire adult size and function of the accessory organs of reproduction, complete secondary sexual characteristics, and adult musculature. They undergo a linear growth spurt, and the epiphyses close when they attain adult height. A composite picture of the measurable and visible portions of this sequence is shown in Fig. 51-17. It must be stressed that this process can start as early as age 8 and as late as age 20, without any evidence of disease. The mechanisms of pubertal onset were described previously (see Fig. 51-7).

Enlargement of the testis is the first and most important clinical sign of puberty. This represents principally an increase in the volume of the seminiferous tubules, and it is preceded by small increases in plasma FSH. Leydig cells appear, and testosterone secretion is stimulated, as plasma LH increases. Plasma testosterone then climbs rapidly over a 2-year period, during which time pubic hair appears, the penis enlarges, and peak velocity in linear growth is achieved (see Fig. 51-17). Sometime during this interval—at a median age of 13 years—sperm production begins. In about one third of boys, breast growth and tenderness appear transiently. This probably reflects increased production of estradiol secondary to LH stimulation. As testosterone levels continue to climb, the breast tissue regresses. One to two years after adult testosterone levels are reached, closure of the epiphyseal growth centers ends puberty.

The Ovaries

The ovaries, fallopian tubes, and uterus comprise the internal reproductive organs of the female and are situated in the pelvis. Each adult ovary weighs approximately 15 g and is attached to the lateral pelvic wall

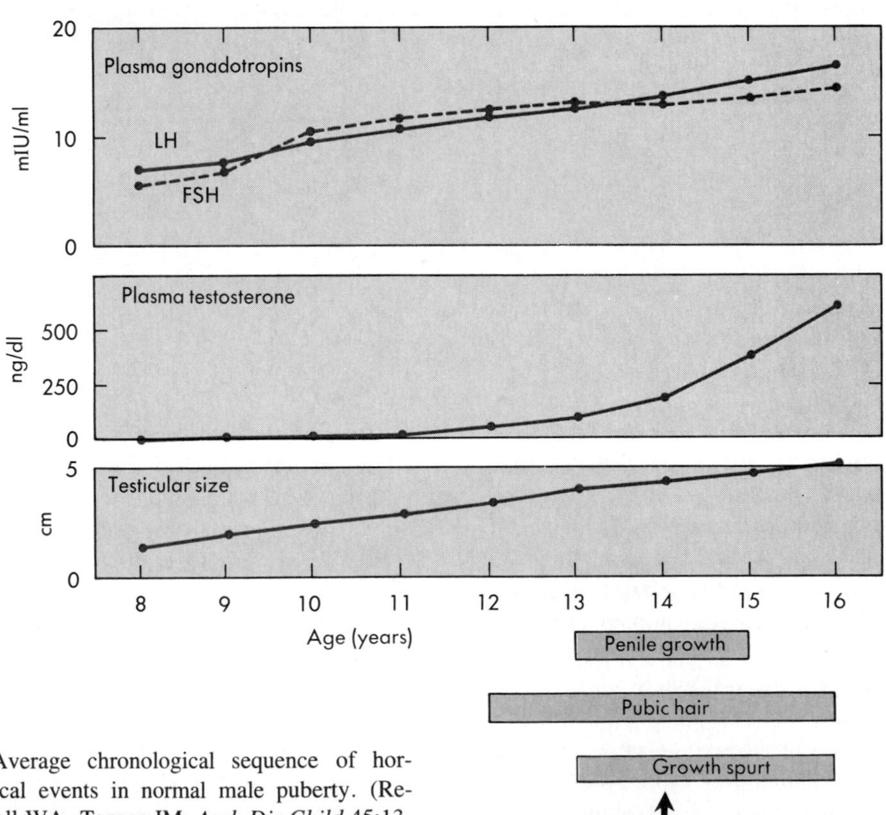

Fig. 51-17 Average chronological sequence of hormonal and biological events in normal male puberty. (Redrawn from Marshall WA, Tanner JM: *Arch Dis Child* 45:13, 1970; and Winter JSD et al: *Pediatr Res* 6:126, 1972.)

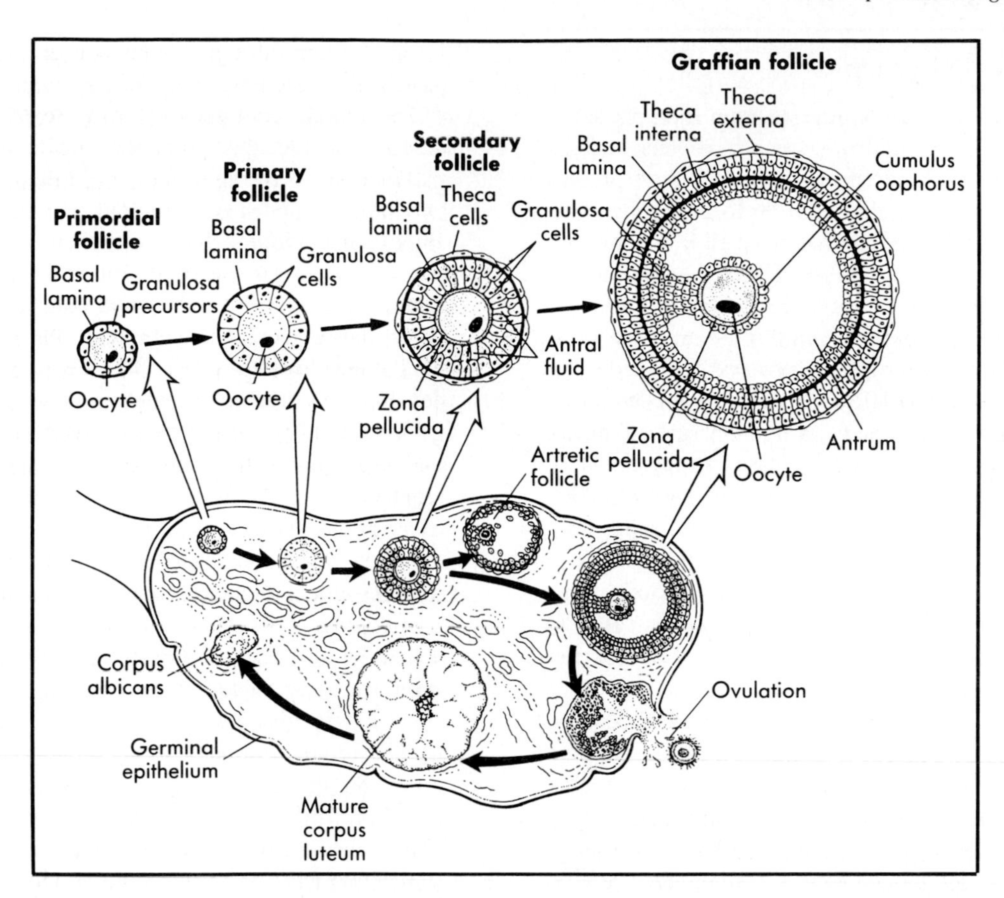

■ **Fig. 51-18** Schematic representation (not to scale) of the structure of the ovary, showing the various stages in the development of the follicle and its successor structure, the corpus luteum. The follicle grows from a primordial size of 25 μ to an ovulatory size of 10 to 20 mm. The oocyte is shielded from indiscriminate exposure to interstitial fluid contents by the basal lamina and the cytoplasm of the surrounding granulosa cells. The hormones and other constituents of the antral fluid are important regulators of follicular development. (Redrawn from Ham AW, Leeson TS: *Histology,* ed 4, Philadelphia, 1968, JB Lippincott.)

and to the uterus by ligaments, through which run the ipsilateral ovarian artery, vein, lymphatic vessels, and nerve supply.

The ovary consists of three zones (Fig. 51-18). The dominant zone is the **cortex**, which is lined by germinal epithelium and contains all the oocytes, each enclosed within a follicle. Follicles in various stages of development and regression can be seen throughout the cortex during the reproductive years (see Fig. 51-18). Interposed between the follicles is the stroma, composed of supporting connective tissue elements and interstitial cells. The other two zones of the ovary are the **medulla**, consisting of a heterogeneous group of cells, and the **hilum**, at which the blood vessels enter. These zones contain scattered steroid-producing cells, whose normal function is unknown. During physical examination the ovaries can be felt through the abdominal wall and also can be well visualized by ultrasonography and computerized axial tomography (CAT scanning).

As a hormone-secreting organ, the ovary functions in two ways. First, the ovarian sex steroids and protein hormones function locally to modulate the complex events in the development and extrusion of the ova. Second, these hormones are secreted into the circulation and act on diverse target organs, including the uterus, fallopian tubes, vagina, breasts, hypothalamus, pituitary gland, adipose tissue, bones, kidney, and liver. Many but not all of the distant effects are closely related to the reproductive sequence.

The fundamental reproductive unit in the female is the single ovarian follicle, which is composed of one germ cell completely surrounded by a cluster of endocrine cells. When fully developed and functional it will (1) maintain and nurture the resident oocyte; (2) mature the oocyte and release it at the right time; (3) prepare the vagina and fallopian tubes to assist in fertilization; (4) prepare the lining of the uterus to accept and implant a zygote; and (5) maintain hormonal support for the fetus until the placenta achieves this capability.

The Biology of Oogenesis

The primordial germ cells migrate from the yolk sac of the embryo to the genital ridge at 5 to 6 weeks of gestation. There, in the developing ovary, they produce oogonia by mitotic division until 20 to 24 weeks, when the total number of oogonia has reached a maximum of 7 million. Beginning at 8 to 9 weeks some oogonia start into the prophase of meiosis, becoming primary oocytes. This process continues until 6 months after birth, when all oogonia have been converted to oocytes. At this time, oocytes are 10 to 25 μm in diameter. They grow to 50 to 120 μm at maturity, the size of the nucleus and cytoplasm having increased proportionately. The first meiotic division is not completed until the time of ovulation; thus primary oocytes have life spans up to 50 years. The lengthy suspension of the oocyte in prophase apparently depends on the hormonal milieu provided by its surrounding sustaining cells. Both X chromosomes are required in the ovary for oocyte meiosis and survival.

From the start, however, there is also a process of oocyte attrition, so that by birth only 2 million primary oocytes remain, and by the onset of puberty the number falls to 400,000. Thus, in contrast to the male, who is continuously producing spermatogonia and primary spermatocytes, the female cannot manufacture new oogonia and must function with a continuously declining number of primary oocytes from which ova can mature. At or soon after menopause there are few, if any, oocytes left, and reproductive capacity ends.

First stage. The first stage of development of the ovarian follicle parallels the prophase of the oocyte and occurs very slowly, over a period that is usually not less than 13 years but that may be as long as 50 years. As an oocyte enters meiosis, it induces a single layer of spindle cells from the stroma to surround it completely. These cells are the precursors of the **granulosa cells.** Cytoplasmic processes from these cells attach to the plasma membrane of the oocyte. In addition, a membrane called the **basal lamina** forms outside the spindle cells, delimiting the complex from the surrounding stroma. This constitutes the **primordial follicle,** which is about 25 μm in diameter (see Fig. 51-18).

Beginning at 21 to 31 weeks of gestation some of these follicles enter the next phase of development. The spindle-shaped cells become cuboidal, and granulosa cells and a **primary follicle** is formed. As the granulosa cells divide and form several layers around the oocyte, a **secondary follicle** is created. The granulosa cells secrete mucopolysaccharides, which form a protective halo, the **zona pellucida,** around the oocyte (see Fig. 51-18). The cytoplasmic processes of the granulosa cells, however, continue to penetrate the zona pellucida, evidently providing nutrients and chemical signals to the maturing primary oocyte within. Thus cytoplasm of the granulosa cells, like that of the male Sertoli cells, forms a filter through which plasma substances must pass before reaching the germ cell.

The follicle continues to grow, reaching a diameter of 150 μm. At this point the oocyte has reached its maximal size, averaging 80 μm in diameter. There are two other concurrent developments. Another layer of spindle interstitial cells is recruited outside the basal lamina and forms the **theca interna**, while the granulosa cells begin to extrude small collections of fluid between themselves. This completes the first stage of follicular development. It is also the maximal degree of development ordinarily found in the prepubertal ovary.

Second stage. In contrast to the first stage, the second stage of follicular development is much more rapid, requiring 70 to 85 days. This stage takes place primarily after menarche, i.e., after the onset of menses. During each menstrual cycle a small cohort of secondary follicles enters the next sequence. The small collections of follicular fluid coalesce into a single central area called the **antrum** (see Fig. 51-18). The fluid in the antrum contains mucopolysaccharides, plasma proteins, electrolytes, glycosoaminoglycans, proteoglycans, gonadal steroid hormones, FSH, inhibin, and other factors. The steroid hormones reach the antrum by direct secretion from granulosa cells and by diffusion from the theca cells outside the basal lamina. A nonsteroidal substance that is capable of inhibiting oocyte meiosis probably also is secreted into the antral fluid and is believed to be of granulosa cell origin. AMH is one candidate to be this factor.

The granulosa cells continue to proliferate and develop gap junctions between them, forming a syncitium of electrical and chemical communication. They displace the oocyte into an eccentric position on a stalk, where it is surrounded by a distinctive granulosa cell layer, two to three cells thick, called the **cumulus oophorus**. The theca cells also proliferate, and those nearest the basal lamina are transformed into cuboidal steroid-secreting cells—the theca interna. Additional peripheral layers of spindle cells from the stroma form around the theca interna and together with an ingrowth of blood vessels comprise the **theca externa**. By the end of this second stage the entire complex, called a graafian follicle or antral follicle (Fig. 57-18), has reached an average diameter of 2,000 to 5,000 μm (2 to 5 mm). Although such follicles may rarely be found in prepubertal ovaries, they are relatively small in size and early in development.

Third stage. The third and final stage of follicular development is the most rapid and occurs only in the postpubertal reproductive ovary. Five to 7 days after the onset of menses, a single graafian follicle from its cohort in that cycle achieves dominance. With rare exceptions, this occurs in only a single ovary each month. In addition to further cellular growth, the production of antral fluid is significantly increased. The colloid osmotic pressure of the fluid also increases because of depolymerization of the mucopolysaccharides. The granulosa cells spread apart, and the cumulus oophorus loosens. At the same time the vascularity of the theca increases greatly. With exponential growth, the total size of this follicle reaches 20 mm in the final 48 hours, corresponding to the midpoint of the menstrual cycle. The portion of the basal lamina adjacent to the surface of the ovary then is subjected to proteolysis. The follicle gently ruptures, releasing the oocyte with its adherent cumulus oophorus into the peritoneal cavity. At this time the initial meiotic division is completed. The resultant secondary oocyte is drawn into the closely approximated fallopian tube, and the first polar body is discarded. In the fallopian tube fertilization causes completion of the second meiotic division, resulting in the haploid (23 chromosome) ovum and the second polar body.

Corpus luteum formation. The residual elements of the ruptured follicle next form a new endocrine structure, the **corpus luteum** (see Fig. 51-18). This unit will provide the necessary balance of gonadal steroids that optimizes conditions for implantation of the ovum, should fertilization occur, and for subsequent maintenance of the zygote until the placenta can assume this function. The corpus luteum is made up of granulosa cells, theca cells, thecal capillaries, and fibroblasts. The granulosa cells comprise 80% of the corpus luteum. They hypertrophy to a diameter of 30 μm, become arranged in rows, and undergo striking changes. The mitochondria develop dense matrices with tubular cristae, numerous lipid droplets form within the cytoplasm, and the smooth endoplasmic reticulum proliferates. This process, called **luteinization,** is precipitated by the exit of the oocyte from the follicle.

The remaining 20% of the corpus luteum consists of theca cells arranged in folds along its outer surface. The theca cells exhibit similar although less dramatic changes of luteinization. The basal lamina between the theca and granulosa cells disappears, allowing direct vascularization of the latter.

The antrum may become temporarily engorged with blood from hemorrhaging thecal vessels, but a clot quickly forms and is subsequently lysed. If fertilization and pregnancy do not ensue, the corpus luteum begins to regress after a 14-day life span. In this process, known as **luteolysis,** the endocrine cells undergo necrosis, and the structure is invaded by leukocytes, macrophages, and fibroblasts. Gradually the former corpus luteum is replaced by an avascular scar known as the **corpus albicans** (see Fig. 51-18).

Atresia of follicles. During the reproductive life span of the average woman only 400 to 500 oocytes (one per month) will undergo the complete sequence of events culminating in ovulation. The remaining millions disappear in a process called **atresia,** which begins almost as soon as the first primordial follicles appear in the fetal ovary.

In first-stage follicles atresia is a relatively simple process precipitated by oocyte degeneration. The oocyte becomes necrotic, its nucleus becomes pyknotic, and the granulosa cells also degenerate. This accounts for the vast majority of oocytes. In more advanced follicles atresia is a more complex process. In some follicles the granulosa cells furthest from the oocyte first undergo necrotic changes. Loss of their function may actually precipitate a resumption of meiosis in the oocyte to the point of extrusion of the first polar body. Eventually the granulosa cells in the cumulus oophorus also die, the protective zona pellucida disappears, and the oocyte degenerates. Eventually fibroblasts invade the follicle, and everything inside the basal lamina collapses into an avascular scar, called a **corpus albicans.** Outside the basal lamina the theca cells dedifferentiate and return to the pool of interstitial cells from which they came.

Hormonal Patterns during the Menstrual Cycle

The menstrual cycle is divided physiologically into three sequential phases. The **follicular phase** begins with the onset of menstrual bleeding and averages 15 days (range is 9 to 23 days). The **ovulatory phase** is of 1 to 3 days' duration and culminates in ovulation. The **luteal phase** has a somewhat more constant length averaging 13 days and ends with the onset of menstrual bleeding. The overall duration of a normal menstrual cycle averages 28 days but can vary from 21 to 35 days, depending mostly on the length of the follicular phase.

A series of cyclic changes in gonadal steroid hormone and inhibin production characterizes adult ovarian function (Fig. 51-19, Table 51-3). This monthly steroid hormone profile results from cyclic changes in pituitary gonadotropins (Fig. 51-19). These, in turn, reflect changing pituitary sensitivity to GnRH and probably changes in the pulsatility of the hypothalamic GnRH generator (Fig. 51-20, *A*). However, the pattern of gonadotropin secretion is also critically regulated by both negative and positive feedback from gonadal steroids and inhibin. These interactions were described in detail in Chapter 48 and presented in Fig. 48-17.

Toward the end of the luteal phase, plasma FSH and LH are at their lowest levels (see Fig. 51-19). The LH/FSH ratio is slightly greater than 1. One to 2 days before the onset of menses FSH levels begin to rise, followed somewhat later by LH levels. The estrogens (estradiol and estrone) increase gradually, stimulated by the rise in FSH in this first half of the follicular phase. Progesterone, 17-hydroxyprogesterone, and the androgens androstenedione and testosterone remain at relatively low, constant levels.

During the second half of the follicular phase FSH

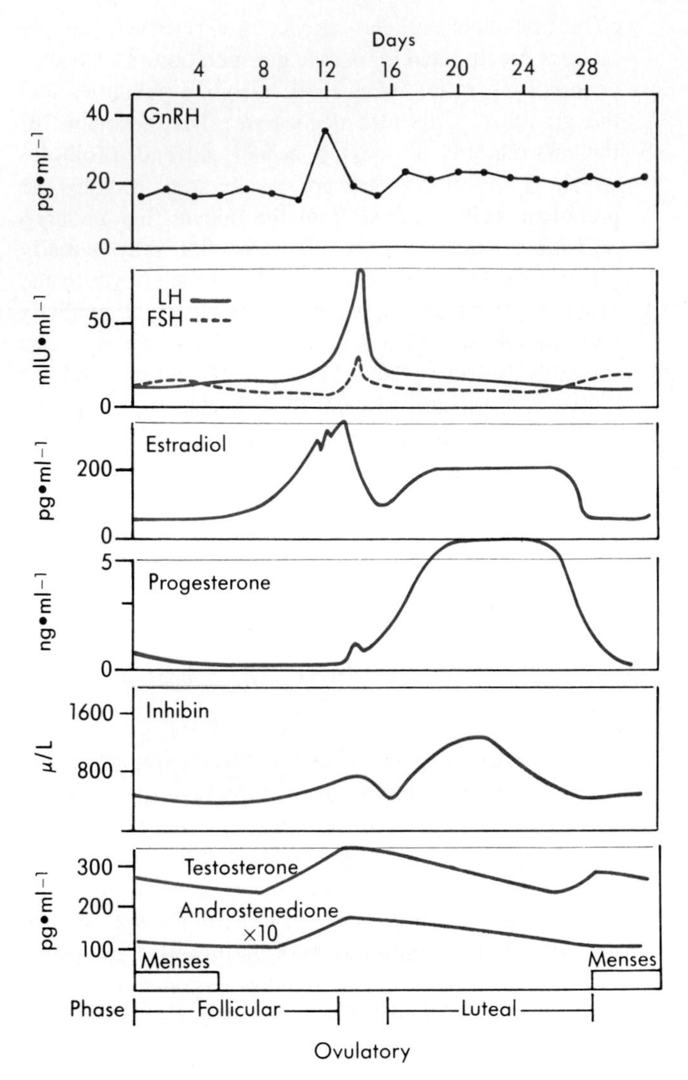

■ **Fig. 51-19** Plasma hormone levels throughout the menstrual cycle. Note increases of estradiol and GnRH preceding ovulatory surges of LH and FSH. The broad peaks of progesterone, estradiol, and inhibin are produced by corpus luteum secretion.

levels fall modestly, whereas LH levels continue to rise very slowly. The LH/FSH ratio therefore increases to about 2. Concurrently, estradiol and estrone production and plasma levels rise sharply, reaching peaks fivefold to ninefold higher just before the ovulatory phase. The estradiol is secreted directly by the dominant follicle, whereas the estrone largely arises by peripheral conversion from estradiol and androstenedione. Plasma progesterone and 17-hydroxyprogesterone remain low until just before the ovulatory phase, when progesterone begins to increase due to ovarian secretion. Androstenedione and testosterone also rise modestly in parallel with 17-hydroxyprogesterone. About half the androgenic steroids are derived from ovarian and half from adrenal secretion. The ovarian contribution results from LH stimulation of the theca cells in the dominant follicle.

The large increment in estradiol secretion from the dominant follicle during the latter half of the follicular phase, augmented by a rise in inhibin (see Fig. 51-19), suppresses FSH secretion by negative feedback.

■ **Table 51-3** Turnover of gonadal steroids in adult women

Steroids	*Plasma concentration (ng/dl)*	*Production rate (µg/day)*	*Metabolic clearance rate (L/day)*
Estradiol			
Early follicular	6	80	1400
Late follicular	50	700	
Middle luteal	20	300	
Estrone			
Early follicular	5	100	2200
Late follicular	20	500	
Middle luteal	10	250	
17-Hydroxy-progesterone			
Early follicular	30	600	2000
Late follicular	200	4000	
Middle luteal	200	4000	
Progesterone			
Follicular	100	2000	2200
Luteal	1000	25000	
Testosterone	40	250	700
Dihydrotestosterone	20	50	400
Androstenedione	150	3000	2000
DHEA	500	8000	1600

Modified from Lipsett MB: In Yen SSC, Jaffe RB, editors: *Reproductive endocrinology,* Philadelphia, 1978, WB Saunders.

The succeeding ovulatory phase is characterized by a very sharp, transient spike in plasma gonadotropin levels. LH increases much more than FSH (Fig. 51-19), so that the LH/FSH ratio rises to about 5. This surge has an ascending limb that averages 14 hours with a doubling time of 5 hours, a plateau that averages 14 hours, and a descending limb that averages 20 hours. Plasma estradiol levels plummet from their peak at the same time that LH and FSH are on their ovulatory upswing. Estrone, 17-hydroxyprogesterone, androstenedione, and testosterone also now decrease but much more gradually than estradiol. In contrast, a small but significant rise in progesterone begins during the ovulatory phase.

After ovulation LH and FSH both continue to decline during the luteal phase, reaching their lowest points in the cycle toward its end, before the onset of menses. The most distinctive and important feature of the luteal phase is a tenfold increase in progesterone, which emanates from the corpus luteum. Estradiol, estrone, 17-hydroxyprogesterone, and inhibin of corpus luteum origin, also increase, providing broad second peaks through the middle of the luteal phase. Androstenedione and testosterone, however, continue to decline during the luteal phase. If pregnancy does not occur, the menstrual cycle ends as the gonadal steroid and inhibin levels decrease dramatically to their lowest values, FSH levels again begin to rise, and bleeding starts. The generally inverse

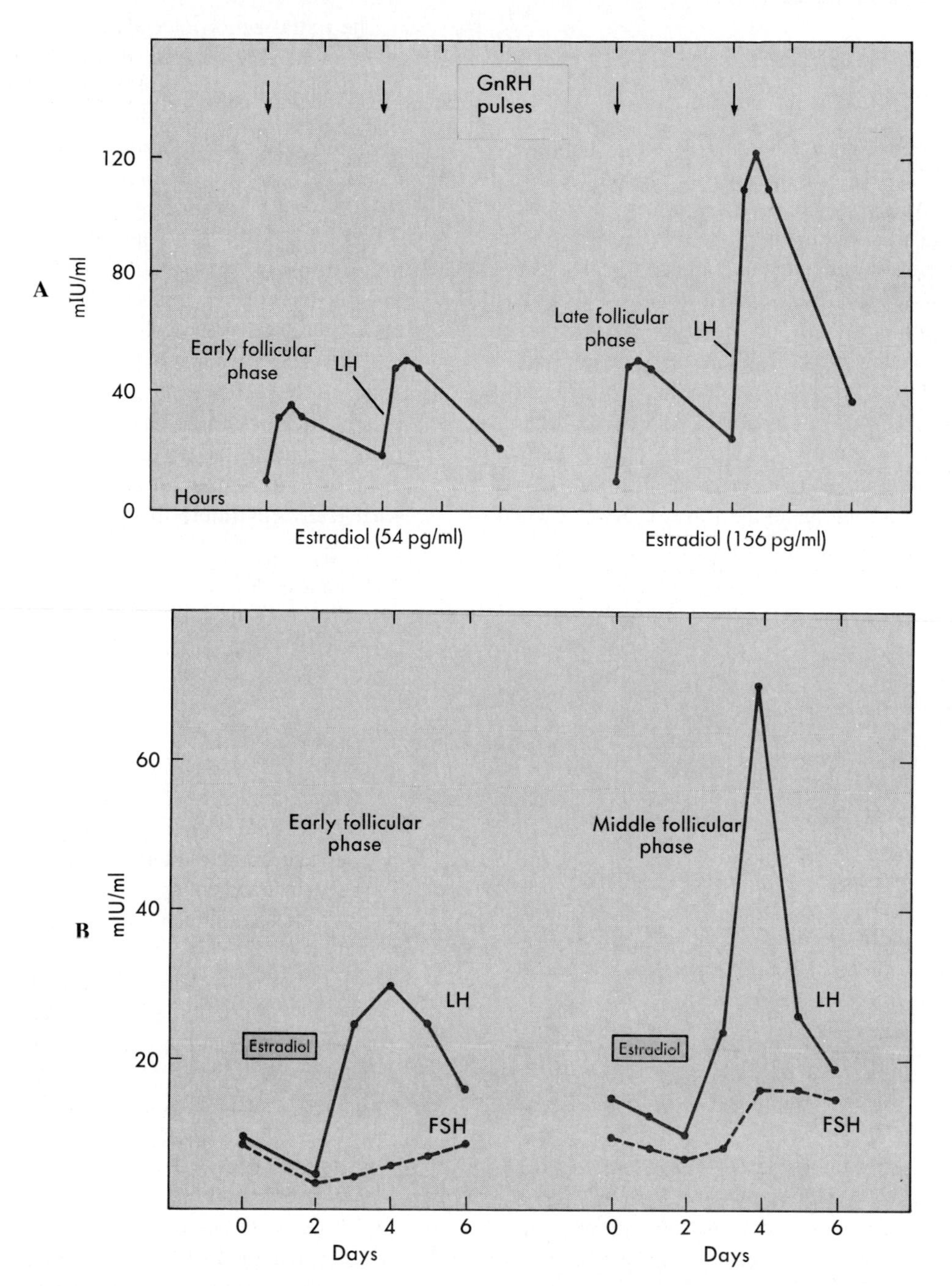

■ **Fig. 51-20** **A,** Increased responsiveness of pituitary gonadotrophs to GnRh in the late follicular phase of the menstrual cycle when endogenous estradiol levels are increased. **B,** Plasma LH and FSH after exogenous estradiol administration. After the initial decrease caused by negative feedback, plasma LH rebounds well above the baseline when estradiol is discontinued. This positive effect of estradiol also is accentuated as the follicular phase of the menstrual cycle progresses. (**A** redrawn from Wang CF et al: The functional changes of the pituitary gonadotrophs during the menstrual cycle, *J Clin Endocrinol Metab* 42:718, 1976. **B** redrawn from Yen SSC et al: *Causal relationship between the hormonal variables*. In Ferin M et al, editors: *Biorhythms and human reproduction*. New York, 1974, John Wiley & Sons.)

relationship between FSH and inhibin throughout the menstrual cycle is congruent with the negative feedback effect of inhibin on FSH secretion.

■ *Hormonal Regulation of Oogenesis*

The development of the primordial follicle is a hormonally independent event. The mechanism by which a particular group of primordial follicles is recruited to descend into the interstitium toward the medulla and to initiate further development into primary follicles is unknown. However, their subsequent growth and further development probably depends on a low constant level of FSH secretion, even before puberty. Once menarche has occurred, recruitment of a cohort may occur in the early luteal phase of one menstrual cycle, with the cohort developing slowly over a period of 60 to 70 days, until the late luteal phase two cycles later. At this point, a total of about 20 follicles in both ovaries have reached the size of 2 to 4 mm and are capable of responding further to the FSH increment in the follicular phase of the next cycle.

The initial action of FSH on the primary follicles is to stimulate growth of the granulosa cells (Fig. 51-21). Additionally, aromatase activity is increased, so that estrogen synthesis from androgen precursors is enhanced. The increasing local estradiol causes proliferation of its own receptors and reinforces FSH actions. FSH action is reinforced by increasing FSH receptors and by synergizing with the gonadotropin to stimulate further granulosa cell hyperplasia and hypertrophy (see Fig. 51-21). This in turn further boosts estradiol production. Thus the initiation of second-stage follicular development may be viewed as a self-propelling mechanism that involves fine coordination between the pituitary gland and ovary and that yields successively increasing rates of follicular growth and estradiol production.

Three other important actions, which develop somewhat later, contribute to this autocatalytic process. (1)

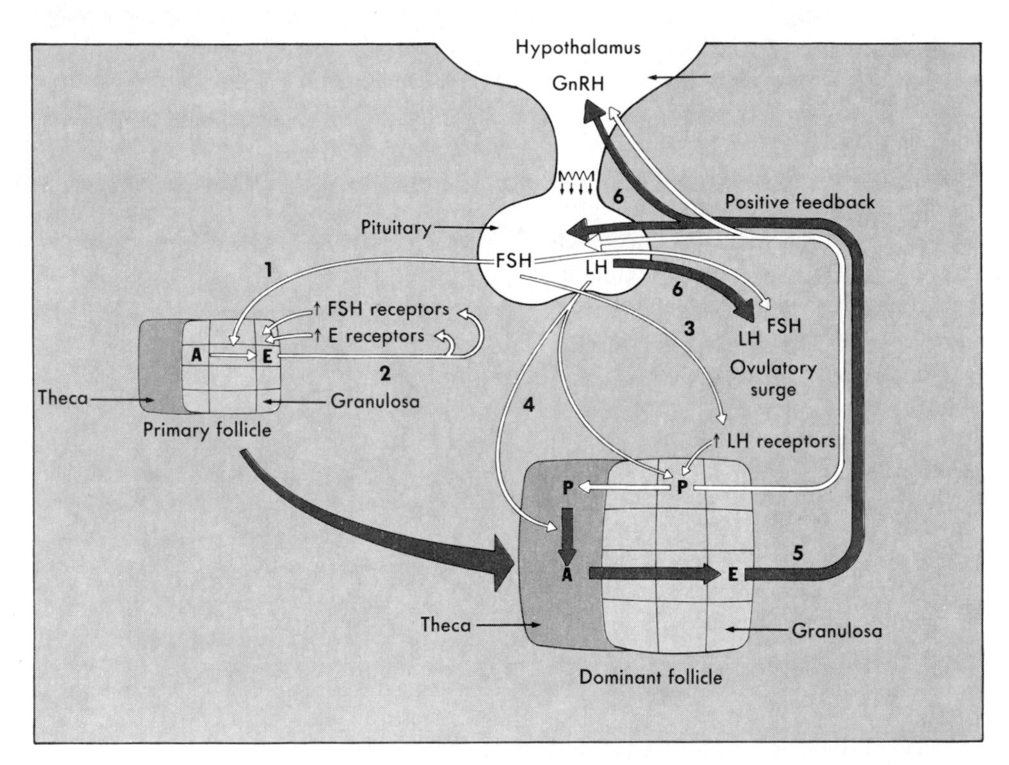

■ **Fig. 51-21** Hormonal regulation of follicular development. *1,* FSH stimulates granulosa cell growth and estradiol *(E)* synthesis in certain primary follicles. *2,* The local estradiol increases its own receptors and FSH receptors, amplifying both hormones' effects. Thus a self-propelling mechanism is set into motion. *3,* FSH later increases LH receptors, augmenting granulosa and theca cell responsiveness to LH. *4,* LH stimulates theca cell growth and androgen *(A)* production. Androgen is then converted to estradiol in the granulosa cells. LH also stimulates progesterone *(P)* production in the granulosa cells. *5,* As a result of two-way steroid traffic, the dominant follicle emerges as a very efficient secretor of estradiol. *6,* Rising estradiol, with late potentiation by progesterone, feeds back positively on the pituitary gland and hypothalamus to evoke the preovulatory surge of LH and FSH.

FSH, along with estradiol, induces LH receptors on the granulosa cells. (2) The slowly rising plasma estradiol levels condition the hypothalamic gonadotropin axis so as to maintain or slightly increase plasma LH while plasma FSH is decreasing. Furthermore, pituitary LH stores are enhanced by estradiol. This is reflected in the fact that administration of exogenous pulses of GnRH produces greater LH responses in the second half of the follicular phase than in the first half (see Fig. 51-20, *A*). (3) Estradiol increases LH receptors in theca cells.

The rising LH stimulates the theca cells to produce increasing amounts of androstenedione and testosterone. These steroids diffuse across the basal lamina, where they serve as substrates for granulosa cell aromatase and sustain the augmented estradiol production (see Figs. 51-4 and 51-21). In addition, LH stimulates the granulosa cells to produce progesterone, some of which diffuses back into the theca cells to serve as a substrate for androgen synthesis (see Fig. 51-21). Thus, although individually granulosa cells and theca cells can synthesize both androgens and estrogens to some extent, their proximity and the two-way traffic of steroids between them increase the overall efficiency of the follicle greatly.

FSH also stimulates production of a variety of molecules by the granulosa cells, which likely have paracrine effects. The situation is analogous to that of the Sertoli cell and spermatogenesis. Thus transferrin and ceruloplasmin pick up iron and copper, respectively, from their plasma-binding analogues and transfer these vital elements to the oocyte. Various granulosa growth factors, such as IGF-1 and transforming growth factors α and β, modulate growth and steroid secretion of neighboring endocrine cells and conceivably of the oocyte itself. For example, IGF-1 potentiates FSH action on granulosa cell differentiation and progesterone synthesis and potentiates LH stimulation of androgen production by theca cells. FSH also stimulates granulosa cell metabolism and provides lactic acid and 2-ketoisocaproic acid as energy sources for the oocyte. Plasminogen activator plays a role in ovulation. Renin and angiotensin, oxytocin, and GnRH-like peptides are all present in elevated concentrations in follicular fluid, but their functions are obscure.

By day five to seven of the follicular phase, only one follicle has reached a size greater than 11 mm. This **dominant follicle** selects itself by outstripping the others. Its key characteristic is increased aromatase activity and therefore more efficient synthesis of estradiol. This may possibly result from a greater density of FSH receptors and therefore less dependency on a waning FSH supply at this point. A temporarily greater production of inhibin and a lesser production of activin at this crucial time may contribute by favoring an increased supply of precursor androgens from the theca cells. Whatever the mechanism, its greater production of estradiol permits the dominant follicle to inhibit further substantive growth of its sister follicles, to prime the GnRH-gonadotropin axis for generating the ovulatory surge, and to alter the tissues of the genital tract so as to favor conception (Fig. 51-22).

An atretic fate for the other follicles is associated with a high androgen/estrogen ratio and a low FSH concentration in their follicular fluid. The fact that cohort follicles are stunted more severely in the ovary that contains the dominant follicle than in the contralateral ovary also suggests secretion of a specific paracrine-acting inhibitor.

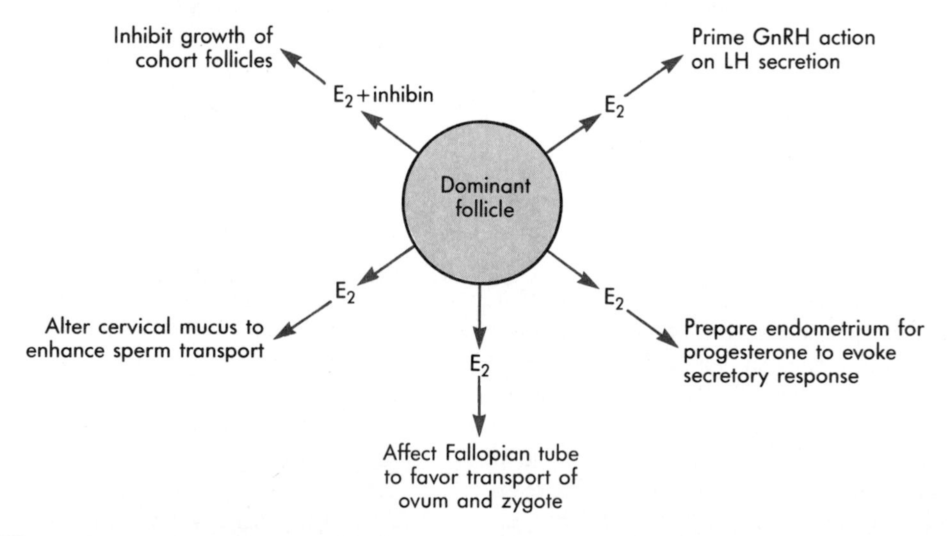

Fig. 51-22 Fuctions of estradiol (E_2) secreted by the dominant follicle. E_2 + inhibin reduce FSH secretion and inhibit cohort follicles.

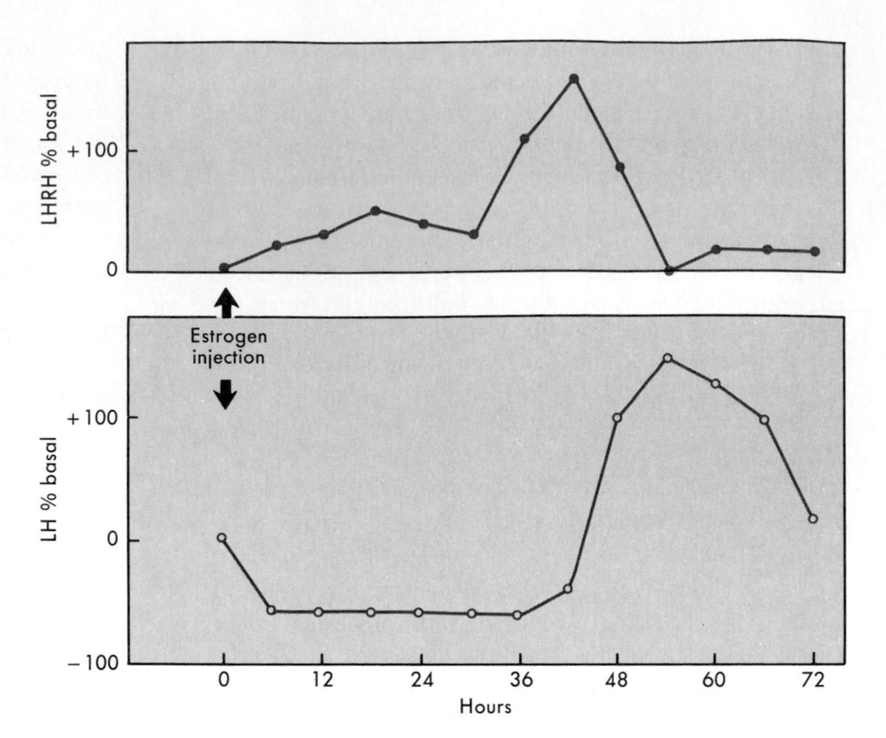

■ **Fig. 51-23** The effect of exogenous administration of estrogen on plasma GnRh and LH levels. After an initial phase of suppression, the positive feedback action of estrogen on LH secretion is seen at 48 hours. This delayed rise in LH levels is preceded by an increase in GnRh levels, suggesting hypothalamic locus of estrogen action. (Redrawn from Miyake A et al: *J Clin Endocrinol Metab* 56:1100, 1983.)

The sharply rising estradiol release from the dominant follicle triggers the ovulatory surge of LH and FSH (Figs. 51-20, *B*, and 51-23). A critical plasma estradiol level of at least 200 pg/ml sustained for at least 2 days is required to elicit this positive feedback effect on LH. The much smaller preovulatory increase in progesterone, although not absolutely required, synergizes with estradiol by amplifying and prolonging the gonadotropin surge. The loci of this positive feedback on gonadotropin secretion are in the pituitary and the hypothalamus. The pituitary gonadotrophs, appropriately primed by the preceding pattern of gonadal steroid exposure, respond at this time to GnRH with heightened sensitivity (see Fig. 51-20, *A*). The LH molecules released are more active, probably because of posttranslational modulation of their sialic acid content by estradiol. In addition, a peak of peripheral plasma GnRH levels precedes the LH/FSH peak (see Fig. 51-19). This suggests an augmented flow of GnRH from the hypothalamus to the pituitary, also attributable to gonadal steroids (see Fig. 51-23). Estradiol and progesterone may be acting in part by decreasing inhibitory endorphinergic activity on the GnRH neurons.

The surge of LH with FSH then triggers ovulation by a multicomponent mechanism, as already described. LH stimulation of the granulosa cell neutralizes the action of an oocyte maturation inhibitor, allowing completion of meiosis. Stimulation of progesterone levels enhances proteolytic enzyme activity and increases distensibility of the follicle. This permits a rapid increase in follicular fluid volume. A pseudoinflammatory response ensues. Local synthesis of prostaglandins, thromboxanes, and leukotrienes lead to follicular rupture. Mucification of the cumulus oophorus and possibly contraction of the follicular wall, stimulated by oxytocin, contribute to extrusion of the oocyte. Plasminogen activator, stimulated by FSH, generates the proteolytic enzyme plasmin, which catalyzes breakdown of the follicular wall. Increases in the concentrations of histamine, bradykinins, platelet-activating factor, and in blood flow ensue. FSH also stimulates the process whereby the oocyte-cumulus complex becomes detached and free-floating just before extrusion. Finally, immediately after the LH surge, LH receptors are temporarily reduced by downregulation. This desensitizes the granulosa and thecal cells to LH. The resultant rapid fall in androgen and estradiol production contributes to loss of integrity of the follicle. The LH surge also neutralizes the activity of the luteinization-inhibiting factor found in preovulatory fluid, thereby stimulating luteinization of the granulosa cells.

The dispatched oocyte is not under any other immediate hormonal influence. However, the organization and growth of the corpus luteum and its secretory pattern are under hormonal control. In the human, LH is essential for luteinization of the granulosa cells and the subsequent high rate of progesterone production by the corpus luteum. Therefore luteinization requires restoration and maintenance of LH receptors; this in turn depends on proper exposure to FSH and LH pulses in the preceding follicular phase. However, during the luteal phase, low frequency/high amplitude LH pulses replace the high frequency/low amplitude pulses of the follicular phase. This alteration may result from conditioning of the GnRH generator by the high progesterone concentrations.

The steadily increasing progesterone, estradiol, and inhibin output of the corpus luteum exerts negative feedback on the pituitary gland, resulting in a gradual decline in both LH and FSH (see Fig. 51-19). If the declining LH levels of the late luteal phase are not replaced by the equivalent placental hormone, human chorionic gonadotropin (HCG), the corpus luteum regresses, and its secretion of progesterone and estradiol ceases completely by 14 days. Prolactin may contribute to sustaining progesterone output by increasing the LH receptors. However, there is relatively little systematic variation in plasma prolactin throughout the menstrual cycle, and corpus luteum function has been observed in prolactin-deficient women. Estradiol also may play a role in maintaining the corpus luteum.

After the eighth postovulatory day the corpus luteum in the nonpregnant female, lacking HCG, begins to regress. This process is marked by increasing corpus luteum concentrations of cholesterol as progesterone synthesis declines. Luteolysis is likely mediated by prostaglandins made within the corpus luteum. By the twelfth postovulatory day progesterone and estradiol levels have fallen low enough to release the pituitary gland from negative feedback inhibition, and the FSH rise of the next cycle begins.

The hormonal regulation of the female reproductive cycle, as just described, has left open an important question: what ultimately determines the monthly cyclicity of the LH/FSH surge and the resultant ovulation? Although the concept of a primary central nervous system clock is inherently attractive, considerable evidence suggests that in the human it is the ovary that determines the basic rhythm. Five observations support this point:

1. No cyclic release of gonadotropins is observed in women whose ovaries never functioned or whose ovaries were removed during the reproductive years or in postmenopausal women after follicular development has ceased.
2. During the reproductive years, the ovulatory gonadotropin surge does not occur until the dominant follicle has reached the appropriate stage of development, whatever number of days that may take.
3. In otherwise normal women who are consistently not ovulating, initiation of follicular development can be achieved by treatment with the antiestrogen, hypothalamic stimulator, clomiphene. After sufficient time has elapsed for a dominant follicle to emerge and grow exponentially—about 10 to 14 days—a spontaneous LH/FSH surge and ovulation can occur.
4. Administration of estrogen with or without progesterone in a format that resembles the normal preovulatory estradiol rise induces an LH surge (see Fig. 51-23), even in postmenopausal women.
5. In a monkey whose pituitary gland has been completely severed from the hypothalamus, central nervous system regulation of gonadotropin secretion is disrupted. However, if GnRH is replaced intravenously in a physiological pulsatile format to reinstitute and sustain basal FSH and LH secretion, subsequent cyclic ovarian function with an LH/FSH surge is automatically observed without any change in the rate or pattern of the GnRH infusion.

Such observations suggest that a GnRH pulse generator in the central nervous system is required to initiate and sustain follicular development. However, it is the pattern of ovarian secretions, most critically from the dominant follicle, that conditions this pulse generator and the pituitary gonadotrophs to respond at the appropriate time with an ovulatory LH/FSH surge. Perhaps no other phenomenon so clearly illustrates the intricate nature of the interactions among endocrine, paracrine, autocrine, and neural mechanisms of regulation.

Ovarian signals can be either overridden or reinforced by other influences on and from the hypothalamus. Loss of cyclic gonadotropin secretion can occur in situations that suggest that the hypothalamus is responding to a caloric, thermal, photic, olfactory, emotional, or inflammatory signal. This is seen in women who are calorically deprived and lose considerable amounts of adipose tissue and lean body mass, and in women who exercise excessively. It is also seen in women who undergo physical translocation, climatic change, or emotional deprivation or who suffer from chronic inflammatory diseases.

Such inhibitory influences may be mediated by hypothalamic endorphins or dopamine or by changes in the levels of adrenal androgen, cortisol, or thyroid hormone. Seasonal variation in reproductive activity suggests possible modulation by melatonin, because conception rates are lowest in the months when darkness is most prevalent and secretion of inhibitory melatonin is highest. It has also been observed that women living in physical proximity can adopt a common timing of their menstrual cycles, possibly because of pheromones. These are chemical signals emitted by one individual that affect another. In the human there is no estrus cycle and no evidence that ovulation is stimulated by sexual behavior. However, female-initiated sexual activity is reportedly increased around the time of ovulation, possibly due to increased androgen levels.

■ *Gonadal Steroid Actions on the Genital Tract*

The cyclic changes in estradiol and progesterone secretion produce effects on the uterus, fallopian tubes, vagina, and breasts (Fig. 51-24). These coordinate precisely with the expectation of conception and the institution of a pregnancy.

Uterus. The function of the uterus is to house and

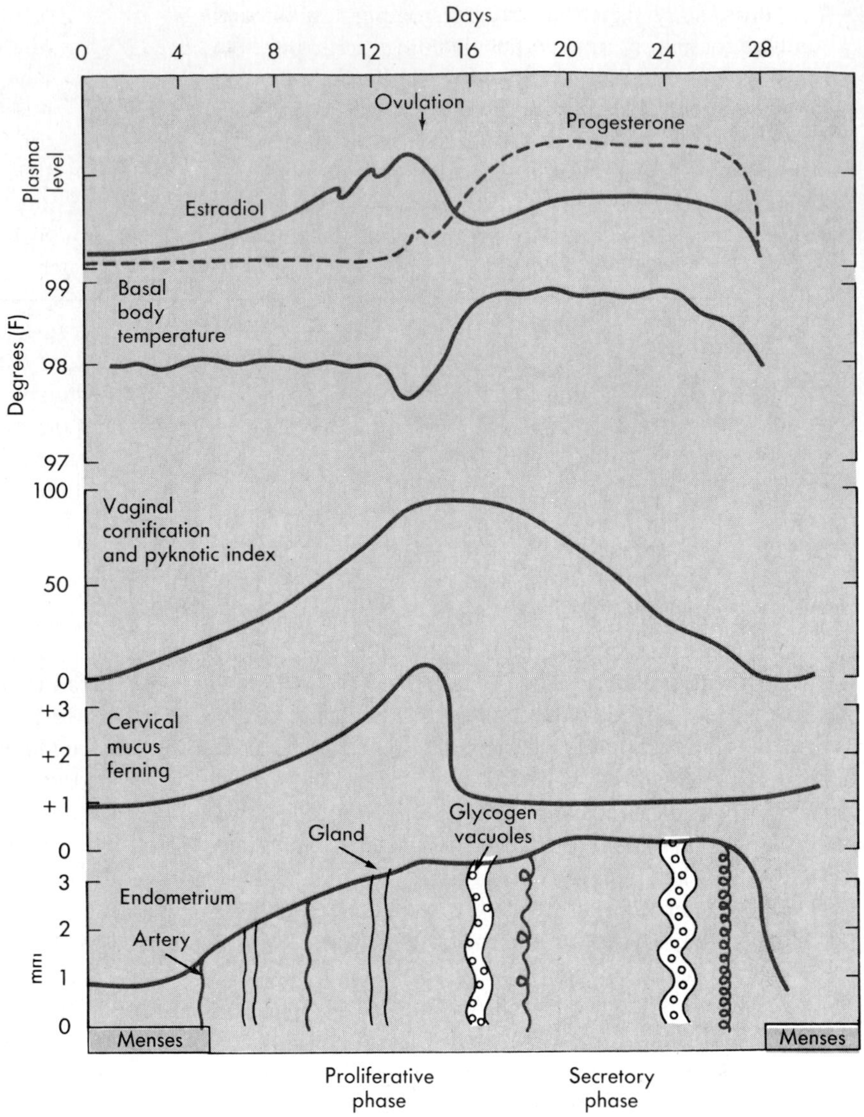

■ **Fig. 51-24** Correlation of biological changes throughout the menstrual cycle with the profiles of plasma estradiol and progesterone levels. (Redrawn from Odell WD: *The reproductive system in women.* In Degroot LJ et al, editors: *Endocrinology,* vol 3, New York, 1979, Grune & Stratton.)

nurture the developing fetus until the appropriate time for evacuation. It is a muscular organ, enclosing a cavity that is lined with stromal cells and a special mucous membrane called the **endometrium.** At the beginning of the follicular phase in each menstrual cycle the uterus is in the process of shedding its lining and is therefore incapable of receiving a conceptus. The endometrium is only 1 to 2 mm thick, and its glands are sparse, straight, and narrow lumened and exhibit few mitoses (see Fig. 51-24). After the menstrual slough has ceased, the increase in estradiol secretion during the follicular phase produces a threefold to fivefold increase in endometrial thickness to a maximum of 8 to 10 mm just before ovulation. Mitoses appear in the glands and stroma, the glands become tortuous, and the spiral arteries that supply the endometrium elongate. This is termed the **proliferative phase** of the endometrium. The mucus elaborated by the cervix also changes dramatically during this phase from a scant, thick, viscous material to a copious, more watery, but more elastic substance that can be stretched into a long, fine thread. It also produces a characteristic fernlike pattern when dried on a glass slide. In this estrogen-stimulated condition the cervical mucus creates a myriad of channels in the opening of the cervix that facilitate the entrance of the sperm and direct their motion forward into the uterine cavity.

Shortly after ovulation the rise in the plasma progesterone level produces marked alterations (see Fig. 51-24). Rapid proliferation of the endometrium is slowed, mitotic activity is reduced, and the thickness is decreased to 5 to 6 mm. The uterine glands become much more tortuous, and they begin to accumulate glycogen in large vacuoles at the base of each cell. As the luteal phase of the cycle progresses, the vacuoles move toward the lumen, and the glands greatly increase their mucus secretions. These contain glycogen, glycoproteins, and glycolipids that sustain and facilitate attachment of a conceptus. The stroma of the endometrium becomes edematous and the originally straight spiral arteries elongate further and become coiled. This is termed the **secretory phase** of the endometrium. At the same time progesterone decreases the quantity of the cervical mucus, causes it to return to its original thick, nonelastic state, and inhibits the ferning pattern.

If pregnancy does not occur and the corpus luteum regresses, lymphocytes and neutrophils appear. The abrupt loss of estradiol and progesterone causes spasmic contractions of the spiral arteries and uterine muscles, probably mediated by locally increased production of prostaglandins. The resultant ischemia produces necrosis, the stroma condenses and degenerates, and the superficial endometrial cells are sloughed along with sludged blood. This comprises the **menstrual period.**

Fallopian tubes. The fallopian tubes are the normal site of fertilization. These bilateral structures are 10 cm long and emerge from the uterus. Each tube ends in fingerlike projections called **fimbriae,** which lie close to the ipsilateral ovary. The fallopian tube consists of a muscular layer surrounding a mucosa lined by an epithelium that contains both ciliated and secretory cells. The cilia beat toward the uterus. During the follicular phase estradiol causes an increase in the number of cilia and in their rate of beating, as well as in the number of actively secreting epithelial cells. Estradiol also stimulates the tubal secretions that provide a mucoid medium in which the sperm may move more efficiently upstream against the ciliary beat. In addition, the fimbria become more vascularized.

As ovulation approaches, tubal contractions increase, and the fimbria undulate so as to draw the shed ovum into the tube. During the luteal phase, progesterone probably maximizes the ciliary beat, enhancing movement of any fertilized ovum toward the uterus. Progesterone also increases the secretion of materials nutritious to the ovum, to any incoming sperm, and to the zygote, should fertilization occur.

Vagina. The vaginal canal is lined with a stratified squamous epithelium that is highly sensitive to estradiol. In the absence of the hormone there is only a thin layer of basal and parabasal cells. For the first few days of the follicular phase of the menstrual cycle the epithelium is relatively thin, and smears taken from the surface show cells with vesicular nuclei from the intermediate layer.

As the cycle progresses to the ovulatory phase, more layers of epithelium are added, and the maturing cells accumulate glycogen. Vaginal smears at this point show many large, eosinophilic-staining, cornified cells with small pyknotic or absent nuclei. The percentage of these cells on a vaginal smear is a sensitive index of estrogenic activity (see Fig. 51-24). Progesterone, on the other hand, reduces the percentage of cornified cells. Vaginal secretions are increased by estradiol, and they too form an important element in the events leading to fertilization.

Breasts. The mammary glands consist of a large series of lobular ducts lined by an epithelium that is capable of secreting milk. These ducts empty into larger milk-conveying ducts that converge at the nipple. These glandular structures are embedded in supporting adipose tissue, and the breasts are separated into lobules by connective tissue.

The development of adult-sized mammary glands is absolutely dependent on estrogens. Before puberty the breasts grow only in proportion to the rest of the body. After the onset of increased estrogen secretion that occurs with pubescence, the growth of the lobular ducts is accelerated, and the area around the nipple (the areola) enlarges. These effects of estrogen on the glandular cells may be mediated by primary action on adjacent stromal cells.

Estrogens also selectively increase the adipose tissue of the breast, giving it its distinctive female shape. The lobular ducts are capable of outpouching to form numerous secretory alveoli. This process is stimulated by progesterone. In various ways, cortisol, growth hormone, prolactin, epidermal growth factor, insulin, IGF-1, and transferrin all contribute to the growth and differentiation of breast tissue. During each menstrual cycle further proliferation of the lobules occurs primarily in parallel with estradiol levels but possibly is fortified by progesterone. This causes swelling of the breasts; however, by the end of the luteal phase breast size and tenderness diminish.

Other tissues. During puberty estradiol is to the female what testosterone is to the male. Estradiol causes almost all the changes that result in the normal adult female phenotype. In addition to stimulating growth of the internal reproductive organs and breasts, estrogens cause pubertal enlargement of the labia majora and labia minora. Linear growth is accelerated by estradiol. However, because the epiphyseal growth centers are more sensitive to estradiol than to testosterone, they close sooner. For this reason the average height of women is less than that of men. The hips enlarge, and the pelvic inlet widens, facilitating future pregnancy. The predominance of estradiol over testosterone as a gonadal steroid in women is responsible for the fact that their total body adipose mass is twice as large as that of men, whereas their muscle and bone mass is only two thirds that of men. The specific deposition of fat about the hips is another effect of estradiol.

The adult skeleton, the kidney, and the liver are also target tissues of estrogens. Estrogen inhibits bone resorption; the loss of this hormone action after menopause contributes to a declining bone mass (osteoporosis) and a resultant increased frequency of fractures. Reabsorption of sodium from the renal tubules is stimulated by estradiol, and this may contribute to the cyclic fluid retention noted by some women. The hepatic synthesis of a number of circulating proteins is increased, including thyroxine-binding globulin, cortisol-binding

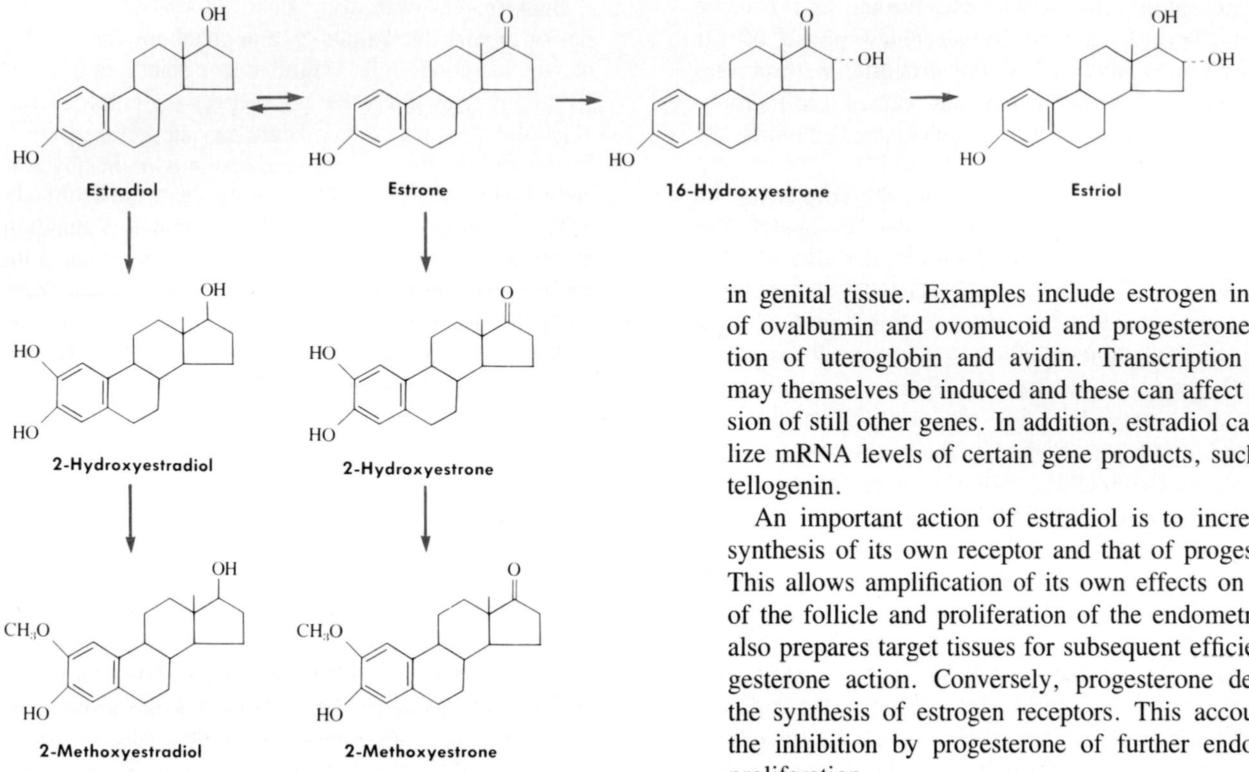

Estradiol ⇄ Estrone → 16-Hydroxyestrone → Estriol

2-Hydroxyestradiol 2-Hydroxyestrone

2-Methoxyestradiol 2-Methoxyestrone

■ **Fig. 51-25** Metabolism of estrogens.

globulin, sex steroid–binding globulin, the renin substrate, angiotensinogen, and very–low density lipoproteins. Estrogens also elevate plasma glucose in susceptible patients.

Only a few systemic actions of progesterone are known. Body temperature is increased by this steroid, accounting for the 0.5° C rise that occurs shortly after ovulation (see Fig. 51-24). Central nervous system actions include an increase in appetite, a tendency to somnolence, and a heightened sensitivity of the respiratory center to stimulation by carbon dioxide. Because progesterone is an aldosterone antagonist, it can induce natriuresis.

■ *Mechanisms of Action of Gonadal Steroids*

Estradiol, estrone, other estrogens, and progesterone all enter cells freely and bind to receptors of the steroid superfamily (see Chapter 50). The estrogen receptor contains 595 amino acids, and the progesterone receptor, 934 amino acids. These receptor proteins are variably phosphorylated. After combination in the cytoplasm or nucleus, the steroid-receptor complex is transformed into an active state. After dimerization, it more efficiently binds to its DNA regulatory element and may also form a complex there with other transcription factors. Gene expression is consequently enhanced or suppressed and various proteins are increased or decreased in genital tissue. Examples include estrogen induction of ovalbumin and ovomucoid and progesterone induction of uteroglobin and avidin. Transcription factors may themselves be induced and these can affect expression of still other genes. In addition, estradiol can stabilize mRNA levels of certain gene products, such as vitellogenin.

An important action of estradiol is to increase the synthesis of its own receptor and that of progesterone. This allows amplification of its own effects on growth of the follicle and proliferation of the endometrium. It also prepares target tissues for subsequent efficient progesterone action. Conversely, progesterone decreases the synthesis of estrogen receptors. This accounts for the inhibition by progesterone of further endometrial proliferation.

Several synthetic compounds have important pharmacological actions by virtue of their ability, as antagonists or as partial agonists, to bind to estrogen and progesterone receptors. **Tamoxifen** binds to estrogen receptors and suppresses the growth of estrogen-responsive breast cancer. **Clomiphene** binds to hypothalamic estrogen receptors in particular. The resulting functional blockade mimics estrogen deficiency at that site and stimulates LH and FSH secretion by negative feedback, which can initiate an ovulatory cycle. **Mifepristone** binds to the progesterone receptor and induces early abortion by loss of progesterone support for the conceptus.

■ *Metabolism of Gonadal Steroids*

Estradiol and estrone bind to SSBG, but their affinities are much lower than that of testosterone. Therefore the estrogens circulate bound loosely to albumin, and they have relatively high metabolic clearance rates (see Table 51-3). In menstruating women most of the circulating estradiol is derived from ovarian secretion; a minor fraction is formed from testosterone in adipose tissue, liver, and other sites. Most of the circulating estrone is derived from estradiol by peripheral 17-hydroxysteroid dehydrogenases. In postmenopausal women estrone is the dominant circulating estrogen, and it is formed from adrenal and theca cell androgens. It can be 16-hydroxylated to estriol (Fig. 51-25).

Sulfated and glucuronidated derivatives of all three estrogens are excreted in the urine. Values range from

20 μg during the early follicular phase to 65 μg at the preovulatory peak. An additional pathway of estrogen metabolism involves 2-hydroxylation and produces the so-called catechol estrogens (see Fig. 51-25). These compounds resemble the catecholamine neurotransmitters norepinephrine and dopamine in their hydroxylated benzene rings. Because 2-hydroxylase activity is present in the hypothalamus, it has been suggested that catechol estrogens generated within the brain might modulate estradiol effects on GnRH release. They bind to estradiol receptors but do not have estradiol actions; in effect, they are natural antiestrogen agents that could increase GnRh by negative feedback.

Progesterone can bind to cortisol-binding globulin, but this is largely prevented by the much higher plasma cortisol concentration. Therefore progesterone circulates loosely bound to albumin. It is reduced to the urinary metabolite, pregnanediol. During the follicular phase of the cycle about half the circulating progesterone is secreted by the ovary and half by the adrenal glands. During the luteal phase, however, the vast majority originates in the ovary.

In women 70% to 80% of circulating testosterone is derived from peripheral conversion of DHEA and androstenedione. About half the daily production stems from adrenal activity and half from ovarian precursors. In pathological situations, ovarian cells can secrete sufficient testosterone to cause virilizing effects.

Female Puberty

The general process of initiation of puberty has already been described (see Fig. 51-7). Reproductive function begins after an increase in gonadotropin secretion from the low levels of childhood (Fig. 51-26). Females differ from males in more clearly demonstrating an earlier rise in FSH than in LH (compare Figs. 51-17 and 51-26). Budding of the breasts is the first observable physical sign of puberty and coincides with the first detectable increase in plasma estradiol, as ovarian secretion commences. The onset of menses (menarche) occurs approximately 2 years later, after LH levels have risen more sharply. Menarche is correlated with both body and bone maturation. It can be delayed by undernutrition or strenuous exercise and occurs later in large sibships. Menarche is accelerated by obesity and blindness.

Because development of the positive feedback effect of estradiol necessary to provoke a preovulatory LH burst is the last step in the maturation of the hypothalamic pituitary ovarian unit, ovulation usually does not occur in the first few cycles. Initial irregularity of

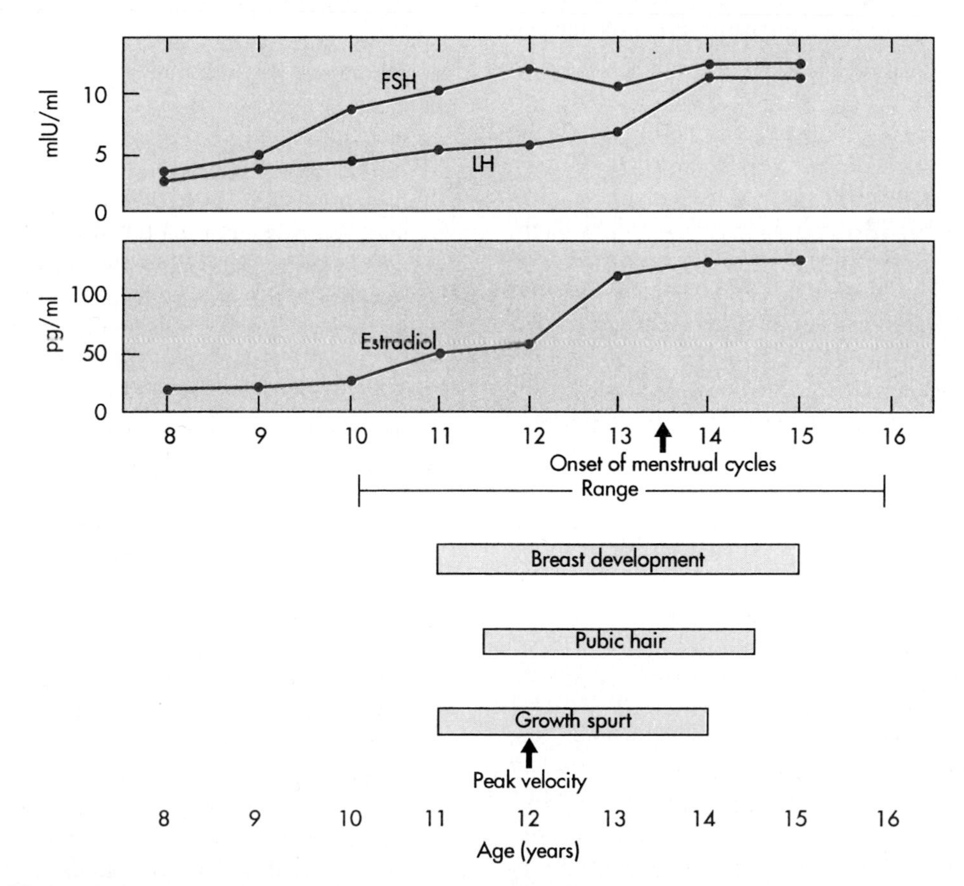

Fig. 51-26 Average chronological sequence of hormonal and biological events in normal female puberty. (Redrawn from Lee PA et al: Puberty in girls, *J Clin Endocrinol Metab* 43:775, 1976; and Marshall WA, Tanner JM: *Arch Dis Child* 45:13, 1970.)

menstrual cycles is therefore common, as the menstrual bleeding is induced by withdrawal of estrogen from graafian follicles undergoing atresia.

The growth spurt and the peak velocity of growth are characteristically earlier in girls than in boys. Further increase in height usually ceases 1 to 2 years after the onset of menses. The development of pubic hair precedes menses, and it correlates best with rising levels of adrenal androgens, especially DHEA-S.

Sexual Functioning

The desire for sexual activity is increased by androgens. During sexual intercourse vascular erectile tissue beneath the clitoris is activated by parasympathetic impulses. This causes the introitus to be tightened around the penis. Simultaneously these impulses stimulate copious secretion of mucus by glands located beneath the labia minora and in the vagina. The secretions lubricate the vagina and help it produce a massaging effect on the penis.

Female orgasm results from spinal cord reflexes similar to those involved in male ejaculation. Involuntary contractions of the skeletal muscle of the perineum, of the musculature of the vagina, uterus, and tubes, and of the rectal sphincter occur. The clitoris retracts against the symphysis pubis. After orgasm the cervix remains widely patent for 20 to 30 minutes, permitting entrance of sperm into the uterus. The first wave of sperm may reach an ovum in the fallopian tube within 10 minutes. However, lacking capacitation, these sperm are unlikely to accomplish fertilization.

Many spermatozoa are trapped and eventually destroyed in the vagina within a few hours. The remainder reach the cervix, where they dwell in storage sites formed by the convoluted mucosa (cervical crypts) and its mucus, and are capacitated. From this reservoir, spermatozoa migrate into the uterine cavity and fallopian tubes over 24 to 48 hours, undergoing tremendous losses along the way. Fewer than one in every 100,000 eventually reach an ovum.

Menopause

The reproductive capacity of women begins to wane in the fifth decade of life, and menses terminate at an average age of 50. For several years before menopause, the frequency of ovulation decreases. The menses occur at variable intervals and with decreased flow, caused by irregular peaks of estradiol without adequate secretion of progesterone in the luteal phase. With the disappearance of virtually all follicles, ovarian secretion of estrogens essentially ceases. From then on, maintenance of the lowered plasma estradiol concentrations (characteristic of menopause) depends on peripheral conversion of androgen precursors secreted by the adrenal glands, with some contribution from ovarian stromal cells. The dominant estrogen becomes estrone rather than estradiol (estrone 3 to 7 ng/dl versus estradiol 1 to 2 ng/dl; compare with Table 51-3).

During the last few years of reproductive life follicular sensitivity to gonadotropin stimulation diminishes, and plasma FSH and LH gradually increase in compensation. Once menopause occurs, the loss of negative feedback from estradiol and inhibin causes gonadotropin levels to average four to ten times those of the normal follicular phase, and the LH/FSH ratio falls to less than 1. Although the cyclicity of gonadotropin secretion is lost, pulsatile secretion persists.

The programmed decline in available estrogenic biological activity causes thinning of the vaginal epithelium and loss of its secretions, a decrease in breast mass, and an accelerated loss of bone. Phenomena such as vascular flushing, which is entrained with LH pulses, emotional lability, and an increase in the incidence of coronary vascular disease also are related to estrogen deficiency. Because adipose tissue contains aromatase activity, it is an important site of production of estrogen from stromal and adrenal androgens. Obese women may therefore suffer somewhat less from estrogen deprivation. Ovarian secretion of androgens may mildly stimulate hair growth in a male pattern.

Pregnancy

Fertilization and Implantation

After ovulation, the ovum is captured by the widened proximal portion of the fallopian tube (ampulla). This is aided by adherence of the "sticky" cumulus oophorus to the cilia in the fimbria. Muscle contractions produce a to-and-fro motion, which mixes the contents and increases the chance of a random encounter between the ovum and sperm, thereby facilitating fertilization. For the ovum, this must take place within 12 to 24 hours. The sperm must reach the ovum within about 48 hours of ejaculation.

Once it is very close to the ovum, the capacitated sperm undergo the acrosomal reaction described previously. As a result of Ca^{++} stimulation, the acrosomal cap releases a corona-penetrating enzyme, a trypsin-like enzyme (known as acrosin), a neurominidase, and hyaluronidases. Together, these disperse and digest the granulosa cells of the cumulus oophorus and the corona radiata and permit attachment to and penetration of the zona pellucida. The latter action is facilitated by sperm receptors for zona proteins, in particular one termed ZP_3.

Penetration of the zona pellucida by the first sperm creates a block to entry by other sperm. This barrier is generated by sequential uptake of Ca^{++} into the ovum, release of materials contained in granules within the ovum, and a resultant alteration in zona surface glycoproteins such as ZP_3. This alteration makes them now reject additional sperm. This important step prevents **polyploidy,** the production of an organism with more than two sets of homologous chromosomes. The polar body resulting from the second reduction-division of meiosis is then released, leaving the ovum in a haploid state, i.e., with 23 chromosomes. After fusion of their respective membranes, the chromatin material of the sperm head is engulfed by the ovum and forms the haploid male pronucleus. The two pronuclei generate a spindle on which the chromosomes are arranged, and a new diploid individual with 46 chromosomes is created.

The zygote, now in the blastocyst stage, traverses the tube in about 3 days. Most of this time is spent in the ampulla, where a delay at the junction with the isthmus may allow time for the endometrium to become better prepared. With the early rise in luteal phase progesterone, transit time through the tube rapidly increases. After another 2 to 3 days in the uterus the zygote initiates implantation. The requisite dissolution of the zona pellucida is brought about by alternate contraction and expansion of the blastocyst, as well as by lytic substances in the uterine secretions. Although their nature is unknown, these and other factors necessary for implantation depend on adequate luteal phase progesterone levels. Very early paracrine signals from the zygote likely induce receptive endometrial responses at the site.

From the initial solid mass of blastocyst cells, a layer of trophoblasts separates. Microvilli of these cells interdigitate with those of endometrial cells, and junctional complexes form between the respective cell membranes. Once firmly attached, trophoblast cells intrude between and burrow beneath endometrial cells, lysing the intercellular matrix with a variety of enzymes. In addition, the trophoblasts phagocytize and digest dead endometrial cells. Prostaglandins, probably of endometrial cell origin, and histamine may also participate in the implantation process.

Penetration by the trophoblasts is limited by concurrent changes in the stroma of the uterus. Late in the normal luteal phase, fibroblast-type stromal cells near uterine blood vessels enlarge and accumulate glycogen and lipid. These decidual cells disappear unless pregnancy supervenes and the corpus luteum is maintained. In the latter case, however, continuing estrogen and especially progesterone stimulate widespread decidualization, rapidly changing the entire stroma into a sheet of compact decidual cells. At the same time, the endometrial glands progressively atrophy. The decidua functions initially as a source of essential nutrients for the embryo until the central circulation has been established. Thereafter, the decidua may provide a mechanical and an immunological barrier to further invasion of the uterine wall. Finally, the decidua also functions as an endocrine organ, producing prolactin, relaxin, prostaglandins, and 1,25-$(OH)_2$-vitamin D.

■ *Functions of the Placenta*

Pregnancy is marked by the development of a unique organ, the placenta, with a limited life span. This organ serves as the fetal gut in supplying nutrients, as the fetal lung in exchanging oxygen and carbon dioxide, and as the fetal kidney in regulating fluid volumes and disposing of waste metabolites. In addition the placenta functions as a versatile endocrine gland, capable of synthesizing many hormones that can affect both maternal and fetal metabolism. Some exhibit regular concentration profiles in maternal plasma (Fig. 51-27), and some also can be found in fetal plasma and amniotic fluid.

Fetal trophoblasts are differentiated very early into two types: an inner layer of cytotrophoblasts and, under the influence of epidermal growth factor and other stimuli, an outer layer of fused syncytiocytotrophoblasts. Both cell types synthesize peptide and protein hormones, many of which are identical or very similar to hypothalamic, pituitary, and gonadal products. The syncytiocytotrophoblasts also synthesize steroid hormones from steroid precursors (Fig. 51-28). The adjacent arrangement of cytotrophoblast and syncytiocytotrophoblast layers forms a placental hypothalamus-pituitary–like unit. The cytotrophoblasts secrete primarily stimulatory and inhibitory hypothalamic-like peptides and gonadal growth factors, which in paracrine manner, regulate output of pituitary-like hormones from the syncytiocytotrophoblast layer.

■ *Human Chorionic Gonadotropin*

Human chorionic gonadotropin (HCG) is the first key hormone of pregnancy. Secreted by the syncytiocytotrophoblast cells, it can be detected in maternal plasma and urine within 9 days of conception. The secretion of HCG may be stimulated by GnRH produced in adjacent cytotrophoblasts. HCG is a glycoprotein of 39,000 molecular weight with two subunits. The α subunit is identical to that of thyroid-stimulating hormone (TSH), LH, and FSH, whereas the β subunit has 80% homology with that of LH. However, specific radioimmunoassays, developed to measure the β subunit of HCG permit reliable discrimination of it from LH. Detection of HCG is the most commonly employed and most specific test for pregnancy. Maternal plasma levels of HCG increase at an exponential rate, reach a peak at 9 to 12 weeks, and then decline to a stable plateau for the remainder of pregnancy (see Fig. 51-27). After delivery, HCG disappears from maternal plasma with a half life of 12 to 24 hours.

HCG acts to maintain the function of the corpus luteum beyond its usual life span of 14 days. The placental gonadotropin stimulates ovarian secretion of progesterone and estrogens by mechanisms essentially identical to those previously described for LH. When the pla-

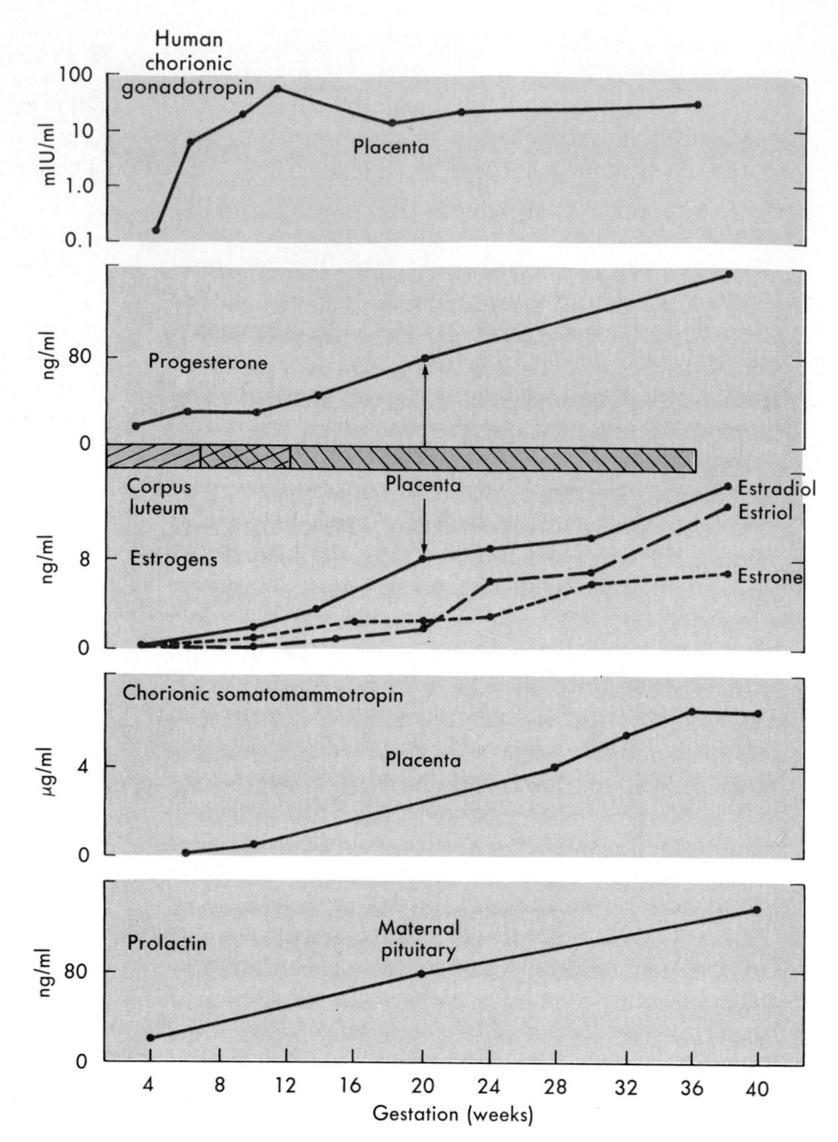

■ **Fig. 51-27** Profile of plasma hormone changes during normal human pregnancy. Note logarithmic scale for HCG. Also note shift from corpus luteum to placenta as the source of estrogens and progesterone between 6 and 12 weeks of gestation. (Redrawn from Goldstein DP et al: *Am J Obstet Gynecol* 102:110, 1968; Rigg LA et al: *Am J Obstet Gynecol* 129:454, 1977; Selenkow HA et al: *Measurement and pathophysiologic significance of human placental lactogen.* In Pecile A, Finzi C: *The foetoplacental unit,* Amsterdam, 1969, Excerpta Medica; and Tulchinsky D et al: *Am J Obstet Gynecol* 112:1095, 1972.)

centa assumes the synthesis of these steroids, thereby relieving the fetus of its dependence on the corpus luteum, HCG secretion declines. The role of the gonadotropin during the remainder of pregnancy is less certain, but other actions are attributed to it. HCG has an inhibitory effect on maternal pituitary LH secretion. Because of its structural overlap with TSH, the plasma concentrations of HCG in pregnancy are high enough to stimulate an increase in maternal thyroid gland activity. In pathological excess HCG can even induce hyperthyroidism. HCG that reaches the fetus may stimulate DHEA-S production by the fetal zone of the adrenal gland and testosterone production by the Leydig cells of the testis. HCG also may stimulate the production of relaxin (see later discussion).

■ *Progesterone*

Progesterone is the hormone most directly responsible for the establishment and sustenance of the fetus in the uterine cavity. As just noted, during the first 2 weeks of pregnancy, progesterone stimulates the tubal and endometrial glands to secrete nutrients on which the free-floating zygote depends. Thereafter, it maintains the decidual lining of the uterus. Progesterone produced by the placenta is the principal substrate for synthesis of cortisol and aldosterone by the fetal adrenal gland (see Fig. 51-28). Lacking the 3β-OL dehydrogenase-$\Delta^{4,5}$ isomerase enzyme complex (see Fig. 50-3), the fetal adrenal gland cannot itself make progesterone. Progesterone may also modulate the secretion of HCG and human chorionic somatomammotropin (HCS).

Other actions of progesterone are important during pregnancy. The steroid inhibits uterine contractions, in part by inhibiting production of prostaglandins and in part by decreasing sensitivity to oxytocin. Progesterone thus prevents premature expulsion of the fetus. It also stimulates the development of the alveolar pouches of the mammary glands, greatly magnifying their eventual capacity to secrete milk. Progesterone may participate in the inhibition of maternal immune responses to antigens from the fetus and thereby help prevent its rejection. Finally, progesterone stimulates the maternal respiratory center to increase ventilation, which helps to dispose of the increased carbon dioxide that results from pregnancy.

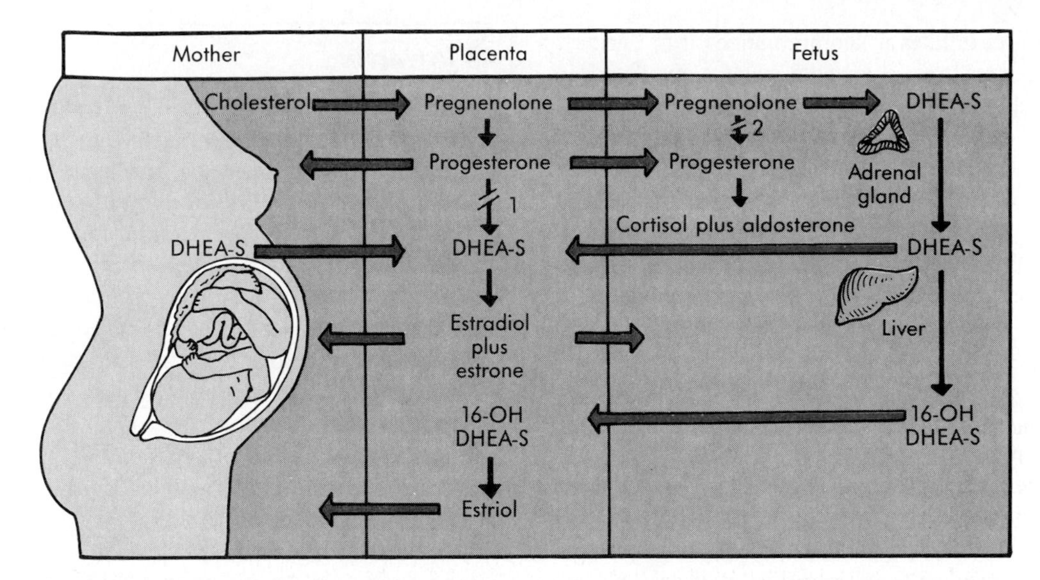

Fig. 51-28 The maternal-fetal-placental unit in steroid hormone synthesis. Progesterone is synthesized in the placenta from maternal cholesterol. In turn this progesterone acts on the mother and also serves as the precursor to fetal cortisol and aldosterone synthesis. Estradiol and estrone are synthesized in the placenta from maternal and fetal DHEA-S and estriol from fetal 16-OH-DHEA-S. *DHEA-S,* Dehydroepiandrosterone sulfate; *16-OH-DHEA-S,* 16α-hydroxydehydroepiandrosterone sulfate; 1, 17-hydroxylase/17,20-desmolase; 2, 3β-OL-dehydrogenase-$\Delta^{4,5}$-isomerase.

The placenta begins to synthesize progesterone at about 6 weeks, and by 12 weeks it is producing enough to replace the corpus luteum for this purpose. During this transition period the otherwise progressive rise in plasma progesterone reaches a temporary plateau (see Fig. 51-27). Cholesterol, present in low-density lipoproteins, is extracted from maternal plasma and serves as the major precursor for placental progesterone. The synthetic pathway is like that of the adrenal gland and the ovary. By term, progesterone production reaches a level of 250 mg/day, which is tenfold greater than the peak rates seen during the luteal phase of the menstrual cycle. About 90% of the progesterone goes to the mother and 10% to the fetus. Maternal urinary pregnanediol excretion also rises markedly, reflecting this huge increase in production of progesterone.

Estrogens

Augmented production of estrogens (estradiol, estrone, and estriol) also occurs throughout pregnancy (see Fig. 51-27), resulting in several important actions. Estrogens stimulate (1) the continuous growth of the uterine myometrium, preparing it for its role in labor, (2) the further growth of the ductal system of the breast, out of which the alveoli will develop, and (3) the enlargement of the external genitalia. In addition, estrogen, along with relaxin, causes relaxation and softening of the pelvic ligaments and the symphysis pubis of the pelvic bones, allowing better accommodation of the expanding uterus.

Estrogens play a paracrine role in placental function. They augment progesterone synthesis by increasing LDL cholesterol uptake and the activity of the $P\text{-}450_{SCC}$ enzyme. They also enhance placental conversion of cortisol to inactive cortisone. This inactivation may relieve inhibition by maternal cortisol of fetal pituitary corticotropin; the latter can then stimulate the production of essential fetal adrenal androgen and later, of fetal cortisol.

Like progesterone, estrogens are initially produced by the corpus luteum under stimulation by HCG. The placenta then assumes this role, but requires steroid hormone precursors from both the maternal and fetal compartments to complete the synthesis of estrogens. This unique example of coordinated maternal-placental-fetal function is depicted in Fig. 51-28. The placenta lacks significant 17-hydroxylase and 17-20 desmolase activity and therefore cannot generate the androgens that serve as substrates for aromatization (see Fig. 51-4). Instead, the placenta extracts DHEA-S (derived from the maternal and fetal adrenal glands), removes the sulfate, and synthesizes estradiol and estrone. The fetal source of DHEA-S predominates as pregnancy progresses. Placental synthesis of estriol, a 16-hydroxylated estrogen, almost entirely depends on precursors from the fetus (see Fig. 51-25). The fetal adrenal gland synthesizes DHEA-S from placental pregnenolone, the fetal liver hydroxylates it in the 16 position, and the placenta then desulfates it and aromatizes it to estriol.

One third of the unconjugated estrogens circulating in maternal plasma at term is accounted for by estriol (see Fig. 51-25). In the form of sulfate and glucuronic acid conjugates, estriol represents 90% of the total estrogen excreted in maternal urine. Because estriol is derived almost entirely from the fetal placental unit, measurement of this estrogen in maternal plasma or urine provides an index to the state of well-being of the fetus.

Human Chorionic Somatomammotropin

Another protein hormone, unique to pregnancy, is **human chorionic somatomammotropin (HCS),** also called human placental lactogen (HPL). The synthesis of HCS by the syncytiocytotrophoblasts can be detected at about 4 weeks of gestation. The maternal plasma concentration rises steadily to a peak of 6 μg/ml at term (see Fig. 51-27). The HCS production rate of 1 to 2 g/day far exceeds that of any other human protein hormone. HCS concentrations correlate well with placental weight; therefore measurement of maternal plasma HCS provides another indicator of placental function. After delivery the hormone rapidly disappears from maternal plasma with a half-life of 20 minutes.

HCS is synthesized by direction of a gene in the growth hormone (GH) family. Although HCS has 96% structural homology with GH, it has only 3% of the latter's growth promoting activity. Nonetheless, its very high plasma concentrations can contribute to anabolism in the pregnant woman. HCS also has lactogenic activity, but this action may be unneeded because of the high concentration of prolactin itself during pregnancy.

HCS stimulates lipolysis and, like GH, antagonizes insulin actions on carbohydrate metabolism; this tends to raise plasma glucose. GnRH and somatostatin produced in neighboring cytotrophoblasts likely stimulate and inhibit HCS synthesis by syncytiocytotrophoblasts. Maternal fasting and hypoglycemia raise plasma HCS levels. As is detailed later, HCS appears to direct maternal metabolism to maintain a continuous flow of substrates, especially glucose, to the fetus. HCS levels in fetal plasma are far below those in maternal plasma, and there is no known direct role for HCS in the fetus. However, somatomedins (IGF-2), produced in the placenta as the result of HCS action, may help to stimulate fetal growth.

Recently a placental human growth hormone variant, closely related to but distinct from both GH and HCS, has been characterized. This variant is by far the major form of growth hormone in maternal plasma, and it reaches elevated levels of 15 ng/ml by term. It is likely regulated by placental GhRH and somatostatin and, because of its high concentrations, may prove to be as important metabolically as HCS.

Prolactin

Another hormone secreted in excess during normal pregnancy is maternal pituitary prolactin. Plasma levels rise linearly, by term reaching values eightfold to tenfold higher than those of nonpregnant women (see Fig. 51-27). The prolactin is largely in the more bioactive nonglycosylated form (see Chapter 48). Prolactin is essential for expression of the mammotropic effects of estrogen and progesterone. More specifically, prolactin stimulates the lactogenic apparatus (Chapter 48, Fig. 48-25). Minimal formation of milk begins at about 5 months of gestation, but significant lactation is inhibited by the great excess of estrogen and progesterone. Lactation is initiated after delivery by the precipitous drop in estrogen and progesterone levels. Although basal prolactin concentrations gradually decline to normal over the next 4 to 8 weeks, they are acutely elevated during each period of suckling (Chapter 48, Fig. 48-23), helping to sustain milk secretion.

Prolactin also suppresses reproductive function in the nursing mother. During the first 7 to 10 days postpartum, plasma FSH and LH levels remain low. FSH then rises to above-normal follicular phase levels, but LH still remains low. The responsiveness of the ovaries to FSH probably is reduced by prolactin, and LH secretion by the pituitary probably is inhibited by prolactin.

A decrease in circulating prolactin, because of either cessation of nursing or administration of a dopaminergic agonist, plays a role in triggering LH release and initiating cycling. Prolactin is also synthesized by the decidual cells of the pregnant uterus. This is the source of the amniotic fluid prolactin, the function of which is currently unknown, but it may relate to regulation of osmolarity in fetal fluids.

Relaxin

In addition to gonadal steroids, the corpus luteum of pregnancy secretes a polypeptide hormone called relaxin. Its structure resembles proinsulin (Fig. 46-3). Plasma levels of relaxin rise early, peak in the first trimester, and then decline somewhat. The production of relaxin by the corpus luteum is stimulated by HCG. Relaxin is also produced by decidual cells. The hormone relaxes pelvic bones and ligaments, inhibits myometrial contractions, and softens the cervix. Thus it may function early to ensure uterine quiescence and prevent abortion but later to facilitate passage out of the uterus.

Other Hormones

The cytotrophoblast-syncytiocytotrophoblast complex reproduces locally several other recognizable hormone axes. One is the CRH-proopiomelanocortin axis, which results in production of a chorionic adrenocorticotropin, β endorphin, lipotropins, and melanocyte-stimulating hormone. Another is the TRH-chorionic thyrotropin pair. The exact roles of these units in regulating maternal or fetal metabolism have yet to be clarified.

Other Maternal Hormonal Changes

The pregnant state induces a characteristic series of changes in a number of other maternal hormones. One of the most significant alterations is in pancreatic islet β-cell function. Insulin secretion, in response to glucose challenge or to meals, increases after the third month of pregnancy. This hypersecretion reaches its peak during the last trimester, coinciding with the peak of plasma HCS. Because maternal sensitivity to insulin is greatly diminished during this same period, insulin hypersecretion may be considered compensatory. In contrast, basal glucagon levels and responses to stimulation do not change significantly.

Aldosterone secretion increases significantly throughout pregnancy, reaching a sixfold to eightfold elevation by term. This is because of increased function of the renin angiotensin system; plasma renin and the renin substrate (angiotensinogen) are both augmented by the high estrogen levels of pregnancy. Aldosterone hypersecretion may be stimulated by a reduction in *effective* circulating blood volume that results from the large placental blood pool. Hyperaldosteronism contributes to the positive sodium balance that is needed to maintain a high total maternal plasma volume and to build the extracellular fluid of the fetus. Another mineralocorticoid, desoxycorticosterone, is also present in thousandfold excess in plasma during pregnancy. It is synthesized in the maternal kidneys by 21-hydroxylation of progesterone originating from the placenta.

The plasma total cortisol level is elevated because of the estrogen-induced increase in cortisol-binding globulin. However, plasma and urinary free cortisol also rises modestly, possibly due to stimulation of the maternal adrenal gland by chorionic adrenocorticotropin and/or stimulation of the maternal pituitary by CRH of placental origin. Maternal plasma levels of ACTH and CRH both increase during pregnancy. The enhanced glucocorticoid activity may contribute to maternal adipose tissue gain and to mammary gland development. It also may be responsible for the plethoric face, thin skin, and susceptibility to bruising of the pregnant woman.

Calcium absorption from the diet increases during pregnancy and offsets the continuing drain created by the growing fetal skeleton. This increase in calcium absorption is accounted for by increased maternal levels of 25-OH-vitamin D and 1,25-(OH_2)-vitamin D. The latter active metabolite in part originates from decidual and placental production. As a result, ionized calcium levels are maintained at normal levels and maternal parathyroid hormone (PTH) secretion is partially suppressed, as manifested by a 50% reduction in maternal PTH levels in the plasma.

Total plasma thyroid hormones, thyroxine (T_4) and triiodothyronine (T_3), are elevated because of estrogen-induced increases in thyroid-binding globulin. Plasma free T_4 concentration usually remains within the normal range. Nonetheless, there are increases during pregnancy in maternal thyroid gland size, radioactive iodine uptake, basal metabolic rate, and resting pulse rate, all of which are compatible with some augmentation of thyroid gland activity. This may be caused by the thyrotropic activity of HCG or secretion of chorionic thyrotropin.

Maternal pituitary GH secretion in response to various stimuli decreases during pregnancy, probably because its anabolic functions are carried out by HCS and/or placental growth hormone variant. Maternal LH and FSH secretion also is suppressed by the high levels of estrogen, progesterone, and inhibin from the corpus luteum and later the placenta.

Maternal-Fetal Metabolism

During normal pregnancy the average gain in maternal weight is 11 kg. About half of this is attributable to changes in maternal tissues and half to the conceptus. The typical distribution of the excess weight is shown in Fig. 51-29. Approximately 75,000 extra calories (250 to 300 kcal/day) must be ingested to support this weight gain; 65,000 kcal support fetal metabolism and growth, and 10,000 kcal are stored in maternal fat. An extra protein intake of 30 g/day ensures adequate supplies for maternal needs and for the accumulation of fetal protoplasm. At birth the protein content of the fetus has reached 400 to 500 g.

From the metabolic standpoint, pregnancy can be divided into two phases. For approximately the first half the mother herself is in an anabolic phase, and the conceptus represents an insignificant nutritional drain. During the second half, and especially the final third, fetal and placental weight increases at an accelerated rate. These demands cause the mother to shift into a state aptly described as "accelerated starvation."

The initial anabolic phase is characterized by normal or even increased sensitivity to insulin. Progesterone and estrogen may facilitate insulin actions at this stage. Maternal plasma levels of glucose, amino acids, free fatty acids, and glycerol are normal or slightly reduced. Carbohydrate and amino acid loads are readily assimilated. Lipogenesis is favored, and lipolysis is braked in maternal adipose tissue, glycogen stores are increased in liver and muscle, and protein synthesis is enhanced.

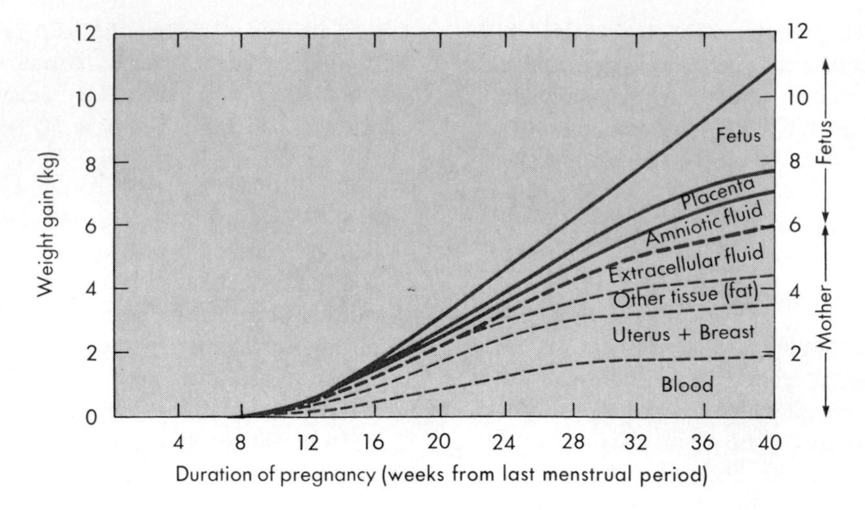

■ **Fig. 51-29** Pattern and components of maternal weight gain during normal pregnancy. (Redrawn from Pitkin RM: *Obstetrics and gynecology.* In Schneider HA, Anderson CE, Coursin DB: *Nutritional support of medical practice,* New York, 1977, Harper & Row.)

The net effect is to stimulate growth of the breasts, uterus, and essential musculature in the mother, while preparing her to withstand the metabolic demands of later fetal growth.

During the later catabolic phase of pregnancy the metabolism of the mother shifts into a mode that effectively accommodates the accelerating needs of the fetus. Insulin sensitivity is replaced by insulin resistance, the locus of which is beyond the insulin receptor. The assimilation of dietary carbohydrate, protein, and fat by maternal tissues is slowed, so that postprandial plasma levels of glucose and amino acids are elevated. This increases the rate of glucose diffusion and of facilitated amino acid transport across the placenta into the fetus. Glucose is the major fuel of the fetus, and the amino acids are required for fetal protein synthesis. By term, the fetus, who is using glucose at a rate of 5 mg/kg/minute compared with the maternal rate of 2.5 mg/kg/minute, must be supplied with up to 25 g/day. During fasting intervals maternal plasma glucose and amino acid levels fall more rapidly than in nonpregnant women, because of continued fetal siphoning of these substances. Conversely, lipolysis is accelerated, and maternal plasma free fatty acid, glycerol, and ketoacid levels rise more rapidly than in nonpregnant women. This ensures alternate oxidative fuels for the mother; in particular, ketoacids and, to a lesser extent, free fatty acids also cross to the fetus, where they may be used instead of some glucose. Placental HCS and/or growth hormone variant is probably the key hormone responsible for insulin resistance and for facilitating lipid mobilization during fasting in the latter stage of pregnancy. The increase in plasma free cortisol also may contribute to these actions.

Along with the other changes in maternal metabolism, plasma cholesterol and triglyceride levels rise throughout pregnancy. The cholesterol is partly used for estrogen and progesterone synthesis. The increased circulating triglycerides are largely the result of an increased hepatic synthesis of very-low density lipoprotein, which is stimulated by estrogens. Some of the triglycerides are stored in the breasts in preparation for milk production. They are shifted away from less specific storage elsewhere by a marked reduction in adipose tissue lipoprotein lipase levels.

■ *Parturition*

Just as the maintenance of the pregnant state depends on a unique hormonal milieu, its termination probably also depends on specific hormonal changes. However, the exact mechanism remains unclear. Roles for cortisol, estrogen, progesterone, relaxin, oxytocin, prostaglandins, and catecholamines in the initiation and maintenance of labor and the final uterine evacuation have been suggested by various lines of evidence. Because much species variation exists, it is difficult to extrapolate the results of studies in subprimates to humans. Fig. 51-30 shows current notions of the endocrine regulation of parturition.

Once the conceptus has reached a critical size, distension of the uterus itself and stretching of the muscle fibers increase their contractility. In the human, uncoordinated uterine contractions begin at least 1 month before the end of gestation. The inherent contractility of the uterus probably in itself would cause eventual evacuation of the conceptus. However, some signal from the fetus probably initiates active labor. In sheep, fetal cortisol has been strongly implicated. This agrees with the fact that fetal cortisol production rises sharply during the last few weeks of gestation. In addition in humans

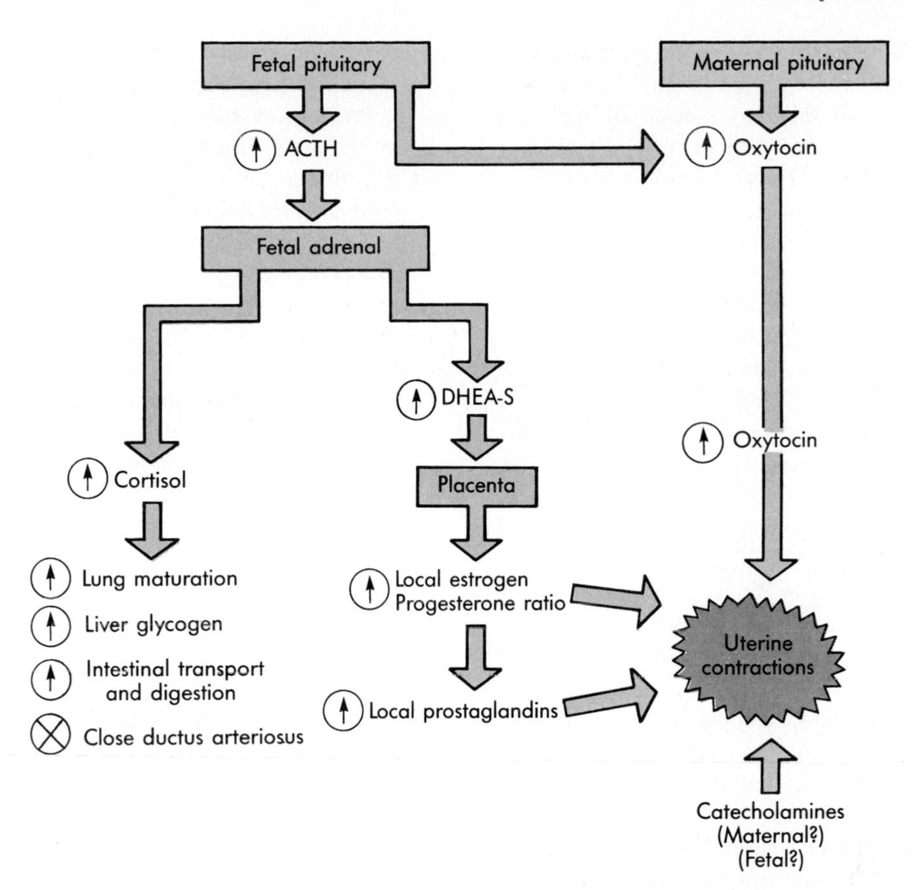

■ **Fig. 51-30** Endocrine regulation of parturition. The fetal pituitary-adrenal axis initiates signals that decrease the ratio of effective progesterone to estrogen in the myometrium. This leads to uterine contractions, which are mediated by prostaglandins. Oxytocin may contribute to labor but is not essential. However, oxytocin sustains uterine contractions after expulsion of the fetus so as to minimize maternal loss of blood. Cortisol prepares the fetus to adapt to extrauterine life successfully. *ACTH,* Adrenocorticotropic hormone; *DHEA-S,* dehydroepiandrosterone sulfate.

maternal plasma CRH levels rise sharply during labor. Although gestation is prolonged in women when the fetus lacks an intact hypothalamic pituitary adrenal unit, the evidence for a surge of fetal cortisol immediately preceding human parturition is weak at best. Nonetheless, the late gestational increase in cortisol secretion is important for preparing the fetus for the abrupt transition to extrauterine life. Cortisol stimulates lung maturation, increases stores of liver glycogen, induces intestinal transport systems and digestive enzymes, and promotes closure of the ductus arteriosus.

The relatively high ratio of progesterone to estrogens is important in maintaining uterine quiescence throughout pregnancy. In some species, but not in humans, a sharp drop in maternal plasma progesterone precedes parturition. Some evidence suggests the existence of a placental progesterone-binding protein whose concentration may be increased by estrogen near term, causing an effective removal of local progesterone from the myometrium and thereby decreasing uterine quiescence.

A key role for prostaglandins PGE_2 and PGF_2-α as intermediates is very likely. The production of these and other prostaglandins in the uterus is augmented by estrogens and suppressed by progesterone. These prostaglandins are found in high concentration in both maternal plasma and amniotic fluid (originating in the fetal membranes) during labor. Prostaglandins increase Ca^{++} within the myometrial cells and thus trigger uterine contractions. Their great effectiveness underlies their use as medical abortofacients. Although prostaglandins are not likely to be the signal for labor to begin, they probably do act as requisite second messengers.

The role of oxytocin in human labor is enigmatic. Unquestionably, this hormone causes myometrial contractions. Furthermore, its secretion can be reflexly increased by stretching of the lower genital tract. However, maternal plasma levels of oxytocin show no consistent increase just before or during early labor. On the other hand, oxytocin receptors in the uterus do increase throughout gestation and dramatically so at term. (This is one additional action of the decidual hormone, relaxin.) Thus stable levels of oxytocin from the mother, perhaps augmented by oxytocin from the fetus and from local syncytiocytotrophoblast production, may have a much greater effect on the musculature of the term uterus. Nevertheless, labor can proceed in the absence of this hormone. Its most important role may be to cause maximal uterine contractions after delivery of the fetus and thereby to minimize blood loss.

Both α- and β-adrenergic receptors are present in the myometrium; α-stimuli cause contraction, and β-stimuli cause relaxation. Therefore circulating catecholamines also may contribute to the final hormonal cascade of parturition.

Once labor has begun, it proceeds in three clinically recognized stages. In the first stage, lasting a variable number of hours, the uterine contractions, which originate at the fundus and sweep downward, force the head of the fetus against the cervix. This progressively widens and thins the opening to the vaginal canal. In the second stage, lasting less than 1 hour, the fetus is forced out of the uterine cavity and through the cervix and is delivered from the vagina. In the third stage, lasting 10 minutes or less, the placenta is separated from the decidual tissue of the uterus and forcefully evacuated. Myometrial contractions at this point act to constrict the uterine vessels and prevent excessive bleeding. Once the placenta has been removed, all its hormonal products disappear from the maternal plasma according to their characteristic half-lives. In general, by 48 to 72 hours the steroid and protein hormone concentrations have reached nonpregnant levels.

■ *Endocrine State of the Fetus*

No maternal protein or peptide hormone effectively traverses the placenta, with the exception of thyrotropin-releasing hormone (TRH) and possibly antidiuretic hormone (ADH). The same is essentially true for maternal thyroid hormones. Steroid hormones, in contrast, can move readily from the mother to the fetus, and vice versa. Catecholamines also cross to the fetus.

Although fetal pancreatic islets are functional by 14 weeks, insulin and glucagon secretion are relatively low. Neither is critically needed for substrate metabolism, because glucose and amino acids are in plentiful supply from the mother. Fetal β cells and α cells respond to their usual stimulators and suppressors in blunted fashion until birth, when responsiveness rapidly increases. Fetal insulin may contribute to anabolism and to deposition of adipose tissue.

Fetal GH is not essential for linear growth; although GH levels are high in plasma, GH receptors are deficient in the fetus. Instead, HCS, placental GH variant, and prolactin may subserve the function of GH prenatally. These may be responsible for the ubiquitous presence of IGF-1 and IGF-2 in fetal plasma and tissues. IGF-2 concentration is especially high, and IGF-2 may be the most important paracrine and autocrine growth factor in fetal life, with IGF-1 assuming this role after birth.

Prolactin concentrations are high in fetal plasma and in amniotic fluid. Prolactin may contribute to fetal growth, to osmotic regulation, and to the production of cortisol and DHEA-S.

The role of fetal thyroid hormone in fetal development is not completely defined. A very small transfer of maternal T_4 or T_3 to the fetus may be important for early fetal maturation. During the last two thirds of gestation, neither maternal nor fetal thyroid hormone may be essential for some developmental processes. At birth, the newborn's thyroid hormone becomes essential for central nervous system development and somatic growth.

Active transport of calcium across the placenta from the mother to the fetus is probably stimulated by fetal parathyroid and placental PTH_{rp} (see Chapter 47), and it keeps fetal plasma calcium levels high. In turn, this slight hypercalcemia inhibits fetal PTH secretion and stimulates fetal calcitonin secretion. Calcitonin, combined with 1,25-$(OH)_2$-vitamin D produced in the fetal kidney and the placenta, promotes bone formation. PTH secretion increases soon after birth, and assumes its regulatory role in calcium metabolism.

Fetal ACTH probably is not essential for the first 12 to 20 weeks, although later it definitely stimulates production of steroids by the fetal zone of the adrenal cortex. The newborn mounts an immediate stress response, as shown by high levels of cortisol in umbilical cord plasma. If endogenous ACTH and cortisol cannot be secreted at this time, death will ensue unless replacement therapy is provided.

■ *Summary*

1. Differences in gonadal function between the genders derive from the process of sexual differentiation. Genetic material on the Y chromosome determines development of a testis, and androgen hormone induces masculinization of the genital ducts and external genitalia. Without this positive input, a female (the neutral) pattern results. Two active X chromosomes are, however, required for oogenesis.

2. Testosterone and estradiol derived from testosterone are synthesized by common pathways and enzymes in homologous cell lines in the testis and ovary. Both genders exhibit gonadotropin surges in fetal life, quiescence in childhood, activation in puberty, and increases in negative feedback late in life caused by gonadal failure. The monthly cyclic ovulatory burst of LH/FSH is unique to the female.

3. Spermatogenesis proceeds within the seminiferous tubules in a locally conditioned hormonal environment behind a blood-testis barrier. Sertoli cells, stimulated by FSH, provide growth factors, binding proteins for tes-

tosterone and trace metals, inhibin, and other factors to nurture and launch spermatozoa.

4. LH stimulates testosterone secretion by Leydig cells. A high local concentration of testosterone is essential to spermatogenesis, either indirectly by affecting Sertoli cells or by direct access to germ cells. Its exact mechanism of action is unknown.

5. In the ovary, oocytes stimulate formation of a follicle, which is a secluded environment analogous to the seminiferous tubules. Under the influence of FSH, a cohort of follicles, with their oocytes suspended in meiosis, begin development each month. The surrounding granulosa cells secrete estradiol synthesized from androgen precursors, which are provided by neighboring theca cells under LH stimulation. Other granulosa cell products similar to those of the Sertoli cell condition the oocyte.

6. A single dominant follicle emerges each month. It grows exponentially and secretes sufficient estradiol to (a) inhibit cohort follicles, (b) prepare the uterus and fallopian tubes for fertilization, and (c) condition the GnRH-gonadotroph axis to provide an ovulatory LH/FSH surge at the appropriate time. After ovulation, the endocrine cells form a corpus luteum, which secretes predominantly progesterone. The latter alters the uterus to favor implantation of a zygote.

7. During puberty, testosterone in the male and estradiol in the female stimulate linear growth and skeletal maturation and enlargement and maturation of the accessory tissues of reproduction.

8. Male and female sexual functioning are stimulated by the autonomic nervous system. An ejaculate of 200 to 400 million sperm requires conditioning (capacitation) in the female genital tract to permit one sperm to fertilize the ovum in the fallopian tube. The zygote is subsequently maintained and protected by secretions from altered uterine cells (decidua) and from placental fetal trophoblastic cells.

9. The placenta initially produces human chorionic gonadotropin, which stimulates the corpus luteum; later the placenta produces its own estrogen and progesterone. In addition, a variety of placental peptides and proteins, including human chorionic somatomammotropin, somatomedins, and pituitary-like ACTH, TSH, and GH, are synthesized. These hormones affect maternal and possibly fetal metabolism. The mother is in an anabolic state early in pregnancy, a condition that facilitates growth of her energy stores and reproductive tissues. The later catabolic phase is marked by insulin resistance and facilitates flow of fuels to the growing fetus.

10. The exact endocrine mechanism of human parturition is unclear, but it likely includes contributions from an increased estrogen/progesterone ratio within the uterus, from relaxin, and from oxytocin. A local increase in prostaglandins is the immediate second messenger that causes uterine contractions. A late gestational rise in fetal cortisol enhances extrauterine survival.

■ Bibliography

Journal articles

Ascoli M, Segaloff DL: On the structure of the luteinizing hormone/chorionic gonadotropin receptor, *Endocr Rev* 10:27, 1989.

Belchetz PE et al: Hypophysial responses to continuous and intermittent delivery of hypothalamic gonadotropin-releasing hormone, *Science* 202:631, 1978.

Bryant-Greenwood GD: Relaxin as a new hormone, *Endocr Rev* 3:62, 1982.

Espey LL, BenHalim IA: Characteristics and control of the normal menstrual cycle, *Obstet Gynecol Clin North Am* 17:275, 1990.

George FW, Wilson JD: Hormonal control of sexual development, *Vitam Horm* 43:145, 1986.

Goebelsmann U: Protein and steroid hormones in pregnancy, *J Reprod Med* 23:166, 1979.

Harman SM et al: Reproductive hormones in aging men. 1. Measurement of sex steroids, basal luteinizing hormone, and Leydig cell response to human chorionic gonadotropin, *J Clin Metab* 51:33, 1980.

Jansen RPS: Endocrine response in the fallopian tube, *Endocr Rev* 5:525, 1984.

Johnson MD: Genes related to spermatogenesis: molecular and clinical aspects, *Semin Reproduct Endocrinol* 9:72, 1991.

Keyes PL, Wiltbank MD: Endocrine regulation of the corpus luteum, *Annu Rev Physiol* 50:465, 1988.

Liu JH, Yen SI: Induction of midcycle gonadotropin surge by ovarian steroids in women: a critical evaluation, *J Clin Endocrinol Metab* 57:797, 1983.

McLachlan RI et al: Circulating immunoreactive inhibin levels during the normal human menstrual cycle, *J Clin Endocrinol Metab* 65:954, 1987.

McNatty KP et al: The microenvironment of the human antral follicle: interrelationships among the steroid levels in antral fluid, the population of the granulosa cells, and the status of the oocyte in vivo and in vitro, *J Clin Endocrinol Metab* 49:851, 1979.

McNatty KP et al: The production of progesterone, androgens, and estrogens by granulosa cells, thecal tissue, and stromal tissue from human ovaries in vitro, *J Clin Endocrinol Metab* 49:687, 1979.

Muller U, Lattermann U: H-Y antigens, testis differentiation, and spermatogenesis, *Exp Clin Immunogenet* 5:176, 1988.

Papadopoulos V: Identification and purification of a human Sertoli cell-secreted protein (HSCSP-80) stimulating Leydig cell steroid biosynthesis, *J Clin Endocrinol Metab* 72:1332, 1991.

Pohl CR, Knobil E: The role of the central nervous system in the control of ovarian function in higher primates, *Annu Rev Physiol* 44:583, 1982.

Qu J et al: Circulating bioactive inhibin levels during human pregnancy, *J Clin Endocrinol Metab* 72:862, 1991.

Rabinovici J, Jaffe RB: Development and regulation of growth and differentiated function in human and subhuman primate fetal gonads, *Endocr Rev* 11:532, 1990.

Rich BH et al: Adrenarche: changing adrenal response to adrenocorticotropin, *J Clin Endocrinol Metab* 52:1129, 1981.

Richards JS, Hedin I: Molecular aspects of hormone action in ovarian follicular development, ovulation, and luteinization, *Annu Rev Physiol* 50:441, 1988.

Rories C, Spelsberg TG: Ovarian steroid action on gene expression: mechanisms and models, *Annu Rev Physiol* 51:653, 1989.

Roseff SJ et al: Dynamic changes in circulating inhibin levels during the luteal-follicular transition of the human menstrual cycle, *J Clin Endocrinol Metab* 69:1033, 1989.

Rossmanith WG et al: Pulsatile cosecretion of estradiol and progesterone by the midluteal phase corpus luteum: temporal link to luteinizing hormone pulses, *J Clin Endocrinol Metab* 70:990, 1990.

SanFilippo S, Imbesi RM: Is the spermatogonium an androgen target cell? An histochemical, immunocytochemical and ultrastructural study in the rat, *Prog Clin Biol Res* 296:177, 1989.

Scott RT Jr, Hodgen GD: The ovarian follicle: life cycle of a pelvic clock, *Clin Obstet Gynecol* 33:551, 1990.

Simpson ER, McDonald PC: Endocrine physiology of the placenta, *Annu Rev Physiol* 43:163, 1981.

Skinner MK: Cell-cell interactions in the testis, *Endocrine Rev* 12:45, 1991.

Skinner MK et al: Stimulation of Sertoli cell inhibin secretion by the testicular paracrine factor PModS, *Molec Cell Endocrinol* 66:239, 1989.

Steer PJ: The endocrinology of parturition in the human, *Bailliere's Clin Endocrinol Metab* 4:333, 1990.

Tonetta SA, DiZerega GS: Intragonadal regulation of follicular maturation, *Endocr Rev* 10:205, 1989.

Tsang BK et al: Androgen biosynthesis in human ovarian follicles: cellular source, gonadotropic control, and adenosine 3′,5′-monophosphate mediation, *J Clin Endocrinol Metab* 48:153, 1979.

Beldhuis JD: The hypothalamic pulse generator: the reproductive core, *Clin Obstet Gynecol* 33:538, 1990.

Ying S-Y: Inhibins, activins, and follistatins: gonadal proteins modulating the secretion of follicle-stimulating hormone, *Endocr Rev* 9:267, 1988.

Books and monographs

Adashi EY: The ovarian cycle. In Yen SSC, Jaffe RB, editors: *Reproductive endocrinology,* Philadelphia, 1991, WB Saunders.

Fawcett DW: Ultrastructure and function of the Sertoli cell. In Hamilton DW, Greep RO, editors: *Handbook of physiology,* sect 7, vol 5, Bethesda, Md, 1975, The American Physiology Society.

Fisher DA: Endocrinology of fetal development. In Wilson JD, Foster DW, editors: *Williams textbook of endocrinology,* Philadelphia, 1992, WB Saunders.

Konigsberg D et al: The ovary development and control of follicular maturation and ovulation. In Degroot LJ et al, editors: *Endocrinology,* vol 3, New York, 1989, Grune & Stratton.

Marshall JL, and Odell WD: The menstrual cycle—hormonal regulation, mechanisms of anovulation, and responses of the reproductive tract to steroid hormones. In Degroot LJ et al, editors: *Endocrinology,* vol 3, New York, 1989, Grune & Stratton.

Veldhuis JD: The hypothalamic-pituitary-testicular axis. In Yen SSC, Jaffe RB, editors: *Reproductive endocrinology,* Philadelphia, 1991, WB Saunders.

Yamamoto M, Turner TT: Epididymis, sperm maturation, and capacitation. In Lipshultz LI, Howards SS, editors: *Infertility in the male,* St Louis, 1991, Mosby–Year Book.

Yen SSC: The human menstrual cycle: neuroendocrine regulation. In Yen SSC, Jaffe RB, editors: *Reproductive endocrinology,* Philadelphia, 1991, WB Saunders.

Index

Entries referring to tables are denoted by a t following the page number. An italic number indicates an illustration.

C

E

G

H

J

K

P

S